Complex Analysis with Applications to Flows and Fields

MATHEMATICAL AND PHYSICS FOR SCIENCE AND TECHNOLOGY

Series Editor

L.M.B.C. Campos

Director of the Center for Aeronautical and Space Science and Technology
Lisbon Technical University

Complex Analysis with Applications to Flows and Fields
L.M.B.C. Campos

MATHEMATICS AND PHYSICS FOR SCIENCE AND TECHNOLOGY

Complex Analysis with Applications to Flows and Fields

L.M.B.C. Campos

Director of the Center for Aeronautical
and Space Science and Technology
Lisbon Technical University

CRC Press
Taylor & Francis Group
Boca Raton London New York

CRC Press is an imprint of the
Taylor & Francis Group, an **informa** business

CRC Press
Taylor & Francis Group
6000 Broken Sound Parkway NW, Suite 300
Boca Raton, FL 33487-2742

Printed in the United States of America on acid-free paper
10 9 8 7 6 5 4 3 2 1

International Standard Book Number: 978-1-4200-7118-4 (Hardback)

Library of Congress Cataloging-in-Publication Data

Braga de Costa Campos, Luis Manuel.
Complex analysis with applications to flows and fields / by Luis Manuel Braga de Costa Campos.
p. cm. -- (Mathematics and physics for science and technology)
Includes bibliographical references and index.
ISBN 978-1-4200-7118-4
1. Functions of complex variables. I. Title. II. Series.

QA331.7.B73 2010
515'.9--dc22 2009036654

Visit the Taylor & Francis Web site at
http://www.taylorandfrancis.com

and the CRC Press Web site at
http://www.crcpress.com

To Bia

Contents

List of Tables, Notes, Diagrams, Classifications, and Lists

Tables

Notes

Diagrams

Classifications

Lists

Series Preface

The aim of the Mathematics and Physics for Science and Technology series is to describe the mathematical methods as they are applied to model natural physical phenomena and solve scientific and technological problems. The primary emphasis is on the application, including formulation of the problem, detailed solution, and interpretation of results. The mathematical methods are presented in sufficient detail to justify every step of solution, and avoid superfluous assumptions.

The main areas of physics are covered, namely:

- Mechanics of particles, rigid bodies, deformable solids and fluids
- Electromagnetism, thermodynamics, and statistical physics as well as their classical, relativistic, and quantum formulations
- Interactions and combined effects (e.g., thermal stresses, magnetohydrodynamics, plasmas, piezoelectricity, and chemically reacting and radiating flows)

The examples and problems chosen include natural phenomena in our environment, geophysics, and astrophysics; the technological implications in various branches of engineering; and other mathematical models, in biological, economic, and social sciences.

The coverage of areas of mathematics and branches of physics is sufficient to lay the foundations of all branches of engineering, namely:

- *Mechanical*—including machines, engines, structures, and vehicles
- *Civil*—including structures and hydraulics
- *Electrical*—including circuits, waves, and quantum effects
- *Chemical*—including transport phenomena and multiphase media
- *Computer*—including analytical and numerical methods and associated algorithms

Particular emphasis is given to interdisciplinary areas, such as electromechanics and aerospace engineering. These require combined knowledge of several areas and have an increasing importance in modern technology.

Analogies are applied in an efficient and concise way, across distinct disciplines, but also stressing the differences and aspects specific to each area, for example:

- Potential flow, electrostatics, magnetostatics, gravity field, steady heat conduction, and plane elasticity and viscous flow
- Acoustic, elastic, electromagnetic, internal, and surface waves
- Diffusion of mass, electricity and momentum.

In each case the analogies are exploited by common mathematical methods with distinct interpretations in each context.

The series is organized as a sequence of mathematical methods, each with a variety of applications. As the mathematical methods progress, the range of applications widens. For example, complex functions are used to study potential flows and electrostatics in the plane. The three-dimensional extension uses generalized functions. The latter are used with differential equations to describe vibrations and waves. The series and integral transforms are applied to initial and boundary-value problems. Tensor calculus is used for elasticity, viscous fluids, and relativity. Thus each method is consolidated with diverse applications before proceeding to the next.

The presentation of the material is intended to remain accessible to the university student. The subjects are introduced at a basic undergraduate level. The deductions and intermediate steps are detailed. Extensive illustrations and detailed legends promote visual and intuitive memory and understanding. The material is presented like a sequence of lectures and can be used to construct the subjects or disciplines of a university curriculum. It is possible to adjust the level of the curriculum by retaining the basic theory and simpler examples, and using the rest as background material for further reading.

The adaptation of the material for specific lecture courses can be made by selecting the basic theory and examples used as applications. Some topics chosen for greater insight may be included according to the motivation. The bibliography gives a choice of approaches to the subject and the possibility to focus more in specific subareas. The presentation follows a logical rather than historical sequence; some references to the original sources are used to give a historical perspective. The notes at the ends of chapters hint at the broader scope of the subject. The contents of each chapter are previewed in an introduction.

The present series embodies a concept of "interdisciplinary education in science and technology". The traditional approach is to study each area of mathematics separately (analysis, geometry, differential equations, etc.) as well as each branch of physics (classical mechanics, heat and thermodynamics, electromagnetism, etc.). The student is then expected to "merge" all these sources of information, for example, know all that is needed about partial differential equations for the Maxwell equations of electromagnetism, calculus of variations for the minimum energy methods of elasticity, thermodynamics for the dynamics of compressible fluids, and so on. The time gaps and disjoint nature of this teaching implies a careful sequence of subjects to ensure each subject starts with the required background. Also there is considerable duplication in that similar methods and analogous problem recur in different contexts. Furthermore, the student discovers the utility of most of the mathematics much later, when it is applied to physical and engineering problems. The combined interdisciplinary study aims to resolve these issues.

Although the course starts at undergraduate level it gradually proceeds to research level and to the frontiers of current knowledge. The presentation of each subject takes into account from the very beginning not only the fundamentals but also the major topics of subsequent use. For example, the treatment of complex functions lays the basis for differential equations, integral transforms, asymptotics, and special functions. Linear algebra and analytic geometry lead to tensor analysis, differential geometry, variational calculus, and relativity. An introduction to fluid mechanics via the potential flow is followed by vortical, compressible, viscous, and multiphase flows. Electrostatics and magnetostatics are followed by unsteady electromagnetic fields and waves, magnetohydrodynamics, and plasmas. These apparently disparate subjects are treated at an early stage, analogies are presented, and at subsequent stages they are combined into multidisciplinary applications.

For example, a fluid may be subject to four restoring forces associated with pressure, gravity, rotation, and magnetic fields. The corresponding wave motions are respectively acoustic, internal, inertial, and magnetic waves. They appear combined as magneto–acoustic–gravity–inertial (MAGI) waves in a compressible, ionized, stratified rotating fluid. The simplest exact solutions of the MAGI wave equation require special functions. Thus the topic of MAGI waves combines six subjects: gravity field, fluid mechanics, and electromagnetism and uses a complex analysis, differential equations, and special functions. This is not such a remote subject, since many astrophysical phenomena do involve this combination of several of these effects, as does the technology of controlled nuclear fusion. The latter is the main source of energy in stars and in the universe; if harnessed, it would provide a clean and inexhaustible source of energy on earth. Closer to our everyday experience there is a variety of electromechanical and control systems that use modern interdisciplinary technology. The ultimate aim of the present series is to build up knowledge seamlessly from undergraduate to research level, across a range of subjects, to cover contemporary or likely interdisciplinary needs. This requires a consistent treatment of all subjects so that their combination fits together as a whole.

The approach followed in the present series is a combined study of mathematics, physics, and engineering, so that the practical motivation develops side by side with the theoretical concepts: the mathematical methods are applied without delay to "real" problems, not just to exercises. The electromechanical and other analogies simulate the ability to combine different disciplines, which is the basis of much of modern interdisciplinary science and technology. Starting with the simpler mathematical methods, and consolidating them with the detailed solutions of physical and engineering problems, gradually widens the range of topics that can be covered. The traditional method of separate monodisciplinary study remains possible, selecting mathematical disciplines (e.g., complex functions) or sets of applications (e.g., fluid mechanics). The combined multidisciplinary study has the advantage of connecting mathematics, physics, and technology at an earlier stage. Moreover, preserving that link provides a broader view of the subject and the ability to innovate. Innovation requires an understanding of technological aims, the physical phenomena that can implement them, and the mathematical methods that quantify the expected results. The combined interdisciplinary approach to the study of mathematics, physics, and engineering is thus a direct introduction to a professional experience in scientific discovery and technological innovation.

Preface

The present volume, *Complex Analysis with Applications to Flows and Fields*, consists of four parts; they present the theory of functions of a complex variable, starting with the complex plane (Part 1) and proceeding through the calculus of residues (Part 2) and power series (Part 3) to the conformal mapping (Part 4). The detailed applications cover twice as much space as the mathematical theory and concern potential flows, gravity field, electro- and magnetostatics, steady heat conduction, and other problems. The physical and engineering problems that cover about two-thirds of the text are main motivation; the mathematical results that occupy the remaining about one-third of the material are sufficient to fully justify the solution of problems without additional external references. The self-contained nature of the book concerns both the mathematical background and physical principles needed to formulate problems, justify the solutions, and interpret the results. It is backed by bibliography and other indices.

Organization of the Book

In each part the mathematical theory (physical and engineering applications) appear in alternating odd (even) numbered chapters, for example, Chapters 11, 13, 15, and 17 (12, 14, 16, and 18) in Part 2. The penultimate chapter, for example, Chapter 29 in Part 3, deals with some fundamental mathematical concepts. The last chapter of each part is a collection of 20 detailed examples, for example, Chapter 10 at the end of Part 1 consists of 20 worked out Examples 10.1 to 10.20. The chapters are numbered sequentially (Chapters 1 to 40). The formulae are numbered sequentially in a chapter between curved brackets, for example, (15.20) means formula 20 of Chapter 15. A chapter (e.g., 24) is divided into nine sections (e.g., 24.1 to 24.9); the section may be divided into subsections (e.g., 24.5.1 to 24.5.3). The figures are numbered by chapter (e.g., Figures 12.1 to 12.7 in Chapter 12). The conclusion of each chapter includes references to: (i) the figures as a kind of visual summary; (ii) the note(s), list(s), table(s), diagram(s) and classification(s) as additional support. The latter (ii) apply at the end of each chapter, and are numbered within the chapter (e.g., Note 24.1, List 29.1, Table 24.1, Classification 24.1, Diagram 21.1); if there is more than one they are numbered sequentially (e.g., Notes 24.1 to 24.3). The chapter starts with an introductory preview, and related topics may be mentioned in the notes at the end. The lists of mathematical symbols and physical quantities appear before the main text, and the index of subjects and bibliography at the end of the book.

About the Author

Luis Manuel Braga da Costa Campos was born in Lisbon, Portugal in 1950. He graduated in 1972 as a mechanical engineer from the Instituto Superior Tecnico (IST) of Lisbon Technical University. The tutorials as a student (1970) were followed by a career at the same institution (IST) through all levels: assistant on probation (1972), assistant (1974), auxiliary professor (1978), assistant professor (1982), chair of applied mathematics and mechanics (1985). He has been the coordinator of undergraduate and postgraduate degrees in aerospace engineering since their inception in 1991. He is coordinator of the Applied and Aerospace Mechanics Group in the Department of Mechanical Engineering. He is director and founder of the Center for Aeronautical and Space Science and Technology.

He completed his doctorate on "waves in fluids" at the engineering department of Cambridge University, England (1977). It was followed by a Senior Rouse Ball Scholarship at Trinity College, while on leave from IST. His first sabbatical was as a senior visitor at the Department of Applied Mathematics and Theoretical Physics of Cambridge University, England (1984). His second sabbatical (1991) was as a Alexander von Humboldt Scholar at the Max Planck Institute for Aeronomy in Katlenburg-Lindau, Germany. He could not pursue further sabbaticals abroad owing to major commitments at his home institution, which included extensive travels related to participation in scientific meetings, representation at individual or national levels in international institutions, and involvement in collaborative research projects.

He received the von Karman medal from the Advisory Group for Aerospace Research and Development (AGARD) and Research and Technology Organization (RTO). His participation in AGARD/RTO has been as vice-chairman of the System Concepts and Integration Panel, and as chairman of the Flight Mechanics Panel and of the Flight Vehicle Integration Panel. He has been a member of the Flight Test Techniques Working Group, which is related to the creation of an independent flight test capability active in Portugal for the past 20 years, and which has been used in national and international projects, including those from Eurocontrol and the European Space Agency. He has participated in various committees in the European Space Agency (ESA) as a national representative at the Council and Council of Minister levels.

The author has participated in several activities sponsored by the European Union. He has been involved in 27 research projects with industry, research, and academic institutions. He has been a member of various committees and has been vice-chairman of the Aeronautical Science and Technology Advisory Committee. He has been in the Space Advisory Panel on the future role of the EU in space. He has also been a member of the Space Science Committee of the European Science Foundation, and has been in close coordination with the Space Science Board of the National Science Foundation of the United States. He has been a member of the Committee for Peaceful Uses of Outer Space (COPUOS) of the United Nations. He has been working with these and other institutions as a consultant and advisor.

Regarding his contribution to professional societies, he has been a member and vice-chairman of the Portuguese Academic of Engineering; a fellow of the Royal Aeronautical Society, Astronomical Society, and Cambridge Philosophical Society; an associate fellow of the American Institute of Aeronautics and Astronautics; a founding member of the European Astronomical Society. He has been a member of various other professional associations in aeronautics, engineering, mechanics, acoustics, physics, astronomy, and mathematics. He is or has been a member of Editorial or Honorary Board of *Progress of Aerospace Sciences, Air & Space Europe, International Journal of Aeroacoustics, Revue d'Acoustique, Integral Transforms and Special Functions.* He is a reviewer in *Mathematical Reviews*, and has reviewed two dozen journals.

The author's publications include 4 books, 110 papers in 82 journals, and 160 communications in symposia. His areas of research center on four topics: acoustics, magnetohydrodynamics, special functions, and flight dynamics. His work on acoustics is related to the generation, propagation, and refraction of sound in flows with mostly aeronautical applications. His work on magnetohydrodynamics is related to magneto-acoustic-gravity-inertial waves in solar-terrestrial and stellar physics. The developments on special functions have been mostly based on differintegration operators, generalizing the ordinary derivative, and primitive to complex order. His work on flight dynamics is related to aircraft and rockets, including trajectory optimization, performance, stability, control, and atmospheric disturbances.

The author's interest in topics ranging from mathematics to physics and engineering fits in with the aims and content of the present series; his university teaching and scientific and industrial research relates to the build-up of the series from undergraduate to research level. His professional activities on the technical side are balanced by other cultural and humanistic interests. These are not reflected in publications, except for one book, which is a literary work. His complementary nontechnical interests include classical music (mostly orchestral and choral), plastic arts (painting, sculpture, architecture), social sciences (psychology and biography), history (classical, renaissance, and overseas expansion), and technology (automotive, photo, audio). He speaks four languages (Portuguese, English, French, and Spanish) and reads six (Italian and German). He is listed in various biographical publications, including *Who's Who in the World* since 1986.

Acknowledgments

The present book would require a long list of acknowledgments if all contributions were to be duly recorded. Those which follow are a mere selection based on subjective memory. I would like to first acknowledge the successive generations of university students to whom various parts of the course were taught over the years at Instituto Superior Tecnico; it is hoped that the experience they provided to me is reflected in this book for the benefit of future generations of students. I owe it Mr. Henrique Nuno for the drawings and to the following for typing my manuscript over a period of time: Fernanda Proença, Irene Patriarca, Fernanda Venâncio, Ana Monteiro, Bruno de Souza, Martinha de Sousa, Sónia Marques, Lurdes de Sousa and Sofia Pernadas. The members of my group who have taught similar subjects have made contributions in various ways; in alphabetical order of surname: A.J.M.N. Aguiar, F.S.R.P. Cunha, A.R.A. Fonseca, P.J.S. Gil, M.H. Kobayashi, F.J.P. Lau, J.M.G.S. Oliveira, and P.G.T.A. Serrâo. The final form of the present volume owes most to four persons: Mr. Jorge Coelho for the drawings; Professor J.M. André Junior for very helpful criticisms; Emeritus Professor A.G. Portela for several pages of written general and specific comments and suggestions. At last but not least, to my wife who more than deserves the dedication as the companion of the author in preparing this work.

Mathematical Symbols

The mathematical symbols are those of more common use in the context of (i) sets, quantifiers, and logic; (ii) numbers, ordering, and vectors; (iii) functions, limits, and convergence; (iv) derivatives, integrals and operators. It concludes with a list of functional spaces, most but not all of which appear in the present volume. The section where the symbol first appears may be indicated after a semicolon, for example, "10.2" means Section 10.2.

Sets, Quantifiers, and Logic

Sets

$A \equiv \{x : \dots\}$ – set A whose elements x have the property.

$A \cup B$ – union of sets A and B.

$A \cap B$ – intersection of sets A and B.

$A \supset B$ – set A contains set B.

$A \subset B$ – set A is contained in set B.

Quantifiers

$\forall_{x \epsilon A}$ – for all x belonging to A holds ...

$\exists_{x \epsilon A}$ – there exists at least one x belonging to A such that ...

$\exists^{1}_{x \epsilon A}$ – there exists one and only one x belonging to A such that ...

$\exists^{\infty}_{x \epsilon A}$ – there exist infinitely many x belonging to A such that ...

Logic

$a \wedge b$ – a and b.

$a \vee b$ – or (inclusive): a or b or both.

$a \dot{\vee} b$ – or (exclusive): a or b but not both.

$a \Rightarrow b$ – implication: a implies b.

$a \Leftrightarrow b$ – equivalence: a implies b and b implies a.

Constants

$e = 2.7182\ 81828\ 45904\ 52353\ 60287.$

$\pi = 3.1415\ 92653\ 58979\ 32384\ 62643.$

$\gamma = 0.5772\ 15664\ 90153\ 28606\ 06512.$

$\log 10 = 2.3025\ 85092\ 99404\ 56840\ 179915.$

Numbers, Ordering, and Vectors

Types of numbers

$|C$ – complex numbers: 1.2.
$|C^n$ – ordered sets of n complex numbers.
$|F$ – transfinite numbers.
$|H$ – hypercomplex numbers.
$|I$ – irrational numbers: real nonrational numbers: 1.2.
$|L$ – rational numbers: ratios of integers: 1.1.
$|N$ – natural numbers: positive integers: 1.1.
$|N_0$ – nonnegative integers: zero plus natural numbers: 1.1.
$|P$ – prime numbers: natural numbers without divisors.
$|Q$ – quaternions: 1.9.
$|R$ – real numbers: 1.2.
$|R^n$ – ordered sets of n real numbers.
$|Z$ – integers: 1.1.

Complex numbers

$|\ldots|$ – modulus of complex number...: 1.4.
arg $(\ldots)$ – argument of complex number...: 1.4.
Re $(\ldots)$ – real part of complex number...: 1.3.
Im $(\ldots)$ – imaginary part of complex number...: 1.3.
$\ldots^*$ – conjugate of complex number...: 1.6.

Ordering of numbers

sup $(\ldots)$ – supremum: smallest number larger or equal than all numbers in the set.
max $(\ldots)$ – maximum: largest number in set.
min $(\ldots)$ – minimum: smallest number in set.
inf $(\ldots)$ – infimum: largest number smaller or equal than all numbers in set.

Vectors

$\vec{A} \cdot \vec{B}$ – inner product.
$\vec{A} \wedge \vec{B}$ – outer product.
$\vec{A} \cdot \left(\vec{B} \wedge \vec{C}\right)$ – mixed product.
$\vec{A} \wedge \left(\vec{B} \wedge \vec{C}\right)$ – double outer product.
$\left|\vec{A}\right|$ – modulus.
ang $\left(\vec{A}, \vec{B}\right)$ – angle of vector $\vec{B}$ with vector $\vec{A}$.

Functions, Limits, and Convergence

Values of functions

$f(a)$ – value of function f at point a.
$f(a+0)$ – right-hand limit at a.
$f(a-0)$ – left-hand limit at a.
$f_{(n)}(a)$ – residue at pole of order n at a: 15.8.
$\overline{B}$ – upper bound: $|f(z)| \leq \bar{B}$ for z in ....
$\underline{B}$ – lower bound: $|f(z)| \geq \underline{B}$ for z in ....

Iterated sums and products

$\sum_a$ – sum over a set.

$\sum_{n=a}^{b}$ – sum from $n=a$ to $n=$ b.

$\sum_{n,m=a}^{b}$ – double sum over $n,m=a, \ldots, b$.

$\prod_a$ – product over a set.

$\prod_{n=a}^{b}$ – product from $n=a$ to $n=b$.

Limits

lim – limit when x tends to a: $x \to a$: 11.2.
l.i.m. – limit in the mean.
$a \sim O(b)$ – a is of order b: lim $b/a \neq 0,\infty$: 19.7.
$a \sim o(b)$ – b is of lower order than a: lim $b/a = 0$: 19.7.

Convergence

A.C. – absolutely convergent: 21.2.
A.D. – absolutely divergent: 21.2.
C. – convergent: 21.2
C.C. – conditionally convergent: 21.2.
Cn – converges to class n: $C0 \equiv C$.
D. – divergent: 21.1.
N.C. – nonconvergent: divergent or oscillatory: 21.1.
O. – oscillatory: 21.1.
T.C. – totally convergent: 21.7.
U.C. – uniformly convergent: 21.5.
applies to:
- power series: 21.1.
- series of fractions: 27.9.
- infinite products: 27.9.
- continued fractions.

Derivatives, Integrals, and Operators

Differentials and Derivatives

$\mathrm{d}\Phi$ – differential of Φ.

$\mathrm{d}\Phi/\mathrm{d}t$ – derivative of Φ with regard to t.

$\partial\Phi/\partial t \equiv \partial_t\dot{\Phi} \equiv \dot{\Phi}$ – partial derivative of Φ with regard to t.

$\partial\Phi/\partial x_i \equiv \partial_i\Phi \equiv \Phi_{,i}$ – partial derivative of Φ with regard to x_i.

$\partial^n\Phi/\partial x_{i_1}\ldots\partial x_{i_n} \equiv \partial_{i_1}\ldots\partial_{i_n}\Phi \equiv \Phi_{,i_1,\ldots i_n}$ – nth-order partial derivative of Φ with regard to $x_{i_1},\ldots,x_{i_n}$.

Directed Derivatives

$\partial\Phi/\partial\ell \equiv \vec{\ell}.\nabla\Phi \equiv \ell_i\partial_i\Phi \equiv \ell_i\Phi_{,i}$ – derivative of Φ in the direction of the unit vector $\vec{\ell}$: 18.1

$\partial\Phi/\partial n \equiv \vec{n}\cdot\nabla\Phi$ – derivative in the direction normal to a curve: 18.1.

$\partial\Phi/\partial N \equiv \vec{N}\cdot\nabla\Phi$ – derivative in the direction normal to a surface: 28.1.

$\partial\Phi/\partial s \equiv \vec{s}\cdot\nabla\Phi$ – derivative in the direction of the unit tangent to a curve: 18.1

Invariant Operator

$\nabla\Phi$ – gradient of a scalar Φ: 11.7.

$\nabla.\vec{A}$ – divergence of a vector $\vec{A}$: 11.7.

$\nabla\wedge\vec{A}$ – curl of a vector $\vec{A}$: 11.7.

$\nabla^2\Phi \equiv \Delta\Phi$ – Laplacian of a of scalar Φ: 11.7.

Integrals

$\int\ldots\mathrm{d}x$ – primitive of ... with regard to x: 13.1.

$\int^{y}\ldots\mathrm{d}x$ – indefinite integral of ... at y: 13.2.

$\int_a^b \mathrm{d}x$ – definite integral of ... between a and b: 13.2.

$⨍_a^b\cdots\mathrm{d}x$ – principal value of integral: 17.8.

$\int^{(z+)}$ – integral along a loop around z in the positive (counterclockwise) direction: 13.5.

$\int^{(z-)}$ – idem in the negative (clockwise) direction: 13.5.

$\int_L$ – integral along a path L: 13.2.

$\int_C^{(+)}$ – integral along a closed path or loop C in the positive direction: 13.5.

$\int_C^{(-)}$ – integral along a closed path or loop C in the negative direction: 13.5.

Functional Spaces

The sets of numbers and spaces of functions are denoted by calligraphic letters, in alphabetical order:
(a, b) – set of functions over interval from a to b.
omission of interval: set of function over real line $)-\infty,+\infty($.
$\mathcal{A}(\ldots)$ – analytic functions in ...: 27.1.
$\bar{\mathcal{A}}(\ldots)$ – monogenic functions in ...: 31.1.
$\mathcal{B}(\ldots)$ – bounded functions in ...: $\mathcal{B} \equiv \mathcal{B}^0$: 13.3.
$\mathcal{B}^n(\ldots)$ – functions with bounded n-th derivative in ....
$\mathcal{C}(\ldots)$ – continuous functions in ...: $\mathcal{C} \equiv \mathcal{C}^0$: 11.2.
$\mathcal{C}^n(\ldots)$ – functions with continuous n-th derivative in ....
$\bar{\mathcal{C}}(\ldots)$ – piecewise continuous functions in ...: $\bar{\mathcal{C}} \equiv \bar{\mathcal{C}}^0$.
$\bar{\mathcal{C}}^n(\ldots)$ – functions with piecewise continuous n-th derivative in ....
$\widetilde{\mathcal{C}}(\ldots)$ – uniformly continuous function in...: 13.4.
$\widetilde{\mathcal{C}}^n(\ldots)$ – function with uniformly continuous n-th derivative in....
$\mathcal{D}(\ldots)$ – differentiable functions in ...:$\mathcal{D} \equiv \mathcal{D}^0$: 11.2.
$\mathcal{D}^n(\ldots)$ – n-times differentiable functions in ....
$\mathcal{D}^\infty(\ldots)$ – infinitely differentiable functions or smooth in ... 27.1.
$\bar{\mathcal{D}}(\ldots)$ – piecewise differentiable functions in ...:$\bar{\mathcal{D}} \equiv \bar{\mathcal{D}}^0$.
$\bar{\mathcal{D}}^n(\ldots)$ – functions with piecewise n-th derivative in ....
$\mathcal{E}(\ldots)$ – Riemann integrable functions in ...: 13.2.
$\bar{\mathcal{E}}(\ldots)$ – Lebesgue integrable functions in ...
$\mathcal{F}(\ldots)$ – functions of bounded fluctuation (or bounded variation) in ...; $\mathcal{F} \equiv \mathcal{E} \equiv \mathcal{F}^0$.
$\mathcal{F}^n(\ldots)$ – functions with n-th derivative of bounded fluctuation (variation) in ....
$\mathcal{G}(\ldots)$ – generalized functions (or distributions) in ....
$\mathcal{H}(\ldots)$ – harmonic functions in ...: 11.6.
$\mathcal{I}(\ldots)$ – integral functions in ...: 27.9.
$\mathcal{I}_m(\ldots)$ – rational-integral functions of degree m in ... $\mathcal{I} \equiv \mathcal{I}_0$: 27.9.
$\mathcal{J}(\ldots)$ – square integrable functions with a complete orthogonal set of functions – *Hilbert space.*
$\bar{\mathcal{K}}(\ldots)$ – Lipshitz functions in ....
$\mathcal{K}^n(\ldots)$ – homogeneous functions of degree n in ....
$\mathcal{L}^1(\ldots)$ – absolutely integrable functions in ....
$\mathcal{L}^2(\ldots)$ – square integrable functions in ....
$\mathcal{L}^p(\ldots)$ – functions with power p of modulus integrable in ... - *normed* space: $\mathcal{L}^p \equiv \mathcal{W}^p_0$.
$\mathcal{M}^+(\ldots)$ – monotonic increasing functions in ....
$\mathcal{M}^+_0(\ldots)$ – monotonic nondecreasing functions in ....
$\mathcal{M}^-_0(\ldots)$ – monotonic nonincreasing functions in....
$\mathcal{M}^-(\ldots)$ – monotonic decreasing functions in ....
$\mathcal{N}(\ldots)$ – null functions in ....
$\mathcal{O}(\ldots)$ – orthogonal systems of functions in ....
$\widetilde{\mathcal{O}}(\ldots)$ – complete orthogonal systems of functions in ....
$\mathcal{P}(\ldots)$ – polynomials in ...: 27.7.
$\mathcal{P}_n(\ldots)$ – polynomials of degree n in ...: 27.7.
$\mathcal{Q}(\ldots)$ – rational functions in...: 27.7.
$\mathcal{Q}^m_n(\ldots)$ – rational functions of degrees n, m in...: 27.7.
$\mathcal{R}(\ldots)$ – real functions, that is, with the real line as range.
$\mathcal{S}(\ldots)$ – complex functions, that is, with the complex plane as range.

$\mathcal{T}(\ldots)$ – functions with compact support, that is, which vanish outside a finite interval.
$\mathcal{T}^n(\ldots)$ – temperate functions of order n: n-times differentiable functions with first $(n-1)$ derivatives with compact support.
$\mathcal{T}^\infty(\ldots)$ – temperate functions: smooth or infinitely differentiable functions with compact support.
$\mathcal{U}(\ldots)$ – single-valued functions in…: 9.1.
$\widetilde{\mathcal{U}}(\ldots)$ – injective functions in…: 9.1.
$\bar{\mathcal{U}}(\ldots)$ – surjective functions: 9.1.
$\tilde{\bar{\mathcal{U}}}(\ldots)$ – bijective functions: 9.1.
$\mathcal{U}_n(\ldots)$ – multivalued functions with n branches in…: 6.1.
$\mathcal{U}_\infty(\ldots)$ – many-valued functions in…:6.2.
$\mathcal{U}^1(\ldots)$ – univalent functions, in …: 37.4.
$\mathcal{U}^m(\ldots)$ – multivalent functions taking m values in …: 37.4.
$\mathcal{U}^\infty(\ldots)$ – manyvalent functions in …: note 37.4.
$\mathcal{U}_n^m(\ldots)$ – multivalued multivalent functions with n branches and m values in …: note 37.4.
$\mathcal{V}(\ldots)$ – good functions, that is, with decay at infinity faster than some power.
$\mathcal{V}^N(\ldots)$ – good functions of degree N, that is, with decay at infinity faster than the inverse of a polynomial of degree N.
$\widetilde{\mathcal{V}}(\ldots)$ – fairly good functions, that is, with growth at infinity slower than some power.
$\widetilde{\mathcal{V}}^N(\ldots)$ – fairly good functions of degree N, that is, with growth at infinity slower than a polynomial of degree N.
$\overline{\mathcal{V}}(\ldots)$ – very good or fast decay functions, that is, with faster decay at infinity than any power.
$\mathcal{W}_q^p(\ldots)$ – functions with generalized derivatives of orders up to q such that for all the powers p of the modulus is integrable …—*Sobolev* space.
$\mathcal{X}_0(\ldots)$ – self-inverse linear functions in …: 37.5.
$\mathcal{X}_1(\ldots)$ – linear functions in …: 35.2.
$\mathcal{X}_2(\ldots)$ – bilinear, homographic, or Mobius functions in …: 35.4.
$\mathcal{X}_3(\ldots)$ – self-inverse bilinear functions in …: 37.5.
$\mathcal{X}_a(\ldots)$ – automorphic functions in …: 37.6.
$\mathcal{X}_m(\ldots)$ – isometric mappings in …: 35.1.
$\mathcal{X}_r(\ldots)$ – rotation mappings in …: 35.1.
$\mathcal{X}_t(\ldots)$ – translation mappings in …: 35.1.
$\mathcal{Y}(\ldots)$ – meromorphic functions in …: 37.9.
$\mathcal{Z}(\ldots)$ – polymorphic functions in …: 37.9.

Physical Quantities

The physical quantities are denoted by lower- or uppercase arabic or greek letters. Some are also represented by mathematical symbols. Calligraphic (Gothic or Old English) uppercase letters are reserved for functional spaces (geometries and associated coordinate transformation groups). The subscripts may be omitted to simplify and lighten the text and equations when there is no ambiguity or risk of confusion; they are introduced again when a distinction is necessary. For example the complex potential for the electrostatic field f_e is denoted by f, except when in comparison with others, e.g., the complex potential for the velocity of a potential flow f_v. The latter is also denoted by f when there is no risk of ambiguity.

Lowercase Arabic Letters

a – radius of cylinder: 24.6.
$\vec{\mathrm{a}}$ – acceleration: 2.1.
b – distance of point multipole from cylinder axis: 24.6.
c – phase speed of waves: 22.1.
c_* – speed of light: 26.1.
c_0 – speed of sound: 22.4.
e – electric charge: 6.1.
f – complex potential: 12.3, for example, for the velocity of a potential flow f_v.
f_e – complex potential of an electrostatic field: 24.3.
f_g – complex potential of a gravity field: 18.4.
f_m – complex potential of a magnetostatic field: 26.3.
f_v – complex potential for the velocity of a potential flow: 12.3.
$\vec{\mathrm{f}}$ – force density per unit volume, area or length: 14.2.
$\vec{\mathrm{f}}_b$ – dilatation force density: 28.3.
$\vec{\mathrm{f}}_e$ – electrical force density: 28.3.
$\vec{\mathrm{f}}_{em}$ – electromagnetic force density: 28.3.
$\vec{\mathrm{f}}_g$ – gravity force density: 28.3.
$\vec{\mathrm{f}}_l$ – vortical force density or Lamb vector: 28.3.
$\vec{\mathrm{f}}_m$ – magnetic force density: 28.3.
$\vec{\mathrm{f}}_n$ – stagnation force density: 28.3.
$\vec{\mathrm{f}}_p$ – hydrodynamic force density: 28.3.
$\vec{\mathrm{g}}$ – acceleration of gravity: 18.3.
g^* – complex conjugate gravity field: 18.4.
h_i – scale factors: 11.9.
i – imaginary unit $i \equiv \sqrt{-1}$: 1.2.
$\vec{\mathrm{j}}$ – electric current per unit area: 26.5.

k – wavenumber: 22.1.
– factor in the induced drag coefficient: 34.9.

$\vec{k}$ – wavevector: 12.1.

m – mass: 2.9.

m_0 – added mass: mass of fluid extrained by a body in motion: 28.6.

$\overline{m}$ – total mass: mass of a body plus added mass: 28.6.

n – coordinate normal to a curve: 18.1.

$\vec{n}$ – unit normal vector to a curve: 12.1.

p – pressure: 14.2.

p_0 – stagnation pressure: 14.5.

p_n – multipole strength per unit area: 26.4.

q – flow rate of source/sink per unit area or length: 18.1.
– electric charge density per unit area: 24.1.

q_n – multipole strength per unit length: 26.4.

r – polar coordinate: 1.4.

s – arc length: 11.9.
– coordinate tangent to a curve: 18.1.

$\vec{s}$ – unit tangent vector to a curve: 12.1.

t – time: 2.1.

v – complex velocity: 6.3.

v^* – complex conjugate velocity: 12.3.

$\vec{v}$ – velocity vector: 6.1.

w – density per unit area of heat source/sink: 32.1.
– downwash velocity behind wing: 34.7.

x – Cartesian coordinate: 1.2.

$\vec{x}$ – position vector of observer: 6.5.

y – Cartesian coordinate: 1.2.

z – complex number: 1.2
– Cartesian coordinate: 6.5.

Capital Arabic Letters

A – admittance: 4.4.
– amplitude of a wave: 22.1.

$\vec{B}$ – magnetic induction vector: 26.1.

C_D – drag coefficient: 28.4, 34.6.

C_L – lift coefficient: 28.5, 34.6.

C_M – pitching moment coefficient: 28.7, 34.6.

D – drag force: 28.2.

$\vec{D}$ – electric displacement vector: 24.1.

E – energy: 8.5.

E_e – electrical energy: 24.2.

E_k – kinetic energy: 8.5.

E_m – magnetic energy: 26.2.

$\vec{E}$ – electric field vector: 24.1.

E^* – complex conjugate electric field: 24.3

F – force vector: inertia force: 2.9.

F_b – dilatation force: 28.3.
F_e – electric force: 24.3, 28.3.
F_{em} – electromagnetic or Laplace–Lorentz force: 6.1, 28.3.
F_g – gravity force: 18.5, 28.3.
F_l – vortical force: 28.3.
F_m – magnetic force: 26.3, 28.3.
F_n – stagnation force: 28.3.
F_p – hydrodynamic force: 28.2.
F_s – suction force: 34.1.
G – gravitational constant: 18.3.
$\vec{\mathrm{G}}$ – heat flux: 32.1.
H – enthalpy: 14.3.
$\vec{\mathrm{H}}$ – magnetic field vector: 26.1.
$\vec{\mathrm{J}}_e$ – electric current: 26.1.
$\vec{\mathrm{J}}_v$ – mass flux: 24.1.
L – lift: 28.2.
M – pitching moment: 28.2.
$\vec{\mathrm{M}}$ – moment of forces: 28.2.
N – coordinate normal to a surface: 28.1.
$\vec{\mathrm{N}}$ – unit vector normal to a surface: 18.1.
P_n – moment of 2^n-multipole: 12.9 (e.g., monopole P_0, dipole P_1, quadrupole P_2).
Q_v – volume flow rate: 12.2.
$\vec{\mathrm{Q}}_e$ – electric polarization vector: 24.1.
$\vec{\mathrm{Q}}_m$ – magnetic polarization vector: 26.1.
R_N – remainder of a series after N terms: 21.1.
S – area element of a surface: 28.1.
– entropy density: 14.3.
$\overline{S}$ – total entropy in a domain: 32.1.
S_N – sum of the first N terms of a series: 21.1.
T – temperature: 32.1.
X – resistance: 4.4.
Y – reactance: 4.4.
Z – impedance: $Z = X + iY$: 4.4.

Lowercase Greek Letters

α – angle-of-attack of a flow: 14.8.
β – internal angle in a corner: 24.9.
χ_e – electric susceptibility: 24.1.
χ_m – magnetic susceptibility: 26.1.
ε – dielectric permittivity: 24.1.
ϕ – phase of a wave: 22.9.
γ – circulation density per unit length: 18.2.
– external angle in a corner: 33.5.
– adiabatic exponent: 14.6.
φ – polar angle: 1.4.
κ – thermal conductivity scalar in isotropic medium: 22.1.

λ – damping: 2.3.
– wavelength: $\lambda = 2\pi/k$: 12.1.
μ – mass density per length: 18.4.
– magnetic permeability: 26.1.
ω – angular frequency: 2.2, 22.1.
ρ – mass density per unit area: 14.1.
σ – electric charge density per unit length: 24.2.
τ – period $\tau = 2\pi/\omega$: 2.2.
$\vec{\varpi}$ – vorticity: 14.3.

Capital Greek Letters and Others

Φ – scalar potential: 12.1.
Φ_e – electric potential: 24.2.
Φ_g – gravity potential: 18.3.
Φ_m – magnetic potential: 26.7.
Φ_v – velocity potential: 12.1.
Γ – circulation: 12.1.
ϑ – electric current per unit length: 26.2.
Λ – dilatation: 12.2.
$\vec{\Omega}$ – angular velocity: 6.1.
Ψ_e – field function of electric field: 24.6.
Ψ_g – field function of gravity field: 18.3.
Ψ_m – field function of magnetic field: 26.2.
Ψ_v – stream function: 22.2.

Part 1
Complex Domain: Circuits and Stability

The complex numbers are the simplest for which all direct (sum, product, power) and inverse (subtraction, division, root) operations are closed (Chapter 1), that is, when applied to complex numbers these operations always lead to complex numbers (Chapters 3 and 5). Since a complex number is an ordered pair of real numbers, it can be represented on the plane (Chapter 1), and the corresponding geometry has a number of features: (i) the complex plane maps one-to-one to a sphere (Chapter 9), and has only one point-at-infinity; (ii) a multivalued function (Chapter 7) can be represented with each branch on a sheet of a Riemann surface, with the sheets connected at branch-points, and separated by branch-cuts. Since a complex number involves two real numbers, it allows some two-dimensional motions to be represented by one variable, for example, an electron moving transversely to a magnetic field (Chapter 6); it also allows the combination of several properties of mechanical or electrical circuits into a single complex impedance (Chapter 4). Likewise a complex function involves two real functions, and thus can represent in a single expression several kinds of motion (Chapter 2), for example, oscillatory, damped, or unstable regimes (Chapter 8).

1

Complex Numbers and Quaternions

The complex numbers appear at the end of a hierarchy (Sections 1.1 and 1.2) formed by the positive integers, integers, rationals, irrationals, and real numbers; the complex numbers are the simplest for which all three direct (sum, product, power) and inverse (subtraction, division, root) operations are closed, that is, when applied to a complex number the result is also a complex number. A complex number is an ordered pair of real numbers, which can be represented as a point on the plane (Sections 1.3–1.6); this provides a graphical illustration of some properties of elementary real and complex functions (Sections 1.7 and 1.8). A quaternion (Section 1.9) is a generalization of complex number in four dimensions for which the product is noncommutative. Further generalizations of the concept of number (transfinite, hypercomplex) may have less properties. The operations like sum, subtraction, product, division, power, and root may be applied not only to numbers but also to other entities (functions, multiplicities, sets, rings) as long as they retain similar properties.

1.1 Peano (1889, 1891) Postulates for Natural Numbers

An operation $\otimes$ is ***closed*** with regard to a set of numbers iff (if and only if) when applied to any two elements x, y of the set, the result is an element of the set:

$$\otimes \text{ closed in } A\text{: } \forall_{x,y\in A} \Rightarrow x \otimes y \in A. \tag{1.1}$$

Next, consider the three **direct operations** (sum, product, and power) and three **inverse operations** (subtraction, division, and root) to ascertain whether they are closed with regard to sets of numbers (natural, integer, rational, real, complex, and quaternions) in Table 1.1.

The **natural numbers** correspond to the usual counting of objects:

$$|N \equiv \{1, 2, 3, \ldots, \infty\}\,. \tag{1.2}$$

The natural numbers $|N$ can be introduced by the five **Peano postulates** (1889, 1891), via a **successor function**, S, as follows: (i) the number “1” is an integer $1 \in |N$; (ii) the successor of an integer is an integer $n \in |N \Rightarrow Sn \in |N$; (iii) the successor is not unity: $n \in |N \Rightarrow Sn \neq 1$; (iv) identical successors correspond to the same number: $n, m \in |N \wedge Sn = Sm \Rightarrow n = m$; and (v) if a property is true for “one” and every successor it is true for all integers: $1 \in M\ \forall n (n \in M \Rightarrow Sn \in M) \Rightarrow |N \subseteq M$. The last property (v) is the **axiom of induction**, *viz. if a property holds (i) for* $n = 1$, *and if (ii) holding for* n *implies holding for* $n+1$, *then it holds for all* n *natural numbers.* The five axioms are **independent**, because it is possible to define a system which satisfies all but one, viz. (a) the set {2,3,...} satisfies all properties (ii–v) except (i); (b) the set $|N \cup (1/2)$ with $S1 = 1/2$ and $S{}^1\!/_2 = 2$ only fails (ii); (c) the set {1} fails only (iii); (d) the set {1,2} with $S1 = S2 = 1$ fails only (iv); and (e) the set $|N \cup (-1)$ fails only (v).

TABLE 1.1
Sets of Numbers and Algebraic Operations

Set	Natural	Integers	Rational	Real	Complex	Quaternions
Symbol	$\vert N$	$\vert Z$	$\vert L$	$\vert R \equiv \vert L \cup \vert I$	$\vert C$	$\vert Q$
Sum	×	×	×	×	×	×
Subtraction	–	×	×	×	×	×
Product	×	×	×	×	×	× (1)
Division	–	–	×	×	×	×
Power	×	×	–	×	×	×
Root	–	–	–	–	×	–

Note: Successively larger sets of numbers: positive integers or natural $|N$, integers $|Z$, rational $|L$, irrational $|I$, real $|R$, complex $|C$, and quaternions $|Q$. The operations closed (not closed) with regard to a set are marked with a cross x (slash –). The direct operations (sum, product, power) are always closed. The inverse operations (subtraction, division, root) become closed enlarging the set of numbers. The complex numbers are the simplest for which all operations indicated are closed. Beyond the complex numbers, for example, for quaternions, some properties are lost. x: operation is closed with regard to the set; –: operation is not closed with regard to the set; (1): product is noncommutative.

The idea of the Peano postulates is to start with a number and define the remaining through the successor function. The natural numbers are closed with regard to the direct operations: sum, product, power. The number **zero (one)** is the **neutral element** of the sum (product), viz. $n + 0 = n = 0 + n\,(1 \times n = n = n \times 1)$ for all $n \in |N$. Starting the Peano postulate (i) with one (zero) leads to the **positive (nonnegative) integers** $|N$ in (1.2) [$|N_0$ in (1.3)]:

$$|N_0 \equiv \{0, 1, 2, 3, \ldots, \infty\} \equiv \{0\}\, U\, |N. \tag{1.3}$$

Although the natural numbers are closed with regard to the sum, they are not closed with regard to the subtraction. The difference of two positive integers $n - m$ is a positive integer if n is **larger** $(n > m)$ than m, but has no solution in the set of positive integers if n is **smaller** $(n < m)$ than m. This difficulty is overcome by extending the set of natural numbers to the set of ***integers***:

$$|Z \equiv \{0, \pm 1, \pm 2, \ldots, \pm\infty\}, \tag{1.4}$$

consisting of **positive** $n \in |N$ and **negative** $-n \in |N$ integers that have same *modulus* $|n| = -|n|$ and opposite signs, plus zero.

The integers are closed with respect to the sum, subtraction, and product, but not with regard to the division: the ratio n/m of two integers, apart from sign $|n|\,/\,|m|$, is an integer iff $|n|$ is a **multiple** of $|m|$ or $|m|$ is **a submultiple** of $|n|$, viz.: $|n| = |m|\,p$ with $p \in |N$; otherwise, the ratio does not exist as an integer number. This difficulty is resolved by extending further the set of integers to the set of **rational** numbers:

$$|L \equiv \{x, y \in |N : x/y \in |L\}, \tag{1.5}$$

which are ordered pairs of integers, specified by their ratio. The rational numbers are closed with regard to the three direct operations (sum, product, and power) and also with regard to two of the three *inverse* operations (subtraction and division). However, the rational numbers are not closed with regard to the third inverse operation, viz. the root, for which three cases arise: (i) the Nth root of a rational number is a rational number $q = \sqrt[N]{p}$ iff the Nth power of q is $p = q^N$; (ii) there are positive rational numbers $p > 0$ whose roots are not rational numbers, for example, $\sqrt{2}$ is not rational as will be proved in the sequel

(Section 1.2); and (iii) a negative rational number never has a square root as a rational number. The lifting of the limitation (ii) leads to irrational and real numbers (Section 1.2) and (iii) is resolved by the introduction of complex numbers (Sections 1.3–1.8, Chapter 3).

1.2 Irrational Numbers (Pythagoras, VI b.c.) and Dedekind (1858) Section

In the rational numbers can be introduced a **strict order relation**: of two distinct numbers one must be larger than the other:

$$\forall p, q \in |L : p \neq q \Rightarrow p > q \dot{\vee} q > p; \tag{1.6}$$

the condition (1.6) is (i) **exclusive**, if $p > q$ excludes the possibility $q > p$; and (ii) **exhaustive**, if $p \neq q$ there are no possibilities other than $p > q$ or $q < p$. The proof that $\sqrt{2}$ is irrational (Pythagoras VI b.c.) can be made by "reduction ad absurdum": (i) suppose that $\sqrt{2}$ is rational and specified by the **irreducible fraction** (1.7a), that is, in its lowest terms:

$$\sqrt{2} = \frac{p}{q}; \quad \sqrt{2} = \frac{2q - p}{p - q}; \tag{1.7a,b}$$

(ii) then it is also given by (1.7b) for any p, q, since

$$0 = (2q - p)^2 - 2(p - q)^2 = 2q^2 - p^2; \tag{1.7c}$$

(iii) choosing q such that $0 < p - q < q$ it follows from (1.7b) that the denominator in (1.7a) is not in its lowest terms. This contradiction proves that $\sqrt{2}$ cannot be a rational number. This leads to the introduction of the irrational numbers, in order to make the square root a closed operation for positive arguments.

One way to introduce irrational numbers is through the **Dedekind section** (1858): given a number p, the **upper (lower) class** are sets such that (i) each class has at least one number; and (ii) all numbers of the upper class $y \in p^+$ exceed all numbers of the lower class $x \in p^-$, viz. $y > x$. *A rational number* p *specifies a Dedekind section: it is the largest (or smallest) member, that is,* **supremum (infimum)**, *of the lower (upper) class; if a section is not specified by a rational number, then it defines an* **irrational** *number, for example,* $\sqrt{2}$. The operations on rational numbers may be extended to irrational numbers, as in the example of the sum: (i) let p, q be two numbers, and (p^+, p^-) and (q^+, q^-) their respective (upper, lower) classes; (ii) their sums form a lower class $p^- + q^-$ whose elements are all less than those of the upper class $p^+ + q^+$; and (iii) the resulting Dedekind section is the sum $p + q$ of the two numbers.

Thus, the set of **real** numbers $|R$ can be defined as the union of sets of rational $|L$ and irrational $|I$ numbers:

$$|R \equiv |L \cup |I = \{x : -\infty < x < +\infty\}. \tag{1.8}$$

The set of real numbers establishes a one-to-one correspondence with the points on a real line: some of these points correspond to rational numbers, others to irrational numbers. There are in a given finite interval $a \leq x \leq b$: (i) a **finite** number $M - N + 1$ integers $a \leq N \leq x \leq M \leq b$, where $N - 1 = f(a)$ and $M = f(b)$ is the **integer part** of a real number, obtained by omitting the decimals; (ii) there is **denumerable infinity** of rational numbers x/p, actually less than the number of ordered pairs of integers (x, p), because some may not be irreducible fractions, for example, $2x/2p = x/p$; and (iii) there is a **nondenumerable infinity** of irrational numbers.

1.3 Cartesian Parts: Real and Imaginary (Argand, 1806; Descartes, 1637a; Gauss, 1797)

The extension from the rational to the real numbers makes the square roots $\sqrt{p}$ of a positive number $p > 0$ closed, but there remains another problem: a negative number $p = -|p| < 0$ has no real square root $\sqrt{-|p|}$. For operation square root to be closed, the real numbers have to be extended to the complex numbers, defined (1.9a) as an ordered pair of real numbers:

$$|C \equiv \{x, y \in |R : z = x + iy\}, \quad i^2 = -1, \tag{1.9a,b}$$

where the symbol "i" satisfies (1.9b) as justified subsequently in Section 3.4. Note that (i) the direct operations (sum, product, power) are closed for all sets of numbers, viz. natural $|N$, integer $|Z$, rational $|L$, and real $|R$; (ii) the inverse operations (subtraction, division, root) may not be closed, and their closure may require the extension of the concept of number. Three examples of the process (ii) were given: (i) the subtraction becomes closed (1.1) by extending the natural numbers (1.2, 1.3) to the integers (1.4); (ii) the division becomes closed extending the integers (1.4) to the rational numbers (1.5); and (iii) the square root becomes closed extending the rational (1.5) to the real (1.8) and complex numbers (1.9a,b). Thus, the *complex numbers are the simplest for which all direct (sum, product, power) and inverse (subtraction, division, roots) operations are closed*, as shown in Table 1.1. This justifies the emphasis on the properties of (Sections 1.3–1.8) and operations on (Chapter 3) complex numbers.

A complex number is defined as an ordered pair (x,y) of real numbers, for which the notation usually adopted is (1.10a):

$$z = x + iy : \quad x \equiv \mathrm{Re}\,(z), \quad y \equiv \mathrm{Im}\,(z), \tag{1.10a–c}$$

where Re (Im) denote the **real (imaginary) parts** (1.10b) [(1.10c)]. The real and imaginary parts can be taken as coordinates, respectively, x,y in a Cartesian frame of reference (Descartes, 1637a), so that complex number z is represented by a point in Argand's plane (1806, used before by Gauss in 1797). There exists a one-to-one correspondence between the set $|C$ of complex numbers z, the set $|R \times |R$ of ordered pairs (x,y) of real numbers, and the points P on the complex plane.

1.4 Polar Coordinates: Modulus and Argument

The distance from the point P in Figure 1.1, which represents the complex number z, from the origin defines the **modulus**:

$$z = x + iy : \; r = |z| \equiv \left|x^2 + y^2\right|^{1/2}, \tag{1.11}$$

where $|\ldots|^{1/2}$ means the square root taken with plus sign before the radical, that is, the positive square root. The angle between the straight line OP and the real axis OX, measured

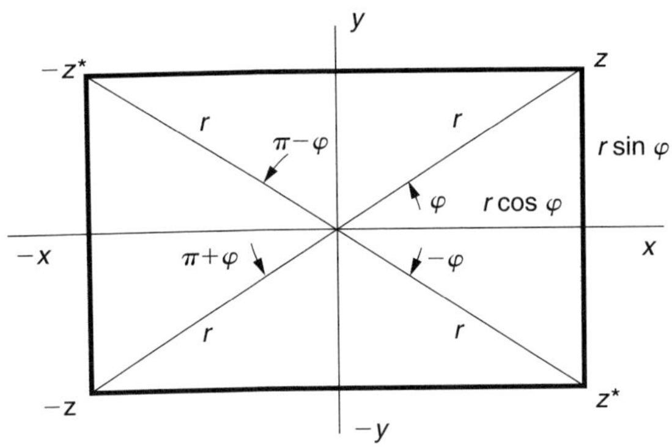

FIGURE 1.1
A complex z is transformed by symmetry relative to the (i) real axis to its conjugate z^*; (ii) the origin to the symmetric $-z$; and (iii) to the imaginary axis to the conjugate symmetric $-z^*$.

from the latter, in the **positive direction**, that is, *counterclockwise*, defines the **argument** of the complex number z:

$$\varphi \equiv \arg(z) = \begin{cases} \operatorname{arc\,tan}(y/x) & \text{for } x \geq 0, \quad (1.12a) \\ \pi + \operatorname{arc\,tan}(y/x) & \text{for } x \leq 0; \quad (1.12b) \end{cases}$$

in (1.12a,b) it has been taken into account that the inverse circular tangent arc tan is one-to-one in the right-hand half circle $-\pi/2 < \varphi \leq +\pi/2$, corresponding to the upper line (1.12a); the lower line (1.12b) adds left-hand half circle $\pi/2 < \varphi \leq 3\pi/2$ to build-up the full circle $-\pi/2 < \varphi \leq 3\pi/2$, or, equivalently, $-\pi < \varphi \leq +\pi$ or $0 \leq \varphi < 2\pi$. The latter range is denoted by $(0, 2\pi($, and it will henceforth be assumed that the function arc tan will be extended to this range in the form (1.12a,b) whenever necessary. *Thus a complex number can be uniquely represented by (1.11) its modulus* $r \geq 0$ *and (1.12a,b) argument* $0 \leq \varphi < 2\pi$, *except at the origin and infinity* r $\neq 0, \infty$, because (i) at origin $r = 0$ the argument φ is undetermined; (ii) at infinity $r = \infty$ there is only one point (as shown in Section 9.2), and the argument φ is again undetermined, because the "point at infinity" can be reached in any direction. The modulus, argument, and real and imaginary parts of complex numbers can be used to represent regions of the plane (Example 10.1).

1.5 Moivre's Formula, Origin and Infinity

The failure of the polar representation (r, φ) at the origin and infinity can be justified analytically by starting from the Cartesian representation (x,y), which is valid everywhere. They are related (Figure 1.1) by

$$x = r\cos\varphi, \quad y = r\sin\varphi; \tag{1.13a,b}$$

this leads to Moivre's formula,

$$z = x + iy = r(\cos\varphi + i\sin\varphi) = re^{i\varphi}, \tag{1.14}$$

for the complex number z, where the notation used was

$$e^{i\varphi} \equiv \cos\varphi + i\sin\varphi; \tag{1.15}$$

this can be justified in the context of the theory of elementary transcendental functions. *The polar (r, φ) and Cartesian (x,y) representations of a complex number (1.14) are related by (1.13a,b) and its inverse (1.11;1.12a,b) for $r \neq 0, \infty$.* The inversion fails if the Jacobian is zero or infinity:

$$\begin{vmatrix} \partial x/\partial r & \partial x/\partial\varphi \\ \partial y/\partial r & \partial x/\partial\varphi \end{vmatrix} = \begin{vmatrix} \cos\varphi & -r\sin\varphi \\ \sin\varphi & r\cos\varphi \end{vmatrix} = r \neq 0, \infty, \tag{1.16}$$

that is, at the origin $r = 0$ and point at infinity $r = \infty$, where the argument φ is undetermined. The complex numbers gain relative to the real numbers by the property that the root, or inverse of power, is a closed operation; they loose the strict ordering property (1.6), which (i) holds for rational numbers; (ii) extends to real numbers, by the Dedekind section; and (iii) does not extend to complex numbers. For example, (a) complex numbers can be partially ordered by their modulus, argument, real, or imaginary part; (b) they are not strictly ordered, for example, two complex numbers may have the same modulus or argument or real or imaginary part and still not coincide.

1.6 Conjugate and Reflection on the Origin and Axis

The point obtained (Figure 1.1) by reflection of P upon the real axis defines the complex **conjugate**:

$$z = x + iy\text{: } z^* = x - iy = re^{-i\varphi}. \tag{1.17}$$

The reflection on the origin defines the **symmetric** complex

$$-z = -x - iy = re^{i(\pi+\varphi)}, \tag{1.18}$$

and the symmetric conjugate,

$$-z^* = -x + iy = re^{i(\pi-\varphi)}, \tag{1.19}$$

specifies the reflection in the imaginary axis. Making the convention that *the quadrants follow each other cyclically in the positive direction 1,2,3,4,1,2,..., then the complex numbers z in (1.14), $-z^*$ in (1.19), $-z$ in (1.18), and z^* in (1.17) lie on successive quadrants of the complex plane (Table 1.2 and Figure 1.1).*

TABLE 1.2
Reflection on the Axis and at the Origin

Transformation	**Conjugate**	**Symmetric**	**Symmetric conjugate**	**Name**
Reflection upon	Real axis	Origin	Imaginary axis	Representation
$(x, y) \rightarrow$	$(x, -y)$	$(-x, -y)$	$(-x, y)$	Cartesian
$z \rightarrow$	z^*	$-z$	$-z^*$	Complex
$(r, \varphi) \rightarrow$	$(r, -\varphi)$	$(r, \pi + \varphi)$	$(r, \pi - \varphi)$	Polar

Note: Reflections of a complex number on the origin and coordinate axis, using Cartesian and polar representations, as shown in Figure 1.1.

1.7 Power with Integral Exponent and Logarithm

The modulus and argument of the polar representation of a complex number, z, can also be applied to any complex function, for example, the **integral power** (1.20b) whose exponent (1.20a) is a positive or negative integer:

$$n \in |Z: \quad z^n = \left(r e^{i\varphi}\right)^n = r^n e^{in\varphi}; \tag{1.20a,b}$$

from (1.20b) follow the formulas:

$$n \in |Z; \quad z \in |C : |z^n| = |z|^n, \quad \arg(z^n) = n \arg(z) \tag{1.21a,b}$$

for *the modulus (1.21a) and argument (1.21b) of an integral power*. The power can be extended to nonintegral exponent (Sections 5.7 and 5.8).

Similarly, the real and imaginary parts of a complex number z can be applied to a function $f(z)$, for example, the Neperian logarithm or logarithm of base e:

$$\log z = \log\left(r e^{i\varphi}\right) = \log r + \log e^{i\varphi} = \log r + i\varphi, \tag{1.22}$$

where the polar representation was used; it leads to the formulas:

$$\operatorname{Re}(\log z) = \log|z|, \quad \operatorname{Im}(\log z) = \arg(z), \tag{1.23a,b}$$

for *the real (1.23a) and imaginary (1.23b) parts of the natural logarithm.*

1.8 Real, Imaginary, and Complex Exponential

The real and imaginary parts, and modulus and argument are four functions that can be applied to any complex expression. For example, in the case of the real $x \neq 0 = y$, imaginary $x = 0 \neq y$ or complex $x \neq 0 \neq y$ exponential:

$$e^z = e^{x+iy} = e^x e^{iy} = e^x \cos y + i e^x \sin y, \tag{1.24}$$

the modulus (1.25a), argument (1.25b), and the real (1.25c) and imaginary (1.25d) parts are, respectively,

$$|e^z| = e^x, \quad \arg(e^z) = y, \quad \operatorname{Re}, \operatorname{Im}(e^z) = e^x(\cos y, \sin y); \tag{1.25a–d}$$

alternatively

$$|e^z| = \exp\{\operatorname{Re}(z)\}, \quad \arg(e^z) = \operatorname{Im}(z), \tag{1.26a,b}$$

$$\operatorname{Re},\ \operatorname{Im}(e^z) = \exp\{\operatorname{Re}(z)\}\left(\cos\{\operatorname{Im}(z)\}, \sin\{\operatorname{Im}(z)\}\right), \tag{1.26c,d}$$

hold for the *modulus (1.25a) ≡ (1.26a), argument (1.25b) ≡ (1.26b), real and imaginary (1.25c) ≡ (1.26c) and imaginary (1.25d) ≡ (1.26d) parts of the complex exponential.* Two more cases are considered in Example 10.2.

1.9 Noncommutative Product of Quaternions (Hamilton, 1843)

A generalization of the complex number from 2 to 4 dimensions is the **quaternion** *(Hamilton, 1843) defined by*

$$\tilde{x} \equiv x_0 + x_1 i + x_2 j + x_3 k, \tag{1.27}$$

where i, j, k satisfy the following multiplication rules:

$$ii = jj = kk = -1, \quad ij = k = -ji, \quad ki = j = -ik, \quad jk = i = -kj, \tag{1.28a–d}$$

which fall into three groups: (i) squares of i, j, or k equal -1 as in (1.9b); (ii) products are skew-symmetric, for example, $ij = -ji$; and (iii) products follow in cyclic permutations (i,j,k), for example, $jk = i$. The latter two symbolic rules (ii, iii) are similar to the outer product (1.28c,d) of orthogonal base vectors (i, j, k) in three dimensions. The sum of quaternions is commutative:

$$\begin{aligned} \tilde{x} + \tilde{y} &= (x_0 + x_1 i + x_2 j + x_3 k) + (y_0 + y_1 i + y_2 j + y_3 k) \\ &= (x_0 + y_0) + i\,(x_1 + y_1) + j\,(x_2 + y_2) + k\,(x_3 + y_3) = \tilde{y} + \tilde{x}; \end{aligned} \tag{1.29}$$

the product of quaternions is not commutative:

$$\begin{aligned} \tilde{x} \times \tilde{y} &\equiv (x_0 + x_1 i + x_2 j + x_3 k) \times (y_0 + y_1 i + y_2 j + y_3 k) \\ &= (x_0 y_0 - x_1 y_1 - x_2 y_2 - x_3 y_3) + i\,(x_0 y_1 + x_1 y_0 + x_2 y_3 - x_3 y_2) \\ &\quad + j\,(x_0 y_2 - x_1 y_3 + x_2 y_0 + x_3 y_1) + k\,(x_0 y_3 + x_1 y_2 - x_2 y_1 + x_3 y_0) \\ &\neq \tilde{y} \times \tilde{x}, \end{aligned} \tag{1.30}$$

as follows interchanging $(x^i,\ y^i)$ with $i = 0, 1, 2, 3$. A quaternion represents a point (x_1, x_2, x_3, x_4) in a four-dimensional space.

Subtracting the **scalar part** x_0 from the quaternion, $\vec{x}$ specifies the **vector part** in (1.31a), which represents a point in three-dimensional space:

$$\vec{x} \equiv \tilde{x} - x_0 = x_1\, i + x_2\, j + x_3\, k: \quad \tilde{x} = x_0 + \vec{x}, \quad \tilde{x}^* \equiv x_0 - \vec{x}, \tag{1.31a–c}$$

and the **conjugate** (1.31c) of a quaternion (1.31b) is defined by the same (opposite) scalar (vector) part. This is similar to the conjugate (1.17) of a complex number taking the real x (imaginary y) part as the scalar x_0 (vector $\vec{x}$). *The product of a quaternion by its conjugate is commutative, real, and positive and specifies the* **norm** $\|\ldots.\|$ *and the* **modulus** $|\ldots.|$:

$$\begin{aligned} \|\tilde{x}\| \equiv |\tilde{x}|^2 \equiv \tilde{x} \times \tilde{x}^* &= \left(x_0 - \vec{x}\right) \times \left(x_0 + \vec{x}\right) = (x_0)^2 + (x_1)^2 + (x_2)^2 + (x_3)^2 \\ &= \tilde{x}^* \times \tilde{x}, \end{aligned} \tag{1.32}$$

and the **inverse** $\tilde{x}^{-1}$:

$$\tilde{x}^{-1} \equiv \tilde{x}^* \left|\tilde{x}\right|^{-2}, \quad \tilde{x}^{-1} \times \tilde{x} = \left|\tilde{x}\right|^{-2} \tilde{x}^* \times \tilde{x} = 1 = \tilde{x} \times \tilde{x}^{-1}. \tag{1.33a,b}$$

This is similar to the modulus (1.11) and inverse (3.19) of a complex number.

Note 1.1. *Types of Numbers*: The extension from complex numbers (1.9a,b) in two dimensions $N = 2$ to quaternions (1.27, 1.28a–d) in four dimensions $N = 4$ "loses" the commutativity of the product, and there is no square root (Table 1.1). Additional properties would be "lost" extending complex numbers to **hypercomplex** numbers, other than quaternions $N = 4$, viz. in higher dimensions $N = 3, 5, 6, \ldots$. Quaternions were used historically to develop vector algebra in three dimensions, until it became clear that it is simpler to do without them. The N-dimensional geometry with $N \geq 3$ is based on N-tuples, or ordered sets of N real numbers, viz. x_n with $n = 1, \ldots, N$, rather than any extension of complex numbers. Complex coordinates $z_n = x_n + iy_n$ are used in spinors in quantum mechanics, both nonrelativistic and relativistic. Although quaternions can be used as coordinates in a four-dimensional space, they are not suitable for the space-time of general relativity because (i) the norm (1.33) of a quaternion corresponds to a Cartesian metric which is positive-definite, that is, has signature $+ + + +$ viz. all signs $+$ in (1.32); (ii) the interval in space-time corresponds to an indefinite metric with signature $- - - +$ or $+ + + -$, that is, one reversed sign in (1.32). Quaternions were invented (Hamilton, 1843) before relativity (Einstein, 1904) and were conceived in connexion with the representation of rotations and rigid displacements in three-dimensional space, which remains their main application. The complex number $e^{i\varphi}$ [quaternion (1.27) $\equiv$ (1.31a)] can be used to represent a rotation in the plane (in space). The rotations in the plane (space) are (are not) commutative, that is, the final position does not (does) depend on the order, that is, it is not (is) changed if the two rotations are interchanged. Two successive rotations in the plane (space) correspond to the product of complex numbers (quaternions), implying that the product is (is not) commutative. Properties of spinors relate to quaternions. The theory of numbers concerns many other topics, for example, **transfinite (prime) numbers** that lie beyond natural numbers (have no natural numbers as divisors). The preceding account suggests that *complex numbers are not only the simplest (Section 1.2) but also the most general, which satisfy all the usual properties of sum and product (Chapter 3) as a continuous set.*

Conclusion 1: The polar and Cartesian representations (Figure 1.1) on the plane, of a complex number z, its conjugate z^*, symmetric $-z$, and symmetric conjugate $-z^*$, also represented in Table 1.2. Table 1.1 lists types of numbers and operations between them.

2

Stability of an Equilibrium Position

When a body (or system) is in a position (or state) of equilibrium, and is subjected to a perturbation, three cases can arise: (i) if it remains in the new, disturbed position, the equilibrium is said to be *indifferent* (Section 2.1); (ii) if it always returns to the old, initial position, the equilibrium is *stable*; and (iii) if it can get farther from the equilibrium position, that is, the perturbation increases, the equilibrium is *unstable*. The return to equilibrium in the case of stability, or deviation from equilibrium in the unstable case, can be monotonic or oscillatory. The words stable, indifferent, and unstable are sometimes reserved for the monotonic case (Section 2.3); the oscillatory case (Sections 2.2 and 2.4) is called *damped*, *oscillatory*, or *overstable*, depending on whether the amplitude of the oscillation decays, stays constant, or grows. All these six cases of motion are included in a single complex expression (Sections 2.5–2.7), which is discussed (Sections 2.8 and 2.9) for a system of arbitrary order. The existence of an equilibrium is not sufficient to ensure that it actually occurs; usually only stable equilibria are found in nature or assure the correct functioning of an engineering device, bearing in mind that perturbations are almost inevitably present. The question of existence of equilibria, and their stability, applies to various kinds of systems, for example, mechanical, electrical, chemical, and so on, both in static and in dynamic conditions.

2.1 Trajectory Following a Perturbation of Equilibrium

A body is in an equilibrium position $x = 0$. At a time $t = 0$ it is displaced to a position x_0, and its subsequent motion for $t > 0$ is specified by the position as a function of time by (2.1b):

$$\zeta = \omega + i\lambda: \quad x(t) = \mathrm{Re}\left(x_0\, \mathrm{e}^{i\zeta t}\right), \tag{2.1a,b}$$

where (2.1a) is generally complex. Since x must be real, the real part of the complex expression in parentheses in (2.1b) is taken. The first and second derivatives of the **displacement** (2.1b) with regard to time specify, respectively, the **velocity** v and **acceleration** a:

$$v(t) \equiv \frac{\mathrm{d}x}{\mathrm{d}t} = \mathrm{Re}\left(i\zeta x_0 \mathrm{e}^{i\zeta t}\right), \quad a(t) \equiv \frac{\mathrm{d}v}{\mathrm{d}t} = \frac{\mathrm{d}^2x}{\mathrm{d}t^2} = \mathrm{Re}\left(-\zeta^2 x_0 \mathrm{e}^{i\zeta t}\right); \tag{2.2a,b}$$

here was used $i^2 = -1$ in (1.9b), because $i = \mathrm{e}^{i\pi/2}$ and $i^2 = \mathrm{e}^{i\pi} = -1$. The expressions (2.1, 2.2a,b) are examples of the **complex representation** of a real quantity, and sometimes the function real part of Re is omitted.

The motion (2.1, 2.2a,b) is analyzed next for all possible values of ζ to assess the nature of the equilibrium position $x = 0$. The trivial case of (2.1a,b) is (2.3a):

$$\zeta = 0: \quad x(t) = x_0, \quad v(t) = 0 = a(t), \tag{2.3a,b}$$

that is, *the equilibrium is* **indifferent** *at* x = *0, since the body, after being displaced to* x_0, *remains at rest in the new position (Figure 2.1a).*

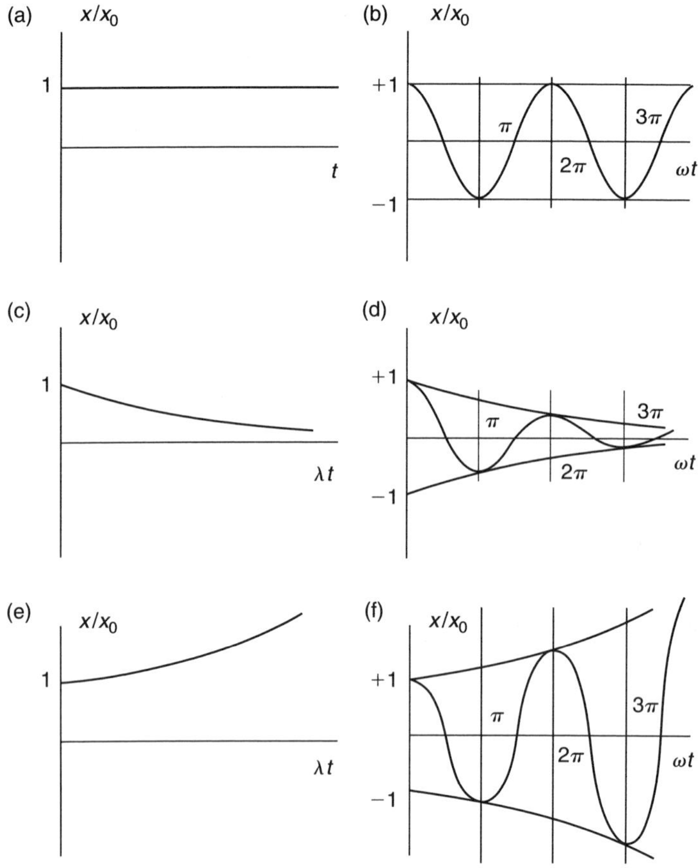

FIGURE 2.1
The six types of motion specified by (2.1a,b) fall in three cases. The first set is (i) indifferent equilibrium, with rest at displaced position (a). The second set is (ii) stability with (ii-1) oscillation around equilibrium position (b) with frequency ω; (ii-2) monotonic decay (c) toward equilibrium position with damping λ; (ii-3) the combination (d) of damped amplitude (ii-2) and harmonic oscillation (ii-1). The third set is (iii) instability with (iii-a) monotonic (e) [(iii-b) oscillatory (f)] growth. The latter (f) is called overstability.

2.2 Oscillatory Motion with Constant Amplitude

The case (II) of real (2.4a), the motion

$$\zeta = \omega: \quad x(t) = \operatorname{Re}\left(x_0 e^{i\omega t}\right) = x_0 \cos(\omega t), \tag{2.4a,b}$$

$$v(t) = \operatorname{Re}\left(i x_0 \omega e^{i\omega t}\right) = -x_0 \omega \sin(\omega t), \tag{2.4c}$$

$$a(t) = \operatorname{Re}\left(-x_0 \omega^2 e^{i\omega t}\right) = -x_0 \omega^2 \cos(\omega t), \tag{2.4d}$$

is **oscillatory**, *with* **frequency** ω, *and* **period** $\tau = 2\pi/\omega$, *with constant amplitudes* $x_0, x_0\omega, x_0\omega^2$, *respectively, for the displacement (2.4b), velocity (2.4c), and acceleration (2.4d) that are out-of-phase by* $\pi/2, \pi$, *that is, a quarter- and a half-period* $\tau/4, \tau/2$. The latter statement corresponds to

$$v(t) = \omega x\left(t + \frac{\pi}{2\omega}\right) = \omega x\left(t + \frac{\tau}{4}\right), \quad a(t) = \omega^2 x\left(t + \frac{\pi}{\omega}\right) = \omega^2 x\left(t + \frac{\tau}{2}\right), \tag{2.5a,b}$$

which follows from (2.4b–d) using

$$\cos\left(\omega t + \frac{\pi}{2}\right) = -\sin(\omega t), \quad \cos(\omega t + \pi) = -\cos(\omega t). \tag{2.6a,b}$$

The displacement from the equilibrium position $x = 0$ never exceeds the initial displacement $|x(t)| \leq x_0$ and the body passes through the equilibrium position every half-period $x((n + 1/2)\tau) = 0$ as shown in Figure 2.1b.

2.3 Attenuation or Amplification and Stability or Instability

In the case (III) of pure imaginary (2.7a)

$$\zeta = i\lambda: \quad x(t) = \mathrm{Re}\left(x_0 e^{i^2\lambda t}\right) = x_0 e^{-\lambda t}, \tag{2.7a,b}$$

there are two possibilities:

$$v(t) = \mathrm{Re}\left(x_0 i^2 \lambda^2 e^{i^2\lambda t}\right) = -x_0 \lambda e^{-\lambda t}, \quad a(t) = \mathrm{Re}\left(x_0 i^4 \lambda^2 e^{i^2\lambda t}\right) = x_0 \lambda^2 e^{-\lambda t}, \tag{2.7c,d}$$

(IIIA) if $\lambda > 0$ *there is* **attenuation**, *and the displacement (2.7b), velocity (2.7c), and acceleration (2.7d) decay monotonically from their initial values, respectively,* $x_0, -\lambda x_0$, *and* $\lambda^2 x_0$, *to zero, as the body returns to the position of equilibrium* $x(\infty) = 0$, *which is* stable (Figure 2.1c); (IIIB) if $\lambda < 0$ *there is* **amplification**, *and the displacement, velocity, and acceleration increase monotonically as the body gets progressively farther from the position of equilibrium that is* **unstable** (Figure 2.1e). The evolution is monotonic, and,

$$x(t) = -\lambda^{-1} v(t) = \lambda^{-2} a(t), \tag{2.8}$$

in *both cases the acceleration has the same sign as the displacement, whereas the velocity has the same (opposite) sign in the case* $\lambda < 0\,(\lambda > 0)$ *of amplification (attenuation)*, that is, the deviation (approach) to equilibrium involves continuing (reversing) the initial motion from $x(0) = x_0$ to $x(\infty) = \infty\,[x(\infty) = 0]$.

2.4 Damped Oscillation or Overstable Growth

The general case (IV) of complex (2.9a) corresponds to

$$\zeta = \omega + i\lambda: \quad x(t) = \mathrm{Re}\left(x_0 e^{i\omega t} e^{i^2\lambda t}\right) = x_0 e^{-\lambda t}\cos(\omega t), \tag{2.9a,b}$$

a *motion which is oscillatory with frequency* ω *and period* $\tau = 2\pi/\omega$, *and has*

$$v(t) = \mathrm{Re}\left(x_0 i(\omega + i\lambda)\right) e^{i\omega t} e^{i^2\lambda t} = -x_0 e^{-\lambda t}\{\omega\sin(\omega t) + \lambda\cos(\omega t)\}, \tag{2.9c}$$

$$a(t) = \mathrm{Re}\left(x_0 i^2(\omega + i\lambda)^2\right) e^{i\omega t} e^{i^2\lambda t} = x_0 e^{-\lambda t}\left\{\left(\lambda^2 - \omega^2\right)\cos(\omega t) + 2\omega\lambda\sin(\omega t)\right\}, \tag{2.9d}$$

nonconstant amplitude: (IVA) if $\lambda > 0$ *there is an exponential decay* (Figure 2.1d) *corresponding to a* **damped** oscillation that *takes the body back to the equilibrium position* $x(\infty) = 0$, *with progressively smaller overshoots on either side; (IVB) if* $\lambda < 0$ *there is exponential growth* (Figure 2.1f), *corresponding to an amplified oscillation designated* **overstability**, *because although the body goes through the equilibrium position every half-period, it overshoots progressively more on either side.* The sequence of maxima and minima can be used to define a logarithmic decrement (Example 10.3).

2.5 General Relations for Amplitudes and Phases

The frequency ω and attenuation/amplification factor λ have the same dimensions, namely, the inverse of time, and when they are comparable $\lambda \sim \omega$ there is no simplification of the general formulas (2.9a–d), which specify the following *relations between the amplitudes (2.10a,b) and phases (2.11a,b) of the displacement (2.9b), velocity (2.9c), and acceleration (2.9d):*

$$|v| = |x|\,|\zeta| = |x|\sqrt{\omega^2+\lambda^2}, \qquad |a| = |x|\left|\zeta^2\right| = |x|\left(\omega^2+\lambda^2\right), \tag{2.10a,b}$$

$$\arg(v) = \arg(x) + \arg(i\zeta) = \arg(x) + \frac{\pi}{2} + \operatorname{arc\,tan}\left(\frac{\lambda}{\omega}\right), \tag{2.11a}$$

$$\arg(a) = \arg(x) + \arg\left(-\zeta^2\right) = \arg(x) - \pi + 2\operatorname{arc\,tan}\left(\frac{\lambda}{\omega}\right). \tag{2.11b}$$

The formulas (2.9c,d) have been deduced from (2.9a,b), and (2.7c,d) from (2.7a,b), and (2.4c,d) from (2.4a,b), using the complex representations; they can be checked by direct derivation of the real expressions, for example, to derive (2.9d) from (2.9b):

$$\begin{aligned}\frac{\mathrm{d}^2\left\{x_0 \mathrm{e}^{-\lambda t}\cos(\omega t)\right\}}{\mathrm{d}t^2} &= \frac{\mathrm{d}\left\{-x_0\mathrm{e}^{-\lambda t}\left\{\lambda\cos(\omega t)+\omega\sin(\omega t)\right\}\right\}}{\mathrm{d}t} \\ &= x_0\mathrm{e}^{-\lambda t}\left\{\left(\lambda^2-\omega^2\right)\cos(\omega t) + 2\lambda\omega\sin(\omega t)\right\}.\end{aligned} \tag{2.12}$$

The deduction of the algebraic relations (2.10a,b and 2.11a,b) from the real representation would require determination of the amplitudes from the maxima of x, v, $a(t)$, and of the phases as the time lags at these maxima; the use of the complex representation is much simpler.

2.6 Predominantly or Weakly Oscillatory Motion

If the frequency ω and attenuation/amplification factor λ are of dissimilar magnitudes, the general results in Sections 2.5 and 2.6 simplify, in the two extremes of (i) motion that is **predominantly oscillatory** (2.13a):

$$\omega \gg |\lambda|: \quad x(t), v(t), a(t) = x_0\mathrm{e}^{-\lambda t}\left\{\cos(\omega t), -\omega\sin(\omega t), -\omega^2\cos(\omega t)\right\}, \tag{2.13a–d}$$

the amplitude and phase relations are the same as for the oscillatory motion (2.4b–d) but the amplitude grows (decays) for $\lambda < 0\,(\lambda > 0)$ *slowly on the time scale of a period*

$1/|\lambda| \gg 1/\omega = \tau/2\pi$; *(ii) motion that is* **predominantly attenuated/amplified** (2.14a):

$$|\lambda| \gg \omega: \quad x(t), v(t), a(t) = x_0 \cos(\omega t)\, e^{-\lambda t} \left\{1, -\lambda, \lambda^2\right\}, \tag{2.14a–d}$$

the relations between amplitudes are the same as for the purely attenuated/amplified motion, (2.7a–d), but there is a long-period $\tau = 2\pi/\omega \gg 2\pi/|\lambda|$ of oscillation in phase, which implies small overshoots in the damping, and a large-scale inversion of motion at long intervals of time in the case of overstability.

2.7 Frequency and Attenuation/Amplification Factor

It is clear from the preceding account that the type of motion specified by (2.1b) and the nature of the equilibrium position $x = 0$ are determined solely by the position of (2.1a) in the complex ζ-plane, which acts as a diagnostic diagram in Figure 2.2 and Table 2.1. All six cases are determined by the combination of (i) zero $\omega = 0$ or nonzero $\omega \neq 0$ frequency, for monotonic or oscillatory motion [the sign of the frequency does not matter since $\mathrm{Re}\left(e^{i\omega t}\right) = \cos(\omega t) = \mathrm{Re}\left(e^{-i\omega t}\right)$, that is, it can be taken always positive]; (ii) constant $\lambda = 0$, increasing $\lambda > 0$ or decreasing $\lambda < 0$ amplitude, since the sign of the imaginary part of ζ is important to distinguish attenuation from amplification $\mathrm{Re}(e^{i^2\lambda t}) = e^{-\lambda t} \neq e^{\lambda t} = \mathrm{Re}(e^{-i^2\lambda t})$. The position (Figure 2.2) of the ζ-plane specifies (i) rest at the origin (Figure 2.1a); (ii) oscillation with constant amplitude (Figure 2.1b) on the real axis; (iii/iv) monotonic decay (growth) on [Figure 2.1c (2.1e)] the positive (negative)

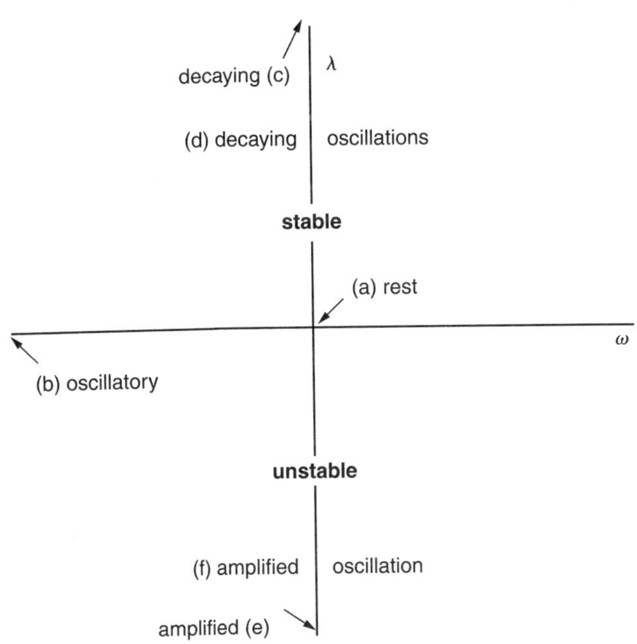

FIGURE 2.2
The position in the (ω, λ)-plane of frequency ω and damping/amplification ($\lambda > / < 0$) indicates the six types of motion and three cases of stability in Table 2.1 and Figure 2.1a–f.

TABLE 2.1
Amplification/Attenuation Factor

Parameters Meaning	$\lambda > 0$ **Attenuation**	$\lambda = 0$ **Constant amplitude**	$\lambda < 0$ **Amplification**
$\omega = 0$ Monotonic	*stable* Figure 2.1c	*indifferent* Figure 2.1a	*unstable* Figure 2.1e
$\omega \neq 0$ Nonzero Frequency	*damped* Figure 2.1d	*oscillatory* Figure 2.1b	*overstable* Figure 2.1f

Note: The equation of motion (2.1b) as the real representation of a complex quantity (2.1a) leads to the six cases illustrated in Figure 2.1a–f. The parameters are initial amplitude x_0, oscillation frequency ω, and damping λ. The six cases are indicated in the diagnostic plane $\omega + i\lambda$ in Figure 2.2.

imaginary axis; and (v/vi) oscillatory decay or stability (oscillatory growth or overstability) in [Figure 2.1d (2.1f)], the upper (lower) half-plane.

2.8 Differential Equation and Stability Criteria

The nature of the equilibrium, and the motion in its neighborhood, depends only on the value of (2.1a) in (2.1b), and it is indicated next how it can be determined in the case of a system described by a linear ordinary differential equation of order N with constant coefficients:

$$\sum_{n=0}^{N} B_n \frac{\mathrm{d}^n x}{\mathrm{d}t^n} = 0, \tag{2.15}$$

without forcing, that is, homogeneous in the sense the r.h.s. is zero. Bearing in mind that (2.1b) satisfies

$$\frac{\mathrm{d}^n x}{\mathrm{d}t^n} = \frac{\mathrm{d}^n \left(x_0 \mathrm{e}^{i\zeta t}\right)}{\mathrm{d}t^n} = (i\zeta)^n x_0 \mathrm{e}^{i\zeta t} = (i\zeta)^n x(t), \tag{2.16}$$

it follows that (2.15) is met by

$$x(t)\, Q_N(\zeta) = 0, \qquad Q_N(\zeta) \equiv \sum_{n=0}^{N} B_n (i\zeta)^n = (i)^N B_N \prod_{n=0}^{N} (\zeta - \zeta_n), \tag{2.17a,b}$$

for all t in (2.17a), provided that ζ be a root ζ_n of the complex polynomial (2.17b) of degree N.

To each distinct ζ_n root of the **characteristic polynomial** (2.17b) corresponds to a particular solution (2.18a):

$$x_n(t) = \exp(i\zeta_n t); \quad x(t) = \sum_{n=1}^{N} A_n x_n(t), \tag{2.18a,b}$$

their linear combination (2.18b) involving N arbitrary constants $A_1, \ldots, A_N$ specifies the general solution. If a root ζ_0 is of multiplicity q, then the q "identical solutions" (2.18a) are replaced by (2.18d):

$$k = 0, \ldots, q-1: \quad x_k(t) = t^k \exp(i\zeta_0 t), \tag{2.18c,d}$$

which are q distinct solutions (2.18c).

From the receding results follow **the stability criteria**: *the equilibrium $x = 0$ of the system described by the differential equation (2.15) with characteristic polynomial (2.17b) is specified by the nature of the roots (2.18a–d) as follows (Figure 2.2): (i) if all roots have positive imaginary parts the equilibrium is stable, and the motion monotonic attenuated if all real parts are zero, or damped oscillatory if at least one real part is nonzero; (ii) if at least one root has negative imaginary part the equilibrium is unstable, and the motion monotonic amplified if all such roots have zero real part, and oscillatory overstable if at least one such root has nonzero real part; and (iii) in the remaining case of some roots with zero imaginary part and none with negative imaginary part, there are three subcases: (a) indifferent equilibrium if the only root is zero (both real and imaginary parts); (b) oscillation with constant amplitude if at least one of the roots with zero imaginary part has nonzero real part; and (c) if there are multiple roots there is instability if their real parts are all zero, and overstability if at least one real part is nonzero.* The arbitrary constants in the general solution (2.18b) are determined from initial conditions, as shown next in the particular case of the harmonic oscillator.

2.9 Initial Conditions for Harmonic Oscillator

The application of initial conditions to render unique the solution of a differential equation describing the stability of a system is illustrated next in the simplest case of a harmonic oscillator (Subsection 2.9.1) leading to three interchangeable pairs of arbitrary constants (Subsection 2.9.2).

2.9.1 Equation of Motion of a Harmonic Oscillator

As a simple example suppose that the displacement satisfies the differential equation

$$m > 0, \; k > 0: \quad m\ddot{x} = -kx, \tag{2.19}$$

which states a balance between (i) the **inertia** force, equal to **mass** m times acceleration; and (ii) the force of a linear **spring** that is proportional to the displacement, through the **resilience** k, and pulls the particle back to the origin $x = 0$; a large (small) resilience means (i) for same displacement x, a strong (weak) pulling force; and (ii) for the same mass m, a higher (lower) acceleration. The mechanical system (2.19) consisting of a mass and a spring omits the damper (included in Section 4.1, Figure 4.1a, and Example 10.4). For the present purpose of application of initial conditions, this case is sufficient. The subsequent solutions of (2.19) assume that there is a spring $k \neq 0$. In the absence of the spring (2.20a), the acceleration is zero (2.20b):

$$k = 0: \quad \ddot{x} = 0, \quad \dot{x}(t) = v_0, \quad x(t) = x_0 + v_0 t, \tag{2.20a–d}$$

and hence the motion is uniform (2.20c,d). The two arbitrary constants in (2.20d) are the initial velocity (2.20c) and initial position $x_0 = x(0)$. The equation of motion (2.19) applies to the harmonic oscillator that involves six assumptions: (i) one-dimensional since there is a single displacement x; (ii) vibration since it depends only on time t; (iii) linear since it involves the displacement x and time derivatives in (2.15), but no powers or cross-products; (iv) second-order since the highest order of derivation is two, corresponding to the acceleration; (v) without damping or amplification, which is considered in Examples 10.4 and 10.5; and (vi) it is a free oscillation due to initial conditions, that is, without applied external forces. The relaxation of these assumptions is considered in Note 2.1.

The equation of motion (2.19) can be written in the form (2.21a):

$$\ddot{x} + \omega^2 x = 0, \quad \omega \equiv \sqrt{\frac{k}{m}}, \tag{2.21a,b}$$

where (2.21b) is the **natural frequency**, because it will be shown next (2.23b) that the motion is an oscillation with this frequency. To prove this, the equation of motion (2.21a) is put into the form

$$0 = \left\{ \frac{\mathrm{d}^2}{\mathrm{d}t^2} + \omega^2 \right\} x(t) = \left\{ Q_2 \left(\frac{\mathrm{d}}{\mathrm{d}t} \right) \right\} x(t), \tag{2.22a}$$

showing (2.1b) that characteristic polynomial (2.17b) in the case is

$$Q_2(\zeta) = (i\zeta)^2 + \omega^2 = -\zeta^2 + \omega^2 = -(\zeta - \omega)(\zeta + \omega); \tag{2.22b}$$

it has real symmetric roots (2.23a):

$$\zeta_\pm = \pm\omega: \quad x_\pm(t) = \exp(i\zeta_\pm t) = \exp(\pm i\omega t), \tag{2.23a,b}$$

corresponding to solutions (2.23b) that are indeed oscillations with frequency (2.21b).

2.9.2 Three Interchangeable Pairs of Arbitrary Constants

A linear combination (2.18b) of (2.23b), viz.:

$$x(t) = A_+ x_+(t) + A_- x_-(t) = A_+ \mathrm{e}^{i\omega t} + A_- \mathrm{e}^{-i\omega t}, \tag{2.24}$$

where $A_\pm$ are arbitrary constants, is also a solution of the equation of motion (2.19) $\equiv$ (2.22a), viz. it is the general solution. Using the identity (1.15) in the form

$$\exp(\pm i\omega t) = \cos(\omega t) \pm i \sin(\omega t), \tag{2.25}$$

the law of motion (2.24) becomes

$$x(t) = A_1 \cos(\omega t) + A_2 \sin(\omega t), \tag{2.26}$$

where A_1, A_2 are given by

$$A_1 \equiv A_+ + A_-, \quad A_2 \equiv i(A_+ - A_-). \tag{2.27a,b}$$

That is, a second pair of arbitrary constants related to the first pair.

The velocity is specified from (2.26) by

$$v(t) = \frac{\mathrm{d}x}{\mathrm{d}t} = \omega\{-A_1 \sin(\omega t) + A_2 \cos(\omega t)\}. \tag{2.28}$$

The arbitrary constants may be determined from **initial conditions**, for example, of **single-point** type, specifying initial displacement (2.29b) and velocity (2.29c) at time (2.29a):

$$t = 0: \quad x_0 = x(0) = A_1, \quad v_0 = v(0) = \omega A_2. \tag{2.29a–c}$$

Thus, the displacement (2.26) [velocity (2.28)] are given by (2.30a) [(2.30b)]:

$$x(t) = x_0 \cos(\omega t) + \frac{v_0}{\omega}\sin(\omega t), \quad v(t) = -x_0\omega\sin(\omega t) + v_0\cos(\omega t), \tag{2.30a,b}$$

confirming the initial values (2.29b) [(2.29c)] for (2.29a).

A third pair (A, α) of arbitrary constants is

$$A\mathrm{e}^{i\alpha} \equiv A_1 + iA_2 = 2A_-, \tag{2.31}$$

where (2.27a,b) was used. Real A and α imply

$$A_1 = A\cos\alpha, \quad A_2 = A\sin\alpha; \tag{2.32a,b}$$

thus, the displacement (2.26) is given by

$$x(t) = A[\cos(\omega t)\cos\alpha + \sin(\omega t)\sin\alpha] = A\cos(\omega t - \alpha) = \mathrm{Re}\left\{A\mathrm{e}^{i(\omega t-\alpha)}\right\}. \tag{2.33}$$

This shows that the **amplitude** (2.34a) and **phase** (2.34b) of the oscillation are given by

$$A = |A_1 + iA_2| = \left|(A_1)^2 + (A_2)\right|^{1/2} = \left|{x_0}^2 + \left(\frac{v_0}{\omega}\right)^2\right|^{1/2}, \tag{2.34a}$$

$$\tan\alpha = \frac{A_2}{A_1} = \frac{v_0}{x_0\,\omega}, \tag{2.34b}$$

in terms of the initial conditions (2.29a–c). If at time $t = 0$ the velocity (displacement) is zero $v_0 = 0\,(x_0 = 0)$, the phase is $\alpha = 0 \ (\alpha = \pi/2)$, and the amplitude is $x_0\,(v_0/\omega)$. If at time $t = 0$ neither the displacement nor the velocity is zero, then (2.34a,b) do not simplify. It is possible to eliminate the phase in (2.33) by a suitable choice of origin of time:

$$\bar{t} \equiv t - \frac{\alpha}{\omega} = t - \frac{\alpha\tau}{2\pi}: \quad x(t) = A\cos(\omega\bar{t}) \equiv \bar{x}(\bar{t}), \tag{2.35a,b}$$

where the period τ was used.

Note 2.1. *Analysis of the Stability of Systems*: An example of harmonic oscillator appears in the rolling oscillations of a ship (Chapter 8). The six assumptions (i) to (vi) stated in Subsection 2.9.1 can be relaxed as indicated next. Forced oscillations (vi) are considered (Chapter 4) for a second-order system with damping (v), and the corresponding initial conditions appear in Example 10.5; this problem is an application of the solution ordinary differential equations; the multidimensional (i) case involves the solution of coupled systems of ordinary differential equations. Both linear and nonlinear (ii) oscillations can be considered for mechanical systems with one or several degrees of freedom; the nonlinear oscillations (iii) lead to elliptic functions. If the displacement were to depend on position and time, this would (iv) lead to waves described by partial differential equations.

Note 2.2. *Alternate Forms of the Stability Theorem*: The solution of the linear homogeneous ordinary differential equation with constant coefficients (2.15) can be written in several alternate forms:

$$\frac{x(t)}{x_0} = \mathrm{e}^{i\zeta t} = \mathrm{e}^{-st} = \mathrm{e}^{\vartheta t}, \tag{2.36a–c}$$

for example, (i) the kernel of the Fourier transform uses (2.36a) ≡ (2.1b); (ii) the kernel of the Laplace transform uses (2.36b); and (iii) the characteristic polynomial of the differential equation is usually (2.36c) put in the form

$$P_n(\vartheta) \equiv \sum_{n=0}^{N} B_n \vartheta^n = \sum_{n=0}^{N} B_n (-s)^n = \sum_{n=0}^{N} B_n (i\zeta)^n \equiv Q_N(\zeta). \qquad \text{(2.37a–c)}$$

The stability theorem (Section 2.8) was stated in terms of ζ, viz.:

$$\vartheta = -s = i\zeta = i(\omega + i\lambda) = -\lambda + i\omega, \qquad (2.38)$$

and can be restated in terms of ϑ or s, with ω (λ) having the same meaning of frequency (decay rate).

Conclusion 2: The stability conditions for the motion (2.1a,b) about an equilibrium position $x = 0$, are indicated in Table 2.1, viz. the position in the complex ζ-plane indicates (Figure 2.2) the cases (Figure 2.1) of: (a) indifferent equilibrium and body at rest in the perturbed position; (b) oscillation with constant amplitude about the equilibrium position; (c) monotonic attenuation towards the position of stable equilibrium; (e) monotonic amplification away from the position of unstable equilibrium; (d) oscillation with decreasing amplitude, i.e., damping towards the equilibrium position; (f) oscillation with increasing amplitude, i.e., overstable oscillation, which is an oscillatory form of instability.

3

Addition, Product, and Inverses

An operation upon complex numbers is a relation between two ordered pairs of real numbers and a third ordered pair, and it can be represented graphically on the complex plane. The sum (Sections 3.1 and 3.2) is most conveniently performed in the Cartesian representation, and the product (Sections 3.3 and 3.4) in the polar representation. The operations sum and product, and their inverses subtraction and division, can be used (Sections 3.5 and 3.6) together with the conjugate operator to determine the real and imaginary parts, and the modulus and argument, of complex expressions. The algebra of complex numbers is related to the geometry of the Cartesian plane (Sections 3.8 and 3.9), and implies a number of trigonometrical relations (Section 3.7). The algebra of complex numbers extends to functions of one or more variables, and thus to complex spaces whose real dimension is twice the number of complex variables.

3.1 Complex Addition and Rule of the Parallelogram

Starting with two complex numbers in the Cartesian representation:

$$u = a + ib, \quad v = c + id, \tag{3.1a,b}$$

the operator addition is defined by the sum of each element of the pair,

$$u + v \equiv (a + c) + i\,(b + d), \tag{3.2}$$

and corresponds (Figure 3.1) to the addition of u and v as vectors, according to the **rule of the parallelogram**. Hence, the real (imaginary) part is the sum of the real (imaginary) parts:

$$\operatorname{Re}(u + v) = \operatorname{Re}(u) + \operatorname{Re}(v)\,, \quad \operatorname{Im}(u + v) = \operatorname{Im}(u) + \operatorname{Im}(v)\,. \tag{3.3a,b}$$

The addition of complex numbers has the following properties: (i) it is closed because the sum of complex numbers is a complex number, that is, (1.1) holds for the operation $+$ in $|C$; (ii) it is commutative (3.4a) and associative (3.4b):

$$u + v = v + u, \quad (u + v) + w = u + (v + w)\,; \tag{3.4a,b}$$

(iii) it has a neutral element, namely, zero $0 + 0i$; (iv) each complex number $u = a + ib$ has a symmetric $-u = -a - ib$, given by (1.18). Hence, the ordered pair $(|C, +)$, where $|C$ is the set of complex numbers and $+$ the operation addition defined by (3.1a,b; 3.2), is a **commutative group**.

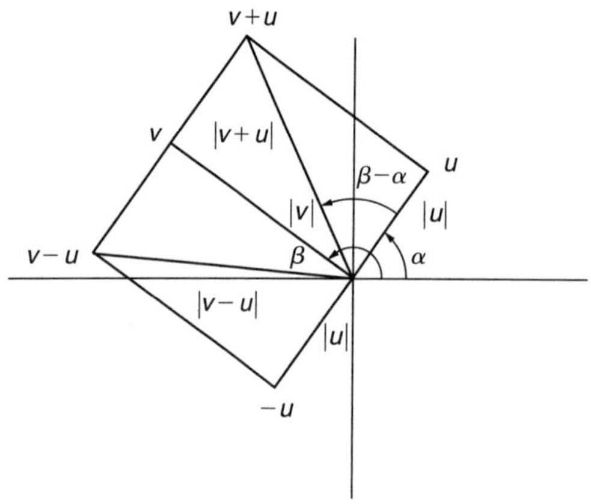

FIGURE 3.1
Taking two complex numbers u, v as the sides of a rectangle their sum is the diagonal. Their difference $u - v$ is the diagonal of the rectangle obtained by reflection on the v-direction.

3.2 Modulus, Argument, and Triangular Equalities (Pythagoras, VI b.c.)

The modulus (1.11) of the sum (3.2) is given by

$$\begin{aligned} |u+v|^2 &= (a+c)^2 + (b+d)^2 = \left(a^2+d^2\right) + 2\,(ac+bd) \\ &= |u|^2 + |v|^2 + 2\,|u|\,|v|\,(\cos\alpha\cos\beta + \sin\alpha\sin\beta)\,, \end{aligned} \tag{3.5}$$

where α, β are the arguments of u,v. The expression (3.5) simplifies

$$\alpha \equiv \arg(u)\,, \beta \equiv \arg(v): \quad |u+v| = \Big||u|^2 + |v|^2 + 2\,|u|\,|v|\cos(\alpha-\beta)\Big|^{1/2}, \tag{3.6}$$

which *specifies the length* $|u+v|$ *of one side of a triangle, from the lengths* $|u|, |v|$ *of the other two sides, and the angle* $\alpha - \beta$ *between them* (Figure 3.2a). In the case $\alpha - \beta = \pm\pi/2$, the two perpendicular sides lead to **Pythagoras theorem (VI b.c.)** *for the rectangular triangle (Figure 3.2b):*

$$\arg(u) - \arg(v) = \pm\frac{\pi}{2}: \quad |u+v|^2 = |u|^2 + |v|^2\,. \tag{3.7}$$

From (1.12a) and (3.2) follows the expression:

$$\begin{aligned} \arg(u+v) &= \text{arc }\tan\left(\frac{b+d}{a+c}\right) \\ &= \text{arc }\tan\left(\frac{|u|\sin\alpha + |v|\sin\beta}{|u|\cos\alpha + |v|\cos\beta}\right), \end{aligned} \tag{3.8}$$

for *the argument of the sum of complex numbers (3.8);* other alternative expressions are given in Example 10.6.

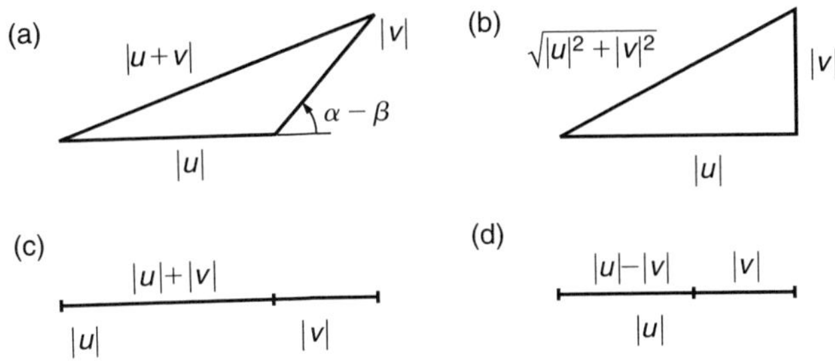

FIGURE 3.2
The sum of complex numbers by the rule of the diagonal of the rectangle (Figure 3.1) is equivalent to summing complex numbers as the sides of a triangle (a). A particular case is a rectangular triangle (b) with orthogonal sides for which holds the Pythagoras theorem. The degenerate cases are collapsed triangles with parallel (c) or antiparallel (d) sides.

3.3 Complex Product, Homothety, and Rotation

Starting with complex numbers in the polar representation:

$$u = |u|\, e^{i\alpha}, \qquad v = |v|\, e^{i\beta}, \tag{3.9a,b}$$

the operation product is defined by

$$uv \equiv |u|\, |v|\, e^{i(\alpha+\beta)}, \tag{3.10}$$

and corresponds (Figure 3.3) to (i) multiplying the moduli, that is, multiplying $|u|$ by $|v|$,which yields an extension if $|v| > 1$, a contraction if $|v| < 1$, and leaves the length unchanged if $|v| = 1$; (ii) adding the arguments, that is, adding α to β, which yields a counterclockwise (clockwise) rotation if $\beta > 0\,(\beta < 0)$. Thus, *the product of complex numbers corresponds algebraically to the product (sum) of moduli (arguments):*

$$|uv| = |u|\, |v|, \quad \arg(u+v) = \arg(u) + \arg|v|, \tag{3.11a,b}$$

and geometrically to a **homothety** (i) [**rotation** (ii)]; these are discussed in more detail subsequently (Section 35.1). Bearing in mind that $|uv|\,/\,|v| = |u|\,/1$ and $\arg(u) = \arg(uv) - \arg(v)$, the *triangles with vertices at* $(0,\ 1,\ u)$ *and at* $(0, v, uv)$ *are* **similar**, *that is, have equal angles between the sides, and the lengths of the sides are in the same ratio;* the triangles $(0,\ 1,\ v)$ and $(0,\ u,\ uv)$ are also similar, and both pairs are illustrated in Figure 3.4. The product of complex numbers is closed, has neutral element (viz. unity 1), and each element u has an inverse $1/u$ (as shown in Section 3.5). It is commutative (3.12a) and associative (3.12b), and distributive (3.12c) with regard to the sum:

$$uv = vu, \quad (uv)\, w = u\,(vw), \quad (u+v)\, w = uw + vw. \tag{3.12a–c}$$

It follows that *the ordered pair* $(|C, \times)$ [*triad* $(|C, +, \times)$], *where* $|C$ *is the set of complex numbers, and* $+, \times$ *are the operations addition and product, is a commutative group* **(ring)**.

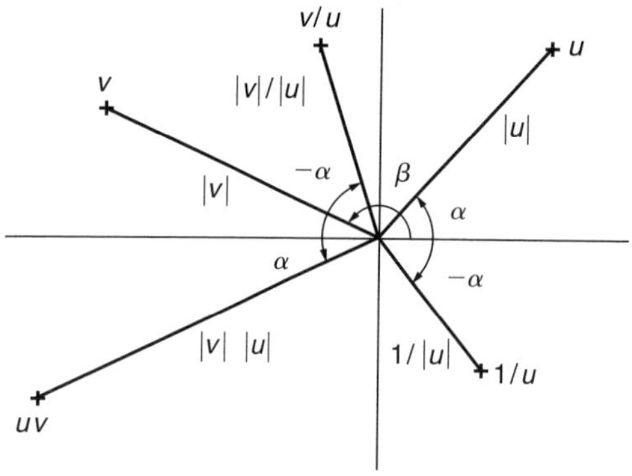

FIGURE 3.3
The inverse $1/u$ of a complex number has the inverse modulus and argument with opposite sign. The product u, v of complex numbers multiplies the moduli and adds the arguments. It follows that the ratio of complex numbers divides the moduli and subtracts the arguments.

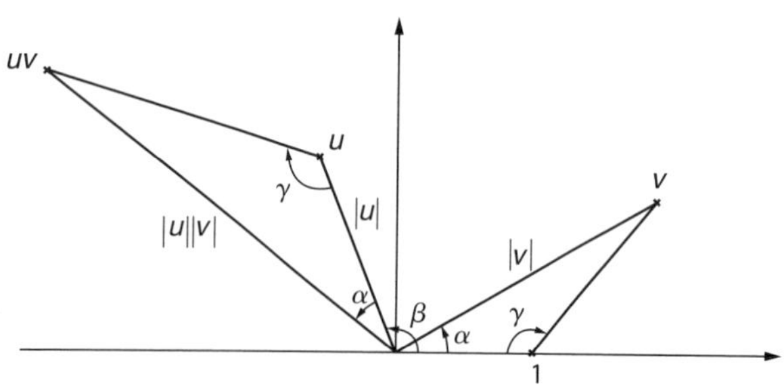

FIGURE 3.4
The triangles $(0, 1, v)$ and $(0, u, uv)$ are similar because the latter is obtained from the former by (i) rotating through $\beta = \arg(u)$ and (ii) multiplying all sides by $|u|$.

3.4 Meaning of the Imaginary Symbol "i"

The real (1.13a) [imaginary (1.13b)] parts of the product of complex numbers in polar form (3.10) are given, respectively, by

$$\operatorname{Re}(uv) = |u|\,|v| \cos(\alpha + \beta) = |u| \cos\alpha\, |v| \cos\beta - |u| \sin\alpha\, |v| \sin\beta = ac - bd; \tag{3.13a}$$

$$\operatorname{Im}(uv) = |u|\,|v| \sin(\alpha + \beta) = |u| \sin\alpha\, |v| \cos\beta + |u| \cos\alpha\, |v| \sin\beta = bc + ad; \tag{3.13b}$$

hence, the product of complex numbers in the Cartesian representation is specified by

$$(a + ib)(c + id) = \left(ac + i^2 bd\right) + i\,(ad + bc) = (ac - bd) + i\,(ad + bc)\,; \tag{3.14}$$

this corresponds to the ordinary rules of multiplication, and agrees with (3.13a,b) ≡ (3.14), provided that

$$i^2 = -1, \quad \sqrt{-1} = \pm i. \tag{3.15a,b}$$

The *condition (3.15a,b) specifying the imaginary symbol "i"* coincides with (1.9b); it can also be justified by noting that lies on the positive imaginary axis at distance unity from the origin in the complex plane $i = \mathrm{e}^{i\pi/2}$, so that $i^2 = \mathrm{e}^{i\pi} = -1$.

3.5 Conjugate of the Sum, Product, and Inversion

From (3.2) [(3.9a,b; 3.10)] and (1.17), it follows that *the conjugate of the sum (product) is the sum (product) of the conjugates:*

$$(u+v)^* = u^* + v^*, \quad (uv)^* = u^* v^*; \tag{3.16a,b}$$

from the expressions of a complex number and its conjugate (1.17), follow the *formulas for the real and imaginary parts:*

$$x \equiv \mathrm{Re}(z) = \frac{z+z^*}{2}, \quad y = \mathrm{Im}\,(z) = \frac{z-z^*}{2i}, \tag{3.17a,b}$$

and for the modulus and argument:

$$r \equiv |z| = (z^* z)^{1/2}, \quad \varphi \equiv \arg\,(z) = \frac{1}{2}\arg\left(\frac{z}{z^*}\right). \tag{3.18a,b}$$

The **inverse** $1/z$ *of a complex number* z *has the inverse modulus and symmetric argument:*

$$\frac{1}{z} = r^{-1}\mathrm{e}^{-i\varphi} = \frac{z^*}{z^* z} = z^* \left|z\right|^{-2}; \tag{3.19}$$

using the conjugate leads to

$$\mathrm{Re},\mathrm{Im}\left(\frac{1}{z}\right) = r^{-1}\left\{\cos\varphi, -\sin\varphi\right\} = \{x, -y\}\left|x^2+y^2\right|^{-1/2}, \tag{3.20a,b}$$

for the real and imaginary parts of the inverse.

3.6 Complex Representation of Real Quantities

The formulas (3.17a,b) are examples of the complex representation, using z^*, z of real quantities x,y. The sum of complex representations is the complex representation of the sum (3.3a,b), but the product does not commute with the complex representation because $\mathrm{Re}\,(uv) \neq \mathrm{Re}\,(u)\,\mathrm{Re}\,(v)$ in (3.13a) and $\mathrm{Im}\,(uv) \neq \mathrm{Im}\,(u)\,\mathrm{Im}\,(v)$ in (3.13b). From (3.13a)

$$\mathrm{Re}\,(uv + uv^*) = (ac - bd) + (ac + bd) = 2ac = 2\mathrm{Re}\,(u)\,\mathrm{Re}\,(v), \tag{3.21a}$$

which can also be proved in the reverse direction from (3.17a):

$$\begin{aligned} \mathrm{Re}\,(u)\,\mathrm{Re}\,(v) &= \frac{1}{2}\left(u+u^*\right)\frac{1}{2}\left(v+v^*\right) = \frac{1}{4}\left[(uv+uv^*) + (uv+uv^*)^*\right] \\ &= \frac{1}{2}\mathrm{Re}\,(uv+uv^*). \end{aligned} \tag{3.21b}$$

Thus, the *complex representation of real quantities (3.17a,b) satisfies the rules (3.3a,b) for the sum, and for the product:*

$$\mathrm{Re}\,(uv + uv^*) = 2\,\mathrm{Re}\,(u)\,\mathrm{Re}\,(v) = \mathrm{Re}\,(uv + u^*v), \tag{3.22a}$$

$$\mathrm{Im}\,(uv - uv^*) = 2\,\mathrm{Re}\,(u)\,\mathrm{Im}\,(v) = \mathrm{Im}\,(uv + u^*v), \tag{3.22b}$$

$$\mathrm{Re}\,(uv^* - uv) = 2\,\mathrm{Im}\,(u)\,\mathrm{Im}\,(v) = \mathrm{Re}\,(u^*v - uv); \tag{3.22c}$$

all formulas can be proved as (3.21a,b). The complex representation is used in Section 4.5.

As a check on the formulas of complex representation of real quantities, for example, u,v in (3.22a) are put in polar form:

$$u \equiv e^{i\alpha}, v \equiv e^{i\beta}: \quad \operatorname{Re}\left(e^{i(\alpha+\beta)} + e^{i(\alpha-\beta)}\right) = 2\operatorname{Re}\left(e^{i\alpha}\right)\operatorname{Re}\left(e^{i\beta}\right), \tag{3.23}$$

this leads to the **trigonometric multiplication formulas**:

$$2\cos\alpha\cos\beta = \cos(\alpha+\beta) + \cos(\alpha-\beta), \tag{3.24a}$$
$$2\cos\alpha\sin\beta = \sin(\alpha+\beta) - \sin(\alpha-\beta), \tag{3.24b}$$
$$2\sin\alpha\sin\beta = \cos(\alpha-\beta) - \cos(\alpha+\beta), \tag{3.24c}$$

where (3.24b,c) can be deduced similarly from (3.22b,c) as (3.24a) was deduced (3.23) from (3.22a). Adding and subtracting among (3.24a–c) leads the **trigonometric addition formulas:**

$$\cos(\alpha\pm\beta) = \cos\alpha\cos\beta \mp \sin\alpha\sin\beta, \tag{3.25a}$$
$$\sin(\alpha\pm\beta) = \sin\alpha\cos\beta \pm \cos\alpha\sin\beta, \tag{3.25b}$$

these were used in (3.13a,b). These formulas (3.24a–c; 3.25a,b) are generalized next from two to any number of arguments.

3.7 Trigonometric Addition and Multiplication Formulas

The starting point is the identity:

$$\exp\{i(z_1 + \cdots + z_N)\} = \exp(iz_1)\cdots\exp(iz_N), \tag{3.26a}$$

in the form

$$\cos(z_1 + \cdots + z_N) + i\sin(z_1 + \cdots + z_N) = (\cos z_1 + i\sin z_1)\cdots(\cos z_N + i\sin z_N). \tag{3.26b}$$

The cosine (sine) of the sum on the l.h.s. of (3.26a) [(3.26b)] is the real (imaginary) part of the r.h.s., which is obtained as

$$\cos(z_1 + \cdots + z_N) = \sum_{p=0}^{\leq N/2}{}'(-)^p \sin z_1 \ldots \sin z_{2p} \cos z_{2p+1} \ldots \cos z_N, \tag{3.27a}$$

$$\sin(z_1 + \cdots + z_N) = \sum_{p=1}^{\leq N/2}{}'(-)^p \sin z_1 \ldots \sin z_{2p+1} \cos z_{2p+2} \ldots \cos z_N, \tag{3.27b}$$

by taking (i) all terms where sin appears an even $2p$ (odd $2p+1$) number of times, with factor $i^{2p} = (-)^p$ $\left[i^{-1}i^{2p+1} = (-)^p\right]$; (ii) the remaining $N-2p$ $(N-2p+1)$ factors are the cosines; and (iii) all permutations of $z_1, \ldots, z_N$ are included in the sum as indicated by $\sum'$. This proves (3.27a,b), and setting (3.28a) leads to (3.28b,c):

$$z_1 = \cdots = z_N \equiv z: \quad \cos(Nz) = \sum_{p=0}^{\leq N/2} \binom{N}{2p} (-)^p \sin^{2p} z \cos^{N-2p} z, \tag{3.28a,b}$$

$$\sin(Nz) = \sum_{p=0}^{\leq N/2} \binom{N}{2p+1} (-)^p \sin^{2p+1} z \cos^{N-2p-1} z, \tag{3.28c}$$

where the number of permutations is

$$\binom{N}{2p} \equiv \frac{N!}{(2p)!\,(N-2p)!}, \quad \binom{N}{2p+1} = \frac{N!}{[(2p+1)!\,(N-2p-1)!]}, \tag{3.29a,b}$$

respectively, in (3.28b, 3.29a) and (3.28c, 3.29b), and thus a normal sum $\sum$ is used in (3.28b,c).

The **addition (multiplication) formulas** *for the circular cosine (3.27a) [(3.28b, 3.29a)] and sine (3.27b) [(3.28c, 3.29b)] with any number of arguments* are simplest for $N = 2$ and $N = 3$. The former are the **double addition** formulas:

$$\cos(z_1 + z_2) = \cos z_1 \cos z_2 - \sin z_1 \sin z_2, \tag{3.30a}$$

$$\sin(z_1 + z_2) = \sin z_1 \cos z_2 + \cos z_1 \sin z_2, \tag{3.30b}$$

confirming (3.25a,b) ≡ (3.30a,b); these include the **duplication formulas** (3.31a–c):

$$z_1 = z_2 \equiv z: \quad \cos(2z) = \cos^2 z - \sin^2 z = 2\cos^2 z - 1 = 1 - 2\sin^2 z, \tag{3.31a,b}$$

$$\sin(2z) = 2\cos z \sin z, \quad \cos^2, \sin^2(z) = \frac{1 \pm \cos 2z}{2}, \tag{3.31c,d}$$

from which follow (3.31d) ≡ (3.31b).

For $N = 3$, the following **triple addition** formulas are obtained:

$$\begin{aligned} \cos(z_1 + z_2 + z_3) = {} & \cos z_1 \cos z_2 \cos z_3 - \cos z_1 \sin z_2 \sin z_3 \\ & - \cos z_2 \sin z_1 \sin z_3 - \cos z_3 \sin z_1 \sin z_2, \end{aligned} \tag{3.32a}$$

$$\begin{aligned} \sin(z_1 + z_2 + z_3) = {} & -\sin z_1 \sin z_2 \sin z_3 + \sin z_1 \cos z_2 \cos z_3 \\ & + \sin z_2 \cos z_1 \cos z_3 + \sin z_3 \cos z_1 \cos z_2, \end{aligned} \tag{3.32b}$$

which include the **triplication** formulas:

$$z_1 = z_2 = z_3 \equiv z: \quad \cos(3z) = \cos^3 z - 3\cos z \sin^2 z = \cos z\left(4\cos^2 z - 3\right), \tag{3.33a,b}$$

$$\sin(3z) = 3\sin z \cos^2 z - \sin^3 z = \sin z\left(3 - 4\sin^2 z\right). \tag{3.33c}$$

These formulas can also be deduced by recurrence, for example, applying twice (3.30a,b) to

$$\begin{aligned} \cos(z_1 + z_2 + z_3) &= \cos(z_1)\cos(z_2 + z_3) - \sin(z_1)\sin(z_2 + z_3) \\ &= \cos(z_1)\left[\cos(z_2)\cos(z_3) - \sin(z_2)\sin(z_3)\right] \\ &\quad - \sin(z_1)\left[\sin(z_2)\cos(z_3) + \sin(z_3)\cos(z_2)\right], \end{aligned} \tag{3.34}$$

which leads to (3.34) ≡ (3.32a). The quadruple addition and quadruplicating formulas are indicated in Example 10.7.

3.8 Conjugate Complex and Triangular Inequalities

Squaring (3.17a,b) leads to

$$(z+z^*)^2 \geq 0 \geq (z-z^*)^2, \tag{3.35a,b}$$

the *complex conjugate inequalities, where the equality sign in (3.35a)* [(3.35b)] *holds only for imaginary (real)* z. The conjugate can be used to calculate the modulus of the sum:

$$\begin{aligned} |u+v|^2 = (u+v)(u^*+v^*) &= u^*u + v^*v + u^*v + v^*u \\ &= |u|^2 + |v|^2 + u^*v + (u^*v)^*, \end{aligned} \tag{3.36a}$$

where the real representation (3.22a) may be used:

$$u^*v + (u^*v)^* = \operatorname{Re}(u^*v) = \operatorname{Re}\left\{|u|\,|v|\,e^{i(\beta-\alpha)}\right\} = |u|\,|v|\cos(\alpha-\beta); \tag{3.36b}$$

substitution of (3.36b) in (3.36a) yields (3.6).

Bearing in mind that $-1 \leq \cos(\alpha-\beta) \leq +1$, from (3.6) ≡ (3.36a,b)

$$|u|^2 + |v|^2 - 2|u|\,|v| \leq |u+v|^2 \leq |u|^2 + |v|^2 + 2|u|\,|v|, \tag{3.37a}$$

follows the relation:

$$|u| > |v|: \quad |u| - |v| \leq |u+v| \leq |u| + |v|, \tag{3.37b}$$

which is designated the **triangular inequality**, *since it shows* (Figure 3.2a) *that the length of one side of a triangle cannot exceed the sum of the lengths of the other two sides, and cannot be less than their difference; the right- (left-) hand side equality holds in the case in* Figure 3.2c (3.2d) *of parallel (antiparallel) collinear sides.*

3.9 Generalized Schwartz (1890) or Polygonal Inequality

The result (3.37b) can be generalized to finite sum:

$$|z_k| - \sum_{\substack{n=1 \\ n\neq k}}^{N} |z_n| \leq \left|\sum_{n=1}^{N} z_n\right| \leq \sum_{n=1}^{N} |z_n|, \tag{3.38}$$

which is the generalized Schwartz or (1890) **polygonal inequality**, *showing that* (Figure 3.5a) *the length of a straight line between two points cannot exceed the length of any polygonal line joining the two points, and equality on the r.h.s. can only occur for collinear parallel segments* (Figure 3.5b); *the equality on the l.h.s. applies to collinear segments with z_k antiparallel to all others* (Figure 3.5c). *Also the length of one side of the polygon cannot be less than the longest side with the lengths of all remaining sides subtracted from it.* The proof of (3.38) is made by induction: (i) since (3.38) applies for $N = 2$, in which case it

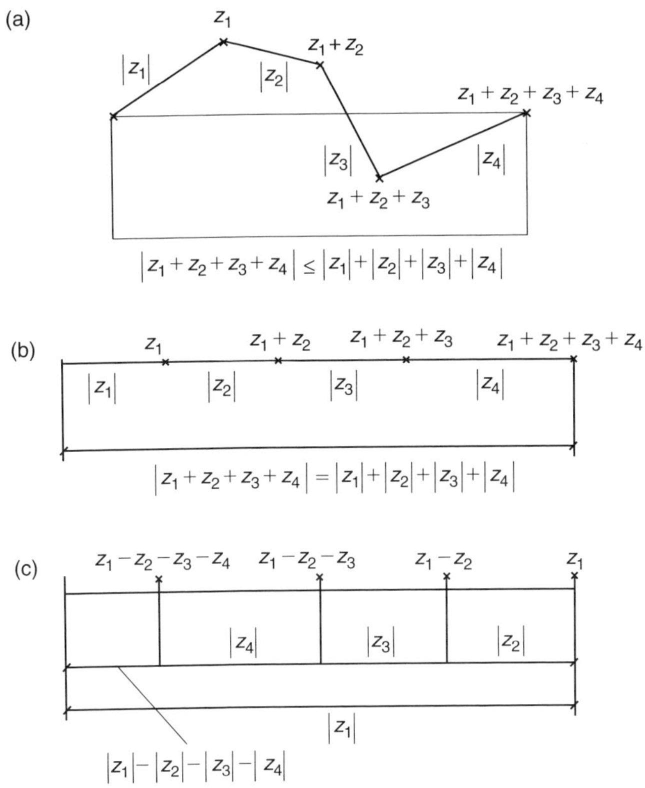

FIGURE 3.5
The polygonal inequality (3.38) states that the length of a polygonal line cannot be less than the length of a straight line joining the ends (a); it is equal only if all sides are parallel (b); if some segments are antiparallel, they are subtracted from the total length (c).

coincides with (3.37b) for $u \equiv z_1$, $v \equiv z_2$; (ii) besides if (3.38) holds for N, then it also holds for $N + 1$. To prove the latter statement: (ii-1) choose (u,v) in (3.39a):

$$u \equiv z_{N+1}, \; v \equiv \sum_{n=1}^{N} z_n : \quad \left| \sum_{n=1}^{N+1} z_n \right| = |u+v| \leq |u| + |v| = |z_{N+1}| + \left| \sum_{n=1}^{N+1} z_n \right| \leq \sum_{n=1}^{N+1} |z_n| \, , \tag{3.39a,b}$$

where (3.37b) and (3.38) were used to prove the right-hand inequality (3.39b); (ii-2) for the left-hand inequality choose (u,v) in (3.40a):

$$u \equiv z_k, \; v \equiv \sum_{\substack{n=1 \\ n\neq k}}^{N+1} z_n : \quad |z_k| - \sum_{\substack{n=1 \\ n\neq k}}^{N+1} |z_n| = |u| - |v| \leq |u+v| = \left| \sum_{n=1}^{N+1} z_n \right| \leq \sum_{n=1}^{N+1} |z_n| \, , \tag{3.40a,b}$$

where (3.37b) and (3.38) were used to prove (3.40b). Note that the left- (3.40b) and right-hand (3.40a) inequalities operate independently.

Note 3.1. *Formal Inequalities and Equalities*: The Schwartz inequality applies not only to numbers (3.38) but also to the integrals of functions. It can also be extended to exponents other than $1/2 = p$ in the modulus, by the Holder and Minkowski inequalities; the latter arise in the context of functional analysis with application to generalized functions, and Fourier series and integrals, and other orthogonal expansions and integral representations. The addition and multiplication formulas in Section 3.7 can be extended from the circular cosine and sine to the circular tangent and to the corresponding hyperbolic functions. Some higher transcendental functions also have addition and/or multiplication formulas, for example, (i) the gamma and polygamma functions; (ii) the elliptic functions of Jacobi and Weierstrass; and (iii) some special functions, for example, Bessel.

Conclusion 3: Graphical representation of operations and relations between complex numbers: (Figure 3.1) sum, difference, and symmetry; (Figure 3.3) product, division, and inverse. The product of complex numbers specifies (Figure 3.4) two pairs of similar triangles, with corresponding angles γ and δ. The (Figure 3.2) triangular inequality (a), has the orthogonal (b), parallel (c), and antiparallel (d) subcases; it is generalized by (Figure 3.5) the polygonal inequality (a), which also has a parallel (b) and antiparallel (c) subcases.

4

Impedance of Associations of Circuits

A mechanical circuit consists (Section 4.1) of masses, dampers, and springs, and is analogous to an electrical circuit consisting (Section 4.3) of inductors, resistors, and capacitors. A mechanical (electrical) circuit is characterized by the impedance (Section 4.2), which is the ratio of the mechanical (electromotive) force to the velocity (electric current). The impedance is generally complex (Section 4.4), since it appears as a coefficient of proportionality between two quantities that may have different amplitudes and phases (Section 4.5). Several circuits, electrical or mechanical, may be associated in series, in parallel (Section 4.6) or in hybrid arrangements (Section 4.9), and the total impedance is specified by the partial impedances of each separate element. The calculation of impedances of complex circuits (Section 4.7) is an application of the algebra of complex numbers, and the laws of association are different (Section 4.8) for mechanical and electrical circuits. The mechanical circuits are used in the suspensions of vehicles (e.g., cars, trains, etc.) and in the "soft" or "shock absorbing" mountings of machinery; the electrical circuits are used in a variety of electrical devices, for example, radios, computers, appliances, and so on. The electromechanical circuits also have applications like control systems, actuators, and so on.

4.1 Inertia, Friction, and Elastic Forces

Consider a basic **mechanical circuit**, represented in Figure 4.1a and consisting of a mass, damper, and spring. When the mass is displaced from its equilibrium position, three **forces** arise: (i) the **inertia** force, equal to the product of mass m and acceleration; (ii) the **friction** force, which, for weak damping, is proportional to the velocity through the **damping coefficient** b; (iii) the **elastic** force, which, for small displacement, is proportional to the latter through the **resilience** k of the spring. Adding all forces leads to the **mechanical force**, which is specified by

$$m > 0, b > 0, k > 0: \quad F = m\ddot{x} + b\dot{x} + kx, \tag{4.1}$$

where dot denotes derivative with regard to time, viz., $x, \dot{x}, \ddot{x}$ are, respectively, the displacement, velocity, and acceleration. The friction force was not considered in the harmonic oscillator (Section 2.9), that is, (2.19) corresponds to (4.1) with $b = 0$. There may be two causes for the motion: free motion if there is no mechanical force $F = 0$, and the initial conditions are a nonequilibrium position (as in Section 2.9); (ii) forced motion, which is independent of initial conditions, and specified by the mechanical force (Section 4.2).

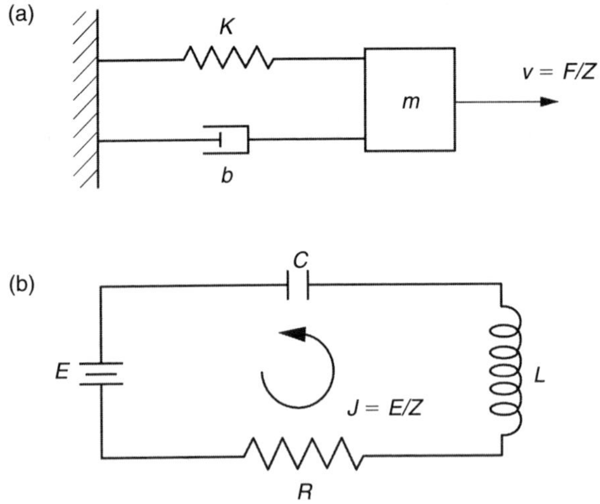

FIGURE 4.1
The analogy between mechanical (a) [electrical (b)] circuits replaces the (i) mass m (induction L) of body (self); (ii) damping b (resistance R) of a damper (resistor); and (iii) resilience k (inverse capacity $1/C$) of a spring (capacitor). These specify the mechanical Z_m (electrical Z_e) impedance of the circuit relating the mechanical force F (electromotive force E) to the velocity v (electric current J).

4.2 Free and Forced Motion of Circuit

In the **free motion** of a circuit, for which the mass is released from a disturbed position, in the absence of mechanical force $F = 0$, the relation (4.1) becomes a homogeneous or unforced differential equation:

$$m\frac{\mathrm{d}^2x}{\mathrm{d}t^2} + b\frac{\mathrm{d}x}{\mathrm{d}t} + kx = 0; \tag{4.2}$$

it is linear with constant coefficients (2.15) and specifies the nature of the equilibrium (Section 2.8) and the type of motion in its vicinity (2.1a,b). Given this type of displacement x, it is related to the velocity v and acceleration a by (2.2a,b), so that

$$v \equiv \dot{x}, \qquad x = -i\frac{v}{\zeta}, \qquad a = i\zeta v. \tag{4.3a–c}$$

Given a velocity $v\,(t)$, imposed upon the circuit by an external influence, the associated mechanical force (4.1) is specified as a function of time by (4.4a):

$$F\,(t) = Z_m v\,(t)\,, \quad Z_m \equiv b + i\left(\zeta m - \frac{k}{\zeta}\right), \tag{4.4a,b}$$

as a linear function of velocity $v(t)$, through a complex constant (4.4b), designated **mechanical impedance**. For example, the wheel of a car travels over an undulating ground, by measuring its velocity v, and knowing the impedance Z_m of the suspension, the mechanical force F being exerted can be determined. In the forced motion, the frequency ω and decay or amplification λ in (4.5a) $\equiv$ (2.1a) are imposed by the mechanical force (4.1) and lead to the mechanical impedance (4.4b) $\equiv$ (4.5b):

$$\zeta \equiv \omega + i\lambda : \quad Z_m = b + m(i\omega - \lambda) - k\frac{\lambda + i\omega}{\omega^2 + \lambda^2}. \tag{4.5a,b}$$

Using (4.3a–c) the mechanical force (4.1) can also be expressed in terms of the acceleration (4.6a) or displacement (4.6b):

$$F(t) = \left(m - i\frac{b}{\zeta} - \frac{k}{\zeta^2}\right) a(t) = \left(-m\zeta^2 + ib\zeta + k\right) x(t). \tag{4.6a,b}$$

In all cases the coefficient is independent of time, and is specified by (i) the mass m of the body, the coefficient b of friction, and the resilience k of the spring; and (ii) the frequency ω and decay/growth λ of the forcing.

4.3 Electrical Induction, Resistance, and Capacity

The **electromechanical analogy** (Table 4.1) replaces a mechanical (Figure 4.1a) by an electrical circuit, represented in Figure 4.1b, with the displacement x replaced by the electric charge q, and the velocity $v \equiv \dot{x}$ by the electric current $j = \dot{q}$. *The devices forming the basic electrical (mechanical) circuit are a capacitor (spring), resistance (damper), and induction (mass), and the circuit is driven by* **the electromotive** *(mechanical)* **force** *E (F), applied externally, for example, through a battery (engine). The total electromotive force is the sum of the contribution of the capacitor $1/C$, resistance R, and self L, multiplying, respectively, the charge, current, and its time derivative:*

$$E = L\ddot{e} + R\dot{e} + \frac{e}{C}. \tag{4.7}$$

The expression (4.7) is analogous to the relation (4.1), exchanging mechanical F for electromotive E force, and velocity $v \equiv \dot{q}$ for current, so that

$$J \equiv \dot{e} : \quad E(t) = Z_e\, J(t), \quad Z_e \equiv R + i\left(\zeta L - \frac{1}{\zeta C}\right), \tag{4.8a,b}$$

TABLE 4.1
Analogy of Mechanical and Electrical Circuits*

Circuit	Mechanical	Electrical
Force	Mechanical: F	Electromotive: E
Variable	Velocity: v	Current: j
Zero phase	Damper: b	Resistor: R
Phase $+\pi/2$	Mass: m	Induction: L
Phase $-\pi/2$	spring: k	Capacitor: $1/C$
Impedance	Z_m: (4.4b)	Z_e: (4.8b)

* Corresponds to Figure 4.1.

the **electrical impedance**, relating electromotive force to current (4.8a), is specified by (4.8b) ≡ (4.8c):

$$Z_e = R + L\left(i\omega - \lambda\right) - \frac{\lambda + i\omega}{C\left(\omega^2 + \lambda^2\right)}. \tag{4.8c}$$

For example, measuring the current J along an electrical circuit, and knowing the impedance Z_e, the electromotive force E being used to drive the circuit is specified.

4.4 Decomposition of Impedance into Inductance and Reactance

The mechanical (electrical) impedance, denoted by Z_m in (4.4b) [Z_e in (4.8b)], consists for real ζ, that is, an imposed frequency without decay or amplification,

$$\zeta = \omega, \lambda = 0: \quad Z_m = b + i\left(m\omega - \frac{k}{\omega}\right), \quad Z_e = R + i\left(L\omega - \frac{1}{\omega C}\right), \tag{4.9a,b}$$

of the damping coefficient b (resistance R) for the real part, plus an imaginary part involving the mass m (induction L of a self) and resilience k of a spring (inverse capacity $1/C$ of a capacitor). A circuit is said to be reactive if the impedance is real, that is, it consists only of a damper (resistance), and the mechanical (electromotive) force is in phase with the velocity (current). The opposite case is the inductive circuit, for which the impedance is imaginary, and which can include a mass m (induction L) and spring of resilience k (capacitor of capacity C), implying that the mechanical (electromotive) force is out-of-phase by $\pm\pi/2$ relative to the velocity (current). Table 4.1 summarizes the electromechanical analogy, in the general case, when the impedance can be decomposed into a reactance and inductance that are, respectively, its real and imaginary parts. If the mechanical force has a frequency and a decay/amplification (4.5a), the real (imaginary) part of the impedance (4.10a), called **reactance** X (**inductance** Y) is (i) for a mechanical circuit (4.5b):

$$Z_m \equiv X_m + iY_m: \quad X_m = b - m\lambda - \frac{k\lambda}{\omega^2 + \lambda^2}, \quad Y_m = m\omega - \frac{k\omega}{\omega^2 + \lambda^2}, \tag{4.10a–c}$$

(ii) for an electrical circuit (4.8c):

$$Z_e \equiv X_e + iY_e: \quad X_e = R - \lambda L - \frac{\lambda/C}{\omega^2 + \lambda^2}, \quad Y_e = \omega L - \frac{\omega/C}{\omega^2 + \lambda^2}, \tag{4.11a–c}$$

The inverse of the impedance:

$$\frac{1}{A} = Z = X + iY: \quad (\text{admittance})^{-1} \equiv \text{impedence} \equiv (\text{resistance}) + i(\text{reactance}), \tag{4.12}$$

is called **admittance**.

4.5 Activity in Terms of the Velocity, Force, and Impedance

The product of the force (4.13a) by the velocity (4.13b),

$$\frac{f(t)}{f_0} = \cos(\omega t) = \operatorname{Re}\left(e^{i\omega t}\right) = \frac{v(t)}{v_0}, \tag{4.13a,b}$$

is the **activity** or work per unit time, which is specified (i) in the real representation by

$$A(t) \equiv f(t)\, v(t) = f_0 v_0 \cos^2(\omega t); \tag{4.14a}$$

(ii) in the complex representation (3.22a) by

$$\begin{aligned} A(t) &= \operatorname{Re}\{f(t)\} \operatorname{Re}\{v(t)\} = \frac{1}{2} \operatorname{Re}(fv + fv^*) = \frac{1}{2} f_0 v_0 \operatorname{Re}\left(e^{i2\omega t} + 1\right) \\ &= \frac{1}{2} f_0 v_0 \left[1 + \cos(2\omega t)\right]. \end{aligned} \tag{4.14b}$$

These are equivalent (4.14a) = (4.14b) on account of (3.31d). The activity may be expressed in terms of the velocity alone using the impedance (4.4a), which is complex. Thus, the real representation is applied in two stages: (i) calculation of the force:

$$\begin{aligned} Z = X + i\mathrm{Y}: \quad f(t) &= \operatorname{Re}\{Zv(t)\} = v_0 \operatorname{Re}\left\{(X + i\mathrm{Y})\, e^{i\omega t}\right\} \\ &= v_0 \left[X\left(\cos(\omega t) - \mathrm{Y} \sin(\omega t)\right)\right], \end{aligned} \tag{4.15a}$$

which shows that the resistance is in phase and the reactance out-of-phase by $\pi/2$, viz. (2.6a); (ii) the activity follows from

$$A(t) = f(t)\, v(t) = v_0^2 \left[X \cos^2(\omega t) - \frac{Y}{2} \sin(2\omega t)\right]. \tag{4.15b}$$

The activity can also be calculated in terms of the velocity and impedance in the complex representation:

$$\begin{aligned} A(t) &= \operatorname{Re}\{f(t)\} \operatorname{Re}\{v(t)\} = \operatorname{Re}\{Zv(t)\} \operatorname{Re}\{v(t)\} \\ &= \frac{1}{2} \operatorname{Re}\left(Zv^2 + Zvv^*\right) = \frac{1}{2} |v|^2 \operatorname{Re}(Z) + \frac{1}{2} \operatorname{Re}\left(Zv^2\right); \end{aligned} \tag{4.16a}$$

the r.h.s. of (4.16a) ≡ (4.16c) is evaluated in terms of the real and imaginary components of the velocity (4.16b):

$$\begin{aligned} \{v_x, v_y\} &\equiv v_0 \cos, \sin(\omega t): \\ A(t) &= \frac{X}{2} |v|^2 + \frac{1}{2} \operatorname{Re}\left[(X + i\mathrm{Y})\left(v_x^2 - v_y^2 + 2iv_x v_y\right)\right] \\ &= \frac{X}{2}\left(v_x^2 + v_y^2\right) + \frac{X}{2}\left(v_x^2 - v_y^2\right) - \mathrm{Y} v_x v_y \\ &= X v_x^2 - \mathrm{Y} v_x v_y = v_0^2 \left[X \cos^2(\omega t) - \frac{\mathrm{Y}}{2} \sin(2\omega t)\right], \end{aligned} \tag{4.16b,c}$$

in agreement with (4.16c) ≡ (4.15b).

The averages over a period,

$$\tau = \frac{2\pi}{\omega}, \theta = \omega t: \qquad \omega \int_0^{2\pi/\omega} \sin(2\omega t) \mathrm{d}t = \int_0^{2\pi} \sin 2\theta \, \mathrm{d}\theta = \left[\frac{-\cos(2\theta)}{2}\right]_0^{2\pi} = 0, \tag{4.17a}$$

$$\omega \int_0^{2\pi/\omega} \cos^2(\omega t)\, \mathrm{d}t = \int_0^{2\pi} \cos^2\theta \, \mathrm{d}\theta = \int_0^{2\pi} \frac{1 + \cos(2\theta)}{2} \mathrm{d}\theta = \left[\frac{\theta}{2} + \frac{\sin(2\theta)}{4}\right]_0^{2\pi} = \pi, \tag{4.17b}$$

applied to (4.15b) ≡ (4.16c) lead the conclusion that

$$\overline{A(t)} \equiv \frac{1}{\tau}\int_0^{\tau} A(t)\,dt = \frac{\pi}{\omega\tau} X v_0^2 = \frac{1}{2} X v_0^2 = \frac{1}{2} A(0) = \frac{f_0 v_0}{2}, \tag{4.18}$$

the average activity over a period is one-half of the square of the amplitude of the velocity multiplied by the resistance, that is, one-half of the initial value, which is the peak value (4.14a) ≡ (4.14b) for the resistive term in (4.17b); the contribution (4.17a) of the reactance to the activity cancels over a period, and thus has zero average. Similar reasonings apply to the energy (activity) associated with the electric current $J(t)$ [velocity $v(t)$] and electromotive force $E(t)$ [mechanical force $F(t)$] with the electrical (mechanical) impedance playing the same role.

4.6 Mechanical Circuits in Parallel or Series

When N mechanical circuits are associated in parallel, for example, $N = 2$ in Figure 4.2a, the velocity v is the same for all of them (4.19a) and if the impedances z_n are generally distinct, so will be the mechanical forces F_n:

$$F_n = z_n v, \qquad F_{mp} = \sum_{n=1}^{N} F_n = v \sum_{n=1}^{n} z_n, \tag{4.19a,b}$$

which add up to the total force (4.19b). The latter specifies the total impedance (4.20a):

$$Z_{mp} = \frac{F_{mp}}{v}, \qquad Z_{mp} = \sum_{n=1}^{N} z_n, \tag{4.20a,b}$$

so that *the total impedance of mechanical circuits associated in parallel is the sum of the impedances (4.20b).* For example, if a wheel is suspended from two parallel springs, the force required to move it at a given velocity increases, and so the inductance (imaginary part of the impedance, apart from sign) is greater.

In contrast, if the N mechanical circuits are associated in series, for example, $N = 2$ in Figure 4.2b, the forces exerted by each of them must be equal, in order that there may be equilibrium at the junctions. The same force F with generally different impedances z_n leads to distinct velocities v_n in (4.21a):

$$v_n = \frac{F}{z_n}, \qquad v = \sum_{n=1}^{N} v_n = F \sum_{n=1}^{N} \frac{1}{z_n}, \tag{4.21a,b}$$

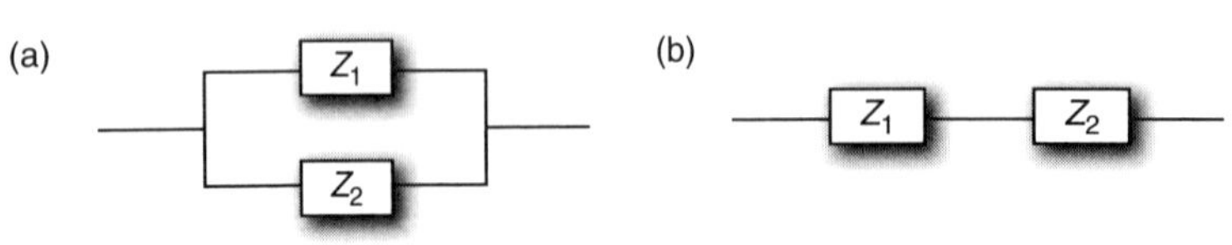

FIGURE 4.2
Two circuits can be associated in parallel (a) or in series (b). The laws of association of impedances are opposite for mechanical and electrical circuits: (i) the impedances add for mechanical (electrical) circuits in parallel (a) [series (b)]; and (ii) the inverse impedances add for mechanical (electrical) circuits in series (b) [parallel (a)].

which add up to the total velocity (4.21b). The latter specifies the total impedance (4.14a):

$$v = \frac{F}{Z_{ms}}, \quad \frac{1}{Z_{ms}} = \sum_{n=1}^{N} \frac{1}{z_n}, \tag{4.22a,b}$$

and it follows that *for mechanical circuits associated in series, the inverses of the impedances add up to the inverse of the total impedance (4.22b).*

4.7 Electromechanical Analogy and Contrasting Laws

In the case of N electrical, instead of mechanical, circuits associated in parallel (series), for example, $N = 2$ in Figure 4.2a (4.2b), the currents add up (are the same) and the electromotive forces are the same (add up):

$$\frac{E}{Z_{ep}} = J = \sum_{n=1}^{N} J_n = E \sum_{n=1}^{N} \frac{1}{z_n}, \quad Z_{es} = \frac{E}{J} = \frac{\sum_{n=1}^{N} E_n}{J} = \sum_{n=1}^{N} z_n. \tag{4.23a,b}$$

It follows that

$$\frac{1}{Z_{ep}} = \sum_{n=1}^{N} \frac{1}{z_n}, \quad Z_{es} = \sum_{n=1}^{N} z_n \tag{4.24a,b}$$

for electrical circuits associated in parallel (series) the inverse impedances (impedances) are added (4.24a) [(4.24b)].

The laws of association of electrical circuits (4.24a,b) are the inverse of those (4.19b, 4.20b) applying to mechanical circuits; the electromechanical analogy (Table 4.1) is followed by a contrast: *the association of mechanical circuits in parallel is equivalent to the association of electrical circuits in series, and vice versa viz:* (i) the impedances add for mechanical (electrical) circuits in parallel (series) in (4.25a):

$$Z_{mp} = \sum_{n=1}^{N} z_n = Z_{es}; \quad Z_{ms} = \left\{ \sum_{n=1}^{N} z_n^{-1} \right\}^{-1} = Z_{ep}, \tag{4.25a,b}$$

(ii) the inverse impedances add for mechanical (electrical) circuits in series (parallel) in (4.25b). The latter law (4.25b) can be rewritten:

$$Z_{ms} = \frac{\prod_{n=1}^{N} z_n}{\sum_{n=1}^{N} \prod_{\substack{m=1 \\ m \neq n}}^{N} z_m} = Z_{ep}, \tag{4.26}$$

where (i) the numerator is the product of all impedances; (ii) the denominator is a sum of N terms that are all the distinct products of $N - 1$ impedances. The simplest cases are

$$N = 2: \quad Z_{ms}^{-1} = z_1^{-1} + z_2^{-2} \equiv Z_{ep}^{-1}, \quad Z_{ms} = \frac{z_1 z_2}{z_1 + z_2} \equiv Z_{ep}, \tag{4.27a,b}$$

or for two impedances and

$$N = 3: \quad Z_{ms}^{-1} = z_1^{-1} + z_2^{-2} + z_3^{-2} \equiv Z_{ep}^{-1}, \quad Z_{ms} = \frac{z_1 z_2 z_3}{z_1 z_2 + z_2 z_3 + z_1 z_3} \equiv Z_{ep}, \tag{4.28a,b}$$

for three impedances.

4.8 Comparison of Two Circuits in Parallel and in Series

Considering two circuits, of impedances z_1, z_2 that can be associated either in parallel $Z_{mp}\,(Z_{ep})$ or in series $Z_{ms}\,(Z_{es})$ in a mechanical (electrical) circuit:

$$Z_{mp} = z_1 + z_2 = Z_{es}, \quad Z_{mp} = \left(z_1^{-1} + z_2^{-1}\right)^{-1} = \frac{z_1 z_2}{z_1 + z_2} = Z_{ep}. \tag{4.29a,b}$$

In the case of identical impedances (4.30a) the total impedance is (4.30b,c)

$$z_1 = z_2 \equiv z_0: \quad Z_{mp} = 2z_0 = Z_{es}, \quad Z_{ms} = \frac{z_0}{2} = Z_{ep}. \tag{4.30a–c}$$

Generalizing to N identical subcircuits in (4.31a) in (4.25a,b) leads to (4.31b,c):

$$z_1 = z_2 = \cdots = z_N \equiv z_0: \quad Z_{mp} = Nz_0 = Z_{es}, \quad Z_{ms} = \frac{z_0}{N} = Z_{ep}. \tag{4.31a–c}$$

Thus *association of* N *identical mechanical (electrical) circuits, multiplies the impedance by* N *if they are in parallel (series), and divides it by* N *if they are in series (parallel).* In the case of two resistances, if they are placed in parallel, there are two paths, so that a larger total current flows for the same electromotive force, implying a reduced resistance (real part of the impedance); if the resistances are placed in series, for the same current, a larger electromotive force is required, and the resistance is greater.

4.9 Hybrid Associations of Three Circuits

Starting with three circuits of impedances z_1, z_2, z_3, there are four distinct types of associations illustrated in Figure 4.3, ranging from all in parallel (4.3a) to all in series (4.3d), through two hybrid cases (4.3b,c). In each case, the total mechanical (electrical) impedances are distinct: (i) all in parallel (series) in Figure 4.3a (4.3d):

$$Z_{mA} = z_1 + z_2 + z_3 = Z_{eD}; \tag{4.32}$$

(ii) two in series (parallel) and one in parallel (series) in Figure 4.3b (4.3c):

$$Z_{mC} = \left(z_1^{-1} + z_2^{-1}\right)^{-1} + z_3 = z_3 + \frac{z_1 z_2}{z_1 + z_2} = Z_{eB}; \tag{4.33}$$

(iii) two in parallel (series) and one series in (parallel) in Figure 4.3c (4.3b):

$$Z_{mB} = \left\{(z_1 + z_2)^{-1} + z_3^{-1}\right\}^{-1} = \frac{z_3\,(z_1 + z_2)}{z_1 + z_2 + z_3} = Z_{eC}; \tag{4.34}$$

(iv) all in series (parallel) in Figure 4.3d (4.3a):

$$Z_{mD} = \left(z_1^{-1} + z_2^{-1} + z_3^{-1}\right)^{-1} = \frac{z_1 z_2 z_3}{z_1 z_2 + z_1 z_3 + z_2 z_3} = Z_{eA}. \tag{4.35}$$

In the case of three identical mechanical (electrical) circuits of impedances (4.36a):

$$z_1 = z_2 = z_3 \equiv z_0: \quad Z_{mA}, Z_{mB}, Z_{mC}, Z_{mD} = \left\{3, \frac{3}{2}, \frac{2}{3}, \frac{1}{3}\right\} z_0 = Z_{eD}, Z_{eB}, Z_{eC}, Z_{eA}, \tag{4.36a–e}$$

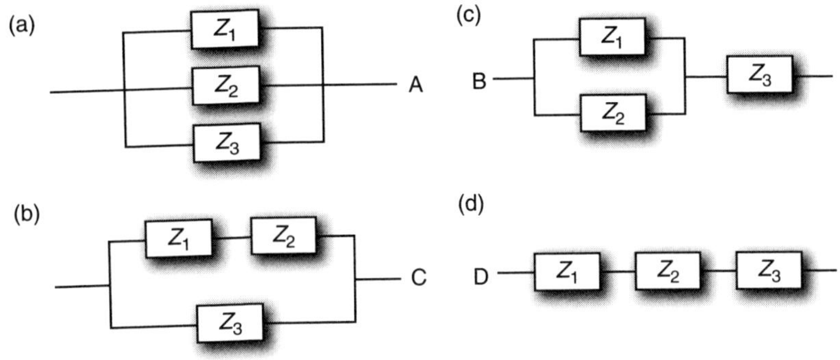

FIGURE 4.3
There are four possible associations of three circuits from all in parallel (a) to all in series (d). The intermediate cases are two in parallel and one in series (c) or two in series and one in parallel (b). The total impedances differ for mechanical (electrical) circuits, for example, are added for (a) [(d)].

it follows that *for mechanical force (electrical) circuits the impedance increases (decreases) as more are put in parallel (series),* viz. (Figure 4.3) it is (i) maximum (4.36a,b) ≡ (4.32) for all impedances in parallel (series) in Figure 4.3a (4.3d); (ii) minimum (4.36e) ≡ (4.35) for all impedances in series (parallel) in Figure 4.3d (4.3a); and (iii) in the intermediate cases, it is larger for (4.36c) ≡ (4.33) than for (4.36d) ≡ (4.34) because there is one impedance in parallel (series) in Figure 4.3b (4.3c) with the other two. There are (i) two distinct associations of two impedances (Section 4.8); (ii) four distinct associations of three impedances (Section 4.9); and (iii) ten distinct associations of four impedances (Example 10.8).

Note 4.1. *Impedances of Circuits and Waves*: The impedance is defined by the ratio between the mechanical (electromotive) force and velocity (electric) current assuming both are complex exponential functions of time, that is, have constant frequency and damping/amplification. The impedance is defined for waves, for example, sound (Subsection 22.4.1) as the ratio of pressure to velocity,

$$p(\omega) = -Z(\omega)v(\omega), \tag{4.37}$$

in the frequency domain with opposite sign. In this case, a Fourier transform (Subsection 17.5.2) leads to a convolution integral in the time domain. The main point of the preceding analysis, that the impedance is complex, remains valid, as do the laws of association of circuits.

Note 4.2. *Electromechanical Analogies and Controls*: The laws of association of impedances are extensible to the association of control systems; they relate also to the time evolution of mechanical and electrical circuits. They are an instance of electromechanical analogies that extend to potential fields and waves. These in turn open the way to interactions between electromagnetism and matter, for example, fluids and solids; also ionized fluids or plasmas. These links extend to large-scale relativistic and small-scale quantum phenomena.

Conclusion 4: Analogy between (Figure 4.1) mechanical (a) [electrical (b)] circuits consisting of mass (induction), damper (resistance) and spring (capacitor). The association of two impedances (Figure 4.2) in parallel (a) and in series (b). With three impedances (Figure 4.3) there are four possible distinct associations, ranging from all in parallel (a) to all in series (d), through two hybrid cases, with two circuits in series and one in parallel (b) or vice-versa (c).

5

Power, Root, and Logarithm

The power with positive integer order n, is an example (Section 5.1) of a single-valued function, since to each z corresponds only one z^n; the corresponding points for successive n lie on a logarithmic spiral (Section 5.2). The inverse function (Section 5.3), namely the root of order n, is **n-valued**, since for each z there are n distinct possible values of $\sqrt[n]{z}$; the corresponding points lie at the vertices of a regular polygon (Section 5.4) with n sides, leading to a number of trigonometric identities (Section 5.5). There are also **many-valued** functions, like the logarithm (Section 5.6), such that to each value of z corresponds an infinite number of values of $\log z$. The function logarithm appears in the definition (Section 5.7) of z^a complex power z with complex exponent a that generalizes both the ordinary nth power and nth root that correspond respectively to $a = n$ and $a = 1/n$. It can be shown that the behavior of the complex power z^a at the origin $z \to 0$ and infinity $z \to \infty$ depends only on the modulus of the base and (Section 5.8) on the real part of the exponent, and not on the imaginary part; this is a particular case of the exponential of a power having different asymptotic behavior as $z \to \infty$ in different angular sectors (Section 5.9). Selecting of the "correct" branch of a function, and taking into account "jump conditions" between branches, may be necessary to arrive at correct results; it is also an indication that the problem in question has several solutions that may or may not be equivalent for a specific purpose.

5.1 n-th Power as the $(n-1)$-Times Iterated Product

The **n-th power**, with n a positive integer, of a complex number z is defined as the product by itself iterated $(n-1)$ times, and is most conveniently calculated in the polar representation:

$$z = re^{i\varphi}: \quad z^n \equiv z \ldots z\,[(n-1)\,\text{times}] = r^n\, e^{in\varphi}; \tag{5.1a,b}$$

the modulus and argument of the power are given by (1.21a,b), and the real and imaginary parts by

$$\operatorname{Re}(z^n) = r^n \cos(n\varphi), \quad \operatorname{Im}(z^n) = r^n \sin(n\varphi); \tag{5.2a,b}$$

the plane may be divided into n angular sectors, in which the real (imaginary) parts of the power alternate in sign according to Table 5.1; the sectors where each sign is constant have an angular width π/n and are rotated by $\pi/2n$ between the real and imaginary parts of z, as shown by the null-lines $k = 0, \ldots, n-1$:

$$k = 0, \ldots, 2n-1: \quad \varphi_k = \left(k + \frac{1}{2}\right)\frac{\pi}{n}, \frac{k\pi}{n}, \tag{5.3a,b}$$

respectively, for the real (5.3a) and imaginary (5.3b) parts of the power.

TABLE 5.1
Signs of Real and Imaginary Parts of Integral Power

z^n	$\mathrm{Re}(z^n)$	$\mathrm{Im}(z^n)$
Positive	$\left(k-\frac{1}{2}\right)\frac{\pi}{n}<\varphi<\left(k+\frac{1}{2}\right)\frac{\pi}{n}$	$\frac{2k\pi}{n}<\varphi<(2k+1)\frac{\pi}{n}$
Zero	$\varphi=\left(k+\frac{1}{2}\right)\frac{\pi}{n}$	$\varphi=\frac{k\pi}{n}$
Negative	$\left(k+\frac{1}{2}\right)\frac{\pi}{n}<\varphi<\left(k+\frac{3}{2}\right)\frac{\pi}{n}$	$(2k-1)\frac{\pi}{n}<\varphi<\frac{2k\pi}{n}$

$k=0,\ldots,2n-1$.
Note: The signs of the real and imaginary parts of z^n depend on the angular sector through the origin.

5.2 Discrete Set of Points on Logarithmic Spiral

Graphically (Figure 5.1) the power z^n of z to the order n (i) raises the modulus to the power n, that is, yields points progressively farther from (closer to) the origin for $|z|>1$ ($|z|<1$), and points on the unit circle for the intermediate case of modulus unity $|z|=1$; (ii) multiplies the argument by n, and thus rotates counterclockwise (clockwise) around the origin for $\arg(z)>0(\arg(z)<0)$, leaving the points on a straight line only in the case of a real base $\arg(z)=0$:

$$\frac{\log|z^n|}{\log|z|}=n=\frac{\arg(z^n)}{\arg(z)}; \tag{5.4a,b}$$

the powers of a complex number z number lie (Figure 5.1) on a logarithmic spiral:

$$\log r=k\varphi, \quad k\equiv\frac{\log|z|}{\arg(z)}, \tag{5.5a,b}$$

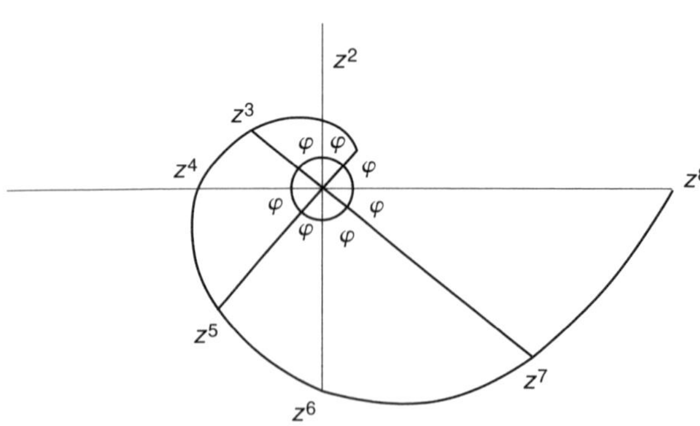

FIGURE 5.1
The successive powers z^n of a complex number lie at equal angles $\varphi=\arg(z)$ along a logarithmic spiral $\log|z^n|=n\log|z|$, and (i/ii) diverge (converge) toward infinity (the origin) as $n\to\infty$ for $|z|>1$ ($|z|<1$); (iii) in the intermediate case $|z|=1$ the spiral degenerates into a circle and z^n divide it in equal angular sectors, for example, three in Figure 5.2.

that is right- (left-) handed for positive arg(z) > 0 [negative arg(z) < 0] argument, and unfolds to infinity (curls up to the origin) for modulus $|z| > 1(|z| < 1)$ greater (smaller) than unity. The degenerate cases $\arg(z) = 0$ and $|z| = 1$ the successive powers lie, respectively, along the real axis (unit circle) that have zero (unit) curvature.

5.3 Inversion of the Power: Roots of Order n

The n-th root is defined as any inverse of the n-th power:

$$w \equiv \sqrt[n]{z} \quad \Leftrightarrow \quad w^n \equiv z \equiv r e^{i\varphi}. \tag{5.6}$$

Using the polar form (5.7a) of w:

$$w = \sigma e^{i\psi}, \quad \sigma^n e^{in\psi} = r e^{i\varphi}, \tag{5.7a,b}$$

it follows that σ, ψ have to be determined satisfying (5.7b) for given r, φ. The number of distinct solutions of (5.7b) specifies the number of roots: (i) there is a single solution for σ:

$$\sigma = \sqrt[n]{r}, \quad \left|\sqrt[n]{z}\right| = \sqrt[n]{|z|}, \tag{5.8a,b}$$

that is, all roots of order n have the same modulus that is the root of order n of the modulus; (ii) the solution for ψ is

$$e^{in\psi} = e^{i\varphi} = e^{i\varphi + i2k\pi}, \quad \psi_k = \frac{\varphi + 2k\pi}{n}, \tag{5.9a,b}$$

where k is any integer; so that the phase (5.9b) proceeds in steps of $2\pi/n = \psi_{k+1} - \psi_k$, and repeats itself, that is, adds 2π, after n steps $\psi_{k+n} = \psi_k + 2\pi$. A set of n distinct values is obtained letting

$$k = 0, 1, \ldots, n-1: \quad \sqrt[n]{z} = r^{1/n} \exp\left\{ i\left(\frac{\varphi}{n} + \frac{2k\pi}{n}\right)\right\}, \tag{5.10}$$

so that *the n roots of order n of a complex number (5.6) are specified by (5.10).* Using (5.10) it follows that:

$$\sqrt[3]{-8i} = \left(2^3 e^{i3\pi/2}\right)^{1/3} = 2\left\{e^{i\pi/2}, \ e^{i7\pi/6}, \ e^{i11\pi/6}\right\} = 2i, -i \pm \sqrt{3}, \tag{5.11}$$

$$\sqrt[6]{-64} = \left(2^6 e^{i\pi}\right)^{1/6} = 2\left\{e^{\pm i\pi/6}, \ e^{\pm i\pi/2}, \ e^{\pm i5\pi/6}\right\} = \pm 2i, \pm i \pm \sqrt{3}, \tag{5.12}$$

where the six roots of (5.12) coincide with the three roots of (5.11) and their complex conjugates, because $\sqrt[6]{-64} = \sqrt[3]{\pm 8i}$ and $\sqrt[3]{8i} = \left(\sqrt[3]{-8i}\right)^*$. Example 10.9 also uses (5.10).

5.4 Regular Polygon Contained in a Circle

The n roots of order n of a complex number can be represented graphically by the construction, illustrated for $n = 3$ in Figure 5.2, by noting that: (i) all roots lie on the circle

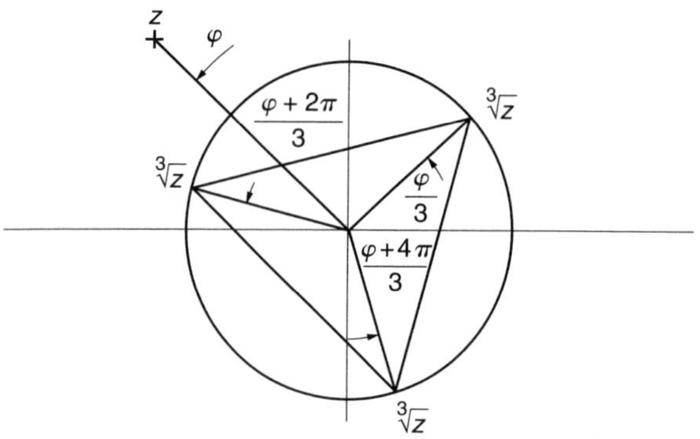

FIGURE 5.2
The roots of order n ($n = 3$ in Figure 5.2) lie on the circle of radius $|z|^{1/n}$ and center at the origin, at equal angular intervals $2\pi/n$, starting in the direction arg $(z)/n$.

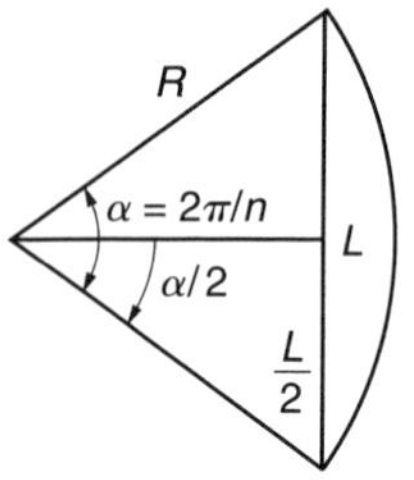

FIGURE 5.3
The n roots of a complex number lie on the vertices of a regular polygon with n sides and center at the origin, for example, an equilateral triangle for $n = 3$ in Figure 5.2. The polygon is made of n triangles, and the perimeter nL can be calculated relating the length of the side L to the radius R. The limit $n \to \infty$ is the perimeter of the circle.

with (5.9b) center at the origin and radius (5.8a) equal to the n-th root of the modulus; (ii) the first root has argument equal to $1/n$ of that of z; (iii) the remaining $n-1$ roots divide the circle into n equal angular sectors measuring $2\pi/n$ each. *Hence the roots of order n of z determine a* ***regular polygon*** *with n sides, contained within the circle of radius $|z|^{1/n}$ and center at the origin, with one vertex lying in the direction $\arg(z)/n$.*

Since the polygon has n sides, the angle of two vertices is $\alpha = 2\pi/n$, and the length of one side $L = 2R\sin(\alpha/2)$, where (Figure 5.3) the radius is R. Hence, *the length of the perimeter of a regular polygon with n sides, with vertices lying on a circle of radius R, is given by:*

$$u \equiv \frac{\pi}{n}: \quad P_n = n2R\sin\left(\frac{\pi}{n}\right) = 2\pi R u^{-1}\sin u. \tag{5.13}$$

Since the perimeter of the polygon cannot exceed that of the circle, follows (5.14a):

$$u^{-1}\sin u \le 1, \quad \lim_{u \to 0} u^{-1}\sin u = 1; \tag{5.14a,b}$$

the equality holds only (5.14b) in the limit $u \to 0$, corresponding to a polygon with $n \to \infty$ sides that coincides with the circle. The result (5.14a) that u exceeds sin u except at the origin, where they coincide (5.14b), can be proved analytically (19.39a).

5.5 Multiple Sums of Sines or Cosines of Equal Angles

As another application consider the set of N-th order roots of $\exp(iN\alpha)$, viz.:

$$n=0,\dots,N-1: \quad \{\exp(iN\alpha)\}^{1/N} = \exp\left(i\alpha + i\frac{2\pi n}{N}\right); \tag{5.15}$$

they are the roots of the polynomial of degree N:

$$z^N - e^{iN\alpha} = \prod_{n=0}^{N-1}\left\{z - \exp\left(i\alpha + i\frac{2\pi n}{N}\right)\right\} \tag{5.16}$$

that can be expanded into the N factors indicated. The power z^{N-1} is missing from the l.h.s. of (5.16), and so the coefficient of z^{N-1} on the r.h.s. is zero:

$$N \geq 2: \quad 0 = -z^{N-1}\sum_{n=0}^{N-1}\exp\left\{i\left(\alpha + \frac{2\pi n}{N}\right)\right\}; \tag{5.17}$$

equating real and imaginary parts to zero leads to

$$N \geq 2: \quad \sum_{n=0}^{N-1}\cos\left(\alpha + \frac{2\pi n}{N}\right) = 0 = \sum_{n=0}^{N-1}\sin\left(\alpha + \frac{2\pi n}{N}\right), \tag{5.18a,b}$$

showing that: *if the sector 2π is divided in N equal parts, the sum of the cosines (sines) is zero; the result still holds if to all angles is added an arbitrary value α.* The geometrical interpretation is that if a regular polygon is contained in a circle of center at the origin, the sum of the cosines (sines) at the vertices is zero. This result also has a mechanical interpretation: (i) consider a **star of forces**, that is, n forces with the same magnitude, applied at the same point (e.g., the origin), and making equal angles $2\pi/n$ (e.g., for $n = 3$ with end points on the vertices of the equilateral triangle in Figure 5.2 as shown in Figure 5.4a); (ii) this system of forces is equivalent to zero, that is, the total or resultant force must be zero (this is most obvious when $n = 2p$ is even, because then there are p pairs of opposite forces that cancel, for example, $n = 4$ in Figure 5.4b); (iii) the x- and y-components of the total force are zero, leading to (5.18a,b). This is generalized next to (15.22a,b).

If $N \geq 3$ the same reasoning can be applied to the coefficient of z^{N-2} that is zero on the l.h.s. of (5.16), and on the r.h.s. is given by

$$N \geq 3: \quad 0 = z^{N-2}\sum_{n=0}^{N-1}\exp\left\{i\left(\alpha + \frac{2\pi n}{N}\right)\right\}\sum_{m=0}^{N-1}\exp\left\{i\left(\alpha + \frac{2\pi m}{N}\right)\right\}; \tag{5.19}$$

equating real and imaginary parts leads to

$$N \geq 3: \quad 0 = \sum_{n,m=0}^{N-1}\begin{matrix}\cos\\ \sin\end{matrix}\left\{2\alpha + 2\pi\frac{(n+m)}{N}\right\} = 0. \tag{5.20a,b}$$

This result (5.20a,b) is geometrically less obvious than (15.18a,b), and can be extended to $1 \leq p \leq N-1$ sets of numbers $n_1,\dots,n_p$, by equating in (15.16) the coefficients of z^{N-p}:

$$1 \leq p \leq N-1: \quad 0 = (-)z^{N-p}\prod_{q=1}^{p}\sum_{n_q=0}^{N-1}\exp\left\{i\left[\alpha + \left(\frac{2\pi}{N}\right)n_q\right]\right\}; \tag{5.21}$$

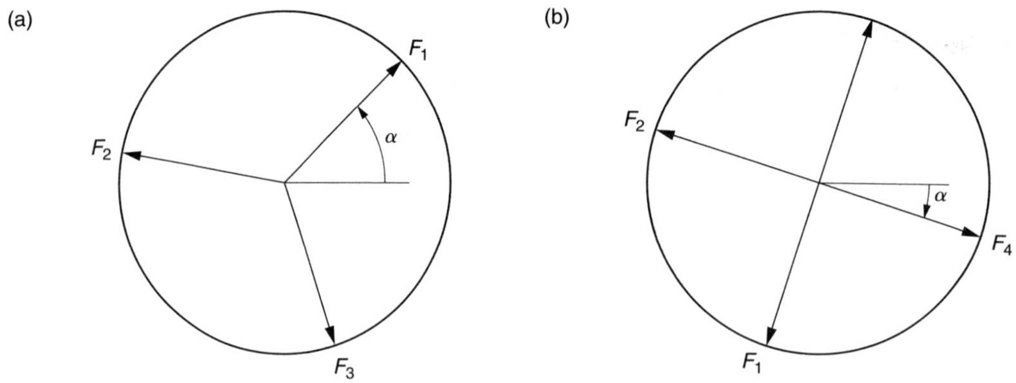

FIGURE 5.4
Dividing the circle in equal sectors, for example, via the roots of a complex number, leads to a system of forces equivalent to zero, for example, three in (a) ≡ (Figure 5.2) and four in (b). The system of forces is equivalent to zero because (i) all forces pass through the same point, the center, so there is no torque; (ii) the components of the forces in any direction cancel, regardless of rotation angle α.

The real and imaginary parts of (5.20) yield:

$$1 \le p \le N-1: \quad 0 = \sum_{n_1,\ldots,n_p=0}^{N-1} \begin{matrix}\cos \\ \sin\end{matrix} \left(p\alpha + \frac{2\pi}{N} \sum_{q=1}^{p} n_q \right), \tag{5.22a,b}$$

showing that: *if* $0 \le \alpha < 2\pi$ *is any angle, the identities (5.22a,b) hold for any natural number* $N \ge 1+p$, *and any* $1 \le p \le N-1$, *for example, the particular cases* $p=1$ $(p=2)$ *are (5.18a,b) [(5.20a,b)] that hold for* $N \ge 2$ $(N \ge 3)$. The only remaining results from (5.16) are two checks: (i) an obvious identity for the power z^N; (ii) for the power $z^0 \equiv 1$, viz.:

$$\begin{aligned} e^{-iN\alpha} &= (-)^N \prod_{n=0}^{N-1} \exp\left[i\left(\alpha + \frac{2\pi n}{N}\right)\right] = (-)^N e^{iN\alpha} \exp\left(\frac{2\pi}{N}\sum_{q=1}^{p} n_q\right) \\ &= (-)^N e^{iN\alpha} e^{i\pi(N-1)} = (-)^N e^{iN\alpha} (-)^{N-1} = -e^{iN\alpha}, \end{aligned} \tag{5.23}$$

another identity that proves nothing new.

5.6 Single-, Multi-, and Many-Valued Functions

The power z^n is a single-valued function, the square root $\sqrt{z}$ a two-valued function, the n-th root $\sqrt[n]{z}$ an n-valued or multivalued function, and many-valued functions, with infinitely many values, also exist, for example, the logarithm. Considering:

$$k = 0, \pm 2, \ldots: \quad z = re^{i\varphi} = re^{i\varphi + i2\pi k}, \tag{5.24}$$

where k is any integer; taking logarithms leads to

$$k \in |Z: \quad \log z = \log r + i\varphi + i2\pi k; \tag{5.25}$$

to each z there correspond infinitely many values of $\log z$, all having the same real part, and differing only in the imaginary part, by multiples of 2π. The reason is that the logarithm is the inverse of the exponential:

$$\exp(z) = \exp(i2\pi k)\exp(z) = \exp(z + i2\pi k), \tag{5.26}$$

and the latter is periodic with period 2π; when inversion is performed, the phase is undetermined to within a multiple of 2π, leading to a many-valued function, viz.:

$$\log(1 \pm i) = \log\left(\sqrt{2}e^{\pm i\pi/4}\right) = \frac{1}{2}\log 2 \pm i\frac{\pi}{4} + i2k\pi, \tag{5.27}$$

is an example.

5.7 Power with Complex Base and Exponent

The n-th power and root of order n are particular cases, respectively $a = n$ and $a = 1/n$ with n a positive integer, of the power with base z and exponent a both complex:

$$k \in |Z: \quad z^a \equiv \exp(a \log z) = \exp(a \log z + i2\pi ka), \tag{5.28}$$

and generally a many-valued function, because its definition involves the logarithm. The latter introduces the factor $\exp(i2\pi ka)$, with k an arbitrary integer: (i) for $a = p$ an integer, the factor is unity, and the function reduces to the integral power z^p that is single-valued; (ii) for $a = 1/q$ the inverse of an integer, the factor takes q distinct values, and the function reduces to the root $\sqrt[q]{z}$ of order q that is q-valued; (iii) for $a = p/q$ a rational number in its lowest terms, the factor takes q distinct values, and the function $\sqrt[q]{z^p}$ is q-valued, as should be expected from (i) and (ii); (iv) for all other values of a, viz. real irrational, imaginary, or complex, the factor takes an infinite number of distinct values, and function z^a is many-valued. An example is:

$$k \in |Z: \quad i^i \equiv \exp(i \log i) = \exp\left\{i\left(i\frac{\pi}{2} + i2\pi k\right)\right\} = \exp\left[-\left(2k + \frac{1}{2}\right)\pi\right], \tag{5.29}$$

where k is arbitrary integer.

5.8 Limiting Behavior at the Origin and Infinity

As another example, with u,v real and positive, consider:

$$\begin{aligned} u,v > 0: \quad (iu)^{-iv} &\equiv \exp\{iu \log(-iv)\} = \exp\left\{iu \log v + \frac{\pi u}{2}\right\} \\ &= \exp\left(\frac{\pi u}{2}\right)\{\cos(u \log v) + i \sin(u \log v)\}, \end{aligned} \tag{5.30}$$

where the principal value was taken, viz. $k = 0$ in (5.28). The function z^a, with complex $z(a)$ in the polar (5.31a) [Cartesian (5.31b)] form, is specified by (5.31c):

$$\begin{aligned} a = \alpha + i\beta: \quad z^a &= \exp\{(\alpha + i\beta)(\log r + i\varphi)\} = \exp(\alpha \log r - \beta\varphi)\exp\{i(\beta \log r + \alpha\varphi)\} \\ z = re^{i\varphi}: \quad &= r^\alpha e^{-\beta\varphi}[\cos(\beta \log r + \alpha\varphi) + i \sin(\beta \log r + \alpha\varphi)]; \end{aligned}$$

(5.31a–c)

TABLE 5.2
Limits of z^a as $z \to 0, \infty$ for All a

z^a	Re(a) > 0	Re(a) < 0
as $z \to 0$:	$z^a \to 0$	$z^a \to \infty$
as $z \to \infty$:	$z^a \to \infty$	$z^a \to 0$

Note: The limits of z^a at the origin $z \to 0$ and infinity $z \to \infty$ depend only on the sign of the real part of the exponent Re(a).

the power with complex base and exponent generally involves [e.g., (5.30) and (5.31a–c)] the real functions cosine and sine with logarithmic arguments. The second factor in (5.31c) has modulus unity and $\beta\varphi$ is bounded, so that the behavior at the origin $r = 0$ and infinity $r = \infty$ is determined solely by $\alpha \equiv \mathrm{Re}(a)$ viz.:

$$\lim_{z \to 0,\infty} |z^a| \sim \lim_{|z| \to 0,\infty} \exp\{\mathrm{Re}(a) \log|z|\}, \tag{5.32}$$

where $\sim$ means that there may be a finite factor. From (5.32) follow the results in Table 5.2, viz.: *the limit of the complex power z^a as the base z tends to the origin $z \to 0$ (to infinity $z \to \infty$) is determined solely by the real part of the exponent* a, *and is* $0\ (\infty)$ *for* $\mathrm{Re(a)} > 0$, *and conversely* $\infty\,(0)$ *for* $\mathrm{Re(a)} < 0$. The limits in Table 5.2 are isotropic, that is, independent of direction arg(z) for the function z^a.

5.9 Vanishing and Divergence on Alternate Sectors

For other functions the limit may be anisotropic, for example, depend on direction, such as:

$$\exp(z^2) = \exp\left[(x+iy)^2\right] = \exp(x^2 - y^2)\exp(i2xy), \tag{5.33}$$

that as $z \to \infty$, is oscillatory with unit modulus along $x^2 = y^2$ the diagonals of the quadrants $y = \pm x$, and in the four sectors separated by these (Figure 5.5) alternatively diverges and vanishes, according to Table 5.3. This is the particular case $n = 2$ of the exponential of an integral power (5.34a):

$$|\exp(z^n)| = \exp\{r^n \cos(n\varphi)\}, \quad \lim_{z \to 0} \exp(z^n) = 1, \tag{5.34a,b}$$

that (i) tends to unity (5.34b) as $z \to 0$ tends to the origin in any direction; (ii) as $z \to \infty$ tends to infinity it vanishes or diverges on alternate sectors of width π/n, viz. from (5.2a):

$k = 0, \ldots, 2n-1:$

$$\lim_{z \to \infty} \exp(z^n) = \begin{cases} \infty & \text{for } (k-1/2)\pi/n < \varphi < (k+1/2)\pi/n, & (5.35\text{a}) \\ \exp\{i|z|^n(-)^n\} & \text{for } \varphi = (k+1/2)\pi/n, & (5.35\text{b}) \\ 0 & \text{for } (k+1/2)\pi/n < \varphi < (k+3/2)\pi/n; & (5.35\text{c}) \end{cases}$$

(iii) along the separation lines (5.5a) it oscillates in phase with modulus unity. Note that (i) the first column of Table 5.1 coincides with the conditions (5.35a–c), viz. both specify the sign of cos($n\varphi$); (ii) Table 5.3 is the particular case $n = 2$ of (5.35a–c).

Note 5.1. *Balancing of Forces*: The results (5.18a,b) hold if the star of forces is rotated by an angle α, which may vary with time $\alpha(t)$, leading to an angular velocity $\omega \equiv d\alpha/dt$. Thus

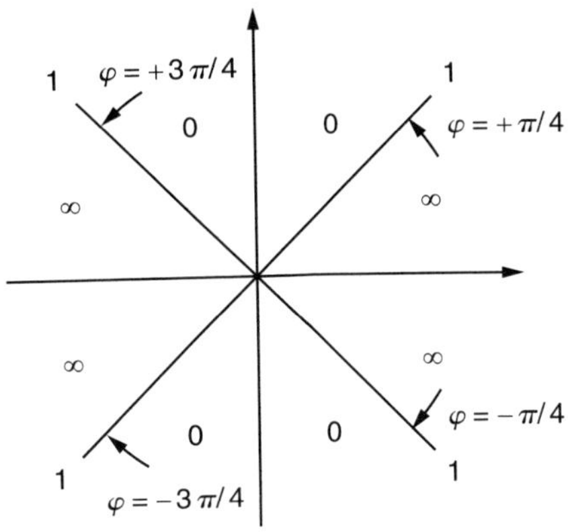

FIGURE 5.5
The function $\exp(z^2)$ in Table 5.3 diverges (vanishes) at infinity in the angular sectors between the diagonals of quadrants which contain the real (imaginary) axis. The boundary between these sectors is the diagonals of quadrants, along which the function oscillates with unit modulus.

TABLE 5.3
Behavior of $\exp(z^2)$ at Infinity in All Directions

$\exp(z^2) \to$	∞	$\exp(\pm i2x^2)$	0
(x,y)	$x^2 > y^2$	$y = \pm x$	$y^2 > x^2$
$r \to \infty$	$-\frac{\pi}{4} < \varphi < +\frac{\pi}{4}$	$\varphi = \pm\frac{\pi}{4}$	$\frac{\pi}{4} < \varphi < \frac{3\pi}{4}$
	or		or
	$+\frac{3\pi}{4} < \varphi < +\frac{5\pi}{4}$		$-\frac{3\pi}{4} < \varphi < -\frac{\pi}{4}$

Note: The signs of the real part of z^2 change (Table 5.1) across the diagonals of quadrants. This implies that $\exp(z^2)$ diverges (tends to zero) at infinity in alternate sectors between the diagonals of quadrants in Figure 5.5. It oscillates with unit modulus along the diagonals which separate the sectors of decay and divergence.

the result (5.18a,b), and its generalizations (5.20a,b) and (5.22a,b) apply to the balancing of multicylinder piston engines; these have a single crankshaft, to which are linked at equal angles the rod and crank mechanisms of each piston, through which the forces are exerted. This is one of the many practical cases of balancing of forces in classical mechanics.

Conclusion 5: The successive integral powers of a fixed complex number lie on a logarithmic spiral (Figure 5.1), whereas the n roots of order n of a complex number specify a regular polygon with the n vertices on a circle of center at the origin (Figure 5.2); one side is represented (Figure 5.3), to calculate its length, and hence the perimeter of the polygon. The diagonals of the quadrants divide the complex z-plane into four sectors (Figure 5.5) containing the directions for which the exp (z^2) function diverges (∞) or vanishes (0) as $z \to \infty$ tends to infinity; it oscillates with unit modulus on the diagonals of quadrants, which are the boundaries of the four sectors. The divisions (Figure 5.4) in three (four) sectors correspond to a star of three (a) [four (b)] forces.

6

Electron in an Electromagnetic Field

A charged particle (Section 6.1) placed in an electric field, experiences an acceleration (Section 6.2) that increases as the electric charge increases and the mass decreases. In the presence of a magnetic field (Sections 6.3 and 6.4), the charged particle rotates around it, with an angular velocity that increases with increasing charge and decreasing mass. The superposition of these two motions (Sections 6.6 and 6.7) specifies the trajectory of the charged particle in an external electromagnetic field, uniform or not. Electromagnetic fields may be used to "focus" a beam of charged particles (Section 6.8), distinguish between positive and negative charges, and separate particles or atoms with different charges and/or masses (Section 6.9). The motion of charged particles in electromagnetic fields can be observed in laboratory experiments and occurs in astrophysics, for example, in the ionosphere of the earth and interplanetary space; some electrical devices, like cathode ray tubes and television screens, are based on deflection of the trajectories of charged particles.

6.1 Electromagnetic or Laplace–Lorentz Force

A particle, for example, an electron of **charge** e placed in an electric **field** $\vec{\mathrm{E}}$ is subject (Section 24.3) to an **electric force** (6.1a); if the charged particles moves with velocity $\vec{\mathrm{v}}$ it carries an **electric current** (6.1b):

$$\vec{\mathrm{F}}_e = e\vec{\mathrm{E}}, \quad \vec{\mathrm{J}} = e\vec{\mathrm{v}}, \quad \vec{\mathrm{F}}_m = \frac{\vec{\mathrm{J}} \wedge \vec{\mathrm{B}}}{c}, \tag{6.1a–c}$$

and in the presence of a magnetic field of induction $\vec{\mathrm{B}}$, it is acted upon (Section 26.3) by a magnetic force $\vec{\mathrm{F}}$ given by (6.1c), where c is an universal constant, namely, the speed of light in vacuo. The sum of the electric and magnetic forces is (Section 28.3) the Laplace–Lorentz or total **electromagnetic force** that balances the inertia force, equal to mass times acceleration:

$$\vec{\mathrm{F}}_e + \vec{\mathrm{F}}_m = \vec{\mathrm{F}}_{em} = e\vec{\mathrm{E}} + \frac{e}{c}\vec{\mathrm{v}} \wedge \vec{\mathrm{B}} = m\frac{d\vec{\mathrm{v}}}{dt}; \quad \frac{d\vec{\mathrm{v}}}{dt} + \vec{\Omega} \wedge \vec{\mathrm{v}} = \vec{\mathrm{a}} \equiv \frac{\vec{\mathrm{F}}}{m}, \tag{6.2a,b}$$

the equation (6.2a) is similar to the Euler equation (6.2b), describing the combined translation and rotation of a spherical rigid body with an angular velocity $\vec{\Omega}$, under a mechanical force $\vec{\mathrm{F}}$ corresponding per unit mass m to an acceleration $\vec{\mathrm{a}}$. Comparison of (6.2a) with (6.2b) shows that

$$\vec{\mathrm{a}} = \frac{e}{m}\vec{\mathrm{E}}, \quad \vec{\Omega} = \frac{e\vec{\mathrm{B}}}{mc}, \tag{6.3a,b}$$

the electric field $\vec{\mathrm{E}}$*(magnetic induction* $\vec{\mathrm{B}}$*) causes a translation (rotation) with acceleration (6.3a) [angular velocity (6.3b)] proportional to the ratio of charge to mass.*

6.2 Uniform Fields and Larmor (1897) Frequency

The integration of the equation of motion (6.2a) specifies the velocity and the trajectory, as shown next in the case of uniform electric and magnetic fields. A Cartesian reference frame is chosen with OZ-axis aligned with the magnetic induction, and OY-axis such that the electric field lies on the YOZ-plane:

$$\vec{\mathrm{B}} = (0,0,B), \quad \vec{\mathrm{E}} = (0, E_\perp, E_{//}); \tag{6.4a,b}$$

thus $E_\perp$ ($E_{//}$) is the component of the electric field orthogonal (parallel) to the magnetic induction. The vector system (6.2a) of three differential equations takes the form:

$$\dot{v}_x = \Omega v_y, \quad \dot{v}_y = -\Omega v_x + a_\perp, \quad \dot{v}_z = a_{//}, \tag{6.5a–c}$$

where (i) dot means derivative with regard to time; (ii) $a_\perp$ ($a_{//}$) is the acceleration (6.3a) due to the component of the electric parallel (6.6a) [orthogonal (6.6b)] to the induction:

$$a_{//} = \frac{eE_{//}}{m}, \quad a_\perp = \frac{eE_\perp}{m}, \quad \Omega \equiv \left|\vec{\Omega}\right| = \frac{eB}{mc}; \tag{6.6a–c}$$

(ii) Ω is associated with the magnetic induction (6.3b) $\equiv$ (6.6c), has the dimensions of inverse time, and is designated **Larmor frequency (1897)**.

6.3 Longitudinal Translation and Transverse Rotation

The system of three differential equations (6.5a–c) separates into two subsets, viz. (6.5c) that specifies alone the longitudinal motion, and (6.5a,b) that are coupled and determine the transversal motion. The motion along the magnetic induction (6.5c) is independent of the transversal motion, and is determined (6.6a) by:

$$v_z(t) = \frac{eE_{//}}{m}t + v_{//}, \quad z(t) = \frac{eE_{//}}{2m}t^2 + v_{//}t, \tag{6.7a,b}$$

where $v_{//}$ is the initial longitudinal velocity and z obtained by integrating $v_z = \dot{z}$ and is given by (6.7b), choosing the XOZ-plane so as to pass through the particle at initial time $t = 0$. From (6.7a,b) it follows that the longitudinal motion is uniformly accelerated (uniform) in the presence $E_{//} \neq 0$ (absence $E_{//} = 0$) of a longitudinal electric field.

To describe the transverse motion that is coupled between the directions OX and OY, a complex velocity is introduced (6.8a):

$$v \equiv v_x + iv_y: \quad \frac{\mathrm{d}v}{\mathrm{d}t} = \dot{v} = \dot{v}_x + i\dot{v}_y = -i\Omega(v_x + iv_y) + ia_\perp = -i\Omega v + ia_\perp, \tag{6.8a,b}$$

that replaces the pair of coupled real equations (6.5a,b) by a single complex differential equation (6.8b); the integration of the latter is elementary:

$$-i\Omega t = -i\Omega \int_0^t \mathrm{d}t = \int_{v_0}^{v} \left(v - \frac{a_\perp}{\Omega}\right)^{-1} \mathrm{d}v = \log\left(\frac{v - a_\perp/\Omega}{v_0 - a_\perp/\Omega}\right), \tag{6.9a}$$

where v_0 denotes the initial complex velocity at time $t = 0$.

6.4 Components of the Velocity and Trajectory of the Particle

The initial complex velocity $v_0 = v_{0x} + iv_{0y}$ may be taken as real $v_0 = v_{0x} \equiv v_\perp$ provided that initial time $t = 0$ is chosen when the particle velocity lies in the XOZ-plane:

$$v(t) = \left(v_\perp - \frac{a_\perp}{\Omega}\right)\exp(-i\Omega t) + \frac{a_\perp}{\Omega}; \tag{6.9b}$$

separating real and imaginary parts of (6.9b) $\equiv$ (6.8a) specifies (6.3b,6.6b) the two transverse components of the velocity:

$$v_x(t) = \left(v_\perp - \frac{cE_\perp}{B}\right)\cos(\Omega t) + \frac{cE_\perp}{B}, \quad v_y(t) = -\left(v_\perp - \frac{cE_\perp}{B}\right)\sin(\Omega t); \tag{6.10a,b}$$

integrating with regard to time, leads to the equations of the trajectory:

$$x(t) = \int_0^t v_x(t)\,\mathrm{d}t = \left(\frac{v_\perp}{\Omega} - \frac{cE_\perp}{B\Omega}\right)\sin(\Omega t) + \frac{cE_\perp}{B}t, \tag{6.11a}$$

$$y(t) = \int_0^t v_y(t)\,\mathrm{d}t = \left(\frac{v_\perp}{\Omega} - \frac{cE_\perp}{B\Omega}\right)\cos(\Omega t), \tag{6.11b}$$

in which the origin of the reference frame was placed at the position of the particle at time $t{=}0$. Thus *a particle of charge e and mass m placed in an uniform electromagnetic field, undergoes a motion specified by the velocity (6.7a; 6.10a,b) and trajectory (6.7b; 6.11a,b).* An alternative derivation of these results using only real instead of complex variables is presented in Example 10.10. The motion will be analyzed first in the cases of (Section 6.5) zero electric field $E_\perp = 0 = E_{//}$, and (Section 6.6) zero initial velocity $v_\perp = 0 = v_{//}$, before proceeding to the general case (Section 6.7) when both are nonzero; all cases are included in the Table 6.1.

6.5 Linear, Circular, and Helical Motion

In the absence of an electric field (6.12a,b), the velocity (6.7a; 6.10a,b) $\equiv$ (6.12c–e):

$$E_{//} = 0 = E_\perp: \quad v_x(t) = v_\perp\cos(\Omega t), \quad v_y(t) = -v_\perp\sin(\Omega t), \quad v_z(t) = v_{//}, \tag{6.12a–e}$$

and coordinates (6.7b; 6.11a,b):

$$x(t) = \frac{v_\perp}{\Omega}\sin(\Omega t), \quad y(t) = \frac{v_\perp}{\Omega}\cos(\Omega t), \quad z(t) = v_{//}t \tag{6.13a–c}$$

of the electron show that, in the presence of uniform magnetic induction, the following trajectories are possible: (I) if it is initially at rest $v_{//} = 0 = v_\perp$ it remains at rest, that is, the magnetic field alone cannot start a motion since the magnetic force $(e/c)\vec{v}\wedge\vec{B} = 0$ is zero for zero velocity $\vec{v} = 0$; (II) if the initial velocity is longitudinal $v_{//} \neq 0 = v_\perp$ the motion is uniform along the magnetic induction, and is unaffected by it because the magnetic force $(e/c)\vec{v}\wedge\vec{B} = 0$ is zero for $\vec{v}$ parallel to $\vec{B}$; (III) if the initial velocity is transverse (6.14a,b)

TABLE 6.1
Trajectories of an Electron in a Uniform Electromagnetic Field

Section	Case	Formulas (6.)	Initial velocity		Electric field		Trajectory: motion	Figure
			Longitudinal*	Transversal*	Longitudinal*	Transversal*		
6.5	I	—	$v_{//} = 0$	$v_\perp = 0$	$E_{//} = 0$	$E_\perp = 0$	At rest	—
	II	(12e, 13c)	$v_{//} \neq 0$	$v_\perp = 0$	$E_{//} = 0$	$E_\perp = 0$	Linear: uniform	—
	III	(12c,d; 13a,b; 14c)	$v_{//} = 0$	$v_\perp \neq 0$	$E_{//} = 0$	$E_\perp = 0$	Circular	6.1
	IV	(12c–e; 13a–c; 14c; 15c,d)	$v_{//} \neq 0$	$v_\perp \neq 0$	$E_{//} = 0$	$E_\perp = 0$	Cylindrical helix	6.2
6.6	V	(16c, 17c)	$v_{//} = 0$	$v_\perp = 0$	$E_{//} \neq 0$	$E_\perp = 0$	Linear uniformly accelerated	—
	VI	(16a,b; 17a,b; 18c,d)	$v_{//} = 0$	$v_\perp = 0$	$E_{//} = 0$	$E_\perp \neq 0$	Cycloid	6.3
	VII	(16a–c; 17a–c; 18c,d)	$v_{//} = 0$	$v_\perp = 0$	$E_{//} \neq 0$	$E_\perp \neq 0$	Elongated cycloid	—
6.7	VIII	(12c,d; 13a,b; 14c; 16c; 17c; 19a,b)	$v_{//} \neq 0$	$v_\perp \neq 0$	$E_{//} \neq 0$	$E_\perp = 0$	Elongated cylindrical helix	6.4
	IX	(10a,b; 11a,b; 18c,d; 20a,b)	$v_{//} \neq 0$	$v_\perp \neq 0$	$E_{//} = 0$	$E_\perp \neq 0$	Trochoid: curled elongated	6.6/6.5
	X	(16c; 17c; 10a,b; 11a,b)	$v_{//} \neq 0$	$v_\perp \neq 0$	$E_{//} \neq 0$	$E_\perp \neq 0$	Distorted trochoid	—

*Relative to the direction of magnetic induction.

Note: These ten cases depend on the nonzero components of the electric field $\vec{E}$ and initial velocity $\vec{v}$ in the direction parallel $E_{//}, v_{//}$ and perpendicular $(E_\perp, V_\perp)$ to the magnetic field as shown in Figures 6.1–6.6.

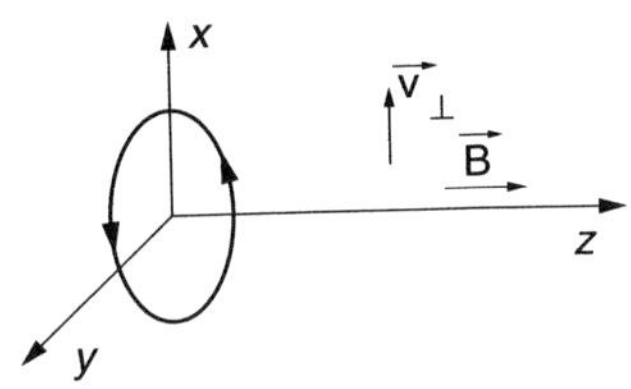

FIGURE 6.1
An electron with initial velocity transverse to an external magnetic field moves in a circle around the latter with constant velocity.

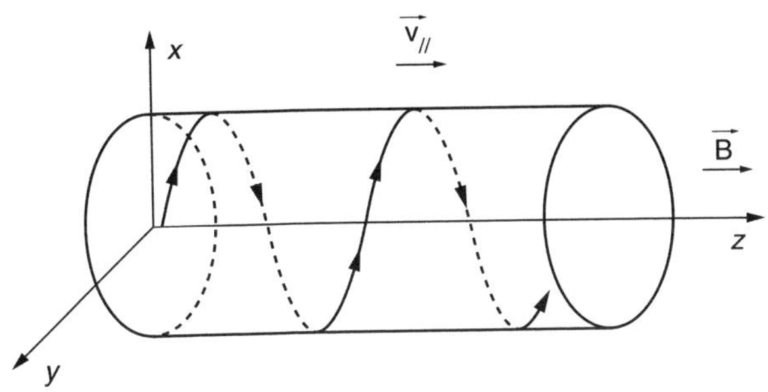

FIGURE 6.2
An electron with initial velocity oblique to an external magnetic field moves along a cylindrical helix with constant azimuthal (axial) velocity, equal to the component of the initial velocity orthogonal (parallel) to the magnetic field.

the motion takes place in the plane passing through the initial velocity and orthogonal to the magnetic induction, and the trajectory (6.12a,b; 6.13a,b) is circular (Figure 6.1):

$$v_{//} = 0 \neq v_\perp : \quad \left| x(t)^2 + y(t)^2 \right|^{1/2} = R \equiv \frac{v_\perp}{\Omega} = \frac{mev_\perp}{cB}; \tag{6.14a–c}$$

it is described with constant tangential velocity equal to the initial velocity and constant angular velocity Ω equal to the Larmor frequency (6.6c) so that the radius (6.14c) increases for larger electric charge and mass, and smaller magnetic induction; (IV) if the initial velocity has both longitudinal and transversal components nonzero (6.15a,b), the superposition of cases (II) and (III) shows that the trajectory is a **cylindrical helix** (Figure 6.2), of radius (6.14c), taken at the angular velocity (6.6c), with linear tangential velocity (6.15c):

$$v_{//} \neq 0 \neq v_\perp : \quad v_0 = \sqrt{v_\perp^2 + v_{//}^2}, \quad \alpha = \arctan\left(\frac{v_{//}}{v_\perp}\right), \tag{6.15a–d}$$

making a constant angle (6.15d) with the direction of the magnetic induction.

6.6 Linear Acceleration and Cycloid in the Plane

In the case where the particle is initially at rest $v_{//} = 0 = v_\perp$, the motion can only be started by the electric field, although it is then modified by the magnetic induction. The components of the velocity (6.7a; 6.10a,b):

$$v_\perp = 0: v_x(t) = \frac{cE_\perp}{B}\left[1 - \cos(\Omega t)\right], \quad v_y(t) = \frac{cE_\perp}{B}\sin(\Omega t), \quad v_z(t) = \frac{eE_{//}}{m}t, \tag{6.16a–c}$$

and the coordinates of the particle (6.7b; 6.11a,b):

$$x(t) = \frac{E_\perp c}{B\Omega}\left[\Omega t - \sin(\Omega t)\right], \quad y(t) = -\frac{E_\perp c}{B\Omega}\cos(\Omega t), \quad z(t) = \frac{E_{//}e}{2m}t^2, \tag{6.17a–c}$$

show that three trajectories (V–VII) are possible. If (V) the electric field and magnetic induction are parallel $E_{//} \neq 0 = E_\perp$, the motion is uniformly accelerated along their common direction (6.16c; 6.17c) because only the electric force and not the magnetic force acts $e\vec{E} \neq 0 = (e/c)\,\vec{v} \wedge \vec{B}$. If (VI) the electric field and magnetic induction are orthogonal (6.18a,b), the trajectory lies in the plane (x, y) orthogonal to the magnetic field, on which lies the electric field (6.16a,b; 6.17a,b); the trajectory corresponds to an uniform circular motion with angular velocity equal to the Larmor frequency Ω in (6.6c) and linear velocity (6.18c) and radius (6.18d):

$$E_{//} = 0 \neq E_\perp: \quad U = \frac{E_\perp c}{B}, \quad S = \frac{U}{\Omega} = \frac{E_\perp c^2 m}{B^2 e}, \tag{6.18a–d}$$

upon which is superimposed an uniform motion with velocity U along the OX axis; thus the trajectory is a **cycloid** (Figure 6.3) that is also the curve described by a point on a circle rolling without sliding on a straight line. If (VII) the electric field and magnetic induction are oblique $E_{//} \neq 0 \neq E_\perp$, the superposition of cases (V) and (VI) leads to a trajectory that projects as the cycloid on the XOY-plane, and is elongated by an uniform acceleration along the OZ axis.

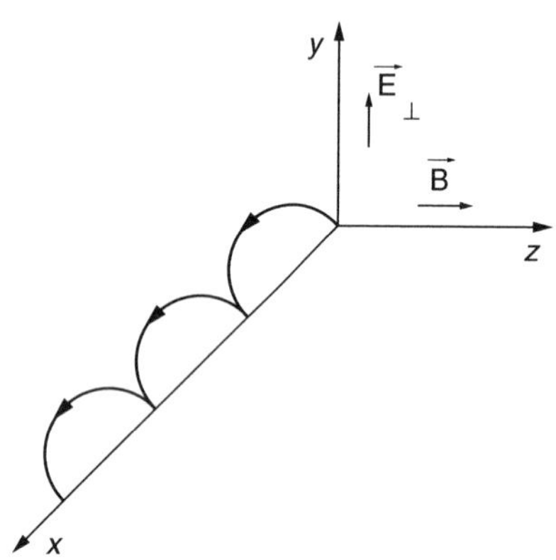

FIGURE 6.3
If the electron starts at rest it will only move in the presence of an electric field. If the latter is orthogonal to the magnetic field the trajectory is a cycloid. It corresponds to a fixed point on a circle rolling without sliding on a plane. It has a cusp where the trajectory touches the axis and inverts the direction changing the tangent between $\pm\infty$.

6.7 Elongated Helix and Plane Trochoid

In the general case of a charged particle with initial velocity in the presence of uniform electric and induction fields, three more (VIII–X) trajectories are possible, in addition to the seven particular cases considered earlier. If (VIII) the electric field and magnetic induction are parallel $E_\perp = 0 \neq E_{//}$, the superposition of the cases (III) and (V) shows that the motion is circular and uniform (6.12c,d; 6.13a,b) in the transversal plane and uniformly accelerated (6.16c; 6.17c) in the direction parallel to the induction; thus the trajectory is a **cylindrical helix progressively elongated** (Figure 6.4), with constant radius (6.14c), angular velocity (6.6c), tangential velocity (6.19a), and inclination (6.19b) increasing with time:

$$v(t) = \left| v_\perp^2 + \left(\frac{eE_{//}}{m} t + v_{//} \right)^2 \right|^{1/2}, \quad \alpha(t) = \text{arc tan} \left(\frac{eE_{//}}{mv_\perp} t + \frac{v_{//}}{v_\perp} \right). \tag{6.19a,b}$$

If (IX) the electric field and magnetic induction are orthogonal $E_{//} = 0 \neq E_\perp$, the motion along OZ is uniform, and the trajectory in the XOY-plane is the superposition (6.10a,b; 6.11a,b) of a circular motion with tangential velocity (6.20a) and radius (6.20b):

$$V = \frac{E_\perp c}{B} - v_\perp, \quad R = \frac{|V|}{\Omega} = \left| \frac{E_\perp c}{B\Omega} - \frac{v_\perp}{\Omega} \right| = \left| \frac{E_\perp mc^2}{eB^2} - \frac{v_\perp mc}{eB} \right|, \tag{6.20a,b}$$

and a translation along OX at velocity U in (6.18b); thus the trajectory would be a cycloid (Figure 6.3) if $U = V$ or $v_\perp = 0$ in case VI, and in the present case IX is an **elongated (curled) trochoid** illustrated in Figure 6.5 (6.6); it corresponds to $v_\perp > 0$ ($v_\perp < 0$) translation velocity larger $U > V$ (smaller $U < V$) than the tangential velocity, as for a circle rolling on a plane and sliding in the direction of (direction opposite to) the translation. In the most general case (X) of oblique electric field and magnetic induction, the trajectory projects on the XOY-plane as a trochoid, and is elongated in the OZ direction; thus the particle rotates around the magnetic induction $\vec{B}$ and is simultaneously accelerated in the oblique direction of the electric field $\vec{E}$, changing its initial velocity that is generally oblique to both $\vec{B}$ and $\vec{E}$.

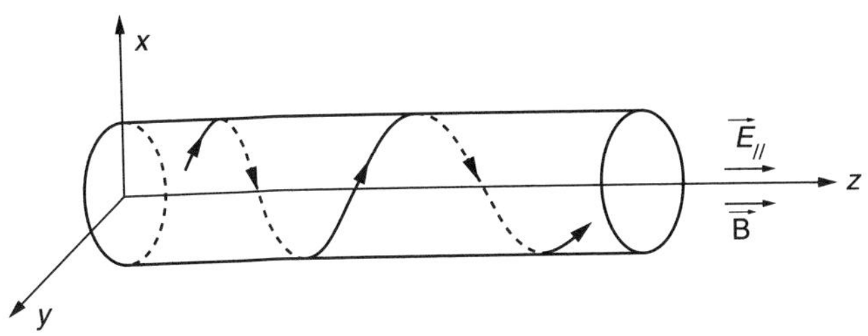

FIGURE 6.4
If the electric and magnetic fields are parallel, and the electron starts with arbitrary oblique initial velocity, the cylindrical helix in Figure 6.2 is deformed by the acceleration along the electric field, that is, increasingly elongated in the axial direction.

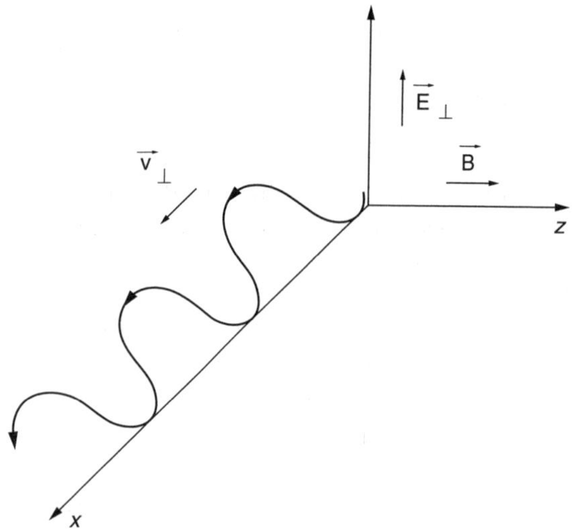

FIGURE 6.5
If the electric and magnetic fields are orthogonal, apart from a uniform motion along the magnetic field, the trajectory is a deformation of the cycloid in Figure 6.3. If the transverse component of the initial velocity is along the cycloid, the latter becomes an elongated trochoid without cusps, because a positive horizontal velocity has been added.

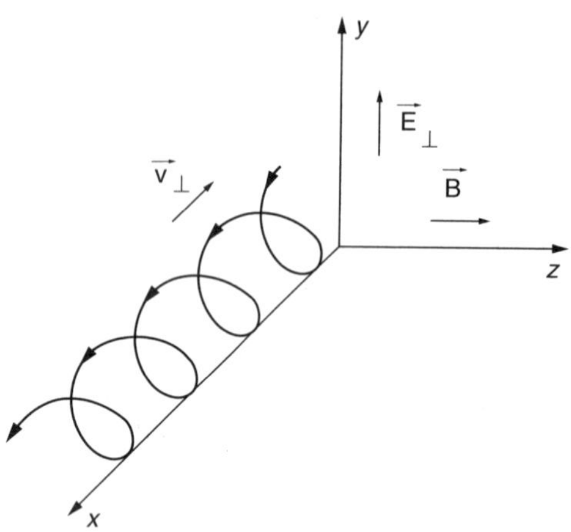

FIGURE 6.6
As in Figure 6.5: if the transverse component of the initial velocity is negative it inverts the motion at the cusps of the cycloid in Figure 6.3, leading to a curled trochoid with double points, where the electron passes again through the same point.

6.8 Oval and Conical Helices and Magnetic Focusing

In the absence of an electric field, the trajectory of the electron can be deformed if the magnetic induction is nonuniform: (XI) if the initial velocity of particle is transverse to $\vec{B}$, the trajectory is circular (Figure 6.1), provided that $\vec{B}$ be constant or varies only longitudinally;

(XI) if $\vec{B}$ varies transversely to itself, the side on which it is larger the radius of curvature is smaller (6.14c) and the angular velocity larger (6.6c), so that the electron describes with approximately constant velocity $\Omega R = v_\perp$ an **oval** trajectory (Figure 6.7); (XII) if $\vec{B}$ varies transversely to itself and the initial velocity is oblique, the cylindrical helix (Figure 6.2) is deformed into an **oval helix** (Figure 6.8) on a cylinder whose directrix is the oval of case (XI); (XIII) if $\vec{B}$ varies longitudinally to itself and the electron has an oblique initial velocity, the radius of the circular motion decreases as B increases, and the trajectory is a **conical helix** (Figure 6.9) on a tapering surface. The case (XIII) demonstrates the effect of **magnetic focusing** of electrically charged particles: as the magnetic induction increases, the magnetic field lines approach each other and trap electrons in their interior, in a tightening spiral (Figure 6.9) with an increasing angular velocity. Table 6.2 contains a few of the many possible cases of electron motion in nonuniform electromagnetic fields, viz: (i) the three cases XI–XIII of nonuniform magnetic induction (Section 6.8); (ii) one case of uniform magnetic and nonuniform electric field (Example 10.11).

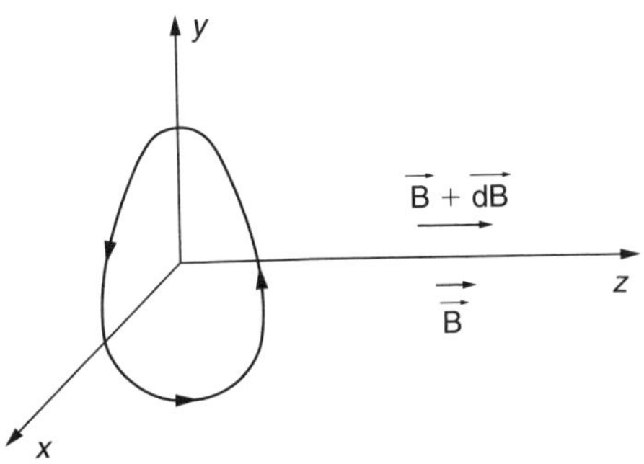

FIGURE 6.7
If the initial velocity of the electron is transverse to a transversely nonuniform magnetic field with constant direction, the circular path in Figure 6.1 is deformed into an oval, with smaller (larger) radius on the side of stronger (weaker) magnetic field, where the angular velocity is larger (smaller).

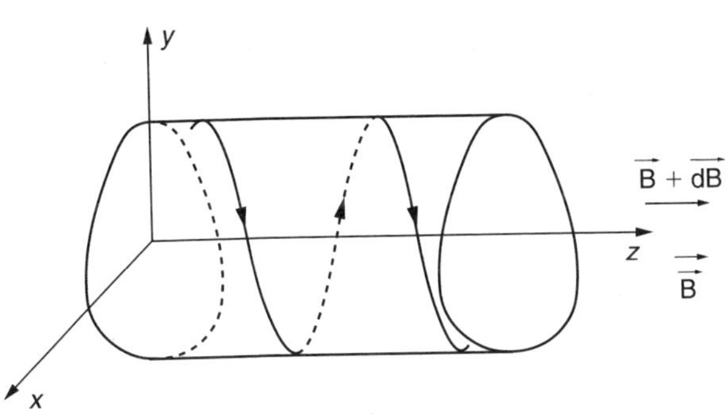

FIGURE 6.8
In the magnetic field is transversely nonuniform with constant direction and the electron starts with an oblique velocity, the oval in Figure 6.7 becomes the directrix of an oval cylinder, upon which lies the oval helix which is the trajectory of the electron.

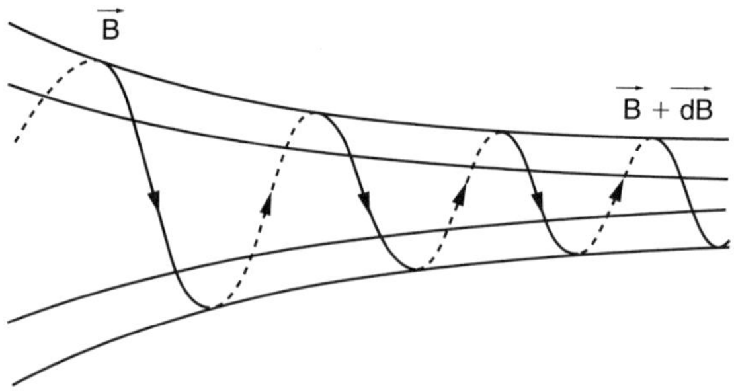

FIGURE 6.9
If the magnetic field has constant direction with axially increasing strength, and the electron starts with oblique velocity, the trajectory is tightening spiral with decreasing radius, corresponding to "magnetic focusing" with an increasing angular velocity.

6.9 Separation of Isotopes in a Mass Spectrograph

Consider a beam of elementary particles, for example, electrons, protons, and neutrons, or a **plasma**, that is, ionized matter in an external electromagnetic field: (a) the **neutrons** that are electrically neutral are the only ones whose trajectory is unaffected by the electromagnetic field; (b) the **electrons (protons)** have a positive (negative) charge and are accelerated along (opposite to) the direction of the electric field, and rotate in the counterclockwise (clockwise) direction around the magnetic induction; (c) the protons have the opposite charge and a much greater mass than the electrons, by a factor 1836, and thus are less accelerated by electric field (6.6a,b) and rotate around the magnetic induction with smaller angular velocity (6.6c) and on helices of larger radius (6.14c), but with the same inclination (6.15d) determined by the initial velocity (6.15c). In this way, the passage through an electromagnetic field separates electrons, protons, and neutrons. The electrons have the same charge (with opposite sign) as the protons, and much smaller mass, they have much larger linear (angular) velocity in the same external electric (magnetic) field, and dominate the **Ohm (Hall) electric current**; the neutrons can have an effect only through collisions, that is, **ambipolar diffusion** in the plasma.

The simplest model of the **atom** consists of a **nucleus** of protons and neutrons, surrounded by a "cloud" of electrons. Since the atom, in a free nonionized form is electrically neutral, the number of electrons is equal to that of protons, and specifies the **atomic number** of the **element**, for example, 8 for oxygen and 92 for uranium. A given **element** may have several **isotopes** that can be distinguished by the number of neutrons, whose mass is comparable to that of protons, and thus affect the mass of the nucleus; this determines the **mass number** of the atom, since electrons are much lighter; for example, uranium-235 has 92 protons + 143 neutrons and uranium-238 has 92 protons + 146 neutrons. The isotopes of a given element are difficult to separate or isolate by chemical processes because they have similar electrical characteristics; often all but a few isotopes are unstable, and have short lives. If a natural sample, containing several isotopes of the same element is placed in a **mass spectrometer**, the magnetic field separates the heavier from the lighter isotopes, by increasing radius of turn (6.14c) and decreasing angular velocity (6.6c). The measurement of the Larmor frequency (or the radius) allows the mass of the isotope to be determined if the charge or element is known, or vice-versa. Another method of separation is a high-speed centrifuge, using the centrifugal force to concentrate heavier isotopes outward.

TABLE 6.2
Some Electron Trajectories in Nonuniform Electromagnetic Fields

Section	Case	Initial velocity		Field		Trajectory	Figure
		Longitudinal	Transversal	Electric	Induction		
6.8	XI	$v_{//} = 0$	$v_{\perp} \neq 0$	$\vec{E} = 0$	$\vec{B} \neq 0$ varies transversely	Oval	6.7
	XII	$v_{//} \neq 0$	$v_{\perp} \neq 0$	$\vec{E} = 0$	$\vec{B} \neq 0$ varies transversely	Helix on oval cylinder	6.8
	XII	$v_{//} \neq 0$	$v_{\perp} \neq 0$	$\vec{E} = 0$	$\vec{B} \neq 0$ varies longitudinally*	Magnetic focusing	6.9
Example 10.11	XII	$v_{//} \neq 0$	$v_{\perp} \neq 0$	$\vec{E}$ varies linearly	$\vec{B}$ uniform parallel to $\vec{E}$	Sinusoid on circular cylinder	10.3

*Relative to direction of magnetic induction.

Note: In contrast with the exhaustive listing in Table 6.1 of ten cases of trajectories of an electron in uniform electric and magnetic fields, only a sample of four cases appears in this table for some nonuniform fields; these correspond to Figures 6.7–6.9 and Figure 10.3 of Example 10.11.

Note 6.1. *Cycloidal and Trochoidal Motions*: As implied by (6.2a,b) the motion of an electron in an external magnetic field is analogous to the rotation of a symmetrical rigid body; the equations (6.5a,b) are the same, and can be solved using: (a) either a complex velocity (Section 6.3); (b) or by eliminating among real quantities (Example 10.10). Besides the 13 cases considered (Chapter 6) for uniform or slowly varying electric field and magnetic induction, there are many other cases with strongly nonuniform fields, for example, one is considered in Example 10.11. The analogy with the rotation of a rigid body breaks down for an unsymmetrical top. The motions of charged particles in nonuniform electromagnetic fields are relevant to geophysics and astrophysics, for example, the ionosphere of the earth. These are other examples of cycloidal and trochoidal motions, for example, a cylinder with circulation in a potential flow (Section 28.6). Classical mechanics gives examples not only of cycloidal and trochoidal motions, but also of generalizations, for example, the hypo (epi) cycloids described by a point on a circle rolling inside (outside) another circle. The cycloid family of plane curves has properties related to the calculus of variations, for example, the cycloid is: (i) the brachistocrone, that is, curve of quickest frictionless descent between two points; (ii) the tautochrone, that is, the pendulum whose period of oscillation is independent of the amplitude, making it an error-free clock.

Conclusion 6: Trajectories of a charged particle subject to electromagnetic fields: (i/ii) circle (cylindrical helix) in a constant magnetic field in the case [Figure 6.1 (6.2)] of transverse (oblique) initial velocity; (iii/iv) elongated helix (cycloid) in constant parallel (orthogonal) electric and magnetic fields [Figure 6.4 (6.3)], with zero initial velocity; (v/vi) curled (elongated) trochoid with double points (points of tangency) for [Figure 6.6 (6.5)] orthogonal electric and magnetic fields with initial velocity opposite to (along) the electric field. For a nonuniform magnetic field: (vii/viii) oval trajectory (oval helix) for [Figure 6.7 (6.8)] magnetic field varying transversely, and initial transverse (oblique) velocity; (ix) conical helix for (Figure 6.9) oblique initial velocity and magnetic field varying longitudinally. The latter (Figure 6.9) illustrates the trapping and focusing of charged particles by converging magnetic field lines.

7

Multivalued Functions, Branch-Points, and Branch-Cuts

A function $w = f(z)$ is multivalued (Section 7.1) if there are at least some values of z to which correspond several values of w, viz. each of the values specifies one branch of the function. One branch may be selected as the principal branch, for example, the branch with smallest phase in modulus. A multivalued function may be treated as single-valued if the same branch is always chosen, for example, the principal branch (Section 7.3). A complex function $w = f(z)$ transforms curves in the z-plane into curves in the w-plane for each of the branches; each of the branches lies on sheet, and different sheets may be connected to form a Riemann surface (Sections 7.1 and 7.2). Some paths in the z-plane will pass from one sheet to another in the Riemann surface of w, leading to a change of branch of the function. To avoid changing to another branch, the lines joining the different sheets in the Riemann surface are never to be crossed; these lines correspond to branch-cuts (Section 7.4) in the z-plane. The branch-cuts join between themselves (Section 7.8) or to infinity (Section 7.5), branch-points where the branches coincide, and in whose neighborhood they separate. Thus, picking the principal branch of a multivalued function, and performing appropriate branch-cuts, will always retain the same branch if the cuts are not crossed. To draw the branch-cuts (Section 7.9) it is necessary to locate the branch-points (Sections 7.6 and 7.7), and join them between themselves or to infinity so as to avoid passing to other branches; the principal branch is usually discontinuous across the two sides of a branch-cut (Section 7.4), since continuity could only be preserved by going over to another sheet of the Riemann surface. Whereas the representation of a single-valued function requires only one sheet, the connection between sheets of a multivalued function may have a meaning: if the branches of a multivalued function represent distinct solutions of the same problem or different states of the same system, then their connections correspond to possible or impossible choices or transitions.

7.1 Riemann (1857) Surface of a Multivalued Function

A function $w = f(z)$ is called **single-valued** if to each value of z corresponds only one value of w, for example, $w = z^3$ or $w = e^z$. The set of values of z (w) is called **domain** (**range**) of the function $w = f(z)$, and it is region of the z-plane (w-plane). Thus a single-valued function maps a region, its domain in the z-plane, into another region, its range, in the w-plane. A function $w = f(z)$ is **multivalued** if to one value of z can correspond several (two or more, even infinitely many) values w_n of w; each value defines one **branch** $w_n(z)$ of the function, where $n = 1, \dots, N$; if the number of branches is N, the function is ***N*-valued**. The domain of an N-valued function lies in the z-plane, and the range consists of N **sheets**, one for each branch, forming a **Riemann surface** (1857); the sheets of a Riemann surface are connected to each other at the points where different branches coincide. Thus *for an N-valued function with complex domain z, a Riemann surface with N connected sheets can* *be constructed as its range w. The branches coincide at branchs-points (Subsection 7.1.1), from which issue branch-cuts (Subsection 7.1.2).*

7.1.1 Coincidence of Branches at a Branch-Point

The simplest instance of a multivalued function is the case $N=2$ of a two-valued function, for example (Section 5.3) the square root (5.10 with $n = 2$, and $k = 0, 1$) that has two branches:

$$w \equiv \sqrt{z - z_1}: \quad w_0 = |z - z_1|^{1/2} e^{i\varphi/2}, \quad w_1 = |z - z_1|^{1/2} e^{i\varphi/2 + i\pi} = -w_0, \qquad (7.1a\text{–}c)$$

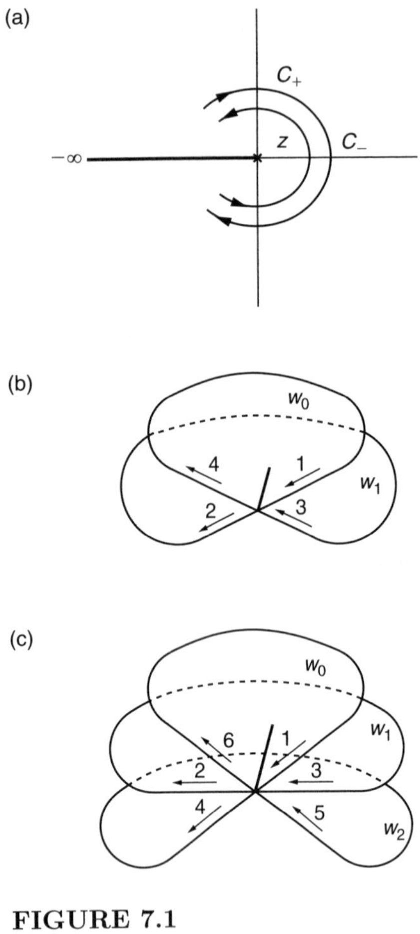

FIGURE 7.1
For a multivalued complex function with a branch-point at the origin a loop in the counter clockwise C_+ (clockwise C_-) direction (a) will pass to the next (preceding) sheet of the Riemann surface. This can be prevented by a semiinfinite branch-cut joining the branch-point to infinity. In the case of a two-valued complex function (b) each branch is mapped into one sheet with the two sheets joined at the branch-cut: an odd (even) number of loops in the same direction around the branch-point leads to the other (same) branch. If the complex function is triple valued, the three sheets are still joined at the same branch-cut (c); successive loops in the same direction around the branch-point change to the next sheet, returning to the same sheet after three loops.

that coincide at the **branch-point** z_1. The function square-root maps the complex z-plane into the sheets w_0, w_1 of a Riemann surface for w. If a path L is drawn in the z-plane starting and ending at z and not enclosing the point z_1, then $|z - z_1|$ and $\arg(z - z_1)$ return to the original value at the end, and the corresponding paths w_0 and w_1 each lie on a distinct sheet. If however the path C_- (C_+) in Figure 7.1a in the z-plane goes round the point z_1 once in the positive (negative) direction, the modulus $|z - z_1|$ returns to the original value, but the phase $\arg(z - z_1)$ is increased (decreased) by 2π:

$$z - z_1 \to (z - z_1)\,\mathrm{e}^{\pm 2i\pi}, \quad \sqrt{z - z_1} \to \sqrt{z - z_1}\,\mathrm{e}^{\pm i\pi}, \quad w_0 \to w_1, \tag{7.2a–c}$$

and the path in the w_0-sheet crosses over to the w_1-sheet.

7.1.2 Branch-Cut Issuing from a Branch-Point

The transformation $\arg(z - z_1) \to \arg(z - z_1) + 2\pi$ that leaves the z-plane unchanged, interchanges the w_0- and w_1-sheets of Riemann surface w:

$$\arg(z - z_1) \to \arg(z - z_1) + 2\pi: \quad w_0 \to w_1, \quad w_1 \to w_0. \tag{7.3a,b}$$

It follows that the function square root can be made single-valued by preventing *the passage between the two sheets of the Riemann w-surface for the square root, viz. by drawing in the* z*-plane a line* or **branch-cut** *joining the branch-point* z_1 *to infinity, for example, along the negative real axis if the branch-point is at the origin* $z_1 = 0$ (Figure 7.1a). Crossing the branch-cut once, or an odd number of times passes from one sheet to the other; not crossing the branch-cut, or crossing it an even number of times, leads back to the same sheet. The Riemann surface with two sheets is sketched in Figure 7.1b that gives the impression that they are joined along a line; this is a defect of attempting a the representation in three-dimensional space that is not possible (a more suitable representation on the Riemann sphere is given in Section 9.7). In fact the two branches (7.1b,c) coincide only at one point $z = z_1$, so the two sheets of the Riemann surface touch only at the branch-point. The branch-cut in the z-plane does not correspond to an intersection of the two sheets of the Riemann surface, but rather to a jump (Section 7.4) needed to avoid going to the other sheets, and to remain on the same sheet.

7.2 Denumerable Infinity of Connected Sheets

The Riemann surface can be constructed for function with two (Section 7.1), a finite number (Subsection 7.2.1) or a denumerable infinity (Subsection 7.2.2) of branches, leading to a classification of branch-points (Subsection 7.2.3).

7.2.1 Multivalued Function with a Finite Number of Branches

The square-root is the particular case $N = 2$ of the N-th root (Section 5.3), with N any positive integer that is an **multivalued function** or N-valued function (5.10) with N branches:

$$n = 0, \ldots, N-1: \quad w_n = \left(\sqrt[N]{z - z_1}\right)_n = |z - z_1|^{1/n} \exp\left[i\frac{\arg(z - z_1)}{N} + \frac{2\pi i n}{N}\right]; \tag{7.4}$$

the z-plane is mapped into a Riemann w-surface with N sheets, all connected at the branch-point $z = z_1$, so that passage from one sheet to the next or preceding is prevented by the branch-cut joining the branch-point z_1 to infinity, for example, as in Figure 7.1a. The Riemann surface with $N = 3$ sheets is sketched in Figure 7.1c; the $(N - 1)$ sheets only touch at one point, viz. $z = z_1$ the branch-point, where all branches coincide $w_0 = w_1 = \cdots = w_{N-1}$. Performing a circuit L that does not enclose the branch-point, that is, does not cross the branch-cut, leads to a path on a single sheet; if the path encloses the branch-point or crosses the branch-cut $k < N$ times in the positive (negative) direction, it ends k sheets forward (rearward):

$$\arg(z - z_1) \to \arg(z - z_1) \pm k2\pi : \quad w_n \to w_{n\pm k}. \tag{7.5}$$

To return to the same sheet it is sufficient to follow a similar path in opposite direction, or alternatively:

$$w_{n\pm N} = |z|^{1/N} \exp\left[i\frac{\arg(z - z_1)}{N} + \frac{i2\pi n}{N} \pm i2\pi\right] = w_n, \tag{7.6}$$

it is sufficient to enclose the origin (or cross the branch-cut) N times in the same direction.

7.2.2 Many-Valued Function with a Denumerable Infinity of Branches

An example of a **many-valued** function (Section 5.6) is the logarithm (5.25) that has a denumerable infinity of branches:

$$n \in |Z : \quad w_n = \log(z - z_1) = \log|z - z_1| + i\arg(z - z_1) + i2\pi n, \tag{7.7}$$

one corresponding to each integer value of n. The Riemann w-surface consists of a denumerable infinity of sheets, represented in Figure 7.2 as a helicoid; all sheets have one, and only one, common point, namely the branch-point $z = z_1$ corresponding to the origin $w = 0$ of the w-plane. The passage between adjoining sheets is prevented in the same way as with the N-th root, viz. by (Figure 7.1a) a branch-cut joining the branch-point z_1 to infinity for example, along the negative real axis for $\log z$. To move k sheets forward [(7.7) with $k = n$)] the branch-cut should be crossed k times in the positive direction, that is, the branch-point

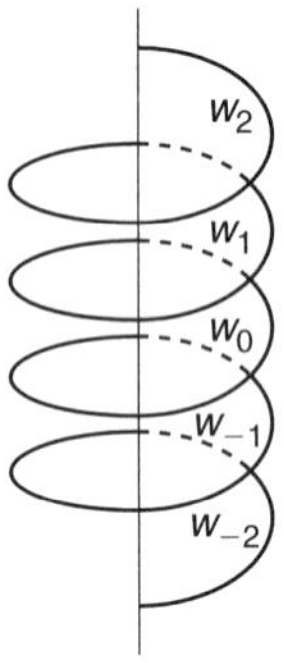

FIGURE 7.2
For a many-valued function with infinitely many branches represented on infinitely many sheets of the Riemann surface proceeding with loops in the same direction around the origin there is never a return to the same sheet. The only way to return to the same sheet is to perform the same number of loops around the branch point in the opposite direction.

z_1 is encircled k times in the counterclockwise direction in the z-plane; to return to the same sheet, it is necessary to reverse the path, since by encircling the branch-point always in the same direction new sheets always appear, and the original sheet is never regained.

7.2.3 Classification of Branch-Points

The logarithm is a many-valued function (Section 5.6) because it is the inverse of the exponential that is periodic with period $2\pi \equiv \tau$. In fact:

$$n \in |Z: \quad w = f(t) = f(t+\tau) = f(t+n\tau) \Rightarrow t = f^{-1}(w) - n\tau, \tag{7.8a,b}$$

any periodic function (7.8a) has a many-valued inverse (7.8b); the result is extensible to two doubly periodic functions:

$$n, m \in |Z: \quad w = f(t) = f(t+n\tau_1+m\tau_2) \Rightarrow t = f^{-1}(\omega) - n\tau_1 - n\tau_2. \tag{7.9a,b}$$

A complex function cannot have more than two fundamental periods, since two is the dimension of the complex plane. The **order** of branch-point of a function is $N-1$, where N is the number of branches of the function in its neighborhood. A branch-point is **algebraic (logarithmic)** if its order is finite (infinite). Some examples are (i) a one-valued function has all points of order zero; (ii/iii) a two-valued function like the square root (N-th root) has an algebraic branch-point of order 1 ($N-1$); (iv) any periodic function has at least one logarithmic branch-point.

7.3 Principal Branch in the Cut-Plane

A multivalued function can be replaced by a single-valued function provided that: (i) one branch is chosen as the **principal branch**; *(ii) branch-cuts are performed in the domain so as to prevent crossing over to other branches in the range.* The principal branch of a multivalued function can be chosen as the branch with smallest argument in modulus, or some other convention. This particular choice corresponds to taking $n=0$ in the case of the N-th root (7.4) or logarithm (7.7):

$$\sqrt[N]{z-z_1} \equiv |z-z_1|^{1/N} \exp\left\{i \arg \frac{(z-z_1)}{N}\right\}, \quad \log(z-z_1) = \log|z-z_1| + i \arg(z-z_1); \tag{7.10a,b}$$

whenever a multivalued function is mentioned, for example, the complex power (Section 5.7), the principal branch is considered, viz. (5.28) with $\log(z)$ given by (7.10b).

The choice of a principal branch is not sufficient, since paths in the z-plane exist that could pass to other sheets in the Riemann w-surface, and lead to different branches of the multivalued function. To prevent this *the branch-cuts in the domain must be considered as impassable barriers that cannot be crossed by any path, and restrict the range to the principal sheet of the Riemann surface, corresponding to the principal branch of the multivalued function.* For example, the N-th root (7.4) [logarithm (7.7)] are N-valued (many-valued) functions, but taken as their principal branches (7.10a) [(7.10b)], become single-valued functions in the complex z-plane with a cut joining (Figure 7.1) infinity to the branch-point.

7.4 Jump Discontinuity across a Branch-Cut

A multivalued function is continuous across a branch-cut if passage from one sheet of the Riemann surface to the next is allowed, that is, changing over to another branch. To retain the principal branch, it is necessary to "jump" from the adjoining sheet back to the principal sheet, implying a discontinuity in the principal branch. The branch-point can be brought to the origin $z_1 = 0$ by a translation, and the branch-cut drawn along the negative real axis, with: (i) the origin excluded $z \notin) - \infty, 0($ from the branch-cut if the function is defined at the branch-point, for example, $z^{1/N} = 0$ at $z = 0$; (ii) the origin included $z \notin) - \infty, 0)$ in the branch-cut if the function is singular at the branch-point, for example, $\log z = -\infty$ for $z = 0$. Bearing in mind that $z = -x + i0 = xe^{i\pi}$ above ($z = -x - i0 = xe^{-i\pi}$ below) the branch-cut along the negative real axis, to the N-th root and logarithm correspond, respectively:

$$0 < x \in |R: \quad (-x \pm i0)^{1/N} = x^{1/N} \mathrm{e}^{\pm i\pi/N}, \quad \log(-x \pm i0) = \log x \pm i\pi, \qquad (7.11\text{a,b})$$

with x real and positive; it follows that *the N-valued (many-valued) function root of order* N *(7.4) [natural logarithm (7.7)], taken as the principal branch (7.10a) [7.10b] with branch-point $z_1 = 0$ at the origin, takes the values (7.11a) [(7.11b)] with upper/lower sign respectively above/below the branch-cut along the negative real axis from $-\infty$ to the origin, the origin being excluded (included) in the branch-cut; thus there is a discontinuity across the cut specified by:*

$$-x \in (-\infty, 0): \quad (-x + i0)^{1/N} - (-x - i0)^{1/N} = x^{1/N} \left(\mathrm{e}^{i\pi/N} - \mathrm{e}^{-i\pi/N} \right) = x^{1/N} 2i \sin\left(\frac{\pi}{N}\right), \qquad (7.12\text{a})$$

$$-x \in (-\infty, 0(: \quad \log(-x + i0) - \log(-x - i0) = 2\pi i. \qquad (7.12\text{b})$$

The discontinuity (7.12a,b) means that if a path C_+ is drawn enclosing the origin once in the positive direction (Figure 7.1a) the N-th root (logarithm) changes by adding (7.12a) [(7.12b)] that is the phase change due to the transformation $\varphi \to \varphi + i2\pi$; reversing the path to C_- would subtract the same value, and taking k loops would imply adding the same term k times.

7.5 Semiinfinite Cut Joining a Branch-Point to Infinity

Considering a multivalued function $w = f(z)$ the **branch-cuts** are lines in the domain or z-plane, preventing passage between distinct sheets of the Riemann w-surface that is the range; the *branch-cuts join between themselves and infinity the* ***branch-points****, where different branches of the multivalued function coincide.* For example, the N-th root (7.4) and natural logarithm (7.7), have a single branch-point at $z = z_1$, where all branches coincide, that is, take same value, finite $\sqrt[N]{0} = 0$ or not $\log 0 = -\infty$; the branch-cut joins the branch-point at the origin to infinity, and any direction is adequate to prevent the branch-point from being encircled, for example, the negative real axis was chosen in (7.12a,b) for a branch-point at the origin. Another example is (Section 5.7) *the function complex power z defined by (5.28) that is single-valued for $a \in |Z$ an integer (positive, negative, or zero), q-valued for $a = p/q$ a rational fraction in its lowest terms and many-valued for a real irrational,* *pure imaginary, or complex; the function becomes single-valued in all cases by choosing the principal branch:*

$$z^a \equiv \exp\{a\log(z)\} = \exp\{a[\log|z| + i\arg(z)]\} = |z|^a \exp[i\arg(z)], \tag{7.13a}$$

on the complex z-plane with a branch-cut along the negative real axis from $-\infty$ to the origin; the jump across the branch-cut is:

$$(-x+i0)^a - (-x-i0)^a = |x|^a \left(e^{i\pi a} - e^{-i\pi a}\right) = 2i|x|^a \sin(\pi a). \tag{7.13b}$$

The origin is included in (excluded from) the branch-cut if a is nonintegral and $\operatorname{Re}(a) < 0$ $[\operatorname{Re}(a) > 0]$ because $z^a \to \infty\,(z^a \to 0)$ as $z \to 0$ a shown in Section 5.8 and Table 5.2. For $a = 1/N$ the (7.13b) agrees with (7.12a). The branch-cut could be any line, straight or curved, joining the origin to infinity, in any direction, since it serves only to prevent circuits round the origin; the choice of the branch-cut along the negative real axis in Figure 7.1a is a convenient convention.

7.6 Infinite Derivative of a Function at a Branch-Point

It is convenient to have a method of locating the branch-points of an arbitrary multivalued function, since once these have been found, the branch-cuts joining them or infinity can be drawn; then the principal branch taken without risk of passing over to other branches, provided that the cuts be regarded as barriers that the paths in domain cannot cross. Taking as examples the functions N-th root and logarithm, and taking their derivatives:

$$\frac{d\left\{(z-z_1)^{1/N}\right\}}{dz} = N^{-1}(z-z_1)^{1/N-1}, \quad \frac{d\{\log(z-z_1)\}}{dz} = \frac{1}{z-z_1}, \tag{7.14a,b}$$

they are both infinity at the branch-point $z = z_1$; this suggests the following **theorem of identification of branch-points**: *if $z = b$ is a branch-point of a multivalued function $f(z)$, then the derivative of the latter is infinite there:*

$$z = b \text{ is a branch point} \Rightarrow f'(b) = \infty; \tag{7.15}$$

with two qualifications: (i) if the function has a factor $f_0(z)$ that is single-valued in the neighborhood of $z = b$, and vanishes there (7.16a):

$$f_0(b) = 0: \quad f(z) = f_0(z) f_1(z), \quad f_1'(b) = \infty, \tag{7.16a–c}$$

then this factor must be removed (7.16b) before calculating the derivative (7.16c); (ii) the condition is necessary but not sufficient viz. a function may be single-valued in the neighborhood of point, have an infinite derivative that the point, and yet it is not a branch-point.

Before proving the theorem, two examples are given, concerning its applicability viz. to explain the need for the two qualifications made. The function $f(z) = z^n$, with n an integer, is single-valued, and thus has no branch-points; its derivative $f'(z) = nz^{n-1}$ is infinite at the origin for n a negative integer, but $z = 0$ is not a branch-point, since the theorem applies only to multivalued functions. The complex power $f(z) = z^a$, with nonintegral exponent a is (Section 7.4) a multivalued function with branch-point at the origin $z = 0$; its derivative $f'(z) = az^{a-1}$ is infinite at the origin for $\operatorname{Re}(a) < 1$, and the criterion (7.15) identifies the branch-point only in this case. If $\operatorname{Re}(a) \geq 1$ then $f'(z) = az^{a-1}$ can be written $a = b + n$, where n is an integer such that $n \le \mathrm{Re}(a) < n+1$, and thus $0 \le (\mathrm{Re}(b)) < 1$; it follows that the function $f(z) = z^a = z^n f_1(z)$ splits into: (i) a single-valued factor $f_0(z) = z^n$, with n an integer that vanishes at the origin as in (7.16a,b); (ii) the remaining factor a multivalued function $f_1(z) = z^b$, whose derivative $f_1'(z) = bz^{b-1}$ is infinite at the origin $f_1' = \infty$ showing that it is a branch-point as in (7.16c). It follows that $f(z) = z^a$ has a branch-point at the origin for all nonintegral values of the exponent a.

7.7 Theorem and Method for the Identification of Branch-Points

To prove the theorem (7.15), assume that the function $w = f(z)$ is multivalued and $z = b$ is a branch-point, where $f(b) = B$. Thus the equation $f(z) = B + \varepsilon$ has multiple roots in the neighborhood of b, one for each branch of the function, and all roots coincide at $z = b$. It follows that the inverse function $z = f^{-1}(w)$ is of the type:

$$g(B) \neq 0: \quad z - b = (w - B)^a g(w) \equiv f^{-1}(w), \tag{7.17a}$$

with real a such that $a > 1$, and the function $g(w)$ does not vanish at $w = B$, for example, $a = N$ a positive integer if there are N branches. On derivation:

$$a > 1: \quad \frac{\mathrm{d}f^{-1}}{\mathrm{d}w} = \frac{\mathrm{d}z}{\mathrm{d}w} = (w - B)^{a-1} \{ag(w) + (w - B) g'(w)\}; \tag{7.17b}$$

bearing in mind that $a - 1 > 0$, it follows that at a branch-point $z - b$, $w = B$ the inverse (direct) function $f^{-1}(f)$ has zero (7.18b) [infinite (7.18c)] derivative:

$$g'(B) \neq \infty: \quad \left.\frac{\mathrm{d}z}{\mathrm{d}w}\right|_{w=B} = 0 \Leftrightarrow \left.\frac{\mathrm{d}w}{\mathrm{d}z}\right|_{z=b} = \left\{\left.\frac{\mathrm{d}z}{\mathrm{d}w}\right|_{w=B}\right\}^{-1} = \infty, \tag{7.18a–c}$$

as stated in (7.15) = (7.18c) where the condition (7.18a) was used. Since $g(w)$ is nonzero at $w = B$, and differentiable in its neighborhood, it has bounded logarithmic derivative (7.19a):

$$\infty \neq C \equiv \frac{\mathrm{d}\{\log[g(w)]\}}{\mathrm{d}w} = \frac{g'(w)}{g(w)}, \quad g'(B) = Cg(B) \neq \infty, \tag{7.19a,b}$$

implying (7.19b), so that the condition (7.18a) is met. The reasoning (7.17a) to (7.19b) relies on the assumption that all needed derivatives exist in a neighborhood of $z = B$ (and $w = B$) excluding the branch-cut.

The qualifications (i) and (ii) in the theorem of localization of branch-points suggest that (i) before the derivative is calculated at any possible branch-point, all factors that are single-valued and vanish at the point be removed, because by multiplying the "infinite" derivative they might make it zero instead; (iii) if the derivative is infinite, it must be checked that it is indeed a branch-point, and not some other singularity, by confirming that (a) the function is multivalued in its neighborhood and (b) all branches coincide at the point. Thus the **method of search for branch-points** *can be stated as follows: (i) consider in turn every point b where the derivative of the function is zero or infinity $f'(b) = 0, \infty$; (ii) for each point remove (7.16b) from the function any factor $f_0(z)$ that is (7.16a) single valued in the neighborhood of b and zero at b; (iii) find if the derivative at b, after the factor $f_0(z)$ is removed, is infinite (7.16c); (iv) check that the function is multivalued in the neighborhood of b, and that all branches coincide at b.*

7.8 Elementary Functions with Two Branch-Points

The usefulness of the theorem is that it limits the search for the branch-points of the multivalued function $f(z)$ to those where $\mathrm{d}f/\mathrm{d}z$ is infinite (plus those where it is zero, if there is a single-valued vanishing factor); all points $z=c$ with finite (and nonzero) derivative are excluded. The points with zero derivative need not be considered if all vanishing single-valued factors like $(z-c)^n$ are suppressed. As an example consider the functions:

$$f_1(z) = \sqrt{(z-z_1)(z-z_2)} = \sqrt{(z_1-z)(z_2-z)}, \tag{7.20a}$$
$$f_2(z) = \sqrt{(z_1-z)(z-z_2)} = \sqrt{(z-z_1)(z_2-z)}, \tag{7.20b}$$

that are the two distinct cases with two branch-points $N=2$ of (7.42a). Both functions have infinite derivative at z_1 and z_2, for example, for (7.20a):

$$2\frac{\mathrm{d}f_1}{\mathrm{d}z} = \sqrt{\frac{z-z_1}{z-z_2}} + \sqrt{\frac{z-z_2}{z-z_1}}, \quad f_1'(z_1) = \infty = f_1'(z_2); \tag{7.21a–c}$$
$$2\frac{\mathrm{d}f_2}{\mathrm{d}z} = \sqrt{\frac{z_1-z}{z-z_2}} - \sqrt{\frac{z-z_2}{z_1-z}}, \quad f_2'(z_1) = \infty = f_2'(z_2). \tag{7.21d–f}$$

It can be checked that these are indeed branch-points, viz. a loop around $z_1(z_2)$ multiplies $z-z_1(z-z_2)$ by $\mathrm{e}^{i2\pi}$, and hence multiplies the function by $\mathrm{e}^{i\pi} = -1$, changing branch to the "other" square root. Thus there are three possibilities for branch-cuts: (I) a single semiinfinite branch-cut passing through z_1 and z_2 and going to infinity, viz. Figure 7.1a with z_2 between z_1 and ∞, so that the second branch-point z_2 lies in the branch-cut (Subsection 7.8.3); (II) two semiinfinite branch-cuts joining z_1 and z_2 separately (Figure 7.3a) to infinity (Subsection 7.8.1); (III) a finite branch-cut (Subsection 7.8.2) joining z_1 to z_1 in Figure 7.4a, with the corresponding Riemann surface in Figure 7.4b. These three cases will be illustrated next by elementary transcendental functions with branch-points at $z_1 = 1 = -z_2$.

A branch-cut is an interval, which may be open (closed) at each end:

$$(x_1, x_2) \equiv \{x:\ x_1 \le x \le x_2\}, \quad (x_1, x_2) \equiv \{x:\ x_1 < x < x_2\}, \tag{7.22a,b}$$
$$)x_1, x_2) \equiv \{x:\ x_1 < x \le x_2\}, \quad (x_1, x_2(\equiv \{x:\ x_1 \le x < x_2\}, \tag{7.22c,d}$$

leading two four cases for real $x_2 > x_1$, namely: (i/ii) closed (open) interval (7.22a) [(7.22b)]; (iii/iv) open (closed) at the lower end $x = x_1$ and vice-versa at the upper end $x = x_2$ in (7.22c) [7.22d].

7.8.1 Two Semiinfinite Branch-Cuts

The function $z = \sin w$ for real w takes real values $z \equiv x$ with $-1 \le x \le +1$ or $|x| \le 1$; this suggests that real $z = x$ with $|x| > 1$ should be excluded from the inverse function $w = \arg\sin z$, corresponding to two semiinfinite branch-cuts from ± 1 to $\pm\infty$, viz:

$$z \notin (-\infty, -1(\ \cup\ ) +1, +\infty): \quad w = \arg\sin z; \tag{7.23a,b}$$

the branch-points $z_1 = 1 = -z_2$ are not included in the branch-cut because the function is defined there $\arg\sin(\pm 1) = \pm\pi/2$. The derivative of (7.23b) can be calculated from the

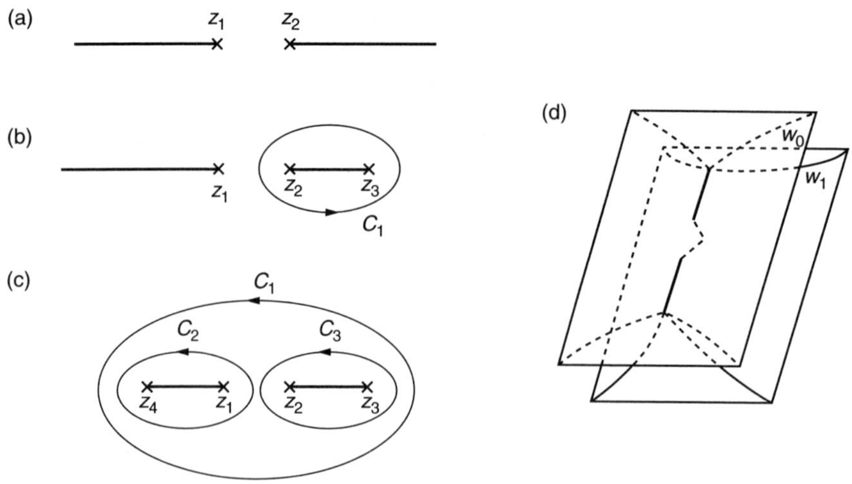

FIGURE 7.3
For a complex function with two branch-points there are three possibilities: (i) a finite branch-cut joining the branch-points (Figure 7.4); (ii) two semiinfinite branch-cuts joining each branch-point separately to infinity (a); (iii) a single semiinfinite branch-cut joining one branch-point to infinity and "hiding" or containing the other branch-point (Figure 8.1h). In the case of a two valued complex function with two branch-points (ii) connected to infinity by semiinfinite branch cuts (d) they prevent a loop around either of the branch-points (a). If two branch-points are connected by a finite branch-cut (c) the latter can be enclosed by a loop C_1, avoiding other branch-points and branch-cuts. In the case of four branch-points joined by two finite branch-cuts (d) three types of loops are possible: (i/ii) enclosing one branch-cut only: C_2 or C_3; (iii) enclosing both branch-cuts: C_1.

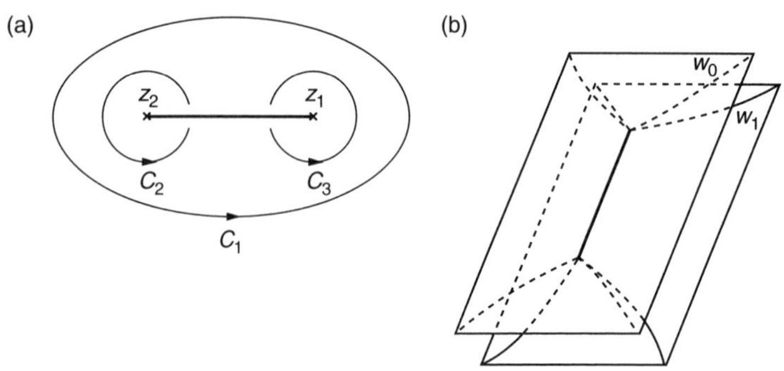

FIGURE 7.4
A two valued complex function with two branch-points joined by a finite branch-cut is represented on a Riemann surface with two sheets joined between the branch-points (b). The finite branch-cut prevents loops C_2 and C_3 around one branch-point but not a loop C_1 around both branch points enclosing the branch-cut.

inverse function:

$$\begin{aligned}\frac{\mathrm{d}w}{\mathrm{d}z} &= \left(\frac{\mathrm{d}z}{\mathrm{d}w}\right)^{-1} = \left[\frac{\mathrm{d}\,(\sin w)}{\mathrm{d}w}\right]^{-1} = \cos^{-1} w \\ &= \left(1-\sin^2 w\right)^{-1/2} = \left(1-z^2\right)^{-1/2} \equiv (1-z)^{-1/2}(1+z)^{-1/2};\end{aligned} \tag{7.23c}$$

the derivative is infinite at $z = \pm 1$, which are the branch-points:

$$z_1 = 1 = -z_2: \quad \frac{\mathrm{d}(\arg\sin z)}{\mathrm{d}z} = \left(1 - z^2\right)^{-1/2} = \frac{1}{f_2(z)}, \tag{7.24a,b}$$

corresponding to the inverse of the function (7.20b). The square root is to be taken in the sense of (7.23), and thus the function becomes single-valued function with cuts in the complex z-plane for $1 \pm z = -x$ real negative with $x > 0$, viz. the branch-cuts are $z = \pm(1 + x)$, that is, are the union of $(-\infty, -1($ with $) + 1, +\infty)$ as indicated in (7.22a).

Using the definition of sine in (7.22b) ≡ (7.25a):

$$2iz = 2i\sin w = \mathrm{e}^{iw} - \mathrm{e}^{-iw}, \quad \mathrm{e}^{2iw} - 2iz\mathrm{e}^{iw} - 1 = 0 \tag{7.25a,b}$$

leads to a quadratic equation (7.25b); its roots (7.26a) show that

$$\mathrm{e}^{iw} = iz \pm \sqrt{1 - z^2}, \quad w^{\pm} = -i\log\left(iz \pm \sqrt{1 - z^2}\right) \equiv \arg\sin z, \tag{7.26a,b}$$

the function $\arg\sin z$ is expressible as logarithm (7.26b). There is no branch-point when the argument of the logarithm vanishes $iz \pm \sqrt{1 - z^2} = 0$ because $iz = \mp\sqrt{1 - z^2}$, has no solution, as can be seen by squaring $0 = 1$; thus the only branch-points arise from $\sqrt{1 - z^2}$, and are $z_1 = 1 = -z_2$, and the two semiinfinite branch-cuts (7.23a) correspond (Figure 7.3a ≡ Figure 8.1g) to the case II.

7.8.2 One Finite Branch-Cut

In order to obtain a case III of a finite branch-cut $) - 1, +1($ it is sufficient to consider the algebraic inverse function $z = \sin^{-1} w = \csc w$; for real w it follows $z \equiv x$ real with $|x| \geq 1$, so it may be expected that $|x| < 1$ is excluded from the inverse function:

$$z \notin (-1, +1): \quad w = \arg\csc z \equiv \arg\sin^{-1} z. \tag{7.27a,b}$$

Bearing in mind $z = \sin w = 1/(\csc w)$, substituting $1/z$ for z in (7.26b) specifies the function $\arg\csc z$ in terms of logarithms:

$$w^{\pm} = -i\log\left[\frac{i}{z} \pm \sqrt{1 - \frac{1}{z^2}}\right] = \arg\csc(z). \tag{7.28}$$

The arguments of the logarithm in (7.26b) are the roots of (7.25b), whose product is -1; thus the product of the two arguments in (7.28) is also -1:

$$-\left(\frac{i}{z} - \sqrt{1 - \frac{1}{z^2}}\right)^{-1} = \frac{i}{z} + \sqrt{1 - \frac{1}{z^2}} \equiv r\mathrm{e}^{i\varphi}. \tag{7.29a}$$

as can be checked directly:

$$\left(\frac{i}{z} - \sqrt{1 - \frac{1}{z^2}}\right)\left(\frac{i}{z} + \sqrt{1 - \frac{1}{z^2}}\right) = \left(\frac{i}{z}\right)^2 - \left(1 - \frac{1}{z^2}\right) = -1. \tag{7.29b}$$

Thus the arg csc has a double infinity of branches:

$$n \in |Z: \quad w_n^+ = -i\log\left(r\mathrm{e}^{i\varphi}\right) = -i\log r + \varphi + 2n\pi, \tag{7.30a}$$

$$w_n^- = -i\log\left(-r^{-1}\mathrm{e}^{-i\varphi}\right) = i\log r - \varphi + (2n + 1)\pi; \tag{7.30b}$$

the same would apply to arg sin in (7.26a,b) with a different value of r, φ, viz.

$$r\mathrm{e}^{i\varphi} = iz + \sqrt{1 - z^2} = -\left(iz - \sqrt{1 - z^2}\right)^{-1}, \tag{7.31}$$

in (7.26b).

7.8.3 One Semiinfinite Branch-Cut with Hidden Branch-Point

For the remaining case I an example is the inverse of $z = \cosh w$, because for real w it follows $z \geq 1$, that is, the inverse function should exclude the values $z \equiv x$ real with $x < 1$, leading to a branch-cut from $z = 1$ to infinity:

$$z \notin (-\infty, 1(: \quad w = \arg\cosh z. \tag{7.32a,b}$$

The derivative of (7.32b) is:

$$\begin{aligned}\frac{\mathrm{d}w}{\mathrm{d}z} &= \left(\frac{\mathrm{d}z}{\mathrm{d}w}\right)^{-1} = \left[\frac{\mathrm{d}(\cosh w)}{\mathrm{d}w}\right]^{-1} = \sinh^{-1} w \\ &= \left(\cosh^2 w - 1\right)^{-1/2} = \left(z^2 - 1\right)^{-1/2} \equiv (z-1)^{-1/2}(z+1)^{-1/2};\end{aligned} \tag{7.33}$$

this corresponds to the function (7.20a), viz.:

$$z_1 = 1 = -z_2 : \quad \frac{\mathrm{d}(\arg\cosh z)}{\mathrm{d}z} = \left(z^2 - 1\right)^{-1/2} = \frac{1}{f_1(z)}. \tag{7.34}$$

The square roots in (7.33) lead to a single-valued function with cuts in the complex z-plane for $z \pm 1 = -x$ negative real with $x > 0$, viz. for real $z = -x \mp 1$; there are "two" branch-cuts $(-\infty; -1)$ and $(-\infty, +1)$, but the former is included in the latter that is the only one indicated in (7.32a) and shown in Figure 8.1h. The branch-point $z = -1$ is an end-point of a branch-cut and the branch-point $z = +1$ is "hidden" within the branch-cut.

The definition of hyperbolic cosine (7.35a):

$$2z = 2\cosh w = \mathrm{e}^{w} + \mathrm{e}^{-w}, \quad \mathrm{e}^{2w} - 2z\mathrm{e}^{w} + 1 = 0, \tag{7.35a,b}$$

leads to the quadratic equation (16.35b); its roots (16.36a):

$$\mathrm{e}^{w} = z \pm \sqrt{z^2 - 1}, \quad w^{\pm} = \log\left(z \pm \sqrt{z^2 - 1}\right) = \arg\cosh z, \tag{7.36a,b}$$

express the inverse function arg cosh as a logarithm (7.36b). There is no branch-point for vanishing argument of the logarithm $z = \mp\sqrt{z^2 - 1}$, because squaring it is clear that this condition cannot be satisfied $0 = -1$; the branch-points and branch-cuts that remain in the argument of the logarithm in (7.36b) are the same that have been considered before in (7.32a) $\equiv$ (7.33).

The derivative of (7.36b), viz.:

$$\frac{\mathrm{d}w}{\mathrm{d}z} = \left(z \pm \sqrt{z^2 - 1}\right)^{-1}\left(1 \pm \frac{z}{\sqrt{z^2 - 1}}\right) = \pm\left(z^2 - 1\right)^{-1/2}, \tag{7.37}$$

agrees with (7.37) $\equiv$ (7.34), since the square root in (7.34) also allows the choice of $\pm$ sign. The two arguments of (7.36b), or roots of (7.35b) are inverse:

$$\left(z - \sqrt{z^2 - 1}\right)^{-1} = z + \sqrt{z^2 - 1} \equiv s\,\mathrm{e}^{i\psi}, \tag{7.38}$$

and specify the double infinity of branches of arg cosh z, viz.:

$$n \in |Z : \quad \arg\cosh z = w_n^{\pm} = \log\left(s^{\pm 1}\mathrm{e}^{\pm i\psi}\right) = \pm\log s \pm i\psi + i2\pi n. \tag{7.39}$$

The condition of minimum imaginary part in modulus specifies two branches $w_0^{\pm}$, and an extra condition is needed to choose a principal branch, viz. either w_0^{+} or w_0^{-}. In the case (7.30a,b) the condition of minimum imaginary part in modulus would not choose a principal branch, for example, using instead the minimum real part in modulus would leave two branches $w_0^{\pm}$, one of which could be taken as principal branch. Since there is only one point-at-infinity (Chapter 9), the two semiinfinite branch-cuts in Figure 7.3a qualify as a single branch-cut, that is, the complement of the finite branch-cut in Figure 7.4a, on the line joining z_1 and z_2 to infinity; the "difference" between the branch-cuts in Figure 7.4a (7.3a) is that it does not (does) pass through the point-at-infinity. Further cases of branch-points, branch-cuts, and principal branches are presented in Example 10.12.

7.9 Functions with Several Branch-Points and Branch-Cuts

The preceding ideas can be extended to functions with several branch-points as end-points of branch-cuts or "hidden" within branch-cuts (Subsection 7.9.1) and relate to possible paths around branch-cuts (Subsection 7.9.2).

7.9.1 Branch-Cuts Connecting Multiple Branch-Points

For functions of the form (7.42a) with three branch-points there are two distinct cases, viz.:

$$f_3(z) = \sqrt{(z-z_3)(z-z_2)(z-z_1)}, \quad f_4(z) = \sqrt{(z_3-z)(z-z_2)(z-z_1)}. \tag{7.40a,b}$$

The new possibility (IV) is a finite branch-cut [e.g., joining (z_2, z_3)] and a semiinfinite branch-cut (e.g., joining z_1 to infinity) as shown in Figure 7.3b with the corresponding Riemann surface with two sheets in Figure 7.3d for a two-valued function. For a function of the form (7.42a) with four branch-points there are three distinct cases:

$$f_5(z) = \sqrt{(z-z_4)(z-z_3)(z-z_2)(z-z_1)}, \tag{7.41a}$$

$$f_6(z) = \sqrt{(z_4-z)(z-z_3)(z-z_2)(z-z_1)}, \tag{7.41b}$$

$$f_7(z) = \sqrt{(z_4-z)(z_3-z)(z-z_2)(z-z_1)}. \tag{7.41c}$$

There are two new possibilities, both with one finite branch-cut and one difference: (V) there is besides (z_2, z_3) another finite branch-cut (z_1, z_4) as shown in Figure 7.3c with the corresponding Riemann surface in Figure 7.3d; (VI) besides the finite branch-cut (z_3, z_4) there are two semiinfinite branch cuts joining z_1 and z_2 to infinity in Figure 7.5.

The functions:

$$f_8(z) = \left\{ \prod_{n=1}^{N} (z-z_n) \right\}^{1/2}, \tag{7.42a}$$

$$\frac{\mathrm{d}f_8}{\mathrm{d}z} = \frac{1}{2} \left\{ \sum_{n=1}^{N} (z-z_n)^{-1} \left[\prod_{\substack{m=1 \\ m\neq n}}^{N} (z-z_m) \right] \right\}^{1/2}, \tag{7.42b}$$

have N branch-points at $z = z_1, \ldots, z_N$, and include as particular cases (7.20a,b; 7.21a–f) for $N=2$, (7.40a,b) for $N=3$ and (7.41a–c) for $N=4$. The inverse of (7.42a) appears in the derivative of the function (7.43) that equals the inverse of the square root of a polynomial of degree N:

$$\frac{\mathrm{d}w}{\mathrm{d}z} = \{P_N(z)\}^{-1/2} = \left\{ A_N \prod_{n=1}^{N} (z-z_n) \right\}^{-1/2}. \tag{7.43}$$

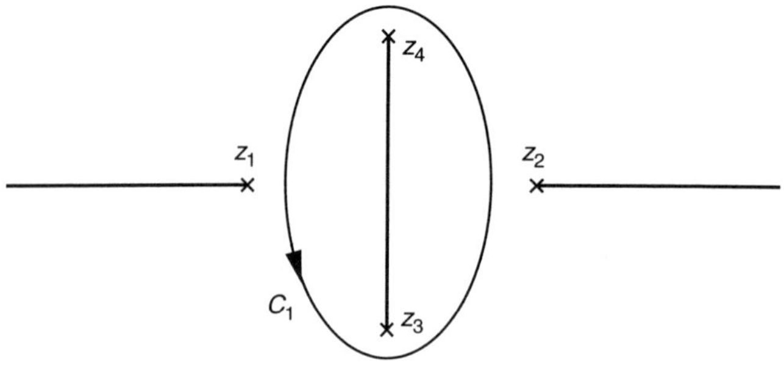

FIGURE 7.5
In the case of four branch-points there are several possibilities, for example, two finite branch-cuts (Figure 7.3c), or one finite and two semiinfinite branch cuts (Figure 7.5).

The functions inverse $z(w)$ to (7.43) lead to: (i) for $N = 2$ circular and hyperbolic functions, for example (7.23c) $\equiv$ (7.24b) and (7.33) $\equiv$ (7.34); (ii/iii) for $N = 3\,(N = 4)$ the Weierstrass (Section 39.8) [Jacobi (Section 39.9)] elliptic functions; (iv) for $N \geq 5$ the hyperelliptic functions (Section 39.7). Note that in some cases, for example (7.33) $\equiv$ (7.34) the derivative of the function may be infinite at two points, but there is only one (7.32a) semiinfinite branch-cut joining the actual branch-point to infinity, and passing through the other "hidden" branch-point. In this case the **hidden branch-point**, lying in a branch-cut is not considered when counting the branch-cuts, whether finite or not. Thus only branch-points which are end-points of a branch-cut are considered next; whether the branch-point is included in (excluded from) the branch-cut, because the function is singular (regular) there, has no effect on the counting of branch-cuts, only that it is an end-point of the branch-cut, and not a hidden branch-point in the interior of the branch-cut. The preceding this qualification should be borne in mind when considering a function with any number N of branch-points.

7.9.2 Paths around Finite and Infinite Branch-Cuts

Two branch-cuts cannot cross, because the jumps across them might not match; the case of three or more semiinfinite branch-cuts is also excluded, because this would be equivalent to the crossing of two distinct branch-cuts at infinity. Considering not more than two semiinfinite branch-cuts, it is necessary to distinguish cases of even and odd numbers of nonhidden branch-points. *A function with an odd $N = 2p+1$ number of nonhidden branch-points has p finite branch-cuts and one semiinfinite branch-cut,* for example $(p = 0, N = 1)$ *in* Figure 7.1a, $(p = 1, N = 3)$ *in* Figure 7.3b. *For a function with an even number of branch-cuts, there are two possibilities: (i) p finite branch-cuts, for example,* $(p = 1, N = 2)$ *in* Figure 7.4a, *and* $(p = 2, N = 4)$ in Figure 7.3c; *(ii) $(p-1)$ finite branch-cuts and two semiinfinite branch-cuts, for example* $(p = 0, N = 2)$ *in* Figure 7.3a *and* $(p = 1, N = 4)$ *in* Figure 7.5.

The allowable paths in the complex plane cannot cross branch-cuts. Thus a path around a semiinfinite branch-cut must start and end at infinity. Finite paths are possible around finite branch-cuts. *The value of the function does not change when z describes a loop around a finite branch-cut, viz.: (i) one loop C_1 in the case (Figures 7.4a and 7.5) with one finite branch-cut; (ii) three loops C_1, C_2, C_3 in the case (Figure 7.3c) with two-finite branch-cuts; (iii) for p finite branch-cuts there number of distinct permissible loops is:*

$$q \equiv p + p\frac{(p-1)}{2} + \cdots = \sum_{q=1}^{p} \binom{p}{q} = 2^p - 1, \tag{7.44}$$

where the number of arrangements of p cuts in q groups is:

$$q=1,\ldots,p: \quad \binom{p}{q} = \frac{p!}{q!\,(p-q)!} = p\,(p-1)\cdots\frac{(p-q+1)}{q!}. \tag{7.45}$$

The binomial theorem (25.38) with integer exponent p, viz:

$$(u+v)^p = \sum_{q=0}^{p} \binom{p}{q} u^p v^{p-q}, \tag{7.46a}$$

in the particular case:

$$u=1=v: \quad 2^p = \sum_{q=0}^{p} \binom{p}{q} = \sum_{q=0}^{p} \frac{p!}{(p-q)!q!}, \tag{7.46b}$$

was used to calculate the sum (7.44). The simplest cases are one (two) finite branch-cuts $p=1 (p=2)$ and $q=1 (q=3)$ loops around them in Figure 7.4a (7.3c). The loops that can be continuously deformed into each other are topologically equivalent (Section 9.7).

Note 7.1. *Separation between Branches of a Function*: Since all circular and hyperbolic function derive from the exponential, they are periodic, and hence have many-valued inverses. The elliptic functions are doubly-periodic (Sections 39.7–39.9), and have doubly many-valued inverses. The special functions that involve logarithms are also many valued, for example, Neumann and Hankel functions, confluent hypergeometric functions of the second kind, associated Legendre and Laguerre functions of the second kind, etc. The solutions of a linear ordinary differential equation with variable coefficients:

$$\sum_{n=0}^{N} A_n(z) \frac{d^n w}{dz^n} = 0, \tag{7.47}$$

often have branch-points for at least some values of the parameters in the coefficients, and thus branch-cuts occur for most special functions. The integral of a multivalued function can be performed (Section 17.8) deforming the path of integration around the branch-cut(s) even semiinfinite (Figures 7.1a and 17.4). The need to avoid crossing branch-cuts can lead to particular paths of integration, like the Hankel contour (Pochhammer loop) around a semiinfinite (finite) branch-cut in that applies to Euler's gamma (beta) function, and also the confluent (Gaussian) hypergeometric function. Other loops, like the "teardrop" or "airfoil" loops pass through a branch-point, requiring restrictions on the exponent.

Note 7.2. *A Real Function with Complex Derivative*: Thus multivalued functions, branch-points, and branch-cuts occur frequently, and care should be taken not to pass inadvertently to another branch, otherwise unexpected results may appear, for example, the annihilation of derivatives. A simpler example is the apparently absurd case of a real function with complex derivative. The function (7.48a):

$$f(z) = z^z, \quad f'(z) = z^z(1+\log z) = f(z)(1+\log z), \tag{7.48a,b}$$

has derivative (7.48b), as follows from (5.28), viz.:

$$\frac{d(z^z)}{dz} \equiv \frac{d[\exp(z\log z)]}{dz} = \exp(z\log z)\frac{d(z\log z)}{dz} = z^z(1+\log z). \tag{7.49}$$

Considering negative real z, viz.:

$$x>0, z=-x: \quad f(-x) = (-x)^{-x} = \exp[-x(\log x - i\pi - i2n\pi)] = x^{-x} e^{i\pi x(1+2n)}, \tag{7.50}$$

the function (7.48a) is (i) real for x rational of the form $-z = x = p/(1+2n)$ with p any integer and n corresponding to a particular branch of the function; (ii) complex otherwise, viz. for other rational x, and all irrational x. The derivative (7.48b):

$$n, m \in |Z: \quad f'(-x) = x^{-x} e^{i\pi x(1+2n)} \left[1 + \log x - i(2m+1)\pi\right], \tag{7.51}$$

is not real for $x = p/(1+2n)$. The explanation is (i) the derivative of the function at x involves the value at neighboring points; (ii) every neighborhood contains an infinite number of points $x = p/(1+2n)$ where the function is real, but p or n are distinct and correspond to another branch n of the function; (iii) at the points $x \neq p/(1+2n)$ the function is not real. Thus the differentiation made always involves a change of branch of the function.

Conclusion 7: The semiinfinite branch-cut (Figure 7.1) from the origin to infinity, prevents (a) circuits C_+ (C_-) in the positive (+) [negative (−)] direction; crossing the branch-cut corresponds to passing to another sheet of the Riemann w-surface, which has (b) two sheets w_0, w_1 for the square-root, (c) a finite number N sheets $w_0, \ldots, w_{N-1}$ for the N-th root. There is (Figure 7.2) a denumerable infinity of sheets $w_0, \ldots, w_{\pm 1}, \ldots$ for the logarithm. If there are (Figure 7.4) two branch-points, joined by a finite branch-cut (b), circuits C_2, C_3 around one branch-point are prevented by the cut (a) but a circuit C_1 around both branch-points is allowed. In the case (Figure 7.3) of four (three) branch-points, with two finite branch-cuts (c) [one finite and one semiinfinite branch-cut (b)], there are three (one) allowable loops, namely C_2, C_3 (C_1), around two branch-points and C_1 (none) around four (more than two) branch-points. As the number N of nonhidden branch-points increases: (i) for $N = 1$ there is a semiinfinite branch-cut (Figure 7.1a); (ii) for $N = 2$ there is one finite (Figure 7.4a) or two semiinfinite (Figure 7.3a) branch-cuts; (iii) for $N = 3$ there is one finite and one semiinfinite branch-cut (Figure 7.3b); (iv) for $N = 4$ there are either two finite branch-cuts (Figure 7.3c) or one finite and two semiinfinite branch-cuts (Figure 7.5).

8

Motion of a Pendulum and a Ship

The stability of a ship depends on the relative position of (i) the center of gravity of the ship or center of mass, where the weight is applied; (ii) the center of gravity of the displaced fluid or center of flotation, where the flotation force, equal and opposite to the weight, is applied. If the center of gravity lies below (above) the center of flotation (Section 8.3), the analogy with a suspended (inverted) pendulum (Section 8.2) shows that (Section 8.1) the equilibrium position, with weight and flotation force opposite along the same straight line, is stable (unstable). The implication is that, if the ship or pendulum is displaced from the position of stable (unstable) equilibrium, the weight and flotation force remain equal and opposite, but act on distinct parallel lines, giving rise to a torque that causes a return toward (further deviation from) the position of equilibrium. Thus the motion following an initial disturbance, will be an oscillation (Section 8.8) [(monotonic divergence Section 8.9)], with a frequency (rate-of-growth in time) that is calculated most simply in the case of motion of small amplitude (Section 8.6), for which the differential equation of the motion can be linearized. For motions with large amplitude, the equation of motion is nonlinear, but the conservation of energy holds in both cases (Section 8.5), since all forces present are associated with gravity that is conservative. The analogy of the ship and pendulum allows the choice of the length of the latter so that it reproduces the motion of the former (Section 8.4); the length of the equivalent pendulum (Section 8.7) is independent (dependent) of angular displacement for small (large) motions. The monotonous divergence (oscillation) is typical of unstable (stable) systems, of which the simple pendulum and transverse motion of a ship are just two examples; various kinds of mechanical, electrical, and other vibrations and waves are also perturbations of an equilibrium state.

8.1 Stability of a Suspended or Inverted Pendulum

Consider (Figure 8.1a) a pendulum, consisting of a point mass m suspended from a fixed point by a rod (or string), of mass negligible compared with m and inextensible, that is, of fixed length, ℓ. The position of the pendulum is specified by the angle φ of the rod with the vertical, and since the potential of gravity $\Phi(\varphi)$ is defined to within a constant, it can be chosen to be zero at the level of the suspension point $\Phi(\pi/2) = 0$. The **potential energy** (8.1a) is specified by $\Phi = mgy$, in which g is the acceleration of gravity, taken to be uniform (constant g), and $y = -\ell\cos\varphi$ is the vertical displacement from the suspension point:

$$\Phi(\varphi) = -mg\ell\cos\varphi; \quad F(\varphi) = -\ell^{-1}\frac{\mathrm{d}\Phi}{\mathrm{d}\varphi} = -mg\sin\varphi, \tag{8.1a,b}$$

the gravity force is specified by (8.1b), that is, the component of the weight mg projected along the tangent to the circle of radius ℓ and center at the origin, along which the mass m moves.

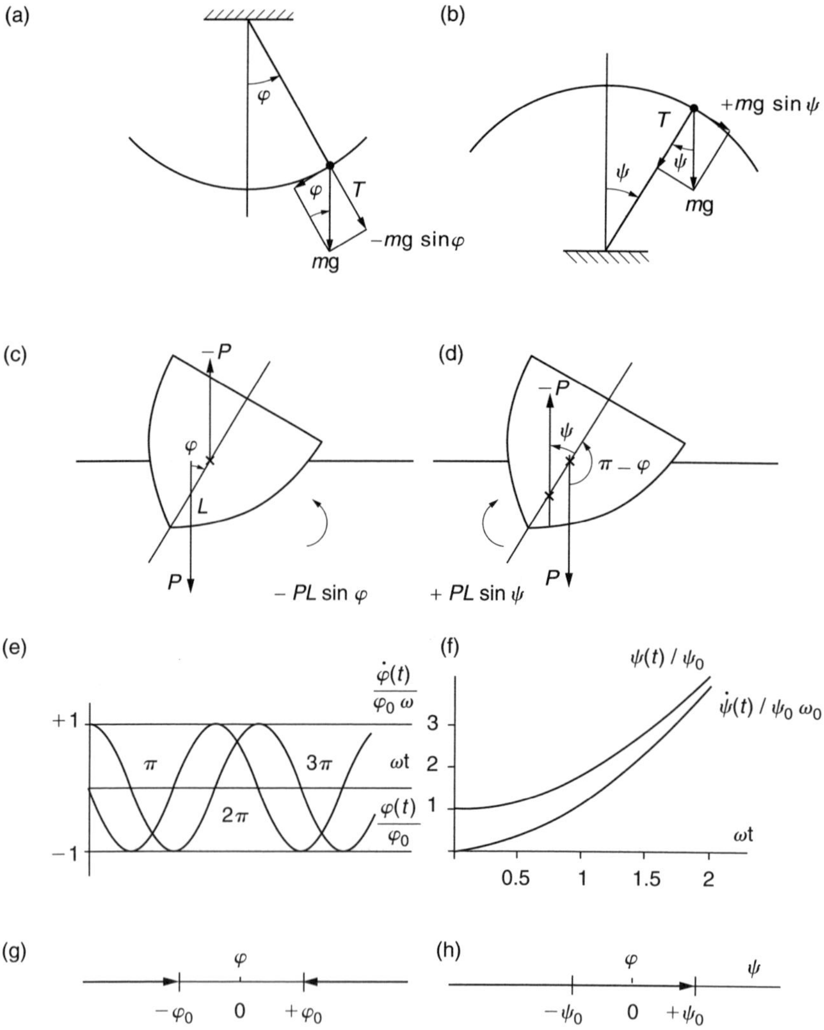

FIGURE 8.1
The stable (unstable) equilibrium position of a pendulum is suspended (a) [inverted (b)], so that the traction (compression) on the rod holding the mass lies on the vertical line of the weight. The stable (unstable) equilibrium position of a ship is with the center of mass and the metacenter on the same vertical line with the latter above (c) [below (d)] so that the flotation force lies on the vertical line of the weight. In the stable case, both for the suspended pendulum (a) [ship with metacenter above the center-of-mass (c)] a deviation from equilibrium leads to a restoring force (moment) toward equilibrium; the inertia leads to an "overshoot" of the equilibrium position, until a stop at the opposite angular deviation, in the absence of damping; then the restoring force (moment) is inverted, again toward equilibrium, resulting in a periodic motion (e) within a range of angles limited (g) by the initial displacement without velocity. In the unstable case, both for the inverted pendulum (b) [ship with metacenter below the center-of-mass (d)] a deviation from equilibrium leads to a force (moment) away from equilibrium; thus the equilibrium is never regained, and instead the angular displacement increases monotonically with time (f) beyond the initial displacement with zero velocity. For large displacements, either in the stable, oscillatory case (a, c, e, g) [or unstable, divergent case (b, d, f, h)] nonlinear effects should be considered.

TABLE 8.1
Motion near Equilibrium of a Simple Pendulum and a Ship Hull

Position	$\varphi = 0$	$\varphi = \pi$
$F, \mathrm{d}\Phi/\mathrm{d}\varphi$	0	0
$\mathrm{d}F/\mathrm{d}\varphi = -mg\cos\varphi$	$-mg < 0$	$+mg > 0$
$\mathrm{d}^2\Phi/\mathrm{d}\varphi^2 = mg\ell\cos\varphi$	$mg\ell > 0$	$-mg\ell < 0$
Potential	Minimum	Maximum
Equilibrium	Stable	Unstable
Figures	8a, c, d, g	8b, d, f, h
Equation of motion—nonlinear	(8.17a)	(8.17b)
—linear	(8.4)	(8.4)
frequency/growth: ω	(8.5a,b)	(8.5a,b)
period/instability time:	(8.26b)	(8.26b)
Linear solutions—initial	(8.25a,b)	(8.29a,b)
conditions: $\varphi_0 \neq 0 = \dot{\varphi}_0$	Section 8.8	Section 8.9
$\varphi_0 = 0 \neq \dot{\varphi}_0$: Example 10.13	(10.56a,b)	(10.57a,b)
$\varphi_0 \neq 0 \neq \dot{\varphi}_0$: Example 10.14	(10.62a,b)	(10.63a,b)

Note: The stable and unstable motions of a pendulum or a ship hull near an equilibrium position are described by the same equations with distinct parameters. The parameters can be made to coincide by introducing the pendulum of length equivalent to the ship (8.7). This leads to the same cases of motion illustrated in Figure 8.1.

The positions of static equilibrium correspond to zero force $F(\varphi) = 0$ or stationary potential $\mathrm{d}\Phi/\mathrm{d}\varphi = 0$, that is, $\sin\varphi = 0$, to a suspended $\varphi = 0$ (inverted $\varphi = \pi$) pendulum, both cases being considered in Table 8.1. *The suspended (inverted) pendulum is stable (unstable)* in Figure 8.1a (8.1b) because, when displaced from the equilibrium position $\varphi = 0\,(\varphi = \pi)$ the force changes from zero $F = 0$ to negative $\mathrm{d}F/\mathrm{d}\varphi < 0$ (positive $\mathrm{d}F/\mathrm{d}\varphi > 0$) for $\mathrm{d}\varphi > 0$, so that the mass is pulled back toward (pushed further away from) the equilibrium position. *The stable (unstable) equilibrium corresponds to minimum (maximum) of the potential energy,* that is, $\mathrm{d}\Phi/\mathrm{d}\varphi = 0$ with $\mathrm{d}^2\Phi/\mathrm{d}\varphi^2 > 0\,(\mathrm{d}^2\Phi/\mathrm{d}\varphi^2 < 0)$.

8.2 Motion of the Pendulum and Force along the Rod

In the position of stable (unstable) equilibrium, illustrated in Figure 8.1a (8.1b), the weight mg of the suspended (inverted) pendulum is balanced by a **traction (compression)** of the rod $T = -mg\,(T = +mg)$. If the pendulum is displaced away from the equilibrium position $\varphi \neq 0, \pi$, a motion ensues, in which the weight mg has two components: (i) the motion is due entirely to the tangential force (8.1b), or tangential component of the weight that is balanced (8.2a) by the **inertia force** that is equal to the mass m multiplied by the **tangential acceleration** $\ell\ddot{\varphi}$ that for circular motion, is the product of the radius ℓ, by the **angular acceleration** $\ddot{\varphi}$:

$$m\ell\ddot{\varphi} = -mg\sin\varphi; \quad T = -mg\cos\varphi - m\ell\dot{\varphi}^2, \tag{8.2a,b}$$

the tension force along the rod is balanced (8.2b) by the radial component of the weight, to which must be added the centrifugal force $m\ell\dot{\varphi}^2$, equal to the product of mass m by radius ℓ and square of angular velocity $\dot{\varphi}$. *The centrifugal force causes always traction*

$-m\ell\dot{\varphi}^2 < 0$ along the rod, whereas the radial component of the weight $-mg\,\ell\cos\varphi$ causes traction (compression) in the lower $|\varphi| < \pi/2$ (upper $|\varphi| > \pi/2$) half of the circle.

8.3 Metacentric Distance and Rolling Torque

Consider a ship, whose transverse cross-section, illustrated in Figure 8.1c and d, is constant over most of its length; if the length of the ship is much greater than the transverse dimensions viz. vertical (**draught**) and horizontal (**beam**), the motion will be mainly transversal, and will consist of a vertical translation (**heave**) and a rotation (**roll**). The forces present consist of (i) the **weight** of the ship $P = Mg$, equal to the product of the mass of the ship M by the acceleration of gravity g, applied at the **center of mass**; (ii) the **flotation force**, equal to the weight of the displaced fluid $G = \rho V$, where ρ is its density and V the volume, applied at its center of gravity, called the **center of flotation**. The flotation force is always antiparallel to the weight, and if it is greater $G > P$ (smaller $G < P$), the difference is balanced by the vertical inertia force, $m\ddot{y}$, that is, the ship is accelerated upward (downward) and a heave motion results.

In the following, the vertical motion is neglected, that is, it is assumed that the flotation force exactly balances the weight at all times $G = P = Mg$, viz. there is no heave, although roll could occur. If the ship is rotated around the center of mass, the section immersed into the water will generally change, and so will the center of flotation that describes the locus of flotation. Take as the **axis of the section** the line in the vertical direction (of gravity, weight, and flotation force), passing through the center of flotation in the equilibrium position. If the ship is rotated around the center of gravity, the center of flotation is displaced along the **locus of flotation**, and the vertical through it intersects the axis of the section at a point called the **metacenter**. The importance of the metacenter is that, since it lies along the direction of the flotation force, the latter can be applied at the metacenter for the purpose of calculating the torque relative to the center of gravity; the latter is given by $-PL\sin\varphi$, where P is the flotation force, equal to the weight of the ship, φ is its angle with the axis of the section, and L is the **metacentric distance** from the metacenter to the center of gravity:

$$-PL\sin\varphi = I\ddot{\varphi}, \tag{8.3}$$

and the applied torque is balanced by the product of the angular acceleration $\ddot{\varphi}$ by the **moment of inertia** I of its transverse section. Thus (8.3) is the equation specifying the transverse rolling motion of the ship, in the particular case when it decouples from all other motions.

8.4 Length of the Pendulum Equivalent to the Ship

The equations of motion of the pendulum (8.2a) and rolling of the ship (8.3) are both of the form:

$$\ddot{\varphi} + \omega^2\sin\varphi = 0, \tag{8.4}$$

where ω has the dimensions inverse of time, and is specified by:

$$\omega = \sqrt{\frac{g}{\ell}}, \quad \sqrt{\frac{PL}{I}}, \tag{8.5a,b}$$

respectively, for the pendulum (8.5a) and ship (8.5b). The weight P of the ship and its moment of inertia I are given, respectively, by:

$$P = Mg, \quad I = MR^2, \tag{8.6a,b}$$

where M is the mass and R the giration radius. It follows that the ship is equivalent, as far as rolling motion is concerned, to a pendulum of length ℓ obtained equating (8.5a) and (8.5b), viz.:

$$\ell_{\mathrm{eq}} = \frac{gI}{PL} = \frac{I}{ML} = \frac{R^2}{L}, \tag{8.7}$$

where (8.6a,b) were used. From (8.7) it follows that *the product of the length of the equivalent pendulum by the metacentric distance is equal to the square of the giration radius.* This relation determines the length of the equivalent pendulum $\ell_{\mathrm{eq}}(\varphi)$ that for large roll angles φ, varies like the inverse of the metacentric distance $L(\varphi)$; note that the radius of giration R is a constant, but the metacentric distance L depends on the roll angle φ, for all shapes of transverse section other than circular. A pendulum of length $\ell(\varphi)$ varying with the angular displacement φ can be obtained by using a suspension string of fixed length, $\ell_0 \geq \max\{\ell(\varphi)\}$, enveloping a suspension cam of radius $r(\varphi) = \ell_0 - \ell(\varphi)$.

8.5 Kinetic, Potential, and Total Energy

Multiplying (8.4) by $\dot{\varphi}$ and integrating leads to

$$0 = \ddot{\varphi}\dot{\varphi} + \omega^2\dot{\varphi}\sin\varphi = \frac{d\left(\frac{\dot{\varphi}^2}{2} - \omega^2\cos\varphi\right)}{dt}, \tag{8.8}$$

assuming that ω^2 is constant, that is, for: (i) a pendulum (8.5a) of constant length ℓ; (ii) the rolling motion of a ship (8.5b) whose metacentric distance $L(\varphi)$ is approximately independent of φ. In these conditions the quantity in curly brackets in (8.8) is a constant, and multiplying it by $m\ell^2$, and using $m\omega^2\ell^2 = mg\ell$ from (8.5a), the constant becomes:

$$E = \frac{1}{2}m\ell^2\dot{\varphi}^2 - mg\ell\cos\varphi = K(\dot{\varphi}) + \Phi(\varphi); \tag{8.9}$$

the constant E is the total energy, since it is the sum of the potential energy (8.1a), and the kinetic energy $K(\dot{\varphi})$; the latter is half the product of the mass m by the square of the tangential velocity $\ell\dot{\varphi}$.

In the case $\dot{\varphi} = 0$ of static equilibrium, the kinetic energy is zero, $K = 0$, and all energy is potential $E = \Phi$, that is, the stable (unstable) case $\varphi = 0\,(\varphi = \pi)$ for the suspended (inverted) pendulum corresponds to negative (positive) total energy $E = -mg\ell < 0\,(E = mg\ell > 0)$. For the ship in the equilibrium position, $\varphi = 0$, the metacentric distance $L_0 \equiv L(0)$ is the distance from the center of gravity to the center of flotation, taken positive upward. Since the total energy at the equilibrium position is $E = MgL_0$, it follows *that the equilibrium position $\varphi = 0$*

of the ship is stable (unstable) in Figure 8.1c and d, like the suspended (inverted) pendulum in Figure 8.1a and b, if the total energy E or metacentric distance L_0 is negative (positive), that is, for center of gravity below (above) the center of flotation, so that the torque resulting from an angular displacement causes a return to (further displacement from) the equilibrium position.

In the case of motion $\varphi(t)$ the total energy remains constant, that is, it is a conserved quantity, as is generally true in the presence of conservative forces. The motion involves an exchange of kinetic and potential energy, between two extremes: (i) all energy is potential $E = \Phi$, when the kinetic energy and angular velocity vanish $K = 0 = \dot{\varphi}$, at the maximum displacement (8.10a):

$$\varphi_{\max} = \arccos\left(\frac{E}{mg\ell}\right); \quad \dot{\varphi}_{\max} = \sqrt{\frac{2E}{m\ell^2} + \frac{2g}{\ell}}, \qquad (8.10\text{a,b})$$

(ii) the maximum angular velocity (8.10b) corresponds to all energy being kinetic $E = K$, at the moment the equilibrium position $\varphi = 0$ is crossed.

8.6 Linearization in the Vicinity of the Equilibrium Position

The equation of motion (8.4) is generally nonlinear, both, for pendulum and ship, for large angular displacements φ. To study the motion in the vicinity of the position of stable $\varphi = 0$ (unstable $\varphi = \pi$) equilibrium small displacements are considered, so that the equations may be linearized by neglecting the squares of angles compared with unity viz. $\varphi^2 \ll 1$ ($\psi^2 \ll 1$ with $\psi \equiv \pi - \varphi$). The angular displacement does not exceed $\varphi \leq 17°$, bearing in mind that a radian corresponds to $180°/\pi = 57°$, the linearization condition is met since $\varphi^2 \leq (17/57)^2 = 0.3^2 = 0.09 \ll 1$. Thus the linearization for the stable case can be performed by the approximations (Section 19.8) corresponding to the first two terms of the series expansions:

$$\varphi^2 \ll 1: \quad \sin\varphi = \varphi\left(1 - \frac{\varphi^2}{6}\right), \quad \cos\varphi = 1 - \frac{\varphi^2}{2}, \qquad (8.11\text{a,b})$$

where (i) the approximation (8.11a) ≡ (19.39a) for the sine agrees with the limit (5.14b), the inequality (5.14a) and the series (27.75) ≡ (30.47); (ii) in the approximation for the cosine (8.11b) ≡ (19.39b) is consistent with $\cos\varphi = \mathrm{d}(\sin\varphi)/\mathrm{d}\varphi$ using (8.11a) and the series (30.45). The linearization for the unstable case is similar:

$$\psi \equiv \pi - \varphi, \quad \psi^2 \ll 1: \quad \sin\varphi = \sin\psi = \psi\left(1 - \frac{\psi^2}{6}\right), \quad \cos\varphi = -\cos\psi = -1 + \frac{\psi^2}{2}, \qquad (8.12\text{a,b})$$

using the angular deviation from the upward vertical $\psi = \pi - \varphi$ in Figures 8.1b and 8.1d instead of the angular deviation φ from the downward vertical in (8.11a,b).

Substituting (8.11a,b) in the equations for the pendulum (8.2a,b), the linearized equation of motion becomes (8.13a):

$$\varphi^2, \dot{\varphi}^2 \ll 1: \quad \ddot{\varphi} + \frac{g}{\ell}\varphi = 0, \quad T = -mg, \qquad (8.13\text{a,b})$$

and the traction on the rod (8.2b) ≡ (8.13b) differs negligibly from the weight, that is, the value in the position of stable equilibrium. The linearization about the position of unstable equilibrium (8.12a,b) leads to a compression of the rod equal to the weight (8.14b):

$$\psi^2, \dot{\psi}^2 \ll 1: \quad \ddot{\psi} - \frac{g}{\ell}\psi = 0, \quad T = mg, \tag{8.14a,b}$$

and an equation of motion (8.14a) that differs from (8.13a) only in the sign of the second term. The tangential force causing the motion in the stable (8.13a) [unstable (8.14a)] case is not constant, and though small compared with the traction (8.13b) [compression (8.14b)], it is not negligible. In the case of the ship, the metacentric distance $L(\varphi)$ as function of the angular displacement φ is approximated by

$$L(\varphi) = \pm|L_0| + \beta\varphi, \quad \beta \equiv \left.\frac{dL}{d\varphi}\right|_{\varphi=0}, \tag{8.15a,b}$$

where $L_0 = \pm|L_0|$ is the distance from the center of flotation to the center of gravity in the equilibrium position, with the upper + (upper −) sign for the stable (unstable) case. The coefficient (8.15b) does not appear in the linearized form (8.11a) of the equation of rolling motion (8.3) ≡ (8.16a,b):

$$\psi = \pi - \varphi: \quad \frac{\ddot{\varphi}}{\varphi} = -\frac{PL_0}{I} = \mp g\frac{|L_0|}{R^2} = \frac{\ddot{\psi}}{\psi}, \tag{8.16a,b}$$

where (8.6a,b) were used.

8.7 Branch-Points as Boundaries of the Angular Motion

The equation of motion is (i) in the stable case (8.17a) for the pendulum (8.13a) [ship (8.16a)]; (ii) in the unstable case (8.17b) for the pendulum (8.14a) [ship (8.16b)]:

$$\ddot{\varphi} + \omega^2\varphi = 0, \quad \ddot{\psi} - \omega^2\psi = 0; \tag{8.17a,b}$$

the parameter ω is the same both in the (8.17a) and unstable (8.17b) cases, and is specified by (8.18a) ≡ (8.5a) for the pendulum and (8.18b) ≡ (8.5b) ≡ (8.16b) for the ship:

$$\omega = \sqrt{\frac{g}{\ell}}, \quad \frac{\sqrt{g|L_0|}}{R}. \tag{8.18a,b}$$

The corresponding equation of conservation of energy can be obtained in the stable case multiplying (8.17a) by $\dot{\varphi}$ and integrating with regard to time (as in Section 8.5), or using (8.11b) in (8.9):

$$\dot{\varphi}^2 + \omega^2\varphi^2 = C, \quad C \equiv \frac{2E}{m\ell^2} + \frac{2g}{\ell} = \dot{\varphi}^2_{max} = \frac{g^2\varphi^2_{max}}{\ell^2}; \tag{8.19a,b}$$

the energy equation (8.19a) is quadratic in the linear approximation for the motion, and the constant C is specified alternatively by the total energy E, maximum angular displacement (8.10a) or angular velocity (8.10b). The energy equation in the case of unstable equilibrium is obtained similarly multiplying (8.17b) by $\dot{\psi}$ and integrating with regard to time (as in Section 8.5) or using (8.12b) in (8.9):

$$\dot{\psi}^2 - \omega^2\psi = D, \quad D \equiv \frac{2E}{m\ell^2} - \frac{2g}{\ell} = \dot{\psi}^2_{max} = -\frac{g^2\psi^2_{max}}{\ell^2}. \tag{8.20a,b}$$

The length ℓ of the pendulum in (8.19–8.20a,b) may be replaced by the equivalent length (8.7) for the ship. The use of the conservation of energy reduces the order of the differential equation from two to one in the nonlinear (8.4) $\rightarrow$ (8.9) case and also in the linear cases, both stable (8.17a) $\rightarrow$ (8.19a) and unstable (8.17b) $\rightarrow$ (8.20a). The motion is specified at all times if two initial conditions are given, for example, the system (pendulum or ship) is displaced from the equilibrium position (stable or unstable) at time $t=0$, and released without velocity:

$$\dot{\varphi}(0)=0=\dot{\psi}(0), \quad \varphi(0)\equiv\varphi_0\neq0\neq\psi_0\equiv\psi(0); \tag{8.21a,b}$$

other initial conditions are considered in Examples 10.13 and 10.14. It will be shown next that the initial conditions (8.21a) [(8.21b)] for the stable (Section 8.8) [unstable (Section 8.9)] cases specify branch-points that by excluding the motion from the branch-cuts, set the boundaries for the angular motion.

8.8 Oscilation about a Position of Stable Equilibrium (Galileo, 1583)

The constant C in (8.19a) is specified by φ_0 in the stable case (8.17a) $\equiv$ (8.22a) with initial conditions (8.21a,b), viz.:

$$\dot{\varphi}^2+\omega^2\varphi^2=\omega^2\varphi_0^2; \quad \dot{\varphi}\equiv\frac{d\varphi}{dt}=\left\{\omega^2\left(\varphi_0^2-\varphi^2\right)\right\}^{1/2}; \tag{8.22a,b}$$

the derivative of the function $\varphi(t)$ specifying the motion has branch-points at $\varphi=\pm\varphi_0$, that is, the initial angular displacement (8.21b) and its symmetric $-\varphi_0$. Thus $|\varphi|=\varphi_0$ sets the limits of the motion in the stable case, for which it will be shown in the sequel that $|\varphi_0|$ is an upper bound on $|\varphi|$, that is, $|\varphi|\leq|\varphi_0|$.

In the case of motion in the vicinity of a position of stable equilibrium (8.22b) leads to the integral (8.23b):

$$\begin{aligned}\alpha\equiv\frac{\varphi}{\varphi_0}: \quad \omega t=\omega\int_0^t dt&=\int_{\varphi_0}^{\varphi}\left(\varphi_0^2-\varphi^2\right)^{-1/2}d\varphi=\int_1^{\varphi/\varphi_0}\left(1-\alpha^2\right)^{-1/2}d\alpha\\&=\arc\sin\alpha\Big|_1^{\varphi/\varphi_0}=\arc\sin\left(\frac{\varphi}{\varphi_0}\right)-\frac{\pi}{2},\end{aligned} \tag{8.23a,b}$$

where was used (8.22b), the change of variable (8.23a) and also (7.24b); inverting and using (8.24a):

$$\frac{\sin}{\cos}\left(\omega t+\frac{\pi}{2}\right)=\pm\frac{\cos}{\sin}(\omega t), \quad \frac{\sin}{\cos}(\omega t)=-\frac{\sin}{\cos}(\omega t+\pi), \tag{8.24a,b}$$

leads to the angular displacement (8.25a) [velocity (8.25b)] as a function of time for an oscillatory motion (Figure 8.1e), with frequency ω in (8.18a,b):

$$\varphi(t)=\varphi_0\cos(\omega t), \quad \dot{\varphi}(t)=-\varphi_0\omega\sin(\omega t). \tag{8.25a,b}$$

and amplitude $\varphi_0\,(\varphi_0\omega)$ and phase $0\,(\pi)$.

The **period** τ is the minimum time after which both the displacement and velocity take again the same value (8.26a):

$$\frac{\sin}{\cos}(\omega t)=\frac{\sin}{\cos}\left\{\omega\left(t+\frac{2\pi}{\omega}\right)\right\}, \quad \tau\equiv\frac{2\pi}{\omega}=2\pi\sqrt{\frac{\ell}{g}},\ \frac{2\pi R}{\sqrt{g|L_0|}}, \tag{8.26a,b}$$

is specified by **Galileo's** (1583) **law** (8.26b) for the pendulum, and its analogue for the ship. Thus *the small oscillations are* **isocronous**, *that is, have a period independent of the amplitude; the period is proportional to (i) the square root of the length of the suspended pendulum; (ii) the giration radius and inverse square root of the metacentric distance, for the rolling motion of the ship with center of gravity below the center of flotation.* The oscillatory motion is confined to $|\varphi| \leq |\varphi_0|$ in (8.25a) or $|\alpha| \leq 1$ in (8.23a), that is, between the branch-points of arc $\sin\alpha$ in Figure 8.1g, viz. the branch-cuts $|\alpha| > 1$ or $|\varphi| > |\varphi_0|$ are inaccessible regions for the motion.

8.9 Divergence Away from the Position of Unstable Equilibrium

In the other case, of motion in the vicinity of the position of unstable equilibrium, the equation of motion (8.17b) ≡ (8.27a) with initial conditions (8.21a,b), viz.:

$$\dot{\psi}^2 - \omega^2\psi^2 = -\omega^2\psi_0^2, \quad \dot{\psi} \equiv \frac{d\psi}{dt} = \left\{\omega^2\left(\psi^2 - \psi_0^2\right)\right\}^{1/2}, \tag{8.27a,b}$$

differs from (8.22a) only in one sign that affects the integration of (8.27b) compared with (8.22b), viz. instead of circular (8.23b) appear (8.28b) hyperbolic (7.34) functions:

$$\beta \equiv \frac{\psi}{\psi_0}: \quad \omega t = \omega\int_0^t dt = \int_{\psi_0}^{\psi}\left(\psi^2 - \psi_0^2\right)^{-1/2} d\psi = \int_1^{\psi/\psi_0}\left(\beta^2 - 1\right)^{-1/2} d\beta$$
$$= \arg\cosh\beta\big|_1^{\psi/\psi_0} = \arg\cosh\left(\frac{\psi}{\psi_0}\right); \tag{8.28a,b}$$

this leads to *the angular displacement (8.29a) [velocity (8.29b)] increasing monotonically (Figure 8.1f) from the initial value* ψ_0 *(zero initial value with slope* $\psi_0\omega$*)*:

$$\psi(t) = \psi_0\cosh(\omega t), \quad \dot{\psi}(t) = \psi_0\omega\sinh(\omega t); \tag{8.29a,b}$$

for large values of the argument the growth is exponential (17.30b):

$$e^{2\omega t} \gg 1: \quad \begin{matrix}\cosh\\ \sinh\end{matrix}(\omega t) = \frac{1}{2}\left(e^{\omega t} \pm e^{-i\omega t}\right) = \frac{1}{2}e^{\omega t}\left[1 \pm O\left(e^{-2\omega t}\right)\right], \tag{8.30a,b}$$

if the approximation (8.30a) holds: (i) in dimensionless terms $e^{2\omega t} \gg 10$ for $\omega t = \frac{1}{2}\log 10 = 1.15$; (ii) for an elapsed time $t \gg 1.15/\omega = 1.15\tau/2\pi = 0.18\tau$ greater than 18% of the period of oscillation in the stable case (8.26b).

The approximation (8.30b) simplifies (8.29a,b) to:

$$t \gg 0.18\tau: \quad \frac{\psi(t)}{\psi_0} = \frac{1}{2}e^{2\pi t/\tau} = \frac{\dot{\psi}(t)}{\psi_0\omega}, \tag{8.31a,b}$$

in the unstable case, of the inverted pendulum, or ship with center of flotation below the center of gravity, the angular displacement (velocity) grow monotonically in proportion to $\psi_0/2\,(\psi_0\omega/2)$ *after a time* 0.18τ*, at an exponential rate with e-folding time equal to* $\tau/2\pi = 0.15\tau$*, where* τ *denotes the period (8.26b) in the stable case.* The same quantity specifies, in the stable (unstable) case, the frequency of the oscillation (rate of exponential growth in time) ω in (8.18a,b), and the period of oscillation τ (e-folding time for the increase of the disturbance 0.15τ) with τ given by (8.26b), exactly (approximately) from time $t > 0$

$(t > 0.18\tau)$. Note that in the unstable case of monotonic divergence $\psi \geq \psi_0$ in (18.29a) or $\beta \geq 1$ in (18.28a), so that the motion is above the positive branch-point of $\arg\cosh\alpha$ in Figure 8.1h, and again the branch-cut $\beta < 1$ or $\psi < \psi_0$ is an inaccessible region for the motion.

Note 8.1. *Oscillations and Stability of Mechanical Systems*: The motions considered (Chapter 8) correspond to a simple, circular, one-dimensional, linear pendulum, because: (i) only one pendulum is considered; (ii/iii) the motion lies in a plane at a constant distance from the suspension point so it is one-dimensional and circular; (iv) the displacements must be small for the linear approximation to hold. Other cases include (i) the multiple pendulum, for example, one suspended from the preceding, or coupled penduli, for example, linked by springs; (ii) a particle rolling and/or sliding in a cylinder is analogous to a pendulum that is circular (noncircular, e.g., cicloidal) if the directrix is (is not) a circle (for example, a cycloid); (iii) the simple pendulum may move in two dimensions on a sphere, for example, the spherical pendulum; (iv) the nonlinear motion may be a large amplitude oscillation or a circulatory motion (viz. the pendulum goes round without stopping) specified by Jacobian elliptic functions; the latter generalize the circular and hyperbolic functions that describe small amplitude motions.

The motion of the ship considered was a pure roll, around the longitudinal axis; other possible angular motions include **pitch** (keel-up or -down) and **yaw** (keel left-or-right); in addition to the three rotations there are three translations, for example, heave (Section 8.3) is the vertical displacement. These six degrees-of-freedom apply not only to a surface ship, but also to a submarine that moves in three-dimensions, like an aircraft does in the atmosphere and a rocket or satellite in space. The motions corresponding to the six degrees-of-freedom of a rigid body may be coupled, nonlinear, etc., for example, for aerospace vehicles. They may also be coupled to continuous deformation (elasticity) or fluid loads (aeroelasticity).

Conclusion 8: Comparison of the motion (Figure 8.1) in the vicinity of the position of stable (unstable) equilibrium, illustrated on the left-hand side (right–hand side), for: (c/d) the rolling motion of a long ship of constant cross-section with weight P, flotation force $-P$ and metacentric distance L, positive (c) [negative (d)] for center of gravity below (above) the center of flotation; (a/b) the analogous motion of a suspended (a) [inverted (b)] pendulum, consisting of a mass m in the gravity field, connected to a fixed point by a rod of negligible weight, able to resist traction (compression) with negligible extension; (e/f) periodic oscillations (e) [monotonic growth (f)] of the angular displacement and velocity, with frequency (rate of exponential growth in time) ω, and initial amplitude $\varphi_0(\psi_0)$ at time $t = 0$ of release with zero velocity; (g/h) branch-cuts showing the finite (g) [infinite (h)] range of stable (unstable) motion.

9

Stereographic Projection and Genus of a Surface

Part 1 has been concerned with the representation of numbers on the complex plane (Chapters 1 and 3) that is the domain of a function (Chapters 3 and 5), and also with the range of the function (Chapters 5 and 7), in particular when it is multivalued (Chapters 7 and 9). The representation of a multivalued function on plane sheets is limited (Chapter 7), and an alternative is a one-to-one map between the plane and a sphere, viz. using the Riemann sphere as a Riemann surface. Projecting a sphere from one pole onto a plane passing through the equator, a one-to-one correspondence (Section 9.1) is obtained between the points on the sphere and the points on the plane (Section 9.2); the pole itself is projected to infinity, implying that there is only one "point at infinity" (Section 9.2). The stereographic projection (Section 9.3) has been used since the times of Ptolemy to make plane maps of the spherical surface of the earth; these maps are reasonably accurate near the equator, but become significantly distorted near the pole of projection. The section of the sphere by a plane is always a circle, and if the plane passes (does not pass) through the pole of projection, the circle is projected onto the plane as a straight line (Section 9.4) [another circle (Section 9.5)]. A particular case of the stereographic projection is obtained by taking a perpendicular plane through the pole of projection that makes an ordered projection of the circle onto a straight line (Section 9.6). A generalization of the stereographic projection is obtained by considering a sphere with branch-cuts (Section 9.7), for example, for a multivalued function. In the case of multivalued functions, taking two or more spheres joined by tubes along the branch-cuts leads, by continuous deformation, to spheres with one or more handles (Section 9.8) that can also be deformed into a torus with one or more holes (Section 9.9). Although the mapping into a plane is the simplest, it can be misleading if there are several sheets or the point-at-infinity is part of the domain or range; in such situations mapping into a closed surface may be more illuminating.

9.1 Injective, Surjective, and Bijective Mappings

A complex function $w = f(z)$ specifies a **mapping** from a **domain** A or region of the z-plane $z \in A$ to a region or **range** $w \in B$ of the w-plane. If the function is **single-valued** each point of the domain $z \in A$ is mapped to one point $w \in B$:

$$f \text{ is one-valued:} \quad \mathcal{V} \equiv \left\{ f(z) : \quad \forall_{z\in A} \quad \exists^1_{w\in B} : \quad w = f(z) \right\}, \tag{9.1a}$$

(Figure 9.1a) whereas if it is **multivalued** at least some points $z \in A$ are mapped to several distinct points of $w \in B$ (Figure 9.1b):

$$f \text{ is multivalued:} \quad \mathcal{V}_n \equiv \left\{ f(z) : \exists_{z\in A; w_1, w_2 \in B} : w_1 = f(z) \wedge w_2 = f(z) \wedge w_1 \neq w_2 \right\}. \tag{9.2}$$

In both cases it may happen that (i) some points $w \in B$ are not images of any point $z \in A$, that is, the mapping does not cover B; (ii) a point $w \in B$ may be the **image** of several

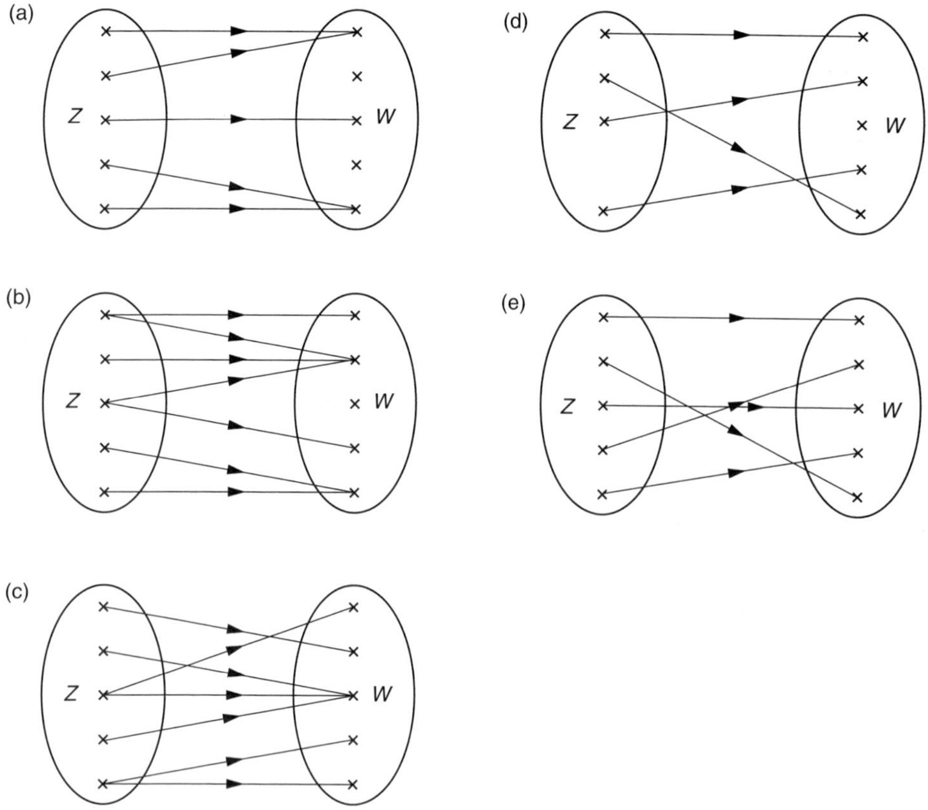

FIGURE 9.1
A single-valued function (a) maps each point of the domain to only one point in the range; a multivalued function (b) maps at least one point in the domain to more than one point in the range. It is possible that: (i) different points of the domain go to the same point of the range, unless the function is injective (d); (ii) some points of the range are not mapped into, unless the function is subjective (c). A surjective function (c) maps into every point of the range at least once; it may be single or multivalued. An injective function (d) maps into the each point of the range from only one point in the domain; some points of the range may not be mapped into, unless it is surjective (c). If the function is is both injective (d) and surjective (c) then it must be single-valued; moreover it is bijective (c), that is, it maps to one-to-one the domain to the range.

points $z \in A$. The case (i) is excluded for a **surjective** function, for which (Figure 9.1c) every point $w \in B$ is the image of at least one point $z \in A$, viz:

$$f \text{ is surjective}: \quad \bar{\mho} \equiv \{f(z): \quad \forall_{w \in B} \exists_{z \in A}: \quad w = f(z)\}, \tag{9.3}$$

that is, the mapping f from A covers B. The case (ii) cannot arise if the function is **injective**, that is, (Figure 9.1d) every $w \in B$ cannot be the image of more than one $z \in A$, viz.:

$$f \text{ is injective}: \quad \tilde{\mho} \equiv \{f(z): \quad f(z_1) = w = f(z_2) \Rightarrow z_1 = z_2\}. \tag{9.4}$$

Neither (i) nor (ii) can occur if the mapping is both injective and surjective, that is, a **bijective** mapping (Figure 9.1e) makes every point $w \in B$ of the range the image of a single point $z \in A$ of the domain:

$$f \text{ is bijective}: \quad \tilde{\tilde{\mathcal{U}}} \equiv \left\{ f(z): \quad \forall_{w \in B} \exists^1_{z \in A}: \quad w = f(z) \right\}; \tag{9.5}$$

a *bijective mapping is single-valued, and has a single-valued inverse.* A single- (multi) valued complex function $w = f(z)$ is a mapping from the z-plane to the w-plane (a Riemann p-surface with N sheets). The classification of functions (Section 9.1) has been made based mostly based on their domain; a classification based on the range leads to univalent/multivalent/many valent functions (Note 37.4). Together (Figures 9.1a–e and 37.8a–d) they clarify generally the question of direct and inverse functions (Table 37.2), not just in the particular case (9.5) of bijective functions in Diagram 37.

9.2 Unit Sphere and the Complex Plane

The geometry of the **Riemann p-surface (1853)** can be clarified by mapping instead the z-plane onto a sphere, for example, via a stereographic projection. Consider a sphere of radius unity, whose points have Cartesian coordinates (r,s,t) satisfying:

$$r^2 + s^2 + t^2 = 1. \tag{9.6}$$

Take a plane passing through the center of the sphere, with Cartesian axis OX, OY coincident with respectively OR, OS, as the complex z-plane with $z = x + iy$. Thus, the origin of the z-plane coincides with the center of the unit sphere, and a straight line through it, perpendicular to the complex plane, coincides with the OT axis, and intersects the sphere at two diammetrically opposite points, taken as north (south) pole $N\,(S)$ in Figure 9.2a. A family of straight lines issued from the north pole satisfies the equations:

$$\frac{r}{x} = \frac{s}{y} = \frac{\sqrt{r^2 + s^2}}{|z|} = 1 - t, \tag{9.7}$$

that *specify (Figure 9.2a) the* ***stereographic projection*** *of the point $Q = (r, s, t)$ on the unit sphere on the point $P \equiv (x, y)$ on the complex z-plane, with $z = x + iy$.*

From (9.6) and (9.7) it follows that

$$z^* z = |z|^2 = x^2 + y^2 = \frac{r^2 + s^2}{(1-t)^2} = \frac{1 - t^2}{(1-t)^2} = \frac{1+t}{1-t}, \tag{9.8}$$

that implies that (i) the north pole $t = 1$ corresponds to the “point at infinity” $|z| = \infty$ of the complex z-plane; (ii) the south pole $t = -1$, corresponds to the origin $z = 0$; (iii) the northern (southern) hemisphere $1 > t > 0$ $(0 > t > -1)$ is projected in the exterior (interior) of the unit circle $|z| > 1\,(|z| < 1)$ of the complex z-plane; (iv) the boundary of the two hemispheres is the equator of the unit sphere $t = 0$, consisting of all the points lying on the complex z-plane, and specifying the unit circle, that is, the circle of radius unity and center at the origin. It follows that *the stereographic projection is a bijective mapping of the unit sphere onto the complex plane, with a single “**point at infinity**” corresponding to pole of projection.*

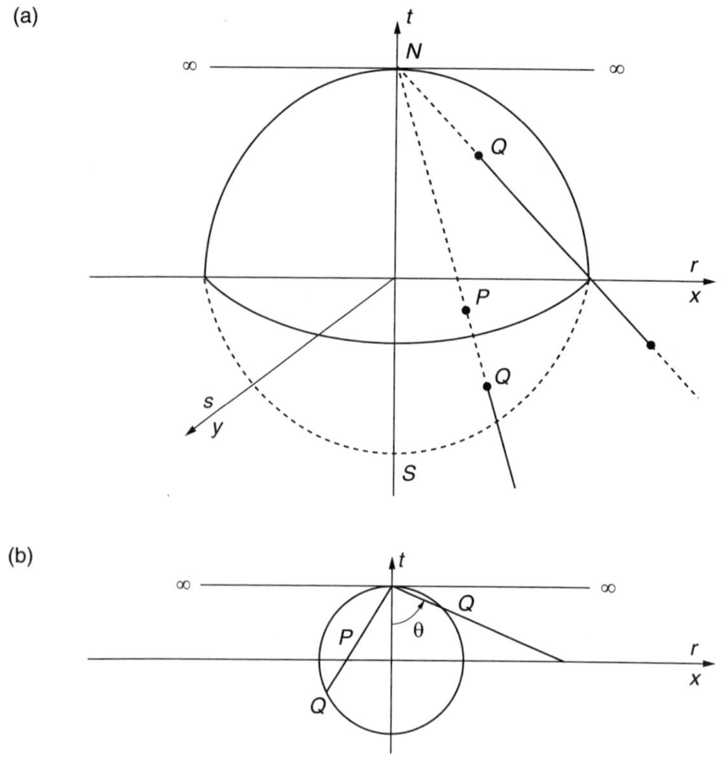

FIGURE 9.2
The stereographic projection of the surface of the sphere (a) from the north pole to the equatorial plane: (i) leaves only the equator unchanged; (ii/iii) the upper (lower) hemisphere is mapped outside (inside) the equatorial circle; (iv/v) the south (north) pole is mapped to the origin (infinity). The mapping is bijective if it is assumed that there is only one point-at-infinity regardless of the direction in longitude. A cross-section through the polar axis (b) maps the circle to a real line through the center orthogonal to the radial line through the projection point; the mapping is again bijective if $\pm\infty$ are understood as the same point. The latter mapping can be strictly ordered, i.e., is an isomorphism.

9.3 Direct and Inverse Stereographic Transformation

The preceding statement shows that the z-plane has a single point at infinity that is obtained as $|z|\to\infty$ for arbitrary direction $0\le\arg(z)<2\pi$. *The "point at infinity" is the inverse of the origin, that is,* $\zeta=1/z\to\infty$ as $z\to 0$, in analogy with the south and north poles of the unit sphere being diametrically opposite. From (9.7) it follows that the point (r,s,t) on the sphere projects stereographically on the point:

$$z\equiv x+iy:\quad x=\frac{r}{1-t},\quad y=\frac{s}{1-t}, \tag{9.9a,b}$$

on the complex z-plane. Using (9.8) to invert (9.9a,b) leads to

$$t=\frac{z^*z-1}{z^*z+1},\quad 1-t=\frac{2}{1+z^*z}, \tag{9.10a,b}$$

that specifies the point (x,y) on the complex z-plane the corresponding point on the unit sphere:

$$r = \frac{2x}{1+x^2+y^2}, \quad s = \frac{2y}{1+x^2+y^2}, \quad t = \frac{x^2+y^2-1}{x^2+y^2+1}, \tag{9.11a–c}$$

It follows that *the points on the unit sphere $Q \equiv (r,s,t)$ and on the complex plane $P \equiv (x,y)$ are mapped onto each other by the direct $Q \to P$ in (9.9a,b) and inverse $P \to Q$ in (9.11a–c) stereographic projections (Figure 9.2a).*

9.4 Mapping of a Circle onto a Straight Line

A plane passing through the north pole (0,0,1) of the unit sphere has equation (9.12a):

$$Ar + Bs + t = 1, \quad \vec{\mathrm{N}} = \left|A^2 + B^2 + 1\right|^{-1/2} \{A, B, 1\} \tag{9.12a,b}$$

where the constants A,B specify the unit normal vector (9.12b). It intersects the unit sphere (9.6) in a circle that is projected stereographically (9.11a–c) into a straight line with: (i) implicit equation (9.13a,b):

$$t = 0: \quad Ax + By = 1, \quad \vec{\mathrm{n}} = \{A, B\} \quad \left|A^2 + B^2\right|^{-1/2}, \tag{9.13a,b}$$

and unit normal vector (9.13b); (ii) explicit equation (9.14a):

$$y = mx + b, \quad m \equiv -\frac{A}{B}, \quad b \equiv \frac{1}{B}, \tag{9.14a–c}$$

with slope (9.14b) and ordinate at the origin (9.14c). Substituting (3.17a,b) in (9.13a) it follows that *the equation of a* ***straight line*** *in the complex z-plane is (9.15a):*

$$\alpha z + \alpha^* z^* = 1, \quad 2\alpha \equiv A - iB = -\frac{m+i}{b}, \tag{9.15a,b}$$

where the complex constant (9.15b) is specified in terms of the components of the normal (9.13b), or slope (9.14b) and ordinate (9.14c).

9.5 Projection of a Circle into Another Circle

A plane not passing through the north pole:

$$C < 1: \quad Ar + Bs + t = C, \tag{9.16a,b}$$

has unit normal vector (9.12b). The distance of the plane (9.16b) from the origin is the projection of the position vector (r,s,t) on the normal $\vec{\mathrm{N}}$ in (9.12b), viz. (9.17):

$$d = N_x r + N_y s + N_z t = \left|1 + A^2 + B^2\right|^{-1/2} (Ar + Bs + t) = \left|1 + A^2 + B^2\right|^{-1/2} C; \tag{9.17}$$

for the plane to cut the unit sphere (9.6), it cannot pass at a distance more than unity from the origin (9.18a):

$$d \leq 1: \quad C^2 \leq 1 + A^2 + B^2, \tag{9.18a,b}$$

that implies (9.18b).

The plane (9.16b) intersects the unit sphere (9.6) along a circle, whose projection (9.11a–c) on the complex z-plane is given by:

$$(1-C)\left(x^2+y^2\right) + 2Ax + 2By = 1 + C; \tag{9.19}$$

this is the equation of a circle (9.20a):

$$(x-x_0)^2 + (y-y_0)^2 = R^2: \quad (x_0, y_0) \equiv \frac{\{A, B\}}{C-1}, \tag{9.20a,b}$$

of center at (9.20b) and radius (9.20c):

$$R^2 = \frac{A^2+B^2}{(1-C)^2} + \frac{1+C}{1-C} = \frac{1+A^2+B^2-C^2}{(1-C)^2}. \tag{9.20c}$$

The radius of the circle projected on the plane (9.20a) is real $R \geq 0$ in (9.20c) on account of (9.18b). Using (3.17a,b) in (9.19) it follows that *the equation of a circle in the complex plane is (9.21a):*

$$\beta z^* z + \alpha z + \alpha^* z^* = 1, \quad (1+C)\{\beta, \alpha\} = \{1-C, A-iB\}, \tag{9.21a,b}$$

where α is a complex and β a real constant, specified by (9.21b) in terms of the constants A,B,C; *the latter determine the center (9.20b) and radius (9.20c).*

9.6 Isomorphism of Circle and Real Line

The rational numbers satisfy (Section 1.2) a strict order relation (Section 1.6). The real numbers are also strictly ordered, but the complex numbers cannot be strictly ordered by modulus, because $z_1 \neq z_2$ can happen for $|z_1| = |z_2|$ if $\arg(z_1) \neq \arg(z_2)$. Thus the strict ordering is a property of the real numbers, lost in the extension to the complex numbers; it is the counterpart for having gained the advantage of performing the square root always that is, either there is strict ordering (real numbers) or there are closed algebraic operations (complex numbers). An **isomorphism** is a bijective mapping that preserves a strict order (1.6). Clearly, isomorphisms may be defined for real, but not for complex, numbers.

In order to obtain an isomorphism that is bijective, from the stereographic projection it is necessary to introduce a strict ordering, by restricting from complex to real numbers. This is done by a section of the unit sphere and complex z-plane by a perpendicular plane passing through the center of the sphere (Figure 9.2b), for example, though the XOZ plane, so that $y = 0 = s$ and (9.9a,b) reduces to first expression:

$$\tan\theta = x = \frac{r}{1-t}, \tag{9.22}$$

where θ is the angle of the line of projection with the vertical; the inverse projection (9.11a–c) with $y = 0$ reduces to two formulas:

$$r = \frac{2x}{1+x^2} = \sin(2\theta), \quad t = \frac{x^2-1}{x^2+1} = -\cos(2\theta), \tag{9.23a,b}$$

Thus if the points on the circle are ordered in the interval $(-\pi/2, +\pi/2)$ the points on the real line become ordered one-to-one in the range $(-\infty, +\infty)$ where $x = \pm\infty$ corresponds to $\theta = \pm\pi/2$, and the mapping remains bijective. It follows that *there is an isomorphism (9.22; 9.23a,b) between the points on the unit circle* $r^2 + t^2 = 1$ or $-\pi/2 < \theta < +\pi/2$ *and the points on the real line* $-\infty < x < +\infty$ *where* $+\infty\,(-\infty)$ *is the infimum (supremum) of all real numbers, and both correspond to the "point at infinity"* that can be reached only in two opposite directions.

Additional properties of the stereographic projection are indicated in Examples 10.15 and 10.16; the stereographic projection can be generalized from the sphere to an ellipsoid (Example 10.17 and Figure 10.4). Any projection of a sphere on a plane causes a distortion, because a sphere has two nonzero principal curvatures, and cannot be developed onto a plane; a surface with only one nonzero principal curvature is developable, for example, a cylinder or a cone. The Mercator (1569) [conical] map is obtained in Example 10.18 (10.19) by projecting the sphere from the center onto a tangent cylinder (cone), and then developing onto a plane. The Mercator projection (Figure 10.5) does not distort the equator, but the distortion increases toward the poles that are mapped to the point-at-infinity. The conical projection (Figure 10.6a,b) does not distort one parallel, and has less distortion near the north pole that is mapped to a finite distance; the conical projection does not cover the whole earth, viz. the south pole and nearby regions. The whole earth can be covered by two conical projections, for example, with cones with opposite vertices relative to the center of the sphere; in this case the equator is projected from the center to the intersection of the two cones. For mapping limited areas of the earth some of the preceding difficulties are reduced.

9.7 Continuous Deformation and Topological Sphere

Returning to the general stereographic projection, it is clear that a function $w = f(z)$ single-valued in the complex z-plane, is also single-valued on the unit sphere. If the function $w = f(z)$ is multivalued with a single branch-point, for example, at the origin of the z-plane as the square root $\sqrt{z}$ (in Section 7.2), the branch-cut joining origin to infinity maps unto the sphere as a cut joining the north to the south pole along half-a-meridian. A multivalued function with one branch-point has a semiinfinite branch-cut in the z-plane (Section 7.5) that maps into a finite cut on the unit sphere. If the multivalued function has two branch-points (Section 7.8) it may have (i) finite branch-cut joining them, as in (7.27a,b) and Figure 7.4a; (ii) two semiinfinite branch-cuts joining the branch-points to infinity, as in (7.23a,b) and Figure 7.3a; (iii) two overlapping branch-cuts that reduce to one, as in (7.32a,b) and Figure 8.1h. Since there is only one point-at-infinity at the north pole the branch-cut (cuts) in all cases (i) to (iii) always map to one finite branch-cut on the sphere. If the function is multivalued each of the branches can be mapped: (i) on a plane sheet, joining all the sheets in a semiinfinite cut, as illustrated for the Riemann surface in (Figures 7.1b,c, 7.3d, and 7.4b); (ii) on a sphere, joining the cuts on each sphere by a tube that crosses all spheres. A representation resulting in the Riemann surface with two spheres and one tube for a two-valued function is given in Figure 9.3a. A mapping is called **topological** if it is continuous, injective, and the inverse mapping is also continuous; a topological mapping can be viewed as a continuous deformation. For example, any number of spheres connected by the same tube can be continuously deformed into a single sphere, and thus are **topologically equivalent** to a sphere. It follows that *a multivalued complex function* $w = f(z)$ *with* $N = 1, 2$ *branch-points can be represented single-valued function on*

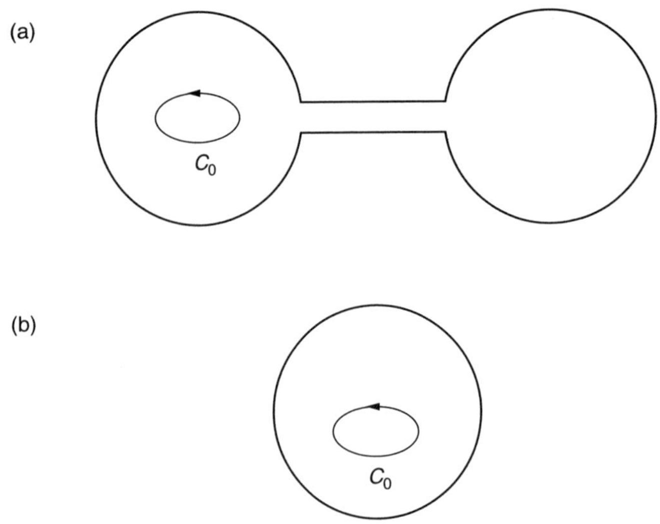

FIGURE 9.3
A single-valued complex function has a bijective mapping onto a sphere (b). A multivalued complex function with one branch-cut can be mapped into one sphere for each branch with a tube joining all spheres at the branch-cuts, for example, two spheres and one tube (a) for a two-valued function with one branch-cut. The spheres with one tube can be continuously deformed into one sphere. They are topologically equivalent to a single sphere; the latter is a surface of genus zero (b) because any loop C_0 can be continuously shrunk to zero.

a Riemann surface consisting of several spheres all joined by the same tube (Figure 9.3a for two spheres) that is topologically equivalent to a single sphere (Figure 9.3b). The number of branches of the function, viz. two-valued (two spheres joined by a tube in Figure 9.4a), N-valued (N spheres joined by the same tube) or many-valued ($N \to \infty$), does not alter the topology, since all cases can be continuously deformed onto one sphere with the same branch-cut. The topology is determined only by the number of branch-points and geometry of branch-cuts joining them.

9.8 Sphere with One Handle and Toroidal Topology

Consider next a two-valued function with four (three) branch-points, for example, the functions (7.40a,b) [(7.41a–c)] corresponding to (7.42a) with $N = 4\,(N = 3)$ that may have two finite cuts (one finite and one semiinfinite cut) in the complex z-plane in Figure 7.3c (7.3b), and can be represented by a Riemann surface with two sheets represented in Figure 7.3d. In the unit sphere the two branch-cuts are finite, and thus the two-valued (multivalued) function with three or four branch-points can be represented by two spheres joined by two tubes (Figure 9.4a); this can be deformed continuously into a **torus**, or doughnut-shape surface in Figure 9.4b. The torus is topologically distinct from the sphere because: (i) on a sphere any closed curve C_0 can be shrunk continuously to zero; (ii) on a torus some closed curves C_0 can be shrunk continuously to zero, but two classes of curves cannot, namely, the **meridional** C_1 and the **equatorial** C_2 curves. The torus can be further deformed into a sphere with a handle in Figure 9.4c, with the two families of unshrinkable curves remaining associated

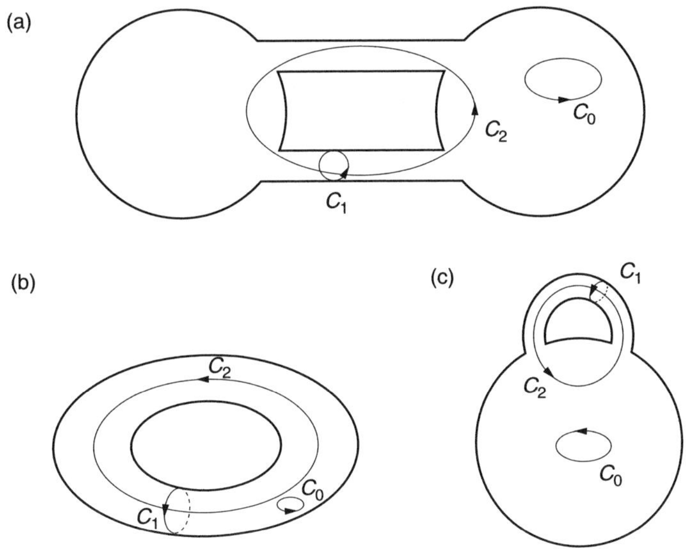

FIGURE 9.4
A multivalued complex function with two branch-cuts can be mapped one-to-one onto two or more spheres joined by two tubes (a). They can be continuously deformed into: (i) a sphere with one handle (c); (ii) a torus with one hole (b). Both are surfaces of genus one, which admit three kinds of loops: (i/ii) the equatorial C_2 (meridional C_1) which cannot be shrunk to zero because they enclose the "hole" (span the cross-section) of the torus; (iii) the loops C_0 which do neither (i) nor (ii) can be shrunk to zero.

with the handle. No continuous deformation will remove the handle or shrink the curves C_1 and C_2. It has been shown that *a multivalued function with three (four) branch-points in the complex* z*-plane, can be represented on a Riemann surface with multiple sheets [illustrated in (Figure 7.3d) for two sheets for two-valued function], or alternatively, on a class of topologically equivalent single-sheet surfaces, including two spheres connected by two tubes (Figure 9.4a), a simple torus with one hole (Figure 9.4b) or a sphere with one* ***handle*** *(Figure 9.4c).*

9.9 Torus with Holes or Sphere with Handles

Proceeding to consider the two-valued function (7.42a) with an arbitrary number N of nonhidden branch-points: (i) if $N = 2M$ is even $M = N/2$ it has either M finite or $(M-1)$ finite and two semiinfinite branch-cuts in the complex z-plane; (ii) if $N = 2M + 1$ is old it has $M = (N-1)/2$ finite branch-cuts plus one semiinfinite branch-cut, for a total of $M + 1 = (N + 1)/2$ cuts in the complex z-plane. These map as $G = N/2$ $[G = (N + 1)/2]$ finite branch-cuts on the unit sphere for a function with N even (odd) branch-points. If the function is two-valued (multivalued), it can be represented as two (more) spheres, each with G cuts joined by G tubes; a continuous deformation collapses the spheres and one tube into a sphere as in (Figure 9.3a,b), leaving $G-1$ handles corresponding to remaining $G-1$ tubes.

A **surface of genus** $\boldsymbol{g}$ is defined as a surface topologically equivalent to a sphere with g handles, or a torus with g holes. Thus it has been proved that a *multivalued function with* *an even (odd) number N of nonhidden branch-points, can be represented as a single-valued function on a Riemann surface of genus $g = (N/2) - 1$ [$g = (N-1)/2$] that is topologically equivalent to (i) two spheres connected by $g+1$ tubes, (ii) a sphere with g handles, or (iii) a torus with g holes.* The simplest cases are the two-valued function with one or two (three or four) branch-points, considered in Section 9.7 (9.8) that are single-valued on a sphere (sphere with one handle, or torus with one hole) that is a surface of genus zero (one), and corresponds to circular and hyperbolic (elliptic) in Section 7.8 (39.8–39.9).

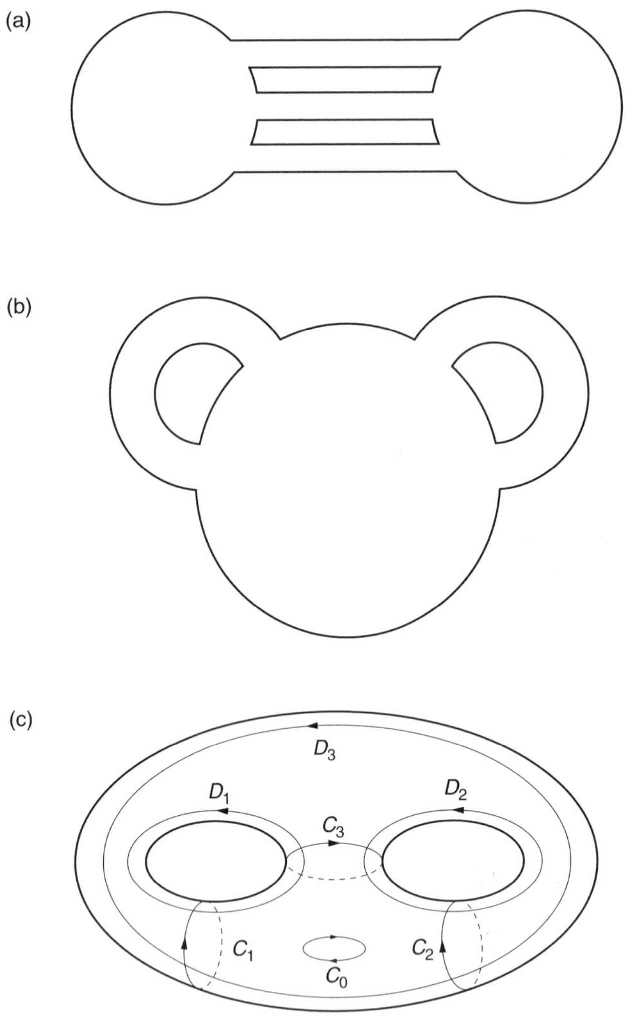

FIGURE 9.5
A multivalued function with three branch-cuts can be mapped one-to-one onto spheres joined by three tubes (a). These can be continuously deformed into: (i) a sphere with two handles (b); (ii) a torus with two holes (c). Both (i) and (ii) are surfaces of genus two, with seven types of loops: (i,ii) two equatorial (D_1, D_2) around each hole; (iii) one equatorial D_3 around both holes; (iv–v) two meriodional (C_1, C_2) joining a hole to the boundary; (vi) one meridional C_3 joining the two holes; (vii) the loops C_0 which do not fall under (i–vi) can be shrunk to zero.

After would come hyperelliptic functions (Section 39.7), viz. the next case in increasing order of complexity is a multivalued function with five or six branch-points that can be represented as single-valued on two spheres connected by three tubes. These are topologically equivalent surfaces of genus two, for example, two spheres joined by three tubes (Figure 9.5a), a sphere with two handles (Figure 9.5b) or a torus with two holes (Figure 9.5c). The double (Figure 9.5c) torus has six classes of **unshrinkable curves**, three meridional (C_1, C_2, C_3) and three equatorial (D_1, D_2, D_3), with the first two connected to one hole each, and the third to both holes together. The sets of unshrinkable curves form the **cohomology class**, and their number is calculated for a surface of genus g in Example 10.20.

Note 9.1. *Geometry of Complex Functions*: The geometry associated with complex functions involves both topology (Sections 9.7–9.9) and the theory of surfaces and curves lying on them. These topics have been covered here only to the extent necessary to represent functions of a complex variable in the subsequent development of the theory and immediate applications. Since the theory of functions is the basis of analysis, its results are used over the rest of the rest of the book as needed. For example: (i/ii) the logarithm (Section 5.6) and complex power (Section 5.7) [inverse circular and hyperbolic functions, e.g., (Subsections 7.8.1–7.8.3)] have one (two) branch-points, and can be mapped onto a sphere; (iii/iv) the inverse Weierstrass (Jacobi) elliptic functions section 39.9 (39.8) have three (four) branch-points and can be mapped on a torus; (v) the inverse hyperelliptic functions (Section 39.7) that have five or more branch-points, can be mapped unto a torus with more than one hole.

Conclusion 9: Examples (Figure 9.1) of single-valued (a) [multivalued (b)] functions; also of surjective (c) [injective (d)] mappings, whose combination is injective (e). The stereographic projection (Figure 9.2) of the unit sphere (r, s, t) on the complex z-plane with $z \equiv x + iy$ is a bijective mapping with a single 'point at infinity' (a); the section by the XOZ-plane, leads to (b) an isomorphism between the unit circle (r, t) and the real line x, with the infimum $+\infty$ and supremum $-\infty$ corresponding the single 'point-at-infinity'. The sphere (Figure 9.3), like two spheres connected by a tube (b), is a surface of genus zero (a), which has one class C_0 of shrinkable curves; it can be used to represent functions with one (two) branch-points, e.g., the elementary transcendental functions logarithm and complex power (inverse circular and hyperbolic functions). The Weierstrass (Jacobi) inverse elliptic functions have three (four) banch-points, and can be represented (Figure 9.4) on a Riemann surface of genus one, e.g., (a) two spheres connected by two tubes, a sphere with one handle (c), or a simple torus with a single hole (c); there all have two classes of unshrinkable curves, one equatorial C_2 and one meridional C_1. The simplest inverse hyperelliptic functions, involving five or six branch-points, are represented (Figure 9.5) on Riemann surfaces of genus two, which are topologically equivalent to (a) two spheres connected by three tubes, (b) a sphere with two handles, or (c) a double torus, with two holes; these all have six classes of unshrinkable closed curves, three meridional C_1, C_2, C_3 and three equatorial D_1, D_2, D_3.

10

Examples 10.1 to 10.20

Example 10.1. *Regions of the Complex Plane.*

Determine the region of the complex z-plane specified: (I) by $|z-a|>|z-b|$ for real $b>a$; (II) by $|z|\geq a\geq|z+a|$ for real a, and find its angular points.

Concerning (I) the meaning of $|z-a|>|z-b|$ is that z is farther from a than from b, so it must lie (Figure 10.1a) to the left of the bisector or the segment (a,b), viz.:

$$b>a: \quad |z-a|>|z-b| \Leftrightarrow \operatorname{Re}(z)>\frac{a+b}{2}. \tag{10.1}$$

Concerning (II) the point z must lie inside (outside) the circle of radius a and center at $(-a,0)$ [the origin], hence it lies in a crescent shaped region (Figure 10.1b), which has two angular points $z_{\pm}$; these are at distance a from the origin and the point $(a,0)$, and thus define equilateral triangles, with sides a and internal angles $\pi/3$, so:

$$z_{\pm}=a\exp\left(\pm\frac{i\pi}{3}\right)=\frac{a}{2}\left(-1\pm i\sqrt{3}\right), \tag{10.2}$$

are the locations of the angular points.

Example 10.2. *Real and Imaginary Parts in Polar Coordinates.*

Obtain the real and imaginary parts of (10.3b):

$$z=x+iy: \quad \begin{matrix}\operatorname{Re}\\ \operatorname{Im}\end{matrix}\left\{\exp\left(\mathrm{e}^{z}\right)\right\}=\exp\left(\mathrm{e}^{x}\cos y\right)\begin{matrix}\cos\\ \sin\end{matrix}\left(\mathrm{e}^{x}\sin y\right), \tag{10.3a,b}$$

using the Cartesian representation (10.3a) and of (10.4b):

$$z=re^{i\varphi}z=re^{i\varphi}: \quad \begin{matrix}\operatorname{Re}\\ \operatorname{Im}\end{matrix}\left\{\sqrt[z]{e}\right\}=\pm\exp\left(r^{-1}\cos\varphi\right)\begin{matrix}\cos\\ \sin\end{matrix}\left(r^{-1}\sin\varphi\right), \tag{10.4a,b}$$

using the polar representation (10.4a).

The former (10.3a,b) results from (1.24):

$$\exp\{\exp(z)\}=\exp\{\mathrm{e}^{x}(\cos y+i\sin y)\}=\exp(\mathrm{e}^{x}\cos y)\exp(i\mathrm{e}^{x}\sin y), \tag{10.5a}$$

and the latter (10.4a,b) uses instead (3.19):

$$\exp\left(\frac{1}{z}\right)=\exp\left\{r^{-1}(\cos\varphi-i\sin\varphi)\right\}=\exp\left(r^{-1}\cos\varphi\right)\exp\left(-ir^{-1}\sin\varphi\right), \tag{10.6b}$$

followed in both cases separation of real and imaginary parts.

Example 10.3. *Logarithmic Decrement of Damped Motion.*

Consider a damped or amplified oscillatory motion:

$$x(t)=A\mathrm{e}^{-\lambda t}\cos(\omega t), \tag{10.7}$$

and determine the ratio of successive maxima and minima.

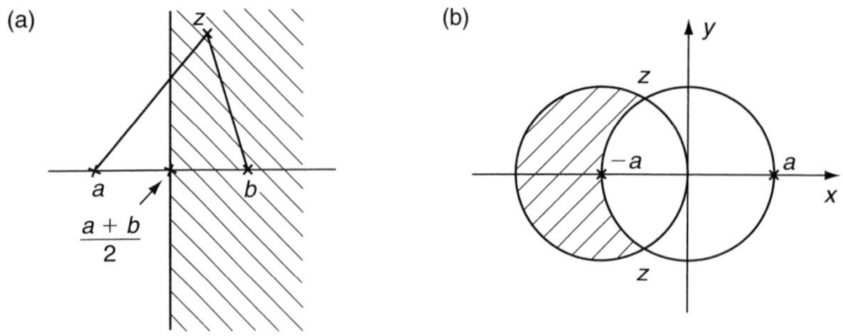

FIGURE 10.1
The locus of points closer to b than to a lies (a) on the r.h.s. of the bisector of the line joining a to b. The locus of points inside (outside) a circle with radius a and center at (0,a) [the origin in (0,0)] is the hatched crescent moon.

The maxima correspond to $\omega t = 2\pi n$ and the minima to $\omega t = (2n+1)\pi$; regardless of sign they are represented by:

$$\omega t = n\pi: \quad x_n = x\left(\frac{n\pi}{\omega}\right) = A(-)^n \exp\left(-\frac{n\pi\lambda}{\omega}\right). \tag{10.8a}$$

The ratio of successive extrema is:

$$-\frac{x_{n+1}}{x_n} = \left|\frac{x_{n+1}}{x_n}\right| = \exp\left(-\frac{\pi\lambda}{\omega}\right), \tag{10.8b}$$

and thus the **logarithmic decrement** of successive extrema is specified by the frequency/period and the decay/growth:

$$\tau = \frac{2\pi}{\omega}: \quad \log\left|\frac{x_{n+1}}{x_n}\right| = -\frac{\pi\lambda}{\omega} = -\frac{\lambda\tau}{2}. \tag{10.9a,b}$$

For damped (amplified) motion $\lambda > 0\,(\lambda < 0)$ the successive extrema reduce (increase) in modulus $|x_n| > |x_{n+1}|\,(|x_n| < |x_{n+1}|)$ as seen in Figure 2.1d (2.1f).

Example 10.4. *Classification of Equilibria from Perturbations.*
Consider the motion of a system in the vicinity of the equilibrium position described by the linear second-order differential equation with constant coefficients (10.10a):

$$\frac{\mathrm{d}^2x}{\mathrm{d}t^2} + 2b\frac{\mathrm{d}x}{\mathrm{d}t} + \omega^2 x(t) = 0, \quad x(t) = A\mathrm{e}^{i\zeta t}, \tag{10.10a,b}$$

where the natural frequency ω is positive real, and the parameter b can take any real values, and (i) classify the possible motions; (ii) determine the corresponding equilibria.

The substitution of (10.10b) in (10.10a) leads (Section 2.8) to the characteristic polynomial (2.17b) $\equiv$ (10.11a):

$$\zeta^2 - 2ib\zeta - \omega^2 = 0, \quad \zeta_\pm = ib \pm \sqrt{\omega^2 - b^2}, \tag{10.11a,b}$$

whose roots are (10.11b). The radical in (10.11b) implies that there are three cases and six subcases: (I) if $b^2 > \omega^2$ the roots are distinct and imaginary (10.12a,b):

$$|b| > \omega: \quad \zeta_\pm = i(b \pm \alpha), \quad \alpha \equiv \left|b^2 - \omega^2\right|^{1/2}, \quad x_\pm(t) = A\mathrm{e}^{-(b\pm\alpha)t}, \tag{10.12a–c}$$

and since $b > \alpha$ in (10.12b), the motion is monotonic with decaying (growing) amplitude for $b > 0\,(b < 0)$ in case IA (IB); (II) if $b^2 = \omega^2$ there is a double imaginary root (10.13a):

$$|b| = \pm\omega: \quad \zeta_\pm = ib, \quad x_+(t) = A\mathrm{e}^{-bt}, \quad x_-(t) = At\mathrm{e}^{-bt}, \tag{10.13a–c}$$

and the motion is monotonic for long time (10.13b,c), with decaying (growing) amplitude for $b > 0\,(b \le 0)$ in case IIA (IIB); (III) if $b^2 < \omega^2$ the roots are distinct complex (10.14a):

$$|b| < \omega: \quad \zeta_\pm = ib \pm \alpha, \quad x_\pm(t) = A\mathrm{e}^{-bt}\mathrm{e}^{\pm i\alpha t}, \tag{10.14a,b}$$

and the motion is oscillatory (10.14b) with effective frequency α in (10.12b), and amplitude decaying (growing) for $b > 0\,(b < 0)$, in the case IIIA (IIIB). The equilibrium is stable (unstable) for $b > 0\,(b < 0)$, corresponding to a force $-b\dot{x}$ with direction opposite (parallel) to the velocity. The intermediate case $b = 0$ is an oscillation with constant amplitude considered in Section 2.9.

Example 10.5. *Amplitude and Phase of Oscillatory Motion.*

Consider the motion specified by:

$$x(t) = A\mathrm{e}^{\lambda t}\cos(\omega t - \alpha), \tag{10.15}$$

and determine the amplitude A and phase α from the initial displacement x_0 and velocity v_0.

Note that (10.15) $\equiv$ (2.33) if $\lambda = 0$, and (10.15) generalizes (2.33) to account for decay $\lambda < 0$ or growth $\lambda > 0$ with time. The velocity associated with (10.15) is:

$$v(t) \equiv \frac{\mathrm{d}x}{\mathrm{d}t} = A\mathrm{e}^{\lambda t}\left[\lambda\cos(\omega t - \alpha) - \omega\sin(\omega t - \alpha)\right]. \tag{10.16}$$

The initial displacement (10.17a) and velocity (10.17b):

$$x_0 \equiv x(0) = A\cos\alpha, \quad v_0 \equiv v(0) = A(\lambda\cos\alpha + \omega\sin\alpha) = \lambda x_0 + A\omega\sin\alpha, \tag{10.17a,b}$$

specify:

$$A = \left|(x_0)^2 + \left(\frac{v_0 - \lambda x_0}{\omega}\right)^2\right|^{1/2}, \quad \tan\alpha = \frac{v_0/x_0 - \lambda}{\omega}, \tag{10.18a,b}$$

the amplitude (10.18a) and phase (10.18b). These reduce to (2.34a,b) for $\lambda = 0$.

Example 10.6. *Alternative Expressions for the Argument of a Complex Number.*

Show that the argument of a complex number can be specified not only by using the inverse tangent (1.12a) but also using other inverse circular functions, such as:

$$x, y > 0: \quad \varphi = \arg(z) = \arg\cos\left\{\left|1 + \left(\frac{y}{x}\right)^2\right|^{-1/2}\right\} = \arg\sin\left\{\left|1 + \left(\frac{x}{y}\right)^2\right|^{-1/2}\right\}, \tag{10.19a,b}$$

in the first quadrant (10.19a,b), and in the other three quadrants using:

$$\{x < 0 < y; \quad x, y < 0; \quad y < 0 < x\} = \{\pi - \varphi; \quad \pi + \varphi; \quad 2\pi - \varphi\}. \tag{10.20a–c}$$

Using the notations:

$$z \equiv x + iy \equiv r\mathrm{e}^{i\varphi}, \; u \equiv a + ib = |u|\mathrm{e}^{i\alpha}, \; v \equiv c + id = |v|\mathrm{e}^{i\beta}, \tag{10.21a–c}$$

show that the argument of the sum of complex numbers is given by:

$$\arg(u+v) = \text{arc sin}\left\{\left|1+\left(\frac{a+c}{b+d}\right)^2\right|^{-1/2}\right\} = \text{arc sin}\left\{\left|1+\left(\frac{\cos\alpha+|v/u|\cos\beta}{\sin\alpha+|v/u|\sin\beta}\right)^2\right|^{-1/2}\right\}, \tag{10.22a,b}$$

$$\arg(u+v) = \text{arc cos}\left\{\left|1+\left(\frac{b+d}{a+c}\right)^2\right|^{-1/2}\right\} = \text{arc cos}\left\{\left|1+\left(\frac{\sin\alpha+|v/u|\sin\beta}{\cos\alpha+|v/u|\cos\beta}\right)^2\right|^{-1/2}\right\}, \tag{10.22c,d}$$

as an alternative to (3.8).

The inversion of (1.13a,b) and (1.11):

$$\{\cos\varphi, \sin\varphi\} = \frac{\{x,y\}}{r} = \left|x^2+y^2\right|^{-1/2}\{x,y\}, \tag{10.23a,b}$$

leads to (10.19a,b) and hence (10.20a,b,c); for $u = a+ib, v = c+id$ then $u+v = x+iy$ with $x = a+c, y = b+d$ leading from (10.19a,b) to (10.22a, 10.22c); the arguments α,β of u,v are used in (10.22b, 10.22d).

Example 10.7. *Quadruple Summation Formulas for Circular Functions.*

Obtain the **trigonometric quadruple summation** formulas for the circular cosine and sine:

$$\begin{aligned}\cos(z_1+z_2+z_3+z_4) = {} & \cos z_1\cos z_2\cos z_3\cos z_4 + \sin z_1\sin z_2\sin z_3\sin z_4 \\ & - \cos z_1\cos z_2\sin z_3\sin z_4 - \cos z_1\cos z_3\sin z_2\sin z_4 \\ & - \cos z_1\cos z_4\sin z_2\sin z_3 - \cos z_2\cos z_3\sin z_1\sin z_4 \\ & - \cos z_2\cos z_4\sin z_1\sin z_3 - \cos z_3\cos z_4\sin z_1\sin z_2,\end{aligned} \tag{10.24a}$$

$$\begin{aligned}\sin(z_1+z_2+z_3+z_4) = {} & \sin z_1\cos z_2\cos z_3\cos z_4 + \sin z_2\cos z_1\cos z_3\cos z_4 \\ & + \sin z_3\cos z_1\cos z_2\cos z_4 + \sin z_4\cos z_1\cos z_2\cos z_3 \\ & - \sin z_1\sin z_2\sin z_3\cos z_4 - \sin z_1\sin z_2\sin z_4\cos z_3 \\ & - \sin z_1\sin z_3\sin z_4\cos z_2 - \sin z_2\sin z_3\sin z_4\cos z_1,\end{aligned} \tag{10.24b}$$

and the corresponding **trigonometric quadruplication** formulas:

$$\cos(4z) = \cos^4 z + \sin^4 z - 6\cos^2 z\sin^2 z = 8\cos^4 z - 8\cos^2 z + 1 = 8\sin^4 z - 8\sin^2 z + 1, \tag{10.25a}$$

$$\sin(4z) = 4\sin z\cos z\left(\cos^2 z - \sin^2 z\right) = 2\sin(2z)\cos(2z). \tag{10.25b}$$

These formulas follow from (3.27a,b) ≡ (10.24a,b) and (3.28b,c) ≡ (10.25a,b) for $N = 4$; in particular (10.25b) coincides with (3.31c) replacing $2z$ by z.

Example 10.8. *Ten Distinct Associations of Four Impedances.*

Consider all distinct associations of four impedances in mechanical (electrical) circuits, and calculate the total impedance; consider also the particular case of equal impedances.

There are ten cases in Figure 10.2: (i) all in parallel (series) in Figure 10.2a (10.2j):

$$Z_{mA} = Z_{eJ} = z_1 + z_2 + z_3 + z_4; \tag{10.26}$$

(ii) two in series (parallel) with others in parallel (series) in Figure 10.2b (10.2g):

$$Z_{mB} = Z_{eG} = \left(z_1^{-1} + z_2^{-1}\right)^{-1} + z_3 + z_4 = \frac{z_1 z_2}{z_1+z_2} + z_3 + z_4; \tag{10.27}$$

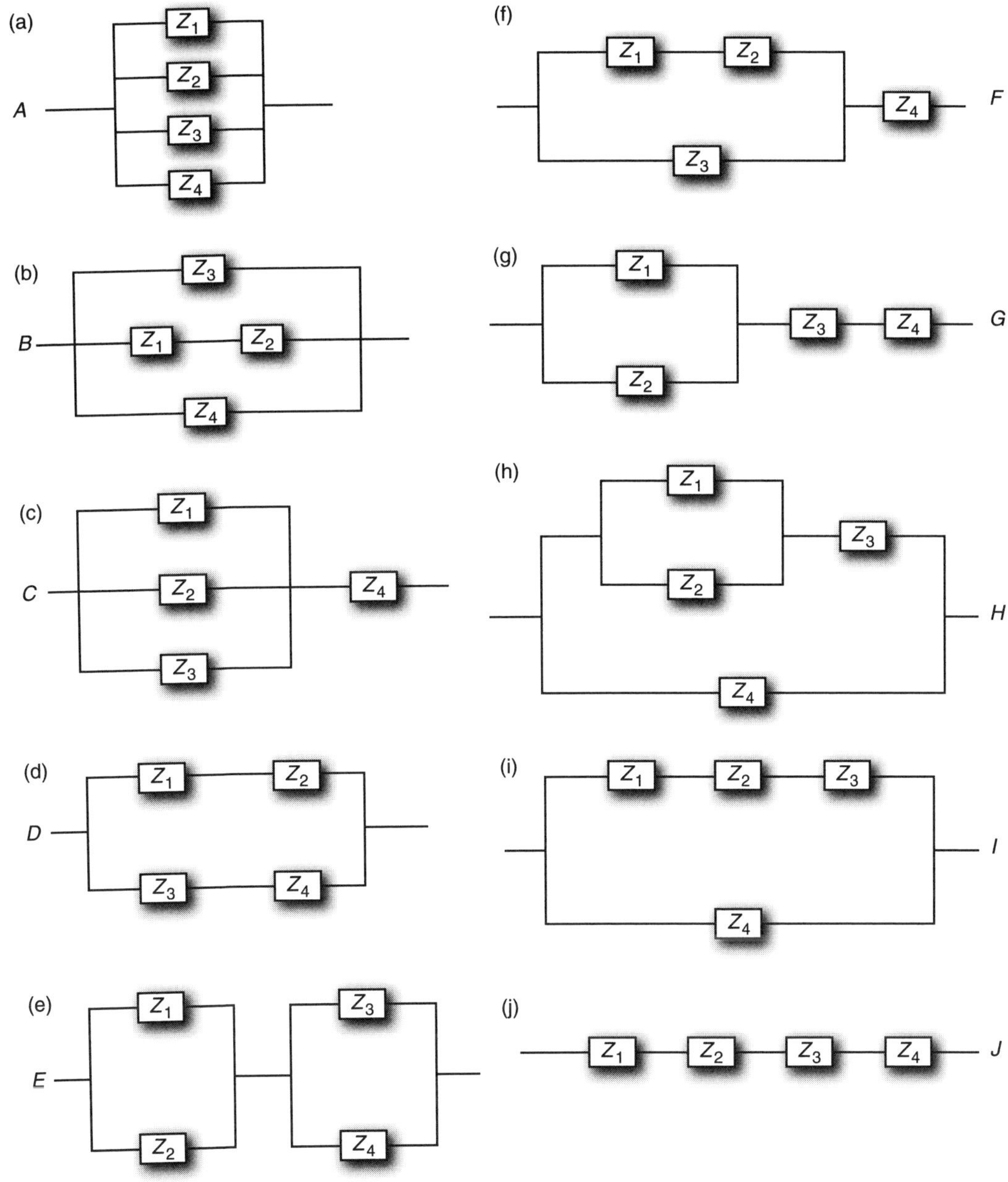

FIGURE 10.2
There are 2, 4, and 10 distinct associations of 2, 3, and 4 circuits in Figures 4.2, 4.3, and 10.2. Concerning the latter case, the ten arrangements of four circuits can be found exhaustively as follows: (i) start with all four circuits in series (j); (ii) put one in parallel and leave the other three in series (i); (iii) keep one in parallel and of the remaining three put two in parallel (h); (iv) keep two in parallel and the other two in series (g); (v) shift one of the circuits in series into one of the parallel branches (f); (vi) shift each of the series circuits into one parallel branch (e); (vii) merge the two pairs of consecutive parallel branches into a single pair (d); (viii) pass one series circuit to parallel (b); (ix) pass one series circuit outside the three parallel branches (c); (x) put all circuits in parallel (a). The exhaustive listing also works in reverse.

(iii) three in parallel (series) plus one in series (parallel) in Figure 10.2c (10.2i):

$$Z_{mC} = Z_{eI} = \left[(z_1 + z_2 + z_3)^{-1} + z_4^{-1}\right]^{-1} = \frac{z_4(z_1 + z_2 + z_3)}{z_1 + z_2 + z_3 + z_4}; \tag{10.28}$$

(iv) two pairs in parallel (series) in Figure 10.2d (10.2e):

$$Z_{mD} = Z_{eE} = \left(z_1^{-1} + z_2^{-1}\right)^{-1} + \left(z_3^{-1} + z_4^{-1}\right)^{-1} = \frac{z_1 z_2}{z_1 + z_2} + \frac{z_3 z_4}{z_3 + z_4}; \tag{10.29}$$

(v) two pairs in series (parallel) in Figure 10.2e (10.2d):

$$Z_{mE} = Z_{eD} = \left[(z_1 + z_2)^{-1} + (z_3 + z_4)^{-1}\right]^{-1} = \frac{(z_1 + z_2)(z_3 + z_4)}{z_1 + z_2 + z_3 + z_4}; \tag{10.30}$$

(vi) one in series (parallel) with one and pair in parallel (series) in Figure 10.2f (10.2h):

$$Z_{mF} = Z_{eH} = \left\{\left[(z_1^{-1} + z_2^{-1})^{-1} + z_3\right]^{-1} + z_4^{-1}\right\}^{-1} = \frac{z_4\left[z_1 z_2 + z_3(z_1 + z_2)\right]}{z_1 z_2 + (z_3 + z_4)(z_1 + z_2)}; \tag{10.31}$$

(vii) a pair in parallel (series) with two in series (parallel) in Figure 10.2g (10.2b):

$$Z_{mG} = Z_{eB} = \left[(z_1 + z_2)^{-1} + z_3^{-1} + z_4^{-1}\right]^{-1} = \frac{z_3 z_4 (z_1 + z_2)}{z_3 z_4 + (z_3 + z_4)(z_1 + z_2)}; \tag{10.32}$$

(viii) one in parallel (series) with one and a pair in series (parallel) in Figure 10.2h (10.2f):

$$Z_{mH} = Z_{eF} = \left[(z_1 + z_2)^{-1} + z_3^{-1}\right]^{-1} + z_4 = z_4 + \frac{z_3(z_1 + z_2)}{z_1 + z_2 + z_3}; \tag{10.33}$$

(ix) one in parallel (series) with three in series (parallel) in Figure 10.3i (10.2c):

$$Z_{mI} = Z_{eC} = \left(z_1^{-1} + z_2^{-1} + z_3^{-1}\right)^{-1} + z_4 = z_4 + \frac{z_1 z_2 z_3}{z_1 z_2 + z_1 z_3 + z_2 z_3}; \tag{10.34}$$

(x) all in (series) parallel in Figure 10.2j (10.2a):

$$Z_{mJ} = Z_{eA} = \left(z_1^{-1} + z_2^{-1} + z_3^{-1} + z_4^{-1}\right)^{-1} = \frac{z_1 z_2 z_3 z_4}{z_1 z_2 z_3 + z_1 z_2 z_4 + z_1 z_3 z_4 + z_2 z_3 z_4}. \tag{10.35}$$

These simplify to:

$$\begin{aligned} &\{Z_{mA}, Z_{mB}, Z_{mC}, Z_{mD}, Z_{mE}, Z_{mF}, Z_{mG}, Z_{mH}, Z_{mI}, Z_{mJ}\} \\ &= \left\{4z_0, \frac{5z_0}{2}, \frac{3z_0}{4}, z_0, z_0, \frac{3z_0}{5}, \frac{2z_0}{5}, \frac{5z_0}{3}, \frac{4z_0}{3}, \frac{z_0}{4}\right\} \\ &= \{Z_{eJ}, Z_{eG}, Z_{eI}, Z_{eE}, Z_{eD}, Z_{eH}, Z_{eB}, Z_{eF}, Z_{eC}, Z_{eA}\}, \end{aligned} \tag{10.36}$$

in the case of equal impedances $z_1 = z_2 = z_3 = z_4 = z_0$.

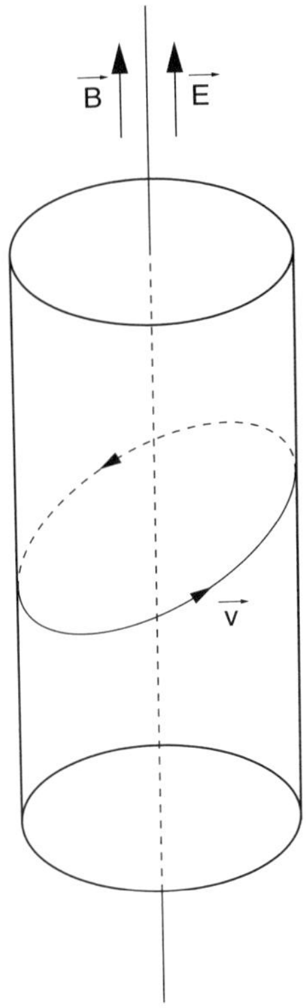

FIGURE 10.3
The trajectory of an electron in: (i) a constant magnetic field lies on a cylinder of which it is the axis; (ii) a parallel nonuniform electric field with strength proportional to the displacement from a point on the axis causes a vertical oscillation. The combination of (i) and (ii) is a closed trajectory on the cylinder with radius (height range) determined by the magnetic (electric) frequency.

Example 10.9. *Roots of Plus/Minus Unity and of the Imaginary Unit.*

Obtain the roots:

$$\sqrt[2]{i} = \pm\frac{1+i}{\sqrt{2}}; \quad \sqrt[3]{i} = -i, \frac{\pm\sqrt{3}+i}{2}; \tag{10.37a,b}$$

$$\sqrt[3]{-1} = -1, \frac{1 \pm i\sqrt{3}}{2}; \quad \sqrt[4]{-1} = \frac{\pm 1 \pm i}{\sqrt{2}}; \tag{10.38a,b}$$

$$\sqrt[4]{1} = \pm 1, \pm i; \quad \sqrt[6]{1} = \pm 1, \frac{\pm 1 \pm i\sqrt{3}}{2}; \quad \sqrt[8]{1} = \pm 1, \pm i, \frac{\pm 1 \pm i}{\sqrt{2}}. \tag{10.39a,b}$$

All the results follow from (5.10) in Section 5.3.

Example 10.10. *Alternative Calculation of Trajectory of an Electron in a Uniform Electromagnetic Field.*

Prove that (6.10a,b) are the solutions of (6.5a,b) using real velocity components, instead of the complex velocity in Sections 6.2–6.4.

Concerning the trajectories of the electron (Chapter 6), elimination between the coupled first-order ordinary differential equations (6.5a,b) can be made, for example, for v_x by differentiating (6.5a) and substituting (6.5b), leading to:

$$\ddot{v}_x = \Omega \dot{v}_y = -\Omega^2 v_x + \Omega a_\perp, \tag{10.40}$$

which is a second-order ordinary differential equation (10.40) for v_x alone. A particular solution of (10.40) is the first term on the r.h.s. of (10.41a):

$$v_x(t) = \frac{a_\perp}{\Omega} + C_1 \cos(\Omega t) + C_2 \sin(\Omega t). \tag{10.41a}$$

and to this can be added the solution of $\ddot{v}_x + \Omega^2 v_x = 0$, which consists (2.26) of the last two terms on the r.h.s. of (10.41a), where C_1, C_2 are arbitrary constants. The solution (10.41a) is the complete integral which satisfies (10.40) and involves two arbitrary constants C_1, C_2. The same arbitrary constants appear in the other component of the velocity (10.5a) $\equiv$ (10.41b):

$$v_y(t) = \Omega^{-1} \dot{v}_x(t) = -C_1 \sin(\Omega t) + C_2 \cos(\Omega t). \tag{10.41b}$$

If the origin of time $t = 0$ is chosen such that the initial velocity $v_\perp$ is parallel to the OX-axis, the arbitrary constants are specified by:

$$t = 0: \quad v_\perp = v_x(0) = \frac{a_\perp}{\Omega} + C_1, \quad 0 = v_y(0) = C_2. \tag{10.42a,b}$$

Substituting (10.42a,b) in (10.41a,b) yields:

$$v_x(t) = \frac{a_\perp}{\Omega} + \left(v_\perp - \frac{a_\perp}{\Omega}\right) \cos(\Omega t), \qquad v_y(t) = -\left(v_\perp - \frac{a_\perp}{\Omega}\right) \sin(\Omega t), \tag{10.43a,b}$$

which coincide with (6.10a,b) $\equiv$ (10.43a,b), bearing in mind that $a_\perp / \Omega = E_\perp c / B$ as follows from (6.6b,c).

Example 10.11. *Trajectory of an Electron in a Nonuniform Electromagnetic Field.*

Consider an electron moving in parallel magnetic (electric) fields, which are constant (attractive and proportional to the distance). Determine the trajectory using cylindrical coordinates, that is, polar coordinates in the transverse direction and a Cartesian longitudinal coordinate.

If direction of the constant external magnetic field is taken as z-axis, the corresponding motion (6.13a,b) is circular in the transverse plane (10.44b,c):

$$\Omega \equiv \frac{eB}{mc}: \quad x(t) = \frac{v_\perp}{\Omega} \sin(\Omega t), \quad y(t) = \frac{v_\perp}{\Omega} \cos(\Omega t), \tag{10.44a–c}$$

with the Larmor or **magnetic frequency** (6.6c) $\equiv$ (10.44a), and radius $v_\perp / \Omega$ where $v_\perp$ is the initial transversal velocity (6.14c). The longitudinal motion (10.45c) is specified by (6.5c; 6.6a) the electric field (10.45a,b) which is proportional to distance and attractive:

$$E(z) = -Az; \; A > 0: \quad \ddot{z} = \dot{v}_z = a_{//} = \frac{eE}{m} = -\frac{eAz}{m}. \tag{10.45a–c}$$

The longitudinal motion corresponds (10.45c) $\equiv$ (10.46a), to a harmonic oscillator:

$$\ddot{z} + \omega^2 z = 0, \quad \omega \equiv \sqrt{\frac{eA}{m}}, \tag{10.46a,b}$$

with **electrical frequency** ω in (10.46b). If the initial longitudinal displacement is zero and velocity is $v_{//}$, the longitudinal motion is specified (2.30a) by:

$$z(t) = \frac{v_{//}}{\omega} \sin(\omega t). \tag{10.47}$$

The trajectory is circular in the transverse plane (10.44b,c), and corresponds to a constant radius (10.48a):

$$R(t) = \left|[x(t)]^2 + [y(t)]^2\right|^{1/2} = \frac{v_\perp}{\Omega}, \quad \varphi(t) = \Omega t, \tag{10.48a,b}$$

and angle (10.48b) proportional to the magnetic frequency and time; substituting (10.48b) in (10.47) it follows that:

$$z(t) = \frac{v_{//}}{\omega}\sin\left[\frac{\omega}{\Omega}\varphi(t)\right], \quad \frac{\omega}{\Omega} = \frac{c}{B}\sqrt{\frac{mA}{e}}, \tag{10.49a,b}$$

the trajectory of the electron lies: (i) on the cylinder with radius (10.48a) and axis along the magnetic and electric fields; (ii) the axial distance (10.49a) is proportional to the sine of the angle of rotation, multiplied by the ratio (10.49b) of electrical (10.46b; 10.45a,b) to magnetic (10.44a) frequency: (iii) the range of longitudinal displacement $z_{\max} - z_{\text{mix}} = 2v_{//}/\bar{\omega}$ depends on the transverse initial velocity and electrical frequency (Figure 10.3).

Example 10.12. *Principal Branches of Multivalued Functions.*

Show that the functions below are single-valued as principal values in the complex z-plane:

$$z \notin (0,\infty): \quad \log(1 - \mathrm{e}^z) = \frac{1}{2}\log\left|1 + \mathrm{e}^{2x} - 2\mathrm{e}^x\cos y\right| + i\operatorname{arccot}\left[\left(\cot y - \mathrm{e}^{-x}\csc y\right)^{-1}\right]; \tag{10.50a,b}$$

$$z \notin (e,\infty): \quad \{1 - \log z\}^{-1/2} = \left|(1-\log r)^2 + \varphi^2\right|^{-1/2}\exp\left[\frac{i}{2}\arctan\left(\frac{\varphi}{1-\log r}\right)\right]; \tag{10.51a,b}$$

The branch-points are identified from the rule (7.15), viz.:

$$\frac{\mathrm{d}\{\log(1-\mathrm{e}^z)\}}{\mathrm{d}z} = \frac{1}{1-e^{-z}} \to \infty \text{ as } z \to 0, \tag{10.52}$$

$$\frac{\mathrm{d}\left\{(1-\log z)^{-1/2}\right\}}{\mathrm{d}z} = (2z)^{-1}(1-\log z)^{-3/2} \to \infty \text{ as } z \to e, 0; \tag{10.53}$$

the branch-cuts for (10.50a) [(10.51a] correspond to real $x > 0$ with $1 - \mathrm{e}^z = -x\,(1 - \log z = -x)$, that is $z = \log(1+x)\,[z = \exp(1+x)]$, which corresponds to $z \geq 0$ ($z \geq e$), including the branch-point, where the function is singular. The principal branches arise from:

$$\log(1-\mathrm{e}^z) = \log\left[1 - \mathrm{e}^x\cos y - i\mathrm{e}^x\sin y\right], (1-\log z)^{1/2} = (1 - \log r - i\varphi)^{-1/2}, \tag{10.54a,b}$$

for (10.50b) $\equiv$ (10.54a) [(10.51b) $\equiv$ (10.54b)].

Example 10.13. *Motion of Suspended/Inverted Pendulum.*

Show that the linear motion of the suspended (inverted) pendulum, or of the ship with center of flotation above (below) the center of gravity, following a disturbance with nonzero velocity in the stable (10.55a) [unstable (10.55b)] equilibrium position:

$$\varphi(0) = 0 \neq \dot{\varphi}(0) \equiv \dot{\varphi}_0, \quad \psi(0) = 0 \neq \dot{\psi}(0) \equiv \dot{\psi}_0, \tag{10.55a,b}$$

is given by:

$$\varphi(t) = \frac{\dot{\varphi}_0}{\omega}\sin(\omega t), \qquad \dot{\varphi}(t) = \dot{\varphi}_0\cos(\omega t), \tag{10.56a,b}$$

$$\psi(t) = \frac{\dot{\psi}_0}{\omega}\sinh(\omega t), \quad \dot{\psi}(t) = \dot{\psi}_0\cosh(\omega t), \tag{10.57a,b}$$

with (10.55a,b) [(10.56a,b)] involving of circular (hyperbolic) functions.

It is sufficient to obtain (10.56a; 10.57a) since (10.56b, 10.57b) follow by differentiation. The initial conditions (10.55a) [(10.55b)] for the stable (unstable) case substituted in (8.19a) [8.20a)] lead to (10.58a) [(10.58b)]:

$$\dot{\varphi}^2 + \omega^2\varphi^2 = \dot{\varphi}_0^2, \quad \dot{\psi}^2 - \omega^2\psi^2 = \dot{\psi}_0^2. \tag{10.58a,b}$$

These may integrated:

$$\alpha \equiv \frac{\varphi\omega}{\dot{\varphi}_0}: \quad \omega t = \int_0^t \omega \mathrm{d}t = \int_0^\varphi \left|\left(\frac{\dot{\varphi}_0}{\omega}\right)^2 - \varphi^2\right|^{-1/2} \mathrm{d}\varphi$$

$$= \int_0^{\varphi\omega/\dot{\varphi}_0} \left|1-\alpha^2\right|^{-1/2} \mathrm{d}\alpha = [\arcsin\alpha]_0^{\varphi\omega/\dot{\varphi}_0} = \arcsin\left(\frac{\varphi\omega}{\dot{\varphi}_0}\right), \tag{10.59a,b}$$

$$\beta \equiv \frac{\psi\omega}{\dot{\psi}_0}: \quad \omega t = \int_0^t \omega \mathrm{d}t = \int_0^\psi \left|\left(\frac{\dot{\psi}_0}{\omega}\right)^2 + \psi^2\right|^{-1/2} \mathrm{d}\psi$$

$$= \int_0^{\psi\omega/\dot{\psi}_0} \left|1+\beta^2\right|^{-1/2} \mathrm{d}\beta = [\operatorname{arcsinh}\beta]_0^{\psi\omega/\dot{\psi}_0} = \operatorname{arcsinh}\left(\frac{\psi\omega}{\dot{\psi}_0}\right), \tag{10.60a,b}$$

and coincide with (10.59b) ≡ (10.56a) [(10.60b) ≡ (10.57a)].

Example 10.14. *Motion of Pendulum for Arbitrary Initial Conditions.*

Compare the stable (10.56a,b) [unstable (10.57a,b)] trajectories for an initial velocity (10.55a) [(10.55b)], with the stable (8.25a,b) [unstable (8.29a,b)] trajectories for an initial displacement (8.21a,b), and by superposition, or otherwise, show that following arbitrary initial displacement and velocity in (10.61a) [(10.61b)]:

$$\varphi(0) \equiv \varphi_0 \neq 0 \neq \dot{\varphi}_0 \equiv \dot{\varphi}(0), \quad \psi(0) \equiv \psi_0 \neq 0 \neq \dot{\psi}_0 \equiv \dot{\psi}(0), \tag{10.61a,b}$$

the angular displacement and velocity are given in the linear approximation as a function of time by:

$$\varphi(t) = \varphi_0\cos(\omega t) + \frac{\dot{\varphi}_0}{\omega}\sin(\omega t), \quad \dot{\varphi}(t) = -\varphi_0\omega\sin(\omega t) + \dot{\varphi}_0\cos(\omega t), \tag{10.62a,b}$$

$$\psi(t) = \psi_0\cosh(\omega t) + \frac{\dot{\psi}_0}{\omega}\sinh(\omega t), \quad \dot{\psi}(t) = \psi_0\omega\sinh(\omega t) + \dot{\psi}_0\cosh(\omega t), \tag{10.63a,b}$$

where (10.62a,b) [(10.63a,b)] involving circular (hyperbolic) functions correspond to the oscillatory (monotonic) motion in Section 8.8 (8.9). In the oscillatory case, show that:

$$\varphi_{\max} = \left|\varphi_0^2 + \left(\frac{\dot{\varphi}_0}{\omega}\right)^2\right|^{1/2}, \quad \dot{\varphi}_{\max} = \left|\dot{\varphi}_0^2 + \omega^2\varphi_0^2\right|^{1/2}, \tag{10.64a,b}$$

specify the maximum angular displacement (10.64a) and velocity (10.64b).

All these problems reduce to linear ordinary differential equations with constant coefficients soluble by the methods of Sections 2.8 and 2.9. Since (10.56–10.57a,b) are a particular case of (10.62–10.63a,b), the following method of solution also applies to the preceding Example 10.13. The stable case (10.62a,b) ≡ (2.30a,b) coincides with Section 2.9 replacing (10.65a) angular φ (linear x) displacement and angular $\dot{\varphi}$ (linear v) velocity; also for the maximum displacement (10.64a) ≡ (2.34a) or amplitude. The maximum velocity is obtained from (2.30b) ≡ (10.65b):

$$(x, v) \to (\varphi, \dot{\varphi}): \quad v_{\max} = \left|(v_0)^2 + (x_0\omega)^2\right|^{1/2}, \tag{10.65a,b}$$

which is analogous to (10.64b). The unstable case in (10.63a,b) corresponds to the differential equation (10.17b) ≡ (10.66a):

$$0 = \ddot{\psi} - \omega^2\psi = \left\{Q_2\left(\frac{\mathrm{d}}{\mathrm{d}t}\right)\right\}\psi(t), \quad -Q_2(\zeta) = \zeta^2 + \omega^2 = (\zeta - i\omega)(\zeta + i\omega) \tag{10.66a,b}$$

whose characteristic polynomial (2.17b) ≡ (10.66b) has complex conjugate $\zeta = \pm i\omega$ roots corresponding to the solution (2.1a,b) ≡ (10.67a):

$$\psi_\pm(t) = \exp\left(\pm i^2\omega t\right) = \exp(\mp\omega t). \tag{10.67a}$$

The general solution is thus:

$$\psi(t) = B_+\exp(\omega t) + B_-\exp(-\omega t), \tag{10.67b}$$

where $B_\pm$ are arbitrary constants. Using the relations with the hyperbolic sine and cosine:

$$\exp(\pm\omega t) = \cosh(\omega t) \pm \sinh(\omega t), \tag{10.68a,b}$$

the angular displacement (10.67b) ≡ (10.69a) and velocity (10.69b) are given by:

$$\psi(t) = B_1\cosh(\omega t) + B_2\sinh(\omega t), \qquad \dot{\psi}(t) = \omega\{B_1\sinh(\omega t) + B_2\cosh(\omega t)\}, \tag{10.69a,b}$$

where B_1, B_2 are new arbitrary constants, related to the preceding by:

$$B_1 = B_+ + B_-, \quad B_2 = B_+ - B_-. \tag{10.70a,b}$$

The arbitrary constants are determined by the initial conditions:

$$t = 0: \quad \psi(0) \equiv \psi_0 = B_1, \quad \dot{\psi}(0) \equiv \dot{\psi}_0 = \omega B_2; \tag{10.71a,b}$$

substitution of (10.71a,b) in (10.69a,b) yields (10.63a,b). Introducing the new constants: (B,β) by:

$$\psi_0 > 0: \quad B_1 = B\cosh\beta, \quad B_2 = -B\sinh\beta, \tag{10.72a,b}$$

the angular displacement (10.69a) becomes:

$$\psi(t) = B[\cosh(\omega t)\cosh\beta - \sinh(\omega t)\sinh\beta] = B\cosh(\omega t - \beta). \tag{10.73}$$

The amplitude B and phase β are specified by the initial conditions (10.71a,b):

$$B = \left|(B_1)^2 - (B_2)^2\right|^{1/2} = \left|\psi_0{}^2 - \left(\frac{\dot{\psi}_0}{\omega}\right)^2\right|^{1/2}, -\tanh\beta = \frac{B_2}{B_1} = \frac{\dot{\psi}_0}{\psi_0\omega}. \tag{10.74a,b}$$

The phase β can be eliminated by a suitable choice of time:

$$\bar{t} = t - \frac{\beta}{\omega}: \quad \psi(t) = B\cosh(\omega\bar{t}). \tag{10.75a,b}$$

The oscillatory (monotonic divergent) motion near the position of stable (unstable) equilibrium can be compared for the linear (2.33) [angular (10.73)] displacement, and also:

$$\dot{x}(t) = -A\omega\sin(\omega t - \alpha), \quad \dot{\psi}(t) = B\omega\sinh(\omega t - \beta), \tag{10.76a,b}$$

for the linear (angular) velocity (10.76a) [(10.76b)].

Example 10.15. *Reflection on the Origin and on the Unit Sphere.*

Relate through the stereographic projection the reflection on the origin for the unit sphere and complex plane.

The points diametrically opposite (r, s, t) and $(-r, -s, -t)$ on the unit sphere are mapped by the stereographic projection (9.9a,b) respectively to the points z and ζ:

$$z = \frac{r + is}{1 - t}, \quad \zeta = -\frac{r + is}{1 + t}, \quad \zeta^* z = -\frac{r^2 + s^2}{1 - t^2} = -1, \tag{10.77a–c}$$

which satisfy (10.77c). Thus the inversion with regard to the center of the sphere $(r, s, t) \to (-r, -s, -t)$ corresponds to the inversion on the complex plane relative to the origin $z \equiv \sigma e^{i\varphi} \to \zeta = -\frac{1}{z^*} = -\sigma^{-1}e^{i\varphi} = \sigma^{-1}e^{i(\varphi+\pi)}$, that is, inverse modulus $1/\sigma$ and opposite phase $\pi + \varphi$. It follows that the inversion maps the plane (sphere) as follows: (i) the origin (south pole) is inverted into the "point-at-infinity" (north pole); (ii) the interior of the unit circle (southern hemisphere) is inverted into the exterior of the unit circle (northern hemisphere); (iii) only the points on the unit circle (equator) are mapped on the same circle, in diametrically opposite positions.

Example 10.16. *Stereographic Projection of Parallel of a Sphere.*

Consider the stereographic projection of a parallel of the sphere.

The parallel (10.78a) of co-latitude θ on the unit sphere (Figure 9.2b) is mapped by the stereographic projection (Section 9.2) on a circle of center at the origin and radius R given (9.8) by (10.78b):

$$t = \cos\theta: \quad R^2 = \frac{1 + \cos\theta}{1 - \cos\theta} = \frac{(1 + \cos\theta)^2}{\sin^2\theta}; \tag{10.78a,b}$$

thus the arc lying between the meridians $\varphi_1 \le \varphi \le \varphi_2$ is projected into a circular arc of length $(R/2\pi)\,|\varphi_2 - \varphi_1|$.

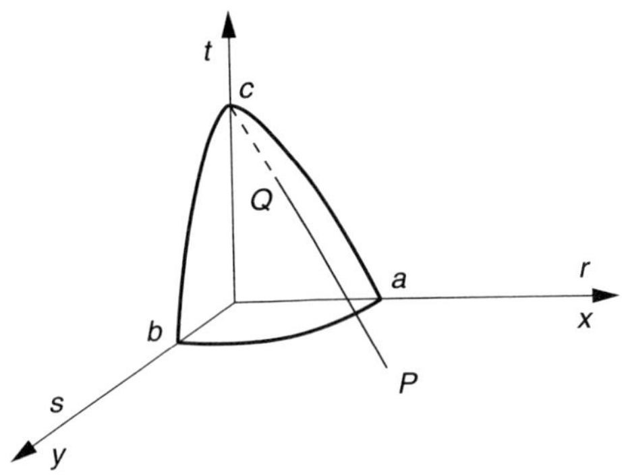

FIGURE 10.4
The stereographic projection of an ellipsoid from one pole to a plane through the center orthogonal to the polar axis, reduces to that of a sphere (Figure 9.2a) with distinct scale changes along each axis.

Example 10.17. *Stereographic Projection of an Ellipsoid.*

Consider the projection of an ellipsoid (Figure 10.4) from the end of one half-axis to the plane of the other two half-axis, and determine the direct and inverse coordinate transformations. Check that the results agree with the particular case of stereographic projection of the unit sphere onto the complex plane.

The ellipsoid of center at the origin, and half-axis (a,b,c) along the (r,s,t)-axis (10.79b):

$$(r',s',t') \equiv \left(\frac{r}{a},\frac{s}{b},\frac{t}{c}\right): \quad 1 = \frac{r^2}{a^2} + \frac{s^2}{b^2} + \frac{t^2}{c^2} = r'^2 + s'^2 + t'^2, \qquad \text{(10.79a–c)}$$

is topologically equivalent to the unit sphere (10.79c) via the transformation (10.79a); thus its projection from the point ($t = c$, $r = 0 = s$) to the complex-z plane is a bijective mapping. The direct stereographic projection, from a point on the ellipsoid (r,s,t) to a point on the plane $z \equiv x + iy$, is given by:

$$x = \frac{r/a}{1-t/c}, \quad y = \frac{s/b}{1-t/c}, \quad z = \frac{r/a + is/b}{1-t/c}, \quad |z|^2 = \frac{c+t}{c-t}, \qquad \text{(10.80a–d)}$$

which coincide with (9.9a,b) using the transformation (10.79a); this applies also to the inverse mapping (9.11a–c):

$$r = \frac{2ax}{1+x^2+y^2}, \qquad s = \frac{2by}{1+x^2+y^2}, \qquad t = c\frac{x^2+y^2-1}{x^2+y^2+1}. \qquad \text{(10.81a–c)}$$

The case $y = 0 = s$ leads to the isomorphism of the ellipse (10.82a) and the real axis, which corresponds to the inverse mapping (10.82b):

$$y = 0 = s: \qquad \frac{r^2}{a^2} + \frac{t^2}{c^2} = 1: \quad r = \frac{2ax}{1+x^2}, \quad t = c\frac{x^2-1}{x^2+1}, \qquad \text{(10.82a–c)}$$

and direct mapping (10.80a).

Example 10.18. *Mercator Mapping of a Sphere onto a Cylinder.*

The **Mercator map** (1569) projects the sphere from its center on a tangent cylinder (Figure 10.5) and then develops the cylinder on a plane. Find the relation between the latitude and longitude on the sphere and the Cartesian coordinates on the plane.

The Mercator map projects the a point P of longitude φ and latitude ψ on the surface of a sphere of radius R, from the center to a point P on a cylinder tangent to the equator, that is, with radius R and axis passing through the north and south poles. After developing the cylinder on a plane, the projection goes to horizontal x and vertical y Cartesian coordinates:

$$x = R\varphi, \quad y = R\tan\psi. \tag{10.83a,b}$$

There is no distortion on the equator, but the distortion increases with latitude, as shown by: (i) the Cartesian area element:

$$\mathrm{d}x\,\mathrm{d}y = R^2\sec^2\psi\,\mathrm{d}\psi\,\mathrm{d}\varphi, \tag{10.84a}$$

compared to the spherical area element:

$$\frac{\mathrm{d}x\,\mathrm{d}y}{R^2\sin\psi\,\mathrm{d}\psi\,\mathrm{d}\varphi} = \sec^2\psi\csc\psi = 2\sec\psi\csc(2\psi); \tag{10.84b}$$

(ii) the arc length (11.17a):

$$(\mathrm{d}s)^2 = (\mathrm{d}x)^2 + (\mathrm{d}y)^2 = R^2\left[(\mathrm{d}\varphi)^2 + \sec^4\psi\,(\mathrm{d}\psi)^2\right], \tag{10.85}$$

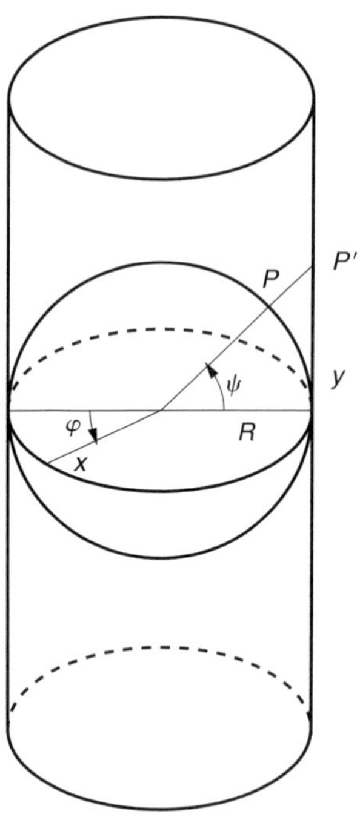

FIGURE 10.5
The sphere has a double curvature and cannot be developed onto a plane without distortion. The Mercator map projects the sphere from the center onto a tangent cylinder. The cylinder can be "unrolled" onto a plane. There is no distortion only at the equator; the parallels become horizontal straight lines. The distortion increases without bound with latitude, and the meridians become parallel lines "converging" to the "poles" at infinity.

The parallels $\psi=$const (meridians $\varphi=$ const) are mapped (10.83b) [(10.83a)] to horizontal $y=$ const (vertical $x=$const) lines, and the poles $\psi=\pm\pi/2$ to $y=\pm\infty$. The polar distortion of the earth (on the Mercator map) is shown by: (i) the distance between meridians which decreases toward the poles (is constant); (ii) the distance between parallels which is constant (increases toward the poles).

Example 10.19. *Mapping of a Sphere onto a Cone.*

Consider the **conical projection** (Figure 10.6a) of a sphere on a tangent cone, followed by the development of the latter on a plane (Figure 10.6b), and relate the latitude and longitude on the sphere to the polar coordinates on the plane.

If the tangent cone has the vertex at distance H from the center of the sphere of radius R the aperture α is given by:

$$H>R: \quad \sin\alpha=\frac{R}{H}, \quad \cos\alpha=\left|1-\left(\frac{R}{H}\right)^2\right|^{1/2}, \quad \tan\alpha=\left|\left(\frac{H}{R}\right)^2-1\right|^{-1/2}. \tag{10.86a–c}$$

The line from the center of the sphere to the vertex of the cone intersects the sphere at the north pole $\theta=0$, and the cone is tangent to the sphere at the parallel of co-latitude $\theta=\alpha$, where there is no distortion. An arbitrary point P of co-latitude θ and longitude φ on the sphere is projected (Figure 10.6b) on the point P' on the cone, whose distance from the vertex r satisfies the condition:

$$r\sin\alpha=|P'Q|=(H-r\cos\alpha)\tan\theta, \tag{10.87a}$$

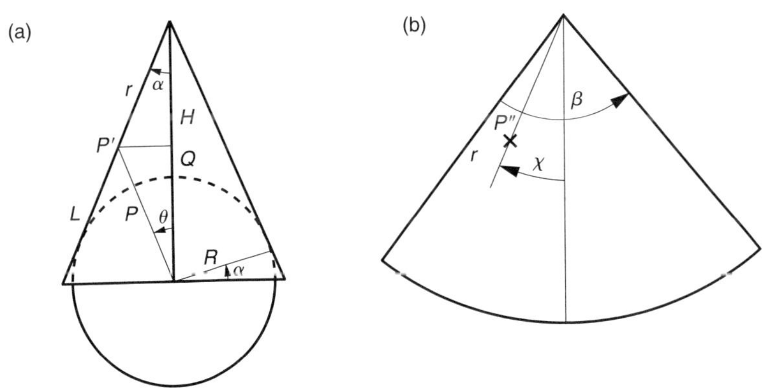

FIGURE 10.6
The conical projection maps a sphere from the center to a tangent cone (a); the cone is a developable surface which can be "unrolled" onto the plane as a circular sector (b). There is no distortion only along the parallel of tangency of the cone to the sphere (instead of the equator in the Mercator map in Figure 10.5). The parallels are mapped to circles, and the distortion increases with bound toward the vertex, corresponding to the north pole where converge the meridians mapped into radial lines. The conical projection does not cover the lower hemisphere unless the cone is extended below the equatorial plane; then the distortion increases without bound and the south pole is not mapped at all. To map the south pole as the north pole and have the same distortion in the upper and lower hemispheres two opposite half-cones with vertices symmetric relative to the center of the sphere may be used enveloping the whole sphere.

where $|P'Q|$ is the length of the segment from P' to its projection Q on the axis. The distance from the vertex depends only on the latitude:

$$r = \frac{H \tan\theta}{\sin\alpha + \cos\alpha \tan\theta} = \frac{(H^2/R)\tan\theta}{1 + \tan\theta\left|(H/R)^2 - 1\right|^{1/2}}, \tag{10.87b}$$

and is preserved when the cone is developed on the plane (Figure 10.6b).

The cone is developed on the plane (Figure 10.6b) as an arc of circle with: (i) radius (10.88a) equal to the length of the directrix of the cone up to the equator; (ii) perimeter equal to the perimeter of the cone (10.88b):

$$L = H\sec\alpha, \quad a = 2\pi H \tan\alpha. \tag{10.88a,b}$$

Hence the angle at the vertex is:

$$\beta = \frac{a}{L} = \frac{2\pi H \tan\alpha}{H\sec\alpha} = 2\pi\sin\alpha = \frac{2\pi R}{H}. \tag{10.88c}$$

The point P is developed onto the point P'' whose polar angle χ with the axis of the circular sector depends only on the longitude:

$$\chi = \frac{\beta\varphi}{2\pi} = \varphi\sin\alpha = \frac{R\varphi}{H}. \tag{10.88d}$$

Thus *the point of co-latitude θ and longitude φ on a sphere of radius R, after projection on a tangent cone with the vertex at a distance H from the center, is developed onto the plane with polar coordinates (10.87b; 10.88d).*

Example 10.20. *Cohomology Classes on a Surface of Genus* g.

Determine the number of **cohomology classes** of unshrinkable curves on a surface of genus g and: (i) check it graphically for the torus with one, two and three holes; (ii) tabulate it up to $g = 8$. In all cases distinguish the various classes of curves, for example, meridional and equatorial.

A multivalued function with $N = 2\,(N = 3)$ branch-cuts is of genus $g = 1\,(g = 2)$, and hence equivalent to a torus with one (two) holes, and has already been considered graphically in Figure 9.4a–c (9.5a–c). The next case is multivalued function with $N = 4$ branch-cuts which has a Riemann surface of genus three, such as a torus with three holes (Figure 10.7), so that it has 13 classes of unshrinkable curves, 6 meridional, and 7 equatorial, with 3 each for single and twin holes, plus one equatorial curve for three holes. Considering the general surface of genus g, for example, a torus with g holes the number of **cohomology classes**, that is, unshrinkable curves, is: (i) for **meridional** curves linking a hole to the boundary, and linking pairs of holes, as the number of independent terms of a symmetric $g \times g$ matrix (10.89a):

$$G_1 = g + \binom{g}{2} = \frac{g(g+1)}{2}; \quad G_2 = \binom{g}{1} + \binom{g}{2} + \cdots \binom{g}{g} = 2^g - 1, \tag{10.89a,b}$$

the number of **equatorial** curves, is $\binom{g}{1}$ around one hole, $\binom{g}{2}$ around two holes up well $\binom{g}{g}$ around all holes, where the arrangements of g holes in n groups are specified by (7.45), and the sum (10.89b) $\equiv$ (7.44). The number of classes of unsinkable curves of a surface of genus g is the sum of the meridional (10.89a) and equatorial (10.89b) numbers, viz.:

$$G \equiv G_1 + G_2 = 2^g + \frac{g^2}{2} + \frac{g}{2} - 1, \tag{10.90}$$

TABLE 10.1
Number of Classes of Unshrinkable Curves on a Surface of Genus g

g	1	2	3	4	5	6	7	8
G_1	1	3	6	10	15	21	28	36
G_2	1	3	7	15	31	63	127	255
$G = G_1 + G_2$	2	6	13	25	46	84	155	291
Figures	9.4	9.5	10.7	–	–	–	–	–

Note: A surface of genus g, for example, a sphere (torus) with g handles (holes), has G_1 (G_2) types of unshrinkable meridional (equatorial) loops. The cases of genus 1, 2, and 3 corresponding to a sphere (torus) with 1, 2, and 3 handles (holes) appear in Figures 9.4, 9.5, and 10.7.

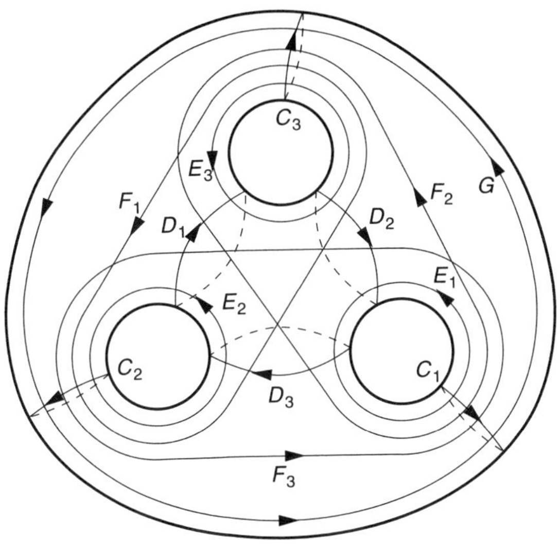

FIGURE 10.7
A surface of genus three is topologically equivalent to a sphere with three handles or a torus with three holes. There are seven distinct sets of unshrinkable equatorial loops: (i) three $\binom{3}{1} = 3$ around each hole (E_1, E_2, E_3); (ii) three $\binom{3}{2} = 3$ around two holes (F_1, F_2, F_3): (i) one $\binom{3}{3} = 1$ around all three holes G. There are six meridional loops; (iv) three $\binom{3}{1} = 3$ joining each hole to the boundary (C_1, C_2, C_3); (v) three $\binom{3}{2} = 3$ joining two holes (D_1, D_2, D_3). The total (i) to (v) is $7 + 6 = 13$ in agreement with Table 10.1 for genus $g = 3$. The loops C_0 other than (i–v) can be shrunk to zero, and do not count toward the cohomology class.

and appears in Table 10.1, for genus up to $g = 8$, including the torus with one (Figure 9.4a–c), two (Figure 9.5a–c), and three (Figure 10.7) holes. For large genus the equatorial classes are much more numerous due to the larger number of combinations, that is, growth is like 2^g (g^2) for equatorial (meridional) classes.

Note 10.1. *From Complex Algebra to Analysis*: The initial Part 1 of the present volume has concentrated on the geometry of the complex plane, as a representation of an ordered pair of real numbers by a single complex number with the advantage that all basic arithmetic operations are closed. The use of complex formulas to represent several real expressions and to simplify the solution some two-dimensional problems is a natural sequel to basic operations and functions with complex variables. The consideration of multivalued functions, branch-points, branch-cuts, and the Riemann sphere are aspects of the geometry of the complex plane linking to the complex analysis in Part 2. The definition of derivative of a complex holomorphic function opens two major paths: (i) the calculus of integrals by the method of residues in several variants; (ii) the use of differentiable functions to obtain solutions of the Laplace and Poisson equations, for example, describing potential flows and gravity fields. These two aspects will progress alongside each other in subsequent developments of (i) complex analysis and (ii) potential fields.

Conclusion 10: Regions of the plane (Figure 10.1) defined by complex inequalities: (a) distance from b less than from a; (b) exterior of circle of radius a and center at the origin and interior of circle of radius a tangent at left to the imaginary axis at the origin. Ten associations of four distinct impedances (Figure 10.2), from all in parallel (a) to all in series (j) arrangement, with 8 intermediate cases. The trajectory of an electron (Figure 10.3) under parallel uniform (nonuniform linear function of distance and attractive) magnetic (electric) field is a sinusoid on a cylinder. The stereographic projection (Figure 10.4) of the ellipsoid of half-axis (a,b,c) on the complex z-plane with $z = x + iy$ is a bijective mapping, and its section by the $s = 0$ plane, an isomorphism of the ellipse of half-axis (a,c) onto the real line. The Mercator (conical) map projects [Figure 10.5 (10.6a,b)] the sphere from its center to a tangent cylinder (cone), and then develops it on the plane. The torus with three holes (Figure 10.7) is an example of a surface of genus three, and has 13 cohomology classes, or families of unshrinkable loops, namely the 3 meridional loops C_1, C_2, C_3, connecting each hole to the outside, the 3 meridional loops D_1, D_2, D_3, connecting two holes, the 3 equatorial loops E_1, E_2, E_3, around one hole each, the three equatorial loops F_1, F_2, F_3, around pairs of holes, and the single equatorial loop G around all three holes.

Part 2
Integrals and Residues: Flows and Gravity

A complex number is an ordered pair of real numbers; similarly a complex function is a pair of real functions. Thus a complex function of a complex variable is a pair of real functions of two real variables, and yields a convenient way of treating plane problems using a single variable. The two real functions of two real variables involve four first-order partial derivatives, and all these are contained in a single complex derivative if the function is holomorphic (Chapter 11); also a holomorphic function has derivatives and primitives of all orders (Chapters 13 and 15). A holomorphic function can be integrated not only along the real axis but also suitable curves in the complex plane (Chapters 17 and 19). If the complex function fails to be holomorphic at an isolated point, that is, has a singularity, the path of integration can be deformed so as to by-pass or surround it; this is a significant advantage over real functions, since if a real function has a singularity in the real axis there is no way of connecting it over the two sides. Thus the integral calculus is more powerful in the complex than in the real domain, viz. it allows the evaluation of integrals involving functions with singularities, for example, using their residues (Chapters 15 and 17). The real and imaginary parts of a holomorphic function satisfy Laplace's equation, except at singularities, where Poisson's equation may be satisfied instead. Thus complex functions with a single variable may be used to obtain two-dimensional solutions of the Laplace and Poisson equations, for example, describing plane potential flows (Chapters 12, 14, and 16) and gravity fields (Chapter 18).

11

Differentiation and Holomorphic Functions

Since the domain of a complex function is a plane (Section 11.1) its derivative at any given point may be obtained by approaching the point from any direction or along any curve (Section 11.2). The derivative only exists if it is independent of the direction of approach, and for this reason a differentiable complex function is called holomorphic; the greek word "holos" stands for symmetry, for example, the most symmetric class in each chrystalographic group is called holoedric. The derivative of a complex function of a complex variable involves all four real partial derivatives and it is independent of direction only if these satisfy the Cauchy–Riemann conditions (Section 11.3). The Cauchy–Riemann conditions can be interpreted as stating that the real and imaginary parts of a complex holomorphic function: (i) specify two families of orthogonal curves (Section 11.8); (ii) satisfy Laplace's equation (Section 11.6). The former (i) can be used to introduce orthogonal curvilinear coordinates in the plane (Section 11.9); these can be used together with (ii) to specify the invariant differential operators (gradient, divergence, curl, and Laplacian) in plane orthogonal curvilinear coordinates, including as a particular case polar coordinates (Section 11.7). The Cauchy–Riemann conditions (Section 11.4) and hence the derivative of a holomorphic function (Section 11.5) can also be expressed not only in Cartesian (Section 11.3), but also in polar (Section 11.4) and other orthogonal curvilinear coordinates (Section 11.9). A non-holomorphic function may have specific properties in particular directions (Example 20.1) that are excluded by the "isotropy" of holomorphic functions; some of the consequences of isotropy are explored next.

11.1 Function, Neighborhood, Limit, and Uniformity

The **neighborhood** of order $\varepsilon > 0$ (real and positive) of a point z_0 (or complex number), is the set of all points z whose distance from z is less than ε:

$$V_\varepsilon(z_0) \equiv \{z : |z - z_0| < \varepsilon\} \subset \bar{V}_\varepsilon(z_0) = \{z : |z - z_0| \le \varepsilon\}; \tag{11.1}$$

the neighborhood $V_\varepsilon(z_0)$ is an **open region** that coincides with the **interior** of the circle of radius ε and center at z_0; its **closure** $\bar{V}_\varepsilon(z_0)$ adds the circle itself, to form a **closed region**. A single-valued function $w = f(z)$ is (Section 9.1) a relation between two sets of points namely, the domain D and the range E, such that to each point of the domain corresponds only one point of the range:

$$z \xrightarrow{f} w : \quad \forall_{z \in D} \quad \exists^1_{w \in E} : \quad w = f(z); \tag{11.2}$$

the function (11.2) maps the neighborhood (11.1) of z_0 in the domain $V_\varepsilon(z_0) \subset D$ into a region $G \subset E$ of the range that may not be contained in any neighborhood, for example, if two close points $|z_1 - z_2| < 2\varepsilon$ are mapped into two distant points $|w_1 - w_2| \gg a$.

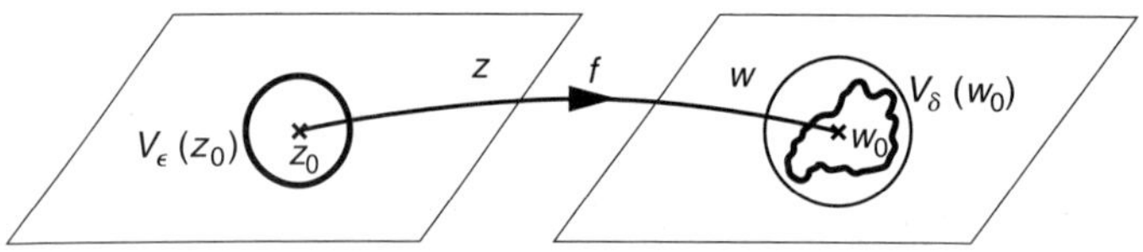

FIGURE 11.1
A complex function $w = f(z)$ maps the z-plane to the w-plane. In particular the point z_0 is mapped to the point $w_0 = f(z_0)$. A neighborhood $V_\varepsilon(z_0)$ of z_0 is mapped into some region of the w-plane. If the function is continuous at z_0 the region is always within a neighborhood $V_\varepsilon(w_0)$ of w_0. Moreover that neighborhood of w_0 can be made as small as we please by choosing the neighborhood $V_\varepsilon(z_0)$ of z_0. If the function were discontinuous at z_0 some points in $V_\varepsilon(z_0)$ would be mapped outside $V_\delta(w_0)$ when δ is small enough no matter how small ε is.

If the neighborhood (11.1) of z_0 is mapped by the function (11.2) into a region G contained in a neighborhood $V_\delta(w_0)$ of w_0 in Figure 11.1, and if δ can be made arbitrarily small by choosing ε, then the **limit** of the function $f(z)$ at z_0 exists and equals w_0:

$$\lim_{z\to z_0} f(z) = w_0: \quad \forall_{\delta>0}\exists_{\varepsilon>0}: \quad z \in V_\varepsilon(z_0) \Rightarrow f(z) \in V_\delta(w_0); \tag{11.3}$$

generally, the choice of $\varepsilon(\delta, z_0)$ depends both on the degree of approximation δ of $f(z)$ to w_0, and on the point z_0. Considering two points z_1, z_2 at which $f(z)$ has limits respectively w_1, w_2, the functions $\varepsilon(\delta, z_1)$ and $\varepsilon(\delta, z_2)$ are usually different, and may be incompatible. If for all z_0 in a region H, an $\varepsilon(\delta)$ can be found which is independent of z_0, then:

$$\forall_{z\in H,\delta>0} \quad \exists_{\varepsilon>0}: \quad |z - z_0| < \varepsilon \Rightarrow |f(z) - w_0| < \delta, \tag{11.4}$$

the limit (11.3) is **uniform** (11.4) in the region H.

11.2 Continuity, Incremental Ratio, and Derivate (Newton, 1670; Leibnitz, 1684)

The limit (11.3) may fail to coincide with $f(z_0) \neq w_0$, for example, if the neighborhoods of z_0 are mapped to the neighborhoods of w_0, but z_0 itself is mapped elsewhere into $f(z_0)$. If $f(z_0) = w_0$, then the point z_0 is mapped together with its neighborhood, and the function $f(z)$ is **continuous**:

$$f(z) \in \mathcal{C}(|C): \quad \lim_{z\to z_0} f(z) = f(z_0) \tag{11.5}$$

where, $\mathcal{C}$ denotes the set of continuous functions of complex variable $z \in |C$. If the function $f(z)$ is continuous at z_1, then $z - z_1 \to 0$ implies that $f(z) - f(z_1) \to 0$, and their ratio, if it exists, specifies (Newton, 1670; Leibnitz, 1684) the **derivate** (or derivative) of the function f at the point z:

$$\Delta z \equiv z - z_0: \quad f'(z) \equiv \lim_{z\to z_0} \frac{f(z) - f(z_0)}{z - z_0} = \lim_{\Delta z\to 0} \frac{f(z+\Delta z) - f(z)}{\Delta z} \equiv \frac{\mathrm{d}f}{\mathrm{d}z}; \tag{11.6}$$

both the Newton–Lagrange $f'(z)$ and Leibnitz $\mathrm{d}f/\mathrm{d}z$ notations for the derivative can be used.

The definition of derivate (11.6) requires the limit to exist, and have the same value, regardless of the way in which $\Delta z \to 0$. Using the Cartesian representation:

$$\Delta z = \Delta x + i\Delta y: \quad f(x+iy) = \Phi(x,y) + i\Psi(x,y), \tag{11.7}$$

it follows that the derivative (11.6) exists iff (if and only if) the limit:

$$\frac{\mathrm{d}f}{\mathrm{d}z} = \lim_{\Delta x, \Delta y \to 0} \frac{\Phi(x+\Delta x, y+\Delta y) + i\Psi(x+\Delta x, y+\Delta y) - \Phi(x,y) - i\Psi(x,y)}{\Delta x + i\Delta y}, \tag{11.8}$$

exists and is independent of the way in that Δx and Δy each tends to zero. In this case the function $f(z)$ belongs to the set $\mathcal{D}(|C)$ of differentiable functions on the complex plane, and is called holomorphic. It will be shown subsequently that a complex holomorphic function, that is, with first-order derivate, has not only derivates of all orders (Chapter 15) that is, is a smooth function $\mathcal{D}^\infty$ but also has a convergent Taylor series (Chapter 23), that is, is an analytic function $\mathcal{A}$. This is generally not the case for a real function, viz: (i) the first-order derivative may exist but higher order derivatives may not exist; (ii) the derivatives of all orders may exist but the Taylor series may fail to converge.

11.3 Holomorphic Function and Continuous Derivatives

If the derivate (11.8) is calculated (Figure 11.2) by approaching the point z along a horizontal line $(\Delta y = 0, \Delta z = \Delta x)$, it is specified by:

$$\frac{\mathrm{d}f}{\mathrm{d}z} = \lim_{\Delta x \to 0} \left\{ \frac{\Phi(x+\Delta x, y) - \Phi(x,y)}{\Delta x} + i\frac{\Psi(x+\Delta x, y) - \Psi(x,y)}{\Delta x} \right\} = \frac{\partial \Phi}{\partial x} + i\frac{\partial \Psi}{\partial x}; \tag{11.9a}$$

if instead the derivate is calculated along a vertical line $(\Delta x = 0, \Delta z = i\Delta y)$, it is specified by.

$$\frac{\mathrm{d}f}{\mathrm{d}z} = \lim_{\Delta y \to 0} \left\{ \frac{\Phi(x+\Delta x, y) - \Phi(x,y)}{i\Delta y} + i\frac{\Psi(x+\Delta x, y) - \Psi(x,y)}{i\Delta y} \right\} = \frac{\partial \Psi}{\partial y} - i\frac{\partial \Phi}{\partial y}. \tag{11.9b}$$

Assuming that the function $f(z)$ is holomorphic, that is, belongs to the set $\mathcal{D}$ of differentiable functions, its derivative exists, is unique, and can be calculated in any direction, for example, horizontal (11.9a) or vertical (11.9b); equating these two expressions leads to the **Cauchy–Riemann conditions:**

$$f(z) \in \mathcal{D}(|C): \frac{\partial \Phi}{\partial x} = \frac{\partial \Psi}{\partial y}, \quad \frac{\partial \Phi}{\partial y} = -\frac{\partial \Psi}{\partial x}. \tag{11.10a,b}$$

It follows that *a necessary condition that the function (11.7) be holomorphic (11.6), is that its real Φ and imaginary Ψ parts satisfy the Cauchy–Riemann conditions (11.10a,b).*

The Cauchy–Riemann **conditions** are necessary, but not sufficient for the function $f(z)$ to be holomorphic: (i) they state that the limit (11.8) is equal along horizontal (11.9a) and vertical (11.9b) directions that must be true if the function $f(z)$ is holomorphic; (ii) it could happen that the limit (11.8) taken along some other curve or direction, for example, a line $\Delta y = \alpha \Delta x \to 0$ of slope $\theta = \arctan \alpha$ in Figure 11.3, would be different from (11.9a) $\equiv$ (11.9b), in which case the function $f(z)$ would not be holomorphic. Example 20.1 concerns a function that meets the Cauchy–Riemann conditions, but is not holomorphic, because direction(s) exist for which the limit (11.8) is different from the horizontal (11.9a) or vertical (11.9b)

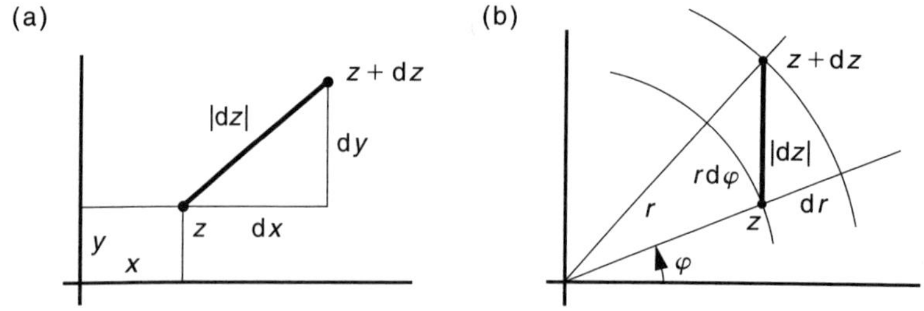

FIGURE 11.2
The infinitesimal displacement dz in the z-plane can be decomposed in several ways, for example: (i) parallel to the Cartesian coordinate axis forming a rectangular triangle with sides dx and dy in (a); (ii) as a radial dr and an azimuthal displacement dφ the latter multiplied by the radius r to yield the length rdφ of the curved side (b). In both cases the triangles are rectangular, because they lie along orthogonal coordinate axis, viz. Cartesian (a) [polar (b)]. Thus the Pythagoras theorem holds for an infinitesimal rectangular triangle even with "curved" sides; it would hold for a finite rectangular triangle only with straight sides.

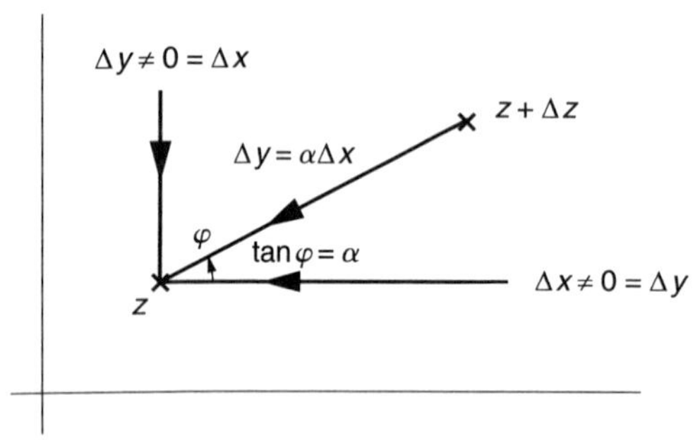

FIGURE 11.3
The derivative of a function at a point z is the limit of the ratio $\Delta f/\Delta z$ where the displacement tends to zero $\Delta z \to 0$ and $\Delta f = f(z+\Delta z) - f(z)$ is the corresponding variation of the function. For the derivative to exist the limit must be the same whatever way Δz tends to zero, for example, along vertical, horizontal, or oblique lines. This is why the function is called holomorphic: it has the maximum symmetry, from the greek "holos" or whole. If it differs for any two directions (Example 20.1) the derivative does not exist, and the function is not holomorphic.

limit. This situation can be avoided if the partial derivates in (11.9a,b) are continuous, because then the limit (11.8) cannot change between horizontal, vertical, or other directions, and the function is holomorphic; it follows that:

$$\Phi, \Psi(x,y) \equiv \mathrm{Re}, \mathrm{Im}\{f(z)\} \in C^1(|R) \wedge \left\{\frac{\partial \Phi}{\partial x}, \frac{\partial \Phi}{\partial y}\right\} = \left\{\frac{\partial \Psi}{\partial y}, -\frac{\partial \Psi}{\partial x}\right\} \Rightarrow f(z) \in \mathcal{D}(|C), \tag{11.11}$$

a sufficient condition that the function (11.7) be holomorphic (11.8) is that its real Φ and imaginary Ψ parts have continuous first-order derivates satisfying the Cauchy–Riemann conditions (11.10a,b) $\equiv$ (11.11).

To prove the statement (11.11), note that since Φ has continuous first-order derivative, it satisfies (Section 19.2) the mean value theorem:

$$\Delta\Phi \equiv \Phi(x+\Delta x, y+\Delta y) - \Phi(x,y) = \frac{\partial\Phi}{\partial x}\Delta x + \frac{\partial\Phi}{\partial y}\Delta y + \varepsilon_1\Delta x + \delta_1\Delta y, \tag{11.12a}$$

where, $\varepsilon_1, \delta_1 \to 0$ as $\Delta x, \Delta y \to 0$; similarly, since Ψ has continuous first-order derivative, its increment is given by:

$$\begin{aligned}\Delta\Psi &= \frac{\partial\Psi}{\partial x}\Delta x + \frac{\partial\Psi}{\partial y}\Delta y + \varepsilon_2\Delta x + \delta_2\Delta y \\ &= -\frac{\partial\Phi}{\partial y}\Delta x + \frac{\partial\Phi}{\partial x}\Delta y + \varepsilon_2\Delta x + \delta_2\Delta y,\end{aligned} \tag{11.12b}$$

where, the Cauchy–Riemann conditions (11.10a,b) were used, and ε_2, δ_2 are infinitesimals in (11.12b):

$$\lim_{\Delta x, \Delta y \to 0} \{\varepsilon_1, \varepsilon_2, \delta_1, \delta_2\} = \{0,0,0,0\}, \tag{11.12c}$$

as ε_1, δ_1 in (11.12a).

Adding (11.12a) to i times (11.12b):

$$\Delta\Phi + i\Delta\Psi = \left(\frac{\partial\Phi}{\partial x} - i\frac{\partial\Phi}{\partial y}\right)(\Delta x + i\Delta y) + \varepsilon\Delta x + \delta\Delta y, \tag{11.13a}$$

specifies the increment of f:

$$\Delta f = \left(\frac{\partial\Phi}{\partial x} - i\frac{\partial\Phi}{\partial y}\right)\Delta z + \eta\Delta z, \tag{11.13b}$$

where, $\varepsilon, \delta, \eta$ are complex quantities defined by:

$$\varepsilon \equiv \varepsilon_1 + i\varepsilon_2, \quad \delta \equiv \delta_1 + i\delta_2, \quad \eta\Delta z \equiv \varepsilon\Delta x + \delta\Delta y; \tag{11.13c}$$

noting that

$$|\Delta x|, |\Delta y| \le |\Delta x + i\Delta y| \equiv |\Delta z|, \quad |\eta||\Delta z| \le |\varepsilon||\Delta x| + |\delta||\Delta y|, \tag{11.14a,b}$$

it follows that η is an infinitesimal:

$$\lim_{\Delta z \to 0} |\eta| \le \lim_{\Delta x, \Delta y \to 0} \left\{\frac{|\varepsilon||\Delta x| + |\delta||\Delta y|}{|\Delta z|}\right\} \le \lim_{\Delta x, \Delta y \to 0} \{|\varepsilon| + |\delta|\} = 0; \tag{11.14c}$$

from (11.13b) it follows that the limit:

$$\lim_{\Delta z \to 0} \frac{\Delta f}{\Delta z} = \frac{\partial\Phi}{\partial x} - i\frac{\partial\Phi}{\partial y} = \frac{\mathrm{d}f}{\mathrm{d}z} \equiv f'(z), \tag{11.15}$$

is unique and independent of the way in which $\Delta z \to 0$, and specifies the derivative $f'(z)$ of $f(z)$ that is thus a holomorphic function, QED.

11.4 Cauchy (1821)–Riemann (1851) Conditions in Cartesian and Polar Coordinates

The Cauchy–Riemann conditions (11.10a,b) in Cartesian coordinates (x, y), can be restated in polar coordinates (r, φ), by taking the corresponding representation of the complex differential:

$$\mathrm{d}x + i\mathrm{d}y = \mathrm{d}z = \mathrm{d}\left(r\mathrm{e}^{i\varphi}\right) = \mathrm{e}^{i\varphi}\left(\mathrm{d}r + ir\mathrm{d}\varphi\right); \tag{11.16}$$

its modulus:

$$|\mathrm{d}z|^2 = (\mathrm{d}x)^2 + (\mathrm{d}y)^2 = (\mathrm{d}r)^2 + r^2(\mathrm{d}\varphi)^2, \tag{11.17a,b}$$

specifies *the* **distance element** *or* **arc-length** *in Cartesian (11.17a) [polar (11.17b)] coordinates, illustrated in Figure 11.2a (11.2b).* The polar representation of a complex function can be separated into real and imaginary parts:

$$f\left(r\mathrm{e}^{i\varphi}\right) \equiv \Phi(r, \varphi) + i\Psi(r, \varphi); \tag{11.18}$$

if the function is holomorphic, its derivates:

$$\left(\frac{\mathrm{d}f}{\mathrm{d}z}\right)_\varphi = \frac{\partial(\Phi + i\Psi)}{\mathrm{e}^{i\varphi}\partial r} = \mathrm{e}^{-i\varphi}\left(\frac{\partial\Phi}{\partial r} + i\frac{\partial\Psi}{\partial r}\right), \tag{11.19a}$$

$$\left(\frac{\mathrm{d}f}{\mathrm{d}z}\right)_r = \frac{\partial(\Phi + i\Psi)}{i\mathrm{e}^{i\varphi}r\partial\varphi} = \frac{\mathrm{e}^{-i\varphi}}{r}\left(\frac{\partial\Psi}{\partial\varphi} - i\frac{\partial\Phi}{\partial\varphi}\right), \tag{11.19b}$$

calculated at constant $\varphi(r)$ and varying $r(\varphi)$ in (11.19a) [(11.19b)] must be equal:

$$\frac{\partial\Phi}{\partial r} = \frac{1}{r}\frac{\partial\Psi}{\partial\varphi}, \quad \frac{\partial\Phi}{\partial\varphi} = -r\frac{\partial\Psi}{\partial r}, \tag{11.20a,b}$$

leading to *the Cauchy–Riemann conditions in polar coordinates (11.20a,b).* As before (Section 11.3), the conditions (11.20a,b) are necessary for the function (11.18) to be holomorphic, and if the four derivates in (11.20a,b) are continuous then the conditions are also sufficient for holomorphy.

11.5 Formulas for the Derivative and Its Modulus and Argument

From (11.9a,b; 11.10a,b; 11.15) [11.19a,b; 11.20a,b)] follow four equivalent expressions for the derivative of a holomorphic function, in the Cartesian (11.7) [polar (11.18)] representation:

$$\frac{\mathrm{d}f}{\mathrm{d}z} = \frac{\partial\Phi}{\partial x} + i\frac{\partial\Psi}{\partial x} = \frac{\partial\Phi}{\partial x} - i\frac{\partial\Phi}{\partial y} = \frac{\partial\Psi}{\partial y} - i\frac{\partial\Phi}{\partial y} = \frac{\partial\Psi}{\partial y} + i\frac{\partial\Psi}{\partial x}; \tag{11.21a}$$

$$\mathrm{e}^{i\varphi}\frac{\mathrm{d}f}{\mathrm{d}z} = \frac{\partial\Phi}{\partial r} + i\frac{\partial\Psi}{\partial r} = \frac{\partial\Phi}{\partial r} - \frac{i}{r}\frac{\partial\Phi}{\partial\varphi} = \frac{1}{r}\frac{\partial\Psi}{\partial\varphi} - \frac{i}{r}\frac{\partial\Phi}{\partial\varphi} = \frac{1}{r}\frac{\partial\Psi}{\partial\varphi} + i\frac{\partial\Psi}{\partial r}. \tag{11.21b}$$

Taking the modulus of (11.21a) and (11.21b) yields:

$$\begin{aligned}|f'(z)|^2 &= \left(\frac{\partial\Phi}{\partial x}\right)^2 + \left(\frac{\partial\Psi}{\partial x}\right)^2 = \left(\frac{\partial\Phi}{\partial x}\right)^2 + \left(\frac{\partial\Phi}{\partial y}\right)^2 \\ &= \left(\frac{\partial\Psi}{\partial y}\right)^2 + \left(\frac{\partial\Phi}{\partial y}\right)^2 = \left(\frac{\partial\Psi}{\partial y}\right)^2 + \left(\frac{\partial\Psi}{\partial x}\right)^2;\end{aligned} \tag{11.22a}$$

$$|f'(z)|^2 = \left(\frac{\partial\Phi}{\partial r}\right)^2 + \left(\frac{\partial\Psi}{\partial r}\right)^2 = \left(\frac{\partial\Phi}{\partial r}\right)^2 + \frac{1}{r^2}\left(\frac{\partial\Phi}{\partial\varphi}\right)^2$$
$$= \frac{1}{r^2}\left\{\left(\frac{\partial\Phi}{\partial\varphi}\right)^2 + \left(\frac{\partial\Psi}{\partial\varphi}\right)^2\right\} = \left(\frac{\partial\Psi}{\partial r}\right)^2 + \frac{1}{r^2}\left(\frac{\partial\Psi}{\partial\varphi}\right)^2. \tag{11.22b}$$

Concerning the argument of the derivative, the following simplifications are possible:

$$\arg\left(\frac{\partial\Phi}{\partial x} - i\frac{\partial\Phi}{\partial y}\right) = \operatorname{arc\,cot}\left(-\frac{\partial\Phi/\partial x}{\partial\Phi/\partial y}\right) = \operatorname{arc\,cot}\left(\frac{\mathrm{d}y}{\mathrm{d}x}\right)_\Phi, \tag{11.23}$$

where, rule of implicit derivation (11.38a) was used; in a similar way can be obtained:

$$\arg\{f'(z)\} = \operatorname{arc\,cot}\left\{\left(\frac{\mathrm{d}y}{\mathrm{d}x}\right)_\Phi\right\} = -\operatorname{arc\,tan}\left\{\left(\frac{\mathrm{d}y}{\mathrm{d}x}\right)_\Psi\right\}; \tag{11.24a}$$

$$\arg\{f'(z)\} = \operatorname{arc\,tan}\left\{\frac{1}{r}\left(\frac{\mathrm{d}r}{\mathrm{d}\varphi}\right)_\Phi\right\} = -\operatorname{arc\,cot}\left\{\frac{1}{r}\left(\frac{\mathrm{d}r}{\mathrm{d}\varphi}\right)_\Psi\right\}. \tag{11.24b}$$

The preceding expressions specify *the derivate of a holomorphic function (11.21a) [(11.21b)], its the modulus (11.22a) [11.22b] and argument (11.24a) [(11.24b)] in Cartesian (11.7) [polar (11.18)] coordinates.*

11.6 Cartesian and Polar Laplace (1825) Equation

If the functions Φ, Ψ have continuous second-order derivates, these are symmetric:

$$\Phi, \Psi \in C^2\left(|R^2\right): \quad \frac{\partial}{\partial y}\left(\frac{\partial\Phi}{\partial x}\right) - \frac{\partial}{\partial x}\left(\frac{\partial\Phi}{\partial y}\right) = 0 = \frac{\partial}{\partial x}\left(\frac{\partial\Psi}{\partial y}\right) - \frac{\partial}{\partial y}\left(\frac{\partial\Psi}{\partial x}\right); \tag{11.25a,b}$$

substituting the Cauchy–Riemann conditions (11.10a,b), it follows that both satisfy the two-dimensional Laplace equation:

$$\frac{\partial^2\Psi}{\partial x^2} + \frac{\partial^2\Psi}{\partial y^2} = 0 = \frac{\partial^2\Phi}{\partial x^2} + \frac{\partial^2\Phi}{\partial y^2}, \tag{11.26}$$

in Cartesian coordinates. To obtain Laplace's equation in polar coordinates, Φ, Ψ are assumed to have continuous second-order derivates, and elimination using the Cauchy–Riemann conditions in polar coordinates (11.20a,b) leads to

$$0 = \frac{\partial}{\partial\varphi}\left(\frac{\partial\Phi}{\partial r}\right) - \frac{\partial}{\partial r}\left(\frac{\partial\Phi}{\partial\varphi}\right) = \frac{1}{r}\frac{\partial^2\Psi}{\partial\varphi^2} + \frac{\partial}{\partial r}\left(r\frac{\partial\Psi}{\partial r}\right), \tag{11.27a}$$

$$0 = \frac{\partial}{\partial r}\left(\frac{\partial\Psi}{\partial\varphi}\right) - \frac{\partial}{\partial\varphi}\left(\frac{\partial\Psi}{\partial r}\right) = \frac{\partial}{\partial r}\left(r\frac{\partial\Phi}{\partial r}\right) + \frac{1}{r}\frac{\partial^2\Psi}{\partial\varphi^2}. \tag{11.27b}$$

Since the Laplacian operator ∇^2 must have the dimensions $(\text{length})^{-2}$ in (11.27a,b) these are divided by r:

$$\nabla^2 A \equiv \frac{\partial^2 A}{\partial x^2} + \frac{\partial^2 A}{\partial y^2} = \frac{1}{r}\frac{\partial}{\partial r}\left(r\frac{\partial A}{\partial r}\right) + \frac{1}{r^2}\frac{\partial^2 A}{\partial\varphi^2} = \frac{\partial^2 A}{\partial r^2} + \frac{1}{r}\frac{\partial A}{\partial r} + \frac{1}{r^2}\frac{\partial^2 A}{\partial\varphi^2}. \tag{11.28a–c}$$

It has been shown that, *if the function* f(z) *is holomorphic and its real and imaginary parts have continuous second-order derivates, then they satisfy Laplace' s equation:*

$$f(z)\in \mathcal{D}(|C)\wedge\Phi,\Psi(x,y)\equiv \mathrm{Re},\mathrm{Im}\{f(z)\}\in C^2\left(|R^2\right):\nabla^2\Phi=0=\nabla^2\Psi, \qquad (11.29a)$$

where, the two-dimensional **Laplace operator** *is specified by (11.28a) [(11.28b,c)] in Cartesian* (x,y) *[polar* (r,φ)*] coordinates.* A solution of Laplace's equation is called a **harmonic** function $\mathcal{H}$. Thus:

$$f(z)\in \mathcal{D}(|C)\Rightarrow f(z)\in A\subset D^\infty\wedge\Phi,\Psi\equiv \mathrm{Re},\mathrm{Im}(f)\in \mathcal{H}, \qquad (11.29b)$$

a complex holomorphic function has real and imaginary parts that are harmonic functions.

11.7 Gradient, Divergence, Curl, and Laplacian

The modulus of the derivate of a holomorphic function (11.22a,b) is the square of the gradient of its real and imaginary parts:

$$\Phi+i\Psi=f(z)\in \mathcal{D}(|C):|\nabla\Phi|^2=|f'(z)|^2=|\nabla\Psi|^2, \qquad (11.30)$$

where, the operator **gradient** is given by

$$\nabla A=\vec{e}_x\frac{\partial A}{\partial x}+\vec{e}_y\frac{\partial A}{\partial y}=\vec{e}_r\frac{\partial A}{\partial r}+\frac{\vec{e}_\varphi}{r}\frac{\partial A}{\partial\varphi}, \qquad (11.31a,b)$$

in Cartesian (11.31a) [polar (11.31b)] coordinates, with unit vectors $\vec{e}_x,\vec{e}_y\left(\vec{e}_r,\vec{e}_\varphi\right)$ along the coordinate axis x,$y\,(r,\varphi)$. The Laplacian of a scalar is the divergence of its gradient:

$$\nabla^2A=\nabla\cdot(\nabla A)=\nabla\cdot\vec{B},\quad \vec{B}\equiv\nabla A; \qquad (11.32a,b)$$

from (11.28a) and (11.31a) [(11.28b) and (11.31b)] follows:

$$\nabla\cdot\vec{B}=\frac{\partial B_x}{\partial x}+\frac{\partial B_y}{\partial y}=\frac{1}{r}\frac{\partial(rB_r)}{\partial r}+\frac{1}{r}\frac{\partial B_\varphi}{\partial\varphi}, \qquad (11.33a,b)$$

that is the expression of the **divergence** in Cartesian (11.33a) [polar (11.33b)] coordinates.
The curl of a gradient is zero:

$$\frac{\partial(\partial A/\partial y)}{\partial x}-\frac{\partial(\partial A/\partial x)}{\partial y}=0=r\nabla\wedge\nabla A=\frac{\partial\left[r\left(r^{-1}\partial A/\partial\varphi\right)\right]}{\partial r}-\frac{\partial(\partial A/\partial r)}{\partial\varphi}; \qquad (11.34a–c)$$

substituting (11.31a) [(11.31b)] and noting that the curl must have dimensions $(\text{length})^{-1}$ in (11.34c), it is not divided (is divided by r), leads to:

$$\nabla\wedge\vec{B}=\vec{e}_3\left(\frac{\partial B_y}{\partial x}-\frac{\partial B_x}{\partial y}\right)=\frac{\vec{e}_3}{r}\left[\frac{\partial(rB_\varphi)}{\partial r}-\frac{\partial B_r}{\partial\varphi}\right], \qquad (11.35a,b)$$

the operator **curl** in Cartesian (11.35a) [polar (11.35b)] coordinates; the curl is a pseudoscalar on the plane, and a vector with the direction orthogonal to the (x,y)-plane or z-plane, hence the unit vector $\vec{e}_3$. Thus have been obtained *the four invariant differential operators, namely the gradient (11.31a) [(11.31b)] and Laplacian (11.28a) [(11.28b,c)] of scalar* A, *and the divergence (11.33a) [(11.33b)] and curl (11.35a) [(11.35b)] of a vector $\vec{B}$, in Cartesian* x,y *(polar* r,φ*) coordinates, and components* $B_x, B_y\,(B_r, B_\varphi)$ *for the vector $\vec{B}$.*

11.8 Families of Plane Orthogonal Curves

The Cauchy–Riemann conditions, specified by (11.10a,b) in Cartesian and (11.20a,b) in polar coordinates, for the pair of functions Φ,Ψ have the following interpretations: (physical) if Φ,Ψ have continuous second-order derivates they are solutions of the Laplace equation (11.29a,b); (analytical) if Φ,Ψ have continuous first-order derivates, the function $\mathrm{f}=\Phi+\mathrm{i}\Psi$ *is holomorphic (11.11); (geometrical) the families of curves:*

$$\Phi(x,y)=\alpha\equiv\text{const},\quad \Psi(x,y)=\beta\equiv\text{const}, \tag{11.36a,b}$$

are mutually orthogonal (Figure 11.4), that is, cut each other at right-angles. There is an implication that *two solutions of Laplace's equation Φ,Ψ specify families of curves (11.36a,b) that are parallel if $\Psi=g(\Phi)$, and otherwise are orthogonal.*

To prove the geometrical property, note that a displacements $\mathrm{d}\tilde{\mathrm{x}}\equiv(\mathrm{d}x,\mathrm{d}y)$ along the curves (11.36a,b) satisfy:

$$0=\mathrm{d}\Phi=\frac{\partial\Phi}{\partial x}\mathrm{d}x+\frac{\partial\Phi}{\partial y}\mathrm{d}y=\nabla\Phi\cdot\mathrm{d}\tilde{\mathrm{x}}=\tilde{\mathrm{n}}_\Phi\cdot\mathrm{d}\tilde{\mathrm{x}}, \tag{11.37a}$$

$$0=\mathrm{d}\Psi=\frac{\partial\Psi}{\partial x}\mathrm{d}x+\frac{\partial\Psi}{\partial y}\mathrm{d}y=\nabla\psi\cdot\mathrm{d}\tilde{\mathrm{x}}=\tilde{\mathrm{n}}_\Psi\cdot\mathrm{d}\tilde{\mathrm{x}}; \tag{11.37b}$$

this implies that the curves of the each family have slopes:

$$\left(\frac{\mathrm{d}y}{\mathrm{d}x}\right)_\Phi=-\frac{\partial\Phi/\partial x}{\partial\Phi/\partial y},\quad \left(\frac{\mathrm{d}y}{\mathrm{d}x}\right)_\Psi=-\frac{\partial\Psi/\partial x}{\partial\Psi/\partial y}; \tag{11.38a,b}$$

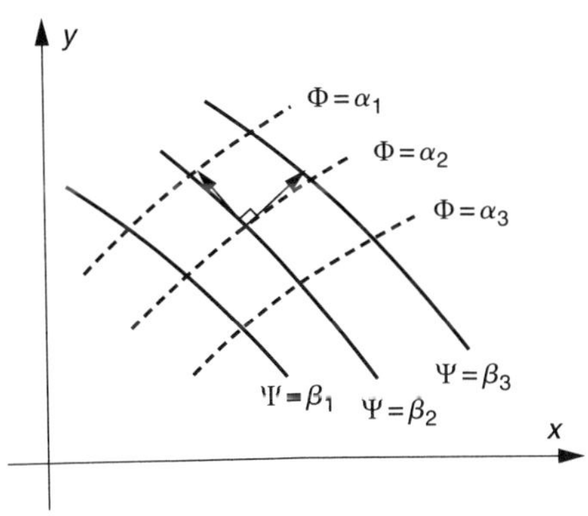

FIGURE 11.4
A complex function $f(z)$ is a pair of real functions $f=\Phi+i\Psi$ of two real variables $z=x+iy$. Thus it specifies two families of curves $\Phi(x,y)=C_1$ and $\Psi(x,y)=C_2$. If the complex function is holomorphic its real (imaginary) parts specify [solid (dotted) lines] orthogonal families of curves, that is: (i) each curve of one family cuts all the curves of the other family at right angles; (ii/iii) the tangent (normal) vectors to one family are orthogonal to the tangent (normal) vectors of the other family; (iv) the tangents to one family lie along the normals to the other family and vice-versa. The real and imaginary parts of a complex function can be used to define conformal coordinates (Subsection 33.3.1), which are plane orthogonal coordinates, all curvilinear except for the Cartesian. Since the function is holomorphic the derivative is the same in every direction, and the scale factors are equal along both axes of conformal coordinates.

the Cauchy–Riemann conditions (11.10a,b) show that the product of slopes is minus unity:

$$\left(\frac{\mathrm{d}y}{\mathrm{d}x}\right)_\Phi \left(\frac{\mathrm{d}y}{\mathrm{d}x}\right)_\Psi = \frac{(\partial\Phi/\partial x)(\partial\Psi/\partial x)}{(\partial\Phi/\partial y)(\partial\Psi/\partial y)} = -1, \tag{11.39}$$

and so the curves (20.36a,b) are orthogonal.

The result can also be proved by considering the normals to the curves:

$$\vec{\mathrm{n}}_\Phi = \nabla\Phi = \left\{\frac{\partial\Phi}{\partial x}, \frac{\partial\Phi}{\partial y}\right\}, \quad \vec{\mathrm{n}}_\Psi = \nabla\Psi = \left(\frac{\partial\Psi}{\partial x}, \frac{\partial\Psi}{\partial y}\right), \tag{11.40a,b}$$

that are orthogonal to the displacements along the curves in (11.37a,b); their inner product:

$$\vec{\mathrm{n}}_\Phi \cdot \vec{\mathrm{n}}_\Psi = \frac{\partial\Phi}{\partial x}\frac{\partial\Psi}{\partial x} + \frac{\partial\Phi}{\partial y}\frac{\partial\Psi}{\partial y} = 0, \tag{11.41}$$

is zero on account of the Cauchy–Riemann conditions (11.10a,b); thus the normals are orthogonal and so are the curves. Thus has been proved in two distinct ways the property that *a pair of functions Φ,Ψ satisfying the Cauchy–Riemann conditions (11.10a,b), specifies two families of curves (11.36a,b) that are mutually orthogonal, because the product of slopes is −1 in (11.39), or equivalently the inner product of normals (11.40a,b) is zero in (11.41).* Calculating (11.41) also in polar coordinates:

$$0 = \nabla\Phi \cdot \nabla\Psi = \frac{\partial\Phi}{\partial r}\frac{\partial\Psi}{\partial r} + \frac{1}{r^{-2}}\frac{\partial\Phi}{\partial\varphi}\frac{\partial\Phi}{\partial\varphi}, \tag{11.42}$$

it follows that *two differentiable curves $\Phi = \text{const}$ and $\Psi = \text{const}$ are orthogonal if the functions Φ,Ψ satisfy (11.41) in Cartesian [(11.42) in polar] coordinates.*

11.9 Orthogonal Plane Curvilinear Coordinates

The orthogonal curves lead to orthogonal curvilinear coordinates (Subsection 11.9.1) that may be used (Subsection 11.9.2) in the invariant differential operators: gradient, curl, divergence, and Laplacian.

11.9.1 Scale Factors for Orthogonal Curvilinear Coordinates

Two families of orthogonal curves may be used to define orthogonal curvilinear coordinates (x_1,x_2); the corresponding **arc length** is:

$$|\mathrm{d}z|^2 = (h_1)^2(\mathrm{d}x_1)^2 + (h_2)^2(\mathrm{d}x_2)^2 = |h_1\mathrm{d}x_1 + ih_2\mathrm{d}x_2|^2, \tag{11.43}$$

where, $h_1\,(h_2)$ is the **scale factor** in the $x_2-(x_2-)$ direction, for example, in Cartesian (11.17a) [polar (11.17b)] coordinates the scalar factors are (11.44a) [(11.44b)]:

$$h_x = 1 = h_y, \quad \{h_r, h_\varphi\} = \{1, r\}. \tag{11.44a,b}$$

The first three scale factors in (11.44a,b) are unity because the corresponding coordinates (x,y,r) are distances; the fourth coordinate φ is an angle, and the scale factor r is such that $r\mathrm{d}\varphi$ is the length of the arc of the circle of radius r. If the coordinates (x_1,x_2) were not orthogonal a cross-term $\mathrm{d}x_1\mathrm{d}x_2$ would appear in the arc length (11.43).

Calculating the derivative of a holomorphic function at constant $x_2(x_1)$ in (11.45a) [(11.45b)]:

$$x_2 = \text{const}: \quad \frac{\mathrm{d}f}{\mathrm{d}z} = \frac{\partial(\Phi + i\Psi)}{h_1 \partial x_1} = \frac{1}{h_1}\left(\frac{\partial \Phi}{\partial x_1} + i\frac{\partial \Psi}{\partial x_1}\right), \tag{11.45a}$$

$$x_1 = \text{const}: \quad \frac{\mathrm{d}f}{\mathrm{d}z} = \frac{\partial(\Phi + i\Psi)}{i h_2 \partial x_2} = \frac{1}{h_2}\left(\frac{\partial \Psi}{\partial x_2} - i\frac{\partial \Phi}{\partial x_2}\right), \tag{11.45b}$$

and equating real and imaginary parts, leads to: (i) *the Cauchy–Riemann conditions in orthogonal curvilinear coordinates:*

$$\Phi + i\Psi \in \mathcal{D}: \qquad h_2 \frac{\partial \Phi}{\partial x_1} = h_1 \frac{\partial \Psi}{\partial x_2}, \quad h_1 \frac{\partial \Phi}{\partial x_2} = -h_2 \frac{\partial \Psi}{\partial x_1}; \tag{11.46a,b}$$

(ii) four expressions for the derivative of a holomorphic function, viz. (11.45a,b) and (11.47a,b):

$$f = \Phi + i\Psi: \qquad \frac{\mathrm{d}f}{\mathrm{d}z} = \frac{1}{h_1}\frac{\partial \Phi}{\partial x_1} - \frac{i}{h_2}\frac{\partial \Phi}{\partial x_2} = \frac{1}{h_2}\frac{\partial \Psi}{\partial x_2} + \frac{i}{h_1}\frac{\partial \Psi}{\partial x_1}; \tag{11.47a,b}$$

(iii) also the modulus (11.48a,b) [argument (11.49a,b)] of the derivative:

$$\left|\frac{\mathrm{d}f}{\mathrm{d}z}\right| = \left|h_1^{-2}\left(\frac{\partial \Phi}{\partial x_1}\right)^2 + h_2^{-2}\left(\frac{\partial \Phi}{\partial x_2}\right)^2\right|^{-1/2} = \left|h_1^{-2}\left(\frac{\partial \Psi}{\partial x_1}\right)^2 + h_2^{-2}\left(\frac{\partial \Psi}{\partial x_2}\right)^2\right|^{-1/2}, \tag{11.48a,b}$$

$$\arg\left(\frac{\mathrm{d}f}{\mathrm{d}z}\right) = \operatorname{arc\,tan}\left\{\frac{h_1}{h_2}\left(\frac{\mathrm{d}x_1}{\mathrm{d}x_2}\right)_\Phi\right\} = -\operatorname{arc\,cot}\left\{\frac{h_1}{h_2}\left(\frac{\mathrm{d}x_1}{\mathrm{d}x_2}\right)_\Psi\right\}. \tag{11.49a,b}$$

Using the Cartesian (11.44a) [polar (11.44b)] scale factors the expressions (11.46a,b; 11.45a,b, 11.47a,b; 11.48a,b; 11.49a,b) lead back to (11.10a,b; 11.21a; 11.22a; 11.24a) [11.20a,b; 11.21b; 11.22b; 11.24b)].

11.9.2 Invariant Operators in Orthogonal Curvilinear Coordinates

The expressions (11.48a,b) are the modulus of the gradient in orthogonal curvilinear coordinates:

$$\nabla A = \frac{\vec{\mathrm{e}}_1}{h_1}\frac{\partial A}{\partial x_1} + \frac{\vec{\mathrm{e}}_2}{h_2}\frac{\partial A}{\partial x_2}. \tag{11.50}$$

Using the Cauchy–Riemann conditions in orthogonal curvilinear coordinates (11.46a,b) in the identity (11.51a):

$$0 = \frac{\partial}{\partial x_1}\left(\frac{\partial \Psi}{\partial x_2}\right) - \frac{\partial}{\partial x_2}\left(\frac{\partial \Psi}{\partial x_1}\right) = \frac{\partial}{\partial x_1}\left(\frac{h_2}{h_1}\frac{\partial \Phi}{\partial x_1}\right) + \frac{\partial}{\partial x_2}\left(\frac{h_1}{h_2}\frac{\partial \Phi}{\partial x_2}\right), \tag{11.51a,b}$$

it follows that (11.51b) must be the Laplace operator apart from a factor with the dimensions $(\text{length})^{-2}$, involving the scale factors, for example, $h_1^{-2}, (h_1 h_2)^{-1}$. Since the Laplacian must be invariant exchanging (x_1, h_1) by (x_2, h_2), the factor $(h_1 h_2)^{-1}$ is chosen in (11.51b), leading to:

$$\nabla^2 A = \frac{1}{h_1 h_2}\left[\frac{\partial}{\partial x_1}\left(\frac{h_2}{h_1}\frac{\partial A}{\partial x_1}\right) + \frac{\partial}{\partial x_2}\left(\frac{h_1}{h_2}\frac{\partial A}{\partial x_2}\right)\right], \tag{11.52}$$

as the Laplacian in orthogonal curvilinear coordinates. Recalling (11.32a,b), and the gradient operator (11.50), follows:

$$\nabla \cdot \vec{\mathrm{B}} = \frac{1}{h_1 h_2}\left[\frac{\partial}{\partial x_1}(h_2 B_1) + \frac{\partial}{\partial x_2}(h_1 B_2)\right], \tag{11.53}$$

that is the divergence in orthogonal curvilinear coordinates.

The identity (11.54b):

$$\vec{\mathrm{B}} = \nabla A: \quad 0 = \frac{\partial}{\partial x_1}\left(\frac{\partial A}{\partial x_2}\right) - \frac{\partial}{\partial x_1}\left(\frac{\partial A}{\partial x_1}\right) = \frac{\partial}{\partial x_1}(h_2 B_2) - \frac{\partial}{\partial x_2}(h_1 B_1), \tag{11.54a,b}$$

applies to a vector $\vec{\mathrm{B}}$ which is the gradient (11.54a) of a scalar; since the curl of gradient is zero (11.34b), the operator (11.54b) must coincide with the curl, apart from a factor with the dimensions $(\text{length})^{-2}$; as in the passage from (11.51b) to (11.52) this factor is $(h_1 h_2)^{-1}$, viz:

$$\nabla \wedge \vec{\mathrm{B}} = \frac{\vec{\mathrm{e}}_3}{h_1 h_2}\left[\frac{\partial}{\partial x_1}(h_2 B_2) - \frac{\partial}{\partial x_2}(h_1 B_1)\right]. \tag{11.55}$$

Thus have been *obtained the gradient (11.50), and Laplacian (11.52) of a scalar* A, *and the divergence (11.53) and curl (11.55) of a vector* $\vec{B}$, *in plane orthogonal curvilinear coordinates* $(\mathrm{x}_1, \mathrm{x}_2)$, *with scale factors* $(\mathrm{h}_1, \mathrm{h}_2)$ *in the arc length (11.43); also the condition of orthogonality of the curves (11.36a,b), viz.*:

$$0 = \nabla\Phi \cdot \nabla\Psi = \frac{1}{h_1^2}\frac{\partial\Phi}{\partial x_1}\frac{\partial\Psi}{\partial x_1} + \frac{1}{h_2^2}\frac{\partial\Phi}{\partial x_2}\frac{\partial\Psi}{\partial x_2}. \tag{11.56}$$

Using the Cartesian (11.44a) [polar (11.44b)] scale factors, leads from the gradient (11.50), divergence (11.53), curl (11.55), Laplacian (11.52), and orthogonality condition (11.56) in plane orthogonal curvilinear coordinates to the corresponding Cartesian (11.31a, 11.33a, 11.35a, 11.28a, 11.41) [polar (11.31b, 11.33b, 11.35b, 11.28b, 11.42)] forms.

Note 11.1. *The Four Invariant Differential Operators:* The four invariant differential operators (gradient and Laplacian of a scalar, and divergence and curl of a vector), plus the Laplacian of a vector, and their generalizations for tensors, can be introduced for oblique curvilinear coordinates in N dimensions. The particular case of plane orthogonal curvilinear coordinates is sufficient to address plane: (i) potential flow (Chapters 12, 14, and 16); (ii) gravity field (Chapter 18); (iii) electrostatics (Chapter 24); (iv) magnetostatics (Chapter 26); (v) steady heat conduction (Chapter 32); (vi) elasticity. The orthogonal property of the real and imaginary parts of a complex holomorthic function can be used to introduce plane orthogonal curvilinear coordinates, for example confocal, coaxal, parabolic, spiral, and dipolar; the preceediing are all conformal coordinates (Subsection 33.3.1). The orthogonal coordinates can be extended to space and to any number of dimensions.

Conclusion 11: For the limit of the function $f(z)$ at z_0 to be w_0, the neighborhoods $V_\varepsilon(z_0)$ of z_0 must transform (figure 11.1) into subsets G of neighborhoods $V_\delta(w_0)$ of w_0; if $w_0 = f(z_0)$ the function is continuous at z, and it has a derivative at z, (figure 11.3) at z iff the ratio $\Delta f/\Delta z$ of increments $\Delta f \equiv f(z+\Delta z) - f(z)$ and $\Delta z = \Delta x + i\Delta y$ is independent of the way in which $\Delta x, \Delta y \to 0$. The increments or differentials dz can be calculated (figure 11.2) either in (a) Cartesian x, y [or (b) polar r, φ] coordinates. The real Φ and imaginary Ψ parts of a holomorphic function $f = \Phi + i\Psi$, or any pair of functions satisfying the Cauchy–Riemann conditions, specify two families of curves $\Phi = \alpha \equiv \text{const.}$ and $\Psi = \beta \equiv \text{const.}$, which are mutually orthogonal (figure 11.4).

12

Potential Flow and Multipoles

A flow is two-dimensional if the velocity lies on a plane, and does not depend on the coordinate normal to the plane. The volume flux across (circulation around) a loop is zero for an incompressible (irrotational) flow that has a stream function (Section 12.1) [potential (Section 12.2)]. A potential flow (Section 12.3) is both irrotational and incompressible, and thus has both potential (and stream function) that is the real (imaginary) part of the complex potential; the latter is a holomorphic function, whose derivative specifies the components of velocity, for example, for a flow source or sink (Section 12.4), a vortex (Section 12.5), or their superposition in a monopole (Section 12.6). The limit of opposing monopoles whose strength increases inversely with mutual distance leads in the coincidence limit to a dipole (Section 12.7); a similar limiting process may be repeated, to lead to a quadrupole (Section 12.8) and hydrodynamic multipoles (Section 12.9) of any order. The potential flow is one of the simplest, in that it is devoid of the effects of compressibility, vorticity, viscosity, turbulence, heat transfer, etc. As such it is the baseline flow against that all these effects can be compared at a subsequent stage.

12.1 Circulation, Potential, Curl, and Vorticity

A flow involves a **velocity field** specified by the velocity vector $\vec{v}$ as a function of position $\vec{x}$ and time t. In a **plane flow** the velocity components lie in a plane (x,y) and do not depend on the coordinate orthogonal to the plane:

$$\vec{v} = \vec{e}_x v_x(x,y,t) + \vec{e}_y v_y(x,y,t), \tag{12.1}$$

where, $\vec{e}_x(\vec{e}_y)$ is the unit vector along the x-axis (y-axis). A two-dimensional flow excludes that (i) there is a velocity component v_z orthogonal to the plane (x,y), viz. $v_z = 0$; (ii) that the velocity components(v_x, v_y) in the plane depend on the orthogonal coordinate z, viz. $\partial v_x/\partial z = 0 = \partial v_y/\partial z$. An example is the flow orthogonal (oblique) to the axis of a static cylinder that is (is not) two-dimensional. The flow around a static sphere is not a plane flow, though it may be **axisymmetric**, that is, independent of the azimuthal coordinate, and thus **two-dimensional**. The flow around a static ellipsoid with three unequal half-axis is always **three-dimensional**. A **one-dimensional** flow would apply to a tube if the velocity varies only longitudinally, that is, it is uniform transversely. A flow is **steady (unsteady)** if the velocity does not (12.2) [does (12.1)] depend on time.

$$\frac{\partial v_x}{\partial t} = 0 = \frac{\partial v_y}{\partial t}: \quad \vec{v} = \vec{e}_x v_x(x,y) + \vec{e}_y v_y(x,y). \tag{12.2}$$

For example if a cylinder is kept static (moves) in a uniform stream, the flow is steady (unsteady), because at any given point the velocity does not (does) vary with time.

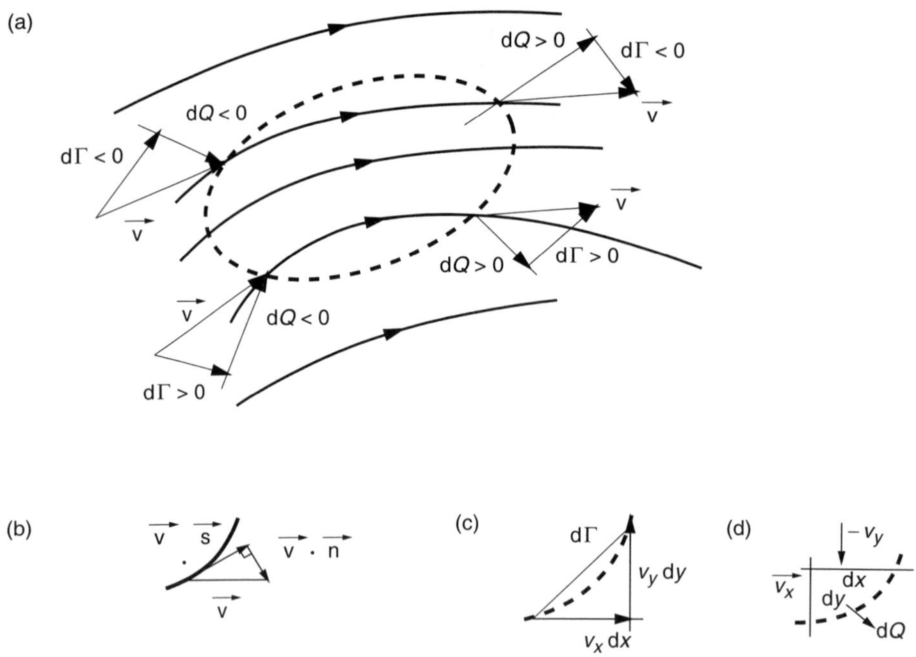

FIGURE 12.1
If a closed curve (dotted line) is drawn in a plane flow it crosses the streamlines (a) to which the velocity is tangent. The flow velocity has (b) components tangent (normal) to the curve, whose integral along the loop specifies the circulation (c) [flow rate (d)]. In an (i) incompressible flow the volume occupied by each fluid element is constant, so the flow rate in balances the flow rate out, and the total is zero along the closed loop. If the flow is (ii) irrotational, that is, there are no vortices (b) inside the loop, then the circulation also balances along the loop. Both (i) and (ii) hold for a potential flow which is incompressible and irrotational.

Consider a (closed) curve or loop in a two-dimensional flow (Figure 12.1a), and let ($\mathrm{d}x$,$\mathrm{d}y$) be the components of the infinitesimal displacement along the tangent. The **circulation** (Figure 12.1c) along the loop is the integral of the tangential component of velocity:

$$\Gamma \equiv \oint_C \vec{v}\cdot \mathrm{d}\vec{x} = \oint_C (v_x\,\mathrm{d}x + v_y\,\mathrm{d}y). \tag{12.3}$$

The flow is **irrotational** if the circulation is zero on every static loop, that is, the arcs with positive tangential velocity are compensated by arcs with negative tangential velocity in Figure 12.1b. In an irrotational flow the integrand (12.3) must be an exact differential:

$$\Gamma = 0: \quad \vec{v}\cdot \mathrm{d}\vec{x} \equiv v_x\,\mathrm{d}x + v_y\,\mathrm{d}y = \mathrm{d}\Phi, \tag{12.4a,b}$$

of a function Φ called the **potential**. *In an irrotational flow the circulation along an open curve is specified by the difference of potential at the end points:*

$$\Gamma_{12} \equiv \int_{P_1}^{P_2} (v_x\,\mathrm{d}x + v_y\,\mathrm{d}y) = \Phi(P_2) - \Phi(P_1), \tag{12.5}$$

and is independent of the path; it vanishes for a closed loop $P_2 \equiv P_1$, because the potential is a single-valued function.

Comparing (12.4b) with:

$$d\Phi = \frac{\partial \Phi}{\partial x} dx + \frac{\partial \Phi}{\partial y} dy, \tag{12.6}$$

it follows that *in an irrotational flow the velocity is the gradient of the potential (12.7a):*

$$\{v_x, v_y\} = \left\{ \frac{\partial \Phi}{\partial x}, \frac{\partial \Phi}{\partial y} \right\} \equiv \nabla \Phi, \quad \vec{v} \cdot d\vec{x}_\Phi = \nabla \Phi \cdot d\tilde{x}_\Phi = d\Phi = 0, \tag{12.7a,b}$$

and is orthogonal to the **equipotentials**, *that is, curves of constant potential* (12.7b); the latter (12.7b) follows from $d\Phi = 0$ in (12.6) for any displacement $d\vec{x}_\Phi$ along an equipotential, so that $\vec{v} \cdot d\vec{x} = 0$ implies that the velocity is orthogonal to the equipotential. If the potential has continuous second-order derivatives, the identity (12.8a):

$$0 = \frac{\partial^2 \Phi}{\partial y \, \partial x} - \frac{\partial^2 \Phi}{\partial x \, \partial y} = \frac{\partial v_y}{\partial x} - \frac{\partial v_x}{\partial y}, \tag{12.8a}$$

implies that:

$$0 = \nabla \wedge \vec{v} = \vec{e}_3 \left(\frac{\partial v_y}{\partial x} - \frac{\partial v_x}{\partial y} \right) \equiv \vec{\varpi}, \tag{12.8b}$$

in an irrotational flow the **vorticity**, *that is, curl of the velocity is zero.* It can be shown that the preceding four conditions are equivalent, viz: (i) the circulation is zero along a static closed loop (12.3; 12.4a); (ii) the circulation along an open curve depends only on the end-points (12.5); (iii) the velocity is the gradient of a potential (12.7a), and hence orthogonal to the stream lines (12.7b); (iv) the vorticity or curl of the velocity is zero (12.8b).

12.2 Flow Rate, Stream Function, Divergence, and Dilatation (Lagrange, 1781; Rankine, 1864)

The integral along a loop of the normal (tangential) velocity specifies (Figure 12 1a) **volume flux** (Figure 12.1d) [circulation (Figure 12.1c)] in (12.9) [(12.3)], viz.:

$$Q \equiv \oint_C \left| \vec{v} \wedge d\vec{x} \right| = \oint_C (v_x \, dy - v_y \, dx). \tag{12.9}$$

A flow is **incompressible** if the volume flux across every fixed loop is zero, because the fluid that enters through some arcs must flow out through the remaining arcs, so the volume (to be more precise, the area) of fluid inside is constant (Figure 12.1d). If a flow is incompressible, the integrand in (12.9) must be an exact differential:

$$Q = 0: \quad \left| \vec{v} \wedge d\vec{x} \right| = v_x \, dy - v_y \, dx = d\Psi, \tag{12.10a,b}$$

of a function Ψ called (Lagrange, 1781; Rankine, 1864) the **stream function**, *because the velocity is tangent to the streamlines, or lines along which the stream function is constant:*

$$\frac{v_y}{v_x} = \left(\frac{dy}{dx} \right)_\Psi = -\frac{\partial \Psi / \partial x}{\partial \Psi / \partial y}; \tag{12.11a,b}$$

the rule of implicit differentiation was used in (12.11b), as follows from:

$$d\Psi = \frac{\partial \Psi}{\partial x} dx + \frac{\partial \Psi}{\partial y} dy, \tag{12.12}$$

for $\Psi = \text{const}$ or $d\Psi = 0$:

$$\Psi = \text{const}: \quad \frac{\partial \Psi}{\partial x} dx_\psi + \frac{\partial \Psi}{\partial y} dy_\Psi = 0. \tag{12.13}$$

Comparing (12.10a,b) with (12.12) it follows that:

$$\{v_x, v_y\} = \left\{ \frac{\partial \Psi}{\partial y}, -\frac{\partial \Psi}{\partial x} \right\}, \tag{12.14}$$

in an incompressible plane flow the stream function specifies the velocity through the two-dimensional curl (12.14).

From (12.10b) also follows that

$$Q_{12} \equiv \int_{P_1}^{P_2} (v_x \, dy - v_y \, dx) = \Psi(P_2) - \Psi(P_1), \tag{12.15}$$

in an incompressible flow the volume flux along a curve is specified by the difference of the stream function at the end-points, and is independent of the curve, because, if another curve is chosen, the difference is the zero volume flux across a loop. If the stream function has continuous second-order derivatives, the identity:

$$0 = \frac{\partial^2 \Psi}{\partial y \, \partial x} - \frac{\partial^2 \Psi}{\partial x \, \partial y} = \frac{\partial v_x}{\partial x} + \frac{\partial v_y}{\partial y} = \nabla \cdot \vec{v} \equiv \Lambda, \tag{12.16}$$

shows that *in an incompressible flow, the* ***dilatation****, that is, divergence of the velocity is zero.* It can be shown that the preceding four conditions are equivalent: (i) the volume flux across a closed loop is zero (12.9; 12.10a); (ii) the volume flux along an open curve depends only on the end-points (12.15); (iii) the velocity is derived from a stream function (12.14), and is tangent (12.11a,b) to the streamlines; (iv) the dilatation or divergence of the velocity is zero (12.16).

12.3 Complex Potential and Conjugate Velocity

A **potential flow** is both irrotational (12.8b) [and incompressible (12.16)], and thus a potential (12.7a) [and a stream function (12.14)] exists:

$$v_x = \frac{\partial \Phi}{\partial x} = \frac{\partial \Psi}{\partial y}, \qquad v_y = \frac{\partial \Phi}{\partial y} = -\frac{\partial \Psi}{\partial x}. \tag{12.17a,b}$$

The Cauchy–Riemann (Section 11.3) conditions (11.10a,b) ≡ (12.17a,b) are thus met, and they are necessary conditions for the **complex potential**:

$$\Phi(x,y) + i\Psi(x,y) = f(x+iy) \in \mathcal{D}, \tag{12.18}$$

whose real (imaginary) part is the potential (stream function) to be a holomorphic function, that is, differentiable. The sufficient condition (11.11) for the complex potential to be a holomorphic function, is that it has continuous first-order derivatives, that is, the velocity (12.17a,b) is continuous; thus impulsive motions are excluded. Since the complex potential is a holomorphic function it has a complex derivative:

$$v^*(z) \equiv \frac{df}{dz} = \frac{\partial \Phi}{\partial x} + i\frac{\partial \Psi}{\partial x} = v_x - iv_y, \tag{12.19}$$

that specifies the **complex conjugate velocity** (12.19) in Cartesian components.

Since the velocity is the gradient (12.7a) of the scalar potential Φ, its polar components (12.20a,b) are given (11.31b) by:

$$v_r = \frac{\partial \Phi}{\partial r} = \frac{1}{r}\frac{\partial \Psi}{\partial \varphi}, \qquad v_\varphi = \frac{1}{r}\frac{\partial \Phi}{\partial \varphi} = -\frac{\partial \Psi}{\partial r}, \tag{12.20a,b}$$

in terms of the potential; and also in terms of the stream function, using the Cauchy–Riemann conditions in polar coordinates (11.20a,b). The derivative of the complex potential (11.18) in polar coordinates (11.19a) specifies:

$$v^*(z) = \frac{df}{dz} = e^{-i\varphi}\left(\frac{\partial \Phi}{\partial r} + i\frac{\partial \Psi}{\partial r}\right) = e^{-i\varphi}(v_r - iv_\varphi), \tag{12.21}$$

the polar components of the velocity. Substituting the potential (12.7a) [stream function (12.14)] in the condition of incompressibility (12.16) ≡ (12.22a) [irrotational flow (12.8b) ≡ (12.22b)] it follows that

$$0 = \frac{\partial v_x}{\partial x} + \frac{\partial v_y}{\partial y} = \frac{\partial^2 \Phi}{\partial x^2} + \frac{\partial^2 \Phi}{\partial y^2} = \nabla^2 \Phi, \tag{12.22a}$$

$$0 = \frac{\partial v_x}{\partial y} - \frac{\partial v_y}{\partial x} = \frac{\partial^2 \Psi}{\partial x^2} + \frac{\partial^2 \Psi}{\partial y^2} = \nabla^2 \Psi, \tag{12.22b}$$

both satisfy Laplace's equation:

$$0 = \nabla^2 \Phi = \nabla^2 \Psi = \left\{\frac{\partial^2}{\partial x^2} + \frac{\partial^2}{\partial y^2}\right\} \Phi, \Psi(x, y), \tag{12.23a}$$

$$0 = \left\{\frac{1}{r}\frac{\partial}{\partial r} r \frac{\partial}{\partial r} + \frac{1}{r^2}\frac{\partial^2}{\partial \varphi^2}\right\} \Phi, \Psi(r, \varphi), \tag{12.23b}$$

for example, in Cartesian (12.23a) ≡ (11.28a) [polar (12.23b) ≡ (11.28b)] coordinates.

In conclusion: *a complex holomorphic function specifies a potential flow, for which: (i) the complex potential (12.18) has the scalar potential (stream function) as the real (imaginary) part; (ii) the complex derivative of the complex potential specifies the complex conjugate velocity, for example, in Cartesian (12.19) [polar (12.21)] components; (iii) the velocity components can also be obtained from either the scalar potential or stream function, for example, in Cartesian (12.17a,b) [polar (12.20a,b)] components; (iv) both the scalar potential and stream function satisfy Laplace's equation, for example, in Cartesian (12.23a) [polar (12.23b)] coordinates; (v) the velocity is tangent (orthogonal) to the streamlines (equipotentials) that are two families of orthogonal curves.*

12.4 Irrotational Flow due to a Source or Sink

As a first example consider the case of a potential that depends only on the radius (12.24a):

$$\Phi = \Phi(r): \quad v_\varphi = 0, \quad 0 = \frac{d}{dr}\left(r\frac{d\Phi}{dr}\right) = \frac{d}{dr}(v_r r), \tag{12.24a–c}$$

so that: (i) the tangential velocity (12.20b) is zero (12.24b); (ii) the radial velocity (12.20a) satisfies (12.23b) ≡ (12.24c); thus the velocity v_r multiplied by the perimeter of a circle is a constant, that is, independent of the radius (12.25a):

$$Q = 2\pi r v_r, \quad v^* = \mathrm{e}^{-i\varphi} v_r = \frac{Q}{2\pi r}\mathrm{e}^{-i\varphi} = \frac{Q}{2\pi z}, \tag{12.25a,b}$$

viz. it coincides with the **flow rate** that is positive $Q > 0$ (negative $Q < 0$) for a **source (sink)** in Figure 12.2a. The corresponding conjugate velocity (12.21) is (12.25b), and its primitive is the complex potential (12.26a):

$$f(z) = \frac{Q}{2\pi}\log z; \quad \Phi(r) = \frac{Q}{2\pi}\log r, \quad \Psi(\varphi) = \frac{Q\varphi}{2\pi}. \tag{12.26a–c}$$

using (5.25) it is confirmed (12.18) that the scalar potential depends only on the radius (12.26b); since the stream function depends only on the angle (12.26c), the streamlines

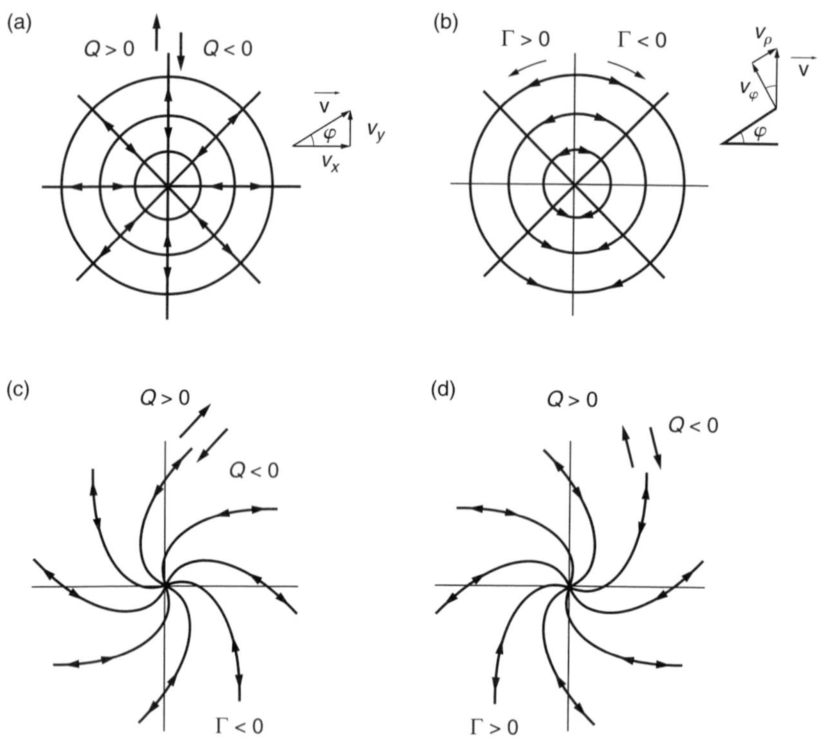

FIGURE 12.2
A source $Q > 0$ (sink $Q < 0$) produces a radial outward (inward) isotropic flow, that is, with the same velocity in every direction, and constant flow rate (a), so that the velocity varies inversely with the distance from the center. Exchanging the radial by an azimuthal velocity leads to a positive $\Gamma > 0$ (negative $\Gamma < 0$) vortex with counterclockwise (clockwise) rotation (b). Combining the source/sink with a positive (c) [negative (d)] vortex leads to the four cases of monopole or spiral flow. The flow is not incompressible (irrotational) for the source/sink (a) [vortex (b)] only at the center. For the spiral flow (c, d) both the irrotational and incompressible conditions fail only at the center. Thus the center of a monopole is always a singularity of a potential flow.

$\Psi = \text{const}$ are radial $\varphi = \text{const}$:

$$\Phi(x, y) = \frac{Q}{4\pi} \log|x^2 + y^2|, \quad \Psi(x, y) = \frac{Q}{2\pi} \arctan\left(\frac{y}{x}\right). \tag{12.27a,b}$$

The Cartesian (12.29a,b) [polar (12.28a,b)] components of the velocity of the potential flow due to a source or sink of flow rate Q are:

$$v_\varphi = 0, \quad v_r = \frac{Q}{2\pi r} = \frac{Q}{2\pi} |x^2 + y^2|^{-1/2}, \tag{12.28a,b}$$

$$\{v_x, v_y\} = \frac{Q}{2\pi r} \{\cos\varphi, \sin\varphi\} = \frac{Q}{2\pi} |x^2 + y^2|^{-1} \{x, y\}, \tag{12.29a,b}$$

in agreement with the scalar potential (12.26b) ≡ (12.27a), stream function (12.26c) ≡ (12.27b), complex potential (12.26a) and conjugate velocity (12.25b), using the relations (12.17a,b; 12.18; 12.19, 12.20a,b; 12.21). The origin is a singularity of the complex potential (12.26a) and conjugate velocity (12.25b), and thus the flow is not potential there, because the condition of incompressibility is violated.

12.5 Incompressible Flow due to a Vortex

The vortex (source or sink) corresponds to a potential that depends (12.30a) [12.24a)] only on the polar angle (radius), so that: (i) the radial velocity (12.20a) is zero (12.30b):

$$\Phi = \Phi(\varphi): \quad v_r = 0; \quad v_\varphi = \frac{\mathrm{d}}{\mathrm{d}\varphi}\left(\frac{\mathrm{d}\Phi}{\mathrm{d}\varphi}\right) = \frac{\mathrm{d}}{\mathrm{d}\varphi}(r v_\varphi), \tag{12.30a–c}$$

(ii) the tangential velocity (12.20b) satisfies (12.23b) ≡ (12.30c), that is, varies inversely with the radius; thus the circulation around a circle is a constant (12.31a), viz. independent of the radius:

$$\Gamma = 2\pi r v_\varphi; \quad v^* = -i \frac{\Gamma}{2\pi r} \mathrm{e}^{-i\varphi} = -i \frac{\Gamma}{2\pi z}, \tag{12.31a,b}$$

the corresponding complex velocity (12.21) is (12.31b). The complex potential (12.32a):

$$f(z) = -i \frac{\Gamma}{2\pi} \log z, \quad \Phi(\varphi) = \frac{\Gamma\varphi}{2\pi}, \quad \Psi(r) = -\frac{\Gamma}{2\pi} \log r, \tag{12.32a–c}$$

is the same as for a source (12.26a), apart from the imaginary factor that interchanges the potential (12.32b) and stream function (12.32c) relative to (12.26b,c); thus the streamlines are circles (Figure 12.2b), and the flow rotates in the counterclockwise (clockwise) direction for a right- (left-) handed vortex with positive $\Gamma > 0$ (negative) $\Gamma < 0$ circulation. *The complex potential (12.32a) and conjugate velocity (12.31b), and potential (12.32b) and stream function (12.32c) of a* **vortex** *with circulation Γ lead to the polar (12.33a,b) and Cartesian (12.34a,b) components of velocity:*

$$v_r = 0, \quad v_\varphi = \frac{\Gamma}{2\pi r} = \frac{\Gamma}{2\pi} |x^2 + y^2|^{-1/2}, \tag{12.33a,b}$$

$$\{v_x, v_y\} = \frac{\Gamma}{2\pi r} \{-\sin\varphi, \cos\varphi\} = \frac{\Gamma}{2\pi} |x^2 + y^2|^{-1} \{-y, x\}. \tag{12.34a,b}$$

The origin is a singularity of the complex potential (12.32a) and conjugate velocity (12.31b), viz. the flow is not potential at the vortex center, because the condition of irrotational flow is violated. Solutions of Laplace's equation depending both on r and φ are considered in Example 20.2.

12.6 Superposition as a Monopole and Spiral Flow

The potential (or stream function) of a potential flow both satisfy Laplace's equation (12.23a,b) that is linear, so that the **principle of superposition** holds: the sum of the velocities two potential flows is the velocity a potential flow, for example, the superposition of a vortex (12.32a) with a source or sink (12.26a), both at the origin, is a **monopole** (12.35a,b):

$$f(z) = \frac{P_0}{2\pi}\log z, \quad v^* = \frac{P_0}{2\pi z}, \quad P_0 \equiv Q - i\Gamma, \tag{12.35a–c}$$

whose **moment** (12.35c) is specified by the flow rate Q and circulation Γ. The scalar potential (12.36a) [stream function (12.36b)]:

$$2\pi\Phi = Q\log r + \Gamma\varphi, \quad 2\pi\Psi = Q\varphi - \Gamma\log r, \tag{12.36a,b}$$

show that the equipotentials (12.37b) [streamlines (12.37a)]

$$\left(\frac{1}{r}\frac{\mathrm{d}r}{\mathrm{d}\varphi}\right)_\Psi = \frac{Q}{\Gamma} \equiv \cot\alpha, \quad \left(\frac{1}{r}\frac{\mathrm{d}r}{\mathrm{d}\varphi}\right)_\Phi = -\frac{\Gamma}{Q} = -\tan\alpha = \cot\left(\frac{\pi}{2}+\alpha\right), \tag{12.37a,b}$$

are orthogonal spirals, making an angle $\alpha(\pi/2+\alpha)$ with the radial direction. Thus (Note 12.2) *a monopole (12.35a,b) corresponds to a spiral flow, with counterclockwise (Figure 12.2d) [clockwise (Figure 12.2c)] rotation depending on the combination of positive* $\Gamma > 0$ *(negative* $\Gamma > 0$*) circulation, and outflow for a source* $Q > 0$ *or inflow for a sink* $Q < 0$*; the moment (12.35c) of the monopole appears in the complex potential (12.35a) and conjugate velocity (12.35b), and is specified by the flow rate* Q *and circulation* Γ *that also appear in the potential (12.36a) and stream function (12.36b); the velocity is the superstition those for the source/sink (12.28–12.29a,b) and vortex (12.33–12.34a,b). The pressure field (14.37c) associated with a spiral flow (12.35b,c) is given (14.27c) by:*

$$p(r) = p_0 - \frac{1}{2}\rho|v^*|^2 = p_0 - \frac{\rho}{2}\frac{|P_0|^2}{(2\pi r)^2} = p_\infty - \rho\frac{Q^2+\Gamma^2}{8\pi^2 r^2}, \tag{12.38}$$

where: (i) ρ_0 *is the mass density, which is constant in an incompressible flow; (ii)* p_0 *is the stagnation pressure in the Bernoulli equation (14.27c); (iii) it coincides* $p_0 = p_\infty$ *with the pressure at infinity, where the velocity is zero.* The monopole has a singularity at the origin, where it is not a potential flow, because both the conditions of irrotational and incompressible flow are violated. The possibility of "nonisotropic" sources/sinks or vortices is considered in Example 20.12.

12.7 Dipole as the Limit of Two Opposing Monopoles

Consider two monopoles with opposite moments $\pm P_0$ symmetrically placed (Figure 12.3) relative to the origin at $\pm\varepsilon$, whose complex conjugate velocity (12.35b) is:

$$v^*(z) = \frac{P_0}{2\pi}\left[\frac{1}{z-\varepsilon} - \frac{1}{z+\varepsilon}\right]. \tag{12.39a}$$

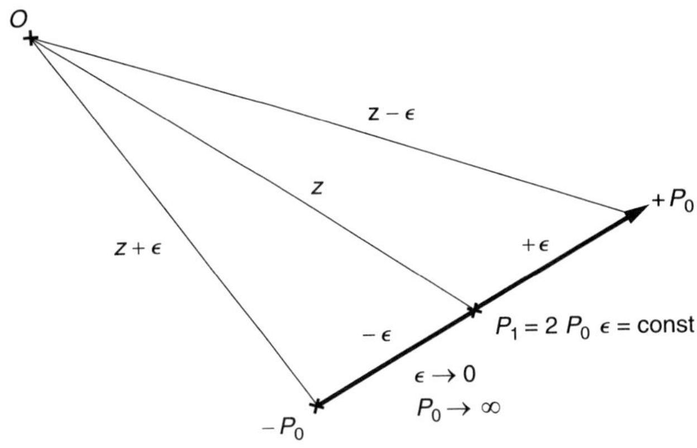

FIGURE 12.3
The monopole (Figure 12.2) is not the only singularity of a potential flow and can be used to build-up others. The next by order of complexity is the dipole: (i) it consists of two opposite monopoles with moments $\pm P_0$ which are at a distance 2ε; (ii) they converge to the same point $\varepsilon \to 0$ increasing (Figure 12.2) the monopole moment $P_0 \to \infty$ so that the product $P_1 = 2P_0\varepsilon$ is the constant dipole moment. The dipole has an axis, and is not isotropic, like the monopole.

For a distant observer the first two terms of the geometric series (21.62b) in (12.39b) are

$$\frac{1}{z \mp \varepsilon} = \frac{z^{-1}}{1 \mp \varepsilon/z} = \frac{1}{z} \pm \frac{\varepsilon}{z^2} + O\left(\frac{\varepsilon^2}{z^3}\right), \tag{12.39b}$$

where, $O(\dots)$ means that terms like ... or higher powers (Section 19.7) are omitted. Substituting (12.39b) in (12.39a) yields:

$$v^*(z) = \frac{P_0}{2\pi z^2}\left[2\varepsilon + O\left(\frac{\varepsilon^2}{z}\right)\right]. \tag{12.40}$$

As the opposite monopoles coincide $\varepsilon \to 0$, they cancel $v^* \to 0$ unless the moment $P_0 \to \infty$ diverges to keep a constant product (12.41a):

$$P_1 = \lim_{\substack{\varepsilon \to 0 \\ P_0 \to \infty}} 2\varepsilon P_0, \quad v^*(z) = \frac{P_1}{2\pi z^2}, \tag{12.41a,b}$$

this leads to (12.41b), since the terms $O\left(P_0\varepsilon^2/z^3\right) = O\left(P_1\varepsilon/z^3\right)$ in (12.40) vanish as $\varepsilon \to 0$. Thus *the limit of opposite monopoles as they approach and the moment increases inversely to the distance (12.41a) is a* **dipole** *(12.41b); its moment:*

$$P_1 = 2\varepsilon P_0 = 2(\varepsilon_x + i\varepsilon_y)(Q - i\Gamma) = 2(Q\varepsilon_x + \Gamma\varepsilon_y) + i2(\varepsilon_y Q - \varepsilon_x \Gamma), \tag{12.42}$$

corresponds to a horizontal dipole (Figure 12.4) *for: (i) horizontally placed source* $Q > 0$ at $-\infty$ *and sink* $-Q$ at $+\infty$; *(ii) vertically placed vortices, viz. right-handed* $\Gamma > 0$ *at* $+i\infty$ *and left-handed* $-\Gamma$ at $-i\infty$. *Conversely a vertical dipole corresponds to a vertical source-sink pair or a horizontal pair of opposite vortices.*

The complex conjugate velocity (12.41b) corresponds to the complex potential (12.43a) for a dipole:

$$f(z) = -\frac{P_1}{2\pi z}; \quad \{\Phi, \Psi\} = \frac{P_1}{2\pi r}\{-\cos\varphi, \sin\varphi\}, \tag{12.43a–c}$$

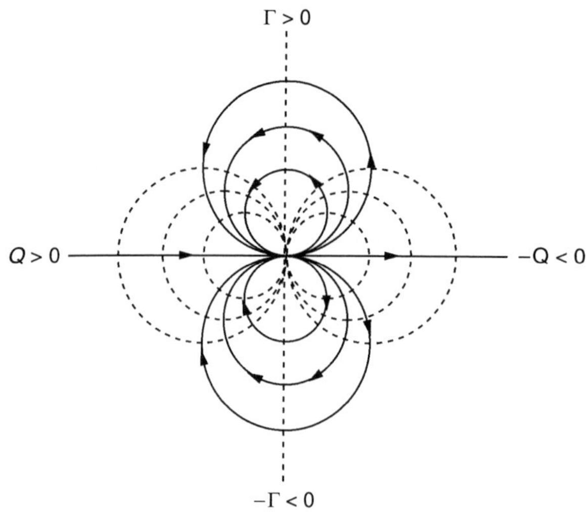

FIGURE 12.4
A dipole has an axis: (i) from the source towards the sink; (ii) perpendicular to the line joining the counterclockwise to the clockwise vortex. The streamlines are the real axis and circles on each side tangent at the dipole location. The equipotentials are the imaginary axis and orthogonal circles obtained by rotation through ninety degrees.

and taking the OX-axis along the dipole axis, the dipole moment P_1 is real, leading to the scalar potential (12.43b) and stream function (12.43c). The equipotentials (streamlines) have equations (12.44a) [(21.44b)] in polar coordinates:

$$r^2 = -\frac{P_1}{2\pi\Phi} r\cos\varphi, \quad r^2 = \frac{P_1}{2\pi\Psi} r\sin\varphi, \tag{12.44a,b}$$

or in Cartesian coordinates:

$$x^2 + y^2 = -\frac{P_1}{2\pi\Phi} x, \quad x^2 + y^2 = \frac{P_1}{2\pi\Psi} y, \tag{12.45a,b}$$

It follows that

$$A \equiv \frac{P_1}{4\pi\Phi}: \quad (x + A)^2 + y^2 = A^2, \tag{12.46a,b}$$

$$B \equiv \frac{P_1}{4\pi\Psi}: \quad x^2 + (y - B)^2 = B^2, \tag{12.47a,b}$$

the streamlines (12.43c) [equipotentials (12.43b)] of a dipole (12.43a), are circles tangent (12.47b) [orthogonal (12.46b)] to the dipole axis in Figure 12.4, viz:

$$r = 2B\sin\varphi, \quad r = -2A\cos\varphi, \tag{12.48a,b}$$

are the equations of circles with center on the imaginary (12.48a) [real (12.48b)] axis, and radius (12.47a)[(12.46a)] such that they are tangent to the real (imaginary) axis. These two families of circles are orthogonal and coincide by a rotation of 90°. The dipole has complex potential (12.43a) and conjugate velocity (12.41b), and polar velocity components:

$$\{v_r, v_\varphi\} = \frac{P_1}{2\pi r^2}\{\cos\varphi, \sin\varphi\} = \frac{P_1}{2\pi}\left|x^2 + y^2\right|^{-3/2}\{x, y\}. \tag{12.49a,b}$$

The origin is a singularity of the velocity of higher order $O\left(1/z^2\right)$ than for a monopole $O(1/z)$ because the source and sink and opposite vortices lead to zero flow rate and zero circulation.

12.8 Quadrupole Moment and Rule of Differentiation

The passage from monopoles to dipoles (Section 12.7) can be continued to quadrupoles (Subsection 12.8.2) and multipoles (Section 12.9) using a differentiation rule (Subsection 12.8.1).

12.8.1 Rule of Differentiation for Multipoles

The limiting process that leads from the monopole (Section 12.6) to the dipole (Section 12.7) can be continued indefinitely, as follows: (i) consider two opposite singularities, leading to the complex potential:

$$F(z) = P_1 \{f(z-\varepsilon) - f(z+\varepsilon)\}; \tag{12.50}$$

(ii) since $f(z)$ is the complex potential of a flow, it is a holomorphic function (Section 12.3), and hence it is differentiable and the mean value theorem can be applied:

$$f(z \mp \varepsilon) = f(z) \mp \varepsilon \frac{\mathrm{d}f}{\mathrm{d}z} + o(\varepsilon), \tag{12.51a}$$

where, the symbol $o(\varepsilon)$ means (Section 19.8) an infinitesimal of order higher than ε; (iii) substituting (12.51a) in (12.50) yields:

$$F(z) = -2P_1\varepsilon \frac{\mathrm{d}f}{\mathrm{d}z} + o(P_1\varepsilon); \tag{12.51b}$$

(iv) taking the limit $\varepsilon \to 0$ and $P_1 \to \infty$ so as to preserve the quadrupole moment (12.52a):

$$P_2 = \lim_{\substack{\varepsilon\to 0\\ P_1\to\infty}} 2P_1\varepsilon, \quad F(z) = -P_2\frac{\mathrm{d}f}{\mathrm{d}z}, \tag{12.52a,b}$$

leads to (12.52b), since the term $o(P_1\varepsilon)$ in (12.52) tends to zero as $\varepsilon \to 0$. Thus *the passage from a dipole* P_1 (2^{n} *− pole* P_{n}) *to a quadrupole* P_2($2^{\mathrm{n}+1}$*-pole of moment* $\mathrm{P}_{\mathrm{n}+1}$), *is made by differentiation, and replacing of the moment* P_1 (P_{n}) *by minus the next moment* $-\mathrm{P}_2$ ($-\mathrm{P}_{\mathrm{n}+1}$), *multiplied by 1*(n) in (12.53a), viz.:

$$n = 1,2,\ldots: \quad P_{n+1} = \lim_{\substack{\varepsilon\to 0\\ P_n\to\infty}} 2n\varepsilon P_n = \lim_{\substack{\varepsilon_1\ldots\varepsilon_n\to 0\\ P_0\to\infty}} 2^n(n!)\varepsilon_1\ldots\varepsilon_n P_0, \tag{12.53a,b}$$

thus the passage (12.53b) from the monopole P_0, *to the* **dipole** $\mathrm{P}_1 = 2\varepsilon_1\mathrm{P}_0$, *to the quadrupole* $\mathrm{P}_2 = 2\varepsilon_2\mathrm{P}_1$, *to octupole* $\mathrm{P}_3 = 4\varepsilon_3\mathrm{P}_2$, *and so on up to the* 2^n -pole of order n, *introduces* n *axis* $\varepsilon_1,\ldots,\varepsilon_{\mathrm{n}}$ *with possibly distinct directions* $\arg(\varepsilon_1),\ldots,\arg(\varepsilon_{\mathrm{n}})$. The need for the factor $1(n)$ will be seen when applying the rule next (subsequently in Section 12.9) to a quadrupole (arbitrary multipole).

12.8.2 Two Axes and Four Lobes of a Quadrupole

The rule holds when passing from the monopole (12.35a,b) to the dipole (12.43a, 12.41b), viz.:

$$f(z) = -P_1\left(\frac{\mathrm{d}}{\mathrm{d}z}\right)\left\{(2\pi)^{-1}\log z\right\} = -\frac{P_1}{2\pi z}, \tag{12.54a}$$

$$v^*(z) = -P_1\left(\frac{\mathrm{d}}{\mathrm{d}z}\right)\left\{(2\pi z)^{-1}\right\} = \frac{P_1}{2\pi z^2}. \tag{12.54b}$$

It can be used to pass from the dipole (12.54a,b) to the **quadrupole**:

$$f(z) = -P_2\left(\frac{\mathrm{d}}{\mathrm{d}z}\right)\left\{(-2\pi z)^{-1}\right\} = -\frac{P_2}{2\pi z^2}, \tag{12.55a}$$

$$v^*(z) = -P_2\left(\frac{\mathrm{d}}{\mathrm{d}z}\right)\left\{\left(2\pi z^2\right)^{-1}\right\} = \frac{P_2}{\pi z^3}. \tag{12.55b}$$

Along the real $z = x$ or imaginary $z = iy$ axis $z^2 = x^2, -y^2$ is real, so the complex potential (12.55a) is real for real quadrupole moment, and the stream function is zero $\Psi = 0$. Thus *the real and imaginary axes are streamlines of a quadrupole with real moment that corresponds* (Figure 12.5) *to: (i) sources* $\mathrm{Q} > 0$ *at* $\mathrm{z} = \pm\mathrm{i}\infty$ *and sinks* $-\mathrm{Q}$ *at* $\mathrm{z} = \pm\infty$*; (ii) positive vortices* $\Gamma > 0$ *at* $\mathrm{z} = \pm(1+\mathrm{i})\infty$ *and negative vortices* $-\Gamma$ *at* $\mathrm{z} = \pm(1-\mathrm{i})\infty$*. The quadrupole has complex potential (12.55a) and conjugate velocity (12.55b) and scalar potential (12.56a) and stream function (12.56b):*

$$\Phi(r,\varphi) = -\frac{P_2}{2\pi r^2}\cos(2\varphi), \quad \Psi(r,\varphi) = \frac{P_2}{2\pi r^2}\sin(2\varphi) = \Phi\left(\varphi + \frac{\pi}{4}\right), \tag{12.56a,b}$$

such that (i) the equipotentials and streamlines lead to a **directivity** *pattern with four* **lobes**, *like a flower with four petals (Figure 12.5) that are orthogonal, and coincide by rotation over* $\pi/4 = 45°$*; (ii) the polar components of velocity are:*

$$\{v_r, v_\theta\} = \frac{P_2}{\pi r^3}\{\cos(2\varphi), \sin(2\varphi)\}. \tag{12.57a,b}$$

In (12.56–12.57a,b) it was assumed that P_2 is real, that is, the quarupole is horizontal. In general the passage from a monopole to a dipole (12.42) ≡ (12.58a) and from a dipole to a quadrupole (12.52a) ≡ (12.58b) introduces (Figure 12.6) two axes in the quadrupole (12.58c):

$$\begin{aligned} P_2 = 2\varepsilon_2 P_1 = 4\varepsilon_1\varepsilon_2 P_0 &= \frac{4(\varepsilon_{1x} + i\varepsilon_{1y})(\varepsilon_{2x} + i\varepsilon_{2y})(Q - i\Gamma)}{2\pi} \\ &= 2\frac{Q - i\Gamma}{\pi}\{(\varepsilon_{1x}\varepsilon_{2x} - \varepsilon_{1y}\varepsilon_{2y}) + i(\varepsilon_{1x}\varepsilon_{2y} + \varepsilon_{1y}\varepsilon_{2x})\}. \end{aligned} \tag{12.58a–c}$$

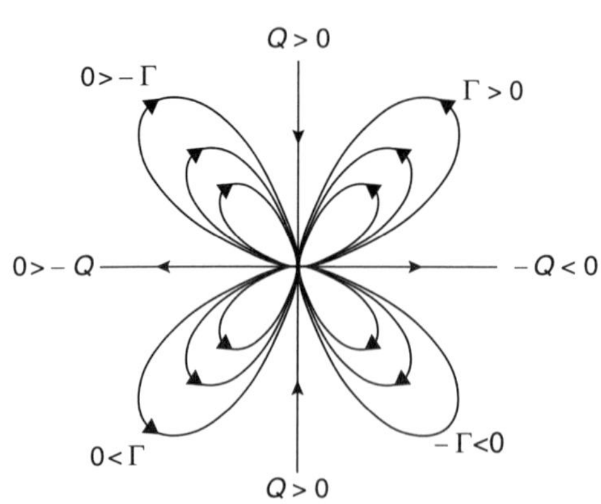

FIGURE 12.5
Whereas the dipole has one axis (Figure 12.4) the quadruple has two axes to which the streamlines in each quadrant are tangent at the origin. The quadruple corresponds to two opposite dipoles made of: (i) one source and one sink along the coordinate axis; (ii) one counterclockwise and one clockwise vortex along the diagonal of the quadrants.

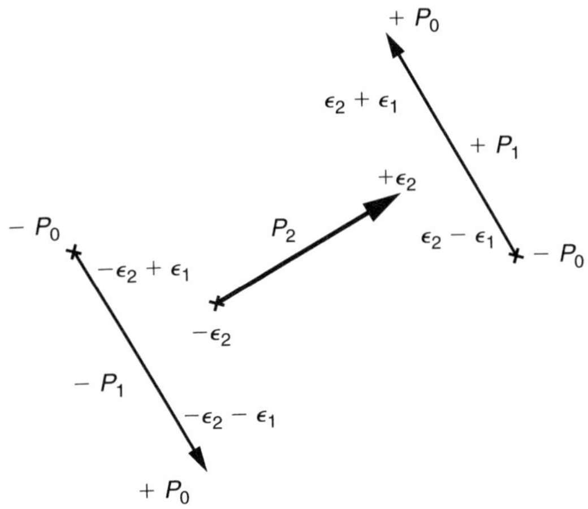

FIGURE 12.6
The process passing from the monopole to the dipole (Figure 12.3) can be extended from the dipole to the quadrupole (Figure 12.6), and so on. Two opposite dipoles of moments $\pm P_1$ at a distance $2\varepsilon_2$ along an orthogonal direction approach $\varepsilon_2 \to 0$ increasing the dipole moment $P_1 \to \infty$ so that the product $P_2 = 2P_1\varepsilon_2$ is the constant quadrupole moment. This leads to the two axes of the quadrupole in Figure 12.5.

If $\varepsilon_{1x}\varepsilon_{2x} = \varepsilon_{1y}\varepsilon_{2y}\,(\varepsilon_{1x}\varepsilon_{2y} = -\varepsilon_{1y}\varepsilon_{2x})$ the quadrupole is horizontal for vortices $\Gamma \neq 0 = Q$ (sources or sinks $\Gamma = 0 \neq Q$), and vertical for sources or sinks (vortices), otherwise it is oblique; in particular it is oblique along the odd (even) quadrants if $\varepsilon_{1x}\varepsilon_{2x} - \varepsilon_{1y}\varepsilon_{2y} = (\pm)(\varepsilon_{1x}\varepsilon_{2y} + \varepsilon_{1y}\varepsilon_{2x})$ for sources or sinks, and vice-versa for vortices. In all cases the origin is a singularity for the velocity of order $O\left(z^{-3}\right)$ higher than $O\left(z^{-2}\right)$ for the dipole.

12.9 Arbitrary Multipole and Directivity Lobes

Proceeding with the limiting process (12.50) to (12.53a,b), *the* **2^n-pole** or **multipole of order** *n (e.g., octupole for $n = 3, 2^3 = 8$), has complex conjugate velocity (12.59a)*:

$$v^*(z) = \frac{P_n}{2\pi}\frac{(-)^n}{(n-1)!}\left(\frac{\mathrm{d}}{\mathrm{d}z}\right)^n \frac{1}{z} = \frac{P_n}{2\pi} n z^{-n-1}, f(z) = -\frac{P_n}{2\pi} z^{-n}, \tag{12.59a,b}$$

and complex potential (12.59b); this corresponds to the scalar potential (12.60a) and stream function (12.60b):

$$\{\Phi, \Psi\}(r, \varphi) = \frac{P_n}{2\pi} r^{-n} \{-\cos(n\varphi), \sin(n\varphi)\}; \tag{12.60a,b}$$

the equipotentials and streamlines lead to a directivity pattern with 2n *lobes, like a flower with* 2n *petals [six lobes or petals* 2n = 6 *for the octupole* ($\mathrm{n} = 3, 2^{\mathrm{n}} = 8$) *in Figure 12.7] that are (i) orthogonal; (ii) coincide by a rotation over* $\pi/2\mathrm{n}$ *in (12.61a):*

$$\Phi\left(r, \varphi + \frac{\pi}{2n}\right) = \Psi(r, \varphi); \quad \Psi\left(r, \varphi = \frac{k\pi}{n}\right) = 0 = \Phi\left(r, \varphi = \frac{(k + 1/2)\pi}{n}\right); \tag{12.61a,b}$$

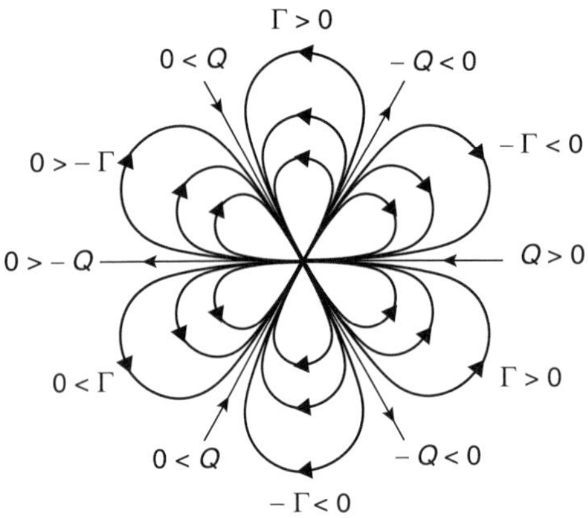

FIGURE 12.7
In the case of an octupule (Figure 12.5) joining two quadrupoles (Figures 12.5 and 12.6) two monopoles cancel [see (12.63a)] leaving six monopoles, namely sources/sinks (positive/negative vortices) along six equally spaced directions starting with the positive real axis (30° below). Besides the six radial lines, the streamlines form "flowers" with six petals in Figure 12.7 for the octupole, versus four (two) petals for the quadrupole (dipole) in Figure 12.5 (12.4).

(iii) have 2n *separating radial lines (12.61b). The velocity is given by:*

$$\{v_r, v_\varphi\} = \frac{P_n}{2\pi} n r^{-n-1} \cos\{(n\varphi), \sin(n\varphi)\}, \tag{12.62a,b}$$

in polar components. The pressure field corresponding to a multipolar flow (12.59a) is given (14.27c) by:

$$p(r) = p_0 - \frac{1}{2}\rho |v^*(z)|^2 = p_\infty - \frac{n^2 \rho}{8\pi^2} |P_n|^2 r^{-2-2n}, \tag{12.63}$$

where ρ *is the mass density and* $p_0 = p_\infty$ *the stagnation pressure at infinity, as in (12.38).*

The complex potential (12.59b) and conjugate velocity (21.59a), the scalar potential (12.60a) and stream function (12.60b), and the polar components of the velocity (12.62a,b), include as particular case $n = 2$ the quadrupole $2^2 = 4$ in (12.55a,b; 12.56a,b; 12.57a,b), and $n = 1$ the dipole $2^1 = 2$ in (12.43a, 12.41b; 12.43b,c; 12.49a,b); for $n = 0$ the monopole $2^0 = 1$ has a logarithmic potential (12.35a), and the complex conjugate velocity (12.35b) satisfies (12.59a) without the factor n in the numerator. The next case $n = 3$ is $2^3 = 8$ the octupole in Figure 12.7 that has six lobes or petals instead of eight for two quadrupoles due to the cancellation (12.64a):

$$\begin{matrix} & + & \\ - & & - \\ & + & \end{matrix} \quad + \quad \begin{matrix} & - & \\ + & & + \\ & - & \end{matrix} \quad = \quad \begin{matrix} & + & - & \\ - & & & + \\ & + & - & \end{matrix}; \qquad \begin{matrix} - & + \\ + & - \end{matrix}, \tag{12.64a,b}$$

this cancellation does not occur for a quadrupole (12.64b) as a limit of two dipoles. The relations between polar and Cartesian components of the velocity for an arbitrary multipole can be used to obtain the transformation between Cartesian and polar components of any vector (Example 20.3).

Note 12.1. *Directivity Patterns for Radiation and Jamming:* It can be shown (Chapters 27 and 28) than an isolated singularity of the complex potential can be represented by a superposition of multipoles, for example, for the gravity field (Chapter 18). The decomposition into multipoles applies to: (i) plane multipoles, which are line multipoles in space; (ii) point multipoles in space and higher dimensions. The representations of an arbitrary source by a superposition of multipoles extends from potential fields satisfying the Poisson (i.e., the forced Laplace) equation, to the forced solutions of the wave equation with sources. In the latter case: (i) the directivity pattern concerns the spatial dependence of the multipole; (ii) the time dependence of the "signal" leads to the spectrum.

As an example, the radiation pattern emitted by antenna consists of directivity lobes, usually with a "main lobe" where most energy is concentrated, to increase the strength of the signal and sensivity to the echo, and obtain maximum range; also the main lobe should be narrow to give better angular discrimination and reject or reduce sensivity to off-axis incoming signals, for example, noise. The ECM (Electronic Counter Measures) or "jamming" are techniques to interfere with the correct emission of the emitter, for example, using the less powerful secondary lobes to send a disrupting signal, for example, blanking noise, out-of-phase cancellation even a false or deceptive "echo" from a nonexisting object. The ECCM (Electronic Counter Counter Measures) or "anti-jam" are techniques to restore correct operation in the presence of jamming, or even use the jamming to identify or collect information on the jammer; one ECCM technique is to point a "null" of the radiation pattern at the "jammer," so that the antena becomes insensitive to signals from that direction. Thus secondary lobe jamming (null steering) are ECM (ECCM) techniques whose effectivness depends on knowledge of the directivity pattern of the emitter.

The exposure to the noise of an emitter also depends on its directivity pattern, that is, it is maximum in the direction of the peak of main lobe and minimum along nulls if they exist, or in the direction of least amplitude. One way to reduce the noise exposure is to turn the emitter "away" from the observer direction that depends on the directivity pattern. The peak of the main lobe may (may not be) on axis, for example, example is a isolated dipole (quadrupole) that has peak on axis (45° off axis). Active noise reduction acts like jamming, using additional source(s) of sound to try to cancel the existing noise. The radiation patterns of practical emitters can be complex superpositions of multipoles and are measured experimentally in: (i) anechoic chambers, that is, rooms with walls absorbing the echo; (ii) test ranges, that is, wide areas without obstacles or reflectors.

Note 12.2. *Complex Potential and Conjugate Velocity of a Monopole:* The monopole of moment (12.35c) serves as a example of the calculation of: (i) the complex potential (12.35a):

$$\begin{aligned} 2\pi(\Phi + i\Psi) &= 2\pi f = 2\pi P_0 \log z = (Q - i\Gamma)(\log r + i\varphi) \\ &= (Q - i\Gamma)\left[\frac{1}{2}\log\left|x^2 + y^2\right| + i\text{arc}\tan\left(\frac{y}{x}\right)\right], \end{aligned} \tag{12.65}$$

and hence the scaler potential (12.36a) ≡ (12.66a) and stream function (12.36b) ≡ (12.66b):

$$2\pi\Phi = Q\log r + \Gamma\varphi = \frac{Q}{2}\log\left|x^2 + y^2\right| + \Gamma\text{arc}\tan\left(\frac{y}{x}\right) \tag{12.66a}$$

$$2\pi\Psi = Q\varphi - \Gamma\log r = Q\text{arc}\tan\left(\frac{y}{x}\right) - \frac{\Gamma}{2}\log\left|x^2 + y^2\right|; \tag{12.66b}$$

(ii) the complex conjugate velocity (12.35b):

$$\begin{aligned} 2\pi(v_x - iv_y) = 2\pi v^*(z) &= \frac{P_0}{z} = \frac{Q - i\Gamma}{r} = e^{-i\varphi} = \frac{Q - i\Gamma}{r}(\cos\varphi - i\sin\varphi) \\ &= \frac{Q - i\Gamma}{x + iy} = \frac{Q - i\Gamma}{x^2 + y^2}(x - iy), \end{aligned} \tag{12.67}$$

$$2\pi(v_r - iv_\varphi) = e^{i\varphi}2\pi v^*(z) = \frac{P_0}{z}e^{i\varphi} = \frac{Q - i\Gamma}{r} = \left|x^2 + y^2\right|^{-1/2}(Q - i\Gamma), \tag{12.68}$$

and hence the polar (12.68) ≡ (12.69a,b) [Cartesian (12.67) ≡ (12.70a,b)] components of the velocity:

$$2\pi \{v_r, v_\varphi\} = r^{-1}\{Q,\Gamma\} = \left|x^2+y^2\right|^{-1/2}\{Q,\Gamma\}, \tag{12.69a,b}$$

$$\begin{aligned} 2\pi \{v_r, v_\varphi\} &= r^{-1}\{Q\cos\varphi - \Gamma\sin\varphi,\ Q\sin\varphi + \Gamma\cos\varphi\} \\ &= \left|x^2+y^2\right|^{-1/2}\{Qx - \Gamma y,\ Qy + \Gamma x\}. \end{aligned} \tag{12.70a,b}$$

The Cartesian (12.69a,b) [polar (12.70a,b)] components of the velocity also follows from the scalar potential (12.66a) or stream function (12.66b) using (12.17a,b) [(12.20a,b)]. The Cartesian (polar) components of the velocity are expressed in terms of: (i) Cartesian coordinates by (12.69b) [12.70b]; (ii) polar coordinates by (12.69a) [(12.70a)]. The expressions (12.65; 12.66a,b; 12.67; 12.68; 12.69a,b; 12.70a,b) simplify for a source/sink $Q \neq 0 = \Gamma$ (vortex $\Gamma \neq 0 = Q$) to (12.26a; 12.26b,c ≡ 12.27a,b; 12.25b; 12.28a,b; 12.29a,b) [(12.32a; 12.32b,c; 12.31a,b; 12.33a,b; 12.34a,b)]. The Cartesian (12.70a,b) and polar (12.69a,b) components of the velocity are related by (20.11–12a,b).

Conclusion 12: The integral (figure 12.1) of the normal (tangential) velocity **(b)** specifies the flow rate **(d)** [circulation **(c)**] along an open curve or around a loop **(a)**. The velocity is (figure 12.2) radial (tangential) for a source/sink **(a)** [vortex **(b)**], viz. outward/inward (counterckwise/clockwise) for positive/negative flow rate (circulation); the superposition of a source/sink with a left-handed/right-handed vortex is an outward/inward spiral flow, rotating as clockwise **(c)** [counterclockwise **(d)**] monopole. The dipole (figure 12.3) is the limit of two monopoles, viz. a source and sink (opposite vortices) along (orthogonal) to the dipole axis, and the streamlines (equipotentials) are circles tangent (orthogonal) to the dipole axis, and passing through the center (figure 12.4). The quadrupole is the limit of two dipoles (figure 12.6) and has two axes. The quadrupole (octupole) [figure 12.5 (12.7)] correspond to four (six) alternating source/sinks or vortices, and the equipotentials and streamlines lead to orthogonal directivity patters withy 4(6) lobes, which coincide by rotation over $\pi/4(\pi/6)$.

13

Primitive and Contour Integrals

A holomorphic function $f(z)$ is defined by the existence of a first-order derivative, and its real and imaginary parts necessarily satisfy the Cauchy–Riemann conditions. It can be shown that (Section 13.1) the latter imply the existence of a primitive F, defined to be a function whose derivative is $f = F'$. Hence a holomorphic functions f always has a primitive F such that $F' = f$, and since the latter is also holomorphic, F too has a primitive $\ldots$, that is, given a holomorphic function an infinite number of successive primitives can be found. A real function can integrated only along the real axis (Section 13.2), whereas a complex function can be integrated (Sections 13.3–13.5) along any contour in the complex plane, where contour means a suitable curve. The integral of a complex function along a contour generally depends on the function itself, the end-points and the contour or path joining them. If the function is holomorphic, it has a primitive, so the integral is the difference of the primitive at the end-points, and is independent of the path joining them; the converse (Section 13.6) can also be proved: if an integral does not depend on the contour joining the end-points, the integrand is a holomorphic function. The operations derivation and integration have several useful properties when applied to holomorphic functions, for example (Section 13.7) the rule of integration by parts and the chain rule of derivation. The derivative and primitive are almost inverse operations, for example, the derivate of an integral with a free upper bound coincides with the integrand. This result may be generalized to cases in which (i) the integrand depends not only on the variable of integration but also on a parameter (Section 13.8); (ii) the limits of integration are functions rather than constants, that is, the end-points that are not fixed but may lie on given curves (Section 13.9). There will be frequent subsequent use of the integration along curved paths, and of the derivation of an integral with regard to a parameter appearing in the integrand, in the limits of integration or in both; the latter occur in problems with variable limits, where the end-points are not fixed, but can "slide" along a curve or surface. These appear in optimization and other problems, for example, the shortest bridge between two banks of a river, a constrained path around an obstacle, etc.

13.1 Existence and Properties of the Primitive of a Function

The **primitive** $F(z)$ of a function $f(z)$ is defined as a function whose derivate is $\mathrm{d}F/\mathrm{d}z \equiv f(z)$, that is:

$$\mathrm{d}F = f(z)\,\mathrm{d}z: \quad f(z) = \Phi + i\Psi, \quad F = U + iV, \tag{13.1a,b}$$

where the function f is (or Φ, Ψ are) given, and the primitive F is (or U,V are) to be found. The primitive:

$$\mathrm{d}U + i\,\mathrm{d}V = (\Phi + i\Psi)(\mathrm{d}x + i\,\mathrm{d}y), \tag{13.2}$$

exists iff the expressions:

$$dU = \Phi\, dx - \Psi\, dy, \quad dV = \Phi\, dy + \Psi\, dx, \tag{13.3a,b}$$

are exact differentials. If the function $f(z) \in C$ is continuous, its primitive $F(z) \in C^1$ if it exists has continuous first-order derivates:

$$dU = \frac{\partial U}{\partial x} dx + \frac{\partial U}{\partial y} dy, \quad dV = \frac{\partial V}{\partial x} dx + \frac{\partial V}{\partial y} dy; \tag{13.4a,b}$$

comparison with (13.3a,b) shows that

$$\frac{\partial U}{\partial x} = \Phi = \frac{\partial V}{\partial y}, \quad \frac{\partial U}{\partial y} = -\Psi = -\frac{\partial V}{\partial x}, \tag{13.5a,b}$$

the real U and imaginary V parts of the primitive (13.1b) satisfy the Cauchy–Riemann conditions (13.5a,b); since $U, V \in C^1$ its follows (Section 11.3) that (13.1b) is holomorphic. It has been shown that: *(i) the primitive (13.1b) of a complex function (13.1a), if it exists, satisfies the Cauchy–Riemann conditions (13.5a,b); (ii) if the function (13.1a) is continuous, and has a primitive (13.1b), the latter is holomorphic, as would result from the definition of primitive* $dF/dz \equiv f(z) \in C$.

In the case where the function (13.1b) is holomorphic, the Cauchy–Riemann conditions (Section 11.3) for its real and imaginary parts are met:

$$\frac{\partial \Phi}{\partial x} = \frac{\partial \Psi}{\partial y}, \quad \frac{\partial \Phi}{\partial y} = -\frac{\partial \Psi}{\partial x}; \tag{13.6a,b}$$

these conditions imply that (13.3a,b) are exact differentials, and hence the primitive (13.1b) exists, is holomorphic, and its real and imaginary parts satisfy the Laplace equation, from (13.5a,b; 13.6a,b):

$$\frac{\partial^2 U}{\partial x^2} = -\frac{\partial^2 U}{\partial y^2}, \quad \frac{\partial^2 V}{\partial y^2} = -\frac{\partial^2 V}{\partial x^2}. \tag{13.7a,b}$$

In conclusion *a holomorphic function (13.1b) has a holomorphic (13.1c) primitive (13.1a), with real* U *and imaginary* V *parts satisfying Laplace's equation (13.7a,b). Since the primitive is holomorphic, it too has a primitive, and thus a holomorphic function has an infinite number of successive primitives. A holomorphic function also has derivatives of all orders (Section 25.4).*

13.2 Riemann Integral of a Complex Function (Cauchy, 1825; Riemann, 1851)

The **Riemann integral** of a complex function $f(z)$ along a curve C between the points a and b, is defined as follows: (i) the **curve** consists of the points $z(t) = x(t) + iy(t)$, where the parameter t belongs to the interval $\alpha \leqq t \leqq \beta$, whose limits specify the end-points $a \equiv z(\alpha)$ and $b \equiv z(\beta)$; (ii) a **partition** $\alpha \equiv t_0 < t_1 < t_2 < \cdots < t_{N-1} < t_N \equiv \beta$ of the interval (α, β) is performed such that successive straight segments of the tangent (Figure 13.1a) have a length not exceeding δ:

$$n = 0, \ldots, N-1: \quad |z_{n+1} - z_n| \equiv |z(t_{n+1}) - z(t_n)| \leqq \delta; \tag{13.8a}$$

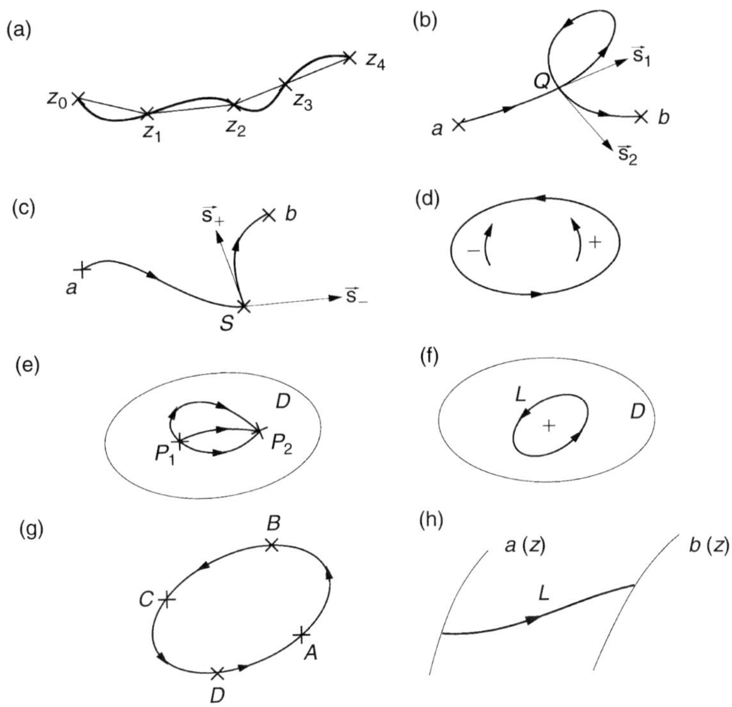

FIGURE 13.1
A curve may be partitioned by a succession of points, joined by secants (a): if for any partition the sum of the lengths of the secants is bounded, the curve is rectifiable, that is, has finite length. At a double (b) [multiple] point a curve has two (several) distinct tangents ($\tilde{s}_1, \tilde{s}_2$) where it crosses itself; a simple curve has no multiple points, so it does not cross itself anywhere. At an angular point (c) the left $\tilde{s}_-$ and right $\tilde{s}_+$ hand tangents do not coincide; in particular they are opposite at cusp (Figure 20.3c,d). A regular curve has a continuous tangent and no angular points. A curved is closed (d) if the start and end points coincide; a closed curve can be described in the positive (negative) direction, that is, counterclockwise (clockwise). Describing a closed curve in the positive or counterclockwise direction the interior (exterior) is to the left (right); vice-versa in the negative direction. A contour is a simple regular rectifiable curve; if it is also closed it is a loop. The integral of a function along two distinct contours joining the same points (e) can be different (must be equal) if the function is not (is) holomorphic between the curves; it follows that the integral around a loop (f) may be nonzero (must be zero) if the function is not (is) holomorphic in the interior. An integral along a loop (g) is the sum of the integrals along any partition of the loop into successive contours. The integral can be taken along a contour (h) whose end-points can move along curves.

(iii) in each segment is taken the arithmetic mean of the function $f(z)$ at the two ends $\{f(z_{n+1}) + f(z_n)\}/2$, and it is multiplied by $(z_{n+1} - z_n)$; if the sum over all $n = 1, \ldots, N-1$ tends to a limit as $\delta \to 0$, it defines:

$$\lim_{\delta \to 0} \sum_{n=0}^{N-1} \frac{1}{2} \{f(z_{n+1}) + f(z_n)\}(z_{n+1} - z_n) \equiv \int_a^b f(z)\,\mathrm{d}z, \tag{13.8b}$$

the Riemann integral of the function $f(z)$ along the curve $z(t)$ from $a \equiv z_0$ to $b \equiv z_N$.

In the case of a real function $f(x)$, the integration is performed along a segment of the real axis $z(t) \equiv x$, and the integral is defined by:

$$a \equiv x_1 < x_2 < \cdots < x_{N-1} < x_N \equiv b: \int_a^b f(x)\,\mathrm{d}x$$
$$\equiv \lim_{\max|x_{n+1}-x_n| \to 0} \sum_{n=0}^{N-1} \frac{1}{2} \{f(x_{n+1}) + f(x_n)\}(x_{n+1} - x_n). \tag{13.9}$$

Both in (13.8b) and (13.9) it has been required that the longest of the segments tend to zero; that implies $N \to \infty$, since infinitely many segments will be needed to cover the finite interval (a,b). The limit in (13.8b) or (13.9) cannot be replaced by $N \to \infty$, because this would allow an infinite partition of some intervals, while leaving others finite, and in this case the sum would not generally yield the integral.

13.3 Rectifiable Curves and Bounded Functions

To consider the existence of the Riemann integral as defined before, the starting point is to take the integrand $f(z) = 1$ to be unity, and define a **rectifiable curve** as one whose sum of the length of secants, for an arbitrary partition, is bounded:

$$t_0 < t_1 < \cdots < t_{N-1} < t_N: \quad \sum_{n=0}^{N-1} |z(t_{n+1}) - z(t_n)| \leqq L \equiv \int_a^b |\mathrm{d}z|, \tag{13.10}$$

where L is the length of the curve between $a \equiv z(t_0)$ and $b \equiv z(t_N)$; thus a rectifiable curve is one of finite length. A function $f(z)$ is **bounded** in a set of points X, for example, along a curve, or in a region, if its modulus lies below a fixed (positive) constant at all points:

$$f(z) \in B(X): \quad \exists_{M>0} \forall_{z \in X}: \quad |f(z)| \leqq M, \tag{13.11}$$

for example, *a continuous function is bounded* C(X) ⊂ B(X). It is clear that *if the function* f(z) *is continuous along a curve* C *that is rectifiable from* a *to* b, *its Riemann integral exists:*

$$\left| \int_a^b f(z)\,\mathrm{d}z \right| \leqq |f(z)|_{\max} \sum_{n=0}^{N-1} |z_{n+1} - z_n| \leqq ML, \tag{13.12}$$

because it has an upper bound, specified (13.12) by the product of (13.10) and (13.11): *the modulus of an integral along a curve cannot exceed the product of the length of the curve* L *by the maximum of the modulus of* M *the function along it, and the equality only holds if the function is a constant* f(z) = M.

In the case of a real integral (13.9), can be obtained both upper and lower bounds:

$$b > a: \quad (b-a)\{f(x)\}_{\min} \leqq \int_a^b f(x)\mathrm{d}x \leqq (b-a)\{f(x)\}_{\max}, \tag{13.13}$$

viz. *the integral of a continuous real function along the interval* (a,b) *lies between the minimum and maximum of the function in* (a,b), *multiplied by* (b − a), *with the equality holding only if the function is a constant* $\mathrm{f(x)} = \{\mathrm{f(x)}\}_{\max} = \{\mathrm{f(x)}\}_{\min} = \mathrm{const}$. By taking the limit

$b \to a$, and recalling that the function $f(x)$ is continuous, $\{f(x)\}_{\min} = f(a) = \{f(x)\}_{\max}$, so that (13.13) implies:

$$f(z) \in C(a,b): \quad \lim_{b \to a} (b-a)^{-1} \int_a^b f(z)\,\mathrm{d}z = f(a), \tag{13.14}$$

that *if the function* f(z) *is continuous along a segment* (a,b)*, then the integral divided by the distance tends to the value of the function as* b → a. The theorem (13.14) was proved for real integrals, but it holds as well for a complex integral:

$$\int f(z)\mathrm{d}z = \int (\Phi + i\Psi)(\mathrm{d}x + i\mathrm{d}y) = \int (\Phi\,\mathrm{d}x - \Psi\,\mathrm{d}y) + i\int (\Phi\,\mathrm{d}y + \Psi\,\mathrm{d}x), \tag{13.15}$$

because it is the sum of two real integrals. If a function has one (or a finite number) of discontinuities at $z = c$, with limit $c + 0$ at right and $c - 0$ at left, then the integral along a segment $c \equiv (a,b)$ containing c, can be determined as the sum of integrals from a to $c - 0$ and from $c + 0$ to b.

13.4 Parametric Limits and Uniform Continuity

If a function $f(z:\zeta)$ of two variables is integrated in one variable along a rectifiable curve, it defines a function of the other variable:

$$G(z) \equiv \int_a^b f(\zeta; z)\,\mathrm{d}\zeta; \quad \lim_{z \to c} G(z) = G(c), \tag{13.16a,b}$$

the latter function (13.16b) is continuous at a point $z = c$, if the limit can be interchanged with the integral:

$$\lim_{z \to c} \int_a^b f(\zeta : z)\,\mathrm{d}\zeta = \int_a^b f(\zeta; c)\,\mathrm{d}\zeta. \tag{13.17}$$

A function $f(\zeta; z)$ of two variables is **continuous** at $z = c$, **uniformly** in $\zeta \in (a,b)$, iff its limit at $z = c$ can be calculated in the same way regardless the value of ζ in (a,b):

$$f(\zeta; z) \in C(c; a,b): \quad \forall_{\zeta \in (a,b) \wedge \delta > 0} \quad \exists_{\varepsilon(\delta,c) > 0} : |z - c| < \varepsilon \Rightarrow |f(\zeta; z) - f(\zeta; c)| < \delta, \tag{13.18}$$

where $\varepsilon(\delta, c)$ depends on δ and c, but not on ζ, provided that $\zeta \in (a,b)$; in the case of ordinary, nonuniform continuity $\varepsilon(\delta, c, \zeta)$ would depend on ζ. The independence of δ on ζ, allows the estimate:

$$\left| \lim_{z \to c} \int_a^b f(\zeta; z)\,\mathrm{d}\zeta - \int_a^b f(\zeta; c)\,\mathrm{d}\zeta \right| \leqq \lim_{z \to c} \int_a^b |f(\zeta; z) - f(\zeta; c)|\,|\mathrm{d}\zeta| \leqq \lim_{z \to c} L\delta(\varepsilon, c) = 0, \tag{13.19}$$

where L is the length (13.10) of the rectifiable curve. It has been proved that: *if the function* f(ζ;z) *is: (i) integrable in* ζ *along the rectifiable curve* (a,b)*; (ii) continuous at* z=c*, uniformly in* ζ ∈ (a,b)*; then: (a) the limit can be taken under integral sign in (13.17); (b) the integral (13.16a) in* ζ *is a continuous function (13.16b) of* z *at* z=c.

13.5 Complex Loop and Contour Integrals

A curve $z(t)$ with $\alpha \leqq t \leqq \beta$ has a **double point** Q in Figure 13.1b iff $z(t_1) = z(t_2) = c$ for $\alpha \leqq t_1 < t_2 \leqq \beta$, that is, it crosses itself at $z = c$; a curve is **simple** iff it has no double points, that is, it never crosses itself. A **contour** is a simple, rectifiable curve (Figure 13.1c), that is, it has finite length and does not cross itself. A curve $z(t)$ with $\alpha \leqq t \leqq \beta$ is **closed** iff $z(\beta) = z(\alpha)$, viz. it starts at the point where it ends, and its closure would make the first point the only double point; a **loop** is a closed contour, that is, a simple, closed, rectifiable curve (Figure 13.1d). A **loop integral** is the limit of a contour integral when the end-points coincide

$$\oint f(z)\mathrm{d}z = \lim_{P_2 \to P_1} \int_{P_1}^{P_2} f(z)\,\mathrm{d}z, \tag{13.20}$$

where $\oint \equiv \int^{(+)} (\int^{(-)})$ indicates the **positive (negative) direction** of rotation, that is, counterclockwise (clockwise). The interchange of the end-points is equivalent to replacing $\mathrm{d}z$ by $-\mathrm{d}z$:

$$\int_a^b f(z)\,\mathrm{d}z = -\int_b^a f(z)\,\mathrm{d}z, \quad \int^{(+)} f(z)\,\mathrm{d}z = -\int^{(-)} f(z)\,\mathrm{d}z, \tag{13.21a,b}$$

showing that *if the contour (13.21a) [loop (13.21b)] is taken in opposite direction the integral changes sign.*

The final value of the contour integral (13.21a) depends on: (i) the integrand $f(z)$; (ii) the end-points a,b; (iii) the contour or path of integration. Hence the integral of a given function between the same end-points along distinct paths (Figure 13.1e) may be different; it follows that the loop integral (13.20) may not be zero (Figure 13.1f), for example, if the integrand $f(z)$ has no primitive (Example 20.4), or if the primitive exists but is multivalued (Chapters 7 and 9), and changes branch after describing the loop. As an example, consider the function z^{-1}, whose primitive, the logarithm, is a many-valued function that varies by $2\pi i$ after describing a loop around the origin:

$$\int_{|z|=R} z^{-1}\mathrm{d}z = \{\log z\}_{Re^{i0}}^{Re^{i2\pi}} = \log\left\{\frac{Re^{i2\pi}}{R}\right\} = 2\pi i, \tag{13.22}$$

such as a circle of a radius R. In the preceding example, the integrand z^{-1}, is not holomorphic at the origin $z = 0$ that lies within the loop of integration. If a loop integral had been taken not enclosing the origin the result would have been zero. This suggests the theorem:

$$f(z) \in \mathcal{D}(A) \Leftrightarrow \oint_{\partial A} f(z)\,\mathrm{d}z = 0 \Leftrightarrow \int_a^b f(z)\,\mathrm{d}z = J(a,b), \tag{13.23a–c}$$

stating that *the following three statements are equivalent in a region* A*: (i) the function* f(z) *is holomorphic (13.23a) at all points* z ∈ A*; (ii) the integral of* f(z) *along the* ∂A *boundary of a (13.23b), or any other loop contained in* A*, is zero; (iii) the contour integral of* f(z) *between end-points in* a, b ∈ A *in* A*, is the same for any contour lying entirely within* A.

13.6 Reciprocal Theorems of Cauchy (1825) and Morera (1886)–Osgood (1896)

The proof of (13.23a–c) starts by showing that (13.23b,c) are equivalent. To prove this consider four points A, B, C, D in counterclockwise sequence along a loop (Figure 13.1g):

$$\oint f(z)\mathrm{d}z \equiv \int_{ABCDA} f(z)\,\mathrm{d}z = \int_{ABC} f(z)\,\mathrm{d}z - \int_{ADC} f(z)\,\mathrm{d}z; \tag{13.24}$$

the loop integral vanishes iff the contour integrals between the same end-points A,C along different paths ABC, ADC are equal:

$$\oint f(z)\mathrm{d}z = 0 \Leftrightarrow \int_{ABC} f(z)\,\mathrm{d}z = \int_{ADC} f(z)\,\mathrm{d}z. \tag{13.25}$$

The proof of the theorem now reduces to the equivalence of (13.23a,b).

The first implication, from (13.23a) to (13.23b) is **Cauchy's first theorem** (1825): *if the function* f(z) *is holomorphic in a region, its contour integral is independent of the path within the region, between the same end-points.* This follows from the **fundamental theorem of analysis:**

$$\int_a^b f(z)\mathrm{d}z = \int_{F(a)}^{F(b)} \mathrm{d}F(z) = F(b) - F(a) \equiv J(a,b), \tag{13.26}$$

stating that *the integral of a function $f(z)$ is the difference of the primitive $F(z)$ at the end-points*; this follows from the definition of Riemann integral (13.8b) with integrand unity $f(z) = 1$ and variable $z \equiv F$, as in (19.1). Since a holomorphic function $f(z)$ has (Section 13.1) a single-valued holomorphic, hence continuous, primitive $F(z)$, the result (13.26) shows that the contour integral depends only on the end-points, and specifies its value $J(a,b)$.

The proof of the theorem is completed by the second implication, from (13.23b) to (13.23a) that is **Morera (1889) – Osgood (1896) theorem**, *namely, the reciprocal of Cauchy's first theorem: if a contour integrals depends only on the end-points and not on the path of integration, the integrand is holomorphic.* To prove the theorem, assume that the integrals:

$$\int_a^z f(\zeta)\mathrm{d}\zeta = J(z), \quad \int_a^{z+\Delta z} f(\zeta)\,\mathrm{d}\zeta = J(z+\Delta z), \tag{13.27a,b}$$

depend only on the upper end-point, the lower end-point being considered fixed (a procedure could be applied to the lower end-point, similar to that that follows for the upper end-point). The incremental ratio for the function $J(z)$ defined by the integrals (13.27a,b) has a well-defined limit:

$$\frac{\mathrm{d}J}{\mathrm{d}z} = \lim_{\Delta z \to 0} \frac{J(z+\Delta z) - J(z)}{\Delta z} = \lim_{\Delta z \to 0} \frac{1}{\Delta z} \int_z^{z+\Delta z} f(\zeta)\,\mathrm{d}\zeta = f(z), \tag{13.28}$$

where (13.14) was used, with $z \equiv a, b \equiv z + \Delta z, b - a = \Delta z$. This proves that the function $J(z)$ defined by the integral is holomorphic, with derivate $J'(z) = f(z)$; since it can be proved (Section 15.4) that the derivate of a holomorphic function is holomorphic, it follows that the integrand $f(z)$ is holomorphic. QED.

13.7 Integration by Parts and Chain Rule (Leibnitz, 1864)

The derivation (Section 11.2) and primitivation (Section 13.1) or integration (Section 13.2) are **almost-inverse operations**, since if F is one primitive of f, then: (i) $F+C$ where C is an arbitrary constant is also a primitive; (ii) f is the derivate of F, unique for all $F+C$. The derivate being unique and the primitive defined to within an additive constant, implies that *(i) the derivate of the integral of a continuous function, with regard to the upper limit of integration, is the value of the function at the upper limit; (ii) the integral of the derivate of a holomorphic function, is the difference of its values at the upper and lower end-points:*

$$f \in C: \quad \frac{\mathrm{d}}{\mathrm{d}z}\int_a^z f(\zeta)\mathrm{d}\zeta = f(z); \quad f \in D: \quad \int_a^z f'(\zeta)\mathrm{d}\zeta = f(z) - f(a). \tag{13.29a,b}$$

This almost-inverse property is maintained for other properties, for example:

$$f, g \in D: \quad (fg)' = f'g + g'f, \quad \int_a^b fg'\mathrm{d}z = \{fg\}_a^b - \int_a^b f'g\,\mathrm{d}z, \tag{13.30a,b}$$

the rules of derivation of the product (13.30a) and of **integration by parts** *(13.30b).*

The rule of derivation of the product (13.30a) can be generalized to **Leibnitz's (1864) chain rule**, *expressing the derivate of order* n *of the product of functions, as a binominal theorem (25.38) of derivatives:*

$$\frac{\mathrm{d}^n\{f(z)g(z)\}}{\mathrm{d}z^n} = \sum_{p=0}^{n} \frac{n!}{p!(n-p)!} f^{(p)}(z)g^{(n-p)}(z); \tag{13.31}$$

the formula (13.31) can be proved by induction: (i) it holds for $n=1$, since then it coincides with the rule of derivation of the product (13.30a); (ii) if it holds for integer n such that $n \geq 1$, then it also holds for $n+1$:

$$\begin{aligned}\frac{\mathrm{d}^{n+1}}{\mathrm{d}z^{n+1}}\{f(z)\} &= \sum_{p=0}^{n} \frac{n!}{p!(n-p)!}\left\{f^{(p+1)}(z)g^{(n-p)}(z) + f^{(p)}(z)g^{(n-p+1)}(z)\right\} \\ &= \sum_{p=0}^{n+1} \frac{(n+1)!}{p!(n+1-p)!} f^{(p)}(z)g^{(n-p+1)}(z),\end{aligned} \tag{13.32}$$

where

$$\frac{n!}{p!(n-p)!} + \frac{n!}{(p-1)!(n-p+1)!} = \frac{n!}{(p-1)!(n-p)!}\left(\frac{1}{p} + \frac{1}{n-p+1}\right) = \frac{(n+1)!}{p!(n+1-p)!}, \tag{13.33}$$

was used to simplify (13.32). QED.

The formula (13.33) can be restated as the following *property (13.34b) of arrangements (13.34a)*:

$$\binom{n}{p} \equiv \frac{n!}{p!(n-p)!}: \quad \binom{n}{p} + \binom{n}{p-1} = \binom{n+1}{p}. \tag{13.34a,b}$$

An example of the use of Leibnitz's rule (13.31) is:

$$\frac{\mathrm{d}^n\left(z^{-1}\mathrm{e}^{az}\right)}{\mathrm{d}z^n} = \sum_{p=0}^{n}\binom{n}{p}\frac{\mathrm{d}^{n-p}\left(\mathrm{e}^{az}\right)}{\mathrm{d}z^{n-p}}\frac{\mathrm{d}^p(1/z)}{\mathrm{d}z^p} = \mathrm{e}^{az}\sum_{p=0}^{n}(-)^p\frac{n!}{(n-p)!}a^{n-p}z^{-1-p}; \tag{13.35}$$

that can be checked, for example, for $n=3$:

$$\left(z^{-1}e^{az}\right)''' = e^{az}\left(\frac{a^3}{z} - \frac{3a^2}{z^2} + \frac{6a}{z^3} - \frac{6}{z^4}\right); \tag{13.36}$$

further cases appear in Example 20.5.

13.8 Derivation of an Integral with Regard to a Parameter

Reconsider the case (13.16a) of an integrand $f(\zeta;z)$ that depends not only on the variable of integration ζ but also on a parameter z, so that the integral is a function of z; in the present instance the aim is to go beyond the sufficient conditions for continuity (13.16b), and consider derivation via the incremental ratio:

$$\frac{d}{dz}\int_a^b f(\zeta;z)\,d\zeta = \lim_{\Delta z\to 0}\frac{1}{\Delta z}\int_a^b \{f(\zeta;z+\Delta z) - f(\zeta;z)\}\,d\zeta, \tag{13.37}$$

that should tend to a limit independent of the way which $\Delta z \to 0$. Assuming that $f(\zeta;z)$ has continuous first-order derivate with regard to z, the mean value theorem can be used:

$$\frac{d}{dz}\int_a^b f(\zeta;z)\,dz = \lim_{\Delta z\to 0}\int_a^b \left\{\frac{\partial f(\zeta;z)}{\partial z} + \varepsilon(\zeta;z)\right\}d\zeta, \tag{13.38}$$

where $\varepsilon(\zeta;z)\to 0$ as $\Delta z \to 0$; from (13.38) follows that derivation under integral sign is permissible:

$$\left|\frac{d}{dz}\int_a^b f(\zeta;z)\,dz - \int_a^b \frac{\partial f(\zeta,z)}{\partial z}d\zeta\right| \leqq \lim_{\Delta z\to 0}\int_a^b |\varepsilon(\zeta;z)|\,|dz| \leqq L \lim_{\Delta z\to 0}|\varepsilon(z)|_{\max} = 0, \tag{13.39}$$

where L is the length of the contour of integration between a and b, and it was assumed that $\varepsilon(z)$ depends on z but not on ζ, that is, $\partial f/\partial z$ is continuous in z, uniformly in ζ. The continuity of the integrand $f(\zeta;z)$ in the variable of integration ζ, uniformly in the parameter z, thus allows both taking limits (13.16b) $\equiv$ (13.17) and differentiating (13.40) under the integral sign in (13.16a).

It has been proved that *if: (i) the function $f(\zeta;z)$ is continuous in ζ along the contour (a,b); (ii) $\partial f/\partial z$ is uniformly continuous in $\zeta \in (a,b)$; then: (a) the derivate with regard to the parameter may be taken under the integral:*

$$\frac{d\left\{\int_a^b f(\zeta;z)\,d\zeta\right\}}{dz} = \int_a^b \frac{\partial f(\zeta;z)}{\partial z}d\zeta; \tag{13.40}$$

(b) the integral specifies a holomorphic function:

$$G(z) \equiv \int_a^b f(\zeta;z)\,d\zeta, \quad G'(z) = \int_a^b \frac{\partial f(\zeta;z)}{\partial z}d\zeta. \tag{13.41a,b}$$

As an example consider the function $f(\zeta;z)=\exp(\zeta z)$, that is (i) continuous along any finity contour (a,b) and (ii) whose derivative $\partial f/\partial z=\zeta e^{\zeta z}$ is uniformly continuous for all finite ζ:

$$\int_a^b e^{\zeta z}\,d\zeta=\left[z^{-1}e^{\zeta z}\right]_a^b=\frac{e^{bz}-e^{az}}{z}, \tag{13.42a}$$

$$\int_a^b \zeta e^{\zeta z}\,d\zeta=\frac{e^{az}-e^{bz}}{z^2}+\frac{be^{bz}-ae^{az}}{z}, \tag{13.42b}$$

where (13.42b) can be obtained: (i) either by differentiating (13.42a) with regard to z; (ii) or integrating (13.42b) by parts (13.30b):

$$\int_a^b \zeta e^{\zeta z}d\zeta=\int_a^b \frac{\zeta}{z}\frac{d}{d\zeta}\left(e^{\zeta z}\right)d\zeta=\left[\frac{\zeta}{z}e^{\zeta z}\right]_a^b-\frac{1}{z}\int_a^b e^{\zeta z}d\zeta=\frac{be^{bz}-ae^{az}}{z}-\frac{1}{z^2}\left[e^{\zeta z}\right]_a^b. \tag{13.42c}$$

Other cases are considered in Examples 20.6 and 20.7.

13.9 Parametric Integral with Variable End-Points

A regular curve is specified by a complex function $z(t)\in C^1$, or pair of real functions $x(t)$, $y(t)\in C^1$ implying that the tangent is continuous, that is, the curve has no cusps or angular points such as the point S in Figure 13.1c. A generalization of (13.26) is the integral with end-points on two regular curves (Figure 13.1h):

$$\int_{a(z)}^{b(z)} f(\zeta)d\zeta=F(b(z))-F(a(z)), \tag{13.43}$$

where F is the primitive of f. Derivation with regard to z yields:

$$\frac{d}{dz}\int_{a(z)}^{b(z)} f(\zeta)d\zeta=f(b(z))\frac{db}{dz}-f(a(z))\frac{da}{dz}, \tag{13.44}$$

where have been used $dF/db=f(b)$, $dF/da=f(a)$ and db/dz, da/dz exist because the curves $b(z)$, $a(z)$ are regular. It has been shown that *if the function* f(z) *is continuous along a contour with end-points on two regular curves* a,b(z)$\in \mathcal{D}$, *it satisfies the derivation formula (13.44)*. A trivial example of (13.44), with $b(z)=z$ and $a(z)\equiv a=\text{const}$ is (13.29a); a not so trivial example is:

$$\frac{d}{dz}\int_z^{z^2} e^{\zeta}\,d\zeta=2ze^{z^2}-e^z=\frac{d\left(e^{z^2}-e^z\right)}{dz}, \tag{13.45}$$

that can be checked directly.

The formulas (13.40) and (13.44) are generalized together in the following theorem: *if the function $f(\zeta;z)$ is continuous in ζ and $\partial f/\partial z$ uniformly continuous in ζ along a contour $\zeta\in(a(z),b(z))$, with end-points lying on regular curves* a, b(z) $\in\mathcal{D}$, *then the function:*

$$a,b(z)\in\mathcal{D},\quad f\in C,\frac{\partial f}{\partial z}\in\tilde{C}:\quad G(z)\equiv\int_{a(z)}^{b(z)} f(\zeta;z)d\zeta, \tag{13.46a}$$

is holomorphic with derivate specified by:

$$\frac{\mathrm{d}G}{\mathrm{d}z} = \int_{a(z)}^{b(z)} \frac{\partial f(\zeta;z)}{\partial z}\mathrm{d}\zeta + f(b(z);z)\frac{\mathrm{d}b}{\mathrm{d}x} - f(a(z);z)\frac{\mathrm{d}a}{\mathrm{d}z}. \tag{13.46b}$$

An example combining (13.42b) and (13.45), is:

$$F(z) \equiv \frac{\mathrm{d}}{\mathrm{d}z}\int_z^{z^2} \mathrm{e}^{\zeta z}\,\mathrm{d}\zeta = \left(3z - \frac{1}{z^2}\right)\mathrm{e}^{z^3} - \left(2 - \frac{1}{z^2}\right)\mathrm{e}^{z^2}; \tag{13.47a}$$

it corresponds to:

$$\begin{aligned} F(z) &= \int_z^{z^2} \zeta \mathrm{e}^{\zeta z}\,\mathrm{d}\zeta + \mathrm{e}^{\zeta z}\Big|_{\zeta=z^2}\frac{\mathrm{d}(z^2)}{\mathrm{d}z} - \mathrm{e}^{\zeta z}\Big|_{\zeta=z}\left(\frac{\mathrm{d}z}{\mathrm{d}z}\right) \\ &= \frac{\mathrm{e}^{z^2} - \mathrm{e}^{z^3}}{z^2} + \frac{z^2\mathrm{e}^{z^3} - z\mathrm{e}^{z^2}}{z} + 2z\mathrm{e}^{z^3} - \mathrm{e}^{z^2}, \end{aligned} \tag{13.47b}$$

where (13.42b) was used with $a = z$ and $b = z^2$. Other cases are considered in Example 20.7.

Note 13.1. *Two Dirichlet Functions*: The contour of integrals (Chapter 13) are the basis of the calculus of residues (Chapter 15) which also applies to real integrals (Chapter 17); other uses of contour integrals include the establishment of series (Chapters 23 and 25) and other infinite (Section 36.6) representations. The Riemann integral applies to functions with isolated discentinuities, viz. by adding the integrals along the intervals of continuity. The **Dirichlet functions**, equal to zero (unity) at rational (irrational) points (13.48a) or vice-versa (13.48b):

$$D(x) = \begin{cases} 0 & \text{if } x \in |L, \\ 1 & \text{if } x \in |I, \end{cases} \quad D_2(x) = \begin{cases} 1 & \text{if } x \in |L, \\ 0 & \text{if } x \in |I, \end{cases} \tag{13.48a,b}$$

have an infinite number of discontinuities in any interval. Neither has a Riemann integral. They have a Lebesgue (1928) integral that is a generalization of the Riemann integral. An earlier extension of the Riemann integral allowing for some singularities is the Stieltjes (1894) integral. The Leibnitz chain rule of differentiation (13.31) corresponds to the binomial formula (25.38): the binomial series (25.37b,c) corresponds to the rule of differentegration of the product (Campos, 1993), that is, derivatives of complex order.

Conclusion 13: Some types (figure 13) of plane curves: **(a)** a rectifiable curve has finite lengh, and may have **(b)** a double point at Q, or **(c)** an angular point at S. Complex integrals may be calculated along contours **(c)**. i.e., simple rectifiable curves, with angular but not double points allowed, and around loops **(d)**, which are closed contours, and can be described in the positive + or negative − direction. If a function is holomorphic in a region D, then: **(e)** its contour integrals along different paths with the same end-points P_1, P_2 are equal; **(f)** any integral along a loop contained entirely in D is zero. A loop integral $ABCDA$ in **(g)** is the difference of two contour integrals, taken between the same end-points, e.g., A and C, along different paths ABC and ADC. An integral can be calculated along a contour L whose end-points lie **(h)** on two curves $a(z)$, $b(z)$, and it can be differentiated if the curves are regular.

14

Pressure and Corner Flows

The potential flow (Chapter 12) is based on the two assumptions that it is (a) irrotational (Section 12.1); (b) incompressible (Section 12.2). To identify the physical conditions under that these approximations hold, it is necessary to consider the fundamental equations of fluid mechanics, viz.: (i) the equation of continuity (Section 14.1), stating the conservation of mass, implies that a flow is incompressible (a) if the mass density is conserved along streamlines, or equivalently the divergence of the velocity is zero (Section 12.1); (ii) the momentum equation shows that the fluid is accelerated in the direction of decreasing pressure, in the inviscid case in the absence of external forces (Section 14.2); (iii) in the absence of viscosity and thermal conduction and other dissipative processes an adiabatic condition (Section 14.3) holds, excluding heat exchanges; (iv) the temperature can be specified by the equation of state (Section 14.3), for example, a perfect gas is a reasonable approximation for air at ambient temperatures. It can be shown that the circulation is conserved along streamlines (a) for an steady irrotational flow (Section 14.4) with two restrictions: (i) homentropic, that is, with constant entropy; (ii) under external forces that are conservative, that is, derive from a potential, like gravity. Also for a homentropic steady irrotational flow the assumption of incompressibility (b) holds if (Section 14.5) the flow velocity is small compared with the sound speed. For an incompressible irrotational (that is, potential) steady flow the difference between the stagnation pressure (at a stagnation point of zero velocity) and the free stream pressure, is the dynamic pressure (Section 14.6) that is equal to the kinetic energy per unit volume; this form of Bernoulli's equation has many applications, for example, to measure the flow velocity using a Pitot (Venturi) tube in a free stream (Section 14.6) [ducted flow (Section 14.7)], from the pressure at two points. The simplest potential flow is uniform and corresponds to a dipole at infinity; the higher order multipoles at the origin (Sections 12.4–12.9) by inversion with regard to the origin lead to multipoles at infinity, corresponding to corner flows (Section 14.8), for example, a quadrupole (higher order 2^n-pole) at infinity corresponds to the flow in a rectangular (acute) corner with angle $\pi/4\,(\pi/2n)$ at the vertex. The extension to a corner with arbitrary angle $0 < \beta \leq 2\pi$ includes the sharp edge for $\beta = 2\pi$ or $n = 1/2$; this allows the consideration of a stream incident on a wedge at an angle-of-attack (Section 14.9). The equations of fluid mechanics are deduced with the aim to particulize to potential flows, while indicating the restrictions involved; the simple properties of potential flows will be modified as the various restrictions are lifted in the sequel, for example, effects of vorticity, compressibility, viscosity, turbulence, and energy and mass transfer, chemical reactions, are introduced.

14.1 Mass Conservation and Equation of Continuity

Consider the conservation of mass (Figure 14.1a) in an infinitesimal rectangle of sides $dx\,dy$. The mass flux, that is, mass per unit time entering the side AB is $\rho v_x dy$, where ρ is the mass density; subtracting the mass flux exiting the opposite side DC yields the net mass

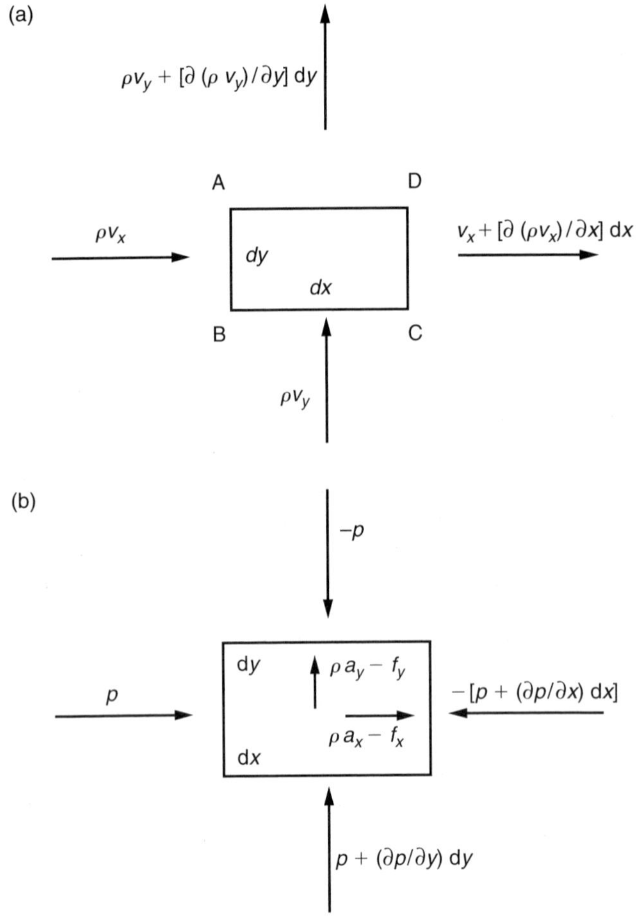

FIGURE 14.1
The conservation of mass (a) requires that the excess (default) on inflow relative to outflow trough the boundary leads to mass accumulation (depletion) in the interior. The conservation of momentum (b) for an inviscid fluid balances resultants of (i) the inertia force in the interior and of (ii) the pressure forces acting along the inward normal to the boundary; the inertia force equals the mass times acceleration.

flux in the x-direction:

$$\Delta \dot{m}_x = \rho v_x \mathrm{d}y - \left\{\rho v_x + \frac{\partial(\rho v_x)}{\partial x}\right\}\mathrm{d}y = -\frac{\partial(\rho v_x)}{\partial x}\mathrm{d}x\,\mathrm{d}y. \tag{14.1}$$

Adding the net mass flux $\Delta \dot{m}_y$ in the y-direction yields the accumulation of mass per unit time $\partial\rho/\partial t$ in the area $\mathrm{d}x\,\mathrm{d}y$, viz.:

$$\frac{\partial \rho}{\partial t}\mathrm{d}x\,\mathrm{d}y = \Delta \dot{m}_x + \Delta \dot{m}_y = -\left[\frac{\partial(\rho v_x)}{\partial x} + \frac{\partial(\rho v_y)}{\partial y}\right]\mathrm{d}x\,\mathrm{d}y. \tag{14.2}$$

Thus the **equation of continuity** *or conservation of mass states that the mass density* ρ *and velocity* $\vec{\mathrm{v}}$ *satisfy:*

$$\frac{\partial \rho}{\partial t} = -\frac{\partial(\rho v_x)}{\partial x} - \frac{\partial(\rho v_y)}{\partial y} = -\nabla\cdot(\rho\vec{\mathrm{v}}); \tag{14.3}$$

this is equivalent to:

$$\rho\left(\frac{\partial v_x}{\partial x}+\frac{\partial v_y}{\partial y}\right)=\frac{\partial \rho}{\partial t}+v_x\frac{\partial \rho}{\partial x}+v_y\frac{\partial \rho}{\partial y}; \tag{14.4}$$

it can be stated more compactly:

$$-\rho\nabla\cdot\vec{\mathrm{v}}=\frac{\partial \rho}{\partial t}+\vec{\mathrm{v}}\cdot\nabla\rho=\frac{\mathrm{d}\rho}{\mathrm{d}t}, \tag{14.5}$$

in terms of the **material derivative**:

$$\frac{\mathrm{d}}{\mathrm{d}t}\equiv\frac{\partial}{\partial t}+\frac{\mathrm{d}x}{\mathrm{d}t}\frac{\partial}{\partial x}+\frac{\mathrm{d}y}{\mathrm{d}t}\frac{\partial}{\partial y}=\frac{\partial}{\partial t}+v_x\frac{\partial}{\partial x}+v_y\frac{\partial}{\partial y}=\frac{\partial}{\partial t}+\vec{\mathrm{v}}\cdot\nabla, \tag{14.6}$$

that is the total derivative with regard to time along the streamlines, consisting of: (i) the ***local time derivative*** *$\partial/\partial t$ at fixed position (x,y); (ii) the* ***convected derivative*** *due to transport by the flow velocity $\vec{\mathrm{v}}$ at fixed time. An* ***incompressible*** *flow has constant mass density along streamlines (14.7a):*

$$\frac{\mathrm{d}\rho}{\mathrm{d}t}=0\Leftrightarrow 0=\nabla\cdot\vec{\mathrm{v}}\equiv\Lambda, \tag{14.7a,b}$$

and hence the dilatation or divergence of the velocity is zero (14.7b), in agreement with (14.5). The condition of incompressibility (14.7b) ≡ (12.16) leads to the vanishing (12.10a) of the mass flux across a fixed loop (12.9). Since the equation of continuity for a general compressible flow (14.3) ≡ (14.4) ≡ (14.5; 14.6) is a scalar equation involving a scalar (the mass density ρ) and a vector (the velocity $\vec{\mathrm{v}}$), another vector equation is needed, viz. the momentum equation (Section 14.2) specifying the force balance.

14.2 Inviscid Momentum Equation (Euler, 1752, 1759)

The normal component of the velocity of a fluid must vanish at a **rigid wall**, *whereas the tangential component may be nonzero (is also zero) for a* **inviscid (viscous)** *fluid,* that is, a **no-slip condition** does not apply (applies). In the case of an inviscid fluid the only internal stresses correspond to the pressure that is a force per unit area, acting in the inward direction; its magnitude is isotropic, that is, independent of direction. The **equation of momentum** balances the inertial force per unit volume (area) in three (two) dimensions, that is, mass density ρ times acceleration $\vec{\mathrm{a}}$, against the resultant of the density of **external forces** $\vec{\mathrm{f}}$ and pressure forces, leading (Figure 14.1b) in the x-direction for an infinitesimal rectangle to:

$$(\rho a_x-f_x)\,\mathrm{d}x\,\mathrm{d}y=p\,\mathrm{d}y-\left(p+\frac{\partial p}{\partial x}\mathrm{d}x\right)\mathrm{d}y=-\frac{\partial p}{\partial x}\mathrm{d}x\,\mathrm{d}y. \tag{14.8}$$

A similar force balance in the y-direction leads to **Euler's equation** (1752, 1759) stating that *in an inviscid fluid the inertial force is balanced by the external force minus the pressure gradient:*

$$\frac{\vec{\mathrm{f}}-\nabla p}{\rho}=\vec{\mathrm{a}}\equiv\frac{\mathrm{d}\vec{\mathrm{v}}}{\mathrm{d}t}=\frac{\partial\vec{\mathrm{v}}}{\partial t}+\left(\vec{\mathrm{v}}\cdot\nabla\right)\vec{\mathrm{v}}; \tag{14.9}$$

in the absence of external forces the flow is in the direction of decreasing pressure. The acceleration is the material derivative (14.6) of the velocity, and thus consists of local $\partial\vec{v}/\partial t$ plus convective $(\vec{v}\cdot\nabla)\vec{v}$ terms. The scalar equation of continuity (14.5) and vector (inviscid) momentum equation (14.9) involve one vector (velocity) and two scalars (mass density and pressure), so another scalar equation is needed, viz. the equation of energy (Section 14.3).

14.3 Adiabatic Condition and Equation of State

For an inviscid fluid there is no dissipation by viscosity; in the absence of thermal conduction (Chapter 32) and all other dissipative processes, the energy equation can be replaced by a statement that no heat is exchanged. *The absence of dissipation corresponds to an adiabatic condition (14.10a) relating the pressure to density:*

$$p\rho^{-\gamma} \equiv \text{const} \equiv k, \quad \gamma \equiv \frac{C_p}{C_v}, \tag{14.10a,b}$$

*where the **adiabatic exponent** γ, is the ratio of **specific heats at constant pressure** C_p (**volume** C_v); the latter depend only on **temperature** that is specified by the equation of state (14.11a):*

$$p = p(\rho, T): \quad p = \rho RT, \quad R = 2.87 \times 10^6\,\text{cm}^2\,\text{s}^{-2}\,\text{K}^{-1}, \tag{14.11a–c}$$

*for example, for a **perfect gas** the equation of state is (14.11b) where (14.11c) is the* **gas constant** in degrees Kelvin K for air. For other substances, for example, liquids, the equation of state (14.11a) may take forms other than (14.11b).

The preceding are the fundamental equations of the adiabatic flow of an inviscid fluid in a given external force field $\vec{f}$, specifying one vector and three scalar quantities, viz. the velocity $\vec{v}$, mass density ρ, pressure p, and temperature T that satisfy four equations, one vector and three scalar, viz. the: (i) equation of continuity (14.5) stating mass conservation; (ii) inviscid momentum equation (14.9) stating force balance; (iii) adiabatic condition (14.10a,b) excluding heat exchanges; (iv) equation of state (14.11a), taking the particular form (14.11b) for a perfect gas. The perfect gas law (14.11b) is a good approximation for atmospheric air (14.11c) at ambient conditions; the adiabatic exponent is $\gamma = 7/5 = 1.4$ for a diatomic gas, viz. air consists mostly of diatomic nitrogen (80%) and oxygen (20%) molecules. The equations of fluid mechanics, in the particular adiabatic form (Section 14.3), are used next to establish the conditions of validity and the implication of the assumptions of (a) irrotational and (b) incompressible flow.

The **enthalpy** is a thermodynamic function of state specified by (14.12a):

$$\mathrm{d}H = \rho^{-1}\,\mathrm{d}p + T\,\mathrm{d}S, \quad \nabla H = \rho^{-1}\nabla p + \mathrm{T}\nabla S, \tag{14.12a,b}$$

where S is the **entropy**, implying equivalently (14.12b). The convective component of the acceleration can be written (14.13b,c):

$$\varpi \equiv \frac{\partial v_y}{\partial x} - \frac{\partial v_x}{\partial y}: \quad a_x - \frac{\partial v_x}{\partial t} = v_x\frac{\partial v_x}{\partial x} + v_y\frac{\partial v_x}{\partial y} = v_x\frac{\partial v_x}{\partial x} + v_y\frac{\partial v_y}{\partial x} - v_y\varpi,$$

$$a_y - \frac{\partial v_y}{\partial t} = v_x\frac{\partial v_y}{\partial x} + v_y\frac{\partial v_y}{\partial y} = v_y\frac{\partial v_y}{\partial y} + v_x\frac{\partial v_x}{\partial y} + v_x\varpi, \tag{14.13a–c}$$

in terms of the vorticity (14.13a); substitution in (14.9) leads to the inviscid momentum equation in the form:

$$f_x - \frac{\partial H}{\partial x} + T\frac{\partial S}{\partial x} = \frac{f_x - \partial p/\partial x}{\rho} = a_x = \frac{\partial v_x}{\partial t} + \frac{\partial\left(v^2/2\right)}{\partial x} - v_y\varpi, \tag{14.14a}$$

$$f_y - \frac{\partial H}{\partial y} + T\frac{\partial S}{\partial y} = \frac{f_y - \partial p/\partial y}{\rho} = a_y = \frac{\partial v_y}{\partial t} + \frac{\partial\left(v^2/2\right)}{\partial y} + v_x\varpi; \tag{14.14b}$$

recalling that the vorticity is orthogonal to (x,y)-plane (12.8b) ≡ (14.13a; 14.15a):

$$\vec{\varpi} \equiv \nabla \wedge \vec{v} = \varpi \vec{e}_3 : \left(\nabla \wedge \vec{v}\right) \wedge \vec{v} = \varpi \vec{e}_3 \wedge \left(v_x \vec{e}_x + v_y \vec{e}_y\right) = \varpi v_x \vec{e}_y - \varpi v_y \vec{e}_x, \quad (14.15a,b)$$

the last terms in (14.14a,b) coincide with (14.15b); this leads to a vector form of the *inviscid (14.9) ≡ (14.16) momentum equation:*

$$\frac{\vec{f} - \nabla p}{\rho} = -\nabla H + T\nabla S = \frac{d\vec{v}}{dt} = \frac{\partial \vec{v}}{\partial t} + \nabla\left(\frac{v^2}{2}\right) + \left(\nabla \wedge \vec{v}\right) \wedge \vec{v}; \quad (14.16)$$

this also specifies *the acceleration or material derivative (14.6) applied to the velocity, as a sum of a local time derivative, a derivative of the modulus of the velocity and a vortical term.*

14.4 Homentropic Flow and Conservation of Circulation (Helmholtz, 1858; Kelvin, 1869)

The distinction should be made:

$$S = \text{const}, \quad 0 = \frac{dS}{dt} = \frac{\partial S}{\partial t} + \vec{v} \cdot \nabla S, \quad (14.17a,b)$$

between: (i) a **homentropic** flow, for that the entropy is constant (14.17a); (ii) an **adiabatic** flow, for which the entropy is conserved along the streamlines (14.17b), and thus can vary transversely, and need not be constant. Thus the homentropic flow is a particular case of isentropic flow. For a homentropic flow (14.18a), the second term in the enthalpy (14.12b) that is the **heat** exchanged, vanishes, leaving only the first term (14.18b,c):

$$dS = 0: \quad dH = \frac{dp}{\rho}, \quad \nabla H = \frac{\nabla p}{\rho}; \quad \nabla H + \frac{d\vec{v}}{dt} = \frac{\vec{f}}{\rho}, \quad (14.18a\text{–}d)$$

this can be substituted in (14.9) leading to the homentropic momentum equation (14.18d). An adiabatic flow in general, and homentropic flow in particular, excludes all forms of dissipation, meaning not only viscosity, but also thermal conduction, for example. Thus:

$$\frac{\vec{f}}{\rho} - \nabla H = \vec{a} = \frac{d\vec{v}}{dt} = \frac{\partial \vec{v}}{\partial t} + \left(\vec{v} \cdot \nabla\right)\vec{v} = \frac{\partial \vec{v}}{\partial t} + \nabla\left(\frac{v^2}{2}\right) + \left(\nabla \wedge \vec{v}\right) \wedge \vec{v}. \quad (14.19)$$

the inviscid momentum equation (14.9) ≡ (14.16) is more general than the homentropic momentum equation (14.18d) ≡(14.19).

In the case of the gravity field (14.20a) the force per unit mass is the gradient of minus the gravity potential Φ_g in (14.20b):

$$\frac{\vec{f}_g}{\rho} = \vec{g} = -\nabla \Phi_g, \quad d\Phi_g = \nabla \Phi_g \cdot d\vec{x} = -\vec{g} \cdot d\vec{x} = -\frac{\vec{f}_g \cdot d\vec{x}}{\rho}, \quad (14.20a,b)$$

not to be confused with the velocity potential $\Phi \equiv \Phi_v$ of an irrotational flow. The *homentropic momentum equation for an irrotational flow in the gravity field:*

$$\nabla \wedge \vec{v} = 0: \quad \frac{\partial \vec{v}}{\partial t} = -\nabla\left(H + \frac{v^2}{2} + \Phi_g\right), \quad (14.21a,b)$$

implies that the circulation along a loop convected with the flow:

$$\frac{\mathrm{d}}{\mathrm{d}t}\oint \vec{\mathrm{v}}\cdot \mathrm{d}\vec{\mathrm{x}} = \oint \frac{\partial \vec{\mathrm{v}}}{\partial t}\cdot \mathrm{d}\vec{\mathrm{x}} = -\oint \nabla\left(H + \frac{v^2}{2} + \Phi_g\right)\cdot \mathrm{d}\vec{\mathrm{x}} = -\oint \mathrm{d}\left(H + \frac{v^2}{2} + \Phi_g\right) = 0, \tag{14.22}$$

vanishes because the integrand is a single-valued function. This is **Kelvin's theorem** (1869; Helmholtz, 1858): *the homentropic momentum equation (14.19), for an irrotational flow (14.23a) in the gravity field (14.23b) implies that the circulation (14.23d) is conserved along streamlines (14.23c):*

$$\nabla\wedge\vec{\mathrm{v}} = 0 = \nabla\wedge\vec{\mathrm{g}}: \quad \frac{\mathrm{d}\Gamma}{\mathrm{d}t} = 0, \quad \Gamma \equiv \oint \vec{\mathrm{v}}\cdot\mathrm{d}\vec{\mathrm{x}}; \tag{14.23a–d}$$

hence, if a flow is irrotational at a time $t = 0$, it will remain irrotational for all time, as long as the streamlines are not cut by a body. If the streamlines are cut by a body, a vortical or turbulent wake may form; also a viscous flow is generally vortical. Thus the maintenance of irrotational flow supposes homentropic conditions and allows for a gravity field. Having clarified the physical conditions for irrotational flow (a), it remains to consider (b) incompressibility.

14.5 Hydrostatic, Dynamic, and Stagnation Pressures (Torricelli, 1643; Bernoulli, 1738)

In an irrotational flow the velocity is the gradient of a potential (14.24a), and thus the homentropic momentum equation (14.19) reduces to (14.24b):

$$\vec{\mathrm{v}} = \nabla\Phi: \quad \nabla\left(\frac{\partial\Phi}{\partial t} + \frac{v^2}{2} + H + \Phi_g\right) = 0, \tag{14.24a,b}$$

thus quantity in curved brackets cannot depend on position, that is, it is a function of time only:

$$\mathrm{d}S = 0, \vec{\mathrm{v}} = \nabla\Phi: \quad \frac{\partial\Phi}{\partial t} + \frac{v^2}{2} + H - \int \vec{\mathrm{g}}\cdot\mathrm{d}\vec{\mathrm{x}} = f(t). \tag{14.25a–c}$$

This is the **Bernoulli equation** (1738): *the homentropic (14.25a) irrotational (14.25b) flow in a gravity field (14.20b) satisfies the conservation equation (14.25c); the absence of a gravity field (14.26a) for a* **steady flow**, *that is, independent of time (14.26b), leads to (14.26c):*

$$\vec{\mathrm{g}} = 0 = \frac{\partial\Phi}{\partial t}: \quad \frac{v^2}{2} + H = H_0 \equiv \text{const}; \tag{14.26a–c}$$

here the constant is the **stagnation enthalpy**, *viz. the enthalpy at a stagnation point, where the velocity is zero. In the case of an irrotational incompressible (that is, potential) homentropic flow the enthalpy is given by (14.18b) ≡ (14.27b) leading to (14.27c):*

$$\rho = \text{const} \equiv \rho_0: \quad H = \frac{p}{\rho_0}, \quad p + \frac{\rho_0 v^2}{2} = p_0 = \text{const}. \tag{14.27a–c}$$

This states the constancy of the **stagnation pressure** *(14.27c) that is the sum of: (i) the* **free stream pressure** *p; (ii) the* ***dynamic pressure*** *(14.28b) that coincides with the kinetic energy per unit volume.*

If the gravity field (14.25c) is taken into account there is in addition (iii) the **hydrostatic pressure**, *that is, weight of the column of fluid lying above the measurement point (14.28a):*

$$p_0 = p + q - \rho_0 \int \vec{g} \cdot d\vec{x}, \quad q \equiv \frac{\rho_0}{2} v^2; \tag{14.28a,b}$$

in the case of a uniform gravity field (14.29a,b) the hydrostatic pressure simplifies (14.29c) to:

$$\vec{g} = -g_0 \vec{e}_z, \quad g_0 = \text{const}: \quad \Phi_g = \rho_0 g_0 z = p_0 - p - \frac{1}{2}\rho_0 v^2. \tag{14.29a–c}$$

If the pressure is uniform (14.30a) this simplifies to the **Torricelli theorem** (1643):

$$p = p_0: \quad v = \sqrt{-2gz} = \sqrt{-\frac{\Phi_g}{\rho_0}}, \tag{14.30a,b}$$

where (14.30b) corresponds to: (i) the kinetic energy per unit volume or dynamic pressure (14.28b) being equal to minus the gravity potential (14.29c); (ii) if a liquid fills a reservoir, then opening an orifice a distance $-z$ *from the free surface, the fluid will exit with velocity (14.30b), if there are no losses; (iii) the latter (14.30b) is also the velocity of impact on the ground* $-z$ *of a body released in free fall from the altitude zero.* The minus sign in (14.30b) implies that a fluid particle gains velocity as the altitude reduces $z < 0$ below its initial level $z = 0$; a rise above the initial level $z > 0$, would lead to an "imaginary" velocity, and is not possible if the fluid particle was initially at rest at the level $z = 0$.

14.6 Compressibility Effects and the Pitot Tube (1732)

The difference between incompressible (14.5) [compressible (Section 14.6)] flow can be illustrated by the influence of the sound speed and Mach number (Subsection 14.6.1) on the use of the Pitot tube to measure the velocity in a flow (Subsection 14.6.2).

14.6.1 Sound Speed and Mach Number

The ratio of pressure by density has the dimensions of the square of a velocity, and the ratio of their differentials specifies the **sound speed** (14.31b):

$$S = \text{const}: \quad c^2 = \left(\frac{\partial p}{\partial \rho}\right)_s = \frac{\gamma p}{\rho} = \gamma R T, \tag{14.31a–c}$$

in adiabatic (14.31a) conditions with constant (14.32a) adiabatic exponent (14.10b); the sound speed depends only on temperature (14.31c) for a perfect gas (14.11b). In (14.31c) ≡ (14.32b) was used the adiabatic condition (14.10a) with constant adiabatic exponent (14.32a,b):

$$\gamma = \text{const}: \quad c^2 = \frac{dp}{d\rho} = k\rho^{\gamma-1}\gamma = \frac{\gamma p}{\rho}. \tag{14.32a,b}$$

In an incompressible fluid $d\rho = 0 \neq dp$, the sound speed is infinite $c = \infty$, implying that perturbations at one point reach instantaneously all points in the fluid; in a compressible fluid the sound speed is finite $c < \infty$, and a perturbation at a point reaches another point at a distance R after a propagation time $t = R/c$. A flow is called **barotropic** if the pressure depends only on the mass density $p = p(\rho)$, for example, a particular case is the adiabatic flow (14.10a). The enthalpy is specified by (14.27b) for an incompressible flow; the latter is replaced in homentropic conditions (14.31a) by (14.18b) $\equiv$ (14.33b) viz.:

$$\gamma = \text{const}: \quad H = \int \rho^{-1}\,\mathrm{d}p = k\gamma \int \rho^{\gamma-2}\,\mathrm{d}\rho = \frac{k\gamma\rho^{\gamma-1}}{\gamma-1} = \frac{c^2}{\gamma-1}, \tag{14.33a,b}$$

in terms of the adiabatic sound speed in (14.32b).

Substituting (14.33b) in (14.26c), thus neglecting the gravity field (14.34a), specifies the relation between the velocity v, sound speed c, and stagnation sound speed c_0 in (14.34b):

$$\vec{g} = 0: \quad \frac{c^2}{\gamma-1} + \frac{v^2}{2} = \frac{c_0^2}{\gamma-1}; \tag{14.34a,b}$$

for the homentropic (14.10a) flow of a perfect gas (14.11b); *the ratios of the pressure, mass density, temperature and sound speed to their stagnation values, are given in a homentropic flow of a perfect gas by:*

$$\frac{p}{p_0} = \left(\frac{\rho}{\rho_0}\right)^{\gamma} = \left(\frac{T}{T_0}\right)^{\gamma/(\gamma-1)} = \left(\frac{c}{c_0}\right)^{2\gamma/(\gamma-1)}. \tag{14.35a–c}$$

The sound speed (14.35c) is constant in incompressible conditions, viz. $\rho = \rho_0 = \text{const}$ implies $c = c_0 = \text{const}$. From (14.34b) in the form:

$$c^2 = c_0^2\left(1 - \frac{\gamma-1}{2}M^2\right), \quad M = \frac{v}{c_0}, \tag{14.36a,b}$$

it follows that *a flow is incompressible iff the sound speed is approximately constant (14.37b), that is, if the* **Mach number** *(14.36b) is small; this implies that the flow velocity is small compared to the sound speed (14.37a):*

$$v^2 \ll c_0^2: \qquad c = c_0, \quad \rho = \rho_0, \quad T = T_0, \quad p = p_0. \tag{14.37a–e}$$

For air $\gamma = 7/5$ at ambient conditions $T = 15\text{C} = 288\,\text{K}$, the gas constant $R = 2.87 \times 10^6\,\text{cm}^2\,\text{s}^{-2}\,\text{K}^{-1}$, leads to the sound speed $c_0 = \sqrt{\gamma R T} = 340\,\text{m}\,\text{s}^{-1}$; if the flow velocity is up to one-third of the sound speed $v \le 0.3\ c_0$, it meets the incompressibility condition $M^2 \equiv (v/c_0)^2 \le 0.09 \ll 1$. Thus air is an incompressible fluid for speeds up to $v = 0.3\ c_0 = 102\,\text{m}\,\text{s}^{-1} = 367\,\text{km}\,\text{h}^{-1}$, including the take-off and landing of aircraft, but not high-speed cruise flight; in water the sound speed is $c_0 = 1550\,\text{m}\,\text{s}^{-1}$, and the incompressibility condition is met up to high speeds $v \le 465\,\text{m}\,\text{s}^{-1}$. The propagation of sound is always a compressibility phenomenon; thus the assumption of incompressibility excludes sound waves.

14.6.2 Measurement of the Velocity in Incompressible and Compressible Flows

From the preceeding account it follows that the potential flow applies with six restrictions: (i) the flow velocity is small compared with the sound speed, to neglect compressibility (Subsection 14.6.1); (ii) the wakes of bodies or other regions of rotational flow are excluded (Section 14.4); (iii) dissipative effects are not accounted for (Section 14.3), for example, thermal; (iv) it does not apply in regions where viscous stresses are significant, for example, in the boundary layer near a wall (Section 14.3); (v) the entropy must be constant, for example, excluding stratification in an atmosphere, for that the entropy increases with altitude in stable conditions; (vi) the external forces per unit mass should derive from a potential (Section 14.4), viz. gravity is included but nonconservative force fields are excluded. These six restrictions are not independent, because a viscous flow (iv) is dissipative (iii) and the latter is nonhomentropic (v) and rotational (ii), that is, does not satisfy the conservation of circulation (Section 14.4). In spite of these restrictions, the potential flow has many applications, of that the **Pitot tube** (1732) is considered next (Figure 14.2a): it is a tube in the flow for that the forward face measures the stagnation pressure p_0 and a side port measures the free stream pressure p, so that the free stream velocity can be calculated (14.27c) by (14.38b):

$$M^2 \equiv \frac{v^2}{c_0^2} \ll 1: \quad v = \sqrt{\frac{2(p_0 - p)}{\rho_0}}, \tag{14.38a,b}$$

if the mass density is known, viz. the latter is constant in incompressible flow (14.38a), and thus coincides with the atmospheric density at the altitude of flight. The Pitot tube should be sufficiently long to measure the pressure undisturbed by the presence of the aircraft, otherwise correction or calibration is necessary.

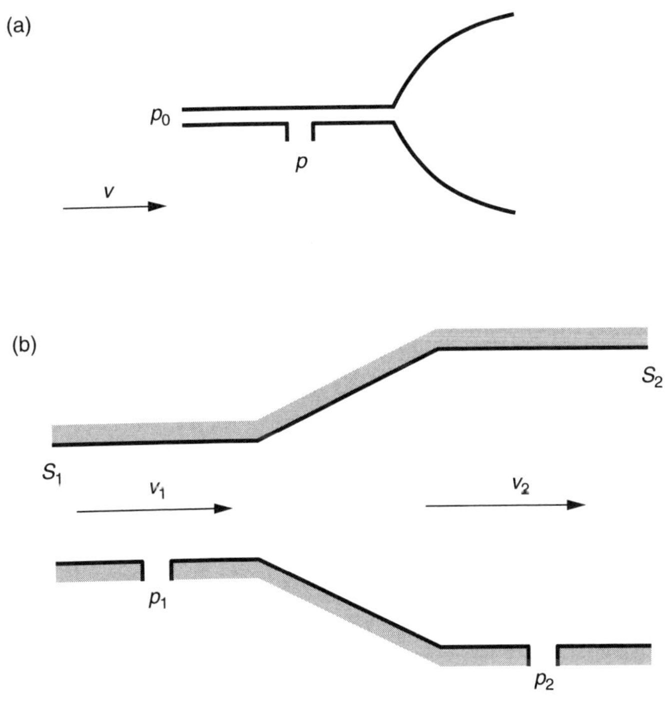

FIGURE 14.2
The velocity of a flow can be measured (a) by comparing the stagnation p_0 (static p) pressure at the front (side) of a boom (Pitot tube) in the flow direction; for accurate results the boom must be long enough so that the body upon which it is mounted does not disturb the flow (Figure 36.12b) or change the velocity at the two pressure taps or ports. In the case of a contraction of a duct (b) the velocities far upstream v_1 and downstream v_2 are related to the difference of static pressures $p_2 - p_1$, for example, measured by a Venturi tube (Figure 14.4).

The result (14.38b) uses the incompressible Bernoulli equation (14.27c) and is restricted to low Mach number (14.38a). In a homentropic compressible flow (14.27c) is replaced by (14.34b; 14.35a–c), viz.:

$$1-\frac{\gamma-1}{2}\frac{v^2}{c_0^2}=\left(\frac{c}{c_0}\right)^2=\frac{T}{T_0}=\left(\frac{\rho}{\rho_0}\right)^{\gamma-1}=\left(\frac{p}{p_0}\right)^{1-1/\gamma}. \tag{14.39a–d}$$

Using (14.39d) specifies the velocity in the compressible case:

$$v^2=c_0^2\frac{2}{\gamma-1}\left[1-\left(\frac{c}{c_0}\right)^2\right]=\frac{p_0}{\rho_0}\frac{2\gamma}{\gamma-1}\left[1-\left(\frac{p}{p_0}\right)^{1-1/\gamma}\right], \tag{14.40}$$

where (14.32b) was used for the stagnation sound speed and also (14.35c) in the last factor. If the relative pressure change is small, the binomial theorem [(25.37b)] may be used:

$$\begin{aligned}\left(\frac{p}{p_0}\right)^{1-1/\gamma}&=\left(1-\frac{p_0-p}{p_0}\right)^{1-1/\gamma}\\&=1-\left(1-\frac{1}{\gamma}\right)\frac{p_0-p}{p_0}\left[1+\frac{1}{2\gamma}\frac{p_0-p}{p_0}\right]+O\left(\left(\frac{p_0-p}{p_0}\right)^3\right),\end{aligned} \tag{14.41}$$

in (14.40), to specify:

$$(p_0-p)^3\ll p_0^3:\quad v^2=2\frac{p_0-p}{\rho_0}\left(1+\frac{p_0-p}{2\gamma p_0}\right); \tag{14.42a,b}$$

this is the first-order compressibility correction beyond (14.38b).

Thus *the measurement of the stagnation* p_0 *(free stream* p*) pressure in the front (side) port of a Pitot tube (Figure 14.2a), specifies the free stream velocity of an irrotational flow in the compressible case (14.40); this reduces at low Mach number (14.38a) to the incompressible case (14.38b), with first-order (14.42a) compressibility correction (14.42b).* The first-order compressibility correction also follows from (14.39d) the adiabatic pressure:

$$p=p_0\left[1-\frac{\gamma-1}{2}\left(\frac{v}{c_0}\right)^2\right]^{\gamma/(\gamma-1)}=p_0\left(1-\frac{\gamma-1}{2}M^2\right)^{\gamma/(\gamma-1)}, \tag{14.43}$$

using again the binomial theorem (25.37b)

$$\begin{aligned}p&=p_0\left[1-\frac{\gamma-1}{2}\frac{v^2}{c_0^2}\frac{\gamma}{\gamma-1}+\frac{(\gamma-1)^2}{8}\frac{v^4}{c_0^4}\frac{\gamma}{\gamma-1}\left(\frac{\gamma}{\gamma-1}-1\right)\right]\\&=p_0-\frac{v^2}{2}\frac{p_0\gamma}{c_0^2}+\frac{v^4}{8c_0^2}\frac{p_0\gamma}{c_0^2};\end{aligned} \tag{14.44a}$$

using (14.32b) for c_0 simplifies (14.44a) to (14.44b);

$$p=p_0-\frac{1}{2}\rho_0v^2+\frac{1}{8}\rho_0\frac{v^4}{c_0^2}=p_0-\frac{1}{2}\rho_0v^2\left(1-\frac{M^2}{4}\right). \tag{14.44b}$$

Thus *the homentropic relations (14.35a–c) apply in particular to the stagnation* p_0 *and free stream* p *pressures that are related in an irrotational steady flow by the compressible (14.43) [incompressible (14.27c)] Bernoulli equation, where (14.44b) specifies the first-order compressibility correction.*

14.7 Venturi Tube (Herschel, 1887) and Variable-Area Duct

The Pitot (Subsection 14.6.2) [Venturi (Subsection 14.7.3)] tube is one method of measuring the velocity of a free stream (flow in a duct). The duct or nozzle can be of varying cross-section and contain a potential (Subsection 14.7.1) [vortical (Subsection 14.7.2)] flow.

14.7.1 Flow in a Duct of Varying Cross-Section

Assuming that the flow through the change of cross-section of the nozzle is potential, the Bernoulli equation (14.27c) ≡ (14.45a) can be applied with equal stagnation pressure far upstream (p_1, v_1) and downstream (p_2, v_2) of the junction (Figure 14.2b) whether it is a contraction $S_1 > S_2$ or an expansion $S_1 < S_2$, viz.:

$$p_1 + \frac{1}{2}\rho_0 v_1^2 = p_2 + \frac{1}{2}\rho_0 v_2^2; \quad v_1 S_1 = v_2 S_2, \tag{14.45a,b}$$

the conservation of volume of fluid (equivalent to conservation of mass in an incompressible flow) yields (14.45b). If two of the four quantities (p_1, p_2, v_1, v_2) are measured or known, the other two can determined from equations (14.45a,b). For example, if the pressures p_1, p_2 are measured at the sections of areas S_1, S_2, the velocities are determined by:

$$\frac{2(p_2 - p_1)}{\rho_0} = v_1^2 - v_2^2 = v_1^2\left\{1 - \left(\frac{S_1}{S_2}\right)^2\right\} = v_2^2\left\{\left(\frac{S_2}{S_1}\right)^2 - 1\right\}, \tag{14.46a,b}$$

that is, *in an potential flow in a duct the pressure is higher* $p_2 > p_1$ *where the velocity is lower* $v_1 < v_2$ *and the cross-section larger* $S_2 > S_1$ in (14.46a,b).

The flow of the fluid along the nozzle causes a reaction force R on its mountings due to the balance of pressure forces and change of momentum:

$$R = p_2 S_2 - p_1 S_1 + \rho_0\left(S_2 v_2^2 - S_1 v_1^2\right), \tag{14.47}$$

where the momentum is the mass flux per unit time $\rho_0 S v$ times the velocity v, that is, $\rho_0 S v^2$. The reaction force (14.47) may be simplified using (14.45b) and (14.46a):

$$R - (p_2 S_2 - p_1 S_1) = \rho_0 S_1 v_1 (v_2 - v_1) = -\rho_0 S_1 v_1^2\left(1 - \frac{S_1}{S_2}\right) = -\frac{2S_1(p_2 - p_1)}{S_1/S_2 + 1}; \tag{14.48a,b}$$

on rearrangement this leads to:

$$R = \bar{p}(S_1 - S_2), \quad \bar{p} = \frac{p_1 S_1 + p_2 S_2}{S_1 + S_2}, \tag{14.49a,b}$$

that shows that *in a potential flow (14.45a,b) the pressure difference (14.46a,b) across of contraction* $S_1 > S_2$ *(expansion* $S_1 < S_2$*) of a duct, leads to a reaction force that acts* $R > 0 (R < 0)$ *opposite to the direction of increase of cross-section, and is proportional to the latter (14.49a), through the arithmetic mean of the pressure at the two sections (14.49b).*

14.7.2 Potential versus Vortical Flow in a Nozzle

In (14.49a,b) it was assumed that the flow remains potential along the entire length of the nozzle that would be the case for very gradual change of cross-section (Figure 14.3a). If the cross-section were to vary abruptly (Figure 14.3b), the potential flow theory would predict

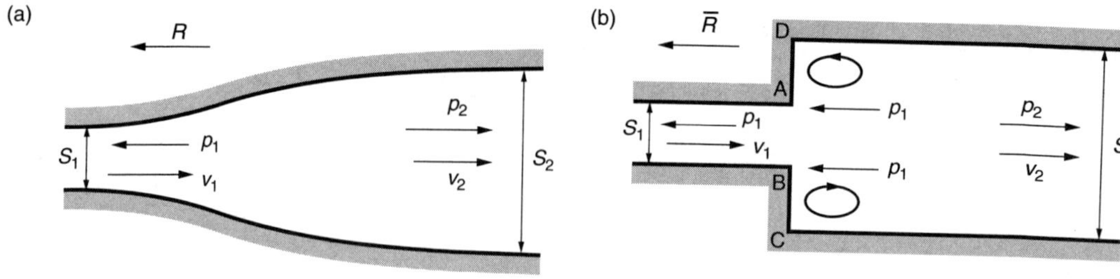

FIGURE 14.3
The flow in a duct of gradually varying cross-section is potential (a) if the area changes slowly enough along its length to avoid flow separation, formation of vortices, and transition to turbulence. The latter are likely to occur in an abrupt change of cross-section (b), which creates a recirculation bubble near the corners.

(Section 14.8) infinite velocity at the salient edges A, B, and zero velocity at the recessed corners C, D (see Section 14.8); in reality, this rapid acceleration and deceleration of the flow would cause the appearance of a separation bubble near each comer, with recirculating, vortical flow. In the case of potential flow, the stagnation pressure is conserved (14.45a) leading to the pressure difference (14.46a,b) and reaction force (14.49a,b). In the case of rotational flow, the pressure in the annulus region is approximately p_1, and thus the reaction force is given by (14.50a,b):

$$p_1 (S_2 - S_1) = \bar{R} = \left(p_2 + \rho_0 v_2^2\right) S_2 - \left(p_1 + \rho_0 v_1^2\right) S_1, \tag{14.50a,b}$$

or alternatively by (14.50b) $\equiv$ (14.47). The pressure difference can be calculated from (14.50a,b), as well as the loss of head or reduction in stagnation pressure:

$$h \equiv p_{01} - p_{02} = \left(p_1 + \frac{1}{2}\rho_0 v_1^2\right) - \left(p_2 + \frac{1}{2}\rho_0 v_2^2\right). \tag{14.51}$$

The only equations common to the two cases of irrotational (14.45a; 14.47) [rotational (14.50a;14.51)] incompressible flow are those stating the conservation of volume flux (14.45b) and momentum (14.50b) $\equiv$ (14.47), the latter involving the reaction force.

The pressure rise as the fluid is slowed down in the wider section in the rotational flow (14.50a,b) is given by:

$$r \equiv \frac{S_1}{S_2} = \frac{v_2}{v_1}: \quad p_2 - p_1 = \rho_0 v_1^2 \frac{S_1}{S_2} - \rho_0 v_2^2 = \rho_0 v_1^2 r (1 - r). \tag{14.52a,b}$$

There is no pressure rise $p_2 = p_1$ *if the cross-section does not change* $r = 1, S_2 = S_1$, *or if the duct is a tube exhausting into a large reservoir* $r = 0, S_2 = \infty$; *the maximum pressure rise* $d^2 (p_2 - p_1)/dr^2 < 0$ *occurs* $d(p_2 - p_1)/dr = 0$ *for* $r = 1/2$, *that is, an expansion to twice the cross-section (14.53a):*

$$S_2 = 2S_1: \quad (p_2 - p_1)_{\max} = \frac{\rho_0 v_1^2}{4} \equiv \frac{p_*}{2}, \quad p_* \equiv \frac{1}{2}\rho_0 v_1^2, \tag{14.53a–c}$$

for that the pressure rise (14.53b) is one-half the initial dynamic pressure (14.53c); the latter is the maximum pressure rise for a potential flow into a reservoir. The loss of head (14.51) is given by:

$$h = p_1 - p_2 + \frac{1}{2}\rho_0 v_1^2 \left(1 - r^2\right) = \frac{1}{2}\rho_0 v_1^2 \left[1 - r^2 - 2r(1 - r)\right] = \frac{1}{2}\rho_0 v_1^2 (1 - r)^2, \tag{14.54}$$

that is, is zero for constant cross-section $r = 1$ *and small for a modest expansion* $r \sim 1$; *there is a total loss of head in (14.54) for a small tube exhausting in a large reservoir* $r = 0$; *in the case of maximum pressure recovery (14.53a–c), the loss of head:*

$$r = \frac{1}{2}: \quad h = \frac{1}{8}\rho_0 v_1^2 = \frac{1}{2}(p_2 - p_1)_{\max} = \frac{p_*}{4}, \tag{14.55}$$

is one-half the gain in pressure, and one-quarter the initial dynamic pressure. The gain in pressure in the potential case (14.46b) would have been $p_2 - p_1 = (3/8)\rho_0 v_1^2 = 3p_*/4$, that is, the sum of (14.53b) and (14.55). Since $p_2 > p_1$ implies $\bar{p} > p_1$ in (14.49b), the reaction force $\bar{R}$ in (14.53b) is smaller in the vortical than in the potential (14.49a) flow case $R > \bar{R}$.

14.7.3 Venturi Tube in Incompressible and Compressible Flow

The Pitot (1732) [Venturi (Herschel, 1887)] tube measures the velocity in a free stream (ducted flow) in Figure 14.2a (14.2b) from the pressure at two points. The formula relating the free stream velocity v to the pressure difference $p_0 - p$ has been given for the Pitot tube in incompressible (14.38a,b), conditions; the formula relating the pressure difference $p_1 - p_2$ to the flow velocities in the duct v_2 and Venturi tube v_1 are obtained next, also in the incompressible case (14.57a,b). The first order compressibility correction (14.43) [(14.62a)] already requires the individual pressures (p_0, p) $[(p_1, p_2)]$, as does the exact compressibility expression for the Pitot (Venturi) tube, viz. in (14.40) [(14.61)]. The pressure difference is thus sufficient in the incompressible case. It can be determined in both cases (Figure 14.4) connecting each pressure port to one side of a U-tube filled with a liquid; when the liquid stabilizes and is at rest $v = 0$ in (14.29c), the difference of height Δz on the two sides of the tube specifies the hydrostatic pressure:

$$\{p_0 - p, p_1 - p_2\} \equiv \Delta p = \rho_* g \Delta z, \tag{14.56}$$

where ρ_* is the density of the liquid in the tube, for example, $\rho_* = 1\,g\,\text{cm}^{-3}$ for water, and $g = 981\,\text{cm}\,\text{s}^{-2}$ is the acceleration of gravity at sea level. To measure the individual pressures

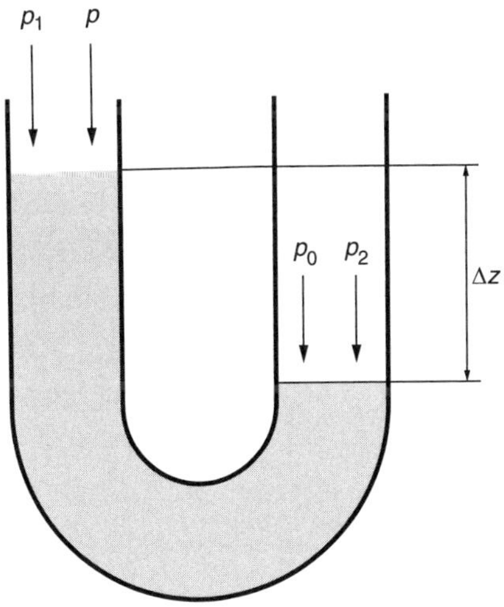

FIGURE 14.4
A U-tube partially filled with a fluid provides via the difference of heights on the two branches a measurement of pressure difference; this may be used together with the Pitot (Venturi) tube in Figure 14.2a(b) to measure the flow velocity.

rather than the pressure difference, as required in the compressible case, two instead of one U-tubes in Figure 14.4 are used each comparing one pressure (p_1, p_0) $[(p_1, p_2)]$ with the atmospheric pressure p_a that is known. From (14.46b) follows (14.57b):

$$M^2 \ll 1: \quad v_2 = \left| \frac{2(p_2 - p_1)/\rho_0}{(S_2/S_1)^2 - 1} \right|^{1/2}, \tag{14.57a,b}$$

that specifies the velocity (14.57b) an incompressible flow (14.57a) in a nozzle (Figure 14.2b) by using a **Venturi tube** *to measure the pressure difference between two sections.*

To extend the velocity in the Venturi tube from the incompressible (14.57a,b) to the compressible case, the starting point must be: (i) the equation of conservation of mass (14.58a) instead of volume (14.45b), viz.:

$$\rho_1 S_1 v_1 = \rho_2 S_2 v_2; \quad v_2^2 + \frac{2\gamma p_2}{(\gamma - 1)\rho_2} = v_1^2 + \frac{2\gamma p_1}{(\gamma - 1)\rho_1}, \tag{14.58a,b}$$

the conservation of stagnation pressure for adiabatic (14.58b) instead of incompressible (14.27b) flow; the latter follows from (14.34b) and (14.32b):

$$\text{const} = \frac{v^2}{2} + \frac{c^2}{\gamma - 1} = \frac{v^2}{2} + \frac{\gamma}{\gamma - 1}\frac{p}{\rho}. \tag{14.59}$$

The velocity v_1 may be eliminated from (14.58b) with the help of (14.58a), viz.:

$$\left[\left(\frac{\rho_2 S_2}{\rho_1 S_1}\right)^2 - 1\right] v_2^2 = v_1^2 - v_2^2 = \frac{2\gamma}{\gamma - 1}\left(\frac{p_2}{\rho_2} - \frac{p_1}{\rho_1}\right) = \frac{2\gamma}{\gamma - 1}\frac{p_2}{\rho_2}\left(1 - \frac{p_1 \rho_2}{p_2 \rho_1}\right); \tag{14.60}$$

using the adiabatic relation (14.10a) leads to:

$$v_2^2 = \frac{[2\gamma/(\gamma - 1)](p_2/\rho_2)\left[1 - (p_1/p_2)^{1-1/\gamma}\right]}{(\rho_2 S_2/\rho_1 S_1)^2 - 1}. \tag{14.61}$$

This *specifies the velocity in the Venturi tube for adiabatic compressible flow (14.61); it simplifies to (14.57b) in the incompressible case (14.57a) with:*

$$v_2^2 = \frac{2(p_2 - p_1)/\rho_2}{(\rho_2 S_2/\rho_1 S_1)^2 - 1}\left(1 + \frac{p_2 - p_1}{2\gamma p_2}\right), \tag{14.62a}$$

as the first-order compressibility correction (14.62a). The latter (14.62a) follows from (14.61) using the first three terms of the binomial expansion (25.37b), as in (14.41):

$$\begin{aligned}\left[\left(\frac{\rho_2 S_2}{\rho_1 S_1}\right)^2 - 1\right] v_2^2 &= \frac{2\gamma}{\gamma - 1}\frac{p_2}{\rho_2}\left[1 - \left(1 - \frac{p_2 - p_1}{p_2}\right)^{1-1/\gamma}\right] \\ &= \frac{2\gamma}{\gamma - 1}\frac{p_2}{\rho_2}\left[\left(1 - \frac{1}{\gamma}\right)\frac{p_2 - p_1}{p_2} + \frac{1}{2\gamma}\left(1 - \frac{1}{\gamma}\right)\left(\frac{p_2 - p_1}{p_2}\right)^2\right] \\ &= \frac{2(p_2 - p_1)}{\rho_2} + \frac{(p_2 - p_1)^2}{\gamma \rho_2 p_2}.\end{aligned} \tag{14.62b}$$

In the incompressible case $\rho_2 = \rho_1 = \rho_0$ the first factor of (14.62a) reduces to (14.57b),

14.8 Corner Flows and Multipoles at Infinity

The flow in the nozzles (Figures 14.2b and 14.3a,b) is uniform in the parts of constant cross-section, if viscosity in the wall boundary layers and vorticity shed from the expansion are ignored; the case of abrupt change of cross-section (Figure 14.23b) includes local flows inside (outside) the square corner at C,D (A,B). The uniform (Subsection 14.8.1) [corner (Subsection 14.8.2)] flows are represented by the simplest forms of complex potential, viz. the (linear) power function.

14.8.1 Uniform Flow and Dipole at Infinity

The potential flow is specified by the complex potential $f(z) + \text{const}$ to within an added constant, since the latter does not affect the complex conjugate velocity, $v^* = \mathrm{d}f/\mathrm{d}z$. Thus the simplest flow is specified by a complex potential that is a linear function (14.63a):

$$f(z) = U\mathrm{e}^{-i\alpha}z, \quad v^*(z) = \frac{\mathrm{d}f}{\mathrm{d}z} = U\mathrm{e}^{-i\alpha}, \tag{14.63a,b}$$

where the complex conjugate velocity (14.63b) shows that:

$$v_x = U\sin\alpha, \quad v_y = U\sin\alpha, \tag{14.64a,b}$$

the complex multiplying constant has: (i) modulus equal to the uniform stream velocity U; (ii) phase equal to $-\alpha$ minus the **angle-of-attack** (Figure 14.5). From (14.63a) it follows that

$$\Phi + i\Psi = U(\cos\alpha - \sin\alpha)(x + iy), \tag{14.65}$$

the potential Φ and stream Ψ functions are both linear in the variables x,y:

$$\Phi(x,y) = U(x\cos\alpha + y\sin\alpha), \quad \Psi(x,y) = U(x\sin\alpha - y\cos\alpha); \tag{14.66a,b}$$

thus the streamlines $\Psi = \text{const}$ (equipotentials $\Phi = \text{const}$):

$$y = x\tan\alpha - \frac{\Psi}{U}\sec\alpha, \quad y = -x\cot\alpha + \frac{\Phi}{U}\csc\alpha, \tag{14.67a,b}$$

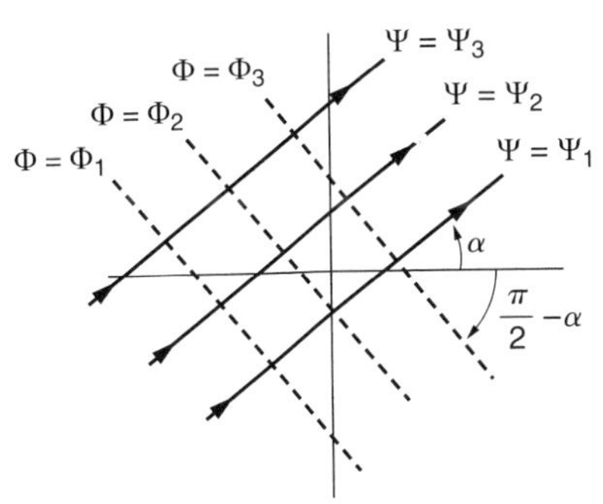

FIGURE 14.5
In a uniform flow the stream lines (equipotentials) are [solid (dotted) lines] are orthogonal straight lines.

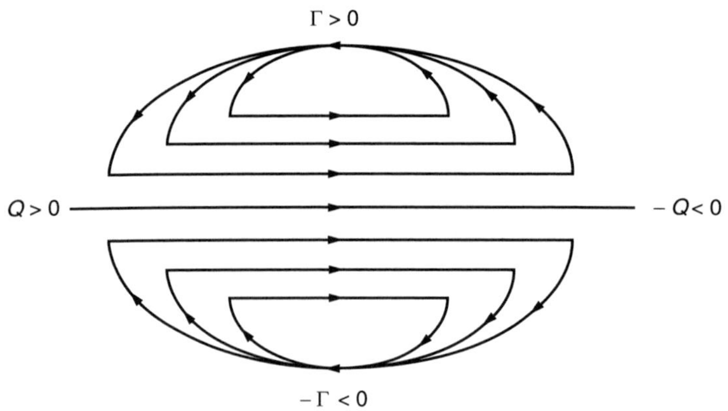

FIGURE 14.6
A uniform flow (Figure 14.5) corresponds to a "dipole at infinity" with: (i) source (sink) in the upstream (downstream) direction; (ii) counterclockwise (clockwise) vortex above (below). It either case (i) or (ii) the axis of the "dipole at infinity" (Figures 12.3 and 12.4) is in the direction of the uniform flow.

are straight lines (14.68a):

$$y = mx + b: \quad m = \tan\left\{\alpha, \frac{\pi}{2} + \alpha\right\}, \quad b = \frac{\{-\Psi \sec\alpha, \Phi \csc\alpha\}}{U}, \tag{14.68a–c}$$

all with the same slope (14.68b) or inclination $\alpha\,(\pi/2 + \alpha)$, and ordinate at the origin (14.68c) determined by the value of $\Psi\,(\Phi)$. Since the flow is uniform with angle-of-attack, α, the streamlines are straight lines of slope $\tan\alpha$, and the equipotentials the orthogonal straight lines of slope $\tan(\pi/2 + \alpha) = -\cot(\alpha)$ in Figure 14.5. Taking the OX-axis along the flow velocity, and performing an inversion (14.69a) that sends the origin $z = 0$ of the z-plane to the point-at-infinity $\zeta = \infty$ of the ζ-plane:

$$\zeta = \frac{1}{z}: \quad e^{-i\alpha} f(z) = Uz = \frac{U}{\zeta} = -\frac{P_1}{2\pi\zeta} = F(\zeta), \quad P_1 = -2\pi U e^{i\alpha}, \tag{14.69a–c}$$

it follows that *the uniform flow of velocity U and angle-of-attack α is (14.69b) due to a dipole of moment (14.69c) at infinity (14.69a), because the velocity (14.63b; 14.64a,b), and streamlines and equipotentials (14.66–67a,b; 14.68a–c) are the same. They correspond (Figure 14.6), with a rotation by α, to: (i) a source $Q > 0$ at $-\infty$ and a sink $-Q$ at $+\infty$; (ii) a positive $\Gamma > 0$ (negative $-\Gamma$) vortex at $+i\infty\,(-i\infty)$; both lead to a negative dipole with axis in the flow direction $\varphi = \alpha$.*

14.8.2 Multipole at Infinity and Corner Flow

A multipole of order n at infinity:

$$F_n(\zeta) = -\frac{P_n}{2\pi\zeta^n} = Uz^n = f_n(z), \quad U \equiv -\frac{P_n}{2\pi}, \tag{14.70a,b}$$

corresponds to a potential flow specified by a power function; for a horizontal axis the potential and stream function:

$$\alpha = 0: \quad \Phi = Ur^n \cos(n\varphi), \quad \Psi = Ur^n \sin(n\varphi), \tag{14.71a,b}$$

and velocity:

$$v^* = Unz^{n-1}, \quad \{v_r, v_\varphi\} = nUr^{n-1}\{\cos(n\varphi), -\sin(n\varphi)\}, \tag{14.72a–c}$$

show that (14.71b) the streamline $\Psi = 0$ corresponds to $\varphi = 0, \pi/n, 2\pi/n, \ldots$, along that the velocity is radial since $v_\varphi = 0$ in (14.72c). Thus *the flow in a corner of aperture* $\beta = 2\pi/\mathrm{n}$ *corresponds to a multipole of order* n *at infinity, and is specified by the complex potential (14.70a,b) and conjugate velocity (14.72a), scalar potential (14.71a) and stream function (14.71b), and polar components (14.72b,c) of the velocity, including: (i) the uniform flow (Figure 14.6) for* $\mathrm{n} = 1$; *(ii) the flow in a square corner (Figure 14.7) for* $\mathrm{n} = 2$ *that corresponds to a quadrupole at infinity (Figure 14.8). In the case of a square corner (14.73a,b):*

$$n = 2: \quad f(z) = Uz^2, \quad \Phi = U\left(x^2 - y^2\right), \quad \Psi = 2Uxy, \tag{14.73a–d}$$

the equipotentialls (streamlines) $\Phi = const\,(\Psi = const)$ *in (14.73c) [(14.73d)] are equilateral hyperbolas (Figure 14.7) with asymptotes along the diagonals of quadrants* $\mathrm{y} = \pm\mathrm{x}$ *(the real* $\mathrm{y} = 0$ *and imaginary* $\mathrm{x} = 0$*) axis; they are mutually orthogonal and coincide by a rotation over* $\pi/4$. *The latter property is a particular case* $\mathrm{n} = 2$ *of the coincidence of equipotentials (14.71a) and streamlines (14.71b) by a rotation over* $\pi/2\mathrm{n} = \beta/2$ *for a flow in a corner of angle* $\beta = \pi/\mathrm{n}$, *viz.:*

$$\beta = \frac{\pi}{n}: \quad \Phi\left(r, \varphi + \frac{\beta}{2}\right) = \Phi\left(r, \varphi + \frac{\pi}{2n}\right) = Ur^n \cos\left(n\varphi + \frac{\pi}{2}\right) = -Ur^n \sin(n\varphi) = -\Psi(r, \varphi). \tag{14.74}$$

This property is shared with multipoles (12.61a), with one difference: for the multipole of order n (flow in a corner of angle, π/n) the streamlines and equipotentials are closed (open)

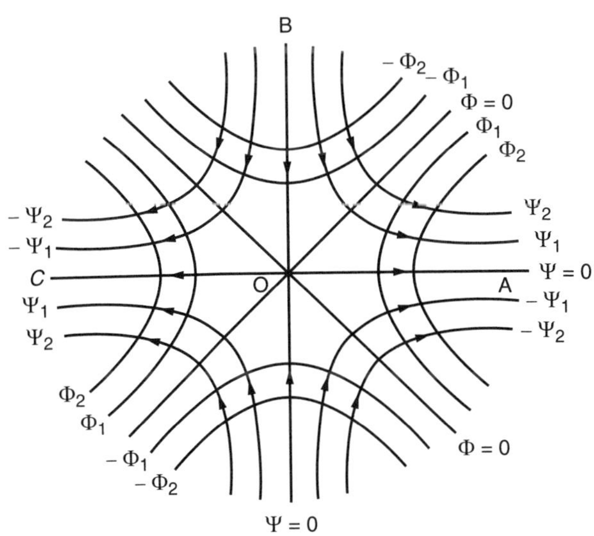

FIGURE 14.7
The potential flow in a rectangular corner has streamlines (equipotentials) which are equilateral hyperbolas with asymptotes along the coordinate axis (diagonals of quadrants). They coincide by rotation through 45°.

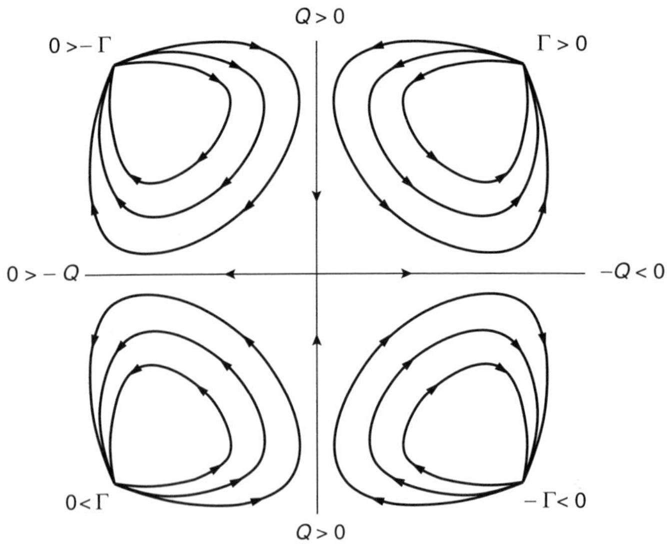

FIGURE 14.8
The uniform (rectangular corner) flow [Figure 14.5(14.7)] corresponds to a "dipole (quadrupole) at infinity" [Figure 14.6 (14.8)]. In the case of the corner flow the sources/sinks (counter clockwise/clockwise vortices) alternate at infinity in the direction of the walls (the diagonals of the walls).

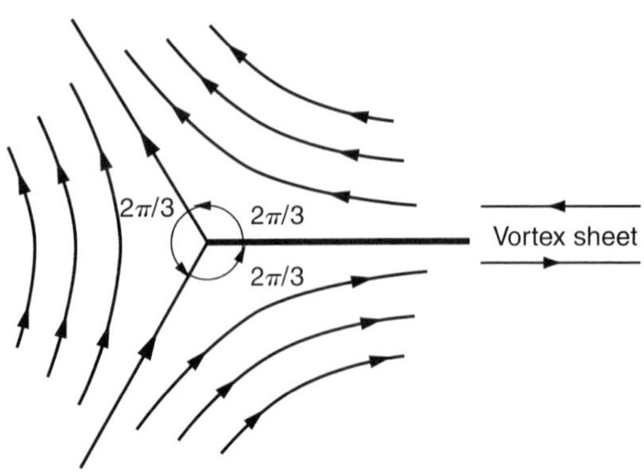

FIGURE 14.9
The "tripole" at infinity would correspond to a flow in a corner of angle $2\pi/3$. Since there are three sectors the velocities must be opposite on the two sides of a separating line; the latter is then a vortex sheet. Thus a "tripole" cannot be a potential flow.

curves, since the center is the origin (infinity). If the corner angle is an even fraction $n = 2p$ of 2π, viz. $\beta = \pi/2p$ then the tangential velocity is continuous along all separating streamlines $\varphi_k = 0, \pi/2p, \pi/p, \ldots, \frac{(2p-1)\pi}{2p}$; for example, $\varphi_k = 0, \pi/2, \pi, 3\pi/2$, for the flow in a square corner (Figure 14.7) corresponding to $n = 2, p = 1$. If the corner aperture is an odd fraction $n = 2p + 1$ of 2π, viz. $\beta = 2\pi/(2p+1) = \pi/(p+1/2)$ then the exponent is a half-integer, and one separating streamline $\varphi_k = 0, \pi/(2p+1), 2\pi/(2p+1), \ldots, (2p-1)\pi/(2p+1)$, will have discontinuous tangential velocities, that is, with opposite signs, leading to the appearance of a **vortex sheet**; for example, $\varphi_k = 0, \pi/3, 2\pi/3$, in Figure 14.9 for $\beta = 2\pi/3$ corresponding

TABLE 14.1
Scaling of Flow with Power-Type Complex Potential $f(z) \sim z^n$

Type of flow	Potential, stream functions	Velocity field	Dynamic pressure
Symbol	Φ, Ψ	$\vec{v} = \nabla\Phi$	$q \equiv \frac{1}{2}\rho_0 v^2$
Corner of angle $\beta = \pi/n$	r^n	r^{n-1}	r^{2n}
Internal orthogonal corner	r^2	r	r^2
Uniform flow	r	r^0	r^0
External orthogonal corner	$r^{2/3}$	$r^{-1/3}$	$r^{-2/3}$
Sharp edge	$r^{1/2}$	$r^{-1/2}$	r^{-1}
Monopole: source/sink	$\log r, \varphi\,(1)$	r^{-1}	r^{-2}
Monopole: vortex	$\varphi, \log r\,(1)$	r^{-1}	r^{-2}
Dipole	r^{-1}	r^{-2}	r^{-4}
Quadrupole	r^{-2}	r^{-3}	r^{-6}
2^n–pole	r^{-n}	r^{-n-1}	r^{-2n-2}

(1) Respectively, for potential, stream function.
Note: Data given for ten potential flows: (i–v) corner with arbitrary angle $0 \le \beta \le 2\pi$, including uniform flow $\beta = \pi$, sharp edge $\beta = 2\pi$ and internal $\beta = \pi/2$ (external $\beta = 3\pi/2$) orthogonal corner; (vi–x) multipole or 2^n-pole of any order, including quadrupole ($n = 2$, $2^2 = 4$), dipole ($n = 1$, $2^1 = 2$) and monopole ($n = 0$, $2^0 = 1$), distinguishing in the latter case source/sink from vortex.

to $n = 3, p = 1$; this would correspond to a "tripole" at infinity and is not a potential flow due to the vortex sheet. Other corner angles or wedge apertures, viz. not submultiples of 2π are considered next; the cases of positive (negative) exponent, that is, corner flows (Section 14.8) [multipoles (Sections 12.4–12.9)] are summarized in Table 14.1.

14.9 Stream past a Wedge and Sharp Edge

The corner flow specifies the potential flow past an edge (Subsection 14.9.1) and the effect of a wedge in a uniform stream (Subsection 14.9.2), of that the semiinfinite plate is a particular case (Subsection 14.9.3).

14.9.1 Potential Flow past an Edge

If n in (14.71b) is taken as an arbitrary real number, the streamline $\Psi = 0$ corresponds to $n\varphi = 0, \pi, \ldots$, and thus represents the flow in a corner of aperture β if $n\beta = \pi$. This specifies the exponent (14.75a) in *the complex potential (14.70a) ≡ (14.75b) for a flow in a corner of angle* β:

$$n \equiv \frac{\pi}{\beta}: \quad f(z) = U z^{\pi/\beta}, \quad v^*(z) = \frac{\pi U}{\beta} z^{\pi/\beta - 1}, \tag{14.75a–c}$$

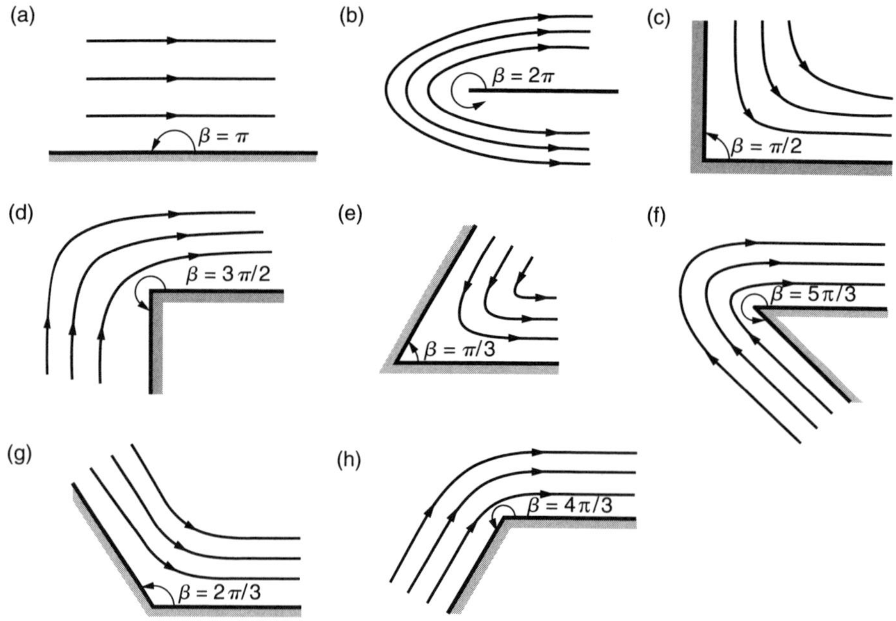

FIGURE 14.10
Potential flow in a corner for eight particular angles: (i/ii) uniform (edge) flow past an infinite wall (a) [semiinfinite plate (b)]; (iii/iv) inside (c) [outside (d)] a rectangular corner; (v/vi) inside (e) [outside (f)] a corner of angle 60°; (vii/viii) inside (g) [outside (h)] a corner of double angle 120°. The vertex is a stagnation point (a singularity of the velocity) for internal (c,e,g) [external (d,f,h)] flows in convex (concave) corners.

and the corresponding complex conjugate velocity (14.75c), scalar potential (14.71a), stream function (14.71b), and polar components of velocity (14.72b,c)≡(14.87b,c). This includes (14.75a) the following cases: (i) for $n=1$ or $\beta=\pi$ the uniform flow (14.63–14.68) in Figure 14.5 ≡ 14.6 ≡ 14.10a; (ii) for $\beta=2\pi, n=1/2, f=U\sqrt{z}$ the flow past the trailing-edge of a semiinfinite plate in Figure 14.10b; (iii) for $\beta=\pi/2, n=2$ the flow (14.73a–d) in the interior of a square corner in Figures 14.7 ≡ 14.8 ≡ 14.10c; (iv) for $\beta=3\pi/2, n=2/3, f=\sqrt[3]{z^2}$ the flow past the edge of orthogonal walls in Figure 14.10d; (v) for $\beta=\pi/3, f=z^3$ the flow in an acute corner of angle 60° in Figure 14.10e; (vi) for $\beta=5\pi/3$, $n=3/5, f=\sqrt[5]{z^3}$ the flow outside the acute corner of angle 60° in Figure 14.10f; (vii) for $\beta=2\pi/3, n=3/2, f=\sqrt{z^3}$ the flow inside an obtuse corner of angle 120° in Figure 14.10g; (viii) for $\beta=4\pi/3, n=3/4, f=\sqrt[4]{z^3}$ the flow outside an obtuse corner of angle 120° in Figure 14.10h. The **corner flows** may be classified into: (i) **acute** $0<\beta<\pi/2$ for exponent $n>2$, and **obtuse** $\pi/2<\beta<\pi$ for exponent $1<n<2$ separated by the **orthogonal** case $\beta=\pi/2$ of exponent $n=2$; (ii) the preceding are convex corners $0<\beta<\pi$ for exponent $n>1$, the corresponding **concave** corners $\pi<\beta<2\pi$ of exponents $1>n>1/2$, being separated by the uniform flow over a flat wall $\beta=\pi$ for $n=1$. The flow in a convex (concave) corner may be designated **interior (exterior) flow**, and the velocity is zero (infinite) at the vertex and infinite (zero) at infinity, as indicated in Table 14.2. The intermediate case is the uniform flow.

The exponent (14.75a) specifies the singularity of the velocity of a potential flow at the vertex of a wedge of internal (external) angle, $\chi(\beta)$:

$$\beta+\chi=2\pi: \quad v^*(z)\sim U z^{\pi/\beta-1}\sim U z^{(\chi-\pi)/(2\pi-\chi)}, \tag{14.76}$$

TABLE 14.2
Stagnation Points and Singularities of Corner Flows

Corner		Concave	Plane	Convex
Angle		$0<\beta<\pi$	$\beta=\pi$	$2\pi>\beta>\pi$
Exponent		$\infty>n>1$	$n=1$	$1>n>1/2$
Velocity	At the vertex	0	U	∞
	At infinity	∞	U	0

Note: In a convex corner the stagnation point is at the corner and the velocity diverges at infinity; the reverse applies to a concave corner. The intermediate case is a plane wall with uniform flow.

for example,: (i) $z^{-1/2}$ at the trailing-edge of (Figure 14.10b) a semiinfinite plate $\beta=2\pi$, $\chi=0$; (ii) $z^{-2/5}$ at (Figure 14.10f) a 60°-wedge $\beta=5\pi/3, \chi=\pi/3$; (iii) $z^{-1/3}$ at (Figure 14.10d) an orthogonal wedge $\beta=3\pi/2, \chi=\pi/2$; (iv) $z^{-1/4}$ at (Figure 14.10h) an 120°-wedge $\beta=4\pi/3, \chi=2\pi/3$; in general the exponent goes from $-1/2$ to zero as the edge becomes less sharp. *In the corner flow (14.71b) all streamlines (14.75a) ≡ (14.77a):*

$$n\equiv\frac{\pi}{\beta}: \quad \Psi=Ur^{\pi/\beta}\sin\left(\frac{\pi\varphi}{\beta}\right); \quad \lim_{r\to\infty}\varphi=0,\beta; \quad r_{\min}=r\left(\frac{\beta}{2}\right)=\sqrt[n]{\frac{\Psi}{U}}, \qquad (14.77\text{a–d})$$

have (i) asymptotes along the walls (14.77c); (ii) point of closest approach to the origin (14.77d) at the diagonal of the corner.

14.9.2 Uniform Stream Incident on a Wedge

The flow (14.75a–c) in the corner of internal angle β was taken with U real, implying an incident stream along the positive x-axis (Figures 14.6–14.10), whereas in the uniform flow (14.63a,b) an angle-of-attack α was considered (Figure 14.5). The factor $e^{-i\alpha}$ is reintroduced in the case of a *stream with angle-of-attack* α, *incident on a* **wedge** *with* **internal angle** χ, *corresponding (Figure 14.11a) to an* **external angle** $\beta=2\pi-\chi$ *in (14.76), leading to: (i) the complex potential (14.78a) and conjugate velocity (14.78b):*

$$f(z)=Ue^{-i\alpha}z^{\pi/(2\pi-\chi)}, \quad v^*(z)=\frac{U}{2-\chi/\pi}e^{-i\alpha}z^{-(\pi-\chi)/(2\pi-\chi)}; \qquad (14.78\text{a,b})$$

(ii) the scalar potential (14.79a) and stream function (14.79b):

$$\Phi,\Psi(r,\varphi)=Ur^{\pi/(2\pi-\chi)}\cos,\sin\left(\frac{\varphi}{2-\chi/\pi}-\alpha\right); \qquad (14.79\text{a,b})$$

(iii) the polar components of velocity:

$$v_r,v_\varphi(r,\varphi)=\frac{U}{2-\chi/\pi}r^{-(\pi-\chi)/2\pi-\chi}\cos,\sin\left(\alpha+\varphi\frac{\pi-\chi}{2\pi-\chi}\right); \qquad (14.80\text{a,b})$$

(iv) these correspond to walls at (14.81c) viz. for that (14.81b) the stream function vanishes (14.81a):

$$\Psi=0: \quad \frac{\varphi}{2-\chi/\pi}-\alpha=0,-\pi, \quad \varphi=\alpha\left(2-\frac{\chi}{\pi}\right),(\alpha-\pi)\left(2-\frac{\chi}{\pi}\right); \qquad (14.81\text{a–c})$$

for example, if the stream is incident from the negative x-direction $\alpha=\pi$, then the walls are at $\varphi=0,2\pi-\chi$. The case:

$$\alpha=-\frac{\chi}{2}: \quad f(z)=Ue^{i\chi/2}z^{1/(2-\chi/\pi)}, \qquad (14.82\text{a,b})$$

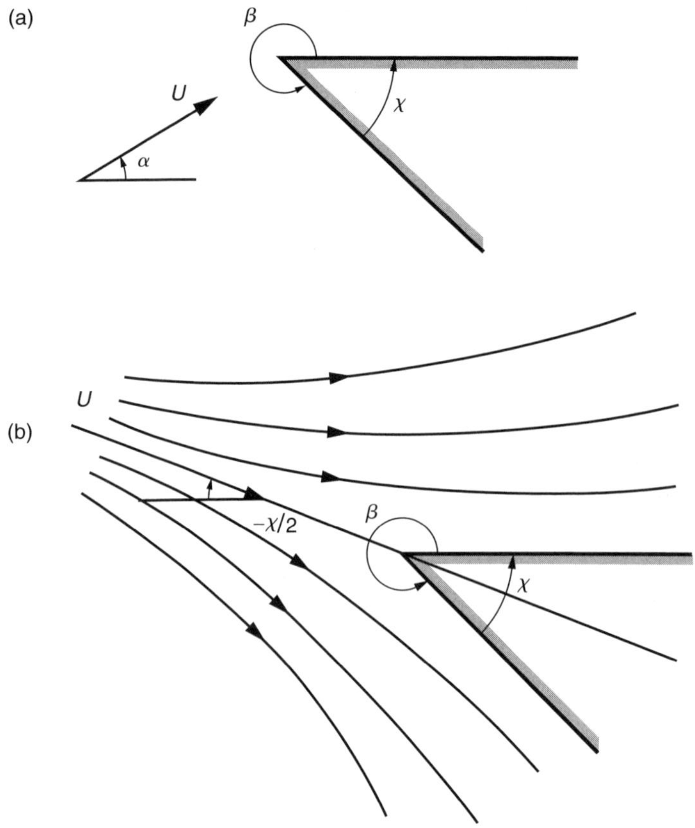

FIGURE 14.11
Uniform stream of velocity U incident (a) at angle-of-attack α on a wedge of internal (external) angle $\chi(\beta)$. The particular case of symmetric incidence (b) corresponds to the angle of attack minus half the internal angle of the wedge.

of angle-of-attack equal to one-half the interior angle of the wedge (14.82a) corresponds (14.82b) the **symmetric flow** *past the wedge* (Figure 14.11b).

14.9.3 Flow past a Semiinfinite Plate

The potential flow past a semiinfinite plate [(14.75b) with $\beta = 2\pi$] has a complex potential (14.83b) that has a branch-cut (14.83a) on the plate:

$$z \notin (0, \infty(: \quad f(z) = U\sqrt{z}; \quad v^*(z) = \frac{U}{2\sqrt{z}}; \tag{14.83a–c}$$

the corresponding complex conjugate velocity (14.83c) leads to the polar components of the velocity:

$$v_r(r, \varphi) = \frac{U}{2\sqrt{r}} \cos\left(\frac{\varphi}{2}\right), \quad v_\varphi(r, \varphi) = -\frac{U}{2\sqrt{r}} \sin\left(\frac{\varphi}{2}\right), \tag{14.84a,b}$$

that show (Figure 14.10b) that: (i) the velocity is tangent to the plate (14.84a,b) and has opposite directions on the two sides (14.84c,d):

$$v_\varphi(r,0)=0=v_\varphi(r,2\pi),\quad v_r(r,0)=\frac{U}{2\sqrt{r}}=-v_r(r,2\pi); \tag{14.85a–d}$$

(ii) outside the plate in the same direction:

$$v_r(r,\pi)=0,\quad v_\varphi(r,\pi)=-\frac{U}{2\sqrt{r}}, \tag{14.86a,b}$$

the velocity has the same modulus (14.86b) = (14.85c,d) but is vertical (14.86a) instead of horizontal (14.85a,b). For a corner of arbitrary angle (14.87a) the velocity is given (14.72b,c) ≡ (14.87b,c) by:

$$n=\frac{\pi}{\beta}:\quad \{v_r,v_\varphi\}=\frac{\pi U}{\beta}r^{\pi/\beta-1}\left\{\cos\left(\frac{\pi\varphi}{\beta}\right),-\sin\left(\frac{\pi\varphi}{\beta}\right)\right\}; \tag{14.87a–c}$$

it is also purely azimuthal (14.88b) at the diagonal angle (14.88a) viz.:

$$\frac{\beta}{2}=\frac{\pi}{2n}:\quad v_r\left(r,\frac{\beta}{2}\right)=0,\quad v_\varphi\left(r,\frac{\beta}{2}\right)=-\frac{\pi U}{\beta}r^{\pi/\beta-1}; \tag{14.88a–c}$$

these (14.88b,c) reduce to (14.86a,b) for $\beta=2\pi$. The pressure field (14.27c) is given in a corner flow (14.72a) ≡ (14.75c) by:

$$p(r)=p_0-\frac{1}{2}\rho_0U^2n^2r^{2n-2}=p_0-\frac{\rho_0}{2}\left(\frac{\pi U}{\beta}\right)^2r^{2\pi/\beta-2}; \tag{14.89}$$

it is constant for a uniform steam $n=1$ or $\beta=\pi$. The superposition of corner flows is considered in Example 20.8.

Note 14.1. *Corners and Sharp Edges in Potential Flows*: In a corner flow the velocity is zero (infinite) at one point, implying that the free stream pressure is the stagnation pressure (minus infinity). For an interior (exterior) flow in a concave (convex) corner of external angle $\beta<\pi$ $(\beta>\pi)$ the: (i) stagnation point of zero velocity is the vertex (infinity); (ii) conversely the singular point where the velocity is infinite is the point at infinity (vertex). In both cases the total force on the infinite walls is infinite, because $|v|^2\sim|z|^{2n-2}\sim r^{2n-2}$ is not integrable over $0\le r<\infty$ for any n (Example 20.17). The singularities of the velocity may in some cases be eliminated by making them coincide with a stagnation point (Chapter 34). The hydrodynamic forces become finite for finite bodies (Chapter 28) and also some cases of semiinfinite bodies. The singularities of a potential corner flow can be useful as checks, viz. a potential flow around any body with a sharp edge, must scale near the edge as:

$$z\to0:\quad f(z)\sim z^{1/2},\quad v^*(z)\sim z^{-1/2}, \tag{14.90a,b}$$

for the complex potential (14.90a) and conjugate velocity (14.90b). For example the potential flow past a flat plate $(-a,+a)$ must have scalings as:

$$\lim_{z\to\pm a}f(z)\sim(z\mp a)^{1/2},\quad \lim_{z\to\pm a}v^*(z)\sim(z\mp a)^{-1/2}, \tag{14.91a,b}$$

near the edges $z=\pm a$; these scalings will be confirmed subsequently (Section 34.1). The product of these scalings leads to factors:

$$z\notin(-a,+a):\quad f(z)\sim\sqrt{(z-a)(z+a)}=\sqrt{z^2-a^2},\quad v^*(z)\sim\frac{z}{\sqrt{z^2-a^2}}, \tag{14.92a–c}$$

in the complex potential (14.92b) and conjugate velocity (14.92c) corresponding to a branch-cut (Section 8.8) on the plate (14.92a). The scaling (14.90a) [(14.90b)] should be satisfied by the complex potential (conjugate velocity) for a body of any shape having a sharp edge, in the neighborhood of the edge. Similarly in the neighborhood of any corner of internal angle β the complex potential (conjugate velocity) must scale as (14.75b) [(14.75c)] as a local limit.

Conclusion 14: The balance (figure 14.1) mass (a) [force (b)] leads to the equation of continuity (section 14.1) [momentum (section 14.2)]. These show that for a potential flow the velocity (figure 14.2) can be determined: (i) from the stagnation p_0 and free stream p pressures using a Pitot tube (a) in a free stream; (ii) from the pressures p_2 in the duct and p_1 at the Venturi (b) associated with a change of cross-section of a nozzle. The pressure difference may be measured by connecting the pressure taps to the two branches of an U-tube field with a liquid (figure 14.3) and determing the difference of height of the liquid between the two sides. The flow in a nozzle (figure 14.4) of varying cross-section (a) remains potential if the area change is very gradual, whereas (b) at an abrupt change of cross-section flow separation occurs and a recirculating, vortical flow causes a loss of stagnation pressure, and reduces the reaction force R on the supports. Some potential flows in corners, viz. of angle β for (i): $\beta = \pi$ infinite (figure 14.5) $\equiv$ (figure 14.6) $\equiv$ (figure 14.10a) and (figure 14.10b) semiinfinite $\beta = 2\pi$ flat plate; (ii/iii) $\beta = \pi/2$ interior (figure 14.7) $\equiv$ (figure 14.8) $\equiv$ (figure 14.10c) and (figure 14.10d) exterior $\beta = 3\pi/2$ of perpendicular walls; (iv/v); interior of (figure 14.10e) acute wedge $\beta = \pi/3$ and (figure 14.10g) obtuse faces $\beta = 2\pi/3$; (vi/vii) exterior of (figure 14.10h) obtuse sides $\beta = 4\pi/3$ and (figure 14.10f) wedge $\beta = 5\pi/3$. The velocity is zero (infinite) at the edge in the interior (figure 14.10c,e,f,g) [exterior (figure 14.10b,d,f,h)] cases. If the sector is an odd fraction of 2π, e.g., $2\pi/3$ in (figure 14.9), the corner flows are separated by streamlines plus one vortex sheet across which the velocity changes sign; this is not a potential flow, and would correspond to a 'tripole' at infinity. The uniform (figure 14.6) $\equiv$ (figure 14.10a) [square corner (figure 14.7) $\equiv$ (figure 14.10c)] flow corresponds to a dipole (figure 14.7) [quadrupole (figure 14.9)] at infinity. A particular case (figure 14.11) of a stream (a) with angle-of-attack α incident on a wedge of internal (external) angle χ ($\beta = 2\pi - \chi$), is the symmetric $\alpha = \chi/2$ flow (b).

15

Loop Integrals and Poles

It was shown (Section 13.5) that if a function is holomorphic in a region, its loop integral in that region is zero. It follows that if the function is holomorphic in a doubly-connected (or ring-shaped) region, the integrals along the inner and outer loops are equal. This implies that a loop may be contracted (Section 15.2), without changing the value of the integral, as long as it crosses only regions where the function is holomorphic. Thus a loop can be shrunk to zero if the function is holomorphic in its interior, implying that the integral is zero (Section 15.1). A point where the function is not holomorphic is called a singularity. If a function is holomorphic in the interior of a loop, except at an isolated singularity, the loop can be shrunk to a small circle around it. If the singularity is of the type called simple pole, the value of the integral will be determined by the value at the pole of an associated function (Section 15.3), called residue (Section 15.7); the derivate of any order of the function can be obtained if the singularity is (Section 15.8) a pole of order N. It follows that if a function is holomorphic in a region, its value, and that of all its derivates, at any interior point, may be determined by loop integrals around the boundary; thus the values of a holomorphic function on the boundary determine the values of the function and all its derivatives at all interior points. This implies that a holomorphic function that is defined as having a first-order derivate also has derivates of all orders (Section 15.4), as well as primitives of all orders (Section 13.1). The singularities of a function, either in the interior (Section 15.6) or boundary (Section 15.5) of a region, if they are of the type called pole, provide a convenient way of evaluating loop integrals (Section 15.9) using residues. The loop integrals with poles will be used to calculate forces on bodies (Chapter 28), for example, hydrodynamic and electromagnetic; the integrals along loops (the real axis) integrals also specify the fields due to source distributions (Chapters 18,24,26), along curves (straight segments).

15.1 Cauchy (1821) First Theorem on Integrals

It was shown (13.23b) that if the function $f(z)$ is holomorphic in a closed region D, with boundary loop ∂D, its integral is zero:

$$f(z) \in \mathcal{D}(D): \quad \int_{\partial D} f(z)\,dz = 0. \tag{15.1}$$

The proof given before (Section 13.1) relied on the existence of primitive, and used indirectly the Cauchy–Riemann conditions (Section 11.3). The theorem (15.1) can also be proved in a different way (Sections 19.1–19.4) that does not involve the Cauchy–Riemann conditions, and uses directly the definition of holomorphic function, and allows a sequence of generalizations: (i) the function need not be holomorphic everywhere in the region D, and it is sufficient that it be holomorphic in the interior $D - \partial D$ of D, and uniformly continuous on the boundary ∂D (Section 19.5); (ii) the function may have singularities if they lie on the boundary, have order less than unity:

$$n=1,\ldots,N: \lim_{\zeta\to\zeta_n} (\zeta-\zeta_n) f(\zeta)=0, \tag{15.2}$$

and are finite in number (Section 19.6).

Thus can be stated **Cauchy's first integral theorem (1821)**: *the integral of a complex function along a loop* $\partial\mathrm{D}$ *(defined as a rectifiable, simple, closed curve in Section 13.5), is zero (15.1), for a function* $\mathrm{f(z)}$ *satisfying one of the following* ***Cauchy conditions****, each of that generalizes the preceding: (i) the function is holomorphic in the closed region* D *bounded by* $\partial\mathrm{D}$*; (ii) the function is holomorphic in the interior (or open region)* $\mathrm{D}-\partial\mathrm{D}$*, and uniformly continuous on the boundary (or closure)* $\partial\mathrm{D}$*; (iii) the function is holomorphic in the interior* $\mathrm{D}-\partial\mathrm{D}$*, and uniformly continuous on the boundary* $\partial\mathrm{D}-\sum_{\mathrm{n}=1}^{\mathrm{N}}\mathrm{V}_{\varepsilon}(\zeta_{\mathrm{n}})$*, excluding the neighborhoods of a finite number* N *of points* ζ_{n}*, where* f *can have ignorable singularities* $\mathrm{f}(\zeta_{\mathrm{n}})=\infty$ *whose order is less than unity (15.2).*

15.2 Doubly-Connected Region and "Shrinking" of a Loop

The first theorem of Cauchy (15.1) ≡ (13.23b) implies that the value of a contour integral (13.23c) is not changed if the path of integration is deformed within the region of holomorphy of the function, without changing the end-points. The analogous theorem for loop integrals states that a loop may be "contracted" (Figure 15.1b) without changing the value of the integral, provided that only regions where the function is holomorphic be crossed. This can be restated: *if the function* $\mathrm{f(z)}$ *satisfies the Cauchy conditions in doubly-connected (or ring-shaped) region* E *lying between the loops* $\underline{\partial\mathrm{E}}$ *and* $\overline{\partial\mathrm{E}}$*, its integral is the same along the two loops:*

$$f(z)\in\mathcal{D}(E): \quad \oint_{\overline{\partial E}} f(\zeta)\,\mathrm{d}\zeta = \oint_{\underline{\partial E}} f(\zeta)\,\mathrm{d}\zeta, \tag{15.3}$$

where the outer (inner) loop is $\overline{\partial\mathrm{E}}\,(\underline{\partial\mathrm{E}})$.

To prove the theorem consider a loop L (Figure 15.1a), consisting of the outer ABC and inner DFG loops, joined by segments CD and GA contained in E. The Cauchy conditions are met for the loop L, and thus the integral is zero:

$$0=\int_{ABCDEFA} f(\zeta)\,\mathrm{d}\zeta = \left(\int_{ABC}+\int_{DEF}+\int_{CD}+\int_{FA}\right) f(\zeta)\,\mathrm{d}\zeta, \tag{15.4}$$

and consists of four terms: (i) letting the connecting segments coincide ($A\to C$ and $D\to F$ in Figure 15.1a), the corresponding integrals cancel each other, since they are taken in opposite directions:

$$\lim_{C\to A,D\to F}\int_{CD} f(\zeta)\,\mathrm{d}\zeta = -\int_{FA} f(\zeta)\,\mathrm{d}\zeta; \tag{15.5}$$

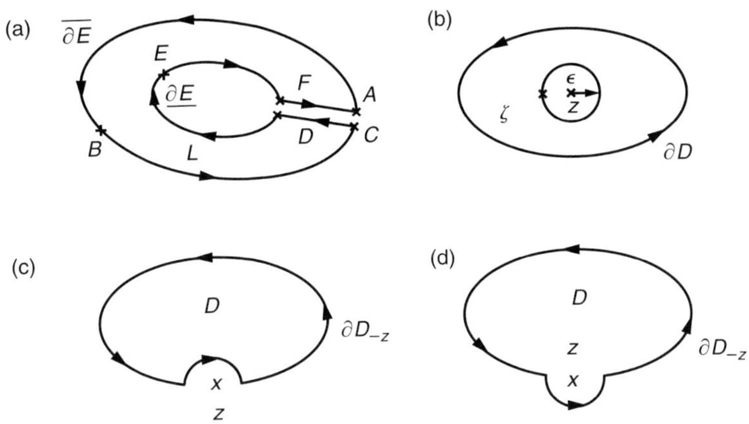

FIGURE 15.1
If a function is holomorphic (a) in a in a ring shaped region with outer $\overline{\partial E}$ (inner $\underline{\partial E}$) boundary its integrals along both boundaries coincide because: (i) it is holomorphic within the loop $ABCDEFA$ so the total integral zero; (ii) the segments $CD \equiv FA$ coincide and are taken in opposite directions, hence cancel; (iii) the remaining loop integrals FED and ABC add to zero, and since they are taken in opposite directions, they have the same value. Thus a loop can be shrunk without changing the value of the integral as long as it crosses only regions where the function is holomorphic. If a function is holomorphic in the whole interior of a loop, the loop integral is zero, because it can be shrunk into a point; if the function has an isolated singularity inside the loop, the loop can be shrunk into an infinitesimal circle around the singularity (b). If the function has a singularity on the boundary it may be enclosed (d) [excluded (c)] by deforming the loop, in which case it may (it cannot) affect the value of the loop integral.

(ii) the remaining two integrals in (15.4) add to zero in the limit and $D \to F$, when they become loop integrals:

$$0 = \lim_{C \to A, D \to F} \left(\int_{ABC} + \int_{DEF} \right) f(\zeta) - \left(\overset{(+)}{\oint_{\overline{\partial E}}} + \overset{(-)}{\oint_{\underline{\partial E}}} \right) f(\zeta)\, \mathrm{d}\zeta = \left(\overset{(+)}{\oint_{\overline{\partial E}}} - \overset{(+)}{\oint_{\underline{\partial E}}} \right) f(\zeta)\, \mathrm{d}\zeta; \quad (15.6)$$

(iii) bearing in mind that they are taken in opposite directions it follows from (15.6) that they are equal (15.3). QED.

15.3 Second Cauchy (1821) Theorem: Value of the Function

The theorem (15.3) is equivalent to Cauchy's first theorem (15.1) because: (i) the latter (15.3) was proved using (15.1); (ii) from (15.3) can be proved (15.1), by noting that if $f(z)$ satisfies the Cauchy conditions at all points of the region D, the boundary $\partial D \equiv \overline{\partial E}$

becomes the outer loop, and the inner loop $\underline{\partial E}$ can be shrunk to zero:

$$0 = \lim_{\underline{\partial E} \to 0} \int_{\underline{\partial E}} f(\zeta)\,\mathrm{d}\zeta = \int_{\overline{\partial E} \equiv \partial D} f(\zeta)\,\mathrm{d}\zeta, \tag{15.7}$$

so that (15.3) leads to (15.7) ≡ (15.1). The result (15.3), besides being equivalent to Cauchy's first theorem, is also useful in proving:

$$z \in D - \partial D: \quad f(z) = (2\pi i)^{-1} \oint_{\partial D} f(\zeta)(\zeta - z)^{-1}\,\mathrm{d}\zeta, \tag{15.8}$$

Cauchy's second theorem (1821): *if the function* f(z) *is holomorphic, or satisfies the Cauchy conditions (Section 15.1) in a region* D, *then its value at any interior point* z ∈ D − ∂D *can be calculated by evaluating an integral (15.8) along the boundary* ζ ∈ ∂D.

To prove (15.8) note that the integrand is holomorphic at all interior points except $\zeta = z$, so that the loop ∂D may be shrunk without crossing $\zeta = z$, for example, into a circle of center at z and radius ε, such that it lies wholly within D in Figure 15.1b:

$$\zeta = z + \varepsilon \mathrm{e}^{i\varphi}, \quad \mathrm{d}\zeta = \varepsilon \mathrm{e}^{i\varphi} i\,\mathrm{d}\varphi; \tag{15.9a,b}$$

substituting into the r.h.s. of (15.8):

$$\oint_{\partial D} f(\zeta)(\zeta - z)^{-1}\,\mathrm{d}\zeta = \int_{|\zeta - z| = \varepsilon} f\left(z + \varepsilon \mathrm{e}^{i\varphi}\right) \varepsilon^{-1} \mathrm{e}^{-i\varphi}\mathrm{d}\left(\varepsilon \mathrm{e}^{i\varphi}\right)$$

$$= \lim_{\varepsilon \to 0} \int_0^{2\pi} f\left(z + \varepsilon \mathrm{e}^{i\varphi}\right) i\,\mathrm{d}\varphi = f(z) \int_0^{2\pi} i\,\mathrm{d}\varphi = 2\pi i f(z), \tag{15.10}$$

where the limit of zero radius for the circle was taken inside the integral (Section 13.4) since the integrand is uniformly continuous in z. The result (15.10) coincides with (15.8). QED. An example is:

$$\int^{(\pi/2+)} \mathrm{e}^{i\zeta} \left(\zeta - \frac{\pi}{2}\right)^{-1} \mathrm{d}\zeta = \mathrm{e}^{i\pi/2}(2\pi i) = -2\pi, \tag{15.11}$$

where the path of integration is any loop taken in the positive direction around the point $\zeta = \pi/2$.

15.4 Third Cauchy (1821) Theorem: All the Derivates

The integrand in (15.8) is a function of ζ and z, such that: (i) it is integrable along the loop ∂D, provided that $z \neq \partial D$ does not lie on it; (ii) the derivate with regard to z is $(\zeta - z)^{-2}$, that is uniformly continuous in $\zeta \in \partial D$, for z in the interior $|\zeta - z| \geq \delta > 0$. Hence both conditions (Section 13.8) of theorem (13.41a,b) are met, showing that $f(z)$ defined by (15.8) is a holomorphic function, whose derivate may be calculated by applying d/dz under integral sign:

$$f'(z) = (2\pi i)^{-1} \oint_{\partial D} f(\zeta) \frac{\partial\left[(\zeta - z)^{-1}\right]}{\partial z} \mathrm{d}\zeta = (2\pi i)^{-1} \oint_{\partial D} f(\zeta)(\zeta - z)^{-2}\,\mathrm{d}\zeta; \tag{15.12}$$

the integral (15.12) has again the properties required for derivation with regard to z under integral sign, and applying the process $(n-1)$-times:

$$n \in |N: \quad f^{(n)}(z) = \frac{n!}{2\pi i} \oint_{\partial D} f(\zeta)\{(\zeta - z)\}^{-n-1} d\zeta, \tag{15.13}$$

leads to the nth derivate of $f(z)$ expressed as a loop integral.

The proof of (15.13) can be made rigorously by induction (Section 1.1), because: (i) it holds for $n = 1$, in that case it reduces to (15.12); (ii) if it holds for n, then it also holds for $n + 1$, since (15.13) can be differentiated with regard to z under the integral:

$$f^{(n+1)}(z) = \frac{n!}{2\pi i} \oint_{\partial D} f(\zeta) \frac{\partial \left[(\zeta - z)^{-n-1}\right]}{\partial z} d\zeta = \frac{(n+1)!}{2\pi i} \oint_{\partial D} f(\zeta)(\zeta - z)^{-n-2} d\zeta, \tag{15.14}$$

leading to a formula (15.14) identical with (15.13) replacing n by $n + 1$. QED. Thus has been proved **Cauchy's third theorem (1821)**: if *the function* f(z) *is holomorphic or satisfies the Cauchy conditions (Section 15.1) in a region* D, *its derivates of any order* $n \in |N$ *can be calculated at any interior point* $z \in D - \partial D$ *from the values on the boundary* $\zeta \in \partial D$ *by means of the loop integral (15.13)*. The preceding theorem implies that if a complex function has first-order derivate in a closed region D, then it is holomorphic in D, and has derivates of all orders at all interior points:

$$f(z) \in \mathcal{D}(D): \quad \forall_{n \in |N \wedge z \in D - \partial D}: \quad f(z) \in \mathcal{D}^{\infty}. \tag{15.15}$$

There is a fourth Cauchy integral (Section 31.4) concerning the number of zeros and poles of a function. An example of theorem (15.13) is:

$$\int^{(\pi/2+)} e^{i\zeta}\left(\zeta - \frac{\pi}{2}\right)^{-2n-1} d\zeta = \frac{2\pi i}{(2n)!} i^{2n} e^{i\pi/2} = (-)^{n-1} \frac{2\pi}{(2n)!}, \tag{15.16}$$

of that (15.11) is the particular case $n = 0$.

15.5 Inclusion or Exclusion of Singularities on the Boundary

A point $z = \zeta$ at that a function $f(z)$ fails to be holomorphic is called a **singularity**. In Section 15.3 (15.4) were considered singularities of the type simple pole (pole of order n) in (15.8) [(15.13)] at a point z in the interior $D - \partial D$ of the region D where $f(z)$ is holomorphic or satisfies the Cauchy conditions. If the singularity $\zeta = z$ lies on the boundary loop $z \in \partial D$, then two possibilities arise: (a) if the function is not holomorphic beyond the boundary, it can be deformed only to exclude the pole, that is, by **indenting** the loop as in Figure 15.1c; (b) if the function is holomorphic beyond the boundary, there is the choice of deforming the loop to exclude the pole by indenting (Figure 15.1c) or including the pole by **enclosing** it within the loop (Figure 15.1d). Thus the two following ways of the deformation of the loop need to be considered: (i) if the pole is **excluded** (Figure 15.1c), by indenting the loop so as to leave z on the outside, the integrals (15.8) and (15.13) are zero; (ii) if the pole is **included** (Figure 15.1d) by deforming the loop so as to enclose z on the inside, and then the integrals (15.8) and (15.13) are calculated as before, except that in (15.10) the circle

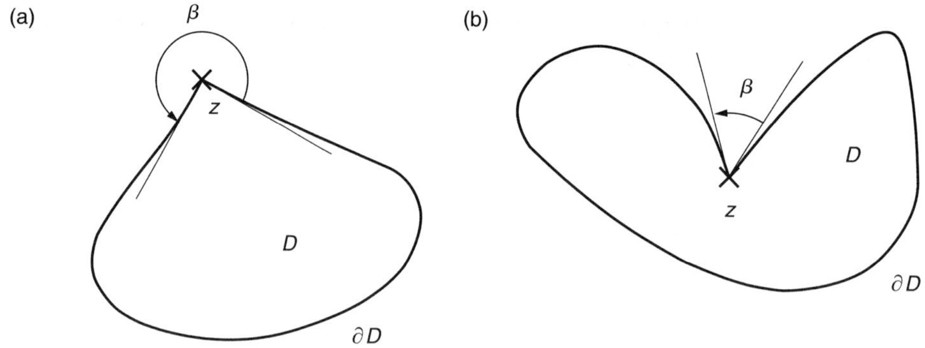

FIGURE 15.2
If only one singularity lies inside a loop, the loop can be shrunk to a circle (Figure 15.1b) around the singularity, with a phase change of 2π. If the singularity is at a regular point on the boundary (Figure 15.1d) it can be enclosed by a half-circle with a phase change of π. If the singularity is on the boundary at a convex (concave) angular point [Figure 15.2a(b)] the circular arc between the tangents (Figure 13.1c) corresponds to a change of argument $\beta > \pi$ ($\beta < \pi$).

$0 \leqq \varphi < 2\pi$ is replaced by a half-circle $0 \leqq \varphi < \pi$, so that the constant factor $2\pi i$ is replaced πi, for a regular point on the boundary. For an **angular points** (Figure 15.2a,b) then 2π is replaced by the **exterior angle** β between the right- and left-hand tangents, that is, $\pi < \beta < 2\pi$ ($0 < \beta < \pi$) for an **salient (reentrant) angular point** in Figure 15.2a (15.2b).

It has been shown that *if the function* $f(z)$ *is holomorphic or satisfies the Cauchy conditions (Section 15.1) in a closed region* D, *the value of the function* f *as well as its derivates* $f^{(n)}$ *of all orders* $n \in N$ *can be calculated at a point in the interior* $z \in D - \partial D$ *or on the boundary* $z \in \partial D$, *if necessary by deforming the latter into a shape* ∂D_z *enclosing* z *(Figure 15.2d), and performing a loop integral along* ∂D_z:

$$z \in D: f^{(n)}(z) = \frac{n!}{i\beta} \oint_{\partial D_z} f(\zeta)(\zeta - z)^{-n-1} \mathrm{d}\zeta. \tag{15.17}$$

where: (i) $\beta = 2\pi$ *for a pole in the interior* $z \in D - \partial D$ *in Figure 15.1b; (ii)* $\beta = \pi$ *for an enclosed pole (Figure 15.1d) at a regular point on the boundary; (iii) for a pole at an angular point (Figure 15.2a,b) on the boundary* β *is the exterior angle between right- and left-hand tangents.* The Riemann integral (Section 13.2) can de extended to a contour with isolated angular points, by splitting the path of integration into arcs between each successive pair of angular points; the Cauchy's first (Section 15.1) second (Section 15.3) and third (Section 15.4) theorems apply to region whose boundary is a loop (Sections 13.5 and 15.5), that is, a simple closed rectifiable curve that can have isolated angular points, that is, a sectionally continous tangent. An example of (15.17) is the integral:

$$\int_{|\zeta|=\pi/2} \mathrm{e}^{i\zeta} \left(\zeta - \frac{\pi}{2}\right)^{-1} \mathrm{d}\zeta = \mathrm{e}^{i\pi/2} i\pi = -\pi, \tag{15.18}$$

whose value is half that of (15.11), because the singularity $\zeta = \pi/2$, which is a simple pole, lies at a regular point on the boundary in (15.18) instead of the interior in (15.11), and has been enclosed by deforming the loop (Figure 15.1d); if the singularity had been excluded (Figure 15.1c) the integral would have been zero. Some loop integrals with poles on the boundary are considered in Example 20.9.

15.6 Holomorphic Function in a Multiply-Connected Region

The region D used (Figure 15.1b) in proving the theorems in Section 15.3–15.5 has only one boundary loop, and thus is simply-connected. The ring-shaped region E in Figure 15.1a used in proving the theorem in Section 15.2, has a boundary $\partial E \equiv \underline{\partial E} + \overline{\partial E}$ consisting of an inner $\underline{\partial E}$ and an outer $\overline{\partial E}$ loop, and is doubly-connected. In general, a **multiply-connected region** D has a boundary D consisting of Figure 15.3a of an outer L and M inner loops L_m, with $m = 1, \ldots, M$. To apply the preceding theorems, consider (Figure 15.3a) the loop $\bar{L}$ consisting of all inner loops $L_1, \ldots, L_M$ joined in turn to the outer loop L by two-way segments, $A_m B_m$:

$$\oint_{\partial D} = \oint_{L}^{(+)} + \sum_{m=1}^{M} \left(\oint_{L_m}^{(-)} + \int_{A_m B_m} + \int_{B_m A_m} \right), \tag{15.19}$$

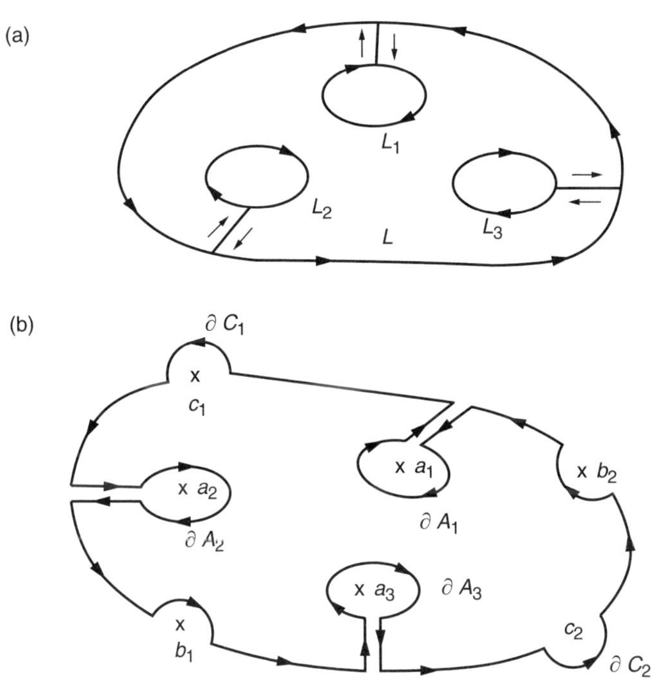

FIGURE 15.3
If a function is holomorphic in a multiply-connected region (a) consisting of one outer boundary L and several inner boundaries L_1, L_2, L_3, ... its integral vanishes around a loop: (i) taking the outer boundary in opposite direction to the inner boundaries; (ii) joining the inner boundaries to the outer boundary by two-way segments. If the function is holomorphic in a loop except for isolated singularities, the loop integral can be calculated by deforming the loop without crossing any singularities. This implies (b) it equals the sum of: (i) loop integrals $\partial A_1, \ldots$ around singularities $a_1, \ldots$ in the interior; (ii) half-loop integrals $\partial C_1, \ldots$ around enclosed $c_1, \ldots$ singularities on the boundary. The (iii) excluded $b_1, \ldots$ singularities on the boundary do not affect the value of the loop integral.

so that the Cauchy (Section 15.1) conditions apply within D if they apply within L. The segments (A_m, B_m) are taken twice in opposite directions, so the corresponding integrals cancel, and the integrals along the inner loops are taken in opposite direction to the outer loop:

$$\oint_{\partial D} = \oint_{L}^{(+)} + \sum_{m=1}^{M} \left(-\oint_{L_m}^{(+)} + \int_{A_m B_m} - \int_{A_m B_m} \right) = \oint_{L}^{(+)} - \sum_{n=1}^{M} \oint_{L_m}^{(+)}, \tag{15.20}$$

or in the same direction if the sign before the integral is reversed.

Using in the formula (15.20), respectively (15.1; 15.8; 15.13) **the first, the second, and the third Cauchy theorems**, *leads to the corresponding generalizations from singly to multiply connected regions: if the function* f(z) *is holomorphic or satisfies the Cauchy conditions (Section 15.1) in a multiply-connected region* D, *whose boundary* ∂D *consists of one outer loop* L *and* M *inner loops* $\mathrm{L_m}$ *with* m = 1,. . . ,M, *then:*

$$\oint_{L} f(\zeta)\, \mathrm{d}\zeta = \sum_{m=1}^{M} \oint_{L_m} f(\zeta)\, \mathrm{d}\zeta, \tag{15.21}$$

$$\frac{i\beta}{n!} f^{(n)}(z) = \left(\oint_{L} - \sum_{m=1}^{M} \oint_{L_m} \right) f(\zeta)(\zeta - z)^{-n-1} \mathrm{d}\zeta, \tag{15.22}$$

in that: (i) $\beta = 2\pi$ *if the pole lies (Figure 15.1b) in the interior* $\mathrm{z} \in \mathrm{D} - \partial\mathrm{D}$*; (ii)* $\beta = 0$ *if the pole lies at a regular point on the boundary* $\mathrm{z} \in \partial\mathrm{D}$ *and is excluded (Figure 15.1c) by indentation of the contour; (iii)* $\beta = \pi$ *if the pole lies at a regular point on the boundary and is included (Figure 15.1d) by deformation of the contour; (iv)* β *is the exterior angle between the right- and left-hand tangents if the pole lies on an angular point on the boundary and the contour is deformed to enclose it, for example, a salient (reentrant) angular point in Figure 15.2a (15.2b).*

15.7 Residue of a Function at a Simple Pole

The simplest type of singularity of a function, that is, point $\zeta = z$ where it fails to be holomorphic is the **pole**. The function $f(\zeta)$ has a **simple pole** at z if it is of the form (15.23b):

$$g \in D(V_\varepsilon(z)): \quad f(\zeta) = (\zeta - z)^{-1} g(\zeta), \tag{15.23a,b}$$

where g is holomorphic (15.23a) in a neighborhood of z. Considering a closed subregion $D \subset V_\varepsilon(z)$ of the neighborhood, the integral over the boundary can be evaluated using (15.22) [(15.21) with $n = 1$] if $z \in D - \partial D\,(z \in \partial D)$, leading to the factor $2\pi i\,(i\beta)$:

$$D \subset V_\varepsilon(z): \quad \int_{\partial D} f(\zeta)\, \mathrm{d}\zeta \equiv \int^{(z+)} f(\zeta)\, \mathrm{d}\zeta = \{2\pi i, i\beta\} f_{(1)}(z), \tag{15.24a}$$

multiplying the **residue at the simple pole** $f_{(1)}(z)$ of the function $f(\zeta)$ at $\zeta = z$:

$$f_{(1)}(z) \equiv g(z) = \lim_{\zeta \to z} (\zeta - z) f(\zeta). \tag{15.24b}$$

The residue function $f_{(1)}(z)$ associated to $f(\zeta)$ is holomorphic at the pole that is a singularity of $f(\zeta)$. There are four methods of calculation of the residue at a simple pole: (i) method I uses (Subsection 15.7.1) the limit (15.24b); (ii) method II uses (Subsection 15.7.2) a partial fraction decomposition (Section 31.8); (iii) method III uses a differentiation rule (Subsection 19.9.1); (iv) method IV uses the Laurent series (Section 27.4).

15.7.1 Calculation of the Residue as a Limit (Method I)

Consider as an example the function:

$$f(\zeta)=\left\{\zeta^2-\zeta(a+b)+ab\right\}^{-1}=\left\{(\zeta-a)(\zeta-b)\right\}^{-1}, \tag{15.25}$$

that has simple poles at $\zeta=a,b$, with residues:

$$f_1(a)=\lim_{\zeta\to a}(\zeta-a)\left\{(\zeta-a)(\zeta-b)\right\}^{-1}=(a-b)^{-1}=-f_1(b). \tag{15.26a,b}$$

The residues become obvious performing the decomposition into partial fractions:

$$f(\zeta)=\frac{1}{a-b}\left\{\frac{1}{\zeta-a}-\frac{1}{\zeta-b}\right\}=\frac{f_{(1)}(a)}{\zeta-a}+\frac{f_{(1)}(b)}{\zeta-b}. \tag{15.27a,b}$$

In the case of rational function, such as (15.25), the residue $f_{(1)}(a)$ at a simple pole $\zeta=a$ may be obtained by suppressing the factor $\zeta-a$ in the denominator, and setting $\zeta=a$ in the remaining terms, to obtain $\left[(\zeta-b)^{-1}\right]_{\zeta=a}=(a-b)^{-1}$, as indicated in (15.26a,b) and (15.27a,b).

15.7.2 Calculation of Residue from a Partial Fraction Decomposition (Method II)

The partial fraction decomposition (15.27a) can be obtained by purely algebraic means in four steps, as follows: (i) the function on the l.h.s. of (15.25) has simple poles at $z=a,b$, and thus must be of the form:

$$\frac{1}{(\zeta-a)(\zeta-b)}=\frac{A}{\zeta-a}+\frac{B}{\zeta-b}=\frac{A(\zeta-b)+B(\zeta-a)}{(\zeta-a)(\zeta-b)}, \tag{15.28a,b}$$

where A, B are constants, to be determined next; (ii) from (15.28b) follows the identity (15.29a):

$$1=(A+B)\zeta-(Ab+Ba), \quad -B=A=\frac{1}{a-b}, \tag{15.29a–c}$$

whose solution (iii) for arbitrary ζ is (15.29b,c); (iv) substituting (15.29b,c) in (15.28a) yields the partial fraction decomposition (15.27b); (v) the function (15.27a) is rational and vanishes as $\zeta\to\infty$. The latter can also be obtained from (15.27b) $\equiv$ (15.28a) noting that $A(B)$ are (15.26a,b) the residues $A\equiv f_{(1)}(a)\left[B\equiv f_{(1)}(b)\right]$ at the simple poles $z=a\,(z=b)$. The values of residues can (method III) also be calculated (19.40–19.41) using the auxiliary function (15.30a):

$$h(\zeta)=(\zeta-a)(\zeta-b), \quad h'(\zeta)=2\zeta-a-b, \tag{15.30a,b}$$

as follows:

$$\frac{1}{(\zeta-a)(\zeta-b)}=\frac{1}{h'(a)(\zeta-b)}+\frac{1}{h'(b)(\zeta-a)}=\frac{1}{a-b}\left(\frac{1}{\zeta-a}-\frac{1}{\zeta-b}\right), \tag{15.30c}$$

leading directly to (15.30c) $\equiv$ (15.27a). Thus the partial fraction decomposition (15.27a) can be obtained in four equivalent ways: (method II) by the algebraic procedure (15.28a,b; 15.29a–c) that can be extended to any rational function (Section 31.8); (method I) from the residues (15.26a,b) at the simple poles in (15.27a,b) that applies to all functions (15.24b); (method III) calculating the residues in (15.30c) with the help of the auxiliary function (15.30a,b) that applies to some irrational functions; (method IV) a more general method is presented in Section 37.4 that applies to any isolated singularity not just poles. The residues (15.26a,b) allow the evaluation of the following integrals:

$$\int^{(a+)}, \int^{(b+)}, \int^{(a+,b+)} \left[\zeta^2 - (a+b)\zeta + ab\right]^{-1} \mathrm{d}\zeta = \frac{2\pi i}{a-b}, -\frac{2\pi i}{a-b}, 0, \tag{15.31a–c}$$

using (15.21) and (15.24a).

15.8 Multiple Pole or Pole of Order n

The simple pole is the particular case $n = 1$ of the **multiple pole of order n** at the point z, where $f(z)$ has the form (24.32b):

$$g \in D(V_\varepsilon(z)): \quad f(\zeta) = (\zeta - z)^{-n} g(\zeta), \tag{15.32a,b}$$

in that the function $g(z)$ is holomorphic (15.32a) in a neighborhood of z. Considering a closed subregion $D \subset V_\varepsilon(z)$, the integral of the function along the boundary can be evaluated by (15.13) [(15.17)] if z lies inside $z \in D - \partial D$ (on the boundary $z \in \partial D$), leading the factor $2\pi i\,(i\beta)$:

$$\oint_{\partial D} f(\zeta)\,\mathrm{d}\zeta \equiv \int^{(z+)} f(\zeta)\,\mathrm{d}\zeta = \{2\pi i, i\beta\}\, f_{(n)}(z), \tag{15.33a}$$

multiplying by the **residue at the pole of order n**, viz. $f_{(n)}(z)$, of the function $f(\zeta)$ at $\zeta = z$:

$$f_{(n)}(z) \equiv \frac{g^{(n-1)}(z)}{(n-1)!} = \frac{1}{(n-1)!}\lim_{\zeta\to z}\frac{\mathrm{d}^{n-1}}{\mathrm{d}\zeta^{n-1}}\left\{(\zeta - z)^n f(\zeta)\right\}; \tag{15.33b}$$

this reduces to (15.24b) for a simple pole $n = 1$. The four methods of calculation of the residue at a simple pole (Section 15.7) can all be extended to multiple pole of any order: (i) method I uses (Subsection 15.8.1) the limit (15.33b): (ii) method II uses (Subsection 15.8.2) a partial fraction decomposition (Section 31.9) for rational functions; (iii) method III uses a rule of differentiation (Subsection 19.9.2); (iv) method IV uses a Laurent series (Section 27.4).

15.8.1 Residue at a Multiple Pole as a Limit (Method I)

Consider as an example the function:

$$f(\zeta) = \left\{\zeta^4 - 2\zeta^2 a^2 + a^4\right\}^{-1} = \left(\zeta^2 - a^2\right)^{-2} = (\zeta - a)^{-2}(\zeta + a)^{-2}, \tag{15.34}$$

that has double poles at $\zeta = \pm a$, with residues:

$$f_{(2)}(\pm a) = \lim_{\zeta \to \pm a} \frac{\mathrm{d}}{\mathrm{d}\zeta} \frac{(\zeta \mp a)^2}{(\zeta - a)^2 (\zeta + a)^2} = \lim_{\zeta \to \pm a} \frac{\mathrm{d}}{\mathrm{d}\zeta} \frac{1}{(\zeta \pm a)^2} = \mp \frac{1}{4a^3}. \tag{15.35}$$

The decomposition of (15.34) into partial fractions:

$$f(\zeta) = \frac{1}{4a^2} \left\{ \frac{1}{(\zeta - a)^2} + \frac{1}{(\zeta + a)^2} \right\} - \frac{1}{4a^3} \left\{ \frac{1}{\zeta - a} - \frac{1}{\zeta + a} \right\}, \tag{15.36}$$

features the residues $f_{(2)}(\pm a)$:

$$f(\zeta) - \frac{(\zeta - a)^{-2} + (\zeta + a)^{-2}}{4a^2} = (\zeta - a)^{-1} f_{(2)}(a) + (\zeta + a)^{-1} f_{(2)}(-a). \tag{15.37}$$

as the coefficients of $(\zeta \mp a)^{-1}$.

15.8.2 Residue at a Multiple Pole from a Partial Fraction Decomposition (Method II)

The partial fraction decomposition (15.36) can be obtained by a purely algebraic method, using a sequence of four steps, viz.: (i) the function is rational, vanishes as $\zeta \to \infty$ and has double poles at $z = \pm a$, and thus can be represented in the form:

$$\frac{1}{(\zeta - a)^2 (\zeta + a)^2} = \frac{A}{(\zeta - a)^2} + \frac{B}{\zeta - a} + \frac{C}{(\zeta + a)^2} + \frac{D}{\zeta + a}, \tag{15.38a}$$

where A, B, C, D are constants, to be determined next; (ii) from (15.38a) follows the identity:

$$1 = A(\zeta + a)^2 + B(\zeta - a)(\zeta + a)^2 + C(\zeta - a)^2 + D(\zeta + a)(\zeta - a)^2, \tag{15.38b}$$

that is a cubic polynomial, from whose coefficients can be determined A, B, C, D; (iii) as an alternative, setting $\zeta = \pm a$ in (15.38b) determines immediately two coefficients (15.39a,b):

$$1 = 4a^2 A = 4a^2 C, \qquad \frac{A}{(\zeta - a)^2} + \frac{C}{(\zeta + a)^2} = \frac{1}{2a^2} \frac{\zeta^2 + a^2}{(\zeta^2 - a^2)^2}, \tag{15.39a–c}$$

so that (iv) the corresponding terms in (15.38a) are (15.39c); (iv) substituting (15.39c) in (15.38a) leaves only two terms:

$$\begin{aligned} \frac{B}{\zeta - a} + \frac{D}{\zeta + a} &= \frac{1}{(\zeta - a)^2 (\zeta + a)^2} - \frac{1}{4a^2} \left[\frac{1}{(\zeta - a)^2} + \frac{1}{(\zeta + a)^2} \right] = \frac{1}{(\zeta^2 - a^2)^2} \left(1 - \frac{\zeta^2 + a^2}{2a^2} \right) \\ &= \frac{1}{2a^2} \frac{a^2 - \zeta^2}{(\zeta^2 - a^2)^2} = -\frac{1}{2a^2} \frac{1}{\zeta^2 - a^2} = -\frac{1}{2a^2} \frac{1}{2a} \left(\frac{1}{\zeta - a} - \frac{1}{\zeta + a} \right), \end{aligned} \tag{15.40}$$

whose decomposition in partial fractions has been considered before (15.27a), leading to (15.40) ≡ (15.36). The residue at the double pole can be calculated [(19.48; 19.49) in Subsection 19.9.2] using the method III of auxiliary function (19.50a,b), extending (15.30a–c). The residues at the multiple pole supply only two coefficients B, D in (15.38a), and a similar but generalized procedure can be used (Section 31.9) to determine A,C. Thus the partial fraction decomposition can be obtained in two equivalent ways: (a) by the algebraic procedure (15.38a,b; 15.39a–c; 15.40); (b) by a general method (Section 31.9) for rational functions. A function with simple and multiple poles is considered in Example 20.10. Other methods of calculation of residues of rational and other functions are considered in Sections 19.9 and 27.4. The residues (15.35) allow the evaluation of the integrals:

$$\int^{(a+)}, \int^{(-a+)}, \int^{(a+,-a+)} \left(\zeta^4 - 2a^2\zeta^2 + a^4\right)^{-1} d\zeta = -\frac{i\pi}{2a^3}, \frac{i\pi}{2a^3}, 0, \tag{15.41a–c}$$

using (15.21) and (15.33a).

15.9 Loop Integral with Poles in the Interior and on the Boundary

The residues can be used to calculate the integral of a function $f(\zeta)$ along the boundary ∂D of a closed region D containing several poles, in the interior $D - \partial D$ and on the boundary ∂D, by using the following method illustrated in Figure 15.3b: (i) the poles $a_m \in D - \partial D$ of order α_m lying in the interior of D are enclosed by loops A_m, for example, small circles; (ii) the poles b_k on the boundary are excluded, and the poles $c_n \in \partial D$ of order γ_n on the boundary are included by deforming ∂D in half-loops C_n, the corresponding integrals being:

$$\int_{\partial A_m} f(\zeta)\, d\zeta = 2\pi i f_{(\alpha_m)}(a_m), \quad \int_{\partial C_n} f(\zeta)\, d\zeta = \pi i f_{(\gamma_n)}(c_n); \tag{15.42a,b}$$

(ii) the function $f(\zeta)$ is holomorphic or satisfies the Cauchy conditions (Section 15.1) in the multiply-connected region:

$$E = D - \sum_{m=1}^{M} A_m - \sum_{n=1}^{N} C_n, \tag{15.43}$$

so that (15.21) the integral along the outer loop ∂D, is equal to the sum of the integrals along the M inner loops ∂A_m with $m = 1, \ldots, M$, and N boundary half-loops ∂C_n with $n = 1, \ldots, N$:

$$\oint_{\partial D} f(\zeta)\, d\zeta = \left(\sum_{m=1}^{M} \oint_{\partial A_m} + \sum_{n=1}^{N} \oint_{\partial C_n} \right) f(\zeta)\, d\zeta; \tag{15.44}$$

(iii) the terms on the r.h.s. are specified by (15.42a,b), allowing the evaluation of the integral on the l.h.s. from the residues at the poles.

Thus has been proved the **theorem of residues**: (Figure 15.3b): *if the function* $\mathrm{f}(\zeta)$ *is holomorphic or satisfies the Cauchy conditions (Section 15.1) in a region* D, *except at: (i)* M *poles* $\mathrm{a_m}$ *of order* α_m *in the interior* $\mathrm{a_m} \in \mathrm{D} - \partial\mathrm{D}$ *with* $\mathrm{m} = 1, \ldots, \mathrm{M}$*; (ii)* N *poles* $\mathrm{c_n}$ *of order* γ_n *on the boundary* $\mathrm{C_n} \in \partial\mathrm{D}$ *with* $\mathrm{n} = 1, \ldots, \mathrm{N}$, *that are included by suitable deformation of* ∂D*; (iii) a finite number of other poles by on the boundary* ∂D, *excluded by suitable indentation of the contour. Then the integral along the loop* ∂D *deformed as stated, is equal to the sum of: (i)* $2\pi\mathrm{i}$ *times the sum of the residues at the poles in the interior; (ii)* $\mathrm{i}\beta_\mathrm{n} = \mathrm{i}\pi$ *times the sum of the residues at the poles at regular points on the boundary that were included; (iii)* $\mathrm{i}\beta_\mathrm{n}$ *times the residue at each pole at an angular point on the boundary,* *where, β_n is the angle between right and left tangents:*

$$\oint\limits_{\partial D} f(\zeta)\,d\zeta = 2\pi i \sum_{m=1}^{M} f_{(\alpha_m)}(a_m) + i \sum_{n=1}^{N} \beta_n f_{(\gamma_n)}(c_n). \tag{15.45}$$

The evaluation of the integrals (15.31a–c) and (15.41a–c) were already examples of the theorem of residues. As another example, consider the function (15.25) multiplied by e^ζ, and integrated along a loop containing both poles that are simple, viz.:

$$\int^{(a+,b+)} e^\zeta \left\{\zeta^2 - (a+b)\zeta + ab\right\}^{-1} d\zeta = 2\pi i \left\{e^a f_{(1)}(a) + e^b f_{(1)}(b)\right\} = 2\pi i \frac{e^a - e^b}{a-b}, \tag{15.46}$$

where the residues (15.26a,b) were used.

As an example with double poles, consider the function (15.34) along a circle with center at one pole $z = a$, and radius $|z - a| = 2a$ such that the other pole $z = -a$ lies on the boundary, and is enclosed by deformation of the loop:

$$\int\limits_{|z-a|=2a} \left(\zeta^4 - 2a^2\zeta^2 + a^4\right)^{-1} d\zeta = 2\pi i f_{(2)}(a) + \pi i f_{(2)}(-a) = -\frac{i\pi}{4a^3}, \tag{15.47}$$

where the residues (15.35) were used. Setting $a = b$ in (15.46) leads to a loop integral enclosing a double pole that can be evaluated using (15.33a,b):

$$\int^{(a+)} e^\zeta (\zeta - a)^{-2} d\zeta = 2\pi i \lim_{\zeta\to a} \frac{d\left(e^\zeta\right)}{d\zeta} = 2\pi i e^a; \tag{15.48a}$$

this result can alternatively be obtained by taking the limit $b \to a$ in (15.46):

$$\lim_{b\to a} 2\pi i \frac{e^a - e^b}{a-b} = 2\pi i e^a \lim_{b\to a} \frac{1 - e^{b-a}}{a-b} = 2\pi i e^a, \tag{15.48b}$$

using (Subsection 27.9.3) the limit $e^{b-a} \simeq 1 + (b-a)$, or $1 - e^{b-a} = a - b$ as $b \to a$. The result (15.48b) is an example of a residue at a double pole being taken as the limit of two coincident simple poles. More generally:

$$\alpha \equiv \sum_{m=1}^{N} \alpha_n: \qquad \lim_{a_1,\ldots,a_n\to a} f(\zeta) \prod_{n=1}^{N} (\zeta - a_n)^{-\alpha_n} = f(\zeta)(\zeta - a)^{-\alpha}, \tag{15.49}$$

if N *poles* $a_1,\ldots,a_N$ *of orders* $\alpha_1,\ldots,\alpha_N$ *coincide at a they are equivalent to a single pole order* $\alpha = \alpha_1 + \cdots + \alpha_N$, for example, a pole of order N can be obtained by the coincidence of N simple poles.

Note 15.1. *Integration by Residues versus Primitives:* The method of residues (Chapter 15) has over conventional integration (Chapter 13) the advantage that it does not require knowledge of the primitive. If the primitive is known the result is the same, for example:

$$\int^{(a+)} (\zeta - a)^{-1} d\zeta = [\log(\zeta - a)]^{(+a)} = [\log|\zeta - a| + i \arg(\zeta - z)]^{(+a)} = 2\pi i, \tag{15.50}$$

since when describing a loop around $z = a$ in the positive (i.e., counterclockwise) direction the logarithm does not change except by adding 2π to the imaginary part. The preceding result (15.50) can be used to evaluate the integral:

$$\int^{(a+)} \left(\zeta^2 - a^2\right)^{-1} d\zeta = (2a)^{-1} \int^{(a+)} \left[(\zeta - a)^{-1} - (\zeta + a)^{-1}\right] d\zeta = (2a)^{-1} 2\pi i = \frac{i\pi}{a}, \tag{15.51a}$$

bearing in mind that the loop encircles $+a$ but not $-a$; the use of the residues (15.26a,b), viz.:

$$b = -a: \quad \int^{(a+)} \left(\zeta^2 - a^2\right)^{-1} d\zeta = \frac{2\pi i}{2a} = \frac{i\pi}{a}, \tag{15.51b}$$

leads to the same result (15.51a) ≡ (15.51b). The method of residues is simpler than ordinary integration for less simple integrands, and applies to arbitrary functions viz. nonsingular terms in the integrand are just evaluated at the poles. The fourth Cauchy integral theorem (Section 31.4) concerns the zeros and poles of a function within a contour. A generalization of Cauchy's integral (15.13) can be used as definition of derivative of complex order (Campos, 1984).

Note 15.2. *Calculus of Residues and Its Applications:* Contour integrals and the calculus of residues can be applied to potential flow not only to calculate the circulation and flow rate (Chapter 12), but also the to calculate the kinetic energy (Section 28.1) of the flow, and forces and torques on bodies (Section 28.2); analogous methods apply to the force and energy of other potential fields (Section 28.3), for example, electro- or magnetostatic. When a pole lies on the boundary, and the integrand is holomorphic in its neighborhood, it is valid mathematically either to include or exclude the pole: the choice will be dictated by the nature of the problem. For example, when considering a wave problem there may be two poles on the real axis, representing waves propagating in opposite directions: including (excluding) one pole is equivalent to considering (not considering) one wave, and thus corresponds to a radiation type of boundary condition. There are cases where the inclusion of a pole on the boundary is both mathematically possible and physically necessary, for example: (i) the Cauchy principal value of on integral (Sections 17.8 and 17.9) relates to a pole on the path of integration; (ii) it occurs when calculating the potential (e.g., gravity) field of a source (e.g., mass) distribution at a point inside the distribution (Example 20.20). A singularity may occur at an angular point on the boundary, for example, for a source/sink at the vertex of a corner flow (Section 36.8); the use of numerical methods discretizing a domain with a mesh consisting of polygons may lead to balances (e.g., of mass) at the vertices, and to the corresponding integrals with singularities at corner points.

Conclusion 15: A loop $\overline{\partial E}$ may (figure 15.1) be (a) shrunk up to $\underline{\partial E}$ without changing the value of the integral, provided that the integrand be holomorphic in the annulus or doubly-connected region between the two loops. If there is only one singularity z in (b) the interior of D, the loop along the boundary ∂D can be shrunk to a small circle of radius ε and center at z, for purpose of calculating the loop integral. If the singularity lies on the boundary, it is necessary to decide whether it should be deformed so as to exclude (c) or include (d) the singularity. The singularity may lie at a regular point on the boundary (d), or at an angular point which may be salient (a) or reentrant (b). Besides singly connected regions, with one boundary loop (figure 15.1b), and doubly-connected regions, ring-shaped with outer $\overline{\partial E}$ and inner $\underline{\partial E}$ loops (figure 15.1c), can also be considered multiply-connected regions (figure 15.3a), with one outer L and several inner L_1, L_2, L_3 loops. The concept of multiply-connected region can be used to calculate by means of residues the integral along a loop ∂D containing (figure 15.3b) poles a_1, a_2, a_3 the interior, as well as poles on the boundary $b_1, b_2\,(c_1, c_2)$ which are included (excluded) by deformation of ∂D.

16

Images on Plane Walls

The effect of a rigid plane wall on the potential flow due to a source/sink is equivalent (Section 16.1) to the introduction of an identical image, leading to tangential velocity at the wall, and doubling of the velocity at large distance; in the case a vortex (Section 16.2) the condition of tangential velocity at the wall, leads to an image with opposite circulation; the flow at large distance appears like due to a dipole perpendicular to the wall, and decays faster than it would for the isolated vortex. For a monopole seen in the far-field, the source/sink part is doubled by the wall effect, and unless absent, dominates the vortex part that decays faster as a dipole (Section 16.3). For a dipole (multipole) the effect (Section 16.4) in the far-field is similar, doubling the real part and causing the imaginary part to decay faster as a quadrupole (multipole one order higher). In the case of a source/sink (Section 16.5) [vortex (Section 16.6)], at any position in a rectangular corner there are three identical (alternating) images, so that in the far-field the strength is multiplied by four [is reduced to a quadrupole]. The superposition of the preceding specifies the flow field due to a monopole in a rectangular corner (Section 16.7), for example, along the diagonal. A source/sink in a corner of angle $\beta = 2\pi/n$ is multiplied in strength by n in the far-field (Section 16.8), whereas a vortex in a corner of angle $\beta = \pi/n$ becomes (Section 16.9) a multipole of order n in the far-field. The representation of wall effects by images applies not only to potential flows (Chapters 16, 28, 34, 36, 38) but also to electrostatic (Chapter 24) and magnetostatic (Chapter 26) fields and various types of waves.

16.1 Identical Image on a Rigid Wall (Rankine, 1864)

Consider the potential flow due to a source $Q > 0$ (sink $Q < 0$) at a distance a from a rigid impermeable wall (Figure 16.1a). The condition that the velocity must be tangent at the wall is met by replacing the wall with an identical image (Figure 16.1b), as will be shown next. Taking the OX-axis along the wall, and the OY-axis through the source/sink Q at $+ia$ and its identical image Q at $-ia$, the complex potential (12.26a) is specified by (16.1a):

$$f(z) = \frac{Q}{2\pi}\left[\log(z - ia) + \log(z + ia)\right] = \frac{Q}{2\pi}\log\left(z^2 + a^2\right), \tag{16.1a}$$

$$v^*(z) = \frac{Q}{2\pi}\left\{(z - ia)^{-1} + (z + ia)^{-1}\right\} = \frac{Q}{\pi}\frac{z}{z^2 + a^2} = \frac{df}{dz}, \tag{16.1b}$$

and the complex conjugate velocity (12.25b) by (16.1b). For real $z = x$ the complex potential (16.1a) is real $f(x) = \Phi(x,0)$, and thus coincides with the scalar potential, implying that the stream function is zero $\Psi(x,0) = 0$, that is, the wall coincides with the streamline $\Psi = 0$; an alternative way to reach the same conclusion is to note that for real $z = x$ the complex conjugate velocity (16.1b) is real $v^*(x) = v_x(x,0)$, and thus coincides with the tangential velocity, implying that the normal velocity is zero $v_y(x,0) = 0$. The flow due to the source/sink exerts a force on the wall (Subsection 16.1.1), simplifies in the far-field

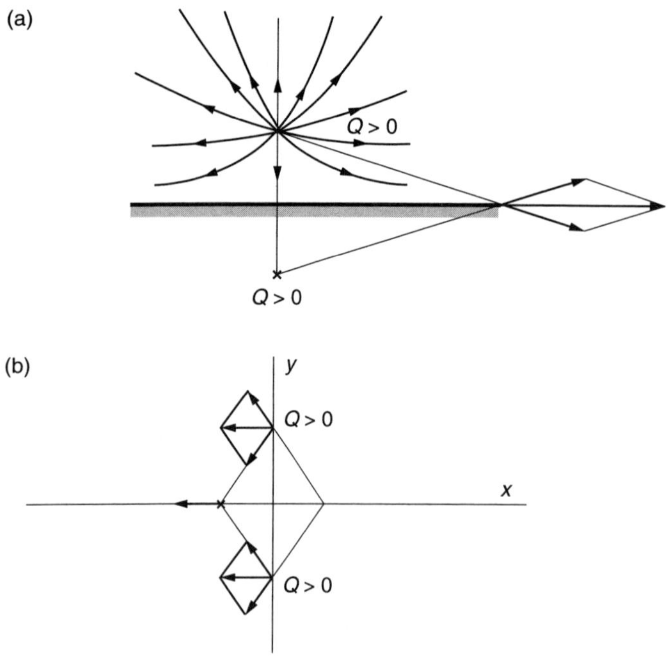

FIGURE 16.1
The potential flow due to a source $Q > 0$ (or sink $Q < 0$) near a straight rigid impermeable wall is unchanged replacing the wall by an identical image source (sink) at a symmetric position, because the sum of the flow velocities due to them is tangent to the wall.

(Subsection 16.1.2) and serves as an example of analytical calculation and geometrical plotting of streamlines (Subsection 16.1.3).

16.1.1 Induced Velocity and Force on the Wall

The image source (sink) induces on the real source (sink) a velocity $Q/2\pi d$ orthogonal to wall, where $d = 2a$ is the mutual distance; thus *a source (sink) of flow rate* Q *at a distance* a *from a wall moves toward (away) from the wall with velocity (16.2a):*

$$v_0^* = -\frac{iQ}{4\pi a} = -iv_y, \quad F^* = \rho Q v_0^* = -\frac{i\rho Q^2}{4\pi a} = -iF_y, \tag{16.2a,b}$$

corresponding to a force (16.2b) of attraction on the wall. The force on the wall is minus the force on the source/sink that is specified by Blasius theorem (28.28a). From (16.2a) it follows that the velocity is away from (toward) the wall $v_y > 0\,(v_y < 0)$ for a source $Q > 0$ (sink $Q < 0$) and (16.2b) implies that the in both case the force $F_y > 0$ is upward, that is, an repulsion. It can be checked (Example 20.11) that the dynamic pressure exerted by the source on the wall, integrated over its length) $-\infty, +\infty$ (adds to attractive force equal in modulus to (16.2b) and of opposite sign. The velocity of the source/sink (16.2a) leads (16.3):

$$t = \int \left(\frac{\mathrm{d}y}{\mathrm{d}t}\right)^{-1} \mathrm{d}y = \int v_y^{-1}\,\mathrm{d}y = \frac{4\pi}{Q}\int_a^y \xi\,\mathrm{d}\xi = \frac{2\pi}{Q}\left(y^2 - a^2\right), \tag{16.3}$$

to the trajectory (16.4a):

$$z_{\pm}(t) = \pm iy = \pm i\sqrt{a^2 + \frac{Qt}{2\pi}}, \quad t_0 = -\frac{2\pi a^2}{Q} = \frac{2\pi a^2}{|Q|}, \tag{16.4a,b}$$

showing that: (i) the source z_+ (its image z_-) moves away from the wall, reaching infinity after an infinite time [$z \to \infty$ as $t \to \infty$ for $Q > 0$ in (16.4a)]; (ii) the sink z_+ (its image z_-) moves toward the wall, reaching it after a finite time [$z = 0$ for $t = t_0$ with $Q < 0$ in (16.4b)]. The calculation of the trajectories of singularities has excluded arbitrary functions of time (Note 16.1).

16.1.2 Flow in the Near-Field and Far-Field

For an observer in the far-field, that is, at a distance large compared with the distance of the source/sink from the wall (16.5a):

$$|z|^2 \gg a^2: \quad f(z) = \frac{Q}{\pi}\log z, \quad v^*(z) = \frac{Q}{\pi z}, \tag{16.5a–c}$$

the flow (16.1a,b) is equivalent (16.5b,c) to a source/sink of double strength $2Q$ at the origin, because: (a) the identical image adds to the source, considering the flow in all space; (b) considering that the wall deflects the flow to a half-space, it doubles the field. Thus have been obtained, for *a source/sink of flow rate Q at a distance a from a rigid wall, the exact [asymptotic (16.5a)] complex potential (16.1a) [16.5b)] and conjugate velocity (16.1b) [(16.5c)]. Also the force (16.2b) of attraction exerted on the wall involving the induced velocity (16.2a) of the source sink; the latter specifies the trajectory (16.4a,b) of the source/sink due to the wall effect. The scalar potential (16.6a) and stream function (16.6b) [polar components of velocity (16.7a,b)]:*

$$\Phi = \frac{Q}{4\pi}\log\left|a^4 + r^4 + 2a^2r^2\cos(2\varphi)\right|, \tag{16.6a}$$

$$\Psi = \frac{Q}{2\pi}\operatorname{arc\,cot}\left\{\cot(2\varphi) + \left(\frac{a}{r}\right)^2\csc(2\varphi)\right\}, \tag{16.6b}$$

$$\{v_r, v_\varphi\} = \frac{Q}{\pi}r\left[a^4 + r^4 + 2a^2r^2\cos(2\varphi)\right]^{-1}\left\{r^2 + a^2\cos(2\varphi), -a^2\sin(2\varphi)\right\}, \tag{16.7a,b}$$

follow from the complex potential $f = \Phi + i\Psi$ in (16.1a) [complex conjugate velocity $v_r - iv_\varphi = e^{i\varphi}v^*$ in (16.1b)]. The derivation of (16.6a,b) from (16.1a) is the particular case $n = 2$ of:

$$\begin{aligned}
\frac{4\pi}{Q}(\Phi + i\Psi) &= \frac{4\pi}{Q}f(z) = \frac{4\pi}{Q}f\left(re^{i\varphi}\right) = 2\log(z^n + a^n) \\
&= 2\log\left(a^n + r^n e^{in\varphi}\right) = 2\log\left[a^n + r^n\cos(n\varphi) + ir^n\sin(n\varphi)\right] \\
&= \log\left|\left[a^n + r^n\cos(n\varphi)\right]^2 + r^{2n}\sin^2(n\varphi)\right| + 2i\ \operatorname{arc\,cot}\left[\frac{a^n + r^n\cos(n\varphi)}{r^n\sin(n\varphi)}\right] \\
&= \log\left|a^{2n} + r^{2n} + 2a^nr^n\cos(n\varphi)\right| + 2i\ \operatorname{arc\,cot}\left[\cot(n\varphi) + \left(\frac{a}{r}\right)^n\csc(n\varphi)\right];
\end{aligned} \tag{16.8}$$

the derivation of (16.7a,b) from (16.1b) is the particular case $n = 2$ of:

$$\frac{2\pi}{nQ}(v_r - iv_\varphi) = e^{i\varphi}\frac{2\pi}{nQ}v^*(z) = \frac{e^{i\varphi}z^{n-1}}{z^n + a^n} = \frac{r^{n-1}e^{in\varphi}}{a^n + r^n e^{in\varphi}} = \frac{r^{n-1}}{r^n + a^n e^{-in\varphi}}$$
$$= \frac{r^{n-1}}{r^n + a^n\cos(n\varphi) - ia^n\sin(n\varphi)} = r^{n-1}\frac{r^n + a^n\cos(n\varphi) + ia^n\sin(n\varphi)}{Y}, \tag{16.9a}$$

where

$$Y \equiv |r^n + a^n\cos(n\varphi) - ia^n\sin(n\varphi)|^2 = [r^n + a^n\cos(n\varphi)]^2 + a^{2n}\sin^2(n\varphi)$$
$$= r^{2n} + a^{2n} + 2a^n r^n\cos(n\varphi). \tag{16.9b}$$

The polar components of the velocity (16.9a,b) $\equiv$ (16.7a,b) can also be obtained from the potential or stream function (16.8) $\equiv$ (16.6a,b) using (12.20a,b). The real axis $\varphi = 0, \pi$ is the streamline $\Psi = 0$ in (16.6b), and along it the normal velocity is zero $v_\varphi = 0$ in (16.7b).

16.1.3 Method of Calculation and Plotting of the Streamlines

The stream function in Cartesian coordinates follows as in (16.8) from (16.1a), viz.:

$$\frac{2\pi}{Q}\Psi = \arg(z^2 + a^2) = \arg(x^2 - y^2 + 2ixy + a^2) = \operatorname{arc\,cot}\left(\frac{x^2 - y^2 + a^2}{2xy}\right); \tag{16.10a}$$

it can be written:

$$a^2 = y^2 - x^2 + 2xy\cot\left(\frac{2\pi\Psi}{Q}\right) = \left[y - x\tan\left(\frac{\pi\Psi}{Q}\right)\right]\left[y + x\cot\left(\frac{\pi\Psi}{Q}\right)\right]; \tag{16.10b}$$

here was used the identity:

$$\beta \equiv \frac{\pi\Psi}{Q}: \quad 2\cot(2\beta) = \cot\beta - \tan\beta, \tag{16.11a}$$

that follows from (3.31a,c):

$$\frac{2\cos(2\beta)}{\sin(2\beta)} = \frac{\cos^2\beta - \sin^2\beta}{\sin\beta\cos\beta}. \tag{16.11b}$$

The streamlines (16.10b) are hyperbolas, with center on the wall at the midpoint between the source and its image, and asymptotes making an angle $\beta = \pi\Psi/Q$ *with the wall;* it can be confirmed from (16.6b) that as $r \to \infty$ the streamline $\Psi = Q\varphi/\pi$ is a radial line through the origin.

The stream function corresponding to (16.1a) is the sum of two terms:

$$\varphi_\pm \equiv \arg(z \pm ia) \equiv \operatorname{arc\,tan}\left(\frac{y \pm a}{x}\right), \quad \Psi = \frac{Q}{2\pi}(\varphi_+ - \varphi_-). \tag{16.12a,b}$$

The streamlines when the stream function is the sum of two terms can be constructed graphically by **Rankine's method (1864)**: (i) each term is plotted in equal steps (16.13a):

$$\frac{2\pi\Psi_\pm}{Q} = \varphi_\pm = \pm n_\pm\varphi_0; \quad \frac{2\pi\Psi}{\varphi_0 Q} = \frac{2\pi}{\varphi_0 Q}(\Psi_+ + \Psi_-) = n_+ + n_- = n = \text{const}, \tag{16.13a,b}$$

(ii) the streamline of the total flow is obtained joining the points $n_+ + n_- = \text{const}$ *in (16.13a,b).* An example is given in Figure 16.2, for: (i) the original source/sink the

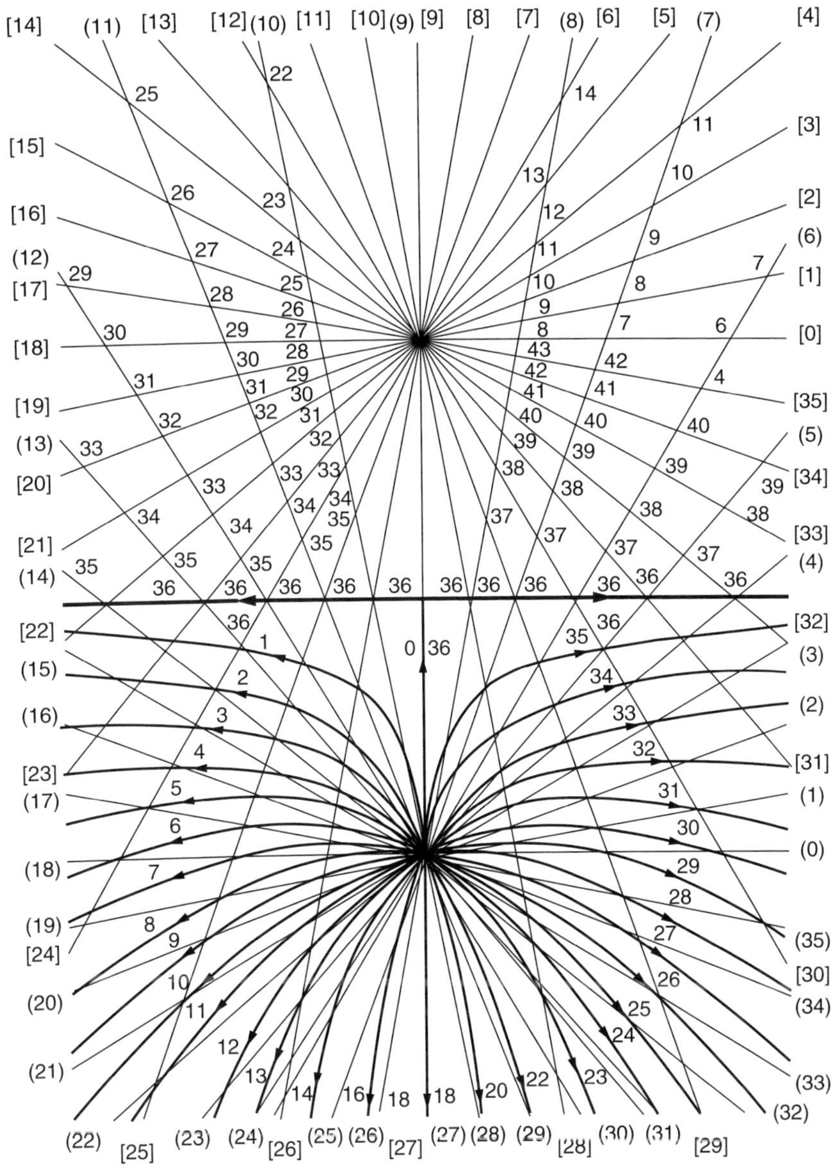

FIGURE 16.2
The streamlines of the potential flow due to a source near a straight rigid impermeable wall can be obtained by the Rankine method as follows: (i) take equally spaced radial lines through the source, for example, 36 radial lines at 10° intervals marked (0) to (35); (ii) take the same number of equally spaced radial lines through the image, viz. 36 radial lines at 10° intervals marked [0] to [35]; (iii) both (i) and (ii) start in the same direction, for example, the positive real axis; (iv) sum the values $(x) + [y]$ on each set (i) + (ii) of lines at all crossing points so that $0 \le x + y \le 72$, though only $4 \le x + y \le 40$ fall into the drawing; (iv) join all the points with the same value of $x + y$ along the same streamline. In Figure 16.2 are drawn the 36 equally spaced streamlines issuing from the source at 10° intervals. The last $0 = 36$ is the only one to hit the wall; it does so at the only stagnation point, where it can turn $\pm\, 90^\circ$ to follow along both sides of the wall.

streamlines are radial lines through $z = ia$, drawn at $10°$ intervals $n_+ = [0-35]$; (ii) the streamlines of the image source/sink are radial lines through $z = -ia$, also drawn with the same $10°$ intervals $n_- = (0-35)$; (iii) the crossing points are labeled $n = n_+ + n_-$, corresponding to the values of $n_\pm$ indicated at the edges; (iv) the crossing points with a given constant value of n lie on the same streamline, for example, $n = 36$ corresponds to the streamline $\Psi = 0$ along the wall and the streamline $\Psi = Q$ joining the source $\varphi_- = -\pi/2$ to its image $\varphi_+ = \pi/2$; (v) the crossing points are labeled on the top half of Figure 16.2, and the corresponding streamlines are drawn on the bottom half [for clarity, since the streamlines are mirror images in the upper half].

16.2 Image Vortex with Opposite Circulation

The potential flow (12.32a) of a vortex (Figure 16.3b) meets the condition of tangent velocity at the wall for Figure 16.3a an image vortex with opposite circulation:

$$f(z) = -\frac{i\Gamma}{2\pi}\left[\log(z-ia) - \log(z+ia)\right] = -\frac{i\Gamma}{2\pi}\log\left(\frac{z-ia}{z+ia}\right); \tag{16.14a}$$

$$v^*(z) = -\frac{i\Gamma}{2\pi}\left[(z-ia)^{-1} - (z+ia)^{-1}\right] = \frac{\Gamma}{\pi}\frac{a}{z^2+a^2}. \tag{16.14b}$$

The complex conjugate velocity (16.14b) is real for real $z \equiv x$, showing that the normal velocity is zero at the wall; an alternative way to reach the same conclusion is to note that on the wall $z \equiv x$ the numerator and denominator of the logarithm (16.14a) have equal moduli $|x - ia| = |x + ia|$, hence the real part of the logarithm is zero, and the imaginary part of the complex potential is zero, that is, it corresponds to the streamline $\Psi = 0$. The flow due to the vortex exerts a force on the wall (Subsection 16.2.1) and simplifies to a dipole in the far-field (Subsection 16.2.2).

16.2.1 Induced Velocity and Force on the Wall

The image vortex induces on the real vortex a horizontal velocity $\Gamma/2\pi d$, where $d = 2a$ is their mutual distance corresponding to the complex conjugate velocity (16.15a); thus *a vortex with circulation Γ at a distance a from a wall moves parallel to the wall with velocity (16.15a) in the positive (negative) x-direction for positive (negative) circulation:*

$$v_0^* = \frac{\Gamma}{4\pi a}, \qquad F^* = -i\rho\Gamma v_0^* = -\frac{i\rho\Gamma^2}{4\pi a} = -iF_y, \tag{16.15a,b}$$

corresponding to a force (16.15b) of attraction on the wall. The force on the wall is minus the force on the singularity that is specified by Blasius theorem (28.28a). The force exerted by a vortex (16.15b) [source/sink (16.2b)] on a wall at a distance a is a similar interchanging the circulation Γ (flow rate Q) that is, it is an attraction in both cases; the same transformation applies to the velocity induced on the vortex (16.15a) [source/sink (16.2a)], except that the velocity has a different direction, viz. parallel (orthogonal) to the wall. Thus the trajectory

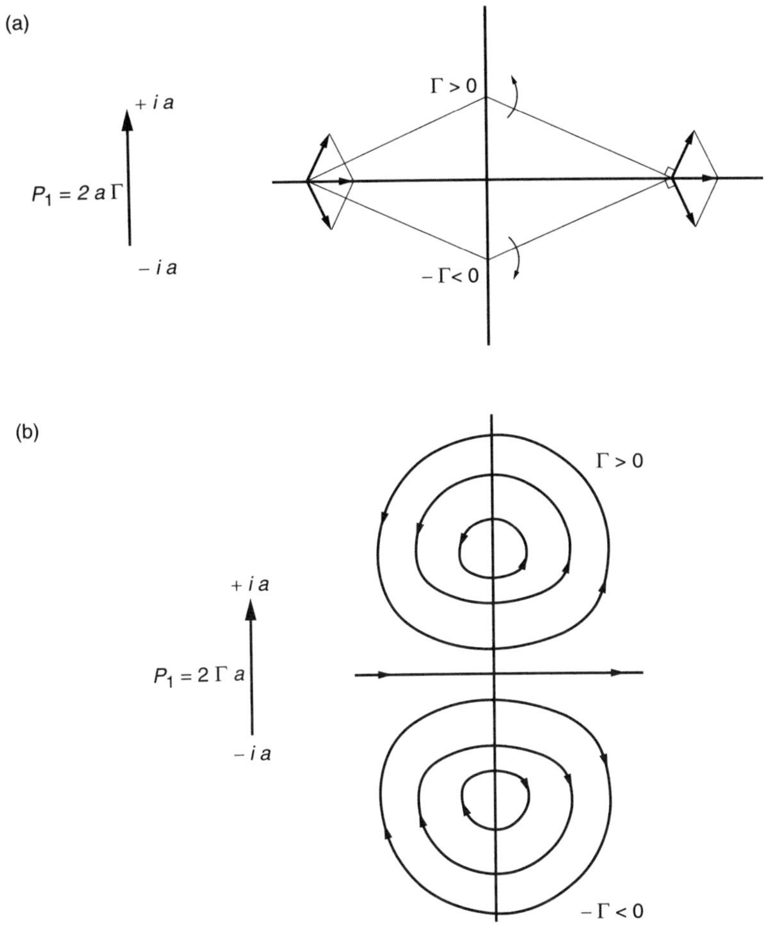

FIGURE 16.3
The condition that the velocity must be tangent to a rigid impermeable wall implies that in the case of a vortex (source/sink) in Figure 16.3 (16.1) the wall can be replaced by an opposite (identical) vortex (source/sink) at the symmetric position; the difference arises because the vortex (source/sink) has an azimuthal (radial) velocity [Figure 12.2b(a)].

(16.4a,b) of the source/sink is replaced for the vortex z_+ (its image z_-) by:

$$z_{\pm}(t) = \pm ia + \int_0^t v_0^* \, \mathrm{d}t = ia + \frac{\Gamma t}{4\pi a}, \tag{16.15c}$$

a uniform motion parallel to the wall.

16.2.2 Dipole Moment in the Far-Field

For an observer in the far-field (16.16a) the vortex with its opposite image (Figure 16.3a) corresponds (Figure 16.3b) to a dipole (16.16b,c):

$$|z|^2 \gg a^2: \quad v^*(z) = \frac{\Gamma a}{\pi z^2} = \frac{P_1}{2\pi z^2}, \quad f(z) = -\frac{P_1}{2\pi z}, \tag{16.16a–c}$$

with (12.54a,b) moment:

$$P_1 = 2\Gamma a = (-i\Gamma)\left[ia - (-ia)\right], \tag{16.16d}$$

showing that: (i) the dipole moment is the circulation times the distance between the original vortex and its image on the wall; (ii) the dipole axis is horizontal (real P_1) because it is a vertical dipole of vortices (Section 12.7). Thus *a vortex of circulation Γ at a distance a from a wall has a complex potential (16.14a) and conjugate velocity (16.14b), and exerts on the wall a force of attraction (16.15b); the induced velocity (16.15a) implies a uniform motion parallel to the wall (16.15c). The opposite image leads to a dipole flow (16.16b,c) in the far-field (16.16a) with moment (16.16d). The scalar potential (16.17a) and stream function (16.17b) [polar components of velocity (16.18a,b)]:*

$$\Phi = \frac{\Gamma}{2\pi}\left\{\arctan\left[\tan\varphi - \left(\frac{a}{r}\right)\sec\varphi\right] - \arctan\left[\tan\varphi + \left(\frac{a}{r}\right)\sec\varphi\right]\right\}, \tag{16.17a}$$

$$\Psi = \frac{\Gamma}{4\pi}\left\{\log\left|a^2 + r^2 + 2ar\sin\varphi\right| - \log\left|a^2 + r^2 - 2ar\sin\varphi\right|\right\}, \tag{16.17b}$$

$$\{v_r, v_\varphi\} = \frac{\Gamma a}{\pi}\left[a^4 + r^4 + 2a^2r^2\cos(2\varphi)\right]^{-1}\left\{\left(r^2 + a^2\right)\cos\varphi, \left(r^2 - a^2\right)\sin\varphi\right\}, \tag{16.18a,b}$$

follow from the complex potential (16.14a) [conjugate velocity (16.14b)] as in (16.8) [16.9)].

The passage from (16.14a) to (16.17a,b) is the particular case $n = 1$ of:

$$\begin{aligned}
\frac{4\pi i}{\Gamma}(\Phi + i\Psi) &= \frac{4\pi i}{\Gamma} f(z) = \frac{4\pi i}{\Gamma} f\left(r\,e^{i\varphi}\right)\\
&= 2\log\left(\frac{z^n - ia^n}{z^n + ia^n}\right) = 2\log\left(\frac{r^n e^{in\varphi} - ia^n}{r^n e^{in\varphi} + ia^n}\right)\\
&= 2\log\left(\frac{r^n\cos(n\varphi) + ir^n\sin(n\varphi) - ia^n}{r^n\cos(n\varphi) + ia^n\sin(n\varphi) + ia^n}\right)\\
&= \log\left|\left[r^n\cos(n\varphi)\right]^2 + \left[r^n\sin(n\varphi) - a^n\right]^2\right| + 2\arctan\left(\frac{r^n\sin(n\varphi) - a^n}{r^n\cos(n\varphi)}\right)\\
&\quad - \log\left|\left[r^n\cos(n\varphi)\right]^2 + \left[r^n\sin(n\varphi) + a^n\right]^2\right| - 2\arctan\left(\frac{r^n\sin(n\varphi) + a^n}{r^n\cos(n\varphi)}\right)\\
&= \log\left|r^{2n} + a^{2n} - 2a^nr^n\sin(n\varphi)\right| + 2i\arctan\left[\tan(n\varphi) - \left(\frac{a}{r}\right)^n\sec(n\varphi)\right]\\
&\quad - \log\left|r^{2n} + a^{2n} + 2a^nr^n\sin(n\varphi)\right| - 2i\arctan\left[\tan(n\varphi) + \left(\frac{a}{r}\right)^n\sec(n\varphi)\right];
\end{aligned} \tag{16.19a}$$

The passage from (16.14b) to (16.18a,b) is the particular case $n = 1$ of:

$$\begin{aligned}
\frac{\pi}{n\Gamma a^n}\left(v_r - iv_\varphi\right) &= e^{i\varphi}\frac{\pi}{n\Gamma a^n}v^*(z) = e^{i\varphi}\frac{\pi}{n\Gamma a^n}v^*\left(re^{i\varphi}\right)\\
&= \frac{e^{i\varphi}z^{n-1}}{z^{2n} + a^{2n}} = \frac{r^{n-1}\,e^{in\varphi}}{r^{2n}\,e^{i2n\varphi} + a^{2n}} = \frac{r^{n-1}}{r^{2n}\,e^{in\varphi} + a^{2n}e^{-in\varphi}}
\end{aligned}$$

$$= \frac{r^{n-1}}{\left(r^{2n}+a^{2n}\right)\cos\left(n\varphi\right)+i\left(r^{2n}-a^{2n}\right)\sin\left(n\varphi\right)}$$
$$= r^{n-1}\frac{\left(r^{2n}+a^{2n}\right)\cos\left(n\varphi\right)-i\left(r^{2n}-a^{2n}\right)\sin\left(n\varphi\right)}{X}, \tag{16.19b}$$

where:

$$X \equiv \left[\left(r^{2n}+a^{2n}\right)^2\cos^2\left(n\varphi\right)+\left(r^{2n}-a^{2n}\right)^2\sin^2\left(n\varphi\right)\right]$$
$$= r^{4n}+a^{4n}+2r^{2n}a^{2n}\left[\cos^2\left(n\varphi\right)-\sin^2\left(n\varphi\right)\right] = r^{4n}+a^{4n}+2r^{2n}a^{2n}\cos\left(2n\varphi\right). \tag{16.19c}$$

It can be confirmed that the wall $\varphi = 0,\ \pi$ is the streamline $\Psi = 0$ in (16.17b), along that the normal velocity is zero $v_\varphi = 0$ in (16.18b). A source/sink (vortex) has a radial (tangential) velocity, and the effect of the wall in the far-field is to double the velocity (16.5c) [weaken the velocity field into a dipole (16.16b)].

16.3 Effect of Wall on Monopole or Spiral Source

For the potential flow of a monopole the source/sink (vortex), has identical (Section 16.1) [opposite (Section 16.2)] image on the wall, and thus the image of the spiral source of strength (16.20a) is the complex conjugate monopole in (16.20b):

$$P_0 \equiv Q - i\Gamma: \quad f\left(z\right) = \frac{1}{2\pi}\left[P_0\log\left(z-ia\right)+P_0^*\log\left(z+ia\right)\right]$$
$$= \frac{Q}{2\pi}\log\left(z^2+a^2\right)-\frac{i\Gamma}{2\pi}\log\left(\frac{z-ia}{z+ia}\right); \tag{16.20a,b}$$

$$v^*\left(z\right) = \frac{1}{2\pi}\left(\frac{P_0}{z-ia}+\frac{P_0^*}{z+ia}\right) = \frac{Qz+\Gamma a}{\pi}\,\frac{1}{z^2+a^2}, \tag{16.20c}$$

thus (12.35a,b) the complex potential (16.20b) [conjugate velocity (16.20c)] are the sum of (16.1a + 16.14a) [(16.1b + 16.14b)].

The force on the wall due to a spiral source is given (28.28a) by the sum of (16.2b) and (16.15b), viz. (16.21a):

$$F^* = \rho\left(Q-i\Gamma\right)v_o^* = \rho\left(Q-i\Gamma\right)\left(-\frac{iQ}{4\pi a}+\frac{\Gamma}{4\pi a}\right); \tag{16.21a,b}$$

the velocity due to the image monopole has source (16.2a) and vortex (16.15a) components (16.21b), leading to the total force:

$$F^* = -i\frac{\rho}{4\pi a}\left(Q-i\Gamma\right)\left(Q+i\Gamma\right) = -i\frac{\rho}{4\pi a}\left(Q^2+\Gamma^2\right) = -iF_y. \tag{16.21c}$$

The two terms correspond to the vertical attractive force between the identical source/sinks (16.2b) [opposite vortices (16.15b)]. Thus a spiral source at a distance a from a wall exerts an attractive force (16.21c).

Using the geometric series (21.62b), the complex conjugate velocity (16.20c) is approximated by:

$$\pi v^*(z) = \frac{\Gamma a}{z^2 + a^2} + \frac{Q/z}{1 + a^2/z^2} = \frac{\Gamma a}{z^2} + \frac{Q}{z} - \frac{Qa^2}{z^3} + O\left(\frac{a^4}{z^5}\right). \tag{16.22}$$

Thus in the far-field (16.23a) the flow (16.23b) is due: (i) at leading order to (16.5c) a double source/sink:

$$|z|^4 \gg a^4: \quad v^*(z) = \frac{Q}{\pi z} + \frac{P_1}{2\pi z^2} + \frac{P_2}{\pi z^3} + O\left(z^{-4}\right), \quad P_2 = -Qa^2; \tag{16.23a–c}$$

(ii) at next order to a vortex effect acting as a dipole (16.16b,d); (iii) the multipolar expansion of the source leads to the next quadrupole term (12.55b) with moment (16.23c). Thus *a monopole of moment (16.20a) at a distance a from a wall has complex potential (16.20b) and conjugate velocity (16.20c), and exerts on the wall an attractive force (16.21c); the induced velocity is the sum of (16.2a) and (16.15a), and the trajectory the sum of (16.4a) and (16.15c). The far-field (16.23a) is dominated by the source/sink component (16.5c) that is a monopole, followed by the vortex dipole (16.16b,d) and a quadrupole (16.23c) due to the source/sink. The sum of (16.6–16.7a,b) and (16.18–16.19a,b) specify, the scalar potential (16.6a + 16.18a), stream function (16.6b + 16.18b), and polar components of velocity (16.7a,b + 16.19a,b);* the wall is a common streamline $\Psi = 0$ with zero normal velocity $v_\varphi = 0$.

16.4 Far-Field of Multipole near a Hard Wall

For the potential flow of an arbitrary multipole (12.59b) the image on the wall has conjugate moment:

$$\begin{aligned} f(z) &= \frac{1}{2\pi}\left[P_n (z - ia)^{-n} + P_n^* (z + ia)^{-n}\right] \\ &= \frac{1}{2\pi}\left(z^2 + a^2\right)^{-n}\left\{\mathrm{Re}(P_n)\left[(z + ia)^n + (z - ia)^n\right] + i\,\mathrm{Im}(P_n)\left[(z + ia)^n - (z - ia)^n\right]\right\} \end{aligned} \tag{16.24}$$

leading to (16.24) for the complex potential, and likewise for the complex conjugate velocity. *In the far-field (16.25a) the flow (16.24) due to a multipole of moment P_n at a distance a from a wall corresponds to: (i) a multipole of the same order with double strength (16.25c) for the real part (16.25b) of the moment:*

$$|z|^2 \gg a^2, \mathrm{Re}(P_n) \neq 0: \quad f(z) = \frac{\mathrm{Re}(P_n)}{\pi} z^{-n}, \tag{16.25a–c}$$

with the imaginary part having a negligible effect; (ii) unless the real part of the moment is zero (16.26b), and the imaginary part specifies the far-field (16.26a):

$$\left|z^2\right| \gg a^2, \quad \mathrm{Re}(P_n) = 0: \quad f(z) = -\frac{na}{\pi}\mathrm{Im}(P_n)\, z^{-n-1}, \tag{16.26a–c}$$

as (16.26c) ≡ (16.27a) a multipole of order n + 1:

$$f(z) = \frac{P_{n+1}}{2\pi} z^{-n-1}, \quad P_{n+1} = -2na \operatorname{Im}(P_n), \tag{16.27a,b}$$

with moment (16.27b). The case (i) [(ii)] includes the source/sink (vortex) near a wall in Sections 16.1.2 (16.2.2). The derivation from (16.24) to (16.26a–c)

$$a^2 \ll |z|^2, \operatorname{Re}(P_n) = 0: \frac{\pi f(z)}{\operatorname{Im}(P_n)} = \frac{1}{2} z^{-2n} i z^n \left[\left(1 + \frac{ia}{z}\right)^n - \left(1 - \frac{ia}{z}\right)^n\right]$$
$$= \frac{i}{2} z^{-n} \left[\left(1 + \frac{ina}{z}\right) - \left(1 - \frac{ina}{z}\right)\right] = \frac{i}{2} z^{-n} 2in \frac{a}{z} = -naz^{-n-1}, \tag{16.28a,b}$$

uses the binomial series (25.37b,c), instead of the geometric series (21.62b) in (16.22).

16.5 Monopole in a Hard-Walled Rectangular Corner

The potential flow due to a source/sink at arbitrary position $z_0 = ae^{i\alpha}$ with $0 < \alpha < \pi/2$ in a rectangular impermeable hard-walled corner corresponds to three identical images at $-z_0, z_0^* - z_0^*$ (in Figure 16.4), leading (12.26a) to the complex potential:

$$f(z) = \frac{Q}{2\pi} [\log(z - z_0) + \log(z + z_0) + \log(z - z_0^*) + \log(z + z_0^*)]$$
$$= \frac{Q}{2\pi} \log\left[\left(z^2 - z_0^2\right)\left(z^2 - z_0^{*2}\right)\right]$$
$$= \frac{Q}{2\pi} \log\left[z^4 + |z_0|^4 - 2z^2 \operatorname{Re}\left(z_0^2\right)\right]; \tag{16.29}$$

this is real both for real $z = x$ and imaginary $z = iy$, showing that the real and imaginary axes correspond to the streamline $\Psi = 0$. In the far-field (16.30a) the flow (16.30b) corresponds to source with four times the strength:

$$|z|^2 \gg \left|z_0^2\right|: \quad f(z) = \frac{2Q}{\pi} \log z, \tag{16.30a,b}$$

because: (a) the flow in all space is due to four source/sinks, viz. the original and three identical images; (b) the flow in the corner is deflected by the orthogonal walls from the source/sink to one-quarter of the space. Thus (16.29) [(16.30b)] *is exact [asymptotic (16.30a)] complex potential of a source/sink of flow rate* Q *at an arbitrary position* z_0 *in a rectangular corner.*

The potential flow due to a vortex in a rectangular corner corresponds Figure 16.4 to images (image) on the walls (vertex) that have opposite (equal) circulation, leading (12.31b) to the complex conjugate velocity:

$$v^*(z) = -\frac{i\Gamma}{2\pi}\left[(z - z_0)^{-1} + (z + z_0)^{-1} - (z - z_0^*)^{-1} - \{z + z_0^*\}^{-1}\right]$$
$$= -\frac{i\Gamma z}{\pi}\left[\left(z^2 - z_0^2\right)^{-1} - \left(z^2 - z_0^{*2}\right)^{-1}\right] = -\frac{i\Gamma z}{\pi} \frac{z_0^2 - z_0^{*2}}{\left(z^2 - z_0^2\right)\left(z^2 - z_0^{*2}\right)}$$
$$= \frac{2\Gamma z}{\pi} \frac{\operatorname{Im}\left(z_0^2\right)}{z^4 + |z_0|^4 - 2z^2 \operatorname{Re}\left(z_0^2\right)}. \tag{16.31}$$

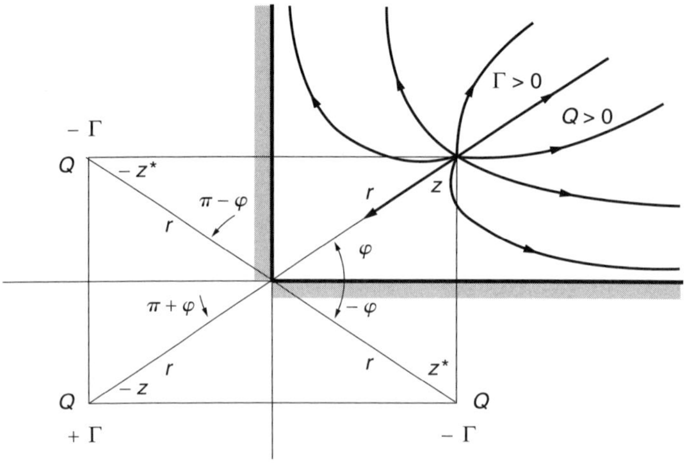

FIGURE 16.4
In the case of a source (sink) in a rectangular corner the walls of the latter may be replaced by identical images obtained by reflection on the walls and corner (Figures 16.1 and 16.3); each pair of sources (sinks) normal to a wall ensures a tangential velocity along that wall (Figure 16.1). In the case of a vortex the images are at the same points (Figure 16.3); the sign of the circulation alternates from one quadrant to the next, since two opposite vortices are needed to ensure a tangential velocity at the wall at equal distance from them.

For real $z = x$ (imaginary $z = iy$) the term in the last denominator is real, and thus the complex conjugate velocity is real $v^*(x) = v_x(x,0)$ [imaginary $v^*(iy) = -iv_y(0,y)$], showing that the velocity is horizontal (vertical) on the horizontal (vertical) wall, that is, always tangent, or with zero normal component.

In the far-field (16.32a) the flow (16.31) is (12.55b) due to:

$$|z|^2 \gg |z_0|^2: \quad v^*(z) = \frac{2\Gamma}{\pi z^3}\operatorname{Im}\left(z_0^2\right) = \frac{P_2}{\pi z^3}, \quad P_2 \equiv 2\Gamma \operatorname{Im}\left(z_0^2\right), \tag{16.32a–c}$$

a quadrupole (16.32b) with moment (16.32c), since: (i) each pair of opposite vortices orthogonal to the horizontal (vertical) wall acts as dipole of moment (16.32d):

$$P_1 = 2\Gamma \,{}^{\operatorname{Im}}_{\operatorname{Re}}(z_0); \quad P_2 = 2P_1 \,{}^{\operatorname{Re}}_{\operatorname{Im}}(z_0) = 4\Gamma \operatorname{Re}(z_0)\operatorname{Im}(z_0) = 2\Gamma \operatorname{Im}\left(z_0^2\right) \tag{16.32d,e}$$

(ii) the two dipoles are opposite and at a distance $2\operatorname{Re}(z_0)\,[2\operatorname{Im}(z_0)]$ leading to the quadrupole strength (16.32e) ≡ (16.32c). Thus *a vortex of circulation* Γ *at an arbitrary position* z_0 *in a rigid rectangular corner has complex conjugate velocity (16.31), and appears in the far-field (16.32a) as a quadrupole (16.32b) of moment (16.32c).*

16.6 Trajectories of a Vortex or Source/Sink in a Corner

The velocity induced on a multipole by its images (Section 16.6.1) implies that it is generally not at rest, for example, the paths are calculated next for a source, sink, and vortex in a rectangular corner (Section 16.6.2); also the associated acceleration of the monopole and force exerted on the walls (Subsection 16.6.3).

16.6.1 Velocity Induced by Images on Walls

The source/sink (Section 16.1) [vortex (Section 16.2)] near a wall are acted by a normal attractive force, and have a normal (parallel) induced velocity, due to the image on the wall. In the case of a monopole (i.e., source/sink combined with vortex) in a rectangular corner (Section 16.5) there are three images (Figure 16.4) that determine the trajectory (Figure 16.5). The monopole has no effect on itself, and thus the complex conjugate velocity at the monopole is that due to the three images (16.33a,b), that is, (16.31) where (i) the first-term an the r.h.s. that is singular for $z = z_0$ is omitted; (ii) in the remaining terms is set $z = z_0$:

$$\begin{aligned} v_0^* &= \frac{Q+i\Gamma}{2\pi}\left[\left(z_0 - z_0^*\right)^{-1} + \left(z_0 + z_0^*\right)^{-1}\right] + \frac{Q-i\Gamma}{2\pi}\frac{1}{2z_0} \\ &= \frac{Q+i\Gamma}{\pi}\frac{z_0}{z_0^2 - z_0^{*2}} + \frac{Q-i\Gamma}{4\pi z_0} = \dot{z}_0^* \equiv \frac{\mathrm{d}z_0^*}{\mathrm{d}t}, \end{aligned} \qquad (16.33\text{a,b})$$

where (i) the circulation is Γ for the image at $-z_0$ and $-\Gamma$ for the images at $(-z_0^*, z_0^*)$; (ii) the flow rate is Q for all images. The position of the monopole (16.34a):

$$z_0 = r\mathrm{e}^{i\varphi}: \quad v_0^* = \frac{\mathrm{d}z_0^*}{\mathrm{d}t} \equiv \dot{z}_0^* = \left[(\dot{r} + ir\dot{\varphi})\mathrm{e}^{i\varphi}\right]^* = (v_r - iv_\varphi)\mathrm{e}^{-i\varphi}, \qquad (16.34\text{a,b})$$

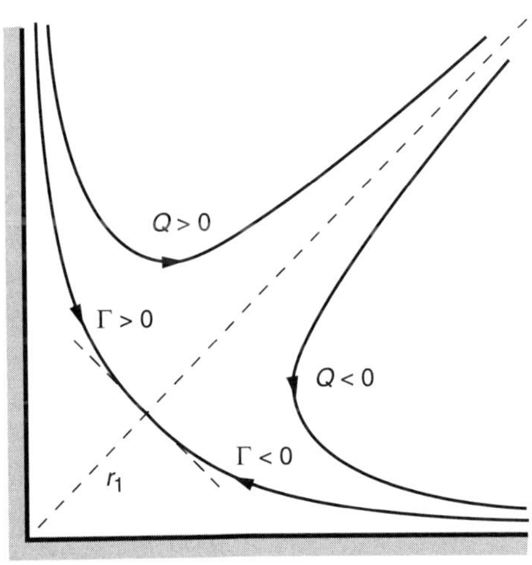

FIGURE 16.5
A source/sink (vortex) in a rectangular corner flow is subject to the sum of velocities induced by its identical (alternating) images in Figure 16.4. As a result: (i) a source $Q > 0$ is "repelled" to infinity along the diagonal; (ii) a sink is "attracted" from infinity along the diagonal and turned again toward infinity along a wall; (iii) a vortex is "attracted" from infinity along one wall and "repelled" to infinity along the other wall. The trajectories are the same in opposite directions for: (i) a source/sink; (ii) a positive/negative vortex. In contrast with the rectangular corner (Figure 16.5), for a flat wall the induced velocity due to one image: (i/ii) repels (attracts) a source (sink) perpendicular to the wall (Figure 16.1); (ii) induces a vortex motion parallel to the wall (Figure 16.3) to the right (left) for counterclockwise (clockwise) vortex.

corresponds (12.21) to the polar velocity components (16.34b); this leads from (16.33a,b) to (16.36a,b):

$$\begin{aligned} v_r - iv_\varphi = e^{i\varphi} v^*\left(re^{i\varphi}\right) &= \frac{Q+i\Gamma}{\pi}\frac{re^{i2\varphi}}{r^2e^{2i\varphi}-r^2e^{-2i\varphi}} + \frac{Q-i\Gamma}{4\pi r} \\ &= \frac{Q-i\Gamma}{4\pi r} + \frac{Q+i\Gamma}{\pi r}\frac{\cos(2\varphi)+i\sin(2\varphi)}{2i\sin(2\varphi)} \\ &= \frac{Q-i\Gamma}{4\pi r} + \frac{Q+i\Gamma}{2\pi r}\left[1-i\cot(2\varphi)\right]; \end{aligned} \tag{16.35}$$

that specifies the polar components of the velocity of the monopole:

$$\dot r \equiv \frac{dr}{dt} = v_r = \frac{3Q}{4\pi r} + \frac{\Gamma}{2\pi r}\cot(2\varphi), \tag{16.36a}$$

$$r\dot\varphi = r\frac{d\varphi}{dt} = v_\varphi = \frac{Q}{2\pi r}\cot(2\varphi) - \frac{\Gamma}{4\pi r}, \tag{16.36b}$$

for the polar components of the velocity of the monopole.

16.6.2 Path of a Vortex, Source, and Sink

The path is obtained eliminating time, for example, by dividing (16.36a) by (16.36b), viz. for a vortex (16.37a–c) [source/sink (16.38a–c)]:

$$\Gamma \neq 0 = Q: \quad \frac{1}{r}\frac{dr}{d\varphi} = -2\cot(2\varphi), \quad r\sin(2\varphi) = \text{const} \equiv r_1, \tag{16.37a–c}$$

$$\Gamma = 0 \neq Q: \quad \frac{1}{r}\frac{dr}{d\varphi} = \frac{3}{2}\tan(2\varphi), \quad r\cos^{3/4}(2\varphi) = \text{const} \equiv r_2, \tag{16.38a–c}$$

where (16.37c) [(16.38c)] follows from (16.37b) [(16.38b)] by separation of variables and integration, viz. (16.39a) [(16.39b)]:

$$0 = \frac{dr}{r} + 2\frac{\cos(2\varphi)}{\sin(2\varphi)}d\varphi = d\{\log r + \log[\sin(2\varphi)]\} = d\{\log[r\sin(2\varphi)]\}, \tag{16.39a}$$

$$0 = \frac{dr}{r} - \frac{3}{2}\frac{\sin(2\varphi)}{\cos(2\varphi)}d\varphi = d\left\{\log r + \frac{3}{4}\log[\cos(2\varphi)]\right\} = d\left\{\log\left[r\cos^{3/4}(2\varphi)\right]\right\}. \tag{16.39b}$$

Thus a vortex (16.37a) in a rectangular corner has a trajectory (16.37c) with the walls as asymptotes $r \to \infty$ as $\varphi \to 0, \pi/2$, whose closest distance to the origin is on the diagonal $r = r_1$ for $\varphi = \pi/4$; if the circulation is positive $\Gamma > 0$ (negative $\Gamma < 0$) the vortex moves to the right $v_\varphi < 0$ (left $v_\varphi > 0$) in (16.36b) $\equiv$ (16.42b) and Figure 16.5. A source/sink (16.38a) in a rectangular corner has a trajectory (16.38c) that goes to infinity along the diagonal $r \to \infty$ for $\varphi = \pi/4$, and stays on one side of the diagonal at a distance not less than $r = r_2$ for $\varphi = 0$; a source $Q > 0$ (sink $Q < 0$) moves away from (toward the) wall $v_r > 0\,(v_r < 0)$ in (16.36a) $\equiv$ (16.43b) and Figure 16.5. Even if the induced velocities on the source/sink (vortex) have equal magnitude, their distinct direction along (across) the relative position, leads to different trajectories, going to infinity along the wall (the diagonal) in Figure 16.5.

16.6.3 Acceleration Associated with the Motion

The velocity (16.33b) specifies by differentiation with regard to time (Section 2.1) the acceleration.

$$a_0 \equiv \ddot z_0 = \dot v_0 = \left[(v_r + iv_\varphi)e^{i\varphi}\right] = \left[\dot v_r + i\dot v_\varphi + i\dot\varphi(v_r + iv_\varphi)\right]e^{i\varphi}; \tag{16.40}$$

thus *the components of the acceleration in polar coordinates are given by:*

$$a_r = \dot{v}_r - \dot{\varphi} v_\varphi = \dot{v}_r - \frac{(v_\varphi)^2}{r}, \quad a_\varphi = \dot{v}_\varphi + \dot{\varphi} v_r = \dot{v}_\varphi + \frac{v_\varphi v_r}{r}, \tag{16.41a,b}$$

in terms of the polar velocity components and their time derivatives. In the case of a vortex (16.42a) the polar components of the velocity (16.36a,b) ≡ (16.42b,c):

$$\Gamma \neq 0 = Q: \quad \{v_r, v_\varphi\} = \frac{\Gamma}{2\pi r}\left\{\cot(2\varphi), -\frac{1}{2}\right\}, \quad \{a_r, a_\varphi\} = \left\{\frac{3\Gamma^2}{16\pi^2 r^3}, 0\right\}, \tag{16.42a–e}$$

lead to an acceleration (16.42d,e) outward from the origin scaling like the inverse cube of distance. In the case of a source/sink (16.43a) from the velocity (16.36b,c) ≡ (16.43a,b) follows:

$$\Gamma = 0 \neq Q: \quad \{v_r, v_\varphi\} = \frac{Q}{2\pi r}\left\{\frac{3}{2}, \cot(2\varphi)\right\}, \tag{16.43a–c}$$

$$\{a_r, a_\varphi\} = -\frac{Q^2}{2\pi^2 r^3}\left\{\frac{9}{8} + \frac{1}{2}\cot^2(2\varphi), \cot(2\varphi)\csc^2(2\varphi)\right\}, \tag{16.43d,e}$$

an acceleration (16.43d,e) that is radial $a_\varphi = 0$ only on the diagonal $\varphi = \pi/4$, that is, the direction the source/sink goes to infinity.

An example of the calculation of the polar component of the acceleration (16.42d,e) [(16.43d,e)] from those of the velocity (16.42b,c) [(16.43b,c)] using (16.41a,b) is:

$$\begin{aligned} a_\varphi = \dot{v}_\varphi + \frac{v_\varphi v_r}{r} &= -\frac{\mathrm{d}(\Gamma/4\pi r)}{dt} - \frac{\Gamma}{4\pi r^2}\frac{\Gamma}{2\pi r}\cot(2\varphi) \\ &= \frac{\Gamma}{4\pi r^2}\left[\frac{\mathrm{d}r}{\mathrm{d}t} - \frac{\Gamma}{2\pi \mathrm{r}}\cot(2\varphi)\right] = 0; \end{aligned} \tag{16.44}$$

this proves (16.42e) from (16.42b,c). Similar calculations apply to the other cases (16.42d; 16.43d,e). Thus a vortex (16.42a) in a corner is repelled from the vertex (16.42e) with a radial acceleration (16.42d) scaling on the inverse cube of distance; for a source/sink (16.43a) the acceleration has the same scaling on distance from the origin, depends on the angle (16.43d,e) and is not radial except along the diagonals of quadrants. The velocity field (12.57a,b) of a quadrupole at the origin also scales like r^{-3} but has a distinct angular dependence. From (16.43b,c) [(16.42b,c)] follows:

$$\Gamma = 0 \neq Q: \quad \vec{v} \cdot \vec{x} = r v_r = \frac{3Q}{4\pi} - \text{const}, \tag{16.45a,b}$$

$$\Gamma \neq 0 = Q: \quad \left|\vec{v} \wedge \vec{x}\right| = v_\varphi r = -\frac{\Gamma}{4\pi} = \text{const}, \tag{16.45c,d}$$

the distinction between the source/sink (16.45a) [vortex (16.45c)] in a rectangular corner, for that the product of the radius by the radial (16.45b) [tangential (16.45d)] velocity is constant.

16.7 Flow and Forces for a Monopole in a Rectangular Corner (Gröbli, 1877; Greenhill, 1878)

The force exerted on a rectangular corner by a monopole can be calculated geometrically (Subsection 16.7.1) [analytically (Subsection 16.7.2)] for example, for a position along the diagonal (an arbitrary position), using Cartesian or polar components (Subsection 16.7.3). The transformations between the Cartesian and polar components of the velocity (Subsection 16.7.3) lead to the invariance of the work and activity (Subsection 16.7.4). The diagonal position leads to simpler analytical formulas for the flow variables (Subsection 16.7.6) that are the solution of a kinematical rather than dynamical problem (Subsection 16.7.5).

16.7.1 Forces due to Identical and Conjugate Images

The force exerted by the monopole on the walls of the rectangular corner is minus the force due to its images (Figure 16.4). The latter force is given (28.28a) by (16.46a) for a monopole at a distance a along the diagonal, where the induced velocity is due to the three images, viz.: (i) one identical image $z = r e^{i5\pi/4}$ at a distance $2r$ in the direction of unit vector $(1+i)/\sqrt{2}$; (ii/iii) two complex conjugate images at $z = r e^{-i\pi/4}\left(e^{i3\pi/4}\right)$, at a distance $2r/\sqrt{2} = r\sqrt{2}$, in the direction of unit vector i (1), viz.:

$$F^* = \rho(Q - i\Gamma)\, v^* = \rho \frac{Q - i\Gamma}{2\pi r\sqrt{2}} \left[Q\left(\frac{1-i}{2} + 1 - i\right) + \Gamma\left(\frac{-i-1}{2} + 1 + i\right)\right]; \qquad (16.46a)$$

the source/sink and vortex contributions to the induced velocity were grouped together in (16.46a) that can be simplified:

$$F^* = \frac{\rho}{4\pi r\sqrt{2}} (Q - i\Gamma)\,[3Q(1-i) + \Gamma(1+i)]; \qquad (16.46b)$$

separated into horizontal and vertical components:

$$F^* = \frac{\rho}{4\pi r\sqrt{2}} \left[3Q^2 + \Gamma^2 - 2Q\Gamma - i\left(3Q^2 + \Gamma^2 + 2Q\Gamma\right)\right], \qquad (16.46c)$$

the source/sink and vortex contribution to the force have: (i/ii) equal signs for the double source/sink (vortex) interaction, but with distinct magnitude $3Q^2$ (Γ^2); (iii) equal magnitude and opposite signs for the source/sink and vortex cross-interaction $\pm 2Q\Gamma$.

16.7.2 Interaction between a Singularity and a Wall

The force on the monopole is specified by (16.47a):

$$F^* = -\rho(Q - i\Gamma)\, v_0^* = -\rho(Q - i\Gamma)\left(\frac{Q - i\Gamma}{4\pi z_0} + \frac{Q + i\Gamma}{\pi} \frac{z_0}{z_0^2 - z_0^{*2}}\right), \qquad (16.47a,b)$$

where the velocity induced by the images is given by (16.33b). In the case of a monopole at the diagonal of corner (16.48a), the force (16.47b) simplifies to (16.48b,c):

$$z_0 = r e^{i\pi/4}: \quad \{F_x, F_y\} = -\frac{\rho}{4\pi r\sqrt{2}} \left\{3Q^2 + \Gamma^2 - 2Q\Gamma, 3Q^2 + \Gamma^2 + 2Q\Gamma\right\}, \qquad (16.48a–c)$$

in agreement with (16.46c) with reversed sign, because the forces on the monopole (16.48b,c) and on the wall (16.46c) are equal and opposite. For a monopole at arbitrary position (16.49a):

$$z_0 = r e^{i\varphi}: \quad F^* \equiv F_x - iF_y = (F_r - iF_\varphi)\, e^{-i\varphi}, \qquad (16.49a–c)$$

using instead of Cartesian (16.49b) ≡ (16.48b,c) the polar components (16.49c) of the force (16.47b) leads to:

$$\{F_r, F_\varphi\} = -\frac{\rho}{4\pi r}\left\{3Q^2 + \Gamma^2, 2Q\Gamma + 2\left(Q^2 + \Gamma^2\right)\cot(2\varphi)\right\}; \tag{16.50a,b}$$

the passage from (16.47b;16.49a,c) to (16.50a,b) is:

$$\begin{aligned} F_r - iF_\varphi &= -e^{i\varphi}\rho(Q - i\Gamma)\left(\frac{Q - i\Gamma}{4\pi r}e^{-i\varphi} + \frac{Q + i\Gamma}{\pi r}\frac{e^{i\varphi}}{e^{2i\varphi} - e^{-2i\varphi}}\right) \\ &= -\rho(Q - i\Gamma)\left[\frac{Q - i\Gamma}{4\pi r} + \frac{Q + i\Gamma}{\pi r}\frac{e^{2i\varphi}}{2i\sin(2\varphi)}\right] \\ &= -\frac{\rho}{4\pi r}(Q - i\Gamma)\{Q - i\Gamma + 2(Q + i\Gamma)[1 - i\cot(2\varphi)]\} \\ &= -\frac{\rho}{4\pi r}(Q - i\Gamma)[3Q + i\Gamma + 2(\Gamma - iQ)\cot(2\varphi)] \\ &= -\frac{\rho}{4\pi r}\left[3Q^2 + \Gamma^2 - 2i\Gamma Q - 2i\left(\Gamma^2 + Q^2\right)\cot(2\varphi)\right]. \end{aligned}$$

A similar calculation applies to (16.46a–c) and (16.48a–c). In particular for a source/sink (16.51a–c) [vortex (16.52a–c)] the forces:

$$\Gamma = 0 \neq Q: F_r = -\frac{3\rho Q^2}{4\pi r} = -\rho v_r Q, \quad F_\varphi = -\frac{\rho Q^2}{2\pi r}\cot(2\varphi) = -\rho v_\varphi Q, \tag{16.51a–c}$$

$$\Gamma \neq 0 = Q: \quad F_r = -\frac{\rho\Gamma^2}{4\pi r} = \rho v_\varphi \Gamma, \quad F_\varphi = -\frac{\rho\Gamma^2}{2\pi r}\cot(2\varphi) = -\rho v_r \Gamma, \tag{16.52a–c}$$

are proportional to the velocities (16.43b,c) [(16.42b,c)]; this agrees with the Blasius theorem (28.28a) ≡ (16.53a), viz.:

$$F_r - iF_\varphi = -\rho(Q - i\Gamma)(v_r - iv_\varphi) = -\rho(v_r Q - v_\varphi \Gamma) + i\rho(v_r \Gamma + v_\varphi Q). \tag{16.53a–c}$$

The force exerted by the monopole on the wall has been expressed in Cartesian (16.46a–c; 16.48a–c) [polar (16.50–53a–c) components], for a diagonal (arbitrary) position. The transformations between Cartesian and polar components are considered next.

16.7.3 Transformation between Cartesian and Polar Components of a Vector

For a monopole on the diagonal (16.54a), the polar components of the force (16.50a,b) simplify to (16.54b):

$$\varphi = \frac{\pi}{4}: \ \{F_r, F_\varphi\} = -\frac{\rho}{4\pi r}\left\{3Q^2 + \Gamma^2, 2Q\Gamma\right\}; \sqrt{2}\{F_x, F_y\} = \{F_r - F_\varphi, F_r + F_\varphi\}; \tag{16.54a–c}$$

they relate to the Cartesian components (16.48a–c) by (16.54c). The latter (16.54a–c) is the particular case (16.54a) of the transformation from polar to Cartesian components of the force vector (16.55a,b):

$$F_x = F_r\cos\varphi - F_\varphi\sin\varphi, \quad F_y = F_r\sin\varphi + F_\varphi\cos\varphi; \tag{16.55a,b}$$

the inverse transformation, from the Cartesian to polar coordinates of the force (16.56a,b) is obtained replacing $+\varphi$ by $-\varphi$ in (16.55a,b):

$$F_r = F_x \cos\varphi + F_y \sin\varphi, \quad F_\varphi = -F_x \sin\varphi + F_y \cos\varphi. \tag{16.56a,b}$$

It can be checked that substitution of (16.56a,b) in (16.55a,b), or vice-versa, leads to an identity; thus the two relations are inverse, and only one need be considered. The transformation between polar and Cartesian components of the complex force (16.55a,b) corresponds to

$$F \equiv F_x + iF_y = (F_r + iF_\varphi)\,e^{i\varphi}, \quad F^* = F_x - iF_y = (F_r - iF_\varphi)\,e^{-i\varphi}; \tag{16.57a,b}$$

either (16.57a) or (16.57b) lead to (16.55a,b), and hence also to (16.56a,b). *The transformation from the polar to the Cartesian components (vice-versa) of a vector, for example, either for the force or the velocity (16.55a,b) ≡ (20.11a,b) [their inverse (16.56a,b) ≡ (20.12a,b)] correspond to a rotation by $-\varphi$ $(+\varphi)$, that is, $|\varphi|$ in the negative (positive) direction in Figure 20.1, that is, clockwise (counter clockwise).* The relations (16.55–16.56a,b) can be obtained (20.11–20.12a,b) by an alternative equivalent method (Example 20.3). The relations between the polar and Cartesian components of a vector are illustrated graphically in Figure 20.1, for example, projections the polar components (v_r, v_φ) on the (x,y)-axis yield (16.55a,b) ≡ (20.11a,b).

16.7.4 Force, Velocity, and Invariance of the Work and Activity

The inner product of the velocity and force vectors is (Section 4.5) the activity (16.58a) or work per unit time (16.55a,b):

$$A \equiv \vec{F}\cdot\vec{v} = F_x v_x + F_y v_y, \quad dW = \vec{F}\cdot d\vec{x} = \left(\vec{F}\cdot\vec{v}\right) dt = A\,dt. \tag{16.58a,b}$$

The activity may be calculated from the complex velocity and force, both in Cartesian (16.59a) [polar (16.59b)] coordinates:

$$A = F_x v_x + F_y v_y = \operatorname{Re}\{(F_x - iF_y)(v_x + iv_y)\} = \operatorname{Re}(F^* v) = \frac{F^* v + F v^*}{2}, \tag{16.59a}$$

$$A = \operatorname{Re}(F v^*) = \operatorname{Re}(F^* v) = \operatorname{Re}\left\{(F_r - iF_\varphi)\,e^{-i\varphi}(v_r + iv_\varphi)\,e^{i\varphi}\right\} = F_r v_r + F_\varphi v_\varphi. \tag{16.59b}$$

The activity (16.58a) and work (16.58b) are **invariants**, *that is, have the same form in all coordinate systems, for example, Cartesian (16.59a) and polar (16.59b) coordinates.* This is consistent with the relation between the polar and Cartesian components (16.55a,b) of the force and velocity:

$$\begin{aligned} A = F_x v_x + F_y v_y &= [(F_r \cos\varphi - F_\varphi \sin\varphi)]\,(v_r \cos\varphi - v_\varphi \sin\varphi) \\ &+ [(F_r \sin\varphi + F_\varphi \cos\varphi)]\,(v_r \sin\varphi + v_\varphi \cos\varphi) = F_r v_r + F_\varphi v_\varphi. \end{aligned} \tag{16.60}$$

The work or activity in one-dimension (4.16a) is given by the product (3.22a) of the two real quantities; in two dimensions the product of two real (3.22a) [imaginary (3.22c)] quantities:

$$\operatorname{Re}(F)\operatorname{Re}(v) = \frac{1}{2}\operatorname{Re}(Fv + Fv^*) = \frac{1}{2}\operatorname{Re}[(F_x + iF_y)\,2v_x] = F_x v_x, \tag{16.61a}$$

$$\operatorname{Im}(F)\operatorname{Im}(v) = \frac{1}{2}\operatorname{Re}(Fv^* - Fv) = \frac{1}{2}\operatorname{Re}[(F_x + iF_y)(-2iv_y)] = F_y v_y, \tag{16.61b}$$

specifies the first (16.61a) [second (16.61b)] term of the activity (16.58a). The sum of (16.61a,b) specifies:

$$F_x v_x + F_y v_y = \operatorname{Re}(F)\operatorname{Re}(v) + \operatorname{Im}(F)\operatorname{Im}(v)$$
$$= \frac{1}{2}\operatorname{Re}(Fv + Fv^*) + \frac{1}{2}\operatorname{Re}(Fv^* - Fv) = \operatorname{Re}(Fv^*), \tag{16.61c}$$

the total activity (16.61c) ≡ (16.59a) ≡(16.59b).

16.7.5 Comparison of a Dynamical Motion with a Kinematical Problem

The forces (16.51a–c) [(16.52a–c)] are not proportional to the accelerations (16.43d,e) [(16.42d,e)], that is, in both cases it is not possible to define an "equivalent mass" of the singularity. Even if the forces were proportional to the accelerations, an "equivalent mass" per unit length could not be formed with just two quantities: mass density ρ and flow rate Q (circulation Γ). An "equivalent mass" per unit length would have the dimensions $m = \rho\ell^2$, where the length ℓ could only be a coordinate of the singularity. This would contrast with the mass of a material particle that is independent of position. Thus the problem is not a **dynamical motion** specified by masses and inertia forces, and integration for the velocity. *The motion of singularities in a potential flow is a* **kinematical problem**, *specified from induced velocities, from that accelerations can be calculated.* By the principle of equal action and reaction, (i) the force exerted by the monopole on the walls (16.46c) is (ii) equal and opposite to the force exerted by the wall or images on the monopole (16.48b,c). However, the force (ii) cannot be related to the acceleration as an "inertia force," since generally an "equivalent mass" does not exist. On the other hand the force (i) exerted by the monopole on the walls has a clear physical meaning: it is the integral of the pressure exerted by the flow on the wall. Thus (ii) could be seen as a device to calculate (i) with reversed sign, and henceforth *the interaction between a singularity and a wall is to be interpreted in terms of the resultant force and/or moment of the pressure distribution on the wall associated with the flow generated by the singularity.*

16.7.6 Flow due to a Monopole on the Diagonal of a Quadrant

The potential flow due to a source/sink (vortex) in a corner (Figure 16.4), simplifies for (16.62a), a diagonal position: (i) to (16.62b,c) for the source (16.29):

$$z_0 = a e^{i\pi/4}: \quad f(z) = \frac{Q}{2\pi}\log\left(z^4 + a^4\right), \quad v^*(z) = \frac{2Q}{\pi}\frac{z^3}{z^4 + a^4}; \tag{16.62a–c}$$

(ii) to (16.63a,b) for the vortex (16.31):

$$v^*(z) = \frac{2\Gamma}{\pi}\frac{a^2 z}{z^4 + a^4}, \quad f(z) = -\frac{i\Gamma}{2\pi}\log\left(\frac{z^2 - ia^2}{z^2 + ia^2}\right). \tag{16.63a,b}$$

A monopole or spiral source is the superposition, viz. of a source/sink (16.62b) [(16.62c)] and a vortex (16.63b) [(16.63c)] for the complex potential (16.64a) [conjugate velocity (16.64b)]:

$$f(z) = \frac{Q}{2\pi}\log\left(z^4 + a^4\right) - \frac{i\Gamma}{2\pi}\log\left(\frac{z^2 - ia^2}{z^2 + ia^2}\right), \tag{16.64a}$$

$$v^*(z) = \frac{2z}{\pi}\frac{Qz^2 + \Gamma a^2}{z^4 + a^4}. \tag{16.64b}$$

In the far-field (16.65a):

$$|z|^4 \gg a^4, Q \neq 0: \quad f(z) = \frac{2Q}{\pi}\log z, \quad v^*(z) = \frac{2Q}{\pi z}, \tag{16.65a–c}$$

the source/sink that is a monopole dominates the vortex term (16.65b,c); the latter is a quadrupole:

$$|z|^4 \gg a^4, \quad Q = 0: \quad f(z) = -\frac{\Gamma a^2}{\pi z^2}, \quad v^*(z) = \frac{2\Gamma a^2}{\pi z^3}, \tag{16.66a–d}$$

and is relevant (16.66c,d) asymptotically (16.66a) only in the absence of the source term (16.66b).

A monopole consisting of a source/sink of flow rate Q and vortex of circulation Γ in a square corner at a distance a from the vertex along the diagonal, has exact [asymptotic (16.65a)] complex potential (16.64a) [(16.65b, 16.66c)], conjugate velocity (16.64b) [16.65c, 16.66d)] and exerts upon the walls a force (16.46c)≡(16.48b,c). The scalar potential (16.67a) and stream function (16.67b) [polar components of velocity (16.67a,b)]:

$$\Phi = \frac{Q}{4\pi}\log\left|a^8 + r^8 + 2a^4r^4\cos(4\varphi)\right| + \frac{\Gamma}{2\pi}\left\{\operatorname{arc\,tan}\left[\tan(2\varphi)\right] - \left(\frac{a}{r}\right)^2 \sec(2\varphi)\right]$$

$$-\operatorname{arc\,tan}\left[\tan(2\varphi) + \left(\frac{a}{r}\right)^2 \sec(2\varphi)\right]\Bigg\}, \tag{16.67a}$$

$$\Psi = \frac{Q}{2\pi}\operatorname{arc\,cot}\left[\cot(4\varphi) + \left(\frac{a}{r}\right)^4 \csc(4\varphi)\right]$$

$$+\frac{\Gamma}{4\pi}\left\{\log\left|a^4 + r^4 + 2a^2r^2\sin(2\varphi)\right| - \log\left|a^4 + r^4 - 2a^2r^2\sin(2\varphi)\right|\right\}, \tag{16.67b}$$

$$\{v_r, v_\varphi\} = \frac{2}{\pi}r\left[r^8 + a^8 + 2a^4r^4\cos(4\varphi)\right]^{-1}\left\{Qr^2\left[r^4 + a^4\cos(4\varphi)\right]\right.$$

$$\left.+\Gamma a^2\left(r^4 + a^4\right)\cos(2\varphi), -Qr^2a^4\sin(4\varphi) + \Gamma a^2\left(r^4 - a^4\right)\sin(2\varphi)\right\}, \tag{16.68a,b}$$

follow from (16.64a) [(16.64b)] as in (16.8; 16.19a) [(16.9a,b; 16.19b,c)] with $n = 2(n = 4)$ for the source/sink Q (vortex) terms. They confirm that the walls $\varphi = 0, \pi, \pm\pi/2$ are the streamline $\Psi = 0$ in (16.67b) with zero normal velocity $v_\varphi = 0$ in (16.68b).

16.8 Multiple Identical Source/Sink Images

The original and images are identical for a source/sink in a corner of angle $\beta = 2\pi/n$, for that there are $(n-1)$ images; the case $n = 2$ is the plane wall $\beta = \pi$ in (Section 16.1), and the case $n = 4$ is the rectangular corner $\beta = \pi/2$ in (Section 16.4). If the source lies on the diagonal of the corner of angle $\beta = 2\pi/n$ at a distance a from the vertex:

$$\beta \equiv \frac{2\pi}{n}, m = 0, \ldots, n-1: \quad z_m = ae^{i\beta/2 + im\beta} = ae^{i\pi/n + i2\pi m/n} = \sqrt[n]{a^n e^{i\pi}} = \sqrt[n]{-a^n}, \tag{16.69}$$

the original and image sources lie on the vertices of a regular polygon with n sides, and vertices on the circle of radius a and center at the origin; these points (16.69) correspond (5.10) to the n roots of $-a^n$, leading to the factorization of the polynomial:

$$\prod_{m=0}^{n-1}(z - z_m) = z^n + a^n. \tag{16.70}$$

The complex potential (12.26a) due to the source/sink and its $(n-1)$ images is specified by:

$$f(z) = \frac{Q}{2\pi}\sum_{m=0}^{n-1}\log(z-z_m) = \frac{Q}{2\pi}\log\left\{\prod_{m=0}^{n-1}(z-z_m)\right\}; \tag{16.71}$$

this leads by (16.70) to (16.72a) and corresponding complex conjugate velocity (16.72b):

$$f(z) = \frac{Q}{2\pi}\log(z^n + a^n), \quad v^*(z) = \frac{nQ}{2\pi}\frac{z^{n-1}}{z^n + a^n}; \tag{16.72a,b}$$

in the far-field (16.73a):

$$|z|^n \gg a^n: \quad f(z) = \frac{nQ}{2\pi}\log z, \quad v^*(z) = \frac{nQ}{2\pi z}, \tag{16.73a–c}$$

the flow corresponds (16.73b,c) to a source/sink of flow rate nQ, that is, n source/sinks (one source/sink) in all space (in $1/n$ of all space). Thus *a source/sink of flow rate* Q *on the diagonal of a corner of angle* $\beta = 2\pi/\mathrm{n}$ *at a distance* a *from the vertex has exact [asymptotic (16.73a)] complex potential (16.72a) [(16.73b)] and conjugate velocity (16.72b) [(16.73c)]; the exact scalar potential (16.74a) and stream function (16.74b) [polar components of velocity (16.75a,b)]:*

$$\Phi = \frac{Q}{4\pi}\log\left|a^{2n} + r^{2n} + 2a^n r^n \cos(n\varphi)\right|, \tag{16.74a}$$

$$\Psi = \frac{Q}{2\pi}\operatorname{arc\,cot}\left\{\cot(n\varphi) + \left(\frac{a}{r}\right)^n \csc(n\varphi)\right\}, \tag{16.74b}$$

$$\{v_r, v_\varphi\} = \frac{nQ}{2\pi}r^{n-1}\left[a^{2n} + r^{2n} + 2a^n r^n \cos(n\varphi)\right]^{-1}\{r^n + a^n\cos(n\varphi), -a^n\sin(n\varphi)\}, \tag{16.75a,b}$$

follow, respectively, from (16.72a) [(16.72b)], as in (16.8) [(16.9a,b)]. They confirm that the walls $\varphi = 0, 2\pi/n, 4\pi/n, \ldots, 2\pi(n-1)/n$ are the streamline $\Psi = 0$ in (16.74b) along that the normal velocity is zero $v_\varphi = 0$ in (16.75b). The potential flow (16.72a,b; 16.73a–c; 16.74a,b; 16.75a,b) due to a source/sink in a corner of angle $\beta = 2\pi/n$ includes the particular cases of a plane wall $\beta = \pi$ (rectangular corner $\beta = \pi/2$) for $n = 2\,(n = 4)$ in (16.1a,b; 16.5a–c; 16.6a,b; 16.7a,b) [first terms on the r.h.s. of (16.64a,b; 16.65a–c; 16.67a,b; 16.68a,b), that is, with $Q \neq 0 = \Gamma$]. *The preceding results (16.72–16.75) extend (16.84a–c) to a source/sink in a corner of arbitrary angle* $0 < \beta \leq 2\pi$, *choosing* $\mathrm{n} = 2\pi/\beta \geq 1$, for example, for $(\beta = 2\pi, n = 1)$ and the source/sink is $z = -a$ aligned with a semiinfinite plate; (ii) for $(\beta = 4\pi/3n = 3/2)$ the source is at an angle 120° in a corner of angle 240°.

16.9 Alternating Vortices in a Rigid Corner

The potential flow due to a vortex in a corner of angle $\beta = 2\pi/p$ with $p = 2n+1$ odd, would have p images with opposite circulation, plus $(p+1)$ images + original with the same circulation, leading to the appearance of a vortex sheet, as in Figure 14.9. There will be no vortex sheet in the case of a corner of angle $\beta = 2\pi/p$ with even $p = 2n$, that is, $\beta = \pi/n$, and in this case a vortex of circulation Γ will have $(2n-1)$ alternating images. If the vortex is at a distance a from the vertex on the diagonal $\beta/2 = \pi/2n$ the images with equal Γ

(opposite $-\Gamma$) circulation lie on:

$$\beta=\frac{\pi}{n}: \quad z_{2m}=ae^{i\beta/2+i2m\beta}=ae^{i\pi/2n+i2m\pi/n}=\sqrt[n]{ia^n}, \tag{16.76a}$$

$$m=0,\ldots,n-1: \quad z_{2m+1}=ae^{-i\beta/2+i2m\beta}=ae^{-i\pi/2n+i2m\pi/n}=\sqrt[n]{-ia^n}; \tag{16.76b}$$

these are the roots (5.10) of the polynomials:

$$\prod_{m=0}^{n-1}(z-z_{2m})=z^n-ia^n, \quad \prod_{m=0}^{n-1}(z-z_{2m+1})=z^n+ia^n. \tag{16.77a,b}$$

The complex potential due to a vortex of circulation Γ in (12.32a) a corner of angle $\beta=\pi/n$ is given by:

$$f(z)=-\frac{i\Gamma}{2\pi}\sum_{m=0}^{n-1}\left[\log(z-z_{2m})-\log(z-z_{2m+1})\right]=-\frac{i\Gamma}{2\pi}\log\left(\prod_{m=0}^{n-1}\frac{z-z_{2m}}{z-z_{2m+1}}\right); \tag{16.78}$$

this may be simplified using (16.77a,b), leading to (16.79a):

$$f(z)=-\frac{i\Gamma}{2\pi}\log\left(\frac{z^n-ia^n}{z^n+ia^n}\right), \quad v^*(z)=\frac{\mathrm{d}f}{\mathrm{d}z}=\frac{n\Gamma}{\pi}\frac{a^nz^{n-1}}{z^{2n}+a^{2n}}, \tag{16.79a,b}$$

and the conjugate velocity (16.79b). In the far-field (16.80a):

$$|z|^n\gg a^{2n}: \quad v^*(z)=\frac{n\Gamma}{\pi}a^nz^{-n-1}=n\frac{P_n}{2\pi}z^{-n-1}, \tag{16.80a,b}$$

the flow (16.80b) corresponds to a multipole (16.81a) $\equiv$ (12.59b) of order n:

$$f(z)=-\frac{P_n}{2\pi z^n}, \quad P_n\equiv 2\Gamma a^n, \tag{16.81a,b}$$

with moment (16.81b).

A vortex of circulation Γ at a distance a from the vertex on the diagonal of a corner of angle, $\beta=\pi/n$, has an exact complex potential (16.79a) [conjugate velocity (16.79b)], corresponding to (16.81a) [(16.80b)] in the far-field (16.80a) to a multipole of order n *with moment (16.81b). The exact scalar potential (16.82a) and stream function (16.82b) [polar components of velocity (16.83a,b)]:*

$$\Phi=\frac{\Gamma}{2\pi}\left\{\mathrm{arc}\left[\tan(n\varphi)-\left(\frac{a}{r}\right)^n\sec(n\varphi)\right]-\arctan\left[\tan(n\varphi)+\left(\frac{a}{r}\right)^n\sec(n\varphi)\right]\right\}, \tag{16.82a}$$

$$\Psi=\frac{\Gamma}{4\pi}\left\{\log\left|a^{2n}+r^{2n}+2a^nr^n\sin(n\varphi)\right|-\log\left|a^{2n}+r^{2n}-2a^nr^n\sin(n\varphi)\right|\right\}, \tag{16.82b}$$

$$\{v_r,v_\varphi\}=\frac{n\Gamma}{\pi}a^nr^{n-1}\frac{\left\{\left(r^{2n}+a^{2n}\right)\cos(n\varphi),\left(r^{2n}-a^{2n}\right)\sin(n\varphi)\right\}}{a^{4n}+r^{4n}+2a^{2n}r^{2n}\cos(2n\varphi)}, \tag{16.83a,b}$$

follow from (16.79a) [(16.79b)] as in (16.19a) [(16.19b,c)]. This confirms that the walls $\varphi=0,\pi/n,2\pi/n,\ldots,2\pi(2n-1)\pi/n$ are the streamline $\Psi=0$ in (16.82b), and have zero normal velocity $v_\varphi=0$ in (16.83b). The potential flow (16.79a,b; 16.80a,b; 16.81a,b; 16.82a,b; 16.83a,b) of a vortex in a corner in a corner of angle $\beta=2\pi/n$ includes the plane wall of $\beta=\pi$ (rectangular corner $\beta=\pi/2$) as the particular case $n=1\,(n=2)$ in (16.14a,b;

16.16a–c; 16.17a,b; 16.18a,b) [the second terms on the r.h.s. of (16.64a,b;16.66a–d; 16.67a,b; 16.68a,b), that is, with $\Gamma \neq 0 = Q$]. The results (16.79–16.83a,b)) extend to a vortex in a corner of arbitrary angle $0 < \beta \le 2\pi$, choosing $n = \pi/\beta \ge 1/2$. *In the case of a monopole consisting of a source/sink (plus a vortex) the superposition of (16.72a,b) [(16.79a,b)] with* $n = 2\pi/\beta\,(n = \pi/\beta)$ *leads to the complex potential (16.84a) [conjugate velocity (16.84c)]:*

$$f(z) = \frac{Q}{2\pi}\log\left(z^{2\pi/\beta} + a^{2\pi/\beta}\right) - i\frac{\Gamma}{2\pi}\log\left(\frac{z^{\pi/\beta} - ia^{\pi/\beta}}{z^{\pi/\beta} + ia^{\pi/\beta}}\right), \tag{16.84a}$$

$$n = \frac{\pi}{\beta}: \quad v^*(z) = \beta^{-1} z^{\pi/\beta - 1}\frac{Qz^{\pi/\beta} + \Gamma a^{\pi/\beta}}{z^{2\pi/\beta} + a^{2\pi/\beta}}. \tag{16.84b,c}$$

in a corner of (16.84b) angle β*; these specify the corresponding potential (16.67a + 16.82a), stream function (16.67b + 16.82b) and polar components of the velocity (16.68a,b + 16.83a,b). The multipole analogy holds only for* $\pi/\beta = n$ *an integer.*

Note 16.1. *Images, Induced Velocities, Paths, and Trajectories*: The images of singularities give rise to induced velocities, unless the velocities add to zero; in the cases of an equilibrium, for example, at rest or in motion, arises the question of stability. The complex potential remains a solution of Laplace's equation if a function of time is added (16.85a) [multiplied (16.85b)]:

$$f_1(x,y,t) = f(x+iy) + g(t); \quad f_2(x,y,t) = h(t)\,f(x+iy). \tag{16.85a,b}$$

These functions appear in the unsteady incompressible Bernoulli equation (14.25c), and thus affect the forces on walls. Adding an arbitrary function of time in (16.85a) does not change the velocity (16.86a):

$$v_1^*(z) = \frac{df_1}{dz} = \frac{df}{dz} = v^*(z), \quad v_2^*(z,t) = \frac{df_2}{dz} = h(t)\frac{df}{dz} = h(t)\,v^*(z), \tag{16.86a,b}$$

multiplying by a function of time in (16.85b) carries over to the conjugate velocity in (16.86b). Thus it is possible to multiply the velocity by any function of time.

It is also possible to add to the velocity a function of time (16.87a):

$$\vec{v}_3\left(\vec{x},t\right) = \vec{v}\left(\vec{x}\right) + \vec{j}(t), \qquad \nabla\cdot\vec{v}_3 - \nabla\cdot\vec{v} = 0 = \nabla\wedge\vec{v}_3 - \nabla\wedge\vec{v}, \tag{16.87a–c}$$

this affects neither the divergence (16.87b) nor the curl (16.87c), so a potential flow remains potential. The preceding calculations of the paths or trajectories of singularities, for example, sources, sinks, vortices, and multipoles was based on induced velocities, and have excluded such arbitrary functions of time in Sections 16.1, 16.2, and 16.5; this exclusion will be retained subsequently in (e.g., Sections 28.8 and 36.3). The choice of arbitrary functions of time can lead to any trajectory. This arbitrariness is due to the assumption of incompressibility that implies an infinite sound speed, and instantaneous propagation of disturbances to all space. In a compressible flow unsteady disturbances would give rise to sound waves, and propagate at sound speed. The wave speed can be modified by other effects, for example, stratification/rotation/ionization combined with gravity/inertial/magnetic forces.

Note 16.2. *Images on Planes, Cylinders, and Spheres*: The effect of a plane wall, a rectangular corner or a corner with arbitrary angle on a source/sink, vortex, monopole, or multipole on any order can discussed via a conformal mapping, viz. the simplest Schwarz–Christofell transformation (Chapter 33) with one angular point (Section 36.1). The approach adopted in this Chapter 15 builds up on the preceding discussion of multipoles (Chapter 12) and corner flows (Sections 14.8 and 14.9). The images on plane walls are an example of how monopoles may generate dipoles, quadrupoles, or multipoles. The method of images applies not only to plane walls (Chapter 16) but also to cylinders (Sections 24.6–24.8, 26.7–26.9, 28.6–28.9); it extends from irrotational (Chapter 1) to rotational flow in two dimensions. As the multipoles extend from two to three-dimensions, so do the images on planes (Sections 16.1–16.4) and cylinders (Sections 24.6, 24.8, 26.7–26.9, 28.6–28.9) extend to point-images on planes and spheres and also hyperplanes and hyperspheres in higher dimensions. The method of images also extends from the Laplace/Poisson equation in the steady case of infinite propagation speed (Section 12.5) to the unsteady case of the wave equation with finite propagation speed. The images on a plane and circle are both instances of inverse points and relate to the Mobius transformation and bilinear conformal mapping (Chapter 35). The relations between Cartesian and polar components of a vector correspond to plane rotations; three-dimensional rotations are specified by the vector calculus or quaternions (Section 1.9). The transformations between coordinate systems lead to the tensor calculus.

Conclusion 16: A source/sink near (figure 16.1) a plane wall (a), has a tangent velocity in the presence of an identical image (b), i.e., with the same flow rate and the same distance, leading to an attraction force; the streamlines of the total flow can be constructed graphically from those of the source and its image using (figure 16.2) Rankine's method. A vortex near (figure 16.3) a plane wall has tangential velocity (a) in the presence of an opposite image; with opposite circulation appearing in the far-field as a dipole (b). A source/sink (vortex) in a corner of angle $\beta = 2\pi/n\,(\beta = \pi/n)$, e.g., $n = 4\,(n = 2)$ in (figure 16.4) has n identical (n alternating opposite) images. The trajectory of a vortex (souce/sink) in a rectangular corner (figure 16.5) goes to infinity along a wall (the diagonal).

17

Improper Integrals and Principal Value

A real integral is a particular case of a contour integral, in that the path of integration lies along the real axis, that is, it consists of one or several straight segments. Complex integrals along open contours in general, and real integrals in particular, can be calculated by means of residues, by two methods: (i) using a change of variable to transform the open contour into a closed loop (Section 17.2); (ii) extension of the open loop until it becomes a closed loop, in such a way (Section 17.3) that the contribution to the integral of the additional path can be calculated (or made to vanish). In both cases (i) and (ii), a loop integral results that may be calculated by residues (Chapter 15). The Riemann integral defined before (Chapter 13) is a proper integral, as it applies to a bounded function over a finite path (Section 17.1). The integral is improper of the first (second) kind if the path is infinite (function is unbounded), and of mixed kind is both features occur. A improper integral of first kind (Sections 17.4 and 17.5) is unilateral (bilateral) if only one (both) of the end-points are at infinity; an improper integral of the second kind (Section 17.6) is unilateral (bilateral) if the singularity of the integrand occurs at the end-points (in the interior of the path of integration). Improper integrals are calculated as limits of proper integrals, for example, as the end-point(s) tend to infinity (the integral approaches its singularity) for the first (second) kind. If the integrand is a many-valued function, with branch-point on the path of integration, the integral may become indeterminate, unless a principal value is chosen (Section 17.7). Even for a single-valued integrand, the primitive may be multi-valued, leading to an improper integral of the second kind whose value is indeterminate (Section 17.8), and can be specified uniquely as the Cauchy principal value (Section 17.9). The improper integrals of first (second) kind apply to source distributions that have infinite extent (singularities); an improper integral of third kind appears if the source distribution has infinite extent and contains singularities. A particular kind of singularity occurs when calculating the field due to a source distribution at point not outside the source distribution but within it; in this case a source element should not create a field upon itself, leading to the Cauchy principal value of the integral.

17.1 Improper Uni(bi)lateral Integrals of the Three Kinds

A **proper integral** is defined (17.1b) as a Riemann's integral (Section 13.2):

$$f \in \mathcal{B}(a,b): \qquad \int_a^b f(z)\,\mathrm{d}z, \tag{17.1a,b}$$

where (i) the integrand is bounded (17.1a) along the path of integration, that is, $|f(z)| \leq M$ for some $M > 0$ and all $z \in (a,b)$; (ii) the path of integration is a rectifiable curve, that is, has finite length $\int_a^b |\mathrm{d}z| = L < \infty$. If one or both of the conditions stated are not met then arises an **improper integral** that can be of three kinds. If (ii) is not met then (17.1b) is an improper integral of the **first kind** that is **unilateral (bilateral)** if one (both) of the end-points a, b tends to infinity; examples with real integrals are

$$\int_a^\infty f(x)\,\mathrm{d}x \equiv \lim_{b\to\infty}\int_a^b f(x)\,\mathrm{d}x, \qquad \int_{-\infty}^{+\infty} f(x)\,\mathrm{d}x = \lim_{-a,b\to\infty}\int_a^b f(x)\,\mathrm{d}x, \tag{17.2a,b}$$

that is an upper unilateral (17.2a) [bilateral (17.2b)] improper integral of first kind. If (i) is not met, for example, $f(z)$ is unbounded $f(c)=\infty$ at a point $z=c$ on the path of integration $c\in(a,b)$, the integral is improper of the **second kind**, and is **bilateral** if $c\neq a,b$ lies on the interior, and **lower (upper) unilateral** if the singularity coincides $c=a\,(c=b)$ with the lower (upper) end-point; examples with real integrals are:

$$\int_a^b f(x)\,\mathrm{d}x \equiv \lim_{\varepsilon\to 0+}\int_{c+\varepsilon}^b f(x)\,\mathrm{d}x, \qquad \int_a^b f(x)\,\mathrm{d}x = \lim_{\varepsilon,\delta\to 0+}\int_a^{c-\varepsilon} + \int_{c+\delta}^b f(x)\,\mathrm{d}x, \tag{17.3a,b}$$

that is a lower unilateral (17.3a) [bilateral (17.3b)] improper integral of the second kind. If neither (i) nor (ii) are met then it is a **mixed improper integral**, or improper integral of the **third kind**, for example:

$$\int_{-\infty}^{+\infty} f(x)\,\mathrm{d}x \equiv \lim_{-a,b\to\infty}\ \lim_{\varepsilon,\delta\to 0+}\left(\int_a^{c-\varepsilon} + \int_{c+\delta}^b\right) f(x)\,\mathrm{d}x, \tag{17.4}$$

that is a mixed real improper twice bilateral integral. The improper integrals of the first (17.2a,b), second (17.3a,b), and third (17.4) kinds, are defined iff the stated limits exist, and are independent of the way in that $-a,b\to\infty$ and $\varepsilon,\delta\to 0+$; the limit $\varepsilon\to 0+$ means that ε tends to zero through positive values. For example:

$$\int_{-\infty}^{+\infty} \exp\{-|x|\}\,\mathrm{d}x = 2, \qquad \int_0^a x^{-1/2}\,\mathrm{d}x = 2\sqrt{a}, \tag{17.5a,b}$$

are real improper bilateral (unilateral) integrals of the first (17.5a) [second (17.5b)] kind. The improper integrals of first, second, and third kind with the power function as integrand are considered in Example 20.17.

17.2 Transformation of a Straight Segment into a Circle

As an example of the calculation of a proper integral, consider a trigonometric integral over 2π radians:

$$J = \int_0^{2\pi} f(\cos\varphi, \sin\varphi)\,\mathrm{d}\varphi. \tag{17.6}$$

The segment of integration $(0,\,2\pi)$ is transformed (Figure 17.1a) into the circle of radius unity and center at the origin (Figure 17.1b) by the change of variable:

$$z = \mathrm{e}^{i\varphi}, \qquad \mathrm{d}z = i\mathrm{e}^{i\varphi}\,\mathrm{d}\varphi = iz\,\mathrm{d}\varphi, \tag{17.7a,b}$$

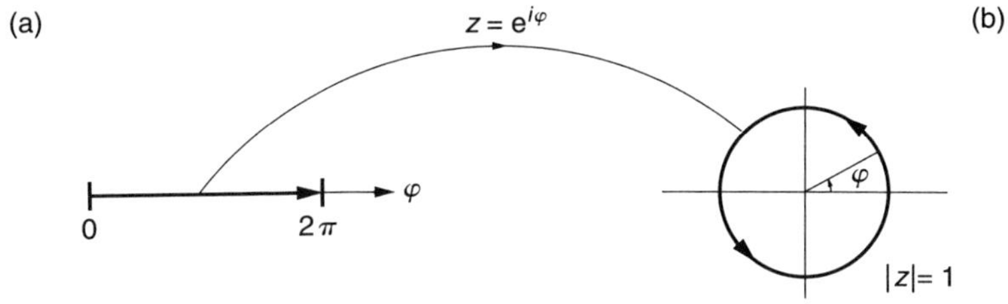

FIGURE 17.1
The complex mapping of a finite interval on the real axis into a circle transforms a real integral into a complex loop integral, that is, when the integrand is a trigonometric function. The loop integral can be evaluated (Figure 15.3b) from the residues at the: (i) poles inside the circle; (ii) enclosed poles on the circle.

this renders the factor in the integrand algebraic:

$$2\cos\varphi, 2i\sin\varphi \equiv \mathrm{e}^{i\varphi} \pm \mathrm{e}^{-i\varphi} = z \pm \frac{1}{z}. \tag{17.8}$$

The integrand in (17.6) as function of z is:

$$F(z) \equiv z^{-1} f\left(\frac{z+z^{-1}}{2}, \frac{z-z^{-1}}{2i}\right); \tag{17.9}$$

assuming that $F(z)$ meets the Cauchy conditions in the unit circle, except at poles in the interior $|z| < 1$ or on the boundary $|z| = 1$, the integral may be calculated by (15.45):

$$J = -i \int_{|z|=1} F(z)\,\mathrm{d}z = 2\pi \sum_{|z|<1} \mathrm{Re}\,s\{F(z)\} + \pi \sum_{|z|=1} \mathrm{Re}\,s\{F(z)\}, \tag{17.10}$$

from the residues at those poles.

It has been shown that *the trigonometric integral (17.6), in that the function (17.9) is analytic within and on the unit circle* $|z| \leq 1$ *except at poles, is evaluated by (17.10) using the residues at those poles.* As an example, consider:

$$\begin{aligned} J &= \int_0^{2\pi} \left(\frac{5}{2} + 2\cos\varphi\right)^{-1} \mathrm{d}\varphi = \int_{|z|-1} (iz)^{-1} \left(\frac{5}{2} + z + z^{-1}\right)^{-1} \mathrm{d}z \\ &= i^{-1} \int_{|z|=1} \left(z^2 + \frac{5z}{2} + 1\right)^{-1} \mathrm{d}z = i^{-1} \int_{|z|=1} \left\{(z+2)\left(z+\frac{1}{2}\right)\right\}^{-1} \mathrm{d}z; \end{aligned} \tag{17.11}$$

the integrand has simple poles at $z = -2, -1/2$, and the former lies outside the unit circle, so only the residue at the latter is needed to evaluate the integral:

$$J = (2\pi i)\, i^{-1} \lim_{z \to -1/2} \frac{z+1/2}{(z+1/2)(z+2)} = \frac{2\pi}{3/2}, \tag{17.12}$$

leading to the result:

$$\int_0^{2\pi} \left(\frac{5}{2} + \cos\varphi\right)^{-1} \mathrm{d}\varphi = \frac{4\pi}{3}. \tag{17.13}$$

Six more cases of the evaluation of trigonometric integrals by residues are given in Example 20.13.

17.3 Closing a Straight Line by a Half-Circle

The integral along an (open) contour L can be calculated by prolonging it into a (closed) loop:

$$\int_L f(z)\,\mathrm{d}z = \oint_{\partial D} f(z)\,\mathrm{d}z - \int_{\partial D - L} f(z)\,\mathrm{d}z$$

$$= 2\pi i \sum_{z \in D - \partial D} \mathrm{Res}\{f(z)\} + \pi i \sum_{z \in \partial D} \mathrm{Res}\{f(z)\} - \int_{\partial D - L} f(z)\,\mathrm{d}z, \tag{17.14}$$

provided that: (i) the loop integral can be evaluated by residues, that is, $f(z)$ be holomorphic or meets the Cauchy conditions in D, except at poles; (ii) the integral along the added path of integration $\partial D - L$ can be evaluated more easily, for example, vanishes. A real proper integral can always be taken, using a translation if necessary, with symmetric end-points:

$$\int_{-a}^{+a} f(x)\,\mathrm{d}x = \oint_{L_\pm} f(z)\,\mathrm{d}z - \int_{C_\pm} f(z)\,\mathrm{d}z, \tag{17.15}$$

where the straight segment $(-a, +a)$ was closed using a semicircle C_+ (C_-) of center at the origin and radius $|z| = a$, lying in the upper (lower) half-plane in Figure 17.2a (17.2b), to form the loop L_+ (L_-), that has edges at $z = \pm a$ and is rectifiable. The integral (17.15) is proper if $f(x)$ is bounded for $-a \le x \le +a$, and is improper of the second kind if $f(x)$ has singularities in this range, viz. unilateral if $f(a) = \infty$ or $f(-a) = \infty$ and bilateral if $f(c) = \infty$ for $-a < c < +a$. The integral along the loop L_+/L_- may be calculated by residues, at the poles in the interior (on the boundary), with the factor $2\pi i$ (πi) having the $+$ $(-)$ sign, if

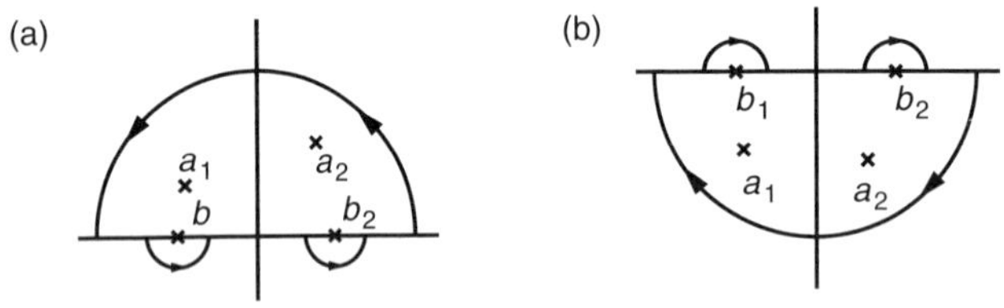

FIGURE 17.2
The path of integration along a symmetric segment $(-a, +a)$ of the real axis can be closed by a half-circle in the upper (a) [lower (b)] complex half-plane. If the integrand decays faster than $1/z$ at infinity in the upper (lower) half plane, the integral along the half-circle vanishes as the radius tends to infinity. This specifies the integral along the whole real axis in terms of the residues at the: (i) poles of the integrand in the upper (lower) half-plane; (ii) the enclosed poles on the real axis. It is assumed that the improper integral of the second or third kind along the real axis exists: (i) the preceding method is not a proof of existence; (ii) it is a method of calculation of the integral, assuming that it exists.

the path is taken in the counterclockwise (clockwise) direction:

$$\int_{-a}^{+a} f(x)\,\mathrm{d}x = \pm 2\pi i \sum \operatorname{Res}\{f(z)\}_{|z|<a,\operatorname{Im}(z)><0}$$

$$\pm\,\pi i \sum \operatorname{Res}\{f(z)\}_{|z|=a,\operatorname{Im}(z)><0;\operatorname{Im}(z)=0,|z|\leq a} - \int_{0}^{\pi} f\left(a\mathrm{e}^{i\varphi}\right) a\mathrm{e}^{i\varphi} i\,\mathrm{d}\varphi. \quad (17.16)$$

It has been shown that *if the function* f(z) *is holomorphic, except for poles, in the upper (lower) semicircle of center at the origin and radius a, its integral along the straight segment* $-\mathrm{a} \leq \mathrm{x} \leq +\mathrm{a}$ *is specified by (17.16), with the upper (lower) signs, both for a proper or improper integral of the second kind.*

17.4 Connecting the Real Axis in the Upper/Lower Half-Plane

The preceding result can be extended to the calculation by residues of improper integrals of the first or third kind along the real axis (Subsection 17.4.1) as an alternative to the calculation via the primitive (Subsection 17.4.2).

17.4.1 Evaluation of Integrals along the Real Axis

If (a) the function $f(x)$ is bounded on the real axis and the improper bilateral integral of the first kind (17.2b) exists, or (b) the function $f(x)$ has poles on the real axis and the improper integral of the third kind (17.4) exists, then, in both cases, the integrals can be calculated by taking the symmetric limits:

$$\int_{-\infty}^{+\infty} f(x)\,\mathrm{d}x: \qquad \int_{-\infty}^{+\infty} f(x)\,\mathrm{d}x = \lim_{a\to\infty} \int_{-a}^{+a} f(x)\,\mathrm{d}x; \quad (17.17)$$

note that the existence of (17.2b) or (17.4) implies that of (17.17), by setting $b=a$, but the inverse may not be true; if the limit (17.17) with $+a$ replaced by some $b \neq a$ had a different value, then the integral (17.2b) or (17.4) would not exist. In other words, the existence of the integral (17.17) will not be proved; it will be evaluated on the assumption that it exists. Taking the limit $a \to \infty$ in (17.16), the poles in the interior (on the boundary) are those in the upper/lower half-plane (on the real axis), leading to

$$\int_{-a}^{+a} f(x)\,\mathrm{d}x = \pm 2\pi i \sum_{\operatorname{Im}(z)><0} \operatorname{Res}\{f(z)\} \pm \pi i \sum_{\operatorname{Im}(z)=0} \operatorname{Res}\{f(z)\} \quad (17.18)$$

where the last term in (17.16) was omitted.

If the function $f(z)$ satisfies the asymptotic condition:

$$\operatorname{Im}(z) >< 0: \quad \lim_{z\to\infty} z f(z) = 0, \quad (17.19)$$

in the upper/lower half-plane, it is indeed legitimate to omit the integral along the half-circle of radius a, since the latter vanishes as $a \to \infty$:

$$\left| \int_0^{\pm\pi} f\left(ae^{i\varphi}\right) ae^{i\varphi} i\, d\varphi \right| \leq |zf(z)|_{\max} \int^{\pm\pi} |d\varphi| \leq \pi\varepsilon, \tag{17.20}$$

because (17.19) implies $|zf(z)| < \varepsilon$, for $|z| > a$, where $\varepsilon \to 0$ as $a \to \infty$. Hence: *if (i) the improper integral of the first (17.2b) or third (17.4) kind exists; (ii) the function* f(z) *is holomorphic in the upper (lower) half-plane, including the real axis* $Im(z) \geq 0 \,(Im(z) \leq 0)$, *except at poles; (iii) the function* f(z) *has the asymptotic property (17.19) in the upper (lower) half-plane; then the integral may be evaluated by (17.18), with the upper (lower) sign.*

17.4.2 Alternative Methods of Calculation of Integrals

As an example consider the integral:

$$\xi = \frac{x}{a}: \quad I(a) \equiv \int_{-\infty}^{+\infty} \left(x^2 + a^2\right)^{-1} dx = a^{-1} \int_{-\infty}^{+\infty} \left(1 + \xi^2\right)^{-1} d\xi, \tag{17.21a,b}$$

It may be evaluated using inverse trigonometric functions:

$$I(a) = a^{-1} \arctan \xi \,\Big|_{-\infty}^{+\infty} = a^{-1} \left\{ \frac{\pi}{2} - \left(-\frac{\pi}{2}\right) \right\} = \frac{\pi}{a}; \tag{17.22a}$$

the integral can be calculated in an elementary manner using residues:

$$I(a) = \pm 2\pi i \lim_{z \to \pm ia} \frac{z \mp ia}{z^2 + a^2} = \pm 2\pi i \lim_{z \to \pm ia} \frac{1}{z \pm ia} = \frac{\pm 2\pi i}{\pm 2ia} = \frac{\pi}{a}, \tag{17.22b}$$

since the integrand $f(z) = \left(z^2 + a^2\right)^{-1}$ meets the asymptotic condition (17.19), and is holomorphic in the whole plane, except at the poles $z = \pm ia$, where the residues are $\pm 1/(2ia)$.

The integral:

$$J(a,b) \equiv \int_{-\infty}^{+\infty} \left\{ \left(x^2 + a^2\right)\left(x^2 + b^2\right) \right\}^{-1} dx, \tag{17.23}$$

involves a quartic polynomial, and can also be calculated by the method of residues that applies to (17.23), because the three preceding conditions are met: (i) the integrand is positive and the integral:

$$J < b^{-2} \int_{-\infty}^{+\infty} \left(x^2 + a^2\right)^{-1} dx, \quad a^{-2} \int_{\infty}^{\infty} \left(x^2 + b^2\right)^{-1} dx = \frac{\pi}{ab^2}, \frac{\pi}{ba^2}, \tag{17.24a,b}$$

is (17.22a,b) bounded by (17.24a,b) and hence exists; (ii) the integrand in (17.23) decays as z^{-4} at infinity, and thus meets the asymptotic condition (17.19) in all directions; (iii) the integrand is holomorphic in the whole complex plane except for poles at $z = \pm ia$, with residues:

$$f_{(1)}(\pm ia) = \lim_{z \to \pm ia} \frac{z \mp ia}{\left(z^2 + b^2\right)(z - ia)(z + ia)} = \pm \left[2ia\left(b^2 - a^2\right)\right]^{-1},$$

$$f_{(1)}(\pm ib) = \pm \left[2ib\left(a^2 - b^2\right)\right]^{-1}. \tag{17.25a,b}$$

The integral (17.23) may be evaluated by closing the real axis with an infinite half-circle in the upper (lower) half-plane:

$$J(a,b) = \pm 2\pi i \left\{ f_{(1)}(\pm ia) + f_{(1)}(\pm ib) \right\} = \frac{\pi}{a^2 - b^2}\left(\frac{1}{b} - \frac{1}{a}\right); \tag{17.26a}$$

both methods yield the same result:

$$J(a,b) = \int_{-\infty}^{+\infty} \left\{ x^4 + \left(a^2 + b^2\right) x^2 + a^2 b^2 \right\}^{-1} \mathrm{d}x = \frac{\pi}{ab(a+b)}, \tag{17.26b}$$

that complies with the inequalities (17.24a,b). The particular case $a = b$ leads to the integral:

$$J(a,a) = \int_{-\infty}^{+\infty} \left(x^4 + 2a^2x^2 + a^4\right)^{-1} \mathrm{d}x = \int_{-\infty}^{+\infty} \left(x^2 + a^2\right)^{-2} \mathrm{d}x = \frac{\pi}{2a^3}; \tag{17.27a}$$

this can also be evaluated directly using the residues at the double poles at $z = \pm ia$, viz.:

$$J(a,a) = \pm 2\pi i \lim_{z \to \pm ia} \frac{\mathrm{d}}{\mathrm{d}z}\left[(z \mp ia)^2 (z - ia)^{-2} (z + ia)^{-2}\right] = \pm 2\pi i \frac{-2}{(\pm 2ia)^3} = \frac{\pi}{2a^3}. \tag{17.27b}$$

Four cases of the evaluation of real improper bilateral integrals by residues are given in Example 20.14.

17.5 Integrals with an Oscillating Factor (Jordan, 1894)

The asymptotic condition (17.19) can be relaxed to (17.29) in the case of an integral along the real axis with an oscillating factor (Subsection 17.5.1), that is, a direct or inverse (Subsection 17.5.2) Fourier transform.

17.5.1 Integral along the Real Axis with Oscillating Factor

The theorem (17.18) also applies to the function (17.28b):

$$g(z) \in \mathcal{D}(|C): \quad f(z) = g(z)\exp(\pm i\omega z), \tag{17.28a,b}$$

assuming that (17.28a) is holomorphic. Bearing in mind that:

$$\omega > 0: |\exp(\pm i\omega z)| = |\exp(\pm i\omega x)| \exp\left(\pm i^2 \omega y\right) = \exp(\mp \omega y), \tag{17.28c–d}$$

the function (17.28b) with upper (lower) sign, is holomorphic for $y \equiv \mathrm{Im}(z) > 0 (< 0)$, that is, in the upper (lower) half-plane, but not in the lower (upper) half-plane, because (17.28d) diverges as $y \to -\infty$ $(y \to +\infty)$. Thus the real axis must be closed by a half-circle Figure 17.2a (17.2b) in the upper (lower) complex plane, as the integral along it vanishes if $g(z)$ decays at infinity:

$$\mathrm{Im}(z) >< 0: \quad \lim_{z \to \infty} g(z) = 0. \tag{17.29}$$

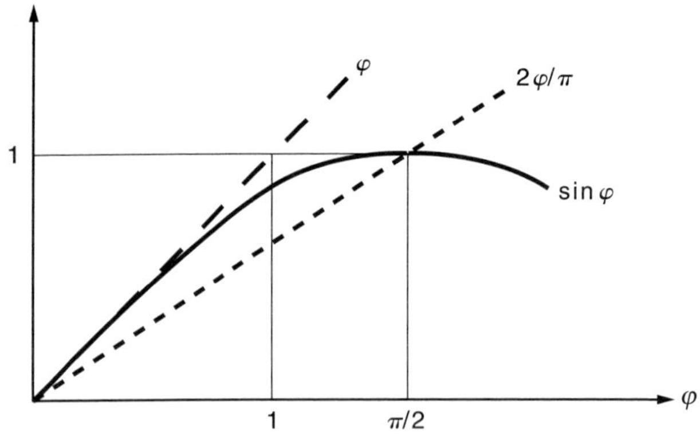

FIGURE 17.3
In the evaluation of the integrals in Figure 17.2 for an integrand of the form $f(z) = g(z)\exp(\pm i\omega z)$ with $\omega > 0$, the asymptotic condition that $f(z)$ decays faster than $1/z$ may be replaced by the condition that $g(z)$vanishes at infinity in the upper (lower) half z-plane. The proof of this relaxed asymptotic condition relies on bounds for the function $\sin\varphi$ in the interval $(0, \pi/2)$, namely: (i) it lies above the secant $\sin\varphi > (2/\pi)\varphi$; (ii) it lies below the tangent at the origin $\sin\varphi < \varphi$ which has slope unity $\mathrm{d}(\sin\varphi)/\mathrm{d}\varphi = \cos\varphi = 1$ for $\varphi = 0$.

It will be proved that if (17.29) is met then the integral of (17.28a,b) over a semicircle with center of center at the origin and infinite radius vanishes. The proof uses one of the inequalities:

$$0 \le \varphi \le \frac{\pi}{2}: \quad \frac{2\varphi}{\pi} \le \sin\varphi \le \varphi. \tag{17.30a,b}$$

The inequalities for the sine function (17.30b) in the interval (17.30a) of the real axis (Figure 17.3) follow from: (i) the slope of the sine $(\sin\varphi)' = \cos\varphi$ at the origin is unity, $\cos 0 = 1$, and the sine lies below the tangent; (ii) the sine lies above the straight line joining the origin to its closest maximum $\sin(\pi/2) = 1$, that is, it lies above the straight line $\varphi/(\pi/2) = 2\varphi/\pi$. The asymptotic condition (17.29) is less restrictive than (17.19), for example, $f(z) = 1/z$ does not meet (17.19) but $g(z) = 1/z$ meets (17.29). The reason for this is the decay of the factor in (17.28b,d):

$$|\exp(\pm i\omega z)| = \exp(\mp\omega y) = \exp(\pm\omega a\sin\varphi), \tag{17.31a}$$

that tends to zero (for $\omega > 0$) on the circle of radius a, as $a \to \infty$, for $\sin\varphi > 0\,(\sin\varphi < 0)$ that is, on the upper $0 < \varphi < \pi$ (lower $\pi < \varphi < 2\pi$) half-plane. It can be confirmed that the integral (17.20) evaluated along the half-circle of infinite radius vanishes:

$$\begin{aligned}
&\left|\int_0^{\pm\pi} g\left(ae^{i\varphi}\right)\exp\left(\pm i\omega a e^{i\varphi}\right) a e^{i\varphi} i\,\mathrm{d}\varphi\right| \le |g(z)|_{\max}\int_0^{\pm\pi} a\exp(\mp\omega a\sin\varphi)\,\mathrm{d}\varphi \\
&\le 2\varepsilon a\int_0^{\pi/2} \mathrm{e}^{-\omega a\sin\varphi}\,\mathrm{d}\varphi \le 2\varepsilon a\int_0^{\pi/2} \mathrm{e}^{-2\omega a\varphi/\pi}\,\mathrm{d}\varphi = 2\varepsilon a\left(-\frac{\pi}{2\omega a}\right)\mathrm{e}^{-2\omega a\varphi/\pi}\Big|_0^{\pi/2} \\
&= \frac{\pi\varepsilon}{\omega}\left(1-\mathrm{e}^{-\omega a}\right) \le \frac{\pi\varepsilon}{\omega},
\end{aligned} \tag{17.31b}$$

where (i) $\varepsilon \equiv |g(z)|_{\max} \to 0$ as $a \to \infty$ on account of the asymptotic condition (17.29); (ii) was used the following property:

$$0 \le \varphi \le \frac{\pi}{2}: \quad |\exp(i\omega z)| = \exp(-\omega a \sin\varphi) \le \exp\left(-\frac{2\omega a \varphi}{\pi}\right), \tag{17.31c,d}$$

arising from the first inequality in (17.30b). It has been proved that *if: (i) the improper integral (17.32), of the first (17.2b) or third (17.4) kinds, exists; (ii) the function* g(z) *is holomorphic in the upper (lower) complex half-plane, except for poles; (iii) the function* g(z) *vanishes at infinity (17.29), in the upper (lower) complex half-plane. Then the integral:*

$$\omega > 0: \int_{-\infty}^{+\infty} g(x) \exp(\pm i\omega x)\, \mathrm{d}x = \pm 2\pi i \sum_{\mathrm{Im}(z) >< 0} \mathrm{Res}\left(g(z)\, \mathrm{e}^{\pm i\omega z}\right) \pm \pi i \sum_{\mathrm{Im}(z)=0} \mathrm{Res}\left(g(z)\, \mathrm{e}^{\pm i\omega z}\right), \tag{17.32}$$

with the upper (lower) sign, can be evaluated by residues at the poles.

17.5.2 Some Direct and Inverse Fourier Transforms

As an example consider the integral (17.33b):

$$g(z) \equiv \frac{1}{z^2 + a^2}, \quad \tilde{g}_\pm(\omega) \equiv \frac{1}{\sqrt{2\pi}} \int_{-\infty}^{+\infty} \left(x^2 + a^2\right)^{-1} \mathrm{e}^{\pm i\omega x}\, \mathrm{d}x, \tag{17.33a,b}$$

that is the **direct** $\tilde{g}_+$ (**inverse** $\tilde{g}_-$) **Fourier transform** of the function (17.33a). The integral exists, since it has the upper bound:

$$\sqrt{2\pi}\, |\tilde{g}_\pm(\omega)| \le \int_{-\infty}^{+\infty} \left(x^2 + a^2\right)^{-1} \left|e^{\pm i\omega x}\right| \mathrm{d}x = \int_{-\infty}^{+\infty} \left(x^2 + a^2\right)^{-2} \mathrm{d}x = \frac{\pi}{a}, \tag{17.34}$$

where (17.22b) was used; the function (17.33a) meets the asymptotic condition (17.29); it is holomorphic except for simple poles at $z = \pm ia$ with residues $1/(\pm 2ia)$, allowing the evaluation the integral:

$$\sqrt{2\pi}\tilde{g}_\pm(\omega) \equiv \int_{-\infty}^{+\infty} \left(x^2 + a^2\right)^{-1} \mathrm{e}^{\pm i\omega x}\, \mathrm{d}x = \frac{\pm 2\pi i}{\pm 2ia} \mathrm{e}^{i^2 \omega a} = \frac{\pi}{a} \mathrm{e}^{-\omega a}; \tag{17.35}$$

this reduces to (17.22b) for $\omega = 0$. A similar extension of (17.27a,b) to a direct and inverse Fourier transform is:

$$j(z) = \left(z^2 + a^2\right)^{-2}, \qquad \tilde{j}_\pm(\omega) \equiv \frac{1}{\sqrt{2\pi}} \int_{-\infty}^{+\infty} \left(x^2 + a^2\right)^{-2} \mathrm{e}^{\pm i\omega x}\, \mathrm{d}x. \tag{17.36a,b}$$

The asymptotic condition (17.29) is met by (17.36a), and the integral exists:

$$\sqrt{2\pi}\, \left|\tilde{j}_\pm(\omega)\right| \le \int_{-\infty}^{+\infty} \left(x^2 + a^2\right)^{-2} \left|\mathrm{e}^{\pm i\omega x}\right| \mathrm{d}x = \int_{-\infty}^{+\infty} \left(x^2 + a^2\right)^{-2} \mathrm{d}x = \frac{\pi}{2a^3}, \tag{17.37}$$

where (17.27a) was used; the integrand is analytic except for double poles at $z = \pm ia$ with residues:

$$\lim_{z\to\pm ia} \frac{\mathrm{d}}{\mathrm{d}z} \frac{\mathrm{e}^{\pm i\omega z}}{(z\pm ia)^2} = \lim_{z\to\pm ia} \frac{\mathrm{e}^{\pm i\omega z}}{(z\pm ia)^2}\left(\pm i\omega - \frac{2}{z\pm ia}\right)$$
$$= \pm i\left(\omega + \frac{1}{a}\right)\frac{\mathrm{e}^{-\omega a}}{-4a^2}; \tag{17.38}$$

these specify the value of the integral:

$$\sqrt{2\pi}\,\tilde{j}_\pm(\omega) = \pm 2\pi i.\pm i\left(\omega + \frac{1}{a}\right)\frac{\mathrm{e}^{-\omega a}}{-4a^2} = \frac{\pi}{2a^3}\mathrm{e}^{-\omega a}(1+\omega a); \tag{17.39}$$

this reduces to (17.27b) for $\omega = 0$. Both Fourier transforms (17.35) [(17.39)] are real, showing that they reduce to **cosine transforms** (17.40a) [(17.41a)] because the function (17.33a) [(17.36a)] is even:

$$\sqrt{2\pi}\,\tilde{g}_\pm(\omega) = \int_{-\infty}^{+\infty} \left(x^2+a^2\right)^{-1}\{\cos,\sin(\omega x)\}\,\mathrm{d}x = \left\{\frac{\mathrm{e}^{-\omega a}\pi}{a}, 0\right\}, \tag{17.40a,b}$$

$$\sqrt{2\pi}\,\tilde{j}_\pm(\omega) = \int_{-\infty}^{+\infty} \left(x^2+a^2\right)^{-2}\{\cos,\sin(\omega x)\}\,\mathrm{d}x = \left\{\frac{\pi}{2a^3}\mathrm{e}^{-\omega a}(1+\omega a), 0\right\}; \tag{17.41a,b}$$

the corresponding **sine transforms** (17.40b) [(17.41b)] are zero, because the corresponding integrands in are odd functions. Three more cases of calculation of Fourier, cosine and sine transforms by residues are given in Example 20.15.

17.6 The Localization Lemma for Holomorphic Functions

Consider the function (17.42b):

$$h(z) \in \mathcal{D}(|C): \qquad g(z) = z^{-1}h(z), \qquad g_{(1)}(0) = h(0), \tag{17.42a–c}$$

where $h(z)$ is holomorphic (17.42a) at the origin $z = 0$, has a simple pole (17.42b) there, with residue (17.42c); the theorem (17.32) can be applied to (17.42a) provided that $h(z)$ be holomorphic in the upper (lower) half-plane, and meets there the asymptotic condition (17.43c) arising from (17.29):

$$\operatorname{Im}(z) >< 0: \qquad \lim_{z\to\infty} z^{-1}h(z) = 0. \tag{17.43a–c}$$

Note that (17.43c) is less stringent than (17.29) that in turn is less stringent than (17.19), for example, the latter do not hold for $h(z) \equiv c = \text{const}$, whereas (17.43a–c) does. The formula (17.32) corresponds to:

$$\int_{-\infty}^{+\infty} h(x)\,\mathrm{e}^{\pm i\omega x}x^{-1}\,\mathrm{d}x = \pm i\pi h(0); \tag{17.44a}$$

equivalently, setting $x = \xi - a$:

$$\int_{-\infty}^{+\infty} h(\xi - a)\, e^{\pm i\omega(\xi - a)} (\xi - a)^{-1}\, d\xi = \pm i\pi h(a). \tag{17.44b}$$

The imaginary part of (17.44b) specifies the **localization lemma** *for holomorphic functions:*

$$h(a) = \frac{1}{\pi} \int_{-\infty}^{+\infty} h(\xi - a) \frac{\sin[\omega(\xi - a)]}{\xi - a} d\xi, \tag{17.45a}$$

stating that if (i) the integral (17.45a) exists; (ii) the function h(z) *is holomorphic in the upper or lower half-plane (17.43a,b); (iii) it meets there the asymptotic condition (17.43c); then the integral specifies the value of the function at* ξ = a, or equivalently:

$$h(0) = \pi^{-1} \int_{-\infty}^{+\infty} h(x) \frac{\sin x}{x} dx, \tag{17.45b}$$

since the integral is specified by the value of the function for $a = 0$ in (17.45a). The real part of (17.45a,b) is zero, because the integrand involves the function cos that is even. The simplest instance of the localization lemma in the form (17.45a) is obtained for $h(x) = 1$, and concerns the evaluation of the bilateral or unilateral **sine-integral**:

$$\int_{-\infty}^{-\infty} \frac{\sin x}{x} dx = \pi, \qquad \int_{0}^{\infty} \frac{\sin x}{x} dx = \frac{\pi}{2}. \tag{17.46a,b}$$

The integrals (17.46a,b) are improper of the first kind, because the integrand is finite everywhere, including at the origin, where it takes the value unity (5.14b).

17.7 Surrounding a Semiinfinite Branch-Cut

The method of residues can be used to calculate integrals along the whole real axis (Sections 17.4–17.6) [positive real axis (Section 17.7)]. These integrals [Subsection 17.5.1 (17.7.1)] relate to the Fourier (Mellin) transform [Subsection 17.5.2 (17.7.2)].

17.7.1 Integral along the Positive Real Axis with a Branch-Point at the Origin

Consider the function (17.47b):

$$\alpha \in |R - |Z: \qquad f(z) = (-z)^{\alpha - 1} g(z), \tag{17.47a,b}$$

that is multivalued (Section 5.7) for real α not an integer (17.47a); the function becomes single-valued by taking the principal branch, and performing a branch-cut (Section 7.5) to exclude the negative real axis for $-z \notin) - \infty, 0)$, that is, the cut lies along the positive real half-axis. Consider the integral of (17.47b) along a loop L surrounding the branch-cut, and

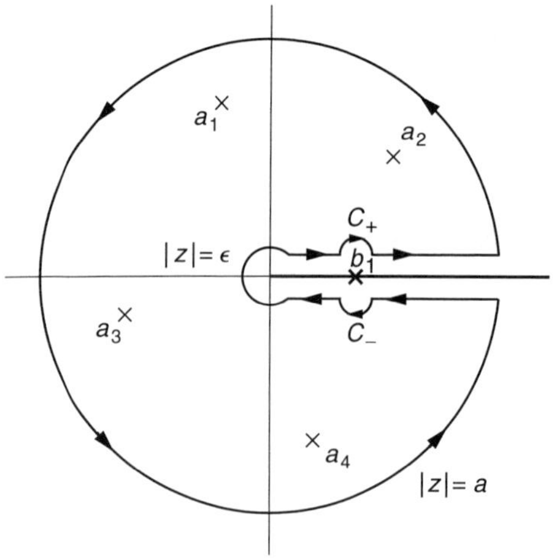

FIGURE 17.4
Consider an integral along the positive real axis, whose integrand has a branch-point at the origin with a semiinfinite branch-cut joining to infinity along the positive real axis. The existence of the improper integral of first or third kind can be proved, and its value calculated, considering a small and a large circle with centers at the origin, connected by segments above and below the branch-cut. The method requires an asymptotic (limit) condition on the large (small) circle as the radius tends to infinity (zero). If both conditions are met the integral can be evaluated in terms of the residues at the poles: (i) outside the positive real axis $a_1, \ldots$; (ii) on the positive real axis $b_1, \ldots$ with half-circle indentations $C_\pm, \ldots$.

consisting (Figure 17.4) of: (i) a circle of large radius $a \to \infty$ that cannot cross the cut, and must enclose it; (ii/iii) two semiinfinite straight lines passing below $(\infty - i0, \varepsilon - i0)$ and above $(\varepsilon + i0, \infty + i0)$; (iv) joining them a circle of radius $\varepsilon \to 0$ taken in the direction opposite to the outer circle. The integral of (17.47b) along this loop:

$$\oint_L (-z)^{\alpha-1} g(z)\,\mathrm{d}z = \left(\int_{|z|=a} - \int_{|z|=\varepsilon} \right) (-z)^{\alpha-1} g(z)\,\mathrm{d}z + \int_{\varepsilon+i0}^{a+i0} (-x-i0)^{\alpha-1} g(x+i0)\,\mathrm{d}x + \int_{a-i0}^{\varepsilon-i0} (-x+i0)^{\alpha-1} g(x-i0)\,\mathrm{d}x. \tag{17.48}$$

is thus the sum of four terms.

As before with (17.19; 17.20) the conditions:

$$\arg(z) \neq 0: \quad \lim_{z\to\infty} z^\alpha g(z) = 0 = \lim_{z\to 0} z^\alpha g(z), \tag{17.49a–c}$$

imply that the integrals along the outer (inner) circle of radius $a \to \infty\,(\varepsilon \to 0)$ vanish:

$$\lim_{a\to\infty} \int_{|z|=a} (-z)^{\alpha-1} g(z)\,\mathrm{d}z = 0 = \lim_{\varepsilon\to 0} \int_{|z|=\varepsilon} (-z)^{\alpha-1} g(z)\,\mathrm{d}z; \tag{17.50a,b}$$

the remaining two integrals can be collected together by setting $-x \pm i0 = x\mathrm{e}^{\pm i\pi}$:

$$\lim_{a\to\infty,\varepsilon\to 0}\left\{\mathrm{e}^{-i\pi(\alpha-1)} - \mathrm{e}^{i\pi(\alpha-1)}\right\}\int_{\varepsilon}^{a} x^{\alpha-1} g(x)\,\mathrm{d}x$$

$$= 2\pi i \sum_{\arg(z)\neq 0} \mathrm{Res}\left\{(-z)^{\alpha-1} g(z)\right\} - \pi i\left(\mathrm{e}^{i\pi\alpha} + \mathrm{e}^{-i\pi\alpha}\right)\sum_{\arg(z)=0}\mathrm{Res}\left\{z^{\alpha-1} g(z)\right\}, \quad (17.51)$$

where the loop integral was evaluated using the residues at the poles: (i) in the interior of L, that is, outside the positive real half-axis; (ii) on the boundary of L, that is, on the positive real half-axis that requires the loop to be deformed both in the upper/lower paths, introducing the factors, $\mathrm{e}^{\pm i\pi\alpha}$, with the corresponding half-circles taken in the negative direction.

The formula (17.51) involves the factors:

$$\mathrm{e}^{i\pi\alpha} \pm \mathrm{e}^{-i\pi\alpha} = 2\cos(\pi\alpha), 2i\sin(\pi\alpha); \quad (17.52\text{a,b})$$

it can be divided by the latter $\sin(\pi\alpha) \neq 0$ since α is not an integer (17.47a), to express the integral along the positive real half in terms of residues at poles outside it and on it:

$$\int_0^{\infty} x^{\alpha-1} g(x)\,\mathrm{d}x = \pi\csc(\pi\alpha)\sum_{\arg(z)\neq 0}\mathrm{Res}\left\{(-z)^{\alpha-1} g(z)\right\} - \pi\cot(\pi\alpha)\sum_{\arg(z)=0}\mathrm{Res}\left\{z^{\alpha-1} g(z)\right\}. \quad (17.53)$$

In the present instance (Section 17.7) was indeed proved, rather than assumed (as in Sections 17.4–17.6), the existence of the integral, since the limits in (17.51) coincide with the definition (17.2a) [(17.4)] improper integral of the first (third) kind, if there are no (there are) poles on the positive real half-axis. It has been proved that: *(i) if the function* g(z) *is holomorphic in the whole complex plane except for poles; (ii) if it satisfies the conditions (17.49a–c) at the origin and infinity, for nonintegral* α *in (17.47a). Then the improper integral of first or third kind (17.53) exists, and can be evaluated by residues at the poles.* The integrals of type (17.53) correspond to Mellin transforms.

17.7.2 Evaluation of Mellin Transforms by Residues

The conditions (17.49a,b) may restrict the value of α, for example, in the case:

$$g_{\pm}(z) \equiv \left(z^2 \pm a^2\right)^{-1}, \quad \lim_{z\to 0} z^{\alpha} g(z) = \lim_{z\to 0} a^{-2} z^{\alpha} = 0, \quad \lim_{z\to\infty} z^{\alpha} g(z) = \lim_{z\to\infty} z^{\alpha-2} = 0, \quad (17.54\text{a–c})$$

the condition (17.49b) [(17.49c)] at the origin (infinity) for the function (17.54a), requires $\mathrm{Re}(\alpha) > 0\,[\mathrm{Re}(\alpha) < 2]$ in (17.54b) [(17.54c)] leading to (17.55a). Take as example the first function in (17.54a) ≡ (17.55b):

$$0 < \mathrm{Re}(\alpha) < 2: \quad g_{-}(z) = \{(z-a)(z+a)\}^{-1}, \quad g^{-}_{(1)}(\pm a) = \pm\frac{1}{2a}, \quad (17.55\text{a–c})$$

that has residue (17.55c) at the pole $z = a\,(z = -a)$ on (outside) the positive real half-axis; this corresponds to the second (first) term of (17.53):

$$a > 0, 0 < \mathrm{Re}(\alpha) < 2: \quad I_{-}(\alpha) \equiv \int_0^{\infty}\frac{x^{\alpha-1}}{x^2 - a^2}\mathrm{d}x = -\frac{\pi}{2}a^{\alpha-2}\{\cot(\pi\alpha) + \csc(\pi\alpha)\}; \quad (17.56\text{a–c})$$

the second function in (17.54a) ≡ (17.57a) is obtained from the first changing a to ia:

$$g_+(z) = \{(z-ia)(z+ia)\}^{-1}, \quad g^+_{(1)}(\pm ia) = \mp\frac{i}{2a}, \tag{17.57a,b}$$

and thus the poles $z = \pm ia$ with residues (17.57b) now lie both outside the positive real half-axis; they both appear in the first term of (17.53):

$$a > 0, 0 < \mathrm{Re}(\alpha) < 2: \quad I_+(\alpha) \equiv \int_0^\infty \frac{x^{\alpha-1}}{x^2+a^2}\,\mathrm{d}x = \frac{\pi}{2}a^{\alpha-2}\csc\left(\frac{\pi\alpha}{2}\right), \tag{17.58a–c}$$

where the following simplifications have been used:

$$(ia)^{\alpha-1}g_{(1)}(-ia) + (-ia)^{\alpha-1}g_{(1)}(+ia) = a^{\alpha-2}\frac{i^\alpha + (-i)^\alpha}{2} = a^{\alpha-2}\frac{\mathrm{e}^{i\pi\alpha/2} + \mathrm{e}^{-i\pi\alpha/2}}{2}, \tag{17.59a}$$

$$a^{2-\alpha}I_+(\alpha) = \pi\csc(\pi\alpha)\cos(\pi\alpha/2) = \frac{\pi\cos(\pi\alpha/2)}{2\sin(\pi\alpha/2)\cos(\pi\alpha/2)} = \frac{\pi}{2}\csc(\pi\alpha/2). \tag{17.59b}$$

As examples, (17.56a,b) ≡ (17.58a,b) are met by $\alpha = 3/2$ in (17.56c) [(17.58c)] leading to (17.60a) [(17.60b)]:

$$I_-\left(\frac{3}{2}\right) \equiv \int_0^\infty \frac{\sqrt{x}}{x^2-a^2}\,\mathrm{d}x = \frac{\pi}{2\sqrt{a}} < I_+\left(\frac{3}{2}\right) \equiv \int_0^\infty \frac{\sqrt{x}}{x^2+a^2}\,\mathrm{d}x = \frac{\pi}{\sqrt{2a}} = \sqrt{2}I_-\left(\frac{3}{2}\right), \tag{17.60a,b}$$

thus the first integral is $1/\sqrt{2}$ times the second. The integrals (17.56c) and (17.58c) would not exist for $\alpha = 2\,(\alpha = 0)$ since in that case the integrand would tend to x^{-1} at infinity (the origin), that is, have a logarithmic singularity. Two more cases of integrals with branch-points appear in Example 20.16.

17.8 Branch-Point within the Path of Integration

Following from the integral with a branch-point type of singularity at one end, the final case is an integral with a singularity of logarithmic type within the path of integration. The integral (17.61b):

$$ab > 0: \quad \int_a^b x^{-1}\,\mathrm{d}x = \log x|_a^b = \log\left(\frac{b}{a}\right), \tag{17.61a,b}$$

is proper (17.61a) if b, a have the same sign $b, a > 0$ or $b, a < 0$, since then the singularity at $x = 0$ lies outside the path of integration. If $ba < 0$ then $x = 0$ lies within the path of integration, and leading (17.62a) to a bilateral improper integral of the second kind (17.62b):

$$a < 0 < b: \quad \int_a^b x^{-1}\,\mathrm{d}x = \lim_{\varepsilon,\delta\to 0+}\left(\int_a^{-\varepsilon} + \int_\delta^b\right)x^{-1}\,\mathrm{d}x$$

$$= \lim_{\varepsilon,\delta\to 0+}\log\left(-\frac{b\varepsilon}{a\delta}\right) = \log\left(-\frac{b}{a}\right) + \lim_{\varepsilon,\delta\to 0+}\log\left(\frac{\varepsilon}{\delta}\right); \tag{17.62a,b}$$

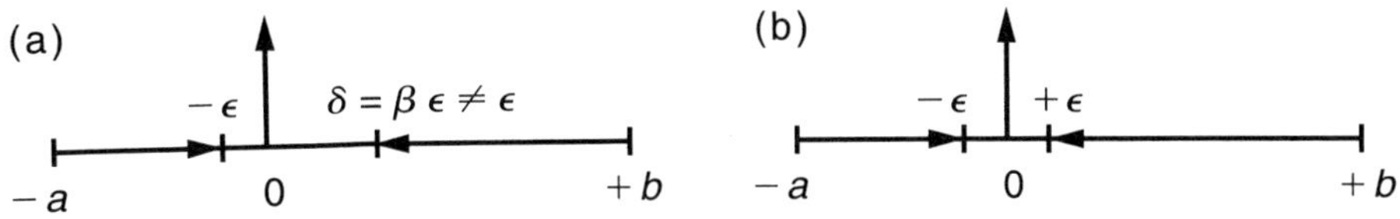

FIGURE 17.5
The real integral along a segment of the real axis with a pole in the interior requires "carving a hole" around the singularity. The value of the integral then depends on the "shape of the hole": if the pole is at the origin and the hole (a) is $-\varepsilon \le x < \beta\varepsilon$ then by choosing β the integral can be given any value from $-\infty$ to $+\infty$. To give the integral a definite value, a symmetric hole (b) is chosen $\beta = 1$ or $-\varepsilon \le x \le +\varepsilon$. This leads to the Cauchy principal value of the integral, and applies to poles of any order.

since ε, δ tend to zero independently, choosing $\delta = \beta\varepsilon \to 0$ for (Figure 17.5a) any real positive $\beta > 0$, adds to $\log|b/a|$ in (17.62b) an arbitrary real constant $-\infty < -\log\beta < +\infty$. Hence the limit (17.62b) depends on the shape of the interval $(-\varepsilon, +\beta\varepsilon)$ around the branch-point, and can be given any value by choosing β; thus the improper integral of the second kind is totally indeterminate, and does not exist. A definite value can be obtained for the integral value by settling on a $\beta = 1$, that is (Figure 17.5b) using a symmetric interval $(-\varepsilon, +\varepsilon)$ in the definition of **Cauchy principal value** of the integral (17.62b). It has been shown that the expression:

$$\fint_a^b x^{-1}\,\mathrm{d}x = \log\left|\frac{b}{a}\right|, \tag{17.63}$$

specifies both *the proper integral (17.61a,b) [Cauchy principal value (17.63)] of the improper integral of first kind (17.62a,b) with a singularity at the origin:*

$$\fint_a^b x^{-1}\,\mathrm{d}x = \lim_{\varepsilon\to 0+}\left(\int_a^{-\varepsilon} + \int_{+\varepsilon}^{b}\right) x^{-1}\,\mathrm{d}x = \log\left|\frac{b}{a}\right|. \tag{17.64}$$

along a path of integration not including (including) the origin.

17.9 Cauchy (1821) Principal Value of an Integral

The result (17.63) can be generalized to:

$$\fint_a^b x^{-1} f(x)\,\mathrm{d}x \equiv \lim_{\varepsilon\to 0+}\left(\int_a^{-\varepsilon} + \int_{+\varepsilon}^{b}\right) f(x)\,\mathrm{d}(\log x) = -f(a)\log|a| + f(b)\log b$$
$$+ \lim_{\varepsilon\to 0+}\{f(-\varepsilon) - f(\varepsilon)\}\log\varepsilon - \lim_{\varepsilon\to 0+}\left(\int_a^{-\varepsilon} + \int_{+\varepsilon}^{b}\right)\log|x|\,\mathrm{d}f(x), \tag{17.65}$$

where integration by parts was used; since the function is differentiable, the mean value theorem (28.5) may be used to cancel one term:

$$f \in \mathcal{D}(a,b): \quad \lim_{\varepsilon\to 0+}[f(-\varepsilon) - f(\varepsilon)]\log\varepsilon = -\lim_{\varepsilon\to 0+} f'(\varepsilon)\,2\varepsilon\log\varepsilon = -f'(0)\lim_{\varepsilon\to 0+} 2\varepsilon\log\varepsilon = 0. \tag{17.66a,b}$$

It follows that

$$b>0>a:\quad \int_a^b f(x)\,x^{-1}\,\mathrm{d}x = f(b)\log b - f(a)\log|a| + \lim_{\varepsilon\to 0+}\left(\int_a^{-\varepsilon} + \int_{+\varepsilon}^{b}\right)\log|x|\,f'(x)\,\mathrm{d}x, \tag{17.67a,b}$$

the Cauchy principal part of the integral (17.67a) of the product a first-order inverse power by a differentiable function (17.66a), can be evaluated by (17.67b).

An integration by parts as in (17.65) applied to:

$$\begin{aligned} b>0>a:\quad \int_a^b x^{-n} f(x)\,\mathrm{d}x &= \int_a^b f(x)\,\mathrm{d}\left(\frac{x^{1-n}}{1-n}\right) \\ &= \left\{\frac{x^{1-n}}{1-n} f(x)\right\}_a^b + \int_a^b \frac{x^{1-n}}{n-1} f'(x)\,\mathrm{d}x, \end{aligned} \tag{17.68a,b}$$

and repeated $(n-1)$ times leads to:

$$\begin{aligned} f\in \mathcal{D}^{n-1}(a,b):\quad \int_a^b x^{-n} f(x)\,\mathrm{d}x &= \{(n-1)!\}^{-1}\int_a^b x^{-1} f^{(n-1)}(x)\,\mathrm{d}x \\ &\quad + \sum_{\mathrm{k}=2}^{\mathrm{n}} \{(1-n)\cdots(1-k)\}^{-1}\left\{x^{1-k} f^{(n-k)}(x)\right\}_a^b; \end{aligned} \tag{17.69a,b}$$

this *evaluates the Cauchy principal value of the integral of the product an* n*-th order inverse power (17.69b) by an* $(\mathrm{n}-1)$ *times differentiable function (17.69a), in terms of the same integral with 1st order inverse power (17.67b), plus terms at the end-points (17.67a) involving derivates up to order* $\mathrm{n}-2$. The formula (17.69b) is proved by induction: (i) it holds for $n=1$, when it reduces to an identity; (ii) if it holds for n, then:

$$\begin{aligned} \int_a^b x^{-n-1} f(x)\,\mathrm{d}x &= -n^{-1}\{x^{-n} f(x)\}_a^b + \frac{1}{n}\int_a^b x^{-n} f'(x)\,\mathrm{d}x \\ &= \frac{1}{n!}\int_a^b x^{-1} f^{(n)}(x)\,\mathrm{d}x + \sum_{k=2}^{n+1}[(-n)\ldots(1-k)]^{-1}\left\{x^{1-k} f^{(n+1-k)}(x)\right\}_a^b, \end{aligned} \tag{17.70}$$

by performing one integration by parts, and using (17.69b) for n, the same formula is obtained for $n+1$. QED. The integral of the power with arbitrary exponent involves (Example 20.17) several of the cases dealt with before (Chapter 17). Cauchy principal values of integrals are evaluated in Section 18.5 and Subsection 34.7.4.

Note 17.1. *Evaluation of Real and Complex Integrals by Residues:* The preceding examples show that the method of residues succeeds in evaluating integrals, like (17.18), (17.32), and (17.53) that would be more difficult to establish by other means; these integrals are relevant to the Fourier (17.33a,b, 17.35; 17.36a,b, 17.39) [Mellin 17.56a–c; 17.58a–c)] transforms in Subsection 17.5.2 (17.7.2). Some of the integrals obtained apply to arbitrary functions, for example, the localization lemma (17.45a,b) and Cauchy principal value (17.69a,b; 17.67a). Loop integrals specify the flow rate and circulation (Chapter 12) in a potential flow, and also the kinetic energy, forces, and moments (Chapter 28); there are analogous results for the gravity field (Chapter 18), electrostatics (Chapter 24), and magnetostatics (Chapter 26). Not only these but many other integrals will be evaluated subsequently by residues and related methods. An example is the gravity field created by a mass distribution that is considered next (Chapter 18).

Conclusion 17: Transformation (Figure 17.1) of (**a**) the interval $0 \leqq \varphi < 2\pi$ of the real axis (**a**) into (**b**) the unit circle $|z| = 1$ used to evaluate trigonometric integrals. Closing a symmetric segment of the real axis by a semicircle (Figure 17.2) in upper (**a**) [lower (**b**)] half-plane L_+ (L_-) taken in the positive (negative) direction, with poles at a_1, a_2 in the interior, and indentations to include the poles at b_1, b_2 on the boundary evaluates the real integral. The integration can be extended to the whole real axis if an asymptotic condition is met; the lower inequality (Figure 17.3) proves that the direct (inverse) Fourier transform can be evaluated from the residues at poles in the upper (lower) complex plane with a less stringent asymptotic condition. Enclosing the branch-cut along the positive real axis (Figure 17.4) between large (small) circles, taken in the positive (negative) direction, having poles at a_1, a_2, a_3 in their interior, and connected by straight segments above and below the positive real half-axis, with indentations on both sides for each pole, e.g., b_1; this allows the evaluation by residues of an integrand with a branch point at the origin. The (Figure 17.5) asymmetric (symmetric) interval $(-\varepsilon, \beta\varepsilon)$ [$(-\varepsilon, +\varepsilon)$] for (**a**) [(**b**)] leads to an indeterminate (unique) improper integral of second kind (Cauchy principal value).

18

Mass and the Gravity Field

An irrotational and compressible flow has a potential generated by mass source/sinks (Section 18.1), and an incompressible rotational flow has a stream function generated by vortices (Section 18.2); across a distribution of sources/sinks (vortices) on a curve the normal (tangential) velocity has a jump. The gravity field (Section 18.3) is always irrotational and is generated by a mass distribution; the mass is positive and creates an attractive field; thus it is analogous to the irrotational flow due to sinks. The normal (tangential) component of the gravity field across a mass distribution on a curve is discontinuous (continuous). The gravity field due to a line-mass (mass distribution) is analogous to a sink (Section 18.4), and this specifies the gravity force (Section 18.5) between line-masses and/or mass distributions. Two examples of gravity fields are those due to a finite (infinite) homogeneous (inhomogeneous) slab [(Section 18.6 (18.8)] for that the integral is nonsingular at an external point; at an internal point (Section 18.7) the integrals must be taken as Cauchy principal values (Section 17.8), to exclude the singularity, since a mass does not create a gravity field on itself. The gravity field of an arbitrary mass distribution can be represented as a superposition of multipoles (Section 18.9). The creation of a gravity field is an intrinsic property of a mass distribution, be it the fall of a body on the earth, the motion of planets around the sun, or the massive objects like galaxies in the universe.

18.1 Irrotational Flow due to Sources or Sinks

For an irrotational (Section 12.1) flow (18.1a):

$$\nabla \wedge \vec{v} = 0, \qquad \vec{v} = \nabla\Phi, \tag{18.1a,b}$$

the velocity is the gradient of a potential (18.1b); in the presence (Section 12.4) of a source $\Lambda > 0$ (sink $\Lambda < 0$) of dilatation (18.2a):

$$\nabla \cdot \vec{v} = \Lambda, \qquad \Lambda = \nabla \cdot \nabla\Phi = \nabla^2\Phi, \tag{18.2a,b}$$

the potential satisfies **Poisson's equation** (18.2b). Consider orthogonal curvilinear coordinates, and denoting by $s \equiv x_1\,(n \equiv x_2)$ the coordinate in (Figure 18.1a,b) tangent (normal) to the curves $x_1 = \text{const}$, and hence normal (tangent) to the curves $x_2 = \text{const}$; the Laplacian operator (11.52) takes the form:

$$\frac{1}{h_s h_n}\left[\frac{\partial}{\partial s}\left(\frac{h_n}{h_s}\frac{\partial\Phi}{\partial s}\right) + \frac{\partial}{\partial n}\left(\frac{h_s}{h_n}\frac{\partial\Phi}{\partial n}\right)\right] = \Lambda = \frac{1}{h_n h_s}\frac{\partial^2 Q}{\partial n\,\partial s}, \tag{18.3}$$

where $h_s\,(h_n)$ is the scale factor in the tangential (normal) direction, and the dilatation Λ is the density per unit area of the flow rate Q.

Suppose that the source/sink lies on the curve x_1 with density per unit length (18.4a):

$$q \equiv \frac{\partial Q}{\partial s}; \quad \Lambda = \frac{1}{h_n h_s}\frac{\partial q}{\partial n}; \tag{18.4a,b}$$

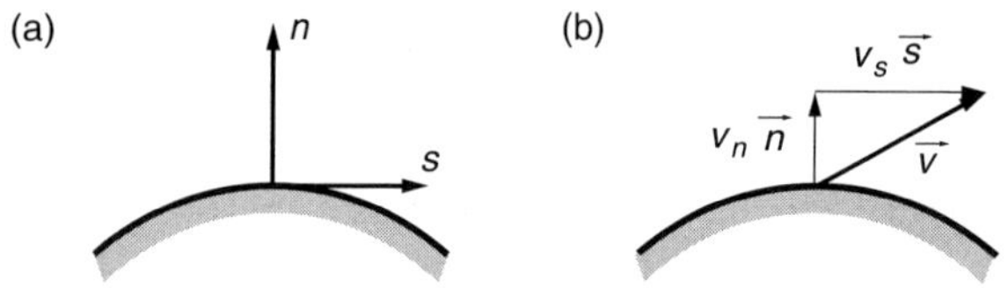

FIGURE 18.1
A regular curve has at each point (a) a normal n (tangential s) direction; a vector $\vec{v}$ with arbitrary direction can be decomposed (b) into normal v_n (tangential v_s) components along the unit normal $\vec{n}$ (tangent $\vec{s}$) vector.

then (18.4b) leads in (18.3) to:

$$\frac{\partial}{\partial s}\left(\frac{h_n}{h_s}\frac{\partial \Phi}{\partial s}\right)+\frac{\partial}{\partial n}\left(\frac{h_s}{h_n}\frac{\partial \Phi}{\partial n}\right)=\frac{\partial q}{\partial n}. \tag{18.5}$$

The contrast between the concentrated source/sinks that vary abruptly along the normal and the smooth variation along the tangent allows the approximation (18.6a) that simplifies (18.5) to the second term on the l.h.s., leading to (18.6b):

$$\frac{\partial}{\partial n} \gg \frac{\partial}{\partial s}: \quad \frac{h_n}{h_s}q=\left[\frac{\partial \Phi}{\partial n}\right]=[v_n]\equiv v_n(+0)-v_n(-0), \tag{18.6a,b}$$

where $[v_n]$ is the jump of the normal velocity. In the case of a distribution of source/sinks on a circle of radius r, the scale factors in polar coordinates (11.44b) $\equiv$ (18.7a,b] lead to (18.7c):

$$h_s=r, h_n=1: \quad r^{-1}q(\varphi)=[v_r]\equiv v_r(r+0,\varphi)-v_r(r-0,\varphi). \tag{18.7a–c}$$

A particular case is a uniform source/sink distribution (18.8a) along the circle of radius r, leading to (18.8b):

$$q(\varphi)\equiv \text{const}=\frac{Q}{2\pi}: \quad 2\pi v_r=r^{-1}\int_0^{2\pi} q(\varphi)\,d\varphi=\frac{Q}{r}, \tag{18.8a,b}$$

that is, equivalent to a line source (12.25a) $\equiv$ (18.8b) of flow rate Q.

For a distribution of source/sinks with density q unit length along the real axis (Figure 18.2a), specifies the jump of the normal velocity:

$$\frac{\partial Q}{\partial y}\equiv q(x): \quad q(x)=\left[\frac{\partial \Phi}{\partial y}\right]=[v_y(x,0)]\equiv v_y(x,+0)-v_y(x,-0). \tag{18.9a,b}$$

Thus *in an irrotational flow (18.1a), the velocity (18.1b) is the gradient of a potential that satisfies Poisson's equation (18.2b)* $\equiv$ *(18.3) forced by the dilatation* Λ*; if the flow sources/sinks are concentrated along a curve with density* q *per unit length (18.4a), the normal velocity is discontinuous (18.6a,b), for example, in polar (18.7c) [Cartesian (18.9a,b) coordinates (Figure 18.2a)].* The distribution of source/sinks may be a fictitious way to generate or simulate an irrotational flow, or it may be the a porous body with wall blowing (suction) acting as a flow source (sink).

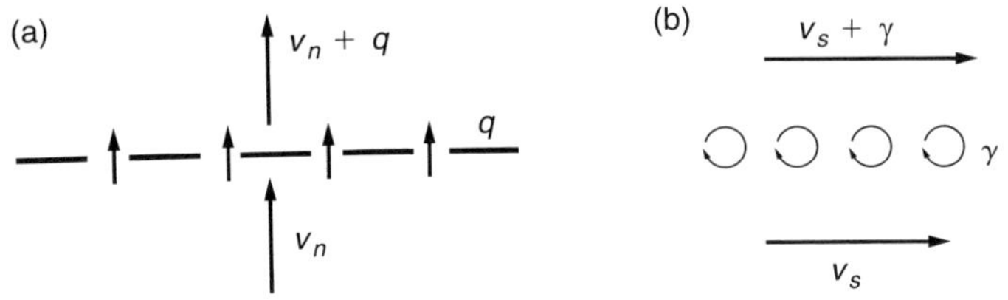

FIGURE 18.2
A porous surface (a) across which there is blowing (suction) corresponds to a surface distribution of mass sources (sinks) with flow rate per unit length q, across which there is a jump q of normal velocity, from v_n below to $v_n + q$ above. A surface distribution of vortices (b) with circulation density per unit length γ corresponds to a vortex sheet across which there is a jump of the tangential velocity γ from v_s below to $v_s + \gamma$ above. In the case of the "vortex sheet" or "tripole" (Figure 14.9) the opposive tangential velocities $-v$ below ($+v$ above) corresponds to a vorticity density $\gamma = 2v$.

18.2 Incompressible Flow due to a Vorticity Distribution

In the case of an incompressible (Section 12.2) flow (18.10a):

$$\nabla \cdot \vec{v} = 0, \quad \vec{v} = \nabla \wedge \Psi \equiv \left\{ \frac{\partial \Psi}{\partial y}, -\frac{\partial \Psi}{\partial x} \right\} = \{v_x, v_y\}, \tag{18.10a,b}$$

the velocity is (Section 12.2) the curl (18.10b) of a stream function (or two-dimensional analogue of the vector potential); the two-dimensional vorticity (Section 12.1) is orthogonal to the (x,y)-plane (18.11a), and forces with minus sign (18.11b):

$$\vec{\varpi} = \vec{e}_3 \Gamma \equiv \frac{\partial v_y}{\partial x} - \frac{\partial v_x}{\partial y}; \quad -\varpi = \nabla^2 \Psi, \tag{18.11a,b}$$

the Poisson equation (18.11b) for the stream function (18.12):

$$-\frac{1}{h_s h_n}\left[\frac{\partial}{\partial s}\left(\frac{h_n}{h_s}\frac{\partial \Psi}{\partial s}\right) + \frac{\partial}{\partial n}\left(\frac{h_s}{h_n}\frac{\partial \Psi}{\partial n}\right)\right] = \varpi = \frac{1}{h_n h_s}\frac{\partial^2 \Gamma}{\partial n \partial s}; \tag{18.12}$$

where the vorticity is minus the density of circulation per unit area. In the case of a vorticity distribution along a curve the density satisfies the analogue (18.13a,b) of (18.4a,b):

$$\gamma = \frac{\partial \Gamma}{\partial s}, \quad \varpi = \frac{1}{h_n h_s}\frac{\partial \gamma}{\partial n}. \tag{18.13a,b}$$

Thus the condition (18.13c) reduces the l.h.s. of (18.12) to the first term (18.13d):

$$\frac{\partial}{\partial n} \gg \frac{\partial}{\partial s}; \quad -\frac{\partial \gamma}{\partial n} = \frac{\partial}{\partial n}\left(\frac{h_s}{h_n}\frac{\partial \Psi}{\partial n}\right); \tag{18.13c,d}$$

the latter (18.13d) leads to:

$$-\frac{h_n}{h_s}\gamma = \left[\frac{\partial \Psi}{\partial n}\right] = [v_s] \equiv v_s(+0) - v_s(-0), \tag{18.14}$$

a discontinuity in the normal component of the stream function that corresponds (18.10b) to the tangential velocity.

In the case of vorticity distributed on a circle of radius r, the tangential velocity is discontinuous:

$$h_s = r, h_n = 1: \quad r^{-1}\gamma(\varphi) = \left[-\frac{\partial\Psi}{\partial r}\right] = [v_\varphi] = v_\varphi(r+0,\varphi) - v_\varphi(r-0,\varphi); \qquad \text{(18.15a–c)}$$

for a constant vorticity (18.16a):

$$\gamma(\varphi) = \text{const} \equiv \frac{\Gamma}{2\pi}: \quad 2\pi v_\varphi = r^{-1}\int_0^{2\pi}\gamma(\varphi)\,d\varphi = \frac{\Gamma}{r}, \qquad \text{(18.16a,b)}$$

this leads to (18.6b) $\equiv$ (12.31a) the line vortex of circulation Γ. For a vorticity distribution with density (18.17a) per unit length along the real axis (Figure 18.2b) the tangential velocity has a jump (18.17b):

$$\gamma(x) = \frac{\partial\Gamma}{\partial y}: \quad -\gamma(x) = \left[\frac{\partial\Psi}{\partial y}\right] = [v_x] = v_x(x,+0) - v_x(x,-0). \qquad \text{(18.17a,b)}$$

Thus *in an incompressible flow (18.10a) the velocity (18.10b) derives from a stream function that satisfies Poisson's equation (18.11b)* $\equiv$ *(18.12) forced by vorticity (18.11a); if the vorticity is concentrated along a curve (18.13a), it causes a jump in the tangential velocity (18.14), for example, in polar (18.15a–c) [Cartesian (18.17a,b)] coordinates* (Figure 18.2b). The distribution of vorticity may be a fictitious way to generate or simulate an incompressible flow, or it may represent a vortex sheet separating two jets with distinct tangential velocities. Besides the jump of normal (tangential) velocity in Figure 18.2a (18.2b) due to a source/sink (vortex) distribution along a curve, there is also the issue of determining the flow velocity inside a domain with sources/sinks (vortices), that is, the self-induced field. This issue will be dealt with in the analogous case (Section 18.8) of the gravity field (Sections 18.4–18.9) that is considered next (Section 18.3).

18.3 Gravity Field and Gravitational Constant

The **gravity field** is specified by the **acceleration of gravity** $\vec{g}$ that is always irrotational (18.18a):

$$\nabla \wedge \vec{g} = 0, \qquad \vec{g} = -\nabla\Phi_g, \qquad \text{(18.18a,b)}$$

and hence derives from a gravity potential (18.18b); the convention on the sign implies that the gravity field lies in the direction of decreasing potential [the inverse of the velocity in (18.1b) that lies direction of increasing scalar potential]. The divergence of the gravity field equals (18.19a) minus the product of the mass density ρ by the **gravitational constant** G:

$$\nabla\cdot\vec{g} = -\rho G, \qquad \rho G = \nabla\cdot\nabla\Phi_g = \nabla^2\Phi_g, \qquad \text{(18.19a,b)}$$

thus the latter appears in the Poisson equation (18.19b) for the gravity potential. The comparison of (18.19a) with (18.2a) shows that the mass that is positive, creates a gravity field analogous to the potential flow of sink:

$$\vec{g} \Leftrightarrow \vec{v}, \qquad -\rho G \Leftrightarrow \Lambda < 0. \qquad \text{(18.20a,b)}$$

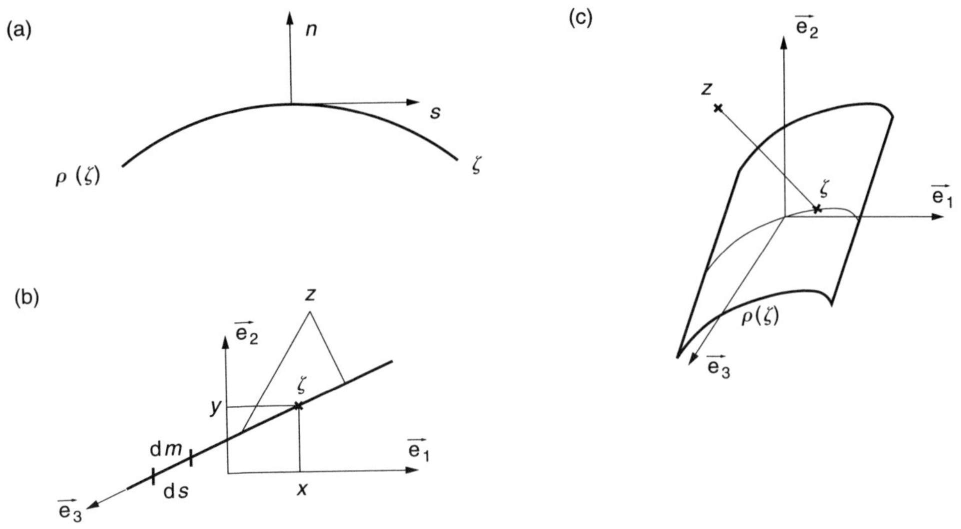

FIGURE 18.3
A mass distribution along a curve (a), like (unlike) [Figure 18.2a(b)] a flow source (vortex sheet) causes (does not cause) a jump in the normal (tangential) component of the gravity field. A point mass (curve) in the plane (b) [(c)] corresponds in space to a mass distribution along a line (on a surface) normal to the plane, obtained by the translation of the original point (curve).

There no "source" in the gravity field $\Lambda = -\rho G > 0$ because a "negative" mass $\rho < 0$ does not exist; there are no "vortices" (18.11a) in the gravity field, because it is always irrotational (18.18a).

The same deduction as (18.11a,b) to (18.1a,b) in, starting from (18.19a,b) leads to analogous results: *the gravity field is specified by the acceleration $\vec{g}$ that is irrotational (18.18a), and hence derives from a potential (18.18b), satisfying Poisson's equation (18.19b), where G is the gravitational constant and ρ the mass density per unit area. A mass distribution (Figure 18.3a) of density μ per unit length along a curve (18.21a):*

$$\mu = \frac{\partial \rho}{\partial s}: \quad [g_s] = 0, \quad -\frac{h_n}{h_s} G\mu = \left[-\frac{\partial \Phi_g}{\partial n} \right] = [g_n], \qquad (18.21\text{a–c})$$

causes a jump (18.21c) in the normal gravity field, and leaves the tangential component continuous (18.21b); for example:

$$g_x(x, +0) = g_x(x, -0), \quad g_y(x, +0) - g_y(x, -0) = -G\mu(x). \qquad (18.22\text{a,b})$$

a mass distribution on the real axis leads to (18.22a,b).

18.4 Line, Surface, and Volume Mass Distributions

Consider a mass distribution $\mu(\varphi)$ on a small circle of radius r; using (18.21b,c) in polar coordinates (18.7a,b) ≡ (12.44b) it follows that the tangential gravity field is zero (18.23a):

$$g_\varphi = 0, \quad g_r = -\frac{G}{r}\mu, \qquad (18.23\text{a,b})$$

and the normal component has a jump (18.23b), that is, if it is zero inside (18.23a), its value outside is (18.23b). The total mass is:

$$m = \int_0^{2\pi} \mu(\varphi)\,\mathrm{d}\varphi = -\frac{2\pi r g_r}{\mathrm{G}}; \tag{18.24}$$

collapsing it to a point specifies the gravity field (18.24; 18.23a,b) ≡ (18.25a,b):

$$\{g_r, g_\varphi\} = \left\{-\frac{m\mathrm{G}}{2\pi r}, 0\right\}; \qquad \Phi_g = \frac{m\mathrm{G}}{2\pi}\log r, \tag{18.25a–c}$$

from (18.18b) follows the gravity potential (18.25c) of a point mass in the plane that corresponds in space (Figure 18.3b) to a line-mass orthogonal to the plane. The gravity field (18.25a,b) of a line-mass is analogous to the velocity (12.25a) of a sink of flow rate $Q = -m\mathrm{G}$. Thus *the gravity field of a line-mass* m *at position* ζ *has complex potential (18.26a) and conjugate gravity (18.26b):*

$$f_g(z) = f(x+iy) = \Phi(x,y) + i\Psi(x,y) = \frac{m\mathrm{G}}{2\pi}\log(z-\zeta), \tag{18.26a}$$

$$g^*(z) \equiv -\frac{\mathrm{d}f_g}{\mathrm{d}z} = g_x(x,y) - i g_y(x,y) = \frac{\mathrm{G}}{2\pi}\frac{m}{\zeta - z}, \tag{18.26b}$$

that add by superposition to: (i) for a mass distribution of density $\mu \equiv \mathrm{d}m/\mathrm{d}s$ *per unit length along a curve* C *in Figure 18.3c that is a surface in space*:

$$f_g(z) = \frac{\mathrm{G}}{2\pi}\int_C \mu(\zeta)\log(z-\zeta)\,\mathrm{d}\zeta, \quad g^*(z) = \frac{\mathrm{G}}{2\pi}\int_C \mu(\zeta)(\zeta - z)^{-1}\,\mathrm{d}\zeta; \tag{18.27a,b}$$

(ii) similarly with the mass density $\rho(\zeta)$ *per unit area over a two-dimensional domain* D *in the plane that is the directrix of a cylinder in space.* The complex conjugate gravity field (18.26b) is given by minus the derivate of the complex potential (18.26a) in agreement with (18.18b).

18.5 Gravity Force of Attraction (Newton, 1687)

The gravity field (18.26b) due to a line-mass at ζ exerts (Section 28.3) on another line mass m' at z a **gravity force:**

$$F_g^* = m' g^* = \frac{\mathrm{G}}{2\pi}\frac{m'm}{\zeta - z}, \tag{18.28}$$

corresponding to **Newton' law of gravity attraction (1687)** *in two-dimensions (Figure 18.4a): two line-masses* m,m', *respectively, at* ζ,z, *attract each other in proportion to: (i) the product of masses; (ii) the inverse of mutual distance. The force per unit length in the direction joining the masses is multiplied by* $-G/2\pi$ *where* G *is the gravitational constant (18.32a,b). In the case of a line-mass and (18.29a) a mass distribution* $\rho(\zeta)$ *in a domain* D *the gravity force is (18.29b):*

$$\mathrm{d}m = \rho\,\mathrm{d}\zeta: \; F_g^* = \frac{m'\mathrm{G}}{2\pi}\int_D (\zeta - z)^{-1}\rho(\zeta)\,\mathrm{d}\zeta; \tag{18.29a,b}$$

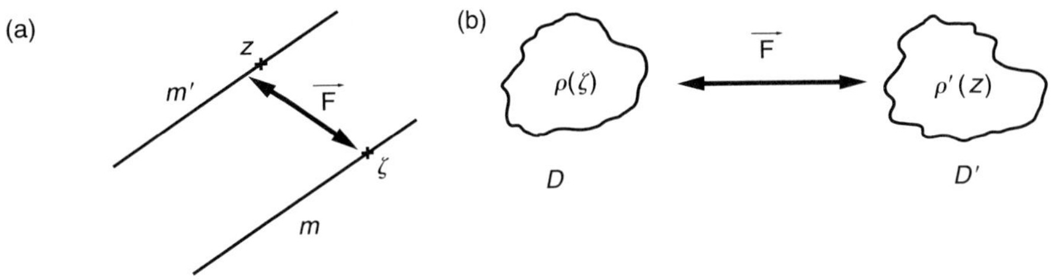

FIGURE 18.4
The gravity force of attraction between to point masses in the plane (a) corresponds in space (Figure 18.3b) to the force of attraction between two line masses. The gravity force of attraction (b) between two mass distributions on curves (domains) in the plane correspond in space to forces between hollow (solid) tubes in the normal direction and with the same cross-section. In all cases the forces between the two mass distributions are equal and opposite.

in the case (Figure 18.4b) of two mass distributions the gravity force is given by (18.30b):

$$dm' = \rho' \, d\zeta' : \; F_g^* = \frac{G}{2\pi} \int_{D'} dz \, \rho'(z) \int_{D} d\zeta \, \rho(\zeta) (\zeta - z)^{-1}, \tag{18.30a,b}$$

where $\rho'(z)\,[\rho(\zeta)]$ *is the mass density(18.30a) [(18.29a)] in* D' (D). *In all cases the gravity field (18.26b; 18.27a,b) and force (18.18; 18.29–30a,b) attract toward the mass creating the field,* since:

$$\arg(g) = \arg(F_g) = -\arg(F_g^*) = -\arg\left[(\zeta - z)^{-1}\right] = \arg(\zeta - z), \tag{18.31}$$

is directed from the field observation point z to the position ζ of the mass source.

The three-dimensional law of attraction between point masses is distinct from (18.28) in that: (i) the perimeter of the unit circle 2π is replaced by the area of the unit sphere 4π; (ii) the inverse distance is replaced by the inverse square distance. The combination of (i) and (ii) replaces the inverse of the perimeter of the circle by the inverse of the area of a sphere. In two dimensions the domains in (18.29–18.30a,b) could be replaced by curves as in (18.27a,b). The two-dimensional forces and masses are per unit length orthogonal to the (x,y)-plane. The law of gravity attraction (18.28) shows that the gravitational constant G has the dimensions of force per unit length F'^* times length [viz. $\zeta - z$] divided by mass per unit length square $(1/m'm)$; its value is:

$$G = (6.6732 \pm 31) \times 10^{-11} \, \mathrm{N\,m^2\,kg^{-2}} \left(\equiv \mathrm{m^3\,kg^{-1}\,s^{-2}}\right); \tag{18.32a,b}$$

in m.k.s. units, viz. kilogram kg for mass, meter m for length, seconds for time and Newton N for force; the force is mass times acceleration $N = \mathrm{kg\ m\ s^{-2}}$, leading from (18.32a) to (18.32b).

18.6 Gravity Field of a Homogeneous Slab

A curve C in two dimensions corresponds (Figure 18.3c) to a surface in three-dimensions, for example, the interval $(-a, +a)$ is a flat plane plate of width $2a$ perpendicular to the z-plane

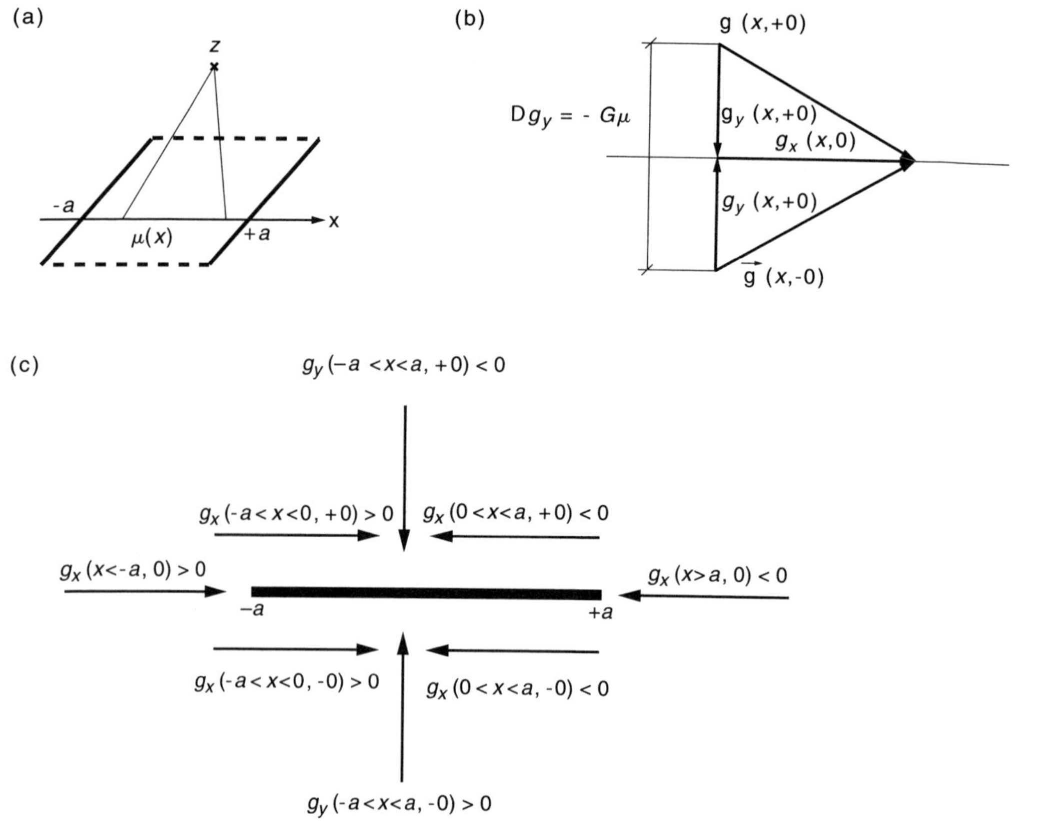

FIGURE 18.5
The gravity field due to (a) a mass distribution of density $\rho(x)$ along a segment $(-a, +a)$ of the real axis in the z-plane corresponds in space (Figure 18.3c) to a strip or an infinite plate with parallel sides normal to the z-plane. The gravity field is always attractive (c) and points: (i) at the edges horizontally toward the plate; (ii) on the sides the horizontal component of the gravity field is continuous across the plate, and points toward the center, where it changes sign, and hence vanishes; (iii) on the sides the vertical component of the gravity fields is downwards above (upwards below), so it changes sign across the plate and is zero there; (iv) both components of the gravity field vanish only at the center of the plate. Comparing the gravity field (b) on each side of the plate: (i) the tangential component is continuous; (ii) the normal component is reversed with a jump equal to the product of the mass density by the gravitational constant; (iii) the angle of the gravity field with the plate is opposite on each side.

(Figure 18.5a). If the mass density μ is constant, the homogenous slab creates (18.27b) a gravity field:

$$g^* \equiv g_x - i g_y = \frac{\mathrm{G}\mu}{2\pi} \int_{-a}^{+a} (\zeta - z)^{-1} \, \mathrm{d}\zeta = \frac{\mathrm{G}\mu}{2\pi} \log\left(\frac{z-a}{z+a}\right); \tag{18.33a}$$

the components parallel g_x (orthogonal g_y) to the plate are:

$$g_x(x,y) = \frac{G\mu}{2\pi}\log\left|\frac{z-a}{z+a}\right| = \frac{G\mu}{4\pi}\log\left|\frac{(x-a)^2+y^2}{(x+a)^2+y^2}\right|, \tag{18.33b}$$

$$g_y(x,y) = -\frac{G\mu}{2\pi}\arg\left(\frac{z-a}{z+a}\right) = \frac{G\mu}{2\pi}\left\{\arg\tan\left(\frac{y}{x+a}\right) - \operatorname{arc}\tan\left(\frac{y}{x-a}\right)\right\}. \tag{18.33c}$$

Thus a *mass distribution of density* $\mu(\zeta)$ *along a curve* C, *creates at an external point* z, *a gravity field (18.27b); in the case of a homogeneous* $\mu \equiv const$ *plate* $(-a,+a)$ *the gravity field (18.33a) has components (18.33b) parallel [(18.33c) orthogonal] to the plate.*

The gravity field calculated before (18.33a–c) applies only to points outside the mass distribution $z \notin C$. To calculate the gravity field due to the mass distribution of density μ along the curve $\zeta \in C$ at a point also on the curve $z \in C$, the integration (18.27b) should exclude z, since a point mass does not exert a force on itself. Thus the path of integration $\zeta \in C-(z-\varepsilon, z+\delta)$ must exclude an interval around z, leading to an improper integral of the second kind; however, the integral is indeterminate (Section 17.8), that is, depends on the shape of the hole $(z-\varepsilon, z+\delta)$ cut in the curve around C. A definite value is obtained by digging a symmetric hole $(z-\varepsilon, z+\varepsilon)$, viz.:

$$z \in C: f_g(z) = \frac{G}{2\pi}\fint_C \mu(\zeta)\log(z-\zeta)\,d\zeta, \quad g^*(z) = \frac{G}{2\pi}\fint_C \mu(\zeta)(\zeta-z)^{-1}\,d\zeta, \tag{18.34a,b}$$

the gravitational potential (18.34a) [gravity field (18.34b)] due to a mass distribution of density μ *along a curve* C, *measured at an internal point* $z \in C$ *is given by the same expression (18.27a) [(18.27b)] as for an external point* $z \notin C$, *replacing the integrals by Cauchy principal values.* For example, in the case of the homogeneous flat plate (18.33a), the gravity field at a point $-a \le x \le +a$ on the plate, is given by (Section 17.8) the Cauchy principal value of the integral:

$$\begin{aligned}
-a \le x \le +a: \quad g_x - i g_y &= \frac{G\mu}{2\pi}\fint_{-a}^{+a}\frac{d\xi}{\xi - x} \\
&= \frac{G\mu}{2\pi}\lim_{\varepsilon\to 0+}\left(\int_{-a}^{x-\varepsilon} + \int_{x+\varepsilon}^{+a}\right)\frac{d\xi}{\xi-x} \\
&= \frac{G\mu}{2\pi}\lim_{\varepsilon\to 0+}\log\left[\frac{(-\varepsilon)(a-x)}{(-a-x)\varepsilon}\right] = \frac{G\mu}{2\pi}\log\left(\frac{a-x}{a+x}\right).
\end{aligned} \tag{18.35}$$

The expressions for the gravity field outside (18.33a) and inside (18.35) the plates are not identical, and the implications are discussed next.

18.7 Gravity Field inside and outside the Mass

The preceding results can be used to show that the gravity field is always one of attraction toward the plate (Subsection 18.7.2), so that the normal component must reverse sign across the plate, and be discontinuous (Subsection 18.7.1).

18.7.1 Tangential Continuity and Normal Discontinuity

Since the argument of the logarithm in (18.35) is real and positive $(a-x)/(a+x)>0$ for $-a<x<a$ the expression is real; thus the gravity field at the plate has zero normal component (18.36a), and the tangential component given by (18.36b):

$$g_y(|x|<a,0)=0, \quad g_x(|x|<a,0)=\frac{G\mu}{2\pi}\log\left(\frac{a-x}{a+x}\right). \tag{18.36a,b}$$

The gravity field (18.33a) just above (below) the plate:

$$g^*(|x|<a,\pm 0)=\frac{G\mu}{2\pi}\log\left(\frac{x-a}{x+a}\right)=\frac{G\mu}{2\pi}\left[\log\left(\frac{a-x}{a+x}\right)\mp i\pi\right], \tag{18.37}$$

has continuous tangential component at the plate:

$$g_x(|x|<a;\pm 0)=\frac{G\mu}{2\pi}\log\left(\frac{a-x}{a+x}\right)=g_x(|x|<a,0), \tag{18.38a,b}$$

in agreement with (18.36b) ≡ (18.38b).

Concerning the component of the gravity field normal to the plate in (18.37) it has opposite signs on both sides:

$$g_y(|x|<a,+0)=-\frac{G\mu}{2}=-g_y(|x|<a,-0), \tag{18.39a,b}$$

that is, it points downward above and upward below, viz. the field is always one of attraction towards the plate. Thus the gravity field normal to the plate is discontinuous, and: (i) its zero value there (18.36a) is the arithmetic mean of the opposite values (18.39a,b) above and below, viz. (18.40a):

$$0=g_y(|x|<a,0)=\frac{g_y(|x|<a,+0)+g_y(|x|<a,-0)}{2}, \tag{18.40a}$$

$$[g_y]\equiv\Delta g_y(|x|<a,0)\equiv g_y(|x|<a,+0)-g_y(|x|<a,-0)=-G\mu; \tag{18.40b}$$

(ii) the jump across the plate agrees with (18.22b) ≡ (18.40b). The limit of the tangential gravity field outside the plate (18.33a) as $y\to 0$ tends to the value (18.36b) on the plate without ambiguity. The normal gravity field outside the plate (18.33c) as $y\to 0$ has the ambiguity $\arctan 0=0,\pm\pi$; this ambiguity is resolved by the approach (18.37) to calculate (18.39a,b). The *continuity of the tangential gravity field at the plate (18.38a,b) and the opposite signs of the normal component above and below (18.39a,b), imply that the total gravity field vectors are mirror images on the plate* (Figure 18.5b).

18.7.2 Attraction on and near the Plate

On the plane of the plate outside it, that is, for $y=0$ and $x>a$ or $x<-a$ the logarithm (18.33a) has real argument so the gravity field is horizontal (18.41a):

$$g_y(|x|>a,0)=0, \quad g_x(|x|>a,0)=\frac{G\mu}{2\pi}\log\left(\frac{x-a}{x+a}\right); \tag{18.41a,b}$$

the horizontal component has opposite signs on the right- and left-hand sides (18.41b), viz:

$$0>g_x(x>a,0)=\frac{G\mu}{2\pi}\log\left(\frac{x-a}{x+a}\right), \quad \frac{G\mu}{2\pi}\log\left(\frac{|x|+a}{|x|-a}\right)=g_x(x<-a,0)>0; \tag{18.42a,b}$$

this is also true (18.36b) of the gravity field on the plate that reverses sign across the origin:

$$0 > g_x(0 < x < a, 0) = \frac{G\mu}{2\pi}\log\left(\frac{a-x}{a+x}\right), \quad \frac{G\mu}{2\pi}\log\left(\frac{a+|x|}{a-|x|}\right) = g_x(-a < x < 0, 0) > 0. \tag{18.42c,d}$$

Thus the gravity field is horizontal in the plane of the plate both inside (18.36a) and outside (18.41a) it; the horizontal component is (18.42a,b) negative (positive) to the right (left) of center:

$$0 > g_x(x > 0, 0) = -g_x(x < 0, 0), \quad g_x(x < 0, 0) > 0, \tag{18.43a,b}$$

that is, the gravity field is always an attraction to the center of the plate $x = 0 = y$.

At the center of the plate the gravity field must reverse its horizontal direction (18.43a,b) and thus the horizontal component must be zero; (18.33a) ≡ (18.44a) shows that at the center of the plate the gravity field:

$$g^*(0 \pm i0) = \frac{G\mu}{2\pi}\log(-1) = \mp\frac{iG\mu}{2}, \quad g_x(0,0) = 0, \quad g_y(0, \pm 0) = \mp\frac{G\mu}{2}, \tag{18.44a–c}$$

is locally vertical (18.44b); it points downward above and upward below (18.44c) at the center, in agreement with (18.39a,b) ≡ (18.44c). The gravity field is discontinuous across the center of the plate, because: (i) the vertical component does change sign (18.39a,b) ≡ (18.44c), either above or below but does not vanish, so is discontinuous; (ii) the horizontal component changes sign (18.43a,b) going continuously through zero (18.44b). The gravity field (18.33a; 18.35) is singular only at $z = \pm a = x$ the edges of the plate; crossing the singularities along the real axis, from outside to inside the plate, the vertical component of the gravity field is discontinuous, jumping from zero outside the plate (18.41a) to nonzero across the edges of the plate (18.39a,b), with different signs above and below. Thus *the gravity field everywhere around the plate points toward the plate (Figure 18.5c) as should be expected of an attractive field.* The gravity field of a number of identical homogeneous parallel plates, with the same mass density, at equal distances is considered in Example 20.18; Example 20.19 concerns two unequal homogeneous plates with the same mass density crossing at any angle at their mid point.

18.8 Field due to a Distribution of Infinite Extent

If the plate has infinite extent the mass density must decay with distance so that the total mass (Subsection 18.8.1) and gravity field (Subsection 18.8.2) are finite; the tangential (normal) continuity (discontinuity) of the gravity field applies both to finite (Subsection 18.7.1) and infinite (18.8.3) plates.

18.8.1 Mass Distribution on an Infinite Plate

A mass distribution on an infinite curve in two dimensions corresponds to a mass distribution on an infinite surface in three-dimensional space, with constant cross-section parallel to z. For example, a mass distribution on the real axis corresponds (Figure 18.3c) in three dimensions to a mass distribution on a plane:

$$g^*(z) = \frac{G}{2\pi} \oint_{-\infty}^{+\infty} \frac{\mu(\zeta)}{\zeta - z} d\zeta, \tag{18.45}$$

where $\mu(\zeta)$ is the mass density per unit length on the real axis, and unit width in the direction perpendicular to the z-plane. The total mass along the real axis must be finite (18.46a):

$$\infty > m \equiv \int_{-\infty}^{+\infty} \mu(x) dx, \qquad \lim_{x \to \pm\infty} \mu(x) = 0, \tag{18.46a,b}$$

and since $\mu(x)$ is positive, the mass density must decay to zero at infinity, in both directions (18.46b). If the latter condition is satisfied for all directions in the upper (lower) complex plane:

$$\operatorname{Im}(\zeta) >< 0: \quad \lim_{\zeta \to \infty} \zeta(\zeta - z)^{-1} \mu(\zeta) = \lim_{\zeta \to \infty} \mu(\zeta) = 0, \tag{18.47}$$

then is also satisfied the asymptotic condition (17.19) that allows the integral (18.45) to be evaluated by residues using (17.18).

The calculation of (18.45) by residues (Section 17.4) was proved assuming that: (iii) the asymptotic condition (17.19) $\equiv$ (18.47) holds; (i) the integral exists, that is, the gravity field is finite and determinate at each point that must be physically true; (ii) the integrand in (18.45) is holomorphic, except for poles, in the upper or lower complex half-plane. If the mass density $\mu(\zeta)$ has the property stated in (ii), so does the integrand in (18.45), at an external point $\operatorname{Im}(z) \neq 0$: in this case the integral (18.45) is improper, bilateral of the first kind, since the mass density $\mu(x)$ must be finite on the axis $|\mu(x)| \leqq M < \infty$ for a continuous distribution. The gravity field at a point z on the plate $\operatorname{Im}(z) = 0$, leads to the appearance of an extra pole at $\zeta = z$ on the real axis and the integral (18.45) becomes improper bilateral of the third kind (Section 17.1); it is uniquely defined if it taken as a Cauchy principal value (Sections 17.8 and 17.9); this is what the calculus of residues does for a pole on the boundary (Section 17.4) by deforming the contour to enclose it. Thus *consider a distribution of mass on the real axis, with density $\mu(x)$ such that: (i) the total mass is finite (18.46a); (ii) the condition (18.47) holds asymptotically in the upper (lower) complex plane; (iii) the function $\mu(\zeta)$ is holomorphic, except for poles, in the upper (lower) complex plane; (iv) the gravity field due to the mass distribution is finite and determinate. Then the gravity field at a point z can be obtained by evaluating the integral (18.45) by residues.*

18.8.2 Calculation of the Gravity Field by Residues

Take as example the mass distribution of density:

$$-\infty < x < +\infty: \quad \mu(x) = \frac{b}{x^2 + a^2}, \tag{18.48}$$

along the real axis. The function $\mu(\zeta)$ satisfies the asymptotic condition (18.47) in all directions that is consistent with the total mass along the real axis being finite:

$$m \equiv \int_{\infty}^{\infty} \mu(x) dx = b \int_{\infty}^{\infty} (x^2 + a^2)^{-1} dx = \frac{\pi b}{a}, \tag{18.49}$$

where (17.21–22a,b) in was used. The function $\mu(\zeta)$ also meets the other condition required to evaluate (18.45) by residues:

$$g^*(z) = \frac{Gb}{2\pi}\int_{\infty}^{\infty}\left(x^2+a^2\right)^{-1}(x-z)^{-1}\,dx, \tag{18.50}$$

namely, $\mu(\zeta)$ is holomorphic in the upper (lower) complex plane, except for a pole at $\zeta = +ia\,(\zeta = -ia)$. To calculate the gravity field, by evaluation of the integral (18.50) by residues, three cases have to be distinguished, depending on where the point z lies: (i/ii) if the gravity field is to be evaluated at a point z below (above) the mass distribution $\operatorname{Im}(z) < 0\,(\operatorname{Im}(z) > 0)$, it is most convenient to close the real axis in the opposite half-plane, that is, the upper (lower), so that the integrand has only one pole in the interior at $\zeta = +ia\,(\zeta = -ia)$, and its residue multiplied by $2\pi i\,(-2\pi i)$ determines the integral (18.50):

$$\operatorname{Im}(z) <> 0: \quad g^*(z) = \frac{Gb}{2\pi}\frac{\pm 2\pi i}{(\pm 2ia)(\pm ia - z)} = \frac{Gb}{2a}\frac{1}{\pm ia - z} = -\frac{Gm}{2\pi}\frac{1}{z \mp ia}; \tag{18.51}$$

(iii) if the gravity field is to be evaluated on the real axis $\operatorname{Im}(z) = 0$, then to (18.51) must be added $\pi i\,(-\pi i)$ times the residue at the simple pole at $x = z$:

$$\operatorname{Im}(z) = 0: \quad g^*(z) = -\frac{Gb/(2a)}{x \mp ia} \pm \pi i\frac{Gb/(2\pi)}{x^2+a^2} = -\frac{Gb}{2a}\frac{x}{x^2+a^2}. \tag{18.52}$$

Thus the gravity field is given: (i) by (18.52) on the mass distribution; (ii) by (18.51) with the *lower* + (*upper*−) sign above (below) it.

18.8.3 Jump and Mean Value of the Normal Gravity Field

The gravity field due to the distribution of mass of density (18.48) on the real axis is given by:

$$g_x - ig_y = g^*(z) = -\frac{Gb}{2a}j^*(z) = -\frac{Gm}{2\pi}j^*(z), \tag{18.53a}$$

where m is the total mass (18.49), and $j^*(z)$ the gravity field per unit mass; the latter is specified outside (18.51) [inside (18.52)] the mass by (18.53b) [(18.53c)]:

$$y <> 0: \; j^*(x+iy) = \frac{1}{x+i(y\mp a)} = \frac{x - i(y\mp a)}{x^2+(y\mp a)^2}, \tag{18.53b}$$

$$y = 0: \; j^*(x) = \frac{x}{x^2+a^2}, \tag{18.53c}$$

above $y > 0$, on $y = 0$ and below $y < 0$ the mass distribution (Table 18.1). It follows that the tangential component of the gravity field j_y is continuous across the mass distribution:

$$g_x(x,\pm 0) = -\frac{Gb}{2a}\frac{x}{x^2+a^2} = -\frac{Gm}{2\pi}\frac{x}{x^2+a^2} = g_x(x,0), \tag{18.54}$$

in agreement with (18.22a); it is nonzero on it, except at origin like for a finite plate (18.44b). The normal component is discontinuous, and takes values above (below) given by:

$$g_y(x,\pm 0) = \mp\frac{Gb}{2}\frac{1}{x^2+a^2} = \mp\frac{G}{2}\mu(x), \tag{18.55}$$

TABLE 18.1
Gravity Field above, below, and in a Mass Distribution

Region	j_x	j_y
$y>0$	$x/\left\{x^2+(y+a)^2\right\}$	$(y+a)/\left\{x^2+(y+a)^2\right\}$
$y=0$	$x/\left(x^2+a^2\right)$	0
$y<0$	$x/\left\{x^2+(y-a)^2\right\}$	$(y-a)/\left\{x^2+(y-a)^2\right\}.$

Note: The mass distribution with density (18.48) on the real axis creates a gravity field (18.53a–c) for which the tangential (normal) component [second (third) column] is continuous (discontinuous). The discontinuous normal component takes on the mass distribution the arithmetic mean of values above and below; since the latter are equal and opposite above (below) the mass distribution, the value on the mass distribution is zero. This corresponds to the change of direction of the normal component of the gravity field (Figure 18.5c) which always points toward the mass distribution, because the latter causes an attraction.

that is, equal and with opposite signs, viz. is always negative above and positive below, so that there is always attraction toward the plate. The implications for the normal component of the gravity field are that (Figure 18.5b): (i) the arithmetic mean is zero that is the value of the field at the mass distribution:

$$\bar{g}_y(x,0) \equiv \frac{g_y(x,+0)+g_y(x,-0)}{2} = 0 = g_y(x,0); \tag{18.56a}$$

(ii) the jump is proportional to minus the product of the gravitational constant by the mass density:

$$[g_y] \equiv \Delta g_y(x,0) \equiv g_y(x,+0) - g_y(x,-0) = -\mathrm{G}\mu(x). \tag{18.56b}$$

The boundary conditions (18.22a,b) at a mass distribution, corresponding to a monopole, are confirmed for: (i) the homogeneous finite plate (18.38a,b; 18.39a,b; 18.40a,b) in Section 18.7; (ii) for an infinite plate (18.54; 18.55; 18.56a,b) with mass density (18.48) in Section 18.8; (iii) also for another mass distribution (20.115) on an infinite plane in Example 20.20. The boundary conditions (18.22a,b) can be generalized from monopoles to multipoles (26.19a,b) and are confirmed for an infinite dipole distribution on a plane (Section 26.9).

18.9 Multipolar Representation of the Gravity Field

Any gravity field has a multipolar representation (Subsection 18.9.1) for example, the (Section 18.6) homogeneous slab (Subsection 18.9.2).

18.9.1 Multipolar Expansion of the Gravity Field

Using the geometric series (30.62a–c):

$$|z| > |\zeta|: \quad (z-\zeta)^{-1} = z^{-1}\left(1-\frac{\zeta}{z}\right)^{-1} = \sum_{n=0}^{\infty} \zeta^n z^{-n-1}, \tag{18.57a,b}$$

in (18.27b) shows that *the gravity field (18.58a) [potential (18.58b)] of an arbitrary mass distribution* $\mu(\zeta)$, *can be represented by a series:*

$$-\frac{2\pi}{G} g^*(z) = z^{-1}\int_C \mu(\zeta)\,d\zeta + \sum_{n=1}^{\infty} z^{-n-1}\int_C \zeta^n \mu(\zeta)\,d\zeta, \tag{18.57c}$$

consisting of: (i) a monopole (12.35a–c); (ii) multipoles (12.59a,b) of all orders:

$$2\pi g^*(z) = -\frac{P_0}{z} - \sum_{n=1}^{\infty} P_n n z^{-n-1}, \qquad 2\pi f(z) = P_0 \log z - \sum_{n=1}^{\infty} P_n z^{-n}; \tag{18.58a,b}$$

the moments are specified by (18.59a) [(18.59b)] for the monopole $n=0$ (multipoles $n = 1,2,\ldots$):

$$P_0 \equiv G\int_C \mu(\zeta)\,d\zeta, \quad P_n \equiv \frac{G}{n}\int_C \zeta^n \mu(\zeta)\,d\zeta. \tag{18.59a,b}$$

Since the gravity field is due to masses that are always positive, the limit process leading to dipoles (Section 12.7), quadrupoles (Section 12.8), and multipoles (Section 12.9) does not apply; this does not prevent the validity general multipole representation (18.58a,b; 18.59a,b).

In the multipole representation (18.58b) of the gravity field (18.58a), the successive moments (18.59a,b) divided by the gravitational constant specify: (i) for the monopole (18.59a) the **total mass** *(18.60):*

$$P_0 = G\int_C \mu(\zeta)\,d\zeta = mG; \tag{18.60}$$

(ii) for the dipole [(18.59b) with n = 1] the mass times (18.61a) the position of the **center of mass** *(18.61b):*

$$P_1 = G\int_C \zeta\mu(\zeta)\,d\zeta = mG\zeta_G, \quad \zeta_G = \frac{\int_C \zeta\mu(\zeta)\,d\zeta}{\int_C \mu(\zeta)\,d\zeta}; \tag{18.61a,b}$$

(iii) for the quadrupole [(18.59b) with n = 2] one-half the mass times (18.62a) the **radius of gyration** *(18.62b):*

$$P_2 = \frac{G}{2}\int_C \zeta^2\mu(\zeta)\,d\zeta = \frac{mG}{2}\zeta_R, \quad \zeta_R \equiv \left|\frac{\int_C \zeta^2\mu(\zeta)\,d\zeta}{\int_C \mu(\zeta)\,d\zeta}\right|^{1/2}; \tag{18.62a,b}$$

(iv) for the multipole of order n in (18.59b) ≡ (18.63a):

$$P_n = \frac{G}{n} m\zeta_n, \qquad \zeta_n \equiv \left|\frac{\int_C \zeta^2\rho(\zeta)\,d\zeta}{\int_C \rho(\zeta)\,d\zeta}\right|^{1/n}, \tag{18.63a,b}$$

the **centroid** *of order n in (18.63b).*

18.9.2 Calculation of the Moments of All Orders

As an example the dipole moments (18.59a,b) of the homogeneous slab $(-a,+a)$ with constant density (18.64a) are (18.64b):

$$\mu = \text{const}: \quad P_n = \mu \frac{\mathrm{G}}{n} \int_{-a}^{+a} \zeta^n \mathrm{d}\zeta = \frac{\mu \mathrm{G}}{n(n+1)} \left[a^{n+1} - (-a)^{n+1} \right]; \tag{18.64a,b}$$

it follows that the multipole moments of odd order vanish (18.65b), because the mass distributions is symmetric:

$$P_0 = 2\mu \mathrm{G} a = m\mathrm{G}: \quad P_{2n+1} = 0, \quad P_{2n} = \frac{\mu \mathrm{G}}{n(2n+1)} a^{2n+1}, \tag{18.65a–c}$$

and only the multipoles of even order remain, with moments (18.65c); the moment of the monopole (18.60) is given by (18.65a). Substitution of (18.65a–c) in (18.58a) specifies the gravity field of the slab:

$$g^*(z) = -\frac{P_0}{2\pi z} - \frac{1}{\pi} \sum_{n=1}^{\infty} P_{2n} n z^{-2n-1} = -\frac{\mathrm{G}\mu}{\pi} \sum_{n=0}^{\infty} (2n+1)^{-1} \left(\frac{a}{z} \right)^{2n+1}, \tag{18.66}$$

as a multipolar series.

The multipolar series can also be obtained from the gravity field (18.33a), viz.:

$$g^*(z) = \frac{\mathrm{G}\mu}{2\pi} \left[\log(z-a) - \log(z+a) \right], \tag{18.67}$$

using the power series for the logarithm (21.64). The latter follows from arithmetic series like (18.57), viz.:

$$(z \mp a)^{-1} = z^{-1} \left(1 \mp \frac{a}{z} \right)^{-1} = \sum_{n=0}^{\infty} (\pm)^n a^n z^{-n-1}, \tag{18.68}$$

by integration:

$$\log(z \mp a) = \log z + \sum_{n=1}^{\infty} (\pm)^n (-n)^{-1} \left(\frac{a}{z} \right)^n. \tag{18.69}$$

Subtracting the two preceding expressions yields:

$$\log\left(\frac{z-a}{z+a} \right) = -2 \sum_{n=0}^{\infty} (2n+1)^{-1} \left(\frac{a}{z} \right)^{2n+1}; \tag{18.70}$$

substitution of (18.70) in (18.67) confirms (18.66).

Note 18.1. *Conventions for Potentials and Fields*: The definition of the gravity (18.18b) [velocity (18.1b)] potential by equating minus (plus) its gradient to the gravity (velocity) vector follows the usual practice in gravity theory (fluid mechanics); since the complex potential (18.26a) is related in the same way to the scalar potential and field function, it follows that the complex conjugate gravity field (18.26b) [velocity (12.21a)] is minus (plus) the derivative of the complex potential. If the gravity field vector had been chosen as plus the gradient of the potential in (18.18b), then the complex conjugate gravity field should be plus the derivative of the complex potential in (18.26b). Thus the choice of sign of the potential affects neither the gravity field nor the force expressed in terms of the mass distribution. The comparison of the complex gravity potential (18.26a) with the complex velocity potential (12.26a), bearing in mind their opposite signs, implies that a mass $m > 0$ corresponds to a sink $Q < 0$, in agreement with (18.20a,b), as can be confirmed comparing the complex conjugate velocity (12.25b) and gravity field (18.26b). This agrees with the fundamental equations involving the dilatation (18.2a) and the mass density (18.19a). Thus the complex gravity (velocity) potential of a monopole or multipole have opposite signs, but the complex conjugate velocity (gravity) have the same sign.

Note 18.2. *Multipolar Expansion in Any Spatial Dimension and Including Time:* The multipolar expansion in the plane applies not only to the gravity field (Section 18.9) but also to other potential fields, for example, electrostatics (Chapter 24) magnetostatics (Chapter 26), potential flow (Chapter 28) and steady heat conduction (Chapter 32), and any potential field satisfying the Poisson equation. For the same potential fields it extends from the plane to space and higher dimensions involving the original and hyperspherical Legendre polynomials. These multipolar expansions for the Poisson or forced Laplace equations, extend to the forced wave equation; the finite propagation speed then introduces a retarded time, or time delay between emission by the source and reception by the observer. Thus the multipolar expansion extends from the plane to space and to space-time.

Note 18.3. *The Classical Gravity Field and the Relativistic Gravitational Field:* The gravity field has been considered in a classical case, in the sense that time is explicitly absent, that is, acts as an independent parameter that does not affect spatial properties. This is still true in restricted relativity, with differences in the dynamics of a particle, for example, in the gravity field, but only for velocities close to that of light. General relativity brings a more fundamental and surprising change: there is no "external gravity field", the gravity field causes an acceleration, and leads to a nonlinear relation between position and time. This corresponds to a curved space-time, due to the local acceleration determined by gravity, and hence by the global mass distribution. Thus the distribution of mass determines the curvature of space-time. The "gravity field" is the local manifestation of the global gravitational field: the latter causes space-time to become curved, and locally this can be replaced by the flat tangent space-time of restricted relativity plus a local acceleration due to the "external gravity field." Thus gravitation is connected at fundamental level with curved geometry in four space-time dimensions.

Conclusion 18: A distribution (figure 18.2) of sources/sinks (**a**) [vortices (**b**)] along (figure 18.1) a curve (**a**) causes a jump in normal (tangential) velocity (**b**); likewise a mass distribution along a curve (**a**) leads to a continuous tangential (discontinuous normal) gravity field. A point in the (x, y)-plane is (figure 18.3) a line mass in three-dimensions (**b**), and a curve in the plane is a surface in three-dimensions (**c**). Thus the force of attraction (figure 18.4) between two points (**a**) [domains (**b**)] in the plane, is in three dimensions the force per unit length between two line-masses (cylinders). A mass distribution (figure 18.5) along a segment the real axis (**a**) corresponds to a plate in space. Across a mass distribution (**c**) the tangential (normal) gravity field is continuous (has a jump proportional to the mass density and gravitational constant) so that the total vector gravity field is a mirror reflection on the two sides (b). Since the gravity field is attractive, in the case of a homogeneous plate (**c**) it points everywhere towards it.

19

Cauchy Conditions and Infinitesimals

The proof of Cauchy's first theorem (1821) given before (Section 13.6) uses the concept of primitive and the Cauchy–Riemann conditions, and makes only indirect use of the definitions of holomorphic function (Section 19.2) and Riemann integral (Sections 19.1). It is possible to give a totally distinct proof of the same theorem (Goursat, 1900) that uses directly the fundamental concepts of derivate and integral, without other intermediaries (Sections 19.3 and 19.4). This proof has the advantage of allowing two extensions, that is, the condition that the function be holomorphic in a closed region for Cauchy's theorem to apply, can be relaxed (Section 19.5) on the boundary to uniform continuity (Goursat, 1900); then (Section 19.6) isolated singularities of order less than a pole are also allowed (Littlewood, 1944) on the boundary. These conditions were called Cauchy conditions (Section 15.1), since they are sufficient conditions for Cauchy's theorem to hold; it can be shown that they are minimal conditions, since violation of any of them will cause the theorem to fail. Obtaining minimal conditions for Cauchy's theorem to hold is important, since it extends the range of cases in which the many useful consequences of this theorem, for example, calculus of residues (part 2) and power series (part 3) can be applied; the notion of singularity with order less than a pole, is an instance of the indication of order of an infinitesimal (Section 19.7). This notion is useful in resolving indeterminate limits of the type 0/0, for example, by L' Hôspital rule (Section 19.8). The latter provide the method III of calculation of the residues of certain classes of functions at poles (Section 19.9). The infinitesimals, are "smaller than anything else, but not zero, and could only be an invention of the devil," according to bishop Berkeley, in his critique of Newton's method of fluxions (1670). In science rather than theology they are the basis of the calculus and a convenient way to calculate limits, for example, a velocity, a direction, or a density, when they exist; and otherwise, to prove that a tentative concept is zero or infinity.

19.1 Calculation of Riemann Integrals Using the Definition

The definition (13.8a,b) of Riemann integral is applied most simply to the unit function $f(z)=1$ in the interval $(z_0 \equiv a, z_N \equiv b)$ for an arbitrary partition $|z_{n+1}-z_n|<\delta$:

$$\int_a^b \mathrm{d}z = \lim_{\delta\to 0}\sum_{n=0}^{N-1}(z_{n+1}-z_n) = z_N - z_0 = a-b = z\big|_a^b, \tag{19.1}$$

that shows that the primitive of 1 is z. The integral of the latter can also be evaluated simply from the definition:

$$\int_a^b z\,\mathrm{d}z = \lim_{\delta\to 0}\sum_{n=0}^{N-1}\frac{1}{2}(z_n+z_{n+1})(z_{n+1}-z_n) = \lim_{\delta\to 0}\sum_{n=0}^{N-1}\frac{1}{2}\left(z^2_{n+1}-z^2_n\right) = \frac{1}{2}\left(b^2-a^2\right) = \frac{1}{2}\,z^2\big|_a^b, \tag{19.2}$$

and its primitive is $z^2/2$. The evaluation of integrals from the definition is usually more tedious than the use of the primitive. Letting $b \to a$ in (19.1, 19.2) leads to a loop ∂D, and proves

$$\oint_{\partial D} \mathrm{d}z = 0 = \oint_{\partial D} z\,\mathrm{d}z, \tag{19.3a,b}$$

that are particular cases of the Cauchy first theorem (15.1), for the functions $f(z) = 1$ and $f(z) = z$.

19.2 Mean Value Theorem and Bounds

To proceed to prove more general cases of Cauchy's theorem in a closed region D, consider a holomorphic function that satisfies (11.13b) in a neighborhood of every point $z_0 \in D$:

$$f'(z_0) - \frac{f(z) - f(z_0)}{z - z_0} \equiv -\varepsilon(z_0, z), \qquad \lim_{z \to z_0} \varepsilon(z_0, z) = 0. \tag{19.4a,b}$$

The expression (19.4a) corresponds to the **mean value theorem:**

$$f(z) - f(z_0) = (z - z_0) f'(z_0) + (z - z_0)\,\varepsilon(z_0, z); \tag{19.5}$$

the difference of a holomorphic function at two neighboring points is proportional to the relative position multiplied by the derivate (19.5), with an error that is an infinitesimal of higher order (19.4b). Integrating (19.5) along the boundary ∂D of D, and using (19.3a,b), leads to:

$$\begin{aligned}\oint_{\partial D} f(z)\,\mathrm{d}z &= \{f(z_0) - z_0 f'(z_0)\} \oint_{\partial D} \mathrm{d}z + f'(z_0) \oint_{\partial D} z\,\mathrm{d}z \\ &\quad + \oint_{\partial D} \varepsilon(z_0, z)(z - z_0)\,\mathrm{d}z = \oint_{\partial D} (z - z_0)\,\varepsilon(z_0, z)\,\mathrm{d}z,\end{aligned} \tag{19.6}$$

showing that only the higher order infinitesimal remains; this suggests that the integral will be ultimately zero. As an intermediate step (19.6) implies

$$z_0 \in D - \partial D, \quad z \in \partial D: \qquad \left| \oint_{\partial D} f(z)\,\mathrm{d}z \right| \le |z - z_0|_{\max} L\, |\varepsilon(z_0, z)|_{\max}, \tag{19.7}$$

that specifies *an upper bound (19.7) for the modulus of the integral of a holomorphic function along a loop* ∂D *of length* L*, where* $|z - z_0|_{\max}$ *is the maximum distance between a fixed point* z_0 *in the interior and a variable point* z *on the boundary, and* $\varepsilon(z_0, z)$ *is the infinitesimal in (19.4a,b).*

19.3 Division into Internal and Boundary Regions

The region may be divided D into a network (Figure 19.1a) of subregions, by considering two families of straight lines parallel to each coordinate axis and intersecting the boundary. Thus

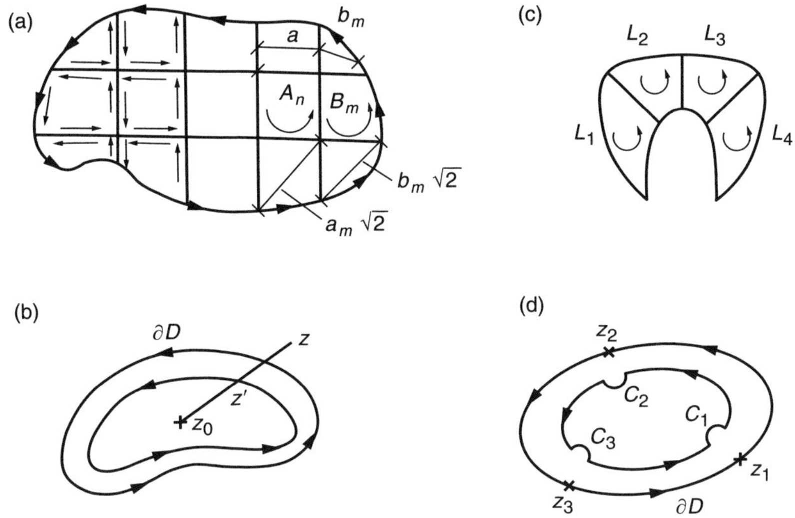

FIGURE 19.1
Division of a domain by a net of horizontal and vertical lines (a) creating: (i) internal cells; (ii) cells with boundary points. Choosing the same direction around all cells (counter clockwise is shown) the sum of all loop integrals equals the loop integral along the boundary, because all inner lines are taken twice in opposite directions. A loop is star-like (b) if there exists at least one central point z_0 in the interior from which all points an the boundary loop can be "seen," that is, a straight line through the central point in any direction crosses the loop only at one point; a star-like loop can be shrunk homotetically $[z \to z' \to z_0$ in (19.17a) as $1 \to \lambda \to 0]$ to any central point. If the loop is not star-like (c), its interior may divided into star-like regions, whose whole boundary is "visible" from at least one central point in each region; taking the same direction in each region as in (a) the sum of all loop integrals is again a loop integral along the boundary. The singularities of a function on the boundary may be excluded by half-circle indentations (Figure 15.1c); if the singularities are ignorable, that is, of order n satisfying $0 > n > -1$, they do not affect (d) the loop integral of the function along the boundary, regardless of whether they enclosed or excluded.

the region D is subdivided into: (i) N interior regions a_n, with $n = 1, \ldots, N$, of rectangular shape, with larger side a_n for which (19.8c) holds:

$$L \le 4a_n; |z - z_0|_{\max} \le a_n\sqrt{2}: \quad \left|\oint_{A_n} f(z)\,\mathrm{d}z\right| \le 4a_n\sqrt{2}a_n\,|\varepsilon_n|_{\max}, \tag{19.8a–c}$$

since: $(\mathrm{i}-1)$ in (19.7) the perimeter cannot exceed (19.8a); $(\mathrm{i}-2)$ the maximum distance between two points is the length of the diagonal between corners that cannot exceed (19.8b); (ii) M boundary regions B_m, with $m = 1, \ldots, M$, each consisting of straight sides of length not exceeding b_m, plus a part of the boundary ∂D, of length L_m, for which holds (19.9c):

$$L \le 3b_m + L_m; |z - z_0|_{\max} \le b_m\sqrt{2}: \quad \left|\oint_{B_m} f(z)\,\mathrm{d}z\right| \le (3b_m + L_m)\sqrt{2}b_m\,|\delta_m|_{\max}, \tag{19.9a–c}$$

since: (ii-1) in (19.7) there are at most three straight sides of length not exceeding b_m, besides the part of the boundary curve L_m; (ii-2) provided that the latter be contained within a square of sides b_m (this is always possible by using, if necessary, a finer subdivision), the maximum distance between two points cannot exceed (19.9b). Denoting by c the largest of the lengths of sides (19.10a), and by ε the largest of the infinitesimals (19.10b):

$$c \equiv \sup(b_m), \quad \varepsilon \equiv \sup(\varepsilon_n, \delta_m), \tag{19.10a,b}$$

an upper bound for the sum of all moduli of integrals (19.8c; 19.9c) is:

$$\sum_{n=1}^{N} \left| \int_{A_n} f(z)\,\mathrm{d}z \right| + \sum_{m=1}^{M} \left| \oint_{B_m} f(z)\,\mathrm{d}z \right|$$
$$\le 4\sqrt{2}\varepsilon \left(\sum_{m=1}^{N} a_n^2 + \sum_{m=1}^{M} b_m^2 \right) + \varepsilon\sqrt{2}c \sum_{m=1}^{M} L_m \le \varepsilon\sqrt{2}(4A + cL), \tag{19.11}$$

where A is the area of polygon made of squares and containing the region D, and L is the total length of the boundary loop.

19.4 Function Holomorphic in the Interior and on the Boundary

If the boundary loops of each internal A_n and boundary B_m region are taken in the same direction, for example, counter clockwise in Figure 19.1a, then: (i) each straight segment in the interior is taken twice in opposite directions, and the corresponding integrals cancel; (ii) the integrals along the boundary arcs L_m are taken all in the same direction, and add up to the loop integral around the boundary:

$$\oint_{\partial D} f(z)\,\mathrm{d}z = \sum_{n=1}^{N} \oint_{A_n} f(z)\,\mathrm{d}z + \sum_{n=1}^{N} \oint_{B_m} f(z)\,\mathrm{d}z; \tag{19.12}$$

an upper bound to the modulus of the latter is given:

$$\left| \oint_{\partial D} f(z)\,\mathrm{d}z \right| \le \sum_{n=1}^{N} \left| \oint_{A_n} f(z)\,\mathrm{d}z \right| + \sum_{m=1}^{N} \left| \oint_{B_m} f(z)\,\mathrm{d}z \right|; \tag{19.13}$$

using (19.11) leads to (19.14b):

$$M \equiv (4A + cL)\sqrt{2}: \qquad \left| \oint_{\partial D} f(z)\,\mathrm{d}z \right| \le \lim_{\varepsilon \to 0} \varepsilon M = 0, \tag{19.14a,b}$$

where (19.14a) is a finite number, and ε an infinitesimal as the largest subregion tends to zero. Thus has been proved:

$$f(z) \in \mathcal{D}(D): \qquad \oint_{\partial D} f(z)\,\mathrm{d}z = 0, \tag{19.15a,b}$$

Cauchy's first theorem that *for a holomorphic function in a closed region (19.15a), the loop integral along the boundary* ∂D *is zero (19.15b).*

19.5 Uniform Continuity on the Boundary (Goursat, 1900)

It is shown next that *Cauchy's first theorem (19.15b) ≡ (19.16c) remains valid if the function is uniformly continuous on the boundary (19.16b), and holomorphic in the interior (19.16a):*

$$f(z) \in \mathcal{D}(D - \partial D) \quad f(z) \in \tilde{C}(\partial D): \qquad \oint_{\partial D} f(z)\,\mathrm{d}z = 0. \tag{19.16a–c}$$

Consider first the case in that the boundary is a **star-like loop** $L \equiv \partial D$ defined by the property that (Figure 19.1b) an interior point z_0 exists, such that any straight line passing through z_0 intersects the boundary L only at one point. The point z_0 can be called **central point,** since from it the whole of the boundary is within the line of sight; from a noncentral point some parts of the boundary would stand in the line of sight of others, that is, a straight line would intersect the boundary at more than one point. If *a region* D *is convex all internal points are central points, whereas a concave region has a subset of central points, or even none. It is always possible to decompose a concave region in the sum of several convex ones, or nonconvex ones with at least one central point.*

If z_0 is a central point for the region within the loop L, then:

$$0 < \lambda < 1: \quad z' \equiv z_0 + \lambda(z - z_0), \quad \lim_{\lambda \to 1} z' = z, \tag{19.17a–c}$$

when $z \in \partial D \equiv L$ lies on the boundary, $z' \in L'$ lies on a homothetic loop L' contained (Figure 19.1b) within the interior $D - \partial D$ of D. Hence the function $f(z)$ is holomorphic on L' and within it, so that the Cauchy theorem in original form is valid:

$$0 = \oint_{L'} f(z')\,\mathrm{d}z' = \lambda \oint_{L} f(z_0 + \lambda(z - z_0))\,\mathrm{d}z; \tag{19.18}$$

since by hypothesis, the function is uniformly continuous as $z' \to z$, the limit (19.17c) or $\lambda \to 1$ can be taken (Section 13.4) under integral sign and (19.18) yields (19.16c). If the region D does not have a central point, then it can be subdivided by a finite number $N - l$ of straight lines into N regions (Figure 19.1c), with one central point each, from which the whole boundary L_n can be seen. The integral along the loop equals as before the sum of the integrals along the boundaries of each subregion:

$$\oint_{\partial D} f(z)\,\mathrm{d}z = \sum_{n=1}^{N} \oint_{L_m} f(z)\,\mathrm{d}z = 0, \tag{19.19}$$

and since the latter vanish, (19.16c) is regained. QED.

19.6 Isolated Ignorable Singularities on the Boundary (Littlewood, 1944)

It is shown next that *Cauchy's first theorem remains valid the function* f(z) *has singularities* z_0 *on the boundary, provided that they are of order lower than unity (19.20a) ≡ (19.20b):*

$$f(z) = o\left[(z - z_n)^{-1}\right] \quad \Leftrightarrow \quad \lim_{z \to z_n} (z - z_n) f(z) = 0. \tag{19.20a,b}$$

The integral of the function along an infinitesimal neighborhood of z_0 is zero (19.21b):

$$M \equiv \sup\left\{1+\frac{\varepsilon}{\delta};1+\frac{\delta}{\varepsilon}\right\}: \lim_{\varepsilon,z\to 0+}\int_{z_n-\varepsilon}^{z_n+\delta} f(z)\,\mathrm{d}z = \lim_{\varepsilon,\delta\to 0+}\frac{1}{2}\{f(z_n+\delta)+f(z_n-\varepsilon)\}(\varepsilon+\delta)$$
$$\le M \lim_{\varepsilon,\delta\to 0+}\{\delta f(z_n+\delta)+\varepsilon f(z_n+\varepsilon)\}=0, \qquad (19.21a,b)$$

since M in (19.21a) is bounded. *A singularity of the type (19.20a) ≡ (19.20b) is thus called an* **ignorable singularity**, *since adding it to a function does not change the value of its integral.* It follows that Cauchy's theorem should remain valid if the function $f(z)$ has a finite number N of singularities z_n with $n = 1,\ldots,N$, on the boundary. The singularities are **isolated**, that is, a neighborhood exits for each singularity that contains no other singularity viz.:

$$\varepsilon < \frac{1}{2}|z_n - z_m|_{\min}: \qquad \forall_{m\neq n}: \quad z_m \notin V_\varepsilon(z_n). \qquad (19.22)$$

The boundary ∂D can be deformed so as to exclude the singularities z_n by half-circles C_n of radius ε_n (Figure 19.1d), leading to a loop L, that differs from ∂D:

$$\oint_{\partial D} f(z)\,\mathrm{d}z - \int_L f(z)\,\mathrm{d}z = \sum_{n=1}^{N}\left\{\int_{z_n-\varepsilon_n}^{z_n+\varepsilon_n} f(z)\,\mathrm{d}z - \int_{C_n} f(z)\,\mathrm{d}z\right\}, \qquad (19.23)$$

only by integrals along paths lying in the neighborhoods of the singular points; these integrals vanish as $\varepsilon_n \to 0$ for all n, that is,

$$\varepsilon \equiv |\varepsilon_n|_{\max}: \qquad \lim_{\varepsilon\to 0}\oint_{\partial D} f(z)\,\mathrm{d}z = \oint_L f(z)\,\mathrm{d}z = 0, \qquad (19.24)$$

where the integral along the loop L is zero because it meets the conditions stated before in Section 19.5. Thus has been proved the result (stated in Section 15.1) concerning the **Cauchy conditions** *that are minimal conditions for Cauchy's theorem to hold: (i) the function* f(z) *is holomorphic in the interior* D − ∂D *of* D; *(ii) is uniformly continuous on the boundary loop* ∂D; *(iii) except at finite number* z_0 *of ignorable singularities (19.20a,b); then the Cauchy first theorem (19.16c) stating that the integral along the boundary is zero holds.* The Cauchy conditions indicated above, assure the validity of Cauchy theorem, and bear no further significant extension, since: (i) if the singularity (19.20a) is of higher order, that is, of order unity, then it could be a simple pole (15.23a,b), in which case (19.24) no longer holds if the path is deformed so as to enclose the pole (15.24a), because the residue is nonzero; (ii) if the function is not uniformly continuous on the boundary, then there may be no extension (19.18) from the interior, where it is holomorphic.

19.7 Infinitesimals of the Same or Higher Order

If two functions $f(z)$, $g(z)$ vanish at point $z = a$, then by taking the limit of their ratio $f(z)/g(z)$, three cases can arise:

$$f(a)=0=g(a): \quad \lim_{z\to a}\frac{f(z)}{g(z)}\begin{cases} =0: & f(z)=o(g(z)), & (19.25a)\\ \neq 0,\infty: & f(z)=O(g(z)), & (19.25b)\\ =\infty: & g(z)=o(f(z)), & (19.25c)\end{cases}$$

where the **Landau symbol** $o(O)$ means **equivalent (higher) order**. The same "O" (higher "o") order relation for infinitesimals (19.25a–c) is a strict order relation (1.4), since *given two functions continuous at a given point:*

$$f,g \in C: \quad f(z) = O(g(z)) \ \dot{\vee} \ f(z) = o(g(z)) \ \dot{\vee} \ g(z) = o(f(z)), \tag{19.26a–c}$$

either they are of the same order or one is of higher order than the other; the three possibilities are mutually exclusive.

It is often convenient to choose for comparison the power function $g(z) = (z-a)^n$, viz.:

$$f(z) = O((z-a)^n) \quad \Leftrightarrow \quad \lim_{z \to a} (z-a)^{-n} f(z) = M \neq 0, \infty, \tag{19.27a}$$

$$f(z) = o((z-a)^n) \quad \Leftrightarrow \quad \lim_{z \to a} (z-a)^{-n} f(z) = 0. \tag{19.27b}$$

Thus *increasing the real part of the exponent increases the order of the infinitesimal:*

$$\operatorname{Re}(\alpha) > 0: \quad O\left((z-a)^{n+\alpha}\right) = o((z-a)^n); \tag{19.28a}$$

this can be proved using (19.27a):

$$\begin{aligned} f(z) = O\left((z-a)^{n+\alpha}\right): \quad &\lim_{z \to a} (z-a)^{-n} f(z) \\ = \lim_{z \to a} (z-a)^{-n-\alpha} f(z)(z-a)^\alpha &= M \lim_{z \to a} (z-a)^\alpha = 0, \end{aligned} \tag{19.28b}$$

where (19.28b) implies (19.27b).

In (19.28a,b) only the real part of the exponent matters (Section 5.8), and it can be positive or negative. The examples which appeared before include (i) in (5.14b) it implies $\sin\varphi = O(\varphi)$ that is confirmed by (8.11a) which also shows that $\sin\varphi - \varphi = O(\varphi^3)$; (ii) at (19.20b) an integrable singularity $f(z) = o((z-a)^{-1})$ the order is lower (19.20a) than at a simple pole, and $f(z) = O((z-a)^{-1})$ for a simple pole (15.23a,b), and $f(z) = O((z-a)^{-n})$ for a multiple pole of order n (15.32a,b); (iii) the asymptotic conditions for the evaluation of real improper integrals by residues are $f(z) = o(z^{-1})$ in (17.19), $g(z) = o(1)$ in (17.29) and $h(z) = o(z)$ in (17.43c), as $z \to \infty$. For example $g(z) = O(1)$ implies $g(0) \neq 0, \infty$, whereas $g(z) = o(1)$ implies $g(0) = 0$ as $z \to 0$.

19.8 Zero of Order *n* and L'Hôspital's (1696; Bernoulli, 1691) Rule

From (19.5) it follows that if a function $f(z)$ is holomorphic in a neighborhood of a, and is zero at a with nonzero derivate, then it is an infinitesimal of order one at a:

$$f(a) \neq 0 \neq f'(a): \quad f(z) = f(a) + f'(a)(z-a) + o(z-a) = O(z-a); \tag{19.29}$$

setting $f(z) \equiv g'(z)$ leads to

$$g'(a) = 0 \neq g''(a): \qquad g'(z) = g''(a)(z-a) + o(z-a); \tag{19.30}$$

after an integration, and setting $f(z) \equiv g(z)$:

$$f(a) = f'(a) = 0 \neq f''(a): \qquad f(z) = \frac{1}{2}(z-a)^2 f''(a) + o\left((z-a)^2\right) = O\left((z-a)^2\right). \tag{19.31}$$

Similarly, by induction, *in the neighborhood of a* **zero of order n**, *where the function and its derivates of order up to* n−1 *vanish, but the derivate of order* n *is nonzero, the function is an infinitesimal of order* n:

$$f(a) = \cdots = f^{(n-1)}(a) = 0 \neq f^{(n)}(a): \quad f(z) = \frac{(z-a)^n}{n!} f^{(n)}(a) + o((z-a)^n) = O((z-a)^n), \tag{19.32}$$

and the inverse of the function $1/\mathrm{f}(\mathrm{z}) = \mathrm{O}\left((\mathrm{z}-\mathrm{a})^{-\mathrm{n}}\right)$ *has a pole of order* n.

If the functions $f(z), g(z)$ both have a zero of order n at $z = a$:

$$\{f(z), g(z)\} = \frac{(z-a)^n}{n!}\left\{f^{(n)}(a), g^{(n)}(a)\right\} + o(1), \tag{19.33}$$

the limit as $z \to a$ of their ratio is given by the ratio of n-th derivates:

$$\lim_{z\to a} \frac{f(z)}{g(z)} = \lim_{z\to a} \frac{f^{(n)}(a) + o(1)}{g^{(n)}(a) + o(1)} = \frac{f^{(n)}(a)}{g^{(n)}(a)}. \tag{19.34}$$

The result (19.34) applies not only to infinitesimals of the same order, but also to infinitesimals of different orders, viz. $n(m)$ for $f(z)\,[g(z)]$, because: (i) if $n > m$ after m differentiations in the numerator and the denominator, the result is zero; (ii) if $n < m$ after n differentiations in the numerator and denominator, the result is infinity. In conclusion is obtained **L'Hôspital's rule** *(1696, due to Bernoulli, 1691)*:

$$\lim_{z\to a} \frac{f(z)}{g(z)} = \lim_{z\to a} \frac{f'(z)}{g'(z)} = \lim_{z\to a} \frac{f^{(n)}(z)}{g^{(n)}(z)}, \tag{19.35}$$

stating that the ratio of two infinitesimals, viz. holomorphic functions vanishing at z = a, *is not changed if the denominator and numerator are differentiated separately, any equal number of times.* The L'Hôspital's rule succeeds in lifting 0/0 type indeterminacy, when at least one of the functions is an infinitesimal of positive integer order.

Applying the rule (19.35) to the integral of a function which is an infinitesimal of order n:

$$\frac{\int_a^z f(\zeta)\,d\zeta}{(z-a)^{n+1}/(n+1)} = \frac{f(z)}{(z-a)^n} = \frac{f'(z)}{n(z-a)^{n-1}}, \tag{19.36}$$

it follows that *if the function* f(z) *is an infinitesimal of order* n *(order higher than* n*), then the order of its integral (primitive) is increased (decreased) by unity as shown in Table 19.1.* The results in Table 19.1 show that order of an infinitesimal can be integrated and differentiated. Some examples of L'Hôspital's rule (19.35) are

$$\lim_{z\to 0} \frac{\sin z}{z} = \lim_{z\to 0} \frac{\cos z}{1} = 1; \ \sin z = z + o(z); \tag{19.37a,b}$$

$$\lim_{z\to 0} \frac{\sin z - z}{z^3} = \lim_{z\to 0} \frac{\cos z - 1}{3z^2} = \lim_{z\to 0} -\frac{\sin z}{6z} = -\frac{1}{6}; \tag{19.38}$$

these lead to the leading terms of the series expansion for $\sin z$ in (19.39a) to $O\left(z^3\right)$:

$$\sin z = z - \frac{z^3}{6} + o\left(z^3\right); \quad \cos z = 1 - \frac{z^2}{2} - \frac{z^4}{24} + o\left(z^4\right), \tag{19.39a,b}$$

integrating between 0 and z yields the expansion (19.39b) of $\cos z$ to $\mathrm{O}(z^4)$, bearing in mind that $\cos 0 = 1$.

$$\cos z = \cos 0 - \int_0^z \sin u\,du = 1 - \int_0^z \left[u - \frac{u^3}{6} + 0(u^3)\right] du = 1 - \frac{z^2}{2} + \frac{z^4}{24} + O(z^4), \tag{19.39c}$$

The result (19.39a) proves analytically (5.14b) and (8.11a); also (19.39b) proves (8.11b).

TABLE 19.1
Infinitesimal Order of a Function, and Its Derivatives and Primitives

Function	**Order equal to**	**Order higher than**
$f(z)$	$O((z-a)^n)$	$o((z-a)^n)$
$f'(z)$	$O\left((z-a)^{n-1}\right)$	$o\left((z-a)^{n-1}\right)$
$f^{(m)}(z)$ with $m=0,\ldots,n$	$O\left((z-a)^{n-m}\right)$	$o\left((z-a)^{n-m}\right)$
$\int_a^z f(\zeta)\,d\zeta \equiv P\{f(\zeta)\}$	$O\left((z-a)^{n+1}\right)$	$o\left((z-a)^{n+1}\right)$
$P^k\{f(z)\}$ with $k=0,\ldots,\infty$	$O\left((z-a)^{n+k}\right)$	$o\left((z-a)^{n+k}\right)$

Note: If it is known that a function is an infinitesimal of order n (order higher than n) at a point, then the order is decreased (increased) by unity by each differentiation (integration).

19.9 Calculation of the Residues of Ratios of Functions

The L'Hôspital' rule provides the method III of calculation of residues at simple (Section 15.7; Subsection 19.9.1) [multiple (Section 15.8; Subsection 19.9.2)] poles.

19.9.1 Calculation of the Residue at a Simple Pole (Method III)

The ratio of a nonzero (19.40b) function $h(\zeta)$ *to a function* $g(\zeta)$ *having a simple zero (19.40a) at* $\zeta = z$ *is a function* $f(\zeta)$ *having a simple pole (19.40c) at* $\zeta = z$*:*

$$g(z) = 0 \neq g'(z), \quad h(z) \neq 0: \quad f(\zeta) \equiv \frac{h(\zeta)}{g(\zeta)} = \frac{(\zeta - z)^{-1} h(\zeta)}{g'(z) + o(1)}, \tag{19.40a–c}$$

whose residue:

$$f_{(1)}(z) = \lim_{\zeta \to z} (\zeta - z) f(\zeta) - \frac{h(z)}{g'(z)}, \tag{19.41}$$

is the ratio of the numerator to the derivate of the denominator at the pole. This method of calculation of the residues (19.42b) at simple poles can be applied to rational functions (19.42a) for example,

$$f(\zeta) = \left\{\zeta^2 - (a+b)\zeta + ab\right\}^{-1}, \quad f_{(1)}(a) = (2\zeta - a - b)^{-1}|_{\zeta=a} = \frac{1}{a-b}; \tag{19.42a,b}$$

this agrees with the result (15.26a) of applying the definition of residue (Subsection 15.7.1), and dispenses with factorizing the polynomials. This method becomes more convenient as the degree of the polynomial increases, for example,

$$f(\zeta) = \left\{(\zeta^2 + a^2)^{-1} (\zeta^2 + b^2)\right\}^{-1} = \left\{\zeta^4 + (a^2 + b^2)\zeta^2 + a^2 b^2\right\}^{-1}, \tag{19.43}$$

has simple poles at $\zeta = \pm ia, \pm ib$, with residues:

$$f_{(1)}(\pm ia) = \left\{4\zeta^3 + 2\left(a^2 + b^2\right)\zeta\right\}^{-1}|_{z=\pm ia} = \left\{\pm 2ia\left(b^2 - a^2\right)\right\}^{-1}, \tag{19.44}$$

in agreement with (17.25a).

The rule (19.40a–c; 19.41) is most useful for the calculation of residues of transcendental functions. For example, the sine has simple zeros $\sin z = 0$ at $z = n\pi$ with integer $n \in |\mathrm{N}$ so that they are simple poles of the circular cosecant $\csc z \equiv 1/\sin z$. The residues of $\csc z$ at its simple poles, $z_n = n\pi$, can be calculated by the definition (15.24b):

$$\begin{aligned} w = \zeta - n\pi: \quad b_n &\equiv \lim_{z \to n\pi} (\zeta - n\pi)\csc(\zeta) = \lim_{w \to 0} \frac{w}{\sin(w + n\pi)} \\ &= (-)^n \lim_{w \to 0} \frac{w}{\sin w} = (-)^n, \end{aligned} \tag{19.45}$$

where (19.37a) was used. The rule (20.40a–c; 20.41) allows the residues of $\csc z$ at the poles $z = n\pi$ to be determinated as follows:

$$f(\zeta) \equiv \csc\zeta \equiv \frac{1}{\sin\zeta}, \qquad b_n \equiv f_{(1)}(n\pi) = \left\{\cos\zeta\right\}^{-1}\Big|_{\zeta = n\pi} = (-)^n, \tag{19.46a,b}$$

in agreement with (19.45) $\equiv$ (19.46b). The calculation of the residues of the combination of a rational and a transcendental function:

$$f(\zeta) = \frac{a^2 + \zeta^2}{\sin\zeta}, \qquad f_{(1)}(n\pi) = \lim_{\zeta \to n\pi} \left(a^2 + \zeta^2\right)\frac{\zeta - n\pi}{\sin\zeta} = (-)^n\left(a^2 + n^2\pi^2\right), \tag{19.47a,b}$$

is given as a final example for a simple poles, using (19.45).

19.9.2 Calculation of the Residue at a Multiple Pole (Method III)

The preceding method of calculation of residues extends from simple (19.40–19.41) to multiple poles: *the ratio of a nonzero (19.48b) function* $\mathrm{h}(\zeta)$ *to a function* $\mathrm{g}(\zeta)$ *with (19.48a) a zero of order* n *at* $\zeta = \mathrm{z}$ *is a function* $\mathrm{f}(\zeta)$ *with (19.48c) a pole of order* n *at* $\zeta = \mathrm{z}$:

$$g(z) = \cdots = g^{(n-1)}(z) = 0 \neq g^{(n)}(z), \; 0 \neq h(z): f(\zeta) \equiv \frac{h(\zeta)}{g(\zeta)} = \frac{n!\,(\zeta - z)^{-n}\,h(\zeta)}{g^{(n)}(z) + o(1)}, \tag{19.48a–c}$$

whose residue (15.32b):

$$f_{(n)}(z) = \lim_{\zeta \to z} \frac{1}{(n-1)!}\frac{\mathrm{d}^{n-1}}{\mathrm{d}\zeta^{n-1}}(\zeta - z)^n\, n!\,(\zeta - z)^{-n}\frac{h(\zeta)}{g^{(n)}(z)} = \frac{n h^{(n-1)}(z)}{g^{(n)}(z)}, \tag{19.49}$$

is n *times the ratio of the* $(\mathrm{n}-1)$*-the order* $[\mathrm{n}$*-the order*$]$ *derivative of the numerator (denominator)*; a simple pole $n = 1$ leads from (19.49) to (19.41). As examples, the preceding rule can be applied to the calculation of the residue at: (i) a double pole of a rational function:

$$f(\zeta) = \left(\zeta^4 - 2\zeta^2 a^4 + a^2\right)^{-1} = (\zeta - a)^{-2}(\zeta + a)^{-2}, \tag{19.50a}$$

$$f_{(2)}(a) = 2\lim_{\zeta \to a} \frac{\left\{\mathrm{d}\left[(\zeta + a)^{-2}\right]/\mathrm{d}\zeta\right\}}{\left\{\mathrm{d}^2\left[(\zeta - a)^2\right]/\mathrm{d}\zeta^2\right\}} = \frac{-4(\zeta + a)^{-3}\Big|_{\zeta = a}}{2} = -\frac{1}{4a^3}, \tag{19.50b}$$

in agreement with (15.35); (ii) a double pole of a transcendental function:

$$f(\zeta)=\csc^2\zeta, \quad f_{(2)}(n\pi)=2\lim_{\zeta\to n\pi}\frac{[\mathrm{d}(1)/\mathrm{d}\zeta]}{[\mathrm{d}^2(\sin^2\zeta)/\mathrm{d}\zeta^2]}=0, \tag{19.51a,b}$$

that can be confirmed from:

$$\begin{aligned} w\equiv\zeta-n\pi: \qquad f_{(2)}(n\pi) &= \lim_{\zeta\to n\pi}\frac{\mathrm{d}}{\mathrm{d}\zeta}(\zeta-n\pi)^2\csc^2\zeta=\lim_{w\to 0}\frac{\mathrm{d}}{\mathrm{d}w}\left(w^2\sin^{-2}w\right)\\ &= \lim_{w\to 0} 2w\csc^3 w\,(\sin w - w\cos w)\\ &= \lim_{w\to 0}\frac{2}{w^2}\left[w-\frac{w^3}{6}-w\left(1-\frac{w^2}{2}\right)\right]=\lim_{w\to 0}\frac{2w}{3}=0, \end{aligned} \tag{19.52}$$

where (19.39a,b) were used; (iii) a triple pole of a combination of rational and transcendental functions:

$$f(\zeta)=\zeta^{-3}\cos\zeta, \quad f_{(3)}(0)=3\lim_{\zeta\to 0}\frac{[\mathrm{d}^2(\cos\zeta)/\mathrm{d}\zeta^2]}{[\mathrm{d}(\zeta^3)/\mathrm{d}\zeta^3]}=-3\lim_{\zeta\to 0}\frac{1}{6}\cos\zeta=-\frac{1}{2}, \tag{19.53a,b}$$

that can be checked from (19.39b):

$$f(\zeta)\equiv\zeta^{-3}\cos\zeta=\frac{1}{\zeta^3}-\frac{1}{2}\zeta+o(\zeta), \qquad f_{(3)}(0)=-\frac{1}{2}. \tag{19.54a,b}$$

The rules of calculation of residues at simple (19.40a–c; 19.41) [multiple (19.48a–c; 19.49)] poles relate to the decomposition of rational functions into partial fractions [Section 31.8 (31.9)].

Note 19.1. *Minimal Conditions of Validity of Cauchy's Theorem*: It is worthwhile to establish the Cauchy conditions (Sections 19.1–19.6) that are minimal for the validity of the first Cauchy theorem, because they also apply to the second and third (Sections 15.1–15.4) and fourth (Section 31.4) Cauchy theorems. This set of four Cauchy theorems are important for: (i) the calculation of integrals by residues (Chapters 13 and 15); (ii) the establishment of ascending (Chapter 23) and ascending-descending (Chapter 25) power series representations; (iii) the location of zeros and poles of functions (Chapters 31 and 39); (iv) many other properties, for example, of special functions and related differential equations; (v) in particular the extension to differintegrations or derivatives of complex order. The infinitesimals (Sections 19.7–19.9) also appear often, for example, in series (part 3) and asymptotic approximations. The infinitesimals with nonintegral exponents lead to branch-points (Chapters 7 and 9). The Goursat proof of the Cauchy theorem (Sections 19.1–19.4) [proof of the divergence theorem)] are similar in dividing in dividing a two- (three-) dimensional region, using parallel lines (planes) into rectangles (tubes); in both cases the integrals over the straight sides cancel, leaving only the integral over the boundary.

Conclusion 19: The division (figure 19) of a region D by (**a**) two families of orthogonal straight lines, into a network of internal A_n, and boundary B_m, subregions, can be used to prove Cauchy's theorem in the original form, for a function holomorphic in D. The extension to a function uniformly continuous on the boundary applies a star-like region (**b**) with a central point z_0, in relation to which the boundary may be contracted homotetically without crossing itself; in order to find central points it may be necessary to divide (**c**) the region into subregions L_n. The final extension (**d**) of Cauchy's theorem allows a finite number of isolated ignorable singularities z_n on the boundary, of order lower than a pole; simple or multiple poles are allowed only if excluded by indenting the boundary loop.

20

Examples 20.1 to 20.20

Example 20.1. *Nonholomorphic Function Satisfying the Cauchy–Riemann Conditions.* Consider the function defined by:

$$z \neq 0: \quad f(x,y) = \frac{x^3(1+i) - y^3(1-i)}{x^2+y^2}, \tag{20.1}$$

and $f(0) = 0$. Show that it satisfies the Cauchy–Riemann conditions at the origin but is not holomorphic there (Pollard, 1928).

The real and imaginary parts are

$$\Phi(x,y) = \frac{x^3 - y^3}{x^2+y^2}, \quad \Psi(x,y) = \frac{x^3+y^3}{x^2+y^2}; \tag{20.2a,b}$$

using the definition of partial derivatives at the origin:

$$\frac{\partial \Phi(0,0)}{\partial x} = \lim_{x\to 0} \frac{\Phi(x,0) - \Phi(0,0)}{x} = \frac{x}{x} = 1, \tag{20.3a}$$

$$\frac{\partial \Phi(0,0)}{\partial y} = \lim_{y\to 0} \frac{\Phi(0,y) - \Phi(0,0)}{y} = -\frac{y}{y} = -1; \tag{20.3b}$$

doing likewise for Ψ it follows that

$$x = 0 = y: \quad \frac{\partial \Phi}{\partial x} = 1 = \frac{\partial \Psi}{\partial y}, \quad \frac{\partial \Phi}{\partial y} = -1 = -\frac{\partial \Psi}{\partial x}, \tag{20.4a,b}$$

so that the Cauchy–Riemann conditions (11.10a,b) are satisfied. The incremental ratios of the function (20.1) along the two diagonals of quadrants coincide:

$$x = y: \quad \lim_{x\to 0} \frac{1}{x(1+i)} \frac{2ix^3}{2x^2} = \frac{i}{1+i} = \frac{1+i}{2}, \tag{20.5a}$$

$$x = -y: \quad \lim_{x\to 0} \frac{1}{x(1-i)} \frac{2x^3}{2x^2} = \frac{1}{1-i} = \frac{1+i}{2}; \tag{20.5b}$$

the incremental ratios of the function (20.1) along the coordinate axis also coincide and have the same value:

$$y = 0: \quad \lim_{x\to 0} \frac{f(x,0)}{x} = \lim_{x\to 0} \frac{1+i}{x} \frac{x^3}{x^2} = 1+i, \tag{20.5c}$$

$$x = 0: \quad \lim_{y\to 0} \frac{f(0,y)}{iy} = \lim_{y\to 0} \frac{i-1}{iy} \frac{y^3}{y^2} = 1+i, \tag{20.5d}$$

but differ from (20.5a) ≡ (20.5b) ≠ (20.5c) ≡ (20.5d).

Thus *the incremental ratio of the function (20.1) has: (i) the same limits along the horizontal (vertical) directions (20.6a) = (20.6b)]:*

$$\lim_{z\to 0} z^{-1} f(z) = 1+i \quad \text{for } y=0, x\to 0 \tag{20.6a}$$

$$1+i \quad \text{for } x=0, y\to 0 \tag{20.6b}$$

$$\frac{1+i}{2} \quad \text{for } x=y\to 0 \tag{20.6c}$$

$$\frac{1+i}{2} \quad \text{for } y=-x\to 0, \tag{20.6d}$$

(ii) also the same values along the diagonals (20.6c) = (20.6d); (iii) the two sets of values are distinct (20.6a) = (20.6b) ≠ (20.6c) = (20.6d). This proves that the function is not holomorphic. This result could be predicted by using (3.17a,b) to write (20.1) as a function of z *and* z***:

$$f\left(\frac{z+z^*}{2}, \frac{z-z^*}{2i}\right) = \frac{(1+i)\left(3z+z^{*2}/z\right)}{4} \equiv g(z,z^*); \tag{20.7}$$

since g(z,z*) *does not depend on* z *alone, the function cannot be holomorphic* (compare with Example 20.4).

Example 20.2. *Solution of Laplace Equation in Polar Coordinates.*
Determine the relation between the constants a and b so that the potential:

$$\Phi(r,\varphi) = Ar^b \cos, \sin(a\varphi), \tag{20.8a}$$

is a solution of Laplace's equation, and identify the potential flows corresponding to integer values.

Substitution of (20.8a) in (11.28b) yields (20.8b) leading to (20.8c):

$$b^2 = a^2: \quad \Phi_\pm(r,\varphi) = Ar^{\pm a} \cos, \sin(a\varphi). \tag{20.8b,c}$$

The particular case $a=n$ a positive integer is: (i) for Φ_- a multipole of integer order $a=-n$ in (12.60a); (ii) for Φ_+ the potential flow in a corner of angle $\beta=\pi/a$ in (14.71a).

Example 20.3. *Relation between Polar and Cartesian Components of the Velocity.*
Use the velocity field of a multipole (Section 12.9), to obtain the relation between the Cartesian and polar components of an arbitrary (contravariant) vector (Figure 20.1).

From (12.59a) follow the Cartesian components of the velocity:

$$\{v_x, v_y\} = \frac{P_n}{2\pi} n r^{-n-1} \{\cos[(n+1)\varphi], \sin[(n+1)\varphi]\}. \tag{20.9a,b}$$

Using the trigonometric relations (3.25a,b) follows:

$$v_x = \frac{P_n}{2\pi} n r^{-n-1} [\cos(n\varphi)\cos\varphi - \sin(n\varphi)\sin\varphi], \tag{20.10a}$$

$$v_x = \frac{P_n}{2\pi} n r^{-n-1} [\sin(n\varphi)\cos\varphi + \cos(n\varphi)\sin\varphi]. \tag{20.10b}$$

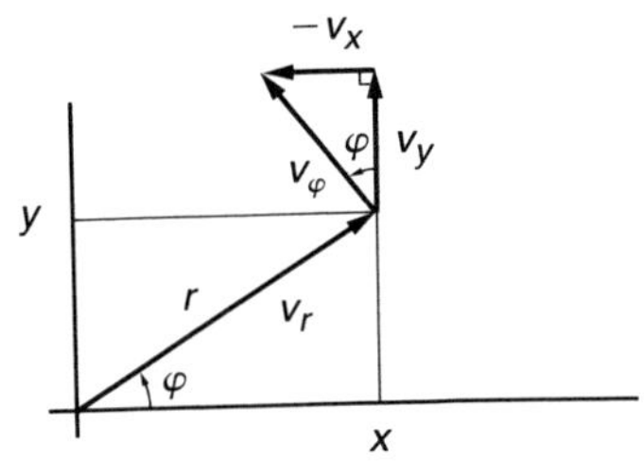

FIGURE 20.1
Transformation from the Cartesian components (v_x, v_y) of the velocity which lie along the coordinate axis and the polar components which are: (i) radial outward v_r; (ii) azimuthal counterclockwise v_φ. Thus passing from the Cartesian to the polar components corresponds to a rotation by φ in the positive direction; vice-versa leads to a rotation by $-\varphi$.

Substitution of (12.62a,b) leads to the transformation from polar to Cartesian components of a vector:

$$v_x = v_r \cos\varphi - v_\varphi \sin\varphi, \quad v_y = v_r \sin\varphi + v_\varphi \cos\varphi, \tag{20.11a,b}$$

that is equivalent to a plane rotation by an angle $-\varphi$, that is, φ in the negative or clockwise direction. Substituting φ by $-\varphi$ leads to the transformation from Cartesian to polar components:

$$v_r = v_x \cos\varphi + v_y \sin\varphi, \quad v_\varphi = -v_x \sin\varphi + v_y \cos\varphi, \tag{20.12a,b}$$

as can be checked writing (12.62a,b) in terms of (20.9a,b); this is equivalent to plane rotation by an angle $+\varphi$ viz. (20.11–20.12a,b) transform into each other by changing the sign of $\pm\varphi$. These formulas apply *to arbitrary vectors, converting Cartesian into polar components (20.12a,b) [vice-versa (20.11a,b)], and correspond to a plane rotation by* $+\varphi(-\varphi)$, *that is,* $|\varphi|$ *in the counterclockmise (clockwise) direction.*

Example 20.4. *Integrals of Holomorphic and Nonholomorphic Functions.*
Consider the functions:

$$f_\pm(x,y) = x^2 - y^2 \pm 2ixy = z^2, z^{*2}, \tag{20.13a,b}$$

and show that $f_+(f_-)$ is (is not) holomorphic, and its integral between the same points along the two distinct paths in Figure 20.2 are equal (unequal).

The function f_+ satisfies the Cauchy–Riemann conditions (11.10a,b) and its real and imaginary parts have continuous first-order derivatives; hence it is holomorphic and can be written (20.14b) as a function of (20.14a) alone:

$$z \equiv x + iy: \quad f_+(x,y) = x^2 - y^2 + 2ixy = (x+iy)^2 = z^2; \tag{20.14a,b}$$

Its primitive specifies the integral from the origin to the point $1+i$,

$$\int_0^{1+i} z^2 \mathrm{d}z = \left[\frac{1}{3}z^3\right]_0^{1+i} = \frac{(1+i)^3}{3} = \frac{1}{3} + i + i^2 + \frac{i^3}{3} = \frac{2}{3}(i-1), \tag{20.14c}$$

that is the same for any path; it will be confirmed in the sequel that the value (20.14c) $\equiv$ (20.16a) holds for the two paths in Figure 20.2. The function f_- does not satisfy the Cauchy–Riemann conditions (11.10a,b), and cannot be holomorphic. Using (3.17a,b) it follows it is

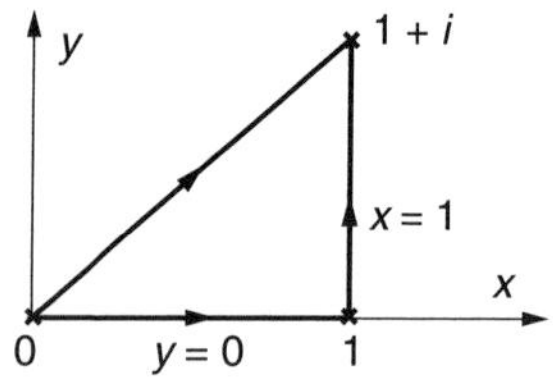

FIGURE 20.2
The integral of a holomorphic (nonholomorphic) function between the same points along different paths must be equal (may be different). This is shown by the integration of the holomorphic (nonholomorphic) function (20.13a) [(20.13b)] from the origin to the point $1 + i$ comparing as paths of integration: (i) a straight line from the origin to the point $1 + i$; (ii) the sides of a rectangular triangle with an edge at (1,0). This leads to the same (a different) result for the two paths.

a function of z^*:

$$f_\pm\left(\frac{z^2+z^*}{2},\frac{z-z^*}{2i}\right)=\frac{z^2+z^{*2}}{2}\pm\frac{z^2-z^{*2}}{2}=z^2,z^{*2}=f_+(z),f_-(z^*); \tag{20.14d}$$

since is cannot be written in terms of z alone it does not have a primitive which is a function of z alone, and its integral between two points may depend on the path; this is confirmed next (20.16b) for the two paths in Figure 10.2.

The integrals:

$$I_\pm=(1+i)\int_0^1 f_\pm(x,x)\,\mathrm{d}x, \tag{20.15a}$$

$$J_\pm=\int_0^1 f_\pm(x,0)\,\mathrm{d}x+i\int_0^1 f_\pm(1,y)\,\mathrm{d}y, \tag{20.15b}$$

joining the origin to the point $1 + i$ (Figure 20.2), either along a straight line (20.15a), or along the real axis first and then parallel to the imaginary axis (20.15b), yield:

$$I_+=\frac{2(i-1)}{3}=J_+,\quad I_-=\frac{2(1-i)}{3}\neq\frac{2(2+i)}{3}=J_-, \tag{20.16a,b}$$

showing that the integral of the holomorphic f_+ (nonholomorphic f_-) function is (is not) the same for both paths.

Example 20.5. *Derivatives of Arbitrary Order and the Chain Rule.*
Prove the identities:

$$\frac{\mathrm{d}^n\{z^m\mathrm{e}^{az}\}}{\mathrm{d}z^n}=\mathrm{e}^{az}\sum_{p=0}^{\leq n,m}\frac{n!m!}{p!(n-p)!(m-p)!}a^{n-p}z^{m-p}, \tag{20.17}$$

$$n\geq m:\ \frac{\mathrm{d}^n\{(z^2-a^2)^n\}}{\mathrm{d}z^n}=\sum_{p=n-m}^{m}\frac{n!(m!)^2}{p!(n-p)!(m-p)!(m-n+p)!}\times(z-a)^{m-p}(z+a)^{m-n+p}; \tag{20.18}$$

also the identities:

$$\frac{\mathrm{d}^n\left\{z^m \begin{matrix}\cos\\ \sin\end{matrix} z\right\}}{\mathrm{d}z^n} = P(z) \begin{matrix}\cos\\ \sin\end{matrix} z \mp Q(z) \begin{matrix}\cos\\ \sin\end{matrix} z, \tag{20.19a}$$

where the polynomial coefficients are

$$P(z) \equiv \sum_{p\geq 0,(n-m)/2}^{\leq n/2} \frac{n!m!}{(2p)!(n-2p)!(m-n+2p)!} z^{m-n+2p}, \tag{20.19b}$$

$$Q(z) \equiv \sum_{p\geq 1,(n-m+1)/2}^{s} \frac{n!m!}{(2p-1)!\,(n-2p+1)!\,(m-n+2p-1)!} z^{m-n+2p-1}; \tag{20.19c}$$

in (20.19c) the sum is up to $s = n/2$ for n even and $s = (n+1)/2$ for n odd.

The identities (20.17, 20.18) follow from (13.31) the chain rule:

$$\frac{\mathrm{d}^n\left\{z^m \mathrm{e}^{az}\right\}}{\mathrm{d}z^n} = \sum_{p=0}^{n} \binom{n}{p} \frac{\mathrm{d}^p\left\{z^m\right\}}{\mathrm{d}z^p} \frac{\mathrm{d}^{n-p}\left\{\mathrm{e}^{az}\right\}}{\mathrm{d}z^{n-p}}; \tag{20.20}$$

$$\frac{\mathrm{d}^n\left\{\left(z^2-a^2\right)^m\right\}}{\mathrm{d}z^n} = \sum_{p=0}^{n} \binom{n}{p} \frac{\mathrm{d}^p\left\{(z-a)^m\right\}}{\mathrm{d}z^p} \frac{\mathrm{d}^{n-p}\left\{(z+a)^m\right\}}{\mathrm{d}z^{n-p}}; \tag{20.21}$$

bearing in mind that:

$$\frac{\mathrm{d}^p\left(z^m\right)}{\mathrm{d}z^p} = \begin{cases} \dfrac{m!}{(m-p)!} z^{m-p} & \text{if } p \leq m \\ 0 & \text{if } p > m \end{cases} \tag{20.22}$$

limits the sum: (i) to $p \leq m$ in (20.20) $\equiv$ (20.17); (ii) to $p \leq m$ and $n - p \leq m$ in (20.21) $\equiv$ (20.18).

Concerning (20.19a) the chain rule is separated into derivatives of even and odd order:

$$\begin{aligned} \frac{\mathrm{d}^n\left\{z^m \begin{matrix}\cos\\ \sin\end{matrix} z\right\}}{\mathrm{d}z^n} &= \sum_{p=0}^{\leq n/2} \binom{n}{2p} \frac{\mathrm{d}^{2p}\left\{\begin{matrix}\cos\\ \sin\end{matrix} z\right\}}{\mathrm{d}z^{2p}} \frac{\mathrm{d}^{n-2p}\left\{z^m\right\}}{\mathrm{d}z^{n-2p}} \\ &\quad + \sum_{p=1}^{s} \binom{n}{2p-1} \frac{\mathrm{d}^{2p-1}\left\{\begin{matrix}\cos\\ \sin\end{matrix} z\right\}}{\mathrm{d}z^{2p-1}} \frac{\mathrm{d}^{n-2p+1}\left\{z^m\right\}}{\mathrm{d}z^{n-2p+1}} \\ &= \begin{matrix}\cos\\ \sin\end{matrix} z \sum_{p=0}^{\leq n/2} (-)^p \frac{n!}{(2p)!\,(n-2p)!} z^{m-n+2p} \frac{m!}{(m-n+2p)!}, \\ &\quad \pm \begin{matrix}\sin\\ \cos\end{matrix} z \sum_{p=0}^{s} (-)^p \frac{n!}{(2p-1)!\;(n-2p+1)!} \\ &\quad \times z^{m-n+2p-1} \frac{m!}{(m-n+2p-1)!}, \end{aligned} \tag{20.23}$$

where (20.22) was used; the coincidence of (20.19a) $\equiv$ (20.23) specifies the polynomial coefficients (29.19b,c) where only positive powers appear $m - n + 2p \geq 0$ and $m - n + 2p - 1 \geq 0$, respectively. For $p > n$ or $p > m$, this (20.17) or (20.18) would lead to factorials of negative integers, for example, $(-1)! = (-1)(-2)\ldots = \infty$ that diverge and suppress the corresponding terms; likewise the factorials suppress in (20.19b) the terms $p > n/2$ and $p < (n-m)/2$ and in (20.19c) the terms $p > (n+1)/2$ and $p < (n-m+1)/2$.

Example 20.6. *Existence and Calculation of Laplace Transforms.*
Show that if the integral defining the Laplace transform:

$$\int_0^\infty f(z)\,\mathrm{e}^{-sz}\,\mathrm{d}z = f\left(-\frac{\mathrm{d}}{\mathrm{d}s}\right)\int_0^\infty \mathrm{e}^{-sz}\,\mathrm{d}z = f\left(-\frac{\mathrm{d}}{\mathrm{d}s}\right)\frac{1}{s}, \tag{20.24}$$

is uniformly convergent with regard to s, it can be evaluated by the rule (20.24); hence apply (20.24) to the following examples:

$$\int_0^\infty z^n \mathrm{e}^{-sz}\,\mathrm{d}z = n!s^{-n-1}, \tag{20.25}$$

$$\int_0^\infty (az+b)^n\,\mathrm{e}^{-sz}\,\mathrm{d}z = \sum_{p=0}^{n}\frac{n!}{(n-p)!}a^p b^{n-p} s^{-p-1}, \tag{20.26}$$

$$\int_0^\infty \left(z^2+a^2\right)^n \mathrm{e}^{-sz}\,\mathrm{d}z = \sum_{p=0}^{n}\frac{n!\,(2p)!}{p!\,(n-p)!}a^{2n-2p}s^{-2p-1}, \tag{20.27}$$

$$\int_0^\infty (z^m-1)^n\,\mathrm{e}^{-sz}\,\mathrm{d}z = \sum_{p=0}^{n}(-)^{n-p}\frac{n!\,(pm)!}{p!\,(n-p)!}s^{-pm-1}, \tag{20.28}$$

of Laplace transforms expanded as polynomials of $1/s$ and indicate the range of values of s for which these hold.

The Laplace integral of any power z^m whose exponent has nonnegative real part $\mathrm{Re}(m) \geq 0$ is uniformly convergent for $\mathrm{Re}(s) \geq \varepsilon > 0$, viz:

$$\begin{aligned}
&M \geq \mathrm{Re}(m) \geq 0, \quad \mathrm{Re}(s) \geq \varepsilon > 0: \quad \left|\int_0^\infty z^m \mathrm{e}^{-sz}\,\mathrm{d}z\right| \leq \int_0^\infty |x^m|\left|\mathrm{e}^{-sx}\right|\mathrm{d}x \\
&= \int_0^\infty x^{\mathrm{Re}(m)}\mathrm{e}^{-x\mathrm{Re}(s)}\,\mathrm{d}x \leq \int_0^\infty x^M \mathrm{e}^{-\varepsilon x}\,\mathrm{d}x \\
&= (-)^M \frac{\mathrm{d}^M}{\mathrm{d}\varepsilon^M}\int_0^\infty \mathrm{e}^{-\varepsilon x}\,\mathrm{d}x = (-)^M\frac{\mathrm{d}^M}{\mathrm{d}\varepsilon^M}\left(\frac{1}{\varepsilon}\right) = M!\varepsilon^{-1-M} < \infty,
\end{aligned} \tag{20.29}$$

where M is any integer larger than $\mathrm{Re}(m) \geq 0$. Thus *all the Laplace transforms (20.25) to (20.28) hold in the interior of the right-hand complex s-plane* $\mathrm{Re}(s) \geq \varepsilon > 0$. The evaluation of the integral (20.25) is similar to the last three steps of (20.29); the application of the chain rule (13.31) to the integrands in (20.26) to (20.28) leads to a sum of terms like (20.25), viz.:

$$\left(b - a\frac{d}{ds}\right)^n \frac{1}{s} = \sum_{p=0}^{n} \binom{n}{p} b^{n-p} a^p \left(-\frac{d}{ds}\right)^p \frac{1}{s}, \tag{20.30}$$

$$\left(a^2 + \frac{d^2}{ds^2}\right)^n \frac{1}{s} = \sum_{p=0}^{n} \binom{n}{p} a^{2n-2p} \left(-\frac{d}{ds}\right)^{2p} \frac{1}{s}, \tag{20.31}$$

$$\left[\left(-\frac{d}{ds}\right)^m - 1\right]^n \frac{1}{s} = \sum_{p=0}^{n} \binom{n}{p} (-)^{n-p} \left(-\frac{d}{ds}\right)^{pm} \frac{1}{s}, \tag{20.32}$$

where (20.30) ≡ (20.26), (20.31) ≡ (20.27), and (20.32) ≡ (20.28).

Example 20.7. *Differentiation of Parametric Integrals.*
Obtain the derivates of the following integrals with variable end-points:

$$I_1 \equiv \frac{d}{dz} \int_{az}^{2az} \log\zeta \, d\zeta = a \log(4az), \tag{20.33}$$

$$I_2 \equiv \frac{d}{dz} \int_{1}^{z^n} \zeta^{-1} \log\zeta \, d\zeta = n^2 z^{-1} \log z, \tag{20.34}$$

$$I_3 \equiv \frac{d}{dz} \int_{\sin z}^{\cos z} \left(1 - \zeta^2\right)^{-1} \zeta \, d\zeta = -\tan z - \cot z, \tag{20.35}$$

$$I_4 \equiv \frac{d}{dz} \int_{0}^{\cosh z} \left(\zeta^2 - 1\right)^{-1} \zeta^3 \, d\zeta = \cosh^2 z \coth z, \tag{20.36}$$

$$J_1 \equiv \frac{d}{dz} \int_{z}^{1/z} \log(\zeta z) \, d\zeta = z^{-2} - 1 - 2\log z, \tag{20.37}$$

$$J_2 \equiv \frac{d}{dz} \int_{z}^{z^2} \zeta^{-1} \log(\zeta z) \, d\zeta = 5z^{-1} \log z, \tag{20.38}$$

$$J_3 \equiv \frac{d}{dz} \int_{0}^{1/z} \arctan(\zeta z) \, d\zeta = \frac{\log 2 - \pi/2}{2z^2}, \tag{20.39}$$

$$J_4 \equiv \frac{d}{dz} \int_{0}^{1/z} \arg\sinh(\zeta z) \, d\zeta = z^{-2} \left(\sqrt{2} - 1 - \arg\sinh 1\right), \tag{20.40}$$

where the limits of integration are such that the integrand has no singularities on the path of integration.

The first (last) four (20.33–20.36) [(20.37–20.40)] integrals follow from (13.44) [(13.46b)]:

$$I_1 = 2a\log(2az) - a\log(az) = a\log\left[\frac{(2az)^2}{az}\right] = a\log(4az), \tag{20.41}$$

$$I_2 = nz^{n-1}(z^n)^{-1}\log(z^n) = n^2 z^{-1}\log z, \tag{20.42}$$

$$I_3 = -\sin z\left(1-\cos^2 z\right)^{-1}\cos z - \cos z\left(1-\sin^2 z\right)^{-1}\sin z = -\cot z - \tan z, \tag{20.43}$$

$$I_4 = \sinh z\left(\cosh^2 z - 1\right)^{-1}\cosh^3 z = \cosh^2 z\coth z, \tag{20.44}$$

$$J_1 = -z^{-2}\log 1 - \log\left(z^2\right) + z^{-1}\int_z^{1/z} \mathrm{d}\zeta = -2\log z + \frac{1}{z^2} - 1, \tag{20.45}$$

$$J_2 = 2z\left(z^2\right)^{-1}\log\left(z^3\right) - z^{-1}\log\left(z^2\right) + z^{-1}\int_z^{z^2}\zeta^{-1}\mathrm{d}\zeta = (6-2+1)z^{-1}\log z, \tag{20.46}$$

$$J_3 = -z^{-2}\arctan 1 + \int_0^{1/z}\left(1+\zeta^2 z^2\right)^{-1}\zeta\,\mathrm{d}\zeta = -\frac{\pi}{4z^2} + \left(2z^2\right)^{-1}\log\left(1+\zeta^2 z^2\right)\Big|_0^{1/z}, \tag{20.47}$$

$$J_4 = -z^{-2}\operatorname{arg\,sinh} 1 + \int_0^{1/z}\left(1+\zeta^2 z^2\right)^{-1/2}\zeta\,\mathrm{d}\zeta = -z^{-2}\operatorname{arg\,sinh} 1 + z^{-2}\left(1+\zeta^2 z^2\right)^{1/2}\Big|_0^{1/z}, \tag{20.48}$$

proving the coincidence of (20.33–20.40) ≡ (20.41–48).

Example 20.8. *Superposition of Corner Flows.*
Consider the superposition of potential flows in corners of angles $\beta = \pi/n$ and $\beta/2 = \pi/2n$ and determine: (i) the location of the stagnation points; (ii) the meaning, if any, of the radial lines $\varphi = k\pi/n$ and $\varphi = (k+1/2)\pi/n$.

The complex potential is specified (Section 14.8) by (20.49b):

$$AB > 0, n = 1,2,\ldots: \quad f(z) = z^n\left(A + Bz^n\right); \quad v^*(z) = nz^{n-1}\left(A + 2Bz^n\right); \tag{20.49a–c}$$

the roots of the conjugate velocity (20.49c) specify the location of the stagnation points, viz. the origin and:

$$k = 0,\ldots,n-1: \quad z_k = \sqrt[n]{\frac{A}{2B}}\exp\left(\frac{i\pi}{n} + \frac{i2k}{n}\right). \tag{20.49d}$$

It was assumed that A, B are real and have the same sign (20.49a), otherwise, if they have opposite signs $AB < 0$ the factor $\exp(i\pi/n)$ is omitted from (20.49d). The streamlines $\varphi = k\pi/n$ that are common to both the z^n and z^{2n} flows, remain streamlines of the flow obtained by superposition. The polar components of the velocity:

$$v_r(r,\varphi) = nr^{n-1}\left[A\cos(n\varphi) + 2Br^n\cos(2n\varphi)\right], \tag{20.50a}$$

$$v_\varphi(r,\varphi) = -nr^{n-1}\left[A\sin(n\varphi) + 2Br^n\sin(2n\varphi)\right], \tag{20.50b}$$

imply that the radial lines (20.51a) are streamlines of both flows (20.51b,c):

$$\varphi = k\pi/n: \quad v_r = nr^{n-1}\left[(-)^k A + 2Br^n\right], \quad v_\varphi = 0; \tag{20.51a–c}$$

$$\varphi = (k+1/2)\pi/n: \quad v_r = -2nr^{2n-1}B, \quad v_\varphi = -nr^{n-1}(-)^k A, \tag{20.52a–c}$$

the lines (20.52a): (i) are equipotentials (20.52c) of z^n; (ii) are streamlines (20.52b) of z^{2n}; (iii) are neither equipotentials nor streamlines of the flow obtained by super position

Example 20.9. *Evaluation of Integrals with Pole(s) on the Boundary.*
Use residues to evaluate the following integrals each with a simple pole at an angular point on boundary:

$$\left(\int_{-1}^{1} + \int_{1}^{i} + \int_{i}^{-1}\right)(\zeta - i)^{-1}\,\mathrm{d}\zeta \equiv \oint_{L_1} (\zeta - i)^{-1}\,\mathrm{d}\zeta = \frac{i3\pi}{2}, \tag{20.53}$$

$$\left(\int\limits_{\substack{|\zeta|=2R\\ 0\le\arg(\zeta)\le\pi/2}} + \int\limits_{\substack{|\zeta-iR|=R\\ \pi/2\le\arg(\zeta)\le\pi}} + \int\limits_{\substack{|\zeta-R|=R\\ 3\pi/2\le\arg(\zeta)\le 2\pi}}\right)\zeta^{-1}\,\mathrm{d}\zeta \equiv \oint_{L_2} \zeta^{-1}\mathrm{d}\zeta = \frac{i\pi}{2}, \tag{20.54}$$

$$\left(\int\limits_{\substack{|\zeta|=2R\\ 0\le\arg(\zeta)\le\pi}} + \int\limits_{\substack{|\zeta+R|=R\\ \pi\le\arg(\zeta)\le 3\pi/2}} + \int\limits_{\substack{|\zeta-R|=R\\ 3\pi/2\le\arg(\zeta)\le 2\pi}}\right)\zeta^{-1}\,\mathrm{d}\zeta \equiv \oint_{L_3} \zeta^{-1}\mathrm{d}\zeta = 0, \tag{20.55}$$

$$\left(\int\limits_{\substack{|\zeta|=2R\\ 0\le\arg(\zeta)\le\pi}} + \int\limits_{\substack{|\zeta+R|=R\\ \pi\le\arg(\zeta)\le 3\pi/2}} + \int\limits_{\substack{|\zeta-R|=R\\ \pi\ge\arg(\zeta)\ge 0}}\right)\zeta^{-1}\,\mathrm{d}\zeta \equiv \oint_{L_4} \zeta^{-2}\mathrm{d}\zeta = i\pi, \tag{20.56}$$

and check the result by Riemann integration along the paths (L_1, L_2, L_3, L_4) in Figures 20.3a–d.

The integrals (20.53–29.56) each has a simple pole with residue unity on the boundary at: (i) a salient angular point at $z = i$ with $\beta = 3\pi/2$ for the integral (20.53) along the path L_1 in Figure 20.3a; (ii/iii) a reentrant angular point at $z = 0$ with $\beta = \pi/2\,(\beta = 0)$ for the integral (20.54) [(20.55)] along the path $L_2\,(L_3)$ in Figure 20.3b (20.3c); (iv) a regular $\beta = \pi$ point at $z = 0$ for the integral (20.56) along the path L_4 in Figure 20.3d. As a check, the integral (20.56) is taken as example for evaluation by Riemann integration along the path L_4 in Figure 20.3d, viz.:

$$\begin{aligned}\oint_{L_4} \zeta^{-1}\mathrm{d}\zeta &= \int_0^{\pi} \left(2R\mathrm{e}^{i\varphi}\right)^{-1} 2R\mathrm{e}^{i\varphi} i\,\mathrm{d}\varphi + \int_{\pi}^{2\pi} \left(R\mathrm{e}^{i\varphi} + R\right)^{-1} R\mathrm{e}^{i\varphi} i\,\mathrm{d}\varphi + \int_{\pi}^{0} \left(R\mathrm{e}^{i\varphi} - R\right)^{-1} R\mathrm{e}^{i\varphi} i\,\mathrm{d}\varphi \\ &= i\pi + \log\left(\mathrm{e}^{i\varphi} + 1\right)\Big|_{\pi}^{2\pi} + \log\left(\mathrm{e}^{i\varphi} - 1\right)\Big|_{\pi}^{0} = i\pi + \log\left(\frac{2}{-i\varepsilon}\right) + \log\left(\frac{i\varepsilon}{-2}\right) \\ &= i\pi + \log\left[\left(\frac{2i}{\varepsilon}\right)\left(-\frac{i\varepsilon}{2}\right)\right] = i\pi,\end{aligned} \tag{20.57}$$

where the same value of $\varepsilon \to 0$ was used at two limits, as in the Cauchy principal value of an integral (Sections 17.8 and 17.9).

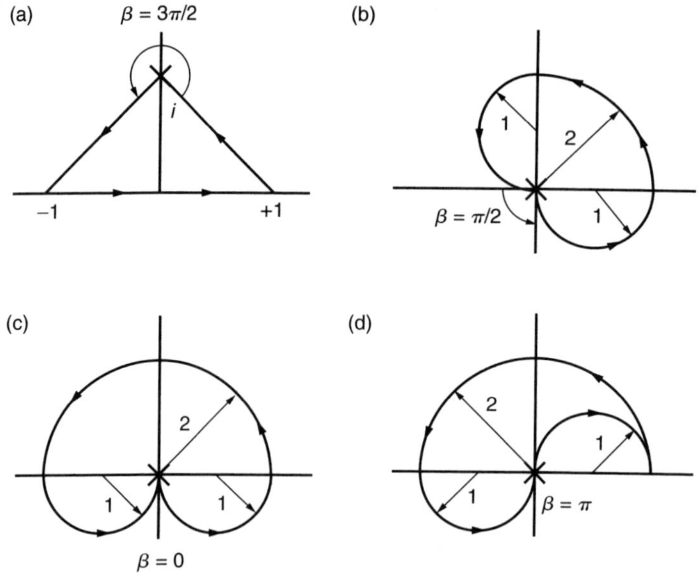

FIGURE 20.3
Examples of loop integrals with a singularity at a point on the boundary: (i/ii) angular point with a sector $\beta = 3\pi/2\,(\pi/2)$ between the tangents before and after the point (a) [(b)]; (iii) a cusp for which $\beta = 0$ because the tangents have opposite directions before and after (c); (iv) a regular point for which $\beta = \pi$ because the tangent is continuous (d).

Example 20.10. *Residues at Simple, Double and Triple Pole.*
Consider the function:

$$f(z) = \left[(z-a)^2 \sin z\right]^{-1}, \tag{20.58}$$

and obtain its residues at simple, double, or triple poles.

If a/π is not an integer (20.59a,b), the function (20.58) has simple poles (a double pole) at $z = n\pi\,(z = a)$ with residues (20.59c) [(20.59d)], for example, calculated from (15.24b) [(15.33b)], viz.:

$$a \neq n\pi; n \in |Z: \quad f_{(1)}(n\pi) = \frac{(-)^n}{(a-n\pi)^2}, \quad f_{(2)}(a) = -\csc a \cot a; \tag{20.59a–d}$$

if $a = m\pi$ for one integer value m, then (20.60a,b) the simple poles remain at $z = n\pi$ with $n \neq m$ with residues (20.59c), and $z = m\pi$ becomes a triple pole, with residue (20.60c):

$$\begin{aligned} a = m\pi; m \in |Z: \quad f_{(3)}(m\pi) &= \lim_{z\to m\pi} \frac{1}{2}\frac{\mathrm{d}^2}{\mathrm{d}z^2}\frac{z-m\pi}{\sin z} = \lim_{w\to 0}\frac{1}{2}\frac{\mathrm{d}^2}{\mathrm{d}w^2}\frac{w}{\sin(w+m\pi)} \\ &= \frac{(-)^m}{2}\lim_{w\to 0}\frac{\mathrm{d}^2}{\mathrm{d}w^2}\frac{w}{\sin w} = \frac{(-)^m}{2}\lim_{w\to 0}\frac{\mathrm{d}^2}{\mathrm{d}w^2}\left[1-\frac{w^2}{6}+o\left(w^2\right)\right]^{-1} \\ &= \frac{(-)^m}{2}\lim_{w\to 0}\frac{\mathrm{d}^2}{\mathrm{d}w^2}\left[1+\frac{w^2}{6}+o\left(w^2\right)\right] = \frac{(-)^m}{6}, \end{aligned} \tag{20.60a–c}$$

where the substitution $w = z - n\pi$ was made and (19.39a) was used.

Example 20.11. *Force Exerted on a Wall by a Monopole.*
Calculate the integral of the dynamic pressure exerted on a wall by the potential flow due to a source or sink and show that it coincides with (Section 16.1) minus the force of repulsion from its identical image (16.2b).

The potential flow due to a source or sink of flow rate Q at a distance a from a wall (Section 16.1) has complex conjugate velocity (16.1b), whose value on the wall:

$$z \equiv x;\ y = 0: \quad v_x(x,0) = \frac{Q}{\pi}\frac{x}{x^2+a^2}, \quad v_y(x,0) = 0, \tag{20.61a,b}$$

shows that it is tangential. From Bernoulli' s equation (14.27c) it corresponds to the distribution of pressure on the wall (20.62a):

$$p(x) = p_0 - q(x), \quad q(x) = \frac{1}{2}\rho v^2 = \rho\frac{Q^2}{2\pi^2}\frac{x^2}{(x^2+a^2)^2}, \tag{20.62a,b}$$

where the dynamic pressure (20.62b): (i) decays fast enough at infinity to meet the condition (17.19), viz. (20.63a); (ii) has double poles at $z = \pm ia$, with residues (20.63b):

$$\lim_{z\to\infty} zq(z) \sim 0\left(z^{-1}\right) \to 0: q_{(2)}(\pm ia) = \rho\frac{Q^2}{2\pi^2}\lim_{z\to\pm ia}\frac{\mathrm{d}}{\mathrm{d}z}\frac{z^2}{(z\pm ia)^2} = \mp i\rho\frac{Q^2}{8\pi^2 a}. \tag{20.63a,b}$$

The force on the wall is the integral of the pressure acting downward, where the constant stagnation pressure can be omitted:

$$F = -i\int_{-\infty}^{+\infty}[p(x)-p_0]\,\mathrm{d}x = i\int_{-\infty}^{+\infty} q(x)\,\mathrm{d}x = \pm i2\pi i q_{(2)}(\pm ia) = \frac{i\rho Q^2}{4\pi a}; \tag{20.64}$$

the integral (20.64) can be evaluated by residues (20.63b), closing the real axis (17.18) either on the upper- or lower half-z-plane, since the asymptotic condition (20.63a) is met in all directions. The result (20.64) ≡ (16.2b) coincides with the force exerted on a wall by a source/sink. It is similar to the electric (magnetic) force exerted by a line electric charge (current) on a plane conductor (24.22) [(26.30b)] in a medium of dielectric permittivity ε (magnetic permeability μ, with c the speed of light in vacuo).

Example 20.12. *Nonisotropic Source/Sink and Vortex.*
Consider a spiral flow with nonisotropic source/sink (vortex) of volume flux $q(\varphi)$ [circulation $\gamma(\varphi)$]:

$$v_r(r,\varphi) = \frac{q(\varphi)}{2\pi r}, \quad v_\varphi(r,\varphi) = \frac{\gamma(\varphi)}{2\pi r}, \tag{20.65a,b}$$

and establish under which conditions it is a potential flow.

Applying to (20.65a,b) the curl (11.35b) [divergence (11.33b)] in polar coordinates leads to

$$\nabla\wedge\vec{\mathrm{v}} = \frac{\vec{\mathrm{e}}_3}{r}\left[\frac{\partial(rv_\varphi)}{\partial r} - \frac{\partial v_r}{\partial\varphi}\right] = -\frac{\vec{\mathrm{e}}_3 q'(\varphi)}{2\pi r^2}, \tag{20.66a}$$

$$\nabla\cdot\vec{\mathrm{v}} = \frac{1}{r}\left[\frac{\partial(rv_r)}{\partial r} + \frac{\partial v_\varphi}{\partial\varphi}\right] = \frac{\gamma'(\varphi)}{2\pi r^2}; \tag{20.66b}$$

it follows that the flow is potential, that is, irrotational (20.66a) [incompressible (20.66b)] only if $q|\varphi| \equiv Q$ [$\gamma(\varphi) \equiv \Gamma$] is constant; this leads to the monopole (Section 22.6) as a superposition of source/sink (12.25a) [vortex (12.31a)]. Thus a potential flow excludes anisotropic source/sinks (vortices), as could be expected from Section 12.4 (12.5).

Example 20.13. *Evaluation of Trignometric Integrals along the Unit Circle.* Evaluate the following trignometric integrals, with real p:

$$I(p) \equiv \int_0^{2\pi} \left(1 - 2p\cos\varphi + p^2\right)^{-1} \mathrm{d}\varphi = \begin{cases} \dfrac{2\pi}{1-p^2} & \text{if } |p| < 1, \quad (20.67\text{a}) \\ \dfrac{2\pi}{p^2-1} & \text{if } |p| > 1, \quad (20.67\text{b}) \end{cases}$$

$$J(p) \equiv \int_0^{2\pi} \left\{\left(1+p^2\right)\sec\varphi - 2p\right\}^{-1} \mathrm{d}\varphi = \begin{cases} \dfrac{2\pi p}{1-p^2} & \text{if } |p| < 1, \quad (20.68\text{a}) \\ \dfrac{2\pi}{p\left(p^2-1\right)} & \text{if } |p| > 1. \quad (20.68\text{b}) \end{cases}$$

Obtain the integrals of even powers of the circular cosine and sine:

$$H(n) \equiv \int_0^{2\pi} \cos^{2n}\varphi \,\mathrm{d}\varphi = \pi 2^{1-2n} \frac{(2n)!}{(n!)^2} = \int_0^{2\pi} \sin^{2n}\varphi \,\mathrm{d}\varphi, \tag{20.69}$$

and prove that the integrals for odd powers are zero. Evaluate also the integrals:

$$K(n) \equiv \int_0^{2\pi} \exp(\cos\varphi) \begin{matrix}\cos\\ \sin\end{matrix} (\sin\varphi) \begin{matrix}\cos\\ \sin\end{matrix} (n\varphi)\,\mathrm{d}\varphi = \frac{\pi}{n!}, \tag{20.70a,b}$$

$$\int_0^{2\pi} \exp(\cos\varphi)\cos(\sin\varphi - n\varphi)\,\mathrm{d}\varphi = \frac{2\pi}{n!} = 2K(n), \tag{20.71}$$

and prove that the combinations cos sin, sin cos in (20.70a,b) and sin in (20.71) would lead to zero.

The integrals may be evaluated by the method of residues (Section 17.2) transforming via the change of variable (17.7a) $\equiv$ (20.72a) into an integral along the unit circle (20.72b):

$$z = \mathrm{e}^{i\varphi}: \quad I(p) = \int_{|z|=1} \left[1 - p(z + 1/z) + p^2\right]^{-1} \frac{\mathrm{d}z}{iz}, \tag{20.72a,b}$$

in the case (20.67a,b); the latter (20.72b) $\equiv$ (20.72c,d) takes the form,

$$I(p) = i \int_{|z|=1} f(z)\,\mathrm{d}z; \quad \frac{1}{f(z)} = pz^2 - \left(1+p^2\right)z + p. \tag{20.72c,d}$$

The integrand (20.72d) $\equiv$ (20.73a):

$$\frac{1}{f(z)} = p(z-p)\left(z - \frac{1}{p}\right), \quad f_{(1)}(p) = \frac{1}{p^2-1} = -f_{(1)}\left(\frac{1}{p}\right), \tag{20.73a–c}$$

has simple poles at $z = p, 1/p$, with residues (20.73b,c). The integral (20.72c) equals $2\pi i^2 = -2\pi$ times the residue at the pole inside the unit circle, viz. (20.67a) [(20.67b)] follows from (20.73b) [(20.73c)] if $|p| < 1\,(|p| > 1)$. The method of residues should not be applied for (20.74a) in (20.67a,b) $\equiv$ (20.74b), since the integral diverges:

$$2I(\pm 1) \equiv \int_0^{2\pi} (1 \mp \cos\varphi)^{-1} \mathrm{d}\varphi = \infty, \tag{20.74a,b}$$

because (19.39a) there is a nonintegrable singularity, viz.:

$$I(+1) :\sim \frac{2}{1-\cos\varphi} \sim \frac{1}{\varphi^2}; \quad I(-1) :\sim \frac{2}{1+\cos\varphi} \sim \frac{1}{(\varphi-\pi)^2}, \tag{20.74c,d}$$

respectively, at $\varphi = 0\,(\varphi=\pi)$ for (20.74c) [(20.74d)]; the integral (20.74a) diverges like (20.67a,b) for $p \to \pm 1$.

The same change of variable (20.72a) also transforms (20.68a,b) into an integral along the unit circle (20.75a) of the function (20.75b):

$$J(p) = i \int_{|z|=1} g(z)\,\mathrm{d}z: \quad g(z) = -z^{-1}\left(2\frac{1+p^2}{z+1/z} - 2p\right)^{-1} = \frac{1}{2z}\frac{z^2+1}{pz^2-(1+p^2)z+p}$$
$$= \frac{z^2+1}{2p}\frac{1}{z(z-p)(z-1/p)}. \tag{20.75a,b}$$

The latter has simple poles at $z = 0,\ p,\ 1/p$ with residues:

$$g_{(1)}(0) = \frac{1}{2p}, \quad g_1(p) = \frac{p^2+1}{2p(p^2-1)} = -g_1\left(\frac{1}{p}\right). \tag{20.75c–e}$$

The integral (20.75a) is evaluated as $2\pi i$ times the residues at the poles in the unit disk, namely: (i) simple poles at $z=0,\ p$ for (20.76a) in (20.76b) $\equiv$ (20.68a):

$$|p|<1: \quad J(p) = 2\pi i^2\left[g_{(1)}(0) + g_{(1)}(p)\right] = -\frac{\pi}{p}\left(1+\frac{p^2+1}{p^2-1}\right) = \frac{2\pi p}{1-p^2}, \tag{20.76a,b}$$

$$|p|>1: \quad J(p) = 2\pi i^2\left[g_{(1)}(0) + g_{(1)}\left(\frac{1}{p}\right)\right] = -\frac{\pi}{p}\left(1-\frac{p^2+1}{p^2-1}\right) = \frac{2\pi}{p(p^2-1)}, \tag{20.76c,d}$$

(ii) simple poles at $z = 0, 1/p$ for (20.76c) in (20.76d) $\equiv$ (20.68b). The case $p = \pm 1$ would lead in (20.75a,b) to the similar singularities (20.74a–d) as (20.72c,d). In the remaining case $|p| = 1$ with $p \neq \pm 1$, the poles $z = p, 1/p$ lie on the path of integration outside the real axis; their residues are multiplied by πi (zero) if the pole is included (excluded) by stretching (indenting) the unit circle, implying for $I(p)$ [$J(p)$] that: (i) it is zero ($1/p$) if both poles are included or excluded; (ii) it is one-half of (20.67a) [it is (20.76e)] if the pole at $\zeta = p$ is included and the pole at $\zeta = 1/p$ is excluded:

$$J(p) = 2\pi i^2\left(\frac{1}{2p} \pm \frac{1}{4p}\frac{p^2+1}{p^2-1}\right) = \frac{\pi}{2p}\frac{\{3p^2-1, p^2-3\}}{1-p^2}; \tag{20.76e,f}$$

(iii) vice-versa leads to one-half of (20.76b) [to (20.76f)].

The result:

$$z = e^{i\varphi}: \quad \int_{|z|=1} z^{-n}\,\mathrm{d}z = i\int_0^{2\pi} e^{i(1-n)\varphi}\,\mathrm{d}\varphi = \begin{cases} 2\pi i & \text{if } n=1 \\ 0 & \text{if } n \neq 1, \end{cases} \tag{20.77a,b}$$

can be used to evaluate the integral:

$$H(n) \equiv \int_0^{2\pi} \cos^{2n\varphi}\,\mathrm{d}\varphi = \int_{|z|=1} 2^{-2n}\left(z+\frac{1}{z}\right)^{2n}(iz)^{-1}\,\mathrm{d}z; \tag{20.78a}$$

using the binomial theorem (20.78b):

$$\left(z+\frac{1}{z}\right)^{2n}=\sum_{p=0}^{2n}\binom{2n}{p}z^{2n-2p};\quad A=\binom{2n}{n}=\frac{(2n)!}{(n!)^2}, \tag{20.78b,c}$$

the coefficient of z^{-1} in the integrand of (20.78a) involves only the term $p=n$, viz. (20.78b); this multiplied by $-i2^{-2n}$ is the residue of the integrand at the pole at the origin, and thus the integral is given by:

$$H(n)=2\pi i.-i2^{-2n}\;A=2^{1-2n}\;\pi A=2^{1-2n}\pi\frac{(2n)!}{(n)!}, \tag{20.78d}$$

in agreement with (20.69). The same result applies replacing $\cos^{2n}\varphi$ by $\sin^{2n}\varphi$. The odd powers of cos, sin change sign under the transformation $\varphi\to\varphi+\pi$, and lead to zero integrals in (20.69).

The integral (20.71) results from the sum of two integrals (20.70a,b) using the addition formula for the cosine (3.25a). The combinations cos sin and sin cos in (20.70a,b) [sin in (20.71)] change sign under the transformation $\varphi\to\varphi+\pi$ and lead to zero integrals. The proof of (20.70b) is similar to (20.70a); thus it remains to prove the latter, which is the real part of the integral:

$$z=e^{i\varphi},\quad d\varphi=-\frac{i}{z}dz:\quad 2K(n)=-i\int_{|z|=1}e^{z}\left(z^{n}+z^{-n}\right)z^{-1}\,dz. \tag{20.79a,b}$$

The pole of order $n+1$ the origin allows evaluation of the integral from its residues (15.33a,b):

$$K(n)=2\pi i.\left(-\frac{i}{2}\right)\lim_{z\to0}\frac{1}{n!}\frac{d^n}{dz^n}\left\{e^z\left(z^{2n+1}+1\right)\right\}=\pi\lim_{z\to0}\frac{1}{n!}\sum_{p=0}^{n}\binom{n}{p}e^z\frac{d^p\left(1+z^{2n+1}\right)}{dz^p}=\frac{\pi}{n!} \tag{20.80}$$

where the chain rule (13.31) was used, and the only non vanishing power is $p=0$, so that (20.80) $\equiv$ (20.70a).

Example 20.14. *Evaluation of Improper Integrals along the Real Line.*
Evaluate the following integrals for a real and positive a:

$$a>0:\quad\int_{-\infty}^{+\infty}\left(x^4-a^4\right)^{-1}dx=-\frac{\pi}{2a^3}, \tag{20.81}$$

$$a>0:\quad\int_{-\infty}^{+\infty}\left(x^4+a^4\right)^{-1}dx=\frac{\pi}{a^3\sqrt{2}}, \tag{20.82}$$

$$a>0:\quad\int_{-\infty}^{+\infty}\left(x^2+a^2\right)^{-3}dx=\frac{3\pi}{8a^5}, \tag{20.83}$$

where (20.82; 20.83) are improper integrals of the first kind, and (20.81) is of the third kind with poles on the real axis included.

The integrand (20.84a) in (20.81) has four simple poles with residues (20.84c,d):

$$\frac{1}{f(z)}=z^4-a^4;\quad f_{(1)}(z)=\frac{1}{4z^3},\quad f_{(1)}(\pm a)=\pm\frac{1}{4a^3},\quad f_{(1)}(\pm ia)=\pm\frac{i}{4a^3}; \tag{20.84a–d}$$

the integrand (20.84a) satisfies the asymptotic condition (17.19) allowing closing the path of integration (17.18) either on the upper or lower half-plane; evaluating the integral using the residues at the poles in the interior (20.84c) and on the boundary (20.84b) leads to

$$\int_{-\infty}^{+\infty} f(x)\,\mathrm{d}x = \pm 2\pi i f_{(1)}(\pm ia) \pm i\pi\left[f_{(1)}(a) + f_{(1)}(-a)\right] = \pm 2\pi i . \pm \frac{i}{4a^3} = -\frac{\pi}{2a^3}, \tag{20.85}$$

that coincides with (20.81) ≡ (20.85). The integrand in (20.86a) has simple poles with residues (20.86c,d):

$$\frac{1}{g(z)} = z^4 + a^4; \quad g_{(1)}(z) = \frac{1}{4z^3}, \quad g_{(1)}\left(\frac{a(1\pm i)}{\sqrt{2}}\right) = -\frac{1\pm i}{4a^3\sqrt{2}}, \quad g_{(1)}\left(\frac{a(-1\pm i)}{\sqrt{2}}\right) = \frac{1\mp i}{4a^3\sqrt{2}}; \tag{20.86a–d}$$

since all poles are outside the real axis the integral is evaluated by

$$\int_{-\infty}^{+\infty} g(x)\,\mathrm{d}x = \pm 2\pi i\left[g_{(1)}\left(\frac{a(1\pm i)}{\sqrt{2}}\right) + g_{(1)}\left(\frac{a(-1\pm i)}{\sqrt{2}}\right)\right] = \pm 2\pi i . \mp \frac{2i}{4a^3\sqrt{2}} = \frac{\pi}{a^3\sqrt{2}}, \tag{20.87}$$

in agreement with (20.82) ≡ (20.87). The integrand of (20.83) is (20.88a) that has triple poles (15.33b) with residues (20.88b):

$$\frac{1}{h(z)} = \left(z^2 + a^2\right)^{-3}, \quad h_{(3)}(\pm ia) = \mp\frac{3i}{16a^5}; \quad \int_{-\infty}^{+\infty} h(x)\,\mathrm{d}x = \pm 2\pi i . \mp \frac{3i}{16a^5} = \frac{3\pi}{8a^5}; \tag{20.88a–c}$$

it meets the asymptotic condition (17.19) allowing closing the path of interaction in the upper or lower half of the complex z-plane, and leading to the integral (20.88c) ≡ (20.83) In (20.88c) were used:

$$h_{(3)}(\pm ia) = \frac{1}{2}\lim_{z\to\pm ia}\frac{\mathrm{d}^2}{\mathrm{d}z^2}(z\pm ia)^{-3} = 6\lim_{z\to\pm ia}(z\pm ia)^{-5} = \mp\frac{3i}{16a^5}, \tag{20.88d}$$

the residues (20.88d) ≡ (20.88b) at the triple roles.

Example 20.15. *Evaluation of Fourier Cosine and Sine Transforms.* Evaluate the following integrals with a and b real and positive:

$$a,b>0: \quad \int_{-\infty}^{+\infty} (x-a)^{-1}\cos(bx)\,\mathrm{d}x = -2\pi\sin(ab), \tag{20.89}$$

$$a,b>0: \quad \int_{-\infty}^{+\infty} (x\pm a)^{-2}\cos(bx)\,\mathrm{d}x = -\pi b\cos(ab), \tag{20.90}$$

$$a,b>0: \quad \int_{-\infty}^{+\infty} (x\pm a)^{-2}\sin(bx)\,\mathrm{d}x = \pm\pi b\sin(ab), \tag{20.91}$$

$$a,b>0: \quad \int_{-\infty}^{+\infty} \left(x^2-a^2\right)^{-2}\cos(bx)\,\mathrm{d}x = -\frac{\pi b}{2a^2}\cos(ab) + \frac{\pi}{2a^3}\sin(ab), \tag{20.92}$$

$$a,b>0: \quad \int_{-\infty}^{+\infty} \left(x^2+a^2\right)^{-2}\cos(bx)\,\mathrm{d}x = \frac{\pi}{2a^2}\left(b+\frac{1}{a}\right)\mathrm{e}^{-ba}, \tag{20.93}$$

that are all improper integrals of the first or third kind. Prove that (20.92; 20.93) with sin instead of cos is zero.

All integrands satisfy the condition (17.29), and since $b > 0$ can be evaluated by the method (17.32) of residues closing the path of integration in the upper half of the complex plane, viz.: (i) for (20.89) using the residue at the simple pole at a:

$$\mathrm{Re}[2\pi i\ \exp(iba)] = -2\pi \sin(ba); \tag{20.94}$$

(ii/iii) for (20.90) [20.91)] using the residues at the double poles at $\mp a$:

$$\mathrm{Re}, \mathrm{Im}\left\{\pi i \lim_{z\to\mp a} \frac{\mathrm{d}}{\mathrm{d}z} \mathrm{e}^{ibz}\right\} = \mathrm{Re}, \mathrm{Im}\left\{(\pi i)(ib)\exp[ib(\mp a)]\right\} = -\pi b\{\cos(ab),\ \mp\sin(ab)\}, \tag{20.95}$$

and taking the real (imaginary) part; (iv) for (20.92) using the residues at the double poles at $\pm a$:

$$i\pi.\frac{1}{4a^2}\cdot\left[\left(ib-\frac{1}{a}\right)\mathrm{e}^{iba}+\left(ib+\frac{1}{a}\right)\mathrm{e}^{-iba}\right] = -\frac{\pi b}{2a^2}\cos(ba) + \frac{\pi}{2a^3}\sin(ba); \tag{20.96}$$

(v) for (20.93) using the double pole at ia:

$$2\pi i \lim_{z\to ia}\frac{\mathrm{d}}{\mathrm{d}z}\left\{(z+ia)^{-2}\mathrm{e}^{ibz}\right\} = (2\pi i)\,i\left(b+\frac{1}{a}\right)\left(-\frac{1}{4a^2}\right)\exp[ib(ia)] = \frac{\pi}{2a^2}\mathrm{e}^{-ab}\left(b+\frac{1}{a}\right); \tag{20.97}$$

(vi) for (20.92; 20.93) with sin instead of cos the integrand would be an odd function of x, and the integral would vanish.

Example 20.16. *Evaluation of Mellin Transforms.*
Evaluate the integrals (20.98c,d), with the conditions (20.98a,b):

$$a > 0 < \mathrm{Re}(\alpha) < 1: \quad \int_0^\infty x^{\alpha-1}\left\{(x+a)^{-1}, (x-a)^{-1}\right\}\mathrm{d}x = \pi a^{\alpha-1}\left\{\csc(\pi\alpha), -\cot(\pi\alpha)\right\}, \tag{20.98a–d}$$

and obtain:

$$\int_0^\infty x^{-1/2}(x+a)^{-1}\,\mathrm{d}x = \frac{\pi}{\sqrt{a}}, \quad \int_0^\infty x^{-3/4}(x-a)^{-1}\,\mathrm{d}x = -\frac{\pi}{\sqrt[4]{a^3}}, \tag{20.99a,b}$$

as particular cases.

The integrals (20.98c) [(20.98d)] are of the type (17.53) where the functions (20.100b):

$$0 < \mathrm{Re}(\alpha) < 1: \quad \frac{1}{g^\pm(z)} = z \pm a, \quad \lim_{z\to 0,\infty} z^\alpha g^\pm(z) = \lim_{z\to 0,\infty}\frac{z^\alpha}{z\pm a} = 0, \tag{20.100a–d}$$

satisfy the conditions (17.49b,c) $\equiv$ (20.100c,d); this allows the evaluation of the integral (20.98a) [(20.98b)]:

$$g^\pm_{(1)}(\mp a) = 1, \quad \int_0^\infty x^{\alpha-1}(x\pm a)^{-1}\,\mathrm{d}x = \pm\pi a^{\alpha-1}\,\begin{matrix}\csc\\ \cot\end{matrix}\,(\pi\alpha), \tag{20.101a,b}$$

using the residue (20.101a) at the pole on the negative (positive) real axis. The particular case of (20.98c) [(20.98d)] for $\alpha = 1/2\,(\alpha = 1/4)$ is (20.99a) [(20.99b)].

Example 20.17. *Proper and Improper Integrals of a Power.*
Consider all convergent cases of the integral of a power along an interval of or the whole real line.

Of all integrals of the function x^α with real x and complex α, only the following converge: (i) the integral over a finite interval:

$$\int_a^b x^\alpha \,\mathrm{d}x = \frac{b^{\alpha+1} - a^{\alpha+1}}{\alpha+1}, \tag{20.102}$$

that is a proper integral for all a, b if $\mathrm{Re}(\alpha) > 0$, and for all α with a, b having the same sign $ab > 0$; (ii) if a, b have opposite signs $ab < 0$, then (20.102) is an improper integral of second kind for $0 > \mathrm{Re}(\alpha) > -1$, bilateral if $a \neq 0 \neq b$, and lower (upper) unilateral if $a = 0 \neq b\,(a \neq 0 = b)$; (iii) letting $b \to \infty\,(a \to \infty)$ leads to an upper (lower) unilateral improper integral of first kind:

$$\int_a^\infty x^\alpha \,\mathrm{d}x = -\frac{a^{\alpha+1}}{\alpha+1}, \qquad \int_{-\infty}^b x^\alpha \,\mathrm{d}x = \frac{b^{\alpha+1}}{\alpha+1}, \tag{20.103a,b}$$

for $\mathrm{Re}(\alpha) > -1$ and $a > 0\,(b < 0)$. The bilateral case $-a, b \to \infty$ in (20.102), is a special case of the improper integral of third kind, viz. $a < 0, b = \infty$, or $a = -\infty, b > 0$, which never converges, because: (i) $x^{\alpha+1}$ diverges either at 0 or at $\pm\infty$ for all $\alpha \neq -1$; (ii) for $\alpha = -1$ then $x^{\alpha+1} = 1$ has a divergent integral for $a \to -\infty$ or $b \to \infty$.

Example 20.18. *Gravity Field of Parallel Plates.*
Obtain the gravity field due to identical homogeneous plates, of density per unit length μ and width $2a$, placed parallel to each other, in the cases of: (i) two plates (Figure 20.4a) at distance $2b$, placed at $(-a \pm ib, +a \pm ib)$, with positive real a, b; (ii) $2n + 1$ plates (Figure 20.4b) at distance b, placed at $(-a \pm imb, +a \pm imb)$ with $m = 0, 1, \ldots, n$. Obtain in the former case (i) the Cartesian components of the gravity field in the whole plane including the limiting form at the plates.

The principle of superposition is applied complex conjugate gravity field (18.33a) in for a plate of width $2a$: (i) at two positions:

$$g^*(z) = \frac{G\mu}{2\pi}\left[\log\left(\frac{z - a - ib}{z + a - ib}\right) + \log\left(\frac{z - a + ib}{z + a + ib}\right)\right], \tag{20.104}$$

that simplifies to:

$$g_x - ig_y = g^*(z) = \frac{G\mu}{2\pi}\log\left[\frac{(z-a)^2 + b^2}{(z+a)^2 + b^2}\right]; \tag{20.105}$$

(ii) at $(2n+1)$ positions:

$$g^*(z) = \frac{G\mu}{2\pi}\sum_{m=-n}^{+n}\int_{-a+imb}^{a+imb}\frac{\mathrm{d}\zeta}{\zeta - z} = \frac{G\mu}{2\pi}\sum_{m=-n}^{+n}\log\left(\frac{z - a - imb}{z + a - imb}\right), \tag{20.106}$$

that simplifies to:

$$g^*(z) = \frac{G\mu}{2\pi}\log\left[\frac{z-a}{z+a}\prod_{m=1}^{n}\frac{(z-a)^2 + m^2b^2}{(z+a)^2 + m^2b^2}\right], \tag{20.107}$$

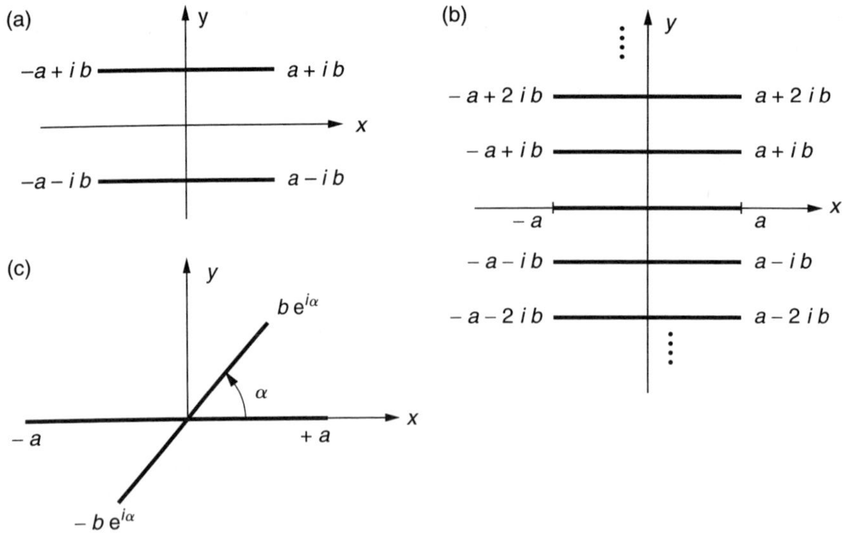

FIGURE 20.4
The gravity field due to several mass distributions: (i) two identical parallel plates (a); (ii) two plates with different lengths crossing at the middle at an arbitrary angle (c); (iii) an odd number of equally spaced identical parallel plates (b).

The Cartesian components of the gravity field (29.105) are

$$\begin{aligned} g_x(x,y) = \frac{G\mu}{4\pi}\Big\{&\log\left|\left[(x-a)^2+b^2-y^2\right]^2+4y^2(x-a)^2\right| \\ &-\log\left|\left[(x+a)^2+b^2-y^2\right]^2+4y^2(x+a)^2\right|\Big\}, \end{aligned} \tag{20.108a}$$

$$\begin{aligned} g_y(x,z) = \frac{G\mu}{2\pi}\Big\{&(+,+,-)\operatorname{arc\,cot}\left[\frac{(x+a)^2+b^2-y^2}{2y(x+a)}\right] \\ &+(-,-,+)\operatorname{arc\,cot}\left[\frac{(x-a)^2+b^2-y^2}{2y(x-a)}\right]\Big\}, \end{aligned} \tag{20.108b}$$

where (20.108a) holds at all positions, and (20.108b) has the upper/middle/lower signs, respectively, above/between/below the plates. It follows that on each plate:

$$g_x(x,\pm b) = \frac{G\mu}{2\pi}\log\left|\frac{x-a}{x+a}\right| + \frac{G\mu}{4\pi}\log\left[\frac{(x-a)^2+4b^2}{(x+a)^2+4b^2}\right], \tag{20.109a}$$

$$g_y(x,\pm b) = \frac{G\mu}{2\pi}\left[\operatorname{arc\,cot}\left(\pm\frac{x+a}{2b}\right) - \operatorname{arc\,cot}\left(\pm\frac{x-a}{2b}\right)\right]; \tag{20.109b}$$

the tangential gravity field is the same in both plates, whereas the normal gravity field has opposite signs on the two plates such that is always *an* attraction.

Example 20.19. *Gravity Field of Two Crossed Plates.*
Consider the gravity field due to two homogeneous plates with the same mass density μ per unit length, with lengths $2a$ and $2b$, and crossing at their centers at an angle α (Figure 20.4c). In the case of orthogonal plates $\alpha = \pi/2$ obtain the Cartesian components of the

gravity field at all points in the plane, including the limiting form at the plates and their crossing point.

The complex conjugate gravity fields due to plates with ends at $\pm a$ and $\pm b e^{i\alpha}$ and density μ is given (18.33a) summed for each plate.

$$g^*(z) = \frac{G\mu}{2\pi}\left[\log\left(\frac{z-a}{z+a}\right) + \log\left(\frac{z-be^{i\alpha}}{z+be^{i\alpha}}\right)\right]. \tag{20.110}$$

In the case of orthogonal plates this simplifies to

$$\alpha = \frac{\pi}{2}: \quad g^*(z) = \frac{G\mu}{2\pi}\log\left[\frac{(z-a)(z-ib)}{(z+a)(z+ib)}\right]. \tag{20.111}$$

The Cartesian components of the gravity field are

$$g_x(x,b) = \frac{G\mu}{4\pi}\log\left|\frac{\left[(x-a)^2+y^2\right]\left[x^2+(y-b)^2\right]}{\left[(x+a)^2+y^2\right]\left[x^2+(y+b)^2\right]}\right|, \tag{20.112a}$$

$$g_y(x,y) = \frac{G\mu}{2\pi}\left[\operatorname{arc\,cot}\left(\frac{x+a}{y}\right) + \operatorname{arc\,tan}\left(\frac{y+b}{x}\right) - \operatorname{arc\,cot}\left(\frac{x-a}{y}\right) - \operatorname{arc\,tan}\left(\frac{y-b}{x}\right)\right]; \tag{20.112b}$$

in particular at the plates:

$$g_x(|x|<a,0) = \frac{G\mu}{2\pi}\log\left(\frac{a-x}{a+x}\right), \quad g_y(|x|<a,0) = \frac{G\mu}{\pi}\operatorname{arc\,tan}\left(\frac{b}{x}\right), \tag{20.113a,b}$$

$$g_x(0,|y|<b) = \frac{G\mu}{2\pi}\log\left(\frac{b-y}{b+y}\right), \quad g_y(0,|y|<b) = \frac{G\mu}{\pi}\operatorname{arc\,tan}\left(\frac{a}{y}\right), \tag{20.114a,b}$$

and vanishes at the crossing point as follows from $g^*(0) = 0$ in (20.111).

Example 20.20. *Gravity Field of an Infinite Mass Distribution.*
Consider an infinite plate with mass density:

$$a > b > 0 < c: \quad \mu(x) = \frac{c}{(x^2+a^2)(x^2+b^2)}, \tag{20.115}$$

and determine the total mass, and gravity field on the plane, including in the plate.

The total mass follows from (17.26b) viz. (20.116):

$$m = \int_{-\infty}^{+\infty}\mu(x)\,dx = \frac{\pi c}{ab(a+b)} = \pi c\frac{1/b-1/a}{a^2-b^2}. \tag{20.116}$$

The complex conjugate gravity field is given (18.27b) by (20.117a,b):

$$g^*(z) = \frac{Gc}{2\pi}\int_{-\infty}^{+\infty} h(x)\,dx, \quad \frac{1}{h(\zeta)} \equiv (\zeta^2+a^2)(\zeta^2+b^2)(\zeta-z); \tag{20.117a,b}$$

the integrand (20.117b) has simple poles at $\zeta = \pm ia, \pm ib, z$, with residues:

$$h_{(1)}(z) = \left[(z^2+a^2)(z^2+b^2)\right]^{-1}, \quad h_{(1)}(\pm ia) = \mp\frac{i/(2a)}{(b^2-a^2)(\pm ia - z)}, \tag{20.118a,b}$$

$$h_{(1)}(\pm ib) = \pm\frac{i/(2b)}{(b^2-a^2)(\pm ib - z)}. \tag{20.118c}$$

Since the asymptotic condition (17.19) is met in all directions, the path of integration in (20.117a) a can be closed (Section 17.4) either in the upper or lower complex plane, for example, (i) for an observer below $y<0$ (above $y>0$) the mass distribution, closing above (below) leaves only two poles $(+ia,\ +ib)$ $[(-ia,\ -ib)]$ in the interior whose residues (20.118b,c) evaluate the integral (20.117a) ≡ (20.119a):

$$y<>0:\quad g^*(z)=\frac{\mathrm{G}c}{2\pi}(\pm 2\pi i)\left[h_{(1)}(\pm ia)+h_{(1)}(\pm ib)\right]$$
$$=\frac{\mathrm{G}c/2}{a^2-b^2}\left[\frac{1}{a(z\mp ia)}-\frac{1}{b(z\mp ib)}\right]\equiv k_\pm(x);\tag{20.119a}$$

(ii) at the plate itself must be added $\pm\pi i$ times the residue (20.118a) at the field point, when closing the real axis above (below):

$$y=0:\quad g^*(x)=k_\pm(x)\pm\frac{i\mathrm{G}c/2}{(x^2+a^2)(x^2+b^2)}.\tag{20.119b}$$

Using (20.116) the complex conjugate gravity field is given by:

$$g^*(z)=\frac{\mathrm{G}c/2}{a^2-b^2}j(z)=\frac{\mathrm{G}m/(2\pi)}{1/b-1/a}j(z),\tag{20.120a}$$

where (20.119a) [(20.119b)] lead to (20.120b) [(20.120c)]:

$$y<>0:\quad j(x+iy)=\frac{1}{a[x+i(y\mp a)]}-\frac{1}{b[x+i(y\mp b)]},\tag{20.120b}$$

$$y=0:\quad j(x)=\frac{1}{a(x\mp ia)}-\frac{1}{b(x\mp ib)}\pm i\frac{a^2-b^2}{(x^2+a^2)(x^2+b)^2}.\tag{20.120c}$$

The latter (20.120c) may be simplified:

$$\left(x^2+a^2\right)\left(x^2+b^2\right)j(x)=\left(x^2+b^2\right)\left(\frac{x}{a}\pm i\right)-\left(x^2+a^2\right)\left(\frac{x}{b}\pm i\right)\pm i\left(a^2-b^2\right)$$
$$=\left(x^2+b^2\right)\frac{x}{a}-\left(x^2+a^2\right)\frac{x}{b},\tag{20.120d}$$

showing that it is real; hence the normal component of the gravity field is zero at the plate.

Thus (20.120b) applies above $y>0$ and below $y<0$ the mass distribution, leading to: (i) the tangential gravity field:

$$j_x(x,y<>0)=\frac{x/a}{x^2+(y\mp a)^2}-\frac{x/b}{x^2+(y\mp b)^2},\tag{20.121a}$$

$$j_x(x,\mp 0)=\frac{x/a}{x^2+a^2}-\frac{x/b}{x^2+b^2},\tag{20.121b}$$

that is continuous across the plate; (ii) the normal gravity field:

$$j_y(x,y<>0)=\frac{y/a\mp 1}{x^2+(y\mp a)^2}-\frac{y/b\mp 1}{x^2+(y\mp b)^2},\tag{20.122a}$$

$$j_y(x,\mp 0)=\pm\left(\frac{1}{x^2+b^2}-\frac{1}{x^2+a^2}\right)=\pm\frac{a^2-b^2}{(x^2+a^2)(a^2+b^2)}\neq 0=j_y(x,0),\tag{20.122b}$$

has opposite signs above and below the plate and hence is discontinuous across the plate. From (20.120d) follows the gravity field on the plate, viz.: (i) the horizontal component is continuous (20.123a) ≡ (20.121a):

$$j_x(x,0)=\frac{x/a}{x^2+a^2}-\frac{x/b}{x^2+b^2}=j_x(x,\pm 0);\tag{20.123a}$$

(ii) the vertical component, which is discontinuous (20.122b), vanishes on the plate:

$$0 = j_y(x,0) = \frac{j_y(x,0) + j_y(x,-0)}{2}, \tag{20.123b}$$

that is, takes the arithmetic mean (20.123b) of the values (20.122b) above and below. The opposite values of the normal component of the gravity field (20.120a; 20.122b) across the plate:

$$g_y(x,\pm 0) = \mp \frac{Gc/2}{(x^2+a^2)(x^2+b^2)} = \mp \frac{G\mu(x)}{2}, \tag{20.124}$$

where $\mu(x)$ is the mass density (20.115), imply in agreement with Section 18.8, that: (i) the mean normal gravity field on the plate is zero (20.123b) ≡ (18.40a) and it is negative above (positive below), that is, there is always attraction towards the plate (20.124) ≡ (18.39a,b); (ii) the jump of the normal gravity field across the plate is (18.40b). The result (20.124) is independent of the particular mass distribution (18.22b), as well as the continuity (20.124a) ≡ (18.22a) of the tangential gravity field.

Note 20.1. *Series Expansions and Potential Fields.*
The two main parallel lines of advance in Part 2 are taken further along Part 3: (i) the calculus of residues used for integration in Part 2, serves in Part 3 to establish general representations of a function in power series, allowing a classification of singularities and functions; (ii) the analogies for potential fields are extended from potential flows and gravity fields in Part 2 to electrostatic and magnetostatic fields in Part 3. In both cases there is more than a build-up of results, there is also the emergence of a coherent view of: (i) classes of functions and their properties, including convergence of series and behavior near singularities; (ii) additional aspects of potential fields, like forces and moments on bodies. These more general mathematical methods (i) enlarge the range of potential fields problems (ii) that can be addressed.

Conclusion 20: The transformation (figure 20.1) from Cartesian (v_x, v_y) to polar $(v_r,\, v_\varphi)$ components of the velocity is a positive rotation by φ and the inverse a negative rotation by $-\varphi$. The integral of a holomorphic (nonholomorphic) function between the same end-points $0, 1+i$, along different paths (figure 20.2), has always (generally does not have) the same value. Paths of integration with one singularity at an angular point (figure 20.3) viz. a salient point (**a**), and two reentrant points (**b**) and (**c**); the path (**d**) has a singularity at a regular point. The gravity field (figure 20.4) due to two (**a**) $[2n+1$ (**b**)$]$ homogeneous plates of width $2a$ at a distance $2b$ (**b**) is a superposition of that due to each plate separately, and likewise for two plates (**c**) crossing at an angle.

Part 3
Power Series: Electricity and Magnetism

After considering the geometry of the complex plane (Part 1) and the basic properties of differentiation and integration of complex functions (Part 2), the latter are used to construct general exact representations of functions (Part 3). A function may be approximated by a polynomial of degree N, that coincides with it at N points; to obtain an arbitrarily high order of approximation at all points, a power series with $N = \infty$ terms is needed. The power series is a particular case of the series of functions, that is, an infinite sum whose terms may be functions other than powers. Series differ from finite sums in that operations with them are subject to conditions of convergence (Chapter 21), to find out if the sum of an infinite number of terms has a finite and definite value at each point of the complex plane (Chapter 29). Power series give convenient representations of functions; for example, in the neighborhood of a regular (singular) point, a holomorphic function can be represented as an ascending (ascending and descending) power series of another function, viz. the Lagrange–Burmann, Taylor and Stirling–Maclaurin (Teixeira, Laurent, and Laurent–Maclaurin) series (Chapter 23 [25]). Series expansions can be used to calculate functions with an arbitrarily high order of accuracy, by truncating the sum, and estimating the error as an upper bound of the remainder. Power series can be used first to identify types of singularities, and then the singularities serve to classify the functions (Chapter 27). Series occur in connection with infinite sequences or chains of processes, for example, multiple reflections of light in a lens (Chapter 22). The representation of a holomorphic function in a power series also serves to decompose a potential flow into a superposition of corner flows. The flow fails to be potential at the singularities of the function that represent points where it is rotational and/or compressible, for example, vortices, sources, or sinks. The latter are monopoles, and their combination leads to dipoles, quadrupoles, and so on; the superposition of multipoles can be used to construct any isolated singularity of a holomorphic function. The multipole representation, as well as other properties of complex functions, applies to the electrostatic (Chapter 24) (magnetostatic [Chapter 26]) field due to electric charges (currents) that have analogies with the potential flow due to sources/sinks (vortices), as well as differences, for example, induced electric charges (surface electric currents) in a conductor. The analogies also include the gravity field (Chapter 18), and apply (Chapter 28) to forces and moments on a body (e.g., the lift and drag forces, the electromagnetic and gravity forces) and to energies (e.g., kinetic, electric, magnetic, and gravity energies).

21

Convergence of and Operations on Series

Since a series is a sum with an infinite number of terms (Section 21.1), it can be (i) convergent, (ii) divergent, or (iii) oscillating depending on whether its sum is, respectively (i) finite and unique, (ii) infinite, or (iii) takes more than one finite value. The sum with a finite number of terms has the association (commutation) properties, and these extend to convergent (Section 21.1) [absolutely convergent (Section 21.3)] series, whose terms may be associated (Section 21.2) [deranged (Section 21.1)] without changing the sum; a series is absolutely convergent (Section 21.3) if the series of moduli converges, this being a stricter requirement than simple convergence. The sum (Section 21.1) [product (Section 21.4)] of a series that involve association (rearrangement) of terms, applies both to simply and absolutely convergent series. Another concept of convergence, stronger than simple, and distinct from absolute, is uniform convergence (Section 21.5); the limit, derivate, and integral operators can be applied term-by-term (Section 21.6) to a uniformly convergent series. The strongest concept of convergence, is the totally convergent series (Section 21.7) that is, both, absolutely and uniformly convergent so that all relations and operations indicated above, and listed in Table 21.1, apply. A series of complex functions, for example, a geometric series (Section 21.8), may have different types of behavior, viz., divergence, oscillation, and simple, absolute, uniform, or total convergence, at distinct points, curves, or regions of the complex plane (Section 21.9). The power series can be seen as the successive approximations of a function by polynomials of increasing degree to obtain an arbitrary small error, provided that the series be convergent. For many problems that have no solution in finite form, the series provides a convenient approach, provided that its convergence properties allow the required operations.

21.1 Convergent, Divergent, and Oscillating Series

A **series** is defined as an infinite sum:

$$\sum_{n=0}^{\infty} f_n(z) \equiv \lim_{N\to\infty} \sum_{n=0}^{N} f_n(z) = f_0(z) + f_2(z) + \cdots + f_n(z) + \cdots, \tag{21.1}$$

where $f_n(z)$ is called the **general term**. The **sum** of the first N terms is given by (21.2a):

$$S_N(z) \equiv \sum_{n=0}^{N-1} f_n(z) = f_0(z) + \cdots + f_{N-1}(z), \quad R_N(z) \equiv \sum_{n=N}^{\infty} f_n(z) = f_N(z) + f_{N+1}(z) + \cdots \tag{21.2a,b}$$

and the **remainder** after N terms (21.2b) adds the terms of (21.1) not present in (21.2a). Taking the limit of partial sums of the series (21.2a) as the number of terms tends to infinity

TABLE 21.1
Transformation of and Operations between Series

Types of series: section 21.2	**Oscillating**	**Divergent**	**Conditionally convergent**	**Absolutely convergent**	**Uniformly convergent**	**Totally convergent**
Classes of convergence	O.	D.	C.C.	A.C.	U.C.	T.C.
Transformations of one series:						
Association of terms: 21.2	–	×	×	×	×	×
Permutation of terms: 21.3; 21.4	–	–	–	×	–	×
Limit term-by term: 21.5; 21.6	–	–	–	–	×	×
Integration term-by-term: 21.5; 21.6	–	–	–	–	×	×
Derivation term-by-term: 21.5; 21.6	–	–	–	–	×	×
Operations on two series:						
Addition: 21.2; 21.7	–	–	×	×	×	×
Product or multiplication : 21.4; 21.7	–	–	×	×	×	×

Note: The crosses × (dashes –) indicate the transformations and operations, which when performed on series, do not (may) affect the convergence or the sum of the series: (i/ii) the association of terms does not (may) affect the divergence (oscillation) of a series; (iii) the association of terms does not affect the convergence of a series and does not change its sum; (iv/v/vi) the limit/differentiation/integration term-by-term of a uniformly convergent series leads to a series whose sum is the limit/derivative/integral of the sum of the original series; (vii/viii) the addition (product) of two convergent series is a series whose sum is the addition (product) of the sums of the original series; (ix) all properties (ii) to (viii) hold for a totally convergent series, which is both uniformly and absolutely convergent.

$N \to \infty$ three cases can arise:

$$\lim_{N\to\infty} S_N(z) = \begin{cases} f(z) \neq \infty & \text{C.} \\ \infty & \text{D.} \\ a, b & \text{O.,} \end{cases} \tag{21.3a–c}$$

namely: (i) if the limit exists, that is, is unique and finite (21.3a), the series (21.1) is **convergent**, with **sum** $f(z)$; (ii) if the sum in unbounded (21.3b), the series is **divergent**, regardless of whether a single limit exists or not; (iii) if the limit does not exist, but the sum is bounded (21.3c), then the series is **oscillatory**. This classification applies both to real and complex series and to operations on series, like the sum and the product (Example 30.1).

Adapting the definition of limit in (11.3) to the case (21.3a), it follows that *the series (21.1) is convergent (21.3a) with sum equal to* f(z), *iff for any positive real* ε, *there exists a positive integer* M, *such that for all* N ≥ M, *the sum (21.2a) of the first* N *terms differs from* f(z) *in modulus by less than* ε:

$$\forall_{\varepsilon>0} \exists_{M\in|N}: \qquad N \geq M \Rightarrow \left| f(z) - \sum_{n=0}^{N-1} f_n(z) \right| = |R_N(z)| \leq \varepsilon(N, z); \tag{21.4}$$

the accuracy or error ε *is an upper bound for the modulus of the remainder (21.2b), and generally depends on the point* z *and the number of terms* N:

$$\lim_{N\to\infty} |R_N(z)| = \lim_{N\to\infty} |f(z) - S_N(z)| \leq \lim_{N\to\infty} \varepsilon(z, N) = 0; \tag{21.5}$$

thus the error can be made as small as needed, by choosing a sufficiently large N. Some examples of series are:

$$O.: \qquad 1-1+1-1-1+1-1+\cdots = 1, 0; \tag{21.6a}$$

$$C.: \qquad 1 - \frac{1}{2} + \frac{1}{3} - \frac{1}{4} + \frac{1}{5} - \frac{1}{6} + \cdots + \frac{(-)^{n-1}}{n} + \cdots = \log 2, \tag{21.6b}$$

$$D.: \qquad 1 + \frac{1}{2} + \frac{1}{3} + \frac{1}{4} + \frac{1}{5} + \frac{1}{6} + \cdots + \frac{1}{n} + \cdots = \infty, \tag{21.6c}$$

where: (i) the first series (21.6a) is oscillating, since its sum alternates between the values 1 and 0; (ii) the second series (21.6b) is convergent and specifies log 2; that is, it can be obtained from the series (21.64b) for $\log(1+z)$ by setting $z = 1$, which is permissible since it converges at this point; (iii) if instead of alternating signs in the convergent series (21.6b), the same sign is taken in all terms, the series (21.6c) is obtained that is divergent, as follows from:

$$\begin{aligned} &1 + \frac{1}{2} + \frac{1}{3} + \frac{1}{4} + \frac{1}{5} + \frac{1}{6} + \frac{1}{7} + \frac{1}{8} + \cdots > 1 + \frac{1}{2} + \left(\frac{1}{4} + \frac{1}{4}\right) + \left(\frac{1}{8} + \frac{1}{8} + \frac{1}{8} + \frac{1}{8}\right) + \cdots \\ &> 1 + \frac{1}{2} + \frac{1}{2} + \frac{1}{2} + \frac{1}{2} + \cdots = \infty; \end{aligned} \tag{21.7}$$

thus **the arithmetic series** *(21.6c) is divergent;* this can be confirmed as a particular case of the harmonic series (29.67a).

21.2 Association of Terms and Sum of a Series

Starting from the oscillating series (21.6a), and associating the terms: (i) in pairs, leads to a convergent series:

$$(1-1)+(1-1)+(1-1)+\cdots=0+0+0+\cdots=0; \tag{21.8a}$$

(iii) in sets of three, regains the same oscillating series:

$$(1-1+1)+(-1+1-1)+(1-1+1)+\cdots=1-1+1-\cdots=1,0. \tag{21.8b}$$

Hence *the association of terms in an oscillating series may change its sum or convergence.* Consider the opposite case of a nonoscillating series f_n, and let g_n be a series obtained from it by association of terms, viz. the first $N+1$ terms of g_n arise from the association of the first $N+K+1$ terms of f_n in (21.9a):

$$\sum_{n=0}^{N} g_n(z)=\sum_{n=0}^{N+K} f_n(z); \qquad \sum_{n=0}^{\infty} g_n(z)=\sum_{n=0}^{\infty} f_n(z), \tag{21.9a,b}$$

since the series f is nonoscillating, the limit of the r.h.s. of (21.9a) exists as $N\to\infty$, and thus the series g_n has the same sum, that is, it converges (diverges) if f_n converges (diverges). It follows that*: (i) the association of terms of a divergent series leads to a divergent series; (ii) if a series is convergent, its sum is unaltered by association of terms.*

The sum of two series is an association of terms, and it may be expected from the last statement that: *the sum of two convergent series* $\mathrm{f_n},\mathrm{g_n}$ *is a series* $\mathrm{f_n}+\mathrm{g_n}$ *converging to their sum*:

$$\sum_{n=0}^{\infty} f_n(z)=f(z), \quad \sum_{n=0}^{\infty} g_n(z)=g(z): \quad \sum_{n=0}^{\infty}\{f_n(z)+g_n(z)\}=f(z)+g(z). \tag{21.10}$$

By way of contrast the following are examples of two divergent series whose sum is: (i) a convergent series (21.11a):

$$(1+1+\cdots)+(-1-1+\cdots)=(1-1)+(1-1)+\cdots=0+0+\cdots=0; \tag{21.11a}$$

$$(1+1+\cdots)+(0-2+0-2+\cdots)=(1+0)+(1-2)+\cdots=1-1+1-\cdots=1,\ 0, \tag{21.11b}$$

(ii) an oscillating series (21.11b). To prove the theorem (21.10) that the convergence of series is preserved by addition, note from (21.4) follows that if the series f_n, g_n converge to f, g, then, for any $\varepsilon>0$, positive integers M_1, M_2 exist such that for all equal or larger N, the sum is within ε of f and g:

$$N\geq M_1: \quad \left|\sum_{n=0}^{N} f_n(z)-f(z)\right|<\varepsilon; \quad N\geq M_2: \quad \left|\sum_{n=0}^{N} g_n(z)-g(z)\right|<\varepsilon; \tag{21.12a,b}$$

choosing N not smaller than M_1 and M_2, both (21.12a,b) can used:

$$\left|\sum_{n=0}^{N}\{f_n(z)+g_n(z)\}-f(z)-g(z)\right|\leq\left|\sum_{n=0}^{N} f_n(z)-f(z)\right|+\left|\sum_{N=0}^{N} g_n(z)-g(z)\right|<2\varepsilon; \tag{21.13}$$

since $\varepsilon\to0$ as by (21.5), it follows that the series f_n+g_n converges to $f+g$ as stated in (21.10). QED.

21.3 Absolute and Conditional Convergence (Dirichlet, 1837)

A series is defined to be **absolutely convergent** iff the series of moduli converges:

$$\sum_{n=0}^{\infty} f_n(z) \quad \text{A.C.} \qquad \Leftrightarrow \qquad \sum_{n=0}^{\infty} |f_n(z)| \ \text{C.} \tag{21.14}$$

The absolute convergence is stronger than ordinary convergence, because: (i) an absolutely convergent series is always convergent:

$$\lim_{N\to\infty} \left| \sum_{n=0}^{N} f_n(z) \right| \le \lim_{N\to\infty} \sum_{n=0}^{N} |f_n(z)|, \tag{21.15}$$

since the existence (*A.C.*) of the limit on the r.h.s. of (21.15) implies the existence (C.) of the limit on the l.h.s.; (ii) a series can converge, for example, (21.6b), and the series of moduli viz. (21.6c) diverge, thus combining ordinary convergence with absolute divergence. Thus can be defined a **conditionally convergent** *series, as a series combining ordinary convergence with absolute divergence:*

$$\sum_{n=0}^{\infty} f_n(z) \ \text{C.} \wedge \sum_{n=0}^{\infty} |f_n(z)| \ \text{D.} \Leftrightarrow \sum_{n=0}^{\infty} f_n(z) \ \text{C.C.} \tag{21.16}$$

Thus *the convergent* (C.) *series may be classified into absolutely (conditionally) convergent series, depending on whether the series of moduli converges* (A.C.) *[diverges* (C.C.)*]. Conversely, the absolutely divergent* (A.D.) *series can be classified into conditionally convergent (simply divergent) series, depending on whether the series without moduli converge* (C.C.) *[diverge* (D.)*]. The relation between the various types of convergence and divergence of series is indicated in Diagram 21.1.*

As the example (21.6b) shows a conditionally convergent series is the difference of two divergent series; deranging the order of the terms of the series, by changing the proportion of positive and negative terms, will thus alter the sum. It can be shown that *if a series of real numbers is conditionally convergent, then its sum can, by suitable reordering of terms, be made to converge to any a priori given number, including* $\pm\,\infty$. Instead of the proof of this theorem (due to Euler) here is given an example of change of the sum of a conditionally convergent series, by derangement of its terms. The series (21.6b) is conditionally convergent, because if instead of alternating the signs of successive terms, they are all given the same sign, it diverges (21.6c). The series (21.6b) consists of equal numbers of positive and negative terms; by rearranging the series so that two positive terms precede each negative term, its sum is changed to:

$$n \in |N_0: \ 1 + \frac{1}{3} - \frac{1}{2} + \frac{1}{5} + \frac{1}{7} - \frac{1}{4} + \cdots + \frac{1}{4n+1} + \frac{1}{4n+3} - \frac{1}{2n+2} + \cdots = \frac{3}{2}\log 2; \tag{21.17}$$

the increase by a factor of 3/2 over (21.6b) is due to the larger proportion of positive terms in (21.17). To prove (21.17) introduce the partial sums:

$$A_n \equiv 1 - \frac{1}{2} + \frac{1}{3} - \frac{1}{4} + \cdots + \frac{(-)^n}{n+1}; \quad B_n \equiv 1 + \frac{1}{2} + \frac{1}{3} + \frac{1}{4} + \cdots + \frac{1}{n+1}, \tag{21.18a,b}$$

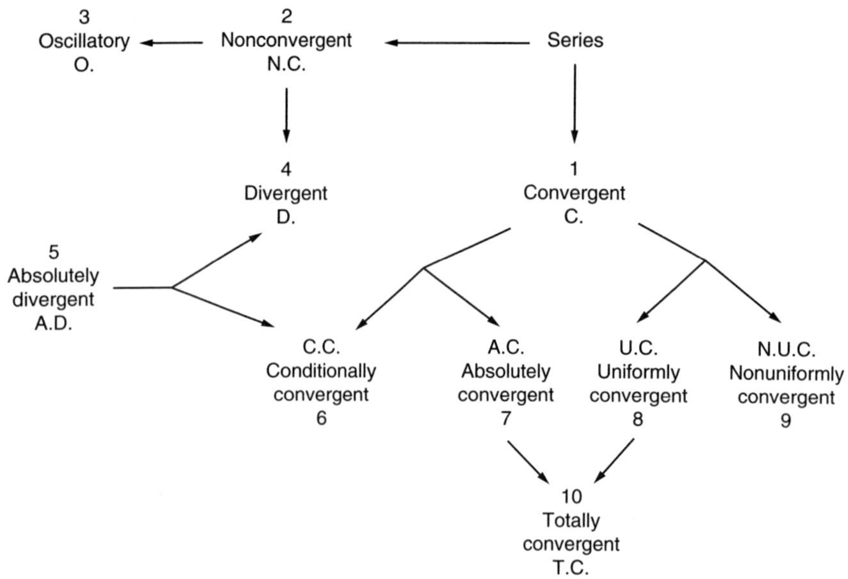

DIAGRAM 21.1
Ten cases of convergence of series. The classification of series by the limit of the sum as the number of terms tends to infinity: (i) convergent if there is a unique finite limit; (ii) divergent if it is unbounded; (iii) oscillatory if it neither (i) nor (ii) hold, that is, it is bounded without a single limit; (iv) the convergent series are conditionally (absolutely) convergent if the series of moduli diverges (converges); (v) this implies that an absolutely divergent series is either divergent or conditionally convergent; (v) a criterion independent from (iv) is uniform (nonuniform) convergence in a region; (vi) the combination of (iv) and (v) leads to the strongest convergence: absolute + uniform = total convergence.

which correspond, respectively, (21.18a) [(21.18b)] to the conditionally convergent (21.6b) [divergent (21.6c)] series. The partial sum of the first $3n$ terms of the rearranged series (21.17) is:

$$C_{3n} \equiv 1 + \frac{1}{3} + \frac{1}{5} + \cdots + \frac{1}{4n-1} - \frac{1}{2} - \frac{1}{4} - \cdots - \frac{1}{2n}$$
$$= B_{4n} - \frac{B_{2n}}{2} - \frac{B_n}{2} = B_{4n} - B_{2n} + \frac{B_{2n} - B_n}{2}; \tag{21.19}$$

this can be evaluated in terms of (21.18a,b):

$$A_{2n} = B_{2n} - B_n, \qquad C_{3n} = A_{4n} + \frac{A_{2n}}{2}; \tag{21.20a,b}$$

bearing in mind that (21.18b) ≡ (21.21a) tends to log 2 as $n \to \infty$ by (21.6b), it follows that

$$\lim_{n\to\infty} A_n = \log 2, \qquad \lim_{n\to\infty} C_n = \frac{3}{2}\log 2, \tag{21.21a,b}$$

that proves (21.21b) ≡ (21.17). QED.

21.4 Permutation of Terms and Product of Series (Cauchy, 1821)

The change of the sum of a series by derangement of its terms cannot occur if it is absolutely convergent: *the sum of an absolutely convergent series is unaffected by derangement or permutation of its terms.* To prove the theorem consider an absolutely convergent series $f_n(z)$, and a series $g_n(z)$ obtained by derangement of its terms, so that the first N terms of f_n are contained among the first $N+K+1$ terms of g_n; it follows that $K+1$ of the first $N+K+1$ terms of g_n correspond to f_n with $n>N$, so that

$$\left|\sum_{n=0}^{N+K} g_n(z) - \sum_{n=0}^{N-1} f_n(z)\right| \le \sum_{n=N}^{\infty} |f_n(z)|; \tag{21.22}$$

since the series f_n is absolutely convergent, the series of moduli converges, and its remainder (21.5) tends to zero:

$$\sum_{n=0}^{\infty} f_n(z)\ A.C.: \qquad 0 = \lim_{N\to\infty} \sum_{n=N}^{\infty} |f_n(z)| = \left|\sum_{n=0}^{\infty} g_n(z) - \sum_{n=0}^{\infty} f_n(z)\right|, \tag{21.23}$$

and this proves that the two series f_n, g_n that differ only in the ordering of terms, have the same sum. QED. An operation that may involve the rearrangement of terms is the product of two series, which involves summing all terms in the matrix or table:

$$f(z)g(z) \equiv \left\{\sum_{n=0}^{\infty} f_n(z)\right\}\left\{\sum_{m=0}^{\infty} g_m(z)\right\} = \begin{bmatrix} f_0g_0 & f_0g_1 & f_0g_2 & \cdots & f_0g_m & \cdots \\ f_1g_0 & f_1g_1 & f_1g_2 & \cdots & f_1g_m & \cdots \\ f_2g_0 & f_2g_1 & f_2g_2 & \cdots & f_2g_m & \cdots \\ \vdots & \vdots & \vdots & & \vdots & \\ f_ng_0 & f_ng_1 & f_ng_2 & \cdots & f_ng_m & \cdots \\ \vdots & \vdots & \vdots & & \vdots & \end{bmatrix} \tag{21.24}$$

The product of series can be summed in several ways: the simplest being by lines (21.25a), columns (21.25b) or diagonals (21.25c):

$$f(z)g(z) = \sum_{n=0}^{\infty} f_n(z)\left\{\sum_{m=0}^{\infty} g_m(z)\right\} = \sum_{m=0}^{\infty} g_m(z)\left\{\sum_{n=0}^{\infty} f_n(z)\right\} = \sum_{n=0}^{\infty}\sum_{k=0}^{n} f_k(z)\,g_{n-k}(z); \tag{21.25a–c}$$

the method of summation by diagonals (21.25c) is known as **Cauchy's Rule (1821)**.

The product of two series is a particular instance of a double or series $h_{n,m}(z)$, in which the general term is $f_n(z)\ g_m(z)$. In analogy with (21.4) or (21.3a), a **double series converges** to $h(z)$ if the limit:

$$\text{C.}: \quad h(z) = \lim_{N,M\to\infty} \sum_{n=0}^{N}\sum_{m=0}^{M} h_{n,m}(z) \equiv \lim_{N,M\to\infty} S_{N,M}(z), \tag{21.26}$$

exists and is independent of the way in which N, M tend to infinity. In the case of *the product of two convergent series (21.24) the resulting double series converges and can be summed by lines (21.25a), columns (21.25b), or diagonals (21.25c).* To prove the rule of summation by lines, the convergence of the single series f_n, g_m can be used:

$$\begin{aligned} 0 &= \lim_{N\to\infty}\left\{f(z)-\sum_{n=0}^{N}f_n(z)\right\}\lim_{M\to\infty}\left\{g(z)-\sum_{m=0}^{M}g_m(z)\right\} \\ &= \lim_{N,M\to\infty}\sum_{n=0}^{N}f_n(z)\left\{\sum_{m=0}^{M}g_m(z)\right\}-f(z)\sum_{m=0}^{\infty}g_m(z)-g(z)\sum_{n=0}^{\infty}f_n(z) \\ &\quad +f(z)g(z)=\sum_{n=0}^{\infty}f_n(z)\left\{\sum_{m=0}^{\infty}g_m(z)\right\}-f(z)g(z), \end{aligned} \tag{21.27}$$

to prove the convergence of the double series for $f_n g_m$. The proof of the product rule by columns (21.25b) is similar to that by lines (21.25a) given above (21.27). Concerning the proof of **Cauchy's Rule of diagonals**, note that the upper left-hand diagonal corresponds to the square less the lower left-hand diagonal:

$$\sum_{n=0}^{N}\sum_{k=0}^{N}f_k(z)g_{n-k}(z)=\sum_{n,m=0}^{N}f_n(z)g_m(z)-\sum_{\substack{n=0\\n+m\geq N+1}}^{N}\sum_{m=0}^{N}f_n(z)g_m(z); \tag{21.28}$$

since the double series converges, the last term on the r.h.s of (21.28) that is its remainder as $N\to\infty$, vanishes; it then follows the equality of the two remaining terms as $N\to\infty$ showing that the sum by diagonals equals that by rows:

$$\begin{aligned} 0 &= -\lim_{N\to\infty}\sum_{n+m\geq N+1}^{N,N}f_n(z)g_m(z)=\lim_{N\to\infty}\sum_{n=0}^{N}\sum_{k=0}^{N}f_n(z)g_{n-k}(z)-\sum_{n,m=0}^{\infty}f_n(z)g_m(z) \\ &= \sum_{n=0}^{\infty}\sum_{k=0}^{\infty}f_n(z)g_{n-k}(z)-f(z)g(z); \end{aligned} \tag{21.29}$$

the last term is the product of the sums, which has already been proven (21.27), and thus (21.25c) follows. QED.

21.5 Uniform Convergence and Series of Functions

Reconsidering the series (21.1) of functions $f_n(z)$ its error $\varepsilon(N,z)$generally depends on the point z; iff for all z in a region D, the error $\varepsilon(N)$ in (21.4) or modulus of the remainder on (21.5), depends only on the number of terms N, the series is **uniformly convergent** in D:

$$\forall_{z\in D}:\quad \sum_{n=0}^{\infty}f_n(z)\ U.C.\Leftrightarrow\varepsilon=\varepsilon(N)\geq\left|\sum_{n=N}^{\infty}f_n(z)\right|. \tag{21.30}$$

It may be expected, as for integrals [Section 13.4 (13.8)], that the limit (13.17) [derivative 13.40)] can be interchanged with the sum sign in a series, if it is uniformly convergent. Before proving this (Section 21.6) three examples are given next of nonuniformly convergent series, for which the sum cannot be interchanged with the operators (Subsection 31.5.1) limit

(21.31a), (Subsection 31.5.2) derivative (21.31b) and (Subsection 31.5.3) integral (21.31c), viz. the three series with real variable x, whose sum after N terms are given by:

$$S_n(x) = 1 - x^N, \quad T_N(x) = \frac{Nx}{1+N^2x^2}, \quad U_N(x) = \frac{N}{1+N^2x^2}. \tag{21.31a–c}$$

The three examples that follow are not essential, and the reader may proceed directly to the proof of the general theorems (Section 21.6) that is very simple, viz. simpler than the examples of cases where they do not apply.

21.5.1 Discontinuous Series of Continuous Functions

The general or n-th term of the series with partial sum (21.31a) may be obtained subtracting the sums after $n+1$ and n terms:

$$f_n(x) = \sum_{k=0}^{n} f_k(x) - \sum_{k=0}^{n-1} f_k(x) = S_{n+1}(x) - S_n(x) = x^n - x^{n+1}; \tag{21.32}$$

in the case of the series (21.31a) the general term is (21.32). It is shown next that the series (21.31a) of continuous functions (21.32) is discontinuous at the point $x = 1$ of nonuniform convergence. The proof follows: (i) the functions (21.32) are continuous for all finite x and n, but the sum of the series (21.31a) is (Figure 21.1) discontinuous at $x = 1$ as $N \to \infty$, viz.:

$$\lim_{N\to\infty}\left(1 - x^N\right) = \begin{cases} \infty & \text{if } x > 1 & (21.33\text{a}) \\ 0 & \text{if } x = 1 & (21.33\text{b}) \\ 1 & \text{if } x < 1; & (21.33\text{c}) \end{cases}$$

(ii) the series converges to unity (21.33c) for (21.34a) and the remainder of the series is (21.34b):

$$0 \le x < 1: \qquad \varepsilon > |S_N(x) - 1| \ge x^N; \qquad N \ge \frac{\log(1/\varepsilon)}{\log(1/x)}, \tag{21.34a–c}$$

(iii) it follows (21.34c) that the convergence is nonuniform that is, $N \to \infty$ for $x = 1$, viz. at the point of discontinuity of the sum (21.33a–c). Thus the remainder of the series: (a) is independent of x for $x \neq 1$, in which case $x < 1$ the series is uniformly convergent (U.C.), and its sum is continuous; (b) at $x = 1$, then $N = \infty$ in (21.34c), and the series is not uniformly convergent (N.U.C.) and its sum is discontinuous. The representation of $S_N(x)$ as a function (21.31a) of x, for $N = 1, 2, \ldots$ in Figure 21.1 shows the reason: as $N \to \infty$ the sum $S_N(x)$ changes from continuous to a downward step function; thus the limit of the sum does not coincide with the sum of the limits at $x = 1$ that is the point of nonuniform convergence.

21.5.2 Nondifferentiable Series of Differentiable Functions

The second example (21.31b) is a series of differentiable functions whose term-by-term derivative does not converge to the derivative of the sum at a point of nonuniform convergence. The proof follows: (i) the partial sums (21.31b) are differentiable and have slope:

$$T'_N(x) = \frac{N\left(1+N^2x^2\right) - 2N^3x^2}{\left(1+N^2x^2\right)^2} = \frac{N\left(1-N^2x^2\right)}{\left(1+N^2x^2\right)^2}; \tag{21.35}$$

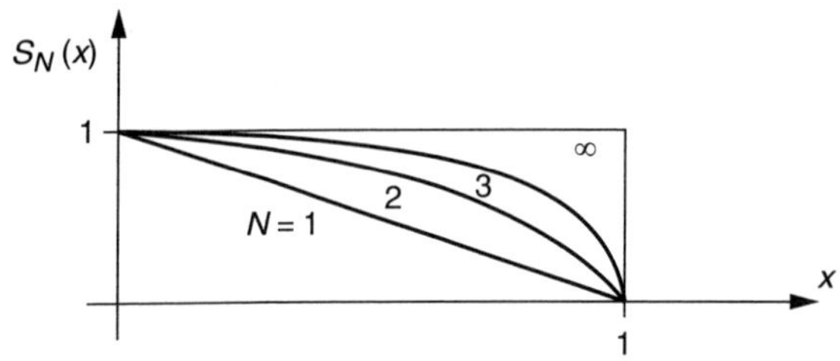

FIGURE 21.1
A first apparent anomaly with nonuniformly convergent series is a series of continuous functions whose sum is discontinuous: (i) all partial sums are continuous and vanish at $x = 1$: (ii) as the order N increases the sums tend to a step; (iii) in limit at $N \to \infty$ there is a discontinuity or downward jump from 1 to 0 at $x = 1$.

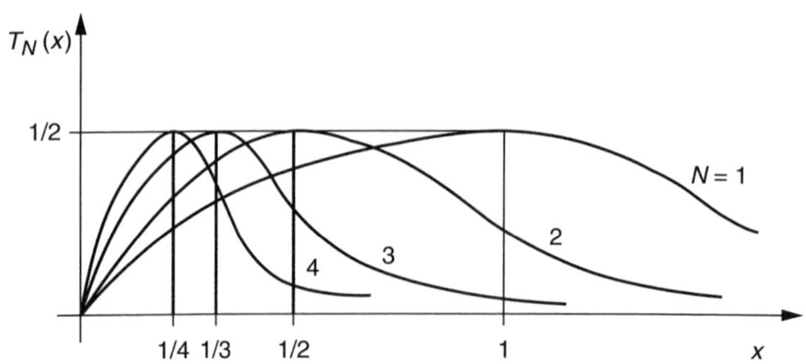

FIGURE 21.2
A second apparent anomaly with nonuniformly convergent series is a series of differentiable functions whose sum is not differentiable: (i) the partial sums are differentiable with slope at the origin increasing as N increases; (ii) as $N \to \infty$ the slope diverges at the origin, so the sum of the series is not differentiable there.

(ii) this implies (21.36a) a maximum (21.36c) at (21.36b):

$$T'_N(x_N) = 0: \quad x_N = \frac{1}{N}, \quad T_{N\max} = T_N\left(x_N = \frac{1}{N}\right) = \frac{1}{2}; \quad T'_N(0) = N, \qquad \text{(21.36a–d)}$$

(ii) the maximum occurs (Figure 21.2b) closer to the origin $x_N = 1/N \to 0$ as $N \to \infty$; (iv) thus the slope (21.35) diverges at the origin (21.36d); (v) the sum of the series (21.31b) is zero $T_N(0) = 0$ for $x = 0$, and for $x \neq 0$ it tends to zero $T_N \to 0$ as $N \to \infty$;

$$x \neq 0: \qquad T(x) = \lim_{N\to\infty} T_N(x) = \lim_{N\to\infty} O\left(N^{-1}x^{-1}\right) = 0 = T_N(0); \quad T'(x) = 0, \qquad \text{(21.37a–c)}$$

(vi) thus the sum of the series (21.31b) is continuous at the origin (21.37a,b), and since it vanishes everywhere its derivative (21.37c) is zero; (vii) the derivative of the series has sum (21.35), and vanishes everywhere (21.38a):

$$x \neq 0: \qquad \lim_{N\to\infty} T'_N(x) = \lim_{N\to\infty} O\left(N^{-1}x^{-2}\right) = 0 \neq \infty = \lim_{N\to\infty} N = \lim_{N\to\infty} T'_N(0), \qquad \text{(21.38a,b)}$$

except at the origin $x=0$ where it diverges (21.38b) by (21.36d); (viii) since the sum of the series is zero (21.37b), it coincides with the remainder (21.39a), and specifies the error (21.39b):

$$\frac{Nx}{1+N^2x^2} \le \varepsilon : \qquad N \ge \frac{1+\sqrt{1-4\varepsilon^2}}{2\varepsilon x}; \tag{21.39a,b}$$

(ix) this shows that (21.39b) it is uniformly convergent for $x \neq 0$, but nonuniformly convergent for $x=0$. The inequality (21.39a) $\equiv$ (21.39c) is quadratic with roots (21.39d):

$$0 \le N^2 - \frac{N}{x\varepsilon} + \frac{1}{x^2} = (N-N_+)(N-N_-), \qquad N_\pm = \frac{1\pm\sqrt{1-4\varepsilon^2}}{2x\varepsilon}; \tag{21.39c,d}$$

thus (21.39a) is satisfied (21.39c) if $N \ge N_+ > N_-$ in (21.39d), leading to (21.39b). Thus the derivative of the series (21.38b) does not converge to the derivative of the sum (21.37b) at the origin, where (Figure 21.2) the slope (21.36d) of the sum diverges, causing nonuniform convergence.

21.5.3 Nonintegrable Series of Integrable Functions

The third example (21.31c) concerns a series of integrable functions whose sum is not integrable at a point of nonuniform convergence although the series of integrals converges. The proof follows: (i) the partial sums (21.31c) of the series (Figure 21.3a) are even and decay outside the origin (21.40a) faster for larger N:

$$x \neq 0: \qquad U(x) \equiv \lim_{N\to\infty} U_N(x) = \lim_{N\to\infty} O\left(N^{-1}x^{-2}\right) = 0 \neq \infty = \lim_{N\to\infty} N = \lim_{N\to 0} U_N(0); \tag{21.40a,b}$$

(ii) the initial value (21.40b) diverges as $N \to \infty$, so the integral or area under the curve will be discontinuous at the origin, causing nonuniform convergence; (iii) the sum after N terms (21.31c) is integrable:

$$V_N(x) \equiv \int_0^x U_N(\zeta)\,d\zeta = \int_0^x \left(1+N^2\zeta^2\right)^{-1} N\,d\zeta = \arctan(Nx); \tag{21.41}$$

(iv) the partial sums of the series of integrals (21.41) are odd functions (Figure 21.3b) and have a finite discontinuity at $x=0$, with a jump from $-\pi/2$ before to $+\pi/2$ after:

$$\lim_{N\to\infty} V_N(x) = \begin{cases} -\pi/2 & \text{if } x<0 & (21.42\text{a}) \\ 0 & \text{if } x=0 & (21.42\text{b}) \\ +\pi/2 & \text{if } x>0; & (21.42\text{c}) \end{cases}$$

(v) since the function vanishes (21.40a) for $x \neq 0$ it coincides with the remainder (21.43a) that has upper bound:

$$\varepsilon \ge \frac{N}{1+N^2x^2}; \qquad N \ge \frac{1+\sqrt{1-4x^2\varepsilon}}{2x^2\varepsilon}, \tag{21.43a,b}$$

(vi) this shows (as in 21.39a–d) that (21.43b) the convergence is uniform everywhere except at $x=0$; (vii) the series of integrals converges nonuniformly at $x=0$ because the slope there (21.44a) $\equiv$ (21.40b) diverges as $N \to \infty$; (viii) this is due to the functions $V_N(x)$ rising more rapidly from zero at the origin (21.44b) to their asymptotic value as $\pi/2$ in (21.44c) as $x \to \infty$ in (21.42c), for increasing N:

$$V_N'(0) = U_N(0) = N, \qquad V_N(0) = 0, \qquad V_N(\infty) = \frac{\pi}{2}, \tag{21.44a–c}$$

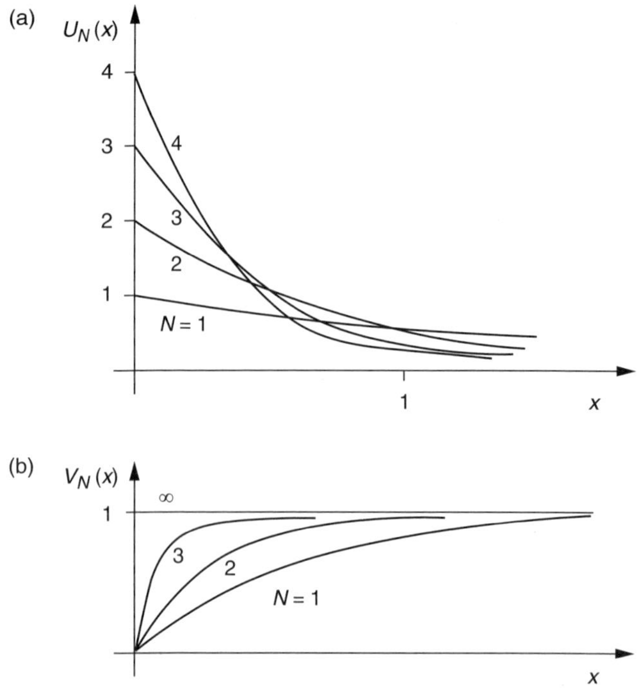

FIGURE 21.3
A third apparent anomaly with nonuniformly convergent series is a series of integrable functions for which the sum is nonintegrable: (i) the sum of the first N terms of the series (a) has a maximum at the origin increasing with N and faster decay afterwards, leading always to the same integral over the positive real line; (ii) the primitive is zero at the origin and rises toward the value unity at infinity more rapidly has N increases (b); (iii) as $N \to \infty$ the primitive has a unit step discontinuity at the origin, jumping up from zero to unity (b); (iv) the sum has an infinite discontinuity at the origin jumping down from infinity to zero (a); (v) thus both (a) and (b) show nonintegrability at the origin, which is the point of nonuniform convergence in Figure 21.3. The point of nonuniform convergence in Figure 21.1 (21.2) is $x = 1$ ($x = 0$).

as represented in Figure 21.3b. Thus the series of integrals (21.41) converges to a discontinuous function (21.42a–c), the convergence is nonuniform at $x = 0$, where the limit for $N \to \infty$ is an upward step in Figure 21.3b; the sum of the series converges (Figure 21.3a) to zero everywhere (21.40a), except at the origin where it diverges (21.40b); its Riemann integral (Section 13.2) is not defined even as an improper integral (Section 17.1) of second kind i.e., the sum of the series is a nonintegrable function.

21.6 Limit, Differentiation, and Integration Term-by-Term

The preceding examples show that at a point of nonuniform convergence, the limit, derivate, and integral of a series may differ from the result of taking term-by-term respectively the limits, derivates and integrals. The same examples suggest that this cannot happen at points where the convergence is uniform: *if: (i) a series of functions (21.1) is uniformly (21.30) convergent (21.4) ≡ (21.5) in a region* D*; (ii) the functions* $f_n(z)$ *are continuous/integrable/differentiable in* D*. Then: (i) the sum* $f_n(z)$ *is also a continuous/integrable/differentiable function; (ii) its limit/integral/derivate can be evaluated by summing the series of limits/integrals/derivates:*

$$\sum_{n=0}^{\infty} f_n(z) = f(z) \text{ U.C.}; \quad f \in \mathcal{C}(V_\varepsilon(a)): \quad \lim_{z\to a} f(z) = \lim_{z\to a} \sum_{n=0}^{\infty} f_n(z) = \sum_{n=0}^{\infty} f_n(a), \tag{21.45a}$$

$$f \in \mathcal{E}(a,b): \quad \int_a^b f(z)\,\mathrm{d}z = \int_a^b \left\{\sum_{n=0}^{\infty} f_n(z)\right\} \mathrm{d}z = \sum_{n=0}^{\infty} \int_a^b f_n(z)\,\mathrm{d}z, \tag{21.45b}$$

$$f \in \mathcal{D}(V_\varepsilon(z)): \quad \frac{\mathrm{d}f(z)}{\mathrm{d}z} = \frac{\mathrm{d}\left\{\sum_{n=0}^{\infty} f_n(z)\right\}}{\mathrm{d}z} = \sum_{n=0}^{\infty} \frac{\mathrm{d}f_n(z)}{\mathrm{d}z}. \tag{21.45c}$$

The theorem implies that permutation of the limit, integral, and derivate with a sum that holds for finite sums, also applies to infinite sums, in the case of uniformly convergent series. The proof of (21.45a) follows from:

$$\begin{aligned} \lim_{z\to a} |f(z) - f(a)| \le \lim_{N\to\infty} \left| f(z) - \sum_{n=0}^{\infty} f_n(z) \right| + \lim_{N\to\infty} \left| f(a) - \sum_{n=0}^{N} f_n(a) \right| \\ + \lim_{z\to a} \lim_{N\to\infty} \left| \sum_{n=0}^{N} \{f_n(z) - f_n(a)\} \right| \le \lim_{N\to\infty} 3\varepsilon(N) = 0; \end{aligned} \tag{21.46}$$

each term is bounded by $\varepsilon(N)$ and independent of (z, a) by (21.30), and $\varepsilon \to 0$ as $N \to \infty$ by (21.5). QED. The interchange of the limit and the infinite sum for an uniformly convergent series (21.45a), is analogous with the interchange (13.17) of a limit and the integral of an uniformly continuous function.

A series of functions $f_n(z)$ uniformly convergent in a region D may be treated, as far as convergence is concerned, as a series of constants or numbers, independent of the point z in D. This is apparent in the proof of the theorem on integration term-by-term:

$$\begin{aligned} \lim_{N\to\infty} \left| \int_a^b f(z)\,\mathrm{d}z - \sum_{n=0}^{N} \int_a^b f_n(z)\,\mathrm{d}z \right| \le \lim_{N\to\infty} \int_a^b \left| f(z) - \sum_{n=0}^{N} f_n(z) \right| \mathrm{d}z \\ \le \lim_{N\to\infty} \varepsilon(N) \int_a^b \mathrm{d}z = L(a,b) \lim_{N\to\infty} \varepsilon(N) = 0, \end{aligned} \tag{21.47}$$

where $\varepsilon(N)$ was taken out of the integral because it does not depend on z, and the length $L(a,b)$ of the path of integration between a, b is finite. QED. The Cauchy Theorem, expressing the derivate of order k of a holomorphic function as a loop integral (15.13) implies that the permutation of integrals with uniformly convergent series (21.45b), also extends to derivates, for example, for the first derivative (15.12):

$$\sum_{n=0}^{\infty}\frac{\mathrm{d}f_n(z)}{\mathrm{d}z}=\sum_{n=0}^{\infty}(2\pi i)^{-1}\int^{(z+)}(\zeta-z)^{-2}f_n(\zeta)\,\mathrm{d}\zeta=(2\pi i)^{-1}\int^{(z+)}(\zeta-z)^{-2}\left\{\sum_{n=0}^{\infty}f_n(\zeta)\right\}\mathrm{d}\zeta$$

$$=\frac{\mathrm{d}\left\{\sum_{n=0}^{\infty}f_n(z)\right\}}{\mathrm{d}z}; \tag{21.48}$$

from (21.48) it follows that if all $f_n(z)$ with $n=0,1,\ldots$ are differentiable, so is their sum, and the derivate is the sum of derivates. QED. This holds for of derivatives of any order of finite sums and uniformly convergent series.

21.7 Total Convergence and Weierstrass M-Test (1876)

The relationship between the various types of convergence is illustrated in Diagram 21.1, where an arrow indicates a particular case, or a subclassification. The absolute (Section 21.3) and uniform (Sections 21.5 and 21.6) convergence are stronger requirements than ordinary convergence (Section 21.1); the absolute and uniform convergence are independent requirements, that is, a series may satisfy one of them and not the other. It follows that the strongest requirement for convergence is a combination of these two: a series (21.1) is **totally convergent** by definition iff it is absolutely (21.14) and uniformly (21.30) convergent:

$$\sum_{n=0}^{\infty}f_n(z)\text{ T.C. }\Leftrightarrow\sum_{n=0}^{\infty}|f_n(z)|C.\wedge\sum_{n=0}^{\infty}f_n(z)\text{ U.C.}; \tag{21.49}$$

thus the totally convergent ($T.C.$) series allows all the transformations of and operations on series discussed before.

A simple method of establishing the total convergence of a series is the **Weierstrass's M-test** (1880): *a sufficient condition that the series of functions (21.1) be totally convergent in a region* D, *is that the functions* $\mathrm{f_n(z)}$ *have an upper bound in modulus* $\mathrm{M_n}$ *at all points* z *in* D, *and the series* $\mathrm{M_n}$ *be convergent:*

$$\forall_{z\in D}:\quad |f_n(z)|\le M_n\wedge\sum_{n=0}^{\infty}M_n\,C.\Leftrightarrow\sum_{n=0}^{\infty}f_n(z)\text{ T.C.} \tag{21.50}$$

The proof of total convergence based on the series of upper bounds is justified by the inequality:

$$\lim_{N\to\infty}\sum_{n=0}^{N}|f_n(z)|\le\lim_{N\to\infty}\sum_{n=0}^{N}M_n, \tag{21.51}$$

showing that: (i) the convergence of the series M_n implies the convergence of $|f_n(z)|$, and the absolute convergence of $f_n(z)$; (ii) the convergence is also uniform for $z\varepsilon D$, since the bounds M_n are independent of z in this region. QED.

A useful extension of this test, reminiscent of the Abel–Dirichlet test (Section 29.6), is the **extended Weierstrass M-test**: *a sufficient condition that the series* $\mathrm{f_n(z)\,g_n(z)}$ *be* *totally convergent in a region* D, *is that: (i) the series* $\mathrm{f_n(z)}$ *be totally convergent; (ii) the functions* $\mathrm{g_n(z)}$ *bounded, in the region* D:

$$\forall_{z\in D}: \quad |g_n(z)| \le M \wedge \sum_{n=0}^{\infty} f_n(z)\, T.C. \Leftrightarrow \sum_{n=0}^{\infty} \{g_n(z)\, f_n(z)\}\, T.C. \tag{21.52}$$

The proof follows from:

$$\sum_{n=0}^{\infty} |g_n(z)\, f_n(z)| = \sum_{n=0}^{\infty} |g_n(z)||f_n(z)| \le M \sum_{n=0}^{\infty} |f_n(z)|, \tag{21.53}$$

that shows that (i) if the series f_n is absolutely convergent then the series $f_n\, g_n$ is absolutely convergent; (ii) since the bound M on $g_n(z)$ does not depend on z, the uniform convergence of f_n extends to $f_n\, g_n$. QED. *The definition of totally convergent series (21.49), shows that it has all the properties listed in Table 21.1, in connection with relations (association and permutation do not change the sum, and limits, derivates and integrals may be taken term-by-term) and operations [the series sum (product) converge to the sum (product) of the series].* Concerning the operations sum (product) of series, their validity has been proved for ordinary convergence in Section 21.2 (21.4), but it also applies for other, stronger types of convergence: *if two series* $\mathrm{f_n}$, $\mathrm{g_m}$ *are absolutely/uniformly/totally convergent their sum (product) also is absolutely/uniformly/totally convergent:*

$$\begin{aligned} &\left\{\sum_{n=0}^{\infty} f_n(z), \sum_{n=0}^{\infty} g_n(z)\right\} \text{A.C., U.C., T.C.} \\ &\qquad \Rightarrow \left\{\sum_{n=0}^{\infty} f_n(z) + g_n(z),\ \sum_{n,m=0}^{\infty} f_n(z)\, g_m(z)\right\} A.C., U.C., T.C. \end{aligned} \tag{21.54}$$

The proof of uniform convergence for the sum (product) coincides with simple convergence (21.13) [(21.25a–c)] with ε independent of z. The proofs for absolute convergence of the sum (product) are also similar, and are based on the inequality (21.55) [identity (21.56)]:

$$\lim_{N\to\infty} \sum_{n=0}^{N} |f_n(z) + g_n(z)| \le \lim_{N\to\infty} \left\{\sum_{n=0}^{N} |f_n(z)| + \sum_{n=0}^{N} |g_n(z)|\right\}, \tag{21.55}$$

$$\begin{aligned} \lim_{N,M\to\infty} \left|\sum_{n=0}^{N}\sum_{m=0}^{M} f_n(z)\, g_m(z)\right| &\le \lim_{N,M\to\infty} \sum_{n=0}^{N}\sum_{m=0}^{M} |f_n(z)||g_m(z)| \\ &= \left\{\lim_{N\to\infty} \sum_{n=0}^{N} |f_n(z)|\right\}\left\{\lim_{M\to\infty} \sum_{m=0}^{M} |f_m(z)|\right\}. \end{aligned} \tag{21.56}$$

The proof for the total convergence of the sum (product) of series follows from the proofs for absolute and uniform convergence. QED.

21.8 Geometric, Logarithmic, and Inverse-Power Series

The type of convergence of a series of functions generally depends on the point z, and this illustrated is next by considering convergence at all points of the plane z of the **geometric**

series; the latter is defined by the constant ratio of successive terms (21.57a):

$$\frac{f_{n+1}(z)}{f_n(z)} = z, \quad f_0(z) \equiv a \neq 0, \tag{21.57a,b}$$

with the first term (21.57b) an arbitrary nonzero constant a. The general term is (21.58a):

$$f_n(z) = z f_{n-1}(z) = z^n f_0(z) = a z^n; \quad S_N(z) = a \sum_{n=0}^{N-1} z^n, \tag{21.58a,b}$$

the sum after N terms (21.58b) can be simplified using:

$$(1-z) S_N(z) = a \sum_{n=0}^{N-1} \left(z^n - z^{n+1}\right) = a\left(1 - z^N\right); \tag{21.59}$$

this leads to

$$z \neq 1: \quad S_N(z) = a\frac{1-z^N}{1-z} = \frac{a}{1-z} + R_N(z), \quad R_N(z) = -\frac{a z^N}{z-1}, \tag{21.60a,b}$$

as the sum (21.60a) of the first N *terms of the geometric series (21.58b), with* $a/(1-z)$ *being the sum of the series for* $|z|<1$, *and* $R_N(z)$ *the remainder (21.60b) after* N *terms.*

The last statement follows from (21.61a,b):

$$|z|<1: \quad \lim_{N\to\infty} R_N(z) = 0, \quad \lim_{N\to\infty} S_N(z) = \frac{a}{1-z}; \tag{21.61a–c}$$

recalling (21.58a) and replacing z by $-z$ leads to the **ascending geometric series** (21.62b,c):

$$|z|<1: \quad \sum_{n=0}^{\infty} a z^n = \frac{a}{1-z}, \quad \sum_{n=0}^{\infty} a(-)^n z^n = \frac{a}{1+z}, \tag{21.62a–c}$$

that converge with in the unit circle (21.62a); the change of variable $z \to 1/z$ leads to convergence $|1/z|<1$ outside the unit circle (21.63a):

$$|z|>1: \quad \sum_{n=0}^{\infty} a z^{-n} = \frac{az}{z-1}, \quad \sum_{n=0}^{\infty} a(-)^n z^{-n} = \frac{az}{z+1}, \tag{21.63a–c}$$

of the **descending geometric series** (21.63b,c). The series (21.62b; 21.63b) [(21.62c; 21.63c)] are **monotonic (alternating)** geometric series since they have fixed (alternating) sign. Thus *there are four geometric series: (i) ascending monotonic (21.62b); (ii) ascending alternating (21.62c); (iii) descending monotonic (21.63b); (iv) descending alternating (21.63c). The convergence of these four series is indicated in* Table 21.2, and is discussed in Section 21.9.

The geometric series may be used to derive other series, as shown in the next two examples; the series (21.62b,c) are uniformly convergent for $|z| \leq 1-\varepsilon$ with $1>\varepsilon>0$ and thus can be (Section 21.6) integrated and differentiated term-by-term. Taking (21.62b) [(21.62c)] with $a=1$ and integrating term-by-term from 0 to z, leads to (21.64a) [(2164b)] the **logarithmic series**:

$$-\log(1-z) = \sum_{n=0}^{\infty} \frac{z^{n+1}}{n+1} = \sum_{n=1}^{\infty} \frac{z^n}{n}; \tag{21.64a}$$

$$\log(1+z) = \sum_{n=0}^{\infty} \frac{(-)^n z^{n+1}}{n+1} = \sum_{n=1}^{\infty} \frac{(-)^{n-1} z^n}{n}; \tag{21.64b}$$

TABLE 21.2
Convergence of Geometric Series in the Complex Plane

Case	**(21.62b)**	**(21.62c)**	**(21.63b)**	**(21.63c)**
Geometric series	Ascending monotonic	Ascending alternating	Descending monotonic	Descending alternating
Divergent (D.)	$\|z\| > 1$	$\|z\| > 1$	$\|z\| < 1$	$\|z\| < 1$
Divergent (D.)	$z = 1$	$z = -1$	$z = 1$	$z = -1$
Oscillating (O.)	$\|z\| = 1 \neq z$	$\|z\| = 1 \neq -z$	$\|z\| = 1 \neq z$	$\|z\| = 1 \neq -z$
Absolutely convergent (A.C.)	$\|z\| < 1$	$\|z\| < 1$	$\|z\| > 1$	$\|z\| > 1$
Totally convergent (T.C.)	$\|z\| \leq 1 - \varepsilon$	$\|z\| \leq 1 - \varepsilon$	$\|z\| \geq 1 + \delta$	$\|z\| \geq 1 + \delta$
where	$0 < \varepsilon < 1$	$0 < \varepsilon < 1$	$\delta > 0$	$\delta > 0$

Note: The region of total convergence is a closed subset of the open region of absolute convergence. The latter is disjoint from the open region of divergence; both are closed by the same boundary of convergence. On the boundary of convergence (always the unit circle in the present case) there may be special points where the convergence differs from the other boundary points, that is, there are at most two subsets of the boundary as concerns convergence.

the particular case $z = 1$ of (21.64b) is (21.6b). Differentiating N times with regard to z, leads from (21.62c) to *the* **binomial series** *with negative integral exponent* $-N$:

$$m = n - N: \quad (1+z)^{-N-1} = \frac{(-)^N}{N!} \sum_{n=N}^{\infty} (-)^n \frac{n!}{(n-N)!} z^{n-N}$$
$$= \sum_{m=0}^{\infty} (-)^m \frac{(m+N)!}{m!N!} z^m = \sum_{m=0}^{\infty} \binom{m+N}{N} (-z)^m; \quad \text{(21.65a,b)}$$

a change of dummy summation variable (21.65a) was used to simplify the series (21.65b). The series (21.65b) is the particular case $a = 1$ of (25.40a–c); the latter is the particular case $\nu = N$ of the binomial series (25.37a–c). The preceding series can be used to sum numeric expansions (Example 30.2) and to expand rational functions in power series (Example 30.3).

21.9 Convergence inside, outside, and on the Unit Circle

The series (21.65b) is the particular case of negative integer exponent $v = -N$, of the binomial series (Section 25.9), whose convergence is established by a general test (Subsection 29.1.3). The same test applies (Subsection 29.2) to the logarithmic series (21.64a,b), as shown in Subsection 29.8.2. Next will be considered the behavior at all points of the complex z-plane of the geometric series starting with the ascending monotonic case (21.62b). The latter is (i) absolutely convergent within the unit circle, because the series of moduli converges:

$$|z| < 1: \quad \sum_{n=0}^{\infty} |a| \, |z|^n = \frac{|a|}{\{1 - |z|\}}; \quad \text{(21.66)}$$

in a closed subcircle of the unit circle:

$$|z| \le 1 - \varepsilon: \quad \sum_{n=0}^{\infty} |a| \, |z|^n \le \frac{|a|}{\varepsilon}, \tag{21.67}$$

the geometric series is (ii) uniformly convergent (21.67), and hence (21.66) is totally convergent.

Outside the unit circle the (21.60b) shows that (iii) the geometric series diverges:

$$|z| > 1: \quad \lim_{N\to\infty} \sum_{n=0}^{\infty} a z^N = \lim_{N\to\infty} \frac{a\left(1 - z^N\right)}{1 - z} = \infty; \tag{21.68}$$

also (21.60b) holds at all points z except $z = 1$, where (21.58b) shows (iv) that it also diverges:

$$z = 1: \quad \sum_{n=0}^{\infty} a z^n = \sum_{n=0}^{\infty} a = \infty. \tag{21.69}$$

The boundary between the regions of convergence $|z| < 1$ in (21.62a,b; 21.66) and divergence $|z| > 1$ in (21.68) is the unit circle. Excluding the point $z = 1$ where the series diverges (21.69), at all other points (21.70a,b) the sum is (21.70c):

$$z = e^{i\varphi}, \ \varphi \ne 0: \quad S_N(z) = \frac{a\left(1 - e^{iN\varphi}\right)}{1 - e^{i\varphi}}; \tag{21.70a–c}$$

the limit of (21.70c) as $N \to \infty$ does not exist, and hence (v) the series oscillates in the range:

$$|z| = 1 \ne z: \quad 0 \le |S_N(z)| \le \frac{2|a|}{|1 - z|}; \tag{21.71}$$

thus the maximum and minimum differ at most by the double of the modulus of (21.62b); for example, for $z = -1$, leads to the oscillating series:

$$z = -1: \quad \sum_{n=0}^{\infty} a z^n = \sum_{n=0}^{\infty} (-)^n a = a - a + a - a + \cdots = a, \ 0; \tag{21.72}$$

this is a multiplied by (21.6a), and alternates be between the values 0, a that are the extremes in (21.71). The five cases (i) to (v) of behavior of the geometric series (21.62b) in distinct regions, curves, or points, also apply to the series (21.62c; 21.63b; 21.63c) via the changes of variable $z \to z$, $1/z$, $-1/z$, leading to the results listed in Table 20.2.

A series may fail to converge everywhere on the complex plane. For example the series:

$$\begin{aligned} S(z) = \sum_{n=-\infty}^{+\infty} z^n &= \cdots + z^{-n} + \cdots + z^{-1} + 1 + z + \cdots + z^n + \cdots = \lim_{N,M\to\infty} \sum_{n=-M}^{+N} z^n \\ &= \lim_{N,M\to\infty} \left[S_N(z) + S_M\left(\frac{1}{z}\right) - 1 \right], \end{aligned} \tag{21.73}$$

is the sum an ascending (21.62b) and a descending (21.63b) series: (i) the ascending series (21.62b) converges (21.66) for $|z| < 1$ but the descending series (21.63b) diverges; (ii) the descending series (21.63b) converges for $|z| > 1$ but the ascending series (21.62b) diverges (21.68); (iii) to prove nonconvergence in the remaining case $|z| = 1$, it is sufficient to show for a particular method of summation:

$$z = e^{i\varphi}: \quad S(z) = 1 + \sum_{n=1}^{\infty} \left(z^n + z^{-n}\right) = 1 + \sum_{n=1}^{\infty} \left(e^{in\varphi} + e^{-in\varphi}\right) = 1 + 2\sum_{n=1}^{\infty} \cos(n\varphi), \quad (21.74)$$

that the series due not converge; this is the case because the general term cos $(n\varphi)$ in (21.74) does not tend to zero as $n \to \infty$, violating a necessary condition for convergence (29.16a). Thus the series (21.73) converges nowhere on the complex plane. If the sums of the geometric ascending (21.62b) and descending (21.63b) series were substituted in (21.73) the result would be zero:

$$S_\infty(z) + S_\infty\left(\frac{1}{z}\right) - 1 = \frac{1}{1-z} + \frac{z}{z-1} - 1 = \frac{1-z}{1-z} - 1 = 0; \qquad (21.75)$$

this is absurd since the series (21.73) clearly diverges for $z = 1$, viz. $\cdots + 1 + \cdots + 1 + \cdots = \infty$; thus the sum (21.75) is invalid, because the regions of convergence of the two series (21.62b) and (21.63b) do not overlap, that is, they are not both valid at any point. This example shows the need to establish the convergence of series, viz. using tests (Chapter 29), both for analytical (Example 30.3) and numerical (Examples 30.2 and 30.19) purposes.

Note 21.1. *Convergence of Infinite Processes:* The initial discussion of the convergence of series (Chapter 21) aims to clarify the relation between the types of convergence (Diagram 21.1) and the associated properties of series (Table 21.1); these are illustrated by simple examples concerning convergence at any point of the complex plane, for example, of the four geometric series in Table 21.2. The convergence properties will be used frequently, not only for specific functions (e.g., elementary transcendental) but also to establish general series representations in the neighborhood of regular (Chapter 23) [singular (Chapter 25)] points. The notions of simple, conditional, absolute, uniform, and total convergence apply to: (i) series, including power series (Chapters 23 and 25) and series of fractions (Subsection 29.5.2); (ii) improper integrals of three kinds (Chapter 17); (iii/iv) infinite products (Sections 35.6 and 35.7) and continued fractions. To each type of convergence may be associated a number of convergence tests, for example, the Weierstrass M-test for total convergence (Section 21.7) is just one of a wide variety; a selected combination of tests may be chosen to determine the type of convergence at every point of the complex plane (Chapter 29) for a significant number of cases. The convergence of an infinite process, for example, a series may be: (i) simple in same cases and difficult in others; (ii) straightforward in same regions (harder in others), for example, within (at the boundary of) the region of convergence. Similarly: (i) some theorems are difficult to prove, though the exceptions are easy to find; (ii) some theorems have straightforward proofs, but the examples of exceptions are less immediate. An instance of (ii) was (a) the straightforward proof that for a uniformly convergent series the limit, derivative, and integration can be performed term-by-term (Section 21.6); (b) the examples of discontinuous/nondifferentiable/nonintegrable series whose terms are continuous/differentiable/integrable functions are less immediate (Subsections 21.5.1–21.5.3).

Conclusion 21: Sums of nonuniformly convergent series as a function of x, as the number of terms N increases: (Figure 21.1) the series with partial sums $S_N(x)$ fails to converge uniformly at $x = 1$ because the downward jump causes a discontinuity, i.e., the series of continuous functions has a discontinuous sum at $x = 1$; (Figure 21.2) the series with partial sums $T_N(x)$ fails to converge uniformly at $x = 0$ because of the infinite slope, i.e., the series of differentiable functions has a sum which is not differentiable at $x = 0$; (Figure 21.3) the series $U_N(x)$ diverges (a) at $x = 0$, and the series of integrals $V_N(x)$ converges nonuniformly (b) at $x = 0$, where it has an upward jump, i.e., the series of integrals exists but not the integral of the sum at $x = 0$.

22

Multiple Reflections in a Lens

A wave incident upon an interface separating two media gives rise (Section 22.1) to two waves, that is, a reflected (transmitted) wave propagating in the first (second) medium (Section 22.2). In the case of three media, for example, separated by two interfaces, waves incident from the two outer media can be reflected or transmitted once only, whereas waves propagating in the inner medium, between the two interfaces, can have multiple internal reflections. The calculation of the exact wave field, in the presence of multiple internal reflections, can be made by summing the reflections of all orders that lead to series (Section 22.6). The series will always converge provided that the process of transmission or reflection conserves or decreases energy, that is, the convergence is more rapid in the presence of adsorption at the interfaces, and/or absorption in the interior of the media (Section 22.5). The scattering of waves by a double interface, for example, a parallel-sided lens (Section 22.8), can also be studied by decomposing the wave field in the interior into two components propagating in opposite directions, orthogonal to the interfaces; the laws of reflection and transmission (Section 22.4) at the two interfaces then specify a system of algebraic equations, whose solution is the same as the sum of the series of multiple reflections. Both methods lead to global reflection and transmission coefficients for a lens consisting of two or more interfaces (Section 22.7), allowing it to be replaced by a single "equivalent" interface. These expressions include as particular instances the cases of transparent or opaque interfaces (Section 22.8), zones of darkness or of total reflection (Section 22.3), and constructive or destructive interference (Section 22.9). Multiple reflections occur for other geometries, and not only for light or electromagnetic waves, but also for acoustic, elastic, and other waves; in the case of elastic waves in solids, the reflections and/or focusing can lead to stress concentrations, crack formation and growth, and eventual structural failure or ejection of material (spalling).

22.1 Period, Frequency, Wavelength, and Wavevector

A **sinusoidal wave** is a function (Figure 22.1) of the position vector $\vec{x}$ and time t specified in the complex representation by (22.1b):

$$\vec{n}.\vec{n} = 1: \quad f(\vec{x}, t) = A(\vec{k}, \omega) \exp\{i(\vec{k}\cdot\vec{x} - \omega t)\} = A(\vec{k}, \omega) \exp\left\{2\pi i\left(\frac{\vec{x}\cdot\vec{n}}{\lambda} - \frac{t}{\tau}\right)\right\}, \tag{22.1a,b}$$

where the **amplitude** A does not depend an position $\vec{x}$ or time t, but generally depends on two parameters: (i) namely the **period** τ (**wavelength** λ) which is the minimum time t (distance $\vec{x}\cdot\vec{n}$ along the direction (22.1a) of unit vector $\vec{n}$) after which the wave field takes

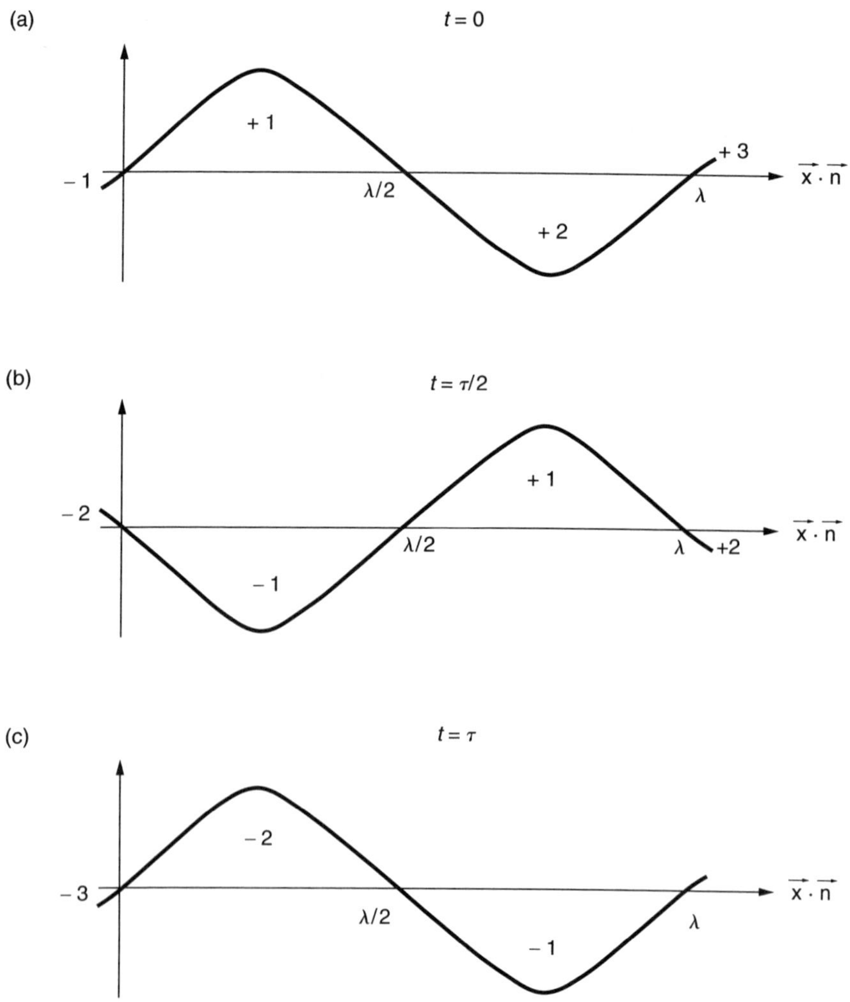

FIGURE 22.1
A wave of permanent shape propagating to the right is "photographed" at three times: (i) when the upward node passes the origin (a): (ii) half-a-period later when the next or downward node passes the origin (b); (iii) one period later when the next upward node passes the origin, and the picture is exactly the same (c) $\equiv$ (a) for the first time; (iv) it will repeat itself every multiple of a period.

the same value in (22.2a) [(22.2b)]:

$$f(\vec{x}+\lambda\vec{n},\, t) = A\exp\left\{i2\pi\left[\frac{\vec{n}\cdot(\vec{x}+\lambda\vec{n})}{\lambda}-\frac{t}{\tau}\right]\right\}$$

$$= A\exp\left[i2\pi\left(\frac{\vec{n}\cdot\vec{x}}{\lambda}-\frac{t}{\tau}\right)\right] = f(\vec{x},t)$$

$$f(\vec{x},t+\tau) = A\exp\left\{i2\pi\left[\frac{\vec{n}\cdot\vec{x}}{\lambda}-\frac{t+\tau}{\tau}\right]\right\} = A\exp\left[i2\pi\left(\frac{\vec{n}\cdot\vec{x}}{\lambda}-\frac{t}{\tau}\right)\right] = f(\vec{x},t); \tag{22.2a,b}$$

(ii) alternatively the inverses multiplied by 2π specify the **frequency** ω (**wavenumber** k) in (22.3a) [(22.3c)]:

$$\omega \equiv \frac{2\pi}{\tau}, \quad \vec{k} \equiv \frac{2\pi}{\lambda}\vec{n}, \quad k = \left|\vec{k}\right| = \frac{2\pi}{\lambda}; \tag{22.3a–c}$$

where (22.3c) follows from (22.3b) because $\vec{n}$ is a unit vector (22.2a). The waveform does not change (22.4a) for constant **phase** (22.4b):

$$f\left(\vec{x}, t\right) = \text{const}: \quad \phi\left(\vec{x},\ t\right) \equiv \vec{k}\cdot\vec{x} - \omega t = \text{const}, \quad \vec{u}\cdot\vec{n} = \vec{n}\cdot\frac{d\vec{x}}{dt} = \frac{\omega}{k} \equiv c; \tag{22.4a–c}$$

thus, the wave does not evolve for an observer moving at the velocity $\vec{u}$ in the **direction of propagation** such that (22.4c) is constant; the latter defines the **phase speed** c in (22.4c) $\equiv$ (22.5a):

$$\omega = k\,c; \quad \lambda = c\,\tau, \tag{22.5a,b}$$

thus the phase speed is the speed of propagation of the wave fronts (Figure 22.1) and using (22.3a) and (22.3c) in (22.5a) it follows (22.5b) that *the wavelength* λ *is the distance travelled by the wave at the phase speed* c *during one period* τ. The vector $\vec{k}$ in (22.3b) $\equiv$ (22.6), equal to the wavenumber k in modulus (22.3c), and in the direction $\vec{n}$ of propagation (22.2a), is designated **wavevector**:

$$\vec{k} = k\vec{n} = \frac{\omega}{c}\vec{n} = \frac{2\pi}{\lambda}\vec{n} = \frac{2\pi}{c\tau}\vec{n}. \tag{22.6}$$

Since the **wavenormal** $\vec{n}$ is constant, the wave front is a plane, and thus (22.1b) represents a **plane wave** *with: (i) frequency* ω *[period* τ *in (22.3a)]; (ii) wavevector* $\vec{k}$*, that is, wavenumber* k *[wavelength* λ *in (22.3c)], and wavenormal* $\vec{n}$ *[in (22.1a)]; (iii) these are related by the phase speed* c *[in (22.5a,b; 22.6)].* The velocity of particles and/or energy is the group velocity that may coincide or not with the phase velocity depending on the type of wave (acoustic, elastic, electromagnetic, on the surface of a liquid).

22.2 Reflection, Transmission, and Inacessible Regions (Snell, 1626; Descartes, 1637b; Fresnel, 1823)

Consider an **interface** separating two media that are characterized in the preceding expressions (22.4c; 22.5a,b; 22.6) by the propagation speeds c_1, c_2, for example, the adiabatic speed of sound (14.31a–c) for acoustic waves. An incident wave will give rise to scattered (i.e., reflected and transmitted) waves (Subsection 22.2.1) that may not reach everywhere, for example, have a total reflection or leave a zone of silence/darkness (Subsection 22.2.2).

22.2.1 Laws of Scattering: Reflection and Transmission

If the interface has a local radius of curvature R large on the scale of wavelength $R^2 \gg \lambda^2$, its irregularities may be neglected, and it can be treated as a plane **scattering** an **incident** plane wave into **reflected** and **transmitted** plane waves. The properties of the medium change in direction of the normal $\vec{N}$ to the interface, and are uniform in the transverse direction. Thus the wavevectors for the incident $\vec{k}_i \equiv (\vec{k}, k_{in})$, reflected $\vec{k}_r \equiv (\vec{k}, k_{rn})$ and transmitted $\vec{k}_t \equiv (\vec{k}, k_{tn})$ waves have equal transverse component $\vec{k}$, whereas the normal components (k_{in}, k_{rn}, k_{tn}) are generally distinct (22.19a,b). This implies: (i) that (I) *the directions of propagation of the incident, reflected, and transmitted waves lie on the same plane (22.7a)*; (ii) denoting (Figure 22.2a) by θ_i, θ_r , θ_t respectively the **angles of incidence, reflection, and transmission** between the directions of propagation $\vec{n}_i, \vec{n}_r, \vec{n}_t$ and the unit normal $\vec{N}$ to the interface leads to (22.7b–d):

$$\left(\vec{n}_i \wedge \vec{n}_r\right) \cdot \vec{n}_t = 0: \quad \vec{n}_i \cdot \vec{N} = \cos\theta_i, \quad \vec{n}_r \cdot \vec{N} = \cos\theta_r, \quad \vec{n}_t \cdot \vec{N} = \cos\theta_t; \qquad \text{(22.7a–d)}$$

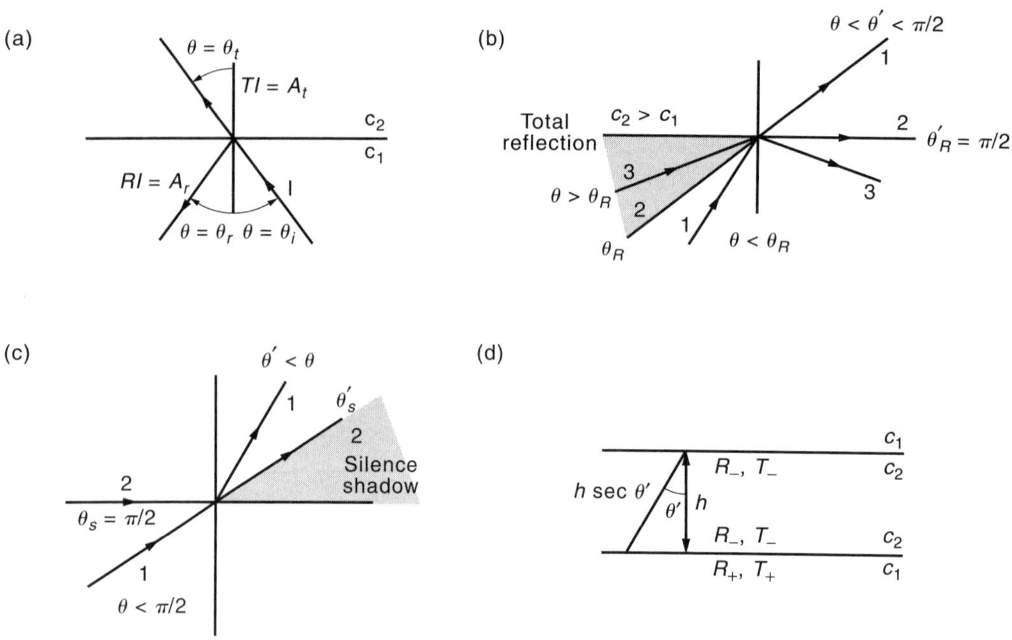

FIGURE 22.2
A plane wave with amplitude I and direction of propagation making an angle of incidence θ_i with the normal to the plane interface between two media with distinct phase speeds (c_1, c_2) gives rise (a) in the same (other) medium to: a reflected (transmitted) wave propagating at an equal $\theta_r = \theta_i$ (unequal $\theta_t \neq \theta_i$) angle with normal; its amplitude $A_r(A_t)$ divided by the amplitude I of the incident wave specifies the reflection R (transmission T) factor. If the phase speed is higher in the medium of transmission (b), the transmitted wave propagates at an angle to the normal larger than the incident wave. This "moving away" from the normal implies that a transmitted wave in the grazing direction $\theta' = \pi/2$ corresponds to an incident wave at the angle of total reflection $\theta = \theta_R$; beyond $\theta > \theta_R$ no transmitted wave exists (see Example 30.4) and there is total reflection. In the case (c) opposite to (b), when the phase speed is higher in the medium of incidence, the transmitted wave makes a smaller angle with the normal than the incident wave. This "moving toward" the normal implies that for a grazing incident wave $\theta = \pi/2$ for transmitted wave is at the angle of silence $\theta' = \theta_s$; beyond $\theta' > \theta_s$ is the zone of silence (shadow) where no sound (light) wave can be received (see Example 30.4). If there are several interfaces, for example, a parallel sides lens (d), the preceding phenomena (a–c) can occur at each interface.

the equality of the transverse components of the wavevector (22.6) yields:

$$k \equiv \frac{\omega}{c_1}\sin\theta_i = \frac{\omega}{c_1}\sin\theta_r = \frac{\omega}{c_2}\sin\theta_t. \qquad \text{(22.8a–c)}$$

From (22.8a,b) it follows that (II) *the angles of incidence and reflection are equal (22.9a):*

$$\theta_i = \theta_r \equiv \theta, \quad \theta' \equiv \theta_t : \quad \frac{c_1}{c_2} = \frac{\sin\theta}{\sin\theta'}. \qquad \text{(22.9a–c)}$$

and from (22.8b,c) it follows that (III) *the angles of incidence θ and transmission $\theta' \equiv \theta_t$ have sines in the same ratio as the propagation speeds in the two media (22.9c).* The results (I, II, III) are the three **laws of wave scattering** (Snell, 1626; Descartes, 1637b; Fresnel, 1823), for reflection and transmission (Figure 22.2a).

22.2.2 Total Reflection and Zone of Silence/Darkness

The most significant of the three Fresnel laws is that of transmission (22.8c) that allows three cases: (i) in the case of media with the same propagation speed $c_1 = c_2$, the wave is not reflected $\theta = \theta'$, that is, the transmitted wave coincides with the incident wave and the interface may be taken as nonexistent; (ii) if the *medium of transmission is* **denser**, *for example, transmission from air to water, that is, has higher propagation speed (22.10a), the wave is shifted away from the normal $\theta' > \theta$, and it takes a grazing direction $\theta' = \pi/2$ for* the **angle of total reflection** (22.10b):

$$c_2 > c_1 : \quad \theta < \arcsin\left(\frac{c_1}{c_2}\right) = \theta_R, \qquad \text{(22.10a,b)}$$

beyond which the wave is totally reflected (Figure 22.2b); (iii) if the medium of incidence is denser, for example, transmission from water to air, then (22.11a) the wave is shifted closer to the normal $\theta' < \theta$, so that for grazing incidence $\theta = \pi/2$, the transmission occurs at the **limit angle** (22.11b):

$$c_1 > c_2 : \quad \arcsin\left(\frac{c_2}{c_1}\right) \equiv \theta'_S, \qquad \text{(22.11a,b)}$$

beyond which no wave is received, so that $\theta' > \theta'_S$ is a ***zone of darkness (silence)*** *for electromagnetic (acoustic) waves (Figure 22.2c).* Thus the transmission from (to) a denser medium, for example, from water to air (air to water), involves the existence of a zone of silence or darkness (total reflection) in Figure 22.2c (22.2b). The three laws of wave scattering apply to a single (Figure 22.2a–c) or multiple (Figure 22.2d) interfaces. A wave incident beyond the angle of total reflection gives rise to a transmitted wave that is not zero, but is **evanescent**, that is, decays exponentially with distance from the interface, faster for shorter wavelengths (Example 30.4); thus only long waves penetrate partially in the medium of transmission in the condition of total reflection. Conversely, the propagation into zone of silence would imply evanescent incident and reflected waves in the medium of incidence (Example 30.4); thus radiation into zone of silence is not possible, except if the constraints of a plane interface at rest are lifted.

22.3 Wave Scattering and Fastest Path (Fermat, 1657)

The total reflection (22.10a,b) and zone of silence (22.11a,b) are both consequences of the third law of scattering (22.9c) that can also be derived from **Fermat's principle**

(1657): *a wave travels between two points along the path with minimum propagation time.* In a homogeneous medium, the wave speed is constant, and the minimum propagation time is along the shortest path, that is, a geodesic, viz. a straight line in flat space, leading to a plane wave (22.1a,b). The scattering by a plane interface between to media (Figure 22.3) with different wave speeds c_1 and c_2 does not take place in line-of-sight from source to observer, that is, the wave path is not one straight line, but rather two straight lines, with the longest segment in the medium of higher phase speed; thus the angle with the normal is larger in the medium with higher phase speed in agreement with the third law of scattering (22.9c). The latter can be obtained by considering the propagation time (22.12a) between a source at (x_s, y_s) and observer at (x_0, y_0), as a function of the position x of the scattering point on the interface $y = 0$, viz.:

$$\psi = \frac{D_s}{c_1} + \frac{D_0}{c_2}, \quad D_s = \left|(x_s - x)^2 + y_s^2\right|^{1/2}, \quad D_0 = \left|(x_0 - x)^2 + y_0^2\right|^{1/2}, \qquad (22.12\text{a–c})$$

where, $D_s(D_0)$ is the distance from the source (observer) to the scattering point (22.12b) [(22.12c)]. The condition of stationary time of propagation is:

$$0 = \frac{\partial\psi}{\partial x} = \frac{1}{c_1}\frac{dD_s}{dx} + \frac{1}{c_2}\frac{dD_0}{dx} = \frac{|x_s - x|}{c_1 D_s} sgn\,(x - x_s) + \frac{|x_0 - x|}{c_2 D_0} sgn\,(x - x_0); \qquad (22.13)$$

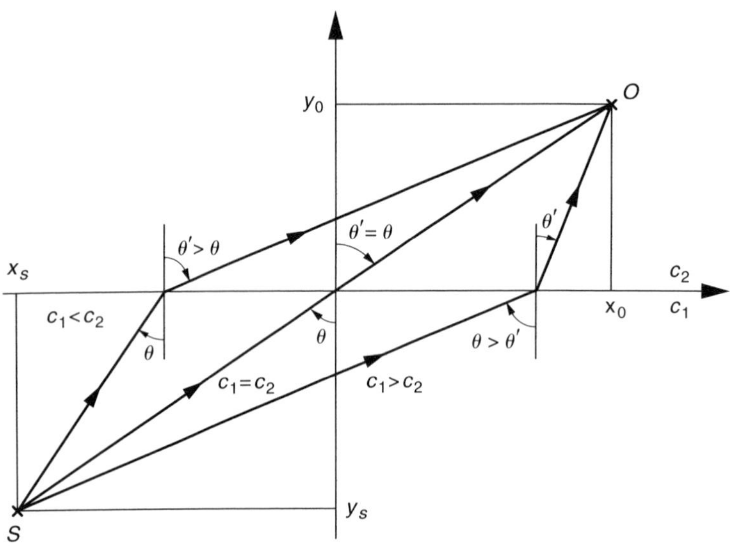

FIGURE 22.3
The principle of Fermat states that a wave travels from a source S to an observer O along the quickest path, that is, with shorter propagation time or smallest time delay. If the medium is homogeneous, that is, has the same phase speed everywhere, the quickest path is the shortest, that is, the straight line from the source to the observer. If the source S and (observer O) are in distinct media with different phase speeds $c_1(c_2)$ the straight line joining S to O no longer is the quickest path: the quickest path is shorter (longer) in the medium of lower (higher) phase speed. Thus the ray path of the wave consists of two straight segments, one in each medium of incidence (transmission) meeting at the same point on the interface between the two media. The position of that point is determined by the shortest travel time, and establishes the law of refraction relating the angles of incidence and transmission.

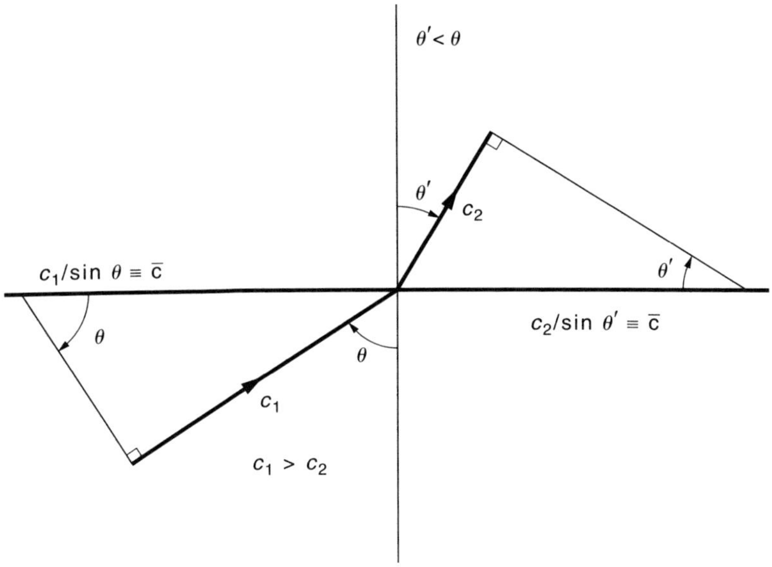

FIGURE 22.4
The law of refraction of a wave at the interface between two media can be established in two ways: (i) by the minimum travel time between a source in one medium and an observer in the other (Figure 22.3); (ii) by projecting the phase speeds on the direction tangent to the interface and noting that they must be equal in order for the phase of the wave to be continuous across the interface. This implies that: (i) the wavenumber transverse to the interface is continuous; (ii) the angle of the direction of propagation with the normal is larger (smaller) in the medium where the phase speed is larger (smaller); (iii) the wavenumber normal to the interface is smaller (larger) in the medium where the phase speed is larger (smaller). This can be seen in Figures 22.2b, 22.3, and 22.4.

there is a solution of (22.13) only if the two terms have opposite signs (22.14a):

$$\operatorname{sgn}(x_s - x) = -\operatorname{sgn}(x_0 - x); \quad x_s > x > x_0 \quad or \quad x_0 > x > x_s, \tag{22.14a,b}$$

that implies (22.14b) that *the scattering point must lie between the normal from observer and source to the interface* (Figure 22.3). Substituting (22.14b) in (22.13) follows:

$$\bar{c} \equiv \frac{c_1}{\sin\theta} = \frac{c_1 D_s}{|x_s - x|} = \frac{c_2 D_0}{|x_0 - x|} = \frac{c_2}{\sin\theta'}, \tag{22.15}$$

the third law of scattering (22.15) ≡ (22.9c); thus *the wave path is longer in the medium of larger phase speed, because the phase speed of wave fronts along the interface is constant* (Figure 22.4), *that is, the incident, reflected, and transmitted waves must all be in phase at the scattering point.*

22.4 Reflection and Transmission of Acoustic Waves

The **reflection (transmission) factor** is defined locally at a point on the interface as the ratio of the amplitude of the reflected A_r (transmitted A_t) wave to the amplitude I of the incident wave:

$$R \equiv \frac{A_r}{I}, \quad T \equiv \frac{A_t}{I}. \tag{22.16a,b}$$

The reflection R (transmission T) factors generally depend on the angle-of-incidence, θ, and are determined by applying boundary conditions at the interface, relevant to the type of wave under consideration for example, electromagnetic, acoustic, or elastic; the scattering (i.e., reflection and transmission) factors are real if only the amplitude changes in (22.16a,b), and are complex if a phase shift is involved. The reflection and transmission factors (Subsection 22.4.1) are determined next for an acoustic plane waves (Subsection 22.4.2), in the simplest case of a plane interface between media at rest. To determine two scattering factors (22.16a,b) two boundary conditions are needed, viz. continuity of (i) pressure and (ii) displacement normal to the interface.

22.4.1 Continuity of Pressure and Normal Displacement at the Interface

The velocity $\vec{v}$ is the time derivative of the displacement $\vec{\zeta}$, and thus the latter is related to the pressure p by the inviscid momentum equation [(14.9) without external forces $\vec{f} = 0$], viz.:

$$-\frac{\nabla p}{\rho} = \frac{d\vec{v}}{dt} = \frac{d^2\vec{\zeta}}{dt} = \left(\frac{\partial}{\partial t} + \vec{v} \cdot \nabla\right)^2 \vec{\zeta}, \tag{22.17}$$

where, ρ is the mass density and the material derivative (14.6) was used. The incident p_i, reflected p_r, and transmitted p_t plane waves are represented (21.1b) by their respective acoustic pressures:

$$p_i,\ p_r,\ p_t(\vec{x}, t) = e^{i(kx-\omega t)} \left\{I \exp(ik_{in}y),\, A_r \exp(ik_{rn}y),\, A_t \exp(ik_{tn}y)\right\}, \tag{22.18a–c}$$

that have: (i) all the same frequency ω and transverse wave number in (22.8a–c), where the x-direction is chosen along $\vec{k} = k\vec{e}_x$; (ii) distinct amplitudes (22.16a,b) and wave numbers normal to the interface, that is, in the y-direction:

$$k_{in} \equiv \vec{k}_i \cdot \vec{N} = \frac{\omega}{c_1}\cos\theta = -\vec{k}_r \cdot \vec{N} \equiv -k_{rn}, \qquad k_{tn} \equiv \vec{k}_t \cdot \vec{N} = \frac{\omega}{c_2}\cos\theta; \tag{22.19a,b}$$

here as before the $c_1(c_2)$ is the sound speed in the medium of incidence (transmission) and $\theta(\theta')$ the angle-of-incidence (transmission) with the normal to the interface.

22.4.2 Acoustic Reflection and Transmission Factors

In a medium at rest the inviscid momentum equation (22.17) simplifies to

$$\frac{\nabla p}{\rho} = -\frac{\partial^2\vec{\zeta}}{\partial t^2} = -(-i\omega)^2\vec{\zeta} = \omega^2\vec{\zeta}, \tag{22.20}$$

where at time derivative (gradient) is equivalent (22.21a) to multiplication by $-i\omega(i\vec{k})$ for a plane wave (22.1a,b), viz.:

$$\left\{\frac{\partial}{\partial t}, \frac{\partial}{\partial \vec{x}}\right\} = \left\{-i\omega,\ i\vec{k}\right\};\ 0 = \vec{\zeta} \cdot \vec{N} = \frac{\vec{N} \cdot \nabla p}{\rho\omega^2} = \frac{i}{\omega^2}\frac{p}{\rho}\left(\vec{k} \cdot \vec{N}\right), \tag{22.21a,b}$$

thus the displacement (22.20) normal to the interface satisfies (22.21b). The frequency ω is continuous (22.22a) and thus the boundary conditions stating the continuity of (i) pressure (22.22b) and (ii) displacement normal to the interface (22.21b) $\equiv$ (22.22c) are

$$[\omega] = 0: \quad [p] = 0 = \left[\frac{p}{\rho}\left(\vec{\mathrm{k}} \cdot \vec{\mathrm{N}}\right)\right]. \qquad (22.22\text{a–c})$$

Substituting (22.18a–c) with (22.23a) and (22.19a,b) in (22.22b,c) leads to the relations (22.23b,c):

$$y = 0: \quad I + A_r = A_t, \quad \frac{I - A_r}{\rho_1 c_1}\cos\theta = \frac{A_t}{\rho_2 c_2}\cos\theta', \qquad (22.23\text{a–c})$$

where $\rho_1(\rho_2)$ is the mass density in the medium of incidence (transmission). Dividing (22.23b,c) by I leads:

$$1 + R = T, \quad (1 - R)\,\rho_2 c_2 \cos\theta = T \rho_1 c_1 \cos\theta'. \qquad (22.24\text{a,b})$$

Solving for the reflection (22.25a) and transmission (22.25b) factors yields:

$$R = \frac{1 - D}{1 + D}, \quad T = \frac{2}{1 + D}, \quad D \equiv \frac{\rho_1 c_1 \cos\theta'}{\rho_2 c_2 \cos\theta}, \qquad (22.25\text{a–c})$$

in terms of the **scattering coefficient** (22.25c) $\equiv$ (22.26a,b):

$$D(\theta) \equiv \frac{\rho_1 c_1}{\rho_2 c_2}\left|1 - \left(\frac{c_2}{c_1}\right)^2 \sin^2\theta\right|^{1/2} \frac{1}{\cos\theta},$$

$$= \frac{\rho_1 c_1}{\rho_2 c_2}\left|1 - \left(\frac{c_1}{c_2}\right)^2 \sin^2\theta'\right|^{-1/2} \cos\theta' \equiv D(\theta'); \qquad (22.26\text{a,b})$$

here (22.9c) was used to show the dependence on the angle-of-incidence, θ in (22.26a) [transmission θ' in (22.26b)] and properties (ρ_1, c_1) $[(\rho_2, c_2)]$ of the media of incidence and transmission.

It has been shown that a *the conditions of continuity of pressure (22.22b) and displacement (22.20) normal to a plane interface (22.22c), for an incident plane wave (22.18a), leads to reflected (22.18b) [transmitted (22.18c)] plane waves, with reflection (22.16a) [transmission (22.16b)] factor specified by (22.25a) [(22.25b)] in terms of the scattering coefficient (22.25c) by: (i) the angle-of-incidence, θ (transmission θ') in (22.26a) [(22.26b)]; (ii) the mass density and sound speed the medium of incidence (ρ_1, c_1) [transmission (ρ_2, c_2)]. The energy is conserved during the scattering (22.23a,b) $\equiv$ (22.24a). In particular for total reflection (22.10a,b) [zone of silence (22.11a,b)] it follows that*

$$c_2 > c_1: \quad \sin\theta_R = \frac{c_1}{c_2}, \quad \theta'_R = \frac{\pi}{2}, \quad D = 0, \quad R = +1, \quad T = 2, \qquad (22.27\text{a–f})$$

$$c_1 > c_2: \quad \sin\theta'_S = \frac{c_2}{c_1}, \quad \theta_S = \frac{\pi}{2}, \quad D = \infty, \quad R = -1, \quad T = 0, \qquad (22.28\text{a–f})$$

the reflection coefficient is R $= +1\,(\mathrm{R} = -1)$, *and there is a double transmission factor* T $= 2$ *(no transmission* T $= 0$) *in agreement with (22.24a).* The double transmission factor is of no consequence, because the transmitted wave is evanescent for total reflection (Example 30.4).

22.5 Adsorption at an Interface and Internal Absorption

If the interface is passive, it cannot amplify the waves, and the sum of the amplitudes A_r (A_t) of the reflected (transmitted) waves cannot exceed the amplitude I of the incident wave (22.29a):

$$|A_r| + |A_t| \leq |I|, \qquad |R| + |T| \leq 1, \tag{22.29a,b}$$

implying that the sum the reflection R and transmission T coefficients cannot exceed unity. In both formulas (22.29a,b) there is: *(i) equality if there is no adsorption of waves at the interface; (ii) inequality if the interface is* **adsorvent**, *that is, withdraws energy from the wave.* Another type of energy loss is **absorption** by the medium, at a rate with regard to distance x proportional (22.30b) to the amplitude A through the **absorption factor** (22.30a):

$$\nu(x) > 0: \qquad \frac{\mathrm{d}A}{\mathrm{d}x} = -\nu(x)\,A; \quad A(x) = A(0)\exp\left\{-\int_0^x \nu(\eta)\;\mathrm{d}\eta\right\}; \tag{22.30a–c}$$

the absorption factor (22.30a) leads to (22.30c) an exponential decay of the amplitude with distance on a scale $1/\nu$; the absorption factor depends on the physical process responsible for wave dissipation, for example, fluid viscosity and thermal conduction for an acoustic wave.

When there are several interfaces separating distinct media its is possible to calculate, from the local reflection, transmission, and **damping** (i.e., absorption + adsorption) **factors**, a single set of **total reflection, transmission, and damping coefficients** that allow the set of interfaces to be replaced by a single equivalent interface. Take as example a **double interface**, consisting of a strip thickness h of a medium 2 lying between two parallel interfaces, forming a parallel-sided lens (Figure 22.5a) with the same unbounded medium 1 above and below. Since the upper and lower media are identical (medium 1) and distinct from the intermediate medium (medium 2), the following physical properties have to be considered: (i) the pairs of reflection and transmission factors (R_-, T_-) for scattering from the medium 2 to 1, and (R_+, T_+) for scattering from the medium 1 to 2; (ii) if there is absorption in the interior of the medium 2, with constant absorption coefficient ν, it introduces (22.30c) an amplitude decay $\mathrm{e}^{-\delta}$ where, $\delta = \nu x = \nu h \sec\theta'$, since (22.31a) a wave making an angle θ' with the normal travels a distance $x = h\sec\theta'$ between the interfaces spaced h. Thus there are *two processes of amplitude decay: (i) gradual absorption in the lens, introducing a factor* $\mathrm{e}^{-\delta}$ *applied (22.31b) to the reflection and transmission factors of the waves in medium 2:*

$$\delta \equiv \nu h \sec\theta': \quad (R_+,\, T_+,\, R_-,\, T_-,) \to \left(R_+,\, T_+\,\mathrm{e}^{-\delta},\, R_-\,\mathrm{e}^{-\delta},\, T_-\right); \tag{22.31a,b}$$

(ii) abrupt adsorption at the interfaces if the inequality (22.29b) applies to the reflection $\mathrm{R}_\pm$ *and transmission* $\mathrm{T}_\pm$ *factors.*

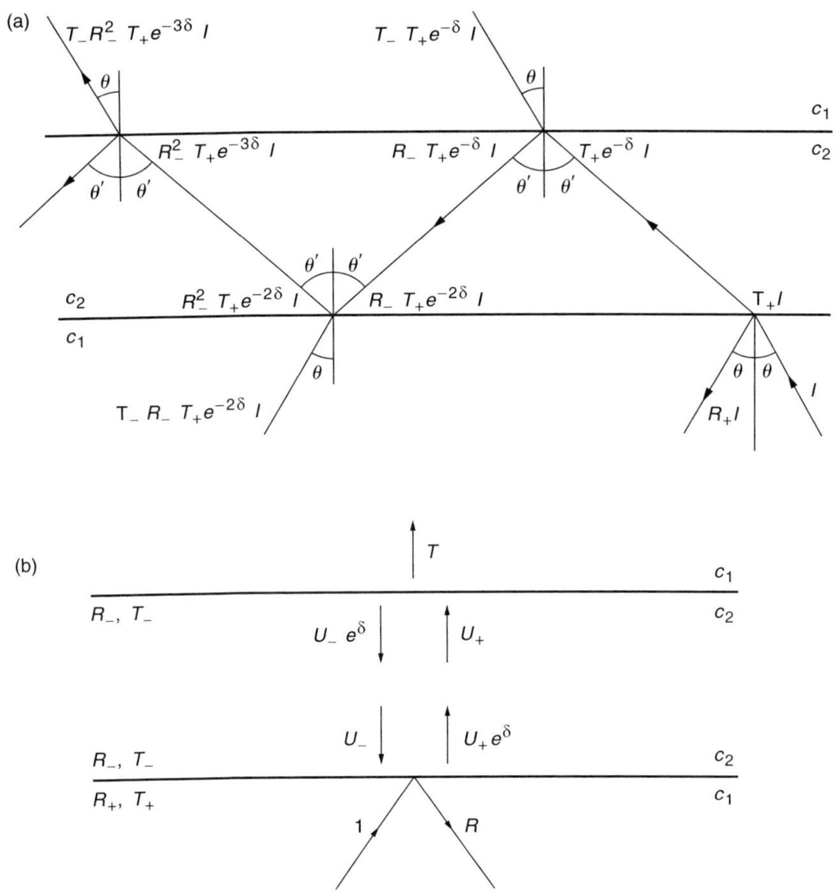

FIGURE 22.5
The scattering of an incident wave by a parallel-sided lens between identical media (distinct media in Figure 30.1) can be considered by two methods. The first method (a) follows the incident wave (from right to left): (i) the first scattering at the lower interface with a reflected (transmitted) wave in the lower medium (the lens); (ii) the wave transmitted into the less undergoes partial absorption until it is scattered at the upper interface, into a reflected (transmitted) wave in the lens (the upper medium); (iii) the doubly-reflected wave is partially absorbed in the lens until it is scattering at the lower interface into a reflected (transmitted) wave in the lens (the lower medium); (iv) the triply reflected wave is again partially absorbed until being scattered at the upper interface into a reflected (transmitted) wave in the lens (upper medium); (v) the process of double successive reflections at the lower and upper interfaces goes on indefinitely with progressively less residual amplitude; (vi) the wave field in the lower (upper) medium is the sum of all reflections (transmissions) without and with double internal reflections of all orders in the lens; (vii) the energy absorbed in the lens during the double reflections of all orders is the energy in the incident wave in excess of the sum of the energies of the waves reflected into (transmitted to) the lower (upper) medium.

The second method (b) relates four fields: (i/ii) the reflected (transmitted) wave in the upper (lower) medium; (iii/iv) the upward (downward) propagating waves in the lens. There are four relations between the four fields: (i) the upward wave in the lens is due to transmission at the lower interface of the wave incident from the lower medium plus reflection of the downward wave in the lens; (ii/iii) the upward wave in the lens is transmitted (reflected) at the upper interface as the transmitted wave in the upper medium (the downward wave in the lens); (iv) the reflected wave in the lower medium is due to the reflection of the incident wave at the lower interface plus transmission of the downward wave in the lens. When applying (i) to (iv) it must be taken into account that both the upward and downward waves in the lens differ at the two interfaces by the internal absorption. Both methods (a) and (b) allow for: (i) internal absorption in the lens; (ii) surface adsorption at the interfaces. Both methods lead to the same total reflection/transmission (damping) coefficients defined as the fraction of the incident wave amplitude which goes into the amplitudes of the reflected /transmitted wave/ (is missing from both).

22.6 Multiple Reflections between Parallel Interfaces

Consider (Figure 22.5a) a wave of amplitude I incident from the lower medium: (i) the scattering at the lower interface yields a reflected wave R_+I in the lower medium, and a transmitted wave T_+I in the lens; (ii) the latter is incident at the upper interface with amplitude $e^{-\delta}T_+I$ giving rise to a wave of amplitude $T_-e^{-\delta}T_+I$ transmitted to the upper medium, and to a wave of amplitude $R_-e^{-\delta}T_+I$ reflected downward into the lens; (iii) the latter is incident on the lower interface with amplitude $e^{-\delta}R_-e^{-\delta}T_+I$, giving rise to a double reflection $R_-e^{-\delta}R_-e^{-\delta}T_+I$ into the lens, and a first order transmission $T_-e^{-\delta}R_-e^{-\delta}T_+I$ into the lower medium; (iv) proceeding in the same way can be considered double reflections of all higher orders, as well as the associated higher order transmissions upward and downward. This leads to the reflected A_r (transmitted A_t) wave amplitudes above (below) the lens specified the scattering series (22.32a)[(22.32b)]:

$$A_r = R_+I + T_-e^{-\delta}R_-e^{-\delta}T_+I + T_-e^{-\delta}\left(R_-e^{-\delta}\right)^3 T_+I + T_-e^{-\delta}\left(R_-e^{-\delta}\right)^5 T_+I + \cdots \tag{22.32a}$$

$$A_t = T_-e^{-\delta}T_+I + T_-e^{-\delta}\left(R_-e^{-\delta}\right)^2 T_+I + T_-e^{-\delta}\left(R_-e^{-\delta}\right)^4 T_+I + \cdots. \tag{22.32b}$$

It is assumed that all waves are in phase, otherwise interference effects occur (Section 22.9). Dividing throughout $A_r(A_t)$ by the amplitude I of the incident wave, specifies **reflection (transmission T) coefficient** $R(T)$ for the lens, as geometric series (22.32a) [(22.32b)]; these are convergent because their ratio $z \equiv R_-^2 e^{-2\delta} < 1$ does not exceed unity (21.62a,b) and thus can be summed (22.33a) [(22.33b)]:

$$R \equiv \frac{A_r}{I} = R_+ + \frac{T_+T_-R_-e^{-2\delta}}{1 - R_-^2 e^{-2\delta}}, \tag{22.33a}$$

$$T \equiv \frac{A_t}{1} = \frac{T_+T_-e^{-\delta}}{1 - R_-^2 e^{-2\delta}}; \tag{22.33b}$$

$$S \equiv 1 - T - R = 1 - R_+ - \frac{T_+T_-e^{-\delta}}{1 - R_-e^{-\delta}}, \tag{22.33c}$$

the **damping or loss coefficient** S specifies (22.33c) the fraction of the amplitude of the incident wave missing from the reflected and transmitted waves. Thus have been obtained *the total reflection (22.33a), transmission (22.33b), and damping (22.33c) coefficients for a lens with adsorption (22.29b) at the two interfaces and absorption (22.31a,b) in the interior.* In the particular case when: (i) absorption in the interior is absent $\delta = 0$, and only adsorption at the interface remains:

$$\delta = 0: \quad R = R_+ + \frac{T_+T_-R_-}{1 - R_-^2}, \quad T = \frac{T_+T_-}{1 - R_-^2}, \quad S = 1 - R_+ - \frac{T_+T_-}{1 - R_-}; \tag{22.34a–c}$$

(ii) if besides there is no adsorption at the interface, the reflection and transmission coefficients simplify to

$$T_\pm = 1 - R_\pm: \quad R = R_+ + \frac{R_-(1 - R_+)}{1 + R_-}, \quad T = \frac{1 - R_+}{1 + R_-}, \quad S = 0, \tag{22.35a–c}$$

implying that the total amplitude is conserved $R + T = 1$, and the damping coefficient (22.35c) vanishes. In the latter case the total reflection R and transmission T coefficients can be expressed in terms of the partial reflection factors alone.

22.7 Total Reflection, Transmission, and Damping Coefficients

The total scattering coefficients for a system of interfaces may be calculated without summing series, by using a different algebraic method, illustrated in Figure 22.5b for a double interface: (i) an incident wave of unit amplitude is considered, so that the amplitude of the reflected (transmitted) wave below (above) the lens is the total reflection R (transmission T) coefficient; (ii) the wave field inside the lens is decomposed into upward U_+ (downward U_-) components, calculated at the upper (lower) interface, so that their value is $U_+e^\delta (U_-e^\delta)$ at the lower (upper) interface. The four quantities $R.T, U_\pm$ are related by four scattering relations, two at each interface, stating that: (i) the upward field U_+ is refracted at the upper interface into a transmitted T_- (reflected R_-) field above T (downward U_- into) the lens (22.36a) [(22.36b)]:

$$T = T_- U_+, \quad U_- e^\delta = R_- U_+; \tag{22.36a,b}$$

(ii) the upward field a U_+ in the lens is due to the transmission T_+ of the incident field, plus reflection R_- of the downward field U_-, both at the lower interface (22.37a):

$$U_+ e^\delta = T_+ + R_- U_-; \quad R = R_+ + T_- U_-, \tag{22.37a,b}$$

(iii) the reflected field R below the lens is due (22.37b) to reflection R_+ of the incident wave, plus transmission $T_-$of the downward field U_-, both at the lower interface.

The coupled system of equations (22.36b; 22.37a) for $U_\pm$ can be solved either for the upward (22.38a) or downward (22.38b) field:

$$U_+ = T_+ e^{-\delta} + R_-^2 e^{-2\delta} U_+, \quad U_- = R_- T_+ e^{-2\delta} + R_-^2 e^{-2\delta} U_-, \tag{22.38a,b}$$

showing that: (i) initially the upward field (22.38a) is due to transmission of the incident wave T_+, at the lower interface, with the factor $e^{-\delta}$ because it crosses the lens once; (ii) initially the downward field (22.38b) is due to transmission at the lower interface T_+ followed by reflection at the upper interface R_-, with factor $e^{-2\delta}$ because it crosses the lens twice; (iii) both fields (22.38a,b) regenerate themselves after a double internal reflector R_-^2 that has the factor $e^{-2\delta}$ because it involves crossing the lens twice. Solving (22.38a,b) specifies explicitly the upward (22.39a) [downward (22.39b)] fields:

$$U_+ = \frac{T_+ e^\delta}{e^{2\delta} - R_-^2}; \quad U_- = \frac{R_- T_+}{e^{2\delta} - R_-^2}, \quad U_+ + U_- = \frac{T_+}{e^\delta - R_-}, \tag{22.39a–c}$$

and their sum (22.39c). Substitution of (22.39b) into (22.37b) [(22.39a) into (22.36a)] yields the total reflection R (22.40a) [transmission T (22.40b)] coefficient:

$$R = R_+ + \frac{R_- T_+ T_-}{e^{2\delta} - R_-^2}, \quad T = \frac{T_- T_+ e^\delta}{e^{2\delta} - R_-^2}. \tag{22.40a,b}$$

There is agreement between (22.33a) ≡ (22.40a) [(22.33b) ≡ (22.40b)] the two methods of solution of the multiple scattering problem: (I) summation of infinite scattering series (Section 22.6; Figure 22.5a); (II) decomposition into upward and downward fields (Section 22.7; Figure 22.5b). The agreement of the reflection and transmission coefficients implies that of the damping coefficient; the latter is given by (22.36a; 22.37b):

$$S \equiv 1 - R - T = 1 - R_+ - T_- (U_- + U_+) = 1 - R_+ - \frac{T_- T_+}{e^{\delta} - R_-}; \qquad (22.40c)$$

here (22.39c) was used, leading to (22.40c) ≡ (22.33c).

22.8 Multiple Media, Transparency, and Opaqueness

It has been shown that the total reflection R, transmission T, and damping S coefficients for a double interface can be calculated by two distinct methods, viz. summation of multiple scattering series (Section 22.6), and solution of system of scattering equations (Section 22.7), with identical results (22.33a–c) ≡ (22.40a–c). Consider next particular types of interface, noting that if the amplitude $A_r (A_t)$ of the reflected (transmitted) waves cannot exceed that I of the incident wave (22.41a):

$$0 \le |A_r|, |A_t| \le I, \quad 0 \le |R|, \quad |T| \le 1, \qquad (22.41a,b)$$

the reflection (transmission) coefficients lie between zero and unity in modulus (22.41b). The *extreme cases of nonadsorptive interfaces* R+T = 1, *are the* **transparent (opaque) interface**, *which transmits (reflects) without loss of amplitude* $|\mathrm{T}| = 1, \mathrm{R} = 0\,(|\mathrm{R}| = 1, \mathrm{T} = 0)$. The opaque interface is realized in the case of incidence into a denser medium, for angles $\theta > \theta_R$ exceeding that of total reflection. The transparent interface is the opposite case. Returning to the double interface (22.33a–c) ≡ (22.40a–c), there are two pairs of particular cases to consider: (i) the passage $1 \to 2$ is opaque (22.42a,b):

$$R_+ = 1, \quad T_+ = 0: \quad R = 1, \quad T = 0 = S, \qquad (22.42a–e)$$

the incident wave is totally reflected below the lens (22.42a), and no wave exists above (22.42d), and no absorption takes place (22.42e); (ii) the passage $1 \to 2$ is transparent (22.43a,b):

$$T_+ = 1, \quad R_+ = 0: \quad R = \frac{T_- R_-}{e^{2\delta} - R_-^2}, \qquad T = \frac{T_- e^{\delta}}{e^{2\delta} - R_-^2}, \qquad S = 1 - \frac{T_-}{e^{\delta} - R_-}, \quad (22.43a–e)$$

there is no zero-order reflection in (22.43c), but there is transmission (22.43d) and absorption (22.43e) to all orders.

The remaining two cases are: (iii) the passage $2 \to 1$ is opaque (22.44a,b):

$$T_- = 0, \quad R_- = 1: \quad R = R_+, \quad T = 0, \quad S = 1 - R_+, \qquad (22.44a–c)$$

the wave entering the lens is dissipated (22.44e), because it cannot be transmitted above (22.44d), or reflected below, that is, only zero-order reflection (22.44c) occurs; (iv) if the passage $2 \to 1$ is transparent (22.45a,b):

$$T_- = 1, \quad R_- = 0: \quad R = R_+, \quad T = T_+ \, e^{-\delta}, \quad S = 1 - R_+ - T_+ e^{-\delta}, \qquad (22.45a–e)$$

the reflection is of zero-order (22.45c), since all transmitted energy transverses the lens (22.45d), apart from absorption (22.45e), and multiple reflections do not occur. In the absence of absorption $\delta = 0$ there is: (i) a trivial simplification of (22.45a–e) for transparent $2 \to 1$ passage:

$$T_- = 1, \quad R_- = 0 = \delta: \quad R = R_+, \quad T = T_+, \quad S = 1 - R_+ - T_+; \tag{22.46a–f}$$

(ii) a not so trivial simplification of (22.43a–e):

$$T_+ = 1, \quad R_+ = 0 = \delta, \quad T_- = 1 - R_-: \quad R = \frac{R_-}{1 + R_-}, \quad T = \frac{1}{1 + R_-}, \quad S = 0, \tag{22.47a–f}$$

for lossless transparent $1 \to 2$ passage.

The angle of transmission (22.9c) between three media separated by parallel interfaces (Figure 30.1) is unaffected by the intermediate medium:

$$\sin\theta' = \frac{c_2}{c_1}\sin\theta, \quad \sin\theta'' = \frac{c_3}{c_2}\sin\theta' = \frac{c_3}{c_1}\sin\theta, \tag{22.48a–c}$$

viz.: (i) the angle-of-incidence from the first medium with wave speed c_1 is θ, and second medium with wave speed c_2 has angle of transmission θ' in (22.48a); (ii) the latter θ' is the angle-of-incidence from the second medium in (22.48b), where θ'' is the angle of transmission to the third medium c_3; (iii) the final relation (22.48c), between the angle-of-incidence, θ in the first medium c_1 and the angle of transmission θ'' to the third medium c_3, is unaffected by the second medium, viz. θ' and c_2 do not appear in (22.48c). *If the first and last media are the same (22.49a), the incident and transmitted waves are parallel (22.49b):*

$$c_1 = c_3: \quad \theta'' = \theta; \quad d_0 = h\sin\theta', \quad d_n = (2n+1)\,h\sin\theta' = (2n+1)\,d_0, \tag{22.49a–d}$$

in all cases of similar or distinct first and third media, the deviation (Figure 30.1) between the point of incidence in the first medium and the point of transmission to the third medium, depends on the angle of transmission to the second medium θ' and distance h between the interfaces, for the first transmission (22.49c), and the double is added for each double reflection (22.49d). The reflection, transmission, and damping coefficients for two parallel interfaces between three distinct media (Figure 30.1) are considered in Example 30.5.

22.9 Constructive and Destructive Interference (Bragg, 1912)

In the preceding expressions (Sections 22.6–22.8) *the amplitudes were added that implies that all wave components are in phase, and there is constructive interference, leading to the largest possible amplitudes*; in *case the transmission through the lens causes a phase shift α, it can be incorporated in the exponential factor:*

$$\exp(-\delta) \quad \to \quad \exp(-\delta + i\alpha)\,. \tag{22.50}$$

This would lead to complex reflection R, transmission T, and damping S coefficients, consisting of amplitude ($|R|$, $|T|$, $|S|$) and phase $[\arg(R, T, S)]$ changes; they would also be complex if the local reflection and transmission factors ($R_\pm$, $T_\pm$) had phase shifts. The **interference** *of two waves with the same amplitude but out-of-phase by α, is specified in the total wave:*

$$\phi \equiv \vec{k}\cdot\vec{x} - \omega t: \quad F\left(\vec{x}, t\right) = Ae^{i\phi} + Ae^{i(\phi-\alpha)} = Ae^{i(\phi-\alpha/2)}\left(e^{i\alpha/2} + e^{-i\alpha/2}\right)$$
$$= 2A\cos\left(\frac{\alpha}{2}\right)e^{i(\phi-\alpha/2)}; \tag{22.51}$$

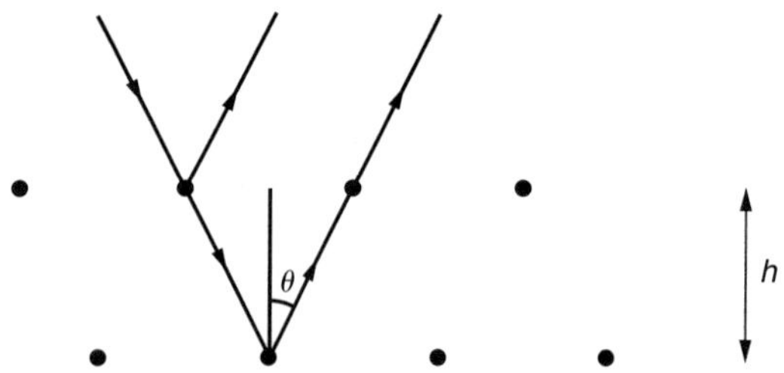

FIGURE 22.6
The waves incident on a crystal are reflected from successive rows of atoms. The strongest reflections are from the first two rows, and the phase shift indicates the interatomic distance.

that has (22.52a) a phase shift $\alpha/2$ and an amplitude (22.52b):

$$F\left(\vec{x},t\right) = B(\alpha)\,e^{i(\phi-\alpha/2)}, \quad B(\alpha) = 2A\cos\left(\frac{\alpha}{2}\right); \tag{22.52a,b}$$

the latter varies between the double (zero) as the phase difference ranges from 0 to π :

$$|B(\alpha)|_{\max} = B(0) = 2A \geq |B(\alpha)| \geq |B(\alpha)|_{\min} = B(\pi) = 0, \tag{22.53}$$

that corresponds to perfect addition (cancellation).

An experimental instance is an x-ray beam incident upon two parallel rows of atoms, a distance h apart (Figure 22.6). Assuming that the angle of reflection is equal to that of incidence, θ, a wave scattered by the second row travels a distance $2h\sec\theta$ in excess of that travelled by a wave scattered from the first row. There is constructive (destructive) interference if the additional distance is equal to one (one-half) a wavelength, that is, *the maximum (minimum) intensity of the x-ray diffraction occurs for the wavelength $\lambda_{\mathrm{M}}\,(\lambda_{\mathrm{m}})$:*

$$\lambda_M = 2h\sec\theta, \quad \lambda_m = 4h\sec\theta, \tag{22.54a,b}$$

where h *is the interatomic distance, and θ is the angle-of-incidence.* The **Bragg law (1912)** stated by (22.54a,b) was obtained considering the first interference from the first two rows of atoms. In addition, there is a doubly-infinite set of additional interferences due to: (i) excess distances an even (odd) multiples of half-a-wavelength, for constructive (destructive) interference; (ii) reflection from the n-th row that interferes with reflection from the m-th row, with an excess distance $2\ |n-m|\,h\sec\theta$, leading to additional sets of maxima and minima. For many crystals the shortest wavelength for x-ray scattering by the first two rows of atoms gives the strongest maximum of interference, and Bragg's law (32.54a) is sufficient to determine interatomic distances.

Note 22.1. *Scattering of a Variety of Waves:* The phase speed and scattering factors depend on the properties of the medium, for example, for acoustic waves: (i) the adiabatic sound speed in a fluid depends (14.31a,b) on the pressure, mass density and adiabatic exponent; (ii) for a perfect gas (14.11b) it depends only on the adiabatic exponent and temperature (14.31c); (iii) the reflection and transmission factors depend on the angle-of-incidence and sound speeds and mass densities of the two media, for a plane interface at rest. Other dependences apply to each specific type of wave, for example (i) electromagnetic waves in vacuo or matter; (ii) acoustic, gravity, and inertial waves respectively in compressible, stratified and rotating fluids; (iii) longitudinal and transversal elastic waves in bars, plates, shells, and three-dimensional solids; (iv) surface waves for example, hydraulic waves in liquids and seismic waves in solids; (v) torsion, bending, and longitudinal waves in elastic rods;

(vi) waves in sheared and swirling flows; (vii) shear waves in ionized fluids in an external magnetic field; (viii) couplings of the preceding, for example, a magneto-acoustic-gravity-inertial waves; (ix) gravitational waves in general relativity at cosmic scales; (x) probability waves in quantum mechanics at microscopic scales. The preceding list is indicative of the variety of waves, but not exhaustive; all these waves exhibit scattering effects. The reflection and transmission coefficients are simplest (Section 22.4) for a plane interface between media at rest, and are modified by: (i) moving media on one or both sides of the interface; (ii) curved, irregular, or random interfaces. The randomly irregular moving interfaces are associated with the "clutter" in the reflection of radar signals, and change not only the directivity but also the spectrum of waves, for example, sound.

Conclusion 22: Propagation in the direction $\vec{n}$ of a sinusoidal wave (Figure 22.1) of wavelength λ and period τ at phase speed c. A wave (Figure 22.2) of amplitude I incident (**a**) upon a plane interface separating to media with wave speeds c_1 (c_2) gives rise to a reflected (transmitted) wave of amplitude $A_r = IR\,(A_t = IT)$ where $R(T)$ is the reflection (transmission) factor. The angles with the normal of the incident θ_i and reflected θ_r waves are equal $\theta_i = \theta_r = \theta$, and generally distinct from the angle $\theta \neq \theta' \equiv \theta_t$ of the transmitted wave. If the medium of transmission (incidence) is the denser of the two (**b**) [(**c**)], i.e., has the higher propagation speed $c_2 > c_1$ $(c_1 > c_2)$, the incident wave is transmitted with deflection away $\theta' > \theta$ (towards $\theta' < \theta$) the normal, leading, for grazing transmission $\theta' = \pi/2$ (incidence $\theta = \pi/2$) to the angle of total reflection θ_R (silence or darkness θ'_S), beyond which $\theta > \theta_R$ there is total reflection ($\theta' > \theta'_S$ no waves exist). These phenomena can be explained in two equivalent ways: (Figure 22.3) a wave follows the fastest path from source S to observer O and thus instead of following a straight line for a homogeneous medium $c_1 = c_2$, it follows a broken line which is longer, i.e., farther from the normal, in the medium with larger phase speed; (Figure 22.4) because the phase is continuous, the wave fronts must travel along the interface with the same speed in both media, and since they are orthogonal to the phase speed they deviate more from the normal in the medium with larger phase speed. In the case of two plane parallel interfaces (Figure 22.2d) a distance h apart, the amplitude loss can occur by adsorption on scattering at the interface between medium 1 and 2 (2 and 1), viz. $|R_+| + |T_+| < 1\,(|R_-| + |T_-| < 1)$ or by absorption in the interior, which depends on the distance of propagation $h \sec \theta'$ between two faces. The total reflection R and transmission T coefficients can be calculated (Figure 22.5) by: (**a**) following the successive scatterings, and summing the series corresponding to reflections or transmissions of all orders; (**b**) decomposing the waves inside the lens in to upward U_+ and downward U_- propagating components, and solving the algebraic system of scattering equations. The amplitudes can be added if the waves are in phase, and subtracted if they are out-of-phase; the maxima and minima of intensity are used (Figure 22.6) in x-ray scattering to determine the interatomic distance h, if the angle of incidence $\equiv$ reflection θ is known.

23

Analytic Series of Ascending Powers

It was shown that if a function $F(z)$ is holomorphic in a region or satisfies the Cauchy conditions (stated in Section 15.1 and proven in Sections 19.1–19.6), then it has derivates of all orders at all interior points (Sections 15.1–15-4). Further it can be shown that the holomorphic function also has a convergent Taylor series, that is, a complex differentiable function is also an analytic function (Section 23.7). This is not true of real functions, since the existence of first-order derivate does not imply the existence of higher-order derivates; nor the convergence of the Taylor series for infinitely differentiable (or smooth) functions. The Taylor (Maclaurin) series allows the calculation, from the value of the function and all its derivates at a point $z = a$ (the origin $z = 0$) its value at all other points z in its circle of convergence; the region of convergence of a Taylor (Maclaurin) series is a circle, because the expansion proceeds in integral ascending powers of $z - a(z)$, that is, holds for $|z - a| > R(|z| < R)$ where R is the radius of convergence. If instead of $f(z) = z - a$ or $f(z) = z$ another holomorphic function $f(z)$ is used, then the region of convergence $|f(z)| < R$ has a noncircular shape, and expanding the analytic function $F(z)$ in powers of $f(z)$ leads to a Lagrange–Burmann series (Section 23.6) valid in this region. Thus the Taylor series is the particular case of the Lagrange–Burmann series, in which the expansion proceeds in powers of the auxiliary function, $f(z) = z - a$. If the series is summed with N terms, the accuracy of the approximation to the function is specified by its remainder, for which more than one form may be found (Sections 23.8 and 23.9). A holomorphic function is not only analytic, but also harmonic (Section 23.1), since its value at one point is the average value on a circle with center at that point. This defining property of harmonic functions can also be extended to noncircular regions: in this case the function has maximum modulus on the boundary (Section 23.2). The proof of all the results stated (Sections 23.1–23.2 and 23.6–23.9) ultimately rests on the harmonic series (Sections 21.8 and 21.9) applied to an auxiliary holomorphic function (Section 23.3) that is used for the ascending power series expansion (Section 23.4) in the Lagrange–Burmann series; the latter is more general than Taylor or Maclaurin series since the region of convergence need not be a circle, and can have any shape specified by (Section 23.5) the auxiliary function. The ascending power series or Taylor's series shows that any potential flow without singularity at a point can be represented by a superposition of corner flows of submultiples of π, and likewise for all potential fields; and the Lagrange–Burmann series allows the replacement of the corners by regions with other shapes.

23.1 Harmonic Function and Mean Value on a Circle

Consider a circle of radius R and center at a:

$$\zeta - a = R\,e^{i\varphi}, \quad d\zeta = R\,e^{i\varphi} i\,d\varphi = i\,(\zeta - a)\,d\varphi, \tag{23.1a,b}$$

and apply Cauchy's second theorem (15.8) to a function $f(z)$ holomorphic in the circle $|z-a| \le R$ or meeting the Cauchy conditions (Section 15.1) there:

$$f(z) = (2\pi i)^{-1} \int\limits_{|\zeta-a|=R} (\zeta-z)^{-1} f(\zeta)\,\mathrm{d}\zeta = \frac{1}{2\pi}\int_0^{2\pi} \frac{\zeta-a}{\zeta-z} f(\zeta)\,\mathrm{d}\varphi; \tag{23.2a,b}$$

in (23.2b) the point z is interior to the circle $|z-a| < |\zeta-a| \le R$, and choosing $z=a$, it follows that

$$f(z) \in \mathcal{D}\left(|z-a| \le R\right): \quad f(a) = (2\pi)^{-1}\int_0^{2\pi} f\left(a + R\,\mathrm{e}^{i\varphi}\right)\,\mathrm{d}\varphi, \tag{23.3}$$

if a function f(z) *holomorphic [(or satisfies the Cauchy conditions Section 15.1)] in a circle, its value at the center* f(a) *is the mean of the values on the boundary of the circle.*

The preceding property can be used to define a **harmonic function** in the plane; it follows that a complex differentiable function is a harmonic function in the plane. The converse is also true for: (i) if the function $f(z)$ is harmonic in the plane, then (23.3) implies that Cauchy's theorem (23.2a) holds for z an arbitrary point and ζ on a circle centered at z; (ii) since $|\zeta - z| = R \ge \varepsilon > 0$, where R is the radius of the circle, the integral converges uniformly with regard to z; (iii) thus (23.3) can be (Section 23.8) differentiated with regard to z under integral sign, and the resulting expression:

$$\mathcal{H}\left(|\mathrm{R}^2\right) \equiv \mathcal{D}\left(|\mathrm{C}\right): \qquad f'(z) = (2\pi i)^{-i} \int\limits_{|\zeta-a|=R} (\zeta-z)^{-2} f(\zeta)\,\mathrm{d}\zeta, \tag{23.4a,b}$$

specifies $f'(z)$ uniquely (23.4b), proving that $f(z)$ is holomorphic. It has been shown that *the classes of harmonic functions in the plane* $\mathcal{H}(|\mathrm{R}^2)$ *and complex differentiable functions* $\mathcal{D}(|\mathrm{C})$ *coincide (23.4a).*

23.2 Lemmas of Constancy and Maximum Modulus

A particular case of the harmonic property (23.3) is that if the function $f(\zeta) = c$ is a constant on the circle, it has the same value at the center $f(a) = c$. This result extends to all interior points and a region of any noncircular shape as the **lemma constancy**: *a function* f(z) *holomorphic in a closed region* D [*or satisfying the Cauchy conditions (Section 15.1) in* D], *is constant in the interior* D − ∂D *iff it is constant on the boundary* ∂D:

$$f(z) \in \mathcal{D}(D): \quad \forall_{\zeta\in\partial D}: \quad f(\zeta) = c \Leftrightarrow \forall_{z\in D-\partial D}: \quad f(z) = c. \tag{23.5}$$

To prove the implication ⇒ the function is assumed to be constant on the boundary, and Cauchy's second theorem is used:

$$f(\zeta) = c: \quad f(z) = (2\pi i)^{-1} c \int^{(z+)} (\zeta - z)^{-1}\,\mathrm{d}\zeta = c, \tag{23.6}$$

to prove that it has the same value c at all interior points. The converse implication ⇐ is due to the Cauchy conditions for holomorphy (Section 19.5), implying that the function is continuous as the boundary is approached from the interior: thus if it is constant in the interior, it takes the same value on the boundary, QED.

The constancy lemma, is a limiting case of the **lemma of maximum modulus**: *a function* f(z) *holomorphic [or satisfying the Cauchy conditions (Section 15.1)], in a closed region D takes the maximum value in modulus at the boundary:*

$$f(z) \in \mathcal{D}(D): \quad \forall_{\zeta \in \partial D}: |f(\zeta)|_{\max} \equiv M \Rightarrow \forall_{z \in D - \partial D}: |f(z)| \leq M, \tag{23.7}$$

where the equality applies iff the function is a constant f (z) = c *with* |c| = M. To prove the theorem note that if z is an interior point, a circle C_ε exists, with radius $\varepsilon < |z_0 - \zeta|_{\min}$ and center at z_0, which is contained within $D - \partial D$. Applying to (23.3) the mean-value theorem for integrals (13.12), it follows $|f(z_0)| \leq M$, where M is the maximum of $|f(z)|$ on the circle C_ε, and equality only holds if $f(z) = c$ is a constant. If $f(z)$ is not constant, then $|f(z_0)| < M$, and the interior point z_0 is not a maximum. Since z_0 is an arbitrary interior point, the maximum of $f(z)$ must lie on the boundary, QED.

23.3 Monotonic Chain of Regions and Loops

The constancy (23.5) and maximum modulus (23.7) lemmas, can be interpreted physically (Section 32.3) or geometrically (Section 23.3). The geometric interpretation is that *a non-constant holomorphic function* f(z) *can be used to define a curve by requiring its modulus to be constant:*

$$R > 0: \qquad \partial D = \{\zeta : |f(\zeta)| = R\}; \tag{23.8}$$

if the curve is closed its interior D − ∂D *in (23.9a) [the whole domain* D *in (23.9b)] is specified by the function in modulus being less than (less or equal to) the bound:*

$$D - \partial D = \{z: \; |f(z)| < R\}, \quad D \equiv \{z : |f(z)| \leq R\}. \tag{23.9a,b}$$

This follows from the maximum modulus lemma (Section 23.2) stating that $|f(z)|$ is: (i) maximum, that is, equal to R, if z lies on the boundary; (ii) smaller, that is, less than R, in the interior. *The region* D_ε *defined by:*

$$0 < \varepsilon < 1: \quad D_\varepsilon \equiv \{z : |f(z)| \leq R(1 - \varepsilon)\}, \tag{23.10}$$

is a closed subregion of (23.8); the subregions are contained in each other monotonically as the parameter increases:

$$0 < \varepsilon < \delta < 1: \quad D \equiv D_0 \supset D_\delta \supset D_\varepsilon, \tag{23.11}$$

as illustrated in Figure 23.1a.

From (23.8; 23.9a; 23.10) it follows that if the region D is defined by the holomorphic function f, and if ζ lies on the boundary (23.12a), z in the interior (23.12b) and η in a subregion (23.12c) then the inequalities (23.12d,e) hold:

$$\zeta \in \partial D, \quad z \in D - \partial D, \quad \eta \in D_\varepsilon: \quad \left|\frac{f(z)}{f(\zeta)}\right| < 1, \quad \left|\frac{f(\eta)}{f(\zeta)}\right| \leq 1 - \varepsilon. \tag{23.12a–e}$$

The regions illustrated in Figure 23.1a become (Figure 23.1b) circles of center at a, choosing as the holomorphic function (23.13a) in which case: (i) the closed region (23.9b) is a circle of

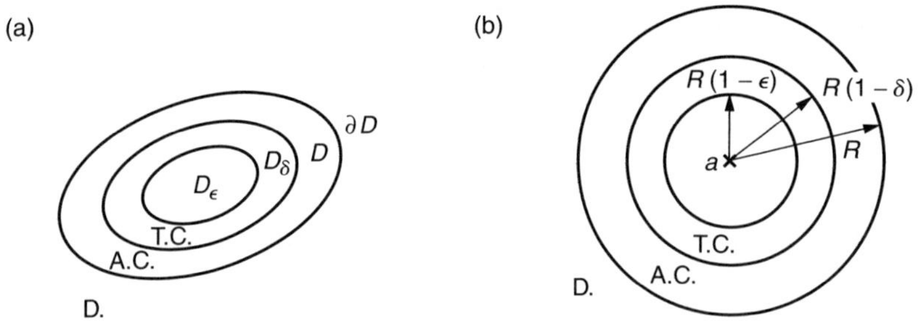

FIGURE 23.1
Expanding an analytic function $F(z)$ in power series of an auxiliary analytic function $f(z)$ $[f(z) = z - a]$ leads to a sequence of noncircular (a) [circular (b)] regions such that: (i) there is divergence (convergence) outside (inside) the boundary ∂D; (ii) the convergence inside the boundary is absolute; (iii) the convergence is total in a closed subregion D_δ; (iv) further closed subregions $D_\varepsilon \subset D_\delta$ can be defined. The convergence on the boundary ∂D is not included in (i) to (iv), and is specified by other methods (Figure 29.1).

radius R, and (23.8) its boundary and (23.9a) its interior; (ii) the closed subregion (23.10) is a circle with the same center and smaller radius $R(1-\varepsilon)$, and the successive subregions (23.11) have decreasing radii $R(1-\varepsilon) > R(1-\delta)$ for $\delta > \varepsilon$; (iii) the conditions (23.12d,e) have the interpretation (23.13a,b):

$$f(z) \equiv z - a: \quad |z-a| < |\zeta - a|, \quad |\eta - a| \le (1-\varepsilon)|\zeta - a|, \tag{23.13a–c}$$

where ζ lies on the circle of radius R, z in its interior, and η is on or within the circle of radius $R(1-\varepsilon)$ with ε satisfying (23.13a), all circles with center at a.

23.4 Geometric Series of Holomorphic Functions

The geometric series (21.62b) converges in the unit circle $|z| < 1$, and to obtain a series converging in (23.9a), z is replaced by (23.12d), and a by $1/f(\zeta)$, so that the sum of the geometric series is:

$$\frac{1}{f(\zeta) - f(z)} = \frac{1/f(\zeta)}{1 - f(z)/f(\zeta)} = \frac{1}{f(\zeta)} \sum_{n=0}^{\infty} \left[\frac{f(z)}{f(\zeta)}\right]^n; \tag{23.14}$$

recalling the conditions of convergence in Section 21.9, it follows that

$$\{f(\zeta) - f(z)\}^{-1} = \sum_{n=0}^{\infty} \{f(z)\}^n \{f(\zeta)\}^{-n-1}, \tag{23.15}$$

the geometric series (23.15) of functions $f(z)$ holomorphic in a region D in (23.12d), converges absolutely (totally) in the interior D − ∂D *in (23.9a) [closed subregion* D_ε *in (23.10)].*

Outside D the series (23.15) diverges. An expression valid in the whole z-plane may be obtained separating (21.60a,b) the first N terms of the geometric series (23.15) of ratio $f(z)/f(\zeta)$ and first term $1/f(\zeta)$:

$$\{f(\zeta) - f(z)\}^{-1} = \sum_{n=0}^{N-1} \{f(z)\}^n \{f(\zeta)\}^{-n-1} + \frac{\{f(z)\}^N \{f(\zeta)\}^{-N}}{f(\zeta) - f(z)}, \tag{23.16}$$

from the remainder after N terms, which is the last term on the r.h.s. When the latter term vanishes as $N \to \infty$, viz. in (23.12d), then the geometric progression (23.16) leads to the series (23.15).

23.5 Regions of Absolute and Uniform Convergence

A function F(z) *holomorphic in the region* D *in (23.9b) is determined by the* **modified Cauchy integral:**

$$F(z) = (2\pi i)^{-1} \int_{\partial D} F(\zeta) \{f(\zeta) - f(z)\}^{-1} f'(\zeta) \mathrm{d}\zeta, \tag{23.17}$$

with the auxiliary holornorphic function f(ζ) *as variable*, so that $df = f'(\zeta)\, d\zeta$ in (15.8) the second Cauchy theorem with (f, z) replaced by (F, f). The loop ∂D can be shrunk to the boundary ∂D_ε of the region D_ε in (23.10) of uniform convergence of the geometric series of functions (23.15); the latter can be integrated term-by-term (Section 21.6) after substitution in (23.17), leading to the series:

$$F(z) = \sum_{n=0}^{\infty} A_n \{f(z)\}^n, \tag{23.18}$$

of integral ascending powers, with coefficients:

$$A_n \equiv (2\pi i)^{-1} \int_{\partial D} F(\zeta) \{f(\zeta)\}^{-n-1} f'(\zeta) \mathrm{d}\zeta; \tag{23.19}$$

in (23.19) the loop ∂D_ε is stretched back to ∂D.

Having proved that the series (23.18) is uniformly convergent in the closed subregion D_ε in (23.10), its convergence considered is next in the larger open region that is the interior of D in (23.9a). The coefficients (23.19) have and upper bound:

$$|A_n| \le (2\pi)^{-1} \int_{\partial D} |F(\zeta) f'(\zeta)| \, |f(\zeta)|^{-n-1} \, \mathrm{d}\zeta \le \frac{ML}{2\pi} |f(\zeta)|^{-n-1}, \tag{23.20}$$

where: (i) since $F(\zeta)f'(\zeta)$ is holomorphic it is bounded in modulus by M; (ii) the finite length of the loop ∂D is L and $|f(\zeta)| = R$ is a constant on ∂D by (23.8). Thus the series of moduli of (23.18) has an upper bound:

$$\left|\frac{f(z)}{f(\zeta)}\right| < 1 : \sum_{n=0}^{\infty} |A_n| \, |f(z)|^n \le \frac{ML}{2\pi} \sum_{n=0}^{\infty} |f(z)|^n \, |f(\zeta)|^{-n-1} = \frac{ML/2\pi}{|f(\zeta)| - |f(z)|}, \tag{23.21}$$

so that the series (23.18) is absolutely convergent in $D - \partial D$ in (23.9a).

Since absolute convergence implies ordinary convergence, the geometric progression of functions obtained substituting (23.16) into (23.17):

$$F(z) = \sum_{n=0}^{N-1} A_n \{f(z)\}^n + R_N(z), \qquad \lim_{N \to \infty} R_N(z) = 0, \tag{23.22a,b}$$

has vanishing (23.22b) remainder, given by

$$R_N(z) = (2\pi i)^{-1} \{f(z)\}^N \int_{\partial D} \{f(\zeta)\}^{-N} \{f(\zeta) - f(z)\}^{-1} F(\zeta) f'(\zeta) \mathrm{d}\zeta. \tag{23.23}$$

Thus it has been shown that *a holomorphic function $F(z)$ can be expanded in a power series progression (23.22a) of an auxiliary holomorphic function $f(z)$, with remainder after N terms (23.23); the latter vanishes (23.22b) as $N \to \infty$ in the open region (23.9a), where the series (23.18) is absolutely convergent; the series is totally (i.e., absolutely and uniformly) convergent in the closed subregion (23.10). The coefficients of the series are given by (23.19) or alternatively by:*

$$A_0 = (2\pi i)^{-1} \int_{\partial D} F'(\zeta) \mathrm{d} \{\log (f(\zeta))\}, \tag{23.24a}$$

$$n = 1, 2, \ldots: \quad A_n = (2\pi i n)^{-1} \int_{\partial D} F'(\zeta) \{f(\zeta)\}^{-n} \mathrm{d}\zeta, \tag{23.24b}$$

where (23.24b) is obtained from (23.19) via an integration by parts, and (23.24a) coincides with (23.19) for $n = 0$.

23.6 Lagrange (1770)–Burmann (1799) Series and Implicit Derivation

To evaluate the integrals specifying the coefficients (23.19) of the Lagrange (1770) series (23.18), choose an auxiliary holomorphic function $f(z)$ having a single zero $z = a$ in (23.9a), assumed to be simple:

$$\exists^1_{a \in D - \partial D}: \quad f(a) = 0 \neq f'(a). \tag{23.25}$$

Hence the function:

$$g(\zeta) \equiv F(\zeta) \left[\frac{\zeta - a}{f(\zeta)} \right]^{n+1} f'(\zeta), \tag{23.26}$$

is holomorphic in D, because the single zero at $\zeta = a$ of $f(\zeta)$ cancels with $\zeta - a$; the integral (23.19) can be calculated by the (15.17) Cauchy third theorem:

$$A_n = (2\pi i)^{-1} \int^{(a+)} (\zeta - a)^{-n-1} g(\zeta) \mathrm{d}\zeta = (n!)^{-1} g^{(n)}(a); \tag{23.27}$$

hence (23.27; 23.26) evaluate explicitly Burmann (1799) the coefficients (23.19) of the Lagrange (1770) series (23.18).

The preceding results (Sections 23.3–23.6) may be summarized as the **Lagrange (1770)–Burmann (1799) theorem**: *a function $F(z)$ holomorphic in a region D in (23.9b) [or satisfying the Cauchy conditions (Section 15.1) in D], can be expanded in series of ascending integral powers of an auxiliary holomorphic function $f(z)$, with a single, simple zero (23.25) in D. The expansion (23.28a):*

$$F(z) = \sum_{n=0}^{N-1} A_n \{f(z)\}^n + R_N(z) = \sum_{n=0}^{\infty} A_n \{f(z)\}^n, \tag{23.28a,b}$$

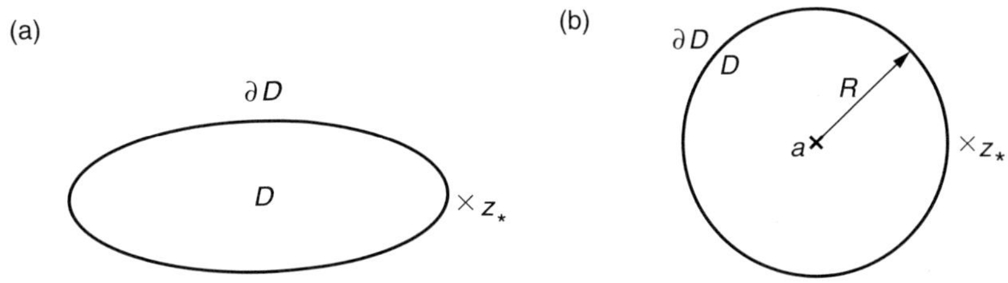

FIGURE 23.2
As is Figure 23.1 the expansion of an analytic function $F(z)$ in power series of an auxiliary function $f(z)$ $[f(z) = z - a]$ is valid a noncircular (a) [circular (b)] region whose extent is limited by the exclusion of the nearest singularity z_* of the expanded function $F(z)$.

has coefficients given by:

$$n!A_n \equiv \lim_{\zeta \to a} \frac{\mathrm{d}^n \left\{ F(\zeta) \left[(\zeta - a)/f(\zeta) \right]^{n+1} f'(\zeta) \right\}}{\mathrm{d}\zeta^n}, \tag{23.29}$$

and remainder after N terms given by (23.23). The latter vanishes as $N \to \infty$ in the open interior (23.9a) of the region D that excludes any singularities z_* of $F(z)$ in Figure 23.2a. The convergence of the series (23.28b) is absolute (total) in the open interior (23.9a) [closed subregion (23.10)] of D in Figure 23.1a (23.2a). A Lagrange–Burmann series may represent different functions inside and outside the boundary of convergence, if it does not converge on the boundary except at isolated points (Example 30.6).

23.7 Taylor (1715) and Stirling (1717)–Maclaurin (1742) Theorems

The simplest auxiliary function $f(\zeta)$ that is holomorphic and has a single, simple zero at $\zeta = a$, is (23.13a) ≡ (23.30a):

$$f(\zeta) = \zeta - a, \quad f'(\zeta) = 1. \tag{23.30a,b}$$

The choice (23.30a,b) simplifies the calculation of the coefficients (23.29) to the ordinary explicit n-th derivative:

$$n!A_n \equiv \lim_{\zeta \to a} \frac{\mathrm{d}^n F}{d\zeta^n} = F^{(n)}(a); \tag{23.31}$$

this substituted into (23.22a; 23.23) leads to the **Taylor theorem** (1715): *a function* F(z) *holomorphic in the neighborhood of a point* z = a, *has the expansion (23.32a):*

$$F(z) = \sum_{n=0}^{N-1} \frac{F^{(n)}(a)}{n!} (z - a)^n + R_N(z) = \sum_{n=0}^{\infty} (n!)^{-1} F^{(n)}(a) (z - a)^n, \tag{23.32a,b}$$

with remainder after N terms:

$$R_N(z) \equiv (2\pi i)^{-1} (z - a)^N \int_{|\zeta - a| = R} F(\zeta) (\zeta - a)^{-N} (\zeta - z)^{-1} \,\mathrm{d}\zeta; \tag{23.33}$$

if the singularity of $\mathrm{F(z)}$ *nearest to* $\mathrm{z = a}$ *is* $\mathrm{z = z_*}$ *(Figure 23.2b), the remainder (23.33) vanishes* $\mathrm{R_N} \to 0$ *as* $\mathrm{N} \to \infty$ *within the circle* $|\mathrm{z - a}| < \mathrm{R} \equiv |\mathrm{z_* - a}|$ *of center at* a *and radius* R *less than the distance of the nearest singularity. The series (23.32b) is absolutely convergent within this circle* $|\mathrm{z - a}| < \mathrm{R}$ *and totally convergent in (Figure 23.1b) a closed subcircle* $|\mathrm{z - a}| \le \mathrm{R}(1 - \varepsilon)$*, with* $1 > \varepsilon > 0$.

In the case $a = 0$ leads to the **Stirling (1717)–Maclaurin (1742) theorem**: *if the singularity of the function* $\mathrm{f(z)}$ *nearest to the origin is* $\mathrm{z = z_*}$*, then it can be expanded in a series (23.34b):*

$$F(z) = \sum_{n=0}^{N-1} \frac{F^{(n)}(0)}{n!} z^n + R_N(z) = \sum_{n=0}^{\infty} (n!)^{-1} F^{(n)}(0) z^n; \tag{23.34a,b}$$

the Maclaurin expansion (23.34a) has remainder after N *terms:*

$$R_N(z) \equiv (2\pi i)^{-1} z^N \int_{|\zeta|=R} F(\zeta)\, \zeta^{-N} (\zeta - z)^{-1}\, \mathrm{d}\zeta, \tag{23.35}$$

that vanishes $\mathrm{R_N} \to 0$ *as* $\mathrm{N} \to \infty$ *for* $|z| < \mathrm{R}$ *within the circle of radius* $\mathrm{R} \equiv |\mathrm{z_*}|$ *and center at the origin; the convergence of the series is absolute (total) in the open interior of this circle* $|\mathrm{z}| < \mathrm{R}$ *(in a closed subcircle* $|\mathrm{z}| < \mathrm{R}(1 - \varepsilon)$ *with* $1 > \varepsilon > 0$). The simplest Stirling–Maclaurin series is the exponential (23.60–23.64), and eleven other instances of Stirling–Maclaurin series are considered in Examples 30.7 and 30.8. If in (23.32a) the Taylor series $z - a$ is replaced by $f(z)$, then the Lagrange series (23.28a) is obtained with coefficients:

$$n! A_n = \lim_{\zeta \to a} \frac{\mathrm{d}^n \{F(\zeta)\}}{\{\mathrm{d}f(\zeta)\}^n}. \tag{23.36}$$

This result may be compared with (23.29) to yield:

$$\frac{\mathrm{d}^n \{F(z)\}}{\{\mathrm{d}f(z)\}^n} = \lim_{\zeta \to z} \frac{\mathrm{d}^n \left\{F(\zeta) \left[(\zeta - z)\, f(\zeta)\right]^{n+1} f'(\zeta)\right\}}{\mathrm{d}\zeta^n}; \tag{23.37}$$

this is the **rule of implicit derivation of order** $\boldsymbol{n}$, and assumes (23.25) that $f(\zeta)$ has a simple zero at $\zeta = z$.

23.8 Implicit Derivatives and Mean-Value Theorem

If $f(\zeta)$ does not vanish at $\zeta = z$, it can be replaced it by $f(\zeta) - f(z)$ that will have a simple zero provided that $f(\zeta)$ is single-valued. Thus is obtained the **rule of implicit** $\boldsymbol{n}$**-th order derivation** *of a holomorphic function* $\mathrm{F(z)}$ *with regard to another holomorphic function,* $\mathrm{f(z)}$:

$$\left.\frac{\mathrm{d}^n F}{\mathrm{d}f^n}\right|_{\zeta = z} = \lim_{\zeta \to z} \frac{\mathrm{d}^n \left\{F(\zeta) \left[(\zeta - z)/(f(\zeta) - f(z))\right]^{n+1} f'(\zeta)\right\}}{\mathrm{d}\zeta^n}. \tag{23.38}$$

The expression (23.38) corresponds to the evaluation, by the residue at a pole of order n, of the integral:

$$\frac{\mathrm{d}^n F}{\mathrm{d}f^n} = \frac{n!}{2\pi i} \int^{(z+)} F(\zeta) \left[\frac{(\zeta - z)}{(f(\zeta) - f(z))}\right]^{n+1} f'(\zeta)\,(\zeta - z)^{-n-1}\,\mathrm{d}\zeta \tag{23.39a}$$

$$= \frac{n!}{2\pi i} \int^{(f(z)+)} F(\zeta)\,\{f(\zeta) - f(z)\}^{-n-1}\,\mathrm{d}f; \tag{23.39b}$$

the latter is Cauchy's integral (15.13) for the n-th derivate of F as a function of f, and serves as a check on the rule (23.38) of implicit derivation of n-th order. Indeed, the latter can be deduced in the inverse way, that is: (i) starting from the third Cauchy theorem (23.39b); (ii) rewriting the integral in the form (23.39a); (iii) then evaluating it by the third Cauchy theorem (15.13), bearing in mind that the function in square brackets is holomorphic at $\zeta = z$.

The Taylor series for real functions $F(x)$ can be deduced from successive applications of the **mean-value theorem for derivates**: if *the function* $\mathrm{f(x)}$ *has continuous first-order derivate in the interval* $\mathrm{a \leq x \leq b}$, *then there is at least one interior point* $\mathrm{a < c < b}$ *where the slope equals that of the straight line joining* $(\mathrm{a, f(a)})$ *to* $(\mathrm{b, f(b)})$:

$$f(x) \in \mathcal{C}^1(a, b): \quad \exists_{a<c<b}: \quad f'(c) = \frac{f(b) - f(a)}{(b - a)} \equiv X. \tag{23.40}$$

The theorem is proved by reduction ad absurdum: suppose it is not true. Then for all $a < x < b$ holds $f'(x) \neq X$, and since $f'(x)$ is continuous, either $f'(x) > X$ or $f'(x) < X$ everywhere. Hence:

$$f(b) - f(a) = \int_a^b f'(x)\,\mathrm{d}x <> X \int_a^b \mathrm{d}x = X(b - a), \tag{23.41}$$

that contradicts (23.40). It follows that $f'(c) = X$ at least at one point $a < c < b$, QED. The geometric interpretation is that a curve without edges joining, $A \equiv (a, f(a))$, ta $B = (B, f(b))$, must have tangent parallel to the straight line joining A to B, at least one point $x = c$ in Figure 23.3a, or possibly more, for example, four $x = c_1$, c_2, c_3, and c_4 in Figure 23.3b. The theorem can fail if the function has discontinuous derivate at one point, for example, $x = c$ in Figure 23.3c, and the slope of the tangent is nowhere equal to that of the line joining A to B. An immediate consequence is **Rolle's theorem**: *if a function* $\mathrm{f(x)}$ *has continuous first-order derivative at an interval,* $\mathrm{a \leq x \leq b}$, *and takes the same value at the end-points, then its derivative vanishes at least at one point in the interior:*

$$f(x) \in \mathcal{C}^1\,(a,\, b): \quad f(b) = f(a) \quad \Rightarrow \quad \exists^1_{a<c<b}: \ f'(c) = 0; \tag{23.42}$$

this result (23.42) corresponds to $X = 0$ in (23.40). QED.

23.9 Darboux Expansion (1876) and Lagrange/Cauchy Remainders

Choosing $b \equiv x$ to be a variable point, the mean-value theorem (23.40) is restated:

$$a < c < x: \quad f(x) = f(a) + (x - a) f'(c), \tag{23.43}$$

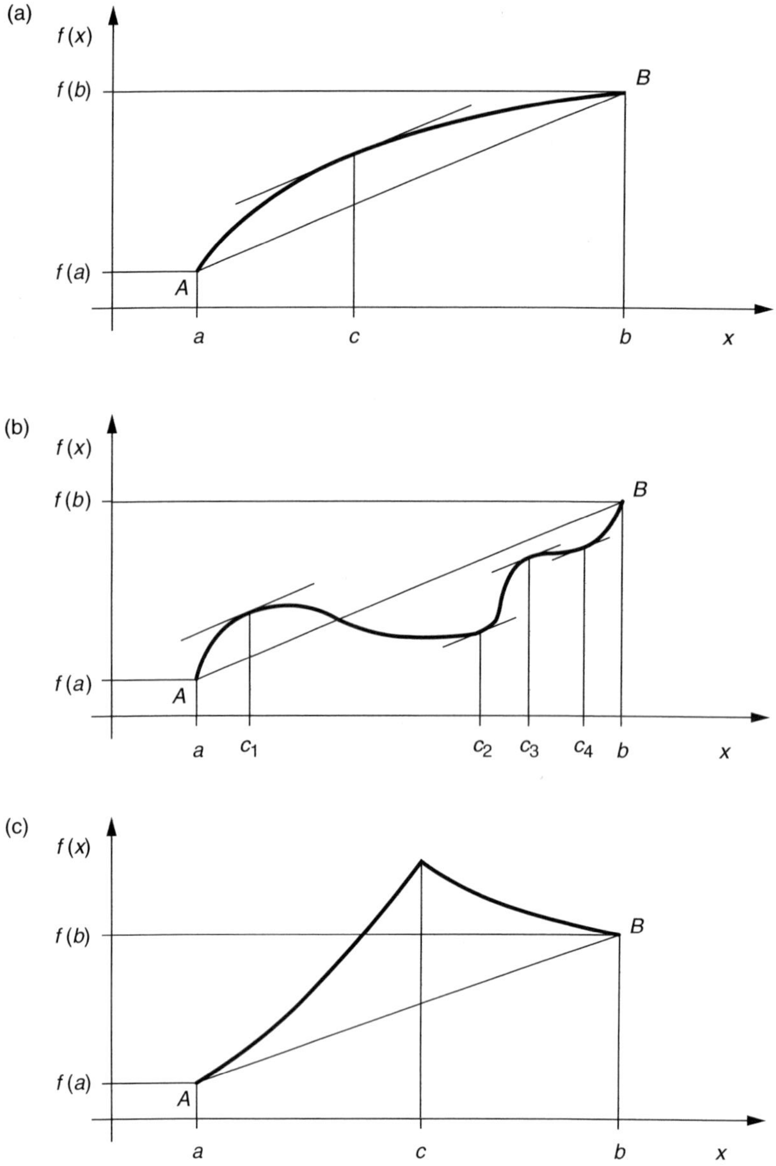

FIGURE 23.3
If function has continuous derivative in an interval (a,b) the tangent to the curve must be parallel to the line joining the two ends at least one internal point (a), possibly several (b). That may not be the case for a function with discontinuous derivative in the interval (a,b), for example, an angular point the interior (c).

for a function having a continuous first-order derivate. If $f(x)$ has continuous second-order derivate, (23.43) can be applied to $f'(x)$:

$$f(x) \in \mathcal{C}^2(a,b): \quad f'(x) = f'(a) + (x+a)\, f''(c); \tag{23.44a}$$

integrating from a to x, yields:

$$a < c < x: \quad f(x) = f(a) + (x-a)f'(a) + \frac{(x-a)^2}{2} f''(c). \tag{23.44b}$$

Proceeding by iteration proves by induction (Section 1.1) that *if the real function $f(x)$ has continuous derivates up to order n in the interval $a \le x \le b$, it has a* **Taylor expansion**:

$$f(x) \in \mathcal{C}^N(a,b): \quad f(x) = \sum_{n=0}^{N-1} \frac{(x-a)^n}{n!} f^{(n)}(a) + R_N(x), \tag{23.45a,b}$$

with the **Lagrange remainder**:

$$a < c < x, \quad 0 < \theta \equiv \frac{c-a}{x-a} < 1: \quad N!(x-a)^{-N} R_N(x) = f^{(N)}(c) = f^{(N)}(a + \theta(x-a)). \tag{23.46a–c}$$

The latter form of the remainder for the real Taylor series has the advantage of being explicit, whereas (23.33) for the complex Taylor series involves an integration. Besides (23.46a–c) and (23.33) other forms of the remainder of a Taylor series can be obtained (Subsection 23.9.2) via the Darboux expansion (Subsection 23.9.1). The latter provides a third method of establishing the Taylor series: (method I) using (Section 23.7) Cauchy's theorem (Sections 23.1–23.5), as a particular case of the Lagrange–Burmann series (Section 23.6); (method II) applying iteratively the mean-value theorem (Section 23.8–23.9); (method III) as a particular case (Subsection 23.9.2) of the Darboux expansion (Subsection 23.9.1). The three methods obviously lead to the same Taylor series, but do not overlap because the remainders are different. In the case of complex holomorphic functions the convergence is assured, and the remainder can serve as an upper bound for the error of truncation of the series at some finite order. In the case of real functions the convergence of the Taylor series is not assured, and it is necessary to prove that the remainder tends to zero as the order tends to infinity (Subsection 23.9.3); an example of a the Taylor series with nonvanishing remainder is given in Section 27.1.

23.9.1 Darboux Expansion and the Taylor Series

The Taylor series for a real function (23.45a,b), with remainder in a form distinct from (23.46a–c), can be obtained from the identity (32.47c):

$$\begin{aligned} f(x) \in \mathcal{C}^N(a,b);\, g(t) \in \mathcal{C}^{N-1}(0,1): &\ \frac{\mathrm{d}\left\{\sum\limits_{n=1}^{N-1} (-)^n (x-a)^n g^{(N-1-n)}(t) f^{(n)}(a + t(x-a))\right\}}{\mathrm{d}t} \\ &= \sum_{n=1}^{N-1} (-)^n \Big\{ (x-a)^n g^{(N-n)}(t) f^{(n)}(a + t(x-a)) \\ &\quad + (x-a)^{n+1} g^{(N-1-n)}(t) f^{(n+1)}(a + t(x-a)) \Big\} \\ &= (-)^{N-1}(x-a)^N g(t) f^{(N)}(a + t(x-a)) - (x-a) g^{(N-1)}(t) f'(a + t(x-a)), \end{aligned} \tag{23.47a–c}$$

where $f(x)\,[g(t)]$ has (32.47a) [(32.47b)] continuous derivates of order up to $N\,(N-1)$ for $a \le x \le b\,(0 \le t \le 1)$. Choosing $g(t)$ to be a polynomial of degree $N-1$, so that

$g^{(N-1)}(t) = \text{const} \equiv g^{(N-1)}(0)$, and integrating (32.47c) between $t = 0$ and $t = 1$, leads to

$$\begin{aligned} f(x) \in \mathcal{C}^N(a,b), a \le x \le b,\ 0 < t < 1 : (-)^{N-1}(x-a)^N \int_0^1 g(t) f^{(N)}(a + t(x-a))\ \mathrm{d}t \\ + \sum_{n=1}^{N-1} (-)^{n-1}(x-a)^n \left\{ g^{(N-1-n)}(1) f^{(n)}(x) - g^{(N-1-n)}(0) f^{(n)}(a) \right\} \\ = g^{(N-1)}(0)\left\{ f(x) - f(a) \right\}, \end{aligned} \tag{23.48a–d}$$

that is the **Darboux expansion** *(1876) for a real function with continuous derivates of order N for $a \le x \le b$, with $g(t)$ a polynomial of degree $N-1$.* This expansion is not explicit because it involves besides $f(x)$ its derivates of order up to N.

The latter can be eliminated if the polynomial $g(t)$ of degree N is chosen such that it vanishes, together with its derivates:

$$g(t) = (t-1)^{N-1}, \quad g^{(N-1-n)}(t) = \frac{(N-1)!}{n!}(t-1)^n, \tag{23.49a,b}$$

at (23.50a) the point $t = 1$:

$$g^{(N-1-n)}(1) = 0; \quad g^{(N-1-n)}(0) = (-)^n \frac{(N-1)!}{n!}; \tag{23.50a,b}$$

thus only the terms at $t = 0$ that involve factorials (23.50b), appear in:

$$f(x) = f(a) + \sum_{n=1}^{N-1} \frac{(x-a)^n}{n!} f^{(n)}(a) + R_N(x); \tag{23.51}$$

this is the Taylor expansion (23.51) $\equiv$ (23.45b), with remainder:

$$R_N(x) = \frac{(x-a)^N}{(N-1)!} \int_0^1 (1-t)^{N-1} f^{(N)}(a + t(x-a))\,\mathrm{d}t. \tag{23.52}$$

Next choose a real number $0 < p \le N$, and note that the function $h(t)$:

$$0 < p \le N, \quad B > h(t) \equiv (1-t)^{N-p} f^{(N)}(a + t(x-a)) > A, \tag{23.53a,b}$$

is continuous in $0 < t < 1$, and thus has upper B and lower A bounds; it follows that

$$\frac{A}{p} = \int_0^1 (1-t)^{p-1} A\mathrm{d}t < \int_0^1 (1-t)^{p-1} h(t)\mathrm{d}t < \int_0^1 (t-1)^{p-1} B\mathrm{d}t = \frac{B}{p}; \tag{23.54}$$

thus $0 < \theta < 1$ exists such that

$$\begin{aligned} p^{-1}h(\theta) = \int_0^1 (1-t)^{p-1} h(t)\ \mathrm{d}t = \int_0^1 (1-t)^{N-1} f^{(N)}(a + t(x-a))\ \mathrm{d}t \\ = (N-1)!(x-a)^{-N} R_N(x). \end{aligned} \tag{23.55}$$

The formula (23.55) together with (23.53a,b) specifies the remainder of Taylor's series.

23.9.2 Four Remainders for the Taylor Series (Taylor, Lagrange, Darboux, Cauchy)

Thus the **Darboux remainder** *of Taylor's expansion (23.45b) ≡ (23.51) after N terms, is given by (23.56c):*

$$0 < \theta < 1; \quad 0 < p \le N: \; R_N^p(x) = \frac{(x-a)^N}{(N-1)!p}(1-\theta)^{N-p} f^{(N)}(a+\theta(x-a)), \tag{23.56a–c}$$

for arbitrary p in (23.56b) and some θ in (23.56a). The particular case $p = N$ is the Lagrange remainder (23.46a–c); $p = 1$ yields the **Cauchy remainder:**

$$0 < \theta < 1: \quad R_N(x) = \frac{(x-a)^N}{(N-1)!}(1-\theta)^{N-1} f^{(N)}(a+\theta(x-a)). \tag{23.57a,b}$$

The integral remainder (23.33) [(23.52)] for the complex (real) Taylor series (23.32a,b) is the fourth (fifth) remainder, adding to the three explicit remainders for the real Taylor series (23.45a,b) ≡ (23.51) due to Darboux (23.56a–c), Lagrange (23.46a–c), and Cauchy (23.57a,b). The Taylor series expansion, and its remainder, have been obtained by three distinct methods: (i) for a complex (Section 23.7) holomorphic function from the Cauchy theorem, as a particular case (23.32a,b) of the Lagrange series (23.28a,b); (ii) for a real function (Section 23.8) by iterative application (23.45a,b) of the mean-value theorem (23.43); (iii) also for a real function as a particular case (23.51) of the Darboux's expansion (Subsection 23.9.1). The three methods lead to five distinct forms (23.33; 23.52; 23.56a–c; 23.46a–c; 23.57a,b) of the remainder after N terms of which two (three) integral (23.33; 23.52) [explicit (23.46a–c; 23.56a–c; 23.57a,b)]; these are useful to estimate the accuracy of the numerical computation of a function using a truncated Taylor expansion.

If the function $f(z)$ is **smooth**, that is, infinitely differentiable (23.58a), then the Taylor theorem (23.51) leads to the Taylor series iff (23.58b) the remainder vanishes $R_N(x) \to 0$, as $N \to \infty$ uniformly in (a, x):

$$f(x) \in \mathcal{D}^\infty(a,x): \quad f(x) = \sum_{n=0}^{\infty} \frac{f^{(n)}(a)}{n!}(x-a)^n \Leftrightarrow \lim_{N\to\infty} R_N(x) = 0. \tag{23.58a,b}$$

For a holomorphic function (23.59a), the derivatives of all orders exist (Section 15.4) and are bounded (23.59b), so the remainder always vanishes (23.59c):

$$f(x) \in \mathcal{D}: \quad \left|f^{(n)}(z)\right| \le M, \quad \lim_{N\to\infty} |R_N(z)| \le \frac{M\,|z-a|^N}{(N-1)!} = 0; \tag{23.59a,b}$$

thus, as the number of terms tends to infinity it has a convergent Taylor series, and is an **analytic function**. Thus *a complex holomorphic function, that is, with first-order derivative (Section 11.3) has: (i) derivatives of all orders (Section 15.4), that is, is a smooth function; (ii) has a convergent Taylor series, that is, an analytic function (Subsection 23.9.2); (iii) has mean value on a circle equal to its value at the center (Section 23.1), that is, it is a harmonic function.* The coincidence of holomorphic, smooth, harmonic, and analytic functions is true for complex functions, but not for real functions. A real function (iii) is generally not harmonic; (ii) if it has derivatives up to order n it may not have derivatives of higher order; (i) if it is infinitely differentiable its Taylor series may not converge, if the remainder does not tend to zero. An example of (i) is given in Section 27.1. Thus for a real (complex) function the vanishing of the remainder is assumed (proved), that is, a smooth real (complex holomorphic) function may fail to have (always has) a convergent Taylor series.

23.9.3 Exponential Series with Remainders

The simplest analytic ascending power series is the Stirling–Maclaurin series (23.34a) for the exponential:

$$|z| < \infty : \quad \mathrm{e}^z = \sum_{\infty}^{\infty} \frac{z^n}{n!} = 1 + z + \frac{z^2}{4} + \frac{z^3}{6} + \frac{z^4}{24} + \frac{z^5}{120} + \cdots; \tag{23.60}$$

the coefficients (23.61a):

$$c_n \equiv \frac{f^{(n)}(0)}{n!} = \frac{\mathrm{e}^0}{n} = \frac{1}{n!}, \quad \lim_{n\to\infty} \frac{c_{n+1}}{c_n} = \lim_{n\to\infty} \frac{z}{n+1} = 0, \tag{23.61a,b}$$

show it has (23.61b) an infinite radius of convergence, by D'Alembert's ratio test (29.31c). The convergence of the complex exponential series (23.60) is proved most simply by noting that the function is holomorphic in the finite complex plane, because it coincides with its derivative; the convergence of the real exponential series is then the particular case of finite real variables. The convergence is absolute excluding the point at infinity, $|z| < \infty$, and total in a closed circle with finite radius $|z| \leq M < \infty$. No convergence issues arise in connection with the exponential expansion (33.62a):

$$\mathrm{e}^z = \sum_{n=0}^{N-1} \frac{z^n}{n!} + R_N(z); \quad R_N(z) = (2\pi i)^{-1} z^N \int_{|\zeta|=R} \zeta^{-N} \mathrm{e}^{\zeta} (\zeta - z)^{-1} \, \mathrm{d}\zeta, \tag{23.62a,b}$$

it has: (i) the Maclaurin remainder (23.62b) ≡ (23.35) that is specified by an integral; (ii) the Darboux's remainder (23.56a–c) ≡ (23.63a–c):

$$0 < \theta < 1; \;\; p = 1, \ldots, N: \quad R_N^p(x) = \frac{x^N}{(N-1)!p} (1-\theta)^{N-p} \exp\left[a + \theta(x-a)\right], \tag{23.63a–c}$$

$$\left\{R_N^N(x), R_N^1(x)\right\} = \frac{x^N}{N!} \exp\left[a + \theta(x-a)\right] \left\{1, N(1-\theta)^{N-1}\right\}, \tag{23.64a,b}$$

that is explicit; (iii) the latter includes the Lagrange (23.64a) [Cauchy (23.64b)] remainder as the particular cases $p = N(p = 1)$. Two sets of five other Maclaurin series appear in Examples 30.7 and 30.8. The convergence of series, with particular reference to power series, is addressed in Chapter 29.

Note 23.1. *Power Series for Regular and Singular Functions:* The simplest power series for an analytic function is Stirling–Maclaurin series and is usually the first choice. If the origin is a singularity, or the expansion should be made around another point, the Taylor series can be used. The Taylor series allows expansion about any regular point on the finite plane; the expansion around the point at infinity is obtained by inversion relative to the origin (Chapter 27). The Lagrange–Burmann series extends the Taylor series to noncircular region of convergence, and which may include the point at infinity (Example 30.6). The proof of the Taylor series shows the difference between: (i) real functions, for which the existence of derivative of order n does not imply the existence of higher order derivatives, and a smooth (i.e., infinitely differentiable) function may have a Taylor series expansion which is not a series, because the remainder does not tend to zero (Section 27.1); (ii) complex functions for which holomorphy, that is, existence of first-order derivative, implies smoothness (existence of derivatives of all orders) and analyticity (convergent Taylor series). Concerning the Taylor, Lagrange–Burmann and Stirling–Maclaurin series of a holomorphic function, it has been proved that: (i) they converge absolutely in an open region; (ii) converge totally in a closed subregion; (iii) diverge in the exterior of the boundary of convergence. The behavior of a series on the boundary of convergence requires finer convergence tests (Chapter 29) than those based only (Chapter 23) on the geometric series (Sections 21.8 and 21.9). The Lagrange–Burmann, Taylor, and Stirling–Maclaurin series involve only ascending powers, and apply only to holomorphic functions. If a complex function has singularities in a bounded domain, these may be enclosed by a loop, so that the function is holomorphic in a doubly-connected region (Sections 15.1 and 15.2); then similar methods (Chapter 23) lead to ascending–descending power series (Chapter 25), viz. ascending (descending) powers associated with the outer (inner) boundary where the function is larger (smaller) in modulus. The regular ascending (Lagrange–Burmann, Taylor, and Stirling–Maclaurin) series can be extended from ordinary derivatives to derivatives of complex order or differentegrations, as can the singular ascending-descending series (Campos, 1990).

Note 23.2. *Solution of a Differential Equation by a Series of Exponentials* (Campos, 1985): Whereas the Taylor series appears in almost every textbook on analysis its extension to the Lagrange–Burmann series appears in just a few. The Lagrange–Burmann series can be obtained as a change of variable $z \to f(z)$ in the Taylor series, though it is perhaps preferable to do the reverse, that is, to go from the general to the particular (Section 23.7). There are problems for which the Taylor series is not a convenient solution, and the more general Lagrange–Burmann series is more appropriate. An example is the solution of a linear ordinary differential equation (23.65a) whose coefficients (23.65b) are linear functions of exponentials:

$$\sum_{n=0}^{N} A_n(z)\frac{\mathrm{d}^n w}{\mathrm{d}z^n} = 0, \quad A_n(z) \equiv a_n - b_n \mathrm{e}^{-z}. \tag{23.65a,b}$$

The solution is conveniently obtained as a power series of e^{-z}:

$$w(z) = \sum_{m=0}^{\infty} C_m \left(\mathrm{e}^{-z}\right)^{m+\nu} = \mathrm{e}^{-z\,\nu} \sum_{m=0}^{\infty} C_m \mathrm{e}^{-mz}, \tag{23.66}$$

with a branch-point of exponent ν. Substituting (23.66) in (23.65a,b) leads to:

$$\begin{aligned} 0 &= \sum_{n=0}^{N} \left(a_n - b_n\,\mathrm{e}^{-z}\right) \sum_{m=0}^{\infty} C_m \frac{\mathrm{d}^n}{\mathrm{d}z^n} \mathrm{e}^{-\nu z - mz} \\ &= \mathrm{e}^{-\nu z} \sum_{m=0}^{\infty} \mathrm{e}^{-mz} \sum_{n=0}^{N} (-)^n \,(m+\nu)^n\, C_m \left(a_n - b_n\,\mathrm{e}^{-z}\right) \\ &= \mathrm{e}^{-\nu z} \sum_{m=0}^{\infty} \mathrm{e}^{-mz}\,[C_m P_m(\nu) - C_{m-1} Q_m(\nu)], \end{aligned} \tag{23.67}$$

where P_m, Q_m are polynomials of degree N:

$$P_m(\nu) = \sum_{n=0}^{N} (-)^n\, a_n\, (m+\nu)^n, \quad Q_{m-1}(\nu) = \sum_{n=0}^{N} (-)^n\, b_n\, (m+\nu-1)^n. \tag{23.68a,b}$$

Since the coefficients of e^{-mz} in square brackets in (23.67) must vanish, the coefficients of the series expansion (23.66) satisfy the recurrence formula:

$$m = 1, 2, \ldots: \quad C_m P_m(\nu) = C_{m-1} Q_m(\nu); \tag{23.69a,b}$$

this recurrence formula specifies all coefficients in terms of the first (23.70b):

$$0 \neq P_1(\nu), P_2(\nu), \ldots: \qquad C_m = C_0 \prod_{k=0}^{m} \left[\frac{Q_k(\nu)}{P_k(\nu)} \right], \tag{23.70a,b}$$

provided that the polynomials in the denominator do not vanish (23.70a). The first coefficient C_0 is an arbitrary constant, because if $w(z)$ in (23.66) satisfies the differential equation (23.65a,b) then $C_0 w(z)$ is also a solution, for any constant C_0. Setting (23.71a), since (23.71b) because there is no such term in (23.66), follows (23.71c):

$$m = 0; \quad C_{-1}(\nu) = 0: \quad 0 = C_0\, P_0(\nu) = C_0 \sum_{n=0}^{N} (-)^n a_n\, \nu^n; \tag{23.71a–c}$$

if $C_0 = 0$ then by (23.69b) all coefficients $C_1 = C_2 = \ldots = 0$ vanish, and a trivial solution $w(z) = 0$ results in (23.66). For a nontrivial solution $C_0 \neq 0$, and thus the exponent ν is a root of the polynomial (23.70c) of degree N. Each distinct root of (23.71c) yields a solution (23.66; 23.70a,b):

$$w(z) = C_0 \mathrm{e}^{-\nu z} \sum_{m=0}^{\infty} \mathrm{e}^{-mz} \prod_{k=0}^{m} \frac{Q_{k-1}(\nu)}{P_k(\nu)}, \tag{23.72}$$

of the differential equation (23.65a,b).

Note 23.3. *Solution in Terms of the Generalized Hypergeometric Function* (Campos, 1987): The polynomials P_0, Q_0 are of degree N and may be factorized:

$$P_0(\nu) \equiv \sum_{n=0}^{N} (-)^n a_n \nu^n = (-)^N a_N \prod_{n=1}^{N} (\nu - \alpha_n) \tag{23.73a}$$

$$Q_0(\nu) \equiv \sum_{n=0}^{N} (-)^n\, b_n\, \nu^n = (-)^N\, b_N \prod_{n=1}^{N} (\nu - \beta_n) \tag{23.73b}$$

where α_n (β_n) are the roots of $P_0(\nu)$ [$Q_0(\nu)$] in (23.73a) [(23.73b)]. Substituting ν by $\nu + m$ ($\nu + m - 1$) in (23.73a) [(23.73b)] specifies (23.68a) [(23.68b)]:

$$P_m(\nu) = P_0(\nu + m) = (-)^N a_N \prod_{n=1}^{N} (\nu + m - \alpha_n), \tag{23.74a}$$

$$Q_{m-1}(\nu) = Q_0(\nu + m - 1) = (-)^N b_N \prod_{n=1}^{N} (\nu + m - 1 - \beta_n); \tag{23.74b}$$

these appear in the recurrence formula for the coefficients (23.69a,b) ≡ (23.70a,b) and hence in the solution (23.66):

$$w(z){=}C_0 \mathrm{e}^{-\nu z} \sum_{m=0}^{\infty} \left(\mathrm{e}^{-z} \frac{a_N}{b_N} \right)^m \prod_{k=0}^{N} \frac{(\nu + k - \alpha_1)\ldots(\nu + k - \alpha_N)}{(\nu + k - 1 - \beta_1)\ldots(\nu + k - 1 - \beta_N)}, \tag{23.75}$$

of the differential equation (23.65a,b). The solution can be expressed in terms of the generalized hypergeometric function (30.112):

$$w(z) = C_0\, \mathrm{e}^{-\nu z}\; {}_{N+1}F_N \left(1, \nu - \alpha_1, \ldots, \nu - \alpha_N; \nu - 1 - \beta_1, \ldots, \nu - 1 - \beta_N; \frac{a_N}{b_N} \mathrm{e}^{-z} \right), \tag{23.76}$$

with: (i) $N+1$ upper parameters namely 1 and $\nu-\alpha_1,\ldots,\nu-\alpha_N$; (iii) N lower parameters $\nu-1-\beta_1,\ldots,\nu-1-\beta_N$; (ii) variable $\mathrm{e}^{-z}a_N/b_N$. The solutions (23.75) $\equiv$ (23.76) of the differential equation (23.65a,b) include explicit formulas for all the coefficients of the series (23.66). The latter is actually a power series (23.77a) of variable e^{-z} with branch-point:

$$w(z)=\sum_{m=0}^{\infty}C_m\left(\mathrm{e}^{-z}\right)^{\nu+m};\quad w(z)=\sum_{m=0}^{\infty}D_m z^{\nu+m}. \tag{23.77a,b}$$

if the solution of the differential equation had be sought as a power series (23.77b) in z with branch-point, this would lead to:

$$O=\sum_{n=0}^{N}\left(a_n-\mathrm{e}^{-z}b_n\right)\sum_{m=0}^{\infty}D_m\frac{\mathrm{d}^n\left(z^{\nu+m}\right)}{\mathrm{d}z^n}=\sum_{m=0}^{N}\left\{a_n-b_n\sum_{k=0}^{\infty}\frac{(-z)^k}{k!}\right\}$$
$$\times\sum_{m=0}^{\infty}D_m(\nu+m)\,(\nu-+m-1)\ldots(\nu+m-n+1)\,z^{\nu+m-n}, \tag{23.78}$$

using the exponential series (23.60). Equating to zero the coefficients of all powers of z, would lead to a systems of equations specifying the coefficients D_m of the series solution (23.76b) in terms of the (a_n,b_n), at least in principle; in practice it would be difficult to obtain explicit formulas for all D_m. This was not the case with the closed form solution (23.75) $\equiv$ (23.76). The two solutions (23.77a) and (23.77b) correspond to a change of variable $z\to e^{-z}$; this change of variable could also be made in the differential equation (23.65a,b). In conclusion, the Lagrange–Burmann series (23.77a) $\equiv$ (26.76) has led to explicit solutions (23.75) $\equiv$ (23.76) of the differential equation (23.65a,b), where the Taylor series (23.77b) fails to do so. The solution as a generalized hipergeometric series (23.75) converges absolutely (Example 30.20) when the variable is less than unity (23.79a) in modulus:

$$1>\left|\frac{a_N}{b_n}e^{-z}\right|=\left|\frac{a_N}{b_N}\right|exp\left[-Re\,(z)\right]\Leftrightarrow Re\,(z)>log\left(\frac{a_N}{b_N}\right)\equiv c; \tag{23.79a,b}$$

this implies absolute convergence for $\mathrm{Re}(z)>c$ in (23.79b); for real z that always (never) includes $z=+\infty(z=-\infty)$. A complete discussion of all possible solutions of (23.65a,b) can be made in the context of the theory of differential equations. The present account has emphasized the advantages in some cases of using series of functions other than power series.

Conclusion 23: A holomorphic function $f(z)$ can (Figure 23.1) be used to specify (**a**) a closed region D, with generally noncircular boundary ∂D, and a chain of regions D_ε, D_δ each contained in the following for $0<\delta<\varepsilon<1$; choosing as function $f(z)=z-a$ these regions become (**b**) circles with center at a and decreasing radii. A function $f(z)$ holomorphic (Figure 23.2) in the region D can (**a**) be expanded in Lagrange-Burmann series of ascending integral powers of the auxiliary function $f(z)$, such that all singularities z_* of $F(z)$ lie outside the boundary ∂D. If the auxiliary function is chosen to be $f(z)=z-a$, then (**b**) the holomorphic function is expanded in Taylor series, in a circle with center at a, and radius R determined by the exclusion of the nearest singularity z_* of $F(z)$. A real function (Figure 23.3) with continuous derivate in $a\le x\le b$, has at least one internal point $a<c<b$ in (**a**), or possibly several $a<c_1<c_2<c_3<c_4<b$ in (**b**), where the tangent is parallel to the straight line joining the end-points $A\equiv(a,f(a))$ and $B\equiv(b,f(b))$. The statement can become false (**c**) if the function has discontinuous derivate at one point only, for example, in the case illustrated no tangent to the curve is parallel to the straight line AB.

24

Electrostatics, Charges, and Dielectrics

The electromagnetic field in the steady case, that is, when if is time independent, splits into decoupled electrostatic (Chapter 24) and magnetostatic (Chapter 26) fields. The electrostatic field is created by electric charges (Section 24.1) that specify the electric field through the potential; also the electric displacement and polarization that depend on the constitutive properties of the material, viz. the dielectric permittivity and electric susceptibility (Section 24.2). The electrostatic field due to positive (negative) charges is analogous to (Chapter 12) the potential flow due to sources (sinks); the gravity field (Chapter 18) due to masses is analogous to sinks or negative charges; thus the Newton's law of attraction leads to (Section 24.3) the Coulomb's law of attraction (repulsion) for charges with opposite (same) signs. The method of images (Chapter 16) applies to charges or electrical multipoles near a plate (Section 24.4) or a corner (Section 24.5) with conducting or insulating walls; it is extended by the first circle theorem (Section 24.8) to the image on a cylinder. The circle theorem specifies (Section 24.7) the effect of introducing a cylindrical conductor or insulator in an arbitrary electric field, for example, the simplest case in a uniform electric field (Section 24.6). The problems of electrostatics with conducting or insulating boundaries may be extended to the interface between two media with distinct dielectric permittivities (Section 24.9). Electrostatics is the simplest model of electric fields and forces in media bounded by insulators, and of the electric charge distributions induced in conductors; it is also a building block to understand unsteady interactions that couple the electric and magnetic fields, for example, in electric circuits and electromagnetic waves.

24.1 Electric Field, Displacement, and Polarization (Maxwell, 1873)

The **Maxwell equations** (1873) *for the steady (i.e., time independent)* ***electrostatic field*** *(24.1a):*

$$\frac{\partial}{\partial t}=0: \quad \nabla\wedge \vec{\mathrm{E}}=0, \quad \nabla\cdot \vec{\mathrm{D}}=q, \tag{24.1a–c}$$

specify that (i) the curl of the **electric field** $\vec{\mathrm{E}}$ *is zero (24.1b), that is, it is an* ***irrotational*** *field; (ii) the* **electric charge** *density* q *forces the divergence of the* **electric displacement** *in* $\vec{\mathrm{D}}$ *in (24.1c).* The system of fundamental Maxwell equations (24.1b,c) must be closed by a **constitutive relation** relating the electric field and displacement that depends on the properties of matter. The relation may be nonlinear for strong electric fields, and **ferroelectric** materials may have spontaneous electric fields. For **dielectrics** the relation between the electric displacement and field is linear; they may have different directions in a crystal. In *an* **isotropic** *dielectric, whose properties must be independent of direction, the* *electric displacement is parallel to the electric field (24.2a):*

$$\vec{D} = \varepsilon \vec{E}, \quad \vec{Q}_e \equiv \vec{D} - \vec{E} = (\varepsilon - 1)\vec{E} = \left(1 - \frac{1}{\varepsilon}\right)\vec{D} = \chi_e \vec{E}, \quad \chi_e \equiv \varepsilon - 1, \tag{24.2a–c}$$

and the coefficient is the **dielectric permittivity** ε. It appears also in the **electric polarization** $\vec{Q}_e$ defined as the difference (24.2b) of the electric displacement and field; it is proportional to the electric field through (24.2c) the **electric susceptibility,** χ_e. *The electric polarization of an isotropic dielectric has the same direction as the electric field, implying that the magnetic susceptibility (dielectric permittivity) is nonnegative (24.3a) [not less than unity (24.3b)]:*

$$\chi_e = \varepsilon - 1 \geq 0 = \chi_{e0}, \quad \varepsilon \geq 1 = \varepsilon_0; \tag{24.3a,b}$$

the minimum value corresponds to vacuo, for which the electric field and displacement coincide, and the polarization is zero.

24.2 Dielectric Permittivity and Electric Susceptibility

The electric field (24.1a) is always minus the gradient (24.4a) of an ***electric potential****:*

$$\vec{E} = -\nabla\Phi_e, \quad q = \nabla \cdot \vec{D} = \nabla \cdot \left(\varepsilon \vec{E}\right) = \nabla \cdot \left(\frac{\varepsilon}{\varepsilon - 1}\vec{Q}_e\right), \tag{24.4a,b}$$

that for an isotropic dielectric satisfies (24.4b); for a ***homogeneous*** *dielectric (24.5a), that is, with dielectric permittivity independent of position:*

$$\varepsilon = \text{const}: \quad \frac{q}{\varepsilon} = \nabla \cdot \vec{E} = -\nabla^2\Phi_e, \tag{24.5a,b}$$

the electric potential satisfies a Poisson equation (24.5b) forced by minus the electric charge density q that acts as the "source" of the electric field, divided by the dielectric permittivity. The electric potential specifies the electric field (24.4a), displacement (24.2a) ≡ (24.6a) and polarization (24.2b) ≡ (24.6b):

$$\vec{D} = -\varepsilon\nabla\Phi_e, \quad \vec{Q}_e = -(\varepsilon - 1)\nabla\Phi_e = -\chi_e\nabla\Phi_e. \tag{24.6a,b}$$

The ***electric energy*** *(Section 28.1) is a quadratic function of the fields:*

$$E_e \equiv \frac{\vec{E} \cdot \vec{D}}{2} = \frac{\varepsilon E^2}{2} = \frac{D^2}{2\varepsilon} = \frac{\varepsilon Q_e^2}{2(\varepsilon - 1)^2} = \frac{\vec{D} \cdot \vec{Q}_e}{2(\varepsilon - 1)} = \frac{\varepsilon \vec{E} \cdot \vec{Q}_e}{2(\varepsilon - 1)}. \tag{24.7}$$

The method of Sections 18.1 and 18.2 shows that (24.1b,c) lead to the following *boundary conditions at the interface between two dielectrics:*

$$0 = [E_s] = \left[\varepsilon^{-1}D_s\right] = \left[\frac{Q_s^e}{\varepsilon - 1}\right] = \left[-\frac{\partial\Phi_e}{\partial s}\right], \tag{24.8a}$$

$$\sigma = [D_n] = [\varepsilon E_n] = \left[\frac{\varepsilon Q_n^e}{\varepsilon - 1}\right] = \left[-\varepsilon\frac{\partial\Phi_e}{\partial n}\right], \tag{24.8b}$$

viz.: (i) the tangential electric field is continuous (24.8a); (ii) the electric charge per unit length σ specifies the jump of the normal component of the electric displacement (24.8b). The boundary conditions (24.8a,b) can be restated:

$$\Phi_1 = \Phi_2, \quad -\sigma = \varepsilon_2 \frac{\partial \Phi_2}{\partial n} - \varepsilon_1 \frac{\partial \Phi_1}{\partial n}, \tag{24.9a,b}$$

in terms of the electric potential Φ, its normal derivative $\partial\Phi/\partial n$ and the dielectric permittivity ε on the two sides (1,2) of an interface. In two-dimensions $q(\sigma)$ in (24.1c) [(24.8b)] specifies the density of electric charge per unit area (length).

24.3 Potential due to Charges and Electric Force (Coulomb, 1785)

The Maxwell equations (24.1b,c) for the electrostatic field (24.1a) of an isotropic (24.2a), homogeneous (24.5a) dielectric:

$$\nabla \wedge \vec{\mathrm{E}} = 0, \quad \nabla \cdot \vec{\mathrm{E}} = \frac{q}{\varepsilon}, \tag{24.10a,b}$$

establish two analogies: (a) first with the irrotational flow (18.1a,b), replacing (24.11a) the electric field by the velocity:

$$\vec{\mathrm{E}} \Leftrightarrow \vec{\mathrm{v}}, \quad \frac{q}{\varepsilon} \Leftrightarrow \Lambda, \tag{24.11a,b}$$

and the dilatation of a source $\Lambda > 0$ (sink $\Lambda < 0$) by (24.11b) a positive (negative) charge q, divided by the dielectric permittivity ε; (b) second with the gravity field (18.18a; 18.19a), replacing (24.12a) the electric by the gravity field:

$$\vec{\mathrm{E}} \Leftrightarrow \vec{\mathrm{g}}, \quad -m\rho \Leftrightarrow \frac{q}{\varepsilon}, \tag{24.12a,b}$$

and the mass density times the gravitational constant (24.12b) by a negative electric charge density divided by the dielectric permittivity. It follows from (12.25b; 24.11a,b) or (18.26b; 24.12a,b) that the complex conjugate electric field due to a line-charge at $z = \zeta$ is (24.13b):

$$f_e(z) = -\frac{e}{2\pi\varepsilon}\log(z-\zeta), \quad E^*(z) \equiv E_x - iE_y = -\frac{\mathrm{d}f_e}{\mathrm{d}z} = \frac{e}{2\pi\varepsilon(z-\zeta)}, \tag{24.13a,b}$$

corresponding to the potential (24.13a), and is attractive (repulsive) for a negative (positive) charge; for a distribution (24.14a,b) of charges $q(\zeta)$ in a domain D the complex potential (24.14a) conjugate electric field (24.14b):

$$f_e(z) = -(2\pi\varepsilon)^{-1}\int_D q(\zeta)\log(z-\zeta)\,\mathrm{d}\zeta, \; E^*(z) = (2\pi\varepsilon)^{-1}\int_D q(\zeta)(z-\zeta)^{-1}\,\mathrm{d}\zeta, \tag{24.14a,b}$$

follows from the **principle of superposition** that applies to the electrostatic field, since the Maxwell' equations (24.1b,c) are linear.

It can be shown (28.37c) in analogy with gravity field (Chapter 18), that the **electric force** *per unit length on a line charge e' is equal to its product by the external electric field (24.15a):*

$$F_e^* = e'E^* = \frac{e'e}{2\pi\varepsilon}\frac{1}{z-\zeta}, \tag{24.15a,b}$$

*leading to **Coulomb's law (1785)**, viz.: the force (per unit length perpendicular to the z-plane) between two line charges, is (i) inversely proportional to the distance r between them multiplied by 2 π (viz. yielding the length of the circle of radius r); (ii) inversely proportional to the dielectric permittivity of the medium; (iii) lies in the direction from one charge to the other; (iv) is proportional to the product of charges, that is, repulsive (attractive) for charges with the same (opposite) sign(s), that is, both positive or negative (one positive and one negative). It is generalized:*

$$F_e^*(z) = \frac{e'}{2\pi\varepsilon} \int_D q(\zeta)(z-\zeta)^{-1} \, d\zeta, \tag{24.16a}$$

$$F_e^* = (2\pi\varepsilon)^{-1} \int_{D'} dz \, q'(z) \int_D d\zeta \, q(\zeta)(z-\zeta)^{-1}, \tag{24.16b}$$

to the electric force between a line-charge and a charge distribution (24.16a) or two charge distributions (24.16b). The electric field (24.13b; 24.14b) and force (24.15b; 24.16a,b) have the common factor:

$$\arg(E) = \arg(F_e) = -\arg(F_e^*) = -\arg\left[(z-\zeta)^{-1}\right] = \arg(z-\zeta), \tag{24.16c}$$

that has the direction from ζ to z, that is, represents a repulsion for equal charges [the opposite sign of the Newton's law (18.31) of attraction in the gravity field]. The following notations (Table 28.1) are used: (i) e for the total electric charge per unit length perpendicular to the ζ-plane of a line-change in (24.13a,b); (ii) $q(\zeta)$ for the electric change density per unit area of the ζ-plane in (24.14a,b; 24.16a,b); (iii) $\sigma(s)$ for the electric change density per unit length along a curve $\zeta(s)$ in the ζ-plane in the boundary condition (24.8b; 24.9b).

24.4 Multipole near Insulating or Conducting Wall

The electric field of a multipole specifies (Subsection 24.4.1) the force it exerts on an insulating or conducting wall (Subsection 24.4.3), and the induced-electric charges in the latter case (Subsection 24.4.2).

24.4.1 Identical or Complex Conjugate Images

The analogy (24.11a,b) with the potential flow due to a source/sink leads (12.59b) to *the complex electric potential (24.17b) and conjugate electric field (24.17c) due to a (21.17a) electric line-multipole of order n:*

$$n = 1, 2, \ldots: \; f_e(z) = \frac{P_n}{2\pi\varepsilon} z^{-n}, \quad E^*(z) = \frac{nP_n}{2\pi\varepsilon} z^{-n-1}; \tag{24.17a,b}$$

and also: (i) the corresponding scalar potential (12.60a) and field function (12.60b) have reversed sign; (ii) the polar components of the electric field (12.62a,b) have the same sign; (iii) in addition the multipole moment P_n is replaced by P_n/ε. If the electrical multipole is placed at a distance, a, from an infinite wall, the effect of the latter may be represented (Section 16.4) by adding or subtracting the image field, specified by the complex conjugate multipole moment:

$$f_{\pm}^{e}(z) = \frac{1}{2\pi\varepsilon}\left[\frac{P_n}{(z-ia)^n} \pm \frac{P_n^*}{(z+ia)^n}\right], \tag{24.18a}$$

$$E_{\pm}^{*}(z) = \frac{n}{2\pi\varepsilon}\left[\frac{P_n}{(z-i\,a)^{n+1}} \pm \frac{P_n^*}{(z+i\,a)^{n+1}}\right], \tag{24.18b}$$

where (24.18a,b) (i) applies to a dipole ($n = 1$), quadrupole ($n = 2$) or multipole on any order (24.17a,b); (ii) for a monopole $n = 0$, then (24.18b) $\equiv$ (24.13b) coincide, and (24.18a) is replaced by (24.13a).

24.4.2 Electric Charge Distribution Induced on a Conducting Wall

On the wall (24.18b) simplifies to

$$y = 0: \quad E_{+}^{*}(x) = \frac{n}{\pi\varepsilon}\text{Re}\left\{P_n (x-ia)^{-n-1}\right\} = E_x(x,\,0), \tag{24.19a,b}$$

$$y = 0: \quad E_{-}^{*}(x) = \frac{in}{\pi\varepsilon}\text{Im}\left\{P_n (x-ia)^{-n-1}\right\} = -iE_y(x,\,0), \tag{24.20a,b}$$

showing that: (i) the identical image electrical multipole [(24.19a,b) and + sign in (24.18a,b)] corresponds to zero normal electric field, hence an **insulating** wall, without surface electric charges, coinciding with a field line; (ii) the opposite image electrical multipole [(24.20a,b) and – sign in (23.18a,b)] with complex conjugate moment, corresponds to a normal electric field, viz. the wall is an equipotential or **conductor** with surface electric charges (24.20c), specified by (24.8b), viz:

$$\sigma(x) = \varepsilon E_y^{-}(x,\,0) = -\frac{n}{\pi}\text{Im}\left[P_n (x-ia)^{-n-1}\right]. \tag{24.20c}$$

For (24.21c) a line charge, e, at a distance a from a conducting wall the complex potential (24.13a) [conjugate field (24.13b)] is given by (24.21a) [(24.21b)]:

$$f_{\pm}^{e}(z) = -\frac{e}{2\pi\varepsilon}\left[\log(z-ia) \pm \log(z+ia)\right], \tag{24.21a}$$

$$F_{\perp}^{e}(z) = \frac{e}{2\pi\varepsilon}\left[(z-ia)^{-1} \pm (z+ia)^{-1}\right]; \tag{24.21b}$$

the corresponding (Figure 24.1) **induced charge distribution** (24.21c):

$$\sigma(x) = -\frac{e}{\pi}\text{Im}\left(\frac{1}{x-ia}\right) = -\frac{e}{\pi}\frac{a}{x^2+a^2}, \tag{24.21c}$$

has: (i) opposite sign to the original electric charge; (ii) peak value in modulus at the closest point (24.21d):

$$|\sigma|_{\max} = |\sigma(0)| = \frac{|e|}{\pi a} \geq |\sigma(x)| \geq |\sigma|_{\min} = |\sigma(\pm\infty)| = 0; \tag{24.21d}$$

$$\int_{-\infty}^{+\infty} \sigma(x)\;\mathrm{d}x = -\frac{ea}{\pi}\int_{-\infty}^{+\infty}\left(x^2+a^2\right)^{-1}\;\mathrm{d}x = -e; \tag{24.21e}$$

(iii) the total induced charge (24.21e) is opposite $-e$, as follows from (17.21a,b; 17.22a,b).

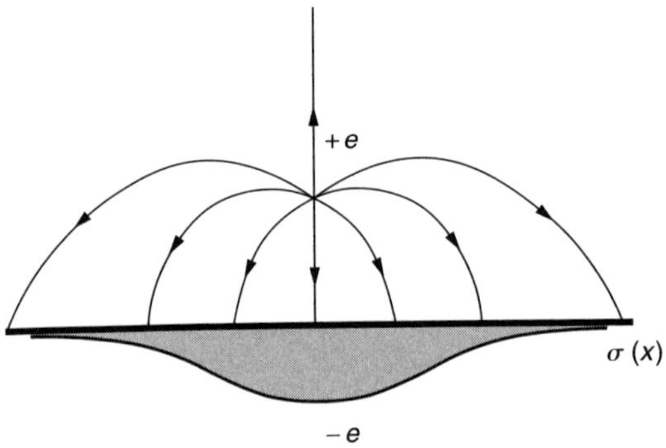

FIGURE 24.1
A line electric charge near a plane compactor has an opposite image; this leads to an electric field normal to the conductor, which specifies the distribution of induced electric charges. The latter peaks closest to the original electric charge and decays away along the plane conductor toward infinity; the electric charges on the conductor are of opposite sign to the original and have the same total value. The opposite induced charges cause the conductor to be subject to a force of attraction toward the original charge. The case of a line positive (negative) charge near an insulating plane is similar to a flow source (sink) near a plane rigid impermeable wall (Figures 16.1 and 16.2); there are no induced electric charges on the insulator because the electric field is tangential. The identical image implies that the insulator is repelled away from the electric charge.

24.4.3 Electric Force on the Wall and Asymptotic Scaling in the Far-Field

In conclusion *an electrical multipole at a distance a from a plane insulator (conductor) has complex conjugate electric field specified by* E_+^* (E_-^*) *in (24.18b) corresponding to the complex potential (24.18a); in the case of a conducting wall, the induced charge distribution is (24.20c). For a line-charge e at a distance, a, from a conducting wall, with potential (24.21a) and conjugate field (24.21b) the induced charge distribution has opposite sign (24.21d) and peaks (24.21e) at the closest point. The force exerted on the boundary:*

$$e' = \pm e: \quad F_e^{\pm} = e' E_* = \pm e \frac{-ie/2\pi\varepsilon}{2a} = \mp \frac{ie^2}{4\pi a\varepsilon} = -iF_y^{\pm}, \tag{24.22}$$

is repulsive (attractive) for a plane insulator (conductor), viz. (24.22) is obtained from (24.15a), bearing in mind that: (i) the distance from the image $e' = \pm e$ to the source is $z - \zeta = 2\,a$; (ii) its electric field is vertical $+i$, and the force exerted on the wall has opposite sign $-i$.

At a large distance an insulating (conducting) wall doubles the strength of a horizontal (14.23a) [vertical (14.23b)] multipole in the complex electrostatic potential (24.18a):

$$|z|^{n+1} \gg a^{n+1}: \quad f_{\pm}^{e}(z) = z^{-n}\frac{P_n \pm P_n^*}{2\pi\varepsilon} = \frac{z^{-n}}{\pi\varepsilon}\left\{\operatorname{Re}(P_n),\; i\operatorname{Im}(P_n)\right\}; \tag{24.23a,b}$$

in the case of a vertical (24.24a)[horizontal(24.25a)] multipole near an insulating (conducting) wall (24.23a) [(24.23b)] vanishes, the far-field decays faster, like (24.24b) [(24.25b)]:

$$P_n + P_n^* = 2\operatorname{Re}(P_n) = 0: \quad f_+^e(z) = -\frac{an}{\pi\varepsilon} z^{-n-1}\operatorname{Im}(P_n) = \frac{z^{-n-1}}{2\pi\varepsilon}P_{n+1}^{+}, \tag{24.24a,b}$$

$$P_n - P_n^* = 2i\operatorname{Im}(P_n) = 0: \quad f_-^e(z) = i\frac{an}{\pi\varepsilon} z^{-n-1}\operatorname{Re}(P_n) = \frac{z^{-n-1}}{2\pi\varepsilon}P_{n+1}^{-}, \tag{24.25a,b}$$

a horizontal (vertical) multipole of order $n+1$ with moment (24.26a) [(24.26b)].

$$P^{+}_{n+1} \equiv -2an\,\mathrm{Im}\,(P_n)\,, \quad P^{-}_{n+1} \equiv i2an\,\mathrm{Re}\,(P_n)\,. \tag{24.26a,b}$$

The proof (24.23–24.26a,b) uses the binomial series (21.65b):

$$(z \mp ia)^{-n} = z^{-n}\left(1 \mp \frac{ia}{z}\right)^{-n} = z^{-n} \pm inaz^{-n-1} + O\left(z^{-n-2}\right); \tag{24.27}$$

substituting (24.27) in (24.18a) yields:

$$\pi\varepsilon f^{e}_{+}(z) = z^{-n}\,\mathrm{Re}\,(P_n) - z^{-1-n} na\,\mathrm{Im}\,(P_n) + O\left(z^{-2-n}\right), \tag{24.28a}$$

$$\pi\varepsilon f^{e}_{-}(z) = iz^{-n}\,\mathrm{Im}\,(P_n) + z^{-1-n} nai\,\mathrm{Re}\,(P_n) + O\left(z^{-2-n}\right). \tag{24.28b}$$

The first terms on the r.h.s. of (24.28a,b) specify in the far-field to leading order $O\left(z^{-n}\right)$ of a multipole of order n in (24.23a,b); if the leading-order vanishes (24.24a; 24.25a), then the next order $O\left(z^{-n-1}\right)$ is specified (24.24b; 24.25b) by the second-term on order $n+1$. The higher order multipoles $O\left(z^{-n-2}\right)$ in (24.27) are not needed.

24.5 Identical or Alternating Images in a Corner

From the preceding account it may be expected that a line-charge, e, in a dielectric ε, placed in an insulating corner of angle (24.29a) at a distance, a, from the origin, will have (16.72b) a complex conjugate electric field (24.29c):

$$\beta = \frac{2\pi}{n}: \quad f^{+}_{e}(z) = -\frac{e}{2\pi\varepsilon}\log\left(z^n + a^n\right), \quad E^{*}_{+}(z) = \frac{ne}{2\pi\varepsilon}\frac{z^{n-1}}{z^n + a^n}, \tag{24.29a–c}$$

and complex potential (24.29b). Since (24.29b) is real for $z = re^{i\varphi}$ with $\varphi = 0, 2\pi/n, 4\pi/n, \ldots, (n-1)2\pi/n$, these radial lines correspond to the field lines $\Psi = 0$; thus the electric field (24.29c) is tangent to the walls that have no surface electric charges, and correspond to an insulating corner with angle $\beta = 2\pi/n$. The corner multiplies the strength of the electric potential (24.30c) and field (24.30d) field:

$$|z|^n \gg a^n; \quad n = \frac{2\pi}{\beta}: \quad f^{+}_{e}(z) \sim -\frac{ne}{2\pi\varepsilon}\log z, \quad E^{*}_{+}(z) \sim \frac{ne}{2\pi\varepsilon z}, \tag{24.30a,b}$$

by a factor (24.30b) in the far-field (24.30a).

In the case of a conducting walls, the image electric charges would be alternating $\pm e$, for example, four alternating charges in a rectangular corner in Figure 24.2; if placed in a corner of angle (24.31a) $\equiv$ (24.31c) on the diagonal $\varphi = \beta/2 = \pi/2n$ at distance a from the origin, would lead (16.79a) to the complex electric potential (24.31b):

$$\beta = \frac{\pi}{n}: \quad f^{-}_{e}(z) = -\frac{e}{2\pi\varepsilon}\log\left(\frac{z^n - ia^n}{z^n + ia^n}\right), \tag{24.31a,b}$$

$$n = \frac{\pi}{\beta}: \quad E^{*}_{-}(z) = \frac{ne}{2\pi\varepsilon} z^{n-1}\left[\left(z^n - ia^n\right)^{-1} - \left(z^n + ia^n\right)^{-1}\right] = i\frac{ne}{\pi\varepsilon}\frac{a^n z^{n-1}}{z^{2n} + a^{2n}}, \tag{24.31c,d}$$

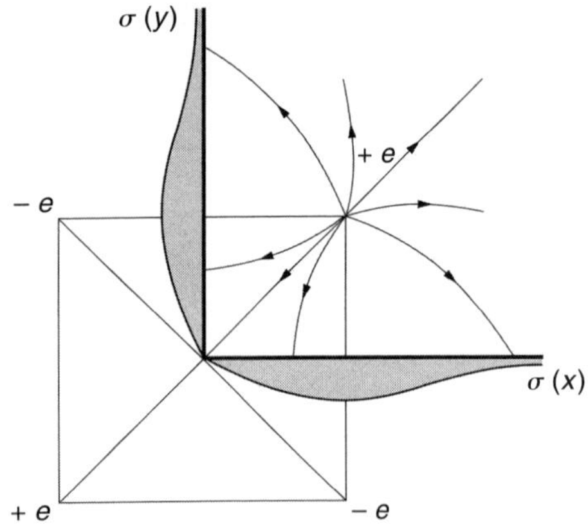

FIGURE 24.2
A line electric charge in a conducting corner has opposite (identical) images on the walls (origin), as for a vortex in a corner (Figure 16.4). The induced electric charges on the conducting corner are of opposite sign to the original line electric charge, vanish at the corner and infinity, and peak in between closest to the original charge. The two opposite images are closer to the original line electric charge than the one equal image; thus there is a net force of attraction of the conducting corner toward the charge. The case of a positive (negative) electric line-charge near an insulating rectangular corner would be similar to a flow source (sink) near a rigid impermeable corner (Figure 16.4): (i) the images are identical; (ii) the electric field is tangent to the insulating walls; (iii) there are no surface electric charges; (iv) the insulating corner is repelled away from the line electric charge.

and corresponding complex conjugate electric field (24.31d). The walls $\varphi = \pi/n$ correspond to zero real part of (24.31b), that is, to the equipotential $\Phi = 0$, viz. the electric field (24.31d) is orthogonal and specifies the distribution of surface electric charges:

$$\sigma(x) = \varepsilon E_y^-(x, 0) = -\frac{ne}{\pi} x^{n-1} \operatorname{Im}\left[(x^n - ia^n)^{-1}\right] = -\frac{ne}{\pi} \frac{a^n x^{n-1}}{x^{2n} + a^{2n}}, \tag{24.32}$$

that have the opposite sign. The surface electric charge peaks:

$$0 = \frac{\mathrm{d}\left[x^{n-1}/(x^{2n} + a^{2n})\right]}{\mathrm{d}x} = \left(x^{2n} + a^{2n}\right)^{-2} x^{n-2} \left[(n-1)\, a^{2n} - (n+1)\, x^{2n}\right], \tag{24.33a}$$

at a distance from the vertex:

$$n = 1, 2, \dots: \quad x_* = a\left(\frac{n-1}{n+1}\right)^{1/(2n)} < a; \tag{24.33b}$$

this leads to the peak charge:

$$|\sigma|_{\max} = |\sigma(x_*)| = \frac{|e|}{2\pi a}\sqrt{n^2 - 1}\left(\frac{n+1}{n-1}\right)^{1/(2n)} \geq |\sigma|_{\min} = 0 = \sigma(0) = \sigma(\infty). \tag{24.33c}$$

The induced charges vanish at the vertex (24.32) for $n \geq 2$, but not for $n = 1$ a plane (24.21c,d). In the case $n = 2$ of a line-charge on the diagonal of corner (Figure 24.2) the peak charge occurs (24.33b) at $x_*/a = \sqrt[4]{1/3} = 0.760$. In all cases the induced charges decay to zero at infinity, and the total induced charge:

$$\int_{-\infty}^{+\infty} \sigma(x)\mathrm{d}x = -\frac{e}{\pi}\int_{-\infty}^{+\infty} n\left(\frac{x}{a}\right)^{n-1}\left[1+\left(\frac{x}{a}\right)^{2n}\right]^{-1}\mathrm{d}\left(\frac{x}{a}\right) = -\frac{e}{\pi}\operatorname{arc\,tan}\left(\frac{x}{a}\right)^{n}\Bigg|_{-\infty}^{+\infty} = -e, \tag{24.34}$$

balances the original charge.

Using the geometric series (21.62a–c):

$$(z^n \mp ia^n)^{-1} = z^{-n}\left[1 \mp i\left(\frac{a}{z}\right)^n\right]^{-1} = z^{-n} \pm ia^n z^{-2n} + O\left(a^{2n}z^{-3n}\right), \tag{24.35}$$

in (24.31d) leads to the asymptotic field (24.36c) at large distance (24.36a) in a corner of angle (24.36b):

$$|z|^{n+1} \gg a^{n+1}; \quad n = \frac{\pi}{\beta}: \quad E_-^*(z) \sim \frac{ine}{\pi\varepsilon}a^n z^{-n-1} = \frac{P_n}{2\pi\varepsilon}nz^{-n-1}; \quad P_n \equiv 2iea^n, \tag{24.36a–d}$$

it corresponds to a vertical multipole of order n and moment (24.36d). Thus *a line-charge e in an insulating (conducting) corner of angle (24.29a) [(24.31a)] on the diagonal* $\varphi = \beta/2 = \pi/n$ ($\varphi = \beta/2 = \pi/2n$) *at a distance, a, from the origin, has a complex electric potential (24.29b) [(24.31b)] and conjugate electric field (24.29c) [(24.31c)]; also the exact scalar potential (16.67a) [(16.74a)] and field function (16.67b) [(16.74b)] with opposite (same) sign, and polar components of the electric field (16.68a,b) [16.75a,b)] with the same signs, plus the substitutions (24.11a,b); the asymptotic scaling (24.30a,b) [(24.36a–d)] at large distance corresponds to a charge n e [vertical multipole of order, n, with moment (24.36d)]. In the case of conducting walls, the induced distribution of electric charges (24.32), vanishes at the vertex and infinity (24.33c) and peaks (24.33c) at the distance (24.33b) from the vertex.*

24.6 Cylinder in a Uniform Electric Field

Consider the superposition of a dipole with real moment P_1 on a uniform electric field E_0, that is, the dipole axis is parallel $P_1 E_0 > 0$ (antiparallel $P_1 E_0 < 0$) to the real axis along E_0, corresponding the complex potential (24.37a):

$$f_e(z) = -E_0 z + \frac{P_1}{2\pi\varepsilon z}, \qquad E^*(z) = E_0 + \frac{P_1}{2\pi\varepsilon z^2}, \tag{24.37a,b}$$

and conjugate electric field (24.37b). The corresponding scalar potential (24.38a) and field function (24.38b):

$$\Phi_e(r,\varphi) = -\left(E_0 r - \frac{P_1}{2\pi\varepsilon r}\right)\cos\varphi, \quad \Psi_e(r,\varphi) = -\left(E_0 r + \frac{P_1}{2\pi\varepsilon\, r}\right)\sin\varphi, \tag{24.38a,b}$$

show that: (i) the dipole antiparallel to the external electric field the circle of radius $r = a$ leads to a streamline $\Psi_+ = 0$ for a dipole moment $P_1 = -2\pi\varepsilon a^2 E_0$, corresponding to a **cylindrical insulator**; (ii) the dipole parallel to the external electric field with moment $P = 2\pi\varepsilon a^2 E_0$ would make the circle $r = R$ the equipotential $\Phi_- = 0$, and hence a **cylindrical conductor**.

Thus *the insertion of a cylindrical insulator (conductor) of radius a in a uniform electric field E_0 is specified by a dipole of moment (24.39a) along (opposite) to the field:*

$$P_1 = \mp 2\pi\varepsilon a^2 E_0 : \quad f_e^{\pm}(z) = -E_0\left(z \pm \frac{a^2}{z}\right), \quad E_{\pm}^{*}(z) = E_0\left(1 \mp \frac{a^2}{z^2}\right), \tag{24.39a–c}$$

in the complex electric potential (24.39b) and conjugate electric field (24.39c); also:

$$\Phi_e^{\pm}(r,\varphi) = -E_0\left(r \pm \frac{a^2}{r}\right)\cos\varphi, \quad \Psi_e^{\pm}(r,\varphi) = -E_0\left(r \mp \frac{a^2}{r}\right)\sin\varphi, \tag{24.40a,b}$$

in the scalar potential (24.40a) and field function (24.40b), and:

$$E_r^{\pm}(r,\,\varphi) = E_0\left(1 \mp \frac{a^2}{r^2}\right)\cos\varphi, \quad E_\varphi^{\pm}(r,\,\varphi) = -E_0\left(1 \pm \frac{a^2}{r^2}\right)\sin\varphi, \tag{24.41a,b}$$

the polar components (24.41a,b) of the electric field.

The electric field (24.41a,b) is tangential on the cylinder $r = a$ in the insulating case $E_r^{+}(a,\,\varphi) = 0$, and in the conducting case is normal $E_\varphi^{-}(a,\,\varphi) = 0$. The *charge distribution induced on the conducting cylinder* r = a *by the external uniform electric field:*

$$\sigma(\varphi) = \varepsilon\, E_r^{-}(a,\,\varphi) = 2\,E_0\,\varepsilon\cos\varphi, \tag{24.42}$$

is (Figure 24.3) positive (negative) along (opposite) to the electric field, and zero orthogonal to it $\varphi = \pi/2$. The total charge on the cylinder is zero (24.43a) and its moment is horizontal

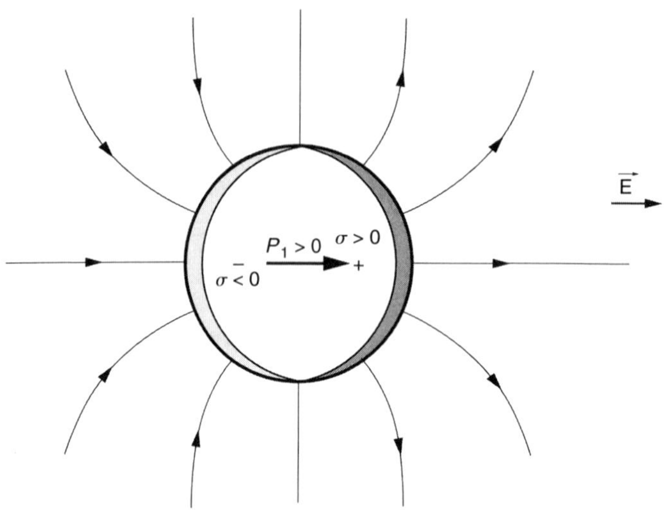

FIGURE 24.3
If a conducting cylinder is placed in a uniform electric field the positive (negative) electric charges accumulate toward the end along (opposite) to the field, in such a way that the total electric field becomes everywhere orthogonal to the surface of the cylinder. Since the electric charges at the two ends of the cylinder balance, the total electric charge on the cylinder is zero. The symmetry (asymmetry) of the electric charge distribution relative to the direction of (direction orthogonal to) the external electric field implies that there is a longitudinal (there is no transversal) dipole moment. Thus the introduction of a conducting (insulating) cylinder in an external uniform electric field is equivalent to a dipole with moment parallel (antiparallel) to the external field.

(24.43b):

$$a\int_0^{2\pi}\sigma(\varphi)\,\mathrm{d}\varphi = 0 = a\int_0^{2\pi} y\sigma(\varphi)\,\mathrm{d}\varphi = a^2\int_0^{2\pi}\sigma(\varphi)\,\sin\varphi\,\mathrm{d}\varphi, \tag{24.43a,b}$$

and equal to the dipole moment:

$$a\int_0^{2\pi} x\sigma(\varphi)\,\mathrm{d}\varphi = a^2\int_0^{2\pi}\sigma(\varphi)\cos\varphi\,\mathrm{d}\varphi = 2E_0\varepsilon a^2\int_0^{2\pi}\cos^2\varphi\,\mathrm{d}\varphi = 2\pi E_0\varepsilon a^2 \equiv P_1, \tag{24.43c}$$

in agreement with (24.43c) ≡ (24.39a).

In the case of the conducting cylinder the condition of zero tangential electric field is still met (24.44a):

$$E_\varphi\,(a,\,\varphi) = 0, \quad \sigma\,(\varphi) = \varepsilon\,E_\varphi\,(a,\,\varphi) = 2\,E_0\,\cos\varphi + \frac{e}{2\pi a}, \tag{24.44a,b}$$

superimposing upon (24.42) the field due to a line charge along the axis that adds a term [(24.13a) with $\zeta = 0$] to (24.45a), viz.:

$$f_e(z) = f_e^-(z) - \frac{e}{2\pi\varepsilon}\log z = -E_0 z + E_0\frac{a^2}{z} - \frac{e}{2\pi\varepsilon}\log z, \tag{24.45a}$$

$$E^*(z) = E_-(z) + \frac{e}{2\pi\varepsilon z} = E_0 + E_0\frac{a^2}{z^2} + \frac{e}{2\pi\varepsilon z}. \tag{24.45b}$$

Thus *a conducting cylinder of radius a in an uniform external electric field* E_0 *with a line charge, e, on the axis, has the complex potential (24.45a) and conjugate electric field (24.45b), corresponding to the induced charge distribution (24.44b).* There is no force on the cylinder in the uniform external electric field, because it is represented by a dipole, and the forces at the equal and opposite (i.e., positive and negative) charges at the same point cancel; there is no torque either, because (Section 28.2) the dipole axis is aligned with the external field, viz. antiparallel to it.

24.7 Reciprocal Point and First Circle Theorem (Kirchhoff, 1845)

The complex electric potential of a uniform electric field is given by (24.46a):

$$f_e(z) = -E_0 z, \quad f_e^\pm(z) = -E_0 z \mp \left[E_0\left(\frac{a^2}{z^*}\right)\right]^*, \tag{24.46a,b}$$

and the insertion (24.39b) ≡ (24.46b) of an insulating (conducting) cylinder of radius a corresponds to the transformation:

$$f_\pm(z) = f(z) \pm f^*\left(\frac{a^2}{z^*}\right). \tag{24.47}$$

This results holds not only for the uniform field (24.39b), but also for any potential field, as shown by the **first circle theorem (Kirchhoff, 1845)**: *if* f(z) *is the complex potential in* *free space, then the insertion of an insulating (conducting) cylinder of radius a corresponds to the potential* f_+ (f_-) *in (24.47).* The proof requires the verification of three properties, namely: (i) the far-field is the same for a distant observer $|z| \gg a$; (ii) the complex potential $f_\pm(z)$ has the same singularities outside the circle $|z| > a$ as $f(z)$; (iii) the circle $|z| = a$ is a field line (equipotential).

Concerning (i) the potential at infinity differs by a constant (24.48a):

$$\lim_{z\to\infty} f_\pm(z) = \lim_{z\to\pm\infty} f(z) \pm f^*(0), \quad \lim_{z\to\infty} E^*_\pm(z) = -\lim_{z\to\pm\infty} \frac{\mathrm{d}f_\pm}{\mathrm{d}z} = -\lim_{z\to\pm\infty} \frac{\mathrm{d}f}{\mathrm{d}z} = E^*(z). \tag{24.48a,b}$$

and thus the field is the same (24.48b). Concerning (ii) the transformation $\zeta = a^2/z^*$ maps the singularities of $f(\zeta)$ outside the circle $|z| > a$ into the circle $|\zeta| < a$ so $f_\pm$ in (24.47) has the same singularities as $f(z)$ outside the circle. Concerning (iii) the complex potential f_+ (f_-) in (24.47) is real (24.49b) [imaginary (24.50a)] on the circle (24.49a):

$$z = ae^{i\varphi}: \quad f_+\left(ae^{i\varphi}\right) = f\left(ae^{i\varphi}\right) + f^*\left(ae^{i\varphi}\right) = 2\,\mathrm{Re}\left[f\left(ae^{i\varphi}\right)\right] = \Phi_+(a,\varphi), \quad \Psi_+(a,\varphi) = 0, \tag{24.49a–c}$$

$$f_-\left(ae^{i\varphi}\right) = f\left(ae^{i\varphi}\right) - f^*\left(ae^{i\varphi}\right) = 2i\,\mathrm{Im}\left[f\left(ae^{i\varphi}\right)\right] = i\Psi_-(a,\varphi), \Phi_-(a,\varphi) = 0; \tag{24.50a,b}$$

thus it coincides there with the field function (potential), and the circle is the field line (24.49c) [equipotential (24.50b)]. There is a second circle theorem that generalizes the first circle theorem to rotational fields. The first circle theorem (Section 24.7) was inferred from the case of a cylinder in a uniform field (Section 24.6), and is applied next to an electric line-charge near a cylinder. The circle theorem (Sections 24.6–24.8; 26.7–26.9; 28.6–28.9) involves the reciprocal points relative to the circle (Sections 35.7–35.9), as shown next.

The image on a circle corresponds to (Figure 24.4) the following geometrical construction: (i) the line joining the field point S to the origin O has length $b > a$ larger than the radius; (ii) since S is outside the circle a tangent to the circle can be drawn AP, and its projection on OS is the image I or reciprocal point; (iii) the other tangent SP' would lead to the same image (24.51a):

$$\sin\psi = \frac{a}{b}, \qquad b' \equiv a\sin\psi = \frac{a^2}{b}, \tag{24.51a,b}$$

whose distance from the center (24.51b) appears in (24.47). The image (24.51b) on a circle of radius a is a **reciprocal point** since $b'b = a^2$. *The points z and* a^2/z^* *are called reciprocal points, or inverse points with regard to the circle of radius* a *(Section 35.7), because the reciprocal of the reciprocal is the original point:*

$$z \to \frac{a^2}{z^*} \to \frac{a^2}{\left(a^2/z^*\right)^*} = z; \tag{24.52a}$$

also, if the original and reciprocal points are interchanged in (24.47), the complex potential changes to its conjugate:

$$f\left(\frac{a^2}{z^*}\right) \pm f^*(z) = \pm\left[f(z) \pm f^*\left(\frac{a^2}{z^*}\right)\right]^* = \pm f^*_\pm(z); \tag{24.52b}$$

in the case of a conducting cylinder f_- *the sign is reversed as well, that is, positive and negative electric charges are interchanged.* Other properties of the reciprocal points are considered in Sections 26.8 and 35.7–35.9.

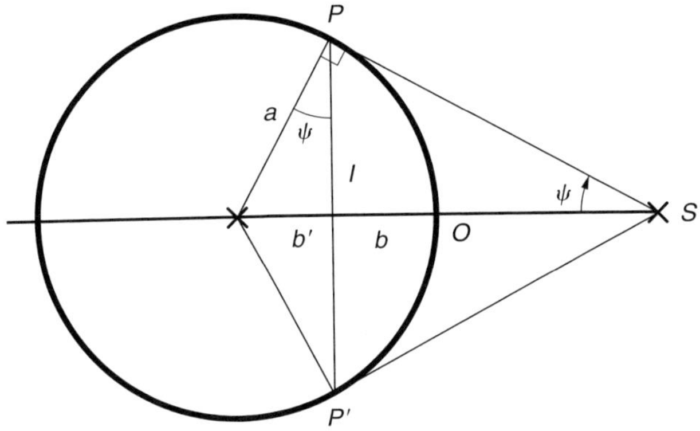

FIGURE 24.4
The image of a line electric charge/current/flow source/sink/vortex on a plane (cylinder) lies at the reciprocal point (Figure 16.1 (24.4)]. The reciprocal point relative to a straight line (Figure 16.1) is at the same distance along the normal. The reciprocal point relative to a circle (Figure 24.4) is on the radial line joining to the center; if the original is at a distance b from the center of a cylinder of radius a, the image is at a distance $c = a^2/b$. Thus if the original is outside the cylinder $b > a$ the image $c = a^2/b < a$ is inside; and vice-versa. The image is at the center $c = 0$ for an original at infinity $b = a$ and vice-versa. If the original is outside the cylinder the location of the image can be found by the following geometric construction: (i) take from the original two tangents to the circle at the points P and P'; (ii) the line joining $P'P$ cuts the radial line to the original OS at the image I. If the original is inside the cylinder it exchanges position with the image and the geometrical construction is reversed: (i) through the original I take a line perpendicular to the radial line OI; (ii) this line intersects the circle at two points P and P'; (iii) the tangents to the circle at these points intersect at the image S, which lies on the same radial line as the original I.

24.8 Induced Electric Charges on a Cylinder

The insertion (24.47) of an insulating f_+ (conducting f_-) cylinder of radius a in the electrostatic field due to a line-charge (24.13a) outside $b > a$ (or inside $b < a$) the cylinder at $z = b$, is specified [Figure 24.5b (24.5a)] by the potential:

$$f_e^{\pm}(z) = -\frac{e}{2\pi\varepsilon}\left[\log(z-b) \pm \log\left(\frac{a^2}{z} - b\right)\right]; \tag{24.53a}$$

this corresponds, apart from a constant which does not affect the field, to:

$$\log\left(\frac{a^2}{z} - b\right) = \log\left[\left(-\frac{b}{z}\right)\left(z - \frac{a^2}{b}\right)\right] = \log\left(z - \frac{a^2}{b}\right) - \log z + \log(-b), \tag{24.53b}$$

the superposition of: (i) an opposite (equal) image charge at the reciprocal point $\zeta = a^2/b$; (ii) an second image at the origin, opposite to the first:

$$E_{\pm}^{*}(z) = \frac{e}{2\pi\varepsilon}\left\{\frac{1}{z-b} \pm \frac{a^2}{z(bz-a^2)}\right\} = \frac{e}{2\pi\varepsilon}\left(\frac{1}{z-b} \pm \frac{1}{z-a^2/b} \mp \frac{1}{z}\right); \tag{24.54a,b}$$

(iii) the opposite images (i) and (ii) add to a zero total electric charge inside the cylinder. The opposite images add to the asymptotic (24.55a) field corresponding to: (i) the original

charge in the first term on the r.h.s. of (24.55b):

$$|z|^2 \gg b^2: \quad E^*_\pm = \frac{e}{2\pi\varepsilon z}\left(1+\frac{b\pm a^2/b}{z}\right) = \frac{e}{2\pi\varepsilon z} + \frac{P_1^\pm}{2\pi\varepsilon z^2}, \quad P_1^\pm \equiv e\left(b\pm\frac{a^2}{b}\right), \tag{24.55a–c}$$

(ii) a dipole of moment (24.55c) in the second term on the r.h.s. of (24.55b). The dipole moment P_1^- is product of the original charge by its distance from the image on the cylinder, for the conductor; for the insulator the sign is reversed in the term in curved brackets, corresponding to the distance to the reciprocal point on the other side of the origin. The passage from (24.54b) to (24.55b) uses the geometric series (21.62a,b):

$$\frac{2\pi\varepsilon}{e}E^*_\pm = \frac{1/z}{1-b/z} \pm \frac{a^2/(bz^2)}{1-a^2/(bz)} = \frac{1}{z} + \frac{b}{z^2} + O\left(\frac{b^2}{z^3}\right) \pm \frac{a^2}{bz^2} + O\left(\frac{a^4}{b^2z^3}\right); \tag{24.56}$$

this justifies the far-field approximation (24.55a) assuming also $O(a^4/b^4) = 1$ in $O\left(a^4/b^2\,z^3\right) = O\left(\left(b^2/z^3\right)\,\left(a^4/b^4\right)\right)$.

In the case of the conducting cylinder, viz.–sign in (24.47) or (24.54a) the electric charge distribution induced on the cylinder (Figure 24.5a) is:

$$\sigma(\varphi) = \varepsilon E_r^-(a,\,\varphi) = -\frac{e}{\pi}\frac{b\cos\varphi - a}{a^2+b^2-2ab\cos\varphi}; \tag{24.57a}$$

it is maximum (minimum) in modulus with opposite (same) sign at the closest (farthest) point:

$$\frac{|e|/\pi\varepsilon}{b+a} = |\sigma(\pi)| = |\sigma|_{\min} \le |\sigma(\varphi)| \le |\sigma|_{\max} = |\sigma(0)| = \frac{|e|/\pi}{b-a}. \tag{24.57b}$$

The electric charges with the opposite (same) sign are stronger and closer (weaker and farther) and occupy a smaller (larger) angular sector (Figure 24.5a), leading to a zero the total electric charge (24.77; 24.81a).

$$z = ae^{i\varphi}: \quad \frac{2\pi\varepsilon}{e}\left(E_r^- - iE_\varphi^-\right) = e^{i\varphi}\left\{\left(ae^{i\varphi}-b\right)^{-1} - \left[e^{i\varphi}\left(be^{i\varphi}-a\right)\right]^{-1}\right\}$$
$$= \left(a-be^{-i\varphi}\right)^{-1} + \left(a-be^{i\varphi}\right)^{-1} = \frac{2\left(a-b\cos\varphi\right)}{a^2+b^2-2ab\cos\varphi}, \tag{24.58a,b}$$

where it is radial (24.58b). The force on the cylinder is equal with reversed sign to the force on the electric charge due to its image; the latter is e times the second term of (24.54a) evaluated at $z = b$, viz.:

$$F^*_\pm = \mp\frac{e^2}{2\pi\varepsilon}\frac{a^2/b}{b^2-a^2} = \mp\frac{e^2}{2\pi\varepsilon}\frac{(a/b)^2}{b-a^2/b}. \tag{24.59}$$

The force (24.59) applies to an electric line-charge outside the cylinder $b > a$, and its sign is reversed for an electric line-charge inside the cylinder $b < a$, so that an insulator (conductor) in Figure 24.5b (24.5a) is always subject to a force of repulsion (attraction) in the presence of an electric line-charge. The remaining expressions, viz. for the complex potential (24.53a,b), and conjugate electric field (24.54a,b), and their asymptotic forms (24.55a–c), as well as the induced charges on the cylinder (24.57a,b) apply to a line charge at a distance b from the center of cylinder of radius a, regardless of whether it lies inside

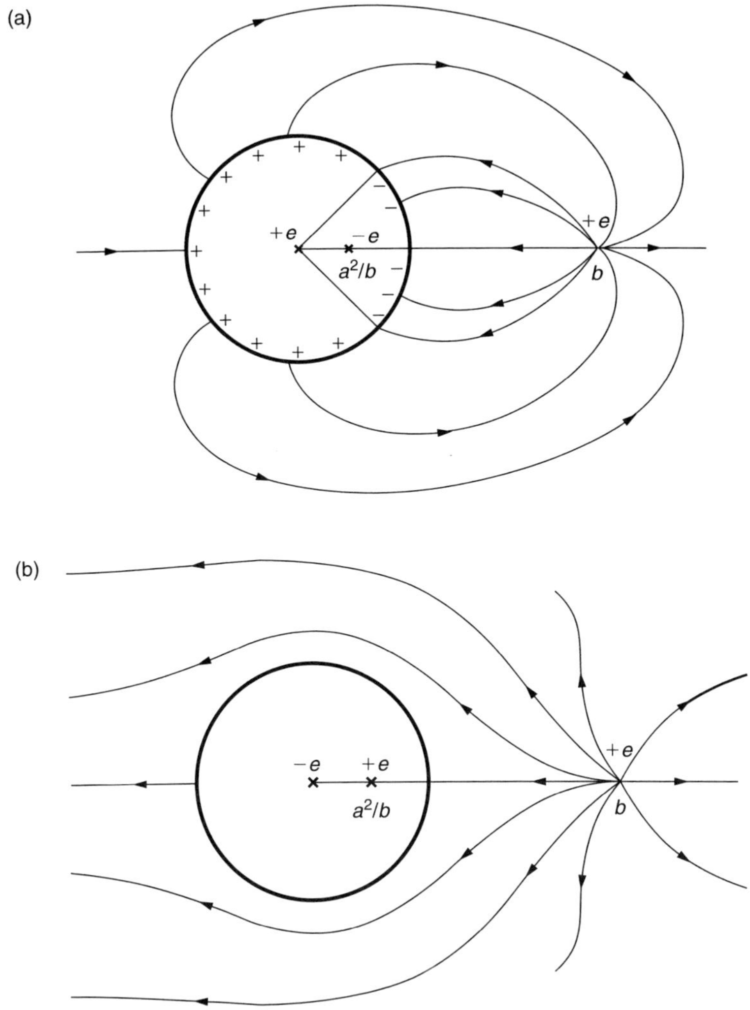

FIGURE 24.5
A line electric charge e per unit length at a distance b from the center of a conducting (a) [insulating (b)] cylinder of radius $a < b$, creates an electric field equivalent to adding two images: (i) an opposite (identical) image $-e$ $(+e)$ at the reciprocal point $c = a^2/b < a$; (ii) the image at the reciprocal point is compensated by the opposite image $+e$ $(-e)$ at the axis of the cylinder, so that the total electric charge inside the cylinder is zero. The system of two images in the conducting (insulating) cylinder implies that: (i) the total electric field is orthogonal (tangent) to the surface of the cylinder; (ii) it determines the distribution of (implies there are no) surface electric charges; (iii) since the closest image is opposite (identical) it implies that there is an attractive (repulsive) force on the cylinder.

$b < a$ or outside $b > a$. The total electric charge on the conducting cylinder is given by (24.81a–c), as shown in Note 24.1.

The reason for the different signs of the dipole moments $P_1^{\pm}$ in (24.55c) is (i) a line-charge attracts opposite charges to the closest side of a conductor (Figure 24.5a), and thus the image is an opposite charge $-e$ at the reciprocal point a^2/b on the same side of the center, leading to the dipole moment P_1^-; (ii) an insulator corresponds to an equal $+e$ instead of opposite charge, and the charge of sign in the second term of the dipole

TABLE 24.1
Electrical Conductors and Insulators

Boundary	Insulator	Conductor
Electro(Magneto)statics – complex potential:	f_e^+, f_m^+	f_e^-, f_m^-
– conjugate field:	E_+^*, H_+^*	E_-^*, H_-^*
Monopole image:	equal	opposite
Electric – charge:	+ e	– e
– current:	+ J	– J
Multipole – image:	equal	conjugate
– moment:	P_n	P_n^*
Cylinder in uniform field – dipole axis:	antiparallel	parallel
Reciprocal point relative to cylinder:	a^2/b	a^2/b
Dipole moment	$b - a^2/b$	$b + a^2/b$
Charge/current/multipole – reciprocal point:	e, J, P_n	$-e, -J, P_n^*$
– center	$-e, -J, -P_n$	$e, J, -P_n^*$

a – radius of cylinder; b – distance of multipole from center.
Note: Comparison of electrical conductors and insulators both for the electro (magneto) static field [Chapters 24 (26)]. The plane electric insulator (conductor) leads to: (i) identical (opposite) images for electric charges and currents, that is, monopoles; (ii) identical (conjugate) moments for multipole images. These distinctions extended to images on a cylinder both at the center and at the reciprocal point (see Figure 24.5).

moment P_1^+ corresponds to the distance from the reciprocal point $-a^2/b$ on the other side of the center. *The images of multipoles on planes and cylinders are indicated in Table 24.1 that applies both to the electrostatic (Chapter 24) [magnetostatic (Chapter 26)] field: (i) the complex potential is* f^+ (f^-) *for an insulator (conductor), for example,* $\mathrm{f}_\mathrm{e}^\pm$ $(\mathrm{f}_\mathrm{m}^\pm)$ *for the electric* $\mathrm{E}_\pm^*$ *(magnetic* $\mathrm{H}_\pm^*$*) conjugate field; (ii) the line-charge (line-current) for the electro (magneto) static field has equal (opposite) image in an insulator (conductor), and a multipole has an image with equal (conjugate) moment, for example, for a plane wall; (iii) for a line-charge (current) near a cylinder (ii) applies to the image at the reciprocal point, and the opposite of (ii) applies to the image at the center; (iv) the opposite signs of images for a conductor (insulator) correspond to a reciprocal point is* $\mathrm{a}^2/\mathrm{b}\left(-\mathrm{a}^2/\mathrm{b}\right)$ *on the same (opposite) side of the center, when calculating the dipole moment.* The electric field in the presence of insulators or conductors is compared in Table 24.1; the extension to dielectric interfaces is considered next.

24.9 Charge near Interface between Two Dielectrics

In the preceding problems with one (Section 24.4) or several (Section 24.5) plane walls, and with cylindrical "obstacles" (Sections 24.6 and 24.8), the latter were taken either as perfect conductors or perfect insulators in a medium of dielectric permittivity ε. An intermediate case between a conductor and insulator is an interface between two media with distinct dielectric permittivities $\varepsilon_1 \neq \varepsilon_2$, for which the boundary conditions (24.8a,b) apply in the general form (24.9a,b). The simplest interfaces between the media with distinct properties are plane (Section 24.9) [cylindrical (Section 26.8)]. They are considered by similar methods in electrostatics (Chapter 24) [magnetostatics (Chapter 26)], considering real, image and virtual electric charges (Subsection 24.9.1) [currents (Subsection 26.8.1)]; these include the particular cases of conducting and insulating walls (Subsection 24.9.2) [(26.8.3)], and specify the forces exerted on the walls and the effective electric charge (Subsection 24.9.3) [current (Subsection 26.8.4)] in the far-field.

24.9.1 Real Image and Virtual Electric Charges

The simplest example (Figure 24.6) is a line-charge, e, at a distance a from the interface, for which the complex potential is due: (i) in the first medium, that is, above the interface (24.60a) to an **image charge** e' at $z = -\, i\, a$, viz. (24.60b), besides the original real charge e:

$$y \geq 0: \quad f_{e1}(z) = -\frac{1}{2\pi\varepsilon_1}\left[e \log(z - ia) + e' \log(z + ia)\right], \tag{24.60a,b}$$

$$y \leq 0: \quad f_{e2}(z) = -\frac{e''}{2\pi\varepsilon_2} \log(z - ia)\,; \tag{24.61a,b}$$

(ii) in the second medium, that is, below the interface (24.61a) by a **virtual charge** e'' alone at the original position (24.61b). The potential (24.60b; 24.61b) is continuous (24.9a)

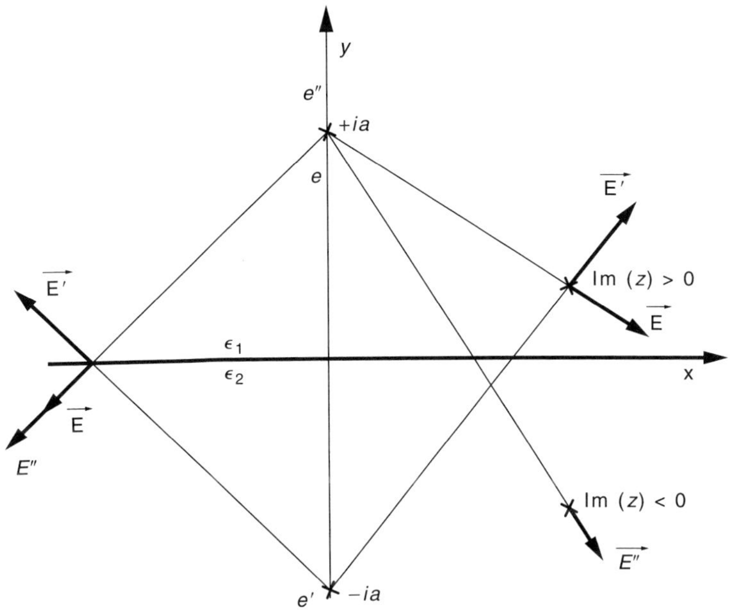

FIGURE 24.6
A joint generalization of the line electric charge near a conducting (insulating) plane [Figure 24.1 (16.1)] is a line electric charge near the plane interface between two dielectrics with distinct permittivities above ε_1 (below ε_2), that is, on the same (opposite) side of the interface as the original charge e. In this case there are two images, at the original (symmetric) point namely the line-charge $e'(e'')$. The image charges (e', e'') are related to the original by the continuity of the tangential (normal) electric field across the interface, which holds always (if there are no surface electric charges). The electric line charge near a insulating (conducting) plane [Figure 24.1 (16.1)] is the particular limiting case of Figure 24.6 when the permittivity of the second, lower medium is zero (infinite), implying that the image charge e' is identical $e' = e$ (opposite $e' = -e$) in (24.70a–d) [(24.71a–d)]. The intermediate cases $0 < \varepsilon_2 < \infty$ lead to a continuous transition between the line charge repelling $\varepsilon_1 > \varepsilon_2$ (attracting $\varepsilon_2 > \varepsilon_1$) the interface, as for the insulating $\varepsilon_1 > \varepsilon_2 = 0$ (conducting $\infty = \varepsilon_2 > \varepsilon_1$) wall, with no force for identical permittivities of the two media $\varepsilon_2 = \varepsilon_1$. The electric field $\vec{E}$ / $\vec{E}'$/ $\vec{E}''$ of the original/image/virtual line-charge $e/e'/e''$ lies along the position vector from $ia/-ia/ia$.

across the interface $y = 0$, but not:

$$y > 0: \quad E_1^*(z) = \frac{1}{2\pi\varepsilon_1}\left(\frac{e}{z - i\,a} + \frac{e'}{z + ia}\right), \tag{24.62a,b}$$

$$y < 0: \quad E_2^*(z) = \frac{1}{2\pi\varepsilon_2}\frac{e''}{z - ia}, \tag{24.63a,b}$$

the normal component of the electric field (24.9b), even in the absence of surface charges.

The boundary conditions (24.9a,b) $\equiv$ (24.64b,c), for an interface without surface charges (24.64a) are

$$\sigma = 0: \quad E_{1x}(x,\,0) = E_{2x}(x,\,0), \quad \varepsilon_1\,E_{1y}(x,\,0) = \varepsilon_2\,E_{2y}(x,\,0)\,; \tag{24.64a–c}$$

they are applied to the electric fields above (24.62a,b) [below (24.63b,c)] the interface:

$$y = +0: \quad \{E_{1x},\,E_{1y}\} = \left[\frac{1}{2\pi\varepsilon_1}\frac{1}{x^2 + a^2}\right]\{x\,(e + e'),\,a\,(e' - e)\}, \tag{24.65a–c}$$

$$y = -0: \quad \{E_{2x},\,E_{2y}\} = \left[\frac{1}{2\pi\varepsilon_2}\frac{1}{x^2 + a^2}\right]e''\,\{x, -a\}. \tag{24.66a–c}$$

Substitution of (24.65–24.66b,c) in (24.64b,c) leads to the relations:

$$\sigma = 0: \quad \frac{e + e'}{\varepsilon_1} = \frac{e''}{\varepsilon_2}, \quad e - e' = e''; \tag{24.67a–c}$$

these specify:

$$\{e', e''\} = \frac{e}{\varepsilon_1 + \varepsilon_2}\{\varepsilon_1 - \varepsilon_2,\,2\varepsilon_2\}, \tag{24.68a,b}$$

the image, e', and virtual, e'', charges in terms of the original electrical charge and dielectric permittivities of the media on the two sides of the interface.

24.9.2 Homogeneous Media or Conducting or Insulating Walls

Two particular cases of (24.68a,b) are mentioned besides the degenerate case. The latter occurs when the dielectric permittivities are the same (24.69a):

$$\varepsilon \equiv \varepsilon_1 = \varepsilon_2: \quad e' = 0, \quad e'' = e, \tag{24.69a–c}$$

and: (i) there is no image (24.69b) or interface; (ii) the virtual charge coincides with the real charge (24.69c); (iii) the potential (24.60b) $\equiv$ (24.61b) $\equiv$ (24.13a) coincides with that of an electric line-charge in free space. One extreme is the second medium with zero dielectric permittivity (24.70a), so that there is no electric displacement (24.70b), although there may be an electric field:

$$\varepsilon_2 = 0: \quad D_2(z) = 0 \neq E_2(z)\,; \quad e' = e, \quad e'' = 0, \tag{24.70a–d}$$

there is no virtual charge (24.70d) and the image is identical (24.70c), that is, the interface is a perfect insulator. The opposite extreme is the second medium with an infinite dielectric permittivity (24.71a), so that there is no electric field (24.71b), although there may be an electric displacement:

$$\varepsilon_2 = \infty: \quad D_2(z) \neq 0 = E_2(z)\,; \quad e' = -e, \quad e'' = 2e, \tag{24.71a–d}$$

the image charge is opposite (24.71c), that is, the interface is a perfect conductor, and the virtual charge (24.71d) is the double of the real charge, because the electric field is expelled, and doubles at infinity. The latter case corresponds to the **Faraday cage**: *the electric field is zero is in the interior of a conductor*, and thus a metal case can act as a "shield."

24.9.3 Force on the Interface and Effective Electric Charge in the Far-Field

The field in the first (24.62b) [second (24.63b)] medium is specified in terms of the original charge (24.68a) [(24.68b)] by (24.72a) [(24.72b)]:

$$\pi\varepsilon_1\left(z^2+a^2\right)E_1^*(z) = \frac{e(z+ia)+e'(z-ia)}{2} = \frac{e}{2}\left[z+ia+(z-ia)\frac{\varepsilon_1-\varepsilon_2}{\varepsilon_1+\varepsilon_2}\right] = e\frac{\varepsilon_1 z+i\varepsilon_2 a}{\varepsilon_1+\varepsilon_2}, \tag{24.72a}$$

$$\pi\left(z^2+a^2\right)E_2^*(z) = e''\frac{z+ia}{2\varepsilon_2} = e\frac{z+ia}{\varepsilon_1+\varepsilon_2}. \tag{24.72b}$$

A line-charge e at distance a from a plane interface between media with distinct dielectric permittivities $\varepsilon_1(\varepsilon_2)$ *creates electric fields above (24.62a,b) [below (24.63a,b)] specified by image (and virtual) charges (24.68a,b), implying for arbitrary dielectric permittivities that: (i) the electric field in the first medium (24.72a) ≡ (24.73a) is:*

$$E_1^*(z) = \frac{e_1 z+ie_2 a}{2\pi\varepsilon_1\left(z^2+a^2\right)}, \qquad \{e_1,\, e_2\} = \frac{2e}{\varepsilon_1+\varepsilon_2}\{\varepsilon_1,\varepsilon_2\}, \tag{24.73a–c}$$

involving the **effective line-charges** *(24.73b,c); (ii) the effective line charge in the first medium (24.3b) specifies the far-field (24.74a) in (24.73a) ≡ (24.74c):*

$$|z|^2 \gg a^2; \quad y>0: \quad E_1^*(z) = \frac{e_1}{2\pi\varepsilon_1 z} = \frac{e}{\pi z(\varepsilon_1+\varepsilon_2)}; \tag{24.74a–c}$$

(iii) the effective line-charge in the second medium (24.73c) specifies the exact electric field (24.63b) ≡ (24.72b) ≡ (24.75b) in the second medium (24.75a):

$$y<0: \quad E_2^*(z) = \frac{e}{\pi(\varepsilon_1+\varepsilon_2)(z-ia)} = \frac{e_2}{2\pi\varepsilon_2(z-ia)}. \tag{24.75a,b}$$

The force exerted by the charge on the interface is minus that due to its image:

$$\tilde{F}_e^* = -\frac{iee'}{4\pi a\varepsilon_1} = -i\frac{e^2}{4\pi a\varepsilon_1}\frac{\varepsilon_1-\varepsilon_2}{\varepsilon_1+\varepsilon_2} = -\tilde{F}_e, \tag{24.76}$$

is repulsive (attractive) if the first medium has larger (smaller) dielectric permittivity. The force vanishes for identical media, and reduces to (24.22) with + (−) sign for $\varepsilon_2=\infty$ $(\varepsilon_2=0)$ corresponding to conducting (24.71a,d) [insulating (24.70a–d)] interface. A combination of the present method (Section 24.9) and circle theorems (Sections 24.6–24.8) can be used to consider an electric charge (current) near a cylindrical interface separating media with distinct [Section 24.9(26.8)] dielectric permittivities (magnetic permeabilities). The plane and circle may be combined for the potential field in a corner with a cylindrical indentation (Section 36.2; Examples 40.10 and 40.11).

Note 24.1. *Total Electric Charge Induced on a Conducting Cylinder by a Line-Charge:* Consider the electric charge distribution induced on a cylinder of radius a by a line-charge e

at a distance b from the axis, lying outside $b > a$, inside $b < a$ or on the surface $b = a$. The induced surface electric charge density (24.57a) integrated over a circle of radius a specifies the total induced charge:

$$\bar{e} \equiv a \int_0^{2\pi} \sigma(\varphi)\, d\varphi = -\frac{ea}{\pi} \int_0^{2\pi} \frac{b \cos\varphi - a}{a^2 + b^2 - 2ab\cos\varphi} d\varphi = \frac{e}{\pi} I\left(\frac{b}{a}\right), \tag{24.77}$$

where the integral:

$$I(p) \equiv \int_0^{2\pi} \frac{1 - p\cos\varphi}{1 + p^2 - 2p\cos\varphi} d\varphi = \begin{cases} 2\pi & \text{if } |p| < 1, & (24.78a) \\ \pi & \text{if } |p| = 1, & (24.78b) \\ 0 & \text{if } |p| > 1. & (24.78c) \end{cases}$$

The integral (24.77) depends only on one parameter (24.79a). The path of integration along the real interval $(0,2\pi)$ is mapped (section 17.2) via the change of variable (24.79b) to the unit circle (24.79c):

$$p \equiv \frac{b}{a}; \ \zeta \equiv e^{i\varphi}: \quad I(p) = \int_{|\zeta|=1} \frac{1 - p(\zeta + 1/\zeta)/2}{1 + p^2 - p(\zeta + 1/\zeta)} \frac{d\zeta}{i\zeta} = \frac{1}{2i} \int_{|\zeta|=1} h(\zeta)\, d\zeta, \tag{24.79a–c}$$

with integrand (24.79d):

$$h(\zeta) \equiv \frac{1}{\zeta} \frac{2\zeta - p(1 + \zeta^2)}{(1 + p^2)\zeta - p(1 + \zeta^2)} = \frac{1}{\zeta} \frac{p\zeta^2 - 2\zeta + p}{p\zeta^2 - (1 + p^2)\zeta + p} = \frac{1}{\zeta p} \frac{p\zeta^2 - 2\zeta + p}{(\zeta - p)(\zeta - 1/p)} \tag{24.79d}$$

The integrand has simple poles at $\zeta = 0,\ p,\ 1/p$ with residues (24.80a–c):

$$h_{(1)}(0) = h_{(1)}(p) = 1 = -h_{(1)}(1/p): \quad I(p) = \pi\, \mathrm{h}_{(1)}(0) + \pi \times \begin{cases} h_{(1)}(p) & \text{if } |p| < 1; \\ h_{(1)}(1/p) & \text{if } |p| > 1. \end{cases} \tag{24.80a–e}$$

The integral (24.79c) equals 2πi times the sum of the residues at the poles inside the unit circle, leading to (24.80d,e) because: (i) the pole $\zeta = 0$ is at the center of the circle; (ii) $\zeta = p$ is inside and $\zeta = 1/p$ is outside the unit circle for $|p| < 1$ in (24.80d); (iii) vice-versa for $|p| > 1$ in (24.80e).

Substitution of (24.80a–c) in (24.80d,e) evaluates the integrals (24.78a,c), and substitution in (24.77) leads to:

$$\bar{e} = \begin{cases} 0 & \text{if } b > a, & (24.81a) \\ e & \text{if } b = a, & (24.81b) \\ 2e & \text{if } b < a. & (24.81c) \end{cases}$$

showing that *a line charge e at a distance b from the axis of a conducting cylinder of radius a leads to a total induced charge that is: (i) zero if the line charge e lies outside the cylinder $b > a$, because the opposite image $-e$ at the reciprocal point $c \equiv a^2/b < a$ inside the cylinder cancels with the identical image $+e$ at the center (24.81a); (ii) if the line charge e lies on the cylinder $b = a$ it coincides with the opposite image $-e$ at the same reciprocal point $c = a$, leaving only the identical image $+e$ at the center (24.81b); (iii) if the line charge e lies inside the cylinder $b < a$, the opposite image $-e$ is at the reciprocal point outside $c > a$, and the identical image $+e$ at the center leads to a double $2e$ total induced electric charge (24.81c).* Consider the cylindrical conductor starting with the line-charge at infinity. As the line-charge approaches it attracts (repels) on the surface of the cylinder opposite (equal) electric charges, that are concentrated towards the closed (farthest) side; the charges of opposite signs cancel (24.81a) since the conducting cylinder has received no net electric charge. When the line charge touches the cylinder (24.81b) it specifies the total net charge. When the line charge moves inside the cylinder it attracts (repels) opposite (equal) charges to the inner (outer) surface; this appears from the outside as a double charge (24.81c). A similar result holds for the surface electric current (circulation) induced by a line-current (line-vortex) near conducting (rigid) cylinder. [Subsection 26.7.2 (Section 28.8)]. The case (iii) follows (24.82a) from (24.77) ≡(24.82b):

$$\mathrm{a} = b : \bar{e} = \frac{e}{2\pi}\int_0^{2\pi} d\varphi = \frac{e}{2}; \quad |p| = 1 : I(p) = \pi h_{(1)}(0) + \frac{\pi}{2}\left[h_{(1)}(p) + h_{(1)}(1/p)\right] = \pi, \tag{24.82a–d}$$

it is a particular case $p = 1$ of the evaluation of the integral (24.82d) ≡ (24.78b) for p on the unit circle (24.82c): (i) the residue at the center (24.80a) is multiplied by $2\pi i$; (ii) the poles on the boundary are included multiplying their residues (24.80b,c) by πi. There are four alternatives for the evaluation of the integral (24.79c) with $|p| = \pm 1 \neq p$: (i/ii) if both poles on the boundary (24.80b,c) are included or excluded the result is (24.78b); (ii) if the pole at $\zeta = p$ is included and that at $\zeta = 1/p$ is excluded the result (24.78b) is replaced by $-\pi - \pi/2 = -3\pi/2$; (ii) vice-versa leads to $-\pi + \pi/2 = -\pi/2$.

Note 24.2. *Decoupled and Coupled Electric and Magnetic Fields:* In the steady (unsteady) case the electromagnetic field decouples (couples) into electrostatics (Chapter 24) and magnetostatics (Chapter 26) [electrical circuits and electromagnetic waves]. The multipole representation applies in two (Chapters 12, 16, 16, 24, 26, and 28) and three or more dimensions both for steady electric and magnetic fields, and for electromagnetic waves. The method of images extends from the plane to space and higher dimensions, and the circle theorem extends from cylinders to spheres and hyperspheres. The constitutive properties of dielectrics are quite distinct from ferroelectric and ferromagnetic materials. The magnetic field couples to fluid motion in magnetohydrodynamics and the electric field also in plasmas.

Conclusion 24: Distribution of electric charges induced by a line charge on a plane conductor (Figure 24.1) [conducting rectangular corner (Figure 24.2)] corresponds to one opposite (three alternating) images. A conducting cylinder in an external uniform electric field (Figure 24.3) has induced charges of opposite (same) sign in the direction along (opposite to) the field, leading to a zero total charge and nonzero dipole moment with axis along the external electric field. The charge distribution induced (Figure 24.5) on a conducting cylinder by a nearby line-charge also corresponds to a dipole (a); the image system on the cylinder has reversed sign for an insulator (b) relative to a conductor. These problems involve the image on the electric charge on a circle (Figure 24.4), which is analogous to the image on a plane (Figure 24.6). The interface between distinct dielectrics (Figure 24.6) includes intermediate cases between perfect conductors and insulators, e.g., specifying the relation between a line-charge e, its image charge e' in the second medium and the virtual charge e'' in the same medium.

25

Singular Series of Ascending–Descending Powers

It was shown (Chapter 23) that a function holomorphic in a closed singly-connected region can be represented by an ascending series of integral powers. If the function has singularities in the interior of the region, they can be surrounded by an inner loop or boundary that together with the outer loop or boundary specifies a ring-shaped or doubly-connected region; within this region the function is holomorphic and hence takes extreme (maximum or minimum) values of the modulus on the boundary (Section 25.1). The function can then be expanded into a series of: (i) ascending powers, associated with the outer loop as for the Lagrange–Burmann series; (ii) descending powers, associated with the inner loop that extend the Lagrange–Burmann to the Teixeira series (Section 25.5). The Teixeira (Lagrange–Burmann) series for a singular (holomorphic) function, proceeds in ascending and descending (ascending only) powers of an auxiliary function $f(z)$, and converge in a doubly- (singly-) connected region, of generally noncircular shape; choosing the auxiliary function $f(z) = z - a$ leads to the Laurent (Taylor) series [Section 23.7 (25.6)], valid in an annulus (circle). Taking the center at the origin $a = 0$ yields the Laurent–Maclaurin (Stirling–Maclaurin) series; this completes the hierarcity of series expansions (Diagram 25.1), from the most general (Teixeira's) is the most restricted (Stirling–Maclaurin's). The series, whether ascending or ascending–descending (Section 25.2) converge absolutely (Section 25.3) [uniformly (Section 25.4)] in the interior of the region (a closed subregion). The ascending power series for holomorphic functions are a particular case of the ascending–descending series for singular functions, obtained in the absence of singularities, when the inner loop can be shrunk to zero, so that the coefficients of descending powers vanish. The examples of series include the geometric (Section 25.2) and binomial (Section 25.9) series. It is possible to define a function of a series, for example, the series (Section 25.8) for the inverse function. The ascending–descending power series represent the potential flow near an isolated singularity, and show that: (i) in addition to the superposition of corner flows of angles submultiples of π that represent the regular part (Chapter 23); (ii) there is a singular part (Chapter 25) corresponding to a multipole expansion. These general representations of nonsingular (Chapter 23) [singular (Chapter 25)] fields also apply to higher dimensions than the plane; the latter is the simplest instance beyond the straight line, and more illuminating.

25.1 Lemma of the Extrema and Doubly-Connected Chain

The lemma of the maximum modulus (Section 23.2) states that a function $f(z)$ holomorphic in a region, either is a constant or takes the maximum value in modulus at the boundary; if in addition the function has no zeros in the region, $f(z) \neq 0$, then $1/f(z)$ is also holomorphic, and if it is nonconstant, the modulus $1/|f(z)|$ is maximum on the boundary, so $|f(z)|$ is minimum on the boundary. This is the extension of the lemma of maximum modulus to minimum modulus as well, viz.: the **lemma of the extrema**: *if the function* f(z) *is holomorphic and nonzero in a closed region, then it takes the extreme, that is, maximum*

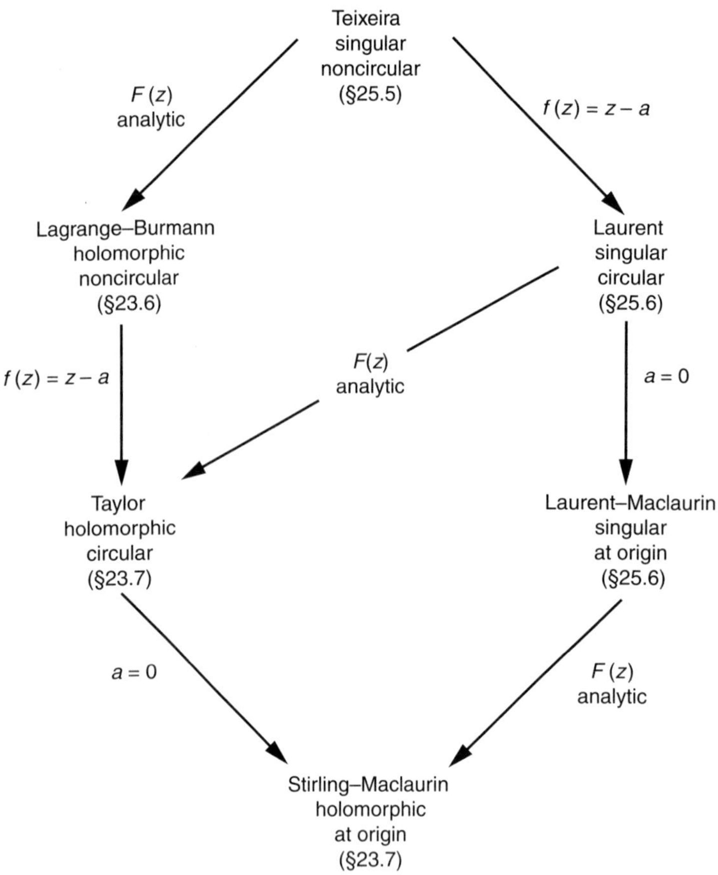

DIAGRAM 25.1
Hierarchy of power series. The hierarchy can be started either from the simplest (Stirling–Maclaurin) or most general (Teixeira): (i) the Maclaurin series expands a function holomorphic at the origin $a = 0$ in ascending powers of z; (ii) if it is holomorphic at another point $a \neq 0$ it can be expanded in Taylor series of ascending powers of $z - a$; (iii) if $f(z) = z - a$ is replaced by another holomorphic auxiliary function $f(z)$ the expansion in ascending powers of the latter leads to the Lagrange–Burmann series valid in a noncircular region; (iv) if the function is holomorphic in an annulus not including the origin $a = 0$, which may be a singularity of $F(z)$, the Laurent–Maclaurin series involves ascending and descending powers of z; (v) if the function is holomorphic in an annulus not including a point $a \neq 0$ other than the origin, the Laurent series involves ascending and descending powers of $z - a$; (vi) replacing $f(z) = z - a$ by another holomorphic auxiliary function leads to the Teixeira series involving ascending and descending powers of $f(z)$, for function $F(z)$ holomorphic in a noncircular annulus $r \leq |f(z)| \leq R$.

and minimum, values in modulus on the boundary:

$$f(z) \in \mathcal{D}(D): \quad f(z) \neq 0; \quad \forall_{\zeta \in \partial f D}: \quad |f(\zeta)|_{\max} = R \geq r \equiv |f(\zeta)|_{\min}: \quad r \leq |f(z)| \leq R. \tag{25.1a–e}$$

where either of the equalities is possible only for constant $f(z) = const \equiv r = R$.

From the preceding theorem it follows than a *region* E *defined by:*

$$f(z) \in \mathcal{D}(E); \quad R > r > 0: \quad E \equiv \{z : r \leq |f(z)| \leq R\}, \tag{25.2}$$

is doubly-connected with boundary:

$$\underline{\partial E} = \{\eta : |f(\eta)| = r\}, \quad \overline{\partial E} = \{\zeta : |f(\zeta)| = R\}, \tag{25.3a,b}$$

and interior:

$$E - \overline{\partial E} - \underline{\partial E} = \{z : r < |f(z)| < R\}; \tag{25.4}$$

the two parts of the boundary may be designated ***inner (outer) boundary*** $\underline{\partial \mathrm{E}}$ *in (25.3a)* [$\overline{\partial \mathrm{E}}$ *in (25.3b)] if they are closed curves (Figure 25.1). The region* $\mathrm{E}_{\varepsilon,\delta}$ *defined by:*

$$0 < \delta < 1, \quad 0 < \varepsilon < \frac{R}{r}(1-\delta) - 1: \quad E_{\varepsilon,\delta} \equiv \{z : (1+\varepsilon)r \leq |f(z)| \leq (1-\delta)R\}, \tag{25.5a–c}$$

is a closed, doubly-connected subregion of (25.2), and a chain of monotonic inclusion (Figure 25.1) is obtained by increasing ε *and decreasing* δ*:*

$$0 < \varepsilon < \varepsilon', \quad \delta' < \delta < 1: \quad E_{\varepsilon',\delta'} \subset E_{\varepsilon,\delta} \subset E_{0,0} \equiv E. \tag{25.6}$$

The constraint (25.5b) arises from the condition $r(1+\varepsilon) < R(1-\delta)$ that the angular region does not collapse to zero. Choosing the function $f(z) = z - a$, all boundaries (Figure 25.2) are circles of center at $\zeta = a$, that is, the doubly-connected region (25.2) is a circular ring or annulus of inner (outer) radius $r(R)$, and (25.5a–c) a subregion of larger inner $r\ (1+\varepsilon)$ [smaller outer $R\ (1-\delta) > r\ (1+\varepsilon)$] radius.

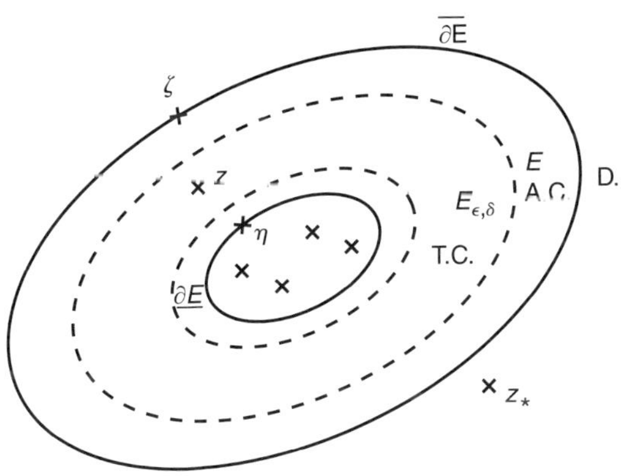

FIGURE 25.1
Let $f(z)$ be an auxiliary function which is holomorphic and nonzero in an annulus $r \leq |f(z)| \leq R$, with inner $\underline{\partial E}$ (outer $\overline{\partial E}$) boundary $|f(z)| = r\,(|f(z)| = R)$. Let $F(z)$ be a holomorphic function in the annulus, that may have singularities both: (i) within the inner boundary; (ii) outside the outer boundary. Then $F(z)$ can be expanded in a series of ascending and descending powers of the auxiliary function $f(z)$, which converges: (i) absolutely in the open interior of the annulus $r < |f(z)| < R$; (ii) totally, that is, also uniformly, in a closed subannulus $r(1+\varepsilon) \leq |f(z)| \leq R(1-\delta)$ with $r(1+\varepsilon) < R(1-\delta)$.

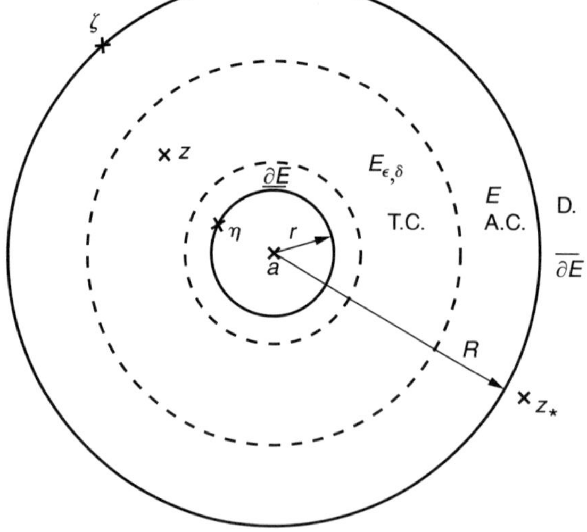

FIGURE 25.2
Choosing the auxiliary function $f(z) = z - a$ in the Teixeira series leads to the Laurent series, and replaces the noncircular (Figure 25.1) by a circular (Figure 25.2) annular region of convergence.

25.2 Ascending and Descending Geometric Series

From (25.2; 25.3a,b; 25.4) the following relations hold, for $\zeta(\eta)$ on the outer (inner) boundary, z within the open region E and w in a closed subregion:

$$\left|\frac{f(z)}{f(\zeta)}\right|, \quad \left|\frac{f(\eta)}{f(z)}\right| < 1; \quad \left|\frac{f(w)}{f(\zeta)}\right| < 1 - \delta; \quad \left|\frac{f(\eta)}{f(w)}\right| \leq \frac{1}{1+\varepsilon}. \tag{25.7a–d}$$

In the circular case (25.8a), the inequalities (25.7a–d) become:

$$f(z) = z - a: \quad |\eta - a| < |z - a| < |\zeta - a|,$$
$$|\eta - a|(1+\varepsilon) \leq |w - a| \leq (1-\delta)|\zeta - a|, \tag{25.8a–c}$$

since (Figure 25.2) the point $\zeta(\eta)$ lies on the circle of radius R (r) with $R > r$, and z lies in between the circles, whereas w is at a distance no less than $r\,\varepsilon(R\,\delta)$ from the closest point on the inner (outer) circle.

The generalized ascending (23.15) geometric series converges absolutely (globally) in (25.4) [(25.5a–c)] on account of (25.7a) [(25.7c)], and similarly the generalized descending geometric series:

$$\frac{1}{f(z) - f(\eta)} = \frac{1/f(z)}{1 - f(\eta)/f(z)} = \sum_{n=0}^{\infty} \{f(z)\}^{-n-1}\{f(\eta)\}^n = \sum_{m=1}^{\infty} \{f(z)\}^{-m}\{f(\eta)\}^{m-1}, \tag{25.9}$$

that also converges absolutely (totally) in (25.4) [(25.5a–c)] on account of (25.7b) [(25.7d)]. The expansion (25.9) truncated (21.60a,b) after the first M terms:

$$\{f(z) - f(\eta)\}^{-1} = \sum_{m=1}^{M} \{f(z)\}^{-m} \{f(\eta)\}^{m-1} + \{f(z)\}^{-M} \{f(\eta)\}^{M} \{f(z) - f(\eta)\}^{-1}, \tag{25.10}$$

is valid for all z, and the remainder after M terms that is the last term in (25.10), vanishes as $M \to \infty$ in (25.4). The geometric expansion (21.60a,b) with first term $1/f(z)$ and ratio $f(\eta)/f(z)$ coincides with (25.10).

25.3 Total Convergence in a Closed Subregion

A function $F(z)$, holomorphic in the doubly-connected region (25.2), or satisfying the Cauchy conditions (Section 15.1) there, can be specified by the ***modified Cauchy integral:***

$$F(z) = (2\pi i)^{-1} \int_{\overline{\partial E}} F(\zeta) \{f(\zeta) - f(z)\}^{-1} f'(\zeta)\, d\zeta - (2\pi i)^{-1} \int_{\underline{\partial E}} F(\eta) \{f(\eta) - f(z)\}^{-1} f'(\eta)\, d\eta, \tag{25.11}$$

using as variable the auxiliary holomorphic function f(z), so that $df = f'(\zeta)\, d\zeta$ [$df = f'(\eta) d\eta$] on the outer $\zeta \in \overline{\partial E}$ (inner $\eta \in \underline{\partial E}$) loop; the case $f(z) = z$ corresponds to the usual form of Cauchy's theorem in a doubly-connected region (Section 15.2). Since both functions $F(z)$, $f(z)$ are holomorphic in the region (25.4), and hence in the two annuli between (25.2) and in the closed subregion (25.5a–c), the inner and outer loops can be deformed from one region to the other.

For example, contracting the region E to $E_{\varepsilon,\delta}$ the uniformly convergent series (23.15) [(25.9)] can be integrated term-by-term (Section 21.7) after substitution in respectively the first (second) term on the r.h.s. of (25.11):

$$F(z) = \sum_{n=0}^{\infty} A_n \{f(z)\}^n + \sum_{m=1}^{\infty} A_{-m} \{f(z)\}^{-m}, \tag{25.12}$$

leading to a series of ascending (descending) powers, with coefficients determined by integrals along the outer (inner) loops:

$$n = 0, 1, \ldots: \quad A_n \equiv (2\pi i)^{-1} \int_{\overline{\partial E}} f(\zeta)^{-n-1} F(\zeta) f'(\zeta)\, d\zeta, \tag{25.13a,b}$$

$$m = 1, 2, \ldots: \quad A_{-m} \equiv (2\pi i)^{-1} \int_{\underline{\partial E}} f(\eta)^{m-1} F(\eta) f'(\eta) d\eta; \tag{25.14a,b}$$

the loops can be stretched back to $\overline{\partial E}$ ($\underline{\partial E}$). The series (25.12) is uniformly convergent in (25.5a–c); since it will be proven (Section 25.4) that it is absolutely convergent in the larger region (25.4), it follows that it is totally convergent in the smaller region (25.5a–c) that is the intersection of the two.

25.4 Absolute Convergence in an Open Region

In order to examine the absolute convergence of the series (25.12), upper bounds are given for the moduli of the coefficients (25.13a,b) [(25.14a,b)] viz. (25.15a,b) [(25.16a,b)].

$$|F(\zeta)f'(\zeta)| \le B: \quad |A_n| \le (2\pi)^{-1} \int_{\partial E} |F(\zeta)f'(\zeta)|\,|f(\zeta)|^{-n-1}\, d\zeta \le \frac{BL}{2\pi}\,|f(\zeta)|^{-n-1}, \tag{25.15a,b}$$

$$|F(\zeta)f'(\eta)| \le b: \quad |A_{-m}| \le (2\pi)^{-1} \int_{\overline{\partial E}} |F(\eta)f'(\eta)|\,|f(\eta)|^{m-1}\, d\eta \le \frac{b\ell}{2\pi}\,|f(\eta)|^{m-1}; \tag{25.16a,b}$$

here $F(z)f'(z)$ is a holomorphic function, and hence bounded in modulus by $B(b)$ on the outer (inner) loop of length $L(\ell)$, where $|f(\zeta)| = R\,(|f(\eta)| = r)$ is a constant. The series of moduli of (25.12), viz.:

$$\begin{aligned}
\left|\sum_{n=-\infty}^{+\infty} A_n\,\{f(z)^n\}\right| &\le \sum_{n=0}^{\infty} |A_n|\,|f(z)|^n + \sum_{n=0}^{\infty} |A_{-n-1}|\,|f(z)|^{-n-1} \\
&\le \frac{BL}{2\pi}\sum_{n=0}^{\infty} |f(z)|^n\,|f(\zeta)|^{-n-1} + \frac{b\ell}{2\pi}\sum_{n=0}^{\infty} |f(\eta)|^n\,|f(z)|^{-n-1} \\
&= \frac{BL/2\pi}{|f(z)| - |f(\zeta)|} + \frac{b\ell/2\pi}{|f(z)| - |f(\eta)|},
\end{aligned} \tag{25.17}$$

has by (23.15)[(25.9)] an upper bound (25.17), and so (25.12) is absolutely convergent in open region (25.4) that contains the region (25.5a–c) of total, that is, absolute and uniform convergence.

Substituting (23.16) [(25.10)] in the first (second) term of the r.h.s. of (25.11), leads to an expansion with N ascending (M descending) powers:

$$F(z) = \sum_{n=-M}^{N-1} A_n\,\{f(z)\}^n + R_N(z) + R_{-M}(z), \tag{25.18}$$

and two remainders:

$$R_N(z) = \{f(z)\}^N (2\pi i)^{-i} \int_{\partial E} F(\zeta)\,\{f(\zeta)\}^{-N}\,\{f(\zeta) - f(z)\}^{-1} f'(\zeta)\, d\zeta, \tag{25.19a}$$

$$R_{-M}(z) = \{f(z)\}^{-M} (2\pi i)^{-1} \int_{\overline{\partial E}} F(\eta)\,\{f(\eta)\}^{M}\,\{f(z) - f(\eta)\}^{-1}\, f'(\eta)\, d\eta, \tag{25.19b}$$

that vanish in (25.4) as $N \to \infty (M \to \infty)$, because absolute convergence implies simple convergence.

25.5 Series of Teixeira (1900): Coefficients and Remainder

The coefficients (25.13–25.14a,b) of the series as an infinite sum over all integers (25.12), as well as the remainders (25.19a,b) for the finite expansion (25.18), may be integrated by parts, leading to the **Teixeira theorem** (1900): *a function* F(z), *holomorphic or satisfying the*

Cauchy conditions (Section 15.1) in a closed doubly-connected region (25.2), can be expanded in a bilateral series of ascending and descending powers of an auxiliary holomorphic function (23.15) without zeros in E:

$$F(z) = \sum_{n=-\infty}^{+\infty} A_n \{f(z)\}^n = \sum_{n=-M}^{N-1} A_n \{f(z)\}^n + R_N(z) + R_{-M}(z); \tag{25.20a,b}$$

the coefficients of the constant A_0 plus ascending A_n (descending A_{-n}) powers are specified (25.13a,b) [(25.14a,b)] by integrals along the outer (inner) loops (25.21a,b) [(25.21c)]:

$$n \in |N : A_n = (2\pi i n)^{-1} \int_{\partial E} F'(\zeta) \{f(\zeta)\}^{-n} \, d\zeta, \tag{25.21a}$$

$$A_0 = -(2\pi i)^{-1} \int_{\partial E} F(\zeta) \, d\{\log f(\zeta)\}, \tag{25.21b}$$

$$n \in |N : A_{-n} = -(2\pi i n)^{-1} \int_{\underline{\partial E}} F'(\eta) \{f(\eta)\}^n \, d\eta. \tag{25.21c}$$

The series (25.20a) is absolutely (totally) convergent in the open interior (25.4) [closed subregion (25.5a–c)] of the region (Figure 25.1). The finite expansion (25.20b) is valid in the whole z-plane, with remainder (25.19a) [(25.19b)] after N ascending (M descending) terms given by (25.22a) [(25.22b)]:

$$R_N(z) = \{2\pi i (N-1)\}^{-1} \{f(z)\}^N \int_{\partial E} \{f(\zeta)\}^{1-N} \left\{ \frac{F(\zeta)}{f(\zeta) - f(z)} \right\}' d\zeta, \tag{25.22a}$$

$$R_{-M}(z) = -\{2\pi i (M+1)\}^{-1} \{f(z)\}^{-M} \int_{\underline{\partial E}} \{f(\eta)\}^{M+1} \left\{ \frac{F(\eta)}{f(z) - f(\eta)} \right\}' d\eta. \tag{25.22b}$$

The theorem allows the existence, in the interior of $\underline{\partial E}$, that is, in the region contained within the inner loop, of: (i) singularities of the function $F(z)$ to be represented; (ii) zeros of the auxiliary function $f(z)$.

25.6 Restriction to Laurent (1843)–Weierstress (1841) and Laurent–Maclaurin Series

A particular case leading to circular regions of convergence is the choice of (25.8a) as an auxiliary function, so that $z = a$ is (23.25) a simple zero $f(a) = 0 \neq f'(a)$, and can be a singularity of $F(z)$, excluded from the circular annulus (25.8a). This leads to the **Weierstress (1841)–Laurent (1843) theorem**: *if the function* F(z) *has an isolated singularity at* z = a, *and the nearest singularity is at* $z = z_*$, *then (Figure 25.2) it can be represented by a Laurent series (25.23a):*

$$F(z) = \sum_{n=-\infty}^{+\infty} A_n (z-a)^n = \sum_{n=-M}^{N} A_n (z-a)^n + R_N(z) + R_{-M}(z), \tag{25.23a,b}$$

with constant term and A_0*, coefficients* $A_n(A_{-n})$ *of ascending (descending) powers given by integrals along circles of radius* $R < |z_* - a|$ $(r > 0)$ *and center at a:*

$$R < |z_* - a|, n \in |N_0: \quad A_n = (2\pi i)^{-1} \int\limits_{|\zeta - a| = R} (\zeta - a)^{-n-1} F(\zeta)\, d\zeta, \tag{25.24a}$$

$$r > 0, m \in |N: \quad A_{-m} = (2\pi i)^{-1} \int\limits_{|\eta - a| = r} (\eta - a)^{m-1} F(\eta)\, d\eta; \tag{25.24b}$$

the series is absolutely (totally) convergent in the open annulus $r < |z - a| < R$ *closed subannulus* $r(1 + \varepsilon) \le |z - a| \le R\,(1 - \delta)$*. The finite expansion (25.23b) with* N *ascending (*M *descending terms), has remainders:*

$$R_N(z) = \frac{(z-a)^N}{2\pi i} \int\limits_{|\zeta - a| = R} F(\zeta)(\zeta - a)^{-N}(\zeta - z)^{-1}\, d\zeta, \tag{25.25a}$$

$$R_{-M}(z) = \left\{2\pi i (z-a)^M\right\}^{-1} \int\limits_{|\eta - a| = r} F(\eta)(\eta - a)^M (z - \eta)^{-1}\, d\eta. \tag{25.25b}$$

If the singularity is located at the origin $a = 0$, the Laurent (25.23a,b) reduces to the **Laurent–Maclaurin series:**

$$F(z) = \sum_{n=-\infty}^{+\infty} A_n z^n = \sum_{n=-M}^{N-1} A_n z^n + R_N(z) + R_{-M}(z), \tag{25.26a,b}$$

with coefficients and (remainders) specified by (25.24a,b)[(25.25a,b)] with $a = 0$. The convergence of three Laurent–Maclaurin series is considered in Example 30.9, and Example 30.10 includes 6 more cases.

25.7 Hierarchy of Power Series Expansions

In this chapter have been considered the expansions, in bilateral power series, that is, ascending and descending, of a function $F(z)$ holomorphic in a doubly-connected region (25.2), so that it can have singularities in the region:

$$E_* = \{z : \quad |f(z)| < r\}, \tag{25.27}$$

that lies within the inner loop. If $F(z)$ has no singularities in E_*, then the inner loop can be shrunk to zero, and: (i) the coefficients (25.21c) [(25.24b)] of descending powers in the Teixeira (25.20a) [Laurent (25.23a) and Laurent–Maclaurin (25.26a)] series vanish; (ii) the coefficients (25.21a,b) [(25.24a)] of ascending powers can be evaluated by Cauchy's theorem, and lead to the coefficients (23.19) ≡ (23.29) [(23.31)] of the Lagrange–Burmann (23.28a) [Taylor (23.32a) and Stirling–Maclaurin (23.34a)] series; (iii) the remainder of the negative powers (25.22b) [(25.25b)] vanishes as the inner loop shrinks to zero, and the remaining remainder, of positive powers coincides with that for the Lagrange (23.23) ≡ (25.19a) Teixeira [Taylor (23.33) ≡ (25.25a) Laurent] series. Thus holds the following **hierarchy of power series expansions** *in Diagram 25.1; (i) the Teixeira series (Section 25.5) is the most general, since it applies to singular functions (25.20a,b; 25.21a–c; 25.22a,b) in noncircular annuli* *of convergence (25.2; 25.3a,b; 25.4; 25.5a–c); (ii) choosing (25.8a) as the auxiliary function leads to the Laurent series (Section 25.6) for singular functions (25.23–25a,b) and circular annuli of convergence (25.8b,c); (iii) assuming instead a holomorphic function (23.28a,b; 23.29; 23.23), leads to the Lagrange–Burmann series (Section 23.6) in a noncircular (23.8; 23.9a,b; 23.10) region of convergence; (iv) the Taylor series (Section 23.7) is the holomorphic (circular) case of Laurent (Lagrange–Burmann) series (23.32a,b; 23.33), that is, applies to holomorphic functions in circular regions (23.13b, c); (v) if the center of the circle is the origin, the Stirling–Maclaurin series (Section 23.7) results (23.34a,b; 23.25); (vi) this is also the holomorphic case of the singular (Section 25.6) Laurent–Maclaurin series (25.26a,b) that is the case of Laurent series with annulus of convergence centered at the origin.*

25.8 Coefficients of Reversion of Series to Third-Order

The various series expansions, from the most general (Teixeira) to the most restricted (Stirling–Maclaurin), and including intermediate cases like the Lagrange–Burmann series, have many applications. The Lagrange–Burmann series allows, for two functions $F(z)$, $f(z)$ both holomorphic in the neighborhood of a point, each to be expanded in power series of the other, in two regions:

$$D_1: \quad F(z) - F(a) = \sum_{n=1}^{\infty} A_n \left\{ f(z) - f(a) \right\}^n, \tag{25.28a,b}$$

$$D_2: \quad f(z) - f(a) = \sum_{n=1}^{\infty} B_n \left\{ F(z) - F(a) \right\}^n. \tag{25.29a,b}$$

The two regions overlap, because they have in common at least one neighborhood of the point $z = a$; in their intersection (25.30a) both series (25.28b; 25.29b) hold, leading to the identity (25.30b):

$$D \equiv D_1 \cap D_2: \quad \zeta = \sum_{n=1}^{\infty} A_n \left(\sum_{m=1}^{\infty} B_m \zeta^m \right)^n, \quad \zeta \equiv F(z) - F(a); \tag{25.30a–c}$$

equating in (25.30b) the coefficients of successive powers of (25.30c), yields a system of algebraic equations that can be solved to specify the coefficients of the inverse series (25.29b) as functions (25.31a) of those of the direct series (25.28b), and vice-versa (25.31b); thus:

$$B_n = B_n(A_1, \ldots, A_n), \quad A_n = A_n(B_1, \ldots, B_n), \tag{25.31a,b}$$

are the formulas of inversion of series.

As an example the formulas of inversion of series are calculated up to the third-order, that is, neglecting powers of the fourth and higher orders:

$$\zeta = A_1(B_1\,\zeta + B_2\,\zeta^2 + B_3\,\zeta^3) + A_2(B_1\,\zeta + B_2\,\zeta^2)^2 + A_3(B_1\,\zeta + B_2\,\zeta^2)^3 + O(\zeta^4); \tag{25.32}$$

equating coefficients of ζ, ζ^2, ζ^3 leads to a system of three equations:

$$1 = A_1\,B_1, \quad 0 = A_1\,B_2 + A_2\,B_1^2, \quad 0 = A_1\,B_3 + 2A_2\,B_1\,B_2 + A_3\,B_1^3, \tag{25.33a–c}$$

that can be solved for A_n in (25.31a) or B_n in (25.31b) with $n = 1, 2, 3$. The triangular system (25.33a–c) can be solved by successive substitutions, viz.: (i) B_1 is specified by (25.33a) ≡ (25.34a):

$$B_1 = \frac{1}{A_1}; \quad B_2 = -\frac{A_2 B_1^2}{A_1} = -\frac{A_2}{A_1^3}, \tag{25.34a,b}$$

$$B_3 = -\frac{2A_2 B_1 B_2}{A_1} - \frac{A_3 B_1^3}{A_1} = -\frac{A_3}{A_1^4} + \frac{2A_2^2}{A_1^5}; \tag{25.34c}$$

(ii) B_2 is specified by (25.33b), where (25.34a) can be substituted leading to (25.34b); (iii) B_3 is specified by (25.33c), where (25.34a,b) can be substituted leading to (25.34c). Thus *if the functions* $\mathrm{f(z)}, \mathrm{F(z)}$ *are holomorphic in a neighborhood of* $\mathrm{z = a}$, *then there is a region (25.30a) in this neighborhood in that both* ***inverse series*** *hold (25.28a,b; 25.29a,b), and their coefficients are related (25.30b,c) by the inversion formulas (25.31a,b), specified to third order by* (25.34a–c). The implication is that the expansions:

$$\zeta = A_1\,\xi + A_2\,\xi^2 + A_3\,\xi^3 + O(\xi^4), \tag{25.35a}$$

$$\xi = \frac{1}{A_1}\zeta - \frac{A_2}{A_1^3}\zeta^2 + \frac{2A_2^2 - A_3 A_1}{A_1^5}\zeta^3 + O(\zeta^4), \tag{25.35b}$$

are inverse, that is, either one replaced into the other leads to an identity to third-order. Five other operations on or functions of series are considered in Example 30.7.

25.9 Binomial Expansion and Series and Inverse Powers

The formulas of inversion of series become increasing complicated at higher orders, but other series expansions can be obtained explicitly to all orders; for example, functions that are infinitely differentiable in a simple form lead in a straightforward manner to Taylor (23.32b) or Maclaurin (23.34b) series. For example, from:

$$\lim_{z\to 0} \frac{\mathrm{d}^n\left\{(z+a)^\nu\right\}}{\mathrm{d}z^n} = \nu(\nu-1)\dots(\nu-n+1)\lim_{z\to 0}(z+a)^{\nu-n} = \frac{\nu!}{(\nu-n)!}a^{\nu-n}, \tag{25.36}$$

the function $F(z) = (z+a)^\nu$ has a Maclaurin series:

$$|z| < |a|: \quad (z+a)^\nu = \sum_{n=0}^{\infty} \binom{\nu}{n} a^{-\nu-n} z^n, \quad \binom{\nu}{n} \equiv \frac{\nu!}{n!(\nu-n)!}, \tag{25.37a–c}$$

designated **binomial series**; its radius of convergence is determined by the exclusion of the singularity $z = -a$ that for complex ν other than an integer is a branch-point (Section 7.6). The series (25.37a) terminates only in the case $\nu = N$ with N a positive integer, since in that case the coefficients (25.37b) with $n > N$ vanish $(-1)! = (-2)! = \ldots = \infty$, and leading to the **binomial formula**:

$$(z+a)^N = \sum_{n=0}^{N} \binom{N}{n} a^{N-n} z^n = \sum_{n=0}^{N} \frac{N!}{n!(N-n)!} a^{N-n} z^n, \tag{25.38}$$

valid in the whole complex plane. The case of negative integral power for the exponent $\nu = -N$;

$$|z| < |a|: \quad (z+a)^{-N-1} = \sum_{m=0}^{\infty} \frac{(-N-1)(-N-2)\ldots(-N-m)}{m!} a^{-N-1-m} z^m \tag{25.39a,b}$$

leads to the ***series for the inverse power:***

$$|z| < |a|: \quad (z+a)^{-N-1} = \sum_{m=0}^{\infty} \binom{N+m}{N} a^{-N-m-1}(-z)^m = \sum_{n=N}^{\infty} \binom{n}{N} a^{-n-1}(-z)^{n-N}, \tag{25.40a–c}$$

where was made the substation $n = m + N$; the radius of convergence is determined by the nearest singularity $z = -a$ that is a pole of order $N + 1$; the particular case $a = 1$ was considered before (25.40b) $\equiv$ (21.65b). The series (25.40a,b) holds replacing $N + 1 \equiv \mu$ by a complex number, viz:

$$|z| < |a|: \quad (z+a)^{-\mu} = \sum_{n=0}^{\infty} \binom{n+\mu}{n} a^{-\mu-n}(-z)^n, \tag{25.41a,b}$$

and is an equivalent expression to the binomial series (25.37a–c). The convergence of the latter is addressed (Subsection 29.1.3) using the combined convergence test. Thus *the binomial series (25.37a–c) is equivalent to (25.41a,b) since both have a branch-point at $z = -a$ limiting the radius of convergence for $\nu(\mu)$ not an integer. The case $\mu = N + 1 = -\nu$ simplifies to the series (25.40a–c) that has the same radius of convergence limited by the pole of order N at $z = -a$. The only case of terminating series is $\nu = N = -\mu$ that leads to the binomial formula (25.38), valid in the whole complex plane. The convergence of the binomial series (25.37b,c) is detailed in Table 29.2 and* Subsection *29.1.3 at all points of the complex plane z, for all values of the parameters a, ν.*

Note 25.1. *Two Parallel Hierarchies of Power Series (Campos, 1990):* The hierarchy of power series in Diagram 25.1 includes six series, from the most general (Teixeira) to the most restricted (Stirling–Maclaurin); the more restricted series are used very often, when more general forms are not needed. There is a similar hierarchy of extended series using instead derivatives of complex order (Campos, 1990). The six expansions and series are established by similar methods, concerning a function holomorphic in a singly (doubly) connected region for ascending (Chapter 23) [ascending–descending (Chapter 25)] power series. Three cases can be distinguished: (i) if the function is holomorphic within a domain, its ascending power series (Chapter 23) can be extended as long as the boundary of the domain does not touch any 'external' singularity, for example, Figure 23.2; (ii) if the function has singularities in the domain, these may be enclosed by an inner loop, leading to a doubly-connected region of holomorphy (e.g., Figure 25.1) and ascending–descending power series (Chapter 25); (iii) if the singularity is isolated within the domain, the inner loop may be shrunk around it, for example, to a circle. The regions of convergence and divergence of ascending (Chapter 23) [ascending–descending (Chapter 25)] power series can be established by comparison with the geometric series (Sections 21.8 and 21.9). The convergence on the boundary (Chapter 29) requires finer convergence tests, including comparison with other series (e.g., the harmonic series) and integrals (viz. improper of second kind). Whereas the Taylor series exists both for real and complex functions (Chapter 23), the Laurent series exists only for complex functions (Chapter 25). The Laurent series allows a classification of singularities and functions (Chapter 27) that is possible in the complex but not in the real domain.

Note 25.2. *The Fourier (1818) Series as a Particular Teixeira (1900) Series:* The Laurent series is discussed in almost every textbook on complex analysis, but its generalization to the Teixeira series appears explicitly in just a few. A common implicit example of Texeira series, that is, a series of ascending and descending powers that uses the variable (25.42a), is (25.42b):

$$\zeta = e^{iz} : \quad F(\zeta) = \sum_{n=-\infty}^{+\infty} C_n \zeta^n = \sum_{n=-\infty}^{+\infty} C_n \, e^{inz} \equiv f(z); \tag{25.42a,b}$$

this corresponds to the complex form of the **Fourier series (1818)**. The region of convergence (25.43a,b):

$$b > 1 > a : \quad b > \left|e^{iz}\right| = \left|e^{ix}\, e^{-y}\right| = \left|e^{ix}\right| \left|e^{-y}\right| = e^{-y} > a \tag{25.43a,b}$$

is a horizontal strip (25.44b):

$$\log\, b > 0 > \log\, a : \quad -\log\, b \le y \equiv Im(z) \le -\log\, a, \tag{25.44a,b}$$

that may include the real axis (25.44a). The Teixeira theorem as stated in Section 25.5 requires that the curves $|f(z)| = r, R$ be closed; that is not the case for $r, R = \left|e^{iz}\right| = e^{-y}$, that is, $y = -\log\, r, -\log\, R$. In spite of these differences the connection between Teixeira (1900) and Fourier (1818) series is worth mentioning.

Note 25.3. *The Complex Fourier Series of Real Variable:* If the horizontal strip (25.43a,b) includes the real axis (25.44a,b) the complex Fourier series (25.42b) $\equiv$ (25.45b) can be considered for real variable (25.45a):

$$z \equiv \theta \in |R : \quad f(\theta) = \sum_{n=-\infty}^{+\infty} C_n \exp(in\theta) = \sum_{n=-\infty}^{+\infty} C_n \exp\left[in(\theta + 2\pi)\right] = f(\theta + 2\pi); \tag{25.45a,b}$$

it specifies a periodic function with period 2π. The period can be changed from 2π to X in (25.46b) by the change of variable (25.46a):

$$\frac{x}{X} = \frac{\theta}{2\pi} : \qquad f(\theta) = \sum_{n=-\infty}^{+\infty} C_n \exp(in\theta) = \sum_{n=-\infty}^{+\infty} C_n \exp\left(\frac{i2\pi n x}{X}\right) = g(x) = g(x + X). \tag{25.46a,b}$$

The coefficients are generally complex (25.47a):

$$C_n = A_n + iB_n : \quad f(\theta) = \sum_{n=-\infty}^{+\infty} (A_n + iB_n)\left[\cos(n\theta) + i\sin(n\theta)\right]$$

$$= \sum_{n=-\infty}^{+\infty} \left[A_n \cos(n\theta) - B_n \sin(n\theta)\right] + i \sum_{n=-\infty}^{+\infty} \left[A_n \sin(n\theta) + B_n \cos(n\theta)\right]; \tag{25.47a,b}$$

thus the complex Fourier series of real variable (25.45b) consists of Fourier cosine and sine (25.47b) both for the real and imaginary part.

Note 25.4. *Some Convergence Properties of Fourier Series:* The complex Fourier series of real variable (25.45a,b) ≡ (25.46a,b) ≡ (25.47a,b) with coefficients (25.48a) has the following convergence properties:

$$C_n \sim (n^{-g}): \quad \sum_{n=-\infty}^{+\infty} C_n e^{in\theta} \begin{cases} \text{A. C.} & \text{for } \mathrm{Re}(g) > 1, \\ \text{N. A. C.} & \text{for } 0 < \mathrm{Re}(g) \le 1, \\ \text{N. C.} & \text{for } \mathrm{Re}(g) < 0. \end{cases} \tag{25.48a–d}$$

The nonconvergence (25.48d) for $\mathrm{Re}(g) < 0$ arises because the general term $C_n\ e^{in\theta}$ does not tend to zero as $n \to \infty$, as required by the Cauchy necessary condition of convergence (29.17a). The results (25.48b,c) follow from the series of moduli of the coefficients:

$$\left| \sum_{n=-\infty}^{+\infty} C_n e^{in\theta} \right| \le \sum_{n=-\infty}^{+\infty} \left| C_n e^{in\theta} \right| = \sum_{n=-\infty}^{+\infty} |C_n|. \tag{25.49}$$

If $\mathrm{Re}(g) > 1$ [$\mathrm{Re}(g) \le 1$] then the series of moduli (25.49) converges (does not converge) as the harmonic series (29.40d) [(29.40a–c)] implying that the Fourier series (25.45b) is (is not) absolutely convergent (25.48b) [(25.48c)], QED. For $0 < \mathrm{Re}(g) \le 1$ there is the possibility that the Fourier series be conditionally convergent. For example, if $C_n \sim O(1/n^2)$ the Fourier series (25.45b) is absolutely convergent (25.48b); if $C_n \sim O(1/n)$ then the Fourier series (25.45b) is not absolutely convergent (25.48c), but the general term tends to zero $C_n e^{in\theta} \to 0$ as $n \to \infty$, so it satisfies the necessary (30.17a) but not sufficient condition for convergence. The possibility of conditional convergence of the Fourier serves (25.45b) with $C_n \sim O(1/n)$ requires specific investigation, since it is not covered by the usual general convergence tests.

Conclusion 25: The region of convergence of an ascending (descending) geometric series associated with a function $f(z)$ is the interior (exterior) of a loop $\overline{\partial E}$ ($\underline{\partial E}$), so that both series converge absolutely (totally) the interior of E (in a closed subregion $E_{\varepsilon,\delta}$); the same (Figure 25.1) type of convergence applies to the Teixeira series, which represents a function $F(z)$, holomorphic in E, but with singularities within $\underline{\partial E}$, i.e., outside E, as a bilateral series of powers, viz. ascending and descending, of the auxiliary function $f(z)$ without zeros in E. The particular case $f(z) = z - a$, leads to (Figure 25.2) the Laurent series, in ascending and descending powers of $z - a$, for a function $F(z)$ with an isolated singularity at $z = a$; if the nearest singularity is at $z = z_*$,then the series converges absolutely (totally) in an open circular annulus $0 < r < |z - a| < R \equiv |z_* - a|$ [closed subannulus $r(1 + \varepsilon) \le |z - a| \le R(1 - \delta)$].

26

Magnetostatics, Currents, and Permeability

The magnetostatic field (Chapter 26) like the electrostatic field (Chapter 24) is specified by three related vectors, viz. the magnetic field, induction, and polarization (Section 26.1); there are differences in the constitutive properties relating them that are specified by the magnetic permeability and susceptibility (Section 26.2). These differences ultimately arise from Maxwell's equations: the electric (magnetic) field is created by electric charges (currents), and is analogous to the potential flow due to source/sinks (vortices), leading (Section 26.3) to a Coulomb (Biot–Savart force) that acts along (across) the relative position vector. In follows that magnetic multipoles interchange the field lines and equipotentials of electric multipoles (Section 26.4), and reverse the alternating and identical images on planes (Section 26.5), corners (Section 26.6), and cylinders (Section 26.7). Bearing in mind these differences of detail, the general methods are similar, for example, for a cylindrical interface between media with distinct magnetic permeabilities (Section 26.8) and the generation of the field by multipole distributions with finite or infinite extent (Section 28.9). The two building blocks to be combined in the electromagnetic phenomena like waves and circuits are best studied separately in electrostatics (Chapter 24) and magnetostatics (Chapter 26). The latter have analogies and differences, for example, the magnetic field is always solenoidal and the electric field is irrotational in steady conditions.

26.1 Magnetic Field, Induction, and Polarization (Maxwell, 1873)

The **Maxwell equations (1873)** *for the steady (26.1a)* ***magnetostatic field:***

$$\frac{\partial}{\partial t} = 0: \quad \nabla \cdot \vec{\mathrm{B}} = 0, \quad \nabla \wedge \vec{\mathrm{H}} = \frac{\vec{\mathrm{j}}}{c}, \tag{26.1a–c}$$

state that (i) the **magnetic induction** $\vec{B}$ *is divergence free (26.1b) that is, is a* ***solenoidal field;*** *(ii) the curl of the* **magnetic field** $\vec{H}$ *equals (26.1c) the* **electric current density** $\vec{\mathrm{j}}$ *divided by the* **speed of light** *in vacuum* $\mathrm{c} = 3.00 \times 10^{10}\, cm\ s^{-1}$. The system of equations (26.1b,c) must be closed by a constitutive relation between the magnetic induction and field. The relation may be nonlinear for strong magnetic fields and **ferromagnetic** materials; for weak magnetic field and nonferromagnetic materials the magnetic induction is a linear function of the magnetic field. The magnetic field and induction may be nonparallel in a crystal; in an isotropic medium they must be parallel, specifying the **magnetic permeability** (26.2a):

$$\vec{\mathrm{B}} = \mu \vec{\mathrm{H}}; \quad \vec{Q}_m \equiv \vec{\mathrm{B}} - \vec{\mathrm{H}} = (\mu - 1)\, \vec{\mathrm{H}} = \left(1 - \frac{1}{\mu}\right) \vec{\mathrm{B}} = \chi_m \vec{\mathrm{H}}, \quad \chi_m \equiv \mu - 1, \tag{26.2a–c}$$

hence the **magnetic polarization**, defined as the difference (26.2b) between the magnetic induction and field, is proportional to the magnetic field through the **magnetic susceptibility** (26.2c). The magnetic (electric) polarization is usually weak (can be strong), and thus the magnetic (electric) susceptibility (26.2c) [(24.2c)] has no fixed sign (is nonnegative). This difference between the magnetic (electric) properties of matter is due to the nature of the magnetic (Section 26.3) [electric (Section 24.3)] force: (i) it acts along (across) the position vector; (ii) it is (is not) divided by the speed of light in vacuo that is a large quantity. *The magnetic induction and field have the same direction in a linear, isotropic material, and thus the magnetic permeability is positive (26.3a):*

$$\mu \geq 0; \quad \mu_0 = 1, \tag{26.3a,b}$$

the magnetic permeability is unity in vacuum (26.3b), when the magnetic field and induction coincide, and the magnetic susceptibility and polarization vanish.

26.2 Magnetic Permeability, Susceptibility, and Field Function

The Maxwell equation (26.1b) implies that the magnetic induction always derives from a field function:

$$\vec{\mathrm{B}} = \nabla \wedge \Psi_m = \vec{\mathrm{e}}_x \frac{\partial \Psi_m}{\partial y} - \vec{\mathrm{e}}_y \frac{\partial \Psi_m}{\partial x}, \tag{26.4a}$$

where the two-dimensional curl was used; thus the field function satisfies (26.1c), viz:

$$-\frac{\vec{\mathrm{j}}}{c} = -\nabla \wedge \left(\frac{\vec{\mathrm{B}}}{\mu} \right) = \vec{\mathrm{e}}_3 \left[\frac{\partial (B_x/\mu)}{\partial y} - \frac{\partial (B_y/\mu)}{\partial x} \right], \tag{26.4b}$$

showing that the electric current $\vec{\mathrm{j}}$ is orthogonal (26.5a) to the (x,y)-plane:

$$\vec{\mathrm{j}} \equiv j \vec{\mathrm{e}}_3 : \quad -\frac{j}{c} = \frac{\partial \left(\mu^{-1} \partial \Psi_m / \partial y \right)}{\partial y} + \frac{\partial \left(\mu^{-1} \partial \Psi_m / \partial x \right)}{\partial x}. \tag{26.5a,b}$$

The magnetic field function satisfies (26.5b) that for a linear, isotropic (26.2a) homogeneous (26.6a) medium reduces to the Poisson equation (26.6b) forced by the electric current density j multiplied by $-\mu/c$:

$$\mu = \text{const}: \quad -\frac{j\mu}{c} = \frac{\partial^2 \Psi_m}{\partial x^2} + \frac{\partial^2 \Psi_m}{\partial y^2} = \nabla^2 \Psi_m; \tag{26.6a,b}$$

this specifies the magnetic induction (26.7a), field (26.7b), and polarization (26.7c):

$$\left\{ \frac{\partial \Psi_m}{\partial y}, \, -\frac{\partial \Psi_m}{\partial x} \right\} = \{B_x, \, B_y\} = \mu \{H_x, \, H_y\} = \frac{\mu}{\mu - 1} \{Q_x^m, \, Q_y^m\}, \tag{26.7a–c}$$

where were used (26.4a) and (26.2a,b). The **magnetic energy** *is a quadratic function:*

$$2E_m = \vec{\mathrm{B}} \cdot \vec{\mathrm{H}} = \mu H^2 = \frac{B^2}{\mu} = \frac{\mu Q_m^2}{(\mu - 1)^2} = \frac{\vec{\mathrm{B}} \cdot \vec{Q}_m}{\mu - 1} = \frac{\mu \vec{\mathrm{H}} \cdot \vec{Q}_m}{\mu - 1}. \tag{26.8}$$

The boundary conditions for the magnetic field are that the normal (tangential) component is continuous (26.9a) [has a jump associated with the electric current density per unit length ϑ

in (26.9b)]:

$$0 = [B_n] = [\mu H_n] = \left[\left(1 - \frac{1}{\mu}\right) Q_n^m\right] = \left[-\frac{\partial \Psi_m}{\partial s}\right], \tag{26.9a}$$

$$\frac{\vartheta}{c} = [H_s] = \left[\mu^{-1} B_s\right] = \left[\frac{Q_s^m}{\mu - 1}\right] = \left[\mu^{-1} \frac{\partial \Psi_m}{\partial n}\right], \tag{26.9b}$$

where (26.9a) [(26.9b)] follows from (26.1b) [(26.1c)] as in Section 18.1 (18.2). In two-dimensions $j(\vartheta)$ denotes the electric current density per unit area (length).

26.3 Electric Current and Magnetic Force (Biot–Savart)

The magnetostatic field in a linear, isotropic, homogeneous medium:

$$\mu = \text{const}: \quad \nabla \cdot \vec{\mathrm{B}} = 0, \quad \nabla \wedge \vec{\mathrm{H}} = \frac{\vec{\mathrm{j}}}{c}, \tag{26.10a,b}$$

is analogous to the velocity (26.11a) field of a incompressible flow [(18.10a; 18.11b)]:

$$\left\{\vec{\mathrm{H}}, H^*, \frac{\vec{\mathrm{j}}}{c}, \frac{\mathrm{J}}{c}\right\} \Leftrightarrow \left\{\vec{\mathrm{v}} = v^*, -\vec{\varpi}, -\Gamma\right\}, \tag{26.11a,b}$$

replacing the vorticity (26.11b) by minus the electric current density divided by the speed of light in vacuum. The minus sign in (26.11b) arises because: (i) the velocity is a polar vector, and thus its two-dimensional curl is a pseudoscalar orthogonal to its plane (26.12a):

$$\vec{\varpi} \equiv \nabla \wedge \vec{v} = \vec{\mathrm{e}}_3 \left(\frac{\partial v_y}{\partial x} - \frac{\partial v_x}{\partial y}\right); \tag{26.12a}$$

$$-\frac{\vec{\mathrm{j}}}{c} = -\nabla \wedge \vec{\mathrm{H}} = -\left(\vec{\mathrm{e}}_x \frac{\partial}{\partial y} - \vec{\mathrm{e}}_y \frac{\partial}{\partial x}\right) \wedge H \vec{\mathrm{e}}_3 = \vec{\mathrm{e}}_y \frac{\partial H}{\partial y} + \vec{\mathrm{e}}_x \frac{\partial H}{\partial x} = \nabla H, \tag{26.12b}$$

the magnetic field is an axial vector, that is, a pseudoscalar perpendicular to the plane (28.12b), and hence its two-dimensional curl is minus the gradient. The comparison of the magnetic (vortical) force in the three-dimensional momentum equation confirms that the electric current (vorticity) enter with opposite signs (28.52). Thus there is an analogy between the magnetic field (velocity) due to an electric current J/c (vortex of circulation, Γ), one right-handed $J > 0\,(\Gamma < 0)$ and the other left-handed $J < 0\,(\Gamma > 0)$. It follows from (12.31b) replacing $-\Gamma$ by J/c that *a line-current* J *has conjugate magnetic field (26.13b):*

$$f_m(z) = -i\frac{J}{2\pi c} \log(z - \zeta), \quad H^*(z) = -\frac{\mathrm{d}f_m}{\mathrm{d}z} = \frac{iJ}{2\pi c} \frac{1}{z - \zeta}, \tag{26.13a,b}$$

corresponding to the complex potential (26.13a); for a distribution $j(\zeta)$ of electric currents in a domain D:

$$f_m(z) = -\frac{i}{2\pi c} \int_D j(\zeta) \log(z - \zeta)\, \mathrm{d}\zeta, \tag{26.14a}$$

$$H^*(z) = \frac{i}{2\pi c} \int_D j(\zeta) (z - \zeta)^{-1}\, \mathrm{d}\zeta, \tag{26.14b}$$

follow (26.14a,b) from (26.13a,b) using the principle of superposition that applies to the electrostatic field, because the Maxwell equations (26.10a,b) are linear. The following notation (Table 28.1) is used: (i) J for the total electric current per unit length perpendicular to the ζ-plane for a line-current (26.13a,b); (ii) j(ζ) for the density of electric current per unit area in the ζ-plane in (26.14a,b); (iii) ϑ(s) for the density of electric current per unit length of a curve in the boundary condition (26.9b).

It can be shown (28.36b) that the **magnetic force** on a line-current J' equals its product by the induction field divided by the speed of light in vacuum:

$$F_m^* = -\frac{iJ'B^*}{c} = -\frac{i\mu J'H^*}{c} = \frac{\mu}{2\pi c^2}\,\frac{J'J}{z-\zeta}, \tag{26.15}$$

leading to **Biot–Savart law**: *the magnetic force per unit length orthogonal to the plane between two line-currents is (i) proportional to the product of the currents; (ii) inversely proportional to the distance multiplied by 2 (that is the perimeter of the circle of radius r); (iii) proportional to the inverse square of the speed of light in vacuo; (iv) it acts perpendicular to the position vector joining the two currents; (v) it has positive (negative), that is, counter-clockwise (clockwise) direction, for currents with equal (opposite) signs. The magnetic force is extended to a line-current and current distribution (26.16a) [two current distributions (26.16b)]:*

$$F_m^* = \frac{\mu J'}{2\pi c^2}\int_D j(\zeta)\,(z-\zeta)^{-1}\,\mathrm{d}\zeta, \tag{26.16a}$$

$$F_m^* = \frac{\mu}{2\pi c^2}\int_D \mathrm{d}z\; j'(z)\int_D j(\zeta)\,(z-\zeta)^{-1}\,\mathrm{d}\zeta, \tag{26.16b}$$

The magnetic field (26.13b; 26.14b) acts in the direction (26.17a) from source to observer plus a positive (i.e., counter clockwise) rotation of 90°, and thus is orthogonal to the relative position vector:

$$\arg(H) = -\arg(H^*) = -\arg\left(\frac{i}{z-\zeta}\right) = -\frac{\pi}{2} + \arg(z-\zeta); \tag{26.17a}$$

$$\arg(F_m) = -\arg(F_m^*) = -\arg\left[(z-\zeta)^{-1}\right] = \arg(z-\zeta), \tag{26.17b}$$

the magnetic force (26.15; 26.16a,b) acts in the direction (26.17b) from the source current to the observer current. Since the electric (magnetic) force acts along (across) the relative position vector, it causes a linear acceleration (angular rotation), as for the trajectories (Section 6.1) of a charged particle in an electromagnetic field; since the speed of light in vacuo c is large, the small factor c^{-2} present (absent) in the magnetic (26.15) [electric (24.15b)] force leads to magnetic polarization generally much weaker than electric polarization, as mentioned in Section 26.1. The exception is ferroelectric (ferromagnetic) materials, for that polarization is strong, and leads to a nonlinear relation between electric displacement (magnetic induction) and electric (magnetic) field.

26.4 Hydrodynamic, Electromagnetic, and Gravity Multipoles

Comparing the forces (Sections 28.2 and 28.3) with the complex potentials and conjugate fields for a potential flow (Chapter 12), gravity (Chapter 18), electro (Chapter 24), and magnetostatic (Chapter 26) fields:

$$F_p^* = -\rho\,(Q - i\Gamma)\,v^*, \quad v^* = \frac{Q - i\Gamma}{2\pi z} \equiv \frac{\mathrm{d}f}{\mathrm{d}z}, \quad f_v = \frac{Q - i\Gamma}{2\pi} \log z, \tag{26.18a–c}$$

$$F_g^* = m' g^*, \quad g^* = -\frac{mG}{2\pi z} \equiv -\frac{\mathrm{d}f_g}{\mathrm{d}z}, \quad f_g = \frac{mG}{2\pi} \log z, \tag{26.19a–c}$$

$$F_e^* = e' E^*, \quad E^* = \frac{e}{2\pi\varepsilon z} \equiv -\frac{\mathrm{d}f_e}{\mathrm{d}z}, \quad f_e = -\frac{e}{2\pi\varepsilon} \log z, \tag{26.20a–c}$$

$$F_m^* = -i\frac{\mu}{c} J' H^*, \quad H^* = \frac{iJ}{2\pi c z} \equiv -\frac{\mathrm{d}f_m}{\mathrm{d}z}, \quad f_m = -\frac{iJ}{2\pi c} \log z, \tag{26.21a–c}$$

it follows that *the line-monopole is specified by: (i) the flow rate Q of a source/sink and circulation Γ of a vortex for the complex potential (26.18c) and conjugate velocity (26.18b) of a potential flow (12.35a–c), with drag plus lift force (26.18a) $\equiv$ (28.28a); (iii) the mass m times the gravitational constant G, with minus sign (18.26a,b) for a sink, for the complex potential (26.19c) and conjugate gravity field (26.19b) with gravity force (26.19a); (iii) the electric charge, e, divided by the dielectric constant, ε, for (24.13a,b) the complex potential (26.20a) and electrostatic field (26.20b), with electric force (26.20a); (iv) the electric current J divided by the speed of light in vacuum c for (26.13a,b) the complex potential (26.21c) and conjugate magnetic field (26.21b), with magnetic force (26.21a) involving the magnetic permeability μ.*

From (26.18–26.21a–c) follow two sets of analogies: (i) for ***irrotational fields*** *(26.22a) between a flow sink $Q < 0$ (source $Q > 0$), negative $e < 0$ (positive $e > 0$) charge, and mass $m > 0$ (no "negative mass" or analogy with source, only with sink, because the gravity field is always attractive); (ii) for* ***solenoidal fields*** *(26.22b) between the circulation Γ of a vortex and an electric current J, with opposite sign and divided by the speed of light in vacuo:*

$$Q \leftrightarrow -m\mathrm{G} \leftrightarrow \frac{e}{\varepsilon}, \quad \Gamma \leftrightarrow -\frac{J}{c}; \tag{26.22a,b}$$

these analogies appear in the complex potential (26.23a) and field (26.23b):

$$f(z) = \frac{P_0}{2\pi} \log(z - \zeta), \quad \frac{\mathrm{d}f}{\mathrm{d}z} = \frac{P_0}{2\pi} \frac{1}{z - \zeta}, \tag{26.23a,b}$$

of a monopole with moment (26.24b):

$$\frac{\mathrm{d}f}{\mathrm{d}z} \equiv \{v^*, \ -g^*, -E^*, -H^*\}, \quad P_0 \equiv \left\{Q - i\Gamma, \ m\mathrm{G}, \ -\frac{e}{\varepsilon}, \ -\frac{iJ}{c}\right\}, \tag{26.24a,b}$$

The negative sign for the gravity, electric, and magnetic potentials, in contrast with the positive sign for the velocity potential in (26.24a), have been taken into account in the (i) and (ii) analogies. The sign convention affects (does not affect) the potentials (26.24a) [the fields (26.24b)]. The two types of potential fields (26.24a,b), viz. irrotational (i) and solenoidal (ii) are in agreement with comparison of the potential flow with: (i) the gravity field (18.20a,b); (ii) the electrostatic field (24.11a,b); (iii) the magnetostatic field (26.11a,b). These analogies are detailed further in: (i) Table 24.1 in the case of electro and magnetostatics; (ii) Table 28.1 adding the gravity field, and irrotational, incompressible and potential flow; (iii) Table 32.1 adding steady heat conduction.

As for the monopole (28.18b,c), the multipole representations in terms of moments apply to all fields (hydrodynamic, gravity, electrostatic, and magnetostatics), viz. they are obtained (Section 12.7) applying $(-\mathrm{d}/\mathrm{d}z)^n$ to the monopole (26.14a,b) and replacing the current J by the dipole moment; this leads to *the complex potential (26.25a) and conjugate field (26.25b) due a multipole distribution of order* n *with moment* $\mathrm{p_n}(\zeta)$:

$$n = 1, 2, \dots: \quad f_m(z) = -\frac{i}{2\pi c} \int_D p_n(\zeta) (z - \zeta)^{-n} \, \mathrm{d}\zeta, \tag{26.25a}$$

$$n = 1, 2, \dots: \quad H^*(z) = -\frac{\mathrm{d}f_m}{\mathrm{d}z} = \frac{in}{2\pi c} \int_D p_n(\zeta) (z - \zeta)^{-n-1} \, \mathrm{d}\zeta; \tag{26.25b}$$

the boundary conditions (26.9a,b) for the interface electric current become:

$$0 = [B_n] = [\mu H_n], \quad [H_t] = \frac{(-)^n}{c} \frac{\mathrm{d}^n q_n}{\mathrm{d}x^n}, \tag{26.26a,b}$$

since a multipole of order n corresponds (Sections 12.8 and 12.9) the application of the operator $-\mathrm{d}/\mathrm{d}x$ to a monopole n times. This will be confirmed for a particular infinite dipole distribution on a plane (Section 26.9). The case $n = 0$ in (26.26a,b) reduces to (26.9a,b) with interface monopole $q_0 \equiv \vartheta$ coinciding with the electric current density per unit length ϑ. The following notation consistent with Table 28.1 is used: (i) P_n for the moment of a line-multipole of order n per unit length orthogonal to the ζ-plane in (26.31a,b); (ii) $p_n(\zeta)$ for the multiplole density per unit area of the ζ-plane in (26.25a,b); (iii) $q_n\,(s)$ for the multipole density per unit length of a curve in the boundary condition (26.26b).

26.5 Current near Conducting or Insulating Plane

Considering an electric current J at a distance a from a wall, the conjugate magnetic field is given (26.13b) for identical (+ sign) [opposite (−sign)] image by:

$$H_{\pm}^*(z) = i\frac{J}{2\pi c}\left[(z - ia)^{-1} \pm (z + ia)^{-1}\right], \tag{26.27a,b}$$

that simplify:

$$H_+^*(z) = \frac{iJ}{\pi c} \frac{z}{z^2 + a^2}, \qquad H_-^*(z) = -\frac{J}{\pi c} \frac{a}{z^2 + a^2}. \tag{26.28a,b}$$

The wall ($y = 0$, $z = x$) is insulating (16.9b) if the magnetic field is normal, that is, the tangential component or real part of H_* is zero; that is the case (26.28a) for a H_+^* equal image; *whereas an equal image current corresponds to an insulating wall, the opposite image current (26.28b) leads to a tangential magnetic field that specifies the* **surface electric current** *distribution on the conducting wall:*

$$\vartheta(x) = cH_x^-(x, 0) = -\frac{Ja/\pi}{x^2 + a^2}; \tag{26.29}$$

the surface current is opposite to the original current, peaks at the closest point, and leads to a total current $-J$, as in (24.21a–e) for the induced electric charge in a conductor. The surface electric current arises from the boundary condition (26.9b) for a zero magnetic field on the other side of a perfect conductor (Note 26.2).

The complex conjugate magnetic force on the boundary is minus the force of the image on the line-current, and is given by (26.21a) ≡ (26.30a):

$$F_{\pm}^{*} = i\frac{\mu J}{c}\left[H_{+}^{*} - i\frac{\mu J}{2\pi c}\frac{1}{z-ia}\right]_{z=ia} = i\frac{\mu J}{c}\left[\pm\frac{iJ}{2\pi c}\frac{1}{z+ia}\right]_{z=ia}$$
$$= i\frac{\mu J}{c}\frac{\pm iJ}{2\pi c}\frac{1}{2ia} = \pm\frac{i\mu J^2}{4\pi c^2 a} = -F_{\pm}, \qquad (26.30\text{a,b})$$

where in (26.30b) only the second term on the r.h.s. of (26.27b) appears because: (i) the line-current does not exert a force on itself; (ii) the force on the line-current is entirely due to its image. Thus *the line-current exerts on insulating (26.27a; 26.28a) [conducting (26.27b; 26.28b)] wall a normal repulsive (attractive) force that is, upper − (lower +) sign for* $F_{\pm}$ *in (26.30b)*. Generalizing (26.27a) like (12.59a) to *a magnetic multipole of moment* P_n, *the complex conjugate magnetic field due to an insulating* H_+*(conducting* H_-*) wall is:*

$$H_{\pm}^{*}(z) = i\frac{nP_n}{2\pi c}\left[(z-ia)^{-n-1} \pm (z+ia)^{-n-1}\right]; \qquad (26.31\text{a,b})$$

it corresponds in the far-field (26.32a) to:

$$|z|^{n+1} \gg a^{n+1}: \quad H_{+}^{*}(z) = i\frac{nP_n}{\pi c}z^{-n-1}, \qquad (26.32\text{a,b})$$

$$H_{-}^{*}(z) = -n(n+1)\frac{P_n}{\pi c}az^{-n-2} = i(n+1)\frac{P_{n+1}}{2\pi c}z^{-n-2}, \quad P_{n+1} \equiv i2naP_n, \qquad (26.33\text{a,b})$$

a double multipole (26.32b) [multipole of higher order (26.33a) with moment (26.33b)]. The passage from (26.31a,b) to (26.33a,b) is similar to (24.27; 24.28a,b).

In (26.31a,b) the multipole moment was assumed to be real, corresponding to a horizontal multipole; the general case of an oblique multipole, when the multipole moment is complex, is similar to the electrostatic case (24.23–24.26a,b). The induced-electric charges (24.32) are similar to the interface electric currents:

$$\vartheta(x) = cH_x^{-}(x,0) = i\frac{nP_n}{2\pi}\left[(x-ia)^{-n-1} - (x+ia)^{-n-1}\right]$$
$$= -\frac{nP_n}{\pi}\left(x^2+a^2\right)^{-n-1}\operatorname{Im}\left\{(x+ia)^{n+1}\right\}, \qquad (26.34\text{a})$$

where a real dipole moment is taken; the complex case would be treated similarly, viz. using the binomial theorem (25.38):

$$\vartheta(x) = -\frac{nP_n}{\pi}\left(x^2+a^2\right)^{-n-1}\operatorname{Im}\left\{\sum_{k=0}^{n+1}\binom{n+1}{k}(ia)^k x^{n+1-k}\right\}$$
$$= -\frac{nP_n}{\pi}\left(x^2+a^2\right)^{-n-1}\sum_{p=0}^{\leq n/2}\binom{n+1}{2p+1}x^{n-2p}\operatorname{Im}\{(ia)^{2p+1}\}$$
$$= -\frac{nP_n}{\pi}\left(x^2+a^2\right)^{-n-1}\sum_{p=0}^{\leq n/2}\frac{(-)^p(n+1)!}{(2p+1)!(n-2p)!}a^{2p+1}x^{n-2p}, \qquad (26.34\text{b})$$

since only the odd powers are imaginary. Thus *the multipole of order n at a distance a from a conducting plane corresponds to an interface electric current distribution (26.34a,b); its* *value at the closest point $x = 0$ is zero for n odd and nonzero for $n = 2m$ even:*

$$n = 2m: \ \vartheta(0) = -\frac{nP_n}{\pi} a^{-2n-2} (-)^m a^{2m+1} = (-)^{m+1} \frac{2m}{\pi} P_{2m} a^{-2m-1}, \tag{26.35a}$$

$$\begin{aligned}\vartheta(0) = cH_-^*(0) &= i\frac{nP_n}{2\pi}\left\{(-ia)^{-n-1} - (ia)^{-n-1}\right\} \\ &= i\frac{nP_n}{2\pi} a^{-n-1}\left\{e^{i\pi(n+1)/2} - e^{-i\pi(n+1)/2}\right\} \\ &= -\frac{nP_n}{\pi} a^{-n-1} \sin\left[\frac{(n+1)\pi}{2}\right] = (-)^{m+1}\frac{2m}{\pi} P_{2m} a^{-2m-1}.\end{aligned} \tag{26.35b}$$

It can be calculated in two ways: (i) from the term in (26.34b) without the x^{n-2p} factor, that is, for $p = n/2 = m$ leading to (26.35a); (ii) from $H_-^*(0)$ in (26.31b) leading to (26.35b) that is nonvanishing only for even, $n = 2m$.

26.6 Image Electric Currents in a Corner

Identical image electric currents in a corner of angle (26.36a) placed on the diagonal $\varphi = \beta/2 = \pi/n$ at a distance a from the vertex lead (16.72a,b) to the complex potential (26.36b) and conjugate magnetic field (26.36c):

$$\beta = \frac{2\pi}{n}: f_m^+(z) = -i\frac{J}{2\pi c} \log(z^n + a^n), \ H_+^*(z) = i\frac{nJ}{\pi c}\frac{z^{n-1}}{z^n + a^n}; \tag{26.36a–c}$$

the latter is normal to the walls that thus are insulating, and asymptotically (26.37a) multiply the field by n, viz.(26.37b,c):

$$|z|^n \gg a^n: \quad f_m^+(z) \sim -i\frac{nJ}{2\pi c}\log z, \quad H_+^*(z) \sim \frac{inJ}{2\pi c z}. \tag{26.37a–c}$$

Alternating images in a corner of angle $\varphi = \beta/2 = \pi/2n$ in (26.38a) at a distance, a, from the origin, lead (16.79a,b) to the complex potential (26.38b) and conjugate magnetic field (26.38c):

$$\beta = \frac{\pi}{n}: \qquad f_m^-(z) = -i\frac{J}{2\pi c}\log\left(\frac{z^n - ia^n}{z^n + ia^n}\right), \tag{26.38a,b}$$

$$H_-^*(z) = i\frac{nJ}{2\pi c} z^{n-1}\left[(z^n - ia^n)^{-1} - (z^n + ia^n)^{-1}\right] = -\frac{nJ}{\pi c}\frac{a^n z^{n-1}}{z^{2n} + a^{2n}}; \tag{26.38c}$$

the magnetic field (26.38c) is parallel to the conducting wall, and specifies the surface electric current distribution:

$$\vartheta(x) = cH_x^-(x, 0) = -\frac{nJ}{\pi}\frac{a^n x^{n-1}}{x^{2n} + a^{2n}}; \tag{26.39}$$

it vanishes at the closest point and at infinity, peaks (24.32) as in (24.33a–c) and has total value $-J$ as in (24.34). Asymptotically (26.40b) the magnetic field (26.38c) corresponds to:

$$|z|^{2n} >> a^{2n}: \ H_-^*(z) \sim -\frac{nJ}{\pi c} a^n z^{-n-1} = \frac{inP_n}{2\pi c} z^{-n-1}, \quad P_n \equiv i2Ja^n, \tag{26.40a–c}$$

a multipole (26.40b) of moment (26.40c). Thus *a line-current on the diagonal at the distance a from the vertex of a corner, with angle (26.36a) [(26.38a)] and insulating (conducting) walls, creates a magnetostatic field with complex potential (26.36b) [(26.38b)] and conjugate field (26.36c) [(26.38c)] that scales asymptotically (26.37a) [(26.40a)] as n times the current (26.37b,c) [a multipole (26.40b) of order n with moment (26.40c)]. In the conducting case the surface current distribution on the wall is (26.39), and has the properties (24.33a–c; 24.34).*

26.7 Cylinder in a Magnetic Field or near a Line-Current

The insertion of a insulating (conducting) cylinder of radius a is specified by the first circle theorem (24.47) for any potential field, for example, a uniform magnetic field (Subsection 26.7.1) or the field of a line-current (Subsection 26.7.2).

26.7.1 Cylinder in a Uniform External Magnetic Field

The complex potential (26.41a) [conjugate magnetic field (26.41b)] of a cylinder of radius a in a uniform external magnetic field H_0 is specified by:

$$f_m^{\pm}(z) = -H_0\left(z \mp \frac{a^2}{z}\right), \quad H_{\pm}^*(z) = H_0\left(1 \pm \frac{a^2}{z^2}\right). \tag{26.41a,b}$$

The corresponding scalar potential (26.42a) and field function (26.42b):

$$\Phi_m^{\pm}(r,\, \varphi) = -H_0\left(r \mp \frac{a^2}{r}\right)\cos\varphi, \quad \Psi_m^{\pm}(r, \varphi) = -H_0\left(r \pm \frac{a^2}{r}\right)\sin\varphi, \tag{26.42a,b}$$

$$H_r^{\pm}(r,\, \varphi) = H_0\left(1 \pm \frac{a^2}{r^2}\right)\cos\varphi, \quad H_{\varphi}^{\pm}(r,\, \varphi) = -H_0\left(1 \mp \frac{a^2}{r^2}\right)\sin\varphi, \tag{26.43a,b}$$

and polar components (26.43a,b) of the magnetic field show that: (i) the upper sign for that the cylinder is an equipotential (26.44a) and corresponds to an insulator, that is, zero tangential magnetic field:

$$\Phi_m^{+}(a, \varphi) = 0 = H_{\varphi}^{+}(a,\, \varphi), \quad \Psi_m^{+}(a,\, \varphi) = -2H_0 a\sin\varphi, \quad H_r^{+}(a,\, \varphi) = 2H_0\cos\varphi; \tag{26.44a–d}$$

(ii) the lower sign for that the cylinder is a field line (26.45a) corresponds to a conductor, with the nonzero tangential magnetic field:

$$\Psi_m^{-}(a,\, \varphi) = 0 = H_r^{-}(a,\, \varphi), \ \Phi_m^{-}(a,\, \varphi) = -2H_0 a\cos\varphi, \ H_{\varphi}^{-}(a,\, \varphi) = -2H_0\sin\varphi = \frac{\vartheta(\varphi)}{c}, \tag{26.45a–d}$$

specifying the interface electric currents (26.45d). Thus *an insulating (conducting) cylinder of radius a in an external uniform magnetic field has complex potential (26.41a) leading to scalar potential (26.42a) and field function (26.42b); the complex magnetic field (26.41b) has polar components (26.43a,b) with upper (lower) signs, leading to (26.44a–d) [(26.45a–d)] on the cylinder.*

Concerning the latter alone, that is, *the electric surface current (Figure 26.1) on a conducting cylinder in an uniform external magnetic field (26.45d) vanishes on axis, and*

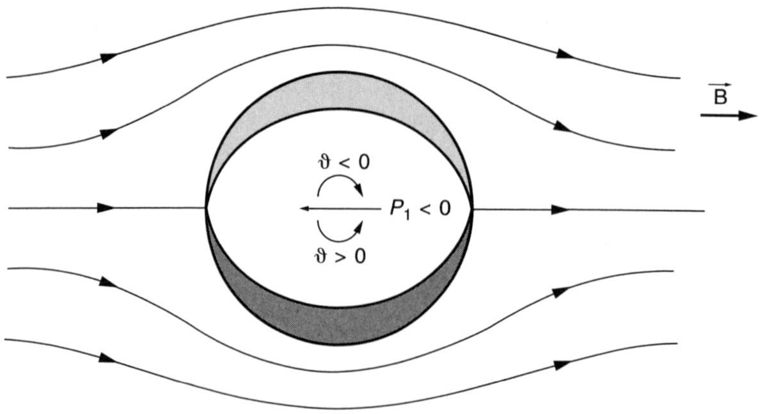

FIGURE 26.1
The introduction of a conducting cylinder in a uniform magnetic (electric) field [Figure 26.1 (24.3)] implies a distribution of surface electric currents (charges) such that the total magnetic (electric) field is everywhere tangent (orthogonal) to the cylinder. The positive and negative electric currents (charges) concentrate at opposite ends of the cylinder in the direction through the axis orthogonal to (along) the external magnetic (electric) field. The total electric current (charge) is zero and there is a dipole moment along the external magnetic (electric) field. The introduction of a conducting (insulating) cylinder in an external magnetic or electric field is equivalent in both cases to a dipole parallel (antiparallel) to the external field. Besides the preceding analogies and differences between the electro (magneto) static field due to electric charges (currents), in the case of a conductor: (i) a nearby electric charge induces a distribution of surface electric charges, and the dielectric displacement is zero in the interior; (ii) a nearby electric current does not induce a distribution of surface electric currents, but such a fictitious electric current distribution is needed to match a nonzero (zero) tangential magnetic field outside (inside).

peaks in the transverse direction with opposite signs; the total current is zero (26.46a) and its moment horizontal (26.46b) and equal (26.46c) to:

$$\int_0^{2\pi} \vartheta(\varphi)\, a\, \mathrm{d}\varphi = -2H_0 ca \int_0^{2\pi} \sin\varphi\, \mathrm{d}\varphi = 0, \tag{26.46a}$$

$$\int_0^{2\pi} x\vartheta(\varphi)\, a\, \mathrm{d}\varphi = -2H_0 ca^2 \int_0^{2\pi} \sin\varphi\cos\varphi\, \mathrm{d}\varphi = 0, \tag{26.46b}$$

$$\int_0^{2\pi} y\vartheta(\varphi)\, a\, \mathrm{d}\varphi = -2H_0 ca^2 \int_0^{2\pi} \sin^2\varphi\, \mathrm{d}\varphi = -2\pi c H_0 a^2 \equiv P_1, \tag{26.46c}$$

the dipole moment in (26.46c) ≡ (26.47b) in (26.41b) ≡ (26.47a):

$$H_-^*(z) - H_0 = -\frac{H_0 a^2}{z^2} = \frac{P_1}{2\pi c z^2}, \quad P_1 \equiv -2H_0 a^2 \mathrm{c}. \tag{26.47a,b}$$

A conducting cylinder in a uniform external electric (magnetic) field has Figure 24.3 (26.1) induced-electric charges (surface electric currents) along (across) the field, leading to a horizontal dipole moment, (24.43c) [(26.47b)] along to the field in both cases. In both cases there is no force and no torque on the cylinder, because: (i) the electric (magnetic) force on the colocated, identical, and opposite electric charges (currents) cancel; (ii) the dipole axis is parallel to at the uniform external field (Section 28.2). In the latter case of a conductor it is possible to add (26.13a,b) a line-current along the axis of the cylinder:

$$f(z) = f_-(z) - i\frac{J}{2\pi c}\log z = -H_0\left(z + \frac{a^2}{z}\right) - i\frac{J}{2\pi c}\log z, \tag{26.48a}$$

$$H^*(z) = H_-^*(z) - \frac{iJ}{2\pi cz} = H_0\left(1 - \frac{a^2}{z^2}\right) + \frac{iJ}{2\pi cz}; \tag{26.48b}$$

this preserves the boundary condition (26.45b) ≡ (26.49a):

$$H_r(a,\ \varphi) = 0, \quad \vartheta(\varphi) = cH_\varphi(a, \varphi) = -\frac{J}{2\pi a} - 2H_0\sin\varphi, \tag{26.49a,b}$$

and adds a constant term to the current (26.49b) relative to (26.45d).

26.7.2 Line-Current near an Insulating or Conducting Cylinder

A cylinder of radius a near a line-current J at a distance b corresponds to the complex potential (26.50a) [conjugate magnetic field (26.50b)] specified by (26.13a,b) and the circle theorem (24.47) viz.:

$$f_m^{\pm}(z) = -i\frac{J}{2\pi c}\left[\log(z - b) \pm \log\left(\frac{a^2}{z} - b\right)\right], \tag{26.50a}$$

$$H_{\pm}^*(z) = i\frac{J}{2\pi c}\left[\frac{1}{z - b} \pm \frac{a^2}{z(bz - a^2)}\right]. \tag{26.50b}$$

On the cylinder (24.49a) the arguments of the logarithms in (26.50a) have the same modulus (opposite phases), and thus the term in square brackets is imaginary (real) for the lower (upper) *sign*− (+) and thus f_- (f_+) is real (imaginary), so the cylinder is a the field line $\Psi = 0$ (equipotential $\Phi = 0$), corresponding to a conductor (insulator). Thus *a line-current* J *at a distance* b *from the center of an insulating (conducting) cylinder of radius,* a, *has complex magnetostatic potential (26.50a) and complex conjugate magnetic field (26.50b) with upper (lower) sign. In the conducting case the corresponding surface electric currents are*

$$\vartheta(\varphi) = cH_\varphi^-(a,\ \varphi) = -\frac{J}{\pi}\frac{b\cos\varphi - a}{a^2 + b^2 - 2ab\cos\varphi}; \tag{26.51}$$

these are largest (smallest) in modulus closest $\varphi = 0$ *(farthest* $\varphi = \pi$*) from the line-current,* as in (24.57a). The passage from (26.50b) to (26.51) is similar to (24.58a,b); similarly (24.56) is relevant in the far-field as shown next.

In both the conductor and insulator cases the far-field is due to leading order to the electric current alone:

$$|z|^2 \gg b^2: \quad H_{\pm}^*(z) = \frac{iJ}{2\pi cz}\left(1 + \frac{b \pm a^2/b}{z}\right), \tag{26.52}$$

because there are two opposite images on the cylinder:

$$\log\left(\frac{a^2}{z} - b\right) = \log\left[-\frac{b}{z}\left(z - \frac{a^2}{b}\right)\right] = \log\left(z - \frac{a^2}{b}\right) - \log z + \log(-b); \tag{26.53}$$

these appear at next order only as a dipole of moment (26.54a):

$$P_1 = J\left(b \pm \frac{a^2}{b}\right): \quad H_{\pm}^*(z) = \frac{iJ}{2\pi cz} + \frac{iP_1}{2\pi cz^2} + O\left(\frac{b^2}{z^3}\right), \tag{26.54a,b}$$

in the second term of (26.52) ≡ (26.54b); the dipole moment (26.54a) is horizontal, and for the conductor (insulator) is the distance from the external current to the reciprocal point on the same (opposite) side of the origin, as in (24.55c). The complex conjugate magnetic force exerted on the cylinder by the line-current is minus that due to its image, that the second term on the r.h.s. of (26.50b) evaluated at $z = b$*, that is, it is,*

$$F_{\pm}^* = \frac{i\mu J}{c}\frac{\pm iJa^2}{2\pi cb\left(b^2 - a^2\right)} = \mp\frac{\mu J^2}{2\pi c^2}\frac{(a/b)^2}{b - a^2/b} = F_{\pm}, \tag{26.55}$$

so that there is attraction (repulsion) on a conducting (insulating) cylinder, that is, lower + (upper −) sign in (26.55); the force changes sign if the line-current is inside the cylinder, so that it is always a force of attraction (repulsion) for a conductor (insulator). The force (26.55) involves: (i) the inverse of the distance from the line-current at b *to its image at* a^2/b*; (ii) a factor* $(a/b)^2$ *smaller for line charge farther* b *from the center of the cylinder of radius* a*. The expression for the force (26.55), as well as those for complex potential (26.50a), conjugate magnetic field (26.50b), their asymptotic forms (26.52) ≡ (26.54a, b) and surface electric currents (26.51), apply to a line-current at distance,* b*, from the center of the cylinder of radius* a*, regardless of whether it lies inside* $b < a$ *or outside* $b > a$*.*

26.8 Current near Cylindrical Magnetic Interface

An electric line-charge e in (Section 24.9) [line-current J in (Section 26.8)] near a plane (Section 24.9) or cylindrical (Section 26.8) interface between media with distinct electrical permittivity (magnetic permeability), induces: (i) as concerns the first medium, an image charge e' (current J') at the reciprocal point (24.60a,b); (ii) in the second medium, a virtual charge e'' (current J'') at the position of the original (24.61a,b). The line-charge near a plane interface was considered before (in Section 24.9), and the line-current near a cylindrical interface is considered next (Section 26.8); the other two combinations, viz. line-charge near a cylindrical interface and line-current near a plane interface can be considered using similar methods. Since the dielectric permittivity varies much more than the magnetic permeability (Sections 26.1 and 26.3), the problem is of greater practical interest for the electric rather than for the magnetic field; the case of the magnetic field does highlight both similarities and differences with the electric field (Table 26.1). The calculation of electric (Section 24.9) [magnetic (Section 26.8)] fields near plane (Section 24.9) [cylindrical (Section 26.8)] interfaces is generally similar with differences of detail (Subsection 26.8.5). The use of real, image, and virtual electric currents (Subsection 26.8.1) includes both conducting and insulating walls (Subsection 26.8.3) and specifies the forces on the walls and effective dipole moment in the far-field (Subsection 26.8.4). The main difference between the plane (Section 24.9) [cylindrical (Section 24.8))] interfaces lies on the image on the plane being replaced by the reciprocal point (Subsection 26.8.2) on the cylinder; the cylindrical case leads to two opposite cancelling images (subsection 26.7.2).

26.8.1 Real, Image, and Virtual Electric Currents

An electric line-current J at a distance b from the center of a cylindrical interface of radius a between media of magnetic permeability $\mu_1\,(\mu_2)$ for $r > a\,(r < a)$, leads to a complex

TABLE 26.1
Comparison of Electro- and Magnetostatics

Statics	Electro	Magneto
In vacuo Equation type Representation by existence of	Electric field $\nabla \wedge \vec{E} = 0$ Irrotational $\vec{E} = -\nabla\Phi$ Scalar potential	Magnetic induction $\nabla \cdot \vec{B} = 0$ Solenoidal $\vec{B} = \nabla \wedge \Psi$ Field function (2–D)
Electric Field Equation	Charge density: q Electric displacement $\nabla \cdot \vec{D} = q$	Current density: $\vec{j}$ Magnetic field $\nabla \wedge \vec{H} = \vec{j}/c$
Boundary Normal Tangential	Charge : σ $[D_n] = \sigma$ $[E_t] = 0$	Current : ϑ $[B_n] = 0$ $[H_t] = \vartheta/c$
Constitutive parameter	Dielectric permittivity: ϵ	Magnetic permeability: μ
Constitutive equation Poisson equation	$\vec{D} = \varepsilon\vec{E}$ $\nabla^2\Phi = -q/\varepsilon$	$\vec{B} = \mu\vec{H}$ $\nabla^2\Psi = -\mu j/c$
Insulator Charge/current Field Displacement/induction	$\varepsilon = 0$ $\sigma = 0$ $\vec{E} \neq 0$ $\vec{D} = 0$	$\mu = \infty$ $\vartheta = 0$ $\vec{H} = 0$ $\vec{B} \neq 0$
Conductor Charge/current Field Displacement/induction	$\varepsilon = \infty$ $\sigma \neq 0$ $\vec{E} = 0$ $D \neq 0$	$\mu = 0$ $\vartheta \neq 0$ $\vec{B} = 0$ $\vec{H} \neq 0$

Note: The comparison of electro- and magnetostatics concerns: (i) the electric field (magnetic induction) vectors and potential (field) function: (ii) the electric charge (current) and electric displacement (magnetic field) vectors; (iii) the dielectric permittivity (magnetic permeability) as the constitutive parameter in the Poisson equation; (iv/v) the extreme cases of a conductor/insulator.

conjugate magnetic field specified (Figure 26.2) by: (i) in the medium (26.56a) of the line-current J by the superposition with the **image current** J' on the circle (24.47) as in (26.50b), viz. (26.56b):

$$r > a: \quad H_1^*(z) = \frac{i}{2\pi c}\left[\frac{J}{z-b} + \frac{J'a^2}{z\,(b\,z - a^2)}\right]; \tag{26.56a,b}$$

$$a < r: \quad H_2^*(z) = \frac{i}{2\pi c}\,\frac{J''}{z-b}, \tag{26.57a,b}$$

in the medium (26.57a) without line-current J, the magnetic field is due (26.57b) to a **virtual current** J'' at the original point. The boundary conditions (26.9a,b) at the interface without surface electric currents (26.58a):

$$\vartheta = 0: \quad H_{1\varphi}\,(a,\varphi) = H_{2\varphi}\,(a,\,\varphi), \quad \mu_1\,H_{1r}\,(a,\,\varphi) = \mu_2\,H_{2r}\,(a,\varphi), \tag{26.58a–c}$$

are used (26.58b,c) next to relate the original J, image J' and virtual J'' currents.

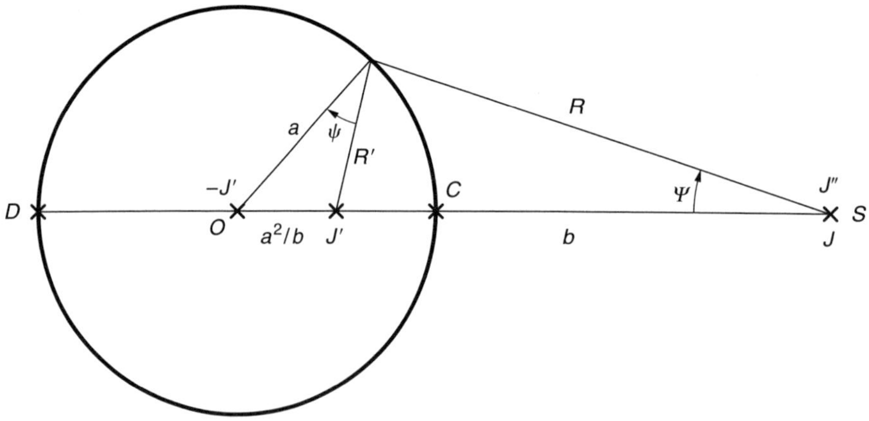

FIGURE 26.2
There is an analogy between the magnetic (electric) field of a line current (charge) near a cylindrical (plane) interface between media with distinct [Figure 26.3 (24.6)] magnetic permeabilities (dielectric permittivities). The image system consists of currents $J''(\pm J')$ at the original point (reciprocal point and center of the cylinder). The relation between the original J and image J', J'' currents is specified by the continuity of the magnetic field components orthogonal (tangent) to the cylinder, which holds always (in the absence of surface electric currents). The extreme cases of a conducting (insulating) cylinder correspond to zero $\mu_2 = 0$ (infinite $\mu_2 = \infty$) magnetic permeability of the second medium and opposite (identical) image current $J' = -J (J' = J)$. There is a continuous variation between these extremes, allowing an attraction $\mu_2 < \mu_1$ (repulsion $\mu_2 > \mu_1$) of the cylinder by the line current, and no force for identical media.

The components of the magnetic field on the interface are

$$H_{r1},\ H_{r2}\ (a,\ \varphi) = \frac{1}{2\pi c}\left\{\frac{J}{R} + \frac{J'a}{bR'},\ \frac{J''}{R}\right\}\sin\psi, \tag{26.59a,b}$$

$$H_{\varphi 1},\ H_{\varphi 2}\,(a,\ \varphi) = \frac{1}{2\pi c}\left\{\frac{J}{R} - \frac{J'a}{bR'},\ \frac{J''}{R}\right\}\cos\psi; \tag{26.60a,b}$$

where $R\,(R')$ is distance from an arbitrary point on the circle to the original current at $(b{,}0)$ [reciprocal point at $\left(a^2/b,\ 0\right)$] and ψ is the angle with the real axis (Figure 26.2). Substituting (26.59–26.60a,b) in the boundary conditions (26.58a,b) yields:

$$\frac{J}{R} - \frac{J'a}{bR'} = \frac{J''}{R}, \qquad \mu_1\left(\frac{J}{R} + \frac{J'a}{bR'}\right) = \mu_2\frac{J''}{R}. \tag{26.61a,b}$$

The magnetic filed $\vec{\mathrm{H}}\big(\vec{\mathrm{H}}''\big)$ due to the original J (virtual J'') line-current (Figure 26.3) is orthogonal to the line from the original current J to the arbitrary point P on the circle, in the positive direction; the magnetic field H' to the image line-current J' has components due to a line-current $J'\,(-J')$ at the reciprocal point (center). It can be seen that $\vec{\mathrm{H}}'$ has radial (tangential) component with the same (26.61b) [opposite (26.61a)] sign to $\vec{\mathrm{H}}$ and $\vec{\mathrm{H}}''$.

26.8.2 Reciprocal Point as Image on the Cylinder

In (26.61a,b) the distance from an arbitrary point (P in Figure 26.3) on the circle $(a,\ \varphi)$ to the original and virtual (image) current $J, J''\,(J')$ at $(b,\,0)$ [at the reciprocal point $\left(a^2/b,\ 0\right)$]

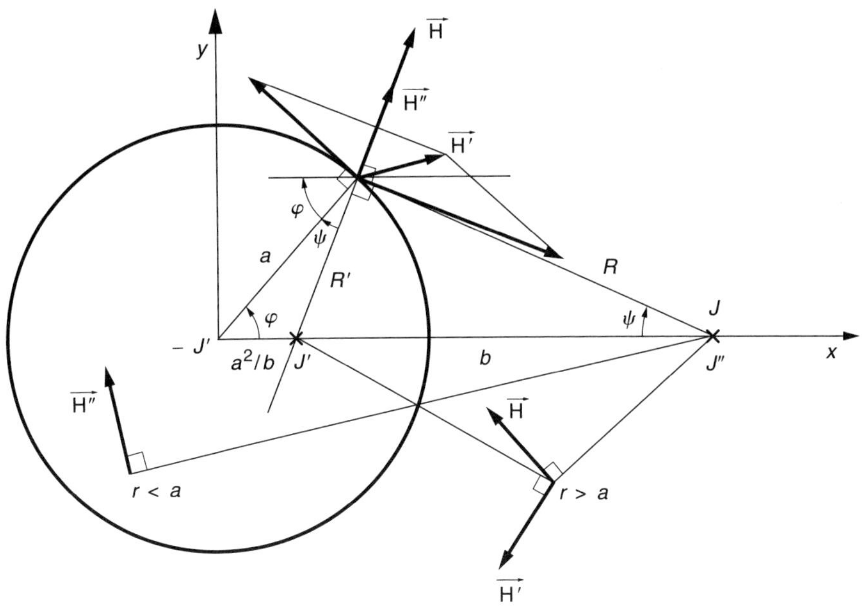

FIGURE 26.3
The image at the reciprocal point in a circle applies to: (i) the electric (magnetic) field of a line charge (current) near a cylinder [Figure 24.4 (26.2)]; (ii) the potential flow of a line monopole (dipole) near a cylinder [Figure 28.12 (28.13)]. The original (reciprocal) point have the property that the ratio of the distance from them of any point on the circle $R'/R = b/a$ equals the distance of the original point from the center of the circle divided by the radius of the circle. This can be checked most readily for the two points on the circle farthest D (closest C) to the reciprocal points $(a \pm a^2/b)/(b \pm a) = a/b$. The magnetic fields $\vec{\mathrm{H}}(\vec{\mathrm{H}}'')$ due to the original J (virtual J'') line-current are orthogonal to the position vector from $(b, 0)$ to the circle; the magnetic field H' is the sum of the contributions due to the opposite image line-currents $J'(-J')$ at the reciprocal point $(a^2/b, 0)$ [center (0,0)].

is denoted by $R\,(R')$ in (26.62a) [(26.62b)]:

$$R^2 \equiv \left|ae^{i\varphi} - b\right|^2 = (b - a\cos\varphi)^2 + (a\sin\varphi)^2 = a^2 + b^2 - 2ba\cos\varphi, \tag{26.62a}$$

$$R'^2 \equiv \left|ae^{i\varphi} - \frac{a^2}{b}\right|^2 = \left(a\cos\varphi - \frac{a^2}{b}\right)^2 + (a\sin\varphi)^2 = a^2 + \frac{a^4}{b^2} - 2\frac{a^3}{b}\cos\varphi. \tag{26.62b}$$

These are related by (26.63a):

$$a^2R^2 = a^4 + b^2a^2 - 2ba^3\cos\varphi = b^2R'^2, \quad \frac{R'}{R} = \frac{a}{b}, \tag{26.63a,b}$$

that is equivalent to (23.63b). This leads to the **reciprocal point theorem** *(26.63b) in Figure 26.2: the distances from an arbitrary point on a circle of radius a to an external point $(b, 0)$ in (26.62a) [the reciprocal point $\left(a^2/b,\ 0\right)$ relative to the circle of radius, a, in (26.62b)] are in a constant ratio (26.63b), equal to a/b. This can be checked in particular for the points on the circle that are closest (farthest) from source:*

$$R = b \mp a, \quad R' = a \mp \frac{a^2}{b}: \quad \frac{R'}{R} = \frac{a}{b}, \tag{26.64a–c}$$

corresponding to C (D) in Figure 26.2, with upper (lower) sign in (26.64a,b), both leading to the same result in (26.64c) $\equiv$ (26.63b).

26.8.3 Homogeneous Medium and Conducting or Insulating Walls

Substituting the reciprocal point theorem (26.63b) in (26.61a,b) leads to

$$J - J' = J'', \quad J + J' = \frac{\mu_2}{\mu_1} J''; \tag{26.65a,b}$$

these specify the image J' and virtual J'' currents in terms of the original current J, viz.:

$$J' = J\,\frac{\mu_2 - \mu_1}{\mu_1 + \mu_2}, \quad J'' = J\,\frac{2\mu_1}{\mu_1 + \mu_2}, \tag{26.66a,b}$$

and magnetic permeabilities of the media on the two sides of the cylindrical interface. As for the plane interface between two dielectrics (24.69a–c; 24.70–24.71a–d), two particular cases are considered beside the degenerate case of identical magnetic permeabilities. In the degenerate case of identical media (26.67a)

$$\mu_1 = \mu_2: \quad J' = 0, \quad J'' = J, \tag{26.67a–c}$$

when there is no image current (26.67b) or interface, and the virtual current coincides with the real current (26.67c), so that the magnetic field (26.56a,b) $\equiv$ (26.57a,b) $\equiv$ (26.13b) is that of a line-current in free space. One extreme case is the interior of cylinder with zero magnetic permeability (26.68a), implying zero magnetic induction (26.68b), though the magnetic field may be nonzero:

$$\mu_2 = 0: \quad B_2(z) = 0 \neq H_2(z); \quad J' = -J; \quad J'' = 2J, \tag{26.68a–d}$$

the image current is opposite (26.68c), corresponding to an conducting cylinder (26.56b) $\equiv$ [(26.50b) with $-$ sign], and the virtual current is double (26.68d) in agreement with the nonzero magnetic field. The opposite extreme case is when magnetic permeability of the interior of the cylinder is infinite (26.69a), so the magnetic field is zero (26.69b), though the magnetic induction may not be:

$$\mu_2 = \infty: \quad H_2(z) = 0 \neq B_2(z), \quad J' = J, \quad J'' = 0, \tag{26.69a–d}$$

the image current is the identical (26.69c) corresponding to an insulating cylinder (26.56b) $\equiv$ (26.50b) with +sign, and virtual current vanishes (26.59d) in agreement with the nonexistence of magnetic field.

26.8.4 Magnetic Force on the Wall and Effective Dipole Moment in the Far-Field

A line-current J *at a distance* b *from the center of a cylindrical interface of radius* a *between media with magnetic permeability* μ_1 (μ_2), *creates a magnetic field outside (26.56a,b) [inside (26.57a,b)] corresponding to an image (26.66*a*) [virtual (26.66*b*)] current; the far-field (26.70a) in the first medium outside the cylinder corresponds to the original line-charge plus a dipole (26.70b):*

$$|z|^2 \gg b^2: \quad H_1^*(z) \sim \frac{iJ}{2\pi c z} + \frac{iP_1}{2\pi c z^2}, \tag{26.70a,b}$$

$$P_1 \equiv Jb\frac{\mu_1\left(1 - a^2/b^2\right) + \mu_2\left(1 + a^2/b^2\right)}{\mu_1 + \mu_2}, \tag{26.70c}$$

whose ***effective dipole moment*** *(26.70c) simplifies to* Jb *for line-charge at large distance from the cylinder* $a^2 << b^2$. The result (26.70a–c) follows from (26.56b):

$$\frac{2\pi c}{iJ} H_1^*(z) - \frac{1}{z} + \frac{b}{z^2} + O\left(\frac{b^2}{z^3}\right) = \frac{J'}{J} \frac{a^2/\left(bz^2\right)}{1 - a^2/\left(bz\right)} = \frac{\mu_2 - \mu_1}{\mu_2 + \mu_1}\left[\frac{a^2}{bz^2} + O\left(\frac{a^4}{b^2 z^3}\right)\right]; \tag{26.71}$$

this simplifies to (26.70b) with dipole moment (26.72a) ≡ (26.70c);

$$P_1 = Jb\left(1 + \frac{a^2}{b^2}\frac{\mu_2 - \mu_1}{\mu_2 + \mu_1}\right); \; O\left(\frac{a^4}{b^2 z^3}\right) \sim O\left(\frac{b^2}{z^3}\right) O\left(\frac{a^4}{b^4}\right), \tag{26.72a,b}$$

since the current lies outside the cylinder $b > a$ the term (26.72b) can be neglected together with $O\left(\mathrm{b}^2/z^3\right)$, so the asymptotic condition (26.70a) is sufficient to omit both. The complex conjugate magnetic force exerted by the line-current on the cylindrical interface is minus the force due to its images, and is specified by the second term on the r.h.s. of (26.56b) evaluated at the line-current $z = b$, viz.:

$$\tilde{F}_m^* = \frac{i\mu_1 J}{c}\frac{iJ'}{2\pi c}\frac{a^2/b}{b^2 - a^2} = -\frac{\mu_1 J^2}{2\pi c^2}\frac{\left(a/b\right)^2}{b - a^2/b}\frac{\mu_2 - \mu_1}{\mu_2 + \mu_1} = \tilde{F}_m, \tag{26.73a,b}$$

where (26.66b) was used. *The force exerted by the line-current on the cylindrical interface is attractive (repulsive) if the first medium has larger (smaller) magnetic permeability than the second. It vanishes for identical media, and in the particular case* $\mu_2 = \infty$ $(\mu_2 = 0)$ *it reduces to (26.55) with upper* − *(lower* + *) sign, corresponding to an insulator (conductor) in (26.69a–d) [(26.68a–d)].*

26.8.5 Analogies and Differences between Electrostatic and Magnetostatic Fields

The image and virtual currents (26.66a,b) can be rewritten:

$$\{J', \, J''\} = \frac{J}{1/\mu_1 + 1/\mu_2}\left\{\frac{1}{\mu_1} - \frac{1}{\mu_2}, \, \frac{2}{\mu_2}\right\}, \tag{26.74a,b}$$

and thus are specified by the same equations as the image (virtual) electric charges (24.68a,b), exchanging the dielectric permittivity by the inverse of the magnetic permeability (26.75a) as in:

$$\frac{1}{\mu} \Leftrightarrow \varepsilon: \quad \nabla^2 \Phi_e = -\frac{q}{\varepsilon}, \quad \nabla^2 \Psi_m = -\frac{\mu j}{c}, \tag{26.75a–c}$$

the Poisson equations for the magnetic field function (26.6b) ≡ (26.75c) [electric potential (24.5b) ≡ (26.75b)]; the corresponding electro(magneto) static analogies *in the Poisson equations (26.75b) [(26.75b)]:*

$$\Phi_e \leftrightarrow \Psi_m, \quad q \leftrightarrow \frac{j}{c}, \quad \frac{1}{\varepsilon} \leftrightarrow \mu, \tag{26.75d–f}$$

concern: (i) the electrostatic potential and magnetostatic field function (26.75d); (ii) the electric charge and current densities (26.75e), the latter divided by the speed of light in vacuo to have the same dimensions; (iii) the inverse of the dielectric permittivity and the magnetic permeability (26.75f). The substitutions (26.75a) ≡ (26.75f) transform the (i) relations (26.65–66a,b; 26.67a–c; 26.69–70a–d) for electric currents near an interface between media with distinct magnetic permeabilities; (ii) to the relations (24.67–68a,b; 24.69a–c; 24.70–71a–d) for electric charges near an interface between media with distinct dielectric permeabilities. *The virtual and image electric charges (currents) are the same for plane (Section 24.9) or cylindrical (Section 26.8) interfaces between distinct media; for both types of interface the dielectric permittivity (magnetic permeability) is replaced by the inverse magnetic permeability (26.75a) ≡ (26.75f) when exchanging magnetic for electric fields.*

Applying the transformations (26.75e,f) to (24.76) [(26.73b)] it follows *that the magnetic (electric) force exerted by a line-current (charge) at a distance a from a plane interface (distance b from a cylindrical interface with radius a), between media with magnetic permeabilities* μ_1, μ_2 *(dielectric permittivities* $(\varepsilon_1, \varepsilon_2)$ *is given by (26.76a) [(26.76b)]:*

$$F_m^* = -i\frac{(J/c)^2}{2\pi a/\mu_1}\frac{1/\mu_1 - 1/\mu_2}{1/\mu_1 + 1/\mu_2} = -i\frac{\mu_1 J^2}{2\pi c^2 a}\frac{\mu_2 - \mu_1}{\mu_2 + \mu_1}, \tag{26.76a}$$

$$\tilde{F}_e^* = -\frac{e^2/\varepsilon_1}{2\pi}\frac{(a/b)^2}{b - a^2/b}\frac{1/\varepsilon_2 - 1/\varepsilon_1}{1/\varepsilon_2 + 1/\varepsilon_1} = -\frac{e^2}{2\pi\varepsilon_1}\frac{(a/b)^2}{b - a^2/b}\frac{\varepsilon_1 - \varepsilon_2}{\varepsilon_1 + \varepsilon_2}. \tag{26.76b}$$

Comparing the plane (cylindrical) interface both in the electrostatic (24.76) [(26.76b)] and magnetostatic (26.73b) [(26.76a)] cases the curvature effect is the same in the ratio between a line-monopole at a distance: (i) b from a plane interface, viz. (24.76; 26.76a) with a replaced by b; (ii) b from the center of a cylindrical interface of radius a in (26.76b; 26.73b):

$$\begin{aligned} i\frac{\tilde{F}_e^*}{F_e^*} &= -i\frac{\tilde{F}_e}{F_e} = \left|\frac{\tilde{F}_e}{F_e}\right| = \left|\frac{\tilde{F}_m}{F_m}\right| = -i\frac{\tilde{F}_m}{F_m} = i\frac{\tilde{F}_m^*}{F_m} \\ &= \frac{(a/b)^2}{b^2 - a^2/b} : \frac{1}{2b} = \frac{2a^2}{b^2 - a^2} = \frac{2}{b^2/a^2 - 1} \equiv k. \end{aligned} \tag{26.76c}$$

The curvature effect: (i) increases the force $k > 1$ *for close monopole* $a < b < a\sqrt{3}$*; (ii) cancels* $k = 1$ *for a monopole at a distance* $b = a\sqrt{3}$*; (iii) decreases the force for* $b > a\sqrt{3}$ *causing a decay* $\sim 2a^2/b^2$ *at large distance* $b^2 \gg a^2$*.*

The electric and magnetic and magnetostatic fields are compared in Table 26.1: (i) the Maxwell equations in vacuo show the first corresponding pair is the electric field (magnetic induction) that is irrotational (solenoidal), implying the existence of a scalar potential (field function in two dimensions); (ii) the second corresponding pair is the electric displacement (magnetic field) due to the electric charge (current density). The designations electric and magnetic field are a bit misleading, since these are not a corresponding pair; it is a historical error mixing (i) with (ii) that remains in posterity. The correct designation would have been magnetic field for $\vec{\mathrm{B}}$ (as electric field for $\vec{\mathrm{E}}$), and magnetic induction for $\vec{\mathrm{H}}$ (as electric displacement for $\vec{\mathrm{D}}$); the current designations are retained, as they are unlikely to change in the future. The boundary conditions (iii) follow from the preceding: (i) continuous tangential (normal) electric field (magnetic induction); (ii) jump of normal (tangential) electric displacement (magnetic field) equal to the interface electric charge (current). The properties of the medium (iv) are specified by the dielectric permittivity (magnetic permeability) that appear in inverse form in the Poisson equation, viz. dividing (multiplying) the electric charge (current density); this is due to their definition as the ratio of the electric displacement (magnetic induction) to the electric (magnetic) field that mixes the two pairs of corresponding vectors in (i,ii). This leads to the contrast seen in Sections 24.9 and 26.8, viz.: (v) an insulator does not support electric charges (currents), so $e = 0 \quad (J = 0)$, and the scalar potential (stream function) in the Poisson equation can be finite and nonzero only if $\varepsilon = 0$ ($\mu = \infty$) that corresponds to zero electric displacement (magnetic field), though the electric field (magnetic induction) may be nonzero; (vi) the reverse happens for a conductor that supports electric charges (currents), so that for a perfect conductor $\varepsilon = \infty$ ($\mu = 0$), and the electric field (magnetic induction) are zero, though the electric displacement (magnetic field) may be nonzero. Thus the **Faraday cage**, *that is, a conductor, excludes not only the electric field but also the magnetic induction.*

26.9 Infinite Magnetic Dipole Distribution

The magnetic (electric) field is generated by electric current (Sections 26.1–26.3) [charge (Sections 24.1–24.3)] distributions, as the gravity field is due to mass distributions (Sections 18.3–18.6) and the potential flow due to, source/sink (Section 12.4) or vorticity (Section 12.5) distributions. The source or forcing distributions that may have finite (Section 18.7) or infinite (Section 18.8) extent, they may be monopoles (26.14a,b) or multipoles (26.25a,b), along a curve (Figure 26.4a), for example, dipoles (Figure 26.4b) along the real axis (Figure 26.4c). A monopole source distribution along real axis was considered before (Section 18.8 and Example 20.20), and a dipole is considered next (Subsection 26.9.1) leading to a different type of normal discontinuity across the multipole distribution (Subsection 26.9.2).

26.9.1 Magnetic Field due to a Dipole Distribution on a Plane

The complex conjugate magnetic field is specified by:

$$H^*(z) = \frac{i}{2\pi c} \int\limits_{-\infty}^{+\infty} q_1(\zeta)\,(z-\zeta)^{-2}\,\mathrm{d}\zeta, \tag{26.77}$$

for a distribution of dipole currents with density $q_1(\zeta)$ along the real axis [$n = 1$ in (26.25b)]. A uniform dipole moment would lead to an infinite total moment, a dipolar moment (26.78a) decaying (Figure 26.4c) sufficiently fast:

$$q_1(x) = \frac{P_1 a/\pi}{x^2 + a^2}, \quad P_1 = \int\limits_{-\infty}^{+\infty} q_1(x)\,\mathrm{d}x, \tag{26.78a,b}$$

leads to a constant total dipole moment (26.78b), where (17.21b; 17.22a,b) was used. The conjugate magnetic field (26.77), is given by an integral:

$$H^*(z) = \frac{iP_1 a}{2\pi^2 c} \int\limits_{-\infty}^{+\infty} h(z,\,\zeta)\,\mathrm{d}\zeta, \quad h(z,\,\zeta) \equiv (\zeta - z)^{-2}\left(\zeta^2 + a^2\right)^{-1}; \tag{26.79a,b}$$

this can be evaluated by the residues of the integrand (26.79b), at its poles:

$$h_{(1)}(z,\,\pm i a) = \left\{\pm 2\,i\,a\,(\pm\,i\,a - z)^2\right\}^{-1}, \tag{26.80a}$$

$$h_{(2)}(z,\,z) = \lim_{\zeta\to z} \frac{\mathrm{d}\left\{\left(\zeta^2 + a^2\right)^{-1}\right\}}{\mathrm{d}\zeta} = -\frac{2z}{\left(z^2 + a^2\right)^2}, \tag{26.80b}$$

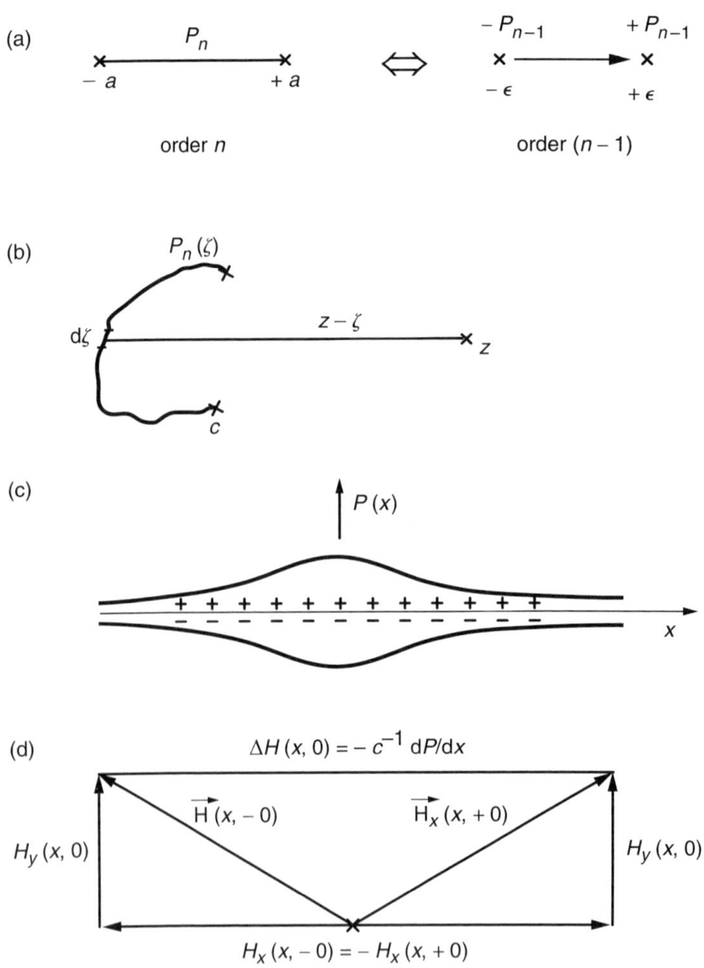

FIGURE 26.4
The multipole of order n is obtained (a) from two opposite multipoles of order $n-1$ reducing their distance $2\varepsilon \to 0$ and increasing the 2^{n-1}-multipole moment $P_{n-1} \to \infty$ so that the product $2P_{n-1}\varepsilon = P_n$ is the constant 2^n-multipole moment; this limiting process was applied in the transition of flow monopoles to dipoles and quadrupoles (Figures 12.3 and 12.6). An electric line-current multipole distribution (b) creates a magnetic field like a monopole mass distribution (Figures 18.3 and 18.5) creates a gravity field. A dipole distribution along the real line (c), like a mass distribution (Figure 18.3) corresponds in space to a distribution on a plane. For the magnetic (d) [gravity (Figure 18.5b)] field due to a dipole (monopole) electric current (mass) distribution along the real line the: (i) normal (tangential) component of the magnetic (gravity) field is continuous across the distribution; (ii) the tangential (normal) component of the magnetic (gravity) field is discontinuous, with: (ii-1) opposite values on the two sides; (ii-2) zero value on the distribution; (ii-3) a jump across the distribution proportional to the derivate of the surface dipole moment (proportional to the surface mass density). The properties (i, ii-1,2,3) apply to solenoidal (irrotational) fields, for example, the magnetostatic field/potential flow (gravity/electrostatic field/potential flow) due to a distribution of electric currents/vortices (masses/electric charges/sources-sinks).

that are (i) two simple poles (15.24b) at $\zeta = \pm\, i\, a$ with residues (26.80a); (ii) a double pole [(15.33b) with $n = 2$] at $\zeta = z$ in (26.80b).

Since (26.79b) meets the asymptotic condition (17.19) in all directions, the complex conjugate magnetic field (26.79a) is calculated using a path of integration closed in upper (lower) complex plane for an observer below $\operatorname{Im}(z) < 0$ (above $\operatorname{Im}(z) > 0$) the real axis, so as to leave only the pole $z = +\, i\, a$ $(z = -\, i\, a)$ inside, whose residue is multiplied by $\pm 2\,\pi\, i$:

$$\operatorname{Im}(z) <> 0: \quad H^*(z) = \frac{iP_1 a}{2\pi^2 c}\,(\pm 2\pi i)\, h_{(1)}(z,\ \pm ia) = \frac{iP_1}{2\pi c}\frac{1}{(z \mp ia)^2}. \tag{26.81}$$

For an observer on the real axis, the Cauchy principal value of the integral should be taken (Sections 17.8–17.9 and 18.9) adding to (26.81) also $\pm\,\pi\, i$ times the residue at the double pole that lies on the real axis and is enclosed by deformation of the contour:

$$\begin{aligned}\operatorname{Im}(z) = 0: \quad H^*(x) &= \frac{iP_1 a}{2\pi^2 c}\,(\pm\pi i)\left\{2h_{(1)}(x, \pm ia) + h_{(2)}(x,\ x)\right\} \\ &= \frac{iP_1}{2\pi c}\frac{1}{(x \mp ia)^2} \mp \frac{iP_1}{\pi c}\frac{ax}{(x^2+a^2)^2};\end{aligned} \tag{26.82a}$$

The latter simplifies to:

$$H^*(x) = \frac{iP_1}{2\pi c}\frac{1}{(x^2+a^2)^2}\left[(x \pm ia)^2 \mp 2iax\right] = \frac{iP_1}{2\pi c}\frac{x^2-a^2}{(x^2+a^2)} = -iH_y(x), \tag{26.82b}$$

showing that it is imaginary; thus the component of the magnetic field parallel to the electric current dipole distribution vanishes. This completes the calculation of the magnetic field outside (26.81) [inside (26.82b)] the distribution of electric current dipoles along the real axis.

26.9.2 Jump of the Normal Derivative across the Dipole Distribution

Separating real and imaginary parts in (26.81) leads to the Cartesian components of the magnetic field above $y > 0$ (below $y < 0$) the plane distribution of currents:

$$y >< 0: \{H_x,\ H_y\}(x, y) = \frac{P_1}{2\pi c}\left[x^2 + (y \pm a)^2\right]^{-2}\left\{2x\,(y \pm a)\,, (y \pm a)^2 - x^2\right\}; \tag{26.83a,b}$$

from (26.82) follows the magnetic field on the dipolar distribution:

$$H_x(x,\ 0) = 0, \quad H_y(x,\ 0) = \frac{P_1}{2\pi c}\frac{a^2 - x^2}{(x^2+a^2)^2}, \tag{26.84a,b}$$

showing that it is tangential. The passage from (28.81) to (26.83a,b) is:

$$\begin{aligned}\frac{2\pi c}{iP_1}(H_x - iH_y) &= [x + i\,(y \pm a)]^{-2} = \left[x^2 - (y \pm a)^2 + 2ix\,(y \pm a)\right]^{-1} \\ &= D\left[x^2 - (y \pm a)^2 - 2ix\,(y \pm a)\right],\end{aligned} \tag{26.85a}$$

$$\begin{aligned}\frac{1}{D} &\equiv \left|x^2 - (y \pm a)^2 + 2ix\,(y \pm a)\right|^2 \\ &= \left[x^2 - (y \pm a)^2\right]^2 + 4x^2\,(y \pm a)^2 = \left[x^2 + (y \pm a)^2\right]^2;\end{aligned} \tag{26.85b}$$

the corresponding calculation on the dipole distribution (26.82b) leads from (26.82a) to (26.84a,b).

The tangential magnetic field (26.83a) just above (below) the dipole distribution takes opposite positive (negative) values (26.86a); that is it is discontinuous:

$$H_x(x,\pm 0) = \pm\frac{p_1}{\pi c}\frac{ax}{\left(x^2+a^2\right)^2}, \quad H_y(x,\pm 0) = \frac{P_1}{2\pi c}\frac{a^2-x^2}{\left(x^2+a^2\right)^2}, \tag{26.86a,b}$$

whereas the normal component (26.86b) ≡ (26.84b) is continuous. Comparing with the magnetic field on the dipole distribution on the real axis, it follows that: (i) the vertical component (26.84b) ≡ (26.86b) is continuous, that is, takes the same value (26.87a):

$$H_y(x,+0) = H_y(x,0) = H_y(x,-0)\,; H_x(x,0) = \frac{H_x(x,+0)+H_x(x,-0)}{2} = 0, \tag{26.87a,b}$$

(ii) the horizontal component (26.84a) that is discontinuous (26.86a), takes the arithmetic mean of the values above and below (26.87b). The jump in the tangential component (26.86a) of the magnetic field across the dipolar current distribution:

$$\begin{aligned}\Delta H_x(x,0) \equiv H_x(x,+0) - H_x(x,-0) &= \frac{2P_1}{\pi c}\frac{ax}{\left(x^2+a^2\right)^2} \\ &= -\frac{P_1 a}{\pi c}\frac{\mathrm{d}\left\{\left(x^2+a^2\right)^{-1}\right\}}{\mathrm{d}x} = -\frac{1}{c}\frac{\mathrm{d}q_1}{\mathrm{d}x},\end{aligned} \tag{26.88}$$

is equal to minus the derivate of the dipolar moment (26.78a), divided by the speed of light in vacuum, in agreement with (26.26b). *Figure 25.4d shows the magnetic field near a surface dipole current distribution: (i) the normal component (26.83b) is continuous (26.84b) ≡ (26.86b) ≡ (26.87a); (ii) the tangential component (26.83a) is discontinuous (26.84a), viz. its value on the current distribution (26.84a) is the arithmetic mean (26.87b) of the values above and below (26.86a) and thus vanishes; (iii) the jump (26.88) relates to the multipole distribution (26.26b), viz. n = 1 for a dipole. The reverse applies to the gravity field due to a mass (Section 18.8 and Example 20.20) distribution, for that the tangential (normal) component is continuous (discontinuous) in Figure 18.5b. A distribution of flow sources/sinks or electric charges (vortices) would, like the gravity (magnetic) field, and any other irrotational (solenoidal) field [Section 18.1 (18.2)], lead to tangential continuity (discontinuity) and normal discontinuity (continuity), across the distribution, regardless of whether it is monopolar [multipolar]* for example, monopole in Section 18.8 and Example 20.20 (dipolar in Section 26.9).

Note 26.1. *Irrotational and Solenoidal Potential Fields:* The last statement reflects the two sets of analogies: (i) the electrostatic field due to electric charges in a dielectric (Chapter 24), the irrotational flow due to mass source/sinks (Chapters 12, 14, 16, and Section 18.1) and the gravity field due to masses (Chapter 18), noting that a sink corresponds to a negative charge and to a mass, since the field is directed inward, and a source corresponds to a positive charge (no gravity analogies since the field is directed outward); (ii) a magnetostatic field due to electric currents in a medium of given magnetic permeability (Chapter 26) is analogous to the incompressible flow due to vortices (Chapters 12, 14, 16, and Section 18.2). These two sets are the irrotational (i) [solenoidal (ii)] potential fields for that the divergence (curl) is zero. These analogies extend to boundary conditions, energies and forces (Sections 18.3–18.9), and apply to a variety of problems (Chapters 34, 36, and 38). The detailed differences between irrotational (solenoidal) potential fields appear in the case of: (i) electrostatics (magnetostatics) in Table 24.1; (ii) also for gravity fields, irrotational and incompressible flows in Table 28.1; (iii) also for steady heat conduction in Table 32.1.

Note 26.2. *Induced-Electric Charges and Interface Electric Currents:* When an electric charge is placed near a conductor it attracts opposite charges, creating an induced-electric charge distribution. When a current is placed near a perfect conductor it creates a magnetic field that does not penetrate the conductor; the discontinuity in the tangential magnetic field is equivalent to an interface electric current. Thus the "induced" electric current is to be understood as a boundary condition for the magnetic due to an electric current near a perfect conductor. If the conductivity is not infinite, or the resistivity is nonzero, the boundary condition without interface electric currents, states that the tangential component of the magnetic field must be continuous across the two sides of the imperfect conductor.

Conclusion 26: Conducting cylinder in an external magnetic field, showing (Figure 26.1) the transverse surface electric currents of opposite signs, leading to a dipole in the direction of the external magnetic field. A line current J near a cylindrical interface between media with distinct magnetic permeabilities gives rise (Figure 26.3) to: (i) to a virtual comment J'' at the same point; (ii) an image current J' at the reciprocal point plus $-J'$ at the center. The reciprocal point satisfies the reciprocal point theorem (Figure 26.2) concerning the constant ratio $R'/a = R/b$ of distances to an arbitrary point on the circle of the original multipole and its image. The magnetic field can be created (Figure 26.4) by a multipole (a) distribution (b) e.g., on an infinite plane (c) decaying fast enough to lead to a finite total dipole moment; the normal magnetic field is continuous across the dipole electric current distribution (d), but not the tangential component; the latter has opposite signs on the two sides, zero arithmetic mean on the surface and a jump across it specified by the derivate of the dipolar moment.

27

Classification of Singularities and Functions

A continuous function (Chapter 21) real or complex, may fail to be differentiable. For a real function, the existence of derivate of a given order does not (Section 27.1) imply the existence, or continuity, of derivates of higher order, and the existence of derivates of all orders does not guarantee the convergence of the Taylor series; in contrast, for a complex function, if it is holomorphic, that is, has first-order derivate (Chapter 21), then (Section 27.2) it has derivates of all orders (Chapter 13), and a convergent Taylor series (Section 25.7 and 27.2), and it is a harmonic function (Section 23.1) as well. Another difference for complex functions is that when they fail to be holomorphic at a point, $z = a$, the associated Laurent series allows the classification of the singularity (Section 27.3) into two types: (i) the pole that is always an isolated singularity; (ii) the essential singularity that may or may not be isolated. The concept of residue at a pole (Chapter 15) may be extended to essential singularities (Section 27.4), while retaining its usefulness for the evaluation of loop integrals. A complex function may have a singularity at infinity (Section 27.5), that is, either a pole or an essential singularity, just as at any other point of the complex plane. The type of singularities of a function (Table 27.1) on the finite part of the complex plane and at infinity, can be used to classify (Table 27.2) the function as (i) a constant (Section 27.6); (ii) a polynomial, or a rational function (Section 27.7); (iii) an integral, a rational-integral, a meromorphic or a polymorphic function (Section 27.9). The latter set (iii) has at least one essential singularity (Section 27.8). The classification of functions in classes allows a listing of the properties associated to each class; it may also serve for the reconstruction of a function from partial data, thus completing the solution of a problem by specifying some desirable properties (e.g., the hodograph method with singularities in Section 38.8 and Example 40.16).

27.1 Chain of Inclusion of Real Functions

The classes of functions of interest in the real domain include the differentiable and continuously differentiable (Subsection 27.1.1), plus the smooth and analytic (Subsection 27.1.2) that form a chain of inclusions.

27.1.1 Continuous, Differentiable, and Continuously Differentiable Functions

A function, real or complex, is differentiable (27.1a) if the limit of the incremental ratio (27.1b) exists:

$$f(z) \in \mathcal{D}: \quad \lim_{z \to a} \frac{f(z) - f(a)}{z - a} = f'(a) \neq \infty; \qquad (27.1\text{a,b})$$

this implies that it is continuous:

$$\lim_{z\to a} f(z) = f(a) + \lim_{z\to a}(z-a)f'(a) = f(a) \Rightarrow f(z) \in C. \qquad (27.1c))$$

The converse need not be true, since $f(z) - f(a) \to 0$ as $z - a \to 0$ does not guarantee the existence of a unique, finite limit for the ratio (27.1b); as an example, the function (27.2a) is continuous (27.2b), but not differentiable at the origin (27.2c):

$$f(z) = \sqrt{z}\colon \lim_{z\to 0} f(z) = \lim_{z\to 0}\sqrt{z} = 0 = f(0), f'(0) = \lim_{z\to 0}\frac{\sqrt{z}}{z} = \lim_{z\to 0}\frac{1}{\sqrt{z}} = \infty; \qquad (27.2a–c)$$

thus the class $\mathcal{D}$ of differentiable functions is contained $\mathcal{D} \subset C$ in the class C of continuous functions.

Concerning the class $\mathcal{D}^n(C^n)$ of functions with derivate (continuous derivate) of order n, in the case of real functions $f(x)$ with $x \in R$: (i) if the derivate of order $n+1$ exists, then the derivates of lower orders must also exist and be continuous $\mathcal{D}^{n+1} \subset C^n \subset \mathcal{D}^n$; (ii) the continuity of the n-th derivate does not guarantee the existence of the $(n+1)$-th derivate, that is, $\mathcal{D}^{n+1} \subset C^n$, because the continuity of $g(x) = f^{(n)}(x)$ does not imply differentiability $g'(x) = f^{(n+1)}(x)$; (iii) the piecewise existence of the n-th derivate, for example, at the left $f^{(n)}(a-0)$ and right $f^{(n)}(a+0)$ of a point $x = a$, does not guarantee continuity, for example, if $f^{(n)}(a-0) \neq f^{(n)}(a+0)$, so that $C^n \not\subset \bar{\mathcal{D}}^n$, where $\bar{\mathcal{D}}^n$ is the set of functions with piecewise continuous n-th derivate; (iv) conversely continuity does not imply piecewise differentiability $\bar{\mathcal{D}}^n \not\subset C$. The property (i) follows from (27.1c), and simple examples are given of (ii) and (iii). Concerning (ii) the function (27.3a) equal to the N-th power (zero) for x positive (negative):

$$f_N(x) \equiv \begin{cases} x^N & \text{if } x \geq 0 \\ 0 & \text{if } x \leq 0 \end{cases}, \quad f_N^{(n)}(x) = \begin{cases} \frac{N!}{(N-n)!}\, x^{N-n} & \text{if } x > 0 \\ 0 & \text{if } x < 0, \end{cases} \qquad (27.3a,b)$$

has (27.3b) derivates: (i) of orders $n = 1, \ldots, N-1$ that are zero and continuous at the origin

$$n = 0, 1, \ldots, N-1: \quad f_N^{(n)}(+0) = 0 = f_N^{(n)}(-0); \quad f_N^{(N)}(+0) = N! \neq 0 = f_N^{(N)}(-0), \qquad (27.4a–c)$$

the N-th derivate (27.4c) is discontinuous. Concerning (iii) the function (27.5a):

$$g_N(x) \equiv \begin{cases} x^{N+1/2} & \text{if } x > 0 \\ 0 & \text{if } x < 0, \end{cases} \qquad g_N^{(n)}(x) = \begin{cases} a_{N,n}x^{N-n+1/2} & \text{if } x > 0 \\ 0 & \text{if } x < 0, \end{cases} \qquad (27.5a,b)$$

has derivates (27.5b), with the coefficients given by (27.5c):

$$\begin{aligned} N, n \in |N: \quad a_{N,n} &\equiv \left(N + \frac{1}{2}\right)\left(N - \frac{1}{2}\right)\ldots\left(N - n + \frac{3}{2}\right) \\ &= 2^{-n}(2N+1)\ldots(2N-2n+3) \equiv 2^{-n}\frac{(2N+1)!!}{(2N-2n+1)!!}; \end{aligned} \qquad (27.5c)$$

in (27.5c) is used the **double factorial**:

$$N \in |N: \quad (2N+1)!! \equiv (2N+1)(2N-1)(2N-3)\ldots 5.3.1. \tag{27.5d}$$

The function (27.5a) has: (i) derivates at the origin of all orders up to $n = N$ in (27.6a) that are zero and continuous (27.6b):

$$n \le N: \quad g_N^{(n)}(+0) = 0 = g_N^{(0)}(-0); \quad n > N: \quad g_N^{(n)}(+0) = \infty \ne 0 = g_N^{(n)}(-0), \tag{27.6a–d}$$

(ii) the derivates beyond (27.6c) the order n do not exist (27.6d).

27.1.2 Smooth and Analytic Functions and Chain of Inclusions

Letting $n \to \infty$ the classes C^n, $\mathcal{D}^n$ become the set of infinitely differentiable or smooth functions $C^\infty \equiv \mathcal{D}^\infty$; these are related to the class $\mathcal{A}$ of analytic functions that can be represented by a Taylor series. An analytic function is smooth $\mathcal{A} \subset \mathcal{D}^\infty$, because its Taylor series involves derivates of all orders; the inverse may not be true $\mathcal{D}^\infty \not\subset \mathcal{A}$, that is, a function may be infinitely differentiable, but its Taylor expansion may fail to converge to it, if the remainder does not vanish. An example is the function (27.7a) in Figure 27.1:

$$f(x) = \exp\left(-\frac{1}{x^2}\right); \quad f^{(n)}(x) = P_{3n}\left(\frac{1}{x}\right)\exp\left(-\frac{1}{x^2}\right), \tag{27.7a,b}$$

its n-th derivative corresponds (27.7b) to the multiplication by a polynomial of degree $3n$ in $1/x$. The singularity of the polynomial at the origin $x = 0$ is dominated by the exponential, so that the derivates of all orders vanish (27.7c) there

$$f^{(n)}(0) = 0; \quad f(x) = R_N(x) = \exp\left(-\frac{1}{x^2}\right), \tag{27.7c,d}$$

thus the Maclaurin expansion (27.34a) reduces to the remainder (27.7d). Since the function coincides with the remainder, the latter is independent of N, and does not tend to zero as $N \to \infty$; it follows that the Maclaurin series does not converge to the function, that is, the series (remainder) is zero everywhere (nonzero for $x \ne 0$) according to (27.7c) [(27.7a)].

Thus has been obtained:

$$\begin{aligned} &C(|R) \supset \mathcal{D}(|R) \supset C^1(|R) \supset \mathcal{D}^2(|R) \supset C^2(|R) \supset \ldots \\ &\supset \mathcal{D}^n(|R) \supset C^n(|R) \supset \mathcal{D}^{n+1}(|R) \supset C^{n+1}(R) \supset \ldots \supset D^\infty(|R) \supset \mathcal{A}(|R), \end{aligned} \tag{27.8}$$

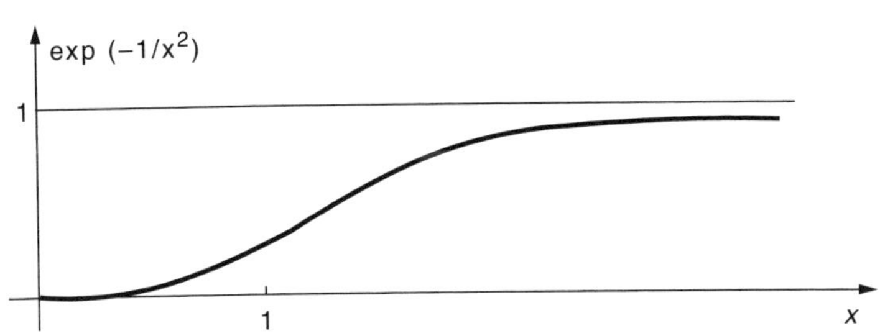

FIGURE 27.1
The complex function $\exp(-1/z^2)$ has a Laurent–Maclaurin series consisting only of inverse powers, corresponding to the essential singularity at the origin. As a real function it vanishes with all its derivatives at the origin. The Maclaurin series: (i) does not hold; (ii) would have all terms zero except the remainder, that coincide with the function.

the **chain of inclusion** *for real functions, stating that the following classes are progressively more restricted: continuous, differentiable, continuously differentiable, twice differentiable, continuously twice differentiable,* n*-times differentiable,* n*-times continuously differentiable,* (n + 1) *times differentiable, smooth, and analytic.*

For real functions continuity (differentiability) is generalized by piecewise continuity (27.9a) [piecewise differentiability (27.9b)]:

$$C(|R) \subset \bar{C}(|R) \equiv \left\{ f(x) : \quad \forall_{x \in |R} : \quad \exists^1 \lim_{\varepsilon \to 0} f(x \pm \varepsilon) \equiv f(x \pm 0) \right\}; \tag{27.9a}$$

$$\mathcal{D}(|R) \subset \bar{\mathcal{D}}(|R) \equiv \left\{ f(x) : \quad \forall_{x \in |R} : \quad \exists^1 \lim_{\varepsilon \to 0} \frac{f(x \pm \varepsilon) - f(x)}{\varepsilon} \equiv f'(x \pm 0) \right\}; \tag{27.9b}$$

piecewise differentiability of any order (27.9d) does not imply continuity:

$$\bar{\mathcal{D}}^n(|R) \not\subset C(|R) : \quad C(|R) \not\subset \bar{\mathcal{D}}^n(|R) \equiv \left\{ f(x) : \quad \forall_{x \in |R} \forall_{m=1,\ldots,n} : \quad \exists^1 f^{(m)}(x \pm 0) \right\}, \tag{27.9c,d}$$

nor the converse (27.9c), even for:

$$\bar{\mathcal{D}}^\infty(|R) \equiv \left\{ f(x) : \quad \forall_{x \in |R} \forall_{n \in |N} \exists^1 f^{(n)}(x \pm 0) \right\}, \tag{27.9e}$$

piecewise smooth functions (27.9e).

27.2 Set of Coincidences for Complex Functions

In the case of complex functions all the preceding classes beyond and including differentiable functions (Subsection 27.2.1) coincide with harmonic functions (Subsection 27.2.2).

27.2.1 Continuous and Holomorphic or Differentiable Complex Functions

The chain of inclusion for real functions (Subsection 27.1.2), involves the following classes that can be defined for functions over any set X, in particular real $X \equiv |R$ (complex $X \equiv |C$) numbers: (i/ii) set $C(\mathcal{D})$ of **continuous (differentiable) functions** (27.10) [(27.11)]:

$$C(X) \equiv \left\{ f(z) : \quad \forall_{a,z \in X} : \quad \lim_{z \to a} f(z) = f(a) \right\}, \tag{27.10}$$

$$\mathcal{D}(X) \equiv \left\{ f(z) : \forall_{z \in X} : \quad \exists^1 f'(z) = \lim_{z \to a} \frac{f(z) - f(a)}{z - a} \right\}; \tag{27.11}$$

(iii/iv) set $\mathcal{D}^n$ (C^n) of **n-times differentiable (n-times continuously differentiable) functions** (27.12) [(27.13)]:

$$\mathcal{D}^n(X) \equiv \{ f(z) : \quad \forall_{z \in X} \quad \exists^1 f^{(n)}(z) \}, \tag{27.12}$$

$$C^n(X) \equiv \{ f(z) : \quad \forall_{z \in X} \quad f^{(n)}(z) \in C(X) \} \tag{27.13}$$

(v/vi) set $\mathcal{D}^\infty(\mathcal{A})$ of **smooth (analytic)** functions (27.14) [(27.15)], that is, infinitely differentiable (with convergent Taylor series):

$$\mathcal{D}^\infty(X) \equiv \left\{ f(z): \quad \forall_{z\in X} \quad \forall_{n\in N}\, \exists^1 f^{(n)}(z) \right\} \tag{27.14}$$

$$\mathcal{A}(X) \equiv \left\{ f(z): \quad \forall_{z,a\in X}: \quad f(z) = \sum_{n=0}^{\infty} \frac{(z-a)^n}{n!} f^{(n)}(a) \right\}. \tag{27.15}$$

The set X is the real (complex) numbers $X \equiv |R$ ($X \equiv |C$) for functions of real (complex) variable; the function is a **real (complex) function** if the range is real (complex) $f(z) \in |R$ $[f(z) \in |C]$.

The arguments (27.1a,b; 27.2a–c) leading to $\mathcal{C}(|R) \supset \mathcal{D}(|C)$ hold for complex as well as for real functions, but the remaining examples work out differently. Extending (27.3a) to a complex function equal to the N-th power, except on the real axis:

$$f(z) \equiv \begin{cases} z^N & \text{if } y \neq 0, \\ 0 & \text{if } y = 0 \end{cases} \qquad \lim_{\substack{y\to 0\\ x\neq 0}} f(x+iy) = x^N \neq 0 = f(x), \tag{27.16a,b}$$

it follows that the function is discontinuous on the real axis, except at the origin; thus it is not differentiable, or holomorphic, unlike (27.3a) that has derivates up to order N. The complex function corresponding to (27.7a):

$$f(z) = \exp\left(-\frac{1}{z^2}\right) = \exp\frac{1}{y^2 - x^2 - 2ixy}, \tag{27.17a}$$

$$\lim_{x\to 0}\lim_{y\to 0} f(z) = \lim_{x\to 0}\exp\left(-\frac{1}{x^2}\right) = 0 \neq \infty = \lim_{y\to 0}\exp\left(\frac{1}{y^2}\right) = \lim_{y\to 0}\lim_{x\to 0} f(z), \tag{27.17b}$$

is also discontinuous at the origin, hence it does not have a single derivate, whereas its real counterpart has infinitely many.

27.2.2 Smooth, Analytic, and Harmonic Complex Functions

The preceding examples are instances of the general fact that the requirement for differentiability is stricter for complex than for real functions. It follows that complex differentiable functions, that is, holomorphic functions, have more properties than real differentiable functions. A holomorphic function (Section 11.2) has derivates of all orders (Section 15.4), that is, is smooth, and has a convergent Taylor series (Section 23.7), that is, is analytic. Thus the chain of inclusion (27.8) for real functions is replaced for complex functions, by:

$$\mathcal{C}(|C) \supset \mathcal{D}(|C) \equiv \mathcal{D}^n(|C) \equiv C^n(|C) \equiv \mathcal{D}^\infty(|C) \equiv \mathcal{A}(|C) \equiv \mathcal{H}(|C), \tag{27.18}$$

the **set of coincidences** *for complex functions (27.18), stating that continuity contains holomorphy but the latter coincides with differentiability and continuous differentiability to any order, smoothness, analyticity and harmonicity.* The set $\mathcal{H}$ of **harmonic functions:**

$$\mathcal{H}(X) \equiv \left\{ f(z): \quad \forall_{z\in X,\, a\in |R}: \quad f(z) = (2\pi a)^{-1} \int_{|\zeta - z| = a} f(\zeta)\, d\zeta \right\} \tag{27.19}$$

is defined by the value of a function at a point being the mean of the values on any circle centered at that point; for complex functions harmonicity coincides with holomorphy

(Section 23.1), and implies:

$$\Phi(x,y) + i\Psi(x,y) = f(x+iy) \in \mathcal{D}(|C): \quad \nabla^2\Phi = 0 = \nabla^2\Psi, \nabla \equiv \frac{\partial^2}{\partial x^2} + \frac{\partial^2}{\partial y^2}, \quad (27.20\text{a–c})$$

that the real and imaginary parts (27.20a) satisfy (27.20b,c) the Laplace equation, provided that second-order continuous partial derivatives exist.

27.3 Ordinary Points, Zeros, and Singularities

A complex function is (is not) differentiable at a regular (irregular) point, and thus can be represented by a Taylor (Laurent) series (Subsection 27.3.1) in its neighborhood. The difference between a Taylor and Laurent series is that the latter includes a set of powers with negative exponents that form its principal part (Subsection 27.3.2).

27.3.1 Taylor (Laurent) Series near a Regular Point (Singularity)

Although the requirements for a complex function to be holomorphic may fail (27.17a,b), even when the corresponding real function is infinitely differentiable (27.7–27.8a,b), the complex function has an important advantage: (i) the real function cannot "pass" over a point on the axis where it is not defined and has no limit; (ii) a complex function can surround a point or region where it is not holomorphic, viz. Laurent series holds in a doubly-connected region enclosing within the inner loop the set of points where it is not holomorphic. This allows the classification of the points of complex functions. For example, the real functions have no general power series representation if differentiability breaks down at a finite order, because then the Taylor series does not exist. For complex functions, the first derivate assures the existence of a Taylor series, and if differentiability breaks down at an isolated point, the Laurent series provides a power series representation that generalizes the Taylor series. A **regular point** of a function is defined (Classification 39.1) by the function being single-valued and finite in its neighborhood. A regular point is a **zero (ordinary point)** if the function does (does not) vanish. An **extraordinary point** is not an ordinary point and can be either (i) a zero or (ii) a **special point**. The latter arises if: *(i) the function is unbounded in its neighborhood, that is, has a* **singularity**, *for example, a pole (Sections 15.7 and 15.8); (ii) the function is multivalued in its neighborhood, that is, has a branch-point (Sections 7.1–7.4).* It is important to stress the neighborhood of the point, not just the point itself, for example, the function $f(z) = \sqrt{z}$ is single-valued and finite at $z = 0$ (actually zero), but $z = 0$ is not a regular point, viz. it is a branch-point because the function is two-valued in any neighborhood of $z = 0$. Thus for a complex function: (i) if it is holomorphic in a region D, every point is a **regular point**, and a Taylor series is valid in a circle with center at the point and radius R such that it is contained in the region (Figure 23.2b); (ii) if it is holomorphic in a region D except for a subregion E that contains **singularities**, where the function is not holomorphic, then there is a Laurent series valid in a doubly connected region, viz. a ring with inner radius r such that the region E lies inside, and outer radius R such that the ring lies inside D (Figure 25.2). A regular point is not isolated, since every neighborhood of it contains other regular points.

In the neighborhood of a regular point (27.21a), the Taylor series (23.32b) ≡ (27.21b) holds

$$f(z) \in \mathcal{D}(V_\varepsilon(a)): \quad f(z) = \sum_{n=0}^{\infty} \frac{f^{(n)}(a)}{n!}(z-a)^n. \qquad (27.21\text{a,b})$$

The regular points can be classified into: (i) **ordinary points**, if the function does not vanish $f(a) \neq 0$, that is, the Taylor series (27.21b) in the neighborhood of $z = a$ is $O(1)$; (ii) if the function and its derivatives up to order $(N-1)$ vanish (27.22a), the point is a **zero of order N:**

$$f(a) = \cdots = f^{(N-1)}(a) = 0 \neq f^{(N)}(a) : f(z) = (z-a)^N \, g(z) = O\left((z-a)^N\right); \tag{27.22a,b}$$

$$0 \neq g(a) = \frac{f^{(N)}(a)}{N!} : \quad g(z) = \sum_{n=0}^{\infty} \frac{f^{(N+n)}(a)}{(N+n)!}(z-a)^n, \tag{27.22c,d}$$

this implies that the function is of order N in its neighborhood (27.22b), with $g(z)$ a nonzero (27.22c) holomorphic (27.22d) function, in agreement with Section 19.8.

27.3.2 Principal Part of a Laurent Series

The **extraordinary points** consist of zeros (27.22a–d) that are regular points, and **special points** that include (i) the branch-points where a function is multivalued (Chapters 7 and 9); (ii) the **singularities** of a single-valued function, where it fails to be holomorphic. The singularities of single-valued functions will be considered next, as contribution to a subsequent (Chapter 39) classification of extraordinary points. If $z = a$ is an isolated singularity of $f(z)$ that is, a neighborhood $V_\varepsilon(a)$ exists (27.23a) such that $f(z)$ is holomorphic at all points $z \neq a$, then the Laurent series (25.26a) $\equiv$ (27.23b) holds:

$$f(z) \in \mathcal{D}\left(V_\varepsilon(a) - \{a\}\right) : \quad f(z) = \sum_{n=-\infty}^{+\infty} A_n (z-a)^n ; \tag{27.23a,b}$$

the **principal part** is defined (27.24b) as the sequence of descending powers:

$$O(1) = f(z) - P(z) \in \mathcal{D}\left(V_\varepsilon(a)\right) : \quad P\{f(z)\} \equiv \sum_{n=1}^{\infty} A_{-n}(z-a)^{-n}, \tag{27.24a,b}$$

so that *the function becomes holomorphic at (27.24a) the singularity $z = a$ if the principal part is subtracted out.*

The nonzero principal part of a complex function at a the singularity can be either terminating (or nonterminating) thus specifying a **pole** (27.25) [**essential singularity** (27.26)]:

$$\text{pole of order } N : \quad \forall_{n \in |N} : \quad n > N \Rightarrow A_{-n} = 0 \neq A_{-N}, \tag{27.25}$$

$$\text{essential singularity} : \quad \forall_{n \in |N} : \quad \exists_{m>n} : \quad A_{-m} \neq 0. \tag{27.26}$$

At a **pole of order, N,** *the function (27.27a) is of $O\left((z-a)^{-N}\right)$ in (27.27b):*

$$A_{-N} \neq 0 = A_{-N-1} = \ldots : f(z) = \sum_{n=-N}^{\infty} A_n (z-a)^n = (z-a)^{-N} \, h(z); \tag{27.27a,b}$$

$$h(a) = A_{-N} \neq 0 : \quad h(z) = \sum_{n=0}^{\infty} A_{n-N} (z-a)^n, \tag{27.27c,d}$$

and $h(z)$ is holomorphic (27.27d) and nonzero (27.27c). Comparison of (27.22a–d) and (27.27a–d) shows that *if the function $f(z)$ has a pole of order N at a, then $1/f(z)$ has a zero of order N at a, and vice-versa.* From (27.27b) it follows that *a pole of order* N *can be removed multiplying by a power with exponent* N. *An essential singularity, in contrast with a pole, cannot be "removed" by multiplication by any power.*

27.4 Residues at Poles and Essential Singularities

The residue of a complex function at a singularity can be defined both for essential singularities (Subsection 27.4.1) and poles (Subsection 27.4.2); of the four methods of calculation of residues at poles only one applies to essential singularities (subsection 27.4.3), so it is the general definition of residue.

27.4.1 Definition of Residue from the Laurent Series

If the function $f(z)$ has an isolated singularity at $z = a$, its Laurent series (27.23b) is uniformly convergent in a closed subset X of a neighborhood of a, and can be integrated term-by-term along a loop contained in X and surrounding, $z = a$:

$$\int^{(a+)} f(z)\,\mathrm{d}z = \sum_{n=-\infty}^{+\infty} A_n \int_{(a+)} (z-a)^n\,\mathrm{d}z. \tag{27.28}$$

The loop can be deformed to (27.29a) circle of radius ε and center at a, because (Section 15.2) the function is holomorphic between the boundary ∂X of X and the circle:

$$z = a + \varepsilon e^{i\varphi}: \quad \int_{|z-a|=\varepsilon} (z-a)^n\,\mathrm{d}z = i\varepsilon^{n+1}\int_0^{2\pi} e^{i(n+1)\varphi}\,\mathrm{d}\varphi = \begin{cases} 0 & \text{if} \quad n \neq -1, \\ 2\pi i & \text{if} \quad n = -1; \end{cases} \tag{27.29a–c}$$

the result (27.29b,c) coincides with (20.77a,b) in because it is independent of the center of the circle, so the integrals must be equal. Substituting (27.29b,c) into (27.28) it follows that

$$\int_{\partial D} f(z)\,\mathrm{d}z = 2\pi i A_{-1}, \qquad A_{-1} = \operatorname{Res}\{f(z)\}_{z=a}; \tag{27.30a,b}$$

if the function $\mathrm{f}(\mathrm{z})$ *is holomorphic in the closed region* D, *except at an isolated singularity* $\mathrm{z} = \mathrm{a}$, *its integral along the boundary is equal to* $2\pi\mathrm{i}$ *times the coefficient* A_{-1} *of the first inverse power* z^{-1} *of the Laurent series that specifies the* ***residue*** *of the function at the singularity.* This form of the **theorem of residues** applies both to isolated essential singularities (Section 27.4) and to poles (Sections 15.9 and 17.4).

27.4.2 Residues at Simple and Multiple Poles

It is checked next that the residue at a pole, as defined by the general theorem (27.30a,b), agrees with the rules of calculation of residues used earlier (Chapters 15 and 19). At a simple pole, the Laurent series (27.23b) simplifies to

$$f(z) = \sum_{n=-1}^{\infty} A_n (z-a)^n = \frac{A_{-1}}{z-a} + O(1), \tag{27.31}$$

where the principal part consists of a single term, whose coefficient A_{-1} is the residue; it may be calculated by the limit:

$$f_{(1)}(a) \equiv \lim_{z\to a} (z-a) f(z) = A_{-1} + \lim_{z\to a} O(z-a) = A_{-1}, \tag{27.32}$$

in agreement with the rule used before (15.24b) $\equiv$ (27.32). At a pole of order N: (i) the principal part of the Laurent series (27.23) has N terms:

$$f(z) = \sum_{n=-N}^{\infty} A_n (z-a)^n = \sum_{n=1}^{N} A_{-n} (z-a)^{-n} + O(1); \tag{27.33}$$

(ii) the singularity at the pole of order N can be "removed" (27.27b) by multiplication by the N-th power of $z-a$:

$$(z-a)^N f(z) = \sum_{n=1}^{N} A_{-n} (z-a)^{N-n} + O\left((z-a)^N\right); \tag{27.34}$$

(iii) the coefficient A_{-1}, becomes the leading term differentiating $(N-1)$ times with regard to z:

$$\frac{\mathrm{d}^{N-1}\left\{(z-a)^N f(z)\right\}}{\mathrm{d}z^{N-1}} = (N-1)!A_{-1} + O(z-a); \tag{27.35}$$

(iv) the limit $z \to a$ cancels the terms beyond the leading term that specifies the residue:

$$A_{-1} = \frac{1}{(N-1)!} \lim_{z\to a} \frac{\mathrm{d}^{N-1}\left\{(z-a)^N f(z)\right\}}{\mathrm{d}z^{N-1}} \equiv f_{(N)}(a), \tag{27.36}$$

in agreement with the rule used earlier (15.33b) $\equiv$ (27.36).

27.4.3 Four Methods of Calculation of Residues

Four methods have been presented for the calculation of residues at simple (multiple) poles: (method I) the limit (27.32) $\equiv$ (15.24b) [(27.36) $\equiv$ (15.32b)] can be applied to any function [Subsection (15.7.2)] [Section 15.7 (15.8)] for that it exists; (method II) the decomposition into partial fractions [Subsection 15.7.1 (15.8.1)] applies to rational functions [Section 31.8 (31.9)] and their combination with functions nonsingular at the pole; (method III) the rule (19.40a–c; 19.41) [(19.48a–c; 19.49)] of differentiation [Subsection 19.9.1 (19.9.2)] is not restricted to rational functions, like method II; (method IV) the coefficient A_{-1} of (27.30a,b) the Laurent series (27.23a,b) that exists in the neighborhood of any isolated singularity (Subsection 27.4.1). The methods I to III fail calculate the residue at an essential singularity, because (method I) an essential singularity (27.26) dominates any power, so the limit (27.36) diverges for any exponent N; (method II) there is no finite partial fraction decomposition near an essential singularity; (method III) an essential singularity cannot be removed by division by a holomorphic function. Thus of the four methods (I to IV) that apply to the calculation of the residue at a pole only method IV applies as well to an essential singularity; it is therefore the general definition of residue at an isolated singularity (27.30a,b).

27.5 Inversion of the Origin and Singularity at Infinity

The point at infinity $\zeta = \infty$ is defined as in the stereographic projection of a sphere (Section 9.3) by inversion (27.37a) of the origin, $z = 0$; thus the behavior of the function $f(z)$ at infinity is specified by the preceding power series (27.23b) with $z - a$ replaced by $1/\zeta$. The function $f(z)$ has an **ordinary point at infinity** if:

$$\zeta = \frac{1}{z}: \ f(z) = \sum_{n=0}^{\infty} A_{-n} z^{-n} = \sum_{n=0}^{\infty} A_{-n}\zeta^{n} = f\left(\frac{1}{\zeta}\right) \equiv F(\zeta), \ A_{-n} = \frac{F^{(n)}(0)}{n!}, \quad (27.37a–c)$$

it is specified (27.37b,c) by a descending Maclaurin (23.34b) series at infinity; there are only nonpositive exponents, thus the function is nonzero and holomorphic, hence finite. The case $A_{-1} \equiv F(0) = \dot{f}(\infty) = 0$ is the instance $N = 1$ of the **zero of order N at infinity** (27.38a):

$$A_{-1} = A_{-2} = \cdots = A_{-N+1} = 0 \neq A_{-N}: \quad f(z) = \sum_{n=N}^{\infty} A_{-n} z^{-n} = z^{-N} g(z) = O\left(z^{-N}\right), \quad (27.38a,b)$$

$$0 \neq g(\infty) = A_{-N}: \quad g(z) \equiv \sum_{n=0}^{\infty} A_{-N-n} z^{-n}, \quad (27.38c,d)$$

for that the function (27.38b) is $O\left(z^{-N}\right)$; the nonvanishing (27.38c) coefficient function (27.38d), has an ordinary point at infinity.

At a **singularity at infinity** the function is specified by a Laurent series (27.23b) ≡ (27.39b) of variable (27.37a) ≡ (27.39a):

$$\zeta = \frac{1}{z}: \quad f(z) = \sum_{n=-\infty}^{+\infty} A_n z^n = \sum_{n=-\infty}^{+\infty} A_n \zeta^{-n} = f\left(\frac{1}{\zeta}\right) \equiv F(\zeta). \quad (27.39a,b)$$

If the highest power (27.40b) is z^N, the function $f(z)$ has a **pole of order N at infinity** (27.40a):

$$h(0) = A_N \neq 0 = A_{N+1}, A_{N+2}, \ldots: \quad f(z) = \sum_{n=-\infty}^{N} A_n z^n = z^N h(z) = O\left(z^N\right),$$

$$h(z) = \sum_{n=-\infty}^{\infty} A_n z^{n-N}, \quad (27.40a–c)$$

implying that it is $O(z^N)$; thus "removing" the singularity z^N leads to the function (27.40c) that is holomorphic (27.40b) and nonzero (27.40a) at infinity, that is, has an ordinary point at infinity. In the case, $N = 1$, the residue at a simple pole at infinity, is given by

$$f_{(1)}(\infty) = A_1 = \lim_{z\to\infty} z^{-1} f(\zeta) = \lim_{\zeta\to\infty} \zeta f\left(\frac{1}{\zeta}\right); \quad (27.41a)$$

the residue at a pole of order N at infinity is specified by

$$f_{(1)}(\infty) = A_1 = \{(N-1)!\}^{-1} \lim_{z\to\infty} f^{(N-1)}(z). \quad (27.41b)$$

In general, the coefficient A_1 specifies the residue at a simple pole (27.41a), pole of order N in (27.41b), or at an **essential singularity at infinity** *(27.39a,b) with* (27.41c):

$$\forall_{n\in|N}\ \exists_{m\in|N}: A_m \neq 0. \quad (27.41c)$$

The exponential (27.42b) has an essential singularity at infinity:

$$\zeta = \frac{1}{z}: \quad \exp(z) \equiv \sum_{n=0}^{\infty} \frac{z^n}{n!}; \quad \exp\left(\frac{1}{\zeta}\right) = \sum_{n=0}^{\infty} \frac{\zeta^{-n}}{n!}, \tag{27.42a–c}$$

the inversion of its argument (27.42a) leads to an essential singularity at the origin (27.42c); the residue is unity $A_{-1} = 1$, so that (27.30a) can be used to evaluate the loop integral (27.43a):

$$\int_{(0+)} \exp\left(\frac{1}{\zeta}\right) \mathrm{d}\zeta = 2\pi i; \qquad \int_{(0+)} \zeta^N \exp\left(\frac{1}{\zeta}\right) \mathrm{d}\zeta = \frac{2\pi i}{(N+1)!}, \tag{27.43a,b}$$

multiplying (27.42c) by any power, the essential singularity at the origin remains (27.44a), and leads to the generalization (27.43b); the latter is obtained by identifying the coefficient of ζ^{-1} in (27.44a):

$$\zeta^N \exp\left(\frac{1}{\zeta}\right) = \sum_{n=0}^{\infty} \frac{\zeta^{N-n}}{n!}, \qquad A_{-1} = \left\{(N+1)!\right\}^{-1}, \tag{27.44a,b}$$

that specifies by $1/(n!)$ with $N - n = -1$ the residue (27.44b); the latter is multiplied by $2\,\pi$ i to evaluate the integral (27.43b).

27.6 Identification of Constants (Cauchy, 1844; Liouville, 1847)

The simplest holomorphic function is one: (i) without any singularity in the finite part of the plane, that is, with Maclaurin series (23.34b) $\equiv$ (27.21a,b) with $a = 0$; (ii) without singularity at infinity, so that all coefficients in (27.21a,b) beyond the first vanish $0 = f'(0) = \ldots$ Thus (27.21a,b) is reduced to $f(z) = f(0) = \text{const}$, which is a constant. Thus a function without singularities, neither at infinity nor in the finite part of the plane, is a constant (Cauchy, 1844). An alternative statement is that $f(z)$ be holomorphic for all finite z, and bounded in modulus by M at infinity; applying Cauchy's theorem (15.12) to express $f'(z)$ by an integral along a circle of radius R and center at the origin, leads to

$$|f'(z)| \le (2\pi)^{-1} \int_{|\zeta|=R} |\zeta - z|^{-2} |f(\zeta)| \,\mathrm{d}\zeta \le \frac{B}{2\pi}\left(\frac{R}{2}\right)^{-2} 2\pi R = \frac{4B}{R}, \tag{27.45}$$

where the radius R is chosen so that $R > 2\,|z|$, and hence $|\zeta - z| > R/2$; taking the limit $R \to \infty$, it follows that $f'(z) = 0$, and $f(z)$ is a constant (Liouville, 1847).

The criteria of Cauchy 1844 (Liouville, 1847) for the identification of a constant "function" are equivalent, since both require the function to be holomorphic in the finite part of the plane, and to be holomorphic (bounded in modulus) at infinity. The condition of being bounded in modulus by B implies that both the real and imaginary parts have upper and lower bounds:

$$|f(z)| < B > 0: \quad -B < \mathrm{Re}\left\{f(z)\right\}, \mathrm{Im}\left\{f(z)\right\} < +B. \tag{27.46}$$

Thus a genuine extension of the theorem is obtained by proving it, while requiring only one of the four conditions (27.46) to be met: (i) to do this consider the auxiliary function:

$$g(z) = \exp\{\pm f(z), \pm i f(z)\}, \quad |g(z)| = \exp\{\pm \operatorname{Re}[f(z)], \mp \operatorname{Im}[f(z)]\}; \qquad (27.47\text{a,b})$$

(ii) if one of the conditions (27.46) is met, then one of the functions (27.47a) is bounded in modulus (27.47b); (iii) since $f(z)$ is holomorphic for finite z, so is $g(z)$, and as $g(z)$ is also bounded in modulus, it reduces to a constant; (iv) if $g(z)$ is a constant, then (27.47a) implies that $f(z)$ is also a constant, and the theorem is also true QED.

Thus can be stated the **theorem of identification of constants** *in two equivalent (i) ≡ (ii) and one generalized (iii) form; a function* f(z) = c, *is a constant iff (i) it has no singularities, either for finite variable* z *or at infinity; (ii) it is holomorphic in the finite part of the plane and bounded in modulus at infinity:*

$$F(z) \in \mathcal{D}(|C) \wedge \exists_{B>0} \quad \forall_{z \in |C} \; |f(z)| < B \Rightarrow f(z) = \text{const}; \qquad (27.48)$$

(iii) it is holomorphic in the finite part of the plane, and either the real or imaginary part has an upper or lower bound (27.46). These conditions are necessary (sufficient) because a constant has the stated properties (i) to (iii) [the properties (i) to (iii) imply that the function $g(z)$ reduces to a constant, as proved before]. The theorem of identification of constants yields a useful method of proving equations that will be used often (e.g., Subsections 27.9.2, 31.8, and 31.9): (i) prove that the difference of the two sides is a function holomorphic everywhere, hence a constant; (ii) the constant is evaluated at a particular point, and shown to vanish, for an exact identity.

27.7 Definition of Polynomial and Rational Function

A rational function (Subsection 27.7.2) is the ratio of two polynomials (Subsection 27.7.1) having at least one noncommon root.

27.7.1 Polynomial of Degree N

The constant is a polynomial of degree 0, and the preceding theorem can be extended to arbitrary degree N, as the **theorem of identification of polynomials**: *the function* f(z) *is a polynomial of degree* N:

$$P_N(z) = \sum_{n=0}^{N} a_n z^n, \quad a_n \equiv \frac{P_N^{(n)}(0)}{n!}, \qquad (27.49\text{a,b})$$

with coefficients (27.49b) iff: (i) it is holomorphic for finite z, *and has a pole of order* N *at infinity; (ii) it is holomorphic for finite* z *and satisfies the asymptotic condition requiring its product by* z^{-N} *to be constant and nonzero at infinity:*

$$\lim_{z \to \infty} z^{-N} f(z) = A \neq 0, \infty; \qquad (27.50)$$

(iii) it is holomorphic for finite z, and meets one of the four asymptotic conditions stating that the real or imaginary part of (27.50) has a lower or upper bound at infinity:

$$\exists^1_{b>0}: \quad \pm \lim_{z \to \infty} \left\{\operatorname{Re}, \operatorname{Im}\left(z^{-N} f(z)\right)\right\} < B. \qquad (27.51)$$

Note that (27.50) implies all four conditions (27.51), but only one of them is needed for the theorem to hold.

The necessary conditions are obvious, since a polynomial of degree N has the stated properties. The sufficient conditions are proved next. Concerning (i) the function $f(z)$ holomorphic in the finite part of the plane has Maclaurin series, (27.21a,b) with $a = 0$, that must terminate at the N-th term, in order that the singularity at infinity be a pole of order N; then (27.21a,b) with $n = 0, \ldots, N$ coincides with (27.49a,b). Concerning (ii) the function $f(z)$, holomorphic in the complex plane for finite z, has derivate of order $N+1$ specified by Cauchy's theorem (15.14); using as loop the circle of radius R and center at the origin, leads to the upper-bound (27.52b):

$$|f(z)| \le BR^N : \left|f^{(N+1)}(z)\right| \le \frac{(N+1)!}{2\pi} \int\limits_{|\zeta|=R} |f(\zeta)| \, |\zeta - z|^{-N-2} \, d\zeta$$

$$\le \frac{(N+1)!}{2\pi} \left(\frac{R}{2}\right)^{-N-2} BR^N 2\pi R = (N+1)! 2^{N+2} \frac{B}{R}, \tag{27.52a,b}$$

where: (a) the condition (27.50) implies the upper bound (27.52a); (b) the radius $R > 2|z|$ was chosen so that $|\zeta - z| > \frac{R}{2}$; in the limit $R \to \infty$ the function has vanishing derivate of order $N+1$, and $f^{(N+1)}(z) = 0$ implies that $f(z)$ is a polynomial of degree N. Concerning (iii) if $f(z)$ is holomorphic for finite z, and satisfies one of the asymptotic conditions (27.51), then the function $g(z) = z^{-N} f(z)$ satisfies the conditions (27.46) to be a constant, as stated in (iii) of Section 27.6, and it follows that $f(z) \sim O(z^N)$ is a polynomial of degree N. QED.

27.7.2 Rational Function of Degrees $\mathcal{N}, \mathcal{M}$

A function $Q_{N,M}(z)$ ***is rational of degrees N, M****, if it satisfies two conditions: (i-1) it has a pole of order N at infinity:*

$$\lim_{z \to \infty} z^{-N} Q_{N,\,M}(z) = c \ne 0, \infty; \tag{27.53a}$$

(i-2) alternatively it satisfies one of the four asymptotic conditions:

$$\exists_{B>0} : \quad \lim_{z \to \infty} \pm \operatorname{Re}, \operatorname{Im}\left\{z^{-N} Q_{N,M}(z)\right\} < B; \tag{27.53b}$$

(ii) in addition it has L poles at $z = a_\ell$ with $\ell = 1, \ldots, L$, whose orders α_ℓ add to M:

$$\ell = 1, \cdots, L \quad Q_{N,M}(z) \in \mathcal{D}\left(|C - \sum_{\ell=1}^{L} \{a_\ell\}\right), \tag{27.54a,b}$$

$$M = \sum_{\ell=1}^{L} \alpha_\ell : \quad \lim_{z \to a_\ell} (z - a_\ell)^{\alpha\ell} Q_{N,M}(z) = b_\ell \ne 0, \infty. \tag{27.54c,d}$$

A rational function $Q_{N,M}(z)$ of degrees N, M may be obtained dividing two polynomials: (i) $P_N(z)/P_M(z)$ of degrees (N, M) without common roots; (ii) $P_R(z)/P_S(z)$ of degrees $(R = N + Q, S = M + Q)$ with Q common roots.

27.8 Essential Singularity as an Accumulation of Poles

Having considered the functions with a finite number of poles, the next case or class of functions either have at least one essential singularity, and/or have an infinite number of poles. Denote by X the set of singular points of a function $f(z)$ that is holomorphic at all other points $z \in |C - X$. A point of X is **isolated** if at least one neighborhood exists that contains no other point of X:

$$a \text{ isolated in } X: \quad \exists_{\varepsilon>0}\forall_{z\in V_\varepsilon(a)-\{a\}}: \quad z \notin X; \tag{27.55}$$

the converse is an **accumulation point** b, that is, a point such that every neighborhood contains at least one other point of X:

$$b \text{ accumulation point of} X: \quad \forall_{\varepsilon>0}\exists_{z\in X\wedge z\neq b}: \quad z \in V_\varepsilon(b). \tag{27.56a}$$

If b *is an accumulation point of* X *then every neighborhood of* b *contains an infinite number of points of* X: this lemma restates the definition (27.55) of accumulation point, with $\exists$ replaced by $\exists^\infty$, viz.:

$$b \text{ accumulation point of } X: \quad \forall_{\varepsilon>0}\exists^\infty_{z\in X,\, z\neq b}: \quad z \in V_\varepsilon(b). \tag{27.56b}$$

The lemma is proved by reduction of ad absurdum: if a neighborhood of b has a finite number $z_1, \dots, z_n$ of points X, then it is possible to choose $0 < \varepsilon < |z_n - b|$; the neighborhood $V_\varepsilon(b)$ contains no points of X other than b; that contradicts b being an accumulation point. QED.

A region D is **compact** if it is contained in a circle of finite radius:

$$D \text{ is compact}: \quad \exists_{R>0}\forall_{z\in D}: |z| < R. \tag{27.57}$$

A compact region D *containing an infinite number of points of* X, *has at least one accumulation point of* X. To prove this note that since D has an infinite number of points of X, if it is divided in two parts, one of them D_1 must have an infinite number of points; proceeding with the division in two parts, and picking always one with an infinite number of points of X, leads to a sequence of regions $D_1 \supset D_2 \supset \dots \supset D_n$ contained in each other. A point $b \in D_n$ as $n \to \infty$ is an accumulation point, since every neighborhood of it will contain at least one region, D_n, and hence an infinite number of points of X. The converse of the theorem states that: *if a set* X *with an infinite number* $\mathrm{z_n}$ *of points has no accumulation point, then the points when ordered by nondecreasing modulus* $|\mathrm{z_1}| \le |\mathrm{z_2}| \le \dots \le |\mathrm{z_n}|$, *must tend to infinity* $|z_n| \to \infty$; for, if $|z_n| < R$ as $n \to \infty$, the circle of radius R that is a compact region, would contain an infinite number of points of X, and at least one accumulation point would have to exist. QED.

The preceding results can be applied to the singularities of an otherwise holomorphic function: (i) *a pole is an isolated singularity,* since there is a neighborhood of it in that the Laurent series (27.27b) has no other singularity; (ii) *a point of accumulation of poles is an essential singularity,* because a point of accumulation of poles is a nonisolated singularity, that is, it is not a pole, so it can only be an essential singularity; (iii) *an essential singularity can be isolated or not,* for example, in the Laurent series (27.23a,b) it is isolated, and if it results from an accumulation of poles, it is not isolated singularity; (iv) *if a function has an infinite number of poles in a compact region, it has at least an essential singularity,* viz. the point of accumulation of poles; (v) *if a function has an infinite number of poles* $\mathrm{z_n}$, *and no essential singularity, then the poles, when ordered by nondecreasing modulus* $|\mathrm{z_1}| \le |\mathrm{z_2}| \le \dots$, *must tend to infinity* $|\mathrm{z_n}| \to \infty$ *as* $\mathrm{n} \to \infty$, otherwise they would have an accumulation point, that is, an essential singularity. This completes the classification of points from the Laurent series (Table 27.1).

TABLE 27.1
Classification of Points from the Laurent Series

Laurent series in the neighborhood of finite point $z = a$:								
$F(z) =$	$\cdots$	$+A_{-n}/(z-a)^n$	$+\cdots+$	$A_{-1}/(z-a)+$	A_0+	$A_1\,(z-a)+\cdots+$	$A_n\,(z-a)^n$	$+\cdots$
	Essential singularity	Pole of order n		Simple pole	Ordinary point	Simple zero	Zero of order n	Vanish identically
$F(z) = F(1/\zeta)$	$\cdots$	$+A_n\,\zeta^n$		$A_1\,\zeta+$		$A_{-1}/\zeta+\cdots+$	A_{-n}/ζ^n	$+\cdots$
Laurent series in the neighborhood of point infinity $z = \infty$:								

The series is truncated at left where indicated.

Note: The Laurent series can be used to classify the point $z = a$ around which is made the expansion in powers $(z-a)^n$ from the lowest exponent n present: (i/ii) zero (pole) of order n for $n(-n)$ positive integer; (iii/iv) simple zero (pole) for $n = 1$ ($n = -1$); (v/vi) ordinary point (essential singularity) for $n = 0$ ($n = -\infty$). The point-at-infinity corresponds to an expansion in $z \to 1/\zeta$ thus exchanging $+n$ by $-n$ and, leading to a classification based on the largest n: (i/ii) zero (pole) of order n at infinity for $-n$ ($+n$) positive integer; (iii/iv) simple zero (pole) for $n = -1$ ($n = 1$); (v/vi) ordinary point (essential singularity) for $n = 0$ ($n = \infty$).

27.9 Integral, Meromorphic, and Polymorphic Functions

After completing (Subsection 27.9.1) the classification of functions (Table 27.2) by their singularities (Table 27.1), a few examples are given of two classes: (i) the meromorphic functions (Subsection 27.9.2) that have an infinite number of poles accumulating only at infinity, and are associated with series of fractions and infinite products; (ii) the polymorphic functions that have essential singularities, serve to demonstrate some of the unusual properties (Subsection 27.9.3) found earlier (Sections 27.1 and 27.2). An essential singularity (Sections 27.3–27.8) can be combined with a branch-point (Sections 7.1–7.6), leading (Subsection 27.9.4) to function with bounded or unbounded fluctuation (Subsection 27.9.5).

27.9.1 Classification of Functions by Their Singularities

The classification of complex functions by their singularities (Table 27.2), can now be completed starting with those that are holomorphic in the finite part of the complex plane: (i) if the function is holomorphic at infinity, it is a **constant** (Section 27.6); (ii) if it has pole of order n at infinity, it is $\mathcal{P}_n$ a **polynomial of degree n** (Subsection 27.7.1); (iii) if it has an essential singularity at infinity, it is an **integral function** I, whose Taylor series (27.21a,b) has an infinite radius of convergence. Proceeding to the functions having singularities in the finite part of the plane: (iv) if the function has a finite number of poles, but no essential singularities, it is Q_n^m **rational function of degrees n, m**, where m is the sum of the orders of the poles in the finite part of the plane, and n is the order or the pole at infinity (Subsection 27.7.2); (v) if it has a finite number of poles with orders adding to m for finite z and an essential singularity at infinity, it is I_m a **rational-integral function of degree m**; (vi) if it has an infinite number of poles, and no essential singularity, it is **meromorphic function** Y, and the poles tend to infinity; (vii) if it has an essential singularity in the finite part of the plane it is **polymorphic function** Z, regardless of what other singularities it may have in the finite plane or at infinity.

The constant, polynomial and rational functions have already been considered, so examples are given of the other four types. The exponential (27.42b) in an integral function, and multiplying it by z^{-N} specifies a rational-integral function (27.58a), with a pole of order N at the origin:

$$z^{-N}\exp(z) = \sum_{n=0}^{\infty} \frac{z^{n-N}}{n}; \quad A_{-1} = \frac{1}{(N-1)!}, \quad \int^{(0+)} z^{-N} \mathrm{e}^{z}\, \mathrm{d}z = \frac{2\pi i}{(N-1)!}, \tag{27.58a–c}$$

the residue is (27.58b) so that its loop integral is given by (27.58c). The circular cotangent:

$$\cot z = \frac{\cos z}{\sin z} = \frac{\cos z}{\sin\left[(z-n\pi)+n\pi\right]} = (-)^n \frac{\cos z}{\sin(z-n\pi)}, \tag{27.59a}$$

$$\lim_{z\to n\pi}(z-n\pi)\cot z = \lim_{z\to n\pi} \frac{z-n\pi}{\sin(z-n\pi)}(-)^n \cos z = (-)^n \cos(n\pi) = 1, \tag{27.59b}$$

has an infinite number of simple poles at $z_n = n\pi$, that is (27.59a) the simple zeros of $\sin z_n = \sin(n\pi) = 0$; since $\cos z$ takes the value $\cos(n\pi) = (-)^n$ at the poles, the residues (27.59b) are all $+1$. The poles $z_n = n\pi \to \infty$ tend to infinity as $n \to \infty$, so the circular cotangent is a meromorphic function. The functions (27.42c) and (27.44a) are polymorphic, since they both have an isolated essential singularity at the origin; the function:

$$\exp\left(-\frac{1}{z^2}\right) = \sum_{n=0}^{\infty} \frac{\left(-z^{-2}\right)^n}{n!} = \sum_{n=0}^{\infty} (-)^n \frac{z^{-2n}}{n!} \tag{27.60}$$

TABLE 27.2
Classification of Complex Functions by Their Singularities

Symbol	Function	Finite complex plane			Point-at-infinity		
		Holomorphic	Pole	Essential singularity	Holomorphic	Pole	Essential singularity
–	Constant	X	–	–	X	–	–
$\mathcal{P}_n$	Polynomial of degree n	X	–	–	–	Order n	–
Q_n^m	Rational of degrees n,m	X	Sum of orders $= m$	–	–	Order n	–
I	Integral	X	–	–	–	–	X
I_m	Rational-integral of degree m	X	Sum of orders $= m$	–	–	–	X
Υ	Meromorphic	X	Infinite isolated	–	?	?	?
Z	Polymorphic	X	?	X	?	?	?

– does not have; X does have; ? this criterion does not affect the classification.

Note: The Laurent series around each singularity is used classify that singularity according to Table 27.1; then the ensemble of all singularities of a function can be used in Table 27.2 to classify the function as: (i) a constant if it is holomorphic in the whole complex plane including infinity; (ii) a polynomial of degree n if the only singularity is a pole of order n at infinity; (iii) a rational function of degrees (n, m), for example, the ratio of polynomials of degrees (n, m) in the numerator (denominator) without common roots, if: (iii–1) there is a pole of order n at infinity; (iii–2) there are poles in the finite plane whose orders add to m; (iv) integral if the only singularity is an essential singularity at infinity, for example, exponential and circular/hyperbolic sine/cosine; (v) rational-integral of degree m if the singularities are (iv) plus (iii–2), for example, an exponential divided by a polynomial of degree m; (vi) meromorphic if the singularities are an infinite number of poles which accumulate at infinity, for example, the circular/hyperbolic secant/cosecant/tangent/cotangent; (viii) polymorphic if there is at least one essential singularity in the finite plane, for example, an exponential or circular/hyperbolic sine/cosine of an inverse power has an essential singularity at the origin.

as a complex function also has an essential singularity at the origin. As a function of real variable $z \equiv x$, was shown to be infinitely differentiable everywhere (27.7a,b) and to have vanishing derivatives of all orders at the origin (27.8a); yet it is not zero (Figure 27.1), and its Taylor series does not converge, viz. its remainder (27.8b) coincides with the function. *The classification of points of a function (a function by its singularities) is summarized in Table 27.1 (27.2),* and further cases appear in Example 30.10 (30.11).

27.9.2 Series of Fractions (Infinite Product) for the Circular Cotangent (Sine)

The circular cotangent (27.59a,b) is a meromorphic function since it has infinite number of poles at $z_n = n\pi$ with residue $+1$; hence subtracting the principal part (27.24b) at all the poles leads to:

$$g(z) \equiv \cot z - \sum_{n=-\infty}^{+\infty} (z - n\pi)^{-1} = \cot z - \frac{1}{z} - \sum_{n=1}^{\infty} \left[(z - n\pi)^{-1} + (z + n\pi)^{-1} \right]$$
$$= \cot z - \frac{1}{z} - 2z \sum_{n=1}^{\infty} \left(z^2 - n^2\pi^2 \right)^{-1}. \tag{27.61}$$

The series on the r.h.s. of (27.61) has general term $O(n^{-2})$, and hence is absolutely convergent, as the harmonic series (29.40d). The function $g(z)$ on the l.h.s. of (27.61) has no singularities, because the principal parts at all the poles have been subtracted out of $\cot z$; also it is bounded for finite z because the series converges. Hence (27.48) it is a constant; the constant can be calculated at any point, for example, the origin using (19.39a,b):

$$\text{const} = g(z) = \lim_{z\to 0} g(z) = \lim_{z\to 0} \cot z - \frac{1}{z} = \lim_{z\to 0} \left(\frac{\cos z}{\sin z} - \frac{1}{z} \right)$$
$$= \lim_{z\to 0} \left(\frac{1 + O(z^2)}{z + O(z^3)} - \frac{1}{z} \right) = \lim_{z\to 0} \frac{1}{z} \left[\frac{1 + O(z^2)}{1 + O(z^2)} - 1 \right]$$
$$= \lim_{z\to 0} \frac{1}{z} \left[1 + O(z^2) - 1 \right] = \lim_{z\to 0} O(z) = 0. \tag{27.62}$$

Thus follows (27.63a) implying (27.63b) from (27.61):

$$g(z) = 0: \qquad \cot z = \frac{1}{z} + 2z \sum_{n=1}^{\infty} \left(z^2 - n^2\pi^2 \right)^{-1}; \tag{27.63a,b}$$

the latter is the **series of fractions** *for the circular cotangent, and converges absolutely (totally) for* $|z| < \infty (|z| \le M < \infty)$, *except in the neighborhood of the poles* $|z - n\pi| > 0$ $(|z - n\pi|) \ge \varepsilon > 0)$.

The series of fractions (27.63b) can be written:

$$\frac{d\left[\log(\sin z)\right]}{dz} = \frac{\cos z}{\sin z} = \cot z = \frac{d(\log z)}{dz} + \sum_{n=1}^{\infty} \frac{d\left[\log\left(z^2 - n^2\pi^2\right)\right]}{dz}. \tag{27.64}$$

Since the series is uniformly convergent for (27.65a) it can be integrated term-by-term (Section 21.6) from O to z leading to (27.65b):

$$|z| \le M < \infty: \quad 0 = \left[\log\left(\zeta^{-1}\sin\zeta\right)\right]_0^z - \sum_{n=1}^{\infty}\left[\log\left(\zeta^2 - n^2\pi^2\right)\right]_0^z$$
$$= \log\left(z^{-1}\sin z\right) - \lim_{z\to 0}\log\left(z^{-1}\sin z\right) - \sum_{n=1}^{\infty}\log\left(\frac{z^2 - n^2\pi^2}{-n^2\pi^2}\right). \tag{27.65a,b}$$

The second term on the r.h.s. of (27.65a) vanishes by (19.37a) ≡ (5.14b) ≡ (27.66a), and the remaining terms lead to (27.66b):

$$\lim_{N\to\infty}\log\left(z^{-1}\sin z\right) = \log 1 = 0: \ \log\left(z^{-1}\sin z\right) = \lim_{N\to\infty}\log\left[\prod_{n=0}^{N}\left(1 - \frac{z^2}{n^2\pi^2}\right)\right]. \tag{27.66a,b}$$

This is the **infinite product** *for the circular sine:*

$$\sin z = z\prod_{n=1}^{\infty}\left(1 - \frac{z^2}{n^2\pi^2}\right), \tag{27.67}$$

that is absolutely (totally) convergent for $|z| < \infty(|z| \le M < \infty)$. The infinite product (27.67) involves the factors z, $1 \pm z/n\pi$ that vanish at the simple zeros $z = 0, n\pi$ of the sine. Other series of fractions and infinite products can be obtained by similar and extended methods. The infinite product (27.67) was derived from the series of fractions (27.63b). The latter was established (27.63a) using the theorem of identification of constants (Section 27.6). The series of factions (infinite products) for a meromorphic function (function with an infinite number of zeros) can also be established by alternate approach (Section 36.7) comparing the method of images (Chapters 16, 24, 26, 28) with the conformal mapping (Chapters 33–35). These infinite representations like series, raise issues of convergence (Chapter 29).

27.9.3 Polymorphic Functions and Essential Singularities

After the meromorphic functions (Subsection 27.9.2), next are considered some polymorphic functions besides (27.42c; 27.44a; 27.60). The three functions shown in Figures 27.1 and 27.2 as functions of real variable, are polymorphic as functions of a complex variable, since they all have an essential singularity at the origin. The function in Figure 27.1 was considered before (27.7a–d; 27.60). The complex (27.68a) function (27.68b):

$$z \in |C: \quad f(z) = \sin\left(\frac{1}{z}\right); \quad \sin\left(\frac{1}{z_n}\right) = 0, \quad z_n = \frac{1}{n\pi}, \quad \lim_{n\to 0} z_n = 0, \tag{27.68a–e}$$

it has an infinite number (27.68d) of zeros (27.68c) on the real axis, with accumulation point (27.68e) at the origin; it is bounded for real variable $|\sin(1/x)| \le 1$, but has unbounded fluctuation at the origin, because it varies between ±1 in ever shorter intervals (Figure 27.2a). This leads for real variable (27.69a) to:

$$x \in |R: \quad f'(x) = x^{-2}\sin\left(\frac{1}{x}\right), \quad f'(0) = \infty. \tag{27.69a–c}$$

an infinite (27.69c) derivative (27.69b) at the origin.

The function (27.68b) is has no limit at the origin for real $z = x$; it has a continuous limit (27.70b) if multiplied by x in (27.70a):

$$h(z) \equiv z\, f(z) = z \sin\left(\frac{1}{z}\right): \qquad \lim_{x \to 0} x \sin\left(\frac{1}{x}\right) = 0; \tag{27.70a,b}$$

its derivative (27.71a):

$$h'(z) = \sin\left(\frac{1}{z}\right) - z^{-1} \cos\left(\frac{1}{z}\right), \qquad h'(0) = \infty, \tag{27.71a,b}$$

does not exist at the origin (27.71b). The function (27.70a) provides an example of **extension by continuity**: (i) the function x sin $(1/x)$ is not defined at the origin, that is, its domain is the set $X = |R- \{0\} \equiv)-\infty,\ 0\ (U)\ 0,\ \infty$ consisting of the real line minus the origin; (ii) the limit (27.70b) at the origin extends the function to the whole real line while preserving continuity.

The function (27.72a) is the algebraic inverse of (27.68a):

$$g(z) \equiv \frac{1}{f(z)} = \csc\left(\frac{1}{z}\right); \qquad z_n = \frac{1}{n\pi} = x_n, \qquad \csc\left(\frac{1}{x_n}\right) = (-)^n\, \infty, \tag{27.72a–c}$$

it has an infinite number of simple poles at the points (27.72b) ≡ (27.68d) where its inverse has simple zeros, with the same accumulation point (27.68e) at the origin; it is unbounded (Figure 27.2b) jumping between $\pm\infty$ at ever closer intervals (27.72c), while lying outside the unit band $|\csc\,(x)| \geq 1$. The three real functions illustrated in Figure 27.1 (27.2a,b), namely, $\exp\left(-1/x^2\right)$, [sin $(1/x)$ and csc $(1/x)$] have "unusual" behavior at the origin that is an essential singularity for the corresponding complex function.

27.9.4 Coincidence of an Essential Singularity with a Branch-Point

The function of a complex variable (27.68b) has an essential singularity at the origin; the origin is a branch-point (Sections 7.1–7.3) of the power with complex exponent z^α. The essential singularity and branch-point at the origin coincide for the product function (27.73b):

$$\zeta = \frac{1}{z}: \quad j(z) = z^\alpha \sin\left(\frac{1}{z}\right) = \zeta^{-\alpha} \sin(\zeta) = j\left(\frac{1}{\zeta}\right); \tag{27.73a,b}$$

this also specifies the behavior at infinity ($\zeta \to \infty$ as $z \to 0$) of the product of the sine by an inverse power. If $\alpha \in |Z$ is an integer the origin $z = 0$ (point at infinity $z = \infty$ or $\zeta = 0$) is an essential singularity only; if α is not an integer $\alpha \notin |Z$ then it is also a branch-point. Using the series of the exponential (27.42a) in the definition of sine leads to the Maclaurin series:

$$\sin z = \frac{e^{iz} - e^{-iz}}{2i} = \frac{1}{2i} \sum_{n=0}^{\infty} \frac{(iz)^n - (-iz)^n}{n!} = \sum_{n=0}^{\infty} \frac{(-)^n}{(2n+1)!} z^{2n+1}, \tag{27.74}$$

that consists of integral powers; the branch-point of the function (27.73b) introduces:

$$j(z) = \sum_{n=0}^{\infty} \frac{(-)^n}{(2n+1)!} z^{\alpha+2n+1} = \sum_{n=0}^{\infty} \frac{(-)^n}{(2n+1)!} \zeta^{-\alpha-2n-1} = j\left(\frac{1}{\zeta}\right), \tag{27.75}$$

nonintegral powers.

As a function of real variable (27.73b) is infinitely oscillating with the same zeros (27.68d) ≡ (27.72b) accumulating at the origin and: (i) constant amplitude (Figure 27.2a)

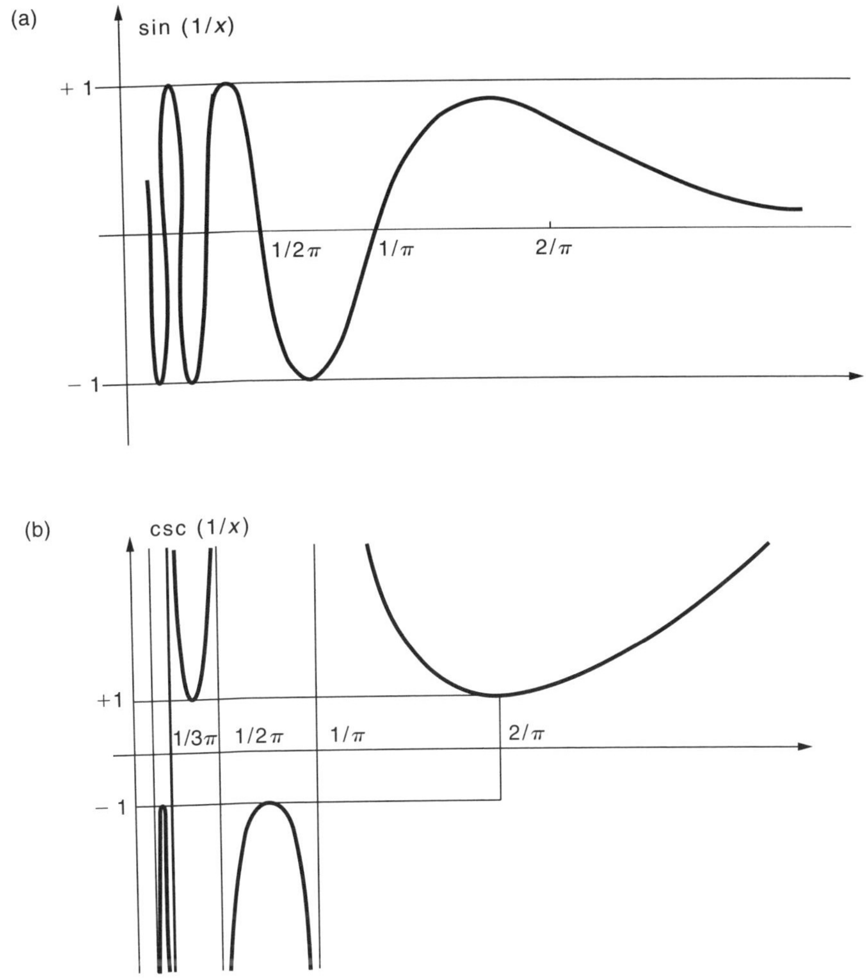

FIGURE 27.2
The function sin $(1/x)$ [its inverse csc $(1/x)$] both have an essential singularity at the origin, and their ranges of variation $|\sin(1/x)| \leq 1[|\csc(1/x)| \geq 1]$ overlap only at the points $1/x = n\pi + \pi/2$ with n integer; they differ most at the simple zeros (poles) $1/x = n\pi$. Although (a) is bounded, unlike (b), it oscillates infinitely as the origin is approached, leading to an unbounded derivative there.

for $\alpha = 0$; (ii) a power-law amplitude (Figure 27.3) for $\alpha \neq 0$. The latter includes (ii-1) linear amplitude growth (Figure 27.4b) for $\alpha = 1$; (ii-2/3) faster (slower) amplitude growth for $\alpha > 1$ $(\alpha < 1)$ in Figure 27.4c (27.4a). The derivative (27.76a) vanishes (27.76b) at the points (27.76c):

$$j'(x) = x^{\alpha-1}\left[\alpha \sin\left(\frac{1}{x}\right) - \frac{1}{x}\cos\left(\frac{1}{x}\right)\right], \quad j'(x_n) = 0, \quad \cot\left(\frac{1}{x_n}\right) = x_n\alpha. \qquad (27.76\text{a–c})$$

that are the local maxima and minima of the function. These are determined by the points of intersection (Figure 27.5) of (27.77b) with (27.7a) that tend to (27.77c) for large n:

$$y_n \equiv \frac{1}{x_n}: \qquad \cot y_n = \frac{\alpha}{y_n}, \qquad \lim_{n\to\infty} y_n = \left(n + \frac{1}{2}\right)\pi. \qquad (27.77\text{a–c})$$

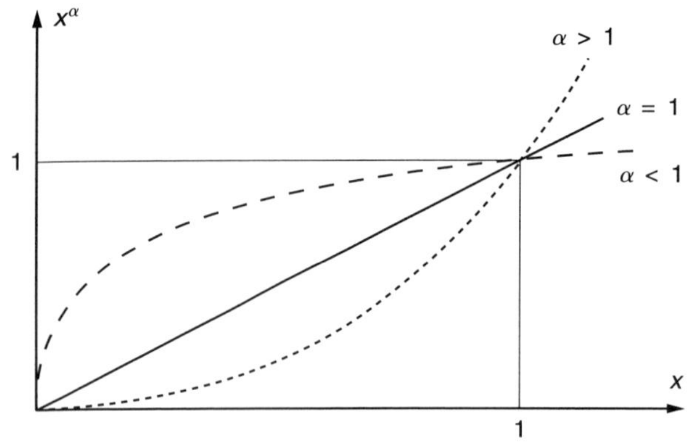

FIGURE 27.3
In the unit interval $0 \leq x \leq 1$ the function x^α lies above (below) the straight line x for $\alpha < 1 (\alpha > 1)$. This is used as amplitude to multiply sin $(1/x)$ in Figures 27.4a,b,c for respectively $\alpha < 1$, $\alpha = 1$ and $\alpha > 1$.

Thus the extrema of the function (27.73b) occur for large n for (27.77a,c) $\equiv$ (27.78a):

$$\lim_{n\to\infty} x_n = \frac{1}{(n+1/2)\pi}; \qquad \lim_{n\to\infty} j_n \equiv \lim_{n\to\infty} j(x_n) = (-)^n \left[\left(n+\frac{1}{2}\right)\pi\right]^{-\alpha}, \qquad (27.78\text{a,b})$$

at the extrema the function takes the value (27.78a,b). Thus *the function (27.73a,b) of complex variable has for α not an integer an irregular singularity, that is, a coincident branch-point and essential singularity at the origin (27.75). As function of real variable it oscillates infinitely (Figures 27.2; 27.4a–c) as the origin is approached, with decreasing amplitude if $\alpha > 0$.* Thus arises the question of whether the function is of bounded fluctuation; this depends on α as shown next.

27.9.5 Infinitely Oscillating Functions with Bounded or Unbounded Fluctuation

The function (27.68b) with real variable is bounded near the origin, with unbounded fluctuation because it oscillates infinitely with unit amplitude. Before proceeding to consider the function (27.73a,b), a rigorous definition of fluctuation is given; this starts with the partition of an interval, as used in the definition of Riemann integral (Section 13.2). Consider: (i) a partition of the interval (a, b) defined by (27.79a); (ii) the corresponding fluctuation of the function is (27.79b) defined as the sum of moduli of differences:

$$a \equiv x_0 < x_1 < \ldots < x_n < \ldots < x_N \equiv b: \quad F(f(x); x_0, \ldots, x_N) = \sum_{n=0}^{N-1} |f(x_{n+1}) - f(x_n)|; \qquad (27.79\text{a,b})$$

(iii) the total **fluctuation** or variation is the supremum of (27.79b) for all possible partitions of the interval. *If a real function has successive maxima (minima) at* $x_{2n}(x_{2n+1})$ *its fluctuation is given by the sum of the differences:*

$$f(x_{2n}) \equiv f_{\max}, \; f(x_{2n+1}) \equiv f_{\min}: \quad F(f(x);\; a, b) = \sum_{n=0}^{N-1} |f(x_{2n}) - f(x_{2n+1})|. \qquad (27.80)$$

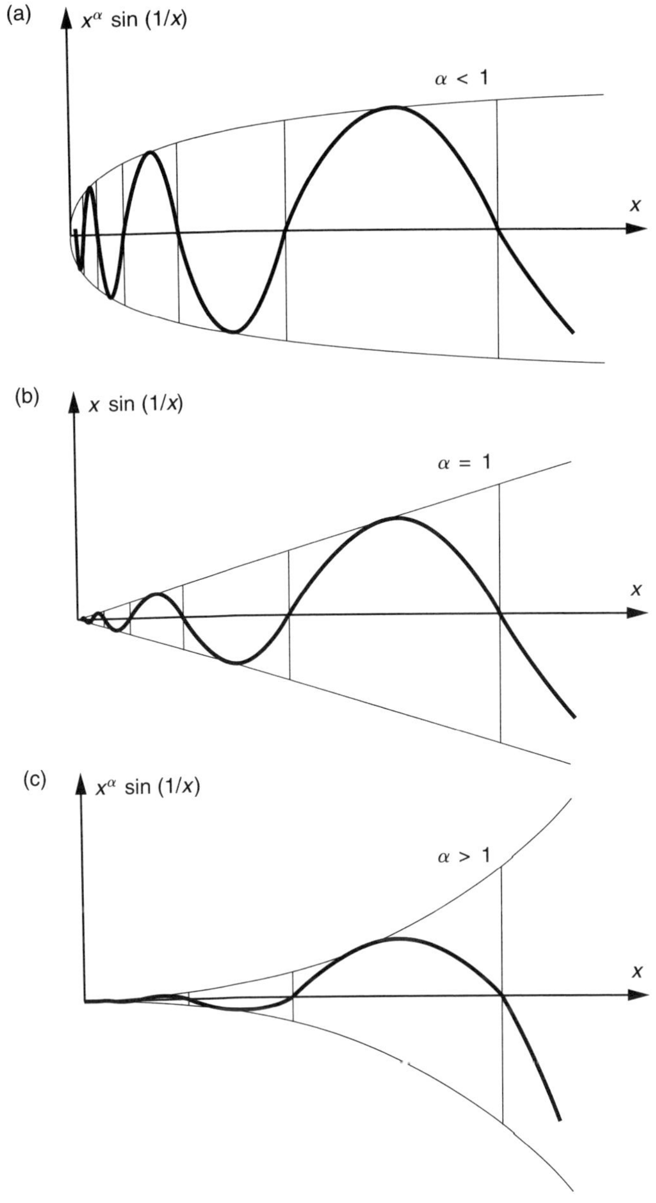

FIGURE 27.4
The function sin $(1/x)$ oscillates infinitely (Figure 27.2a) as the essential singularity at the origin is approached. If multiplied by x^α with $\alpha > 0$ in Figure 27.3 the amplitude decays toward the origin, fast (not fast) enough to be of bounded fluctuation for $\alpha > 1(\alpha \leq 1)$ in (c) [(a,b)]. The fluctuation of the function is the sum of the moduli of the difference between successive maxima and minima. For example (Figure 27.2a): (i) the maxima (minima) are $+1(-1)$; (ii) the modulus of their difference 2; (iii) there is an infinite number of extrema approaching the origin; (iv) thus the function is of bounded (unbounded) fluctuation in any interval not including (including) the origin.

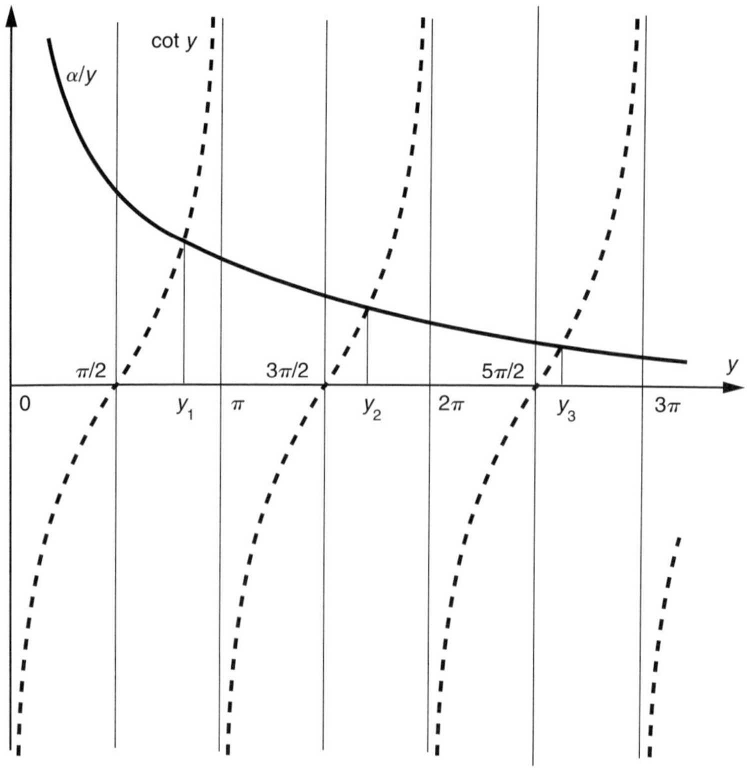

FIGURE 27.5
In the case of the function in Figure 27.4a–c the extrema lie at the points $x_n = 1/y_n$ of intersection of $\cot y$ and α/y. These tend as $n \to \infty$ to $y_n \to n\pi + \pi/2$ so $x_n \to 0$ accumulate toward the origin. The difference of the extrema scales like the series $(x_n)^\alpha \sim (n\pi + \pi/2)^{-\alpha}$ corresponding to a harmonic series $\sum n^{-\alpha}$. The latter converges (diverges) for $\alpha > 1$ ($\alpha \le 1$) in (29.40a–d), so the function is of bounded (unbounded) fluctuation near the origin in (c) [(a,b)].

In the case of the function (27.68b): (i) the difference between successive maxima and minima is 2; (ii) the fluctuation in any interval including the origin is a series of constant terms equal to 2; (iii) the series diverges, so the function has unbounded fluctuation.

The case of the function (27.73b) of real variable: (i) the extrema approaching the origin are (27.78b); (ii) the fluctuation in any interval containing the origin scales as (27.81):

$$a > 0\text{: } F\left(x^\alpha \sin\left(\frac{1}{x}\right);\ 0, a\right) \sim 2\sum_{n=1}^{\infty} j_n \sim \frac{2}{\pi}\sum_{n}^{\infty}\left(n + \frac{1}{2}\right)^{-\alpha} \sim \sum_{n=1}^{\infty} n^{-\alpha}; \qquad (27.81)$$

(iii) the first terms of the series have a finite sum, so they do not affect its convergence (Subsection 29.3.1), and the asymptotic limits (27.78a,b) can be used in (27.81). The following cases can be considered: (i) for $\alpha = 0$ the series (27.81) of constant terms diverges and the function (27.68a) has unbounded fluctuation, as found before; (ii) for $\alpha = 1$ the series (27.81) is the arithmetic series (21.6c) that is divergent, so a linear amplitude (Figure 27.4b) leads to unbounded fluctuation (27.82a); (iii/iv) for faster (slower) amplitude decay $\alpha > 1$ ($\alpha < 1$) in Figure 27.3, the harmonic series (27.81) converges (diverges) by (29.40d) [(29.40b)], and the function in Figure 27.4c (27.4a) has bounded (unbounded) fluctuation

(27.82b) [(27.82a)] in any finite interval including the origin:

$$a > 0: \quad F\left(x^\alpha \sin\left(\frac{1}{x}\right);\ 0, a\right) \begin{cases} = \infty & \text{if } \alpha \geq 1 \\ < \infty & \text{if } \alpha > 1. \end{cases} \tag{27.82a,b}$$

It has been shown that *the function of real variable (27.73b) oscillates infinitely with zeros (27.68d) accumulating at the origin (27.68e). Its amplitude decays toward the origin (Figure 27.4), leading to maxima and minima decreasing in modulus (27.78b) at (27.78a). The total fluctuation (27.80) in any finite interval including the origin (27.81) is bounded (27.82b) for faster than linear amplitude decay (Figure 27.4c), and unbounded (27.82a) otherwise (Figures 27.1 and 27.4a,b).*

Note 27.1. *"Removable" or "Apparent" Singularities:* There is sometimes mention of "removable singularities" in expressions such as

$$f(z) = \frac{e^z - 1}{z}, \qquad g(z) = \frac{z - \sin z}{z^3}, \tag{27.83a,b}$$

because the denominator "appears" to be singular, viz. a simple (triple) "pole" in (27.83a) [(27.83b)]. In fact the simple (triple) "pole" in the denominator cancels with the simple (triple) zero in the numerator, and there is no singularity at all, viz. the point $z = 0$ is an ordinary point where the functions are finite and nonzero:

$$\lim_{z\to 0} \frac{e^z - 1}{z} = 1, \qquad \lim_{z\to 0} \frac{z - \sin z}{z^3} = \frac{1}{6}; \tag{27.84a,b}$$

this can be confirmed from: (i) the leading terms of (23.60) [(19.39a)]; (ii) the Maclaurin series for the exponential (27.42a) [sine (27.74)] that lead to (27.85a) [(27.85b)]:

$$\frac{e^z - 1}{z} = z^{-1} \sum_{n=1}^{\infty} \frac{z^n}{n!} = \sum_{n=0}^{\infty} \frac{z^n}{(n+1)!} = 1 + \frac{z}{2} + \frac{z^2}{6} + \frac{z^3}{24} + \cdots \tag{27.85a}$$

$$\frac{z - \sin z}{z^3} = z^{-3} \sum_{n=1}^{\infty} (-)^n \frac{z^{2n+1}}{(2n+1)!} = \sum_{n=0}^{\infty} (-)^{n+1} \frac{z^{2n}}{(2n+3)!} = \frac{1}{6} - \frac{z^5}{120} + \cdots. \tag{27.85b}$$

The cases of ratios of functions that are finite, zeros, or infinite can be dealt with by the L'Hôpital rule (Sections 19.8 and 19.9). The ignorable singularities (Section 19.6) are not removable singularities in the same sense of (27.83a,b); the ignorable singularities are singularities that can be removed by integration whereas the "removable" singularities are not singularities at all. For this reason the latter do not appear in the classification of singularities in Table 27.1 that is a part of the classification of points of a function in Diagram 39.1. The classification of functions by their singularities in Table 17.2 is relevant to various families of functions. Besides the poles and essential singularities the other important class of special points is branch points (Chapters 7, 9 and 39). These play an essential role in the extension of derivatives to complex order (Campos, 1984).

Note 27.2. *Classification of Points and Families of Functions:* There will be further consideration of the properties of functions in the neighborhood of zeros and poles (Chapter 31) and essential singularities (Chapter 39), besides the earlier discussion of poles and zeros (Chapters 15, 17, 19, and 27). The Laurent power series expansion can be used to classify the singularities of a function in the finite complex plane and at the point-at-infinity. Then these singularities can be used to classify the functions. Although the functions are classified from power series expansions, their properties may suggest the consideration of other infinite representations, besides parametric integrals (Chapter 13), such as series of fractions, infinite products (Subsection 27.9.2 and Section 36.6) and continued fractions. If from a meromorphic function are subtracted the principal parts at all poles, the result is a function holomorphic everywhere that must be a constant: this suggest the existence of a representation of meromorphic functions (like the cosecant) by series of fractions, if the latter converge. Another example is a polynomial that can be factorized from its zeros; similarly a function with an infinite number of zeros (like the sine) could be factorized as infinite product, if it converges. There were the particular examples of series of fractions and infinite products given in Subsection 27.9.2. Like power series (Chapters 23, 25, and 27), the other infinite representations also raise issues of convergence (Chapters 21 and 29).

Note 27.3. *"Pathologic" Functions as Proofs and Counterproofs:* Although the functions with "unusual" properties may appear as "pathologic" cases they can be useful as: (a) counterproofs to dismiss invalid conjectures; (b) as tools used to prove important theorems. The examples of (a) include (i) the real function with complex derivative (Note 7.2) because it is many valued; (ii) the Dirichlet functions (Note 13.1) that are discontinuous everywhere and have no Riemann integral; (iii) the series of continuous/differentiable/integral functions that have a sum that is discontinuous/nondifferentiable/nonintegrable because of nonuniform convergence (Section 21.5); (iv) the bounded infinitely oscillating functions with bounded or unbounded fluctuation in a finite interval containing an essential singularity (Subsections 27.9.3–27.9.5); (v) the series (Example 30.6) that converges to different functions in distinct regions of the plane; (vi) the Osgood function (Section 31.2) that cannot be extended beyond the unit circle, because the singularities are dense throughout the boundary. An example of (b) is the elliptic modular function. It has the following properties: (i) it is holomorphic in the upper half complex plane and cannot be extended; (ii) it is singular only at three points $z = 0, 1, \infty$. The mere existence of such a function (without any further details) is sufficient to prove a number of fundamental theorems on essential singularities and roots of transcendental equations (Sections 35.7 and 39.3–39.6).

Conclusion 27: The function (Figure 27.1) $\exp\left(-1/x^2\right)$ vanishes at the origin with all its derivates; hence it has a null Taylor series, and coincides with its remainder. The function $\sin(1/x)$ has (Figure 27.2a) zeros at $x_n = 1/n\pi$, and thus oscillates in the range ± 1 ever faster as the origin is approached, where its derivate is infinite; the function $\csc(1/x)$ is (Figure 27.2b) singular at the same points $x_n = 1/n\pi$, across which it jumps between $\pm\infty$, between curves lying above $+1$ or below -1, becoming more infinitely more numerous as the origin is approached. Considering the infinitely oscillating functions (Figure 27.4a–c) with power law amplitude (Figure 27.3), the local maxima and minima lie at the intersection (Figure 27.5) of the cotangent with a hyperbola. The difference between the successive maxima and minima approaching the origin, can lead to unbounded fluctuation (Figures 27.1; 27.4a; 27.4b); if the amplitude decays fast enough the fluctuation in bounded (Figure 27.4c).

28

Forces and Moments on Bodies

The kinetic energy of a potential flow contained in a domain can be expressed as an integral over the boundary (Section 28.1) and likewise for the electric and magnetic energies. A body in a stream (Section 28.2) may be acted by: (i) two force components, viz. a drag (thrust) force along (opposite) to the motion, and a transverse lift (down force); (ii) a component of the moment orthogonal to the plane, viz. the pitching moment. The drag/thrust (lift/down force) is associated with source/sinks (vortices), and is analogous (Section 28.3) to the electric (magnetic) force on electric charges (currents). The gravity field has differences as regards the energy (Section 28.3) and similarities as regards the force (Section 28.1). A source (sink) in a uniform stream specifies (Section 28.4) a Rankine semiinfinite body (fairing) that is subject to a drag (thrust); a source-and-sink pair aligned with the flow specifies a Rankine finite body with an oval shape that is subject (Section 28.5) to a pitching moment in a flow when at angle-of-attack. Reversing the position of the source and sink can lead to (i/ii) a valley between mountains (i) and a throated nozzle (ii); (iii) a gap between two semiinfinite bodies. In the dipole limit of coincidence of source-and-sink the Rankine oval becomes a cylinder, whose motion in a potential flow entrains an added mass (Section 28.6) equal to the mass of fluid it displaces. The cylinder as a streamline is preserved by adding a vortex, whose circulation produces a lift or down force (Section 28.7), and displaces the stagnation points where the velocity is zero. The cylinder as a streamline is preserved by a nearby source/sink and/or vortex in the presence of two opposite images; the system of three singularities is not at rest (Section 28.8) because of the forces between the images and the monopole. The forces are modified in the case of a dipole (Section 28.9) near a cylinder. The calculation of forces and energies can be made by similar methods for gravity (Chapter 18), electrostatic (Chapter 24), and magnetostatic (Chapter 26) fields, and potential flows (Chapter 28); the latter receive more emphasis in the present section as a continuation of Chapters 12, 14, and 16.

28.1 Kinetic, Electric, Magnetic, and Gravity Energies

The calculation of the energy in a domain can be reduced to a boundary integral (Subsection 28.1.1), for example, for a corner flow or a multipole (Subsection 28.1.3). This applies to the kinetic energy, as well as to the energies associates with electric, magnetic, and gravity fields (Subsection 28.1.2).

28.1.1 Energy in a Domain as a Boundary Integral

When deriving the equation of continuity specifying the mass balance, it was shown that the mass flux (28.1a) across an infinitesimal rectangle (Figure 14.1a) was equal to the total

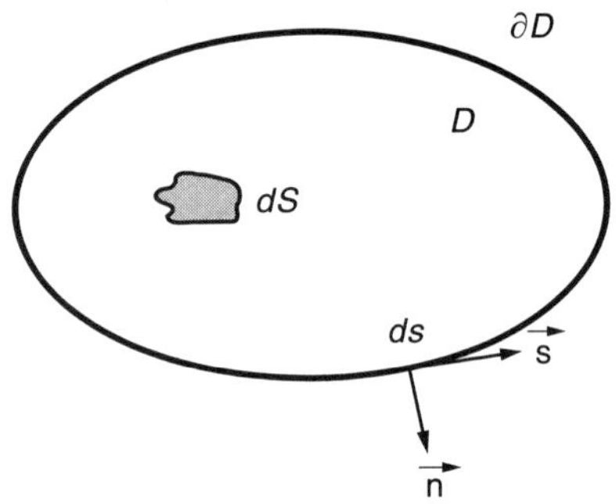

FIGURE 28.1
The divergence theorem applies in the plane to a domain D which is the interior of a closed regular boundary ∂D curve of arc length ds with continuous unit tangent $\vec{s}$ (normal $\vec{n}$) vector at all points.

divergence in the interior (28.1b):

$$\vec{J}_v \equiv \rho\vec{v} \equiv \vec{G}: \qquad \int_D (\nabla\cdot\vec{G})\,dS = \int_{\partial D} (\vec{G}\cdot\vec{n})\,ds; \tag{28.1a,b}$$

it can be proved that (28.1b) holds for an arbitrary domain, and corresponds to the **two-dimensional divergence theorem**: *(i) given a two-dimensional domain D of area dS with a closed regular boundary, that is, where (Figure 28.1) the unit normal (tangent) vector $\vec{n}$ ($\vec{s}$) along the boundary curve ∂D of arc length ds are continuous (28.2a):*

$$\vec{s}\cdot d\vec{x} = ds = |\vec{n}\wedge d\vec{x}| \in C; \quad \vec{G}\in C^1(R^2) \tag{28.2a,b}$$

(ii) a vector field $\vec{G}$ with continuous first order derivative (28.2b); then the flux of the vector through the boundary equals the integral of the divergence in the interior (28.1b). As an example consider the kinetic energy of a fluid in a domain (28.3c):

$$\rho = \text{const}, \quad \vec{v} = \nabla\Phi: \quad E_k = \frac{1}{2}\int \rho|\vec{v}|^2\,dS = \frac{\rho}{2}\int (\nabla\Phi)^2\,dS, \tag{28.3a–d}$$

in the case of a potential flow that is, both irrotational (Section 12.1) and incompressible (Section 12.2). The kinetic energy simplifies to (28.3d) because: (i) the mass density is constant (28.3a) in a incompressible flow; (ii) the velocity is the gradient of a potential (28.3b) in an irrotational flow.

To transform the integrand of (28.3d) into a divergence, the following identity is used

$$\frac{\partial\left[\Phi\left(\partial\Phi/\partial x\right)\right]}{\partial x} - \Phi\frac{\partial^2\Phi}{\partial x^2} = \left(\frac{\partial\Phi}{\partial x}\right)^2; \tag{28.4}$$

adding a similar relation in y, leads to the **Green (1828) first identity** *for (28.5b) a twice differentiable potential (28.5a)*:

$$\Phi\in\mathcal{D}^2: \quad \nabla\cdot(\Phi\,\nabla\Phi) - \Phi\,\nabla^2\,\Phi = \nabla\Phi\cdot\,\nabla\Phi \equiv (\nabla\,\Phi)^2. \tag{28.5a,b}$$

Since the potential satisfies the Laplace equation (28.6a), substitution of (28.5b) in (28.3d) simplifies the latter to (28.6b):

$$\nabla^2\Phi = 0: \qquad E_k = \frac{\rho}{2}\int_D \nabla\cdot(\Phi\nabla\Phi)\,dS = \frac{\rho}{2}\int_{\partial D}\Phi(\vec{n}\cdot\nabla\vec{\Phi})ds, \tag{28.6a,b}$$

where the divergence theorem (28.1b) was also used. The Cauchy–Riemann conditions show that the normal derivative of the potential is the tangential derivative of the stream function [(11.46a) with $h_1 = h_2$ and $x_1 \equiv n, x_2 \equiv s$], and thus:

$$(\vec{n} \cdot \nabla\Phi)\ \mathrm{d}s = \frac{\partial\Phi}{\partial n}\mathrm{d}s = \frac{\partial\Psi}{\partial s}\mathrm{d}s = \mathrm{d}\Psi, \tag{28.7}$$

since the displacement is along a streamline; substitution of (28.7) simplifies (28.8b) to

$$E_k = \frac{\rho}{2}\int\limits_{\partial D} \Phi\ \mathrm{d}\Psi, \tag{28.8}$$

stating that *the kinetic energy of a potential flow in a domain D, is specified by the integral (28.8) along its boundary, involving the potential Φ and stream function Ψ.*

28.1.2 Electric, Magnetic, and Gravity Energies

The electric (24.7) [magnetic (26.8)] energy in a domain D is specified by the integral (28.9b) [(28.10b)] along its boundary:

$$\varepsilon = \text{const}: \quad E_e \equiv \frac{1}{2}\int\limits_D \varepsilon E^2\ \mathrm{d}S = \frac{\varepsilon}{2}\int\limits_{\partial D} \Phi\ \mathrm{d}\Psi, \tag{28.9a–c}$$

$$\mu = \text{const}: \quad E_m \equiv \frac{1}{2}\int\limits_D \mu H^2\ \mathrm{d}S = \frac{\mu}{2}\int\limits_{\partial D} \Phi\ \mathrm{d}\Psi, \tag{28.10a–c}$$

that simplifies to (28.9c)[(29.10c)] in the case of uniform dielectric permittivity (24.5a) [magnetic permeability (26.6a)]; this follows from the kinetic energy (28.8) using the analogy:

$$(\vec{E},\ \varepsilon) \Leftrightarrow (\vec{v},\ \rho) \Leftrightarrow (\vec{H},\ \mu), \tag{28.11a–c}$$

between: (i) the velocity in a potential flow (28.11b) of mass density ρ; (ii/iii) the electrostatic (28.11a) [magnetostatic (28.11c)] field in a medium of dielectric permittivity ε (magnetic permeability μ) in Section 24.3 (26.3). There is no analogy with the gravity field, because the work of the gravity field per unit mass equals minus the gravity potential (28.12a):

$$\mathrm{d}E_g = -\mathrm{d}W = -m\vec{g} \cdot \mathrm{d}\vec{x} = m\,\mathrm{d}\Phi_g; \quad E_g = \int m\ \mathrm{d}\Phi_g; \tag{28.12a,b}$$

thus the energy of the gravity field equals the potential energy (28.12b) that is the integral of the mass times the variation of the gravity potential.

28.1.3 Kinetic Energy of Corner and Multipolar Flows

The preceding results are illustrated next for the kinetic energy of a potential flow in a circle of radius R, in two cases: (i) for a flow in a corner of angle $\beta = \pi/n$, the potential (14.71a) [stream function [(14.71b)] in Section 14.8 lead to the energy inside a circle of radius a:

$$n \equiv \frac{\pi}{\beta}: \quad E_n(r \le a) = \frac{\rho}{2}\int\limits_{r=a} Ur^n \cos(n\varphi)\ \ \mathrm{d}\left[Ur^n \sin(n\varphi)\right]$$

$$= \frac{n}{2}\rho U^2 a^{2n}\int\limits_0^{2\pi} \cos^2(n\varphi)\ \mathrm{d}\varphi = \frac{n\pi}{2}\rho U^2 a^{2n} = \frac{\pi^2}{2\beta}\rho U^2 a^{2n}, \tag{28.13a,b}$$

that vanishes (diverges) for $a \to 0$ $(a \to \infty)$; (ii) the inversion of the flow in a corner of angle $\beta = \pi/n$ with n positive integer is the flow due to the multipole of order n and moment P_n in Section 12.9, for that the potential (12.60a) [stream function (12.60b)] lead to kinetic energy outside a circle of radius a:

$$n = 1, 2, 3, \ldots \quad E_{-n}(r \geq a) = -\frac{\rho}{2} \int_{r=a} \left[-\frac{P_n}{2\pi} r^{-n} \cos(n\varphi) \right] \mathrm{d}\left[\frac{P_n}{2\pi} r^{-n} \sin(n\varphi) \right]$$

$$= n\rho \frac{P_n^2}{8\pi^2} a^{-2n} \int_0^{2\pi} \cos^2(n\varphi)\, \mathrm{d}\varphi = n\rho \frac{P_n^2}{8\pi} a^{-2n}, \qquad (28.14\text{a,b})$$

that vanishes (diverges) as $a \to \infty$ $(a \to 0)$. The kinetic energy has the sign + (−) before the integral in (28.13b) [(28.14b)] because the domain $r \leq a$ $(r \geq a)$ is inside (outside) the boundary that is thus described in the positive (negative) direction. The case (iii) of a monopole (Section 12.6), consisting of a source/sink of flow rate Q and a vortex of circulation Γ is not included in (28.14a), and the kinetic energy in an annulus of inner (outer) radius r_1 (r_2), is best calculated from the radial (12.28b) and tangential (12.33b) components of velocity:

$$E_{-1}(r_1 \leq r \leq r_2) = \frac{\rho}{2} \int_{r_1}^{r_2} r\, \mathrm{d}r \int_0^{2\pi} \mathrm{d}\varphi \left[(v_\varphi)^2 + (v_\varphi)^2 \right]$$

$$= \frac{\rho}{2} \frac{Q^2 + \Gamma^2}{4\pi^2} \int_0^{2\pi} \mathrm{d}\varphi \int_{r_1}^{r_2} r^{-1}\, \mathrm{d}r = \rho \frac{Q^2 + \Gamma^2}{4\pi} \log\left(\frac{r_2}{r_1} \right); \qquad (28.14\text{c})$$

it diverges both as $r_2 \to \infty$ and as $r_1 \to 0$. Thus has been obtained *the kinetic energy: (i/ii) inside (outside) a circle of radius a for a corner flow (28.13b) with arbitrary angle (28.13a) [a multipole (28.14b) excluding a monopole (28.14a)]; (iii) in an annalus for a monopole (28.14c).* The kinetic energy can also be expressed in terms of the complex potential (Example 30.12).

28.2 Drag/Thrust, Lift/Downforce, and Pitching Moment (Kutta, 1902a; Joukowski, 1906; Blasius, 1910)

The calculation of forces and moments on a body (Subsection 28.2.3) due to a pressure distribution on its surface (Subsection 28.2.1) can be made (Subsection 28.2.2) using the calculus of residues (Section 17.1). This leads to the Kutta–Joukowski theorem specifying the lift due to vortices and to analogue results (Subsection 28.2.4) for: (i) the drag (thrust) due to sources (sinks); (ii) the pitching moment due to monopoles and dipoles.

28.2.1 Pressure Distribution on a Body

Since a potential flow is inviscid, the only forces on a body (Figure 28.2a) are a pressure normal to the surface and directed inward. In two dimensions this leads (Figure 28.2b) to a force on a boundary element of length ds, which can be calculated in three ways: (i) along

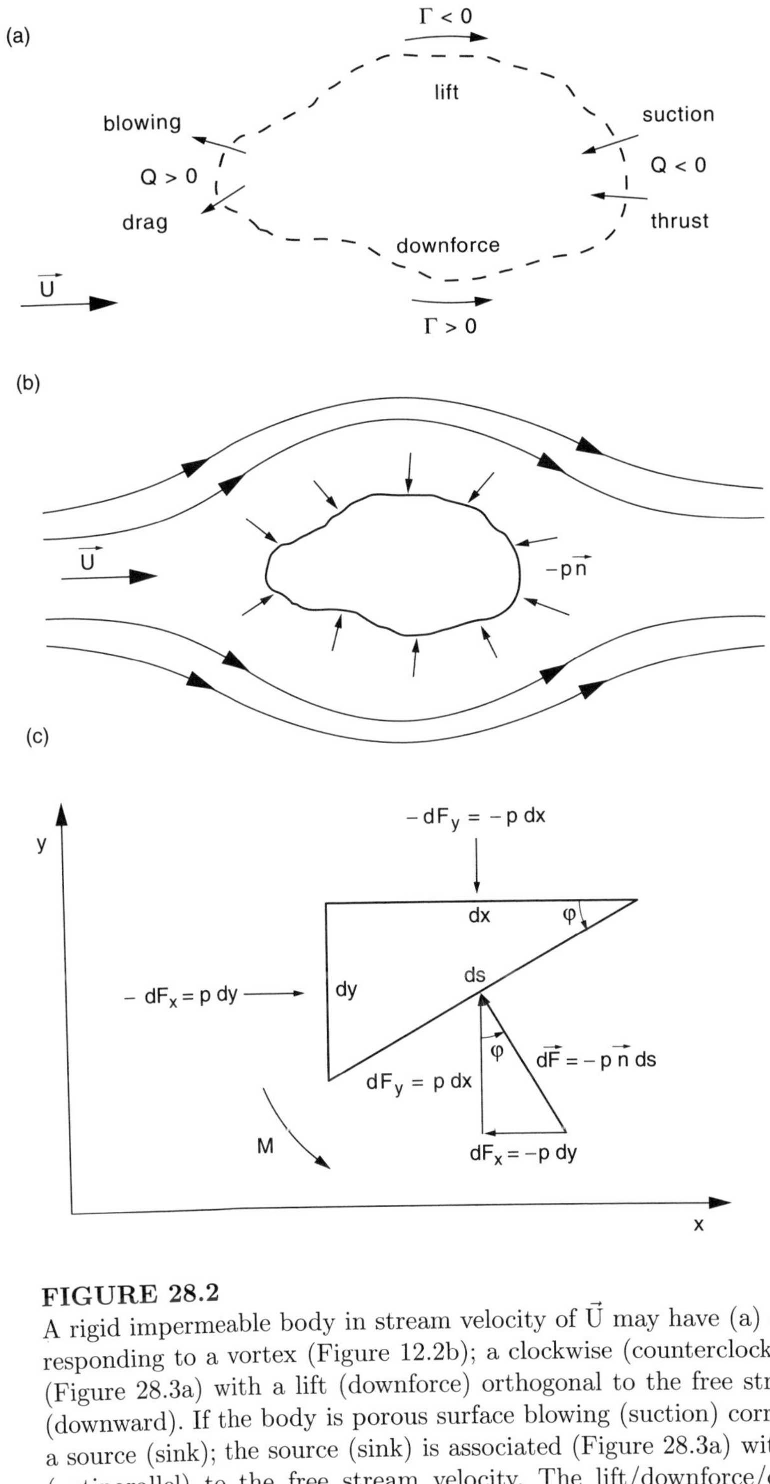

FIGURE 28.2
A rigid impermeable body in stream velocity of $\vec{U}$ may have (a) circulation around it, corresponding to a vortex (Figure 12.2b); a clockwise (counterclockwise) vortex is associated (Figure 28.3a) with a lift (downforce) orthogonal to the free stream velocity and upward (downward). If the body is porous surface blowing (suction) corresponds (Figure 12.2a) to a source (sink); the source (sink) is associated (Figure 28.3a) with a drag (thrust) parallel (antiparallel) to the free stream velocity. The lift/downforce/drag/thrust may arise as the result (b) of the inward pressure exerted along the normal at all points of the body in an

the normal (28.15b) to the tangent (28.15a) vector:

$$\vec{t} = \vec{e}_x \frac{\mathrm{d}x}{\mathrm{d}s} + \vec{e}_y \frac{\mathrm{d}y}{\mathrm{d}s} : \ \mathrm{d}\vec{F} = p\left(\vec{e}_3 \wedge \vec{t}\right)\mathrm{d}s$$
$$= p\vec{e}_3 \wedge \left(\vec{e}_x \mathrm{d}x + \vec{e}_y \mathrm{d}y\right) = p\vec{e}_y \mathrm{d}x - p\vec{e}_x \mathrm{d}y; \qquad (28.15\text{a,b})$$

(ii) opposite (28.15d) to the outward normal (28.15c,d):

$$\vec{n} = \vec{t} \wedge \vec{e}_3 = \vec{e}_x \frac{\mathrm{d}y}{\mathrm{d}s} - \vec{e}_y \frac{\mathrm{d}x}{\mathrm{d}s} : \ \mathrm{d}\vec{F} = -p\vec{n}\mathrm{d}s = -p\vec{e}_x \mathrm{d}y + p\vec{e}_y \mathrm{d}x; \qquad (28.15\text{c,d})$$

(iii) from horizontal (28.15e) and a vertical (28.15f) force components (Figure 28.2c):

$$\mathrm{d}F_x = -p\sin\varphi\,\mathrm{d}s = -p\mathrm{d}y, \quad \mathrm{d}F_y = p\cos\varphi\,\mathrm{d}s = p\mathrm{d}x; \qquad (28.15\text{e,f})$$

The equivalent forces (28.15a,b) ≡ (28.15c,d) ≡ (28.15e,f) are associated with a moment (Figure 28.3b) that is orthogonal to the plane and specified by (28.16a):

$$M = \left(\vec{e}_x x + \vec{e}_y y\right) \wedge \left(\vec{e}_x F_x + \vec{e}_y F_y\right) = \vec{e}_3\left(xF_y - yF_x\right); \qquad (29.16\text{a})$$

in the case of pressure forces on the boundary (28.15e,f) the corresponding moment is:

$$\mathrm{d}M = \begin{vmatrix} x & y \\ \mathrm{d}F_x & \mathrm{d}F_y \end{vmatrix} = p\begin{vmatrix} x & y \\ -\mathrm{d}y & \mathrm{d}x \end{vmatrix} = p\left(x\mathrm{d}x + y\mathrm{d}y\right). \qquad (28.16\text{b})$$

The complex variable (28.17a) and its conjugate (28.17c) may be introduced in the complex conjugate force (28.17b) and real moment (28.17d):

$$z \equiv x + iy : \mathrm{d}F^* = \mathrm{d}F_x - i\mathrm{d}F_y = p\left(-\mathrm{d}y - i\mathrm{d}x\right) = -ip\left(\mathrm{d}x - i\mathrm{d}y\right) = -ip\mathrm{d}z^*, \quad (28.17\text{a,b})$$
$$z^* = x - iy : \mathrm{d}M = p\,\mathrm{Re}[(x+iy)\left(\mathrm{d}x - i\mathrm{d}y\right)] = p\,\mathrm{Re}\left(z\mathrm{d}z^*\right). \qquad (28.17\text{c,d})$$

The free stream pressure p in a steady potential flow is specified by the Bernoulli equation (14.27c), leading in (28.17b,d) after integration over the boundary of the body to

$$F^* = F_x - iF_y = -i\int_L \left(p_0 - \frac{1}{2}\rho v^2\right)\mathrm{d}z^*, \quad M = \mathrm{Re}\left\{\int_L \left(p_0 - \frac{1}{2}\rho v^2\right) z\,\mathrm{d}z^*\right\}; \qquad (28.18\text{a,b})$$

Caption for Figure 28.2 Continued
inviscid flow. The total force results from an integration along the surface of the body, consisting of infinitesimal facets (c). The pressure in the inward normal direction is: (i) decomposed into horizontal (vertical) components, that is, parallel (orthogonal) to the free stream velocity; (ii) integrating over the surface of the body specifies the total horizontal (vertical) force; (iii) the latter is a lift/downforce (drag/thrust) depending on sign, that is, upward/downward (along/opposite to the free stream velocity); (iv) taking into account the position relative to the origin, the integration over the surface of the body specifies a pitching moment orthogonal to the plane of the flow; (v) a positive (negative) pitching moment in the counterclockwise (clockwise) direction corresponds to a nose down (up) relative to the direction of the free stream. These methods of calculation of hydrodynamic forces and moments also apply to electric/magnetic/gravity forces and moments (Chapters 18, 24, and 26).

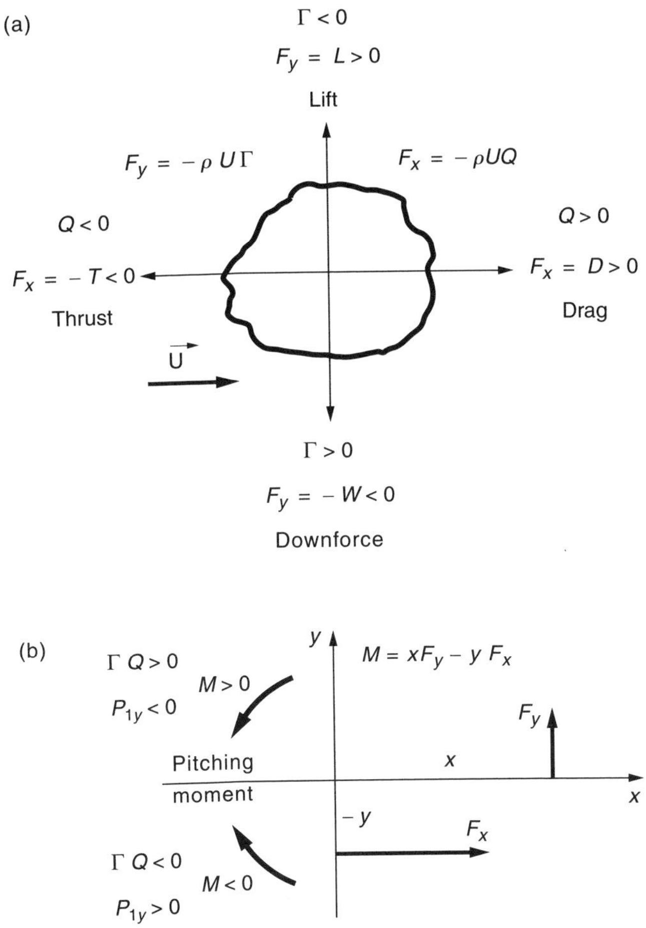

FIGURE 28.3
A source (Figure 12.2a) blowing fluid out of the body into a flow (Figure 28.2a) decreases (increases) the velocity forward (aft); by the Bernoulli law (14.27c) this increases (decreases) the pressure forward (aft), leading to a net drag force along the flow (Figure 28.3a); in the same way it is concluded that a sink (Figure 12.2a) sucking air duct of the flow into a body (Figure 28.2a) causes a net thrust force opposite to the free stream velocity (Figure 28.3a). A clockwise circulation (Figure 12.2b) around the body (Figure 28.2a) increases (decreases) the velocity above (below); this causes an under pressure above (overpressure below) and hence a net upward lift force (Figure 28.3a); a similar reasoning shows that counterclockwise vorticity (Figure 12.2b) around a body (Figure 28.2a) causes a net downward force (Figure 28.3a). These forces have multiple effects: (i) the lift balances the weight of an aircraft at cruise; (ii) the downforce adds to the weight of a car to increase friction, traction and grip in a bend; (iii) the thrust powers a vehicle forward against aerodynamic drag or friction on the ground; (iv) the drag decelerates a vehicle unless it is overcome by a propulsion system. The horizontal and vertical forces, depending on their direction and point of application, cause a pitching moment (Figure 28.3b); the sign of the pitching moment, that is, a pitch-up or down (Figure 28.3b) depends on the monopole and dipole moments (Figures 28.6, 28.8, and 28.9).

the constant stagnation pressure p_0 makes no contribution when integrated along the boundary L of the body. The modulus velocity can be expressed in terms of the complex potential by (28.19a):

$$v^2 = |v^*|^2 = v^* v = \frac{\mathrm{d}f}{\mathrm{d}z}\frac{\mathrm{d}f^*}{\mathrm{d}z^*}; \quad v^2 \mathrm{d}z^* = \frac{\mathrm{d}f}{\mathrm{d}z}\frac{\mathrm{d}f^*}{\mathrm{d}z^*}\mathrm{d}z^* = \left(\frac{\mathrm{d}f}{\mathrm{d}z}\right)^2 \mathrm{d}z; \qquad (28.19\text{a,b})$$

from (28.19a) follows (28.19b), because the surface of the body is a streamline (28.19c), leading to (28.19d):

$$\Psi = \text{const}: \quad \frac{\mathrm{d}f^*}{\mathrm{d}z^*}\mathrm{d}z^* = \mathrm{d}f^* = \mathrm{d}(\Phi - i\Psi) = \mathrm{d}\Phi = \mathrm{d}(\Phi + i\Psi) = \mathrm{d}f = \frac{\mathrm{d}f}{\mathrm{d}z}\mathrm{d}z. \quad (28.19\text{c,d})$$

Substitution of (28.19b) in (28.18a,b) leads to the **Blasius theorem**: *a body with boundary L in a flow of mass density ρ and complex potential $f(z)$ is subject to:*

$$F^* = F_x - iF_y = \frac{i}{2}\rho \int_L \left(\frac{\mathrm{d}f}{\mathrm{d}z}\right)^2 \mathrm{d}z, \quad M = -\mathrm{Re}\left\{\frac{\rho}{2}\int_L z\left(\frac{\mathrm{d}f}{\mathrm{d}z}\right)^2 \mathrm{d}z\right\}. \qquad (28.20\text{a,b})$$

a horizontal F_x and vertical F_y force (28.20a) [pitching moment M in (28.20b)]. The imaginary part of (28.20b) is considered in Example 30.13.

28.2.2 Singular Perturbation of Uniform Flow at Infinity

Consider the body in a free stream with velocity U at infinity (Figure 28.2a or 28.3): (i) the longitudinal force along (opposite) to the motion is the **drag** $D = |F_x| > 0$ (**thrust** $T = -|F_x| < 0$); (ii) the transverse force in the upward (downward) direction is the **lift** $L = |F_y| > 0$ (**downforce** $W = -|F_y| < 0$); (iii) the moment orthogonal to the plane is the **pitching moment** that is positive $M > 0$ (negative $M < 0$) counterclockwise (clockwise), that is, nose-down (nose-up). The design of an air (land) vehicle, for example, an aircraft (car): (i) maximizes lift (downforce) to compensate the weight (provide traction); (ii) while minimizing drag that consumes engine power and fuel to overcome it. For an air vehicle the pitching moment provides attitude control. *The Blasius theorem (28.20a,b) expresses the forces and moments on a body in terms of the complex conjugate velocity. The potential flow due to an arbitrary singularity at the origin is specified by a single-valued complex conjugate velocity that has a Laurent series expansion:*

$$\frac{\mathrm{d}f}{\mathrm{d}z} = v^*(z) = \sum_{n=-\infty}^{+\infty} A_n z^n = \sum_{m=1}^{\infty} A_m z^m + U\mathrm{e}^{-i\alpha} + \frac{Q - i\Gamma}{2\pi z} + \sum_{n=1}^{\infty}\frac{nP_n}{2\pi} z^{-n-1}, \qquad (28.21)$$

that may be interpreted as follows: (i) $A_1/2$, $A_2/3$, $\dots$, $A_m/(m+1)$, correspond to flows in corners (14.72a) of angle $\beta = 2\pi/(m+1)$ in (28.23) for the potential (14.70a); (ii) the constant term A_0 corresponds (28.22a) to an uniform stream (14.63a) of velocity U and angle-of-attack α; (iii) the residue A_{-1} corresponds to sources/sinks and vortices (28.22b), that is, a monopole of moment P_0 in (12.35a–c) with flow rate Q and circulation Γ; (iii) $A_{-2}, A_{-3}, \dots, A_{-n-1}\dots$ correspond to multipoles (12.59a,b) of order n and moment P_n in (28.22c):

$$n = 1, 2, \dots: \qquad A_0 = U\mathrm{e}^{-i\alpha}, \quad A_{-1} = \frac{Q - i\Gamma}{2\pi}, \quad A_{-n-1} = \frac{nP_n}{2\pi}. \qquad (28.22\text{a–c})$$

The corresponding complex potential:

$$f(z) = \sum_{m=1}^{\infty} \frac{A_m}{m+1} z^{m+1} + U e^{-i\alpha} z + \frac{Q - i\Gamma}{2\pi} \log z - \sum_{n=1}^{\infty} \frac{P_n}{2\pi} z^{-n}, \tag{28.23}$$

is not single-valued due to the logarithmic term involving the volume flow rate of the source or sink and circulation of the vortex.

Recalling the integrals (20.77a,b) $\equiv$ [(27.29b,c) with $a = 0$], viz.:

$$n \neq 1: \qquad \int^{(0+)} z^{-1}\, \mathrm{d}z = 2\pi i, \qquad \int^{(0+)} z^{-n}\, \mathrm{d}z = 0, \tag{28.24a–c}$$

it follows that *the residue A_{-1} in (28.22b) of the complex conjugate velocity (28.21) specifies the flow rate and circulation:*

$$\int^{(0+)} v^*(z)\, \mathrm{d}z = 2\pi i A_{-1} = \frac{2\pi i (Q - i\Gamma)}{2\pi} = \Gamma + iQ. \tag{28.25}$$

It will be shown next that the coefficients A_0, A_{-1} (A_0, A_{-1}, A_{-2}) specify the forces (moments) in the case of a uniform flow at infinity. The complex velocity (28.21) with (28.26a) specifies a potential flow with a constant velocity U at infinity, as a constant term plus a superposition of multipoles (28.26b):

$$0 = A_1 = A_2 = \ldots: \qquad \frac{\mathrm{d}f}{\mathrm{d}z} = U + \frac{Q - i\Gamma}{2\pi z} + \frac{P_1}{2\pi z^2} + O\left(z^{-3}\right); \tag{28.26a,b}$$

in (28.26b) the angle-of-attack was taken to be zero [this restriction will be lifted in (28.30a,b)]. In (28.26b) only the monopole (i.e., source/sink plus vortex) and dipole terms have been written explicitly; the reason is that the integrals (28.24c) with (28.24a) vanish, leaving only (28.24b) for $n = -1$. Thus the terms $O\left(z^{-3}\right)$ in (28.26b) and in its square can be omitted:

$$\left(\frac{\mathrm{d}f}{\mathrm{d}z}\right)^2 = U^2 + \frac{U(Q - i\Gamma)}{\pi z} + \left[\left(\frac{Q - i\Gamma}{2\pi}\right)^2 + \frac{P_1 U}{\pi}\right] \frac{1}{z^2} + O\left(\frac{1}{z^3}\right), \tag{28.27}$$

because: (i) the force (28.20a) involves only the coefficient of z^{-1} in (28.27), viz. (28.28a):

$$F^* = F_x - iF_y = \frac{i}{2}\rho\, 2\pi i U \frac{Q - i\Gamma}{\pi} = -\rho U (Q - i\Gamma); \tag{28.28a}$$

$$M = -\mathrm{Re}\left\{\frac{\rho}{2} 2\pi i \left[\frac{(Q - i\Gamma)^2}{4\pi^2} + \frac{P_1 U}{\pi}\right]\right\} = -\mathrm{Re}\left\{i\rho U P_1 + \rho \frac{i}{4\pi}(Q - i\Gamma)^2\right\}, \tag{28.28b}$$

(ii) the pitching moment (28.20b) involves only the coefficient of z^{-2} in (28.27), viz. (28.28b). The imaginary part of (28.28b) is considered in Example 30.13.

28.2.3 Lift in an Incident Stream with Angle-of-Attack

The preceding account proves the **forces and moment theorem**: *a body in a steady potential flow with velocity U at infinity is subject to: (i) a drag (thrust) due to a source (sink) equal to its volume flux times the velocity and mass density (28.29a); (ii) a down* *force (lift) due to a positive (negative) vortex rotating counterclockwise (clockwise) equal to the circulation times the mass density and velocity (28.29b):*

$$F_x = -\rho U Q, \quad F_y = -\rho U \Gamma; \quad M = \rho U P_{1y} - \frac{\rho Q \Gamma}{2\pi}, \tag{28.29a–c}$$

(iii) a nose-down (nose-up) pitching moment due to: (iii-1) a upward (downward) dipole moment P_{1y} transverse to the velocity multiplied by the velocity and mass density; (iii-2) a cross-term due to a spiral source (Section 12.6), involving the product by $\rho/2\pi$ of the flow rate Q of a source/sink times the circulation Γ of a vortex with the opposite (same) signs. If the mean flow has angle-of-attack α:

$$F_x = -\rho U (Q \cos\alpha - \Gamma \sin\alpha), \quad F_y = -\rho U (Q \sin\alpha + \Gamma \cos\alpha), \tag{28.30a,b}$$

$$P_1 \equiv P_{1x} + iP_{1y}: \quad M = \rho U (P_{1y} \cos\alpha - P_{1x} \sin\alpha) - \frac{\rho Q \Gamma}{2\pi}; \tag{28.30c}$$

the horizontal (28.30a) and vertical (28.30b) forces and pitching moment (28.30c) follow from:

$$F^* = F_x - iF_y = -\rho U e^{-i\alpha}(Q - i\Gamma), \quad M = \mathrm{Re}\left\{-i\rho U e^{-i\alpha} P_1\right\} - \frac{\rho Q \Gamma}{2\pi}, \tag{28.31a,b}$$

obtained (28.22a) by substituting $U \to U e^{-i\alpha}$ in (28.28a,b). The transformation from (28.29a,b) to (28.30a,b) corresponds (16.55a,b) to a rotation by an angle, α. This corresponds to (28.28a,b) with U replaced by $U e^{-i\alpha}$ leading to (28.31a,b). The original form of the Kutta–Joukowski theorem (28.29a) concerns only the lift (down force) due to a clockwise (counterclockwise) vortex; the same method (Subsection 28.2.2) using the Blasius theorem (Subsection 28.2.1) specifies (28.29b) the drag (thrust) due to a source (sink).

28.2.4 Analogue Theorems for the Lift, Drag, Thrust, and Pitching Moment

The result (28.29b) is the **Kutta (1902a)–Joukowski (1906) theorem** that may be interpreted (Figure 28.4b) as follows: (i) a negative circulation decreases (increases) the velocity below (above) the body; (ii) by the Bernoulli equation (14.27c) the pressure increases (decreases) below (above) the body; (iii) this results in a lift, viz. $F_y > 0$ for $\Gamma < 0$ in (28.29b); (iv) conversely a positive circulation $\Gamma > 0$ would produce a down force $F_y < 0$. The result (28.29a) is analogous to the Kutta–Joukowski theorem, but applies to a source/sink instead of a vortex, viz.: (i) a source in a potential flow causes a drag, for example, the Rankine semiinfinite body (Section 28.4) in Figure 28.5a; (ii) conversely a sink in a potential flow is associated with a thrust, for example the Rankine semiinfinite fairing in Figure 28.5b. A source in a uniform stream: (i) decreases (increases) the velocity in the forward (aft) direction; (ii) thus the pressure is larger (smaller) in front (behind); (iii) the net force in the free stream direction is a drag. Conversely a sink in a free stream is associated with a net force in the opposite direction, that is, a thrust, as indicated in Figure 28.3a. *An engine acts as a sink, by sucking air into the inlet and ejecting it out of the exhaust; the thrust of an engine (28.29b) is (28.32a) the velocity* U *times the mass rate (29.32b) that is the flow rate times the mass density*:

$$-F_x = \mathrm{T} = \rho U Q = \dot{m} U, \quad \dot{m} = \rho Q. \tag{28.32a,b}$$

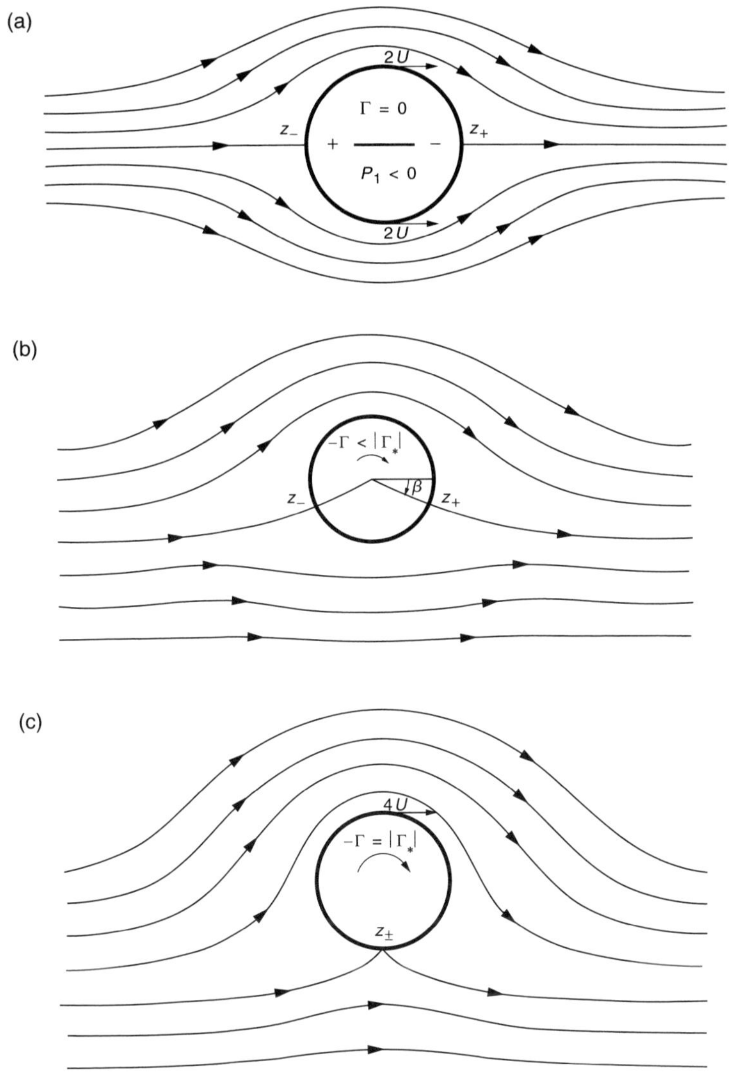

FIGURE 28.4
For a cylinder in a uniform stream there are four cases of potential flows. In the case without circulation it is represented by a dipole opposite to the free stream velocity; this is the limit of the Rankine oval (Figure 28.7) when the line source and sink converge to the same point (the axis of the cylinder) with increasing strength to keep constant the dipole moment (as in Figure 12.3). The stagnation points lie on the axis of symmetry passing through the center of the circular cross-section in the direction of the free stream velocity. Superimposing a line-vortex at the axis of the cylinder does not disturb the circular cross-section as a streamline of the flow. A negative or counterclockwise circulation below the critical value (b) moves the stagnation points downwards to symmetric positions $z_\pm$ relative to the vertical axis; it also destroys the symmetry of the flow relative to the horizontal axis, as required to produce lift. The circulation reaches the critical value $\Gamma_* = 4\pi U a$ when the stagnation points coincide $z_\pm = -ia$ at the lowest point on the cylinder (c), where the velocity induced by the vortex $\Gamma/2\pi a$ equals the velocity $2U$ due to the free stream of velocity U. For supercritical circulation one stagnation point would move into the flow, and the other inside the cylinder; the streamline through the external stagnation point z_+ separates a region of circulating flow around the cylinder (d). The stagnation point is a

(d)

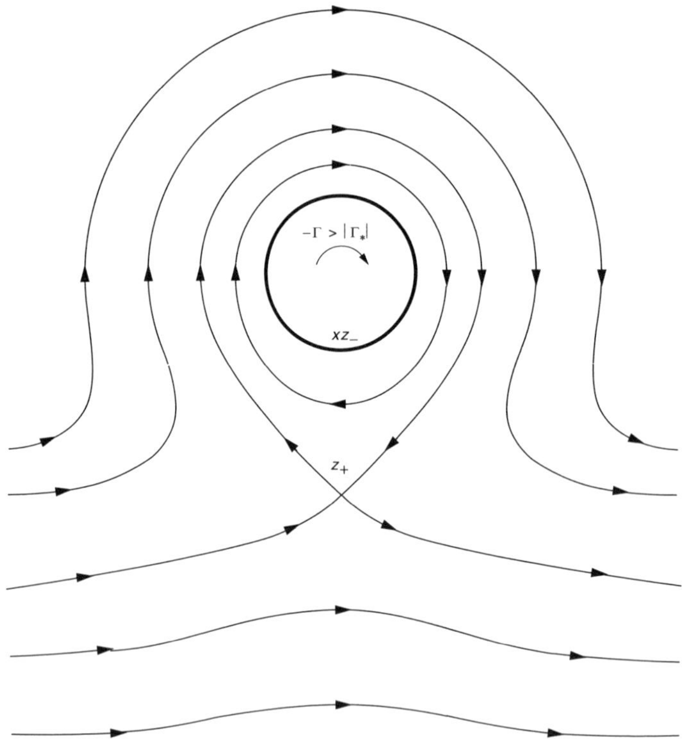

Figure 28.4 Continued
double point (Figure 13.1b, where streamlines cross at right angles, as for a corner flow in Figure 14.7). The symmetry of the flow relative to the vertical axis remains in all cases (a–d), implying that there is no drag. The potential flows without (a) and with subcritical circulation (b) occur in nature and in lifting devices like wings; a critical (c) or supercritical (d) circulation is not achievable in practice. As the circulation increases, so does the velocity below the cylinder, leading to a low pressure; the pressure rises downstream as the influence of the vortex decays. The result is a "reverse" pressure gradient, opposite to the flow velocity, which causes flow separation, for example, vorticity or turbulence. Thus the flow around the cylinder ceases to be potential for values of the circulation much lower than the critical value; the highest achievable value of the circulation determines the maximum lift before flow separation.

The forces (Figure 28.3a) are associated with (Figure 28.3b) a pitching moment (28.29c) that can be due to (i) a spiral flow (Section 12.6) that causes a torque whose direction depends on the signs of the flow rate and circulation, for example, a source and positive vortex lead to a pitch-up (Figure 28.6a); (ii) a dipole (Figure 28.6b) if it is not aligned with the free stream, for example, like the source-and-sink pair in the Rankine oval (Figure 28.7), is acted upon by a pitching moment if the line joining the source to the sink is not aligned with the flow. An external electric (Section 24.6) [magnetic (Section 26.7)] field causes in a conducting cylinder an electric charge (surface current) distribution corresponding to a dipole antiparallel to the field, so there is no moment.

Concerning the second term on the r.h.s. of the pitching moment (28.29c), Figure 28.6a shows the combination of a source and positive vortex that leads to a pitch-up. There are (Figure 28.8) four possible combinations: (i/ii) source (Figure 28.8a) or sink (Figure 28.8b) with counterclockwise vortex (Figure 12.2d); (iii/iv) source (Figure 28.8c) or sink

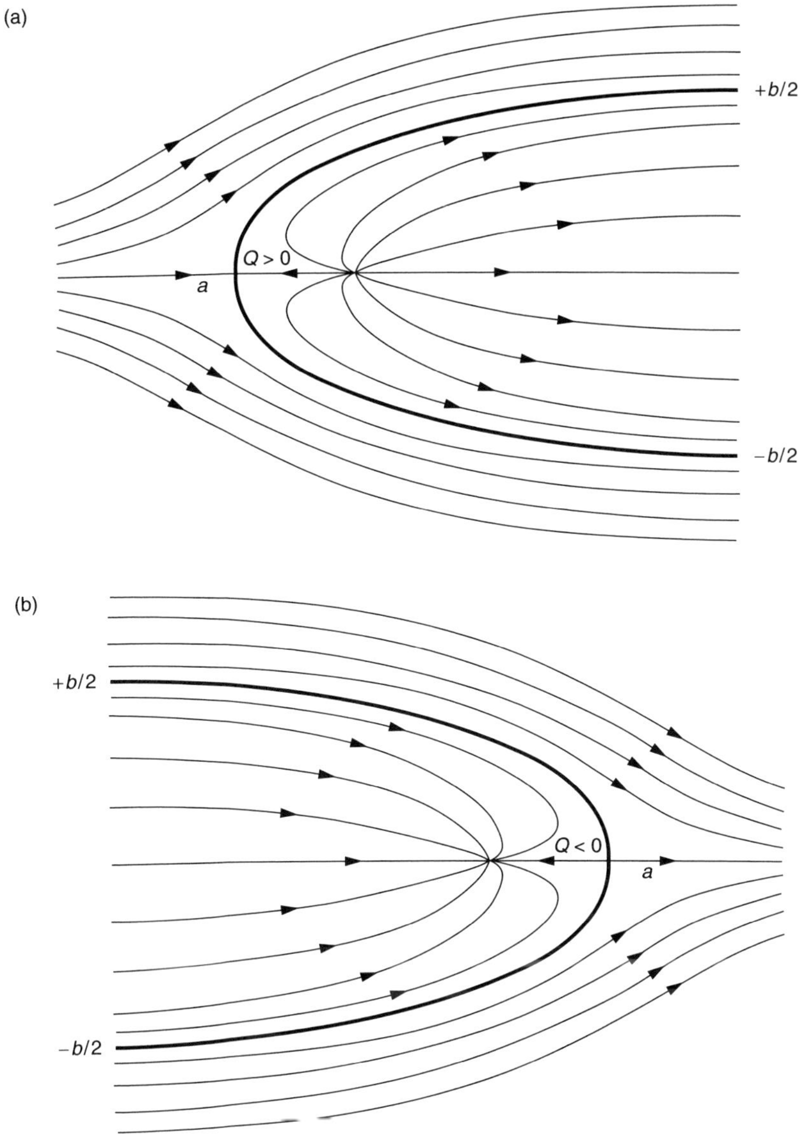

FIGURE 28.5
A source $Q > 0$ (sink $Q < 0$) in a uniform stream velocity U leads to a downstream (upstream) semiinfinite body (a) [fairing (b)] passing trough the stagnation point; the latter lies on the line through the source (sink) parallel to the free stream velocity at a distance a such that the two velocities cancel $U = |Q|/2\pi a$. The streamline through the stagnation point specifies the shape of the Rankine semiinfinite body (a) [fairing (b)]; they are identical by reflection on a line through the source (sink) perpendicular to the direction of the free stream. The Rankine semiinfinite body (fairing) separates the external from the internal flow and either of those flows can be suppressed. Since all the volume flow of the source (sink) lies inside the semiinfinite body (a) [fairing (b)] its asymptotic width b at infinity, where the free stream velocity U is undisturbed must be $b = |Q|/U = \pi a$. The pressure distribution on the Rankine semiinfinite body (a) [faring (b)] is: (i) symmetric transversely, so there is no lift or downforce; (ii) unsymmetric longitudinally, leading to a drag (thrust), as expected for a source (sink) in Figure 28.3a.

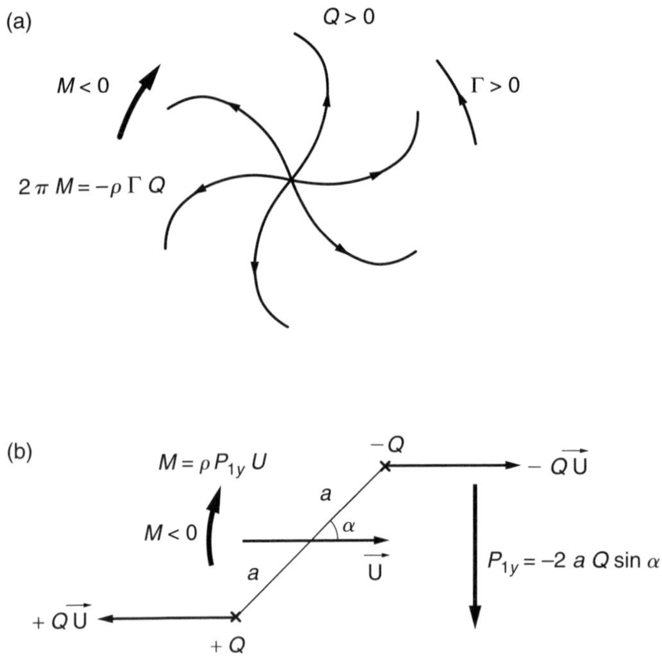

FIGURE 28.6
Two distinct ways to produce a pitching moment are illustrated by particular cases: (i) a monopole or spiral flow consisting of a source combined with a counterclockwise vortex causes a pitch up (a); (ii) a dipole not aligned with the free stream and with the vertical component of the dipole moment upward causes a pitch-down (b). The exhaustive listing of cases (i) [(ii)] appears in Figures 28.8 (28.9).

(Figure 28.8d) with clockwise vortex (Figure 12.2c). The two combinations split into two pairs as concerns the pitching moment: (i/iv) the source with positive vortex (Figure 28.8a) or a sink with negative vortex (Figure 28.8d) lead to a pitch-up; (ii/iii) a sink with a counterclockwise vortex (Figure 28.8b) or a source with a clockwise vortex (Figure 28.8c) lead to a pitch-down. Concerning the other way to cause a pitching moment, viz. the first term on the r.h.s. of (28.29c), a pitch-up is associated with a transverse downward dipole (Figure 28.9a), leaving two possibilities: (left) a sink (source) above (below), associated with a thrust (drag) causes a pitch-up; (right) a counterclockwise (clockwise) vortex aft (forward), associated with a down force (lift) also causes a pitch-up. Conversely a pitch-down is associated with a transverse upward dipole (Figure 28.9b), leaving also two possibilities: (left) a source (sink) above (below), associated with a drag (thrust) causes a pitch-down; (left) a clockwise (counterclockwise) vortex aft (forward), associated with a lift (down force), also causes a pitch-down. The calculation of forces and moments is modified if: (i) the source/sink, vortex, and dipole are at distinct locations (Example 30.14); (ii) for two dipoles, for example, a dipole and its image (Section 24.9). The forces and moments theorem can be extended to a flow with arbitrary singularities at the origin or infinity (Example 30.15), viz. using all terms of the complex conjugate velocity (28.21) in Blasius theorem. Several examples of the calculation of forces and moments appear in the following Sections 28.3–28.9. Other instances of evaluation of forces and moments on bodies are considered in Example 30.16, together with the calculation of flow rate and circulation.

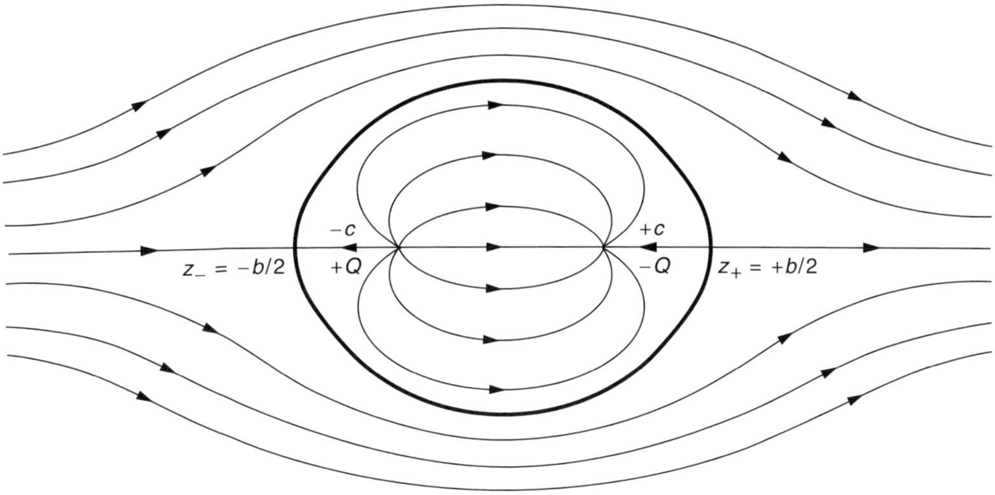

FIGURE 28.7
Combining a source (sink) with opposite flow rates (Figure 28.5a,b), with a uniform stream in the direction from the source to the sink (Figure 28.7) leads to the Rankine oval or finite body. It is symmetric relative to both: (i) the line joining the source to the sink in the flow direction; (ii) the bisector or orthogonal line at equal distance from the source and sink. This double symmetry of the of the pressure distribution implies that there are no net forces on the Rankine oval: (i) the drag of the forward half is balanced by the thrust of the rear half; (ii) the lift on the lower half is balanced by the downforce on the upper half. There is no pitching moment either, because the source and sink correspond to a dipole antiparallel to the free stream; thus the component of the dipole moment transverse to the free stream velocity is zero, excluding a pitching moment (Figures 28.6b and 28.9a,b).

28.3 Hydrodynamic, Electromagnetic, and Gravity Forces

The forces and moments theorem (Subsection 28.2.3) can be applied to gravity, hydrodynamic, and electromagnetic forces (Subsection 28.3.1) that appear in the momentum equation (Subsection 28.3.3), leading to three sets of analogies: (i/ii) the dilatation/electric, and vortical/magnetic forces (Subsection 28.3.2); (iii) the pressure/gravity forces (Subsection 28.3.4).

28.3.1 Analogies of Complex Conjugate Forces

The hydrodynamic force (28.28a) ≡ (28.33a) can be expressed in terms (12.35c) ≡ (28.33b) of the monopole moment:

$$F^* = F_x - iF_y = -2\pi\rho U P_0, \quad P_0 \equiv \frac{Q - i\Gamma}{2\pi}. \tag{28.33a,b}$$

Analogous expressions hold for the gravity (28.34c) ≡ (18.28), electric (28.35c) ≡ (24.15b), and magnetic (28.36c) ≡ (26.15) forces using the corresponding monopole moments

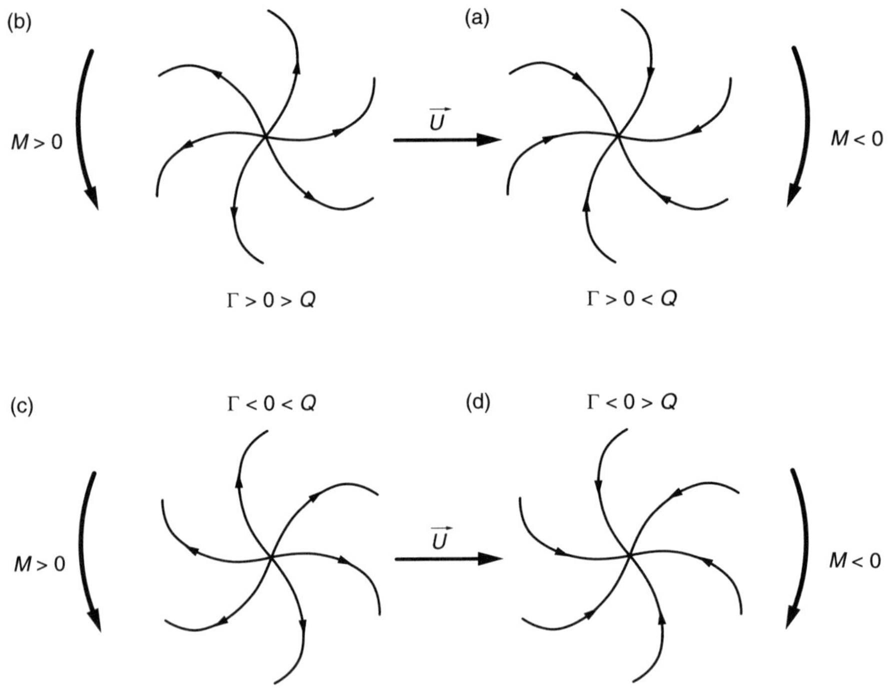

FIGURE 28.8
The source/sink (Figure 12.2a) and counterclockwise/clockwise vortex (Figure 12.2b) can be combined in four ways (Figure 12.2c,d) into a spiral flow (Figure 28.8) or monopole: (i) the source with counterclockwise vortex causes a pitch-up (a), as shown before (Figure 28.6a) = (Figure 28.8a); (ii) the first of the remaining three cases is to reverse both the source (counterclockwise vortex) into a sink (clockwise vortex) leading again to a pitch-up (d); (iii/iv) the remaining two cases reverse only one, that is, source with clockwise vortex (c) [sink with counterclockwise vortex (b)] and both lead to a pitch-down.

(28.34–28.36a) and material properties (28.34–28.36b):

$$2\pi P_0 = -mG; \quad \rho \to -\frac{1}{G}: \quad F_g^* = F_x - iF_y = \frac{1}{G}mGg^* = mg^*, \tag{28.34a–c}$$

$$2\pi P_0 = \frac{e}{\varepsilon}; \quad \rho \to \varepsilon: \quad F_e^* = F_x - iF_y = \varepsilon\frac{e}{\varepsilon}E^* = eE^*, \tag{28.35a–c}$$

$$2\pi P_0 = \frac{iJ}{c}; \quad \rho \to \mu: \quad F_m^* = -\mu\frac{iJ}{c}H^* = -\frac{i\mu H^* J}{c}, \tag{28.36a–c}$$

where are used the following analogies for: (i) the complex conjugate velocity (12.35b,c) ≡ (28.37a), gravity (18.26b) ≡ (28.37b), electrostatic (24.13b) ≡ (28.37c) and magnetostatic (26.13b) ≡ (28.37d) fields:

$$v^* = \frac{Q - i\Gamma}{2\pi z}, \quad g^* = -\frac{mG}{2\pi z}, \quad E^* = \frac{e}{2\pi\varepsilon z}, \quad H^* = \frac{iJ}{2\pi c z}; \tag{28.37a–d}$$

(ii) in all cases the monopole is at the origin with real (imaginary) moment (28.38a) [(28.38b)]for the flow rate of the source/sink, minus the mass times the gravitational constant and electric charge divided by dielectric permittivity (minus the circulation of the

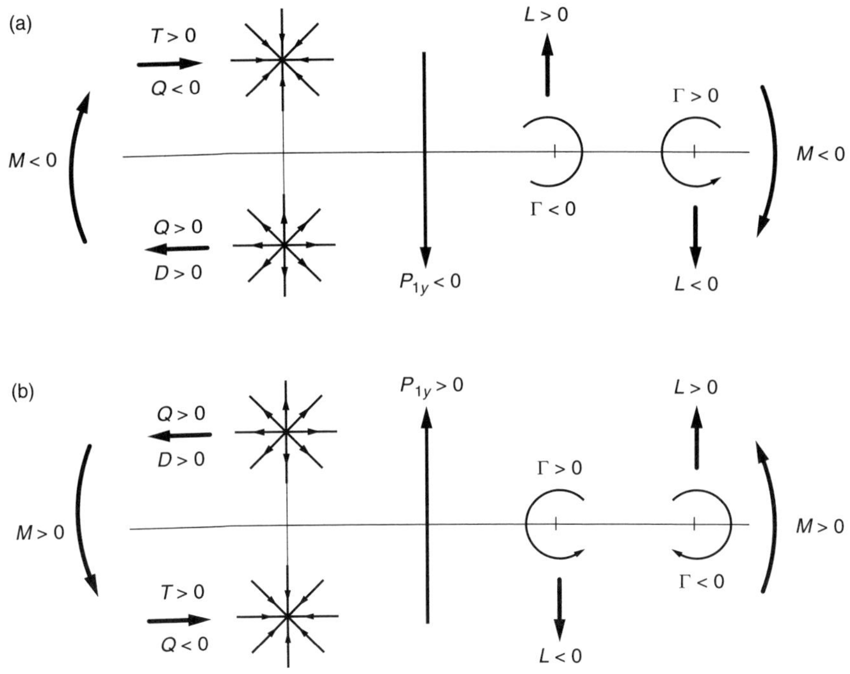

FIGURE 28.9
A dipole not aligned with the mean flow has a nonzero transverse component of the dipole moment: if it points upward (downward) it causes a pitch-down (b) [pitch-up (a)]. The pitch-up due to a downward transverse dipole moment (a) occurs in two cases: (i) a sink above (source below) associated with a thrust (drag) and hence a clockwise moment or pitch-up; (ii) a counterclockwise (clockwise) vortex behind (forward), associated with downforce (lift), causing again a clockwise moment or pitch-up. The pitch-down due to an upward transverse dipole moment (b) occurs in the cases opposite to (i-ii): (iii) a source above (sink below) is associated with a drag (thrust) and hence a counter clockwise moment or pitch-down; (iv) a clockwise (counterclockwise) vortex behind (forward) is associated with a lift (down-force) and hence again a counterclockwise pitching moment or pitch-down.

vortex and plus the electric current divided by the speed of light in vacuo):

$$\operatorname{Re}(P_0) = Q = -m\mathrm{G} = \frac{e}{\varepsilon}, \quad \operatorname{Im}(P_0) = -\Gamma = \frac{J}{c}; \tag{28.38a,b}$$

(ii) the kinetic (28.3d) ≡ (28.39a), electric (28.9c) ≡ (28.39b), and magnetic (28.10c) ≡ (28.39c) energy densities:

$$2\{E_k, E_e, E_m\} = \{\rho v^2, \varepsilon E^2, \mu H^2\}, \quad \rho \Leftrightarrow \varepsilon \Leftrightarrow \mu, \tag{28.39a–d}$$

lead to the analogy (28.39d) between mass density, and dielectric permittivity and magnetic permeability. In the case of the gravity field the monopole moment $P_0 = -mG$ involves (18.24) the total mass m together with the gravitational constant G, so $\rho = m/P_0 = -1/G$ in (28.34b). In (28.34–28.36a–c) the sign of the force is reversed relative to (28.33a), that is $-\rho$ is replaced by ρ because: (i) in (28.33a) it is the force on the body due to the monopole; (ii) in (28.34c; 28.35c; 28.36c) it is the force exerted by the external field on the monopole.

The analogies (28.38a,b) correspond to (28.33b; 28.34a; 28.35a; 28.36a), and (28.39d) correspond to (28.35b; 28.36b). Substitution of (28.34–28.36a,b) in (28.33a) with reversed sign yield (28.34–28.36c), specifying the gravity (28.34c), electric (28.35c), and magnetic forces (28.36c); the latter are discussed next (Subsection 28.3.2) in comparison with the hydrodynamic force (28.33a). The complex conjugate magnetic force (28.36c) ≡ (28.40d) corresponds (28.40c) to

$$\mu\vec{\mathrm{H}} = \vec{\mathrm{B}} = B\vec{\mathrm{e}}_x, \ \vec{\mathrm{J}} = J\vec{\mathrm{e}}_3 :$$
$$c\vec{\mathrm{F}}_m = \vec{\mathrm{J}} \wedge \vec{\mathrm{B}} = J\vec{\mathrm{e}}_3 \wedge B\vec{\mathrm{e}}_x = JB\vec{\mathrm{e}}_y = \mu JH\vec{\mathrm{e}}_y, \quad F_m = \frac{i\mu JH}{c}, \qquad (28.40\text{a–d})$$

the outer vector product of the induction field (28.40a) by the electric current (28.40b) normal to the plane. The **gravity force** (28.34c) *in complex (28.41a) or vector (28.41b) form is the product of the mass by the gravity field:*

$$F_g^* = mg^*, \quad \vec{\mathrm{F}}_g = m\vec{\mathrm{g}} = -m\nabla\Phi_g = -\nabla E_g, \quad \vec{\mathrm{f}}_g = \rho\vec{\mathrm{g}}, \qquad (28.41\text{a–c})$$

and relates to the gradient of the gravity potential and energy (28.12a,b); the ***gravity force density*** *(28.41c) replaces mass m by the mass density ρ.*

28.3.2 Dilatation/Electric and Vortical/Magnetic Forces

From (28.35c) and (28.36c) ≡ (28.40c) follow *the electric and magnetic forces appearing as the first (second) term (28.43a) ≡ (28.43b) in the* **electromagnetic or Laplace–Lorentz force**:

$$F_{em}^* = eE^* - \frac{i\mu JH^*}{c}, \quad \vec{\mathrm{F}}_{em} = e\vec{\mathrm{E}} + c^{-1}\vec{\mathrm{J}} \wedge \vec{\mathrm{B}} \equiv \vec{\mathrm{F}}_e + \vec{\mathrm{F}}_m; \qquad (28.42\text{a,b})$$

in the case of an electron with electric charge e the corresponding electric current is (28.43a):

$$\vec{\mathrm{J}} = e\vec{\mathrm{v}}, \quad \vec{\mathrm{F}}_{em} = e(\vec{\mathrm{E}} + c^{-1}\vec{\mathrm{v}} \wedge \vec{\mathrm{B}}), \qquad (28.43\text{a,b})$$

and the electromagnetic force takes the form (28.43b); the latter was used before to study electron trajectories (Chapter 6), for example, (28.43b) ≡ (6.2a). Using the Maxwell equations for the electrostatic (24.1c) [magnetostatic (26.1c)] field, *the* ***electric [magnetic] force density*** *is given by (28.44)[(28.45)]:*

$$\vec{\mathrm{f}}_e = q\vec{\mathrm{E}} = \vec{\mathrm{E}}(\nabla \cdot \vec{\mathrm{D}}) = \vec{\mathrm{E}}[\nabla \cdot (\varepsilon\vec{\mathrm{E}})], \qquad (28.44)$$
$$\vec{\mathrm{j}}_m = \frac{\vec{j} \wedge \vec{\mathrm{B}}}{c} = (\nabla \wedge \vec{\mathrm{H}}) \wedge \vec{\mathrm{B}} = \mu(\nabla \wedge \vec{\mathrm{H}}) \wedge \vec{\mathrm{H}}, \qquad (28.45)$$

in terms of the electric displacement $\vec{\mathrm{D}}$ *(magnetic induction* $\vec{\mathrm{B}}$*), and electric charge (current) density* $q(\vec{\mathrm{j}})$*, and the dielectric permittivity ε (24.2a) [magnetic permeability μ (26.2a)]; the electromagnetic force density simplifies to (28.46b):*

$$\varepsilon = \text{const}: \qquad \vec{\mathrm{f}}_{em} = \vec{\mathrm{f}}_e + \vec{\mathrm{f}}_m = \varepsilon\vec{\mathrm{E}}(\nabla \cdot \vec{\mathrm{E}}) + \mu(\nabla \wedge \vec{\mathrm{H}}) \wedge \vec{\mathrm{H}}, \qquad (28.46\text{a,b})$$

if the dielectric permittivity is constant (28.46a) in the electric force density (28.44).

The gravity (28.41c) and electromagnetic (28.46b) force densities (per unit volume) may be introduced as external forces in the inviscid momentum equation (14.9), viz.:

$$\rho\vec{\mathrm{g}} + q\vec{\mathrm{E}} + \frac{\vec{\mathrm{j}} \wedge \vec{\mathrm{B}}}{c} = \vec{\mathrm{f}}_g + \vec{\mathrm{f}}_{em} \equiv \vec{\mathrm{f}} = \nabla p + \rho\frac{d\vec{\mathrm{v}}}{dt}, \qquad (28.47\text{a})$$

showing the distinction between the irrotational (gravity, pressure, electric) and solenoidal (magnetic) fields:

$$\rho \frac{d\vec{v}}{dt} = -\nabla p + \rho \vec{g} + q\vec{E} + \frac{\vec{j} \wedge \vec{B}}{c}. \tag{28.47b}$$

The latter are compared next with the **hydrodynamic force** *consisting of drag/thrust (28.29a) and lift/downforce (28.29b) that can be written in the complex form in the plane (28.48a):*

$$F^* = F_x - iF_y = -\rho U (Q - i\Gamma); \quad \vec{f}_p = -\rho\vec{v}(\nabla \cdot \vec{v}) - \rho\vec{v} \wedge (\nabla \wedge \vec{v}) \equiv \vec{f}_b + \vec{f}_\ell, \tag{28.48a,b}$$

this is equivalent to the vector form of the force density (28.48b), where: (i) the second term on the r.h.s. is ***Lamb's vector*** *or* ***vortical force density*** $\vec{f}_\ell$*, associated with vortices or vorticity; (ii) the first term is the analogous* ***dilatation force density*** $\vec{f}_b$ *associated with source/sinks or volume changes.* The first term on the r.h.s. of (28.48b) corresponds to the first term on the r.h.s. of (28.48a), and using instead of the flow rate, the dilatation (28.49a), which is its density (18.3); this specifies the dilatation force (28.49b) density (28.49c):

$$Q \to \nabla.\vec{v} \equiv \Lambda : F_x = -\rho U Q \equiv F_b; \quad \vec{f_b} = -\rho\vec{v}\Lambda = -\rho\vec{v}\left(\nabla.\vec{v}\right) = -\vec{v}\left[\nabla.\left(\rho\vec{v}\right)\right]; \tag{28.49a–d}$$

for a potential flow that is incompressible $\rho =$ const, the two forms (28.49c,d) are equivalent. It will be shown in the sequel that the second is more general, that is, applies to compressible flow (28.60). Replacing the circulation by minus the vorticity (28.50a) that is its density (18.12), specifies the vortical force density $\vec{f_l}$ or Lamb's vector (28.48b) $\equiv$ (28.50b):

$$\varpi \equiv \nabla \wedge \vec{v} \to -\Gamma\vec{e}_3; \quad \vec{f}_\ell = -\rho\vec{v} \wedge (\nabla \wedge \vec{v}); \quad \vec{F}_\ell = \rho U \vec{e}_x \wedge \Gamma\vec{e}_3 = -\rho U\Gamma\vec{e}_y, \tag{28.50a–c}$$

corresponding (28.50c) to the second term of (28.48a).

28.3.3 Momentum Equation with Applied Forces

The vortical force, viz. the second term on r.h.s. of (28.48b) appears in the rotational component of the acceleration (14.16) in viz.:

$$\vec{a} \equiv \frac{d\vec{v}}{dt} = \frac{\partial\vec{v}}{\partial t} + \nabla\left(\frac{v^2}{2}\right) + (\nabla \wedge \vec{v}) \wedge \vec{v} = \frac{\partial\vec{v}}{\partial t} + \nabla\left(\frac{v^2}{2}\right) - \frac{\vec{f_\ell}}{\rho}; \tag{28.51}$$

thus the inviscid momentum equation (28.47b) can be written

$$\rho\frac{\partial\vec{v}}{\partial t} = -\rho\nabla\left(\frac{v^2}{2}\right) - \nabla p + \rho\vec{g} + q\vec{E} + \mu(\nabla \wedge \vec{H}) \wedge \vec{H} - \rho(\nabla \wedge \vec{v}) \wedge \vec{v}. \tag{28.52}$$

This shows the analogy between the vortical force (28.53a) and the magnetic force (28.53b) densities:

$$\vec{f}_b = -\rho\vec{\varpi} \wedge \vec{v}, \quad \vec{f}_m = \frac{\mu}{c}(\vec{j} \wedge \vec{H}) : \quad \vec{\varpi} \equiv \nabla \wedge \vec{v} \leftrightarrow -\vec{j} = -c\nabla \wedge \vec{H}, \quad \rho \leftrightarrow \mu, \tag{28.53a–d}$$

where the vorticity, $\vec{\varpi}$ corresponds to minus the electric current density $\vec{\mathrm{j}}$ in (28.53c); the reason for the minus sign is (Section 26.3) that the vorticity (electric current) is the curl of the velocity (magnetic field) that is a polar (axial) vector. The mass density ρ corresponds to the magnetic permeability μ, in agreement with (28.53d) $\equiv$ (28.39d) $\equiv$ (28.36c).

The inertia force *can be written* $\rho\,\mathrm{d}\vec{\mathrm{v}}/\mathrm{d}t$ *as the mass times acceleration, for a body with constant mass; for a body with variable mass the inertia force is the derivative of (28.54a) the linear momentum per unit volume that (i) is the product of the mass density by the velocity; (ii) coincides with the mass flux in the continuity equation (14.3):*

$$\vec{\mathrm{P}} \equiv \rho\vec{\mathrm{v}}: \quad \frac{\mathrm{d}\tilde{\mathrm{P}}}{\mathrm{d}t} \equiv \frac{\mathrm{d}(\rho\vec{\mathrm{v}})}{\mathrm{d}t} = \vec{\mathrm{v}}\frac{\mathrm{d}\rho}{\mathrm{d}t} + \rho\frac{\mathrm{d}\vec{\mathrm{v}}}{\mathrm{d}t} = -\rho\vec{\mathrm{v}}(\nabla\cdot\vec{\mathrm{v}}) + \rho\frac{\mathrm{d}\vec{\mathrm{v}}}{\mathrm{d}t}; \qquad (28.54\mathrm{a,b})$$

the equation of continuity (14.5) was substituted in the first term an the r.h.s. of (28.54b). Substituting the inviscid momentum equation (28.47b) leads to:

$$\frac{\mathrm{d}(\rho\vec{\mathrm{v}})}{\mathrm{d}t} = -\nabla p + \rho\vec{\mathrm{g}} - \rho\vec{\mathrm{v}}(\nabla\cdot\vec{\mathrm{v}}) + \vec{\mathrm{E}}[\nabla\cdot(\varepsilon\vec{\mathrm{E}})] + \frac{\vec{\mathrm{j}}\wedge\vec{\mathrm{B}}}{c}. \qquad (28.55)$$

This shows the analogy between the dilatation (28.56a) and the electric (28.56b) force densities:

$$\vec{\mathrm{f}}_b = -\rho\vec{\mathrm{v}}\Lambda = -\rho\vec{\mathrm{v}}(\nabla\cdot\vec{\mathrm{v}}), \quad \vec{\mathrm{f}}_e = q\vec{\mathrm{E}} = \vec{\mathrm{E}}[\nabla\cdot(\varepsilon\vec{\mathrm{E}})]: \qquad \vec{\mathrm{v}} \leftrightarrow \vec{\mathrm{E}}, \quad -\rho(\nabla\cdot\vec{\mathrm{v}}) \leftrightarrow q, \qquad (28.56\mathrm{a–d})$$

with: (i) the mass density ρ times the dilatation corresponding (28.56c) to minus the electric charge density $-\,q$; (ii) the mass density ρ corresponding (28.56d) to minus the dielectric permittivity $-\,\varepsilon$. The minus sign arises because the electric force is positive for a positive charge and the dilatation force is: (i) positive for a contraction, $\nabla\cdot\vec{\mathrm{v}} < 0$; (ii) negative for an expansion $\nabla\cdot\vec{\mathrm{v}} > 0$. The closest analogue of the electric force density (28.56b) is not the first (28.56a) $\equiv$ (28.49c) but rather the second form (28.49d) of the dilatation force density that will appear in the sequel in (28.60) $\equiv$ (28.61a,b). Using the decomposition (28.51) of the acceleration in (28.54b) leads to:

$$\frac{\mathrm{d}(\rho\vec{\mathrm{v}})}{\mathrm{d}t} = -\rho\vec{\mathrm{v}}(\nabla\cdot\vec{\mathrm{v}}) - \rho\vec{\mathrm{v}}\wedge(\nabla\wedge\vec{\mathrm{v}}) + \rho\frac{\partial\vec{\mathrm{v}}}{\partial t} + \rho\nabla\left(\frac{v^2}{2}\right) = \vec{\mathrm{f}}_p + \rho\left[\frac{\partial\vec{\mathrm{v}}}{\partial t} + \nabla\left(\frac{v^2}{2}\right)\right], \qquad (28.57)$$

that involves the hydrodynamic force density (28.48b), plus a term associated with the Bernoulli equation (14.24a,b) and the conservation of circulation (14.22). The derivation of (28.57) used the equation of continuity in the first term on the r.h.s. and decomposition of the acceleration (28.51) for the remaining terms; it did not use the momentum equation, and thus is not restricted to inviscid fluids.

28.3.4 The Gravity and Stagnation Forces as Gradients

Comparing the electromagnetic (28.44; 28.45; 28.46a,b), hydrodynamic (28.48b; 28.49c; 28.50b), and gravity (28.41c) force densities follow the analogies between (i) the potential fields, that is, irrotational potential flow due to source/sinks, electrostatic field due to charges, and gravity field due to masses:

$$\{\vec{v}, \nabla.\vec{v} \equiv \Lambda, Q\} \leftrightarrow \left\{\vec{E}, \nabla.\vec{E} = \frac{q}{\varepsilon}, e\right\} \leftrightarrow \{\vec{g}, \nabla.\vec{g} = -\rho g, m\}, \qquad (28.58)$$

(ii) the solenoidal fields, that is, the incompressible flow due to vortices and magnetic filed due to electric currents:

$$\{\vec{v}, \nabla \wedge \vec{v} \equiv \vec{\varpi}, -\Gamma\} \leftrightarrow \left\{\vec{H}, \nabla \wedge \vec{H} = \frac{\vec{j}}{c}, J\right\}, \tag{28.59}$$

The hydrodynamic force (28.48a) [density (28.48b)] consists of: (i) the dilatation force (28.49b) [density (28.49c)] equal to the mass density times velocity times minus the flow rate (dilatation); (ii) the vortical force (28.50c) [density (28.50b)] is equal to mass density times velocity times minus the circulation (vorticity, using the outer vector product). The hydrodynamic force density, with the dilatation term in the form (28.49c), appears in the equation (28.57) for the material derivative of the mass flux or linear momentum that is based on the equation of continuity, and makes no assumption on the constitutive properties of the fluid that may be in inviscid or viscous. The inviscid momentum equation, with external gravity and electromagnetic forces, can be put into a form: (i) showing (28.55) the analogy between the dilatation and electrical forces (28.56a–d); (ii) showing (28.52) the analogy between vortical and magnetic forces (28.53a–d). Including the gravity field as well leads to the analogies among: (i) the ***irrotational*** *fields (28.58), that is, with zero curl, including gravity, electrostatics, and irrotational flow due to source/sinks; (ii) the* ***solenoidal*** *fields (28.59), that is, with zero divergence, viz. the magnetic field and incompressible flow due to vortices.*

These analogies are all embodied in the equation:

$$\frac{\partial(\rho\vec{\mathrm{v}})}{\partial t} = \rho\frac{\partial\vec{\mathrm{v}}}{\partial t} + \vec{\mathrm{v}}\frac{\partial\rho}{\partial t} = -\vec{\mathrm{v}}[\nabla\cdot(\rho\vec{\mathrm{v}})] - \rho\nabla\left(\frac{v^2}{2}\right) - \nabla p + \rho\vec{\mathrm{g}} + q\vec{\mathrm{E}}$$
$$+ \mu(\nabla\wedge\vec{\mathrm{H}})\wedge\vec{\mathrm{H}} - \rho(\nabla\wedge\vec{\mathrm{v}})\wedge\vec{\mathrm{v}}, \tag{28.60}$$

that combines the equations of continuity (14.3) and inviscid momentum (28.52); its terms:

$$\frac{\partial(\rho\vec{\mathrm{v}})}{\partial t} = -\vec{\mathrm{v}}[\nabla\cdot(\rho\vec{\mathrm{v}})] + \vec{\mathrm{E}}[\nabla\cdot(\varepsilon\vec{\mathrm{E}})] + \mu(\nabla\wedge\vec{\mathrm{H}})\wedge\vec{\mathrm{H}}$$
$$-\rho(\nabla\wedge\vec{\mathrm{v}})\wedge\vec{\mathrm{v}} + \rho\vec{\mathrm{g}} - \nabla p - \rho\nabla\left(\frac{v^2}{2}\right), \tag{28.61a}$$

may be interpreted as follows: (i) the l.h.s. is the local time rate of change of linear momentum (28.54a); (ii/iii) the first pair terms on the r.h.s. are the dilatation (28.49d) and electrical (28.44) force densities, with the analogies (28.58); (iv/v) the second pair of terms is the vortical (28.50a,b) and magnetic (28.45) force densities, with the analogies (28.59); (vi/vii) the last pair of forces appearing in (28.61a) ≡ (28.61b):

$$\frac{\partial(\rho\vec{v})}{\partial t} = \left(\vec{f}_b + \vec{f}_e\right) + \left(\vec{f}_m + \vec{f}_\ell\right) + \left(\vec{f}_g + \vec{f}_n\right), \tag{28.61b}$$

are the gravity (28.41c) and stagnation force densities:

$$p_0 - p = \frac{1}{2}\rho v^2 q: \qquad \vec{\mathrm{f}}_n = -\nabla p - \rho\nabla\left(\frac{v^2}{2}\right) = -\nabla p_0 + \frac{v^2}{2}\nabla\rho. \tag{28.62a,b}$$

The latter designation arises because stagnation pressure (28.62a) ≡ (14.27c) can be brought into the pressure force (28.62b) that reduces to its gradient for a constant mass density; this is analogous with the gravity force (28.41b) that is also the gradient of a potential energy.

The ***stagnation force density*** *can be expressed (28.62b) in terms of: (i) the enthalpy (28.62d) using (14.12b); (ii) the stagnation enthalpy (28.62a)*

$$H_0 \equiv H + \frac{v^2}{2}: \quad \vec{\mathrm{f}}_n = -\frac{\rho}{2}\nabla(v^2) - \rho\nabla H + \rho T\nabla S = -\rho\nabla H_0 + \rho T\nabla S; \tag{28.62c,d}$$

the last term in (28.62d) vanishes for a homentropic flow. Thus in a homentropic flow (28.63a) the stagnation force density is the product of minus the mass density by the gradient of the stagnation enthalpy (28.63b):

$$S = \text{const}: \quad \vec{f}_n = -\rho\nabla H_0 = -\rho\nabla\left(H + \frac{v^2}{2}\right); \tag{28.63a,b}$$

$$\rho = \text{const}: \quad \vec{f}_n = -\nabla(\rho H + p - p_0), \tag{28.63c,d}$$

if the flow is incompressible (28.63c), the stagnation force is (28.63d) minus the gradient of the mass density times the enthalpy, plus the dynamic pressure (28.62a). For an incompressible flow (28.64a) the stagnation force density (28.63b) reduces to the gradient of the stagnation pressure (28.64b):

$$\rho = \text{const}: \quad \vec{f}_n = -\nabla p_0 = -\nabla\left(p + \frac{1}{2}\rho v^2\right); \quad \nabla \wedge \vec{v} = 0: \quad \tilde{f}_n = 0, \tag{28.64a–d}$$

if the flow is also irrotational (28.64c) the stagnation pressure is constant (14.27c), and the stagnation force vanishes (28.64d). Thus there is no stagnation force density (28.64d) in a potential flow (28.64a,c); in an incompressible (28.64a) rotational flow there is a stagnation force density (28.64b) that gains an extra term in the compressible case (28.62b). The analogies between the three pairs of dilatation/electric (28.56a–d), vortical/magnetic (28.53a–d), and gravity/stagnation (28.41c; 28.62–64a–d) force densities appears in four forms of the momentum equation for (i/ii) mass density times total (28.47b) [local (28.52)] acceleration; (iii)/(iv) material (local) time derivative (28.57) [(28.60)] of the linear momentum per unit volume or mass flux (28.54a). Of these four (i–iv) only (iii) makes no assumption on the constitutive properties of the fluid, for example, viscous or inviscid. The remaining three (i,ii,iv) apply to a inviscid fluid, and can be extended to a viscous fluid by adding the viscous stresses. The momentum equations is most often written in the form (i); the alternatives (ii,iv) are equally valid, and (iii) is more general since it is actually a restatement of the equation of continuity. The following examples (Sections 28.4–28.9) of forces and moments concern the hydrodynamic case that includes all the preceding analogies.

28.4 Fairing due to a Source or Sink in a Stream (Rankine, 1871)

A source (sink) in a uniform flow represents a semiinfinite body or fairing (Subsection 28.4.1) that experiences (Subsection 28.4.2) a drag (thrust) force.

28.4.1 Flow past a Semiinfinite Body

The spiral flow (12.35a–c) of source/sink of volume flux Q and a vortex of circulation Γ in a uniform stream of velocity U is specified by the complex potential (28.65a) and conjugate velocity (28.65b):

$$f(z) = Uz + \frac{Q - i\Gamma}{2\pi}\log z, \quad v^*(z) = U + \frac{Q - i\Gamma}{2\pi z}; \tag{28.65a,b}$$

the latter (28.65b) vanishes (28.65c) at one stagnation point (28.65d):

$$v^*(z_0) = 0: \qquad z_0 = -\frac{Q - i\Gamma}{2\pi U}, \tag{28.65c,d}$$

The stagnation point of a spiral flow due to monopole is located: (i) on the imaginary axis in the absence of source or sink, and on the right-hand (left-hand) half-plane in the presence of a sink (source); (ii) on the real axis in the absence of a vortex and in the upper (lower) half-plane in the presence of a positive (negative), that is, counterclockwise (clockwise) vortex. In the case of a source/sink without vortex, the complex potential (28.65a) [conjugate velocity (28.65b)] specifies the scalar potential (28.66a) stream function (28.66b) [Cartesian components of velocity (28.68a,b)]:

$$\Phi(r,\varphi) = Ur\cos\varphi + \frac{Q}{2\pi}\log r, \quad \Psi(r,\varphi) = Ur\sin\varphi + \frac{Q\varphi}{2\pi}; \tag{28.66a,b}$$

$$v_x(r,\varphi) = U + \frac{Q}{2\pi r}\cos\varphi, \quad v_y(r,\varphi) = \frac{Q}{2\pi r}\sin\varphi; \tag{28.67a,b}$$

the pressure follows from Bernoulli's equation (14.27c) for a steady flow:

$$p(r,\varphi) = p_\infty + \frac{\rho U Q}{2\pi r}\cos\varphi - \frac{\rho Q^2}{8\pi^2 r^2}, \quad p_\infty \equiv p_0 - \frac{\rho}{2}U^2, \tag{28.68a,b}$$

where p_0 (p_∞) is the stagnation pressure (pressure of the incident stream of uniform velocity U).

The complex potential (28.65a) and conjugate velocity (28.65b), scalar potential (28.66a) and stream function (28.66b), Cartesian components of velocity (28.67a,b) and pressure distribution (28.68a,b) correspond to the potential flow of a source (28.69a) in a uniform stream that has (28.65c,d) a stagnation point (28.69b) at (28.69c):

$$\Gamma = 0 \neq Q; \quad v^*(z_0) = 0: \quad -z_0 = \frac{Q}{2\pi U} \equiv a, \tag{28.69a–c}$$

this leads (28.66b) ≡ (28.70c) to the streamline passing through the stagnation point (28.69c) ≡ (28.70a,b):

$$r_0 = a, \varphi_0 = \pi: \quad Ur\sin\varphi + \frac{Q\varphi}{2\pi} = \text{const} = \frac{Q\varphi_0}{2\pi} = \frac{Q}{2}. \tag{28.70a–c}$$

The streamline (28.70c) specifies the shape of a ***Rankine fairing (1871)*** *in Figure 28.5a, whose half-width at infinity is (28.71b):*

$$Q > 0: \quad \lim_{r\to\infty} y = \lim_{r\to\infty} r\sin\varphi = \lim_{\varphi\to 0}\left(\frac{Q}{2U} - \frac{Q\varphi}{2\pi U}\right) = \frac{Q}{2U} \equiv a\pi \equiv \frac{b}{2}; \tag{28.71a,b}$$

this corresponds to a uniform flow velocity U at infinity, and the flow rate Q of the source (28.71a) passing within the width of the body, viz. $Q/U = 2\,a\,\pi = b$ coincides with (28.71b).

28.4.2 Drag/Thrust Force and Coefficient

The drag force (28.29a) on the body is (28.72a):

$$D = -F_x = \rho U Q = \rho 2\pi U^2 a = \rho U^2 b; \quad C_D \equiv \frac{D}{(\rho/2)U^2 b} = 2, \tag{28.72a,b}$$

it specifies the ***drag coefficient*** *defined as the ratio of drag force to the dynamic pressure of the free stream times the width of the body (28.72b). In the case of a sink (Figure 28.5b) the position of the Rankine semiinfinite body is reversed (related to Figure 28.5a) and (28.72a)*

[(28.72b)] with $Q < 0$ becomes a thrust (thrust coefficient). In the case of the sink (28.73a) the stagnation point (28.69c) is at (28.73b):

$$Q = -|Q| < 0: \quad z_0 = -\frac{Q}{2\pi U} = \frac{|Q|}{2\pi U} \equiv a, \; r_0 - a = 0 = \varphi_0, \; r \sin\varphi = -\frac{Q\varphi}{2\pi U} = a\varphi, \tag{28.73a–e}$$

and the streamline (28.66b) passing through this point (28.73c,d) and specifying the shape of the body is (28.73e). Substituting (28.71b; 28.73e) in (28.68a) specifies the pressure distribution on the Rankine semiinfinite body:

$$p(\varphi) = p_\infty - \frac{\rho}{2}\frac{U^2 a^2}{r^2} + \rho U^2 \frac{a}{r}\cos\varphi = p_\infty - \frac{\rho}{2}U^2[\varphi^{-2}\sin^2\varphi - \varphi^{-1}\sin(2\varphi)]. \tag{28.74}$$

There is no lift on the Rankine semiinfinite body, because: (a) the pressure distribution (28.74) is symmetrical $p(\varphi) = p(-\varphi)$; (b) there is no circulation, viz. $\Gamma = 0$ implies $F_y = 0$ in (28.29b). The longitudinal component of the pressure distribution (28.74) for the sink causes a thrust specified by (28.72b) as $T = -D = \rho U Q = \rho U^2 b$. The Rankine fairing represents the flow near the forward (Figure 28.5a) [aft end (Figure 28.5b)] of a blunt body, like an airship, and leads to a very high drag or thrust coefficient (28.72b); together with the large volume and low density this explains why airships are so sensitive to strong winds, and tend to break away from mooring points, unless stored indoors.

28.5 Oval/Valley/Throat due to a Source and Sink Pair

A flow source ahead of a sink in a uniform stream leads to a finite body (Subsection 28.5.1) subject to a pitching moment at an angle-of-attack (Subsection 28.5.2). Interchanging the source and sink, so that the sink faces the flow (Subsection 28.5.3) leads to (i) either a gap between fairings; (ii) or an internal (external) flow in a throated nozzle (over a valley between hills).

28.5.1 Finite Body in a Stream

A pair of equal and opposite source $Q > 0$ (sink, $-Q < 0$) symmetrically placed at $z = -c$ ($z = c$) in a uniform stream of velocity U have a complex potential (28.75a) [conjugate velocity (28.75b)]:

$$f(z) = Uz + \frac{Q}{2\pi}[\log(z+c) - \log(z-c)] = Uz + \frac{Q}{2\pi}\log\left(\frac{z+c}{z-c}\right), \tag{28.75a}$$

$$v^*(z) = U + \frac{Q}{2\pi}\left(\frac{1}{z+c} - \frac{1}{z-c}\right) = U - \frac{Qc/\pi}{z^2 - c^2}; \tag{28.75b}$$

these correspond to the scalar potential (28.76a) and stream function (28.76b) [Cartesian components of the velocity (28.77a,b)]:

$$\Phi = Ux + \frac{Q}{4\pi}\log\left[\frac{(x+c)^2 + y^2}{(x-c)^2 + y^2}\right], \tag{28.76a}$$

$$\Psi = Uy - \frac{Q}{2\pi}\operatorname{arc\,cot}\left(\frac{x^2 + y^2 - c^2}{2cy}\right), \tag{28.76b}$$

$$\{v_x - U, v_y\} = -\frac{Qc}{\pi}\left|\left(x^2 - y^2 - c^2\right)^2 + 4x^2y^2\right|^{-1}\{x^2 - y^2 - c^2,\; 2xy\}; \tag{28.77a,b}$$

this leads by Bernoulli theorem (14.27c) for a steady flow, to the pressure distribution:

$$p_\infty \equiv p_0 - \frac{\rho}{2}U^2: \quad p = p_0 - \frac{\rho}{2}\left|v^*\right|^2 = p_\infty + \rho\frac{QUc}{\pi}\frac{x^2-y^2-c^2-Qc/(2\pi U)}{\left(x^2-y^2-c^2\right)^2+4x^2y^2}, \quad (28.78\text{a,b})$$

where p_0 (p_∞) is the stagnation pressure (free stream pressure at infinity).

The passage from (28.75a) to (28.76a) [(28.76b)] involves (28.79a) [(28.79b)]:

$$\operatorname{Re}\left[\log\left(\frac{z+c}{z-c}\right)\right] = \log\left|\frac{z+c}{z-c}\right| = \log\left|\frac{x+c+iy}{x-c+iy}\right| = \frac{1}{2}\log\left[\frac{(x+c)^2+y^2}{(x-c)^2+y^2}\right], \quad (28.79\text{a})$$

$$\begin{aligned}\operatorname{Im}\left\{\log\left(\frac{z+c}{z-c}\right)\right\} &= \operatorname{Im}\left\{\log\left(\frac{x+c+iy}{x-c+iy}\right)\right\} = \operatorname{Im}\left\{\log\left[\frac{(x+c+iy)(x-c-iy)}{(x-c)^2+y^2}\right]\right\}\\ &= \arg[(x+c+iy)(x-c-iy)] = \arg\left(x^2-c^2+y^2-i2cy\right)\\ &= -\operatorname{arc\,cot}\left(\frac{x^2+y^2-c^2}{2cy}\right); \end{aligned} \quad (28.79\text{b})$$

the passage from (28.75b) to (28.77a,b) involves:

$$\frac{1}{z^2-c^2} = \frac{1}{x^2-y^2-c^2+2ixy} = \frac{x^2-y^2-c^2-2ixy}{\left(x^2-y^2-c^2\right)^2+4x^2y^2}; \quad (28.79\text{c})$$

the pressure distribution (28.78b) involves (28.75b):

$$\begin{aligned}\left|v^*(z)\right|^2 = v^*v &= \left(U - \frac{Qc/\pi}{z^2-c^2}\right)\left(U - \frac{Qc/\pi}{z^{*2}-c^2}\right)\\ &= U^2 + \frac{(Qc/\pi)^2}{\left|z^2-c^2\right|^2} - \frac{UQc}{\pi}\left(\frac{1}{z^2-c^2}+\frac{1}{z^{*2}-c^2}\right)\\ &= U^2 + \frac{Qc}{\pi}\left[\frac{Qc/\pi}{|z^2-c^2|^2} - 2U\operatorname{Re}\left(\frac{1}{z^2-c^2}\right)\right]\\ &= U^2 + \frac{Qc}{\pi}\frac{Qc/\pi - 2U\left(x^2-y^2-c^2\right)}{\left(x^2-y^2-c^2\right)^2+4x^2y^2}, \end{aligned} \quad (28.79\text{d})$$

where was used (28.79c).

The complex velocity (28.75a) vanishes (28.80a) at the stagnation points (28.80b) on the real axis:

$$v^*(z_\pm) = 0: \quad \pm z_\pm = c\left|1+\frac{Q}{\pi U c}\right|^{1/2} \equiv \frac{b}{2}, \quad (28.80\text{a,b})$$

whose distance $b = z_+ - z_-$ is the length of a **Rankine oval** (Figure 28.7). The streamline (28.76b) passing through the stagnation points (28.80b), other than the incident free stream along the real axis $\Psi = 0$, is specified by $y = 0$ or $\infty = \cot\left(2\pi\Psi/Q\right)$ that leads to $2\pi\Psi_\pm/Q = \pm\pi$ or $\Psi_\pm = \pm Q/2$; this implies that half of the volume flux goes in each of the upper and lower half-planes, and $\Psi_+ - \Psi_- = Q$ is the whole volume flux of the source. The streamline passing through the stagnation points (28.81a) specifies the shape of the oval (28.81b):

$$\Psi = 0: \quad x^2+y^2-c^2 = 2cy\cot\left(\frac{2\pi U y}{Q}\right); \quad (28.81\text{a,b})$$

the roots $y_\pm$ of (28.81b) for $x = 0$ specify the width $|y_+ - y_-|$ of the oval. There is no force on the oval because: (a) the pressure distribution (28.78a,b) is symmetric relative to both the real and imaginary axes $p(-x, y) = p(x, y) = p(x, -y)$; (b) there is no vortex (28.29b) and the opposite source and sink cancel in (28.29a).

28.5.2 Pitching Moment, Coefficient, and Slope

The absence of forces on a body in a potential flow is **D'Alembert's paradox**, *resulting from the balance of pressure forces that is violated: (i) in a potential flow in the presence of sources/sinks (28.29a) or vortices (28.29b); (ii) in a viscous flow due to viscous stresses. Also there is an angle-of-attack of the incident stream for that the pitching moment exerted on a body vanishes,* as shown next for the Rankine oval. The source and sink in (28.75a) attract each other (28.29a) with a force:

$$F^* = -\rho Q U = (-\rho Q)\left(-\frac{Q}{2\pi}\right)(2c)^{-1} = \frac{\rho Q^2}{4\pi c}. \tag{28.82}$$

The far-field of the complex velocity (28.75b) is:

$$v^*(z) = U - \frac{Qc}{\pi z^2} - \frac{Qc^3}{\pi z^4} + O(z^{-6}) = U + \frac{P_1}{2\pi z^2} + O(z^{-4}); \tag{28.83a}$$

the leading term besides the free stream is a dipole (12.41b) of moment (28.83b):

$$P_1 = -2cQ; \quad M = \mathrm{Re}(2i\rho U e^{-i\alpha} cQ) = 2\rho U c Q \sin\alpha, \tag{28.83b,c}$$

if the incident stream has an **angle-of-attack** α this leads by (28.31b) to the pitching moment (28.83c); this corresponds to a pitch-down $M > 0$ for a source $Q > 0$ at the front facing the free stream (Figure 28.7) and vice-versa a pitch-up $M < 0$ for a sink $Q < 0$ at the front. The **pitching moment coefficient** is defined (28.84a) as the ratio of the pitching moment to the dynamic pressure of the free stream times the square of the length of the body:

$$C_M \equiv \frac{2M}{\rho U^2 b^2} = 4\frac{cQ}{Ub^2}\sin\alpha = -\frac{Q/(Uc)}{1 + Q/(\pi U c)}\sin\alpha, \tag{28.84a,b}$$

where (28.80b) was used for the Rankine oval (28.84b); the **pitching moment slope** $C_{M\alpha}$ is given by (28.85a) in the limit small angle-of-attack in (28.85b):

$$C_M(\alpha) = C_{M\alpha}\sin\alpha, \quad C_{M\alpha} = \lim_{\alpha\to 0}\frac{dC_M}{d\alpha} = \frac{1}{1/\pi + Uc/Q} < \pi. \tag{28.85a,b}$$

Thus *an opposite source-and-sink pair of volume flux Q in a free stream of velocity U, specify the complex potential (28.75a) and conjugate velocity (28.75b), scalar potential (28.76a) and stream function (28.76b), Cartesian components of the velocity (28.77a,b) and pressure distribution (28.78a,b) of the potential flow past a Rankine oval (28.81b), whose length b is given by (28.80b); there are no forces on the oval, and if the free stream has an angle-of-attack α there is a pitching moment (28.83c), corresponding to the pitching moment coefficient (28.84a,b), and pitching moment slope (28.85a,b) at small angle of attack.* The presence of angle-of-attack α changes Uz to $Uze^{-i\alpha}$ in (28.75a), U to $Ue^{-i\alpha}$ in (28.75b) and $U(x, y)$ to $U(x\cos\alpha, y\sin\alpha)$ in (28.76a,b), and hence affects the location of the stagnation points and the shape of the body. The shape of the body is not changed by the angle-of-attack for source and sink converging to a dipole and leading to a cylinder (Sections 28.6–28.9).

28.5.3 Throated Nozzle and Valley between Hills

Reversing the position of the source and sink in Rankine's oval (Figure 28.7), so that the sink faces the flow, changes the complex conjugate velocity (28.75b) to (28.86a):

$$v_-^*(z) = U + \frac{Qc/\pi}{z^2 - c^2}; \quad v_-^*(z^\pm) = 0: \quad (z^\pm)^2 = c^2 - \frac{Qc}{\pi U}, \tag{28.86a–c}$$

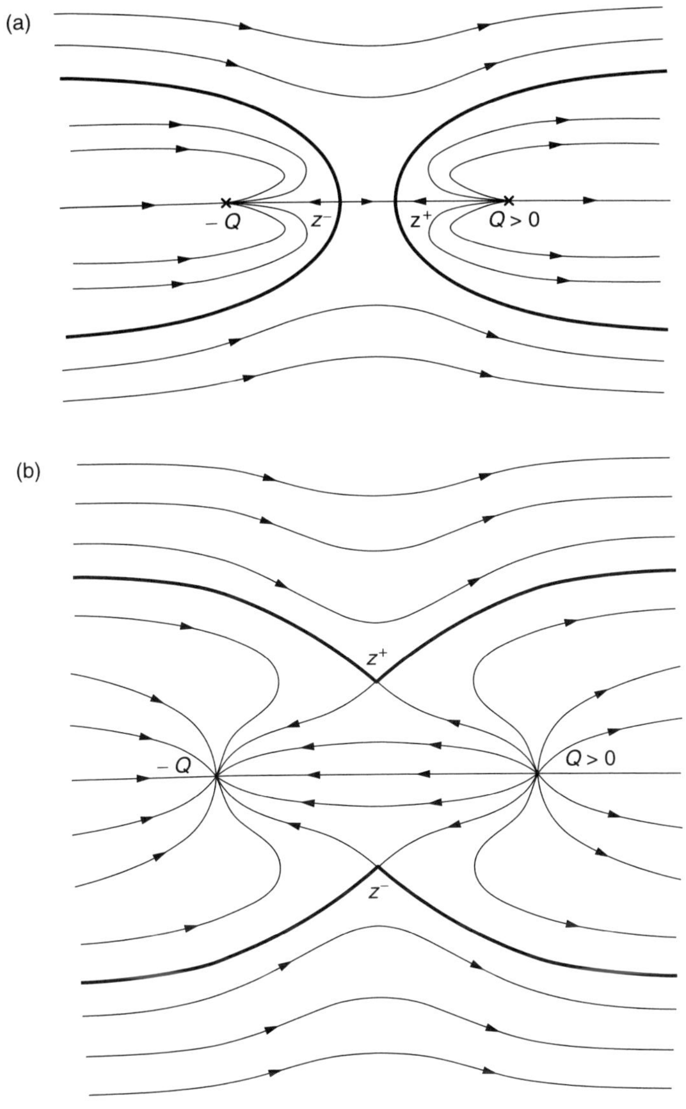

FIGURE 28.10
If in the Rankine oval (Figure 28.7) the position of the source (sink) is reversed, from upstream (downstream) to downstream (upstream) while still keeping the free stream velocity in the source-sink direction, the potential flow is completely changed (Figure 28.10). Two possibilities arise: (i) if the source (equal to minus the sink) flow rate is strong enough relative to the free stream velocity, two symmetric stagnation points appear at $z^- < 0 < z^+$ corresponding (a) to a semiinfinite fairing (body) broadly similar to Figure 28.5a (b), leaving a gap in between; (ii) in the opposite case the semiinfinite body and fairing merge into a throated nozzle narrowest between z^+ and z^- with imaginary $z^\pm$, leading outside to a flow over a "valley between hills" (b). The transition case between (i) and (ii) would have both stagnation points at the origin $z^\pm = 0$, where are joined a pointed semiinfinite body and a fairing; the four streamlines must cut at an angle $\pi/2$ as in a corner flow (Figure 14.7).

this specifies the stagnation points (28.86c) as the roots of (28.86b), viz.

$$z^{\pm} = \begin{cases} \pm c\left|1 - \dfrac{Q}{\pi U c}\right|^{1/2} & \text{if } Q \leq \pi U c, \quad (28.87\text{a}) \\ \pm ic\left|\dfrac{Q}{\pi U c} - 1\right|^{1/2} & \text{if } Q \geq \pi U c, \quad (28.87\text{b}) \end{cases}$$

Thus there are two cases, for *the potential flow with complex velocity (28.86a) that corresponds to: (i) for (28.87a) to a passage between two fairings (Figure 28.10a) at a distance $z^+ - z^-$; (ii) for (28.87b) there are (Figure 28.10b) two subcases: (ii-a) the exterior flow over symmetric hills with a valley in between; (ii-b) the interior flow in a duct with a throat of width* $|z^+ - z^-|$. In both cases (i) and (ii) the flow does not cross the real axis, because the complex velocity (28.86a) is real for real z. Also in both cases (i) [(ii)] four streamlines cross at each stagnation point $z_{\pm}(z^{\pm})$, making angles $\pi/2$ as for a square corner flow (Figure 14.10c); in the case (ii) [(i)] the two streamlines representing the body are (are not) consecutive, leading to an angular (regular) point at each stagnation point in Figure 28.10b (28.10a). In the intermediate (iii) case $Q = \pi U c$ between (28.77a,b) there is a single stagnation point at the origin $z_{\pm} = 0 = z^{\pm}$, where meet six streamlines, viz.: (i) two for the body on each side; (ii) one coming from the source and one going to the sink. The six streamlines meet at equal angles $2\pi/6 = \pi/3$ as for a hexagonal corner flow (Figure 14.10e). Taking the intermediate case (iii) as reference: (i) if the ratio $Q/Uc < \pi$ reduces the stagnation point at the origin opens a gap (Figure 28.10a) between two symmetric stagnation points on the real axis, each with four streamlines making an angle $2\pi/4 = \pi/2$; (ii) if the ratio $Q/Uc > \pi$ increases the stagnation point at the origin divides into two symmetric stagnation points on the imaginary axis, each with (Figure 28.10b) four streamlines and again an angle $2\pi/4 = \pi/4$, that a square corner (Figure 14.10c). Thus there is a discontinuous change of the angle of the streamlines: (i) from $\theta = \pi/4$ for $Q < \pi U c$ to $\theta = \pi/3$ for $Q = \pi U c$ back to $\theta = \pi/4$ for $Q > \pi U c$; (ii) as the stagnation points move from the real axis (Figure 28.10a) passing through the origin and splitting again an the imaginary axis (Figure 28.10b). This corresponds to a change of topology (Section 9.8) of the flow.

28.6 Virtual Mass of a Cylinder and Cavitation

The flow past a cylinder (Section 28.6) may be obtained as a limit of the Rankine oval (Section 28.5) as an alternative to the circle theorem (Section 24.6); both lead to a dipole (Subsection 28.6.1). The pressure distribution on the cylinder specifies the condition of cavitation when its minimum is zero (Subsection 28.6.2). The pressure distribution leads to a pitching moment for a flow at angle-of-attack (Subsection 28.6.3). The entrainment of fluid by the cylinder specifies an added mass (Subsection 28.6.4); the latter together with vortical (Subsection 28.6.5) and applied forces determine the path of cylinder (Subsections 28.6.6 and 28.6.7) in the flow.

28.6.1 Dipole Moment and Flow past a Cylinder

In the limit of coincidence of source and sink $c \to 0$ with flow rate diverging $Q \to \infty$ such that the product is constant (28.88a):

$$P_1 = \lim_{\substack{Q\to\infty \\ c\to 0}} -2cQ: \quad a^2 \equiv x^2 + y^2 = \lim_{\substack{Q\to\infty \\ c\to 0}} 2cy \cot\left(\frac{2\pi U y}{Q}\right) = 2cy\left(\frac{Q}{2\pi U y}\right) = \frac{cQ}{\pi U} = -\frac{P_1}{2\pi U}, \quad (28.88\text{a,b})$$

the Rankine oval (28.81b) becomes a circle (28.88b) with radius a specified by (28.89a):

$$P_1 = -2\pi U a^2 : \qquad v^*(z) = U + \frac{P_1}{2\pi z^2} = U\left(1 - \frac{a^2}{z^2}\right), \tag{28.89a,b}$$

thus the potential flow past a cylinder (Figure 28.10a) without circulation has a complex conjugate velocity (28.89b). The latter coincides with the same limit applied to (28.75b):

$$\lim_{\substack{Q\to\infty\\ c\to 0}} v^*(z) - U = -\lim_{\substack{Q\to\infty\\ c\to 0}} \frac{Qc}{\pi z^2} = -\lim_{Q\to\infty} \frac{Q}{\pi z^2}\left(-\frac{P_1}{2Q}\right) = \frac{P_1}{2\pi z^2}, \tag{28.90a}$$

$$\lim_{\substack{Q\to\infty\\ c\to 0}} v^*(z) - U = -\lim_{\substack{Q\to\infty\\ c\to 0}} \frac{Qc}{\pi z^2} = -\lim_{Q\to\infty} \frac{Q}{\pi z^2}\frac{\pi U a^2}{Q} = -U\frac{a^2}{z^2} = \frac{P_1}{2\pi z^2}, \tag{28.90b}$$

where (28.88a) [(28.88b)] was used in (28.90a) [(28.90b)]; the latter (28.90a) ≡ (28.90b) agree with (28.89a). The conjugate velocity (28.90a) ≡ (28.90b) ≡ (28.89a,b) corresponds to the complex potential:

$$f(z) = U\left(z + \frac{a^2}{z}\right) = Uz - \frac{P_1}{2\pi z^2} = Uz + \frac{Ua^2}{z}, \tag{28.91a}$$

that is due to a free stream plus a dipole (28.91a) ≡ (12.43a) of moment (28.89a); alternatively the complex potential could be obtained (24.47) ≡ (28.91c) from the circle theorem (Section 24.7).

$$f_0(z) = Uz : \; f(z) = f_0(z) + f_0^*\left(\frac{a^2}{z^*}\right) = Uz + U\left(\frac{a^2}{z^*}\right)^* = Uz + \frac{Ua^2}{z^2}. \tag{28.91b,c}$$

The complex potential (28.91a) of a cylinder of radius a in a free stream of velocity U specifies the scalar potential (28.92a) and stream function (28.92b):

$$\Phi(r,\varphi) = U\left(r + \frac{a^2}{r}\right)\cos\varphi, \quad \Psi(r,\varphi) = U\left(r - \frac{a^2}{r}\right)\sin\varphi, \tag{28.92a,b}$$

for example, the circle is the streamline $\Psi(a,\varphi) = 0$. The polar components of the velocity (28.93a,b) follow from the complex conjugate velocity (28.89b):

$$v_r(r,\varphi) = U\left(1 - \frac{a^2}{r^2}\right)\cos\varphi, \quad v_\varphi(r,\varphi) = -U\left(1 + \frac{a^2}{r^2}\right)\sin\varphi; \tag{28.93a,b}$$

this confirms that the velocity is tangent to the cylinder (28.94a,b):

$$v_r(a,\varphi) = 0, \quad v_\varphi(a,\varphi) = -2U\sin\varphi, \quad p(a,\varphi) = p_0 - 2\rho U^2\sin^2\varphi, \tag{28.94a–c}$$

and leads (12.27c) to the pressure distribution (28.94c). The drag (lift) on the forward (lower) hemisphere is (28.95a) [(28.96a)]:

$$D = 2\rho U^2 a \int_{-\pi/2}^{\pi/2} \sin^2\varphi\cos\varphi\, d\varphi = \frac{4}{3}\rho U^2 a, \quad C_D \equiv \frac{D}{\rho U^2 a} = \frac{4}{3}, \tag{28.95a,b}$$

$$L = -2\rho U^2 a \int_{\pi}^{0} \sin^3\varphi\, d\varphi = \frac{8}{3}\rho U^2 a, \quad C_L \equiv \frac{L}{\rho U^2 a} = \frac{8}{3} = 2C_D, \tag{28.96a–c}$$

and corresponds to a drag (lift) coefficient (28.95b) [(28.96b)] and lift-to-drag ratio of two (28.96c); the force on the forward (lower) hemisphere is balanced by the force on the rear (upper) hemisphere, so there is zero total drag (lift) on the cylinder, in agreement with (Subsection 28.5.2) the D'Alembert paradox (in the absence of circulation and viscosity).

28.6.2 Pressure Coefficient and Cavitation Velocity

The **pressure coefficient** *is defined by the difference between stagnation pressure* p_0 *and pressure* p *divided by the dynamic pressure (28.97a), and is given (28.94c) for the cylinder in the free stream by (28.97b):*

$$C_p(\varphi) \equiv \frac{p_0 - p}{(1/2)\rho U^2} = 4\sin^2\varphi; \quad C_{p\max} = C_p\left(\pm\frac{\pi}{2}\right) = 4. \tag{28.97a–c}$$

it has the maximum value (28.97c) at the sides of the cylinder, that is the direction through the center orthogonal to the free stream. The maximum (minimum) pressure (28.94c) is in the longitudinal (transverse) direction (28.98a):

$$p_{\max} = p(a,0) = p(a,\pi) = p_0 > p(a,\varphi) > p_{\min} = p\left(a, \pm\frac{\pi}{2}\right) = p_0 - 2\rho U^2 = p_\infty - \rho U^2; \tag{28.98a}$$

the maximum coincides with the stagnation pressure. The minimum: (i) coincides with stagnation pressure minus four times dynamic pressure of the free stream at infinity; (ii) this equals (28.68b) the pressure minus twice the dynamic pressure calculated for the free stream. The maximum (minimum) corresponds to the minimum (maximum) velocity on the cylinder; the latter is zero (twice the free stream velocity) in the (29.98b) upstream/downstream (in the lateral direction):

$$0 = v_{\min} = v(a,0) = v(a,\pi) \le |v(a,\varphi)| = |v_\varphi(a,\varphi)| \le \left|v\left(a, \pm\frac{\pi}{2}\right)\right| = |v|_{\max} = 2U. \tag{28.98b}$$

Since the pressure must be nonnegative (28.99a) the velocity cannot exceed the **cavitation velocity** (28.99b):

$$p_{\min} \equiv p_0 - 2\rho U^2 \ge 0, \quad U \le U_{\text{cav}} = \sqrt{\frac{p_0}{2\rho}}; \tag{28.99a,b}$$

for velocities higher than the cavitation speed forms a two-phase flow consisting of gas bubbles in a liquid, for example, air bubbles in water; cavitation occurs when the flow past a body is accelerated beyond the cavitation speed; cavitation reduces the efficiency of ship's propellers, and causes stresses and radiates noise due to the collapse of gas bubbles. In the case of the cylinder cavitation occurs first at the sides $\varphi \sim \pm\pi/2$ where the velocity is maximum (28.98b) and the pressure (28.98a) is minimum. The cavitation velocity is reduced in the presence of circulation (Subsection 28.7.2). There are no forces on the cylinder because (a) there is no circulation (28.29b) and the opposite sources and sinks in the dipole (28.88a) cancel (28.29a); (b) *the pressure distribution on the cylinder (28.94c) is symmetric relative to the real and imaginary axes* $\mathrm{p}(-\varphi) = \mathrm{p}(\varphi) = \mathrm{p}(\pi - \varphi)$; *the same symmetry applies to the pressure distribution in the flow (28.93a,b) that specified by*

$$p(r,\varphi) = p_\infty - \rho\left(\frac{Ua}{r}\right)^2\left[\frac{a^2}{2r^2} - \cos(2\varphi)\right], \tag{28.100a}$$

where p_0 (p_∞) *is the stagnation pressure [free stream pressure at infinity (28.68b)].* From the Bernoulli equation (14.27c) by substitution of the velocity (28.93a,b) follows:

$$p(r,\varphi) = p_\infty - \frac{\rho}{2}U^2\left[\left(1-\frac{a^2}{r^2}\right)^2\cos^2\varphi + \left(1+\frac{a^2}{r^2}\right)^2\sin^2\varphi\right]$$

$$= p_0 - \frac{\rho}{2}U^2 - \frac{\rho}{2}U^2\left[\left(\frac{a}{r}\right)^4 - 2\left(\frac{a}{r}\right)^2\left(\cos^2\varphi - \sin^2\varphi\right)\right] \qquad (28.100\text{b})$$

the pressure in the flow (28.100b) ≡ (28.100a).

28.6.3 Pitching Moment at an Angle-of-Attack

For a free stream with an angle-of-attack α*, the pressure distribution on the cylinder (28.94c) is not symmetric relative to this direction* $p(\varphi-\alpha) \neq p(\varphi+\alpha)$*, and a pitching moment (28.31b) corresponding to the dipole moment (28.89a) results (28.101a) in a pitch-up:*

$$M = \mathrm{Re}[-i\rho U e^{-i\alpha}(-2\pi U a^2)] = \mathrm{Re}(i2\pi\rho U^2 a^2 e^{-i\alpha}) = 2\pi\rho U^2 a^2\sin\alpha, \qquad (28.101\text{a})$$

$$C_M \equiv \frac{M}{(\rho/2)U(2a)^2} = \pi\sin\alpha = C_{M\alpha}\sin\alpha, \quad C_{M\alpha} \equiv \lim_{\alpha\to 0}\frac{\mathrm{d}C_M}{\mathrm{d}\alpha} = \pi, \qquad (28.101\text{b,c})$$

this corresponds to a pitching moment coefficient (28.101b), and a pitching moment slope at small angle-of-attack (28.101c). The kinetic energies of an uniform flow of velocity U (dipole of moment, P_1) inside (outside) the sphere of radius a is given by (28.13b) ≡ (28.102a) [(28.14b) ≡ (28.102b)] with $n = 1$:

$$E_{+1}(r \le a) = \frac{\pi}{2}\rho a^2 U^2, \quad E_{-1}(r \ge a) = \frac{\rho P_1^2}{8\pi a^2}. \qquad (28.102\text{a,b})$$

The two kinetic energies are equal (28.103a) for a dipole of moment (28.103b) equal to plus or minus the area of the circle (28.103c) times the velocity:

$$E_{+1}(r \le a) = E_{-1}(r \ge a): \qquad P_1 = \pm\, 2\,\pi\, a^2\, U = \pm\, 2A\, U, \quad A = \pi\, a^2; \qquad (28.103\text{a–c})$$

this dipole moment coincides with that (28.89a) ≡ (28.103b) that represents the introduction in an uniform stream of velocity U of a cylinder of radius a and area (28.103c). *The uniform flow of velocity* U *past a cylinder of radius* a *corresponds to a dipole (28.89b) with moment (28.89a) ≡ (28.103b,c) such there is equality (28.103a) of the kinetic energy associated with the free stream outside (28.102a) [dipole flow inside (28.102b)] the cylinder.*

28.6.4 Virtual Mass and Force of Resistance

The kinetic energy of the fluid, inside and outside the cylinder (28.102a) ≡ (28.102b) ≡ (28.103a) ≡ (28.104a):

$$E_k = \frac{\rho}{2}\pi a^2 U^2 = \frac{1}{2}m_0 U^2, \quad m_0 \equiv \pi\rho a^2 = \rho A, \qquad (28.104\text{a,b})$$

corresponds to a ***virtual mass*** *(28.104b) of fluid entrained by the body that coincides with the* ***displaced mass*** *of fluid. Adding the kinetic energy of the cylinder of mass* m *to the kinetic energy of the entrained fluid specifies the total kinetic energy (28.105a):*

$$E = E_k + \frac{1}{2}mU^2 = \frac{1}{2}\overline{\mathrm{m}}U^2, \quad \overline{\mathrm{m}} \equiv m + m_0, \qquad (28.105\text{a,b})$$

where (28.105b) is the **total mass**, *that is, the mass of the cylinder plus that of the entrained fluid.* The rate of change of the kinetic energy with time specifies the activity of the applied force that is the product of the force by the velocity (28.106a):

$$FU = \frac{dE}{dt} = \overline{m}U\frac{dU}{dt}; \quad m\frac{dU}{dt} - F = -m_0\frac{dU}{dt}, \tag{28.106a,b}$$

the equation of motion (28.106b) of the cylinder of mass m states that the **inertia force**, *that is, mass times acceleration, equals: (i) the applied force F; (ii) plus a* **force of resistance** *equal to minus the virtual mass times the acceleration.* The virtual mass and force of resistance represents the effect of flow entrainment on the motion of a body through the fluid, viz. a cylinder in the present case. A dot is used to denote time derivative, for example, if $\vec{x}$ is a position, $\vec{v} \equiv \dot{\vec{x}}$ is the velocity and $\vec{a} = \dot{\vec{v}} = \ddot{\vec{x}}$ is the acceleration (Section 2.1). *The force of resistance corresponds to a* **buoyancy force** *in the case of a free fall from altitude, x_3, and acts against gravity (28.107a):*

$$\overline{m}\ddot{x}_3 = mg - m_0 g \equiv F_0; \quad \ddot{x}_3 = g\frac{m - m_0}{\overline{m}} = g\frac{m - m_0}{m + m_0}, \tag{28.107a,b}$$

this reduces the acceleration (28.107b) relative to that g of gravity. If the buoyancy force exceeds the weight $m_0\, g > mg$ the acceleration is upward, for example, for an ascending balloon.

28.6.5 Force on the Cylinder with Circulation

If the cylinder has velocity components, (u_x, u_y), in the presence of circulation Γ, it is acted upon by a (Figure 28.11) transverse force given (28.28a) by

$$F_x - iF_y = \frac{i}{2}\rho 2\pi i\,(v_x - iv_y)\left(-\frac{i\Gamma}{\pi}\right) = \rho\Gamma\,(v_y + iv_x)\,; \tag{28.108}$$

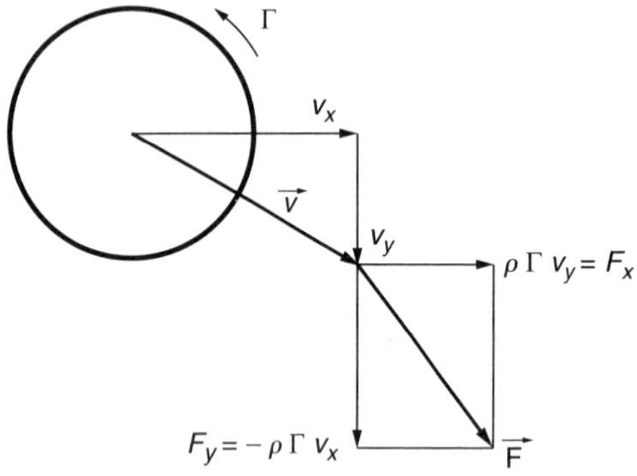

FIGURE 28.11
A cylinder with circulation in a uniform stream is subject to a vortical force orthogonal to the velocity (Figure 28.11); this is similar to an electron in an uniform external magnetic field (Chapter 6), for which the magnetic force is also orthogonal to the velocity. Thus the possible trajectories of a cylinder with circulation in uniform stream include: (i) the cycloid (Figure 6.3) for a cylinder with initial velocity not subject to other forces; (ii/iii) the elongated (curled) trochoid [Figure 6.5 (6.6)] for a cylinder with initial velocity subject to a constant force, for example, gravity or buoyancy, depending on whether it has a component parallel (antiparallel) to the initial velocity.

this is balanced by the inertia force, equal to acceleration $(\dot{v}_x, \dot{v}_y)$ times the total mass:

$$\overline{m}\dot{v}_x = F_x = \rho\Gamma v_y, \quad \overline{m}\dot{v}_y = F_y = -\rho\Gamma v_x. \tag{28.109a,b}$$

The system of equations describing the motion of the cylinder of radius, a, with circulation Γ relative to a potential flow of density ρ is (28.110b,c):

$$\omega \equiv \frac{\rho\Gamma}{\overline{m}}: \quad \dot{v}_x = \omega v_y, \quad \dot{v}_y = -\omega v_x, \tag{28.110a–c}$$

where ω in (28.110a) has the dimensions of frequency. The coincidence of (28.110b–c) $\equiv$ (6.2a,b) shows that ω in (28.110a) plays the role of Larmor frequency (Section 6.2). This is due to the analogy between the motion of a cylinder with circulation (a moving electron, corresponding to an electric current) in an external velocity (magnetic) field, with the vortical (magnetic) force being orthogonal to the velocity (28.50c) [second term in (28.43b)].

28.6.6 Path of a Cylinder in a Flow

Recalling the results of Section 6.5, it follows that *the equations of motion (28.110b,c) imply that for a cylinder of radius a with circulation Γ in a potential flow: (i) if the cylinder is initially at rest, it remains at rest; (ii) if the cylinder is set in motion at initial velocity U in a potential flow it describes a circular path, with constant tangential (angular) velocity U [ω in (28.110a)], and radius R given by:*

$$\frac{R}{a} = \frac{U}{\omega a} = \frac{\overline{m}U}{\rho\Gamma a} = \frac{mU}{\rho\Gamma a} + \frac{UA}{\Gamma a} = \frac{mU}{\rho\Gamma a} + \frac{\pi U a}{\Gamma}, \tag{28.111}$$

where the total mass (28.105b; 28.104b) was used; if the mass of the cylinder is negligible compared with the added mass (28.112a), then (28.111) simplifies to the second term, implying that the radius of the circular trajectory (28.112b):

$$m \ll m_0 \equiv \rho\pi a^2: \quad \frac{R}{a} = \frac{UA}{\Gamma a} = \frac{\pi U a}{\Gamma} = \frac{U}{2v_\varphi}, \quad v_\varphi = \frac{\Gamma}{2\pi a}, \tag{28.112a–c}$$

is independent of the density of the fluid; it is determined by the radius of the cylinder a and circulation Γ, so that the ratio of the radius of the path R to the a radius of the cylinder, equals (28.112b) the ratio of the tangential velocity U to twice the circulation velocity (28.112c) at radius a. The mean position of the cylinder does not change, because there is no drag or thrust in a potential flow, and the circular motion results from the lift or down force transverse to the velocity.

28.6.7 Cylinder in Motion with Applied Force

The application of a constant force to the cylinder is equivalent to adding an electric field to the motion of an electron in the magnetic field, that is, the superposition of a linear and a circular motion leads to a cycloid (Section 6.6) or a trochoid (Section 6.7). *If a constant force F_0 is applied to the cylinder, for example, the weight minus buoyancy force for a cylinder in the gravity field (28.113a), the equations of motion (28.110b,c) become (28.113b,c):*

$$f \equiv \frac{F_0}{\overline{m}} - g: \quad \dot{v}_x = v_y\omega + f, \quad \dot{v}_y = -\omega v_x; \tag{28.113a–c}$$

these imply that the trajectory is a trochoid, resulting from the superposition on the circular motion due to lift, of an uniform translation due to the applied thrust, with velocity (28.114a):

$$V = \frac{f}{\omega} = \frac{F_0}{\rho\Gamma} = -\frac{UF_0}{F_y}; \quad F_y \equiv -\rho\Gamma U, \tag{28.114a,b}$$

the translation velocity is proportional to the initial velocity of the cylinder, multiplied by the ratio of the applied force to the lift force (28.114b) ≡ (28.29b); if the vortical and the applied forces are equal $F_0 = Y$, the trajectory is a cycloid.

28.7 Flow past a Cylinder with Circulation

A rigid cylinder with impermeable walls cannot have a source or sink, since the latter would imply a normal velocity; the absence of sources or sinks excludes a drag force (28.29a), but lift is possible (Subsection 28.7.6) in the presence of circulation (Subsection 28.7.1). The latter affects the position of the stagnation points (Subsection 28.7.3), deflects the streamlines (Subsection 28.7.4) and determines the topology of the flow (28.7.5); the circulation also reduces the cavitation velocity (Subsection 28.7.2). The maximum circulation and lift (Subsection 28.7.6) are limited both: (i) by cavitation (Subsection 28.7.2); (ii) and by flow separation (Subsection 28.7.7).

28.7.1 Velocity Field of Cylinder with Circulation

A lift force (28.29b) on a rigid impermeable cylinder is possible, since the circulation of a vortex at the center preserves the circle as a streamline; thus a vortex with circulation Γ can be added (12.32a) to the complex potential (28.91a) of the flow of free stream velocity U past a cylinder of radius a, leading to (28.115a):

$$f(z) = U\left(z + \frac{a^2}{z}\right) - \frac{i\Gamma}{2\pi}\log z; \quad v^* = U\left(1 - \frac{a^2}{z^2}\right) - \frac{i\Gamma}{2\pi z}, \tag{28.115a,b}$$

corresponding to the complex conjugate velocity is (28.115b). *The complex potential (28.115a) specifies the scalar potential (28.116a) and stream function (28.116b) of the uniform stream incident on a cylinder with circulation:*

$$\Phi(r,\varphi) = U\left(r + \frac{a^2}{r}\right)\cos\varphi + \frac{\Gamma\varphi}{2\pi}, \tag{28.116a}$$

$$\Psi(r,\varphi) = U\left(r - \frac{a^2}{r}\right)\sin\varphi - \frac{\Gamma}{2\pi}\log r; \tag{28.116b}$$

(28.116b) confirms that the circle $r = a$ is the streamline $\Psi(a,\varphi) - (\Gamma/2\pi)\log a =$ const; the complex conjugate velocity (28.115b) specifies the polar components of the velocity (28.117a,b):

$$v_r(r,\varphi) = U\left(1 - \frac{a^2}{r^2}\right)\cos\varphi, \quad v_\varphi(r,\varphi) = -U\left(1 + \frac{a^2}{r^2}\right)\sin\varphi + \frac{\Gamma}{2\pi r}; \tag{28.117a,b}$$

(28.117a) confirms that the velocity is tangential on the cylinder:

$$v_r(a,\varphi) = 0, \quad v_\varphi(a,\varphi) = -2U\sin\varphi + \frac{\Gamma}{2\pi a}; \tag{28.118a,b}$$

thus, the circulation of the vortex adds a term to the tangential velocity in (28.118a,b) relative to (28.94a,b). This affects the cavitation velocity as shown next.

28.7.2 Effect of the Circulation on the Cavitation Velocity

The velocity is tangential on the cylinder (28.118a,b), with extrema given in modulus by (28.119b):

$$|\Gamma| \le 4\pi U a \equiv \Gamma_* : \quad |v(a,\varphi)|_{\max} = 2U + \frac{|\Gamma|}{2\pi a} \ge |v(a,\varphi)| \ge 2U - \frac{|\Gamma|}{2\pi a} \equiv |v(a,\varphi)|_{\min}, \tag{28.119a,b}$$

assuming that the circulation does not exceed the critical value (28.119a); the condition (28.119a) implies the existence of stagnation point(s) on the cylinder, as discussed in Subsection 28.7.3. The minimum (maximum) pressure occurs at the point of highest (lowest) velocity:

$$\begin{aligned} p_{\min} &= p_0 - \frac{\rho}{2}(v_{\max})^2 = p_0 - 2\rho U^2\left(1 + \frac{|\Gamma|}{4\pi U a}\right)^2 = p_0 - 2\rho U^2\left(1 + \frac{|\Gamma|}{\Gamma_*}\right)^2 \\ &\ge p(a,\varphi) \ge p_0 - \frac{\rho}{2}(v_{\min})^2 = p_0 - 2\rho U^2\left(1 - \frac{|\Gamma|}{\Gamma_*}\right)^2 \equiv p_{\max}. \end{aligned} \tag{28.119c}$$

The constraint (28.119a) applies only to the maximum pressure that does not affect cavitation. The cavitation velocity is specified by zero minimum pressure (28.120a), leading for arbitrary circulation to (28.120b):

$$0 = p_{\min} = p_0 - 2\rho\left(U + \frac{|\Gamma|}{4\pi a}\right)^2 : \quad U \le \sqrt{\frac{p_0}{2\rho}} - \frac{|\Gamma|}{4\pi a} \equiv U_{\text{cav}}. \tag{28.120a,b}$$

Thus *the presence of circulation, regardless of whether it is positive(negative), that is counterclockwise (clockwise), and causes a down force (lift), always: (i) increases the maximum tangential velocity (28.119b); (ii) decreases the minimum pressure (28.119c); (iii) lowers the free stream velocity for that cavitation occurs (28.120b); (iv) cavitation occurs at zero velocity (28.121a), in the absence (presence) of a free stream velocity for a circulation (28.121b) [(28.121c)]:*

$$U_{\text{cav}} = 0; \quad |\Gamma| \le \begin{cases} 4\pi a\sqrt{\dfrac{p_0}{2\rho}} & \text{if } U = 0, \\ 4\pi a\left[\sqrt{\dfrac{p_0}{2\rho}} - U\right] & \text{if } U \ne 0; \end{cases} \tag{28.121a–c}$$

(v) the condition (28.121b) leads (28.122a,b) to a vortical force (lift or down force) that takes the maximum value (28.122a):

$$|F_y| = \rho U|\Gamma| \le 4\pi U a\sqrt{\frac{p_0\rho}{2}}, \quad |C_y| = \frac{|F_y|}{\rho U^2 a} \le \frac{2\pi}{U}\sqrt{\frac{2p_0}{\rho}}, \tag{28.122a,b}$$

(vi) the latter corresponds to a maximum transverse force coefficient (28.122b). The cavitation first occurs at the sides of the cylinder $\varphi = \pm\pi/2$, where the tangential velocity is largest; the presence of circulation does not change this position, but diminishes the cavitation velocity. Thus the maximum achievable lift can be limited by: (i) cavitation (Subsection 28.7.2), for example, in water; (ii) flow separation (Subsection 28.7.7), for example, in air.

28.7.3 Position of the Stagnation Points

The circulation affects the position of the **stagnation points** (28.123a) where the velocity (28.115b) vanishes (28.123b):

$$v^*(z_\pm) = 0: \quad Uz^2 - \frac{i\Gamma}{2\pi}z - a^2U = 0; \quad z_\pm = \frac{i\Gamma}{4\pi U} \pm \sqrt{a^2 - \left(\frac{\Gamma}{4\pi U}\right)^2}, \tag{28.123a–c}$$

the roots (28.123c) of (28.123b) specify the position of the stagnation points:

$$\Gamma_* \equiv 4\pi U a: \qquad z_\pm = a \begin{cases} \dfrac{i\Gamma}{\Gamma_*} \pm \sqrt{1 - \dfrac{\Gamma^2}{\Gamma_*^2}} & \text{if } |\Gamma| \le \Gamma_*, \qquad (28.124a) \\ i\left\{\dfrac{\Gamma}{\Gamma_*} \pm \sqrt{\dfrac{\Gamma^2}{\Gamma_*^2} - 1}\right\} & \text{if } |\Gamma| \ge \Gamma_*; \qquad (28.124b) \end{cases}$$

here Γ_* is the **critical circulation** *(28.119a) for that the two stagnation points coincide at the lowest position (28.127a–c). The uniform potential flow past a cylinder with circulation is illustrated in four cases by order of increasing modulus of the circulation. In the (Case I) absence of circulation (28.125a), the flow is symmetric with regard to the incident velocity (Figure 28.4a), and the stagnation points (28.125b) in*

$$\Gamma = 0: \qquad z_\pm = \pm a, \quad v_\varphi(a, 0) = 0 = v_\varphi(a, \pi), \quad v_\varphi\left(a, \pm\frac{\pi}{2}\right) = \mp 2U, \qquad (28.125\text{a–c})$$

(28.125a) are diametrically opposed on the midline (28.125c,d), with the maximum velocity (28.125d) occurring on the orthogonal line, $z = \pm\, i\, a$. If (Case II) the circulation in modulus is less than the critical value (28.126a), the stagnation points (28.124a) $\equiv$ (28.126c) lie on the circle:

$$|\Gamma| \le \Gamma_*, \quad \frac{|\Gamma|}{\Gamma_*} \equiv \sin\beta: \quad z_\pm = a\,(\pm\cos\beta + i\sin\beta)\,\mathrm{sgn}\,(\Gamma) = \pm\, a\, \mathrm{e}^{\pm i\beta}\mathrm{sgn}\,(\Gamma), \qquad (28.126\text{a–c})$$

at an angle β in (28.126b); depending an the sign of the circulation sign (Γ), the stagnation points lie below (above) the midline for negative (positive) circulation (Figure 28.4b). The critical circulation (28.127a) $\equiv$ (28.119a) is (Case III):

$$|\Gamma| = \Gamma_* = 4\pi U a: \qquad v_{\varphi *} = \frac{\Gamma_*}{2\pi} = 2U, \quad z_\pm = i\, a\, \mathrm{sgn}\,(\Gamma), \qquad (28.127\text{a–c})$$

that for that the tangential velocity (28.127b) at the radius a, exactly cancels the maximum velocity $-2U$ due (28.98b) to the deformation of the incident stream at the transverse point on the circle, $z_\pm = +i\, a$, where the stagnation points (28.124a) $\equiv$ (28.124b) $\equiv$ (28.127c) coincide; the coincident stagnation points (28.127c) lie at the transverse position at the bottom (top) of the cylinder for negative (positive) circulation (Figure 28.4c). The (Case IV) supercritical circulation (28.128a) leads to stagnation points on the transverse line through the center (28.128c):

$$|\Gamma| \ge \Gamma_*, \quad \frac{|\Gamma|}{\Gamma_*} \equiv \cosh\beta: \quad z_\pm \mathrm{sgn}\,(\Gamma) = ia\,(\cosh\beta \pm \sinh\beta) = i\, a\, \mathrm{e}^{\pm\beta}, \qquad (28.128\text{a–c})$$

one $|z_-| = a\mathrm{e}^{-\beta} < a$ inside and the other outside $|z_+| = a\mathrm{e}^{\beta} > a$ the circle with β given by (28.128b); the stagnation points lie below (above) the center for negative (positive) circulation, and in both cases there is one stagnation point in the flow (Figure 28.4d), and another inside the cylinder. The flow inside the cylinder (Figure 28.12) is considered in the same four cases I to IV as the flow outside (Figure 28.4): (Case I) in the absence of circulation (Figure 28.12a) there is a dipole (Figure 12.4) "squeezed" inside a circle with a tangential velocity on the boundary; (Case II) as the circulation increases below the critical value (Figure 28.12b) the dipole flow is distorted by the tangential velocity of the vortex that has no effect at the center, and has an increasing effect outward; (Case III) for the critical circulation (Figure 28.12c) there is a streamline with stagnation point on the circle; (Case IV) beyond the critical circulation there is a stagnation point inside the cylinder (Figure 28.12d) with streamlines inside (outside) that pass (do not pass) through the center.

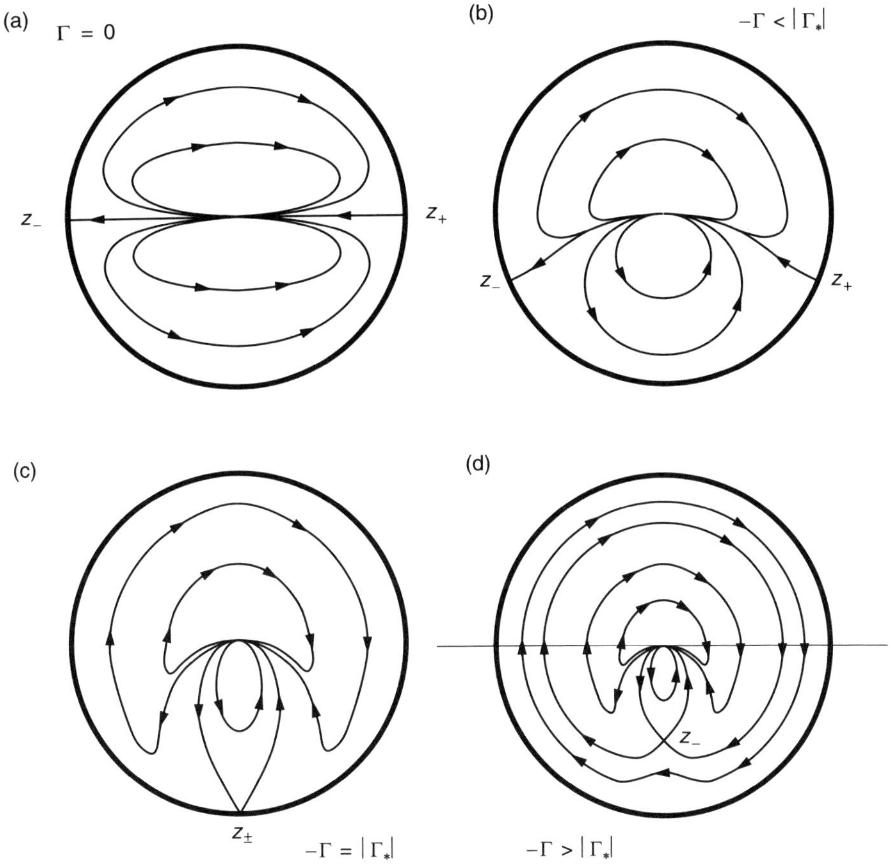

FIGURE 28.12
As for the Rankine semiinfinite (finite) bodies [Figure 28.5 (28.7)] for the cylinder there are also external (internal) flows [Figure 28.4 (28.12)], divided by the "separating stream line" used as a wall. The internal flow in the cylinder (Figure 28.12) has the same four cases as the external flow (Figure 28.4). In the absence of circulation (a) the dipole with axis antiparallel to the free stream is constrained to lie within the circular cross-section; the streamlines are no longer circles as for a free dipole (Figure 12.4) and are "squashed" in the vertical direction while remaining symmetric relative to the horizontal axis. The symmetry relative to the horizontal axis is destroyed by the presence of circulation, for example, if it is negative and subcritical the stagnation points move downward to symmetric positions $z_\pm$; this implies (b) that the streamlines above the separating streamline occupy a larger part of the circular cross-section than those below. All streamlines are tangent to the horizontal dipole axis at the center, including in the cases of: (c) critical circulation when the stagnation points coincide at the lowest position on the cylinder; (d) overcritical circulation when there is a stagnation point in the cylinder. The latter is a double point (Figure 13.1b) where four streamlines cross at right-angles as for a corner flow (Figure 14.7). Between the streamlines passing through the stagnation point and the cylinder there is a region of circulating flow. Only the cases (a,b) (not the cases (c,d)] are achievable in practice both for the internal (external) potential flow past a cylinder [Figure 28.12 (28.4)].

28.7.4 Topology of the Internal and External Flow

The topology of the potential flow around the cylinder is determined by the position of the stagnation points and the streamlines issuing from them. In the (Case I) absence of circulation (Figures 28.4a and 28.12a) the line through the center in the direction of the free stream cuts the cylinder at the stagnation points $z_{\pm;}$*; the latter are diametrically opposite and the flow is symmetric about the line joining them. The streamlines outside (inside) the cylinder in Figure 28.4a (28.12a) are open (closed) and tangent (parallel) to the axis of symmetry at the center (infinity). For (Case II) a subcritical circulation (Figures 28.4b and 28.12b) the stagnation points are symmetric relative to the line through the center orthogonal to the free stream. Four streamlines issue from the stagnation points: the circle and the streamline inside and outside that cut a right-angles* $2\pi/4 = \pi/2$ *like a square corner flow (Figure 14.10c); thus the streamlines are perpendicular to the circle, that is, radial at the stagnation points. The symmetry relative to the line through the center in the free stream direction is lost in all cases with circulation, whereas the symmetry relative to the orthogonal line through the center remains. The streamlines remain closed (open) inside (outside) the cylinder in Figure 28.4b (28.12b), and are stretched above (shrunk below). For (Case III) the critical circulation (Figures 28.4c and 28.12c) the stagnation points coincide at the point on the circle on the line through the center orthogonal to the free stream direction. At the stagnation point converge six streamlines: the sides of the circle and two streamlines inside and two outside the cylinder. The six streamlines cut at equal angles* $2\pi/6 = \pi/3$ *as in a hexagonal corner (Figure 14.10e); thus the streamlines outside and inside the cylinder hit the stagnation point at an angle* $\pi/3$ *with the tangent to the circle. The angle between the streamlines outside (inside) the cylinder at the stagnation point in Figure 28.4c (28.12c) is thus* $2\pi/3$*. For (Case IV) a supercritical circulation (Figures 28.12d and 29.11d) the stagnation points separate along the line through the center orthogonal to the free stream direction. One stagnation point lies outside (Figure 28.4d) and the other inside (Figure 28.12d) the cylinder. Four streamlines issue from each stagnation point, cutting at right-angles* $2\pi/4 = \pi/2$*. The outer (inner) streamline through the stagnation point inside (outside) the cylinder separates a region of circulatory flow between itself and the cylinder. At infinity the free stream becomes uniform because the dipole and vortex effects decay the former more rapidly. Conversely at the center the dipole* $O(r^{-2})$ *has a stronger singularity than the vortex* $O(r^{-1})$ *and for this reason all streamlines inside the cylinder passing through the center are tangent to each other, and to the diameter in the free stream direction.*

28.7.5 Aiming Distance and Separating Streamline

The **separating streamline** is defined as the streamline that starts and ends at the stagnation points. In the absence of circulation (28.129a) the separating streamline that divides the flow that passes above and below the cylinder (28.125a–c), is the real axis outside the circle (Figure 28.4a). All streamlines (28.129b) are parallel to the separating streamline at infinity, and can be identified by the distance from the latter, designated **aiming distance** (28.129b):

$$\Gamma = 0: \quad h \equiv \lim_{r\to\infty} r\sin\varphi = \frac{\Psi}{U}; \quad h = \left(r - \frac{a^2}{r}\right)\sin\varphi, \tag{28.129a–c}$$

the aiming distance h *can replace the stream function* Ψ *in the equation of streamlines (28.92b)* $\equiv$ *(28.129c). The meaning of (28.129b) is that the volume flux between the real axis and the streamline of aiming distance* h *is* $Uh = \Psi$*. The streamline of aiming distance* h *passes closest to the circle at (28.130a), for that the radial distance in (28.129c) takes the* *minimum value $r_{\min}$ in (28.130b):*

$$\varphi = \frac{\pi}{2}: \quad h = r_{\min} - \frac{a^2}{r_{\min}}; \quad r_{\min} = r\left(\frac{\pi}{2}\right) = \frac{h}{2} \pm \left|\frac{h^2}{4} + a^2\right|^{1/2} > a, h, \quad \mathrm{d}_{\min} = r_{\min} - a, \tag{28.130a–d}$$

the minimum distance of the streamline from the center (circle) is (28.130c) [28.130d)]. In the presence of circulation the aiming distance can still be defined by (28.129b), but if (28.116b) is used instead of (28.92b), it is minus infinity (28.131b), implying that all streamlines start asymptotically at infinite distance from the real axis:

$$h \equiv \lim_{r\to\infty} r \sin\varphi = \lim_{r\to\infty} \frac{\Psi + (\Gamma/2\pi)\log r}{U} = \infty \operatorname{sgn}(\Gamma); \tag{28.131a,b}$$

$$\lim_{r\to\infty} \{v_r(r, \varphi), v_\varphi(r, \varphi)\} = U\{\cos\varphi, \sin\varphi\}, \tag{28.131c,d}$$

the asymptotic velocity (28.117a,b) is U parallel to the real axis (28.131c,d) because the vortex has zero velocity at infinity. Thus it is the vortex effect at intermediate distances that accumulates to an infinite asymptotic lateral deflection of the streamlines. For subcritical circulation (28.124a; 28.126a–c) streamlines are open (Figure 28.4b); for supercritical circulation (28.124b) the separating a streamline has a double point, coincident with the stagnation point z_+ in (28.128a–c), and the streamlines (Figure 28.4d) inside (outside) are closed (open). In the intermediate case (28.124a) ≡ (28.124b) of (Figure 28.4c) critical circulation (28.119a), the stagnation point $z_\pm = -i\,a$ in (28.127a–c) corresponds to $r = a$ and $\varphi = -\pi/2$ ($\varphi = \pi/2$) for negative (positive) circulation; the separating streamline (28.116b) corresponding to the critical circulation (28.132a) ≡ (28.119a) has equation (28.132b):

$$\Gamma = \Gamma_* = \pm 4\pi U a: \quad \left(\frac{r}{a} - \frac{a}{r}\right)\sin\varphi \mp 2\log r = \frac{\Psi(a, \pm\pi/2)}{Ua}; \tag{28.132a,b}$$

this is consistent with (28.131b) that is valid for arbitrary circulation.

28.7.6 Pressure Distribution and Lift Force on the Cylinder

The velocity (28.118b) on the cylinder is tangential (28.118a) and is: (i) symmetric fore and aft $v_\varphi(a, \varphi) = v_\varphi(a, \pi - \varphi)$, with or without circulation; (ii) skew-symmetric above and below $v_\varphi(a, \varphi) = -v_\varphi(a, -\varphi)$, in the absence of circulation $\Gamma = 0$. A clockwise vortex $\Gamma < 0$, increases (decreases) the velocity above (below), causing a reduced (increased) pressure, and hence a next upward force, that is, lift. This is confirmed from the Bernoulli equation (14.27c) for the pressure; using the polar components (28.117a,b) of velocity to calculate *the pressure at an arbitrary point of the flow*:

$$p(r, \varphi) = p_\infty - \frac{\rho}{2}U^2\left(\frac{a}{r}\right)^2\left[\frac{a^2}{r^2} - 2\cos(2\varphi)\right] + \frac{\rho U \Gamma}{2\pi r}\left(1 + \frac{a^2}{r^2}\right)\sin\varphi - \frac{\rho\Gamma^2}{8\pi^2 r^2}, \tag{28.133a}$$

where p_∞ is (28.68b) the free stream pressure at infinity; this simplifies to the pressure distribution on the cylinder to:

$$p(a, \varphi) = p_0 - 2\rho U^2 \sin^2\varphi + \frac{\rho U\Gamma}{\pi a}\sin\varphi - \frac{\rho\Gamma^2}{8\pi^2 a^2}; \tag{28.133b}$$

this can alternatively be obtained directly from (28.118a,b). In the absence of circulation the pressure distribution in the flow (28.133a) [on the cylinder (28.133b)] simplify to (28.100a) [(29.94c)]. The first two terms coincide in (28.133a) ≡ (28.100a) ≡ (28.100b) and (28.133b) ≡ (28.94c), and the last two terms add the effect of circulation. The passage from (14.27c; 28.117a,b) to (28.133a) is similar:

$$p(r,\varphi) = p_0 - \frac{\rho}{2}\left\{U^2\left(1-\frac{a^2}{r^2}\right)^2\cos^2\varphi + \left[\frac{\Gamma}{2\pi r} - U\left(1+\frac{a^2}{r^2}\right)\sin\varphi\right]^2\right\}$$

$$= p_0 - \frac{\rho}{2}U^2 - \frac{\rho}{2}U^2\frac{a^4}{r^4} + \rho U^2\frac{a^2}{r^2}\left(\cos^2\varphi - \sin^2\varphi\right) + \frac{\rho U\Gamma}{2\pi r}\left(1+\frac{a^2}{r^2}\right)\sin\varphi - \frac{\rho\Gamma^2}{8\pi^2 r^2} \qquad (28.133c)$$

to (28.100a) to (28.100b), including in addition the effect of circulation, and the free stream pressure (28.68b).

In the absence of circulation $\Gamma = 0$, the pressure is symmetric fore and aft, and above and below $p(a,\varphi) = p(a,\pi-\varphi) = p(a,-\varphi)$, like the modulus of the velocity, and the flow exerts no force on the cylinder; in the presence of circulation $\Gamma \neq 0$, the pressure is symmetric fore and aft $p(a,\varphi) = p(a,\pi-\varphi)$ [unsymmetric above and below $p(a,\varphi) \neq p(a,-\varphi)$] so that there is no drag (there is lift). The horizontal F_x (vertical F_y) force may be calculated integrating (28.15e) [(28.15f)] over the circle of radius a. The drag vanishes in agreement (28.29a) with D'Alembert's paradox:

$$y = a\sin\varphi: \quad F_x = -\int p\,\mathrm{d}y = -\int_0^{2\pi} p(a,\varphi)\,a\cos\varphi\,\mathrm{d}\varphi = 2\rho U^2 a\int_0^{2\pi}\sin^2\varphi\cos\varphi\,\mathrm{d}\varphi$$

$$-\frac{\rho U\Gamma}{\pi}\int_0^{2\pi}\sin\varphi\cos\varphi\,\mathrm{d}\varphi + \frac{\rho\Gamma^2}{8\pi^2 a}\int_0^{2\pi}\cos\varphi\,\mathrm{d}\varphi = 0, \qquad (28.134a,b)$$

since the velocity is symmetric fore and aft, hence the pressures are identical. The vertical force is proportional to minus the circulation:

$$x = a\cos\varphi: \quad F_y = \int p\,\mathrm{d}x = -\int_0^{2\pi} p(a,\varphi)\,a\sin\varphi\,\mathrm{d}\varphi = 2\rho U^2\int_0^{2\pi}\sin^3\varphi\,\mathrm{d}\varphi$$

$$-\frac{\rho U\Gamma}{\pi}\int_0^{2\pi}\sin^2\varphi\,\mathrm{d}\varphi + \frac{\rho\Gamma^2}{8\pi^2 a}\int_0^{2\pi}\sin\varphi\,\mathrm{d}\varphi = -\rho U\Gamma, \qquad (28.135a,b)$$

in agreement with the Joukowski theorem (28.29b) ≡ (28.135b).

28.7.7 Flow Separation and Maximum Lift Coefficient

The pressure gradient on the surface of cylinder is given (28.133b) by:

$$\frac{\mathrm{d}p(a,\varphi)}{\mathrm{d}\varphi} = -2\rho U^2\sin(2\varphi) + \frac{\rho U\Gamma}{\pi a}\cos\varphi. \qquad (28.136)$$

In the absence of circulation the pressure gradient is specified by the first term on the r.h.s. that is positive $\mathrm{d}p/\mathrm{d}\varphi > 0$ (negative $\mathrm{d}p/\mathrm{d}\varphi < 0$) in the forward $\pi/2 < \varphi < \pi$ (aft $0 < \varphi < \pi/2$) part, as shown on Figure 28.4a: (i) on approach to (passing beyond) the cylinder, the streamlines diverge (converge); (ii) since a constant volume flux passes between them, the velocity decreases (increases); (iii) hence the pressure increases (decreases); (iv) thus the pressure gradient is positive (negative) in the forward (aft) part. In the boundary layer near to the cylinder where the velocity is small, a favorable (adverse), that is, positive (negative) pressure gradient, will accelerate (reverse) the flow, avoiding (leading to) **flow separation**. Thus the potential flow is maintained on approach to a blunt body like a cylinder, but in the aft part there is a tendency for boundary layer separation to lead to the break-down of potential flow, and formation of a vortex or turbulent wake. The second term on the r.h.s. of (28.136) shows that negative (positive) circulation, reinforces (reduces) the tendency for reverse flow and break-down of potential conditions in the aft part $0 < \varphi < \pi/2$. Thus increasing lift that requires a large (in modulus) negative circulation, also increases the tendency for flow separation; since flow separation destroys or reduces lift, the condition of flow separation specifies the maximum attainable lift. Only subcritical circulations are usually achievable in practice so that: (i) the flow pattern in Figure 28.4a,b are observed in real flows; (ii) the cases in Figure 28.4c,d were included for completeness of the theory. The maximum circulation attainable without boundary layer separation, determines the maximum lift (28.135b) ≡ (28.29b) that may be obtained, for a given incident stream and body shape. The critical value (28.119a) for negative circulation corresponds to a lift (28.137a):

$$L_* = -\rho U\left(-\Gamma_*\right) = 4\pi\rho U^2 a; \quad C_{L*} \equiv \frac{L_*}{(1/2)\,\rho U^2 2a} = 4\pi, \tag{28.137a,b}$$

the **lift coefficient**, defined as the ratio of the lift to the dynamic pressure multiplied by the length of the body (in the present case the diameter of the cylinder) would take the value (28.137b) for the critical circulation. This value is much larger than the maximum lift coefficient (28.138a) achievable in practice:

$$C_{L\,\max} \sim 2.0 - 2.8, \qquad \frac{\Gamma_{\max}}{\Gamma_*} \sim \frac{C_{L\,\max}}{4\pi} \sim 0.16 - 0.22, \tag{28.138a,b}$$

and shows that the maximum circulation achievable without flow separation is about 20% of the critical circulation. The maximum lift coefficient can be increased: (i) for a cambered airfoil (Chapter 34); (ii) by using flaps, with one more slots to energize the boundary layer; (iii) by blowing air out of holes to delay flow separation. Even with these extra means the achievable lift coefficient hardly reaches $C_L \sim 3$; thus the case of stagnation point in the flow does not occur in real flows that cease to be potential for lower circulation. The flow past a cylinder with two trailing vortices represents the simplest case of separated wake; the vortex wake may consist of a "vortex street" (Section 36.9) with a succession of vortices. Other related problems include (i) cylinder moving in a cylindrical cavity; (ii) a cylinder near a wall; (iii) two static eccentric cylinders; (iv) two cylinders in relative motion; (v) a cylinder near a plane in a stream. Next are considered a monopole (dipole) near a cylinder [Section 28.8 (28.9)].

28.8 Moving Vortex and Source/Sink Image System

A monopole near a cylinder (Subsection 28.8.1) exerts a force and a moment (Subsection 28.8.2); thus, together with its image system, it is not at rest (Subsection 28.8.3), viz. the vortex (source/sink) moves (Subsection 28.8.4) azimuthaly (radially).

28.8.1 Images at the Reciprocal Point and at the Center

The potential flow due (14.35a–c) to a monopole (i.e., source/sink combined with vortex) at a position $z = b$ near a cylinder of radius a is specified by (24.47) the circle theorem:

$$f(z) = \frac{Q - i\Gamma}{2\pi} \log(z - b) + \frac{Q + i\Gamma}{2\pi} \log\left(\frac{a^2}{z} - b^*\right), \tag{28.139a}$$

where (i) the reciprocal point is at a^2/z; (ii) it corresponds to a monopole with the same (opposite) flow rate (circulation), that is, an identical source/sink (an opposite vortex). Apart (24.53b) for a nonessential constant (28.139a) can be rewritten:

$$f(z) = \frac{Q - i\Gamma}{2\pi} \log(z - b) + \frac{Q + i\Gamma}{2\pi} \left\{\log\left(z - \frac{a^2}{b^*}\right) - \log z\right\}, \tag{28.139b}$$

showing that (Figure 28.13) there are two images of a monopole on a cylinder: (i) a complex conjugate image, with the same flow rate and opposite circulation at the conjugate point $c = a^2/b^*$; *(ii) an opposite complex conjugate image, with opposite flow rate and same*

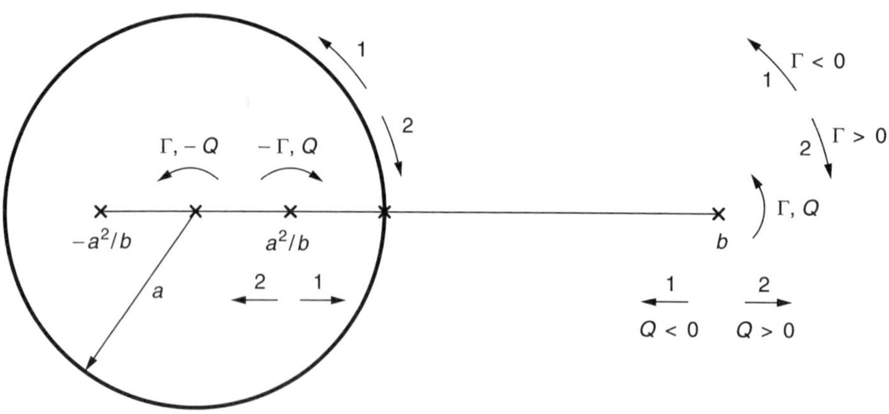

FIGURE 28.13
The influence of a cylinder on the flow due to a source/sink (vortex) is equivalent to two images: (i) an identical (opposite) image at the reciprocal point (Figures 24.4 and 26.2); (ii) an image at the center of the cylinder opposite to (i), that is opposite (identical) to the original, to ensure that there is no net mass flow (circulation) inside the cylinder. The system of images (i+ii) together with the original source/sink (vortex) ensure a tangential velocity at the cylinder. The dominant image, closest to the source (sink), is also a source (sink), and induces a radial outward (inward) velocity; thus the cylinder goes farther from (closer to) the source (sink), which exerts upon it a repulsive (attractive) force. The dominant image, closest to a counterclockwise (clockwise) vortex is an opposite vortex, and induces a tangential clockwise (counterclockwise) velocity; thus a vortex near a cylinder moves in a circular path, in the direction opposite to its circulation, that is, clockwise for a counterclockwise vortex and vice-versa. A similar analysis applies to a monopole inside or outside the cylinder, exchanging the original and reciprocal points (Figures 24.4 and 26.2). There is an analogy between the hydrodynamic monopole (Figure 28.13), that is, a source/sink (counterclokwise/clockwise vortex) near a cylinder for a potential flow, and the electro(magneto) static field of an positive/negative electric charge (counterclockwise/clockwise current) near a cylindrical insulator (conductor) in Figure 24.5 (26.2).

circulation at the center. The complex conjugate velocity:

$$v^*(z) = \frac{Q - i\Gamma}{2\pi}\frac{1}{z-b} + \frac{Q + i\Gamma}{2\pi}\left[\left(z - \frac{a^2}{b*}\right)^{-1} - z^{-1}\right]$$
$$= \frac{Q - i\Gamma}{2\pi}\frac{1}{z-b} + \frac{Q + i\Gamma}{2\pi}\frac{a^2/z}{b^* z - a^2}, \tag{28.140}$$

follows from either (28.139a) or (28.139b). It corresponds in the far-field (28.141b) to the original monopole as the leading term, second in the l.h.s. of (28.141a):

$$2\pi v^*(z) - \frac{Q - i\Gamma}{z} = \frac{(Q - i\Gamma)\, b}{z^2} + \frac{(Q + i\Gamma)\, a^2}{b^* z^2} + O\left(\frac{1}{z^3}\right) = \frac{P_1}{z^2}; \tag{28.141a}$$

$$|z|^2 \gg a^2, \frac{a^4}{|b|^2} : P_1 \equiv Q\left(b + \frac{a^2}{b^*}\right) - i\Gamma\left(b - \frac{a^2}{b^*}\right) = P_{1x} + iP_{1y}; \tag{28.141b,c}$$

the following term on the r.h.s. of (28.141a), that is, the leading term for the influence of the cylinder, is a dipole whose moment (28.141c) for the source/sink (vortex) involves the distance from the original monopole to the reciprocal point on the opposite (same) side of the center of the circle $-a^2/b^*$ $(+a^2/b^*)$ *in Figure 28.13; the reciprocal point* $-a^2/b^*$ $(+a^2/b^*)$ *on the opposite (same) side of the center appears because the image of the source/sink (vortex) has flow rate* Q *(circulation* $-\Gamma$*) equal (opposite) to the original* $Q(\Gamma)$. The passage from (28.140) to (28.141a):

$$v^*(z) = \frac{Q - i\Gamma}{2\pi z}\left[1 + \frac{b}{z} + O\left(\frac{b^2}{z^2}\right)\right] + \frac{Q + i\Gamma}{2\pi z}\left[\frac{a^2}{b^* z} + O\left(\frac{a^4}{b^{*2} z^2}\right)\right], \tag{28.142a}$$

uses the geometric series (21.64a–c):

$$2\pi v^*(z) - \frac{Q - i\Gamma}{z} = \frac{b}{z^2}\left[Q - i\Gamma + \frac{(Q + i\Gamma)\, a^2}{b^* b}\right] + O\left(\frac{b^2}{z^3}, \frac{a^4}{b^{*2} z^3}\right); \tag{28.142b}$$

the identity (28.141a) $\equiv$ (28.142b) also justifies the asymptotic approximation (28.141b) in (28.141a)

28.8.2 Forces and Moments Exerted on the Cylinder

The complex potential (28.139a) is real on the cylinder:

$$2\pi f\left(ae^{i\varphi}\right) = Q \log\left[\left(ae^{i\varphi} - b\right)\left(ae^{-i\varphi} - b^*\right)\right] - i\Gamma \log\left(\frac{ae^{i\varphi} - b}{ae^{-i\varphi} - b^*}\right)$$
$$= Q \log\left[a^2 + |b|^2 - 2a\,\mathrm{Re}\left(be^{-i\varphi}\right)\right] + 2\Gamma \arg\left(ae^{i\varphi} - b\right) \equiv \Phi(a, \varphi). \tag{28.143}$$

The two logarithms involve complex conjugate variables, and hence the same (opposite) modulus and real part (argument and imaginary part); it follows that complex potential on the cylinder is real, and coincides with the scalar potential. Thus the cylinder is the streamline, $\Psi(a, \varphi) = 0$, for any position b of the monopole of inside or outside. *The force exerted by the monopole on the cylinder is specified (28.145) by the velocity (28.140) induced by the images (28.144) at the monopole:*

$$v_0^*(z) \equiv v^*(z) - \frac{Q - i\Gamma}{2\pi(z-b)} = \frac{(Q+i\Gamma)a^2}{2\pi z(b^*z - a^2)}, \tag{28.144}$$

$$F^*(b) = F_x - iF_y = \rho(Q - i\Gamma)v_0^*(b) = \frac{\rho(Q-i\Gamma)(Q+i\Gamma)a^2}{2\pi b\left(|b|^2 - a^2\right)} = \frac{\rho(Q^2+\Gamma^2)a^2}{2\pi b\left(|b|^2 - a^2\right)}, \tag{28.145}$$

and is attractive for a source, sink or vortex; the reverse sign applies for monopole inside, for the same reasons as in (24.59) [(26.55)] for an electric charge (current) near a cylinder. In (24.59) [(26.55)] the real axis was taken through the monopole, leading to real b; in (28.145) the monopole is at any position relative to the real axis, so b is complex. Using the same induced velocity as in (28.144), and the dipole moment (28.141c) in (28.31b) specifies the pitching moment:

$$\begin{aligned} M + \frac{\rho Q\Gamma}{2\pi} &= -\mathrm{Re}\left\{i\rho P_1 v^*(b)\right\} = \mathrm{Re}\left\{-i\rho\left[Q\left(b + \frac{a^2}{b^*}\right) - i\Gamma\left(b - \frac{a^2}{b^*}\right)\right]\frac{(Q+i\Gamma)a^2}{2\pi b(b^*b - a^2)}\right\} \\ &= \frac{\rho a^2}{2\pi\left(|b|^2 - a^2\right)}\mathrm{Re}\left\{(\Gamma - iQ)\left[Q\left(1 + \frac{a^2}{|b|^2}\right) - i\Gamma\left(1 - \frac{a^2}{|b|^2}\right)\right]\right\}; \end{aligned} \tag{28.146}$$

thus a pitching moment exists only if both the sources/sink and vortex (28.147a) are present (28.147b):

$$\begin{aligned} Q \neq 0 \neq \Gamma: M &= -\frac{\rho Q\Gamma}{2\pi} + \frac{\rho Q\Gamma a^4}{\pi|b|^2\left(|b|^2 - a^2\right)} = \frac{\rho Q\Gamma}{2\pi|b|^2}\frac{2a^4 - |b|^4 + a^2|b|^2}{|b|^2 - a^2} \\ &= \frac{\rho Q\Gamma}{2\pi|b|^2}\frac{(2a^2 - |b|^2)(|b|^2 + a^2)}{|b|^2 - a^2}. \end{aligned} \tag{28.147a,b}$$

The pitching moment (28.147b) is nonzero only for a combination of source/sink and vortex (28.147a); it vanishes for two monopole positions $|b| = \pm\, 2\sqrt{a}$ *that lie outside the cylinder; the pitching moment has opposite signs on either side, that is positive (negative) or counterclockwise (clockwise) for* $|b| < a\sqrt{2}$ $\left(|b| > a\sqrt{2}\right)$, *assuming* $\Gamma\, Q > 0$, *and* $|b| > a$; *for a monopole inside the cylinder* $|b| < a$ *the pitching moment is positive (negative) or counterclockwise (clockwise) for opposite* $Q\Gamma < 0$ *(equal* $Q\Gamma > 0$*) signs of the flow rate and circulation.*

28.8.3 Rotating Vortex and Translating Source/Sink

The complex potential (28.139a) ≡ (28.139b) and conjugate velocity (28.140) hold at all points except at the monopole; since the monopole does not act upon itself, the induced field is due only to the images, and corresponds to the omission of the first term in the velocity (28.140), as in (28.144) ≡ (28.148b):

$$b = |b|\,e^{i\varphi} : v_0^*(b) \equiv \left(\dot{b}_r - i\dot{b}_\varphi\right)e^{-i\varphi} \equiv \frac{db}{dt} \equiv \dot{b} = \frac{(Q+i\Gamma)a^2}{2\pi b\left(|b|^2 - a^2\right)}. \tag{28.148a,b}$$

The polar components (28.148a) of the velocity of the monopole:

$$\dot{b}_r = \frac{Qa^2}{2\pi|b|\left(|b|^2 - a^2\right)}, \quad \dot{b}_\varphi = -\frac{\Gamma a^2}{2\pi|b|\left(|b|^2 - a^2\right)}, \tag{28.149a,b}$$

show that the monopole has a radial (28.149a) [azimuthal (28.149b)] velocity due to the source/sink (vortex) in Figure 28.13: (i) a source $Q > 0$ (sink $Q < 0$) has a radial outward $\dot{b}_r > 0$ (inward $\dot{b}_r < 0$) motion; (ii) a vortex with positive $\Gamma > 0$ (negative $\Gamma < 0$) circulation moves azimuthally around the cylinder in the negative $\dot{b}_\varphi < 0$ (positive $\dot{b}_\varphi > 0$) direction, that is, clockwise (counterclockwise); (iii) all directions are reversed between monopole outside $|b| > a$ (inside $|b| < a$) the cylinder.

The velocity of monopole (28.148b) ≡ (28.149a,b) consists of two terms: (i) the first is due to the image with equal (opposite) flow rate (circulation) at the reciprocal point:

$$\dot{b}_1 = \frac{Q + i\Gamma}{2\pi b\left(1 - a^2/|b|^2\right)}, \qquad \left\{\dot{b}_{1r}, \dot{b}_{1\varphi}\right\} = \frac{\{Q, -\Gamma\}}{2\pi\left(|b| - a^2\Big/|b|\right)}; \tag{28.150a,b}$$

(ii) the second is due to the image at the center with opposite (equal) flow rate (circulation) that induces the velocity:

$$\dot{b} - \dot{b}_1 = \dot{b}_2 = -\frac{Q + i\Gamma}{2\pi b}, \qquad \left\{\dot{b}_{2r}, \dot{b}_{2\varphi}\right\} = \frac{\{-Q, \Gamma\}}{(2\pi\ |b|)}. \tag{28.151a,b}$$

The reciprocal point rotates with the vortex and moves in the radial direction in opposition to the source/sink, specifying the motion of the monopole and the whole image system (Figure 28.13). The opposite images at the center and at the reciprocal point mean that (i) the original line monopole prevails (28.141a) in the far-field (28.141b) as if it was at the axis of the cylinder; (ii) the original monopole and image with their respective positions add dipole term (28.141c) at the next order. Both images, at the center (28.151a,b) and at the reciprocal point (28.150a,b) affect the velocity of the monopole (28.149a,b) and the force (28.145) on the cylinder that exist independently for the source/sink and vortex; in contrast the pitching moment (28.147b) is a coupling effect exists only if both the source/sink and vortex are present (28.147a); it has opposite sign to $Q\ \Gamma$ as b becomes large $b^2 >> 2\,a^2$ and the two opposite images coincide at the center, weakening the dipole effect that is first (second) term and the r.h.s. of (28.29c) [(28.147)]. There is no net circulation in the flow (28.139a) because the original and image monopole cancel; however, the vortex moves circumferentially (28.149b) relative to the cylinder. This shows that the observation of an unsteady flow can be misleading: a vortex can move around a cylinder without circulation.

28.8.4 Trajectories outside or inside the Cylinder

The velocities (28.149a,b) apply both to a monopole outside $|b| > a$ or inside $|b| < a$ the cylinder. In the case of a vortex (28.152a) the velocity is azimuthal (28.149b) that is, it describes a circle of radius equal to the initial position (28.152b) with tangential velocity (28.149b) and angular velocity (28.152c):

$$Q = 0 \neq \Gamma: \quad r = |b|, \quad \omega \equiv \frac{\dot{b}_\varphi}{|b|} = -\frac{\Gamma a^2}{2\pi\,|b|^2\left(|b|^2 - a^2\right)}; \tag{28.152a–c}$$

the motion can go on indefinitely in all cases. In the case of a source or sink (28.153a) the velocity (28.149a) ≡ (28.153b) is radial and the time elapsed along the trajectory is given by (28.153a–c):

$$Q \neq 0 = \Gamma: \quad \frac{\mathrm{d}r}{\mathrm{d}t} = \frac{Qa^2}{2\pi r\left(r^2 - a^2\right)}, \quad \frac{2Qa^2t}{\pi} = 4\int r\left(r^2 - a^2\right)\mathrm{d}r = \left[r^4 - 2a^2r^2\right]_{r_1}^{r_2}, \tag{28.153a–c}$$

leading to four cases: (I) for a source (28.154a) outside the cylinder (28.154b), the image is a source inside the cylinder, hence the induced velocity radial outward (28.153b), and the time (28.153c) elapsed between the initial position $r_1 = b$ and a position $r_2 = d$ is (28.154c):

$$Q > 0, \quad a < r_1 \equiv b < d \equiv r_2 : \quad t_I = \frac{\pi}{2Qa^2}\left(d^2 - b^2\right)\left(d^2 + b^2 - 2a^2\right), \qquad \text{(28.154a–c)}$$

so it reaches infinity after an infinite time, that is, $d \to \infty$ as $t_I \to \infty$; (II) for a sink (28.155a) outside the cylinder (28.155b), the image sink induces a radial inward velocity, so it takes the finite time (28.155c) to hit the cylinder at $r_2 = a$ starting at $r_1 = b$, viz.:

$$Q = -\left|Q\right| < 0, \quad r_1 \equiv b > a \equiv r_2 : \quad t_{II} = \frac{\pi}{2\left|Q\right|}\left(\frac{b^2}{a} - a\right)^2; \qquad \text{(28.155a–c)}$$

(III) for a sink (28.156a) inside the cylinder (28.156b), the image sink outside induces a radial outward velocity, and it takes a finite time (28.155c) to hit the cylinder at $r_2 = a$ starting at $r_1 = b$:

$$Q = -\left|Q\right| < 0, \quad r_1 \equiv b < a \equiv r_2 : \quad t_{III} = \frac{\pi}{2\left|Q\right|}\left(a - \frac{b^2}{a}\right)^2 = t_{II}; \qquad \text{(28.156a–c)}$$

(IV) for a source (28.157a) inside the cylinder (28.157b) the image source outside induces a radial inward velocity, and it takes a finite time (28.157c) to reach the center:

$$Q > 0, \quad a > b \equiv r_1 > 0 \equiv r_2 : \quad t_{IV} = \frac{\pi}{2Qa^2}\left[r^4 - 2a^2r^2\right]_a^0 = \frac{\pi a^2}{2Q}. \qquad \text{(28.157a–c)}$$

In (28.154c) was made the simplification:

$$\begin{aligned}\frac{2Qa^2}{\pi}t_I &= \left[r^4 - 2a^2r^2\right]_b^d = d^4 - 2a^2d^2 - b^4 + 2a^2b^2 \\ &= \left(d^2 - b^2\right)\left(d^2 + b^2\right) - 2a^2\left(d^2 - b^2\right) = \left(d^2 - b^2\right)\left(d^2 + b^2 - 2a^2\right), \qquad \text{(28.158a)}\end{aligned}$$

similar to the factorization of the numerator of (28.147b). The simplification:

$$\begin{aligned}\frac{2\left|Q\right|}{\pi}t_{II} &= \frac{2\left|Q\right|}{\pi}t_{III} = -a^{-2}\left[r^4 - 2a^2r^2\right]_b^a \\ &= -\frac{-a^4 - b^4 + 2a^2b^2}{a^2} = \frac{\left(a^2 - b^2\right)^2}{a^2} = \left(a - \frac{b^2}{a}\right)^2, \qquad \text{(28.158b)}\end{aligned}$$

applies to (28.154c) $\equiv$ (28.155c). Thus *there is a distinction between the motion of monopoles (Figure 28.13) at a distance b from a cylinder of radius a: (i) a vortex (28.152a) moves an infinite time along a circle of radius of radius (28.152b) with angular velocity (28.152c), with opposite directions if it is outside $|b| > a$ (inside $|b| < a$); (ii/iii) a sink (28.155a) [(28.156a)] moves toward the cylinder and reaches it in the same finite time (28.155c) [$\equiv$ (28.156c)], from the outside (28.155b) [inside (28.156b)]; (iv/v) a source (28.154a) [(28.157a)] moves away from the cylinder, and comes to rest at the center (infinity) in a finite (28.157c) [infinite (28.154c) with $d \to \infty$] time, if it is inside (28.157b) [(outside (28.154b)]. The trajectory of a monopole is a spiral specified by the superposition of the radial (azimuthal) motion of the source/sink (vortex) in (28.153–157a–c) [(28.152a–c)].* The distinct motions, that is, paths or trajectories of vortices, sinks and sources are demonstrated: (i) near a wall (Sections 16.1 and 16.2); (ii) in a rectangular corner (Section 16.6); (iii) near a cylinder (Section 28.8); (iv) near a semiinfinite plate (Section 36.3).

28.9 Dipole outside or inside a Cylinder

The circle theorem (24.47) applies to any multipole, for example, to a dipole inside or outside a cylinder (Subsection 28.9.1); the forces on the cylinder are not specified by the Kutta–Joukowski theorem (Subsection 28.2.3), and require an alternate derivation (Subsection 28.9.2) from the Blasius theorem (Subsection 28.2.1).

28.9.1 Dipole outside or inside a Cylinder

The complex potential (12.43a) due to a dipole of moment P_1 at b:

$$f(z) = -\frac{P_1}{2\pi (z-b)} - \frac{P_1^*}{2\pi \left(a^2/z - b^*\right)}, \tag{28.159a}$$

is modified (Figure 28.14) in the presence of cylinder of radius a by (24.47) the second term in (28.159a). The conjugate velocity corresponding to the complex potential (28.159a) is:

$$v^*(z) = \frac{\mathrm{d}f}{\mathrm{d}z} = \frac{P_1}{2\pi (z-b)^2} - \frac{P_1^* a^2}{2\pi b^{*2} \left(z - a^2/b^*\right)^2}, \tag{28.159b}$$

showing that the image dipole has moment $-P_1^*\, a^2/b^{*2}$ and is located at the reciprocal point $c = a^2/b$. The dipole may be outside $|b| > a$ (inside $|b| < a$) the cylinder, in that case the reciprocal point will be inside $|c| = a^2/\,|b| < |b|$ (outside $|c| = a^2/\,|b| > |b|$) the circle of radius a. The calculation of the forces and moment exerted by the dipole on the cylinder can be made via Blasius theorems (28.20a,b); the baseline force and moments theorem (28.29–28.30a–c) assume that the singularity is at the origin, and does not apply in the present case when the singularity is located elsewhere.

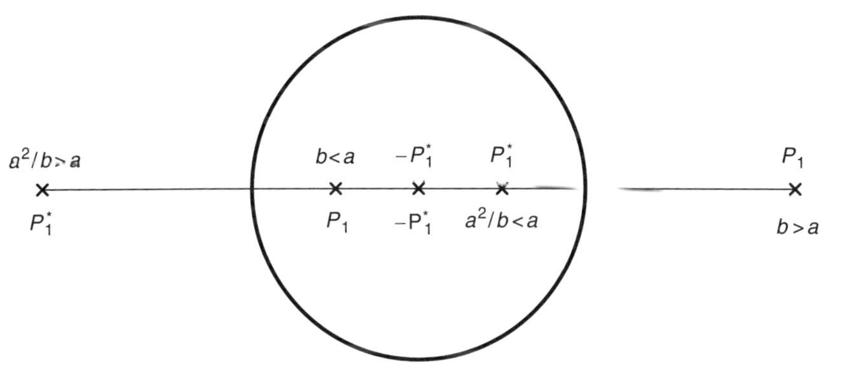

FIGURE 28.14
The images on a cylinder apply to monopoles for the electrostatic/magnetostatic field/potential flow (Figures 24.4–24.5, 26.2, and 28.13) and dipoles (Figure 28.14) and multipoles. The effect of a cylinder on the flow of a dipole of moment P_1 is represented by two images: (i) a conjugate dipole of moment P_1^* at the reciprocal point; (ii) an opposite conjugate dipole of moment $-P_1^*$ at the center of the cylinder. The original dipole may be outside or inside the cylinder and: (i) exchanges the position with the image conjugate dipole at the reciprocal point; (ii) leaves the same opposite conjugate image dipole at the center. In both cases the original and two image dipoles enforce the boundary condition of tangential velocity at the surface of the cylinder.

For a dipole outside the cylinder $|b| > a$, the Blasius theorem applied to the cylinder is affected only by the image singularity inside at the reciprocal point. Thus the square (28.160b) of the complex conjugate velocity (28.159b) for a dipole on the real axis (28.160a):

$$b \in |R: \quad [v^*(z)]^2 = \left(\frac{df}{dz}\right)^2 = \left[\frac{P_1}{2\pi(z-b)^2}\right]^2 + \left[\frac{P_1^* a^2}{2\pi b^2 (z - a^2/b)^2}\right]^2 - \frac{|P_1|^2 a^2}{2\pi^2 b^2 (z-b)^2 (z-a^2/b)^2}, \tag{28.160a,b}$$

is expanded in Laurent series around the conjugate point, corresponding to the change of variable (28.161a):

$$w = z - \frac{a^2}{b}: \quad \left(\frac{df}{dz}\right)^2 = \frac{A_{-2}}{w^2} + \frac{A_{-1}}{w} + \cdots; \tag{28.161a,b}$$

the only terms of interest in (28.161b) are (28.162b) those:

$$\begin{aligned} c \equiv b - \frac{a^2}{b}: \quad -\frac{2\pi^2 b^2}{|P_1|^2 a^2}\left(\frac{df}{dz}\right)^2 &= (z-b)^{-2}\left(z - \frac{a^2}{b}\right)^{-2} = w^{-2}(w-c)^{-2} \\ &= (wc)^{-2}\left(1 - \frac{w}{c}\right)^{-2} = (wc)^{-2}\left[1 + \frac{2w}{c} + O\left(\frac{w^2}{c^2}\right)\right] \\ &= \frac{1}{c^2 w^2} + \frac{2}{c^3 w} + O(1), \end{aligned} \tag{28.162a,b}$$

that depend only on the distance (28.162a) from the dipole to its image at the reciprocal point. The binomial series (25.40a–c) was used in (28.162b) to second-order to determine (28.163b,c) the coefficients (28.161b):

$$b > a: \quad \{A_{-1}, A_{-2}\} = -\frac{|P_1|^2 a^2}{2\pi^2 b^2}\left\{\frac{2}{c^3}, \frac{1}{c^2}\right\}, \tag{28.163a–c}$$

that are sufficient to calculate the forces and moment as shown next. The first (second) term on the r.h.s. of (28.160b) correspond to the dipole (its image) and exert no force or moment, because each consists of poles of order four, with zero residue A_{-1} and coefficient A_{-2} in the Laurent series, which lead to a zero integral (20.77a,b) ≡ (27.29b,c) ≡ (28.24a–c); this is not the case for the third term on the r.h.s. of (28.160b) is a cross-term between the dipole and its conjugate, implying that the monopoles are reversed, and have an effect proportional to the square of the modulus of the dipole moment, and independent of the direction of the dipole axis. Since this term is real, the dipole will not exert a pitching moment and the force will be in the x-direction, as shown next.

28.9.2 Flow Singularities, Forces, and Moments

Substituting (28.163b,c) in the Blasius theorem (28.20a,b) and recalling (28.24a–c) it follows that:

$$F^* = \frac{i\rho}{2}\int^{(0+)}\left(\cdots + \frac{A_{-1}}{w} + \cdots\right) dw = \frac{i\rho}{2}(2\pi i) A_{-1} = -\rho\pi A_{-1}, \tag{28.164a}$$

$$M = \mathrm{Re}\left\{-\frac{\rho}{2}\int^{(0+)}\left(\frac{A_{-2}}{w^2} + \cdots\right) w\, dw\right\} = \mathrm{Re}\left\{-\frac{\rho}{2}(2\pi i) A_{-2}\right\} = \mathrm{Re}(-i\rho\pi A_{-2}). \tag{28.164b}$$

Since A_{-2} is real in (28.163c) there is no pitching moment (28.165b); the real negative A_{-1} in (28.163b) also implies (28.165c,d) a force along the x-axis:

$$b > a: \quad M = 0 = F_y, \quad F_x = \frac{\rho \left|P_1\right|^2 a^2 b}{\pi \left(b^2 - a^2\right)^3}. \tag{28.165a–d}$$

Thus (Figure 28.14) *a dipole of moment P_1 at a distance b from a cylinder of radius a, exerts no moment (28.165b) but only to a force of attraction (28.165c,d) that depends on the modulus of the dipole moment, but not on the direction of the dipole axis. The force applies for a dipole outside the cylinder (28.165a); in the opposite case of a dipole inside the cylinder $b < a$ the sign in (28.165d) is reversed, so that the force on the cylinder is still positive, and it is repelled away from the dipole:*

$$b >< a: \quad F^* = \rho \left|P_1\right|^2 a^2 \frac{b}{\pi} \left|b^2 - a^2\right|^{-3}. \tag{28.166a,b}$$

Thus there are two cases: (i) for a dipole outside the circle (28.165a) then (28.166b) $\equiv$ (28.165d) is positive real, that is, a force of attraction of the cylinder toward the dipole, since the image at the reciprocal point $a^2/b < a$ inside the cylinder implies $c > 0$ in (28.162a); (ii) for a dipole inside the cylinder $b < a$, then (28.165d) is real negative but (28.166b) is real positive, and the cylinder is repelled by the dipole, corresponding to the reciprocal point $a^2/b > a$ now outside the cylinder, so that $c < 0$ in (28.162a).

In the latter case of the dipole inside the cylinder (28.161a,b) should be replaced by an expansion around the pole at $z = b$ in the cylinder:

$$s \equiv z - b: \quad \left(\frac{\mathrm{d}f}{\mathrm{d}z}\right)^2 = \frac{B_{-2}}{s^2} + \frac{B_{-1}}{s} + \cdots; \tag{28.167a,b}$$

this corresponds (28.162b) to:

$$\begin{aligned} -\frac{2\pi^2 b^2}{\left|P_1\right|^2 a^2} \left(\frac{\mathrm{d}f}{\mathrm{d}z}\right)^2 &= (z-b)^{-2} \left(z - \frac{a^2}{b}\right)^{-2} = s^{-2}(s+c)^{-2} \\ &= (sc)^{-2}\left(1 + \frac{s}{c}\right)^{-2} = (sc)^{-2}\left[1 - \frac{2s}{c} + 0\left(\frac{s^2}{c^2}\right)\right] \\ &= \frac{1}{c^2 s^2} - \frac{2}{sc^3} + O(1). \end{aligned} \tag{28.168}$$

Comparing (28.168) with (28.167b) it follows (28.168c,d):

$$b < a: \quad \{B_{-1}, B_{-2}\} = -\frac{\left|P_1\right|^2 a^2}{2\pi^2 b^2}\left\{-\frac{2}{c^3}, \frac{1}{c^2}\right\} = \{-A_{-1}, A_{-2}\}; \tag{28.169a–c}$$

the first (second) coefficients in (28.169b,c) and (28.163b,c) have the opposite (same) sign. Thus: (i) the pitching moment (28.165b) is still zero; (ii) the sign is (28.164a) $\equiv$ (28.165c) changed for dipole inside instead of outside the cylinder preserving (28.166b). The Blasius theorem (28.20a,b) specifies the lift, drag, and pitching moment due to pressure forces for any potential flow or distribution of singularities. The Kutta–Joukowski theorem (28.29a–c) and related results (28.31a–c) are a particular application of the Blasius theorems with two assumptions: (i) the flow velocity is finite at infinity; (ii) all singularities are collocated at the origin. Thus the Kutta–Joukowski theorem requires generalization, within the context of the Blasius theorems if either of the conditions (i) or (ii) is not met. The forces on a dipole near a cylinder could not be obtained from the forces and moment theorem, neither

(i) in the original form (28.29–28.30a–c) for a uniform flow at infinity, nor (ii) in generalized form (Example 30.15) for an arbitrary flow at infinity, because in both cases the only (other) singularity is assumed to be at the origin. Another extension of the original forces and moments theorem is (iii) to allow for a source/sink, vortex and dipole with distinct locations (Example 30.14); it also does not cover the present case of two dipoles. Other cases of forces and moments (Example 30.16) are also distinct. The forces and moments theorems apply to the potential flow (Chapters 12, 14, 18, 28, 34, 36, 38), electro and magnetostatics (Chapters 24, 26, 36) and to the gravity field (Chapter 18); the relevant notation is summarized next.

Note 28.1. *Comparison of Two-Dimensional Potential Fields*: The gravity field has analogies to other potential fields, as shown in two dimensions in the Table 28.1; it includes the conventions and notation as well. The gravity field $\vec{g}$ is associated with minus the gradient of the gravity a potential Φ_g in (18.18b); the latter satisfies the Poisson equation (18.19b) involving the gravitational constant G and mass density ρ per unit area; the mass density per unit length μ causes jump (18.21c) in the normal gravity field involving also the gravitational constant. The analogies and differences with an irrotational (incompressible) flow [Section 18.1 (18.2)] are: (i) the velocity is plus the gradient (the two-dimensional curl) of a velocity potential (18.1b) [stream function (18.10b); (ii) the velocity potential (stream function) satisfies a Poisson Equation (18.2b) [(18.11b)] forced by the dilatation Λ [minus the vorticity $\vec{\varpi}$], with is the flow rate (18.3) [circulation [(18.12)] per unit area; (iii) the flow rate q (circulation γ) per unit length causes a jump in the normal (18.6b) [tangential [18.14)] velocity. There are two additional analogies of the irrotational (incompressible) flow with the electro (magneto) static field [Chapter 24 (26)]; (i) the electric field (magnetic induction) is minus the gradient of an electrostatic potential (24.4a) [plus the two-dimensional curl of a magnetic field function (26.4a)]; (ii) the electrostatic potential (magnetostatic field function) satisfies a Poisson equation (24.5b) [(26.6b)] forced by the density q (j) of electric charge e (current J) per unit area; (iii) the density of electric charge σ (current ϑ) per unit length causes a jump in the normal electric field (24.8b) [tangential magnetic field (26.9b)]. The latter two (ii, iii) involve: (a) for the electrostatic field the electric permittivity ε of the medium; (b) for the magnetostatic field the magnetic permeability of the medium μ divided by the speed of light in vacuo c.

Note 28.2. *Forces and Moments on Bodies in Potential Fields: For an arbitrary body moving in a potential flow of velocity U_i the kinetic energy is a quadratic function (28.170a):*

$$E_k = \frac{1}{2}\sum_{i,j=1}^{2,3} m_{ij}\, U_i U_j, \quad m_{ij} = m_0 \delta_{ij}, \tag{28.170a,b}$$

with a two- (three-) dimensional mass matrix m_{ij}; the mass matrix for the cylinder is diagonal (28.170b), with the displaced mass (28.104b) on the diagonal, where δ_{ij} is the identity matrix. Likewise the electric (magnetic) energy of a body with electric charges (currents) in an external electric E_i (magnetic, H_j) field is a quadratic function (28.171a) [(28.171b)]:

$$E_e = \frac{1}{2}\sum_{i,j=1}^{2,3} P_{ij}^{e} E_i E_j, \quad E_m = \frac{1}{2}\sum_{i,j=1}^{3} P_{ij}^{m} H_i H_j, \tag{28.171a,b}$$

where the electric [magnetic] polarization matrix reduces to a diagonal or scalar form like (28.170b) only in particular cases, for example, the circular cylinder or the sphere.

The methods of calculation of forces (lift, down force, drag, and thrust) and pitching moment for a body in a plane potential flow are general; the first examples considered simple shapes, like cylinders (Sections 28.6–28.9) and Rankine finite (Section 28.5) and semiinfinite

TABLE 28.1
Two-Dimensional Potential Fields

Case	Irrotational Flow	Incompressible Flow	Gravity Field (1)	Electrostatic Field (2)	Magnetostatic Field (3)
Poisson equation	$\nabla^2 \Phi_v = \Lambda$	$\nabla^2 \Psi_v = -\varpi$	$\nabla^2 \Phi_g = \rho G$	$\nabla^2 \Phi_e = \frac{q}{\varepsilon}$	$\nabla^2 \Psi_e = -\frac{\mu}{c} j$
Variable	Φ Velocity potential	Ψ_v Stream function	Φ_g Gravity potential	Φ_e Electric potential	Ψ_m Magnetic field function
Forcing	Λ Dilatation	ϖ Vorticity	ρ Mass density per unit area	q Electric charge per unit area	j Electric current per unit area
Normal jump	$[v_n] = q$	$[v_n] = 0$	$[g_n] = -G\mu$	$[E_n] = \frac{\sigma}{\varepsilon}$	$[H_n] = 0$
Tangential jump	$[v_s] = 0$	$[v_s] = -\gamma$	$[g_s] = 0$	$[E_s] = 0$	$[H_s] = \frac{\mu}{c}\vartheta$
Forcing	q Flow rate per unit length	γ Circulation per unit length	μ Mass density per unit length	σ Electric charge density per unit length	ϑ Electric current density per unit length
Monopole	Line source/sink	Line vortex	Line mass	Line electric charge	Line electric current
Total strength	Q Flow rate	Γ Circulation	m Total mass	e Electric charge	J Electric current
Field	$\vec{v}$ Velocity	$\vec{v}$ Velocity	$\vec{g}$ Gravity	$\vec{E}$ Electric	$\vec{H}$ Magnetic
Complex conjugate	$v^* = \frac{Q}{2\pi z}$	$v^* = -\frac{i\Gamma}{2\pi z}$	$g^* = \frac{mG}{2\pi z}$	$E^* = \frac{e}{2\pi\varepsilon z}$	$H^* = \frac{iJ}{2\pi c z}$
Complex potential	$f_v = \frac{Q}{2\pi}\log z$	$f_v = -\frac{i\Gamma}{2\pi}\log z$	$f_g = \frac{mG}{2\pi}\log z$	$f_e = -\frac{e}{2\pi\varepsilon}\log z$	$f_m = -\frac{iJ}{2\pi c}\log z$
Relation	$v^* = df_v/dz$	$v^* = df_v/dz$	$g^* = -df_g/dz$	$E^* = -df_e/dz$	$H^* = -df_m/dz$

(1) G – gravitational constant (2) ε – dielectric permittivity μ – magnetic permeability
(3) c – speed of light in vacuo

(Section 28.4) bodies. The values obtained for dimensionless parameters like the pressure, lift, drag, and pitching moment coefficients are representative of engineering practice in some cases; in other cases they are too large or not achievable in practice or do not correspond to desirable aerodynamic properties. The conformal mapping (Part 4) can be used to transform simple shapes into many others, for example, the cylinder into airfoil shapes that are efficient to generate high lift with low drag (Chapter 34). The Schwartz–Christoffel transformation (Chapter 33) of the cylinder into polygons with finite or infinite sides extends the present results to other flows of interest (Chapter 36) for example, channels, slits, wells, etc. The drag can be calculated without recourse to source or sinks using the hodograph method (Chapter 38) applied to free jets. Some of the preceding methods extend to vortical and unsteady flows. The combination of these methods provides a relatively simple but representative first approach to efficient aerodynamic design.

Conclusion 28: The divergence theorem applies to a domain whose boundary is a closed regular curve which has continuous normal and tangent (Figure 28.1). A body in a potential flow is subject (Figure 28.2) to the pressure acting in the inward normal direction (b), resulting in a horizontal and vertical force in the plane and a pitching moment normal to the plane (c) associated with (a) the circulation, blowing or suction; if the body is in a stream with uniform velocity U at infinity (Figure 28.3), the force along (opposite) to the velocity is the drag (thrust), and the force upward (downward) is the lift (down force). There is also a pitching moment normal to the plane, which is negative (positive) for a pitch-up (pitch down), i.e., in the clockwise (counter clockwise) direction. Concerning (a) the forces: (i) the lift (down force) is associated with a negative (positive) i.e., clockwise (counter clockwise) vortex; (ii) the drag (thrust) is associated with a source (sink). Concerning (b) the pitching moment it is due (Figure 28.6) to: (i) either a monopole consisting of source/sink and vortex (a); (ii) or a dipole with moment having a component transverse to the incident stream (b). The pitching moment due to a monopole (Figure 28.8) is: (i) a pitch- up for a source and counter clockwise vortex (a) or a sink and clockwise vortex (d); (ii) a pitch-down for a sink and a counter clockwise vortex (b) or a source and a clockwise vortex (c). A pitch- up (Figure 28.9) is also caused by (a) a transverse downward dipole, due to a sink (source) above (below) or a counter clockwise (clockwise) vortex behind (ahead); inverting the positions leads to a transverse upward dipole and a pitch- down (b).

A source (**a**) [sink (b)] in an uniform flow represents a semiinfinite body (Figure 28.5) (fairing) subject to a drag (thrust). A pair of opposite source-and-sink in a uniform flow represents an oval body (Figure 28.7), which is not subject to forces, but is subject to a pitching moment if the flow has angle-of-attack. Reversing the position of the source and sink, so that the latter faces the flow, there are Figure 28.10 two possibilities. (i) for weak sink and source there is a gap between two semiinfinite bodies (a); (ii) a stronger sink-source pair closes the gap and leads (b) to an inner (outer) flow which is a throated nozzle (valley between mountains) with two angular points. A body with circulation in a stream (Figure 28.11) is subject to a force orthogonal to the velocity, i.e., lift or downforce. The potential flow (Figure 28.4) past a cylinder (a) can be obtained as the dipolar limit of the oval (Figure 28.7) when the distance between the source and sink decreares in inverse proportion to the increase of the flow rate. If the cylinder has circulation it is subject to a force (Figure 28.11) orthogonal to the velocity, i.e., lift or downforce, causing a motion along a circle, cycloid or trochoid. A negative circulation increasing in modulus on a cylinder (Figure 28.4) moves the stagnation points from the mid-line (a), to symmetric positions on the lower half (b), to the lowest point (c) or into the flow (d). The corresponding flow inside the cylinder (Figure 28.12) is a dipole (a) more distorted by increasing circulation (b) until (c) streamlines form which do not pass through the center (d). The monopole near a cylinder, creates a related flow field whether it is inside or outside, because it exchanges position with the reciprocal point (Figure 28.13); the images consist of: (i) a conjugate monopole at the reciprocal point, with the same flow rate and opposite circulation; (ii) an opposite conjugate monopole at the center with opposite flow rate and the same circulation. The image system rotates with the vortex and moves radially in opposition to the source/sink. The image system is similar for other singularities, e.g., (Figure 28.14) a multipole of moment P_n has an image with conjugate moment P_n^* at the reciprocal point and another image with opposite conjugate moment $-P_n^*$ at the center. The monopole (Figure 28.11), dipole (Figure 28.12) or multipole images apply to the two cases when the original multipole is outside $b > a$ (inside $b < a$) the circle, and thus its reciprocal point $a^2/b < a$ $\left(a^2/b > a\right)$ is inside (outside) the circle.

29

Combined Test of Convergence

The deductions of the Lagrange–Burmann, Taylor, and Stirling–Maclaurin (Chapter 23) and Teixeira and Laurent (Chapter 25) expansions of holomorphic functions, included indication of the regions of divergence and of convergence, but the behavior of the series on the boundary of convergence was not addressed. In the case of the Taylor series of a real function the convergence was not proved generally, and must be established on a case-by-case basis, using convergence tests. If a function is defined by a series, then the tests of convergence can be used to establish its properties, for example: (i) it is bounded if the series is conditionally convergent, and the order of the terms is not changed; (ii) it is continuous, differentiable, or integrable if the series is uniformly convergent, and consists respectively of continuous, differentiable, and integrable functions; (iii) the value of the function at a point can be obtained by summing the terms of the series in any order, if it is absolutely convergent. Next is presented a combined test of convergence (Section 29.1) that (i) applies to a wide class of series of functions commonly encountered in applications, including power and harmonic series; (ii) indicates the behavior of the series, that is, divergent, oscillating, or conditionally, absolutely, uniformly, or totally convergent (Chapter 21); (iii) applies to all points of the complex plane, including the whole of the boundary of convergence. The theorem (Section 29.1) combines the tests of D'Alembert (Section 29.3), Gauss (Section 29.5), and Abel–Dirichlet (Section 29.6) plus the Weierstrass K-test (Section 29.7), and is used to specify the behavior of series of functions at all points of the complex plane, for example, for the Gaussian hypergeometric series (Section 29.9). The establishment of the combined convergence test uses as intermediate steps alternative convergence criteria (Section 29.2), the comparison of convergence of series and integrals (Section 29.4) and the concepts of radius and exponent of a series (Section 29.8). The combined convergence test establishes the convergence properties at every point of the complex plane for simple series, for binomial (Sections 23.9 and 29.1), logarithmic (Sections 21.8 and 29.9), and Gaussian (Section 29.9), and generalized (Example 30.30) hypergeometric; it also applies to most series whose general term is of a not too "pathologic" form. It thereby indicates the extent to that can be used the solution of for example, a problem in terms of series.

29.1 Behavior of Series at All Points of the Complex Plane

The convergence of a series is addressed (Subsection 29.9.1) by a combined test applying inside, outside and at the boundary of convergence (Subsection 29.1.2), taking as first example the binomial series (Subsection 29.1.3).

29.1.1 Interior, Exterior, and Boundary of Convergence

The following theorem applies to a series of functions (29.1), and it indicates its behavior at all points of the complex plane, so that it may designated a **combined test of**

convergence: *the series of functions*:

$$\sum_{n=0}^{\infty} F_n(z) \equiv \lim_{N\to\infty} \sum_{n=0}^{N} F_n(z) = F(z), \tag{29.1}$$

whose terms, beyond a certain order (29.2a), are in the ratio (29.2b):

$$\exists_{m\in|N} \forall_{n>m} : \quad \frac{F_{n+1}(z)}{F_n(z)} = f(z)\left\{1 - n^{-1}g(z) + O\left(n^{-\lambda}\right)\right\}, \quad \text{Re}\,(\lambda) < 1, \qquad \text{(29.2a–c)}$$

with parameter λ *satisfying (29.2c) is: (i) divergent if* $|f(z)| > 1$ *in (29.3a) and Figure 29.1a and is absolutely (totally) convergent if* $|f(z)| < 1\,(< 1-\varepsilon,$ *with* $0 < \varepsilon < 1)$ *in (29.3b) [(29.3c)] and Figure 29.1b (29.1c):*

$$0 < \varepsilon < 1: \quad |f(z)| \begin{cases} > 1 & \text{D.,} & \text{(29.3a)} \\ < 1 & \text{A.C.,} & \text{(29.3b)} \\ \le 1-\varepsilon & \text{T.C.,} & \text{(29.3c)} \end{cases}$$

so that (29.6b) [(29.6a,c)] specifies the ***open (closed) region of absolute (total) convergence****; (ii) at all points (Figure 29.1d) on (29.6d) the* ***boundary of convergence*** $|f(z)| = 1$*, excluding the particular points (29.6e) for that* $f(z) = 1$*, the series converges absolutely (conditionally) for* $E \equiv Re\{g(z)\} > 1\,(0 < E \le 1)$*, in (29.4d) [(29.4c)] and diverges (oscillates) for* $E < 0\,(E = 0)$ *in (29.4a) [(29.4b)]:*

$$|f(z)| = 1 \neq f(z): \quad \begin{cases} E \equiv \text{Re}\,\{g(z)\} < 0 & \text{D.,} & \text{(29.4a)} \\ \text{Re}\,\{g(z)\} = 0 & \text{O.,} & \text{(29.4b)} \\ 0 < \text{Re}\,\{g(z)\} \le 1 & \text{C.C.,} & \text{(29.4c)} \\ \text{Re}\,\{g(z)\} > 1 & \text{A.C.;} & \text{(29.4d)} \end{cases}$$

(iii) at (29.6e) the ***particular points*** z *satisfying* $f(z) = 1$*, that lie (Figure 29.1e) on the boundary of convergence, the series diverges (converges absolutely) for* $E \equiv Re\ \{g\,(z)\} < 1$ $(E > 1)$*, in (29.5a) [(29.5d)] and for* $E = 1$ *it diverges (oscillates) for* $b \equiv Im\ \{g\,(z)\} = 0$ $(b \neq 0)$ *in (29.5b) [(29.5c)]*

$$f(z) = 1: \quad \begin{cases} E \equiv \text{Re}\,\{g(z)\} < 1 & & \text{D.,} & \text{(29.5a)} \\ \text{Re}\,\{g(z)\} = 1 \begin{cases} \text{Im}\,\{g(z)\} = 0 \\ \text{Im}\,\{g(z)\} \neq 0 \end{cases} & & \begin{matrix}\text{D.,} \\ \text{O.,}\end{matrix} & \begin{matrix}\text{(29.5b)} \\ \text{(29.5c)}\end{matrix} \\ \text{Re}\,\{g(z)\} > 1 & & \text{A.C.,} & \text{(29.5d)} \end{cases}$$

There are $3+4+4 = 11$ cases in all (29.3a–c; 29.4a–d; 29.5a–d) that can proved in the four groups indicated next.

29.1.2 General and Particular Points on the Boundary of Convergence

The series (29.1), with terms in the ratio (29.2b,c) beyond a certain order (29.2a), converges absolutely (totally) in (29.6b) [(29.6a,c)] an open region D (closed subregion D_ε):

$$0 < \varepsilon < 1: \quad D \equiv \{z : |f(z)| < 1\} \quad \supset D_\varepsilon \equiv \{z : |f(z)| \le 1-\varepsilon\}, \qquad \text{(29.6a–c)}$$

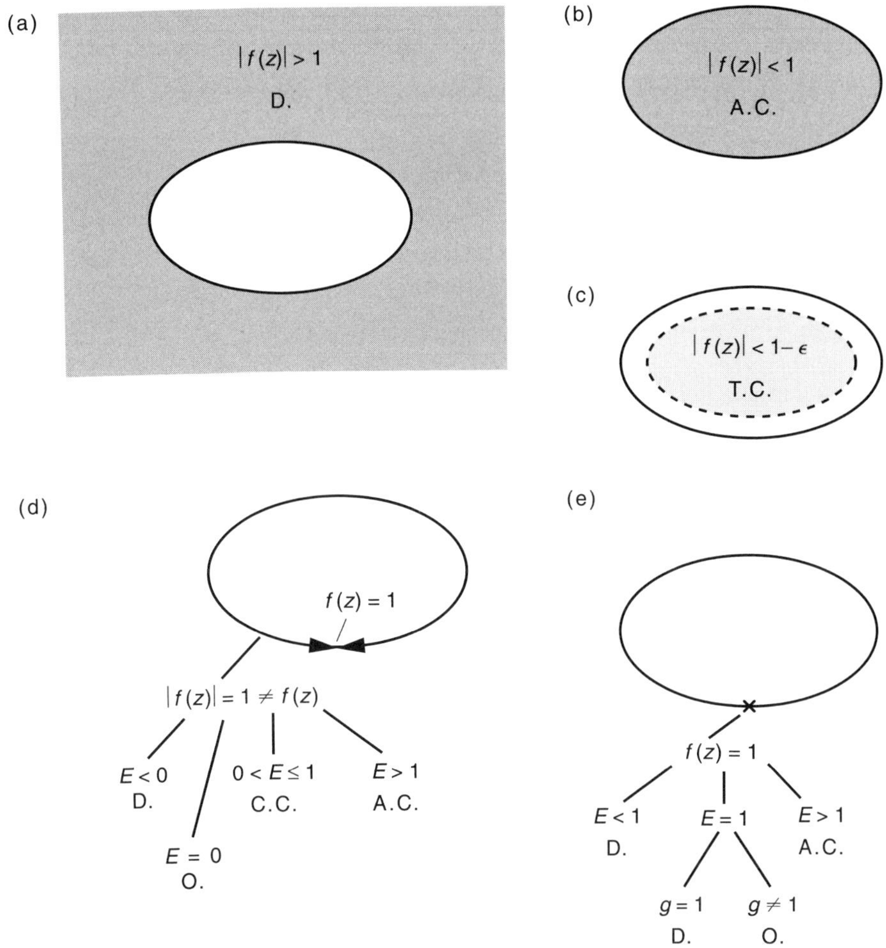

FIGURE 29.1
Consider a series of complex functions (29.1) for which, beyond a certain order, the ratio of successive terms satisfies (29.2a–c). The convergence of the series in the whole complex plane can be established as follows in three criteria. The modulus or the ratio of successive terms $R \equiv |f(z)|$ or **radius** specifies the regions of convergence: (i) divergence (a) for $R > 1$; (ii) absolute convergence (b) for $R < 1$; (iii) also uniform convergence (c) for $R \leq 1-\varepsilon$ with $0 < \varepsilon < 1$. Thus (iii) specifies a closed subregion of the open region where (ii) holds; the regions of divergence (i) and convergence (ii) are separated by a boundary of convergence $R = 1$. To establish the convergence on the boundary it is necessary: (i) to go beyond $R \equiv |f(z)|$ and to consider the next term $g(z)$ in (29.2b) with $E \equiv \{g(z)\}$ the **exponent**; (ii) on the boundary of convergence$|f(z)| = 1$ should be distinguished the special points for which $f(z) = 1$. At the special points $f(z) = 1$ on the boundary of convergence (e) the series: (iv/v) diverges (converges absolutely) for $E < 1$ $(E > 1)$; (vi/vii) in the intermediate case $E = 1$ it diverges (oscillates) for real $g(z) = 1$ [complex $g(z) \neq 1$]. At the remaining, nonspecial points $f(z) \neq 1 = |f(z)|$ on the boundary of convergence (d) the series: (viii) converges absolutely for $E > 1$ as in (v); (ix) diverges for $E < 0$ as in (iv); (x) oscillates for $E = 0$ as in (vii); (xi) for $0 < E \leq 1$ it is conditionally convergent, that is, convergent but not absolutely as in (v). The proof of the eleven cases (i) to (xi) requires a combination of four convergence tests: D'Alembert, Gauss, Weierstrass-K, and Abel–Dirichlet tests.

and diverges outside; on the boundary of convergence (29.6d), including the particular points (29.6e):

$$\partial D \equiv \{z : |f(z)| = 1\} = (\partial D - E) \cup E, \quad E \equiv \{z : f(z) = 1\}, \tag{29.6d,e}$$

the series is absolutely convergent (absolutely divergent) if $E \equiv \operatorname{Re}\{g(z)\} > 1\,(E \leq 1)$ in (29.7b) [(29.7a)]:

$$|f(z)| = 1: \quad \begin{cases} E \equiv \operatorname{Re}\{g(z)\} \leq 1 & \text{A.D.}, \\ E \equiv \operatorname{Re}\{g(z)\} > 1 & \text{A.C..} \end{cases} \tag{29.7a,b}$$

In order to distinguish in the absolutely divergent series between conditionally convergent and nonconvergent, the case (29.7a) is subdivided into two: (i) at the particular points satisfying $f(z) = 1$ in (29.6e), where absolute divergence is actually nonconvergence (29.8a):

$$f(z) = 1: \quad \begin{cases} \operatorname{Re}\{g(z)\} \leq 1 & \text{N.C.}, \\ \operatorname{Re}\{g(z)\} > 1 & \text{A.C.}; \end{cases} \tag{29.8a,b}$$

(ii) on the boundary of convergence except at the particular points $f(z) \neq 1$, the absolute divergence is (29.7a) nonconvergence (conditional convergence) for $E \equiv \operatorname{Re}\{g(z)\} < 0$ $(0 < E \leq 1)$ in (29.9a) [(29.9b)]:

$$|f(z)| = 1 \neq f(z) \quad \begin{cases} E \equiv \operatorname{Re}\{g(z)\} \leq 0 & \text{N.C.}, \\ 0 < \operatorname{Re}\{g(z)\} \leq 1 & \text{C.C.}, \\ \operatorname{Re}\{g(z)\} > 1 & \text{A.C..} \end{cases} \tag{29.9a–c}$$

In order to distinguish nonconvergence (29.9a) into divergence and oscillation on the boundary of convergence: (i) the oscillation is single-out for (29.10a) ≡ (29.4b) and (29.10b) ≡ (29.5c):

$$|f(z)| = 1: \ 0. \quad \begin{cases} E \equiv \operatorname{Re}(g(z)) = 0 & \text{if } f(z) \neq 1, \\ \operatorname{Re}(g(z)) - 1 = 0 \neq \operatorname{Im}(g(z)) & \text{if } f(z) = 1; \end{cases} \tag{29.10a,b}$$

(ii) leaving divergence for the remaining cases (29.11a) ≡ (29.4a) and (29.11b) ≡ (29.5a,b):

$$|f(z)| = 1: \ D. \quad \begin{cases} \operatorname{Re}(g(z)) < 0 & \text{if } f(z) \neq 1, \\ \operatorname{Re}(g(z)) < 1 \quad \text{or} \quad g(z) = 1 & \text{if } f(z) = 1. \end{cases} \tag{29.11a,b}$$

The tables (29.3a–c; 29.4a–d; 29.5a–d) ≡ (29.7–8a, b; 29.9a–c; 29.10–11a,b) are equivalent, and provide a detailed discrimination of all possible cases of convergence at all points (29.6a–e) of the plane (Figure 29.1).

29.1.3 Convergence of the Binomial Series

Before proceeding to prove (Sections 29.2–29.7) the combined convergence test (Subsections 29.1.1 ≡ 29.1.2) an example of its application is given, viz. the binomial series (25.37a–c) that (29.12a) has coefficients (29.12b):

$$F(z) \equiv (z+a)^{\nu} = \sum_{n=0}^{\infty} C_n z^n : \quad C_n \equiv a^{\nu-n} \binom{\nu}{n} = a^{\nu-n} \frac{\nu(\nu-1)\dots(\nu-n+1)}{n!}; \qquad (29.12\text{a,b})$$

the ratio of successive coefficients to $O(1/n)$ is:

$$\frac{C_{n+1}}{C_n} = \frac{1}{a}\frac{\nu-n}{n+1} = -\frac{1}{a}\frac{1-\nu/n}{1+1/n} = -\frac{1}{a}\left[1 - \frac{\nu+1}{n} + O\left(\frac{1}{n^2}\right)\right]. \qquad (29.12\text{c})$$

Comparing with (29.1) and (29.2a,b) the order of approximation is (29.13a), and the general term of the series (29.1) ≡ (29.12a) is (29.13b):

$$\lambda = 2\text{: } F_n(z) = C_n z^n, \quad \frac{F_{n+1}(z)}{F_n(z)} = \frac{C_{n+1}\, z}{C_n} = -\frac{z}{a}\left[1 - \frac{\nu+1}{n} + O\left(\frac{1}{n^2}\right)\right]; \qquad (29.13\text{a–c})$$

the ratio of successive terms leads to (29.13c) ≡ (29.2b) with nonconstant (29.14a) [constant (29.14b)]:

$$f(z) = -\frac{z}{a}, \quad g = \nu + 1. \qquad (29.14\text{a,b})$$

Concerning the function (29.12a) three cases arise: (i) if $\nu = N$ is a positive integer, the binomial series (29.12b) terminates (25.38) as polynomial of degree N that is a rational function analytic in the whole complex z-plane except at infinity where it has a pole of order N; (ii) if N is an a negative integer the series does not terminate (25.40a–c) and the sum has a pole of order N at $z = -a$; (iii) if ν is not an integer then the series does not terminate (25.37a–c; 25.41a–b) and $z = -a$ is a branch-point (Chapter 7) of (29.12a). In the cases (ii) and (iii), the radius of convergence of the binomial series is determined by the singularity at $z = -a$, viz. there is: (i) divergence for $|f(z)| > 1$ or $|z| > |a|$ in (29.3a); (ii) absolute convergence for $|z| < |a|$ in (29.3b); (iii) uniform convergence for $|z| < |a|(1-\varepsilon)$ with $0 < \varepsilon < 1$ in (29.3c). On the boundary of convergence $|z| = |a|$ the particular point $z = -a$ is considered separately, viz.: (i) if Re $(\nu) > 0$ then (29.12a) has a zero at $z = -a$, and since Re(g) = Re$(1+\nu) > 1$ in (29.5d) the binomial series is absolutely convergent; (v) if Re$(\nu) < 0$ then $z = -a$ is a singularity of (29.12a) and for Re(g) = Re$(1+\nu) < 1$ in (29.5a) and the binomial series diverges; (vi) if Re$(\nu) = 0 \neq \nu$ then $(z+a)^{\nu} = O^{i\text{Im}(\nu)}$ at $z = -a$, and Re$(g) = 1 \neq g(z)$ and the binomial series oscillates (29.5c); (vii) if $\nu = 0$ then $(z+a)^{\nu} = 0^{\circ}$ at $z = -a$ and $g = 1$, and the binomial series diverges (29.5b). On the boundary of convergence excluding the singularity $|z| = |a|, z \neq -a$ the binomial series: (viii) converges absolutely (29.4d) for Re$(g) > 1$, so that Re $(\nu) > 0$ and the function (29.12a) is not singular anywhere on the boundary not even near $z = -a$; (ix) converges conditionally (29.4c) for $1 \geq$ Re$(g) > 0$, so that $0 \geq$ Re$(\nu) > -1$ and the function (29.12a) has a singularity at $z = -a$ weaker that a pole, viz. an ignorable (Section 19.6) or integrable singularity; (x) oscillates (29.4b) for Re$(g) = 0$ so that Re$(\nu) = -1$ and (29.12a) has a simple pole at $z = -a$; (xi) diverges (29.4a) for Re$(g) < 0$ so that Re$(\nu) < -1$ and (29.12a) has a singularity at $z = -a$ stronger that a pole. The eleven cases (i) to (xi) are indicated in Table 29.2. They agree with Table 29.1 of convergence of power series in terms of: (i) the radius of convergence (29.15a) ≡ (29.71a):

$$R = \lim_{n\to\infty} \left|\frac{C_n}{C_{n+1}}\right| = a; \quad E = \lim_{n\to\infty} n\left[1 + \text{Re}\left(a\frac{C_{n+1}}{C_n}\right)\right] = \text{Re}\,(1+\nu); \qquad (29.15\text{a,b})$$

TABLE 29.1
Convergence of General Power Series

Behavior of series	Inside or outside boundary of convergence	On the boundary of convergence	
		Except at the particular points	At the particular points
	$\lvert z\rvert \neq R \neq \lvert\alpha\rvert$	$\lvert z\rvert = R, z \neq \alpha$	$z = \alpha$
Divergent (D.)	$\lvert z\rvert > R$	$E < 0$	$E < 1$ or $\beta = 1$
Oscillatory (O.)	–	E = 0	$E = 1 \neq \beta$
Conditionally convergent (C.C.)	–	$0 < E \leq 1$	–
Absolutely convergent (A.C.)	$\lvert z\rvert < R$	$E > 1$	$E > 1$
Totally convergent (T.C.)	$\lvert z\rvert \leq R(1-\varepsilon)$ where $0 < \varepsilon < R$.	–	–

R, E: radius and exponent (29.71a,b) of convergence.
Note: The convergence at all points of the complex plane is indicated for a power series (29.68a,b; 29.69a–c). This is a particular case of the combined convergence test for series of functions (29.1; 29.2a–c) illustrated in Figure 29.1. The criteria for convergence of power series in Table 29.1 are applied to three specific series in Table 29.2.

(ii) the exponent of convergence (29.15b) ≡ (29.71b). The latter result (29.15b) ≡ (29.15c):

$$E = \lim_{n\to\infty} n\left[1 + Re\left(\frac{\nu - n}{n+1}\right)\right] = \lim_{n\to\infty} \frac{n}{n+1}\left[1 + Re(\nu)\right] = Re(1+\nu), \tag{29.15c}$$

follows from (29.71b) with $\alpha = -a$ and (29.12c).

29.2 Cauchy (1821) Necessary and Sufficient Conditions

Before proving in sequence (Sections 29.2–29.7) the various propositions that constitute the combined convergence test (Section 29.1), some preliminary lemmas and theorems are presented (Section 29.2), both because are useful in their own right, and will be applied in the subsequent proofs. The first two such results are a necessary (Subsection 29.21) [necessary and sufficient (Subsection 29.2.2)] condition for convergence.

29.2.1 Necessary Condition for Convergence

The necessary condition is **Cauchy's lemma (1821)**: *a necessary, but not sufficient, condition of convergence of the series (29.1) is that the general term decays to zero:*

$$\sum_{n=1}^{\infty} F_n(z) C. \Rightarrow \lim_{n\to\infty} F_n(z) = 0. \tag{29.16a}$$

The condition is not sufficient, as shown by the divergent series (21.6c), whose general term $F_n = 1/n \to 0$ decays to zero as $n \to \infty$. To prove that the condition is necessary, it is shown that: *if the general term of a series does not decay to zero, the series diverges or*

TABLE 29.2
Convergence of Specific Power Series: Binomial, Logarithmic, Gaussian Hypergeometric

Series	Binomial	Logarithmic	Gaussian Hypergeometric
Subsection(s)	25.9.3; 29.1.3	21.8.2, 29.8.2	29.9
Formula(s)	(25.37b,c) $\equiv$ (29.12a,b)	(25.64a,b) $\equiv$ (29.73)	(29.74)
Sum: $F(z)$	$(z+a)^{\nu}$	$\log(1+z)$	$F(\alpha,\beta;\gamma;z)$
General term: $F_n(z)$	$(z^n/n!)\,\nu\,(\nu-1)\ldots(\nu-n+1)$	$(-)^{n-1}n^{-1}z^n$	$(\alpha)_n\,(\beta)_n/[(n+1)!\,(\gamma)_n]$
Ratio (29.2a) with: $f(z)\equiv\ldots$	$-z/a$	-1	1
$g(z)\equiv\ldots$	$\nu+1$	1	$\gamma+1-\alpha-\beta$
To order $O\left(x^{-\lambda}\right)$ with	$\lambda=2$	$\lambda=2$	$\lambda=2$
Convergence:	a	1	1
Radius: $R=$	$\mathrm{Re}\,(\nu+1)$	1	$\mathrm{Re}\,(\gamma+1-\alpha-\beta)$
Exponent: $E=$	$a\in C$	$a=1$	$a=1$
Region $\|z\|>a$:	D.	D.	D.
$\|z\|<a$:	A.C.	A.C.	A.C.
$\|z\|\le a\,(1-\varepsilon)$; $0<\varepsilon<1$;	T.C.	T.C.	T.C.
Boundary:	$\|z\|=a\neq -z$	$\|z\|=1\neq -z$	$\|z\|=1\neq z$
A.C.	$\mathrm{Re}(\nu)>0$	–	$\mathrm{Re}\,(\alpha+\beta-\gamma)<0$
D.	$\mathrm{Re}(\nu)<-1$	–	$\mathrm{Re}\,(\alpha+\beta-\gamma)>1$
O.	$\mathrm{Re}(\nu)=-1$	–	$\mathrm{Re}\,(\alpha+\beta-\gamma)=1$
C.C.	$-1<\mathrm{Re}(\nu)\le 0$	$\times$	$0\le\mathrm{Re}\,(\alpha+\beta-\gamma)<1$
Special points:	$z=a$	$z=-1$	$z=1$
A.C.	$\mathrm{Re}(\nu)>0$	–	$\mathrm{Re}\,(\alpha+\beta-\gamma)<0$
D.	$\mathrm{Re}(\nu)<0$	$\times$	$\mathrm{Re}\,(\alpha+\beta-\gamma)>0$
D.	$\nu=0$	–	$\alpha+\beta=\gamma$
O.	$\mathrm{Re}(\nu)=0\neq\mathrm{Im}(\nu)$	–	$\mathrm{Re}\,(\alpha+\beta-\gamma)=0$
			$0\neq\alpha+\beta-\gamma$

Note: Convergence at all points of the plane of three specific power series: (i) binomial; (ii) logarithmic; (iii) Gaussian hypergeometric. This is an application of the convergence criteria for general power series in Table 29.1. The latter are a particular case of the convergence tests for series of complex functions of the (29.1; 29.2a–c) illustrated in Figure 29.1.

oscillates:

$$\lim_{n\to\infty} F_n(z) \neq 0 \Rightarrow \sum_{n=1}^{\infty} F_n(z) \quad \text{D.} \quad \text{or} \quad \text{O.} \tag{29.16b}$$

To prove (29.16b) introduce the sum $S(z)$ of the first n of the series (29.17a):

$$S_n\,(z) \equiv \sum_{k=0}^{n-1} F_k\,(z); \quad \lim_{n\to\infty} \quad \{S_{n+1}\,(z) - S_n\,(z)\} = \lim_{n\to\infty} F_n\,(z) \neq 0, \tag{29.17a,b}$$

since the limit of (29.17a) depends on the way in that $n \to \infty$, the series does not converge (29.17b), that is, it diverges or oscillates (29.16b). QED.

29.2.2 Necessary and Sufficient Condition for Convergence

A more precise statement, that is, a necessary and sufficient condition, is obtained replacing a single term $F_n\,(z)$ of the series by a sum of m terms from n to $n + m -\ 1$, with arbitrary m. This is **Cauchy's convergence principle (1821)**: *the series (29.1) converges iff the sum of an arbitrary number m of terms following* $\mathrm{F_n\,(z)}$, *tends to zero as* $\mathrm{n} \to \infty$:

$$\sum_{n=0}^{\infty} F_n(z) C. \qquad \Leftrightarrow \qquad \forall_{m \in |N} : \quad \lim_{n\to\infty} \sum_{k=n}^{n+m-1} F_k(z) = 0. \tag{29.18}$$

Introduce (29.19a) as the sum of the m terms of the series (29.1) following, $F_{n-1}\,(z)$;

$$S_{n,m}(z) \equiv \sum_{k=n}^{n+m-1} F_k(z) - \sum_{k=0}^{n-1} F_k(z) = S_{n+m}(z) - S_n(z); \quad S_n(z) \equiv S_{0,n}(z) \tag{29.19a,b}$$

this implies (29.19b) for (29.17a). Using the notation (29.19a), *the principle of convergence (19.18), states that the series (29.1) converges iff (if and only if):*

$$\sum_{n=0}^{\infty} F_n(z) C. : \ \forall_{\varepsilon>0} \ \forall_{m\in|N} \ \exists_{n\in|N} \ \ |S_{n,m}(z)| < \varepsilon; \tag{29.20}$$

thus $S_{n+m}(z)$ *can be made arbitrarily small, for all* m, *by choosing sufficiently large* n.

To prove the necessary condition, assume that the series (29.1) converges (29.21a,b):

$$\forall_{\varepsilon>0} \ \ \exists_{n\in N} : \quad |S_n(z) - F(z)| < \frac{\varepsilon}{2}, \quad |S_{n+m}(z) - F(z)| < \frac{\varepsilon}{2}; \tag{29.21a,b}$$

it follows from (29.19a) that:

$$|S_{n,m}(z)| \leq |S_{n+m}(z) - F(z)| + |F(z) - S_n(z)| < \varepsilon, \tag{29.21c}$$

that proves (29.20). Concerning the sufficient condition, assume that (29.20) holds: then the sequence $S_{n,1}(z), S_{n,2}(z), \ldots,$ has an infinite number of points in a neighborhood of order ε of $F(z)$, that is, $S_{n,m}(z) \in V_\varepsilon\,(F(z))$. It follows that the sequence has (Section 28.8) at least one accumulation point in $V_\varepsilon\,(F(z))$, and since ε can made arbitrary small the accumulation point is unique; for, suppose $F(z), G(z)$ are two accumulation points associated with the sequences:

$$|S_{n,p}(z) - F(z)| < \varepsilon, \quad |S_{n,q}(z) - G\,(z)| < \varepsilon; \tag{29.22a,b}$$

their difference can be made arbitrarily small:

$$|G(z) - F(z)| \leq |S_{n,p}(z) - F(z)| + |S_{n,q}(z) - G(z)| + |S_{n,p}(z)| + |S_{n,q}(z)| < 4\varepsilon, \tag{29.22c}$$

that is, $G(z) = F(z)$ as $\varepsilon \to 0$. Thus $S_{n,m}(z)$ has an unique limit as $n \to \infty$, that is, the series converges. QED.

29.3 Region of Convergence and D'Alembert's Ratio (1768)

The comparison of a general series with the geometric series leads (Subsection 29.3.1) to the ratio test (Subsection 29.3.2); the latter specifies the region of convergence, that is, part I of (29.3a–c) the combined convergence test in Subsection 29.1.1.

29.3.1 Comparison of the Convergence of Two Series

The Cauchy's lemma (29.16a) and convergence principle (29.19a; 29.20) are concerned with the terms of the series (29.18) beyond a certain order N, and suggests that the first N terms of a series do not affect its convergence: *if the terms of two series* $\mathrm{F_n}(\mathrm{z}), \mathrm{G_n}(\mathrm{z})$ *coincide beyond a certain order, the difference of the series is a bounded function, and one series converges, diverges or oscillates if, respectively, the other does the same:*

$$\exists_{n\in|\mathrm{N}} \forall_{m\geq n}: F_m(z) = G_m(z) \Rightarrow \sum_{n=0}^{\infty} F_n(z)\mathrm{C., D., \ O.} \Leftrightarrow \sum_{n=0}^{\infty} G_n(z)\mathrm{C., D., O.} \qquad (29.23)$$

The result follows from:

$$B(z) \in (|C): \sum_{n=0}^{\infty} F_n(z) - \sum_{n=0}^{\infty} G_n(z) = \sum_{n=0}^{N-1} \{F_n(z) - G_n(z)\} \equiv B(z), \qquad (29.24)$$

where the function $B(z)$ is bounded because it is a finite sum of bounded terms. QED.

To start proving the combined convergence test consider the series (29.1), whose terms are in the ratio (29.2b), implying that

$$\lim_{n\to\infty} \frac{F_{n+1}(z)}{F_n(z)} = f(z). \qquad (29.25)$$

If the ratio of successive terms is in modulus less than unity (29.26a):

$$r \equiv |f(z)| < 1: \quad \exists_{n\varepsilon|N} \ \forall_{m\geq n}: \quad |F_{m+1}(z)| < r\,|F_m(z)|, \qquad (29.26\text{a,b})$$

the series of moduli beyond the n-th term is bounded (29.24b) by a geometric series:

$$\sum_{m=n}^{\infty} |F_m(z)| \leq |F_n(z)| \sum_{k=0}^{\infty} r^k = \frac{|F_n(z)|}{1-r}, \qquad (29.26\text{c})$$

and converges (21.64a–c); since the first N terms do not affect convergence, it follows that the series of moduli $|F_n(z)|$ converges for $|F_n(z)| < 1$, and the series (29.1) is absolutely convergent in the open region D in (29.6b), as stated in (29.3b). In the closed subregion D_ε in (29.6a,c) $\equiv$ (29.27a):

$$|f(z)| \leq 1-\varepsilon; |F_n(z)| \leq B: \quad \sum_{m=n}^{\infty} |F_m(z)| \leq |F_n(z)| \sum_{k=0}^{\infty} (1-\varepsilon)^k \leq \frac{B}{\varepsilon}, \qquad (29.27\text{a–c})$$

since the bound (29.27b) is independent of z, the convergence (29.27c) is also independent of z, that is, it is uniform. Since $D_\varepsilon \subset D$ the convergence is also absolute, so the series converges totally in (29.6a,c), as stated in (29.3c).

29.3.2 Ratio Test and Region of Convergence

In the case (29.3a) $\equiv$ (29.28a) it follows from (29.25) that

$$|f(z)| > 1: \quad \exists_{n\in|N} \quad \forall_{m\varepsilon|N}: \quad |F_{m+1}(z)| > |F_m(z)|, \qquad (29.28\text{a,b})$$

and hence the general term (29.28b) does not tend to zero (29.16b), implying that the series does not converge, that is, it diverges or oscillates. In the present case, the series diverges, as stated in (29.3a), since: *if the general term of a series tends to a nonzero limit, the series diverges:*

$$\lim_{n\to\infty} F_n(z) = c \neq 0 \quad \Rightarrow \quad \sum_{n=1}^{\infty} F_n(z) \quad D. \tag{29.29}$$

The condition $c = \alpha + i\beta \neq 0$ implies that at least one of α, β is nonzero, that is, is positive or negative; the proof is similar in all four cases, and so real positive $\alpha > 0$ may be chosen. Thus follows:

$$\exists_{n \in |N} \forall_{m \geq n} : \quad \operatorname{Re}\{F_m(z)\} > \alpha > 0, \quad \operatorname{Re}\left\{\sum_{m=n}^{\infty} F_m(z)\right\}$$

$$= \lim_{N\to\infty} \sum_{m=n}^{n+N} \operatorname{Re}\{F_m(\alpha)\} \geq \lim_{N\to\infty} N\alpha = \infty. \tag{29.30a,b}$$

QED. The preceding results: (i) complete the proof of (29.3a–c); (ii) prove the **D'Alembert ratio test (1768)**:

$$\lim_{n\to\infty} \left|\frac{F_{n+1}(z)}{F_n(z)}\right| \begin{cases} > 1 & \text{D.,} & (29.31a) \\ = r(z) < 1, & \text{A.C.,} & (29.31b) \\ \leq 1 - \varepsilon & \text{T.C.,} & (29.31c) \end{cases}$$

if the ratio of succeeding terms of the series, in modulus, as, $n \to \infty$: (i) tends to a value above unity the series diverges (29.31a) in Figure 29.1a; (ii) if the limit is a number $r(z)$, less than unity in modulus, the series converges absolutely (29.31b) in Figure 29.1b; (iii) if r does not depend on z the series converges totally (29.31c) in Figure 29.1c. Bearing in mind (29.25), the D'Alembert ratio test (29.31a–c) proves (29.3a–c). The ratio test does not clarify the behavior of the series on the boundary of convergence; for this further tests are needed (Sections 29.4–29.6).

29.4 Convergence Integrals and Harmonic Series

The comparison of the convergence of series and integrals (Subsection 29.4.1) establishes the convergence of the harmonic series (Subsection 29.4.2); the latter is used subsequently (Subsection (29.5.2) to prove part III or (29.5a–d) of the combined convergence test in Subsection 29.1.1.

29.4.1 Comparison of Convergence of Series and Integrals

The determination (29.3a–c) of the region of convergence of the series relied on the D'Alembert ratio test (29.31a–c); an alternative is the **Cauchy test** in Example 30.17. Both tests are based on a comparison (29.25; 29.26–29.27a–c; 29.28a,b) with the geometric series (21.62a–c). To consider the behavior of the series (29.1) on its boundary of convergence (29.4a–d), the Gauss test will be used (Section 29.5) that is based on the harmonic series:

$$\sum_{n=1}^{\infty} n^{-g} \leftrightarrow \int_1^{\infty} x^{-g}\, dx. \tag{29.32}$$

The arithmetic series (21.6c) is the particular case $g = 1$ of the harmonic series (29.32), and is divergent (21.7). The convergence of the harmonic series (29.32) for general g is investigated next, by comparison with the unilateral improper integral of the first kind (Section 17.1) with a similar integrand; the main difference between the series (integral) is the discrete (continuous) sum over $n\,(x)$. The behavior of the series and integral will be similar, if their difference specifies a bounded function (29.33b):

$$E \equiv \operatorname{Re}(g) > 0: \quad C(g) \equiv \lim_{N\to\infty} \left\{ \sum_{n=1}^{N} n^{-g} - \int_{1}^{N} x^{-g}\,\mathrm{d}x \right\} = \sum_{n=1}^{\infty} A_n; \tag{29.33a,b}$$

this will be the case if the series (29.33b) $\equiv$ (29.33c) converges:

$$A_n = n^{-g} - \int_{n}^{n+1} x^{-g}\,\mathrm{d}x = \int_{n}^{n+1} \left(n^{-g} - x^{-g}\right)\,\mathrm{d}x; \tag{29.33c}$$

thus A_n compares one term of the series with one step of the integral. Noting that:

$$n^{-g} - x^{-g} = y^{-g}\big|_x^n = \int_x^n \mathrm{d}\left(y^{-g}\right) = -g\int_x^n y^{-1-g}\,\mathrm{d}y, \tag{29.34a}$$

the general term (29.33c) can rewritten as a double integral:

$$A_n = \int_n^{n+1} \mathrm{d}x \int_n^x g y^{-1-g}\,\mathrm{d}y. \tag{29.34b}$$

Using (29.35a) $\equiv$ (29.33a):

$$g \equiv E + ib: \ \left|y^{-1-g}\right| = \left|y^{-1-E}\right|\left|y^{-ib}\right| = y^{-1-E}, \quad \left|y^{-ib}\right| \equiv \left|\exp\left(-ib\log y\right)\right| = 1, \tag{29.35a–c}$$

in the general term (29.34b) leads to upper bound:

$$\begin{aligned} |A_n| &\le \int_n^{n+1} \mathrm{d}x \int_n^x |g|\, y^{-1-E}\,\mathrm{d}y = \frac{|g|}{E}\int_n^{n+1} \left(n^{-E} - x^{-E}\right)\mathrm{d}x \\ &\le \frac{|g|}{E}\left\{n^{-E} - (n+1)^{-E}\right\}\int_n^{n+1}\mathrm{d}x, \end{aligned} \tag{29.36a}$$

since $x < n + 1$; hence the series of moduli of (29.34b) is also bounded:

$$\sum_{n=1}^{N} |A_n| \le \frac{|g|}{E}\sum_{n=1}^{N}\left\{n^{-E} - (n+1)^{-E}\right\} = \frac{|g|}{E}\left\{1 - (N+1)^{-E}\right\} < \frac{|g|}{E}; \tag{29.36b}$$

this proves that *the series (29.33b) is absolutely convergent for (29.33a), and its sum with the terms in any order, specifies the bounded function $C\,(g)$.*

29.4.2 Convergence of the Harmonic Series

The preceding result can be used to investigate the convergence of the harmonic series:

$$\lim_{N\to\infty}\sum_{n=1}^{N} n^{-g} - C(g) = \lim_{N\to\infty}\int_1^N x^{-g}\mathrm{d}x = -\lim_{N\to\infty}\frac{1-N^{1-g}}{1-g}, \tag{29.37}$$

where (29.33b) was used. This leads to four cases: (i) the limit on the r.h.s. of (29.37) exists for:

$$\mathrm{Re}\,(1-g) < 0: \quad \sum_{n=1}^{\infty} n^{-g} = C(g) - \frac{1}{1-g}, \tag{29.38a}$$

proving that the series converges, and since the series of moduli also converges:

$$E \equiv \mathrm{Re}(g) > 1: \quad \sum_{n=1}^{\infty}\left|n^{-g}\right| = \sum_{n=1}^{\infty} n^{-E} = C\,(E) - \frac{1}{1-E}, \tag{29.38b}$$

the convergence is absolute; (ii) in the opposite case Re $(1-g) > 0$ or $E < 1$, the series diverges: (ii-1) for $1 > E > 0$ because $\mathrm{Re}\,(1-g) = 1-E > 0$ and (29.37) diverge; (ii-2) for $E \le 0$ because $\left|n^{-g}\right| = n^{-E}$ does not tend to zero as $n \to \infty$; (iii) the intermediate case $E \equiv \mathrm{Re}\,(g) = 1$ corresponds to (29.39a):

$$\begin{aligned} g = 1+ib,\ b \neq 0: \quad \sum_{n=1}^{\infty} n^{-g} - C\,(g) &= -\frac{i}{b}\lim_{N\to\infty}\left(1-N^{-ib}\right) \\ &= -\frac{i}{b}\lim_{N\to\infty}\left[\bar{1} - \exp\left(-ib\log N\right)\right], \end{aligned} \tag{29.39a,b}$$

and the series (29.39b) oscillates between $C\,(g) \pm 2\,i/b$, with a range of variation $4/b$; (iv) only one case was left out viz. $b = 0,\ E = 1$, in that $g = 1$, and the series n^{-1} diverges by (21.6c). It has been proved that the **harmonic series**:

$$g = E + ib: \quad \sum_{n=1}^{\infty} n^{-g} \begin{cases} E < 1 & \text{D.}, & (29.40a) \\ E = 1,\ b = 0 & \text{D.}, & (29.40b) \\ E = 1,\ b \neq 0 & \text{O.}, & (29.40c) \\ E > 1 & \text{A.C.}, & (29.40d) \end{cases}$$

of general term n^{-g} with $g = E + i\,b$, converges absolutely (diverges) for $E > 1\,(E < 1)$, and for $E = 1$ diverges (oscillates in a range $4/b$) for $b = 0\,(b \neq 0)$. This convergence test may be extended to the **generalized harmonic series** (Example 30.18).

29.5 Gauss Test (1812a) and Euler (1735)–Mascheroni (1790) Constant

The harmonic series (Subsection 29.42) leads to a comparison test (Subsection 29.5.1) that is used to: (i) prove the existence of Euler's constant and digamma function (Subsection 29.5.2); (ii) establish part III that is, (29.5a–d) of the combined convergence test (Subsection 29.1.1) concerning convergence at particular points on the boundary.

29.5.1 Comparison Test for Harmonic Series

The harmonic series suggests the following **comparison theorem**: *if beyond a certain order (29.41b) the general terms of the series $F_n(z), G_n(z)$ differ by $\left(n^{-\lambda}\right)$ in (29.41a) with Re $(\lambda) > 1$:*

$$\operatorname{Re}(\lambda) > 1; \quad \exists_{m \in |N} \quad \forall_{n \geq m}: \quad F_n(z) - G_n(z) = O\left(n^{-\lambda}\right), \tag{29.41a–c}$$

then two series have the same behavior:

$$\sum_{n=1}^{\infty} F_n(z) \quad \text{C., O., D.} \Leftrightarrow \sum_{n=1}^{\infty} G_n(z) \quad \text{C., O., D.} \tag{29.41d}$$

The proof follows from:

$$B(z) \equiv \sum_{k=1}^{n} \{F_k(z) - G_k(z)\} \in B(|C): \quad \sum_{n=1}^{\infty} F_n(z) - \sum_{n=1}^{\infty} G_n(z) = B(z) + \sum_{n=m}^{\infty} O\left(n^{-\lambda}\right), \tag{29.42a,b}$$

where (29.42a) is a bounded function, and in (29.42b) the last term is a harmonic series (38) with exponent λ such that Re $(\lambda) > 1$, so it converges by (29.40d); it follows that the two series have the same behavior, that is, if one converges, oscillates, diverges, the other does respectively the same. Before applying these results to the series (29.1), some consequences of the comparable behavior of the series and integral in (29.32) are mentioned.

29.5.2 Euler's Constant and Digamma Function

Setting $g = 1$ in (29.33b) is allowed by (29.33a), and shows it follows that $C\ (1) \equiv \gamma$ defines a constant:

$$\gamma \equiv \lim_{N\to\infty} \sum_{n=1}^{N} n^{-1} - \log N = \lim_{N\to\infty} 1 + \frac{1}{2} + \cdots + \frac{1}{N} - \log N, \tag{29.43}$$

namely the **Euler (1735) – Mascheroni (1790) constant**; this is an irrational number, whose first 50 digits are

$$\gamma = 0.\ 57721\ 56649\ 01532\ 86060\ 65120\ 90082\ 40243\ 10421\ 59335\ 939992; \tag{29.44}$$

its meaning is that *although the Leibnitz or arithmetic series (21.6c) and the logarithm diverge as $N \to \infty$, their difference (29.43) tends to a constant value (29.44) that is Euler's constant.* The **digamma function** is defined by (29.45a):

$$\psi\,(1+z) \equiv \lim_{N\to\infty} \left\{\log N - \sum_{n=1}^{N} \frac{1}{n+z}\right\}; \ \psi\,(0) = -\gamma, \tag{29.45a,b}$$

its value at the origin (29.45b) is minus the Euler constant (29.44). The sum of the two:

$$\psi\,(1+z) + \gamma = \lim_{N\to\infty} \sum_{n=1}^{N} \left\{\frac{1}{n} - \frac{1}{n+z}\right\} = z \sum_{n=1}^{N} \frac{1}{n\,(n+z)}, \tag{29.46}$$

leads to a series of general term $O\left(n^{-2}\right)$ that is absolutely convergent (29.40d); this implies by the comparison theorem (29.41a–d), that (29.45a) as well as (29.46) converges, except for $z = -1, -2, \ldots,$ that are poles with residue -1:

$$n \in |N: \quad \psi_{(1)}\,(-n) = \lim_{z\to -n} (z+n)\,\psi(z) = \lim_{z\to -n} \frac{z}{n} = -1. \tag{29.47a,b}$$

Thus it has been proved that *the digamma function, defined by the convergent series of fractions (29.45a) ≡ (29.46; 29.43), is bounded in the complex z-plane excluding negative integers* $z \neq -1, -2, \ldots$, *that (29.47a,b) are simple poles with residue* −*1.*

29.5.3 Gauss Test and Convergence at Particular Points

The series (29.1) at the particular points z such that (29.48a) has ratio (29.2b) ≡ (29.48b):

$$f(z) = 1: \quad \frac{F_{n+1}(z)}{F_n(z)} = 1 - \frac{g}{n} + O\left(n^{-\lambda}\right) = \left(1 + \frac{1}{n}\right)^{-g} + O\left(n^{-\mu}\right)$$
$$= \left(\frac{n+1}{n}\right)^{-g} + O\left(n^{-\mu}\right), \qquad (29.48\text{a,b})$$

where $\operatorname{Re}(\mu)$ is (29.49a) the infinum, or the least of 2 and $\operatorname{Re}(\lambda)$; thus (29.2b) leads to (29.49b). It follows from the comparison theorem (29.41a–d) applied to (29.49b):

$$\mu \equiv \inf\{2, \operatorname{Re}(\lambda)\} > 1: \quad \sum_{n=1}^{\infty} F_n(z) \leftrightarrow \sum_{\infty}^{\infty} \left\{n^{-g} + O\left(n^{-\mu}\right)\right\}, \qquad (29.49\text{a,b})$$

that the behavior of the series (29.1) at (29.48a), follows that of the harmonic series (29.40a–d). This leads to the **Gauss test (1812a)**:

$$\operatorname{Re}(\lambda) > 1: \quad \sum_{n=1}^{\infty} C_n: \quad \frac{C_{n+1}}{C_n} = 1 - \frac{g}{n} + O\left(n^{-\lambda}\right), \qquad (29.50\text{a–c})$$

the series (29.50b) with terms in the ratio (29.50c) with (29.50a), has the same convergence properties as the harmonic series (29.40a–d). This proves (29.5a–d) ≡ (29.40a–d) the part III of the combined convergence test in Subsection 29.1.1.

29.6 Criteria and Sums of Abel (1826, 1839)–Dirichlet (1862)

The D'Alembert ratio (29.29a–c) ≡ (29.3a–c) [Gauss (29.50a–c) ≡ (29.40a–d) ≡ (29.5a–d)] tests specify the behavior of the series (39.1) outside the boundary of convergence [at the particular points z satisfying $f(z) = 1$ on the boundary of convergence $|f(z)| = 1$] that are parts I (III) of the combined convergence test in Subsection 29.1.1. The remaining part II concerns (29.4a–d) all other points z on the boundary of convergence $f(z) \neq 1 = |f(z)|$, for that the factor $f(z)$ cannot be omitted from the ratio (29.2b) so that it does not reduce to (29.48b). Thus whereas the D'Alembert (Gauss) test in Subsection 29.3.2 (29.5.3) specifies the behavior of the series (29.1) resulting from the first (second) factor in (29.2b), next is needed a test applying to a series whose terms are products of two factors, for example, the **Abel (1839)–Dirichlet (1862) test**:

$$|S_N(z)| \equiv \left|\sum_{n=1}^{N} U_n(z)\right| < A, \quad \sum_{n=1}^{\infty} \{V_n(z) - V_{n-1}(z)\} \ A.C., \quad \lim_{n\to\infty} V_n(z) = 0:$$
$$\Rightarrow \sum_{n=1}^{\infty} \{U_n(z)\, V_n(z)\} \quad C. \qquad (29.51\text{a–d})$$

a sufficient condition that the series $U_n V_n$ *converges (29.51d), is that: (i)* U_n *has bounded partial sums (29.51a); (ii) the series* $V_n - V_{n-1}$ *is absolutely convergent (29.51b); (iii)* $V_n \to 0$ *tends to zero as* $n \to \infty$ *in (29.51c).*

The proof of the theorem uses the **Abel sums (1826)**:

$$\begin{aligned} S_0 = 0: \quad \sum_{n=1}^{N} U_n V_n &= \sum_{n=1}^{N} (S_n - S_{n-1})\ V_n \\ &= \sum_{n=1}^{N} (S_n V_n - S_{n-1} V_{n-1}) + \sum_{n=1}^{N} S_{n-1} (V_{n-1} - V_n) \\ &= S_N V_N - \sum_{n=1}^{N} S_{n-1} (V_n - V_{n-1}); \end{aligned} \tag{29.52a,b}$$

taking the modulus leads to

$$\begin{aligned} \lim_{N\to\infty} \left| \sum_{n=1}^{N} U_n V_n \right| &\le \lim_{N\to\infty} |V_N| \, |S_N| + \lim_{N\to\infty} \sum_{n=1}^{N} |S_{n-1}| \, |V_n - V_{n-1}| \\ &\le A \left\{ \lim_{N\to\infty} |V_N| + \sum_{n=1}^{N} |V_n - V_{n-1}| \right\} \le AB, \end{aligned} \tag{29.53}$$

where the bound (29.51a) is independent of N. The series (29.51d) converges since in (29.53): (i) the first term in curly brackets vanishes because $V_N \to 0$ as $N \to \infty$ by (29.51c); (ii) the second term is bounded by B, because the series $V_n - V_{n-1}$ converges absolutely (29.51b), so $|\, V_n - V_{n-1}|$ converges. QED.

In the case (29.54a) the partial sums of U_n never exceed unity, and the first condition (29.51a) is satisfied with $A = 1$ in (29.54c); choosing V_n to be a real positive decreasing sequence (29.54b) simplifies the term (29.51b) $\equiv$ (29.54d):

$$U_n \equiv (-)^n; \ V_n > V_{n+1} > \ldots > 0: \quad A = 1, \quad \sum_{n=0}^{N-1} (V_n - V_{n-1}) = V_{N-1} - V_0; \tag{29.54a–d}$$

if further $V_N \to 0$ as $N \to \infty$ in (29.51c) $\equiv$ (29.55a), the series (29.55c) is absolutely convergent (29.55b):

$$\lim_{N\to\infty} V_n = 0: \quad \sum_{n=0}^{\infty} |\, V_n - V_{n-1}| = V_0, \quad \sum_{n=1}^{\infty} \{V_n - V_{n-1}\} \quad A.C. \tag{29.55a–c}$$

because (29.55b) $\equiv$ (29.55d):

$$\sum_{n=1}^{\infty} |V_n - V_{n-1}| = -\lim_{N\to\infty} \sum_{n=1}^{N} (V_{n-1} - V_n) = -\lim_{N\to\infty} (V_0 - V_N) = -V_0. \tag{29.55d}$$

Thus all conditions of the Abel–Dirichlet test (29.51a–d), have been met for the convergence of (29.51d; 29.54a,b); this is the series, $(-)^n \, V_n$, as stated in the **Abel Test**: *a real alternating series whose terms decrease monotonically to zero, is convergent:*

$$f_n > f_{n+1} > \ldots > 0: \quad \lim_{n\to\infty} f_n = 0 \Rightarrow \sum_{n=1}^{\infty} (-)^n f_n C. \tag{29.56}$$

As an example, the Leibnitz **alternating arithmetic series** (27.57b):

$$f_n \equiv \frac{1}{n}: \sum_{n=1}^{\infty} f_n = \sum_{n=1}^{\infty} \left\{ \frac{(-)^n}{n} \right\} = 1 - \frac{1}{2} + \frac{1}{3} - \frac{1}{4} + \frac{1}{5} - \cdots = \log 2, \tag{29.57a,b}$$

obtained by choosing (29.57a) in (29.56), converges, as stated in (21.6b); the nonalternating series (21.6c) diverges, so the convergence of (29.57b) is conditional, that is, the sum may be changed by derangement of the terms (Section 21.3). Alternating series may be used to bound the sum of series from above and below providing an estimate of the numerical error of truncating the series at a certain order (Example 30.19).

29.7 Boundary of Convergence and Weierstrass K-Test (1856)

The final part II of the combined convergence test (Subsection 29.1.1), concerning (29.4a–d) all points on the boundary of convergence (other than the particular points), is based (Subsection 29.7.1) on the Abel test (Section 29.6) and leads to the Weierstrass K-test (Subsection 29.7.2).

29.7.1 Convergence on the Boundary of Convergence

Returning to the series (29.1), the ratio of terms (29.2b) consists of two factors, neither of them can be omitted at the nonparticular points (28.58a) on the boundary of convergence (28.58b):

$$f(z) \neq 1 = |f(z)|: \quad \frac{F_{n+1}(z)}{F_n(z)} = f(z) \left[\left(\frac{n+1}{n} \right)^{-g} + O\left(n^{-\mu}\right) \right]; \tag{29.58a–c}$$

in (29.58c) $\equiv$ (29.2b) have been applied the same transformations as in (29.48b) and μ is given by (29.49a) $\equiv$ (29.65a,b). From (29.58c) it follows that

$$\sum_{n=1}^{\infty} F_n(z) \leftrightarrow \sum_{n=1}^{\infty} n^{-g} \{f(z)\}^n, \tag{29.59}$$

the series (29.1) has the same convergence properties as a series (29.59) of harmonic n^{-g} and geometric f^n products; this can be used to establish the remaining part II of (29.4a–d) the combined convergence test (Subsection 29.1.1). The general term of the series (29.59) has modulus (29.60c):

$$g \equiv E + ib, \quad |f(z)| = 1: \qquad \left|n^{-g}\right| |f(z)|^n = n^{-E}, \tag{29.60a–c}$$

that does not tend to zero as $n \to \infty$ for $E = \mathrm{Re}\,(g) < 0$ in (29.60a) and thus (29.29) the series diverges, proving (29.4a). If $E > 1$ then the series of moduli of (29.59), viz.:

$$\sum_{n=1}^{\infty} \left|n^{-g}\right| |f(z)|^n = \sum_{n=0}^{\infty} n^{-E}, \tag{29.60d}$$

converges (29.40d) for $E \equiv \mathrm{Re}(g) > 1$ in (29.60a), and thus the series (29.59) is absolutely convergent, proving (29.4d). In the case (29.4b) $\equiv$ (29.61a) the series (29.59) with (29.60b)

$\equiv$ (29.61b):

$$g = ib; \quad f = e^{i\varphi}: \quad \sum_{n=1}^{\infty} f^n n^{-g} = \sum_{n=1}^{\infty} e^{in\varphi} n^{-ib} = \sum_{n=1}^{\infty} \exp\{i[n\varphi - b\log n]\}$$
$$= \lim_{N\to\infty} \exp\left\{ i \prod_{n=1}^{N} (n\varphi - b\log n) \right\}, \qquad (29.61a\text{–}c)$$

oscillates infinitely (29.61c), proving (29.4b) for $\varphi \neq 0$, and confirming (29.5c) for $\varphi = 0 \neq b$.

The last (29.4c) of the 11 cases of the combined convergence test (Subsection 29.1.1) states that for $0 < E = \mathrm{Re}\,(g(z)) \leq 1$ the series (29.59) is conditionally convergent, that is, it is convergent (ii) but absolutely divergent (i). The latter part (i) follows from the series of moduli (29.60d) that diverges (29.40a) for $E < 1$, and for $E = 1$ also diverges because it coincides with the arithmetic series (21.6b). The remaining proof (ii) that the series (29.59) converges for $0 < E \leq 1$, follows by showing it meets all three conditions (29.51a–c) of the Abel–Dirichlet test for the convergence of (29.51d), viz.: (ii-1) choosing (29.62a,b) it has (29.51a) bounded partial sums (29.62c):

$$U_n(z) \equiv [f(z)]^n, \quad |f(z)| = 1 \neq f(z): \quad \left|\sum_{n=1}^{\infty} f^n\right| = \lim_{N\to\infty} \left|\sum_{n=1}^{N} f^n\right|$$
$$= \lim_{N\to\infty} \left| \frac{f\left(1 - f^N\right)}{(1-f)} \right| \leq \frac{2}{|1-f|}, \qquad (29.62a\text{–}c)$$

on account of (29.61b) with $\varphi \neq 0$, hence $f \neq 1$; (ii) choosing (29.63a) then (29.51b) is absolutely convergent:

$$V_n(z) \equiv n^{-g(z)}: \ |V_n - V_{n-1}| \equiv \left|n^{-g} - (n-1)^{-g}\right| = \left|n^{-g}\right| \left|1 - \left(1 - \frac{1}{n}\right)^{-g}\right|$$
$$= n^{-E}\left|-\frac{g}{n} + O\left(n^{-2}\right)\right| = O\left(n^{-1-E}\right), \qquad (29.63a,b)$$

because (29.63b) is a harmonic series (29.40d) with exponent $1 + E > 1$, and hence convergent; (iii-3) using (29.63a) it follows that

$$E \equiv \mathrm{Re}\,(g(z)) > 0: \quad \lim_{n\to\infty} |V_n(z)| = \lim_{n\to\infty} \left|n^{-g(z)}\right| = \lim_{n\to\infty} n^{-E} = 0, \qquad (29.64a,b)$$

proving (29.51c). This completes the proof of the last of all the 11 cases of the combined convergence test in Subsection 29.1.1.

29.7.2 Weierstrass Second or K-Test

In the proof of some of the 11 cases in the (Section 29.1) combined test of convergence, $\mathrm{Re}\,(\lambda)$ on (29.2a) was replaced by μ according to:

$$\mu \equiv \inf\{2, \mathrm{Re}\,(\lambda)\} \Rightarrow \begin{cases} \mu = \mathrm{Re}\,(\lambda) & \text{if } 1 < \mathrm{Re}\,(\lambda) \leq 2 \\ \mu = 2 & \text{if } \mathrm{Re}\,(\lambda) \geq 2. \end{cases} \qquad (29.65a,b)$$

For example, this was used in (29.48b; 29.49a,b) and (29.58c). The preceding results lead to the **Weierstrass K-test (1856)**:

$$\operatorname{Re}(\lambda) > 1: \quad \sum_{n=1}^{\infty} C_n z^n, \quad \frac{C_{n+1}}{C_n} = 1 - \frac{g}{n} + O\left(n^{-\lambda}\right), \tag{29.66a–c}$$

the series (29.66b) with coefficients satisfying the ratio (29.66c) with (29.66a) has on its circle of convergence $|z| = 1 \neq z$ *excluding the point* $z = 1$*, the following behavior it: (i) diverges for* $E \equiv \operatorname{Re}(g) < 0$ *in (29.67a); (ii) oscillates for* $E = 0$ *in (29.67b); (iii) converges conditionally* $0 < E \leq 1$ *in (29.67c); (iv) converges absolutely for* $E > 1$ *in (29.67d):*

$$\operatorname{Re}(\lambda) > 1, \quad E \equiv \operatorname{Re}(g): \quad \sum_{n=1}^{\infty} C_n z^n \begin{cases} E < 0 & \text{D.}, & (29.67\text{a}) \\ E = 0 & \text{O.}, & (29.67\text{b}) \\ 0 < E \leq 1 & \text{C.C.}, & (29.67\text{c}) \\ E > 1 & \text{A.C.}, & (29.67\text{d}) \end{cases}$$

This is designated Weierstrass K-test (Section 29.7) to distinguish from M-test (Section 21.7). The Gauss test (29.40a–d) ≡ (29.50a–c) [Weierstrass K-test (29.67a–d)] are complementary, in that they specify the behavior of the power series with coefficients C_n satisfying (29.50c) ≡ (29.66c), at the particular point $z = 1$ (general points $z \neq 1$) on the boundary of convergence. Also the D'Alembert test (29.31a–c) ≡ (29.3a–c), Gauss test (29.41a–d) ≡ (29.5a–d) and Weierstrass K-test (29.67a–d) ≡ (29.4a–d) complement each other to complete the proof of the combined convergence test stated in Subsection 29.1.1.

29.8 Radius and Exponent of a Power Series

In the case of power series the combined convergence test (Subsection 29.1.1) reduces to a radius and exponent of convergence (Subsection 29.8.1); examples include the binomial (Subsection 29.1.3) and logarithmic (Subsection 29.8.2) series.

29.8.1 Radius and Exponent of Convergence

The two most important particular cases of the series (29.1) are the generalized harmonic type (Example 30.18), and the power series (26.68a):

$$\sum_{n=0}^{\infty} C_n z^n: \quad \frac{C_{n+1}}{C_n} = \alpha^{-1}\left\{1 - \frac{\beta}{n} + O\left(n^{-\lambda}\right)\right\}, \tag{29.68a,b}$$

whose coefficients are assumed to have the ratio (29.68b); it corresponds to the series (29.1) ≡ (29.69a) of general term (29.69b) with ratio of successive terms (29.69c):

$$\sum_{n=0}^{\infty} F_n(z): \quad F_n(z) \equiv C_n z^n, \quad \frac{F_{n+1}(z)}{F_n(z)} = \frac{z}{\alpha}\left\{1 - \frac{\beta}{n} + O\left(n^{-\lambda}\right)\right\}; \tag{29.69a–c}$$

this coincides with (29.2b) setting (29.70b,c):

$$\operatorname{Re}(\lambda) > 1: \qquad f(z) = \frac{z}{\alpha}, \quad g(z) = \beta, \tag{29.70a–b}$$

and keeping (29.2a) $\equiv$ (29.70a). It follows that the combined convergence test (Subsection 29.1.1) can be restated in a simplified form: *the power series (29.69a) $\equiv$ (29.68a), with* **radius of convergence** *(29.71a) and* **exponent of convergence** *(29.71b):*

$$R \equiv |\alpha| = \lim_{n\to\infty} \left|\frac{C_n}{C_{n+1}}\right|, \quad E \equiv \operatorname{Re}(\beta) = \lim_{n\to\infty} n\left\{1 - \operatorname{Re}\left(\frac{C_{n+1}\alpha}{C_n}\right)\right\}, \tag{29.71a,b}$$

has the following behavior indicated in Table 29.1: (i) diverges for $|z| > R$; (ii) converges absolutely for $|z| < R$; (iii) converges totally for $|z| \le R - \varepsilon$ with $0 < \varepsilon < R$. On the circle of convergence $|z| = R$, excluding the particular point $z = \alpha$, the series: (iv) diverges for $E < 0$; (v) oscillates if $E = 0$; (vi) converges conditionally if $0 < E \le 1$; (vii) converges absolutely if $E > 1$. At the particular point $z = \alpha$, the series: (viii) diverges if $E < 1$; (ix) converges absolutely if $E > 1$; (x) oscillates if $E = 1 \ne \beta$; (xi) diverges if $\beta = 1$.

29.8.2 Convergence of Logarithmic Series

The convergence of the binomial series (25.37a–c) $\equiv$ (29.12a,b) has already been discussed in Subsection 29.1.3 in agreement with Tables 29.1 and 29.2. Since the convergence of the geometric series (Section 21.8) was discussed (Section 21.9) in Table 21.2, the only other series given as example in the present part 3, for that the convergence is not yet fully discussed, is the logarithmic series (21.64a–c). The latter is a power series with coefficients:

$$C_n = \frac{(-)^{n-1}}{n}, \quad \frac{C_{n+1}}{C_n} = -\frac{n}{n+1} = -1 + \frac{1}{n} + O\left(\frac{1}{n^2}\right), \quad -\alpha = 1 = \beta, R = 1 = E, \tag{29.72a–f}$$

corresponding to (29.72b) $\equiv$ (29.68b) with (29.72c,d) that is, radius and exponent of convergence both unity (29.72e,f). It follows as indicated in the second column of Table 29.2, in agreement with Table 29.1, that:

$$\log(1+z) = z - \frac{z^2}{2} + \frac{z^3}{3} - \frac{z^4}{4} + \cdots + \frac{(-)^{n-1}}{n} z^n + \cdots, \tag{29.73}$$

the logarithmic series (29.73) $\equiv$ (21.64b) diverges (converges absolutely) outside (inside) the unit circle $|z| > 1$ $(|z| < 1)$, and converges totally in a closed subcircle $|z| \le 1 - \varepsilon$, with $0 < \varepsilon < 1$. On the circle of convergence $|z| = 1$, it is conditionally convergent at all points $|z| = -1$, except at $z = -1$ where it diverges. For real, $z \equiv x$, the boundary of convergence $|x| = 1$ consists of two points $x = \pm 1$, and at $x = -1 = \alpha$ $(x = 1 \ne \alpha)$ the series (29.73) coincides with minus (21.6c) [with (21.6b)] and diverges (converges conditionally).

29.9 Gaussian or Three-Parameter Hypergeometric Series (Gauss, 1812b)

The combined convergence test applies (Subsections 29.1.1 and 29.1.2) to much more general functions than the elementary transcendentals, for example, to the Gaussian hypergeometric function. As final example of application of the combined convergence test consider the

hypergeometric series (Gauss, 1812b):

$$F(a,b;c;z) = 1 + \frac{ab}{c}z + \frac{a(a+1)b(b+1)}{c(c+1)}\frac{z^2}{2!} + \cdots$$

$$= 1 + \sum_{n=1}^{\infty} \frac{z^n}{n!} \prod_{m=0}^{n-1} \frac{(a+m)(b+m)}{c+m}, \qquad (29.74)$$

with parameters a, b, c, and variable z, generally complex, excluding c negative integer $-c \in |N$*; in the latter case (29.75a) has a simple pole with residue (29.75b):*

$$-p \in |N: \quad F_{(1)}(a,b;-p;z) = \lim_{c \to -p} (p+c) F(a,b;c;z)$$

$$= \sum_{n=p+1}^{\infty} \frac{z^n}{n!} \left\{ \prod_{\substack{m=0 \\ m \neq p}}^{n-1} \frac{(a+m)(b+m)}{c+m} \right\} (a+p)(b+p),$$

$$= \frac{z^{p+1}}{(p+1)!} \left\{ \prod_{m=0}^{p-1} \frac{(a+m)(b+m)}{c+m} \right\}$$
$$\times (a+p)(b+p) F(a+p+1, b+p+1; p+2; z). \qquad (29.75\text{a,b})$$

The hypergeometric series includes as particular cases many others, for example: (i) the logarithmic series (29.73) ≡ (21.64a):

$$F(1,b;b;z) = 1 + \sum_{n=1}^{\infty} \frac{(n-1)!}{n!} z^n = 1 - \sum_{n=1}^{\infty} \frac{(-)^{n-1}}{n} (-z)^n = 1 - \log(1-z); \qquad (29.76)$$

$$a \neq 1: \quad F(a,b;b;z) = 1 + \sum_{n=1}^{\infty} \frac{z^n}{n!} a(a+1)\ldots(a+n-1) = (1-z)^{-a}, \qquad (29.77)$$

(ii) the binomial series (29.77) ≡ (25.38).

The hypergeometric series (29.75) is a power series (29.68a), with coefficients:

$$C_n = \frac{a\ldots(a+n-1)\,b\ldots(b+n-1)}{(n+1)!c(c+1)\ldots(c+n-1)} = \frac{(a)_n (b)_n}{(c)_n (n+1)!}, \qquad (29.78)$$

where the **Pochammer symbol** (28.79a):

$$(a)_n \equiv a(a+1)\ldots(a+n-1), \quad (a)_{n+1} = (a)_n (a+n), \qquad (29.79\text{a,b})$$

has the property (29.79b). The ratio of coefficients (29.78) of the hypergeometric series:

$$\frac{C_{n+1}}{C_n} = \frac{(a)_{n+1} (b)_{n+1} (c)_n\, n!}{(a)_n (b)_n (c)_{n+1} (n+1)!}, \qquad (29.80\text{a})$$

simplifies using (29.79b) to

$$\frac{C_{n+1}}{C_n} = \frac{(a+n)(b+n)}{(1+n)(c+n)} = 1 - \frac{c+1-a-b}{n} + O(n^{-2}). \qquad (29.80\text{b})$$

This coincides with (29.68b) ≡ (29.80b) for

$$\alpha = 1 = R, \quad \beta = c+1-a-b, \quad E = \text{Re}\,(c+1-a-b), \qquad (29.81\text{a–c})$$

so that the radius of convergence is unity. It follows from Table 29.1, as stated in Table 29.2, that *the hypergeometric series (29.74): (i) converges absolutely (totally) for* $|z| < 1\,(|z| \leq 1 - \varepsilon$, *with* $0 < \varepsilon < 1)$, *and diverges for* $|z| > 1$; *(ii) for* $|z| = 1 \neq z$ *it diverges (oscillates) if* $Re\,(a+b-c) > 1 \quad [Re\,(a+b-c) = 1]$, *and converges absolutely (conditionally) for* $Re\,(a+b-c) < 0 \quad [1 > Re\,(a+b-c) \geq 0]$; *(iii) at the particular point* $z = 1$ *it diverges (converges absolutely) for* $Re\,(a+b-c) > 0$ $[Re\,(a+b-c) < 0]$, *and if* $Re\,(a+b-c) = 0$ *it diverges (oscillates) for* $a+b = c\,(a+b \neq c)$. The convergence of the generalized hypergeometric series can be considered in a similar way (Example 30.20).

Note 29.1. *Convergence of Infinite Expansions.* The establishment of convergence tests is often based on three complementary approaches: (i) obtaining the sum (proving the boundedness) of particular series, for example, the geometric (Sections 21.8 and 21.9) [harmonic (Sections 29.4 and 29.5)] series; (ii) use of these particular series for comparison with more general series to prove convergence (divergence) by upper (lower) bounds, viz. the d'Alembert and Cauchy (Gauss) test use the comparison with geometric (Section 29.3 and Example 30.17) [harmonic (Section 29.5)] series; (iii) comparison with other infinite sequences, for example, improper integrals of the first kind (Section 29.4). The preceding methods can be combined to specify the convergence of a series in the whole complex plane, including the boundary of convergence (Section 29.1), for example, for power series (Sections 29.8 and 29.9); the combined convergence test (Sections 29.1 and 29.8) is actually a selection among the variety of convergence tests available that applies most simply to all points of the complex plane for the classes of series that will be found in various subsequent applications. The subject of convergence of series is quite extensive, in particular as concerns the more difficult aspect of convergence on the boundary of convergence, for example, Tauberian theorems. Most convergence tests apply not only to (i) power series (Chapters 23 and 25) but also to improper (ii) integrals (Chapter 13), (iii) series of fractions (iv) infinite products (Subsection 27.9.2), and (v) continued fractions because these (i–v) may be transformed into each other. Also the extension of the concept of summability to higher orders, allowing the summation of some series that are divergent at zero order, extends to other infinite processes.

LIST 29.1 Power Series

General:

Stirling–Maclaurin: 23.34a,b; 23.35.
Taylor: 23.32a,b; 23.33; 23.52; 23.56a–c; 23.57a,b.
Lagrange–Burmann: 23.22a; 23.23; 23.24a,b; 23.29.
Laurent–Maclaurin: 25.26a,b.
Laurent–Weierstrass: 25.23a,b; 25.24a,b; 25.25a,b.
Teixeira: 25.12; 25.13a,b; 25.18; 25.19a,b; 25.21a–c.

Particular:

Monotonic arithmetic: 21.6c.
Alternating arithmetic: 21.6b.
Ascending monotonic geometric: 21.62b.
Ascending alternating geometric: 21.62c.
Descending alternating geometric: 21.63c.
Descending monotonic geometric: 21.63b.
Logarithmic: 21.64a,b.
Binomial: 25.37a–c; 25.38; 25.40a–c; 25.41a,b.

LIST 29.1 Continued

Exponential: 23.60; 30.43.
Circular cosine: 30.45.
Sine: 27.74; 30.47.
Harmonic: 29.40a–d.
Generalized harmonic: 30.105a,b.
Confluent hypergeometric: 30.116a.
Gaussian hypergeometric: 29.74.
Generalized hypergeometric: 30.112.
Of fractions: 27.63b, 29.45a, 36.142, 36.149.
Infinite products: 27.67, 36.141.

LIST 29.2 Convergence Tests

Cauchy necessary condition: 29.16a.
Cauchy necessary and sufficient condition: 29.18.
Cauchy root test: 30.98a–c.
D'Alembert ratio test: 29.31a–c.
Comparison with integral: 29.32.
Comparison with harmonic series: 29.41a–c.
Gauss test: 29.50a–c.
Weierstrass M-test: 21.50.
Extended Weierstrass M-test: 21.52.
Weierstrass K-test: 29.67a–d.
Abel test: 29.56.
Abel–Dirichlet test: 29.51a–d.
Power series test: 29.68a,b; 29.69a–c.
Combined test: 29.1; 29.2a–c; 29.3a–c; 29.4a–d; 29.5a–d.

Conclusion 29: The series (Figure 29.1), with ratio of succeeding terms given by (29.2a,b), has a region of convergence D, determined by the modulus of $f(z)$, which is the O(1) coefficient in the ratio (29.2b), viz.: (a) the series diverges for $|f(z)| > 1$, in the exterior of D; (b) it converges absolutely for $|f(z)| < 1$, in the interior of D; (c) it converges totally in a D_ε closed subregion $|f(z)| \leq 1-\varepsilon$, with $0 < \varepsilon < 1$. The behavior on the boundary ∂D of convergence $|f(z)| = 1$, depends on the coefficient $-g(z)$ of the $O\left(n^{-1}\right)$ term in the ratio (29.2b), through $E \equiv Re\{g(z)\}$, and is distinct on two sets of points: (d) at all points z for which $f(z) \neq 1$, the series diverges if $E < 0$, oscillates if $E = 0$, converges conditionally if $0 < E \leq 1$, and converges absolutely if $E > 1$; (e) at the particular points z for which $f(z) = 1$, the series diverges for $E < 1$, converges absolutely for $E > 1$, and for $E = 1$ diverges if $g = 1$, and oscillates if $g \neq 1$.

Tables 29.1 and 29.2: The combined convergence test for power series in Table 29.1 is applied to several particular series in Table 29.2, viz. the logarithmic, binomial, and Gaussian hypergeometric.

Lists 29.1 and 29.2: List 29.1 includes the general and particular series, and List 29.2 the convergence tests considered in Chapters 21, 23, 25, 27, and 29. The elementary transcendental functions provide examples of power series, series of fractions, infinite products, and continued fractions. The convergence tests apply also to the latter.

30

Examples 30.1 to 30.20

Example 30.1. *Convergence, Oscillation, and Divergence of Complex Series.*

Indicate the conditions of convergence (C.), oscillation (O.), and divergence (D.) of a real series; combine these for the real and imaginary parts of a complex series, and conclude on the convergence, oscillation, and divergence of the latter. Also relate the convergence of the sum or product of two series to that each of the series.

For a real series, depending on whether the limit [(21.1)] exists or not, and the sums (21.2a) are bounded or not, four cases arise in the definition (21.3a–c) of convergence (C.), divergence (D.), and oscillation (O.) in Table 30.1. In the case of a complex series $f_n = \Phi_n + i\, \Psi_n$, depending on the real Φ_n and imaginary Ψ_n parts, there are nine cases in Table 30.2, viz.: (i) it converges only if both the real and imaginary parts converge (one case); (ii) diverges if at least one of the real and imaginary parts diverges (five cases); (iii) oscillates if of the real and imaginary parts none diverges and at least one oscillates (three cases). For the sum of two series there are six cases in Table 30.3, viz.: (i/ii/iii) if one converges the other is the deciding factor; (iii/iv) if only one diverges, then the result is divergence; (v) if both oscillate the result could be oscillation or convergence; (vi) if both diverge any result is possible (divergence, oscillation or convergence). For the product of two series there are also six cases in Table 30.3, viz.: (i/ii/iii) if one series converges the other is the deciding factor; (v,vi) if one series diverges the result is divergence; (iv) if both the series oscillate the result is oscillation or divergence. The last case is illustrated by the product of two oscillating series in Table 30.4, whose sum by diagonals (1a) following Cauchy's Rule (21.25c):

$$h_n \equiv \sum_{m=0}^{n} f_m g_{n-m} = 1, -(1+1), (1+1+1), -(1+1+1+1)$$
$$= 1, -2, 3, -4, \quad \sum_{n=1}^{\infty} h_n = 1, -1, 2, -2, \ldots \tag{30.1a,b}$$

diverges (30.1b).

Example 30.2. *Sum of Geometric and Binomial Series.*

Sum the following sixteen series:

$$1 \pm \frac{1}{2} + \frac{1}{4} \pm \frac{1}{8} + \frac{1}{16} \pm \cdots = 2, \frac{2}{3}; \tag{30.2a,b}$$

$$1 \pm \frac{1}{3} + \frac{1}{9} \pm \frac{1}{27} + \frac{1}{81} \pm \cdots = \frac{3}{2}, \frac{3}{4}; \tag{30.2c,d}$$

$$\frac{1}{2} \pm \frac{1}{8} + \frac{1}{24} \pm \frac{1}{64} + \frac{1}{260} \pm \cdots = \log 2, \log\left(\frac{3}{2}\right); \tag{30.3a,b}$$

$$\frac{1}{3} \pm \frac{1}{18} + \frac{1}{81} \pm \frac{1}{324} + \frac{1}{1215} \pm \cdots = \log\left(\frac{3}{2}\right), \log\left(\frac{4}{3}\right); \tag{30.3c,d}$$

TABLE 30.1
Convergence of Real Series

Limit	Bounded	Unbounded
Exists	C.	D.
does not exist	O.	D.

Note: Concerning real series (i) convergence requires the existence of a finite limit for the sum. Nonconvergence is caused either by (ii) inexistence of a limit (oscillation) or (iii) infinite limit (divergence). In the case (iv) both occur, that is, infinite oscillation, the criterion divergence is taken as dominant. Thus divergent series means unbounded sum, with or without limit [cases (iii) and (iv)]. Oscilatory means bounded without limit [case (ii)].

TABLE 30.2
Convergence of Complex Series

Complex series $f_n = \Phi_n + i\,\Psi_n$		**Real part: Φ_n**		
		Φ_n C.	Φ_n O.	Φ_n D.
Imaginary part: Ψ_n	Ψ_n C.	C.	O.	D.
	Ψ_n O.	O.	O.	D.
	Ψ_n D.	D.	D.	D.

Note: A complex series converges if both the real imaginary parts converge. If at least one of the real or imaginary parts is unbounded the complex series diverges. The remaining case of oscillatory complex series requires: (i) both the real and imaginary parts to be bounded; (ii) at least one to have no limit.

TABLE 30.3
Convergence of Sums and Products of Series

$\sum_n f_n$	C.	C.	C.	O.	O.	D.
$\sum_n g_n$	C.	O.	D.	O.	D.	D.
$\sum_n (f_n + g_n)$	C.	O.	D.	C. or O.	D.	C. or O. or D.
$\left(\sum_n f_n\right)\left(\sum_m g_m\right)$	C.	O.	D.	O. or D.	D.	D.

Note: Convergence of the addition and product of two complex series for all six combinations of convergence/oscillation/divergence of each series. The simplest case is all convergent. The other five cases require individual consideration, and can be established by construction of suitable examples. If one series converges and the other oscillates (diverges) the latter dictates an oscillatory (divergent) addition and product. Two oscillatory series may add to an oscillatory series; also to a convergent series, for example, zero for two symmetric oscillatory series with opposite signs. The product of oscillatory series is oscillatory, or may be divergent (Table 30.4). The addition or product of an oscillatory and a divergent series always diverges. The product of divergent series always diverges, and the addition may: (i) diverge; (ii) oscillate, for example, one series differs from the other by ± 1 in successive terms; (iii) converge, for example, to zero if the series have opposite signs.

$$1 \pm 1 + \frac{3}{4} \pm \frac{1}{2} + \frac{5}{16} + \cdots = 4, \frac{4}{9}; \tag{30.4a,b}$$

$$1 \pm \frac{2}{3} + \frac{1}{3} \pm \frac{4}{27} + \frac{5}{81} \pm \cdots = \frac{9}{4}, \frac{9}{16}; \tag{30.4c,d}$$

$$1 \pm \frac{N}{2} + \frac{N(N+1)}{8} \pm \frac{N(N+1)(N+2)}{48} + \frac{N(N+1)(N+2)(N+3)}{384} \pm \cdots = 2^N, \left(\frac{2}{3}\right)^N; \tag{30.5a,b}$$

TABLE 30.4
Divergent Product of Oscillating Series

		f_1	f_2	f_3	f_4
		+1	**−1**	**+1**	**−1**
g_1	+1	+1	−1	+1	−1
g_2	−1	−1	+1	−1	+1
g_3	+1	+1	−1	+1	−1
g_4	−1	−1	+1	−1	+1

Note: Of the cases of addition and products of series in Table 30.3 perhaps the only one not immediately obvious is a divergent product of oscillatory series. It is shown in Table 30.4 that multiplying the series $+1, -1, +1, -1, \ldots$ by itself the products of increasing order are read summing the diagonals: $+1, -2, +3, -4, \ldots$. The sum $+1, -1, +2, -2, \ldots$ diverges.

$$1 \pm \frac{N}{3} + \frac{N\,(N+1)}{18} \pm \frac{N(N+1)(N+2)}{162} + \frac{N(N+1)(N+2)(N+3)}{1944} \pm \cdots = \left(\frac{3}{2}\right)^N, \left(\frac{3}{4}\right)^N. \tag{30.5c,d}$$

Note that two distinct series, viz. monotonic (30.3c) and alternating (30.3b) have the same sum $\log(3/2)$.

The series (30.2a) [(30.2b)] is geometric monotonic (21.62a) [alternating (21.62b)] with first term $a = 1$ and ratio, $z = 1/2$, hence sum $1/(1 \mp 1/2) = 2, 2/3$; likewise for (30.2c,d) with ratio $z = 1/3$ and sum $1/(1 \mp 1/3) = 3/2, 3/4$. The series (30.3a,b) is logarithmic (21.64a,b) with $z = \pm 1/2$, hence sum $\mp \log(1 \mp 1/2) = \log 2, \log(3/2)$; likewise for (30.3c,d) with $z = \pm 1/3$ and sum $\mp \log(1 \mp 1/3) = \log(3/2), \log(4/3)$. The series (30.4a,b) are (21.65b) inverse binomials $(1 \mp 1/2)^{-2} = 4, 4/9$, and (30.4c,d) are inverse binomials $(1 \mp 1/3)^{-2} = 9/4, 9/16$. The series (30.5a,b) are (21.65b) inverse binomials $(1 \mp 1/2)^{-N} = 2^N, (2/3)^N$, and (30.5c,d) are $(1 \mp 1/3)^{-N} = (3/2)^N, (3/4)^N$. Setting $N = 1$ in (30.5a–d) yields (30.2a–d).

Example 30.3. *Stirling–Maclaurin Expansions via Geometric Series.*
Obtain the expansions:

$$|z| < |z+a|^{-1}: \quad \left(1 + az + z^2\right)^{-1} = 1 - az + z^2\left(a^2 - 1\right) + az^3\left(2 - a^2\right) + O\left(z^4\right); \tag{30.6a,b}$$

$$|\exp(z)| > |az|: \quad \{\exp(z) - az\}^{-1} = 1 + z\,(a-1) + z^2\left(a^2 - 2a + \frac{1}{2}\right) + z^3\left(a^3 - 3a^2 + 2a - \frac{1}{6}\right) + O\left(z^4\right); \tag{30.7a,b}$$

$$|\log(1+z)| < \frac{1}{|a|}: \quad \{1 + a\log(1+z)\}^{-1} = 1 - az + a\left(a + \frac{1}{2}\right)z^2 - az^3\left[\frac{1}{3} + a\,(a+1)\right] + O\left(z^4\right), \tag{30.8a,b}$$

up to the third order.

All the expansions are Maclaurin series (35.34a) whose coefficients can be determined from the derivatives of the function at the origin. An alternative is to use the geometric series, for example, for (30.6b) with ratio (30.9a):

$$r \equiv z(z+a): \quad \left(1+az+z^2\right)^{-1} = 1 - z(a+z) + z^2(a+z)^2 - z^3(a+z)^3 + O\left(z^4\right); \tag{30.9a,b}$$

the ratio (30.9a) specifies the region of convergence $|r| < 1$ in (30.6a); the expansion (30.9b) keeping only powers of z up to z^3 leads to (30.6b). The case of (30.7b) is another geometric series:

$$r \equiv aze^{-z}: \quad \left(e^z - az\right)^{-1} = e^{-z}\left[1 + aze^{-z} + a^2z^2e^{-2z} + a^3z^3e^{-3z} + O\left(z^4\right)\right], \tag{30.10a,b}$$

whose ratio (30.10a) specifies the region of convergence $|r| < 1$ in (30.7a); substituting in (30.10b) the exponential series to the same order:

$$\exp(-nz) = 1 - nz + \frac{n^2z^2}{2} - \frac{n^3z^3}{6} + O\left(z^4\right), \tag{30.10c}$$

and keeping powers of z up to z^3 leads to (30.7b). The geometric series for (30.8b) has ratio (30.11a):

$$\begin{aligned} r \equiv a\log(1+z): \quad & \{1 + a\log(1+z)\}^{-1} \\ &= 1 - a\log(1+z) + a^2\log^2(1+z) - a^3\log^3(1+z) + O\left(z^4\right), \end{aligned} \tag{30.11a,b}$$

that specifies the region of convergence $|r| < 1$ in (30.8a); substitution of the logarithmic series (21.64b) in (30.11b) leads to (30.8b).

Example 30.4. *Wave Scattering into the Zone of Silence or for Total Reflection.*

Express the normal component of the wavevector of incidence, reflection, and transmission in terms of the transverse wavenumber, k, and of the angle of incidence, θ, or transmission, θ'. Use the results to analyze wave scattering by a plane interface near the conditions of: (i) total reflection; (ii) zone of silence.

Consider a wave of frequency, ω, incident from (transmitted to) a medium of wave speed $c_1(c_2)$, with an angle of incidence θ (transmission θ'). The horizontal wavenumber (22.8a–c) is conserved:

$$\frac{\omega}{c_1}\sin\theta = k = \frac{\omega}{c_2}\sin\theta'; \tag{30.12}$$

thus the vertical components of the incident k_i and reflected k_r (transmitted k_t) wavevector are given by

$$-k_r = k_i = \frac{\omega}{c_1}\cos\theta = \left\{\left(\frac{\omega}{c_1}\right)^2 - k^2\right\}^{1/2} = \left\{\left(\frac{\omega}{c_1}\right)^2 - \left(\frac{\omega}{c_2}\right)^2\sin^2\theta'\right\}^{1/2}, \tag{30.13a–c}$$

$$k_t = \frac{\omega}{c_2}\cos\theta' = \left\{\left(\frac{\omega}{c_2}\right)^2 - k^2\right\}^{1/2} = \left\{\left(\frac{\omega}{c_2}\right)^2 - \left(\frac{\omega}{c_1}\right)^2\sin^2\theta\right\}^{1/2}, \tag{30.14a–c}$$

in terms of the angle of incidence (30.13a) [(30.14c)], or transmission (30.13c) [(30.14a)].

Example 30.4.1. *Total reflection of waves.*

In the case of the speed of propagation larger in the medium of transmission the vertical component of the transmitted wavevector (30.14c) is given by:

$$c_2 > c_1: \quad k_t(\theta) = \frac{\omega}{c_2}\left\{1 - \left(\frac{c_2}{c_1}\right)^2\sin^2\theta\right\}^{1/2} = \frac{\omega}{c_2}\left\{1 - \left(\frac{\sin\theta}{\sin\theta_R}\right)^2\right\}^{1/2}, \tag{30.15a,b}$$

in terms of the angle of total reflection (22.10a). For an incident wave, with wavevector $\vec{k} = (k, k_i)$, propagating in the tangential x- and normal y-directions:

$$f_i(x, y, t) = A \exp\left[i\omega\left(\frac{x\cos\theta + y\sin\theta}{c_1} - t\right)\right], \tag{30.16a}$$

the transmitted wave:

$$f_t(x, y, t) = AT \exp\left\{i\frac{\omega}{c_2}x\cos\theta - i\omega t\right\} g(y), \tag{30.16b}$$

has: (i) amplitude a modified by the transmission factor T, for example, specified by (22.25b,c) for acoustic waves; (ii) propagates in the x-direction; (iii) in the y-direction:

$$g(y) = \begin{cases} \exp(ik_t y) & \text{if} \quad \theta < \theta_R, & (30.17\text{a}) \\ 1 & \text{if} \quad \theta = \theta_\text{R}, & (30.17\text{b}) \\ \exp(-|k_t|y) & \text{if} \quad \theta < \theta_R, & (30.17\text{c}) \end{cases}$$

the wave: (a) propagates upward (30.17a) for incidence below the angle of total reflection when k_t is real in (30.15b); (b) does not propagate nor decay at (30.17b) the angle of total reflection when $k_t = 0$ in (30.15b); (c) is evanescent beyond the angle of total reflection (30.17c) when the vertical wavenumber (30.15b) is imaginary $k_t = i|k_t|$. In the latter case, its amplitude decays exponentially with distance, faster for larger vertical wavenumber or shorter wavelength, $\lambda \equiv 2\pi/k_t$, for example, a decay $1/e$ at a distance $y = \lambda_t/2\pi = 1/k_t$. Thus the condition of total reflection back into the medium of incidence gives rise to a transmitted wave that is not zero but whose amplitude decays exponentially with distance into the medium of transmission.

Example 30.4.2. *Zone of silence for waves.*

If the speed of propagation is larger in the medium of incidence the vertical component (30.13c) of the reflected wavenumber can be written:

$$c_1 > c_2: \quad -k_r(\theta') = \frac{\omega}{c_1}\left\{1 - \left(\frac{c_1}{c_2}\right)^2 \sin^2\theta'\right\}^{1/2} = \frac{\omega}{c_1}\left\{1 - \left(\frac{\sin\theta'}{\sin\theta'_S}\right)^2\right\}^{1/2}, \tag{30.18a,b}$$

in terms of the angle θ'_S of the zone of silence (22.11b). The incident wave (30.16a) gives rise to a reflected wave:

$$f_r(r, y, t) = AR \exp\left\{i\frac{\omega}{c_1}x\cos\theta - i\omega t\right\} h(y), \tag{30.19}$$

with: (i) amplitude modified by the reflection factor R, for example specified by (22.25a,c) for acoustic waves; (ii) propagates in the x-direction; (iii) in the y-direction:

$$h(y) = \begin{cases} \exp(ik_r y) & \text{if } \theta' < \theta'_S, & (30.20\text{a}) \\ 1 & \text{if } \theta' = \theta'_S, & (30.20\text{b}) \\ \exp(-|k_r|y) & \text{if } \theta' > \theta'_S, & (30.20\text{c}) \end{cases}$$

the wave: (a) propagates downward (30.20a) for real k_r when the transmitted wave is outside the zone of silence (30.18b); (b) does not propagate or decay (30.18b) at angle of silence (30.20b); (c) a transmitted wave in the zone of silence (30.18b) would imply an imaginary vertical wavenumber $k_r = i|k_r|$ for reflection (30.18b), hence a reflected wave decaying exponentially with distance from the interface toward the medium of incidence (30.20c), the decay being faster for larger wavelength, $\lambda_r = 2\pi/k_r$. The limit of **rays** that follow paths tangent to the wave vector, corresponds to zero wavelength, $\lambda \to 0$; hence both in the case of total reflection (30.17c) [zone of silence (30.20c)] the vertical wave number is infinite $k_t \to \infty (k_r \to \infty)$, and there is no evanescent wave $g(y) = 0$ [$h\ (y) = 0$]. The existence of an evanescent wave is an oscillatory wave phenomenon that implies that the wave cannot propagate, but still penetrates the medium with amplitude (energy) decaying like (30.21a) [(30.21b)]:

$$A \sim \exp\left(-\left|k\right| y\right) = \exp\left(-\frac{2\pi y}{\lambda}\right), \quad E \sim A^2 \sim \exp\left(-2\left|k\right| y\right) = \exp\left(-\frac{\pi y}{\lambda}\right), \quad (30.21\text{a,b})$$

that is, the penetration increases with the wavelength, and only disappears for $\lambda \to 0$.

Example 30.5. *Parallel Sided Lens Separating Three Distinct Media.*

Consider a parallel sided lens separating distinct media (Figure 30.1) and calculate the total reflection, R, transmission, T, and damping coefficients, S, allowing for interfacial adsorption, and including or not internal absorption.

The a parallel sided lens, illustrated in Figure 30.1, is made of a material 2 separating the materials 1 and 3, with wave speeds, respectively, c_2, c_1, c_3. A wave is incident from medium 1, and the reflection and transmission factors are by denoted (R_1, T_1) for the scattering $1 \to 2, (R_2, T_2)$ for $2 \to 1$, and (R_3, T_3) for $2 \to 3$. Either the method of series summation (Section 22.6) or the method of composition into upward and downward fields (Section 22.7) can be used to show that the global reflection, R, transmission, T, and damping, S, factors for the lens, are given by

$$R = R_1 + \frac{T_1 T_2 R_3}{1 - R_2 R_3}, \quad T = \frac{T_1 T_3}{1 - R_2 R_3}, \quad (30.22\text{a,b})$$

$$S = 1 - R - T = 1 - R_1 - \frac{T_1 (T_2 R_3 + T_3)}{1 - R_2 R_3}, \quad (30.22\text{c})$$

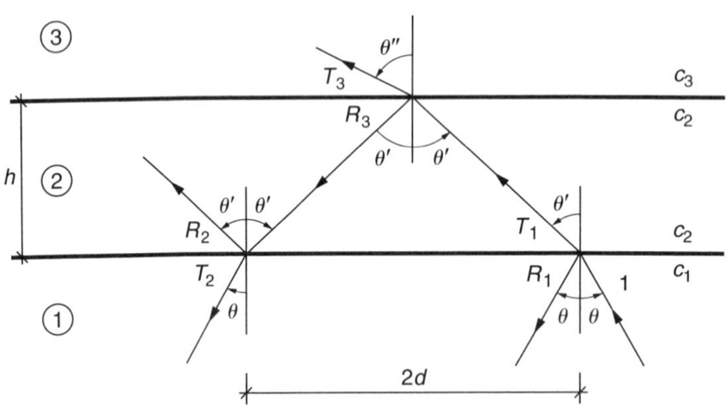

FIGURE 30.1
The scattering of waves by a parallel-sided lens between distinct (identical) media [Figure 30.1 (22.5)]. The generalization from identical to different media below and above the lens adds a third pair of reflection and transmission factors: (i) for identical media below and above the lens (Figure 22.5) the two pairs of reflection and transmission factors $(R_\pm, T_\pm)$ concern scattering by the interface $1 \to 2$ $(2 \to 1)$ for (R_+, T_+) $[(R_-, T_-)]$; (ii) for distinct media below and above the lens (Figure 30.1) the three pairs of reflection and transmission factors (R_{1-3}, T_{1-3}) concern scattering by the interfaces $1 \to 2/2 \to 1/2 \to 3$ respectively for $(R_1, T_1)/(R_2, T_2)/(R_3, T_3)$; (iii) thus the former (i) is the particular case of the latter (ii) with $(R_1, T_1) \equiv (R_+, T_+)$ and $(R_2, T_2) \equiv (R_-, T_-) \equiv (R_3, T_3)$.

allowing for adsorption at the interfaces. If in addition there is also linear absorption in the interior with coefficient, ν, then the substitutions:

$$\{R_1, T_1, R_2, T_2, R_3, T_3\} \to \{R_1, T_1 e^{-\delta}, R_2 e^{-\delta}, T_2, R_3 e^{-\delta}, T_3\}, \tag{30.23}$$

lead from (30.22a–c) to

$$R = R_1 + \frac{T_1 T_2 R_3}{e^{2\delta} - R_2 R_3}, \quad T = \frac{T_1 T_3 e^{\delta}}{e^{2\delta} - R_2 R_3}, \tag{30.24a,b}$$

$$\delta \equiv \nu h \sec\theta' : \quad S = 1 - R_1 - \frac{T_1 \left(T_2 R_3 + T_3 e^{\delta}\right)}{e^{2\delta} - R_2 R_3}; \tag{30.24c,d}$$

here h is the thickness of the lens and θ' the angle with the vertical of the direction of propagation in the interior. If the media outside the lens are identical then

$$1 \equiv 3 : \quad R_1 = R_+, \quad T_1 = T_+, \quad R_2 = R_- = R_3 \quad T_2 = T_- = T_3; \tag{30.25a–d}$$

the substitutions (30.25a–d) in (30.24a,b,d) lead to (22.33a–c) $\equiv$ (22.40a–c).

Example 30.6. *The Same Lagrange–Burmann Series Representing Two Distinct Functions in Nonoverlapping Regions.*

Expand the function (30.26a) in power series of an auxiliary function (30.26b):

$$F(z) \equiv z, \quad f(z) \equiv \frac{2}{z + 1/z} = f\left(\frac{1}{z}\right), \tag{30.26a,b}$$

and discuss the convergence in the whole complex plane.

Bearing in mind that $f(0) = 0 \neq 2 = f'(0)$ meets (23.25) with $a = 0$, the Lagrange–Burmann series (23.28b) has coefficients (23.29) $\equiv$ (30.27a):

$$\begin{aligned} n!A_n &= \lim_{\zeta \to 0} \frac{d^n}{d\zeta^n} \left\{ F(\zeta) \left[\frac{\zeta}{f(\zeta)}\right]^{n+1} f'(\zeta) \right\} = 2^{-n} \lim_{\zeta \to 0} \frac{d^n}{d\zeta^n} \left[\left(\zeta^2 + 1\right)^{n-1} \left(1 - \zeta^2\right) \zeta \right] \\ &= 2^{-n} \lim_{\zeta \to 0} \frac{d^n}{d\zeta^n} \left\{ \sum_{m=0}^{n-1} \binom{n-1}{m} \left(\zeta^{2m+1} - \zeta^{2m+3}\right) \right\}. \end{aligned} \tag{30.27a}$$

Since the term in curly brackets in (30.27a) involves only odd powers, the even coefficients are zero (30.27b):

$$A_{2p} = 0, \quad (2p+1)!\, A_{2p+1} = 2^{-1-2p}(2p+1)! \left[\binom{2p}{p} - \binom{2p}{p-1}\right], \tag{30.27b,c}$$

and the odd coefficients are given by (30.27c), with $2p+1 = n = 2m+1 (2p+1 = n = 2m+3)$ for the first $p = m$ (second $p = m + 1$) term; the latter can be simplified:

$$A_{2p+1} = 2^{-1-2p}(2p)! \left[\frac{1}{p!p!} - \frac{1}{(p-1)!\,(p+1)!}\right] = 2^{-1-2p} \frac{(2p)!}{p!(p+1)!}. \tag{30.28a}$$

This corresponds to the Lagrange–Burmann series:

$$z = F(z) = \sum_{p=0}^{\infty} A_{2p+1} \left[f(z)\right]^{2p+1} = \sum_{p=0}^{\infty} A_{2p+1} \left(\frac{2}{z + 1/z}\right)^{2p+1}. \tag{30.28b}$$

Since the function (30.26b) has poles at $z = \pm\, i$, the unit circle is the radius of convergence, that is, the series (30.28b) converges for $|z| < 1$ in (30.29a). The change of variable $z \to 1/z$

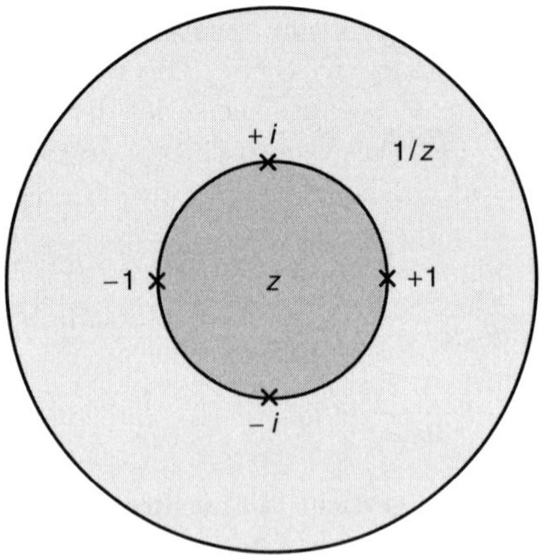

FIGURE 30.2
The same Lagrange–Burmann series (30.29a–d) represents distinct functions z $(1/z)$ inside $|z| < 1$ (outside $|z| > 1$) the unit circle. On the unit circle $|z| = 1$ the series: (i) converges at the points $z = \pm 1$, where the functions coincide $z = 1/z$; (ii) at all other points $z \neq \pm 1$ on the boundary of convergence $|z| = 1$ the functions differ $z \neq 1/z$ and the series diverges. The Lagrange–Burmann series representing the function z $(1/z)$ inside (outside) the unit circle is lacunary (Figure 31.2c) because: (i) the boundary of convergence is densely covered with singularities, that is, all points on the unit circle $|z| = 1$ are singularities except $z = \pm 1$; (ii) thus analytic continuation (Section 31.1) beyond the boundary of convergence is not possible.

specifies a series converging for $|z| > 1$; it actually the same series, because (30.26b) is unchanged by the transformation $z \to 1/z$, but now it converges to $1/z$, in (30.29d):

$$\sum_{n=0}^{\infty} \frac{(2n)!}{n!(n+1)!}\left(z + \frac{1}{z}\right)^{-2n-1} = \begin{cases} z & \text{if } |z| < 1, & (30.29\text{a}) \\ \pm 1 & \text{if } z = \pm 1, & (30.29\text{b}) \\ \infty & \text{if } |z| = 1 \neq \pm z, & (30.29\text{c}) \\ \dfrac{1}{z} & \text{if } |z| > 1; & (30.29\text{d}) \end{cases}$$

the Lagrange–Burmann series (30.28a,b) represents (Figure 30.2) different functions z in (30.29a) inside $|z| < 1$ [$1/z$ in (30.29d) outside $|z| > 1$] the unit disk; that is possible since the series does not converge (30.29c) on the unit disk $|z| = 1$, except (30.29b) at the points $z = \pm 1$, where the two functions coincide. The latter aspects (30.29b,c) are proved using the combined test on the unit circle:

$$\begin{aligned} z = e^{i\varphi} : \quad \frac{A_{2n+1}}{A_{2n-1}}\left(z + \frac{1}{z}\right)^{-2} &= (2\cos\varphi)^{-2}\frac{(2n+2)(2n+1)}{(n+1)(n+2)} \\ &= \sec^2\varphi \frac{n+1/2}{n+2} = \sec^2\varphi\left[1 - \frac{3}{2n} + O\left(\frac{1}{n^2}\right)\right]; \end{aligned} \qquad (30.30\text{a,b})$$

comparing (30.30a) ≡ (30.31a) with real coefficients in (29.68b; 29.70a,b) specifies the radius (30.31b) and exponent (30.31c) of convergence:

$$\frac{A_{2n+1}}{A_{2n-1}}\left(z + \frac{1}{z}\right)^{-2} = R\left[1 - \frac{E}{n} + O(n^{-2})\right]: \quad R = \sec^2\varphi, \quad E = \frac{3}{2}. \qquad (30.31\text{a–c})$$

Thus two cases arise: (i) $R > 1$ at all points on the unit circle except $\varphi = 0, \pi$, implying divergence (30.29c); (ii) at $\varphi = 0, \pi$ then $R = 1$ and since $E = 3/2 > 1$ there is absolute convergence (30.29b). Thus the series: (i) converges on the unit circle only at the two points $\varphi = 0, \pi$ or $z = \pm 1$ where the functions $z = 1/z$ coincide; (ii) there is no possible analytic continuation (Section 31.1) from z inside or $1/z$ outside because the series diverge on the boundary at a dense set of points. Thus (30.29a–d) concerns two distinct functions represented by the same series in regions that overlap only at two isolated points where the two functions coincide.

Example 30.7. *Operations on and Functions of Power Series.*

Deduce from the Maclaurin series to order three:

$$F(z) = a_0 + a_1\, z + a_2\, z^2 + a_3\, z^3 + O\left(z^4\right), \tag{30.32}$$

for the function $F(z)$, the first four terms of the Maclaurin series for the following five related functions:

$$|F(z) - a_0| < |a_0| : \quad \{F(z)\}^{-1}\, a_0 = 1 - \frac{a_1}{a_0} z - \left[\frac{a_2}{a_0} - \left(\frac{a_1}{a_0}\right)^2\right] z^2$$
$$- \left[\frac{a_3}{a_0} - 2\frac{a_1 a_2}{a_0^2} + \left(\frac{a_1}{a_0}\right)^3\right] z^3 + O\left(z^4\right), \tag{30.33}$$

$$|z| < \infty : \quad \exp\{F(z)\}$$
$$= \exp(a_0)\left\{1 + a_1 z + \left(a_2 + \frac{a_1^2}{2}\right) z^2 + \left(a_3 + a_1 a_2 + \frac{a_1^3}{6}\right) z^3 + O(z^4)\right\}, \tag{30.34}$$

$$|F(z) - a_0| < |a_0| : \quad \{F(z)\}^{-2}\, a_0^2 = 1 - 2\frac{a_1}{a_0} z - \left[2\frac{a_2}{a_0} - 3\left(\frac{a_1}{a_0}\right)^2\right] z^2$$
$$- 2\left[\frac{a_3}{a_0} - 3\frac{a_1 a_2}{a_0^2} + 2\left(\frac{a_1}{a_0}\right)^3\right] z^3 + O\left(z^4\right), \tag{30.35}$$

$$|F(z) - a_0| < |a_0| : \quad \log\{F(z)\} = \log(a_0) + \frac{a_1}{a_0} z + \left[\frac{a_2}{a_0} - \frac{1}{2}\left(\frac{a_1}{a_0}\right)^2\right] z^2$$
$$+ \left[\frac{a_3}{a_0} - \frac{a_1 a_2}{a_0^2} + \frac{1}{3}\left(\frac{a_1}{a_0}\right)^3\right] z^3 + O\left(z^4\right), \tag{30.36}$$

$$|F(z) - a_0| < |a_0| : \quad \{F(z)\}^{-n}\, a_0^n = 1 - n\frac{a_1}{a_0}\, z - n\left[\frac{a_2}{a_0} - \frac{n+1}{2}\left(\frac{a_1}{a_0}\right)^2\right] z^2$$
$$- n\left\{\frac{a_3}{a_0} - (n+1)\left[\frac{a_1 a_2}{a_0^2} - \frac{n+2}{6}\left(\frac{a_1}{a_0}\right)^3\right]\right\} z^3 + O\left(z^4\right), \tag{30.37}$$

and establish the regions of convergence. From (30.37) with $n = 1$ ($n = 2$) follows (30.33) [(30.35)].

The series (30.33) is (21.62c) a geometric series:

$$\{F(z)\}^{-1} = \frac{a_0^{-1}}{1 + (z/a_0)\left(a_1 + a_2 z + a_3 z^2\right)} + O\left(z^4\right)$$
$$= a_0^{-1}\left[1 - \frac{z}{a_0}\left(a_1 + a_2 z + a_3 z^2\right) + \left(\frac{z}{a_0}\right)^2 \left(a_1 + a_2 z\right)^2 - \left(\frac{z}{a_0}\right)^3 a_1^3 + O\left(z^4\right)\right], \tag{30.38a}$$

whose ratio specifies the region of convergence:

$$1 > \left| \left(\frac{z}{a_0} \right) \left(a_1 + a_2 z + a_3 z^2 + \cdots \right) \right| = \left| \frac{F(z) - a_0}{a_0} \right|. \tag{30.38b}$$

The exponential (30.34) has (23.60) an infinite radius of convergence:

$$\begin{aligned} \exp\{F(z)\} &= \exp(a_0) \exp\left[z \left(a_1 + a_2 z + a_3 z^2 \right) \right] + O\left(z^4\right) \\ &= \exp(a_0) \left\{ 1 + z\left(a_1 + a_2 z + a_3 z^2\right) + \frac{z^2}{2}\left(a_1 + a_2 z\right)^2 + \frac{z^3}{6} a_1^3 + O\left(z^4\right) \right\}. \end{aligned} \tag{30.39}$$

The series (30.35) is (21.65b) an inverse binomial series:

$$\begin{aligned} \{F(z)\}^{-2} &= a_0^{-2} \left[1 + \frac{z}{a_0}\left(a_1 + a_2 z + a_3 z^2\right) \right]^{-2} + O\left(z^4\right) \\ &= a_0^{-2} \left\{ 1 - \frac{2z}{a_0}\left(a_1 + a_2 z + a_3 z^2\right) \right. \\ &\quad \left. + 3\left(\frac{z}{a_0}\right)^2 \left(a_1 + a_2 z\right)^2 - 4\left(\frac{z}{a_0}\right)^3 a_1^3 + O\left(z^4\right) \right\}, \end{aligned} \tag{30.40}$$

whose region of convergence is (30.38b). The series (30.36) is (21.64b) the logarithmic series:

$$\begin{aligned} \log\{F(z)\} &= \log(a_0) + \log\left[1 + \frac{z}{a_0}\left(a_1 + a_2 z + a_3 z^3\right) \right] + O\left(z^4\right) \\ &= \log(a_0) + \frac{z}{a_0}\left(a_1 + a_2 z + a_3 z^2\right) \\ &\quad - \frac{(z/a_0)^2}{2}\left(a_1 + a_2 z\right)^2 + \frac{(z/a_0)^3}{3}\left(a_1\right)^3 + O\left(z^4\right), \end{aligned} \tag{30.41}$$

whose radius of convergence is (30.38b). The series (30.37) is (21.65b) $\equiv$ (25.40a–c) an inverse binomial:

$$\begin{aligned} \{F(z)\}^{-n} &= (a_0)^{-n} \left\{ 1 + \frac{z}{a_0}\left(a_1 + a_2 z + a_3 z^2\right) \right\}^{-n} + O\left(z^4\right) \\ &= (a_0)^{-n} \left\{ 1 - \frac{nz}{a_0}\left(a_1 + a_2 z + a_3 z^2\right) + \frac{n(n+1)}{2}\left(\frac{z}{a_0}\right)^2 \left(a_1 + a_2 z\right)^2 \right. \\ &\quad \left. - \frac{n(n+1)(n+2)}{6}\left(\frac{z}{a_0}\right)^3 \left(a_1\right)^3 + O\left(z^4\right) \right\}, \end{aligned} \tag{30.42}$$

with region of convergence (30.38b). The series (30.33–30.37) could also be obtained as Maclaurin series (23.34a,b) using the derivatives of the function at the origin, but this would be more laborious. The results are *five operations on series (30.33–30.37), in addition to the reversion of series (25.35a,b) in Section 25.8.*

Example 30.8. *Stirling–Maclaurin Series to Fifth-Order.*

Obtain the Stirling–Maclaurin series to fifth-order:

$$\exp(z) = \sum_{n=0}^{\infty} \frac{z^n}{n!} = 1 + z + \frac{z^2}{2} + \frac{z^3}{6} + \frac{z^4}{24} + \frac{z^5}{120} + \frac{z^6}{720} + \cdots, \tag{30.43}$$

$$\frac{\exp z - 1}{z} = \sum_{n=0}^{\infty} \frac{z^n}{(n+1)!} = 1 + \frac{z}{2} + \frac{z^2}{6} + \frac{z^3}{24} + \frac{z^4}{120} + \frac{z^5}{720} + \cdots, \tag{30.44}$$

$$\cos z = \sum_{n=0}^{\infty} \frac{(-)^n \, z^{2n}}{(2n)!} = 1 - \frac{z^2}{2} + \frac{z^4}{24} - \frac{z^6}{720} + \cdots, \tag{30.45}$$

$$\frac{2\,(1 - \cos z)}{z^2} = 2\sum_{n=0}^{\infty} \frac{(-)^n \, z^{2n}}{(2n+2)!} = 1 - \frac{z^2}{12} + \frac{z^4}{360} + \cdots, \tag{30.46}$$

$$\sin z = \sum_{n=0}^{\infty} \frac{(-)^n \, z^{2n+1}}{(2n+1)!} = z - \frac{z^3}{6} + \frac{z^5}{120} - \frac{z^7}{5040} + \cdots, \tag{30.47}$$

$$\frac{6\,(z - \sin z)}{z^3} = 6\sum_{n=0}^{\infty} \frac{(-)^n \, z^{2n}}{(2n+3)!} = 1 - \frac{z^2}{20} + \frac{z^4}{840} + \cdots, \tag{30.48}$$

and prove that the radius of convergence is infinity, that is, all functions are integral.

All the series can be obtained from the Stirling–Maclaurin series (33.34b), using the derivatives of the function, as for the exponential series (33.60). A shorter alternative approach is: (i) from (1.15) is obtained in the complex form (30.49a):

$$\exp(\pm iz) = \cos z \pm i \sin z: \qquad 2\cos z = e^{iz} + e^{-iz}, \qquad 2i\sin z = e^{iz} - e^{-iz}, \tag{30.49a–c}$$

and thus the cosine (30.45) [sine (30.47)] series is the sum (30.49b) [difference (30.49c)] of two exponential series (30.43); (ii) the series (30.44; 30.46; 30.48) coincide, respectively, with (30.43; 30.45; 30.49) suppressing the first term, and multiplying the rest of the series by a constant factor so that the first term becomes unity.

Example 30.9. *Convergence of Laurent–Maclaurin Series.*

Obtain the Laurent–Maclaurin series:

$$\left\{z^N (1-z)\right\}^{-1} = \sum_{n=-N}^{\infty} z^{n-N} = z^{-N} + z^{-N+1} + z^{-N+2} + \cdots \tag{30.50}$$

$$z^{-N} \log(1+z) = \sum_{n=1-N}^{\infty} \frac{(-)^{n+N-1}}{n+N} z^n = z^{1-N} - \frac{z^{2-N}}{2} + \frac{z^{3-N}}{3} - \frac{z^{4-N}}{4} + \cdots, \tag{30.51}$$

$$\begin{aligned}\{z(1+z)\}^{-N-1} &= \sum_{n=0}^{\infty} \binom{-N-1}{N} z^{n-N-1} = \sum_{n=0}^{\infty} (-)^n \binom{n+N}{N} z^{n-N-1} \\ &= z^{-N-1} - (N+1)\,z^{-N} + \frac{1}{2}(N+1)(N+2)\,z^{1-N} \\ &\quad + \frac{1}{6}(N+1)(N+2)(N+3)\,z^{2-N} + \cdots\end{aligned} \tag{30.52}$$

and find their behavior at all points of the complex plane (Table 30.5), where $0 < \varepsilon < 1$ and $0 < \delta < 1 - \varepsilon$.

TABLE 30.5
Convergence of Power Series (30.50–30.52)

Series	(30.50)	(30.51)	(30.52) with $N \geq 1$
T.C.	$\varepsilon \leq \lvert z\rvert \leq 1-\delta$	$\varepsilon \leq \lvert z\rvert \leq 1-\delta$	$\varepsilon \leq \lvert z\rvert \leq 1-\delta$
A.C.	$0 < \lvert z\rvert < 1$	$0 < \lvert z\rvert < 1$	$0 < \lvert z\rvert < 1$
O.	$\lvert z\rvert = 1 \neq z$	–	–
C.C.	–	$\lvert z\rvert = 1 \neq -z$	–
D.	$z = 1$	$z = -1$	$\lvert z\rvert = 1$
D.	$\lvert z\rvert > 1$	$\lvert z\rvert > 1$	$\lvert z\rvert > 1$,

$0 < \varepsilon < 1$, $0 < \delta < 1 - \varepsilon$.
Note: This convergence follows from Table 29.1.

TABLE 30.6
Singularities and Residues of Functions

$f(z)$	z_n	N	A_{-1}
$\sec z$	$(n+1/2)\,\pi$	1	$(-)^{n-1}$
$\tan z$	$(n+1/2)\,\pi$	1	-1
$\cot^2 z$	$n\,\pi$	2	0
$\csc^3 z$	$n\,\pi$	3	$(-)^n/2$
$z^N \exp(1/z)$	0	∞	$1/(N+1)\,!$
$z^{2N+1}\cos(1/z)$	0	∞	$(-)^{N+1}/(2N+2)\,!$

z_n – singular point; N – order of pole; $N=\infty$: essential singularity; A_{-1} – residue.
Note: For each of the six complex functions is indicated: (i) the location z_n of the singularities; (ii) the type $N = 1/n/\infty$ for a simple pole/pole of order n, essential singularity; (iii) the residue A_{-1}.

The series (30.51; 30.52; 30.53) are obtained most simply multiplying by z^{-N} respectively the geometric (21.62b), logarithmic (21.64b), and inverse binomial (21.65b) series. Their convergence properties are indicated in Table 30.5, and follow from: (i) the geometric series (Table 21.2) for (30.50); (ii) the logarithmic series (Table 29.2) for (30.51); (iii) for (30.52) taken as a series of functions with general term (30.53a) in the ratio (30.53b):

$$f_n(z) \equiv z^{n-N-1}(-)^n \binom{n+N}{N} = z^{n-N-1}\frac{(-)^n}{N!}\frac{(n+N)!}{n!}, \tag{30.53a}$$

$$\frac{f_{n+1}(z)}{f_n(z)} = -z\frac{1+(N+1)/n}{1+1/n} = -z\left[1+\frac{N}{n}+O\left(\frac{1}{n^2}\right)\right], \tag{30.53b}$$

from the combined convergence test (Subsection 29.1.1), in agreement with the binomial series in Table 29.2 since $N = 2, \ldots$. The series (30.50–30.52) could also be obtained from the Laurent–Maclaurin expansion (25.24a).

Example 30.10. *Location of Singularities of Functions and Identification of Their Type.*

Locate on the complex plane the singularities z of the functions indicated in Table 30.6 and determine their residue, A_{-1}; also indicate the type of point N, where $N = n$ $(N = \infty)$ means that it is a pole of order n (essential singularity).

The secant is the inverse of the cosine (30.54a) and has simple poles at its simple zeros (30.54b) with residues (30.54c):

$$f(z) \equiv \sec z = \frac{1}{\cos z}, \quad \cos\left[\left(n+\frac{1}{2}\right)\pi\right] = 0, \quad f_{(1)}\left(\left(n+\frac{1}{2}\right)\pi\right) = (-)^{n-1}, \tag{30.54a–c}$$

calculated from

$$w \equiv z - n\pi - \frac{\pi}{2}: \qquad \lim_{z \to n\pi + \pi/2} \frac{z - n\pi - \pi/2}{\cos z}$$
$$= \lim_{w \to 0} \frac{w}{\cos(w + n\pi + \pi/2)} = (-)^n \lim_{w \to 0} \frac{w}{-\sin w} = (-)^{n-1}. \qquad (30.55\text{a,b})$$

The tangent (30.56a) has simple poles at the simple zeros of the cosine in the denominator:

$$f(z) \equiv \tan z \equiv \frac{\sin z}{\cos z}, \quad f_{(1)}\left(n\pi + \frac{\pi}{2}\right) = (-)^{n-1} \sin\left(n\pi + \frac{\pi}{z}\right) = (-)^{n-1}(-)^n = -1, \qquad (30.56\text{a,b})$$

with residues (30.56b) calculated from (30.55b). The cotangent square (30.57a) has double poles from the simple zeros of the sine (30.57b) that appears in the denominator:

$$f(z) = \cot^2 z = \left(\frac{\cos z}{\sin z}\right)^2, \quad \sin(n\pi) = 0, \quad f_{(2)}(n\pi) = 0, \qquad (30.57\text{a–c})$$

with residues (30.57c) specified by

$$w \equiv z - n\pi: \qquad \lim_{z \to n\pi} \frac{\mathrm{d}}{\mathrm{d}z} (z - n\pi)^2 \frac{\cos^2 z}{\sin^2 z} = \lim_{w \to 0} \frac{\mathrm{d}}{\mathrm{d}w} \frac{w^2 \cos^2(w + n\pi)}{\sin^2(w + n\pi)}$$
$$= \lim_{w \to 0} \frac{\mathrm{d}}{\mathrm{d}w} w^2 \left[\frac{1 + O(w^2)}{w^2 + O(w^4)}\right]^2 = \lim_{w \to 0} \frac{\mathrm{d}}{\mathrm{d}w} \left[1 + O(w^2)\right] = \lim_{w \to 0} O(w) = 0. \qquad (33.58\text{a,b})$$

The cube of the cosecant (30.59a) has triple poles at the zeros of the sine (30.59b):

$$f(z) \equiv \csc^3 z = (\sin z)^{-3}, \quad f_{(3)}(n\pi) = \frac{(-)^n}{2}, \qquad (33.59\text{a,b})$$

with residues (30.59b) ≡ (30.60b):

$$w \equiv z - n\pi: \qquad \frac{1}{2} \lim_{z \to n\pi} \frac{\mathrm{d}^2}{\mathrm{d}z^2} \left(\frac{z - n\pi}{\sin z}\right)^3 = \frac{1}{2}(-)^{3n} \lim_{w \to 0} \frac{\mathrm{d}^2}{\mathrm{d}w^2} \left(\frac{w}{\sin w}\right)^3$$
$$= \frac{(-)^n}{2} \lim_{w \to 0} \frac{\mathrm{d}^2}{\mathrm{d}w^2} \left[1 - \frac{w^2}{6} + O(w^4)\right]^{-3}$$
$$= \frac{(-)^n}{2} \lim_{w \to 0} \frac{\mathrm{d}^2}{\mathrm{d}w^2} \left[1 + \frac{w^2}{2} + O(w^4)\right] = \frac{(-)^n}{2}. \qquad (30.60\text{a,b})$$

The function (30.61a) has Laurent series derived from (23.60) ≡ (30.43) with an essential singularity at the origin whose coefficient of z^{-1} corresponds to $N - n = -1$ or $n = N + 1$ and specifies the residue (20.61b):

$$z^N \exp\left(\frac{1}{z}\right) = \sum_{n=0}^{\infty} \frac{z^{N-n}}{n!} \equiv \sum_{n=0}^{\infty} A_{N-n} z^{N-n}, \quad A_{-1} = \frac{1}{(N+1)!}. \qquad (30.61\text{a,b})$$

The function (30.62a) has Laurent series derived from (30.45)

$$z^{2N+1}\cos\left(\frac{1}{z}\right)=\sum_{n=0}^{\infty}\frac{(-)^n}{(2n)!}z^{2N+1-2n}=\sum_{n=0}^{\infty}A_{2N+1-2n}z^{2N+1-n},\quad A_{-1}=\frac{(-)^{N+1}}{(2N+2)!},\tag{30.62a,b}$$

whose coefficient of z^{-1} corresponds to $2N+1-2n=-1$ or $n=N+1$ and specifies the residue (30.62b).

Example 30.11. *Type of Singularities of Functions at Infinity.*

Classify the point at infinity of the functions in Table 30.7 indicating the residue there, in case it is a singularity; also indicate the **type of point** N, where $N=+n$ ($N=-n$) means a pole (zero of order n), and $N=\infty$ an essential singularity.

The type of singularity N at infinity is specified by the highest exponent, z^n, and the residue at infinity A_1 is the coefficient of z:

$$f(z)=A_N z^N+\cdots+A_1 z+\cdots.\tag{30.63}$$

viz.: (i) the first three functions in Table 30.7 are rational:

$$\frac{(z+a)^2}{z+b}=z+O(1):\quad N=1,\quad A_1=1,\tag{30.64a–c}$$

$$\frac{z+a}{(z+b)^2}=z^{-1}+O\left(z^{-2}\right):\quad N=-1,\quad A_1=0,\tag{30.65a–c}$$

$$a\left(z+\frac{1}{z}\right)=az+O\left(\frac{1}{z}\right):\quad N=1,\quad A_1=a,\tag{30.66a–c}$$

(ii) the binomial expansion (23.38) is used in the fourth and fifth functions in Table 30.7:

$$(a+z)^n=\sum_{m=0}^{n}\binom{n}{m}z^m a^{n-m}=z^n+\cdots+na^{n-1}z+\cdots:\quad N=n,\quad A_1=na^{n-1},\tag{30.67a–c}$$

TABLE 30.7
Point at Infinity of Some Functions

$f(z)$	N	$A_{-1}=$
$(z+a)^2/(z+b)$	1	1
$(z+a)/(z+b)^2$	-1	–
$a(z+1/z)$	1	a
$(a+z)^n$	n	$n\,a^{n-1}$
$a(z+1/z)^{2n+1}$	$2n+1$	$(?n+1)\,a$
$z^{-2n}\sin z$	∞	$(-)^n/(2n+1)!$

$n\in|N$ – pole of order n; $-n\in|N$ – zero of order n; $N=\infty$: essential singularity: A_{-1} – residue at singularity.

Note: For each of the six complex functions is indicated the type of point at infinity: (i/ii) $N=+1/-1$: simple pole/zero; (iii/iv) $N=+n/-n$ with n positive integer: pole/zero of order n; (v) $N=\infty$: essential singularity. In the case of a singularity, that is, pole or essential singularity, the residue A_{-1} is indicated.

$$a\left(z+\frac{1}{z}\right)^{2n+1}=a\sum_{p=0}^{2n+1}\binom{2n+1}{p}z^{2n+1-p}=az^{2n+1}+\cdots+a\binom{2n+1}{2n}z+\cdots, \tag{30.68a}$$

$$N=2n+1, \qquad A_1=a\binom{2n+1}{2n}=(2n+1)\,a; \tag{30.68b,c}$$

(ii) the sixth and last case in Table 30.7 is the function:

$$z^{-2n}\sin z=\sum_{p=0}^{\infty}\frac{(-)^p}{(2p+1)!}z^{2p+1-2n}=\cdots+\frac{(-)^n}{(2n+1)!}z+\cdots, \tag{30.69a}$$

$$N=\infty, \quad A_{-1}=\frac{(-)^n}{(2n+1)!} \tag{30.69b,c}$$

where was used the series (30.47).

Example 30.12. *Alternative Forms of the Kinetic Energy of a Potential Flow.*

Express the kinetic energy of a potential flow in six forms, all in terms of the complex potential.

The kinetic energy E_k of a potential flow in a region is given (28.8) by (30.70b):

$$\begin{aligned} f=\Phi+i\Psi: \quad E_k&=\frac{\rho}{2}\int\Phi\,\mathrm{d}\Psi=\frac{\rho}{2}\int\mathrm{Re}\,(f)\mathrm{Im}\,(\mathrm{d}f)\\ &=\frac{\rho}{8i}\int_{\partial D}(f+f^*)\,(\mathrm{d}f-\mathrm{d}f^*)=\frac{\rho}{4i}\int_{\partial D}\mathrm{Re}\,(f)\,(\mathrm{d}f-\mathrm{d}f^*)\\ &=\frac{\rho}{4}\int_{\partial D}(f+f^*)\,\mathrm{Im}\,(\mathrm{d}f)=\frac{\rho}{4}\mathrm{Im}\left\{\int_{\partial D}(f\,\mathrm{d}f+f^*\,\mathrm{d}f)\right\}\\ &=\frac{\rho}{4}\mathrm{Im}\left\{\int_{\partial D}(f\,\mathrm{d}f-f\,\mathrm{d}f^*)\right\}, \end{aligned} \tag{30.70a–b}$$

in terms of the complex potential (30.70a). The complex representation (3.17a,b) of the scalar potential Φ and stream function Ψ was used in (30.70b–f), and (3.22b) in (30.70g,h).

Example 30.13. *Imaginary Part of the Blasius Integral for the Pitching Moment.*

Use complex potential to evaluate the integral

$$N\equiv\int_{\partial D}p\,(y\,\mathrm{d}x-x\,\mathrm{d}y), \tag{30.71}$$

where p is the pressure on a body in a stream with uniform velocity at infinity.

The expression (30.71) is the imaginary part of the integral (28.17d) whose real part specifies the pitching moment:

$$\begin{aligned} M+iN&=\int_{\partial D}p\,z\,\mathrm{d}z^*=\int_{\partial D}p\,(x+iy)\,(\mathrm{d}x-i\mathrm{d}y)\\ &=\int_{\partial D}p\,[(x\,\mathrm{d}x+y\,\mathrm{d}y)+i\,(y\,\mathrm{d}x-x\,\mathrm{d}y)]. \end{aligned} \tag{30.72}$$

Using the same derivation as for Blasius theorem [(28.17d) to (28.28b) in Subsection 28.2.1] leads from (30.72) to

$$N = -\mathrm{Im}\left\{\frac{\rho}{2}(2\pi i)\left[\frac{(Q-i\Gamma)^2}{4\pi^2}+\frac{P_1 U}{\pi}\right]\right\}; \tag{30.73a}$$

this simplifies to

$$N = -\rho U P_{1x} + \frac{\rho}{4\pi}\left(\Gamma^2 - Q^2\right), \tag{30.73b}$$

where ρ is the mass density, U the mean flow velocity, P_{1x} the horizontal component of the dipole moment, Γ the circulation, and Q the volume flux. Replacing U by $Ue^{-i\alpha}$ generalizes (30.73a,b) to:

$$\begin{aligned} N &= -\mathrm{Im}\left\{i\rho\left[P_1 U e^{-i\alpha} + \frac{(Q-i\Gamma)^2}{4\pi}\right]\right\} \\ &= -\rho U\left(P_{1x}\cos\alpha + P_{1y}\sin\alpha\right) + \frac{\rho}{4\pi}\left(\Gamma^2 - Q^2\right), \end{aligned} \tag{30.74}$$

for nonzero angle-of-attack

Example 30.14. *Noncollocated Forces and Moment Theorem.*

Obtain the noncollocated forces and moments theorem, specifying the forces and moments on a body in a uniform stream at infinity, with a vortex, a source/sink and a dipole each at a different location. Compare with the Kutta–Joukowski and related theorems, and identify the extra terms.

The complex potential is specified by

$$f(z) = Uz - i\frac{\Gamma}{2\pi}\log z + \frac{Q}{2\pi}\log(z-a) - \frac{P_1}{2\pi(z-b)} + \cdots \tag{30.75}$$

where: (i) the OX-axis is taken along the free stream velocity at infinity so that U is real; (ii) the origin is chosen at the center of the vortex of circulation, Γ; (iii) the source/sink of flow rate Q is located at $z = a$; (iv) the dipole of moment P_1 is located at $z = b$; (v) it will be shown that the higher-order multipoles do not appear in the force and moments, as could be expected from Section 28.2. The complex conjugate velocity corresponding to (30.75) is:

$$\frac{\mathrm{d}f}{\mathrm{d}z} = U - \frac{i\Gamma}{2\pi z} + \frac{Q}{2\pi(z-a)} + \frac{P_1}{2\pi(z-b)^2} + \cdots \tag{30.76a}$$

$$= U + \frac{Q - i\Gamma}{2\pi z} + \frac{Qa + P_1}{2\pi z^2} + O\left(z^3\right), \tag{30.76b}$$

where the flow rate of the source/sink adds to the dipole moment due to its position distinct from the vortex. Substituting:

$$\left(\frac{\mathrm{d}f}{\mathrm{d}z}\right)^2 = U^2 + \frac{U(Q-i\Gamma)}{\pi z} + \frac{(Q-i\Gamma)^2/2\pi + 2U(Qa+P_1)}{2\pi z^2} + O\left(z^{-3}\right), \tag{30.76c}$$

in the Blasius theorem (28.20a,b), and using (28.24a,b) shows by comparison of (30.71b) with (28.27) that: (i) the coefficient of z^{-1} is the same, so nothing changes regarding forces, viz. horizontal (28.29a) and vertical (28.29b); (ii) the coefficient of z^{-2} in (30.71b) adds an extra term in (30.73a) the pitching moment relative to (28.28b),

$$\text{viz.}: M = -\mathrm{Re}\left\{\frac{\rho}{2}2\pi i\left[\frac{(Q-i\Gamma)^2}{4\pi^2} + U\frac{Qa+P_1}{\pi}\right]\right\} = -\frac{\rho Q\Gamma}{2\pi} + \rho U\bar{P}_{1y}. \tag{30.77}$$

Thus a source/sink not collocated with the vortex changes the dipole moment P_1 to the **effective dipole moment** (39.78a):

$$\bar{P}_{1y} \equiv P_{1y} + Q \text{ Im } (a) ; \quad \bar{P}_{1x} \equiv P_{1x} + Q \text{ Re } (a) , \tag{30.78a,b}$$

The corresponding real part (30.78b) does not appear for zero angle-of-attack. In the case of incident flow at an angle-of-attack, α, the change $U \to U\mathrm{e}^{-i\alpha}$ in (30.77) leads to the pitching moment:

$$M + \frac{\rho Q \Gamma}{2\pi} = -\text{Re}\left\{\frac{\rho}{2} 2\pi i U \, \mathrm{e}^{-i\alpha} \frac{Qa + P_1}{\pi}\right\} = \rho U \left(\bar{P}_{1y} \cos \alpha - \bar{P}_{1x} \sin \alpha\right), \tag{30.79}$$

This is the **noncollocated forces and moments theorem**: *a uniform flow at infinity with vortex at the origin, source/sink at $z = a$, and dipole and higher-order multipoles at any locations (30.75), either without (with) angle-of-attack, has (i) the same forces (28.29a,b) [(28.31a)] as if the singularities were colocated; (ii) a distinct pitching moment, with extra terms (30.78a,b) in the noncollocated (30.77) [(30.79)] relative to the collocated (28.29c) [(28.31c)] case in the absence (presence) of angle-of-attack.*

Example 30.15. *Potential Flow with Arbitrary Singularities at the Origin and Infinity.*

Consider a potential flow whose complex conjugate velocity is represented by arbitrary singularities at the origin and infinity, and determine: (i) the flow rate Q and circulation, Γ; (ii) the horizontal, F_x and vertical, F_y, forces and pitching moment, M. State the latter result (ii) as the generalized forces and moments theorem. Check that the results agree with Kutta–Joukowski theorem and extend the latter to the next order.

Example 30.15.1. *Flow rate and circulation in a potential flow.*

A potential flow with an arbitrary singularity at origin and infinity corresponds to the single-valued complex conjugate velocity specified by a Laurent series:

$$v^*(z) = \frac{\mathrm{d}f}{\mathrm{d}z} = \sum_{n=-\infty}^{+\infty} A_n z^n = A_0 + \sum_{n=1}^{\infty} \left(A_n z^n + A_{-n} z^{-n}\right), \tag{30.80}$$

with infinite radius of convergence. The corresponding complex potential has a singularity at the origin (infinity) if $A_{-n} \neq 0$ ($A_n \neq 0$) for at least one $n = 1, 2, \ldots$:

$$f(z) = \sum_{\substack{n=-\infty \\ n \neq -1}}^{\infty} A_n \frac{z^{n+1}}{n+1} + A_{-1} \log z; \tag{30.81}$$

the complex potential has a logarithmic singularity if the residue is not zero $A_{-1} \neq 0$; in that case it is not single-valued, because the origin is a branch-point. The integral of the complex velocity around a loop:

$$\int^{(0+)} v^*(z) \,\mathrm{d}z = \int (v_x - iv_y)(\mathrm{d}x + i\mathrm{d}y) = \int [(v_x \mathrm{d}x + v_y \mathrm{d}y) + i (v_x \mathrm{d}y - v_y \mathrm{d}x)] = \Gamma + iQ, \tag{30.82}$$

specifies through its real (imaginary) part the circulation (12.3) [volume flux (12.9)]. Bearing in mind (28.24a–c), *the circulation, Γ and volume flux, Q are specified by the residue A_{-1} (minus the residue $-A_{-1}$) of the complex velocity at the origin (infinity):*

$$\zeta \equiv \frac{1}{z}: \quad \Gamma + iQ = \int^{(0+)} v^*(z)\,\mathrm{d}z = -\int^{(\infty+)} v^*\left(\frac{1}{\zeta}\right)\zeta^{-2}\,\mathrm{d}\zeta = 2\pi i A_{-1}; \tag{30.83a,b}$$

thus both are conserved adding the monopoles at the origin and infinity. The different signs in (30.83a) [(30.83b)] arise because the interior (exterior) directions relative to the boundary of domain have distinct signs (Section 31.3), for example, a positive exterior (interior) direction is counterclockwise (clockwise) in Figure 31.5. The multipoles do not contribute to the circulation (mass flux) because they consist of equal and opposite vortices (sources/sinks).

Example 30.15.2. *Generalized forces and moment theorem.*

The Blasius theorem (28.20a,b) specifies the forces and moments in terms of the complex conjugate velocity, for example, if it is singular only at the origin or infinity (30.80), its square is specified by a double Laurent series:

$$n+m=k: \quad \left(\frac{\mathrm{d}f}{\mathrm{d}z}\right)^2 = \sum_{n,m=-\infty}^{+\infty} A_n A_m z^{n+m} = \sum_{n,k=-\infty}^{+\infty} A_k A_{n-k} z^n; \qquad (30.84\text{a,b})$$

here Cauchy's rule (21.25a–c) for the product of series was used

$$\left(\frac{\mathrm{d}f}{\mathrm{d}z}\right)^2 = \sum_{n=-\infty}^{+\infty} B_n z^n, \qquad B_n \equiv \sum_{k=-\infty}^{+\infty} A_k A_{n-k}. \qquad (30.85\text{a,b})$$

Recalling (28.24a,b) and substituting (30.85a,b) in the Blasius theorem (28.20a,b) leads to

$$F^* = F_x - iF_y = \frac{i}{2}\rho 2\pi i B_{-1} = -\pi\rho \sum_{k=-\infty}^{+\infty} A_k A_{-1-k}, \qquad (30.86\text{a})$$

$$M = -\mathrm{Re}\left\{\frac{\rho}{2} 2\pi i B_{-2}\right\} = \pi\rho\,\mathrm{Im}\left\{\sum_{k=-\infty}^{+\infty} A_k A_{-2-k}\right\}, \qquad (30.86\text{b})$$

the **generalized forces and moments theorem**: *a body in a potential flow with complex conjugate velocity having an arbitrary singularity at the origin and infinity (30.80), is acted upon by horizontal, F_x and vertical, F_y forces (pitching moment M) specified by (30.86a) [(30.86b]).*

Example 30.15.3. *Next order beyond the Kutta–Joukowski theorem.*

In the case of a velocity bounded at infinity (30.87a):

$$0 = A_1 = A_2 = \cdots: \quad F^* = F_x - iF_y = -2\pi\rho A_0 A_{-1}, \quad M = \pi\rho\,\mathrm{Im}\left[2A_0A_{-2} + (A_{-1})^2\right], \qquad (30.87\text{a–c})$$

the only relevant coefficients:

$$A_0 = U\mathrm{e}^{-i\alpha}, \quad A_{-1} = \frac{Q - i\Gamma}{2\pi}, \quad A_{-2} = \frac{P_1}{2\pi}, \qquad (30.88\text{a–c})$$

are: (i) the free stream velocity U and angle-of-attack α in (30.88a); (ii) the flow rate Q and circulation Γ of the monopole (30.88b); (iii) the moment P_1 of a dipole (30.88c). Substitution of (30.88a–c) in (30.87b,c) without (with) angle-of-attack leads to (28.28a,b) (28.31a,b), and hence to the original form of the Kutta–Joukowski theorem and associated results (28.29a–c) [(28.30a–c)]. The next approximation beyond the Kutta–Joukowski theorem and related results is to allow the velocity to have a simple pole at infinity, as for the flow in a rectangular corner:

$$0 = A_2 = A_3 = \cdots: \quad F^* = F_x - iF_y = -2\pi\rho\left(A_0A_{-1} + A_1A_{-2}\right), \qquad (30.89\text{a,b})$$

$$M = \pi\rho\,\mathrm{Im}\left[2A_0A_{-2} + (A_{-1})^2 + 2A_1A_{-3}\right], \qquad (30.89\text{c})$$

in that case: (i) the dipole moment (30.88c) affects the forces (30.89b) together with the corner flow $v^* = A_1 z$; (ii) the pitching moment (30.89c) is affected by a quadrupole (30.90a):

$$A_{-3} \equiv \frac{P_2}{2\pi}; \quad 0 = A_{-4} = A_{-5} = \cdots, \tag{30.90a,b}$$

the restriction (30.90b) to a quadrupole at the origin is equivalent to the restriction (30.89a) to a rectangular corner flow at infinity that is also a quadrupole (Section 12.8); substitution of (30.88a–c; 30.90a) in (30.81) [(30.80)] leads to the complex potential (30.91a) [conjugate velocity (30.91b)]:

$$f^*(z) = A_1 z^2 + U e^{-i\alpha} z + \frac{Q - i\Gamma}{2\pi} \log z - \frac{P_1}{2\pi z} - \frac{P_2}{2\pi z^2} + \cdots, \tag{30.91a}$$

$$v^*(z) = 2A_1 z + U e^{-i\alpha} + \frac{Q - i\Gamma}{2\pi z} + \frac{P_1}{2\pi z^2} + \frac{P_2}{\pi z^3} + \cdots. \tag{30.91b}$$

The case of a simple pole at infinity, or triple pole at the origin, is included in the next Example 30.16.

Example 30.16. *Volume Flux, Circulation, Forces and Pitching Moment in a Potential Flow.*

Consider the potential flows in Table 30.8 that have at most a triple pole at the origin, and determine for a loop round the origin: (i) the coefficients of the Laurent–Maclaurin series up to A_1; (ii/iii) the volume flux, Q and circulation, Γ; (iv/v) the horizontal, F_x and vertical, F_y force; (vi) the torque, M in the direction normal to the plane of the flow.

The complex conjugate velocity for the flows in Table 30.8 has singularity at the origin of order not higher than a triple pole in:

$$v^*(z) = U\left(\frac{c}{z}\right)^2 \exp\left(-\frac{z}{c}\right) = \frac{Uc^2}{z^2} - \frac{Uc}{z} + \frac{U}{2} - \frac{Uz}{6c} + O\left(z^2\right), \tag{30.92}$$

$$v^*(z) = U\frac{c}{z}\cos\left(\frac{z}{c}\right) = \frac{Uc}{z} - \frac{Uz}{2c} + O\left(z^3\right), \tag{30.93}$$

$$v^*(z) = U\left(\frac{c}{z}\right)^2 \sin\left(\frac{z}{c}\right) = \frac{Uc}{z} - \frac{Uz}{6c} + O\left(z^3\right), \tag{30.94}$$

$$v^*(z) = U\left(1 + \frac{c}{z}\right)^2 = \frac{Uc^2}{z^2} + \frac{2Uc}{z} + U, \tag{30.95}$$

$$v^*(z) = U\frac{z}{c}\left(1 + \frac{c}{z}\right)^2 = \frac{Uc}{z} + 2U + \frac{Uz}{c}, \tag{30.96}$$

$$v^*(z) = U\left(1 + \frac{c}{z}\right)^3 = \frac{Uc^3}{z^3} + \frac{3Uc^2}{z^2} + \frac{3Uc}{z} + U, \tag{30.97}$$

where the series for the exponential (30.43), cosine (30.45), and sine (30.47) were used, respectively, in (30.92; 30.93; 30.94). The results in Table 30.8 may be obtained as follows: (i) comparison of (30.92–30.97) with the general Laurent series (30.80) for the complex conjugate velocity specifies the coefficients A_n with $n = 1, 0, -1, -2, -3$; (ii) the residue A_{-1} specifies (30.83b) the flow rate and circulation; (iii) together with (A_0, A_1, A_2) specifies the forces (30.89b); (iv) together with, A_{-3} specifies the pitching moment (30.89c).

Example 30.17. *Cauchy Test (1821) on the Convergence of Series.*

Prove the **Cauchy convergence test (1821)** *for a series of functions*:

$$\sum_{n=0}^{\infty} F_n(z) = F(z): \quad \lim_{n\to\infty} \left|F_n(z)\right|^{1/n} \begin{cases} > 1 & \text{D.} \\ \equiv r(z) < 1 & \text{A.C.} \\ r(z) \le 1 - \varepsilon & \text{T.C.}; \end{cases} \tag{30.98a–c}$$

TABLE 30.8
Circulation, Flow Rate, Drag, Lift, and Pitching Moment in a Potential Flow

$v^*(z) = df/dz$	$U(c/z)^2 e^{-z/c}$	$U(c/z)\cos(z/c)$	$U(c/z)^2 \sin(z/c)$	$U(1+c/z)^2$	$U(z/c)(1+c/z)^2$	$U(1+c/z)^3$
A_{-3}	0	0	0	0	0	Uc^3
A_{-2}	Uc^2	0	0	Uc^2	0	$3Uc^2$
A_{-1}	$-Uc$	Uc	Uc	$2Uc$	U c	$3Uc$
A_0	$U/2$	0	0	U	2 U	U
A_1	$-U/(6c)$	$-U/(2c)$	$-U/(6c)$	0	U/c	0
Q	$-2\pi U a$	$2\pi U a$	$2\pi U a$	$4\pi U a$	$2\pi U a$	$6\pi U a$
Γ	$2\pi U b$	$-2\pi U b$	$-2\pi U b$	$-4\pi Ub$	$-2\pi Ub$	$-6\pi Ub$
F_x	$4\pi\rho U^2 a/3$	0	0	$-4\pi\rho U^2 a$	$-4\pi\rho U^2 a$	$-6\pi\rho U^2 a$
F_y	$-4\pi\rho U^2 b/3$	0	0	$4\pi\rho U^2 b$	$4\pi\rho U^2 b$	$6\pi\rho U^2 b$
M	$4\pi\rho U^2 ab$	$2\pi\rho U^2 ab$	$2\pi\rho U^2 a b$	$12\pi\rho U^2 ab$	$2\pi\rho U^2 ab$	$30\pi\rho U^2 ab$

Where $c \equiv a + ib$. $A_{-3}, \ldots, A_1$ - coefficients of Laurent series for complex conjugate velocity.
Q – flow rate; Γ – circulation; M – pitching moment; F_x – drag; F_y – lift.
Note: The complex conjugate velocity of six potential flows is expanded in Laurent–Maclaurin series around the origin, recording the five coefficients ($A_{-3}, \ldots, A_1$) which are needed to calculate the following set of data: (i) flow rate Q; (ii) circulation Γ; (iii/iv) horizontal F_x and vertical F_y forces; (v) pitching moment M.

compare the Cauchy (30.98a–c) [D'Alembert (29.31a–c)] tests, and the related formulas (30.99b) [(29.71a)] for:

$$F(z) = \sum_{n=0}^{\infty} C_n \, z^n, \quad |z| < R = \lim_{n\to\infty} |C_n|^{-1/n}, \tag{30.99a,b}$$

the radius of convergence (30.99b) of a power series (30.99a).

There are three cases: (i) if (30.98a) holds then $F_n(z)$ tends to infinity as $n \to \infty$ and the series diverges; (ii) if the limit (30.100a) exists:

$$\lim_{n\to\infty} |F_n(z)|^{1/n} \equiv r(z), \quad \lim_{n\to\infty} |F_n(z)| = r^n, \tag{30.100a,b}$$

then the remainder of the series of moduli (30.100b) can be summed as a geometric series:

$$|r(z)| < 1: \; |R_N(z)| \equiv \left| \sum_{n=N}^{\infty} F_n(z) \right| \le \sum_{n=N}^{\infty} |F_n(z)| \le B \sum_{n=N}^{\infty} r^n = \frac{B r^N}{(1-r)}; \tag{30.101a,b}$$

this tends to zero as $N \to \infty$ (30.101a), so the series is absolutely convergent (30.98b); (iii) if (30.102a,b) then the upper bound is (30.102c) independent of z:

$$R \le 1 - \varepsilon \quad 0 < \varepsilon < 1: \quad |R_N(z)| \le B \sum_{n=N}^{\infty} (1-\varepsilon)^N = \frac{B(1-\varepsilon)^N}{\varepsilon}, \tag{30.102a–c}$$

and the series is totally convergent (30.98c). An alternative proof of (ii) [(iii)] is to use instead of the remainder (30.101b) [(30.102c)] after N terms the sum of the series of moduli:

$$r \le 1 - \varepsilon: \quad \sum_{n=0}^{\infty} |F_n(z)| \le B \sum_{n=0}^{\infty} r^n = \frac{B}{1-r} \le \frac{B}{\varepsilon}. \tag{30.103a,b}$$

For a power series (30.99a) the general term is (30.104a):

$$F_n(z) = C_n \, z^n, \quad 1 > |r(z)| = \lim_{z\to\infty} |z| \; |C_n|^{1/n}, \tag{30.104a,b}$$

and the condition (30.100a) ≡ (30.104b) specifies the radius of convergence (30.99b).

Example 30.18. *Convergence of the Generalized Harmonic Series.*

Discuss the convergence of the **generalized harmonic series**:

$$c \equiv a + ib: \quad F(z) = \sum_{n=1}^{\infty} n^{-c-z}, \tag{30.105a,b}$$

Compare with the original harmonic series (29.40a–d) that is the particular case $z \equiv x + i\,y = 0$.

The coefficients satisfy:

$$C_n \equiv n^{-c-z}: \quad \frac{C_{n+1}}{C_n} = \left(1 + \frac{1}{n}\right)^{-c-z} = 1 - \frac{c+z}{n} + O\left(n^{-2}\right); \tag{30.106a,b}$$

the series has the same behavior as the harmonic series (29.40a–d) with $g = c + z$, viz. *the generalized harmonic series (30.105a,b), with coefficients (30.106a,b) has the following convergence properties (Figure 30.3): (i) absolutely (totally) convergent for* $(x > 1 - a)$

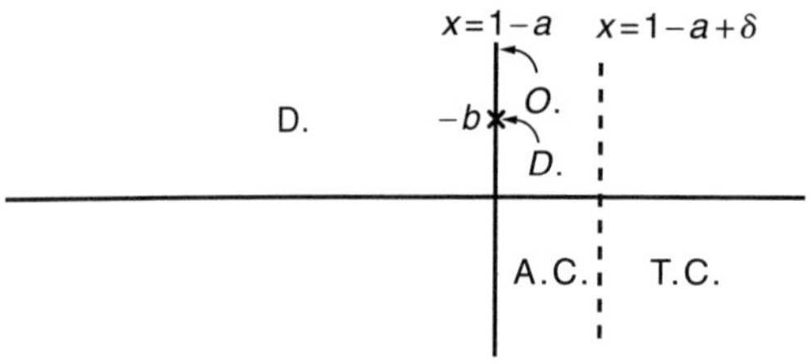

FIGURE 30.3
The convergence of the generalized harmonic series (30.105a,b) is determined by vertical slices of the z-plane: (i/ii) diverges (converges absolutely) to the left (right) of $x = 1 - a$; (iii) converges also uniformly in a closed subregion to the right $x \geq 1 - a + \delta$ with $\delta > 0$; (iv/v) on the boundary of convergence $x = 1 - a$ oscillates (diverges) for $y \neq -b$ $(y = -b)$.

($x \geq 1 - a + \delta$ with $\delta > 0$); (ii) for $x = 1 - a$ it is oscillatory (divergent) for $y \neq -b$ ($y = -b$); (iii) it diverges for $x < 1 - a$.

Example 30.19. *Error of Truncation of Conditionally/Absolutely Convergent Series.*
Show that *if*

$$\lim_{n\to\infty} f_n = 0, \quad f_n > f_{n+1} > \ldots > 0, \tag{30.107a,b}$$

for the real alternating series:

$$S_N \equiv \sum_{n=1}^{N} (-)^{n-1} f_n, \quad \lim_{N\to\infty} S_N = S, \tag{30.108a,b}$$

the odd (even) sums form a decreasing (increasing) sequence, with the limit lying between them

$$S_1 > S_3 > \ldots > S_{2n-1} > S_{2n+1} > \ldots > S > \ldots > S_{2n} > S_{2n-2} > S_4 > S_2. \tag{30.109}$$

Use the result to calculate the sum of series:

$$\sum_{n=1}^{\infty} \frac{(-)^{n-1}}{n} = 1 - \frac{1}{2} + \frac{1}{3} - \frac{1}{4} + \frac{1}{5} \cdots = \log 2 = 0.693, \tag{30.110a}$$

$$\sum_{n=1}^{\infty} \frac{(-)^{n-1}}{n^2} = 1 - \frac{1}{2^2} + \frac{1}{3^2} - \frac{1}{4^2} + \frac{1}{5^2} - \cdots = \frac{\pi^2}{12} = 0.822, \tag{30.110b}$$

with three digit accuracy. Compare the rate of convergence of the two series, noting that (30.110a) [(30.110b)] is conditionally (absolutely) convergent.

The inequalities (30.109) follow from (30.111a):

$$S_{2n} - S_{2n-2} = -f_{2n} + f_{2n-1} > 0 > f_{2n+1} - f_{2n} = S_{2n+1} - S_{2n-1}; \tag{30.111a}$$

$$\varepsilon_n \equiv |S - S_n| < |S_{n+1} - S_n| < f_{n+1} \tag{30.111b}$$

also *(30.111b) provides an upper bound for the* ***error of truncation*** *of the series.* The conditionally convergent series (30.110a) needs 20 terms for a 4% accuracy as shown in Table 30.9; the absolutely convergent series (30.110b) with half the terms gives a similar accuracy of 4% in Table 30.10. This comparison shows the improved numerical accuracy of an absolutely relative to a conditionally convergent series, whose terms scale as $O\left(n^{-1}\right)$ $\left[O\left(n^{-2}\right)\right]$:

TABLE 30.9

Sums of Alternating Conditionally Convergent Series

n	1	2	3	4	5	6
S_{2n-1}	1.000	0.833	0.783	0.760	0.746	0.737
S_{2n}	0.500	0.583	0.617	0.635	0.646	0.653
n	**7**	**8**	**9**	**10**	∞	**error**
S_{2n-1}	0.730	0.725	0.722	0.719	0.693	+0.026
S_{2n}	0.659	0.663	0.666	0.669		−0.024

Note: The sums of the alternating conditionally convergent series (30.110a) show convergence, with one accurate digit after 10 terms.

TABLE 30.10

Sums of Alternating Absolutely Convergent Series

n	1	2	3	4	5	∞	Error
S_{2n-1}	1.000	0.861	0.839	0.831	0.828	0.822	+0.06
S_{2n}	0.750	0.799	0.811	0.816	0.818		−0.04

Note: The sums of the alternating absolutely (conditionally) convergent series (30.110b) [(30.110a)] showing faster (slower) convergence, for example, two (one) accurate digits (digit) after five (ten) terms; thus absolute convergence can lead to higher accuracy with less terms than conditional convergence. For the numerical calculation of a function a representation as absolutely convergent series with terms $O\left(n^{-2}\right)$ is preferable to a conditionally convergent series with terms $O\left(n^{-1}\right)$.

(i) it requires fewer terms for the same accuracy; (ii) it accumulates fewer round-off errors. The latter (ii) can invalidate a high-accuracy calculation with many terms.

Example 30.20. *Convergence of Non-Gaussian Hypergeometric Series.*

Consider the **generalized hypergeometric series** of type, ${}_pF_q$, defined by

$$ {}_pF_q\left(a_1,\ldots,a_p;b_1,\ldots,b_q;z\right) = 1 + \sum_{n=1}^{\infty}\frac{z^n}{n!}\prod_{m=0}^{n-1}\frac{(a_1+m)\ldots(a_p+m)}{(b_1+m)\ldots(b_q+m)}; \tag{30.112} $$

prove that its radius of convergence is zero (infinity) if $p > q+1\,(p \le q)$, *and is unity if* $p = q+1$, *according to Table 30.11. In the case* $p = q+1$ *of radius of convergence unity, show that the behavior of the generalized hypergeometric series* ${}_{q+1}F_q$ *is specified by*

$$ g(z) = E + i\,D \equiv 1 + \sum_{s=1}^{p} b_s - \sum_{r=1}^{p+1} a_r, \tag{30.113} $$

at all points on the complex z-*plane, according to Table 30.11.* Compare with the Gaussian hypergeometric series (Section 29.9) and simpler series.

Example 30.20.1. *Confluent hypergeometric series.*

The Gaussian hypergeometric series (29.74), has two upper and one lower parameter. The simpler cases are: (i) no upper or lower parameter:

$$ |z| < \infty: \quad {}_0F_0(z) = \sum_{n=0}^{\infty}\frac{z^n}{n!} = \exp(z), \tag{30.114} $$

TABLE 30.11
Radius of Convergence of Generalized Hypergeometric Series

Radius of convergence	***R*:**	**0**	**1**	∞
Type of series	${}_pF_q$:	$p > q+1$	$p = q+1$	$p \leq q$
Example of function		Diffluent	Gaussian	Confluent
Formula		(30.117a)	(29.74)	(30.116a)
Symbol		${}_2F_0$	${}_2F_1$	${}_1F_1$
Parameters	*Upper* : $a_1, \ldots, a_p$	a,b	a,b	a
	Lower : $b_1, \ldots, b_q$	–	c	c

Note: The radius of convergence of generalized hypergeometric (30.112) series ${}_pF_q$ takes only three values: (i) zero for $p > q+1$ meaning it converges nowhere except at the origin; (ii) infinite for $p < q+1$ implying it converges everywhere in the finite complex plane; (iii) in the intermediate case $p = q+1$ the radius of convergence is unity, as detailed in Table 30.12, which includes the boundary of convergence.

leads to the exponential series (30.114) $\equiv$ (23.60) that has infinite radius of convergence; (ii) one upper parameter:

$$|z| < 1 : {}_1F_0\left(a; z\right) = \sum_{n=0}^{\infty} \frac{z^n}{n!} a\left(a+1\right)\ldots\left(a+n-1\right) = \sum_{n=0}^{\infty} \binom{-a}{n} \left(-\right)^n z^n = \left(1-z\right)^{-a}, \tag{30.115}$$

leads to the binomial series (30.115) $\equiv$ (25.41a,b) that has radius of convergence unity; (iii) one upper and one lower parameter, leads (Kummer, 1836) to the **confluent hypergeometric series**:

$$|z| < \infty : \quad {}_1F_1\left(a; c; z\right) = 1 + \sum_{n=1}^{\infty} \frac{z^n}{n!} \prod_{m=0}^{n-1} \frac{a+m}{b+m}, \tag{30.116a}$$

that has infinite radius of convergence:

$$\frac{C_{n+1}}{C_n} = \frac{z\left(a+n\right)}{\left(b+n\right)\left(n+1\right)} = \frac{z}{n} + O\left(\frac{1}{n^2}\right); \quad R = \lim_{n\to\infty}\left|\frac{C_n}{C_n+1}\right| = \lim_{n\to\infty}\frac{n}{|z|} = \infty, \tag{30.116b,c}$$

and specifies an integral function; (iv) the diffluent hypergeometric series:

$$z = 0 : \quad {}_2F_0\left(a, b; z\right) = 1 + \sum_{n=1}^{\infty} \frac{z^n}{n!} \prod_{m=0}^{n-1} \left(a+m\right)\left(b+m\right), \tag{30.117a}$$

diverges

$$\frac{C_{n+1}}{C_n} = \frac{\left(a+n\right)\left(b+n\right)}{n+1} = O\left(n\right), \tag{30.117b}$$

everywhere except at the origin. These examples agree with Table 30.11.

Example 30.20.2. *Generalized hypergeometric series.*

The generalized hypergeometric series (30.112) has ratio of coefficients:

$$\frac{C_{n+1}}{C_n} = \frac{z}{n+1} \frac{\left(a_1+n\right)\ldots\left(a_p+n\right)}{\left(b_1+n\right)\ldots\left(b_q+n\right)} \sim z n^{p-q-1}, \tag{30.118}$$

and thus: (i) if $p \geq q+1$ the series diverges everywhere except at the origin, for example, the diffluent case (30.117a,b) for that $p = 2, q = 0, p-q-1 = 1$; (ii) if $p \leq q$ the series

TABLE 30.12
Convergence of Generalized Hypergeometric Series

Region:	$\lvert z\rvert \neq 1$	$\lvert z\rvert = 1 \neq z$	$z = 1$
Behavior:			
T.C.	$\lvert z\rvert \le 1-\varepsilon$	–	–
A.C.	$\lvert z\rvert < 1$	$E > 1$	$E > 1$
C.C.	–	$0 < E \le 1$	–
O.	–	$E = 0$	$E = 1, D \neq 0$
D.	–	–	$E = 1, D = 0$
D.	$\lvert z\rvert > 1$	$E < 0$	$E < 1,$

E, D defined by (30.113).
Note: Convergence of the generalized hypergeometric series ${}_pF_{p+1}$ in the whole complex plane, including the unit circle, which is the boundary of convergence. The convergence depends only on two parameters (E, D) in (30.113).

has infinite radius of convergence, for example. the confluent series (30.116a,b) for that $p = 1 = q, p - q - 1 = -1$ and the exponential (30.114), for that $p = 0 = q, p - q - 1 = -1$; (iii) if $p = q + 1$ the series has radius of convergence unity, for example, the Gaussian case (29.74) for that $p = 2, q = 1, p - q - 1 = 0$ and the binomial series (30.115), for that $p = 1, q = 0, \quad p - q - 1 = 0$. To complete the proof of Table 30.12, it is shown: (iii) that (30.118) implies (30.119):

$$p = q + 1: \quad \frac{C_{n+1}}{C_n} = z\frac{(1 + a_1/n)\ldots(1 + a_{p+1}/n)}{(1 + b_1/n)\ldots(1 + b_p/n)\,(1 + 1/n)}$$
$$= z\left\{1 + \left[-1 + \sum_{r=1}^{p+1} a_r - \sum_{s=1}^{q} b_s\right] n^{-1} + O\left(n^{-2}\right)\right\}$$
$$\equiv z\left[1 - \frac{E + iD}{n} + O\left(\frac{1}{n^2}\right)\right], \tag{30.119}$$

with $E + iD$ defined by (30.113). From (30.119) it follows that the radius of convergence is unity; applying to (30.113) the combined test (Subsection 29.1.1) proves the whole of Table 30.12. In the Gaussian case: (i) the parameters are $q = 1,\ p = 2,\ \ a_1 = a,\ \ a_2 = b,\ \ b_1 = c$; (ii) the ratio of successive coefficients (30.119) becomes (30.120);

$$p = 2; \quad q = 1: \quad \frac{C_{n+1}}{C_n} = \frac{z\,(1 + a/n)\,(1 + b/n)}{(1 + c/n)\,(1 + 1/n)} = z\left[1 + \frac{a + b - c - 1}{n} + O\left(n^{-2}\right)\right]; \tag{30.120a–c}$$

(iii) the parameters (E,D) are given by (30.113) $\equiv$ (30.121a,b):

$$E = \operatorname{Re}\ (c + 1 - a - b)\,, \quad D = \operatorname{Im}\,(c + 1 - a - b)\,; \tag{30.121a,b}$$

(iv) Table 30.12 follows from the combined convergence tests (Subsection 29.1.1) and agrees with Table 29.2.

Note 30.1. *Analytic Continuation in a Plane and Conformal Mapping into Another Plane:* Part 3 already gives a reasonable overview of the two lines of reasoning started in Part 2, and that gain in Part 4 additional tools extending their reach: (i) a further exploration on the properties of certain classes of complex functions, for example, factorization of polynomials, partial fraction decomposition of rational functions, roots and periods of transcendental functions; (ii) these are developed further, viz. the analytic continuation of functions in the complex plane and the conformal mapping into another complex plane; (iii) the conformal mapping of the upper half-plane and unit disk into a variety of regions, extends substantially the variety and scope of solutions of potential field problems; (iv) the latter include for potential flows: airfoils, wings, ducts, channels, wells, wakes, jets, slits, etc.; (v) the other potential field analogies lead to similar problems, for example, gravity, electric and magnetic fields of plates, condensers, etc.; (vi) an additional potential field concerns steady heat conduction, with a variety of boundary conditions, viz. isothermal, heated or convective processes, for example, in heat exchangers.

Conclusion 30: Refraction of waves by a parallel sided lens of thickness h, of material 2, separating media of materials 1 and 3, with wave speeds respectively c_1, c_2, c_3; the (Figure 30.1) first three refractions involve all reflection and transmission coefficients, viz. R_1, T_1 for the passage $1 \to 2$, and R_2, T_2 for $2 \to 1$, and R_3, T_3 for $2 \to 3$. The same series (30.29a–d), of powers of an auxiliary function with poles at $\pm i$, represents (Figure 30.2) two different functions, viz., $z(1/z)$ inside (outside) the unit disk; on the unit circle the series converges only at ± 1, where the two functions coincide. The region of convergence (Figure 30.3) of the generalized harmonic series (30.100a,b) is a right-hand half-plane, open (closed) for absolute (total) convergence; there is divergence on the left half plane, and on the boundary line it diverges (oscillates) at one point $-b$ (all other points $\neq -b$).

Part 4
Conformal Mapping: Heat and Aerodynamics

A complex function establishes a correspondence or mapping between points on the two planes; if the function is holomorphic (Chapter 11) it maps smooth curves into smooth curves, and thus preserves angles, that is, has the conformal property (Chapter 33). A conformal transformation generally does not preserve lengths, that is, it can map regions of different extent one onto the other, for example, a wide variety of regions can be mapped into or from a half-plane or the inside or outside of a circle. Since a holomorphic function yields a solution of the Laplace equation, that is, is a harmonic function, this property is preserved under conformal mapping. The solution of the Laplace equation for a given region, for example, a half-plane or a circle, can be used to obtain the solution in a variety of other regions, for example, airfoils (Chapter 34), channels (Chapter 36), and so on, into which the half-plane or circle can be mapped conformally. The latter, for example, channels, are instances of the mapping into polygons, viz. with infinite sides (Chapter 36); these involve critical points (Chapter 35) that are singularities, for example, branch-points, where the conformal property is not preserved, viz. leading to the appearance of edges. The method of conformal mapping applies to all potential fields, viz. potential flow, gravity, electrostatic, and magnetostatic field, and also (Chapter 32) steady heat conduction; a variant of conformal mapping, viz. the hodograph method (Chapter 38) applies to problems with free boundaries, that is, jets out of reservoirs, along walls or around bodies. The conformal mapping uses a function holomorphic in a part of the complex plane (Chapter 37); the function may be extended to other regions of the complex plane, by matching with analytic functions defined over those regions, such that their values coincide in overlapping subregions. This process of analytic continuation leads to monogenic functions (Chapter 31) that besides ordinary points can, have special points, viz.: (i) zeros which allow the factorization of polynomials (Chapter 31); (ii) poles that lead to the partial fraction decomposition of rational functions; (iii) essential singularities that relate to the roots of transcendental functions and equations (Chapter 39).

31

Analytic Continuation and Rational Functions

The analysis of functions of a complex variable can be envisaged at three successive levels. At the first local level (part 2) a holomorphic function is defined by the existence of derivate at a point (Chapter 11); this definition involves only the neighborhood of the point, and allows the consideration of the primitive and the deduction of solutions of the Laplace equation in the plane, that is, harmonic functions (Section 23.1). At the second level (part 3) of a region, an analytic function is defined as a function holomorphic at all points of the region; by means of contour and loop integrals, it can be shown that a such a function is infinitely differentiable (Chapter 15), that is, smooth, and that it has a Taylor series (Chapter 25) with radius of convergence determined by the nearest singularity. At the third or global level (part 4), a set of functions analytic in overlapping regions defines a single monogenic function, if the functions coincide in the overlapping subregions; a monogenic function provides the analytic continuation of each of its component functions beyond the region where it was originally defined. The analytic continuation allows the study of a function over the maximum possible extent of the complex plane, including special points, such as zeros and singularities; the singularities of complex analytic function are either poles (Chapter 31) or essential singularities (Chapter 39). The analytic continuation of a monogenic function (Section 31.1) can be used to extend its domain, for example, if it is real on the real axis by reflection (Section 31.2). The analytic continuation is not possible for lacunary functions, whose singularities are dense on the boundary of convergence (Section 31.1); analytic extension may be possible across an arc for a function analytic on both sides with a finite discontinuity (Section 31.3). The analytic continuation depends on the location of poles that can be determined, together with zeros via Cauchy's fourth integral theorem (Section 31.4). The zeros relate to: (i) the factorization of polynomials (Sections 31.5–31.7); (ii) the decomposition of rational function into partial fractions (Sections 31.8 and 31.9) based on their poles. The location of the zeros of a polynomial, even without their precise determination, can be sufficient to ensure the stability of a system, for example, if all zeros of the characteristic polynomial are on the r.h.s. half-plane (Chapter 2).

31.1 Theorem of Monodromy and Lacunary Functions (Osgood, 1929)

An analytic function can generally be extended beyond its domain (Subsection 31.1.1) unless the boundary is densely covered by singularities (Subsection 31.1.2), that is, it is a lacunary function.

31.1.1 Elements of a Monogenic Function

The starting point is an almost self-evident **theorem of monodromy**: *let $f_n(z)$ be a set of analytic functions defined in the regions, D_n, with $n = 1, \ldots, N$. Suppose that each region*

D_n intersects at least one other region D_m, with $n \neq m = 1, \ldots, N$ over a continuous arc or subdomain and that in the overlapping subregion the functions coincide:

$$\forall_{z \in D_n \cap D_m}: \quad f_n(z) = f_m(z). \tag{31.1a}$$

Then the ensemble of the analytic functions $f_n(z)$ for $n = 1, \ldots, N$, specifies a single monogenic function $f(z)$, defined over the union of the regions $D_1, \ldots, D_n$:

$$z \in \bigcup_{n=1}^{N} D_n \equiv D: \quad z \in D_n \Rightarrow f(z) \equiv f_n(z) \in \bar{A}. \tag{31.1b}$$

The function $f_n(z)$ defined for $z \in D_n$ has for ***analytic continuation*** *into the region $z \in D_m - (D_n \cap D_m)$ the function $f_m(z)$; the* ***monogenic function*** *$f(z)$ is the analytic continuation into $z \in D - D_n$ of each of its* ***element functions*** *$f_n(z)$ defined for $z \in D_n$. The element or* ***component function*** *$f_n(z)$ is the* ***representation*** *of the monogenic function $\bar{A}$ in the region D_n.* In Figure 31.1a, a set of three analytic functions is defined over a chain of three regions, specifying a single monogenic function over the union of the regions.

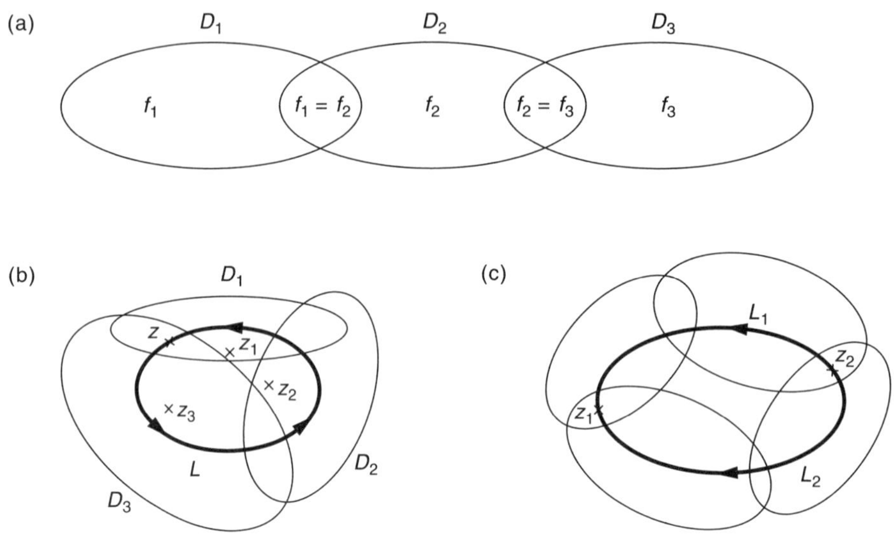

FIGURE 31.1
Consider (a) three functions f_{1-3} analytic in the regions D_{1-3}, such they coincide in the overlapping subregions; then: (i) each pair of functions is the analytic continuation of the other function beyond its original domain, for example, (f_2, f_3) perform the analytic continuation of f_1 from its original domain D_1 to the extended domain $D_1 \cup D_2 \cup D_3$; (iii) the three functions f_{1-3} are the elements or component functions of a single monogenic function analytic over the union $D_1 \cup D_2 \cup D_3$ of the three domains. If the analytic continuation is performed along a closed loop (b) the value of the monogenic function will be increased by $2\pi i$ times the sum of the residues at the singularities in the interior. Likewise for of the monogenic function when analytic continuation is performed by two distinct paths between the same points (c): the difference is $2\pi i$ times sum of the residues at the singularities between the two paths. If there are no singularities within the loop (b) [between the two paths (c)] the value of the monogenic function is unchanged.

An analytic function $f_n(z)$ is specified by its Taylor series (31.2b) about a regular point $z_n = a_n$, with radius of convergence r_n determined (31.2a) by the nearest singularity b_n:

$$|z - a_n| < r_n \equiv |b_n - a_n| : \quad f_n(z) = \sum_{m=0}^{\infty} \frac{(z - a_n)^m}{m!} f_n^{(m)}(a_n). \tag{31.2a,b}$$

In this case the regions D_n would be circles, overlapping in subregions not containing the singularities, for example in Example 40.1. The subregions form either an **open (closed) chain**, for example, in Figure 31.1a (31.1b,c). Performing the analytic continuation along a closed loop, for example, L in Figure 31.1b, then (i) the monogenic function returns to the same value, if no singularities are enclosed within the loop; (ii) if singularities are enclosed, the function is increased by $2\pi i$ times the sum of the residues. If a monogenic function is obtained by analytic continuation between the same points z_1, z_2 along different paths, for example, L_1, L_2 in Figure 31.1c, then (i) the same value is obtained, if there are no singularities between the paths; (ii) if there are singularities between the paths, the two values differ by $2\pi i$ times the sum of the residues at those singularities.

The preceding statements form the **theorem of monodromy** that restates, for a monogenic function, the properties of the analytic functions (Chapter 23) that are its components. These properties extend from analytic to monogenic functions, and can be proved for several component functions as for two component functions: *let* $\mathrm{f}_1, \mathrm{f}_2$ *two functions analytic in the regions* $\mathrm{D}_1, \mathrm{D}_2$ *that overlap in subregion* L, *either a domain* D *or a continuous arc* A, *where they coincide* $\mathrm{f}_1(\mathrm{z}) = \mathrm{f}_2(\mathrm{z})$ *for* $\zeta \in \mathrm{L}$. *Then there exists a monogenic function* f(z) *that coincides with* $\mathrm{f}_1\,(\mathrm{f}_2)$ *in* $\mathrm{D}_1\,(\mathrm{D}_2)$ *and is analytic in the union* $\mathrm{D}_1 \cap \mathrm{D}_2$:

$$\forall_{z_1 \in D_1,\, z_2 \in D_2} : \quad f(z_1) = f_1(z_1), \quad f(z_2) = f_2(z_2). \tag{31.3}$$

The theorem extends to any number of regions and component functions (31.1a,b); it is illustrated in Figure 31.2 in the case of two regions overlapping along a curve L. For a point $z_1\,(z_2)$ within $D_1\,(D_2)$ Cauchy's second theorem (15.8) implies:

$$z_1 \in D_1 - \partial D_1 : \quad (2\pi i)^{-1} \int_{\partial D_1, \partial D_2} (\zeta - z_1)^{-1} f(\zeta)\, d\zeta = f(z_1),\ 0, \tag{31.4a}$$

$$z_2 \in D_1 - \partial D_2 : \quad (2\pi i)^{-1} \int_{\partial D_1, \partial D_2} (\zeta - z_2)^{-1} f(\zeta)\, d\zeta = 0, f(z_2). \tag{31.4b}$$

Adding (31.4a,b) and noting that $\partial D_1 + \partial D_2 \equiv \partial D$, because the overlapping arc A is described twice in opposite directions by the same function, leads to

$$z \in D - \partial D : \quad f(z) = (2\pi i)^{-1} \int_{\partial D_1 + \partial D_2 \equiv \partial D} (\zeta - z)^{-1} f(\zeta)\, d\zeta; \tag{31.5}$$

this is Cauchy's theorem for $f(z)$ in $D \equiv D_1 \cup D_2$, and proves that the monogenic function is analytic over the union of the two domains. QED.

Two examples of analytic continuation over a set of three (four) regions are given in Example 40.1.

31.1.2 Impossibility of Analytic Continuation

The preceding theorem shows that (Figure 31.2b) if a function is analytic in a region D_1, whose boundary ∂D_1 has a finite arc without singularities, then analytic continuation into

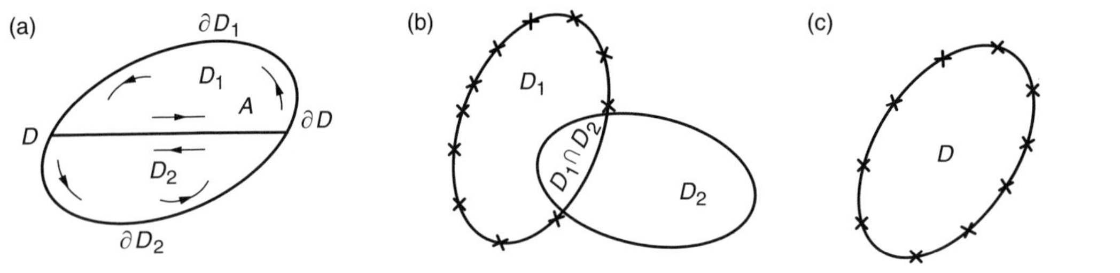

FIGURE 31.2
For the analytic continuation of two functions (f_1, f_2) between two regions (D_1, D_2) they must coincide in the (Figure 31.1a) overlapping region. The overlapping region need not be a subdomain like in Figure 31.1a–c, and could be only an arc of the boundary as in Figure 31.2a. Thus a function analytic in a domain D can be continued analytically if there is a finite arc on the boundary without singularities (b) for overlap with coincidence. On the other hand analytic continuation is impossible for a lacunary function (c) which is analytic a domain D whose boundary is densely filled with singularities, that is, there is no finite arc without singularities. Two examples of lacunary functions were given before (Figure 30.2): (i) the Lagrange–Burmann series (30.29a–d) is analytic inside (outside) the unit circle and represents the function $z(1/z)$; (ii) it is singular everywhere on the boundary unit circle $|z| = 1$ except the points $z = \pm 1$; (iii) thus analytic continuation is impossible either inward or outward across the boundary densely covered with singularities.

another region D_2 is possible, if a function f_2 analytic in D_2 and coinciding with f_2 in $D_1 \cap D_2$, can found. The impossibility of analytic continuation was found earlier, in Example 30.6 of a series (30.29a–d) converging to different function $z\,(1/z)$ inside ($|z| < 1$) [outside $|z| > 1$] the unit circle; on the unit circle $|z| = 1$ the two series coincide only at the points $z = \pm 1$ where the functions coincide $z = 1/z$. At every other point $|z| = 1 \neq \pm z$ of the boundary of convergence $|z| = 1$ the series do not converge; thus the region of overlap is just two points $z = \pm 1$ in Figure 30.2, and not an arc, so analytic continuation is not possible. This just means that $f(z) = z$ and $g(z) = 1/z$ are different functions, not two representations of the same monogenic function. Thus *the analytic continuation is impossible for a function specified by a series that diverges at the whole of its boundary of convergence, or converges only at a set of points not containing a continuous arc.* A class of functions for that analytic continuation is impossible is **lacunary functions**: *analytic continuation is impossible (Figure 31.2c) for a lacunary function, whose boundary is densely covered with singularities,* that is, *every arc of the boundary, however, small, contains singularities of the function.* As evidence that lacunary functions exist, and another example of impossible analytic continuation, consider the **Osgood function** (1929) defined by the series:

$$|z| < 1: \quad f(z) \equiv \sum_{n=0}^{\infty} z^{n!} = 1 + z + z^2 + z^6 + z^{24} + \cdots. \tag{31.6}$$

The ratio of succeeding terms (31.7b) of the series (31.7a) is (31.7c):

$$f(z) = \sum_{n=0}^{\infty} f_n(z), \quad f_n(z) = z^{n!}, \quad \frac{f_n(z)}{f_{n-1}(z)} = z^n; \tag{31.7a–c}$$

this shows that the series is absolutely convergent (29.31b) within the unit disc $|z| < 1$, and totally convergent (29.31c) in a closed subdisk $|z| \le 1 - \varepsilon$, with $0 < \varepsilon < 1$; since in the latter region it is an uniformly convergent series of analytic functions $z^{n!}$, it specifies an analytic function $f(z)$. The series (31.6) diverges (29.31a) for $|z| > 1$ and the function is not analytic on the boundary of convergence, viz. every arc of the unit circle has singularities of $f(z)$. To prove this, consider the complex numbers (31.8a) with unit modulus (31.8b) and argument (31.8c) a rational $p/q \in |Z$ multiple of 2π:

$$z_{p,q} = \exp\left(\frac{i2\pi p}{q}\right): \quad |z_{p,q}| = 1, \quad \arg(z_{p,q}) = \frac{2\pi p}{q}; \tag{31.8a–c}$$

they form a set (31.9a) dense on $|z| = 1$:

$$p, q \in |N, \frac{p}{q} \in |L : (z_{p,q})^{(q+n)!} = \exp\{i2\pi p(q+n)\cdots(q+1)(q-1)\cdots 1\} = 1, \tag{31.9a,b}$$

and lead to a general term of the series (31.6) equal to unity (31.9b), beyond the order q in (31.10a):

$$n \ge q: \quad f(z_{p,q}) = \sum_{n=0}^{q-1} (z_{p,q})^{n!} + \sum_{n=q}^{\infty} 1 = \infty; \tag{31.10a,b}$$

thus the series (31.6) diverges (31.10b) and the function is singular at all points (31.9a); since these form a set dense set on $|z| = 1$, the Osgood function (31.6) is lacunary and has no analytic continuation beyond the unit circle. QED.

31.2 Conjugate Property and Reflection Principle (Riemann, 1863; Schwartz, 1890)

Two particular instances of impossible analytic continuation (Example 20.6 and the Osgood function (31.6)] where given, before proceeding to two general methods that show that (i) analytic continuation can be performed via a the reflection (Section 31.2) principle (Schwartz) on the real axis (Figure 31.4); (ii) nonanalytic extension (Section 31.3) can be made (Plemelj, 1908) across an arc (Figure 31.5a,b). The method of analytic continuation (i) by reflection on the real axis (Subsection 31.2.2) uses the properties (Figure 31.3) of complex conjugate functions (Subsection 31.2.1).

31.2.1 Incompatible Pairs of Holomorphic Functions

It has been shown that analytic continuation is possible between two regions overlapping only in a contour, for example, a segment of the real axis. Taking the conjugate z^* of a complex number z is equivalent (Figure 31.3) to performing a reflection on the real axis (31.11a,b), and the range (31.11c) has for reflection (31.11d):

$$z = x + iy, z^* = x - iy: \; f(z) = \Phi(x,y) + i\Psi(x,y), f^*(z^*) = \Phi(x,-y) - i\Psi(x,-y). \tag{31.11a–d}$$

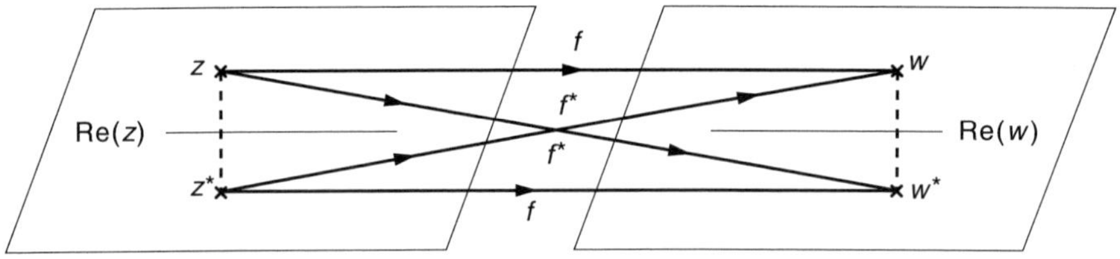

FIGURE 31.3
A complex function $w = f(z)$ is a mapping from z-plane to the w-plane. The reflection on the real axis in the z-plane (w-plane) corresponds to the conjugate $z \to z^*(f \to f^*)$. This leads to two pairs of mappings: (i) the basic pair consisting of the original $w = f(z)$ and double conjugate $w^* = f(z^*)$ mappings; (ii) the alternative pair including one conjugate, viz. either in the z-plane $w = f(z^*)$ or in the w-plane $w = f^*(z)$. It can be shown that if the first pair (i) is analytic, the alternate pair (ii) is not; and vice-versa. The reason is that a reflection $z \to z^*$ or $f \to f^*$ is not an analytic mapping (see Diagram 33.1), but two reflections $z \to z^*$ and $f \to f^*$ lead to an analytic mapping of symmetrical points in both planes.

If the function $w = f(z)$ is holomorphic, the function $f^*(z^*)$ in (31.11c,d), obtained by reflection on both the z-plane (31.11a,b) [w-plane (31.11c,d)], is also holomorphic:

$$\forall_{z\in D}: \quad f(z) \in \mathcal{D}(D) \Leftrightarrow f^*(z^*) \in \mathcal{D}(D^*). \tag{31.12}$$

The proof (Sections 11.2 and 11.3) is that: (i) the Cauchy–Riemann conditions (11.10a,b) are the same for both functions (31.11c,d):

$$f(x) \equiv \Phi(x,y) + i\Psi(x,y): \quad \frac{\partial\Phi}{\partial x} = \frac{\partial\Psi}{\partial y}, \frac{\partial\Phi}{\partial y} = -\frac{\partial\Psi}{\partial x}. \tag{31.13a,b}$$

$$f^*(z^*) \equiv \Phi(x,-y) - i\Psi(x,-y): \quad \frac{\partial\Phi}{\partial x} = \frac{\partial(-\Psi)}{\partial(-y)}, \frac{\partial\Phi}{\partial(-y)} = -\frac{\partial(-\Psi)}{\partial x}; \tag{31.14a,b}$$

(ii) if one pair has continuous first-order derivatives:

$$\{\Phi(x,y), \Psi(x,y)\} \in C^1 \Leftrightarrow \{\Phi(x,-y), \mp\Psi(x,-y)\} \in C^1. \tag{31.15}$$

so has the other pair. QED.

The function $z^* = x - iy$ is not holomorphic, because $\Phi = x, \Psi = -y$ does not satisfy the Cauchy–Riemann conditions, for example, $\partial\Phi/\partial x = 1 \neq -1 = \partial\Psi/\partial y$. Hence $f(z^*)$ and $f^*(z)$ are not holomorphic if (31.12) are, as could be checked from relations like (31.13a,b) not being satisfied. This can be proved as follows: (i) the necessary (Section 21.2) and sufficient (Section 21.3) conditions for $f = \Phi + i\Psi$ to be holomorphic differ in the latter's requirement Φ, Ψ to have continuous first order derivates, and coincide in demanding that the Cauchy–Riemann conditions (31.13a,b) be satisfied; (ii) since the requirement of continuous first-order derivates goes over (31.15) between all functions $\mathrm{Re}, \mathrm{Im}\{f(z), f^*(z), f(z^*), f(z^*)\}$, only the Cauchy–Riemann conditions have to be considered; (iii) for $f^*(z)\,[f(z^*)]$ the Cauchy–Riemann conditions are (31.16a,b) [(31.17a,b)]:

$$f^*(z) \equiv \Phi(x,y) - i\Psi(x,y): \quad \frac{\partial\Phi}{\partial x} \neq \frac{\partial(-\Psi)}{\partial y}, \frac{\partial\Phi}{\partial y} \neq -\frac{\partial(-\Psi)}{\partial x}; \tag{31.16a,b}$$

$$f(z^*) \equiv \Phi(x,-y) + i\Psi(x,-y): \quad \frac{\partial\Phi}{\partial x} \neq \frac{\partial\Psi}{\partial(-y)}, \frac{\partial\Phi}{\partial(-y)} \neq -\frac{\partial\Psi}{\partial x}. \tag{31.17a,b}$$

The conditions (31.13a,b) coincide with (11.10a,b) in Section 11.2, and are consistent with (31.14a,b) $\equiv$ (31.13a,b), showing that $f^*(z^*)$ is analytic; the conditions (31.16a,b) $\equiv$ (31.17a,b) are consistent with each other, and incompatible with (31.13a,b) $\equiv$ (31.14a,b), and show that if $f(z)$ is analytic then $f^*(z), f(z^*)$ cannot be analytic. However, if $f^*(z)$ is holomorphic then $f^*(z)$ is also holomorphic:

$$\forall_{z\in D}: \quad f^*(z) \in \mathcal{D}(D) \Leftrightarrow f(z^*) \in \mathcal{D}(D^*). \tag{31.18}$$

The two statements (31.12) and (31.18) are equivalent, because

$$g(z) \equiv f^*(z) \in \mathcal{D}(D) \Leftrightarrow f(z^*) = \{f^*(z^*)\}^* = g^*(z^*) \in \mathcal{D}(D^*). \tag{31.19}$$

Thus follows **the conjugate property**: *the four functions (31.12; 31.18) illustrated in Figure 31.3, form two incompatible holomorphic pairs that is, if one is holomorphic, so is the other of the same pair, but not those of the other pair.*

31.2.2 Complex Function Real on the Real Axis

The pair of functions (31.12) or (31.18) can perform the analytic continuation between the domain D and its image D^* on the real axis (Figure 31.4) if the two functions coincide for real $z = x$, for example, if they are both real $f^*(x) = f(x)$. This suggests the **Riemann (1863) Schwartz (1890) reflection principle**: *(i) if the function* f(z) *is analytic in a domain with a segment* L *of the real axis as part of the boundary, and (ii) the function is (31.20a) real* f = f* *on the real axis* z = x; *then the conjugate property (31.20b) $\equiv$ (31.20c)*:

$$f(x) = f^*(x): \quad f(z) = f^*(z^*), \quad f(z^*) = f^*(z), \tag{31.20a–c}$$

holds (Figure 31.3) and specifies the analytic continuation from the domain D into its reflection D^* on the real axis (Figure 31.4). To prove these statements consider a function $f(z)$ analytic in the domain $z \in D$, and real for real $z \equiv x$; its Taylor series (23.32b) about a point on the real axis $z \in x$ has real coefficients:

$$f(z) = \sum_{n=0}^{\infty} a_n (z-x)^n, \quad a_n \equiv \frac{f^{(n)}(x)}{n!} \in |R; \tag{31.21a,b}$$

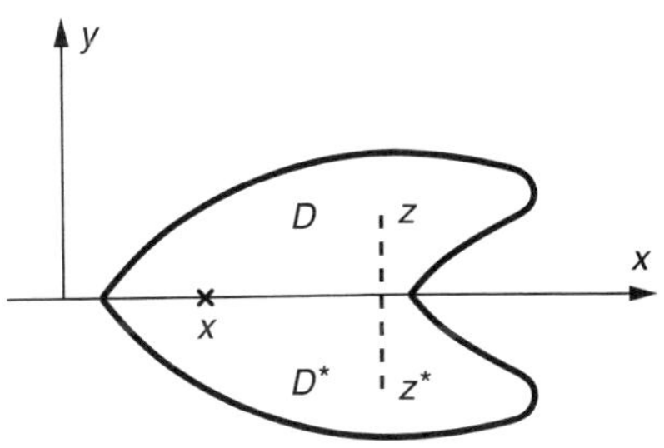

FIGURE 31.4
A function analytic in a domain D on the upper complex z-half-plane and real on the real axis can be continued analytically to the image domain D^* on the lower complex z-half-plane, using the same Taylor series with real coefficients around a point on the real axis. This "Schwartz reflection principle" passes from the function $w = f(z)$ to $w = f^*(z^*)$; this is possible because if one of them is analytic the other is also analytic (Figure 31.3). It would not be possible with $w = f^*(z)$ or $w = f(z^*)$ because these are not analytic at the same as time $w = f(z)$.

it follows that:

$$a_n^* = a_n: \quad f^*(z^*) = \left\{ \sum_{n=0}^{\infty} a_n (z^* - x)^n \right\}^* = \sum_{n=0}^{\infty} a_n (z - x^*) = \sum_{n=0}^{\infty} a_n (z - x)^n = f(z); \tag{31.21c}$$

thus the reflection principle holds in the form (31.20b), and because of

$$f(z^*) = \{f^*(z^*)\}^* = \{f(z)\}^* = f^*(z), \tag{31.21d}$$

it also holds (31.21d) in the form (31.20c). QED.

The reflection principle when applied to a holomorphic function, $w = f(z)$, implies that reflection in the z-plane is equivalent to reflection in the w-plane (Figure 31.3); thus the pairs of functions (31.12) or (31.18) are not only simultaneously holomorphic, but also provide the analytic continuation of each other. The circle theorem (24.47) uses a pair of functions like (31.12). The conjugate property holds for the functions:

$$f^*(z) = f(z^*): \quad f(z) = z, \ \mathrm{e}^z, \sin z, \tanh z, \tag{31.22a}$$

but not for the functions:

$$f^*(z) \neq f(z^*): \quad f(z) = iz, \mathrm{e}^{iz}, \sin(iz), \tanh(iz), \tag{31.22b}$$

because although all are analytic functions, the former (31.22a) [latter (31.22b)] set is (is not) real for real z. As an example, the following two cases are checked directly:

$$f(z) \equiv \mathrm{e}^z: \ f^*(z) = (\mathrm{e}^z)^* = (\mathrm{e}^{x+iy})^* = (\mathrm{e}^x \mathrm{e}^{iy})^* = \mathrm{e}^x \mathrm{e}^{-iy} = \mathrm{e}^{x-iy} = \mathrm{e}^{z*} = f(z^*), \tag{31.23a}$$

$$\begin{aligned} f(z) \equiv iz: \quad f^*(z) &= (iz)^* = (ix - y)^* = -ix - y \\ &= -i(x - iy) = -iz^* = -f(z^*) \neq f(z^*) = iz^*, \end{aligned} \tag{31.23b}$$

leaving some of the others for Example 40.2. The preceding cases are instances of the following property: *a pair of real functions of two real variables can always be written (3.17a,b) as the complex function:*

$$\Phi(x, y) + i\Psi(x, y) = \Phi\left(\frac{z + z^*}{2}, \frac{z - z^*}{2i}\right) + i\Psi\left(\frac{z + z^*}{2}, \frac{z - z^*}{2i}\right) \equiv f(z, z^*); \tag{31.24}$$

if (31.24) does not depend on z^*, and is differentiable with regard to z, then it specifies a holomorphic function.

31.3 Analytic Extension with Jump across an Arc (Plemelj, 1908)

The Schwartz reflection principle provides analytic continuation across the real axis (31.20a–c), and it can be extended to an arc A. Next is presented a method of analytic extension with a jump across an arc. Given a closed contour or loop C, the **positive (negative)** direction is such that the area being bounded is on the left: (i) **outside** *is counterclockwise (clockwise), as in Section 23.5 and Figure 13.1d; (ii)* ***inside*** *is the reverse* that

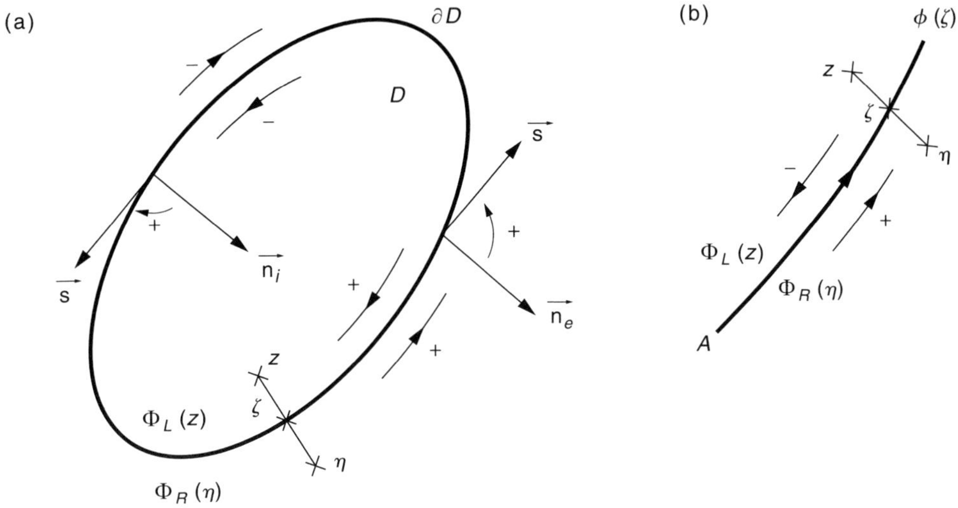

FIGURE 31.5
It is possible to perform nonanalytic continuation of a function across a closed boundary (a) [an open arc (b)], using: (i) two functions $\Phi^L(z)$ and $\Phi^R(\eta)$ analytic on each side; (ii) a specified jump $\Phi^R(\zeta)$ - $\Phi^L(\zeta)$ across the boundary loop (open arc). Describing the ∂D loop (a) on: (i) the outside the positive (negative) direction is counterclockwise (clockwise) relative to the interior $D - \partial$D; (ii) on the inside, the interior is the interior of the domain $|R^2 - D$, and the positive (negative) direction is reversed to clockwise (counterclockwise). For (b) the open arc the positive (negative) direction is like (i) opposite on both sides corresponding to a counterclockwise rotation. The unit normal toward the exterior $\vec{n}_e$ ($\vec{n}_i$) in the case (i) [(ii)] rotates toward the tangent $\vec{s}$ in the counterclockwise (clockwise) direction.

is, *clockwise (counterclockwise) in (Figure 31.5a). In both cases the normal vector pointing to the exterior of the region being bounded $\bar{n}_i$ ($\bar{n}_e$) always forms a right-handed dyad with the tangent vector, $\bar{s}$, in the sense it turns in the positive direction toward it. Assume that the integral (31.25a) along a loop ∂D(open arc A) in Figure 31.5a (31.5b) exists at least as (Sections 17.8 and 17.9) as a Cauchy principal value:*

$$\eta, \zeta \in \partial D, A\colon (2\pi i)^{-1} ⨍_{\partial D,A} [\phi(\zeta) - \phi(\eta)] (\zeta - \eta)^{-1}\, d\zeta \in C; \tag{31.25a}$$

$$\zeta \in \{A, \partial D\}; z \in \{|C - A, |C - \partial D\} : \Phi(z) \equiv (2\pi i)^{-1} \int_{A,\partial D} \phi(\zeta)(\zeta - z)^{-1}\, d\zeta; \tag{31.25b}$$

then the function defined by (31.25b) vanishes at infinity (31.26c) and is analytic on the whole complex plane (31.26a):

$$\eta \in \partial D, A;\ \ \Phi(z) \in \mathcal{D}(|C - A, |C - \partial D) : \quad \Phi_L(\eta) - \Phi_R(\eta) = \phi(\eta), \tag{31.26a,b}$$

$$\Phi(\infty) = 0, \quad \Phi_R(\eta) + \Phi_L(\eta) = (\pi i)^{-1} \int_{\partial D,A} \phi(\zeta)(\zeta - \eta)^{-1}\, d\zeta, \tag{31.26c,d}$$

excluding the loop (arc), where it has a jump (31.26b) [mean value (31.26d)] between right Φ_R and left Φ_L values. Also the functional equation (31.26b) has the unique solution (31.25b).

The proof is started by considering the function is represented by the integral (31.25b) that on account of the existence of (31.25a) can be written as:

$$2\pi i\Phi(z) = \int_{\partial D} \phi(\zeta)(\zeta - z)^{-1} \mathrm{d}\zeta = \int_{\partial D} [\phi(\zeta) - \phi(z)](\zeta - z)^{-1} \mathrm{d}\zeta + \phi(z) \int_{\partial D} (\zeta - z)^{-1} \mathrm{d}\zeta. \tag{31.27}$$

The function Φ is denoted Φ_L at (31.28b) an interior point (31.28a):

$$z \in D - \partial D: \quad \Phi_L(z) = \phi(z) + (2\pi i)^{-1} \int_{\partial D} [\phi(\zeta) - \phi(z)](\zeta - z)^{-1} \mathrm{d}\zeta, \tag{31.28a,b}$$

where the second integral on the r.h.s. of (31.27) was evaluated by Cauchy's theorem (15.8); for z also in the interior (31.29a) holds (31.29b):

$$z \in D - \partial D: \quad \int_{\partial D} (\zeta - z)^{-1} \mathrm{d}\zeta = 2\pi i; \quad \eta \notin D: \int_{\partial D} (\zeta - \eta)^{-1} \mathrm{d}\zeta = 0, \tag{31.29a–d}$$

for η an exterior point (31.29c) the integral with the same integrand vanishes (31.29d). The latter leads to (31.30b) for an exterior point (31.30a):

$$\eta \notin D: \quad \Phi_R(\eta) = (2\pi i)^{-1} \int_{\partial D} [\phi(\eta) - \phi(\eta)](\zeta - \eta)^{-1} \mathrm{d}\zeta. \tag{31.30a,b}$$

Letting the points z inside (31.28a,b) and η outside (31.30a,b) tend (Figure 31.5a) to the same point ζ on the boundary (31.31a):

$$\eta, z \to \zeta \in \partial D: \ \Phi_R(\eta) = (2\pi i)^{-1} \int_{\partial D} [\phi(\zeta) - \phi(\eta)](\zeta - \eta)^{-1} \mathrm{d}\zeta = \Phi_L(\eta) - \phi(\eta), \tag{31.31a,b}$$

proves (31.31b), implying: (i) directly (31.26b) form (31.31b); (ii) adding (31.28b) to (31.31b) yields:

$$\begin{aligned} \Phi_L(\eta) + \Phi_R(\eta) &= (\pi i)^{-1} \int_{\partial D} \phi(\zeta)(\zeta - \eta)^{-1} \mathrm{d}\zeta + \lim_{z \to \eta} \phi(z) \left\{ 1 - (2\pi i)^{-1} \int_{\partial D} (\zeta - z)^{-1} \mathrm{d}\zeta \right\} \\ &\quad - (2\pi i)^{-1} \phi(\eta) \int_{\partial D} (\zeta - \eta)^{-1} \mathrm{d}\zeta = (\pi i)^{-1} \int_{\partial D} \phi(\zeta)(\zeta - \eta)^{-1} \mathrm{d}\zeta, \end{aligned} \tag{31.32}$$

where (31.29a–d) were used to obtain (31.26d).

Theorem (31.26a–d) applies as well to an open arc A in Figure 31.5b, by closing it along a loop ∂D in Figure 31.5a, and specifying $\phi(\zeta) = 0$ on the extension of the arc $\zeta \in \partial D - A$. The solution (31.25b) of the functional equation (31.26b) vanishes (31.26c) at infinity as $z \to \infty$. To show that it is a unique solution, Φ, suppose there are two solutions Φ_1, Φ_2 and let $\Phi_0 = \Phi_1 - \Phi_2$ be the difference: (i) then by (31.26a) it is continuous on A; (ii) hence it is analytic in the whole complex plane, and must be a constant, by Liouville's theorem; (iii) on account of (31.26c) the constant must be zero $\Phi_0 = 0$ so $\Phi_1 = \Phi_2$ and the solution is unique. This completes the proof all properties (31.26a–d). QED.

The possibility (Section 31.2) or impossibility (Section 31.1) of analytic continuation of a single-valued function depends on its behavior at the boundary points that may be regular (irregular) points, the former (latter) being ordinary points or zeros (poles or essential singularities). The properties of an analytic function in the neighborhood of zeros and poles are considered next (Sections 31.4–31.9) [essential singularities in Chapter 29].

31.4 The Cauchy (1821) Fourth Integral Theorem

The first three integral theorems of Cauchy concern functions that are analytic (Sections 15.1 and 15.2) or have a simple (Section 15.3) or multiple (Section 15.4) pole, and lead to calculus of residues (Sections 15.7–15.9). The fourth theorem uses the calculus of residues to calculate the excess of number of zeros over the number of poles in a region. It is based on the remark that if a function $f(z)$ has a zero (pole) of order N at point (31.33a):

$$f(z) \sim O\left((z-a)^{\pm N}\right), \qquad \frac{\mathrm{d}\{\log[f(z)]\}}{\mathrm{d}z} = \frac{f'(z)}{f(z)} \sim \pm\frac{N}{(z-a)} + O(1), \tag{31.33a,b}$$

then its logarithmic derivate (31.33b) has a simple pole at the same point with residue $\pm N$. The theorem of residues (31.33b) then suggests the **Cauchy fourth integral theorem (1821)**: *let* $\mathrm{f(z)}$ *function be on analytic in a region* D, *except at* m *poles* $\mathrm{b_m}$ *of order* β_m *with* $\mathrm{m} = 1, \dots, \mathrm{M}$, *and with* N *zeros of order* α_n *at* $\mathrm{a_n}$ *with* $\mathrm{n} = 1, \dots, \mathrm{N}$ *in its interior; if* $\mathrm{g(z)}$ *is analytic in* D, *then:*

$$(2\pi i)^{-1} \int_{\partial D} g(z)\left[\frac{f'(z)}{f(z)}\right] \mathrm{d}z = \sum_{n=1}^{N} \alpha_n g(a_n) - \sum_{m=1}^{M} \beta_m g(b_m). \tag{31.34}$$

The theorem of residues (15.45) can be used to evaluate the integral (31.34) as $2\pi i$ times the sum of the residues $\alpha_n g(a_n)\,[-\beta_m g(b_m)]$ at the simple poles at a_n with $n = 1, \dots, N$ (b_m with $m = 1, \dots, M$) of the integrand; these correspond to zeros (poles) of order α_n (β_m) of the function $f(z)$, that is, the logarithmic derivate $f'(z)/f(z)$ has residue α_n ($-\beta_m$) in (31.33b) multiplied by the value $g(a_n)\,[g(b_m)]$ of the analytic function $g(z)$ at $z = a_n$ ($z = b_m$). All zeros and poles lie inside the region d, so only the first term on the r.h.s. of (15.45) is relevant. QED.

Choosing the auxiliary function to be unity (31.36a) it follows that *the excess of the sum Z of (31.35a) the orders α_n of the zeros $n = 1, \dots, N$ over the sum P of (31.35b) the orders β_m of the poles $m = 1, \dots, M$ of a function, $f(z)$, analytic in a region D, except for poles, is given by (31.35c):*

$$\begin{aligned} Z \equiv \sum_{n=1}^{N} \alpha_n, \quad P = \sum_{m=1}^{M} \beta_m: \quad Z - P &= (2\pi i)^{-1} \int_{\partial D} \frac{f'(z)}{f(z)}\,\mathrm{d}z \\ &= (2\pi i)^{-1} \Delta_{\partial D}\{\log[f(z)]\} = (2\pi)^{-1} \Delta_{\partial D}\{\arg[f(z)]\}, \end{aligned} \tag{31.35a–c}$$

where $\Delta_{\partial D}$ denotes the variation along the loop ∂D taken once in the positive, that is, counterclockwise, direction. The result (31.35c) follows from (31.34) with (31.36a) noting that (31.36b):

$$g(z) = 1: \quad Z - P = \int_{\partial D} \frac{f'(z)}{f(z)}\,\mathrm{d}z = \int_{\partial D} \{\log [f(z)]\}'\,\mathrm{d}z = \Delta_{\partial D} \{\log [f(z)]\}$$
$$= \Delta_{\partial D} \{\log |f(z)|\} + i\Delta_{\partial D} \{\arg [f(z)]\} = i\Delta_{\partial D} \{\arg [f(z)]\}, \quad \Delta_{\partial D} \{\log |f(z)|\} = 0, \qquad \text{(31.36a–c)}$$

where (i) $\log |f(z)|$ returns to the same value after the loop (31.36c); (ii) $\arg (f(z))$ may not do so. For a function analytic in D, there are no poles $P = 0$, and (31.35c) specifies the number of zeros, that is, *the number* Z *of zeros of a function* f(z) *analytic in a region* D *is given by (31.35c) with* P = *0.* For example, the exponential $f(z) = \mathrm{e}^z$, is defined as the function that equals it derivative (31.37a) and hence is analytic everywhere. It follows from (31.37b):

$$(\mathrm{e}^z)' = \mathrm{e}^z: \quad Z = (2\pi i)^{-1} \int_{\partial D} \left[\frac{(\mathrm{e}^z)'}{(\mathrm{e}^z)}\right] \mathrm{d}z = (2\pi i)^{-1} \int_{\partial D} \mathrm{d}z = 0, \qquad \text{(31.37a,b)}$$

that the exponential has no zeros anywhere

$$z \in |C, \quad |z| < \infty: \quad \exp(z) \neq 0. \qquad \text{(31.38a,b)}$$

on the finite complex plane; also the point-at-infinity is an essential singularity (27.42a).

31.5 Number of Zeros and Poles of a Function

If the function has zeros (poles), their number $Z(P)$ will depend on the region D considered in (31.35c). Henceforth the number $Z(P)$ of zeros (poles) is counted according to their multiplicity (31.35a) [(31.35b)], that is, a zero (pole) of multiplicity α_n (β_m) is counted α_n (β_m) times. As an example, consider a function $f(z)$ analytic (31.39a) in disk of radius R, and with a zeros a_n with $n = 1, \ldots, Z$ in a subdisk (31.39b) of radius εR with $\varepsilon < 1/2$:

$$f(z) \in \mathcal{A}(|z| \leq R), \; |a_1| \leq \cdots \leq |a_N| \leq \varepsilon R: \; F(z) \equiv \frac{f(z)}{\prod_{n=1}^{Z} (1 - z/a_n)} \in \mathcal{A}(|z| < R); \qquad \text{(31.39a–c)}$$

the function (31.39c) is analytic for $|z| \leq R$; since (31.39a) is analytic it has as upper bound (31.40a,b), leading for (31.39c) to (31.40b):

$$|f(z)| \leq B(R): \quad |F(z)| \leq \frac{B(R)}{\prod_{n=1}^{Z} (1/\varepsilon - 1)}; \qquad \text{(31.40a,b)}$$

by the lemma of maximum (Section 25.1) its value at the center cannot be greater:

$$|f(0)| = |F(0)| \leq B(R) \left(\frac{1}{\varepsilon} - 1\right)^{-Z}; \qquad \text{(31.41)}$$

thus the number of zeros within $|z| \le \varepsilon R$ satisfies:

$$\varepsilon < \frac{1}{2}: \quad Z \le \frac{\log\{B(R)/|f(0)|\}}{\log(1/\varepsilon - 1)}. \tag{31.42a,b}$$

Replacing z by $z - a$, it follows that *if the function $f(z)$ is analytic in (31.39a) the circle $|z - a| \le R$, with (31.40a) maximum modulus $B(R)$, the number Z of zeros in the subcircle $|z - a| \le \varepsilon R$ with $\varepsilon < 1/2$, cannot exceed (31.42a,b).*

For example, the function $f(z) = \sin z$ has zeros at $a_n = n\pi$, so that a circle $|z| \le N\pi + \varepsilon$ with $0 < \varepsilon < \pi$ contains $2N+1$ zeros. To check this from (31.35c), note that the logarithmic derivate of the sine is the cotangent:

$$f(z) = \sin z, \quad \frac{f'(z)}{f(z)} = \frac{\cos z}{\sin z} = \cot z; \tag{31.43a,b}$$

it has poles at $a_n = n\pi$ with residue $+1$ in (31.44b):

$$\begin{aligned} \zeta = z + n\pi: \quad \alpha_n &\equiv \lim_{z->n\pi} (z - n\pi)\cot z = \lim_{\zeta->0} \zeta \cot(\zeta + n\pi) \\ &= \lim_{\zeta->0} \zeta \cot \zeta = \lim_{\zeta->0} \frac{\zeta}{\sin\zeta} \cos\zeta = 1, \end{aligned} \tag{31.44a,b}$$

where the change of variable (31.44a) was made; an alternative calculation is (27.59a,b) $\equiv$ (31.44b). Thus the circle $|z| \le R$ of radius $N\pi < R < (N+1)\pi$ contains $2N+1$ poles of (31.43b) with residues unity (31.44b), that is, it contains $Z = 2N + 1$ zeros of $\sin z$. Whereas, the theorem (31.35c) gives exact the number of zeros, (31.42a,b) supplies only an upper bound. For example, the function $f(z) = \cos z$ has simple zeros at $a_n = (n + 1/2)\pi$, that is, it has $2N$ zeros within the circle of radius, $|z| \le N\pi$. Choosing $\varepsilon = 1/3$, a bound for the modulus of $\cos z$ within a circle of tripe radius is:

$$|z| \le \frac{N\pi}{\varepsilon} = 3N\pi: \quad |\cos z| \le \frac{(|e^z| + |e^{-z}|)}{2} \le |e^z| \le e^{3N\pi}. \tag{31.45}$$

Applying the formula (31.42b) with $a = 0, \cos 0 = 1, \varepsilon = 1/3$, yields:

$$Z \le \frac{\log\left(e^{3N\pi}\right)}{\log 2} = \frac{3N\pi}{\log 2} = 13.597\, N, \tag{31.46}$$

that is a somewhat gross overestimate of $Z = 2N$.

31.6 Theorem of Rouché (1858) and Fundamental Theorem of Algebra

The comparison of the number of zeros of two functions (Subsection 31.6.1), for example, a polynomial and its leading power, can be used to prove that the number of roots of a polynomial equals its degree (Subsection 31.6.2); this leads (Subsection 31.6.3) to the factorization of polynomials and roots of rational functions.

31.6.1 Comparison of the Number of Zeros of Two Functions

It is possible to use inequalities to determine the exact number of zeros or poles of a function, as shown by **Rouche Theorem (1858):** *two functions* f(z), g(z) *analytic except for poles* *in a region D, have the same excess of the number of zeros over the number of poles in D, provided one of the following conditions is met: (a) the variation of their arguments along a loop of the boundary ∂D differs by less than 2π:*

$$0 \le \Delta_{\partial D} \{\arg [f(z)]\} - \Delta_{\partial D} \{\arg [g(z)]\} < 2\pi; \tag{31.47}$$

(b) the ***relative error*** *(31.48b) ≡ (31.48c) is less than unity in modulus (31.48a):*

$$|k(z)| < 1: \quad k(z) \equiv \frac{f(z) - g(z)}{g(z)}, \quad f(z) = g(z)\{1 + k(z)\}. \tag{31.48a–c}$$

The statement (a) or (31.47) of the theorem follows noting that the argument of a function varies by multiples of 2π along a loop:

$$\Delta_{\partial D} \log \{f(z)\} - \Delta_{\partial D} \log \{g(z)\} = \Delta_{\partial D} \log \left\{ \frac{f(z)}{g(z)} \right\} = n 2\pi; \tag{31.49}$$

the only integer n in (31.49) consistent with (31.42) is $n = 0$, so that the value of (31.35c) is the same for the two functions $f(z)$, $g(z)$; the statement (b) relies on the property (31.48a) that implies all four relations (31.50a–d):

$$|\, k| < 1, -1 < Re\,(k) < +1, \quad 0 < 1 + Re\,(k)\,, \quad 0 \le \Delta_{\partial D}\{\arg\,(1 + k)\} < 2\pi; \tag{31.50a–d}$$

substituting (31.50d) into (31.48c):

$$\Delta_{\partial D}\{\arg\,[f\,(z)]\} - \Delta_{\partial D}\{\arg\,[g\,(z)]\} = \Delta_{\partial D}\,\{\arg[1 + k(z)]\} < 2\pi, \tag{31.51}$$

leads back to (31.47), that is, the statement (a) that $f(z)$, $g(z)$ have the same excess of the number of zeros over the number of the poles. QED.

31.6.2 Roots and Factorization of a Polynomial

As an example compare a polynomial (31.52a) of degree N with its leading term (31.52b):

$$f(z) \equiv P_N(z) = \sum_{n=0}^{N} a_n z^n, \quad g(z) = a_N z^N. \tag{31.52a,b}$$

The polynomial will be dominated by its leading term provided that the relative error:

$$k(z) \equiv \frac{P_N(z)}{a_N z^N} - 1 = \sum_{n=0}^{N-1} \frac{a_n}{a_N} z^{n-N}, \tag{31.53}$$

has modulus less than unity:

$$|k(z)| \le \sum_{n=0}^{N-1} \left| \frac{a_n}{a_N} \right| |z|^{n-N} < \frac{1}{N} \sum_{n=0}^{N-1} \left\{ \frac{|z|}{R} \right\}^{n-N} < 1; \tag{31.54a}$$

this is always the case for sufficiently large z:

$$n = 0, \ldots, N-1: \quad |z| > R \equiv \max_{n=0,\ldots,N-1} \left| \frac{a_n N}{a_N} \right|^{1/(N-n)}. \tag{31.54b}$$

Hence *outside the circle of radius* R *given by (31.54b), the polynomial (31.52a) is dominated by the term of highest degree (31.52b).* As both $f(z)$, $g(z)$ are analytic, Rouché's theorem (31.48a–c), implies that a polynomial has the same number of zeros as its term of highest degree; the number of zeros of (31.52b) is (31.55) given by (31.35c):

$$Z = (2\pi i)^{-1} \int^{(0+)} \frac{\left(a_N z^N\right)'}{\left(a_N z^N\right)} \mathrm{d}z = (2\pi i)^{-1} \int^{(0+)} \frac{N}{z} \mathrm{d}z = N, \tag{31.55}$$

and thus coincides with the degree of the polynomial.

31.6.3 Roots of a Rational Equation

The preceding results prove the **fundamental theorem of algebra**: *a polynomial of degree N has N roots, and can be* **factorized**:

$$a_N \neq 0: \qquad P_N(z) \equiv \sum_{n=0}^{N} a_n z^n = a_N \prod_{n=1}^{N} (z - z_n), \tag{31.56a,b}$$

where z_n are the roots; if there are L roots z_ℓ of **multiplicity** *α_ℓ the factorization becomes (31.57b):*

$$\sum_{\ell=1}^{L} \alpha_\ell = N, \quad P_N(z) = a_N \prod_{\ell=1}^{L} (z - z_\ell)^{\alpha_\ell}, \tag{31.57a,b}$$

where L is the number of distinct roots (31.57a). A extension is the **theorem of roots of a rational equation**: *consider the roots of the irreducible rational equation (31.58b), where c is a constant and $f_{M,N}(z)$ is the ratio of two polynomials (31.58a,c) and (31.56a,b) without common roots*:

$$b_M \neq 0: \quad c = f_{M,N}(z) \equiv \frac{Q_M(z)}{P_N(z)}, \quad Q_M(z) = \sum_{m=0}^{M} b_m z^m; \tag{31.58a–c}$$

then for each value of c the number of roots is (31.59a) the largest of M,N, because: (i) the rational equation (31.58b) is equivalent to (31.59b):

$$J = \sup(M, N): \quad 0 = Q_M(z) - c\, P_N(z) = A \prod_{j=1}^{J} (z - z_j); \tag{31.59a,b}$$

(ii) the coefficient A given by

$$A = \begin{cases} b_M & \text{if } M > N, \quad (31.60a) \\ b_N - a_N c & \text{if } M = N, \quad (31.60b) \\ -a_N c & \text{if } N > M, \quad (31.60c) \end{cases}$$

depending on degrees $N(M)$ of the polynomials (31.56a,b) [(31.58a,c)]; (iii) thus (31.59b) is a polynomial of degree (31.59a), and this specifies the number of roots. QED.

31.7 Legendre's Theorem and Roots of Polynomials

Some problems require the precise determination of the roots of a polynomial. For other problems, like stability (Chapter 2), it may suffice to locate the roots on the complex plane, for example, real, imaginary, complex conjugate pair, positive real part, etc. The location of roots of a polynomial in the complex plane (Subsection 31.7.2) relates to the coefficients through a number of "search theorems" (Subsection 31.7.1).

31.7.1 Relation between Roots and Coefficients of a Polynomial

If a polynomial (31.61b) has real (31.61a) coefficients a_n and z_0 is a root, then its conjugate z_0^* is also a root (31.61c):

$$a_n = a_n^* : \quad \sum_{n=0}^{N} a_n \, z_0^n = 0, \quad \sum_{N=0}^{N} a_n^* \, z_0^{*n} = \left\{ \sum_{n=0}^{N} a_n z_0^n \right\}^* = 0; \tag{31.61a–c}$$

hence *the roots of a polynomial with real coefficients are either real or complex conjugate pairs.* It follows that a polynomial of even $N = 2p$ (odd $N = 2p+1$) degree can have $2q$ pairs of complex conjugate roots with $0 \leq q \leq p$, and the remaining $2\,(p-q)$ $[2\,(p-q)+1]$ roots must be real; in particular, *a polynomial of odd degree with real coefficients must have at least one real root.* Also, if a polynomial is real on a line, and has a root it changes sign if the multiplicity is odd, and keeps the sign if the multiplicity is even. Thus *if a polynomial is real along a line (e.g., a polynomial with real coefficients along the real axis), and changes sign* p *times, it has at least* p *roots on the line* (p *real roots*).

If the polynomial (31.52a) with real coefficients (31.61a) has (31.61c) real roots (real roots with fixed sign), then the polynomial:

$$\varepsilon P_N(z) + z \frac{dP_N}{dz} = \sum_{n=0}^{N} (n + \varepsilon)\, a_n z^n \tag{31.62}$$

with $\varepsilon > 0$ also has real coefficients and real roots (with fixed sign). Inserting iteratively more coefficients, leads to a real polynomial of degree M in n, with negative roots $-\varepsilon_m$:

$$\varepsilon_1, \ldots, \varepsilon_m > 0 : \quad S_M(n) \equiv \prod_{m=1}^{M} (n + \varepsilon_m). \tag{31.63}$$

This proves **Legendre's theorem:**

$$Q_N(z) = \prod_{m=1}^{M} a_n z^n S_M(n), \tag{31.64}$$

if $S_M(n)$ is (31.63) a polynomial with real coefficients and negative roots, and $P_N(z)$ is (31.52a) a polynomial with real coefficients (31.61a) and real roots (real roots with fixed sign), then (31.64) also has real roots (real roots with fixed sign).

31.7.2 Location of the Roots of a Polynomial in the Complex Plane

As an example of the use of the preceding theorems to locate the roots of a polynomial, it will be proven that the following polynomial has two roots each in the second and third quadrants:

$$P_4(z) = z^4 + z^3 + 4z^2 + 2z + 3 = 0. \tag{31.65}$$

A polynomial of the fourth degree with real coefficients can have: (i) four real roots; (ii) two real roots and a complex conjugate pair; (iii) two complex conjugate pairs. The statement is that to the polynomial (31.65) applies the case (iii), with the two pairs of complex conjugate roots having negative real parts. The roots may coincide or not. The equation has (i) no positive real root, because all terms are positive for $z \equiv x > 0$. Concerning $z \equiv x < 0$ negative real roots:

$$z = -x, \quad x > 0: \quad x^4 - x^3 + 4x^2 - 2x + 3 = 0, \tag{31.66a–c}$$

there are (ii) none for $x > 1$ because the first two and last three terms are positive, and none for $0 < x \leq 1$ because the first three and last two terms are positive. Thus (31.65) has no real roots; (iii) it has no pure imaginary roots either:

$$z = iy: \quad 0 = y^4 - iy^3 - 4y^2 + 2iy + 3 = \left(y^2 - 3\right)\left(y^2 - 1\right) - iy\left(y^2 - 2\right), \tag{31.67a,b}$$

because the real (imaginary) parts of (31.67b) do not vanish together, viz. none of their roots $y = \pm\sqrt{3}, \pm 1$ $\left(y = 0, \pm\sqrt{2}\right)$ coincide. Thus there are no roots on the axis, and the four roots must lie in the quadrants.

The argument of (31.65) varies along any quadrant by $+2\pi$:

$$\begin{aligned} z = re^{i\varphi}; \quad \alpha \leq \varphi < \alpha + \frac{\pi}{2}: \quad & \Delta \arg\{P_4(z)\} = \arg\left\{z^4\left[1 + O\left(z^{-1}\right)\right]\right\} \\ = \Delta \arg\left(r^4 e^{i4\varphi}\right) = 4\frac{\pi}{2} = 2\pi. & \end{aligned} \tag{31.68}$$

The argument of (31.65) does not change along the real axis, because (31.65) and (31.66c) are real positive for real $z \equiv x$. The argument along the positive imaginary axis can be calculated: (i) using (31.67b):

$$\arg\{P_4(iy)\} = \arg\tan[g(y)], \quad g(y) \equiv \frac{-y^3 + 2y}{y^4 - 4y^2 + 3}; \tag{31.69a,b}$$

(ii) the numerator (denominator) has roots $\left(0, \pm\sqrt{2}\right)$ $\left[\left(\pm 1, \pm\sqrt{3}\right)\right]$; (iii) the term in brackets goes through ∞ twice:

y	∞		$\sqrt{3}$		$\sqrt{2}$		1		0
$g(y)$	0	$-$	∞	$+$	0	$-$	∞	$+$	0

(31.69c)

(iv) each time arc tan jumps by $-\pi$, so $\arg\{P_4(iy)\}$ varies by -2π along the positive imaginary axis. Thus the total change is $\Delta_1 = 2\pi - 2\pi = 0$ on the first, and $\Delta_2 = 2\pi + 2\pi = 4\pi$ in the second quadrant (Figure 31.6). This implies (31.35c) that there are zero (two) roots in the first (second) quadrant; since the coefficients of (31.65) are real, the roots are complex conjugate pairs, and the other two roots must lie in the third quadrant. The stability (2.17a) of a linear system with constant coefficients (2.15) is specified by the roots of a characteristic polynomial (2.17b); a sufficient condition (Section 2.8) for stability is that the roots have negative real parts. If (31.65) is the characteristic polynomial, then stability is proved, without calculating its roots. For the linear system specified by the fourth-order differential equation with constant coefficients:

$$\mathrm{d}^4x/\mathrm{d}t^4 + i\,\mathrm{d}^3x/\mathrm{d}t^3 - 4\mathrm{d}^2x/\mathrm{d}t^2 - 2i\,\mathrm{d}x/\mathrm{d}t + 3x(t) = 0 \tag{31.70}$$

Similar methods can be used to locate roots of polynomials with some arbitrary coefficients (Example 40.3).

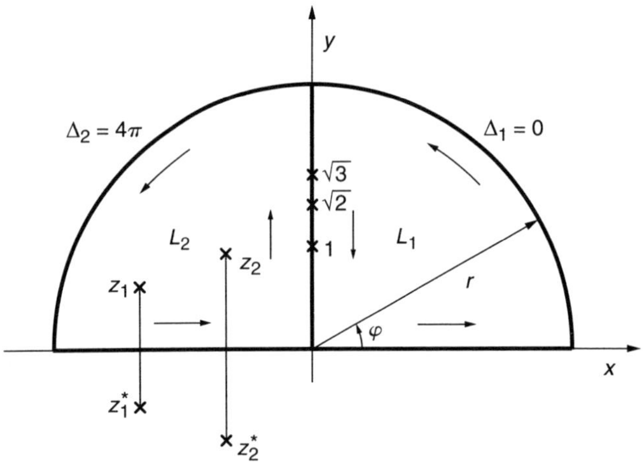

FIGURE 31.6
If a function $f(z)$ is analytic in a domain D except for poles its logarithmic derivative $(\mathrm{d}/\mathrm{d}z)\{\log f(z)\} = f'(z)/f(z)$ integrated along the boundary ∂D specifies $Z - P$ the excess of the number of zeros Z over the number of poles P. When summing the number of zeros Z (poles P) each is counted according to its multiplicity. In the case of a polynomial, which is an analytic function in the finite plane, hence without poles, the preceding method yields the number of zero Z inside the loop ∂D. For example the number of zeros in the upper complex z-half-plane can be found by integration along a half-circle $0 \leq \varphi \leq \pi$ with center at the origin and infinite radius $r \to \infty$. The number of zeros in the first (second) quadrant can be found by integrating over a quarter-circle $0 \leq \varphi \leq \pi/2$ $(\pi/2 \leq \varphi \leq \pi)$ with center at the origin and infinite radius $r \to \infty$. If the polynomial has real efficients its roots are complex conjugate pairs z_1, z_1^*, z_2, z_2^*, etc.; it follows that the number of zeros in the third (fourth) quadrant is the same as in the second (first) quadrant.

31.8 Rational Functions and Simple Fractions

An irreducible rational function is the ratio of two polynomials (Subsection 31.8.1) without common roots; the simple roots of the polynomial in the denominator are simple poles of the rational function, and lead to the decomposition in partial fractions (Subsection 31.8.2).

31.8.1 Irreducible and Simple Rational Functions

A **rational function of degrees** M, N is defined (Section 27.7) as the ratio of two polynomials Q_M, P_N of degrees respectively M, N:

$$N > M: \quad f_{M,N}(z) = \frac{Q_M(z)}{P_N(z)} = \frac{\sum_{m=1}^{M} b_m z^m}{\sum_{n=1}^{N} a_n z_n}; \tag{31.71a}$$

it is **irreducible**, that is there are no roots common to the two polynomials; any rational function may be rendered irreducible by dividing out factors common to the numerator and denominator. A rational function is **simple** if the degree of the denominator is larger than that of the numerator. for example, $N > M$ in (31.71a); if a rational function is not simple, then the numerator is of equal or higher degree than the denominator, and can be **divided** by the latter, with a polynomial of degree $M - N \geq 0$ as ratio, leaving a remainder that is a polynomial of degree $S < N$ lower than N:

$$M \geq N > S: \quad f_{M,N}(z) = P_{M-N}(z) + \frac{R_S(z)}{P_N(z)}; \tag{31.71b,c}$$

then the last term of (31.72) is a simple fraction. Thus *a rational function either is simple (31.71a), or can be split into the sum of a polynomial and a simple rational function (31.71b,c). The examples:*

$$\frac{z^3}{z^2 - (a+b)\,z + ab} = z + a + b + \frac{\left(a^2 + b^2 + ab\right) z - ab\,(a+b)}{(z-a)\,(z-b)}, \tag{31.72a}$$

$$\frac{z^3}{z^2 - 2\,a\,z + a^2} = z + 2a + a^2\,\frac{3z - 2a}{(z-a)^2}, \tag{31.72b}$$

illustrate the latter case of reduction of a nonsimple rational function to a simple one, plus a polynomial whose degree is the difference of the degree of the numerator and denominator of the initial nonsimple rational function.

The method of division of polynomials is illustrated in the case (31.72a): (i) the rational function (31.72a) is the ratio of a cubic by a quadratic polynomial, and tends to z as $z \to \infty$, so the term z is subtracted in:

$$\frac{z^3}{(z-a)\,(z-b)} - z = \frac{z^3 - (z-a)\,(z-b)\,z}{(z-a)\,(z-b)} = -\frac{a\,b\,z}{(z-a)\,(z-b)} + \frac{(a+b)\,z^2}{(z-a)\,(z-b)}; \tag{31.73a}$$

(ii) the first term on the r.h.s. of (31.73a) is the ratio of a linear by a quadratic polynomial, and hence is a simple fraction that is brought to the l.h.s. in:

$$\frac{z^3}{(z-a)\,(z-b)} - z + \frac{abz}{(z-a)\,(z-b)} = (a+b)\,\frac{z^2}{(z-a)\,(z-b)}; \tag{31.73b}$$

(iii) the r.h.s. of (31.73b) is not a simple fraction, because it is the ratio of two quadratic polynomials, and as $z \to \infty$ tends to the constant $a + b$ that is subtracted in:

$$\begin{aligned} \frac{z^3}{(z-a)\,(z-b)} - z + \frac{abz}{(z-a)\,(z-b)} - a - b &= (a+b)\left[\frac{z^2}{(z-a)\,(z-b)} - 1\right] \\ = (a+b)\,\frac{z^2 - (z-a)\,(z-b)}{(z-a)\,(z-b)} &= (a+b)\,\frac{(a+b)\,z - a\,b}{(z-a)\,(z-b)}; \end{aligned} \tag{31.73c}$$

(iv) the l.h.s. of (31.73c) is the ratio of a linear by a quadratic polynomial, hence a simple fraction; (v) this completes the division of the two polynomials:

$$\frac{z^3}{(z-a)\,(z-b)} = z + a + b + \frac{\left[(a+b)^2 - ab\right] z - ab\,(a+b)}{(z-a)\,(z-b)}; \tag{31.73d}$$

and proves (31.72a) $\equiv$ (31.73d). Setting $a = b$ in (31.72a) yields (31.72b). The preceding *method of division of two polynomials applies to any degree* N (M) *of the numerator (denominator) with* $\mathrm{N} \geq \mathrm{M}$*: (i) if* $\mathrm{S} \equiv \mathrm{N} - \mathrm{M} \geq 0$ *then the leading term is subtracted out* $(S+1)$ *times, leading to a polynomial of degree* $\mathrm{N} - \mathrm{M}$*, first on the r.h.s. of (31.71c); (ii) what remains is a simple fraction, second on the r.h.s. of (31.71c).*

31.8.2 Partial Fraction Decomposition Associated with Simple Poles

Consider the rational function (31.71a), in that the polynomial in the denominator has all roots distinct (31.56a,b), that is, simple, z_n: (i) since the fraction is irreducible, the factor $z - z_n$ does not appear in the numerator and the rational function has a simple pole at $z = z_n$, with residue:

$$A_n = \lim_{z \to z_n} (z - z_n) \frac{Q_M(z)}{P_N(z)} = \frac{Q_M(z_n)}{\left\{ a_N \prod_{\substack{k=1 \\ k \neq n}}^{N} (z_n - z_k) \right\}}; \tag{31.74}$$

(ii) because the fraction is simple, it vanishes at infinity:

$$N > M: \quad \lim_{z \to \infty} \frac{Q_M(z)}{P_N(z)} = \lim_{z -> \infty} O\left(z^{M-N}\right) = 0; \tag{31.75}$$

(iii) the function:

$$g(z) \equiv f_{M,N}(z) - \sum_{n=1}^{N} \frac{A_n}{z - z_n} = \text{const} = g(\infty) = 0, \tag{31.76}$$

has no singularities, since all poles of $f_{M,N}(z)$ have been subtracted out, and so, by Liouville's theorem (Section 27.5) it is a constant; (iv) the value of the constant, evaluated at infinity, is zero.

This proves the **decomposition and integration theorem**: *a simple irreducible rational function (31.71a), whose denominator has all roots z_n simple (31.56a,b), can be decomposed (31.77a) into the sum of N simple fractions of degree 1, with coefficients A_n in (31.77b):*

$$f_{M,N}(z) = \sum_{n=1}^{N} \frac{A_n}{z - z_n}, \qquad A_n \equiv \frac{Q_M(z_n)}{a_N \sum_{\substack{k=0 \\ k \neq n}}^{N} (z_n - z_k)}; \tag{31.77a,b}$$

$$\int f_{M,N}(z)\, \mathrm{d}z = \sum_{n=1}^{N} A_n \log(z - z_n) = \log\left\{ \prod_{n=1}^{N} (z - z_n)^{A_n} \right\}, \tag{31.77c}$$

its indefinite integral is specified by (31.77c). The coefficients (31.77b) are obtained suppressing from the denominator the vanishing factor, in agreement with the method II(III) of calculation of residues at simple poles (15.24b) [(19.41)]. The case (15.30c) ≡ (31.78a):

$$\left\{z^2 - (a+b)\,z + a\,b\right\}^{-1} = \left\{(z-a)(z-b)\right\}^{-1} = (a-b)^{-1}\left\{(z-a)^{-1} - (z-b)^{-1}\right\}, \tag{31.78a}$$

$$\int \left\{z^2 - (a+b)\,z + ab\right\}^{-1} \mathrm{d}z = (a-b)^{-1} \log\left(\frac{z-a}{z-b}\right), \tag{31.78b}$$

leads to the primitive or indefinite integral (31.78b). Two examples of partial fraction decompositions are given: (i) the simple rational function (31.78a) specifies the indefinite rational integral (31.78b); (ii) applying (31.77a) to the nonsimple rational fraction (31.72a) leads to the partial fraction decomposition (31.79b):

$$\frac{z^3}{z^2-(a+b)z+ab} = z+a+b+\frac{1}{a-b}\left(\frac{a^3}{z-a}-\frac{b^3}{z-b}\right), \tag{31.79a}$$

$$\int \frac{z^3}{z^2-(a+b)z+ab}\,dz = \frac{z^2}{2}+(a+b)z+(a-b)^{-1}\log\left\{\frac{(z-a)^{a^3}}{(z-b)^{b^3}}\right\}, \tag{31.79b}$$

and rational indefinite integral (31.79b). The latter case (31.79a,b) requires: (i) a division (31.72a) to separate the simple fraction: (ii) the residues at the poles of the latter specify the partial fraction decomposition:

$$\begin{aligned}\frac{z^3}{(z-a)(z-b)}-z-a-b &= \frac{\left(a^2+b^2+ab\right)z-ab(a+b)}{(z-a)(z-b)}\\ &= \frac{1}{z-a}\left[\frac{\left(a^2+b^2+ab\right)z-ab(a+b)}{(z-b)}\right]_{z=a}\\ &\quad+\frac{1}{z-b}\left[\frac{\left(a^2+b^2+ab\right)z-ab(a+b)}{z-a}\right]_{z=b}\\ &= \frac{1}{a-b}\left(\frac{a^3}{z-a}-\frac{b^3}{z-b}\right);\end{aligned} \tag{31.80}$$

(iii) the indefinite integration or primitive of (31.80) $\equiv$ (31.79a) is immediate (31.79b).

31.9 Decomposition into Partial Fractions and Rational Integrals

The partial fraction decomposition of a rational function can be extended from simple (Subsection 31.8.2) to multiple (Subsection 31.9.1) poles. It uses an extended residue rule (Subsection 31.9.2) and leads to rational integrals (Subsection 31.9.3) for poles of any order.

31.9.1 Partial Fraction Decomposition for Multiple Poles

In the decomposition theorem (31.77a–c) it was assumed that the polynomial in the denominator has simple roots (31.56a,b); in the general case (31.57b), for each root (31.57a) of multiplicity, α_ℓ, the irreducible rational function (31.71a) has a pole of order α_ℓ, with principal part:

$$f_{M,N}(z) = \sum_{k=1}^{\alpha_\ell} A_{\ell,-k}\,(z-z_\ell)^{-k} + O(1). \tag{31.81}$$

Subtracting from the function the principal parts at all poles, leads to a function without singularities, that is, a constant; the constant is zero, because all sums (31.81) vanish at infinity, together with the rational function that is simple (31.75). This specifies the decomposition into fractions:

$$f_{M,N}(z) = \sum_{\ell=1}^{L}\sum_{k=1}^{\alpha_\ell} A_{\ell,-k}\,(z-z_\ell)^{-k}; \tag{31.82}$$

the N coefficients $A_{\ell,-k}$, of that the $A_{\ell,-1}$ are the residues at the poles z_ℓ, have to be determined.

31.9.2 Calculation of Coefficients by the Extended Residue Rule

The method for determining all coefficients in (31.82) is an extension of one of the preceding methods [(15.33b) and (19.49)] of calculation of residues at a pole of order n, and may be called the extended residue rule, viz.: (i) multiply the rational fraction by the order of the pole:

$$(z - z_\ell)^{\alpha_\ell} f_{M,N}(z) = \sum_{p=1}^{\alpha_\ell} A_{\ell,-k} (z - z_\ell)^{\alpha_\ell - p} + O\left((z - z_\ell)^{\alpha_\ell}\right); \tag{31.83}$$

(ii) bring the coefficient $A_{\ell,-k}$ to the leading position, by derivation $(\alpha_\ell - k)$ times:

$$\frac{\mathrm{d}^{\alpha_\ell - k}\left\{(z - z_\ell)^{\alpha_\ell} f_{M,N}(z)\right\}}{\mathrm{d}z^{\alpha_\ell - k}} = \sum_{p=1}^{k} A_{\ell,-k} (\alpha_\ell - p) \cdots$$
$$(\alpha_\ell - p - 1) \cdots (k - p + 1) (z - z_\ell)^{k-p} + O\left((z - z_\ell)^k\right); \tag{31.84}$$

(iii) taking the limit $z \to z_\ell$ all terms vanish except the last $k = p$ that specifies $A_{\ell,-k}$:

$$k = 1, \ldots, \alpha_\ell: \qquad \lim_{z \to z_\ell} \frac{\mathrm{d}^{\alpha_\ell - k}\left\{(z - z_\ell)^{\alpha_\ell} f_{M,N}(z)\right\}}{\mathrm{d}z^{\alpha_\ell - k}} = (\alpha_\ell - k)! A_{\ell,-k}. \tag{31.85a,b}$$

*In the case of a simple root $\alpha_\ell = 1 = k$, then (31.85b) leads back to the residue (31.77a) $\equiv$ (31.74) at a simple pole with $A_n \equiv A_{n,-1}$; in the case of a multiple root $\alpha_\ell \geq 2$, then (31.85b) with $k = 1$ specifies the residue at a pole of order α_ℓ in (15.32b); (ii) for other values of k in (31.85a) the **extended residue rule** (31.85b), specifies as well as all the coefficients $(A_{\ell,-1}; A_{\ell,-2}; \ldots; A_{\ell,-\alpha_\ell})$ of the principal part of the rational function (31.82).*

31.9.3 The General Partial Fraction Decomposition and Integration

This proves the **extended partial fraction decomposition and integration theorem**: *the simple irreducible rational function (31.71a), in that the polynomial in the denominator (31.57b) has L roots z_ℓ of (31.57a) multiplicity α_ℓ, can be decomposed into a sum of partial fractions of degrees up to α_ℓ in (31.82), with coefficients specified by the extended residue rule (31.85a,b), leading to the rational integral:*

$$\int f_{M,N}(z) = \sum_{\ell=1}^{L} \left\{ A_{\ell,-1} \log(z - z_\ell) + \sum_{k=2}^{\alpha_\ell} \frac{A_{\ell,-k} (z - z_\ell)^{1-k}}{1-k} \right\}. \tag{31.86}$$

This reduces to the first term (31.77c) if $\alpha_\ell = 1$, with, $L = N$, $A_{\ell,-1} = A_\ell$. An example of rational functions with a double pole at $z = a$ is the decomposition of the simple rational function (31.87a):

$$\frac{z}{z^2 - 2az + a^2} = \frac{1}{z - a} + \frac{a}{(z - a)^2}, \tag{31.87a}$$

$$\int \frac{z}{z^2 - 2az + a^2} \, \mathrm{d}z = \log(z - a) - \frac{a}{z - a}, \tag{31.87b}$$

leads to the rational indefinite integral (31.87b).

Another nonsimple rational function (31.72b) has partial fraction decomposition (31.88a):

$$\frac{z^3}{z^2 - 2az + a^2} = z + 2a + \frac{3a^2}{z-a} + \frac{a^3}{(z-a)^2}, \tag{31.88a}$$

$$\int \frac{z^3}{z^2 - 2az + a^2}\,\mathrm{d}z = \frac{z^2}{2} + 2az + 3a^2 \log(z-a) - \frac{a^3}{z-a}, \tag{31.88b}$$

and corresponding rational indefinite integral (31.88b). The passage from (31.72b) to (31.88a) involves a partial fraction decomposition in the simple fraction:

$$\begin{aligned}\frac{z^3}{z^2 - 2az + a^2} - z - 2a &= a^2\frac{3z-2a}{(z-a)^2} = a^2\frac{3(z-a)+a}{(z-a)^2}\\ &= \frac{3a^2}{z-a} + \frac{a^2}{(z-a)^2} = \frac{A_{-1}}{z-a} + \frac{A_{-2}}{(z-a)^2},\end{aligned} \tag{31.89a}$$

in agreement with (31.88a) ≡ (31.89a). The coefficients could also have been calculated from the extended residue rule (31.85a,b):

$$A_{-1} = \lim_{z\to a}\frac{\mathrm{d}}{\mathrm{d}z}\left\{(z-a)^2 a^2\frac{3z-2a}{(z-a)^2}\right\} = a^2\lim_{z\to a}\frac{\mathrm{d}}{\mathrm{d}z}(3z-2a) = 3a^2, \tag{31.89b}$$

$$A_{-2} = \lim_{z\to a}\left\{(z-a)^2 a^2\frac{3z-2a}{(z-a)^2}\right\} = \lim_{z\to a} a^2(3z-2a) = a^3, \tag{31.89c}$$

in agreement with (31.89a) ≡ (31.89b,c). In the case (31.87a) the simple fraction has a double pole:

$$\frac{z}{(z-a)^2} = \frac{(z-a)+a}{(z-a)^2} = \frac{1}{z-a} + \frac{a}{(z-a)^2}, \tag{31.90}$$

and the algebraic decomposition (31.90) ≡ (31.87a) is equivalent to the extended residue rule. Six more cases of decomposition of simple and nonsimple rational functions, into partial fractions with simple or multiple poles are given in Example 40.4 together with the corresponding rational integrals.

Note 31.1. *Methods for Partial Fraction Decomposition of Rational Functions*: The method of partial fraction decomposition (Sections 31.8 and 31.9) relates to the calculus of residues, either in original form (Sections 15.6 and 15.7) or via l'Hôpital's rule (Sections 19.8 and 19.9), viz.: (i) the definition of the residue at a simple pole is (15.24b), and it can be calculated also by the rule (19.41) for any function that becomes (31.74) ≡ (31.77b) for an irreducible rational function; (ii) this is sufficient to perform the partial fraction decomposition (31.77a) of a rational function with simple poles, and to calculate the corresponding indefinite integral (31.77c); (iii) the residue at a multiple pole is defined by (15.32b), and can be calculated also by the rule (19.49) that becomes (31.85a,b) with $k = 1$; (iv) this not sufficient for the partial fraction decomposition of a rational function with multiple poles (31.82) that involves the coefficients (31.85a,b) for $k = 2, \ldots, \alpha_\ell$, not just the residue $k = 1$; (v) all the coefficients (31.85a,b), not just the residue A_{-1}, are also needed to evaluate the rational indefinite integral (31.86); (vi) if the rational function (31.71a) is simple, the preceding steps (i) to (v) are sufficient, for example, (31.78a,b; 31.87a,b); (vii) if the rational function (31.71b,c) is not simple then the preceding method (i–v) applies to its simple part viz. the second term on the r.h.s. of (31.72a,b), and what remains is a polynomial whose integration is straightforward, for example, in (31.79a,b; 31.88a,b).

It is convenient to have a standard procedure (i–vii) for partial fraction decomposition, not only for the calculation of indefinite integrals, but also because this is an algebraic manipulation of common occurrence, for example, to calculate the sources, forces and moments in potential fields (Chapter 28). The partial fraction decomposition of rational functions can be made by: (i) purely algebraic methods, multiplying out the denominators, an equating coefficients for example, (15.28–15.30) and (15.38–15.40); (ii) by using the properties of complex functions, including the calculus of residues (31.77–31.79) and its extensions (31.81–31.90). The latter approach applies both to (i) the finite partial fraction decomposition of rational functions, that is, a finite number of fractions (Sections 31.8 and 31.9); (ii) the representation of a meromorphic function (Subsection 27.9.3) that has an infinite number of poles, based on the sum of its principal parts at the poles leading to series of fractions and to infinite products. The latter (ii) unlike the former gives rise to convergence issues (Chapters 21 and 29).

Conclusion 31: A monogenic function (figure 31) consisting (a) of three functions f_1, f_2, f_3, analytic in the domains D_1, D_2, D_3, and coinciding in the overlapping regions $f_1 = f_2$ in $D_1 \cap D_2$ and $f_2 = f_3$ in $D_2 \cap D_3$. The analytic continuation along a loop L starting and ending at z, depends (b) on the singularities z_1, z_2, z_3 enclosed; the same applies when comparing the analytic continuation along (c) distinct paths L_1, L_2 with the same starting z_1 and end z_2 points. For the purpose of analytic continuation between two domains D_1, D_2, it is sufficient (figure 32) that they overlap (a) over an arc A of their boundaries ∂D_1, ∂D_2. Thus analytic continuation to a domain D_2 exterior to a domain D_1 is possible, if the boundary of the latter has (b) a finite arc without singularities; conversely, analytic continuation is impossible (c) for a lacunary function with singularities forming a dense set on the boundary of its domain. One form of analytic continuation is (figure 31.4) the reflection on the real axis, which corresponds to taking the conjugate; in the case of a function $w = f(z)$ analytic in D and real on the real axis, this leads to the identities (figure 31.3) associated with reflections in the z- and w-planes. The extension of a function (figure 31.5) across the boundary of a domain (a) or an arc (b) involves a jump if it is not analytic on the boundary, but is possible if its analytic on both sides and a boundary integral exists at least as a principal value. The methods such as factorization of polynomials (partial fraction decomposition) depend on the location of zeros (poles), e.g., the zeros of a quartic polynomial (figure 31.6).

32

Steady Heat Conduction

A material is said to conduct heat if in the presence of a temperature gradient there is a heat flux; their ratio specifies the thermal conductivity (Section 32.1). Thus a body of high thermal conductivity transmits, for a given temperature gradient, a larger heat flux than one of low conductivity. The sign of the thermal conductivity is determined by the condition that heat flows from the hotter to the cooler bodies. In a steady state, that is, time-independent conditions, there is no accumulation or rarefaction of heat, and the heat flux out of closed region must be: (i) zero if it contains no heat sources or sinks; (ii) match the output of heat sources or sinks when they exist. It follows that the steady state temperature satisfies the Laplace (Poisson) equation in the absence (presence) of heat sources or sinks (Section 32.1); thus solutions in terms of complex functions may be obtained, for example, heat source/sinks in a corner with isothermal/adiabatic walls (Section 32.4). The latter are two among more choices of boundary conditions (Section 32.2) that apply also to other irrotational or solenoidal potential fields (Section 32.3). The solution with cylindrical symmetry (Section 32.5) corresponds to the potential field due to a line source, and the two constants of integration are determined from boundary conditions (Section 32.2) at the surface of a body, specifying: (i) the temperature, if it is contact with a heat reservoir; (ii) the temperature gradient, if it is subjected to a heat influx or outflux; (iii) a linear combination of the two in the case of convection, that is, transfer of heat from a solid to a fluid, the latter at rest (in motion) for free (forced) convection (Section 32.7). These boundary conditions apply to: (i) a solid cylinder (Section 32.5); (ii) a cylindrical cavity in an unbounded medium (Section 32.5); (iii) the hollow cylindrical tube with thin or thick walls (Section 32.6); (iv) concentric tubes of different materials (Section 32.8). The latter (iv) corresponds to a thermal conductivity with finite discontinuities at the interfaces between different materials; the thermal conductivity may also be continuously nonuniform in an inhomogeneous material (Section 32.9), just as the heat source/sinks. The applications of steady heat conduction include the heating of wires, the cooling of shafts and heat exchangers using banks of tubes. The steady conduction of heat is analogous to the steady diffusion of electric charges in a resistive medium also (also of a chemical species into another).

32.1 Heat Flux and Thermal Conductivity (Fourier, 1818)

The **heat flux** vector $\vec{\mathrm{G}}$ is defined as the heat that crosses in the unit time the unit area normal to the $\vec{\mathrm{G}}$ direction. Thus the amount of heat that crosses in the unit time an element of length $\mathrm{d}s$ with unit normal $\vec{\mathrm{n}}$ is $(\vec{\mathrm{G}} \cdot \vec{\mathrm{n}})\mathrm{d}s$. Considering (Figure 32.1) a closed domain D whose boundary $\partial\mathrm{D}$ is regular, that is, has a unique unit outward normal $\vec{\mathrm{n}}$ at every point, the heat flux out of D is given by:

$$\int_{\partial D} \left(\vec{\mathrm{G}} \cdot \vec{\mathrm{n}}\right) \mathrm{d}s = \int_{D} w \, \mathrm{d}S; \tag{32.1}$$

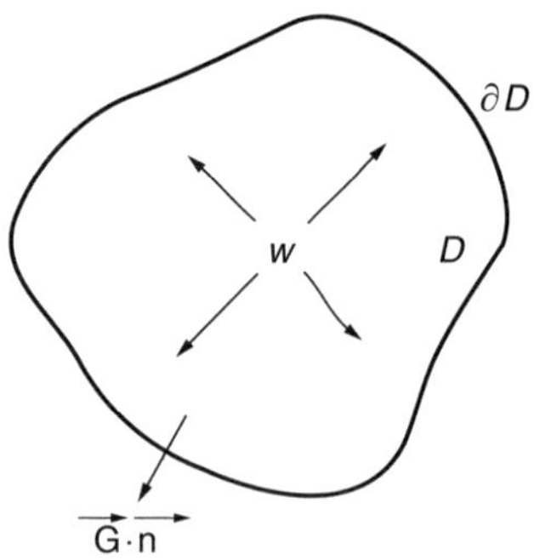

FIGURE 32.1
Steady, that is, time independent heat conduction with heat sources of density w in a domain D and a heat flux $\vec{\mathrm{G}}$ across the outward normal $\vec{\mathrm{n}}$ to the boundary ∂D.

it is balanced, in **steady** conditions, that is, if there is no time dependence, viz. accumulation or rarefaction of heat, by the integral over the domain D of the **heat density** w, that is, the output (input) of heat per unit area $\mathrm{d}S$ and unit time of heat sources $w > 0$ (sinks $w < 0$); if there are no heat sources or sinks the total heat flux through ∂D is zero, that is, the outflow through some parts of the boundary ∂D is exactly balanced by the influx through other parts. Assuming that the heat flux vector $\vec{\mathrm{G}}(\vec{\mathrm{x}})$ has continuous first order derivates, the divergence theorem (28.1b) can be used in (32.1):

$$\int_D \left(\nabla \cdot \vec{\mathrm{G}} - w\right) \mathrm{d}S = 0; \tag{32.2}$$

and since the domain D is arbitrary, this leads to

$$\nabla \cdot \vec{\mathrm{G}} = w, \tag{32.3}$$

the **heat equation** *(32.3), balancing the divergence of the heat flux* $\vec{\mathrm{G}}$ *against the heat density* w *of heat sources* $w > 0$ *or sinks* $w < 0$. This is analogous to the continuity equation expressing mass conservation (14.3) replacing: (i) the mass flux $\rho\vec{\mathrm{v}}$ by the heat flux $\vec{\mathrm{G}}$; (ii) rate-of-change of mass density $\partial\rho/\partial t$ by minus the heat density $-w$; (iii) the minus sign appears since a heat source $w > 0$ (sink $w < 0$) corresponds to an excess of outflow over inflow of heat.

In the heat equation (32.3), the sources or sinks are given functions of position $w(\vec{\mathrm{x}})$ in the steady case, and the heat flux $\vec{\mathrm{G}}$ must be expressed as a function of the temperature T or its derivatives, for example, of first $\nabla T \equiv \partial T/\partial\vec{\mathrm{x}}$ and/or higher order. In an isothermal medium, that is, at uniform temperature, there is no heat flux, and the temperature gradient is zero, $\nabla T = 0$. Heat fluxes occur when the temperature is nonuniform, and thus it may be assumed that: (i) the heat flux is proportional to the temperature gradient, that is, a linear function, although their direction may not be the same in an anisotropic material, for example, a crystal; (ii) if the material is isotropic the heat flux must be parallel to the temperature gradient (32.4b), and have the opposite direction (32.4a) because heat flows from the hotter to the cooler regions:

$$\kappa(\vec{\mathrm{x}}) > 0: \quad \vec{\mathrm{G}} = -\kappa\nabla T; \tag{32.4a,b}$$

(iii) the factor $\kappa(\vec{\mathrm{x}})$ is the **thermal conductivity**, and in the case of a homogeneous material is independent of position. The **Fourier law** (1818) *states that the heat flux is* *antiparallel to the temperature gradient, and proportional to its modulus through the thermal conductivity* κ, *for an isotropic medium*; for moderate temperature changes the thermal conductivity is independent of temperature. The thermal conductivity may depend on the position in the case of an inhomogeneous material, and substituting (32.4b) in (32.3) leads to

$$w = \nabla \cdot (-\kappa \nabla T) = -\kappa \nabla^2 T - \nabla \kappa \cdot \nabla T; \tag{32.5a}$$

the heat equation (32.5a) specifies steady heat conduction in a medium at rest, assumed to be linear and isotropic; if the medium is also homogeneous(32.5b) the temperature satisfies (32.5c) the Laplace (Poisson) equation in the absence (presence) of heat source/sinks forced by the volume density w *of heat divided by the thermal conductivity* κ*:*

$$\kappa = \text{const}: \qquad -\frac{w}{\kappa} = \nabla \cdot \nabla T = \nabla^2 T. \tag{32.5b,c}$$

Regardless of whether it is constant or not and depends or not on temperature, the thermal conductivity is positive (32.4a), because the heat flux (32.4b) is from higher to lower temperatures; this is required by the condition that entropy density grows with time:

$$T^2 \frac{\partial S}{\partial t} = -\vec{\mathrm{G}} \cdot \nabla T = \kappa \left(\nabla T\right)^2 > 0, \tag{32.5d}$$

in agreement with the second principle of thermodynamics applied to irreversible processes.

The latter result is justified next. By analogy with the equation of continuity (14.3) the heat equation (32.3) $\equiv$ (32.6a) involves the heat Q output per unit time:

$$w = \nabla . \vec{G} = -\frac{\partial Q}{\partial t}: \qquad \frac{\mathrm{d}\bar{S}}{\mathrm{d}t} = \int_D \frac{1}{T}\frac{\partial Q}{\partial t} \mathrm{d}A, \tag{32.6a,b}$$

the total entropy $\bar{S}$ production in a domain D with area element dA is the integral of the heat divided by temperature (32.6b), according to the first principle of thermodynamics. Substitution of (32.6a) in (32.6b) yields:

$$\frac{\mathrm{d}\bar{S}}{\mathrm{d}t} = -\int_D \frac{\nabla . \vec{G}}{T} \mathrm{d}s = -\int_D \left[\nabla . \left(\frac{\vec{G}}{T}\right) - \vec{G} . \nabla\left(\frac{1}{T}\right)\right] \mathrm{d}A, \tag{32.6c}$$

where was used a relation similar to (28.5b) with $\Phi \equiv 1/T$ and $\vec{G} = \nabla \Phi$. The divergence theorem transforms the first term on the r.h.s. of (32.6c) into an integral (32.6d) over a surface 'at infinity' or sufficiently far for the heat flux to vanish:

$$\int_D \nabla . \left(\frac{\vec{G}}{T}\right) \mathrm{d}A = \int_{\partial D_\infty} \frac{\vec{G}.\vec{n}}{T} \mathrm{d}s = 0: \quad \frac{\mathrm{d}\bar{S}}{\mathrm{d}t} = -\int_D \frac{\vec{G}.\nabla T}{T^2} \mathrm{d}A = \int_D \frac{\kappa \left(\nabla T\right)^2}{T^2} \mathrm{d}s = \int_D \frac{\partial S}{\partial t} \mathrm{d}A, \tag{32.6d,e}$$

the second term on the r.h.s. of (32.6c) thus specifies the entropy production (32.6e). The second principle of thermodynamics requires the latter to be positive for an arbitrary domain, implying (32.5d) for the entropy density $S = d\bar{S}/dA$ per unit area. In (32.6a–e) the area element was denoted dA (instead of dS) to avoid confusion with the differential of the entropy S.

32.2 Regularity, Asymptotic, and Boundary Conditions

The temperature distribution T satisfies a second-order differential equation, viz. (32.5a) [(32.5c)] for an inhomogeneous (homogeneous) medium, whose solution becomes unique if **two limit conditions** are imposed, for example, (i) **regularity conditions**, for example, the temperature (heat flux) must be finite except possibly at singularities like heat source/sinks; (ii) **asymptotic conditions**, for example, at infinity $r \to \infty$, the temperature must be bounded implying a zero heat flux; (iii) **boundary conditions**, specifying at the boundary ∂D of a domain:

$$T|_{\partial D} = T_0, \quad \left.\frac{\partial T}{\partial n}\right|_{\partial D} = -\frac{G_n}{\kappa}, \quad \{T|_{\partial D} - T_0\}\, h = -\kappa \left.\frac{\partial T}{\partial n}\right|_{\partial D}, \tag{32.7a–c}$$

one of the following: (a) the temperature (32.7a), if the body is in contact with a heat **reservoir** that can supply or withdraw any amount heat required to maintain a fixed temperature T_0 [Dirichlet problem (Sections 37.9 and 38.1)]; (b) the temperature gradient (32.7b), if the surface of the body is subjected to a given heat flux G_n along the normal, for example, it is in contact with a heat **radiator** $G_n > 0$ or **absorber** $G_n < 0$ [Neumann problem (Note 37.1)]; (c) in the case (32.7c) of heat **convection** between the body at temperature T and a fluid at temperature T_0, the normal heat flux $G_n = -\kappa \partial T/\partial n$ is proportional to the temperature difference $T - T_0$, through the **surface thermal conductivity** h, leading to a boundary condition (32.7c) that is a linear, nonhomogeneous relation between the temperature and its gradient [mixed or Robin problem (Note 37.1)]. *The convective boundary condition (32.7c) states that: (i) if the fluid is hotter (colder) than the body it supplies (withdraws) heat to (from) the body; (ii) the temperature gradient is proportional to the temperature difference between the fluid and the body through the ratio of surface and volume thermal conductivities.*

The case of a convection condition (32.7c) with infinite (zero) surface thermal conductivity $h = \infty\, (h = 0)$, leads to the case (32.7a) of a heat reservoir [to the case (32.7b) with $G_0 = 0$, that is, an **adiabatic** wall that allows no heat flux, and imposes a zero temperature gradient normal to it]. The preceding boundary conditions on the temperature (32.7a), heat flux (32.7b), or convection (32.7c) apply at the surface separating the body or domain D from an external medium. If the body consists of two materials with distinct thermal conductivities $\{\kappa_1, \kappa_2\}$ separated by an interface ∂D, the boundary conditions state the continuity of temperature and heat flux:

$$T_1 = T_2, \quad \kappa_1 \frac{\partial T_1}{\partial n} = \kappa_2 \frac{\partial T_2}{\partial n}. \tag{32.8a,b}$$

The latter (32.8b) shows that since the heat flux (32.4b) is continuous, the temperature gradient is discontinuous, being larger (smaller) for the material with smaller (larger) thermal conductivity. This result is the converse of the following: if a given temperature gradient is imposed, a material with high (low) thermal conductivity, will lead to a large (small) heat flux, that is, act as a good **conductor (insulator)** of heat, for example, copper (asbestos).

32.3 Irrotational and Solenoidal Potential Fields

There is the following *analogy between* **potential fields of irrotational (solenoidal) type** $A(B)$ *that have zero curl (divergence) and hence are described by a potential (32.9a, c) [stream (32.10a) or field (32.10b) function]:*

$$\Phi \equiv \Phi_v, \Phi_g, -\Phi_e, T: \quad \nabla^2\Phi = \left\{\Lambda,\, \rho G, -\frac{q}{\varepsilon}, -\frac{w}{\kappa}\right\}, \quad \nabla\Phi = \left\{\vec{\mathrm{v}}, -\vec{\mathrm{g}}, -\vec{\mathrm{E}}, -\frac{\vec{\mathrm{G}}}{\kappa}\right\}, \tag{32.9a–c}$$

$$\Psi \equiv \Psi_v, \Psi_h: \quad \nabla^2\Psi = \left\{-\varpi, -\frac{j\mu}{c}\right\}, \quad \nabla \wedge \vec{\Psi} = \left\{\vec{\mathrm{v}}, \vec{\mathrm{B}} \equiv \mu\vec{\mathrm{H}}\right\}, \tag{32.10a–c}$$

and detailed in Classification 38.1. The irrotational potential fields (32.9a) are (32.9b,c): (i) the velocity $\vec{\mathrm{v}}$ *potential* Φ_v *of an irrotational flow (Sections 12.1 and 18.1) due to the dilatation* Λ *associated with a (18.1b)* $\equiv$ *(32.9c) source* $\Lambda > 0$ *or sink* $\Lambda < 0$ *in (18.2b)* $\equiv$ *(32.9b); (ii) the gravity* $\vec{\mathrm{g}}$ *potential* Φ_g *due (18.18b)* $\equiv$ *(32.9c) to mass density* $\rho > 0$ *in (18.19b)* $\equiv$ *(32.9b), where G is the gravitational constant (Section 18.3); (iii) the electric* $\vec{\mathrm{E}}$ *potential* Φ_e *due (24.4a)* $\equiv$ *(32.9b) to a positive* $q > 0$ *or negative* $q < 0$ *electric charge density (24.5b)* $\equiv$ *(32.9b) in a medium (Section 32.2) with dielectric permittivity* ε*; (iv) the temperature T and associated heat flux (32.4b)* $\equiv$ *(32.9c) for steady heat conduction (32.5c)* $\equiv$ *(32.9b) in a medium at rest (Section 32.1) with thermal conductivity* κ *and heat sources* $w > 0$ *or sinks* $w < 0$*. The solenoidal (32.10a) potential fields (32.10b,c) are: (v) the stream function* Ψ_v *of an incompressible flow (Sections 12.2 and 18.2) due (18.10b)* $\equiv$ *(32.10c) to counterclockwise* $\varpi > 0$ *or clockwise* $\varpi < 0$ *vorticity (18.11b)* $\equiv$ *(32.10b); (vi) the magnetic induction* $\vec{\mathrm{B}}$ *and magnetic field* $\vec{\mathrm{H}}$ *associated (26.4a)* $\equiv$ *(32.10c) with the magnetic field function* Ψ_m *due to counterclockwise* $j > 0$ *or clockwise* $j < 0$ *electric current density* $\vec{j}$ *in (26.6b)* $\equiv$ *(32.10b) in a medium of magnetic permeability* μ*, denoting by c the speed of light in vacuo (Section 26.2). In the cases (iii,iv,vi) the medium was assumed to be linear, isotropic, homogeneous and at rest.*

In the absence of forcing terms, all six fields (i) to (iv) have both potential Φ and stream Ψ functions that both satisfy the Laplace equation, and thus have the properties of harmonic functions for example, the lemmas maximum (maximum and minimum) modulus in Section 24.1 (26.1). It follows that*: consider the following potential fields in a domain* D *with closed regular boundary* ∂D*. (a) potential flow with (i) neither source/sinks nor (v) vortices: (b) electro (magneto) static field without electric (iii) charges [(vi) currents]; (c) gravity field (ii) without masses; (d) temperature due to steady heat conduction (iv) without heat source/sinks. The flow source/sinks (i) and vortices (vi), masses (ii), electric charges (iii) and currents (vi), and heat sources (iv) could exist outside* D*. The following properties hold for the scalar potential, stream or field function, and the velocity, gravity, electro, or magnetostatic field and temperature and heat flux: (a) if they are zero on the boundary* ∂D *they are also zero inside* D*; (b) if they are constant on the boundary* ∂D *they have the same value in the interior of* D*; (c) if they are nonuniform in the interior* D $-$ ∂D *they have maximum value in modulus on the boundary* ∂D*; (d) if they are nonuniform and nonzero in the interior* D$-\partial$D *they have maximum and minimum value in modulus on the boundary* ∂D.

These statements suggest three further results, not proven in here, concerning **unicity**: *there is a unique potential field satisfying Laplace or Poisson equation in a domain D, with satisfies on the boundary* ∂D *one of the three* ***boundary conditions****:*

$$\lambda > 0: \quad \Phi\,|_{\partial D} = c_1, \qquad \left.\frac{\partial\Phi}{\partial n}\right|_{\partial D} = c_2, \qquad \left[\Phi + \lambda\frac{\partial\Phi}{\partial n}\right]_{\partial D} = c_3, \tag{32.11a–d}$$

of the: (i/ii) ***first (second) kind*** *specifying the value* c_1 *in (32.11b) [normal derivative* c_2 *in (32.11c)] of the potential on the boundary, without (with) an arbitrary added constant; (iii)* ***third kind*** *specifying a linear combination* c_3 *in (32.11d) with positive parameter (32.11a), without arbitrary added constant.* The arbitrary added constant appears only in the case (ii) because it involves only the temperature gradient, not the temperature, and thus is unaffected by adding a constant to the temperature. In the case of steady heat conduction the boundary condition of: (i/ii) first (second) kind corresponds to specified temperature (32.7a) ≡ (32.11b) [heat flux (32.7b) ≡ (32.11c)] on the boundary; (iii) third kind corresponds to convection (32.11d) ≡ (32.7c) ≡ (32.12a), viz.:

$$0 = \left[T + \frac{\kappa}{h} \frac{\partial T}{\partial n} \right]_{\partial D} - T_0 \equiv \left[T + \lambda \frac{\partial T}{\partial n} \right]_{\partial D} - T_0, \quad \lambda \equiv \frac{\kappa}{h} > 0, \tag{32.12a,b}$$

where the parameter (32.11a) ≡ (32.12b) must be positive because heat flows from hotter to colder bodies, that is: (i) physically this is the second principle of thermodynamics that entropy must increase (32.5d) for an isolated system undergoing an irreversible process; (ii) mathematically $\lambda < 0$ would lead to failure of the proof of unicity of solution.

32.4 Corner with Isothermal or Adiabatic Walls

Since the temperature distribution for steady heat conduction without (with) sources satisfies Laplace (Poisson) equation, it is specified by the real and imaginary parts of a complex potential that is analytic (singular), for example, in a corner without (with) heat source/sinks [Subsection 32.4.1 (32.4.2)].

32.4.1 Temperature Distribution in a Corner or Wedge

The complex potential (14.71a,b) for a corner of angle (32.13a), leads to the temperature distribution (32.13b):

$$\beta = \frac{\pi}{n}: \quad T_{\pm}(r, \varphi) = T_0 + A r^{\pi/\beta} \left\{ \cos, \sin \left(\frac{\pi \varphi}{\beta} \right) \right\}, \tag{32.13a,b}$$

where a constant temperature T_0 has been added; the latter does not affect the heat flux (32.4b), viz. in polar components:

$$\mathrm{G}_r^{\pm}(r, \varphi) = -\kappa \frac{\partial T}{\partial r} = -\frac{\kappa A \pi}{\beta} r^{\pi/\beta - 1} \left\{ \cos, \sin \left(\frac{\pi \varphi}{\beta} \right) \right\}, \tag{32.14a}$$

$$\mathrm{G}_\varphi^{\pm}(r, \varphi) = -\frac{\kappa}{r} \frac{\partial T}{\partial \varphi} = -\frac{\kappa A \pi}{\beta} r^{\pi/\beta - 1} \left\{ -\sin, \cos \left(\frac{\pi \varphi}{\beta} \right) \right\}. \tag{32.14b}$$

The second case in (32.13b; 32.14a,b) concerns isothermal walls at temperature T_0 in (32.15b,c):

$$0 \le \varphi \le \beta: \quad T_-(r, 0) = T_0 = T_-(r, \beta); \quad \mathrm{G}_\varphi^{+}(r, 0) = 0 = \mathrm{G}_\varphi^{+}(r, \beta), \tag{32.15a–e}$$

the first case corresponds to adiabatic walls, for that the normal heat flux is zero (32.15d,e). In both cases the temperature is T_0 at the corner and diverges at infinity; (ii) the heat flux (32.14a,b) vanishes at an the vertex of a "corner" $0 < \beta < \pi$, and diverges at infinity;

(iii) the reverse happens for a wedge $\pi < \beta < 2\pi$; (iv) the intermediate case $\beta = \pi$ corresponds a uniform heat flux over a flat wall. Thus *the steady heat conduction in a corner of angle (32.13a; 32.15a) without heat source/sinks and with isothermal (adiabatic) walls is specified in polar coordinates by the temperature* T_- (T_+) *in (32.13b), and heat flux* $\vec{\mathrm{G}}^-$ $(\vec{\mathrm{G}}^+)$ *polar components in (32.14a,b).*

32.4.2 Monopole or Multipole Heat Sources/Sinks

A heat line source $w > 0$ (sink $w < 0$) corresponds (12.26b) to the temperature (32.16a):

$$T(r) = T_0 - \frac{w}{2\pi\kappa}\log r, \quad \vec{\mathrm{G}} = -\kappa\nabla T = \frac{w}{2\pi r}\vec{\mathrm{e}}_r, \tag{32.16a,b}$$

and radial heat flux (32.16b); the latter (32.16b) is unaffected by adding a constant T_0 to the temperature in (32.16a). Using (16.74a) [(16.82b)] with $Q(-\Gamma)$ replaced by $-w/\kappa$ according to (32.9b), specifies the temperature (32.17b) [(32.18b)] due to n identical ($2n$ alternating) heat source/sinks, with equal angular spacing (32.17a) [(32.18a)] at a distance, a, from the origin, that is, at the vertices of a regular polygon with n ($2n$) sides on a circle of radius a, viz.:

$$0 \le \varphi \le \beta = \frac{2\pi}{n}: \quad T_-(r,\varphi) = T_0 - \frac{w}{4\pi\kappa}\log\left[r^{2n} + a^{2n} + 2a^n r^n\cos(n\varphi)\right], \tag{32.17a,b}$$

$$\begin{aligned} 0 \le \varphi \le \beta = \frac{\pi}{n}: \quad T_-(r,\varphi) = T_0 - \frac{w}{4\pi\kappa}\left\{\log\left|r^{2n} + a^{2n} - 2a^n r^n\cos(n\varphi)\right|\right. \\ \left. -\log\left|r^{2n} + a^{2n} + 2a^n r^n\cos(n\varphi)\right|\right\}; \end{aligned} \tag{32.18a,b}$$

in both cases the heat flux normal to the walls scales as $\mathrm{G}_\varphi \sim \sin(n\varphi)$ and thus vanishes at the walls (32.15d,e), that is, the walls are adiabatic. In (32.17b; 32.18b) as in (16.67a,b; 16.74a) the arguments of the logarithms are positive so the modulus is not necessary and may be replaced by simple brackets:

$$(r^n - a^n)^2 \ge 0: \; r^{2n} + a^{2n} \ge 2a^n r^n \ge 2a^n r^n\cos,\sin(n\varphi). \tag{37.18c,d}$$

Thus, *the steady heat conduction due to* n *identical (2*n *alternating) heat source/sinks (32.16a,b) at equal angular spacing at a distance,* a*, from the origin, is specified by the temperature (32.17b) [(32.18b)] that corresponds to a corner of angle (32.17a) [(32.18a)] with adiabatic walls (32.15d,e).*

A heat line-multipole of order n with moment P_n, corresponds (12.60a) to a temperature:

$$T_\pm(r,\varphi) = T_0 - \frac{P_n}{2\pi\kappa}r^{-n}\{-\cos,\sin(n\varphi)\}, \tag{32.19}$$

that leads to isothermal (adiabatic) walls at $\varphi = 0,\ \pi/n$ for T_- (T_+) since (32.15b,c) [(32.15d,e)] are satisfied. It was shown in *(12.59a,b) that the complex potential (32.20a) of a multipole of order* $(\mathrm{n}+1)$ *and moment* $\mathrm{P}_{\mathrm{n}+1}$*, is the complex conjugate field (32.20b) of multipole of order* n *and moment (32.20c):*

$$f(z) = \frac{P_n}{2\pi z^n}, \quad \frac{\mathrm{d}f}{\mathrm{d}z} = \frac{P_{n+1}}{2\pi z^{n+1}}, \quad P_{n+1} = -nP_n. \tag{32.20a–c}$$

Thus (16.75a,b) [(16.83a,b)], with $Q(\Gamma)$ replaced by $-w/\kappa$, specify *the temperature for steady heat conduction due to* $n(2n)$ *dipoles [(32.19) with* $n = 1$*] equally spaced at a distance* a

from the origin:

$$0 \le \varphi \le \frac{2\pi}{n} \equiv \beta: \quad T_\pm(r,\varphi) = T_0 - \frac{wn}{2\pi\kappa} r^{n-1}\left[a^{2n} + r^{2n} + 2a^n r^n \cos(n\varphi)\right]^{-1}$$
$$\{r^n + a^n \cos(n\varphi), \quad -a^n \sin(n\varphi)\}, \tag{32.21a,b}$$
$$0 \le \varphi \le \frac{\pi}{n} \equiv \beta: \quad T_\pm(r,\varphi) = T_0 - \frac{wn}{\kappa\pi} r^{n-1} a^n \left[a^{4n} + r^{4n} + 2a^{2n} r^{2n} \cos(2n\varphi)\right]^{-1}$$
$$\left\{\left(r^{2n} + a^{2n}\right)\cos(n\varphi), \left(r^{2n} - a^{2n}\right)\sin(n\varphi)\right\}, \tag{32.22a,b}$$

corresponding to a corner of angle (32.21a) [(32.22a)], with isothermal (adiabatic) walls for T_- (T_+), that satisfy (32.15b,c) [32.15d,e)]. The temperature distributions $T_\pm(r,\varphi) - T_0$ due to dipoles (32.21a,b) [(32.22a,b)] are the derivatives $\{d/dr, rd/d\varphi\}$ of the temperature distributions $T_-(r,\varphi) - T_0$ due to monopoles (32.17a,b) [(32.18a,b)].

32.5 Solid Cylinder and Cylindrical Cavity

The solution of the heat equation with cylindrical symmetry (Subsection 32.5.1), is simplest in the cases of a solid cylinder (cylindrical cavity) of radius a in Subsection 32.5.2 (32.5.3).

32.5.1 Solution of Heat Equation with Cylindrical Symmetry

Not all complex potentials are relevant to steady heat conduction, for example, the uniform field with a cylinder corresponds to a constant heat flux at infinity, and thus diverging temperature, viz. (32.13b) with $\beta = \pi$; also the complex potential is most suited to isothermal (adiabatic) boundary conditions, viz. (32.7a) [(32.7b) with $\mathrm{G}_n = 0$]. Other boundary conditions, for example, convective (32.7c) [heated (32.7b) with $\mathrm{G}_n \neq 0$] boundaries can be considered solving the heat equation (32.5a), for example, polar coordinates (11.31b; 11.33b) with cylindrical symmetry (32.23a), that is, depending only on the radius:

$$\frac{1}{r}\frac{\mathrm{d}}{\mathrm{d}r}\left[r\kappa(r)\frac{\mathrm{d}T}{\mathrm{d}r}\right] = -w(r); \quad \vec{\mathrm{G}} = -\kappa\frac{\mathrm{d}T}{\mathrm{d}r}\vec{\mathrm{e}}_r, \tag{32.23a,b}$$

in this case: (i) the heat flux (32.4b) is (11.31b) radial (32.23b); (ii) the density of heat source/sinks in (32.23a) can depend only on the radius. An example of power law heat source/sink density is considered (32.24):

$$w(r) = w_0\left(\frac{r}{a}\right)^n, \tag{32.24}$$

and the thermal conductivity is taken as a constant (32.5b) for a homogeneous medium; the heat equation (32.23a) also holds for a radially nonuniform thermal conductivity and is a particular case of (32.5a) that will be considered subsequentily (Section 32.9).

For cylindrical symmetry, the density of heat sources/sinks can depend only on the distance from the origin $w(r)$, and integrating (32.23a) once leads to (32.25b):

$$\kappa = \kappa(r): \quad -r\kappa(r)\frac{\mathrm{d}T}{\mathrm{d}r} = A + \int^r \xi w(\xi)\mathrm{d}\xi, \tag{32.25a,b}$$

where A is an arbitrary constant of integration, and the thermal conductivity $\kappa(r)$ may depend on the radius; the meaning of (32.25b) is that the heat flux G through a circle of radius r:

$$2\pi r \mathrm{G}_r = -2\pi r\kappa(r)\frac{\mathrm{d}T}{\mathrm{d}r} = 2\pi A + \int_0^{2\pi} \mathrm{d}\varphi \int_0^r \eta w(\eta)\,\mathrm{d}\eta, \tag{32.26}$$

is the sum of: (i) a constant $2\pi A$ independent of the radius r in the absence of sources/sinks (first term); (ii) it can depend on r only due to the presence of heat sources/sinks with nonzero total output (second term) over the disk of radius r. Integrating once more (32.25b) with constant thermal conductivity (32.27a) specifies the temperature distribution (32.27b):

$$\kappa = \text{const}: \quad T(r) = B - \frac{A}{\kappa}\log r - \kappa^{-1}\int^r \zeta^{-1}\,\mathrm{d}\zeta \int^\zeta \xi w(\xi)\mathrm{d}\xi, \tag{32.27a,b}$$

where: (i) the first term, involving two arbitrary constants of integration, is the temperature due to a line-source at the origin (32.16a), that is, the general solution of the Laplace equation with cylindrical symmetry; (ii) the second term is the temperature distribution imposed by the heat sources/sinks, distributed with density $w(r)$ in the cylinder of radius r, that is, a forced particular solution of the Poisson equation.

32.5.2 Heat Sources/Sinks in a Cylinder

In the case of (Figures 32.2 and 32.3) a solid cylinder of radius, a, the temperature distribution (32.27b) applies in the region, $0 \le r \le a$. The regularity condition that the temperature be finite at the center implies $A = 0$, and in the absence of heat sources/sinks $w = 0$, the temperature is uniform $T(r) = B$, and equal to the value at the surface $T(r) = T_0$, that is, there is no heat flux $\mathrm{G}(r) = 0$. This result is a particular instance of the constancy lemma (Section 23.1 ≡ Section 32.3), because in the absence of heat sources/sinks the temperature satisfies the Laplace equation. that is, is a harmonic function; a function harmonic in a region is constant in its interior if it is constant on the boundary, and hence uniform temperature at the surface of the solid cylinder $T(a) = T_0$ implies uniform temperature in the interior $T(r) = T_0$ in steady conditions (Figure 32.2). This result holds not only for circles, but also for arbitrary regions: *the steady heat conduction in a compact region* D, *without heat sources of sinks, leads to a uniform temperature in the interior* $\mathrm{D} - \partial\mathrm{D}$ *iff the temperature is uniform on the boundary.*

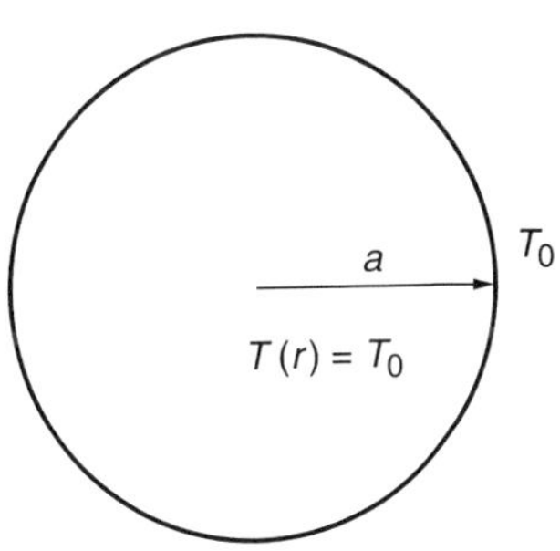

FIGURE 32.2
The steady heat conduction in a compact, that is, finite body, without heat sources in the interior and with constant temperature T_0 at the surface, leads to the same constant temperature in the interior, for example, for a solid cylinder. The same holds for any other potential field, for example, potential flow (Chapters 12, 14, 16, and 28) electro/magnetostatic field (Chapters 24 and 26), and gravity field (Chapter 18).

The temperature will not be uniform in the presence of heat sources/sinks, for example, for an wire of electrical resistivity R the electric currents J dissipate an amount of heat $J^2R = w_0$ per unit volume and time; in this case the heat density $w_0 = \text{const}$ is uniform, that is, corresponds to $n = 0$ in (32.24b). Substitution of (32.24) in (32.25b) specifies the heat flux:

$$n \neq -2: \quad G(r) = -\kappa(r)\frac{dT}{dr} = \frac{A}{r} + \frac{w_0}{n+2}a^{-n}r^{n+1}, \tag{32.28a,b}$$

where the thermal conductivity may be nonuniform; assuming it is uniform (32.29a) a further integration leads to the temperature profile (32.29b):

$$\kappa = \text{const}: \quad T(r) = B - \frac{A}{\kappa}\log r - \frac{w_0/\kappa}{(n+2)^2}a^{-n}r^{n+2}, \tag{32.29a,b}$$

that coincides with (32.27b; 32.24) ≡ (32.29b). Consider a solid cylinder (32.30a) with a boundary condition:

$$0 \le r \le a: \quad T(a) = T_0, \quad T'(a) = -\frac{G_0}{\kappa}, \quad T(a) + \frac{\kappa}{h}T'(a) = T_0, \tag{32.30a–d}$$

of one the three kinds: (i) isothermal (32.7a) ≡ (32.30b); (ii) constant normal heat flux G_0 in (32.30b) ≡ (32.7b); (iii) convection (32.7c) ≡ (32.30c) from a fluid at temperature T_0 with surface thermal conductivity h.

In all cases the boundedness condition, that the temperature (32.29b) be finite on the axis of the cylinder, requires that: (i) the arbitrary constant A be zero (32.31a): (ii) the exponent n be restricted to $n > -2$ in (32.31b). The remaining constant of integration B in (32.29b) can be determined from one of the boundary conditions (32.30b) [(32.30d)], leading respectively to (32.31c) [(32.31d)]:

$$A = 0,\ n > -2: \quad B - T_0 - \frac{w_0}{\kappa}\left(\frac{a}{n+2}\right)^2 = \left\{0, \frac{w_0 a}{h(n+2)}\right\}. \tag{32.31a–d}$$

The boundary condition (32.30c) cannot be applied, because the heat flux (32.28b; 32.32a) is specified by the heat sources (32.32b):

$$A = 0: \quad G(r) = \frac{w_0}{n+2}a^{-n}r^{n+1}, \quad G(a) = \frac{w_0 a}{n+2}, \tag{32.32a–c}$$

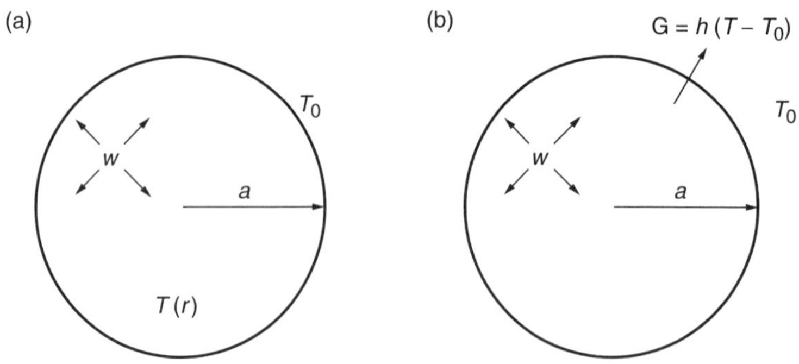

FIGURE 32.3
The steady heat conduction in a solid cylinder, with heat sources w in the interior, is compared for: (i) constant temperature T_0 at the surface (a); (ii) convection (b) with surface thermal conductivity h from a static fluid at temperature T_0. The latter case leads to a larger temperature difference between the outside and the inside of the cylinder.

and takes the value (32.32c) at the surface. Thus: (i) the boundary condition (32.30c) is inconsistent if G_0 does not take the value (32.32c); (ii) if G_0 in (32.30c) does take the value (32.32c), then the boundary condition is redundant, and it fails to determine the constant B.

32.5.3 Cylindrical Cavity in Unbounded Medium

The case of a cylindrical cavity of radius a in an infinite medium is similar as concerns the boundedness condition as $r \to \infty$ requires: (i) again $A = 0$ in (32.29b); (ii) instead of $n > -2$ in (32.33a,b) that $n < -2$ in (32.33c,d):

$$0 \le r \le a: \quad n > -2\,; \quad a \le r < \infty: \quad n < -2\,. \tag{32.33a–d}$$

The isothermal boundary condition (32.30b) ≡ (32.34a) is the same (Figure 32.3a) for the solid cylinder and cylindrical cavity, but in the convective boundary condition (32.30d) the outward normal (Figure 32.3b) is reversed, so the sign is exchanged on the r.h.s. of (32.31d). Substituting (32.31c,d) in (32.29b) specifies:

$$T(a) = T_0: \quad T_1(r) = T_0 + \frac{w_0}{\kappa}\left(\frac{a}{n+2}\right)^2\left[1 - \left(\frac{r}{a}\right)^{n+2}\right], \tag{32.34a,b}$$

$$T(a) = T_0 + \frac{\kappa}{h}T'(a): \quad T_2^{\pm}(r) = T_1(r) \pm \frac{w_0 a}{h(n+2)}. \tag{32.35a,b}$$

The temperature (32.34b) [(32.35b)] inside a cylinder (32.33a,b) [outside a cylindrical cavity (32.33c,b)] of radius a, for a material with constant thermal conductivity (32.27a), due to heat sources $w_0 > 0$ or sinks $w_0 < 0$ given by (32.24), with boundary conditions: (i) isothermal (32.34a) in Figure 32.3a; (ii) convective (32.35a) in Figure 32.3b with upper (lower) sign. In the case (32.34a) of isothermal boundary condition (Figure 32.3a) the temperature (32.34b) is given by the same expression (32.36a) at: (i) the center of the cylinder; (ii) at infinity away from the cavity:

$$T_1(0) = T_0 + \frac{w_0}{\kappa}\left(\frac{a}{n+2}\right)^2 = T_1(\infty), \tag{32.36a}$$

it is higher (lower) than at the surface to for heat sources $w_0 > 0$ (sinks $w_0 < 0$). The (Figure 32.3b) convective boundary condition (32.35a): (i) increases the temperature at the center of the cylinder relative to the surface because (32.36c) in holds in the case (32.36b):

$$n > -2: \quad T_2^{+}(0) = T_0 + \frac{w_0 a}{n+2}\left(\frac{1}{h} + \frac{a/\kappa}{n+2}\right) > T_1(0) = T_0 + \frac{w_0}{\kappa}\left(\frac{a}{n+2}\right)^2; \tag{32.36b,c}$$

$$n < -2: \quad T_2^{-}(\infty) = T_0 + \frac{w_0 a}{n+2}\left(-\frac{1}{h} + \frac{a/\kappa}{n+2}\right) > T_1(\infty) = T_0 + \frac{w_0}{\kappa}\left(\frac{a}{n+2}\right)^2, \tag{32.36d,e}$$

(ii) also increases the temperature at infinity relative to the surface of the cavity because (32.36e) holds in the case (32.36d); (iii) the conclusions (i/ii) hold for heat sources $w_0 > 0$, and are reversed for heat sinks $w_0 < 0$. In all cases convection increases the temperature difference relative to a wall at the same temperature; thus *convection from a hotter (cooler) fluid is a more effective heating (cooling) process than a heat reservoir at the same temperature.*

32.6 Hollow Tube with Thick or Thin Walls

The solution of the steady heat equation or the Laplace equation with cylindrical symmetry (Subsection 32.5.1) involves two arbitrary constants of integration, one of that vanishes by the boundedness condition: (i) either at the center for a solid cylinder (Subsection 32.5.2); (ii) or at infinity for a cylindrical cavity (Subsection 32.5.3); thus only one boundary condition remains to be satisfied. In the case of a hollow cylinder (Section 32.6) both the center and infinity are excluded, and the boundedness condition is met with neither of the two arbitrary constants of integration being zero; thus two compatible, nonredundant boundary conditions are applied at the same boundary (distinct boundaries), that is, single-point (two-point) boundary conditions in Subsection 32.6.1 (32.6.2). The temperature profiles for a thick hollow tube simplify in the case of a thin tube (Subsection 32.6.3), if the thickness is small compared with the radius.

32.6.1 Single-Point Boundary Conditions

A case of common occurrence (Figures 32.4, 32.5a–c, 32.6a–d, and 32.7) is a thick tube of inner a_- (outer a_+) radius (32.37a) at without heat sources/sinks, that is, $w_0 = 0$ in the heat flux (32.28b) [temperature (32.29b)] that simplify to (32.37b) [(32.37c)] as in (32.16a) [(32.16b)]:

$$a_- \leq r \leq a_+ : \quad T(r) = B - \frac{A}{\kappa} \log r, \quad \mathrm{G}(r) = -\kappa \frac{\mathrm{d}T}{\mathrm{d}r} = \frac{A}{r}. \tag{32.37a–c}$$

Since the origin and infinity are excluded, neither of the two constants of integration (A, B) vanishes, and two and independent and compatible boundary conditions are needed to determine them. The **single-point boundary conditions** apply at one interface, for example,

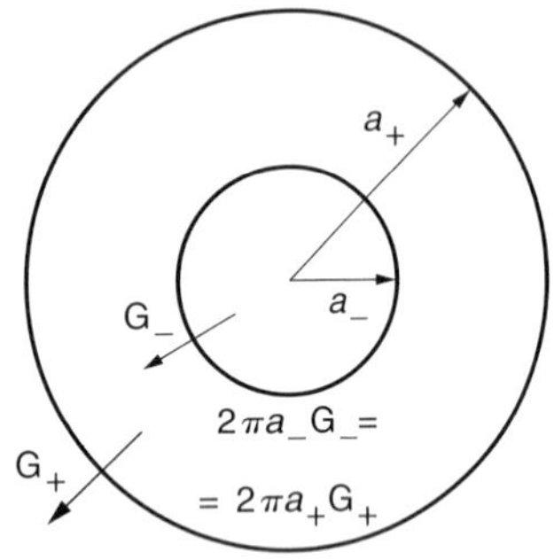

FIGURE 32.4
In the case of steady heat conduction in a doubly-connected region without heat sources, for example, a hollow cylinder, the total heat flux must be equal across the two boundaries, because heat accumulation or loss is not possible in the interior. Thus only one boundary condition should be applied, for example, the heat flux across one boundary. If the heat flux is specified also across the other boundary one of two undesirable cases occurs: (i) the two heat fluxes are consistent, and hence the second boundary condition is redundant; (ii) the two heat fluxes are incompatible, and the second boundary condition has turned a well-posed problem with a unique solution into an ill-posed problem without a solution.

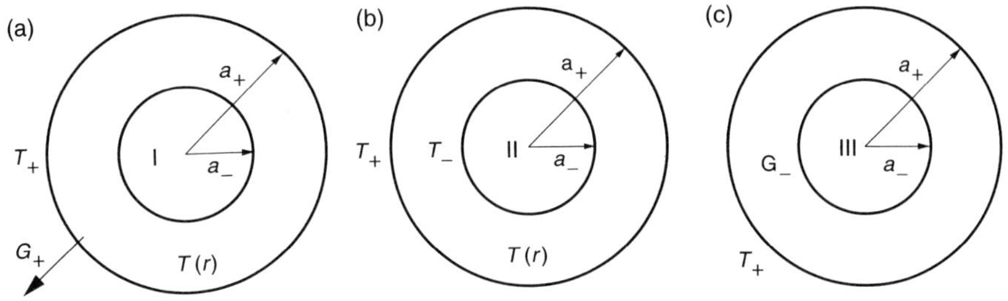

FIGURE 32.5
There are three distinct cases of two compatible, nonredundant boundary conditions for steady heat conduction in a doubly-connected region without heat sources, imposing either one or two temperature(s) or one heat flux (two heat fluxes would be either redundant or incompatible by Figure 32.4). The three distinct cases (Tables 32.1 and 32.2), for example, for a hollow cylindrical tube, are: (i) temperature and heat flux at the same boundary, for example, the outer wall in (a); (ii) temperature at each boundary, that is, both the inner and outer walls in (b); (iii) temperature (heat flux) at one (the other) boundary, for example, the outer (inner) wall in (c).

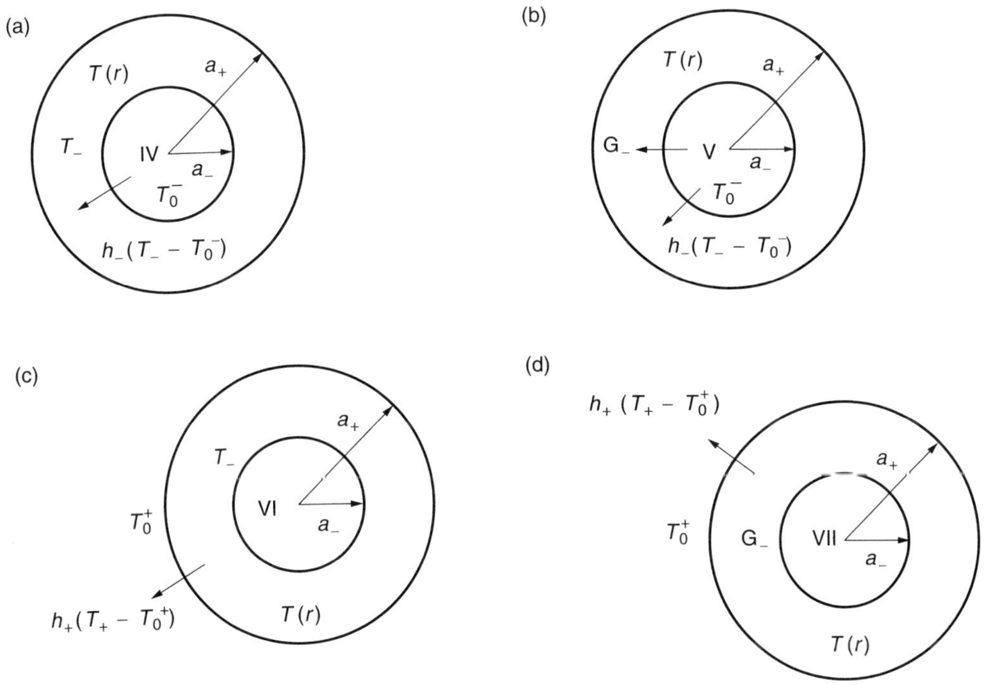

FIGURE 32.6
Allowing one convection boundary condition adds to the three cases in Figure 32.5, four more distinct cases of two compatible, nonredundant boundary conditions for steady heat conduction in a doubly-connected region without heat sources (Figure 32.6). The four new cases are, for example, for a hollow cylindrical tube: (iii/iv) convection and temperature (heat flux) at the same boundary, for example, inner in (a) [(b)]; (v/vi) convection (temperature heat /flux) at one boundary (the other boundary), for example, the outer (inner) wall in (c/d).

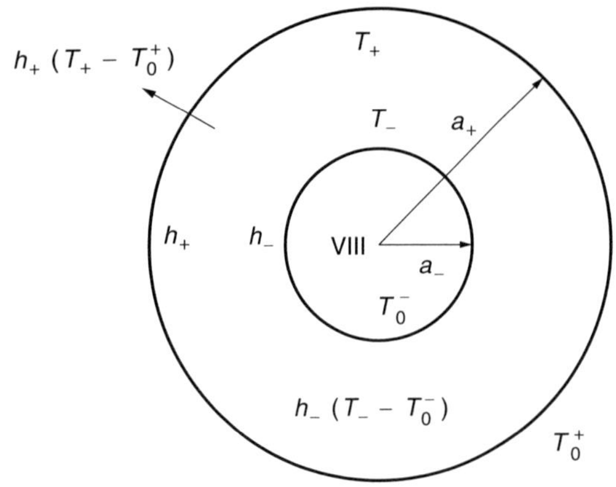

FIGURE 32.7
The last of eight distinct cases of two compatible, nonredundant boundary conditions for steady heat conduction in a doubly-connected region without heat sources is (viii) convection at both boundaries, for example, in a hollow cylindrical tube. The four restrictions common to Figures 32.5–32.7 can be justified as follows: (a) the eight distinct cases (i) to (viii) exclude mere exchanges of the same conditions between two boundaries, as in Tables 32.1 and 32.2; (b) the compatible, nonredundant boundary conditions exclude specifying the heat flux at the two boundaries, which is not possible for steady heat conduction in a medium without heat sources. The specification of the heat flux at two boundaries would be possible if at least one (possibly both) of the following conditions allow for different heat fluxes at the two boundaries: (i) unsteady heat conduction, that is, accumulation (depletion) of heat in excess (default); (ii) heat sources (sinks) in the interior compensating the default (excess) heat flux.

(Figure 32.5a) temperature and heat flux at the outer boundary:

$$I: \quad T(a_+) = T_+, \quad -\kappa T'(a_+) = \mathrm{G}_+. \tag{32.38a,b}$$

Substitution of (32.37c) in (32.38b) ≡ (32.37b) [in (32.38a)] specifies the constants of integration (32.39a) [(32.39b)]:

$$I: \quad \mathrm{G}_+ = \frac{A}{a_+}, \quad B = T_+ + \frac{A}{\kappa}\log a_+ = T_+ + \frac{a_+\mathrm{G}_+}{\kappa}\log a_+; \tag{32.39a,b}$$

these appear in (32.37b) [(32.37c)] the temperature (32.40a) [heat flux (32.40b)]:

$$I: \quad T(r) = T_+ + \frac{a_+\mathrm{G}_+}{\kappa}\log\left(\frac{a_+}{r}\right), \quad \vec{\mathrm{G}}(r) = G_+\frac{a_+}{r}\vec{\mathrm{e}}_r. \tag{32.40a,b}$$

A heat influx $\mathrm{G}_+ > 0$ (outflow $\mathrm{G}_+ < 0$) through the boundary increases (decreases) the temperature inside the tube relative to the outer surface, leading to the value (32.41d) at the inner surface:

$$I: \quad T(a_-) = T_+ + \frac{a_+\mathrm{G}_+}{\kappa}\log\left(\frac{a_+}{a_-}\right); \quad \vec{\mathrm{G}}(a_-) = \mathrm{G}_+\frac{a_+}{a_-}\vec{\mathrm{e}}_r, \tag{32.41a,b}$$

the conservation of the total heat flux through the inner and outer boundary is expressed by (32.41b). In either case the maximum and minimum temperatures occur at the boundaries, in agreement with the general properties of harmonic functions (Sections 24.1 and 32.3).

32.6.2 Two-Point Boundary Conditions

The two-point boundary conditions apply at the two interfaces, and are similar if they specify the same quantities, for example, (Figure 32.5b) temperature (32.42a,b), for example, the **two-point similar boundary conditions:**

$$II: \quad T_\pm = T(a_\pm) = B - \frac{A}{\kappa}\log a_\pm. \tag{32.42a,b}$$

The boundary conditions (32.42a,b) can be solved for the arbitrary constants:

$$II: \quad A = -\kappa\frac{T_+ - T_-}{\log(a_+/a_-)}, \qquad B = T_- + \frac{A}{\kappa}\log a_-; \tag{32.43a,b}$$

they lead by (32.37b) [(32.37c)] to the temperature distribution (32.44a) [heat flux (32.44b)]:

$$II: \quad T(r) = T_- + (T_+ - T_-)\frac{\log(r/a_-)}{\log(a_+/a_-)}, \tag{32.44a}$$

$$II: \quad \vec{G}(r) = \frac{\kappa}{r}\frac{T_- - T_+}{\log(a_+/a_-)}\vec{e}_r. \tag{32.44b}$$

Again the temperature in the interior always lies between the values on the boundaries:

$$II: \quad T(a_-) \equiv T_- > T(r) > T_+ \equiv T(a_+) \quad \text{or} \quad T_- < T(r) < T_+, \tag{32.45a,b}$$

depending on whether $T_- > T_+$ in (32.45a) [$T_+ > T_-$ in (32.45b)] in the boundary conditions (32.42a,b). The two-point similar boundary conditions specifying the heat fluxes $G_\pm$ fail (Figure 32.4) to be independent and consistent, because the heat flux (32.37c) involves only one constant of integration A, and thus: (i) if the heat fluxes $G_\pm$ at the two boundaries $r = a_\pm$ lead to different values of A, the two boundary conditions are incompatible; (ii) if they determine the same value of A, the two boundary conditions are redundant, and the other constant of integration B is left undetermined.

The one-point boundary conditions are always dissimilar, for example, II in (32.42a,b), otherwise they would be either incompatible or redundant. The two-point boundary conditions can be similar, for example, I in (32.38a,b), or dissimilar, for example, specifying (Figure 32.5c) the temperature (32.46b) at the outer boundary and heat flux at the inner boundary, where the outward normal points radially inward (32.46a); this leads to a pair of compatible **dissimilar two-point boundary conditions:**

$$III: \quad G_- \equiv G(a_-) = -\frac{A}{a_-}, \quad T_+ = T(a_+) = B - \frac{A}{\kappa}\log a_+ = B + \frac{G_- a_-}{\kappa}\log a_+. \tag{32.46a,b}$$

The constants of integration (A,B) in (32.46a,b) specify by (32.37b) [(32.37c)] the temperature (32.47a) [heat flux (32.47b)]:

$$III: \quad T(r) = T_+ - \frac{G_- a_-}{\kappa}\log\left(\frac{a_+}{r}\right), \quad \vec{G}(r) = -G_-\frac{a_-}{r}\vec{e}_r; \tag{32.47a,b}$$

in particular their respective values:

$$III: \quad T(a_-) = T_+ - \frac{G_- a_-}{\kappa}\log\left(\frac{a_+}{a_-}\right), \quad \vec{G}(a_+) = -G_-\frac{a_-}{a_+}\vec{e}_r, \tag{32.48a,b}$$

at the other boundary.

32.6.3 Small Thickness Compared with the Radius

The **thickness** e and **mean radius** $\bar{a}$ of the tube are defined by (32.49a,b):

$$e \equiv a_+ - a_-, \qquad 2\bar{a} \equiv a_+ - a_- : \qquad a_\pm = \bar{a} \pm \frac{e}{2}. \tag{32.49a–c}$$

and are an alternative way to specify the inner a_- and outer a_+ radii (32.49c). The tube is **thin** if the thickness is small relative to the mean radius (32.50a), allowing approximations such as (32.50b–e):

$$e^2 \ll \bar{a}^2 : \quad \frac{\bar{a}}{a_\pm} = \frac{1}{1 \pm e/2\bar{a}} = 1 \mp \frac{e}{2\bar{a}}, \tag{32.50a,b}$$

$$\log\left(\frac{a_\pm}{\bar{a}}\right) = \log\left(1 \pm \frac{e}{2\bar{a}}\right) = \pm\frac{e}{2\bar{a}}, \tag{32.50c}$$

$$\log\left(\frac{a_+}{a_-}\right) = \log\left(\frac{a_+}{\bar{a}}\right) - \log\left(\frac{a_-}{\bar{a}}\right) = \frac{e}{\bar{a}}; \tag{32.50d}$$

$$\log\left(\frac{a_\pm}{r}\right) = \log\left[\frac{\bar{a}}{r}\left(1 \pm \frac{e}{2\bar{a}}\right)\right] = \log\left(\frac{\bar{a}}{r}\right) + \log\left(1 \pm \frac{e}{2\bar{a}}\right) = \log\left(\frac{\bar{a}}{r}\right) \pm \frac{e}{2\bar{a}}; \tag{32.50e}$$

the **thin tube approximation** (32.49a–c; 32.50a–e) may be applied to the three cases [I ≡ (32.40a,b), II ≡ (32.44a,b), III ≡ (32.47a,b)] considered before:

$$I: \quad T(r) = T_+ + \frac{\mathrm{G}_+ e}{2\kappa} - \frac{G_+}{\kappa}\left(\bar{a} + \frac{e}{2}\right)\log\left(\frac{r}{\bar{a}}\right), \tag{32.51a}$$

$$\vec{\mathrm{G}}(r) = \mathrm{G}_+ \frac{\bar{a} + e/2}{r}\vec{e}_r; \tag{32.51b}$$

$$II: \quad T(r) = T_- + (T_+ - T_-)\left[\frac{1}{2} + \frac{\bar{a}}{e}\log\left(\frac{r}{\bar{a}}\right)\right], \quad \vec{\mathrm{G}}(r) = \frac{\kappa\bar{a}}{er}(T_- - T_+)\vec{e}_r, \tag{32.52a,b}$$

$$III: \quad T(r) = T_+ - \frac{\mathrm{G}_- e}{2\kappa} + \frac{G_-}{\kappa}\left(\bar{a} - \frac{e}{2}\right)\log\left(\frac{r}{\bar{a}}\right), \tag{32.53a}$$

$$\vec{\mathrm{G}}(r) = -\mathrm{G}_- \frac{\bar{a} - e/2}{r}\vec{e}_r. \tag{32.53b}$$

Thus have been obtained for *steady heat conduction in a thick (32.49a,c) [thin (32.50a–e)] hollow cylindrical tube of thermal conductivity* κ, *for (Figure 32.5a–c) boundary conditions I ≡ (32.38a,b)/II = (32.42a,b)/III ≡ (32.46a,b) the temperature and heat flux I ≡ (32.40a,b)/II ≡ (32.44a,b)/III ≡ (32.47a,b) [I ≡ (32.51a,b)/II ≡ (32.52a,b)/III = (32.53a,b)].*

32.7 Convective Transfer in Heat Exchangers

Besides boundary conditions imposing the temperature and/or heat flux at the wall(s) of a hollow cylinder (Section 32.6), convective boundary conditions can also be applied (Subsection 32.7.1) that are particularly relevant for heat exchangers (Section 32.7). The single-point boundary conditions are always dissimilar (Subsections 32.6.1 and 32.7.2) viz. each specifies a boundary condition of different kind, for example: among the three (32.7a–c). The two-point boundary conditions (Subsections 32.6.2 and 32.7.3) can always be dissimilar, and can also similar if they are compatible and nonredundant, for example: (i) the heat flux at one boundary of the hollow cylinder specifies the heat flux at the other boundary, so heat fluxes at two boundaries are either incompatible or redundant; (ii) the conditions at two boundaries can be both temperature or convective types, because there is neither redundancy nor incompatibility.

32.7.1 Compatible, Nonredundant Boundary Conditions

The possible combinations of boundary conditions of three kinds, viz. temperature $T \equiv$ (32.7a) $\equiv$ (32.11b) $\equiv$ (32.30b)/heat flux $H \equiv$ (32.7b) $\equiv$ (32.11c) $\equiv$ (32.30c)/convective $C \equiv$ (32.11a,d) $\equiv$ (32.30d), at the two boundaries of a hollow tube, are: (i) the single-point boundary conditions (Table 32.1) must be dissimilar, leading to three combinations (TH, TC, HC), doubled to six by interchanging the inner and the outer boundaries; (ii) the two-point similar boundary conditions (Table 32.2) are independent and nonredundant for temperature (T) or convection (C), but not for heat flux (H), adding two more possibilities; (iii) the two-point dissimilar boundary conditions (Table 32.2) allow all possible $3 \times 2 = 6$ combinations, of that half in interchanging the three pairs (TH, TC, HC) between the two boundaries. Thus there is a total of $3 + 2 + 3 = 8(6 + 2 + 6 = 14)$ distinct sets of boundary conditions, not counting (counting) distinct boundaries. Of these three have already been considered, viz. steady heat conduction in a hollow cylindrical tube (Section 32.6), with: (I) single-point boundary conditions (I: 32.39–32.41a,b; 32.51a,b) for the temperature (32.38a) and heat flux (32.38b) at the outer wall (inner wall instead also possible); (II) two-point similar (II: 32.43–32.45a,b; 32.52a,b) boundary conditions (32.42a,b) for temperature (not possible for heat flux); (III) two-point dissimilar (III: 32.47–32.48a,b; 32.53a,b) boundary conditions for heat flux (32.46a) at inner wall and temperature (32.46b) at outer wall (reverse also possible). Thus have been covered three out of five possible combinations of boundary conditions of first (32.7a) $\equiv$ (32.11b) $\equiv$ (32.30b) and second (32.7b) $\equiv$ (32.11c) $\equiv$ (32.30c) kinds viz.: (i) temperature (T) and heat flux (H) at the same boundary, leaves two possibilities; (ii) one at each boundary, gives two more possibilities; (iii) temperature (T) but not heat flux at both boundaries the fifth possibility. In fact the cases I–III are all three distinct combinations of boundary conditions of first and second kinds, not counting interchanges of boundaries that merely exchange, ($a_\pm, T_\pm, \mathrm{G}_\pm$).

The boundary conditions of third kind (32.11a,d), are of particular interest for convection (32.7c) $\equiv$ (32.30d). Convective heat transfer is used in heat exchangers, for example, for

TABLE 32.1
Single-Point Boundary Conditions

Boundary	Inner	Outer
Combination:		
TH	I in reverse	I $\equiv$ (32.38–32.41a,b; 32.51a,b) Figure 32.5a
TC	IV $\equiv$ (32.55a,b; 32.57a,b) Figure 32.6a	IV in reverse
HC	V $\equiv$ (32.56a,b; 32.58a,b) Figure 32.6b	V in reverse

Note: When considering heat conduction in a doubly-connected region, for example, a hollow cylindrical tube (Figures 32.4–32.6) the two boundary conditions can be applied: (i) at the same boundary (Table 32.1); (ii) at distinct boundaries (Table 32.2). In the case (i) of single point boundary conditions it is possible to specify at the same boundary: (i–1) the temperature (T) and heat flux (H); (i–2,3) a convection condition (C) and temperature (T) or heat flux (H); (i–4,5.6) instead the same conditions (i–1, 2, 3) can be applied at the other boundary.

TABLE 32.2
Two-Point Boundary Conditions

Boundary	Inner		
	Temperature: T	**Heat flux: H**	**Convective: C**
Outer			
Temperature	II ≡ (32.42–32.45a,b; 32.52a,b) Figure 32.5b	III ≡ (32.46–32.48a,b; 32.53a,b) Figure 32.5c	VI in reverse
Heat flux	III in reverse	Redundant or incompatible	VII in reverse
Convective	VI ≡ (32.59a; 32.60a; 32.61a,b) Figure 32.6c	VII ≡ (32.59a; 32.60b; 32.62a,b) Figure 32.6d	VIII ≡ (32.59a; 32.60c; 32.63a,b) Figure 32.7

Note: The heat conduction in a doubly-connected region, for example, a hollow cylindrical tube (Figures 32.4–32.7) is uniquely specified by boundary conditions at the same (distinct) boundary(ies) in Table 32.1 (32.2). In the latter case (Table 32.2) it is possible to combine the following boundary conditions: (ii–1, 2, 3) convective (C) at one boundary with temperature (T) or heat flux (H) or convective (C) at the other boundary; (ii–4,5) temperature (T) at one-boundary and temperature (T) or heat flux (H) at the other boundary; (ii–6) heat flux at both boundaries is either redundant (or incompatible), because only one constant of integration is involved, and its specified twice by the same (distinct) values; (ii–7, 8, 9) the remaining three cases exchange the boundaries in one of the preceding cases.

air conditioning systems, thermal power stations, etc. Two typical configurations are: (i) a bank of tubes, with hot water inside cooled by an air flow outside; (ii) a wall with tubes with water heated or vaporized by the gas outside in a chamber. The convective boundary condition at the outer wall of the tube will be combined next with an isothermal (VI), heated (VII) or convective (VIII) boundary condition on the inner wall; interchanging the walls in case VI and VIII would lead to two more cases. Thus the three (five) cases with convection are considered in Section 32.7 out of a possible total of five (nine) not counting (counting) interchanges between boundaries. This adds to: (i) the two cases (CT, CH) at the same boundary that are considered next (Subsection 32.7.2), viz. convection with temperature (IV ≡ CT) and with heat flux (V ≡ CH) at the same boundary, leaving two more possibilities interchanging boundaries; (ii) the three cases ($VI \equiv CT,\;\; VII \equiv CH,\;\; VIII \equiv CC$) at two boundaries in Subsection 32.7.3; (iii) the overall sample of eight (eleven) cases out of a possible eight (fourteen) not counting (counting) boundary interchanges is a representative sample, since it covers all cases apart from interchanges of boundaries, as shown in Table 32.1 (32.2) for single-point (two-point) boundary conditions.

32.7.2 Single-Point Boundary Conditions with Convection

The convective boundary condition (32.30d) at the inner wall, for a fluid at temperature T_0^- and surface thermal conductivity, h_-, leads from (32.37b,c) to a relation between constants of integration:

$$T_0^- = T(a_-) - \frac{\kappa}{h_-} T'(a_-) = B - \frac{A}{\kappa}\log a_- + \frac{A}{h_- a_-}, \tag{32.54}$$

taking into account that the outward normal at the inner surface points inward. Combining with given temperature T_- in (32.42b) [heat flux G$_-$ in (32.46a)] at the same inner boundary

Figure 32.6a (32.6b), specifies the constants of integration:

$$IV: \quad A = h_- a_- \left(T_0^- - T_-\right), \quad B = T_- + \frac{h_- a_-}{\kappa}\left(T_0^- - T_-\right)\log a_-; \tag{32.55a,b}$$

$$V: \quad A = -\mathrm{G}_- a_-, \quad B = T_0^- + \frac{\mathrm{G}_-}{h_-} - \frac{\mathrm{G}_- a_-}{\kappa}\log a_-. \tag{32.56a,b}$$

Substituting (32.55a,b) [(32.56a,b)] back in (32.37b,c) leads to

$$IV: \quad T(r) = T_- + \left(T_- - T_0^-\right)\frac{h_- a_-}{\kappa}\log\left(\frac{r}{a_-}\right), \quad \vec{\mathrm{G}}(r) = \frac{T_0^- - T_-}{r} h_- a_- \vec{\mathrm{e}}_r, \tag{32.57a,b}$$

$$V: \quad T(r) = T_0^- + \frac{\mathrm{G}_-}{h_-} + \frac{\mathrm{G}_- a_-}{\kappa}\log\left(\frac{r}{a_-}\right), \vec{\mathrm{G}}(r) = \mathrm{G}(r)\vec{\mathrm{e}}_r = -\mathrm{G}_-\frac{a_-}{r}\vec{\mathrm{e}}_r. \tag{32.58a,b}$$

The steady heat conduction in a cylinder of thermal conductivity κ, with inner of radius a_- with surface thermal conductivity h_- in contact with a fluid at temperature T_0^-, is specified if (IV) [(V)] the same inner wall is Figure 32.6a (32.6b) kept at temperature T_- (has a heat flux G_-) by the temperature (32.57a) [(32.58a)] and heat flux (32.57b) [(32.58b)].

32.7.3 Two-Point Boundary Conditions with Convection

The convective boundary condition (32.30d) at the outer wall relates the constants of integration (A,B) in (32.37b,c) as in (32.54), with opposite sign in the last term in (32.59a) because the outward normal is radial outward (inward) at the outer (inner) boundary:

$$T_0^+ = T\left(a_+\right) + \frac{\kappa}{h_+}T'\left(a_+\right) = B - \frac{A}{\kappa}\log a_+ - \frac{A}{h_+ a_+}; \tag{32.59a}$$

$$T(r) - T_0^+ = \frac{A}{h_+ a_+} + \frac{A}{\kappa}\log\left(\frac{a_+}{r}\right) = \frac{A}{\kappa}\left[\log\left(\frac{a_+}{r}\right) + \frac{\kappa}{h_+ a_+}\right], \tag{32.59b}$$

this can be used to eliminate one of the constants of integration from the temperature (32.37b), leading to (32.59b). A particular case is (32.58a) at the inner boundary $r = a_-$ with reversed normal $A \to -A$ and $A = Ga$. The surface thermal conductivity at the outer (inner) wall is h_+ (h_-) in contact with external fluid at temperature, T_0^+ $\left(T_0^-\right)$; thus $\left(h_+, T_+^0\right)$ $\left[\left(h_-, T_-^0\right)\right]$ appear at the outer (32.59a), [inner (32.54)] wall in Figure 32.7. Retaining a convective boundary condition (32.59a,b) at the outer wall, the remaining constant of integration A can be determined from three cases at the inner wall: (VI) constant temperature T_- in (32.60a) and Figure 32.6c; (VII) constant heat flux G_- in (32.60b) and Figure 32.6d; (VIII) convection (32.54) in (32.60c) and Figure 32.7:

$$VI: \quad T_- = T\left(a_-\right) = T_0^+ + \frac{A}{\kappa}\left[\log\left(\frac{a_+}{a_-}\right) + \frac{\kappa}{h_+ a_+}\right], \tag{32.60a}$$

$$VII: \quad \mathrm{G}_- = +\kappa T'\left(a_-\right) = -\frac{A}{a_-}, \tag{32.60b}$$

$$VIII: \quad T_0^- = T_0^+ + \frac{A}{\kappa}\left[\log\left(\frac{a_+}{a_-}\right) + \frac{\kappa}{h_+ a_+} + \frac{\kappa}{h_- a_-}\right]; \tag{32.60c}$$

in (32.60b,c) the heat flux at the inner wall is projected on the outward normal that lies radically inward $-\vec{\mathrm{e}}_r$, instead of radially outward $\vec{\mathrm{e}}_r$ at the outer surface.

The convective boundary condition (32.30d) $\equiv$ (32.60d) at the inner surface:

$$T_0^- = T\left(a_-\right) - \frac{\kappa}{h_-}T'\left(a_-\right) = T_- - \frac{\kappa}{h_-}\left(-\frac{A}{a_-\kappa}\right) = T_- + \frac{A}{h_- a_-}, \tag{32.60d}$$

leads to (32.60c) on substitution of (32.59b) with $r = a_-$; alternatively (32.60c) can be obtained subtracting (32.59a) from (32.54). The boundary condition (32.60a)/(32.60b)/(32.60c) specifies the remaining constant of integration A in the temperature (32.59b), viz. (32.61a)/(32.62a)/(32.63a), from that follows (32.61b)/(32.62b)/(32.63b) for the heat flux:

$$\textit{VI}: \quad T(r) = T_0^+ + \left(T_- - T_0^+\right) \frac{\log(a_+/r) + \kappa/(h_+a_+)}{\log(a_+/a_-) + \kappa/(h_+a_+)}, \tag{32.61a}$$

$$\vec{\mathrm{G}}(r) = \frac{T_- - T_0^+}{\log(a_+/a_-) + \kappa/(h_+a_+)} \frac{\kappa}{r} \vec{\mathrm{e}}_r; \tag{32.61b}$$

$$\textit{VII}: \quad T(r) = T_0^+ - \frac{a_-\mathrm{G}_-}{\kappa}\left[\log\left(\frac{a_+}{r}\right) + \frac{\kappa}{h_+a_+}\right], \quad \vec{\mathrm{G}}(r) = -\mathrm{G}_- \frac{a_-}{r}\vec{\mathrm{e}}_r; \tag{32.62a,b}$$

$$\textit{VIII}: \quad T(r) = T_0^+ + \left(T_0^- - T_0^+\right) \frac{\log(a_+/r) + \kappa/(h_+a_+)}{\log(a_+/a_-) + \kappa/(h_-a_-) + \kappa_+/(h_+a_+)}, \tag{32.63a}$$

$$\vec{\mathrm{G}}(r) = \frac{T_0^- - T_0^+}{\log(a_+/a_-) + \kappa/(h_-a_-) + \kappa/(h_+a_+)} \frac{\kappa}{r} \vec{\mathrm{e}}_r. \tag{32.63b}$$

Thus have been obtained *the temperature (heat flux) for steady heat conduction in a hollow cylinder of thermal conductivity* κ *in contact with a fluid of temperature* T_0^+ *at the outer boundary (32.59a) of surface thermal conductivity* h_+, *for three conditions at the inner boundary: (VI) constant temperature* T_- *in (32.60a) leading (Figure 32.6c) to (32.61a) [(32.61b)]; (V) constant heat flux* G_- *in (32.60b) leading (Figure 32.6d) to (32.62a) [(32.62b)]; (vi) inner boundary in contact with a fluid (32.60c) at temperature* T_0^- *with surface thermal conductivity* h_- *leading (Figure 32.7) to (32.63a) [(32.63b)].*

32.8 Concentric Cylinders of Different Materials

The consideration of multiple walls made of materials with distinct thermal conductivities (Subsection 32.8.1) leads to a law of association in series of thermal conductivities (Subsection 32.8.2) for cylindrical or plane walls.

32.8.1 Boundary Conditions at Multiple Interfaces

In order to benefit from the distinct thermal conductivities of different materials, a heat insulation may consist of several layers. In the case (Figure 32.8) of two concentric cylinders of distinct materials of thermal conductivities κ_- (κ_+) for $a_- \le r < a$ ($a < r \le a_+$), the temperature distribution is given by (32.37b) with different constants of integration in each layer, viz.:

$$T(r) = \begin{cases} A\log r + B & \text{for } a_- \le r \le a, & (32.64\text{a}) \\ C\log r + D & \text{for } a \le r \le a_+, & (32.64\text{b}) \end{cases}$$

There are four constants of integration are determined from four boundary conditions. Two of these apply to the inner a_- or outer a_+ surfaces as before (Sections 32.6 and 32.7); the other two concern (32.8a,b) the continuity of temperature (32.65b) and heat flux (32.65c) at the interface (32.65a):

$$r = \mathrm{a}: \quad A\log a + B = C\log a + D, \quad \kappa_- A = \kappa_+ C; \tag{32.65a–c}$$

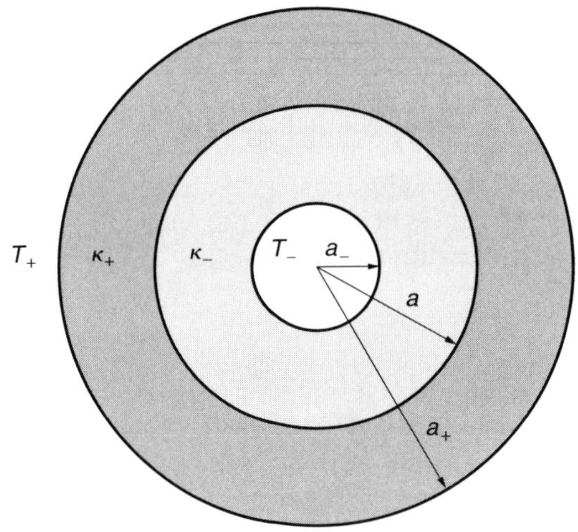

FIGURE 32.8
The steady heat conduction in two coaxial cylindrical tubes without heat sources in the interior with for the inner (outer) tube with: (i) thermal conductivities κ_- (κ_+); (ii) inner/outer radii a_-/a (a/a_+). Two compatible, nonredundant boundary conditions: (i) render the problem unique, for example, given temperatures at the inner a_- and outer a_+ radius. There are another two implicit compatible, nonredundant boundary-conditions: (ii) continuity of temperature and heat flux at the intermediate boundary of radius a; the four boundary conditions (i+ii) specify two pairs of arbitrary constants, one in each of the two coaxial tubes. In the case of several $n \geq 3$ cylindrical layers there would: (a) be $(n-2)$ pairs of boundary conditions at the $(n-2)$ intermediate boundaries; (ii) together with two boundary conditions at the inner and outer surfaces this leads to $2(n-1)$ conditions; (iii) there specify the $(n-1)$ pairs of arbitrary constants in each of the $(n-1)$ coaxal cylindrical layers. In all cases this specifies an equivalent thermal conductivity $\bar{\kappa}$, which leads to the same heat flux for a single layer for the same temperature difference between the innermost and outermost boundary. This equivalent thermal conductivity applies to homothetic geometries, for example, several axial tubes, concentric spheres, or parallel walls.

these relate the four constants of integration, so that two may be eliminated from (32.64a,b), for example: (i) retaining (32.64a); (ii) using (32.65a,b) to express (32.64b) $\equiv$ (32.66a,b):

$$a \leq r \leq a_+: \quad T(r) = C \log\left(\frac{r}{a}\right) + A \log a + B = A\left[\frac{\kappa_-}{\kappa_+}\log\left(\frac{r}{a}\right) + \log a\right] + B, \tag{32.66a,b}$$

in terms of the same constants $(A,\ B)$.

For example, if the inner (outer) wall is kept at temperature, T_- (T_+), this leads in (32.64a,b) to:

$$T(r) = \begin{cases} A \log(r/a_-) + T_- & \text{for } a_- \leq r \leq a, \quad (32.67\text{a}) \\ (\kappa_-/\kappa_+)\, A \log(r/a_+) + T_+ & \text{for } a \leq r \leq a_+, \quad (32.67\text{b}) \end{cases}$$

where (32.65c) was used; the remaining constant of integration A is determined from the continuity of temperature (32.68b) at the intermediate boundary (32.68a):

$$r = a: \quad A = \frac{T_+ - T_-}{\log(a/a_-) + (\kappa_-/\kappa_+)\log(a_+/a)}; \tag{32.68a,b}$$

the heat flux:

$$a_- \le r \le a_+ : \vec{G}(r) = -\frac{\kappa_- A}{r}\vec{e}_r = \frac{\kappa_+ C}{r}\vec{e}_r = \frac{(T_- - T_+)/r}{\kappa_-^{-1}\log(a/a_-) + \kappa_+^{-1}\log(a_+/a)}\vec{e}_r, \tag{32.69a,b}$$

has a single expression (32.69b) for over the whole thickness of the tube (32.69a) and hence is continuous at $r = a$. The temperature gradient is discontinuous at $r = R$:

$$\frac{\mathrm{d}T}{\mathrm{d}r} = \frac{(T_+ - T_-)/r}{\kappa_+\log(a/a_-) + \kappa_-\log(a_+/a)} \times \begin{cases} \kappa_+ & \text{for } a_- \le r \le a, \quad (32.70\text{a}) \\ \kappa_- & \text{for } a \le r \le a_+, \quad (32.70\text{b}) \end{cases}$$

due to the different thermal conductivities on the two sides. *The steady heat conduction in two coaxial cylindrical tubes (figure 32.8) of inner (outer) radii $a_-(a_+)$, with interface $a_- < a < a_+$ separating distinct inner κ_- (outer κ_+) thermal conductivities leads to the inner (32.64a) [outer (32.66b)] temperature, in particular (32.67a) [(32.67b)] for (32.68b) inner (outer) wall at temperature $T_-(T_+)$; the temperature and heat flux (32.69a,b) are continuous across the interface, but not the temperature gradient (32.70a,b).*

32.8.2 Association in Series of Thermal Conductivities

Comparing the heat flux (32.69b) of the composed cylinder (Figure 32.8) with that (32.44b) a cylinder of equal inner and outer radii at the same temperatures (Figure 32.5b) made of a single material of thermal conductivity κ, the heat flux would be the same if:

$$\kappa^{-1}\log\left(\frac{a_+}{a_-}\right) = \kappa_+^{-1}\log\left(\frac{a_+}{a}\right) + \kappa_-^{-1}\log\left(\frac{a}{a_-}\right); \tag{32.71}$$

the equation becomes a trivial identity if a single material is considered $\kappa_- = \kappa = \kappa_+$. Thus have been obtained *the temperature (32.67a,b), its gradient (32.70a,b) and the heat flux (32.69a,b), in (32.68b) two concentric cylinders, of conductivities κ_- (κ_+) for $\mathrm{a}_- \le \mathrm{r} < \mathrm{a}$ $(\mathrm{a} < \mathrm{r} \le \mathrm{a}_+)$, with temperatures T_- (T_+) at the inner (outer) wall (Figure 32.8). The heat flux is the same of a homogeneous cylinder of conductivity κ given by (32.71). This is the* case n = 2 of the **law of association in series** *of the thermal conductivities κ_n with* $\mathrm{n} = 1, \ldots, \mathrm{N}$ *of* $(\mathrm{N}+1)$ *cylindrical walls (32.72a) of radii* $\mathrm{a}_1, \ldots, \mathrm{a}_{\mathrm{N}+1}$:

$$\kappa^{-1}\log\left(\frac{a_{N+1}}{a_1}\right) = \sum_{n=1}^{N}\kappa_n^{-1}\log\left(\frac{a_{n+1}}{a_n}\right); \qquad \frac{x_{N+1} - x_1}{\kappa} = \sum_{n=1}^{N}\frac{x_{n+1} - x_n}{\kappa_n}; \tag{32.72a,b}$$

this corresponds to (32.72b) for $(N+1)$ plane walls at positions $x_1, \ldots, x_{N+1}$.

The result (32.72b) can be proved as follows: (i) for steady heat conduction in one dimension the Laplace equation (11.26) for one Cartesian coordinate (32.73a):

$$0 = \nabla^2 T = \frac{\mathrm{d}^2 T}{dx^2}, \qquad \text{const} = \frac{\mathrm{d}T}{\mathrm{d}x} = -\frac{\mathrm{G}}{\kappa_n}, \tag{32.73a,b}$$

leads to a constant temperature gradient in each layer (32.73b); (ii) since the heat flux must be continuous across all interfaces and the overall temperature difference is a sum of the temperature differences resulting from the linear temperature profiles in each layer:

$$-\frac{(x_{N+1} - x_1)\,\mathrm{G}}{\kappa} = T_{N+1} - T_1 = \sum_{n=1}^{N} T_{n+1} - T_n = -\mathrm{G}\sum_{\kappa=1}^{N}\frac{x_{n+1} - x_n}{\kappa_n}; \tag{32.73c}$$

from (32.73c) follows (32.72b). The latter (32.72b) also follows from (32.72a) with the cylindrical approximation (32.74a) in the limit of infinite radius (32.74b):

$$a_{n+1} = a_n \left(1 + \frac{x_{n+1} - x_n}{R}\right): \quad \lim_{R\to\infty} R \log\left(\frac{a_{n+1}}{a_n}\right)$$
$$= \lim_{R\to\infty} R \log\left(1 + \frac{x_{n+1} - x_n}{R}\right) = \lim_{R\to\infty} R\left[\frac{x_{n+1} - x_n}{R} + O\left(\frac{1}{R^2}\right)\right] = x_{n+1} - x_n. \tag{32.74a,b}$$

Introducing the **conductance** *for the plane (32.75a) [cylindrical (32.75b)] walls leads in both cases (32.72a) [(32.72b)] to the law of association (32.75c) of inverse sums:*

$$\tilde{\kappa}_n \equiv \kappa_n \left\{ (x_{n+1} - x_n)^{-1}, \left[\log\left(\frac{a_{n+1}}{a_n}\right)\right]^{-1} \right\}: \tilde{\kappa}^{-1} = \sum_{n=1}^{N} \tilde{\kappa}_n^{-1} = \left(\sum_{n=1}^{N} \tilde{\kappa}_n\right)^{-1} \sum_{m=1}^{N} \left(\prod_{\substack{p=1\\p\neq m}}^{m} \tilde{\kappa}_p\right); \tag{32.75a–c}$$

the law of association for the thermal conductance (32.75c) is similar to the law of association of impedances in series (4.22b) [parallel (4.24a)] of mechanical (electrical) circuits in Section 4.6 (4.7), that leads to (4.26) ≡ (32.75c).

32.9 Parallel Walls of an Inhomogeneous Substance

There are two kinds of nonuniformities for steady heat conduction, for example, in the solution (32.25a,b; 32.26) of the heat equation (32.23a) with cylindrical symmetry, viz.: (i) nonuniform distribution of heat sources (Section 32.5); (ii) nonuniform thermal conductivity (Section 32.8). These nonuniformities can occur in any geometry, for example, Cartesian (Section 32.9) for parallel walls (Subsection 32.9.1) without heat sources/sinks and with nonuniform thermal conductivity, for example, a linear profile (Subsection 32.9.2).

32.9.1 Parallel Walls Continuously-Varying Thermal Conductivity

The composite cylinder (walls) with several concentric (parallel) layers with distinct thermal conductivities in the law of association in series (32.72a) [(32.72b)] is an example of an inhomogeneous material with stepwise constant properties and finite discontinuities at the junctions. This may be used as an approximation to a continuous variation of thermal conductivity, or the latter may be used directly in the heat equation (32.5a), for example, in Cartesian coordinates (32.76b) for a plane wall of thickness L in (32.76a):

$$0 \le x \le L: \quad -w(x) = \frac{\mathrm{d}}{\mathrm{d}x}\left[\kappa(x)\frac{\mathrm{d}T}{\mathrm{d}x}\right]. \tag{32.76a,b}$$

From (36.76b) follows the heat flux:

$$\vec{\mathrm{G}}(x) = \vec{\mathrm{e}}_x \mathrm{G}(x): \quad \mathrm{G}(x) = -\kappa(x)\frac{\mathrm{d}T}{\mathrm{d}x} = A + \int^{x} w(\eta)\,\mathrm{d}\eta. \tag{32.77a,b}$$

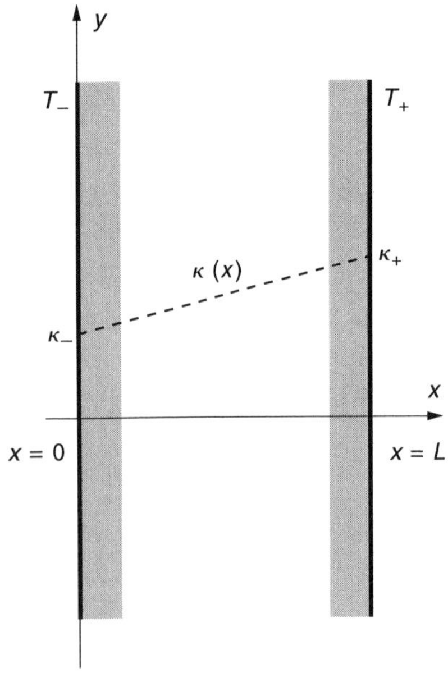

FIGURE 32.9
The steady heat conduction can be considered: (i) with (without) heat sources/sinks [Figures 32.1 and 32.3 (32.2 and 32.4–32.9)]; (ii) with uniform (nonuniform) thermal conductivity [Figures 32.1–32.8 (32.9)]. The steady heat conduction is compared for parallel walls at given temperature, without heat sources/sinks, in the cases of a homogeneous (inhomogeneous) medium, with constant thermal conductivity (nonuniform thermal conductivity a linear function of the distance from a wall).

It is constant (32.78b) in the absence of heat/sources or sinks (32.78a):

$$w(x) = 0: \quad \frac{\mathrm{d}T}{\mathrm{d}x} = -\frac{A}{\kappa(x)}; \qquad \text{(32.78a,b)}$$

that is, in this case the temperature gradient (32.78b) is nonconstant due to the nonuniformity of the thermal conductivity. The temperature is specified by (32.79b):

$$T_- \equiv T(0): \quad T(x) = T_- - A\int_0^x [\kappa(\eta)]^{-1}\,\mathrm{d}\eta, \qquad \text{(32.79a,b)}$$

where T_- is the temperature (32.79a) at the inner wall (Figure 32.9). The remaining constant of integration A can be determined (32.80b) from another boundary condition, for example, the temperature at the outer wall (32.80a):

$$T(L) = T_+: \quad A = \frac{T_- - T_+}{\int_0^L [\kappa(x)]^{-1}\,\mathrm{d}x}. \qquad \text{(32.80a,b)}$$

A parallel-sided wall (32.76a) of a material with nonuniform thermal conductivity has a heat flux (32.77a,b) in presence of heat sources/sinks. In their absence (32.78a), if the boundaries are kept at temperatures (32.79a; 32.80a), the temperature (its gradient) is specified by (32.79b) [(32.78b)], with the constant A determined by (32.80b).

32.9.2 Linear Thermal Conductivity Profile

In the case of uniform thermal conductivity (32.81a), the constant (32.80b) is given by (32.81b):

$$\kappa(x) = \text{const} \equiv \kappa_0 : \ A = \frac{T_- - T_+}{L/\kappa_0}; \ T(x) = T_- + \frac{(T_+ - T_-)\,x}{L}, \quad \mathrm{G}(x) = \frac{\kappa_0 (T_- - T_+)}{L} \equiv A, \tag{32.81a–d}$$

the temperature is a linear function of distance (32.81c), and the heat flux (32.81d) is constant. For a wall with nonuniform thermal conductivity, with temperature T_- (T_+) at the inner (32.78a) [outer (32.80a)] boundary, the temperature in the interior is specified (32.79b; 32.80b) by:

$$T(x) = T_- - (T_- - T_+)\frac{g(x)}{g(L)}, \quad g(x) \equiv \int_0^x [\kappa(\xi)]^{-1}\,\mathrm{d}\xi. \tag{32.82a,b}$$

The average inverse thermal conductivity over the width of the wall (32.76a) is:

$$\frac{1}{\bar{\kappa}} \equiv \frac{1}{L}\int_0^L [\kappa(x)]^{-1}\,\mathrm{d}x = L^{-1} g(L), \quad g(L) = \frac{L}{\bar{\kappa}}; \tag{32.83a,b}$$

the average inverse thermal conductivity (32.83a) is used in agreement with the law of association in series (32.72b) exchanging between continuous and discrete (32.83a) ≡ (32.84a,b) forms:

$$0 = x_1 < x_2 < \cdots < x_n < \cdots < x_n < x_{n+1} \equiv L : \quad \frac{1}{\bar{\kappa}} = \frac{1}{L}\sum_{n=1}^{N}\frac{x_{n+1} - x_n}{\kappa_n}. \tag{32.84a,b}$$

In the case of a linear variation of thermal conductivity between values κ_- (κ_+) at the inner (outer) boundary (32.85a):

$$\kappa(x) = \kappa_- + \left(\frac{\kappa_+ - \kappa_-}{L}\right)x, \quad g(x) = \frac{L}{\kappa_+ - \kappa_-}\log\left[1 + \left(\frac{\kappa_+}{\kappa_-} - 1\right)\frac{x}{L}\right], \tag{32.85a,b}$$

the integral (32.82b) is evaluated by (32.85b) ≡ (32.86):

$$\begin{aligned} g(x) &= \int_0^x \left(\kappa_- + \frac{\kappa_+ - \kappa_-}{L}\xi\right)^{-1} \mathrm{d}\xi = \frac{L}{\kappa_+ - \kappa_-}\log\left[\kappa_- + \frac{\kappa_+ - \kappa_-}{L}\xi\right]_0^x \\ &= \frac{L}{\kappa_+ - \kappa_-}\log\left(1 + \frac{\kappa_+ - \kappa_-}{\kappa_-}\frac{x}{L}\right). \end{aligned} \tag{32.86}$$

Thus *the steady heat conduction in a slab of material (32.76a) with temperatures (32.79a) [(32.80a)] at the inner (outer) wall is specified for arbitrary nonuniform thermal conductivity (32.82b) by the temperature (32.82a), and temperature gradient [heat flux (32.87b)]:*

$$\frac{\mathrm{d}T}{\mathrm{d}x} = \frac{T_+ - T_-}{g(L)\,\kappa(x)} = \frac{T_+ - T_-}{L}\frac{\bar{\kappa}}{\kappa(x)}, \tag{32.87a}$$

$$\bar{G} = -\vec{e}_x \kappa(x)\frac{\mathrm{d}T}{\mathrm{d}x} = \vec{e}_x\frac{T_- - T_+}{g(L)} = \vec{e}_x\frac{T_- - T_+}{L}\bar{\kappa}, \tag{32.87b}$$

where (32.83a) is the average inverse thermal conductivity.

In the case (32.88a) of uniform thermal conductivity (32.88b):

$$\text{const} = \kappa(x) \equiv \kappa_0 = \bar{\kappa}: \quad g(x) = \frac{x}{\kappa_0}, \quad \frac{dT}{dx} = \frac{T_+ - T_-}{L}, \tag{32.88a–c}$$

then (32.82b/32.82a/32.87a/32.87b) reduce respectively to (32.88b/32.81a/32.88a/32.81d). In the case of thermal conductivity varying linearly (32.85a) between the values κ_- (κ_+) *at the wall at temperature,* T_- (T_+)*, leads to: (i) the temperature profile (32.89a) and gradient (32.89b):*

$$T(x) = T_- - \frac{T_- - T_+}{\log(\kappa_+/\kappa_-)} \log\left[1 + \left(\frac{\kappa_+}{\kappa_-} - 1\right)\frac{x}{L}\right], \tag{32.89a}$$

$$\nabla T = \vec{e}_x \frac{dT}{dx} = \frac{T_+ - T_-}{\log(\kappa_+/\kappa_-)} \frac{\kappa_+/\kappa_- - 1}{L + (\kappa_+/\kappa_- - 1)\,x}, \tag{32.89b}$$

$$\vec{G}(x) = -\kappa(x)\nabla T = \vec{e}_x \frac{T_+ - T_-}{L} \frac{\kappa_- - \kappa_+}{\log(\kappa_+/\kappa_-)} \equiv \vec{e}_x G, \tag{32.89c}$$

and constant heat flux (32.89c). The temperature gradient is not constant, and takes the values (32.90a) [(32.90b)] at the inner (outer) wall:

$$T'(0) = \frac{T_+ - T_-}{L} \frac{\kappa_+/\kappa_- - 1}{\log(\kappa_+/\kappa_-)} = -\frac{G}{\kappa_-}, \tag{32.90a}$$

$$T'(L) = \frac{T_+ - T_-}{L} \frac{1 - \kappa_-/\kappa_+}{\log(\kappa_+/\kappa_-)} = -\frac{G}{\kappa_+}; \tag{32.90b}$$

$$\kappa_- T'(0) = \kappa_+ T'(L) = \kappa(x)\,T'(x) = -G = -A, \tag{32.90c}$$

where (32.90a,b) satisfy the condition (32.90c) ≡ (32.78b) of constancy of the heat flux. The heat flux that can be transferred between walls at given temperatures, in one (several) layers, is limited by the thermal conductivity of the material(s).

Note 32.1. *Heat, Mass, and Momentum Transfer:* The steady heat conduction appears together with other irrotational (and solenoidal) potential fields in Classification 38.1. In comparison with other potential fields the boundary conditions are of particular interest, as shown by the combinations in Tables 32.1 and 32.2. Besides the boundary conditions for steady heat condition, there are also initial conditions for unsteady heat conduction. The Fourier series/Fourier integrals/ Laplace transform are used to solve heat conduction problems in finite/infinite/semiinfinite spatial domains, both for steady (unsteady), that is, time independent (dependent) conditions. Heat transfer can occur not only by conduction and convection but also by radiation. Heat, mass, and momentum transfer may occur simultaneously and influence each other in dissipative and turbulent fluids, for example, combustion. In this case the thermal conductivity is only one of the diffusive parameters, with others applying to mass, and momentum transfer, like viscosities and mass diffusivities, and cross-terms; these diffusion parameters satisfy relations generalizing (32.5d) to ensure that entropy must increase in irreversible thermodynamic processes. The diffusion parameters can be deduced from the statistical properties of microscopic matter in statistical physics, kinetic theory and quantum mechanics. The twenty-six problems of steady heat conduction considered in the present section appear in List 32.1.

LIST 32.1 Twenty-Six Problems of Steady Heat Conduction

A – In a corner: Section 32.4
  AA – Without sources: (32.13a,b; 32.14a,b; 32.15a)
    1 – Isothermal walls: (32.15b,c)
    2 – Adiabatic walls: (32.15d,e)
  AB – With heat sources/sinks: (32.16a,b)
    3 – Isothermal walls
    4 – Adiabatic walls: (32.17a,b; 32.18a,b)
  AC – With heat multipole: (32.19; 32.21a,b; 32.22a,b)
    5 – Isothermal walls;
    6 – Adiabatic walls

B – Heat sources/sinks: (32.26; 32.27a,b; 32.24; 32.28a,b; 32.29a,b)
  BA – Solid cylinder: (32.33a,b)
    7 – Isothermal walls: (32.34a,b; 32.32b)
    8 – Convection at wall: (32.35a,b; 32.36b,c)
  BB – Cylindrical cavity in unbounded medium: (32.33c,d)
    9 – Isothermal wall: (32.34a,b; 32.32b)
    10 – Convection at wall: (32.35a,b; 32.36d,e)

C – Hollow cylinder and cylindrical shell: (32.37a–c)
  CA – Nonconvective boundary conditions:
    CAA – Hollow cylinder:
      11 – Inner temperature and heat flux (I): (32.38–32.41a,b)
      12 – Inner and outer temperature (II): (32.42–32.45a,b)
      13 – Inner temperature and outer heat flux (III): (32.46–32.48a,b)
    CAB – Cylindrical shell: (32.49a–c; 32.50a–e)
      14 – Inner temperature and heat flux (I): (32.51a,b)
      15 – Inner and outer temperature (II): (32.52a,b)
      16 – Outer temperature and inner heat flux (III): (32.53a,b)

D – Hollow cylinder: Section 32.7
  DA – Convection at inner wall: (32.54)
    17 – Temperature at inner wall (IV): (32.55a,b; 32.57a,b)
    18 – Heat flux at inner wall (V): (32.56a,b; 32.58a,b)
  DB – Convection at outer wall: (32.59a,b).
    19 – Temperature at inner wall (VII): (32.60a; 32.61a,b).
    20 – Heat flux at outer wall (VII): (32.60b; 32.62a,b).
    21 – Convection at both walls (VIII): (32.60c; 32.63a,b).

E – Multiple walls: Section 32.8.
  22 – Two cylindrical walls: (32.67–32.70a,b; 32.71).
  23 – Multiple cylindrical walls: (32.72a).
  24 – Multiple parallel walls: (32.72b; 32.73a–c).

F – Thermal conductivity for parallel walls: Section 32.9.
  23 – Uniform: (32.81a,d; 32.88a–c).
  24 – Nonuniform, with heat/sources sinks: (32.76–32.77a,b).
  25 – Nonuniform without heat sources/sinks: (32.78–32.80a,b; 32.82–32.83a,b; 32.87a,b).
  26 – Linear profile of thermal conductivity: (32.85a,b; 32.89–32.90a–c).

Conclusion 32: In steady conditions (figure 32.1) the production of heat by sources $w > 0$ (sinks $w < 0$) in a domain is balanced by the heat flux $\vec{G}$ out of its boundary ∂D with unit outward normal $\vec{n}$. One of the simplest domains corresponds to cylindrical symmetry and leads (figure 32.2) to uniform temperature, and no heat flux, in the absence of heat sources/sinks; in the presence of heat sources/sinks, the temperature in the interior is nonuniform, and has a smaller difference (figure 32.3) if (a) a temperature T_0 is imposed at the surface, than if (b) there is convection from a fluid at temperature T_0; thus the effect of the surface thermal conductivity is to enhance heat transfer in the convective relative to the conductivity case. In the case of a cylindrical tube, the heat flux at the two walls cannot be imposed (figure 32.4), because the two conditions would be either (i) incompatible or (ii) redundant. Two compatible and independent boundary conditions can be imposed (figure 32.5) e.g.: (I) the temperature and heat flux at the same wall e.g., the outer wall in (a); (II) the temperature at both walls (b); (III) the heat flux at one wall (e.g., inner) and the temperature at the other (outer) wall (c). The solid cylinders (figure 32.5) model the cooling of a wire or rod, and the hollow cylinders (figure 32.6–32.7) the cooling of tubes viz.: (IV/V) convection at one boundary (e.g., inner) combined with fixed temperature (a) [heat flux (b)] at the same boundary; (VI/VII) convection at one boundary (e.g., outer) and fixed temperature (c) [heat flux (d)] at the other boundary. A particularly important case (VIII) is the heat exchanger between two fluids, with convection conditions at both walls (figure 32.7). Another configuration of interest is the case (figure 32.8) of concentric cylinders made of different materials with distinct thermal conductivities κ_-, κ_+, implying continuous temperature and heat flux and discontinuous temperature gradient at the interface. The thermal conductivity may vary continuously in an inhomogeneous medium, e.g., a fourth configuration of interest is a parallel-sided slab (figure 32.9).

33

Conformal and Critical Points

The mapping from one complex plane to another through a holomorphic function has the conformal property of preserving the modulus and direction of the angles between two curves (Section 33.1); the verification that an isogonal mapping preserves the modulus but reverses the direction of angles, is specified by a nonholomorphic function, leads to the proof that the conformal property is exclusive of holomorphic mappings (Section 33.2). The isogonal and conformal mappings share the property of preserving the modulus of angles that is equivalent to preserving the ratio of lengths (Section 33.3); the latter may vary from one point to another, but is independent of direction at a fixed point. At the poles (zeros) of the function the conformal property is broken, since the ratio of lengths is infinite (zero), and a point becomes a neighborhood (vice-versa); at such critical points of the second (first) kind (Section 33.4), the angles are (Section 33.5) divided (multiplied) by an integer, leading to the appearance of edges (reversals of paths). By selecting suitably the location and exponent of the critical points, it is possible to map conformally a region with a smooth boundary into another with edges, for example, the upper or lower half-plane, and the interior (Section 33.8) or exterior (Section 33.9) of a circle, may be mapped into the interior (Section 33.6) or exterior (Section 33.7) of a polygon. The selection of the location and angles at the vertices, allows the conformal mapping into a variety of regions (Chapters 34 and 36), including: (i) finite polygons, such as plates, triangles, rectangles; (ii) infinite obstacles, such as slits, channels, steps, flaps, ducts, and bends.

33.1 Preservation of Modulus and Direction of Angles

A complex function $\zeta = f(z)$ maps points of the z-plane into the ζ-plane; if it is continuous (Section 11.1) two curves C_1, C_2 that intersect at z_0, and pass through points z_1, z_2 in its neighborhood, are mapped (Figure 33.1) from the z-plane to the ζ-plane, as curves Γ_1, Γ_2 intersecting at $\zeta_0 = f(z_0)$, and passing through the points $\zeta_{1,2} = f(z_{1,2})$ in its neighborhood. If the curves are regular, that is, have continuous tangent, viz. no edges, and the function (33.1a) is holomorphic at z_0, its derivate (33.1b) can be calculated along either of the curves:

$$\zeta = f(z) \in \mathcal{D}\colon \lim_{z_1 \to z_0} \frac{f(z_1) - f(z_0)}{z_1 - z_0} \equiv f'(z_0) \equiv \lim_{z_2 \to z_0} \frac{f(z_2) - f(z_0)}{z_2 - z_0}; \tag{33.1a,b}$$

assuming that the derivate at z_0 is neither zero nor infinite (33.2a), it can be eliminated from (33.1b) by dividing the two terms (33.2b):

$$f'(z_0) \neq 0, \infty: \quad \lim_{z_1 z_2 \to z_0} \frac{f(z_1) - f(z_0)}{z_1 - z_0} \frac{z_2 - z_0}{f(z_2) - f(z_0)} = 1. \tag{33.2a,b}$$

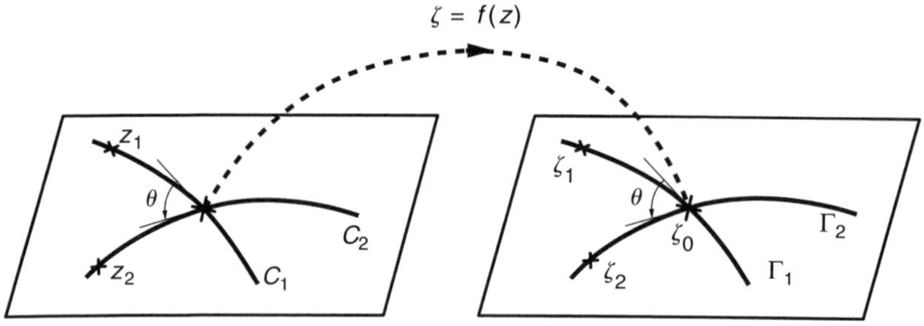

FIGURE 33.1
Let (C_1, C_2) be two regular curves with continuous tangent through the point z_0 in the z-plane; and let these be mapped by an analytic function $\zeta = f(z)$ into two curves (Γ_1, Γ_2) also regular through the point $\zeta_0 = f(z_0)$ in the ζ-plane. Since the ratio of lengths is the same $|d\zeta|/|dz| = |f'(z)|$ in all directions through the same point, the angle θ between the curves is preserved by the conformal mapping both in: (i) modulus $|\theta|$; (ii) direction $\operatorname{sgn}(\theta)$, where sgn is the sign function, equal to $+1$ (-1) for θ positive (negative).

Taking the modulus (argument) of (33.2b) leads to (33.3a) [(33.3b)]:

$$\lim_{z_1, z_2 \to z_0} \left\{ \left| \frac{f(z_1) - f(z_0)}{f(z_2) - f(z_0)} \right| \Big/ \left| \frac{z_1 - z_0}{z_2 - z_0} \right| \right\} = 1. \tag{33.3a}$$

$$\lim_{z_1, z_2 \to z_0} \frac{\arg\{f(z_1) - f(z_0)\} - \arg\{f(z_2) - f(z_0)\}}{\arg(z_1 - z_0) - \arg(z_2 - z_0)} = 1, \tag{33.3b}$$

it follows that: (i) the ratio of lengths along C_1, C_2 in the z-plane, is equal to the ratio of lengths along Γ_1, Γ_2 in the ζ-plane; (ii) the angle between the curves C_1, C_2 in the z-plane, is equal in modulus and direction to the angle between the curves Γ_1, Γ_2 in the ζ-plane. This proves that *a holomorphic function (33.1a) performs at all* **conformal points** *with derivate neither zero nor infinity (33.2a) a mapping from the z-plane to the ζ-plane that preserves (33.3a,b) at each point for all directions, the: (i) ratio of lengths; (ii) modulus of angles; (iii) direction of angles. The ratio of the length in the ζ-plane to the length in the z-plane is the modulus of the derivative of the holomorphic function:*

$$h \equiv \lim_{z \to z_0} \frac{|\zeta - \zeta_0|}{|z - z_0|} = \lim_{z \to z_0} \frac{|f(z) - f(z_0)|}{|z - z_0|} = |f'(z_0)|, \tag{33.4}$$

corresponding to the scale factor in Subsection 33.3.1. Preserving the modulus of angles (ii) does not guarantee that the direction of angles is maintained, as shown next (iii).

33.2 Inversion of Angles and Isogonal Mapping

A conformal mapping transforms (Figure 33.2) a triangle in the z-plane, into another triangle, with the same angles and possibly different size, in the ζ-plane; in contrast the inverted or image triangle, corresponds to an **isogonal mapping,** that: (i) preserves the ratio of lengths, that is, the modulus of angles; (ii) reverses the direction of angles. Hence, the conformal (isogonal) mapping differ only in the property of preserving (reversing) the direction of angles. The simplest conformal (isogonal) mapping is the identity $\zeta = z$ (conjugate

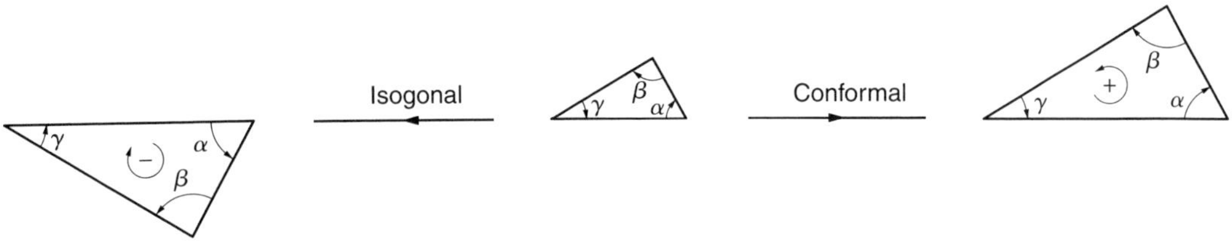

FIGURE 33.2
A conformal mapping preserves the modulus and direction of angles (Figure 33.1) and transforms a triangle into a similar triangle, with the same angles and possibly different size, if the scale factor multiplying the lengths of all sides is not unity, viz. an expansion (contraction) if the scale factor exceeds (is less than) unity. An isogonal mapping preserves the modulus of angles but reverses their direction, hence is not conformal, and maps a triangle into a "reflected image" in a different "position," possibly with different "size" but similar "shape."

$\zeta = z^*$) that is (is not) analytic; in fact the reflection on the real axis that reverses angles and is an isogonal mapping, is not analytic since it violates the Cauchy–Riemann conditions [(11.10a,b)] that are not satisfied by

$$z^* = x - iy: \quad \frac{\partial x}{\partial x} = 1 \neq -1 = \frac{\partial(-y)}{\partial y}, \quad \frac{\partial x}{\partial y} = 0 = \frac{\partial y}{\partial x}. \qquad (33.5a,b)$$

If $\zeta = f(z)$ is a conformal mapping (Figure 33.3a) then (i/ii)$f^*(z)$ $[f(z^*)]$ is an isogonal mapping [Figure 33.3b (33.3c)] obtained by inversion in the ζ-plane (z-plane); (iii) $f^*(z^*)$ is conformal (Figure 33.3d) again, since it involves reflections in both the z-plane and ζ-plane. This relates (Diagram 33.1) to the property (Section 31.2) that: (i) if $f(z)$ is analytic, then $f^*(z^*)$ also is analytic, and both specify conformal mappings; (ii) whereas $\{f(z^*), f^*(z)\}$ are not analytic, and both specify isogonal mappings; (iii) the transformation $f(z^*)$ to $f^*(z)$ is conformal again, because like (i) it involves two successive isogonal mappings (ii). It was shown before (31.11–31.19) that if $f(z)$ is holomorphic or analytic then $f^*(z^*)$ is also analytic, but $f^*(z)$, $f(z^*)$ cannot be analytic. These instances of conformal $\{f(z), f^*(z^*)\}$ [isogonal $\{f(z^*), f^*(z)\}$] mapping for analytic (nonanalytic) functions, suggest that analyticity is not only a sufficient, but also a necessary, condition for conformal mapping.

33.3 Transformation of Angles, Lengths, and Areas

The proof of the converse of the conformal property (Subsection 33.3.2) introduces in passing the plane conformal coordinates (Subsection 33.3.1).

33.3.1 Plane Conformal Coordinates

To consider the necessary (Section 33.3) instead of sufficient (Section 33.1) conditions for conformal and isogonal mappings, the starting point is a complex mapping (33.6a):

$$f(x+iy) = \zeta \equiv \Phi + i\Psi \in \mathcal{C}^1: \quad \mathrm{d}\Phi = \Phi_x\,\mathrm{d}x + \Phi_y\,\mathrm{d}y, \quad \mathrm{d}\Psi = \Psi_x\,\mathrm{d}x + \Psi_y\,\mathrm{d}y, \qquad (33.6a–c)$$

whose real Φ and imaginary Ψ parts have continuous first order derivates (33.6b,c) that are denoted in this Section 33.3 by a suffix, for example, $\Phi_x \equiv \partial\Phi/\partial x$. The arc lengths $|\mathrm{d}z|$,

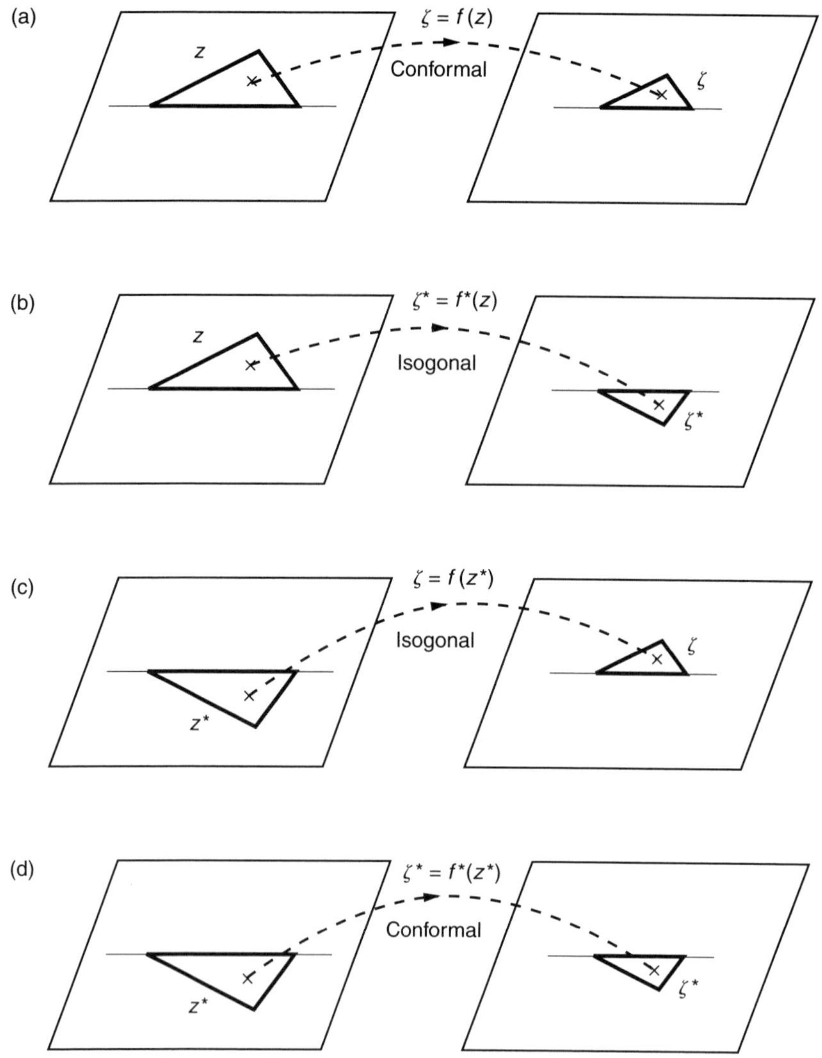

FIGURE 33.3
There are four related possibilities for complex functions $\zeta = f(z)$ [mappings from the z-plane to the ζ-plane], as shown in Figure 31.2 (33.3) and Diagram 33.1: (i) if $\zeta = f(z)$ is an analytic function the mapping is conformal and transforms a triangle into a similar triangle in (a); (ii/iii) the function $f^*(z)[f(z^*)]$ is not analytic, and the nonconformal isogonal mapping transforms the triangle into a similar image reflected in the ζ-plane (z-plane) in (b) [(c)]; (iv) the function $f^*(z^*)$ is analytic, and the double isogonal mapping with reflections both in the z-plane and ζ-plane is again a conformal mapping between similar "reflected" triangles (d).

($|\mathrm{d}\zeta|$) in the $z(\zeta)$-planes are given respectively by (33.7a) [(33.7b)]:

$$|\mathrm{d}z|^2 = |\mathrm{d}x + i\mathrm{d}y|^2 = (\mathrm{d}x)^2 + (\mathrm{d}y)^2 : \quad |\mathrm{d}\zeta|^2 = (\mathrm{d}\Phi)^2 + (\mathrm{d}\Psi)^2$$
$$= \left(\Phi_x^2 + \Psi_x^2\right)(\mathrm{d}x)^2 + 2\left(\Phi_x\Phi_y + \Psi_x\Psi_y\right)\mathrm{d}x\,\mathrm{d}y + \left(\Phi_y^2 + \Psi_y^2\right)(\mathrm{d}y)^2, \qquad (33.7\text{a,b})$$

where (33.6b,c) were used in (33.7b). The equivalent conditions of preservation at each point of (i) the ratio of lengths, or (ii) modulus of angles, in all directions $\mathrm{d}x$, $\mathrm{d}y$, require that

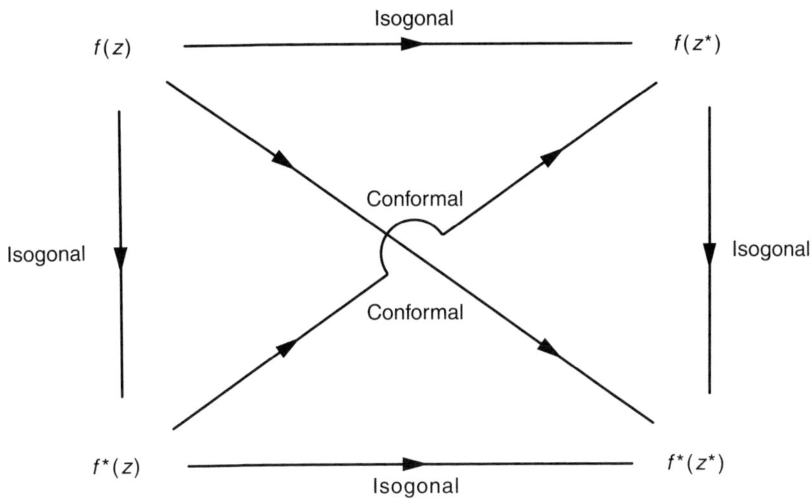

Diagram 33.1
Conformal and isogonal mappings. Starting with a mapping $\zeta = f(z)$ from the z-plane to the ζ-plane performed by an analytic function, which is conformal: (i/ii) the reflection on the real axis $z \to z^*(f \to f^*)$ in the z-plane (ζ-plane) keeps the modulus but reverses the direction of angles (Figure 33.2), and thus is an isogonal but nonconformal mapping; (iii) the reflection on the real axis both in the plane and ζ-plane $f^*(z^*)$ reverses both angles, so both the modulus and direction of angles are preserved, and the mapping is again conformal. The statements (i/ii) [(iii)] relate to the property (Figure 31.3) that if $\zeta = f(z)$ is an analytic function specifying a conformal mapping, then $f(z^*)$ and $f^*(z)[f^*(z^*)]$ is not (is) analytic, and does not (does) specify a conformal mapping.

the coefficients of $(\mathrm{d}x)^2, (\mathrm{d}y)^2$, $\mathrm{d}x\,\mathrm{d}y$ in (33.7a,b) be proportional:

$$h^2 \equiv \left|\frac{\mathrm{d}\zeta}{\mathrm{d}z}\right|^2 = \frac{\Phi_x^2 + \Psi_x^2}{1} = \frac{\Phi_y^2 + \Psi_y^2}{1} = \frac{2\left(\Phi_x\Phi_y + \Psi_x\Psi_y\right)}{0} \tag{33.8}$$

where h specifies the ratio of lengths; it is the **scale factor** (33.4) $\equiv$ (33.90) of the orthogonal curvilinear coordinate system of arc length (33.9b):

$$h = \left|\frac{\mathrm{d}\zeta}{\mathrm{d}z}\right| = |f'(z)|\,, \quad |\mathrm{d}\zeta|^2 = h^2\left[(\mathrm{d}x)^2 + (\mathrm{d}y)^2\right]. \tag{33.9a,b}$$

Comparing with general plane curvilinear orthogonal coordinates (11.43) it follows that **conformal coordinates** *are a plane curvilinear orthogonal coordinates with the same scale factor in both directions specified by the modulus of the derivative of a holomorphic function (33.4) $\equiv$ (33.9a) that is independent of direction.* The examples of conformal coordinates, that is, plane orthogonal curvilinear coordinates with the same scale factor in both directions include confocal coordinates (Subsection 36.4.4) and coaxal coordinates. The scale factor is calculated for six conformal mappings in Example 40.5 and Table 40.1.

33.3.2 Inverse Conformal Property

The conditions (33.8) are quadratic, hence insensitive to sign, so that both preserved and reversed angles are included. From the three equal ratios (33.8), follow the pair of equations:

$$\Phi_x\,\Phi_y + \Psi_x\,\Psi_y = 0, \quad \Phi_x^2 + \Psi_x^2 = \Phi_y^2 + \Psi_y^2; \tag{33.10a,b}$$

these can be solved algebraically:

$$\mu \equiv \frac{\Psi_x}{\Phi_x} = -\frac{\Phi_y}{\Psi_y}, \quad \Phi_x^2\left(1+\mu^2\right) = \Psi_y^2\left(1+\mu^2\right). \tag{33.11a,b}$$

Thus the Riemann metric (33.7b) is: (i) orthogonal [(11.43)] because (33.10a) the coefficient of $dx\ dy$ in (33.7b) vanishes, that is, the curves $\Phi = \text{const}$ and $\Psi = \text{const}$ satisfy the orthogonal condition (11.41) ≡ (33.10a); (ii) besides has the same scale factor h in (33.9a) for both the Φ- and Ψ-directions in agreement with (33.4) ≡ (33.9a,b):

$$|d\zeta|^2 = \left(\Phi_x^2+\Psi_x^2\right)(dx)^2 + \left(\Phi_y^2+\Psi_y^2\right)(dy)^2 = |f'(z)|^2\left[(dx)^2+(dy)^2\right] = |f'(z)|^2|dz|^2. \tag{33.12}$$

There are two solutions of (33.11a,b):

$$\frac{\partial\Phi}{\partial x} \equiv \Phi_x = \pm\Psi_y \equiv \pm\frac{\partial\Psi}{\partial y}, \quad \frac{\partial\Phi}{\partial y} \equiv \Phi_y = \mp\Psi_x \equiv \mp\frac{\partial\Psi}{\partial x}. \tag{33.13a,b}$$

The example of the identity $\zeta = z$ (conjugate $\zeta = z^*$) transformation $\Phi = x$, $\Psi = \pm y$, shows that the upper (lower) sign corresponds to conformal (isogonal) mapping; the expressions (33.13a,b) with upper (lower) sign correspond (Section 31.2) to the Cauchy–Riemann conditions (31.13a,b) for $f(z)$ and (31.14a,b) for $f^*(z^*)$ [(31.17a,b) for $f(z^*)$ and (31.16a,b) for $f^*(z)$] with the original (reversed) signs.

It has been shown that it is a sufficient condition for the continuously differentiable functions $\Phi, \Psi(x,y) \in C^1\left(|R^2\right)$ to specify a conformal (isogonal) mapping, that they satisfy the Cauchy–Riemann conditions with the original (reversed) sign up (down) in (11.10a,b), that is, that $f = \Phi + i\,\Psi$ $(f^* = \Phi - i\Psi)$ be analytic. The two transformations coincide in: (i) the ratio of lengths (33.14b) [(33.15b)] for the conformal (33.14a) [isogonal (33.15a)] mapping:

$$f(z) \in \mathcal{D}: \quad h = \left|\frac{d\zeta}{dz}\right| = |f'(z)| = |\nabla\Phi| = |\nabla\Psi|, \quad \frac{dA_\zeta}{dA_z} = h^2 = |f'(z)|^2 = |\nabla\Phi|^2 = |\nabla\Psi|^2, \tag{33.14a–c}$$

$$f^*(z) \in \mathcal{D}: \quad h = |\nabla\Phi| = |\nabla\Psi|, \quad \frac{dA_\zeta}{dA_z} = -h^2 = -|\nabla\Phi|^2 = -|\nabla\Psi|^2; \tag{33.15a–c}$$

(ii) the ratio of areas is the square (33.14c) [minus the square (33.15c)] of the ratio of lengths (33.14b) [33.15b)] since the angles are preserved in modulus and also (not) in direction in the conformal (33.14a) [isogonal (33.15a)] case. The ratio of areas is specified by the Jacobian of the transformation:

$$\begin{aligned}\frac{dA_\zeta}{dA_z} = \frac{\partial(\Phi,\Psi)}{\partial(x,y)} = \begin{vmatrix}\Phi_x & \Psi_x\\ \Phi_y & \Psi_y\end{vmatrix} &= \Phi_x\Psi_y - \Phi_y\Psi_x = \pm\left(\Phi_x^2+\Phi_y^2\right)\\ &= \pm|\nabla\Phi|^2 = \pm|\nabla\Psi|^2,\end{aligned} \tag{33.15d}$$

where were used (33.13a,b): for the isogonal (33.15a,c) [conformal (33.14a,c)] mapping the lower (upper) sign applies in (33.15d). The derivative $f'(z)$ exists only for the conformal mapping, and can be used in (33.14a–c).

33.4 Critical Points of the First and Second Kinds

The conformal property (Section 33.1) breaks down at a **critical point** (33.16a,b) of the first (second) kind where the derivate is zero (infinity). It follows that *a* **critical point** *of* *the first (second) kind for (33.16b) the direct mapping (33.16a), is a critical point of the second (first) kind for (33.16d) inverse mapping* (33.16c), because if one function has zero derivative its inverse has an infinite derivative:

$$\zeta = f(z): \quad \frac{\mathrm{d}\zeta}{\mathrm{d}z} = f'(z) = 0, \infty; \qquad z = f^{-1}(\zeta): \frac{\mathrm{d}z}{\mathrm{d}\zeta} = \frac{1}{f'(z)} = \infty, 0. \tag{33.16a–d}$$

At a **critical point of the first kind** *the direct (inverse) mapping has zero (infinite) derivative (33.17a) [(33.17b)] and the ratio of lengths and areas is zero:*

$$\frac{\mathrm{d}\zeta}{\mathrm{d}z} = f'(z) = 0: \quad \frac{\mathrm{d}z}{\mathrm{d}\zeta} = \infty; \quad \mathrm{d}z \neq 0, \quad \mathrm{d}\zeta = 0; \tag{33.17a–d}$$

a neighborhood (33.17c) of the critical point of the first kind z_0 in the z-plane is mapped into (33.17d) a point $\zeta_0 = f(z_0)$ in the ζ-plane as shown in on the l.h.s. of Figure 33.4.

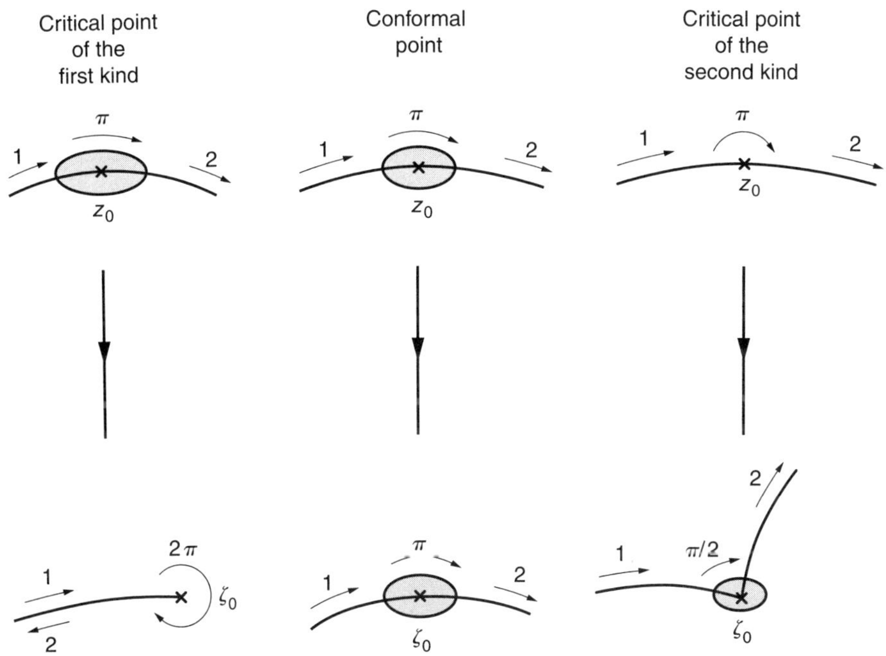

FIGURE 33.4
A regular curve, that is, with continuous tangent through a conformal point z_0 in the z-plane is mapped into a regular curve through the corresponding point ζ_0 in the ζ-plane. If ζ_0 is nonconformal it can be a critical point of the (i) first or (ii) second kind. For a critical point of the first kind: (i–1) the angle (Table 33.1) is multiplied a factor; (i–2) for example, the factor two in $\pi \to 2\pi$ reverses the regular curve in the z-plane into a sharp edge in the ζ-plane; (i–3) an example is the mapping of an uniform flow (Figure 14.10a) into the potential flow past a semiinfinite plate (Figure 14.10b). For a critical point of the second kind: (ii–1) the angle (Table 33.2) is divided by a factor; (ii–2) if the factor is two in $\pi \to \pi/2$ a regular curve is mapped into a rectangular corner; (ii–3) an example is the mapping of an uniform flow (Figure 14.10a) into the potential flow in a rectangular corner (Figure 14.10c).

Consider a critical point of the first kind corresponding to a zero of order, $N \geq 2$; in its neighborhood the function is given by (33.18c):

$$g(z) \in \mathcal{A}; g(z_0) \neq 0, \infty : \quad f(z) = (z - z_0)^N g(z), \tag{33.18a–c}$$

where (33.18a) is holomorphic, finite, and nonzero (33.18b); hence it is continuous and nonzero, so that the ratio of its limit along two paths exists and is given by

$$1 = \lim_{z_1, z_2 \to z_0} \frac{g(z_1)}{g(z_2)} = \lim_{z_1, z_2 \to z_0} \frac{f(z_1) - f(z_0)}{f(z_2) - f(z_0)} \left(\frac{z_2 - z_0}{z_1 - z_0} \right)^N; \tag{33.19a}$$

taking the argument of (33.19a):

$$\lim_{z_1, z_2 \to z_0} \frac{\arg\{f(z_1) - f(z_0)\} - \arg\{f(z_2) - f(z_0)\}}{\arg(z_1 - z_0) - \arg(z_2 - z_0)} = N, \tag{33.19b}$$

it follows that *at a critical point of the first kind that is a zero order N, the angles are multiplied by the order N of the zero of the mapping function; thus a regular curve, in the z-plane ceases to be (remains) regular in the ζ-plane, for $N = 2\,p$ even ($N = 2\,p - 1$ odd), as the direction of travel is maintained (reversed),* because the angle π in the z-plane transforms to $2\,p\pi$ $[(2\,p - 1)\,\pi]$ in the ζ-plane that is congruent to $0(\pi)$. The example $N = 2$ in Figure 33.4 on the l.h.s., shows that an angle π in the z-plane becomes 2π in the ζ-plane, that is, the curve folds back on itself and the direction of travel is reversed at the critical point of the first kind. A stagnation point of a potential flow is a critical point of the first kind, where may be joined streamlines with several directions; likewise field lines with distinct directions can converge at a point of zero field (electric, magnetic, or gravity).

33.5 Multiplication and Division of Angles into Edges

The critical points of first (Section 33.4) [second (Subsection 33.5.1)] kind are considered first for integer N (inverse integer $1/N$) exponents, that is, multiple zeros (an algebraic branch-point), as particular cases, then proceeding to the case of a general branch-point (Subsection 33.5.2).

33.5.1 Algebraic Branch-Point as the Inverse of a Zero

At a **critical point of the second kind** *the derivate is infinite (33.20a), the inverse function is stationary, (33.20b) and the ratio of lengths and areas infinite:*

$$\frac{d\xi}{dz} = f'(z) = \infty : \quad \frac{dz}{d\zeta} = 0; \quad d\zeta \neq 0, \quad dz = 0, \tag{33.20a–d}$$

a critical point (33.20d) of the second kind $z = z_0$ is mapped into (33.20c) the neighborhood of $\zeta_0 = f(z_0)$ as illustrated in the r.h.s. of Figure 33.4. If the stationary point of the inverse mapping $z = f^{-1}(\zeta)$ is a zero of order, $N \geq 2$, the direct mapping $\zeta = f(z)$ has N branches, that is, a branch-point where N roots coincide (33.21c):

$$j(z) \in \mathcal{A}; \quad j(z_0) \neq 0, \infty : \quad f(z) = (z - z_0)^{1/N}\, j(z); \tag{33.21a–c}$$

the function (33.21a) is holomorphic, finite, and nonzero (33.21b); thus it has a well-defined ratio along two paths:

$$1 = \lim_{z_1, z_2 \to z_0} \left\{ \frac{j\,(z_1)}{j\,(z_2)} \right\}^N = \lim_{z_1, z_2 \to z_0} \left\{ \frac{f(z_1) - f(z_0)}{f(z_2) - f(z_0)} \right\}^N \frac{z_2 - z_0}{z_1 - z_0}; \tag{33.22a}$$

the argument of (30.22a):

$$\lim_{z_1, z_2 \to z_0} \frac{\arg\,\{f(z_1) - f(z_0)\} - \arg\{f(z_2) - f(z_0)\}}{\arg\,(z_1 - z_0) - \arg\,(z_2 - z_0)} = \frac{1}{N}, \tag{33.22b}$$

shows that *at a critical point of the second kind obtained by inversion of a zero of order* N*, the angles are divided by the number* N *of branches, so that a regular curve in the z-plane is mapped in the* ζ*-plane into a curve having a corner with internal angle,* π/N, for example, an orthogonal corner $\pi/2$ in the case of $N = 2$ branches on the r.h.s. of Figure 33.4. Thus *(Table 33.1) if a function has a zero of order* N*: (i) the angles of the direct (inverse) mapping are multiplied (divided) by the order of the zero at the critical point of the first (33.19b) [second (33.22b)] kind; (ii) the ratio of length and areas is zero (infinite); (iii) a neighborhood (point) is mapped into a point (neighborhood), as shown in Figure 33.4 on the left- (right-) hand side. The two expressions coincide only in the case* $N = 1$ *in (33.19b)* $\equiv$ *(33.22b)* $\equiv$ *(33.3b) of a conformal point, where: (i) a regular curve is mapped into a regular curve, (iii) and neighborhood into a neighborhood, with finite (ii) nonzero ratio of lengths and areas (Figure 33.4 at the center).*

TABLE 33.1
Conformal and Critical Points

Properties	**Critical point: first kind**	**Conformal point**	**Critical point second kind**
Angles	Multiplied	Preserved	Divided
Scale factor	$h = 0$	$h \neq 0,\ \infty$	$h = \infty$
Transforms point to	Point	Point	Neighborhood
Transforms neighborhood to	Point	Neighborhood	Neighborhood
Example	Zero of order N	Arbitrary point	Algebraic branch-point of order M
Derivative	Zero	Finite and nonzero	Infinite
Angle θ becomes	$N\theta$	θ	θ/M

Note: A complex function $\zeta = f(z)$ performs a mapping from the z- to the ζ-plane, and the points where the function is analytic, that is, has a unique nonzero, finite derivative, are conformal points: (i) the angle between two curves is the same in modulus and direction in the z-plane and in ζ-plane; (ii) because all lengths in any direction are multiplied by the same scale factor $h \equiv |\mathrm{d}\zeta|/|\mathrm{d}z| = |f'(z)|$. If the derivative of the mapping vanishes (is infinite) the nonconformal point is a critical point of the first (second) kind: (i) the scale factor is zero $h = 0$ (infinity $h = \infty$) so a neighborhood (point) in the z-plane $|\mathrm{d}z| \neq 0(\mathrm{d}z = 0)$ is mapped to a point $\mathrm{d}\zeta = 0$ (neighborhood $|\mathrm{d}\zeta| \neq 0$) in the ζ-plane; (ii) the case of a zero (pole) of order N (M) suggests that the argument of the function is multiplied (divided) by N (M), so the angle of two curves is multiplied by N (M) when mapping from the z-plane to the ζ-plane.

33.5.2 Branch-Points as Critical Points of First or Second Kind

The ratio of angles is the exponent at the critical point both for: (i) a critical point of the first kind that is a zero of order N in (33.18a–c), viz. the exponent N is the ratio of angles (33.19a,b); (ii) a critical point of the second kind that is an algebraic branch-point (Subsection 7.2.3) of order $N-1$ in (33.21a–c), viz. the exponent $1/N$ is again the ratio of angles (33.22a,b). The case (ii) is generalized next to from an algebraic to a logarithmic branch-point (Subsection 7.2.2) by allowing the exponent to be any complex number; this also includes the case (i). *A critical point with exponent γ in (33.23a) is of the first (second) kind if the real part of the exponent exceeds (33.23b) [is less than (33.23c)] unity:*

$$f(z) \sim (z - z_0)^{\gamma}: \quad \lim_{z\to z_0} \frac{\mathrm{d}f}{\mathrm{d}z} \sim \lim_{z\to z_0} (z - z_0)^{\gamma-1} = \begin{cases} 0 & \text{if } \mathrm{Re}(\gamma) > 1, \\ \infty & \text{if } \mathrm{Re}(\gamma) < 1. \end{cases} \tag{33.23a–c}$$

If the exponent is real (33.24a) it specifies the ratio of angles (33.24b):

$$\gamma \in |R: \quad \lim_{z_1, z_2 \to z_0} \frac{\arg\{f(z_1) - f(z_0)\} - \arg\{f(z_2) - f(z_0)\}}{\arg(z_1 - z_0) - \arg(z_2 - z_0)} = \gamma, \tag{33.24a,b}$$

as follows taking arguments in:

$$\lim_{z_1, z_2 \to z_0} \frac{f(z_1) - f(z_0)}{f(z_2) - f(z_0)} = \lim_{z_1, z_2 \to z_0} \left(\frac{z_1 - z_0}{z_2 - z_0}\right)^{\gamma}. \tag{33.24c}$$

The choice of the exponent allows the turning of a curve at a critical point by any angle chosen a priori. This property is used in the set of four Schwartz–Christoffel transformations (Sections 33.6–33.9) of that one pair each has critical points of the first (second) kind. The four transformations map the upper half plane or circle into the interior or exterior of a polygons that may have infinite sides. Five cases of multiplication and/or division of angles at critical points of a conformal mapping are given in Example 40.5 and Table 40.2.

33.6 Interior Polygonal Transformation (Christoffel, 1868; Schwartz, 1868)

The exponent at a critical point need not be integral, and the mapping may have several critical points, for example, the mapping $\zeta = f(z)$ with derivate:

$$\gamma_n > 0: \quad \frac{\mathrm{d}\zeta}{\mathrm{d}z} = A \prod_{n=1}^{N} (z - x_n)^{-\gamma_n/\pi}, \tag{33.25}$$

has N critical points on the real axis that are of the second kind $\mathrm{d}\zeta/\mathrm{d}z = \infty$ if $\gamma_n > 0$. The ratio of angles is, γ_n/π, implying that a regular curve with angle π in the z-plane when passing through the critical point x_n maps into a curve a corner with angle γ_n. This can be confirmed by noting that as $z = x$ travels along the real axis between two points $x_{n-1} \le x \le x_n$ the $\arg(\mathrm{d}\zeta/\mathrm{d}z)$ is a constant, implying that the path is a straight line in the ζ-plane. As $z = x$ passes through $x = x_n$ the nth term of (33.25) has a change in the sign of $z - x_n$, and so the expression is multiplied, for passage above (below) the point (33.26a), by a factor (33.26b):

$$\Delta\{\arg(z - x_n)\} = \mp\pi: \; (-)^{-\gamma_n/\pi} = \{\exp(\mp i\pi)\}^{-\gamma_n/\pi} = \exp(\pm i\gamma_n); \tag{33.26a,b}$$

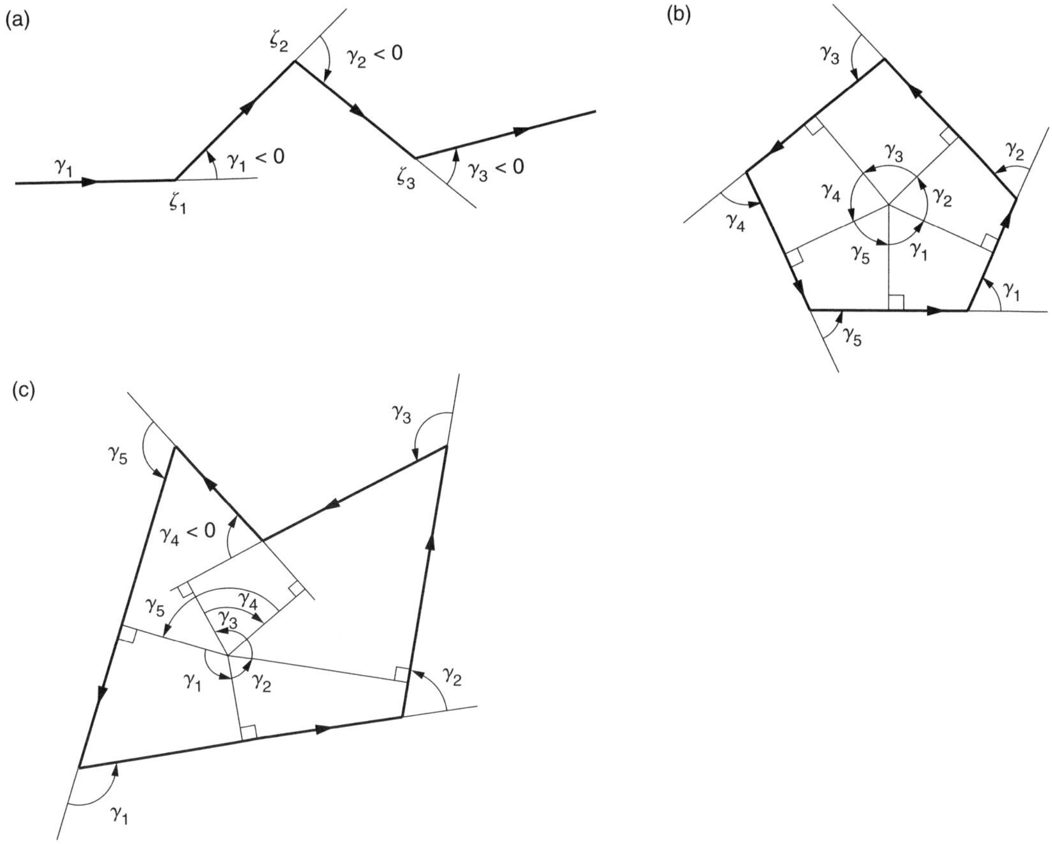

FIGURE 33.5
The critical points of a mapping, by suitable choices of the points and exponents, can be used to map a regular curve with continuous tangent, for example, a straight line or circle, into a polygonal line (a) where the tangent turns by specified angles γ_n at given corners ζ_n any number of times $n = 1, 2, 3, \ldots$. The polygonal line forms a closed polygon (b) if the external angles γ_n, which equal the angles between the normals at their common center, add to the solid angle of a full circular sector of 2π. The polygon will be convex (b) [concave (c)] if all external angles have the same sign (at least one has distinct sign). The polygon may have infinite sides (e.g., Figure 36.11 and many others) if one or more corners or vertices is at infinity; combining the preceding cases leads to a large variety of shapes specified by conformal mapping with critical points.

this implies that at the point $\zeta_n = f(x_n)$ on the ζ-plane corresponding to x_n, there is an edge with external angle $\pm\gamma_n$. Thus if $z = x$ starts before the first critical point $x < x_1$, then ζ follows a straight line, until z goes through $z = x_1$ and ζ turns through $\pm\gamma_1$ at $\zeta_1 = f(x_1)$, keeping the new direction until z passes through $z = x_2$ and ζ turns through $\arg(\zeta) = \pm\,\gamma_2$ at $\zeta_2 = f(x_2)$ and so on, forming a polygonal line (Figure 33.5a). The latter is a closed a polygon, if the external angles add to 2π; the reason (Figure 33.5b,c) is that the external angles are equal to the angles between perpendiculars taken from an internal point, and the latter add to 2π. The angles are positive (negative) at convex (concave) corners, as shown in Figure 33.5b (33.5c).

The mapping (33.25) is holomorphic except at the critical points, and so the function $\zeta = f(z)$ is continuous; thus it may be expected that if the points x_n are passed above (below) that is, in the upper (lower) half-plane, the points $\zeta_n = f(x_n)$ will be passed inside the angles $\pm\gamma_n$; it can be proved (Section 37.7) that a conformal mapping of a curve C in the z-plane into a Γ in curve ζ in the ζ-plane, transforms the interior of C into the interior of Γ. Thus has been shown that **the Schwartz–Christoffel or interior polygonal transformation** $\zeta = f(z)$ *with derivate (33.25), maps the complex-z upper (lower) half-plane* $\text{Im}(z) > 0$ $[\text{Im}(z) < 0]$, *into the interior of a polygon in the* ζ*-plane, with vertices at the points* $\zeta_n = \text{f}(x_n)$ *corresponding to the critical points, with external angles* $\pm\gamma_n$ *determined by the exponents; the two polygons are the mirror image of each other on the side ending at* ζ_1. *The polygon* is **convex (concave)** *if (33.27b) all angles are (at least one angle is) of the same (opposite) sign, for example all (some) angles are positive* $0 < \gamma_n \leq \pi$ *(negative,* $0 > \gamma_n > -\pi$*), as shown in Figures 33.5b (33.5c) and 33.6a (33.6b); it is an (33.27a) open polygonal line (closed polygon), for example, Figures 33.5a (33.5b,c) if the sum of the external angles is not (is)* 2π:

$$\sum_{n=1}^{N}\gamma_n \begin{cases} = 2\pi: & \text{closed polygon,} \\ \neq 2\pi: & \text{open line;} \end{cases} \qquad \text{polygon} \begin{cases} \text{convex } \forall_n: & \gamma_n > 0, \\ \text{concave} \exists_n: & \gamma_n < 0. \end{cases} \tag{33.27a,b}$$

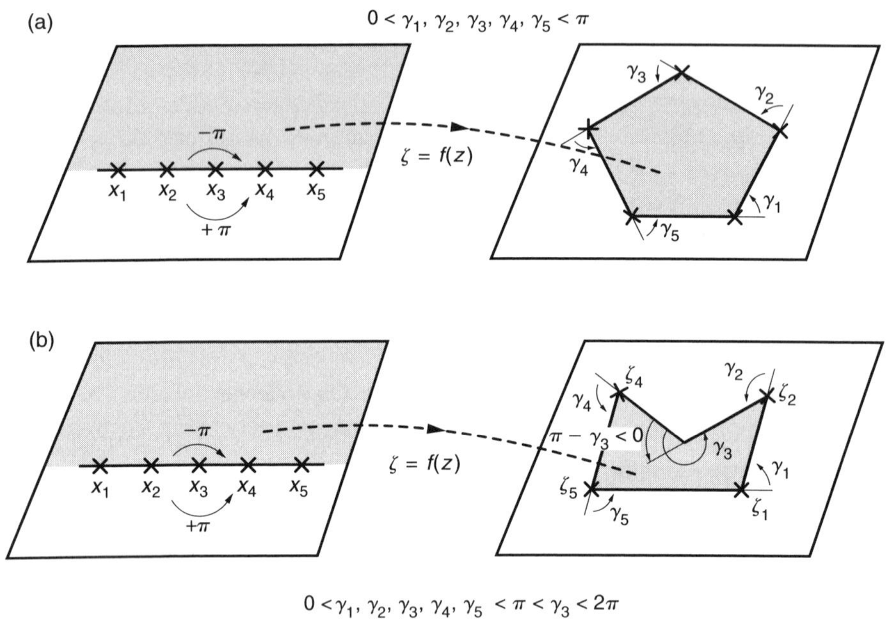

FIGURE 33.6
The first Schwartz–Christoffel transformation or interior polygonal mapping in Table 33.1 maps the upper complex z-half-plane into the interior of a closed polygon, which may be convex (concave) in Figure 33.6a ≡ 33.5b (33.6b ≡ 33.5c). The same first Schwartz–Christoffel transformation in Table 33.2 maps the lower complex half-z-plane into the image polygon (Figure 33.7a); the points outside the polygon and image polygon are not mapped into. The second Schwartz–Christoffel transformation or exterior polygonal mapping in Table 33.2 maps: (i) the upper complex half-z-plane into the exterior of the polygon in Figure 33.7b; (ii) the lower complex half-z-plane is mapped into the exterior of the image polygon in Figure 33.7b; (iii) both regions (i) and (ii) include the point at infinity and overlap also at all points outside the polygon and its image.

For example, the mapping of the upper half z-plane $\mathrm{Im}(z) > 0$ into a triangle with vertices at (33.28a,b), and external angles (33.28c), is given by (33.28d)

$$\zeta_n = \zeta(z_n);\, z_{1-3} = -1, 0, 1;\, \gamma_{1-3} = \alpha, \beta, 2\pi - \alpha - \beta: \\ \frac{\mathrm{d}\zeta}{\mathrm{d}z} = A(z+1)^{-\alpha/\pi} z^{-\beta/\pi} (z-1)^{\alpha/\pi+\beta/\pi-2}; \tag{33.28a–d}$$

in the case particular case (33.29a) of an equilateral triangle, the transformation is (33.28d):

$$\alpha = \frac{2\pi}{3} = \beta\colon \zeta = A \int \left\{z\left(z^2-1\right)\right\}^{-2/3} \mathrm{d}z + B. \tag{33.29a,b}$$

The constant $A(B)$ in (33.28d) and (33.29b) can be used to rotate (translate) the triangle.

33.7 Interior and Exterior Mappings and Point-at-Infinity

The polygonal mapping (33.25) is called **interior** since it maps the upper (lower) z-complex half plane into the interior of a polygon (its image polygon). The image polygon is obtained by reversing all angles or applying a reflection on the first side (Figure 33.7a). In the interior mapping of a finite, closed polygon, the point z_∞ that is mapped to infinity $f(z_\infty) = \infty$, need not be considered because it lies outside the region of the z-plane being transformed. In the case of the **exterior** mapping, the point-at-infinity $\zeta = \infty$ is in the mapped region, and it corresponds to z_∞ such that $f(z_\infty) = \infty$; hence $z = z_\infty$ is a singularity, and in order that the inverse function be single-valued it must be a simple pole (33.30a):

$$\zeta(z) = (z - z_\infty)^{-1} + O(1), \quad \frac{\mathrm{d}\zeta}{\mathrm{d}z} = -(z - z_\infty)^{-2} + O(1), \tag{33.30a,b}$$

corresponding to a double pole of the derivative (33.30b). In order that the $\arg(\mathrm{d}\zeta/\mathrm{d}z)$ does not change at the point at infinity, the factor (33.30b) must be inserted into (33.25); in order that ζ be real for real z, it must be multiplied by a further factor, $(z - z_\infty *)^{\;2}$; in the transformation itself there is a change from external to internal angles, so that $-\gamma_n$ is replaced by $+\gamma_n$ leading to

$$\frac{\mathrm{d}\zeta}{\mathrm{d}z} = A\left\{(z - z_\infty)(z - z_\infty^*)\right\}^{-2} \prod_{n=1}^{N} (z - x_n)^{\gamma_n/\pi}; \tag{33.31}$$

thus the critical points x_n are of first kind for $\gamma_n > 0$. It has been shown that *the* **exterior polygonal transformation** $\zeta = \mathrm{f}(\mathrm{z})$ *with derivate (33.31), maps the upper (lower) complex-*z *half-plane into the exterior of the polygon with vertices at* $\zeta_\mathrm{n} = \mathrm{f}(\mathrm{x_n})$ *corresponding to the critical points, and with internal angles* $\pm\gamma_\mathrm{n}$ *determined by the exponents; the two polygons are mirror images, and have* z_∞ *as critical point mapped to infinity,* $\zeta = \infty$.

The interior (33.25) [exterior (33.31)] polygonal transformation can be used to map the upper complex-z half-plane into the interior (exterior) of a polygon of arbitrary shape and

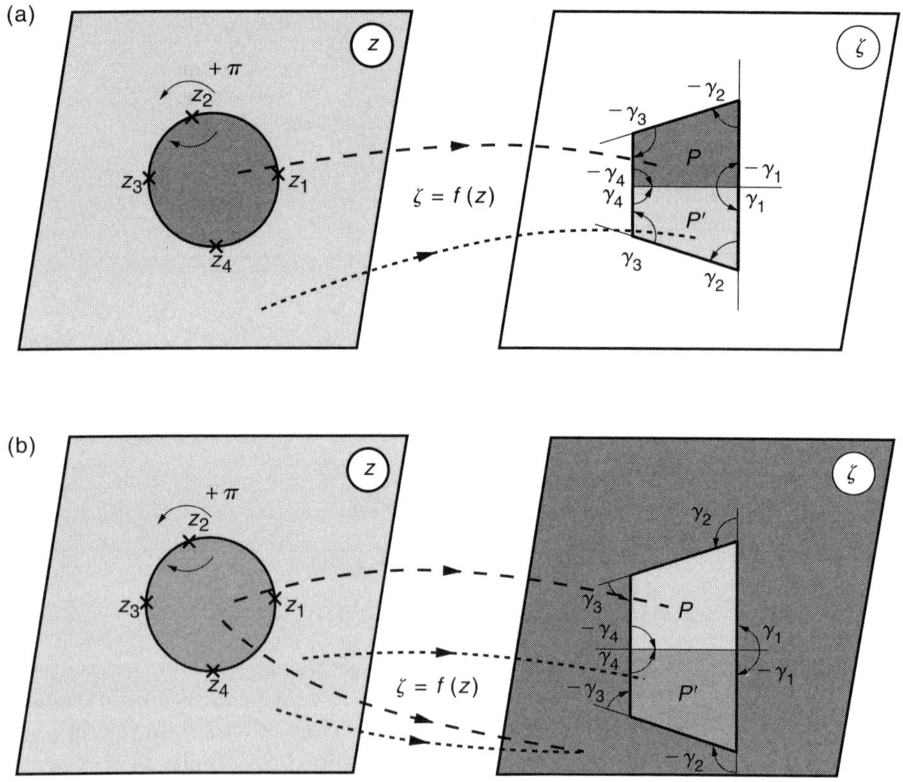

FIGURE 33.7
The first/second (third/fourth) Schwartz–Christoffel transformations or interior/exterior polygonal (disk) mappings map the z-plane divided by the real axis (unit circle) into the upper/lower half-z-plane (interior/exterior of the unit circle) in Figure 33.6 (33.7). The first/third (second/fourth) Schwartz–Christoffel transformation in Table 33.1 are polygonal/disc (i) interior [(ii) exterior] mappings [Figure 33.7a (b)]. The interior mappings **(a)**: (i–1) the (upper) complex half-z-plane or interior (exterior) of the unit circle map to the interior of the polygon (interior of the image polygon); (i–2) the real axis or unit circle is mapped to the edges of the polygon and its image; (c–3) there is no mapping to the outside of the polygon and its image, which is left "uncovered" The exterior mappings **(b)**: (ii–1) map the upper (lower) complex half-z-plane or interior/exterior of the unit disk to the outside of the polygon (image polygon); (ii–2) the real axis or unit circle is again mapped to the edges of the polygon and its image; (ii–3) as for the interior mapping the interior of the polygon (image polygon) is mapped only once, but in reverse from the lower (upper) complex half-z-plane or exterior (interior) of the unit circle; (ii–4) unlike for the interior mapping, for which the exterior of the polygon and its image is not mapped into at all, for the exterior mapping it is mapped into twice, overlapping the maps from the lower and upper complex half-z-plane or interior and exterior of the unit circle.

position in the ζ-plane, by choosing: (i) the exponents γ_n/π that determine the external angles at the vertices $\zeta_n = f(x_n)$; (ii) the critical points x_n that determine the position of the vertices, and the lengths of the sides $|\zeta_n - \zeta_{n-1}| = |f(x_n) - f(x_{n-1})| \equiv L_n$; (iii) the argument of the multiplying constant $\beta \equiv \arg(A)$ can be used to rotate the polygon through an angle β; (iv) the constant of integration B, for example, in (33.28), can be *used to translate one point ζ of the polygon to $\zeta + B$. The modulus $|A|$ of the multiplying constant causes (Sections 3.3 and 35.2) a homothetic* ***magnification (shrinking)*** *if $|A| > 1$ ($|A| < 1$), and in the case $|A| = 1$ does not change the size.* As an example consider the mapping (33.32d) of the upper half z-plane into the exterior of a closed polygon with four sides, with vertices at (33.32a,b) and angles (33.32c):

$$\zeta_n \equiv \zeta(z_n); \; z_{1-4} = \pm 2, \pm 1; \; \gamma_{1-4} = \alpha, \beta, \delta, 2\pi - \alpha - \beta - \delta:$$
$$\frac{d\zeta}{dz} = Az^{-4}(z+2)^{\alpha/\pi}(z+1)^{\beta/\pi}(z-1)^{\delta/\pi}(z-2)^{2-\alpha/\pi-\beta/\pi-\delta/\pi}; \tag{33.32a–d}$$

the case of a rectangle (33.33a) leads to (33.33b):

$$\alpha = \beta = \delta = \frac{\pi}{2}: \quad \zeta = A\int z^{-4}\sqrt{(z^2-4)(z^2-1)}\,dz + B. \tag{33.33a,b}$$

A, B in (33.32) and (33.33b) are constants that can be used to change the size, location, and orientation of the polygon.

33.8 Mapping of a Disk into the Interior of a Polygon

The transformation (33.25), with critical points of the second kind at $x_n \to z_n$ can be written in a slightly modified form:

$$\frac{d\zeta}{dz} = A\prod_{n=1}^{N}\left(1 - \frac{z}{z_n}\right)^{-\gamma_n/\pi}, \tag{33.34}$$

where: (i) the new multiplying constant (33.35a) may change the size and orientation of the polygon, but not its shape; (ii) the critical points x_n on the real axis are replaced (33.35c) by critical points z_n lying (33.35b) on a circle of radius a with arguments φ_n:

$$C \equiv A\prod_{n=1}^{N}(-x_n)^{\gamma_n/\pi}: \quad z = a\exp(i\varphi), \quad z_n = a\exp(i\varphi_n). \tag{33.35a–c}$$

The circle is mapped onto a polygon, because that for $\varphi_{n-1} < \varphi < \varphi_n$, the expression $d\zeta/d\varphi$ has constant argument:

$$\begin{aligned}
\frac{d\zeta}{d\varphi} &= \frac{d\zeta}{dz}\frac{dz}{d\varphi} = iae^{i\varphi}C\prod_{n=1}^{N}\left\{1 - \exp\left[i(\varphi - \varphi_n)\right]\right\}^{-\gamma_n/\pi} \\
&= iae^{i\varphi}C\prod_{n=1}^{N}\left\{\exp\left[\frac{i(\varphi - \varphi_n)}{2}\right] \quad (-2i)\sin\left(\frac{\varphi - \varphi_n}{2}\right)\right\}^{-\gamma_n/\pi} \\
&= iaC\exp\left(\frac{i}{2\pi}\sum_{n=1}^{N}\gamma_n\varphi_n\right)\exp\left[i\varphi\left(1 - \frac{1}{2\pi}\sum_{n=1}^{N}\gamma_n\right)\right]\prod_{n=1}^{N}\left[-2i\sin\left(\frac{\varphi - \varphi_n}{2}\right)\right]^{-\gamma_n/\pi},
\end{aligned} \tag{33.36}$$

where: (i) the second factor depends on φ and can be omitted for a closed polygon (33.27a) $\equiv$ (33.37a); (ii) the first factor is a constant C_- in (33.37b):

$$\sum_{n=1}^{N}\gamma_n = 2\pi, \quad C_- \equiv -\frac{i}{4}aC\exp\left(\frac{i}{2\pi}\sum_{n=1}^{N}\gamma_n\varphi_n\right), \tag{33.37a,b}$$

in that are included all terms independent of φ namely also

$$\prod_{n=1}^{N}(-2i)^{-\gamma_n/\pi} = 2^{-\frac{1}{\pi}\sum\limits_{n=1}^{N}\gamma_n}\prod_{n=1}^{N}\left(\mathrm{e}^{-i\pi/2}\right)^{-\gamma_n/\pi}$$
$$= 2^{-2}\exp\left[i\frac{\pi}{2}\sum_{n=1}^{N}\frac{\gamma_n}{\pi}\right] = 2^{-2}\mathrm{e}^{i\pi} = -\frac{1}{4}; \tag{33.37c}$$

(iii) the dependence on φ is thus specified by

$$\frac{\mathrm{d}\zeta}{\mathrm{d}\varphi} = C_-\prod_{n=1}^{N}\left[\sin\left(\frac{\varphi-\varphi_n}{2}\right)\right]^{-\gamma_n/\pi}, \tag{33.38}$$

that is similar to (33.25).

As φ passes through φ_n on the inside (outside), the factor $\sin\{(\varphi-\varphi_n)/2\}$ changes sign, and a factor $\exp(\pm i\pi)$ is inserted, corresponding to a change of argument (33.26a) so that the whole expression is multiplied by (33.26b), corresponding to a jump $\pm\gamma_n$ in $\arg(\mathrm{d}\zeta/\mathrm{d}\varphi)$, that is, the external angles of a polygon P (its mirror image P'). The mapping (33.34b) is internal since it is holomorphic at the origin (33.39a,b):

$$\lim_{z\to 0}\frac{\mathrm{d}\zeta}{\mathrm{d}z} = C; \quad \lim_{z\to\infty} z^2\frac{\mathrm{d}\zeta}{\mathrm{d}z} = D. \tag{33.39a,b}$$

it has a double zero at infinity (33.39b) $\equiv$ (33.40b) for a closed polygon (33.37a) $\equiv$ (33.40a):

$$\sum_{n=1}^{N}\gamma_n = 2\pi: \quad \lim_{z\to\infty} z^2\frac{\mathrm{d}\zeta}{\mathrm{d}z} = \lim_{z\to\infty}\prod_{n=1}^{N}\left(1-\frac{z}{z_n}\right)^{-\gamma_n/\pi} z^{\gamma_n/\pi}$$
$$= \lim_{z\to\infty}\prod_{n=1}^{N}\left(\frac{1}{z}-\frac{1}{z_n}\right)^{-\gamma_n/\pi} = \prod_{n=1}^{N}(-z_n)^{\gamma_n/\pi} \equiv D. \tag{33.40a,b}$$

It has been shown that the **interior disk transformation** *(33.34) maps the interior (exterior) of the circle, into the interior of a polygon with vertices at* $\zeta_n = f(z_n)$ *and external angles* $\pm\gamma_n$ *; the two polygons are closed (33.37a) mirror images. The whole complex* z*-plane is mapped into the polygon* P *and its image leaving* P' *the rest of the* ζ*-plane inaccessible both for interior polygonal (33.25) [disk (33.34)] transformation (Figure 33.7a).* Thus the internal polygonal (disk) transformation (33.25) [(33.34)] maps a point in the z-plane to a point in the ζ-plane: (i) inside the polygon P (image P') if it originates from the upper (lower) half z-plane or inside (outside) the disc; (ii) the region outside P and P' is not covered by the mapping.

33.9 Finite Interior and Overlapping Exterior

The exterior disk transformation (Subsection 33.9.1) is the last of the four Schwartz–Christoffel mappings (Sections 33.6–33.9); all four involve a number of arbitrary constants (Subsection 33.9.2).

33.9.1 Exterior Disk Transformation

The interior transformation, maps the whole z-plane into a finite, nonoverlapping region of the ζ-plane (Figure 33.7a), whereas for an exterior transformation (Figure 33.7b) the whole ζ-plane is mapped with overlap in the region exterior to the polygons. The external transformation is analogous to (33.34), with critical points of the first kind:

$$\frac{d\zeta}{dz} = C \prod_{n=1}^{N} \left(1 - \frac{z_n}{z}\right)^{\gamma_n/\pi}, \tag{33.41}$$

that is, reversed sign, as from (33.25) to (33.31). It is holomorphic at infinity (33.42a):

$$\lim_{z\to\infty} \frac{d\zeta}{dz} = C; \quad \lim_{z\to 0} z^2 \frac{d\zeta}{dz} = D, \tag{33.42a,b}$$

it has a double pole at the origin (33.42b) $\equiv$ (33.43b) for a closed polygon (33.43a) $\equiv$ (33.40a):

$$\sum_{n=1}^{N} \gamma_n = 2\pi: \quad \lim_{z\to 0} z^2 \frac{d\zeta}{dz} = A \lim_{z\to 0} \prod_{n=1}^{N} \left(1 - \frac{z_n}{z}\right)^{\gamma_n/\pi} z^{\gamma_n/\pi}$$
$$= A \lim_{z\to 0} \prod_{n=1}^{N} (z - z_n)^{\gamma_n/\pi} = A \prod_{n=1}^{N} (-z_n)^{\gamma_n/\pi} \equiv D, \tag{33.43a,b}$$

where the constant D is the same (33.40b) $\equiv$ (33.43b) for the interior (33.34) [exterior (33.41)] disk mappings. With the choice of the constant $C = 1$ in (33.42) the mapping is an identity to leading order at infinity, that is, it preserves the shape, size, and orientation of figures. The behavior at infinity of the transformation (33.41) needs examination beyond leading order:

$$\frac{d\zeta}{dz} = C\left[1 - \frac{1}{\pi z}\sum_{n=1}^{N} \gamma_n z_n + O\left(\frac{1}{z^2}\right)\right], \tag{33.44}$$

because there is a logarithmic singularity; this leads to a many-valued function that can be eliminated by setting (33.45a):

$$\sum_{n=1}^{N} \gamma_n z_n = 0; \quad \sum_{n=0}^{N} \gamma_n = 2\pi, \tag{33.45a,b}$$

in the case of a closed polygon with given angles (33.27a) $\equiv$ (33.45b), the condition (33.45a) can be met by choosing suitably the position of one of the critical points, for example, the last.

It remains to check that (33.41) transforms the circle into a polygon, that is, arg $(d\zeta/d\varphi)$ is constant between two critical points as in (33.35–33.38). In the present case, the proof uses again (33.35b,c), this time in (33.41), leading to:

$$\frac{d\zeta}{d\varphi} = iae^{i\varphi} C \sum_{n=1}^{N} \{1 - \exp[-i(\varphi - \varphi_n)]\}^{\gamma_n/\pi}$$
$$= iae^{i\varphi} C \prod_{n=1}^{N} \left[\exp\left[-\frac{i(\varphi - \varphi_n)}{2}\right] 2i \sin\left(\frac{\varphi - \varphi_n}{2}\right)\right]^{\gamma_n/\pi}$$
$$= iaC \exp\left(\frac{i}{2\pi}\sum_{n=1}^{N} \gamma_n \varphi_n\right) \exp\left[i\varphi\left(1 - \frac{1}{2\pi}\sum_{n=1}^{N} \gamma_n\right)\right] \prod_{n=1}^{N} \left[2i \sin\left(\frac{\varphi - \varphi_n}{2}\right)\right]^{\gamma_n/\pi}, \tag{33.46}$$

where: (i) the second factor that depends on φ is unity for a closed polygon (33.27a) $\equiv$ (33.45b); (ii) the constant:

$$C_+ \equiv -4iCa \exp\left(\frac{i}{2\pi}\sum_{n=1}^{N}\gamma_n\varphi_n\right), \tag{33.47a}$$

includes all factors independent of φ, such as

$$\begin{aligned}\prod_{n=1}^{N}(2i)^{\gamma_n/\pi} &= 2^{\frac{1}{\pi}\sum_{n=1}^{N}\gamma_n}\prod_{n=1}^{N}\left(e^{i\pi/2}\right)^{\gamma_n/\pi} \\ &= 2^2\exp\left(i\frac{\pi}{2}\sum_{n=1}^{N}\frac{\gamma_n}{\pi}\right) = 4e^{i\pi} = -4;\end{aligned} \tag{33.47b}$$

(iii) the remaining term depending on φ is

$$\frac{d\zeta}{dy} = C_+\prod_{n=1}^{N}\left[\sin\left(\frac{\varphi-\varphi_n}{2}\right)\right]^{\gamma_n/\pi}, \tag{33.48}$$

that is similar to (33.38).

As φ passes through φ_n on the outside (inside) a factor:

$$(-)^{\gamma_n/\pi} = \{\exp(\pm i\,\pi)\}^{\gamma_n/\pi} = \exp(\pm i\,\gamma_n), \tag{33.49}$$

appears, implying as for (33.26b) external angles $\pm\gamma_n$ at the vertices $\zeta_n = f(z_n)$. It has been shown that *the* **exterior disk transformation** *(33.41) specifies the mapping of the exterior (interior) of the circle into the exterior of a polygon with vortices at $\zeta_n = f(z_n)$ and external angles, $\pm\alpha_n$; the two polygons are closed (33.45b) mirror images, with the location of the critical points meeting the condition (33.45a), for the mapping to be single-valued. The whole complex z-plane is mapped one-to-one to the polygons $P + P'$, and two-to-one on the outside both for the exterior polygonal (33.31) [disk (33.41)] transformation (Figure 33.7b).* Thus a point in the ζ-plane: (i) inside the polygon P (image P') is mapped from the lower Im $z < 0$ (upper Im $z > 0$) complex-z half-plane or inside $|z| < R$ (outside $|z| > R$ the circle); (ii) outside both of the polygon and its image, is mapped from two points z_1, z_2, such that $\zeta(z_1) = \zeta(z_2)$, and one lies in the upper, and the other in the lower, complex-z half-plane Im $(z_1) > 0 >$ Im (z_2) or one inside $|z_1| < R$ (and the other outside $|z_2| > R$) the disk. For each critical point that is mapped to infinity, the polygon has two infinite sides, that is, straight lines.

33.9.2 Number of Arbitrary Constants for the Four Schwartz–Christoffel Mappings

The four Schwartz–Christoffel mappings (Table 33.2) consist of: (I/III) the interior polygonal (disk) transformation of the upper half z-plane (Figure 33.6a,b and Section 33.6) [interior of the unit circle (Figure 33.7a and Section 33.8)] into the interior of a polygon, and associated mapping of the lower half z-plane (exterior of the circle) into the interior of the image polygon, leaving the rest of the ζ-plane uncovered (Figure 33.7a); (II/IV) the exterior polygonal (disk) transformation of the upper half z-plane (Section 33.7) [interior of the unit circle (Section 33.9)] into the exterior of a polygon, and the associated mapping of the lower half z-plane (exterior of a circle) into the exterior of the image polygon, so that (Figure 33.7b) there is (i) no overlap inside the polygons and (ii) there is overlap outside the

TABLE 33.2
Four Schwartz–Christoffel Transformations

Number from to polygon	I half-plane interior	II half-plane exterior	III circle interior	IV circle exterior
$d\zeta/dz$	(33.25)	(33.31)	(33.34)	(33.41)
Section	33.6	33.7	33.8	33.9.1
Figure	33.6a,b	–	33.7a	33.7b
Interior polygon	$\text{Im}(z) > 0$	–	$\lvert z\rvert < a$	–
Exterior polygon	–	$\text{Im}(z) > 0$	–	$\lvert z\rvert < a$
Interior image	$\text{Im}(z) < 0$	–	$\lvert z\rvert > a$	–
Exterior image	–	$\text{Im}(z) < 0$	–	$\lvert z\rvert > a$
Parameters	$2N+4$	$2N+6$	*	*
Closed polygon (33.50a)	$2N+3$	$2N+5$	$2N+3$	*
Added constraint (33.50b)	$2N+2$	$2N+4$	$2N+2$	$2N+2$

P: polygon with external angles γ_n; P': image polygon with external angles $-\gamma_n$.
* : not valid without constraints.
Note: There are four Schwartz–Christoffel transformations which map from a half-plane (circle) to the interior (exterior) of a polygon, that is, the interior (exterior) polygonal (disk) mappings. This raises several questions: (i) if the upper half z-plane (interior of the unit circle) is mapped to the interior (exterior) of a polygon, to where is mapped the lower half-z-plane (exterior of the unit circle)? (ii) the mapping to the exterior of the polygon includes the point-at-infinity in the range $\infty = f(z_\infty)$ for which point z_∞ in the domain? (iii) the boundary, that is, real axis (unit circle) is always mapped to the boundary of the polygon or also elsewhere?; (iv) how many independent parameters appear in each of the four mappings for a polygon with N sides? Some of the answers to these issues appear in this table and the remaining in Figures 33.5–33.7.

polygons. After completing the discussion of the four polygonal mappings (viz. I $\equiv$ polygonal interior in Section 33.6; II $\equiv$ polygonal exterior in Section 33.7; III $\equiv$ disk interior in Section 33.8; IV $\equiv$ disk exterior in Subsection 33.9.1) and before proceeding to various applications (Chapters 34, 36, and 38), the number of arbitrary constants involved in each mapping is considered (Subsection 33.9.2) in all four cases.

The polygon in the ζ-plane is uniquely specified by the positions $\zeta_n \equiv \zeta(z_n)$ of the N vertices, that is, a total of $2N$ parameters. Thus if a Schwartz–Christoffel transformation has more than $2N$ parameters say $2N + S$, the excess S parameters are dependent on the other $2N$, as indicated in Table 33.2. The interior polygonal (33.25) [disk (33.34)] transformation is of the second kind and has $2\,N+4$ independent parameters, that is, 4 excess parameters, viz.: (i) the N coordinates x_n (angles φ_n) on the real axis (33.25) [on the circle (33.35c)] of the N critical points; (ii) the N angles γ_n at the critical points (33.27a) of second kind; (iii) two parameters for the complex arbitrary multiplying constant A, viz. the modulus (phase) that specifies a homothety $|A|$ [rotation arg (A)]; (iv) two parameters for the arbitrary complex constant of integration B that is added to $\zeta(z)$, and represents a translation. If the polygon is closed the condition (33.27a) eliminates one angle, for example, (33.50a) and reduces total the number of independent parameters to $2N+3$, that is, 3 excess parameters:

$$\gamma_N = 2\pi - \sum_{n=0}^{N-1} \gamma_n; \quad z_N = \frac{\sum_{n=0}^{N-1} \gamma_n z_n}{\gamma_N}, \tag{33.50a,b}$$

the exterior polygonal transformation (33.31) adds the two coordinates of the point z_∞ mapped to infinity, for a total of $2N+6$ $(2N+5)$ independent parameters for a [closed (33.50a)] polygon, that is, 6 (5) excess parameters. The exterior disk transformation (33.41) imposes the unicity condition (33.45a), for example, to specify the position of the last critical point (33.50b), reducing the number of independent parameters from $2N+5$ $(2N+3)$ for an open to $2N+4$ $(2N+2)$ for a closed (33.50a) polygon, that is, 4 (3) excess parameters in the case of the half-plane exterior (interior) transformation. The number of parameters for the circle interior transformation is the same as for the half-plane interior transformation, with the constraint (33.50a) always imposed; for the circle interior transformation the constraint (33.50b) must also be imposed for unicity.

Note 33.1. *Mapping into "Polygons" with Infinite and Collapsed Sides:* The polygonal transformations (Sections 33.6–33.9) are not only a good illustration of the general properties (Sections 33.1–33.5) of the conformal mapping, but also have many applications; of particular interest is the ability to map into "polygons" with infinite or collapsed sides, including single or multiple plates, channels, ducts, bends, wells, steps, etc. (Chapters 34 and 36) as well as jets with free boundaries (Chapter 38). The basic conformal mappings, for example, linear and bilinear are addressed next (Chapter 35), before concluding with general properties, like (Chapter 37) existence of conformal mapping into arbitrary regions, matching of interior and exteriors in the z-plane and ζ-plane, etc. Since the conformal mapping is performed by an analytic function, the properties of the latter (Chapter 31), including singularities (Chapter 39) are also relevant. The conformal mappings, in particular the polygonal and disk interior and exterior mappings, are one of the most powerful tools to solve potential field problems in two dimensions, and will be applied in subsequent Chapters 34, 36, and 38 of the present last part of this volume; these are a sample of the variety applications to the theory complex functions and two-dimensional potential fields. The conformal mapping also specifies various curvilinear orthogonal coordinates, with the same scale factor in the two directions (Subsection 33.3.1). The conformal mapping can be extended to any number of dimensions in the context of metric spaces and the tensor calculus.

Conclusion 33: The conformal mapping (figure 33.1) from the z- to the ζ-plane, maps two curves C_1, C_2 passing through z_1, z_2 and intersecting at z_0 in two curves Γ_1, Γ_2, passing through ζ_1, ζ_2 and intersecting at ζ_0 in an angle equal in modulus and direction; the isogonal mapping (figure 33.2) differs in reversing the angles, i.e., the conformal (isogonal) mapping transforms an infinitesimal triangle into a similar (reversed) triangle with a different scale and thus preserves (reverses) the orientation. As shown in Diagram 33.1, starting (figure 33.3) with a conformal mapping (a): (i/ii) the conjugate in the domain (c) [range (b)] leads to an isogonal mapping; (iii) both conjugates lead back to a conformal mapping (d). At a conformal point (figure 33.4) a regular curve is mapped into another regular curve, and a neighborhood into a another neighborhood (center); at a critical point of the first (second) kind, e.g., of order two [on the left (right) hand side], a neighborhood is mapped into a point (vice-versa), and a regular curve into a curve reversed onto itself (gains a square corner). There are two types of Schwartz-Christoffell polygonal transformations, mapping (figure 33.5) the real axis into a polygonal line (a), which becomes a closed polygon if the external angles add to 2π, viz. convex (b) if all angles have the same sign, and concave (c) if at least one angle has distinct sign. The interior polygonal transformation, e.g., with five critical points, performs the mapping of the upper complex-z half-plane into a pentagon in the ζ-plane, (figure 33.6) convex in (a), and with concave corner in (b). There is also an exterior polygonal mapping, and the disk transformation also has two forms, namely (figure 33.7) interior (a) [exterior (b)] which maps: (i) the interior (exterior) of the disk into the interior (exterior) of the polygon P; (ii) and the exterior (interior) of the disk into the interior (exterior) of the image polygon, (iii) thus the outside of the polygon $P+P'$ is not covered (is covered twice).

34

Wing Sections and Planforms

The simplest external disk transformation (Subsection 33.9.1) has two diametrically opposite critical points, and maps the circle into a flat plate (Section 34.1); since the flow past a flat plate without incidence remains uniform, the inverse transformation is a third way to specify the flow past a circle. Thus the flow past a circular cylinder can be obtained by four methods, viz.: (i) by the circle theorem (Section 24.7); (ii) as a limit of the Rankine oval (Section 28.5); (iii) through the insertion of a dipole (Section 28.6); (iv) via the inversion of the Schwartz–Christoffel transformation (Section 34.1). The simplest case of external disk mapping is the Joukowski transformation, and when applied to a circle it leads to: (i) an ellipse if the center is the origin, viz. the potential flow past an elliptic cylinder (Section 34.2), including the flat plate (circular cylinder) as the particular case of zero smaller half-axis (equal half-axis); (ii) a circular arc if the center is on the imaginary axis; (iii) if the center is on the real axis, a symmetric airfoil with rounded leading-edge and sharp trailing-edge; (iv) if the center is off both axes, the airfoil (v) with the circular arc as midline yields a cambered or unsymmetric airfoil. These Joukowski airfoils (Section 34.3) have a sharp trailing edge, where the velocity is made finite by coincidence with a stagnation point; this Kutta condition of specifies the circulation, and thus the lift on the airfoil, regardless of whether it is: (i) a flat plate (Section 34.1); (ii) a Joukowski airfoil (Sections 34.2–34.4). The trailing edge need not be sharp, and the Kutta conditions may not apply Section 34.5 to: (iii) other parametric families of airfoils, for example, von Karman–Trefftz or von Mises; (iv) an arbitrary airfoil.

The lift, drag, and pitching moment coefficients (Section 34.6) also apply to a three-dimensional wing, of that the airfoil is a two-dimensional section. The two-dimensional airfoil with Joukowski (Sections 34.2–34.4) or other (Section 34.5) profiles corresponds three-dimensionally to a wing of "infinite" span, that is, a section of a wing far from the tips. The "tip" effect of a wing of finite span is represented by replacing the wing by a Prandtl "lifting line" (Section 34.7), that is, a distribution of circulation along the span at the center of each airfoil section; the methods that apply to a lifting-line can also be extended to a lifting-surface (Note 38.1). An arbitrary distribution of circulation for a symmetrically loaded wing can be represented by a Fourier sine series, and the total circulation specifies the lift of the wing. The distribution of circulation with finite span also induces a downwash velocity perpendicular to the incident stream that causes a drag (Section 34.8); the latter is a three-dimensional "wing" effect absent in two-dimensional "airfoil" theory. In the straight, level and steady flight of aircraft or birds the weight is balanced by lift, and the drag is overcome by thrust; thus it is desirable to have the least drag, that is, to need the minimum thrust for a given weight or lift. This condition is met by a wing with uniform downwash along the span that is achieved if the circulation varies elliptically. The case of "elliptic loading" (Section 34.9), is one instance of the aerodynamics of two-dimensional airfoils and three-dimensional wings; these are subjected to forces such as lift and drag, and moments such as pitch, represented by dimensional scaling laws and dimensionless coefficients that depend on wing geometry, that is, airfoil sections and planform shape. For incompressible flow the drag consists of induced drag due to the downwash associated with lift, and form drag due to viscosity that is independent of lift; the induced (form) drag dominates at low (high) speed, and there is an intermediate minimum drag speed at that the total drag is minimum, and the form and induced drag are equal. The long-range flight of aircraft depends on generating enough lift to balance the weight, while incurring in the minimum possible drag, to reduce the required engine thrust and associated fuel consumption; thus lift-to-drag ratio is a measure of aerodynamic efficiency, for example, of airfoils and wings; values of 20 are achieved by airliners in cruise and gliders can exceed 50.

34.1 Flow past a Flat Plate and Kutta (1902b) Condition

The exterior mapping of the circle (Subsection 33.9.1) onto a flat plate (Subsection 34.1.1) specifies the potential flow with an angle-of-attack (Subsection 34.1.2). Placing a stagnation point at the trailing edge ensures a finite velocity there (Subsection 34.1.3) and determines the lift (Subsection 34.1.4); the suction force at the leading-edge (Subsection 34.1.3) balances the horizontal component of the lift. The vertical component of the lift is associated with a pitching moment (Subsection 34.1.6).

34.1.1 Exterior Disk Mapping to a Flat Plate

The simplest external disk transformation (Subsection 33.9.1) with angles $0 < \gamma_n \le \pi$ in (33.41) applies to a polygon (33.45b) with (34.1a) two coincident sides, that is, a flat plate (Figure 34.1a); the two critical points must satisfy (33.45a), that is, $z_1 + z_2 = 0$; thus they must be diametrically opposite, for example, lie on the real axis (34.1b) of a circle of radius a and center at the origin. The transformation (33.41) with (34.1a,b) leads to (34.1c):

$$\gamma_1 = \pi = \gamma_2; \quad z_1 = a = -z_2: \quad \frac{\mathrm{d}\zeta}{\mathrm{d}z} = \left(1 - \frac{a}{z}\right)^{\pi/\pi} \left\{1 - \left[\frac{(-a)}{z}\right]^{\pi/\pi}\right\} = 1 - \frac{a^2}{z^2}. \tag{34.1a–c}$$

It can be checked that the conformal mapping (34.1c) transforms the circle (34.2a) of radius, a, into a flat plate (34.2b):

$$z = a\mathrm{e}^{i\varphi}: \zeta = z + \frac{a^2}{z} = a\left(\mathrm{e}^{i\varphi} + \mathrm{e}^{-i\varphi}\right) = 2a\cos\varphi = x; \ 0 \le \varphi < 2\pi \Leftrightarrow -2a \le x < +2a, \tag{34.2a–d}$$

thus (34.2c) corresponds to the segment (34.2d) on the real axis of the ζ-plane. The flat plate of **chord** or length $4a$ lies parallel to the streamlines, so that it does not disturb (Figure 34.1a) a uniform stream of velocity U in the ζ-plane:

$$g_0(\zeta) = U\zeta = U\left(z + \frac{a^2}{z}\right) = f_0(z); \tag{34.3}$$

thus (34.3) specifies the potential flow past a circular cylinder in the z-plane. The potential flow with uniform velocity U at infinity along the x-axis past a circular cylinder of radius a, has been obtained in four distinct but equivalent ways: (i) by using the circle theorem (Section 24.7) to specify (24.47) the perturbation potential $U\ a^2/z$ due to the cylinder; (iii) as a limit (28.90a,b) of the Rankine oval (Section 28.5); (iii) by inserting a dipole (Section 28.6) of moment $P_1 = -2\pi a^2 U$ in (28.89a,b) an uniform flow of velocity U; (iv) by inverting (34.3) the mapping (34.1c) of the cylinder into a flat plate (Subsection 34.1.1).

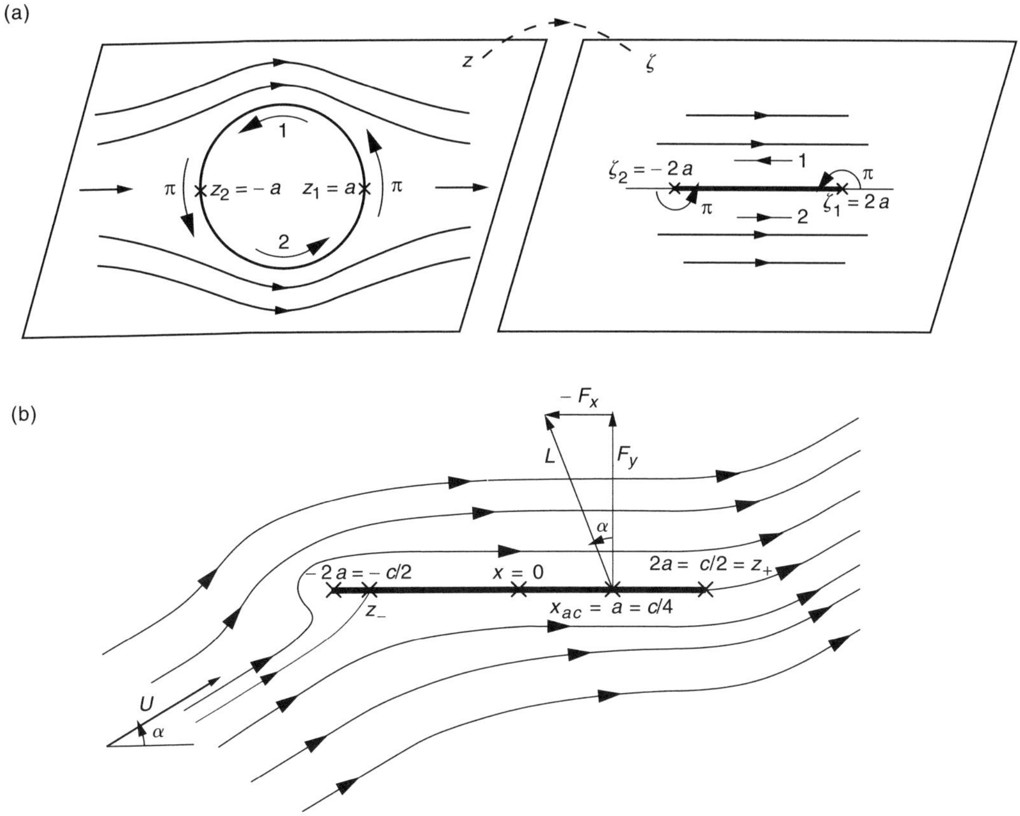

FIGURE 34.1
The Schwartz–Cristoffel exterior polygonal transformation with two critical points with angles $\gamma_1 = \pi = \gamma_2$ maps the circle of radius a onto a flat plate (a) of chord $c = 4a$. A flat plate does not (does) disturb a uniform stream if it is aligned (a) [it is not aligned (b)] with the free stream. In the latter case (b) of a free stream of velocity U incident on a flat plate with angle-of-attack, the velocity has an inverse square root $1/\sqrt{z}$ singularity at the leading and trailing edges, as for a semiinfinite plate (Figure 14.10b). By placing one stagnation point at the trailing-edge, the simple zero $z \sim \sqrt{\zeta}$ compensates for the singularity and leads to a finite velocity at the trailing-edge. The Kutta condition placing the stagnation point at the trailing-edge specifies the circulation around the flat plate; the latter determines by the Joukowski theorem the lift on the flat plate. The lift force is orthogonal to the free stream velocity. The component tangent to the flat plate is compensated by the suction force at the leading-edge; the latter is associated with the inverse-square-root singularity at the leading-edge which remains, since the second stagnation points is located behind on the underside of the flat plate. The lift force is applied at the aerodynamic center, where the pitching moment is zero; for a flat plate the aerodynamic center is located at a distance one-quarter of the chord from the leading edge.

The potential flow past a circular cylinder (34.3), with negative circulation $\Gamma < 0$ to give lift, is taken as reference case for further conformal transformations (34.4b):

$$z \to z\mathrm{e}^{-i\alpha} : f(z) = U\left(\mathrm{e}^{-i\alpha}z + \mathrm{e}^{i\alpha}\frac{a^2}{z}\right) - i\frac{\Gamma}{2\pi}\log z, \qquad \text{(34.4a,b)}$$

the substitution (34.4a) is equivalent to a rotation by α in the positive direction (16.55a,b; 16.57b) and implies that the free stream has an angle-of-attack α; a constant term $-i\alpha\Gamma/2\pi$ was omitted in (34.4b), since it does not affect the conjugate velocity:

$$\frac{\mathrm{d}f}{\mathrm{d}z} = U\left(\mathrm{e}^{-i\alpha} - \mathrm{e}^{i\alpha}\frac{a^2}{z^2}\right) - \frac{i\Gamma}{2\pi z} = U\mathrm{e}^{-i\alpha}\left[1 + O\left(\frac{1}{z}\right)\right]; \qquad (34.4c)$$

it leads at infinity $z \to \infty$ to an uniform flow $U\mathrm{e}^{-i\alpha}$ with velocity U and angle-of-attack α. The stagnation points (34.5a) of (34.4c) satisfy (34.5b):

$$\frac{\mathrm{d}f}{\mathrm{d}z} = 0: \quad 0 = z^2 - \frac{i\Gamma}{2\pi U}\mathrm{e}^{i\alpha}z - a^2\mathrm{e}^{2i\alpha} = (z - z_+)(z - z_-); \qquad (34.5a,b)$$

for (28.126a–c) subcritical circulation (34.6a):

$$\frac{-\Gamma}{4\pi U a} \equiv \sin\beta < 1: \quad z_\pm = \frac{i\Gamma}{4\pi U}\mathrm{e}^{i\alpha} \pm \mathrm{e}^{i\alpha}\sqrt{a^2 - \left(\frac{\Gamma}{4\pi U}\right)^2}$$
$$= -a\mathrm{e}^{i\alpha}(i\sin\beta \mp \cos\beta) = \pm a\exp\{i(\alpha \mp \beta)\}; \qquad (34.6a,b)$$

the stagnation points (34.5b) $\equiv$ (34.6b) lie on the cylinder $|z_\pm| = a$ in the directions arguments $\pm\beta$ as in (28.126a–c), to that must be added the angle-of-attack α of the free stream at infinity. The conjugate velocity (34.4c) $\equiv$ (34.7b) depends: (i) explicitly on the position of the stagnation points (34.5b):

$$\Gamma = -4\pi U a\sin\beta : v^* = \frac{\mathrm{d}f}{\mathrm{d}z} = U\mathrm{e}^{-i\alpha}z^{-2}\left[z^2 - i\frac{\Gamma}{2\pi U}\mathrm{e}^{i\alpha}z - a^2\mathrm{e}^{2i\alpha}\right]$$
$$= U\mathrm{e}^{-i\alpha}\left(1 - \frac{z_+}{z}\right)\left(1 - \frac{z_-}{z}\right) = U\mathrm{e}^{-i\alpha}\left\{1 - \frac{a}{z}\mathrm{e}^{i\alpha - i\beta}\right\}\left\{1 + \frac{a}{z}\mathrm{e}^{i\alpha + i\beta}\right\}; \qquad (34.7a,b)$$

(ii) implicitly on the circulation (34.6a) $\equiv$ (34.7a) through (34.6b). In (34.7b): (i) the flow velocity and angle-of-attack at infinity, is specified by (U, α); (ii) also (a, β) determines the radius of the cylinder and the circulation $-\Gamma$ around it (34.7a); (iii) $c = 4a$ is the chord or length of the flat plate (34.2d).

34.1.2 Flat Plate at Angle-of-Attack in a Stream

The inverse (34.8b) $\equiv$ (34.8c) of (34.2b) $\equiv$ (34.8a):

$$\zeta = z + \frac{a^2}{z}, \quad z^2 - \zeta z + a^2 = 0, \quad 2z = \zeta \pm \sqrt{\zeta^2 - 4a^2}, \quad F(\zeta) = f(z(\zeta)), \qquad (34.8a–d)$$

substituted in (34.4b) specifies the complex potential (34.8d) in the ζ-plane for the potential flow with velocity U and angle-of-attack α past a flat plate (Figure 34.1b) of chord $c = 4a$. The transformation (34.8b) applied to a uniform flow of velocity U in the z-plane (34.9a):

$$g(z) = Uz = \frac{U}{2}\left\{\zeta \pm \sqrt{\zeta^2 - 4a^2}\right\} = G(\zeta), \quad \frac{\mathrm{d}G}{\mathrm{d}\zeta} = \frac{U}{2} \pm \frac{U\zeta}{2}\left(\zeta^2 - 4a^2\right)^{-1/2}, \qquad (34.9a,b)$$

leads to a complex potential (34.9a) [conjugate velocity (34.9b)]; the complex conjugate velocity has branch-points at $z = \pm a$ with exponent $-1/2$ as predicted (14.91a,b; 14.92a–c) for a flat plate the basis of its two sharp edges at $\zeta = \pm 2a$. *The complex potential in the z-plane (34.4b) appears in (34.8d), and the conjugate velocity in the z-plane (34.7b) and external disk mapping (34.1c) specify, the conjugate velocity in the* ζ*-plane:*

$$v^*(\zeta) = \frac{dF}{d\zeta} = \frac{df/dz}{d\zeta/dz} = U e^{-i\alpha} \frac{z - a e^{i\alpha - i\beta}}{z - a} \frac{z + a e^{i\alpha + i\beta}}{z + a}; \qquad (34.10)$$

this shows that: (i) the velocity is unchanged $U e^{-i\alpha}$ *in magnitude* U *and angle-of-attack* α *at infinity* $z \to \infty$*, because the conformal mapping (38.8a) of the cylinder of radius* a *to the flat plate of chord* $c = 4a$ *causes no asymptotic deformation of the flow* $\zeta \to z$ *in (34.2b) and* $d\zeta/dz \to 1$ *in (34.1c); (ii) it is zero at the points* $\zeta_\pm$ *corresponding to the stagnation points* $z_\pm$ *on the cylinder (34.6a,b), unless these coincide with critical points; (iii) at the critical points* $z^\pm = \pm a$ *corresponding to the edges of the plate* $\zeta^\pm = \pm 2a$ *the velocity is infinite, unless they coincide with stagnation points. Thus the singularities in the velocity field, for example, arising from critical points of the first kind of the conformal mapping, may be eliminated by coincidence with stagnation points; this* ***Kutta condition (1902b)*** *determines the circulation around the airfoil, because the latter is specified by the position of the stagnation points. The complex potential (conjugate velocity) in the* ζ*-plane (34.2b) of the flat plate (34.2d) is given by (34.4b; 34.8d) [(34.10)] as a function of the coordinate in the z-plane of the circle, which was mapped conformally onto the flat plate airfoil; for example the center of the airfoil* $\zeta = 0$ *corresponds (34.2b) to* $\varphi = \pm\pi/2$*, and the complex conjugate velocity is given by (34.4c) with* $z = +ia$ *above (*$z = -ia$ *below):*

$$v^*(\pm ia) = U\left(e^{-i\alpha} + e^{i\alpha}\right) - \frac{i\Gamma}{2\pi(\pm ia)} = 2U\cos\alpha \mp \frac{\Gamma}{2\pi a} = v_x^\pm, \quad v_y^\pm = 0. \qquad (34.11a,b)$$

The Cartesian components of the velocity show that it is tangent to the flat plate and differs on the upper and lower side due to the circulation; a clockwise or negative circulation $\Gamma < 0$ *leads to a higher velocity on the upper side* $v^+ > v^-$*, hence a lower pressure* $p^+ < p^-$ *by the Bernoulli equation (14.27b), and an upward vertical force, that is, a lift; conversely a positive circulation would lead to a downforce.*

34.1.3 Suction Force at a Sharp Edge

In the case of the flat plate (34.10), the two critical points coincide with the two stagnation points only in the case $\alpha = 0 = \beta$ of zero incidence and circulation, when the flow will be uniform. The flow will not be uniform (Figure 34.1b), if one critical and one stagnation point are made to coincide (34.12a), viz. at the trailing-edge $z^+ = a = z_+$, where $c = 4a$ is the chord (34.2d). The latter condition $a = z_+$ implies (34.12b) by (34.6b); then the circulation (34.7a) [velocity (34.10)] is given by (34.12c) [(34.12d)]:

$$z^+ = z_+ = a = \frac{c}{4}; \alpha = \beta: \quad \Gamma = -\pi U c \sin\alpha, \quad v^*(z) = U e^{-i\alpha} \frac{z + a e^{i2\alpha}}{z + a}. \qquad (34.12a\text{–}d)$$

The velocity is still singular $df/d\zeta = \infty$ at the leading edge $z = -a$, unless the angle-of-attack is zero $\alpha = 0$; in that case there is no circulation $\Gamma = 0$ and the flow is uniform $df/d\zeta = U$. For positive angle of attack $\alpha > 0$ the forward stagnation point $z_- = a e^{i\beta}$ lies (34.6b) on the underside of the flat plate (Figure 34.1c), leaving the singularity (34.9a,b) at the leading-edge, and associated with a suction force (Subsection 34.1.3). The flat plate corresponds to the circle (34.2a) of radius, a, and center at the origin, and on it

$$v^*\left(a e^{i\varphi}\right) = U \frac{e^{i\varphi - i\alpha} + e^{i\alpha}}{e^{i\varphi} + 1} = U\left(\cos\alpha + \frac{\sin\alpha \sin\varphi}{1 + \cos\varphi}\right) = v_x(\varphi), \qquad (34.13a)$$

the velocity is horizontal. The simplification in (34.13a):

$$\frac{v^*\left(a\mathrm{e}^{i\varphi}\right)}{U}=\frac{\mathrm{e}^{i\varphi-i\alpha}+\mathrm{e}^{i\alpha}}{\mathrm{e}^{i\varphi}+1}=\frac{(\mathrm{e}^{i\varphi-i\alpha}+\mathrm{e}^{i\alpha})(\mathrm{e}^{-i\varphi}+1)}{(\mathrm{e}^{-i\varphi}+1)(\mathrm{e}^{i\varphi}+1)}$$

$$=\frac{\mathrm{e}^{-i\alpha}+\mathrm{e}^{i\alpha-i\varphi}+\mathrm{e}^{i\varphi-i\alpha}+\mathrm{e}^{i\alpha}}{2+\mathrm{e}^{i\varphi}+\mathrm{e}^{-i\varphi}}=\frac{\cos\alpha+\cos(\alpha-\varphi)}{1+\cos\varphi}$$

$$=\frac{\cos\alpha+\cos\alpha\cos\varphi+\sin\alpha\sin\varphi}{1+\cos\varphi}=\cos\alpha+\frac{\sin\alpha\sin\varphi}{1+\cos\varphi}=v_x(\varphi), \qquad (34.13\text{b})$$

confirms (34.13a) ≡ (34.13b) that the velocity is tangent to the plate (34.14a):

$$v_y(\varphi)=0; \quad v_x\left(\varphi=0\right)=U\cos\alpha; \qquad (34.14\text{a,b})$$

on the plate the velocity (34.13a) is horizontal (34.14a) and the Kutta conditional implies that it is finite (34.14b) at the trailing edge, and equal to the horizontal projection of the free stream velocity. The inverse square root singularity (34.9b) of the velocity at the leading edge remains, leading to a suction force as shown next. *A velocity varying (14.75c) ≡ (34.15a) like the inverse power of distance in a corner of angle β leads:*

$$v^*(z)\sim Cz^{\pi/\beta-1}, \quad p\sim\frac{\rho}{2}\left|C\right|^2 z^{2\pi/\beta-2}, \qquad (34.15\text{a,b})$$

to a singularity (14.27c) of the pressure (13.15b); by the Blasius theorem (28.20a) it corresponds to a ***suction force****:*

$$F_s=i\frac{\rho}{2}C^2\int^{(0+)}z^{2\pi/\beta-2}\,\mathrm{d}z=\begin{cases}0 & \text{if } 0<\beta<2\pi, \\ -\rho\pi C^2 & \text{if } \beta=2\pi,\end{cases} \qquad (34.16\text{a}),(34.16\text{b})$$

The suction force vanishes (34.16a) except for a sharp edge when it is finite (34.16b).

34.1.4 Stagnation Point at Trailing Edge (Kutta Condition) and Lift

A flat plate of chord c in an uniform stream of velocity U at angle-of-attack α, meets the Kutta condition by placing one stagnation point at the trailing edge (34.12a), leading to: (i) the circulation (34.12c), and by Kutta–Joukowski theorem (28.29b) to the lift:

$$L=-\rho U\Gamma=\rho\pi U^2 c\sin\alpha=\frac{\rho}{2}U^2 cC_L; \qquad (34.17)$$

(ii) the corresponding to a lift coefficient (34.18a):

$$C_L\left(\alpha\right)=2\pi\sin\alpha, \quad C_{L\alpha}\equiv\frac{\mathrm{d}C_L}{\mathrm{d}\alpha}=2\pi\cos\alpha, \quad C_{L\alpha}\sim 2\pi+O\left(\alpha\right), \qquad (34.18\text{a–c})$$

has (34.18b) slope 2π at small angle-of-attack (34.18c); (iii) the complex conjugate velocity (34.12d) leads to a tangential velocity (34.13a) at the plate; (iv/v) the latter specifies by Bernoulli's theorem (24.27c) to the pressure distribution (34.19a) [pressure coefficient (34.19b)] on the plate:

$$p(\varphi)=p_0-\frac{\rho}{2}[v_x(\varphi)]^2=p_0-\frac{\rho}{2}U^2C_p(\varphi), \; C_p(\varphi)=\frac{1+\cos\left(\varphi-2\alpha\right)}{1+\cos\varphi}, \qquad (34.19\text{a,b})$$

where p_0 is the stagnation pressure and ρ the mass density; (vi) at the trailing edge (34.20a–c), the velocity (Figure 34.1b) is equal to that at infinity U projected in the direction of the plate (34.20d) ≡ (34.14b):

$$\varphi=0, z=a, \zeta=2a: \quad v_x(0)=U\cos\alpha, \quad p(0)=p_0-\frac{\rho}{4}U^2\left(1+\cos 2\alpha\right), \qquad (34.20\text{a–e})$$

and the pressure (34.19a) is given by (34.20e). Using:

$$\left|1+e^{i\varphi}\right|^2 = (1+\cos\varphi)^2 + \sin^2\varphi = 2(1+\cos\varphi), \tag{34.21a}$$

and likewise for $(\varphi - 2\alpha)$ in the conjugate velocity (34.12d):

$$C_p(\varphi) \equiv 2\frac{p_0 - p(\varphi)}{\rho U^2} = \frac{\left|v^*\left(ae^{i\varphi}\right)\right|}{U^2} = \left|e^{i\alpha}\right|^2 \frac{\left|1+e^{i(\varphi-2\alpha)}\right|^2}{\left|1+e^{i\varphi}\right|^2} = \frac{1+\cos(\varphi-2\alpha)}{1+\cos\varphi}, \tag{34.21b}$$

specifies the pressure coefficient (34.21b) ≡ (34.19b) on the flat plate at angle-of-attack.

34.1.5 Singularity (Finite Velocity) at the Leading (Trailing) Edge

The finite velocity (34.20d) is due to the Kutta condition (34.11a) at the trailing edge; (vii) near the leading edge (34.22a,b) the scalings (34.22c,d) ≡ (19.39a,b):

$$\varphi = \pi + \varphi, \ \psi \to 0: \quad \sin\varphi \sim \sin(\pi+\psi) \approx -\psi, \quad 1+\cos\varphi \sim 1-\cos\psi \approx \frac{\psi^2}{2}, \tag{34.22a–d}$$

imply (34.2b) that

$$x(\varphi) + 2a = 2a(1+\cos\varphi) \sim a\psi^2, \quad \psi \sim \left(\frac{x}{a}+2\right)^{1/2}; \tag{34.23a,b}$$

(viii) thus the velocity (34.13a) ≡ (34.24) scales like the inverse of the angle:

$$v_x(\varphi) = v_x(\psi+\pi) = U\left(\cos\alpha - \frac{\sin\alpha\sin\psi}{1-\cos\psi}\right) = -2\frac{U}{\psi}\sin\alpha + O(1); \tag{34.24}$$

(ix) the singularity of the velocity (34.24; 34.23b) ≡ (34.25a,b) like the inverse square root of distance from the edge:

$$v_x(\varphi) \sim -2\frac{U}{\psi}\sin\alpha = -2\mathrm{U}\left(\frac{x}{a}+2\right)^{-1/2}\sin\alpha = C(x+2a)^{-1/2}, \ C \equiv -2U\sqrt{a}\sin\alpha, \tag{34.25a,b}$$

agrees (34.9b) with thc cnd condition for a sharp edge; (x) the corresponding suction force (34.16b) is:

$$F_s \sim -4\rho\pi a\, U^2\sin^2\alpha = -\rho\pi c U^2\sin^2\alpha = -L\sin\alpha = F_x, \tag{34.26}$$

(xi) suction force at the leading edge is a horizontal force that balances (Figure 34.1b) the projection of the lift (34.17) that is orthogonal to the free stream velocity, on the direction of the plate; (xii) the component of the lift normal to the plate is:

$$F_y = L\cos\alpha = \pi\rho c U^2\sin\alpha\cos\alpha = \frac{\pi}{2}\rho c U^2\sin(2\alpha), \tag{34.27}$$

and corresponds to a vertical force; (xiii) concerning (28.31b) the pitching moment (34.28b) relative to the center of the airfoil it is unaffected by the Kutta condition, and depends only on (28.89a) the dipole moment (34.28a):

$$P_1 = -2\pi a^2 U e^{-i\alpha} = -\frac{\pi}{8}c^2 U e^{-i\alpha}, \quad M = \mathrm{Re}\left\{-i\rho U e^{-i\alpha} P_1\right\} = \frac{\pi}{8}\rho U^2 c^2\sin(2\alpha); \tag{34.28a,b}$$

(xiv) the pitching moment (34.28b) corresponds to the component of lift (34.17) normal to the plate (34.27) applied at the aerodynamic center:

$$x_{ac} = \frac{M}{F_y} = \frac{c}{4} = a, \tag{34.29}$$

located at quarter chord (Figure 34.1b); (xv) the pitching moment coefficient (34.30a):

$$C_M(\alpha) = \frac{2M}{\rho U^2 c^2} = \frac{\pi}{4}\sin(2\alpha) = \frac{\pi}{2}\sin\alpha\cos\alpha = \frac{1}{4}C_L(\alpha)\cos\alpha, \tag{34.30a}$$

and slope (34.30b):

$$C_{M\alpha} \equiv \frac{dC_M}{d\alpha} = \frac{\pi}{2}\cos(2\alpha), \quad C_{M\alpha}(0) = \frac{\pi}{2} = \frac{C_{L\alpha}(0)}{4}, \tag{34.30b,c}$$

are related to the lift coefficient (34.18a) [slope (34.18b)]; (xvi) both the lift (34.18a) and pitching moment (34.30a) coefficients vanish at zero angle of attack, and for small angles their ratio (34.30c) is four; (xvii) for positive angle-of-attack $\alpha > 0$ there is an upward force $F_y > 0$ in (34.27) and a pitch-down $M < 0$ in (34.28b); (xviii) the converse, that is, a downward force $F_y < 0$ and a pitch-up $M < 0$ for a negative angle-of-attack; (xix) the pitching moment in both cases causes a return to zero angle-of-attack, that is a statically stable condition of zero pitching moment; (xx) the dynamic 'overshoot' of the statically stable position leads to pitch or angle-of-attack oscillations (Subsection 34.6.4), as for the suspended pendulum (section 8.8).

34.1.6 Qualitative Description of the Potential Flow around a Body

The lift increases with angle-of-attack (34.17) in the same proportion as the circulation (34.12c); an increase in angle-of-attack α projects a larger proportion $\sin\alpha$ of the chord across the direction of the free stream. A positive angle-of-attach leads (Figure 34.1b) to (i) a convergence (divergence) of streamlines above (below) the flat plate or airfoil; (ii) hence to a velocity larger (smaller) than the free stream velocity; (iii) the associated pressure is lower (higher); (iv) the net resultant pressure force is upward, that is, a lift. The aerodynamic center is the point where the lift should be placed to produce the pitching moment; since it is located in the rear arc there is a pitch-down, that is the pitching moment is counterclockwise or positive. This implies that the pressure forces in the forward part of the airfoil, near the suction effect of the leading-edge, have a smaller moment than the pressure forces on the rear part, associated with the tangential flow from the trailing-edge, when the Kutta condition is enforced. The situation is reversed for large negative angle-of-attack, when the lift becomes a downforce, and there is a pitch-up, that is, a clockwise moment. Other combinations can arise (Section 34.6) if the angles of attack for zero lift and zero pitching moment do not coincide, for example, for some of the airfoils considered in the sequel (Sections 34.2–34.6). The lift (34.17; 34.18a–c) and pitching moment (34.28b; 34.30a–c) coefficients for a flat plate are representative, as an order of magnitude, of those of thin airfoils (Sections 34.2–34.5). The pitching moment slope (34.30b) at small angle of attack is $\pi/2$, compared with less than π for the Rankine oval (28.85b); the two values coincide for (34.31a), when the stagnation points (28.80b) are at (34.31b) specifying the length of the Rankine oval:

$$Q = \pi U c, \ \pm z_\pm = c\sqrt{2} = b/2: \qquad x^2 + y^2 - c^2 = 2cy\cot\left(\frac{2y}{c}\right), \tag{34.31a–c}$$

whose shape (28.81b) is (34.31c).

34.2 Joukowski (1910) Transformation and the Elliptic Cylinder

In the case of the flat plate (Figure 34.1b), the singularity at the trailing-edge can be eliminated by the Kutta condition of coincidence of a critical and a stagnation point. The coincidence of the other stagnation point with the singularity at the leading edge would occur only at zero angle-of-attack, $\alpha = 0$, that is, zero lift $\Gamma = 0$ in (34.17); for nonzero lift the singular velocity at the leading edge and the stagnation point on the underside cause the existence of a region of reverse flow that can lead to separation of the boundary layer, that is, undesirable drag and lift loss. These can be avoided by having rounded corners, for example, at the leading and trailing edges by moving the critical points away from the surface to the interior. This suggests an extension of the circle mapping, viz. the Joukowski transformation (Subsection 34.2.1) that leads to an elliptic cylinder (Subsection 34.2.2); the latter has finite velocity on the whole surface (Subsection 34.2.3), except in the limit of the flat plate.

34.2.1 Extension from the Circle to the Joukowski Transformation

The preceding remarks suggest considering the conformal mapping (34.8a) with a replaced by an arbitrary constant b, specifying the (1910) **Joukowski transformation** (34.32a):

$$\zeta = z + \frac{b^2}{z}, \quad \frac{\mathrm{d}\zeta}{\mathrm{d}z} = 1 - \frac{b^2}{z^2} = \frac{(z-b)(z+b)}{z^2}; \tag{34.32a,b}$$

it has (34.32a,b) critical points of the first kind $\mathrm{d}\zeta/\mathrm{d}z = 0$ at $z = \pm b$, and of the second kind $\mathrm{d}\zeta/\mathrm{d}z = \infty$ at the origin $z = 0$. Using

$$\zeta \pm 2b = \frac{z^2 \pm 2bz + b^2}{z} = \frac{(z \pm b)^2}{z}, \tag{34.32c}$$

the Joukowski transformation (34.32a) can be rewritten in the "symmetric" form:

$$\frac{\zeta + 2b}{\zeta - 2b} = \left(\frac{z+b}{z-b}\right)^2; \tag{34.32d}$$

this shows that the critical points of first kind $z = \pm b$ in the z-plane, lead to square-roots near $\zeta = \pm 2b$ in the ζ-plane; as before (14.91a,b) these correspond to inverse square root singularities for the velocity at an end-plate, viz. near the edge of a flat plate (34.25a,b).

If singularities are placed within a body the velocity will be continuous and finite on its surface, that is, there will be no sharp edges; this may be expected to happen for (34.32a) with $b < a$, since in this case (Figure 34.2a) the critical points $z = \pm b$ lie within the circle $|z| < a$. This can be checked by applying the Joukowski transformation (34.32a) with real positive b to a circle (34.33a) of radius, a, and center at the origin leading to (34.33b):

$$z = a\mathrm{e}^{i\varphi}: \quad \zeta = a\mathrm{e}^{i\varphi} + \frac{b^2}{a}\mathrm{e}^{-i\varphi} \equiv x + iy; \tag{34.33a,b}$$

this specifies an ellipse:

$$0 \le b \le a: \quad x(\varphi) = \left(a + \frac{b^2}{a}\right)\cos\varphi, \quad y(\varphi) = \left(a - \frac{b^2}{a}\right)\sin\varphi, \tag{34.34a,b}$$

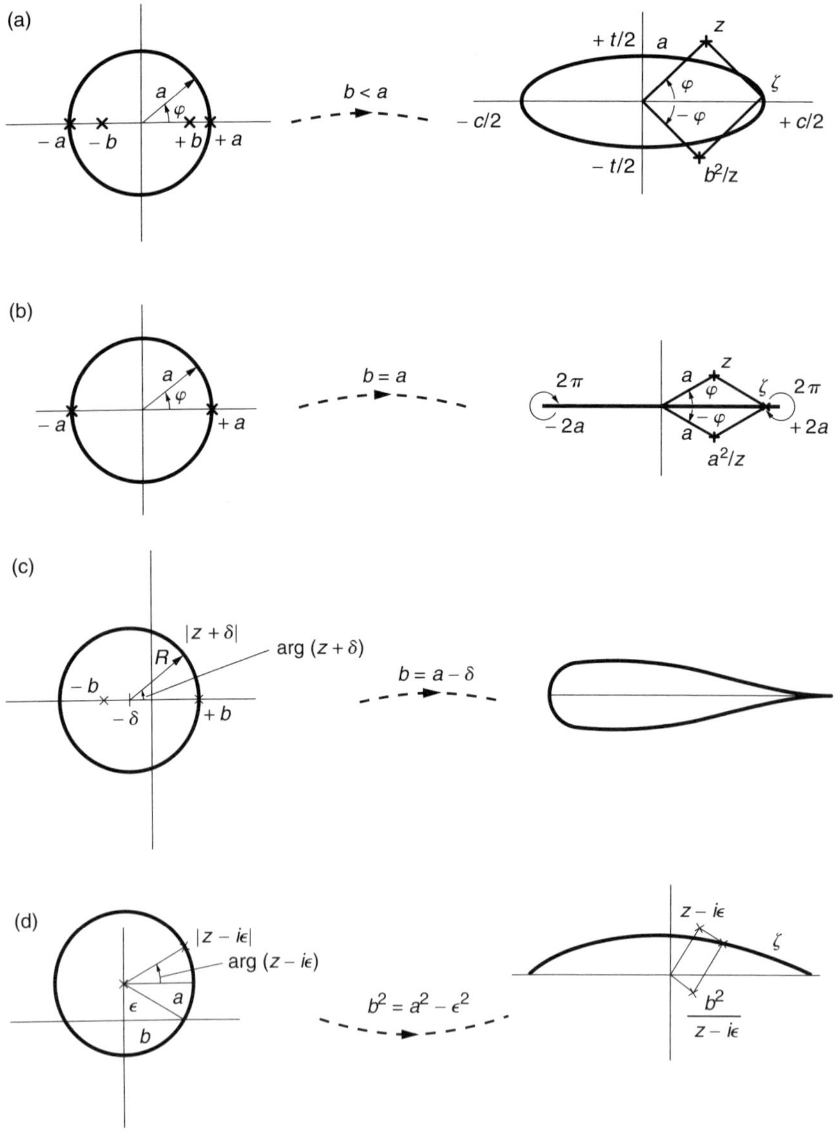

FIGURE 34.2
If in the Schwartz–Christoffel transformation (Figure 34.1) the critical points are moved from the circle to its interior, this leads to the Joukowki transformation, for which there are no sharp edges and the flat plate (Figure 34.2b) becomes a cylinder with elliptic cross-section (Figure 34.2a). The elliptic cylinder (Figure 34.2a) includes as particular cases: (i) the circular cylinder (Figure 34.1a) for zero eccentricity or equal minor and major axis; (ii) the flat plate (Figure 34.2b) for zero minor axis or "thickness." The Joukowki transformation applied to the circle with center on the real axis so that one critical point lies inside (the other on the circle) leads to a symmetric Joukowki airfoil (Figure 34.2c) with rounded leading-edge (sharp trailing-edge). If instead the Joukowki transformation is applied to a circle with center on the imaginary axis, such that both critical points lie on the circle, both the leading and trailing-edges are sharp (Figure 34.2d) and correspond to a "circular arc airfoil" or camber line without thickness.

with longitudinal (transverse) half-axis, equal to half $c/2\,(t/2)$ the **chord** c (**thickness** t):

$$\frac{x^2}{c^2} + \frac{y^2}{t^2} = 1: \quad 2a \geq a + \frac{b^2}{a} \equiv \frac{c}{2} \geq a - \frac{b^2}{a} \equiv \frac{t}{2} \geq 0. \tag{34.35a,b}$$

The case $b = 0$ when (34.31a) reduces to the identity $\zeta = z$ retains the circular cylinder of radius, $c/2 = a = t/2$; the opposite limit $b = a$ leads back (Figure 34.2b $\equiv$ 34.1a) to a flat plate of zero thickness $t = 0$ and chord $c = 4a$.

34.2.2 Circular Cylinder and Flat Plate as Particular Cases

In the general case of the potential flow past an elliptic cylinder of longitudinal (transverse) axis equal to the chord c (thickness t) it follows from (34.35a,b) that

$$a = \frac{c+t}{4}, \qquad b = \frac{\sqrt{c^2 - t^2}}{4}. \tag{34.36a,b}$$

The Joukowski transformation (34.32a,b) with parameter (34.36b), transforms the flow with complex potential (34.4b) and conjugate velocity (34.4c) past the circle of radius (34.36a) into the flow past on elliptic cylinder (34.34a,b) with (34.35a,b) chord c and thickness t in Figure 34.2a. The corresponding conjugate velocity is determined by the ratio of (34.7b) and (34.32b):

$$v^*(z) \equiv \frac{\mathrm{d}f}{\mathrm{d}\zeta} = \frac{\mathrm{d}f/\mathrm{d}z}{\mathrm{d}\zeta/\mathrm{d}z} = U\mathrm{e}^{-i\alpha}\frac{z - a\mathrm{e}^{i\alpha - i\beta}}{z-b}\frac{z + a\mathrm{e}^{i\alpha + i\beta}}{z+b}; \tag{34.37a}$$

it is nonsingular on the surface:

$$v^*\left(a\mathrm{e}^{i\varphi}\right) = U\frac{\left(\mathrm{e}^{i\varphi} - \mathrm{e}^{i\alpha - i\beta}\right)\left(\mathrm{e}^{i\varphi - i\alpha} + \mathrm{e}^{i\beta}\right)}{\mathrm{e}^{2i\varphi} - b^2/a^2}, \tag{34.37b}$$

because $b < a$. The critical points lie within the elliptic cylinder, whereas the stagnation points $\varphi_\pm = \alpha \pm \beta$ lie on the surface. Using:

$$\left|u\mathrm{e}^{ir} \pm v\mathrm{e}^{is}\right|^2 = (u\cos r + v\cos s)^2 + (u \sin \pm v \sin s)^2 = u^2 + v^2 \pm 2uv\cos(r-s), \tag{34.38a}$$

$$\left|\mathrm{e}^{i2\varphi} - \frac{b^2}{a^2}\right|^2 = \left[\cos(2\varphi) - \frac{b^2}{a^2}\right]^2 + \sin^2(2\varphi) = 1 + \left(\frac{b}{a}\right)^4 - 2\left(\frac{b}{a}\right)^2\cos(2\varphi), \tag{34.38b}$$

leads from (34.37b) to the pressure distribution on the elliptic cylinder:

$$p(\varphi) = p_0 - \frac{\rho}{2}\left|\frac{\mathrm{d}f}{\mathrm{d}\zeta}\right|^2 = p_0 - 2\rho U^2\frac{[1-\cos(\varphi-\alpha+\beta)][1+\cos(\varphi-\alpha-\beta)]}{1+(b/a)^4 - 2(b/a)^2\cos(2\varphi)}. \tag{34.39}$$

For (34.40a) the circular cylinder the pressure distribution (34.39) simplifies to (34.40b):

$$b = 0: \quad p(\varphi) = p_0 - 2\rho U^2[1-\cos(\varphi-\alpha+\beta)][1+\cos(\varphi-\alpha-\beta)]. \tag{34.40a,b}$$

In the absence of circulation (34.41a), the pressure distributions on the elliptic (34.39) and circular (34.40a,b) cylinders become (34.41b) symmetric in $\varphi - \alpha$:

$$\beta = 0: \quad p(\varphi) = p_0 - 2\rho U^2\frac{\sin^2(\varphi-\alpha)}{1+(b/a)^4 - 2(b/a)^2\cos(2\varphi)}, \tag{34.41a,b}$$

and there is no lift; the corresponding dimensionless **pressure coefficient** is

$$\beta = 0: \quad C_p \equiv \frac{2(p_0 - p)}{\rho U^2} = \frac{4\sin^2(\varphi-\alpha)}{1+(b/a)^4 + 2(b/a)^2\cos(2\varphi)}; \tag{34.42a,b}$$

it simplifies to (28.97a), in the case of a circular cylinder $b = 0$ at zero angle-of-attack $\alpha = 0$ without circulation.

34.2.3 Peak Velocity, Circulation, and Onset of Cavitation

If the rear stagnation point (34.7b) is placed $z_+ = 0$ at the trailing edge (34.43a), the circulation is given by (34.12c) and the complex conjugate velocity (34.37a) in the flow by

$$\beta = \alpha: \quad v^*(z) = U e^{-i\alpha} \frac{z-a}{z-b} \frac{z+ae^{2i\alpha}}{z+b}; \tag{34.43a,b}$$

it simplifies on the elliptic cylinder to

$$\beta = \alpha: \quad v^*\left(ae^{i\varphi}\right) = U \frac{\left(e^{i\varphi}-1\right)\left(e^{i\varphi-i\alpha}+e^{i\alpha}\right)}{e^{i2\varphi}-b^2/a^2}; \tag{34.44a,b}$$

it has Cartesian components:

$$\begin{aligned}\{v_x(\varphi), v_y(\varphi)\} = U &\left[1+\left(\frac{b}{a}\right)^4 - 2\left(\frac{b}{a}\right)^2 \cos(2\varphi)\right]^{-1} \\ &\left\{\left(1+\frac{b^2}{a^2}\right)\left[\cos\alpha + 2\sin\alpha\sin\varphi - \cos(2\varphi-\alpha)\right],\right. \\ &\left.\left(1-\frac{b^2}{a^2}\right)\left[\sin\alpha - 2\sin\alpha\cos\varphi - \sin(2\varphi-\alpha)\right]\right\}; \end{aligned} \tag{34.45a,b}$$

this follows from

$$\begin{aligned}&\left[1+\left(\frac{b}{a}\right)^4 - 2\left(\frac{b}{a}\right)^2\cos(2\varphi)\right] \frac{v_x(\varphi)-iv_y(\varphi)}{U} \\ &= \left|e^{i2\varphi}-\frac{b^2}{a^2}\right|^2 \frac{v^*\left(ae^{i\varphi}\right)}{U} = \left(e^{i\varphi}-1\right)\left(e^{i\varphi-i\alpha}+e^{i\alpha}\right)\left(e^{-2i\varphi}-\frac{b^2}{a^2}\right) \\ &= \left[e^{i2\varphi-i\alpha}-e^{i\alpha}+e^{i\varphi}\left(e^{i\alpha}-e^{-i\alpha}\right)\right]\left(e^{-2i\varphi}-\frac{b^2}{a^2}\right) \\ &= e^{-i\alpha}-e^{i\alpha-i2\varphi}+2ie^{-i\varphi}\sin\alpha + \left(\frac{b}{a}\right)^2\left[e^{i\alpha}-e^{i2\varphi-i\alpha}-2ie^{i\varphi}\sin\alpha\right], \end{aligned} \tag{34.45c}$$

where was used (34.38b). For a flat plate (34.46a) then (34.45c) ≡ (34.45a,b) simplifies to (34.14a) ≡ (34.46b) and (34.46c) ≡ (34.13a):

$$\begin{aligned} b = a: \quad v_y(\varphi) = 0, \quad \frac{v_x(\varphi)}{U} &= \frac{\cos\alpha + 2\sin\alpha\sin\varphi - \cos(2\varphi-\alpha)}{1-\cos(2\varphi)} \\ &= \cos\alpha + \frac{\sin\alpha\left[2\sin\varphi - \sin(2\varphi)\right]}{2\sin^2\varphi} \\ &= \cos\alpha + \frac{2\sin\alpha\sin\varphi(1-\cos\varphi)}{1-\cos^2\varphi} = \cos\alpha + \frac{2\sin\alpha\sin\varphi}{1+\cos\varphi}. \end{aligned} \tag{34.46a–c}$$

The velocity (34.45a,b) on the elliptic cylinder is: (i) zero at the trailing-edge (34.47a,b):

$$v_x(0) = 0 = v_y(0); \quad v_x(\pi) = 0, \quad v_y(\pi) = 4U\frac{\sin\alpha}{1-b^2/a^2}, \tag{34.47a–d}$$

(ii) vertical (34.47c,d) at the leading edge except for the flat plate $b = a$, when it is singular (34.25a,b); (iii) the velocity is horizontal at the midposition:

$$v_x\left(\pm\frac{\pi}{2}\right) = 2U\frac{\cos\alpha \pm \sin\alpha}{1+b^2/a^2}, \quad v_y\left(\pm\frac{\pi}{2}\right) = 0; \tag{34.48a,b}$$

(iv) in particular (34.48a,b) with $b = a$ coincides with the flat plate (34.13a) for $\varphi = \pm\pi/2$. In the case (34.49a) of the circular cylinder with circulation (34.49b), the velocity (34.37b) is best specified in polar coordinates:

$$b = 0 \neq \beta - \alpha: \quad v_r(\varphi) = 0, \quad v_\varphi(\varphi) = -2U\left\{\sin(\varphi - \alpha) + \sin\beta\right\}, \qquad \text{(34.49a–c)}$$

using

$$\begin{aligned}\frac{v_r(\varphi) - iv_\varphi(\varphi)}{U} &= \mathrm{e}^{i\varphi - i\alpha}\left(1 - \mathrm{e}^{i\alpha - i\beta - i\varphi}\right)\left(1 + \mathrm{e}^{i\alpha + i\beta - i\varphi}\right) \\ &= \mathrm{e}^{i\varphi - i\alpha} - \mathrm{e}^{-i\beta} + \mathrm{e}^{i\beta} - \mathrm{e}^{-i\varphi + i\alpha} = 2i\left[\sin\beta + \sin(\varphi - \alpha)\right]. \qquad \text{(34.49d)}\end{aligned}$$

From (34.49b,c) $\equiv$ (34.49d) it follows that the stagnation points occur for $\varphi = (\alpha - \beta, \alpha + \beta + \pi)$ for arbitrary circulation in agreement with (34.6b).

The velocity at the side of the elliptic cylinder (34.48a) is maximum (34.50a) for a diagonal angle-of-attack (34.50b) when it takes the value (39.50c).

$$0 = \frac{\mathrm{d}(\cos\alpha + \sin\alpha)}{\mathrm{d}\alpha} = \cos\alpha - \sin\alpha: \quad \alpha = \frac{\pi}{4}, \quad v_x\left(\frac{\pi}{2}\right) = \frac{2\sqrt{2}U}{1 + b^2/a^2}. \qquad \text{(34.50a–c)}$$

At zero angle-of-attack (34.51a) and in the absence of circulation (34.51b), the modulus of the velocity (34.42b) on the elliptic cylinder is (34.51a):

$$\alpha = 0 = \beta: \qquad v^2 \equiv \left|v^*\left(ae^{i\varphi}\right)\right|^2 = 2\frac{p_0 - p}{\rho} = U^2\frac{4\sin^2\varphi}{1 + (b/a)^4 - 2(b/a)^2\cos(2\varphi)}. \qquad \text{(34.51a–c)}$$

Its maximum corresponds to:

$$\begin{aligned}0 &= \left[1 + (b/a)^4 - 2(b/a)^2\cos(2\varphi)\right]^2 d\left(v^2\right)/d\varphi \\ &= 4U^2\left\{2\sin\varphi\cos\varphi\left[1 + (b/a)^4 - 2\left(b/a^2\right)\cos(2\varphi)\right] + 4(b/a)^2\sin^2\varphi\sin(2\varphi)\right\} \\ &= 4U^2\sin(2\varphi)\left\{1 + (b/a)^4 + 2(b/a)^2\left[2\sin^2\varphi - \cos(2\varphi)\right]\right\} \\ &= 4U^2\sin(2\varphi)\left[1 + (b/a)^4 + 2(b/a)^2\left(4\sin^2\varphi - 1\right)\right]. \qquad \text{(34.51d)}\end{aligned}$$

The term in square brackts is positive:

$$1 + (b/a)^4 + 2(b/a)^2\left(4\sin^2\varphi - 1\right) = \left[1 - (b/a)^2\right]^2 + 8(b/a)^2\sin^2\varphi \geq 0, \qquad \text{(34.51e)}$$

so (34.52d) only vanishes for $\varphi = \pm\pi/2$. Thus (34.51e) the maximum velocity is at the sides of the elliptic cylinder, and is given (34.48a,b) by:

$$\alpha = 0 = \beta: \quad U \leq |v|_{\max} = \frac{2U}{1 + b^2/a^2} \leq 2U; \qquad \text{(34.52a–c)}$$

on account of the condition (34.34a) it is: (i) smaller than for the corresponding $b = 0$ circular cylinder (28.98b); (ii) larger $b < a$ than the free stream velocity. The maximum velocity corresponds to the minimum pressure (34.52d):

$$p_{\min} = p_0 - \frac{2\rho U^2}{\left(1 + b^2/a^2\right)^2} \geq 0; \; U \leq \left(1 + \frac{b^2}{a^2}\right)\sqrt{\frac{p_0}{2\rho}} \equiv U_{\text{cav}}, \qquad \text{(34.52d,e)}$$

the condition of nonnegative pressure (34.52d) leads to the cavitation velocity (34.52e); it is increased relative to the case $b = 0$ of the circular cylinder (28.99b). Thus *the potential flow with free stream velocity* U *and angle-of-attack* α *past an elliptic cylinder (Figure 34.2a) with chord* c *and thickness* t *in (34.36a,b), is specified for arbitrary sub-critical circulation (34.7a) [circulation (34.12c) such that the trailing edge is a stagnation point,* $\alpha = \beta$*] by the complex conjugate velocity (34.37a) [(34.43b)] in the flow and (34.37b) [(34.44b)* $\equiv$ *(34.45c)* $\equiv$ *(34.45a,b)] on the surface. For general circulation (34.7a) the pressure distribution on the elliptic (circular) cylinder is (34.39) [(34.40a,b)]; in the absence of circulation the pressure distribution on the elliptic cylinder simplifies to (34.41a,b), corresponding to the pressure coefficient (34.42a,b). For the circulation (34.12c) corresponding to a stagnation point at the trailing edge (34.12b), the Cartesian components of the velocity on the elliptic cylinder (34.45a,b), vanish at the trailing edge (34.47a,b), and simplify to (34.47c,d) [(34.48a,b)] at the leading edge (mid position). The velocity at midposition is maximum (34.50c) for a diagonal angle-of-attack (34.50b). For zero angle-of-attack (34.52a) in the absence of circulation (34.52b) the surface velocity is maximum (34.52c) in the direction transverse to the incident stream where the pressure is minimum (34.52d). The cavitation velocity (34.52e) that should not be exceeded if cavitation is to be avoided, is increased relative to a circular cylinder corresponding to* $b = 0$ *in (28.99b).*

34.3 Circular Arc and Symmetric Airfoils

A bluff body like an elliptic cylinder has a negative pressure gradient at the rear, leading to flow separation and lift loss (Subsection 28.7.7). Thus a profile designed to give high lift, that is, and **airfoil**, should have a rounded leading-edge to avoid singularities of the velocity, and a long gently sloping aft end with a pointed trailing-edge to avoid flow separation. The main features of an airfoil are the thickness (Subsection 33.3.1) and curvature of the midline (Subsection 33.3.2).

34.3.1 Thick and Thin Symmetric Airfoils

A finite velocity both at the leading and trailing edges of an airfoil can be obtained by placing the forward critical point within the circle and the rear critical point on the circle. This is done most simply (Figure 34.2c) by placing the center of the circle on the negative real axis at, $z_0 = -\delta$, and choosing a radius $a = b + \delta$ such that $z = +b$ lies on the circle and $z = -b$ within it (34.53a,b):

$$b = a - \delta, z = ae^{i\varphi} - \delta: \quad x_1 + iy_1 \equiv \zeta_1 = z + \frac{b^2}{z} = ae^{i\varphi} - \delta + b^2 \frac{ae^{-i\varphi} - \delta}{\left|ae^{i\varphi} - \delta\right|^2}; \quad \text{(34.53a–c)}$$

the Joukowski transformation (34.32a) leads in this case (34.53a,b) to the **thick symmetric airfoil** (34.53c) $\equiv$ (34.54a,b):

$$x_1(\varphi) = (a\cos\varphi - \delta)\{1 + d_1(\varphi)\} = x_1(-\varphi), \quad y_1(\varphi) = a\sin\varphi\{1 - d_1(\varphi)\} = -y_1(-\varphi), \quad \text{(34.54a,b)}$$

where d_1 is defined by (34.55a):

$$d_1(\varphi) \equiv \frac{b^2}{\left|ae^{i\varphi} - \delta\right|^2} = \frac{(a - \delta)^2}{a^2 + \delta^2 - 2a\delta\cos\varphi} = 1 - 2\frac{\delta}{a}(1 - \cos\varphi) + O\left(\frac{\delta^2}{a^2}\right) = d_1(-\varphi); \quad \text{(34.55a,b)}$$

the profile is symmetric relative to the real axis because the x-coordinate (34.54a) [y-coordinate (34.54b)] is an even (odd) function of φ, bearing in mind that (34.55a) is also an even function of φ. The latter (34.55a) simplifies to (34.55b) if the displacement of the center is small compared with the radius (34.56a) leading to (34.56b,c):

$$\delta^2 \ll a^2 : x_1(\varphi) = 2(a-\delta)\cos\varphi - 2\delta\sin^2\varphi, \quad y_1(\varphi) = 2\delta\sin\varphi\,(1-\cos\varphi)\,; \tag{34.56a–c}$$

the latter case (34.56a–c) corresponds to the **thin symmetric airfoil**. The geometry of the thick symmetric Joukowski airfoil is considered in Example 40.6.

The **chord** (34.57b) is the distance between the leading and trailing edges, and the **thickness** (34.57c) is the largest vertical distance between upper and lower surface. For the thin symmetric airfoil (34.57a) they are given by (34.57b,c):

$$\begin{aligned}\delta^2 \ll a^2 : \quad & c \equiv x_{1\,\max} - x_{1\,\min} = x(0) - x(\pi) = 4(a-\delta) = 4b, \\ & t \equiv y_{1\,\max} - y_{1\,\min} = y\left(\frac{\pi}{2}\right) - y\left(-\frac{\pi}{2}\right) = 4\delta; \quad \frac{t}{c} = \frac{\delta}{b}, \quad c + t = 4a;\end{aligned} \tag{34.57a–e}$$

the thickness-to-chord ratio (34.57d) ≡ (34.58a) is small:

$$\frac{t}{c} = \frac{\delta}{b} \sim \frac{\delta}{a-\delta} \sim \frac{\delta}{a}; \quad c + t = 4(b+\delta) = 4a, \tag{34.58a,b}$$

the relation of the chord to the radius of the cylinder is preserved (34.58b) ≡ (34.11a). From (34.56b,c) follows the slope of the thin symmetric profile:

$$\frac{dy_1}{dx_1} = \frac{dy_1/d\varphi}{dx_1/d\varphi} = -\frac{\delta}{a}\frac{\cos(2\varphi) - \cos\varphi}{\sin\varphi} \sim O\left(-\frac{3\varphi}{2}\right), O\left(\frac{2}{\varphi-\pi}\right), \tag{34.59a,b}$$

that is: (i) proportional to the thickness-to-chord ratio, that is, would vanish for a flat plate; (ii) zero (34.59a) at the trailing-edge $\varphi = 0$ by (19.39a,b), that is, the latter is a "cusp," where the upper and lower surfaces are tangent; (iii) it is infinite at $\varphi = \pi$ the rounded leading edge (34.59b), where the tangent is vertical. The end-plate singularity at the cusp on the trailing-edge is removed by coincidence with a stagnation point, viz. by setting (34.60a):

$$a - \delta = b = z_+ = ae^{i\alpha - i\beta} - \delta; \quad \Gamma = -4\pi U a \sin\alpha = -\pi U(c+t)\sin\alpha, \tag{34.60a,b}$$

since, $\alpha = \beta$, the circulation (34.7a) is given by (34.60b) where (34.58b) was used. *The circulation (34.60b) has a small increase $O(t/a)$ of the order of the thickness to chord ratio of the airfoil, when compared with the case $t = 0$ of the flat plate (34.12c); the thin (34.56a–c; 34.55b) symmetric Joukowski airfoil with chord (34.57b), thickness (34.57c) and slope (34.59a,b), is a particular case of the thick (34.54a,b; 34.55a) symmetric Joukowski airfoil.*

34.3.2 Circular Arc Airfoil and Camber Angle

If instead of displacing the center of the circle to $z_0 = -\delta$ to the negative real axis (Figure 34.2c), it is displaced to $z_0 = i\varepsilon$ on the positive imaginary axis (Figure 34.2d), the critical points $z = \pm b$ lie on the circle for:

$$a^2 = b^2 + \varepsilon^2, \quad z = i\varepsilon + ae^{i\varphi}, \quad |z - i\varepsilon| = a. \tag{34.61a–c}$$

and both the leading and trailing edges are sharp. The Joukowski transformation (34.31a) with (34.61c) leads to

$$x_2 + iy_2 \equiv \zeta_2 = z + \frac{b^2}{z} = ae^{i\varphi} + i\varepsilon + b^2\frac{ae^{-i\varphi} - i\varepsilon}{|ae^{i\varphi} + i\varepsilon|^2}, \tag{34.62}$$

the strongly cambered arc profile:

$$x_2(\varphi) = a\cos\varphi\{1 + d_2(\varphi)\}, \quad y_2(\varphi) = (a\sin\varphi + \varepsilon)\{1 - d_2(\varphi)\}, \tag{34.63a,b}$$

where d_2 is defined by (34.64a):

$$d_2(\varphi) \equiv \frac{b^2}{|ae^{i\varphi} + i\varepsilon|^2} = \frac{a^2 - \varepsilon^2}{a^2 + \varepsilon^2 + 2a\varepsilon\sin\varphi} = 1 - 2\frac{\varepsilon}{a}\sin\varphi + O\left(\frac{\varepsilon^2}{a^2}\right); \tag{34.64a,b}$$

the profile is an arc because it has zero thickness as can be checked from (34.63a,b; 34.64a):

$$x_2\left(\pm\frac{\pi}{2}\right) = 0, \quad y_2\left(\pm\frac{\pi}{2}\right) = 2\varepsilon, \quad t = y\left(+\frac{\pi}{2}\right) - y\left(-\frac{\pi}{2}\right) = 0, \tag{34.65a–c}$$

and in Example 40.6. The zero thickness (34.65c) is an exact result arising from (34.63b) to all orders in ε/a:

$$y_2\left(\pm\frac{\pi}{2}\right) = (\varepsilon \pm a)\left[1 - \frac{a^2 - \varepsilon^2}{(\varepsilon \pm a)^2}\right] = \frac{(\varepsilon \pm a)^2 - a^2 + \varepsilon^2}{\varepsilon \pm a} = \frac{2\left(\varepsilon^2 \pm \varepsilon a\right)}{\varepsilon \pm a} = 2\varepsilon. \tag{34.66}$$

If the displacement from the origin is small (34.67a) to leading order (34.63a,b; 34.64b) simplify to the **weakly cambered circular arc profile** (34.67b,c):

$$\varepsilon^2 \ll a^2: \quad x_2(\varphi) = 2\cos\varphi\,(a - \varepsilon\sin\varphi), \quad y_2(\varphi) = 2\varepsilon\sin^2\varphi; \tag{34.67a–c}$$

since the approximation (34.67a) implies constant curvature, the profile corresponds to a circular arc. The chord (34.68a) and **height** (34.68b) of the weakly cambered circular arc profile:

$$c = x_2(0) - x_2(\pi) = 4a = 4\sqrt{b^2 + \varepsilon^2}, \quad h \equiv y_{2\,\max} = y_2\left(\pm\frac{\pi}{2}\right) = 2\varepsilon, \quad \frac{h}{c} = \frac{\varepsilon}{2a}, \tag{34.68a–c}$$

show that the height-to-chord ratio (34.68c) is small. The **chord line** is defined as a line joining the leading and trailing-edges, that is, for a symmetric (asymmetric) airfoil the upper and lower surfaces are (are not) symmetric relative to the chord line. The profile in (34.54–34.55a,b; 34.56a–c; 34.57a–e) has a symmetric shape (Figure 34.2c), whereas the circular are profile (34.63a,b; 34.65a–c; 34.67a–c; 34.68a–c) is (Figure 34.2d) the extreme case of asymmetry in that the lower and upper surfaces coincide. The geometry of the strongly cambered arc profile is considered in Example 40.6.

In the case of the circular arc profile the **camber angle** η is defined (Figure 34.3a) as the angle of the chord line with a straight line joining the trailing-edge to the highest point (34.69a):

$$\frac{\varepsilon}{a} = \tan\eta \approx \eta; \quad R = \frac{2a}{\sin\eta} \approx \frac{2a}{\eta} \approx \frac{2a^2}{\varepsilon}, \quad \frac{h}{R} \sim \frac{\varepsilon^2}{a^2} \ll 1, \tag{34.69a–c}$$

in the a case of small height (34.67a) the camber angle is small (34.69a), and the radius of curvature (34.69b) large; hence the height (34.68b) is negligible compared with the radius of curvature (34.69c). The slope (34.67a–c) of the weakly cambered circular arc profile:

$$\varepsilon^2 \ll a^2: \quad \frac{dy_2}{dx_2} = \frac{dy_2/d\varphi}{dx_2/d\varphi} = -2\frac{\varepsilon}{a}\cos\varphi, \tag{34.70a,b}$$

vanishes at the midpoint, and takes extreme values (34.70c) at the leading and trailing edges, where the inclination is small and equal to twice the camber angle (34.69a):

$$\left(\frac{dy_2}{dx_2}\right)_{\max,\min} = \pm 2\frac{\varepsilon}{a} = \pm 2\eta \approx \pm\tan(2\eta). \tag{34.70c}$$

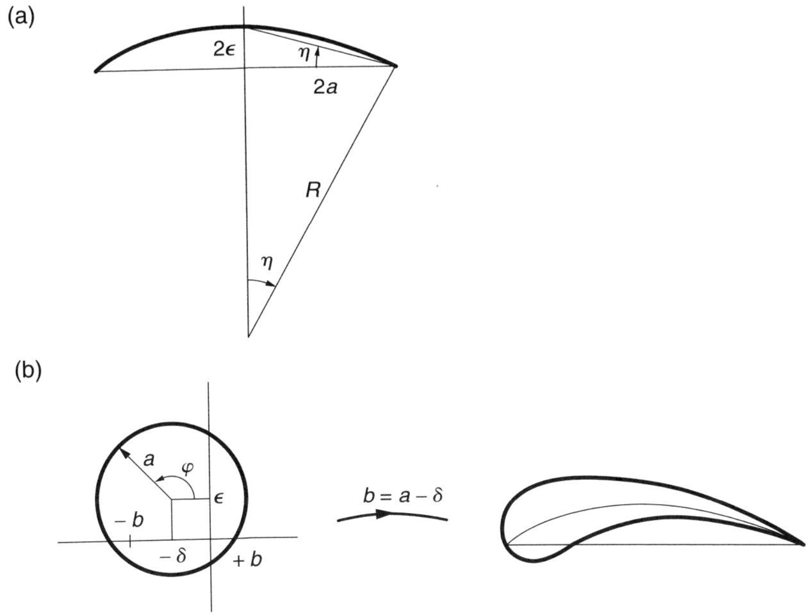

FIGURE 34.3
The Joukowki transformation applied to a circle with the center off both coordinate axes, such that one critical point lies inside (the other on the circle) combines (Figure 34.3b) the features of Figure 34.2c,d, that is, it leads to a symmetric Joukowski airfoil (Figure 34.2c) superimposed on the camber line (Figures 34.2d ≡ 34.3a); the result is the unsymmetric or cambered Joukowski airfoil (Figure 34.3b) with a rounded (sharp) leading (trailing) edge. The superposition of the symmetric Joukowski airfoil (Figure 34.2c) and camber line (Figures 34.2d ≡ 34.3a) is valid for the thin, weakly-cambered Joukowski airfoil (Figure 34.3b), for which the off set of the center of the circle from the origin is much smaller than radius $\varepsilon^2 + \delta^2 \ll a^2$, that is, to first-order in $(\varepsilon + i\delta)/a$. The exact mapping involving all higher orders leads to the thick, strongly-cambered Joukowski airfoil (Figure 40.2).

If the rear stagnation point is placed at the trailing-edge (34.71a) in (34.6b) and Figure 34.2d:

$$z_+ = b; \frac{\varepsilon^2}{a^2} \sim \eta^2 \ll 1 : \mathrm{e}^{i\alpha - i\beta} = \frac{b - i\varepsilon}{a} = 1 - \frac{i\varepsilon}{a} = \exp\left(-\frac{i\varepsilon}{a}\right) = \mathrm{e}^{-i\eta}, \qquad \text{(34.71a–c)}$$

it specifies the Kutta condition (34.71c) in terms of the small (34.71b) camber angle (34.70c); (34.69a) determines the circulation (34.7a) ≡ (34.72b) by (34.71c) ≡ (34.72a):

$$\beta = \alpha + \eta : \quad \Gamma = 4\pi U a \sin\beta = \pi U c \sin(\alpha + \eta) . \qquad \text{(34.72a,b)}$$

The result (34.72b) is valid (34.71b) for small camber:

$$\sin\beta = \sin(\alpha + \eta) = \sin\alpha\cos\eta + \sin\eta\cos\alpha = \sin\alpha\left[1 + \eta\cot\alpha + O\left(\eta^2\right)\right] . \qquad \text{(34.72c)}$$

Thus *the circulation (34.72b) is increased by the camber relative to a flat plate (34.12c) for the (34.67a–c) weakly cambered circular arc profile, with chord (34.68a), height (34.68b), and slope (34.70b); it is a particular case of the strongly cambered arc profile (34.63a,b; 34.64a; 34.65a–c). The increase in circulation (34.72c) is equivalent to a flat plate adding the camber to the angle-of-attack (34.72a,b).*

34.4 Cambered or Unsymmetric Joukowski (1916) Airfoil

To obtain higher circulation and hence lift, the camber (Subsection 34.3.2) can be combined with the thickness (Subsection 34.3.1), that is, by placing the circle (Figure 34.3b) with the center off both axes; the rear critical point lies on the circle to yield a sharp trailing-edge where the velocity is made finite by applying the Kutta condition of coincidence with the rear stagnation point; the forward critical point lies within the circle, so there is no singularity at the leading-edge of the airfoil that is rounded. This leads to the unsymmetric airfoil (Subsection 34.4.1) that increases lift (Subsection 34.4.2) both due to thickness and camber, and specifies the velocity and pressure distributions (Subsection 34.4.3).

34.4.1 Thick and Thin Unsymmetric Airfoil

The conditions:

$$a^2 = \varepsilon^2 + (b+\delta)^2, \quad z = ae^{i\varphi} - \delta + i\varepsilon, \quad |z+\delta-i\varepsilon| = a, \tag{34.73a–c}$$

include the particular cases: (i) of the circular arc $\delta = 0$ in (34.61a–c); (ii) symmetric airfoil $\varepsilon = 0$ in (34.53a–c); (i+ii) the flat plate $\varepsilon = 0 = \delta$ in (34.2a). Substitution of (34.72b) in the Joukowski transformation (34.32a) yields:

$$x_3 + iy_3 \equiv \zeta_3 = z + \frac{b^2}{z} = ae^{i\varphi} - \delta + i\varepsilon + b^2\frac{ae^{-i\varphi} - \delta - i\varepsilon}{|ae^{i\varphi} - \delta + i\varepsilon|^2}. \tag{34.74}$$

Separation real and imaginary parts specifies the shape of the **thick unsymmetric airfoil**:

$$x_3(\varphi) = (a\cos\varphi - \delta)\{1 + d_3(\varphi)\}, \quad y_3(\varphi) = (a\sin\varphi + \varepsilon)\{1 - d_3(\varphi)\}, \tag{34.75a,b}$$

where d_3 is defined by (34.76a):

$$\begin{aligned} d_3(\varphi) &\equiv \frac{b^2}{|ae^{i\varphi} - \delta + i\varepsilon|^2} = \frac{\left[\sqrt{a^2-\varepsilon^2} - \delta\right]^2}{a^2 + \varepsilon^2 + \delta^2 + 2a(\varepsilon\sin\varphi - \delta\cos\varphi)} \\ &= 1 - 2\left[\frac{\varepsilon}{a}\sin\varphi + \frac{\delta}{a}(1-\cos\varphi)\right] + O\left(\frac{(\delta+\varepsilon)^2}{a^2}\right); \end{aligned} \tag{34.76a,b}$$

the approximation (34.76b) for small combined thickness and camber (36.77a), follows from (34.77):

$$\begin{aligned} d_3(\varphi) &= \frac{\left(\sqrt{a^2-\varepsilon^2} - \delta\right)^2}{(a\cos\varphi - \delta)^2 + (a\sin\varphi + \varepsilon)^2} = \frac{(a-\delta)^2\left[1 + O\left(\varepsilon^2/a^2\right)\right]}{a^2+\varepsilon^2+\delta^2+2a(\varepsilon\sin\varphi - \delta\cos\varphi)} \\ &= \left(1 - \frac{2\delta}{a}\right)\left[1 + \frac{2}{a}(\varepsilon\sin\varphi - \delta\cos\varphi)\right] + \mathrm{O}\left(\frac{(\varepsilon+\delta)^2}{a^2}\right) \\ &= 1 - 2\frac{\delta}{a} - \frac{2}{a}(\varepsilon\sin\varphi - \delta\cos\varphi) + O\left(\frac{(\varepsilon+\delta)^2}{a^2}\right). \end{aligned} \tag{34.77}$$

The case of the thick, strongly cambered Joukowski airfoil (34.75a,b; 34.76a) simplifies to: (i) the symmetric airfoil (34.54a,b; 34.55a) in the absence of camber for $\varepsilon = 0$; (ii) circular arc (34.63a,b; 34.64a) for $\delta = 0$; (iii) in general is not the sum of those two expressions.

It becomes the sum of (34.56b,c) and (34.67b,c) in the case (34.76b) of the **thin unsymmetrical weakly cambered airfoil**:

$$(\delta + \varepsilon)^2 \ll a^2 : y_3(\varphi) = 2\delta \sin\varphi\,(1 - \cos\varphi) + 2\varepsilon \sin^2\varphi = y_1(\varphi) + y_2(\varphi), \qquad \text{(34.78a,b)}$$

$$x_3(\varphi) = 2\,(a - \delta)\cos\varphi - 2\delta \sin^2\varphi - \varepsilon \sin(2\,\varphi) = x_1(\varphi) + x_2(\varphi) - 2a\cos\varphi, \qquad \text{(34.78c)}$$

where were neglected: (i) the second-order thickness (34.67a) [curvature (34.70a)] effects; (ii) also their cross-product (34.78a), for example, in (34.76b) ≡ (34.77). The flat plate term $2a\cos\varphi$, that appears in (34.2b) must be subtracted in (34.78c) because it is repeated in (34.56b) and (34.67b). The shape of the thin weakly cambered unsymmetric Joukowski airfoil (34.78b,c) follows from substitution of (34.76b) in (34.75a,b) ≡ (34.74):

$$\begin{aligned} x_3(\varphi) + iy_3(\varphi) &= ae^{i\varphi} - \delta + i\varepsilon + \left(ae^{-i\varphi} - \delta - i\varepsilon\right) d_3(\varphi) \\ &= ae^{i\varphi} - \delta + i\varepsilon + \left(ae^{-i\varphi} - \delta - i\varepsilon\right)\left[1 - 2\frac{\varepsilon}{a}\sin\varphi - 2\frac{\delta}{a}(1 - \cos\varphi)\right] \\ &= 2a\cos\varphi - 2\delta - 2\left[\varepsilon\sin\varphi + \delta(1 - \cos\varphi)\right](\cos\varphi - i\sin\varphi) \\ &= 2a\cos\varphi - 2\delta\left(1 + \cos\varphi - \cos^2\varphi\right) - 2\varepsilon\sin\varphi\cos\varphi \\ &\quad + i2\sin\varphi\left[\varepsilon\sin\varphi + \delta(1 - \cos\varphi)\right]; \end{aligned} \qquad \text{(34.79a,b)}$$

this proves in particular (34.56a–c) [(34.67a–c)] for the symmetric (circular arc) airfoil $\varepsilon = 0$ $(\delta = 0)$. The upper y_3^+ (lower y_3^-) surfaces of the thin weakly cambered unsymmetric airfoil:

$$0 \le \varphi \le \pi: \quad y_3^{\pm}(\varphi) \equiv y_3(\pm\varphi) = \pm 2\delta\sin\varphi\,(1 - \cos\varphi) + 2\varepsilon\sin^2\varphi, \qquad \text{(34.80a–c)}$$

have: (i) for arithmetic mean (34.81) the **camber line** y_2 that is defined as the line at equal distance from both surfaces, and coincides with the circular arc (34.81) ≡ (34.67c):

$$\frac{y_3^+(\varphi) + y_3^-(\varphi)}{2} = 2\varepsilon\sin^2\varphi = y_2(\varphi); \qquad \text{(34.81)}$$

$$\frac{y_3^+(\varphi) - y_3^-(\varphi)}{2} = 2\delta\sin\varphi\,(1 - \cos\varphi) = y_1(\varphi), \qquad \text{(34.82)}$$

(ii) whereas the half-thickness (34.82) coincides with that of the symmetric airfoil (34.56c) ≡ (34.82). The geometry of the thick strongly cambered Joukowski airfoil is considered in Example 40.6.

The unsymmetric airfoil has chord (34.57b) and thickness (34.57c) equal to those of the symmetric shape, but since the latter is superimposed on a camber line, the heights at the midpoints are increased by (34.68b):

$$y_3\left(\pm\frac{\pi}{2}\right) = 2\,(\varepsilon \pm \delta) = h \pm \frac{t}{2}. \qquad \text{(34.83)}$$

The thickness-to-chord ratio is the same (34.57d) as for the symmetric airfoil. From (34.78b,c) follows the slope:

$$\frac{dy_3}{dx_3} = \frac{dy_3/d\varphi}{dx_3/d\varphi} = \frac{\delta}{a}\,\frac{\cos(2\varphi) - \cos\varphi}{\sin\varphi} - 2\frac{\varepsilon}{a}\cos\varphi = \frac{dy_1}{dx_1} + \frac{dy_2}{dx_2}, \qquad \text{(34.84)}$$

to the first order approximation (34.78a) of

$$\frac{\mathrm{d}y_3}{\mathrm{d}x_3} = \frac{2\delta\cos\varphi - 2\delta\cos(2\varphi) + 4\varepsilon\sin\varphi\cos\varphi}{-2a\sin\varphi\,[1 + O(\varepsilon/a, \delta/a)]}. \tag{34.85}$$

The total slope (34.85) ≡ (34.84) of the weakly cambered thin Joukowski airfoil coincides with the sum of those of the symmetric airfoil (34.59a) and of the circular arc (34.70b); at the cusp at the trailing edge the inclination is twice the camber angle (34.70c) as for the circular arc. The Kutta condition takes the same form (34.71c) ≡ (34.86a) as for the circular arc:

$$\beta = \alpha + \eta: \quad \Gamma = -4\pi U a\sin\beta = -\pi U(c+t)\sin(\alpha+\eta), \tag{34.86a,b}$$

and *the circulation (34.7a) is proportional: (i) to the flow velocity; (ii) to the chord-plus-thickness; (iii) and to the sine of the* ***incidence*** *(34.86a) that is the sum of the* ***angle-of-attack*** *α of the flow at infinity, and the* ***geometric camber*** *η of the unsymmetric airfoil. These also appear (28.28b) in the lift (34.87a):*

$$L = -\rho U\Gamma = \frac{\rho}{2}cU^2 C_L, \quad C_L(\alpha; t, \eta) = 2\pi\left(1 + \frac{t}{c}\right)\sin(\alpha+\eta), \tag{34.87a,b}$$

where the lift coefficient (34.87b) is increased relative to that (34.18a) of the flat plate by two factors: (i) the first due to the thickness-to-chord ratio (34.60a); (ii) the second is due to the camber (34.72a) and corresponds to the term in square brackets in (34.72c). Thus the camber has the same effect on lift as the angle-of-attack, that is, they are interchangeable for the same total incidence (34.86a).

34.4.2 Combined Effects of Camber and Thickness

The center of the circle is displaced to the point $a = -\delta, i\varepsilon, i\varepsilon - \delta$ respectively for the symmetric (34.53b), circular arc (34.61b) and cambered (34.73b) airfoils, implying that z must be replaced by (34.73b) in the second and third terms of the complex potential (34.4a), but not in the first term for the free stream

$$f = U\left(z\mathrm{e}^{-i\alpha} + \frac{a^2\mathrm{e}^{i\alpha}}{z+\delta-i\varepsilon}\right) - i\frac{\Gamma}{2\pi}\log(z+\delta-i\varepsilon); \tag{34.88a}$$

hence in the conjugate velocity:

$$\frac{\mathrm{d}f}{\mathrm{d}z} = U\mathrm{e}^{-i\alpha}\frac{(z-z_+)(z-z_-)}{(z+\delta-i\varepsilon)^2}, \tag{34.88b}$$

is given by

$$\begin{aligned}\frac{\mathrm{d}f}{\mathrm{d}z} &= U\mathrm{e}^{-i\alpha} - \frac{a^2\mathrm{e}^{i\alpha}}{(z+\delta-i\varepsilon)^2} - \frac{i\Gamma}{2\pi(z+\delta-i\varepsilon)}\\ &= U\mathrm{e}^{-i\alpha}\frac{(z+\delta-i\varepsilon)^2 - i(\Gamma/2\pi)\mathrm{e}^{i\alpha}(z+\delta-i\varepsilon) - a^2\mathrm{e}^{2i\alpha}}{(z+\delta-i\varepsilon)^2}.\end{aligned} \tag{34.89}$$

Since the same translation (34.73b) applies to z and $z_\pm$, the term in the numerator of (34.89) coincides with (34.5b), leading to (34.89) ≡ (34.88b). The translation by $i\varepsilon - \delta$ in (34.73b) also applies to the stagnation points (34.6b) ≡ (34.90b):

$$\eta = \beta - \alpha: \quad z_\pm + \delta - i\varepsilon = \pm a\exp\{i(\alpha \mp \beta)\}, \tag{34.90a,b}$$

where the camber angle may be introduced as in (34.72a) ≡ (34.90a). The velocity in the ζ-plane of the airfoil follows as the ratio of (34.88b) and (34.32b):

$$v^*(z) \equiv \frac{\mathrm{d}f}{\mathrm{d}\zeta} = \frac{\mathrm{d}f/\mathrm{d}z}{\mathrm{d}\zeta/\mathrm{d}z} = U\mathrm{e}^{-i\alpha}\frac{z^2}{z^2-b^2}\frac{(z-z_+)(z-z_-)}{(z-\delta+i\varepsilon)^2}. \tag{34.91}$$

The Kutta condition places the rear stagnation point at the trailing-edge (34.92a), and thus cancels the factors (34.92b) in (34.91) so that the velocity on the unsymmetric airfoil (34.73b) is given by (34.92c):

$$z_+ = b, z - z_+ = z - b: \quad v^*(z) = \frac{U\mathrm{e}^{-i\alpha}z^2(z-z_-)}{(z+b)(z+\delta-i\varepsilon)^2}. \tag{34.92a–c}$$

Using (34.73a) ≡ (34.93a), (34.73b), (34.90b), and (34.71c) specifies the three factors in (34.92c):

$$b = a-\delta: \frac{z+b}{a} = \mathrm{e}^{i\varphi} - \frac{\delta}{a} + \frac{i\varepsilon}{a} + 1 - \frac{\delta}{a} = \mathrm{e}^{i\varphi} + \mathrm{e}^{i\eta} - \frac{2\delta}{a}; \tag{34.93a,b}$$

$$\mathrm{e}^{-i\alpha}\frac{z-z_-}{a} = \mathrm{e}^{-i\alpha}\left(\mathrm{e}^{i\varphi} - \frac{\delta}{a} + \frac{i\varepsilon}{a} + \frac{\delta}{a} - \frac{i\varepsilon}{a} + \mathrm{e}^{i\alpha+i\beta}\right) = \mathrm{e}^{i\varphi-i\alpha} + \mathrm{e}^{i\beta}, \tag{34.93c}$$

$$\left(\frac{z}{z+\delta-i\varepsilon}\right)^2 = \left(\frac{a\mathrm{e}^{i\varphi}-\delta+i\varepsilon}{a\mathrm{e}^{i\varphi}}\right)^2 = \left(1-\mathrm{e}^{-i\varphi}\frac{\delta-i\varepsilon}{a}\right)^2 = 1 - 2\mathrm{e}^{-i\varphi}\frac{\delta-i\varepsilon}{a}. \tag{34.93d}$$

Substituting (34.93b–d) in (34.92c) specifies *the velocity on the thin weakly cambered Joukowski airfoil:*

$$v^*\left(a\mathrm{e}^{i\varphi}-\delta+i\varepsilon\right) = U\left(\mathrm{e}^{i\varphi-i\alpha}+\mathrm{e}^{i\beta}\right)\frac{1-2\mathrm{e}^{-i\varphi}(\delta-i\varepsilon)/a}{\mathrm{e}^{i\varphi}+\mathrm{e}^{i\eta}-2\delta/a}. \tag{34.94}$$

The pressure distribution corresponding to the velocity (34.94) that satisfies the Kutta condition (34.95a) for a symmetric airfoil (34.95b) is given by (34.95c):

$$\alpha = \beta, \delta = 0: \quad p(\varphi) = p_0 - \frac{\rho}{2}|v^*|^2 = p_0 - \frac{\rho}{2}U^2 C_p, \tag{34.95a–c}$$

where the pressure coefficient:

$$C_p = \frac{[1+\cos(\varphi-2\alpha)](1+4\eta\sin\varphi)}{1+\cos(\varphi-\eta)}; \tag{34.95d}$$

simplifies to (34.19a) for the flat plate $\varepsilon = 0 = \eta$. The passage from the velocity (34.94) to the pressure (34.95c) involves:

$$\begin{aligned}\left|\frac{v^*}{U}\right|^2 &= \frac{\left|\mathrm{e}^{i\varphi-i\alpha}+\mathrm{e}^{i\alpha}\right|^2\left|1+2i(\varepsilon/a)\mathrm{e}^{-i\varphi}\right|^2}{\left|\mathrm{e}^{i\varphi}+\mathrm{e}^{i\eta}\right|^2} = \frac{\left|1+\mathrm{e}^{i\varphi-2i\alpha}\right|^2\left|1+2i\eta\mathrm{e}^{-i\varphi}\right|^2\left|\mathrm{e}^{i\alpha-i\eta}\right|^2}{\left|1+\mathrm{e}^{i\varphi-i\eta}\right|^2}\\ &= \frac{[1+\cos(\varphi-2\alpha)](1+4\eta\sin\varphi)}{1+\cos(\varphi-\eta)},\end{aligned} \tag{34.96}$$

where were used: (i) in the first and last terms (34.38a); (ii) in the middle term:

$$\left|1 + 2i\eta e^{-i\varphi}\right|^2 = (1 + 2\eta \sin\varphi)^2 + (2\eta \cos\varphi)^2 = 1 + 4\eta \sin^2\varphi + O\left(\eta^2\right), \tag{34.97}$$

in the weak camber approximation.

34.4.3 Velocity and Pressure Distributions Including Leading/Trailing Edges

The conjugate velocity (34.94) at the leading edge (34.98a) is given by (34.98b):

$$\varphi = \pi: \quad v^*(\pi) = U\left(e^{i\beta} - e^{-i\alpha}\right)\frac{1 + 2(\delta - i\varepsilon)/a}{e^{i\eta} - 1 - 2\delta/a} = U\left(e^{i\alpha + i\eta} - e^{-i\alpha}\right)\frac{a + 2(\delta - i\varepsilon)}{i\varepsilon - 2\delta}, \tag{34.98a,b}$$

where were used (34.72a; 34.71c). The velocity at the leading edge (34.98b) has Cartesian components:

$$\{v_x(\pi), v_y(\pi)\} = \frac{2Ua}{\varepsilon^2 + 4\delta^2}\sin\alpha\,\{\varepsilon, 2\delta\}, \quad \frac{v_y(\pi)}{v_x(\pi)} = \frac{2\delta}{\varepsilon} = \frac{t}{h}, \tag{34.99a–c}$$

with inclination proportional to the ratio of the thickness (34.75c) to the height (34.68c). The passage from (34.98b) to (34.99a,b):

$$\begin{aligned} v^*(\varphi = \pi) &= \frac{U\left[e^{i\alpha}(1 + i\varepsilon/a) - e^{-i\alpha}\right]\left[a + 2(\delta - i\varepsilon)\right](-i\varepsilon - 2\delta)}{\varepsilon^2 + 4\delta^2} \\ &= -2Uai\sin\alpha\frac{2\delta + i\varepsilon}{\varepsilon^2 + 4\delta^2} = v_x(\pi) - iv_y(\pi), \end{aligned} \tag{34.100}$$

shows that *the velocity (34.100) ≡ (34.99a,b) is large $O(1/\varepsilon, 1/\delta)$ at the leading edge of the unsymmetric airfoil, and has inclination (34.99c). For the symmetric airfoil (34.101a) [circular arc (34.102a)] the velocity is vertical (34.101b,c) [horizontal (34.102b,c)] at the leading-edge:*

$$\varepsilon = 0: \quad v_x(\pi) = 0, \quad v_y(\pi) = \frac{Ua}{\delta}\sin\alpha, \tag{34.101a–c}$$

$$\delta = 0: \quad v_x(\pi) = \frac{2Ua}{\varepsilon}\sin\alpha, \quad v_y(\pi) = 0; \tag{34.102a–c}$$

it vanishes only at zero angle-of-attack, when the leading edge is a stagnation point. For nonzero angle-of-attack $\alpha \neq 0$ the velocity is large at the leading-edge for small thickness (camber), and becomes singular for the flat plate, $\varepsilon = 0 = \delta$.

The trailing-edge corresponds (34.103a) ≡ (34.103b) to the point where

$$a - \delta = b = z = ae^{i\varphi} - \delta + i\varepsilon: \quad e^{i\varphi} = 1 - \frac{i\varepsilon}{a} = e^{-i\eta}, \quad \varphi = -\eta, \tag{34.103a–c}$$

the inclination is minus the camber (34.103c) ≡ (34.104b); at the trailing-edge (34.104a) the conjugate velocity (34.92c) simplifies to (34.104c):

$$\begin{aligned} z = b = z_+, \varphi = -\eta: \frac{df}{d\zeta} &= Ue^{-i\alpha}b^2\frac{\left(ae^{i\varphi} - ae^{i\alpha + i\beta}\right)}{2ba^2e^{i2\varphi}} \\ = e^{-i\alpha}\frac{Ub}{2a}\left(e^{i\varphi} - e^{i\alpha + i\beta}\right)e^{-2i\varphi} &= \frac{Ub}{2a}\left(e^{-i\alpha - i\eta} + e^{i\alpha + i\eta}\right)e^{2i\eta}, \end{aligned} \tag{34.104a–c}$$

where were used (34.73b; 34.90b) and (34.103c; 34.72a). Using also (34.103a) specifies *the velocity at the trailing edge*:

$$\{v_x(-\eta), v_y(-\eta)\} = u\{\cos(2\eta), -\sin(2\eta)\} \quad u \equiv U\left(1 - \frac{\delta}{a}\right)\cos(\alpha + \eta); \quad (34.105a\text{–}c)$$

it has an inclination (34.105a,b) equal to twice the camber angle, that is, is horizontal for the symmetric airfoil; the modulus of the velocity (34.105c) is reduced by the thickness and apart from this factor, corresponds to the velocity U projected on the angle of incidence (34.72a) $\equiv$ (34.90a). The inclination of the velocity at the trailing edge (34.105a,b) is twice the camber 2η because the curvature of the airfoil (Figure 34.3c) turns the flow by an angle η over each of the forward (rearward) part near the leading (trailing) edge and corresponds to the tangent to the circular arc (34.70c). Since (34.105c) $\equiv$ (34.106) applies for small camber, the modulus of the velocity can be approximated:

$$u = U\left(1 - \frac{\delta}{a}\right)(\cos\alpha\cos\eta - \sin\alpha\sin\eta) = U\left(1 - \frac{\delta}{a}\right)\cos\alpha\left[1 - \eta\tan\alpha + O\left(\eta^2\right)\right], \quad (34.106)$$

There is no deflection of the flow for the symmetric airfoil:

$$\eta = 0: \quad v_x(0) = U\left(1 - \frac{\delta}{a}\right)\cos\alpha, \quad v_y(0) = 0, \quad (34.107a\text{–}c)$$

that is, the velocity is horizontal at the trailing-edge, as in the case $\delta = 0$ of the flat plate (34.20d); the thickness reduce the horizontal velocity. The velocity is nonzero at the trailing-edge because it is a cusp; the streamline issuing from it makes angle π with the surface, that is, the flow is locally uniform. The camber affects both the velocity at the trailing-edge (34.106) and the circulation (34.72c).

34.5 Parametric Families and Generic Airfoils (von Karman–Trefftz, 1918; von Mises, 1920; Carafoli)

The Joukowski airfoils (Sections 34.3 and 34.4) have only one-parameter b, not counting the two coordinates of the center of the circle. Two alternative generalizations are the von Karman–Trefftz (von Mises) airfoils [Subsections 34.5.1 (34.5.2)] that form a two-parameter (finite-parameter) family and have trailing-edge with an angular point instead of a cusp (camber line possibly with inflexions instead of curvature with constant sign). These may be modified into Carafoli airfoils (Subsection 34.5.3) with rounded rather than sharp trailing-edge. All of the preceding are particular cases of the generic airfoil (Subsection 34.5.4) used to calculate forces and moments (Section 34.6) on wings (Sections 34.7–34.9). The relation between these airfoil families is indicated in Diagram 34.1.

34.5.1 Airfoils with Angular Edges (von Karman–Trefftz, 1918)

The Joukowski transformation (34.32d) is the particular case $k = 2$ of the **von Karman–Trefftz transformation (1918):**

$$k \in |R: \quad \frac{\zeta + kb}{\zeta - kb} = \left(\frac{z + b}{z - b}\right)^k, \quad (34.108)$$

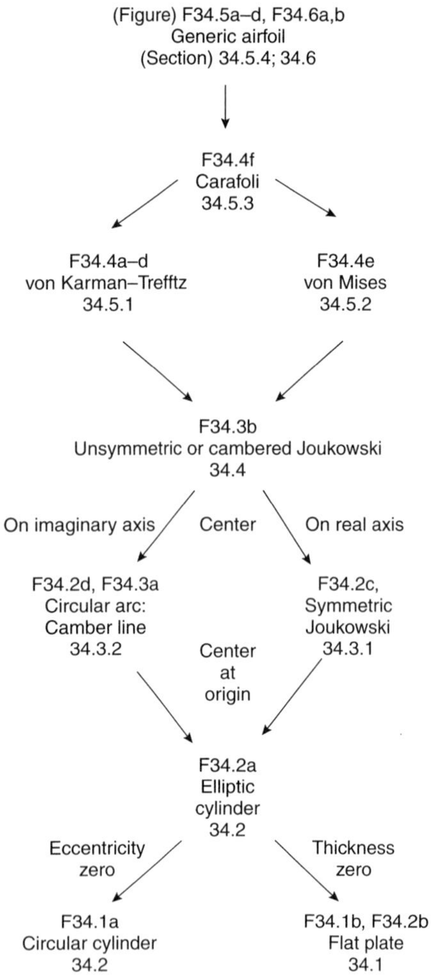

Diagram 34.1

Some airfoil families, starting with a circular cylinder (Figure 34.1a) with (i) center at the origin the Joukowki transformation leads to a cylinder with elliptic cross-section (Figure 34.2a); the particular case of zero minor axis is a flat plate (Figures 34.1 and 34.2b). If the circle has the center offset from the origin to the (ii) real [(iii) imaginary] axis, the Joukowki transformation leads to a symmetric airfoil (circular arc or camber line) in Figure 34.2a(d). Superimposing the symmetric Joukowki airfoil (Figure 34.2c) on the camber line (Figures 34.2d ≡ 35.3a) leads to the unsymmetric or cambered Joukowki airfoil (Figure 35.3b); the latter corresponds to the Joukowki transformation applied to a circle with origin outside both coordinate axes. The generalization of the Joukowki transformation leads to broader families of airfoils. The Joukowki airfoils have a cusp at the trailing edge, that is, the upper and lower surfaces are tangent there; the von Karman–Trefftz airfoils (Figure 34.4a–d) allow a finite angle between the upper and lower surfaces at the trailing-edge. The Joukowki airfoils have a camber line (Figures 34.2d ≡ 34.3d) without inflexions; the von Mises airfoils allow s-shaped camber lines with inflexions (Figure 34.4e). The Carafoli airfoils are a modification of any of the preceding in which the trailing-edge is rounded (Figure 34.4f) instead of sharp. All of these are particular cases of the generic airfoil (Figures 34.5b,d and 34.6). From the point-of-view of forces and moments the "generic airfoil" is characterized by two real parameters, which take particular values for the Joukowki/von Karman–Trefftz/von Mises/Carafoli transformations.

that introduces besides b a second parameter k. It can be written in the two inverse explicit forms:

$$\zeta = kb\frac{(z+b)^k + (z-b)^k}{(z+b)^k - (z-b)^k}, \tag{34.109a}$$

$$z = b\frac{(\zeta+kb)^{1/k} + (\zeta-kb)^{1/k}}{(\zeta+kb)^{1/k} - (\zeta-kb)^{1/k}}; \tag{34.109b}$$

this leads to the asymptotic limits:

$$\zeta = z + \frac{(k^2-1)\,b^2}{3z} + O\left(b^4 z^{-3}\right), \tag{34.110a}$$

$$z = \zeta - \frac{(k^2-1)\,b^2}{3\zeta} + O\left(b^4 \zeta^{-3}\right), \tag{34.110b}$$

showing that there is no deformation at infinity. The binomial series (25.37a–c) must be taken to the third-order with four terms:

$$\begin{aligned} |z| > b: \quad (z \pm b)^k &= z^k\left(1 \pm \frac{b}{z}\right)^k = z^k \sum_{n=0}^{\infty} \binom{k}{n}\left(\pm\frac{b}{z}\right)^n \\ &= z^k\left[1 \pm \frac{kb}{z} + \frac{k(k-1)\,b^2}{2z^2} \pm \frac{k(k-1)(k-2)\,b^3}{6z^3}\right] + O\left(\left(\frac{b}{z}\right)^4\right), \end{aligned} \tag{34.111a,b}$$

in (34.109a) in order to obtain the asymptotic approximation:

$$\begin{aligned} z^4 \gg b^4: \quad \zeta &= kb\frac{(1+b/z)^k + (1-b/z)^k}{(1+b/z)^k - (1-b/z)^k} = \frac{2kb}{2kb/z}\frac{1+k(k-1)\,b^2/(2z^2)}{1+(k-1)(k-2)\,b^2/(6z^2)} \\ &= z\left[1 + (k-1)\frac{b^2}{2z^2}\left(k - \frac{k-2}{3}\right)\right] = z + \left(k^2-1\right)\frac{b^2}{3z}, \end{aligned} \tag{34.111c,d}$$

to second-order (34.111d) = (34.110a); the passage from (34.109b) to (34.110b) is similar replacing $(z,\ k,\ b) \to (\zeta, 1/k, kb)$. From (34.108) follows the scaling (34.112a):

$$\lim_{z\to\pm b}\left[\zeta(z) \mp kb\right] \sim O\left((z \mp b)^k\right): \quad 2\gamma = (2-k)\,\pi, \qquad \beta = \pi - \gamma = \frac{k\pi}{2}, \tag{34.112a–c}$$

thus at the edges the internal angle is (34.112b), corresponding to a corner flow (Section 14.8) with angle (34.112c); thus the trailing edge (Figure 34.4a–d) it is not a cusp ($\gamma \neq 0, \beta \neq \pi$) except for the Joukowski airfoil $k = 2$.

From (34.112a) follows:

$$\lim_{z\to\pm b}\frac{\mathrm{d}\zeta}{\mathrm{d}z} = O\left((z \mp b)^{k-1}\right) = O\left((\zeta \mp 2b)^{1-1/k}\right) = \left\{\lim_{\zeta\to\pm 2b}\frac{\mathrm{d}z}{\mathrm{d}\zeta}\right\}^{-1}, \tag{34.113}$$

implying at the trailing-edge $(z = b, \zeta = 2b)$ that: (i) if it is not a stagnation point (34.114a) the velocity the velocity (34.114b) is singular for $k > 1$, including the Joukowski airfoil $k = 2$,

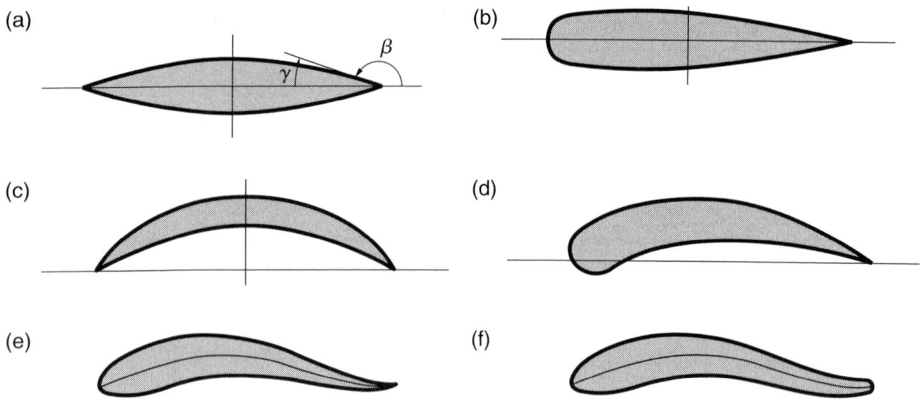

FIGURE 34.4
By generalizing or modifying the Joukowski transformation other airfoil families are obtained. For example, changing the exponent or angle at the critical points from $\gamma = \pi$ to $\gamma < \pi$ in the von Karman–Trefftz transformation eliminates the cusp at the trailing-edge, that is, the upper and lower surfaces are no longer tangent there, but make a finite angle: (i/ii) the flat plate (circular arc airfoil) in Figure 34.2b(d) becomes a convex (concave) lens airfoil in Figure 34.4a(c), with sharp but not cusped leading and trailing edges; (iii/iv) the symmetric (unsymmetric) Joukowski airfoil in Figure 34.2c (34.3a) become symmetric (unsymmetric) von Karman–Trefftz airfoils in Figure 34.4b(d), with rounded leading-edge and sharp or angular but not cusped trailing-edge. The Joukowski transformation with two critical points leads to a circular arc airfoil (Figure 34.2d) or camber line (Figure 34.3a) with curvature of fixed sign; introducing more than two critical points in the von Mises transformation leads to a camber line with inflexion points, and hence an s-shaped airfoil (Figure 34.4e). All of the preceding airfoils (Figures 34.2b–d and 34.4a–e) have sharp trailing-edges, corresponding to a critical point on the circle; changing the radius of the circle so that the critical point lies inside leads to the Carafoli airfoils with rounded trailing-edge (Figure 34.4f).

when the singularity is an inverse square root:

$$\lim_{z\to b}\frac{\mathrm{d}f}{\mathrm{d}z} \neq 0 = O(1): \lim_{z\to 2b}\frac{\mathrm{d}f}{\mathrm{d}\zeta} = \left(\lim_{z\to b}\frac{\mathrm{d}f}{\mathrm{d}z}\right)\left(\lim_{z\to 2b}\frac{\mathrm{d}z}{\mathrm{d}\zeta}\right) = O\left((\zeta-2b)^{1/k-1}\right); \quad (34.114\text{a,b})$$

$$\begin{aligned}\lim_{z\to b}\frac{\mathrm{d}f}{\mathrm{d}z} = O(z-b):\quad &\lim_{z\to 2b}\frac{\mathrm{d}f}{\mathrm{d}\zeta} = \left(\lim_{z\to b}\frac{\mathrm{d}f}{\mathrm{d}z}\right)\left(\lim_{z\to 2b}\frac{\mathrm{d}z}{\mathrm{d}\zeta}\right)\\ &= O(z-b)\,O\left((z-b)^{1-k}\right) = O\left((z-b)^{2-k}\right) = O\left((\zeta-2b)^{2/k-1}\right), \quad (34.115\text{a,b})\end{aligned}$$

(ii) if the Kutta condition is applied by placing the stagnation point at the trailing edge (34.115a), then velocity (34.115b) is zero for $k < 2$, and finite for $k = 2$ the Joukowki airfoil. *The von Karman–Trefftz (Joukowski) transformation (34.108) with* $\mathrm{k} \neq 2\,(\mathrm{k} = 2)$ *maps the circle: (i) with center at the origin into a symmetric "convex lens" (flat plate) with angular point (cusp) at the leading and trailing edges in Figure 34.4a (34.2b); (ii) with center on the negative real axis into a symmetric airfoil with rounded leading edge and with angular point (cusp) at the trailing edge in Figure 34.4b (34.2c); (iii) with center on the imaginary axis into a an "concave lens" (arc) with (without) thickness and with angular points (cusps) at the leading and trailing edges in Figure 34.4c (34.2d); (iv) with origin outside the axis into an unsymmetric concave lens (unsymmetric airfoil) with rounded leading edge and with angular point (cusp) at the trailing edge in Figure 34.4d (34.3b). The von Karman–Trefftz airfoils (34.109a,b) do not deform the flow at infinity (34.110a,b) neither in general nor for the Joukowski airfoil,* k $= 2$. *The leading and trailing edges are not cusped (34.112a,b), and the velocity is: (i) finite at the leading-edge; (ii) singular at the trailing-edge (34.114b) for* k > 1 *if it is not a stagnation point (34.114a); (iii) if the trailing-edge is a stagnation point (34.115a) the velocity is zero (34.115b) for* k < 2.

34.5.2 Airfoils with Camber Line Having Inflexion(s) (von Mises, 1920)

The **von Mises transformation (1920)**:

$$\frac{\mathrm{d}\zeta}{\mathrm{d}z} = \sum_{n=1}^{N}\left(1 - \frac{b_n}{z}\right), \tag{34.116}$$

includes the Joukowski transformation (34.32b) that is the particular case $N = 2, b_1 = b = -b_2$; generally (34.116) it involves N parameters $b_1, \ldots, b_N$ corresponding to N critical points of the first kind. From (34.116) follows the far-field scaling (34.117a):

$$\frac{\mathrm{d}\zeta}{\mathrm{d}z} = 1 - z^{-1}\sum_{n=1}^{N} b_n + z^{-2}\left\{\sum_{n,m=1}^{N} b_n b_m\right\} + O\left(z^{-3}\right); \sum_{n=1}^{N} b_n = 0. \tag{34.117a,b}$$

The integration of (34.117a) will lead to a logarithmic term, unless the condition (34.117b) is met; thus a necessary condition for the mapping (34.117a) to be single-valued is (34.117b). In this case:

$$\zeta = z - \frac{B}{z} + O\left(z^{-2}\right), \quad z = \zeta + \frac{B}{\zeta} + O\left(\zeta^{-2}\right), \quad B \equiv \sum_{\substack{n,\, m = 1 \\ n \neq m}}^{N} b_n b_m, \tag{34.118a–c}$$

the term following the identity (34.118a,b) at infinity $\zeta \to z$ as $z \to \infty$ corresponds to a dipole (34.118c). The unicity condition (34.117b) is met by the Joukowski airfoil $b_1 = b = -b_2$ that has $N = 2$ critical points, that is, the minimum number N for (34.117b) to hold. The simplest von Mises profile of higher complexity than the Joukowski type is (34.116) with $N = 3$ critical points that must satisfy (34.117b) leading to (34.119a–c) in (34.119d):

$$b_1 = a, \quad b_2 = b, \quad b_3 = -a - b: \quad \frac{\mathrm{d}\zeta}{\mathrm{d}z} = \left(1 - \frac{a}{z}\right)\left(1 - \frac{b}{z}\right)\left(1 + \frac{a+b}{z}\right); \tag{34.119a–d}$$

the simplification (34.120a) of (34.119d) is consistent with (34.117a,b):

$$\frac{\mathrm{d}\zeta}{\mathrm{d}z} = 1 - \frac{ab + a^2 + b^2}{z^2} + \frac{ab\,(a+b)}{z^3}; \zeta = z + \frac{a^2 + b^2 + ab}{z} - \frac{ab\,(a+b)}{2z^2}. \tag{34.120a,b}$$

integration of (34.120a) leads to (34.120b) that is consistent with (34.118a,c); in (34.120b) was omitted a constant corresponding to a translation.

The transformation (34.120b) maps the circle (34.121a) into (34.121b):

$$z = a\mathrm{e}^{i\varphi}: \quad x + iy \equiv \zeta = a\mathrm{e}^{i\varphi} + \left(a + b + \frac{b^2}{a}\right)\mathrm{e}^{-i\varphi} - \frac{b}{2}\left(1 + \frac{b}{a}\right)\mathrm{e}^{-2i\varphi}, \tag{34.121a,b}$$

corresponding to the von Mises profile:

$$x_4(\varphi) = \left(2a + b + \frac{b^2}{a}\right)\cos\varphi - \frac{b}{2}\left(1 + \frac{b}{a}\right)\cos(2\varphi), \tag{34.122a}$$

$$y_4(\varphi) = -\left(b + \frac{b^2}{a}\right)\left[\sin\varphi - \frac{1}{2}\sin(2\varphi)\right] = -b\left(1 + \frac{b}{a}\right)\sin\varphi\,(1 - \cos\varphi). \tag{34.122b}$$

The slope of the profile:

$$\frac{\mathrm{d}y_4}{\mathrm{d}x_4} = \frac{\mathrm{d}y_4/\mathrm{d}\varphi}{\mathrm{d}x_4/\mathrm{d}\varphi} = b\left(1 + \frac{b}{a}\right)\csc\varphi\frac{\cos(2\varphi) - \cos\varphi}{2b\left(1 + \frac{b}{a}\right)\cos\varphi - 2a - b - b^2/a}, \tag{34.123a}$$

vanishes (34.123b) at two points (34.123c):

$$\frac{\mathrm{d}y_4}{\mathrm{d}\varphi} = 0 : 2\cos^2\varphi - \cos\varphi - 1 = 0, \cos\varphi = \frac{1}{4} \pm \frac{3}{4} = 1, -\frac{1}{2}, \varphi = 0, \frac{2\pi}{3}, \tag{34.123b–e}$$

corresponding (34.123d) to (34.123e): (i) the trailing edge $\varphi = 0$; (ii) an intermediate angle $\varphi = 2\pi/3$. Thus the profile has a point of inflexion between the trailing-edge and the point in the forward arc (34.124a) below the chord line (34.124b):

$$x_4\left(\frac{2\pi}{3}\right) = -a - \frac{b}{4}\left(1 + \frac{b}{a}\right) < 0, \; y_4\left(\frac{2\pi}{3}\right) = -b\left(1 + \frac{b}{a}\right)\frac{3\sqrt{3}}{4}. \tag{34.124a,b}$$

The von Mises profile (34.120a,b) has an s-shape as shown in Figure 34.4e with camber line having an inflexion point, that is, s-shaped; the shape (34.122a,b), with an inflexion between the trailing-edge and the point (34.124a,b), is a particular two-parameter $2 = N - 1$ case (34.119a–d) of the general von Mises transformation (34.116) with N parameters, and one constraint (34.117b) to ensure unicity (34.118a–c).

34.5.3 Airfoils with Rounded Trailing Edge (Carafoli)

An airfoil with a rounded trailing-edge (Figure 34.4f) is easier to build. Examples include the **Carafoli airfoils** that can be obtained from any of the preceding profiles, Joukowski (34.32a–d; 34.125a), von Karman–Trefftz (34.108; 34.125b) or von Mises (34.116; 34.125c) by enlarging the radius of the circle (34.125a) so that the critical points lie (34.125b,c) inside:

$$z \equiv a e^{i\varphi} + \delta - i\varepsilon : \quad a > b, \quad a > \sup\{|b_n|\}. \tag{34.125a–c}$$

In this case, the mapping of the circle is conformal onto the whole airfoil profile, and there are no edges. Another method to round-off edges (Subsection 36.1.1) is to add a term to the Schwartz–Christoffel transformation (Sections 33.6–33.9).

34.5.4 Generic Airfoil

The one-parameter b family of (Figures 34.2b–d; 34.3b) of Joukowski airfoils airfoils has been extended to: (i) the two parameter family of von Karman–Trefftz airfoils (34.108) that has angular points instead of cusps at the edges (Figures 34.4a–d); (ii) the $(N-1)$ parameter family of von Mises airfoils (34.116;34.117b) that may have a camber line with inflexions (Figure 34.4e); (iii) the Carafoli airfoils (34.125a–c) that modify all of the preceding with a rounded trailing edge (Figure 34.4f). The Joukowski transformation (34.32a) may be inverted:

$$z^2 - \zeta z + b^2 = 0: \quad z = \frac{\zeta}{2}\left\{1 + \sqrt{1 - \frac{4b^2}{\zeta^2}}\right\} \approx \zeta - \frac{b^2}{\zeta} + O\left(\zeta^{-2}\right), \qquad \text{(34.126a,b)}$$

showing that it does not distort the far-field, since $z \to \zeta$ as $\zeta \to \infty$. The von Karman–Trefftz (34.108) [von Mises (34.116; 34.117b)] transformations also preserve the far-field (34.110a) [(34.118a–c)] while changing the coefficients of the following term. A circle in the z-plane may be mapped into **generic airfoil** of arbitrary shape in the ζ-plane by the transformation:

$$A_1 \equiv \bar{\mathrm{b}}^2 \mathrm{e}^{i\vartheta}: \quad z = \zeta + \sum_{n=1}^{\infty} A_n \zeta^{-n} = \zeta - \frac{\bar{\mathrm{b}}^2 \mathrm{e}^{i\vartheta}}{\zeta} + O\left(\zeta^{-2}\right), \qquad \text{(34.127a,b)}$$

consisting the identity, plus a Laurent series (27.37b) converging at infinity. Since the coefficients A_n are generally complex, in the generic airfoil (34.127b) the coefficient of ζ^{-1} is generally complex (34.127a), that is, it involves two real parameters, $\bar{\mathrm{b}}, \vartheta$. In the case of the Joukowski (34.126b)/ von Karman–Trefftz (34.110b)/von Mises (34.118b) airfoils, the parameters (34.127a) of the generic airfoil are (34.128a,b)/(34.128a,c)/(34.128a,d):

$$\vartheta = 0, \quad \bar{\mathrm{b}}^2 = b^2, \frac{\left(k^2 - 1\right) b^2}{3}, - \sum_{\substack{n, m = 1 \\ n \neq m}}^{N} b_n b_m \equiv -B; \qquad \text{(34.128a–d)}$$

the phase term is zero (34.128a) for these particular families of airfoils, but may be nonzero in general. Since by the Blasius theorem (Section 28.2) the forces (28.31a) [pitching moment (28.31b)] depend only on the monopole (and dipole) coefficient of the far-field expansion, it is not affected by the terms omitted in (34.126b; 34.127b), and the Joukowski (34.126a) [generic (34.127a)] airfoils are not distinguishable for $\vartheta = 0$ (are distinguishable only for $\vartheta \neq 0$); the latter remark concerns only the forces and moments that are determined by the leading terms in the far-field, and does not apply in the near field. Thus *the forces (pitching moment) on the generic airfoil (34.127b) are the same (the same apart from the phase factor, ϑ) as for: (i) the Joukowski airfoil (34.32a–d; 34.126a,b) with (34.128b); (ii) the von Karman–Trefftz airfoils (34.108; 34.110b) with (34.128c); (iii) the von Mises airfoils (34.116; 34.117b) with (34.128d) in (34.118a–c).*

34.6 Lift and Pitching Moment Axis and Coefficients

The monopole and dipole moments of a generic airfoil (Subsection 34.6.1) can be used to calculate the lift (pitching moment) and respective coefficients and axis [Subsections 34.6.2 (34.6.3)]. Also to determine: (i) the airfoil focus where the pitching moment is independent of total incidence (Subsection 34.6.4); (ii) the aerodynamic center, where the aerodynamic forces are applied, and the pitching moment vanishes (Subsection 36.6.5).

34.6.1 Monopole and Dipole Moments of Generic Airfoils

The conformal mapping for the generic airfoil (34.127b) corresponds to leading order to the circle transformation (34.2a) replacing a^2 by the coefficient (34.127a) ≡ (34.129a). Substituting (34.129a) in the complex potential (28.115a) for the mean flow with (34.129b) velocity U and angle-of-attack α past a cylinder of radius a with circulation Γ, leads to (34.129c):

$$a^2 \to e^{i\vartheta}\bar{b}^2, \quad U \to e^{-i\alpha}U: \quad f(z) = e^{-i\alpha}U\left(z + e^{i\vartheta}\frac{\bar{b}^2}{z}\right) - \frac{i\Gamma}{2\pi}\log z. \qquad (34.129a–c)$$

The complex potential for a uniform flow (14.63a) with a monopole (12.35b) and dipole (12.43a) at the origin:

$$f(z) = Ue^{-i\alpha}z - \frac{P_1}{2\pi z} + \frac{P_0}{2\pi}\log z, \qquad (34.130a)$$

specifies the monopole (34.130b) [dipole (34.130c)] moments:

$$P_0 = -i\Gamma, \quad P_1 \equiv -2\pi U e^{i\vartheta - i\alpha}\bar{b}^2. \qquad (34.130b,c)$$

The monopole moment (12.35c) reduces (34.130b) to the circulation of a vortex, because a source/sink is incompatible with a rigid impermeable cylinder; the dipole moment (34.130c) reduces to (28.89a)≡(34.131b) using the transformation (34.131a) inverse to (34.129a):

$$e^{i\vartheta}\bar{b}^2 \to a^2: \quad P_1 = -2\pi U e^{-i\alpha}a^2, \qquad (34.131a,b)$$

and allowing for the angle-of-attack of the free stream. For the generic airfoil the horizontal F_x (vertical, F_y) forces, that is, parallel (orthogonal) to the chord in a stream of mass density ρ and velocity U at an angle-of-attack α, are specified (28.31a) by:

$$F^* \equiv F_x - iF_y = i\rho\Gamma U e^{-i\alpha} = -\rho\pi U^2(c+t)\sin(\alpha+\eta)\{\sin\alpha + i\cos\alpha\}, \qquad (34.132)$$

where Γ is the circulation (34.86b) that involves the chord c, thickness t, and camber η of the airfoil. The pitching moment is determined (28.31b) by the dipole moment (34.130c), viz.:

$$M = \mathrm{Re}\left(-i\rho U e^{-i\alpha}P_1\right) = \mathrm{Re}\left(i2\pi\rho U^2\bar{b}^2 e^{i\vartheta - 2i\alpha}\right) = 2\pi\rho U^2\bar{b}^2\sin(2\alpha - \vartheta), \qquad (34.133)$$

where the only parameter not present in (24.132) is the phase ϑ for generic (34.127a,b) airfoils.

34.6.2 Lift Force, Coefficient, and Axis

The horizontal F_x (vertical F_y) force (24.132) is the projection, in the direction of (34.134a) [transverse to (34.134b)] the chord, of a **lift** force (34.134c) orthogonal (Figure 34.5a) to the direction of the flow at infinity (Figure 34.5a):

$$F_y = L\cos\alpha, \quad F_x = -L\sin\alpha: \quad L = \rho\pi U^2 c\left(1 + \frac{t}{c}\right)\sin(\alpha+\eta) = -\rho U\Gamma; \quad (34.134a–c)$$

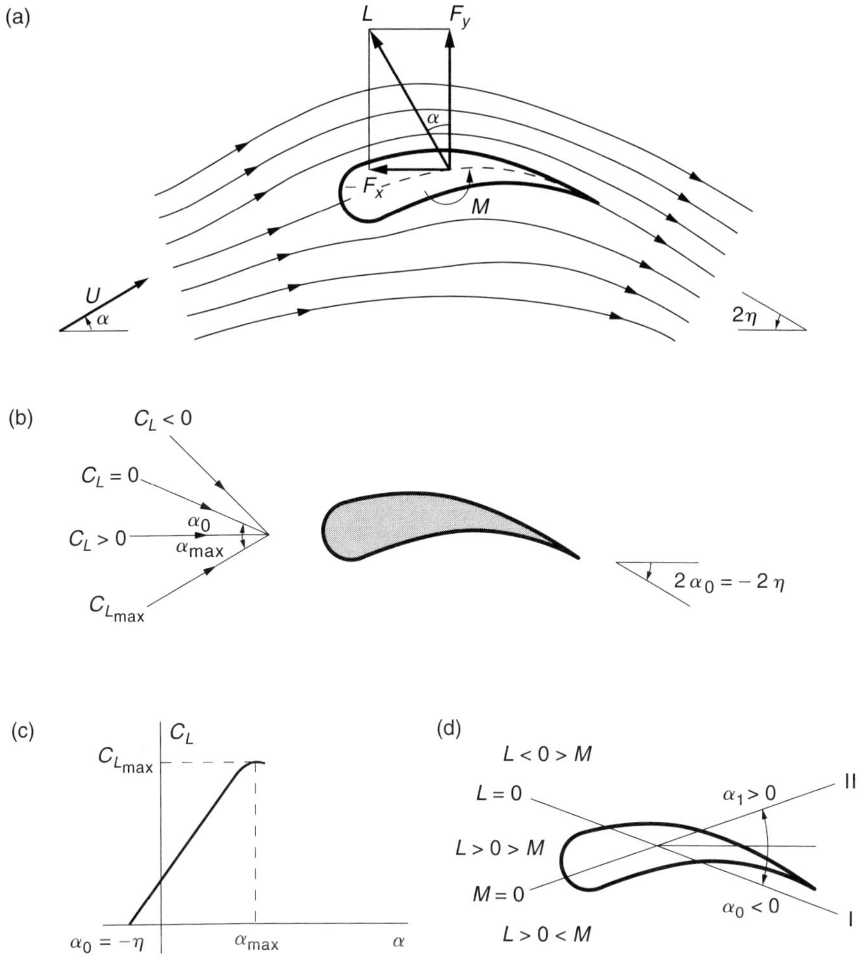

FIGURE 34.5
A generic cambered airfoil (a) in a free stream with velocity U and angle-of-attack α turns the free stream into a downwash at twice the camber angle; this affects: (i) the lift which is orthogonal to the free stream velocity and applied at the aerodynamic center (a.c); (iii) the pitching moment which vanishes only when taken relative to the a.c. An airfoil has two axis: (i) the lift axis I corresponding to the angle-of-attack (b) generally negative $\alpha_0 < 0$ for which there is no lift (c); (ii) the pitching moment axis II for which the pitching moment vanishes (d). Generally the two axes do not coincide, leading (d) to three cases: (i) for an angle-of-attack below both axes there is lift $L > 0$ and pitch-down $M > 0$; (ii) the reverse for an angle-of-attack above both axes, that is, downforce $L < 0$ and pitch-up $M < 0$; (iii) between the two axes there is: (iii–1) lift $L > 0$ and pitch-up $M < 0$ if the lift axis is above the pitch axis; (iii–2) downforce $L < 0$ and pitch-down $M > 0$ if the lift axis is below the pitch axis. The lift L or dimensional lift coefficient C_L increases (c) almost linearly from zero at the angle of zero lift $\alpha_0 = -\eta$ equal to minus half of the downwash angle to the angle-of-attack for maximum lift $\alpha_{\max}$, beyond which flow separation occurs, and the lift coefficient again drops below $C_{L_{\max}}$; flow separation also limits the pitching moment M and dimensionless pitching moment coefficient C_M.

the lift would be perpendicular to the chord $F_x = L, F_y = 0$ *at zero angle-of-attack* $\alpha = 0$. *The forces on the airfoil per unit span, must scale dimensionally as the dynamic pressure of the incident stream (14.28b) ≡ (34.135a) times a length (34.135b), for example, the chord* c:

$$p_0 - p = \frac{\rho}{2}U^2: \quad L = (p_0 - p)\, c\, C_L = \frac{\rho}{2}U^2 c\, C_L; \tag{34.135a,b}$$

the factor is the dimensionless ***lift coefficient*** C_L *given in the present case by (34.136a); if the total incidence* $\beta = \alpha + \eta$, *that is, the angle-of-attack of the flow* α *plus the geometric camber* η, *is small* $\beta^2 \ll 1$, *for example, for* $\beta^2 \le 0.1$ *or* $\beta \le 0.3$ *for* $\beta \le 18^o$, *the lift coefficient is a linear function of the angle-of-attack:*

$$C_L(\alpha) = 2\pi\left(1 + \frac{t}{c}\right)\sin(\alpha + \eta) = 2\pi\left(1 + \frac{t}{c}\right)(\alpha + \eta)\left[1 + O\left((\alpha + \eta)^2\right)\right] = C_{L0} + \alpha C_{L\alpha}; \tag{34.136a}$$

the ***lift constant (slope)*** *(34.136b) [(34. 136c)]:*

$$C_{L0} \equiv C_L(0) = 2\pi\left(1 + \frac{t}{c}\right)\eta, \quad C_{L\alpha} \equiv \frac{\partial C_L}{\partial \alpha} = 2\pi\left(1 + \frac{t}{c}\right), \tag{34.136b,c}$$

is the lift coefficient at zero incidence (34.136b) [its derivative with regard to angle-of-attack (34.136c)]. The lift at zero angle-of-attack (34.136b) is due to the camber, for example, it would be zero for a flat plate or a symmetric airfoil.

The lift slope (34.136c) is 2π *for a flat plate of zero thickness,* $t = 0$, *and for a thin airfoil of thickness* $t^2 \ll c^2$ *compared with the chord* c, *has a modest increase by* $1 + t/c$. *The lift constant (34.136b) shows (Figure 34.5b) that there is lift at zero angle-of-attack for a cambered airfoil, that is, with nonzero curvature. An uncambered airfoil, of zero curvature, experiences no lift at zero angle-of-attack. A cambered airfoil experiences no lift for a negative angle-of-attack equal to minus the camber (34.137a), that is, at the* ***angle of zero lift*** α_0; *this specifies the* ***lift axis***, *or* ***first axis I*** *of the airfoil (Figure 34.5d) such that a flow incident from above* $\alpha < \alpha_0$ *(below* $\alpha > \alpha_0$*) causes a downforce* $L < 0$ *(lift* $L > 0$*) according to (34.137b):*

$$\alpha_0 \equiv -\eta: C_L(\alpha) = C_{L\alpha}\beta = C_{L\alpha}(\alpha + \eta) = C_{L\alpha}(\alpha - \alpha_0), \quad \beta = \alpha + \eta = \alpha - \alpha_0; \tag{34.137a–c}$$

this states that the lift coefficient is proportional, through the lift slope, to the ***incidence*** β; *latter is (34.137c) the difference between the flow angle-of-attack of the flow* α *and the angle of zero lift* α_0; *the latter vanishes for an uncambered airfoil, in that case the angle-of-attack coincides with the incidence, and the lift coefficient is proportional to angle-of-attack. The lift on airfoil increases with the angle-of-attack (Figure 34.5c), until* ***flow separation*** *(Subsection 28.7.7) occurs at the maximum angle-of-attack,* $\alpha_{\max}$. *The preceding formulas cease to apply beyond that point, but they can be used to estimate the* ***maximum lift coefficient***, $C_{L_{\max}} \equiv C_L(\alpha_{\max})$.

34.6.3 Pitching Moment, Coefficient and Axis

The pitching moment (34.133) has the dimensions of a force, for example, lift (34.135b), multiplied by a length, for example, the chord c, *and thus scales dimensionally (34.138a) as*

$$M = C_M(p_o - p)\, c = C_M \frac{\rho}{2}U^2 c^2, \quad C_M = 4\pi\left(\frac{\bar{b}}{c}\right)^2 \sin(2\alpha - \vartheta), \tag{34.138a,b}$$

where (34.138b) is the **pitching moment coefficient**. *At small angle-of-attack $\alpha^2 \ll 1$, for example, $\alpha^2 \le 0.1$, or $\alpha \le 0.3$ or $\alpha \le 18°$, the pitching moment coefficient (34.139a):*

$$C_M(\alpha) = 4\pi \left(\frac{\bar{\text{b}}}{c}\right)^2 \left\{-\sin\vartheta + 2\alpha\cos\vartheta + O\left(\alpha^2\right)\right\} \equiv C_{M_0} + C_{M\alpha}\alpha + O\left(\alpha^2\right), \quad (34.139\text{a})$$

is a linear function of angle-of-attack with **pitch constant** *(34.139b) [(***slope***) 34.139c)]:*

$$\{C_{M0}, C_{M\alpha}\} = 4\pi \left(\frac{\bar{\text{b}}}{c}\right)^2 \{-\sin\vartheta, 2\cos\vartheta\}. \quad (34.139\text{b,c})$$

In the case of the Joukowski airfoil: (i) the chord is given by (34.57b) ≡ (34.128b) ≡ (34.140a), and the phase is zero (34.128a) ≡ (34.140b):

$$\bar{\text{b}} = b = \frac{c}{4}, \quad \vartheta = 0: \quad C_{M0} = 0, \quad C_M(\alpha) = C_{M\alpha}\alpha = \frac{\pi}{2}\alpha, \quad (34.140\text{a–d})$$

implying that: (i) the pitch constant is zero (34.140c); (ii/iii) the pitching moment coefficient (34.140d) is proportional to angle-of-attack α with slope (34.141a); (ii) the pitching moment slope (34.141a) is one quarter of the lift slope (34.136a) for a flat plate (34.141b,c):

$$C_{M\alpha} = \frac{\pi}{2} = \frac{C_{L\alpha}/4}{1 + t/c}; \quad t = 0: \quad C_{M\alpha} = \frac{C_{L\alpha}}{4}, \quad (34.141\text{a–c})$$

with a decrease (34.141a) due to the thickness for a thin airfoil.

The pitching moment vanishes (34.141d) at the **angle of zero moment** *(34.141e):*

$$C_M(\alpha) = C_{M\alpha}(\alpha - \alpha_1), \quad \alpha_1 \equiv -\frac{C_{M0}}{C_{M\alpha}} = \frac{1}{2}\tan\vartheta; \quad (34.141\text{d,e})$$

the latter specifies (Figure 34.5d) the **moment axis** *or* **second axis II** *of the airfoil that is specified by the phase ϑ in the generic airfoil (34.127b). Depending or the relative position of the lift I and moment II axes three situations can arise: (i) if they coincide $\alpha_0 = \alpha_1$ in (34.137a) ≡ (34.141e), then at this angle-of-attack there is no lift and no moment, and above $\alpha > \alpha_0$ (below $\alpha < \alpha_0$) there is a lift (downforce) and positive (negative) moment, that is, pitch-down (pitch-up), for example, this is the case of the uncambered Joukowski airfoil, for that the angles of zero lift and moment are both zero; (ii) if the angle of zero moment is larger than the angle of zero lift $\alpha_1 > \alpha_0$, for example, for a cambered Joukowski airfoil $\alpha_1 = 0 > \alpha_0 = -\eta$, at zero pitching moment there is lift, and at zero lift there is negative pitching moment, that is, a pitch-up, implying that pressure forces on the acting upon the forward part of the airfoil exert a large torque than those acting upon the rear part; (iii) for a non-Joukowski airfoil with phase $\vartheta \neq 0$ such that $\alpha_0 > \alpha_1$ the angle of zero lift exceeds that of zero pitching moment, that is, tan $\vartheta < -2\eta$, there (Figure 34.5d) is a lift at the angle of zero moment, and a pitch-up at zero lift.*

34.6.4 Static Stability and Frequency of Oscillations in Pitch

The pitch oscillations are specified by equating the pitching moment to minus the product of moment of inertia I by the acceleration of angle-of-attack, that has opposite direction (34.142a):

$$-I\text{d}^2\alpha/\text{d}t^2 = M = \frac{1}{2}\rho U^2 c^2 C_M(\alpha) = \frac{\alpha}{2}\rho U^2 c^2 C_{M\alpha}[1 + O(\alpha)]; \quad (34.142\text{a,b})$$

the slope of the pitching moment versus angle-of-attack, calculated at zero angle-of-attack, appears in (34.142b). The pitch oscillations (34.142d) with small angle-of-attack (34.142c) have frequency (34.142e):

$$\alpha^2 \ll 1: \quad d^2\alpha/dt^2 + \omega^2\alpha = 0, \quad \omega \equiv Uc\sqrt{C_{M\alpha}\frac{\rho}{2I}}. \tag{34.142c–e}$$

The small amplitude pitch oscillation of aircraft (34.142a) [ships (8.3)] are similar (34.142c)≡(8.4), with distinct pitching moments (34.142b) [(8.3)] leading to different natural frequencies (34.142e) [(8.5b)]. In aeronautical engineering the pitching moment is taken positive nose-up, that is, clockwise; this the opposite of the general convention that the positive direction is counterclockwise. If the aeronautical convention is used, then the sign of the pitching moment, its coefficient and slope must be reversed; for example, the static stability condition that the pitching moment opposes the deviation of the angle-of-attack from the equilibrium $\alpha = 0$ is $C_{M\alpha} < 0$ in the aeronautical convention, instead of $C_{M\alpha} > 0$ in the general convention. The static pitch stability condition $C_{M\alpha} > 0$ ensures a real natural frequency (34.142e); using the aeronautical convention $C_{M\alpha}$ in (34.142e) is replaced by $-C_{M\alpha}$, so the condition of static stability and real oscillation frequency becomes $C_{M\alpha} < 0$.

34.6.5 Airfoil Axis, Center, and Focus

The real axis is taken as the **airfoil axis** and the point $\varphi = 0\,(\varphi = \pi)$ where it intersects the airfoil as the **trailing (leading) edge**. The lift (34.134c; 34.135a,b) [pitching moment (34.133; 34.138a,b)] and the respective coefficients (34.136a–c; 34.137a–c) [(34.139a–c; 34.140a–d)] are calculated at the **airfoil center** that is (Figure 34.6a) the center of the circle in z-plane superimposed on the ζ-plane. *Since the lift is orthogonal to the* ***wind direction***,that is, *mean flow velocity at infinity, the pitching moment is given along the airfoil axis by (34.143a):*

$$M' = M - xL\cos\alpha; \quad C_M(\alpha, x) = C_M(\alpha) - \frac{x}{c}C_L(\alpha)\cos\alpha, \tag{34.143a,b}$$

this corresponds to (34.143b) in terms of the lift (34.135b) and of the pitching moment (34.138a) coefficients. For small (34.144a) angle of attack: (i) *the pitching moment coefficient (34.143b) is given (34.136a; 34.139a) by (34.144b):*

$$\alpha^2 \ll 1: \quad C_M(\alpha, x) = C_{M0} - \frac{x}{c}C_{L0} + \left(C_{M\alpha} - \frac{x}{c}C_{L\alpha}\right)\alpha; \tag{34.144a,b}$$

(ii) the pitching moment (34.144b) is independent of angle of attack at the ***airfoil focus*** *(34.145a):*

$$\frac{x_m}{c} = \frac{C_{M\alpha}}{C_{L\alpha}}: C_{M1} \equiv C_M(x_m) = C_{M0} - \frac{x_m}{c}C_{L0} = C_{M0} - C_{M\alpha}\frac{C_{L0}}{C_{L\alpha}}, \tag{34.145a,b}$$

where it takes the value (34.145b); (iii) the lift force is applied at the ***lift or aerodynamic center*** *(a.c.) where (36.146b) the pitching moment is zero (34.146a):*

$$C_M(\alpha, x_{ac}) = 0, \quad \frac{x_{ac}}{c} = \frac{C_M(\alpha)}{C_L(\alpha)} = \frac{C_{M\alpha}}{C_{L\alpha}}\frac{\alpha - \alpha_1}{\alpha - \alpha_0}; \tag{34.146a,b}$$

(iv) the lift center (34.146b) and airfoil focus coincide (34.145a) only if the angle of zero lift equals the angle of zero pitching moment:

$$\frac{x_m}{x_{ac}} = \frac{C_{M\alpha}}{C_{L\alpha}}\frac{C_L(\alpha)}{C_M(\alpha)} = \frac{\alpha - \alpha_0}{\alpha - \alpha_1}; \tag{34.147}$$

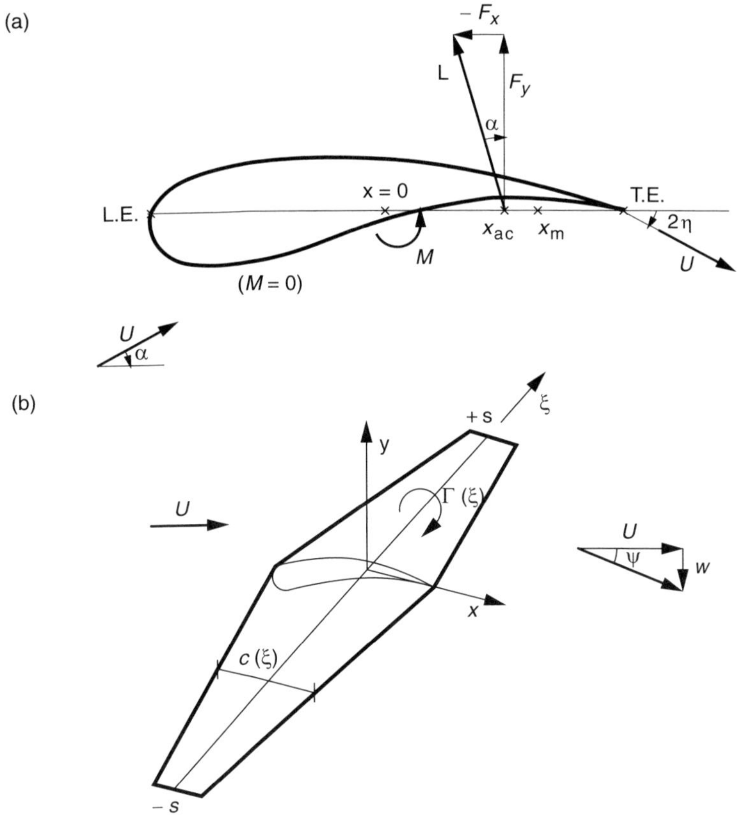

FIGURE 34.6
A generic airfoil (Figure 34.6a) of which the Joukowski/von Karman–Trefftz/von Mises/Carafoli airfoils are particular cases (Diagram 34.1; Figures 34.2b–d; 34.3b; and 34.4a–f): (i) turns the free stream velocity U from the angle-of-attack α upstream to the downwash angle -2η downstream due to the camber η of the airfoil; (ii) has a lift force L applied at the aerodynamic center x_{ac}; (iii) the pitching moment takes the value M at the airfoil center or reference point $x = 0$, vanishes at the aerodynamic center x_{ac} and is independent of the angle-of-attack at the metacenter x_m; (iv) the metacenter coincides with the aerodynamic center for equal angles of zero lift and zero pitching moment (Figure 34.5d), but not otherwise. A wing consists of a constant or more usually a continuously varying airfoil section (Figure 34.6a) along the span (Figure 34.6b), that is, the perpendicular direction. Along the direction perpendicular to the airfoil, that is, along the span may vary: (i) the chord of the airfoil, leading to a nonrectangular wing planform, for example, swept-back, delta, double delta, ogival, cranked, etc.; (ii) the camber of the airfoil, that is, for a twisted wing, which causes the downwash angle to vary along the span; (iii) the thickness of the airfoil, for example, a wing tapering from root to tip; (iv) the plane of the two wings may be the same and horizontal, or they may be inclined upward (downward), that is, the dihedral (anhedral) angle is most visible from the front.

(v) for a Joukowski airfoil (34.57b; 34.128a,b) ≡ (34.148a–c) the airfoil focus (34.145a) is located (34.141a) at (34.148d):

$$\bar{b} = b = \frac{c}{4}, \vartheta = 0: \quad \frac{x_m}{c} = \left[4\left(1 + \frac{t}{c}\right)\right]^{-1}; \qquad (34.148a\text{–}d)$$

(vi) due to the effect of camber it is farther than the aerodynamic center (34.146b) that is located (34.136a; 34.140d) at

$$\frac{x_{ac}}{c} = -\left[4\left(1+\frac{t}{c}\right)\left(1+\frac{\eta}{\alpha}\right)\right]^{-1} = \frac{x_m/c}{1+\eta/\alpha}; \tag{34.149}$$

(vii) for a Joukowski airfoil the focus (34.148d) and aerodynamic center (34.149) coincide (34.150b) only if there is no camber, that is, if the airfoil is symmetric (34.1150a):

$$\eta = 0: x_m = x_{ac} = -\frac{c/4}{1+t/c} = -\frac{c}{4}\left[1-\frac{t}{c}+O\left(\frac{t^2}{c^2}\right)\right] = -\frac{c-t}{4}+O\left(\frac{t^2}{c}\right); \quad \text{(34.150a,b)}$$

(viii) both are located at a quarter chord for a flat plate (34.29), that is, zero thickness in (34.150b), and move a rearward closer to the airfoil center for a thin airfoil. The camber separates the focus and aerodynamic center because it affects the lift (34.134c) but not the pitching moment (34.133). A non-Joukowski airfoil with nonzero phase will also have distinct aerodynamic center and focus, even without camber. The airfoil focus and aerodynamic center lie (do not lie) along the chord line for small (large) angle of attack as shown next.

34.6.6 Lift Center or Aerodynamic Center

The pitching moment is calculated next without restriction to small angle-of-attack (34.144a) and for a generic airfoil. The pitching moment (34.151b) at an arbitrary position (34.151a) is the moment of the lift force (34.134a,b) placed at the aerodynamic center (34.151b)

$$\begin{aligned}\vec{x} = r\left(\vec{e}_x\cos\varphi+\vec{e}_y\sin\varphi\right): \quad \vec{e}_3 M' &= (\vec{x}_{ac}-\vec{x})\wedge\vec{F} \\ &= \vec{e}_3 M - \vec{x}\wedge(-\vec{e}_x\sin\alpha+\vec{e}_y\cos\alpha)L,\end{aligned} \tag{34.151a,b}$$

where M is the pitching moment relative to the center of the airfoil; the pitching moment at all positions lies in the direction of the unit vector $\vec{e}_3$ normal to the (x, y)-plane:

$$\begin{aligned}M' &= M - rL\left(\cos\alpha\cos\varphi+\sin\alpha\sin\varphi\right) = M - rL\cos(\alpha-\varphi) \\ &= \rho\pi U^2\left[2\bar{b}^2\sin(2\alpha-\vartheta) - r(c+t)\sin(\alpha+\eta)\cos(\alpha-\varphi)\right],\end{aligned} \tag{34.152}$$

where (34.133; 34.134c) were used. The pitching moment (34.152) ≡ (34.153b) is independent of the incidence (34.86a) ≡ (34.153a,b):

$$\begin{aligned}\beta = \alpha+\eta: \quad M'(\beta) &= \rho\pi U^2\left[2\bar{b}^2\sin(2\beta-2\eta-\vartheta) - r(c+t)\sin\beta\cos(\beta-\eta-\varphi)\right] \\ &= \frac{\rho\pi U^2}{2}\left[4\bar{b}^2\sin(2\beta-2\eta-\vartheta) - r(c+t)\sin(2\beta-\eta-\varphi) - r(c+t)\sin(\eta+\varphi)\right],\end{aligned} \tag{34.153a,b}$$

if the conditions (34.154a,b) are met:

$$\varphi_m = \eta+\vartheta, \quad r_m = \frac{4\bar{b}^2}{c+t}: \quad M_1 = -2\rho\pi U^2\bar{b}^2\sin(2\eta+\vartheta), \tag{34.154a–c}$$

specifying the pitching moment at the focus (34.154a–c). Thus *the pitching moment is given by (34.152) ≡ (34.153b) at an arbitrary position (34.151a) for an arbitrary airfoil (34.127a,b), with chord c, thickness t, camber η, and phase ϑ, in a stream of velocity U mass density ρ at and angle-of-attack α = β − η; it is independent of the incidence (34.153a) at the airfoil focus (34.154a,b) where it takes the value (34.154c). For a Joukowski airfoil (34.148a,b) ≡ (34.155a,b) these simplify to (34.155c–e):*

$$\bar{\mathrm{b}} = \frac{c}{4}, \quad \vartheta = 0 : \varphi_m = \eta, r_m = \frac{c}{4\left(1 + t/c\right)} = x_m, \quad M_1 = -\frac{\pi}{8}\rho U^2 c^2 \sin\left(2\eta\right), \quad \text{(34.155a–e)}$$

showing that: (i/ii) the airfoil focus lies on the camber line (34.155c) at the position (34.155d) ≡ (34.148d); (iii) the pitching moment (34.155e) leads to a coefficient (34.155f):

$$C_{M1} = \frac{2M_1}{\rho U^2 c^2} = -\frac{\pi}{4}\sin\left(2\eta\right) = -\frac{\pi\eta}{2} + O\left(\eta^2\right); \quad \text{(34.155f)}$$

(iv) this coincides with (34.155f) ≡ (34.145b) on account of (34.140c,d) and (34.136b,c) for small camber. The pitching moment is negative at the airfoil focus because it lies behind the aerodynamic center $x_m > x_{ac}$ in (39.149). The pitching moment at the airfoil focus (34.155e) vanishes for zero camber (34.150a); in that case it is also the aerodynamic center (34.150a,b).

34.7 Spanwise Distribution of Circulation along a Lifting-Line (Prandtl, 1918)

A wing can be represented by the distribution of circulation along its span (Subsection 34.7.1), specifying the lift (Subsection 34.7.2), and through the downwash velocity (Subsections 34.7.3–34.7.4) also the drag (Subsection 34.7.5).

34.7.1 Circulation around the Lifting-Line along the Span

The airfoil is represented as far as aerodynamic forces, that is, lift is concerned, by a vortex of circulation Γ concentrated at its center. For a wing (Figure 34.6b) of span $2s$: (i) the $\zeta = x + iy$-plane is that of the airfoil, that is, the section of the wing normal to the spanwise coordinate $-s \le \xi \le +s$; (ii) at each spanwise coordinate the airfoil may have a different chord $c(\xi)$, thickness $t(\xi)$, and camber $\eta(\xi)$, thus specifying through (34.86b) the circulation $\Gamma(\xi)$ at station ξ. If the span of the wing is much larger than the mean chord $4s^2 \gg \bar{c}^2$, the flow is approximately two dimensional over each section except near the tips, and the wing may be represented by a **lifting-line (Prandtl, 1918)** that is, a distribution of circulation $\Gamma(\xi)$ along the span; the integral along the span of the lift (28.28b) of the airfoil section specifies the total lift (34.156a) in a stream of incident velocity U and mass density ρ:

$$\bar{\mathrm{L}} = -\rho U \int_{-s}^{+s} \Gamma(\xi)\, \mathrm{d}\xi; \quad w(\xi) = -\left(4\pi\right)^{-1} \int_{-s}^{+s} \left(\xi - \chi\right)^{-1} \frac{\mathrm{d}\Gamma}{\mathrm{d}\chi}\mathrm{d}\chi, \quad \text{(34.156a,b)}$$

the variation of circulation along the span also gives rise to a vertical velocity or **downwash** velocity where the integral (34.156b) is a Cauchy principal value (Sections 17.8 and 17.9), because the vortex element $\mathrm{d}\Gamma\left(\chi\right)$ does not induce a velocity on itself. The factor $1/4\pi$ instead of $1/2\pi$ applies in (34.156b) because in three dimensions the area of the unit sphere is 4π instead of the perimeter 2π of the unit circle in two dimensions. The exchange of the factor $(2\pi)^{-1}$ by $(4\pi)^{-1}$ from two to three dimensions also applies to the Newton law of the gravity force (Section 18.5), and to elastic (magnetic) forces [Section 24.3 (26.3)]. A negative circulation produces lift (34.156a).The variation of circulation along the span

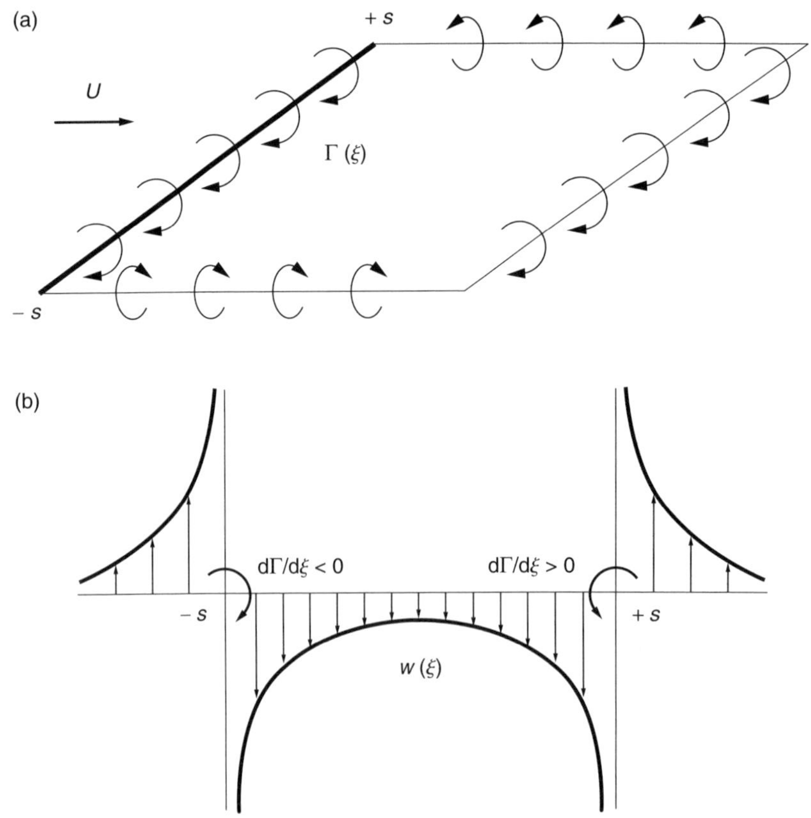

FIGURE 34.7
The airfoil section (Figure 34.6a) corresponds to a plane flow, which is the same in all parallel planes, corresponding to a wing of "infinite span," for which there is no wing tip effect. A real wing of finite span has two wing tips from which issue vortices with opposite circulations (a); together with the circulation around the airfoil sections of the wing this forms a U-shaped or horseshoe vortex. Since the vorticity is conserved in a homentropic flow (Section 14.4), the horseshoe vortex must be "closed" downstream with a "fourth side." The wing tip vortices induce (b) behind the wing: (i) a downwash between them in the wake of the wing; (ii) an upwash outside the wing; (iii) the downwash and upwash are separated by a singular jump of the velocity across the vortex axis. The downwash combined with the variation of circulation along the span of the wing causes a drag force opposite to the free stream, even in an inviscid fluid; this breaks the D'Alembert paradox of zero drag of a body in a potential flow (e.g., Figure 28.8 for Rankine oval), because the flow is rotational behind the wing and the pressure distribution is no longer symmetric forward and aft.

(Figure 34.6b) leads (Figure 34.7a) to wing vortices with axis along the stream and opposite circulations, causing (Figure 34.7b): (i) a downwash between them, that is, behind the wing; (ii) an upwash outside the wing tips.

The circulation lies along the span (34.157b) and the vortical force (28.53a) ≡ (34.157d) associated with a velocity (34.157a) in the plane of the airfoil is given by (34.157d) (34.157e)] as a force density (total force):

$$\vec{v} = \vec{e}_x U - \vec{e}_y w; \ \nabla \wedge \vec{v} \to -\vec{e}_\xi \Gamma; \quad \vec{f}_b \to \vec{F}_b: \quad \vec{f}_b = \rho\left(\nabla \wedge \vec{v}\right) \wedge \vec{v}$$

$$\vec{F}_b = -\vec{e}_\xi \rho \Gamma \wedge \left(\vec{e}_x U - \vec{e}_y w\right) = -\vec{e}_y \rho \Gamma U - \vec{e}_x \rho \Gamma w \equiv \vec{e}_x D + \vec{e}_y L; \qquad (34.157\text{a–e})$$

it has two components, viz.: (i) the lift (34.158a) integrated along the span in (34.156a) that is positive for negative circulation:

$$L(\xi) = -\rho U \Gamma(\xi); \quad D(\xi) = -\rho w(\xi)\Gamma(\xi), \tag{34.158a,b}$$

(ii) the drag (34.158b) integrated along the span in (34.159) specifies the total drag of the wing that is positive since (34.156b) the circulation and downwash velocity have the opposite signs:

$$\overline{\mathrm{D}} = -\rho \int_{-s}^{+s} w(\xi)\Gamma(\xi)\mathrm{d}\xi = -\rho U \int_{-s}^{+s} \Gamma(\xi)\tan\psi(\xi)\mathrm{d}\xi; \tag{34.159}$$

the **downwash angle** $\psi(\xi)$ is (Figure 34.6b) the angle of the total velocity with the incident stream and is given by (34.160a):

$$\frac{w(\xi)}{U} \equiv \tan\psi(\xi) \sim \psi(\xi)\left\{1 + O\left(\left[\frac{w(\xi)}{U}\right]^2\right)\right\}; \tag{34.160a,b}$$

it is small if the downwash velocity is small compared with the free stream velocity (34.160b). Substituting (34.156b) into (34.159), leads to a double-integral formula for drag (34.161):

$$4\pi\overline{\mathrm{D}} = \rho \int_{-s}^{+s} \mathrm{d}\xi \int_{-s}^{+s} \mathrm{d}\chi \, (\xi - \chi)^{-1} \, \Gamma(\xi)\frac{\mathrm{d}\Gamma}{\mathrm{d}\chi} = -\rho \int_{-s}^{+s} \mathrm{d}\xi \int_{-s}^{+s} \mathrm{d}\chi \log|\xi - \chi| \frac{\mathrm{d}\Gamma}{\mathrm{d}\xi}\frac{\mathrm{d}\Gamma}{\mathrm{d}\chi}; \tag{34.161}$$

an integration by parts (34.162b) was performed in (34.161):

$$\Gamma(\pm s) = 0: \quad \int_{-s}^{+s} (\xi - \chi)^{-1}\,\Gamma(\xi)\mathrm{d}\xi = \int_{-s}^{+s} \Gamma(\xi)\,\mathrm{d}\left(\log|\xi - \chi|\right) = -\int_{-s}^{+s} \log|\xi - \chi|\,\mathrm{d}\Gamma(\xi), \tag{34.162a,b}$$

bearing in mind that the circulation vanishes at the wing tips (34.162a). *For a wing of span 2s represented by a lifting-line with circulation distribution $\Gamma(\xi)$ for $-s \le \xi \le s$ in an incident stream of velocity U and density ρ, the lift, downwash and drag are specified by respectively (34.156a), (34.156b), and (34.161).*

34.7.2 Fourier Series for the Circulation and Total Lift

There is no drag (34.161) for a uniform circulation $\mathrm{d}\Gamma/\mathrm{d}\chi = 0$ in agreement with D'Alembert paradox (Subsection 28.5.2). *The drag is due to the variation of circulation along the span $\vec{e}_\xi$ that gives rise to longitudinal vortices in the streamwise direction $\vec{e}_x$. These may be represented by* ***trailing vortices*** *issuing from the wing tips with opposite circulations, resulting in a* ***horseshoe vortex*** *(Figure 34.7a) that causes the downwash velocity behind the wing (Figure 34.7b).* The horseshoe vortex must be closed by a distant wake (Figure 34.7a), since vorticity is conserved along streamlines in a homentropic flow by Kelvin's theorem (Section 14.4). The effect of the closing vortex may be neglected by assuming it lies in a distant part of the wake. In order to apply the preceding formulas the circulation $\Gamma(\xi)$ must be specified for the lifting-line of an arbitrary wing along the span. The spanwise coordinate (34.163b) can be represented using an angular variable (34.163a):

$$0 \le \theta \le \pi: \quad -s \le \xi \equiv s\cos\theta \le s; \quad \Gamma(\theta) = Re\left\{\sum_{n=-\infty}^{+\infty} A_n \mathrm{e}^{in\theta}\right\} \tag{34.163a–c}$$

the circulation is specified by the real part of a Laurent series [(25.42a,b) ≡ (34.163c)]; the series will be real:

$$\Gamma(\xi) = A_0 + \sum_{n=1}^{\infty} \left(A_n e^{in\theta} + A_{-n} e^{-in\theta} \right)$$
$$= A_0 + \sum_{n=1}^{\infty} \left[(A_n + A_{-n}) \cos(n\theta) + i (A_n - A_{-n}) \sin(n\theta) \right], \quad (34.164)$$

if the coefficients are complex conjugate:

$$A_{-n} = A_n^* : \quad \{A_0 \equiv d_0, A_n + A_{-n} = 2\text{Re}(A_n) \equiv d_n, i(A_n - A_{-n}) = -2\,\text{Im}(A_n) \equiv e_n\} \in |R\,; \quad (34.165\text{a,b})$$

this leads (34.165a,b) ≡ (34.165c) to a **Fourier series** (34.165d) specifying the circulation along the span:

$$2A_n = d_n - ie_n : \quad \Gamma(\theta) = d_0 + \sum_{n=1}^{\infty} \left[d_n \cos(n\theta) + e_n \sin(n\theta) \right]. \quad (34.165\text{c,d})$$

The region of convergence of the Fourier series (34.165d) as a function of the complex variable θ is a horizontal strip (25.44a,b) that can include the real axis.

If the series (34.165c) for the circulation is uniformly convergent (Section 21.6 and Note 25.4) it can be integrated term-by-term, viz. to calculate the lift (34.156a), downwash (34.156b), and drag (34.161). For **symmetric loading** the circulation is equal (34.166a) at equal distances from the midline or wing root $\theta = \pi/2$ or $s = 0$ in (34.163b); this condition (34.166a) is met (not met) by the sine (cosine) terms in (34.165c), implying (34.166b) that the circulation simplifies to a **sine series** (34.166c):

$$\Gamma\left(\frac{\pi}{2} + \theta\right) = \Gamma\left(\frac{\pi}{2} - \theta\right) : \quad d_n = 0, \quad \Gamma(\theta) = \sum_{n=1}^{\infty} e_n \sin(n\theta). \quad (34.166\text{a–c})$$

The condition of symmetric loading implies (34.166b) for $n \neq 0$, but would allow $d_0 \neq 0$; the vanishing of the circulation at the wing tips (34.162a) requires $d_0 = 0$. Thus *the circulation is specified along the span (34.163b) for symmetric loading (34.166a) by a sine series (34.166c) that vanishes (34.162a) at the wing tips* $\xi = \pm s$ *or* $\theta = 0, \pi$. Integrating (34.166c) term-by-term, after multiplication by $\sin(m\theta)$, leads to:

$$\int_0^{\pi} \sin(m\theta)\sin(n\theta)\,d\theta = \frac{1}{2}\int_0^{\pi} \{\cos[(n-m)\theta] + \cos[(n+m)\theta]\}\,d\theta$$
$$= \begin{cases} 0 & \text{if } n \neq m, & (34.167\text{a}) \\ \dfrac{\pi}{2} & \text{if } n = m. & (34.167\text{b}) \end{cases}$$

It follows that in the integral (34.168) involving the circulation:

$$\int_0^{+\pi} \Gamma(\theta)\sin(m\theta)\,d\theta = \sum_{n=1}^{\infty} e_n \int_0^{+\pi} \sin(m\theta)\sin(n\theta)\,d\theta = \frac{\pi}{2} e_m, \quad (34.168)$$

only one coefficient (34.167b) appears. Substituting (34.163b) ≡ (34.169a) and (34.166c) and into (34.156a) and using (34.168) leads to (34.169b):

$$d\xi = -s\sin\theta d\theta: \quad \bar{L} = \rho U s\int_0^{+\pi}\Gamma(\theta)\sin\theta d\theta = \rho U s\sum_{n=1}^{\infty}e_n\int_0^{+\pi}\sin\theta\sin(n\theta)\ d\theta = \frac{\pi}{2}\rho U e_1 s, \tag{34.169a,b}$$

showing that *the lift (34.169b) of a symmetrically loaded wing (34.166a) is specified by the first coefficient of the Fourier series for the circulation (34.166c) that vanishes at the wing tips.*

34.7.3 Equivalent Complex Potential and Downwash Velocity

The downwash velocity will be calculated by two methods: (Subsection 34.7.4) the evaluation of the integral (34.156b) as Cauchy principal value using the solution of a finite difference equation; (Subsection 34.7.3) a simpler method that does not require the evaluation of any integrals, and is based on the consideration of an equivalent complex potential. The circulation (34.166c) is the difference of the potential above and below the airfoil (34.170a); since the potential is specified to within an added constant, it can be taken with opposite signs above and below (34.170b) and it equals (34.170c) half the circulation:

$$\Phi^+(\theta) - \Phi^-(\theta) = \Gamma(\theta), \Phi^+(\theta) = -\Phi^-(\theta), \Phi^+(\theta) = \frac{\Gamma(\theta)}{2} = \sum_{n=1}^{\infty}\frac{e_n}{2}\sin(n\theta) \equiv \Phi(\theta). \tag{34.170a–c}$$

The potential is the real part of the complex expression (34.171a):

$$\Phi(\theta) + i\Psi(\theta) = \frac{1}{2}f\left(e^{-i\theta}\right), \quad 2\Psi(\theta) = \operatorname{Im}\left\{f\left(e^{-i\theta}\right)\right\}, \tag{34.171a,b}$$

whose imaginary part specifies a consistent stream function (34.171b); it can be checked that the descending Laurent series (34.172b) with variable $e^{i\theta}$ and coefficients (34.165c) ≡ (34.172a) leads

$$2A_{-n} = 2A_n^* = d_n + ie_n: \ 2f\left(e^{i\theta}\right) = \sum_{n=0}^{\infty}2A_{-n}\left(e^{i\theta}\right)^{-n} = \sum_{n=0}^{\infty}(d_n + ie_n)\, e^{-in\theta} = \Gamma(\theta) + i2\Psi(\theta), \tag{34.172a,b}$$

to: (i) the circulation (34.165d) as its real part; (ii) the stream function (34.173a) as its imaginary part:

$$2\Psi(\theta) = \sum_{n=0}^{\infty}[e_n\cos(n\theta) - d_n\sin(n\theta)]. \tag{34.173a}$$

In the case of symmetric loading (34.166b) ≡ (34.173b) the stream function (34.173a) simplifies to (34.173c):

$$d_n = 0: \quad \Psi(\theta) = \frac{1}{2}\sum_{n=1}^{\infty}e_n\cos(n\theta). \tag{34.173b,c}$$

The corresponding downwash follows from (34.173c; 34.169a):

$$w(\theta) = \frac{1}{2}\frac{\mathrm{d}\Psi}{\mathrm{d}\xi} = \frac{1}{2}\frac{\mathrm{d}\Psi/\mathrm{d}\theta}{\mathrm{d}\xi/\mathrm{d}\theta} = \frac{1}{4s}\csc\theta\sum_{n=1}^{\infty} n e_n \sin(n\theta). \tag{34.174}$$

In three dimensions relative to two dimensions there is the factor $1/2 = 2\pi/4\pi$ that is the ratio of the perimeter (area) of the unit circle (sphere), as mentioned before (Section 18.5 and Subsection 34.7.1). The same downwash velocity can be obtained by: (i) considering instead of the stream function (34.174) the potential (34.180); (ii) alternatively (34.192) by evaluating the integral (34.156b).

The equivalent complex potential (34.172b) was considered only in terms of an angular variable, θ, because its variation is needed only along the span. The downwash velocity (34.174) corresponds to a radius, $r = s/2$, suggesting a complex variable (34.175a):

$$z = \frac{s}{2}\mathrm{e}^{i\theta}; \quad \xi = z + \frac{s^2}{4z} = \frac{s}{2}\left(\mathrm{e}^{i\theta} + \mathrm{e}^{-i\theta}\right) = s\cos\theta, \tag{34.175a,b}$$

the circle theorem (34.47) applied to the $s/2$ leads to the correct spanwise coordinate (34.175b) $\equiv$ (34.163b). The equivalent two-dimensional complex potential is specified by the descending Laurent series (34.176a):

$$2f(z) = \sum_{n=0}^{\infty} A_{-n} z^{-n}, \quad A_{-n} = \frac{1}{2}\left(\frac{s}{2}\right)^n (\mathrm{d}_n + ie_n), \tag{34.176a,b}$$

whose coefficients are written in the form (34.176b) with (34.172a) $\equiv$ (34.165a,c) dimensionless real d_n and imaginary parts e_n. Thus the complex potential is given by

$$f\left(r\mathrm{e}^{i\theta}\right) = \Phi(r,\theta) + i\Psi(r,\theta) = \frac{1}{2}\sum_{n=1}^{\infty}\left(\frac{s}{2r}\right)^n (\mathrm{d}_n + ie_n)\,\mathrm{e}^{-in\theta}. \tag{34.177a}$$

The real (imaginary) part of (34.177a) specifies the Fourier series:

$$\Phi(r,\theta) = \frac{1}{2}\sum_{n=1}^{\infty}\left(\frac{s}{2r}\right)^n \{\mathrm{d}_n\cos(n\theta) + e_n\sin(n\theta)\}, \tag{34.177b}$$

$$\Psi(r,\theta) = \frac{1}{2}\sum_{n=1}^{\infty}\left(\frac{s}{2r}\right)^n \{e_n\cos(n\theta) - \mathrm{d}_n\sin(n\theta)\}, \tag{34.177c}$$

for the potential (34.177b) [stream function (34.177c)]; these reduce to

$$r = \frac{s}{2}: \quad \Phi\left(\frac{s}{2},\theta\right) = \Phi(\theta) = \frac{\Gamma(\theta)}{2}, \quad \Psi\left(\frac{s}{2},\theta\right) = \Psi(\theta), \tag{34.178a–c}$$

to the earlier expressions (34.178b) $\equiv$ (34.165d) [(34.178c) $\equiv$ (34.173a)] for the radius (34.178a). The potential (34.178b) for symmetric loading (34.173b) $\equiv$ (34.180a) is (34.179b):

$$\mathrm{d}_n = 0: \ \Phi(r,\theta) = \frac{1}{2}\sum_{n=1}^{\infty}\left(\frac{s}{2r}\right)^n e_n\sin(n\theta). \tag{34.179a,b}$$

The downwash velocity is given by (34.174), or alternatively (11.20a) replacing $\partial\Psi/\partial\theta$ by $r\partial\Phi/\partial r$ in terms of the potential (34.179b) with $r = s/2$:

$$\begin{aligned} w(\theta) &= \lim_{r\to s/2}\frac{r\,\partial\Phi/\partial r}{2\,\mathrm{d}\xi/\mathrm{d}\theta} = \frac{r}{s\sin\theta}\lim_{r\to s/2}\frac{s}{8r^2}\sum_{n=1}^{\infty} n\lim_{r\to s/2}\left(\frac{s}{2s}\right)^{n-1} e_n\sin(n\theta) \\ &= \frac{\csc\theta}{4s}\sum_{n=1}^{\infty} e_n n\sin(n\theta). \end{aligned} \tag{34.180}$$

The factor $\csc\theta$ in (34.180) ≡ (34.174) at the wing tips $\xi = \pm s$ or $\theta = 0, \pi$ in (34.163a) leads to a divergent series with terms $O(n^2)$, corresponding to the singularity of the downwash velocity at the trailing vortices (Figure 34.7b). Thus the **equivalent complex potential** *(34.175–34.176a,b; 34.177a–c) specifies the circulation in general (34.165d) ≡ (34.178b) and for symmetric loading (34.166a–c); also the lift (34.169b), downwash (34.174) ≡ (34.180) and drag.* Before proceeding to calculate the drag (Subsection 34.7.5), the method of the equivalent potential is checked by evaluating the downwash velocity directly from the integral (34.156b).

34.7.4 Downwash Velocity as Principal Value of an Integral

The downwash velocity is due to the variation of the circulation (34.166c) along the span (34.163b). Introducing the coordinate (34.181a) in analogy with (34.163b) leads to (34.181b):

$$\chi = s\cos\varphi: \quad d\Gamma(\chi) = \sum_{n=1}^{\infty} e_n n \cos(n\varphi)\, d\varphi. \tag{34.181a,b}$$

Substituting (34.181a,b) and (34.163b) specifies the downwash velocity (34.156b) ≡ (34.182):

$$w(\theta) = (4\pi s)^{-1} \sum_{n=1}^{\infty} e_n n \fint_0^{\pi} (\cos\varphi - \cos\theta)^{-1} \cos(n\varphi)\, d\varphi \equiv \frac{1}{4\pi s}\sum_{n=1}^{\infty} e_n n I_n(\theta). \tag{34.182}$$

The integrals in (34.182) are Cauchy principal values (Sections 17.8 and 17.9), since the self-effect of vorticity on itself must be excluded by omitting a symmetric neighborhood of $\theta = \varphi$, viz.:

$$I_n(\theta) \equiv \lim_{\varepsilon\to 0}\left(\int_0^{\theta-\varepsilon} + \int_{\theta+\varepsilon}^{\pi}\right)(\cos\varphi - \cos\theta)^{-1}\cos(n\varphi)\, d\varphi. \tag{34.183}$$

The identity:

$$J(\theta,\varphi) \equiv \left(\frac{d}{d\varphi}\right)\log\left\{\frac{\sin[(\theta+\varphi)/2]}{\sin[\pm(\theta-\varphi)/2]}\right\} = \frac{\sin\theta}{\cos\varphi - \cos\theta}, \tag{34.184a}$$

specifies the primitive for the integral (28.182) with $n = 0$:

$$\begin{aligned} I_0(\theta) &= \csc\theta \lim_{\varepsilon\to 0}\left[\log\left\{\frac{\sin[(\theta+\varphi)/2]}{\sin[(\theta-\varphi)/2]}\right\}\Bigg|_{\varphi=0}^{\varphi=\theta-\varepsilon} + \log\left\{\frac{\sin[(\theta+\varphi)/2]}{\sin[(\theta-\varphi)/2]}\right\}\Bigg|_{\varphi=\theta+\varepsilon}^{\pi=\varphi}\right] \\ &= \csc\theta \lim_{\varepsilon\to 0}\log\left[\frac{\sin(\theta-\varepsilon/2)}{\sin(\varepsilon/2)}\frac{\sin(\theta/2)}{\sin(\theta/2)}\frac{\cos\theta}{-\cos\theta}\frac{\sin(-\varepsilon/2)}{\sin(\theta+\varepsilon/2)}\right] \\ &= \csc\theta \lim_{\varepsilon\to 0}\log\left[\frac{\sin(\theta-\varepsilon/2)}{\sin(\theta+\varepsilon/2)}\right] = 0. \end{aligned} \tag{34.184b}$$

The identity (34.184a) follows from

$$\begin{aligned} 2J(\theta,\varphi) &= \frac{\cos[(\theta+\varphi)/2]}{\sin[(\theta+\varphi)/2]} \pm \frac{\cos[(\theta-\varphi)/2]}{\sin[\pm(\theta-\varphi)/2]} \\ &= \frac{\cos[(\theta+\varphi)/2]\sin[\pm(\theta-\varphi)/2] \pm \sin[(\theta+\varphi)/2]\cos[(\theta-\varphi)/2]}{\sin[(\theta+\varphi)/2]\sin[\pm(\theta-\varphi)/2]} \end{aligned}$$

$$= \frac{2\sin\left[\pm(\theta-\varphi)/2 \pm (\theta+\varphi)/2\right]}{\cos\left[(\theta+\varphi)/2 \mp (\theta-\varphi)/2\right] - \cos\left[(\theta+\varphi)/2 \pm (\theta-\varphi)/2\right]}$$

$$= \frac{\pm 2\sin\theta}{\pm(\cos\varphi - \cos\theta)} = \frac{2\sin\theta}{\cos\varphi-\cos\theta}, \tag{34.185}$$

this (34.185) ≡ (34.184a) evaluates the integral (34.183) for $n = 0$ in (34.184b).

For nonzero order the integrals (34.183) ≡ (34.186a) coincide:

$$I_n = I_n - I_0\cos(n\theta) = \int_0^\pi \frac{\cos(n\varphi)-\cos(n\theta)}{\cos\varphi-\cos\theta}\mathrm{d}\varphi; \quad I_1(\theta) = \pi, \tag{34.186a,b}$$

the second particular case of (34.183) ≡ (34.186a) after $n = 0$ in (34.184b) is $n = 1$ in (34.186b). The identity:

$$\begin{aligned}&\{\cos[(n+1)\varphi] - \cos[(n+1)\theta]\} + \{\cos[(n-1)\varphi] - \cos[(n-1)\theta]\}\\ &\quad = 2\cos(n\varphi)\cos\varphi - 2\cos(n\theta)\cos\theta\\ &\quad = 2\cos(n\varphi)[\cos\varphi - \cos\theta] + 2\cos\theta[\cos(n\varphi)-\cos(n\theta)],\end{aligned} \tag{34.187}$$

leads in (34.186a) to the recurrence formula (34.188a):

$$I_{n+1}(\theta) + I_{n-1}(\theta) = 2\cos\theta I_n(\theta), \qquad \int_0^\pi \cos(n\varphi)\,\mathrm{d}\varphi = 0, \tag{34.188a,b}$$

where was used (34.188b). The integral (34.186a) for general n can be determined by solving the finite difference equation (34.188a) with the initial value (34.184b) for $n = 0$ and (34.186b) for $n = 1$. The finite difference equation (34.188a) has power type solutions with exponent n in (34.189a), where the base satisfies (34.189b):

$$I_n(\theta) = [z(\theta)]^n: \quad 0 = z^2 - 2z\cos\theta + 1 = z^2 - z\left(\mathrm{e}^{i\theta} + \mathrm{e}^{-i\theta}\right) + 1 = \left(z - \mathrm{e}^{i\theta}\right)\left(z - \mathrm{e}^{-i\theta}\right). \tag{34.189a,b}$$

The solutions are (34.189b) ≡ (34.190a), leading (34.189a) to (34.190b):

$$z_\pm(\theta) = \mathrm{e}^{\pm i\theta}, \quad [z_\pm(\theta)]^n = \mathrm{e}^{\pm in\theta} = \cos(n\theta) \pm i\sin(n\theta); \tag{34.190a,b}$$

the general solution of (34.188a) ≡ (34.190a) is a linear combination of (34.190b):

$$I_n(\theta) = B\cos(n\theta) + C\sin(n\theta); \quad 0 = I_0 = B, \quad \pi = I_1 = C\sin\theta, \tag{34.190c}$$

the coefficients in (34.190a) are determined from the initial values (34.184b) ≡ (34.190b) and (34.186b) ≡ (34.190c). Substitution of (34.190b,c) in (34.190a) specifies the integral (34.183) ≡ (34.186a) ≡ (34.191):

$$I_n(\theta) = \int_0^\pi (\cos\varphi - \cos\theta)^{-1}\cos(n\varphi)\,\mathrm{d}\varphi = \pi\csc\theta\sin(n\theta); \tag{34.191}$$

substituting (34.191) in (34.182) specifies:

$$w(\theta) = \frac{1}{4\pi s}\sum_{n=1}^\infty e_n n I_n(\theta) = \frac{1}{4s}\csc\theta\sum_{n=1}^\infty e_n n\sin(n\theta), \tag{34.192}$$

the downwash velocity (34.192) due to a symmetric distribution of circulation (34.166c) along the span (34.163b). The result (34.192) ≡ (34.180) ≡ (34.174) obtained via the integrations (34.181–34.191) coincides with that obtained without integrations via the equivalent potential (34.176–34.179) [stream function (34.170–34.173)].

34.7.5 Induced Drag for Symmetric Loading

The induced drag is specified (34.161) by a double integral that is also a principal value (Sections 17.8 and 17.9) and could also be evaluated, as before (Subsections 34.7.3 and 34.7.4). Since the principal value of the integral (34.182) has already been for the downwash velocity (Subsection 34.7.4), it is simpler to substitute (34.192), (34.166c) and (34.169a) in (34.159):

$$\overline{\mathrm{D}} = \frac{\rho}{4}\int_0^\pi \left(\sum_{n=1}^\infty e_n n \sin(n\theta)\right)\left(\sum_{m=1}^\infty e_m \sin(m\theta)\right) d\theta$$

$$= \frac{\rho}{4}\sum_{n,m=1}^\infty n e_n e_m \int_0^\pi \sin(n\theta)\sin(m\theta)\, d\theta = \frac{\rho\pi}{8}\sum_{n=1}^\infty n e_n^2, \qquad (34.193)$$

where (34.167a,b) was used. *A symmetrically loaded wing with circulation (34.166c) as a function of the spanwise coordinate (34.163b) along a lifting-line, in a stream of velocity U and mass density ρ has total lift $\overline{\mathrm{L}}$ [(drag $\overline{\mathrm{D}}$ and downwash velocity w] and specified respectively by (34.169b) ≡ (34.194a) [(34.174) ≡ (34.180) ≡ (34.192) ≡ (34.194b) and (34.193) ≡ (34.194c)] by the first e_1 (all e_n) Fourier coefficients:*

$$\overline{\mathrm{L}} = \frac{\pi}{2}\rho U s e_1, \quad \left\{w(\theta), \overline{\mathrm{D}}\right\} = \frac{1}{4}\sum_{n=1}^\infty n e_n \left\{s^{-1}\csc\theta \sin(n\theta), \rho\frac{\pi}{2}e_n\right\}. \qquad (34.194a–c)$$

The lift and drag determine the conditions of flight of aircraft and birds, for example, the simplest case, (Figure 34.8) of lift $\overline{\mathrm{L}}$ equal to weight W, and drag $\overline{\mathrm{D}}$ equal to thrust T, corresponds to: (i) level flight, so that there is no change in potential energy, and the weight is transverse to the direction of flight; (ii) straight flight so that there are no centrifugal forces to be balanced; (iii) steady flight so that there is no acceleration or inertia forces. The flight in conditions without the restrictions (i) to (iii) is described by the same aerodynamic forces (lift and drag), thrust and weight balanced against the inertia force involving the

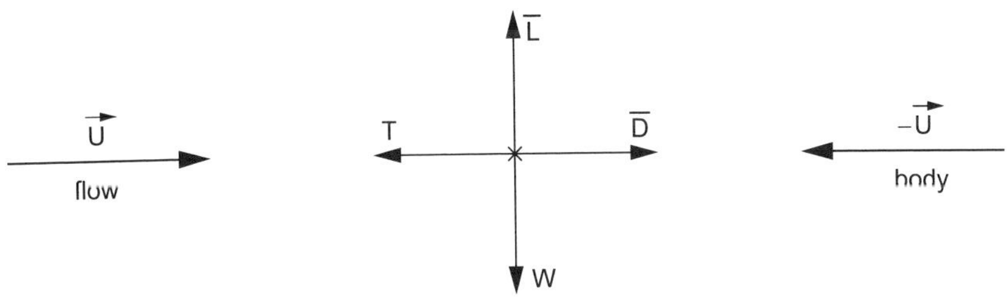

FIGURE 34.8
The lift and drag forces play an essential role in the flight of birds and aircraft. The simplest flight condition is "cruise": (i) straight, that is, no turns or centrifugal forces; (ii) level, that is, no climb or descent to change the potential energy; (iii) steady, that is, no acceleration or deceleration or inertia force. This simplest cruise or nonmanoeuvring flight may occupy must of the flight hours of an aircraft, for example, an airliner. The "cruise" flight: (i) requires that the weight W downward be balanced by the opposite upward lift L; (ii) to generate lift requires a forward velocity, which also causes an opposite drag force D; (iii) the latter must be balanced by the thrust T of the propulsion system in the direction of the velocity. Thus the simplest straight steady level flight requires the balance of two pairs of orthogonal forces: (i) lift versus weight; (ii) drag versus thrust. The drag is along the mean flow velocity and opposite to the body velocity.

acceleration. The straight and level steady flight corresponds to cruise, where an aircraft spends most of the time; thus it is the design condition for the optimization of a transport aircraft.

34.8 Uniform Downwash and Elliptic Loading

The minimization of drag for a given lift (Subsection 34.8.1) specifies the optimum span wise loading, and also the effect on lift of the finite span (Subsection 34.8.2), that is, the wing tip effect.

34.8.1 Spanwise Loading for Minimum Drag

The steady, straight, and level flight, corresponds to the cruise conditions of an aircraft (most of the time spent by a bird in a long migration); it is desirable to have for a given weight or lift $W = \overline{L}$ the minimum drag $\overline{D} = T$ or thrust since this minimize the engine thrust, fuel consumption, and fuel load (physical effort) for a given range. Since the lift fixes the coefficient e_1 in (34.169b) ≡ (34.191a), the minimum drag is obtained in (34.193) ≡ (34.194c) by setting all other coefficients equal to zero (34.195a):

$$e_1 \neq 0 = e_2 = e_3 = \cdots : \quad \overline{D}_{\min} = \frac{\pi}{8}\rho e_1^2 = \frac{\overline{L}^2}{2\pi\rho U^2 s^2}; \tag{34.195a,b}$$

it follows that the minimum lift-induced drag (34.195b) is proportional to the square of the lift. Generally the drag (34.194c) is given by (34.196a):

$$\overline{D} = \overline{D}_{\min}(1+\mu), \quad \mu \equiv e_1^{-2}\sum_{n=2}^{\infty} n e_n^2, \tag{34.196a,b}$$

where μ measures the deviation from minimum drag conditions (34.196b). The condition of minimum drag (34.195a) corresponds by (34.192) to (i) a uniform downwash velocity along the span (34.197a):

$$w = \frac{e_1}{4s}; \quad \overline{L} = 2\pi\rho s^2 U w, \quad \overline{D}_{\min} = 2\pi\rho s^2 w^2, \tag{34.197a–c}$$

(ii/iii) a lift (34.169a) [minimum drag (34.195b)] proportional to the downwash velocity (34.197b) [its square (34.197c)].

It has been shown that the minimum drag for a given lift (34.194a) ≡ (34.197b) is obtained for a uniform downwash velocity along the span (34.197a) that corresponds (34.195a) to a circulation (34.166c) ≡ (34.198b) varying elliptically along the span (34.163b) ≡ (34.198a), from a maximum of (34.198c) at the root to zero at the tip:

$$\xi = s\cos\theta, \quad \Gamma = e_1\sin\theta = 4ws\sin\theta = 4w\sqrt{s^2-\xi^2} = \Gamma_0\sqrt{1-\frac{\xi^2}{s^2}}, \quad \Gamma_0 \equiv 4ws. \tag{34.198a,c}$$

*The condition of **elliptic loading** (34.198b), or any other loading specified a "a priori" $\Gamma(\xi)$, may be obtained by tailoring the wing geometry, that is, choosing the evolution of the airfoil section along the span, viz.: (i) varying the chord $c(\xi)$ specifies the wing **planform**, for example, rectangular, trapezoidal, triangular; (ii) varying the thickness $t(\xi)$ introduces **taper** for example, from root to tip; (iii) varying the camber $\eta(\xi)$ introduces **twist**. For a wing of constant airfoil section, that is, constant chord, and no twist, and constant*

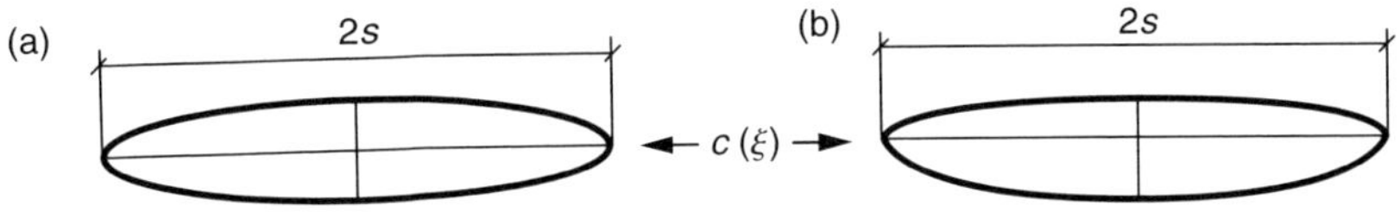

FIGURE 34.9
The most efficient flying machine for the same weight needs the least thrust, for the minimum fuel consummation and emissions, least fuel load and cost. Thus should be maximized the weight to thrust ratio; in cruise this equals (Figure 34.8) the lift-to-drag ratio. Thus maximizing the latter is an objective of aerodynamic design. The wing planform which maximizes lift-to-drag ratio in an incompressible, inviscid flow is elliptical. This implies that the wing planform is: (i) an ellipse (a); (ii) or two half-ellipses with the same major axis along the span (b) and distinct minor axis adding to the root chord, that is, maximum chord at the root of the wing.

thickness-to-chord ratio, the chord $c(\xi)$ reproduces the circulation $\Gamma(\xi)$ distribution, for example, elliptic loading (34.198b) implies that the distance from the leading c_+ to the trailing c_- edge varies (34.199b) elliptically:

$$c(0) = c_+(0) + c_-(0) : c(\xi) = c_+(\xi) - c_-(\xi) = c(0)\sqrt{1 - \frac{\xi^2}{s^2}}; \quad c_\pm(\xi) = c_\pm(0)\sqrt{1 - \frac{\xi^2}{s^2}}, \tag{34.199a–c}$$

the elliptic loading may correspond (Figure 34.9b) to two elliptical plan forms (34.199b) with the same span as one axis (34.199c) and the other two half-axis adding to the chord at the root (34.199a); in the particular case (34.200a):

$$c_+(0) = c_-(0) : \quad \frac{c(\xi)}{c(0)} = \frac{c_\pm(\xi)}{c_\pm(0)} = \sqrt{1 - \frac{\xi^2}{s^2}}, \tag{34.200a,b}$$

then the wing platform is (Figure 34.9a) a single ellipse (34.200b). In both cases the loading remains elliptical and the downwash velocity is uniform along the span.

34.8.2 Effect of Finite Span on Lift

The total lift $\overline{\mathrm{L}}$ of a three-dimensional wing is given by a formula similar to that (34.135b) for the lift per unit span of a two-dimensional airfoil, replacing the chord c by the area S, that is, it is proportional to the area S and incident dynamic pressure (34.135a) in (34.201a):

$$\overline{\mathrm{L}} = (p_0 - p)\, S\overline{\mathrm{C}}_L = \frac{\rho}{2} U^2 S \overline{\mathrm{C}}_L, \quad \overline{\mathrm{C}}_L = \frac{4\pi s^2 w}{US} = \pi\psi\lambda, \tag{34.201a,b}$$

through the total lift coefficient (34.201b); the latter is specified, in the case of elliptic loading (34.197b), as the product of π by the (34.160b) downwash angle ψ and the **aspect ratio** λ; the latter is defined as the square of the wing span $2s$ divided by the area (34.202a):

$$\lambda \equiv \frac{4s^2}{S} = \frac{2s}{\bar{c}}; \quad \bar{c} \equiv \frac{S}{2s} = \frac{1}{2s}\int_{-s}^{+s} c(\xi)\, d\xi, \tag{34.202a,b}$$

it is equal to the **span** $2s$ divided by the **mean chord** (34.202b). The mean chord of a wing with any shape is the chord of a wing with a rectangular plan form and the same

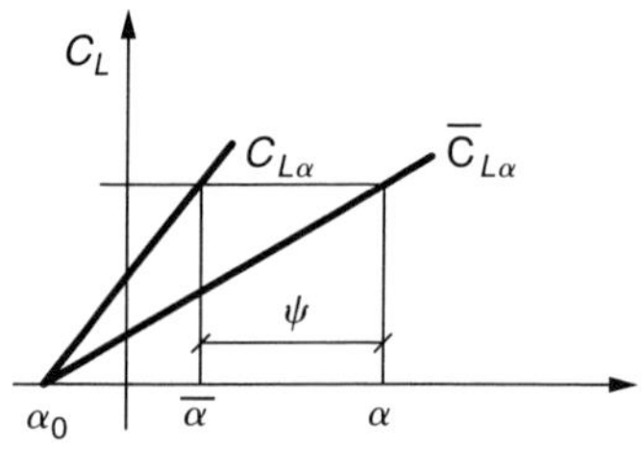

FIGURE 34.10
An airfoil or wing of "infinite span" has the highest lift, because the flow is not disturbed by wing tip vortices. The latter reduce lift more as the aspect ratio reduces, that is, the ratio of span to average chord or mean chord is smaller. This can be seen in the lift-slope or rate-of-change of the lift coefficient with angle-of-attack: (i) it is highest for the airfoil or wing of "infinite span"; (ii) the wing of "finite span" causes a downwash which subtracts from the angle of attack leading to a smaller lift slope; (iii) thus the same lift needs a larger angle-of attack and more drag, as the aspect ratio reduces.

span and area. The aspect ratio λ for a wing of given area is large for large span and small chord, for example, the wings of gliders; the aspect ratio is smaller for the same area for a sweptback or delta (i.e., triangular) plan form that has smaller span and larger mean chord. Considering the induced drag (34.196a,b; 34.197c) associated with the lift (34.197b), their ratio (34.203b) is the ratio of drag (34.203a) and lift (34.202a) coefficients:

$$\overline{\mathrm{D}} = \frac{\rho}{2} U^2 S \overline{C}_D, \quad \frac{\overline{C}_D}{\overline{C}_L} = \frac{\overline{\mathrm{D}}}{\overline{\mathrm{L}}} = (1+\mu)\frac{\overline{\mathrm{D}}_{\min}}{\overline{\mathrm{L}}} = (1+\mu)\frac{w}{U} = (1+\mu)\,\psi; \qquad (34.203\mathrm{a,b})$$

the induced drag is smaller than the lift in the proportion of the downwash angle or ratio of downwash and free stream velocities, with nonelliptic loading increasing the drag-to-lift ratio. In the limit $\lambda \to \infty$ of the wing of "infinite" span or two-dimensional airfoil, since the lift coefficient is finite (34.201b), there is no downwash $\psi = 0$ and no induced drag, $D = 0 = C_D$; this agrees with the verification (Section 34.2) that the only force on a two-dimensional airfoil in a potential flow is lift due to the circulation; in the absence of circulation there would be no force (D'Alembert's paradox in Subsection 28.5.2).

The lift on an airfoil can be compared with that of a wing to assess the "tip" effect of finite span. At small angle-of-attack α the lift coefficient C_L is a linear function with slope $C_{L\alpha} \equiv \mathrm{d}C_L/\mathrm{d}\alpha$, vanishing at the angle-of-zero lift α_0. Assuming that (Figure 34.10) that the angle of zero lift is the same for the airfoil and wing, for the same lift coefficient, the lift slope is smaller for the wing $\overline{C}_{L\alpha}$ than for the airfoil $C_{L\alpha}$ because in the angle-of-attack of the wing $\bar{\alpha}$ is increased relative to that the airfoil α by the downwash angle ψ:

$$\overline{C}_L = \overline{C}_{L\alpha}(\bar{\alpha} - \alpha_0) = C_{L\alpha}(\alpha - \alpha_0), \quad \bar{\alpha} = \alpha + \psi. \qquad (34.204\mathrm{a,b})$$

The ratio of lift slopes:

$$\frac{C_{L\alpha}}{\overline{C}_{L\alpha}} = \frac{\bar{\alpha} - \alpha_0}{\alpha - \alpha_0} = 1 + \frac{\psi}{\alpha - \alpha_0} = 1 + \frac{C_{L\alpha}\psi}{\overline{C}_L} = 1 + \frac{C_{L\alpha}}{\pi\lambda}, \qquad (34.205)$$

simplifies for elliptic loading (34.201b), to

$$\overline{C}_{L\alpha} = \frac{C_{L\alpha}}{1 + C_{L\alpha}/(\pi\lambda)} = \frac{1}{1/C_{L\alpha} + 1/(\pi\lambda)}, \quad C_{L\alpha} = \frac{\overline{C}_{L\alpha}}{1 - \overline{C}_{L\alpha}/(\pi\lambda)} = \frac{1}{1/\overline{C}_{L\alpha} - 1/(\pi\lambda)}. \qquad (34.206\mathrm{a,b})$$

The lift slope (34.136c) for an airfoil of small thickness-to-chord ratio specifies the lift slope of the wing (34.206a) of aspect ratio, λ:

$$\overline{C}_{L\alpha} = \frac{2\pi\,(1+t/c)}{1+2\,(1+t/c)/\lambda} = \frac{2\pi}{1+2/\lambda}\left[1+\frac{t/c}{1+2/\lambda}+O\left(\frac{t^2}{c^2}\right)\right]. \qquad (34.207\text{a,b})$$

The lift slope $\overline{C}_{L\alpha}$ of an elliptically loaded wing of aspect ratio λ is smaller than the lift slope $C_{L\alpha}$ of its airfoil section, and the two coincide (34.206a,b) only in the limit of infinite span. The thickness increases the lift slope both of the airfoil (34.136c) [wing (34.207a,b)], with a moderate effect since it is small relative to the chord. A smaller lift slope implies, for the same maximum angle-of-attack, a lower achievable lift in Figure 34.10. For a flat plate or very thin Joukowski airfoil the lift slope is $C_{L\alpha} = 2\pi$, and the elliptically loaded wing has lift slope $\overline{C}_{L\alpha} = 2\pi/\,(1+2/\lambda)$ that is reduced by 6.25%, 16.7%, 40%, respectively, for aspect ratios $\lambda = 30,\ 10,\ 3$. A thickness-to-chord ratio $t/c = 0.1$ of 10% of the airfoil improves the lift slope by a factor $(t/c)/(1+2/\lambda) = 0.625, 1.67, 4.00\%$ for the same aspect ratios $\lambda = 30, 10, 3$; twice (half) as much for a thick $t/c = 0.2$ (thin $t/c = 0.05$) airfoil or wing section. These results use (34.207b) $\equiv$ (34.207c):

$$\overline{C}_{L\alpha} = \frac{2\pi}{1+2/\lambda}\quad\frac{1+t/c}{1+2t/\left[c\lambda\,(1+2/\lambda)\right]} = \frac{2\pi}{1+2/\lambda}\left[1+\frac{t}{c}\left(1-\frac{2/\lambda}{1+2/\lambda}\right)+O\left(\frac{t^2}{c^2}\right)\right], \qquad (34.207\text{c})$$

that is the approximation of the lift slope (34.207b) of an elliptically-loaded wing to the lowest order of the thickness-to-chord ratio of the airfoil.

34.9 Induced, Form, and Total Drag

The preceding statements suggest that for low-speed flight the most desirable wing for maximum lift and minimum drag should have a large aspect ratio. This conclusion is significantly modified by compressibility effects in high-speed flight, due to wave drag. In flight at velocities U small compared with that c_0 of sound $U^2 \ll c_0^2$, only the two other forms of drag remain: (i) the **induced drag** (34.197c) is proportional (34.195b) to the square of the lift (34.197b); this leads to an induced drag coefficient C_{Di} in (34.203a) proportional to the square of the lift coefficient (34.201a), viz.:

$$C_{Di} = \frac{\overline{D}_i}{(\rho/2)\,U^2 S} = (1+\mu)\,\overline{D}_{\min}\frac{\overline{C}_L}{\overline{L}} = \frac{(1+\mu)\,\overline{C}_L\overline{L}}{2\pi\rho U^2 s^2} = (1+\mu)\,\overline{C}_L^2\frac{S}{4\pi s^2} = (1+\mu)\,\frac{\overline{C}_L^2}{\pi\lambda}, \qquad (34.208)$$

where was used the general correction (34.196a,b) to the elliptic loading (34.195b) and the aspect ratio (34.202a,b); (ii) the **form drag** D_f with coefficient C_{Df} that is lift independent, and is due to frictional forces associated with fluid viscosity within the boundary layer near the wing. Thus:

$$C_D \equiv C_{Df} + C_{Di} = C_{Df} + k\overline{C}_L^2, \qquad k \equiv \frac{1+\mu}{\pi\lambda}, \qquad (34.209\text{a,b})$$

the total drag coefficient (34.209a) of a wing in on incompressible flow, is the sum of: (i) the form drag coefficient due to viscosity that is lift independent; (ii) and the induced drag coefficient that is proportional to the lift coefficient squared, through a factor (34.209b)

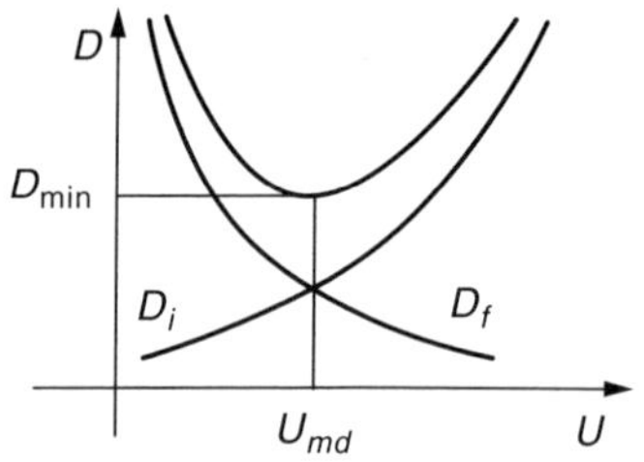

FIGURE 34.11
The drag in an incompressible flow consists of two parts: (i) the friction drag D_f due to the effect of viscosity increases with the square of the velocity; (ii) the induced drag, associated with the lift and downwash increases with the inverse square of the velocity. The reason for (ii) is: (ii–1) the lift must balance the weight at "cruise" (Figure 34.8); (ii-2) the lift-induced-drag involves a product by the downwash velocity and is proportional to the square of the lift; (iii–3) at a constant weight a smaller velocity requires a larger lift coefficient $C_L \sim 1/U^2$ and an even larger drag coefficient $C_{Di} \sim (C_L)^2 \sim 1/U^4$. Thus the induced drag varies like $D_i \sim C_{Di}U^2 \sim 1/U^2$ whereas the form drag varies like $D_f \sim C_{Df}U^2 \sim U^2$. If follows that the total drag increases both at low (high) speeds due to the induced (form) drag. Thus there must exist a minimum drag speed U_{md} for which the total drag is minimum; this corresponds to the minimum thrust, fuel consumption and emissions for "cruise" flight. In the case of "elliptic loading" along the span, that is, an elliptic wing (Figure 34.9) the minimum drag speed (Figure 34.11) corresponds to equal induced and form drag, that is, each is one-half of the total drag. For a compressible flow there is an additional "wave drag" due to the work done to compress the fluid, for example, across shock waves.

that is minimum $\mu = 0$ for an elliptically loaded (34.195a,b) wing of large aspect ratio (34.202a,b).

Substituting (34.209a) into (34.203a), it follows that the *total drag D in (34.210b) in low subsonic flight (34.210a), consists of the form D_f (induced D_i) drag, proportional to the square (inverse square) of velocity:*

$$U^2 \ll c^2 : \overline{D} = C_{Df}\frac{\rho U^2 S}{2} + k\frac{2W^2}{\rho U^2 S} \equiv D_f + D_i; \quad W = \overline{\mathrm{L}} = \frac{\rho}{2}U^2 S\overline{C}_L, \qquad (34.210\text{a–c})$$

the weight (34.210c) was put equal to lift for steady, straight, and level flight (Figure 34.9). The form drag increases with velocity in the same proportion as the dynamic pressure, and dominates at high-speed; at low-speed, to balance the weight W with lift L, high-values of the lift coefficient C_L are needed, leading to large induced drag (34.208) that becomes dominant. Since the form (induced) drag increases at high (low) speed (Figure 34.11), there is an intermediate velocity U_{md} at that (34.211a) the total drag is minimum (34.211b):

$$0 = \frac{d\overline{D}}{dU} = \rho SUC_{Df} - \frac{4W^2 k}{\rho SU^3}, \quad \frac{d^2\overline{D}}{dU^2} = \rho SC_{Df} + \frac{12W^2 k}{\rho SU^4} > 0; \qquad (34.211\text{a,b})$$

at the ***minimum drag speed*** *(34.211a) ≡ (34.212a) the form and induced drag are equal (34.212b):*

$$U_{md} = \sqrt[4]{\frac{k}{C_{Df}}}\sqrt{\frac{2W}{\rho S}}, \quad D_{\text{fmd}} = \frac{\rho}{2}C_{Df}SU_{md}^2 = k\frac{2W^2}{\rho S}U_{md}^{-2} = D_{imd}; \qquad (34.212\text{a,b})$$

then the total drag is twice either of them

$$D_{md} = 2D_{\text{fmd}} = \rho S C_{Df}\, U_{md}^2 = 2D_{imd} = \frac{4kW^2}{\rho S U_{md}^2} = 2W\sqrt{C_{Df}\,k}. \tag{34.213}$$

Thus the lift coefficient for minimum drag (34.214a) is determined by (34.214b):

$$C_{D\,\min} = 2C_{Df} = 2kC_{Lmd}^2, \quad C_{Lmd} = \sqrt{\frac{C_{Df}}{2k}} = \sqrt{\frac{\pi\lambda C_{Df}}{2\,(1+\mu)}}, \tag{34.214a,b}$$

in terms of the form drag coefficient and (34.209b), the aspect ratio (34.202a) and deviation from elliptic loading (34.196a,b). The maximum lift-to-drag ratio or maximum weight-to-thrust ratio in steady, straight and level flight is

$$\left(\frac{W}{T}\right)_{\max} = \left(\frac{L}{D}\right)_{\max} = \frac{C_{Lmd}}{2C_{Df}} = \left[\frac{\pi\lambda}{8\,(1+\mu)\,C_{Df}}\right]^{1/2}, \tag{34.215}$$

the minimum drag (34.214a) specifies the least thrust needed to sustain steady straight and level flight at the minimum drag speed (34.212a); the latter corresponds to the lowest fuel consumption per unit time and longest endurance for a given fuel load (assuming the that specific fuel consumption, per unit thrust, is independent of velocity). For elliptic loading, $\mu = 0$, a laminar flow airfoil $C_{Df} = 0.02$ and moderate aspect ratio $\lambda = 20$, from (34.215) follows $T_{\min} = W/20$; thus cruising flight is possible with a thrust equal to 5% of the weight; this compares with 100% for hover, and more than that for upward motion. The lift-to-drag ratio L/D can attain $L/D \sim 50$ for high-performance cylinder that flies at low speed, for example, Mach number $M < 0.2$; a jet airliner in cruise at $M = 0.8$ achieves about an $L/D \sim 20$. This explains how good aerodynamic design allows birds (aircraft) to fly long distances, with moderate effort (using engines with thrust much smaller than the weight).

Note 34.1. *An Extension from the Lifting-Line to the Lifting-Surface Theory:* In the lifting-line theory of wings (Section 34.7) an airfoil is represented by a point vortex whose circulation varies along the span leading to a one-dimensional distribution; *if the airfoil is represented by a vorticity distribution along the chord (34.216b) then the variation of the circulation along the span (34.216a) leads*

$$-s < \xi < +s, \quad c_-(\xi) \le x \le c_+(\xi); \quad \Gamma = \Gamma(\xi; x), \tag{34.216a–c}$$

to a two-dimensional circulation (34.216c), representing a ***lifting-surface****, where the chord $c(\xi)$ at station ξ specifies the* ***plan form*** *of the wing, for example, rectangular, sweptback, delta, swept forward, etc. In the lifting-surface (lifting-line) theory specifies: (i) the total lift of the wing (34.217) [(34.156a)]:*

$$\overline{\mathrm{L}} = -\rho U \int_{-s}^{+s} \mathrm{d}\xi \int_{c_-(\xi)}^{c_+(\xi)} \mathrm{d}x\, \Gamma(\xi;\, x); \tag{34.217}$$

(ii) the downwash velocity (34.218) [(34.156b)]:

$$-4\pi w(\xi) = \int_{-s}^{+s} \mathrm{d}\chi \;\fint_{c_-(\xi)}^{c_+(\xi)} \mathrm{d}x\, (\xi - \chi)^{-1}\, \frac{\partial \Gamma(\chi;\, x)}{\partial \chi}; \tag{34.218}$$

(iii) the total drag (34.219) [(34.159)]:

$$-\overline{\mathrm{D}} = \rho \int_{-s}^{+s} \mathrm{d}\xi\, w(\xi) \int_{c_-(\xi)}^{c_+(\xi)} \mathrm{d}x\, \Gamma(\xi;\, x)\,, \tag{34.219}$$

or alternatively (34.220b) [(34.161)]:

$$-\overline{\mathrm{D}} = \rho \int_{-s}^{+s} \mathrm{d}\xi \fint_{-s}^{+s} \mathrm{d}\chi \fint_{c_-(\xi)}^{c_+(\xi)} \mathrm{d}x \left[\frac{\partial\Gamma(\xi, x)}{\partial\xi}\right]\left[\frac{\partial\Gamma(\chi;\, x)}{\partial\chi}\right] \log|\xi - \chi|\,; \tag{34.220}$$

in the integrals for the lift (34.217), downwash velocity (34.218) and drag (34.219) ≡ (34.220) the circulation is specified by (34.86b) extended from two to three dimensions:

$$\Gamma(\xi; x) = \pi U\left[c(\xi) + t(\xi; x)\right]\sin\left(\alpha + \eta(\xi)\right], \tag{34.221}$$

so that the chord $c(\xi)$ and camber $\eta(\xi)$ can vary along the span ξ, and the thickness $t(\xi; x)$ along the chord as well, allowing for (i) plan form shape specified by the chord; (ii) taper specified by the thickness; (iii) twist specified by the camber. The expressions (34.217–34.221) are three-dimensional extensions of two-dimensional lifting-line theory; a lifting-surface would use three-dimensional rather than two-dimensional potential theory. The integrals can be evaluated numerically for complex shapes; the downwash (34.218) and drag (34.220) integrals involve Cauchy principal values (Sections 15.8–15.9; Subsection 34.7.6).

Note 34.2. *High-Lift, Low-Drag, and Efficient Aerodynamic Design:* The ability to fly with a thrust of only 5% of the weight, at speeds higher than possible on land or in water, is a demonstration of the importance of efficient aerodynamic design. There is a significant difference between a blunt body like a Rankine oval (Section 28.5) and a wing of high aspect ratio with well designed airfoil sections (Chapter 34). Similar methods apply to slender bodies, whose length is large compared with the cross-section, for example, fuselages and rockets. The aerodynamics can be approached by: (i) analytical methods to understand the physics of the flow around simple shapes; (ii) numerical methods for more complex practical configurations; (iii) wind tunnel tests of scale models; (iv) in-flight measurements in full scale. The potential flow theory can calculate the induced drag associated with lift (Chapter 34) and also due to free jets (Chapter 38). The form drag relates to viscous flows and the wave drag is an effect of compressibility. The aerodynamic forces on flight vehicles for example, lift and drag, are used to calculate performance, for example, speed, range, altitude; the moments relate to stability, and control for example, the pitching moment relates to pitch-up or pitch-down attitude. The aerodynamic loads on structures can couple to deformations, leading to aero elasticity that is a combination of fluid and solid mechanics. The preceding account has concentrated (Chapter 34) on a single airfoil in a steady flow. Extensions include: (i) multiple airfoils, for example, tandem wings or airfoil with flap; (ii) a cascade of parallel airfoils; (iii) a fixed airfoil with separated flow; (iv) an oscillating airfoil in unsteady flow.

Conclusion 34: The simplest disk transformation (Figure 34.1) has two diametrically opposite critical points (a), and maps a circle into a flat plate; thus the flow with angle-of-attack and circulation around a circular cylinder leads to the flow past a flat plate (b), where the Kutta condition is applied at the trailing edge, so that the coincidence of stagnation and critical points there leads to a finite velocity. The same transformation, with the parameter $b \neq a$ distinct from the radius of the circle (Figure 34.2) yields: (a) for $b < a$ and circle with center at the origin, an elliptic cylinder, of which the flat plate is the case (b) of zero minor axis; (c) for center on the real axis, one critical point on and the other inside the circle, a symmetric Joukowski airfoli; (d) for center on the imaginary axis and critical points on the circle, a circular arc profile with camber angle η in (Figure 34.3a); (Figure 34.3b) for center off both axes, the symmetric airfoil (Figure 34.2c) with cambered mean-line (Figure 34.2d $\equiv$ 34.3a) yields an asymmetric airfoil (Figure 34.3c). The flow past these airfoils can be obtained by applying the corresponding transformation to the flow with angle-of-attack α past the circle, and the Kutta condition at the trailing edge specifies the circulation, and determines uniquely the velocity and pressure distributions. The von Karman–Trefftz transformation is a generalization of the Joukowki mapping, which replaces: (i) the flat plate (Figure 34.1b $\equiv$ 34.2b) by a thick airfoil with two axes of symmetry (Figure 34.4a): (ii) the symmetric airfoil with cusp at the trailing-edge (Figure 34.2c) by a symmetric airfoil with an angular point at the trailing edge (Figure 34.4b); (ii) the circular arc airfoil or camber line (Figure 34.2d $\equiv$ 34.3a), by a cambered profile with nonzero thicknes (Figure 34.4c); (iv) the unsymmetric airfoil with nonzero velocity at the cusp at the trailing-edge (Figure 34.3b), by an asymmetric airfoil with a stagnation point at the sharp but not cusped trailing-edge (Figure 34.4d). A distinct generalization of the Joukowski mapping is the von Mises transformation, which can lead to airfoils (Figure 34.4e) a with camber line having one or more inflexions. Any of the preceding airfoils becomes a Carafoli airfoil (Figure 34.4f) with both the leading and trailing-edges rounded by enlarging the circle so that all critical points lie inside. An airfoil in a stream (Figure 34.5) with angle-of-attack α and velocity U experiences (a) a lift L orthogonal to the stream with longitudinal F_x and transverse F_y force components, plus a pitching moment M, which depend on the chord, thickness and camber. The lift coefficient C_L for a cambered airfoil (b) is zero at the angle of zero lift, positive at zero angle-of-attack and attains the maximum value (c) at the angle-of-attack $\alpha_{\max}$ just before boundary layer separation occurs. The angles of zero lift α_0 (zero pitching moment α_1), determine the lift I (pitch II) axis (Figure 34.5d) of the airfoil, e g. in the case illustrated there is lift at zero pitching moment, and there is a pitch-down at zero lift. The system of forces on airfoil (Figure 34.6a) thus consists of a pitching moment applied at the airfoil center $x = 0$, and a lift force applied at the aerodynamic center x_{ac}, with the resulting moment being independent of angle-of-attack at the airfoil focus x_m. The forces on each airfoil section are specified by the circulation, which varies along the span of the wing (Figure 34.6b), causing longitudinal vortices in the stream wise direction (Figure 34.7a). The resulting horseshoe vortex (Figure 34.7) is closed in the far wake (a), and induces a downwash velocity between the wings (b), which causes drag. In straight, level and steady flight (Figure 34.8) the weight W is balanced by the total lift $\overline{L}$, and the total drag $\overline{D}$ is overcome by the thrust T. The minimum thrust for given weight is obtained by spanwise elliptic loading; in the case of a wing of constant cross-section, the plan form is (Figure 34.9) an ellipse if it is symmetrical (a), but if it the leading and trailing edges are not symmetrical elliptic loading leads to a plan form which is not an ellipse (b). The lift slope (Figure 34.10) of a wing of finite span $\overline{C}_{L\alpha}$ is smaller than that of an airfoil section $C_{L\alpha}$, because of the effect of downwash angle ψ, which is the difference of the three $\bar{\alpha}$ and two α dimensional angle-of-attack. To the induced drag D_i which is minimized for elliptic loading, must be added for subsonic flight, the form drag D_f, so that the total drag D is (Figure 34.11) minimum at a velocity U_{md}, at which the form and induced drag are equal; above (below) it the form (induced) drag dominates.

35

Linear and Homographic Transformations

The properties of conformal mappings can be illustrated by considering the simplest transformations, namely, the isometric, linear, and bilinear. A conformal mapping preserves the modulus and direction of angles; if it is also isometric, that is, preserves distances, then it leaves all figures unchanged; the isometric, conformal mapping consists of rotations and translations (Section 35.1). Adding homotheties, that is, projections from the origin, leads to the linear mapping that is the most general transformation without critical points, that is, conformal everywhere (Section 35.2). The ratio of two linear transformation specifies the bilinear, homographic, or Mobius mapping (Section 35.4) that has critical points; it consists of the preceding transformation plus the inversion; it is the most general mapping univalent on the whole complex plane, that is, it maps distinct points on to distinct points, and its domain and image are the whole complex plane (Section 35.5). The bilinear mappings form a group, of that the linear and isometric mappings are subgroups. A point that is mapped onto itself is called (Section 35.1) a fixed point of the transformation: (i) the isometric mapping has one fixed point at infinity (Section 35.1); (ii) the linear mapping has one fixed point at finite distance (Section 35.2); (iii) the bilinear mapping has two fixed points, distinct or coincident, at finite distance or at infinity (Section 35.6). The repeated application of these mappings makes all points tend to (or away from) the limit points that are then called attractive (repulsive); the third possibility, is that of indifferent fixed points, when successive mappings do not change the distance, that is, move the points around a circle. The linear (bilinear) mapping [Section 35.3 (35.6)] have the three types of limit points. The linear (bilinear) mapping leaves invariant a three (four) point cross-ratio [Section 35.3 (35.6)]. This is an instance of the general property of the bilinear transformation of mapping circles and straight lines into circles or straight lines (Sections 35.8 and 35.9); it relates to the inversion (Section 35.7) with regard to the straight line (circle), that is, the image (circle) methods in potential fields with plane walls (Chapters 16, 24, and 26) [cylindrical bodies (Chapters 24, 26, 28, 34, and 36)]. Besides being the most general univalent mapping, and mapping between straight lines and circles, the bilinear transformation appears in a variety of other problems, for example, the composition of spatial rotations.

35.1 Rotation, Translation, and Isometric Mappings

A mapping that preserves the shapes of all figures must: (i) preserve the modulus and direction of all angles, that is, it must be conformal (Section 33.1), and hence be specified (Section 33.3) by a holomorphic function $\zeta = f(z)$; (ii) preserve all distances, that is, be **isometric**, so that the scale factor (33.4) must be unity, that is mapping function must have (33.14b,c) derivate of modulus unity everywhere (35.1a):

$$\left|\frac{\mathrm{d}\zeta}{\mathrm{d}z}\right| = 1, \quad \frac{\mathrm{d}\zeta}{\mathrm{d}z} = \mathrm{e}^{i\alpha}, \quad \chi_m \equiv \{\zeta : \ \zeta = z\mathrm{e}^{i\alpha} + b\}, \tag{35.1a–c}$$

leading to (35.1b,c). Thus a **conformal isometric** *mapping that preserves the shapes of all figures, consists (35.1a–c) of: (ii) a* ***rotation*** $\eta = z e^{i\alpha}$ *through a real angle* α*, in the positive (negative), that is, counterclockwise (clockwise) direction for* $\alpha > 0\,(\alpha < 0)$*; (ii) a* ***translation*** $\zeta = \eta + b$ *by a complex number* b*, of arbitrary modulus and direction.* The isometric mapping (35.1c), involves three real parameters $\alpha \in |R\,, b \in |C$. A point unchanged by the mapping is called a **fixed point,** $\zeta = z$. The rotations $\eta = z e^{i\alpha}$ have the origin $\zeta = z = 0$ as the fixed point, and applied iteratively to any other point $z \neq 0$ describe a circle of radius $|\,z|$ around it, that is, the fixed point is **indifferent**. The translations $\zeta = \eta + b$ have a fixed point at infinity, $\zeta = \eta = \infty$; this property is inherited by the isometric mapping (35.1c): iterated mappings lead to infinity, showing that the fixed point is **attractive**. In the case of an isometric transformation that is not purely a translation (35.2a), the iterated mappings (35.2a,b):

$$\alpha \neq 0: \quad z_N = e^{i\alpha} z_{N-1} + b = e^{i\alpha}\left(e^{i\alpha} z_{N-2} + b\right) + b = e^{i2\alpha} z_{N-2} + b e^{i\alpha} + b, \qquad (35.2\text{a,b})$$

lead, using the geometric sum (21.60a) of ratio (35.3a), to (35.3b):

$$e^{i\alpha} \neq 1: \quad z_N = e^{iN\alpha} z_0 + b \sum_{n=0}^{N-1} e^{in\alpha} = e^{iN\alpha} z_0 + b \frac{e^{iN\alpha} - 1}{e^{i\alpha} - 1}; \qquad (35.3\text{a,b})$$

the latter is bounded by

$$|z_\infty| = \lim_{N\to\infty} |z_N| \leq \lim_{N\to\infty} \left\{ |z_0|\left|e^{iN\alpha}\right| + |b| \frac{\left|e^{iN\alpha} - 1\right|}{\left|e^{i\alpha} - 1\right|} \right\} = |z_0| + \frac{2\,|b|}{\left|e^{i\alpha} - 1\right|}, \qquad (35.3\text{c})$$

where was used:

$$\left|e^{i\psi} - 1\right| = \left|(\cos\psi - 1)^2 + \sin^2\psi\right|^{1/2} = |2 - 2\cos\varphi|^{1/2} \leq \sqrt{4} = 2; \qquad (35.3\text{d})$$

it follows that *an isometric mapping that is not purely a translation, that is, includes a rotation, does not tend iteratively to infinity.*

The ordered pair $(\chi_m, \otimes)$, where χ_m is the set of isometric mappings (35.1c) and $\otimes$ the composition of functions, is a commutative group, since: (i) the composition of two isometric mappings:

$$\eta = \zeta e^{i\beta} + c = z e^{i(\alpha+\beta)} + b e^{i\beta} + c, \qquad (35.4)$$

is an isometric mapping that adds the two rotations, and adds the second translation to the first translation with the second rotation; (ii, iii) it has the associative and commutative properties; (iv) it includes the identity $\zeta = z$, in the case (35.1c) of no rotation $\alpha = 0$ and no translation $b = 0$; (v) the inverse:

$$z = \zeta e^{-i\alpha} - b e^{-i\alpha}, \qquad (35.5)$$

is an isometric mapping, consisting of the symmetric rotation $-\alpha$, plus the symmetric translation $-b$ combined with the symmetric rotation. *The ordered pair* $(\chi_r, \otimes)$ $[(\chi_t, \otimes)]$*, consisting of the rotations (35.6a) [(translations (35.6b)]:*

$$\chi_r \equiv \{\zeta : \ \zeta = z e^{i\alpha}\}, \quad \chi_t \equiv \{\zeta : \ \zeta = z + b\}, \qquad (35.6\text{a,b})$$

and the operation $\otimes$ *composition of functions is also a commutative group, viz. a subgroup of the isometric group;* these correspond to rigid body displacements in two-dimensions, viz. in the complex plane.

35.2 Group of Linear Mappings and Homothety

The isometric mapping has no critical points, because its derivate (35.1a) is unity, and hence is finite and nonzero everywhere. The most general conformal mapping without critical points is specified by an analytic function, $\zeta = f(z)$, whose derivate $d\zeta/dz$ has no singularities, that is, by Liouville's theorem (Section 27.6) is a constant $d\zeta/dz = a \neq 0$ other than zero. Thus a conformal mapping without critical points is a specified by a linear transformation (35.7a–c):

$$X_1 \equiv \{\zeta : \zeta = az + b\}, \quad 0 \neq \frac{d\zeta}{dz} = a = |a|\, e^{i\alpha}, \quad \zeta = az + b = |a|\, e^{i\alpha} z + b, \qquad (35.7a–c)$$

and (35.7b,c) consists (Figure 35.1) of: (i) a rotation $\eta = e^{i\alpha} z$ through an angle $\alpha = \arg(a)$; (ii) a homothety $\zeta = |a|\ \eta$; (iii) a translation $\zeta = \zeta + b$ by b. The **homothety** $\zeta = |a|\, z$, multiplies the modulus of a complex number by a constant and does not affect the argument, that is, corresponds to a projection from the origin that causes a **magnification (reduction)** if $|a| > 1\,(|a| < 1)$; it produces no change if $|a| = 1$; in that case the linear mapping (35.7c) reduces to the isometric mapping (35.1c). Thus a linear mapping involves four real, or two complex a, b, constants.

The ordered pair $(\chi_1, \otimes)$, *where* χ_1 *is the set of linear mappings (35.7c) and* $\otimes$ *the composition, is a commutative group,* because (i) it is closed, since the composition of two linear mappings is a linear mapping:

$$\eta = c\,\zeta + d = c\,(a\ z + b) + d = \ (a\ c)\,z + (b\ c + d), \qquad (35.8)$$

with homothety $|ac| = |a|\ \ |c|$ equal to the product of homotheties, rotation $\arg(ac) = \arg(a) + \arg(c)$ equal to the sum of rotations, and translation $d + c\,b$ equal to the second translation d plus the first translation b with the second homothety $|c|$ and rotation $\arg(c)$;

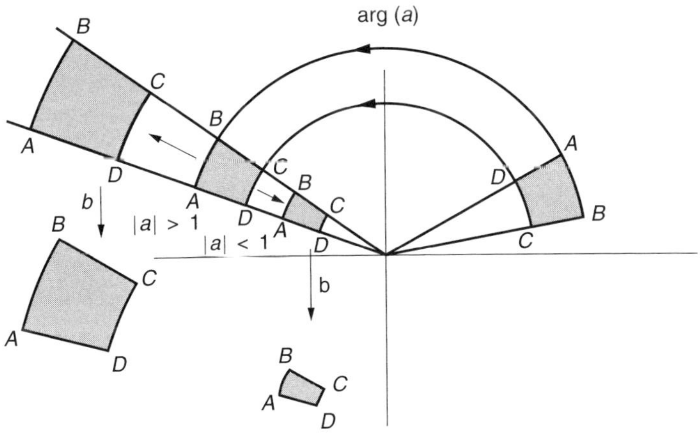

FIGURE 35.1
A linear homogeneous mapping $\zeta = az$ with constant a consists of: (i) a rotation by $\arg(a)$; (ii) no further change if $|a| = 1$, and a homothety with amplification (contraction) if $|a| > 1$ ($\||a| < 1$). The linear nonhomogeneous mapping $\zeta = az + b$ with constant a,b adds: (iii) a translation by b after (i) and (ii). The order of (i) and (ii) may be interchanged without affecting the final result; in contrast interchanging (iii) with (i) or (ii) will generally affect the final result.

(ii–iv) it has the commutative and associative properties, and includes the identity $\zeta = z$ for $a = 1, b = 0$ in (35.7c); (v) the inverse mapping is linear:

$$a \neq 0: \quad z = \frac{\zeta - b}{a} = |a|^{-1} e^{-i\alpha}\zeta - \frac{b}{a}, \tag{35.9a,b}$$

and consists of the translation $-b$ followed by inverse homothety $1/|a|$ and symmetric rotation $-\arg(a)$. *The isometric group* $(\chi_m, \otimes)$ *is (35.1c) a subgroup of (35.7c) the linear group* $(\chi_1, \otimes.)$

35.3 Attractive, Repulsive, and Indifferent Limit Points

The fixed point (35.10a) of the linear mapping (35.7c) is (35.10b):

$$z_0 = \zeta_0 = az_0 + b, \quad z_0 = \frac{b}{1-a}; \tag{35.10a,b}$$

it is placed at finite distance (35.10b) if $a \neq 1$, and lies at infinity if $a = 1$. The translation b in the linear mapping (35.7c) can be eliminated by making the difference of two points (35.11a):

$$\zeta_1 - \zeta_2 = a(z_1 - z_2); \quad \frac{\zeta_1 - \zeta_2}{\zeta_3 - \zeta_2} = \frac{z_1 - z_2}{z_3 - z_2}, \tag{35.11a,b}$$

and the homothety and rotation are eliminated by a division (35.11b). It follows that **the three-point cross-*ratio*** *(35.11b) is an invariant of the linear mapping (35.7c).* A straight line in the z-plane is specified by a linear relation between $\mathrm{Re}(z)$, $\mathrm{Im}(z)$, and a linear mapping transforms it into a straight line in the ζ-plane. Rewriting (35.11b) with the permutation of points (35.12a) and likewise for z it follows (35.12b) that:

$$(\zeta_1, \zeta_2, \zeta_3) \to (\zeta, \zeta_1, \zeta_2): \zeta - \zeta_1 = \frac{\zeta_2 - \zeta_1}{z_2 - z_1}(z - z_1), \tag{35.12a,b}$$

the linear transformation (35.12b) maps the straight line through the points z_1, z_2 *in z-plane, into the straight line through the points* ζ_1, ζ_2 *to the* ζ*-plane.* For example, the transformation of the straight line through (35.13a) to the straight line through (35.13b) is given by (35.13c):

$$z_{1,2} = -1, i;\ \zeta_{1,2} = 1, i:\ \zeta - 1 = \frac{i-1}{i+1}(z+1) = i(z+1);\ \zeta - i = i(z - i), \tag{35.13a–d}$$

this corresponds (Figure 35.2) to (35.13d) a rotation by $\pi/2 = \arg(i)$ around $\zeta_2 = i = z_2$.

The linear transformation (35.12b) with (35.14a) one fixed point is (35.14b):

$$\zeta_1 = z_0 = z_1: \quad \zeta - z_0 = k(z - z_0), \quad k \equiv \frac{\zeta_1 - z_0}{z_1 - z_0} \neq 1; \tag{35.14a–c}$$

the coefficient (35.14c) cannot be unity, because the second point must be distinct, and thus cannot be a fixed point $\zeta_1 \neq z_0$ *in (35.10a). If (35.14b,c) is applied iteratively n-times, as*

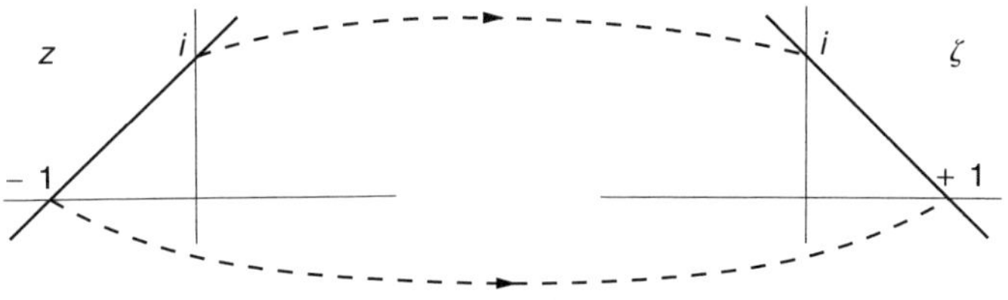

FIGURE 35.2
A linear transformation (35.7a) ≡ (35.12b) with a,b constant maps straight lines in the z-plane onto straight lines in the ζ-plane: two points (z_1, z_2) in (35.12a) locate the straight line in the z-plane and their images $\zeta_{1-2} = f(z_{1,2})$ locate the image straight line. The case illustrated is (35.13a–d).

*$n \to \infty$ it leads to a **limit point** that may be of one of three types:*

$$\begin{aligned} &\zeta \to z_0 && \text{if } |k| < 1 && z_0 \text{ attractive}, && (35.15\text{a})\\ &\zeta \to \infty && \text{if } |k| > 1 && z_0 \text{ repulsive}, && (35.15\text{b})\\ &|\zeta - z_0| = R && \text{if } |k| = 1 && z_0 \text{ indifferent}, && (35.15\text{c}) \end{aligned}$$

*namely: (i) if $|k| < 1$ all points tend to the fixed point z_0 that is **attractive**; (ii) if $|k| > 1$ all points $z \neq z_0$ tend to infinity, that is, away from the fixed point that is **repulsive**; (iii) if $|k| = 1 \neq k$ the successive points $\zeta_1, \ldots, \zeta_n, \ldots$ lie at a constant distance R from z_0, that is, on a circle with center at z_0 and radius $R \equiv |z - z_0|$ determined by the first iteration, and the fixed point is **indifferent**. In the last case the point ζ takes the same position after q iterations if* arg $(k) = p/q$ *is a rational number in its lowest terms, whereas if* arg (k) *is irrational it never takes the same position.*

35.4 Univalent Mapping and Homographic Transformation

The linear mapping (35.7c) is **univalent** because it takes each value in the range only once, that is, it maps (35.11a) distinct points $z_1 \neq z_2$ into distinct points $\zeta_1 \neq \zeta_2$ in the ζ-plane. The most general univalent mapping of the whole z-plane is specified by a function $\zeta = f(z)$ without essential singularities, because in the vicinity of such a singularity the function takes the same value at many distinct points (Section 39.2). A function without essential singularities is rational (Section 27.7), that is, is the ratio of two polynomials, $\zeta = P(z)/Q(z)$. The mapping is univalent iff $P(z)$, $Q(z)$ has a single root, and this is the case (Subsection 31.6.3) iff $P(z)$, $Q(z)$ are both of the first degree. Thus the *most general univalent mapping of the whole complex z-plane, is the ratio of two linear transformation,* that is, *the **bilinear, homographic or Mobius transformation** (35.16a):*

$$\mathcal{X}_2 \equiv \left\{\zeta\colon\ \zeta = \frac{az+b}{cz+d}\right\}, \quad \zeta = \frac{az+b}{cz+d}, \quad \lambda \equiv ad - bc \neq 0, \tag{35.16a–c}$$

that involves (35.16b) four complex, or eight real quantities, whose cross-ratio is nonzero (35.16c); otherwise for a zero (35.17a) cross-ratio (35.16c), the mapping (35.16b) would reduce to a constant (35.17b), and could not be univalent:

$$ad = bc: \quad \zeta = \frac{az+b}{cz+d} = \frac{az+b}{cz+bc/a} = \frac{az+b}{(c/a)(az+b)} = \frac{a}{c} = \frac{b}{d}. \tag{35.17a,b}$$

The linear mapping is the most general mapping without critical points, and hence the bilinear mapping that is more general must have critical points:

$$\frac{d\zeta}{dz} = \frac{a(cz+d) - c(az+b)}{(cz+d)^2} = \frac{ad-bc}{(cz+d)^2}; \tag{35.18}$$

the bilinear mapping has a critical point of the first $d\zeta/dz = 0$ (second $d\zeta/dz = \infty$) kind at $z = \infty$ $(z = -d/c)$, both with exponent 2.

Since the homographic transformation (35.16b) is the ratio of two linear mappings, it must involve an inversion:

$$\eta = cz + d, \quad \xi = \frac{1}{\eta}, \quad \zeta = \xi\left(b - \frac{ad}{c}\right) + \frac{a}{c}; \tag{35.19a–c}$$

this follows from (35.20) ≡ (35.16b):

$$\zeta = \frac{b - ad/c}{\eta} + \frac{a}{c} = \frac{1}{c}\left(\frac{bc - ad}{cz+d} + a\right) = \frac{az+b}{cz+d}. \tag{35.20}$$

The bilinear mapping (35.16b) consists of two linear mappings (35.19a,c), with an inversion (35.19b) in between. The **inversion,** $z \to 1/z$, inverts the modulus $|1/z| = |z|$, and changes the sign of the argument $\arg(1/z) = -\arg(z)$; thus it performs (Figure 35.3) a reflection on the real axis, and a magnification (reduction) for $|z| > 1$ $(|z| < 1)$. The coefficients in the transformation (35.19c) can be obtained as follows: (i) it must be a linear transformation that leads to (35.16b):

$$\frac{az+b}{cz+d} = \zeta = A\xi + B = \frac{A}{cz+d} + B = \frac{cBz + A + Bd}{cz+d}; \tag{35.21}$$

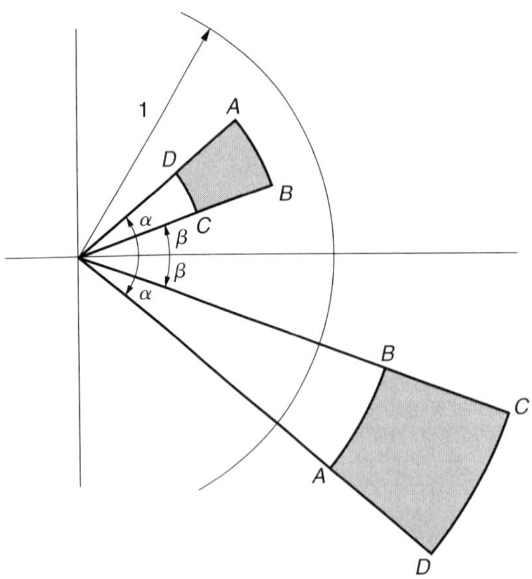

FIGURE 35.3
The inversion $\zeta = 1/z$ with regard to the origin consists of: (i) a reflection on the real axis $\arg(\zeta) = -\arg(z)$; (ii) a homothety which is an amplification (contraction) for $|\zeta| > 1$ $(|\zeta| < 1)$ or $|z| < 1$ $(|z| > 1)$, and no further change for $|\zeta| = 1$ or $|z| = 1$. The order of (i) and (ii) may be interchanged without effect on the final result.

(ii) hence the constants (A, B) must satisfy (35.22a,b):

$$a = cB, \quad b = A + Bd: \quad B = \frac{a}{c}, \quad A = b - \frac{ad}{c}, \tag{35.22a–d}$$

leading to (35.22c,d); (iii) the latter substituted in (35.21) lead to (35.19c).

35.5 Bilinear Group (Mobius) and Self-Inverse Function

The group of bilinear transformations includes as progressively more restricted subgroups (Subsection 35.5.1) the linear and isometric; other particular bilinear and linear transformations, for example, the self-inverse (Subsection 35.5.2) do not form subgroups.

35.5.1 Subgroups of the Group of Bilinear Transformations

The ordered pair $(\chi_2, \otimes)$*, where* χ_2 *is the set of bilinear transformations (35.16a), and* $\otimes$ *the operation composition, is a commutative group,* since: (i) the composition of two homographic transformations is a homographic transformation:

$$\lambda' = a'd' - b'c' \neq 0: \quad \zeta' = \frac{a'\zeta + b'}{c'\zeta + d'} = \frac{(a'a + b'c)\ z + \ a'b + b'd}{(c'a + d'c)\ z + c'b + d'd}, \tag{35.23a,b}$$

$$\lambda'' \equiv (a'a + b'c)(c'b + d'd) - (a'b + b'd)\ (c'a + d'c) = (ad - bc)\,(a'd' - b'c') = \lambda'\lambda \neq 0; \tag{35.23c}$$

(ii, iii) the composition of bilinear mappings is associative and commutative; (iv) the identity $\zeta = z$ is included in (35.16b) for $b = c = 0 \neq a = d$; (v) the inverse of (35.16b) is also a bilinear mapping:

$$\lambda \equiv (-d)\,(-a) - bc \neq 0: \qquad z = \frac{(-d)\,\zeta + b}{c\zeta + (-a)}. \tag{35.24a,b}$$

The composition of two bilinear mappings (35.16c) and (35.23b) is a bilinear mapping:

$$\zeta' = \frac{a'\,(az + b)/(cz + d) + b'}{c'\,(az + b)/(cz + d) + d'} = \frac{a'\,(az + b) + b'\,(cz + d)}{c'\,(az + b) + d'\,(cz + d)} = \frac{(a'a + b'c)\,z + a'b + b'd}{(c'a + d'c)\,z + c'b + d'd}, \tag{35.24c}$$

in the agreement with (35.23b) $\equiv$ (35.24c); the identity (35.23c) follows performing the products on both sides. It follows that *the cross-ratio of the composition of two bilinear mappings is the product of the cross-ratios, and hence nonzero because each factor is nonzero. Of the* **four elementary transformations:** *(i,ii) the rotation (35.6a) [translation (35.6b)] form separate groups* $(\chi_r, \otimes)$ $[(\chi_t, \otimes)]$*, whose union (35.1c) is the isometric group* $(\chi_m, \otimes)$*; (iii, iv) adding the homothety leads to the linear (35.7c) group* $(\chi_1, \otimes)$ *and the inversion to the bilinear (35.16b,c) group* $(\chi_2, \otimes)$*. Each of these groups contains the preceding,* that is,

$$(\chi_r, \otimes) \cup (\chi_t, \otimes) \equiv (\chi_m, \otimes) \subset (\chi_1, \otimes) \subset (\chi_2, \otimes), \tag{35.25}$$

states the relation between the simplest complex transformation groups that correspond to five two-dimensional geometries: (i) congruent $\mathfrak{B}$ for the translation, χ_t; (ii) plane rotation $\mathfrak{J}_0$ for χ_r; (iii) isometric conformal $\mathfrak{M}_0\mathfrak{Q}$ for χ_m; (iv) affine $\mathfrak{A}$ for linear transformations χ_1; (v) rigid body $\mathfrak{J}_3$ for the bilinear transformations χ_2. A subset of the bilinear (linear) transformations that does not form a subgroup consists of the self-inverse transformations; these coincide with their inverses and are considered next (Subsection 35.5.2).

35.5.2 Bilinear and Linear Self-Inverse Transformations

A homographic transformation (35.16b) coincides with its inverse (35.24a) if (35.26a) is met:

$$d = -a: \quad \zeta = \frac{az+b}{cz-a} \Leftrightarrow z = \frac{a\zeta+b}{c\zeta-a}, \quad \lambda \equiv -a^2 - bc; \tag{35.26a–d}$$

this can be checked substituting (35.26b) into (35.26c) leading to the identity:

$$\zeta = \frac{a\,(a\zeta+b)/(c\zeta-a)+b}{c(a\zeta+b)/(c\zeta-a)-a} = \frac{a\,(a\zeta+b)+b\,(c\zeta-a)}{c\,(a\zeta+b)-a\,(c\zeta-a)} = \frac{\left(a^2+bc\right)\zeta}{a^2+bc} = \zeta. \tag{35.26e}$$

Thus *the* ***self-inverse bilinear transformation*** *that coincides with its inverse, is of the form (35.26b) ≡ (35.26c) with cross-ratio (35.26d). It maps the complex* z*-plane back to itself after two successive transformations with the* ζ*-plane (35.26e) as intermediate.*

The composition of two self-inverse bilinear transformations (35.26b; 35.27a):

$$\zeta' = \frac{a'\zeta+b'}{c'\zeta-a'}: \quad \zeta' = \frac{a'\,(a'z+b)+b'\,(cz-a)}{c'\,(az+b)-a'\,(cz-a)} = \frac{(a'a+b'c)\,z+a'b-b'a}{(c'a-a'c)\,z+a'a+c'b}$$
$$= \frac{a''z+b''}{c''z+d''}, \tag{35.27a,b}$$

is generally not a self-inverse bilinear transformation, because the condition (35.26a) is not met (35.27c):

$$d'' = a'a + c'b \neq -a'' = -a'a - b'c. \tag{35.27c}$$

A particular case (35.28a) of (35.26b) is the **self-inverse linear transformation** (35.28b,c):

$$c = 0; \quad \frac{b}{a} \equiv q \in |C: \quad \zeta = q - z, \qquad z = q - \zeta; \tag{35.28a–c}$$

these do not form a group either:

$$\zeta' = q' - \zeta = q' - (q - z) = q' - q + z. \tag{35.28d}$$

Thus the sets of self-inverse bilinear (35.29a) [linear (35.29b)] transformations:

$$X_3 \equiv \left\{\zeta\colon \zeta = \frac{az+b}{cz-a}, \; a^2 + bc \neq 0\right\}, \; X_0 \equiv \{\zeta : \zeta = q - z\}, \tag{35.29a,b}$$

do not form groups with the operation composition.

Applying (35.26b,c) ≡ (35.30d,e) to any function (35.30a,b) leads to a **self-inverse function** (35.30c):

$$\zeta = f(\eta)\,; \; z = f(w)\,; \; f(f(w)) = w\colon \; f(\eta) = \frac{af(w)+b}{cf(w)-a}, f(w) = \frac{af(\eta)+b}{cf(\eta)-a}. \tag{35.30a–e}$$

For example the transformation (35.31a,b) leads to the function (35.31c):

$$\begin{aligned}\exp(2\eta) &= \frac{ae^{2w}+b}{ce^{2w}-a} = \frac{ae^{w}+be^{-w}}{ce^{w}-ae^{-w}}\\ &= \frac{(a+b)\cosh w + (a-b)\sinh w}{(c-a)\cosh w + (c+a)\sinh w},\end{aligned} \qquad (35.31\text{a-c})$$

using:

$$\exp(\pm w) = \cosh w \pm \sinh w, \qquad 2\cosh,\ \sinh w = e^{w} \pm e^{-w}. \qquad (35.31\text{d–f})$$

The function (35.31c) ≡ (35.32e) is self-inverse (35.32b–d) for all values of the parameters (35.32a):

$$g(w) = \frac{1}{2}\log\left\{\frac{a+b+(a-b)\tanh w}{c-a+(c+a)\tanh w}\right\} = \frac{1}{2}\log\left\{\frac{a-b+(a+b)\coth w}{(a+c)+(c-a)\coth w}\right\}; \qquad (35.32\text{a–e})$$

a particular case appears in Subsection 38.9.4. Other self-inverse functions are considered in Note 38.1. A self-inverse function can be used to test the accuracy of a numerical or computer algorithm: (i) if applied twice it should yield exactly the same result; (ii) any deviation is a measure of the error of computation.

35.6 Four-Point Cross-Ratio and Fixed Points

The difference of two points confirms that the bilinear mapping (35.16b) is univalent:

$$\zeta_1 - \zeta_2 = \frac{z_1 a + b}{z_1 c + d} - \frac{z_2 a + b}{z_2 c + d} = \frac{(ad-bc)(z_1 - z_2)}{(z_1 c + d)(z_2 c + d)}, \qquad (35.33\text{a})$$

because $z_1 \neq z_2$ implies $\zeta_1 \neq \zeta_2$. From (35.33a) follows:

$$(\zeta_1 - \zeta_2)(\zeta_3 - \zeta_4) = \frac{(z_1 - z_2)(z_3 - z_4)(ad-bc)^2}{\prod_{n=1}^{4}(cz_n + d)}; \qquad (35.33\text{b})$$

thus all coefficients can be eliminated from (35.33b) by performing

$$\frac{\zeta_1 - \zeta_2}{\zeta_1 - \zeta_4}\frac{\zeta_3 - \zeta_4}{\zeta_3 - \zeta_2} = \frac{z_1 - z_2}{z_1 - z_4}\frac{z_3 - z_4}{z_3 - z_2}, \qquad (35.34)$$

the ***four-point cross-ratio*** *(35.34) that is an invariant of the homogrophic mapping (35.16b), that is, is unchanged by a transformation of the bilinear type.* Rewriting (35.34) with (35.35a) leads to (35.35b):

$$(\zeta_1, \zeta_2, \zeta_3, \zeta_4) \to (\zeta, \zeta_1, \zeta_3, \zeta_2): \quad \frac{\zeta - \zeta_1}{\zeta - \zeta_2} = \frac{(\zeta_3 - \zeta_1)(z_3 - z_2)}{(\zeta_3 - \zeta_2)(z_3 - z_1)}\frac{z - z_1}{z - z_2}, \qquad (35.35\text{a,b})$$

there is only one homographic transformation (35.35b) that maps the points z_1, z_2, z_3 in the z-plane to the points $\zeta_1, \zeta_2, \zeta_3$ the ζ-plane. The transformation (35.35b) is used in Example 40.7 to map between sets of three points in (not in) a straight line, representing a straight line (circle).

The bilinear mapping (35.16b) reduces to the linear mapping for $c = 0$; in that case it has one fixed point (35.10a,b). Excluding this degenerate case, the bilinear mapping in generally has two fixed points (35.36b):

$$z_\pm = \zeta_\pm = \frac{z_\pm a + b}{z_\pm c + d}, \quad z_\pm^2 c + (d - a)\, z_\pm - b = 0, \tag{35.36a,b}$$

that are the roots (35.37b) of (35.36b):

$$(a - d)^2 + 4bc \neq 0 \neq z_+ - z_- : \quad 2cz_\pm = (a - d) \pm \left|(a - d)^2 + 4bc\right|^{1/2}; \tag{35.37a,b}$$

the roots are distinct for nonzero discriminant (35.37a) and coincide for (35.38b) for zero discriminant (35.38a):

$$(a - d)^2 + 4bc = 0 : \quad z_+ = z_- = \frac{a - d}{2c} = \sqrt{-\frac{b}{c}} = -\frac{2b}{a - d}. \tag{35.38a,b}$$

The bilinear mapping (35.35) with distinct fixed points at (35.39a), is given by (35.39b):

$$z_{1,2} \equiv z_\pm = \zeta_\pm = \zeta_{1,2} : \quad \frac{\zeta - z_+}{\zeta - z_-} = k\frac{z - z_+}{z - z_-}, \quad k \equiv \frac{z_+ - \zeta_3}{z_- - \zeta_3}\frac{z_- - z_3}{z_+ - z_3} \neq 1, \tag{35.39a–c}$$

where the coefficient (35.39c) cannot be unity, because the third point must be distinct from $z_\pm$, and thus cannot be a fixed point $\zeta_3 \neq z_3$. *If the fixed points of the bilinear mapping are distinct, then (35.39a) can be applied iteratively n-times, and letting $n \to \infty$ specifies the limit points, leading to three cases:*

$$z \neq z_- : \quad |k| < 1, \quad \zeta \to z_+, \quad z_+ \text{ attractive}, \quad z_- \text{ repulsive}; \tag{35.40a–c}$$

$$z \neq z_+ : \quad |k| > 1, \quad \zeta \to z_-, \quad z_- \text{ attractive}, \quad z_+ \text{ repulsive}; \tag{35.41a–c}$$

$$z \neq z_+ : \quad |k| = 1 \neq k, \quad |\zeta - z_+| = R\,|\zeta - z_-|, \quad z_\pm \text{ indifferent}, \tag{35.42a–c}$$

namely: (i) if $|k| < 1$ then (35.40a–c) all points tend to the first fixed point z_+, except the second fixed point z_- that stays in place, that is, the first (second) fixed point is attractive (repulsive); (ii) if $|k| < 1$ then (35.41a–c) the converse happens, that is, all points except the first fixed point tend to the second fixed point, that is, the latter (former) is attractive (repulsive), (iii) if $|k| = 1 \neq k$, then (35.42a–c) the points lie on a circle or straight line, whose reciprocal points (Section 35.7) are the fixed points so both are indifferent. In the last case (iii), the successive points lie on a circle (straight line) if the first iteration is not (is) equidistant $|z - z_+| \neq |z - z_-|$ ($|z - z_+| = |z - z_-|$) from the fixed points.

35.7 Reciprocal Points with Regard to the Circle and the Straight Line

The circle theorem (Sections 24.6, 26.7, 28.6–28.9, and 36.2) involves the reciprocal point, whose analogue for the real axis is the complex conjugate (Subsection 35.7.1). Generalizing to the reciprocal point for a circle or straight line at any position (Subsection 35.7.2) leads to the bilinear transformations (Subsection 35.7.3) between half-planes and circles (Sections 35.8 and 35.9).

35.7.1 Reciprocal Point Theorems on the Circle and Straight Line

The analogue of the circle theorem (24.47) $\equiv$ (35.43a–c) for the real axis is (35.44a–d):

$$z = r\mathrm{e}^{i\varphi}; \ \frac{a^2}{z^*} = \frac{a^2}{r}\mathrm{e}^{i\varphi}; \ f(z) \in \mathcal{D}(|C):$$

$$F_\pm(z) = f(z) \pm f^*\left(\frac{a^2}{z^*}\right) = \Phi(r,\varphi) \pm \Phi\left(\frac{a^2}{r},\varphi\right) + i\left[\Psi(r,\varphi) \mp \Psi\left(\frac{a^2}{r},\varphi\right)\right], \quad (35.43\text{a–d})$$

$$z = x + iy, \ z^* = x - iy; \ g(z) \in \mathcal{D}(|C):$$

$$\mathrm{G}_\pm(z) = g(z) \pm g^*(z^*) = \Phi(x,y) \pm \Phi(x,-y) + i\left[\Psi(x,y) \mp \Psi(x,-y)\right]. \quad (35.44\text{a–d})$$

If the function (35.43c) [(35.44c)] is holomorphic so are (35.43d) [(35.44d)] as shown by: (i) the reflection principle (31.19); (ii) the conditions of conformal mapping (Section 33.2). From (35.43c) [(35.44c)] it follows that on circle $|z| = a$ (real axis $y = 0$): (i) F_+ (G_+) in (35.45a) [(35.46a)] is real; (ii) F_- (G_-) in (35.45b) [(35.46b)] is imaginary:

$$F_+(a,\varphi) = 2\Phi(a,\varphi) = 2\mathrm{Re}\left[f\left(a\mathrm{e}^{i\varphi}\right)\right], \ F_-(a,\varphi) = 2i\Psi(a,\varphi) = i2\mathrm{Im}\left[f\left(a\mathrm{e}^{i\varphi}\right)\right], \quad (35.45\text{a,b})$$

$$\mathrm{G}_+(x,0) = 2\Phi(x,0) = 2\mathrm{Re}\left[f(x)\right], \ \mathrm{G}_-(x,0) = 2i\Psi(x,0) = i2\mathrm{Im}\left[f(x)\right] \quad (35.46\text{a,b})$$

Thus the **reciprocal point** with regard to the circle (35.43a,b) [real axis (35.44a,b)] appears in the **circle theorem** (35.43d) [**reflection principle** (35.44d)] for a holomorphic function (35.43c) [(35.44c)]. They imply that the circle $|z| = a$ (real axis $y = 0$): (i) is a field line $\Psi = 0$, because F_+ (G_+) is real; (ii) is an equipotential, $\Phi = 0$, because F_- (G_-) is imaginary.

An example is the potential flow due to a monopole with flow rate Q and circulation, Γ, at a position $z = b$, for that the complex potential (conjugate velocity) is (12.35a–c) with $z \to z - b$ in (35.47a) [(35.47b)]:

$$f(z) = \frac{Q - i\Gamma}{2\pi}\log(z - b), \quad v^*(z) = \frac{\mathrm{d}f}{\mathrm{d}z} = \frac{Q - i\Gamma}{2\pi(z - b)}. \quad (35.47\text{a,b})$$

The real axis is a streamline $\Psi = 0$ (equipotential $\Phi = 0$) for (G_+, $\mathrm{dG}_+/\mathrm{d}z$) [($\mathrm{G}_-$, $\mathrm{dG}_-/\mathrm{d}z$)]:

$$2\pi\mathrm{G}_\pm(z) = (Q - i\Gamma)\log(z - b) \pm (Q + i\Gamma)\log(z - b^*), \quad (35.48\text{a})$$

$$2\pi\frac{\mathrm{dG}_\pm}{\mathrm{d}z} = \frac{Q - i\Gamma}{z - b} \pm \frac{Q + i\Gamma}{z - b^*}, \quad (35.48\text{b})$$

in agreement with (16.20a–c) with $b = ia$ for $f = \mathrm{G}_+$, $v^* = \mathrm{dG}_+/\mathrm{d}z$, corresponding to a rigid impermeable wall; for (G_-, $\mathrm{dG}_-/\mathrm{d}z$) the flow would be orthogonal to the real axis. The circle $|z| = a$ is a streamline (equipotential) for

$$2\pi F_\pm(z) = (Q - i\Gamma)\log(z - b) \pm (Q + i\Gamma)\log\left(\frac{a^2}{z} - b^*\right), \quad (35.49\text{a})$$

$$2\pi\frac{\mathrm{d}F_\pm}{\mathrm{d}z} = \frac{Q - i\Gamma}{z - b} \pm \frac{a^2}{z}\frac{Q + i\Gamma}{b^* z - a^2}, \quad (35.49\text{b})$$

in agreement with (28.139a; 28.140) for (F_+, $\mathrm{d}F_+/\mathrm{d}z$) corresponding to a rigid impermeable cylinder; for (F_-, $\mathrm{d}F_-/\mathrm{d}z$) the flow would be normal to the cylinder. The reciprocal point is considered next relative to any straight line (circle), not just the real axis (the circle with center at the origin).

35.7.2 Determination of a Circle and a Straight Line from the Reciprocal Points

The points p, q are said to be **reciprocal** with respect to the circle of center at s and radius R, if (Figure 35.4a) they lie on the same radial line s, and the product of their distances to s is the square of the radius R^2:

$$p = s + \mu e^{i\alpha}, \quad q = s + \frac{R^2}{\mu} e^{i\alpha}; \tag{35.50a,b}$$

the inverse points (35.50a,b) lie in the direction $e^{i\alpha}$ at distances $(\mu,\ R^2/\mu)$ from the center. For example the points $p = z = re^{i\varphi}$ and $q = R^2/z^* = (R^2/r)e^{i\varphi}$ are reciprocal with respect to the circle of radius R and center at the origin, $s = 0$; this was used in the circle theorem (Section 24.7) and its applications to cylinders in potential fields, for example, electrostatics (Sections 24.6–24.8), magnetostatics (Sections 26.7 and 26.8), and potential flows (Sections 26.6–28.9). The points p, q are reciprocal (35.50a,b) with regard to the circle

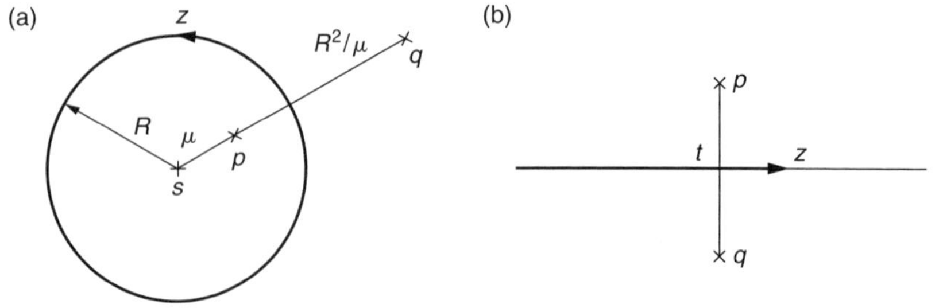

FIGURE 35.4
Two points $(p,\ q)$ are reciprocal (a) with regard to a circle if they lie on the same radial line through the center s at distances whose product is the square of the radius R. If the radius of the circle becomes infinite, that is, for a straight line (b) the reciprocal points are at the same distance along the same normal; thus a straight line is a bisector of its reciprocal points. The name reciprocal points is due to the fact that, both for the circle and straight line, the reciprocal of the reciprocal is the same point, that is: (i) if p is reciprocal to q, then q is reciprocal to p; (ii) applying reciprocity twice leads to an identity. The reciprocal of a complex number z with regard to the real (imaginary) axis is (Figure 1.1) its conjugate z^* (minus its conjugate $-z^*$). The reciprocal of a complex number z relative to the unit circle $1/z^*$ compared with the inverse $1/z$ has: (i) the same modulus $1/|z| = 1/|z^*|$; (ii) opposite argument $\arg(1/z) = -\arg(1/z^*)$. The reciprocal point with respect to a circle with center at the origin was used in the circle theorem applied to electro(magneto)statics [Figure 24.4 (26.2)] and potential flow (Figures 28.13 and 28.14). The reciprocal point in relation to a straight line corresponds to the images: (i) on a rigid plane of line sources/sinks (vortices) in a potential flow [Figure 16.1 (16.3)]; (iii/iv) line charges (currents) near an insulating or conducting plane (Figure 24.1). The images on planes can be extended to: (i) multiple images in corners (Figures 16.4 and 24.2); (ii) infinite images on parallel walls (Figures 36.13a and 40.7a). The reciprocal point on a circle specifies one of the two images on a cylinder (the other is at the axis) for: (i/ii) the electro(magneto)static field of a line charge (current) near a cylinder [Figure 24.5 (26.1)]; (iii/iv) the potential flow of a line source/sink/vortex or monopole (multipole) inside or outside a cylinder [Figure 28.13 (28.14)]. The images on a straight line (circle) also apply to a line electric charge (current) near a plane (cylindrical) interface separating two media [Figure 24.6 (26.3)] with distinct dielectric permittivity (magnetic permeability).

with radius R and center at s if holds:

$$(p-s)(q^*-s^*) = \mu e^{i\alpha}\frac{R^2}{\mu}e^{-i\alpha} = R^2; \tag{35.51}$$

the points on the circle have a constant ratio of distances to the inverse points:

$$z = s + Re^{i\varphi}: \quad \frac{|z-p|}{|z-q|} = \frac{\left|Re^{i\varphi} - \mu e^{i\alpha}\right|}{\left|Re^{i\varphi} - (R^2/\mu)\, e^{i\alpha}\right|} = \frac{\mu}{R}\frac{\left|Re^{i\varphi} - \mu e^{i\alpha}\right|}{\left|\mu e^{i\varphi} - Re^{i\alpha}\right|} = \frac{\mu}{R} \equiv \lambda \tag{35.52a,b}$$

where by (34.38a):

$$\left|Re^{i\varphi} - \mu e^{i\alpha}\right|^2 = R^2 + \mu^2 - 2\mu R\cos(\varphi - \alpha) = \left|\mu e^{i\varphi} - Re^{i\alpha}\right|^2. \tag{35.53}$$

Thus *the equation of the circle with reciprocal points, p, q is (35.54a):*

$$\frac{|z-p|}{|z-q|} = \lambda \begin{cases} \neq 1 & \text{circle,} \\ = 1 & \text{straight line,} \end{cases} \begin{matrix} (35.54a) \\ (35.54b) \end{matrix}$$

for $\lambda \neq 1$ in (Figure 35.4a); if $\lambda = 1$ then (35.54b) specifies the straight line (Figure 35.4b) at equal distance from p, q. Thus two points p, q are reciprocal with regard to a straight line, if they lie on the same perpendicular to it, at equal distances, on opposite sides, for example, the complex conjugate points z, z^* are reciprocal points with regard to the real axis: this was used in the reflection principle (Section 33.2), and applications to walls in potential fields, for example, potential flows (Sections 16.1–16.4), electrostatics (Section 24.4), and magnetostatics (Section 26.5).

In the case $\lambda = 1$, the straight line (35.54b) passes through (35.55b) the midpoint t between the inverse points p and q, and has inclination (35.53c) equal to 90° plus that of the line joining p to q:

$$\lambda = 1: \quad t = \frac{p+q}{2}, \qquad \beta \equiv \arg(t) = \frac{\pi}{2} + \arg(p) - \arg(q) = \frac{\pi}{2} + \arg\left(\frac{p}{q}\right). \tag{35.55a–c}$$

If $\lambda \neq 1$, then (35.54a) specifies a circle, and eliminating α (s) between (35.50a,b) leads to (35.56a) [(35.56b)]:

$$q - \left(\frac{R}{\mu}\right)^2 p = \left(1 - \frac{R^2}{\mu^2}\right)s, \quad p - q = \mu\left(1 - \frac{R^2}{\mu^2}\right)e^{i\alpha}; \tag{35.56a,b}$$

using (35.57a) it follows that: *the circle with inverse points (p, q) has the center s, and radius R, given by:*

$$\lambda \equiv \frac{\mu}{R} \neq 1: \quad s = \frac{p - \lambda^2 q}{1 - \lambda^2}, \quad R = \frac{|p-q|}{\lambda - 1/\lambda}. \tag{35.57a–c}$$

It can be confirmed (35.58a) that the equation (35.54a,b) specifies (35.58b):

$$z \equiv x + i\,y: \quad 0 = |z-p|^2 - \lambda^2|z-q|^2 = \left(1-\lambda^2\right)\left(x^2+y^2\right) + O\,(x,y)\,, \tag{35.58a,b}$$

for: (i) $\lambda \neq 1$ a circle, since it is quadratic, and the coefficients of x^2, y^2 are equal; (ii) for $\lambda = 1$ a straight line, since (35.58b) becomes a linear relation between x and y. *In the case, $\lambda \neq 1$, the circle (35.54a) with inverse points p, q, has radius (35.57c), and center at (35.57b); if $\lambda = 1$, the straight line (35.54b) equidistant (35.55a–c) from p, q, is the degenerate casc of a circle with radius and center at infinity $s = \infty = R$.* The determination of a straight line (family of circles) from a pair of conjugate points is considered in Example 40.8.

35.7.3 Bilinear Mapping between Reciprocal Points

Rewriting the bilinear transformation (35.35b) in the form (35.59a) leads to (35.59b,c):

$$\zeta_{1,2} \equiv P, Q; z_{1,2} \equiv p, q; \zeta_3 \to \zeta_1, z_3 \to z_1: \qquad \frac{\zeta - P}{\zeta - Q} = k\frac{z-p}{z-q}, \qquad k = \frac{P-\zeta_1}{Q-\zeta_1}\frac{q-z_1}{p-z_1}; \tag{35.59a–c}$$

comparing with (35.54a,b), it follows that *the bilinear transformation (35.59b,c), maps the circle $\lambda \neq 1$ (straight line $\lambda = 1$) with reciprocal points p, q in the z-plane (35.54a,b) into the circle $\lambda\,|k| \neq 1$ (straight line $\lambda\,|k| = 1$) in the ζ-plane, with reciprocal points P, Q that are the homographic transforms of p, q:*

$$\frac{|\zeta - P|}{|\zeta - Q|} = \lambda\,|k|, \qquad P = \frac{ap+b}{cp+d}, \qquad Q = \frac{aq+b}{cq+d}. \tag{35.60a–c}$$

Thus there are four possibilities of transformation: (i) between straight lines, $\lambda = 1 = |k|$; (ii) between circles $|k| \neq 1 \neq \lambda$; (iii) from straight line to circle $\lambda = 1 \neq |k|$; (iv) from circle to straight line $\lambda\,|k| = 1 \neq \lambda$. The case (i) corresponds to the linear transformation (35.12b) between straight lines (Section 35.3). Next examples will be given [Section 35.8 (35.9)] of transformations of types (iii, iv) [(ii)]; in all cases the inverse points correspond $(p, q) \to (P, Q)$, but the third point $z_1 \to \zeta_1$ is free. This indeterminacy can be removed by returning to (35.35b): *there is only one bilinear mapping (35.35b) assigning three collinear (noncollinear) points z_1, z_2, z_3 on a straight fine (circle) in the z-plane, to three collinear (noncollinear) points $\zeta_1, \zeta_2, \zeta_3$ on a straight line (circle} in the ζ-plane.* For example, assigning the points (35.61a) in the z-plane to (35.61b) in the ζ-plane, specifies the bilinear transformation (35.61c) ≡ (35.61d):

$$z_{1-3} = -1, 0, 1; \quad \zeta_{1-3} = -1, i, 1: \qquad \frac{\zeta+1}{\zeta-i} = \frac{1+1/z}{1-i}, \qquad \zeta = \frac{z+i}{1+iz}; \tag{35.61a–d}$$

it maps the real axis $\mathrm{Im}(z) = 0$ in the z-plane to the unit circle $|\zeta| = 1$ in the ζ-plane (Figure 35.5).

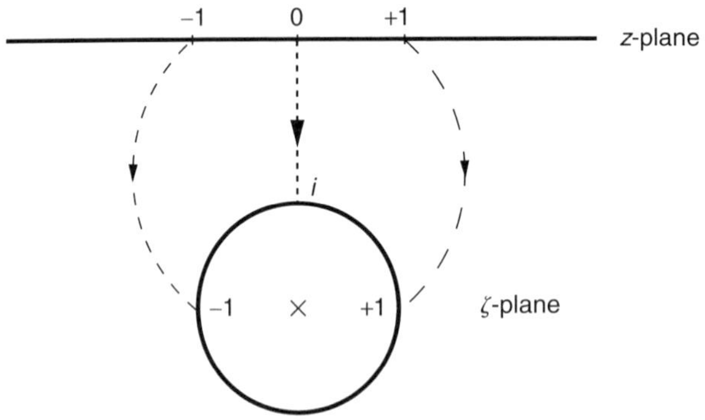

FIGURE 35.5
A bilinear transformation (35.16b) ≡ (35.35b) maps straight lines or circles in the z-plane to straight lines or circles in ζ-plane: the three points z_{1-3} in (35.35a) locate a straight line (circle) if they are (are not) collinear, and the mapped points ζ_{1-3} locate the image; the latter is a straight line (circle) if ζ_{1-3} are (are not) collinear. The example given (31.61a–d) is a bilinear mapping of a straight line in the z-plane (collinear z_{1-3}) onto a circle in the ζ-plane (noncollinear ζ_{1-3}).

35.8 Mapping of a Half-Plane into a Unit Disk

A particular example (35.61a–d) given has been of mapping of the real axis to the unit circle, with a particular assignment of three points. Next is considered the general form of such mappings, for an arbitrary assignment of points. They are bilinear mappings (35.16b), such that two points z, z^* that are reciprocal with regard to the real axis in the z-plane, are mapped to points ζ, $1/\zeta^*$ that are reciprocal with regard to the unit circle in the ζ-plane. The origin $\zeta = 0$ and point at infinity $1/\zeta^* = \infty$ are reciprocal points in the ζ-plane and thus (Figure 35.6a) if $z = \alpha$ maps to $\zeta = 0$, then $z = \alpha^*$ maps to $\zeta = \infty$, leading to the bilinear transformation (35.16b) $\equiv$ (35.62c):

$$a\alpha + b = 0 = c\alpha^* + d: \quad \zeta(z) = \frac{a}{c}\frac{z-\alpha}{z-\alpha^*}, \tag{35.62a–c}$$

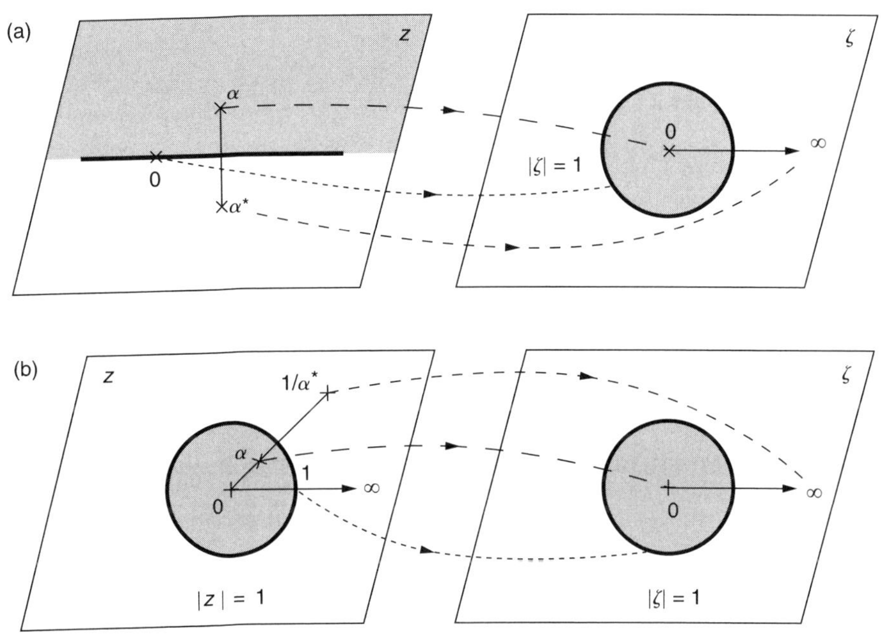

FIGURE 35.6
The bilinear mapping (Figure 35.5) between straight lines and circles can be made unique mapping the real axis (a) [unit circle (b)] in the z-plane to the unit circle in the ζ-plane, with: (i) a given point α in the upper half-z-plane (inside the unit circle in z-plane) mapped to the center of the unit circle in the ζ-plane at the origin; (ii) the reciprocal point α^* $(1/\alpha^*)$ in the lower half-z-plane (outside the unit circle in the z-plane) mapped to the point at infinity in the ζ-plane. In this case: (i/ii) the upper/lower half-z-plane (interior/exterior of the unit circle in the z-plane) is mapped to the interior/exterior of the unit circle in the ζ-plane; (iii) the real axis (unit circle) in the z-plane is mapped to the unit circle in the ζ-plane as indicated in Tables 35.1 (35.2). The tables also include the cases when: (i) the point α is the lower instead of upper half-z-plane (outside instead of inside the unit circle in the z-plane); (ii) the unit circle in the z-plane (ζ-plane) is replaced by a circle of arbitrary radius S (R). All circles have the center at the origin.

with coefficients b (d) specified by (35.62a) [(35.62b)]. The origin (35.63a) must be mapped by (35.62c) to the unit circle (35.63b):

$$z = 0: \ |\zeta(0)| = 1 = \left|\frac{a\alpha}{c\alpha^*}\right| = \left|\frac{a}{c}\right|, \qquad \frac{a}{c} = e^{i\lambda}, \tag{35.63a–c}$$

specifying the condition (35.63c). Thus the four complex constants of the bilinear mapping (35.16b), have determined in terms of one complex, a, and one real λ constant.

It can be checked that the bilinear mapping (35.62c; 35.63c) ≡ (35.64f):

$$\alpha \in |C, \ \lambda \in |R, \ \mathrm{Im}(\alpha) > 0: \quad \mathrm{Im}(z) >< 0, \quad |\zeta| <> 1: \quad \zeta = e^{i\lambda}\frac{z-\alpha}{z-\alpha^*}, \tag{35.64a–f}$$

includes the particular case (35.61b) for (35.65a,b) since (35.65a) ≡ (35.61d):

$$\alpha = -i, \ \lambda = -\frac{\pi}{2}: \zeta = e^{-i\pi/2}\frac{z+i}{z-i} = \frac{z+i}{i(z-i)} = \frac{z+i}{1+iz}. \tag{35.65a–c}$$

More generally, it can be checked that real z implies $|z-\alpha| = |z-\alpha^*|$ and $|\zeta| = |a/c| = 1$ in (35.62c) using (35.63b), that is, the real z-axis maps to the unit ζ-circle; also if $\mathrm{Im}(\alpha) > 0$ then the upper (lower) z-plane $\mathrm{Im}(z) > 0$ ($\mathrm{Im}(z) < 0$) is mapped into the interior $|\zeta| < 1$ (exterior $|\zeta| > 1$) of the unit circle, because z is closer to α than to α^* (vice-versa), that is, $|z-\alpha| < |z-\alpha^*|$ (the reverse). The opposite would apply if $\mathrm{Im}(\alpha) < 0$, that is, the upper (lower) z-half-plane would map into the exterior (interior) of the unit ζ-circle. It has been proved that *the bilinear transformation (35.64f) with (35.64a,b) real λ and complex α is the only univalent mapping of the real axis of the z-plane to the unit circle in the ζ-plane that maps the upper (lower) half-z-plane (35.64d) into the interior (exterior) of the unit circle (35.64e) for* $\mathrm{Im}(\alpha) > 0$ *in (35.64c) and into the exterior (interior) of the unit circle for* $\mathrm{Im}(\alpha) < 0$. *The half-plane to disk mapping (35.64a–f) can be extended (Table 35.1) to a disk of arbitrary radius (Example 40.9).*

35.9 Mapping between Interiors and Exteriors of Circles

Since the bilinear mapping transforms circles into circles, it should map the unit circles into each other; also since the bilinear transformation is the only mapping univalent in the whole

TABLE 35.1
Conformal Mapping of a Half-Plane to a Disk

Case	$\mathrm{Im}(\alpha) > 0$	$\mathrm{Im}(\alpha) < 0$
$\mathrm{Im}(z) > 0$	$\lvert\zeta\rvert < R$	$\lvert\zeta\rvert > R$
$\mathrm{Im}(z) = 0$	$\lvert\zeta\rvert = R$	$\lvert\zeta\rvert = R$
$\mathrm{Im}(z) < 0$	$\lvert\zeta\rvert > R$	$\lvert\zeta\rvert < R$

$R = 1$ – Mobius mapping (35.64a–f).
$R \neq 1$ – extension (40.90a–f) in Example 40.9.
Note: The bilinear transformation (35.64a–f) [it extension (40.90a–f)] maps: (i/ii) the upper/lower complex half-z-plane to the interior/exterior of the unit disk $R = 1$ (disk of radius R) with center at the origin; (iii) the boundary in the z-plane, that is, the real axis, is mapped onto the boundary in the ζ-plane, that is, the circle of radius unity $R = 1$ (radius R) and center at the origin. In (i/ii) it is assumed that the complex parameter α lies in the upper half-α-plane; in the opposite case (i/ii) are interchanged.

complex z-plane, the same transformation will be the only one to map the exteriors of the unit disks as well. In this case (Figure 35.6b) the origin $\zeta = 0$ and point at infinity $\zeta = \infty$ of the ζ-plane correspond to points $z = \alpha$, $1/\alpha^*$ of the z-plane reciprocal with respect to the unit circle that is, the bilinear transformation (35.16b) $\equiv$ (35.66c):

$$a\alpha + b = 0 = c + \alpha^* d: \quad \zeta(z) = \frac{a}{c}\frac{z-\alpha}{z - 1/\alpha^*}, \tag{35.66a–c}$$

has the coefficients $b\,(d)$ specified by (35.66a) [(35.66b)]. The point (35.67a) lies on the unit circle in the z-plane, and must be mapped to some point on the unit circle $|\zeta| = 1$ in the ζ-plane (35.67b):

$$z = 1: \; |\zeta\,(1)| = 1 = \left|\frac{a\,(1-\alpha)}{c\,(1 - 1/\alpha^*)}\right| = \left|\alpha^* \frac{a}{c}\right|, \quad \frac{a}{c} = \frac{e^{i\lambda}}{\alpha^*}, \tag{35.67a–c}$$

leading to (35.67c). Substituting (35.67c) in (35.66c) yields (35.68f):

$$\alpha \in |C\,, \quad \lambda \in |R\,, \quad |\alpha| < 1: \quad |z| <> 1, \;\; |\zeta| <> 1: \quad \zeta = e^{i\lambda}\frac{z-\alpha}{\alpha^* z - 1}; \tag{35.68a–f}$$

the eight real parameters of the bilinear mapping (35.16b) have been determined in terms of three, viz. real λ and complex α.

It can be checked that (35.68f) maps the unit circle $|z| = 1$ in the z-plane (35.68a) to the unit circle $|\zeta| = 1$ in the ζ-plane (35.69c):

$$z \equiv e^{i\varphi}, \alpha \equiv k e^{i\psi} : |z - \alpha| - |\alpha^* z - 1| = \left|e^{i\varphi} - k e^{i\psi}\right| - \left|k e^{i(\varphi-\psi)} - 1\right| = 0, \tag{35.69a–c}$$

since:

$$\left|e^{i\varphi} - k e^{i\psi}\right|^2 = 1 + k^2 - 2k\cos(\varphi - \psi) = \left|k e^{i(\varphi-\psi)} - 1\right|^2, \tag{35.70}$$

using the identity (34.38a). The interior $r \equiv |z| < 1$ (exterior, $r \equiv |z| > 1$) of the unit circle in the z-plane, maps to the interior $|\zeta| < 1$ (exterior $|\zeta| > 1$) of the unit circle in the ζ-plane, because:

$$\begin{aligned} z = r e^{i\varphi}, \alpha = k e^{i\psi} : |z - \alpha|^2 - |\alpha^* z - 1|^2 &= \left|r e^{i\varphi} - k e^{i\psi}\right|^2 - \left|k r e^{i(\varphi-\psi)} - 1\right|^2 \\ &= r^2 + k^2 - 2kr\cos(\varphi-\psi) - k^2 r^2 - 1 + 2kr\cos(\varphi-\psi) = \left(r^2 - 1\right)\left(1 - k^2\right), \end{aligned} \tag{35.71a–c}$$

with $k < 1$ is negative (positive), that is, $|z - \alpha| < |\alpha^* z - 1|$ and $|\zeta| < 1$ (the reverse $|\zeta| > 1$) for $r < 1$ $(r > 1)$. Note that (35.69c) $\equiv$ (35.70) is the particular case $r = 1$ of (35.71c) with common (35.69a,b) $\equiv$ (35.71a,b). It has been proved that *the bilinear transformation (35.68f) with (35.68a,b) complex α with and real λ, is the only univalent mapping of the unit circle in the z-plane to the unit circle in the ζ-plane that maps the interior (exterior) of the former to the interior (exterior) of the latter (35.68e) for $|\alpha| < 1$ in (35.68c); for $|\alpha| > 1$ then (35.68f) maps the interior (exterior) of the former to the exterior (interior) of the latter. The mapping between unit disks (35.68a–f) can be extended (Table 35.2) to disks of arbitrary radii (Example 40.9).*

Note 35.1. *Conformal Mappings and Coordinates:* The general properties of the conformal mapping (Sections 33.1–33.5) were followed by the most useful, viz. the polygonal (Sections 33.6–33.9) transformations (Chapter 33); the present Chapter 35 builds up on the simplest conformal mappings, viz. isometric (Sections 35.1), linear (Sections 35.2 and 35.3)

TABLE 35.2
Conformal Mapping between Disks

Case	$\lvert\alpha\rvert < S$	$\lvert\alpha\rvert > S$
$\lvert z\rvert < S$	$\lvert\zeta\rvert < R$	$\lvert\zeta\rvert > R$
$\lvert z\rvert = S$	$\lvert\zeta\rvert = R$	$\lvert\zeta\rvert = R$
$\lvert z\rvert > S$	$\lvert\zeta\rvert > R$	$\lvert\zeta\rvert < R$

$R = 1 = S$ – bilinear mapping (35.68a–f).
$R \neq 1 \neq S$ – extension (40.93a–f) in Example 40.9.
Note: The bilinear transformation (35.68a–f) [its extension (40.93a–f)] maps: (i/ii) the interior/exterior of the unit circle $S = 1$ (circle of radius S) in the z-plane to the interior exterior of the unit circle $R = 1$ (circle of radius R) in the ζ-plane; (iii) the boundaries are mapped onto each other, that is, one circle onto the other. All circles have center at the origin. In (i–ii) it is assumed that the parameter α lies inside the circle in z-plane $|\alpha| < S$; in the opposite case (i/ii) are interchanged.

and bilinear (Sections 35.4–35.9). Another major application of the conformal mapping is the construction of orthogonal curvilinear coordinate systems (Subsection 33.3.1), with equal scale factors along the two axes; the examples of conformal coordinates include the parabolic, bipolar confocal (Subsections 36.4.4 and 36.4.5), and coaxal; these lead to three-dimensional coordinates by translation or rotation, for example, parabolic cylinder coordinates and prolate and oblate spheroidal coordinates. The curvilinear confocal coordinate systems are as numerous as the holomorphic functions; besides the standard coordinate systems, others can be introduced as needed, for example, spiral and dipolar.

Note 35.2. *Specific Points and Iterated Mappings:* The consideration of the linear and bilinear mappings has involved the three kinds of inverse points:

$$z_0 = 0: \qquad z = re^{i\varphi} \to \zeta = \frac{1}{z} = r^{-1}e^{-i\varphi}, \tag{35.72}$$

$$\operatorname{Im}(z_0) = 0: \qquad z = x + iy \to \zeta = z^* = x - iy, \tag{35.73}$$

$$|z_0| = R: \qquad z = re^{i\varphi} \to \zeta = \frac{R^2}{r}e^{i\varphi}, \tag{35.74}$$

viz.: (i) the inverse point with regard to another point is the algebraic inverse, for example, (35.72) for the inverse relative to the origin; (ii) the inverse point with regard to a line is on the same normal at the same distance on the opposite side, for example, the inverse with regard to the real axis is the complex conjugate (35.73); (iii) the inverse with regard to a circle lies on the same radius on the same side of the center at a distance such that the product of distances is the square of the radius, for example, (35.74) is the inverse point with regard to the circle of radius R and center at the origin, and corresponds to the reciprocal point used in electrostatics (24.52a,b) [magnetostatics (26.62–26.63a,b] and potential flow (28.91b,c; 28.139a,b; 28.159a,b).

The **specific points** of a transformation (Table 35.3) include the reciprocal points (35.72–35.74), for the bilinear mapping, and more generally any mapping may have: (i) a fixed point that is unchanged by the mapping:

$$z_0 \text{ fixed point}: \quad f(z_0) = z_0; \tag{35.75}$$

(ii) the **iterated mapping**, that is, mapping applied (35.76a) successively n-times (35.76b) in the limit (35.76c) as $n \to \infty$:

$$f_1(z) \equiv f(z), \quad f_n(z) \equiv f(f_{n-1}(z)), \qquad f_\infty(z) \equiv \lim_{n\to\infty} f_n(z), \tag{35.76a–c}$$

TABLE 35.3
Specific Points of a Transformation

Type	Subtype	Transformation
Fixed – point		$f(z) = z$
Limit point of $z_{n+1} = f(z_n)$	Attractive: z_1	$\lim\limits_{n\to\infty} \lvert z_n - z_1\rvert = 0$
	Indifferent: z_0	$\lvert z_n - z_0\rvert = \text{const}$
	Repulsive: z_{-1}	$\lim\limits_{n\to\infty} \lvert z_n - z_{-1}\rvert = \infty$
Inverse point relative to $\zeta = f(z)$	Origin	$\zeta = 1/z$
	Real axis	$\zeta = z^*$
	Imaginary axis	$\zeta = -z^*$
	Circle*	$\zeta = a^2/z^*$

* Of radius a with center at the origin.
Note: A class of special points of a complex (bilinear) mapping is the limit (inverse) points. The limit point results from the iterated application of the mapping $z_{n+1} = f(z_n)$ as $n \to \infty$. It can be: (i/ii) an attractive z_1 (repulsive z_{-1}) point if z_n converges to z_1 (diverges away from z_{-1}); (iii) it is an indifferent point z_0 in the intermediate case when z_n keeps a constant distance, that is, moves around a circle with center at z_0. The bilinear mapping (35.16a) includes the reciprocal points relative to: (iv) the origin $1/z$; (v/vi) reflection on a line, for example, the real z^* (imaginary $-z^*$) axis; (vii) a circle with radius a and center at the origin a^2/z^*. The latter (v–vi) [(vii)] can be generalized relative to a straight line (circle) at any position in the complex plane [Figure 35.4b(a)]. A particularly important case of special point is the fixed point $f(z) = z$, which may be an attractive point (i).

specifies the **limit point**:

$$\zeta \text{ is a limit point} \quad |f_\infty(z) - \zeta| \begin{cases} = 0 & \text{attractive}, \\ = \text{const} & \text{indifferent}, \\ = \infty & \text{repulsive}, \end{cases} \qquad \begin{matrix}(35.77a)\\(35.77b)\\(35.77c)\end{matrix}$$

that may be attractive (35.77a), indifferent (35.77b) or repulsive (35.77c). Examples of fixed (limit) points were given for the linear (35.10a,b) [(35.15a–c)] and bilinear (35.36–35.39a,b) [(35.40–35.42a–c)] mappings.

Note 35.3. *Operator Power Series Applied to Self-Inverse Functions:* An **operator power series** is a series (35.78) whose general n-th term involves the function applied to itself $(n-1)$ times:

$$L\{g(z)\} \equiv A_0 + \sum_{n=1}^{\infty} A_n \left(g\left(\cdots\left(g\left(z\right)\right)\right)\right)\cdots\left[(n-1) - \text{times}\right]. \qquad (35.78)$$

An operator power series (35.78) applied to a self-inverse function (35.79a) becomes the linear operator (35.79b):

$$g\left(g(z)\right) = 1: \qquad L\{g(z)\} = \sum_{n=0}^{\infty} A_{2n} + g(z)\sum_{n=0}^{\infty} A_{2n+1}, \qquad (35.79a,b)$$

where the two series are assumed to converge. The examples for elementary transcendental functions include the hyperbolic sine (35.80b) [cosine (35.81)] and exponential (35.82):

$$a \in |C: \sinh\{ag(z)\} = \sum_{n=0}^{\infty} \frac{a^{2n+1}}{(2n+1)!} g(g(\cdots g(z)))\cdots[(2n+1) - \text{times}]\cdots$$

$$= g(z)\sum_{n=0}^{\infty} \frac{a^{2n+1}}{(2n+1)!} = g(z)\sin a, \quad (35.80\text{a,b})$$

$$\cosh\{ag(z)\} \equiv 1 + \sum_{n=0}^{\infty} \frac{a^{2n}}{(2n)!} g(g(\cdots(g(z)))\cdots[(2n) - \text{times}]$$

$$= \sum_{n=0}^{\infty} \frac{a^{2n}}{(2n)!} = \cos a, \quad (35.81)$$

$$\exp\{ag(z)\} = 1 + \sum_{n=1}^{\infty} \frac{a^{n}}{n!} g(g(\cdots(g(z)))\cdots[n - \text{times}]$$

$$= \cosh\{ag(z)\} + \sinh\{ag(z)\} = \cos a + g(z)\sin a, \quad (35.82)$$

where a is an arbitrary constant (35.80a).

Conclusion 35: The linear mapping (Figure 35.1) consists of (i) a rotation $\arg(a)$ and (ii) a translation b, which do not change figures, and form an isometric transformation, plus (iii) the homothety, which is a magnification (reduction) for $|a| > 1$ ($|a| < 1$) and does not change the size or position for $|a| = 1$; the linear mapping allows a straight line to be transformed into another (Figure 35.2), i.e., a rotation (translation) is sufficient if they are convergent (parallel). The bilinear mapping adds a fourth elementary transformation, namely (iv) the inversion (Figure 35.3) which involves a reflection on the real axis and a homothety. The bilinear transformation (Figure 35.4) preserves circles and straight lines, and their reciprocal points: (a) the reciprocal points of a circle lie on the same radius, are at a distance from the center whose product is the square of the radius, and lie on the same side of the center; (b) the straight line is a degenerate circle of radius infinity, and its reciprocal points lie on a normal line, at equal distance from it, on opposite sides. The bilinear transformation can be used to map the real axis into the unit circle, either (Figure 35.5) with a particular assignment of three points, or (Figure 35.6a) generally involving one complex and one real parameter. It then maps the upper (lower) half-plane into the interior (exterior) of the unit circle, or vice-versa; (Figure 35.6b) a different bilinear transformation maps between them the interiors and exteriors of the unit circles.

36

Channels, Condensers, and Wakes

This chapter concerns three sets of configurations for potential flows: (i) corner flows, with sharp (Sections 14.8 and 14.9), rounded (Section 36.1), and indented (Section 36.2) edges that include a semiinfinite plate (Section 36.3); (ii) channels (Section 36.5) that include the Pitot tube (Section 36.5), a slit (Section 36.4), and as its complement, a finite plate (Section 44.1); (iii) a single (Sections 36.6 and 36.7) [double (Sections 36.8 and 36.9)] row of monopoles that in the case of vortices forms a wake and may be divided into parallel walls (rectangular wells). Each of these sets of geometries can be considered for: (a) potential flow due to vortices and sources/sinks; (b/c) electro (magneto) static field due to electric charges (currents) with insulating or conducting walls, for example, condensers. Thus a variety of problems arises from a small set using simple transformations (List 36.1).

The simplest Schwartz–Christoffel transformation (Section 33.6) with one critical point maps the upper half-plane into a corner with arbitrary angle. At the vertex the velocity is zero (infinite) for a convex $0 < \beta < \pi$ (concave, $\pi < \beta < 2\pi$) corner (Section 14.8); this can be avoided by rounding-off the corner (Section 36.1) via a change in the conformal mapping. Another possibility is to use the circle theorem (Section 24.6) to surround the corner by a cylindrical indentation (Sections 36.1 and 36.2). The flow due to a monopole (Section 16.4), that is, source/sink (Section 16.8) [vortex (Section 16.9)] in a corner, can be extended to include a cylindrical indentation (Section 36.2). It corresponds to the electrostatic (Chapter 24) [magnetostatic (Chapter 26)] field due to a line electric charge (current) in an insulating corner with a cylindrical indentation; interchanging field lines, and equipotentials leads to the induced electric charges (surface electric currents) on the conducting corner with cylindrical indentation. A monopole is generally not at rest in a corner, due to the velocity induced by its images, for example, the path (trajectories) of a vortex or source/sink in a rectangular corner (Section 16.6) [past a semiinfinite plate (Section 36.3)].

The next simplest Schwartz–Christoffel transformation (Section 33.9) has two critical points, and maps the circle into a flat plate (Section 34.1). The opposite of a lifting-line (Section 34.7) across that the velocity is normal, is a flat plate on that the velocity is tangential. The Joukowski transformation from the circle to a flat plate, also maps a vortex into a flow circulating around the plate whose streamlines are ellipses (Section 36.4). The equipotentials are the orthogonal curves, that is, confocal hyperbolas that together with the ellipses form a system of plane confocal coordinates that are a particular case of conformal coordinates (Subsection 33.3.1). Interchanging equipotentials and streamlines, leads to the potential flow through a slot between two aligned semiinfinite plates. The latter is a limiting case of the potential flow out of a nozzle with flat convergent sides (Section 36.5). The opposite limiting case is the potential flow out of a channel with semiinfinite parallel sides. Adding a uniform counterflow that cancels the volume flux out of the channel, leads to the potential flow past a Pitot tube. The electrostatic and magnetostatic analogues of the potential flows with flat finite or semiinfinite plates as obstacles are the fields outside insulators or condensers with flat plates.

An example of infinite instead of semiinfinite walls is a point source/sink at equal distance from parallel walls. The problem can be solved by considering the images (Section 36.6) on the two walls that form a discrete infinite set, leading to a series of fractions for the complex velocity; the complex potential is a sum of logarithms, leading to the logarithm of an infinite product. An alternative method to solve the same problem is the conformal mapping (Section 33.5) of a source/sink in a plane to a source/sink in a vertical strip that leads to functions with a real period, that is, circular functions (Section 36.6). Thus comparing two solutions of the same problem, viz. a source between parallel walls by two different methods, that is, infinite images and conformal mapping, leads to series of fractions and infinite products for circular functions; the same result was obtained by a mathematical rather than physical method based on residues (Subsection 27.9.2).

Both methods show (Section 36.6) that the velocity field due to the source is weakened (enhanced) in the direction perpendicular (parallel) to the walls; in the case of a vortex between two parallel walls (Section 36.7) the streamlines are elongated in the direction parallel to the walls. Exchanging equipotentials and field lines, leads to the fan of field lines (cat's eye pattern) for a line of electric charges (currents) between conducting walls. By translation the flow in a parallel-sided well with sources/sinks at the corners is also obtained (Section 36.8). If the singularities, that is, source/sink, vortex or electric charge or current, are located in a rectangular well away from the bottom wall, a single image on the latter, in addition to a denumerable infinity on the side walls, specifies the potential fields (Section 36.9). Thus one vortex between parallel walls (Section 36.8) [in a rectangular well (Section 36.9)] leads to a single (double) row of vortices. The flows around obstacles and in cavities, as well as the electric and magnetic fields with various conducting or insulating boundaries, serve to analyze phenomena in nature and engineering designs.

36.1 Rounded Wedge and Cylindrical Indentation

The simplest Schwartz–Christoffel (Section 36.6) or interior polygonal transformation (33.25) has a single critical point $N = 1$ that may be placed at the origin; the external angle (36.1a) corresponds (Figure 36.1a) to a corner of internal angle β, and the multiplying constant is chosen to be (36.1b):

$$\gamma = \pi - \beta; \quad A \equiv \frac{\beta}{\pi}: \quad \frac{\mathrm{d}\zeta}{\mathrm{d}z} = \frac{\beta}{\pi} z^{-(\pi-\beta)/\pi} = \frac{\beta}{\pi} z^{\beta/\pi - 1}, \quad \zeta = z^{\beta/\pi}; \tag{36.1a–d}$$

thus the uniform flow of velocity U along the x-axis in the z-plane corresponds in the ζ-plane to:

$$f = Uz = U\zeta^{\pi/\beta}, \quad \frac{df}{d\zeta} = \frac{\pi U}{\beta}\zeta^{\pi/\beta - 1}, \tag{36.2a,b}$$

that is the flow (36.2a) $\equiv$ (14.75b) in a **sharp corner** of angle β. For (36.3a) this leads to the flow past a semiinfinite **end-plate**:

$$\beta = 2\pi: \quad f = U\sqrt{\zeta}, \quad \frac{\mathrm{d}f}{\mathrm{d}\zeta} = \frac{U/2}{\sqrt{\zeta}}, \tag{36.3a–c}$$

that has an inverse square-root singularity for the velocity (36.3c) $\equiv$ (14.83c). This type of singularity will occur for other types of flow, whenever there is an end-plate, as an approximation near a sharp edge [(14.90a,b); (34.25a); (36.75b)]. The edge $\zeta = 0$ of a sharp wedge is a singularity $\mathrm{d}f/\mathrm{d}\zeta = \infty$ (stagnation point $\mathrm{d}f/\mathrm{d}\zeta = 0$) for the velocity, if the corner is concave $\pi < \beta < 2\pi$ (convex $0 < \beta < \pi$). The singularity (stagnation) point may be avoided by two methods: (i) rounding-off the corner (Subsection 36.1.1); (ii) enclosing

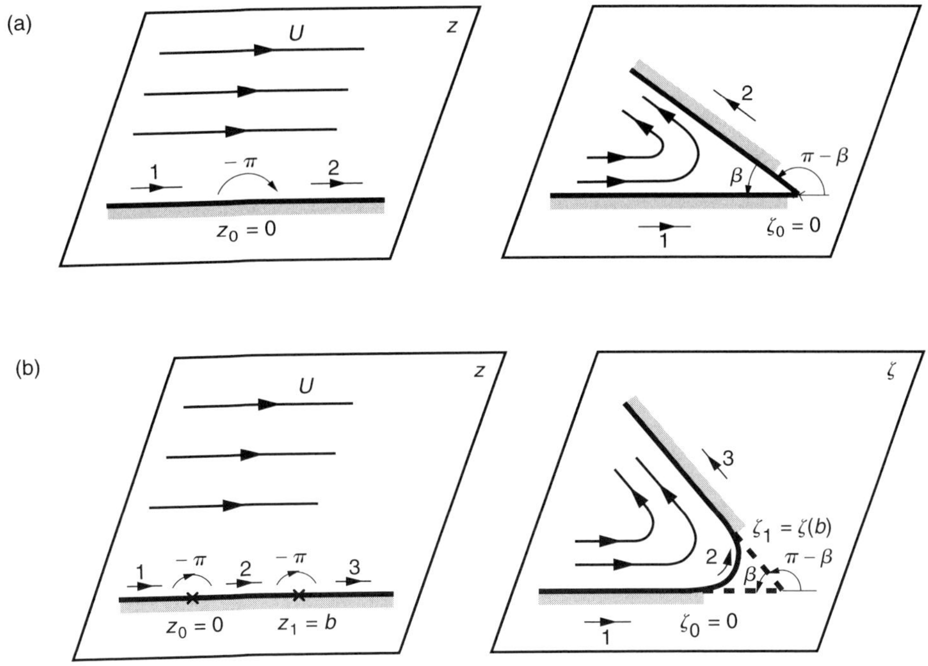

FIGURE 36.1
A Schwartz–Christoffel exterior polygonal with a single critical point (Figure 33.4) maps the upper complex z-plane into a corner (a) with arbitrary angle β; this includes the corner flows (Figure 14.10). The velocity is zero (singular) at the vertex for a convex $\beta < \pi$ (concave $\beta > \pi$) corner. This may be avoided by modifying the Schwartz–Christoffel transformation to round-off the corner (b) so that the velocity turns from one wall to the other without a stagnation point or a singularity. The alternative is to enclose the corner by a circular indentation (Figures 36.2, 36.3 and 36.5); this alternative creates two square corners which are stagnation points as for the potential flow in a rectangular corner (Figure 14.10c).

the corner by an indentation, for example, cylindrical (Subsection 36.1.2) using the circle theorem (Section 24.7). These methods apply to all potential fields, for example, potential flow (Subsection 36.1.2) and electro and magnetostatics (Subsection 36.1.3).

36.1.1 Flow in a Sharp or Rounded Corner

A finite, nonzero velocity can be obtained by rounding-off the corner, that is, using in the Schwartz–Christoffel transformation (33.25) in addition to (36.1c) a second term in (36.4):

$$\frac{d\zeta}{dz} = \frac{\beta}{\pi}\left[z^{\beta/\pi-1} + \varepsilon\,(z-b)^{\beta/\pi-1}\right], \tag{36.4}$$

involving two real constants b, ε. As $z = t$ moves along the real axis (Figure 36.1b) the first term changes sign at $z = 0$, but since its modulus is small the phase change is gradual, until z goes through $z = b$, and the phase change $\pi - \beta$ is complete. This is shown by calculating (36.4) along the real axis (36.5b) between $z = 0$ and $z = b$ in (36.5a):

$$0 < z \equiv t < b: \quad \frac{d\zeta}{dt} = \frac{\beta}{\pi}\left[t^{\beta/\pi-1} + e^{i\pi(\beta/\pi-1)}\varepsilon\,(b-t)^{\beta/\pi-1}\right]$$

$$= \frac{\beta}{\pi}\left[t^{\beta/\pi-1} - e^{i\beta}\varepsilon(b-t)^{\beta/\pi-1}\right]; \tag{36.5a,b}$$

the transformed curve in the $\zeta = x + i\,y-$plane (36.6a) is (36.6b):

$$\zeta = x + iy: \quad \left\{ \frac{\mathrm{d}x}{\mathrm{d}t}, \frac{\mathrm{d}y}{\mathrm{d}t} \right\} = \frac{\beta}{\pi} \left\{ t^{\beta/\pi - 1} - \varepsilon\,(b-t)^{\beta/\pi-1} \cos\beta, -\varepsilon\,(b-t)^{\beta/\pi-1} \sin\beta \right\}; \tag{36.6a,b}$$

it has slope:

$$\frac{\mathrm{d}y}{\mathrm{d}x} = \frac{\mathrm{d}y/\mathrm{d}t}{\mathrm{d}x/\mathrm{d}t} = \left[\cot\beta - \varepsilon^{-1} \left(\frac{b}{t} - 1 \right)^{1-\beta/\pi} \csc\beta \right]^{-1}; \tag{36.6c}$$

thus at $z = t = 0$ the curve is horizontal $\mathrm{d}y/\mathrm{d}x = 0$, and it turns smoothly until $\mathrm{d}y/\mathrm{d}x = \tan\beta$ it reaches the inclination β at $z = t = b$.

It can be confirmed that the velocity:

$$v^*(z) \equiv \frac{\mathrm{d}f}{\mathrm{d}\zeta} = \frac{\mathrm{d}f/\mathrm{d}z}{\mathrm{d}\zeta/\mathrm{d}z} = \frac{U\pi/\beta}{z^{\beta/\pi-1} + \varepsilon\,(z-b)^{\beta/\pi-1}}, \tag{36.7}$$

is small but nonzero at the start (end) $t = 0$ $(t = b)$ of the rounded corner in (36.8a) [(36.8b)]:

$$v^*(0) = \frac{U\pi}{\beta\varepsilon}(-b)^{1-\beta/\pi}, \quad v^*(b) = \frac{U\pi}{\beta} b^{1-\beta/\pi} = -v^*(0)\,\mathrm{e}^{i\beta}\varepsilon. \tag{36.8a,b}$$

The velocity is: (i) horizontal at $z = 0$ and has inclination β at $(t = b)$; (ii) its modulus is $(U/\varepsilon)\,b^{1-\beta/\pi}$ at (36.8a) the former position $z = 0$ and ε times this at (36.8b) the latter position; (iii) at $z = b$ besides there is a change of sign from (36.8a) to (36.8b). The passage from (36.7) to (36.8b) is

$$\begin{aligned} v_x(b) - iv_y(b) = v^*(b) &= \frac{U\pi}{\beta} b^{1-\beta/\pi} = \frac{U\pi}{\beta}\left(-\mathrm{e}^{-i\pi} b\right)^{1-\beta/\pi} \\ &= \mathrm{e}^{-i\pi} \frac{U\pi}{\beta}(-b)^{1-\beta/\pi} \mathrm{e}^{i\beta} = -v^*(0)\,\mathrm{e}^{i\beta}\varepsilon; \end{aligned} \tag{36.8c}$$

the horizontal (36.8d) [vertical (38.80e)] velocity components:

$$\{v_x(b), v_y(b)\} = v^*(0)\,\varepsilon\,\{-\cos\beta, \sin\beta\} = v^*(0)\,\varepsilon\,\{\cos(\pi-\beta), \sin(\pi-\beta)\}, \tag{36.8d,e}$$

correspond to a rotation by $\pi - \beta = \gamma$ in (36.1a) and a reduction in magnitude by ε. *The modification (36.4) of the Schwartz–Christoffel transformation (36.1c) at a corner of angle β involving two real constants b, ε, leads to a small, nonzero velocity (36.7) that turns by an angle (36.1a) between $z = 0$ in (36.8a) and $z = b$ in (36.8b), its magnitude varying by a factor ε in (36.8d,e).*

36.1.2 Vertex Enclosed by Cylindrical Indentation

Another way to avoid the zero or singularity at the vertex (Figure 36.1a), instead of rounding-off the corner (Figure 36.1b), is to enclose it in a circular arc (Figure 36.2a). The circular arc will meet the walls of the corner at right-angles at two points that are

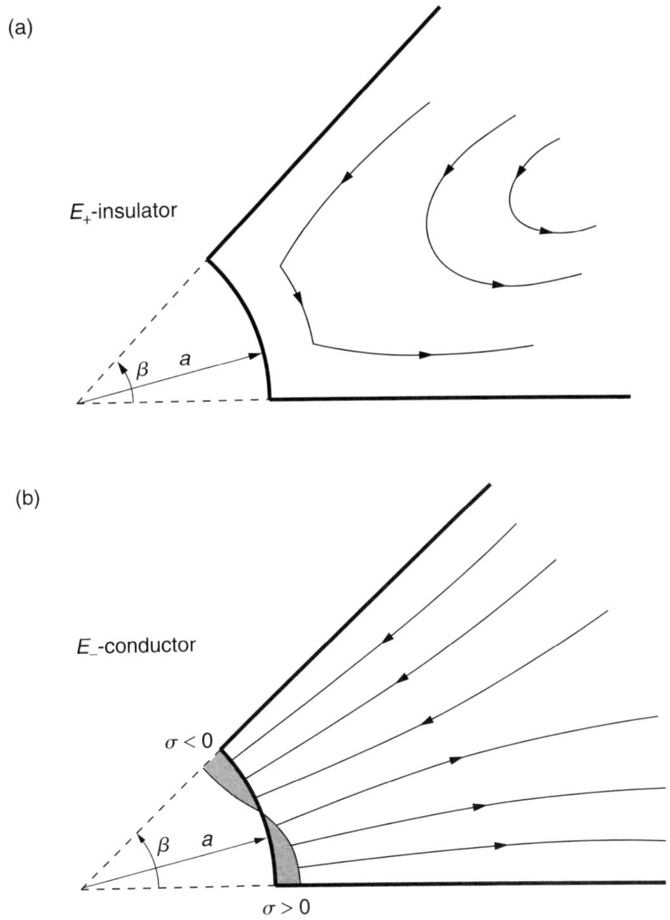

FIGURE 36.2
There is an analogy between (a) the potential flow (electrostatic field) in a rigid-walled (insulating) corner with a circular indentation. A distinct case (b) is the electrostatic field in a corner with: (i) insulating side walls, to which it is tangent; (ii) a conducting circular indentation to which it is orthogonal. It induces an electric charge distribution on the latter which reverses sign at the mid-point.

stagnation points. The complex potential of the flow in a corner of angle β is specified by (36.2a); the circle theorem [(24.47) with + sign] specifies (36.9a) the effect of the introduction of a circle of radius a and center at the origin:

$$f(z) = U\left[z^{\pi/\beta} + \left(\frac{a^2}{z}\right)^{\pi/\beta}\right]; \quad v^*(z) = \frac{df}{dz} = \frac{U\pi}{\beta z} z^{\pi/\beta}\left[1 - \left(\frac{a}{z}\right)^{2\pi/\beta}\right]; \qquad (36.9\text{a,b})$$

from *the complex potential (36.9a) follows the conjugate velocity (36.9b) for the potential flow in a corner of angle β with a cylindrical indentation of radius, a, that corresponds in three-dimensions to the (Figure 36.2a) wedge with cylindrical indentation.* The straight walls $\varphi = n\beta$ with n integer (36.10a) lead to a real potential (36.10b):

$$n \in |Z: \quad f\left(re^{in\beta}\right) = U\left[r^{\pi/\beta}e^{in\pi} + \left(\frac{a^2}{r}\right)^{\pi/\beta} e^{-in\pi}\right] = (-)^n U r^{\pi/\beta}\left[1 + \left(\frac{a}{r}\right)^{2\pi/\beta}\right], \tag{36.10a,b}$$

$$f\left(ae^{i\varphi}\right) = Ua^{\pi/\beta}\left(e^{i\pi\varphi/\beta} + e^{-i\pi\varphi/\beta}\right) = 2Ua^{\pi/\beta}\cos\left(\frac{\pi\varphi}{\beta}\right), \tag{36.10c}$$

as does the circular arc (36.10c); thus both correspond to the streamline $\Psi = 0$. This is confirmed from the velocity (36.9b) in polar (12.21) components:

$$v_\varphi(r,0) = 0 = v_\varphi(r,\beta), \quad v_r(r,0) = \frac{U\pi}{\beta r}r^{\pi/\beta}\left[1 - \left(\frac{a}{r}\right)^{2\pi/\beta}\right] = -v_r(r,\beta), \tag{36.11a–d}$$

$$v_r(a,\varphi) = 0, v_\varphi(a,\varphi) = i\frac{U\pi}{\beta}a^{\pi/\beta-1}\left(e^{i\pi\varphi/\beta} - e^{-i\pi\varphi/\beta}\right) = -\frac{2U\pi}{\beta a}a^{\pi/\beta}\sin\left(\frac{\pi\varphi}{\beta}\right), \tag{36.12a,b}$$

that (i): is radial (36.11a–d) [azimuthal (36.12a,b)] at the plane (circular) walls; (ii) and vanishes at the corners $v^* = 0$ for $r = a, \varphi = 0, \beta$; (iii) for positive U is radial inward (36.11d) [outward (36.11c)] along the wall $\varphi = \beta$ $(\varphi = 0)$, and vice-versa for U negative; (iv) for positive U is clockwise (36.12b) on the circular indentation (Figure 36.2a), and counterclockwise for negative U.

36.1.3 Electrically Conducting or Insulating Walls

Using the analogy between a potential flow (Chapters 12 and 14) and an electrostatic (magnetostatic) field [Chapter 24 (26)] as stated in (26.18–21b,c; 26.24a,b) leads from (36.9a) [(36.9b)] to electrostatic (magnetostatic) potential (36.13a) [(36.14a)] and conjugate the electric (magnetostatic) field (36.13b) [(36.14b)]:

$$f_e^\pm(z) \equiv -E_0\left[z^{\pi/\beta} \pm \left(\frac{a^2}{z}\right)^{\pi/\beta}\right], \quad E_\pm^*(z) = -\frac{df_e^\pm}{dz} = \frac{E_0\pi}{\beta z}z^{\pi/\beta}\left[1 \mp \left(\frac{a}{z}\right)^{2\pi/\beta}\right], \tag{36.13a,b}$$

$$f_m^\pm(z) = -iH_0\left[z^{\pi/\beta} \pm \left(\frac{a^2}{z}\right)^{\pi/\beta}\right], \quad H_\pm^*(z) = -\frac{df_m^\pm(z)}{dz} = i\frac{H_0\pi}{\beta z}z^{\pi/\beta}\left[1 \mp \left(\frac{a}{z}\right)^{2\pi/\beta}\right], \tag{36.14a,b}$$

where: (i) in the electric (36.13b) [magnetic (36.14b)] field E_0 (iH_0) replaces U in (36.9b); (ii) the convention relating the electric (36.13b) [magnetic (36.14b)] field to the electrostatic (36.13a) [magnetostatic (36.14a)] potential uses the −sign instead of +sign in (36.9b); (iii) the ±signs in (36.13–36.14a,b) allow for conductors and insulators as described next. The upper signs (case I) correspond (36.15a) to (36.9b) the corner flow (36.15b):

$$\frac{v^*(z)}{U} = \frac{E_+^*}{E_0} = \frac{H_+^*}{iH_0}: \frac{v_r - iv_\varphi}{U} = \frac{E_r^+ - iE_\varphi^+}{E_0} = \frac{H_r^+ - iH_\varphi^+}{iH_0}, \tag{36.15a,b}$$

leading to

$$\frac{v_r(r,\varphi)}{U} = \frac{E_r^+(r,\varphi)}{E_0} = -\frac{H_\varphi^+(r,\varphi)}{H_0}, \quad \frac{v_\varphi(r,\varphi)}{U} = \frac{E_\varphi^+(r,\varphi)}{E_0} = \frac{H_r^+(r,\varphi)}{H_0}, \tag{36.16a–d}$$

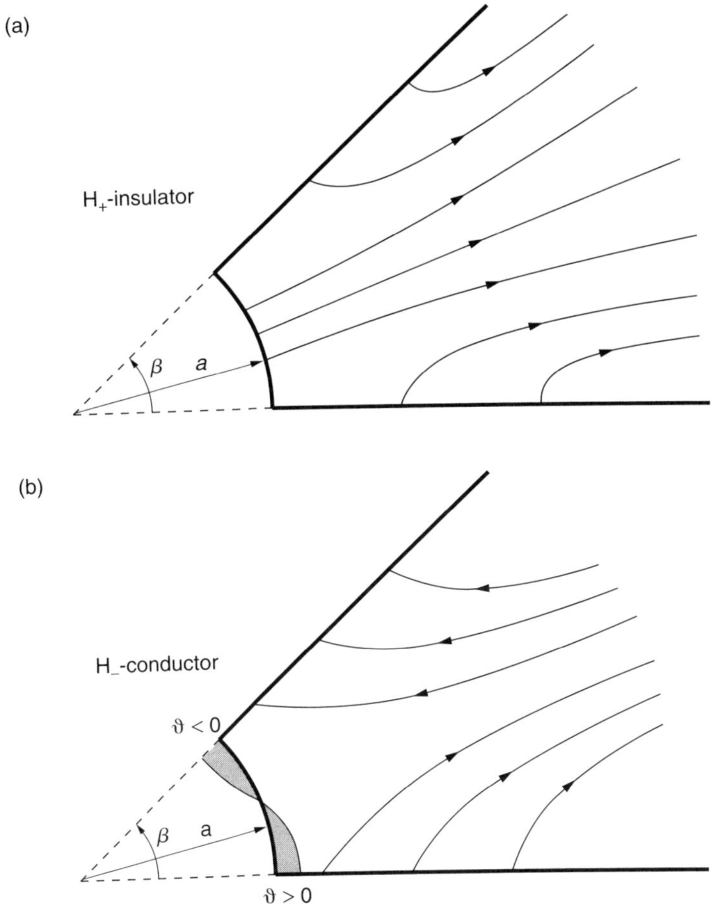

FIGURE 36.3
The magnetostatic field in a corner with: (i) insulating side walls to which it is orthogonal; (ii) an insulating (a) [conducting (b)] circular indentation to which it is orthogonal (tangent). The case (b) of a conducting circular indentation requires, for matching to a zero magnetic field inside the wall, a surface electric current distribution which changes sign at the mid-point. Concerning the electro (magneto) static field [Figure 36.2 (36.3)] there are four possible combinations of insulating/conducting side walls/circular indentation.

two subcases: (IA) the electric field (36.16a,c) is tangent to the walls (36.11a–d) and indentation (36.12a,b) so there are no induced electric charges (Figure 36.2a), corresponding to an insulator ; (IB) the magnetic field (36.16b,d) is normal to the walls (36.11a–d) and indentation (36.12a,b), so there are no surface electric currents (Figure 36.3a), corresponding again to an insulator. The lower signs (cases II) correspond (36.13–36.14a,b) to (36.17a):

$$\frac{E_-^*}{E_0} = \frac{H_-^*}{iH_0} : \frac{E_r^-(r,\varphi)}{E_0} = -\frac{H_\varphi^-(r,\varphi)}{H_0}, \frac{E_\varphi^-(r,\varphi)}{E_0} = \frac{H_r^-(r,\varphi)}{H_0}, \tag{36.17a–c}$$

leading to (36.17b,c).

In the case II of lower signs in the electrostatic (36.13a,b) [magnetostatic (36.14a,b)] fields, the polar components are given by

$$\frac{\{E_r^-(r,\varphi), E_\varphi^-(r,\varphi)\}}{E_0} = \frac{\{-H_\varphi^-(r,\varphi), H_r^-(r,\varphi)\}}{H_0}$$
$$= \frac{\pi}{\beta r} r^{\pi/\beta}\left\{\left[1+\left(\frac{a}{r}\right)^{2\pi/\beta}\right]\cos\left(\frac{\pi\varphi}{\beta}\right), -\left[1-\left(\frac{a}{r}\right)^{2\pi/\beta}\right]\sin\left(\frac{\pi\varphi}{\beta}\right)\right\}; \tag{36.18a–d}$$

this result follows (36.19b) from (36.13b) using (36.19a):

$$z = re^{i\varphi}: \; E_r^- - iE_\varphi^- = E_-^* e^{i\varphi} = \frac{E_0\pi}{\beta r} r^{\pi/\beta} e^{i\pi\varphi/\beta}\left[1+\left(\frac{a}{r}\right)^{2\pi/\beta} e^{-i2\pi\varphi/\beta}\right]$$
$$= \frac{E_0\pi}{\beta r} r^{\pi/\beta}\left[e^{i\pi\varphi/\beta} + \left(\frac{a}{r}\right)^{2\pi/\beta} e^{-i\pi\varphi/\beta}\right]. \tag{36.19a,b}$$

The same result with reversed sign in the second term in square brackets would apply: (i) to the velocity (36.9b) in a potential flow; (ii/iii) to the electrostatic (36.16a,c) [magnetostatic (36.16b,d)] field with insulating walls. From (36.18a–d) on the side walls (38.20a–d) [circular indentation (38.21a–d)]:

$$E_\varphi^-(r,0) = 0 = H_r^-(r,0): \; \frac{E_r^-(r,0)}{E_0} = \frac{\pi}{\beta r} r^{\pi/\beta}\left[1+\left(\frac{a}{r}\right)^{2\pi/\beta}\right] = -\frac{H_\varphi^-(r,0)}{H_0}, \tag{36.20a–d}$$

$$E_\varphi^-(a,\varphi) = 0 = H_r^-(a,\varphi): \; \frac{\sigma(\varphi)}{E_0\varepsilon} = \frac{E_r^-(a,\varphi)}{E_0} = \frac{2\pi}{\beta a} a^{\pi/\beta}\cos\left(\frac{\pi\varphi}{\beta}\right)$$
$$= -\frac{H_\varphi^-(a,\varphi)}{H_0} = \frac{\vartheta(\varphi)}{c}, \tag{36.21a–d}$$

follow the two subcases of case II: (IIA) the lower signs in (36.13a,b) correspond to an electric field tangential (normal) to the side walls (36.20a,c) [indentation (36.21a,c)] that are insulating (conducting) in Figure 36.2b, leading to the induced charge distribution (36.21c) only on the indentation; (IIB) the lower signs in (36.14a,b) correspond to a magnetic field that is normal (tangential) on the side wall (36.20b,d) [indentation (36.21b,d)] that again are insulating (conducting) in Figure 36.3b, leading to surface electric currents (36.21d) only on the indentation.

The induced electric charges (surface electric currents) on the conducting circular indentation [Figure 36.2b (36.3b)] in the electro (magneto) static case (36.21c) [(36.21d)], peak at the corners $\varphi = 0$ with opposite signs and vanish at the middle $\varphi = \beta/2$; the total electric charge is zero. The electro (magneto) static fields have been considered in two cases: (I) all walls insulating in Figure 36.2a (36.3a); (II) conducting cylindrical indentation and insulating side walls in Figure 36.2b (36.3b). The two remaining cases are: (III) conducting side walls and insulating cylindrical indentation that is obtained from case II multiplying the complex potential by i, to interchange equipotentials and field lines; (IV) all walls conducting that is obtained multiplying by i the complex potential of case I. The insulating cases of the electrostatic (36.13a,b) [magnetostatic (36.14a,b)] field correspond (36.15a,b; 36.16a–d) to the potential flow (36.9a,b). The cases of electrostatic (magnetostatic) field (36.17a–c) with *insulating side wall (36.20a–d) and conducting indentation (36.21a–d) specify the induced electric charges (surface electric current) on the latter. The side walls are insulating in both cases. The electric and magnetic filed are specified by (36.18a–d) for a conducting indentation; changing the sign in the second term in square brackets specifies the electric and magnetic fields for an insulating indentation, and also (36.16a–d) the velocity field of a potential flow the complex potential for the flow (36.9a) leads to the scalar potential (36.22a) [stream function (36.22b)]:*

$$\{\Phi(r,\varphi), \Psi(r,\varphi)\} = Ur^{\pi/\beta}\left\{\left[1+(a/r)^{2\pi/\beta}\right]\cos\left(\frac{\pi\varphi}{\beta}\right), \left[1-\left(\frac{a}{r}\right)^{2\pi/\beta}\right]\sin\left(\frac{\pi\varphi}{\beta}\right)\right\}, \tag{36.22a,b}$$

that: (i) also apply to the electrostatic (36.13a,b) [magnetostatic (36.14a,b)] field in the insulating case (36.16a–d); (ii) in the conducting case (36.17a–c) the same expressions apply reversing the sign of the second term in square brackets. The passage from (36.9a) to (36.22a,b) is similar to (36.19a,b) from (36.13b; 36.14b) to (36.18a–d).

36.2 Identical/Alternating Images for Irrotational/Solenoidal Fields

The preceding examples can be extended to a corner without (with) without cylindrical indentation for the irrotational/solenoidal: (i) flows due to source-sinks/(vortices) [Subsection 36.2.1 (36.2.3)]; (ii) the analogue electro/magnetostatic fields due to electric/charges currents [Subsection 36.2.2 (36.2.4)].

36.2.1 Source/Sink or Vortex in a Corner

The transformation of the upper-half z-plane into a corner of angle β in the ζ-plane is (36.1d) $\equiv$ (36.23a); it is applied (12.26a; 12.25b) or (35.47a,b) to a source/sink of volume flux, Q on the diagonal of the corner (Figure 36.4a) at a distance b from the origin that corresponds to the position (36.23b):

$$\begin{aligned}\zeta = z^{\pi/\beta}, \zeta_0 = ib^{\pi/\beta}: \quad & h(\zeta) = \frac{Q}{2\pi}\left[\log(\zeta-\zeta_0) + \log(\zeta-\zeta_0^*)\right] \\ = \frac{Q}{2\pi}\log\left[(\zeta-\zeta_0)(\zeta-\zeta_0^*)\right] = & \frac{Q}{2\pi}\log\left[(z^{\pi/\beta}-ib^{\pi/\beta})(z^{\pi/\beta}+ib^{\pi/\beta})\right] \equiv f_+(z),\end{aligned} \tag{36.23a–c}$$

together with its image (35.48a,b); it leads to the complex potential (36.23c) $\equiv$ (36.24a):

$$f_+(z) = \frac{Q}{2\pi}\log\left(z^{2\pi/\beta}+b^{2\pi/\beta}\right), \quad v_+^*(z) = \frac{\mathrm{d}f_+}{\mathrm{d}z} = \frac{Q/(\beta z)}{1+(b/z)^{2\pi/\beta}}, \tag{36.24a,b}$$

and the corresponding complex conjugate velocity is (36.24b). This agrees with the first terms of (16.84a–c). The potential (36.24a) is real for $z = x, xe^{i\beta}$, and thus the walls correspond to the streamline $\Psi(x,0) = 0 = \Psi(x,\beta)$; the factor in the denominator in the velocity (36.24b) is also real, and thus the complex conjugate velocity $v^* = (v_r - iv_\varphi)\,e^{-i\varphi}$ scales as $c/z = (c/r)\,e^{-i\varphi}$ with real c, showing that it is radial $v_r \neq 0 = v_\varphi$ along walls, that is, tangent. Thus follows the *complex potential (34.24a) and conjugate velocity (34.24b) of the potential flow due to a source/sink of flow rate* Q *in a corner (36.23a) angle* β *along the diagonal at a distance* b *from the origin (Figure* 36.4a).

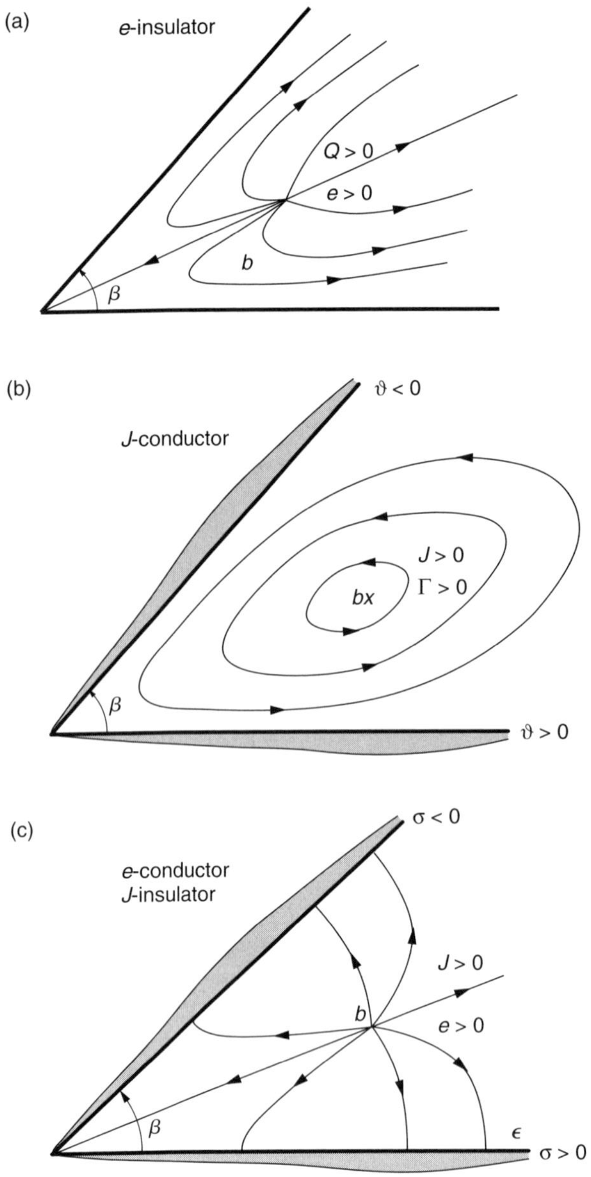

FIGURE 36.4
Analogy (a) between the potential flow (electrostatic (field) due to a line source (positive line electric charge) on the axis in a rigid walled (insulating) corner. Analogy (b) between the potential flow (magnetostatic field) due to a counterclockwise line vortex (line electric current) on the axis a rigid walled (conducting) corner. The tangential magnetic field is associated with (b) an electric current distribution in the conducting walls which: (i) vanishes at the vertex; (ii) peaks at the point on the wall closest to the original current; (iii) decays toward infinity. Analogy (c) between the electro (magneto) static field due to a line positive electric charge (counterclockwise electric current) on the axis in a corner with conducting (insulating) walls to which the field is orthogonal. The orthogonal electric (tangential magnetic) field is associated with a distribution of electric charges (currents) in the conducting walls (c) [(b)].

The same transformation (36.23a,b) would apply to (12.32a; 12.31b) or (35.47a,b) to specify:

$$g(\zeta) = -\frac{i\Gamma}{2\pi}\left[\log(\zeta-\zeta_0) - \log(\zeta-\zeta_0^*)\right] = -\frac{i\Gamma}{2\pi}\log\left(\frac{\zeta-\zeta_0}{\zeta-\zeta_0^*}\right), \tag{36.25}$$

leading to

$$f_-(z) = -\frac{i\Gamma}{2\pi}\log\left(\frac{z^{\pi/\beta}-ib^{\pi/\beta}}{z^{\pi/\beta}+ib^{\pi/\beta}}\right), \tag{36.26a}$$

$$v_-^*(z) = \frac{\mathrm{d}f_-}{\mathrm{d}z} = \frac{\Gamma/(\beta z)}{(z/b)^{\pi/\beta}+(b/z)^{\pi/\beta}}, \tag{36.26b}$$

the complex potential (36.26a) and conjugate velocity (36.26b) of a vortex of circulation Γ *in (Figure 36.4b) a corner of angle* β *at a distance* b *from the origin along the diagonal.* This agrees with the second terms of (16.84a–c). The passage from (36.26a) to (36.26b) follows from:

$$\begin{aligned} v_-^*(z) &= -\frac{i\Gamma}{2\pi}\frac{\pi}{\beta}z^{\pi/\beta-1}\left[\left(z^{\pi/\beta}-ib^{\pi/\beta}\right)^{-1} - \left(z^{\pi/\beta}+ib^{\pi/\beta}\right)^{-1}\right] \\ &= -\frac{(i\Gamma/2\beta z)\,2ib^{\pi/\beta}z^{\pi/\beta}}{z^{2\pi/\beta}+b^{2\pi/\beta}} = \frac{\Gamma/\beta z}{(z/b)^{\pi/\beta}+(b/z)^{\pi/\beta}}. \end{aligned} \tag{36.27}$$

The numerator and denominator in (36.26a) are complex conjugate for real $z = x$, and thus have the same modulus; it follows that the logarithm is imaginary, and thus $f(x)$ is real, implying that the real axis is the streamline $\Psi(x,0) = 0$; likewise for $z = xe^{i\beta}$ along the other wall of the corner. The same conclusion is reached from the complex conjugate velocity (36.26b) since it scales like c/z along the walls with a real coefficient c, and thus is radial.

36.2.2 Induced Electric Charges/Surface Electric Currents in a Corner

Replacing in (36.24b) the flow rate Q of a source/sink (26.18c; 26.20c) [(26.18b; 26.20b)] by $-e/\varepsilon$ (e/ε) specifies the electrostatic complex potential (36.28a) [conjugate field (36.28b)] due to an electric charge e in an insulating corner (Figure 36.4a) filled with a medium of dielectric permittivity, ε:

$$\frac{f_+(z)}{Q} = -\frac{\varepsilon f_e^+(z)}{e}, \quad E_+^*(z) = -\frac{\mathrm{d}f_e^+}{\mathrm{d}z} = \frac{e}{Q\varepsilon}v_+^*(z); \tag{36.28a,b}$$

in both cases the source and image are identical (36.23c). In the case of conducting walls the source and image are opposite (36.25); in the case of a conducting corner (24.31a,b) then (34.28b) is replaced by an electrostatic field similar to (36.26b) where instead of Γ appears $-ie/\varepsilon(ie/\varepsilon)$ in the electrostatic complex potential (36.29a) [conjugate field (36.29b)]:

$$-\frac{\varepsilon}{e}f_e^-(z) = \frac{i}{\Gamma}f_-(z), \quad E_-^*(z) = -\frac{\mathrm{d}f_e^-}{\mathrm{d}z} = \frac{ie}{\Gamma\varepsilon}v_-^*(z). \tag{36.29a,b}$$

It follows from (36.29a,b) that *the normal electric field (36.30c) corresponds to the tangential velocity on the wall (36.30a) due to a vortex in a corner of angle* β *at a position (36.23b):*

$$\varphi = 0, \beta: \quad \frac{v_r^-(r,0)}{\Gamma} = -\frac{\varepsilon}{e}E_\varphi^-(r,0) = \frac{1}{\beta r}\left[\left(\frac{r}{b}\right)^{\pi/\beta} + \left(\frac{b}{r}\right)^{\pi/\beta}\right]^{-1} = -\frac{\sigma(r)}{e}, \tag{36.30a–d}$$

it also specifies the electric charge distribution (36.30d) induced in the same conducting corner of angle β by a point charge e at a distance b from the origin along the diagonal (Figure 36.4c) corresponding the complex potential (36.26a; 36.29a) [conjugate electric field (36.26b; 36.29b)], where ε is the dielectric permittivity of the medium.

An electric current near insulating walls has identical images (36.24a,b) leading to a normal magnetic field; thus replacing in (36.23c) the flow rate Q of the source by $-iJ/c(iJ/c)$ specifies (26.18b; 26.21b) [(26.18c; 26.21c)] the magnetostatic complex potential (36.31a) [conjugate field (36.31b)] due to a current J in an insulating corner that is orthogonal to the wall (Figure 36.4c), so that there are no surface electric currents:

$$\frac{c}{J} f_m^+(z) = \frac{i}{Q} f_+(z), \quad H_+^*(z) = -\frac{df_m^+}{dz} = -\frac{iJ}{Qc} v_+^*(z). \tag{36.31a,b}$$

For conducting walls the images are opposite (36.25); in this case (Figure 36.4b) of a conducting corner Γ in (36.26a,b) should be replaced (26.18c; 26.21c) [(26.18b; 26.21b)] by $-J/c(J/c)$ in the magnetostatic complex potential (36.32a) [conjugate field (36.32b)] due to a line current J:

$$\frac{c}{J} f_m^-(z) = \frac{1}{\Gamma} f_-(z), \quad H_-^*(z) = -\frac{df_m^-}{dz} = -\frac{J}{\Gamma c} v_-^*(z); \tag{36.32a,b}$$

it follows that *the tangential magnetic field (36.33c) relates to the tangential velocity (36.33b) due to a vortex (36.26b) at a distance b from the origin along the diagonal of a corner of angle β:*

$$\varphi = 0, \beta: \quad \frac{v_r^-(r,0)}{\Gamma} = \frac{1}{\beta r}\left[\left(\frac{r}{b}\right)^{\pi/\beta} + \left(\frac{b}{r}\right)^{\pi/\beta}\right]^{-1} = -\frac{\vartheta(r)}{J} = -\frac{cH_r(r,0)}{J}; \tag{36.33a–d}$$

it also specifies the surface electric currents (36.33d) induced on the walls (36.33c) of the same corner of angle β, by an electric current J at the same position (Figure 36.4b) with corresponding complex potential (36.26a; 36.32a) [conjugate magnetic field (36.26b; 36.32b)], where c is the speed of light in vacuo. The distribution of induced electric charges (36.30a–d) [surface electric currents (36.33a–d)] associated with by a line charge (current) in a corner is the same in dimensionless form. The induced electric charges have opposite signs on the two walls, since (36.26b) changes sign between $z = r$ and $z = re^{i\beta}$; the latter multiplies the denominator by $(e^{i\beta})^{\pm\pi/\beta} = e^{\pm i\pi} = -1$. Thus the total electric charge on both walls is zero, corresponding to the zero total electric charges of the alternating images. This completes all cases of the electro (magneto) static field in a corner with: (I) insulating walls in Figure 36.4a (36.4c); (II) conducting walls in Figure 36.4c (36.4b).

36.2.3 Monopole in Corner with Cylindrical Indentation

Applying the circle theorem [(24.47) with +sign] to (36.24a) with an identical image:

$$f_+(z) = \frac{Q}{2\pi} \log\left[\left(z^{2\pi/\beta} + b^{2\pi/\beta}\right)\left(a^{4\pi/\beta} z^{-2\pi/\beta} + b^{2\pi/\beta}\right)\right], \tag{36.34a}$$

$$v_+^*(z) = \frac{Q}{\beta z}\left\{\left[1 + \left(\frac{b}{z}\right)^{2\pi/\beta}\right]^{-1} - \left[1 + \left(\frac{bz}{a^2}\right)^{2\pi/\beta}\right]^{-1}\right\}, \tag{36.34b}$$

specifies *the complex potential (36.34a) and conjugate velocity (36.34b) of a source/sink of circulation Q at a distance b from the origin along the diagonal of a corner of angle β with*

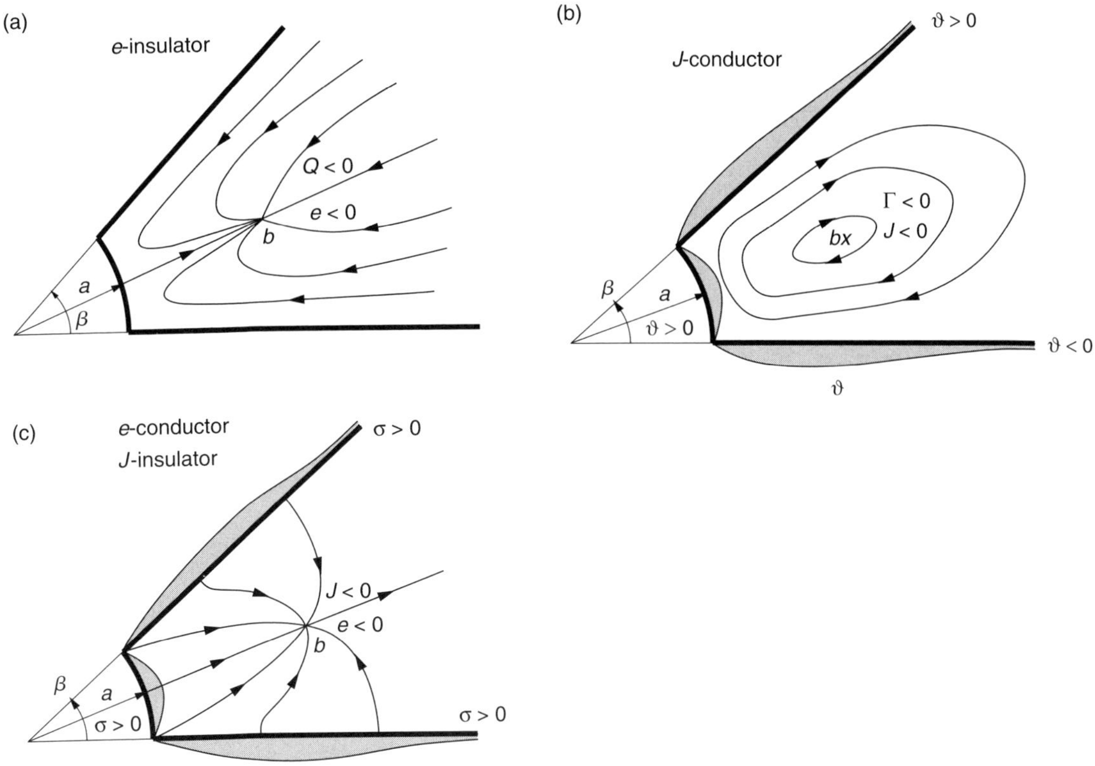

FIGURE 36.5
Analogy (a) between the potential flow (electrostatic field) due to a line sink (negative line electric charge) on the axis of a hard-walled (insulating) corner with a circular indentation. Analogy (b) between the potential flow (magnetostatic field) due to a clockwise line vortex (line electric current) on the axis of a hard-walled (conducting) corner with a circular indentation. Analogy (c) between the electro (magneto) static field due to a line negative electric charge (clockwise electric current) on the axis of a conducting (insulating) corner with a circular indentation. The orthogonal electric (c) [tangent magnetic (b)] field to the conducting corner is associated with a surface electric charge (current) distribution which: (i) vanishes at the corners; (ii) peaks at the middle of the circular indentation closest to the original electric charge (current); (iii) decays to zero at infinity; (iv) peaks at an intermediate position on the side walls. There are for possible combinations of conducting/insulating side walls/circular indentation, for example, Figure 40.4.

a circular indentation of radius a, for example, a sink in Figure 36.5a. In the case of a vortex of circulation Γ *the complex potential (36.35a) and conjugate velocity (36.35b) are (Figure 36.5b) given by*

$$f_-(z) = -\frac{i\Gamma}{2\pi}\left[\log\left(\frac{z^{\pi/\beta} - ib^{\pi/\beta}}{z^{\pi/\beta} + ib^{\pi/\beta}}\right) - \log\left(\frac{a^{2\pi/\beta}z^{-\pi/\beta} + ib^{\pi/\beta}}{a^{2\pi/\beta}z^{-\pi/\beta} - ib^{\pi/\beta}}\right)\right],$$

$$= -\frac{i\Gamma}{2\pi}\log\left(\frac{z^{\pi/\beta} - ib^{\pi/\beta}}{z^{\pi/\beta} + ib^{\pi/\beta}}\,\frac{a^{2\pi/\beta}z^{-\pi/\beta} - ib^{\pi/\beta}}{a^{2\pi/\beta}z^{-\pi/\beta} + ib^{\pi/\beta}}\right), \tag{36.35a}$$

$$v_-^*(z) = \frac{\Gamma}{\beta z}\left\{\left[\left(\frac{z}{b}\right)^{\pi/\beta} + \left(\frac{b}{z}\right)^{\pi/\beta}\right]^{-1} - \left[\left(\frac{bz}{a^2}\right)^{\pi/\beta} + \left(\frac{a^2}{bz}\right)^{\pi/\beta}\right]^{-1}\right\}, \tag{36.35b}$$

as follows from the circle theorem [(24.47) with − sign] applied to (36.26a). The passage from (36.34a) [(36.35a)] to (36.34b) [(36.35b)] is broadly similar to (36.27) from (36.26a) to (36.26b), for example,

$$\begin{aligned} v_-^*(z) = \frac{\mathrm{d}f_-}{\mathrm{d}z} &= -\frac{i\Gamma}{2\pi}\frac{\pi}{\beta z}\left\{z^{\pi/\beta}\left[\left(z^{\pi/\beta} - ib^{\pi/\beta}\right)^{-1} - \left(z^{\pi/\beta} + ib^{\pi/\beta}\right)^{-1}\right]\right.\\ &\left.\quad -a^{2\pi/\beta}z^{-\pi/\beta}\left[\left(a^{2\pi/\beta}z^{-\pi/\beta} - ib^{\pi/\beta}\right)^{-1} - \left(a^{2\pi/\beta}z^{-\pi/\beta} + ib^{\pi/\beta}\right)^{-1}\right]\right\}\\ &= -\frac{i\Gamma}{2\beta z}2ib^{\pi/\beta}\left(\frac{z^{\pi/\beta}}{z^{2\pi/\beta} + b^{2\pi/\beta}} - \frac{a^{2\pi/\beta}z^{-\pi/\beta}}{a^{4\pi/\beta}z^{-2\pi/\beta} + b^{2\pi/\beta}}\right)\\ &= \frac{\Gamma}{\beta z}\left\{\left[\left(\frac{z}{b}\right)^{\pi/\beta} + \left(\frac{b}{z}\right)^{\pi/\beta}\right]^{-1} - \left[\left(\frac{bz}{a^2}\right)^{\pi/\beta} + \left(\frac{a^2}{bz}\right)^{\pi/\beta}\right]^{-1}\right\}. \end{aligned} \tag{36.36}$$

Concerning the boundary conditions, for the potential of the source (36.34a): (i) it is real on both side walls $z = x, x\mathrm{e}^{i\beta}$, that thus correspond to the streamline $\Psi = 0$; (ii) on the circular indentation $z = a\mathrm{e}^{i\varphi}, a^2/z = a\mathrm{e}^{-i\varphi}$ so the two factors in curved brackets in (36.34a) are complex conjugates, their product in square brackets is the square of the modulus, hence real positive, and thus the logarithm is real, leading again to the streamline $\Psi = 0$. Concerning the velocity for the source (36.34b): (i) on the side walls $z = x, x\mathrm{e}^{i\beta}$ the term in curly brackets is real and thus $v^* = (v_r - iv_\varphi)\,\mathrm{e}^{-i\varphi}$ and $1/z \sim r^{-1}\mathrm{e}^{-i\varphi}$, imply a radial velocity $v_r = 0 \neq v_\varphi$ that is, tangent to the side walls; (ii) on the circular indentation $z = a\mathrm{e}^{i\varphi}$ then $b/z = (b/a)\,\mathrm{e}^{-i\varphi}$ and $(bz/a^2) = (b/a)\,\mathrm{e}^{i\varphi}$ are complex conjugates, so the difference of their inverses is imaginary, and imaginary $v_r - iv_\varphi$ implies $v_r = 0$, so the velocity is tangential again. Concerning the potential of the vortex (36.35a): (i) on the side walls the terms in numerator and denominator are complex conjugate pairs appearing as ratios, so the argument of the logarithm has unit modulus, thus the logarithm is imaginary, then multiplying by $-i\Gamma/2\pi$ leads to a real potential, hence the streamline $\Psi = 0$; (ii) on the circular indentation $z = a\mathrm{e}^{i\varphi}$ the terms in numerator and denominator are again complex conjugate pairs appearing as ratios, leading again to the streamline $\Psi = 0$. Concerning the velocity for the vortex (36.35b): (i) on the side walls $z = x, x\mathrm{e}^{i\beta}$ it scales like $v_-^* \sim c/z$ with real c, so it is radial, that is, tangential; (ii) on the circular indentation $z = a\mathrm{e}^{i\varphi}$ the two terms in square brackets are complex conjugate, so their difference is imaginary, and the coefficients of z^{-1} that scales as $v_r - iv_\varphi$ is imaginary implying that the radial velocity is zero, and the flow tangential.

36.2.4 Line Electric Charge/Current in a Corner with Cylindrical Indentation

Replacing Q in (36.34b) by e/ε specifies (36.28b) the electrostatic field of a charge in a corner with insulating walls with a circular indentation (Figure 36.5a), since all images, circumferential on the straight walls, plus radial on the circle, are identical. The case of all walls conducting corresponds to alternating circumferential images and opposite images on the circle, leading to (36.29a,b) applied to (36.35a,b), viz.: *the potential flow (electrostatic field) in a corner of angle* β *with a circular indentation of radius* a *due to a vortex of circulation* Γ *(electric charge* e*) at a distance* b *from the origin along the diagonal is specified [Figure 36.5b (36.5c)] by the complex potential (36.35a) [(36.29a)] and conjugate velocity (36.35b) [conjugate electric field (36.29b)] leading to the tangential velocity (induced* *electric charge distribution): (i) on the straight side walls (36.37a,b) to (36.37c.d):*

$$\varphi = 0, \beta; r > a: \quad v_\varphi^- (r,0) = 0 = E_r^- (r,0), \frac{v_r^- (r,0)}{\Gamma} = -\frac{\sigma (r)}{e} = -\frac{\varepsilon E_\varphi^- (r,0)}{e}$$
$$= \frac{1}{\beta r}\left\{\left[\left(\frac{r}{b}\right)^{\pi/\beta} + \left(\frac{b}{r}\right)^{\pi/\beta}\right]^{-1} - \left[\left(\frac{a^2}{br}\right)^{\pi/\beta} + \left(\frac{br}{a^2}\right)^{\pi/\beta}\right]^{-1}\right\}; \qquad (36.37\text{a–d})$$

$$r = a, 0 \le \varphi \le \beta: \quad v_r^- (a,\varphi) = 0 = E_\varphi^- (a,\varphi), \frac{v_\varphi^- (a,\varphi)}{\Gamma} = \frac{\sigma (\varphi)}{e} = \frac{\varepsilon E_r^- (a,\varphi)}{e}$$
$$= \frac{2}{\beta a}\sin\left(\frac{\pi\varphi}{\beta}\right)\frac{(a/b)^{\pi/\beta} - (b/a)^{\pi/\beta}}{(a/b)^{2\pi/\beta} + (b/a)^{2\pi/\beta} + 2\cos(2\pi\varphi/\beta)}; \qquad (36.38\text{a–d})$$

(ii) on the circular arc (36.38a,b) to (36.38c,d). The passage from (36.35b) to (36.38c) is broadly similar to (36.19a,b) from (36.13b) to (36.18b):

$$v_r^- (a,\varphi) - i v_r^- (a,r) = \mathrm{e}^{i\varphi} v^* \left(a\mathrm{e}^{i\varphi}\right)$$
$$= \frac{\Gamma}{\beta a}\left\{\left[\left(\frac{a}{b}\right)^{\pi/\beta}\mathrm{e}^{i\pi\varphi/\beta} + \left(\frac{b}{a}\right)^{\pi/\beta}\mathrm{e}^{-i\pi\varphi/\beta}\right]^{-1}\right.$$
$$\left. - \left[\left(\frac{b}{a}\right)^{\pi/\beta}\mathrm{e}^{i\pi\varphi/\beta} + \left(\frac{a}{b}\right)^{\pi/\beta}\mathrm{e}^{-i\pi\varphi/\beta}\right]^{-1}\right\}$$
$$= \frac{\Gamma}{\beta a}\frac{\left[(b/a)^{\pi/\beta} - (a/b)^{\pi/\beta}\right]\left(\mathrm{e}^{i\pi\varphi/\beta} - \mathrm{e}^{-i\pi\varphi/\beta}\right)}{(a/b)^{2\pi/\beta} + (b/a)^{2\pi/\beta} + \mathrm{e}^{2i\pi\varphi/\beta} + \mathrm{e}^{-i2\pi\varphi/\beta}}$$
$$= i\frac{2\Gamma}{\beta a}\sin\left(\frac{\pi\varphi}{\beta}\right)\frac{(b/a)^{\pi/\beta} - (a/b)^{\pi/\beta}}{(a/b)^{2\pi/\beta} + (b/a)^{2\pi/\beta} + 2\cos(2\pi\varphi/\beta)}, \qquad (36.39)$$

leads to (36.38c,d). The induced electric charge vanishes at the corners between the side walls and the circular indentation, that is, for $r = a$ in (36.37d) and $\varphi = 0, \beta$ in (36.38d). The induced electric charge peaks at the middle of the circular indentation $\varphi = \beta/2$ in (36.38d); since the induced electric charge on the side walls vanishes at the corner and at infinity, it peaks for an intermediate value.

Applying (36.31a,b) to (36.34a,b) specifies the magnetic field due to a current J in an insulating corner with a circular indentation (Figure 36.5c) that is normal to the walls, so that there are no surface electric currents. Applying (36.32a,b) to (36.35a,b) leads to the *potential flow (magnetostatic field) in a corner of angle β with circular indentation of radius* a *due to a vortex of circulation Γ (electric current* J*) at a distance* b *from the origin along the diagonal that is specified (Figure 36.5b) by the complex potential (36.35a) [(36.32a)] and conjugate velocity (36.35b) [conjugate magnetic field (36.32b)], leading to the tangential velocity (surface electric current distribution): (i) on the straight side walls* *to (36.40a–d) ≡ (36.37–36.38a–d):*

$$\varphi = 0, \beta; r > a: \quad -\frac{\vartheta(r)}{J} = -\frac{cH_r^-(r,0)}{J} = \frac{v_r^-(r,0)}{\Gamma} = -\frac{\varepsilon E_\varphi^-(r,0)}{e} = -\frac{\sigma(r)}{e}, \tag{36.40a–d}$$

$$r = a, 0 \le \varphi \le \beta: \quad \frac{\vartheta(\varphi)}{J} = -\frac{cH_\varphi^-(a,\varphi)}{J} = \frac{v_\varphi^-(a,\varphi)}{\Gamma} = \frac{\varepsilon E_r^-(a,\varphi)}{e} = \frac{\sigma(\varphi)}{e}; \tag{36.41a–d}$$

(ii) on the circular arc (36.40a–d) ≡ (36.38a–d). The induced electric charges (36.37–36.38a–d) on a conductor are the same as the surface electric currents (36.40–41a–d). The following cases were considered for the electro (magneto) static field due to a line-charge (current): (I) all walls insulating in Figure 36.5a (36.5c); (IV) all walls conducting in Figure 36.5c (36.5b). The remaining cases (II) of conducting cylindrical indentation and insulating side walls or (III) vice-versa are obtained (Example 40.10) by considering opposite instead of identical images and vice-versa. The case of a monopole of order n with complex moment in a corner with or without cylindrical indentation is considered in Example 40.11.

36.3 Path of a Monopole past a Sharp Edge

A monopole, that is, a source/sink or a vortex in a corner without or with [Sections 16.5 and 16.9/(36.2)] circular indentation is subject to a velocity induced by its images, and thus describes a path, for example, in a rectangular corner (Section 16.6) [past the edge of a semiinfinite plate (Section 36.3)]. The paths, trajectories and equilibrium positions are discussed separately for the vortex (source/sink) in Subsection 36.3.1 (36.3.2).

36.3.1 Path and Trajectory of a Vortex

The method of calculation of the trajectory proceeds in five steps, illustrated next for a vortex of circulation Γ: (i) it has an opposite image on a wall leading [(16.14a) or (35.48a)] to the potential $h(\zeta)$ in (36.42b):

$$\zeta = \sqrt{-z} = i\sqrt{z}: \quad h(\zeta) = -\frac{i\Gamma}{2\pi}\log\left(\frac{\zeta - \zeta_0}{\zeta - \zeta_0^*}\right) = -\frac{i\Gamma}{2\pi}\log\left[\frac{\sqrt{z} - \sqrt{z_0}}{\sqrt{z} + \sqrt{z_0^*}}\right] = f(z), \tag{36.42a,b}$$

that corresponds to $f(z)$ in the plane of the plate, placed (36.42a) on the negative real axis (Figure 36.6a); (ii) the complex potential (36.42b) applies everywhere except $z \neq b$ at the vortex, where it is singular $f(z = z_0) = -i\infty$, because it includes the "effect of the vortex on itself" that must be subtracted out:

$$\begin{aligned} f_0(z) = f(z) + \frac{i\Gamma}{2\pi}\log(z - z_0) &= -\frac{i\Gamma}{2\pi}\log\left[\frac{\sqrt{z} - \sqrt{z_0}}{(z - z_0)\left(\sqrt{z} + \sqrt{z_0^*}\right)}\right] \\ &= \frac{i\Gamma}{2\pi}\log\left[\left(\sqrt{z} + \sqrt{z_0}\right)\left(\sqrt{z} + \sqrt{z_0^*}\right)\right]; \end{aligned} \tag{36.43a}$$

(iii) the complex potential (36.43a) is nonsingular at the vortex, as can be checked from:

$$f_0(z_0) = \frac{i\Gamma}{2\pi}\log\left[2\sqrt{z_0}\left(\sqrt{z_0}+\sqrt{z_0^*}\right)\right] = \frac{i\Gamma}{2\pi}\log\left[4\sqrt{z_0}\,\mathrm{Re}\left(\sqrt{z_0}\right)\right]; \tag{36.43b}$$

(iv) the complex conjugate velocity associated with the complex potential (36.43a) is

$$v_0^*(z) = \frac{\mathrm{d}f_0(z)}{\mathrm{d}z} = \frac{i\Gamma}{4\pi\sqrt{z}}\left[\left(\sqrt{z}+\sqrt{z_0}\right)^{-1} + \left(\sqrt{z}+\sqrt{z_0^*}\right)^{-1}\right]; \tag{36.44a}$$

(v) its value at z_0 specifies:

$$v_0^*(z_0) = \frac{i\Gamma}{8\pi z_0} + \frac{i\Gamma/(4\pi)}{\sqrt{z_0}\left(\sqrt{z_0}+\sqrt{z_0^*}\right)}, \tag{36.44b}$$

the velocity of the vortex.

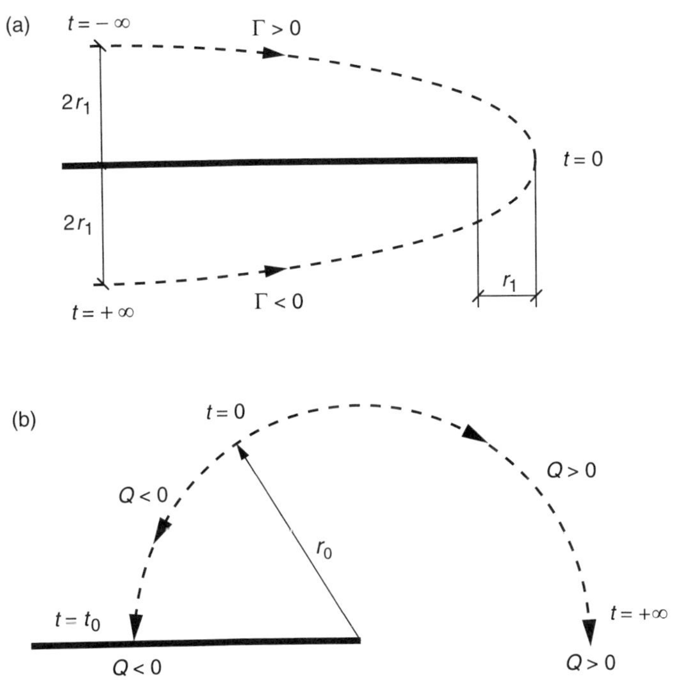

FIGURE 36.6
A vortex is swept by the potential flow past a semiinfinite plate along a trajectory (a) which is parallel to the plate far upstream and downstream, at a distance equal to twice the distance of closest approach to the edge, where the velocity is maximum and equal to the double of the asymptotic velocity (Table 36.1); the trajectory is the same for clockwise $\Gamma < 0$ (counterclockwise $\Gamma > 0$) vortices taken in opposite directions, i.e., counterclockwise (clockwise). A source (sink) is swept by a potential flow past a semiinfinite plate (b) along a circle with center at the edge passing through the initial position; the source (sink) travel in opposite directions toward a point aligned with the plate (toward) the plate, which it reaches after an infinite (finite) time with zero (infinite) velocity [Table 36.2 (36.3)].

The polar coordinates the position of the vortex (36.45a):

$$z_0 = re^{i\varphi}, \quad v_0^* (z_0) = \left(\frac{dz_0}{dt}\right)^* = \frac{d\left(re^{-i\varphi}\right)}{dt} = e^{-i\varphi}\left(\frac{dr}{dt} - ir\frac{d\varphi}{dt}\right), \tag{36.45a,b}$$

and its complex conjugate velocity (36.45b), coincide with (16.34a,b); substituting (36.45b) in (36.44b) leads to

$$v_r \equiv \frac{dr}{dt} = -\frac{\Gamma}{8\pi r}\tan\left(\frac{\varphi}{2}\right), \quad v_\varphi \equiv r\frac{d\varphi}{dt} = -\frac{\Gamma}{4\pi r}, \tag{36.46a,b}$$

that is a system of two coupled nonlinear ordinary first-order differential equations. The passage from (33.44b) to (36.46a,b):

$$\begin{aligned} v_r - iv_\varphi = e^{i\varphi}v_0^*\left(re^{i\varphi}\right) &= \frac{i\Gamma}{8\pi r} + \frac{i\Gamma}{4\pi r}\frac{e^{i\varphi/2}}{e^{i\varphi/2} + e^{-i\varphi/2}} \\ &= \frac{i\Gamma}{8\pi r}\left[1 + \frac{\cos(\varphi/2) + i\sin(\varphi/2)}{\cos(\varphi/2)}\right], \end{aligned} \tag{36.47}$$

is broadly similar to the passage (16.35) from (16.33a,b) to (16.36a,b). In order to obtain the **path** of the vortex, that is, the curve it describes, the time is eliminated from (36.46a,b), viz. by division (36.48a):

$$\frac{dr}{r} = \frac{1}{2}\tan\left(\frac{\varphi}{2}\right)d\varphi, \quad \log r + \log\left[\cos\left(\frac{\varphi}{2}\right)\right] = \text{const} \equiv \log r_1, \tag{36.48a,b}$$

leading to the integral (36.48b). Thus: (i) the path is (36.49), where r_1 is the distance of closest approach to the edge of the plate (Figure 36.6a) at $\varphi = 0$:

$$r(\varphi) = r_1 \sec\left(\frac{\varphi}{2}\right) \geq r_1 \equiv r(0); \tag{36.49}$$

(ii) far from the edge the vortex moves parallel to the plate at a distance $2r_1$ in (36.50):

$$y(\pm\infty) \equiv \lim_{r\to\infty} r\sin\varphi = \lim_{\varphi\to\pm\pi} r_1\sec\left(\frac{\varphi}{2}\right)\sin\varphi = \lim_{\varphi\to\pm\pi} 2r_1\sin\left(\frac{\varphi}{2}\right) = \pm 2r_1; \tag{36.50}$$

(iii) the radial (36.46a) ≡ (36.51b) [azimuthal (36.46b) ≡ (36.51c)] velocity of the vortex:

$$V \equiv \frac{|\Gamma|}{8\pi r_1}: \qquad v_r(\varphi) \equiv \frac{dr}{dt} = -V\sin\left(\frac{\varphi}{2}\right) sgn(\Gamma), \tag{36.51a,b}$$

$$v_\varphi(\varphi) \equiv r\frac{d\varphi}{dt} = -2V\cos\left(\frac{\varphi}{2}\right) sgn(\Gamma), \tag{36.51c}$$

increases (decreases) in modulus from zero ($2V$) at the point closest to the edge to V (zero) at infinity. Thus V in (36.51a) is the modulus velocity of the vortex parallel to the semi-infinite plate far from the edge; the sign function (36.69b) as a factor $sgn(\Gamma) = +1\,(-1)$ both in (36.51b,c) shows that the velocity has opposite signs for counterclockwise $\Gamma > 0$ (clockwise $\Gamma < 0$) vortices.

The **trajectory** of the vortex, that is, the position as a function of time, is obtained substituting the path (36.49) back into one of the equations of motion, for example, (36.46b) $\equiv$ (36.52a):

$$\frac{d\varphi}{dt} = -\frac{\Gamma}{4\pi r^2} = -\frac{\Gamma}{4\pi r_1^2}\cos^2\left(\frac{\varphi}{2}\right); \tag{36.52a}$$

$$\frac{\Gamma}{4\pi r_1^2}t = -\int_0^{\varphi}\sec^2\left(\frac{\psi}{2}\right)d\psi = -2\tan\left(\frac{\varphi}{2}\right), \tag{36.52b}$$

and integrating (36.52b) with initial time $t = 0$ at the position $\varphi = 0$ aligned with the semi-infinite plate. Substituting (36.52b) in the equation of the path (36.49) specifies completely the trajectory of the vortex (36.53b,c):

$$\tau \equiv \frac{t|\Gamma|}{8\pi r_1^2} = \frac{Vt}{r_1}: \qquad \varphi(\tau) = -2\arctan[\tau \mathrm{sgn}(\Gamma)], \quad r(\tau) = r_1\sqrt{1+\tau^2}, \tag{36.53a–c}$$

in terms of the dimensionless time (36.53a). Five points of the trajectory are indicated in Table 36.1. *A vortex with circulation Γ is swept past a semiinfinite plate (Figure 36.6a) along the path (36.49) with velocity (36.51a–c), passing at a minimum distance r_1 from the edge that is one-half the distance (36.50) from the plate when far from the edge; the trajectory has opposite directions for positive $\Gamma > 0$ (negative $\Gamma < 0$) vorticity, that is, clockwise $v_\varphi < 0$ (counterclockwise $v_\varphi > 0$), and is specified by (36.53b,c) as function of the dimensionless time (36.53a) in Table 36.1.* The latter conclusion follows noting that at time $t = -\infty\,(t = +\infty)$ for a positive or counterclockwise vortex $\Gamma > 0$ then $\tau = -\infty\,(\tau = +\infty)$ from (36.53a) and $\varphi = \pi\,(\varphi = -\pi)$ from (36.53b), so the vortex starts above and ends below the plate; vice-versa for a negative or clockwise vortex $\Gamma < 0$.

36.3.2 Path, Trajectory and Equilibrium of Source and Sink

The same five-step method applies to the trajectory of a source/sink, starting: (i) with the potential $h(\zeta)$ for [(16.1a) or (35.48a)] an identical image:

$$\begin{aligned} h(\zeta) &= \frac{Q}{2\pi}\log[(\zeta - \zeta_0)(\zeta - \zeta_0^*)] \\ &= \frac{Q}{2\pi}\log\left[\left(\sqrt{z} - \sqrt{z_0}\right)\left(\sqrt{z} + \sqrt{z_0^*}\right)\right] + \frac{iQ}{2} \equiv f(z), \end{aligned} \tag{36.54}$$

where was used transformation (36.42a) to the plane of the semiinfinite plate (Figure 36.6b). The potential (36.54) applies everywhere except at the source/sink $z \neq z_0$, where (ii) the effect of the source/sink must be subtracted out (36.55a):

$$\begin{aligned} f_0(z) = f(z) - \frac{Q}{2\pi}\log(z - z_0) &= \frac{Q}{2\pi}\log\left[\frac{\left(\sqrt{z} - \sqrt{z_0}\right)\left(\sqrt{z} + \sqrt{z_0^*}\right)}{z - z_0}\right] \\ &= \frac{Q}{2\pi}\log\left(\frac{\sqrt{z} + \sqrt{z_0^*}}{\sqrt{z} + \sqrt{z_0}}\right), \end{aligned} \tag{36.55a}$$

$$f_0(z_0) = \frac{Q}{2\pi}\log\left(\frac{\sqrt{z_0} + \sqrt{z_0^*}}{2\sqrt{z_0}}\right) = \frac{Q}{2\pi}\log\left[\frac{\mathrm{Re}\left(\sqrt{z_0}\right)}{\sqrt{z_0}}\right], \tag{36.55b}$$

TABLE 36.1
Trajectory of a Vortex past a Semiinfinite Plate

Dimensionless time*	Position		Velocity***	
	Radial**	**Angular**	**Radial******	**Azimuthal*******
τ : (36.53a)	r: (36.53c)	φ : (36.53b)	v_r : (36.51b)	v_φ (36.51c)
0	r_1	0	0	$-2V$
$\mp 1/\sqrt{3} = \mp 0.577$	$2r_1/\sqrt{3} = 1.155r_1$	$\pm\pi/3$	$\mp V/2$	$-V\sqrt{3} = -1.732V$
∓ 1	$r_1\sqrt{2} = 1.414r_1$	$\pm\pi/2$	$\mp V/\sqrt{2} = \mp 0.707V$	$-V/\sqrt{2} = -1.414V$
$\mp\sqrt{3} = \mp 1.732$	$2r_1$	$\pm 2\pi/3$	$\mp V\sqrt{3}/2 = \mp 0.866V$	$-V = -1.000V$
$\mp\infty$	∞	$\pm\pi$	$\mp V$	0

*$\tau = Vt/r_1$: dimensionless time; **r_1: distance of closest approach to the edge of the plate; ***$V \equiv |\Gamma|/(8\pi r_1)$: velocity at infinity; **** upper (lower) sign above (below) the plate; ***** trajectory for a counterclockwise vortex $\Gamma > 0$.

Note: The trajectory of a vortex past a semiinfinite plate (Figure 36.6a) is specified in polar coordinates with origin at the edge by the radial distance and azimuthal angle as a functions of dimensionless time: the radial and azimuthal velocities are also indicated. The velocity is purely azimuthal at the point of closest approach aligned with the semiinfinite plate; the velocity is parallel to the plate far upstream and downstream at twice the distance of closest approach.

leading (iii) to a finite value (36.55b) of the complex potential at the source/sink. The (iv) complex conjugate velocity (36.56a) associated with (36.55a):

$$v_0^*(z) \equiv \frac{\mathrm{d}f_0}{\mathrm{d}z} = \frac{Q}{4\pi\sqrt{z}}\left[\left(\sqrt{z}+\sqrt{z_0^*}\right)^{-1} - \left(\sqrt{z}+\sqrt{z_0}\right)^{-1}\right], \tag{36.56a}$$

$$v_0^*(z_0) = -\frac{Q}{8\pi z_0} + \frac{Q/\left(4\pi\sqrt{z}_0\right)}{\sqrt{z}_0+\sqrt{z_0^*}} = -\frac{Q}{8\pi z_0}\left[1 - \frac{\sqrt{z_0}}{\mathrm{Re}(\sqrt{z_0})}\right], \tag{36.56b}$$

is also (v) nonsingular (36.56b) at the source/sink. Substituting (36.45b) in (36.56b) leads via a calculation broadly similar to (36.47) leads to the polar components of the velocity:

$$\begin{aligned} v_r - iv_\varphi = e^{i\varphi}v_0^*\left(re^{i\varphi}\right) &= -\frac{Q}{8\pi r} + \frac{Q}{4\pi r}\frac{e^{i\varphi/2}}{e^{i\varphi/2}+e^{-i\varphi/2}} \\ &= -\frac{Q}{8\pi r}\left[1 - \frac{\cos(\varphi/2)+i\sin(\varphi/2)}{\cos(\varphi/2)}\right] = \frac{iQ}{8\pi r}\tan\left(\frac{\varphi}{2}\right), \end{aligned} \tag{36.57a}$$

$$v_r \equiv \frac{\mathrm{d}r}{\mathrm{d}t} = 0, \quad v_\varphi \equiv r\frac{\mathrm{d}\varphi}{\mathrm{d}t} = -\frac{Q}{8\pi r}\tan\left(\frac{\varphi}{2}\right). \tag{36.57b,c}$$

The corresponding trajectory is (36.57a) a circle (36.58a) and the tangential velocity (36.57b) leads to (36.58b):

$$r = r_0: \quad -\frac{Qt}{8\pi r_0^2} = \int_{\varphi_0}^{\varphi}\cot\left(\frac{\psi}{2}\right)\mathrm{d}\psi = 2\log\left[\sin\left(\frac{\varphi}{2}\right) - \sin\left(\frac{\varphi_0}{2}\right)\right], \tag{36.58a,b}$$

where (r_0, φ_0) is the initial position at time $t = 0$.

Thus *a source/sink describes a circular path (36.58a) ≡ (36.59a) around the edge of a semiinfinite plate (Figure 36.6b) with radius (36.59a) determined by its initial position:*

$$r(\varphi) = r_0, \quad \overline{V} \equiv \frac{|Q|}{8\pi r_0}, \quad v = v_\varphi(\varphi) = -\overline{V}\tan\left(\frac{\varphi}{2}\right)\mathrm{sgn}(Q) = r_0\omega(\varphi), \tag{36.59a–d}$$

and linear (angular) velocity $v(\varphi)$ $[\omega(\varphi)]$ *specified by (36.57c) ≡ (33.59c) [(36.59d)], where (36.59b) is the velocity at an angle* $\varphi = \pi/2$ *corresponding to a position orthogonal to the edge of the plate. The velocity has opposite directions* $\mathrm{sgn}(Q) = +1(-1)$ *for a source* $Q > 0$ *(since* $Q < 0$*). The trajectory (36.58a,b) ≡ (36.60b,c) is specified by (36.60b,c):*

$$\bar{\tau} = \frac{t|Q|}{16\pi r_0^2} - \frac{t\overline{V}}{2r_0}, \quad r(\bar{\tau}) = r_0 = \mathrm{const}, \quad \sin\left(\frac{\varphi}{2}\right) = \sin\left(\frac{\varphi_0}{2}\right)\exp\left[-\bar{\tau}\,\mathrm{sgn}(Q)\right], \tag{36.60a–c}$$

in terms of the dimensionless time (36.60a). The latter (36.60a) is similar to the dimensionless time (36.53a) for the vortex, replacing: (i) the circulation Γ *by half of the flow rate* $Q/2$*; (ii) the minimum distance* r_1 *by the initial distance* r_0 *from the edge of the plate. The trajectory of the vortex (Figure 36.6a) is quite different from those (Figure 36.6b) of a: (a) source (36.61a) that will move to a position of rest (36.61e) aligned with the edge (36.61d) at the initial distance (36.61c), taking an infinite time (36.61b) to get there (Table 36.2):*

$$Q > 0; \bar{\tau} \to \infty: \quad r_\infty = r_0, \quad \varphi_\infty = 0, \quad v_\infty = 0; \tag{36.61a–e}$$

$$Q = -|Q| < 0; \quad \tau = \bar{\tau}_0: \quad r(\bar{\tau}_0) = r_0, \ \varphi(\bar{\tau}_0) = \pi, \ v(\bar{\tau}_0) = \infty, \tag{36.62a–e}$$

$$\frac{t_0|Q|}{16\pi r_0^2} = \frac{t_0\overline{V}}{2r_0} \equiv \bar{\tau}_0 = -\log\left[\sin\left(\frac{\varphi_0}{2}\right)\right] = \log\left[\csc\left(\frac{\varphi_0}{2}\right)\right], \tag{36.62f}$$

TABLE 36.2
Trajectory of a Source past a Semiinfinite Plate

Dimensionless time*	Position		Velocity***	
	Radial**	Angular	Radial	Azimuthal
$\bar{\tau}$: (36.60a) ≡ (36.63)	r: (36.60b)	φ : (36.60c)	v_r : (36.57b)	v_φ : (36.59c)
0	r_0	$\pi/2$	0	$-\bar{V}$
$(1/2)\log 2 = 0.347$	r_0	$\pi/3$	0	$-\bar{V}/\sqrt{3} = -0.577\bar{V}$
0.614	r_0	$\pi/4$	0	$-0.414\bar{V}$
1.005	r_0	$\pi/6$	0	$-0.268\bar{V}$
∞	r_0	0	0	0

*$\bar{\tau} \equiv \bar{V}t/(2r_0)$: dimensionless time **$r_0$: initial distance of source from the edge of the plate; ***$\bar{V} \equiv |Q|/(8\pi r_0)$: initial velocity at $\varphi = \pi/2$; φ_0: initial azimuth of source from the edge of the plate: $\varphi_0 = \pi/2$.

Note: A source has a circular trajectory around a semiinfinite plate (Figure 36.6b), with: (i) the radius equal to the initial value; (ii) the azimuthal angle tending to zero, that is, source aligned with the plate; (iii) the purely azimuthal velocity reduces to zero as the position aligned with the plate is approached; (iv) the latter position of stable equilibrium is reached after an infinite time.

TABLE 36.3
Trajectory of a Sink past a Semiinfinite Plate

Dimensionless time*	**Position**		**Velocity*****	
	Radial**	Angular	Radial	Azimuthal
$\bar{\tau}$: (36.60a) ≡ (36.63)	r : (36.60b)	φ : (36.60c)	v_r : (36.57b)	v_φ : (36.59c)
0	r_0	$\pi/2$	0	$\bar{V}$
$(1/2)\log(3/2) = 0.203$	$2r_0$	$2\pi/3$	0	$\bar{V}\sqrt{3} = 1.732\bar{V}$
0.267	$3r_0$	$3\pi/4$	0	$2.414\bar{V}$
0.312	$5r_0$	$5\pi/6$	0	$3.733\bar{V}$
$(1/2)\log 2 = 0.347$	r_0	π	0	∞

*$\bar{\tau} \equiv \bar{V}t/(2r_0)$: dimensionless time; **$r_0$: initial distance of source from the edge of the plate; ***$\bar{V} \equiv |Q|/(8\pi r_0)$: initial velocity at $\varphi = \pi/2$; φ_0: initial azimuth of source from the edge of the plate: $\varphi_0 = \pi/2$.

Note: The trajectory of a sink near a semiinfinite plate (Figure 36.6b) is a circle with radius equal to the initial distance from the edge, as for the source in Table 36.2. In contrast: (i) the purely azimuthal velocity has the opposite direction, away from the position of unstable equilibrium aligned with the plate; (ii) the velocity increases without bound as the plate is approached; (iii) the collision takes plate at 'infinite velocity' after a finite time.

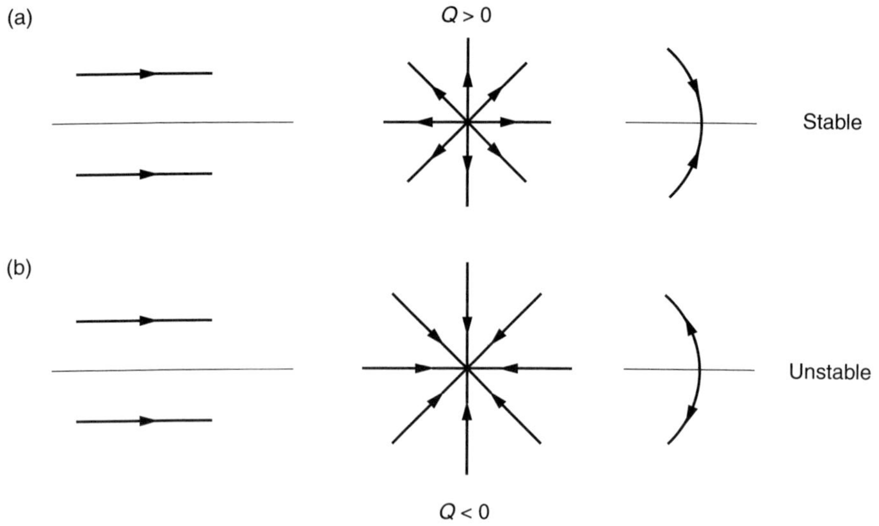

FIGURE 36.7
A vortex (source/sink) has no (has an) equilibrium position in the potential flow past a semiinfinite plate [Figure 36.6a(b); Tables 36.1 (36.2–36.3)]. The equilibrium position is aligned with the plate and is stable (unstable) for the source (sink), since a deviation from equilibrium leads to a motion back to (away from) equilibrium.

(b) sink (36.62a) that travels in the opposite direction, and hits the plate (36.62d) at the distance equal to the initial distance (36.62c), in a finite time (36.62b) ≡ (36.62f) with an "infinite" impact velocity (36.62e) in Table 36.3. The time the sink takes to reach the plate (36.62f) tends to infinity $\bar{\tau}_0 \to \infty$ as the starting position approaches the alignment with the plate $\varphi_0 \to 0$ that is a position of rest. Both of the preceding conclusions for a source (36.61a–e) or a sink (36.62a–f) follow from the dimensionless time (36.60a) as a function of angular position (36.60c) along the trajectory:

$$\bar{\tau} = \operatorname{sgn}(Q) \log \left[\frac{\sin(\varphi_0/2)}{\sin(\varphi/2)} \right]. \tag{36.63}$$

Any position aligned with the plate is an equilibrium position (36.64a) of the source/sink. The equilibrium is: (i) indifferent in the radial direction because the radius of trajectory (36.64a–c) is constant (36.64b):

$$\varphi = 0: \quad \frac{\mathrm{d}^2 r}{\mathrm{d}t^2} = 0; \quad \frac{\mathrm{d}\dot{\varphi}}{\mathrm{d}\varphi} = -\frac{Q}{16\pi r^2} \sec^2\left(\frac{\varphi}{2}\right), \tag{36.64a–c}$$

(ii) in the azimuthal direction (36.57c) the equilibrium is stable (unstable) for a source $Q > 0$ (sink $Q < 0$) because the (36.64c) is negative (positive) as shown in Figure 36.7a (36.7b). The vortex (36.46a,b) has no equilibrium position because the modulus of its velocity never vanishes (36.65a):

$$\left|\vec{v}\right|^2 = (v_\varphi)^2 + (v_r)^2 = \left(\frac{\Gamma}{8\pi r}\right)^2 \left[4 + \tan^2\left(\frac{\varphi}{2}\right)\right] \geq \left(\frac{\Gamma}{4\pi r}\right)^2, \tag{36.65a}$$

$$\left|\vec{v}\right|^2 = (v_\varphi)^2 = \left(\frac{Q}{8\pi r}\right)^2 \tan^2\left(\frac{\varphi}{2}\right), \tag{36.65b}$$

whereas for the source/sink (36.57a,b) the modulus of the velocity (36.65b) vanishes in the direction $\varphi = 0$ aligned with the plate.

36.4 Circulation around a Flat Plate and Flow through a Slit

The flow past a semiinfinite plate (Section 36.3) is a particular case of corner flow (Section 14.8) that corresponds to the interior polygonal Schwartz–Christoffel transformation with one critical point (Section 36.1). The Joukowski transformation has two critical points, and maps a circle into a finite flat plate (Section 34.1); adding a vortex at the center leads to a circulating flow (Subsection 36.4.1) corresponding to the flat plate without free stream. Exchanging the vortex by a source/sink leads to the flow through a slot between two aligned semiinfinite plates (Subsection 36.4.2). The two problems interchange the equipotentials and streamlines that form a system of confocal ellipses and hyperbolas, with foci at the edges of the plates (Subsection 36.4.4); this specifies the confocal coordinates (Subsection 36.4.5) that are a particular case of conformal coordinates (Subsection 33.3.1). The corresponding problem in electro (magneto) statics is (Subsection 36.4.3) a condenser or insulator with one (two) finite (aligned semiinfinite) plate(s).

36.4.1 Circulating Flow around a Flat Plate

The flow due to point vortex of circulation Γ at the origin of the ζ-plane may be represented in direct form, that is, with explicit complex potential f by (12.32a) $\equiv$ (36.66b):

$$\zeta \to \frac{2\zeta}{b}: \ f = -\frac{i\Gamma}{2\pi}\log\left(\frac{2\zeta}{b}\right), \quad \zeta = \frac{b}{2}\exp\left(-\frac{i2\pi f}{\Gamma}\right), \qquad \text{(36.66a–c)}$$

and in inverse form with explicit coordinate ζ by (36.66c); the factor $b/2$ was introduced in (36.66a) so that the circle $|\zeta| = b/2$ is transformed (Figure 36.8a) into the flat plate with edges at $z = \pm b$ by the Joukowski transformation (34.32a) $\equiv$ (36.67b):

$$b \to \frac{b}{2}: \ z = \zeta + \frac{b^2}{4\zeta} = \frac{b}{2}\left(e^{i2\pi f/\Gamma} + e^{-i2\pi f/\Gamma}\right) = b\cos\left(\frac{2\pi f}{\Gamma}\right), \qquad \text{(36.67a,b)}$$

with parameter (36.67a). Thus *the flow with circulation Γ around a flat plate with edges at $\zeta = \pm b$ is specified by the complex potential (36.68a) and conjugate velocity (36.68b):*

$$f(z) = \frac{\Gamma}{2\pi}\operatorname{arc\,cos}\left(\frac{z}{b}\right), \quad \frac{df}{dz} = \frac{\Gamma/(2\pi)}{\sqrt{b^2 - z^2}}; \qquad \text{(36.68a,b)}$$

the velocity (36.68b) has an inverse square-root singularity at the edges: (i) as predicted before (14.92c); (ii) in agreement with end-plate effect (34.9b); (iii) this leads to the suction force (34.16b) that turns the flow around the edges.

In the sequel $|\ldots|^{1/2}$ denotes the square root with positive sign, and sign (...) denotes the sign function:

$$|a|^{1/2} \equiv \left|\sqrt{a}\right|, \quad \operatorname{sgn}(a) = \begin{cases} +1 & \text{if } a > 0, \\ 0 & \text{if } a = 0, \\ -1 & \text{if } a < 0. \end{cases} \qquad \text{(36.69a,b)}$$

The velocity (36.68b) for positive circulation: (i) on the real axis (36.70a) $\equiv$ (36.71a) is horizontal on (36.70b,d) [vertical outside (36.71b,d)] the flat plate, with (36.70c) [(36.71c)] positive velocity, that is, to the right below (upward to the right), and negative velocity,

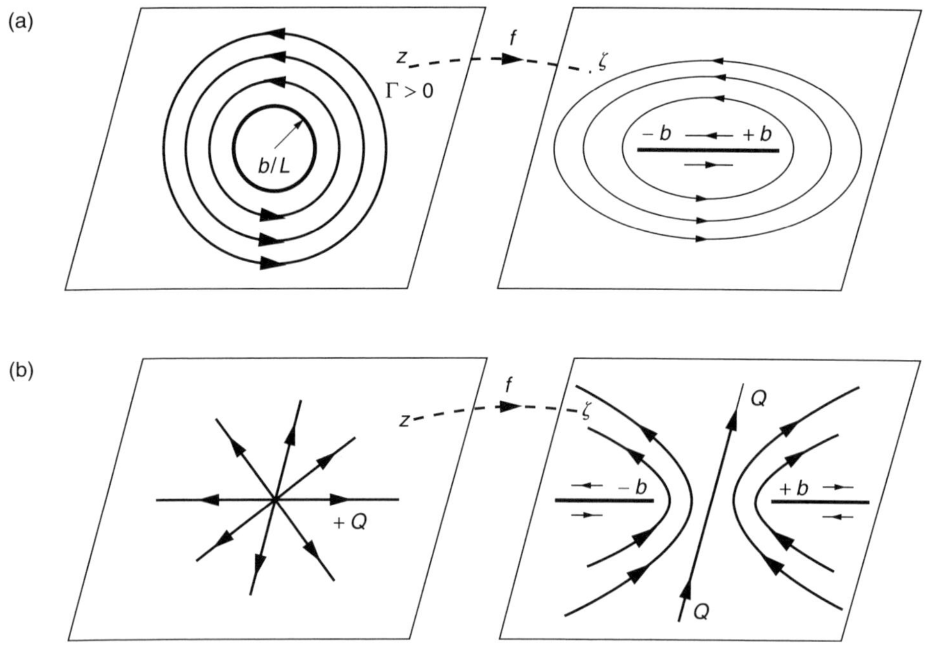

FIGURE 36.8
The circle is a streamline of a vortex with axis at its center (Figure 12.2b). The Joukowski mapping of the circle into a flat plate (Figure 34.1a) leads to the potential flow with circulation around a flat plate (a). Exchanging the streamlines and equipotentials replaces the vortex (Figure 12.2b) by a source (Figure 12.2a); the same transformation then leads to the potential flow through the gap between two aligned semiinfinite plates (b).

that is, to the left above (downward to the left):

$$y = 0, |x| < b: \quad v_x = -\frac{\Gamma}{2\pi}\left|b^2 - x^2\right|^{-1/2} \operatorname{sgn}(y), \quad v_y = 0, \tag{36.70a–d}$$

$$y = 0, |x| > b: \quad v_x = 0, \quad v_y = \frac{\Gamma}{2\pi}\left|x^2 - b^2\right|^{-1/2} \operatorname{sgn}(x); \tag{36.71a–d}$$

$$x = 0: \quad v_x = -\frac{\Gamma}{2\pi}\left|y^2 + b^2\right|^{-1/2} \operatorname{sgn}(y), \quad v_y = 0, \tag{36.72a–c}$$

(ii) on the imaginary axis (36.72a), that is, the line perpendicular to the plate at equal distance from the edges, the velocity is horizontal (36.72b,c) to the left above and to the right below (Figure 36.8a). The reverse would apply for negative circulation. The minimum velocity on the plate is (36.73a) at the middle (36.70a), and it increases to become singular at the edges:

$$v_{\min}(|x| < b; y = 0) = \frac{\Gamma}{2b}; \quad \Gamma = 2\left[f(b) - f(-b)\right] = \frac{\Gamma}{\pi} \operatorname{arc\,cos}\left(\frac{z}{b}\right)\Big|_{-b}^{+b} \tag{36.73a,b}$$

the circulation (36.73b) follows from the difference of potential (36.68a) between the edges of the plate. The volume flux around the flat plate is infinite (36.73c):

$$Q = \int_b^\infty v_y(x,0)\, dx = \frac{\Gamma}{2\pi}\int_b^\infty \left(x^2 - b^2\right)^{-1/2} dx = \frac{\Gamma}{2\pi}\left[\operatorname{arg\,cosh}\left(\frac{x}{b}\right)\right]_0^\infty = \infty, \tag{36.73c}$$

as follows from (36.71d).

36.4.2 Slit between Aligned Semiinfinite Plates

Exchanging the vortex $-i\Gamma$ of circulation Γ is by a source or sink of volume flux Q, then the circular cosine in (36.67b) is replaced by a hyperbolic cosine:

$$z = \frac{b}{2}\left(e^{2\pi f/Q} + e^{-2\pi f/Q}\right) = b\cosh\left(\frac{2\pi f}{Q}\right), \tag{36.74}$$

and the streamlines and equipotentials are interchanged; thus *the potential flow with flow rate Q through a slot between two aligned semiinfinite plates (Figure 36.8b) with edges at $\zeta = \pm b$ has complex potential (36.76a) and conjugate velocity (36.75b):*

$$f(z) = \frac{Q}{2\pi}\arg\cosh\left(\frac{z}{b}\right), \quad \frac{\mathrm{d}f}{\mathrm{d}z} = \frac{Q/2\pi}{\sqrt{z^2-b^2}}. \tag{36.75a,b}$$

The inverse square-root singularity of the velocity or end-plate effect remains in (36.75b) as in (36.68b). The functions (36.68b) [(36.75b)] have: (i) the same branch-points $z = \pm b$; (ii) the branch-cuts [Subsection 7.8.1 (7.8.2)] that correspond to two aligned semi-infinite plates (one finite plate) in Subsection 36.4.2 (36.4.1). Join $z = \pm b$ to $\pm\infty$ in (36.75b) [Join $z = +b$ to $z = -b$ in (36.68b)] as for arc $\sin(z/b)$ in (7.23a,b) [arc $\csc(z/b)$ in (7.27a,b)].

The velocity (36.75b) for a source: (i) on the real axis (36.76a) $\equiv$ (36.77a) is vertical upward (36.76c,d) in the slot (36.76b) and horizontal (36.77c,d) on the plates (36.77b), toward (from) infinity above (below):

$$y = 0, \quad |x| < b: \quad v_x = 0, \quad v_y = \frac{Q}{2\pi}\left|b^2 - x^2\right|^{-1/2}, \tag{36.76a–d}$$

$$y = 0, \quad |x| > b: \quad v_x = \frac{Q}{2\pi}\left|x^2 - b^2\right|^{-1/2}\operatorname{sgn}(y)\operatorname{sgn}(x), \quad v_y = 0; \tag{36.77a–d}$$

$$x = 0: \quad v_x = 0, \quad v_y = \frac{Q}{2\pi}\left|b^2 + y^2\right|^{-1/2}, \tag{36.78a–c}$$

(ii) on the imaginary axis or axis of the slot (36.78a) the velocity (36.78b,c) vertical upward (Figure 36.8b). The reverse would apply for a sink. The minimum (maximum) velocity in the slot (36.76d) [on the midline (36.76c)] is (36.79b) at (36.79a):

$$y = 0: \quad v_{\min} = v_x(0,0) = \frac{Q}{2\pi b}; \quad \bar{v} = \frac{Q}{2b} = \pi v_{\min}, \tag{36.79a–c}$$

the average velocity of the flow through the slot (36.79c) is π times larger. The volume flux (36.76d) through the slot is $Q/2$ in (36.80a) because only one half of the flow rate of the source/sink (Figure 36.11a) goes through the slot (Figure 36.8b):

$$\int_{-b}^{+b} v_y(x,0)\,\mathrm{d}x = \frac{Q}{2\pi}\int_{-b}^{+b}\left(b^2 - x^2\right)^{-1/2}\mathrm{d}x = \frac{Q}{2\pi}\left[\operatorname{arc}\cos\left(\frac{x}{b}\right)\right]_{-b}^{+b} = Q/2; \tag{36.80a}$$

$$\Gamma = f(b) - f(\infty) = \frac{Q}{2\pi}\left[\arg\cosh\left(\frac{z}{b}\right)\right]_{\infty}^{b} = -\infty, \tag{36.80b}$$

the circulation around each plate is infinite (36.80b) and clockwise (counterclockwise) around the right (left) semi-infinite plate, hence negative in (36.80b) [positive for the range $-\infty$ to $-b$].

36.4.3 Insulators or Condensers with Flat Plates

The electrostatic analogue of the flow circulating around a flat plate (Figure 36.8a), is the electric field around an insulating flat plate (Figure 36.9a). Exchanging equipotentials for field line leads to the (Figure 36.9b) electric field near a condenser with a conducting finite flat plate that induces positive (negative) charges above (below), with surface density:

$$y = 0, \quad |x| < b: \quad \sigma(x) = \frac{e}{2\pi}\left|b^2 - x^2\right|^{-1/2} \operatorname{sgn}(y), \tag{36.81a–c}$$

given by (36.70a–c) ≡ (36.81a–c), with $-\Gamma$ replaced by the total charge e; the charge density varies from $e/2\pi b$ at the midposition to ∞ at the edges, and the total charge is e, that is, a mean value $\overline{e} = e/2b$ over the conductor of length $2b$. *The potential flow through the slot*

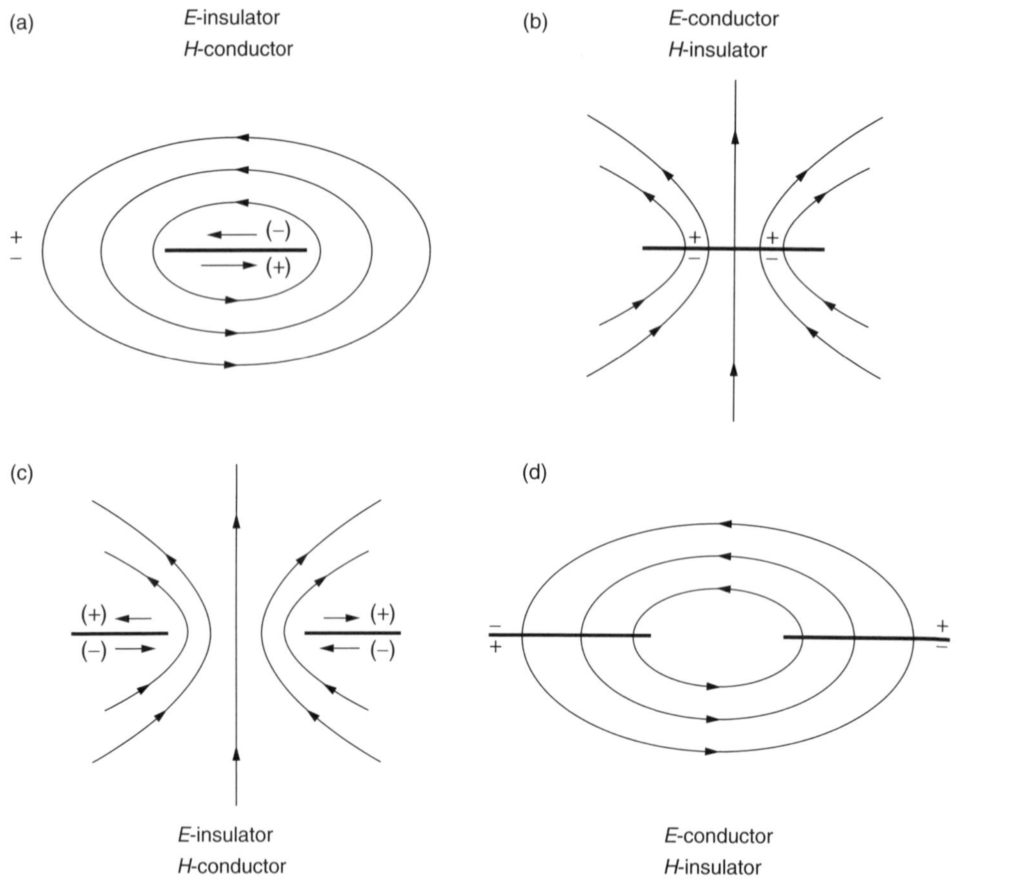

FIGURE 36.9
The streamlines of the potential flow with circulation around a flat plate (Figure 36.8a) coincide with the field lines of the electro (magneto) static field: (i) around (a) an insulating (conducting) flat plate; (ii) between (d) two conducting (insulating) aligned semiinfinite plates; (iii) because the field is tangent (orthogonal) to the plate(s) in the case (i) [(ii)]. The streamlines of the potential flow through the gap between two aligned semiinfinite plates (Figure 36.8b) coincide with the field lines of the electro (magneto) static field: (iv) of a conducting (insulating) flat plate (b); (v) of two insulating (conducting) aligned semiinfinite plates (c); (vi) because the field is orthogonal (tangent) to the plate(s) in the case (iv) [(v)].

(Figure 36.8b) has for electrostatic analogue the electric field outside two an aligned semiinfinite flat insulators (Figure 36.9c). Exchanging the field lines for equipotentials, leads to (Figure 36.9d) the electric field of a condenser with two plane aligned semiinfinite conducting plates that induces positive (negative) charges on opposite sides of each conductor, with surface density:

$$y = 0, |x| > b: \quad \sigma(x) = \frac{e}{2\pi} \left|x^2 - b^2\right|^{-1/2} \operatorname{sgn}(x) \operatorname{sgn}(y), \tag{36.82a–c}$$

given by (36.77a–c) ≡ (36.82a–c) with Q replaced by e.

In the case of the magnetostatic field Figure 36.9a (36.9b) represents the magnetic field near a flat plate conductor (insulator); in the former case there is a surface current distribution:

$$y = 0, |x| < 0: \quad \vartheta(x) = \frac{J}{2\pi} \left|b^2 - x^2\right|^{-1/2} \operatorname{sgn}(y), \tag{36.83a–c}$$

given by (36.70a–c) ≡ (36.83a–c), with Γ replaced by the total current $-J$. In the case of *two flat aligned semiinfinite conductors (insulators) the magnetic field is illustrated in Figure 36.9c (36.9d); there are induced currents in the former case:*

$$y = 0, |x| > b: \quad \vartheta(x) = \frac{J}{2\pi} \left|x^2 - b^2\right|^{-1/2} \operatorname{sgn}(x) \operatorname{sgn}(y), \tag{36.84a–c}$$

given by (36.77a–c) ≡ (36.84a–c) with Q replaced by J. The magneto (electro) static field for the single-plate (36.81a–c) [(36.83a–c)] condenser leads to the same induced electric charge (surface electric current) distribution; the same applies for the condenser with two semiinfinite aligned plates (36.82a–c) [(36.84a–c)].

36.4.4 Confocal, Orthogonal Ellipses, and Hyperbolas

In all of the preceding cases the equipotentials and stream or fields lines are the same set of orthogonal curves that will be shown next to be confocal quadrics (Figure 36.10a); there is more than one parametrization of confocal coordinates, and the choice (36.74) ≡ (36.85a,b) is conformal, that is, leads to the same scale factor along both coordinate curves (Subsection 33.3.1). The starting point is (36.74) ≡ (36.85a):

$$\zeta \equiv \frac{2\pi f}{Q} \equiv \xi + i\eta: \quad x + iy = z = b\cosh\zeta = b\cosh(\xi + i\eta). \tag{36.85a,b}$$

The relations (1.15) between circular and hyperbolic functions:

$$\cosh(i\eta), \sinh(i\eta) \equiv \frac{e^{i\eta} \pm e^{-i\eta}}{2} \equiv \cos(\eta), i\sin(\eta), \tag{36.86a,b}$$

are used when expanding (36.85b):

$$\cosh(\xi + i\eta) = \cosh\xi\cosh(i\eta) + \sinh\xi\sinh(i\eta) = \cosh\xi\cosh\eta + i\sinh\xi\sinh\eta. \tag{36.87}$$

Substitution of (36.87) in (36.85b) leads to the transformation:

$$x = b\cosh\xi\cos\eta, \quad y = b\sinh\xi\sin\eta; \tag{36.88a,b}$$

this shows that the orthogonal families of curves (ζ, η) consist:

$$\cos^2\eta + \sin^2\eta = 1 = \cosh^2\xi - \sinh^2\xi, \tag{36.89a,b}$$

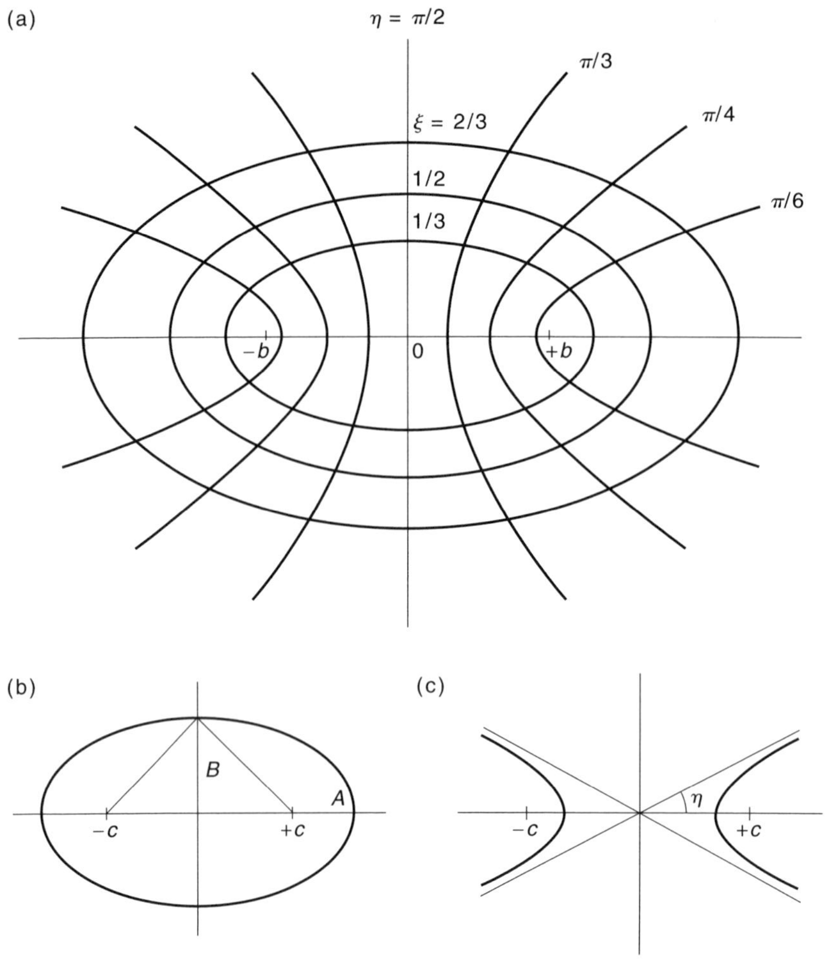

FIGURE 36.10
The streamlines and equipotentials are the same set (a) of orthogonal confocal ellipses (b) and hyperbolas (c) in all ten cases of: (i–v) circulating flow around a finite plate (Figure 36.8a) and electro/magnetostatic field of an insulating/conducting finite plate (Figure 36.9a,b); (vi–x) the potential flow through a gap between two aligned semiinfinite plates (Figure 36.8b) and the electro/magnetostatic field of a two insulating/conducting aligned semiinfinite plates (Figure 36.9c,d). The stream/field lines and equipotentials in cases (i,iii,iv,ix,x) are interchanged in the other cases (ii,v,vi,vii,viii).

for $\zeta \equiv \text{const}$ ($\eta \equiv \text{const}$):

$$\frac{x^2}{(b\cosh\xi)^2} + \frac{y^2}{(b\sinh\xi)^2} = 1 = \frac{x^2}{(b\cos\eta)^2} - \frac{y^2}{(b\sin\eta)^2}, \tag{36.90a,b}$$

of ellipses (36.90a) [hyperbolas (36.90b)].

The ellipses (36.90a) ≡ (36.91a):

$$\frac{x^2}{A^2} + \frac{y^2}{B^2} = 1: \quad A = b\cosh\xi > b\sinh\xi = B, \tag{36.91a–c}$$

have larger (36.91b) and smaller (36.91c) half-axis and are defined by the condition that the sum of the distances to the foci c is a constant, for example, equal for a point on the positive real and imaginary axis in Figure 36.10b:

$$2\sqrt{c^2 + B^2} = (A - c) + (A + c) = 2A, \quad c^2 + B^2 = A^2. \tag{36.92a,b}$$

Thus the foci of the ellipse (36.91a–c) are the points, $\pm b$:

$$c^2 = A^2 - B^2 = b^2 \left(\cosh^2 \xi - \sinh^2 \xi\right) = b^2. \tag{36.93}$$

The property that the points on an ellipse are at a distance from the foci whose sum is constant leads to a simple geometric construction used by gardeners: the ends of a string are attached at the foci, and a marker running through the taut string traces the ellipse; different lengths of string lead to ellipses with distinct eccentricities.

The hyperbolas (36.90b) $\equiv$ (36.94a):

$$y = \sqrt{x^2 \tan^2 \eta - b^2 \sin^2 \eta}; \qquad \lim_{x \to \infty} \frac{y}{x} = \pm \tan \eta, \tag{36.94a,b}$$

have asymptotes (36.94b) making an angle $\pm\eta$ with the real axis (Figure 36.10c). They intersect the real axis at (36.95a,b):

$$y = 0: \quad x = \pm b \cos \eta; \qquad \lim_{\eta \to 0.\pi} x = \pm b; \tag{36.95a–c}$$

the hyperbola of zero slope (36.95c) coincides with the part of the real axis $|x| \geq b$ outside the foci $\pm b$. Thus *(36.88a,b) are the relations between Cartesian* (x, y) *and confocal* (ζ, η) *coordinates, whose orthogonal curves* $\zeta = const\,(\eta = const)$ *are ellipses (36.91a) with half-axis (36.91b,c) [hyperbolas (36.94a) of asymptotes (36.94b)], with common: (i) foci at* $\mathrm{y} = 0, \mathrm{x} = \pm \mathrm{b}$, *as illustrated in Figure 36.10a–c; (ii) scale factors along the two axis (36.96b):*

$$(\mathrm{d}s)^2 \equiv (\mathrm{d}x)^2 + (\mathrm{d}y)^2 = h^2 \left[(\mathrm{d}\xi)^2 + (\mathrm{d}\eta)^2\right], \tag{36.96a}$$

$$\frac{h^2}{b^2} \equiv \cosh^2 \xi - \cos^2 \eta = \sin^2 \xi + \sin^2 \eta. \tag{36.96b}$$

specifying the arc length (36.96a).

36.4.5 Scale Factor for Conformal Coordinates

The arc length in Cartesian coordinates leads by (36.88a,b) to the arc length in confocal coordinates:

$$\begin{aligned} b^{-2} \left[(\mathrm{d}x)^2 + (\mathrm{d}y)^2\right] &= [\mathrm{d}\left(\cosh \xi \cos \eta\right)]^2 + [\mathrm{d}\left(\sinh \xi \sin \eta\right)]^2 \\ &= (\sinh \xi \cos \eta \, \mathrm{d}\xi - \cosh \xi \sin \eta \, \mathrm{d}\eta)^2 + (\cosh \xi \sin \eta \, \mathrm{d}\xi + \sinh \xi \cos \eta \, \mathrm{d}\eta)^2 \\ &= \left(\sinh^2 \xi \cos^2 \eta + \cosh^2 \xi \sin^2 \eta\right) \left[(\mathrm{d}\xi)^2 + (\mathrm{d}\eta)^2\right]. \end{aligned} \tag{36.97}$$

The absence of a cross-term $\mathrm{d}\xi \, \mathrm{d}\eta$ in (36.97) proves that the confocal coordinates are plane orthogonal coordinates (11.43); thus the families of confocal eclipses (36.90a) and hyperbolas (36.90b) are orthogonal. The confocal coordinates are also conformal (Subsection 33.3.1) since the scale factor in (36.97) is the same both directions (ξ, η):

$$h_\xi = h_\eta = b \left|\sinh^2 \xi \cos^2 \eta + \cosh^2 \xi \sin^2 \eta\right|^{1/2} \equiv h. \tag{36.98}$$

For conformal coordinates the scale factor is (33.14b) the modulus of the derivative of the conformal mapping:

$$h = \left|\frac{\mathrm{d}z}{\mathrm{d}\zeta}\right| = \left|\frac{\mathrm{d}\,(b\cosh\zeta)}{\mathrm{d}\zeta}\right| = b\,|\sinh\zeta = b\,|\sinh\zeta|\,. \tag{36.99}$$

Using (36.86a,b) ≡ (36.100a,b):

$$\cosh(i\eta) = \cos\eta, \quad \sinh(i\eta) = i\sin\eta, \tag{36.100a,b}$$

implies

$$\begin{aligned}\sinh\zeta \equiv \sinh(\xi + i\eta) &= \sinh\xi\cosh(i\eta) + \cosh\xi\sinh(i\eta)\\ &= \sinh\xi\cos\eta + i\cosh\xi\sin\eta.\end{aligned} \tag{36.101}$$

The modulus of (36.101) is the scale factor (36.99) and coincides with (36.98). Thus *the scale factor (36.96b) of confocal coordinates (36.96a) is specified by (i) the complex form (36.99); (ii) the real from (36.98); (iii) the simpler real alternate forms:*

$$\begin{aligned}\frac{h^2}{b^2} &= \sinh^2\xi\cos^2\eta + \left(1+\sinh^2\xi\right)\sin^2\eta = \sinh^2\xi + \sin^2\eta,\\ &= \left(\cosh^2\xi - 1\right)\cos^2\eta + \cosh^2\xi\sin^2\xi = \cosh^2\xi - \cos^2\eta,\end{aligned} \tag{36.102a,b}$$

used in (36.96b) ≡ (36.102a,b).

36.5 Convergent Channel (Harris, 1901) and Added Length

In the flow through the slot (Section 36.4) the two semiinfinite plates (Figure 36.9b) are aligned. This is the particular case $\beta = \pi/2$ of the flow out of a duct with flat sides (Subsection 36.5.1), converging at an angle 2β with edges at the distance $2b$. The other extreme particular case is the flow out of a channel with parallel sides (Subsection 36.5.2); cancelling the flow in the duct leads to a Pitot tube (Section 14.6) in a uniform stream (Subsection 36.5.3). The electro/magnetostatic analogues include a condenser with two parallel semiinfinite plates (Subsection 36.5.4).

36.5.1 Convergent Duct with Straight Semiinfinite Sides

The convergent duct (Figure 36.11b) is equivalent to a degenerate polygon with two pairs of coincident sides and four vertices, that is, two at the edges with angles $\gamma_1 = -\pi = \gamma_3$, and two coincident at infinity with angle $\gamma_2 = \pi + 2\beta$ in (36.103a). Choosing the vertices z_1, z_2, z_3 to be the images of (36.103b), the interior Schwartz–Christoffel transformation (33.25) is given by (36.103c):

$$\gamma_{1-3} = -\pi, \pi + 2\beta, -\pi; \quad \zeta_{1-3} = -1, 0, 1:$$
$$\frac{\mathrm{d}z}{\mathrm{d}\zeta} = A\,(\zeta+1)\,\zeta^{-(\pi+2\beta)/\pi}\,(\zeta-1) = A\left(\zeta^{1-2\beta/\pi} - \zeta^{-1-2\beta/\pi}\right); \tag{36.103a–c}$$

this can be integrated:

$$z = \frac{\pi A}{2\,(\pi-\beta)}\zeta^{2-2\beta/\pi} + \frac{\pi A}{2\beta}\zeta^{-2\beta/\pi} + B, \tag{36.104}$$

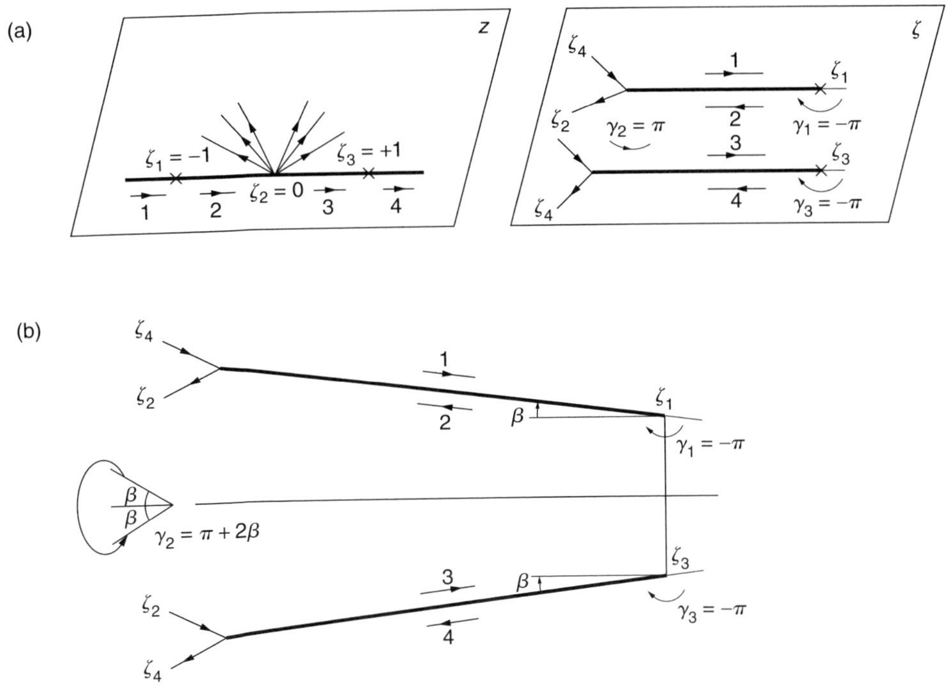

FIGURE 36.11
The real axis can be mapped using a Schwartz–Christoffel interior polygonal transformation with three critical points into a channel with two infinite sides (a); the choice of exponents specifies the angles, for example, a parallel-sided (a) [convergent (b)] channel.

where the constants A, B allow a translation B, a homothety $|A|$ and a rotation arg (A). A source/sink of volume flux $2Q$ at the origin $\zeta = 0$ of the ζ-plane (12.26a) is chosen:

$$f(\zeta) = \frac{Q}{\pi}\log\zeta, \quad \zeta = \exp\left(\frac{\pi f}{Q}\right), \tag{36.105a,b}$$

so that (Figure 36.11a) half the volume flux Q goes into the upper-half z-plane:

$$z = \frac{\pi A}{2\beta}\left\{\exp\left(-\frac{2\beta f}{Q}\right) + \left(\frac{\pi}{\beta} - 1\right)^{-1}\exp\left[\frac{2(\pi-\beta)f}{Q}\right]\right\} + B. \tag{36.106}$$

that is, into the duct at infinity in the ζ-plane.

Thus has been obtained *the complex potential* $f(\zeta)$ *in inverse form (36.106) for the potential flow with volume flux* Q *out of a duct with flat semiinfinite sides converging at an angle* 2β *with* $0 < \beta \leq \pi/2$ *in Figure 36.11b. The extreme cases are: (i) the case (36.107a) of aligned sides:*

$$\beta = \frac{\pi}{2}: \quad z = A\left[\exp\left(\frac{\pi f}{Q}\right) + \exp\left(-\frac{\pi f}{Q}\right)\right] + B = 2A\cosh\left(\frac{\pi f}{Q}\right) + B, \tag{36.107a,b}$$

leads back to the slot between semiinfinite aligned plates (36.74) $\equiv$ *(36.107b) with* $A = b/2$ *and* $B = 0$*; (ii) in the opposite extreme case of parallel sides (36.108a) yields (36.108b):*

$$\beta = 0: \quad z = \frac{A}{2}\left[\exp\left(\frac{2\pi f}{Q}\right) - \frac{2\pi f}{Q}\right] + B. \tag{36.108a,b}$$

The latter (36.108b) is the limit of (36.106) as $\beta \to 0$:

$$\begin{aligned}
&\lim_{\beta\to 0}\left\{\frac{\pi A}{2\beta}\exp\left(-\frac{2\beta f}{Q}\right)+\frac{\pi A}{2(\pi-\beta)}\exp\left[\frac{2(\pi-\beta)f}{Q}\right]+B\right\}\\
&\quad = B+\frac{A}{2}\exp\left(\frac{2\pi f}{Q}\right)+\lim_{\beta\to 0}\frac{\pi A}{2\beta}\left[1-\frac{2\beta f}{Q}+O\left(\beta^2\right)\right]\\
&\quad = B+\frac{A}{2}\exp\left(\frac{2\pi f}{Q}\right)-\frac{\pi A f}{Q}+\lim_{\beta\to 0}\frac{\pi A}{2\beta}; \qquad (36.109)
\end{aligned}$$

this coincides with (36.108b) $\equiv$ (36.109) apart from an infinite constant that can be neglected. The legitimacy of doing so can be checked by starting with (36.108a) $\equiv$ (36.110a) in the Schwartz–Christoffel transformation (36.103c) that is nonsingular in β:

$$\beta = 0: \quad \frac{\mathrm{d}z}{\mathrm{d}\zeta} = A\left(\zeta-\frac{1}{\zeta}\right), \qquad z = A\left(\frac{\zeta^2}{2}-\log\zeta\right)+B; \qquad (36.110\text{a–c})$$

substituting (36.105a,b) into (36.110c) leads to (36.108b). QED.

36.5.2 Flow out of a Channel with Parallel Sides

The constants A, B in (36.106; 36.107b; 36.108b) can be determined by specifying two conditions, for example, the position of the edges in the ζ-plane. Consider (Figure 36.12a), the channel with parallel semiinfinite flat sides (36.110c) at a distance b, so that the edges corresponding to $\zeta = \pm 1$ are placed at $z = 0, ib$, leading in (36.110c) to (36.111a,b):

$$0 = z(+1) = \frac{A}{2}+B, \quad ib = z(-1) = \frac{A}{2}+B-i\pi A: \quad -2B = A = -\frac{b}{\pi}, \qquad (36.111\text{a–d})$$

whose solution is (36.111c,d). Substituting (36.111c,d) and (36.105b) in (36.110c):

$$z = -\frac{b}{\pi}\left(\log\zeta+\frac{1-\zeta^2}{2}\right) = \frac{bf}{Q}+\frac{b}{2\pi}\left[1-\exp\left(\frac{2\pi f}{Q}\right)\right], \qquad (36.112\text{a,b})$$

leads to *the mapping (36.112a) of the real z-axis into two parallel semiinfinite plates at distance b; also the flow (Figure 36.12a) with volume flux Q out of this channel, is specified by the inverse complex potential (36.112b), and conjugate velocity (36.113):*

$$\frac{\mathrm{d}f}{\mathrm{d}z} = \left(\frac{\mathrm{d}z}{\mathrm{d}f}\right)^{-1} = \frac{Q/b}{1-\exp(2\pi f/Q)}; \qquad (36.113)$$

in these inverse forms the complex potential (position) appears as the independent (dependent) variable.

The inverse form (36.112b) is in some respects more convenient than the direct form, in that it specifies the coordinates x, y in terms of the potential Φ and stream Ψ functions as parameters:

$$x = \frac{b\Phi}{Q}+\frac{b}{2\pi}\left[1-\exp\left(\frac{2\pi\Phi}{Q}\right)\cos\left(\frac{2\pi\Psi}{Q}\right)\right], \qquad (36.114\text{a})$$

$$y = \frac{b\Psi}{Q}-\frac{b}{2\pi}\exp\left(\frac{2\pi\Phi}{Q}\right)\sin\left(\frac{2\pi\Psi}{Q}\right); \qquad (36.114\text{b})$$

the equipotentials (streamlines) may be obtained directly by plotting x, y at constant Φ (constant Ψ), as a function $-\infty < \Psi < +\infty\,(-\infty < \Phi < +\infty)$. The upper $x < 0, y = b$

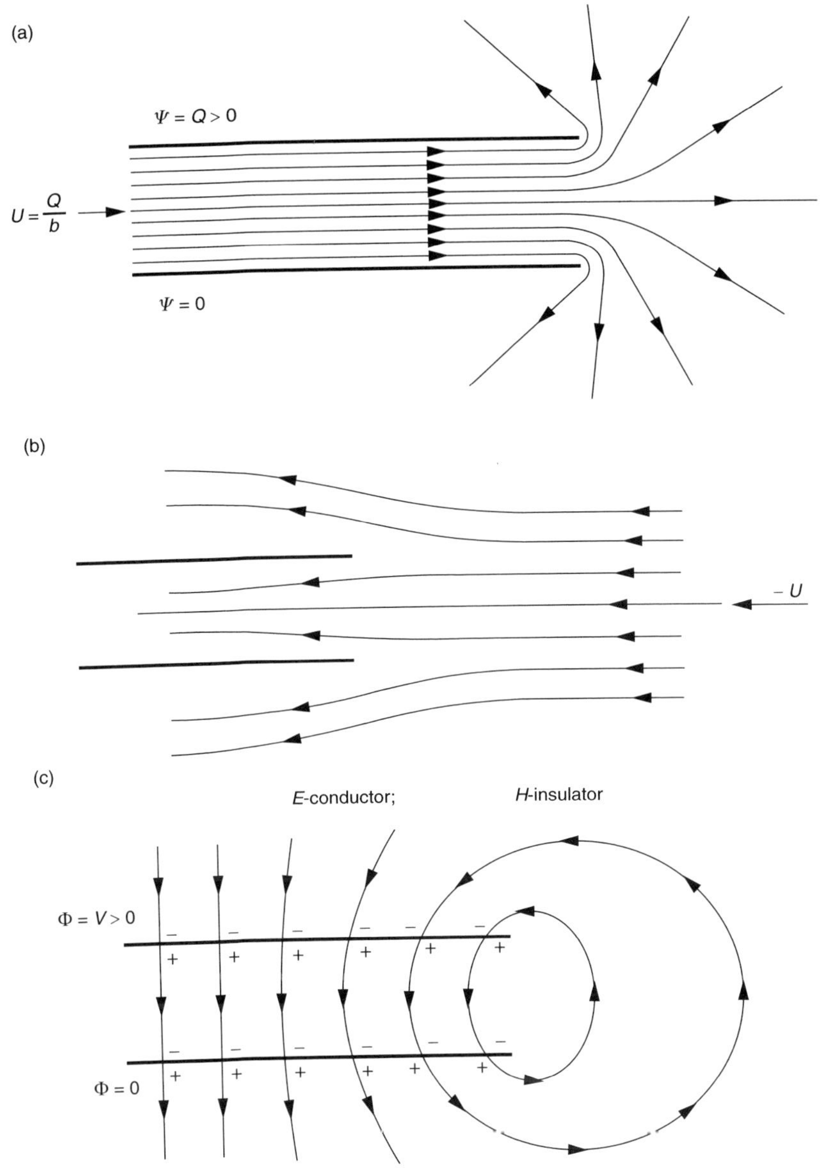

FIGURE 36.12
A source of flow rate Q at the far and of a duct with parallel semiinfinite sides at a distance b causes a uniform flow of velocity $U = Q/b$ which spreads in all directions at the duct exit (a). Superimposing a uniform stream with velocity $-U$ aligned with the duct axis cancels the flow in the duct (b), corresponding to a Pitot tube in a free stream (Figure 14.2a); this determines the pressure distribution on the outer side, and the error in measuring static pressure and free stream velocity due to the perturbation caused by the tube. The analogue electro (magneto) static field concerns two parallel semiinfinite conducting (insulating) plates with aligned edges (c); the induced electric charge distribution on the "parallel plate condenser" has an inverse square root singularity at the edges. This is an integrable singularity, and thus the total electric charge due to the concentration of electric charges near the edge is finite. The total electric charge on the parallel plate condenser is the same as if: (i) the electric charge density was equal to the constant value far from the edges; (ii) the concentration of charges near the edges is compensated by an "added length" of conductor "extending" the plates.

(lower $x < 0, y = 0$) wall corresponds to the streamline $\Psi = Q\,(\Psi = 0)$ in (36.114b), confirming that the volume flux is Q in the channel. With $\Psi = 0, Q$ as Φ goes from $\pm\infty$ to 0, then x goes from $-\infty$ to 0, so that the inner (outer) side of the walls correspond to $\Phi < 0\,(\Phi > 0)$, and the edges to $\Phi = 0$. The velocity (36.113) is horizontal and identical at both the lower $f = \Phi$ (upper $f = \Phi + iQ$) walls:

$$x \le 0: \quad v_x(x,0) = \frac{Q/b}{1 - \exp(2\pi\Phi/Q)} = v_x(x,b), \quad v_y(x,0) = 0 = v_y(x,b). \quad (36.115\text{a,b})$$

Near the edges $f \approx \Phi \approx 0$ it follows from (36.112b) that

$$z = \frac{bf}{Q} + \frac{b}{2\pi}\left[-\frac{2\pi f}{Q} - \frac{(2\pi f/Q)^2}{2} + O\left(f^3\right)\right] = -\frac{b\pi f^2}{Q^2} + O\left(f^3\right), \qquad (36.116)$$

that is, $z \approx f^2$ or $f \approx \sqrt{z}, df/dz \sim 1/\sqrt{z}$ so that the velocity has an inverse square root singularity as should be expected at a sharp edge [(36.68b) and (14.90b)]. As $\Phi \to +\infty\,(\Phi \to -\infty)$ along the outer (inner) side of the walls, $v_x \to 0\,(v_x \to Q/b \equiv U)$ in (36.115a), showing that far outside (inside) the channel the fluid is at rest (there is uniform flow of velocity $U = Q/b$ equal to the volume flux divided by the width).

36.5.3 Pressure Correction for a Pitot Tube

The flow far inside the channel can be suppressed by superimposing a uniform flow of velocity (36.117a) that corresponds to the transformation (36.117b):

$$U = \frac{Q}{b}: \quad g = f - Uz = f - \frac{Qz}{b} = \frac{Q}{2\pi}\left\{\exp\left[\frac{2\pi}{Q}(g + Uz)\right] - 1\right\}, \qquad (36.117\text{a,b})$$

where (36.112b) was used; solving for z leads to the inverse complex potential (36.118a):

$$z = -\frac{g}{U} + \frac{b}{2\pi}\log\left(1 + \frac{2\pi g}{Q}\right), \quad \frac{dg}{dz} = \left(\frac{dz}{dg}\right)^{-1} = -U\left(1 + \frac{Q}{2\pi g}\right), \qquad (36.118\text{a,b})$$

and conjugate velocity (36.118b). The latter follows from

$$\begin{aligned}\frac{dg}{dz} &= \left(-\frac{1}{U} + \frac{b/Q}{1 + 2\pi g/Q}\right)^{-1} = \left[-\frac{1}{U}\left(1 - \frac{1}{1 + 2\pi g/Q}\right)\right]^{-1} \\ &= -U\frac{1 + 2\pi g/Q}{2\pi g/Q} = -U\left(1 + \frac{Q}{2\pi g}\right).\end{aligned} \qquad (36.119)$$

The streamlines $\Psi = \text{const}$ (equipotentials $\Phi = \text{const}$) are given (36.118a) $\equiv$ (36.120a) in parametric form by (36.120b,c):

$$g = \Phi + i\Psi: \quad x = -\frac{\Phi}{U} + \frac{b}{4\pi}\log\left[\left(1 + \frac{2\pi\Phi}{Q}\right)^2 + \left(\frac{2\pi\Psi}{Q}\right)^2\right], \qquad (36.120\text{a,b})$$

$$y = -\frac{\Psi}{U} + \frac{b}{2\pi}\arctan\left[\frac{\Psi}{\Phi + Q/(2\pi)}\right]; \qquad (36.120\text{c})$$

for example, the lower wall $x < 0, y = 0$ is the streamline $\Psi = 0$. At the lower wall (36.121a,b) the velocity (36.118b) is horizontal (36.121d):

$$\Psi = 0, \quad g = \Phi: \; v_x(x,0) = -U\left(1 + \frac{Q}{2\pi\Phi}\right), \quad v_y(x,0) = 0, \qquad (36.121\text{a–d})$$

and: (i) it has a singularity at the edge $\Phi = 0$; (ii) at a large distance outside $\Phi \to \infty$ the velocity tends to that $-U$ of the free stream; (iii) far inside the channel it tends to zero $v_x \to 0$, that is, $\Phi \to -Q/2\pi$, as can be checked from (36.120b) with $\Psi = 0$ and $x \to -\infty$.

Thus has been obtained *the potential flow (36.118a,b) with incident velocity $-U$ aligned with a* ***Pitot tube (1732)*** *with (Figure 36.12b) parallel semiinfinite walls at a distance b, with the static p_∞ (stagnation p_0) pressure measured at large distance on the inside (outside) of the tube. The free stream velocity U is calculated by the Bernoulli equation (14.27c) ≡ (36.122a):*

$$p_0 = p_\infty + \frac{\rho}{2}U^2, \quad U = \sqrt{2\frac{(p_0 - p_\infty)}{\rho}}, \tag{36.122a,b}$$

assuming that p_0 (p_∞) is measured at velocity zero (equal to $-U$) in agreement with [(14.38b) in the incompressible case (14.38a). An error ε in the velocity produces an error 2ε in the pressure, because the latter is quadratic in (36.122a); since the difference of two pressures appears in (36.122b), the total error is 4ε. Along the wall of the tube (36.120b) is approximated by (36.123a) and the relative error in the velocity (36.121c) is (36.123b):

$$\Phi \sim -xU, \quad \varepsilon_0 \sim -\frac{Q}{2\pi\Phi}; \quad \varepsilon \sim 4\varepsilon_0 \sim -\frac{2Q}{\pi\Phi} \sim \frac{2Q}{\pi x U} = \frac{2b}{\pi x}, \tag{36.123a–c}$$

thus the relative error in the pressure is (36.123c). Hence *if the static (stagnation) pressures are measured on the outside (inside) of a Pitot tube of width b at distance x from the edge, there is a relative error (36.123c) in the estimate of the free stream velocity,* for example, ε =3% for a Pitot tube of length $x/b = 2/(\pi\varepsilon) = 1/(0.015\pi) = 21$ times the width. Thus the accurate measurement of the velocity needs a long thin tube (Figure 14.2a), in order to be away (Figure 36.12b) from the disturbed flow, so as not to corrupt the measurement.

36.5.4 Added Length due to Concentration of Charges

The parallel-sided ejector channel (Figure 36.12a) has for electro (magneto) static analogue an insulator (conductor) without (with) surface electric charges (currents), with the complex potential specified by (36.118a) where the flow rate Q is replaced by Φ_0/ε (Ψ_0/c), where Φ_0 (Ψ_0) is the potential (field function) and ε the dielectric permittivity (c the speed of light in vacuo). Exchanging field lines and equipotentials leads to (Figure 36.12c) the electro (magneto) static conductor (insulator) with (without) surface electric charge (currents). In the former case *of a condenser with two parallel flat semiinfinite plates at a distance* b *with potential difference Φ_0 the complex potential in inverse from (36.124a) [conjugate electric field (34.124b)]:*

$$z = \frac{ibg}{\Phi_0} + \frac{b}{2\pi}\left[1 - \exp\left(\frac{i2\pi g}{\Phi_0}\right)\right], \tag{36.124a}$$

$$E^* = -\frac{\mathrm{d}g}{\mathrm{d}z} = -\left(\frac{\mathrm{d}z}{\mathrm{d}g}\right)^{-1} = -\frac{i\Phi_0/b}{1 - \exp(i2\pi g/\Phi_0)}, \tag{36.124b}$$

is obtained from (36.112b) [(36.113)] by replacing: (i) f by $i\,g$ to interchange streamlines and equipotentials; (ii) Q by Φ_0. The equipotentials Φ = const and field lines Ψ = const are given in parametric form by (36.124a) with $g = \Phi + i\Psi$ or $i\,g = -\Psi + i\Phi$:

$$x = -\frac{b\Psi}{\Phi_0} + \frac{b}{2\pi}\left[1 - \exp\left(-\frac{2\pi\Psi}{\Phi_0}\right)\cos\left(\frac{2\pi\Phi}{\Phi_0}\right)\right], \tag{36.125a}$$

$$y = \frac{b\Phi}{\Phi_0} - \frac{b}{2\pi}\exp\left(-\frac{2\pi\Psi}{\Phi_0}\right)\sin\left(\frac{2\pi\Phi}{\Phi_0}\right); \tag{36.125b}$$

the lower (upper) plate $y = 0$ ($y = b$) is at the potential $\Phi = 0$ ($\Phi = \Phi_0$), so that the potential difference between the two plates is Φ_0. On the lower $g = i\Psi$ (upper $g = \Phi_0 + i\,\Psi$) plate (36.126a,b) the electric field (36.124b) is vertical (36.126c):

$$x < 0; y = 0,\ b: \quad E_x = 0, \quad \sigma(x) = \pm\varepsilon E_y = \pm\frac{\Phi_0\varepsilon/b}{1 - \exp(-2\pi\Psi/\Phi_0)}, \qquad (36.126\text{a–d})$$

and specifies the induced surface charges (36.126d); bearing in mind that the unit normal is downward (upward) on the lower (upper) side of the lower plate, the electric charge density has the lower (upper) sign in (36.126d). The surface electric charges are: (i) singular at the edges $\sigma \to \infty$ as $\Psi \to 0$; (ii) far from the edges $|\Psi| \to \infty$ tend to positive uniform $\sigma \to \Phi_0/b$ inside $\Psi \to +\infty$, and to zero $\sigma \to 0$ through negative values outside $\Psi \to -\infty$; the signs of the charges are reversed at the upper plate (Figure 36.12c).

In the case of two infinite plates $-\infty < x < +\infty$ at a distance $y = 0, b$ equal to b with a potential difference, Φ_0, the potential satisfies the Laplace equation (36.127b) between the plates (36.127a), with the boundary conditions (36.127c,d) on the plates:

$$0 \le y \le b: \quad \frac{d^2\bar{\Phi}}{dy^2} = 0, \quad \bar{\Phi}(0) = 0, \quad \bar{\Phi}(b) = \Phi_0; \qquad (36.127\text{a–d})$$

it has for solution: (i) a linear potential (36.128a); (ii) a uniform electric field (36.128a–c):

$$\bar{\Phi}(y) = \Phi_0\frac{y}{b}, \quad \bar{E} = -\frac{d\bar{\Phi}}{dy} = -\frac{\Phi_0}{b}; \quad \sigma_0 = \varepsilon\bar{E} = -\varepsilon\frac{\Phi_0}{b}, \qquad (36.128\text{a–c})$$

(iii) this leads to a uniform surface electric charge (36.128c). The latter (36.128c) corresponds to the numerator of (36.126d), that is the first term of (36.124a) with $z = ibg/\Phi_0$ or $g = -iz\Phi_0/b$, so that $\Phi = \text{Re}\,(g) = \Phi_0 y/b$, and $E_y = \text{Im}\,(dg/dz) = -\Phi_0/b$ in agreement respectively with (36.128a,b). The uniform charge density (36.128c) on the condenser with infinite plates is equal to that (36.126c) in the semiinfinite condenser far from the edges, $\Psi \to \infty$. For the latter the charge density becomes singular at the edges $\sigma \to \infty$ for $\Psi \to 0$; this is an inverse square-root singularity that is integrable, leading to a finite total charge. The total charge e_+ (e_-) on the inner $\Psi > 0$ (outer $\Psi < 0$) side of the condenser, within a distance $x = -L$ from the edge, large (36.129a) compared with the width b, is given by the change in Ψ/ε in (36.125a):

$$L^2 \gg b^2: \quad e_+ \approx \Phi_0\varepsilon\left(\frac{L}{b} + \frac{1}{2\pi}\right), \qquad e_- \approx -\frac{\Phi_0\varepsilon}{2\pi}\log\left(\frac{2\pi L}{b}\right), \qquad (36.129\text{a–c})$$

where: (i) the first term of (36.129b) corresponds to the condenser with infinite plates $\Phi_0\varepsilon/b = -\sigma_0$, with σ_0 given by (36.128c); (ii) the second term in (36.129b) is $\Phi_0\varepsilon/2\pi$ an "edge" effect due to the concentration of charges on the inside; (iii) in the case of the infinite condenser there are no charges on the outside, so the external charges (36.129c) are wholly an "edge" effect arising from the last term in (36.125a). Adding (36.129b,c) it follows that:

$$\bar{e} \equiv (e_+ - e_-)\,L = \frac{\Phi_0\varepsilon}{b}\left(L + \bar{L}\right) = -\sigma_0\left(L + \bar{L}\right), \quad \bar{L} \equiv \frac{b}{2\pi}\log\left(\frac{2\pi e L}{b}\right), \qquad (36.130\text{a,b})$$

with $e = 2.711828\ldots$, implying that *in a condenser consisting of two semiinfinite plates at a distance b and potential difference* Φ_0*, there is a concentration of charges near the edges such that the total charge* $\bar{e}$ *in (36.130a) in a length L large compared with the width (36.129a), is equal to the charge on the corresponding infinite condenser (36.128c), with an added length* $\bar{L}$ *in (36.130b) that accounts for the "edge" effect;* the definition of an "added" length is possible in other geometries, whenever the singularity at the edge is integrable and the total effect is finite.

36.6 Monopole Images on Parallel Walls

The flow out of a parallel-sided duct (Subsection 36.5.1) corresponds to a source (sink) at the deep end of the tube. Instead the source/sink is considered next between two parallel infinite walls (Subsection 36.6.1) leading to: (i) infinite images; (ii) an alternative is mapping the complex plane into a strip (Subsection 36.6.2). Both methods show the wall effect on the flow (Subsection 36.6.3). The comparison of the two methods provides a physical derivation of (Subsection 36.6.4) a series of fractions (infinite product), as an alternative to the mathematical method (Subsection 27.9.1).

36.6.1 Infinite Images of a Source/Sink between Walls

Consider (Figure 36.13a) a line source/sink of mass flux Q at the origin between two vertical parallel walls $\mathrm{Re}\,(z) = \pm b/2$ at a distance b. The boundary condition at the walls that the horizontal velocity is zero, can be met by introducing (Section 16.1) two identical image sources/sinks at $z = \pm b$; the image at $z = \pm b$ alone would preserve the wall at $Re\,(z) = b/2$ but the other image $z = \mp b$ perturbs the wall, unless an extra pair of images is added at $z = \mp 2b$; the latter perturbs the wall at $\mathrm{Re}\,(z) = b/2$, unless another pair of images is added image at $z = \pm 3b$. Proceeding in this way, it follows that the effect of two parallel walls is represented by a infinite set of sources/sinks of equal mass flux at the points $z = \pm nb$ with $n = 1, 2, \ldots$, like visual images in two parallel mirrors. To each image source/sink is associated (12.26a) a complex potential:

$$f_n(z) = \frac{Q}{2\pi}\log(z - nb) = \frac{Q}{2\pi}\left\{\log\left(1 - \frac{z}{nb}\right) + \log(-nb)\right\}, \tag{36.131}$$

where the last term is a constant, that can be omitted. The total flow due to the source plus all the images has: (i) a complex potential:

$$\begin{aligned} f(z) &= \frac{Q}{2\pi}\log z + \sum_{n=1}^{\infty}\left[f_n(z) + f_{-n}(z)\right] \\ &= \frac{Q}{2\pi}\log z + \sum_{n=1}^{\infty}\log\left\{\left(1 - \frac{z}{nb}\right)\left(1 + \frac{z}{nb}\right)\right\} \\ &= \frac{Q}{2\pi}\log\left\{z\prod_{n=1}^{\infty}\left(1 - \frac{z^2}{n^2b^2}\right)\right\}, \end{aligned} \tag{36.132}$$

specified by an infinite product (36.132); (ii) a complex velocity:

$$\begin{aligned} \frac{\mathrm{d}f}{\mathrm{d}z} &= \frac{Q}{2\pi z} + \sum_{n=1}^{\infty}\left\{\frac{\mathrm{d}f_n}{\mathrm{d}z} + \frac{\mathrm{d}f_{-n}}{\mathrm{d}z}\right\} \\ &= \frac{Q}{2\pi z} + \frac{Q}{2\pi}\sum_{n=1}^{\infty}\left(\frac{1}{z - nb} + \frac{1}{z - nb}\right) \\ &= \frac{Q}{2\pi}\left(\frac{1}{z} + 2z\sum_{n=1}^{\infty}\frac{1}{z^2 - n^2b^2}\right) \end{aligned} \tag{36.133}$$

given by a series of fractions (36.133). Since the infinite product (36.132) [series of fractions (36.133)] has general term $1 + O\left(n^{-2}\right)$ $\left[O\left(n^{-2}\right)\right]$, it is absolutely convergent, as the series harmonic series (29.40d) of general term $O\left(n^{-2}\right)$.

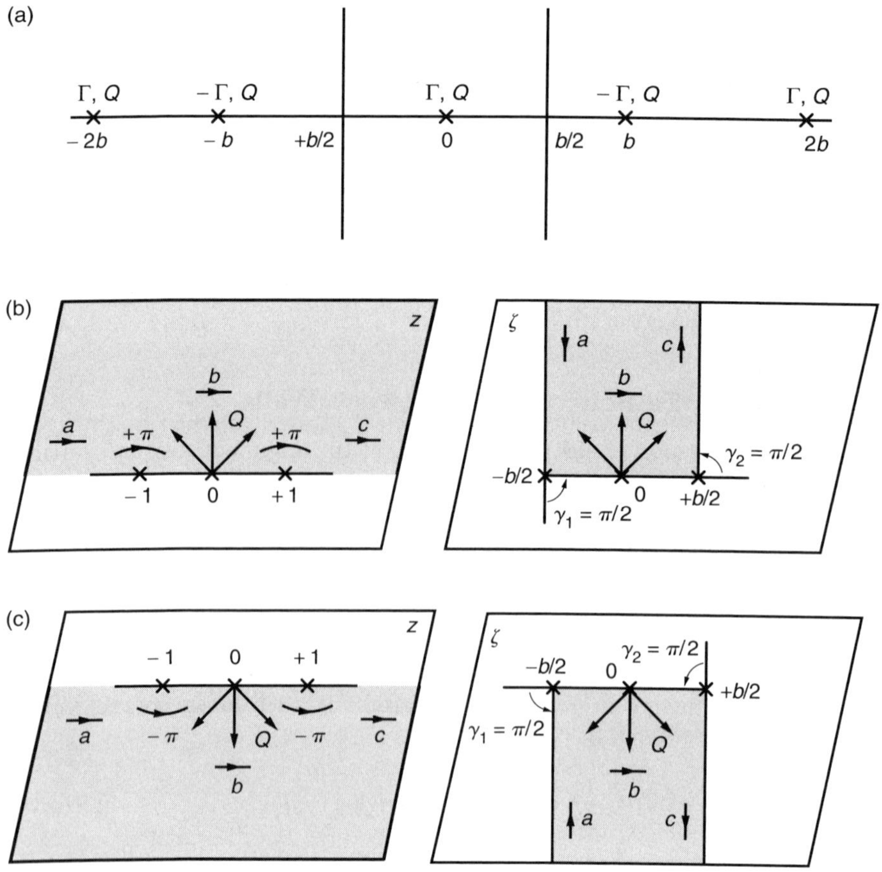

FIGURE 36.13
The potential flow due to a source/sink (vortex) at equal distance from two parallel infinite walls can be obtained in two equivalent ways: (i) considering the infinite identical (alternating) images on the walls (a), instead of just one identical (opposite) image on one wall [Figure 16.1(16.3)], because the second wall leads to an "iteration" of images; (ii) using a Schwartz–Christoffel interior polygonal transformation with two critical points to map the upper (b) [lower (c)] complex half-z-plane with a source at the origin into an upward (downward) square "well." The "upward" (b) and "downward" (c) square wells are the analytic continuation of each other by the Schwartz reflection principle (Figure 31.4), and add together to the vertical strip between the parallel walls (a). In the case of a vortex it may be taken together with its opposite in a vertical strip of double width, to ensure periodic repetition along the horizontal axis. If the source/sink (vortex) are at unequal distance from the two walls the positions of the infinite images change (Figure 40.7a).

36.6.2 Mapping the Complex Plane into a Strip

Another method of obtaining this potential field is to map the complex ζ-plane conformally on the strip $-b/2 < \operatorname{Re}(z) < +b/2$. Starting with the upper-half z-plane, $\operatorname{Im}(z) > 0$, it is mapped into the degenerate triangle $\operatorname{Im}(z) > 0, -b/2 < \operatorname{Re}(z) < +b/2$, using a Schwartz–Christoffel mapping (33.25) with two critical points (Figure 36.13b) at (36.134a), with angles

(36.134b):

$$\zeta_{1-2} = -1, +1; \ \gamma_1 = \frac{\pi}{2} = \gamma_2 : \quad \frac{dz}{d\zeta} = A\pi \left\{ (1-\zeta)(1+\zeta) \right\}^{-1/2} = \frac{A}{\sqrt{1-\zeta^2}}; \tag{36.134a–c}$$

the constants A, B of integration (36.135a) of (36.134c) are determined so that the critical points $\zeta = \pm 1$ are mapped to the edges $z = \pm b/2$ of the triangles (36.135b,c) leading to (36.135d,e):

$$z = A \arcsin \zeta + B : \ z(\pm 1) = \pm \frac{b}{2} = B \pm \frac{A\pi}{2} \Leftrightarrow B = 0, \ A = \frac{b}{\pi}. \tag{36.135a–e}$$

The same transformation maps (Figure 36.13c) the lower-half ζ-plane into the image degenerate triangle $\operatorname{Im}(z) < 0, -b/2 < \operatorname{Re}(z) < +b/2$, with two vertices at $\pm b/2$ and the third at infinity, hence two parallel sides; thus the whole ζ-plane is mapped to the strip (36.136a), with the real axis corresponding to segment $(-b/2, +b/2)$ in the conformal mapping (36.135a,d,e) $\equiv$ (36.136b,c):

$$-\frac{b}{2} < Re(z) < +\frac{b}{2} : \ z = \frac{b}{\pi} \arcsin \zeta, \ \zeta = \sin\left(\frac{\pi z}{b}\right). \tag{36.136a–c}$$

The same conclusion could be obtained applying the reflection principle (Subsection 31.2.2), by reflecting on the real axis in the ζ-plane [on the segment $(-b/2, +b/2)$ in the z-plane], because z is real for real ζ; this performs the analytic continuation from one half-plane (half-strip) to the other. Thus a source/sink of mass flux Q at the origin corresponds (12.26a) by (34.136c) to the complex potential (36.137a):

$$f = \frac{Q}{2\pi} \log \zeta = \frac{Q}{2\pi} \log \left\{ \sin\left(\frac{\pi z}{b}\right) \right\}, \quad \frac{df}{dz} = \frac{Q}{2b} \cot\left(\frac{\pi z}{b}\right), \tag{36.137a,b}$$

and leads to the conjugate velocity (36.137b).

36.6.3 Flow Attenuation and Enhancement by Walls

A source/sink of flow rate Q at the origin, at equal distance from two parallel walls with spacing b, has complex potential (36.137a) $\equiv$ (36.132) and conjugate velocity (36.137b) $\equiv$ (36.133). The latter show that the velocity along the real (36.138a,b) [imaginary (36.139a,b)] axis $z = x$ ($z = iy$) is horizontal (vertical):

$$v_x(x,0) = \frac{Q}{2b} \cot\left(\frac{\pi x}{b}\right) < \frac{Q}{2\pi x}, \quad v_y(x,0) = 0, \tag{36.138a,b}$$

$$v_x(0,y) = 0, \quad v_y(0,y) = \frac{Q}{2b} \coth\left(\frac{\pi y}{b}\right) > \frac{Q}{2\pi y}. \tag{36.139a,b}$$

From the inequalities:

$$\tan \xi > \xi > \tanh \xi \Leftrightarrow \cot \xi < \frac{1}{\xi} < \coth \xi, \tag{36.140a,b}$$

it follows that *the velocity due to a source between walls is smaller (larger) than in free space, in the direction perpendicular (parallel) to the walls, that is, the flow is weakened (enhanced), with the extremes being attained (Figure 36.14a): (i) at the stagnation points $df/dz = 0$ for $z = \pm b/2$ on the walls closest to the source in (36.138a); (ii) the flow far from the source is uniform $df/dz \to Q/2b$ as $z \to \infty$ in (36.137b), that is, it is parallel to the walls, and due to half the volume flux $Q/2$ in a channel of width b.* The potential flow due to a source/sink at an unsymmetrical position between parallel walls is considered in Example 40.12.

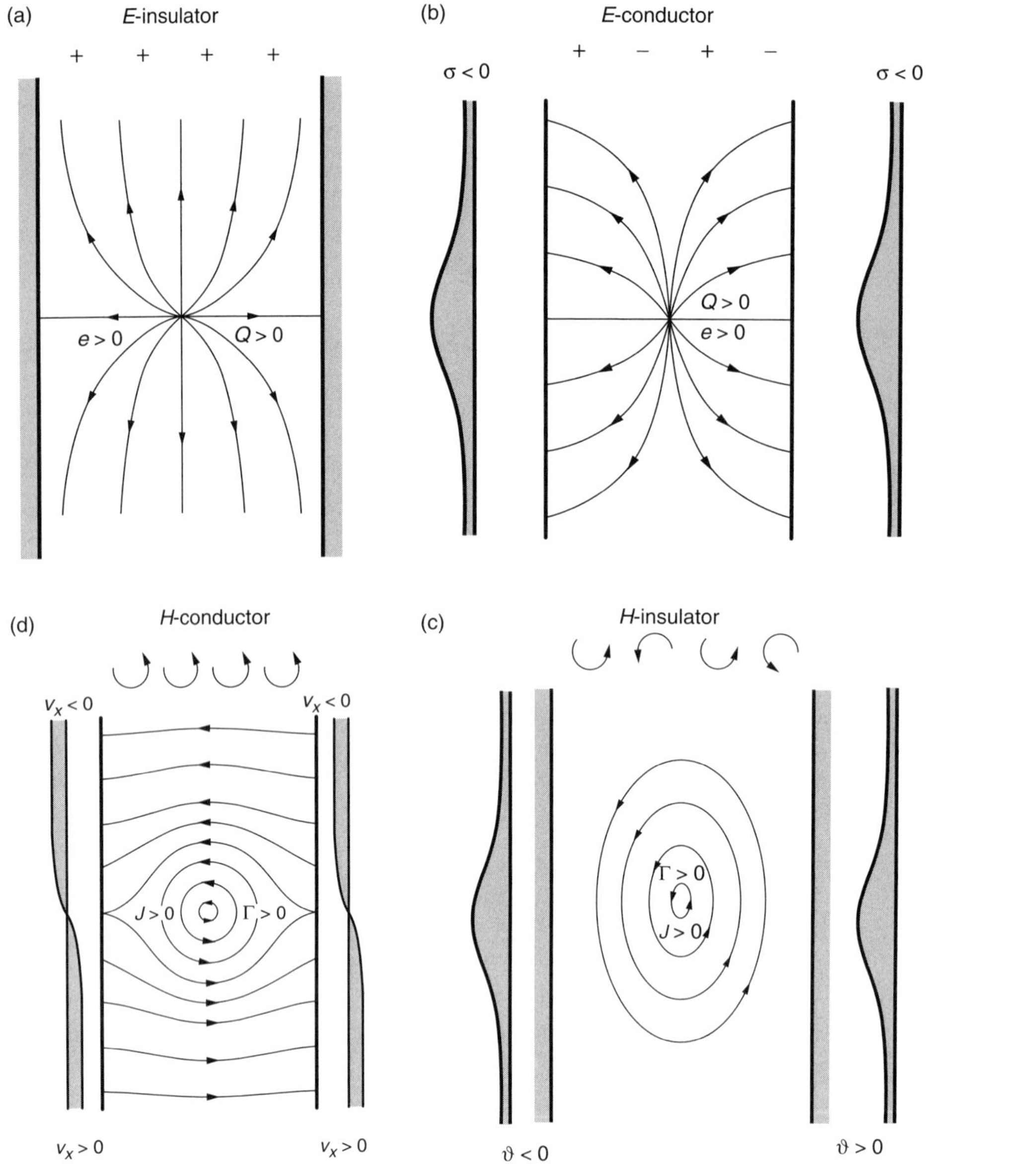

FIGURE 36.14
Analogy (a) between the potential flow (electrostatic field) due to a line source (positive electric charge) at equal distance from two parallel (insulating) walls with identical images; the flow velocity (electric field) is enhanced (reduced) parallel (orthogonal) the walls. Analogy (b) between the potential flow (electrostatic field) of a row of equally distant aligned alternating sources and sinks (positive and negative electric charges); the normal electric field corresponds to a distribution of electric charges on the conducting walls, which peaks closest to the original line-charge and decays toward infinity. Analogy (c) between the potential flow (magnetostatic field) of a counterclockwise vortex (electric current) at equal distance from parallel rigid (conducting) walls, with alternating images; the electric current (charge) distribution of the conducting wall is similar for the magneto (electro) static field (b) [(c)]. Analogy (d) between the potential flow (magnetostatic field) of a row of equally spaced identical counterclockwise vortices (electric currents) leading to the Kelvin cat's eye pattern

36.6.4 Physical and Mathematical Derivation of Series of Fractions and Infinite Products

The same problem, viz. a source/sink between parallel walls has been solved by two distinct methods: (i) infinite images (Subsection 36.6.1; Figure 36.13a) on the walls; (ii) conformal mapping (Subsection 36.6.2; Figure 36.13b,c) of the plane to a strip. Both methods (i) and (ii) must lead to the same result for the complex potential (36.132) ≡ (36.137a), [conjugate velocity (36.133) ≡ (36.137b)]. Thus is obtained through a physical example the infinite product (36.141) [series of fractions (36.142)]:

$$\sin\left(\frac{\pi z}{b}\right) = \frac{\pi z}{b}\prod_{n=1}^{\infty}\left(1-\frac{z^2}{n^2 b^2}\right), \tag{36.141}$$

$$\cot\left(\frac{\pi z}{b}\right) = \frac{b}{\pi z} + \frac{2bz}{\pi}\prod_{n=1}^{\infty}\frac{1}{z^2-n^2b^2}, \tag{36.142}$$

for two circular functions with period b, viz. the sine (36.141) [cotangent (36.142)]. When comparing (36.132) with (36.137a) was added a constant $(Q/2\pi)\log(\pi/b)$, which does not change the flow velocity (36.133) ≡ (36.137b) ≡ (36.142). These coincide for $b=\pi$ with those (27.67) ≡ (36.141) [(27.63b) ≡ (36.142)] obtained using a mathematical method, that is, viz. subtracting from the meromorphic function (36.142) the principal parts at all the poles (Subsection 27.9.2). The change of variable $z \to \pi z/b \equiv \xi$ changes the period $z/2\pi \to \xi/2b$ of circular functions from 2π to $2b$, and leads from (27.67) [(27.63b)] to (36.141) [(36.142)]. Thus the mathematical and physical methods are equivalent. The first term on the r.h.s. of (36.141) [(36.142)] implies that $\sin\xi \sim \xi\,(\cot\xi \sim 1/\xi)$ as $\xi \equiv \pi z/b \to 0$; this implies that $z=0$ is a simple zero (pole) as well as all congruent points $z=2nb$ with n integer.

36.7 Confined Vortex and Single Vortex Row

If the source/sink in Figure 36.13a is replaced by a vortex the images are alternating instead of identical (Subsection 36.7.1); the electric (magneto) static analogues of a source/sink (vortex) between parallel walls are insulators (condensers) with two parallel infinite plates

Caption for Figure 36.14 Continued
(insulating walls). The velocity at the mid position is: (i) horizontal; (ii) changes sign at the closest point $y=0$; (iii) tends to constant opposite values $\pm U$ at infinity $y \to \pm\infty$. This corresponds to the "vortex sheet" (Figure 14.9), that is, a discontinuity of tangential velocity, "smoothed-out" as a continuous velocity profile. There is an electric charge (current) distribution in the case of normal electric (b) [tangential magnetic (c)] field; there are stagnation points at the closest points on the walls for the row of identical sources (a) [vortices (d)]. The streamlines of the source (vortex) are symmetrical (a) [(c)] for equal distance from the walls and become unsymmetrical otherwise [Figure 40.6a(b)].

(Subsection 36.7.2). The images of a vortex on two parallel walls form a vortex row (Subsection 36.7.3) and if equally spaced move at a constant velocity parallel to the walls (36.7.4).

36.7.1 Alternating Images of Vortex between Walls

Consider a vortex (12.32a; 12.31b) in Figure 36.13a at equal distance from the walls, $\operatorname{Re}(z) = \pm b/2$; its images at $\pm nb$ are alternating leading to vortices of circulation $(-)^n \Gamma$, and adding to a total complex potential (36.143) [conjugate velocity (36.144)]:

$$\begin{aligned} f(z) &= -\frac{i\Gamma}{2\pi}\left\{\log z + \sum_{n=1}^{\infty}(-)^n \log\left[\left(1-\frac{z}{nb}\right)\left(1+\frac{z}{nb}\right)\right]\right\} \\ &= -\frac{i\Gamma}{2\pi}\log\left[z\prod_{n=1}^{\infty}(-)^n\left(1-\frac{z^2}{n^2b^2}\right)\right], \end{aligned} \tag{36.143}$$

$$\begin{aligned} \frac{df}{dz} &= -\frac{i\Gamma}{2\pi}\left[\frac{1}{z}+\sum_{n=1}^{\infty}(-)^n\left(\frac{1}{z-nb}+\frac{1}{z+nb}\right)\right] \\ &= -\frac{i\Gamma}{2\pi}\left[\frac{1}{z}+2z\sum_{n=1}^{\infty}\frac{(-)^n}{z^2-n^2b^2}\right], \end{aligned} \tag{36.144}$$

that involve an infinite product (series of fractions). The infinite product (36.143) [series of fractions (36.144)] for the complex potential (conjugate velocity) of a vortex at equal distance from two parallel walls, is similar to the infinite product (36.132) [series of fractions (36.133)] for a source, with two differences: (i) replacing $-i\Gamma$ by Q in the constant factor; (ii) inserting an alternating sign $(-)^n$ that preserves the absolute convergence of the general term $O\left(n^{-2}\right)$.

In the present case the period is $2b$, and the function:

$$z \equiv x + iy: \quad \zeta = \exp\left(\frac{i\pi z}{b}\right) = \exp\left(-\frac{\pi y}{b}\right)\exp\left(\frac{i\pi x}{b}\right), \tag{36.145a,b}$$

maps the ζ-plane into a vertical strip of width $2b$ in the z-plane, for example, $-b/2 < \operatorname{Re}(\zeta) < +b/2$. The pair of vortices of circulation $\pm\Gamma$ at $z = 0, b$, are repeated indefinitely as images; the original and images all correspond to the two points $\zeta = \pm 1$ of the complex potential:

$$\frac{2\pi i f}{Q} = \log\left(\frac{\zeta-1}{\zeta+1}\right) = \log\left(\frac{e^{i\pi z/2b}-e^{-i\pi z/2b}}{e^{i\pi z/2b}+e^{-i\pi z/2b}}\right) = \log\left[i\tan\left(\frac{\pi z}{2b}\right)\right]; \tag{36.146}$$

the complex potential simplifies to (36.147a) apart from a constant that does not affect the conjugate velocity (36.147b):

$$f(z) = -\frac{i\Gamma}{2\pi}\log\left[\tan\left(\frac{\pi z}{2b}\right)\right], \quad \frac{df}{dz} = -\frac{i\Gamma}{2b}\csc\left(\frac{\pi z}{b}\right). \tag{36.147a,b}$$

The simplification:

$$\begin{aligned} -\frac{i\Gamma}{2\pi}d\frac{\{\log\left[\tan\left(\pi z/2b\right)\right]\}}{dz} &= -\frac{i\Gamma}{2\pi}\frac{\pi}{2b}\cot\left(\frac{\pi z}{2b}\right)\sec^2\left(\frac{\pi z}{2b}\right) \\ &= -\frac{i\Gamma/\pi}{4\cos\left(\pi z/2b\right)\sin\left(\pi z/2b\right)} = -\frac{i\Gamma}{2b}\csc\left(\frac{\pi z}{b}\right), \end{aligned} \tag{36.148}$$

was used in the passage from (36.147a) to (36.147b).

Comparing (36.144)≡(36.147b) leads to series of fractions (36.149) for the circular cosecant:

$$\csc\left(\frac{\pi z}{b}\right) = \frac{b}{\pi z} + \frac{2\pi z}{b}\sum_{n=1}^{\infty}\frac{(-)^n}{z^2 - n^2 b^2}; \tag{36.149}$$

this can be simplified:

$$b = \pi: \quad \csc z = \frac{1}{z} + 2z\sum_{n=1}^{\infty}\frac{(-)^n}{z^2 - n^2\pi^2}, \tag{36.150a,b}$$

confirming that it has poles at $z = n\pi$ with residues $(-)^n$, in agreement with (19.46a,b). Comparing (36.143) ≡ (36.147a) leads to a logarithmic identity for the circular tangent:

$$\log\left[\tan\left(\frac{\pi z}{2b}\right)\right] = \log\left[\frac{\pi z}{b}\prod_{n=1}^{\infty}(-)^n\left(1 - \frac{z^2}{n^2 b^2}\right)\right], \tag{36.151}$$

that is equivalent to:

$$b = \pi/2: \quad \log(\tan z) = \log\left[2z\prod_{n=1}^{\infty}(-)^n\left(1 - \frac{4z^2}{n^2\pi^2}\right)\right]. \tag{36.152a,b}$$

An infinite product for the function tan z cannot be obtained by removing log from both sides of (36.152b) because: (i) the function tan z has zeros at $z_n^+ = n\pi$ and poles at $z_n^- = n\pi + \pi/2$; (ii) both $z_m^\pm = m\pi/2$ the zeros (poles) for $m = 2n$ even ($m = 2n + 1 odd$) are singularities when log is applied on the l.h.s. of (36.152a); (iii) the same points $z_m^\pm = m\pi/2$ are zeros of the product in square brackets on the l.h.s. of (36.152b); (iv) thus the r.h.s. of (36.152b) has logarithmic singularities at the same points as the l.h.s., and the equation can be valid outside these points; (v) removing the log on both sides, then tan z on the l.h.s. has poles at $z_n^- = n\pi + \pi/2$ whereas r.h.s. has zeros, so the relation cannot be valid. This difficulty did not arise in the passage from (36.132)≡(36.137a) to (36.141) or (29.67), because there are zeros at $z_n^+ = n\pi$ on both sides, and the sine has no poles, only an essential singularity at infinity. The series of fractions for the circular cotangent (27.63b) [cosecant (36.152b)] differ only on signs, because all the poles are simple and at the same locations $z_n = n\pi$, and only the residues $1\,[(-)^n]$ differ.

The complex potential (36.147a) and conjugate velocity (36.147b) have period 2b, that is, are unchanged by the substitution $z \to z + 2b$ that transforms the sequence of opposite vortices in Figure 36.13a into itself. Considering *the complex potential (36.147a) due to a vortex of circulation* Γ *at the origin between parallel walls at* $\mathrm{Re}\,(\mathrm{z}) = \pm\mathrm{b}/2$, *the complex conjugate velocity (36.147b) shows that the velocity along the real (36.153a,b) [imaginary (36.154a,b)] axis is vertical (horizontal):*

$$v_x(x,0) = 0, \quad \frac{\Gamma}{2b} \ge v_y(x,0) = \frac{\Gamma}{2b}\csc\left(\frac{\pi x}{b}\right) > \frac{\Gamma}{2\pi x}, \tag{36.153a,b}$$

$$\frac{\Gamma}{2b} \le -v_x(0,y) = \frac{\Gamma}{2b}\mathrm{csch}\left(\frac{\pi y}{b}\right) < \frac{\Gamma}{2\pi y}, \quad v_y(0,y) = 0. \tag{36.154a,b}$$

From the inequalities:

$$\sinh\xi > \xi > \sin\xi \Leftrightarrow \mathrm{csch}\,\xi < \frac{1}{\xi} < \csc\xi, \tag{36.155a,b}$$

for $\Gamma > 0$ it follows that *the velocity compared with a vortex in free space, is increased (reduced) parallel (orthogonal) to the walls, so that the stream lines change from circular to elongated parallel (compressed orthogonal) to the walls (Figure 36.14b) with two extremes: (i) at the closest point on the walls $z = \pm b/2$ the velocity is $\pm\Gamma/2b$, instead of $\pm\Gamma/\pi b$ in free space; (ii) far from the vortex there is no net flow $v_x \to 0$ as $y \to \infty$ in (36.154a), because there is no volume flux.*

36.7.2 Induced Charges/Currents on Parallel Wall Condenser

Figure 36.14a represents not only the potential (36.132) ≡ (36.137a) [conjugate velocity (36.133) ≡ (36.137b)] due to a source/sink of volume flux Q at equal distance from infinite parallel rigid walls $\operatorname{Re}(z) = \pm b/2$, but also replacing Q by $-e/\varepsilon$ $(+e/\varepsilon)$, the electrostatic potential (electric field) of a charge, e, in a dielectric of permeability, ε, between insulating walls. The case of conducting, walls corresponds Figure 36.14b to a vortex (36.147a) replacing Γ by e/ε, and exchanging field lines and equipotentials $f \to -if$, that is, the substitution $-\Gamma$ to e/ε:

$$f(z) = -\frac{e}{2\pi\varepsilon}\log\left\{\tan\left(\frac{\pi z}{2b}\right)\right\}, \quad E^*(z) = -\frac{\mathrm{d}f}{\mathrm{d}z} = \frac{e}{2\varepsilon b}\csc\left(\frac{\pi z}{b}\right); \qquad (36.156\text{a,b})$$

the electric field is orthogonal to the walls:

$$z = \pm\frac{b}{2} + iy: \quad E^*(z) = \frac{e}{2\varepsilon b}\csc\left(\frac{i\pi y}{b} \pm \frac{\pi}{2}\right) = \pm\frac{e}{2\varepsilon b}\sec\left(\frac{i\pi y}{b}\right) = \pm\frac{e}{2\varepsilon b}\operatorname{sech}\left(\frac{\pi y}{b}\right); \qquad (36.157\text{a,b})$$

it specifies the induced charges:

$$\sigma(y) = \mp\varepsilon E_x\left(\pm\frac{b}{2}, y\right) = -\frac{e}{2b}\operatorname{sech}\left(\frac{\pi y}{b}\right), \quad E_y\left(\pm\frac{b}{2}, y\right) = 0, \qquad (36.158\text{a,b})$$

that are of sign opposite to of that of the line charge, have maximum value in modulus $|e|/2b$ at the closest point, and decay to zero far away since $\sigma(y) \to 0$ as $y \to \infty$; the total charge on each wall equals the line-charge with opposite sign:

$$\xi = \frac{\pi y}{b}: \quad \int_{-\infty}^{+\infty} \sigma(y)\,\mathrm{d}y = -\frac{e}{2\pi}\int_{-\infty}^{+\infty} \operatorname{sech}\xi\,\mathrm{d}\xi = -\frac{e}{\pi}\left[\operatorname{arc\,tan}\left(\mathrm{e}^{\xi}\right)\right]_{-\infty}^{+\infty} = -e. \qquad (36.159\text{a,b})$$

In (36.159b) was used the primitive inverse to the derivation:

$$\frac{1}{2}\int \operatorname{sech}\xi\,\mathrm{d}\xi = \int \left(\mathrm{e}^{\xi} + \mathrm{e}^{-\xi}\right)^{-1}\mathrm{d}\xi = \int \frac{\mathrm{e}^{\xi}}{1 + \mathrm{e}^{2\xi}}\mathrm{d}\xi = \operatorname{arc\,tan}\left(\mathrm{e}^{\xi}\right). \qquad (36.160)$$

The electrostatic field (36.156a,b) due to a line charge e in a dielectric of permeability ε between parallel conductors, induces the charges (36.158a) and leads to a fan of field lines Figure 36.14c orthogonal to the streamlines of the vortex in Figure 36.14b. The vertical, that is, tangential velocity along the walls in the latter case is given by (36.157a) replacing e by the circulation Γ. The latter can be replaced by Q for in the potential flow due to alternating sources/sinks $(-)^n Q$ at $z = \pm nb$ whose conjugate velocity is given by (36.156b) with e/ε replaced by Q.

Figure 36.14c represents not only the potential flow (36.143–36.144) ≡ (36.147a,b) due to a vortex of circulation Γ at equal distance from two infinite parallel rigid walls, but also replacing Γ by $-J/c$, where c is the speed of light in vacuum, the magnetostatic field due to a current between parallel conductors, with surface electric currents:

$$H_x\left(\pm\frac{b}{2},y\right)=0,\quad \vartheta(y)=H_y\left(\pm\frac{b}{2},y\right)=-\frac{J}{2b}\operatorname{sech}\left(\frac{\pi y}{b}\right),\tag{36.161a,b}$$

similar to (36.157a). The case of the magnetostatic field due to a current between insulators corresponds via the relation $J/c\leftrightarrow-\Gamma$ to the potential flow due to a row of identical vortices:

$$f(z)=-\frac{i\Gamma}{2\pi}\log\left\{\sin\left(\frac{\pi z}{b}\right)\right\},\quad \frac{\mathrm{d}f}{\mathrm{d}z}=-\frac{i\Gamma}{2b}\cot\left(\frac{\pi z}{b}\right),\tag{36.162a,b}$$

that is obtained from (36.137a,b) replacing Q by $-i\Gamma$, and corresponds to exchanging equipotentials and streamlines in Figure (36.14a) leading to Figure 36.14d. The potential flow due to a vortex at unequal distance from parallel walls is considered in Example 40.12.

36.7.3 Cat's Eye Pattern (Kelvin) and Hyperbolic Tangent Profile

The cat's eye pattern (Kelvin) in Figure 36.14d corresponds to the complex potential (36.162a) and conjugate velocity (36.162b); using in the latter the relations (36.100a,b) and their inverses:

$$\begin{aligned}\xi=i\eta:\quad &\cos(i\xi)=\cosh(i^2\xi)=\cosh\xi,\\ &\sin(i\xi)=-i\sinh(i^2\xi)=i\sinh\xi,\end{aligned}\tag{36.163a–c}$$

leads to:

$$\begin{aligned}\cot\left(\frac{\pi z}{b}\right)&=\cot\left[\frac{\pi(x+iy)}{b}\right]=\frac{\cos[\pi(x+iy)/b]}{\sin[\pi(x+iy)/b]}\\ &=\frac{\cos(\pi x/b)\cosh(\pi y/b)-i\sin(\pi x/b)\sinh(\pi y/b)}{\sin(\pi x/b)\cosh(\pi y/b)+i\cos(\pi x/b)\sinh(\pi y/b)}\\ &=\frac{\begin{array}{c}[\cos(\pi x/b)\cosh(\pi y/b)-i\sin(\pi x/b)\sinh(\pi y/b)]\\ \times[\sin(\pi x/b)\cosh(\pi y/b)-i\cos(\pi x/b)\sinh(\pi y/b)]\end{array}}{\sin^2(\pi x/b)\cosh^2(\pi y/b)+\cos^2(\pi x/b)\sinh^2(\pi y/b)}\\ &=\frac{\cos(\pi x/b)\sin(\pi x/b)-i\cosh(\pi y/b)\sinh(\pi y/b)}{\cosh^2(\pi y/b)-\cos^2(\pi x/b)}\\ &=\frac{\sin(2\pi x/b)-i\sinh(2\pi y/b)}{\cosh(2\pi y/b)-\cos(2\pi x/b)};\end{aligned}\tag{36.163d}$$

thus *the velocity field due to an infinite row of vortices with the same circulation Γ with equal spacing b along (Figure 36.14d) along the real axis is given by:*

$$\{v_x(x,y),v_y(x,y)\}=\frac{\Gamma}{2b}\left[\cosh\left(\frac{2\pi y}{b}\right)-\cos\left(\frac{2\pi x}{b}\right)\right]^{-1}\left\{-\sinh\left(\frac{2\pi y}{b}\right),\ \sin\left(\frac{2\pi x}{b}\right)\right\};\tag{36.164a,b}$$

this tends at large distance to equal and opposite horizontal velocities (36.165a,b):

$$\begin{aligned}&v_x(x,\pm\infty)=\mp\frac{\Gamma}{2b},\quad v_y(x,\pm\infty)=0;\\ &\Delta v_x\equiv v_x(x,+\infty)-v_x(x,-\infty)=-\frac{\Gamma}{b},\end{aligned}\tag{36.165a–c}$$

so the jump in velocity (36.165c) is equivalent (18.17) to a vortex sheet of strength $-\Gamma/b$. The velocity at the walls or midline between vortices is horizontal (36.166a):

$$v_y\left(\pm\frac{b}{2},y\right)=0,\quad v_x\left(\pm\frac{b}{2},y\right)=-\frac{\Gamma}{2b}\tanh\left(\frac{\pi y}{b}\right),\tag{36.166a,b}$$

and leads to a hyperbolic tangent velocity profile (36.166b) for the ***unidirection shear flow****, that is, whose variation (in the y-direction) is transverse to the velocity (in the x-direction).* The result (36.165a,b) can be obtained: (i) from (36.164a,b):

$$\begin{aligned}\frac{2b}{\Gamma}v_x\left(\pm\frac{b}{2},y\right)&=-\frac{\sinh(2\pi y/b)}{\cosh(2\pi y/b)+1}=-\frac{2\sinh(\pi y/b)\cosh(\pi y/b)}{2\cosh^2(\pi y/b)}\\&=\tanh(\pi y/b),\end{aligned}\tag{36.166c}$$

(ii) directly from (36.162b):

$$\begin{aligned}v^*\left(\pm\frac{b}{2}+iy\right)&=-\frac{i\Gamma}{2b}\cot\left[\frac{\pi}{b}\left(\pm\frac{b}{2}+iy\right)\right]=-\frac{i\Gamma}{2b}\cot\left(\pm\frac{\pi}{2}+\frac{i\pi y}{b}\right)\\&=\frac{i\Gamma}{2b}\tan\left(\frac{i\pi y}{b}\right)=-\frac{\Gamma}{2b}\tanh\left(\frac{\pi y}{b}\right)=v_x\left(\pm\frac{b}{2},y\right).\end{aligned}\tag{36.166d}$$

The latter also specifies the velocity on a vertical line through the vortex:

$$v^*(iy)=-\frac{i\Gamma}{2b}\cot\left(\frac{i\pi y}{b}\right)=-\frac{\Gamma}{2b}\coth\left(\frac{\pi y}{b}\right)=-v_x(0,y),\tag{36.167a}$$

that is horizontal as at the walls (36.165a,b). Using (36.140a,b) in (36.167a) it follows that

$$|v_x(0,y)|=\left|\frac{\Gamma}{2b}\coth\left(\frac{\pi y}{b}\right)\right|\geq\frac{|\Gamma|}{2\pi y},\tag{36.167b}$$

the horizontal velocity on vertical axis is larger for a vortex at equal distance from parallel walls than it would be in free space at the same distance.

36.7.4 Shear Layer and Static Equilibrium

A flow with velocity U_+ (U_-) at $y\to+\infty$ ($y\to-\infty$) can be represented by:

$$\frac{\mathrm{d}f}{\mathrm{d}z}=\frac{U_++U_-}{2}-\frac{i}{2}(U_--U_+)\cot\left(\frac{\pi z}{b}\right),\tag{36.168a}$$

$$f(z)=\frac{(U_++U_-)z}{2}-i\frac{b}{\pi}(U_--U_+)\log\left[\sin\left(\frac{\pi z}{b}\right)\right],\tag{36.168b}$$

the superposition of a uniform flow at mean velocity (36.169a) with a ***discretized vortex sheet*** *of strength (36.169b):*

$$2\bar{U}=U_++U_-,\quad U_--U_+=\frac{\Gamma}{b},\quad\frac{\mathrm{d}f}{\mathrm{d}z}=\begin{cases}\bar{U}-\Gamma/(2b)=U_+ & \text{as}\quad y\to+\infty\\ \bar{U}+\Gamma/(2b)=U_- & \text{as}\quad y\to-\infty,\end{cases}\tag{36.169a–c}$$

as indicated in (36.169c). Considering the row of vortices in Figure 36.14d, the velocities induced on each vortex n by the next pairs, $n\pm1, n\pm2,\ldots$ all cancel, so the vortices are in equilibrium. This can be confirmed by calculating the induced velocity at one vortex, for example, that at $z = 0$, by subtracting its velocity field from (36.162b) and taking the limit, $z \to 0$, viz.:

$$v_0(0) \equiv \lim_{z\to 0}\left\{-\frac{i\Gamma}{2b}\cot\left(\frac{\pi z}{b}\right)+\frac{i\Gamma}{2\pi z}\right\} = \lim_{z\to 0} -\frac{i\Gamma}{2b}\left\{\cot\left(\frac{\pi z}{b}\right)-\frac{b}{\pi z}\right\} = 0. \qquad (36.170)$$

The result (36.170) can be proved by two methods: (i) either by using (19.39a,b):

$$\eta \equiv \frac{\pi z}{b}: \quad \cot(\eta) \equiv \frac{\cos\eta}{\sin\eta} = \frac{1+O(\eta^2)}{\eta} = \frac{1}{\eta}+O(\eta); \qquad (36.171\text{a,b})$$

(ii) using (36.142) in (36.170):

$$v_0(0) = \lim_{z\to\infty} -\frac{i\Gamma}{2b}\left[\cot\left(\frac{\pi z}{b}\right)-\frac{b}{\pi z}\right] = \lim_{z\to\infty}\left\{-\frac{i\Gamma z}{\pi}\sum_{n=1}^{\infty}\frac{1}{z^2-n^2b^2}\right\} = 0. \qquad (36.171\text{c})$$

In a similar way it can be proved from (36.137b) and (36.133) that the row of sources in Figure 36.14a is an equilibrium at rest.

Thus *the infinite row of alternating or identical equally spaced vortices or sources/sinks is in equilibrium at rest that is unstable* because: (i) displacement of the source/sink outside the real axis will lead to a vertical velocity induced by the pairs of images at the same distance, with decreasing modulus and same or alternating sign, so the total vertical velocity has the sign of the first pair, and hence there will be no return to the equilibrium position; (ii) a displacement of the vortex along the real axis will keep the induced velocities vertical but they become larger for the images on the side of the displacement and the net vertical velocity excludes a return to equilibrium. The stability theory (Chapter 2) can be used to prove that a row of equally spaced identical or alternating sources/sinks or vortices is unstable relative to sinusoidal disturbances with small amplitude. The velocity for identical vortices (36.162b) is vertical (horizontal) on the real (imaginary) axis, and tends to the constant value $U = \Gamma/b$ at large distances. The cat's eye pattern has (Figure 36.14d) a bifurcated field line $\Psi = 0$ passing through the points $z = \pm b/2$, such that: (i) within it the flow is circulating: (ii) outside the flow has opposite directions above and below. The cat's eye pattern may originate as follows: (i) start with two flows with opposite velocities $\mp U$ with $U \equiv \Gamma/2b$ for $y >< 0$, that is, an unstable vortex sheet at $y = 0$; (ii) if the latter breaks up into a sequence of identical vortices it forms a cat's eye pattern, with the hyperbolic tangent velocity profile (36.166b) on the vertical lines through the two stagnation points $\eta = \pm b/2$. The potential flow due to a monopole (dipole) or spiral source combining a source/sink and a vortex (parallel and orthogonal moments), at equal distance from two parallel walls is considered in Example 40.13 (40.14).

36.8 Source/Sink in a Well, on a Wall or at a Corner

The potential fields of monopoles at the origin $\zeta = 0$ between parallel walls $\text{Re}(\zeta) = \pm b/2$ have been considered in the four cases of flow sources/sinks or electric charges with identical (Figure 36.14a) [alternating (Figure 36.14b)] images, and vortices or electric currents with identical (Figure 36.14d) [alternating (Figure 36.14c)] images. The sources/sinks can be displaced to the wall, corresponding (Subsection 36.8.2) to the corner of a rectangular well (Figure 36.15a,b); the source/sink (vortex) can also be placed in a rectangular well [Subsection 36.8.3 (Section 36.9)] away from the walls (Figure 36.16a–d)]. Both problems can be solved by two methods (Subsection 36.8.1): (i) either performing a translation in the preceding solution with source/sink at equal distance from the walls; (ii) or performing the conformal mapping again with displaced initial points.

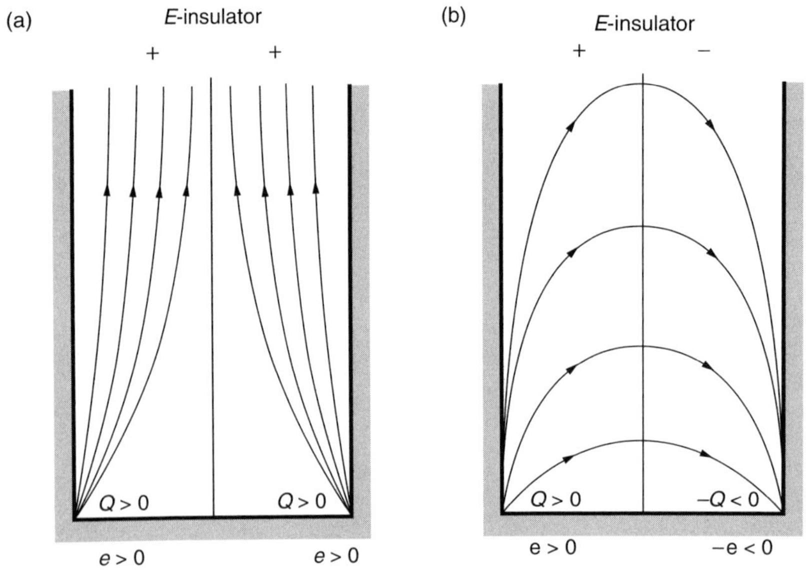

FIGURE 36.15
A translation by half the distance between walls places the sources (electric charges) in Figure 36.14a (b) at the walls in Figure 36.15a(b). The symmetry relative to the real axis allows reversing the Schwartz reflection principle (Figure 31.4), "cutting-off" the lower half by a segment of the real axis to which the velocity (electric field) is tangent. This leads to the potential flow (electrostatic field) of equal sources (a) [alternating sources and sinks (b)] or positive (alternating positive and negative) electric charges at the corners of a rectangular well. There is a stagnation point at the middle of the lower wall for the potential flow in both cases (a,b). In both cases there are no electric charges induced on the insulating walls by the tangential electric filed. The identical (alternating) sources/electric charges lead to a net (a) [no net (b)] flow/electric field asymptotically far into the well; the case of identical sources(sinks) of flow rate $Q > 0\,(Q < 0)$ only $Q/2$ goes into the well of with b, so the uniform velocity in the well at infinity is $U_\infty = Q/2b$.

36.8.1 Translation of Source/Sink or Critical Points of Conformal Mapping

In the former two cases of Figure 36.14a (36.14b) a translation by $b/2$ leads to two equal (opposite) sources at $z = \pm b/2$ on the two parallel infinite walls; closing on the real segment $(-b/2, +b/2)$ that is a field line, results in the **rectangular well** in Figure 36.15a (36.15b). Concerning the first case of *two identical sources/sinks of volume flux* Q *at the corners of a rectangular well* $\operatorname{Im}(z) > 0, |\operatorname{Re}(z)| < b/2$ *in Figure 36.15a the complex potential and conjugate velocity (36.172b):*

$$f(z) = \frac{Q}{2\pi}\log\left\{\cos\left(\frac{\pi z}{b}\right)\right\}, \qquad \frac{df}{dz} = -\frac{Q}{2b}\tan\left(\frac{\pi z}{b}\right), \tag{36.172a,b}$$

may be obtained by two methods: (a) performing a translation $z \to z + b/2$ in (36.137a):

$$f\left(z + \frac{b}{2}\right) = \frac{Q}{2\pi}\log\left\{\sin\left(\frac{\pi z}{b} + \frac{\pi}{2}\right)\right\} = \frac{Q}{2\pi}\log\left\{\cos\left(\frac{\pi z}{b}\right)\right\}; \tag{36.173}$$

(b) mapping into the strip $-b/2 < \operatorname{Re}(z) < b/2$ the complex z-plane through the conformal mapping (36.136c), with sources at $z = \pm b/2$ corresponding to $\zeta = \pm 1$:

$$\begin{aligned} f(z) &= \frac{Q/2}{2\pi} \log\left[(\zeta+1)(\zeta-1)\right] = \frac{Q}{4\pi}\log(\zeta^2-1) \\ &= \frac{Q}{4\pi}\log\left[\sin^2\left(\frac{\pi z}{b}\right)-1\right] = \frac{Q}{2\pi}\log\left[\cos\left(\frac{\pi z}{b}\right)\right] + \frac{iQ}{4}, \end{aligned} \tag{36.174}$$

the volume flux being $Q/2$ at each comer.

In the case of *opposite sources/sinks of volume flux* $\pm Q$ *at the corners* $z = \pm b/2$ *of a rectangular well (Figure 36.15b) the complex potential (36.175a) and conjugate velocity (36.175b):*

$$f(z) = \frac{Q}{2\pi}\log\left\{\tan\left[\left(\frac{\pi z}{2b}+\frac{\pi}{4}\right)\right]\right\}, \quad \frac{\mathrm{d}f}{\mathrm{d}z} = \frac{Q}{2b}\sec\left(\frac{\pi z}{b}\right), \tag{36.175a,b}$$

can be obtained: (a) by a translation through $b/2$ in (36.156a), with $-e/\varepsilon$ replaced by Q:

$$f\left(z+\frac{b}{2}\right) = \frac{Q}{2\pi}\log\left[\tan\left(\frac{\pi z}{2b}+\frac{\pi}{4}\right)\right] = \frac{Q}{2\pi}\log\left[\tan\left(\frac{\pi z}{b}\right)+\sec\left(\frac{\pi z}{b}\right)\right]; \tag{36.176}$$

(b) using a conformal transformation (36.136c) with a monopole $\pm Q/2$ at $z = \mp b/2$, corresponding to $\zeta = \mp 1$, viz.:

$$j = \frac{Q/2}{2\pi}\log\left(\frac{\zeta+1}{\zeta+1}\right) = \frac{Q}{4\pi}\log\left[\frac{\sin(\pi z/b)+1}{\sin(\pi z/b)-1}\right] = \frac{Q}{2\pi}\log\left[\tan\left(\frac{\pi z}{b}\right)+\sec\left(\frac{\pi z}{b}\right)\right] + \frac{iQ}{2}; \tag{36.177}$$

In (36.176) [(36.177)] were used (36.178b) [(36.178d)] in terms of the variable (36.178a) [(36.178c)]:

$$\begin{aligned} \xi = \frac{\pi z}{2b}: \quad \tan\left(\xi+\frac{\pi}{4}\right) &= \frac{\sin(\xi+\pi/4)}{\cos(\xi+\pi/4)} = \frac{\cos\xi+\sin\xi}{\cos\xi-\sin\xi} = \frac{(\cos\xi+\sin\xi)^2}{\cos^2\xi-\sin^2\xi} \\ &= \frac{1+\sin(2\xi)}{\cos(2\xi)} = \tan(2\xi)+\sec(2\xi), \end{aligned} \tag{36.178a,b}$$

$$\begin{aligned} \eta = \frac{\pi z}{2b} = 2\xi: \quad \log\left(\frac{\sin\eta+1}{\sin\eta-1}\right) &= \log\left[\frac{(\sin\eta+1)^2}{-\cos^2\eta}\right] = 2\log\left(\frac{\sin\eta+1}{\cos\eta}\right)+i\pi \\ &= 2\log(\tan\eta+\sec\eta)+i\pi; \end{aligned} \tag{36.178c}$$

the constant in (36.177) relative to (36.176) is a translation that does not affect the conjugate velocity (36.175b). The latter follows from (36.175a):

$$\frac{\mathrm{d}f}{\mathrm{d}z} = \frac{Q}{2\pi}\frac{\pi}{2b}\frac{\sec^2(\xi+\pi/4)}{\tan(\xi+\pi/4)} = \frac{Q/b}{4\sin(\xi+\pi/4)\cos(\xi+\pi/4)} = \frac{Q}{2b}\frac{1}{\sin(2\xi+\pi/2)} = \frac{Q}{2b}\sec(2\xi), \tag{36.178d}$$

using the same variable (36.178a).

36.8.2 Identical/Opposite Sources/Sinks at the Corners of a Well

The electrostatic field, in a rectangular well $\text{Im}(z) > 0, |\text{Re}(z)| < b/2$ *filled with a dielectric of permeability* ε *with equal* e *(opposite* $\pm e$*) electric charges at the corners* $z = \mp b/2$*, is obtained by setting* $Q \to e/\varepsilon$ *in (36.172b) [(36.175b)]. Returning to the case of the potential flow with identical (opposite) sources the velocity (36.172b) [(36.175b)] is: (i) horizontal on the bottom wall:*

$$v_x(x,0) = \frac{Q}{2b}\left\{-\tan\left(\frac{\pi x}{b}\right), \sec\left(\frac{\pi x}{b}\right)\right\}, \quad v_y(x,0) = 0, \tag{36.179a,b}$$

with stagnation point $df/dz = 0$ *in (36.172b) [finite nonzero value* $df/dz = Q/2b$ *in (36.175b)] at the middle* $z = 0$*; (ii) on the midline the velocity tends to uniform (36.180a,c) [decays (36.180b,d)]:*

$$v_x(0,y) = \left\{0, \frac{Q}{4b}\text{sech}\left(\frac{\pi y}{b}\right)\right\}, \quad v_y(0,y) = \left\{\frac{Q}{2b}\tanh\left(\frac{\pi y}{b}\right), 0\right\}, \tag{36.180a–d}$$

and is vertical (horizontal); (iii) vertical at the walls and tending to uniform (decaying):

$$v_x\left(\pm\frac{b}{2}, y\right) = 0, \quad v_y\left(\pm\frac{b}{2}, y\right) = \frac{Q}{2b}\left\{\coth\left(\frac{\pi y}{b}\right), \mp\text{csch}\left(\frac{\pi y}{b}\right)\right\}, \tag{36.181a–c}$$

at infinity $y \to \infty$*, since* $df/dz \to -Q/2b$ *in (36.172b) [*$df/dz \to 0$ *in (36.175b)]. The source/sink in Figure 36.15a may be seen as lying: (i) at the corners of a rectangular well; (ii) by symmetry on the* $\text{Re}(z)$*-axis as lying on the walls of an infinite parallel sided duct. The flow rate is double in the case (ii) relative to (i).* In (36.181b) [(36.181c)] was used (36.172b)[(36.175b)] on the side walls (36.181d) [(36.181e)]:

$$\begin{aligned}\frac{2b}{Q}v^*\left(\pm\frac{b}{2} + iy\right) &= -\tan\left(\frac{i\pi y}{b} \pm \frac{\pi}{2}\right) = \cot\left(\frac{i\pi y}{b}\right)\\ &= -i\coth\left(\frac{\pi y}{b}\right) = -iv_y\left(\pm\frac{b}{2}, y\right),\end{aligned} \tag{36.181d}$$

$$\begin{aligned}\frac{2b}{Q}v^*\left(\pm\frac{b}{2} + iy\right) &= \sec\left(\frac{i\pi y}{b} \pm \frac{\pi}{2}\right) = \mp\csc\left(\frac{i\pi y}{b}\right)\\ &= \pm i\text{csch}\left(\frac{\pi y}{b}\right) = -iv_y\left(\pm\frac{b}{2}, y\right);\end{aligned} \tag{36.181e}$$

the direction of the vertical velocity at the walls in (36.181d) $\equiv$ (36.181b) [(36.181e) $\equiv$ (36.181c)] agrees with Figure 36.16a(b). A vortex or an electric current could not be placed in a corner or on a wall, but can be located in a well away from the walls (Figure 36.16b,d); this location will be considered first for source/sinks (Figure 36.16a,c) and electric charges.

36.8.3 Source on the Axis of a Rectangular Well

Instead of equal sources at the corners of a rectangular well $\text{Im}(z) > 0, -b/2 < \text{Re}(z) < b/2$ as in Figure 36.15; a single source on the axis at a distance, a, from the wall is considered in Figure 36.16a. The effect of the sidewalls is represented by (36.137a) with z replaced by $z - ia$, and the effect of the bottom wall by adding an identical image with $z + ia$, viz.:

$$\begin{aligned}f(z) &= \frac{Q}{2\pi}\left\{\log\left[\sin\left(\pi\frac{z - ia}{b}\right)\right] + \log\left[\sin\left(\pi\frac{z + ia}{b}\right)\right]\right\}\\ &= \frac{Q}{2\pi}\log\left[\sin\left(\frac{\pi z}{b} - \frac{i\pi a}{b}\right)\sin\left(\frac{\pi z}{b} + \frac{i\pi a}{b}\right)\right]\\ &= \frac{Q}{2\pi}\log\left\{\frac{1}{2}\left[\cos\left(\frac{i2\pi a}{b}\right) - \cos\left(\frac{2\pi z}{b}\right)\right]\right\};\end{aligned} \tag{36.182}$$

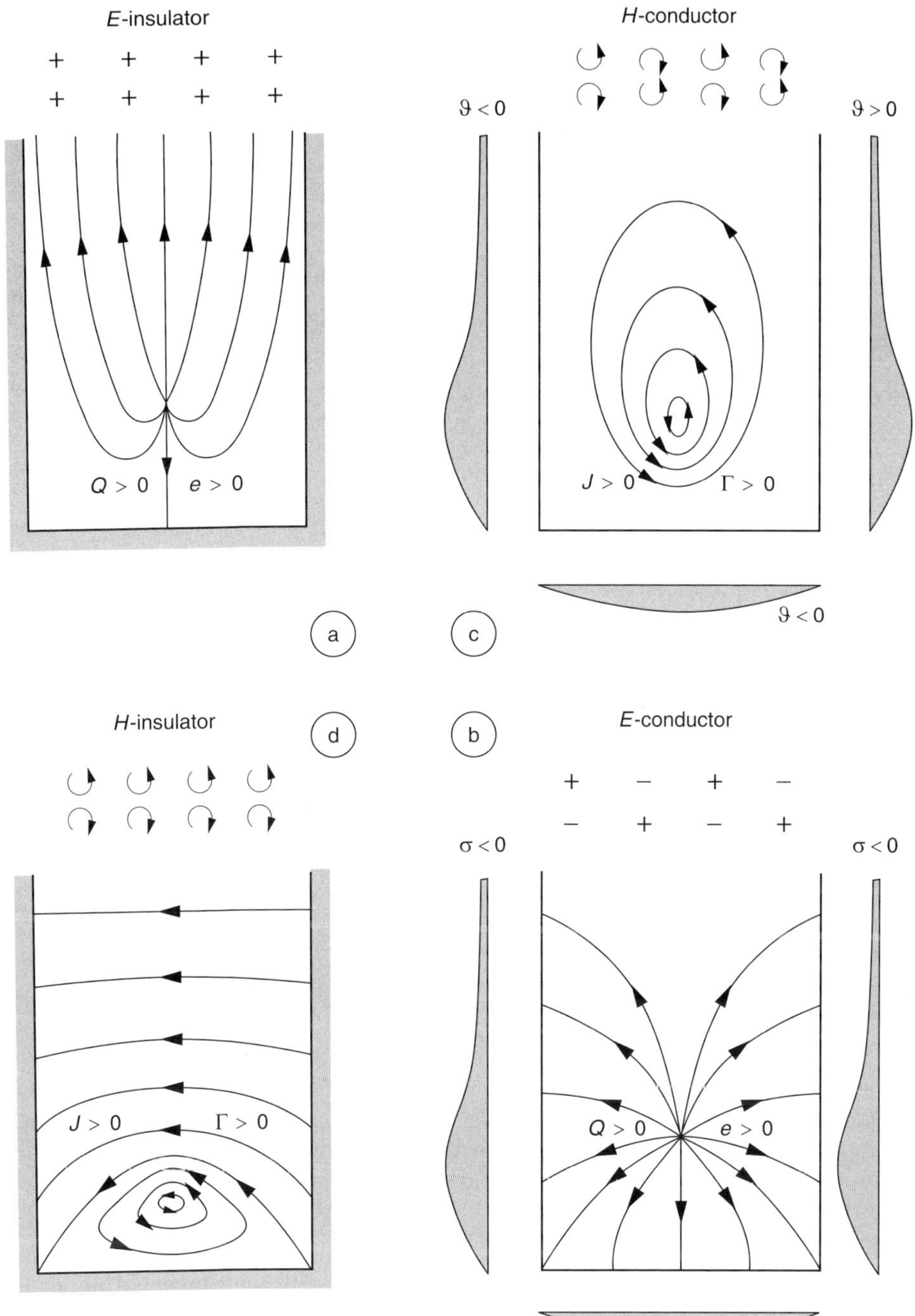

FIGURE 36.16

The case of a monopole in a rectangular well away from the walls or corners (Figure 36.16) is distinct from monopoles at the walls or corners (Figure 36.15) because it involves two sets of images: (i) the infinite images on the side walls leading to a row of identical or alternating monopoles (Figure 36.13a); (ii) the identical or opposite images on the bottom leading to a double parallel rows (as in Figure 36.17b but with zero stagger $d = 0$). This leads to similar analogies for the monopole between walls (in a well) in Figure 36.14a–d (36.16a–d); in the this can be simplified:

$$f(z) = \frac{Q}{2\pi} \log \left[\cosh\left(\frac{2\pi a}{b}\right) - \cos\left(\frac{2\pi z}{b}\right) \right], \tag{36.183a}$$

$$\frac{df}{dz} = \frac{Q}{b} \frac{\sin(2\pi z/b)}{\cosh(2\pi a/b) - \cos(2\pi z/b)}. \tag{36.183b}$$

An added constant was omitted in the potential (36.183a), that does not affect the conjugate velocity (36.183b). Thus all images are identical on the sidewalls and bottom wall and *(36.183a) specifies the complex potential and (36.183b) the conjugate velocity due to a source/sink of volume flux Q on the axis of a rectangular well of width b at a distance a from the bottom (Figure 36.16a).* The flow: (i) has one stagnation point $z = 0$ at the middle of the bottom wall:

$$v_x(0, x) = \frac{Q}{b} \frac{\sin(2\pi x/b)}{\cosh(2\pi a/b) - \cos(2\pi x/b)}, \qquad v_y(0, x) = 0; \tag{36.184a,b}$$

(ii) the velocity is vertical on the midline:

$$v_x(0, y) = 0, \quad v_y(0, y) = \frac{Q}{b} \frac{\sinh(2\pi y/b)}{\cosh(2\pi y/b) - \cosh(2\pi a/b)}, \tag{36.185a,b}$$

with opposite signs above and below the source; (iii) the velocity on the side walls is vertical and increases monotically from the bottom:

$$v_x\left(\pm\frac{b}{2}, y\right) = 0, \quad v_y\left(\pm\frac{b}{2}, y\right) = \frac{Q}{2b}\left[\tanh\left(\pi\frac{y+a}{b}\right) + \tanh\left(\pi\frac{y-a}{b}\right)\right]. \tag{36.186a,b}$$

Caption for Figure 36.16 Continued
latter case of the well it is possible to have different boundary conditions at the bottom and side walls, for example, one conducting and the other insulating (Figure 40.5) as for the corner with circular indentation (Figure 36.3b). Analogy (a) between the potential flow (electrostatic field) of a line source (line electric charge) at equal distance from the side walls of a rigid (insulating) rectangular well, with identical images on the sides and bottom; the flow velocity (electric field) is enhanced parallel to the side walls far from the bottom wall. Analogy (c) between the potential flow (magnetostatic field) of a counterclockwise line vortex (line electric current) at equal distance from the side walls of a rectangular well, with opposite/alternating images on the bottom/side wall(s). The tangential magnetic field corresponds to surface electric currents which vanish at: (i) the corners; (ii) infinity along the side walls. Analogy (d) between the potential flow (magnetostatic field) of a double row of counterclockwise line vortices (electric currents) with unstaggered aligned images, opposite/identical on the bottom/side wall(s); the flow (field) is uniform with the same direction far away for the double vortex row in Figure 36.16d, instead of having opposite direction for the vortex sheet for the single vortex row in Figure 36.14d. Analogy (b) between the potential flow (electrostatic field) of a double row of line source/sinks (line positive/negative electric charges) unstaggered and aligned with alternating images both on bottom and side walls. The electric field normal to the conducting walls corresponds to surface electric distributions which: (i) peak at point on the on the bottom and side walls closest to the original electric line charge; (ii) vanish at the corners and at infinity along the walls.

(iv) both vertical velocities (36.185b) and (36.186b) far from the bottom of the square well $y \to \infty$ tend to a limiting value Q/b, equal to the flow rate of the source divided by the width of the duct; this corresponds to the uniform flow $\mathrm{d}f/\mathrm{d}z \to Q/b$ at the far end as $z \to x + i\infty$ in (36.183b). The velocity on the side-walls is calculated: (i) for a source at equal distance from parallel infinite walls (36.137b) by:

$$\frac{2b}{Q} v^* \left(\pm \frac{b}{2} + iy \right) = \cot \left(\frac{i\pi y}{b} \pm \frac{\pi}{2} \right) = -\tan \left(\frac{i\pi y}{b} \right)$$
$$= -i \tanh \left(\frac{\pi y}{b} \right) = -i v_y \left(\pm \frac{b}{2}, y \right), \tag{36.186c}$$

Implying:

$$v_x \left(\pm \frac{b}{2}, y \right) = 0, \ v_y \left(\pm \frac{b}{2}, y \right) = \frac{Q}{2b} \tanh \left(\frac{\pi y}{b} \right); \tag{36.186d,e}$$

(ii) for a source at a distance a from the bottom of a square well (36.182) the velocity on the side walls (36.186a,b) is the sum of (36.186d,e) with y replaced by $y \pm a$.

36.9 Parallel and Staggered Double Vortex Street (von Karman, 1911; Lamb, 1932)

If the source in Figure 36.16a is replaced by a vortex, the streamlines of the latter become elongated, and instead of having two axes of symmetry between walls (Figure 36.14c), take an oval shape (Subsection 36.9.1) with only one axis of symmetry in the case of a well (Figure 36.16c). The corresponding electro (magneto) static fields (Subsection 36.9.2) have field lines coincident with or orthogonal to (Figure 36.16b) those of the vortex and cause on conducting walls analogous (Subsection 36.9.4) induced electric charge (surface electric current) distributions (Subsection 36.9.3). The vortex in a well corresponds to a double vortex row (Subsection 36.9.5) that is in convective equilibrium (Subsection 36.9.6) if parallel or with symmetric stagger.

36.9.1 Elongated, Nonsymmetric Vortex in a Rigid Well

The effect of the bottom wall on a vortex is represented by taking (36.147a) with an opposite image, that is, subtracting the same expression with z replaced by $z \mp ia$:

$$f(z) = -\frac{i\Gamma}{2\pi} \log \left\{ \frac{\tan [\pi (z - ia)/2b]}{\tan [\pi (z - ia)/2b]} \right\}, \tag{36.187a}$$

$$\frac{\mathrm{d}f}{\mathrm{d}z} = -\frac{i\Gamma}{2b} \left[\csc \left(\pi \frac{z - ia}{b} \right) - \csc \left(\pi \frac{z + ia}{b} \right) \right]. \tag{36.187b}$$

Thus the images are alternating on the sidewalls and opposite on the bottom wall, and *the complex potential (36.187a) and conjugate velocity (36.187b) due to a vortex of circulation* Γ *on the axis of a rectangular well of width* b *at a distance* a *from the bottom (Figure 36.16c),* show that: (i) there are stagnation points $\mathrm{d}f/\mathrm{d}z = 0$ at the corners (36.188a), since by (36.188b):

$$z = \pm \frac{b}{2}: \quad \csc \left(\pm \frac{\pi}{2} \pm \frac{i\pi a}{b} \right) = \pm \sec \left(\pm \frac{i\pi a}{b} \right) = \pm \mathrm{sech} \left(\frac{\pi a}{b} \right), \tag{36.188a,b}$$

and the two terms in (36.187b) cancel; (ii) the velocity on the axis of the well is horizontal:

$$v_x(0,y) = \frac{\Gamma}{2b}\left[\operatorname{csch}\left(\pi\frac{y+a}{b}\right) - \operatorname{csch}\left(\pi\frac{y-a}{b}\right)\right], \quad v_y(0,y) = 0; \tag{36.189a,b}$$

(iii) the value at the midpoint on the bottom wall is less than twice that for a vortex in free space at the same distance a:

$$v_y(0,0) = 0, \quad v_x(0,0) = \frac{\Gamma}{b}\operatorname{csch}\left(\frac{\pi a}{b}\right) < \frac{\Gamma}{\pi a}, \tag{36.190a,b}$$

because of (36.155b); thus although the velocities due to the vortex and image add-up, the side walls tend to reduce horizontal velocities. The velocity at the points on the sidewalls closest to the vortex is vertical:

$$v_x\left(\pm\frac{b}{2},a\right) = 0, \quad \pm v_y\left(\pm\frac{b}{2},a\right) = \frac{\Gamma}{2b}\left\{1 - \operatorname{sech}\left(\frac{2\pi a}{b}\right)\right\} \le \frac{\Gamma}{2b}; \tag{36.191a,b}$$

as the vortex moves to a distance from the bottom wall much larger than the width $2\pi a/b \gg 1$ it tends to the value $\Gamma/2b$ that applied (36.153b) to the vortex between infinite walls at the closest point.

36.9.2 Electro/Magnetostatic Fields in Conducting/Insulating Wells

The electrostatic field due to a charge e on the axis of a rectangular well $\operatorname{Im}(z) > 0, -b/2 < \operatorname{Re}(z) < b/2$ *filled with a dielectric of permeability* ε, *at a distance a from the bottom is given: (a) for insulating walls (Figure 36.16a), by (36.183b) with Q replaced by* e/ε; *(b) for conducting walls (Figure 36.16b) by*

$$f(z) = \frac{e}{2\pi\varepsilon}\log\left\{\frac{\tan[\pi(z-ia)/2b]}{\tan[\pi(z+ia)/2b]}\right\}, \tag{36.192a}$$

$$E^*(z) = -\frac{\mathrm{d}f}{\mathrm{d}z} = -\frac{e}{2b\varepsilon}\left[\csc\left(\pi\frac{z-ia}{b}\right) - \csc\left(\pi\frac{z+ia}{b}\right)\right], \tag{36.192b}$$

that corresponds to exchanging equipotential and field lines in Figure 36.16c, that is, substituting Γ by e/ε and f by $-if$ in (36.187a). The induced charges on the bottom conductor are given by

$$E_x(x,0) = 0: \quad \sigma(x) = \varepsilon E_y(x,0) = -(e/b)\operatorname{Im}\left\{\csc\left(\frac{\pi x}{b} - \frac{i\pi a}{b}\right)\right\}$$
$$= -\frac{e}{b}\frac{\sinh(\pi a/b)\cos(\pi x/b)}{\sinh^2(\pi a/b) + \sin^2(\pi x/b)}, \tag{36.193a,b}$$

that vanish at the corners $x = \pm b/2$. This agrees with the induced charges on the side conductors:

$$E_y\left(\pm\frac{b}{2},y\right) = 0, \quad \sigma(y) = \pm\varepsilon E_x\left(\pm\frac{b}{2},y\right)$$
$$= -\frac{e}{2b}\left[\operatorname{sech}\left(\pi\frac{y-a}{b}\right) - \operatorname{sech}\left(\pi\frac{y+a}{b}\right)\right], \tag{36.194a,b}$$

that also vanishes at the corners. In (36.193b) was used:

$$\begin{aligned}\csc\left(\frac{\pi x}{b}\pm i\frac{\pi a}{b}\right) &= \left[\sin\left(\frac{\pi x}{b}\right)\cos\left(\frac{i\pi a}{b}\right)\pm\cos\left(\frac{\pi x}{b}\right)\sin\left(\frac{i\pi a}{b}\right)\right]^{-1}\\ &= \left[\sin\left(\frac{\pi x}{b}\right)\cosh\left(\frac{\pi a}{b}\right)\pm i\cos\left(\frac{\pi x}{b}\right)\sinh\left(\frac{\pi a}{b}\right)\right]^{-1}\\ &= X^{-1}\left[\sin\left(\frac{\pi x}{b}\right)\cosh\left(\frac{\pi a}{b}\right)\mp i\cos\left(\frac{\pi x}{b}\right)\sinh\left(\frac{\pi a}{b}\right)\right],\end{aligned} \tag{36.194c}$$

where:

$$\begin{aligned}X &\equiv \left|\sin\left(\frac{\pi x}{b}\right)\cosh\left(\frac{\pi a}{b}\right)\pm i\cos\left(\frac{\pi x}{b}\right)\sinh\left(\frac{\pi a}{b}\right)\right|^2\\ &= \sin^2\left(\frac{\pi x}{b}\right)\cosh^2\left(\frac{\pi a}{b}\right)+\cos^2\left(\frac{\pi x}{b}\right)\sinh^2\left(\frac{\pi a}{b}\right)\\ &= \left[1-\cos^2\left(\frac{\pi x}{b}\right)\right]\cosh^2\left(\frac{\pi a}{b}\right)+\cos^2\left(\frac{\pi x}{b}\right)\sinh^2\left(\frac{\pi a}{b}\right)\\ &= \cosh^2\left(\frac{\pi a}{b}\right)-\cos^2\left(\frac{\pi x}{b}\right)=\sinh^2\left(\frac{\pi a}{b}\right)+\sin^2\left(\frac{\pi x}{b}\right)\\ &= \frac{1}{2}\left[\cosh\left(\frac{2\pi a}{b}\right)-\cos\left(\frac{2\pi x}{b}\right)\right];\end{aligned} \tag{36.194d}$$

substitution of (36.193d) in (36.193c) yields (36.192b).

The magnetostatic field due to a current J on the axis of a rectangular well of width b at a distance, a, from the bottom, is given: (a) by (36.192a,b) in the case of conducting walls (Figure 36.16c), with e replaced by J, leading to surface electric currents on the sides (bottom) that coincide with (36.193b) [(36.194b)]; (b) in the case of insulating walls (Figure 36.16d) by:

$$f(z) = -\frac{iJ}{2\pi c}\log\left[\cos\left(\frac{2\pi z}{b}\right)-\cosh\left(\frac{2\pi a}{b}\right)\right], \tag{36.195a}$$

$$\frac{\mathrm{d}f}{\mathrm{d}z} = -\frac{iJ}{bc}\frac{\sin(2\pi z/b)}{\cosh(2\pi a/b)-\cos(2\pi z/b)}, \tag{36.195b}$$

obtained replacing Q by $-iJ$ in (36.183a,b). There are two sets of walls in the cases of: (i) the rectangular well (Figure 36.16a–d); (ii) the corner with a circular indentation (Figures 36.2a,b; 36.3a,b; 36.5a–c). In these cases four combinations are possible: (i) all walls are insulating for electro (magneto) static field in Figures 36.16a; 36.2a; 36.5a (Figures 36.16d; 36.3a; 36.5c); (ii) all walls are conducting for the electro (magneto) static field in Figure 36.16b; 36.5c (Figures 36.16b and 36.5b); (iii) one set of walls conducting (e.g., the circular indentation) and the other insulating (e.g., the side walls) for the electro (magneto) static field in Figure 36.2b (Figure 36.3b); (iv) the reverse of (iii). The cases illustrated are a sample of all possible combinations, and all of the remaining possibilities that are considered in Example 40.10. These and other problems with similar symmetries appear in List 36.1.

36.9.3 Induced Electric Charges (Surface Electric Currents) on the Bottom and Side Walls

The real line-charge e (current J) is balanced (36.196a) by the induced charges at the bottom of the well (36.196b) and at the two side walls (36.196c):

$$-e = e_b + 2e_s, \quad e_b \equiv \int_{-b/2}^{+b/2}\sigma(x)\,\mathrm{d}x, \quad e_s \equiv \int_0^{\infty}\sigma(y)\,\mathrm{d}y. \tag{36.196a–c}$$

LIST 36.1

Potential Flows and Fields: 42 Problems.

Configuration	Potential flow			Insulator/conductor			
	Free stream	Monopole		Electrostatics		Magnetostatics	
Case		Vortex	Source/sink	Free field	Electric charge	Free field	Electric current
CORNER:							
sharp	36.1;F36.1a	36.2.1;F36.4b	36.2.1;F36.4a	–	36.2.2;F36.4a/c	–	36.2.2;F36.4c/b
rounded	36.1.1;F36.1b	–	–	–	–	–	–
with indentation	36.1.2;F36.2a	36.2.3;F36.5b	36.2.3;F36.5a	36.1.3; F36.2a/b	36.2.4; F36.5a/c	36.1.3; F36.3a/b	36.2.4; F36.5c/b
SEMIINFINITE PLATE	–	36.3.1;F36.6a T36.1	36.3.2;F36.6b; F36.7a, b T36.2–3	–	–	–	–
FLAT PLATE	36.4.4–5; F36.10a–c	36.4.1;F36.8a		36.4.3; F36.9a/b	–	36.4.3; F36.9b/a	–
slit	36.4.2;F36.8b	–	36.4.2;F36.8b	36.4.3; F36.9c/d	–	36.4.3; F36.9d/c	–
SEMIINFINITE DUCT – sides: parallel	36.5.2;F36.11a	–	36.5.2; F36.12a	36.5.4; F36.12c	–	36.5.4; F36.12c	–

convergent	36.5.1;F36.11b	–	–	–	–	–	–
Pitot tube	–	–	36.5.3F36.12c	–	–	–	–
PARALLEL WALLS	36.6.1–3; F36.13a–c, 14a	36.7.1; F36.13a 14c, d	36.6.1–3; F36.14a, b	–	36.7.2; F36.14a/b	–	36.7.2; F36.14d/c
Single vortex row	–	36.7.3–4; F36.14d	–	–	–	–	–
WELL: at corner	–	–	36.8.1–2; F36.15a, b	–	36.8.2; F36.15a/b	–	–
on axis	–	36.9.1; F36.16c, d	36.8.3; F36.16a	–	36.9.2–4; F36.16a/b	–	36.9.2–4; F36.16d/c
Double vortex row	–	36.9.5–6; F36.16d, 17a,b	–	–	–	–	–
SUM	8	8	9	4	5	4	4

– The chapters/sections/subsections are indicated, for example, 36/36.4/36.6.1; followed by corresponding figures, for example, F36.15c.

Note: The list includes 42 problems in this chapter making the analogy between: (i) free potential flow and electro (magneto) static field; (ii) vortex and electric current; (iii) source/sink and positive/negative electric charge. The analogies apply to several configurations: (i–iii) a sharp (rounded) corner [Figure 36.1a(b)] and a corner with a circular indentation (Figures 36.2–36.5); (iv) a monopole convected by the flow past a semiinfinite plate (Tables 36.1–36.3; Figures 36.6 and 36.7); (v/vi) the flow around a flat plate or through a slit (Figures 36.8–36.10); (vii–ix) the flow out of convergent or parallel-sided channel or into a Pitot tube (Figures 36.11 and 36.12); (x–xii) monopoles, that is, sources/sinks or vortices between parallel walls, corresponding to a single row (Figures 36.13 and 36.14); (xiii–xvi) monopoles in a rectangular well corresponding to a double row (Figures 36.15–36.17).

The total electric charge induced on each side wall (36.196c) is specified (36.194b) by the difference (36.197a) of two integrals (36.197b):

$$e_s = I_+ - I_-, \quad I_\pm \equiv -\frac{e}{2b}\int_0^\infty \operatorname{sech}\left[\frac{\pi(y\mp a)}{b}\right] dy, \tag{36.197a,b}$$

these can be evaluated (36.198b) via the change of variable (36.198a):

$$\xi \equiv \frac{\pi}{b}(y\mp a): \quad I_\pm = -\frac{e}{2\pi}\int_{\mp\pi a/b}^\infty \operatorname{sech}\xi \, d\xi = -\frac{e}{\pi}\left[\arctan\left(e^\xi\right)\right]_{\mp\pi a/b}^\infty$$
$$= -\frac{e}{2} + \frac{e}{\pi}\arctan\left[\exp\left(\mp\frac{\pi a}{b}\right)\right], \tag{36.198a,b}$$

where was used the primitive (36.160). Substituting (36.198b) in (36.197a) and (36.196a) specifies:

$$-\frac{e}{\pi}\left\{\arctan\left[\exp\left(\frac{\pi a}{b}\right)\right] - \arctan\left[\exp\left(-\frac{\pi a}{b}\right)\right]\right\} = e_s = -\frac{e+e_b}{2}, \tag{36.199a,b}$$

the total electric charge induced at the bottom (36.196b) and on the side walls (36.196c) of a rectangular well of width b by a line-charge e *on the axis at a distance a from the bottom (36.199a,b).*

The preceding result (36.199) can be checked by calculating directly the total electric charge (36.196b) on the bottom wall (36.193b), using the change of variable (36.200a) in the integral (36.200b):

$$\eta \equiv \frac{\sin(\pi x/b)}{\sinh(\pi a/b)}: \quad e_b = -\frac{e}{b}\int_{-b/2}^{+b/2} \frac{\cos(\pi x/b)/\sinh(\pi a/b)}{1+\left[\sin(\pi x/b)/\sinh(\pi a/b)\right]^2} dx$$
$$= -\frac{e}{\pi}\int_{-\operatorname{csch}(\pi a/b)}^{\operatorname{csch}(\pi a/b)} \left(1+\eta^2\right)^{-1} d\eta = -\frac{e}{\pi}\arctan\eta\Big|_{-\operatorname{csch}(\pi a/b)}^{+\operatorname{csch}(\pi a/b)}; \tag{36.200a,b}$$

this simplifies to

$$e_b = -\frac{2e}{\pi}\arctan\left[\operatorname{csch}\left(\frac{\pi a}{b}\right)\right] = -e - 2e_s. \tag{36.201a,b}$$

The ratio of (3.30b) to (3.30a) with $u = z_1$ leads to the identity (36.202a):

$$\tan(u-v) = \frac{\tan u - \tan v}{1+\tan u \tan v}, \tag{36.202a}$$

$$\tan(\arctan u - \arctan v) = \frac{u-v}{1+uv}, \tag{36.202b}$$

and its inverse (36.202b); the latter is used in (36.199a), leading to

$$\tan\left[\left(\frac{-\pi e_s}{e}\right)\right] = \frac{\exp(\pi a/b) - \exp(-\pi a/b)}{2} = \sinh\left(\frac{\pi a}{b}\right) = -\cot\left(\frac{\pi e_b}{2e}\right), \tag{36.203a,b}$$

where (36.201a) was substituted. The identity (36.204a) applied to (36.203a,b) implies (36.204b):

$$\tan\xi = -\cot\left(\xi + \frac{\pi}{2}\right), \qquad -\frac{\pi e_s}{e} = \frac{\pi e_b}{2e} + \frac{\pi}{2}; \tag{36.204a,b}$$

the coincidence of (36.204b) ≡ (36.196a) confirms that the total charge induced on all walls is −e, and proves the consistency of (36.196a) ≡ (36.199b), and (36.201b) ≡ (36.204b), specifying the induced electric charges on the bottom (36.196b) ≡ (36.199b) ≡ (36.201a) ≡ (36.203b) [on the side walls (36.196c) ≡ (36.199a) ≡ (36.201b) ≡ (36.203a)] due to a line-charge e at a distance a from the bottom of a square well and at equal distance b/2 from the side walls (Figure 36.16a).

36.9.4 Similarity of Electric Charge and Current Distributions on Conductors

It has been found that the ratio of the line electric charge e (current J) to the induced electric charge σ (surface electric current ϑ) on a conductor is equal for all shapes of conductor considered so far: (i) for (36.30d) ≡ (36.33d) a corner with arbitrary angle (Subsection 36.2.2); (ii) also (36.37d) ≡ (36.40d) with a cylindrical indentation (Subsection 36.2.3); (iii/iv) also (36.81c) ≡ (36.83c) for a finite flat plate and for (36.82c) ≡ (36.84c) two aligned semiinfinite plates (Subsection 36.4.3); (v) again (36.158a) ≡ (36.161b) for parallel infinite walls (Subsection 36.7.2); (vi) at last (36.193–36.194b) for a rectangular well (Subsection 36.9.3). Also (vii) in the case of electro (magneto) statics (36.21c,d) in a conducting corner (Subsection 36.1.3) the distribution of induced electric charges (currents) is the same to within scaling factors. A comparison of electostatics (magnetostatics) in Chapter 24(26) again shows in every case that the same geometry of conductor leads to similar induced electric charge (surface electric current) distributions, viz. for: (viii) monopole near a wall (24.21c–e) [(26.29)]; (ix) a multipole near a wall (24.32; 24.33a, b; 24.34) [(26.39)]; (x) cylinder in a uniform field (24.42; 24.43a–c) [(26.45d; 26.46a–c)]; (xi) adding a monopole an the axis (24.44b) [(26.49b)]; (xii) monopole near a cylinder (24.57a,b) [(26.51)]. There is a physical difference (Note 26.2) in that the induced electric charges (surface electric currents) in a conductor are real (fictitious), that is, (i) an electric charge near a conductor attracts real opposite charges, whose distribution ensures that the electric field is normal, and is equivalent to the image(s); (ii) an electric current near a conductor must have a tangential magnetic field, represented by a fictitious surface current distribution that is equivalent to the image(s). The image electric charge (i) [current (ii)] can be realized physically, and corresponds to: (i) a line-charge (line-current) in two dimensions, that is, both scalar; (ii) a point charge (point current) in three dimensions, that is, a scalar (vector). Thus there is a greater distinction between the irrotational (solenoidal) electric (magnetic) field due to electric charges (currents) in three and more dimensions, than in the plane (Section 32.3). It is shown next that in the plane case the real (fictitious) induced electric charge (electric current) distribution induced in (equivalent to) a conductor is mathematically the same, for any given geometry.

The preceding conclusion can be proved generally by the following comparisons of the electro (magneto) static field [Chapter 24 (26)]: (i) the relevant Maxwell equation of electro (magneto) statics is (24.5b) ≡ (36.205a) [(26.1c) ≡ (36.205b)]:

$$\nabla\cdot\vec{\mathrm{E}} = \frac{q}{\varepsilon}, \qquad \nabla\wedge\vec{H} = \frac{\vec{j}}{c}, \tag{36.205a,b}$$

involving the electric $\vec{\mathrm{E}}$ (magnetic $\vec{\mathrm{H}}$) field, electric charge q (current $\vec{j}$) density, and the constant dielectric permittivity ε (speed of light in vacuum c); (ii) the corresponding boundary condition is that the surface electric charge σ (current ϑ) density is due to the normal (24.8b) ≡ (36.206a) [tangential (26.9b) ≡ (36.206b)] component of the electric (magnetic) field:

$$\sigma = \varepsilon E_n, \quad \vartheta = c H_s; \tag{36.206a,b}$$

(iii) the complex electrostatic f_e (magneto static f_m) potential due to a line charge e (current J) at the origin is (24.13a) ≡ (36.207a) [(26.13a) ≡ (36.207b)]:

$$2\pi f_e(z) = -\frac{e}{\varepsilon}\log z, \quad 2\pi f_m(z) = -i\frac{J}{c}\log z, \tag{36.207a,b}$$

$$\frac{\varepsilon}{e} f_e(z) = -\frac{1}{2\pi}\log z = -i\frac{c}{J} f_m(z), \tag{36.207c}$$

leading to (36.207c); (iv) the corresponding complex conjugate electric (magnetic) field is (36.208a) [(36.208b)]:

$$E^*(z) = -\frac{\mathrm{d}f_e}{\mathrm{d}z} = \frac{e}{2\pi\varepsilon z}, \quad H^*(z) = -\frac{\mathrm{d}f_m}{\mathrm{d}z} = \frac{iJ}{2\pi c z}, \tag{36.208a,b}$$

$$\frac{\varepsilon}{e} E^*(z) = \frac{1}{2\pi z} = -i\frac{c}{J} H^*(z), \tag{36.208c}$$

leading to (36.208c); (v) the electric (magnetic) field is normal (36.209a) [tangential (36.209b)] to the induced electric charge (surface electric current) distribution, leading from (36.208e) to (36.209e); (vi) substituting (36.206a,b) in (36.209c) implies (36.209d)

$$E^* = -iE_n, \; H^* = H_s: \qquad E_n\frac{\varepsilon}{e} = H_s\frac{c}{J} \quad \Rightarrow \quad \frac{\sigma}{e} = \frac{\vartheta}{J}; \tag{36.209a–d}$$

The induced electric charge σ (surface electric current ϑ) density in a conductor due the to an external line charge e (current J) are in the same ratio $\sigma/e = \vartheta/J$ for any given geometry of conductor. The result extends to multipoles of arbitrary order that are a linear superposition of monopoles. The relation (36.209d) applies directly in 11 of the 12 cases (i) to (xii) mentioned before; it is modified only in the case (x) of a conducting cylinder in an external uniform electric E_0 (magnetic H_0) field, for that the induced electric charge (24.42) ≡ (36.209e) [surface electric current (26.45d) ≡ (36.209f)] distribution:

$$\frac{\sigma(\varphi)}{2E_0\varepsilon} = \cos\varphi = \mathrm{Re}(e^{i\varphi}), \quad \frac{\vartheta(\varphi)}{2H_0 c} = -\sin\varphi = \mathrm{Re}(e^{i\varphi}), \tag{36.209e,f}$$

correspond to a rotation by 90° of the dipole axis (24.43a–c) [(26.46a–c)].

36.9.5 Staggered and Equilateral Double Vortex Street

The complex potential (36.187a) and conjugate velocity (36.187b) apply to the potential flow due to an infinite row of pairs of opposite vortices, leading the parallel **double vortex street** (von Karman, 1911) in Figure 36.16d; the field line $\Psi = 0$ is bifurcated with one branch coincident with the real axis and the other going over the vortex, so that the field lines are closed inside and open outside. If the vortex and image coalesce, since they are opposite they cancel, and the double cat's eye pattern disappears, suggesting that it is unstable. This is confirmed by the fact that the velocity has the same value far away above

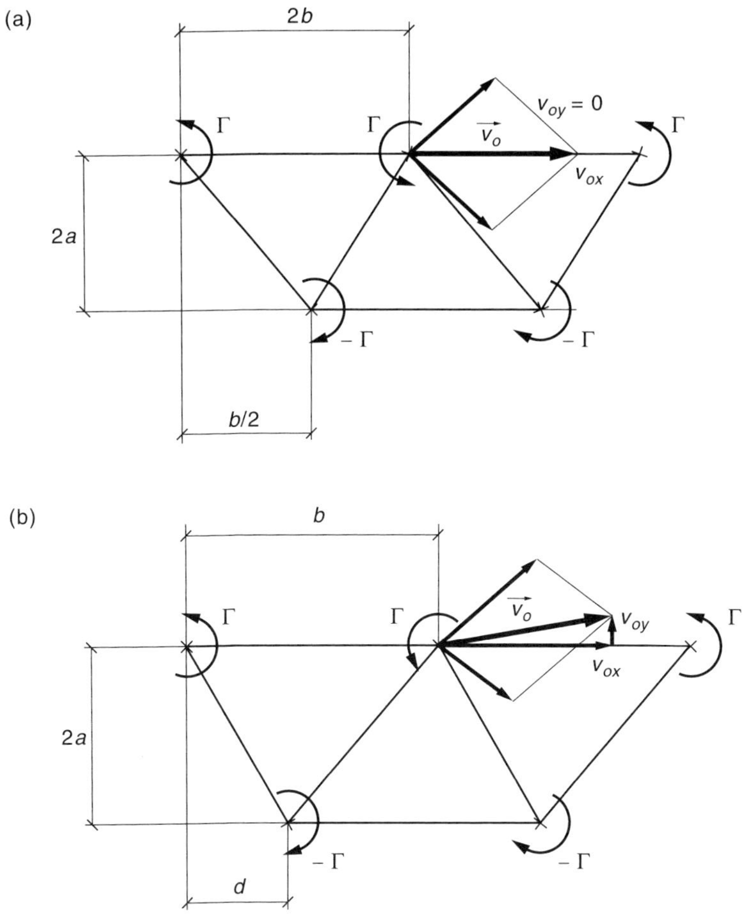

FIGURE 36.17
Considerer a double vortex street, consisting of two rows of identical vortices with opposite circulation in each row (Figure 36.16d), for example, the tip vortices from a wing of finite span (the two longitudinal rows in Figure 34.7a). If the stagger d between the vortices in the two rows is arbitrary (b), the vortices in one row induce an out-of-line velocity upon the vortices in the other row, so the double vortex row cannot be stable. Stability is possible for a von Karman "double vortex street" with stagger equal to half the distance between the vortices in the same row (a), because the vortices in the other row induce a horizontal velocity. This necessary condition for stability is not sufficient; for sufficiency it is necessary to prove that the displacement of one vortex from its equilibrium position leads to an induced velocity from the other vortices restoring equilibrium.

and below $U = \Gamma/b$, so that no vortex sheet need exist unlike in the case (Figure 36.14d) of the cat's eye pattern. The cat's eye pattern and the vortex sheet are also unstable. Thus the parallel double vortex street (Figure 36.16d) consists of two opposite cat's eye patterns (Figure 36.14d). The double vortex street, instead of aligned vortices (Figure 36.16d) at the corners of rectangles with sides $(x = b, y = 2a)$, can be **staggered** (Figure 36.17a) with vortices at the edges of equal isosceles triangles of base $x = b$ and height $h = 2a$. Both are particular cases, with stagger $d = 0\,(d = b/2)$ of *a double vortex street (Figure 36.17b) consisting of: (i) infinite rows of vortices with alternating circulations* $\pm\Gamma$ *at a distance* b *apart; (ii) the two rows are parallel at a distance* 2a, *and one is shifted by* d *relative to* *latter. The corresponding complex potential (36.210a) [conjugate velocity (36.210b)]:*

$$f(z) = -\frac{i\Gamma}{2\pi}\log\left\{\frac{\sin\left[\pi(z-ia)/b\right]}{\sin\left[\pi(z-d+ia)/b\right]}\right\}, \tag{36.210a}$$

$$\frac{\mathrm{d}f}{\mathrm{d}z} = -\frac{i\Gamma}{2b}\left[\cot\left(\pi\frac{z-ia}{b}\right) - \cot\left(\pi\frac{z-d+ia}{b}\right)\right], \tag{36.210b}$$

corresponds to subtracting from (36.162a) [(36.162b)] for a single row with $z - ia$ the same expression for $z - d + ia$.

36.9.6 Convection Velocity of General Vortex Street

The velocity induced at one vortex, for example, that at $z = ia$, by all the others is obtained by subtracting the corresponding term from (26.210b) and letting $z \to ia$, viz:

$$\begin{aligned} v_0^*(ia) &= \lim_{z\to ia}\left(\frac{\mathrm{d}f}{\mathrm{d}z} + \frac{i\Gamma}{2\pi}\frac{1}{z-ia}\right) \\ &= \lim_{z\to ia}\frac{i\Gamma}{2b}\cot\left(\pi\frac{z-d+ia}{b}\right) = \frac{i\Gamma}{2b}\cot\left(\pi\frac{2ia-d}{b}\right), \end{aligned} \tag{36.211}$$

where (36.170) was used to cancel two terms; these correspond to the influence of a row on a vortex in the same row that has been shown by two methods (36.171a,b; 36.171c) to lead to zero velocity. The term that remains in (36.211) is the velocity induced at one vortex by all vortices in the other row. The convection velocity of the double vortex row (36.211) corresponds to (36.162b; 36.163d; 36.164a,b) with reversed sign and $x = -d, y = 2a$:

$$\{v_{0x}, v_{0y}\} = \frac{\Gamma}{2b}\left[\cosh\left(\frac{4\pi a}{b}\right) - \cos\left(\frac{2\pi d}{b}\right)\right]^{-1}\left\{\sinh\left(\frac{4\pi a}{b}\right), \sin\left(\frac{2\pi d}{b}\right)\right\}. \tag{36.212a,b}$$

A vortex street consisting of two rows of vortices (Figure 36.17b), with alternating circulations $\pm\Gamma$, spacing a, distance b between rows, and relative displacement d between vortex positions, moves with uniform velocity (36.212a,b); the velocity is parallel to the rows (36.213a):

$$v_{0y} = 0: \quad \sin\left(\frac{2\pi d}{b}\right) = 0, \quad d = 0, \frac{b}{2}, \tag{36.213a–c}$$

in two (36.213b) cases, namely the parallel (staggered) vortex street with stagger (36.214a) [(36.214c)] in (36.214b) [(36.214d)]:

$$d = 0: \quad v_{0x} = \frac{\Gamma}{2b}\coth\left(\frac{2\pi a}{b}\right); \quad d = \frac{b}{2}: \quad v_{0x} = \frac{\Gamma}{2b}\tanh\left(\frac{2\pi a}{b}\right). \tag{36.214a–d}$$

Thus the vortices are Figure (36.16d) [(36.17a)] at the corners of identical rectangles (36.213c) [isosceles triangles (36.213d)].

The passage from (36.212a) to (36.214b) [(36.214d)] corresponds to (36.215a) [(36.215b)].

$$\begin{aligned} \frac{2b}{\Gamma}v_{0x} &= \frac{\sinh(4\pi a/b)}{\cosh(4\pi a/b) \mp 1} \\ &= \frac{2\sinh(2\pi a/b)\cosh(2\pi a/b)}{2\sinh^2, \cosh^2(2\pi a/b)} = \coth, \tanh(2\pi a/b). \end{aligned} \tag{36.215a,b}$$

It can be checked that in both cases the velocity induced by each pair of vortices in the: (i) same row cancel; (ii) other row add to a horizontal velocity. The sum of all such contributions is the horizontal velocity (36.214a–d). In the particular case $a = 0$, when two vortex rows merge into one, the velocity is: (i) zero in the parallel case (36.214c,d) because opposite vortices coincide and cancel; (ii) infinite for the staggered case (36.214a,b) because for a single row of alternating vorticles the induced velocities in one direction are larger than in the opposite direction. In the case of the parallel vortex street (Figure 36.16d), if a vortex is displaced from its equilibrium position along a row, the vertical velocities induced by pairs of vortices in the same row will be unequal, and it will not return to the same position; this is the same instability as for the single vortex row (Subsection 36.7.4). In the case of the staggered vortex street (Figure 36.17a) a displacement from the equilibrium position along the: (i) vertical has horizontal induced velocities; (ii) horizontal has vertical velocity components induced by vortices in both rows. Thus the staggered vortex street may be stable for a particular choice of geometric parameters. The stability method (Chapter 2) can be used together with Fourier series (Notes 25.2–25.4) to show that the parallel (staggered) double vortex street is unstable (stable) with regard to small wavelike disturbances for all (some) values of the geometric parameters.

Note 36.1 *Potential Flow and Electric, Magnetic and Gravity Fields:* The electro (magneto) static field due to electric charges (currents) in Chapter 24 (26) is analogous to the potential flow due to source/sinks (vortices). Whereas the electric and magnetic field are decoupled in steady conditions, in a potential flow source/sinks and vortices can coexist. The gravity (Chapter 18) field is due to masses that are positive, and corresponds to sinks, because the field is attractive; thus there are no analogues of the source and/or vortex for the gravity field. Thus of the main four potential fields, the potential flow is the one that allows a greater variety on combinations; for this reason the potential flow is taken as the main example of potential field, and studied in detail. The transposition of results to the gravity (Chapter 18), electro (Chapter 24), and magneto (Chapter 26) static field can be made using the respective analogies; distinct properties, like induced electric charges and surface electric currents on conductors, have been mentioned explicitly (Chapter 36). List 36.1 indicates the potential flow (electro/magnetostatic field) problems in the present chapter, to that may be added those in other Chapters [12, 14, 16, 18, 34, and 38 (24, 26)]. These examples of distinct problems relate to the conformal mappings (Chapters 33 and 35) used as solutions.

Conclusion 36: The simplest polygonal transformation (Figure 36.1) has a single critical point (a), and maps an uniform flow into a flow in two-dimensional sharp corner; the corner may be rounded (b), removing the stagnation point of the velocity, by a modification of the transformation involving an additional term with the same exponent. Another way to exclude the corner (Figures 36.2-36.3) is to have a cylindrical indentation, corresponding to: (Figure 36.2a) the streamlines for a potential flow, and field lines for an electrostatic (magneto) field with all walls insulating (conducting); (Figure 36.2b) the field lines for the electro (magneto) static field with conducting (insulating) in indentation and insulating (conducting) walls; (Figure 36.3a) the field lines for the electro (magneto) static field with all walls conducting (insulating); (Figure 36.3b) field lines for electro (magneto) static field with insulating (conducting) circular indentation and conduting (insulating) side walls. The corner with sharp edge (with cylindrical indentation) is reconsidered Figures 36.4 (36.5) for: (Figure 36.4a) [(36.5a)] flow due to a source/sink or electrostatic field due to an electric charge near insulator, i.e., case I; (Figure 36.4b) [(36.5b)] flow due to a vortex or magnetostatic field due to an electric current near conductor (case II); (Figure 36.4c) [(36.5c)] electro (magneto) static field due to an electric charge (current) near conductor (insulator), i.e., case III. The trajectory (Figure 36.6) of a vortex past a semiinfinite plate starts and ends at infinity (a), whereas (b): (i) a source moves to an equilibrium position aligned with the plate; (ii) a sink moves in the opposite direction and collides with the plate; (iii) the equilibrium position aligned with the plate is (Figure 36.7) stable (a) [unstable (b)] for the source (sink), whereas the vortex has no equilibrium position. The Joukowski transformation with parameter b/2, when applied to a vortex of circulation Γ (source/sink of volume flux 2Q) at the origin of the z-plane, yields (Figure 36.8) in the ζ-plane the flow (a) circulating around a flat plate of chord 2 b [(b) going through a slot of width 2b between aligned semiinfinite plates], with edges at $\zeta = \pm\, b$ and circulation Γ (volume flux Q). The electrostatic analogue of the finite plate (Figure 36.8a) is an insulator (a) [conductor (b)] consisting of a finite flat plate; in the magnetostatic case the conductor (a) [insulator (b)] are reversed. The case of aligned semiinfinite plates (Figure 36.8b) leads (Figure 36.9) to (c) electrostatic insulator (magnetostatic conductor), and vice-versa (d). In all of four cases, (Figure 36.9a-d) the equipotentials and field or stream lines are a (Figure 36.10) set of orthogonal (a) ellipses (b) and hyperbolas (c) with common foci at $\pm\, c$, which specify the plane confocal coordinates (ξ, η). The aligned semiinfinite plates (Figure 36.8b), are one extreme $\beta = \pi/2$ of (Figure 36.11) the (b) duct with flat sides converging at an angle 2β, the other extreme being (a) the channel with parallel sides. All of these are obtained by Schwartz-Christoffel transformations (Figure 36.11b) of the real axis starting (Figure 36.12) with (a) the potential flow with volume flux Q out of a channel of width b; superimposing a stream of velocity $-\, Q/b = -\, U$ cancels the flow in the channel, and specifies the potential flow past a Pitot tube (b). The electrostatic analogue of the channel is the condenser with parallel semiinfinite plates leading (c) to a concentration of induced charges near the edges which is equivalent a distribution of uniform charges with added length. The potential flow due to a source/sink (vortex) between parallel walls (Figure 36.13) can be represented (a) by an infinity of identical (alternating) images, leading to infinite products for the complex potential and series of fractions for the conjugate velocity. An alternate method is to map (Figure 36.13) the upper (b) [lower (c)] half plate into degenerate image triangles by the same transformation, i.e., the whole plane is mapped into a strip with each half being the reflection of the other. The methods of infinite images or conformal transformation can be used (Figure 36.14) to determine the: (a) [(b)] potential flow due identical (alternating) source/sinks between walls, or electrostatic field due to a charge between parallel insulators (conductors); (c) [(d)] the potential flow due to a alternating (identical) vortices between walls or magnetostatic field due to a current between conductors (insulators). A translation (Figure 36.14) of (a) [(b)] gives the electrostatic field (Figure 36.15) of equal (a) [opposite (b)] charges at the corners of a well, which coincides with the potential flow due to two sources (a) [a source and a sink (b)] on two parallel infinite walls, or taking one-half on the corners of a rectangular well. Placing the source on the axis of the well away from the bottom, leads (Figure 36.16) to: (a) [(b)] the potential flow identical (alternating) pairs of source/sinks and electrostatic field of a charges in an insulating (conducting) well; (c) [(d)] the potential flow due to alternating (identical) pairs of vortices, or magnetostatic field due to a current in a conducting (insulating) well. The patterns [(Figure 36.14d) (36.16d)] are the cat's eye (vortex street), corresponding to a single row (two parallel rows) of vortices. The two parallel rows of vortices may Figure (36.17) be offset (Figure 36.17b), between the two limits of parallel (Figure 36.16d) and staggered (Figure 36.17a) double vortex streets.

37

Mapping of Domains and Boundaries

The preceding applications of the conformal mapping have shown on several instances two properties: (i) a variety of domains, for example, a half-plane, a circle, or an arbitrary polygon can be mapped into each other; (ii) if the boundary of a region is mapped into a curve, the interior (exterior) of the region is mapped into the interior (exterior) of the curve. Since two regions can be mapped into each other through the unit disk, it is sufficient to address (i) and (ii) for mappings from the unit disc to an arbitrary region. The mappings besides being conformal or analytic should be univalent, that is, take each value in the range only once; in this way the range does not overlap with itself. The question thus arises of the most general conditions in that these properties hold. A first approach is to find necessary conditions, for example, conditions that if not met cause the property to fail; it is then proved that those necessary conditions are also sufficient. For example, the simple, that is, conformal and univalent, mapping into the unit disk does not exist for (a) multiply connected regions, for example, a doubly connected region may be mapped into an annulus rather than a disk; (b) a region with a single boundary point, for example, the punctured plane, that is, the plane less a point, has no simple mapping into a unit disk. Thus two necessary conditions for the existence of simple mapping of a region into a unit disk (Section 37.3) are that it be simply connected and the boundary consists of at least two points; these conditions are also sufficient (Section 37.5). The map is unique if a point is assigned as well as direction through it, that is, the conditions of unicity are the same as for bilinear mappings into the unit disk. Concerning the correspondence of boundaries and interiors (Section 37.7), the property (ii) above holds: (α) for a domain without singularities of the mapping function; (β) if the mapping function has one singularity in the interior that is a simple pole, viz. if it were a multiple pole or essential singularity it would fail to be univalent; (γ) if the mapping function has a singularity on the boundary, then the image region extends to infinity, and some asymptotic conditions have to be imposed for the theorem to hold. The mapping between multiply connected regions cannot be simple, and requires the use of multivalued functions (Section 37.4). The proof of existence of conformal simple (nonsimple) mapping between simply connected (multiply connected) regions involves the refcrence and minimax functions (Section 37.7) [the complete collection of fundamental regions of an automorphishm group (Section 37.6)]. The proof of existence of a conformal mapping from the unit disk to an arbitrary compact region is based on a sequence of holomorphic functions that converges uniformly, implying that the limit function is also holomorphic (Section 37.2); the uniform convergence of the sequence of functions holds if its elements are bounded (Section 37.1). Thus the starting point for all these proofs, as usual with existence and unicity theorems, is two sets of results concerning: (i) some bounds on the modulus, and real or imaginary parts of holomorphic functions and their derivates in disks (Section 37.1); (ii) the existence of uniformly convergent subsequences of sequences of functions in compact domains (Section 37.2). A set of results complementary to the existence and unicity theorems, is the construction of complex functions in a disk (in the complex half-plane) from the values of their real (or imaginary) parts on the boundary circle (Section 37.8) [straight line (Section 37.9)]; the latter specify the potential field inside or outside a cylinder (in the upper or lower half-plane) from the values of the potential or stream function on the boundary circle (real axis). These results (Sections 37.8 and 37.9) are based on an extension of the Cauchy theorems (Chapters 13, 15, and 31) to the exterior (Section 37.8) as well as the interior (Sections 13.6, 15.4, and 31.4) of a boundary. If the main interest is explicit solutions of problems rather than existence and unicity theorems it is possible to proceed directly to Sections 37.8 and 37.9. The latter concerns the field created by a distribution of sources or by a given potential on a boundary; both processes occur in nature and are used in engineering devices.

37.1 Unicity of Mappings and Bounds in the Unit Disk (Schwartz, 1890; Caratheodory, 1912; Borel, 1928)

A univalent conformal mapping can be performed through the unit disk, suggesting as starting point the establishment of upper bounds in the unit disk for the modulus (Subsection 37.1.1) [modulus of the real and imaginary parts (Subsection 37.1.2)] of a holomorphic function, and also for the modulus of its derivatives (Subsection 37.1.3). The first result (Subsection 37.1.1) proves a unicity theorem for the unit disk, and together with the other results will be used in subsequent existence and unicity theorems for general domains (Sections 37.2–37.7).

37.1.1 Unicity of Simple Mappings in the Unit Disc (Schwartz, 1890)

The parameters α, λ in the general bilinear mapping (35.68a–f) between unit discs may be determined by choosing a point and a direction through it, for example, (i) the centers correspond, that is, $z=0$ is equivalent to $\zeta=0$ if $\alpha=0$ since then $\zeta=-\mathrm{e}^{i\lambda}z$; (ii) the real axis correspond, that is $\arg(\zeta)=\arg(z)$ if $0=\arg(\zeta)-\arg(z)=\lambda-\pi$ implying $\lambda=\pi$ and hence $\zeta=-\mathrm{e}^{i\pi}z=z$. *Thus the bilinear mapping between unit disks that keeps the center at the origin and the direction of the real axis is the identity map* $\zeta=\mathrm{z}$. The point $z=a$ can be mapped to the origin (37.1b) and a straight line of inclination α through it can be mapped to the real axis (37.1c) by the linear transformation (37.1a):

$$\zeta(z)=(z-a)\mathrm{e}^{-i\alpha}, \quad \zeta(a)=0, \quad \arg(\zeta)=\arg(z-a)-\alpha; \tag{37.1a–c}$$

since the linear mapping is a particular case (Section 35.3) of the bilinear mapping, and the composition of bilinear mappings is a bilinear mapping (Section 35.5), this suggests the unicity theorem for the unit disk: *the only conformal univalent (37.2c) mapping from the unit disk* $|\mathrm{z}|\leq 1$ *in the z-plane to the unit disk* $|\mathrm{w}|\leq 1$ *in the w-plane (37.2d) that preserves one point (37.2a,e,f) and a direction through it (37.2b,g) is the identity map* $\zeta=\mathrm{z}$ *in (37.2h):*

$$c\in|C, \quad \alpha\in|R: \quad w\equiv f(z)\in\mathcal{D}\cup\mathcal{U}^1(|z|\leq 1), \quad |w|\leq 1, \quad |a|\leq 1:$$
$$\{f(a)=c, \quad \arg[f(a)]=\alpha\}\Rightarrow f(z)=z. \tag{37.2a–h}$$

Note that the theorem was proved before only in the class of bilinear mappings (Section 35.9) that are univalent in the whole plane, whereas the statement (37.2a–h) applies to the wider class of simple **mappings**, that is, conformal and univalent mappings in the unit disk. The conformal mapping that is performed by an analytic function is designated **simple** (**nonsimple**) if the function is univalent (nonunivalent), that is, maps distinct points to distinct points (not always); univalent (Section 35.4) means that the function takes each value in the range only once (37.36b) $\equiv$ (37.28), that is, assigns different values in the range to different values in the domain and its image does not overlap with itself.

Bearing in mind (37.1a–c) the theorem needs only to be proved for a function $f(z)$ analytic in a disk (37.2c) ≡ (37.3b) that has a simple zero at the center (37.3c,d), it follows that function (37.3e) is holomorphic in the disk:

$$R<1: \quad f(z)\in \mathcal{D}(|z|\le 1), \quad f(0)=0\neq f'(0): \quad g(z)=z^{-1}f(z), \quad |g(z)|\le \left|g\left(Re^{i\varphi}\right)\right|_{\max}, \tag{37.3a–f}$$

and by the lemma of the maximum (Section 33.1) takes maximum value in modulus at the boundary (37.3f); the equality in (37.3f) holds only if $g(z)=c$ is a constant, that is, $f(z)=cz$ is a linear function with $|c|=1/R$, or $c=R^{-1}e^{i\lambda}$. This result is **Schwartz's lemma**: *if the function* f(z) *is analytic in the disk (37.3b) and has a simple zero at the center (37.3c,d) with fixed derivative, then it has the upper bound:*

$$|f(z)|\begin{cases} =\dfrac{r}{R} & \text{if } f(z)=\dfrac{z}{R}e^{i\lambda}, \qquad (37.4a)\\[2mm] <\dfrac{r}{R}B(R) & \text{if } f(z)\neq\dfrac{z}{R}e^{i\lambda}, \qquad (37.4b)\end{cases}$$

where the equality (37.4a) holds in the linear case, and in general (37.4b) the constant unity is replaced by the maximum modulus (37.5a) of $f(z)$ in the disk $|z|\le R$:

$$B(R)\equiv\max_{|z|\le R}|f(z)|=\max_{0\le\varphi<2\pi}\left|f\left(Re^{i\varphi}\right)\right|: \quad r<R\Rightarrow B(r)\le B(R); \tag{37.5a,b}$$

the lemma of the maximum implies that the maximum of $|f(z)|$ occurs on the boundary $|z|=R$, and that if the disk is reduced to the radius $r<R$, the bound on the modulus $B(r)\le B(R)$ cannot increase (37.5b).

The Schwartz's lemma is applied next to a holomorphic function $\zeta=f(z)$, mapping the unit disk $|z|\le 1\equiv R$ to the unit disk $|\zeta|\le 1\equiv B$, that is, so that $|\zeta|=|f(z)|\le|z|$ because $R=1=B$ in (37.4b) or $R=1=c$ in (37.4a). If the function $\zeta=f(z)$ is univalent, it has an inverse mapping $z=f^{-1}(\zeta)$ that satisfies the condition $|z|\le|\zeta|$, by the Schwartz lemma. Thus must hold $|\zeta|=|z|$, that is, $\zeta=e^{i\lambda}z$; in order that the real axis correspond $\lambda=0$, that is, $\zeta=z$ and the identity map is obtained. This provides a general proof of the unicity theorem on the simple mapping between units disks, stated before. The Schwartz lemma (37.4a,b), of that further applications will be given subsequently, supplies an upper bound for the modulus of a function univalent in a disk, using its maximum modulus (37.5a). Alternate forms of this upper bound can be obtained by considering instead of the maximum modulus (37.5a), the modulus of the maximum or minimum of the real or imaginary parts:

$$A(R)\equiv|\min,\max\{\mathrm{Re},\mathrm{Im}[f(z)]\}|. \tag{37.6}$$

The auxiliary analytic function (37.7a) ≡ (27.47a):

$$h(z)\equiv\exp\{\pm f(z),\pm if(z)\}, \quad |h(|z|\le R)|\le\exp\{A(R)\}, \tag{37.7a,b}$$

has extreme values on the boundary $|z|=R$; it follows that

$$r<R: \quad A(r)\le A(R), \quad B(r)\le B(R), \quad B(R)\ge A(R); \tag{37.8a–c}$$

thus *for a function holomorphic in the unit disk (37.3b) both the maximum modulus (37.5a) and the maximum of the modulus of the real and imaginary parts (37.6) are nondecreasing functions of the radius (37.8a,b), and (37.8c) the former cannot be lower that the latter.*

37.1.2 Bounds for the Modulus of the Real or Imaginary Part (Caratheodory, 1912; Borel, 1928)

To improve on the Schwartz lemma, consider again a function $f(z)$ analytic in the unit disk (37.3b) and with a simple zero at the center (37.3c,d) so that $A(R) \geq A(0) = 0$. The function $h(z)$ defined by (37.9a):

$$h(z) \equiv \frac{f(z)}{2A(R) - f(z)}, \quad f(z) = \frac{2A(R)\,h(z)}{1 + h(z)}, \tag{37.9a,b}$$

is holomorphic, because either the real or imaginary part of the denominator of (37.9a) does not vanish, and so $h(z)$ cannot have any new singularities. Also $h(z)$ vanishes at the origin (37.10b) because $f(z)$ does, and is of bound unity in the unit disk:

$$f(z) \equiv \Phi + i\Psi: \quad h(0) = 0, \quad |h(z)|^2 = \frac{\Phi^2 + \Psi^2}{[2A(R) - \Phi]^2 + \Psi^2} \leq 1, \tag{37.10a–c}$$

because $|\Phi| \leq A(R)$, and hence $|2A(R) - \Phi| \geq |\Phi|$, in the case where the real part is taken in (37.6); if the imaginary part is taken the same result holds with Ψ replacing Φ. In either case $h(z)$ satisfies the conditions of the Schwartz lemma (37.2a,b):

$$|h(z)| \leq \frac{r}{R}: \quad |f(z)| \leq \frac{2A(R)\,|h(z)|}{1 - |h(z)|} \leq \frac{2r}{R-r} A(R), \tag{37.11a,b}$$

where the equality holds only if $h(z) = cz = z/Re^{i\lambda}$ is a linear function (37.4a), and hence $f(z)$ is a homographic function (37.9b). Thus was proved the **Borel–Caratheodory lemma**:

$$|z| \equiv r < R: \quad |f(z)| \leq \frac{2r}{R-r} A(R) \quad \text{if } f(z) \underset{\neq}{=} \frac{2z}{Re^{i\lambda} + z} A(R), \tag{37.12a,b}$$

if the function $f(z)$ is analytic in the unit disk (37.3b), has a simple zero at the center (37.3c,d) and has a bound (37.6) for its real or imaginary part, then it satisfies (37.12b), where the equality holds only for a homographic transformation of the type (37.9b).

The restriction of the Schwartz (37.2a,b) and Borel–Caratheodory (37.12a,b) lemmas to a function with a zero at the origin can be lifted by replacing $f(z)$ by $f(z) - f(0)$:

$$\begin{aligned} |f(z)| &\leq |f(z) - f(0)| + |f(0)| \leq |f(0)| + \frac{r}{R} \max_{|z| \leq R} |f(z) - f(0)| \\ &\leq |f(0)| + \frac{r}{R} \{B(R) + |f(0)|\}, \end{aligned} \tag{37.13a}$$

$$\begin{aligned} |f(z)| &\leq |f(0)| + \frac{2r}{R-r} \max_{|z| \leq R} |\mathrm{Re}, \mathrm{Im}\{f(z) - f(0)\}| \\ &\leq |f(0)| + \frac{2r}{R-r} \{|f(0)| + A(R)\}. \end{aligned} \tag{37.13b}$$

This yields the **extended Schwartz [Borel–Caratheodory] lemma**: *if the function $f(z)$ is holomorphic is the disk (37.3b), and has an upper bound for the modulus (37.5a) [an upper or lower bound for the real or imaginary part (37.6)], it satisfies (37.14b) [(37.15)] for (37.14a):*

$$|z| \equiv r < R: \quad |f(z)| \le \left(1 + \frac{r}{R}\right) |f(0)| + \frac{r}{R} B(R), \tag{37.14a,b}$$

$$|f(z)| \le \frac{2rA(R) + (R+r)|f(0)|}{R-r} \le \frac{R+r}{R-r}\left[A(R) + |f(0)|\right]; \tag{37.15}$$

this reduces to (37.4b) [(37.12b)] for $f(0) = 0$.

37.1.3 Upper Bounds for the Modulus of Derivates

The second inequality (37.15) can be usefully extended to all derivates of $f(z)$ calculated by the Cauchy third theorem (15.13) ≡ (37.16b) on a subcircle with center at z and radius (37.16a), such that it lies entirely within the disk of radius R:

$$|\zeta - z| = \frac{R-r}{2}: f^{(n)}(z) = \frac{n!}{2\pi i} \int_{|\zeta - z| = (R-r)/2} f(\zeta)(\zeta - z)^{-n-1}\, d\zeta. \tag{37.16a,b}$$

On the circle:

$$|\zeta| \le |\zeta - z| + |z| = r + \frac{R-r}{2} = \frac{R+r}{2} \equiv \delta, \tag{37.17a}$$

the inequality (37.15) yields (37.17c):

$$r < R: \frac{\max|f(\zeta)|}{A(R) + |f(0)|} \le \frac{R+\delta}{R-\delta} = \frac{3R+r}{R-r} < \frac{4R}{R-r}; \tag{37.17b,c}$$

on substitution of (37.17c) in (37.16b):

$$\left|f^{(n)}(z)\right| \le \frac{n!}{2\pi}\left[A(R) + |f(0)|\right] \frac{4R}{R-r}\left(\frac{R-r}{2}\right)^{-n} 2\pi, \tag{37.18}$$

it follows that

$$\left|f^{(n)}(z)\right| \le \frac{2^{2+n} n! R}{(R-r)^{n+1}}\left[A(R) + |f(0)|\right]. \tag{37.19}$$

If the function $f(z)$ is analytic in the unit disk (37.3b), and has a bound (37.6) for its real or imaginary part, then its n-th order derivate has upper bound in modulus specified by (37.19).

37.2 Existence of a Uniformly Convergent Subsequence (Vitali, 1903; Montel, 1910; Osgood, 1929)

Besides the bounds on holomorphic functions (Section 37.1), the proof of the existence and unicity theorems (Sections 37.3–37.7) will require some results on sequences of series of functions (Subsection 37.2.1), in particular on uniformly convergent subsequences (Subsection 37.2.2). These results apply in particular to holomorphic and univalent functions, and lead to some properties of simple mappings (Subsection 37.2.3).

37.2.1 Uniform Convergence in a Subregion (Vitali, 1903; Osgood, 1929)

The consideration of sequences of functions analytic in a disk or in a compact domain as regards uniform convergence, leads to the **Osgood–Vitali theorem**: *let* $f_n(z)$ *be a sequence of functions analytic (37.20a) in compact domain* D, *and with an upper bound* B *for the modulus (37.20a) independent of* z *and* n:

$$f_n(z) \in \mathcal{A}(D): \quad \exists_{B\in|R} \forall_{z\in D, n\in|N}: \quad |f_n(z)| < B \Rightarrow \forall_{z\in E<\partial D}: \quad \sum_{n=1}^{\infty} f_n(z) \text{ U.C.} \tag{37.20a–c}$$

Let $f_n(z)$ *converge at a sequence of points having a limit point in* D. *Then* $f_n(z)$ *converges uniformly to a limit in any region* E *bounded by a contour* ∂E *in the interior of* D*, that is, its limit is a function analytic in* E. It is sufficient to prove the theorem for a circle with center at the limit point at the origin; for if the limit point is z_0 in the interior of D, the theorem can be proved in a circle of center at z_0 and radius R such that it lies entirely within D; then analytic continuation can be used to extend the result to any other point z within D. The sequence of functions $f_n(z)$ analytic in a circle of radius R and center at the origin, has Maclaurin series:

$$\lim_{n\to\infty} f_n(z) \equiv \lim_{n\to\infty} \sum_{p=0}^{\infty} a_{n,p} z^p = \sum_{p=0}^{\infty} a_p z^p = f(z), \tag{37.21}$$

and the aim is to prove that (i) the limit can be taken under the series; (ii) the coefficients $a_{n,p} \to a_p$ converge; (iii) hence an analytic function $f(z)$ is obtained.

Taking the limit point at the origin, that is, at the center of the circle:

$$|f_n(z) - a_{n,0}| = |f_n(z) - f_n(0)| \le |f_n(z)| + |f_n(0)| \le 2B(R), \tag{37.22a}$$

where B denotes the maximum modulus (37.5a). Noting that $f_n(z) - f_n(0)$ vanishes at the origin, the Schwartz lemma (37.4a,b) holds:

$$|z| \le R: \quad |f_n(z) - f_n(0)| \le \frac{2B(R)|z|}{R}. \tag{37.22b,c}$$

Let a be a point where the sequence converges:

$$\begin{aligned} |f_n(0) - f_{n+m}(0)| &\le |f_n(0) - f_n(a)| + |f_{n+m}(a) - f_{n+m}(0)| + |f_n(a) - f_{n+m}(a)| \\ &\le \frac{4B(R)|a|}{R} + |f_n(a) - f_{n+m}(a)|. \end{aligned} \tag{37.23}$$

Since the origin is the limit point $|a|$ can be chosen arbitrary small; hence the first term of (37.23) is arbitrary small; the second also because the sequence converges at a. Thus the sequence $f_n(0) \equiv a_{n,0}$ converges to limit, that is, $a_{n,0} \to a_0$ proving (37.21) for $p = 0$. Next consider the function:

$$g_n(z) \equiv \frac{f_n(z) - a_{n,0}}{z} = \sum_{p=1}^{\infty} a_{n,p} z; \tag{37.24}$$

it has the same properties as $f_n(z) - f_n(0)$, namely: (i) it tends to a limit at $z = a$; (ii) it satisfies a Schwartz type inequality (37.22c) with a different bound $B(R)$. Thus can be deduced as in (37.23) that $g_n(0) = a_{n,1}$ converges, that is, $a_{n,1} \to a_1$, proving (37.22a) for

$p=1$. Proceeding in the same way it is proved that all $a_{n,p} \to a_p$ have limits as $n \to \infty$. The proof of the theorem is completed by taking the limit under the summation that is valid for uniformly convergent series. The series (37.21) is uniformly convergent for $|z| \le R-\varepsilon$ by the Weierstrass M-test (Section 21.7), if a bound for the coefficient $|a_{n,p}| < M/R^p$ exists that is independent of z. The bound just indicated is the Cauchy bound (39.69d) $\equiv$ (37.25a), and it follows from (37.21):

$$|a_{n,p}| = \frac{1}{p!}\left|f_n^{(p)}(0)\right| \le \frac{M}{R^p}; \quad M \ge 2^{2+p}\{A(R)+|f_n(0)|\}, \tag{37.25a,b}$$

using (37.19) with $z=0=r$; this leads to the upper bound (37.25b) that is independent of z and n. QED.

37.2.2 Uniform Subsequences and Normal Families (Montel, 1910)

The Osgood–Vitali theorem implies the following **theorem on uniform subsequences**: *a sequence $f_n(z)$ of analytic functions bounded in a compact region D contains a subsequence $g_n(z)$ uniformly convergent in any region $E \subset D - \partial D$ interior to D:*

$$f_n(z) \in \mathcal{A}(D) \cap \mathcal{B}(D): \quad \exists_{\{g_n\}\subset\{f_n\}} \forall_{z\in E\subset D-\partial D}: \quad \sum_n g_n(z) \text{ U.C.} \tag{37.26}$$

Consider a sequence of functions $f_n(z)$ analytic and bounded in D, and a sequence of points $z_1, \ldots, z_n, \ldots$ with limit point z_0 in D. Then all points $\zeta_n = f_n(z)$ lie in the compact region D, and they have (Section 27.8) at least one accumulation point. Thus there is a sequence $n_1, n_2, \ldots$ such that $f_{n1}, f_{n2}, \ldots$, converges at z_1. From this sequence can be selected a subsequence $f_{m1}, f_{m2}, \ldots$ that converges also at z_2, and so on:

$$f_{n1}, f_{n2}, \ldots \equiv \Phi_{1,1}, \Phi_{1,2}, \ldots, \Phi_{1,n}, \ldots \quad \text{converges at } z_1, \tag{37.27a}$$

$$f_{m1}, f_{m2}, \ldots \equiv \Phi_{2,1}, \Phi_{2,2}, \ldots, \Phi_{2,n}, \ldots \quad \text{converges at } z_1, z_2, \tag{37.27b}$$

$$\Phi_{p,1}, \Phi_{p,2}, \ldots, \Phi_{p,n}, \ldots \quad \text{converges at } z_1, z_2, \ldots, z_p; \tag{37.27c}$$

the diagonal sequence:

$$g_1, g_2, \ldots, g_n \equiv \Phi_{1,1}, \Phi_{2,2}, \ldots, \Phi_{n,n}, \ldots \quad \text{converges at } z_1, z_2, \ldots, z_n, \ldots, \tag{37.27d}$$

converges at z_1 from $n=1$, at z_2 from $n=2$, etc. at all points $z_1, \ldots, z_n$; thus by the Osgood–Vitali theorem it converges uniformly in any region interior to D. QED. A family of functions in a domain is a **normal family** if from every infinite sequence of functions from the family a uniformly convergent subsequence can be formed; this subsequence can converge either (i) to a finite function or (ii) to a constant infinity. A normal family need not be formed by analytic functions, viz. other types of functions can be considered. The Osgood–Vitali theorem can be restated: *a family of analytic functions in a compact region is a normal family.* It contains a subsequence converging uniformly to an analytic function, that is, the case (i) above; only the case (i) is possible, because (ii) is excluded because analytic functions that are bounded.

37.2.3 Some Properties of Simple Mappings

A simple mapping $\zeta = f(z)$ was defined (Subsection 37.1.1) as a conformal and univalent mapping, that is, specified by an analytic function that assigns different values of ζ to distinct values of z:

$$f(z) \in \mathcal{U}^1(D): \quad \forall_{z_1, z_2 \in D}: \quad z_1 \neq z_2 \Rightarrow f(z_1) \neq f(z_2); \tag{37.28}$$

a univalent function takes each value in the range only once (37.36b). *A function* f(z) *simple in a domain has nonvanishing derivate there.* To prove this suppose that the homorphic function has vanishing derivate (37.29a,b) at some point a in D:

$$A \neq 0: \quad f'(z) = A(z-a) + O\left((z-a)^2\right); \quad f(z) = \frac{A}{2}(z-a)^2 + O\left((z-a)^3\right), \quad (37.29\text{a–c})$$

the function (37.29c) has a double root at $z=a$ and hence is not univalent:

$$z_1 = a + \varepsilon \neq a - \varepsilon = z_2: \quad f(z_1) = \frac{A\varepsilon^2}{2} + O\left(\varepsilon^3\right) = \frac{A(-\varepsilon)^2}{2} = \frac{A\varepsilon^2}{2} = f(z_2); \qquad (37.30)$$

this is equivalent to the remark that if the derivate vanishes at a point $z=a$, it is a critical point of first kind (Section 33.4), that is, the mapping is not conformal, hence not simple.

An application of the theorem of uniform subsequences, is the proof of a **lemma on simple mappings**: *a sequence* $f_n(z)$ *of functions, univalent and holomorphic in compact region converges either to a constant or to a simple mapping.* The possibility of the limit being a constant, for example, zero, is illustrated by the case $f_n(z) = z/n \to 0$ as $n \to \infty$. To prove the theorem note that the limit function exists and is analytic by the Osgood–Vitali theorem (Subsection 37.2.1), and thus specifies a conformal mapping. To complete the proof it must be shown that the limit function is univalent, for example, by assuming that it is not and then arrive at a contradiction. If the limit function $f(z)$ is not univalent, it takes the same value $f(z_1) = \zeta_0 = f(z_2)$ at two distinct points $z_1 \neq z_2$; by the theorem on uniformly convergent subsequences, there is a subsequence $g_n(z)$ of $f_n(z)$ converging uniformly at z_1 and z_2. Considering:

$$g_n(z) - \zeta_0 = \{f(z) - \zeta_0\} + \{g_n(z) - f(z)\}, \qquad (37.31)$$

the second term can be made arbitrary small for sufficiently large n. Thus $g_n(z) - \zeta_0$ has the same number of zeros as $f(z) - \zeta_0$, viz. two $z_1 \neq z_2$; it follows that the sequence $f_n(z)$ contains nonunivalent functions that contradicts the hypothesis. QED.

37.3 Simply Connected Region with at Least Two Boundary Points (Riemann, 1863)

Two necessary conditions for the existence of a simple mapping are established first (Subsection 37.3.1) and then it is shown that they are also sufficient (Subsection 37.3.2); before proceeding to the proof (Section 37.5), some additional properties are established, viz. correspondence between interiors and boundaries (Subsection 37.3.3) and conditions of unicity of the mapping (Subsection 37.3.4).

37.3.1 Two Necessary Conditions for the Existence of a Simple Mapping

The preceding results can be used to prove the existence and unicity of simple mappings of an arbitrary region D of the z-plane into the unit disk $|\zeta| \leq 1$ in the ζ-plane by first establishing any restrictions on the region D. First the region must be simply connected. For example, consider a doubly connected region D with outer (inner) boundary $\partial\bar{D}(\partial\underline{D})$, and let L be a contour containing the inner boundary (Figure 37.1a). A simple mapping

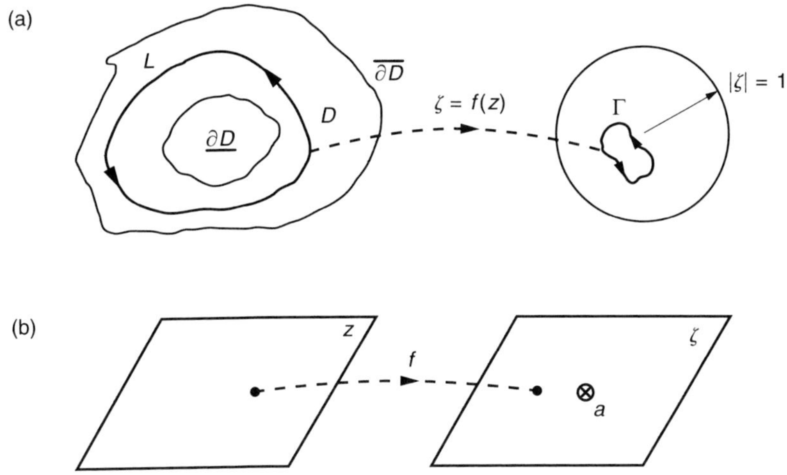

FIGURE 37.1
A simple mapping is conformal and univalent: (i) conformal implies the preservation of the modulus and direction of angles, that is, of the ratio of lengths; (ii) univalent means that each point in the range is mapped from just one point in the domain, that is, the range does not overlap with itself nor is folded over itself. A doubly (or multiply) connected region cannot be mapped conformally to a singly connected region (a), because: (i) a loop containing the hole (one or more holes) cannot be shrunk to zero; (ii) the image loop can be shrunk to zero; (iii) thus the mappings is not continuous, hence it is not differentiable, and cannot be conformal. A simply connected region cannot be mapped conformally to a "punctured" domain, which misses one point (b). If it could a contradiction would be reached because: (i) the punctured domain can be mapped conformally on to a domain without point-at-infinity; (ii) the combined mapping would be conformal and bounded in the whole domain; (iii) then by the Liouville theorem it would reduce to a constant, and would map the whole domain to a point. These two examples show that a simple mapping into a singly connected region requires that: (I) the domain be simply connected; (II) the range has at least two boundary points. These two necessary conditions for the existence of a simple mapping turn out to be also sufficient.

into the unit disk would transform L into a loop Γ that could be shrunk continuously to zero. However, the loop L cannot be shrunk to zero without crossing the inner boundary, showing that the mapping cannot be continuous, hence it is not conformal or simple. Second the region must have at least two boundary points. For suppose that the region D has only one boundary point, that is, it is the **punctured plane**, for example, the (Figure 37.1b) complex plane z less the point a. Then the simple mapping $\zeta = 1/(z-a)$ sends the point $z = a$ to infinity $\zeta = \infty$. Next is made a simple mapping $w = f(\zeta)$ into the w-plane minus infinity; the composition of the two mappings is a function analytic in the whole complex w-plane. By the Liouville theorem (Subsection 27.6.1) the function must be a constant $w = f(\zeta) = \text{const}$ and $\zeta = \text{const}$; thus all points of the z-plane are mapped to a single point of the ζ-plane, and the mapping is not univalent or simple. These two necessary conditions will be shown to be sufficient as well (Section 37.5).

37.3.2 Proof of Sufficiency and the Simple Mapping Theorem

Thus can be stated (Figure 37.2) the **simple mapping theorem**: *a simply connected region D whose boundary ∂D consists of at least two points of $a \neq b$, can be mapped by a simple,*

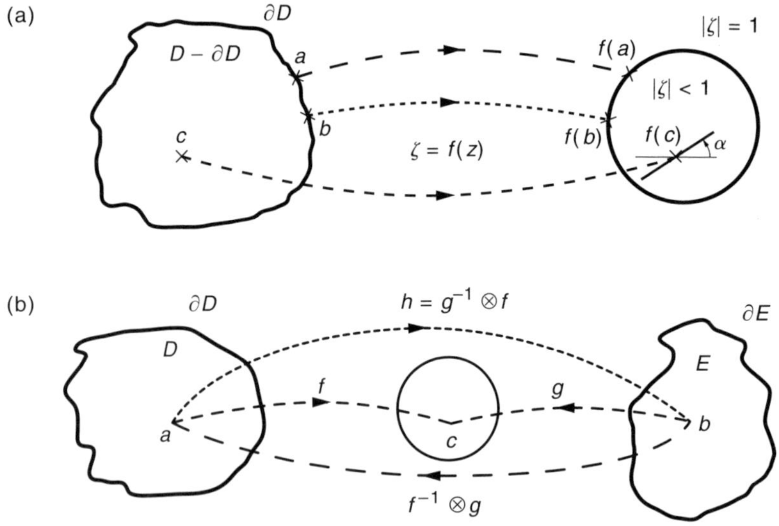

FIGURE 37.2
The simple, that is, conformal univalent, mapping of a simply connected region with at least two boundary points to the unit disk exists, and can be made unique (a) choosing a point and a direction through it, that is, to within a translation and rotation. It follows that any simply connected region D with at least two boundary points can be simply mapped to another simply connected region E with at least two boundary points as follows: (i) both D (E) can be mapped to the unit disk by simple mappings $f(g)$; (ii) a simple mapping is conformal and univalent, hence single-valued and univalent, hence injective, hence invertible; (iii) hence g^{-1} is a simple mapping, and its composition with the simple mapping f is another simple mapping $g^{-1} \circ f$ from D to E.

that is, holomorphic and univalent (37.32c) function $\zeta = f(z)$, from the z-plane *to the unit disk $|\zeta| \leq 1$ in the ζ-plane. The function maps simply the interior $D - \partial D$ (boundary ∂D) of D into (37.32d) [(37.32e)] the interior of the unit circle $|\zeta| < 1$ (the unit circle $|\zeta| = 1$). The mapping is unique if are specified the transforms of: (i) a given (37.32a) point a in D; (ii) a direction α through it (37.32b):*

$$f(a) = c \in |C, \arg\{f(a)\} = \alpha \in |R : \exists^1 f(z) \in \mathcal{D}(D) \cap \mathcal{U}^1(D) : \begin{cases} z \in D - \partial D \Rightarrow |f(z)| < 1 \\ z \in D \qquad\quad\; \Rightarrow |f(z)| = 1. \end{cases} \tag{37.32a–e}$$

The necessary conditions have already been demonstrated, and the sufficient conditions will be proved in Section 37.5. The two statements added to Riemann's theorem are proven next.

37.3.3 Correspondence of Interiors and Boundaries

First concerning the correspondence of boundaries and interiors. If $z_1 \in D - \partial D$ is an interior point of D, there exists a circle $V_\varepsilon(z_1)$ with center at z_1 and radius ε wholly contained in D; it is mapped into a region $W_\varepsilon(\zeta_1)$ contained in $|\zeta| \leq 1$, to that $\zeta_1 \equiv f(z_1)$ belongs: hence ζ_1 is an interior point of $|\zeta| \leq 1$. In addition if $z_2 \in \partial D$ lies on the boundary of D, every neighborhood $V_\varepsilon(z_2)$ contains points internal and external to D, and thus its transform $W_\varepsilon(\zeta_2)$ contains points internal and external to $|\zeta| \leq 1$. Conversely for every neighborhood $W_\delta(\zeta_2)$ there exists a z_2 such that the transform $W_\varepsilon(\zeta_2)$ of $V_\varepsilon(z_2)$ lies within $W_\delta(\zeta_2) \supset W_\varepsilon(\zeta_2)$; since every neighborhood of $\zeta_2 = f(z_2)$ contains points internal and external to $|\zeta| \leq 1$ and $\zeta_2 = f(z_2)$ then z_2 lies on the boundary $|z_2| = 1$. QED.

37.3.4 Conditions of Unicity of the Mapping

Second concerning the unicity of the mapping, the conditions (37.32a,b) are similar to those (37.2a,b,f,g) used to render the homographic mapping between unit circles unique (Subsection 37.1.1). This suggests that the function $f(z)$ be replaced by (37.33a):

$$g(z) \equiv \mathrm{e}^{-i\alpha}[f(z) - c]: \quad g(a) = 0 = \arg\{g(a)\}, \qquad \text{(37.33a–c)}$$

so that the conditions (37.32a,b) are replaced by (37.33b,c). Suppose that two simple mappings exist $g(z), h(z)$ from the region D to the unit disk satisfying (37.33b,c). Since the inverse and composition of simple mappings is simple, the function $g \circ h^{-1}$ maps simply the unit circle into itself, while preserving the origin (37.33b); the theorem for the unit disk (Subsection 37.1.1) states that the only such mapping is the identity $g \circ h^{-1} = 1$, hence $g = h$. Thus the Riemann mapping satisfying the conditions (37.33a–c), or alternatively (37.32a,b), is unique. QED. The fundamental theorem (37.3.2) on the existence of simple mappings into a unit disk (Figure 37.2a) implies through the composition of simple mappings that are invertible, that *two simply connected regions with two boundary points can be mapped into each other by a simple mapping (Figure 37.2b).*

37.4 Mapping between Multiply Connected Regions by Multivalent/Multivalued Functions

The extension of the conformal mapping from simply connected to multiply connected regions (Subsection 37.4.2) implies allowing nonsimple (Subsection 37.4.1) mappings, for which the function may be multivalued (Chapters 7 and 9). A multivalued function can be univalent if there is no overlap in the range (Subsection 37.4.3), for example, the logarithm is a many-valued univalent function that maps on annulus into an infinite strip (Subsection 37.4.5). This leads to the introduction of a reference function (Subsection 37.4.6) mapping between disks with one puncture. The theorem on uniformly convergent subsequences (Subsection 37.2.2) can be extended to nonsimple mappings (Subsection 37.4.4). Some of there results will be used subsequently (Section 37.5) in the proof of the Riemann theorem on the existence of a simple mapping from the unit disk to a simply connected compact region with at least two boundary points.

37.4.1 Univalent/Multivalent Functions or Mappings

An analytic mapping is nonsimple if it is specified by a multivalent function, that is, if there are at least two points in the domain mapped into the same point in the range:

$$f(z) \notin \mathcal{U}^1(D): \quad \exists_{z_1, z_2 \in D}: z_1 \neq z_2 \wedge f(z_1) = w_0 = f(z_2). \qquad \text{(37.34a,b)}$$

Let w_0 be such a point in the range that is mapped from points $z_1, \ldots, z_N$ in the domain (37.35a) of the function, each a zero (37.35b) of order α_n:

$$\exists_{z_1,\ldots,z_N \in D}: \quad f(z) - w_0 \equiv O((z - z_n)^{\alpha_n}); \quad Z(w_0) = \sum_{n=1}^{N} \alpha_n = m, \tag{37.35a–c}$$

then the function is said to **take m-times the value at the point** w_0, where m is (37.35c) the sum of the orders of all zeros. A function is **m-multivalent** if there is at least one point (no points) where it takes that value m (more than m) times:

$$f \text{ is } m\text{-multivalent}: \quad \mathcal{U}^m = \{f(z): \forall_{w \in E}: Z(w) \le m \wedge \exists_{w \in E}: Z(w) = m\}; \tag{37.36a}$$

in particular $Z(w) = 1$ for all w for a **univalent** function, that is, is a function that takes only once every value in its range (37.36b) $\equiv$ (37.28):

$$f \text{ is univalent}: \quad \mathcal{U}^1 \equiv \{f(z): \quad \forall_{w \in E} \exists^1_{z \in D}: \quad w = f(z)\}. \tag{37.36b}$$

If $Z(w) = \infty$ the function is many-valent, that is, there is at least value in the range taken at an infinite number of points in the domain:

$$f \text{ is many valent}: \quad \mathcal{U}^\infty \equiv \{f(z): \quad \exists_{w \in E} \exists^\infty_{z \in D}: \quad f(z) = w\}. \tag{37.36c}$$

A single-valued univalent function is injective (9.2b) that is, maps distinct points in the domain to distinct points in the range; a multivalued univalent function cannot be injective, because at least one point in the domain maps to more than one point in the range. This leads to the question of inversion for single-, multi-, and many-valued functions and uni/multi/many-valent functions. (Note 37.4).

37.4.2 Inversion of Single/Multi/Many Valued and Uni/Multi/ Many Valent Functions

This question can be clarified by the example of the complex exponential: (i) it is single-valued (37.37a) since to each point in the domain corresponds only one point in the range (37.37b):

$$w \equiv \exp(z) \in \mathcal{U}^\infty_1(|C): \quad \forall_{z \in |C} \quad \exists^1_{w \in |C}: \quad w = \exp z; \tag{37.37a,b}$$

$$\forall_{z \in |C} \quad \exists_{w \in |Z}: \quad w = \exp z = \exp(z + i2n\pi) = \exp(z_n), \quad z_n = z + i2n\pi, \tag{37.37c,d}$$

(ii) it is many valent (37.37a) since to each point in the range (37.37b) there exist an infinite number of distinct points in the domain (37.37c), all differing by (37.37d) a multiple of $2\pi i$. Whereas the exponential is a single-valued (37.37a,b) many-valent (37.37b,c) function, its inverse, that is, the logarithm, is a many-valued (37.39a–c) univalent (37.40a,b) function, because: (iii) to each value in the domain the logarithm assigns an infinite number of distinct points in the range (37.39b), differing by multiples of $2\pi i$:

$$z \equiv \log(w) \in \mathcal{U}^1_\infty(|C): \forall_{z \in |C} \quad \forall_{n \in |Z}: \quad \log(w) = \log|w| + i\arg(w) + i2\pi n = w_n, \tag{37.38a,b}$$

hence its is a many-valued function (37.38b), with principal branch (37.39a):

$$w_0 = \log|w| + i\arg(w), \quad w_n = w_0 + i2\pi n, \tag{37.39a,b}$$

all other branches (37.39b) differing by a multiple of $2\pi i$; (iv) it is univalent (37.38a) because each value in the range corresponds to only one value in the domain (37.40b, c):

$$\operatorname{Re}(z) = \log|w|, \quad \exists^1_{n\in|Z}: \quad \operatorname{Im}(z) = \arg(w) + 2\pi n, \tag{37.40a,b}$$

namely: (i) the real part is unique (37.40a); (ii) the imaginary part (37.40b) is matched by the argument adding only one multiple of 2π (other integers $m \neq n$ will not do).

The preceding example suggests the following results: *(i) a single-valued univalent function is injective, that is, maps one-to-one the domain to the range, and has a single-valued univalent inverse, and vice-versa:*

$$w = f(z) \in \mathcal{U}^1_1(D) \equiv \widetilde{\mathcal{U}}(D) \Leftrightarrow z = f^{-1}(w) \in \mathcal{U}^1_1(E) \equiv \widetilde{\mathcal{U}}(E); \tag{37.41}$$

(ii) an n-valued univalent function has n-branches and hence its inverse is a n-valent single-valued function (39.42) and vice-versa:

$$w = f(z) \in \mathcal{U}^1_n(D) \Leftrightarrow z = f^{-1}(w) \in \mathcal{U}^n_1(E); \tag{37.42}$$

(iii) an n-valued m-valent function has an inverse that is a m-valued n-valent function and vice-versa:

$$w = f(z) \in \mathcal{U}^m_n(D) \Leftrightarrow z = f^{-1}(w) \in \mathcal{U}^n_m(D); \tag{37.43}$$

(iv) the preceding statements remain valid if $m = \infty$ and/or $n = \infty$:

$$w = f(z) \in \left\{\mathcal{U}^1_1, \mathcal{U}^n_1, \mathcal{U}^\infty_1, \mathcal{U}^1_m, \mathcal{U}^1_\infty, \mathcal{U}^\infty_\infty\right\} \Leftrightarrow z = f^{-1}(w) \in \left\{\mathcal{U}^1_1, \mathcal{U}^1_n, \mathcal{U}^1_\infty, \mathcal{U}^m_1, \mathcal{U}^\infty_1, \mathcal{U}^\infty_\infty,\right\}, \tag{37.44}$$

for example, (v) an m-valued many-valent function has a m-valent many-valued inverse, and vice-versa; (vi) a many-valued many-valent function has a many-valued many-valent inverse, and vice-versa. The statements (ii) to (iv) assume that the ranges do not overlap.

37.4.3 Multivalued Univalent Function and Nonoverlapping Range

Whether it is single-valued or multivalued, an univalent (37.36b) function: (i) cannot map distinct points in the domain to the same point in the range; (ii) hence a point in the range comes from only one point in the domain; (ii) that point in the domain can (cannot) map to other points in the range if the function is multivalued (single-valued), and in this case the function is not (is) injective. A univalent function maps an open curve into an open curve, that is, if there are no multiple points in the domain there are no multiple points in the range, viz. a curve that does not cross itself in the domain cannot cross itself in the range. *A univalent function maps the domain in a range without overlaps.* Thus a univalent single (multi)-valued function maps the domain into a range without overlaps and consisting of one (several) sheets. The superposition of the sheets in the case of a multi-valued univalent function may lead to: (i) a multiply connected region if they are not all next to each other; (ii) to a simply connected region if the ranges in each are all next to each other without overlaps. An example of (ii) is (Figure 37.3) the logarithmic mapping of the annulus (37.48a–c) into the strip (37.49a–c). In the limit when the inner (outer) radius tends to zero (infinity) in (37.45c), the logarithm (34.45a) maps the whole complex z-plane into: (i) the horizontal strip (37.45d) for its principal branch (37.45b):

$$w = \log z \in \mathcal{U}^1_\infty: \quad w_0 = \log|z| + i\arg(w): \quad 0 < |z| < \infty, 0 \leq \operatorname{Im}(w_0) = \arg(z) < 2\pi; \tag{37.45a–d}$$

(ii) the whole plane univalently for all branches (37.45e):

$$w_n = \log|z| + i[\arg(z) + 2\pi n] = w_0 + i2\pi n: \quad 2\pi n \leq \operatorname{Im}(w_n) < 2\pi(n+1), \tag{37.45e,f}$$

because they are nonoverlapping (37.45f).

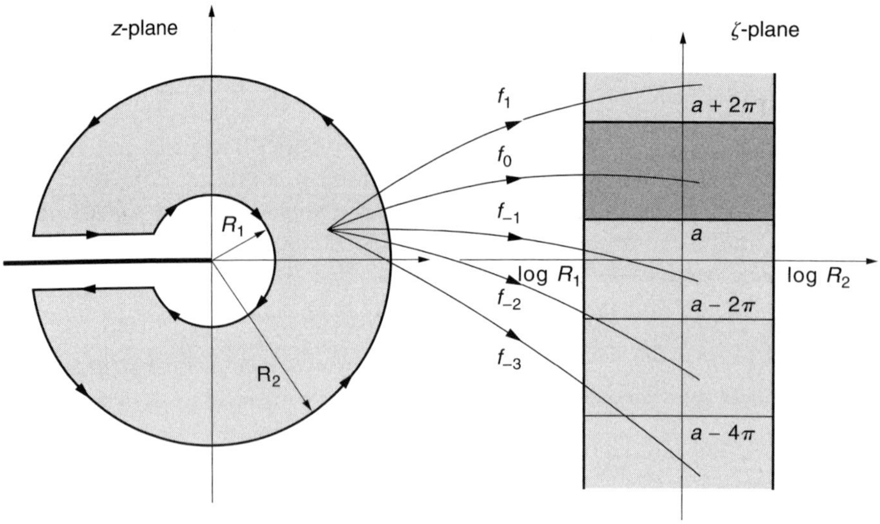

FIGURE 37.3
The logarithm has a branch-point at the origin (Figure 7.1a) with a semiinfinite branch-cut joining to infinity. The branch-cut can be avoided by taking as domain an annulus in the z-plane with the inner and outer circles joined above and below the branch-cut. Since $|z_1| < |z| < |z_2|$ the cut-annulus is mapped by the logarithm $\zeta = \log z$ into a vertical strip $\log|z_1| \leq \mathrm{Re}(\zeta) \leq \log|z_2|$ in the ζ-plane. The logarithm is a many-valued function whose principal branch has imaginary part $0 \leq \mathrm{Im}[\log(z)] = \arg(z) < 2\pi$, and thus occupies one rectangle of the strip. The other branches differ by multiples of $2\pi i$, and fill the whole strip with congruent rectangles of height 2π. One point in the cut-annulus in the z-plane is mapped to an infinite number of points in the ζ-plane, one in each rectangle. Thus the logarithm is a many-valued function; it is also univalent because each point in the strip in the ζ-plane is mapped from only one point in the cut-annulus in the z-plane.

37.4.4 Convergence of a Sequence of Nonsimple Mappings

The uniform convergence of a sequence of analytic functions preserves this m-valued property: *a uniformly convergent sequence of analytic functions that take a value m times at a point in their range, converges to an analytic function that also takes the m times same value at that point:*

$$f_n \in \mathcal{D}(D) \cap \mathcal{U}^m(D) \wedge f_n(z) \text{ U.C.} \Rightarrow \lim_{N\to\infty} f_n(z) = f(z) \in \mathcal{D} \cap \mathcal{U}^m. \tag{37.46}$$

The proof follows from the fourth Cauchy theorem (31.36c) $\equiv$ (37.47b) on the number of zeros of analytic function in a region D:

$$f_n(z) \in \mathcal{U}^m(D): \quad \alpha_m = \lim_{n\to\infty} (2\pi i)^{-1} \int_{\partial D} f_n'(z)\left[f_n(z) - w_0\right]^{-1} \mathrm{d}z$$

$$= (2\pi i)^{-1} \int_{\partial D} f'(z)\left[f(z) - w_0\right]^{-1} \mathrm{d}z = \alpha_m, \tag{37.47a,b}$$

where (37.47a) is the set of functions that take m values w_0 in the range E of D, since: (i) the uniform convergence allows taking the limit under integral sign in (37.47b); (ii) it follows that in every subdomain the number of zeros α_m of the limit function minus w_0 is the same for all the members of the sequence; (iii) hence the limit function takes every value the same number of times. QED. *The Osgood–Vitali theorem (Subsection 37.2.1) and the theorem on uniform subsequences (Subsection 37.2.2) apply to analytic functions, whether they are univalent that is, simple mappings, or multivalued that is, nonsimple mappings, because they hold for each branch.*

37.4.5 Mapping of an Anulus into Infinite Strip

There is distinction between: (i) number n of branches of a multivalued function (multivalent or not) belonging to $\mathcal{U}_n$, that is, number of values it takes at a point z in the domain, viz. how many points in the range $w_1, \ldots, w_n$ the point z is mapped to; (ii) number of times m the function (multivalued or not) belonging to $\mathcal{U}^m$ takes the value w_0, that is, how many points $z_1, \ldots, z_N$ map to w_0 and the order of the zeros α_m there (37.35a–c). An example of this distinction is the logarithm that is: (i) a many-valued function, with an infinite number of branches (Sections 7.2–7.4), requiring a cut along the negative real axis in the z-plane (37.48a,c), so that it can be considered in an annulus (37.48b) of inner (outer) radius $R_1\,(R_2)$:

$$z \in |C\ -) - \infty, 0): \quad R_1 \le |z| \le R_2 \quad -\pi < \arg(z) < +\pi; \qquad (37.48\text{a–c})$$

(ii) its values specify a mapping (Figure 37.3) to

$$\begin{gathered} n \in |N: \quad w_n = \log z + in2\pi = \log|z| + i\arg(z) + i2\pi n, \\ \log R_1 \le \operatorname{Re}(w_n) = \log|z| \le \log R_2, \quad 2\pi n \le \operatorname{Im}(w_n) < 2\pi(n+1), \end{gathered} \qquad (37.49\text{a–d})$$

a vertical strip (37.49c) the complex w-plane, with each branch (37.49b) occupying one rectangle (37.49a) of height 2π in (37.49d). This mapping: (i) is performed by a many-valued function that has infinitely many branches, but takes one value only at each point of the range and hence is univalent; (ii) the domain is a doubly connected region and the range is a singly connected region. Thus the number of branches of a multivalued function is not connected with how many times it takes a given value: (i) a single-valued function may take the same value many times, for example, $\sin z = \sin(z + 2n\pi)$ for all integer values of $n \in |Z$, so it is many valent; (ii) a many-valued function may take each value in the range only once, for example, the logarithm (37.49b) that is univalent. A nonsimple mapping can map a simply or multiply connected region into another simply or multiply connected region, that is, all four combinations are possible. Thus, *a univalent function is restricted to mappings between simply connected regions, whereas the multivalued function, also holomorphic, can perform the mapping of multiply connected regions as well.* There is no simple mapping of a domain with only one boundary point, for example, a punctured plane (Subsection 37.3.1) or disk into the unit disk; next is constructed a nonsimple mapping between punctured unit disks.

37.4.6 Reference Mapping between Punctured Unit Disks

The bilinear transformation (37.50c)[≡ (35.68f) with $\lambda = 0$] maps the unit circle (37.50a) with a deleted point (37.50b) into:

$$|z| < 1, \quad z \ne a: \quad \zeta = \frac{z-a}{a^* z - 1}, \quad 0 < |\zeta| < 1, \qquad (37.50\text{a–d})$$

the unit disk punctured at the origin (37.50d), leading to an annulus as in Figure 37.3. The logarithm (37.51a) then maps the unit disk punctured at the origin (37.50b) into the left-hand half-plane (37.51b):

$$\xi(z) = \log\zeta = \log\left(\frac{z-a}{a^*z-1}\right), \quad \mathrm{Re}(\xi) = \log|\zeta| < 0, \quad \xi(0) = \log a, \tag{37.51a–c}$$

and the origin of the z-plane goes to the point (37.51c). The half-plane is mapped to the unit disk with the point (37.51c) going to the center by the transformation (37.52a):

$$\Theta = \frac{\xi - \log a}{\xi + \log a}, \quad 0 < |\Theta| < 1, \tag{37.52a,b}$$

because $\mathrm{Re}(\log a) < 0$ by (37.51b,c) and (37.52b) states that ξ is closer to log a than to $-\log a$. Thus *the mapping of the unit disk (37.50a) in the z-plane punctured (37.50b) at* $z \neq a$, *into the unit disk (37.52b) in the* Θ*-plane, with corresponding origin* $\Theta = 0$ *for* $z = 0$, *is performed by:*

$$\Theta(z) = \frac{\log[(z-a)/(a^*z-1)] - \log a}{\log[(z-a)/(a^*z-1)] + \log a}, \tag{37.53}$$

*the **reference mapping** (37.53) that is the composition of (37.52a; 37.51a).*

For future use (Subsection 37.5.2) is needed the derivative of the reference function (37.53) $\equiv$ (37.52a; 37.51a; 37.50c):

$$\frac{d\Theta}{d\xi} = 2(\xi + \log a)^{-2}\log a, \quad \frac{d\xi}{dz} = \frac{1}{z-a} - \frac{a^*}{a^*z-1}, \tag{37.54a,b}$$

at the origin:

$$\Theta'(0) = \lim_{z\to 0}\frac{d\Theta}{dz} = \lim_{\xi\to\log a}\frac{d\Theta}{d\xi}\lim_{z\to 0}\frac{d\xi}{dz} = \frac{a^* - 1/a}{2\log a}. \tag{37.55}$$

This implies (37.56b) in unit disk (37.56a):

$$0 < |a| < 1: \quad \Theta'(0) = \frac{|a|^2 - 1}{2a\log a}; \quad |\Theta'(0)| > 1, \tag{37.56a–c}$$

the last property (37.56c) follows from the change of variable (37.57a,b) in (37.56c) $\equiv$ (37.57c):

$$x > 0, \quad 0 < |a| \equiv e^{-x} < 1: \quad |\Theta'(0)| = \frac{1 - |a|^2}{2|a||\log a|} = \frac{|e^x - e^{-x}|}{2|x|} = \frac{|\sinh x|}{|x|} > 1. \tag{37.57a–c}$$

Thus *the derivative of the reference function (37.53) at the origin has modulus greater than unity (37.56c).* This result completes all the preliminaries needed to prove next the existence of a conformal mappings for simply (multiply) connected regions [Section 37.5 (37.6)].

37.5 Minimax, Reference, and Punctured Mappings

The proof of the fundamental theorem of existence of a simple, that is, conformal univalent mapping from simply connected region a region with at least two boundary points in the z-plane to the unit disk $|w| \leq 1$ in the $w = f(z)$-plane is made in two stages: (I) first prove the existence a minimax function (Subsection 37.5.1); (II) then show that this function covers the entire disk (Subsection 37.5.2). The reference and punctured mappings are considered in this context (Subsection 37.5.3).

37.5.1 Existence and Construction of the Minimax Function

It is assumed that the domain D has at least one exterior point that can be mapped to infinity by a linear transformation that also maps an interior point to the origin. The mapping function is sought as the solution of the following **extremal problem**: *consider the set of functions* f_n *holomorphic in the domain (37.58a) that do not exceed modulus unity (37.58b) and vanish at the origin (37.58c):*

$$f_n \in \mathcal{D}(D), \quad |f_n(z)| \leq 1, \quad f_n(0) = 0: \quad f(z) = \lim_{n \to \infty} f_n(z), \tag{37.58a–d}$$

then a limit function (37.58d) exists, and it maps the domain to the unit disk. First is proved that the limit function exists. The set (37.58a–c) is nonempty, because the linear function $f_0(z) = cz$ meets all properties for sufficiently small z. Also all functions have bounded first derivative at the origin, as follows applying the Cauchy theorem [(15.13) with $n = 1$] to a circle of radius R:

$$|f'(0)| = (2\pi)^{-1} \left| \int_{|z|=R} \zeta^{-2} f_n(\zeta)\, d\zeta \right| \leq (2\pi)^{-1} \int_{|z|=R} |f_n(\zeta)|\, |\zeta|^{-2}\, d\zeta \leq \frac{|f_n(\zeta)|_{\max}}{R^{-1}} \leq \frac{1}{R}, \tag{37.59}$$

using (35.58b). Taking B as an upper bound (37.60b) for (37.59):

$$n \in |N, \quad B \equiv \sup |f_n'(0)|: \quad |f_n'(0)| > B - \frac{1}{n}, \tag{37.60a–c}$$

there is a sequence of functions (37.60a) satisfying (37.60c). Hence by the theorem on subsequences (Subsection 37.2.2) there is a uniformly convergent subsequence whose limit function (37.60c) satisfies (37.61a):

$$|f'(0)| = \lim_{n \to \infty} |f_n'(0)| \geq B = |f'(0)|, \tag{37.61a,b}$$

and (37.60b) implies (37.61b). The function (37.58d) is the **minimax function** since, among all functions whose modulus does not exceed unity, it has the highest modulus of the derivative at the origin. Having proved the existence of the minimax function, it remains to show that it maps the interior (boundary) of the domain to the interior (boundary) of the unit circle.

37.5.2 Coverage of the Whole Unit Circle

The uniform limit (37.58d) applied to (37.58b) implies that the limit function is holomorphic and performs a conformal mapping into a subset of the unit circle (37.62a):

$$E_0 \equiv \{z : |f(z)| \leq 1\} \subset E \equiv \{w : |w| < 1\}; \quad E_0 \equiv E, \tag{37.62a,b}$$

it remains to prove that the two sets coincide (37.62b), that is, the mapping covers the whole of the unit circle. This will be proved by "reductio ad absurdum" by assuming that the set E_0 has at least one boundary point a inside the circle, and showing that this leads to a contradiction. The contradiction is built by using the properties of the reference function (Subsection 37.4.6). Applying the reference function (37.53) to the limit function (37.58d) specifies a holomorphic function (37.63a):

$$\Theta(f(z)) \equiv g(z) \in \mathcal{D}(D), \quad g(z) \in E - \{b\}, \tag{37.63a,b}$$

that maps the domain D conformally to the unit disk (37.62a) with one point deleted (37.63b). It satisfies the conditions (37.58a–c), and thus belongs to the set E_0 in (37.62a). However (36.56c) implies that its derivate at the origin (37.64):

$$|g'(0)| = |\Theta'(0) f'(0)| = |\Theta'(0)||f'(0)| > |f'(0)|, \tag{37.64}$$

has a larger modulus that the minimax function; this contradicts the definition of the minimax function and shows that the set E_0 cannot have a boundary point inside the circle. Thus the sets (37.62a) coincide (37.62b) completing the proof of the theorem. QED.

37.5.3 Mapping into the Punctured Disk

It was shown (Subsections 37.5.1 and 37.5.2) that *there is a conformal mapping into the unit disk of a simply connected domain with at least one exterior point.* This theorem can be extended further: *a simply connected domain in the complex plane having more than two boundary points can be mapped into the unit disk.* To prove the theorem it is sufficient to construct a conformal mapping of the complex plane with three deleted points (a,b,c) into the unit disk; one of the deleted points can be infinity, and one of the others could be the origin. This can be done using the elliptic modular function that is constructed in Section 39.4, and has the following properties: (i) it is analytic in the whole complex plane (37.65a); (ii) it never takes three values (37.65b); (iii) its imaginary part is positive (37.65c):

$$E(z;a,b,c) \in \mathcal{A}(|C) \wedge \forall_{z\in|C} : E(z;a,b,c) \neq \{a,b,c\} \wedge \operatorname{Im}\{E(z;a,b,c)\} > 0. \tag{37.65a–c}$$

The elliptic modular function is lacunary (Subsection 31.1.2), because it has no analytic continuation (37.65c) below the real axis, that is, the real axis is densely covered with singularities; the construction of the elliptic modular function (Section 39.3) does not matter here, only that a function with the properties (37.65a–c) does exist. These properties are used to prove the theorem in a sequence of two steps as follows: (i) the domain D with the three boundary points deleted is mapped by the elliptic modular function into the upper complex plane; (ii) the latter can be mapped into the unit disk by a bilinear transformation. QED.

A key role is played in the preceding proof by the existence of a elliptic modular function with the stated properties (37.65a–c); no other details of the elliptic modular function matter, neither here nor in the proof of the Picard and related theorems (Sections 39.3–39.6; Note 39.1). The first Picard theorem (39.37) states than a holomorphic function that does not take three values is a constant; the value ∞ is included in the count of three. Thus *a simply connected region punctured at three (more than three) points can (cannot) be mapped simply to the unit disk.* Recalling that there is no simple mapping from the unit disk to a plane punctured at one point (Subsection 37.3.1), it follows that: *a simply connected region punctured at two or three (one or more than three) points can (cannot) be mapped simply into the unit disk. A simple mapping into the unit disk exists for a compact region with: (a) either one exterior point; (b) or two boundary points.* Having proved the existence of simple mappings for simply connected regions, the conformal mapping of multiply connected regions, that is necessarily nonsimple, is considered next.

37.6 Automorphism Group and Fundamental Regions

It is shown next that the number of one-to-one mappings between two multiply connected regions is the number of nonequivalent automorphism groups (Subsection 37.6.3). This requires the definition of automorphism group (Subsection 37.6.1) and associated fundamental regions (Subsection 37.6.2).

37.6.1 Automorphism Group of a Nonsimple Conformal Mapping

Let $w = f(z)$ be a nonsimple conformal mapping into the unit disk (37.66b) so that the function is analytic and multivalued, and its branches (Chapter 7) are designated f_n in (37.66a):

$$\{f_n(z)\} \equiv f(z) \in \mathcal{D}(D), \quad |f(z)| \leq 1; \quad f_m(z) = A_{mn}(z)\, f_n(z), \tag{37.66a–c}$$

any two branches are connected (37.66c) by a bilinear mapping (37.67a,b) whose cross-ratio is taken to be unity (37.67c) as a normalization condition:

$$A_{mn}(z) \in X_2: \quad A_{mn}(z) \equiv \frac{a_{nm}z + b_{nm}}{c_{nm}z + d_{nm}}, \quad a_{nm}d_{nm} - b_{nm}c_{nm} = 1. \tag{37.67a–c}$$

The ordered set (f, A_n) *formed by the multivalued analytic function (37.66a) with the unit disk as range (37.66b) and all the normalized bilinear mappings (37.67a–c) between its branches (37.66c) form a group, viz.* ***automorphism group*** *of the function,* because (i) the composition of bilinear mappings is another bilinear mapping, as they all relate branches of the same analytic function; (ii) the associative property holds; (iii) the identity map is included; (iv) every map has an inverse.

If A_n is an element of the automorphism group (37.68a) distinct from unity (37.68b), then all its powers are distinct (37.68c):

$$A_n \in (f, A_m) \equiv X \wedge A_n \neq 1: \quad \forall_{p,q \in |N}: \quad p \neq q \Rightarrow (A_n)^p \neq (A_n)^q. \tag{37.68a–c}$$

Hence *every automorphism group has an infinite number of elements.* To prove the theorem suppose that two distinct powers (37.69a) of the same mapping were equal (37.69b):

$$s \equiv p - q > 0: \quad (A_n)^p = (A_n)^q; \quad (A_n)^{p-q} = I, \tag{37.69a–c}$$

then the power (37.69c) is an identity. Next take a loop L in the domain D that is mapped to a loop Γ in the unit disk. When the loop L is described once the loop Γ is described s times; if the loop L is shrunk to zero then loop Γ is also shrunk to zero and it coincides with the same point each of the s times. Thus it is the identity map. This contradicts the hypothesis (39.68b) and proves the theorem. QED.

37.6.2 The Complete Collection of Fundamental Domains

Let $g(z)$ be any other (37.70b) conformal mapping of the same domain (37.70a) into the unit disk (37.70c):

$$g(z) \in \mathcal{D}(D), \quad C \in X_2: \quad |g(z)| < 1; \quad g_0(z) = C(f_0(z)), \quad f_0(z) = C^{-1}(g(z)). \tag{37.70a–e}$$

thus there exists a bilinear mapping C in (37.70d,e) from an element g_0 of g to an element f_0 of f the f-automorphism group (37.66a–c). Applying C to the general element f_n of leads to (37.71a):

$$Cf_n = CA_nf_0 = CA_nC^{-1}g_0 = B_ng_n, \quad B_n \equiv CA_nC^{-1} \in X_2, \tag{37.71a,b}$$

where (37.71b) are bilinear mappings that form a group (B_n, g). This proves the **transformation theorem**: *if* (f, A_n) *an automorphism group (37.66–67a–c), and if* g *is any other holomorphic mapping to the unit disk (37.70a,c), then its automorphism group is* $(\mathrm{g}, \mathrm{B_n})$, *where (37.71a) the transformation* $\mathrm{B_n}$ *is derived from* $\mathrm{A_n}$ *by (37.71b) using the bilinear mapping between two initial elements (37.70d,e).*

Let $f(z)$ be a homorphic function mapping a $\boldsymbol{M+1}$**-multiply connected domain** D into the unit disk. The boundary ∂D of the $(M+1)$-multiply connected region is assumed (Figure 37.4) to consist [as (Figure 15.3) in the residue theorem (Section 13.9)] of: (i) an outer loop C_0; (ii) M inner loops C_m with $m = 1, \ldots, M$ surrounding the holes. The region can be made simply connected $\overline{D}$ by joining each inner boundary to the outer boundary by a branch-cut L_m with two sides $L_m^{\pm}$ such that they do not cross. The mapping of $\overline{D}$ by the function f specifies the **fundamental domain** of the automorphism group. *There are infinitely many fundamental domains, since each element of the automorphism group transforms the fundamental domain (37.67a) into another fundamental domain (37.67b):*

$$E_0 = f(\bar{D}), \quad E_n = A_n f(\bar{D}); \quad E(D) = \bigcup_n E_n(\bar{D}), \tag{37.72a–c}$$

the **complete collection** (37.72c) of fundamental domains is such that (i) all domains are contained in the unit disk (37.73a); (ii) two domains do not intersect (37.73b):

$$E_n \subset \{w : |w| < 1\}, \quad n \neq m \Rightarrow E_n \cap E_m = 0, \quad \forall_w : \quad |w| \leq 1 \Rightarrow \exists_{E_n} : w \in E_n, \tag{37.73a–c}$$

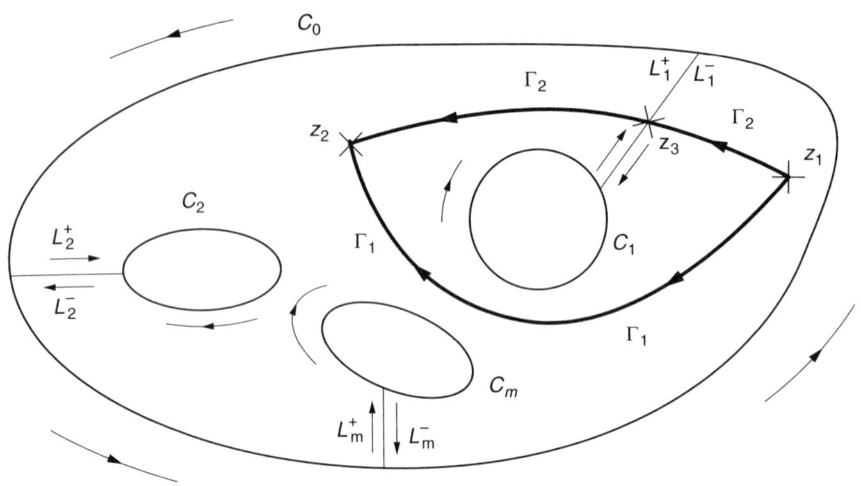

FIGURE 37.4
There is no simple mapping of a multiply connected region to a simply connected region (Figure 37.1a). Two multiply connected regions can be simply mapped into each other if each can be mapped into the unit disk (Figure 37.2b). The simple mapping from a multiply connected region (Figure 37.4) to the unit disk involves a multivalued function. It is analytic and hence specifies a conformal mapping in a region connecting outer to the inner boundaries by segments taken in opposite directions as in Figure 15.3a. It must be ensured that the analytic continuation across all segments leads to single monogenic function (Figure 31.1a). The coverage of the domain (Figure 37.4) is made a complete collection of fundamental domains each associated with an automorphism group. The automorphism groups consist of the branches of multivalued analytic function together with the bilinear mappings with cross-ratio unity. In this way is made a conformal mapping of the multiply connected region into the unit disk consistent with all the branches of the multivalued function.

(iii) every point of the unit disk lies either inside $\mathrm{E_n} - \partial\mathrm{E_n}$ *or on the boundary of the domain (37.73c).* These properties follow from the remarks: (i/ii) since the conformal mapping takes each value once, the images of the mapping by distinct branches of the function f cannot have common points; (iii) each point of the circle is an image of a point of the interior or boundary of $\bar{D}$, and so must be mapped to the interior or boundary of some E_n. QED.

37.6.3 Generation of Mappings by Equivalent Automorphisms

The transformations of the automorphism group can cover the whole of the domain $\bar{D}$ by analytic continuation (Figure 37.4) as follows: (i) let z be the starting point z_1 in any subdomain in the interior of $\bar{D}$ between two branch-cuts and the analytic continuation to another point z_2 is sought; (ii) paths Γ_1 (Γ_2) can be followed avoiding (crossing) branch-cuts, and the final values of the function must be consistent in both cases; (iii) if the path Γ_2 crosses a branch-cut at a point z_3, then the original branch f_0 becomes f_0^+ on the right-side of the cut, represented by a bilinear mapping (37.74a):

$$f_0^+ = A^+ f_0; \quad f_0^- = \left(A^-\right)^{-1} A^+ f_0^+, \quad A^- f_0^- = A^+ f_0^+, \tag{37.74a–c}$$

(iv) on the left side of the mapping it becomes (37.74b), where A^- is the bilinear mapping for the left side of the cut, since (37.74c) must hold; (v) continuing in the same way any number of cuts can be crossed in a consistent analytic continuation.

Since a multivalued homorphic function has a denumerable set (finite or infinite) of branches at any point, the bilinear transformations A_n of its automorphism group $f(D, A_n)$ generate the conformal mapping of the $(M+1)$-multiply connected domain into the unit disk:

$$f \in \mathcal{D}(D): \quad \underset{n}{U} A_n(f(D)) \equiv E\{w : |w| \leq 1\}. \tag{37.75}$$

Two automorphism groups are **equivalent** (37.76a) if a bilinear mapping exists such that for some ordering of the elements (37.71b) $\equiv$ (37.76b) holds:

$$(f, A_n) \sim (g, B_n) \Leftrightarrow \exists_{C \in X_2}: \quad B_n = C A_n C^{-1}. \tag{37.76a,b}$$

Two equivalent automorphism perform a conformal mapping of a $(M+1)$-multiply connected domain into the unit disk; and vice-versa, two conformal mappings of the same $(M+1)$-multiply connected domain into the unit disk must be equivalent on account of the transformation theorem (37.71b). This proves the **theorem of nonsimple mappings**: *a necessary and sufficient condition for the existence of a univalent conformal mapping between two multiply connected domains* D *and* $\tilde{\mathrm{D}}$ *is that their automorphism groups be equivalent:*

$$\tilde{D} \equiv f(D) \in \mathcal{D} \Leftrightarrow (g, D) \sim \left(h, \tilde{D}\right); \tag{37.77}$$

the number of such mapping is the number of distinct equivalent groups.

37.7 Correspondence of Interiors and Boundaries for Compact and Noncompact Regions

After proving the existence of one to-one conformal mappings for simply connected (multiply connected) domains [Section 37.5 (37.6)], next are considered extensions of the theorems on the correspondence of interior and boundaries (Subsection 37.3.3) for simple mappings, specified by a univalent function; they may be applied to any branch of a multivalued function. The simple mapping theorems (Sections 37.3 and 37.5) can be extended from the unit disk to an arbitrary compact domain (Subsection 37.7.1), and also to noncompact domains subject to some restrictions; these restrictions also apply to converse mapping theorem on the correspondence of interiors and boundaries (Subsection 37.7.2) that is proved for all compact (some noncompact) regions [Subsection 37.7.3 (37.7.4)].

37.7.1 Extended Simple Mapping Theorem

The simple mapping theorem (Subsection 37.3.2) concerns the existence and unicity of a simple mapping into the unit disk, and can be restated in an equivalent form, for simple mappings into other regions. Let D, E be two simply connected regions whose boundaries ∂D, ∂E have at least two distinct points each; then by the simple mapping theorem, simple mappings $\zeta = f(z), g(z)$ exist from D, E to the unit disk $|\zeta| \le 1$, and they become unique assigning a point and a direction through it :

$$f(a) = c = g(b), \quad \arg\{f(a)\} = \alpha = \arg\{g(b)\}. \tag{37.78a,b}$$

The function $g^{-1} \otimes f$ maps (Figure 37.2b) the region D into E because: (i) D is mapped to the unit disk by f; (ii) the unit disk is mapped to E by g^{-1}; (iii) the composition $g^{-1} \otimes f$ is taken in the order f first (a) and g^{-1} second (b). The overall mapping $g^{-1} \otimes f$ from D to E: (i) assigns the point $a \in D$ to $b \in E$, and a direction through them; (ii) it is simple, that is, analytic and univalent because it results from the composition and inversion of simple functions; (iii) also the boundaries $\partial D, \partial E$ correspond through the circle $|\zeta| = 1$, and the interiors $D - \partial D, E - \partial D$ through the open disk $|\zeta| < 1$. Thus follows the extended **simple mapping theorem**: *two simply connected regions* D, E *whose boundaries consist of at least two distinct points each, can be mapped simply into each other. The mapping* $\zeta = \mathrm{h}(\mathrm{z})$ *is unique if a point (37.79a–c) and a direction (37.79d) through it are assigned:*

$$a \in D, c \in E: \quad h(a) = c, \quad \arg\{h'(a)\} = \alpha \in |R. \tag{37.79a–d}$$

The boundary ∂D (interior $D - \partial D$) of D is mapped to the boundary ∂E (interior $E - \partial E$) of E.

37.7.2 Converse Theorem on Boundaries and Interiors

Also of interest is the converse theorem: let $\zeta = f(z)$ be a conformal mapping from a region D in the z-plane to a region E in the ζ-plane that maps the simply, that is, univalently, the boundary of the former ∂D to the boundary of the latter ∂E. Does it map the interior $D - \partial D$ of D simply, that is, holomorphic, to the interior $E - \partial E$ of E? The reply is yes, in case I of a function holomorphic in D, that is, without singularities, so that the region E is compact, that is, does not extend to infinity (Figure 37.5a); besides $f(z)$ need only be uniformly continuous on the boundary of D, and holomorphic in the interior according to the Cauchy conditions (Sections 19.1–19.6). Case II arises if $\zeta = f(z)$ has singularities in the interior of D; in order that it remains univalent, it can have only one singularity, and it must be a simple pole $z = a$; if it had a multiple pole or an essential singularity the inverse would have several roots or branches, and the function would not be univalent. In this case II the region E is not compact (Figure 37.5b) for it includes the point at infinity $\infty = f(a)$ as the transform of the simple pole. The last case III arises if the singularity $z = a$ lies on the boundary of D, so that (Figure 37.5c): (i) the boundary ∂E of D may have an **angular point** of argument α at the singularity; (ii) this point is mapped to infinity $\infty = f(a)$, and it is assumed that the boundary ∂E of E has **asymptotes** making an angle β. In this case

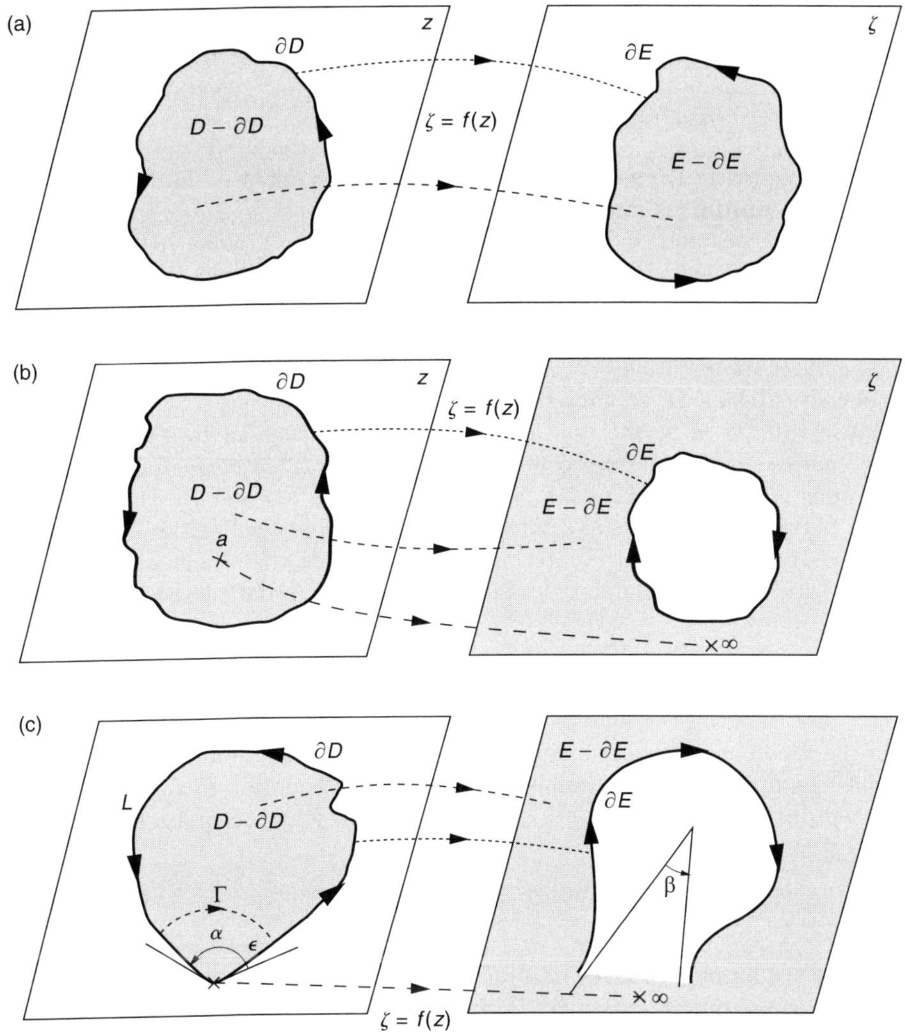

FIGURE 37.5
The simple, that is, conformal univalent, mapping of the interior $D - \partial D$ of a simply connected region D to the interior $E - \partial E$ of another simply connected region E, also maps the boundary ∂D of the former to the boundary ∂E of the latter in the following three cases. In the first case (a) the range is a compact region, that is, is contained in a disk of finite radius and center at the origin, that is, does not contain the point-at-infinity; in this case the boundary of the range ∂E is described in the same direction as the boundary ∂D of the domain. In the second case (b) the range is a noncompact region, which includes the point-at-infinity as a simple pole of the mapping, in order to remain univalent; in this case the boundary of the range ∂E is described in the direction opposite to the boundary ∂D of the domain. The third case (c) is a noncompact region with asymptotes at infinity making an angle β, with the point-at-infinity in the range mapped from an angular point of angle α on the boundary of the domain; in this case a restriction must be placed on the angles (α, β) relative to the order of the singularity (37.80a,b). In this case (c) like the preceding (b), the range is a noncompact region, and its boundary is described in the opposite direction to the boundary of the domain.

the mapping is still univalent, provided that the order of the singularity (37.80a) be limited by (37.80b):

$$f(a) = O\left((z-a)^{-\mu}\right), \quad \mu < \frac{\beta + 2\pi}{\alpha}, \tag{37.80a,b}$$

as will be proved subsequently (Subsection 37.7.4). Thus can be stated the **theorem of correspondence of boundaries and interiors**: *let* $\zeta = \mathrm{f}(\mathrm{z})$ *be a conformal,* that is, *holomorphic, mapping of the interior* $\mathrm{D} - \partial\mathrm{D}$ *of a region* D *in the* z*-plane to the interior* $\mathrm{E} - \partial\mathrm{E}$ *of a region* E *in the* ζ*-plane; let the same function map continuously and univalently the boundary* $\partial\mathrm{D}$ *of* D *to the boundary* $\partial\mathrm{E}$ *of* E*. Then* $\zeta = \mathrm{f}(\mathrm{z})$ *maps simply,* that is, *one-to-one, the interior of* D *to that of* E*, in three cases: (i) if it has no singularities in* D*, in that case* E *is a compact region (Figure 37.5a); (II) if it has a single simple pole* $\mathrm{f}(\mathrm{a}) = \infty$ *in the interior* $\mathrm{a} \in \mathrm{D} - \partial\mathrm{D}$ *of* D*, in that case* E *is noncompact (Figure 37.5b),* that is, *includes the point at infinity in its interior; (III) if (i) it has a singularity* $\mathrm{a} \in \partial\mathrm{D}$ *on the boundary of* $\partial\mathrm{D}$*, and (ii) its transform* $\infty = \mathrm{f}(\mathrm{a})$ *has (Figure 37.5c) asymptotes of angle* β*, and (iii) the order of the function at the singularity satisfies (37.80a,b).* The theorem of correspondence between boundaries and interiors is proved first (Subsection 37.7.3) for regular points (case I) and poles in the interior (case II) and is then (Subsection 7.7.4) extended to singularities on the boundary (case III), including proof of the asymptotic condition (37.80a,b).

37.7.3 Mapping of a Region: Points and Poles in the Interior

To demonstrate the theorem of correspondence of boundaries and interiors in any of the three cases, it must be proved that the number of roots of the equation $\zeta = f(z)$, that is, the **number of** ζ**-points** of the function $f(z)$, is unity (37.81a) for $\zeta \in E$ and zero (37.81b) for $\zeta \notin E$, where $z \in D$:

$$E \equiv \{\zeta : \ \zeta = f(z) \wedge z \in D\}: \quad N(\zeta) = \begin{cases} 1 & \text{if } \zeta \in E, \\ 0 & \text{if } \zeta \notin E. \end{cases} \tag{37.81a,b}$$

Thus the formula for the number of zeros of a function in a region is needed. The number Z of zeros of a function analytic in a domain is given by (31.36b) where the number P of poles is zero:

$$N(b) = (2\pi)^{-1} \Delta_{\partial D}\{\arg(f(z) - b)\} = (2\pi)^{-1} \Delta_{\partial E}\{\arg(\zeta - b)\}; \tag{37.82}$$

thus (37.82) specifies the number of b-points in case I of a mapping analytic in the whole domain D. Two possibilities arise in (Figure 37.5a): (i) b lies within E as ζ goes around ∂E, the argument of $\zeta - b$ changes by 2π, and $N(b) = 1$ in agreement with (37.81a); (ii) if b lies outside E, the argument of $\zeta - b$ does not change after a loop around ∂E, and $N(b) = 0$ in agreement with (37.81b). Thus is proved the case I of the theorem of correspondence.

In the case II the mapping $\zeta = f(z)$ has a single simple pole in the interior of D, so $P = 1$ in (31.36b) and the number of b-points is given by:

$$N(b) - 1 = (2\pi)^{-1} \Delta_{\partial D}\{\arg(f(z) - b)\} = -(2\pi)^{-1} \Delta_{\partial E}\{\arg(\zeta - b)\}; \tag{37.83}$$

here the loop ∂E is taken in direction opposite to ∂D, because (Figure 37.5b) the region E contains the point at infinity and ∂E goes around it, viz. it is an exterior direction (Section 31.3). There are again two possibilities: (i) if b lies inside E, as ζ goes round the loop ∂E, there is no change in $\arg(\zeta - b)$, so $N(b) - 1 = 0$, in agreement with (37.81a); (ii) if b lies outside E, then $\arg(\zeta - b)$ changes by -2π as ζ goes round ∂E, and $N(b) - 1 = -1$, in agreement with (37.81b). Thus (37.81a,b) has so far a been proved in both cases I and II. QED.

37.7.4 Angular Point and Asymptotes to a Curve

The remaining possibility (case III) concerning the theorem of correspondence is (Figure 37.5c) the case of a singularity at $a \in \partial D$ on the boundary of the region D that may be an **angular point**, where the tangents at right and left are distinct and make an angle α. The point a is mapped $f(a) = \infty$ to infinity, and it is assumed that the boundary ∂E has asymptotes making an angle β. Draw a circle of small radius ε about $z = a$ and denote by Γ the part within D, so that the argument of $z - a$ varies: (i) by (37.84a) along Γ:

$$\Delta_\Gamma \{\arg(z-a)\} = -\alpha; \quad \Delta_{L-\Gamma} \{\arg(\zeta - b)\} = 2\pi - \beta, \tag{37.84a,b}$$

(ii) by (47.84b) along $L - \Gamma$ in (37.84b) that is the part of the boundary ∂D outside the circle. Hence the loop L consists of the circular arc $L - \Gamma$ plus the part of ∂D outside it. Since ε can be made small enough that all b-points of $\zeta = f(z)$ in ∂D lie within L, the case I of the theorem applies to L, because (37.82) it contains no singularities:

$$2\pi N(b) = \Delta_L \{\arg[f(z) - b]\} = \Delta_{L-\Gamma} [\arg(\zeta - b)] + \Delta_\Gamma [\arg(\zeta - b)]. \tag{37.85}$$

The first term on the r.h.s. is known (37.84b); concerning the second it is assumed (37.86a) $\equiv$ (37.86b) that $\zeta = f(z)$ is of order μ at the singularity $z = a$:

$$\zeta = f(z) = O\left((z-a)^{-\mu}\right): \quad \lim_{z \to a} (z-a)^\mu f(z) = c \neq 0, \infty, \tag{37.86a,b}$$

so that

$$\Delta_\Gamma [\arg(\zeta - b)] = \Delta_\Gamma \{\arg[f(z) - b]\} = -\mu \Delta_\Gamma [\arg(z-a)] = +\mu\alpha, \tag{37.87}$$

where (37.84a) was used.

Substituting (37.87) and (37.84b) into (37.85) yields:

$$b \in E: \quad 0 < N(b) = \frac{2\pi - \beta + \alpha\mu}{2\pi} < 2, \tag{37.88a,b}$$

where $N(b)$ must be an integer; since $N(b) = 1$ for (37.88a) an interior point $b \in E$ by (37.81a), it must lie in the open range) 0,2 (. Since $\alpha\mu > 0$ and $\beta < 2\pi$, the condition $N(b) > 0$ is trivially met by (37.88b); the second condition $N(b) < 2$ leads in (37.88b) to the bound (37.80b) on the order μ of the function $f(z)$ at $z = a$ in (38.80a) $\equiv$ (38.86a) $\equiv$ (38.86b). Since $N(b) = 1$ in (37.88b), it follows (37.89a) that the order of the mapping (37.86a) $\equiv$ (37.89a) is precisely (37.89b):

$$1 = N(b) = 1 + \frac{\alpha\mu - \beta}{2\pi}: \quad \mu = \frac{\beta}{\alpha}, \quad f(z) = O\left((z-a)^{-\beta/\alpha}\right). \tag{37.89a–c}$$

If the point $b \notin E$ lies (37.90a) outside E, then $2\pi - \beta$ is replaced by $-\beta$ in (37.84b) leading to (37.90b):

$$b \notin E: \quad -1 < N(b) = \frac{\alpha\mu - \beta}{2\pi} < +1; \tag{37.90a,b}$$

the inequalities (37.90b) follow: (i) from (37.80b) for (37.91):

$$\frac{\alpha\mu - \beta}{2\pi} < \frac{\alpha[(\beta + 2\pi)/\alpha] - \beta}{2\pi} = 1; \tag{37.91}$$

(ii) from (37.92a,b) for (37.92c):

$$\alpha\mu > 0, \beta < 2\pi: \quad \frac{\alpha\mu - \beta}{2\pi} > \frac{\alpha\mu}{2\pi} - 1 > -1. \tag{37.92a–c}$$

Since $N(b)$ is an integer, the only value compatible with (37.90b) is $N(b) = 0$, in agreement with (37.81b) and (37.89b). This completes the proof of case III of the correspondence theorem. QED. In order to interpret the condition (37.80b) note that if the boundary ∂D is regular $\alpha = \pi$ at $z = a$, and ∂D if is regular at infinity $\beta = \pi$, then $\mu < 3$. Thus the function $\zeta = z^3$ is not a simple map of the upper half-z-plane into the upper-half ζ-plane; in fact for each ζ such that $\operatorname{Im}(z) > 0$ there are two z such that $\operatorname{Im}(z) > 0$, that is, the mapping is two-to-one instead of one-to-one.

37.8 Interior and Exterior Integral Theorems (Cauchy, 1821; Schwartz, 1890)

The preceding existence and unicity theorems (Sections 37.1–37.7) illustrate the main properties of conformal mappings that had already been found in specific cases, for example, the Schwartz–Christoffel (Chapter 33) and Mobius (Chapter 35) transformations. These allow the mapping of the interior and exterior of boundary curves, suggesting that the Cauchy theorem (15.8) for interior of a loop, can also be applied to the exterior of the loop, assuming that the function is analytic outside the loop; this implies it has a finite unique value at infinity $f(\infty)$. The Cauchy theorem (Subsection 37.8.1) specifies the values of a harmonic (i.e., analytic) function in a domain (e.g., a disk) from its values on the boundary. Since the real and imaginary parts of a complex holomorphic function are related [by the Cauchy–Riemann conditions (11.10a,b)] it is possible to express the values of a harmonic function in a disk in terms of the values on the boundary of only its real or imaginary part (Subsection 37.8.2).

37.8.1 Cauchy Theorem (1821) for the Exterior of a Boundary

The Cauchy second theorem (15.8) applies (Section 15.3) to a doubly connected region (Figure 37.6) consisting of: (i) the loop ∂D as inner boundary, described in the opposite direction (Section 31.3) from the outside:

$$2\pi i f(z) = \left\{ \int_{\partial D} + \int_{|\zeta| = R \to \infty} \right\} f(\zeta)(\zeta - z)^{-1}\, d\zeta, \tag{37.93}$$

plus the integral over a circle with center at the origin and large radius R, taken in the same as the inner direction, and enclosing the loop ∂D. The second integral on the r.h.s. of (37.93) can be taken as the limit as $R \to \infty$:

$$\begin{aligned} \zeta = Re^{i\varphi}: \quad \lim_{R\to\infty} \int_{|\zeta|=R} f(\zeta)(\zeta - z)^{-1}\, d\zeta &= i \lim_{R\to\infty} \int_0^{2\pi} f\left(Re^{i\varphi}\right)\left[1 - \frac{z}{R}e^{-i\varphi}\right]^{-1} d\varphi \\ &= i f(\infty) \int_0^{2\pi} d\varphi = 2\pi i f(\infty), \end{aligned} \tag{37.94}$$

since the function $f(z)$ is holomorphic outside ∂D and thus has a finite value at infinity. Substituting (37.94) in (37.93) and leaving the integration along the loop in the positive

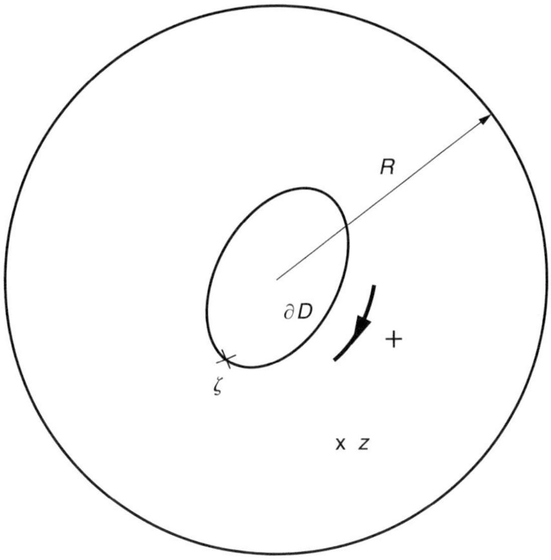

FIGURE 37.6
The Cauchy theorem uses the values of a complex function on a boundary ∂D to specify the values at any point z in: (i) the interior $D - \partial D$, assuming that it is analytic in D, by (Figure 15.1b) shrinking the boundary to a small circle around$z$; (ii) the exterior $|C - D - \partial D$ considering also a circle of large radius $R \to \infty$ (Figure 37.6) so that the value of the function at infinity is involved. The loop is described in opposite directions, for example, clockwise in (i) and counterclockwise in (ii) in agreement with Figure 31.5a.

direction viewed from the outside, that is, negative from the inside, yields:

$$(2\pi i)^{-1} \int_{\partial D} f(\zeta)(\zeta - z)^{-1} \mathrm{d}\zeta = f(z) - f(\infty), \tag{37.95}$$

for a point z outside ∂D; if z is inside ∂D then the r.h.s. of (37.95) reduces to $-f(\infty)$, since the pole $\zeta = z$ of the integrand on the r.h.s. is outside the region of interest.

The preceding result (37.95) can be combined with the original interior Cauchy theorem (15.8) in a single statement of the **Cauchy interior and exterior integral theorem (1821)**: *let the function* $\mathrm{f}(\zeta)$ *be holomorphic (Section 13.3) or satisfy the Cauchy conditions (Section 19.6) on a closed loop* $\partial\mathrm{D}$. *If the function is holomorphic in the interior* $\mathrm{D} - \partial\mathrm{D}$, *then:*

$$f(z) \in \mathcal{D}(D - \partial D): \quad (2\pi i)^{-1} \int_{\partial D+} f(\zeta)(\zeta - z)^{-1} \mathrm{d}\zeta = \begin{cases} f(z) & \text{if } z \in D - \partial D, \quad (37.96a) \\ 0 & \text{if } z \in |C - D, \quad (37.96b) \end{cases}$$

the Cauchy integral is equal the value of the function (37.96a) [zero (37.96b)] for a point inside (outside) the loop. If the function is holomorphic outside the loop:

$$f(z) \in \mathcal{D}(|C - D): \quad (2\pi i)^{-1} \int_{\partial D-} f(\zeta)(\zeta - z)^{-1} \mathrm{d}\zeta$$

$$= \begin{cases} -f(\infty) & \text{if } z \in D - \partial D, \quad (37.97a) \\ f(z) - f(\infty) & \text{if } z \in |C - D, \quad (37.97b) \end{cases}$$

then the Cauchy integral is specified by minus the value at infinity (37.97a) [plus the value of the function (37.97b)] if it is inside (outside) the loop. Both for the interior (37.96a,b) [exterior (37.97a,b)] of the loop: (i) the value of the function appears when it is inside (37.96a) [(37.97b)]; (ii) a constant is added in all cases, viz. zero (minus the value of the function at infinity).

37.8.2 Holomorphic Function in a Disk Determined from the Real Part on the Circle (Schwartz, 1890)

The Cauchy integral (37.96–37.97a,b) specifies the values of a function at all points in its region of holomorphy from the values only the points on the boundary; it was also shown (Section 11.3) that the real and (imaginary) parts of a holomorphic function are related [by the Cauchy–Riemann conditions (11.10a,b)]. Thus it is possible to determine the values of a holomorphic function $f = \Phi + i\Psi$ at all points of D from just the values of the real part $\Phi = \mathrm{Re}(f)$ on the boundary; the same would apply to the imaginary part using the function $\Psi = \mathrm{Re}(-if)$. This can be done explicitly for a disk according to **Schwartz disk theorem (1890)**: *a complex function (37.98a) holomorphic inside (37.99a) [outside (37.100a)] a disk:*

$$f\left(re^{i\varphi}\right) = \Phi(r,\varphi) + i\Psi(r,\varphi), \quad f(z) \in \mathcal{D}(|z| >< a), \tag{37.98a,b}$$

is determined by the values of its real part (37.99b) [(37.100b)] on the boundary circle to within an arbitrary imaginary constant:

$$|z| \le a: \quad f(z) = (2\pi)^{-1} \int_0^{2\pi} \Phi(a,\alpha) \frac{ae^{i\alpha} + z}{ae^{i\alpha} - z} d\alpha + i\Psi(0), \tag{37.99a,b}$$

$$|z| \ge a: \quad f\left(\frac{a^2}{z}\right) = (2\pi)^{-1} \int_0^{2\pi} \Phi(a,-\alpha) \frac{ze^{i\alpha} + a}{ze^{i\alpha} - a} d\alpha + i\Psi(\infty), \tag{37.100a,b}$$

The function may be holomorphic (Section 11.3) or need only meet the Cauchy conditions (Section 9.6) on the boundary: (i) it is uniformly convergent from the interior; (ii) it may have a finite number of discontinuities or ignorable singularities (Sections 19.1–19.6) on the boundary. The inversion on the disk (Section 35.7) or circle theorem (Section 24.7) relate (37.99–37.100a,b), so it is sufficient to prove one of them, for example, (37.99a,b); the terms in the fractions in the integrands (37.99b) and (37.100b) correspond to interchanging (a,z). The theorem is proved next, before proceeding to its consequences for harmonic boundary-value problems in the disk; there is also an analogue for half-planes (Section 37.9). The arbitrary imaginary constant appears in (37.99–37.100b), because it relates to the family of Ψ-curves orthogonal to the Φ-curves that are defined to within a added constant (Section 11.8).

To prove the Schwartz theorem the variable (37.101a) is introduced on the boundary:

$$\zeta = ae^{i\alpha}, \quad \frac{a^2}{\zeta} = ae^{-i\alpha}, \quad G(\zeta) \equiv \Phi(a,\alpha), \tag{37.101a–c}$$

and the corresponding reciprocal point (37.101b) relative to the circle of radius a ; then the boundary function (37.101c) satisfies:

$$2G(\zeta) = 2\mathrm{Re}\left\{f\left(ae^{i\alpha}\right)\right\} = f\left(ae^{i\alpha}\right) + f^*\left(ae^{-i\alpha}\right) = f(\zeta) + f^*\left(\frac{a^2}{\zeta}\right). \tag{37.102}$$

Integrating along the circle leads to

$$2\int\limits_{|\zeta|=a} G(\zeta)(\zeta-z)^{-1}\,\mathrm{d}\zeta = \int\limits_{|\zeta|=a} f(\zeta)(\zeta-z)^{-1}\,\mathrm{d}\zeta + \int\limits_{|\zeta|=a} f^*\left(\frac{a^2}{\zeta}\right)(\zeta-z)^{-1}\,\mathrm{d}\zeta$$
$$= 2\pi i\left[f(z)+f^*(0)\right], \tag{37.103}$$

where: (i) the first term on the r.h.s. is evaluated Cauchy's theorem (37.96a), since z is an interior point and f is holomorphic in the disk; (iii) in the second term on the r.h.s. of (37.103) is used the change of variable (37.104a):

$$\eta \equiv \frac{a^2}{\zeta}: \quad \int\limits_{|\zeta|=a} f^*\left(\frac{a^2}{\zeta}\right)(\zeta-z)^{-1}\,\mathrm{d}\zeta = \int\limits_{|\eta|=a} f^*(\eta)\left(1-\frac{\eta z}{a^2}\right)^{-1}\eta^{-1}\,\mathrm{d}\eta = 2\pi i f^*(0), \tag{37.104a,b}$$

together with the theorem of residues: (ii-a) the pole at $\eta = a^2/z$ lies outside $|\eta| = a$ for z an interior point $|z| < a$ because $a^2/|z| > a$; (ii-b) from $\mathrm{d}\zeta = -(a/\eta)^2\,\mathrm{d}\eta$ it follows that the sign is changed because the η-direction is opposite to the ζ-direction; (ii-c) only the other pole at $\eta = 0$ lies within $|z| < a$ and its residue is $f^*(0)$. Setting $z = 0$ in (37.103) simplifies it to:

$$2\int\limits_{|\zeta|=a} \zeta^{-1}G(\zeta)\,\mathrm{d}\zeta = 2\pi i\left[f(0)+f^*(0)\right] = 4\pi i\Phi(0). \tag{37.105}$$

Using (37.105) and (37.103) leads to

$$f(z) - i\Psi(0) = f(z) + f^*(0) - \Phi(0) = (2\pi i)^{-1}\int\limits_{|\zeta|=a} G(\zeta)\left(\frac{2}{\zeta-z}-\frac{1}{\zeta}\right)\mathrm{d}\zeta$$
$$= (2\pi i)^{-1}\int\limits_{|\zeta|=a} G(\zeta)\,\frac{\zeta+z}{\zeta-z}\,\frac{\mathrm{d}\zeta}{\zeta}; \tag{37.106}$$

this coincides to the Schwartz theorem (37.99b), on substitution of (37.101a,c). QED. The Schwartz theorem will be used to calculate (Section 38.9) the velocity inside a jet from its value at the free surface. It is applied next to the solution of boundary-value problems on a disk.

37.9 Harmonic Functions Defined by Boundary Values (Poisson, 1820; Dirichlet, 1850; Robin, 1886; von Neumann, 1961)

The Cauchy (Schwartz) theorem [Subsection 37.8.1 (37.8.2)] provides the solution [Subsection 37.9.2 (37.9.1)] of the Dirichlet problem for a plane (disk): given the values of the real or imaginary part of a harmonic function on the real axis (a circle), specify the value of the complex holomorphic function in the upper or lower half-plane (interior or exterior of the disk). The complex holomorphic function: (i) extends the harmonic function, viz. its real (imaginary) part from the boundary to the interior or exterior; (ii) also specifies the function determining the orthogonal family of curves, viz. its imaginary (real) part.

37.9.1 Schwartz (1890) Transform and Poisson (1820) Formula

The Schwartz theorem (37.99b) for the circle of radius a involves **Schwartz integral transform (1890)**:

$$f(r,\varphi) = (2\pi)^{-1} \int_0^{2\pi} \Phi(a,\alpha)\, S(r,\varphi;\alpha-\varphi)\, d\alpha, \tag{37.107}$$

with Schwartz kernel:

$$z = re^{i\varphi}: S(r,a,\alpha-\varphi) \equiv \frac{ae^{i\alpha}+re^{i\varphi}}{ae^{i\alpha}-re^{i\varphi}} = \frac{a^2-r^2-2iar\sin(\alpha-\varphi)}{a^2+r^2-2ar\cos(\alpha-\varphi)}. \tag{37.108a,b}$$

that follows from:

$$S(r,a;\alpha-\varphi) = \frac{\left(ae^{i\alpha}+re^{i\varphi}\right)\left(ae^{-i\alpha}-re^{-i\varphi}\right)}{\left(ae^{i\alpha}-re^{i\varphi}\right)\left(ae^{-i\alpha}-re^{-i\varphi}\right)} = \frac{a^2-r^2-ar\left(e^{i\alpha-i\varphi}-e^{i\varphi-i\alpha}\right)}{a^2+r^2-ar\left(e^{i\alpha-i\varphi}+e^{i\varphi-i\alpha}\right)}. \tag{37.108c}$$

Substituting (37.108b) and (37.98a,b) in the Schwartz theorem, and taking real and imaginary parts leads to **Poisson (1820) interior disk formula** *(37.109a,b) and its analogue (37.109c,d):*

$$\Phi(r,\varphi) \in \mathcal{H}(r \le a): \quad \Phi(r,\varphi) = (2\pi)^{-1} \int_0^{2\pi} \Phi(a,\alpha)\left(a^2-r^2\right)\left[a^2+r^2-2ar\cos(\alpha-\varphi)\right]^{-1} d\alpha, \tag{37.109a,b}$$

$$\Phi + i\Psi \equiv f\left(re^{i\varphi}\right) \in \mathcal{H}(r \le a): \quad \Psi(r,\varphi) = -(2\pi)^{-1} \int_0^{2\pi} \Phi(a,\alpha)\, 2ar\sin(\alpha-\varphi) \times \left[a^2+r^2-2ar\cos(\alpha-\varphi)\right]^{-1} d\alpha + \Psi(0), \tag{37.109c,d}$$

stating that (i) a real harmonic function $\Phi(r,\varphi)$ inside (37.109a) the disk $r \le a$ is specified by (37.109b) its values $\Phi(a,\alpha)$ on the boundary $r = a$; (ii) the latter also specifies in the disk (37.109c), to within an added constant $\Psi(0)$, the (37.109d) orthogonal function $\Psi(r,\varphi)$. The first result (37.109b) is the solution of ***Dirichlet interior problem (1850)*** *for the disk (Figure 37.7a): determine the function Φ that satisfies the Laplace equation (37.110b) in the disk (37.110a) and satisfies a boundary condition of the first kind (37.110c) specifying its values on the boundary:*

$$r \le a: \quad \nabla^2\Phi = 0, \quad \Phi(r,\varphi)\big|_{r=a} \equiv \Phi(a,\varphi); \tag{37.110a–c}$$

hence (37.109a,b) is the solution of (37.110a–c). In addition (37.109c,d) is the solution of the ***disk orthogonality problem*** *specifying in the disk (37.111a) the family of curves $\Psi = \text{const}$ or "field lines" orthogonal (37.111b) to the family of curves $\Phi = \text{const}$ or "equipotentials" (37.111c):*

$$r \le a: \quad 0 = \nabla\Psi \cdot \nabla\Phi, \quad \Phi(r,\varphi)\big|_{r=a} = \Phi(a,\varphi); \tag{37.111a–c}$$

hence (37.109c,d) is the solution of (37.111a–c) to within an added constant. The arbitrary constant in (37.109d) comes from the imaginary constant in the Schwartz formula (37.99b), and arises because if Ψ is orthogonal to Φ, then $\Psi + \text{const}$ is also orthogonal, viz. belongs to the same family of orthogonal curves.

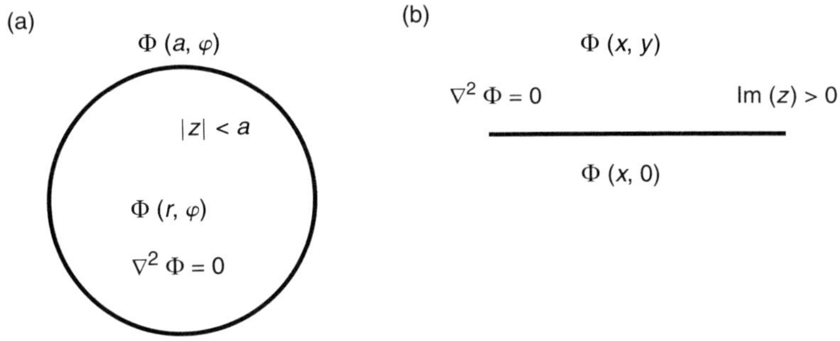

FIGURE 37.7
The Cauchy theorem specifies the values of a function analytic in the interior (exterior) of loop from its boundary values [Figure 15.1b (37.6)]. Also the real and imaginary parts of a complex analytic function are related by the Cauchy–Riemann conditions; there also specify orthogonal families of curves in Figure 11.4. It follows that the potential and stream function inside and outside a cylinder (a) [in the upper and lower complex half-plane (b)] can be determined from the values of just one them (the potential or the stream function) on the cylinder (real axis). To specify both the stream function and the potential would be either redundant or inconsistent.

Proceeding in a similar way from (37.100b) instead of (37.99b) and using the reciprocal point on the circle (37.101a,b) ≡ (37.112a–c):

$$s e^{i\psi} \equiv \eta = \frac{a^2}{z} = \frac{a^2}{r} e^{-i\varphi}, \quad |\eta| = \frac{a^2}{r} \equiv s, \quad \arg(\eta) = -\varphi = \psi, \tag{37.112a–c}$$

leads to:

$$\begin{aligned} \frac{z e^{i\alpha} + a}{z e^{i\alpha} - a} &= \frac{a e^{i\alpha} + a^2/z}{a e^{i\alpha} - a^2/z} = \frac{a e^{i\alpha} + \left(a^2/r\right) e^{-i\varphi}}{a e^{i\alpha} - \left(a^2/r\right) e^{-i\varphi}} \\ &= \frac{a e^{i\alpha} + s e^{i\psi}}{a e^{i\alpha} - s e^{i\psi}} = \frac{a^2 - s^2 - 2ias \sin(\alpha - \psi)}{a^2 + s^2 - 2as \cos(\alpha - \psi)}, \end{aligned} \tag{37.112d}$$

using a simplification similar to (37.108a–c), with (s, ψ) replacing (r, φ). This proves the **exterior Poisson formula** (37.113a,b) [**exterior orthogonality relation** (37.113c,d)]:

$$\Phi(s,\psi) \in \mathcal{H}(s \geq a): \quad \Phi(s,\psi) = (2\pi)^{-1} \int_0^{2\pi} \Phi(a,-\alpha)\left(a^2 - s^2\right)\left[a^2 + s^2 - 2as\cos(\alpha - \psi)\right]^{-1} d\alpha, \tag{37.113a,b}$$

that is similar to (37.109a,b) [(37.109c,d)]:

$$\begin{aligned} \Phi + i\Psi \equiv f\left(s e^{i\psi}\right) \in \mathcal{A}(s \geq a): \quad \Psi(s,\psi) = &-(2\pi)^{-1} \int_0^{2\pi} \Phi(a,-\alpha)\, 2as \sin(\alpha - \psi) \\ &\times \left[a^2 + s^2 - 2as\cos(\alpha - \psi)\right]^{-1} d\alpha + \Psi(\infty). \end{aligned} \tag{37.113c,d}$$

For a real harmonic (complex analytic) function outside the disk $r = a$, the exterior (37.113a,b) [(37.113c,d)] and interior cases (37.109a,b) [(37.109c,d)] transform into each other via the substitutions $(r, \varphi, \alpha) \leftrightarrow (s, \psi, -\alpha)$.

37.9.2 Dirichlet (1850) Problem for a Disk and for the Half-Plane

The Dirichlet problem can also be considered for the half-plane (Figure 35.7b), starting with Cauchy theorem (15.8) or the real axis closed by a large half-circle:

$$2\pi i f(z) = \lim_{R\to\infty}\left\{\int_{-R}^{+R} + \int_{|\zeta|=R\to\infty,\mathrm{Im}(\zeta)>0}\right\} f(\zeta)(\zeta - z)^{-1}\,\mathrm{d}\zeta. \tag{37.114}$$

The function besides being holomorphic in (37.115b) the upper half-plane (37.115a) satisfies the asymptotic condition (37.115c):

$$0 \le \varphi \le \pi: \quad f\left(re^{i\varphi}\right) \in \mathcal{D}(\mathrm{Im}(z) > 0), \quad \lim_{R\to\infty} R\left|f\left(Re^{i\varphi}\right)\right| = 0; \tag{37.115a–c}$$

this implies (Subsection 17.4.1) that the second integral on the r.h.s. of (37.114) is $O(1/R)$ and vanishes as $R \to \infty$. Thus (37.114) reduces to the integration along the real axis:

$$f(x+iy) \equiv \Phi(x,y) + i\Psi(x,y): \quad 2\pi i f(x+iy) = \int_{-\infty}^{+\infty} f(t)(t - x - iy)^{-1}\,\mathrm{d}t. \tag{37.116a,b}$$

Substituting (37.116a) in (37.116b) would specify $\Phi(x,y)$ in terms of $\Phi(x,0)$ and $\Psi(x,0)$; the result sought should involve only $\Phi(x,0)$, since $\Psi(x,0)$ can be determined from $\Phi(x,0)$ as in the case of the disk (Subsection 37.8.2). Thus what is needed is the analogue for the half-plane of the Schwartz theorem (37.99–37.100a,b) or Poisson (37.109a,b) and analogue formula (37.109c,d) for the disk. All of these were obtained applying Cauchy theorem to the disk and considering the reciprocal point (Section 35.7). A similar method can be applied to a half-plane, considering the reciprocal point of $z \equiv x + iy$ relative to the real axis that is the complex conjugate $z^* = x - iy$. The Cauchy formula (37.116b) for the half-plane with z replaced by z^* would be zero, because the pole is outside the upper half-plane:

$$z = x + iy \to z^* = x - iy: \quad 0 = \int_{-\infty}^{+\infty} f(t)(t - x + iy)^{-1}\,\mathrm{d}y. \tag{37.117a,b}$$

Thus to the r.h.s. of (37.116b) can be added or subtracted the same expression (37.117b) with z replaced by z^*:

$$(t - x - iy)^{-1} \pm (t - x + iy)^{-1} = \left[(t-x)^2 + y^2\right]^{-1} 2\{t - x, iy\}; \tag{37.118}$$

using (37.118) in the sum of (37.116b) and (37.117b) leads to

$$2\pi i\left[\Phi(x,y) + i\Psi(x,y)\right] = \int_{-\infty}^{+\infty} \Phi(t,0)\left[(t-x)^2 + y^2\right]^{-1} 2\left[iy + (t-x)\right]\mathrm{d}t, \tag{37.119}$$

where $\Phi(x,y)\,[\Psi(x,y)]$ is specified by the imaginary (minus the real) part of the r.h.s. of (37.119) divided by 2π.

Thus have been proved: *for a real harmonic function (37.120b) in upper half-plane (37.120a) decaying at infinity like (37.120c):*

$$\Phi(x,y) \in \mathcal{H}(y \ge 0): \quad \nabla^2\Phi = 0, \quad \lim_{|x^2+y^2|\to\infty} \left|x^2 + y^2\right|^{-1/2}\Phi(x,y) = 0, \tag{37.120a–c}$$

(i) the ***Poisson half-plane formula*** *(37.121) that specifies the solution of the Dirichlet problem for the half-plane:*

$$\Phi(x,y) = \pi^{-1} \int_{-\infty}^{+\infty} \Phi(t,0) \left[(t-x)^2 + y^2\right]^{-1} y\,\mathrm{d}t; \tag{37.121}$$

$$\Psi(x,y) = \pi^{-1} \int_{-\infty}^{+\infty} \Phi(t,0) \left[(t-x)^2 + y^2\right]^{-1} (x-t)\,\mathrm{d}t; \tag{37.122}$$

(ii) the associated ***half-plane orthogonality formula*** *(37.122) specifies the family of orthogonal functions (37.123a, b):*

$$\Phi + i\Psi \equiv f(x+iy) \in \mathcal{H}(y \geq 0): \quad 0 = \nabla\Psi \cdot \nabla\Phi, \qquad \lim_{|x^2+y^2|\to\infty} \left|z^2 + y^2\right|^{1/2} \Psi(x,y) = 0, \tag{37.123a–c}$$

satisfying the same asymptotic condition (37.123c) ≡ (37.120c). The analogue results for the lower half-plane are obtained replacing y by −y. The corresponding results for *a complex function holomorphic in the upper half-plane (37.115a,b) and meeting the asymptotic condition (37.115c), are: (i) the Cauchy half-plane integral (37.116b); (ii) the* ***Schwartz half-plane formula:***

$$f(z) = \frac{i}{\pi} \int_{-\infty}^{+\infty} \Phi(t,0) \left|z-t\right|^{-2} (z^* - t)\,\mathrm{d}t, \tag{37.124}$$

that also applies to the lower half-plane replacing z by z^*; (ii) the ***Cauchy integral transform*** *(37.125a) involving the* ***Cauchy kernel*** *(37.125b):*

$$f(x+iy) = \pi^{-1} \int_{-\infty}^{+\infty} \Phi(t,0)\, C(x,y;t)\,\mathrm{d}t, \quad C(x,y;t) \equiv \frac{y + i(x-t)}{(x-t)^2 + y^2}. \tag{37.125a,b}$$

The results (37.124) ≡ (37.125a,b) are the sum of (37.121) and (37.122) in (37.116b; 37.118):

$$f(z) = \Phi(x,y) + i\Psi(x,y) = \pi^{-1} \int_{-\infty}^{+\infty} \Phi(t,0) \left|x + iy - t\right|^{-2} \left[y + i(x-t)\right] \mathrm{d}t, \tag{37.126}$$

where (37.126) coincides with (37.119) ≡ (37.125a,b) ≡ (37.126).

Note 37.1. *Dirichlet, Neumann, and Robin Boundary-Value Problems for Harmonic Functions:* The Dirichlet problem on the disk (half-plane) is a particular two-dimensional case of the solution of the Laplace equation that can be obtained for any domain in any dimension using generalized functions, namely, Green's functions; in particular for the three kinds of boundary conditions (Section 32.3), and also the regularity and asymptotic conditions (Section 32.2). The latter approach includes: (i) the interior (exterior) Dirichlet problem with boundary conditions of first kind (37.109a,b; 37.110a–c) [(37.113a,b)]; (ii) the Neumann problem with boundary conditions of the second kind, that is, specifying the normal derivative instead of the values of the function on the boundary (Section 32.3); (iii) the mixed or Robin boundary-value problem of the third kind is a linear combination of the preceding two and can be obtained by superposition, since the Laplace and

Poisson equations are linear. The account on the Dirichlet problem (Section 37.9) is sufficient to calculate (Section 38.1) the field in a cylinder (half-plane) due to a potential on the circle (real axis). The solution of the Dirichlet problem (Sections 37.8 and 37.9) needs only the values of the potential or of the stream function at the boundary; since they are related to specify both would be either redundant or inconsistent. One of them, that is, the potential (stream function) specified on the boundary determines both in the interior or exterior, thus providing: (i) the analytic continuation of the potential (stream function) from the boundary to the interior or exterior; (ii) the orthogonal family of curves, that is, the streamlines from the equipotentials (vice-versa). The latter (ii) involves an arbitrary added constant, specifying any member of a parallel family of orthogonal curves. The electrostatic (magnetostatic) field due to an electric charge (current) distribution would correspond to the Neumann problem for a boundary condition of the second kind; it is an instance of a monopole distribution, considered for the gravity field (Section 18.9). This is a particular case of a multipole distribution, considered for the magnetostatic field (Section 26.3) in the case of a dipole (Section 26.9). The Schwartz theorem (Section 27.8) is used to specify (Section 38.9) complex conjugate velocity in a jet from its value on the free boundary. The Poisson type formulas are applied (Subsection 28.1.2) to determine the potential and stream function in a half-plane (disk) due to a given potential on the real line (boundary circle) [Subsection 28.1.3 (28.1.4)]; also in Example 40.15. These examples are included (Subsection 28.1.1) in a classification (Diagram 38.1) of boundary conditions (Note 38.1).

The general Dirichlet problem considered in Section 37.8 (37.9) for the particular case of the circle (half-plane) is, for an arbitrary domain:

$$\Phi \in \mathcal{H}(D): \quad \nabla^2\Phi = 0, \quad \Phi|_{\partial D} = f; \tag{37.127a–c}$$

it specifies a harmonic function (37.127b) in the domain (37.127a) with given values (37.127c) on the boundary, that is, corresponding to the boundary condition of first kind (37.127c) $\equiv$ (32.7a). The boundary condition of the second kind (32.7b) $\equiv$ (37.128c) leads to the **Neumann problem** (38.128a,b):

$$\Phi \in \mathcal{H}(D): \quad \nabla^2\Phi = 0, \quad \left.\frac{\partial\Phi}{\partial n}\right|_{\partial D} = g. \tag{37.128a–c}$$

The normal derivative corresponds to a surface dipole distribution and hence the solution of the Neumann problem in general uses the logarithmic potential as for an electric charge distribution in vacuo (24.13a):

$$-\frac{\mathrm{e}}{\varepsilon} \equiv 1: \quad f(z) = (2\pi)^{-1} \int_{\partial D} \log(\zeta - z) \left[\left.\frac{\partial\Phi}{\partial n}\right|_{\zeta}\right] \mathrm{d}\zeta + f(\infty), \tag{37.129}$$

to within an arbitrary constant; only half the field is considered $0 \le \arg(z) \le \pi$ for the half-plane:

$$\Phi + i\Psi \equiv f(x+iy) \in \mathcal{D}(y \ge 0):$$

$$f(x+iy) = \pi^{-1} \int_{-\infty}^{+\infty} \log(t - x - iy) \left[\frac{\partial\Phi}{\partial y}(t,0)\right] \mathrm{d}t + f(\infty). \tag{37.130}$$

The solution of the Neumann problem (37.128a–c) is (37.129); in particular in (37.131a,b) is the solution of the plane Neumann problem:

$$\Phi(x,y) \in \mathcal{H}(y \geq 0): \quad \Phi(x,y) = (2\pi)^{-1} \int_{-\infty}^{+\infty} \log\left|(t-x)^2 + y^2\right| \frac{\partial \Phi}{\partial y}(t,0)\mathrm{d}t + \Phi(\infty); \tag{37.131a,b}$$

$$\Phi + i\Psi \equiv f(x+iy) \in D(y \geq 0): \quad \Psi(x,y) = \pi^{-1} \int_{-\infty}^{+\infty} \operatorname{arc\,cot}\left(\frac{t-x}{y}\right) \frac{\partial \Phi}{\partial y}(t,0)\mathrm{d}t + \Psi(\infty), \tag{37.132a,b}$$

also (37.132a,b) specifies is the corresponding orthogonal family (37.133b) of functions satisfying the asymptotic condition (37.133c):

$$\Psi \in \mathcal{H}(D): \quad \nabla\Phi \cdot \nabla\Psi = 0, \qquad \lim_{|x^2+y^2| \to 0} \left|x^2+y^2\right|^{-1/2} \{\Phi(x,y), \Psi(x,y)\} = 0, \tag{37.133a,b}$$

The boundary condition of the third kind (32.7c) $\equiv$ (37.134c,d) leads to the **mixed or Robin problem** (37.134a,b):

$$\Phi \in \mathcal{H}(D): \quad \nabla^2\Phi = 0; \quad \lambda > 0: \quad \left[\Phi + \lambda\frac{\partial \Phi}{\partial n}\right]_{\partial D} = h. \tag{37.134a–d}$$

Comparing with Dirichlet (37.135b,d) [Neumann (37.135c,e)] problems with arbitrary functions $h_1\,(h_2)$:

$$\Phi_1, \Phi_2 \in \mathcal{H}(D): \quad \nabla^2\Phi_1 = 0 = \nabla^2\Phi_2; \quad \Phi_1|_{\partial D} = h_1, \quad \left.\frac{\partial \Phi_2}{\partial n}\right|_{\partial D} = h_2, \tag{37.135a–e}$$

leads by superposition (37.136a) to a harmonic function (37.136b):

$$\Phi \equiv \Phi_1 + \Phi_2; \quad D: \quad \nabla^2\Phi = 0, \quad \Phi_2|_{\partial D} + \lambda\left.\frac{\partial \Phi_1}{\partial n}\right|_{\partial D} = h - h_1 - \lambda h_2, \tag{37.136a–c}$$

and to the boundary condition (37.136c) that (h_1, h_2) must satisfy; the latter is not explicit because $\Phi_1\,(\Phi_2)$ depends on $h_1\,(h_2)$ through (37.135c) [(37.135d)]. Thus *the mixed problem (37.134a–d) is equivalent to the superposition (37.136a) of Dirichlet (37.135b,d) and Neumann (37.135c,e) problems where the arbitrary functions* $(\mathrm{h}_1, \mathrm{h}_2)$ *satisfy (37.136c).*

Note 37.2. *Existence and Unicity Theorems and Methods of Solution:* Whereas the results in Sections 37.8 and 37.9 are explicit solutions of potential field problems, the properties of conformal mappings (Chapters 33 and 35) correspond to existence and unicity theorems (Sections 37.3–37.7). The latter give the assurance that a given problem has at least one solution, and specify the conditions in that it is unique. The methods used to prove existence and unicity theorems are typically: (i) to find upper (or lower) bounds (Section 37.1); (ii) to use these bounds to prove convergence (Section 37.2); (iii) to show that the bounds are violated if certain conditions are not met. The latter approach serves to prove various properties (Sections 37.3–37.7) often by reduction ad absurdum using the bounds established before. The methods of proof of existence and unicity theorems tend to follow one of three lines of reasoning, and their combinations: (i) positive direct proof by using convergence properties, for example, the existence of uniformly convergent subsequences (Section 37.2); (ii) proof by the negative, showing that a result must be true, otherwise some bound (Section 37.1) or constraint (Subsection 37.4.3) is violated (Subsection 35.7.2), for example, the use of the minimax, reference and elliptic modular functions to prove the existence of simple mappings; (iii) introduction of new concepts that illuminate the issue in question (Section 37.6), for example, the complete collection of fundamental domains of the automorphism groups as the necessary and sufficient condition for the existence of one-to-one conformal mappings between multiply connected regions. These methods of proof of existence and unicity of a solution, may give no indication of how the solution can be actually obtained in a specific case. The existence and unicity theorems may cover wide range of cases, for example, simple conformal mapping between arbitrary compact regions, whereas explicitly analytic solutions are available only in a few cases, for example, bilinear mapping between disks or half-planes (Sections 35.8 and 35.9) or polygons (Sections 33.6–33.9). The specific cases for that explicit solutions are available may go beyond the conditions of validity of the existence and unicity theorems, for example, the Schwartz–Christoffel transformation (Sections 33.6–33.9) allow for noncompact regions, branch-points and multivalued functions excluded by the simple conformal mapping, leaving two alternatives: (a) to extend the existence and unicity theorems of the simple mapping (Section 37.3), for example, to noncompact regions and singularities on the boundary (Section 37.7) and to nonsimple mappings of multiply connected regions (Sections 37.4 and 37.6); (b) to prove the required properties for the particular mapping, for example, the mapping of interiors to interiors and boundaries to boundaries was proved for bilinear (Sections 35.6–35.9) and polygonal (Sections 33.6–33.9) conformal transformations. There is no harm in doing both (a) and (b) as a cross-check.

Note 37.3. *Special Points on Curves:* The conformal mapping may involve special points on curves; these are particular instances of points on curves, surfaces and subspaces in differential geometry. Table 37.1 lists a few special points encountered so far from the properties of the tangent: (i) at a **regular point** the curve has continuous first-order derivative, and hence continuous tangent and normal:

$$z(t) \in C^1(t_1, t_2): \quad \dot{z} \equiv \frac{dz}{dt} \in C; \tag{37.137}$$

TABLE 37.1
Classification of Points on Curves

Point	Tangent	Section	Figure
Regular	Continuous	–	–
Cusp	Opposite signs right and left	6.6; E20.9; 34.2–34.6; 39.4	F6.3; F20.3c,d; F34.2c; F34.3b; F39.4
Angular	Distinct right and lift	15.5; E20.9; 33.5–33.9	F15.2a,b; F20.3a,b; F33.5–33.7; F34.4a–e
Double	Two continuous	6.7; 12.7; 28.5; 28.8	F6.6; F12.4; F28.4d; F28.10; F28.12d
Multiple	More than two continuous	12.8–12.9	F12.5; F12.7

E – example; F – figure.
Note: A regular point on a curve has a continuous tangent. An example of nonregular point is an angular point, which has distinct tangents at right and left; a particular case of angular points is a cusp, where the right and left tangents are antiparallel. A curve may cross itself once (n times) at a double (multiple) point with two ($n + 1$) distinct tangents.

(ii) at an **angular point** the right- and left-hand derivatives exist but are distinct, specifying the internal angle β:

$$\varepsilon > 0: \quad z'(t \pm 0) \equiv \lim_{\varepsilon \to 0} z'(t \pm \varepsilon); \quad \beta = \arg[z'(t+0)] - \arg[z'(t-0)]; \qquad (37.138\text{a,b})$$

(ii) a particular case of angular point is the **cusp** (Figure 39.4) for that the right- and left-hand tangents have opposite directions, that is, the angle is $\beta = \pi$:

$$t \text{ is a cusp}: \quad z'(t+0) = -z'(t-0) \Leftrightarrow \arg[z'(t+0)] = \arg[(z'(t-0)] + \pi; \qquad (37.139\text{a,b})$$

(iv) a **multiple point**, where the curve crosses itself, that is,

$$t_1 < t_3 < t_4 < \cdots < t_N < t_2: \quad z(t_3) = z(t_4) = \cdots = z(t_N), \qquad (37.140)$$

that is, the coordinates are the same for a number N of distinct values of the parameter that specifies the **multiplicity** of the point.

The examples of angular points include (i) the residues at a pole at an angular point of a contour (Section 15.5, Figure 15.2a,b; Example 20.9, Figure 20.3a–d); (ii) the vertices of a polygonal mapping (Sections 33.6–33.9, Figures 33.5–33.7); (iii) the angular point (Figure 37.5c) associated with a conformal univalent mapping of a singularity on the boundary to a noncompact region (Section 37.7); (iv) the angles at corners and plates (Chapter 36) and cusp at the trailing edge of some airfoils (Chapter 34), for example, the Joukouski airfoils (Sections 34.3 and 34.4). The examples of cusps include (v) the cycloid corresponding to the trajectory of an electron in a electromagnetic field (Section 6.5) and a cylinder with circulation in a uniform flow (Section 29.7) as shown in Figure 6.3; (vi) the residues at poles of boundary curves with cusps in Example 20.9 and Figure 20.3c,d; (vii) the three vertices of the Schwartz triangle (Figure 39.4) associated with the construction of the elliptic modular function (Section 39.4). The examples of multiple points include: (viii) the trajectory of an electron in an electromagnetic field (Section 6.7) and of a cylinder with circulation in a uniform flow (Section 28.7) viz. a trochoid that has double points (Figure 6.6); (ix) the equipotentials of a multipole of order N that have $2N$ petals or lobes (Section 12.9), and cross at the origin that is a point of multiplicity $4N$, for example, 4,8,12 respectively for a dipole, quadrupole, and octupole (Figures 12.4, 12.5, and 12.7); (x) the stagnation points of potential flows, for example, in Figures 28.4d and 28.12d; (xi) the polhode that is the curve associated with the spinning of a heavy symmetric top, for example, gyroscope; (xii) the epi(hipo) cycloids/trochoids that are curves described by a point on a circle rolling/sliding outside/inside another circle.

Note 37.4. *Classification of Functions as Correspondences or Mappings:* The relation between single-valued (Section 9.1), multivalued (Sections 7.1 and 7.2) and multivalent (Subsections 37.4.1–37.4.4) functions is indicated in Diagram 37.1. A **function** is a relation from a set of points called the **domain** D to a set of points called the **range** E:

$$f : D \to E: \quad \forall_{z \in D} \exists_{w \in E}: \quad w = f(z). \qquad (37.141)$$

A function is: (i) **single-valued** if (Figure 9.1a) to every point in the domain corresponds only one point in the range (9.1); (ii) **multivalued** (Figure 37.8a) $\equiv$ (Figure 9.1b) if there is at least one point in the domain mapped to more than one point in the range (9.2). A function is: (i) **univalent** (Figure 37.8a) if every point in the range is mapped from only one point in the domain (37.36b); (ii) **multivalent** (Figure 35.8b) if there is at least one point in the range to that are mapped more than one point in the domain (37.34a,b). The contrast between multivalued function with n branches $\mathcal{U}_n$ and multivalent function with m values $\mathcal{U}^m$ is illustrated in Figure 37.8a (37.8b).

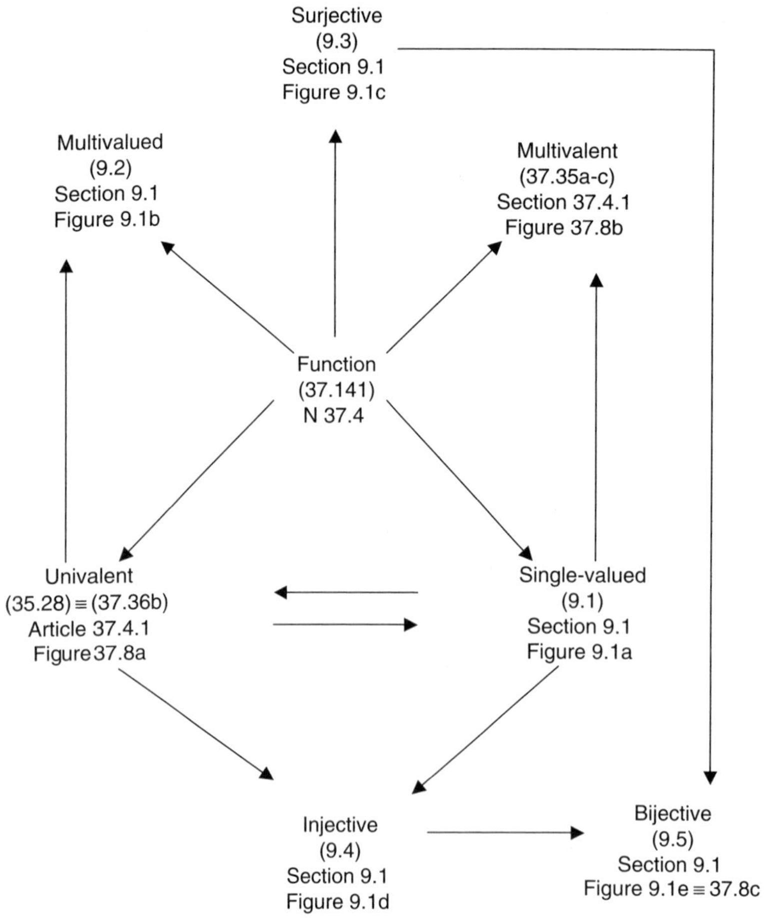

DIAGRAM 37.1
Classification of functions as correspondences or mappings. A function may be classified by two criteria, depending on how the domain (range) are mapped from (into). Concerning the range a single-valued (multivalued) function maps each point (at least one point) in the domain to just one (more than one) point in the range. Concerning the domain a univalent (multivalent) function maps into each point (at least one point) in the range just one (more than one) point in the domain. If follows (Diagram 37.1) that: (i) single-/multivalued are mutually exclusive; (ii) uni-/multivalent are also mutually exclusive; (iii) the four combinations of single/multivalued with uni-/multivalued are possible. The four cases (iii) are considered with their inverses in Table 37.2. There it is noted that a single-valued, univalent function maps between distinct points in the domain and distinct points in the range, and hence is injective; if it is surjective then every point in the range is mapped from one point in the domain. The combination of injective and surjective is a bijective function which maps one-to-one the domain into the range.

The concepts of single-/multivalued and univalent/multivalent functions are independent, that is, all four combinations are possible, for example, (i) the linear function $w = cz$ with nonzero coefficient $c \neq 0$ is single-valued and univalent; (ii) any periodic function (7.8a) is **many-valent**, takes infinitely many times the same value, for example, the sine

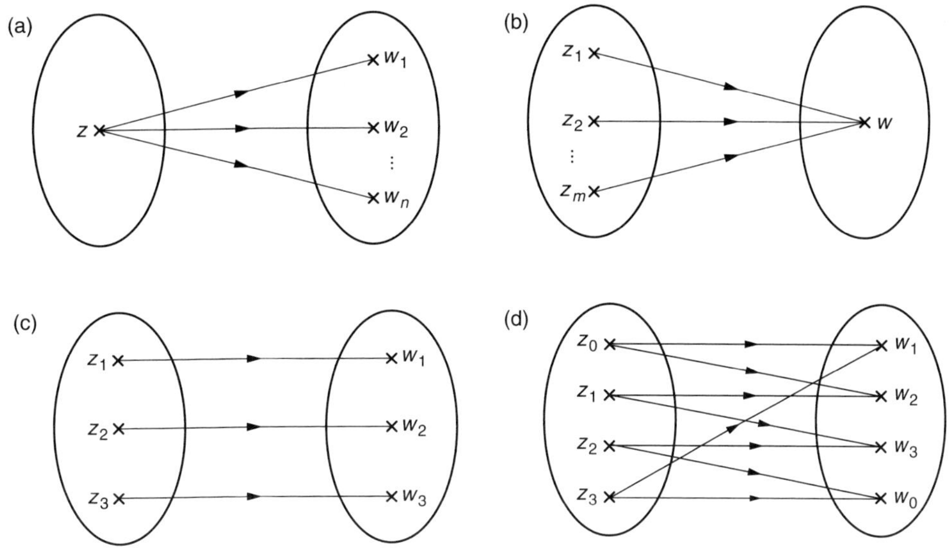

FIGURE 37.8
A multivalued function with n branches maps a single point in the domain to n points in the range (a). A multivalent function with m values maps from m distinct points in the domain to the same point in the range (b). Both (a) and (b) are excluded for a single-valued univalent function (c), which is injective, that is, maps distinct points in the domain to distinct points in the range (c). Both (a) and (b) can occur for a multivalued multivalent function with n branches and m values, so that: (i) m points in the domain can map to the same point in the range; (ii) one point in the domain can map to n points in the range; (iii) the case of two branches $n = 2$ and two valves $m = 2$ is illustrated in (d). Figures 9.1 and 37.8, Table 37.2 and Diagram 37.1 are complementary.

is single-valued and many-valent (37.142b) because it takes the same value (37.142c) for all (37.143a,d):

$$n \in |Z: \quad \sin z = w \in \mathcal{V}^{\infty}, \quad \sin z = \sin z_n, \quad z_n = z + 2n\pi; \qquad (37.142a\text{–}d)$$

(iii) the logarithm is univalent and many-valued (7.7) = (37.49a–d); (iv) the logarithm of the sine (37.143b) is many-valued many-valent function (37.143a):

$$n, m \in |Z: \quad \log(\sin z) = w \in \mathcal{V}_{\infty}^{\infty}, \quad \log(\sin z) = \log[\sin(z + 2n\pi)]$$
$$= \log|\sin z_n| + i \arg(\sin z_n) + i2\pi m, \qquad (37.143a\text{–}d)$$

since it has infinitely many branches (37.143c) and takes infinitely many times (37.142d) the same values (37.143a,b).

A single-valued univalent function is injective (9.4) because it maps between distinct points (Figure 9.1d). *Any function, single- or multivalued can be made* **surjective** *(9.3) by choosing the range to be the image of the domain:*

$$f \quad \text{surjective}: \quad E = f(D) \Leftrightarrow \forall_{w \in E} \exists_{z \in D}: \quad w = f(z), \qquad (37.144)$$

because then every point in the range is (Figure 9.1c) the image of at least one point in the domain. A single-valued univalent function is (37.145a) injective (9.4) and can be made

TABLE 37.2
Properties of Direct and Inverse Functions

Function: direct[*inverse*]	**Single-valued**	**Multivalued**
Univalent	Injective: $\widetilde{\mathcal{U}}$ Figure 35.8c [*injective:* $\widetilde{\mathcal{U}}$]	Multivalued (n) and univalent: $\mathcal{U}^n$ Figure 35.8a [*single-valued and multivalent (n):* $\mathcal{U}^n$]
Multivalent	Single-valued and multivalent (m): $\mathcal{U}^m$ Figure 35.8b [*multivalued (m) and univalent:* $\mathcal{U}_m$]	Multi-valued (n) and multivalent (m): $\mathcal{U}_n^m$ Figure 35.8d [*multivalued (m) and multivalent (n):* $\mathcal{U}_m^n$]

Note: A single-valued (n-valued) function maps each point (some points) in the domain into one point (n points) in the range. A univalent (m-valent) function maps into each point of the range one (up to m points) from the domain. If follows that: (i) a single-valued univalent function maps distinct points between the domain and range, and is injective together with its inverse; (ii) a single-valued m-valent function has for inverse a univalent m-valued function; (iii) an n-valued univalent function has for inverse a single-valued n-valent function; (iv) an n-valued, m-valent function has for inverse an m-valued, n-valent function.

bijective (37.145b) by choosing the range as the image of the domain (Figure 37.8c):

$$\mathcal{U} \cap \mathcal{U}^1 \equiv \widetilde{\mathcal{U}}, \quad \mathcal{U} \cap \mathcal{U}^1 \cap \bar{\mathcal{U}} = \bar{\mathcal{U}} \cap \widetilde{\mathcal{U}} = \bar{\widetilde{\mathcal{U}}}, \tag{37.145a,b}$$

because then it is both injective and surjective (9.5).

The inverse of a single-valued function is a univalent function (37.146a), and vice-versa; this applies regardless of whether the univalent function is single-valued [(37.41); Figure 37.8c] or multivalued [(37.42); Figure 37.8a]:

$$n = 1, 2, \ldots: \quad w = f(z) \in \{\mathcal{U}, \mathcal{U}_n\} \Rightarrow z = f^{-1}(w) \in \{\mathcal{U}^1, \mathcal{U}^n\}, \tag{37.146a,b}$$

the inverse of a multivalued function with n *branches is a multivalent function with* n *values (37.146b), and vice-versa;* this applies regardless of whether the multivalued function is univalent [(37.42); Figure 37.8a] or multivalent [(37.43); Figure 37.8d]. Thus *there are (Table 37.2) four possible combinations:*

$$w = f(z) \in \left\{\widetilde{\mathcal{U}}, \mathcal{U}_n^1, \mathcal{U}_1^m, \mathcal{U}_n^m\right\} \Leftrightarrow z = f^{-1}(w) \in \left\{\widetilde{\mathcal{U}}, \mathcal{U}_1^n, \mathcal{U}_m^1, \mathcal{U}_m^n\right\}, \tag{37.147a–d}$$

viz.: (i) a single-valued univalent function is injective (37.147a) and has an injective inverse (Figure 37.8c); (ii) a multivalued univalent function with n branches (Figure 37.8a) has (37.147b) a single-valued multivalent inverse with n values (Figure 37.8b); (iii) a single-valued multivalent function with m values (Figure 37.8b) has (37.147c) a multivalued univalent inverse with m branches (Figure 37.8a); (iv) a multivalued multivalent function with n branches and m values (37.147d) has a multivalued multivalent inverse with m branches and n values (Figure 37.8d). The statements (ii) to (iv) assume that the ranges do not overlap. The many-valued (many-valent) functions have been considered together with the multivalued (multivalent) functions with $n = \infty$ branches ($m = \infty$ values), in Note 37.4, Table 37.2, and Diagram 37.1.

Conclusion 37: A doubly connected region (Figure 37.1a) cannot be mapped conformally to the unit disk mapping because a loop L around the inner boundary $\underline{\partial D}$ cannot be shrunk to zero, whereas the image loop Γ can be shrunk continuously to zero; thus the mapping is discontinuous, excluding a holomorphic function. The whole complex z-plane (Figure 37.1b) cannot be mapped conformably one univalently to the punctured ζ-plane, i.e., ζ-plane missing one point, because a further simple mapping would lead to a function bounded at infinity; the latter would reduce to a constant, and so would do the first mapping which thus would not be univalent. Thus the necessary conditions for the existence of a simple, i.e., conformal and univalent mapping, into the unit disk are (Figure 37.2) that the region D be (a) simply connected, and has at least two boundary points; these conditions are also sufficient, and unique if a point c and a direction through it are assigned. Using the unit disk as intermediate and noting that the composition and inversion of simple mappings is simple, it follows that two simply connected regions, with at least two boundary points each, can also be mapped simply (b) one on to the other. The conformal mapping of multiply-connected and punctured regions is possible using multivalued functions, e.g., the logarithm maps the annulus (Figure 37.3), which is punctured at the branch-point at the origin, and surrounds the branch-cut along the negative real axis, into a strip in the plane, which is a simply-connected region where it takes each value only once, i.e., there are no overlaps. The one-to-one conformal mapping also exists between multiply-connected domains (Figure 37.4) whose boundaries consists of one outer and several inner loops around holes. The conformal mapping can also be considered as the converse result, viz. a univalent function or branch of a multivalued function on a region D, sends the boundary ∂D to a curve ∂E, and specifies a simple mapping between the interiors $D - \partial D, E - \partial D$ (Figure 37.5) in three cases: (a) the region E is compact, because there are no singularities in D; (b) the region E is noncompact, because it contains the point-at-infinity, as the image of a single simple pole in the interior of D; (c) the boundary ∂E extends to infinity, and has asymptotes making an angle β at the image of a singular point of angle α on the boundary of D, where the mapping is of order $-\alpha/\beta$. The interior (exterior) mappings can be related to Cauchy theorem for a function (Figure 37.6) holomorphic inside or outside a loop. Since holomorphic functions are harmonic, they provide the solutions of the Laplace equation in a domain from its values on the boundary, (Figure 37.7) e.g., specify the harmonic function at all points inside or outside a cylinder (a) [in a upper or lower half-plane (b)], from the values of its real or imaginary part on the boundary circle (real axis); this corresponds to the Dirichlet problem, which has boundary conditions of first kind. The simple functions are single-valued (Figure 37.8c) and univalent, and map conformally between simply-connected domain and range (Figure 37.2a,b). The multivalued univalent functions (Figure 37.8a) can map into a singly-connected (Figure 37.3) or a multiconnected (Figure 37.4) range. The multivalent (Figure 37.8b) [multivalued (Figure 37.8a)] functions assign to at least one point in the range more than one point in the domain (vice-versa). A single-valued univalent function is injective, i.e., maps between distinct points, like its inverse (Figure 37.8c); a multivalued multivalent function has a multivalued multivalent inverse (Figure 37.8d)

38

Hodograph for Free Jets

The preceding potential flow problems have concerned free flows or flows forced by sources, sinks, and vortices, in all space or in regions bounded by rigid boundaries, plus the analogue problems, for example, electro (magnetostatic) fields due to electric charge (current) distributions. Another type of problem (Diagram 38.1) is the field, for example, electric, due to a given surface potential, for example, on a disk or on a plane (Section 38.1), where again the boundaries are fixed. Yet another class of problems are those of free jets, for that the boundary is not known a priori and must be determined as part of the solution to the problem. In the potential flows considered before the boundaries were rigid impermeable walls placed at positions known "a priori" and where the velocity is tangential. The same boundary condition, namely tangential velocity, applies to free jets for which the surface of separation from the surrounding fluid at rest is not determined "a priori," that is, the shape of the free surface is part of the solution of the problem. Since a free surface separates a fluid at rest from a flow, it must be an isobar, that is, a line of constant pressure; by the Bernoulli law the velocity is constant in modulus along an isobar, although the direction generally varies for a curved free surface. If the jet issues from a hole in a reservoir (Sections 38.2–38.4) or impinges on a rigid obstacle (Sections 38.5–38.8) the direction of the velocity is tangential, hence known, although its modulus generally varies because the pressure on the obstacle is usually nonuniform. Thus a free jet with confining walls, or impinging upon obstacles or with given direction (Section 38.9) can be represented using lines of constant modulus or angle of the velocity that are the coordinates in the hodograph plane. Mapping the region occupied by the free jet in the hodograph plane conformally onto the physical plane leads to a relation between the complex velocity and complex potential. The latter would have to be integrated to obtain explicity the complex potential in the physical plane; even if the explicit complex potential is not available it is possible to obtain by integration some results, such as the shape of the free surface or the pressure distribution on obstacles. The problems with free jets considered include: (i/ii) the jet issuing from a slit in a wall, without or with (Section 38.3) a reentrant tube, including the calculation of the contraction ratio (Section 38.2); (iii/v) the jet perpendicular (Section 38.4) [oblique (Section 38.5)] to a flat plate or incident symmetrically on an arrow (Section 38.6) leading to a calculation of the drag; (vi/vii) the attachment (deflection) of a jet to (from) a wall, due the Coanda effect (Section 38.7) [a fluidic source (Section 38.8)]; (viii) the collision or merging of several free jets (Section 38.9). Some of the problems have multiple interpretations, for example, (iii) the deflection of a flow by a plate perpendicular to wall (Section 38.5); (v) the thrust on a surfboard riding on a stream; (viii) a jet incident normal to a wall. The hodograph method specifies both the shape of free jets and the forces and moments they exert on obstacles.

38.1 Fields due to Potentials on Planes and Cylinders

Among the types of potential flow or field problems (Diagram 38.1) are included (i) given potentials on fixed surfaces (Section 38.1); (ii) moving boundaries, for example, free jets

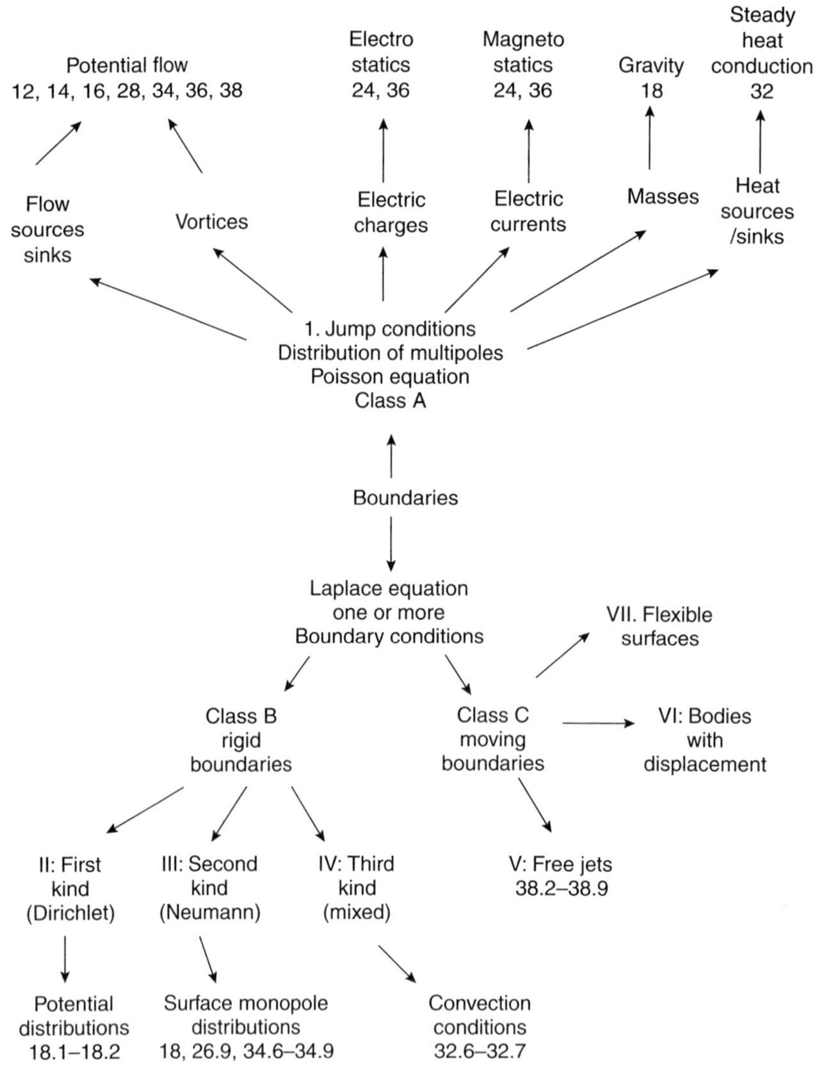

Chapters: 12, 14, 16, 18, 24, 26, 38, 32, 34, 35, 38

Sections: 18.1–18.2, 26.9, 32.6–32.7, 34.6–34.9, 38.2–38.9

DIAGRAM 38.1

Boundary conditions for potential fields. The source "forcing" the potential fields can be distributed: (i) in domains in free space; (ii) along curves or surfaces, as distributions which may also act as boundaries. The "sources" forcing the potential fields (Classification 38.1) are: (i/ii) sources/sinks and vortices in a potential flow; (iii/iv) electric charges (currents) in electro (magneto) statics; (v) mass distributions for the gravity field; (vi) heat sources/sinks for steady heat conduction. The surface distributions (i,iii,v,vi) [(ii,iv)] lead to normal (tangential) jump conditions (Figures 18.1 and 18.2) because [Section 18.1(18.2)] the corresponding fields are irrotational (solenoidal). The boundaries with (i–vi) [without (vii–xii)] forcing relate to the Poisson (Laplace) equation. The boundaries without sources (vii–xii) may be rigid or moving. For the rigid boundaries, steady heat conduction gives the example of boundary conditions three kinds: (vii) imposed temperature (Dirichlet problem); (viii) specified gradient or heat flux (Neumann problem); (ix) combination of (vii) and (viii) in convection (Robin problem). The moving boundaries may be considered for fluids: (x) moving rigid bodies whose displacement determines the fluid velocity at the surface; (xi) flexible surfaces which move in response to pressure or porous surfaces with blowing or suction; (xii) free jets whose free surfaces are constant pressure boundaries not known "a priori," to be found as part of the solution of the problem. Besides the boundary conditions there may be: (xiii) asymptotic conditions at infinity; (xiv) regularity conditions at edges; (xv) other "special" conditions required for the existence of a unique solution.

CLASSIFICATION 38.1

Comparison of seven potential fields.

Field	Nondissipative fluid flow			Electrostatic	Magnetostatic	Gravity	Steady heat conduction
	Irrotational	Incompressible	Potential				
Case	**I**	**II**	**III**	**IV**	**V**	**VI**	**VII**
Chapter		**12,14,16,28,34,36,38**		**24,28,36**	**24,28,36**	**18,28**	**32,36**
Symbol	Q	Γ	$P_0 \equiv Q - i\mathrm{T}$	e	J	m	Q
Generation by: per unit area	Flow rate of source/sink	Circulation of vortex	Monopole or spiral source	Electric charge	Electric current	Mass	Heat source
Per unit length:	q	γ	p_0	σ	ϑ	μ	q
Constitutive parameter		Mass density		Dielectric permittivity	Magnetic permeability	Gravitational constant	Thermal conductivity
Symbol		ρ		ε	μ	G	κ
Potential: Φ	$\nabla^2\Phi = \Lambda$	–	$\nabla^2\Phi = \Lambda$	$\nabla^2\Phi = -q/\varepsilon$	–	$\nabla^2\Phi = -\mathrm{G}\rho$	$\nabla^2 T = -w/\kappa$
Field function: Ψ	–	$\nabla^2\Psi = -\varpi$	$\nabla^2\Psi = -\varpi$	–	$\nabla^2\Psi = -j/\mu$	–	–
Basic vector field	Velocity $\vec{\mathrm{v}} = \nabla\varphi$	Velocity $\vec{\mathrm{v}} = \nabla \wedge \Psi$	Velocity $\vec{\mathrm{v}} = \nabla\Phi = \nabla \wedge \vec{\Psi}$	Electric field $\vec{\mathrm{E}} = -\nabla\Phi$	Magnetic field $\vec{\mathrm{H}} = \nabla \wedge \Psi$	Acceleration of gravity $\vec{g} = -\nabla\Phi$	Heat flux $\vec{\mathrm{G}} = -\kappa\nabla T$
Forcing	Dilatation: Λ	Vorticity: ϖ	Dilatation: Λ vorticity: ϖ	Electric charge density: e	Electric current density: j	Mass density: ρ	Heat density: w
Quadratic function	$2E_k \equiv \vec{\mathrm{p}} \cdot \vec{\mathrm{v}} = v^2 = p^2/\rho$			$2E_e \equiv \vec{\mathrm{D}} \cdot \vec{\mathrm{E}} = \varepsilon E^2 = D^2/\varepsilon$	$2E_h = \vec{\mathrm{B}} \cdot \vec{\mathrm{H}} = \mu H^2 = B^2/\mu$	Potential energy: Φ	$T^2\dot{S} \equiv -\vec{\mathrm{G}} \cdot \nabla T = \kappa(\nabla T)^2 = G^2/\kappa$
Meaning	Kinetic energy			Electric energy	Magnetic energy	(not quadratic)	Rate of entropy production

CLASSIFICATION 38.1

Continued

Field	Nondissipative fluid flow			Electrostatic	Magnetostatic	Gravity	Steady heat conduction
	Irrotational	Incompressible	Potential				
Case	I	II	III	IV	V	VI	VII
Chapter	12,14,16,28,34,36,38			24,28,36	24,28,36	18,28	32,36
Force density: $\vec{f} = \ldots$	$-\vec{v}[\nabla \cdot (\rho\vec{v})]$	$\rho(\nabla \wedge \vec{v}) \wedge v$	$-\vec{v}[\nabla \cdot (\rho\vec{v})]$ $\rho(\nabla \wedge \vec{v}) \wedge \vec{v}$	$q\vec{E} =$ $\vec{E}[\nabla \cdot (\vec{E}/\varepsilon)]$	$c^{-1}\vec{j} \wedge \vec{B} =$ $-\vec{B} \wedge [\nabla \wedge (\vec{B}/\mu)]$	$\rho\vec{g}$	–
Meaning	Dilatation	Vortical	Hydrodynamic	Electric	Magnetic	Gravity	
Pressure	Dynamic pressure $p = p_0 - \frac{1}{2}\rho v^2$			Electric pressure $p_e = \varepsilon E^2/8\pi$	Magnetic pressure $p_m = \mu H^2/8\pi$	–	–
Boundary condition jump [. . .] of	$[v_n] = q, [v_s] = -\gamma$ Components: n-normal, s-tangent to interface			$[E_n] = \sigma/\varepsilon$ $[E_s] = 0$	$[H_n] = 0$ $[H_s] = \vartheta/c$	$[g_n] = \mu,$ $[g_s] = 0$	$[G_n] = -q$ $[G_s] = 0$
Alternative form in terms of potential or field function	$[\partial\Phi/\partial n] = q = [\partial\Psi/\partial s]$ $-[\partial\Phi/\partial s] = -\gamma = [\partial\Psi/\partial n]$			$[\varepsilon\partial\Phi/\partial n] = \sigma$ $[\partial\Phi/\partial s] = 0$	$[\mu\partial\Psi/\partial s] = 0$ $[\partial\Psi/\partial n] = \vartheta$	$[\partial\Phi/\partial n] = \mu$ $[\partial\Phi/\partial s] = 0$	$[\kappa^{-1}\partial T/\partial n] = -q$ $[\partial T/\partial s] = 0$
Condition of validity	No viscosity, no thermal conduction			Steady $\partial\vec{E}/\partial t = 0$	Steady $\partial\vec{H}/\partial t = 0$	Classical non-relativistic	Steady $\partial T/\partial t = 0$

Note: (i–iii) Irrotational or incompressible flow, or both together, that is, potential flow; (iv–v) the electro (magneto) static field; (vi) the gravity field; (vii) steady heat conduction. Not included are: (viii) rotating fluids; (ix) viscous flow in a pipe; (x–xi) deflection of membranes and torsion of prisms; (xii) surface tension and capillarity; (xiii–xiv) biharmonic problems, like plane elasticity and steady plane creeping viscous flow. For the first seven out of fourteen cases are indicated: (i) the "sources" which generate the field; (ii) the relevant constitutive properties of the medium; (iii) the dependent variables which describe the field; (iv–vi) the energy, force, and pressure associated with the field; (vii) the boundary conditions; (viii) some of the conditions of validity or physical restrictions for potential theory to hold; (ix) the chapter where the subject is detailed, including the analogies which allow passage between corresponding problems.

(Sections 38.2–8.9). The consideration of the fields due (Subsection 38.1.2) to given potentials on a plane (Subsection 38.1.3) or a circle (Subsection 38.1.4), is preceded by a classification of potential flows and fields from the point-of-view of boundary conditions (Subsection 38.1.1).

38.1.1 Boundary Conditions for Potential Fields and Flows

The potential fields and flows can be classified from the point-of-view of boundary conditions (Diagram 38.1) in three classes, seven boundary conditions and twelve cases. The first class A (fields with sources) concerns the fields in free space due to singularities, that is: (i/ii) vortices, sources and sinks in a potential flow (Chapters 12, 14, 16, 28, 34, 36, and 38); (iii) gravity field due to mass distributions (Chapter 18); (iv/v) electro (magneto) static field due to electric charges (Chapters 24 and 36) [currents (Chapters 26 and 36)]; (vi) steady heat conduction (Chapter 32). These fields exist in an unbounded medium (viz. free space) as a point/curve/domain in the complex plane, corresponding by adding the third orthogonal dimension, to respectively line/surface/volume distributions in space. The crossing of line or surface distributions leads to jump conditions for monopoles (multipoles) involving (Section 18.8, Example 20.20) [(Sections 26.2 and 26.9)] the potential (its derivatives); the calculation of the field inside the multipole distribution requires Cauchy principal values (Sections 17.8 and 17.9) of field integrals (Sections 18.6, 18.9, and 26.9; Example 20.20). This class A of problems and its six cases (i) to (vi) in free space involves at most jump boundary conditions (type I) across multipoles. The multipole distributions forcing the fields can be combined with rigid (class B) or moving (class C) boundaries; the free or nonforced fields correspond to solutions of the Laplace instead of the Poisson equation, with the same boundary conditions.

The boundary conditions form two classes, viz. rigid (B) [moving (C)] each with three cases, that is, boundary conditions II to IV (V–VII). The presence of boundaries can be represented by: (a) images on planes (Chapters 16, 24, and 26) and cylinders (Chapters 24, 26, and 28); (b) conformal mappings (Chapters 34, 36, and 38). The class B of rigid boundary conditions can be of three kinds (Section 32.2) and lead to three types of problems: (vii) potential specified on the boundary (type II: Dirichlet problem in Sections 37.8 and 37.9), for example, for electrostatic potential (Section 38.1) or specified temperature (Section 32.3), that is, boundary condition of the first kind; (viii) normal derivative specified at the boundary (type III: Neumann problem), that is, heat flux or a surface distribution of singularities, for example, vortices, sources, sinks, mass (Section 18.9), electric charges, or currents (Section 26.9); (ix) a linear combination of potential and normal derivative specified at the boundary (type IV: Robin or mixed problem), for example, heat convection (Sections 32.3, 32.6, and 32.7). The third class C of moving boundaries includes three cases: (x) free jets (Sections 38.2–38.9) having as boundary a free streamline along that the velocity is constant, and whose shape is determined by pressure equilibrium (type V of boundary condition); (xi) bodies moving in a fluid or pulsating bodies, or bodies with suction or blowing, for that the normal velocity is prescribed (type VI of boundary condition), correspond to the boundary condition of the second kind (type III) possibly on a boundary with prescribed motion (unlike type V); (xii) flexible surfaces (type VII of boundary condition) for that the flow velocity couples to the body motion, for example, through the pressure, stresses, heat, mass or momentum transfer, leading to aeroelastic, membrane and other effects. Thus besides the boundary conditions of the "three kinds" (Section 32.3) corresponding to types II to IV, there are four more viz. types I and V to VII. In addition to the boundary conditions there may be asymptotic conditions (Section 32.2) for compact or noncompact domains. Finiteness or boundedness conditions may also apply in compact or noncompact domains.

38.1.2 Discontinuous Potential Distributions

Consider [(Figure 38.1a (38.1b)] a plane (a cylinder) with a strip (half-circle) at a constant potential Φ_0 and the rest at zero potential (38.1a,b) [(38.2a,b)]:

$$\Phi(x,y) = \begin{cases} \Phi_0 & \text{if } |x| < a \\ 0 & \text{if } |x| > a, \end{cases} \qquad \Phi(a,\varphi) = \begin{cases} \Phi_0 & \text{if } 0 < \varphi < \pi, \\ 0 & \text{if } \pi < \varphi < 2\pi. \end{cases} \tag{38.1–2a,b}$$

Both are Dirichlet problems, that is, with boundary condition of first kind (type II in Diagram 38.1). The plane problem (38.1a,b) is solved by Poisson's formula (37.121) ≡ (38.3a) [(37.122) ≡ (38.3b)] for the potential (field function):

$$\{\Phi(x,y), \Psi(x,y)\} = \frac{\Phi_0}{\pi} \int_{-a}^{+a} \left[(t-x)^2 + y^2\right]^{-1} \{y, x-t\}\, dt; \tag{38.3a,b}$$

the disk problem (38.2a,b) is solved by Poisson formula [(37.109a,b) ≡ (38.4a) [(37.109c,d) ≡ (38.4b)] for the scalar potential:

$$\{\Phi(r,\varphi), \Psi(r,\varphi)\} = \frac{\Phi_0}{2\pi} \int_0^{\pi} \left[a^2 + r^2 - 2a\cos(\alpha - \varphi)\right]^{-1} \left\{a^2 - r^2, -2ar\sin(\alpha - \varphi)\right\} d\alpha. \tag{38.4a,b}$$

It remains to evaluate the integrals for the plane (cylindrical) potential [Subsection 38.1.3 (38.1.4)].

38.1.3 Potential in a Half-Plane due to a Finite Strip

In the case of the potential due to a strip (Figure 38.1a), the integrand in the potential (38.3a) [field function (38.3b)] is the real (imaginary) part of

$$z \equiv x + iy: \quad [i(t-z)]^{-1} = [y + i(t-x)]^{-1} = \left[(t-x)^2 + y^2\right]^{-1} [y + i(x-t)]; \tag{38.5a,b}$$

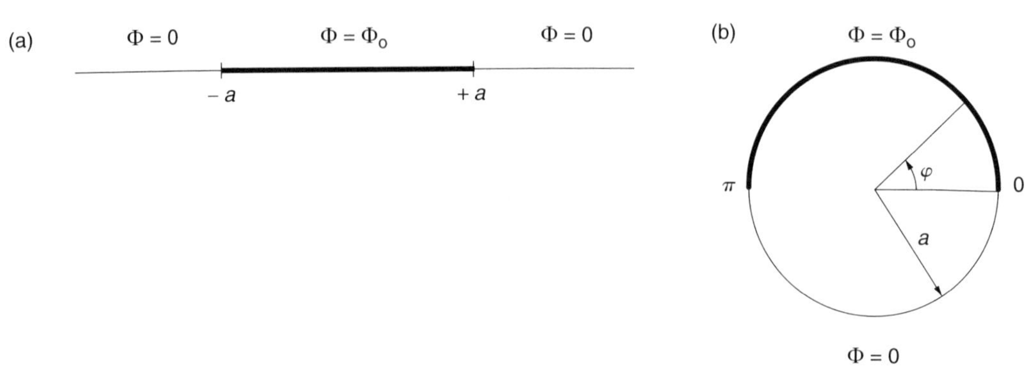

FIGURE 38.1
An example of boundary-value problem is the potential in a half-plane (a) [cylinder (b)] due to a potential distribution on a finite strip (half-hemisphere). This relates to the Cauchy/Poisson/Schwartz theorems using the values [Figure 37.7b(a)] of a function on a boundary to determine its values at other points of a domain where it is analytic (Figures 15.1b and 37.6).

the primitive is given by

$$i^{-1}\int (t-z)^{-1}\,dt = -i\log(t-z) = -\frac{i}{2}\log\left[(t-x)^2+y^2\right] + \text{arc cot}\left(\frac{x-t}{y}\right). \tag{38.6a,b}$$

Both integrals (38.3a,b) are evaluated by (38.5–38.6a,b):

$$\Phi(x,y) = \frac{\Phi_0}{\pi}\left[\text{arc cot}\left(\frac{x-a}{y}\right) - \text{arc cot}\left(\frac{x+a}{y}\right)\right] sgn(y), \tag{38.7a}$$

$$\Psi(x,y) = \frac{\Phi_0}{2\pi}\log\left[\frac{(a+x)^2+y^2}{(a-x)^2+y^2}\right], \tag{38.7b}$$

where: (i) the integration (38.3a,b) is from $-a$ to $+a$ both in (38.7a,b); (ii) in (38.7a) the positive (negative) sign applies above (below) the strip. *The uniform potential (38.1a,b) on a strip of the real axis creates in the plane a field with potential (38.7a) [field function (38.7b)].*

Next is confirmed that the solution for the potential (38.7a) does meet the boundary conditions (38.1a,b). Using identity (38.8):

$$\arctan\alpha + \text{arc}\cot\alpha = \frac{\pi}{2}, \tag{38.8}$$

leads to the equivalent expression:

$$\Phi(x,y) = \frac{\Phi_0}{\pi}\left[\text{arc tan}\left(\frac{x+a}{y}\right) - \text{arc tan}\left(\frac{x-a}{y}\right)\right] sgn(y). \tag{38.9}$$

for the potential:

$$\lim_{y\to\pm 0}\arctan\frac{x+a}{y} = \begin{cases} \pm\dfrac{\pi}{2} & \text{if } x > -a, \quad (38.10a)\\ \mp\dfrac{\pi}{2} & \text{if } x < -a, \quad (38.10b)\end{cases}$$

$$\lim_{y\to\pm 0}\arctan\frac{x-a}{y} = \begin{cases} \mp\dfrac{\pi}{2} & \text{if } x < a, \quad (38.10c)\\ \pm\dfrac{\pi}{2} & \text{if } x > a. \quad (38.10d)\end{cases}$$

Substitution of (38.10a–d) in (38.9) specifies the potential on the real axis:

$$x < -a: \quad \Phi(x,\pm 0) = \pm\frac{\Phi_0}{\pi}\left[\left(\mp\frac{\pi}{2}\right) - \left(\mp\frac{\pi}{2}\right)\right] = 0, \tag{38.11a}$$

$$-a < x < a: \quad \Phi(x,\pm 0) = \pm\frac{\Phi_0}{\pi}\left[\left(\pm\frac{\pi}{2}\right) - \left(\mp\frac{\pi}{2}\right)\right] = \Phi_0, \tag{38.11b}$$

$$x > a: \quad \Phi(x,\pm 0) = \pm\frac{\Phi_0}{\pi}\left[\left(\pm\frac{\pi}{2}\right) - \left(\pm\frac{\pi}{2}\right)\right] = 0, \tag{38.11c}$$

where the upper (lower) sign applies (38.9) above (below) the strip, leading in both cases to the same result in (38.11a–c). Thus *the potential on the real axis is: (i) zero and continuous outside the strip (38.11a,c) ≡ (38.12a):*

$$\Phi(|x| > a, y = \pm 0) = 0; \quad \Phi(|x| < a, y \pm 0) = \Phi_0, \tag{38.12a,b}$$

(ii) on the strip (38.11b) ≡ (38.12b) the potential takes same values on the upper (38.1a) [lower] side. This agrees with the boundary conditions (38.12a,b) ≡ (38.1a,b). The field due to two potential strips in considered in Example 40.15.

38.1.4 Potential inside a Cylinder due to a Half-Circle

In the case of the potential due to a half cylinder (Figure 38.1b), the integrand in (38.4a) [(38.4b)] is the real (imaginary part) of:

$$s = ae^{i\alpha}, \quad z = re^{i\varphi}: \quad \frac{s+z}{s-z} = \frac{ae^{i\alpha}+re^{i\varphi}}{ae^{i\alpha}-re^{i\varphi}} = \frac{a^2-r^2-i2ar\sin(\alpha-\varphi)}{a^2+r^2-2ar\cos(\alpha-\varphi)}; \tag{38.13a–c}$$

where was used (37.108b,c); the complex primitive is given by:

$$\begin{aligned}\int \frac{s+z}{s-z}\mathrm{d}\alpha &= \int \frac{s+z}{s-z}\frac{\mathrm{d}s}{is} = -i\int\left(\frac{2}{s-z}-\frac{1}{s}\right)ds \\ &= -2i\log(s-z)+i\log s = i\log\left(ae^{i\alpha}\right)-2i\log\left(ae^{i\alpha}-re^{i\varphi}\right) \\ &= i\log a-\alpha-2i\log\left|ae^{i\alpha}-re^{i\varphi}\right|+2\arg\left(ae^{i\alpha}-re^{i\alpha}\right) \\ &= i\log a-i\log\left|a^2+r^2-2ar\cos(\alpha-\varphi)\right|-\alpha+2\mathrm{arc}\cot\left(\frac{a\cos\alpha-r\cos\varphi}{a\sin\alpha-r\sin\varphi}\right),\end{aligned} \tag{38.14a,b}$$

where was used (34.38a). Both integrals (38.4a,b) are evaluated (38.13a–c; 38.14a,b) by:

$$\mathrm{sgn}(a-r)\Phi(r,\varphi) = -\frac{\Phi_0}{2}+\frac{\Phi_0}{\pi}\left[\mathrm{arc}\cot\left(\cot\varphi+\frac{a}{r}\csc\varphi\right)-\mathrm{arc}\cot\left(\cot\varphi-\frac{a}{r}\csc\varphi\right)\right], \tag{38.15a}$$

$$\Psi(r,\varphi) = \frac{\Phi_0}{2\pi}\log a-\frac{\Phi_0}{2\pi}\mathrm{sgn}(a-r)\log\left(\frac{a^2+r^2+2ar\cos\varphi}{a^2+r^2-2ar\cos\varphi}\right), \tag{38.15b}$$

an integration from 0 to π inside the cylinder $r<a$; the sign of the potential (38.15a) is reversed $r>a$ outside the cylinder corresponding to an integration in the opposite direction from π to 0, and replacement of α by $-\alpha$; in the field function (38.15b) only the second term on the r.h.s. changes sign. *The uniform potential on the upper half of a cylinder (38.2a,b) creates a field function (38.15b) [potential (38.15a)] where the positive (negative) sign applies inside (outside).* Using (38.8) the potential (38.15a) becomes:

$$\mathrm{sgn}(a-r)\Phi(r,\varphi) = -\frac{\Phi_0}{2}-\frac{\Phi_0}{\pi}\left[\mathrm{arc}\tan\left(\cot\varphi+\frac{a}{r}\csc\varphi\right)-\mathrm{arc}\tan\left(\cot\varphi-\frac{a}{r}\csc\varphi\right)\right]. \tag{38.16}$$

The verification of the consistency of (38.16) with (38.2a,b) is similar to (38.10a–d); starting with (19.39a,b) leads to (38.17a,b):

$$r=a,\ \varphi\to\pm 0: \quad \cot\varphi\pm\frac{a}{r}\csc\varphi = \frac{\cos\varphi\pm 1}{\sin\varphi} \sim \frac{1-\varphi^2/2\pm 1}{\varphi} \sim \frac{2}{\varphi}, -\frac{\varphi}{2}; \tag{38.17a–d}$$

substituting (38.17a–d) in (38.16), viz:

$$\lim_{\varphi\to\pm 0}\Phi(a,\varphi) = \frac{\Phi_0}{2}+\frac{\Phi_0}{\pi}\lim_{\varphi\to\pm 0}\mathrm{arc}\tan\frac{2}{\varphi} = \frac{\Phi_0}{2}+\frac{\Phi_0}{\pi}\left(\pm\frac{\pi}{2}\right) = \frac{\Phi_0}{2}\pm\frac{\Phi_0}{2} = \Phi_0, 0, \tag{38.18a,b}$$

confirms that the boundary condition (38.2a,b) ≡ (38.18a,b) is satisfied outside the cylinder; inside the cylinder all signs would be reversed. The field due to potentials in two angular sectors of a cylinder is considered in Example 40.15.

38.2 Width of the Vena Contracta of a Jet (Borda, 1766)

Consider a free jet issuing from a small orifice in a large reservoir (Figure 38.2a). The width of the orifice is b, and since the stream lines converge toward it, the jet initially becomes narrower until in the absence of gravity they became parallel and form a **vena contracta**

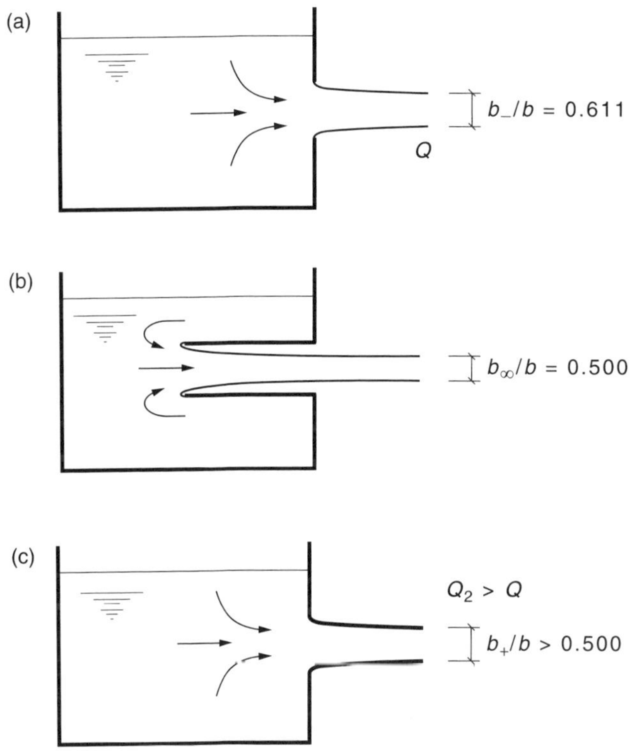

FIGURE 38.2
An example of a free surface is the jet issuing from a slit in the wall of a reservoir (a), that is symmetric if gravity is neglected. If a long reentrant tube is fitted (b) the contraction ratio of the jet is maximized. Defining the contraction ratio as the ratio of the asymptotic width of the jet to the width of the slit in the reservoir: (i) it takes the lowest possible value 0.500 for a long reentrant tube (b); (ii) by comparison it is 0.611 without reentrant tube (a). A third alternative is to fit a rigid outlet nozzle (c) that: (i) imposes the shape of the jet; (ii) reduces pressure losses; (iii) increases the flow rate. The increase in flow rate explains why: (i) a nozzle is put in water reservoirs/fountains since roman times when "time to fill with water" was limited; (ii) an athlete in a long run breathes with lips shaped like a nozzle; (iii) the latter also helps to blow a candle. In contrast with the nozzle (c), the shape of the free jets (a,b) is not known "a priori" and is part of the solution of the problem; the contraction ratios indicated can be obtained by the hodograph method presented next.

of width, b_∞. The force exerted by the fluid on the orifice (38.19b) is approximately equal to the width b multiplied by the dynamic pressure (38.19c):

$$p_a \equiv p_0: \quad F = bq = b(p_a - p) = \frac{\rho}{2}U^2 b, \quad q \equiv p_a - p = \frac{\rho}{2}U^2, \tag{38.19a–c}$$

where: (i) ρ is the mass density; (ii) the variation of velocity U across the jet is neglected; (iii) the Bernoulli equation (14.27c) was used stating that the dynamic pressure equals the stagnation or atmospheric pressure (38.19a) minus the free stream pressure. At the vena contracta the force (38.20a) is equal to the velocity times the mass flux:

$$F_\infty = \rho\, QU = \rho b_\infty U^2; \quad Q = U b_\infty, \tag{38.20a,b}$$

the mass flux is the mass density times the volume flux is (38.20b); the latter is the product of the by the velocity U by the with b_∞ of the vena contract where the velocity is uniform. The force at the vena contracta (38.19a) will be equal to that at the orifice (38.20a) if there are no pressure losses, otherwise F_∞ is an overestimate in (38.21a):

$$F_\infty \geq F: \quad \frac{b_\infty}{b} \geq \frac{1}{2}, \tag{38.21a,b}$$

showing that: *(i) a free jet issuing from an orifice contracts to at most half the width, if there are no pressure losses, for example, it flows out of a long tube (Figure 38.2b); (ii) in the presence of pressure losses, for example, a free jet issuing from a hole in a reservoir (Figure 38.2a), the width of the vena contracta is more than half the width of the hole. The ratio (38.21b) will shown in the sequel [Subsection 38.3.3 (38.3.4)] to be 0.500 (0.611) in the case (i) [(ii)].*

In the case (Figure 38.2a) when there are pressure losses the stagnation pressure is higher than the atmospheric pressure (38.22a):

$$p_0 = p_a + \Delta p: \quad F_- = bq = b(p_0 - p) = b(p_a - p + \Delta p) = b\left(\frac{\rho U^2}{2} + \Delta p\right), \tag{38.22a,b}$$

and the force (38.19b) is replaced by (38.22b); this leads (38.23a) to (38.23b):

$$F_- = \rho b_- U^2: \quad \frac{b_-}{b} \equiv \frac{F_-}{\rho U^2 b} = \frac{1}{2} + \frac{\Delta p}{\rho U^2} > \frac{1}{2}, \tag{38.23a,b}$$

a larger contraction ratio. If a tube is fitted with a nozzle projecting outward (Figure 38.2c), then the vena contracta is fixed at b_+ and the balance of forces (38.19b) $\equiv$ (38.20a) is replaced by (38.24):

$$\rho U^2 b_+ = F_+ = p_0 b - p_a b_+ = p_a(b - b_+) + \frac{\rho U^2 b}{2}, \tag{38.24}$$

that states that the force of the jet results from the difference of pressure forces at the inlet and outlet of the tube. Solving (38.24) for the contraction ratio, it is found to be again larger that one-half (38.21b):

$$\frac{Q_+}{U} = b_+ = \frac{b}{2}\frac{\rho U^2 + 2p_a}{\rho U^2 + p_a} > \frac{b}{2}, \tag{38.25}$$

implying that the flow rate is also increased in the same proportion. Thus *fitting an outlet nozzle to a fountain is a way to collect more water in a given time span;* this is current practice since roman times, when access to public fountains was based on a fixed time allowance. A long distance runner does not breathe with an open month, but rather with semiclosed lips shaped like a "nozzle"; this again increases flow rate of air expelled from the lungs. When blowing a candle shaping the lips helps direct the air flow and increase the flow rate.

38.3 Slit in a Wall and Reentrant Tube in a Reservoir (Helmholtz, 1868)

The contraction ratio (Subsection 38.3.4) as well as the shape of boundary interface (Subsection 38.3.3) can be calculated from the potential flow associated with the free jet (Subsection 38.3.1) by using a representation in the hodograph plane of the velocity (Subsection 38.3.2). The hodograph method (Subsection 38.3.2) for free jets (Subsection 38.3.1) applies to the slit in a reservoir without (or with) a reentrant tube [Subsection 38.3.3 (38.3.4)].

38.3.1 Free Surfaces and Rigid Obstacles

The Bernoulli equation (14.25c) for potential flow is simplified to (14.27c) ≡ (38.26b):

$$gz \ll v^2: \quad p + \frac{\rho}{2}v^2 = p_0 \equiv \text{const}, \tag{38.26a,b}$$

by the omission of the effect of gravity (38.26a) on the shape of the jet compared with that of the dynamic pressure. The jet has two possible kinds of boundaries, namely (i) rigid walls and (ii) free surfaces. The latter (ii) separates the fluid at rest from the fluid in motion, and the nonentrainment of one by the other being possible only if viscosity is neglected, as implied (Sections 14.2 and 14.4) by the assumption of irrotational flow. A **free surface** cannot support forces, so the pressure is continuous across it and equal to the pressure p in the fluid at rest. The pressure on the side of the flow is the same p, and bearing in mind that the stagnation pressure p_0 is also constant, as well as the mass density ρ in an incompressible flow, it follows that *the velocity is constant in modulus on a free surface, and its direction is tangential, that is, follows its curvature.* In the case (i) of an obstacle the velocity is also tangential so that its direction is determined, but the modulus varies so that the pressure force exerted by jet on the rigid wall is generally nonzero and nonuniform. Since the free surfaces (straight rigid walls) are lines of constant modulus (angle) of the velocity, they are represented by straight lines in the hodograph plane, whose coordinates are the modulus v and angle θ of the velocity. The use of the hodograph method involves a conformal mapping between the physical and hodograph plane.

38.3.2 Representation in the Physical and Hodograph Planes

The flow in the physical plane (38.27a) of the Cartesian coordinates x, y, is described by the complex potential $f(z)$ that specifies (12.19) the conjugate velocity:

$$z = x + iy: \quad \frac{\mathrm{d}f}{\mathrm{d}z} = v_x - iv_y = v\mathrm{e}^{-i\theta}, \tag{38.27a–c}$$

both the (i) Cartesian components of the velocity (38.27b); (ii) the modulus v and angle θ of the velocity with the x-axis (38.27c). The latter specify:

$$\zeta \equiv -\log\left(\frac{\mathrm{d}f}{\mathrm{d}z}\right) = -\log v + i\theta, \tag{38.28}$$

the **hodograph plane** *whose coordinates (38.28) are minus the logarithm of the modulus* v *of the velocity, and the angle of the velocity θ.* As an example of the relation between the physical and hodograph planes, consider (Figure 38.3a) a free jet issuing from an orifice in

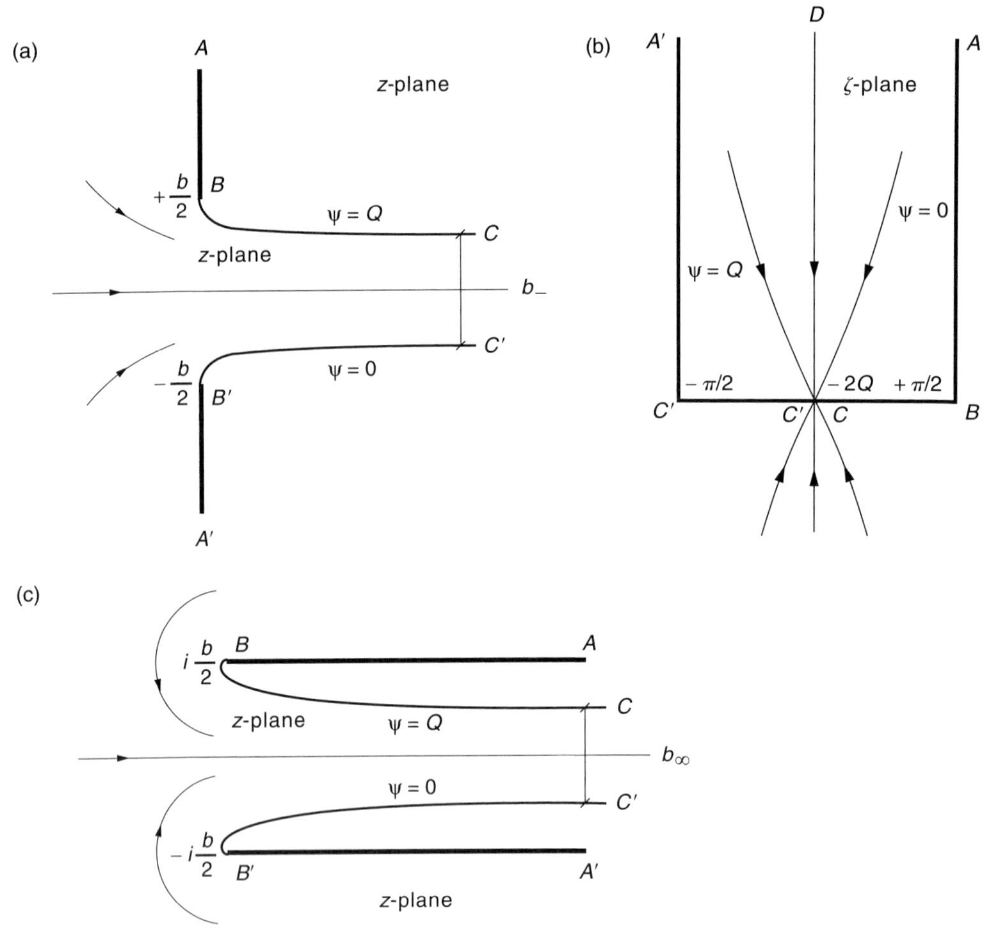

FIGURE 38.3
On the free surfaces $(BC, B'C')$ of the jets the pressure is constant; by the Bernoulli law, in the absence of gravity for an incompressible fluid, the modulus of the velocity is constant along the free surfaces. The direction of the velocity changes along the free streamline, determining its shape. Thus the region occupied by the fluid in the physical plane, either for the slit (Figure 38.2a ≡ 38.3a) [reentrant tube [Figure 38.2b ≡ 38.3c)] is not known "a priori," and is part of the solution of the problem. The region occupied by the fluid can be specified in the hodograph plane (b) whose coordinates are: (i) along the x-axis $-\log(v/U)$ where v is the modulus of the velocity and U the constant asymptotic velocity; (ii) along the y-axis minus the angle $-\theta$ of the velocity. These are the real (i) and imaginary (ii) parts of the hodograph variable (38.29). The region occupied by the flow is bounded by: (i) the walls of the reservoir (a) [reentrant tube (c)] where the velocity makes a constant angle $\theta = \pm\pi/2\,(\theta = \pm\pi)$, corresponding to vertical lines in the hodograph plane (b); (ii) the free surfaces where the velocity is constant and equal to the asymptotic velocity $v = U$, corresponding to the segment of the real axis between the vertical lines. Thus in both cases the fluid occupies a vertical half-strip in the hodograph plane (b). The width of the strip is twice in the case (c) relative to (a), and can be made equal using the hodograph variable $\zeta/2$ instead of ζ. Transforming back from the hodograph plane (b) to the physical plane with different variable, specifies the distinct jet shapes [Figure 38.3a(c)] and contraction ratios [Figure 38.2a(b)] for the exit from a reservoir through a slit on the wall (a reentrant tube).

the wall of an infinite reservoir. The velocity at the free surfaces BC, $B'C'$ of the jet is a constant U, and so is the volume flux Q through any section from the orifice of width b to the vena contracta of width b_∞. At the orifice the flow is not uniform, but it becomes so far away at the vena contracta $Q = b_\infty U$, implying that U is the maximum value of the velocity $v \le U$. The coordinates in the hodograph plane are specified by:

$$\zeta \equiv -i\log\left(\frac{1}{U}\frac{\mathrm{d}f}{\mathrm{d}z}\right) = -i\log\left(\frac{v}{U}\mathrm{e}^{-i\theta}\right) = -\theta - i\log\left(\frac{v}{U}\right), \tag{38.29}$$

showing (Figure 38.3b) that: (i) the upper AB (lower $A'B'$) rigid wall of the reservoir along that the velocity is vertical downward $\theta = -\pi/2$ (upward $\theta = \pi/2$), corresponds to a vertical line $\mathrm{Re}(\zeta) = +\pi/2\,(-\pi/2)$, that is, $AB\,(A'B')$ in the hodograph plane; (ii) the upper BC (lower $B'C'$) free surface, where the velocity is $v = U$, becomes the segment $(-\pi/2, +\pi/2)$ of the real axis, with the origin representing $\theta = 0$ the vena contracta $C'C$; (iii) the region occupied by the flow $v \le U$, is the upper half-strip $\mathrm{Im}(\zeta) \ge 0, -\pi/2 \le \mathrm{Re}(\zeta) \le +\pi/2$, and the volume flux $-Q$ into this upper half corresponds to a source of strength $-2Q$ half-way between parallel walls at distance, whose potential is given (36.137a) by:

$$f = -\frac{Q}{\pi}\log(\sin\zeta) = -\frac{Q}{\pi}\log\left\{\sin\left[-i\log\left(\frac{1}{U}\frac{\mathrm{d}f}{\mathrm{d}z}\right)\right]\right\}. \tag{38.30}$$

Thus has been obtained *a transcendental differential equation (38.30) specifying the complex potential $f(z)$ of a free jet issuing from an orifice with volume flux Q and velocity U at the surface.*

38.3.3 Calculation of the Shape of a Free Surface

It is not necessary to integrate (38.30) to obtain the shape of the free surface. The latter corresponds to $v = U$, that is, (38.31a) in (38.29), so that the complex potential (38.30) becomes (38.31b):

$$\zeta = -\theta: \quad \Phi + i\Psi = -\frac{Q}{\pi}\log(-\sin\theta) = -\frac{Q}{\pi}\log|\sin\theta| - iQ; \tag{38.31a,b}$$

the stream function $\Psi = -Q$ confirms that the volume flux is Q between the two free surfaces. The potential varies along the free surface:

$$\mathrm{d}\Phi = -\frac{Q}{\pi}\mathrm{d}\{\log|\sin\theta|\} = -\frac{Q}{\pi}\cot\theta\,\mathrm{d}\theta; \tag{38.32}$$

since it is a stream line, the velocity is given by

$$\Psi = \mathrm{const}: \qquad U\mathrm{e}^{-i\theta} = v_x - iv_y = \frac{\mathrm{d}f}{\mathrm{d}z} = \frac{\mathrm{d}\Phi}{\mathrm{d}z}; \tag{38.33a,b}$$

the ratio of (38.32) and (38.33b) determines the complex displacement on the free surface:

$$\mathrm{d}x + i\,\mathrm{d}y = \mathrm{d}z = \frac{\mathrm{d}\Phi}{\mathrm{d}\Phi/\mathrm{d}z} = -\frac{Q}{\pi U}\mathrm{e}^{i\theta}\cot\theta\,\mathrm{d}\theta; \tag{38.34}$$

its Cartesian components are

$$\left\{\frac{\mathrm{d}x}{\mathrm{d}\theta}, \frac{\mathrm{d}y}{\mathrm{d}\theta}\right\} = -\frac{Q}{\pi U}\left\{\frac{\cos^2\theta}{\sin\theta}, \cos\theta\right\} = -\frac{Q}{\pi U}\{\csc\theta - \sin\theta, \cos\theta\}. \tag{38.35a,b}$$

Using the primitive (38.74) and integrating from the lower lip $x=0, y=-b/2$, where $\theta=\pi/2$ because the velocity is vertically upward:

$$x=-\frac{Q}{\pi U}\int_{\pi/2}^{\theta}(\csc\varphi-\sin\varphi)\,\mathrm{d}\varphi=-\frac{Q}{\pi U}\left[\log\tan\left(\frac{\varphi}{2}\right)+\cos\varphi\right]_{\pi/2}^{\theta}, \tag{38.36a}$$

$$y+\frac{b}{2}=-\frac{Q}{\pi U}\int_{\pi/2}^{\theta}\cos\varphi\,\mathrm{d}\varphi=-\frac{Q}{\pi U}\left\{\sin\varphi\right\}_{\pi/2}^{\theta}, \tag{38.36b}$$

leads to the parametric equation of the free surface:

$$x(\theta)=-\frac{Q}{\pi U}\left[\log\tan\left(\frac{\theta}{2}\right)+\cos\theta\right],\quad y(\theta)=-\frac{b}{2}+\frac{Q}{\pi U}(1-\sin\theta). \tag{38.37a,b}$$

The vena contracta corresponds to the limit $x\to\infty, \theta\to 0$, and its width is given by

$$-\frac{b_\infty}{2}=y(0)=-\frac{b}{2}+\frac{Q}{\pi U}=-\frac{b}{2}+\frac{b_\infty}{\pi}, \tag{38.38a}$$

using (38.20b); this specifies the contraction ratio:

$$\frac{b_\infty}{b}=\frac{\pi}{2+\pi}=0.611. \tag{38.38b}$$

The parametric equations (38.37a,b) specify the shape of a jet of volume flux Q issuing from a hole of width b in the wall of a reservoir (Figure 38.3a), as well as the width $b_\infty=0.611b$ and velocity $U=Q/b_\infty=1.637Q/b$ at the vena contracta.

38.3.4 Reentrant Tube and Minimum Contraction Ratio (Helmholtz, 1868)

In the preceding case of a jet issuing from a reservoir through a hole (Figure 38.3a), there are pressure losses near the edges leading to a contraction ratio (38.38b) greater than one-half (38.21b). The latter value should be attained if an infinitely long tube is attached to the hole (Figure 38.3c), so that the free jet forms within this tube; the free jet occupies only a part of the space between the walls, and is distinct from the potential flow in a channel (Subsection 36.5.2) that fills the whole space. In the present problem the angle of the velocity along upper AB (lower $A'B'$) wall is $+\pi\,(-\pi)$, suggesting the use as hodograph variable of one-half of (38.29):

$$\zeta\equiv-\frac{i}{2}\log\left(\frac{1}{U}\frac{\mathrm{d}f}{\mathrm{d}z}\right)=-\frac{i}{2}\log\left(\frac{v}{U}\mathrm{e}^{-i\theta}\right)=-\frac{\theta}{2}-\frac{i}{2}\log\left(\frac{v}{U}\right), \tag{38.39}$$

so that these become the same vertical lines $\mathrm{Re}(\zeta)=+\pi/2\,(-\pi/2)$ in the hodograph plane (Figure 38.3b). The free surfaces BC, $B'C'$ map to the segment $(-\pi/2,+\pi/2)$ of the real axis, and the flow region to the same half-strip as before, so that (38.30) is replaced by:

$$f=-\frac{Q}{\pi}\log(\sin\zeta)=-\frac{Q}{\pi}\log\left\{\sin\left[-\frac{i}{2}\log\left(\frac{1}{U}\frac{\mathrm{d}f}{\mathrm{d}z}\right)\right]\right\}, \tag{38.40}$$

as *the transcendental differential equation specifying the complex potential of a free jet of volume flux Q, issuing from a reservoir through a semiinfinite tube with parallel sides (Figure 38.3c) and with maximum velocity U.*

The latter configuration should yield the minimum value of the contraction ratio between the exit hole of width b and the vena contracta of width, b_∞. To prove this consider the free surface where $v = U$, and $\zeta = -\theta/2$ in (38.39), so that (38.40) implies:

$$\Phi + i\Psi = -\frac{Q}{\pi}\log\left\{-\sin\left(\frac{\theta}{2}\right)\right\} = -\frac{Q}{\pi}\log\left|\sin\left(\frac{\theta}{2}\right)\right| - iQ; \tag{38.41}$$

the stream function $\Psi = -Q$ confirms that the volume flux is Q. The variation of the potential:

$$\mathrm{d}\Phi = \mathrm{d}\left\{-\left(\frac{Q}{\pi}\right)\log\left|\sin\left(\frac{\theta}{2}\right)\right|\right\} = -\frac{Q}{2\pi}\cot\left(\frac{\theta}{2}\right)\mathrm{d}\theta, \tag{38.42}$$

specifies the displacement along the free surface:

$$\mathrm{d}x + i\mathrm{d}y \equiv \mathrm{d}z = \frac{\mathrm{d}\Phi}{\mathrm{d}\Phi/\mathrm{d}z} = -\frac{Q}{2\pi U}\mathrm{e}^{i\theta}\cot\left(\frac{\theta}{2}\right)\mathrm{d}\theta, \tag{38.43}$$

where (38.33b) was used again as in (38.34).

The Cartesian displacements:

$$\left\{\frac{\mathrm{d}x}{\mathrm{d}\theta}, \frac{\mathrm{d}y}{\mathrm{d}\theta}\right\} = -\frac{Q}{2\pi U}\left\{\cot\left(\frac{\theta}{2}\right) - \sin\theta, 1 + \cos\theta\right\}, \tag{38.44a,b}$$

follow from

$$\begin{aligned}\mathrm{e}^{i\theta}\cot\left(\frac{\theta}{2}\right) &= (\cos\theta + i\sin\theta)\cot\left(\frac{\theta}{2}\right) = \left[1 - 2\sin^2\left(\frac{\theta}{2}\right) + 2i\sin\left(\frac{\theta}{2}\right)\cos\left(\frac{\theta}{2}\right)\right]\cot\left(\frac{\theta}{2}\right)\\ &= \cot\left(\frac{\theta}{2}\right) - 2\sin\left(\frac{\theta}{2}\right)\cos\left(\frac{\theta}{2}\right) + 2i\cos^2\left(\frac{\theta}{2}\right) = \cot\left(\frac{\theta}{2}\right) - \sin\theta + i(1+\cos\theta).\end{aligned} \tag{38.44c}$$

Integrating (38.44a,b) from the lower lip $x = 0,\ y = -b/2,\ \theta = \pi$ yields:

$$\left\{x, y + \frac{b}{2}\right\} = -\frac{Q}{2\pi U}\left\{2\log\left[\sin\left(\frac{\varphi}{2}\right)\right] + \cos\varphi, \varphi + \sin\varphi\right\}_\pi^\theta; \tag{38.45a,b}$$

this specifies the parametric equations of the free surface:

$$x(\theta) = -\frac{Q}{2\pi U}\left\{2\log\left[\sin\left(\frac{\theta}{2}\right)\right] + \cos\theta - 1\right\}, \tag{38.46a}$$

$$y(\theta) = -\frac{b}{2} + \frac{Q}{2U} - \frac{Q}{2\pi U}(\theta + \sin\theta). \tag{38.46b}$$

The asymptotic width of the free jet as $x \to \infty, \theta \to 0$, is given by

$$-\frac{b_\infty}{2} = y(0) = -\frac{b}{2} + \frac{b_\infty}{2}, \quad \frac{b_\infty}{b} = \frac{1}{2} = 0.500, \quad Q \equiv b_\infty U = \frac{bU}{2}, \tag{38.47a–c}$$

demonstrating (Helmholtz, 1868) the expected contraction ratio of one-half (38.21b) $\equiv$ (38.47b) that had been predicted (Section 38.2) before (Borda, 1766). *The shape of the free surface of a jet issuing from a reservoir with a volume flux* Q *through (Figure 38.3c) a long tube of width b, leading to a vena contracta of width (38.47b) at velocity* U *is specified by (38.46a,b) where the flow rate is (38.47c).*

38.4 Flat Plate Orthogonal to a Jet or to a Wall (Kirchhoff, 1869; Rayleigh, 1876a)

A problem converse to the jet issuing from a hole in a wall (Figure 38.3a) is obtained interchanging the hole and the wall, that is, considering a free jet incident orthogonal to a flat plate of width b in Figure 38.4. This is to be distinguished from the potential flow past a flat plate (Section 34.1) that fills all space. The potential flow past a vertical barrier in a wall (Figure 38.5) [normal to a flat plate (Figure 38.4)] can also be considered by the hodograph method (Subsection 38.4.1); the free jets behind have a parabolic interface (Subsection 38.4.3) with a cavity, resulting in a net drag, that is, it specifies a drag coefficient of 0.880 for a flat plate orthogonal to stream (Subsection 38.4.2). By the reflection principle (Section 31.2) the flow orthogonal to a plate also specifies the effect of a barrier perpendicular to a wall (Figure 38.5); again this flow with a free surface is distinct from the potential flow past the same obstacle occupying all space.

38.4.1 Flow past a Vertical Barrier in a Wall

The angle of the velocity at the upper AB (lower $A'B'$) half of the plate is $\theta = \pi/2\,(\theta = -\pi/2)$, so that the hodograph transformation:

$$\xi \equiv i\log\left(\frac{1}{U}\frac{\mathrm{d}f}{\mathrm{d}z}\right) = i\log\left(\frac{v}{U}\mathrm{e}^{-i\theta}\right) = \theta + i\log\left(\frac{v}{U}\right), \tag{38.48}$$

maps the flow to (Figure 38.4b) the half-strip $-\pi/2 \le \mathrm{Re}(\xi) \le \pi/2, \mathrm{Im}(\xi) \ge 0$. The latter is mapped to the upper half η-plane (Figure 38.4c) by (38.30) $\equiv$ (36.137a) $\equiv$ (38.49a) the transformation $\eta = \sin\xi$. The point at infinity C', C, D where the flow is uniform, lies at the origin in the η-plane, and using an inversion is mapped back to infinity in the $1/\eta$-plane (Figure 38.4d). The upper half $1/\eta$ plane transforms, by the square, $1/\eta^2$, into the flow past a semiinfinite plate (14.83b). Thus from (14.90a) follows $1/\eta \sim \sqrt{f}$ or $f \sim 1/\eta^2$ in the f-plane (Figure 38.4e) as stated in (38.49b):

$$\eta = \sin\xi, \quad f = \frac{k}{\eta^2}. \tag{38.49a,b}$$

The f-plane is the plane of the complex potential $f = \Phi + i\Psi$ since the real axis $\Psi = 0$ in Figure 38.4e corresponds in Figure 38.4d to the stream line $\Psi = 0$ that forms (Figure 38.4c,b,a) the free surfaces $B'C'$, BC, the flat plate $B'A'AB$, and the line $DA'A$ dividing the flow that goes over and under the plate. Thus the complex potential is given from (38.48; 38.49a,b) by

$$f = k\csc^2\xi = k\csc^2\left[i\log\left(\frac{1}{U}\frac{\mathrm{d}f}{\mathrm{d}z}\right)\right], \tag{38.50}$$

where the constant k is related to the width b of the plate.

The latter relation can be obtained noting that on the flat plate (38.51a) the velocity is vertical (38.51b) and the displacement is specified by (38.51c):

$$x = 0: \quad \frac{\mathrm{d}f}{\mathrm{d}z} = -iv, \quad i\mathrm{d}y = \mathrm{d}z = \frac{\mathrm{d}z}{\mathrm{d}f}\frac{\mathrm{d}f}{\mathrm{d}\eta}\mathrm{d}\eta = \frac{2k}{iv\eta^3}\mathrm{d}\eta, \tag{38.51a–c}$$

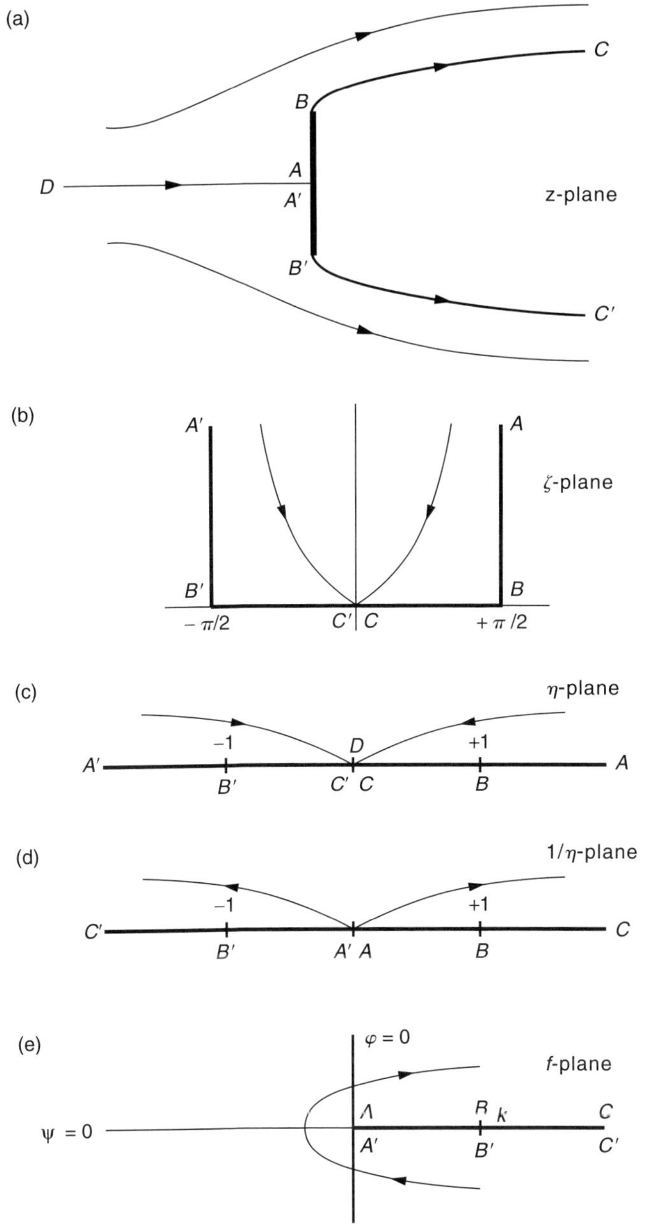

FIGURE 38.4
The flow orthogonal to a flat plate (a) leads to two free streamlines issuing from the edges; these separate a region of "zero" or constant pressure behind the plate. This region may be interpreted as "vacuous," "stagnant fluid" or a "recirculation bubble" or a "detached wake"; the important point is that the lack of a pressure behind the plate to balance the pressure on the forward face leads to a drag force. Thus it is possible to calculate the drag of the flat plate, which would not be the case if the potential flow filled all space. A drag coefficient $C_D = 0.880$ is obtained solving the "free surface problem" by a sequence of steps: (i) map the flow region (a) to the same half-strip in the hodograph plane (b) as before (Figure 38.3b); (ii) using an inverse Schwartz–Christoffel transformation with two critical points (Figure 36.13b) map the half-strip to the upper half complex plane (c); (iii) by inversion relative to the origin (Figure 35.3) bring the stagnation point at the middle of the plate to the origin (d); (iv) using the function square root changes to the flow past a semiinfinite plate (e) which is known (Figure 14.10b). Inverting the whole process ($iv \rightarrow i$) back to the physical plane (e) specifies the flow orthogonal to the flat plate and the drag coefficient $C_D = 0.880$.

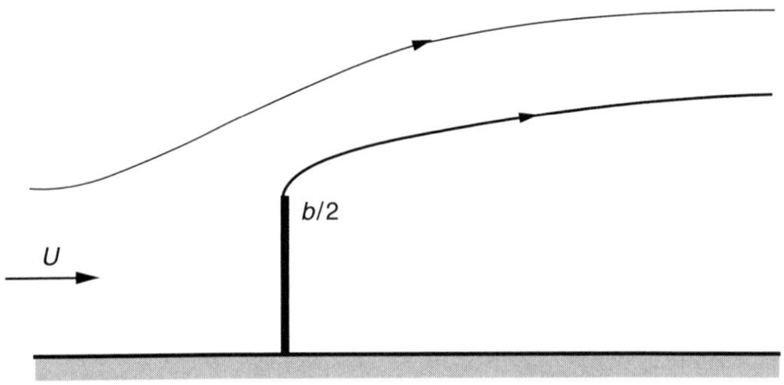

FIGURE 38.5
By the Schwartz reflection principle (Figure 31.4) the solution is the same for the preceding problem (Figure 38.4) taking only the upper half (Figure 38.5). This leads to a free stream of velocity U parallel to a wall with a vertical obstacle; the drag coefficient is the same as for the "complete" flat plate orthogonal to the free stream.

using (38.49b); here η is given by (38.49a; 38.48):

$$\eta = \sin\left\{\frac{\pi}{2} + i\log\left(\frac{v}{U}\right)\right\} = \cos\left\{i\log\left(\frac{v}{U}\right)\right\} = \cosh\left\{\log\left(\frac{v}{U}\right)\right\}$$
$$= \frac{\{\exp[\log(v/U)] + \exp[-\log(v/U)]\}}{2} = \frac{v/U + U/v}{2}. \tag{38.52}$$

The latter relation (38.52) $\equiv$ (38.53a) can be solved (38.53b) for U/v:

$$\left(\frac{U}{v}\right)^2 - 2\eta\left(\frac{U}{v}\right) + 1 = 0: \quad \frac{U}{v} = \eta + \sqrt{\eta^2-1} = \frac{1}{\eta - \sqrt{\eta^2-1}}, \tag{38.53a,b}$$

leading to (38.53c,d):

$$\frac{v}{U} = \eta - \sqrt{\eta^2-1}, \quad \frac{U}{v} - \frac{v}{U} = 2\sqrt{\eta^2-1}. \tag{38.53c,d}$$

Integrating (38.51c) between the center of the plate $y=0, v=0, \eta=\infty$, and the upper edge $y=b/2, v=U, \eta=1$, leads to

$$\frac{b}{2} = -\int_{\infty}^{1} \frac{2k}{\eta^3 v}\,\mathrm{d}\eta = \frac{2k}{U}\int_{1}^{\infty}\left(\eta + \sqrt{\eta^2-1}\right)\eta^{-3}\,\mathrm{d}\eta = \frac{2k}{U}(1+I), \tag{38.54}$$

involving the integral (38.55a); using the identity (38.55a) $\equiv$ (38.60a), to be proved in the sequel (Subsection 38.4.2):

$$I \equiv \int_{1}^{\infty}\sqrt{\eta^2-1}\,\eta^{-3}\,\mathrm{d}\eta = \frac{\pi}{4}; \quad b = \frac{k}{U}(4+\pi), \tag{38.55a,b}$$

specifies the constant k in (38.55b) in terms of the width b of the plate. The latter appears in the complex potential (38.50):

$$-f = \frac{bU}{4+\pi}\operatorname{csch}^2\left[\log\left(\frac{1}{U}\frac{\mathrm{d}f}{\mathrm{d}z}\right)\right] = \frac{4bU/(4+\pi)}{\left[U^{-1}\mathrm{d}f/\mathrm{d}z - U\left(\mathrm{d}f/\mathrm{d}z\right)^{-1}\right]^2}, \tag{38.56}$$

the nonlinear first-order differential equation (38.56) is satisfied by the complex potential of a free jet of velocity U at infinity incident orthogonally upon a flat plate of width b.

38.4.2 Calculation of Drag Force and Coefficient

Before proceeding with the study of free jet around the plate, the integral (38.55a) quoted above is evaluated using the change of variable:

$$\eta = \cosh\alpha : \quad \sqrt{\eta^2 - 1} = \sinh\alpha, \quad d\eta = \sinh\alpha\, d\alpha, \tag{38.57a–c}$$

that simplifies the integration to:

$$I \equiv \int_0^\infty \sinh^2\alpha\, \mathrm{sech}^3\alpha\, d\alpha = \int_0^\infty \left(\mathrm{sech}\alpha - \mathrm{sech}^3\alpha\right) d\alpha. \tag{38.58}$$

The identity:

$$\frac{d(\tanh\alpha\,\mathrm{sech}\,\alpha)}{d\alpha} - \mathrm{sech}^3\alpha = -\mathrm{sech}\,\alpha\tanh^2\alpha = -\mathrm{sech}\,\alpha\left(1 - \mathrm{sech}^2\alpha\right), \tag{38.59a}$$

specifies the integrand of (38.58) in the form:

$$\frac{d(\tanh\alpha\,\mathrm{sech}\,\alpha)}{d\alpha} - \mathrm{sech}\,\alpha = 2\left(\mathrm{sech}^3\alpha - \mathrm{sech}\,\alpha\right); \tag{38.59b}$$

the integral (38.58) is evaluated by

$$2I = -\left[\tanh\alpha\,\mathrm{sech}\,\alpha\right]_0^\infty + \int_0^\infty \mathrm{sech}\,\alpha\, d\alpha = \left[\arctan(\sinh\alpha)\right]_0^\infty = \frac{\pi}{2}, \tag{38.60a}$$

where: (i) the first term on the r.h.s. vanishes since $\tanh 0, \infty = 0{,}1$ and $\mathrm{sech}\, 0, \infty = 1{,}0$; (ii) the integrand in the second term on the r.h.s. has primitive specified by

$$\frac{d[\arctan(\sinh\alpha)]}{d\alpha} = \frac{\cosh\alpha}{1 + \sinh^2\alpha} = \mathrm{sech}\,\alpha. \tag{38.60b}$$

The result (38.60a) ≡ (38.55a) agrees with the earlier statement.

The pressure Δp exerted by the fluid on the plate satisfies:

$$\Delta p + p_\infty + \frac{\rho}{2}v^2 = \mathrm{const} = \frac{\rho}{2}U^2 + p_\infty, \tag{38.61}$$

with p_∞ denoting the pressure on the fluid at rest adjacent to the free surface, where the velocity U equals that of the incident stream, so that p_∞ is also the pressure in the free stream. The velocity v on the plate is generally different (38.53b) from U, resulting in a pressure force:

$$D = 2\int_0^{b/2} \Delta p\, dy = \rho \int_0^{b/2} \left(U^2 - v^2\right) dy; \tag{38.62}$$

this is a drag, calculated from (38.51c; 38.53d):

$$D = -2k\rho U \int_\infty^1 \left(\frac{U}{v} - \frac{v}{U}\right)\eta^{-3}\, d\eta = \frac{4\rho b U^2}{4+\pi}\int_1^\infty \sqrt{\eta^2 - 1}\,\eta^{-3}\, d\eta = \frac{\rho b U^2 \pi}{4+\pi}, \tag{38.63}$$

where (38.55a,b) were used. The general from of the drag is (28.72b) ≡ (38.64a) the product of the width of the plate b, by the dynamic pressure $(\rho/2)U^2$ of the incident stream, by a drag coefficient:

$$D = \frac{\rho}{2}U^2 b C_D, \quad C_D \equiv \frac{2\pi}{4+\pi} = 0.880, \tag{38.64a,b}$$

the latter is given in the present case (38.63) by (38.64b). *A flat plate of width b moving at velocity U orthogonally to a potential flow of density ρ with free surfaces trailing from its edges (Figure 38.4a), experiences a drag force (38.63) ≡ (38.64a), corresponding to the drag coefficient (38.64b). This also applies to the free stream along a wall with an orthogonal flat plate (Figure 38.5).*

38.4.3 Parabolic Cavity Formed by Two Free Surfaces

If the potential flow occupies the space behind the flat plate (Section 39.1) there is no drag, that is, the drag results from the existence of free surfaces, forming a cavity of static fluid behind the flat plate. On the free surfaces the modulus of the velocity is a constant, and hence equal to that of the incident stream, $v = U$, implying (38.65a); also the complex potential for the free streamline that corresponds to zero stream function $\Psi = 0$, is the real potential. The latter varies along the free streamline (38.65b) as a function (38.50; 38.48) of the angle, θ:

$$\frac{\mathrm{d}f}{\mathrm{d}z} = U\mathrm{e}^{-i\theta} = \frac{\mathrm{d}\Phi}{\mathrm{d}z}: \quad \Phi = k\csc^2\theta = \frac{bU}{4+\pi}\csc^2\theta; \tag{38.65a,b}$$

in (38.65b) was used (38.55b). The complex displacement along the free surface satisfies:

$$\mathrm{d}x + i\,\mathrm{d}y = \mathrm{d}z = \frac{\mathrm{d}\Phi}{\mathrm{d}\Phi/\mathrm{d}z} = -\frac{2b}{4+\pi}\mathrm{e}^{-i\theta}\csc^2\theta\cot\theta\,\mathrm{d}\theta, \tag{38.66}$$

where (38.65a,b) were used. The parametric equations of the free surface:

$$\left\{\frac{\mathrm{d}x}{\mathrm{d}\theta}, \frac{\mathrm{d}y}{\mathrm{d}\theta}\right\} = -\frac{2b}{4+\pi}\left\{\csc\theta\cot^2\theta, \csc\theta\cot\theta\right\}, \tag{38.67a,b}$$

may be integrated between $\pi/2$ and θ:

$$x(\theta) = \frac{b}{4+\pi}\left\{\csc\theta\cot\theta + \log\left[\tan\left(\frac{\theta}{2}\right)\right]\right\}, \tag{38.68a}$$

$$y(\theta) = \frac{b}{2} + \frac{2b}{4+\pi}(\csc\theta - 1), \tag{38.68b}$$

where the evaluation of the integrals will be justified subsequently (38.72a,b; 38.73–38.74). Far from the flat plate as (38.69a):

$$\theta \to 0: \quad 2\theta^2 x \sim \frac{2b}{4+\pi} \sim y\theta, \quad y^2 \sim \frac{4bx}{4+\pi}, \tag{38.69a–c}$$

the free surface (38.69b) takes a parabolic shape (38.69c). *The parametric equations (38.68a,b) specify the free surfaces issuing from the edges $x = 0, y = ib/2$ of a flat plate of width b in an orthogonal stream, showing that the asymptotic shape (38.69a,b) is parabolic (38.69c). In the case of a flat plate orthogonal to a wall (Figure 38.5) a free stream parallel to the wall has one free streamline (38.69a–c).*

The calculation of the shape of the free surface involves the evaluation (38.68a,b) of the integrals (38.67a,b) that is justified next:

$$\frac{y(\theta)-b/2}{2b/(4+\pi)}=-\int_{\pi/2}^{\theta}\csc\varphi\cot\varphi\,\mathrm{d}\varphi=\{\csc\varphi\}_{\pi/2}^{\theta},\tag{38.70}$$

$$J(\theta)\equiv-\frac{4+\pi}{2b}x(\theta)=\int_{\pi/2}^{\theta}\csc\varphi\cot^2\varphi\,\mathrm{d}\varphi,\tag{38.71}$$

where: (i) the result (38.70) $\equiv$ (38.68b) is immediate; (ii) concerning (38.71) it may be evaluated by a method similar to that used for (38.55a) or (38.58). Instead of (38.59a) the identity (38.72a):

$$-\frac{\mathrm{d}(\csc\varphi\cot\varphi)}{\mathrm{d}\varphi}-\csc^3\varphi=\csc\varphi\cot^2\varphi=\csc\varphi\left(\csc^2\varphi-1\right),\tag{38.72a}$$

is used in the form:

$$2\left(\csc^3\varphi-\csc\varphi\right)=-\frac{\mathrm{d}(\csc\varphi\cot\varphi)}{\mathrm{d}\varphi}-\csc\varphi;\tag{38.72b}$$

this specifies the integrand in (38.71):

$$\begin{aligned}2J(\theta)&=2\int_{\pi/2}^{\theta}\left(\csc^3\varphi-\csc\varphi\right)\mathrm{d}\varphi=-[\csc\varphi\cot\varphi]_{\pi/2}^{\theta}-\int_{\pi/2}^{\theta}\csc\varphi\,\mathrm{d}\varphi\\&=-\csc\theta\cot\theta-\left\{\log\left[\tan\left(\frac{\varphi}{2}\right)\right]\right\}_{\pi/2}^{\theta};\end{aligned}\tag{38.73a}$$

substitution of (38.73a) in (38.71) leads to

$$x(\theta)=-\frac{2b}{4+\pi}\left(-\frac{1}{2}\right)\left\{\csc\theta\cot\theta+\log\left[\tan\left(\frac{\theta}{2}\right)\right]\right\},\tag{38.73b}$$

that coincides with (38.68a) $\equiv$ (38.73b). The primitive inverse to

$$\frac{\mathrm{d}\{\log[\tan(\varphi/2)]\}}{\mathrm{d}\varphi}=\frac{1}{2}\frac{\sec^2(\varphi/2)}{\tan(\varphi/2)}=\left[2\cos\left(\frac{\varphi}{2}\right)\sin\left(\frac{\varphi}{2}\right)\right]^{-1}=(\sin\varphi)^{-1}=\csc\varphi,\tag{38.74}$$

was used in (38.73a).

38.5 Center of Pressure and Dividing Streamline on a Surfboard (Rayleigh, 1876b, 1891)

The force on a flat plate inclined in a stream (Figure 38.6) also specifies the thrust on a surf board riding the surface of a stream. This is not the same as an oblique barrier on a wall (Figure 38.7a), because the reflection principle applied to the latter (Section 38.6)

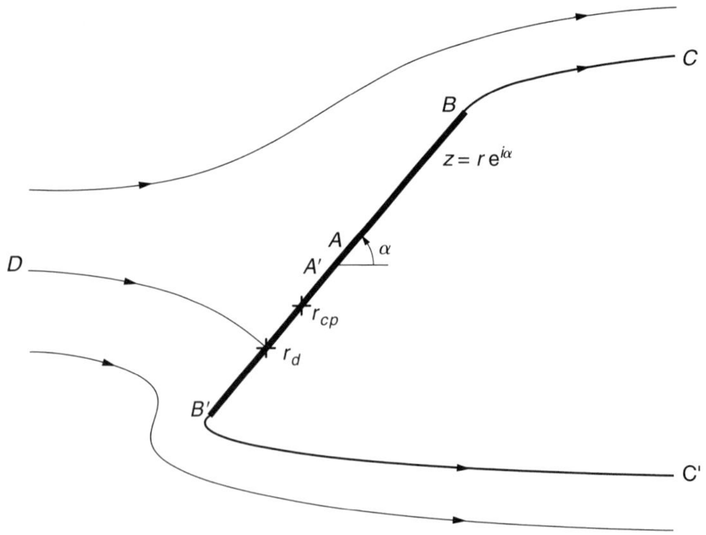

FIGURE 38.6
The drag coefficient of the flat plate (Figure 38.4a) will be reduced if it is inclined to the free stream (Figure 38.6). The torque exerted by the flow on the flat plate orthogonal to the free stream is zero relative to the mid-point which is thus the center of pressure; for the flat plate oblique to the flow the center of pressure moves below the middle point (Figure 38.6). The dividing streamline which separates the flow passing above and below the flat plate orthogonal to the free stream has (Figure 38.4a) an "impact point" on the plate which coincides with the stagnation point and center of pressure at the middle of the plate; in the case of the flat plate inclined to the free stream (Figure 38.6) the dividing streamline has an impact point at the plate which is a stagnation point, but is not at the center of the plate, and lies below the center of pressure. All these conclusions plus the drag and pitching moment coefficients of a flat plate at various inclinations to the free stream are detailed in Table 38.1. They follow from the same sequence of hodograph method and conformal mappings (Figure 38.4a–e) with a prerotation of the flat plate to incline it to the free stream.

would lead to an arrow (Figure 38.7b). The method used for the flow incident normal to a plate (Section 38.4) can be extended to oblique incidence (Subsection 38.5.1), and specifies the drag, the center of pressure and the point where the dividing stream line hits the plate (Subsection 38.5.2).

38.5.1 Dividing Streamline for a Flat Plate at Incidence

The jet incident on a plate (Figure 38.6) at an angle α to the stream velocity generalizes the case $\alpha = \pi/2$ of a normal plate (Figures 38.4a–e and 38.5); the generalization corresponds to a translation in Figure 38.4b of the strip from (38.75a) to (38.75b) implying (38.75c):

$$-\frac{\pi}{2} \le \theta \le \frac{\pi}{2}, \quad \alpha - \pi \le \theta' \le \alpha, \quad \theta' = \theta + \alpha - \frac{\pi}{2}. \tag{38.75a–c}$$

It follows that the complex potential (38.49a,b) is replaced by

$$f = \frac{k}{[\eta + \sin(\alpha - \pi/2)]^2} = \frac{k}{(\eta - \cos\alpha)^2}; \tag{38.76}$$

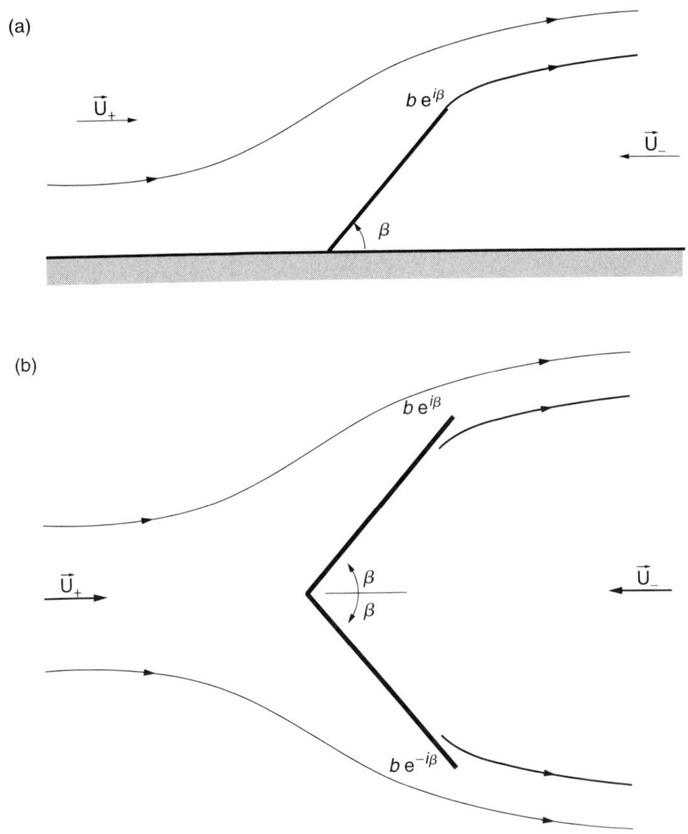

FIGURE 38.7
The Schwartz reflection principle (Figure 31.4) applied to the flat plate orthogonal to the free stream (Figure 38.4a) does lead to the flow along a wall with a vertical obstacle (Figure 38.5). However, the flow along a wall with an oblique obstacle (Figure 38.7a) by reflection on the real axis (Figure 31.4): (i) does not lead to the flat plate inclined to the free stream (Figure 38.6); (ii) leads instead to a bent lamina or symmetric arrow with its axis aligned with the free stream (Figure 38.7b). Comparing the drag coefficient of the straight (bent) lamina [Figure 38.6 (38.7b)] in Table 38.2, it follows that it is lower (higher) for the arrow when the free stream is incident from the front (rear) because: (i) the apex deviates the flow from the front reducing the drag; (ii) the flow from behind is "trapped" in the wedge, increasing the drag; (iii) both effects are more pronounced as the arrow makes a smaller angle at the apex; (iv) for example, compared with the drag coefficient 0.880 of a flat plate orthogonal to the free stream (Figure 38.4a), relative to the same length projected transversely to the flow direction, the arrow with 20° total angle at the apex has a drag coefficient decreased (increased) to 0.199 (4.150) for flow from the front (rear).

from (38.48; 38.49a) it follows that the complex potential satisfies:

$$f = k\left\{\sin\left[i\log\left(\frac{1}{U}\frac{\mathrm{d}f}{\mathrm{d}z}\right)\right] - \cos\alpha\right\}^{-2}. \tag{38.77}$$

The potential flow with velocity U incident at angle α on a plate of width b has complex potential that satisfies the nonlinear first-order differential equation (38.77), where the constant k is related to the width b of the plate by (38.78b):

$$\eta \to \eta - \cos\alpha: \quad \frac{b}{2} = \frac{2k}{U}\int_1^\infty \left(\eta + \sqrt{\eta^2 - 1}\right)(\eta - \cos\alpha)^{-3}\,\mathrm{d}\eta; \tag{38.78a,b}$$

the latter (38.78b) obtained by the substitution (38.78a) in (38.54), as in the passage from (38.49b) to (38.76). The drag force exerted on the plate is given by the same transformation applied to (38.63):

$$D = 2k\rho U\int_1^\infty \sqrt{\eta^2 - 1}\,(\eta - \cos\alpha)^{-3}\,\mathrm{d}\eta. \tag{38.79}$$

The displacement $\mathrm{d}r$ along the free streamline on the plate (38.80a) is given by (38.80b):

$$z = r e^{i\alpha}: \quad \mathrm{d}r = -\frac{2k}{U}\left(\eta + \sqrt{\eta^2 - 1}\right)(\eta - \cos\alpha)^{-3}\,\mathrm{d}\eta, \tag{38.80a,b}$$

as follows from (38.51c; 38.53b).

In order to evaluate the integrals the variable η is changed to β defined by

$$\eta \equiv \frac{1 - \cos\alpha\cos\beta}{\cos\alpha - \cos\beta}, \quad \mathrm{d}\eta = -\frac{\sin^2\alpha\sin\beta}{(\cos\alpha - \cos\beta)^2}\,\mathrm{d}\beta, \tag{38.81a,b}$$

where was used

$$\begin{aligned}(\cos\alpha - \cos\beta)^2\frac{\mathrm{d}\eta}{\mathrm{d}\beta} &= \cos\alpha\sin\beta(\cos\alpha - \cos\beta) - \sin\beta(1 - \cos\alpha\cos\beta)\\ &= \sin\beta\left(\cos^2\alpha - 1\right) = -\sin^2\alpha\sin\beta;\end{aligned} \tag{38.81c}$$

from (38.81a) also follows:

$$\left\{\eta - \cos\alpha, \sqrt{\eta^2 - 1}\right\}(\cos\alpha - \cos\beta) = \sin\alpha\{\sin\alpha, \sin\beta\}, \tag{38.82a,b}$$

using

$$(\eta - \cos\alpha)(\cos\alpha - \cos\beta) = 1 - \cos\alpha\cos\beta - \cos\alpha(\cos\alpha - \cos\beta) = 1 - \cos^2\alpha = \sin^2\alpha, \tag{38.82c}$$

$$\begin{aligned}\left(\eta^2 - 1\right)(\cos\alpha - \cos\beta)^2 &= (1 - \cos\alpha\cos\beta)^2 - (\cos\alpha - \cos\beta)^2\\ &= 1 + \cos^2\alpha\cos^2\beta - \cos^2\alpha - \cos^2\beta = \left(1 - \cos^2\alpha\right)\left(1 - \cos^2\beta\right)\\ &= \sin^2\alpha\sin^2\beta.\end{aligned} \tag{38.82d}$$

Combining (38.82a) with (38.81b) leads to

$$(\eta - \cos\alpha)^{-3}\,\mathrm{d}\eta = -\csc^4\alpha(\cos\alpha - \cos\beta)\sin\beta\,\mathrm{d}\beta. \tag{38.82e}$$

This appears together with (38.81a; 38.82b) in the displacement (38.80b):

$$\sin^4\alpha\frac{\mathrm{d}r}{\mathrm{d}\beta}=2\frac{k}{U}\left(1-\cos\alpha\cos\beta+\sin\alpha\sin\beta\right)\sin\beta; \tag{38.83a}$$

integrating along the plate this yields:

$$-\frac{U}{k}\sin^4\alpha\; r\left(\beta\right)=2\cos\beta+\cos\alpha\sin^2\beta+\sin\alpha\sin\beta\cos\beta-\beta\sin\alpha+C. \tag{38.83b}$$

where C is a constant of integration, and was used

$$-2\int\sin^2\beta\,\mathrm{d}\beta=\int\left[\cos\left(2\beta\right)-1\right]\mathrm{d}\beta=\frac{1}{2}\sin\left(2\beta\right)-\beta=\sin\beta\cos\beta-\beta, \tag{38.83c}$$

The center (upper) edge of plate corresponds in (38.78b; 38.79) as in (38.54; 38.63) to $\eta=\infty\,(\eta=1)$ hence $\beta=\alpha\,(\beta=\pi)$ in (38.81a), and the lower edge $\eta=-1$ corresponds to $\beta=0$. The constant of integration C in (38.83b) is chosen so that the origin is midway between the edges at $\beta=0,\pi$:

$$0\le\beta\le\pi:\quad 0=r\left(0\right)+r\left(\pi\right)=-\frac{k}{U}\csc^4\alpha\left(2C-\pi\sin\alpha\right),\quad 2C=\pi\sin\alpha; \tag{38.84a,b}$$

substituting (38.84b) in (38.83b) specifies the coordinate along the plate:

$$-\frac{U}{k}\sin^4\alpha\; r\left(\beta\right)=2\cos\beta+\cos\alpha\sin^2\beta+\sin\alpha\sin\beta\cos\beta+\left(\frac{\pi}{2}-\beta\right)\sin\alpha. \tag{38.85}$$

The length b of the plate corresponds to (38.86a):

$$b=r\left(\pi\right)-r\left(0\right)=\frac{k}{U}\csc^4\alpha\left(4+\pi\sin\alpha\right),\quad k=\frac{bU\sin^4\alpha}{4+\pi\sin\alpha}, \tag{38.86a,b}$$

thus determining (38.86b) the constant k. The latter reduces to (38.55a) for a mean flow orthogonal $\alpha=\pi/2$ to the flat plate. The value $\beta=\alpha$ specifies:

$$\begin{aligned}r_d=r\left(\alpha\right)&=-\left[2\cos\alpha\left(1+\sin^2\alpha\right)+\left(\frac{\pi}{2}-\alpha\right)\sin\alpha\right]\frac{k}{U}\csc^4\alpha\\&=-b\frac{2\cos\alpha\left(1+\sin^2\alpha\right)+\left(\pi/2-\alpha\right)\sin\alpha}{4+\pi\sin\alpha},\end{aligned} \tag{38.87}$$

the distance from the center of the plate of the **impact point** on the dividing streamline (Figure 38.6) where the stream separates into two flows, one going past each side of the plate.

38.5.2 Center of Pressure and Drag Coefficient

Using (38.82b,e) the drag of the plate (38.79) is given by

$$D=2k\rho U\csc^3\alpha\int_0^\pi\sin^2\beta\,\mathrm{d}\beta=\pi k\rho U\csc^3\alpha=\frac{\rho U^2 b\pi\sin\alpha}{4+\pi\sin\alpha}, \tag{38.88}$$

where (38.86b) was substituted. The pitching moment about the center of the plate or straight lamina is specified by the same integral (38.88) as the drag, with the distance

(38.85) as a factor in the integrand (38.89):

$$M = 2k\rho U \csc^3\alpha \int_0^\pi r(\beta)\sin^2\beta\, d\beta$$

$$= -\frac{2\rho U^2 b\sin\alpha}{4+\pi\sin\alpha}\frac{k}{U}\csc^3\alpha\cos\alpha\int_0^\pi \sin^4\beta\, d\beta; \tag{38.89}$$

here the only nonvanishing term is (38.90a):

$$\int_0^\pi \sin^4\beta\, d\beta = \frac{3\pi}{8}, \quad M = -\frac{3\pi}{4}\frac{\rho U^2 b^2\cos\alpha\sin\alpha}{(4+\pi\sin\alpha)^2}, \tag{38.90a,b}$$

leading to for the moment, (38.90b) that is negative, that is, clockwise or pitch-up in Figure 36.6.

The nonvanishing integral (38.89; 38.85) is evaluated by

$$8\int_0^\pi \sin^4\beta\, d\beta = 2\int_0^\pi [1-\cos(2\beta)]^2\, d\beta = 2\int_0^\pi \left[1-2\cos(2\beta)+\frac{1+\cos(4\beta)}{2}\right] d\beta$$

$$= \int_0^\pi [3-4\cos(2\beta)+\cos(4\beta)]\, d\beta = 3\pi; \tag{38.91a}$$

substitution of (38.85) in (38.89) also leads to other terms that vanish, namely: (i/ii) the first and third:

$$3\int_0^\pi \cos\beta\sin^2\beta\, d\beta = \sin^3\beta\Big|_0^\pi = 0 = \sin^4\beta\Big|_0^\pi = 4\int_0^\pi \sin^3\beta\cos\beta\, d\beta; \tag{38.91b,c}$$

(iv) the fourth (38.91e):

$$\gamma = \frac{\pi}{2}-\beta: \quad 4\int_0^\pi \left(\frac{\pi}{2}-\beta\right)\sin^2\beta\, d\beta = 4\int_{-\pi/2}^{\pi/2} \gamma\sin^2\left(\gamma+\frac{\pi}{2}\right) d\gamma$$

$$= 4\int_{-\pi/2}^{+\pi/2} \gamma\cos^2\gamma\, d\gamma = 2\int_{-\pi/2}^{+\pi/2} \gamma[1+\cos(2\gamma)] = \left[\gamma^2\right]_{-\pi/2}^{\pi/2} + \int_{-\pi/2}^{+\pi/2} \gamma\left\{\frac{d[\sin(2\gamma)]}{d\gamma}\right\} d\gamma$$

$$= [\gamma\sin(2\gamma)]_{-\pi/2}^{\pi/2} - \int_{-\pi/2}^{+\pi/2} \sin(2\gamma)\, d\gamma = \frac{1}{2}[\cos(2\gamma)]_{-\pi/2}^{\pi/2} = 0, \tag{38.91d,e}$$

where was made a change of variable (38.91d) and an integration by parts.

The **center of pressure** is defined (38.92a) as the point where the drag (38.88) should be applied to produce the pitching moment (38.90b):

$$-r_{cp} = -\frac{M}{D} = \frac{3b}{4}\frac{\cos\alpha}{4+\pi\sin\alpha} < -r_d: \quad 0 > r_{cp} > r_d; \tag{38.92a,b}$$

the center of pressure (38.92b) is always closer to the center of the plate than the dividing streamline (38.87), since the term in the numerator is larger than (3/4) $\cos\alpha$:

$$0 \leq \alpha \leq \frac{\pi}{2}: \quad (4+\pi\sin\alpha)\frac{r_{cp}-r_d}{b} = 2\cos\alpha\left(1+\sin^2\alpha\right) + \left(\frac{\pi}{2}-\alpha\right)\sin\alpha - \frac{3}{4}\cos\alpha$$
$$= \cos\alpha\left(\frac{5}{4}+2\sin^2\alpha\right) + \left(\frac{\pi}{2}-\alpha\right)\sin\alpha > 0; \quad (38.92\text{c,d})$$

the center of pressure (point of contract of the dividing streamline) both lie (38.92a) below the middle of plate (Figure 38.6).

It has been shown that a flat plate (Figure 38.6) or straight lamina of width b making an angle α with the velocity U of a stream: (i) separates the flow that passes over from that passes below the plate at the impact point (38.87) below the center of the plate; (ii) has center of pressure at (38.92a) also below the center of the plate and closer to the center of the plate than (i); (iii) experiences a drag (38.88) ≡ (38.93a) corresponding to a drag coefficient per unit length (38.93b) [per unit length projected in the direction transverse to the flow (38.93c)]:

$$2D \equiv \rho U^2 b C_D \equiv \rho U^2 b \overline{C}_D \sin\alpha, \quad \overline{C}_D = \frac{2\pi}{4+\pi\sin\alpha}, \quad C_D = \overline{C}_D \sin\alpha; \qquad (38.93\text{a–c})$$

(iv) is subject to a pitching moment (38.90b) ≡ (38.94a) corresponding to the moment coefficient per square length (38.94b):

$$2M = \rho U b^2 C_M = \rho U b^2 \overline{C}_M \sin\alpha, \quad -C_M = \frac{3\pi}{4}\frac{\sin(2\alpha)}{(4+\pi\sin\alpha)^2},$$
$$-\overline{C}_M = -C_M \csc\alpha = \frac{3\pi}{2}\frac{\cos\alpha}{(4+\pi\sin\alpha)^2}, \qquad (38.94\text{a–c})$$

[projected on the direction normal to the flow (38.94c)].

The preceding quantities are indicated in Table 38.1 for five values of the inclination spaced 15° from zero up to normal incidence 90° of the flow: (i/ii) the drag coefficient normalized to the width of the plate (38.93c) [to the projection of the length of the plate transverse to the flow direction (38.93b)] coincide at normal incidence with the value (38.64b), and as the inclination increases reduces to zero (38.95a) [increases to (38.95b)] when the plate is aligned with the flow:

$$0 = C_D(0) \leq C_D(\alpha) \leq C_D\left(\frac{\pi}{2}\right) = \frac{2\pi}{4+\pi} = 0.880, \qquad (38.95\text{a})$$

$$0.880 = \frac{2\pi}{4+\pi} = \overline{C}_D\left(\frac{\pi}{2}\right) \leq \overline{C}_D(\alpha) \leq \overline{C}_D(0) = \frac{\pi}{2} = 1.571; \qquad (38.95\text{b})$$

(iii) the pitching moment coefficient (38.94b) vanishes at the two extremes of plate aligned with or orthogonal to flow and peaks with a small value 6.6×10^{-2} in (38.96a) at an angle 32.5°:

$$0 = C_M(0) \leq C_{M\,\max} = C_M(32.5°) = 0.066 > C_M\left(\frac{\pi}{2}\right) = 0; \qquad (38.96\text{a})$$

$$0 = \overline{C}_M\left(\frac{\pi}{2}\right) \leq \overline{C}_M(\alpha) \leq \overline{C}_M(0) = \frac{3\pi}{32} = 0.295; \qquad (38.96\text{b})$$

(iv) the pitching coefficient normalized to the length of the plate orthogonal to the flow direction (38.93c) vanishes at normal incidence and increases to a value (30.96b) at grazing

TABLE 38.1
Aerodynamics of a Flat Plate at an Angle to a Stream

α	0°	15°	30°	45°	60°	75°	90°
C_D	0.000	0.338	0.564	0.714	0.810	0.863	0.880
$\overline{C}_D$	1.571	1.305	1.128	1.010	0.935	0.893	0.880
$-C_M\left(\times 10^{-2}\right)$	0.000	5.086	6.575	6.087	4.518	2.381	0.000
$-\overline{C}_M\left(x10^{-1}\right)$	2.945	1.965	1.315	0.861	0.522	0.246	0.000
$-r_{cp}/b\times\left(10^{-1}\right)$	1.875	1.505	1.166	0.852	0.558	0.276	0.000
$-r_d/b$	0.500	0.499	0.483	0.430	0.328	0.178	0.000

α: angle of the incident stream with the plate;
C_D: drag coefficient normalized to the length of the plate (38.93c);
$\overline{C}_D$: idem for the projection of the length of the plate orthogonal to the stream (38.93b);
C_M: pitching moment coefficient (38.94b);
$\overline{C}_M$: idem for the projection of the length of the plate orthogonal to the stream (38.94c);
r_{cp}: position of center of pressure (38.92a);
r_d: impact point of dividing streamline (38.87).
Note: The main aerodynamic parameters are indicated for a flat plate (Figure 38.6) at seven angles of inclination to the free stream, in 15° steps, starting with aligned (0° in Figure 34.1a) and ending with orthogonal (90° in Figure 38.4a): (i/ii) the drag coefficient calculated for the total length of the plate C_D (for the length of the plate projected along the normal to the free stream velocity $\overline{C}_D$); (iii/iv) likewise two values of the pitching moment C_M $(\overline{C}_M)$; (v/vi) the distance from the mid-point divided by the length of the plate, of the center of pressure where the pitching moment vanishes (impact point for the dividing streamline separating the flow that passes above and below the plate).

incidence; (v) the center of pressure (38.92a) is at the center of the plate for normal incidence and deviates most (38.97a) for flow parallel to the plate:

$$0.1875 = \frac{3}{16} = \left(\frac{|r_{cp}|}{b}\right)_{\max} = \frac{|r_{cp}(0)|}{b} > \frac{|r_{cp}(\alpha)|}{b} \geq \frac{r_{ep}(\pi/2)}{b} = 0; \tag{38.97a}$$

$$0.500 = \frac{1}{2} = \left(\frac{|r_d|}{b}\right)_{\max} = \frac{|r_d(0)|}{b} > -\frac{r_d(\alpha)}{b} \geq -\frac{r_d(\pi/2)}{b} = 0, \tag{38.97b}$$

(vi) the impact point (38.87) of the separating streamline moves from the center of the plate at normal incidence to plate, and is already close to the edge (0.483) at 30°.

38.6 Arrow or Bent Lamina in a Stream (Rethy, 1879; Bobyleff, 1881)

The flow impinging on a plate at an angle (Figure 38.6) in a wall (Figure 38.7a) would correspond, by reflection on the midline, to a stream incident on an arrow (Figure 38.7b) from the front; the case of a stream incident upon a bent lamina from behind is also included. Using the corner transformation to straighten the arrow (Subsection 38.6.1), leads to a modification of the integrals (Subsection 38.6.2) for the flat plate normal to flow (Subsection 38.4.1) that is distinct from the inclined flat plate (Subsection 38.5.1) and specifies a drag formula (Subsection 38.6.3) that can be compared with preceding cases (Subsection 38.6.4).

38.6.1 Straightening an Arrow into a Flat Plate

If the arrow is symmetric (Figure 38.7b), it corresponds to the flow past an oblique barrier in a wall (Figure 38.7a). The symmetric arrow with a corner angle β is mapped into a flat plate by a corner transformation (14.75b) $\equiv$ (33.98a) with the constant A and the exponent n are chosen so that $\zeta = \pm 1$ is mapped to the edges at an angle $\pm\beta$ in (38.98b,c):

$$\xi = A\zeta^n: \quad e^{i\beta} = A, \quad e^{-i\beta} = Ae^{-in\pi}, \tag{38.98a–c}$$

from the ratio of (38.98b,c) follows (38.99a) specifying the exponent (38.99b) of the corner flow (38.99c):

$$e^{in\pi} = e^{i2\beta}: \quad \beta = \frac{n\pi}{2}, \quad \xi = e^{i\beta}\zeta^{2\beta/\pi}; \tag{38.99a–c}$$

the exponent (38.99b) is specified by the angle 2β between the two sides of the arrow, that also appears (38.98a) $\equiv$ (38.99c) in the amplitude (38.98b). The arrow is convex (concave) for $0 < \beta < \pi/2\,(\pi/2 < \beta < \pi)$ or $0 < n < 1\,(1 < n < 2)$, corresponding to an incident stream from the front $\vec{U}_+$ (rear $\vec{U}_-$) in Figure 38.7b; this applies also for the straight barrier oblique to the wall Figure 38.7a. The intermediate case between the concave/convex arrow (oblique barrier) is Figure 38.7b (38.7a) is the plate (barrier) orthogonal to the flow ($\beta = \pi/2$ or $n = 1$) in Figure 38.5a (38.6). For stream of velocity U incident symmetrically on the arrow all the arguments concerning the flat plate orthogonal to the flow (Subsection 38.5.1) hold, provided that the arrow together with the tangential velocity is first straightened into a flat plate: (i) in (38.48) holds (38.100a) along the plate; (ii) the transformations (38.100b) lead from the plate to the arrow; (iii) substitution in (38.52) than leads to (38.100c) using (39.98a) and (39.48):

$$\begin{aligned} \theta = 0, \quad \log\left(\frac{v}{U}\right) = |\xi| \to |\zeta|: \quad \eta &= \cosh\left(\log|\zeta|\right) = \cosh\left\{\log\left[|\xi|^{1/n}\right]\right\} \\ = \cosh\left(\frac{1}{n}\log|\xi|\right) &= \cosh\left[\frac{1}{n}\log\left(\frac{U}{v}\right)\right] = \frac{(U/v)^{1/n} - (v/U)^{1/n}}{2}. \end{aligned} \tag{38.100a–c}$$

Solving for U/v yields:

$$\frac{U}{v} = \left(\eta + \sqrt{\eta^2 - 1}\right)^n, \quad \frac{v}{U} = \left(\eta + \sqrt{\eta^2 - 1}\right)^{-n} = \left(\eta - \sqrt{\eta^2 - 1}\right)^n, \tag{38.101a,b}$$

that reduce to (38.53b,c) in the case of a flat plate $\beta = \pi/2$ and $n = 1$ in (38.99b). The drag (38.63) is specified by (38.102a):

$$D = 2k\rho U\,(J_+ - J_-), \quad J_\pm \equiv \int_1^\infty \left(\eta \pm \sqrt{\eta^2 - 1}\right)^n \eta^{-3}\, d\eta; \tag{38.102a,b}$$

one of the integrals (38.102b) appears also in the displacement along the plate (38.54):

$$\frac{b}{2} = \frac{2k}{U} \int_1^\infty \left(\eta + \sqrt{\eta^2 - 1}\right)^n \eta^{-3}\, d\eta = \frac{2k}{U} J_+. \tag{38.103}$$

Eliminating k between (38.102a; 38.103) specifies the relation between the drag and the length of arrow (38.104a):

$$D = \frac{\rho}{2} U^2 b \left(1 - \frac{J_-}{J_+}\right) \equiv \frac{\rho}{2} U^2 b C_D; \quad C_D = 1 - \frac{J_-}{J_+}, \tag{38.104a,b}$$

this shows that the drag coefficient (38.104b) is specified by the integrals (38.102b).

38.6.2 Evaluation of Integrals and Shape Function

The integrals (38.103; 38.102b) [(38.102a,b)] specifying the width (drag) of an arrow of half-angle β are a generalization of those (38.54) [(38.63)] applying to a flat plate orthogonal to flow, and are distinct from the other generalization (38.78b) [(38.79)] for the flat plate at an angle to the flow. An integration by parts of (38.102b) leads to

$$J_\pm = \int_1^\infty \left(\eta \pm \sqrt{\eta^2 - 1}\right)^n \mathrm{d}\left(-\frac{1}{2\eta^2}\right) = \frac{1}{2} \pm \frac{n}{2} \int_1^\infty \left(\eta \pm \sqrt{\eta^2 - 1}\right)^n \frac{\eta^{-2}}{\sqrt{\eta^2 - 1}} \mathrm{d}\eta, \tag{38.105}$$

where (i) the boundary term at $\eta = \infty$ vanishes (38.106c) because $\beta < \pi$ in (38.106a) implies that the exponent (38.99b) does not exceed 2 in (38.106b):

$$0 < \beta < \pi, \quad 0 < n = \frac{2\beta}{\pi} < 2: \quad \lim_{\eta \to \infty} \eta^{-2} \left(\eta \pm \sqrt{\eta^2 - 1}\right)^n \sim \lim_{\eta \to \infty} \eta^{n-2} = 0; \tag{38.106a–c}$$

(ii) the second term on the r.h.s. of (38.105):

$$\mathrm{d}\left[\left(\eta \pm \sqrt{\eta^2 - 1}\right)^n\right] = n \left(\eta \pm \sqrt{\eta^2 - 1}\right)^{n-1} \left(1 \pm \frac{\eta}{\sqrt{\eta^2 - 1}}\right) \mathrm{d}\eta = \pm \frac{n \left(\eta + \sqrt{\eta^2 - 1}\right)^n}{\sqrt{\eta^2 - 1}} \mathrm{d}\eta, \tag{38.106d}$$

uses (38.106d).

The change of variable (38.107a):

$$2\eta = \sqrt{t} + \frac{1}{\sqrt{t}}: \quad J_\pm = \frac{1}{2} \pm n \int_0^1 t^{\mp n/2} (1 + t)^{-2} \,\mathrm{d}t, \tag{38.107a,b}$$

reduces the integral (38.105) to a quadratic denominator in (38.107b) instead of a cubic in (38.102b). The change of variable (38.107a) implies:

$$\eta^2 - 1 = \left(\frac{\sqrt{t} + 1/\sqrt{t}}{2}\right)^2 - 1 = \frac{t + 1/t}{4} - \frac{1}{2} = \left(\frac{1/\sqrt{t} - \sqrt{t}}{2}\right)^2, \tag{38.108a}$$

$$\eta \pm \sqrt{\eta^2 - 1} = \frac{\sqrt{t} + 1/\sqrt{t}}{2} \pm \frac{1/\sqrt{t} - \sqrt{t}}{2} = \frac{1}{\sqrt{t}}, \sqrt{t}, \tag{38.108b}$$

$$\begin{aligned} \left(\eta^2 \sqrt{\eta^2 - 1}\right)^{-1} \frac{\mathrm{d}\eta}{\mathrm{d}t} &= 8 \left[\left(\sqrt{t} + \frac{1}{\sqrt{t}}\right)^2 \left(\frac{1}{\sqrt{t}} - \sqrt{t}\right)\right]^{-1} \frac{1}{4\sqrt{t}} \left(1 - \frac{1}{t}\right) \\ &= \frac{2(t - 1)}{(t + 1)^2 (1 - t)} = -\frac{2}{(1 + t)^2}, \end{aligned} \tag{38.108c}$$

that were used in the passage from (38.105) to (38.107b); the minus sign in (38.108c) is incorporated in (38.107b) as the reversal of limits of integration, since $\eta = 1, \infty$ corresponds (38.107a) to $t = 1, 0$.

Bearing in mind that $n > 0$ in (38.99b), then (i) since $t^{n/2} \to 0$ as $t \to 0$ the integration by parts of J_- in (38.107b) reduces the denominator to a linear expression in (38.109a):

$$J_- - \frac{1}{2} = n \int_0^1 t^{n/2} \, \mathrm{d}\left[(1+t)^{-1}\right] = \frac{n}{2} - \frac{n^2}{2} \int_0^1 t^{n/2-1} (1+t)^{-1} \, \mathrm{d}t; \tag{38.109a}$$

$$J_+ - \frac{1}{2} = n \int_0^1 \left[t^{-n/2} (1+t)^{-1} - t^{1-n/2} (1+t)^{-2}\right] \mathrm{d}t. \tag{38.109b}$$

(ii) for J_+ the integration by parts cannot be applied directly, because $t^{-n/2} \to \infty$ as $t \to 0$. Thus the integrand of J_+ in (38.107b) is split into two terms in (38.109b). Since (38.106b) implies $1 - n/2 > 0$, than $t^{1-n/2} \to 0$ as $t \to 0$, the second term on the r.h.s. of (38.109b) can be integrated by parts:

$$n \int_0^1 t^{1-n/2} \, \mathrm{d}\left[(1+t)^{-1}\right] = \frac{n}{2} - n\left(1 - \frac{n}{2}\right) \int_0^1 t^{-n/2} (1+t)^{-1} \, \mathrm{d}t; \tag{38.110a}$$

$$J_+ - \frac{1}{2} = \frac{n}{2} + \frac{n^2}{2} \int_0^1 t^{-n/2} (1+t)^{-1} \, \mathrm{d}t, \tag{38.110b}$$

substituting (38.110a) back in (38.109b) yields (38.110b).

Both integrals (38.109a) and (38.110b) involve the geometric series (21.62c) that converge uniformly for $0 < t < 1 - \varepsilon$ with $0 < \varepsilon < 1$, absolutely for $t < 1$ and conditionally for $t = 1$, leading to

$$J_- - \frac{1}{2} - \frac{n}{2} = -\frac{n^2}{2} \sum_{m=0}^{\infty} (-)^m \int_0^1 t^{m+n/2-1} \, \mathrm{d}t = -\frac{n^2}{2} \sum_{m=0}^{\infty} \frac{(-)^m}{m + n/2}, \tag{38.111a}$$

$$J_+ - \frac{1}{2} - \frac{n}{2} = \frac{n^2}{2} \sum_{m=0}^{\infty} (-)^m \int_0^1 t^{m-n/2} \, \mathrm{d}t = \frac{n^2}{2} \sum_{m=0}^{\infty} \frac{(-)^m}{m - n/2 + 1}. \tag{38.111b}$$

The width of the arrow (38.103) is specified by

$$\frac{bU}{4k} = J_+ = \frac{1}{2} + \frac{n}{2} + \frac{n^2}{4} \sum_{j=0}^{\infty} \left(j + \frac{1}{2} - \frac{n}{4}\right)^{-1} - \frac{n^2}{4} \sum_{j=0}^{\infty} \left(j + 1 - \frac{n}{4}\right)^{-1}, \tag{38.112}$$

where were separated even $m = 2j$ (odd $m = 2j + 1$) terms. Recalling the definition of the digamma function (29.45a):

$$\psi(z) \equiv \lim_{N \to \infty} \left\{ \log N - \sum_{j=0}^{N} (j+z)^{-1} \right\}, \tag{38.113}$$

the l.h.s. of (38.112) can be rewritten:

$$\frac{bU}{2k} = 1 + n + \frac{n^2}{2} \left[\psi\left(1 - \frac{n}{4}\right) - \psi\left(\frac{1}{2} - \frac{n}{4}\right)\right]; \tag{38.114a}$$

Substitution of (38.99b) leads to

$$\frac{1}{f(\beta)} \equiv \frac{bU}{2k} = 1 + \frac{2\beta}{\pi} + 2\left(\frac{\beta}{\pi}\right)^2 \left[\psi\left(1 - \frac{\beta}{2\pi}\right) - \psi\left(\frac{1}{2} - \frac{\beta}{2\pi}\right)\right], \tag{38.114b}$$

the **shape function** (38.114b); it depends on the half-angle of the arrow, and relates the parameter k to the width b.

38.6.3 Drag of an Arrow with Arbitrary Aperture

The parameter k also appears together with (38.109a; 38.110b) in the integral for the drag (38.102a) leading to

$$D = k\rho U n^2 \int_0^1 \left(t^{n/2-1} + t^{-n/2}\right)(1+t)^{-1}\, dt. \tag{38.115}$$

The change of variable (38.116a):

$$s = \frac{1}{t}: \qquad \int_0^1 t^{n/2-1}(1+t)^{-1}\, dt = \int_1^\infty s^{-n/2}(1+s)^{-1}\, ds, \tag{38.116a,b}$$

combines the second term (38.116b) on the r.h.s. of (38.115) with the first in a single integral:

$$D = \frac{\rho}{2} bU^2 f(\beta)\, n^2 \int_0^\infty t^{-n/2}(1+t)^{-1}\, dt; \tag{38.117}$$

in the coefficient the constant k was expressed in terms of the width of the arrow (38.114b). The integral (38.117) is of the type with branch-point (Section 17.7) and can be evaluated using (17.53), with: (i) the parameter (38.118a); (ii) the function (38.118b) that is analytic except for a simple pole at $t = -1$ with residue unity (38.118c):

$$0 < \alpha = 1 - \frac{n}{2} = 1 - \frac{\beta}{\pi} < 1, \quad g(t) = \frac{1}{1+t}, \quad g_{(1)}(-1) = 1:$$

$$D = \frac{\rho}{2} bU^2 f(\beta)\, n^2 \pi \csc(\alpha\pi) = 2\rho U^2 b \beta^2 \pi^{-1} f(\beta) \csc\beta, \tag{38.118a–d}$$

where was used (38.99b).

Thus *an arrow (an oblique barrier in a wall) with symmetric sides, total width 2b (b) and angle $2\beta\,(\beta)$, placed symmetrically in a stream (in a stream along the wall) of velocity U and density ρ experiences [Figure 38.7b (38.7a)] a drag specified by (38.119a) [one half of (38.119a)]:*

$$2D = \rho U^2 b C_D = \rho U b^2 \overline{C}_D \sin\beta, \quad C_D = \frac{4}{\pi}\beta^2 f(\beta)\csc\beta, \quad \overline{C}_D = C_D \csc\beta, \tag{38.119a–c}$$

corresponding in both cases to the drag coefficient (38.119b) [(38.119c)] per unit width of the arrow (projected transversely to the flow direction), in terms of the shape function (38.114b), involving the digamma function (38.113). In the particular case of a flat plate $\beta = \pi/2$: (i) the drag coefficient is (38.64b) $\equiv$ (38.119b), leading to the value (38.120a) for the shape function; (ii) the shape function is defined by (38.114b) and equals (38.120b):

$$f\left(\frac{\pi}{2}\right) = \frac{2}{4+\pi}; \quad \frac{1}{f(\pi/2)} = 2 + \frac{\psi(3/4) - \psi(1/4)}{2}, \tag{38.120a,b}$$

(iii) comparison of (38.120a,b) implies (38.121a):

$$\psi\left(\frac{3}{4}\right)-\psi\left(\frac{1}{4}\right)=\pi, \quad \psi(1-z)-\psi(z)=\pi\cot(\pi z). \tag{38.121a,b}$$

that is a particular case of the *symmetry formula (38.121b) of the digamma function.* Thus it has been checked that the results for the bent lamina lead back to the flat lamina for $\beta=\pi/2$. The property (38.121b) of the digamma function can be proved from the definition (38.113) as a series:

$$\begin{aligned} m=j+1: \ \psi(1-z)-\psi(z) &= \lim_{N\to\infty}\sum_{j=0}^{N}\left(\frac{1}{j+z}-\frac{1}{j+1-z}\right) \\ &= \frac{1}{z}+\sum_{m=1}^{\infty}\left(\frac{1}{m+z}-\frac{1}{m-z}\right) \\ &= \frac{1}{z}+2z\sum_{m=1}^{\infty}\frac{1}{z^2-m^2}=\pi\cot\pi z, \end{aligned} \tag{38.121c,d}$$

using: (i) the change of dummy summation variable (38.121a); (ii) the series of fractions (27.63b) for the circular cotangent.

38.6.4 Comparison of the Drag of an Arrow with a Flat Plate

To compare the inclined flat plate with the arrow the reference length is taken to be: (i) the distance between the edges for the arrow; (ii) the length of the plate projected on the direction transverse to the flow. For an arrow with each side half the length of the plate, and angle equal to the inclination of the plate these two lengths coincide, so the comparison is relevant. *The ratio drag coefficients for the bent and inclined lamina of the same width is given by (38.122b) [(38.122a)] using the length (38.93c; 38.119c) [length projected in the direction transverse to the flow direction (38.93b; 38.119b)]:*

$$\frac{\overline{C}_D(\beta)}{\overline{C}_D(\alpha)}=2\left(\frac{\beta}{\pi}\right)^2(4+\pi\sin\alpha)f(\beta)\csc^2\beta, \tag{38.122a}$$

$$\frac{C_D(\beta)}{C_D(\alpha)}=2\left(\frac{\beta}{\pi}\right)^2(4+\pi\sin\alpha)f(\beta)\csc\beta\csc\alpha. \tag{38.122b}$$

Using (38.114b) in the limit of small angles (38.123a), it follows that the ratio of drag coefficients of the bent and straight lamina at the same small inclination is (38.123b):

$$\begin{aligned} \lim_{\beta\to 0} f(\beta)=1: \quad \lim_{\beta\to\alpha\to 0}\frac{C_D(\beta)}{C_D(\alpha)} &= \lim_{\beta\to\alpha\to 0}\frac{\overline{C}_D(\beta)}{\overline{C}_D(\alpha)} \\ &= \lim_{\alpha\to 0}\frac{2}{\pi^2}(4+\pi\sin\alpha)\alpha^2\csc^2\alpha=\frac{8}{\pi^2}=0.810; \end{aligned} \tag{38.123a,b}$$

thus *the bent lamina has 19% less drag than the flat lamina of the same length at the same shallow angle of inclination relative to the free stream.*

TABLE 38.2
Drag of a Bent versus a Straight Lamina β Flux

Ratio $\overline{C}_D(\beta)/\overline{C}_D(\alpha)$	Arrow $\overline{C}_D(\beta)$	Angle $\alpha=\beta$	Flat plate $\overline{C}_D(\alpha)$	Angle $\alpha=\beta$	Arrow $\overline{C}_D(\beta)$	Ratio $\overline{C}_D(\beta)/\overline{C}_D(\alpha)$
0.829	0.199	10°	0.240	170°	0.996	4.150
0.848	0.359	20°	0.423	160°	0.990	2.340
0.867	0.489	30°	0.564	150°	0.984	1.745
0.884	0.593	49°	0.671	140°	0.975	1.453
0.892	0.637	45°	0.714	135°	0.970	1.359
0.901	0.677	50°	0.751	130°	0.964	1.284
0.920	0.745	60°	0.810	120°	0.950	1.173
0.942	0.800	70°	0.849	110°	0.931	1.096
0.967	0.844	80°	0.872	100°	0.907	1.040
1.000	0.880	90°	0.880	180°	0.880	1.000

Note: The drag coefficient calculated for the length projected in the direction normal to the free stream velocity is compared for: (i) a flat plate or lamina (Figure 38.6) for which it takes the same value to either side of the normal direction, for example, the drag is the same at 30° (150°) representing an inclinations of 60° forward (aft) of the vertical; (ii) a bent plate or lamina (Figure 38.7b) for which the drag coefficient is smaller (larger) for flow incident from the front (rear), which is deflected by the apex (trapped in the wedge).

Table 38.2 indicates: (i) the drag coefficient of the bent lamina or arrow for half-angles at 10° intervals from 10° to 170°, including the cases of flow incident from the front $\beta < 90°$ (behind $\beta > 90°$); (ii) there is no difference between flow from the front α or behind $\pi - \alpha$ for the flat plate (38.93b,c). The drag coefficients: (i) both coincide with the value (38.64b) for the flat plate orthogonal to the flow for $\alpha = \beta = \pi/2$; (ii) for a flow from the rear of the arrow $\beta > \pi/2$, the trapping of almost stagnant fluid inside the bent lamina leads to a drag coefficient close to unity for $\beta > 150°$, that is, almost total pressure loss; (iii) in the case of a flow from behind the drag coefficient of the bent lamina is larger than for the flat plate, more so the more concave it is; (iv) for flow from the front the drag coefficient is lower for the arrow than for the flat plate inclined at the same angle, with a drag reduction of 19% for the bent relative to the straight lamina at shallow angle (38.123b). Table 38.2 indicates a drag reduction of 17% for the lamina of half-angle 10° compared with a the straight lamina of the same length at the same inclination. This is close to the ultimate value (38.123b) of 19% for "zero" angle that is exact. Thus *a bent lamina has less (more) drag than a straight lamina of the same length set sideways at the same angle as its aperture, if the flow comes from the front (rear).*

38.7 Jet Attachment around a Wall (Coanda Effect)

The preceding problem concerned a vertical (oblique) barrier [Section 38.4 (38.6)] in a wall of infinite extent. Another case of wall flow is a jet parallel to a curved wall or a wall with an edge that experiences the Coanda effect, that is, the tendency to follow the wall or go round the edge rather than continue in a straight line. The Coanda effect is considered for a semiinfinite plate, with the jet of height h, velocity U, and volume flux $Q = hU$ going round the edge (Figure 38.8a). This should be distinguished from the potential flow past a semiinfinite plate (Sections 16.8 and 36.3) that occupies all space and has no free streamline. The

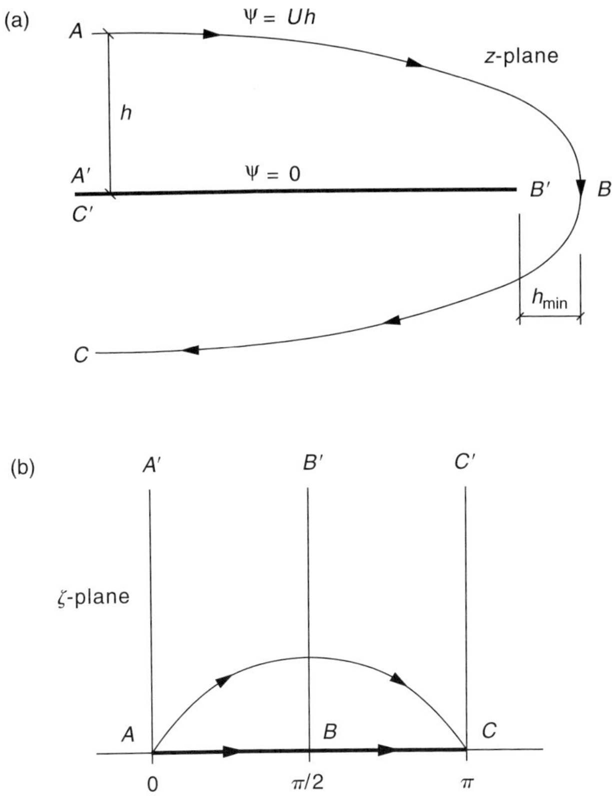

FIGURE 38.8
There are two distinct potential flows past a semiinfinite plate: (i) the corner flow (Figure 14.10b) which fills all space, has no free streamline and is a problem of conformal mapping; (ii) the free jet around a semiinfinite plate (Figure 38.8), which does not occupy all space but only the region between the plate and the free streamline and is a hodograph problem. In the hodograph plane (b) the flow goes from a source (to a sink) far upstream (downstream) above (below) the semiinfinite plate. Both in the corner flow (Figure 14.10b) and free jet (Figure 38.8a) the velocity has an inverse-square-root singularity at the edge of the semiinfinite plate. This is associated with a finite suction force around the edge of the semiinfinite plate; as for the suction force at the leading-edge of a flat plate airfoil (Figure 34.1b), which balances the horizontal component of lift in a flow with angle-of-attack. In the corner flow (Figure 14.10b) this finite suction force is powerless against the infinite momentum flux of the fluid filling all space. In the free jet (Figure 38.8a) the suction force at the edge of the semiinfinite plate exactly balances the momentum flux of the jet, turning it around the edge to reverse its direction. This demonstrates the "Coanda effect"' that a flow tends to remain "attached" to a wall, and turn around it following its curvature. A practical application is the flaps deflected downward behind the airfoil sections of a wing; this turns the flow downwards behind the wing, increasing the downwash velocity and angle (Figure 34.6), and thus the lift. This allows a lower stalling speed, lower take-off and landing speeds, and the use of shorter runways. The increase in drag is countered by increased engine thrust; the higher fuel consumption is of little concern over the short periods of flight, like take-off and landing; fuel consumption is important in cruise which takes most of the flight time of an aircraft (Figures 34.8 and 34.11).

Coanda effect implies that the free streamline curves around the edge (Subsection 38.7.1), where the jet should be narrowest (Subsection 38.7.2) and the velocity highest, as for a vortex going past the edge of a semiinfinite plate at angle-of-attack (Subsection 36.3.1). In the case of the flat plate (Subsection 34.1.3) the suction force at the leading-edge balances (34.26) the horizontal component of the lift specified by the Kutta condition at the trailing-edge; in the present case there is no lift, and the suction force balances the momentum flux (Subsection 36.7.3), that is, turns the jet around the edge of the semiinfinite plate, thereby explaining the Coanda effect of jet attachment to the wall.

38.7.1 Jet Turning around a Sharp Edge

The free streamline in the physical plane (Figure 38.8a) has constant velocity, $v = U$; the angle of the velocity varies from $\varphi = 0$ at infinity, to $\varphi = -\pi/2$ near the edge, $\varphi = -\pi$ back at infinity. The hodograph variable (38.29) covers the strip $0 < \varphi < \pi$ along the real axis (Figure 38.8b). This corresponds to the flow (36.67b) around a finite plate (38.124a):

$$\zeta = -i \operatorname{arg\,cosh} t = -i \log\left(t + \sqrt{t^2-1}\right), \quad \frac{\mathrm{d}f}{\mathrm{d}z} = U\left(t + \sqrt{t^2-1}\right), \tag{38.124a,b}$$

where (7.36b) was used; comparison of (38.124a) with (38.29) gives (38.124b). The transformations (38.125a–c) in (36.67b) lead to (38.125d):

$$\Gamma = 2\pi; \quad t = \frac{z}{b}; \quad f = \zeta: \quad t = \cos\zeta = \cosh\left(i\zeta\right), \tag{38.125a–d}$$

that coincides with (38.124a) $\equiv$ (38.125d). The flow rate is $Q = Uh$, where h is the thickness of the jet at infinity; the flow is due to a source (sink) above (below) the plate at infinity at $t = 1\,(t = -1)$ corresponding to the complex potential (38.126b):

$$Q = 2Uh: \quad f\left(t\right) = \frac{Q}{2\pi}\left[\log\left(t-1\right) - \log\left(t+1\right)\right] = \frac{Uh}{\pi}\log\left(\frac{t-1}{t+1}\right), \tag{38.126a,b}$$

where the source/sink strength was doubled (38.126a) since it lies on the boundary and only half the volume flux goes into the jet.

38.7.2 Narrowest Section of the Free Jet

From the complex potential (38.126b) in the t-plane follows the conjugate velocity:

$$\frac{\mathrm{d}f}{\mathrm{d}t} = \frac{Uh}{\pi}\left(\frac{1}{t-1} - \frac{1}{t+1}\right) = \frac{2Uh}{\pi}\frac{1}{t^2-1}; \tag{38.127a}$$

together with the complex potential in the physical z-plane (38.124b) in terms of the t-variable, it specifies in parametric form the shape of the free streamline:

$$\begin{aligned} \frac{\mathrm{d}z}{\mathrm{d}t} = \frac{\mathrm{d}f/\mathrm{d}t}{\mathrm{d}f/\mathrm{d}z} &= \frac{2h/\pi}{\left(t^2-1\right)\left(t + \sqrt{t^2-1}\right)} \\ &= \frac{2h}{\pi}\frac{t - \sqrt{t^2-1}}{t^2-1} = \frac{2h}{\pi}\left(\frac{t}{t^2-1} - \frac{1}{\sqrt{t^2-1}}\right); \end{aligned} \tag{38.127b}$$

in (38.127) was used (38.53b) $\equiv$ (7.38).

At the edge $z = 0$ the velocity (38.124b) is infinite so $t = \infty$ in (38.128a); integrating (38.127b) from $t = \infty$ to t yields (38.128b):

$$z\left(\infty\right) = 0: \quad z\left(t\right) = \frac{h}{\pi}\left[\log\left(t^2-1\right) - 2\log\left(t + \sqrt{t^2-1}\right) + 2\log 2\right], \tag{38.128a,b}$$

where: (i) the primitive (38.128c) similar to (38.106d) with $n=1$ was used:

$$\frac{\mathrm{d}\left[\log\left(t+\sqrt{t^2-1}\right)\right]}{\mathrm{d}t}=\left(t+\sqrt{t^2-1}\right)^{-1}\left(1+\frac{t}{\sqrt{t^2-1}}\right)=\frac{1}{\sqrt{t^2-1}}; \tag{38.128c}$$

$$\lim_{t\to\infty}\log\left[\frac{t^2-1}{\left(t+\sqrt{t^2-1}\right)^2}\right]=\lim_{t\to\infty}\log\left(\frac{t^2}{4t^2}\right)=-2\log 2, \tag{38.128d}$$

(ii) the constant of integration was chosen (38.128d) to satisfy (38.128a).

The free streamline with the change of variable (38.129a,b) corresponds to (38.129c):

$$\begin{aligned}-1<t\equiv\sin\varphi<+1,\ -\frac{\pi}{2}<\varphi<\frac{\pi}{2}:\quad &\frac{\pi}{h}(x+iy)-2\log 2\\ &=\log\left(-\cos^2\varphi\right)-2\log\left(\sin\varphi+i\cos\varphi\right)=\log\left(-\cos^2\varphi\right)-2\log\left(ie^{-i\varphi}\right)\\ &=i\pi+\log\left(\cos^2\varphi\right)-2\left(\frac{i\pi}{2}-i\varphi\right)=2\log\left(\cos\varphi\right)+2i\varphi;\end{aligned} \tag{38.129a–c}$$

thus its parametric equations (38.129c) are (38.130a,b):

$$y(\varphi)=\frac{2h\varphi}{\pi},\quad x(\varphi)=\frac{2h}{\pi}\log\left(2\cos\varphi\right)=\frac{2h}{\pi}\log\left[2\cos\left(\frac{\pi y}{2h}\right)\right]; \tag{38.130a–c}$$

eliminating the parameter leads to the *explicit formula (38.130c) for the shape of the free streamline of a jet of velocity* U *and thickness* h *at infinity going round a semiinfinite plate (Figure 38.8a). It follows that (i) the minimum width of the jet is (38.131a):*

$$\frac{h_{\min}}{h}=\lim_{y\to 0}\frac{x}{h}=\frac{2}{\pi}\log 2=0.441,\quad v_{\max}=\frac{Uh}{h_{\min}}=2.226U, \tag{38.131a,b}$$

(ii) the corresponding maximum velocity is (38.131b); (iii) both occur at the horizontal section near the edge.

38.7.3 Suction Force at the Edge and Momentum Flux

Near the edge $t\to\infty$ the velocity (38.127a) scales as (38.132a) leading to the complex potential (38.132b):

$$\frac{\mathrm{d}f}{\mathrm{d}t}\sim\frac{2Uh}{\pi t^2},\quad f\sim-\frac{2Uh}{\pi t}. \tag{38.132a,b}$$

The coordinate (38.127b) scales:

$$\frac{\mathrm{d}z}{\mathrm{d}t}=\frac{2h}{\pi}\frac{t\sqrt{t^2-1}-t^2+1}{\left(t^2-1\right)^{3/2}}\sim\frac{2h}{\pi t^3}\left(t^2\sqrt{1-1/t^2}-t^2+1\right)\sim\frac{2h}{\pi t^3}\left(t^2-\frac{1}{2}-t^2+1\right)=\frac{h}{\pi t^3}, \tag{38.132c}$$

as (38.133a,b):

$$\frac{\mathrm{d}z}{\mathrm{d}t}\sim\frac{h}{\pi t^3},\quad z\sim-\frac{h}{2\pi t^2};\quad f\sim-\frac{2Uh}{\pi}\sqrt{-\frac{2\pi z}{h}}, \tag{38.133a–c}$$

substituting (38.133b) in (38.132b) yields (38.133c); this is the usual (Subsections 14.9.3 and 34.1.4) inverse square root singularity for the velocity (38.134a):

$$v^*=\frac{\mathrm{d}f}{\mathrm{d}z}\sim iU\sqrt{\frac{2h}{\pi z}}\equiv\frac{C}{\sqrt{z}},\quad C\equiv iU\sqrt{\frac{2h}{\pi}}, \tag{38.134a,b}$$

with coefficient (38.134b). *The inverse square root singularity of the velocity (38.134a) corresponds to a constant suction force at (34.16b) at a sharp edge:*

$$F_s = -\pi\rho C^2 = -\pi\rho\left(iU\sqrt{\frac{2h}{\pi}}\right)^2 = 2\rho U^2 h = \rho U Q; \tag{38.135}$$

the suction force at the edge of a semiinfinite plate equals the momentum flux of the jet of flow rate Q and density ρ; this is exactly the force needed to turn the jet around the edge, demonstrating the **Coanda effect**.

38.8 Fluidics: Deflection of a Jet by a Small Source

In the absence of wall attraction or Coanda effect (Section 38.7) a jet parallel to a semi-infinite plate (Figure 38.9a) can be deflected (Subsection 38.8.4) by placing a source on the plate near the edge (Subsection 38.8.3); this causes a stagnation point upstream as for the Rankine fairing (Section 28.4). A stagnation point is a singularity for the hodograph variable, as is also the source because in both cases the logarithm of the velocity is singular. The modification of the hodograph method to allow for the presence of monopoles and/or stagnation points (Subsection 38.8.1) is based on the construction of a hodograph function (Subsection 38.8.2) using methods of analytic continuation (Sections 31.1 and 31.2); it can be used to demonstrate the fluidic effect of deflection of a "large" jet by a 'small" source.

38.8.1 Hodograph with Singularities and Stagnation Points (Joukowski, Tchaplygin, Levi–Civita)

Since the hodograph variable:

$$\zeta \equiv \log\left(U\frac{\mathrm{d}z}{\mathrm{d}f}\right) = \log\left(\frac{U}{\mathrm{d}f/\mathrm{d}z}\right) = \log\left(\frac{U}{v}\mathrm{e}^{i\theta}\right) = i\theta + \log\left(\frac{U}{v}\right), \tag{38.136}$$

involves the logarithm of the velocity, it is singular when the velocity is infinite (zero), that is, at monopoles (stagnation points). Among the approaches to use the hodograph method in the presence of singularities and/or stagnation points (Joukowski, Levi–Civita, Tchaplygin), the simplest is to use these points to construct the principal part of the hodograph function, as shown next. The flow in the t-plane (Figure 38.9b) consists of a uniform stream of velocity U plus a source of strength Q at $t = -a$, where the strength must be doubled because only half goes into the jet (38.137a):

$$\frac{\mathrm{d}f}{\mathrm{d}t} = U + \frac{Q}{\pi(t+a)} = \frac{U(t+b)}{t+a}. \tag{38.137a,b}$$

The complex conjugate velocity in the t-plane (38.137a) can be written in the equivalent form (38.137b), where $t = -b$ specifies (38.138a) the location (38.138b,c) of the stagnation point:

$$U(t+a) + \frac{Q}{\pi} = U(t+b): \quad b = a + \frac{Q}{\pi U} = a(1+\varepsilon), \quad \varepsilon \equiv \frac{b-a}{a} = \frac{Q}{\pi U a}. \tag{38.138a–c}$$

Thus *for a jet of velocity U past a semiinfinite plate, with a source of flow rate Q at a distance a from the edge, there is a stagnation point whose: (i) the absolute distance of*

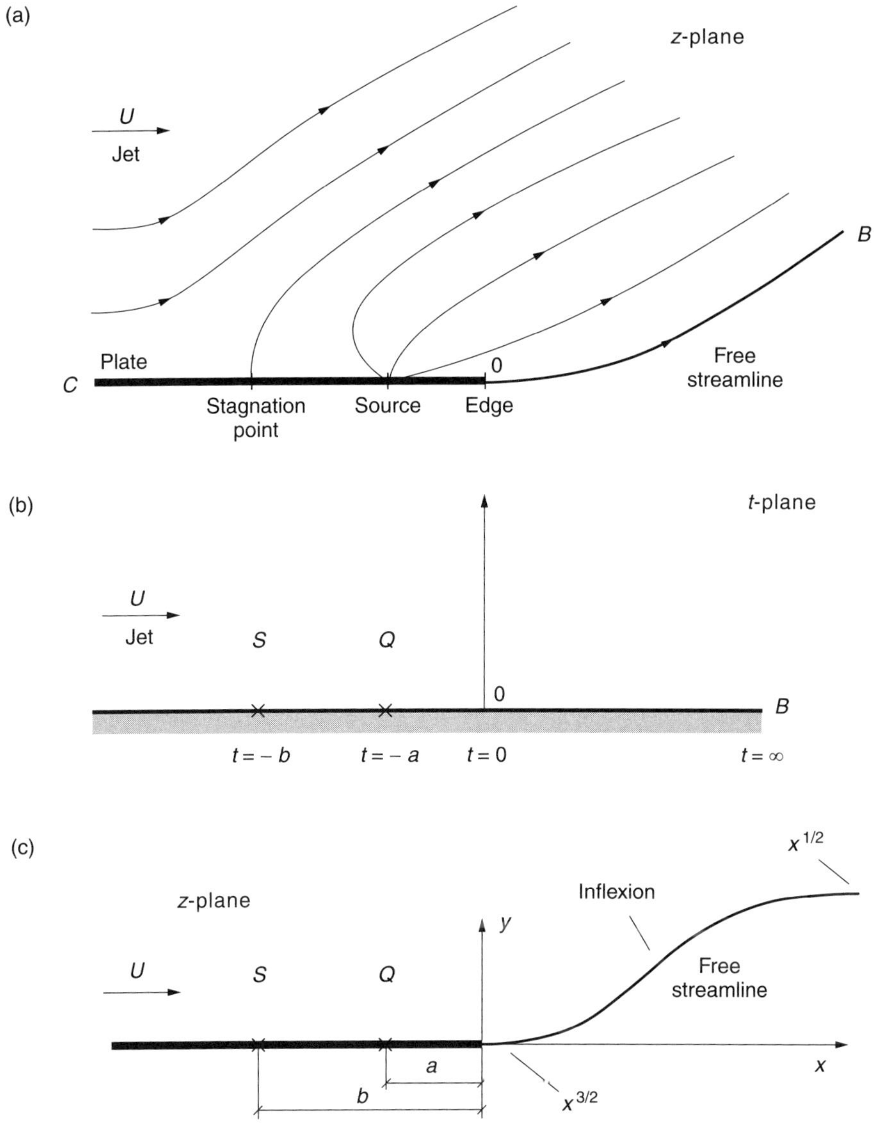

FIGURE 38.9

The free stream problems (Figures 38.3–38.8) represent a first class of hodograph problems for which there are neither stagnation points nor singularities in the flow; thus the hodograph variable (38.28), which is the logarithm of the velocity, is finite. The second class of hodograph problems involves a source in the jet, whose presence also creates a stagnation point(s), as for the Rankine bodies (Figures 28.5 and 28.7). The zero (infinite) velocity at the stagnation point (source/sink/vortex/monopole/multipole) leads to a singularity in the logarithm of the velocity, which is the hodograph variable. In this case (Figures 38.9 and 40.10) may be introduced an auxiliary hodograph function, whose singularities are poles at the locations of the stagnation points and sources/sinks/vortices/monopoles/multipoles. Performing the analytic continuation of the hodograph function over all space (b) together with hodograph method, specifies the flow, for example, a jet past a semiinfinite plate (a), this time with a flow source near the edge. The flow source deflects the free streamline (c) and hence changes the asymptotic direction of the jet. This demonstrates the "fluidic" effect: a small source near the edge of a semiinfinite plate with a jet stream can change its direction, even with a volume flux small relative to that of the jet. If the flow source was inside the jet along a wall without an edge (Figure 40.10) the effect would be much smaller: a modest increase in the volume flux of the jet due to the small relative flow rate of the source and no change of the direction of the jet (Example 40.16).

(38.138b) from the edge is proportional to the distance of the source from the edge; (ii) the relative distance (38.138c) is the ratio of the source strength per radian Q/π by the volume flux of the jet across a distance equal to the spacing of the source from the edge Ua. Since $\mathrm{d}f/\mathrm{d}t$ is already available it suffices to have $\mathrm{d}f/\mathrm{d}z$ to specify $\mathrm{d}z/\mathrm{d}t$, that is, the shape of the free streamline is parametric form; also $\mathrm{d}f/\mathrm{d}z$ is specified by the hodograph variable (38.136) that can be obtained by integration of the hodograph function $\mathrm{d}\zeta/\mathrm{d}t$.

38.8.2 Construction of the Hodograph Function

The **hodograph function** is the derivative of the hodograph variable (38.136):

$$\frac{\mathrm{d}\zeta}{\mathrm{d}t} = \frac{\mathrm{d}\left[\log\left(U\mathrm{d}z/\mathrm{d}f\right)\right]}{\mathrm{d}t} = \frac{\mathrm{d}\left\{\log\left[\left(\mathrm{d}z/\mathrm{d}t\right)/\left(\mathrm{d}f/\mathrm{d}t\right)\right]\right\}}{\mathrm{d}t} = \frac{\mathrm{d}\left[\log\left(\mathrm{d}z/\mathrm{d}f\right)\right]}{\mathrm{d}t} - \frac{\mathrm{d}\left[\log\left(\mathrm{d}f/\mathrm{d}t\right)\right]}{\mathrm{d}t}, \tag{38.139}$$

that consists of two terms: (i) the first is a conformal mapping, hence analytic; (ii) the second is the logarithmic derivative of the complex conjugate velocity (38.137b) in the t-plane that has a simple pole (simple zero) at a source (38.140) [stagnation point (38.141)]:

$$\frac{\mathrm{d}\zeta}{\mathrm{d}t} = -\frac{\mathrm{d}\left\{\log\left[\left(t+a\right)^{-1}\right]\right\}}{\mathrm{d}t} = \frac{1}{t+a}, \tag{38.140}$$

$$\frac{\mathrm{d}\zeta}{\mathrm{d}t} = -\frac{\mathrm{d}\left[\log\left(t+b\right)\right]}{\mathrm{d}t} = -\frac{1}{t+b}. \tag{38.141}$$

Thus *a monopole (stagnation point) corresponds to a pole of the hodograph function with residue* $(+1)-1$; in the present case the hodograph function is

$$\frac{\mathrm{d}\zeta}{\mathrm{d}t} = \frac{1}{t+a} - \frac{1}{t+b} + g_+\left(t\right), \tag{38.142}$$

where: (i) the principal part is specified by the pole with residue $+1\,(-1)$ at the source $t=-a$ (stagnation point $t=-b$); (ii) the function:

$$g_+\left(t\right) \in \mathcal{A}\left(\mathrm{Im}\left(t\right) > 0\right): \qquad \lim_{|t|\to\infty} t g_+\left(t\right) = 0, \tag{38.143a,b}$$

is analytic in the upper t-plane (38.143a) and has at least a simple zero at infinity (38.143b).

To perform the analytic continuation of the hodograph function (38.142) to the whole complex t-plane, the reflection principle (Section 31.2) on the real axis may be used, provided that the function be real for real t. The hodograph variable (38.136) is: (i) real for $t<0$ that corresponds to the plate where $\theta=0$; (ii) imaginary for $t>0$ where corresponds to the free streamline where $v=U$. Thus the function (38.144a) is real on the whole of the real t-axis:

$$j\left(t\right) \equiv i\sqrt{t}\frac{\mathrm{d}\zeta}{\mathrm{d}t};\ h(t) = j(t) - \frac{i\sqrt{-a}}{t+a} + \frac{i\sqrt{-b}}{t+b} = i\frac{\sqrt{t}-\sqrt{-a}}{t+a} - i\frac{\sqrt{t}-\sqrt{-b}}{t+b} + i\sqrt{t}g_+(t) = \frac{i}{\sqrt{t}+\sqrt{-a}} - \frac{i}{\sqrt{t}+\sqrt{-b}} + i\sqrt{t}g_+(t), \tag{38.144a,b}$$

the function (39.144b) obtained from (38.144a) by subtracting the principal parts at the poles, corresponding to the source at $t=-a$ and stagnation at $t=-b$, is analytic in the upper half t-plane. Thus the function (38.144b) is: (i) real on the whole real axis; (ii) analytic in the upper half t-plane; (iii) by reflection it is also analytic in the lower-half t-plane. Thus it is analytic in the whole complex plane, and by Liouville's theorem (Section 27.6) reduces to a constant. The constant can be evaluated at infinity, where: (i) from (38.142; 38.141) follows (38.145a):

$$\lim_{t\to\infty} t\frac{\mathrm{d}\zeta}{\mathrm{d}t} = \lim_{t\to\infty}\left[\frac{1}{1+a/t} - \frac{1}{1+b/t} + tg_+(t)\right] = \lim_{t\to\infty} O\left(\frac{b-a}{t}\right) = 0, \tag{38.145a}$$

$$\lim_{t\to\infty} h(t) = \lim_{t\to\infty} j(t) = \lim_{t\to\infty} i\sqrt{t}\frac{\mathrm{d}\zeta}{\mathrm{d}t} = \lim_{t\to\infty} O\left(\frac{1}{\sqrt{t}}\right) = 0, \tag{38.145b}$$

(ii) from (38.144a,b) follows (38.145b); (iii) the constant value of (38.144b) is zero (38.146a):

$$h(t) = \mathrm{const} = h(\infty) = 0: \quad \frac{\mathrm{d}\zeta}{\mathrm{d}t} = -\frac{i}{\sqrt{t}}j(t) = \frac{i}{\sqrt{t}}\left(\frac{\sqrt{-a}}{t+a} - \frac{\sqrt{-b}}{t+b}\right) = \frac{i\sqrt{a}}{(t+a)\sqrt{t}} - \frac{i\sqrt{b}}{(t+b)\sqrt{t}}; \tag{38.146a,b}$$

(iv) this specifies (38.144a,b) the hodograph function (38.146b).

38.8.3 Shape of the Free Streamline

The hodograph function (38.146b) ≡ (38.147) can be written:

$$\frac{\mathrm{d}\zeta}{\mathrm{d}t} = \frac{1/(2\sqrt{t})}{\sqrt{t}-i\sqrt{a}} - \frac{1/(2\sqrt{t})}{\sqrt{t}+i\sqrt{a}} + \frac{1/(2\sqrt{t})}{\sqrt{t}+i\sqrt{b}} - \frac{1/(2\sqrt{t})}{\sqrt{t}-i\sqrt{b}}. \tag{38.147}$$

The uniform flow $\mathrm{d}f/\mathrm{d}z = U$ at infinity corresponds in (38.136) to zero hodograph variable $\zeta = 0$ and $\mathrm{d}f/\mathrm{d}t = U$ in (38.137b) corresponds to $t = \infty$; thus (38.147) can be integrated from $\zeta = 0$ at $t = \infty$, leading to:

$$\log\left(\frac{\sqrt{t}-i\sqrt{a}}{\sqrt{t}+i\sqrt{a}}\frac{\sqrt{t}+i\sqrt{b}}{\sqrt{t}-i\sqrt{b}}\right) = \zeta = \log\left(U\frac{\mathrm{d}z}{\mathrm{d}f}\right), \tag{38.148a}$$

this specifies the inverse of the complex conjugate velocity:

$$\frac{\mathrm{d}z}{\mathrm{d}f} = \frac{1}{U}\frac{\sqrt{t}-i\sqrt{a}}{\sqrt{t}+i\sqrt{a}}\frac{\sqrt{t}+i\sqrt{b}}{\sqrt{t}-i\sqrt{b}}, \tag{38.148b}$$

that can be compared with (38.137b) the complex conjugate velocity in the t-plane:

$$\frac{\mathrm{d}f}{\mathrm{d}t} = U\frac{\sqrt{t}+i\sqrt{b}}{\sqrt{t}+i\sqrt{a}}\frac{\sqrt{t}-i\sqrt{b}}{\sqrt{t}-i\sqrt{a}}, \tag{38.149}$$

The hodograph function (38.147) specifies by integration the hodograph variable (38.136) ≡ (38.148a) that leads to the inverse of the complex conjugate velocity in the physical z*-plane (38.148b); multiplying by the complex conjugate velocity the* t*-plane (38.137b) ≡ (38.149) specifies the shape of the free streamline:*

$$\frac{\mathrm{d}z}{\mathrm{d}t} = \frac{\mathrm{d}z}{\mathrm{d}f}\frac{\mathrm{d}f}{\mathrm{d}t} = \left(\frac{\sqrt{t}+i\sqrt{b}}{\sqrt{t}+i\sqrt{a}}\right)^2 = \left(1 + i\frac{\sqrt{b}-\sqrt{a}}{\sqrt{t}+i\sqrt{a}}\right)^2, \tag{38.150}$$

after an integration.

The partial fraction decomposition (Section 31.9) of the rational function (38.150) is immediate:

$$\frac{dz}{dt} = 1 + 2i\frac{\sqrt{b}-\sqrt{a}}{\sqrt{t}+i\sqrt{a}} - \left(\frac{\sqrt{b}-\sqrt{a}}{\sqrt{t}+i\sqrt{a}}\right)^2. \tag{38.151}$$

The factors $1/\sqrt{t}$ should be inserted in (38.151) ≡ (38.152a):

$$\frac{dz}{dt} = 1 + \frac{A/\sqrt{t}}{\sqrt{t}+i\sqrt{a}} + \frac{B/\sqrt{t}}{\left(\sqrt{t}+i\sqrt{a}\right)^2} + \frac{C}{\sqrt{t}}, \tag{38.152a}$$

$$z = t + 2A\log\left(\sqrt{t}+i\sqrt{a}\right) - \frac{2B}{\sqrt{t}+i\sqrt{a}} + 2C\sqrt{t}, \tag{38.152b}$$

for an immediate integration (38.152b). Comparing (38.151) ≡ (38.152a) leads to

$$\begin{aligned}\sqrt{t}\left(\sqrt{t}+i\sqrt{a}\right)^2\left[\frac{dz}{dt}-1\right] &= A\left(\sqrt{t}+i\sqrt{a}\right) + B + C\left(\sqrt{t}+i\sqrt{a}\right)^2 \\ &= 2i\sqrt{t}\left(\sqrt{t}+i\sqrt{a}\right)\left(\sqrt{b}-\sqrt{a}\right) - \left(\sqrt{b}-\sqrt{a}\right)^2\sqrt{t}.\end{aligned} \tag{38.153}$$

Equating the coefficients of $\left(t, \sqrt{t}, 1\right)$ yields a system of three algebraic equations:

$$\begin{aligned}C = 2i\left(\sqrt{b}-\sqrt{a}\right): \quad A &= -2iC\sqrt{a} - 2\sqrt{a}\left(\sqrt{b}-\sqrt{a}\right) - \left(\sqrt{b}-\sqrt{a}\right)^2 \\ &= \left(\sqrt{b}-\sqrt{a}\right)\left(4\sqrt{a}-2\sqrt{a}-\sqrt{b}+\sqrt{a}\right) = \left(\sqrt{b}-\sqrt{a}\right)\left(3\sqrt{a}-\sqrt{b}\right), \\ B = -i\sqrt{a}\left(A+iC\sqrt{a}\right) &= -i\sqrt{a}\left(\sqrt{b}-\sqrt{a}\right)\left(3\sqrt{a}-\sqrt{b}-2\sqrt{a}\right) = i\sqrt{a}\left(\sqrt{b}-\sqrt{a}\right)^2;\end{aligned} \tag{38.154a–c}$$

substitution of (28.154a–c) in (38.152b) yields the integral of (38.150):

$$z = t + \left(\sqrt{b}-\sqrt{a}\right)\left[4i\sqrt{t} + 2\left(3\sqrt{a}-\sqrt{b}\right)\log\left(\sqrt{t}+i\sqrt{a}\right) - 2i\sqrt{a}\frac{\sqrt{b}-\sqrt{a}}{\sqrt{t}+i\sqrt{a}}\right], \tag{38.155}$$

that specifies the shape of the free stream line.

38.8.4 Initial and Asymptotic Deflection of the Jet

Far from the edge (38.156a) the shape of the free streamline is specified by the leading real and imaginary terms in (38.155) ≡ (38.156b):

$$t \to \infty: \quad x + iy = z = t + \left(\sqrt{b}-\sqrt{a}\right)4i\sqrt{t}; \quad x \sim t, \quad y \sim 4\left(\sqrt{b}-\sqrt{a}\right)\sqrt{x}, \tag{38.156a–d}$$

this corresponds (38.156c,d) to a parabola with: (i) axis in the direction of the plate (38.157a):

$$x \sim \frac{y^2}{2R}, \quad R \equiv 8\left(\sqrt{b}-\sqrt{a}\right)^2 = 8a\left(\sqrt{1+\varepsilon}-1\right)^2 = 8a\left(\frac{\varepsilon}{2}\right)^2 = 2a\varepsilon^2 = \frac{2Q^2}{\pi^2U^2a}, \tag{38.157a,b}$$

(ii) a radius of curvature (38.157b) at the apex that increases (38.138b,c) for smaller jet velocity and larger mass flux of a source closer to the edge. Near the edge $t \to 0$ the leading terms of (38.150) are:

$$\frac{\mathrm{d}z}{\mathrm{d}t} = \left(\frac{\sqrt{b}}{\sqrt{a}}\frac{1-i\sqrt{t/b}}{1-i\sqrt{t/a}}\right)^2 = \frac{b}{a}\frac{1-i2\sqrt{t/b}}{1-i2\sqrt{t/a}} + O(t);$$
$$= \frac{b}{a}\left[1+i2\sqrt{t}\left(\frac{1}{\sqrt{a}}-\frac{1}{\sqrt{b}}\right)\right] + O(t); \tag{38.158}$$

this leads by (38.159a) to the differential equations (38.158b,c) for the free streamline:

$$\mathrm{d}z = \mathrm{d}x + i\mathrm{d}y: \quad \frac{\mathrm{d}x}{\mathrm{d}t} \sim \frac{b}{a}, \quad \frac{\mathrm{d}y}{\mathrm{d}t} \sim 2\frac{b}{a}\left(\frac{1}{\sqrt{a}}-\frac{1}{\sqrt{b}}\right)\sqrt{t}; \tag{38.159a–c}$$

an integration specifies the shape of the free streamline:

$$x \sim \frac{b}{a}t = \left(1+\frac{Q}{\pi U a}\right)t, \quad y \sim \frac{4}{3}t^{3/2}\frac{b}{a}\left(\frac{1}{\sqrt{a}}-\frac{1}{\sqrt{b}}\right), \tag{38.160a,b}$$

that corresponds to the parabola:

$$y \sim \frac{4}{3}t^{3/2}\frac{1+\varepsilon}{\sqrt{a}}\left(1-\frac{1}{\sqrt{1+\varepsilon}}\right) \sim \frac{2}{3}t^{3/2}\frac{\varepsilon}{\sqrt{a}} = \frac{2}{3}\frac{Q}{\pi U}\left(\frac{x}{a}\right)^{3/2}. \tag{38.160c}$$

The free streamline (38.150) ≡ (38.155) of a jet of velocity U past a semiinfinite plate, with a source (38.137a) of flow rate Q at a distance a from the edge (Figure 38.9a), and hence a stagnation point (38.137b) at (38.138b,c): (i) starts at the edge (38.160a–c) with increasing slope (38.161a); (ii) ends (38.157a,b) with decreasing slope (38.161b) at infinity:

$$\lim_{x\to 0}\frac{\mathrm{d}y}{\mathrm{d}x} \sim x^{1/2}, \quad \lim_{x\to\infty}\frac{\mathrm{d}y}{\mathrm{d}x} \sim \frac{\mathrm{d}\left(\sqrt{2Rx}\right)}{\mathrm{d}x} \sim \sqrt{\frac{R}{2x}}; \tag{38.161a,b}$$

*(iii) hence there must be an inflexion point in between (Figure 38.9c). The deflection of jet is larger for smaller jet velocity, and larger source strength closer to the edge, and reduces (Figure 38.9c) but does not disappear with distance. This demonstrates a **fluidic effect**: a small source can affect significantly the asymptotic deflection of larger jet if located at a critical position.* The fluidic effect may be related at the deeper level to jet instabilities. A source inside a jet also creates a stagnation point, but without the end-plate effect the influence on the jet is weaker (Example 40.16).

38.9 Jets Merging, Splitting, or Colliding with a Wall

The free streamlines can occur without obstacles in the case of collision of jets (Subsection 38.9.1). A line of symmetry of the jets can be replaced by a wall (Subsection 38.9.2), for example, for a jet incident normal to a wall (Subsection 38.9.3), upon that it exerts a force (Subsection 38.9.4). Since the stream function is known on the free streamline of each jet, the Schwartz formula (Subsection 38.8.2) can be used to specify the potential flow, in a way somewhat analogous (Section 38.1) to the field in a half-plane (cylinder) being specified by the potential on a strip (half-circle) [Subsection 38.1.3 (38.1.4)].

38.9.1 Collision, Splitting or Merging of Jets

The case of the collision of two jets (Figure 38.10a) of widths (a_1, a_2) in the direction $(0, \gamma)$, resulting on the emergence of another two jets of widths (b_1, b_2) in direction (α, β) must satisfy: (i) the conservation of the volume flux (38.162a); (ii) the momentum balance (38.162b,c):

$$a_1 + a_2 = b_1 + b_2: \quad a_1 - a_2 \cos\gamma = b_1 \cos\alpha + b_2 \cos\beta$$
$$a_2 \sin\gamma + b_2 \sin\beta = b_1 \sin\alpha. \qquad (38.162\text{a–c})$$

The momentum balance (38.162b,c) is similar to the collision of particles with the same asymptotic velocity because: (i) the velocity is constant along the free streamlines; (ii) at infinity all jets are uniform. As concerns the conservation of the mass flux it is similar to the conservation of mass for a collision of particles; it may not hold for a disintegration of particles. The stream function is given by

$$\frac{\Psi}{U} = \begin{cases} 0 & \text{on } AB, & (38.163\text{a}) \\ b_1 & \text{on } B'C, & (38.163\text{b}) \\ b_1 - a_2 & \text{on } C'D, & (38.163\text{c}) \\ b_1 - a_2 + b_2 & \text{on } D'A'. & (38.163\text{d}) \end{cases}$$

on the four free surfaces.

For the splitting of a jet into two or merging of two jets into one (38.164a), and (38.162a,b) would lead to (38.164b,c):

$$a_2 = 0: \quad b_1 + b_2 = a_1 = b_1 \cos\alpha + b_2 \cos\beta; \quad \alpha = 0 = \beta, \qquad (38.164\text{a–c})$$

there imply that the jets would have all the same direction (38.164d,e). A nontrivial solution is possible for the collision of jets, since there are seven parameters $(a_1, a_2, b_1, b_2; \alpha, \beta, \gamma)$ to satisfy three relations (38.162a–c), for example, four parameters could be chosen. The special cases of collision of two jets include (i) head-on collision (38.165a) in Figure 38.10b when (38.162b,c) simplify to (38.165b,c):

$$\gamma = 0: \quad a_1 - a_2 = b_1 \cos\alpha + b_2 \cos\beta, \quad b_1 \sin\alpha = b_2 \sin\beta; \qquad (38.165\text{a,b})$$

(ii) symmetric (38.157a,b) emergent jets (Figure 38.10c) when (38.165b,c) simplify further to (38.166c,d):

$$\gamma = 0, \quad \alpha = \beta: \quad b_1 = b_2 \equiv b, \quad a_1 - a_2 = 2b\cos\alpha; \qquad (38.166\text{c–d})$$

then (38.162a) $\equiv$ (38.167a) implies (38.167b,c):

$$a_1 + a_2 = 2b, \quad a_1 = b(1 + \cos\alpha), \quad a_2 = b(1 - \cos\alpha); \qquad (38.167\text{a–c})$$

(iii) head-on collision with opposite (38.168a–c) emerging jets (Figure 38.11a) simplifies (38.167a–c) and further to (38.168d–f):

$$\gamma = 0, \quad \alpha = \frac{\pi}{2} = \beta: \quad b \equiv b_1 = b_2 = a_1 = a_2. \qquad (38.168\text{a–f})$$

The last case corresponds to two opposite jets along a wall colliding to form a normal jet (Figure 38.11b) or a jet incident normal to wall (Figure 38.11c). Since the method of solution is the same in all cases, the last simplest case serves as an illustration.

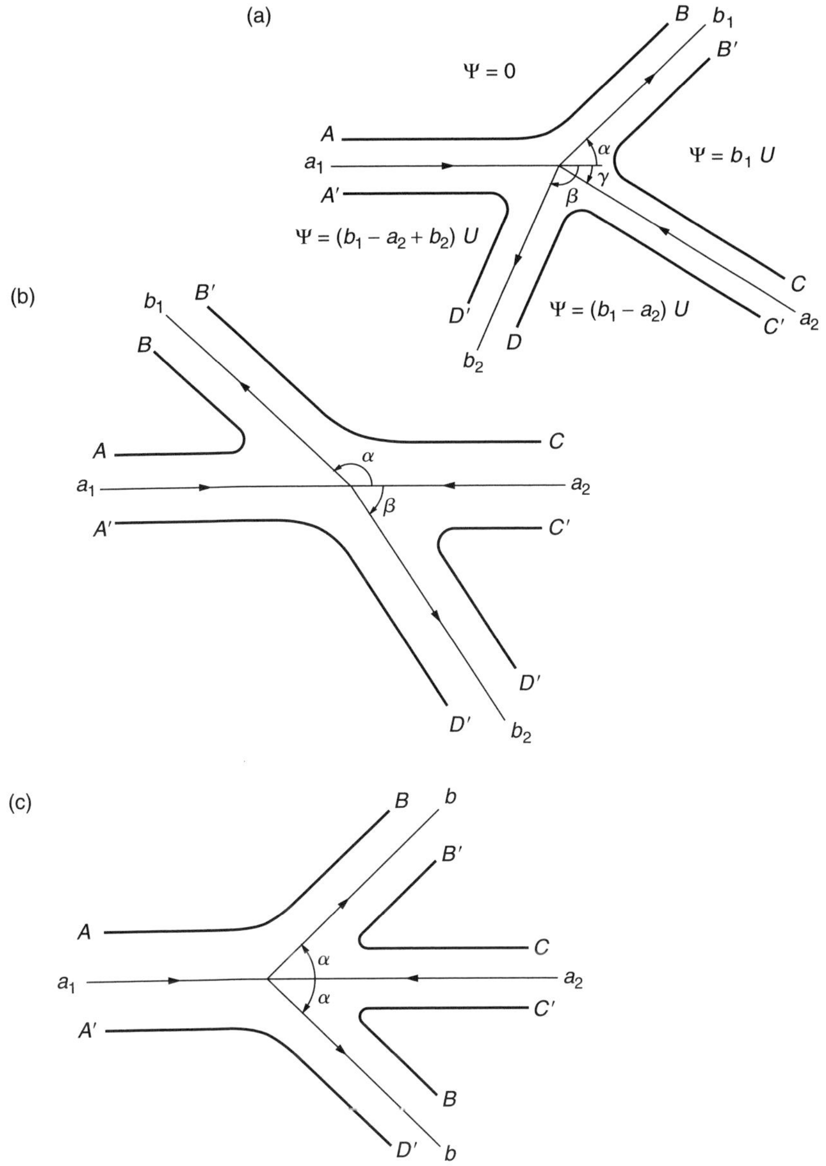

FIGURE 38.10
A third class of hodograph problems, besides those without (with) singularities in the hodograph variable, and not needing (needing) a hodograph function [Figures 38.3–38.8) (38.9 and 40.10)] are the free jet problems without any rigid boundaries (Figures 38.10 and 38.11). The variation of the stream function across a jet specifies the flow rate. Thus in the case of several "colliding," "mixing," "merging," or "separating" jets (Figures 38.10a–c and 38.11a–c) the stream function is known in all directions on a circle. The Schwartz formula, based on the Cauchy theorem, determines the complete complex potential, that is, potential plus stream function, at all points of the region of holomorphy or analicity, from the values of the stream function alone on the "boundary circle" in the hodograph plane. An example is two jets with arbitrary directions and flow rates "colliding" (a) at an angle and "splitting" into another two jets. A particular case is two opposite jets "colliding head-on" (b) and "diverging" in two side jets. A further particularization is two jets colliding head on and "splitting" symmetrically (c). In all cases must be conserved: (i) the flow rate; (ii/iii) the two components of the momentum; like the conservation of mass and momentum in the collision or disintegration of material particles. The three conservation laws (i) to (iii) limit the geometry of colliding/merging/splitting jets.

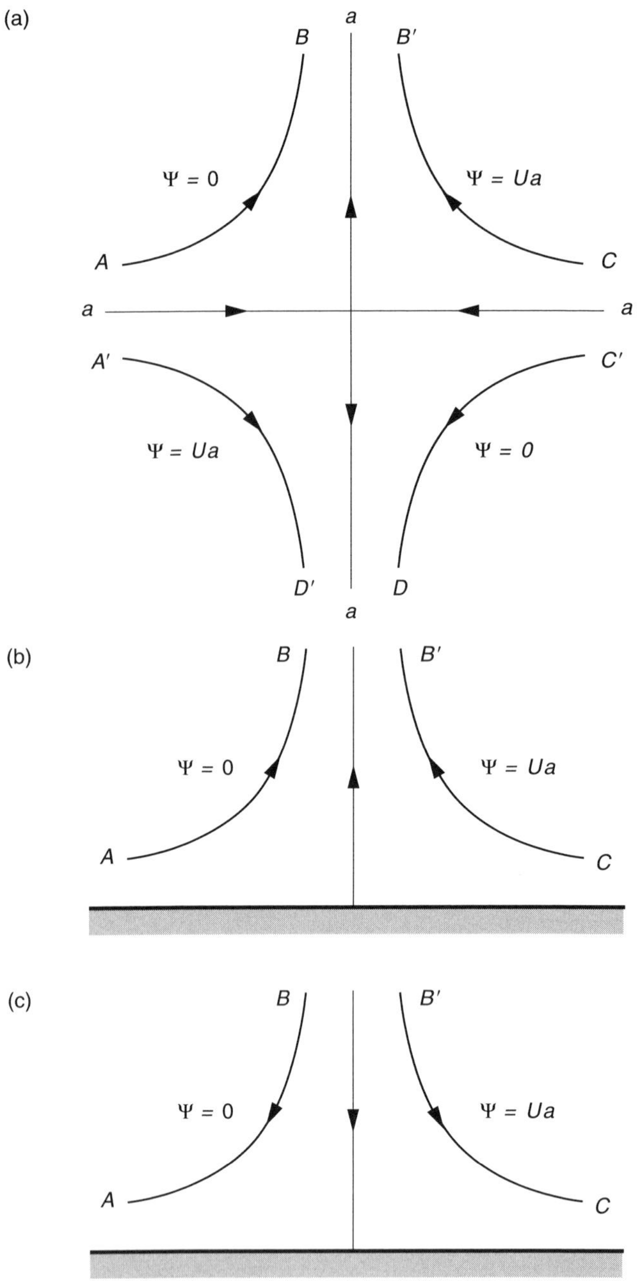

FIGURE 38.11
The simplest example of the third kind of hodograph problem (Figure 38.10) involving the Schwartz theorem in the hodograph plane is the "head-on collision" of two identical jets (a). Since the streamlines are symmetric relative to both the real and imaginary axes, the problem is the same as: (i) two opposite jets along a wall colliding head on and deflected orthogonally to the wall (b); (ii) reversing the flow direction, a jet perpendicular to a wall and splitting symmetrically (c). In all cases the streamlines are given by the same function $y(x)$ and $x(y)$, which must be self-inverse (Figure 38.12).

38.9.2 Jet Incident Normal to a Plane Wall

In this case the jet occupies only a part of the half-plane, unlike the flow in a rectangular corner (Section 14.8). Using as hodograph variable (38.169a):

$$\zeta \equiv \frac{\mathrm{d}f}{\mathrm{d}z} = U\mathrm{e}^{-i\theta}: \quad \Psi\left(\mathrm{on}\,|\zeta| = U\right) = \begin{cases} 0a & \text{on} \quad AB \quad \text{or} \quad 0 < \theta < \dfrac{\pi}{2}, \\ Ua & \text{on} \quad B'C \quad \text{or} \quad \dfrac{\pi}{2} < \theta < \pi, \\ 0a & \text{on} \quad C'D \quad \text{or} \quad \pi < \theta < \dfrac{3\pi}{2}, \\ Ua & \text{on} \quad D'A' \quad \text{or} \quad \dfrac{3\pi}{2} < \theta < 2\pi, \end{cases} \tag{38.169a–e}$$

the free streamlines in the hodograph plane lie on a circle of radius U, and the stream function (38.163a–d) is given by (38.169b–e). Using Schwartz's theorem (37.99b) with the scalar potential $\Phi = f$ replaced by $f = i\Psi$ the stream function, the complex potential is specified by

$$f(\zeta) = \frac{iUa}{2\pi}\left(\int_{\pi/2}^{\pi} + \int_{3\pi/2}^{2\pi}\right)\frac{U\mathrm{e}^{i\theta} + \zeta}{U\mathrm{e}^{i\theta} - \zeta}\,\mathrm{d}\theta; \tag{38.170}$$

the integrand is evaluated by

$$\int \frac{U\mathrm{e}^{i\theta} + \zeta}{U\mathrm{e}^{i\theta} - \zeta}\,\mathrm{d}\theta = \int\left(-1 + \frac{2U\mathrm{e}^{i\theta}}{U\mathrm{e}^{i\theta} - \zeta}\right)\mathrm{d}\theta = -\theta - 2i\log\left(U\mathrm{e}^{i\theta} - \zeta\right). \tag{38.171}$$

Substitution of (38.171) in (38.170) yields:

$$f(\zeta) = -\frac{iUa}{2} + \frac{Ua}{\pi}\log\left(\frac{\zeta^2 - U^2}{\zeta^2 + U^2}\right), \quad \zeta^2 = -U^2\coth\left(\frac{\pi f}{2Ua} + \frac{i\pi}{4}\right); \tag{38.172a,b}$$

the substitution of (38.171) in (38.170) is valid to within an arbitrary constant:

$$\begin{aligned} \mathrm{const} + f(\zeta) &= \frac{iUa}{2\pi}\left(-\pi + \frac{\pi}{2} - 2\pi + \frac{3\pi}{2}\right) + \frac{Ua}{\pi}\log\left[\frac{(-U-\zeta)(U-\zeta)}{(iU \quad \zeta)(\quad iU - \zeta)}\right] \\ &= -\frac{iUa}{2} + \frac{Ua}{\pi}\log\left(\frac{\zeta^2 - U^2}{\zeta^2 + U^2}\right); \end{aligned} \tag{38.173}$$

the constant in the complex potential arises from the initial angle in the stream function (38.169b–e), and is of no consequence in the sequel. The passage from (38.172a) to (38.172b) involves:

$$g \equiv \frac{\pi f}{2Ua} + \frac{i\pi}{4}: \quad \frac{\zeta^2 - U^2}{\zeta^2 + U^2} = \exp(2g); \tag{38.174a,b}$$

solving (38.174b) for ζ^2 leads to

$$\frac{\zeta^2}{U^2} = \frac{1 + \mathrm{e}^{2g}}{1 - \mathrm{e}^{2g}} = \frac{\mathrm{e}^{-g} + \mathrm{e}^{g}}{\mathrm{e}^{-g} - \mathrm{e}^{g}} = -\coth g; \tag{38.174c}$$

substitution of (38.174a) in (38.174c) yields (38.172b).

38.9.3 Shape of Free Streamline

The displacement along the streamlines is given by

$$\frac{\mathrm{d}z}{\mathrm{d}\zeta} = \left(\frac{\mathrm{d}\zeta}{\mathrm{d}z}\right)^{-1} = \left(\frac{\mathrm{d}\zeta}{\mathrm{d}f}\right)^{-1}\left(\frac{\mathrm{d}f}{\mathrm{d}z}\right)^{-1} = \frac{1}{\zeta}\frac{\mathrm{d}f}{\mathrm{d}\zeta}, \tag{38.175}$$

where the two factors are specified by (38.169a) and (38.172a) ≡ (38.176):

$$\frac{\mathrm{d}f}{\mathrm{d}\zeta} = \frac{Ua}{\pi}2\zeta\left[\left(\zeta^2 - U^2\right)^{-1} - \left(\zeta^2 + U^2\right)^{-1}\right]. \tag{38.176}$$

Substitution of (38.176) in (38.175) gives (38.177a):

$$\frac{\pi}{a}\mathrm{d}z = \left\{(\zeta - U)^{-1} - (\zeta + U)^{-1} - i\left[(\zeta + iU)^{-1} - (\zeta - iU)^{-1}\right]\right\}\mathrm{d}\zeta, \tag{38.177a}$$

$$\frac{\pi}{a}z = \log\left(\frac{\zeta - U}{\zeta + U}\right) + i\log\left(\frac{\zeta - iU}{\zeta + iU}\right), \tag{38.177b}$$

whose integral is (38.177b). Substituting the velocity along the free streamlines (38.169a) in (38.177b) leads to:

$$\begin{aligned}\frac{\pi}{a}(x + iy) &= \log\left(\frac{\mathrm{e}^{-i\theta} - 1}{\mathrm{e}^{-i\theta} + 1}\right) + i\log\left(\frac{\mathrm{e}^{-i\theta} - i}{\mathrm{e}^{-i\theta} + i}\right)\\ &= \log\left(\frac{\mathrm{e}^{-i\theta/2} - \mathrm{e}^{i\theta/2}}{\mathrm{e}^{-i\theta/2} + \mathrm{e}^{i\theta/2}}\right) + i\log\left(\frac{\mathrm{e}^{-i\theta/2 - i\pi/4} - \mathrm{e}^{i\theta/2 + i\pi/4}}{\mathrm{e}^{-i\theta/2 - i\pi/4} + \mathrm{e}^{i\theta/2 + i\pi/4}}\right)\\ &= \log\left[-i\tan\left(\frac{\theta}{2}\right)\right] + i\log\left[-i\tan\left(\frac{\theta}{2} + \frac{\pi}{4}\right)\right]\\ &= \log\left[-\tan\left(\frac{\theta}{2}\right)\right] + \frac{i\pi}{2} + i\log\left[\tan\left(\frac{\theta}{2} + \frac{\pi}{4}\right)\right] + \frac{\pi}{2};\end{aligned} \tag{38.178}$$

this specifies the parametric (38.179a) equations of the free streamline (38.179a,b):

$$t \equiv -\tan\left(\frac{\theta}{2}\right): \quad \pi\left(\frac{x}{a} - \frac{1}{2}\right) = \log\left[-\tan\left(\frac{\theta}{2}\right)\right] = \log t, \tag{38.179a,b}$$

$$\pi\left(\frac{y}{a} - \frac{1}{2}\right) = \log\left[\tan\left(\frac{\theta}{2} + \frac{\pi}{4}\right)\right] = \log\left(\frac{1 - t}{1 + t}\right). \tag{38.179c}$$

In (38.179c) was used a particular case of:

$$\tan(\alpha + \beta) = \frac{\tan\alpha + \tan\beta}{1 - \tan\alpha\tan\beta}, \tag{38.180a}$$

that follows from the ratio of (3.25b) by (3.25a); the particular case:

$$\tan\left(\frac{\theta}{2} + \frac{\pi}{4}\right) = \frac{\tan(\theta/2) + 1}{1 - \tan(\theta/2)} = \frac{1 - t}{1 + t}, \tag{38.180b}$$

performs the passage in (38.179c). The parameter can be eliminated from (38.179c) ≡ (38.181c) using the auxiliary variables (38.181a,b):

$$t = \mathrm{e}^X, X \equiv \pi\left(\frac{x}{a} - \frac{1}{2}\right): \quad \pi\left(\frac{y}{a} - \frac{1}{2}\right) = \log\left(\frac{1 - \mathrm{e}^X}{1 + \mathrm{e}^X}\right) = \log\left[-\tanh\left(\frac{X}{2}\right)\right] \equiv F(X); \tag{38.181a–c}$$

this specifies the equation for the free surface:

$$y = \frac{a}{2} + \frac{a}{\pi} \log \left\{ \tanh \left[\frac{\pi}{2} \left(\frac{1}{2} - \frac{x}{a} \right) \right] \right\} \equiv G(x), \tag{38.182}$$

in explicit form.

38.9.4 Force on the Wall

The free surface of a jet normal to a wall is specified by (38.182), corresponding to the free streamline AB in Figure 38.11a implying $0 < \theta < \pi/2$, so that $0 < \theta/2 < \pi/4$ and $\sin(\theta/2) < \cos(\theta/2)$ and thus $t > -1$ in (38.179a), implying that (38.179b,c) are real. *The free streamline (38.182) has asymptotes: (i) in the horizontal direction (38.183a); (ii) in the vertical direction (38.183b):*

$$\lim_{x \to -\infty} y = \frac{a}{2}, \quad \lim_{y \to -\infty} x = \frac{a}{2}; \quad Q = \rho a U, \quad F = QU = \rho a U^2, \tag{38.183a–d}$$

these correspond to the mass flux (38.183c) and to a normal force in the wall equal to the flux of momentum (38.183d). The equation of the free streamline (38.182) can be solved for x as (38.184):

$$x = \frac{a}{2} + \frac{a}{\pi} \log \left\{ \tanh \left[\frac{\pi}{2} \left(\frac{1}{2} - \frac{y}{a} \right) \right] \right\} \equiv G(y), \tag{38.184}$$

implying that x and y are interchangeable, that is, the function G is the same in (38.182) and (38.184), because in Figure 38.11a, the wall be placed alternatively: (i) along the horizontal line $A - C$ as in Figure 38.11b; (ii) along the vertical line $B - D$. Using symmetric variables (38.181b) and (38.185a) it follows that

$$Y \equiv \pi \left(\frac{y}{a} - \frac{\pi}{2} \right): \quad Y = \log \left[-\tanh \left(\frac{X}{2} \right) \right] \equiv F(X), \quad X = F(Y) = F(F(X)), \tag{38.185a–c}$$

the function (38.185b) equals its inverse, and applied twice leads to an identity (38.185c). This is shown in more symmetric from by (38.181c; 38.185b) ≡ (38.186a,b):

$$e^Y = \frac{1 - e^X}{1 + e^X} \Leftrightarrow e^X = \frac{1 - e^Y}{1 + e^Y}: \tag{38.186a, b}$$

this corresponds (38.187a,b):

$$u \equiv e^X = \exp\left(\frac{\pi x}{a} - \frac{\pi}{2} \right), \quad v \equiv e^Y = \exp\left(\frac{\pi y}{a} - \frac{\pi}{2} \right), \tag{38.187a,b}$$

to a bilinear transformation that equals its inverse by (38.188a,b):

$$u = \frac{1 - v}{1 + v} \equiv h(v), \quad v = \frac{1 - u}{1 + u} \equiv h(u). \tag{38.188a,b}$$

The self-inverse transformation (38.188a,b) is a particular case of the general bilinear self-inverse transformation (35.26b) ≡ (38.189a) with parameters (38.189b):

$$u = \frac{av + b}{cv - a} = \frac{1 - v}{1 + v}: \quad \{a, b, c\} = \{-1, 1, 1\}; \tag{38.189a,b}$$

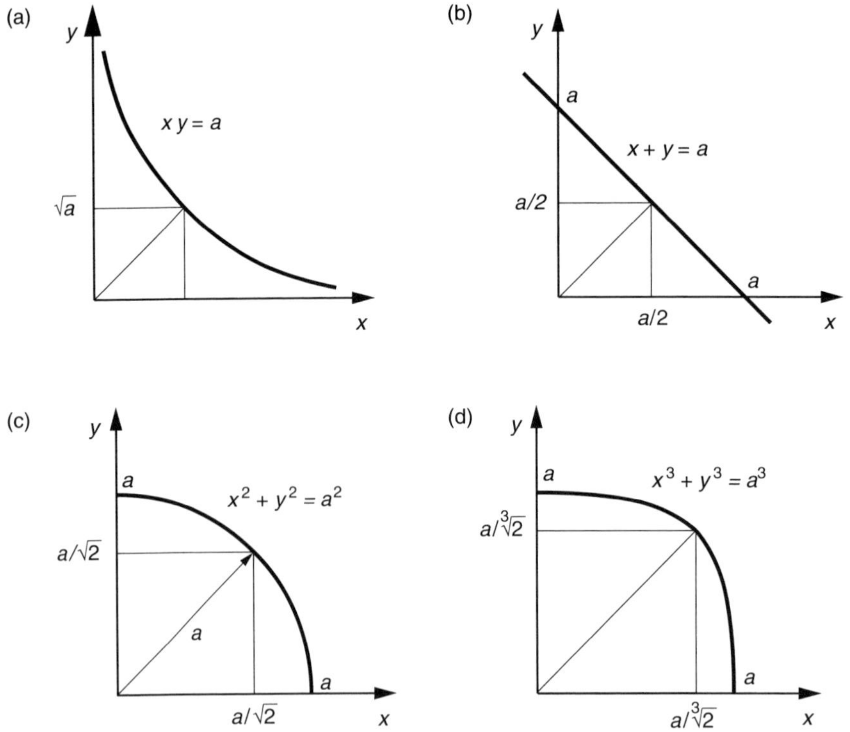

FIGURE 38.12
The stream function for the head-on collision of identical jets (Figure 38.11a) is symmetric in the Cartesian variables $\Psi(x,y) = \Psi(y,x)$, like all the functions in Figure 38.12a–d. Solving for x or y leads to the same function $x(y)$ or $y(x)$. Applying the function twice leads to an identity $x(y(x)) = x$ proving that the function is self-inverse, that is, coincides with its inverse. The self-inverse functions can be obtained solving any symmetric function of two variables, for example: (i) the stream function of the potential flow in a corner (Figure 14.10c); (ii) the stream function of identical jets colliding head on (Figure 38.11a); (iii) the equilateral hyperbola with asymptotes along the coordinate axis (a); (iv) a straight line orthogonal to the diagonal of odd quadrants (b); (v) a circle with center at the origin (c); (vi) a cubic with center at the origin (d). The latter (iii-vi) are some of the simplest examples of self-inverse functions; the streamlines of the head-on collision of identical jets (Figure 38.11a–c) are self-inverse bilinear functions with exponential variables.

the same parameters (38.189b) used in (35.32a) $\equiv$ (38.190):

$$2g(X) = \log\left[\frac{a+b+(a-b)\tanh X}{c-a+(c+a)\tanh X}\right] = \log(-\tanh X) \equiv F(X), \tag{38.190}$$

lead to the function (38.190) $\equiv$ (38.181c) specifying the shape of the free streamline.

Note 38.1. *Self-Inverse (Symmetric) Functions in One (Two) Variable(s):* Any function symmetric in two variables (38.191a) will give a self-inverse function when solved for either of them (38.191b,c):

$$\Psi(x,y) = \Psi(y,x): \quad y = f(x) \Leftrightarrow x = f(y). \tag{38.191a–c}$$

The symmetric functions of the two variables (38.191a) ≡ (38.192a) include functions of the form (38.192b–e):

$$\Psi(x,y) = \Psi(y,x) = F(xy), F(x+y), F\left(r \equiv \sqrt{x^2+y^2}\right), F(x^n+y^n), \qquad (38.192\text{a–e})$$

Simple examples of self-inverse functions include: (i) the (Figure 40.12a) equilateral parabola (38.193a,e); (ii) the (Figure 40.12b) straight line of slope −1 in (38.193b,f); (iii) the (Figure 40.12c) circle of radius a in (38.193c,g); (iv) the (Figure 40.12d) higher-order curve (38.193d,h):

$$0 = \Psi(x,y) = xy - a, \quad x+y-a, \quad x^2+y^2-a^2, \quad x^n+y^n-a^n: \qquad (38.193\text{a–d})$$

$$y = \frac{a}{x}, \ a-x, \sqrt{a^2-x^2}, \ \sqrt[n]{a^n-x^n}. \qquad (38.193\text{e–h})$$

All of these curves are unchanged interchanging the (x,y) axis. Along the diagonal the curves (38.193h) have a distance from the origin (38.194b) increasing with n in (38.194a):

$$n = 1,2,3,4,\ldots,\infty: \ \frac{x}{a} = \frac{y}{a} = \frac{1}{\sqrt[n]{2}} = 0.500, 0, 707, 0.794, 0.841, \ldots, 1. \qquad (38.194\text{a,b})$$

The self-inverse bilinear transformation (35.26b) ≡ (38.195a):

$$y = \frac{ax+b}{cx-a}, \quad 0 = cxy - a(x+y) - b \equiv \Psi(x,y) \equiv \Psi(y,x), \qquad (38.195\text{a–b})$$

is associated with the symmetric function of two variables (38.195b); the latter is a linear combination of (38.192a,b). A self-inverse linear transformation must arise from a linear function of two variables of the form (38.192b), namely (38.196a):

$$0 = \Psi(x+y) = a(x+y) + b; \quad y = q - x, \quad q = -\frac{b}{a}, \qquad (38.196\text{a–c})$$

this leads to (38.196b,c) in agreement with (35.28b) ≡ (38.196b). The case (38.182) ≡ (38.184) is less immediate example of self-inverse function.

Note 38.2. *Operators on the Composition of Self-Inverse Functions:* It has been shown that the composition of the two self-inverse bilinear (35.27a,b) [linear (35.28a–d)] transformations is a bilinear (linear) transformation that is generally not self-inverse:

$$f \circ f = 1 = g \circ g \not\Rightarrow (f \circ g) \circ (f \circ g) = 1. \qquad (38.197)$$

However when the implication (38.197) holds, the composition is commutative: *if the composition of two self-inverse functions (38.198a,b) is self-inverse (38.198c), then it is also commutative (35.198d):*

$$f \circ f = 1 = g \circ g \ \wedge \ (f \circ g) \circ (f \circ g) = 1 \quad \Rightarrow \quad f \circ g = g \circ f. \qquad (38.198\text{a–d})$$

The proof is immediate using the associative property of the composition:

$$g \circ f = (g \circ f) \circ [(f \circ g) \circ (f \circ g)] = [g \circ (f \circ f) \circ g] \circ (f \circ g) = (g \circ g) \circ (f \circ g) = f \circ g. \quad (38.199)$$

QED. A consequence is: *if the ordered pair* (X,o), *where o is the composition and* X *a set of self-inverse functions is a group, then it is also a commutative group.* The properties of self-inverse functions were considered (Subsection 35.5.2) in the context of bilinear mappings, leading to the application of operator power series to self-inverse functions (Note 35.3). The latter subject is reconsidered and developed further next.

Consider an even (38.200a) [odd (38.201a)] analytic operator, specified by a series of even (38.200b) [odd (38.201b)] powers:

$$E(x) = E(-x); \quad E(x) = \sum_{n=0}^{\infty} c_n x^{2n}: \quad E(f) = E(1), \tag{38.200a–c}$$

$$D(x) = -D(-x); \quad D(x) = \sum_{n=0}^{\infty} d_n x^{2n+1}: \quad D(f) = fD(1); \tag{38.201a–c}$$

when applied to a self-inverse function (38.202a) it leads to a constant (38.200a) ≡ (38.202b) [times the function (38.201c) ≡ (38.202e)]:

$$f \circ f = 1: \quad E(f) = \sum_{n=0}^{\infty} c_n (f \circ f)^n = \sum_{n=0}^{\infty} c_n = E(1), \tag{38.202a,b}$$

$$D(f) = \sum_{n=0}^{\infty} d_n f^{2n+1} = f \circ \sum_{n=0}^{\infty} d_n (f \circ f)^n = f \circ \sum_{n=1}^{\infty} d_n = fD(1). \tag{38.202c}$$

An even (38.200a,b) [odd (38.201a,b)] analytic operator applied to a self-inverse function (38.202a) equals its value with argument unity (38.202c) [multiplied by the function (38.302c)]. The examples involving hyperbolic functions include

$$\cosh f = \cosh 1, \quad \sinh f = f \sinh 1, \quad e^f = \cosh 1 + f \sinh 1, \tag{38.203a–c}$$

where: (i/ii) the analytic operator is even (38.203a) ≡ (38.200a–c) [odd (38.203b) ≡ (38.201a–c); (iii) the sum of (38.203a,b) coincides with (38.203c) ≡ (38.207a–c). The latter is proved next.

Any function can be split (38.204a) into the sum of an even (38.204b) and odd (38.204c) function:

$$F(x) = E(x) + D(x); \quad 2\{E(x), D(x)\} = F(x) \pm F(-x) = 2\{E(-x), -D(-x)\}. \tag{38.204a–c}$$

Considering an operator specified by an analytic function (38.206a), applying it to a self-inverse function (38.202a) and using (38.200c; 38.201c) yields:

$$F(f) = E(f) + D(f) = E(1) + fD(1) = \frac{[(1+f)F(1) + (1-f)F(-1)]}{2}; \tag{38.205}$$

an alternative proof of the same result for an analytic operator (38.206a) applied to a self-inverse function (38.202a) is (38.206b):

$$\begin{aligned} F(r) \in \mathcal{A}: \quad F(f) &= \sum_{n=0}^{\infty} a_n f^n = \sum_{n=0}^{\infty} a_{2n} f^{2n} + f \sum_{n=0}^{\infty} a_{2n+1} f^{2n} \\ &= \sum_{n=0}^{\infty} a_{2n} + f \sum_{n=0}^{\infty} a_{2n+1} = \frac{[F(1) + F(-1)]}{2} + \frac{f[F(1) - F(-1)]}{2}. \end{aligned} \tag{38.206a,b}$$

LIST 38.1
Twelve Free Jet and Boundary Problems

Configuration	Problem	Section	Figure
Jet issuing from reservoir	Slit on wall	38.3.1–38.3.3	38.3a,b
	Reentrant tube	38.3.1–38.3.4	38.3b,c
Jet impinging on an obstacle	Orthogonal to plate	38.4	38.4a–e
	Oblique to plate	38.5	38.6
	Aligned with arrow	38.6	38.7b
	Orthogonal to wall	38.9.2–38.9.3	38.12b
Jets along walls	Coanda effect	38.7	38.8a, b
	Fluidic deflection	38.8	38.9a–c
	Vertical barrier	38.4	38.5
	Oblique barrier	38.6	38.7a
Unbounded jets	Internal source	Example 40.16	40.10a–c
	Merging/colliding	38.9.1	38.10a,b, 38.11, 38.12a

Note: All twelve problems involve at least one free boundary along which the pressure and modulus of the velocity are constant, but the direction can change leading to different shapes. In the case of unbounded jets all boundaries are free surfaces (Figures 38.10a–c and 38.11a). For jets issuing from a reservoir the free surfaces start at the edges of the exit (Figures 38.2 and 38.3). For jets impinging on obstacles (Figures 38.4–38.7) or in the presence of walls (Figures 38.8, 38.9, and 38.11b,c) there is a mixture of rigid and free boundaries. The region occupied by the jet may be easier to determine and have a simpler shape in the hodograph plane, whose coordinates are the modulus and direction of the velocity. If the potential flow problem can be solved in the hodograph plane, mapping back to the physical plane describes the flow, including the shape of the free streamlines which separate the jet from the external fluid at constant pressure.

Thus *an analytic operator (38.207a) ≡ (38.206a) of a self-inverse function (38.207b) ≡ (38.202a), is a linear function (38.207c) ≡ (38.206b) ≡ (38.205):*

$$F \in \mathcal{A}; \quad f \circ f = 1: \quad 2F(f) = F(1) + F(-1) + f[F(1) - F(-1)] \\ = (1+f)F(1) + (1-f)F(-1), \qquad (38.207\text{a–c})$$

with coefficients specified by the values at ±1 of the operator.

Note 38.3. *Physical Interpretation of the Hodograph Method:* The hodograph method has been applied to a variety of problems indicated in the List 38.1; this raises the question of what meaning should be given to the "vacuous region" not occupied by the jet, where the pressure is zero. Actually the pressure could be constant and nonzero that is consistent with a constant velocity along streamline; all that matters to calculate forces is that there is a pressure difference rather than the absolute value of the pressure. The region of space not occupied by the jet, for example, behind the flat plate allows the existence of drag without: (i) viscosity (form drag) or lift (lift induced drag); (iii) a source in the Blasius theorem. The regions of recirculating flow behind an obstacle or the wake of a body, may have nearly constant pressure; the boundary of the jet in the hodograph method could be interpreted as the boundary of vortical or turbulent regions, where the potential flow does not apply.

Note 38.4. *Analogies among the Twelve Potential Fields:* Six basic potential fields (Classification 38.1) have been considered so far: (i/ii) the potential flow due to source/sinks (vortices) treated together (Chapters 12, 14, 16, 28, 34, 36, and 38); (iii) the gravity field due to mass distributions (Chapter 18); (iv/v) the electrostatic (magnetostatic) field due to electric charges (Chapter 24) [currents (Chapter 26)] that have similarities and interchanges (also Chapter 36); (vi) steady heat conduction with sources and sinks (Chapter 32). Six additional potential fields are; (vii) potential flow in a rotating vessel; (viii) incompressible viscous flow in a pipe; (ix) torsion of a beam; (x) deflection of a membrane; (xi–xii) surface tension and capillarity. Of the six basic analogies one pair (i/ii) is chosen as the baseline for more thorough discussion, and advantage is taken of this background to present more concisely the remaining four (iii) to (vi) [or ten (ii) to (xiii)] including: (a) the fundamental equations, boundary conditions, energy and forces; (b) the main analogies and differences, as well as specific aspects; (c) a number of typical examples illustrating the preceding points (a,b).

The potential flow was chosen as the baseline because it is the most general in most respects, viz.: (iv) the electrostatic field due to positive (negative) electric charges is similar to sources (sinks), but has no analogue for vortices; (v) the magnetostatic field due to electric currents is analogous to the potential flow due to vortices but has no analogue for sources or sinks; (iii) the gravity field due to masses is always attractive, and thus corresponds to sinks in a potential flow, but has no analogue for sources and vortices; (vi) steady heat conduction has heat sources and sinks, but no analogue for vortices. Concerning the additional six potential fields: (vii) the potential flow in a uniformly rotating vessel has closer analogy to the torsion of a beam; (viii) also the steady viscous flow of an incompressible fluid in a pipe. (viii–x) the remaining four are two special cases viz. torsion (deflection) of beams (ix) [membranes (x)] and two analogues viz. surface tension (xi), and capillarity (xii).

Thus the potential flow (i/ii) is the simplest potential field with most properties of all the others, but still not all, for example: (a) the velocity in a potential flow is tangent to a body, corresponding in electrostatics (magnetostatics) to the tangential electric (magnetic) field at an insulator (conductor); (b) the normal electric (magnetic) field to a conductor (insulator) would correspond to a normal velocity, and does not apply to the potential flow past a rigid, impermeable body at rest; (c) the induced electric charges in a conductor (the insulating properties) associated with the normal electric (magnetic) field, have no direct analogue in a potential flow except for (i) moving (2), porous (3), pulsating, or (4) deformable bodies. These three cases concern: (1) moving bodies, for that the flow velocity must match the normal velocity at the surface; (2) porous bodies with fluid blowing or suction at the surface specifying the normal velocity; (3) pulsating bodies, with a specified motion at surface that generate waves; (4) deformable bodies for that the fluid motion couples to surface deflections, for example, aeroelasticity.

Most of these analogies are discussed but not all, for example, the fluid pressure is considered in detail but neither the viscous stresses nor the electric and magnetic pressure, nor the electromagnetic stresses. The stresses and strains are most relevant in connection with: (ix) the torsion of beams; (x) the deflection of membranes; (xi–xii) surface tension and capillarity are examples of deformations. The latter (ix–xii) are particular cases of elasticity. Plane elasticity (xiii) leads to a biharmonic equation for the stress function; the biharmonic equation is also satisfied by (xiv) the stream function for a plane, viscous, steady, incompressible, creeping flow. The additional six (vii–xii) harmonic [two (xiii–xiv) biharmonic] fields can be addressed by extensions of the present methods, applied here to the six basic harmonic fields (i–vi). The potential flow combines the rotational (solenoidal) singularities, viz. vortices (source/sinks) that is also the case for torques (forces) in plane elasticity. The angular velocity in a rotating flow (vii) has some properties analogous to the vorticity. The potential flow has in addition problems with free surfaces, such as jets (Chapter 38); these have analogues in solids, for example, a melting front advancing in a metal transition from solid to fluid. Overall the choice of the potential flow as baseline is justified by: (i) providing a considerable variety of problems and analogies; (ii) allowing the coverage of the other twelve harmonic (two biharmonic) fields using similar methods.

Note 38.5. *Potential Flows and Fields as an Introduction to Electromechanics of Continua:* The potential fields are the simplest representation of much wider classes of phenomena, viz.: (i) the potential flow excludes compressibility, vorticity, viscosity, and turbulence that are all important aspects of fluid mechanics; (ii) the electric and magnetic fields decouple only in the steady case, that is, electrostatics and magnetostatics do not apply to quasi-steady fields in electrical circuits nor the unsteady electromagnetic fields, like electromagnetic waves and optics; (iii) steady heat conduction excludes not only unsteady or time dependent effects, but also convection and radiation of heat, and its couplings to mass and momentum transport, associated with thermodynamics and statistical properties of microscopic matter; (iv) the gravity field excludes the curved space-time geometry of the gravitational field in general relativity and cosmology of the universe at large scales; (v) plane elasticity is only one of the many aspects of the mechanics of solids (e.g., strings, bars, beams, membranes, plates, shells, crystals) with different constitutive properties (e.g., elasticity, plasticity, viscoelasticity) and phenomena (stresses, strains, cracks, dislocations). In addition there are interactive phenomena like thermal stresses, piezoelectricity, magnetofluidmechanics, etc. In spite of their simplicity potential fields do include some of the basic properties of fluids and solids, gravity and electromagnetic fields, and mass, momentum and energy transfer in matter. The potential fields are thus a suitable introduction to most branches of physics; they also serve as an illustration of the properties of functions of a complex variable that are the basis for subsequent developments of analysis, for example, differential equations, integral transforms and numerical methods. These in turn enable to development to physical models overcoming the limitations of potential fields. The latter remain useful as the standard and as the background case for comparison. Thus the potential fields are a suitable introduction to mathematical modeling of physical phenomena and engineering processes, but they are also no more than an introduction to much wider classes of phenomena that lie beyond.

Conclusion 38: The electric field due to a constant potential (Figure 38.1) on: (a) a strip of the plane; (b) a half-cylinder. Contraction ratio and flow rate for a fluid exiting a large reservoir (Figure 38.2) through: (a) a slit on the wall (Figure 38.3a); (b) a re-entrant tube (Figure 38.3c); (c) an outlet nozzle. Representation (Figure 38.3) in same hodograph plane (b) with different variables of two free jets issuing from a reservoir, either (a) through a hole or (c) in a long tube, both of width b carrying a volume flux Q. The problem converse to (Figure 38.3a) a jet issuing from a hole in a reservoir, is a jet incident (Figure 38.4) on a flat plate (a), which leads through a slightly different change of variable to a hodograph representation (b) similar to Figure 38.3b. The flow is quite different, viz. instead of a sink of volume flux $-2Q$ in (Figure 38.3b), the half-strip (Figure 38.4b) is mapped into a dipole (Figure 38.4c) in the η-plane, then inverted for an uniform flow (Figure 38.4d) in the $1/\eta$ plane, and finally mapped to a flow past a semiinfinite plate (Figure 38.4e) in the f-plane. The flow normal to a finite plate (Figure 38.4a) corresponds by the reflection principle to a vertical barrier on a wall (Figure 38.5). The flow oblique to a plate (Figure 38.6) does not correspond to an oblique barrier on a wall (Figure 38.7a), because the latter is reflected as an arrow (Figure 38.7b). The flow oblique to a plate also specifies the thrust on a surfboard riding on the surface of stream (Figure 38.6). The flow incident on an oblique barrier on a wall corresponds to a bent lamina or arrow (Figure 38.7b), and can come from the front U_+ or behind U_-. The hodograph method applies to jets in partial contact with walls, e.g.: (Figure 38.8) the Coanda effect (a) of jet attachment to a wall around a semiinfinite plate (b); (Figure 38.9) the fluidic effect (a) of a source near the edge of a semiinfinite, causing a stagnation point (b) and deflecting the jet (c). The hodograph method also applies to the collision of jets without boundaries, at any angle (a), head-on (b) or symmetric (c); the simplest case (Figure 38.11) of square collision (a) corresponds to: (b) two opposite jets along a wall colliding head-on; (c) a jet normal to a wall. The free streamline in this case is specified by a self-inverse function, unchanged by interchange of the two Cartesian coordinates. Some examples of self-inverse functions, (Figure 38.12) unchanged by the interchange of the (x, y)-axis: (a) equilateral parabola; (b) straight line of slope-1; (c) circle; (d) higher-order curves.

39

Essential Singularities, Roots, and Periods

The analytic continuation (Chapter 31) allows the study of a complex function over the whole of the complex plane, including the neighborhood of singularities (Section 39.1); the latter include poles (essential singularities) where the function is infinite (indeterminate), that is, exceeds any bound (Section 39.1) [(comes arbitrarily close to any value, however, small or large (Section 39.2)]. This result can be extended to the Picard second theorem (Section 39.3) stating that an integral function takes all values an infinite number of times, possibly excluding one exceptional value; the exponential gives the example of an exceptional value, viz. it is never zero, so $\exp(z) = 0$ has no roots and $\exp(z) = c$ has an infinite number of roots for all $c \neq 0 \equiv b$. This demonstrates the existence of an infinite number of roots for any transcendental complex equation $f(z) = c$, where $f(z)$ is an analytic nonrational function, for all values of c, excluding possibly one exceptional value $b \neq 0$, viz. $b = 0$ for the exponential $f(z) \equiv \exp z$. The first Picard theorem states that an analytic function that has two exceptional values is a constant (Section 39.3); this is a considerable extension of the Liouville theorem (Section 27.6) since it shows that an analytic function reduces to a constant if it fails to take any two distinct values, that is, $f(z) \neq a$, b with $a \neq b$ implies $f(z) = \text{const}$. The two theorems of Picard can be proved via the elliptic modular function of Legendre that relates to coverings of the complex plane by Schwartz triangles (Section 39.4). The first Picard theorem states that a function that does not take two values is a constant, and holds not only over the whole complex plane but also over parts of it, viz.: (i) over a dense ray, that is, an infinitesimal angular sector where, by the Julia theorem (Section 39.5), lie an infinity number of roots of a transcendental equation for every nonexceptional value; (ii) over a disk, with the Landau radius (Section 39.6) that is the maximum value of the radius beyond that the function either ceases to holomorphic or takes all values but one an infinite number of times. These theorems apply to nonrational analytic functions, that is, to transcendental functions. An important class of transcendental functions is related to hyperelliptic integrals that are integrals of the inverse square root of a polynomial, with branch-cuts joining its roots. The integrals along these branch-cuts specify the periods (Section 39.7), that is, quantities that may be added to the integral without changing the inverse function. Considering polynomials of the second degree leads to (Section 39.8) the periods of the exponential, circular, and hyperbolic functions. The polynomials of third (fourth) degree lead (Section 39.9) to the elliptic functions of Weierstrass (Jacobi). The assurance given by the rather abstract theorems of Casorati–Weierstrass and Picard that every complex nonalgebraic or transcendental equation has roots, actually underlies the feasibility of the numerical methods of computation of the roots.

39.1 Classification of Special and Singular Points

Classification 39.1 summarizes a classification of points of a complex function. The simplest special points (Subsection 39.1.1) among the regular (special) points are zeros (poles); the

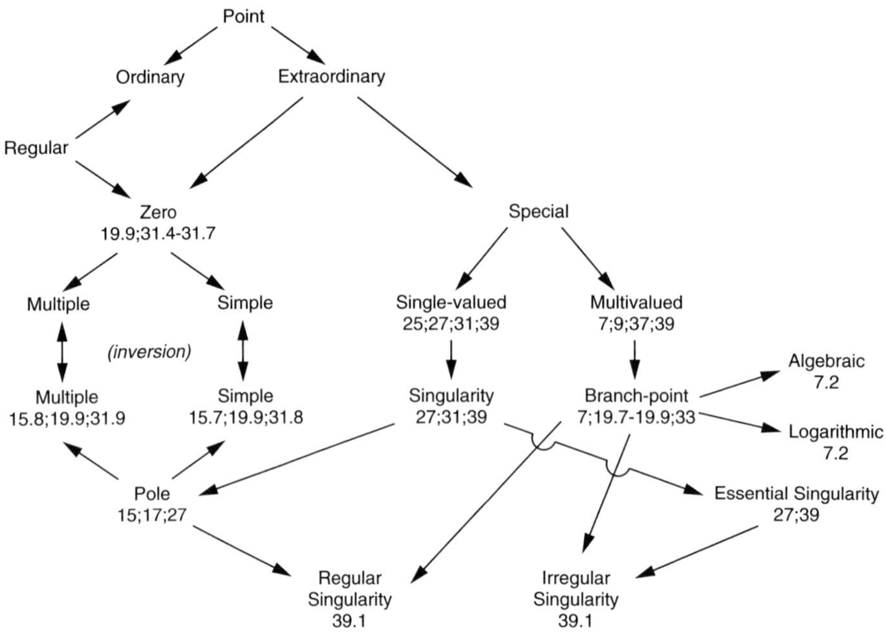

CLASSIFICATION 39.1

Points of a function. The classification of points of an analytic function (Table 27.1) can be included in a wider classification of points of a function (Classification 39.1). For a regular point the function is finite and single-valued in its neighborhood; this includes simple and multiple zeros, and ordinary points (i.e., nonzero regular points). The extraordinary points are zeros if they are regular and are special points if they are not regular. The nonregular or special points can be branch-points (singularities) if the function is multivalued (single-valued). The branch-points of a multivalued function are algebraic (logarithmic) if the number of branches in their neighborhood is finite (infinite). The singularities of a single-valued function are poles (essential singularities) if the principal part of the Laurent series in the neighborhood of the singularity (Table 27.2) is a terminating (nonterminating) series, that is, has a finite (infinite) number of terms. The simple (multiple) poles of a function $f(z)$ are simple (multiple) zeros of the inverse function $1/[f(z)]$ and vice-versa. A branch-point may coincide with a singularity: the coincidence of a branch-point with a pole (essential singularity) leads to a regular (irregular) singularity. For example a Laurent series multiplied by a power with complex exponent leads to a regular (irregular) singularity if the principal part is terminating (nonterminating).

special points include (Subsection 39.1.2) essential singularities and branch-points that may be combined.

39.1.1 Zeros and Poles as Special Points

A **regular point** is a point in whose neighborhood the function is holomorphic and single-valued. This leads two possibilities, viz. a **zero** and an ordinary point; the **ordinary point** is thus a point where the function is holomorphic, single-valued and nonzero. A **extra-ordinary point** is not an ordinary point, and thus is either a zero if the function is holomorphic or an **special point**. The special points can be (i) either a **singularity** if the function is single-valued but not holomorphic; (ii) a **branch-point** if the function is not single-valued. A branch-point of a multivalued or many-valued function (Chapters 7 and 9), is an extraordinary and special point; it is an **algebraic (logarithmic) branch-point** (Subsection 7.2.3) if the function has a finite (infinite) number of branches in its neighborhood, viz. is multivalued (many-valued). The critical points of conformal mappings (Chapters 33 and 35) may be branch-points, for example, powers with nonintegral exponents. A multivalued function need not be singular at a branch-point, for example, $f(z)=z^{\alpha}$ with α noninteger has a branch point at $z=0$ where it vanishes if $\mathrm{Re}(\alpha)>0$; the derivative at a branch-point may (or may not) singular (Section 7.6), for example, $\mathrm{d}f/\mathrm{d}z = z^{\alpha-1} \to \infty$ $(\to 0)$ as $z \to 0$ for $\alpha < 1$ $(\alpha > 1)$. For $f(z)=z^{\alpha}$ with nonintegral $\alpha < 0$ both the function and the derivative would be singular at the branch-point. A special point $z=a$ of a single-valued function is called a **singularity** and is classified (Tables 27.1 and 27.2) from the Laurent series (25.23c) as: (i) a **pole of order n** if the series terminates at the negative power $-n$ in (39.1a,b):

$$n \in |N: \quad f(z) = \sum_{m=-n}^{\infty} A_m (z-a)^m; \quad f_2(z) = \sum_{m=-\infty}^{+\infty} A_m (z-a)^m; \tag{39.1a–c}$$

(ii) **essential singularity** if the series of powers with negative exponents does not terminate (39.1c). There may be relations between different types of points, for example, the function $f(z)$ has a pole of order N at $z=a$ iff $1/f(z)$ has a zero of order n at $z=a$.

39.1.2 Branch-Points, Essential, Regular, and Irregular Singularities

The classification may not be exclusive, for example, and essential singularity (pole of order n) may coincide with a branch-point leading to an **irregular (regular) singularity**; thus the point $z=a$ is a regular (an irregular) singularity of the function defined by the series (39.3a–c) [(39.2a,b)]:

$$\alpha \notin |N: \quad f_3(z) = \sum_{m=-\infty}^{+\infty} A_m (z-a)^{\alpha+m} = (z-a)^{\alpha} \sum_{n=-\infty}^{+\infty} A_m (z-a)^m, \tag{39.2a,b}$$

$$\alpha \notin |N; \quad n \in |N: \quad f_4(z) = \sum_{m=-n}^{+\infty} A_m (z-a)^{\alpha+m} = (z-a)^{\alpha} \sum_{m=-n}^{+\infty} A_m (z-a)^m; \tag{39.3a–c}$$

thus (39.2b) [(39.3c)] is the product of a nonintegral power by a Laurent series at $z=a$ for an essential singularity (pole of order n); in the latter case of a pole of order n, the use of (39.3e) leads (39.3d) to the series (39.3f) near a regular singularity:

$$k=m+n, \ \beta=\alpha-n \notin |N: \quad f_4(z) = \sum_{k=0}^{\infty} A_{k-n} (z-a)^{\beta+k} = (z-a)^{\beta} \sum_{k=0}^{\infty} A_{k-n} (z-a)^k. \tag{39.3d–f}$$

As a further example consider the series of arbitrary powers (39.4a):

$$f_5(z) = \sum_{n=-\infty}^{+\infty} A_n (z-a)^{\alpha_n}; \quad f_6(z) = \sum_{n=-\infty}^{+\infty} A_n (z-a_n)^{\alpha_n}; \tag{39.4a,b}$$

for finite α_n then (39.4b) is a further generalization of (39.4a). The points a_n in (39.4b) are: (i/ii) zeros (poles) for α_n positive (negative) integer; (iii) ordinary points for $\alpha_n = 0$; (iv) branch-points for α_n not an integer. In (39.4a) is included a series of fractions

(Subsection 27.9.2) for α_n a negative integer. The binomial theorem (25.37a–c) can be used to expand each term of (39.4b) in a Maclaurin series (39.5b):

$$|z| < \inf |a_n| \equiv R: \quad (z - a_n)^{\alpha_n} = \sum_{m=0}^{\infty} \binom{\alpha_n}{m} (-a_n)^{\alpha_n - m} z^m, \tag{39.5a,b}$$

that is absolutely convergent within the radius of convergence (39.5a). It is also uniformly convergent in a closed subregion (39.6a,b); thus (39.5b) can be summed term-by-term after substitution in (39.4b) leading to the Maclaurin series (39.6c):

$$|z| \leq R(1-\varepsilon); 0 < \varepsilon < 1: f_6(z) = \sum_{m=0}^{\infty} B_m z^n, B_m \equiv \sum_{n=-\infty}^{+\infty} A_n \binom{\alpha_n}{m} (-a_n)^{n-m}, \tag{39.6a–d}$$

with coefficients (39.6d); the radius of convergence (39.6b) is further limited by the nearest singularity z_* of the function $f_6(z)$ if $|z_*| < R$; in that case $R = |z_*|$ in (39.5a). The convergence of the series (39.6c) can be established from the coefficients (39.6d) that depend on the coefficients of (39.5b). If the double series (39.6c,b) converges it specifies (39.4b) as an analytic function in the disk $|z| < R$ of radius (39.6a,b), and analytic continuation is possible (Sections 31.1 and 31.2) if the boundary of the circle is not densely covered with singularities.

39.2 Zeros, Poles, and Essential Singularities (Casorati, 1868; Weierstrass, 1876; Picard, 1880)

Whereas a function is arbitrarily small (large) near a zero (pole), in the neighborhood of an essential singularity it is both (Subsection 39.2.1), and in fact takes all possible values (Subsection 39.2.2) apart possibly from one exceptional value.

39.2.1 Comparison of Zeros, Poles, and Essential Singularities

Of the extraordinary and special points of a complex function (Section 39.1), all have been discussed in some detail before, except for the essential singularity that was considered only briefly (Sections 27.8 and 27.9). The behavior of a function analytic in the neighborhood of (iv) an essential singularity is considered next, in comparison with an (i) ordinary point, a (ii) a zero and (iii) a pole. The latter give examples of the relevance of the classification of points of a complex functions (Diagram 39.1) to the **behavior of a function** *analytic in the neighborhood of a point* z = a, *viz: (i) for an ordinary point that is not a zero, the function is finite and nonzero,* that is, *has upper* M *and lower* m *bounds in modulus:*

$$\forall_{\varepsilon>0} \quad \exists_{M>m>0} \quad \forall_{z\in V_\varepsilon(a)}: \quad M \geq |f(z)| \geq m; \tag{39.7}$$

(ii) at a zero the function vanishes, and it can be made arbitrarily small by choosing a suitably small neighborhood:

$$\forall_{m>0} \quad \exists_{\varepsilon>0} \quad \forall_{z\in V_\varepsilon(a)}: \quad |f(z)| \leq m; \tag{39.8}$$

(iii) at a pole the function is infinite, that is, it can be made arbitrarily large by choosing a suitable small neighborhood:

$$\forall_{M>0} \quad \exists_{\varepsilon>0} \quad \forall_{z\in V_\varepsilon(a)}: \quad |f(z)| \geq M; \tag{39.9}$$

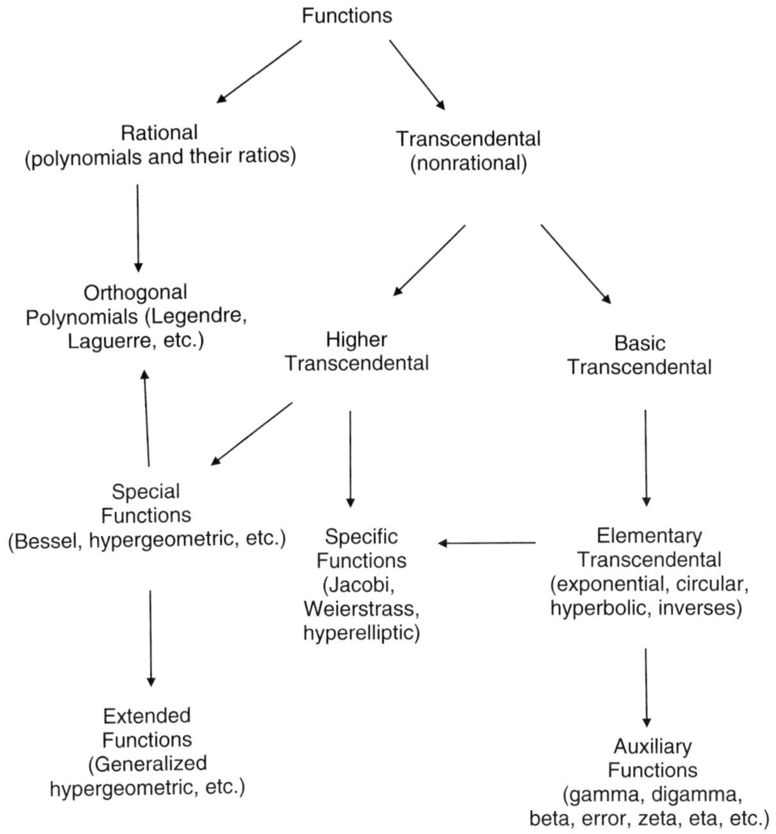

DIAGRAM 39.1
Families of functions. The relations between the families of functions (List 39.1) can be outlined graphically (Diagram 39.1). The functions can be divided into rational (transcendental) which can (cannot) be expressed in finite terms. The rational functions are polynomials and their ratios: they include the orthogonal polynomials which are particular cases of special functions with terminating series. The "other path" toward special functions is through the transcendental functions. The basic transcendental functions start with the simplest: the elementary functions; they provide solutions of linear differential equations with constant coefficients. The elementary functions are used to define and establish properties of the auxiliary functions. The "other branch" of transcendental functions is that of "higher transcendental functions." It includes the specific functions, for example, elliptic which arise in the solution of nonlinear differential equations; the solution of linear differential equations with variable coefficients leads to the special functions, and their generalization, the extended functions.

(iv) for an essential singularity the function is ***unbounded in a dense subset*** *of every neighborhood, that is, an infinite number of points exists where the function exceeds any given bound:*

$$\forall_{\varepsilon>0, M>0} \quad \exists^{\infty}_{z\in V_\varepsilon(a)}: \quad |f(z)| > M. \tag{39.10}$$

Note the distinction between an infinite value (39.9) that implies $|f(z)| > M$ at all points of a neighborhood, and unbounded value (39.10) that implies $|f(z)| > M$ at infinite number of points, but not necessary at all points of the neighborhood.

The behavior indicated in the theorem can be proved as follows: (i) if $z = a$ is an ordinary point that is not a zero (39.11a), then by the mean value theorem or the second-order approximation (39.11b) of the Taylor series (23.32a):

$$f(a) \neq 0: \quad f(z) = f(a) + f'(a)(z-a) + O\left((z-a)^2\right); \tag{39.11a,b}$$

a radius (39.12a) can be found such that hold the upper and lower bounds in modulus:

$$\varepsilon > |z-a| : 0 < m \leq |f(a)| - \varepsilon |f'(a)| \leq |f(z)| \leq |f(a)| + \varepsilon |f'(a)| \leq M, \tag{39.12a,b}$$

where terms of $O(\varepsilon^2)$ were omitted. (ii) If $f(z)$ has (39.13b) a zero of order n at $z = a$, then it is smaller than any bound (39.13b):

$$|g(z)| \leq M: \quad |f(z)| = |z-a|^n |g(z)| \leq B\varepsilon^n \leq m, \quad \varepsilon \leq \sqrt[n]{\frac{m}{B}}, \tag{39.13a–c}$$

where $g(z)$ is analytic and nonzero at $z = a$; thus g(z) has a finite upper bound (39.13a) and this implies that the product (39.13b) can be made arbitrarily small $< m$ by choosing a small ε in (39.13c). (iii) If $f(z)$ has (39.14b) a pole of order N at $z = a$, then it exceeds any bound (39.14b):

$$|g(z)| \geq b: \quad |f(z)| = |z-a|^{-N} |g(z)| \geq b\varepsilon^{-n} \geq M, \quad \varepsilon < \sqrt[n]{\frac{b}{M}}, \tag{39.14a–c}$$

because $g(z)$ is analytic and nonzero; hence $g(z)$ has a nonzero lower bound (39.14a) that divided by ε^N for small ε this can exceed any number M, however large for ε in (39.14c). (iv) In the case of an essential singularity at $z = a$, it is necessary to make sure that the coefficients A_{-n} of the principal part (27.24b) of the Laurent series (27.23b) do not all vanish beyond any order. An upper bound for A_{-n} can be found (25.24b) by integrating along a circle of radius ε center at $z = a$:

$$|A_{-n}| \leq (2\pi)^{-1} \int_{|z-a|=\varepsilon} |f(\zeta)| |\zeta - a|^{n-1} |d\zeta| \leq \varepsilon^n \left| f\left(a + \varepsilon^{i\varphi}\right)\right|_{\max}; \tag{39.15}$$

if $f(\zeta)$ is bounded at all points $|\zeta - a| = \varepsilon$, then $A_{-n} \to 0$ as $\varepsilon \to 0$, and the point $z = a$ is not a singularity. Hence, if $z = a$ is a singularity, then $f(z)$ is not bounded in any neighborhood, and (39.10) must hold at a set z_n of points. If the set is finite, then by choosing $\delta < \inf\{|z - z_n|\}$ the function would be bounded by in $V_\delta(a)$ that contradicts $z = a$ being a singularity. Thus, *if* z = a *is a essential singularity of the function,* f(ζ)*, in every neighborhood it is unbounded at an infinite number of points that form a dense set.*

39.2.2 Casorati (1868)–Weierstrass (1876) and Picard's Second (1880) Theorems

The proof based on (39.14a–c) [(39.15)] shows that a function is unbounded in the neighborhood of any singular point, be it a pole (an essential singularity). From (39.14a–c) it has been proved that a function is unbounded at all points of the neighborhood of a pole; in the case of an essential singularity (39.15) it has only been proved that the set of points where the function is unbounded is dense, and it is shown next that it does not cover the whole neighborhood. To emphasize this difference between poles and essential singularities, consider the function:

$$c \in |C: \quad g(z) \equiv \frac{1}{f(z) - c}, \quad f(z) = c + \frac{1}{g(z)}, \tag{39.16a–c}$$

where c is an arbitrary constant. It will be shown next that: *the point $z=a$ is a pole of order N (essential singularity) of $f(z)$ iff it is a zero of order N (essential singularity) of $g(z)$ defined by (39.16b,c), for all constant c in (36.16a).* To prove the theorem, consider first the case of a pole of order N of $f(z)$ at $z-a$; by definition (39.17a,b) holds:

$$\zeta \equiv z-a, \quad f(\zeta)=O\left(\zeta^{-N}\right): \quad g(z)=\frac{1}{O(\zeta^{-N})-c}=\frac{O\left(\zeta^{N}\right)}{O(1)-O(\zeta^{N})}=O\left((z-a)^{N}\right), \tag{39.17a–c}$$

implying (39.16b) that $\zeta=0$, that is, $z=a$ is (39.17c) a zero of order N of $g(z)$; conversely, if $z=a$ is a zero (39.18a) of order N of $g(z)$, then from (39.16c) it follows that

$$g(z)=O\left(\zeta^{N}\right): \quad f(\zeta)=c+\frac{1}{O(\zeta^{N})}=O(1)+O\left(\zeta^{-N}\right)=O\left((z-a)^{-N}\right), \tag{39.18a,b}$$

and $\zeta=0$, that is, $z=a$, is a pole (39.18b) f order N of $f(z)$. In the case where $z=a$ is an essential singularity of $f(z)$, the reasoning is by exclusion of all other possibilities for $g(z)$, viz. if $z=a$ were: (i) an ordinary point of $g(z)$ that is not a zero, then $f(z)$ would be bounded by (39.16c); (ii) a zero of order N of $g(z)$, then it would be (39.18b) a pole of order N of $f(z)$; (iii) a pole (39.19a) of order N of $g(z)$, then $f(z)$ would have by (39.16c) an ordinary point (39.19b):

$$g(z)=O\left(\zeta^{-N}\right): \quad f(z)=c+\frac{1}{O(\zeta^{-N})}=c+O\left(\zeta^{N}\right)=O(1). \tag{39.19a,b}$$

In any of the three cases (i,ii,iii) the point $z=a$ would not be an essential singularity of $f(z)$; in order that $z=a$ be an essential singularity of $f(z)$, there is only one remaining possibility, viz. that $z=a$ is an essential singularity of $g(z)$, for all values of the constant c in (39.16c). QED.

Bearing in mind that $z=a$ is an essential singularity of $g(z)$, the latter is unbounded (39.20a) by (39.10):

$$M<|g(z)|=\frac{1}{|f(z)-c|}; \quad |f(z)-c|<\frac{1}{M}<\varepsilon, \tag{39.20a,b}$$

thus $f(z)-c$ is arbitrary small (39.20b), that is, $f(z)$ is arbitrary close to any constant c given "a priori"; furthermore, for each c, the number of points z such that $|f(z)-c|<\delta$ is infinite, in every neighborhood of $z=a$. Thus has been proved the **Casorati (1868)–Weierstrass (1876) theorem on essential singularities**: *if $\mathrm{z}=\mathrm{a}$ is an essential singularity of $\mathrm{f}(\mathrm{z})$, then in every neighborhood $\mathrm{V}_{\varepsilon}(\mathrm{a})$ there exists an infinite number of points z_n such that the function comes arbitrary close to any constant value c chosen a priori:*

$$\forall_{\varepsilon,\delta>0,c\in|C}: \quad \exists^{\infty}_{z_n\in V_{\varepsilon}(a)}: \quad |f(z_n)-c|<\delta. \tag{39.21}$$

The Casorati–Weierstrass theorem shows that a function is **indeterminate** in the neighborhood of an essential singularity, but it does not indicate whether it actually takes every value c. If the function took every rational value c but no irrational values then the Casorati–Weierstrass theorem would still be true. This matter is clarified by a more precise statement, namely the **Picard (1880) second theorem**: *if the function $\mathrm{f}(\mathrm{z})$ has an essential singularity, $\mathrm{z}=\mathrm{a}$, then in every neighborhood $\mathrm{V}_{\varepsilon}(\mathrm{a})$ it takes all values c, with one possible exception, at an infinite number points:*

$$\forall_{\varepsilon>0,c\in|C-\{b\}} \quad \exists^{\infty}_{z\in V_{\varepsilon}(a)}: \quad f(z)=c(\neq b); \tag{39.22}$$

the theorem of Picard shows that $f(z)$ actually takes all values c, but leaves the possibility that an **exceptional value** b may exist that the function approaches arbitrarily close but does not equal $\delta \geq |f(z)-b|>0$.

39.3 Exceptional Value (Picard, 1879) and Infinite Number of Roots

Whereas a rational equation has a finite number of roots for every value of the parameters (Subsection 31.6.3) a transcendental equation specified by an integral function that has an essential singularity: (i) may fail to have roots for at most one exceptional value of the parameter that may or may not exist (Subsection 39.3.1); (ii) for every other value it has an infinite number of roots (Subsection 39.3.2).

39.3.1 Possible Existence of an Exceptional Value

An example of exceptional value of an integral function is provided by the exponential that (31.37a,b) never takes the value zero, viz. $e^z = 0$ has no roots anywhere on the complex plane. If follows that

$$g(z) \in \mathcal{A}: \quad \exp\{g(z)\} \neq 0 \equiv b; \tag{39.23a,b}$$

the exponential of an analytic function (39.23a) never takes the exceptional value zero (39.23b), and for all other values the equation (39.23b) has an infinite number of roots (39.23c):

$$\exists^{\infty}_{\zeta \in |C}: \quad \exp(\zeta) = \exp\{g(z)\} = c \neq b = 0. \tag{39.23c}$$

To prove the second statement (39.23c), an inversion (39.24a) is used to place the essential singularity at the origin. The function $\exp(1/z)$ has an isolated essential singularity at the origin, as shown by its Laurent series (27.42c). The equation (39.24b):

$$z = \frac{1}{\zeta}: \quad \exp\left(\frac{1}{z}\right) = c; \quad n \in |Z: \quad \frac{1}{z_n} = \log|c| + i\arg(c) + i2\pi n, \tag{39.24a–d}$$

has roots (39.24d) for all values of the integer (39.24c). Choosing an arbitrarily small neighborhood of the origin (39.25a):

$$|z_n| < \varepsilon: \quad n > \frac{1}{2\pi}\left\{\left|\frac{1}{\varepsilon^2} - (\log|c|)^2\right|^{1/2} - \arg(c)\right\}, \tag{39.25a,b}$$

it contains an infinite number of roots of (39.24b), namely z_n given by (39.24d) with n satisfying (39.25b); the number of roots is infinite as $\varepsilon \to 0$ for all c, except for $c = 0$; in that case the r.h.s. of (39.25b) diverges and no conclusion can be drawn. Thus $b = 0$ is the exceptional value that the function $\exp(1/z)$ never takes in the neighborhood of the essential singularity, $z = 0$. The inequality (39.25b) follows from (39.25a; 39.24d):

$$\frac{1}{\varepsilon^2} < |z_n|^{-2} < (\log|c|)^2 + [\arg(c) + 2\pi n]^2, \tag{39.26}$$

solving for n.QED.

The exceptional value may not exist, in which case the integral function $f(z)$ takes all values, an infinite number of times, in any neighborhood of the essential singularity. As a second example, consider the function $\sin(1/z)$ that has an essential singularity at the origin, as follows (27.74) from (39.27a); it takes every value, without exception, in its neighborhood, for example, the value zero (39.27b):

$$\sin\left(\frac{1}{z}\right) = \sum_{n=0}^{\infty} \frac{(-)^n}{(2n+1)!} z^{-2n-1}: \ \sin\left(\frac{1}{z}\right) = 0, \ z_{n\pi} = \frac{1}{n\pi}, \ |z_n| < \varepsilon \Rightarrow n > \frac{1}{\pi\varepsilon}, \tag{39.27a–d}$$

is taken at the points (39.27c), of that an infinite number exists in every neighborhood (39.27d) of the origin. Five other instances of an infinite number of roots of an integral function are given in Example 40.17. Another case is a nonintegral function, viz. the logarithm is many-valued with a branch-point at the origin (Subsection 7.2.2); it is univalent, and hence has a single root (Subsection 37.4.3).

Before proceeding to prove the Picard second theorem, its consequences regarding the solution of equations are discussed. The fundamental question concerns the roots (39.28c) of a complex equation (39.28b) specified by a holomorphic function (39.28a):

$$f(z) \in \mathcal{D}(|C), \quad c \in |C: \quad f(z_n) = c. \tag{39.28a–c}$$

If $f(z)$ is a rational function (39.29a) of degrees N,M, that is, is the ratio of two polynomials (39.29a–c):

$$f_{M,N}(z) = \frac{Q_M(z)}{P_N(z)}, \quad Q_N(z) - cP_N(z) = 0, \quad S \equiv \sup(N,M), \tag{39.29a–c}$$

and then the equation (39.28c) ≡ (39.29b) for every value of c has a finite number roots (39.29c), by the fundamental theorem of algebra (Subsection 31.6.3); these roots may be multiple, that is, some or all could coincide. If $f(z)$ is analytic but is not a rational function, it must have (Subsection 27.9.1) at least one essential singularity, that is, it is an integral function; then two questions arise: (i) does (39.28c) have roots for all complex values of c? (ii) if so, how many roots z_n? The Casorati–Weierstrass theorem (39.21) gives an answer that is satisfactory from the numerical point-of-view: *for every value of c there is an infinite number of points z_n for that the integral function $f(z_n)$ comes arbitrarily close to c; thus for all complex c in (39.28b) roots of (39.28c) can be found with any numerical accuracy specified a priori (39.30b) for an integral function (39.30a):*

$$f(z) \in \mathcal{I}(|C): \quad \forall_{c \in |C, \varepsilon > 0} \quad \exists^{\infty}_{z \in |C}: \quad |f(z) - c| < \varepsilon. \tag{39.30a,b}$$

However the Weierstrass theorem does not state whether the function $f(z)$ actually takes exactly the value c. It could fail to take an infinite number of values c or even a dense set of values of c. This possibility is excluded by *the second Picard theorem (39.22) that states that (39.28c) where $f(z)$ is an integral function has an infinite number of roots for all c, but allows for the possible existence of one exceptional value b, for that there are no roots;* the existence of one (or several or even an infinite or dense set of) exceptional values would fall out of the scope of Casorati–Weierstrass theorem, because it only covers arbitrarily close values.

The preceding arguments gives the answer to the **fundamental problem of roots** *(Table 39.1): (i) if the function $f(z)$ is (39.29a) rational function of degrees N,M, then the roots of (39.28c) exist far all c, their number is finite (39.29c), and some or all roots may coincide; (ii) if the function $f(z)$ is a holomorphic function that is not rational, that is, is an integral function, then the equation (39.28c) has an infinite number of roots (39.22) for each value of c, except possibly for one exceptional value $b \neq c$.* The roots of a rational equations are all finite, that is, lie within a disk of finite radius. The roots of transcendental equation, for example, (39.31a):

$$\sin z = 0, \quad z_n = n\pi, \quad \lim_{n \to \infty} z_n = \infty, \tag{39.31a–c}$$

may not lie (39.31b) within any disk with finite radius (39.31c). The question of the position or distribution of roots of an integral function in the complex plane will be discussed (Sections 39.5–39.6) after: (i) the proof of both of the Picard theorems (Subsection 39.3.2); (ii) the construction of the elliptic modular function (Subsection 39.4.2). The latter

TABLE 39.1
Roots of an Equation: Rational or Integral

Function		Rational	Analytic
Type		order $(M, N)^*$	nonrational
Roots for		all $c \in \|C$	$c \in \|C - \{b\}^{**}$
Number or roots		$\sup(M, N)$	∞
Location of roots	Finite plane	Yes	Generally not
	Disk radius	$R \equiv \sup\|z_n\|$	Landau radius: 39.3–4; E40.18
	Dense rays: 39.5	No	Yes

* Ratio of two polynomials of degrees M and N without common roots;
** One exceptional value possible.
Note: Concerning the roots of a complex equation $f(z) = c$ there are two cases to consider. If $f(z)$ is a rational function, that is, the ratio of two polynomials of degrees N, M without common roots, then $f(z) = c$ is a rational equation whose degree S is the largest or supremum of (N, M); in this case I for every value of c there are S roots, all in the finite complex plane; some of the roots may be multiple, and are counted toward S by their order of multiplicity. If the function $f(z)$ is integral (case II), that is, analytic in the whole complex plane except for an essential singularity at infinity, there may be at most one exceptional value b for which $f(z) = b$ has no roots (e.g., $e^z = 0$ has no roots, so $b = 0$ is the exceptional value). For all other values $c \neq b$ then $f(z) = c$ has an infinite number of roots. If the exceptional value does not occur then $f(z) = c$ has an infinite number of roots for every c (e.g., sin z has no exceptional value). In the case II every angular sector, however small, with vertex at the essential singularity, contains an infinite number of roots (Figure 39.5a,b) by the Julia theorem. The roots ordered by modulus generally tend to infinity and do not lie within a circle of finite radius. By the Landau theorem (Figure 40.10) if the function doest not take two values up to beyond the Landau radius then if is not holomorphic or it reduces to a constant.

appears: (i) in the proof of the theorems of Picard; (ii) in subsequent theorems on the distribution of roots, viz. the Landau radii (Subsections 39.6.3, 39.6.7, and Example 40.18); (iii) it has been used to prove some theorems on punctured conformal mappings (Subsection 37.5.3). One method of construction of the elliptic modular function (Subsection 37.4.2) involves triangular coverings of the complex plane by conformal transformations of Schwartz–Christoffel type (Chapter 33). Thus there is a close link between all these topics, as discussed next.

39.3.2 Proof of Both of the Theorems of Picard (1879, 1880)

Although the Picard second theorem (39.22) appears as a more precise form of the Casorati–Weierstrass theorem (39.21), it is much more difficult to prove. Two methods appear in the literature: (i) elementary but long-winded proofs using Schottky's and other theorems that can be approached in more than one way, for example, through the theorems of Caratheodory–Borel, Landau, or Bloch; (ii) the original proof of Picard and its subsequent refinements e.g., due to Caratheodory. The latter approach (ii) is shorter and more precise if prior knowledge is assumed of the properties of the elliptic modular function that is not elementary. The elliptic modular function will be constructed in the sequel (Section 39.4), because its explicit form is not needed here but only its properties; those properties have already been used in connection with conformal mapping (Subsection 37.5.3). The **elliptic modular function** E(z): *(i) is analytic except (39.32a) for singularities at* $z = 0, 1, \infty$; *(ii) its imaginary part (39.32b) is positive everywhere:*

$$E(z) \in \mathcal{A}(|C - \{0, 1, \infty\}): \quad \text{Im}\{E(z)\} > 0. \qquad (39.32\text{a,b})$$

The first property is relatively easy to satisfy, for example, the function (39.33a):

$$g(z) = \frac{z^3+1}{z(z-1)} \in \mathcal{A}(|C - \{0,1,\infty\}):$$
$$g_1(0) = -1, g_1(1) = 2, g_1(\infty) = 1; \quad g(-i) = -i, \tag{39.33a–e}$$

is analytic everywhere, except at $z = 0, 1, \infty$ that are simple poles with residues (39.33b–d); it does not satisfy $\operatorname{Im}\{g(z)\} > 0$ the property (39.32b), for example, (39.3e). The latter property (39.32b) is not met by any of the usual elementary functions; it implies that the elliptic modular function cannot be continued analytically to the lower half complex E-plane and thus the real axis is densely covered with singularities (Subsection 31.1.2); hence it is a lacunary function in upper-half complex plane. Thus the existence of an elliptic modular function with the properties stated in (39.32a,b) is the nonelementary assumption (considered in Section 39.4) on the basis of that the proof of both of the Picard theorems will be made next.

The Picard second theorem (39.22) is equivalent to the following statement: *if there exists one neighborhood of the point $z = a$ where the integral function $f(z)$ does not take two values, then $z = a$ is not an essential singularity of $f(z)$:*

$$\exists_{\varepsilon>0; b,d \in |C} \quad \forall_{z \in V_\varepsilon(a)}: \quad f(z) \neq b, d \quad b \neq d \Rightarrow z = a \text{ is not essential singularity off } (z). \tag{39.34}$$

The theorem implies that if $z = a$ is an essential singularity of $f(z)$, then it must take all values, with one exception at most, in every neighborhood. To prove the theorem, note that

$$f(z) \neq b \neq d; \quad g(z) \equiv \frac{f(z) - d}{b - d} \neq 0, 1, \tag{39.35a–c}$$

if $f(z)$ does not take (39.35a,b) the values b, d, then the function $g(z)$ defined by (39.35c) does not take the values $0, 1$; taking it as argument (39.36a) of the elliptic modular function:

$$F(z) \equiv E(g(z)): \quad \forall_{\varepsilon>0} \, \exists_{M>0} \, \forall_{z \in V_\varepsilon(a)}: \quad |F(z)| < M, \tag{39.36a,b}$$

the latter is bounded in the neighborhood of $z = a$; thus $z = a$ cannot be (39.36b) an essential singularity of $F(z)$; since $E(z)$ and $g(z)$ are holomorphic, then $f(z)$ is also bounded, so $z = a$ cannot be an essential singularity of $f(z)$. QED.

A similar argument serves to prove the **Picard first theorem (1879)**: *a holomorphic function that does not take two distinct values reduces to a constant:*

$$f(z) \in \mathcal{D}(|C): \quad f(z) \neq b, d \wedge b \neq d \Rightarrow f(z) \equiv \text{constant}. \tag{39.37}$$

To prove the theorem we note that if $f(z)$ does not take (39.35a) the values b, d, then $g(z)$ does not take the values 0,1 in (39.35b); the elliptic modular function (39.36a) with the argument g is holomorphic (39.38a) with a lower bound (39.38b) for the imaginary part:

$$F(z) \equiv E(g(z)) \in \mathcal{D}(|C): \quad \operatorname{Im}\{F(z)\} > 0; \tag{39.38a,b}$$

a holomorphic function, with a lower bound (viz. zero in this case) for its imaginary part, is a constant by Liouville's theorem (Section 27.6); thus $F(z), g(z)$ and hence $f(z)$ are all constants. QED. The Picard second theorem is a substantial extension of Liouville's theorem: to prove that a holomorphic function reduces to a constant it is not necessary to find any bounds in the whole complex plane, for its modulus or real or imaginary parts; it suffices to show it fails to take two finite distinct values anywhere on the plane. The first Picard theorem (39.37) can be proved from the second (39.22), without using the elliptic modular function, by an elementary reasoning, as follows: (i) a holomorphic function that does not take two values (39.37) cannot have an essential singularity (39.22), and hence must be a rational function; (ii) but a rational function takes all values (39.29a–c), unless it is a constant; (iii) thus the function must be a constant. QED.

39.4 Triangular Coverings (Schwartz, 1890) and the Modular Function (Legendre)

The construction (Subsection 39.4.2) of the elliptic modular function can be made: (i) analytically via the hypergeometric function (Section 29.9) and the Schwartzian derivative; (ii) geometrically via the Schwartz triangles that provide a covering of the plane using conformal mappings (Subsection 39.4.1).

39.4.1 Triangular or Four-Color Coverings of the Plane

The proof of the Picard first (39.37) and second (39.22) theorems depend on the existence of the Legendre modular function (39.32a,b) that will be constructed next using one of the Schwartz triangular mappings that cover the complex plane. The *triangular mapping of the plane show that four colors are the minimum sufficient to cover a plane with polygons (viz. triangles) with different colors on every side, and the same colors only touching at the vertices.* If the polygons were not triangles a larger number of colors would be needed. The Schwartz–Christoffel transformation maps the upper-half complex plane $\operatorname{Im}(z) > 0$ into a triangle with arbitrary vertices $(A,\ B,\ C)$ and internal angles $(\alpha,\ \beta,\ \gamma)$ that correspond to external angles $(\pi-\alpha, \pi-\beta, \pi-\gamma)$. The triangle is closed if (33.27a) the external angles add to $2\pi = 3\pi - \alpha - \beta - \gamma$, that is, the internal angles add to $\pi = \alpha+\beta+\gamma$. The limiting form (14.75b) $\equiv$ (36.2a) of the mapping at a vertex $\zeta = 0$ with internal angle β is $z \sim \zeta^{\pi/\beta}$ or $dz/d\zeta \sim \zeta^{\pi/\beta - 1}$, showing that $\zeta = 0$ is a branch-point unless the angle β is a submultiple of π. Thus the *univalent conformal (i.e., simple) mapping of the upper-half complex z-plane (39.39a) into a triangle in the ζ-plane requires that the internal angles at the vertices (39.39b) be submultiples of π :*

$$\zeta(z) \in A(\operatorname{Im}(z) > 0): \quad \alpha, \beta, \gamma = \frac{\pi}{n}, \frac{\pi}{m}, \frac{\pi}{p}; \tag{39.39a,b}$$

thus the integers n,m,p in (39.39b) $\equiv$ (39.40a) must satisfy (39.40b):

$$n, m, p \in |Z: \quad \frac{1}{n} + \frac{1}{m} + \frac{1}{p} = \frac{\alpha+\beta+\gamma}{\pi} = s; \tag{39.40a,b}$$

in order that the sum of angles be π, for a triangle with flat sides in the ζ-plane, (39.41a) must hold:

$$s = \frac{1}{n} + \frac{1}{m} + \frac{1}{p} \begin{cases} = 1 \text{ for triangle with flat sides in the plane,} & (39.41a) \\ > 1 \text{ for a spherical triangle,} & (39.41b) \end{cases}$$

whereas for a **spherical triangle** *the sum of the internal angles (39.41b) exceeds π.*

To meet the condition (39.41a), the angles can be ordered $\alpha \geq \beta \geq \gamma$ implying $n \leq m \leq p$. The lowest value of n is $n = 2$, so that (39.41a) implies $m > 2$. If $n = 2$ and $m = 3$ then $1/p = 1 - 1/2 - 1/3 = 1/6$, that is, $p = 6$; if $n = 2$ and $m = 4$ then $1/p = 1 - 1/4 = 1/4$, implies $p = 4$. It is not possible to have $n = 2$ and $m \geq 5$ because then $1/p = 1 - 1/2 - 1/m = 1/2 - 1/m \geq 1/2 - 1/5 = 3/10 > 1/4$, and $p < 4$ contradicting $m \leq p$. The next choice for n is $n = 1/3$ and since $n \leq m \leq p$ only $m = p = 1/3$ meets (39.41a). For $n \geq 4$ it is not possible to meet (39.41a) with $n \leq m \leq p$. This exhausts the three possibilities to satisfy (39.41a) with integer (n, m, p). *The condition (39.40b) of simple mapping of the upper-half* z-*plane into a triangle with straight sides (39.41a) is met only in three cases:*

$$\{m, n, p\} = \left\{\frac{1}{3}, \frac{1}{3}, \frac{1}{3}\right\}, \left\{\frac{1}{2}, \frac{1}{4}, \frac{1}{4}\right\}, \left\{\frac{1}{2}, \frac{1}{3}, \frac{1}{6}\right\}, \tag{39.42a–c}$$

that are (Table 39.2) the **three triangular or four-color coverings of the plane:**

$$\alpha = \beta = \gamma = \frac{\pi}{3}, \quad \alpha = \frac{\pi}{2} \neq \frac{\pi}{4} = \beta = \gamma, \quad \frac{\pi}{2} = \alpha \neq \frac{\pi}{3} = \beta \neq \frac{\pi}{6} = \gamma. \tag{39.43a–c}$$

viz. by: (i) equilateral (39.42a) ≡ (39.43a) triangles (Figure 39.1a); (ii) isosceles triangles (39.42b) ≡ (39.43b) with two angles $\pi/4$ and the third double angle $\pi/2$, that is, rectangular isosceles triangles (Figure 39.1b); (iii) irregular triangles (39.42c) ≡ (39.43c) whose angles are $(1, 2, 3) \times \pi/6$ *simple multiples (Figure 39.1c).* The nonoverlapping covering of the plane by equilateral triangles (Figure 39.1a) leads to: (i) larger equilateral triangles by sets of four, for example, 2–5; (ii) hexagons by sets of six (e.g., 11–16); (iii) lozenges by sets of eight, for example, 2–8. The nonoverlapping covering of the plane by isosceles triangles leads (Figure 39.1b) to squares by associating two (e.g., 1–2), four (e.g., 2–5), eight (e.g., 1–8), sixteen (e.g., 1–16), or any multiple 2^n of 2. The nonoverlapping covering of the plane by irregular triangles is possible only for multiple angles $30° - 60° - 90°$, and leads (Figure 39.1c) to: (i) equilateral triangles associating two (e.g., 1–2); (ii) lozenges associating four (e.g., 1–4); (iii) hexagons associating twelve (e.g., 3–14); (v) larger lozenges associating sixteen (e.g., 1–16); (v) David stars of six points associating twenty-four (e.g., 1–24). There are more coverings of the plane by polygons with more than three sides, but only three

TABLE 39.2
Triangular Conformal Coverings of the Plane

Case	**I**	**II**	**III**
Triangle	Equilateral	Isosceles	Irregular
Angle $-\alpha$	$\pi/3$	$\pi/2$	$\pi/2$
Angle $-\beta$	$\pi/3$	$\pi/4$	$\pi/3$
Angle $-\gamma$	$\pi/3$	$\pi/4$	$\pi/6$
Figure	38.1a	38.1b	38.1c

Note: The simplest nonoverlapping covering of the plane by polygons of different colors, such that no two common sides have the same color, uses triangles with four colors (Figure 39.1). If the triangles are obtained by reflections on the sides there are only three nonoverlapping coverings or simple mappings of the plane, using: (i) equilateral triangles (Figure 38.1a), that is, angles of 60° at all corners; (ii) isosceles rectangular triangles (Figure 39.1b), that is, with angles 45°, 45°, 90° at the corners; (iii) rectangular irregular triangles (Figure 39.1c) with multiple angles, that is, 30°, 60°, 90°.

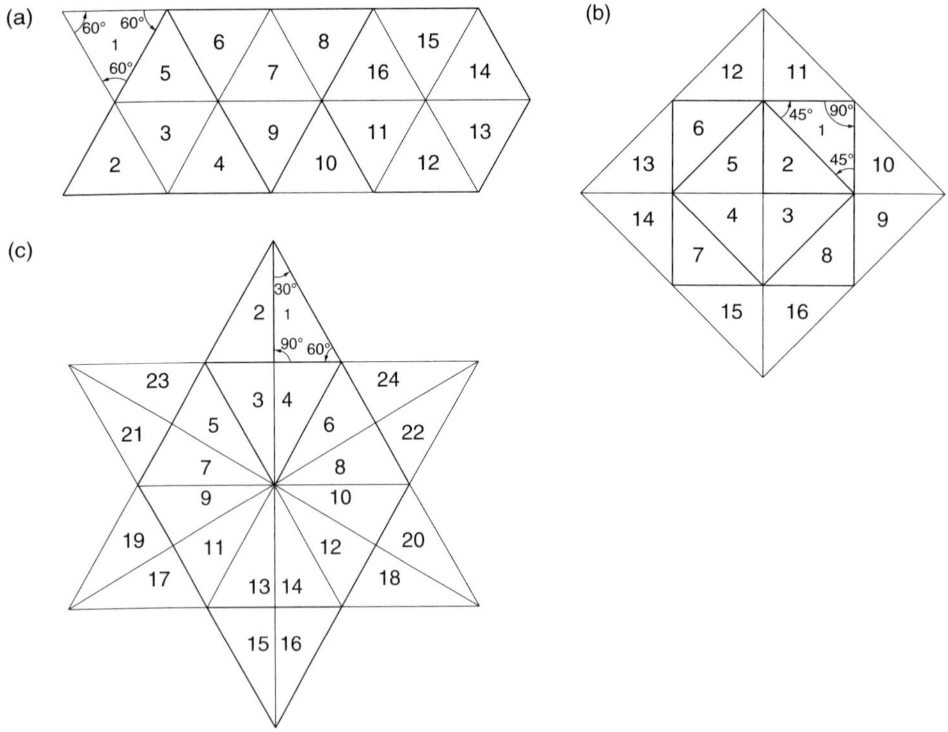

FIGURE 39.1
The Schwartz–Christoffel interior polygonal transformation (Figure 33.6a) with three critical points can be used to map conformally the upped half complex plane into the interior of a triangle. The Schwartz reflection principle (Figure 31.4) can be used to reflect the triangle on its sides to cover the whole complex plane. There are no overlaps, that is, the mapping is both conformal and univalent, that is, is a simple mapping, only in three cases. The first case (I) is equilateral triangles (a) hence with angles 60°. They can be combined: (I-i) in pairs (e.g., 3+5) to form lozenges; (I-ii) in sets of four, for example, (2 + 3 + 5 + 6) to form parallelograms; (I-iii) in sets of six (i.e., 3 to 9) to form hexagons: (I–IV) in sets of sixteen (e.g., 1–16) to form an arrow. They also form larger equilateral triangles in sets of four (e.g., 2 to 5) or 2^{2n} with n a positive integer. The second case (III) is isosceles rectangular triangles (b), hence with angles 45°, 45°, 90°. They form progressively larger squares, in sets of: (II-i) two (e.g., 1 to 2); (II-ii) four (e.g., 2 to 5); (II-iii) eight (e.g., 1 to 8); (II–iv) sixteen (e.g., 1 to 16); (II–V) any power 2^n of two. They also from progressively larger isosceles rectangular triangles, in sets of: (i) two (e.g., 2 + 3); (ii) four (e.g., 5 + 6 + 12 + 13); (iii) eight (e.g., 1 + 2 + 3 + 8 + 9 + 10 + 11 + 16); (iv) any other power 2^n of two. The third case (III) is irregular rectangular triangles with multiple angles: 30°, 60°, 90°. These can be combined in sets of: (III-i) two (e.g., 1 + 2) to form equilateral triangles; (III-ii) four (e.g., 1 to 4) to form lozenges; (III-iii) eight (e.g., 1 to 8) to form larger equilateral triangles; (III-iv) sixteen (e.g., 1 to 16) to form larger lozenges; (III-v) twenty-four (e.g., 1 to 24) to form a David star with six points; (III-vi) larger equilateral triangles in sets of 2^{2n+1}; (III-vii) larger lozenges in sets of 2^{2n}. The three cases I to III are the only four-color mappings of the plane: (i) the plane is covered by identical triangles without overlaps; (ii) using only four colors it is possible to have different colors across each side.

coverings based on reflections of triangles on their sides, and these are the only four color mappings.

39.4.2 Spherical, Curved, and Cusped Triangles

The inequality (39.40b; 39.41b) for spherical triangles has four solutions, of that one is indicated as example:

$$N \in |N: \quad \alpha = \frac{\pi}{2} = \beta, \quad \gamma = \frac{\pi}{N}, \quad \frac{\alpha + \beta + \gamma}{\pi} = 1 + \frac{1}{N} > 1, \tag{39.44a–d}$$

where N is any integer. The sphere is covered by $4N$ spherical triangles (Figure 39.2), with angles $\pi/2$ at the equator, and angle π/N at the poles, obtained by cutting the equator with $2N$ equally spaced meridians. The case of $N = 5$, that is, 20 triangles is illustrated in Figure 39.2, where only ten triangles are visible in the nearby hemisphere; the five northern

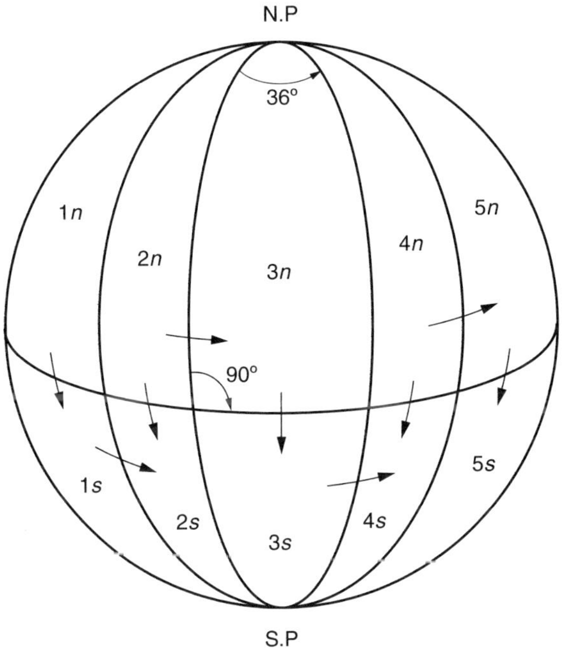

FIGURE 39.2
The Schwartz–Christoffel transformation with three critical points (Figure 33.6a) can also be used to map conformably the upper complex half plane as the domain into a triangle on the Riemann sphere as the range (Figure 9.3b). By applying the reflection principle (Figure 31.4) across the curved sides of the spherical triangles, the whole sphere can be covered. There is a finite number of cases of coverings of the sphere without overlaps, corresponding to conformal univalent, that is, simple maps. One case (Figure 39.2) consists of the upper and lower hemispheres divided by N equally spaced meridians into $2N$ spherical triangles in each hemisphere. All $4N$ spherical triangles can be obtained by reflections of one of them on the equator and meridians. This leads to one of a set of four two-color coverings of the sphere: (i) the sphere is covered by curved triangles without overlaps; (ii) alternating two colors in each hemisphere with stagger across the equator there are always distinct colors across each side.

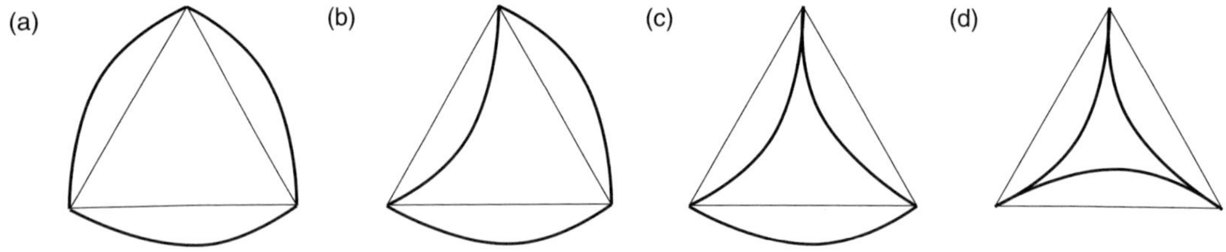

FIGURE 39.3
For the plane (spherical) triangles [Figure 39.1(39.2)] with straight (convex) sides, the sum of internal angles is 180° (more than 180°). In the case of a triangle with concave sides (Figure 39.3d) the sum of internal angles is less than 180°; the sum of internal angles takes the minimum value 0° for three concave sides cusped at the vertices. A triangle with curved sides may have: (i) all sides convex (Figure 39.3a), like the spherical triangles (Figure 39.2); (ii) all sides concave, like the cusped triangle (Figure 39.3d); (iii) two convex and one concave side (Figure 39.3b); (iv) vice-versa (Figure 39.3c).

(1n–5n) and five southern (1s–5s) triangles are shown, with the arrows indicating the a sequence of reflections on the sides that are conformal univalent mappings, covering the Riemann sphere (Chapter 9). The inequality (39.40b; 39.41b) also applies to **circular-arc triangles** *in the plane with three convex sides (Figure 39.3a). For triangles with some concave sides,* for example, *three (Figure 39.3d), two (Figure 39.3c) or one (Figure 39.3b), the sum of internal angles may be less than* π, *and the condition (39.40b) with* $\mathrm{s} < 1$ *has an infinite number of solutions.* In particular, the vertex of a triangle may be a **cusp**, that is, a point where the right- and left-hand-side tangents have the same modulus and opposite signs (39.45a):

$$z'(t+0) = -z'(t-0), \quad \alpha = \pi, \quad \frac{\mathrm{d}\zeta}{\mathrm{d}z} \sim z^{-1} = z^{-\alpha/\pi}, \tag{39.45a–c}$$

corresponding to a zero internal angle and external angle of π in (39.45b) and a simple pole (39.45c) in the Schwartz–Christoffel transformation. A simple pole is consistent with the inverse mapping being univalent; this would not be the case for a multiple pole, whose inverse would have several branches.

A circular-arc triangle may have three cusps (Figure 39.3d). The Legendre modular function, used in the proofs of the two theorems of Picard (39.22) and (39.37) and two theorems (Subsection 37.5.3) on conformal mapping, can be constructed on the basis of the **Schwartz triangle** that *has three cusps at* $\zeta = 0, 1, \infty$, that is, *whose boundary (Figure 39.4) or sides are the half-circle (39.46a) of radius* $1/2$ *and center at (1/2,0), and two vertical lines (39.46b,c) joining the vertices at* $\zeta = 0$ *to the third vertex at infinity* $\zeta = \infty$:

$$S_t \equiv \left\{ \left| \zeta - \frac{1}{2} \right| \geq \frac{1}{2}, \quad 0 \leq \operatorname{Re}(\zeta) < 1 \right\}; \quad f_0(\zeta) \in \mathcal{A}(S_t), \tag{39.46a–d}$$

the upper-half complex z-plane $\operatorname{Im}(z) > 0$ can be mapped (39.46d) simply (i.e., univalently and conformably) to the interior of the Schwartz triangle (39.46a–c) specifying an analytic function, except for singularities at the vertices $\zeta = 0,\ 1,\ \infty$. By successive reflection on the vertical sides $\operatorname{Re}(\zeta) = 0, 1$, the Schwartz triangle covers the whole complex ζ-plane in the x-direction above $\operatorname{Im}(\zeta) \geq 1/2$. The inversion with regard to the half-circle (39.46a) covers

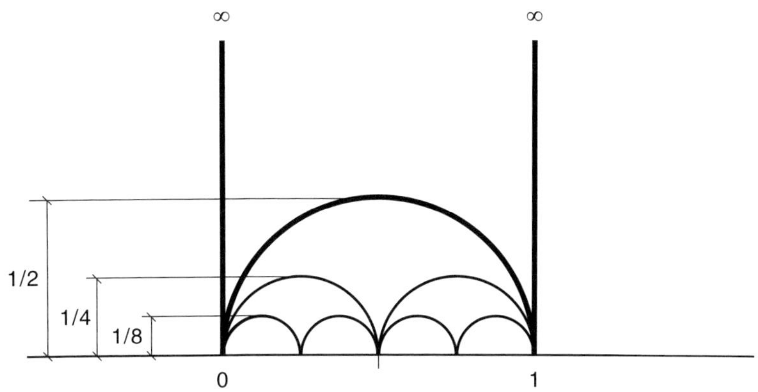

FIGURE 39.4
The Schwartz–Christoffel interior polygonal transformations (Figure 33.6a) with three critical points can be used to map the upper complex half-plane into: (i) a finite triangles in the plane (Figure 39.1); (ii) a spherical triangle (Figure 39.2) which is always finite; (iii) a triangle in the plane which may have infinite sides if one vertex is at infinity (Figure 39.4); (iv) the triangle may have some or all sides straight (Figure 39.1), convex (Figure 39.2) or concave (Figure 39.3). An example combining (iii) and (iv) is the Schwartz triangle (Figure 39.4) consisting of: (i) a half-circle with radius 1/2 and center at the point (0,1/2) joining the origin to the point (1,0) in the upper complex half-plane, as the curved convex side; (ii) two vertical parallel straight lines $\operatorname{Im}(\zeta) = 0, 1$ as two infinite sides meeting at the vertex at infinity. The mapping is conformal except at the three vertices: the point-at-infinity and the cusps at $\zeta = 0, 1$. By applying the Schwartz reflection principle (Figure 31.4) on the sides the ζ-plane is covered without overlaps in the horizontal direction. The reciprocal point relative to the circle (Figure 35.4a) extend the mapping downwards. Considering half-circles with successive half radius 2^{-n} at each iteration, the mapping is extended analytically and univalently up to, but not including, the real axis. Thus has been constructed the Legendre elliptic modular function $E(z)$ which: (i) is analytic in the whole complex plane except for the points $z = 0, 1, \infty$; (ii) has positive imaginary part $\operatorname{Im}\{E(z)\} > 0$. A bilinear transformation can map the points $(0, 1, \infty)$ to any three points (a,b,c) while preserving the simple mapping. It is not difficult to introduce a complex function analytic in the whole complex plane except at three points, for example, $f(z) = [(z-a)(z-b)(z-c)]^{-1}$ but it does not satisfy $\operatorname{Im}[f(z)] > 0$. The remarkable property of the elliptic modular function is that its the imaginary part is everywhere positive. The elliptic modular function is used to prove a number of properties concerning conformal mappings and roots of transcendental equations (Figures 39.5–39.7 and 40.11).

all points of the complex ζ-plane above $\operatorname{Im}(\zeta) \geq 1/4$. The inversion on the two subcircles (39.47a,b):

$$\left|\zeta - \frac{1}{4}\right| < \frac{1}{4} > \left|\zeta - \frac{3}{4}\right|, \quad \operatorname{Im}(\zeta) \geq \frac{1}{8}; \quad \operatorname{Im}(\zeta) \geq 2^{-n}, \tag{39.47a–d}$$

covers the region (39.47c). Proceeding iteratively dividing each half-circle in two half-circles (Figure 39.4) after n iterations the ζ-plane has been covered for (39.47d), and the lower boundary consists of 2^n circles of radius 2^{-n}. As $n \to \infty$: (i) the whole upper-half ζ plane is covered $\operatorname{Im}(z) > 0$ without overlaps specifying function that is holomorphic except at $\zeta = 0, 1, \infty$; (ii) the real axis is densely covered by singularities infinitesimally spaced $2^{-n} < \varepsilon$ for $n > -(\log\varepsilon)/(\log 2) = -1.4427\log\varepsilon$ and thus no analytic continuation (Subsection 31.1.2) is possible for $\mathrm{Im}(\zeta) \le 0$. The properties (i) and (ii) coincide with those assumed for the Legendre modular function (39.32a,b), used in the proof of the Picard first (39.37) and second (39.22) theorems. Further extensions of the Picard theorem are considered next (Sections 39.5 and 39.6); it is possible to proceed directly to the periods of elementary transcendental and elliptic functions (Sections 39.7–39.9).

39.5 Dense Rays and Julia (1924) Theorem

The Picard theorems state that an analytic univalent function has an infinite number of roots but gives no indication of their location. Some examples suggest that an infinite number of roots may lie in a narrow angular sector (Subsection 39.5.1), that is, a dense ray, and this can be proved by the Julia theorem (Subsection 39.5.2). It relates to the asymptotic scaling of complex functions (Subsection 39.5.3); for directions through the origin it leads to the exponential order of a function (Subsection 39.5.4).

39.5.1 Roots Going to Infinity along a Dense Ray

The second theorem of Picard can be stated in two equivalent ways: an analytic function in the whole complex plane z: (i) if it does not take values $b \ne f(z) \ne c \ne b$ then it is a constant; (ii) if it is not a constant $f(z) \ne \text{const}$, then it must take all values $f(z) = c \ne b$ with one possible exception. In the previous statement was emphasized the whole complex plane, because the Picard theorems give no indication about the location of roots. Some indication about possible distribution of roots is given by generalizations of the Picard theorem, by proving that (i) or (ii) hold over a part of the z-plane, for example: (i) over an angular sector $\alpha - \varepsilon < \arg(z) < \alpha + \varepsilon$, for some α and any arbitrarily small ε, by the Julia theorem (Section 39.5); (ii) over a disk $|z| = R_L$, where R_L is (Section 39.6) the Landau radius, that is, is the maximum value for that (i) or (ii) do hold, since they must hold as $|z| \to \infty$. The Julia theorem is proved first, preceded by two examples.

The equation $\mathrm{e}^z = c$ has no roots for a real variable $z = x$ and $c < 0$. The corresponding complex equation (39.48b):

$$n \in |Z: \quad \mathrm{e}^z = c \ne 0, \quad z_n = \log|c| + i\arg(c) + i2\pi n, \tag{39.48a–c}$$

has an infinite number (39.48a) of roots (39.48c) for all values of c, apart from the exceptional value zero $c \ne b = 0$. As $n \to \pm\infty$ the roots tend (39.49a) to infinity $z_n \to \infty$ that is the only essential singularity of the exponential, since the latter is an integral function. All the infinite roots (39.49a) lie on a vertical line (39.49b), and hence their argument (39.49c) tends to $\pm\pi/2$ as $n \to \pm\infty$:

$$\lim_{n->\infty} |z_n| = \infty: \quad \mathrm{Re}(z_n) = \log|c|,$$
$$\lim_{n\to\pm\infty} \arg(z_n) = \lim_{n\to\pm\infty} \arctan\left[\frac{\arg(c) + 2\pi n}{\log|c|}\right] = \arctan(\pm\infty) = \pm\frac{\pi}{2}. \tag{39.49a–c}$$

Thus all roots beyond a certain order lie within an arbitrary small angular sector about the imaginary axis:

$$\forall_{\varepsilon>0} \quad \exists_{n\in|N} \quad \forall_{m\in|N}: \quad m>n \Rightarrow \frac{\pi}{2}-\varepsilon \leq \arg(z_m) \leq \frac{\pi}{2}+\varepsilon, \tag{39.50}$$

that is designated **dense ray** (Figure 39.5a).

The real equation $\cos x = c$, has no roots for $c>1$ or $c<-1$; the corresponding complex equation (39.51a):

$$c=\cos z \equiv \frac{e^{iz}+e^{-iz}}{2}, \qquad e^{2iz}-2ce^{iz}+1=0, \tag{39.51a,b}$$

has roots (39.51b) for all c, that is, there is no exceptional value in this case (39.52b):

$$n\in|Z: \quad e^{iz}=c\pm\sqrt{c^2-1}; \quad z_n=-i\log\left(c\pm\sqrt{c^2-1}\right)+2\pi n,$$

$$\lim_{n\to\infty}\arg(z_n)=\lim_{n\to\infty}\frac{\log\left(c\pm\sqrt{c^2-1}\right)}{2\pi n}=0; \tag{39.52a–d}$$

the roots (39.52c) tend to infinity along a horizontal line (39.52d), and the dense ray

$$\forall_{\varepsilon>0}\ \exists_{n\in|N}\ \forall_{m\in|N}: \quad m>n \Rightarrow -\varepsilon<\arg(z_m)<\varepsilon, \tag{39.53}$$

is the real axis (Figure 39.5b).

39.5.2 The Julia (1924) Theorem and Dense Star of Dense Rays

The preceding examples suggest the **Julia theorem (1924)**: *an integral function, that is, a function with one essential singularity, has at least one ray direction $\varphi=\alpha$ in whose angular neighborhood it takes an infinite number of times every value c, with one possible exception:*

$$\exists_{0\leq\alpha<2\pi} \quad \forall_{c\in|C,\varepsilon>0} \quad \exists_z^{\infty}: \quad f(z)=c\neq b \ \wedge\ \alpha-\varepsilon\leq\arg(z)\leq\alpha+\varepsilon. \tag{39.54}$$

The change of variable $z=a+1/\zeta$ transforms the point at infinity $\zeta=\infty$ to the point, $z=a$, and thus the theorem can be restated: *if* f(z) *has an essential singularity at* z = a, *then there is at least one angular sector in the neighborhood of a where* f(z) *takes every value* c, *with one possible exception, an infinite number of times.* Since both statements are equivalent, only the second statement is proved by reduction ad absurdum; if in every angular neighborhood of a, say $\alpha-\varepsilon\leq\varphi\leq\alpha+\varepsilon$, the function $f(z)$ takes one nonexceptional value a finite number of times $\leq N$, then in the whole neighborhood $0\leq\varphi<2\pi$, that value is taken not more than $\pi N/2\varepsilon$ times, and thus the point $z=a$ cannot be an essential singularity. QED.

The examples given before (39.48a–c; 39.49a–c; 39.50) [(39.51a,b; 39.52a–d; 39.53)] for the exponential (cosine) had only one dense ray along the imaginary (real) axis in the Figure 39.5a(b). In the neighborhood of an essential singularity the dense rays may occur in one, several, or every direction. For example, the function (39.55b):

$$n,m\in|Z, \quad \exp(e^z)=c: \quad (\exp z)_n=\log|c|+i\arg(c)+i2\pi n,$$

$$z_{n,m}=\frac{1}{2}\log\left|(\log|c|)^2+[\arg(c)+2\pi n]^2\right|+i\arctan\left[\frac{\arg(c)+2\pi n}{(\log|c|)}\right]+i2\pi m, \tag{39.55a–d}$$

has a double (39.55d) infinity (39.55c) of roots (39.55a) for each $c\neq0$, so that every ray through the origin contains an infinite number of them and is a dense ray; thus the roots of (39.55d) of (39.55b) form **a dense star of dense rays**. The ordering of zeros toward infinity is one aspect of the asymptotic behavior of complex functions, of which two examples are given next (Subsections 39.5.3 and 39.5.4).

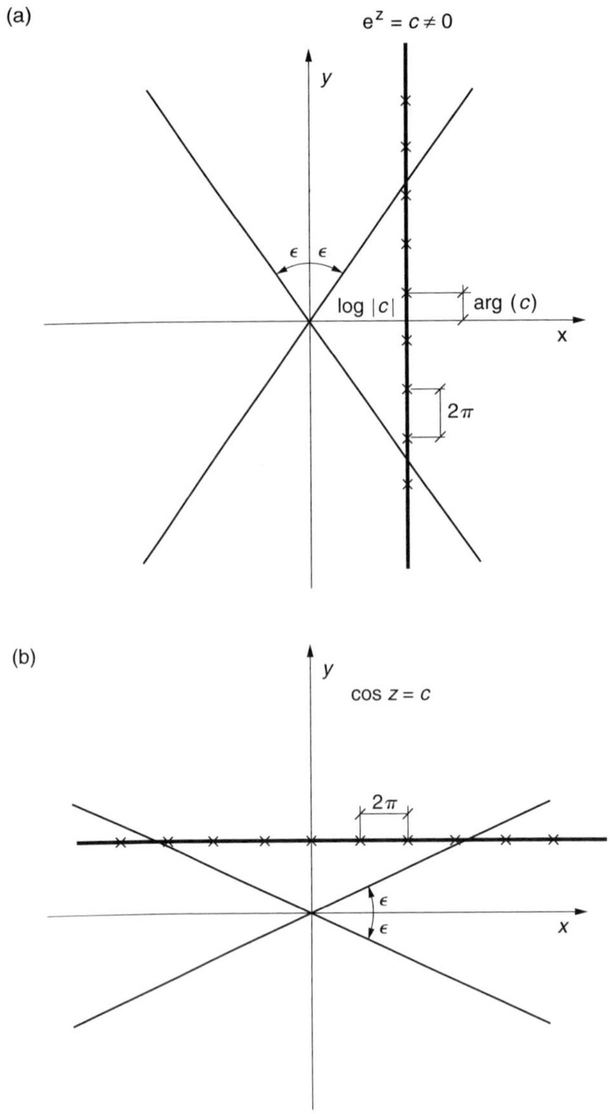

FIGURE 39.5
One of the results obtained via the elliptic modular function is the Picard second theorem on the existence of roots of a transcendental equation (Table 39.1): an integral function takes every value (apart from one possible exceptional value) an infinite number of times. An extension is the Julia theorem: in any angular sector however small of an essential singularity an integral function takes an infinite number of times every possible value (with one possible exception). As examples the exponential (circular cosine) are analytic in the whole finite complex plane with an essential singularity at infinity, and hence integral functions. Concerning the exponential it never vanishes, so the exceptional value is $b = 0$; for every other value the equation (39.48a) has an infinite number of roots (39.48b) lying on a vertical line (a); any angular sector however small with vertex at the origin and including the imaginary axis includes an infinite number of roots. A second example is the circular cosine which takes every value in the complex plane, so that in this case there is no exceptional value. The equation $\cos z = c$ has for every value of c without exception an infinite number of roots (39.52b) lying along a horizontal line (b). An angular sector however small with vertex at the origin and including the real axis contains an infinite number of roots.

39.5.3 Asymptotic Scaling of a Complex Function

Using the Cartesian representation (39.56a) of the function (39.55b) ≡ (39.56b):

$$z \equiv x + iy: \; f(z) \equiv \exp(\mathrm{e}^z) = \exp\left(\mathrm{e}^{x+iy}\right) = \exp(\mathrm{e}^x \cos y + i\mathrm{e}^x \sin y), \tag{39.56a,b}$$

its modulus is given by

$$|f(z)| = |\exp(\mathrm{e}^z)| = \exp(\mathrm{e}^x \cos y) = \exp\{\exp[\mathrm{Re}(z)]\cos[\mathrm{Im}(z)]\}. \tag{39.56c}$$

The function tends to e at the origin (39.57a) and to unity at infinity (39.57c) in the second and third quadrants (39.57b):

$$\lim_{z\to 0} \exp(\mathrm{e}^z) = e; \quad x < 0: \quad \lim_{z\to 0} \exp(\mathrm{e}^z) = 1. \tag{39.57a–c}$$

Along the imaginary axis (39.58a) it oscillates with modulus in the range (39.58b):

$$x = 0: \quad \lim_{z\to\pm\infty} |\exp(\mathrm{e}^z)| = \lim_{z\to\pm\infty} \exp(\cos y) \in \left(\frac{1}{\mathrm{e}}, \mathrm{e}\right). \tag{39.58a,b}$$

In the first two quadrants (39.59a) the modulus (39.56c) of the function (39.56b) tends to zero (infinity) for negative (positive) $\cos y < 0$ ($\cos y > 0$), that is, on the horizontal strips (39.59b) [(39.59d)]:

$$x > 0: \quad \lim_{z\to\pm\infty} |\exp(\mathrm{e}^z)| = \begin{cases} 0 & \text{if} \quad 2n\pi + \dfrac{\pi}{2} < \mathrm{Im}(z) < 2n\pi + \dfrac{3\pi}{2}, \\ 1 & \text{if} \quad \mathrm{Im}(z) = n\pi + \dfrac{\pi}{2}, \\ \infty & \text{if} \quad 2n\pi - \dfrac{\pi}{2} < \mathrm{Im}(z) < 2n\pi + \dfrac{\pi}{2}; \end{cases}. \tag{39.59a–d}$$

On the boundary lines (39.60a,b) between (39.59b,d) it oscillates infinitely (39.60c) with unit modulus (39.59c):

$$x > 0, z = x + in\pi + \frac{i\pi}{2}: \quad \exp(\mathrm{e}^z) = \exp\{i\mathrm{e}^x (-)^n\}. \tag{39.60a–c}$$

The asymptotic scaling (39.57a–c; 39.58a,b; 39.59a–d; 39.60a–c) of the modulus (39.56c) of the function (39.56a,b) with a dense star (39.55a–d) of rays is illustrated its Figure 39.6. In a direction $\varphi =$ const. through the origin it: (i) vanishes on the second and third quadrants $\pi/2 < \varphi < 3\pi/2$; (ii) oscillates in a finite range (39.58b) along (39.58a) the imaginary axis $\varphi = \pm\pi/2$; (iii) in the first and fourth quadrants $|\varphi| < \pi/2$ it oscillates without bound. Next is considered (Subsection 39.5.4) the behavior at infinity in all directions of the class of functions of exponential order.

39.5.4 Function of Exponential Order (Phragmen, Lindelof)

A complex function has **exponential order** (39.61a) if its asymptotic scaling is specified by (39.61d), where (39.61b,c) are complex constants:

$$\vartheta \equiv u + iv, \quad 0, \infty \neq A, B, \in |C: \quad \lim_{z\to 0} g(z) \exp\left(-z^{\vartheta} A\right) = B. \tag{39.61a–d}$$

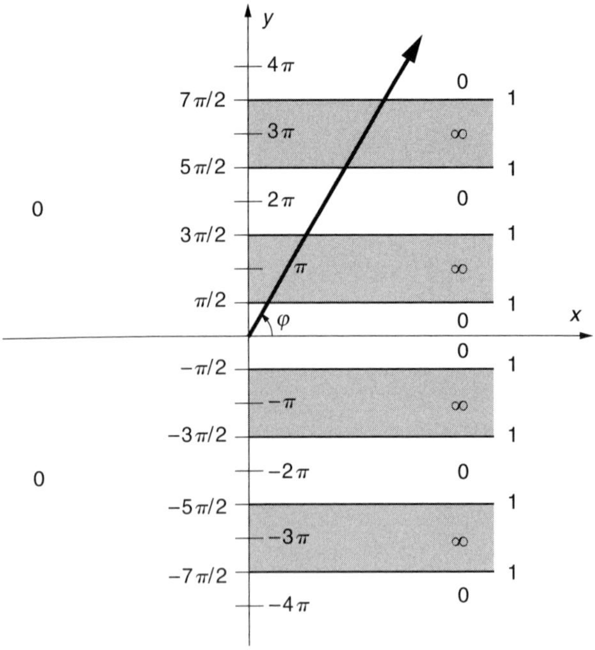

FIGURE 39.6
The examples of the Julia theorem (Figure 39.5a,b) involve the asymptotic behavior of two particular complex functions at infinity. A general class of complex functions is those of exponential order at infinity (39.61a–d), whose asymptotic behavior (Figure 39.6) is: (i) vanish in the left-hand half-plane; (ii) on the right-hand half-plane vanish (diverge) on alternating horizontal strips of width π centered around the real axis; (iii) on the horizontal lines which are the boundaries between the strips it oscillates in phase with unit modulus. Thus taking a straight line through the origin toward infinity it crosses an infinite number of strips of convergence and divergence.

The mixed polar (39.62a) and Cartesian (39.61a) representations lead to the asymptotic scaling (39.62b):

$$A = ae^{i\alpha}: \quad \lim_{z\to\infty} g(z) \sim \lim_{z\to\infty} B\exp\left(z^{\vartheta} A\right) = B \lim_{z\to\infty} \exp\left[ae^{i\alpha}\left(re^{i\varphi}\right)^{u+iv}\right]$$
$$= B \lim_{z\to\infty} \exp\{ar^{u}\exp[-\varphi v + i(\alpha + \varphi u + v\log r)]\}, \tag{39.62a,b}$$

using the definition (5.28) of power with complex exponent:

$$r^{iv} \equiv \exp\left[\log\left(r^{iv}\right)\right] = \exp(iv\log r). \tag{39.62c}$$

The modulus of (39.62b) is (39.63b):

$$0 \le \varphi < 2\pi: \quad \lim_{z\to\infty} |g(z)| \sim |B| \lim_{r\to\infty} \exp\left[ar^{u}\, e^{-\varphi v}\cos(\alpha + \varphi u + v\log r)\right], \tag{39.63a,b}$$

and the angle is taken the range (39.63a). This leads to the following cases: (i) if (39.64a) the limit is a constant (39.64b):

$$u < 0: \lim_{z\to\infty} |g(z)| = |B|; \quad u = 0: \lim_{r\to\infty} |g(z)| \in |B|\exp\left(-ae^{-\varphi v}, ae^{-\varphi v}\right), \tag{39.64a–d}$$

(ii) if (39.64c) the limit oscillates in the finite range (39.64d); (iii) if (39.65a) the term cos $(v \log r)$ oscillates indefinitely while r^u diverges, so the limit oscillates infinitely (39.65b):

$$u > 0 \neq v: \lim_{z \to \infty} |g(z)| \in (0, \infty(; v = 0: \lim_{|z| \to \infty} |g(z)| \sim |B| \exp[a r^u \cos(\alpha + \varphi u)], \quad (39.65\text{a–d})$$

(iv) in the case (39.65c,d) with (39.66a) the limit is zero (infinity) for negative (positive) cos $(\alpha + \varphi u)$ in (33.66b) [(39.66d)]:

$$v = 0 < u: \quad \lim_{z \to \pm\infty} |g(z)| \begin{cases} 0 & \text{if } 2n\pi + \frac{\pi}{2} < \alpha + \varphi u < 2n\pi - \frac{\pi}{2}, \\ 1 & \text{if } \alpha + \varphi u = n\pi + \frac{\pi}{2}, \\ \infty & \text{if } 2n\pi - \frac{\pi}{2} < \alpha + \varphi u < 2n\pi + \frac{\pi}{2}, \end{cases} \quad (39.66\text{a–d})$$

and tends to $|B|$ in modulus in the boundary case (39.66c). *The asymptotic behavior (39.64–39.66a–d) of the function with exponential order (39.61a–d)*, includes as particular cases: (i) for $A = 1 = B$ and $\vartheta = n$ an integer (5.34a,b; 5.35a–c); (ii) for $n = 2$ the case (5.33) in Figure 5.5.

39.6 Cauchy Bounds and Landau (1904) Radius (Schottky, 1904; Caratheodory, 1912; Montel, 1927)

The infinite number of roots of an equation involving an integral function must accumulate somewhere in the complex plane, and may be sought: (i) in a particular direction going to infinity, for example, a dense ray by the Julia theorem (Subsection 39.5.2); (ii) within a circle of bounded Landau radius (Subsection 39.6.3). Because the conditions of the Picard theorem hold for the entire complex plane, they may also hold for a finite radius, viz. consider a circle of radius R and center at the origin, $|z| < R$. As $R \to \infty$ a function either (i) ceases to be analytic or (ii) takes all values but one; this will still be true if $R > R_L$, but not if $R < R_L$, where R_L is the Landau radius. Thus the Landau radius is the maximum radius of a circle with center at the origin, in that a function can be analytic and fail to take two values; beyond the **Landau radius** the function either (i) ceases to be analytic or (ii) takes all values but one. This extends the first Picard theorem from infinite radius to a finite Landau radius R_L. Next is obtained an upper bound $R_1 > R_L$ for the Landau radius, by showing that either (i) or (ii) must happen for $|z| > R_L$. The finite Landau radius may be sought by putting bounds on the function, for example, the Cauchy (Landau) bounds based on the Taylor series (Subsection 39.6.1) [elliptic modular function (39.6.2)]. The upper bound for the Landau radius (Subsection 39.6.3) may be replaced by a precise value due to Caratheodory (Subsection 39.6.4) that also involves the elliptic modular function. An elementary proof of the existence of the Landau radius can be given, not involving the elliptic modular function; it gives no estimative of the value, just a functional form (Subsection 39.6.5). This proof uses the Schottky theorem (Subsection 39.6.5) and leads to further generalizations, due to Montel (Subsection 39.6.6) of the Landau and Julia theorems; four of these generalizations are proven (Subsection 39.6.7).

39.6.1 Cauchy Bounds for the Terms of a Taylor Series

The proof (Subsection 39.2.2) of the Casorati–Weierstrass theorem (39.21) was based on bounds (39.15) for the coefficients of the principal part of the Laurent expansion near an essential singularity; the generalization to the Picard second theorem (39.22) was proved (Subsection 39.3.2) using the properties (39.32a,b) of the elliptic modular function. A further generalization to the Landau theorem (1904) is based both on the elliptic modular function and on Cauchy bounds that are considered first. The coefficients (39.67b) of the Taylor series [(23.32a,b) ≡ (39.67a)] of an analytic function $f(z)$ about a point $z=a$:

$$f(z)=\sum_{n=0}^{\infty} a_n (z-a)^n, \quad a_n \equiv \frac{f^{(n)}(a)}{n!}, \tag{39.67a,b}$$

can be calculated by integrating in the Cauchy third theorem (15.13) over a circle with center at a and radius (39.68a) smaller than the radius of convergence (39.68b):

$$r<R: \quad |a_n| \leq \frac{1}{2\pi} \int_{|z-a|=r} |f(z)| \, |z-a|^{-n-1} \, dz = \frac{B}{r^n}, \tag{39.68a,b}$$

where B is an upper bound (39.69b) for $|f(z)|$. This specifies **Cauchy bounds**: *if $f(z)$ is analytic in the disk (39.69a) it has an upper bound (39.69b), and the coefficients a_n in (39.68b) of its Taylor series (39.67a), and its n-th derivate at the center (39.67b), have upper bounds (39.69c,d):*

$$f \in \mathcal{A}(|z-a| \leq r), \quad |f(z)| \leq B: \quad |a_n| < \frac{B}{r^n}, \quad \left| f^{(n)}(a) \right| < \frac{n! B}{r^n}. \tag{39.69a–d}$$

The bounds hold trivially for $n=0$.

39.6.2 Bounds Using the Elliptic Modular Function

Consider the function $f(z)$ that is holomorphic for $|z|<r$, and does not take (39.35a,b) two values b, d, so that $g(z)$ does not take the values 0, 1 in (39.35c); the elliptic modular function of $g(z)$ is holomorphic (39.32a), and so is the **Landau function**:

$$G(z) \equiv \exp\{iF(z)\} = \exp\{iE[g(z)]\} = \exp\left\{ iE\left(\frac{f(z)-d}{b-d} \right) \right\}; \tag{39.70}$$

it has derivate:

$$G'(z) = \frac{i}{b-d} f'(z) E'\left(\frac{f(z)-d}{b-d} \right) \exp\left\{ iE\left(\frac{f(z)-d}{b-d} \right) \right\}. \tag{39.71}$$

The function (39.71) has bound unity because:

$$|G(z)| = |\exp\{i\mathrm{Re}[F(z)]\}| \exp\{-\mathrm{Im}[F(z)]\} = \exp\{-\mathrm{Im}[E(g(z))]\} \leq 1, \tag{39.72}$$

taking into account (39.32b). Applying to (39.71) the Cauchy's inequality (39.69d) for $a=0$ and $n=1$, yields:

$$a_0 \equiv f(0), a_1 \equiv f'(0): \quad \zeta \equiv \frac{a_0-d}{b-d};$$
$$r < R_1 \equiv |G'(0)|^{-1} = \left| \frac{b-d}{a_1} \right| \left| \frac{\{\exp[-iE(\zeta)]\}}{E'(\zeta)} \right| > R_L; \tag{39.73a–d}$$

it follows that the radius (39.73d) ≡ (39.74a,b):

$$R_1 \equiv \left|\frac{b-d}{a_1}\right| K\left(\frac{a_0-d}{b-d}\right), \quad K(\zeta) \equiv \left|\frac{\exp[-iE(\zeta)]}{E'(\zeta)}\right| = \frac{\exp\{\mathrm{Im}[E(\zeta)]\}}{E'(\zeta)} > |E'(\zeta)|^{-1}, \tag{39.74a,b}$$

depends only on the first two coefficients a_0, a_1, of the Taylor series (39.67a,b) for $f(z)$. The inequality in (39.74b) arises because the elliptic modular function (40.182a) has positive imaginary part (40.182b).

39.6.3 Landau (1904) Radius for the Picard First (1880) Theorem

This proves the **Landau theorem (1904)**: *if the nonconstant function $f(z)$ is (i) analytic (39.75a) in a disk $|z| \le r$, and (ii) does not take the two distinct values b, d, in (39.75b,c) then the radius of the disk must be no greater than the first estimate of the Landau radius (39.75d; 39.74a,b; 39.73a,b):*

$$f(z) \in \mathcal{A}(|z| \le r) \quad \wedge \quad f(z) \ne b, d, b \ne d: \quad r < R_1 \ge R_L, \tag{39.75a–d}$$

that is an upper bound for the exact Landau radius. The theorem can be restated: *if the nonconstant function $f(z)$ is extended over a disk $|z| \equiv r$ of radius $r > R_1$ larger than the Landau radius, then either (i) it ceases to be holomorphic, or (ii) it takes all values except possibly one.* The theorem of Landau extends the Picard first theorem to a finite disk, and reduces to it in the case of an infinite disk $r \to \infty$, whose radius exceeds the Landau radius: if $f(z)$ is holomorphic over an infinite disk $|z| \le \infty$, then either (i) it does not take two values and reduces to a constant, or (ii) it is not a constant and takes all values except possibly one. In the latter case (ii), it is either (a) a rational function of type $Q_{n,m}(z)$, with $N \equiv \sup(n, m)$ roots for each value of c in (39.29a–c), or (b) an integral function, with an infinite number of roots (39.22) for each value of c, apart from the possible exceptional value $c \ne b$. Note that the theorems of Julia (Landau) [Section 39.5 (Subsection 39.6.3)] are refinements of the Picard theorem, specifying the distribution of roots in argument (modulus), that is, in rays (disks). A further extension of the Landau first theorem (39.75a–d; 39.74a,b; 39.73a,b) is possible, viz. the extended Landau theorem (Example 40.18) that uses in the estimate of the Landau radius the second derivative of function at the origin. An alternative to (i) the improvement of the estimate (37.74a,b) the Landau radius is (ii) the calculation of the exact value (Subsection 39.6.4). Since both (i) and (ii) involve the elliptic modular function, the comparison leads to an inequality, that is, (i) always satisfied (Subsection 39.6.4); (ii) satisfied by the elliptic modular function in particular (Example 40.18).

39.6.4 Exact Value of the Landau Radius (Caratheodory, 1912)

Consider a function analytic in a disk (39.76a) and not taking distinct values 0, 1 there (39.76b):

$$g(z) \in \mathcal{A}(|z| \le R) : g(z) \ne 0, 1; \tag{39.76a,b}$$

the elliptic modular function with this argument is analytic in the (39.32a,b) neighborhood of the origin (39.77a) and has positive imaginary part (39.77b):

$$G(z) \equiv E(a, b;\ g(z)) \in \mathcal{A}(|z| \le R); \ \mathrm{Im}\{G(z)\} > 0. \tag{39.77a,b}$$

From (39.77a,b) it follows that $G(z)$ and $G(0)$ $[G^*(0)]$ lie in the upper (lower) complex-G half-plane (39.78a):

$$\mathrm{Im}\{G(z)\} > 0 < \mathrm{Im}\{G(0)\} = -\mathrm{Im}\{G^*(0)\}; \ |G(z) - G(0)| < |G(z) - G^*(0)|, \tag{39.78a,b}$$

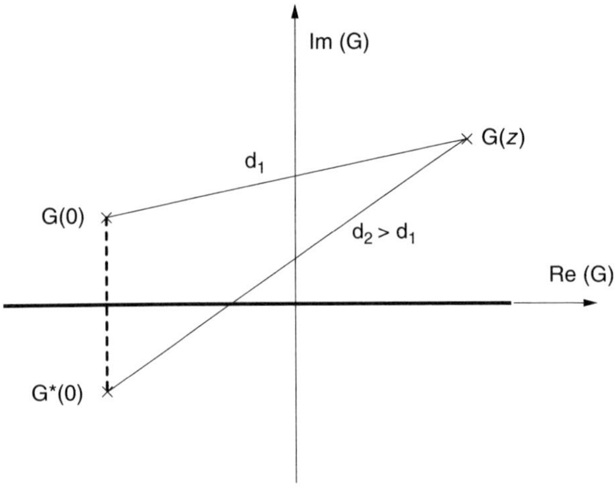

FIGURE 39.7
The second theorem of Picard states that an integral function takes an infinite number of times every value (with one possible exception). The generalization due to Julia (Landau) concerns the angular (radial) distribution of the roots [Figure 39.5 (40.11)]. The Landau theorem states that there is a finite radius beyond which a nonconstant complex function either ceases to be holomorphic or takes all values but one. The Picard second theorem does state the same property over the whole complex plane, that is, for a disk of "infinite radius"; the Landau theorem specifies a finite radius for which upper bounds can be found (Figure 40.11) in terms of the elliptic modular function; the latter is also used in the Caratheodory theorem which states the exact value of the Landau radius. The proof of the theorem relies on the simple property (Figure 39.7): a point on the upper-half complex plane is closer to another point in the upper-half complex plane, than to the conjugate or image of the latter on the real axis, which lies in the lower-half complex plane.

it follows (39.78b) that $G(z)$ is closer to $G(0)$ than to $G^*(0)$ in Figure 39.7. Thus the function (39.79a) does not exceed unity in modulus (39.79b):

$$H(z) \equiv \frac{G(z) - G(0)}{G(z) - G^*(0)}: \quad |H(z)| < 1; \quad R|H'(0)| \le 1, \tag{39.79a–c}$$

the Schwartz theorem (37.4b) the implies (39.79c). To calculate the upper bound for the radius R of the disk in (39.79c) is needed: (i) the derivative (39.80c) of (39.79a):

$$a_0 \equiv g(0), a_1 \equiv g'(z): H'(z) = [G(z) - G^*(0)]^{-2} [G(0) - G^*(0)] G'(z), \tag{39.80a–c}$$

that involves the value (39.80a) [slope (39.80b)] of the function at the center in E' (as a factor) in (39.80d,e):

$$G'(z) = \frac{\mathrm{d}[E(a,b,g(z))]}{\mathrm{d}g(z)} \frac{\mathrm{d}g}{\mathrm{d}z}, \quad G'(0) = a_1 E'(a_0); \tag{39.80d,e}$$

(ii) the value of (39.80c) at the origin:

$$\frac{a_1 E'(a_0)}{H'(0)} = G(0) - G^*(0) = 2\, i \,\mathrm{Im}\{G(0)\} = 2\, i\, \mathrm{Im}\{E(a_0)\}. \tag{39.81}$$

Substituting (39.81) in (39.79c) yields:

$$R \le |H'(0)|^{-1} = \frac{2\,\mathrm{Im}\{E(a_0)\}}{|a_1 E'(a_0)|} \equiv R(a_0, a_1) \equiv R_L, \tag{39.82}$$

as an upper bound for the radius.

The bound (39.82) is the exact value of the Landau radius, because it cannot be improved. To prove the last statement it is sufficient to find one function for that the bound is achieved, that is, the equality holds in (39.82). Such a function is (39.83a) that leads to (39.83b) and hence the equality in (39.82):

$$H(z) \equiv \frac{z}{R}: \quad H'(0) = \frac{1}{R}; \tag{39.83a,b}$$

inverting (39.79a) leads to (39.83c) in general:

$$G(z) = \frac{G(0) - G^*(0)H(z)}{1 - H(z)}: \quad H(z) = \frac{z}{R} \Rightarrow G(z) = \frac{G(0)\,R - G^*(0)\,z}{R - z}, \tag{39.83c–e}$$

and to (39.79e) for the particular case (39.83a) $\equiv$ (39.83d). Thus has been proved the **Caratheodory theorem**: *if a function is analytic in a disk and does not take (39.76a) the values (39.76b) then the radius of the disk cannot exceed the Landau radius, whose exact value is (39.82); it depends only on the value (slope) of the function (39.80a) [(39.80b)] at the origin a_0 (a_1), and involves the elliptic modular function (39.77a). The theorem implies that if the Landau radius is exceeded (i) either the function ceases to be analytic or (ii) it takes one of the values (39.76b).* The comparison of the upper bound (39.74a,b) $\equiv$ (39.84c) for (39.84a,b):

$$b = 1, d = 0: \frac{|\exp[-iE(a_0)]|}{|a_1 E'(a_0)|} \equiv R_1 \ge R(a_1, a_2) = \frac{2\,\mathrm{Im}[E(a_0)]}{|a_1 E'(a_0)|}, \tag{39.84a–c}$$

with the exact value (39.82) leads to the inequality:

$$2\mathrm{Im}[E(a_0)] \le |\exp[-iE(a_0)]| = \exp\{\mathrm{Im}[E(a_0)]\}; \tag{39.85a}$$

this is equivalent to (39.85c):

$$\lambda \equiv \mathrm{Im}[E(a_0)] > 0: \quad 0 \le \mathrm{e}^{\lambda} - 2\lambda = 1 - \lambda + \frac{\lambda^2}{2} + \sum_{n=3}^{\infty} \frac{\lambda^n}{n!} = \left(1 - \frac{\lambda}{2}\right)^2 + \frac{\lambda^2}{4} + \sum_{n=3}^{\infty} \frac{\lambda^n}{n!}, \tag{39.85b,c}$$

that is satisfied for any real positive (39.32b) value of λ in (39.85b). This proves the consistency of the exact Landau radius (39.82) and its upper bound (39.73a–d).

39.6.5 Schottky (1904) Theorem and Three Proofs of the Landau Theorem

A third elementary proof, not using the elliptic modular function, of the Landau theorem, follows from the **Schottky theorem (1904)**: *consider a function analytic in a disk (39.86a) and not taking the values zero and unity (39.86b); in a concentric subdisk (39.86c) of radius k times smaller there is an upper bound for the modulus of the function (39.86d) depending only on its value at the origin (39.80a) and the radius:*

$$g(z) \in \mathcal{A}(|z| < R); \quad g(z) \ne 0,1; \quad 0 < k < 1: \quad |z| \le kR \Rightarrow |g(z)| \le \Omega(a_0, k). \tag{39.86a–d}$$

The proof uses the functions (39.87b) of variable (39.87a) that take on the subdisk the same value as f on the disk:

$$z' = kz: \quad j(z) \equiv g(z') = g(kz); \quad j(0) = g(0) = a_0, \tag{39.87a–c}$$

the two function also take the same value (39.87c) at the center. The functions (39.87b) form a normal family (Subsection 37.2.2) in the subdisk, and thus have an upper bound that can depend only on a_0 and k. QED.

The Schottky theorem (39.86a–d) leads to a simple proof of the Landau theorem, for a function analytic in a disk (39.76a) not taking the values zero and unity (39.76b). The second Cauchy theorem (15.12) yields (39.88b) in a subdisk of radius one-half (39.88a):

$$k = \frac{1}{2}: \quad |a_1| \equiv |g'(0)| = \left| (2\pi i)^{-1} \int_{|\zeta|=kR} \zeta^{-2} g(\zeta)\, d\zeta \right| \\ \le (2\pi)^{-1} \left(\frac{R}{2}\right)^{-2} \Omega\left(a_0, \frac{1}{2}\right) 2\pi \left(\frac{R}{2}\right) = \frac{2\Omega(a_0)}{R}. \tag{39.88a,b}$$

Thus follows the **Montel theorem (1977)**: *if a function is analytic in a disk (39.76a), and does not take the values zero and unity (39.76b), then the radius of the disk has an upper bound (39.89c) that depends only on the value of the function (39.89a) and its first derivative (39.89b) at the center:*

$$a_0 \equiv g(0); \quad a_1 \equiv g'(0): \quad R \le \frac{2\Omega(a_0)}{|a_1|}; \tag{39.89a–c}$$

the Landau (Caratheodory) theorem (39.84c) [(39.82)] theorem specifies:

$$\Omega(a_0) = \frac{\operatorname{Im} E(a_0)}{|E'(a_0)|} \le \frac{|\exp[-iE(a_0)]|}{|2E'(a_0)|}, \tag{39.89d,e}$$

an overestimate (39.89e) [precise value (39.89d)] of the bound in terms of the elliptic modular function.

39.6.6 Successive Generalizations of Theorems on Bounds of Functions

Three proofs were given of the Landau theorem. The simplest (I) is the Montel proof (Subsection 39.6.5) that is elementary, since it does not use the elliptic modular function; it uses instead the Schottky theorem (39.86a–d) to prove the existence of a finite Landau radius, and its dependence only on the value of the function and its first derivative at the origin (39.89a–c). It does not supply any value, either (iii) precise or (ii) an upper bound, for the Landau radius. The latter results involve the elliptic modular function in the Landau (Caratheodory) proof in Subsection 39.6.3 (39.6.4) that specifies an upper bound (39.84c) [precise vale (39.82)] for the Landau radius. The Caratheodory proof (39.82) is consistent with the upper bound provided by the Landau proof (39.94a–c) because the comparison of the two (39.85a) specifies an inequality (39.85b,c) that is always satisfied. The still weaker Montel proof retains its interest as a simpler approach.

The original Montel theorem leads to the following generalization, obtained by replacing $a_1 \neq 0$ by $a_n \neq 0$ in the Montel proof (39.88b):

$$g(z) \in \mathcal{A}(|z| \le R); \quad g(z) \neq 0,1; \quad a_n \equiv g^{(n)}(0) \neq 0: \quad R \le 2\left|\frac{\Omega(a_0)}{a_n}\right|^{1/n}; \tag{39.90a–d}$$

this reduces to (39.89a–c) for $n = 1$ and states the **extended Montel theorem (1927)**: *if a function is analytic in a disk (39.90a) and does not take the values zero and unity (39.91b), and its n-th derivative does not vanish at the origin (39.90c), the radius has an upper bound (39.90d).* The proof is similar to (39.88a,b) using instead of the second (39.88b) the $(n+1)$-th coefficient (39.67b) ≡ (39.91) of the Maclaurin series (39.67a) and the corresponding form of the Cauchy third theorem (15.13) for $g(z)$:

$$|a_n| \equiv \left|\frac{g^{(n)}(0)}{n!}\right| = \left|(2\pi i)^{-1}\right| \int_{\zeta=R/2} \left|\zeta^{-n-1} g(\zeta)\, d\zeta\right|$$

$$\leq (2\pi)^{-1}\left(\frac{R}{2}\right)^{-n-1} 2\pi\left(\frac{R}{2}\right)|g(\zeta)|_{\max} = \left(\frac{R}{2}\right)^{-n}\Omega(a_0); \tag{39.91}$$

solving (39.91) for R yields (39.90d). QED.

A succession of progressively more general theorems can be proved for a complex function that does not take two values in a disk: (i) if the radius is infinity then either the function is a constant or it is not analytic, by the Picard first theorem (Subsection 39.3.2); (ii) if the function is analytic the radius has an upper bound (39.84a–c) by the Landau theorem (Subsection 39.6.3); (iii) the precise value of the Landau radius is (39.82) by the Caratheodory theorem (Subsection 39.6.4); (iv) if the function has nonzero n-th derivative at the center the bound on the radius is (39.90d) by the extended Montel theorem (Subsection 39.6.6); (v) by imposing further restrictions on the function the generalized Montel–Julia theorems are obtained, of that three cases are given next (Subsection 39.6.7).

39.6.7 Three Generalizations of the Landau–Julia Theorems (Montel, 1927)

Imposing on the function in the Schottky–Montel theorems (39.86a–d; 39.89a–c) three distinct sets of additional constraints leads to **three generalized Landau–Julia theorems (Montel, 1927)**: *consider a function holomorphic in a disk (39.76a) and not taking the values zero or unity (39.76b). Assume that the function meets one of the following three sets of conditions: (I) it takes the value* a_0 *at the origin (39.92a) and a distinct value* A_1 *at (39.92b) a point* z_1*:*

$$g(0) = a_0 \neq A_1 - g(z_1): \quad |z| \leq R(z_1; a_0, A_1); \tag{39.92a–c}$$

(II) it does not take the value zero (unity) more than p $(q \geq p)$ *times (39.93a) [(39.93b)]:*

$$N(f(z) = 0) \leq p, N(f(0) = 0) \leq q \geq p: \quad |z| \leq R(a_1, \ldots, a_p); \tag{39.93a–c}$$

(III) the function and its derivates at s points are given (39.94b) or bounded in modulus, and the orders of the derivatives add (39.94a) to p + 1:

$$\sum_{j=1}^{s} n_j = p+1; \quad f^{(n_j)}(z_j) = A_j: \quad |z| \leq R(z_1, \ldots, z_p; a_1, \ldots, a_p). \tag{39.94a–c}$$

Then the radius of the disk cannot exceed a finite value that depends only on the parameters indicated (39.92c–39.94c) respectively in cases I/II/III. Conversely, if radius (39.92–39.94c) is exceeded, the function either (i) ceases to be analytic or (ii) takes at least one of the values zero or unity. The theorem is equally valid in the Landau (Julia) form, for a finite disk (infinitesimal sector) with center (apex) at the origin.

The proof is similar in the three cases, and is given in case I. Suppose that the radius (39.92c) does not exist, and take a circle of radius $R > |z_1|$. There is at least one function $g(z)$ analytic in the disk satisfying (39.92a,b), and also (39.76b). A succession of disks of radii R_n exists each with an analytic function $g_n(z)$ satisfying (39.92a,b) and (39.76b). The sequence $g_1(z)$, $\ldots$, $g_n(z)$ is a normal family (Subsection 37.2.2) and has a uniformly convergent subsequence. The latter converges (39.95a) to an analytic function (39.95b) with the same properties (39.76b).

$$\lim_{n\to\infty} g_n(z) = g(z) \in \mathcal{A}(|C): \quad g(z) = 0,1 \Rightarrow g(z) = \text{const.} \tag{39.95a–c}$$

by the Picard first theorem (39.37) the limit function is a constant (39.95c). The value of the constant would have to be a_0 (A) at $z = 0$ $(z = z_1)$ by (39.92a,b); this is impossible because $a_0 \neq A_1$. The contradiction shows (39.92c) must be finite. Also it can depend only on a_0, A_1, QED. The proofs of (38.93a–c) [(38.94a–c)] are similar, excluding polynomials of degree p by requiring $a_{p+1} \neq 0$ $(a_{p+m} \neq 0$ with $m \geq 1)$. The function not taking the values (0,1) in (39.76b) may be replaced by a function not taking the values (b,d); the center of disk or origin of the ray be shifted to any point by a translation.

The theorems of Julia (Montel) in Subsections 39.5.2 (39.6.5–39.6.7) are qualitative. Quantitative results, for example, in the Schottky theorem (Subsection 39.6.5) use the elliptic modular function. The latter was used eight times in the preceding proofs of the theorems of: (i/ii) Picard (two in Subsection 39.3.2); (iii/iv) Landau (Subsection 39.6.3) plus its extension (Example 40.18); (v/vi) Caratheodory (Subsection 89.6.4) and Schottky (Subsection 39.6.5); (vii/viii) punctured conformal mappings (two in Subsection 37.5.3). The elliptic modular function was constructed geometrically via simple mappings (Section 39.4); the analytical calculation of the elliptic modular function uses the Schwartzian derivative and the hypergeometric function (Section 29.9). The hypergeometric function also specifies (Example 40.20) the periods of the Jacobian elliptic functions (Section 39.9); the latter are a particular case of the hyperelliptic integrals (Section 39.7). These specify for degree: (i) two the elementary transcendental functions (exponential, circular, hyperbolic) in Section 39.8; (ii/iii) three (four) the elliptic functions of Weierstrass (Jacobi) in Section 39.9.

39.7 Periods and Inversion of Hyperelliptic Integrals

The simpler holomorphic functions are the **rational functions** (Section 27.7) that are the ratio of two polynomials, and can be expressed in finite terms using only powers; for rational functions algebraic methods are often sufficient, although the methods of analysis may also be useful for example, for the decomposition in to partial fractions (Chapter 17; Sections 31.8 and 31.9). The nonrational functions are **transcendental functions** that have no finite representation in terms of powers, viz. they may have nonterminating power series, for example, Teixeira, Laurent, Lagrange–Burmann, Taylor, or Stirling–Maclaurin (Chapters 23 and 25); the power series are an infinite extension of polynomials, and raise convergence issues (Chapters 21 and 29). Another approach to some important classes of transcendental functions (Table 39.2) is the inversion of integrals of rational functions; if the integrand has branch-points, this leads to periodic functions (Chapter 7), whose periods are specified by integrals along the segments joining the branch-points (Subsection 39.7.2). The integrals are solution of nonlinear first-order differential equations (Subsection 39.7.1).

39.7.1 Conformal Mapping and Analytic Continuation

Consider a system (e.g., mechanical) with one dependent (e.g., position z) and one independent (e.g., time ζ) variable that satisfies the conservation of the total energy:

$$\text{const} = 2E = \left(\frac{\mathrm{d}z}{\mathrm{d}\zeta}\right)^2 + 2\Phi(\zeta), \tag{39.96a}$$

consisting of: (i) an energy E per unit mass (e.g., kinetic) quadratic on the first derivate (e.g., velocity); (ii) an energy (e.g., potential Φ) per unit mass depending on position. The nonlinear first-order differential equation (39.96) is equivalent to the second-order differential equation:

$$\frac{\mathrm{d}^2 z}{\mathrm{d}\zeta^2} = -\frac{\mathrm{d}\Phi}{\mathrm{d}\zeta} = \frac{F(\zeta)}{m}, \tag{39.96b}$$

where minus the gradient of the potential corresponds to the inertia force per unit mass, equal to the acceleration. The second-order differential equation (39.96b) is linear (nonlinear) if: (i) the force is linear (nonlinear); (ii) the potential is quadratic (has higher-order terms). The passage from (39.96a) to (39.96b) follows from

$$0 = \frac{\mathrm{d}E}{\mathrm{d}\zeta} = \left(\frac{\mathrm{d}^2 z}{\mathrm{d}\zeta^2} + \frac{\mathrm{d}\Phi}{\mathrm{d}\zeta}\right)\frac{\mathrm{d}z}{\mathrm{d}\zeta}, \tag{39.96c}$$

where: (i) the velocity cannot be zero for all time $\mathrm{d}z/\mathrm{d}\zeta \neq 0$; (ii) hence the term in square brackets must vanish, leading to (39.96b). The solution of the problem is the calculation of the function $z(\zeta)$ satisfying (39.96a) $\equiv$ (39.96b); one approach is the application to (39.96a) of **separation of variables**:

$$\zeta = \int \mathrm{d}\zeta = \int \mathrm{d}z \left(\frac{\mathrm{d}z}{\mathrm{d}\zeta}\right)^{-1} = \sqrt{2}\int^{\zeta} |E - \Phi(z)|^{-1/2}\,\mathrm{d}z + B; \tag{39.97}$$

the **solution by quadratures** reduces to an integration and adds an arbitrary constant B. *If the function Φ (i.e., the "potential") is (39.98a) a polynomial of degree N the* ***elliptic integral*** *(39.97) of degree N specifies the solution (39.98b) by inversion:*

$$2E - 2\Phi(z) \equiv P_n(z) = \sum_{n=1}^{N} a_n z^n \sim A^{-2} z^N: \quad \zeta - B = \int \{P_N(z)\}^{-1/2}\mathrm{d}z$$

$$= A\int \left\{\prod_{n=1}^{N} (z - z_n)^{-1/2}\right\}\mathrm{d}z; \tag{39.98a,b}$$

in the (39.98b) polynomial (39.98a) was factorized (31.56a, b) with N roots x_n, and leading coefficient $a_N = 1/A^2$. The transformation (39.98b) coincides with the Schwartz–Christoffel mapping (33.25) of the upper complex z-plane into the interior of a polygon, with right-angles $\gamma_n = \pi/2$ at all vertices $\zeta(z_n)$ if the roots are distinct, and multiples viz. angles $s\pi/2$ for roots of multiplicity s. The function $\zeta(z)$ defined in the polygon, can then be extended by reflection on its sides. The simplest cases are (i) simple zero at the origin leads to the square-root that is a two-valued function (Sections 7.1 and 7.2) with a branch-cut along the real axis that maps the whole complex ζ-plane into the upper-half complex z-plane; (ii) a double zero at the origin leads to the logarithm (Subsection 39.8.1) and the upper half

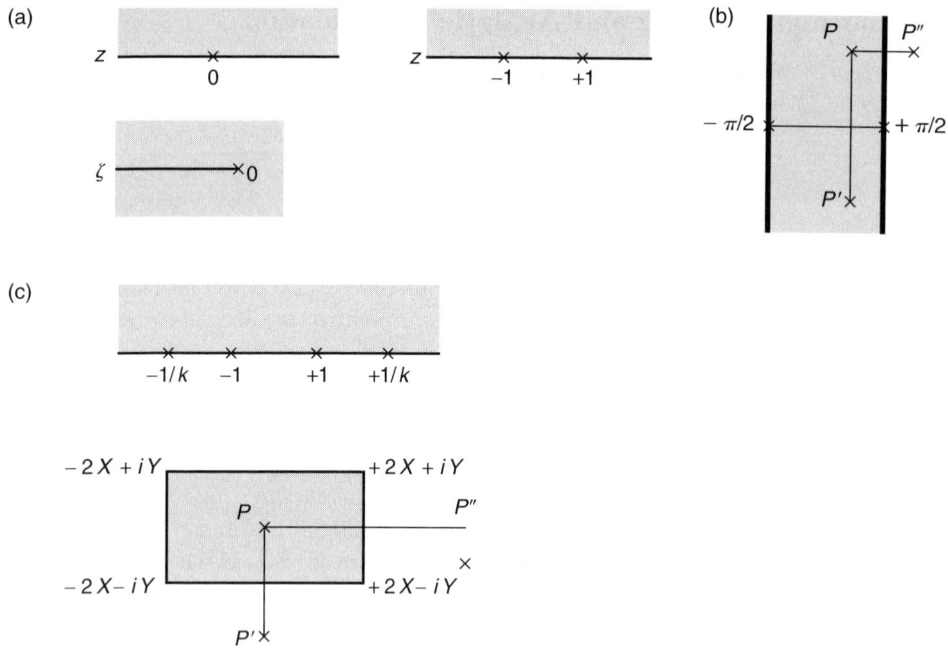

FIGURE 39.8
The Schwartz–Christoffel interior polygonal transformation (Figure 33.6a) can be used to map the upper-half z-complex plane into: (i) the whole complex ζ-plane with one critical point and a semiinfinite branch-cut (a), for example, leading to the logarithm; (ii) a strip in the complex ζ-plane (b) using two critical points, for example, leading to the circular functions; (iii) a rectangle in the complex ζ-plane (c) using four critical points, leading to the Jacobian elliptic functions.

z-plane is mapped (Figure 39.8a) to the whole ζ-plane, with a cut along the negative real axis; (iii) two simple zeros at $z = \pm 1$ lead to a triangle with two parallel sides (Figure 39.8b), that is, by reflection on the third side to a strip that covers the whole ζ-plane by reflection on two sides, that is, specifies one real period of a circular function (Subsection 39.8.2): (iv) four simple zeros at $z = \pm 1, \pm 1/k$, lead to a rectangle (Figure 39.8c), and reflections on two pairs of sides cover the whole ζ-plane with two periods, one real and one pure imaginary, for Jacobian elliptic functions (Section 39.9).

39.7.2 Contour Integration in the Plane with Branch-Cuts

The approach (Subsection 39.7.1) via conformal mapping (Chapter 33) and analytic continuation (Chapter 31), leads to the similar results (Subsection 39.7.2) as branch-cuts (Chapter 7) with contour integration (Chapter 17). The function (39.98a) ≡ (39.99a):

$$\frac{\mathrm{d}\zeta}{\mathrm{d}z} = \{P_N(z)\}^{-1/2} = A(z - z_n)^{-1/2}, \qquad \left.\frac{\mathrm{d}\zeta}{\mathrm{d}z}\right|_{z=z_n} = \infty, \tag{39.99a,b}$$

has branch-points (39.99b) at $z = z_n$ that are the critical points of the second kind (Section 33.5) of the conformal mapping. It can be made single-valued by taking the principal branch of $\zeta(z)$ in the complex z-plane with M branch-cuts, for example, $(z_{m-1}, z_m) \equiv C_m$ with $m = 1, \ldots, M$. If the value ζ_0 is known at a point z_0, the value ζ at another point z may be determined (Figure 39.9a) by: (i) taking a direct path L_1 that does not cross or enclose

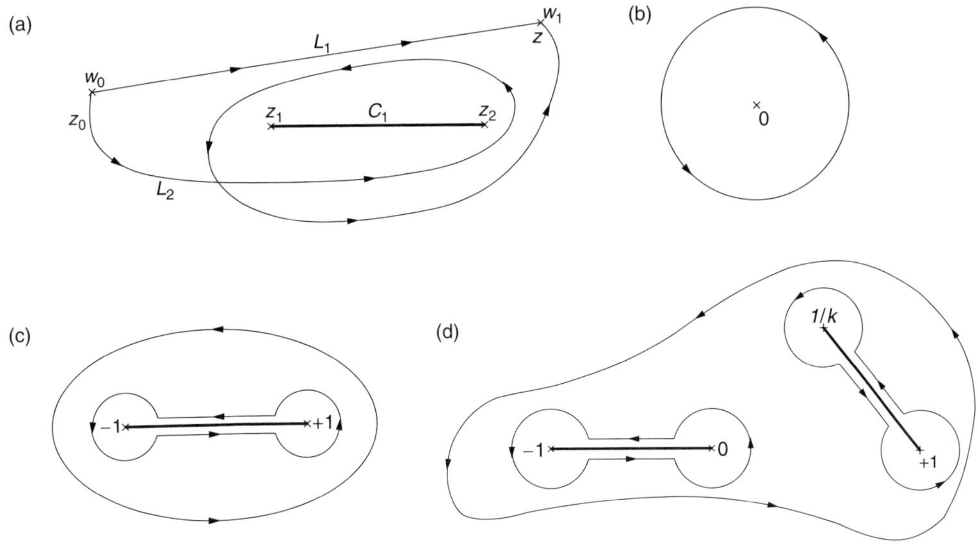

FIGURE 39.9
The Schwartz–Christoffel interior polygonal transformation (33.25) with right-angles $\gamma = \pi/2$ has exponents $-1/2$, leading by integration to elliptic integrals (39.98b). The latter are integrals $\zeta(z)$ of the inverse square root of a polynomial; the inverse function $z(\zeta)$ has periods (39.100b) corresponding to the integrals around the branch-cuts. In the case I of a polynomial of second-degree with a double root the integrand ζ^{-1} leads to the log $z = \log \zeta$ which has a finite branch-cut in the ζ-plane (Figure 39.8a); the integral around the branch-point specifies (Figure 39.9b) the period $2\pi i$ of the inverse function $\zeta = \exp z$. In the case II of a polynomial of second degree with distinct roots the integrand leads to an inverse circular sine (Figure 39.8b); the integral around the finite branch-cut specifies (Figure 39.9c) the period 2π of circular functions. In the case III of a polynomial of the fourth-degree with distinct roots the elliptic integral leaps to the inverse elliptic sine (Figure 39.8c); the integrals around the two finite branch-cuts specify (Figure 39.9d) the two periods of Jacobian elliptic functions. In all of the three cases I to III considered before, the value of the elliptic integral along two paths L_1 and L_2, one of which encircles a branch-cut, differs by the corresponding period (Figure 39.9a).

any branch-cut and leads to a value ζ_1; (ii) taking a path L_2 that goes β_m times round the m-th branch-cut C_m in the positive direction and adds to ζ_1 the value (39.100a):

$$\zeta_2 - \zeta_1 = \sum_{m=1}^{M} \beta_m T_m; \quad T_m \equiv 2 \int_{z_{n-1}}^{z_n} \{P_N(\zeta)\}^{-1/2} \, d\zeta, \tag{39.100a,b}$$

the integrals (39.100b) around the branch-cut may be contracted to the cut, since for single roots the inverse square-root singularities at the edges are integrable. Considering the inverse function, if follows:

$$z(\zeta_1) = z(\zeta_2) = z\left(\zeta + \sum_{m=1}^{M} \beta_m T_m\right), \tag{39.101}$$

that it is periodic, with periods T_m with $m = 1, \ldots, M$ that may not all be independent. Thus *the inversion* $z(\zeta)$ *of the* $\zeta(z)$ *hyperelliptic integral (39.98b) leads to a function (39.101) that has (39.100b) periods* T_m *determined by integrals along the branch-cuts* (z_{n-1}, z_n) *joining the roots* z_n *of the polynomial.*

39.8 Logarithm, Exponential, and Circular/Hyperbolic Functions

For a polynomial of first degree, that is, a linear function, (39.98b) leads to a square-root that is a two-valued function with period, $4\pi i$. The next simplest elliptic integrals, for a quadratic function, lead to the logarithm/circular/hyperbolic functions (Subsection 39.8.1/2/3) for double/distinct roots with negative/positive leading coefficient. The singly periodic functions are defined in a fundamental strip (Subsection 39.8.4) and repeated on congruent strips.

39.8.1 Double Root, One-Branch Point, and Logarithm

The simplest case of a polynomial of second degree is coincident roots (39.102a) that may be placed by translation at the origin:

$$P_2(z) = z^2: \quad \zeta = \int z^{-1}\,\mathrm{d}z = \log z, \quad z = \exp(\zeta), \tag{39.102a–c}$$

the integral (39.98b) $\equiv$ (39.102b) defines the **logarithm** whose inverse is the **exponential** (39.103c). The logarithm is a single-valued function in the complex z-plane with (Figure 39.8a) with (Sections 7.2–7.4) a semiinfinite branch-cut joining the origin to infinity; the exponential is a periodic function, whose fundamental period is obtained integrating (39.102b) in the positive direction around the branch-point (Figure 39.9b) at the origin (39.103b):

$$n \in |N: \quad T = \int\limits_{|z|=R} z^{-1}\,\mathrm{d}z = 2\pi i; \quad \exp(z) = \exp(z + n2\pi i), \tag{39.103a–c}$$

thus it has been shown that the exponential is unchanged (39.103c) by adding a multiple (39.103a) of $2\pi i$ to the argument, in agreement with the geometric representation (Section 1.5).

39.8.2 Two Distinct Roots and Circular Functions

If the polynomial $P_2(z)$ remains of the second degree but has two distinct roots, by means of a rotation and translation they can be placed at $z = \pm 1$: if the leading coefficient $a_2 < 0$ is negative, it can be made equal to $a_2 = -1$ by a homothety. Thus the polynomial (39.104a):

$$P_2(z) = 1 - z^2: \quad \zeta = \int\limits_0^z \left(1 - t^2\right)^{-1/2} \mathrm{d}t = \{\arcsin t\}_0^z, \tag{39.104a,b}$$

specifies by inverting of the elliptic integral (39.104b) $\equiv$ (39.105b) one of the **circular functions** namely, namely the **sine**:

$$n \in |N: \quad \zeta = \arcsin z, \quad z = \sin\zeta = \sin(\zeta + 2n\pi), \tag{39.105a–c}$$

The inverse function arc sin (39.104b) $\equiv$ (39.105b) is single-valued in the complex z-plane with (7.27b) two semiinfinite branch-cuts (Figure 7.3a) joining $z = \pm 1$ to $\pm\infty$ along the real axis. Integrating (Figure 39.9c) between the branch-points $z = \pm 1$, specifies in (39.106) half the period:

$$X = 2\int_{-1}^{+1} \left(1-t^2\right)^{-1/2} \mathrm{d}t = 2\left\{\arcsin t\right\}_{-1}^{+1} = 2\pi. \tag{39.106}$$

Geometrically (Figure 39.8b) the sine is defined first in the vertical strip $-\pi/2 \le \mathrm{Re}(z) < \pi/2$ or $0 \le \mathrm{Re}(z) < \pi$; it is extended (Section 31.2) by reflection to $0 \le \mathrm{Re}(z) < 2\pi$, that is the **fundamental strip**, repeated periodically to cover the whole ζ-plane by analytic continuation.

39.8.3 Reversed Sign of Leading Coefficient and Hyperbolic Functions

If the polynomial of the second degree has distinct roots and positive leading coefficient it can by a linear transformation, be reduced to (39.107a):

$$P_2(z) = z^2 - 1: \quad \zeta = \int_1^z \left(t^2 - 1\right)^{-1/2} \mathrm{d}t = \left\{\arg\cosh t\right\}_1^z, \tag{39.107a,b}$$

inversion of the corresponding elliptic integral (39.107b) $\equiv$ (39.108b) leads to one of the **hyperbolic functions**, namely (39.108c) the **cosine**:

$$n \in |N: \quad \zeta = \arg\cosh z, \quad z = \cosh(\zeta + n2\pi i). \tag{39.108a–c}$$

The arg cosh is single-valued in the complex z-plane, with a semiinfinite branch-cut (7.32a,b) joining infinity to the branch-point $z = +1$, “hiding” (Figure 39.8h) the other branch-point (Subsection 7.8.3); the period is given by twice the integral (39.107b) taken between ± 1, and since it is similar to (39.106) with a change of sign under the radical, it is imaginary (39.109):

$$Y = 2\int_{-1}^{+1} \left(t^2 - 1\right)^{-1/2} \mathrm{d}t = 2i\int_{-1}^{+1} \left(1 - t^2\right)^{-1/2} \mathrm{d}t = 2\pi i, \tag{39.109}$$

as indicated in (39.108c). The hyperbolic functions of complex variable are similar (36.86a,b; 36.136a–c) to the circular functions with a rotation by $i = \mathrm{e}^{i\pi/2}$, as shown the periods (39.106) and (39.109) and relations like:

$$2\cos z \equiv \mathrm{e}^{iz} + \mathrm{e}^{-iz} \equiv 2\cosh(iz). \tag{39.110}$$

The hyperbolic functions are defined in a **fundamental region** that is a horizontal strip $-\pi \le \mathrm{Im}(z) < \pi$ or $0 \le \mathrm{Im}(z) < 2\pi$, and by periodic reflection cover the whole ζ-plane. The vertical (horizontal) position of the fundamental strip implies that the circular (hyperbolic) functions do (39.105c; 39.106) [do not (39.108c; 39.109)] appear to be periodic for real variable. The preceding results can be used to evaluate the elliptic integral (39.98b) in terms of elementary functions for an arbitrary polynomial of the second degree (Example 40.19).

39.8.4 Congruences of the Fundamental Strip for Singly Periodic Many-Valued Function

As an example of complex function with a single period is taken the exponential (39.103a–c). The period is changed from $2\pi i$ to any complex value $2Z$ via the change of independent

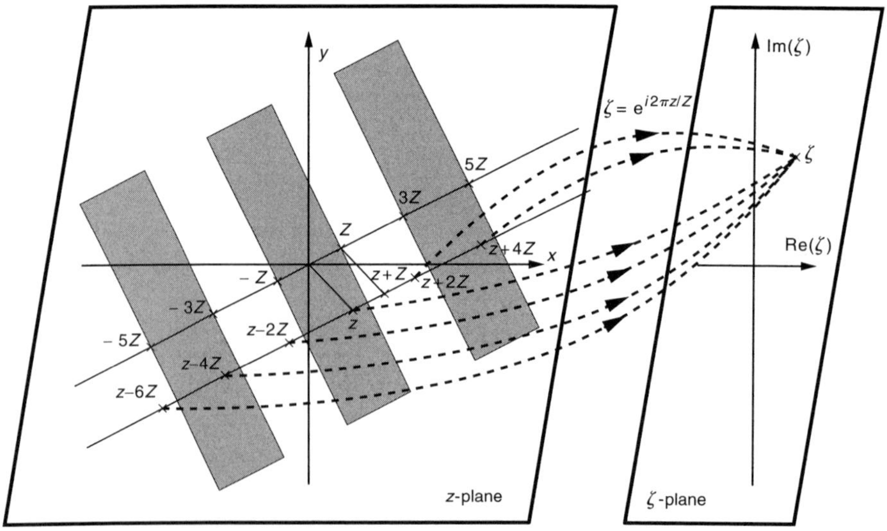

FIGURE 39.10
Consider a complex function $\zeta(z)$ with one complex period $2Z$. Taking two lines through $\pm Z$ perpendicular to the line joining $-Z$ to $+Z$ specifies a fundamental strip. Given any point z in the fundamental strip, all points $z_n = z + 2nZ$ with n integer lie on congruent strips, and lead to the same value of the function. Thus all congruent points z (mod Z) on the z-plane are mapped to the same point in the ζ-plane.

variable (39.111a):

$$z \to \frac{i\pi z}{Z}; \quad n \in |Z: \quad \zeta \equiv f(z) \equiv \exp\left(\frac{i\pi z}{Z}\right); \tag{39.111a–c}$$

as follows from (39.111c) with n integer (39.111b).

$$f(z) = \exp\left(\frac{i\pi z}{Z} + i2\pi n\right) = \exp\left[\frac{i\pi(z+2nZ)}{Z}\right] = f(z+2nZ). \tag{39.111d}$$

The mapping (39.111c) is represented (Figure 39.10) starting with the following construction in the Z-plane: (i) mark the complex number Z that is the fundamental period; (ii) join Z to the origin, and mark points at equal distances Z, $\pm 3Z$, $\pm 5Z, \ldots, Z + 2nZ$ with n integer that are the multiple periods; (iii) through the multiple periods draw perpendicular lines, forming a set of parallel strips; (iv) the fundamental strip passes through Z and $Z- 2Z = -Z$, and all the others are congruent; (v) let z be any point in the fundamental strip; (vi) draw through z a line parallel to the line of periods, and mark the congruent points $z + 2nZ$ spaced by a multiple of the period $2Z$; (vii) the many-valued function (39.111b) repeats itself on all strips congruent to the fundamental strip, since each point z in the fundamental strip, together with all congruent points $z + 2nZ$ is mapped to the same point in the ζ-plane (Figure 39.10). In the case of circular (39.105a–c) [hyperbolic (39.108a–c)]: (i) the period is real (39.106) [imaginary (39.109)]; (ii) the fundamental strip $-\pi < \text{Im}(z) \leq \pi$ ($-\pi < \text{Re}(z) \leq \pi$) is vertical (horizontal) in Figure 39.8b or Figure 39.10 with $Z = \pi$ ($Z = i\pi$); (iii) the congruent strips $-(2n-1)\pi < \text{Im}(z) \leq (2n+1)\pi$ [$-(2n-1)\pi < \text{Im}(z) \leq (2n+1)\pi$] with n integer are also vertical (horizontal); (iv) the function is univalent in the fundamental strip and many-valued in the congruent strips obtained by horizontal (vertical) translation by $2n\pi$ ($i2n\pi$).

39.9 Elliptic Functions of Jacobi (1827) and Weierstrass (1895)

The elliptic integrals of lowest degree beyond the quadratic corresponding to the elementary transcendental functions (Section 39.8) are the cubic (quartic) leading to the elliptic functions (Subsection 39.9.1) of Weierstrass (Jacobi). These are doubly periodic, for example, the Jacobian elliptic functions (Subsection 39.9.2) have one real (one imaginary) period that are related to the circular (hyperbolic) functions in the particular cases with one period and a fundamental strip; (ii) otherwise they determine a fundamental parallelogram where the function is univalent (Subsection 39.9.3), repeating itself by translations as a many-valued function in a double infinity of congruent parallelograms.

39.9.1 Cubic and Quartic Elliptic Integrals

In a cubic polynomial the coefficient of the square $a_2 = 0$ can be made to vanish by a translation, the leading term can be put to unity $a_0 = 1$ by a homothety, yielding (39.112a):

$$P_3(z) = z^3 - az - b: \qquad \zeta = \int \left\{z^3 - az - b\right\}^{-1/2} \mathrm{d}z, \quad z = P(\zeta; a, b), \qquad (39.112\text{a–c})$$

the corresponding elliptic integral (39.112b) specifies by inversion the **Weierstrass elliptic** function (39.112c). The **Jacobian elliptic functions** that correspond to a quartic polynomial, are considered next (Table 39.3). A function whose fundamental region is a

TABLE 39.3
Inversion of Elliptic Integrals

Degree	Polynomial	Function	Section (s)
One	Linear	Square root	7.1–7.4
Two	Equal roots	Exponential, logarithm	7.2–7.4; 23.9; 37.4; 39.8
	Negative*	Circular	3.6–3.7; 7.8; 39.8
	Positive*	Hyperbolic	7.8; 39.8
Three	Cubic	Elliptic: Weierstrass	39.9
Four	Quartic	Elliptic: Jacobi	39.9
More than four	Quintic and higher	Hyperlliptic	7.9; 39.7

* Coefficient of the leading term.

Note: The elliptic integrals are (39.98b) integrals of the inverse square root of a polynomial $P_N(z)$. The roots of the polynomial are branch-points of the integrand $1/\sqrt{P_N(z)} \equiv \{P_N(z)\}^{-1/2}$, and determine the nature of the elliptic integral $\zeta = \zeta(z)$, or by inversion of the "elliptic function" $z = z(\zeta)$. For example, a double root $P_2(z) = z^2$ leads to the logarithm $\zeta = \int z^{-1}\,\mathrm{d}z = \log z$, which has (Figure 7.1a ≡ Figure 39.8a) one branch-point and a branch-cut joining it to infinity; the inverse function is the exponential $z = \exp(\zeta)$. If the polynomial is of the second degree with distinct roots, there are two branch-points. These may be joined: (i) by one finite branch-cut [e.g., Figure 7.4a for the arc csc in (7.27a,b)]; (ii) to infinity by two semiinfinite branch-cuts [e.g., Figure 8.1g for arc sin in (7.22a,b)]; (iii) one infinite branch-cut joining one branch-point to infinity and containing or "hiding" the other branch-point [e.g., Figure 8.1h for arg cosh in (7.32a,b)]. Generally an elliptic integral involving a polynomial of the second degree with distinct roots leads by inversion to circular or hyperbolic functions. If the polynomial is of degree three (four) there are three (four) branch-points in the elliptic integral; its inversion leads to the Weierstrassian (Jacobian) elliptic functions. Polynomials of degrees higher than four, with five or more branch-points in the hyperelliptic integral, lead by inversion to hyperelliptic functions. The elementary (elliptic) functions, namely circular and hyperbolic (Weierstrassian and Jacobian) have one (two) periods, associated with finite branch-cuts; thus their fundamental region is a strip (a rectangle) [e.g., Figure 39.8b(c)], and they cover the whole complex plane by reflection on the sides [Figure 39.10 (39.11)].

parallelogram (Figure 39.8c) with vertices at $(0,\ X,\ iY,\ X + iY)$, can be extended by reflection to the rectangle, $-2X < \operatorname{Re}(z) \leq 2X, -Y < \operatorname{Im}(z) \leq Y$, and by repetition of the real $4X$ and imaginary $i2Y$ period covers the whole ζ-plane. Such a function requires four critical points in the Schwartz–Christoffel transformation (33.25) ≡ (39.98b) say, at (39.113a) with right angles (39.113b) leading to (39.113c):

$$z_{1-4} = \pm 1, \pm\frac{1}{k}, \quad \gamma_{1-4} = \frac{\pi}{2}: \quad \zeta = A\int\left\{\left(t^2-1\right)\left(t^2-\frac{1}{k^2}\right)\right\}^{-1/2} dt + B; \qquad (39.113\text{a–c})$$

choosing $\zeta = 0$ for $z = 0$, so that that $B = 0$ in (39.113c) leads to the inverse elliptic sine function (39.114c):

$$B = 0, \quad A = \frac{1}{k}: \quad \zeta = \int_0^z \left\{\left(1-t^2\right)\left(1-k^2t^2\right)\right\}^{-1/2} dt = \arg sn\,(z;k), \qquad (39.114\text{a–c})$$

taking the value (39.114b) for the other constant. In the plane these can be only two independent fundamental periods. Two fundamental periods are obtained by integrating between two pairs of branch-points (39.113a), where start the branch-cuts: (i) the integral over $(-1,\ +1)$ leads to a quarter period X in (39.115a):

$$-1 < k < +1: \quad X = \int_0^1 \left\{\left(1-t^2\right)\left(1-k^2t^2\right)\right\}^{-1/2} dt = \operatorname{Re}(X), \qquad (39.115\text{a,b})$$

that is real for real k in the unit interval (39.115a); (ii) for the same range of k the integral over $(-1/k,\ 1/k)$ leads to the period (39.116a):

$$-1 < k < +1: \quad X + iY = \int_0^{1/k} \left\{\left(1-t^2\right)\left(1-k^2t^2\right)\right\}^{-1/2} dt, \qquad (39.116\text{a,b})$$

that specifies (39.117a):

$$iY = \int_1^{1/k} \left\{\left(1-t^2\right)\left(1-k^2t^2\right)\right\}^{-1/2} dt, Y = \int_1^{1/k} \left\{\left(t^2-1\right)\left(1-k^2t^2\right)\right\}^{-1/2} dt = \operatorname{Re}(Y), \qquad (39.117\text{a,b})$$

where (39.117b) is real. Thus the **elliptic sine function** (39.118):

$$n, m \in |N: \quad sn\,(z;k) = sn\,(z + 4nX + i2mY;k), \qquad (39.118)$$

has the **real** $4X$ **(imaginary** $i2Y$**) periods** (39.115b) [(39.117b)] *for the parameter (39.115a) ≡ (39.116a).*

39.9.2 Pair of Periods of Jacobian Elliptic Functions

Setting $k = 0$ in (39.114c) [(39.115b)] regains (39.104b) [(39.106)], that is, the elliptic sine function of zero modulus is the circular sine function (39.119b) that has one real period 2π in (39.119a):

$$X = \frac{\pi}{2}: \quad sn\,(z;0) = \sin z = \sin(z + 4X) = \sin(z + 2\pi). \qquad (39.119\text{a,b})$$

Setting $k=1$ in (39.114c) leads to the hyperbolic tangent (39.120b):

$$Y = \frac{i\pi}{2}: \quad \zeta = \int_0^z \left(1-t^2\right)^{-1} dt = \arg\tanh z, \qquad (39.120\text{a,b})$$

that is, the elliptic sine of modulus unity is the hyperbolic tangent that has imaginary period πi corresponding to (39.117a):

$$sn\,(z;1) = \tanh z = \tanh\,(z+i2Y) = \tanh\,(z+i\pi). \qquad (39.120\text{c})$$

Thus the elliptic functions contain both the circular (hyperbolic) types, with one real (imaginary) period for modulus zero $k=0$ (unity $k=1$); for intermediate moduli $0<k<1$ the elliptic function have one real (imaginary) period, corresponding to integrals along two segments (Figure 39.9d). The periods of elliptic functions can be expressed in terms of (Example 40.20) Gaussian hypergeometric series (Section 29.9); this allows analytic continuation for complex values of k.

39.9.3 Congruencies of Fundamental Rectangle for the Elliptic Sine

The analogue of the mapping (Figure 39.10) for a many-valued function with one period (39.111a–d) is indicated next for a function with two periods, taking as example the elliptic sine (39.118) in Figure 39.11: (i) the case $k=0$ ($k=1$) of circular (hyperbolic) functions is excluded, since it corresponds to a single real (imaginary) fundamental period 2π ($i\pi$), and hence vertical (horizontal) strips congruent to $-\pi < \text{Re}\,(z) \leq \pi$ ($-\pi/2 < \text{Im}\,(z) \leq \pi/2$) in

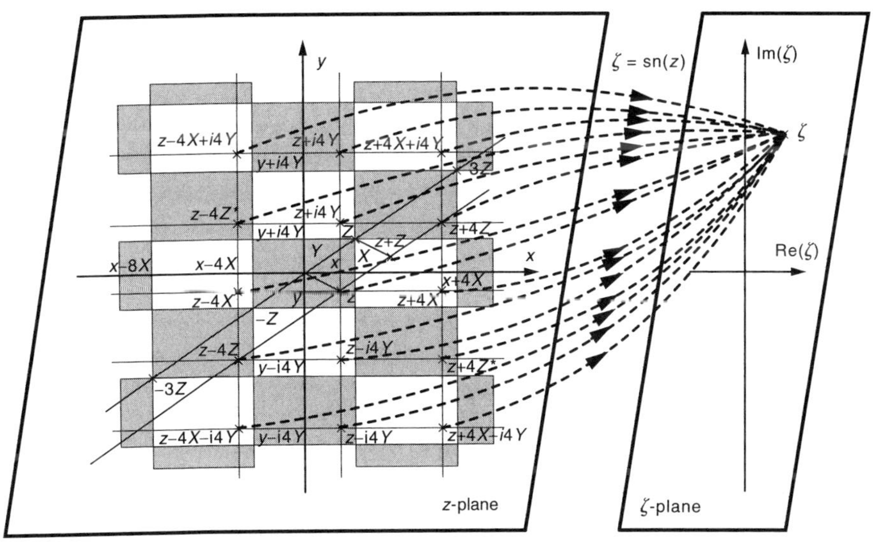

FIGURE 39.11
Consider a complex function $\zeta(z)$ with two periods (Z_1, Z_2). If the periods are complex and have distinct arguments they specify a parallelogram; a particular case is one real $Z_1 \equiv 4X$ and one imaginary $Z_2 \equiv i2Y$ period, leading to a fundamental rectangle with vertices at $(\pm 2X \pm iY)$, sides parallel to the coordinate axis and half-diagonal $Z = (Z_1 + iZ_2)/2 = 2X + iY$. Given a point z in the fundamental parallelogram, all points $z_{n,m} = z + 4nX + i2mY$ lie on congruent parallelograms; the function $\zeta(z)$ takes the same value at all the points $z_{n,m}$, which are mapped to a single point in the ζ-plane.

Figure 39.10; (ii) for $0 < k < 1$ there are one real $4X$ and one imaginary $i2Y$ fundamental period, specifying the fundamental rectangle (39.121a,b) with diagonal (39.121c):

$$-2X < \operatorname{Re}(z) \leq 2X; \quad -Y < \operatorname{Im}(z) \leq Y: \quad Z \equiv 4X + i2Y; \tag{39.121a–c}$$

(iii) drawing parallel vertical (horizontal) lines at distances multiple of $4X(2Y)$ covers the z-plane with rectangles congruent to the fundamental rectangle (39.121a,b); (iv) given a point z in the fundamental rectangle there is a double infinity (39.122a,b) of congruent points (39.122c) at that the function (39.78) takes the same value (39.122d)

$$n, m \in |Z: \quad z_{n,m} = z + 4nX + i2mY, \quad \zeta = g_n(z_{n,m}; k), \tag{39.122a–d}$$

(v) thus all are mapped to same point (39.122d) in the ζ-plane (Figure 39.11).

Note 39.1. *Roots of Rational and Transcendental Equations:* The first motivation for the present Chapter 39 is the question of roots of equations that by combining the results for rational (Subsection 31.6.3) [(transcendental (Sections 39.2 and 39.3)] may be summarized in Table 39.1. The motivations to address the two Picard theorems, and to do so via the elliptic modular function has been to prove two sets of fundamental results: (Section 39.3) the existence of roots of transcendental equations; (Subsection 37.5.3) some the properties of conformal mappings. The importance of these results and their illumination of some essential aspects of complex function theory can be escaped completely only at some loss of insight. Not all textbooks on complex analysis mention the Casorati–Weierstrass theorem (39.21) that was the inspiration for those of Picard; those that may or may not mention the Picard theorems (39.22; 39.37). Some state the Picard theorem(s) without proof, on the grounds that no elementary derivation is available. At the opposite end of the spectrum some works concentrating on the finer aspects of complex analysis, address the Picard theorems at an early stage as a stepping stone to further developments. Most textbooks that give proofs of the Picard theorem, do so after covering must of the basic aspects of complex analysis. The fundamental importance of the Casorati–Weierstrass and the Picard theorems concerning the existence of roots of equations is a reason to include proofs of both. The Casorati–Weierstrass theorem shows that in any neighborhood of an essential singularity an integral function comes arbitrarily close to any value; this sufficient to ensure the existence of approximate numerical solutions of an equation $f(z) = c$ for any c, with arbitrary accuracy of the root(s) z for any integral function $f(z)$. The Weierstrass theorem does not prove that the function takes precisely every value c; the Picard theorem clarifies this showing that there may be one exceptional value for that no roots exist, otherwise for all other values of c exist an infinite number of roots. Although Casorati–Weierstrass theorem is fairly straightforward to prove, this is not the case with Picard theorem. Although the statements made above are very simple to formulate, like some fundamental results elsewhere (e.g., in the theory of numbers), they are less simple to prove. The proofs of these statements in the literature fall into two classes: (i) elementary, but long-winded proofs that can be approached in more than one way, for example, through the theorems of Schottky, Caratheodory–Borel, Landau, or Bloch that are all interrelated; (ii) the original proof of Picard, and its subsequent refinements that is short, if prior knowledge is assumed of the properties of the elliptic modular function that is not elementary. The choice (ii) of using the properties of the elliptic modular function (i) leads to short proofs of the two Picard theorems (Subsection 39.3.2); (ii) is followed by a precise quantitative extension to the Landau (Subsection 39.6.3) and Caratheodory (Subsection 39.6.4) theorems; (iii) serves for comparison with the qualitative theorems of Julia (Subsection 39.5.2), Schottky and Montel (Subsection 39.6.5) and further generalizations (Subsections 39.6.6–7); (iv) also allows the proof of some theorems concerning punctured conformal mappings (Subsection 37.5.3).

Note 39.2. *Branch-Points, Branch-Cuts, and Periodic Functions:* The conformal mapping and contour integration lead from the elementary transcendental functions (circular and hyperbolic) to the Weierstress and Jacobi elliptic functions, and to the hyperlliptic integrals. The latter have branch-points and lead to periods that are integrals along lines joining branch-points. Thus three fundamental aspects of complex function theory are related at a deeper level: (i) branch-points and cuts (Chapter 7) for multivalued functions on the Riemann sphere (Chapter 9); (ii) essential singularities (Chapter 25) of integral functions (Chapter 27) and roots of transcendental equations (Chapters 31 and 39); (iii) conformal mappings between single or multiply connected compact or noncompact regions involving automorphic groups (Chapters 33, 35, and 37). These theoretical and abstract issues are the beginning of a deeper view of the fundamentals of complex analysis that would form an extensive subject. The approach adopted has been to use the minimum necessary to clarify some fundamental aspects of : (i/ii) the representation of multivalued and multivalent functions; (iii/iv) the essential singularities of transcendental functions and the existence of roots of equations and of conformal mappings (v/vi) the periods of elementary transcendental and elliptic functions. The first level of complex analysis, like the four Cauchy theorems, the calculus of residues, the power series, the analytic continuation and conformal mappings, involve more calculations, often related to potential field problems. The second level of complex analysis, like the topology of the Riemann surfaces, the existence of conformal mappings between multiply connected regions, the bounds and asymptotic scalings of functions, tends to involve more the introduction of new concepts and their properties. This division of complex analysis in two levels is an indication of trends, and is not sharp, as there are multiple interrelations. For example, the fundamental region of circular and hyperbolic (elliptic) functions is a strip (rectangle), and thus they specify potential fields between parallel walls (Sections 36.6 and 36.7) [in a box (Section 39.9)]. An example is the wall effects on an aircraft model in a wind tunnel with rectangular cross-section. The question of wall effects is addressed more simply in terms of elementary functions for a cylinder moving or fixed in cylindrical cavity. The close connection between the theory of complex functions and potential fields in the plane implies that the former solves problems of the latter, and the latter provides interpretations for results of the former.

Note 39.3. *Complex Analysis Applied to Families of Functions:* The basic elements of complex analysis developed in the introductory volume were intended to: (i) support immediate application to a variety of potential field problems in fluids, solids, mechanics, heat and electromagnetism with a maximum simplicity; (ii) to lay the ground work for future expansion into various families of functions as the need for them arises. It may thus be appropriate to indicate the relation between families of functions (Diagram 39.1), some of their main elements (List 39.1) and the sequence in that the subject might be addressed. The Laurent series (Chapter 25) was used (Chapter 27) to: (i) classify the singularities of a function (Table 27.1); (ii) classify functions by their singularities (Table 27.2). A broader classification of special points appears in Classification 39.1 that includes and extends Diagram 27.1, by not being restricted to Laurent series. The families of functions in List 39.1 are (i) neither an extension of Diagram 27.2, because the latter is comprehensive; (ii) nor based on the classification of special points in Classification 39.1, because several of the families of functions can have various types of special points. The families of functions in List 39.1 are those of most frequent use in the solution of problems, listed by approximate order of increasing complexity.

The rational functions are those that can be represented in finite terms, and some of their properties have already been addressed (Chapters 7 and 31). The transcendental functions require infinite representations, for example, series; the elementary transcendental functions viz. exponential, circular, hyperbolic, and inverse (logarithmic and cyclometric) are

LIST 39.1
Some Classes of Functions

Higher Transcendental			**Rational**	**Basic transcendental**	
Specific	**Extended**	**Special**	**Polynomials**	**Elementary**	**Auxiliary**
31. Zeta	–	13. Airy	–	1. Exponential 1.5–1.8; 3.5; 17.5; 23.9; 27.1–27.2; 27.9	7. Gamma
	–		–		
32. Theta	15. Extended Bessel	14. Bessel	–	2. Circular 3.5–3.8; 5; 17.5; 23.9; 29.8–29.9; 37.9; 39.8	8. Beta
			–		
33. Weierstrass Elliptic 39.9	18. Extended Confluent Hypergeometric	15. Confluent Hypergeometric E30.20	16. Hermite	3. Hyperbolic: 7.8; 39.8	9. Digamma/ poligamma 29.5.2
				4. Logarithm: 5; 7; 37.4; 39.8	10. Error
			17. Laguerre	5. Inverse circular: 7.8; 39.8	11. Exponential/ sine/cosine integrals
				6. Inverse hyperbolic: 7.8; 39.8	12. Fresnel
34. Jacobian Elliptic 39.8	24. Mathieu 25. Extended Gaussian Hypergeometric	19. Gaussian Hypergeometric 29.9; E30.20; E40.20	20. Legendre 21. Tchebycheff 22. Jacobi 23. Gegenbauer		
35. Hyperelliptic 39.7	28. Extended Heun 30. Extended Generalized Hypergeometric	27. Lamé 29. Generalized Hypergeometric E30.20; N23.3	26. Sonine		

Note: The complex functions may be listed in several families. The rational functions, that is, polynomials and their ratios, can be expressed in finite terms. Of the transcendental functions, which cannot be expressed in finite terms, the simplest are the elementary transcendental functions (Section 39.8), viz. exponential, circular, and hyperbolic, and their inverses (logarithm and cyclometric functions: arc circ and arg hyp). The basic transcendental functions include, besides the elementary transcendental functions, the auxiliary functions; one example is the digamma function (Subsection 29.5.2). Beyond the preceding are the higher transcendental functions, divided into specific, special, and extended. The specific functions include the elliptic (Section 39.9) which are associated with the solution of nonlinear differential equations (Section 39.7). The special functions include the hypergeometric, for example, Gaussian (Section 39.9), confluent and generalized (Example 30.20); they are associated with the solution of linear differential equations with variable coefficients (Notes 23.2 and 23.3). This can lead to further generalizations of the special functions, viz. the extended functions. The series for special functions terminate for particular choices of parameters, for example, the Gaussian hypergeometric series (29.74) terminates at the n-the term if $a = -n$ (or $b = -n$) is a negative integer; thus the special functions contain as particular cases special polynomials, which are rational functions. All of these functions can be treated by the methods of complex analysis presented before, although only a few (mostly the simplest) were used as examples. The generalized functions or distributions, for example, Heaviside unit jump, Dirac unit impulse, and derivatives require a different approach and are not included in this list.

the baseline building block for the solutions of most problems; they have appeared in multiple occasions, and can be treated systematically as the simplest example of many properties of complex functions. The auxiliary functions are defined on the basis of elementary functions (e.g., parametric integrals: Sections 13.8 and 13.9), or by series of fractions, like the digamma function (Subsection 29.5.2); these and other related functions can appear almost as frequently as the elementary functions, for example, in the context of Laplace transforms (Example 20.6), because they allow their calculation with unrestricted parameters in important cases; these auxiliary functions are tabulated, like the elementary functions, and amenable to numerical methods of calculation.

The special functions form an extensive subject, of that the hypergeometric series (Section 29.9; Example 30.20) are but an example; they occur often in the less simple problems. The special functions together with their extensions can be addressed in the context of complex analysis as do the specific functions that include the elliptic functions (Section 39.9); the elliptic functions were of considerable historic interest before the computer age, because reducing the solution of a problem to elliptic functions would allow the use of their numerical tables; in the computer the age the elliptic functions are still of use as exact solutions of nonlinear problems (Section 39.7). The specific (special) functions also appear most often as are solutions of differential equations, in less simple cases viz. nonlinear (Section 39.7) [linear with variable coefficients (Notes 23.3 and 23.4)].

The orthogonal polynomials are particular cases of special functions for that their series expansions terminate after a finite number of terms, for example, a finite or terminating confluent (Gaussian) hypergeometric series [Section 39.9 (Example 30.20)] lead to Laguerre (Legendre) and associated polynomials; the same special function may have several subcases of terminating series, leading to more that on set of orthogonal polynomials, for example, the Legendre, Tchebycheff, Jacobi, and Gegenbauer polynomials are all finite cases of the Gaussian hypergeometric series, not necessarily all mutually independent. The extended special functions go in the opposite direction of the orthogonal polynomials, viz. they are more general than the special functions, for example, a special function may have a series expansion with a branch-point but no essential singularity, and for the corresponding extended special function the branch-point may coincide with an essential singularly leading to an irregular singularity. There are many other relations between the families of functions, for example: (i) the periods of Jacobian elliptic functions (Section 39.9) are specified by particular hypergeometric functions (Example 40.20); (ii) the evaluation of particular hypergeometric functions may involve auxiliary or elementary functions for example, (29.76–29.77); (iii) the auxiliary functions are based on elementary functions, for example, the definition of digamma function (29.43) involves the logarithm. Thus there is a loose hierarchy (Diagram 39.1) from (i) elementary, to (ii) auxiliary, to (iii) special, to (iv) specific functions that is followed in the gradual approaches to the subject; this hierarchy is not strict, for example, the Lame function [viz. a special function (iii)] satisfies a linear-second order differential equation, whose coefficients involve an elliptic function [viz. a specific function (iv)].

The hierarchy from (i) to (iv) of families of functions is loosely related to differential equations: (i) the elementary functions are sufficient to solve linear differential equations with constant coefficients (Section 2.8), for example, the stability problems (Chapter 2); (ii) the special functions are needed to solve some linear differential equations with variable coefficients, for example, of the form in Notes 23.3 and 23.4; (iii) the specific functions are solutions of some nonlinear differential equations, for example, the hyperelliptic integral (39.97) is a solution of the nonlinear first-order differential equation (39.96a). The auxiliary functions (ii) appear as a means of evaluation of special and specific functions, or integrals or series, some of that may also arise from differential equations. The multiple relations between the families of functions in List 39.1 become apparent in many other problems. The application of complex analysis to the families of functions in List 39.1 was kept in mind while selecting the subsets to be covered in the present volume. Thus the treatment of functions of complex variable in the present volume has had three objectives: (i) to establish the basics of function theory of general use; (ii) to allow the applications to various potential fields; (iii) to support subsequent developments, for example, as regards the families of functions in List 39.1 that find application in many other, nonpotential problems. In this sense (iii) List 39.1 (38.1) on complex functions (potential fields) serve as a conclusion to the present volume. A more comprehensive listing would include extra topics, such as: (i) a more systematic account on series of fractions and infinite products, (Subsection 29.9.2; Sections 36.6 and 36.7), as well as other infinite representations for example, continued fractions, applied to elementary and other functions; (ii) complex methods for nonpotential flows, for example, rotational or unsteady, and for the mechanics of deformable solids, for example, plane elasticity, bars, membranes, and surface tension.

Conclusion 39: There are three triangular coverings of the plane (Figure 39.1) using (a) equilateral, (b) isosceles and (c) irregular triangles. The sphere can also be covered by spherical triangles, e.g., (Figure 39.2) whose sides are concave circular arcs. The circular arc triangles can also be considered in the plane (Figure 39.3) with concave or convex sides, e.g., three (a), two (b), one (c) or none (d) convex sides. A particular case of one or two concave side(s) is a cusp at a vertex; a triangle may have cusps at all three vertices, e.g., a finite (Figure 39.3d) or infinite (Figure 39.4) triangle. The latter is the Schwartz triangle (Figure 39.4) with cusps at $z=0,1,\infty$, which leads to the elliptic modular function; the latter is used to prove that every equation $f(z)=c$ where $f(z)$ is an integral function has an infinite number of complex roots z_n for all values of c, excluding possibly one exceptional value $c\neq b$. The roots of the equation $e^z=c\neq 0$ [$\cos z=c$] lie (Figure 39.5) on a vertical [(a) horizontal (b)] line, so that an infinite number of them are contained in an ray about the imaginary (real) axis with with arbitrarily small ε angular range. The asymptotic scaling of a complex function may vary along directions through the origin or strips (Figure 39.6). The property of the elliptic modular function of having a positive imaginary part implies that it is closer on the complex G-plane to another value than to the conjugate of the latter (Figure 39.7). The Schwartz-Christoffell transformation (Figure 39.8) with: (i) one critical point of angle π specifies the logarithmic function (a); (ii) with two right angles at two critical points leads to the circular functions (b), defined initially in a degenerate triangles with parallel sides, and then by reflection on a fundamental strip $-\pi/2\leq \mathrm{Re}(\zeta)<\pi/2$, repeated periodically; (iii) with four right-angles at four critical points (c) elliptic functions in a parallelogagram, extended by reflection to $-2X<\mathrm{Re}(\zeta)\leq +2X$, $-Y<\mathrm{Im}(\zeta)\leq +Y$, and covering the ζ-plane with real $4X$ and imaginary $i2Y$ periods. A hyperelliptic integral (Figure 39.9) of a polynomial with at least two roots z_1, z_2 in (a) may be evaluated along a direct path L_1, or one L_2 surrounding some branch-cuts joining branch-points, which determine the periods of the inverse functions, e.g.: (b) the loop integral around the branch-point of the logarithm determines the period $2\pi i$ of the inverse function, i.e., the exponential; (c) the integral along the line joining the branch-points $z=\pm 1$ specifies the period 2π of the circular functions, which correspond to a polynomial with two distinct roots; (d) a polynomial of fourth degree with distinct roots leads to flow branch-points and two periods for Jacobian elliptic functions. The singly – (doubly –) periodic function have for fundamental region a strip (Figure 39.10) [parallelogram (Figure 39.11)] and repeat their values at congruent points.

40

Examples 40.1 to 40.20

Example 40.1. *Analytic Continuation of Power Series.*
Consider the monogenic function (40.1a) [(40.1b)]:

$$f(a,b;z) = z^a(1-z)^b = f(b,a;1-z), \tag{40.1a}$$
$$g(a,b;z) = (z-a)^{-1}(z-b)^{-1} = g(b,a;z), \tag{40.1b}$$

and obtain a formula for inversion relative to the origin $z \to 1/z$ [linear transformation $z = (z+d)$]. Expand (40.1a) [(40.1b)] in three (four) distinct power series, and based on their regions of convergence consider the possible analytic continuations.

Example 40.1.1. *Analytic continuation of a monogenic function.*
The inversion (linear) property for (40.1a) [(40.1b)] is (40.2a) [(40.26b):

$$f\left(a,b;\frac{1}{z}\right) = e^{i\pi b} f(-a-b,\, b;\, z), \tag{40.2a}$$
$$g(a,b;cz+d) = c^{-2} g\left(\frac{a-d}{c}, \frac{b-d}{c}; z\right). \tag{40.2b}$$

and follows from:

$$\left(\frac{1}{z}\right)^a\left(1-\frac{1}{z}\right)^b = z^{-a-b}(z-1)^{\mathrm{b}} = e^{i\pi b} z^{-a-b}(1-\mathrm{z})^{\mathrm{b}}, \tag{40.3a}$$
$$(cz+d-a)^{-1}(cz+d-b)^{-1} = c^{-2}\left(z-\frac{a-d}{c}\right)^{-1}\left(z-\frac{b-d}{c}\right)^{-1}. \tag{40.3b}$$

The three distinct power series expansions for (40.1a), are

$$|z|<1: \quad f(a,b;z) = \sum_{n=0}^{\infty}(-)^n \binom{b}{n} z^{n+a}, \tag{40.4a,b}$$
$$|1-z|<1: \quad f(a,b;z) = \sum_{n=0}^{\infty}(-)^n \binom{a}{n} (1-z)^{n+b}, \tag{40.5a,b}$$
$$|z|>1: \quad f(a,b;z) = e^{i\pi b}\sum_{n=0}^{\infty}(-)^n \binom{b}{n} z^{a+b-n}, \tag{40.6a,b}$$

obtained from: (i) the binomial series (25.37a–c) applied to (40.1a), leading to (40.4a,b); (ii) the binomial series for (40.7a) [(40.7b)] leads to (40.5a,b) [(40.6a,b)]:

$$f(a,b;z) = (1-z)^b[1-(1-z)]^a, \quad f(a,b;z) = e^{i\pi b} z^{a+b}\left(1-\frac{1}{z}\right)^b. \tag{40.7a,b}$$

Applying successively (40.2a) and (40.1a):

$$f(a,b;z) = e^{i\pi b} f\left(-a-b, b; \frac{1}{z}\right) = e^{i\pi b} f\left(b, -a-b; 1-\frac{1}{z}\right) = e^{i\pi b}\left(1-\frac{1}{z}\right)^{b}\left(\frac{1}{z}\right)^{-a-b}. \tag{40.7c}$$

follows (40.7b) ≡ (40.7d) that can also be obtained from:

$$f(a,b;z) = z^{a+b}\left(\frac{1}{z}-1\right)^{b} = e^{i\pi b} z^{a+b}\left(1-\frac{1}{z}\right)^{b}. \tag{40.7d}$$

The series (40.4b) [(40.6b)] converges inside (40.4a) (outside) [(40.6a)] the unit disk, that is, disjoint regions except for the boundary; the series (40.5b) converges in an overlapping region (40.5a), viz. the disk of radius unity and center at the point unity (Figure 40.1a), and performs the analytic continuation between (40.4b) and (40.6b).

Example 40.1.2. *Analytic continuation over three or four regions.*
The four distinct power series expansions for the function (40.1b):

$$|z-a| > |b-a|: \quad g(a,b;z) = \sum_{n=0}^{\infty} (b-a)^n (z-a)^{-n-2}, \tag{40.8a,b}$$

$$|z-a| < |b-a|: \quad g(a,b;z) = -\sum_{n=0}^{\infty} (b-a)^{-n-1} (z-a)^{n-1}, \tag{40.9a,b}$$

$$|z-b| < |b-a|: \quad g(a,b;z) = -\sum_{n=0}^{\infty} (a-b)^{-n-1} (z-b)^{n-1}, \tag{40.10a,b}$$

$$|z-b| > |b-a|: \quad g(a,b;z) = \sum_{n=0}^{\infty} (a-b)^n (z-b)^{-n-2}; \tag{40.11a,b}$$

are obtained as follows: (i/ii) the geometric series (21.62a,b) around $z = a\,(z = b)$ yield (40.12a) ≡ (40.8a,b) [(40.12b) ≡ (40.11a,b)]:

$$g(a,b;z) = (z-a)^{-2}\left(1-\frac{b-a}{z-a}\right)^{-1} = (z-b)^{-2}\left(1-\frac{a-b}{z-b}\right)^{-1}; \tag{40.12a,b}$$

(iii/iv) similarly the geometric series around $z = a$ $(z = b)$ yield (40.12c) ≡ (40.9a,b) [(40.12d) ≡ (40.10a,b)]:

$$g(a,b;z) = -(z-a)^{-1}(b-a)^{-1}\left(1-\frac{z-a}{b-a}\right)^{-1} = -(z-b)^{-1}(a-b)^{-1}\left(1-\frac{z-b}{a-b}\right)^{-1}. \tag{40.12c,d}$$

The series (40.8b; 40.9b) [(40.10b; 40.11b)] converge in disjoint regions (40.8a; 40.9a) [(40.10a; 40.11a)] that overlap only on the boundaries; any of the former pair overlaps with any of the latter pair (Figure 40.1b), so analytic continuation is possible.

Example 40.2. *Reflection Principle and Complex Conjugate of Functions.*
Show that the function (40.13b) [(40.13c)] of complex variable (40.13a);

$$z = x + iy \quad f(z) = \exp(iz), \quad g(z) = i\sin z, \tag{40.13a–c}$$

that is not real for $z \equiv x$ real, does not satisfy the reflection principle; (ii) indicate another relation that its conjugate does satisfy.

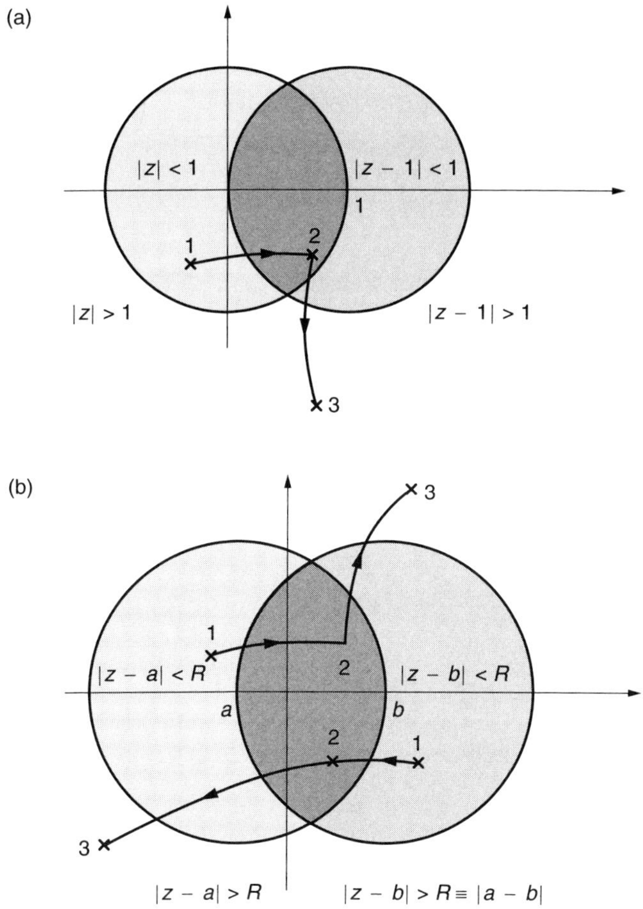

FIGURE 40.1
The same function (40.1a) is represented by three distinct power series (a) converging in three regions: (i/ii) the interior (exterior) of the unit circle; (iii) the interior of the circle of radius unity and center at the point (1, 0). Although the regions (i) and (ii) are disjoint, since (iii) overlaps with both of them, it provides analytic continuation (i–iii–ii). Another function (40.1b) is represented by four distinct power series (b) converging in four regions: the interiors and exteriors of the circles with centers at $z = a$ and $z = b$ with the some radius $R = |a - b|$ equal to the distance between the centers. Although each pair regions is disjoint, each region of one pair overlaps with the two regions of the other pair, so analytic continuation is possible (Figure 31.1a). In both cases (Figure 40.1a,b) analytic continuation extends the functions to the whole complex plane.

From (40.13b) [(40.13c)] follows (40.14a,b) [(40.15a,b)]:

$$f^*(z) = [\exp(iz)]^* = \exp(-iz^*)\,,\; f(z^*) = \exp(iz^*)\,, \tag{40.14a,b}$$

$$g(z^*) = (i\sin z)^* = -i\sin z^*,\; g(z^*) = i\sin z^*; \tag{40.15a,b}$$

this shows that

$$f(z)^* \neq f^*(z) = \frac{1}{f(z^*)}, \quad g(z^*) \neq g^*(z) = -g(z^*)\,. \tag{40.16a–d}$$

The reflection principle is not met (40.16a) [(40.16c)], but the relation (40.16b) [(40.16d)] holds.

Example 40.3. *Location of Roots of Some Polynomials of Even Degree.*
Show that *the polynomial equation:*

$$\alpha, \beta \in |R: \quad 0 = z^{2n} + \alpha^2 z^{2n-1} + \beta^2 \equiv P_{2n}(z), \tag{40.17}$$

has $n - 1$ (n) roots with positive real parts if n is odd (even).

The polynomial (40.17) is an analytic function in the right-hand half-plane $\mathrm{Re}(z) > 0$, and the number of zeros (40.18b) is specified (31.35c) by the variation of the argument of its logarithmic derivative (40.18a):

$$f(z) \equiv \frac{\mathrm{d}\{\log[P_{2n}(z)]\}}{\mathrm{d}z}, \quad Z = (2\pi)^{-1} \Delta\{\arg[f(z)]\} = (2\pi i)^{-1} \int_L f(z)\,\mathrm{d}z = Z_1 + Z_2, \tag{40.18a,b}$$

along a loop surrounding the right-hand half-plane that consists of: (i) a circle of radius $R \to \infty$ taken in the positive, that is, counterclockwise direction $-\pi/2 \le \varphi \le \pi/2$; (ii) the imaginary axis $z = iy$ closing the loop from $+\infty > y > -\infty$. From (40.18a; 40.17) follows:

$$f(z) \equiv \frac{P'_{2n}(z)}{P_{2n}(z)} = \frac{2nz^{2n-1} + (2n-1)\alpha^2 z^{2n-2}}{z^{2n} + \alpha^2 z^{2n-1} + \beta^2} = \frac{2n}{z} + O\left(\frac{1}{z^2}\right); \tag{40.19}$$

thus the first contribution, that is, that of the half-circle, is n:

$$Z_1 \equiv \lim_{R\to\infty} \frac{1}{2\pi i} \int_{-\pi/2}^{+\pi/2} \mathrm{d}\varphi\, i R e^{i\varphi} f\left(Re^{i\varphi}\right) = \frac{1}{2\pi} 2n \int_{-\pi/2}^{+\pi/2} \mathrm{d}\varphi = n; \tag{40.20a}$$

$$Z_2 \equiv \frac{1}{2\pi} \int_{i\infty}^{-i\infty} f(iy)\,\mathrm{d}y, \tag{40.20b}$$

the second contribution (40.20b) is given: (i) by (40.21b) for n even (40.21a):

$$n = 2m\text{: } Z_2 = (2\pi)^{-1} \Delta\{\arg[P_{4m}(iy)]\} = (2\pi)^{-1} \Delta\left\{\arg\left(y^{4m} - i\alpha^2 y^{4m-1} + \beta^2\right)\right\} = 0, \tag{40.21a,b}$$

since the real part has fixed sign and the imaginary part changes sign once at $y = 0$, and thus there is no 2π change of argument; (ii) by (40.22b) for n odd [(40.22a)]:

$$\begin{aligned} n = 2m+1\text{:} \quad Z_2 &= (2\pi)^{-1} \Delta\{\arg[P_{4m+2}(iy)]\} \\ &= (2\pi)^{-1} \Delta\left\{\arg\left(-y^{4m+2} + i\alpha^2 y^{4m+1} + \beta^2\right)\right\} = -1, \end{aligned} \tag{40.22a,b}$$

because both of the real (imaginary) parts change sign as y goes from $+\infty$ to $-\infty$ across $y^{2m-1} = \pm\beta\,(y = 0)$. Adding (40.20a) to (40.21b) [(40.22b)] for even (odd) n leads to (40.23a) [(40.23b)]:

$$\mathrm{Re}(z) > 0\text{:} \quad Z = Z_1 + Z_2 = \begin{cases} n = 2m & \text{if } n \text{ is even}, \quad (40.23a) \\ n - 1 = 2m & \text{if } n \text{ is odd}, \quad (40.23b) \end{cases}$$

for the total number of roots of the polynomial (40.17) in the right-hand half-plane.

Example 40.4. *Decomposition of Rational Functions into Partial Fractions.*

Obtain the decomposition into partial fractions of the following seven rational functions (40.24–40.30a):

$$\frac{(z-c)(z-d)}{z-a} = z + a - c - d + \frac{(a-c)(a-d)}{z-a}, \tag{40.24a}$$

$$\int \frac{(z-c)(z-d)}{(z-a)}\,\mathrm{d}z = [(a-c)(a-d)]\log(z-a) + z\left(\frac{z}{2} + a - c - d\right); \tag{40.24b}$$

$$\frac{z-c}{(z-a)(z-b)} = \frac{(a-c)/(z-a) - (b-c)/(z-b)}{a-b}, \tag{40.25a}$$

$$\int \frac{z-c}{(z-a)(z-b)}\,\mathrm{d}z = \frac{a-c}{a-b}\log(z-a) + \frac{b-c}{b-a}\log(z-b); \tag{40.25b}$$

$$\frac{(z-c)(z-d)}{(z-a)(z-b)} = 1 + \frac{(a-c)(a-d)/(z-a) - (b-c)(b-d)/(z-b)}{a-b}, \tag{40.26a}$$

$$\int \frac{(z-c)(z-d)}{(z-a)(z-b)}\,\mathrm{d}z = z + \frac{(a-c)(a-d)\log(z-a) - (b-c)(b-d)\log(z-b)}{a-b}; \tag{40.26b}$$

$$(z-a)^{-2}(z-b)^{-1} = \frac{1/(z-b) - 1/(z-a)}{(a-b)^2} + (a-b)^{-1}(z-a)^{-2}, \tag{40.27a}$$

$$\int \frac{\mathrm{d}z}{(z-a)^2(z-b)} = (a-b)^{-2}\log\left(\frac{z-b}{z-a}\right) - (a-b)^{-1}(z-a)^{-1}; \tag{40.27b}$$

$$\frac{(z-c)(z-d)}{(z-a)^2} = 1 + \frac{2a-c-d}{z-a} + \frac{(a-c)(a-d)}{(z-a)^2}, \tag{40.28a}$$

$$\int \frac{(z-c)(z-d)}{(z-a)^2}\mathrm{d}z = z + (2a-c-d)\log(z-a) - \frac{(a-c)(a-d)}{z-a}, \tag{40.28b}$$

$$\frac{z-c}{(z-a)^2(z-b)} = \frac{b-c}{(a-b)^2}\left(\frac{1}{z-b} - \frac{1}{z-a}\right) + \frac{a-c}{a-b}\frac{1}{(z-a)^2} \tag{40.29a}$$

$$\int \frac{z-c}{(z-a)^2(z-b)}\,\mathrm{d}z = \frac{b-c}{(a-b)^2}\log\left(\frac{z-b}{z-a}\right) - \frac{a-c}{a-b}\frac{1}{z-a}; \tag{40.29b}$$

$$\begin{aligned}(z-a)^{-3}(z-b)^{-1} &= \frac{1/(z-a) - 1/(z-b)}{(a-b)^3} \\ &\quad - \frac{(z-a)^{-2}}{(a-b)^2} + \frac{(z-a)^{-3}}{a-b},\end{aligned} \tag{40.30a}$$

$$\begin{aligned}\int (z-a)^{-3}(z-b)^{-1}\,\mathrm{d}z &= (a-b)^{-3}\log\left(\frac{z-a}{z-b}\right) \\ &\quad + \frac{(z-a)^{-1}}{(a-b)^2} - \frac{(z-a)^{-2}}{2(a-b)};\end{aligned} \tag{40.30b}$$

and the corresponding (40.24–40.30b) indefinite integrals.

The rational function (40.24a) is not simple and its decomposition can be made by dividing out the simple pole successively two times:

$$\frac{(z-c)(z-d)}{z-a} = (z-c)\left(1 + \frac{a-d}{z-a}\right) = z - c + (a-d)\left(1 + \frac{a-c}{z-a}\right), \tag{40.31}$$

leading to (40.31) ≡ (40.24a). The rational function (40.25a) is simple and the residues at the simple poles specify the partial fraction decomposition:

$$\frac{z-c}{(z-a)(z-b)}=\frac{1}{z-a}\left[\frac{z-c}{z-b}\right]_{z=a}+\frac{1}{z-b}\left[\frac{z-c}{z-a}\right]_{z=b}, \tag{40.32}$$

leading to (40.32) ≡ (40.25a). The rational function (40.26a) is not simple, and since the terms of highest degree in the numerator and denominator are both z^2, the constant term is 1:

$$\frac{(z-c)(z-d)}{(z-a)(z-b)}=1+\frac{(z-c)(z-d)-(z-a)(z-b)}{(z-a)(z-b)}=1+\frac{(a+b-c-d)z+cd-ab}{(z-a)(z-b)}; \tag{40.33}$$

the remainder is the principal part that follows from the residues at the simple poles as in (40.32) leading to (40.26a). The rational function (40.27a) is simple, with: (i) the residue at the simple pole specifying the first term of the partial fraction decomposition:

$$(z-a)^{-2}(z-b)^{-1}=(b-a)^{-2}(z-b)^{-1}+A_{-1}(z-a)^{-1}+A_{-2}(z-a)^{-2}\equiv f(z); \tag{40.34}$$

(ii) the second coefficient is the residue at the double pole (40.35a):

$$A_{-1}=\lim_{z\to a}\frac{\mathrm{d}}{\mathrm{d}z}\left[(z-a)^2 f(z)\right]=\lim_{z\to a}\frac{\mathrm{d}}{\mathrm{d}z}\frac{1}{z-b}=-\frac{1}{(a-b)^2}; \tag{40.35a}$$

$$A_{-2}=\lim_{z\to a}\left[(z-a)^2 f(z)\right]=\lim_{z\to a}\frac{1}{z-b}=\frac{1}{a-b}, \tag{40.35b}$$

(iii) the third coefficient A_{-2} follows (40.35b) from the extended residue rule (31.85b) that also applies to (40.35a); (iv) substitution of (40.35a,b) in (40.34) yields (40.27a).

The same method applies to (40.28a) in three stages: (i) the leading terms in the numerator and denominator are z^2, so the ratio is unity:

$$\begin{aligned}\frac{(z-c)(z-d)}{(z-a)^2}&=1+\frac{(z-c)(z-d)-(z-a)^2}{(z-a)^2}=1+\frac{(2a-c-d)z+cd-a^2}{(z-a)^2}\\&=1+B_{-1}(z-a)^{-1}+B_{-2}(z-a)^{-2}\equiv g(z);\end{aligned} \tag{40.36}$$

(ii) the residue B_{-1} and coefficient B_{-2} at the double pole $z=a$ are given (31.85b) by

$$\begin{aligned}\{B_{-1},B_{-2}\}&=\lim_{z\to a}\left\{\frac{\mathrm{d}}{\mathrm{d}z},1\right\}\left[(z-a)^2 g(z)\right]=\lim_{z\to a}\left\{\frac{\mathrm{d}}{\mathrm{d}z},1\right\}\left[(2a-c-d)z+cd-a^2\right]\\&=\{2a-c-d,(a-c)(a-d)\};\end{aligned} \tag{40.37a,b}$$

(iii) substitution of (40.37a,b) in (40.36) yields (40.28a). The rational function (40.29a) is simple and: (i) the residue at the simple pole specifies the first term of the partial fraction decomposition:

$$\frac{(z-c)}{(z-a)^2(z-b)}=(b-c)(b-a)^{-2}(z-b)^{-1}+C_{-1}(z-a)^{-1}+C_{-2}(z-a)^{-2}\equiv h(z); \tag{40.38}$$

(ii) the residue C_{-1} and coefficient C_{-2} at the double pole $z = a$ given (31.85b) by

$$\{C_{-1}, C_{-2}\} = \lim_{z\to a} \left\{\frac{\mathrm{d}}{\mathrm{d}z}, 1\right\} \left[(z-a)^2 h(z)\right]$$
$$= \lim_{z\to a} \left\{\frac{\mathrm{d}}{\mathrm{d}z}, 1\right\} \frac{z-c}{z-b} = \left\{\frac{c-b}{(a-b)^2}, \frac{a-c}{a-b}\right\}; \tag{40.39a,b}$$

(iii) substitution of (40.39a,b) in (40.38) yields (40.29a). The rational function (40.30a) is simple and: (i) the residue at the simple pole specifies the first term of the partial fraction decomposition:

$$(z-a)^{-3}(z-b)^{-1} = (b-a)^{-3}(z-b)^{-1} + D_{-1}(z-a)^{-1}$$
$$+ D_{-2}(z-a)^{-2} + D_{-3}(z-a)^{-3} \equiv j(z); \tag{40.40}$$

(ii) the coefficients of the principal part at the triple pole are specified (31.85b) by

$$\{D_{-1}, D_{-2}, D_{-3}\} = \lim_{z\to a} \left\{\frac{1}{2}\frac{\mathrm{d}^2}{\mathrm{d}z^2}, \frac{\mathrm{d}}{\mathrm{d}z}, 1\right\} \left[(z-a)^3 j(z)\right]$$
$$= \lim_{z\to a} \left\{\frac{1}{2}\frac{\mathrm{d}^2}{\mathrm{d}z^2}, \frac{\mathrm{d}}{\mathrm{d}z}, 1\right\} \frac{1}{z-b} = \left\{(a-b)^{-3}, -(a-b)^{-2}, (a-b)^{-1}\right\}; \tag{40.41a–c}$$

where D_{-1} in (40.41b) is the residue; (iii) substitution of (40.41a–c) in (40.40) yields (40.3a). The rational integrals (40.24b) to (40.30b) can be checked by multiplying out the denominators on both sides, yielding identical polynomials. The rational integrals (40.24b) to (40.30b) follow respectively from the partial fraction decompositions (40.24a) to (40.30a).

Example 40.5. *Ratio of Lengths and Angles in Complex Mappings.*

Determine the ratio of lengths or scale factor h for the conformal mappings in Table 40.1; and also for the conformal mappings in Table 40.2, the location and type of the critical points, and the ratio of angles.

TABLE 40.1
Scale Factors of Conformal Mappings

Mapping function $\zeta = f(z)$	**Derivative** $f'(z) = \mathrm{d}\zeta/\mathrm{d}z$	**Modulus*** $h = \lvert f'(z)\rvert$
z^n	nz^{n-1}	nr^{n-1}
$\exp(z)$	$\exp(z)$	$\exp(x)$
$\log z$	$1/z$	$1/r$
$z\log z - z$	$\log z$	$\left\lvert(\log r)^2 + \varphi^2\right\rvert^{1/2}$
$z + a^2/z$	$1 - a^2/z^2$	$\left\lvert 1 + (a/r)^4 - 2(a/r)^2\cos(2\varphi)\right\rvert^{1/2}$
$\sin z$	$\cos z$	$\left\lvert\cosh^2 y - \sin^2 x\right\rvert^{1/2}$

* Using $z = x + iy = re^{i\varphi}$.
Note: The table indicates for the six conformal mappings specified by holomorphic functions: (i) the derivatives which exist and are independent of the direction; (ii) the modulus which specifies the scale factor; (iii) the latter would be the same for the two coordinate axes if the mapping was used to define conformal coordinates.

Example 40.5.1. *Ratio of lengths in a conformal mapping.*

The ratio of lengths or scale factor for a conformal mapping (40.42a,b) is the modulus of derivative (33.9a) ≡ (40.42c):

$$z \equiv re^{i\varphi} = x + iy, \quad \zeta = f(z), \quad h \equiv |f'(z)|; \tag{40.42a–c}$$

it is immediate for the functions:

$$\left|\frac{\mathrm{d}(z^n)}{\mathrm{d}z}\right| = n\left|z^{n-1}\right| = nr^{n-1}, \quad \left|\frac{\mathrm{d}(\mathrm{e}^z)}{\mathrm{d}z}\right| = |\mathrm{e}^z| = |\mathrm{e}^x|\left|\mathrm{e}^{iy}\right| = \mathrm{e}^x, \quad \left|\frac{\mathrm{d}(\log z)}{\mathrm{d}z}\right| = \left|\frac{1}{z}\right| = \frac{1}{r}, \tag{40.43–46}$$

$$\left|\frac{\mathrm{d}(z\log z - z)}{\mathrm{d}z}\right| = |\log z| = |\log r + i\varphi| = \left|(\log r)^2 + \varphi^2\right|^{1/2}, \tag{40.47}$$

$$\left|\frac{\mathrm{d}(z + a^2/z)}{\mathrm{d}z}\right| = \left|1 - \frac{a^2}{z^2}\right| = \left|1 - \frac{a^2}{r^2}\mathrm{e}^{-2i\varphi}\right| = \left|1 + \left(\frac{a}{r}\right)^4 - 2\left(\frac{a}{r}\right)^2\cos(2\varphi)\right|^{1/2}, \tag{40.48}$$

$$\begin{aligned}\left|\frac{\mathrm{d}(\sin z)}{\mathrm{d}z}\right| &= |\cos z| = |\cos x\cosh y + i\sin x\sinh y| = \left|\cos^2 x\,\cosh^2 y + \sin^2 x\,\sinh^2 y\right|^{1/2}\\ &= \left|\cosh^2 y - \sin^2 x\right|^{1/2} = \left|\sinh^2 y + \cos^2 x\right|^{1/2} = \frac{1}{\sqrt{2}}\left|\cosh(2y) - \cos(2x)\right|^{1/2},\end{aligned} \tag{40.49}$$

in the Table 40.1; the conformal mapping (40.48) corresponds to the circle transformation (34.3).

Example 40.5.2. *Critical points and ratio of angles.*

The critical points of first (second) kind are the points of vanishing (infinite) derivative; the ratio of angles ψ at a branch point (40.50b) with real exponent (40.50a) is the value (40.50c) of the exponent:

$$\alpha \in |R: \quad \zeta = z^\alpha, \quad \psi = \frac{\arg(\zeta)}{\arg(z)} = \frac{\arg(z^\alpha)}{\arg(z)} = \alpha, \tag{40.50a–c}$$

and thus: (i) the ratio of angles is the order of the zero if the critical point of first kind is a zero, for example, n for the first line in Table 40.2: (ii) if the inverse function has a zero of order N, the ratio of angles is $1/N$, for example, double (triple) zero of (40.51b) [(40.52b)] at (40.51c) [(40.52c)]:

$$\psi = \tfrac{1}{2}: \quad \zeta = \arcsin\sqrt{z}, \quad z = \sin^2\zeta, \quad \zeta_n = n\pi, \quad z = O\left(\zeta - n\pi\right)^2, \tag{40.51a–d}$$

$$\psi = \tfrac{1}{3}: \quad \zeta = \arg\,\coth\left(\sqrt[3]{z}\right), \quad z = \coth^3\zeta, \quad \zeta_n = i\pi\left(n + \frac{1}{2}\right),$$

$$z = O\left(\left[\zeta - i\left(n + \frac{1}{2}\right)\pi\right]^3\right); \tag{40.52a–d}$$

(iii) if the function has a branch point with real exponent α, the latter specifies the ratio of angles, for example, α in for the function (40.53b) with branch-points at (40.53c):

$$\psi = \alpha: \quad \zeta = [1 - \exp(z)]^\alpha = [1 - \exp(z - i2\pi n)]^\alpha \sim O\left(z - z_n\right)^\alpha, \quad z_n = i2\pi n; \tag{40.53a–c}$$

TABLE 40.2
Ratio of Angles at Critical Points

Mapping function $\zeta = f(z)$	Critical point		Ratio of angles
	First kind $f'(z) = 0$	Second kind $f'(z) = \infty$	ψ
$f(z) = z^n$	$z = 0$	–	n
$f(z) = \arcsin(\sqrt{z})$	–	$z_n = n\pi^*$	$1/2$
$f(z) = \arg\coth(\sqrt[3]{z})$	–	$z_n = i(n+{}^1\!/_2)\,\pi^*$	$1/3$
$f(z) = [1 - \exp(z)]^\alpha$	$z_n = i2\pi n^*$	–	α
$f(z) = \left[1 - \exp\left(z^2\right)\right]^\alpha$	$z = 0$	–	2α
	$z = \pm (1+i)\sqrt{\pi m}^{**}$	–	α

$^*\,n \in |Z$: integer; $^{**}m \in |Z$: positive integer.
Note: Table 40.1 (40.2) concerns the scale factors (ratio of angles) for conformal mappings at conformal (critical) points. In the case of Table 40.2 five mappings are considered indicating: (i/ii) the critical points of the first (second) kind where the derivative is zero (infinite); (iii) the ratio of angles in the range and in the domain.

(iv) the function (40.54a) has branch-points at (40.54b):

$$\zeta = \left[1 - \exp\left(z^2\right)\right]^\alpha = \left[1 - \exp\left(z^2 - i2\pi n\right)\right]^\alpha = O\left(z - z_n^\pm\right)^\alpha \tag{40.54a}$$

$$z_n^\pm = \pm\sqrt{i2\pi n} = \pm(1+i)\sqrt{\pi n}; \tag{40.54b}$$

the term in square brackets in (40.54a) has a double (40.55b) [simple (40.55e)] root for n zero (40.55a) [nonzero (40.55d)]:

$$n = 0: \quad \zeta \sim O\left(z^{2\alpha}\right), \quad \psi = 2\alpha; \quad n \neq 0: \quad \zeta \sim O\left(\left(z - z_n^\pm\right)^\alpha\right), \quad \psi = \alpha, \tag{40.55a–f}$$

leading to the exponent (40.55c) [(40.55f)]. This completes the proof of Table 40.2.

Example 40.6. *Geometry of General Joukowski Airfoils.*

Determine the chord, thickness and height of the general unsymmetric thick strongly cambered Joukowski airfoil. Consider the particular cases of the thick symmetric (strongly cambered thin) airfoil, and obtain the approximation to next order beyond the thin symmetric (weakly cambered) airfoil. Discuss possible alternative definitions of chord, thickness and height in the general case of the thick strongly cambered Joukowski airfoil; illustrate one case of these differences with regard to the angular position of the leading-edge defined as the point with vertical tangent.

Example 40.6.1. *Thick and strongly cambered Joukowski airfoil.*

The chord of the general Joukowski airfoil (34.75a,b; 34.76a) is defined as the distance between the leading L.E. (trailing T.E.) edges (Figure 40.2) that correspond to the point φ

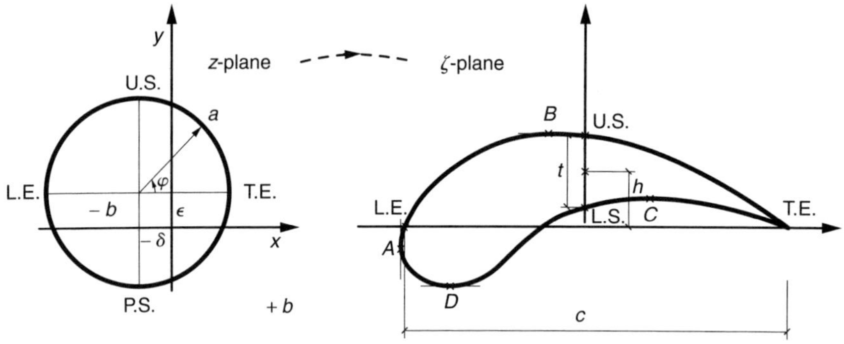

FIGURE 40.2
The Joukowski transformation applied to a circle with center off both coordinate axes leads to the cambered or unsymmetric Joukowski airfoil (Figures 34.3b and 40.2); one critical point is on (inside) the circle corresponding to the sharp (rounded) trailing (leading) edge. An offset of the center of the circle from the origin small (not small) compared with the radius $\varepsilon^2+\delta^2<<R^2$ $\left(\varepsilon^2+\delta^2 \cong R^2\right)$ leads to the thin (thick) weakly (strongly) cambered unsymmetric Joukowski airfoil [Figure 34.3b (40.2)]. In the latter case of the thick strongly-cambered unsymmetric Joukowski airfoil: (i) the position of the trailing edge (T.E.) is clear; (ii) the position of the leading edge (L.E.) at the intersection with the chord line is distinct from the position of the vertical tangent (A); (iii) the choice (ii) of the position of the leading-edge determines the separation between the upper/lower surfaces (U.S./L.S.); (iv) the highest B (lowest D) point on the upper (lower) surface, both with horizontal tangent, do not lie on the same vertical, nor does the third point C with horizontal tangent on the lower surface; (v) the vertical distance between the highest B and lowest D point is larger than thickness t of the airfoil measured as the distance between the upper and lower surfaces along the vertical to the middle of the chord line; (vi) the height h of the camber line can be taken at the mid point between the upper and lower surface in the measurement of the thickness (v); (vii) the chord c of the airfoil depends on the choice (ii) of the position of the leading edge.

$= \pi$ $(\varphi = 0)$ where the real axis or airfoil axis intersects the profile (40.56a):

$$c \equiv x(0) - x(\pi) = 2a\left[1 + \frac{\left(a^2+\varepsilon^2-\delta^2\right)\left(\sqrt{a^2-\varepsilon^2}-\delta\right)^2}{\left(a^2+\varepsilon^2+\delta^2\right)^2 - 4a^2\delta^2}\right]; \tag{40.56a}$$

$$t \equiv y\left(\frac{\pi}{2}\right) - y\left(-\frac{\pi}{2}\right) = 2a\left[1 - \frac{\left(a^2+\delta^2-\varepsilon^2\right)\left(\sqrt{a^2-\varepsilon^2}-\delta\right)^2}{\left(a^2+\varepsilon^2+\delta^2\right)^2 - 4a^2\varepsilon^2}\right], \tag{40.56b}$$

$$h \equiv \frac{1}{2}\left[y\left(\frac{\pi}{2}\right) + y\left(-\frac{\pi}{2}\right)\right] = \varepsilon\left[1 + \frac{\left(a^2-\varepsilon^2-\delta^2\right)\left(\sqrt{a^2-\varepsilon^2}-\delta\right)^2}{\left(a^2+\varepsilon^2+\delta^2\right) - 4a^2\varepsilon^2}\right], \tag{40.56c}$$

the coordinates of the point $\varphi = \pi/2\,(\varphi = -\pi/2)$ on the upper U.S. (lower L.S.) surface (Figure 40.2) specify the thickness (height) through their difference (40.56b) [arithmetic mean (40.56c)]. From (34.75a; 34.76a) follow the extreme horizontal positions:

$$x(0), x(\pi) = (-\delta \pm a)\left[1 + \frac{\left(\sqrt{a^2-\varepsilon^2}-\delta\right)^2}{(a \mp \delta)^2 + \varepsilon^2}\right]; \tag{40.57a}$$

these specify the chord:

$$\begin{aligned}
c &= x(0) - x(\pi) \\
&= 2a + \left(\sqrt{a^2+\varepsilon^2} - \delta\right)^2 \left(\frac{a-\delta}{a^2+\delta^2+\varepsilon^2-2a\delta} + \frac{a+\delta}{a^2+\delta^2+\varepsilon^2+2a\delta}\right) \\
&= 2a + \left(\sqrt{a^2+\varepsilon^2} - \delta\right)^2 \frac{(a-\delta)\left(a^2+\delta^2+\varepsilon^2+2a\delta\right) + (a+\delta)\left(a^2+\delta^2+\varepsilon^2-2a\delta\right)}{\left(a^2+\delta^2+\varepsilon^2\right)^2 - 4a^2\delta^2} \\
&= 2a + 2a\left(a^2+\delta^2+\varepsilon^2-2\delta^2\right) \frac{\left(\sqrt{a^2-\varepsilon^2}-\delta\right)^2}{\left(a^2+\delta^2+\varepsilon^2\right)^2 - 4a^2\delta^2}. \qquad (40.57\text{b})
\end{aligned}$$

From (34.75b; 34.76a) follow the vertical positions:

$$y\left(\pm\frac{\pi}{2}\right) = (\varepsilon \pm a)\left[1 - \frac{\left(\sqrt{a^2+\varepsilon^2}-\delta\right)^2}{(a\pm\varepsilon)^2+\delta^2}\right]; \qquad (40.57\text{c})$$

these specify thought their difference (sum) the thickness (twice the height of the camber line):

$$\begin{aligned}
\{t, 2h\} &= y\left(\frac{\pi}{2}\right) \mp y\left(-\frac{\pi}{2}\right) = \{2a, 2\varepsilon\} \\
&\quad + \left(\sqrt{a^2+\varepsilon^2}-\delta\right)^2 \left(-\frac{\varepsilon+a}{a^2+\varepsilon^2+\delta^2+2a\varepsilon} \pm \frac{\varepsilon-a}{a^2+\varepsilon^2+\delta^2-2a\varepsilon}\right) \\
&= \{2a, 2\varepsilon\} + \left(\sqrt{a^2-\varepsilon^2}-\delta\right)^2 \frac{\pm(\varepsilon-a)\left(a^2+\varepsilon^2+\delta^2+2a\varepsilon\right) - (\varepsilon+a)\left(a^2+\varepsilon^2+\delta^2-2a\varepsilon\right)}{\left(a^2+\varepsilon^2-\delta^2\right)^2 - 4a^2\varepsilon^2} \\
&= \{2a, 2\varepsilon\} + \left(\sqrt{a^2-\varepsilon^2}-\delta\right)^2 \frac{\left\{-2a\left(a^2+\varepsilon^2+\delta^2-2\varepsilon^2\right), -2\varepsilon\left(a^2+\varepsilon^2+\delta^2-2a^2\right)\right\}}{\left(a^2+\varepsilon^2+\delta^2\right)^2 - 4a^2\varepsilon^2}.
\end{aligned}$$

(40.57d)

These agree with the chord (40.57b) ≡ (40.56a) [thickness and height (40.57d) ≡ (40.56b,c)] of the thick strongly cambered Joukowski airfoil.

Example 40.6.2. *Joukowski airfoils either thick or strongly cambered.*

The simplification of (40.56a–c) applies to the thick symmetric (40.58a–d) [strongly cambered thin [(40.59a–d)] airfoil:

$$\varepsilon = 0: \quad h = 0, \quad c = \frac{4a^2}{a+\delta}, \qquad t = \frac{4a^2\delta}{a^2+\delta^2}; \qquad (40.58\text{a–d})$$

$$\delta = 0: \quad t = 0, \quad c = \frac{4a^3}{a^2+\varepsilon^2}, \quad h = 2\varepsilon. \qquad (40.59\text{a–d})$$

In (40.56a) [(40.56b)] were made the simplifications (40.60b) [(40.60c)] for (40.60a) leading to (40.58c) [(40.58d)]:

$$\varepsilon = 0\text{:} \quad \frac{c}{2a} = 1 + \frac{\left(a^2-\delta^2\right)(a-\delta)^2}{\left(a^2+\delta^2\right)^2 - 4a^2\delta^2} = 1 + \frac{(a-\delta)^2}{a^2-\delta^2} = 1 + \frac{a-\delta}{a+\delta} = \frac{2a}{a+\delta}, \qquad (40.60\text{a,b})$$

$$\frac{t}{2a} = 1 - \frac{\left(a^2+\delta^2\right)(a-\delta)^2}{\left(a^2+\delta^2\right)^2} = 1 - \frac{(a-\delta)^2}{a^2+\delta^2} = \frac{a^2+\delta^2-(a-\delta)^2}{a^2+\delta^2} = \frac{2a\delta}{a^2+\delta^2};$$

(40.60c)

in (40.56a) [(40.56b,c)] were made the simplifications (40.61b) [(40.61c,d)] for (40.61a) leading to (40.59c) [(40.59b,d)]:

$$\delta = 0: \frac{c}{2a} = 1 + \frac{\left(a^2+\varepsilon^2\right)\left(a^2-\varepsilon^2\right)}{\left(a^2+\varepsilon^2\right)^2} = 1 + \frac{a^2-\varepsilon^2}{a^2+\varepsilon^2} = \frac{2a^2}{a^2+\varepsilon^2}, \tag{40.61a,b}$$

$$\frac{t}{2a}, \frac{h}{\varepsilon} = 1 \mp \frac{\left(a^2-\varepsilon^2\right)^2}{\left(a^2+\varepsilon^2\right)^2 - 4a^2\varepsilon^2} = 1 \mp 1 = 0{,}2. \tag{40.61c,d}$$

this completes the proof of (40.58–40.59a–d).

The geometric parameters of the thick symmetric (40.58a–d) [strongly cambered this (40.59a–d)] Joukowski airfoil are approximated to the second-order in the thickness (40.62a–d) [camber (40.63a–d)] by:

$$\varepsilon = 0, \quad \delta^3 \ll a^3: \quad h = 0, \quad \frac{c}{4} = a - \delta + \frac{\delta^2}{a}, \quad \frac{t}{4} = \delta - \frac{\delta^3}{a^2}, \tag{40.62a–d}$$

$$\delta = 0, \quad \varepsilon^4 \ll a^4: \quad t = 0, \quad \frac{c}{4} = a - \frac{\varepsilon^2}{a}; \quad h = 2\varepsilon, \tag{40.63a–d}$$

and agree with the first-order approximations (34.58a,b) [(34.68a–c)]. The passage from (40.58c) [(40.58d)] to (40.62c) [(40.61d)] to third-order is (40.64a) [(40.64b)]:

$$c = \frac{4a}{1+\delta/a} = 4a\left[1 - \frac{\delta}{a} + \frac{\delta^2}{a^2} + O\left(\frac{\delta^3}{a^3}\right)\right], \tag{40.64a}$$

$$t = \frac{4\delta}{1+\delta^2/a^2} = 4\delta\left[1 - \frac{\delta^2}{a^2} + O\left(\frac{\delta^4}{a^4}\right)\right], \tag{40.64b}$$

the passage from (40.59c) to (40.63c) is similer to (40.64b):

$$c = \frac{4a}{1+\varepsilon^2/a^2} = 4a\left[1 - \frac{\varepsilon^2}{a^2} + O\left(\frac{\varepsilon^4}{a^4}\right)\right]. \tag{40.64c}$$

From (40.58c) [(40.59c)] it follows that the thickness (camber) reduces the chord, at the first (40.62c) [second (40.63c)] order. The thickness (40.58d) [height (40.59d)] of the symmetric (thin) airfoil is a first-order effect, that is, it has a third-order correction (40.62d) [is exact at first-order (40.63d)].

Example 40.6.3. *Alternative definitions of airfoil chord, thickness, and height.*

For the thick strongly cambered Joukowski airfoil (Figure 40.2) the: (i) point A with smallest x coordinate or vertical tangent $dx/dy = 0$ is not the leading edge; (ii) the points B, C, D of zero slope $dy/dx = 0$ are not those specifying the thickness or height. The thickness of the airfoil, if it were defined as the maximum vertical distance between two points may would be forward of $|\varphi| = \pi/2$; if the chord was defined as the maximum distance from the trailing edge, then the line (L.E.–A) would not be horizontal. To demonstrate these remarks the angular coordinate of the point A is calculated from (39.75a; 34.76a):

$$x_3(\varphi) = (a\cos\varphi - \delta)\left[1 + \frac{f}{g(\varphi)}\right], \tag{40.65a}$$

where

$$f \equiv \left(\sqrt{a^2-\varepsilon^2} - \delta\right)^2, \quad g(\varphi) = a^2 + \varepsilon^2 + \delta^2 + 2a(\varepsilon\sin\varphi - \delta\cos\varphi). \tag{40.65b,c}$$

The point A corresponds to a vertical tangent (40.66a), leading to (40.66b);

$$\frac{\mathrm{d}x_3}{\mathrm{d}\varphi} = 0: \quad 0 = \sin\varphi\,[g(\varphi) + f]\, g(\varphi) + 2(a\cos\varphi - \delta)\, f\,(\varepsilon\cos\varphi + \delta\sin\varphi), \qquad (40.66\text{a,b})$$

the condition (40.66b) corresponds (40.65a) to the vanishing of

$$0 = -\frac{\mathrm{d}x_3}{\mathrm{d}\varphi} = a\sin\varphi\left[1 + \frac{f}{g(\varphi)}\right] + (a\cos\varphi - \delta)\frac{g'(\varphi)f}{[g(\varphi)]^2} \qquad (40.66\text{c})$$

this implies, using (40.65c):

$$a\sin\varphi\,[f + g(\varphi)]\, g(\varphi) = -(a\cos\varphi - \delta)\, f g'(\varphi) = -2(a\cos\varphi - \delta)\, f\,(\delta\sin\varphi + \varepsilon\cos\varphi), \qquad (40.66\text{d})$$

that coincides with (40.66d) $\equiv$ (40.66a). Since the point A is close to the negative real axis, approximation (40.67a,b) may be applied to the lowest order, leading to (40.67c,d):

$$\varphi = \pi + \psi, \psi^2 \ll 1: \quad \sin\varphi = -\sin\psi = -\psi, \quad \cos\varphi = \cos\pi = -1; \qquad (40.67\text{a–d})$$

substituting (40.67c,d) in (40.66b) yields (40.68a):

$$0 = -\psi(\bar{g} + f)\bar{g} + 2(a + \delta)\, f\,(\varepsilon + \psi\delta), \quad \bar{g} = g(\pi) \equiv (a + \delta)^2 + \varepsilon^2, \qquad (40.68\text{a,b})$$

where (40.65c) is evaluated (40.68b) for $\psi =\ 0$ or $\varphi = \pi$. Solving (40.68a) for ψ leads to

$$\psi = \frac{\varepsilon}{(1 + \bar{g}/f)(\bar{g}/2)/(a + \delta) - \delta} = \frac{\varepsilon}{a + 2\delta} = \frac{\varepsilon}{a} - \frac{2\varepsilon\delta}{a^2} + O\left(\frac{\varepsilon\delta^2}{a^3}\right); \qquad (40.69)$$

the approximation (40.70b,c) of (40.65b; 40.68b) to second-order (40.70a):

$$\varepsilon^2, \delta^2, \varepsilon\delta \ll 1: f = a^2 - 2a\delta, \quad \bar{g} = a^2 + 2a\delta, \qquad (40.70\text{a-c})$$

was used in:

$$\frac{\bar{g}}{2}\frac{1 + \bar{g}/f}{a + \delta} = \left(1 + \frac{a^2 + 2a\delta}{a^2 - 2a\delta}\right)\frac{a^2 + 2a\delta}{2a + 2\delta} = \frac{a}{2}\left(1 + \frac{1 + 2\delta/a}{1 - 2\delta/a}\right)\frac{1 + 2\delta/a}{1 + \delta/a};$$
$$= \frac{a}{2}\left(2 + \frac{4\delta}{a}\right)\left(1 + \frac{\delta}{a}\right) = a\left(1 + \frac{3\delta}{a}\right) = a + 3\delta, \qquad (40.70\text{d})$$

and substituted in (40.69). Thus the point on the profile with vertical tangent: (i) lies at angle (40.69) below the real axis; (ii) its deviation from the real axis is due to the camber at lowest order; (iii) the thickness only appears at the second-order.

Example 40.7. *Bilinear Mapping between Circles and Straight Lines.*
Consider the four sets of three points:

$$L_1 \equiv \{2, 2i, 1 + i\}, \quad L_2 = \{1,\ 0,\ -1\}; \quad C_1 \equiv \{1, i, 2i\}, \quad C_2 = \{-1,\ i,\ 1\}. \qquad (40.71\text{a–d})$$

Show that the first (last) two sets (40.71a,b) [(40.71c,d)] correspond to three points that do lie (do not lie) on a straight line. Obtain the univalent mappings between: (i) the two straight lines $L_1 \to L_2$; (ii) the two circles $C_1 \to C_2$; (iii) a straight line and a circle $L_2 \to C_1$.

Three points (z_1, z_2, z_3) *lie in a straight line if the relative arguments have equal modulus (40.72a):*

$$\arg(z_1 - z_2) = \pm\arg(z_1 - z_3) \Leftrightarrow \frac{z_1 - z_2}{z_1 - z_3} = \lambda \in |R, \qquad (40.72\text{a,b})$$

that is equivalent to the three-point cross-ratio being real (40.72b). The set of three points (40.71a,b) [(40.71c,d)] lies (does not lie) on a straight line:

$$\lambda \equiv \frac{z_1 - z_2}{z_1 - z_3} = 2, \frac{1}{2}, \frac{3+i}{5}, \frac{1+i}{2}. \tag{40.73a–d}$$

Geometrically: (i) the mid point between 2 and $2i$ is $1 + i$, so (40.71a) lie on a straight line; (ii) all three points (40.71b) are on the real axis; (iii/iv) of the three points in (40.71c) [(40.71d)] two are on the imaginary (real) axis and one on the real (imaginary) axis, so they could not possibly be on a straight line. *The most general mapping univalent in the whole complex plane is the bilinear transformation that is uniquely determined (35.35a,b) by mapping (40.74a):*

$$\frac{\zeta - \zeta_1}{\zeta - \zeta_2} = k\frac{z - z_1}{z - z_2}, \qquad k \equiv \frac{\zeta_3 - \zeta_1}{\zeta_3 - \zeta_2}\frac{z_3 - z_2}{z_3 - z_1}, \tag{40.74a,b}$$

from three (z_1, z_2, z_3) into another three points ($\zeta_1, \zeta_2, \zeta_3$) in (40.74b). Using (40.74a,b) in (40.71a–d) leads to:

$$\{L_1 \to L_2, C_1 \to C_2, L_2 \to C_1\}: \ k = \left\{-2, \frac{2i}{1+3i}, 1 + \frac{i}{2}\right\}, \tag{40.75a,b}$$

$$\zeta = \left\{\frac{z - 2i}{3z - 4 - 2i}, \frac{z - 3iz - 5 + i}{z + iz + 3 + i}, \frac{2z + 3iz - 2 - i}{z - 1 + 2i}\right\}, \tag{40.75c}$$

for the three transformations indicated.

Example 40.8. *Family of Circles Defined by Its Conjugate Points.*

Obtain the equation of the family of circles of radius R whose conjugate points are (p, q). Consider the particular case $(p = 1, q = i)$ and the limit $R \to \infty$ in Cartesian coordinates.

The equation of the circle with conjugate points (40.76a,b) is (35.54a) $\equiv$ (40.76c) in:

$$p = 1,\ q = i\text{:}\ \lambda = \frac{|z - p|}{|z - q|} = \frac{|z - 1|}{|z - i|} = \left|\frac{(x-1)^2 + y^2}{x^2 + (y-1)^2}\right|^{1/2}, \tag{40.76a–c}$$

that can be rewritten as:

$$0 = \left(1 - \lambda^2\right)\left(x^2 + y^2\right) + 2\left(\lambda^2 y - x\right) + 1 - \lambda^2. \tag{40.77}$$

The case (40.78a) is the straight line (40.78b) that passes through the midpoint (40.78c) between (40.76a,b):

$$\lambda = 1: \quad y = x, \quad \frac{p+q}{2} = \frac{1+i}{2}. \tag{40.78a–c}$$

If $\lambda \neq 1$ the presence of the quadratic term $x^2 + y^2$ and absence of the cross-term xy shows that (40.77) is the equation:

$$\lambda \neq 1\text{:} \quad 0 = x^2 + y^2 + 2\frac{\lambda^2 y - x}{1 - \lambda^2} + 1; \tag{40.79a,b}$$

of a circle:

$$O = \left(x - \frac{1}{1-\lambda^2}\right)^2 + \left(y + \frac{\lambda^2}{1-\lambda^2}\right)^2 - \frac{1+\lambda^4}{\left(1-\lambda^2\right)^2} + 1 \equiv (x-a)^2 + (y-b)^2 - R^2, \tag{40.80a}$$

with center at (40.80c) and radius (40.80d):

$$\lambda > 1: \quad s \equiv a + ib = \frac{1 - i\lambda^2}{1 - \lambda^2}, \quad R = \frac{\lambda\sqrt{2}}{\lambda^2 - 1}; \tag{40.80b–d}$$

the parameter (40.79a) was chosen to satisfy (40.80b), corresponding to a positive radius (40.80d), that follows from the simplification of the last two terms in (40.80a):

$$R^2 = \frac{1 + \lambda^4}{\left(1 - \lambda^2\right)^2} - 1, \quad R\left(\lambda^2 - 1\right) = \left|1 + \lambda^4 - \left(\lambda^2 - 1\right)^2\right|^{1/2} = \lambda\sqrt{2}. \tag{40.81a,b}$$

In the case $\lambda = 1$, the radius is infinite and the center is at infinity corresponding to the straight line (40.78a–c). Thus *the equation of the family of circles (Figure 40.3) with conjugate points (40.76a,b) is (40.80a) ≡ (40.82):*

$$\left[x\left(1 - \lambda^2\right) - 1\right]^2 + \left[y\left(1 - \lambda^2\right) + \lambda^2\right] = 2\lambda^2, \tag{40.82}$$

with the radius (40.80c) and center at (40.80b) specified by the parameter(40.79a); for (40.78a) it degenerates to the straight line (40.78b) orthogonal to the line joining the reciprocal points (40.76a,b) and passing through the midpoint (40.78c) between them.

The preceding results can be obtained from Section 35.7 as follows. The parameter λ in (35.57a) ≡ (40.83a) satisfies (35.57c) ≡ (40.83b):

$$\lambda = \frac{\mu}{R}: \quad |p - q| = R\left(\lambda - \frac{1}{\lambda}\right) = \mu - \frac{R^2}{\mu}, \quad \lambda^2 - \lambda\frac{|p - q|}{R} - 1 = 0, \tag{40.83a–c}$$

that leads to (40.83c). The reciprocal points (36.76a,b) specify (40.84a):

$$|p - q| = |1 - i| = \sqrt{2}, \quad \lambda^2 - \lambda\frac{\sqrt{2}}{R} - 1 = 0, \tag{40.84a,b}$$

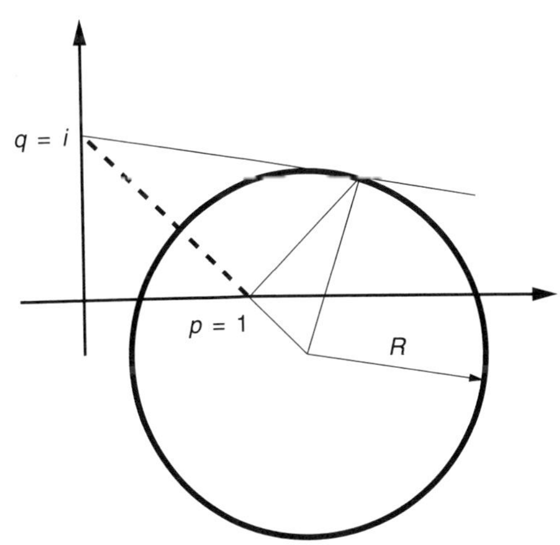

FIGURE 40.3
There is a one-parameter family of circles (Figure 35.4a) with given reciprocal points (p, q), for example, for the points $p = 1$ and $q = i$ in Figure 40.3. The distance of all points on the circle to $p = 1$ and $q = i$ is in a constant ratio λ, which determines the radius of the particular circle of the family.

and lead to (40.84b) ≡ (40.83c) ≡ (40.80d); this is equivalent to

$$\lambda = \frac{1}{R\sqrt{2}} + \left|1 + \frac{1}{2R^2}\right|^{1/2}, \tag{40.85}$$

where the positive real root was chosen. An alternative choice of parameter (40.83a,c) ≡ (40.86a) is (40.86b):

$$0 = \mu^2 - |p - q|\,\mu - R^2 = \mu^2 - \mu\sqrt{2} - R^2, \quad \mu = \frac{1}{\sqrt{2}} + \left|\frac{1}{2} + R^2\right|^{1/2}. \tag{40.86a,b}$$

The coordinates of the center of the circle are specified by (35.57b) ≡ (40.87c) ≡ (40.80c):

$$p = 1, q = i: \quad s = \frac{p - \lambda^2 q}{1 - \lambda^2} = \frac{1 - i\lambda^2}{1 - \lambda^2}, \tag{40.87a–c}$$

in agreement with (40.76a,b) ≡ (40.87a,b).

Example 40.9. *Complex Mappings of Disks with Arbitrary Radius.*

Obtain the mapping that is univalent in the whole complex z-plane and maps (i) the real axis [(ii) the disk of radius S and center at the origin] into the disk of radius R and center at the origin.

Example 40.9.1. *Bilinear mapping between a half-plane and a nonunit disk.*

The only conformal mapping in that is univalent in the whole complex plane is the bilinear mapping (40.88a) ≡ (35.16b):

$$\zeta = \frac{az + b}{cz + d}; \quad \zeta = \frac{a}{c}\,\frac{z - \alpha}{z - \alpha^*}, \tag{40.88a,b}$$

to map the upper-half complex z-plane into the disk of radius R in the ζ-plane it must map the two points $z = (\alpha, \alpha^*)$ reciprocal relative to the real axis into points $(\zeta, R^2/\zeta^*)$ reciprocal with regard to the circle of radius R, and center at the origin, viz. if $z = \alpha$ maps to $\zeta = 0$ then $z = \alpha^*$ maps to $\zeta = \infty$, leading to (40.88b). In addition the point $z = 0$ on the real line must map to the circle $|\zeta| = R$, so that (40.89a):

$$R = \left|\frac{a\alpha}{c\alpha^*}\right| = \left|\frac{a}{c}\right|, \quad \frac{a}{c} = Re^{i\lambda}, \tag{40.89a,b}$$

implies (40.89b). Substituting (40.89b) in (40.88b) yields:

$$\alpha \in |C\,, \quad \lambda \in |R\,, \quad \operatorname{Im}(\alpha) > 0: \quad \operatorname{Im}(z) >< 0, \quad |\zeta| <> R: \quad \zeta = \mathrm{Re}^{i\lambda}\frac{z - \alpha}{z - \alpha^*}, \tag{40.90a–f}$$

the (Table 35.1) concerning the bilinear mapping (40.90a,b,f) of the upper (lower) half z-plane (40.90d) into the interior (exterior) of the disk of radius R and center at the origin (40.90e) in the ζ-plane for $\operatorname{Im}(\alpha) > 0$, *and vice-versa for* $\operatorname{Im}(\alpha) < 0$. For α in the upper-half complex plane $\operatorname{Im}(\alpha) > 0$ in (40.90c), a point in the upper $\operatorname{Im}(z) > 0$ [lower $\operatorname{Im}(z) < 0$] half-complex plane is closer to (farther from) α than from α^*, implying $|z - \alpha| < |z - \alpha^*|$ [$|z - \alpha| > |z - \alpha^*|$] and thus its image is inside $|\zeta| < R$ (outside $|\zeta| > R$) the circle; the reverse applies if α is $\operatorname{Im}(\alpha) < 0$ in the lower-half complex plane. Choosing the unit circle $R = 1$ in the ζ-plane in (40.90a–f) leads to (35.64–f); conversely the substitution $\zeta \to \zeta/R$ in (30.64a–f) leads back to (40.90a–f).

Example 40.9.2. *Mobius Transformation between Nonunit Circles.*

To map the disk of radius S in the z-plane to the disk of radius R in the ζ-plane, the points $z = (\alpha, S^2/\alpha^*)$ inverse with regard to the circle $|z| = S$ must be mapped to the points $(\zeta, R^2/\zeta^*)$ inverse with regard to the circle $|\zeta| = R$. Choosing the inverse points $\zeta = 0, \infty$ in the bilinear mapping (40.88a) leads to (40.91a):

$$\zeta = \frac{a}{c}\frac{z-\alpha}{z-S^2/\alpha^*}; \qquad R \equiv |\zeta| = \left|\frac{a}{c}\frac{S-\alpha}{S-S^2/\alpha^*}\right|, \tag{40.91a,b}$$

in addition the point $z = S$ on the circle $|z| = S$ must map (40.91b) to the circle $|\zeta| = R$. Simplifying (40.91b) to (40.92a):

$$R = \left|\frac{\alpha^* a}{cS}\frac{S-\alpha}{\alpha^*-S}\right| = \left|\frac{\alpha^* a}{cS}\right|, \qquad \frac{a}{c} = e^{i\lambda} R \frac{S}{\alpha^*}, \tag{40.92a,b}$$

leads to (40.92b). Substitution of (40.92b) in (40.91a) yields *(Table 35.2) the bilinear mapping (40.93a,b,f) from (to) the disk (40.93d) [(40.93e)] of radius $R\,(S)$ and center at the origin of the z-plane (ζ-plane):*

$$\alpha \in |C, \quad \lambda \in |R\,, \quad |\alpha| > S: \quad |z| <> S, \quad |\zeta| <> R\!: \zeta = RSe^{i\lambda}\frac{z-\alpha}{\alpha^* z - S^2}; \tag{40.93a–f}$$

for $|\alpha| < 1$ the mapping is from interior (exterior) to interior (exterior) of the disks, and vice-versa for $|\alpha| > 1$. Choosing the unit circles $S = 1\,(R = 1)$ in the z-plane (ζ-plane) in (40.93a–f) leads to (35.68a–f); conversely the changes of variable (40.94a–c) in (35.68f) lead back to (40.94d):

$$\{\zeta, z, \alpha\} = \left\{\frac{\zeta}{R}, \frac{z}{S}, \frac{\alpha}{S}\right\}\!: \quad \frac{\zeta}{R} = e^{i\lambda}\frac{z/S - \alpha/S}{\alpha^* z/S^2 - 1}, \tag{40.94a–d}$$

that coincides with (40.94d) $\equiv$ (40.93f). If α lies inside the circle in the z-plane $|\alpha| < S$, then a point inside $|z| < S$ (outside $|z| > S$) the circle is closer to (farther from) α than to the inverse point S^2/α^*, and thus it is mapped to the interior $|\zeta| < R$ (exterior $|\zeta| > R$) of the circle in the ζ-plane); the reverse applies if α is outside the circle $|\alpha| > S$ in the z-plane.

Example 40.10. *Electric and Magnetic Fields with Mixed Insulating and Conducting Walls.*

Consider the electro (magneto) static field of an electric charge (current) in a corner with cylindrical indentation (rectangular well) for all combinations of conducting or/and insulating walls not covered in Sections 36.1–36.2 (36.7–36.9). Interpret the induced electric change distribution an conducting walls using the analogy with potential flows.

Example 40.10.1. *Electric charges/currents in a corner with a circular indentation.*

Concerning the corner with cylindrical indentation and electrostatic field: (case I) for all walls insulating (Figure 36.5a) all images are identical and there are no induced charges; (case IV) for all walls conducting (Figure 36.5c) the circumferential images are alternating and the radial images opposite leading to the induced charges (36.37–36.38a–d); (case II) if the straight sidewalls (circular indentation) are insulating (is conducting) the images (image)

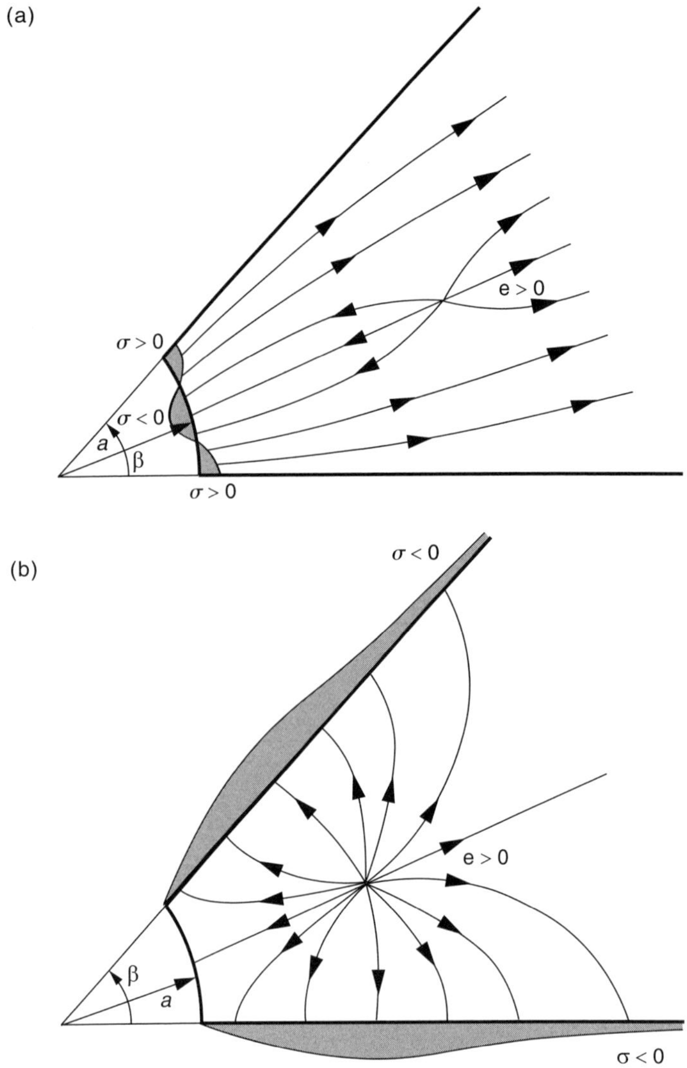

FIGURE 40.4
Electrostatic field of a line electric charge on the axis of a corner with a circular indentation, for which: (i) the circular indentation is conducting, hence the normal electric field and induced surface electric charge distribution; (ii) the side walls are insulating (a) [conducting (b)], hence the tangential (normal) electric field, and absence (presence) of surface electric charges. The surface distributions are represented in Figures 36.2a and 36.5c and also in Figure 40.4a,b: (i) they vanish at the corners and peak at the mid-point for the circular indentation; (ii) on the side wall the peak is closest to original electric charge with decay to zero at the corner and infinity; (iii) they do not vanish at the corner in the case of Figure 40.4a as for a reverse flow.

are (is) identical (opposite) in the circumferential (radial) direction (Figure 40.4a); (case III) if the straight sidewalls (circular indentation) are conducting (is insulating), the image (images) are (is) alternating (identical) in the circumferential (radial) direction (Figure 40.4b). If the sidewalls are insulating the circumferential images are identical leading to the complex electrostatic potential (36.34a) in, where the $+(-)$ sign corresponds to identical (opposite) radial image, and hence to insulating (case I) [conducting (case II)] circular indentation (40.95a):

$$f_{\mathrm{e}}^{+\pm}(z) = -\frac{\mathrm{e}}{2\pi\varepsilon}\left\{\log\left(z^{2\pi/\beta} + b^{2\pi/\beta}\right) \pm \log\left[\left(\frac{a^2}{z}\right)^{2\pi/\beta} + b^{2\pi/\beta}\right]\right\}, \tag{40.95a}$$

$$E_{+\pm}^{*}(z) = -\frac{\mathrm{d}f_{\mathrm{e}}^{+\pm}}{\mathrm{d}z} = \frac{\mathrm{e}}{\varepsilon\beta z}\left\{\left[1+\left(\frac{b}{z}\right)^{2\pi/\beta}\right]^{-1} \mp \left[1+\left(\frac{bz}{a}\right)^{2\pi/\beta}\right]^{-1}\right\}; \tag{40.95b}$$

the corresponding complex conjugate electric field is (40.95b). The case I of upper sign in (40.95b) coincides with (36.34b) replacing Q by e/ε; the case II changes one sign because the image on the circle is opposite instead of identical. The conjugate electric field (40.95b) is real on the sidewalls, that is, it is tangential, showing that they are insulating in both cases I and II. On the circular indentation:

$$E_r^{++}(a,\varphi) = 0, E_\varphi^{++}(a,\varphi) = -\frac{2\mathrm{e}}{\varepsilon\beta a}\frac{\sin(2\pi\varphi/\beta)}{(b/a)^{2\pi/\beta} + (a/b)^{2\pi/\beta} + 2\cos(2\pi\varphi/\beta)}, \tag{40.96a,b}$$

$$\begin{aligned} E_\varphi^{+-}(a,\varphi) = 0, E_r^{+-}(a,\varphi) &= \frac{2\mathrm{e}}{\varepsilon\beta a}\frac{(a/b)^{2\pi/\beta} + \cos(2\pi\varphi/\beta)}{(b/a)^{2\pi/\beta} + (a/b)^{2\pi/\beta} + 2\cos(2\pi\varphi/\beta)} \\ &= \frac{\sigma^{+-}(\varphi)}{\varepsilon} = \frac{e}{\varepsilon}\frac{\vartheta^{+-}(\varphi)}{J}. \end{aligned} \tag{40.97a–c}$$

The passage from (40.95b) to (40.96a,b) [(40.96a,b)] corresponds to the upper (lower) sign in:

$$\begin{aligned} z = a\mathrm{e}^{i\varphi} : E_r^{+\pm}(a,\varphi) - iE_\varphi^{+\pm}(a,\varphi) &= \mathrm{e}^{i\varphi}E_{+\pm}^{*}\left(a\mathrm{e}^{i\varphi}\right) \\ &= \frac{\mathrm{e}}{\varepsilon\beta a}\left\{\left[1+\left(\frac{b}{a}\right)^{2\pi/\beta}\mathrm{e}^{-i2\pi\varphi/\beta}\right]^{-1} \mp \left[1+\left(\frac{b}{a}\right)^{2\pi/\beta}\mathrm{e}^{i2\pi\varphi/\beta}\right]^{-1}\right\} \\ &= \frac{\mathrm{e}}{\varepsilon\beta a}\left[1+\left(\frac{b}{a}\right)^{4\pi/\beta} + \left(\frac{b}{a}\right)^{2\pi/\beta}\left(\mathrm{e}^{i2\pi\varphi/\beta} + \mathrm{e}^{-i2\pi\varphi/\beta}\right)\right]^{-1} \\ &\quad \left\{\left(\frac{b}{a}\right)^{2\pi/\beta}\left(\mathrm{e}^{i2\pi\varphi/\beta} - \mathrm{e}^{-i2\pi\varphi/\beta}\right), 2 + \left(\frac{b}{a}\right)^{2\pi/\beta}\left(\mathrm{e}^{i2\pi\varphi/\beta} + \mathrm{e}^{-i2\pi\varphi/\beta}\right)\right\} \\ &= \frac{2\mathrm{e}}{\varepsilon\beta a}\frac{\left\{i(b/a)^{2\pi/\beta}\sin(2\pi\varphi/\beta), 1 + (b/a)^{2\pi/\beta}\cos(2\pi\varphi/\beta)\right\}}{1 + (b/a)^{4\pi/\beta} + 2(b/a)^{2\pi/\beta}\cos(2\pi\varphi/\beta)}. \end{aligned} \tag{40.98a,b}$$

The electric field is (case I) tangential (40.96a,b), hence (Figure 36.5a) there are no induced electric charges; (case II) radial (40.97a) leading (Figure 40.4a) to the induced electric charge distribution (40.97b). The latter (40.97c) also applies to the surface electric current distribution for insulating side walls and conducting circular indentation (Figure 40.4b).

Example 40.10.2. *Conducting/insulating side walls and circular indentation.*

The case of conducting side walls leads to alternating circumferential images (36.35a), leading (24.47) to the complex electrostatic potential (40.99a) where the + (−) sign applies to identical (opposite) image in the circular indentation:

$$f_e^{-\pm}(z) = -\frac{e}{2\pi\varepsilon}\left[\log\left(\frac{z^{\pi/\beta} - ib^{\pi/\beta}}{z^{\pi/\beta} + ib^{\pi/\beta}}\right) \pm \log\left(\frac{a^{2\pi/\beta}z^{-\pi/\beta} - ib^{\pi/\beta}}{a^{2\pi/\beta}z^{-\pi/\beta} + ib^{\pi/\beta}}\right)\right], \tag{40.99a}$$

$$E_{-\pm}^*(z) = -\frac{df_e^{-\pm}}{dz} = i\frac{e}{\varepsilon\beta z}\left\{\left[\left(\frac{z}{b}\right)^{\pi/\beta} + \left(\frac{b}{z}\right)^{\pi/\beta}\right]^{-1} \pm \left[\left(\frac{bz}{a^2}\right)^{\pi/\beta} + \left(\frac{a^2}{bz}\right)^{\pi/\beta}\right]^{-1}\right\}; \tag{40.99b}$$

the corresponding complex conjugate electric field is (40.99b), using a simplification similar to (30.36). The case IV or lower sign in (40.99b) coincides with (36.35b) replacing Γ by ie/ε; the case III changes one sign because the images on the circle is identical instead of opposite. On the circular indentation:

$$E_r^{-+}(a,\varphi) = 0, \quad E_\varphi^{-+}(a,\varphi) = -\frac{2e}{\varepsilon\beta a}\cos(\pi\varphi/\beta)\frac{(a/b)^{\pi/\beta} + (b/a)^{\pi/\beta}}{(a/b)^{2\pi/\beta} + (b/a)^{2\pi/\beta} + 2\cos(2\pi\varphi/\beta)}, \tag{40.100a,b}$$

$$\begin{aligned} E_\varphi^{-}(a,\varphi) = 0, \quad E_r^{--}(a,\varphi) &= -\frac{2e}{\varepsilon\beta a}\sin\left(\frac{\pi\varphi}{\beta}\right)\frac{(b/a)^{\pi/\beta} - (a/b)^{\pi/\beta}}{(a/b)^{2\pi/\beta} + (b/a)^{2\pi/\beta} + 2\cos(2\pi\varphi/\beta)} \\ &\equiv \frac{\sigma^{--}(\varphi)}{\varepsilon} = \frac{e}{\varepsilon}\frac{\vartheta^{--}(\varphi)}{J}, \end{aligned} \tag{40.101a–c}$$

the electric field is: (case III) tangential (40.100a,b) so there are no induced electric charges; (case IV) normal (40.101a) so there are induced electric charges (40.101b) coincident with (40.101b) ≡ (36.38d). The passage from (40.99b) to (40.100a,b) [(40.101a,b)] corresponds to the upper (lower) sign in:

$$\begin{aligned} E_r^{-\pm}(r,\varphi) - iE_\varphi^{-\pm}(r,\varphi) &= e^{i\varphi}E_{-\pm}^*\left(ae^{i\varphi}\right) \\ &= i\frac{e}{\varepsilon\beta r}\left\{\left[\left(\frac{a}{b}\right)^{\pi/\beta} e^{i\pi\varphi/\beta} + \left(\frac{b}{a}\right)^{\pi/\beta} e^{-i\pi\varphi/\beta}\right]^{-1}\right. \\ &\quad \left.\pm\left[\left(\frac{b}{a}\right)^{\pi/\beta} e^{i\pi\varphi/\beta} + \left(\frac{a}{b}\right)^{\pi/\beta} e^{-i\pi\varphi/\beta}\right]^{-1}\right\} \\ &= i\frac{e}{\varepsilon\beta r}\left[\left(\frac{a}{b}\right)^{2\pi/\beta} + \left(\frac{b}{a}\right)^{2\pi/\beta} + e^{i2\pi\varphi/\beta} + e^{-i2\pi\varphi/\beta}\right]^{-1} \\ &\quad \left[\left(\frac{a}{b}\right)^{\pi/\beta}\left(e^{-i\pi\varphi/\beta} \pm e^{i\pi\varphi/\beta}\right) + \left(\frac{b}{a}\right)^{\pi/\beta}\left(e^{i\pi\varphi/\beta} \pm e^{-i\pi\varphi/\beta}\right)\right] \\ &= i\frac{2e}{\varepsilon\beta r}\frac{\left\{\left[(b/a)^{\pi/\beta} + (a/b)^{\pi/\beta}\right]\cos(\pi\varphi/\beta), i\left[(b/a)^{\pi/\beta} - (a/b)^{\pi/\beta}\right]\sin(\pi\varphi/\beta)\right\}}{(a/b)^{2\pi/\beta} + (b/a)^{2\pi/\beta} + 2\cos(2\pi\varphi/\beta)}. \end{aligned} \tag{40.102}$$

Along the side wall the electric field is normal (40.101b) ≡ (36.38d):

$$E_r^{-\pm}(r,0) = 0 = E_r^{-\pm}(r,\beta)\,, \quad \frac{\mathrm{e}}{\varepsilon}\frac{\sigma^{-\pm}(r)}{J} = \frac{\sigma^{-\pm}(r)}{\varepsilon} = E_\varphi^{-\pm}(r,0) = -E_\varphi^{-\pm}(r,\beta)$$
$$= -\frac{\mathrm{e}}{\varepsilon\beta r}\left\{\left[\left(\frac{r}{b}\right)^{\pi/\beta} + \left(\frac{b}{r}\right)^{\pi/\beta}\right]^{-1} \pm \left[\left(\frac{br}{a}\right)^{\pi/\beta} + \left(\frac{a^2}{br}\right)^{\pi/\beta}\right]^{-1}\right\}, \tag{40.103a–c}$$

and specifies the induced charge distribution, in the case III (IV) with upper (lower) sign in (40.103b). The case IV coincides with (40.103b) ≡ (36.37d). The induced electric charges (surface electric currents) coincide (40.103b) ≡ (40.103c) for the electro (magneto) static field in Figure 40.4b (40.4a).

Example 40.10.3. *Induced electric charges on the side walls.*

In the case of insulating (conducting) circular indentation (40.99b) with upper (cower) sign the induced electric currents on the side walls have opposite signs:

$$\frac{\mathrm{e}}{\varepsilon}\frac{\vartheta^{-\pm}(r)}{J} = \frac{\sigma^{-\pm}(r)}{\varepsilon} = E_\varphi^{-\pm}(r,0) = -E_\varphi^{-\pm}(r,\beta)$$
$$= -\frac{\mathrm{e}}{\beta\varepsilon r}\left\{\left[\left(\frac{r}{b}\right)^{\beta/\pi} + \left(\frac{b}{r}\right)^{\beta/\pi}\right]^{-1} \pm \left[\left(\frac{br}{a^2}\right)^{\pi/\beta} + \left(\frac{br}{a^2}\right)^{\pi/\beta}\right]^{-1}\right\}, \tag{40.104}$$

because (40.99b) changes sign for $z = r$ and $z = r\mathrm{e}^{i\beta}$. The induced electric charges vanish (40.105a) [do not vanish (40.105b)] at the corners for conducting side walls and conducting (insulating) circular indentation:

$$\sigma^{--}(a,0) = 0 = \sigma^{--}(a,\beta)\,, \quad \sigma^{-+}(a,0) = -\sigma^{-+}(a,\beta) = -\frac{2\mathrm{e}}{\beta a}\left[\left(\frac{a}{b}\right)^{\pi/\beta} + \left(\frac{b}{a}\right)^{\pi/\beta}\right]. \tag{40.105a,b}$$

In the case (Figure 40.4b) of side walls and circular indentation both conducting, the electric field must be orthogonal to both; since at the corner the walls are orthogonal, the electric field must vanish, and hence also the induced electric charges; this corresponds to the zero velocity at the corner for the potential flow due to a vortex (Figure 36.5b). In the case (Figure 40.4a) of conducting circular indentation and insulating side walls the electric filed is normal to the former and tangential to the latter, and thus radial in both cases; thus the electric field can be continuous at the corner without vanishing, and the electric charge density, and the electric charge density can be nonzero on the circular indentation while being zero on the side walls. There is (there is no) discontinuity of the induced electric charge (electric field) at the corner between conducting and insulating walls. This would correspond in the case of a potential flow to a source/sink with rigid side walls and a free surface at the circular indentation, which can be crossed by the flow (Figure 36.5c). When calculating the induced electric charges on the side walls it should be noticed that: (i) they are proportional to the normal component of the electric field; (ii) the outward unit normal is opposite on the two side walls; (iii) if the horizontal component of the electric field is opposite (equal) on the two side walls the induced electric charges are equal (opposite). In the case of the magnetic field: (i) it is the tangential component which specifies the surface electric current; (ii) if the tangential magnetic field is equal (opposite) on the side walls the surface electric current component of the are also equal (opposite).

Example 40.10.4. *Induced electric charge on the circular indentation.*

The induced electric charges on the circular indentation (40.97b) [(40.101b)] with insulating (conducting) side walls in Figure 40.4a (36.5c): (ii) peak in the middle (40.106b) [(40.106d)]; (ii) do not vanish (40.106a) [do vanish (40.106c)] at the corners:

$$\begin{aligned}\left[\sigma^{+-}(\varphi)\right]_{\max} &= \sigma^{+-}(0) = \sigma^{+-}(\beta) = \frac{2e/(\beta a)}{(b/a)^{2\pi/\beta}+1} \\ &\geq \sigma^{+-}(\varphi) \geq \left[\sigma^{+-}(\varphi)\right]_{\min} = \sigma^{+-}\left(\frac{\beta}{2}\right) = -\frac{2e/(\beta a)}{(b/a)^{2\pi/\beta}-1},\end{aligned} \tag{40.106a,b}$$

$$\begin{aligned}\left[\sigma^{--}(\varphi)\right]_{\max} &= \sigma^{--}(0) = \sigma^{--}(\beta) = 0 \geq \sigma^{--}(\varphi) \\ &\geq \left[\sigma^{--}(\varphi)\right]_{\min} = \sigma^{--}\left(\frac{\beta}{2}\right) = -\frac{2e/(\beta a)}{(b/a)^{2\pi/\beta}-(a/b)^{2\pi/\beta}}.\end{aligned} \tag{40.106c,d}$$

In (40.106a,b) [(40.106c,d)] were used the simplifications (40.106e) [(40.106f)]:

$$\frac{(a/b)^{2\pi/\beta}\pm 1}{(a/b)^{2\pi/\beta}+(b/a)^{2\pi/\beta}\pm 2} = \frac{\pm\left[(a/b)^{2\pi/\beta}\pm 1\right]}{\left[(a/b)^{2\pi/\beta}\pm 1\right]\left[(b/a)^{2\pi/\beta}\pm 1\right]} = \pm\frac{1}{(b/a)^{2\pi/\beta}\pm 1}, \tag{40.106e}$$

$$\frac{(b/a)^{\pi/\beta}-(b/a)^{\pi/\beta}}{(a/b)^{2\pi/\beta}+(b/a)^{2\pi/\beta}-2} = \frac{(b/a)^{\pi/\beta}-(a/b)^{\pi/\beta}}{\left[(b/a)^{\pi/\beta}-(a/b)^{\pi/\beta}\right]^2} = \frac{1}{(b/a)^{\pi/\beta}-(a/b)^{\pi/\beta}}. \tag{40.106f}$$

In the case of circular indentation and side walls both conducting (Figure 36.5c) the radially and circumferentially alternating images lead to an induced electric charge distribution (40.106d) that is everywhere opposite to the original electric charge.

In the case of insulating walls and conducting circular indentation (Figure 40.4a) the induced surface electric charge density is due to two rings of opposite images (Note 40.1), identical (opposite) to e on the outside (inside), leading to: (i) the opposite sign (40.106b) on axis $\varphi = \beta/2$; (ii) the same sign (40.106a) at the corners $\varphi = 0, \beta$. Thus there are two angles $\varphi_\pm$ for which the induced surface electric charge density (40.97b) vanishes:

$$\left(\frac{a}{b}\right)^{2\pi/\beta} = -\cos\left(\frac{2\pi\varphi_\pm}{\beta}\right) = \cos\left(\mp\pi \pm \frac{2\pi\varphi_\pm}{\beta}\right); \tag{40.107a}$$

the two angles (40.107a) $\equiv$ (40.107b) span an angular sector of width (40.107d) around the diagonal (40.107c):

$$\varphi_\pm = \frac{\beta}{2} \pm \frac{\beta}{2\pi}\operatorname{arc\,cos}\left[\left(\frac{a}{b}\right)^{2\pi/\beta}\right], \quad \varphi_+ + \varphi_- = \beta, \quad \Delta\varphi \equiv \varphi_+ - \varphi_- = \frac{\beta}{\pi}\operatorname{arc\,cos}\left[\left(\frac{a}{b}\right)^{2\pi/\beta}\right]. \tag{40.107b–d}$$

For a monopole outside the circular indentation $b > a$ the angles of zero surface electric charge (40.107b) are real. They would be imaginary for a monopole inside the circular indentation; in that case the surface electric charge does not vanish at any angle. For monopole inside or outside the circular indentation holds the inequality:

$$0 < \left[(b/a)^{\pi/\beta} - (b/a)^{\pi/\beta}\right]^2 = (b/a)^{\pi/\beta} + (b/a)^{\pi/\beta} - 2, \tag{40.108a}$$

it follows that:

$$(b/a)^{\pi/\beta} + (b/a)^{\pi/\beta} > 2 \geq 2\cos(2\pi\varphi/\beta)\,, \tag{40.108b}$$

showing that the denominators of (40.97b) and (40.101c) cannot vanish. Thus the density of electric charge on the circular indentation is finite, regardless of whether: (i) the side walls are conducting or insulating; (ii) the monopole is inside or outside the circular indentation.

Example 40.10.5. *Opposite induced electric charges and reverse potential flow.*

The opposite induced electric charges on the circular indentation of a corner with a insulating walls and containing a line-charge (Figure 40.4a) can be interpreted in terms of a potential flow. In the absence of sources or sinks the flow approaches the corner along one wall and goes back outwards along the other wall (Figure 36.2a), since the incompressible fluid cannot accumulate at the vertex; the velocity is zero at the vertex and normal to the axis (36.12b) that separates inflow from outflow. The case analogous to Figure 40.4a is a flow source (positive electric charge) on the axis with: (i) rigid, impermeable (insulating) side walls; (ii) the circular indentation as a free surface (conductor) to that the velocity (electric field) in normal. Thus the radial inflow along the axis due to the source must be compensated by a radial outflow along the walls (Figure 36.5a), since the incompressible fluid cannot accumulate near the vertex. 'Cutting' the flow by a circular free surface the normal velocity will be radial inward (negative induced electric charges) near the axis and radial outward (positive induced electric charges) near the walls. In the case of a line sink (negative electric charge) the reverse happens: (i) the fluid is 'extracted' from the corner leading to an outward radial velocity (positive induced electric charge) near the axis; (ii) the region near the vertex is refilled with fluid by radial inward flow (negative induced electric charges) near the walls. The reverse flow (opposite induced electric charge) effect is more pronounced near to the vertex, i.e., for smaller a/b in (40.107b) or larger b/a. The angular sector (40.107d) of induced electric charge on the circular indentation opposite to the original line charge becomes narrower for smaller a/b, that is closer to the vertex. The circular indentation must have induced electric charges with opposite signs (both inflow and outflow) because the original line-charge (source or sink) together with all the identical and opposite images leads to a zero total electric charge (flow rate); also the induced electric charge on (flow rate through) the insulating (rigid and impermeable) side walls is zero. In the case of all walls conducting (Figure 36.5c) the induced electric charge on the circular indentation can be of the same sign because there are also induced electric charges on the side walls; the total charge balance was demonstrated (Subsection 36.9.3) for the analogous problem of a rectangular well with all walls conducting (Figure 36.16b). In the case of a rectangular well with the side walls insulating and bottom wall conducting (Figure 40.5a), the induced electric charge can have the opposite sign at the bottom, because there is not the convergence effect of the vertex (Figure 40.4a). The potential flow is incompressible and cannot accumulate at or vacate the region near the vertex, and thus there must be a reverse flow (opposite induced electric charges). In the case of parallel rigid impermeable (insulating) side walls the velocity can have the same direction at the free surface (the induced electric charges can have the same signs) at the bottom. This is demonstrated next.

Example 40.10.6. *Electric charge/current in a well.*

Concerning the rectangular well and electrostatics the four cases are: (case I) all walls insulating (Figure 36.16a) when all images on the side walls and bottom wall are identical and there are no induced charges; (case IV) all walls conducting (Figure 36.16b) when the images on the side walls are alternating and the images on the bottom wall opposite, leading to the induced charges (36.193–36.194a,b); (case III) if the side walls are insulating and the bottom conducting (Figure 40.5a) the lateral images are identical and opposite to

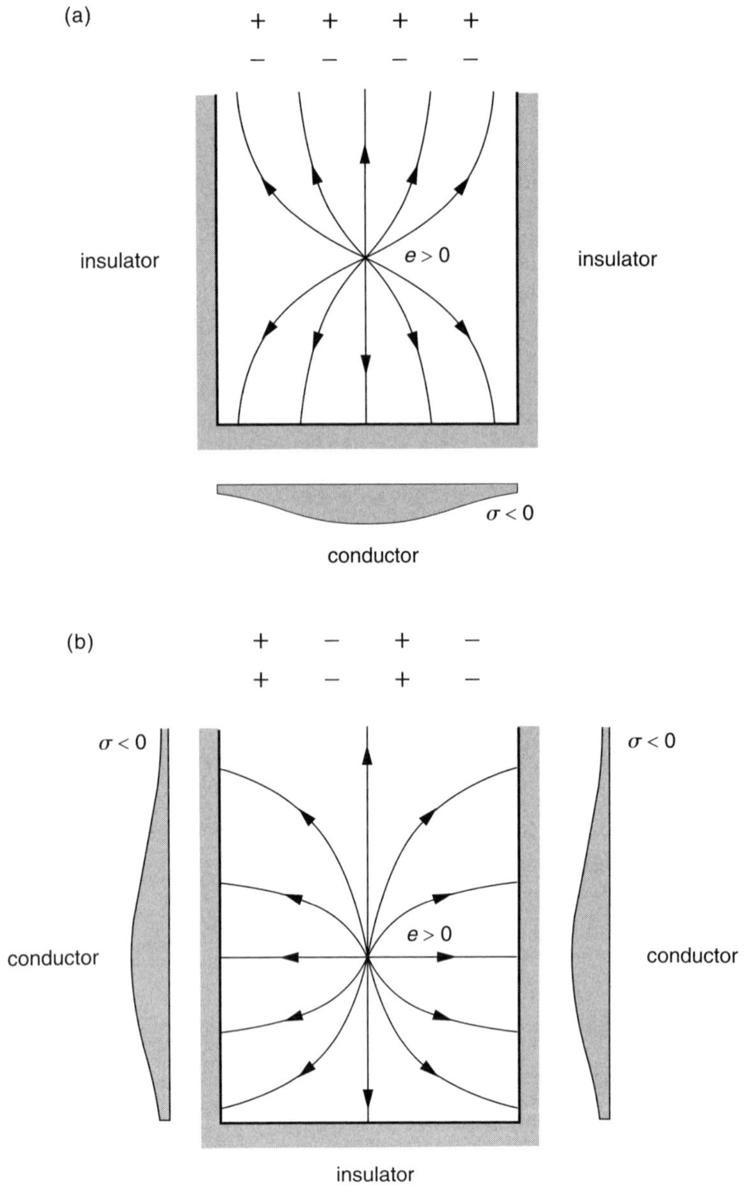

FIGURE 40.5
Electrostatic field of a line electric charge at equal distance of the insulating (a) [conducting (b)] side walls of a rectangular well with a conducting (insulating) bottom. In the case (a) the electric field is tangent (normal) to the side walls (bottom) implying equal (opposite) images; in the case (b) the electric field is orthogonal (tangent) to the side walls (bottom) implying alternating (identical) images. The induced electric charge distributions on conducting surfaces with normal electric field are represented in Figure 36.16d and also in Figure 40.5a,b: (i) they vanish at the corner; (ii/iii) peak on the bottom (sidewall) closest to the original electric charge; (iv) decay to zero at infinity along the sidewalls.

the vertical images; (case III) if the side walls are conducting and the bottom insulating (Figure 40.5b) the lateral images are alternating and the images on the bottom are identical. If the sidewalls are insulating the lateral images are identical, leading (36.182) to the complex electrostatic potential (40.109a), where the upper (lower) sign refers to identical (opposite) image on the bottom wall that is insulating (case I) [conducting (case II)]:

$$f_e^{+\pm}(z) = -\frac{e}{2\pi\varepsilon}\left\{\log\left[\sin\left(\pi\frac{z-ia}{b}\right)\right] \pm \log\left[\sin\left(\pi\frac{z+ia}{b}\right)\right]\right\}, \tag{40.109a}$$

$$E^*_{+\pm}(z) = -\frac{df_e^{+\pm}}{dz} = \frac{e}{2b\varepsilon}\left[\cot\left(\pi\frac{z-ia}{b}\right) \pm \cot\left(\pi\frac{z+ia}{b}\right)\right]; \tag{40.109b}$$

the corresponding complex conjugate electric field is (40.109b). The case I of upper signs in (40.109b) coincides with (36.182) replacing Q by $-e/\varepsilon$; the case II reverses one sign because the image on the bottom wall is opposite instead of identical. The electric field is normal to the side walls that are insulating:

$$E_y^{+\pm}\left(\pm\frac{b}{2}, y\right) = 0, \quad E_x^{+\pm}\left(\pm\frac{b}{2}, y\right) = \frac{e}{2b\varepsilon}\left[\tan\left(\pi\frac{y-a}{b}\right) \pm \tan\left(\pi\frac{y+a}{b}\right)\right], \tag{40.110a,b}$$

the passage from (40.109b) to (40.110a,b) uses:

$$z = \pm\frac{b}{2} + iy: \quad \cot\left(\pi\frac{z \pm ia}{b}\right) = \cot\left(\pm\frac{\pi}{2} + i\pi\frac{y \pm a}{b}\right)$$
$$= -\tan\left(i\pi\frac{y \pm a}{b}\right) = \tanh\left(\pi\frac{y \pm a}{b}\right). \tag{40.111a,b}$$

Using (36.163d) specifies the electric field (40.109d) on the bottom walls:

$$E_x^{++}(x,0) = 0, \quad E_x^{++}(x,0) = \frac{e}{b\varepsilon}\text{Re}\left\{\cot\left(\pi\frac{x-ia}{b}\right)\right\}$$
$$= \frac{e}{b\varepsilon}\frac{\sin(2\pi x/b)}{\cosh(2\pi a/b) - \cos(2\pi x/b)}, \tag{40.112a,b}$$

$$E_x^{+-}(x,0) = 0, \quad E_y^{+-}(x,0) = \frac{e}{b\varepsilon}\text{Im}\left\{\cot\left(\pi\frac{x-ia}{b}\right)\right\}$$
$$= -\frac{e}{b\varepsilon}\frac{\sinh(2\pi a/b)}{\cosh(2\pi a/b) - \cos(2\pi x/b)} \equiv \frac{\sigma(x)}{\varepsilon} = \frac{e}{\varepsilon}\frac{\vartheta(x)}{J}. \tag{40.113a–c}$$

The electric field is: (Case I) tangential (40.112a,b) so there are no induced electric charges; (Case II) normal (40.113a) so there is (Figure 40.5b) the induced electric charge distribution (40.113c). The density of surface electric charges takes the value (40.114a) [(40.114b)] at the middle (corners) on the bottom wall:

$$\sigma(0) = -\frac{e}{b}\frac{\sinh(2\pi a/b)}{\cosh(2\pi a/b) - 1} = -\frac{e}{b}\coth\left(\frac{\pi a}{b}\right), \tag{40.114a}$$

$$\sigma\left(\pm\frac{b}{2}\right) = -\frac{e}{b}\frac{\sinh(2\pi a/b)}{\cosh(2\pi a/b) + 1} = -\frac{e}{b}\tanh\left(\frac{\pi a}{b}\right), \tag{40.114b}$$

the former is larger in modulus:

$$\frac{b}{e}|\sigma(0)| = \coth\left(\frac{\pi a}{b}\right) > 1 > \tanh\left(\frac{\pi a}{b}\right)\frac{b}{e}\left|\sigma\left(\pm\frac{b}{2}\right)\right|. \tag{40.114c}$$

In (40.114a,b) was used:

$$x = 0, \pm\frac{b}{2}: \quad \frac{\sin(2\pi a/b)}{\cosh(2\pi a/b) \mp 1} = \frac{2\sinh(\pi a/b)\cosh(\pi a/b)}{2\sinh^2(\pi a/b), \cosh^2(\pi a/b)} = \left\{\coth\left(\frac{\pi a}{b}\right), \tanh\left(\frac{\pi a}{b}\right)\right\}. \tag{40.115}$$

The latter also applies (36.113c) to the surface current distribution for the magnetostatic field in Figure 40.5b.

Example 40.10.7. *Electrostatic/magnetostatic field in with conducting/insulating side/ bottom walls.*

If the sidewalls are conducting the lateral images are alternating leading (36.187a) to a complex electrostatic potential (40.116a) with upper (lower) sign corresponding to identical (opposite) image on the bottom wall that is insulating (conducting):

$$f_e^{-\pm}(z) = -\frac{e}{2\pi\varepsilon}\left\{\log\left[\tan\left(\pi\frac{z-ia}{2b}\right)\right] \pm \log\left[\tan\left(\pi\frac{z+ia}{b}\right)\right]\right\}; \tag{40.116a}$$

$$E^*_{-\pm}(z) = -\frac{df_e^{-\pm}(z)}{dz} = \frac{e}{2\varepsilon b}\left\{\csc\left[\frac{\pi(z-ia)}{b}\right] \pm \csc\left[\frac{\pi(z+ia)}{b}\right]\right\}, \tag{40.116b}$$

the corresponding complex conjugate electric field is (40.116b). In the case IV of all walls conducting (40.116b) coincides with (36.187b) using the lower sign and replacing $-i\Gamma$ by e/ε; in the case III of insulating bottom wall there is a change of sign since the image on the bottom wall is identical instead of opposite. Using (36.193c, d):

$$\csc\left(\pi\frac{x \pm ia}{b}\right) = 2\frac{\sin(\pi x/b)\cosh(\pi a/b) \mp i\cos(\pi x/b)\sinh(\pi a/b)}{\cosh(2\pi a/b) - \cos(2\pi x/b)}, \tag{40.117}$$

to calculate the electric field (40.116b) at the bottom wall:

$$\begin{aligned}\frac{2\varepsilon b}{e}E^*_{-\pm}(x) &= \frac{2\varepsilon b}{e}\left[E_x^{-+}(x,0) - iE_y^{-+}(x,0)\right] = csc\left(\pi\frac{x-ia}{b}\right) \pm csc\left(\pi\frac{x-ia}{b}\right)\\ &= 2\frac{\sin(\pi x/b)\cosh(\pi a/b) - 2i\cos(\pi x/b)\sinh(\pi a/b)}{\cosh(2\pi a/b) - \cos(2\pi x/b)},\end{aligned} \tag{40.118}$$

leads to:

$$E_y^{-+}(x,0) = 0, \quad E_x^{-+}(x,0) = \frac{e}{b\varepsilon}\frac{\sin(\pi x/b)\cosh(\pi a/b)}{\cosh(2\pi a/b) - \cos(2\pi x/b)}, \tag{40.119a,b}$$

$$E_x^{--}(x,0) = 0, \quad E_y^{--}(x,0) = \frac{e}{b\varepsilon}\frac{\cos(\pi x/b)\sinh(\pi a/b)}{\cosh(2\pi a/b) - \cos(2\pi x/b)}, \tag{40.120a,b}$$

The electric field is: (case III) tangential (40.109a,b) so there no induced charges (case IV) normal (40.120a) so there are the induced charges (40.120b) that coincide with (40.120b) $\equiv$ (36.183b). Using on the side walls:

$$z = \pm\frac{b}{2} + iy: \csc\left(\pi\frac{z \pm ia}{b}\right) = \csc\left(\pm\frac{\pi}{2} + i\pi\frac{y \pm a}{b}\right) = \pm\sec\left(i\pi\frac{y \pm a}{b}\right) = \pm\text{sech}\left(\pi\frac{y \pm a}{b}\right), \tag{40.121c,d}$$

shows that the electric field (40.116b) is normal (40.122a):

$$\begin{aligned}E_y^{-+}\left(\pm\frac{b}{2}, y\right) = 0, E_x^{-\pm}\left(\pm\frac{b}{2}y\right) &= \mp\frac{e}{2b}\left[\text{sech}\left(\pi\frac{y-a}{b}\right) \pm \text{sech}\left(\pi\frac{y+a}{b}\right)\right]\\ &= \frac{\sigma^{-\pm}(y)}{\varepsilon} = \frac{e}{\varepsilon}\frac{\vartheta^{-\pm}(y)}{J}.\end{aligned} \tag{40.122a–c}$$

specifies the induced electric charges where: (case IV) the lower sign coincides with (40.122b) $\equiv$ (36.194b); (case III) the upper sign applies to the induced electric charges (40.122b) [surface electric currents (40.122c)] for the electro (magneto) static field in Figure 40.5b (40.5a).

Example 40.11. *Multipole in a Corner with a Circular Indentation.*

Consider the exact and asymptotic complex potential of the flow due to a multipole of order n at a position b in a corner of angle β with or without cylindrical indentation of radius a.

If the multipole moment P_n is real in (40.123b) the images are identical, and the mapping into a corner of angle β is used (40.123a):

$$\zeta = z^{\pi/\beta}: \quad g(\zeta) = -\frac{P_n}{2\pi}\zeta^{-n} = -\frac{P_n}{2\pi}z^{-n\pi/\beta} \equiv f_1(z). \tag{40.123a,b}$$

The complex potential of a multipole of order n with real moment P_n in a corner of angle β in (40.123b) assumes that it is located at the vertex. If the multipole is at a location outside the vertex the complex potential (40.123b) is modified to:

$$g(\zeta) = -\frac{P_n}{2\pi}[\zeta(z) - \zeta(b)]^{-n} = -\frac{P_n}{2\pi}\left(z^{\pi/\beta} - b^{\pi/\beta}\right)^{-n} \equiv f_2(z). \tag{40.124}$$

The simplest form (40.123b) is regained asymptotically (40.125b) at large distance (40.125a):

$$|z|^{\pi/\beta} \gg |b|^{\pi/\beta}: \quad f_2(z) \sim -\frac{P_n}{2\pi}z^{-n\pi/\beta} = f_1(z). \tag{40.125a,b}$$

If the corner has a circular indentation of radius a, the modification of the complex potential is specified the circle theorem (24.47):

$$f_3(z) \equiv f_2(z) + f_2^*\left(\frac{a^2}{z^*}\right) = -\frac{P_n}{2\pi}\left\{\left(z^{\pi/\beta} - b^{\pi/\beta}\right)^{-n} + \left[\left(\frac{a^2}{z}\right)^{\pi/\beta} - (b^*)^{\pi/\beta}\right]^{-n}\right\}, \tag{40.126}$$

that can be applied to (40.124) but not to (40.123b). If the multipole moment P_n has a nonzero imaginary part, the image with complex conjugate moment must be included in the complex potential (40.124) for a corner of angle 2β in (40.127a), leading to (40.127b):

$$\begin{aligned}\zeta = z^{\pi/2\beta}: \quad h(\zeta) &= -\frac{P_n}{2\pi}[\zeta(z) - \zeta(b)]^{-n} - \frac{P_n^*}{2\pi}[\zeta(z) - \zeta(b^*)]^{-n} \\ &= -\frac{P_n}{2\pi}\left(z^{\pi/2\beta} - b^{\pi/2\beta}\right)^{-n} - \frac{P_n^*}{2\pi}\left[z^{\pi/2\beta} - (b^*)^{\pi/2\beta}\right]^{-n} \equiv f_4(z).\end{aligned} \tag{40.127a,b}$$

Asymptotically (40.128a) this simplifies:

$$|z|^{\pi/2\beta} >> |b|^{\pi/2\beta}: f_4(z) \sim z^{-n\pi/2\beta} \times \begin{cases}\dfrac{1}{\pi}\mathrm{Re}(P_n) & \text{if } \mathrm{Re}(P_n) \neq 0, \\ \dfrac{in}{2\beta}z^{-\pi/2\beta}\,\mathrm{Im}\left(P_n b^{\pi/2\beta}\right) & \text{if } \mathrm{Re}(P_n) = 0,\end{cases} \tag{40.128a–c}$$

in two cases: (i) if the multipole has a nonzero real part, it dominates the far-field, leading to double amplitude (40.128b); (ii) if the multipole has zero real part, the opposite imaginary part of the image leads in the far-field to a multipole of order $n+1$, hence the faster decay in (40.128c). The passage from (40.127b) to (40.128a–c) corresponds to

$$\begin{aligned}
-2\pi f_4 &= z^{-n\pi/2\beta}\left\{P_n\left[1-\left(\frac{b}{z}\right)^{\pi/2\beta}\right]^{-n}+P_n^*\left[1-\left(\frac{b^*}{z}\right)^{\pi/2\beta}\right]^{-n}\right\}\\
&\sim z^{-n\pi/2\beta}\left\{P_n+P_n^*+\frac{n\pi}{2\beta}z^{-\pi/2\beta}\left[P_n b^{\pi/2\beta}+P_n^*\left(b^*\right)^{\pi/2\beta}\right]\right\}\\
&= 2z^{-n\pi/2\beta}\left[\operatorname{Re}\left(P_n\right)+\frac{in\pi}{2\beta}z^{-\pi/2\beta}\operatorname{Im}\left(P_n b^{\pi/2\beta}\right)\right]. \qquad (40.129)
\end{aligned}$$

The insertion of a circular indentation of radius a would add to (40.127b) terms specified by the circle theorem (24.47) as in the passage from (40.124) to (40.126).

Example 40.12. *Monopole at Asymmetric Position between Parallel Walls.*

Consider the potential flow due to a source/sink (vortex) in an asymmetric position between parallel walls, that is, not on the centerline [Figure 40.6a (40.6b)], and determine the complex potential and conjugate velocity.

The potential flow due to a source/sink (vortex) of volume flux Q (circulation Γ) at a position $z = a$ between parallel walls $\operatorname{Re}(z) = \pm b/2$ with $b > 2a$ leads to images in Figure 40.7 that are located as indicated in Table 40.3 that can be justified as follows: (i/ii) the original monopole at $z = a$ has image on the right $z = b/2$ (left $z = -b/2$) wall at $b/2 - a + b/2 = b - a\,[-(a + b/2) - b/2 = -a - b]$; (iii) the image at $b - a$ has a new image on the left wall at $-(b - a + b/2) - b/2 = a - 2b$; (iv) the image at $-a - b$ has a new image on the right wall at $a + b + b/2 + b/2 = 2b + a$; (v) the image at $2b + a$ has a new image on the left wall at $-(2b + a) - b/2 - b/2 = -a - 3b$; (vi) the image at $a - 2b$ has a new image on the right wall at $2b - a + b/2 + b/2 = 3b - a$. Proceeding in this way are obtained the locations of the monopole and of its infinite images on the two walls images indicated

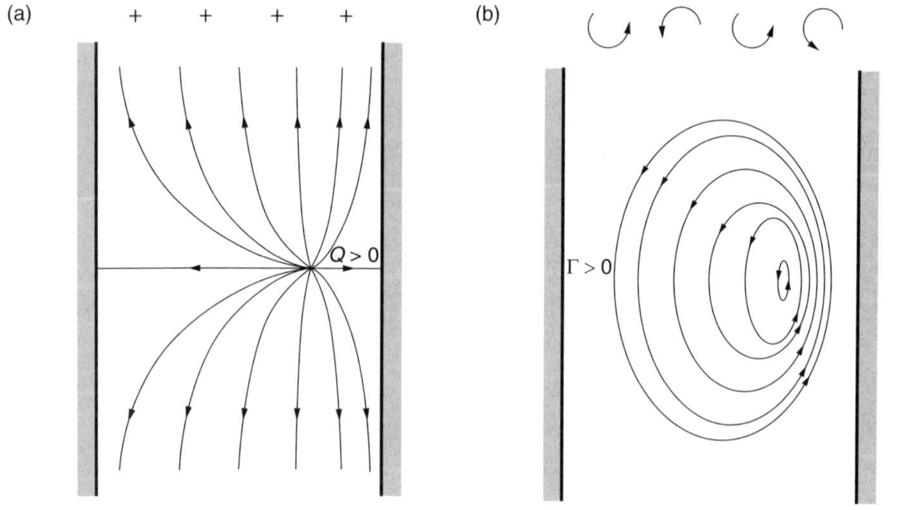

FIGURE 40.6
The potential flow of a source (a) [counterclockwise vortex (b)] at unequal distance from parallel infinite walls is symmetric (unsymmetric) relative to the horizontal (vertical) axis. The symmetry relative to both the horizontal and vertical axis applies in the case of equal distance from both walls in Figure 36.14a(c).

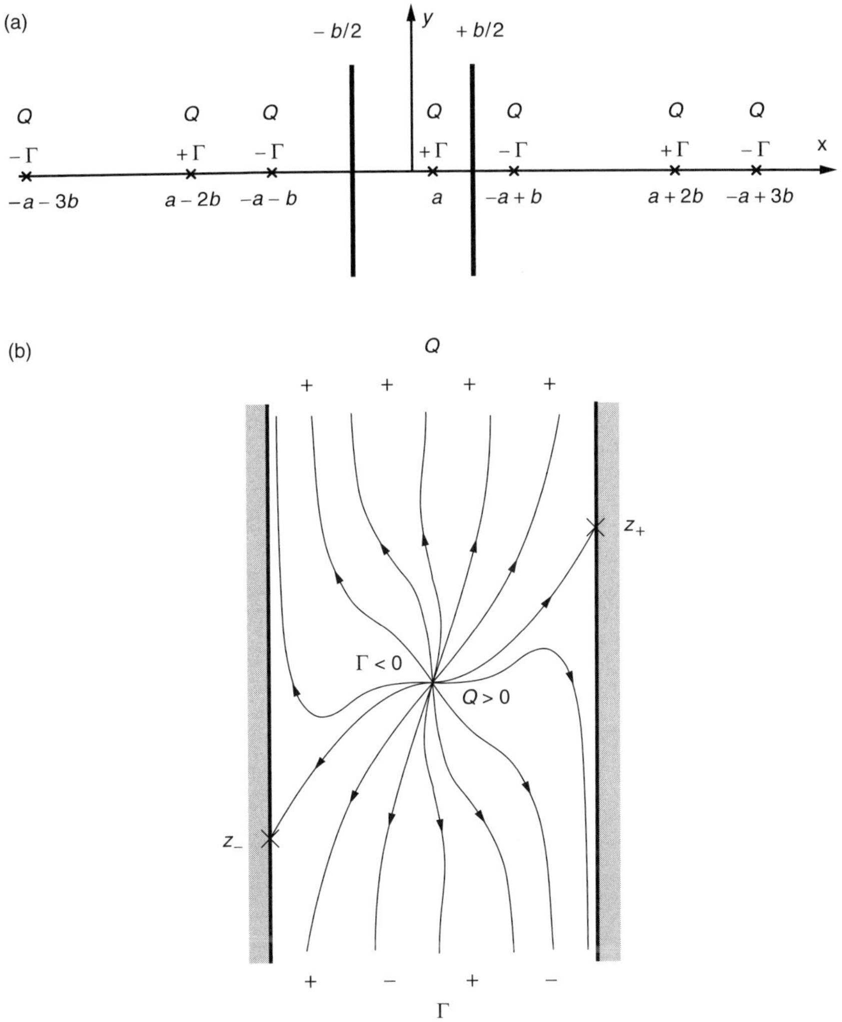

FIGURE 40.7
The identical (alternating) images of a source/sink (vortex) at unequal distance from parallel infinite walls (a) are indicated in Table 40.3; if the monopole is at equal distance from both walls the spacing between images is the same (Figure 36.13a). The monopole at unequal distance from two parallel infinite walls was considered for a source (vortex) in Figure 40.5a(b). The monopole at equal distance from parallel infinite walls is considered for: (i/ii) a source (vortex) in Figure 36.14a(c); (iii) their superposition in a spiral flow (b) in Figure 40.7b; (iv/v) also for a horizontal (vertical) dipole in Figure 40.8a(b).

in Table 40.3 that correspond to the coordinates:

$$n = 0, \pm 1, \pm 2, \dots : \quad z_n^+ = a + 2nb, \quad z_n^- = -a + (2n-1)\,b. \tag{40.130a–c}$$

In the case of a source/sink all images are identical, leading to the complex potential that is specified by an infinite product (40.131a), made totally convergent $O\left(n^{-2}\right)$ by adding an

TABLE 40.3
Images of a Spiral Source between Walls

Monopole	**Left-hand image**		**Original**	**Right-hand images**	
Position:	$a-2nb$	$-a-(2n-1)\,b$	a	$-a+(2n-1)\,b$	$a+2nb$
Source/sink*	Q	Q	Q	Q	Q
Vortex**	Γ	$-\Gamma$	Γ	$-\Gamma$	$+\Gamma$

* Flow rate; ** circulation.
Note: For a monopole at unequal distance between parallel walls (Figure 40.7a) the position of the images is indicated, distinguishing identical (alternating) images for sources/sinks (vortices). This includes the particular case of equal spacing (Figure 38.13a).

infinite constant:

$$\begin{aligned} f(z) &= \frac{Q}{2\pi}\sum_{n=-\infty}^{\infty}\left[\log\left(z-z_n^+\right)+\log\left(z-z_n^-\right)\right] \\ &= \frac{Q}{2\pi}\log\left\{(z-a)\prod_{n=1}^{\infty}(z-a-2nb)(z-a+2nb)\left[z+a+\frac{(2n-1)}{b}\right]\right. \\ &\quad\times\left.\left[z+a-\frac{(2n-1)}{b}\right]\right\} = \frac{Q}{2\pi}\log\left\{(z-a)\prod_{n=1}^{\infty}\left[(z-a)^2-4n^2b^2\right]\right. \\ &\quad\times\left.\left[(z+a)^2-(2n-1)^2b^2\right]\right\}, \\ &= \frac{Q}{2\pi}\log\left\{(z-a)\prod_{n=1}^{\infty}\left[1-\left(\frac{z-a}{2nb}\right)^2\right]\left[1-\left(\frac{z+a}{(2n-1)b}\right)^2\right]\right\}+\text{const}; \end{aligned} \tag{40.131a}$$

$$\begin{aligned} \frac{df}{dz} &= \frac{Q}{2\pi}\left\{(z-a)^{-1}+2(z-a)\sum_{n=1}^{\infty}\left[(z-a)^2-4n^2b^2\right]^{-1}\right. \\ &\quad\left.+2(z+a)\sum_{n=1}^{\infty}\left[(z+a)^2-(2n-1)^2b^2\right]^{-1}\right\}, \end{aligned} \tag{40.131b}$$

the corresponding complex velocity is specified by a series of fractions (40.120b) that is also totally convergent $O\left(n^{-2}\right)$. In the case of a source/sink at equal distance from both walls $a=0$, then (40.131a,b) simplify to (36.132; 36.133).

In the case of a vortex the positions images are alternating (Table 40.3), and positions of identical (40.132b) [opposite (40.132c)] images are:

$$n=1,2,\ldots:\quad z_n^{+\pm}=a\pm 2nb,\quad z_n^{-\pm}=-a\pm(2n-1)\,b. \tag{40.132a–c}$$

For a vortex the identical and opposite images must be distinguished when calculating the complex potential:

$$\begin{aligned} f(z) &= -\frac{i\Gamma}{2\pi}\left\{\log(z-a)+\sum_{n=1}^{\infty}\log\left(z-z_n^{++}\right)+\log\left(z-z_n^{+-}\right)-\log\left(z-z_n^{+-}\right)-\log\left(z-z_n^{--}\right)\right\} \\ &= -\frac{i\Gamma}{2\pi}\log\left\{(z-a)\prod_{n=1}^{\infty}\frac{(z-a-2nb)(z-a+2nb)}{[z+a-(2n-1)\,b]\,[z+a+(2n-1)\,b]}\right\} \end{aligned}$$

$$= -\frac{i\Gamma}{2\pi}\log\left[(z-a)\prod_{n=1}^{\infty}\frac{(z-a)^2-4n^2b^2}{(z+a)^2-(2n-1)^2b^2}\right]$$

$$= -\frac{i\Gamma}{2\pi}\log\left\{(z-a)\prod_{n=1}^{\infty}\left[1-\frac{(z-a)^2}{4n^2b^2}\right]\left[1-\frac{(z-a)^2}{(2n-1)^2b^2}\right]\right\}+\text{const}, \tag{40.133a}$$

$$\frac{\mathrm{d}f}{\mathrm{d}z} = -\frac{i\Gamma}{2\pi}\left\{(z-a)^{-1}+2(z-a)\sum_{n=1}^{\infty}\left[(z-a)^2-4n^2b^2\right]^{-1}\right.$$

$$\left.-2(z+a)\sum_{n=1}^{\infty}\left[(z+a)^2-(2n-1)^2b^2\right]^{-1}\right\}. \tag{40.133b}$$

If the vortex is at the centerline between walls $a = 0$ then (40.133a,b) simplify to (36.143; 36.144).

Example 40.13. *Spiral Monopole at Equal Distance from Parallel Walls.*

Consider a spiral or monopole source consisting of a source/sink and a vortex at equal distance between two parallel walls (Figure 40.7b) and determine: (i) the complex potential and conjugate velocity; (ii) the location of stagnation points; (iii–v) the components of the velocity on the real (iii) and imaginary (iv) axis and on the wall (v).

The potential flow, due to a spiral or monopole singularity of volume flux Q and circulation Γ, at equal distance from two walls at $\mathrm{Re}(z) = \pm b/2$, has the complex potential (40.134a) [conjugate velocity (40.134b)] equal to the sum of (36.137a) [(36.137b] and (36.147a) [(36.147b)]:

$$f(z) = \frac{Q}{2\pi}\log\left[\sin\left(\frac{\pi z}{b}\right)\right] - \frac{i\Gamma}{2\pi}\log\left[\tan\left(\frac{\pi z}{2b}\right)\right], \tag{40.134a}$$

$$v^*(z) = \frac{\mathrm{d}f}{\mathrm{d}z} = \frac{Q\cos(\pi z/b) - i\Gamma}{2b\sin(\pi z/b)}. \tag{40.134b}$$

The roots (40.137a) of (40.134b) specify the location of the stagnation points:

$$i\frac{\Gamma}{Q} = \cos\left(\frac{\pi z_\pm}{b}\right) = \cos\left(\frac{\pi x_\pm}{b}\right)\cosh\left(\frac{\pi y_\pm}{b}\right) - i\sin\left(\frac{\pi x_\pm}{b}\right)\sinh\left(\frac{\pi y_\pm}{b}\right); \tag{40.135}$$

equating real and imaginary parts in (40.135) leads to:

$$\cos\left(\frac{\pi x_\pm}{b}\right) = 0, \quad x_\pm = \pm\frac{b}{2}, \quad \frac{\Gamma}{Q} = \mp\sinh\left(\frac{\pi y_\pm}{b}\right); \tag{40.136a–c}$$

this specifies the positions of (40.137b) of the stagnation points (40.137a):

$$v^*(z_\pm) = 0: \quad z_\pm = x_\pm + iy_\pm = \pm\frac{b}{2} \mp i\,\arg\sinh\left(\frac{\Gamma}{Q}\right), \tag{40.137a,b}$$

that is: (i) at the closest point on the walls $z_\pm = b/2$ for the source/sink in the absence of circulation $\Gamma = 0$ in Figure 36.14a; (ii) at infinity $z_\pm = \pm i\infty$ for the vortex in the absence of sources $Q = 0$ in Figure 36.14c; (iii) at an intermediate position on the walls (40.137b) for a monopole or spiral source (Figure 40.7b) viz. for a source and negative vortex $Q > 0 > \Gamma$ or sink and positive vortex $\Gamma > 0 > Q$, that is, for $Q\Gamma < 0$ the stagnation point moves up (down) on the right (left) wall, and vice-versa for $Q\Gamma > 0$. The Cartesian components of the velocity follow from (40.134b):

$$y = 0: \quad v_x = \frac{Q}{2b}\cot\left(\frac{\pi x}{b}\right), \quad v_y = \frac{\Gamma}{2b}\csc\left(\frac{\pi x}{b}\right), \tag{40.138a-c}$$

$$x = 0: \quad v_x = -\frac{\Gamma}{2b}\operatorname{csch}\left(\frac{\pi y}{b}\right), \quad v_y = \frac{Q}{2b}\coth\left(\frac{\pi y}{b}\right), \tag{40.139a-c}$$

$$x = \pm\frac{b}{2}: \quad v_x = 0, \quad v_y = \frac{Q\tanh(\pi y/b) \pm \Gamma\operatorname{sech}(\pi y/b)}{2b}, \tag{40.140a-c}$$

respectively on the real axis (40.138a–c), imaginary axis (40.139a–c) and on the walls (40.140a–c). The passage from (40.134b) to (40.138–40a–c) is exemplified:

$$\begin{aligned} 2b\,(v_x - iv_y) &= 2b\, v^*\left(\pm\frac{b}{2} + iy\right) = Q\cot\left(\frac{i\pi y}{b} \pm \frac{\pi}{2}\right) - i\Gamma\csc\left(\frac{i\pi y}{b} \pm \frac{\pi}{2}\right) \\ &= -Q\tan\left(\frac{i\pi y}{b}\right) \mp i\Gamma\sec\left(\frac{i\pi y}{b}\right) = -iQ\tanh\left(\frac{\pi y}{b}\right) \mp i\Gamma\operatorname{sech}\left(\frac{\pi y}{b}\right), \end{aligned} \tag{40.141}$$

in the case (40.140a–c).

Example 40.14. *Dipole Transverse to or Aligned with Parallel Walls.*

Consider a dipole with real P_{1x} (imaginary iP_{1y}) moment (Figure 40.8a (40.8b)] at equal distance from parallel walls $\operatorname{Re}(z) = \pm b/2$ and determine: (i) the complex potential: (ii) the conjugate velocity; (iii–v) the nonzero Cartesian components of the velocity on the real (iii) and imaginary (iv) axis and on the walls (v).

A dipole with real moment $P_1 \equiv P_{1x}$ at equal distance from parallel walls $\operatorname{Re}(z) = \pm b/2$, has (Figure 40.8a) alternating images at $\pm b, \pm 2b, \ldots, \pm nb$, leading (36.149) to the complex

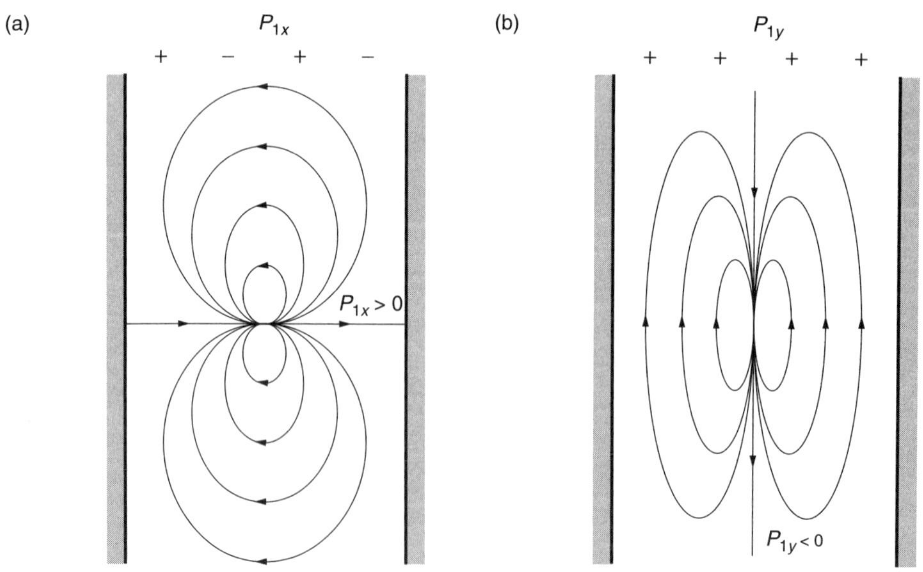

FIGURE 40.8
Potential flow due to a horizontal (a) [vertical (b)] dipole at equal distance between parallel infinite rigid impermeable walls, leading to alternative (identical) images (Figure 36.13a) and the velocity fields in Table 40.4.

TABLE 40.4
Velocity Field of Dipole between Walls

Dipole moment:	**Real** $P_1 = P_{1x}$	**Imaginary** $P_1 = iP_{1y}$	**Dipole moment**
Images*	**Alternating**	**Identical**	**Images****
$v_x(x,0) = A\times$	$\csc(\pi x/b)\cot(\pi x/b)$	$-\csc^2(\pi x/b)$	$\times B = v_y(x,0)$
$v_x(0,y) = A\times$	$-\operatorname{csch}(\pi y/b)\coth(\pi y/b)$	$\operatorname{csch}^2(\pi y/b)$	$\times B = v_y(0,y)$
$v_y(\pm b/2,y) = A\times$	$\pm\operatorname{sech}(\pi y/b)\tanh(\pi y/b)$	$-\operatorname{sech}^2(\pi y/b)$	$\times B = v_y(\pm b/2,y)$

$^*A = P_{1x}\pi/2b^2$; $^{**}B = P_{1y}\pi/2b^2$.
Note: The potential flow due to a horizontal (vertical) dipole at equal distance between parallel walls (Figure 40.8a (b)] leads to identical (alternating) images. In both cases are indicated the Cartesian components of the velocity along: (i) the real axis, which is perpendicular to the walls; (ii) the imaginary axis, which is parallel to the walls at equal distance from them; (iii) the two walls.

potential (40.131a) [conjugate velocity (40.131b)]:

$$f(z) = -\frac{P_{1x}}{2\pi}\sum_{n=-\infty}^{+\infty}(-)^n(z-nb)^{-1}$$
$$= -\frac{P_{1x}}{2\pi}\left\{z^{-1} + 2z\sum_{n=1}^{\infty}(-)^n\left(z^2-n^2b^2\right)^{-1}\right\} = -\frac{P_{1x}}{2b}\csc\left(\frac{\pi z}{b}\right), \tag{40.142a}$$
$$\frac{\mathrm{d}f}{\mathrm{d}z} = \frac{P_{1x}}{2\pi}\frac{1}{z^2} + 2\sum_{n=1}^{\infty}\frac{z^2+n^2b^2}{\left(z^2-n^2b^2\right)^2} = \frac{P_{1x}\pi}{2b^2}\csc\left(\frac{\pi z}{b}\right)\cot\left(\frac{\pi z}{b}\right); \tag{40.142b}$$

in the case of imaginary dipole moment $P_1 \equiv iP_{1y}$ the images (Figure 40.8b) are identical leading (36.142) to the complex potential (40.143a) [conjugate velocity (40.143b)]:

$$f(z) = -i\frac{P_{1y}}{2\pi}\left[\frac{1}{z} + \sum_{n=1}^{\infty}\frac{1}{z-nb} + \frac{1}{z+nb}\right]$$
$$= -i\frac{P_{1y}}{2\pi}\left(\frac{1}{z} + 2z\sum_{n=1}^{\infty}\frac{1}{z^2-n^2b^2}\right) = -i\frac{P_{1y}}{2b}\cot\left(\frac{\pi z}{b}\right), \tag{40.143a}$$
$$\frac{\mathrm{d}f}{\mathrm{d}z} = i\frac{P_{1y}}{2\pi}\left[\frac{1}{z^2} + 2\sum_{n=1}^{\infty}\frac{z^2+n^2b^2}{\left(z^2-n^2b^2\right)^2}\right] = i\frac{P_{1y}\pi}{2b^2}\csc^2\left(\frac{\pi z}{b}\right). \tag{40.143b}$$

The complex potential of a dipole with real P_{1x} (imaginary P_{1y}) moment $P_1 \equiv P_{1x}$ ($P_1 = iP_{1y}$) coincides (40.142a) [(40.143a)] with the conjugate velocity of a vortex (source/sink), viz. (36.147b) [(36.137b)] replacing $i\Gamma$ (Q) by $P_1 \equiv P_{1x}$ ($-iP_{1y} = P_1$) since in both cases the images on the walls are alternating (identical). The nonzero velocity components on the real and imaginary axes, and on the wall:

$$P_1 = P_{1x}: \quad v_y(x,0) = v_y(0,y) = v_x\left(\pm\frac{b}{2},y\right) = 0, \tag{40.144a–d}$$

$$P_1 = iP_{1y}: \quad v_x(x,0) = v_x(0,y) = v_x\left(\pm\frac{b}{2},y\right) = 0, \tag{40.145a–d}$$

are given in Table 40.4, respectively in the case of real and imaginary dipole moment. The case of a complex dipole moment follows by superposition.

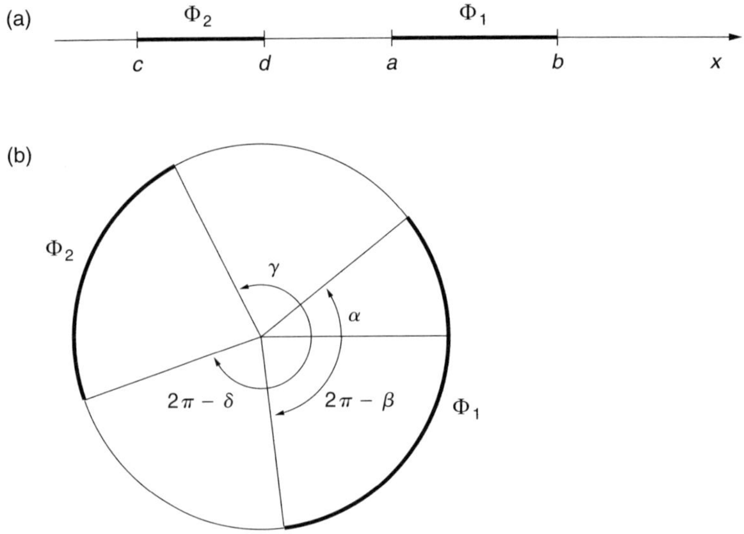

FIGURE 40.9
The potential in the upper/lower half-complex plane (a) [interior/exterior of a cylinder (b)] due to two uniform potentials or finite segments (circular sectors) compared with one finite interval (circular sector) in Figure 39.1a(b).

Example 40.15. *Field due to Potential Distributions Doubly Connected Regions.*

Determine the potential and field function due to two constant potentials on two strips (two sectors of a cylinder of radius a), viz. Figure 40.9a (40.9b). Consider the particular case of equal potentials and symmetric strips (sectors) of the plane (cylinder).

Example 40.15.1. *Field due to two potential strips.*

The potential (field function) due to a given potential Φ_1 (Φ_2) on the strips (a, b) [(c,d)] on the real line (Figure 40.9a) is specified by (38.3a,b):

$$\{\Phi(x,y),\, \Psi(x,y)\} = \frac{1}{\pi}\left(\int_a^b \Phi_1 + \int_c^d \Phi_2\right) dt \left[(t-x)^2 + y^2\right]^{-1} \{y, x-t\}; \qquad (40.146\text{a,b})$$

using the primitives (38.5–38.6a,b):

$$\int \frac{\{y, x-t\}}{(t-x)^2 + y^2} dt = \left\{\operatorname{arc\,cot}\left(\frac{x-t}{y}\right), -\frac{1}{2}\log\left[(t-x)^2 + y^2\right]\right\}, \qquad (40.147\text{a,b})$$

the integrals (40.146a,b) are evaluated by

$$\begin{aligned}\pi\Phi(x,y) \equiv \Phi_1 &\left[\operatorname{arc\,cot}\left(\frac{x-b}{y}\right) - \operatorname{arc\,cot}\left(\frac{x-a}{y}\right)\right] \\ &+ \Phi_2\left[\operatorname{arc\,cot}\left(\frac{x-d}{y}\right) - \operatorname{arc\,cot}\left(\frac{x-c}{y}\right)\right],\end{aligned} \qquad (40.148\text{a})$$

$$2\pi\Psi(x,y) = \Phi_1 \log\left[\frac{(x-a)^2 + y^2}{(x-b)^2 + y^2}\right] + \Phi_2 \log\left[\frac{(x-c)^2 + y^2}{(x-d)^2 + y^2}\right]. \qquad (40.148\text{b})$$

These simplify to

$$d = -a, c = -b, \Phi_1 = \Phi_2 \equiv \Phi_0 : \frac{\pi}{\Phi_0}\Phi(x,y) = \operatorname{arc\,cot}\left(\frac{x-b}{y}\right) - \operatorname{arc\,cot}\left(\frac{x+b}{y}\right) - \operatorname{arc\,cot}\left(\frac{x-a}{y}\right) + \operatorname{arc\,cot}\left(\frac{x+a}{y}\right), \tag{40.149a}$$

$$\frac{2\pi}{\Phi_0}\Psi(x,y) = \log\left[\frac{(x-a)^2+y^2}{(x+a)^2+y^2}\quad\frac{(x+b)^2+y^2}{(x-b)^2+y^2}\right], \tag{40.149b}$$

if the potentials are equal on symmetric strips.

Example 40.15.2. *Field due to two potential angular sectors.*

The constant potential Φ_1 (Φ_2) on two sectors (β,α) [(γ, δ)] of (Figure 40.9b) a sphere of radius a leads (38.4a,b) to the scalar potential (40.150a) [field function (40.150b)]:

$$\{\Phi(r,\varphi), \Psi(r,\varphi)\} = \frac{1}{2\pi}\left(\int_\beta^\alpha \Phi_1 + \int_\gamma^\delta \Phi_2\right) d\alpha\left[r^2+a^2-2ar\cos(\alpha-\varphi)\right]^{-1} \times \left\{a^2-r^2, -2ar\sin(\alpha-\varphi)\right\}; \tag{40.150a,b}$$

using the primitives (38.13–38.14a–c):

$$\int \frac{\left\{a^2-r^2, -2ar\sin(\alpha-\varphi)\right\}}{a^2+r^2-2ar\cos(\alpha-\varphi)} d\alpha = \left\{-\alpha + 2\operatorname{arc\,cot}\left(\frac{a\cos\alpha - r\cos\varphi}{a\sin\alpha - r\sin\varphi}\right), \log a - \log\left[a^2+r^2-2ar\cos(\alpha-\varphi)\right]\right\}, \tag{40.151a,b}$$

leads to

$$\pi\Phi(r,\varphi) = \frac{1}{2}\left[(\beta-\alpha)\Phi_1 + (\gamma-\delta)\Phi_2\right] + \Phi_1\left[\operatorname{arc\,cot}\left(\frac{a\cos\alpha - r\cos\varphi}{a\sin\alpha - r\sin\varphi}\right) - \operatorname{arc\,cot}\left(\frac{a\cos\beta - r\cos\varphi}{a\sin\beta - r\sin\varphi}\right)\right] + \Phi_2\left[\operatorname{arc\,cot}\left(\frac{a\cos\delta - r\cos\varphi}{a\sin\delta - r\sin\varphi}\right) - \operatorname{arc\,cot}\left(\frac{a\cos\gamma - r\cos\varphi}{a\sin\gamma - r\sin\varphi}\right)\right], \tag{40.152a}$$

$$2\pi\Psi(r,\varphi) = \left[\Phi_1(\alpha-\beta) + \Phi_2(\delta-\gamma)\right]\log a + \Phi_1\log\left[\frac{r^2+a^2-2ar\cos(\beta-\varphi)}{r^2+a^2-2ar\cos(\alpha-\varphi)}\right] + \Phi_2\log\left[\frac{r^2+a^2-2ar\cos(\gamma-\varphi)}{r^2+a^2-2ar\cos(\delta-\varphi)}\right] \tag{40.152b}$$

This simplifies in the case of equal potentials in symmetric angular sectors to

$$\Phi_1 = \Phi_2 \equiv \Phi_0, \beta = -\alpha, \gamma = \pi - \alpha, \delta = \pi + \alpha :$$

$$\frac{\pi}{\Phi_0}\Phi(r,\Phi) = 2\alpha + \operatorname{arc\,cot}\left(\frac{a\cos\alpha - r\cos\varphi}{a\sin\alpha - r\sin\varphi}\right) + \operatorname{arc\,cot}\left(\frac{a\cos\alpha - r\cos\varphi}{a\sin\alpha + r\sin\varphi}\right) + \operatorname{arc\,cot}\left(\frac{a\cos\alpha + r\cos\varphi}{a\sin\alpha + r\sin\varphi}\right) + \operatorname{arc\,cot}\left(\frac{a\cos\alpha + r\cos\varphi}{a\sin\alpha - r\sin\varphi}\right) \tag{40.153a}$$

$$\frac{2\pi}{\Phi_0}\Psi(r,\Phi) = 4\alpha\log a + \log\left[\frac{\left(r^2+a^2\right)^2 - 4a^2r^2\cos^2(\alpha+\varphi)}{\left(r^2+a^2\right)^2 - 4a^2r^2\cos^2(\alpha-\varphi)},\right] \tag{40.153b}$$

The sum of $\text{arc cot}\left[(a\cos\alpha \pm r\cos\varphi)\,/\,(a\sin\alpha \pm r\sin\varphi)\right]$ for all four possible combinations of signs in the numerator and denominator appears on the r.h.s. of the potential (40.153a).

Example 40.16. *Deflection and Enlargement of a Jet by a Source.*

Consider a uniform flow of velocity U occupying the upper z-plane $\text{Im}(z) > 0$ with free streamline on the real axis (Figure 40.10a). A source is introduced in the jet (Figure 40.10b)

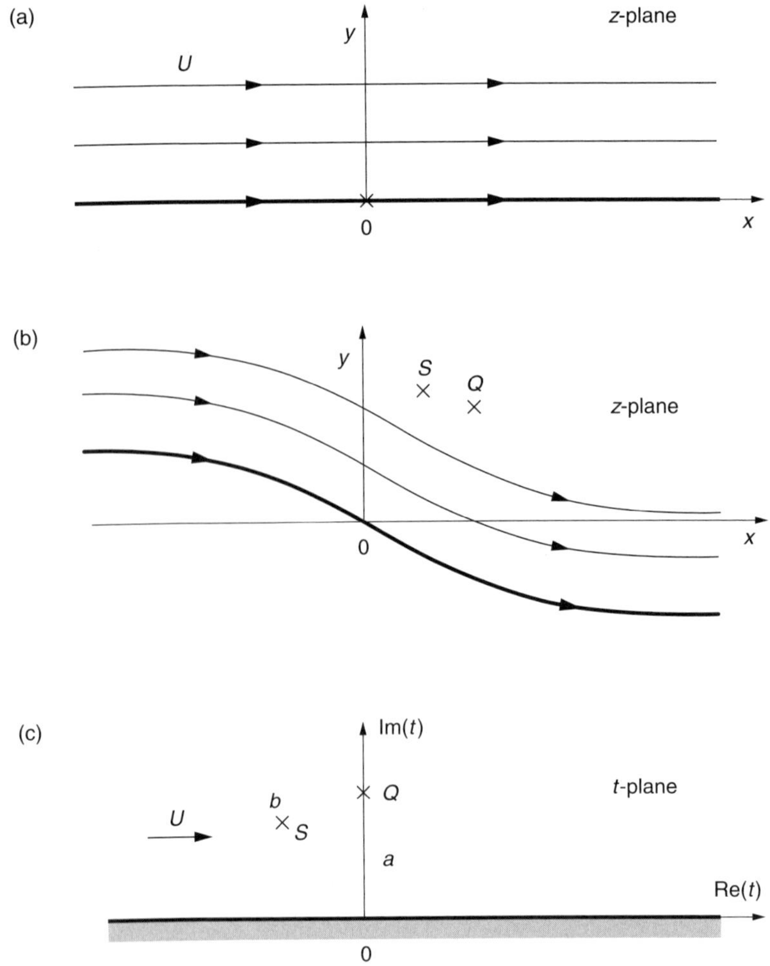

FIGURE 40.10
A flow source in a free stream parallel to a wall (a) causes the appearance of a stagnation point (b). Both the source and stagnation point are singularities of the hodograph variable (38.28) as the logarithm of the complex velocity. This requires the use of the second hodograph method, to construct the hodrograph function in the upper half-plane (c) and extend it analytically to the whole plane, as in Figure 38.9. The two hodograph problems of the second kind use similar methods to reach conclusions with opposite practical significance: (i) the source in a parallel stream over an infinite wall does not change its direction, and increases the width in proportion to the flow rate, which is a small effect for a weak source (Figure 40.10a–c): (ii) even a weak source near the edge of a semiinfinite plate with a free jet has a stronger "fluidic effect" that causes in the near (far) field a change in the direction (a lateral displacement) of the jet (Figure 38.9a–c).

and the deflection of the free streamline is to be determined using the hodograph method (Figure 40.10c). Compare with the fluidic effect on a jet flowing past a semiinfinite plate with a source near the edge (Section 38.8).

Example 40.16.1. *Source/sink in a free jet.*

The flow in the t-plane (Figure 40.10b) consists of a uniform stream with velocity U and source of flow rate Q at a distance a from the free streamline on the real axis, plus its identical image, leading to the complex potential (40.154a,b):

$$\frac{df}{dt} = U + \frac{Q}{2\pi}\left(\frac{1}{t-ia} + \frac{1}{t+ia}\right) = U + \frac{Q}{\pi}\frac{t}{t^2+a^2} = U\frac{(t-b)(t-b^*)}{t^2+a^2}, \qquad (40.154\text{a,b})$$

Rewriting in the form (40.154b), specifies the location b (b^*) of the stagnation point (its image):

$$b = -a\left(\varepsilon + i\sqrt{1-\varepsilon^2}\right), \qquad \varepsilon \equiv \frac{Q}{2\pi U a}, \qquad (40.155\text{a,b})$$

viz.: (i) the coordinates of the stagnation point are proportional to those of the source (40.155a); (ii) they are affected by the dimensionless parameter (40.155b) that compares the flow rate of the source per radian $Q/2\pi$ with the volume flux of the jet between the source and the free streamline Ua. The passage from (40.154a,b) to (40.155a,b) results from (40.156a):

$$U\left(t^2+a^2\right) + \frac{Qt}{\pi} = U\left(t-b\right)\left(t-b^*\right): \frac{Q}{\pi} = -U\left(b+b^*\right),\ a^2 = b^* b, \qquad (40.156\text{a–c})$$

that must be valid for arbitrary t leading to the identities (40.156b,c). From the latter follow (40.157a,b):

$$Q = -2\pi U \operatorname{Re}(b),\ a = |b|\,;\ 0 = b^2 + \frac{Q}{\pi U}b + a^2 = b^2 + 2a\varepsilon b + a^2, \qquad (40.157\text{a–c})$$

eliminating b^* between (40.156b,c) leads to the quadratic equation (40.157c) for b, whose roots are (40.155a,b).

The hodograph function (38.139) has a pole with residue +1 (−1) at the source (stagnation point) as its principal part (38.158b):

$$f_+ \in \mathcal{A}(\operatorname{Im}(t) > 0): \quad \frac{d\zeta}{dt} = \frac{1}{t-ia} - \frac{1}{t-b} + f_+(t), \qquad (40.158\text{a,b})$$

to that is added a function analytic in the upper complex t-plane (38.158a). The hodograph variable:

$$\zeta \equiv \log\left(U\frac{dz}{df}\right) = \log\left(\frac{U}{v}e^{i\theta}\right) = i\theta + \log\left(\frac{U}{v}\right), \qquad (40.159)$$

is purely imaginary along the free streamline, where the velocity $v = U$ is constant. The function:

$$f(t) = \frac{1}{t-ia} - \frac{1}{t+ia} - \frac{1}{t-b} + \frac{1}{t-b^*}, \qquad (40.160)$$

that adds to the principal part of (40.158b) its opposite images is: (i) analytic in the upper complex t-plane except for the poles at $t = ia$, b; (ii) imaginary on the real axis; (iii) it vanishes as $|t| \to \infty$. Thus the function $\pm i\ f(t)$ meets (i) and (iii) and is real on the real axis. Since $i[f(t) - \mathrm{d}\zeta/\mathrm{d}t]$ eliminates the poles it is: (i) analytic in the upper complex, t-plane; (ii) real on the real axis; (iii) vanishes at infinity because (40.158b) and (40.169) must coincide there. On account of (i) and (ii) the reflection principle (Subsection 31.2.2) can be applied, leading to a function analytic in the whole complex t-plane. By the Liouville theorem (Section 25.6) it is a constant; the value of the constant evaluated at infinity, is zero. Implying that $f(t) = \mathrm{d}\zeta/\mathrm{d}t$. Using (40.159):

$$f(t) = \frac{\mathrm{d}\zeta}{\mathrm{d}t} = \frac{\mathrm{d}\left[\log\left(U\mathrm{d}z/\mathrm{d}f\right)\right]}{\mathrm{d}t}, \tag{40.161}$$

it follows that:

$$U\frac{\mathrm{d}z}{\mathrm{d}f} = \exp\left\{\int f(t)\,\mathrm{d}t\right\} = \frac{(t-ia)(t-b^*)}{(t+ia)(t-b)}, \tag{40.162}$$

that specifies the complex conjugate velocity $\mathrm{d}f/\mathrm{d}z$.

Example 40.16.2. *Flow rate of source/sink and jet width.*

The product of the inverse complex conjugate velocity in the physical z-plane (40.162) by the conjugate velocity in the t-plane (40.154b) specifies the shape of the free streamline (40.163):

$$\begin{aligned}\frac{\mathrm{d}z}{\mathrm{d}t} &= \frac{\mathrm{d}z}{\mathrm{d}f}\frac{\mathrm{d}f}{\mathrm{d}t} = \left(\frac{t-b^*}{t+ia}\right)^2 = \left(1 - \frac{b^*+ia}{t+ia}\right)^2 \\ &= 1 - 2\frac{b^*+ia}{t+ia} + \frac{(b^*+ia)^2}{(t+ia)^2} = 1 + \frac{A_{-1}}{t+ia} + \frac{A_{-2}}{(t+ia)^2};\end{aligned} \tag{40.163}$$

the coefficients in the partial fraction decomposition confirm the generalized residue rule (31.85a,b) applied to the double pole at $t = ia$:

$$A_{-1} = \lim_{t\to -a}\frac{\mathrm{d}}{\mathrm{d}t}(t-b^*)^2 = \lim_{t\to -a} 2(t-b^*) = -2(b^*+ia), \tag{40.164a}$$

$$A_{-2} = \lim_{t\to -a}(t-b^*)^2 = (b^*+ia)^2 . \tag{40.164b}$$

A third alternative is to equate the coefficients of powers of t in (40.163) ≡ (40.165a):

$$(t-b^*)^2 = (t+ia)^2 + A_{-1}(t+ia) + A_{-2}; \tag{40.165a}$$

this leads to the system of equations:

$$A_{-1} = -2b^* - 2ia, \quad A_{-2} = (b^*)^2 - (ia)^2 - iaA_{-1} = (b^*+ia)^2 . \tag{40.165b,c}$$

Substituting (40.165b,c) ≡ (40.164a,b) in (40.163) specifies the partial fraction decomposition in three equivalent ways and hence:

$$\frac{\mathrm{d}z}{\mathrm{d}t} = \frac{\mathrm{d}}{\mathrm{d}t}\left\{t - (b^*+ia)\left[2\log(t+ia) + \frac{b^*+ia}{t+ia}\right]\right\}; \tag{40.166a}$$

its primitive.

The integration of (40.166a) from $z = 0$ at $t = 0$ leads to:

$$z = t - (b^*+ia)\left[2\log\left(1-\frac{it}{a}\right) + \frac{it}{a}\frac{b^*+ia}{t+ia}\right], \tag{40.166b}$$

In the passage from (40.166a) to (40.166b) was used (40.166c):

$$(b^* + ia)\left[\frac{1}{t+ia} - \frac{1}{ia}\right] = -\frac{t}{ia}\frac{b^*+ia}{t+ia} = \frac{it}{a}\frac{b^*+ia}{t+ia} = \frac{1}{a}\frac{b^*+ia}{1+ia/t} = \frac{ib^*}{a} - 1 + O\left(\frac{1}{t}\right); \tag{40.166c–d}$$

using (40.166d) at large distance (40.167a) the leading real and complex terms of (40.166b) are (40.167b):

$$t \to \infty: \quad z(t) \sim t - i(b^* + ia)\left(-i\pi + i\frac{b^*}{a} - 1\right); \quad b^* + ia = -a\varepsilon + O\left(\varepsilon^2\right), \tag{40.167a–c}$$

also (40.155a) implies (40.167c) to lowest order in the parameter. From (40.167b,c) follows to lowest order in ε and highest order in t:

$$z \equiv x + iy : x \sim t, y \sim a\varepsilon\left(\frac{b^*}{a} - \pi\right) \sim -a\varepsilon(\pi + \varepsilon) \sim -a\pi\varepsilon + O\left(\varepsilon^2\right) \sim -\frac{Q}{2U}; \tag{40.168a–c}$$

this shows that the source increases the volume flux of the jet in proportion to the flow rate of the source divided by the mean flow velocity $y \sim -Q/2U$, that is, half the volume flux goes into lowering the free stream line; the direction of the jet is not changed. In the case of a jet flowing by a semi infinite plate, a source near the edge is much more effective (Subsection 38.8.4) at deflecting the jet in the far-field.

Example 40.17. *Finite/Infinite Number of Roots of Transcendental Equations.*
Find all the roots of the following six complex equations:

$$\exp(z) = -1: \quad z_n = (2n+1)\pi i; \tag{40.169a,b}$$

$$\log(z) = 2 + i\pi: \quad z = -e^2; \tag{40.170a,b}$$

$$\sin(z) = 5i: \quad z_n, \pi - z_n = 2n\pi + i \arg\sinh 5; \tag{40.171a,b}$$

$$\sec(z) = \tfrac{1}{2}: \quad \pm z_n = 2n\pi - i \arg\cosh 2 \tag{40.172a,b}$$

$$\tanh(z) = 2i: \quad z_n = in\pi + i \arctan 2; \tag{40.173a,b}$$

$$\exp(z) + \exp(-z) = 0: \quad z_n = i\frac{\pi}{2} + in\pi; \tag{40.174a,b}$$

where n is any integer.
The roots ultimately all reduce to the exponential, viz:

$$-1 = \exp(i\pi) = \exp(z) = \exp(z - i2\pi n); \tag{40.175}$$

$$z = \exp(\log z) = \exp(2 + i\pi) = e^2 e^{i\pi} = -e^2; \tag{40.176}$$

$$5 = -i\sin(z) = -i\sin(z - 2n\pi) = -\sinh(iz - i2n\pi); \tag{40.177}$$

$$2 = \cos(z) = \cos(z - 2n\pi) = \cosh(iz - i2n\pi); \tag{40.178}$$

$$2 = -i\tanh(z) = -\tan(iz) = -\tan(iz + n\pi); \tag{40.179}$$

$$0 = \cosh(z) = \cos(iz) = \cos\left(-\frac{\pi}{2} - n\pi\right). \tag{40.180}$$

The logarithm (40.170a) has a single root (40.170b) because it is a univalent function (Subsection 37.4.5); the integral functions (40.169a; 40.171–40.174a) have an infinite number of roots (40.169b; 40.171–40.174b).

Example 40.18. *Extension of Landau Radius and Theorem.*
Elaborate on the first estimate of the Landau radius (39.74a,b) based on the value of the Landau function and its first derivate at the origin (39.73a,b), by considering also the second order derivative at the origin to obtain a second estimate of the Landau radius.

Compare with the exact value (39.82) of the Landau radius to obtain an inequality for the elliptic modular function.

Example 40.18.1. *Improved estimate of the Landau radius.*

The starting point is the Landau function (39.70) that can be constructed in four steps as follows: (i) the function (40.181a) is analytic in a disk of radius r and center at $z = a$, and does not take two distinct values (b, d) in (40.181b,c); (ii) the function (40.181d) is analytic in the same region and does not take the two distinct values (0, 1); (iii) the elliptic modular function (40.182a) is also analytic and has positive imaginary part (40.182b); (iv) the Landau function (40.182c) is analytic and has upper bound unity for the modulus (40.182d):

$$f, g, F, G \in \mathcal{A}(|z-a| < r): f(z) \neq b, d; b \neq d; \quad g(z) \equiv \frac{f(z) - d}{b - d} \neq 0, 1; \tag{40.181a–d}$$

$$F(z) \equiv E(g(z)), \quad \operatorname{Im}\{F(z)\} > 0; \quad G(z) \equiv \exp\{iF(z)\}, \quad |G(z)| < 1 \equiv B. \tag{40.182a–d}$$

The Landau radius R_L is (Figure 40.11) such that for $|z - a| > R_L$ the function (40.169a) either ceases to be analytic or takes at takes at least one of the two values $b \neq d$. If: (i) nothing more is known about the function $f(z)$, the Landau radius is infinite, as for Picard second theorem (39.22); (ii/iii) if the first (second) derivative of the function is known at the center $f'(a)$ $[f''(a)]$ then using the Cauchy bound (39.69d) for the first $n = 1$ (second $n = 2$) derivative of the Landau function in (40.183a) [(40.183b)]:

$$|G'(0)| < \frac{B}{R_1}, \quad |G''(0)| < \frac{2B}{(R_2)^2}, \tag{40.183a,b}$$

leads on account of (40.182d) ≡ (39.72) to the first (40.184b) [second (40.184c)] estimate of the Landau radius:

$$R_L < R_2 < R_1: \quad R_1 = |G'(0)|^{-1}, \quad R_2 = \left|\frac{2}{G''(0)}\right|^{-1/2}. \tag{40.184a–c}$$

The latter (40.184c) is an improvement of the former (40.184b) if (Figure 40.11) it is closer to the Landau radius, that is, smaller (40.184a).

Example 40.18.2. *Extended Landau theorem.*

The second-order derivative of the Landau function (39.70) ≡ (40.181d; 40.182a,c) follows from the first-order derivate (39.71):

$$\begin{aligned} G''(z) = \exp\left[iE\left(\frac{f(z)-d}{b-d}\right)\right]\Bigg\{ &- \left[\frac{f'(z)}{b-d}E'\left(\frac{f(z)-d}{b-d}\right)\right]^2 \\ &+ i\frac{f''(z)}{b-d}E'\left(\frac{f(z)-d}{b-d}\right) + i\left[\frac{f'(z)}{b-d}\right]^2 E''\left(\frac{f(z)-d}{b-d}\right)\Bigg\}. \end{aligned} \tag{40.185}$$

The Cauchy bound (39.69d) for the second derivative yields:

$$\frac{2B}{r^2} > |G''(0)| = |\exp[iE(\zeta)]|\,|b-d|^{-2}\left|-[a_1E'(\zeta)]^2 + i(b-d)a_2E'(\zeta) + iE''(\zeta)(a_1)^2\right|, \tag{40.186}$$

where appear the value of the function and of its first two derivatives at the point $z = a$:

$$a_0 \equiv f(a), \quad a_1 \equiv f'(a), \quad a_2 \equiv \frac{f''(a)}{2}, \quad \zeta \equiv \frac{a_0 - d}{b - d}; \tag{40.187a–d}$$

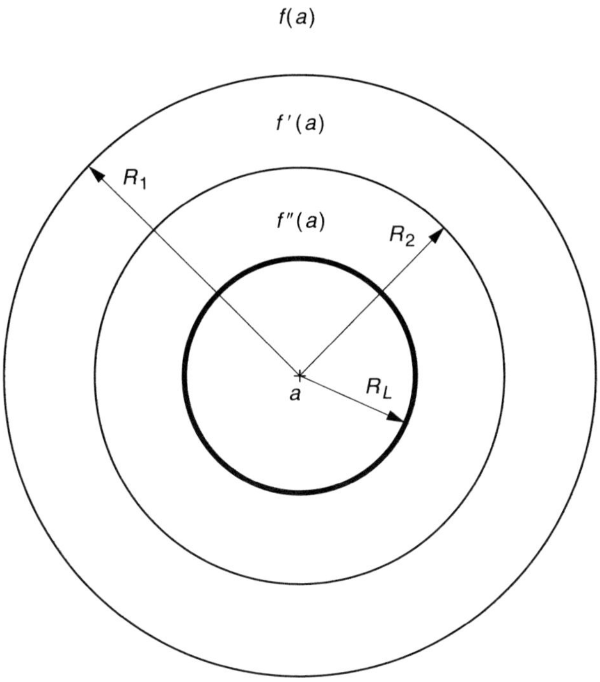

FIGURE 40.11
The two Picard theorems state that an integral function: (i) that does not take two values is a constant; (ii) that is not a constant takes an infinite number of times every value except possibly one. These theorems hold over the whole complex plane, corresponding to a circle of "infinite radius." The extension to finite radius is the Landau theorem: if a nonconstant function does not take two values in a disk in which it is analytic, the radius cannot exceed the Landau radius R_L. Thus beyond the Landau radius the nonconstant function either (i) cases to be analytic or (ii) takes all possible values except one at most. The Landau radius has an upper bound $R_1 > R_L$ determined only by the value of the function and its derivative at the center of the disk (Subsection 39.6.2); this upper bound can be improved R_2 (Example 40.18) to a lower value $R_L < R_2 < R_1$ by considering the second derivative of the function at the center of the disk. The original Landau upper bound R_1 (improved or reduced lower bound R_2) involve the elliptic modular function and its first (first and second) order derivative(s). These bounds have no advantage over the exact value of the Landau radius, established by the Caratheodory theorem (Subsection 39.6.3) which involves: (i) the value of the function and its first-order derivative at the center of the disk; (ii) the elliptic modular function and its first order derivative. The condition that the exact Landau radius is less than the: (i) higher R_1 upper bound, serves as a check on the Catatheodory theorem (Subsection 39.6.4); (ii) lower R_2 upper bound leads to an inequality involving the elliptic modular function and its first two derivatives (Example 40.18).

using the upper bound $B = 1$ in (39.72) $\equiv$ (40.182d) for the modulus of the Landau function leads to the upper bound for the Landau radius:

$$r^2 < \frac{2}{|G''(0)|} = 2\,|b-d|^2\,|\exp[-iE(\zeta)]|$$
$$= \left|\,[a_1 E'(\zeta)]^4 + \left[a_2 (b-d)\, E'(\zeta) + (a_1)^2\, E''(\zeta)\right]^2\right|^{-1/2}. \tag{40.188}$$

This leads to **extended Landau theorem**: *if the nonconstant function $f(z)$ is holomorphic (40.181a) in the disk $|z| < r$, and does not take the values (40.181b,c), then r does not exceed the second estimate of the Landau radius:*

$$r < R_2 \equiv \sqrt{2}\,|b-d|\left|\exp\left[-\frac{i}{2}E(\zeta)\right]\right|\left|[a_1 E'(\zeta)]^4 + \left[a_2(b-d)E'(\zeta) + (a_1)^2 E''(\zeta)\right]^2\right|^{-1/4}, \tag{40.189}$$

where the variable is (40.187d), and (a_0, a_1, a_2) are the first three coefficients of the Maclaurin series for $f(z)$:

$$f(z) = f(0) + zf'(0) + f''(0)\frac{z^2}{2} + O(z^3) = a_0 + a_1 z + \frac{1}{2}a_2 z^2 + O(z^3). \tag{40.190}$$

The theorem can be restated: *if the nonconstant function $f(z)$ is extended to a disc $|z| = r$ of radius $r > R_2$ larger than R_2 then it: (i) either ceases to holomorphic, or (ii) takes at least one of the two values b or d.*

Example 40.18.3. *Inequality for the elliptic modular function.*

Taking (40.191a,b) ≡ (39.84a,b) and comparing the first (second) upper bound for the Landau radius (39.74a,b) ≡ (40.191c) [(40.189) ≡ (40.191d)]:

$$b = 1,\ d = 0: \quad R_1 = \left|\frac{\exp[-iE(a_0)]}{a_1 E'(a_0)}\right|, \tag{40.191a–c}$$

$$\frac{R_2}{R_1} = \sqrt{2}\left|\exp\left[\frac{iE(a_0)}{2}\right]\right|\left|1 + \frac{\left[E''(a_0)/E'(a_0) + a_2\big/(a_1)^2\right]^2}{[E'(a_0)]^2}\right|^{-1/4}; \tag{40.191d}$$

it follows that the second estimate (40.191d) of the upper bound for the Landau radius improves on the first (40.191c):

$$\frac{R_2}{R_1} = \sqrt{2}\exp\left\{-\frac{1}{2}\mathrm{Im}[E(a_0)]\right\}\left|1 + \frac{\left[E''(a_0)/E'(a_0) + a_1^{-2}a_2\right]^2}{[E'(a_0)]^2}\right|^{-1/4} < 1, \tag{40.192}$$

if the factor (40.192) is less than unity. The latter is the case (40.193a):

$$\begin{aligned} R_2 < R_1:\ &\left[(a_1)^2 E''(a_0) + a_2 E'(a_0)\right]^2 + \left[(a_1)^2 E'(a_0)\right]^2 \\ &\geq \left[2(a_1)^2 E'(a_0)\right]^2 \exp\{-2\mathrm{Im}[E(a_0)]\}, \end{aligned} \tag{40.193a,b}$$

if the condition (40.193b) is met. The second bound (40.189) for the Landau radius cannot be less than the exact Landau radius (39.82):

$$\begin{aligned} &4\exp\{2\,\mathrm{Im}[E(a_0)]\}\left|\left[(a_1)^2 E''(a_0) + a_2 E'(a_0)\right]^2 + [a_1 E'(a_1)]^4\right|^{-1} \\ &\equiv (R_2)^4 \geq (R_L)^4 = 16\{\mathrm{Im}[E(a_0)]\}^4 |a_1 E'(a_0)|^{-4}; \end{aligned} \tag{40.194}$$

this leads to:

$$|\mathrm{Im}[E(a_0)]|^{-4}\exp\{2\,\mathrm{Im}[E(a_0)]\} \geq 4\left|1 + \frac{\left[E''(a_0)/E'(a_0) + a_2 a_1^{-2}\right]^2}{[E'(a_0)]^2}\right|. \tag{40.195}$$

This is *an inequality satisfied by the elliptic modular function of a function analytic in a disk, not taking the values (0,1), and with first three terms of the Maclaurin series at the center specified by (40.190).*

Example 40.19. *Evaluation of the Integral of a Rational Quadratic Function.*
Evaluate the indefinite integral:

$$f(x) = \int \left(ax^2 + bx + c\right)^{-1/2} dx, \tag{40.196}$$

in terms of elementary functions, for all values of the parameters a, b, c, real or complex. Obtain real integrals for real a, b, c.

Example 40.19.1. *Indefinite integral of inverse square root of a quadric.*
The binomial in the inverse square root has roots (40.197a):

$$x_\pm = -\frac{b}{2a} \pm d, \qquad d^2 \equiv \left(\frac{b}{2a}\right)^2 - \frac{c}{a}, \tag{40.197a,b}$$

where (40.197b) is the discriminant. As in Section 39.8 there are three cases: (i) a double root:

$$d = 0: \quad x_\pm = -\frac{b}{2a}, \quad y \equiv x + \frac{b}{2a}: \quad \sqrt{a} f(x) = \int y^{-1} \, dy = \log y, \tag{40.198a–d}$$

using (39.102b); (ii) two real distinct roots:

$$\begin{aligned} d^2 > 0;\ x_\pm = -\frac{b}{2a} \pm |d| : \sqrt{a} f(x) &= \int \left[(y - |d|)(y + |d|)\right]^{-1/2} dy \\ &= \int \left(y^2 - |d|^2\right)^{-1} dy = -\text{arc}\sin\left(\frac{y}{|d|}\right), \end{aligned} \tag{40.199a–c}$$

using (39.104b); (iii) a pair of complex conjugate roots:

$$\begin{aligned} d^2 > 0;\ x_\pm = -\frac{b}{2a} \pm i\,|d|\,, \sqrt{a} f(x) &= \int \left[(y - i\,|d|)(y + i\,|d|)\right] \\ &= \int \left(y^2 + |d|^2\right)^{-1/2} dy = \arg\sinh\left(\frac{y}{|d|}\right). \end{aligned} \tag{40.200a–c}$$

The preceding three results can be summarized: *the primitive of the inverse square root of a binomial is given by:*

$$\int \left(ax^2 + bx + c\right)^{-1/2} dx = \frac{1}{\sqrt{a}} \times \begin{cases} -\text{arc}\ \sin\left[p\left(x + \dfrac{b}{2a}\right)\right] & \text{if } b^2 > 4ac, \quad (40.201\text{a}) \\ \log\left(x + \dfrac{b}{2a}\right) & \text{if } b^2 = 4ac, \quad (40.201\text{b}) \\ \arg\ \sinh\left[p\left(x + \dfrac{b}{2a}\right)\right] & \text{if } b^2 < 4ac, \quad (40.201\text{c}) \end{cases}$$

where: (i) the discriminant (40.197b) distinguishes the three cases (40.201a–c); (ii) it appears in the cases (40.201a,c) via:

$$p \equiv |d|^{-1} = \left|\left(\frac{b}{2a}\right)^2 - \left(\frac{c}{a}\right)\right|^{-1/2}, \quad p\left(x + \frac{b}{2a}\right) = \left|b^2 - 4ac\right|^{-1/2}(2ax + b). \tag{40.202a,b}$$

The results (40.201a–c) are written in such a way that all quantities are real for real (a,b) and $a > 0$; in the case (40.201a) also (40.202b) must be less than unity for a real integral. All expressions also hold for complex (a, b, c).

Example 40.19.2. *Evaluation in terms of elementary functions.*

The inverse circular and hyperbolic functions can be expressed (Section 7.8) in terms of logarithms, leading to a single expression to replace all three in (40.201a–c). In (40.201c) appears hyperbolic sine that is defined by (40.203a):

$$w \equiv \sinh z = \frac{e^z - e^{-z}}{2}; \quad e^{2z} - 2we^z - 1 = 0; \tag{40.203a,b}$$

this is equivalent to the quadratic equation (40.203b), whose roots (40.204a):

$$e^z = w + \sqrt{w^2+1}, \quad z = \log\left(w + \sqrt{w^2+1}\right) = \arg\sinh w, \tag{40.204a,b}$$

specify the inverse hyperbolic sine as the logarithm (40.204b) of a rational function. The variable in (40.201c) is (40.205b) ≡ (40.202b):

$$b^2 < 4ac: \quad w = p\left(x + \frac{b}{2a}\right) = \left(4ac - b^2\right)^{-1/2}(2ax+b); \tag{40.205a,b}$$

this leads from (40.201c) to

$$\begin{aligned}\sqrt{a}f(x) &= \arg\sinh\left[\left(4ac-b^2\right)^{-1/2}(2ax+b)\right]\\ &= \log\left\{\left(4ac-b^2\right)^{-1/2}\left[2ax+b+\sqrt{4a^2x^2+4abx+4ac}\right]\right\},\end{aligned} \tag{40.206}$$

where (40.204b) was used. Omitting a constant term that does not affect the indefinite integral (40.206), leads to

$$a > 0: \quad \int \left(ax^2+bx+c\right)^{-1/2} dx = \frac{1}{\sqrt{a}} \log\left|2ax+b+2\sqrt{a}\sqrt{ax^2+bx+c}\right|; \tag{40.207}$$

this *alternative expression (40.207) for the primitive of the inverse square root of a binomial (40.201a–c; 40.202a,b)* can be checked by differentiating (40.207).

Example 40.20. *Periods of Elliptic Functions as Series of the Modulus.*

Express the two periods of Jacobian elliptic functions (Section 39.9) in terms of Gaussian hypergeometric series (Section 29.9).

Example 40.20.1. *Two periods of Jacobian elliptic functions.*

The real quarter period of the Jacobian elliptic functions (39.115a) is a function of the modulus $z \equiv k$:

$$t = \sin\varphi: \quad X(k) = \int_0^1 \left[\left(1-t^2\right)\left(1-k^2t^2\right)\right]^{-1/2} dt = \int_0^{\pi/2} \left(1 - k^2\sin^2\varphi\right)^{-1/2} d\varphi; \tag{40.208a,b}$$

For $|k| \le 1 - \varepsilon$ with $0 < \varepsilon < 1$ the binomial expansion (25.37a–c) is a uniformly convergent, and can be integrated term-by-term:

$$1 > \varepsilon > 0, \ |k| \le 1 - \varepsilon: \quad X(k) = \sum_{n=0}^{\infty} \binom{-1/2}{n} (-)^n k^{2n} \int_0^{\pi/2} \sin^{2n}\varphi \, d\varphi; \tag{40.209a–c}$$

the integral in (40.209a) was obtained before (20.69):

$$a_n \equiv n! \int_0^{\pi/2} \sin^{2n} \varphi d\varphi = \frac{\pi}{4} 2^{1-2n} \frac{(2n)!}{n!} = \frac{\pi}{2} 2^{-n} \frac{(2n)!}{(2n)!!}$$
$$= \frac{\pi}{2} 2^{-n} (2n-1)!! = \frac{\pi}{2} \left(n - \frac{1}{2}\right)\left(n - \frac{3}{2}\right) \cdots \frac{3}{2}\frac{1}{2}. \qquad (40.209d)$$

using the double factorial:

$$(2n)!! \equiv 2n\,(2n-2)\ldots 4.2, \ (2n-1)!! \equiv (2n-1)\,(2n-3)\ldots 3.1; \qquad (40.209e,f)$$

this equivalent to:

$$\frac{2}{\pi} a_n = 2^{-2n} \frac{(2n)!}{n!} = \frac{2n(2n-1)\ldots 2.1}{4n(4n-4)\ldots 4.1}$$
$$= \frac{2n-1}{2}\frac{2n-3}{2} \cdots \frac{3}{2}.\frac{1}{2} = \left(n - \frac{1}{2}\right)\left(n - \frac{3}{2}\right) \cdots \frac{3}{2}\frac{1}{2}. \qquad (40.210)$$

Substituting (40.209d) in (40.209c) the real quarter period (40.196a,b) is specified by:

$$|k| < 1 : X(k) = \frac{\pi}{2}\left[1 + \sum_{n=1}^{\infty} \frac{k^2}{n!}\left(n - \frac{1}{2}\right)\left(n - \frac{3}{2}\right)\cdots \frac{3}{2}\frac{1}{2}\frac{a_n}{n!}\right]$$
$$= \frac{\pi}{2}\left\{1 + \sum_{n=1} \left[\frac{k^n}{n!}\left(n - \frac{1}{2}\right)\left(n - \frac{3}{2}\right)\cdots \frac{3}{2}\frac{1}{2}\right]^2\right\}, \qquad (40.211a,b)$$

and the series converges absolutely for $|k| < 1$. The result (40.211b) is a particular case of (29.74) the Gaussian hypergeometric series:

$$X(k) = \frac{\pi}{2} {}_2F_1\left(\frac{1}{2}, \frac{1}{2}; 1; k^2\right); \qquad (40.212)$$

it can be written (40.211b) $\equiv$ (40.212) explicitly:

$$|k| < 1: \quad X(k) = \frac{\pi}{2}\left\{1 + \frac{k^2}{4} + \left(\frac{3}{8}\right)^2 k^6 + \left(\frac{5}{16}\right)^2 k^4 + \left(\frac{35}{128}\right)^2 k^8 + O\left(k^{10}\right)\right\}. \qquad (40.213a,b)$$

In the case $k = 0$ of circular functions the quarter period is $X = \pi/2$ and the period $4X = 2\pi$; for elliptic functions with real modulus $0 < k < 1$ the real period $4X > 2\pi$ is larger than for circular functions.

Example 4.20.2. *Relation with Gaussian hypergeometric functions.*

The imaginary half period of elliptic functions (39.116b) is given by:

$$t \equiv \frac{1}{k}\sin\psi : \ iY(k) + X(k) = \int_0^{1/k} \left[\left(1 - t^2\right)\left(1 - k^2 t^2\right)\right]^{-1/2} \mathrm{d}t = \int_0^{\pi/2} \left(k^2 - \sin^2\psi\right)^{-1/2} \mathrm{d}\psi. \qquad (40.214a,b)$$

For $|k| \geq 1 + \delta$ with $\delta > 0$ the binomial expansion (25.37a–c) is absolutely convergent and can be integrated term-by-term:

$$0 < \delta, \ |k| \geq 1 + \delta : \quad iY(k) + X(\kappa) = \sum_{n=0}^{\infty} k^{-2n-1} (-)^n \binom{-1/2}{n} \int_0^{\pi/2} \sin^{2n}\psi \, \mathrm{d}\psi. \qquad (40.215a–c)$$

Using again the integral (40.209d):

$$|k| > 1: \quad iY(k) + X(k) = \frac{\pi}{2k}\left\{1 + \sum_{n=1}^{\infty} \frac{k^{-2n}}{(n!)^2}\left[\frac{1}{2}\cdot\frac{1}{3}\cdots\left(n-\frac{3}{2}\right)\left(n-\frac{1}{2}\right)\right]^2\right\}, \tag{40.216a,b}$$

where the series converges absolutely for $|k| > 1$. This is again a particular case of the hypergeometric series (29.74):

$$iY(k) + X(k) = \frac{\pi}{2k}\,{}_2F_1\left(\frac{1}{2};\frac{1}{2};1;\frac{1}{k^2}\right); \tag{40.217}$$

it can be written (40.216b) ≡ (40.217) explicitly:

$$|k| > 1: \quad iY(k) + X(k) = \frac{\pi}{2}\left\{\frac{1}{k} + \frac{k}{4} + \left(\frac{3}{8}\right)^2 k^3 + \left(\frac{5}{16}\right)^2 k^5 + \left(\frac{35}{128}\right)^2 k^7 + O\left(k^9\right)\right\}. \tag{40.218a,b}$$

The hypergeometric series (40.212) [(40.217)] converges inside (outside) the unit disk; they have the same parameters ($a = 1/2 = b$, $c = 1$), and thus the exponent of convergence $E = \operatorname{Re}(c + 1 - a - b) = 1$ implies they: (i) converge conditionally on the boundary of convergence $|k| = 1 \neq k^2$ except at the points $k = \pm 1$; (ii) they do not converge for $k = \pm 1$, when the series (40.213b) = ± (40.218b) coincide with the same (opposite) sign. The case $k = 1$ of hyperbolic functions is not included in (40.212; 40.217) and can be addressed by: (i) analytic continuation of the hypergeometric function; (ii) using the complementary modulus k', defined by $k'^2 + k^2 = 1$ for Jacobian elliptic functions.

Note 40.1. *Electric (Magnetic) Fields: and Electric Charges (Currents) with Insulating and/or Conducting Walls.* It was shown that the induced electric charges (surface electric currents) in a conductor due to an external electric (magnetic) field are the same (Subsection 36.9.4). The terminology of electric charges and fields is adopted concerning three geometries: (i) a corner with circular indentation (Figure 40.12); (ii) a square well (Figure 40.13); (iii) a rectangle (Figure 40.14). The first two (i) and (ii) have been considered before (Sections 36.3 and 36.8–36.9; Example 40.10) in terms of elementary functions (Section 38.8) which have single periods, representing a single infinity of images (Figure 39.10); the third (iii) involves a double infinity of images (Figure 39.11), corresponding to doubly periodic functions, and lead to elliptic functions (Section 38.9). The three geometries (Figure 40.12–40.14) have in common two sets of walls, allowing four combinations: (a/d) both sets insulating or conducting; (b/c) one set conducting and the other insulating. In the case (a) of all walls insulating all images are identical and no surface electric charges are induced. In the case (d) of all surfaces conducting, there are induced electric charges on all walls; since the electric field is orthogonal to all walls, it must vanish at corners, where the induced electric charge is zero. In the cases (b, c) when one (the other) set of walls in conducting (insulating), the electric field is normal (tangential), and thus has the same direction at a square corner. It follows that the electric field can be continuous and nonzero at a square corner between a conducting and an insulating wall; thus the electric charge density on the conducting wall need not vanish at a square corner with a an insulating wall. It follows that at a square corner between a conducting and an insulating wall the electric charge (field) need not vanish and may be discontinuous (continuous). A jump of electric charge (current) possible in electro (magneto) statics because dissipative effects are ignored, as for a potential flow.

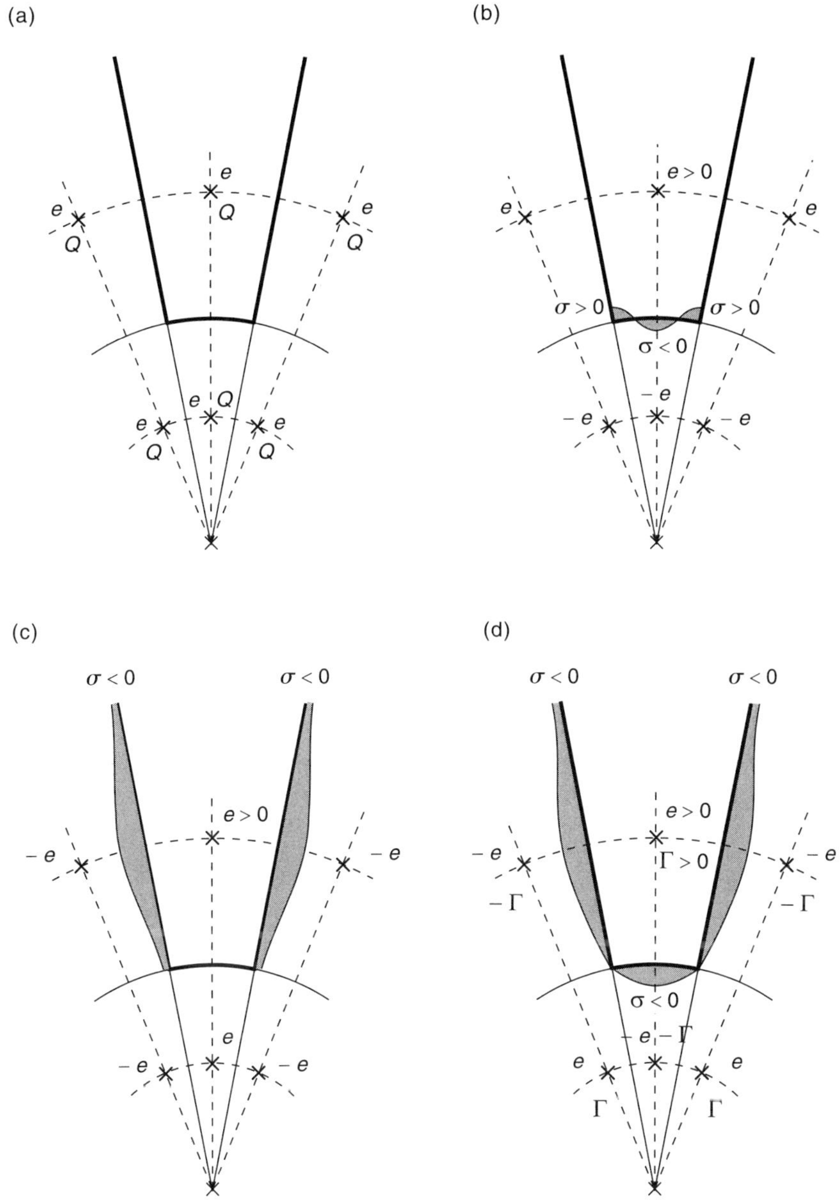

FIGURE 40.12
The corner with cylindrical indentation allows four combinations for the electro (magneto) static field due to a line electric charge (current): (a) all walls insulating (Figure 36.5a(c)]; (b) side walls conducting and cylindrical indentation insulating (Figure 40.4b); (c) cylindrical indentation conducting and side walls insulating (Figure 40.4a); (d) all walls conducting (Figure 36.5c(b)]. The preceding Figures (36.5, 40.4) showed only the monopole in the corner with indentation whereas the Figure 40.12 includes the nearest identical (opposite) images on insulating (conducting) walls.

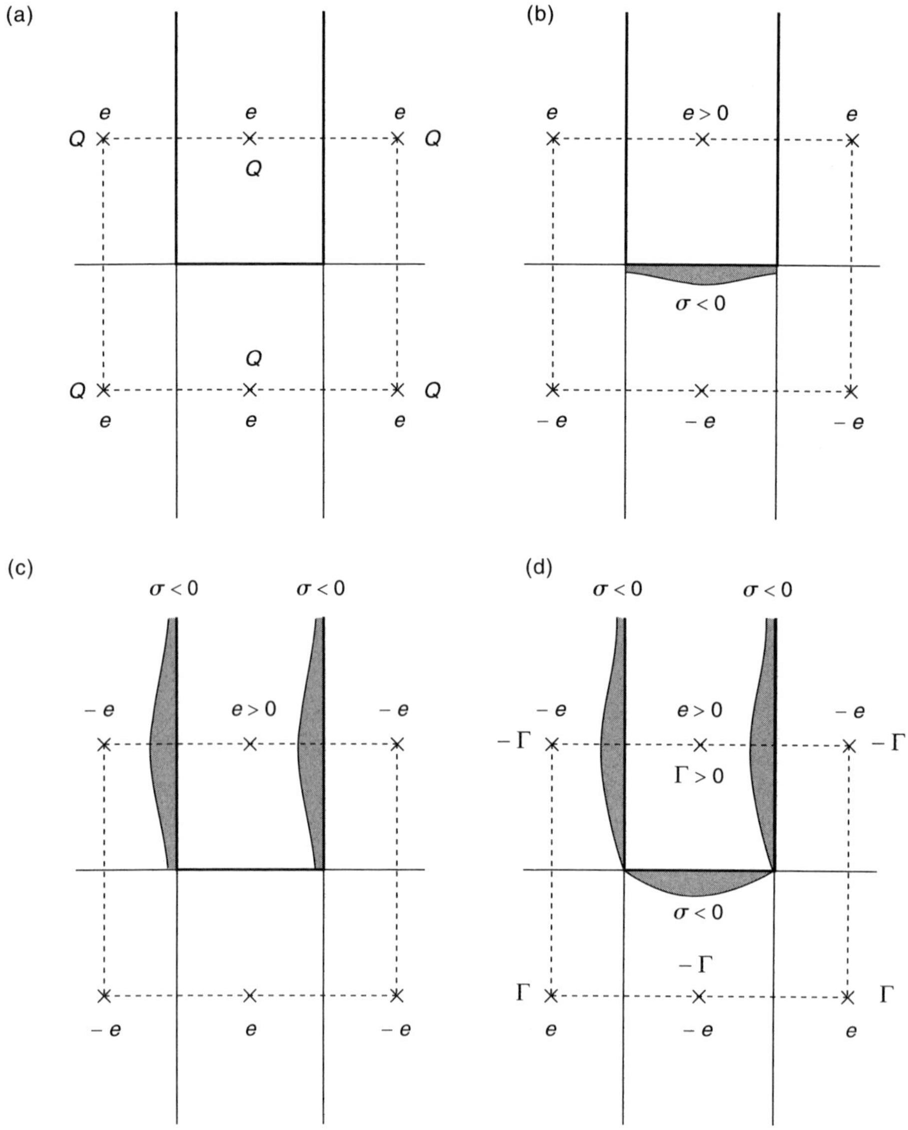

FIGURE 40.13
The square well allows four combination of the electro (magneto) static field due to line electric charges (currents) with: (a) all walls insulating [Figure 36.16a(b)] (b) bottom wall conducting and side walls insulating (Figure 40.5a); (c) side walls conducting and bottom wall insulating (Figure 40.5b); (d) all walls conducting [Figure 36.16b(c)]. The conducting (insulating) walls in electro/magnetostatics) may be replaced for a potential flow by rigid, impermeable walls (free surfaces).

Note 40.2. *Irrotational Incompressible Flow with Sources/Sinks (Vortices) and Rigid and/or Free Surfaces:* The case (a) of an electric charge with insulating walls in electrostatics is analogous to a source or sink in an irrotational flow with rigid walls; the images on all walls are identical. The case (d) of alternating images on all walls corresponds to the incompressible flow dues to a vortex with rigid walls; it corresponds to electrostatics with all walls conducting. The case of coexistent insulating (conducting) walls in electrostatics,

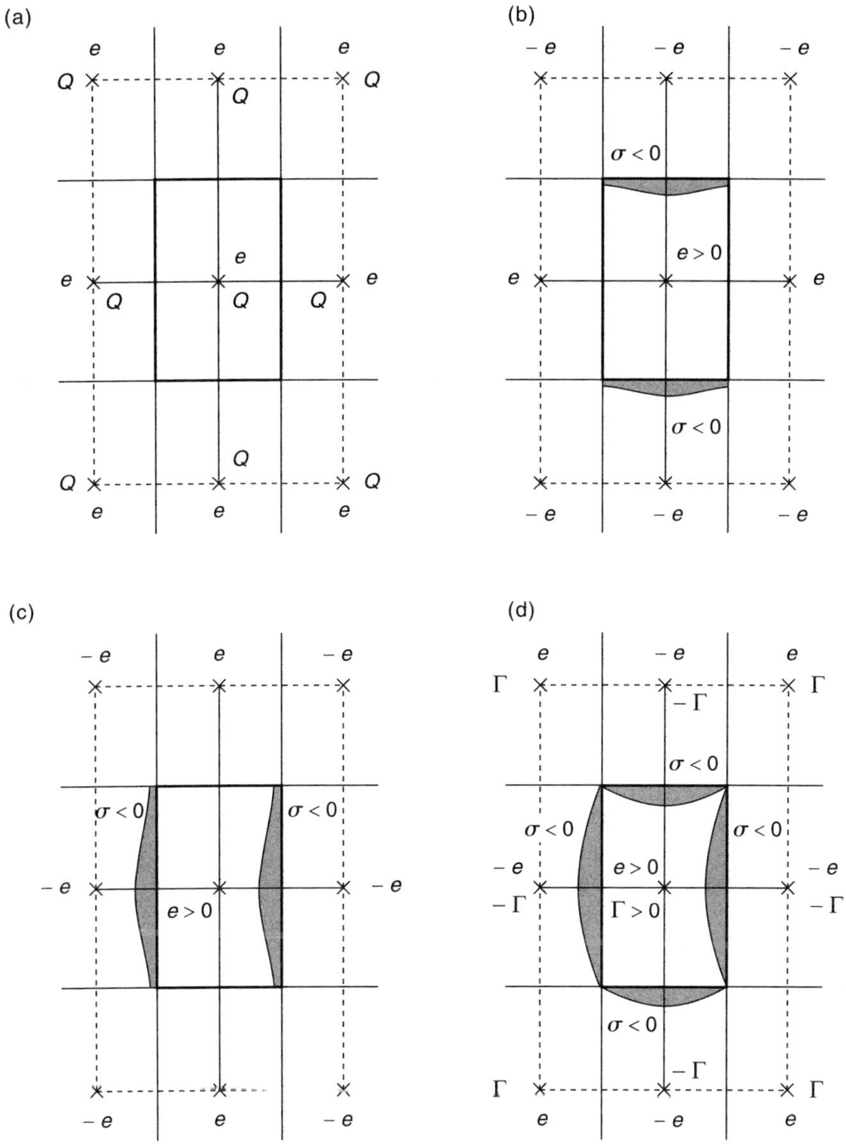

FIGURE 40.14
The tunnel with rectangular cross-section containing a line electric charge (current) allows four combinations of electro (magneto) static field: (a) all walls insulatimg; (b) horizontal walls conducting and vertical walls insulating; (c) vice-versa; (d) all walls conducting. The double infinity (single infinity with two rows) of images in Figure(s) 40.14 (40.12 and 40.13) leads to one (two) periods in Figure 39.10 (39.11), and solutions in terms of elementary (elliptic) functions [Section 39.8 (39.9)].

with tangential (normal) electric field, would correspond in a potential flow to a velocity tangent to a rigid wall (normal to a free surface) which can (cannot) stand a flow pressure. Thus the cases (a, d) correspond to potential flow bounded by rigid, impermeable walls, to which the velocity must be tangent; it follows that the velocity must be zero at corners. The cases (b, c) correspond to free surfaces (conductors) bounded by rigid, impermeable walls (insulators). The velocity is: (i) tangent to the rigid, impermeable walls; (ii) normal to the free surfaces which travel with the flow (no matter crosses them); (iii) at the orthogonal contact of a free surface with a rigid, impermeable wall the velocity need not vanish. The reason is that potential flow ignores dissipation, and there is no viscosity to impose a zero tangential velocity. Thus on a rigid, impermeable wall the velocity is zero (can be nonzero in the tangential direction), for a viscous (inviscid) flow. A normal velocity can exist for: (i) a permeable wall with blowing or suction; (ii) a moving or pulsating wall; (iii) a flexible wall responding with deformations to pressure or stress changes.

Note 40.3. *Synopsis of Complex Analysis and Applications*: The mathematical theory presented (List 40.1) follows five lines of reasoning, mostly sequential: (i) operations of complex numbers (Chapters 1, 3, and 5) and their representation (Chapters 7 and 9) on the plane and sphere (part I); (ii) differentiation (Chapters 11 and 19) and integration (Chapters 13 and 19) of complex functions including (Chapters 15 and 17) the calculus of residues (part II); (iii) convergence of series in general (Chapters 21 and 29) and power series expansions (Chapters 23, 25, and 27) of complex functions (part III); (iv) analytic continuation

LIST 40.1
Five Foundations in Analysis

A – Operations and Representation of Complex Number
- Chapter 1 – Complex numbers and quaternions
- Chapter 3 – Addition product and inverse
- Chapter 5 – Power, root, and logarithm
- Chapter 7 – Multivalued functions, branch-points, and branch-cuts
- Chapter 9 – Stereographic projection and genus

B – Differentiation and Integration
- Chapter 11 – Differentiation and holomorphic functions
- Chapter 13 – Primitive and contour integrals
- Chapter 15 – Loop integrals and poles
- Chapter 17 – Improper integrals and principal value
- Chapter 19 – Cauchy conditions and infinitesimals

C – Convergence and Power Series
- Chapter 20 – Combined test of convergence
- Chapter 21 – Convergence of and operations on series
- Chapter 23 – Analytic series of ascending powers
- Chapter 25 – Singular series of ascending-descending powers
- Chapter 27 – Classification of singularities and functions
- Chapter 29 – Combined test of convergence

D – Analytic Continuation and Conformal Mapping
- Chapter 31 – Analytic continuation and rational functions
- Chapter 33 – Conformal and critical points
- Chapter 35 – Linear and homographic transformations
- Chapter 37 – Mapping of domains and boundaries
- Chapter 39 – Essential singularities, roots, and periods

E – Special Points and Values
- Zeros: Chapters 5, 19, 27, 31, 39
- Poles: Chapters 13, 15, 17, 19, 27, 31, 39
- Essential singularities: Chapters 27, 39
- Branch-points: Chapters 7, 9, 17, 33, 35, 37, 39
- Periods: Chapters 7, 39

LIST 40.2
Five General Applications

F – Oscillations of Bodies and Circuits
  Chapter 2 – Stability of an equilibrium position
  Chapter 4 – Impedance of association of circuits
  Chapter 8 – Motion of pendulum and ship
G – Potential Fields
  Chapter 18 – Mass and the gravity field (also Sections 28.3, 37.8–37.9, 38.1)
  Chapter 24 – Electrostatics, electric charges, and dielectrics (also Sections 28.1, 28.3; 36.1–36.2, 36.4–36.8, 37.8–37.9, 38.1)
  Chapter 26 – Magnetostatics, currents, and permeability (also Sections 28.1, 28.3 36.1–36.2, 36.4–36.8)
H – Potential Flow
  Chapter 12 – Potential flow and multipoles
  Chapter 14 – Pressure and corner flows
  Chapter 16 – Images on plane walls
  Chapter 28 – Forces and moments on bodies
  Chapter 34 – Wing sections and plan forms
  Chapter 36 – Channels, condensers and wakes
  Chapter 38 – Hodograph for free jets
I – Waves and Diffusion
  Chapter 22 – Multiple reflections in a lens
  Chapter 32 – Steady heat conduction
J – Trajectories of Particles and Singularities
  Electron in electromagnetic field: Chapter 6
  Simple linear pendulum: Chapter 8
  Source/sink in a flow: Chapters 12, 16, 28, 36
  Vortices in flow: Chapters 12, 16, 28, 36
  Free streamlines of jets: Chapter 38

(Chapters 31 and 39) and conformal mapping (Chapters 33, 35, and 37) as a synthesis of the preceding (part IV). Some topics have cut across the four lines of reasoning, like special points and values: (i) zeros (Chapters 5, 19, 27, 31 and 39); (ii) poles (Chapters 13, 15, 17, 19, 27, 31 and 39); (iii) essential singularities (Chapters 27 and 39); (iv) branch-points (Chapters 7, 9, 17, 33, 35, 37, and 39); (v) periods (Chapters 7 and 39).

The applications (List 40.2) have been mainly in five areas developed in parallel and often interrelated: (i) solids, that is, oscillations (Chapter 2) of circuits (Chapter 4) and bodies (Chapter 8); (ii) fields, that is, gravity (Chapter 18), electrostatic (Chapters 24, 28, 36, and 38), magnetostatic (Chapters 26, 28, and 36) and electromagnetic (Chapters 6 and 28); (iii) fluid mechanics, as concerns potential flows (Chapters 12, 14, 16, 28, 34, 36, and 38); (iv) heat diffusion (Chapter 32) and waves (Chapter 22). Some topics have cut along several areas like: (v) trajectories of particles and singularities such as: (v–1) electron in electromagnetic field (Chapter 6); (v–2) simple linear pendulum (Chapter 8); (v–3,4) sources/sinks and vortices in a potential flow (Chapters 16, 28, and 36); (v–5) free surface of jets (Chapter 38). The Lists 40.1 (40.2) for the mathematical theory (physical applications) indicate: (i) the logical sequence of topics: (ii) the way they are interwoven in this volume.

Note 40.4. *Extension to Generalized Functions and Differential Equations*: The four parts that constitute the present volume in the context of mathematical physics applied to science and technology cover: (i) most of the basic aspects of the theory of functions of complex variables; (ii) a substantial number of applications including the main potential fields, viz. potential flow, gravity, electrostatic and magnetic fields, and steady heat conduction. There remain a number of topics, both concerning complex functions (i) and potential fields (ii) that are addressed in the second volume of the present series by similar and extended methods. Among the former (i) are the infinite representations of functions other than power series, for example, series of fractions for meromorphic functions; also infinite products (Subsection 27.9.21) and continued fractions; all these have in common convergence issues, lending to the relevance to enhanced summation methods that (i) do not change the limit value of convergent expansions; (ii) can make to converge some oscillatory and divergent expansions. All these properties, often find their simplest illustration with the elementary transcendental functions, viz., the exponential, circular and hyperbolic, and the inverse or cyclometric functions, for example, logarithm and arc circular and arg hyperbolic; they have served as occasional example and can be treated systematically in the complex plane, including elementary and not so elementary properties. When convergence is not possible it remains to classify some of the 'types of infinity; for example, in terms of transfinite numbers.

Besides the aspect complex functions (i) also applications to potential fields (ii) can be expanded, beyond the present volume that has concentrated on potential flows mostly in a strict sense of steady, irrotational, and incompressible. Three extensions are rotational or compressible or unsteady flows. In the present volume were addressed the five basic potential fields: potential flow, gravity, electric, magnetic, and steady heat conduction; six additional potential fields can be considered: rotating fluids, viscous flow in tubes, torsion of beams, deflection of membranes, capillarity and surface tension. All 11 potential fields involve solutions of Laplace and Poisson equations, with the biharmonic equation arising in another two cases: plane elasticity and plane viscous flow. The 80 or so problems solved in the present volume relate to the 5 basic potential fields; about 60 more would cover all 11 harmonic plus 2 biharmonic potential fields. The list of about 140 potential field problems finds a match in a list over 140 topics of complex analysis. This ensemble is the background on the theory of functions for subsequent developments. The generalized functions may be seen as an extension of complex functions to more than two dimensions, in particular as concerns singularities like multipoles; these provide the extension of potential fields from two to three (or more) dimensions.

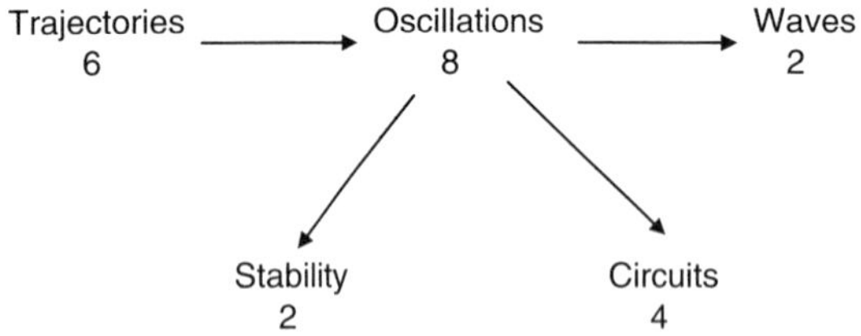

DIAGRAM 40.1
Five nonpotential field problems. The applications of complex analysis (Classification 40.1) chosen besides potential fields (Classification 40.2) are mostly electromechanical problems (Diagram 40.1), such as: (i) trajectories of an electron in an electromagnetic field; (ii–iv) other particle motions include oscillations, which relate to stability and electrical and mechanical circuits; (v) the oscillations of continuous media are waves, for example, in scattering by a parallel-sided lens.

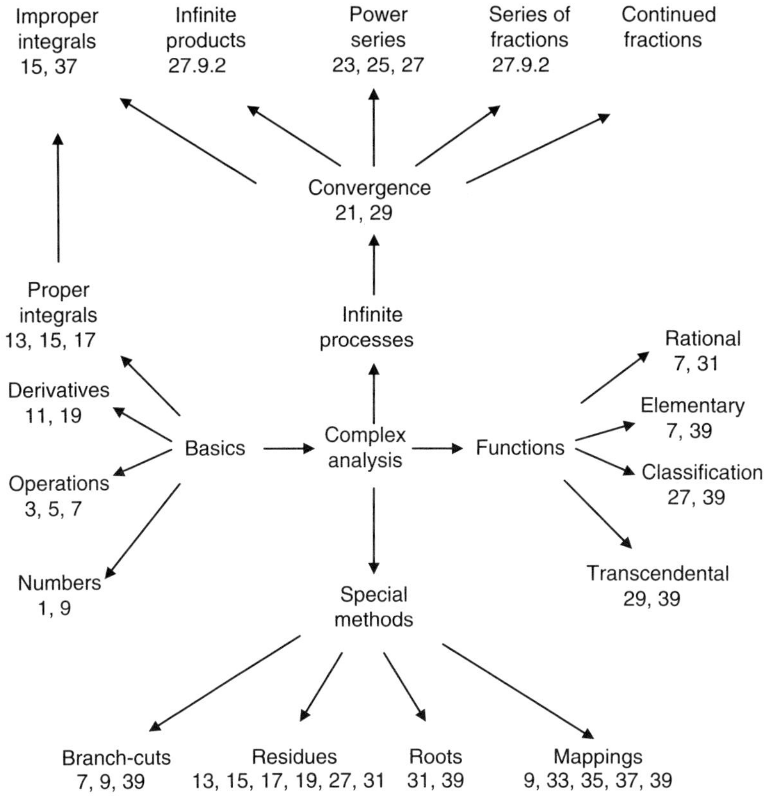

CLASSIFICATION 40.1

Seventeen topics in complex analysis. The main topics in complex analysis can be organized in four groups. The basics (I) include: (I.i) complex numbers; (I.ii) operations between them; (I.iii) complex functions and derivatives; (I.iv) the primitive and proper integrals, that is, integrals of bounded functions along finite contours. The convergence issues (II) arise for: (II.i) improper integrals involving unbounded functions and/or infinite contours; (II.ii) power series; (II.iii) series of fractions; (II.iv) infinite products; (II.v) other infinite representations into which some of the preceding can be transformed, for example, continued fractions. There are a number of associated methods (III) such as: (III.i) the representation of multivalued functions, which involves branch-points, branch-cuts, the Riemann sphere and associated topologies; (III.ii) the residues at singularities, essential or poles; (III.iii) the relation between poles and zeros as roots of algebraically inverse functions; (III.iv) the conformal mapping, which may involve the preceding. These methods may be applied (IV) to complex functions (IV.i): as a classification into classes; (IV.ii–iv) to establish the properties of classes of functions, such as the rational, elementary and higher transcendental functions.

The generalized functions, like Heaviside, Dirac, and Green's functions are linear operators and thus loose the nonlinear aspects of complex analysis, in exchange for the extension from two to any number of dimensions and having derivatives of any order. The elementary transcendental functions together with the generalized functions, are the basis for obtaining solutions of ordinary and partial differential equations; some examples of ordinary differential equations occurred in the present volume, for example, in dynamics, paths and trajectories. The potential fields as solutions of Laplace and Poisson equations

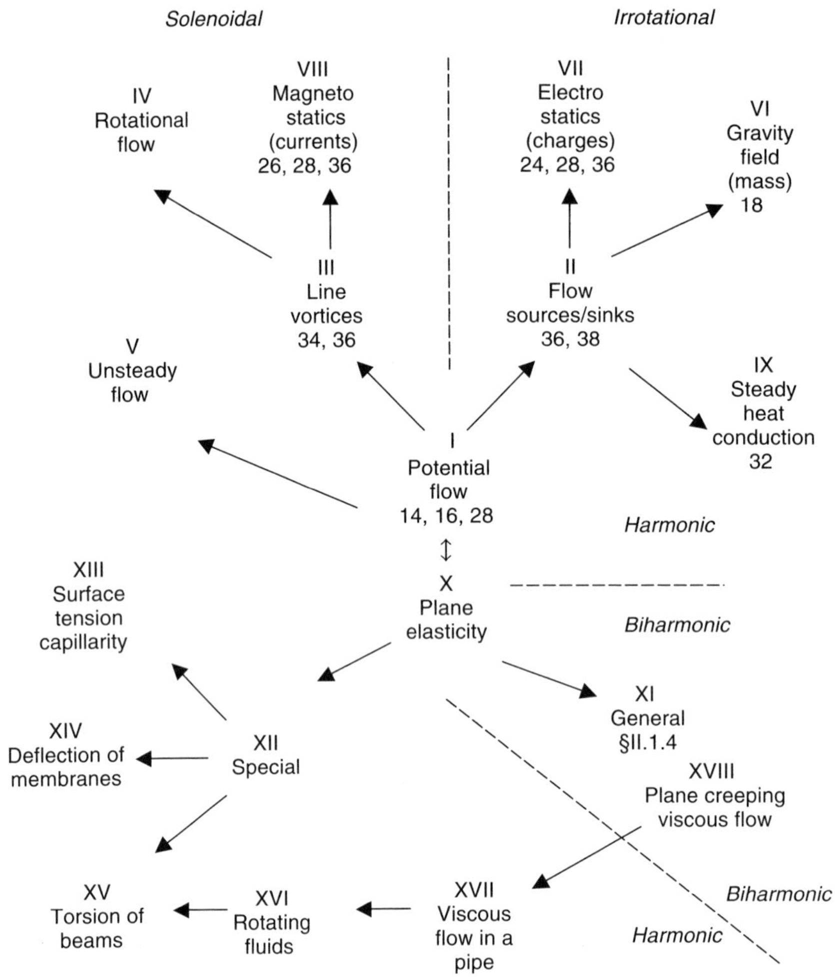

CLASSIFICATION 40.2

Eighteen potential fields. The main application of complex analysis has been potential fields in the plane, which includes two groups: irrotational and solenoidal. The irrotational potential fields include: (i) irrotational flow; (ii) electrostatics; (iii) gravity field; (iv) steady heat conduction. The solenoidal fields include: (v) rotational flow; (vi) magnetostatics. The unsteady flow (vii) relates to time dependence, and may be irrotational and/or potential or neither. The potential fields of second-order, related to solutions of the Laplace and Poisson equations in terms of harmonic functions also apply to: (viii) surface tension and capillarity; (ix) deflection of membranes; (x) torsion of beams; (xi) rotation of fluids; (xii) viscous flow in a tube. The fourth-order problems involving biharmonic functions concern: (xiii) plane elasticity; (xiv) plane steady viscous creeping flow ("creeping" in the sense that the velocity low enough to neglect inertial effects). The methods demonstrated for (i–vi) apply also to (vii–xiii).

have extension to other partial differential equations, for example, wave and diffusion equations, both linear and nonlinear. As a final remark it may be noted that the present volume started at very elementary level with operations between complex numbers, and gradually progressed to moderately advanced topics, such as: (i) the Picard theorems and roots of transcendental equations (Chapter 39) and conformal mapping into multiply connected regions (Chapter 37); (ii) induced charges (Chapter 24) and surface electric currents (Chapter 26) in dielectric media with a plane or cylindrical interface and the hodograph method (Chapter 38) for free jets. Whereas topics like (i) [(ii)] can be found in mathematical (physical) texts, they have not often been combined in the same volume. Diagrams 40.1 (40.2 + 40.3) outline the relation between seventeen topics in complex analysis (15 applications to potential fields plus 5 other problems). Some were covered in the present volume; the rest can be addressed using similar and extended methods, most are covered in the next two volumes of the series.

Conclusion 40: Paths for analytic continuation of a monogenic function (Figure 40.1) between three (a) [four (b)] regions. Geometry (Figure 40.2) of the thick strongly-cambered asymmetric Joukowsky airfoil. Family of circles (Figure 40.3) with conjugate points (p = 1, q = i), which includes the bisector for infinite radius. Electric field due to a line-charge in a corner with circular indentation (rectangular well) with: (Figure 40.4a) [(Figure 40.5a)] insulating sides and conducting bottom; (Figure 40.4b) (40.5b)] vice versa. Potential flow due to a source (a) [vortex (b)] of volume flux Q (circulation Γ) in an off-axis position between parallel walls, showing the position of the images for arbitrary location of the original (Figure 40.7a). Potential flow due to a monopole combining source and vortex (Figure 40.7b) at equal distance from parallel walls. Potential flow due (Figure 40.8) to a horizontal (a) [vertical (b)] dipole at equal distance from parallel walls. Field due to a constant potential (Figure 40.9) on two strips (sectors) of a plane (a) [cylinder (b)]. Parallel uniform (Figure 40.10) stream (a) with free streamline disturbed (b) by a source and a stagnation point (c). The Landau radius (Figure 40.11) is such that when exceeded a nonconstant function either (i) ceases to be holomorphic or (ii) takes all values but one; progressively smaller upper bounds can be determined as more information is available about the function; the comparison with the exact value leads to inequalities satisfied by the elliptic modular function. The four possible combinations of all walls insulating (a) [conducting (d)] or some conducting and some insulating (b, c) apply to: (i) a corner with cylindrical indentation (Figure 40.12); (ii) a square well (Figure 40.13); (iii) a tunnel with rectangular cross section (Figure 40.14). In all cases the insulator (conductor) in electro/magneto statics can be replaced for a potential flow by a rigid impermeable wall (a free surface) with a tangential (normal) velocity.

Bibliography

The general works on mathematics (section 1) include, among other subjects, the theory of functions, which is the basis of the calculus. The present volume is concerned mainly with complex analysis (section 3), which relates also to real functions (section 2). The main application are potential flows, which are a particular area of hydrodynamics (section 4) and aerodynamics (section 5) relating to fluid mechanics. The second main application is electro and magnetostatics, which are the simplest case of electromagnetism (section 6). The bibliography includes different approaches to the subject that support or complement the text such as: (i) distinct or alternative treatment of the same topics; (ii) further developments of these and other related matters. The part of the bibliography that has influenced most the content of the present volume is indicated by one, two or three asterisks (more asterisks indicate greater influence).

1 General Mathematics

**Abramowitz, M. and Stegun, I. *Handbook of Mathematical Functions.* Dover 1964, 2nd ed. 1965.

*Aleksandrov, A. D., Kolmogorov, A. N. and Lavrentiev, M. A. *Mathematics: Its Content, Methods and Meaning.* Massachussets Institute of Technology Press, 1963, 3 vols.

**Bateman, H. *Partial Differential Equations of Mathematical Physics.* Cambridge University Press 1932.

Beyer, W. H. (editor). *Standard Mathematical Tables.* CRC Press 1964, 28th ed. 1987.

Beyer, W. H. (editor) *Handbook of Mathematical Sciences.* CRC Press 1987, 6th ed. 2000.

*Bourbaki, N. (pseudonym). *Élements de Mathématiques.* Hermann, trad. Addison-Wesley, circa 30 vols.

**Erdelyi, A. *Higher Transcenental Functions.* Academic Press, 1953, 3 vols.

*Erdelyi, A. (editor). *Tables of Integral Transforms.* McGraw-Hill 1954, 2 vols.

Glaisher, J. W. L. et al. Tables of Bessel functions, vol. 8, *British Association Mathematical Tables*, 11 vols., Cambridge University Press 1952.

*Gradshteyn, I. S. and Rhyzdnik, I. M. *Tables of Integrals, Series and Products.* Academic Press, 1985, 2nd ed. 1980.

*Jahnke, E. and Emde, F. *Tables of Functions.* G.E. Stechert 1941, reprinted Dover 1945.

James, G. J. & James, R. C. *Mathematics Dictionary.* Van Nostrand 1949, 2nd ed. 1959.

Kant, I. *Logic.* Friedrich Nicolovius, Konigsberg 1800, translation Hartmann, R. S. S & Schwartz, W. 1974, Dover 1988.

Korner, S. *The Philosophy of Mathematics.* Hutchinson 1960, reprinted Dover 1986.

*Luke, Y. L. *Mathematical Functions and Their Approximations.* Academic Press 1975, 2 vols.

*Mansell, W. E. Tables of logarithms to 110 decimals. Vol. 8, *Royal Society Mathematical Tables*, 14 vols., Cambridge University Press 1964.

Marshaal, M. *Bourbaki: A Secret Society of Mathematicians.* American Mathematical Society 2006.

*Prudnikov, A. P., Brishkov, I. A., and Marichev, O. I. *Tables of Functions.* Nauka 1981, Gordon & Breach 1990, 3 vols.

Russel, B. S. and Whitehead, A. N. *Principia Mathematica.* Cambridge University Press, 1st ed. 1910, 3rd ed. 1927.

**Smirnov, V. I. *Cours de Mathématiques Supérieures.* Editions Mir or Pergamon Press, 9 vols.

*Vinogradov, I. M. (editor). *Encyclopedia of Mathematics.* Kluwer 1995, 5 vols.

*Weisstein, E. W. *Encyclopedia of Mathematics.* CRC Press 2009, 3 vols.

Zwillinger, D. (editor). *Standard Mathematical Tables of Formulas.* CRC Press 1966.

2 Real Functions

Apostol, T. M. *Mathematical Analysis.* Addison-Wesley 1963.

*Appell, P. *Cours d'analyse.* Gauthier-Villars 1920.

Appell, P. *Functions algebriques.* Paris 1929–1930, reprinted Chelsea 1976, 2 vols.

Banach, S. *Théorie des operations linéaires.* Lvov 1932, reprinted Chelsea 1978.

Berman, G. *Problémes d'analyse mathematique.* Edition Mir 1976.

Chatterji, S. D. *Cours d'analyse.* Presses polytechniques et universitaires romandes 1997, 4 vols.

*Chilov, G. *Analyse mathematique.* Editions Mir 1975.

*Courant, R. *Differential and Integral Calculus.* Interscience 1st ed. 1934, 2nd ed. 1937, 2 vols.

Ferreira, J. C. *Análise matemática.* Fundação Calouste Gulbernkian 1985.

Franklin, P. *Methods of Advanced Calculus.* McGraw-Hill 1944.

Gauss, C. F. *Untersuchungen uber Hohere Arithmetik.* Berlin 1889 reprinted Chelsea 1965.

Hardy, G. H. *A Course of Pure Mathematics.* Cambridge University Press 1st ed. 1908, 10th ed. 1957.

Hermite, C. *Cours d'annalyse.* Gaulthier-Villars 1973.

Hildebrand, F. *Advanced Calculus for Engineers.* Prentice-Hall 1949.

**Hobson, W. *Functions of Real Variable.* Cambridge University Press 1st ed. 1907, 2nd ed. 1921–6, reprinted Dover, 2 vols.

*Jordan, C. *Cours d'analyse.* Gautheir-Villars, 1909–15, 3 vols.

Klein, F. *Vorlesungen ueber Entwicklung der Mathematik.* Berlin 1926, reprinted Chelsea 1970, 2 vols.

Knopp, K. *Problem Book in the Theory of Functions.* Reprinted Dover 1952, 2 vols.

**Kolmogorov, A. N. and Fomine, S. *Théorie des fonctions et analyse fonctionelle.* Editions Mir 1937.

Landau, E. *Differential and Integral Calculus.* Groningen 1934, reprinted Chelsea 1950.

Landau, E. *Foundations of Analysis.* Berlin 1929, reprinted Chelsea 1951.

Lebedev, N. N., Skalskaya, I. P., and Velyand, Y. S. *Worked Problems in Applied Mathematics.* Dover 1965.

Littewood, D. E. *Theory of Functions.* Cambridge University Press 1926.

*Montel, P. *Familes normales.* Paris 1927, reprinted Chelsea 1974.

Osgood, W. F. *Advanced Calculus.* MacMillan 1928.

*Picard, E. *Traité d'analyse.* Gauthier-Villars, 1903–8, 3 vols.

Picard, E. and Simart, G. *Functions algébriques.* Paris 1897–1906, reprinted Chelsea 1971, 2 vols.

Pólya, G. and Szego, G. *Aufgaben und Lehrsatze aus der Analysis.* Springer 1925, 4. Aufgabe 1970.

Randolph, J. F. *Basic Real and Abstract Analysis.* Academic Press 1968.

Rothe, R., Ollendorf, F., and Pohlausen, K. *Theory of Functions as Applied to Engineering Problems.* MIT Press 1933, reprinted Dover 1961.

Rudin, W. *Principles of Mathematical Analysis.* McGraw-Hill 1953.

Schwartz, L. *Analyse.* Hermann 1970.

Schwartz, L. *Analyse hilbertienne.* Hermann 1979.

Sturm, J. C. F. *Cours d'annalyse.* Gauthier-Villars 1880.

*Teixeira, F. G. *Curso de análise infinitesimal.* Tipografia Ocidental (Porto), 1891–1903, 3 vols.

*Titchmarsh, E. C. *The Theory of Functions.* Oxford University Press 1st ed. 1932, 2nd ed. 1939.

Vallée de Poussin, C. J. de la *Cours d'analyse infinitesimale.* Louvain 1914, 2 vols.

Weyl, H., Landau, E., and Riemann, B. *Der Kontinuum und andere Monographien.* Berlin 1917, reprinted Chelsea 1960.

3 Complex Analysis

Achieser, N. I. *Theory of Approximation.* Frederick Ungar 1956, reprinted Dover 1992.

Ahlfors, L. A. *Complex Analysis.* McGraw-Hill 1953, 3rd ed. 1979.

Ávila, G. *Variáveis complexas.* Livros Técnicos e Científicos 2000.

Bieberbach, L. *Conformal Mapping.* 1949, Chelsea Publications 1953.

Brezinski, C. *Continued Fractions and Padé Approximants.* Springer 1991.

**Bromwich, J. T. I. A. *Introduction to the Theory of Infinite Series.* MacMillan 1908, 2nd ed. 1926.

**Caratheodory, C. *Theory of Functions (Real and Complex).* Verlag Birkhauser 1950, Chelsea Publications 1854, 1+2 vols.

Carrier, G. F., Krook, M., and Pearson, C. E. *Functions of a Complex Variable.* McGraw-Hill 1966.

Cartan, H. *Functions analytiques d'une ou plusieurs variables complexes.* Hermann 1985.

*Chabat, B. *Analyse complexe.* Nauka 1985, Editions Mir 1990, 2 vols.

**Chrystal, G. *Textbook of Algebra.* A. C. Black 1904, Chelsea Publishing 1964, 2 vols.

*Churchill, R. V., Brown, J. W., and Verhey, R. E. *Complex Variables and Applications.* McGraw-Hill 1st ed. 1949, 3rd ed. 1974.

*Copson, E. T. *An Introduction to the Theory of Functions of a Complex Variable.* Oxford University Press 1935.

Courant, R. *Dirichlet's Principle, Conformal Mapping and Minimal Surfaces.* Interscience 1950.

Davis, H. T. *The Summation of Series.* Principia Press 1962.

Dettman, J. W. *Applied Complex Variables.* MacMillan 1965, 5th ed. 1970, reprinted Dover 1984.

Egrafov, M. S. *Recueil de Problemes sur la théorie des fonctions analytiques.* Editions Mir 1974.

Flanigan, F. J. *Complex Variables.* Allyn & Becon 1972, reprinted Dover 1983.

Ford, W. B. *Divergent Series and Asymptotic Expansions.* Michigan 1916, reprinted Chelsea 1960.

*Forsyth, A. R. *Theory of Functions of a Complex Variable.* Cambridge University Press 1st ed. 1893, 3rd ed. 1918, Dover 1965, 2 vols.

Goffman, C. and Pedrick, G. *A First Course in Functional Analysis.* Prentice-Hall 1965.

***Goursat, E. *Cours d'analyse mathematique.* Gauthier-Villars, 1911, 3 vols., Dover 1959, 5 vols.

Hardy, G. H. *Divergent Series.* Oxford 1949, reprinted Chelsea 1991.

Hille, E. *Analytic Function Theory.* Chelsea Publications 1976, 2 vols.

Hormander, L. *Complex Analysis in Several Variables.* North-Holland 1966.

Jolley, L. B. W. *Summation of Series.* Chapman & Hall 1925, reprinted Dover 1961.

Knopp, K. *Allgemeine und Aufgabenzammlung der Funktionentheorie.* Tubingen 1921–1947, 2 vols., Dover 1945, 4 vols.

*Knopp, K. *Elemente des Funktionentheorie.* Dover 1952.

**Knopp, K. *Infinite Sequences and Series.* Dover 1956.

*Knopp, K. *Theory and Applications of Infinite Series.* Tubingen 1921, 4th ed. 1947, reprinted Haffner 1971.

Krantz, S. G. *Handbook of Complex Variables.* Birkhauser 1999.

Lagrange, J. L. *Theorie des functions analytiques.* Imprimerie de la Republique 1794.

Laurent, H. *Théorie des residus.* Gauthier-Villars 1866.

*Lavrentiev, M. and Chabat, B. *Méthodes de théorie des fonctions d'une variable complexe.* Editions Mir 1977.

Lusternik, L. and Sobolev, V. *Précis d'analyse fonctionel.* Nauka 1982, Editions Mir 1989.

*Markushevich, A. I. *Theory of Functions of a Complex Variable.* Reprinted Chelsea 1965, 3 vols.

Michel, A. N. and Herbert, C. J. *Applied Algebra and Functional Analysis.* Prentice-Hall 1981, reprinted Dover 1993.

Miller, K. S. *Advanced Complex Calculus.* Harper 1960, Dover 1970.

Nehari, Z. *Conformal Mapping.* McGraw-Hill 1952, reprinted Dover 1975.

Nevanlinna, R. *Le théoreme de Borel-Picard et la théorie des fonctions méromorphes.* Paris 1926.

Nevanlinna, R. *Analytic Functions.* Springer 1973.

Nevanlinna, R. and Paatero, V. *Complex Analysis.* Birkhauser 1969, reprinted Chelsea 1982.

Osgood, W. F. *Lehrbuch der funktionentheorie.* Cambridge Massachussets 1929–1932, reprinted Chelsea 1965.

Polya, G. and Szego, G. *Isoperimetric Inequalities in Mathematical Physics.* Princeton University Press 1960.

*Riesz, F. and Sz. Nagy, B. *Functional Analysis.* Frederick Ungar 1955, reprinted Dover 1990.

Rivlin, T. J. *Approximation of Functions.* Blaisdell 1969, reprinted Dover 1981.

Schinzinger, R. and Laura, P. A. A. *Conformal Mapping.* Elsevier 1991.

*Sobolev, S. L. *Functional Analysis.* 1950, American Mathematical Society 1963.

Sveschinikov, A. and Tikhonov, A. *The Theory of Functions of a Complex Variable.* Mir Publishers 1978.

Timan, A. F. *Approximation of Functions of a Real Variable.* 1960, Clarendoen 1963, Dover 1994.

Tutschke, W. and Vasudeva, H. L. *An Introduction to Complex Analysis.* Chapman & Hall 2004.

*Valiron, G. *Théorie des fonctions.* Masson, 1ére edition 1942, 3éme edition 1966.

Wall, H. S. *Analytic Theory of Continued Fractions.* Van Nortrand 1948.

***Whittaker, E. T. and Watson, G. N. *Course of Modern Analysis.* Cambridge University Press 1st ed. 1902, 4th ed. 1927.

4 Hydrodynamics

*Abbott, I. H. and Doenhoff, A. E. *Theory of Wing Sections.* McGraw-Hill 1949, reprinted Dover 1959.

Agostini, I. G. *Leonardo da Vinci.* Novara 1956, 2 vols.

Aris, A. *Vectors, Tensors and the Basic Equations of Fluid Mechanics.* Prentice-Hall 1962, reprinted Dover 1989.

Bassett, A. B. *A Treatise on Hydrodynamics.* Deighton & Bell 1888, reprinted Dover 1961, 2 vols.

**Batchelor, G. K. *Fluid Dynamics.* Cambridge University Press 1967.

*Bateman, H., Dryden, H. L., and Murnaghan, F. P. *Hydrodynamics.* Dover 1956.

Birkhoff, G. *Hydrodynamics.* Princeton University Press 1950, reprinted Dover 1955.

Curle, N. and Davies, H. J. *Modern Fluid Dynamics.* Van Nostrand 1968, 2 vols.

D'Alembert, J. R. *Reflexions sur le cause generale des vents.* David 1747.

D'Alembert, J. R. *Traité sur l'equilibre et mouvement des fluides.* David 1744, Briasson 1770.

D'Alembert, J. R. *Essai d'une nouvelle theorie de la resistance des fluides.* David 1752.

Diaz, J. B. and Pai, S. I. (editors). *Fluid Dynamics and Applied Mathematics.* Gordon & Breach 1962.

Duncan, W. J., Thom, A. S., and Young, A. D. *Mechanics of Fluids.* Arnold 1960.

**Durand, W. F. *Aerodynamic Theory.* Springer, 1934–6, 6 vols., Dover 1963, 3 vols.

Fediaevski, C., Voitkounski, I., and Faddeev, K. *Hydromécanique.* Editions Mir 1974.

Gates III, W. H. *Leonardo da Vinci.* Instituto Português do Património Arquitectónico 1998.

*Glauert, H. *Airfoil and Airscrew Theory.* Cambridge University Press 1936.

Golstein, S. *Modern Developments in Fluid Dynamics.* Clarendon Press 1938, Dover 1965.

**Howe, M. S. *Hydrodynamics and Sound.* Cambridge University Press, 2007.

***Lamb, H. *Hydrodynamics.* Cambridge University Press 1st ed. 1879, 6th ed. 1932.

***Landau, L. D. and Lifshitz, E. F. *Fluid Mechanics.* Volume 6 of Course of Theoretical Physics. Pergamon 1953.

*Lavrentiev, M. A. and Chabat, B. *Effets hydrodynamiques et modèles mathématiques.* Nauka 1977, Editions Mir 1980.

*Liepmann, W. H. and Roshko, A. *Elements of Gas Dynamics.* Wiley 1957.

**Lighthill, M. J. *An Informal Introduction to Fluid Mechanics.* Clarendon Press 1986.

*Löitsyanski, L. G. *Mechanics of Fluids and Gases.* Pergamon 1966.

Meyer, R. E. *Introduction to Mathematical Fluid Dynamics.* Wiley 1971, Dover 1982.

***Milne-Thomson, L. M. *Theoretical Hydrodynamics.* MacMillan 1st ed. 1958, 4th ed. 1996, Dover 1973.

Oswatitisch, K. *Gas Dynamics.* Academic Press 1956.

*Prandtl, L. and Tietjens, O. G. *Fundamentals of Hydro and Aeromechanics.* Teubner 1934, reprinted Dover 1957, 2 vols.

Retic, L. *Unknown Leonardo.* McGraw-Hill 1974.

*Rutherford, D. E. *Fluid Dynamics.* Oliver & Boyd 1959.

Sedov, L. I. *Two-Dimensional Problems in Hydrodynamics and Aerodynamics.* 1950, Wiley 1965.

Spurk, J. H. *Fluid Mechanics.* Springer 1997.

Truesdell, C. A. *Mechanical Foundations of Elasticity and Fluid Mechanics.* Gordon & Breach 1966.

*van Dyke, M. *An Album of Fluid Motion.* Parabolic Press 1982.

van Dyke, M. *Perturbation Methods in Fluid Dynamics.* Parabolic Press 1985.

5 Aerodynamics

Anderson, J. D. *A History of Aerodynamics.* Cambridge University Press 1997.

Anderson, J. D. *Modern Compressible Flow with Historical Perspective.* McGraw-Hill 1982, 2nd ed. 1990.

Anderson, J. D. *Hypersonic and High-Temperature Gas Dynamics.* McGraw-Hill 1989.

Anderson, J. D. *Fundamentals of Aerodynamics.* McGraw-Hill 1984, 3rd ed. 2001.

Ashley, H. and Landahl, M. *Aerodynamics of Wings and Bodies.* Addison-Wesley 1965, reprinted Dover 1985.

**Batchelor, G. K. *Homogeneous Turbulence.* Cambridge University Press 1953.

Bers, L. *Subsonic and Transonic Gas Dynamics.* Wiley 1958.

Birkhoff, G. and Zarantonello, E. H. *Jets, Wakes and Cavities.* Academic Press 1957.

Brederode, V. *Aerodinâmica incompressível.* Edição do autor Lisboa 1997.

Carafoli, E. (editor). *Wing Theory in Supersonic Flow.* Pergamon 1969.

**Chandrasekhar, S. *Hydrodynamic and Hydromagnetic Stability.* Oxford University Press 1961.

Courant, R. and Friedrichs, K. O. *Supersonic Flow and Shock Waves.* Interscience 1948, reprinted Springer 1976.

Covert, E. (editor). *Thrust and Drag: Its Prediction and Verification.* American Institute of Aeronautics and Astronautics 1985.

*Drazin, P. G. and Reid, W. H. *Hydrodynamic Stability.* Cambridge University Press 1981.

Freiedrichs, K. O. *Special Topics in Fluid Mechanics.* Gordon & Breach 1966.

Godounov, S. et al. *Resolution numerique des problèmes multidimensionels de la dynamique des gaz.* Nauka 1976, Editions Mir 1979.

Hayes, W. D. and Probstein, R. F. *Hypersonic Flow Theory.* Academic Press 1959.

Hinze, J. O. *Turbulence.* McGraw-Hill 1959, reprinted 1975.

Houghton, E. L. and Boswell, R. P. *Further Aerodynamics for Engineering Students.* Arnold 1970.

Houghton, E. L. and Carpenter, P. W. *Aerodynamics for Engineering Students.* Arnold 1970, 4th ed. 1993.

Howarth, H. L. (editor). *High-Speed Flow.* Oxford University Press 1953.

Joseph, D. D. *Stability of Fluid Motions.* Springer 1976, 2 vols.

Karamcheti, K. *Ideal Fluid Aerodynamics.* Wiley 1966.

Kuethe, A. M. and Chow, C. Y. *Foundations of Aerodynamics.* Wiley 1976, 4th ed. 1986.

Lafita, F. *Aerodinamica Aplicada.* Madrid 1946, 2 vols.

Lin, C. C. *Hydrodynamics Stability.* Cambridge University Press 1955.

*Manwell, A. R. *The Hodograph Equations.* Oliver & Boyd 1971.

*Milne-Thomson, L. M. *Theoretical Aerodynamics.* MacMillan 1958, 4th ed. 1966, reprinted Dover 1973.

*Monin, A. S. and Yaglom, A. M. *Mechanics of Turbulence.* Nauka 1965, M.I.T. Press 1971, 2 tomos.

*Ockendon, H. and J. R. *Viscous Flow.* Cambridge University Press 1995.

*Ockendon, H. and J. R. *Waves and Compressible Flow.* Springer 2004.

*Ockendon, H. and Taylor, A. B. *Inviscid Fluid Flows.* Springer 1983.

Panton, R. L. *Incompressible Flows.* Wiley 1984, 2nd ed. 1996.

Robinson, A. and Laurmann, J. A. *Wing Theory.* Cambridge University Press 1956.

Rosenhead, L. (editor). *Laminar Boundary Layers.* Oxford University Press 1963, reprinted Dover 1988.

*Saffman, P. G. *Vortex Dynamics.* Cambridge University Press 1992.

**Schlichting, H. *Grenzschichttheorie.* Verlag G. Braun 1951, achten ausgabe 1982, translation McGraw-Hill 1955, 3rd ed. 1968.

Schlichting, H. and Gersten, K. *Grenzschich-theorie.* Neun Auflage Springer 1997.

Shapiro, A. H. *The Dynamics and Thermodynamics of Compressible Fluid.* Ronald 1959, 2 vols.

Townsend, A. A. *Structure of Turbulent Shear Flow.* Cambridge University Press 1976.

Truesdell, C. A. *Kinematics of Vorticity.* Indiana University Press 1954.

Twaithes, B. W. (editor). *Incompressible Aerodynamics.* Clarendon Press 1960, reprinted Dover 1987.

*von Mises, R. *Compressible Fluid Flow.* Academic Press 1958.

von Mises, R. and Friederichs, K. O. *Fluid Dynamics.* Springer 1971.

Woods, L. C. *Subsonic Plane Flow.* Cambridge University Press 1961.

Young, A. D. *Boundary Layers.* BSP Professional Books 1989.

*Zeldovich, Y. B. and Raizer, Y. P. *Physics of Shock Waves and High-Temperature Hydrodynamic Phenomena.* 1961, reprinted Dover 2002.

Zeytounian, R. K. *Écoulements rotationels de fluids parfaits.* Springer 1998.

6 Electricity and Magnetism

Akhiezer, A. and Akhiezer, I. *Electromagnetisme et ondes electromagnetiques.* Nauka 1985.

Altman, C. and Suchy, K. *Reciprocity, Spatial Mapping and Time Reversal in Electromagnetics.* Kluwer 1991.

*Bateman, H. *Electrical and Optical Wave Motion.* Cambridge University 1955.

Bolton, W. *Mechatronics: Electronic Control Systems in Mechanical and Electrical Engineering.* Pearson Education 1995, 3rd ed. 2003.

**Born, M. and Wolf, E. *Optics.* Pergamon 1st ed. 1953, 5th ed. 1975.

Boulanger, P. and Hayes, M. *Bivectors and Waves in Mechanics and Optics.* Chapman & Hall 1993.

Bradley, D. A., Dawson, D., Burd, N.C., & Loader, A.J. *Mechatronics: Electronics in Products and Processes.* Chapman & Hall 1991.

Ceninkcent, S. *Mechatronics.* Wiley 2006.

Correia, L. M. *Wireless Flexible Personalized Communications.* Wiley 2001.

Dereniak, E. L. and Boreman, G. D. *Infrared Detectors and Systems.* Wiley 1996.

Fernandes, A. S. C. *Antenas de onda estacionária.* Fundação Calouste Gulbenkian 1979.

Fleury, P. and Mathieu, J. P. *Courants alternatifs et ondes hertziennes.* Eyrolles 1969.

Galejs, J. *Antenas in Inhomogeneous Media.* Pergamon 1969.

Hertz, H. *Electrical Waves.* Teubner 1876.

Histand, M.B. & Alciatore, D.G. *Mechatronics and Measurement Systems.* McGraw-Hill 1999.

Jeans, J. *Theory of Electricity and Magnetism.* Cambridge University Press 1st ed. 1908.

*Jones, D. S. *Electromagnetic Theory.* Pergamon 1964.

Labrique, F. and Santana, J. J. E. *Electrónica de potência.* Fundação Calouste Gulbenkian 1997.

***Landau, L. D. and Lifshitz, E. F. *Field Theory.* Volume 2 of Course of Theoretical Physics. Pergamon 1939.

***Landau, L. D. and Lifshitz, E. F. *Electrodynamics of Continuous Media.* Volume 8 of Course of Theoretical Physics. Pergamon 1956.

Maxwell, J. C. *Matter and Motion.* Cambridge University Press 1877, reprinted Dover 1960.

*Maxwell, J. C. *Treatise on Electricity and Magnetism.* Clarendon Press 1891, Dover 1954.

Mendiratta, S. K. *Introdução ao electromagnetismo.* Fundação Calouste Gulbenkian 1984.

Newton, I. *Optiks.* G. Bell 1931, Reprinted Dover 1952.

Reitz, J. R. and Milford, E. J. *Electromagnetic Theory.* Eddison-Wesley 1967.

Schelkunoff, S. A. *Electromagnetic Waves.* Van Nostrand 1943.

Silva, C.W. de, *Mechatronics: an Integrated Approach.* CRC Press 2005.

Silva, J. F. A. *Electrónica industrial.* Fundação Calouste Gulbenkian 1998.

Silva, M. M. *Circuitos eléctricos e electrónicos.* Fundação Calouste Gulbenkian 1995.

**Silveira, A. *Teoria da electricidade.* IST 1954, 2 vols.

Skilling, H. H. *Electricity Waves.* Wiley 1948.

Slater, L. J. and Frank, N. *Electromagnetism.* McGraw-Hill 1947.

Smythe, W. R. *Static and Dynamic Electricity.* Hemisphere Publishing 1939.

**Sommerfeld, A. *Electrodynamics.* Academic Press 1952.

**Stratton, J. A. *Electromagnetic Theory.* McGraw-Hill 1941.

References

(Authors mentioned in the text, in chronological order, indicating: date, name, publications)

VI b.c. Pythagoras in the *Thirteen Books of Euclids Elements*, T.L. Heath, 1908, Cambridge University Press, rep. Dover 1956.

1569 Mercator, G.K. *Chronologia hoc est Temporum,* 1569; *Atlas Gerardi Mercatonis*, Amesterdam 1609.

1583 Galileo, G. *Discursi*, Firenze 1632; *Dialogus I-IV,* Roma 1699.

1637a Descartes, R. P. *Geometrie*: Open Court 1925, Dover 1956.

1637b Descartes, R. P. *Dioptrique.* Leyden 1637; *Principia Philosophiae*, Amsterdam 1644.

1643 Torricelli, E. *De moto gravium naturalitas accelerato.* Firenze.

1657 Fermat, P. *Oeuvres* 2, 354 (1891).

1670 Newton, I. *Methodis fluxionum.* Published 1736 (posthumous), London; *Isaaci Newtoni opuscula mathematica,* J. Cartillioneus (1744), Lausanne 1823.

1684 Leibnitz, G.W. *Nova methodus pro maximis et minimis.* Paris.

1687 Newton, I. *Philosophiae naturalis principia mathematica*, 2 vols., Cambridge Univesrsity Press, repr. University of California Press 1934; *Exposition des découvertes philosophiques de M. le chevalier Newton*, C. MacLaurin, Royal Society of London, traduction Académic des Sciences de Paris (1746).

1690 Rolle, M. *Traité d'Algébre.* Paris.

1691 Bernoulli, J. *Differentialrechung.* Manuscript found in 1921.

1696 l'Hôspital G.F.A. *Annalyse des infinitement petites pour l'intelligence des courbes. Paris.*

1715 Taylor, B. *Methodus incrementorum.* Doctoral thesis, Cambridge.

1732 Pitot, M.A. Description d'une machine pour la mesure de la vitesse d'un écoulement d'eau. *Mémoires de l'Académie des Sciences*, 172.

1735 Euler L. De progressionibus harmonicus observations. *Commentarii Academia Scientarum Imperialis Petropolitanae* 7, 150–161.

1738 Bernoulli, D. *Hydrodinamica.* Argentorati.

1742 Maclaurin, C. *Treatise on Fluxions.*

1752 Euler, L. Principles géneraux du mouvement des fluides. *Histoire de l'Académie de Berlin 1755*, in C.A. Truesdell, *Rational Fluid Mechanics 1687–1765.*

1759 Euler, L. De Principis motus fluidorum. *Novi Commentari Academia Petropolitana.* 14,1: *Opera Omnia,* Basel Akademie.

1766 Borda, J.C. *Mémoires de L'Académie des Sciences.*

1768 D'Alembert, J.R. *Opuscules* 5, 171–182.

1770 Lagrange, J.L. *Mémoires de l'Académie de Berlin* 24; in *Oeuvres*, 14 vols., Gauthier-Villars (Paris), 1878–1888.

1781 Lagrange, J.L. *Nouvelles Mémoires de l'Academie de Berlin*; *Theorie des functions analytiques*, Imprimerie de la République, 1974.

1790 Mascheroni, L. *Adnotationes ad Calculum Integralem Euleri*, vol. 1–2, Ticino, Italy.

1797 Gauss, C.F. *Mathematische Annalen* 58, 118; *Werke* 3, 20–23.

1799 Bürmann, H. *Mémoires de l'Institut de France*, 2, 17.

1806 Argand, J.R. Essai sur une manière de representer des quantités imaginaires dans le constructions geometriques.

1812a Gauss, C.F *Werke* 1, 340.

1812b Gauss, C.F. Disquisitiones generales circam seriem infinitam... *Commentationes Societones Regiae Scientarium Göttensis Recentiores. Werke* 1, 185.

1818 Fourier, J.B.J. *Theorie analytique de la chaleur.* Paris, repr. Dover 1955.

1820 Poisson, S.D. Mémoire sur les integrales défenies et la sommation de séries. *Journal de l'École Royalle Polytechnique* 11, 295–341 (1820) ; 12, 404–509 (1823).

1821 Cauchy, A.L. *Analyse Algébrique.* Gauthier-Villars (Paris).

1823 Cauchy, A.L. *Resumé des leçons données à L'École Royalle Polytechnique sur le calcus l'infinitesimal.* Paris.

1823 Fresnel, A.J. *Mémoires de l'Académie des Sciences* 11, 393 (1832); *Oeuvres* 2, 25.

1825 Cauchy, A.L. Mémoire sur les integrales prises entre limites imaginaires. *Ouevres*, 14 vols, Gauthier-Villars (Paris), 1881–1895.

1825 Laplace, P.S. *Mécanique Céleste.* 5 vols. , Bachelier (Paris), repr. Chelsea 1982.

1826 Abel, N.H. *Journal fur die reine und angewandte Mathematik* 1, 311–339.

1827 Jacobi, C.G.J. *Astronomische Nachrichten* 6, 123; *Fundamenta nova theoriae functionum ellipticarum,* Konigsberg; *Gesammelte Werke* 1, 49–239, repr. Chelsea 1969.

1828 Green, G. Essay on electricity and magnetism, Nottingham; *Mathematical Papers* p.3, Cambridge University Press 1871.

1836 Kummer, E.E. Ueber die hypergeometrische Reihe... *Journal fur reine und angewandte Mathematik* 15, 39–83, 127–172.

1837 Dirichlet, P.G.L. *Berliner Abhandlungen* 48.

1839 Abel, N.H. *Journal de Mathematiques* 4,397.

1840 Laplace, P.S. *Essai sur les Probabilités.* Gauthier-Villars (Paris).

1841 Weierstrass, K.W.T. *Werke* 1, 51–66.

1843 Hamilton, W.R. *Lectures on quaternions*, Hodges & Smith (Dublin) 1853; *Elements of quaternions*, Longmans & Green (London) 1866; *Mathematical Papers*, Cambridge University Press 1966.

1843 Laurent, P.A. *Comptes rendus de l'Académie des Sciences* 17, 348–349.

1844 Cauchy, A.L. *Comptes Rendus de L'Academie des Sciences* 19, 1377–1378.

1845 Kirchhoff, G.R. *Poggendorf Annalen* 64; *Vorlesungen ueber matematishe Physik: Mechanik*, Teubner (Leipzig) 1877; *Gesammelte Abhandlungen 1.*

1847 Liouville, J. *Leçons*; Borchardt, C.W. *Journal fur reine and angewandte Mathematik* 277–310 (1880).

1847 Riemann, B.: Versuch einer allgemeinem auffassung der Integration und Differentiation. *Werke* 385–398.

1850 Dirichlet, P.G.J. *Abhandlungen der Königliche Preussiche Akademie der Wissenschaften* 99–116.

1851 Riemann, B. Grundlagen fur allgemaine Theorie der Functianen einer veränderlichen complexen Grösse. *Inaugunal dissertation* (Göttingen); *Werke* 1.35–80.

1856 Weierstrass, K.W.T. *Journal fur die reine and angewandte Mathematik* p.29; *Werke* 1, 185.

1857 Riemann, B. Theorie der Abel'schen functionem. *Journal fur die reine and angewandte Mathematik* 54; *Werke* 120–176.

1858 Dedekind, R. *Stetigkeit und irrationale Zahlen,* Braunschweig (5th ed. 1927); *Gesamnelte mathematische Werke* 3, 315–334, Vieweg (Braunschweig).

1858 Helmholtz, H.L. *Crelle* 40.

1858 Rouché, E. *Journal de l'École Polytechnique,* 21.

1862 Dirichlet, P.G.J. *Journal fur reine und anglewandte Mathematik* 7, 253–255.

1863 Riemann, B. *Gesammelte mathematische Werke*, Teubner (Leipzig), repr. Dover 1953.

1864 Rankine, W.J.M. On plane waterlines in two dimensions. *Philosophical Transactions of the Royal Society (Miscellanous scientific papers* p. 495, London 1881).

1868 Casorati, F.: *Teórica della funzioni di variabili complessa.* Pavia.

1868 Christoffel, E.B. *Annali di Matematica Pura Applicata* 11, 89–103 (1868); 4, 1–9 (1871).

1868 Helmholtz, H.L. *Berliner Monatsberichte* 23.4.1868; *Philisophical Magazine* 11.1868; *Wissenschaftliche Abhandlungen* 1,154.

1869 Kelvin, W.T. On vortex motion. *Transactions of the Edinburg Society* 25; *Papers* 4, 13.

1869 Kirchhoff, G.R. Zur theorie freier Flussigkeitsstrahlen. *Crelle* 70; *Gesammelte Abhandlungen*, 416; *Mechanik* 21–22.

1871 Rankine, W.J.M. On the mathematical theory of stream lines, especially those with four foci and upwards. *Philosophical Transactions of the Royal Society* p.167.

1873 Maxwell, J.C. *Treatise of Electricity and Magnetism,* 2 vols., Clarendon Press (Oxford) 3rd ed. 1891, repr. Dover 1954.

1876 Darboux, J.G. *Journal de Mathématiques.* 2, 271; 2, 291.

1876a Rayleigh. J.W.S. Notes on hydrodynamics. *Philosophical Magazine 2, 430–441;* 1, 297–296. *Papers* Cambridge University Press 1899–1920, 6 vols., repr. Dover 1969.

1876b Rayleigh, J.W.S. On the resistance of fluids. *Philosophical Magazine*; *Papers* 1.287–296.

1876 Weierstrass, K. W. T.: *Abhandlungen aus der Funcktionlebre.*

1877 Gröbli. W. *Die die Bewegung paralleler geradliniger* Wirbelflächen, Zurich.

1878 Greenhill, A.G. Plane vortex motion. *Quarterly Journal of Mathematics* 15,10.

1879 Picard, E. Sur une proprieté des functions entières. *Comptes Rendus de l'Academie des Sciences* 88, 1024–1027.

1879 Rethy, M. *Klausenburger Berichte.*

1880 Picard, E. Sur les functions entières, *Comptes Rendus de l'Académie des Sciences* 89, 662–665; Mémoire sur les functions entières, *Annales de l'École Normale Supérieure* 9, 145–166.

1881 Bobyleff, D. *Journal of the Russian Physico-Chemical Society* 13; *Wiedemann Bleiblätter* 6, 163.

1886 Morera, G. Un teorema fundamentale nella teorica delle funzioni di una variabili complessa. *Rendiconti del Instituto Lombardo per Scienze et Lettere* 19, 304–308.

1886 Robin, G. 1886 Sur la distribution d'electricité à la surfaces des conductteurs fermés et des conducteurs ouverts. *Annales Scientifiques de l'École Normale Supérieure* 3, 31–358.

1887 Herschel, C. The Venturi meter. *American Society of Civil Engineers.*

1889 Peano, G. *Arithmetices principia, nova methodo exposita.* Bocca (Torino).

1889 Schwartz, H.A. Ueber der Integration der partiellen Differentialgleichun $\nabla^2 u = 0$ unter vorshrieben grenz- und unstetigsteits bedingungen, *Monatsberichte der koniglichen Akademie der Wissenschaften zu Berlin* 767–795 (*Abhandlungen* 144–171); Zur integration der partieller differential gleichung ∇^2 u = 0, *Journal fur reine und angewandte Mathematik* 74, 218–253 (*Abhandlugen* 175–210).

1890 Schwartz, H.A. *Abhandlungen*, Teubner (Berlin) 1890, repr. Chelsea 1972.

1891 Peano, G. Sul concetto di numero. *Rivista di Matemática* 1, 87–102, 256–257.

1891 Rayleigh, J.W.S. Experiments on aerodynamics. *Nature* 95, 108–109; *Papers* 3, 491–495.

1894 Jordan, C. *Cours d'Annalyse.* 2, 285–286, Gauthier-Villars (Paris).

1894 Stieltjes, T.H. Recherches sur les fractions continues. *Comptes Rendus de l'Académie des Sciences de Paris* 118, 1401–1403.

1895 Weierstrass, K.W.T. *Gesammelte Werke.* 2, 245–255.

1896 Osgood, W.F. *Bulletin of the American Mathematical Society* 2,296–302.

1897 Larmor, J. *Philosophical Magazine* 44,503; *Proceedings of the Royal Society of London* 60,514; *Cambridge Philosophical Transactions* 1900; *Aether and Matter,* p.301 (1904).

1900 Goursat, E. *Transactions of the American Mathematical Society* 1,14.

1900 Teixeira, F.G. Sur les séries ordennnées selon les puissances d'une function donnée. *Journal für reine und angewandte Mathematik* 122, 97–123. 1. 129–161 in *Obras Mathematicas,* 7 vols., Imprensa da Universidade de Coimbra, 1904–1915.

1901 Harris, R.A. Two-dimensional fluid motion through spouts composed of two plane walls. *Annalen der Mathematik,* 2.

1902a Kutta, W.M. Aufstriebskräfte in strömender Flussigkeiten *Illustrierte. aeronautishe Miheilunger*, 138.

1902b Kutta, W.M. *Sitzberachtung der konigliche Bayerischen Academie der Wissenschaften.*

1903 Vitali, G. Rendicenti del Instituto Lambardo per Scienze et Lettere 36, 772–774; *Annali di Matemática Pura Aplicada* 10, 73 (1904).

1904 Landau, E. Ueber eine verallgemeinerung des Picardischen Satzes. *Sitzberichte der Prussiche Akademie der Wissenschaften* 38, 1118–1131.

1904 Schottky, F. Ueber ein Picard'schen Satz und die Borel'shen Ungleichungen, *Sitzberacchtung der Preussichen Akademie des Wissenschaften* 2, 1244–1262.

1905 Einstein, A. Zur Elekrodynamic bewegter Körper. *Annalen der Physik* 17, 891–921.

1906 Joukowski, N.Y. Sur les tourbillons adjoints. *Transactions of the Physical Section of the Imperial Society of Friends of the Natural Sciences (Moscow)* 23,2.

1908 Plemelj. J. *Monatshefte for Mathematik und Physik* 19, 205–210.

1910 Blasius H. Funktiontheoretische methoden in der Hydrodynamik. *Zeitschrift for Mathematik und Physik* 58.

1910 Joukowski, N. Y. *Zeitschrift fur Flugtechnik und Motorluftschiffahrt.*1, 281 (1910); 3, 81 (1912).

1910 Mellin H.J. *Mathematische Annalen* 305–307.

1910 Montel, P. *Leçons sur les series de polynomes à une variable complexe.* Gauthier-Villars (Paris).

1911 von Karman, Th. Flussigkeits und Luftwiderstand. *Physikalische* 13,49; *Nachrichten der Göttinger Gesellschaft fur Wissenschaften*, 547 (1912).

1912 Bragg, W.L. *The crystalline state. Vol. I.: General Survey.* Bell (London) 1933.

1912 Caratheodory, C. *Theory of complex functions.* 2 vols, Birkhauser Velag (Basel) 1950, repr. Chelsea 1954.

1916 Joukowski, N. *Aerodynamique.* Paris.

1918 Prandtl, L. Tragflügeltheorie. *Nachrichten der Gesellschaft der Wissenschaften Gottingen,* 151 (1918), 107 (1919).

1918 von Karman, Th. & Trefftz, E. Potentialströmung um gegebene Tragflächenquerschnitte. *Zeitschrift für Flugwissenchaften* 9,11.

1920 von Mises, R. Zur theorie der Tragflügelauftriebs. *Zeitschrift fur Flugtechnik und Motorluftschiffarht* 11, 68.

1924 Julia, G. *Leçons sur les functions uniformes à un point singulier essential isolé.* Gauthier-Villars (Paris).

1927 Montel, P. *Leçon sur les familles normales de functions analytiques et leurs applications.* Gauthier-Villars (Paris).

1928 Borel, E. *Leçons sur la théorie des functions.* Gauthier. Villars (Paris).

1928 Lebesgue, H. *Leçons sur l'Integration.* repr. Chelsea 1973.

1928 Pollard, S. *Proceedings London Mathematical Society,* 28, 159–160.

1929 Osgood, W.F. *Lehrbuch der Funktionentheorie.* Cambridge Massachusetts 1929, repr. Chelsea 1965.

1932 Lamb, H. *Hydrodynamics.* Cambridge University Press 1879, 6th ed. 1932.

1961 von Neumann, J. *Collected Works.* Pergamon Press, 4 vols., 1961.

1984 Campos, L.M.B.C. On a concept of derivative of complex order with application to special functions. *Institute of Mathematics and Applications Journal of Applied Mathematics,* 33, 109–133.

1985 Campos, L.M.B.C. On vertical hydromagnetic waves in a compressible atmosphere under an oblique magnetic field. *Geophysical and Astrophysical Fluid Dynamics* 32, 217–272.

1987 Campos, L.M.B.C. On waves in gases. Part II: Interaction of sound with magnetic and internal modes. *Reviews of Modern Physics* 59, 363–462.

1990 Campos, L.M.B.C. On generalizations of the series of Taylor, Lagrange, Laurent and Teixeira. *International Journal of Mathematics and Mathematical Sciences* 13, 687–708.

1993 Campos, L.M.B.C. Fractional calculus of analytic and branched functions, p. 65–144 in *Recent Advances in the Fractional Calculus,* ed. R.N. Kalia, Global Press.

Index

Note: *Italicized* page references denote figures, tables, diagrams, classifications and lists.